Trigonometric Formulas

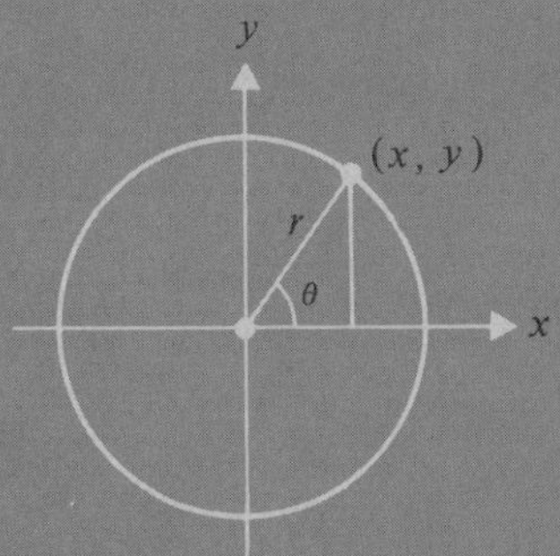

$$\sin\theta = \frac{y}{r} \qquad \cos\theta = \frac{x}{r} \qquad \tan\theta = \frac{y}{x}$$

$$\csc\theta = \frac{r}{y} \qquad \sec\theta = \frac{r}{x}$$

$$\sin^2\theta + \cos^2\theta = 1$$
$$1 + \tan^2\theta = \sec^2\theta$$
$$1 + \cot^2\theta = \csc^2\theta$$

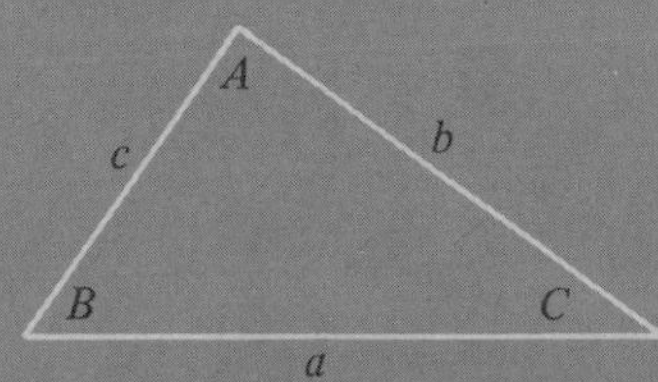

Cosine law: $b^2 = a^2 + c^2 - 2ac\cos B$

Sine law: $\dfrac{\sin A}{a} = \dfrac{\sin B}{b} = \dfrac{\sin C}{c}$

$$\sin(A+B) = \sin A\cos B + \cos A\sin B$$
$$\sin(A-B) = \sin A\cos B - \cos A\sin B$$
$$\cos(A+B) = \cos A\cos B - \sin A\sin B$$
$$\cos(A-B) = \cos A\cos B + \sin A\sin B$$

$$\tan(A+B) = \frac{\tan A + \tan B}{1 - \tan A\tan B}$$

$$\tan(A-B) = \frac{\tan A - \tan B}{1 + \tan A\tan B}$$

$$\sin 2A = 2\sin A\cos A$$
$$\cos 2A = \cos^2 A - \sin^2 A = 2\cos^2 A - 1 = 1 - 2\sin^2 A$$
$$\tan 2A = \frac{2\tan A}{1 - \tan^2 A}$$

$$\sin A\sin B = \tfrac{1}{2}[-\cos(A+B) + \cos(A-B)]$$
$$\sin A\cos B = \tfrac{1}{2}[\sin(A+B) + \sin(A-B)]$$
$$\cos A\cos B = \tfrac{1}{2}[\cos(A+B) + \cos(A-B)]$$

$$\sin A + \sin B = 2\sin\left(\frac{A+B}{2}\right)\cos\left(\frac{A-B}{2}\right)$$

$$\sin A - \sin B = 2\cos\left(\frac{A+B}{2}\right)\sin\left(\frac{A-B}{2}\right)$$

$$\cos A + \cos B = 2\cos\left(\frac{A+B}{2}\right)\cos\left(\frac{A-B}{2}\right)$$

$$\cos A - \cos B = -2\sin\left(\frac{A+B}{2}\right)\sin\left(\frac{A-B}{2}\right)$$

Calculus and Analytic Geometry

Calculus and Analytic Geometry

Donald W. Trim

UNIVERSITY OF MANITOBA
WINNIPEG, MANITOBA, CANADA

ADDISON-WESLEY
PUBLISHING COMPANY
Reading, Massachusetts
Menlo Park, California
London
Amsterdam
Don Mills, Ontario
Sydney

Sponsoring Editor: *Wayne D. Yuhasz*
Developmental Editor: *Jeffrey M. Pepper*
Production Editor: *Doris L. Machado*
Copy Editor: *Martha K. Morong*

Text Designer: *Margaret Ong Tsao*
Illustrator: *Vantage Art, Inc.*
Cover Designer: *Hannus Design Associates, Richard Hannus*
Cover Photograph: *Building Reflections by Geoffrey Gove, The Image Bank*
Art Coordinator: *Dick Morton*

Production Manager: *Karen M. Guardino*
Production Coordinator: *Peter Petraitis*

The text of this book was composed in Monophoto Times Roman
by Composition House Limited.

Library of Congress Cataloging in Publication Data

Trim, Donald W.
Calculus and analytic geometry.
Includes index.
1. Calculus. 2. Geometry, Analytic. I. Title.
QA303.T78 1983 515'.15 82-16287
ISBN 0-201-16270-9

ISBN 0-201-16270-9
ABCDEFGHIJ-DO-89876543

To those students who encouraged the writing of this text, to those who suffered through its many revisions, and to those who will use it in the future.

Preface

This text, like many others, evolved from a set of notes developed by the author over a number of years of teaching freshman and sophomore calculus. It represents that treatment of single and multivariable calculus with which we have had the greatest degree of success. For the most part, our approach is intuitive, making free use of geometry and familiar physical settings to motivate and illustrate concepts. For instance, limits are introduced intuitively, but a brief optional section is included to introduce the more advanced student to the mathematical definition of a limit. Derivatives are motivated and defined algebraically as instantaneous rates of change, and interpreted geometrically as slopes of tangent lines. The definite integral is motivated through two geometric and two physical problems, but its definition is algebraic. This permits us to use the definite integral in a wide variety of applications without fear that it has been associated with any one in particular.

The order of topics for single variable calculus is reasonably standard with a few minor variations. The sine and cosine functions are differentiated earlier than the remaining transcendental functions in order to satisfy the needs of other concurrent courses. Newton's iterative procedure is treated first as an application of differentiation and later as an example of a sequence. Differential equations are introduced briefly in Chapter 5 and used as examples and exercises in Chapters 8, 9, and 10, so that the student is familiar with the concepts before detailed discussions in Chapter 16. L'Hôpital's rule is discussed in two parts. In Chapter 4 it is first applied to simple indeterminate forms ($0/0$ and ∞/∞), then in Chapter 8 to more advanced forms with the help of logarithms.

Chapter 11 (infinite series), Chapters 12–15 (multivariate calculus), and Chapter 16 (differential equations) have been written with a minimum of interdependence and can therefore be taught in any order.

To aid the flow of our discussion we have reduced the theorem-proof, theorem-proof format. This means that "proofs" of some results are essentially plausibility arguments, proofs of others are left to the exercises, and some proofs are completely omitted. Sections of the book that deal with approximations and iterative procedures require the use of an electronic calculator, as do some exercises (marked with a "C") in other sections. It was not our intention to write a text that makes full use of calculators, for then we would have had to deal with too many aspects of numerical analysis. On the other hand, we would certainly have been remiss had we omitted these topics altogether.

Because our approach is intuitive and our emphasis is on problem solving (there are over 4400 problems), we run a considerable risk of having our work labeled "cookbookish." We have endeavored not to let this happen; our emphasis is indeed on problem solving, but problem solving through an understanding of calculus concepts. It is only when the student begins to understand calculus that he or she can also begin to feel it at work. For example, when area, volume, work, lengths of curves, fluid pressure, and centroids, etc., are considered as applications of the definite integral, we constantly remind the student not to memorize formulas. We stress the need to understand the definite integral as a "limit-summation," and then to apply this concept to the particular application at hand. Likewise, we do not discuss curve sketching simply as an application of the theory of maxima and minima, but instead we try to involve the student in a decision-making process of how best to obtain a sketch suitable for a specific problem. We do this by illustrating various useful curve-sketching techniques as calculus unfolds, beginning with tables of values, progressing to building procedures such as addition and multiplication of ordinates, and symmetry principles, using limits near points of discontinuity, and finally adding information determined by relative extrema and points of inflection. In this way, the student sees various aspects of calculus at work on the same problem and begins to appreciate how they complement one another. Integration techniques are approached from an organizational point of view rather than one of memorization. We suggest that there are a few simple integration formulas that must be known, but beyond this it is a matter of organizing an integral so that these elementary formulas can be utilized.

Various supplements have been prepared to aid instructors and students. Available to students is a manual containing detailed solutions to all even-numbered exercises. For instructors there is a manual with answers and selected solutions to odd-numbered exercises and a set of transparencies for the more complicated figures in the text. Any errors in the text or supplements are the responsibility of the author, and we would appreciate having them brought to our attention.

ACKNOWLEDGMENTS

I wish to thank the following individuals who reviewed the manuscript and offered many helpful suggestions:

Edmon L. Adams
University of North Dakota

Charles W. Austin
California State University

C. M. Biles
Humboldt State University

Carlos A. Curley
Wheaton College

Daniel S. Drucker
Wayne State University

Baxter Johns
Baylor University

Joseph F. Krebs
Boston College

Kenneth K. Kubota
University of Kentucky

Philip M. Locks
University of Maine at Orono

Stanley M. Lukawecki
Clemson University

Ronald P. Morash
The University of Michigan at Dearborn

Karl Petersen
University of North Carolina at Chapel Hill

Erik A. Schreiner
Western Michigan University

Fritz Steinhardt
The City College of The City University of New York

David Tischler
CUNY—Queens College

Janet M. Rich
Miami-Dade Community College

It is a pleasure to acknowledge the staff of Addison-Wesley Publishing Company for a superb effort in editing, design, and illustration. Special thanks are due to S. H. Quigley, Wayne Yuhasz, Jeffrey Pepper, and Dick Morton for their encouragement throughout the writing of the text.

Most of all, a sincere thanks to a wife, son, and daughter, who many times must have wondered whether they still had a husband and father.

To the Student

This book has been written for you, the student of calculus. It begins with an introduction to the fundamentals of plane analytic geometry in order to give you a way of visualizing problems geometrically. You will be surprised at the number and variety of problems that you will be able to solve using calculus, but as it so often happens in mathematics, the most difficult part of the solution to any problem is the first line. Once started, the solution frequently unfolds smoothly and easily, but that first step sometimes seems impossible. One of the best ways to start a problem is with a diagram. A picture, no matter how rough, is invaluable in giving you a "feeling" for what is going on. It displays the known facts surrounding the problem; it permits you to see what the problem really is and how it relates to these known facts; and it often suggests that all-important first step. Don't underestimate the value of a diagram; get into the habit of making one at every opportunity—not just to solve problems, but to understand what calculus is all about.

Our study of calculus really begins in Chapter 2 with a discussion of limits. The idea of limits is fundamental to all of calculus, and it is surprising that the founders of calculus, Gottfried Leibniz (1646–1716) and Isaac Newton (1645–1727), had not formulated the idea of a limit. Their development of the subject was based upon the concept of "infinitesimals" —quantities that are extremely close to zero, but not equal to zero. It was the work of such mathematicians as L. Euler (1707–1783), J. D'Alembert (1717–1783), and, later, A. Cauchy (1789–1857) that led to the rigorous formulation of limits and a firm foundation upon which to develop the rest of calculus.

All of calculus originates with two basic ideas—the derivative and the definite integral. If one quantity depends upon another, and the first quantity is in a state of change, so also will be the second. For example, the speed of a car depends upon how far the accelerator is depressed—more depression, more speed. As soon as the accelerator depression is changed, so also is the speed of the car, and the derivative describes how these changes are related. When a furnace is turned on, the amount of heat it produces affects room temperature—more heat, higher temperature. Derivatives relate the rate at which temperature rises to the rate at which heat is produced.

The definite integral is a completely different kind of entity. Suppose, for instance, that we use a pair of scissors to cut some strange shape from a piece of paper and ask for its area. This could be very difficult since there are very few shapes for which we have an area formula. We could, however, estimate its area by drawing rectangles on the piece of paper and adding together their areas. By drawing smaller and smaller rectangles near the curved edges, we could get a better and better estimate of the true area. In essence, this is the idea of the definite integral; it is a sum of terms where the number of terms is increased indefinitely.

Surprising as it may seem, these two concepts which appear so different are really intimately connected, and it is this connection that makes calculus so fascinating. In fact, calculus is a subject wherein every topic depends heavily upon those that went before, and at the same time leads naturally into the next. Once we identify these interdependences, we begin to understand the inner beauty of calculus.

You will find that we place considerable emphasis on problem solving, and we do this because you will never learn calculus simply by reading about it. You must involve yourself in the subject, and the best way to do this is with problems. Only when you solve problems do you begin to think about what you have read, and then you begin to understand. The exercises in each section begin with routine, drill-type problems designed to reinforce fundamentals. It follows, therefore, that if you have difficulties with early exercises, then you have not understood the fundamentals, and you should therefore review material in that section. In later exercises we begin to make you think. Some of the exercises are of the same type as were found earlier, but with more involved calculations; some require you to think about a concept from a different point of view; others may draw upon ideas from previous sections; still others may do all of these. It is in these exercises that you find most of the applications, for only when you have mastered fundamentals can you begin to apply calculus to problems from other fields. Most exercise sets contain problems marked with an asterisk (*). These are challenging problems and should only be attempted after the previous problems have been solved. Certain exercises are also marked with a "C" to indicate the need for an electronic calculator. There are many other exercises for which a calculator could eliminate laborious hand calculations and use of trigonometry and logarithm tables, but exercises marked with a "C" definitely require a calculator. In addition, a calculator is essential for those sections that deal with approximations and iterations (Sections 4.8, 9.9, 11.5, and 11.9).

At the end of each chapter there is a brief summary of the results in that chapter and a set of review exercises. Don't underestimate the value of these exercises; they test your understanding of the material in the entire chapter. Section exercises develop expertise in small areas; chapter exercises test knowledge of major blocks of subject matter.

The key word in our approach to calculus is "think." Above all else, we hope that you will train your mind to think and to think logically. We want you to learn how to organize facts and interpret them mathematically. If there is a problem, we want you to be able to decide exactly what that problem is and finally, to produce a step-by-step procedure by which to solve it. Do not, therefore, read this book expecting formulas that can be used to solve all problems; we won't give them to you. We will give you a few formulas and, we hope, a great deal of insight into the ideas surrounding these formulas. Then you begin to think for yourself. You use what you know to solve the first problem. Having gained from that experience, you then move on to the next problem. In this way, your calculus studies will be a true learning experience—not an exercise in memorization and regurgitation of facts and formulas, but a developmental period during which you both learn calculus and increase your mental capacity. When you reach the end of the road and successfully complete your calculus courses, think back and compare how much you knew then to how much you know now. Congratulate yourself on accomplishing a great deal, and mostly through your own efforts.

Winnipeg, Canada
November 1982

D.W.T.

Contents

Plane Analytic Geometry and Functions

1

In the first three sections of this chapter we introduce the principles of plane analytic geometry, proceeding from coordinates along a line in Section 1.1 to coordinates in a plane in Section 1.2. In Section 1.3 we discuss curves and their representation by use of equations, paying particular attention to straight lines, circles, and parabolas. Students who have previously studied analytic geometry could skip these sections and begin with functions in Section 1.4, but we strongly recommend against this. Even the best-prepared student can benefit by at least a cursory reading of these sections. Not only is it helpful to refresh one's memory on concepts learned some time ago, but it is also wise to become familiar with the terminology and conventions set forth in these sections.

1.1 REAL NUMBERS AND COORDINATES

At an early stage in our mathematical education we learned how the real-number line serves as a geometric representation of the real-number system. In this section we reverse these roles and use the real-number system to "co-ordinatize" or identify points on a line. Essentially the two approaches are one and the same; they set up a correspondence between real numbers and points on a line. To coordinatize a line (which we call the x-axis), we begin by choosing some point of reference O on the line. This point, called the origin, is identified (or coordinatized) by the real number zero. We now start at O and mark off equal distances along the line in both directions (Fig. 1.1). These marks could, for example, be one centimeter apart or one meter apart. The first point to the right of O is one unit of distance from O and is coordinatized by the real number 1; the second point is two units from O and is identified by the number 2, and so on. Unit distances to the left of O are identified by the negative integers -1, -2, -3, and so on. Fractional numbers such as $\frac{3}{4}$ and $-\frac{5}{3}$ identify points that are $\frac{3}{4}$ of our unit distance to the right of O, and $\frac{5}{3}$ of our unit distance to the left of O, respectively. Continuation of this process leads to a coordinatization of the entire x-axis. Every point on the axis is identified by a real number that we denote by x, and this number is the *directed distance* from the origin O to the point. For example, point Q in Fig. 1.1 is two and one-half units of distance to the right of O, and therefore its x-coordinate is $\frac{5}{2}$.

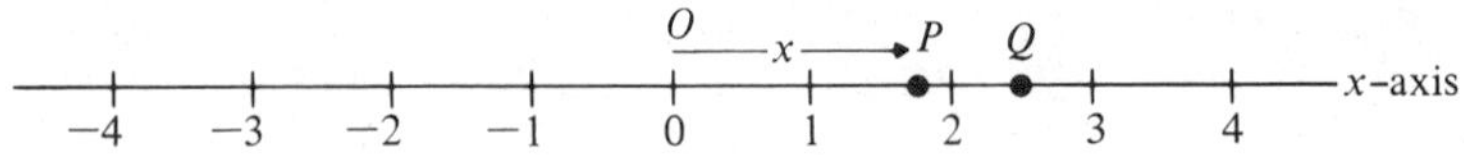

Figure 1.1

To describe Q we should say "the point whose x-coordinate is $\frac{5}{2}$," but because this is somewhat cumbersome, we generally say "the point $x = \frac{5}{2}$." Thus, to refer to a general point P on the x-axis, we simply say "the point x" (meaning, of course, the point with coordinate x).

Figure 1.2 shows a second coordinatization of the x-axis using a unit distance smaller than that in Fig. 1.1. Because the unit of length can be chosen arbitrarily, as can the point of reference O, the number of ways to coordinatize a line is infinite.

To represent sets of points on the x-axis we define open and closed intervals.

▷ **Definition 1.1** The set of all points on the x-axis between two given points $x = a$ and $x = b$ $(a < b)$, but not including these points, is called an *open interval.* It is denoted by (a, b). ◁

The open interval (a, b) is described algebraically as all points on the x-axis whose x-coordinates satisfy the inequalities $a < x < b$. For example, $1 < x < 2$ describes the open interval $(1, 2)$. We have shown this interval in Fig. 1.3, where the open circles at $x = 1$ and $x = 2$ indicate that these points are not included in the set.

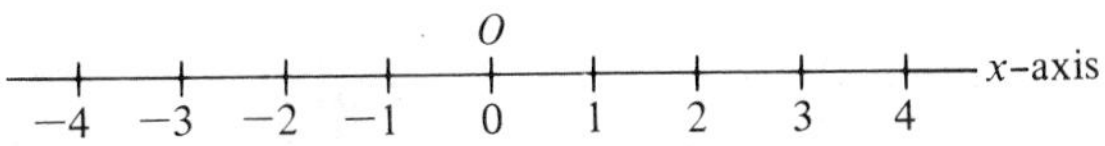

Figure 1.2

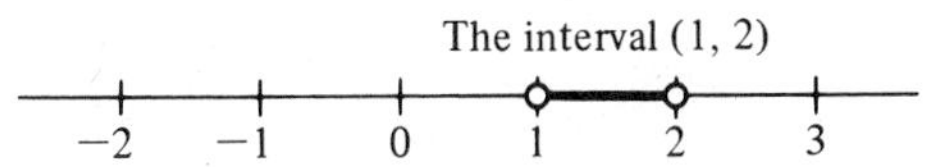

Figure 1.3

▷ **Definition 1.2** The set of all points on the x-axis between two given points $x = a$ and $x = b$ $(a < b)$, including these points, is called a *closed interval.* It is denoted by $[a, b]$. ◁

Algebraically, $[a, b]$ is described as all points whose x-coordinates satisfy $a \leq x \leq b$. The closed interval $[1, 2]$ in Fig. 1.4 is therefore described by $1 \leq x \leq 2$. Note that a closed interval contains its end points, whereas an open interval does not.

Intervals need not be open or closed. For example, the set of points whose coordinates satisfy $1 < x \leq 2$ is denoted by $(1, 2]$. It is said to be a half-open (and half-closed) interval; it contains one end but not the other (Fig. 1.5). Intervals need not be of finite length. The set of points whose coordinates satisfy $x > 2$ has only one end, namely $x = 2$, which is not contained in the set, and we therefore regard it as an open interval (Fig. 1.6). It is denoted by $(2, \infty)$. The interval described by $x \leq 2$ is denoted by $(-\infty, 2]$; it contains its only end point, and is therefore a closed interval (Fig. 1.7).

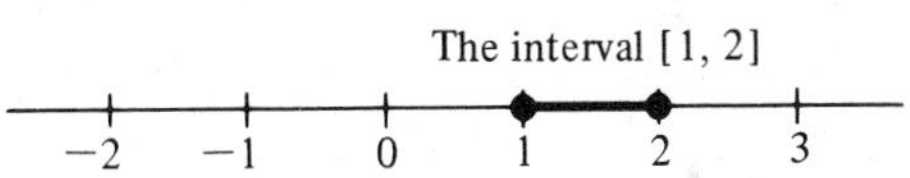

Figure 1.4

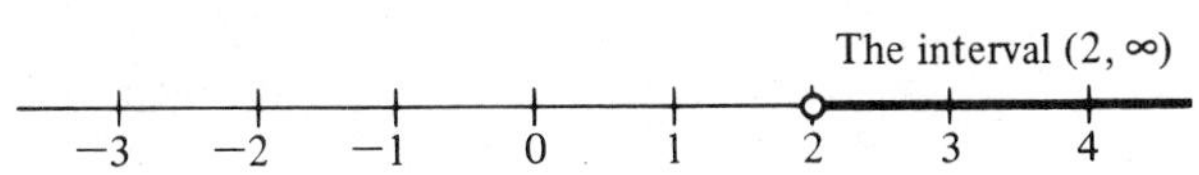

Figure 1.6

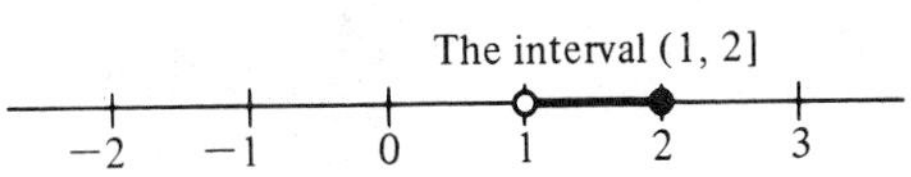

Figure 1.5

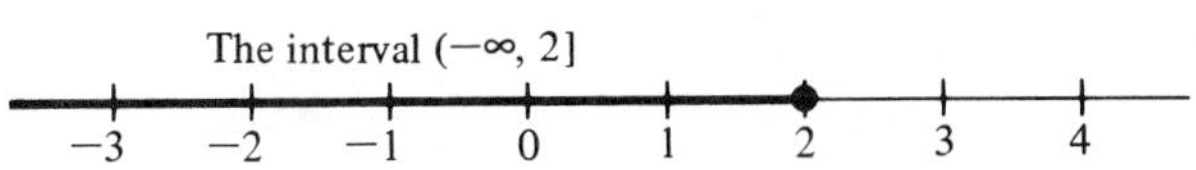

Figure 1.7

Intervals are building blocks for describing more complicated sets of points on the x-axis. For example, the set of points between $x = 1$ and $x = 3$ but not including $x = 1$, $x = 2$, or $x = 3$ consists of all points in the open intervals $1 < x < 2$ and $2 < x < 3$ (Fig. 1.8). The set of all points satisfying the inequality $x^2 \geq 9$ consists of all points in the closed intervals $x \leq -3$ and $x \geq 3$ (Fig. 1.9).

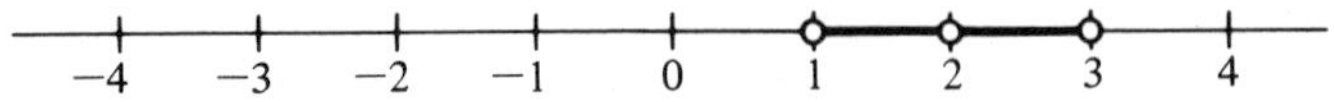

Figure 1.8

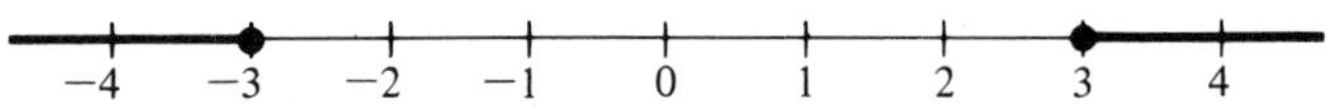

Figure 1.9

In order to determine the set of points described by a complicated inequality such as $2x^2 + 3x \leq x^2 - 2x + 5$, we must first simplify the inequality. The following rules are used to manipulate inequalities:

$$\text{If } a < b, \text{ then} \qquad a + c < b + c \quad \text{for any real } c. \tag{1.1}$$

$$\text{If } a < b, \text{ then} \qquad \begin{cases} ac < bc & \text{if } c > 0, \quad (1.2a) \\ ac > bc & \text{if } c < 0. \quad (1.2b) \end{cases}$$

Each inequality $<$ in these properties may be replaced by $\leq$ provided that $>$ is at the same time replaced by $\geq$.

Example 1 Find all points whose x-coordinates satisfy the following inequalities:

a) $4 - 2x \leq 5$;

b) $1 + 2x^2 < 5$;

c) $\dfrac{2x - 1}{x + 2} > 1$.

Solution a) If we add -4 to both sides of the inequality, we get

$$-2x \leq 1.$$

Multiplication by $-\frac{1}{2}$ then gives us the closed interval

$$x \geq -\tfrac{1}{2}.$$

b) If we subtract 1 from both sides of the inequality,

$$2x^2 < 4,$$

and divide by 2, we have

$$x^2 < 2.$$

Thus we find that points must lie in the open interval $-\sqrt{2} < x < \sqrt{2}$.

c) For fractional inequalities it is often useful to bring all terms to one side of the inequality:

$$0 < \frac{2x - 1}{x + 2} - 1 = \frac{(2x - 1) - (x + 2)}{x + 2} = \frac{x - 3}{x + 2}.$$

A fraction can be positive only if the numerator and the denominator are either both positive or both negative. Figure 1.10 illustrates that $x - 3$ is positive when $x > 3$ and negative when $x < 3$, and that $x + 2$ is positive when $x > -2$ and negative when $x < -2$. Since $x - 3$ and $x + 2$ must be of the same sign in order to have a positive quotient, it follows that for a point to satisfy

$$\frac{x - 3}{x + 2} > 0 \quad \text{or} \quad \frac{2x - 1}{x + 2} > 1,$$

it must lie in either $x < -2$ or $x > 3$. □

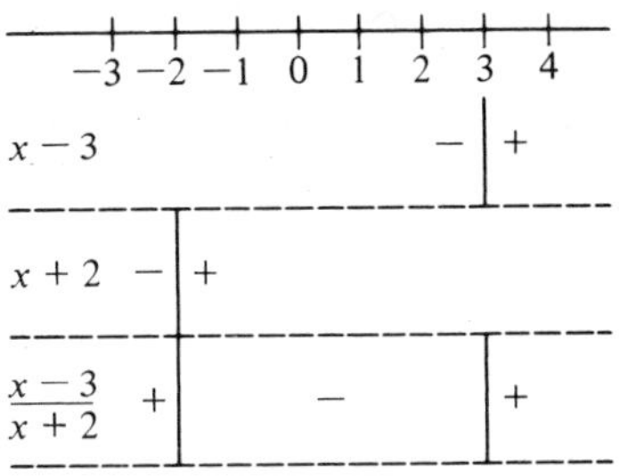

Figure 1.10

In many places throughout this text we find it necessary to discuss the "sign" of a quantity, and schematics such as that shown in Fig. 1.10 are very helpful.

One of the most useful notations for specifying sets of points on the x-axis is as follows.

▷ **Definition 1.3** The *absolute value* of a real number x is

$$|x| = \begin{cases} x & \text{if } x \geq 0, \\ -x & \text{if } x < 0. \end{cases} \tag{1.3}$$ ◁

For example, $|10| = 10$, $|-9| = 9$, and $|2(4)^2 - 9(5)| = |32 - 45| = |-13| = 13$. The absolute value then is an operation that defines the size or "magnitude" of a number without regard for sign.

To see how the absolute value can be used to specify sets of points on the x-axis, consider first the set described by the inequality $|x| < 2$. This inequality states that the magnitude of x must be less than 2. This means that x must lie between -2 and 2; hence $|x| < 2$ describes all points in the open interval

$$-2 < x < 2.$$

Similarly, the inequality $|x - 5| < 2$ states that the magnitude of $x - 5$ must be less than 2, and therefore

$$-2 < x - 5 < 2.$$

Addition of 5 to each of these terms gives the interval

$$3 < x < 7.$$

Alternatively, we could have reasoned that $x - 5$ is the directed distance from 5 to x on the x-axis (Fig. 1.11). The inequality $|x - 5| < 2$ requires this distance to be less in magnitude than 2; in other words, the distance from 5 to x must be less than 2. Hence x must lie between 3 and 7.

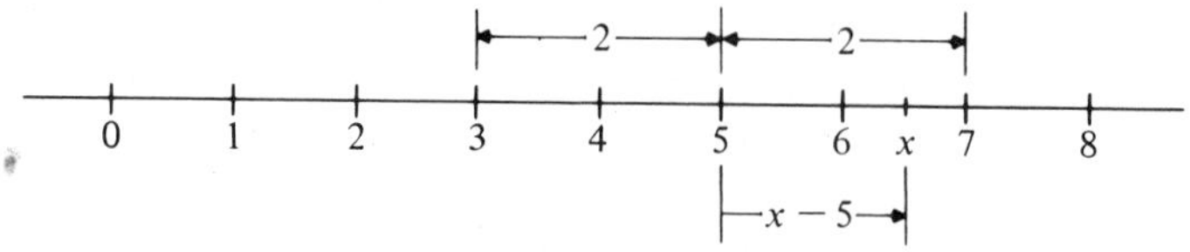

Figure 1.11

The inequality

$$|x^2 + 1| \leq 2$$

requires that $x^2 + 1$ fall between -2 and 2, including ± 2:

$$-2 \leq x^2 + 1 \leq 2.$$

When 1 is subtracted from each of these terms,

$$-3 \leq x^2 \leq 1.$$

Now the left-hand inequality $-3 \leq x^2$ is really not a restriction on x at all—every real x satisfies this inequality. Consequently, x must satisfy only $x^2 \leq 1$ or $-1 \leq x \leq 1$. The original inequality $|x^2 + 1| \leq 2$ is therefore equivalent to $-1 \leq x \leq 1$ or $|x| \leq 1$.

Example 2 Find all points that satisfy each of the following inequalities:

a) $|3 - 5x| \leq 2$;

b) $|4x^2 - 7| > 2$.

Solution a) The absolute value may be replaced by the double inequality

$$-2 \leq 3 - 5x \leq 2,$$

and when we subtract 3, we find

$$-5 \leq -5x \leq -1.$$

Multiplication of both sides of these inequalities by $-\frac{1}{5}$ gives

$$1 \geq x \geq \tfrac{1}{5}$$

or

$$\tfrac{1}{5} \leq x \leq 1.$$

b) The inequality $|4x^2 - 7| > 2$ will be satisfied if x satisfies either $4x^2 - 7 > 2$ or $4x^2 - 7 < -2$. From the first, $4x^2 > 9$ or $x^2 > \frac{9}{4}$. This implies that either $x > \frac{3}{2}$ or $x < -\frac{3}{2}$. From the inequality $4x^2 - 7 < -2$, we find that $x^2 < \frac{5}{4}$ or $-\sqrt{5}/2 < x < \sqrt{5}/2$. Consequently, in order to satisfy $|4x^2 - 7| > 2$, x must lie in one of the three intervals $x < -\frac{3}{2}$, $x > \frac{3}{2}$, or $-\sqrt{5}/2 < x < \sqrt{5}/2$ (Fig. 1.12). These may be stated alternatively as $|x| < \sqrt{5}/2$ and $|x| > \frac{3}{2}$. □

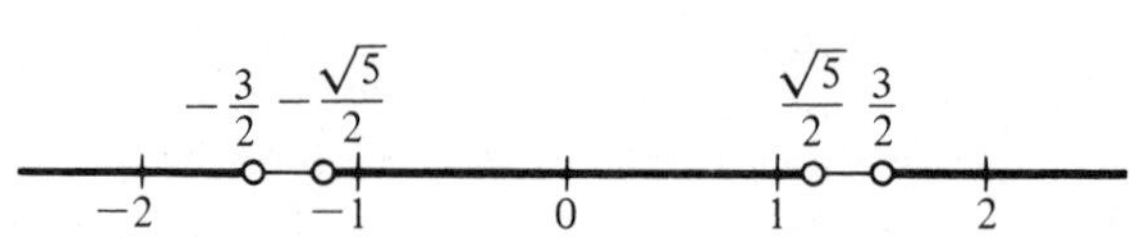

Figure 1.12

Example 3 Describe each of the following sets of points in terms of absolute values:

a) $-4 \leq x \leq 6$;

b) x between 1 and 3 but not equal to 1, 2, or 3.

Solution a) The length of the interval is 10 with the midpoint at $x = 1$. We can state alternatively that x must lie within a distance of 5 on either side of 1; i.e.,

$$|x - 1| \leq 5.$$

b) To specify that x must be within a distance of 1 on either side of 2, we write

$$|x - 2| < 1.$$

The condition that $x \neq 2$ is now easily incorporated:

$$0 < |x - 2| < 1. \square$$

The absolute value of the product of two real numbers is always equal to the product of their absolute values:

$$|ab| = |a|\,|b|. \tag{1.4a}$$

The absolute value of the sum of a and b is equal to the sum of their absolute values *only* when a and b have the same sign; that is,

$$|a + b| = |a| + |b|$$

when a and b are either both positive or both negative. When a and b have opposite signs, the correct statement is

$$|a + b| < |a| + |b|.$$

In general, then, we may state that for any two real numbers a and b,

$$|a + b| \leq |a| + |b|. \tag{1.4b}$$

This inequality finds its way into many applications of mathematics.

EXERCISES 1.1

1. Coordinatize the x-axis with unit distance equal to (a) 2 centimeters; (b) $\frac{3}{4}$ centimeter. Mark on the axis points with coordinates $-3, -2.5, -2, -1.5, -1, -0.5, 0, 0.5, 1, 1.5, 2, 2.5$, and 3.

In Exercises 2–19, find all points that satisfy the given inequality.

2. $3x - 2 < 7$

3. $2x^2 + 12 \geq 19$

4. $\dfrac{3}{x} > 5$

5. $\dfrac{3}{x} > -5$

6. $x^2 - 4 \leq 5$

7. $15 + x^2 < 14$

8. $|3x - 2| < 7$

9. $|2 - 3x| \geq 7$

10. $|2x^2 + 5| \leq 6$

11. $|2x^2 + 6| \leq 5$

12. $|2x + 6| \leq 5$

13. $|x^2 - 5| > 4$

14. $|x^2 - 4| \geq 5$

15. $|x^2 + 5| > 4$

16. $|x^2 + 4| \geq 5$

17. $\dfrac{2x-6}{x-2} > 3$

18. $\dfrac{3x+5}{4x-2} < 3$

19. $\dfrac{9x+5}{3x-2} < 2$

20. Verify that the inequality

$$x^2 + 4x < 6$$

can be written in the form

$$(x+2)^2 < 10,$$

and thereby find all points satisfying the inequality.

Use the technique in Exercise 20 to find all points that satisfy the inequality in each of Exercises 21–24.

21. $x^2 + 6x + 1 \le 0$

22. $2x^2 + 4x - 5 > 0$

23. $-x^2 - 2x \ge 1$

24. $|x^2 + 2x + 5| \ge 3$

25. Use inequality (1.4b) to prove that

$$|a \pm b| \ge ||a| - |b||$$

for any two real numbers a and b.

26. Prove that if $a < b$, then:

$\dfrac{1}{a} > \dfrac{1}{b}$ if a and b have the same sign;

$\dfrac{1}{a} < \dfrac{1}{b}$ if a and b have opposite signs.

In Exercises 27–40, find all points that satisfy the given inequality.

27. $x + \dfrac{1}{x} \ge 2$

28. $\dfrac{1}{x+2} - \dfrac{1}{x-2} < 5$

29. $11 < x^2 + 6x + 4 < 20$

30. $|x^2 + 2x - 3| \le 1$

31. $\dfrac{3x^2+5}{x^2-4} > 1$

32. $\dfrac{4x^2-64}{x^2-9} > 3$

33. $\left|\dfrac{3x+4}{x-2}\right| < 4$

34. $\dfrac{x+5}{x-3} < \dfrac{x+2}{x+1}$

35. $|x^2 + 2x + 3| \ge 5$

36. $|3x + 5| \le 4x$

37. $|3x + 5| \le |4x|$

38. $\dfrac{2x+3}{x-5} > x + 4$

39. $\dfrac{x-3}{x+1} > \dfrac{2x+4}{x}$

40. $|x^2 + 5| + |x - 2| \le 8$

Express the condition in each of Exercises 41–44 algebraically, and find all points that satisfy the condition.

41. The sum of the distances from a point x to $x = 1$ and $x = 5$ must be less than 7.

42. The sum of the distances from a point x to $x = -1$ and $x = 2$ must be greater than 4.

43. The absolute value of the difference of the distances from a point x to $x = 5$ and $x = 12$ must be less than 14.

44. The absolute value of the difference of the distances from a point x to $x = 2$ and $x = -6$ must be less than 3.

45. The arithmetic mean (average) of any two real numbers a and b is defined as $(a + b)/2$. The geometric mean of any two positive numbers a and b is defined as $\sqrt{ab}$. Show that the arithmetic mean of any two distinct positive numbers is always greater than their geometric mean.

46. Find all points that satisfy each of the following inequalities.

a) $x^2 + 12x + 11 < 0$

b) $\dfrac{2x^2 + 11x + 5}{x+2} < x - 3$

c) $2x + 1 < \dfrac{x^2 - x - 6}{x+5}$

d) $\dfrac{2x+1}{x+2} < \dfrac{x-3}{x+5}$

***47.** Find all points that satisfy

$$\frac{|3x+1|}{x+5} < x.$$

1.2 RECTANGULAR COORDINATES IN THE PLANE

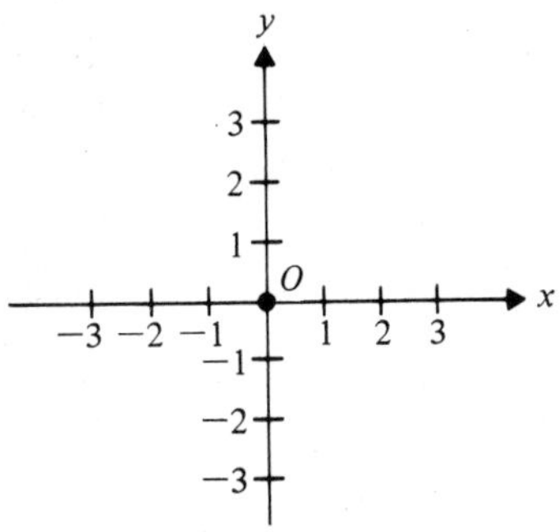

Figure 1.13

The technique of coordinatizing a line by means of real numbers can be extended to define rectangular coordinates in a plane (and, in Chapter 12, rectangular coordinates in space). For a plane, we begin by choosing any point O in the plane, called the origin, and any line in the plane through O, called the x-axis. This line, which we coordinatize as in Section 1.1, is drawn horizontally with a positive direction to the right (denoted by an arrowhead in Fig. 1.13). Next we draw through O a line perpendicular to the x-axis called the y-axis. This line is coordinatized with a positive direction usually chosen upward. In Fig. 1.13, we coordinatized the x- and y-axes with exactly the same unit distance, but we will see in Section 1.3 that it is sometimes more convenient to have different scales along the two axes. Unless we indicate otherwise, however, unit distances along the axes are chosen equal.

These two coordinatizations of the axes can be used to identify uniquely every point in the plane. Through the point P in Fig. 1.14 we draw lines parallel to the x- and y-axes to intersect the axes at points A and B. The x-coordinate of A is $\frac{5}{2}$, and the y-coordinate of B is $\frac{3}{2}$. We call these the x- and y-coordinates of P, and write them together in the form $(\frac{5}{2}, \frac{3}{2})$. The x-coordinate of P (also called the *abscissa* of P) is therefore the distance cut off on the x-axis by the line through P parallel to the y-axis (Fig. 1.15). Because the x- and y-axes are perpendicular, we can state alternatively that the x-coordinate of P is the perpendicular distance from the y-axis to P. Similarly, the y-coordinate of P (also called the *ordinate* of P) is the perpendicular distance from the x-axis to P (or the distance cut off on the y-axis by the line through P parallel to the x-axis).

By this construction, every point in a plane can be identified or coordinatized by an ordered pair of real numbers (x, y) called its x- and y-coordinates. Conversely, given any pair of real numbers x and y, there is one and only one point in the plane with these as x- and y-coordinates. Because coordinates (x, y) are defined in terms of a pair of perpendicular axes, we call them rectangular (or Cartesian)† coordinates. Figure 1.16 shows four points together with their Cartesian coordinates.

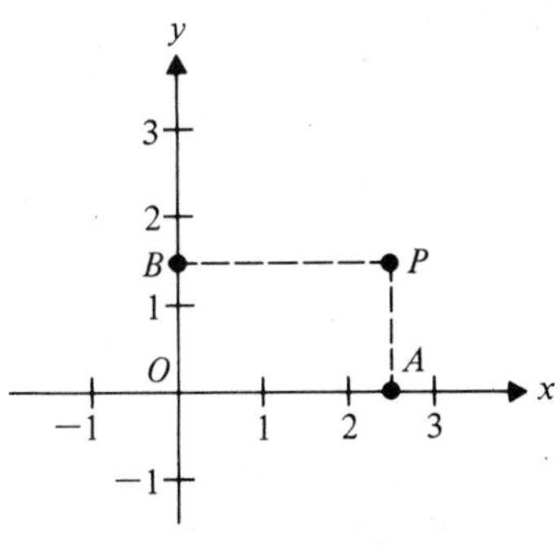

Figure 1.14

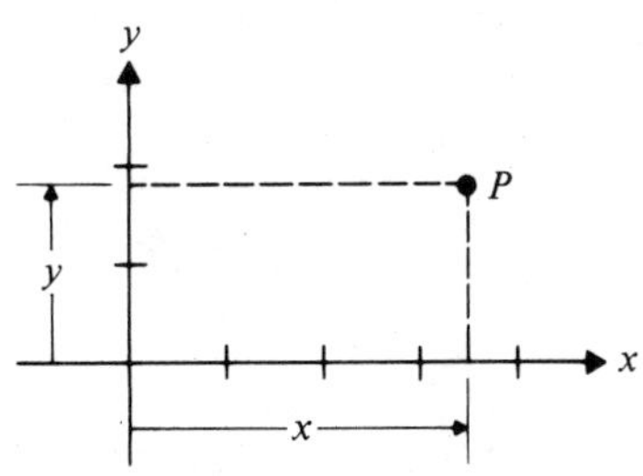

Figure 1.15

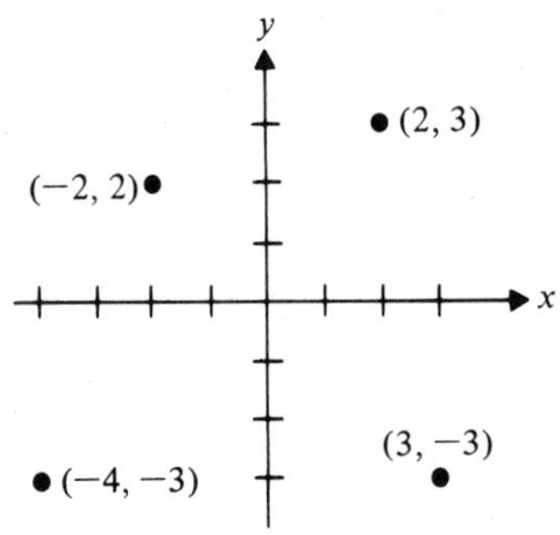

Figure 1.16

† They are called Cartesian coordinates after the French mathematician René Descartes (1596–1650), one of the founders of analytic geometry.

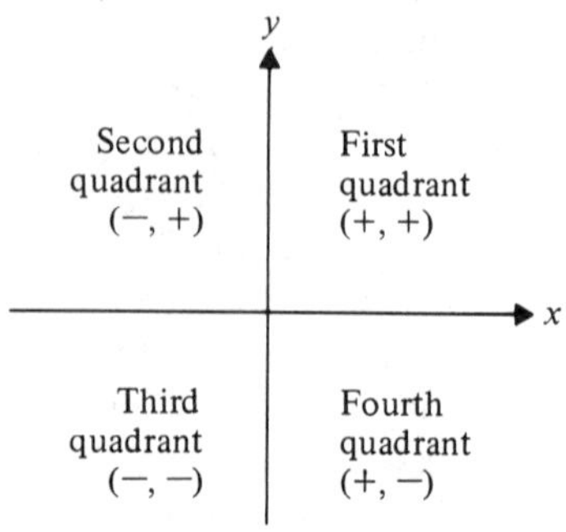

Figure 1.17

The axes divide the plane into four parts which, beginning at the upper right and proceeding counterclockwise, are called, respectively, the first, second, third, and fourth quadrants. The axes themselves are not considered part of any quadrant. Points in the first quadrant have x- and y-coordinates that are both positive, points in the second quadrant have a negative x-coordinate and a positive y-coordinate, and so on (Fig. 1.17).

If P and Q are any two points with coordinates (x_1, y_1) and (x_2, y_2), respectively (Fig. 1.18), then by the Pythagorean relation for right-angled triangle PQS,

$$|PQ|^2 = |PS|^2 + |QS|^2,$$

where $|PQ|$ denotes the length of the line segment joining P and Q, and similarly for $|PS|$ and $|QS|$. But

$$|PS| = |x_2 - x_1| \quad \text{and} \quad |QS| = |y_2 - y_1|;$$

therefore

$$|PQ|^2 = |x_2 - x_1|^2 + |y_2 - y_1|^2$$

or

$$|PQ| = \sqrt{(x_2 - x_1)^2 + (y_2 - y_1)^2}. \tag{1.5}$$

This very important result expresses the length of a line segment joining two points P and Q in terms of their coordinates (x_1, y_1) and (x_2, y_2). For example, if the coordinates of P and Q are $(-3, 4)$ and $(2, -1)$, then

$$|PQ| = \sqrt{(2 + 3)^2 + (-1 - 4)^2} = 5\sqrt{2}$$

(Fig. 1.19).

In Fig. 1.20, point $R(x, y)$ is the midpoint of the line segment joining $P(x_1, y_1)$ and $Q(x_2, y_2)$. Because triangles PTR and RUQ are congruent, it follows that $|RU| = |PT|$, and for P and Q as shown in Fig. 1.20, this condition becomes $x_2 - x = x - x_1$. Consequently,

$$x = \frac{x_1 + x_2}{2}. \tag{1.6a}$$

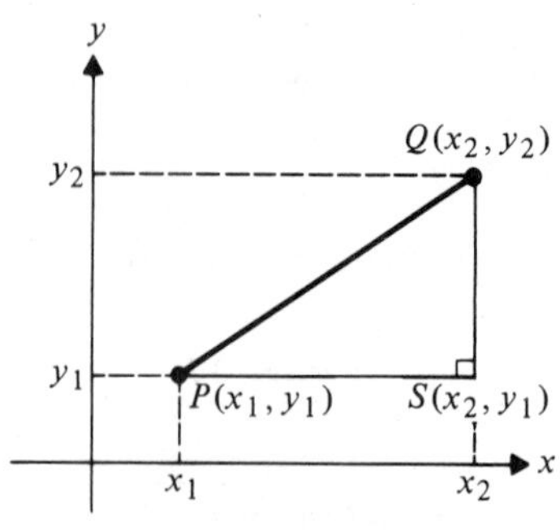

Figure 1.18

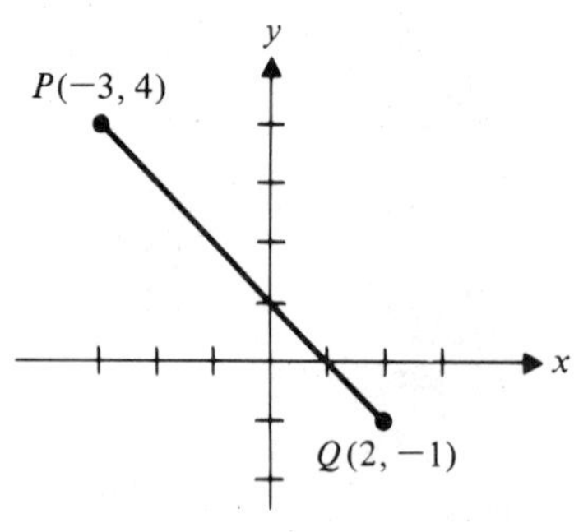

Figure 1.19

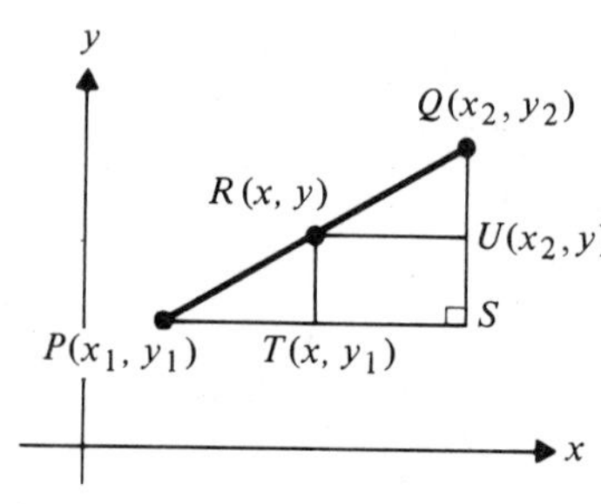

Figure 1.20

Similarly, equality of $|RT|$ and $|QU|$ gives

$$y = \frac{y_1 + y_2}{2}. \tag{1.6b}$$

Derivation of these results assumed that $x_2 > x_1$ and $y_2 > y_1$. The same formulas are valid for any x_1, x_2, y_1, and y_2 whatsoever.

We have shown that the coordinates of the midpoint of a line segment are the averages of the corresponding coordinates of the ends of the line segment.

EXERCISES 1.2

1. Show a plane coordinatized with Cartesian coordinates in each of the following ways:

a) equal unit distances of 1 centimeter along the x- and y-axes;

b) unit distances along the x- and y-axes of $\frac{1}{2}$ centimeter and 1 centimeter, respectively;

c) unit distances along the x- and y-axes of 1 centimeter and 1 millimeter, respectively.

2. On a plane with rectangular coordinates show the points $(2, 5)$, $(-3, 6)$, $(7, -2)$, $(6, -3)$, $(-4, 1)$, $(1, 3)$, $(-\frac{1}{3}, 1)$, $(2, \frac{3}{2})$, $(-\frac{1}{2}, -\frac{3}{4})$.

3. A square has sides of length 4 units. Find the coordinates of its vertices if:

a) one vertex is at the origin, two of its sides lie along the coordinate axes, and one vertex lies in the third quadrant;

b) its diagonals lie along the axes.

4. Show all points in a plane with Cartesian coordinates x and y that have each of the following:

a) an x-coordinate equal to 4;

b) a y-coordinate equal to -1;

c) equal x- and y-coordinates;

d) a y-coordinate equal to the negative of the x-coordinate.

5. Points with Cartesian coordinates $(0, 0)$, $(7, 0)$, and $(3, 5)$ form three vertices of a parallelogram. What are possible coordinates for the fourth vertex?

6. Determine whether each of the triangles with the following vertices is right-angled.

a) $A(2, 9)$, $B(-1, 3)$, $C(6, 7)$

b) $A(-1, 4)$, $B(3, 6)$, $C(1, 1)$

7. A ship sails N 30° E from a pier at 5 kilometers per hour. If coordinates are set up with the pier as origin and the positive x- and y-axes in the east and north directions, respectively, what are the coordinates of the ship after 3 hours?

8. A ladder 10 meters long rests against a wall with its feet 4 meters from the base of the wall (Fig. 1.21). If a man stands on the middle rung of the ladder, what are the coordinates of his feet?

9. What are the coordinates of points A, B, and C in Fig. 1.22 if the angle formed by lines BC and BA is 90°, and the side lengths are as shown?

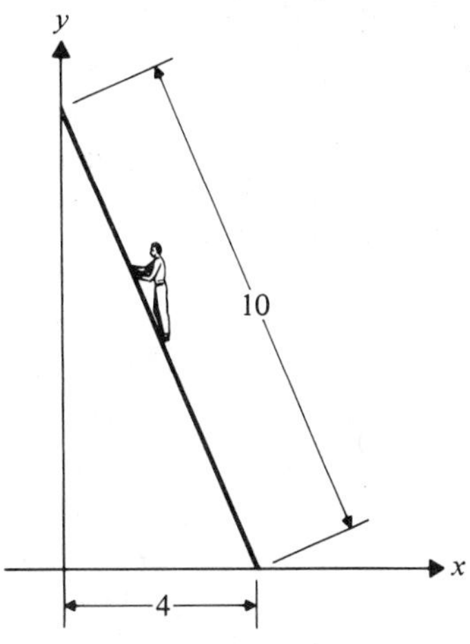

Figure 1.21

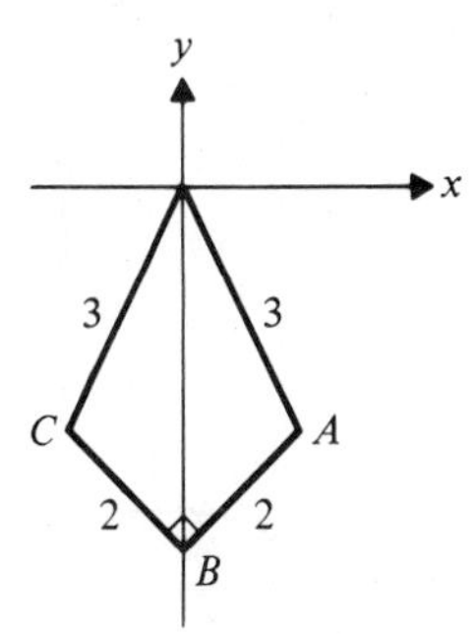

Figure 1.22

10. Lines that intersect at an angle other than 90° (Fig. 1.23) can also be used to coordinatize the plane. If l_1 and l_2 are coordinatized as shown, and P is any point in the plane, we draw through P lines parallel to l_2 and l_1. The lengths x and y cut off on l_1 and l_2 are then called the coordinates (x, y) of P.

a) Show the plane coordinatized with equal unit distances of 1 centimeter along two lines that intersect at 30°.

b) On your diagram mark points $(1, 3)$, $(-2, 3)$, $(4, -5)$, $(-1, -2)$, $(\frac{1}{2}, -\frac{1}{2})$, $(-2, \frac{1}{2})$.

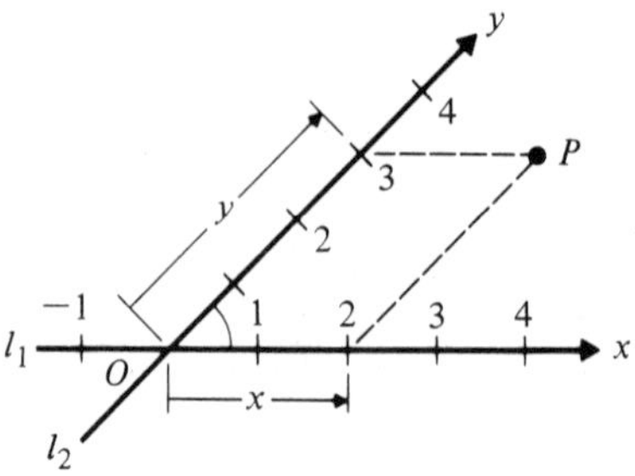

Figure 1.23

11. If the radius of the circle in Fig. 1.24 is 10 centimeters, find the length of line segment AB.

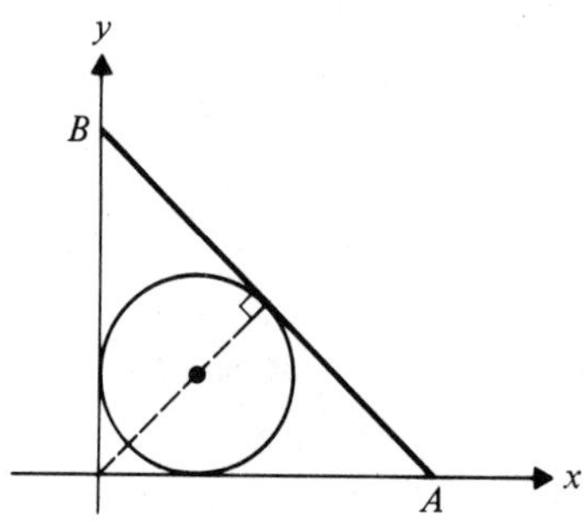

Figure 1.24

12. If the minute and hour hands of a clock are, respectively, 10 centimeters and 7.5 centimeters long (Fig. 1.25), find coordinates for the tips of the hands when the time is:

a) 9:00;

b) 4:00;

c) 7:30;

d) 1:40.

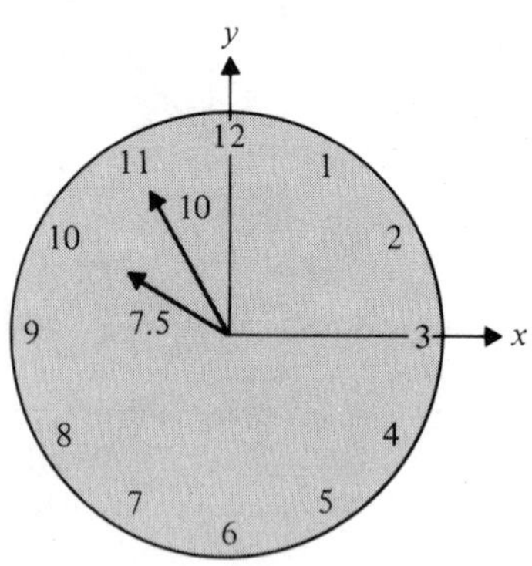

Figure 1.25

13. If a circle passes through three points (0, 0), (4, 0), and (0, 3), what are the coordinates of its center?

14. Plot points $P(2, 1)$, and $Q(5, 5)$ on a Cartesian plane coordinatized with unit distances of 1 centimeter and 0.5 centimeter along the x- and y-axes, respectively. According to distance formula (1.5),

$$|PQ| = \sqrt{(5-2)^2 + (5-1)^2} = 5.$$

Measure $|PQ|$ with a ruler. Is it 5 centimeters? Explain.

*** 15.** A man 2 meters tall walks along the edge of a straight road 10 meters wide (Fig. 1.26). On the other edge of the road stands a streetlight 8 meters high. If rectangular coordinates (with unit distances equal to 1 meter) are set up as shown, find the coordinates for the tip of the man's shadow when he is 10 meters from the point directly opposite the light.

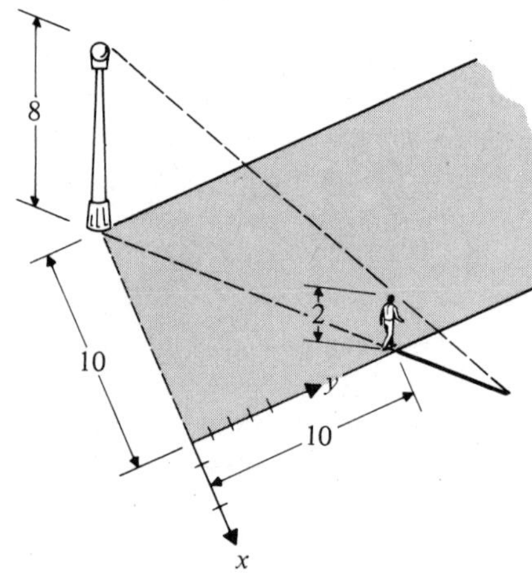

Figure 1.26

*** 16.** Generalize the result of Eq. (1.6) to prove that if a point R divides the length PQ so that

$$\frac{\text{length } PR}{\text{length } RQ} = \frac{r_1}{r_2},$$

where r_1 and r_2 are positive integers, then the coordinates of R are

$$x = \frac{r_1 x_2 + r_2 x_1}{r_1 + r_2}, \qquad y = \frac{r_1 y_2 + r_2 y_1}{r_1 + r_2}.$$

*** 17.** Prove that in any triangle the sum of the squares of the lengths of the medians is equal to three-quarters of the sum of the squares of the lengths of the sides. (A median of a triangle is a line segment drawn from one vertex to the midpoint of the opposite side.)

1.3 EQUATIONS AND CURVES

Analytic geometry is a union of algebra and geometry. On the one hand, it provides a way to describe geometric objects algebraically; on the other hand, it permits a geometric visualization of algebraic statements. In Section 1.2 we began the study of plane analytic geometry by using ordered pairs of real numbers (x, y) to describe the positions of points in a plane, an algebraic description of a geometric object. This led to Eq. (1.5), an algebraic formula that defines the length of the line segment joining two points.

In this section we bring out the other aspect of analytic geometry: that algebraic statements can be visualized geometrically. Consider, for example, the equation $y = x^2$. Algebraically, we speak of pairs of values x and y that satisfy this equation. To illustrate this we have constructed Table 1.1, called a

Table 1.1

x	-6	-5	-4	-3	-2	-1	0	1	2	3	4	5	6
y	36	25	16	9	4	1	0	1	4	9	16	25	36

table of values. Each pair of entries in this table represents a pair of values x and y that satisfy the equation $y = x^2$. If we write these pairs in the form (x, y), and interpret each as the coordinates of a point, we obtain the thirteen points shown in Fig. 1.27(a). (Note that because the values of y range from 0 to 36 and those for x range only from -6 to 6, we have used different scales along the x- and y-axes.) We could add many more points to this illustration by choosing noninteger values of x between -6 and 6, as well as values of x beyond -6 and 6. We would find that all such points lie on the curve shown in Fig. 1.27(b). This curve then is a geometric visualization of solutions of the equation $y = x^2$. Every pair of values x and y that satisfies $y = x^2$ is represented by a point on the curve. Conversely, the coordinates (x, y) of every point on the curve provide a pair of numbers that satisfies $y = x^2$. Algebraic solutions then are represented geometrically as points on a curve; points on the curve provide algebraic solutions. This algebraic–geometric interplay of analytic geometry is vital to our development of calculus; it allows us, on the one hand, to visualize algebraic problems geometrically and, on the other hand, to formulate geometric problems algebraically.

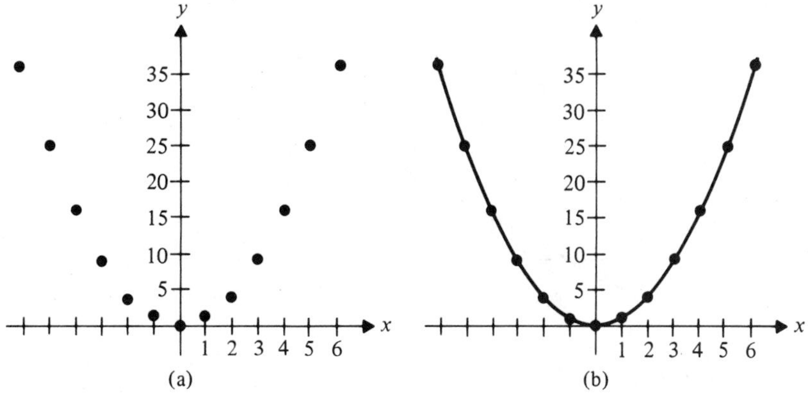

Figure 1.27

If (x, y) are the coordinates of a point in the plane, then the point will be on the curve in Fig. 1.27(b) if and only if x and y satisfy the equation $y = x^2$. In other words, this equation completely characterizes all points on the curve. We therefore call $y = x^2$ the *equation of the curve*. The equation of a curve then is an equation that the coördinates of every point on the curve satisfy, and at the same time is an equation that no point off the curve satisfies. In the remainder of this section we discuss three classes of curves—straight lines, parabolas, and circles—and show that each can be described by a particular type of equation. Once again note the algebraic–geometric interplay; the shape of a curve determines the form of its equation and, conversely, the form of an equation dictates the shape of the curve that it represents.

The Straight Line

The equation of a curve is an equation that completely characterizes all points on the curve. It is a condition that the coordinates of every point on the curve satisfy and, at the same time a condition that no point off the curve satisfies. In particular, every point on the vertical line in Fig. 1.28 has an x-coordinate equal to 2; the equation of the line is therefore $x = 2$. Similarly, the equation of the horizontal line in the same diagram is $y = -\frac{5}{2}$.

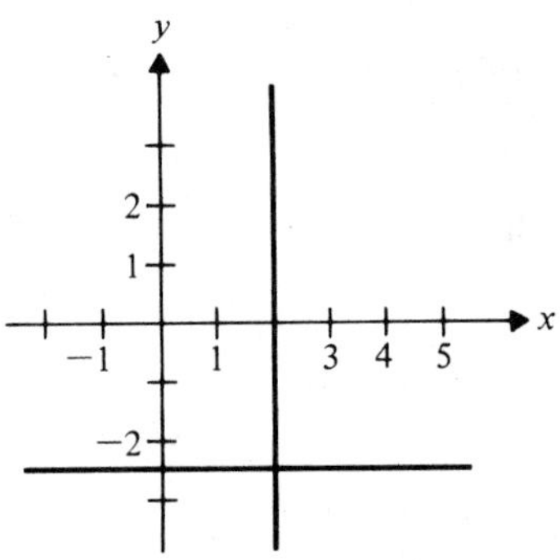

Figure 1.28

To find the equation of a straight line that is not parallel to either axis, we introduce the idea of "slope" (Fig. 1.29a).

▷ **Definition 1.4** The *slope* of a line through points (x_1, y_1) and (x_2, y_2) is

$$m = \frac{y_2 - y_1}{x_2 - x_1}. \quad \triangleleft \tag{1.7}$$

It is customary to call the difference $y_2 - y_1$ the *rise* (since it represents the vertical distance between the points), and $x_2 - x_1$ the *run* (since it is the horizontal distance between the points). It is straightforward to show using similar triangles that m is independent of the two points chosen on the line; that is, no matter what two points we choose on the line to evaluate m, the result will always be the same. The four numbers x_1, x_2, y_1, and y_2 will vary, but the ratio $(y_2 - y_1)/(x_2 - x_1)$ will remain unchanged. Note in Fig. 1.29(b) that a horizontal line has slope zero (since $y_2 - y_1 = 0$), whereas a vertical line has no slope (since $x_2 - x_1 = 0$). In Fig. 1.29(c), line l_1, which leans to the

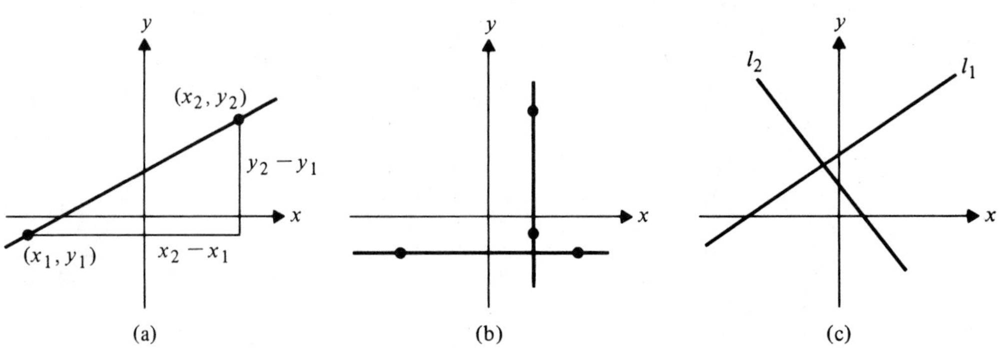

Figure 1.29

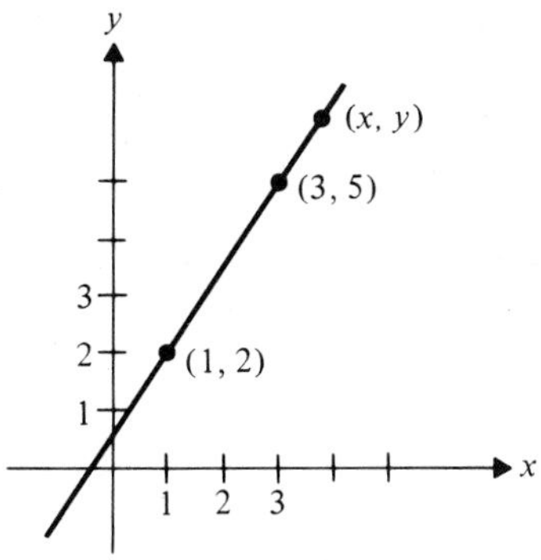

Figure 1.30

right, has a positive slope, and line l_2, which leans to the left, has a negative slope.

In Fig. 1.30 we have shown the line through the points (1, 2) and (3, 5). Its slope is $(5-2)/(3-1) = \frac{3}{2}$. To find the equation for this line, we let (x, y) be the coordinates of any other point on the line. Since the slope of the line must also be given by $(y-2)/(x-1)$, it follows that

$$\frac{y-2}{x-1} = \frac{3}{2},$$

or

$$2y = 3x + 1.$$

If (x, y) are the coordinates of any point not on this line, then they will not satisfy this equation because the slope $(y-2)/(x-1)$ joining (x, y) to (1, 2) will not be equal to $\frac{3}{2}$. Thus $2y = 3x + 1$ is the equation for the straight line through (1, 2) and (3, 5).

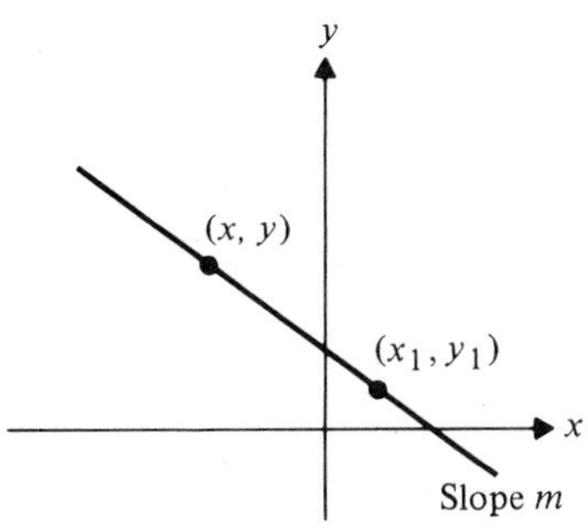

Figure 1.31

We can use this procedure to find the equation for the straight line through a point (x_1, y_1) with slope m (Fig. 1.31). If (x, y) are the coordinates of any other point on the line, then the slope of the line is $(y-y_1)/(x-x_1)$; therefore,

$$\frac{y-y_1}{x-x_1} = m,$$

or

$$y - y_1 = m(x - x_1). \tag{1.8a}$$

This is called the *point-slope* formula for the equation of a straight line; it uses the slope m of the line and a point (x_1, y_1) on the line to determine the equation of the line.

When the coordinates (x_1, y_1) and (x_2, y_2) of two points on the line are known, then the slope of the line can be determined using Definition 1.4.

Example 4 Find the equation of the line through the points $(-1, 1)$ and (2, 3).

Solution Since the slope of the line is $(3-1)/(2+1) = \frac{2}{3}$, we can use the point $(-1, 1)$ and the slope $\frac{2}{3}$ in the point-slope formula to give us

$$y - 1 = \tfrac{2}{3}(x + 1)$$

or

$$3y = 2x + 5.$$

We can verify that the same result is obtained if (2, 3) is used instead of $(-1, 1)$. □

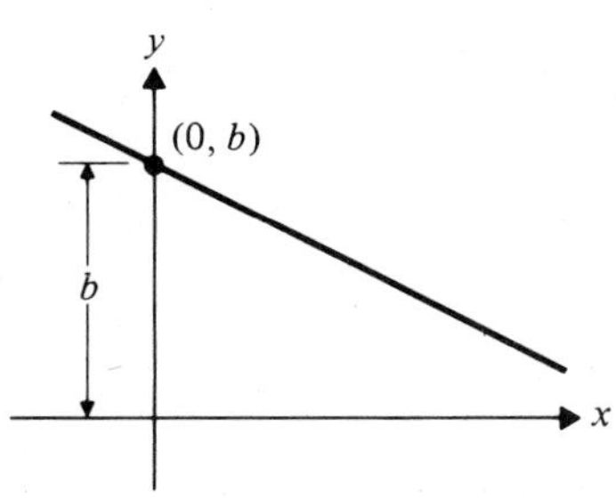

Figure 1.32

When a line intersects the y-axis in a point with coordinates $(0, b)$, the number b is called the *y-intercept* of the line (Fig. 1.32). If the slope of the line is m, then its equation takes the form

$$y - b = m(x - 0)$$

or

$$y = mx + b. \tag{1.8b}$$

This is called the *slope-intercept* formula for the equation of a straight line. Other forms for the equation of a straight line are introduced in the exercises following this section.

Example 5 Find the equation for each of the straight lines in Fig. 1.33.

Solution Since the slope of l_1 is $-\frac{5}{4}$ and its y-intercept is 5, we use the slope-intercept formula (1.8b):

$$y = -\tfrac{5}{4}x + 5$$

or

$$4y = 20 - 5x.$$

The equation of l_2 is clearly $x = -1$. Note that both the slope-intercept and the point-slope formulas fail to give the equation of l_2. □

To find the point of intersection of two straight lines—say, $3y = 2x + 5$ and $4y = 20 - 5x$ (Fig. 1.34)—we must find that point whose coordinates (x, y) satisfy both equations. If we solve each equation for y and equate the expressions, we obtain $(2x + 5)/3 = (20 - 5x)/4$, which immediately yields $x = \frac{40}{23}$. Either of the original equations then gives $y = \frac{65}{23}$, and the point of intersection of the lines is $(\frac{40}{23}, \frac{65}{23})$. Note once again the algebraic–geometric interplay. Geometrically, $(\frac{40}{23}, \frac{65}{23})$ is the point of intersection of two straight lines. Algebraically, it is the solution of the equations $3y = 2x + 5$ and $4y = 20 - 5x$.

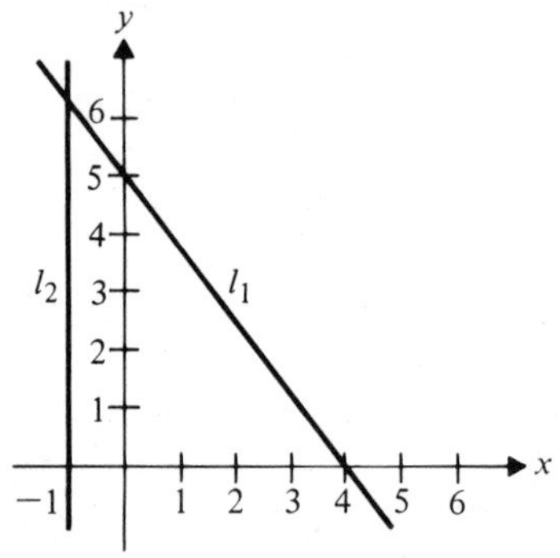

Figure 1.33

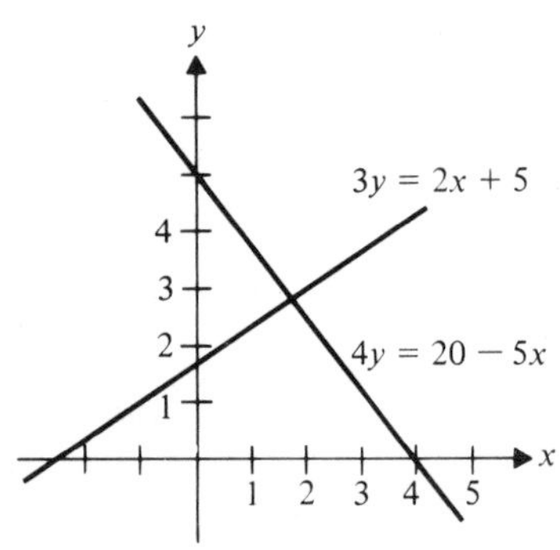

Figure 1.34

Of particular importance to the study of lines are the concepts of parallelism and perpendicularity.

▷ **Definition 1.5** Two lines are said to be *parallel* if they have no point of intersection. ◁

Example 6 Verify that lines with the equations $2x - y = 4$ and $4x - 2y = 7$ are parallel.

Solution If we solve each of these equations for y and equate the resulting expressions, we get

$$2x - 4 = 2x - \tfrac{7}{2},$$

an obvious impossibility. Consequently, the lines do not intersect. □

Geometrically, the following is clear.

▶ **Theorem 1.1** Two lines are parallel if and only if they have the same slope. ◀

For example, if we write the equations of the lines in Example 6 in slope-intercept form (1.8b), we have

$$y = 2x - 4 \quad \text{and} \quad y = 2x - \tfrac{7}{2}.$$

Since the coefficient of the x-term, 2, identifies the slope of each line, then each has slope 2, and they are therefore parallel.

▷ **Definition 1.6** Two lines are said to be *perpendicular* if they intersect at right angles. ◁

The following theorem gives a test for perpendicularity of straight lines in terms of slopes.

▶ **Theorem 1.2** Two lines with nonzero slopes m_1 and m_2 are perpendicular if and only if

$$m_1 = -\frac{1}{m_2} \quad \text{or} \quad m_1 m_2 = -1. \quad ◀ \tag{1.9}$$

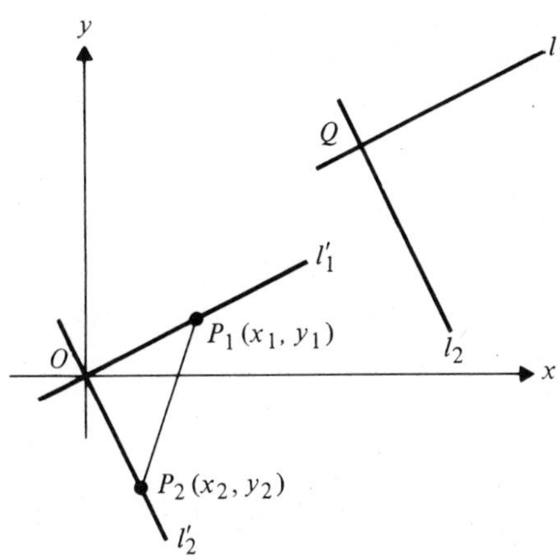

Figure 1.35

Proof Suppose nonparallel lines l_1 and l_2 with nonzero slopes m_1 and m_2 intersect at a point Q (Fig. 1.35). If we draw lines l'_1 and l'_2 through the origin parallel to l_1 and l_2, then we can state that l_1 and l_2 are perpendicular if and only if l'_1 and l'_2 are perpendicular. But if $P_1(x_1, y_1)$ and $P_2(x_2, y_2)$ are any two points on l'_1 and l'_2, then l'_1 and l'_2 are perpendicular if and only if triangle OP_1P_2 is right-angled at O; i.e., if and only if

$$|P_1P_2|^2 = |OP_1|^2 + |OP_2|^2$$

or

$$(x_1 - x_2)^2 + (y_1 - y_2)^2 = (x_1^2 + y_1^2) + (x_2^2 + y_2^2).$$

When the left side is expanded and the equation is simplified, we have

$$x_1x_2 + y_1y_2 = 0.$$

In other words, l_1 and l_2 are perpendicular if and only if

$$\frac{y_1}{x_1} = -\frac{x_2}{y_2}.$$

But $m_1 = y_1/x_1$ is the slope of l'_1, which is also the slope of l_1, and $m_2 = y_2/x_2$ is similarly the slope of l_2. Consequently, l_1 and l_2 are perpendicular if and only if $m_1 = -1/m_2$ or $m_1m_2 = -1$. □

Example 7 Find the equation of the straight line that passes through the point (2, 4) and is perpendicular to the line $3x + y = 5$.

Solution Since the slope of the given line $y = -3x + 5$ is -3, the required line has slope $\frac{1}{3}$. Using the point-slope formula (1.8a), we find that the equation of the required line is

$$y - 4 = \tfrac{1}{3}(x - 2)$$

or

$$x - 3y = -10. \square$$

EXERCISES 1.3A

In each of Exercises 1–10, draw the line described and find its equation.

1. Through the points (1, 2) and $(-3, 4)$

2. Through the points $(3, -6)$ and $(5, -6)$

3. Through the point $(-2, -3)$ with slope 3

4. Through the point $(1, -\frac{3}{4})$ with slope $-\frac{1}{2}$

5. The y-axis

6. The x-axis

7. Through the point (4, 3) and crossing the x-axis at -2

8. Through the point $(-1, -2)$ and crossing the y-axis at 4

9. Crossing the x- and y-axes at 1 and -3, respectively

10. Through the origin and the midpoint of the line segment joining (3, 4) and $(-7, 8)$

Verify the special forms for the equation of a straight line in each of Exercises 11–13.

11. The symmetric or two-point formula for the equation of a straight line is

$$\frac{x - x_1}{x_2 - x_1} = \frac{y - y_1}{y_2 - y_1}$$

if (x_1, y_1) and (x_2, y_2) are two points on the line and provided that $x_2 - x_1 \neq 0$ and $y_2 - y_1 \neq 0$.

12. The slope–x-intercept formula for the equation of a straight line is

$$y = m(x - a)$$

if m is the slope of the line, and it crosses the x-axis at $x = a$ (called the x-intercept).

13. The two-intercept formula for the equation of a straight line is

$$\frac{x}{a} + \frac{y}{b} = 1$$

if its x- and y-intercepts are a and b, respectively ($ab \neq 0$).

Determine whether the pair of lines in each of Exercises 14–18 is perpendicular.

14. $y = 3x + 2,\ y = -\dfrac{x}{2} + 1$

15. $y = -x + 4,\ y = x + 6$

16. $2x + 3y = 1,\ 3x - 2y = 5$

17. $x + y + 2 = 0,\ 3x - y = 4$

18. $x = 0,\ y = 5.$

Find the point of intersection of each pair of straight lines given in Exercises 19–23.

19. $x + y = 0, x - 2y = -3$

20. $x = 1, y = 2$

21. $3x + 4y = 6, x - 6y = 3$

22. $y = 2x + 6, x = y + 4$

23. $\frac{x}{2} + \frac{y}{3} = 1, 2x - \frac{y}{4} = 15$

Determine whether the lines in each of Exercises 24–27 are parallel.

24. $x - y = 5, 2x + 3y = 4$

25. $x + 3y = 4, 2x + 6y = 7$

26. $x = 6, y = 2$

27. $x = 3y + 4, y = \frac{x}{3} - 2$

28. Show that an equation represents a straight line if and only if it can be written in the form $Ax + By + C = 0$, where A and B are not both zero.

In Exercises 29–36, find the equation for the line described.

29. Parallel to $x + 2y = 15$, and through the point of intersection of $2x - y = 5$ and $x + y = 4$

30. Perpendicular to $x - y = 4$, and through the point of intersection of $2x + 3y = 3$ and $x - y = 4$

31. Parallel to the line through (1, 2) and (−3, 0), and through the point (5, 6)

32. Perpendicular to the line through (−3, 4) and (1, −2), and through the point (−3, −2)

33. Crossing through the first quadrant to form an isosceles triangle with an area of 8 square units

34. Through (3, 5) and crossing through the first quadrant to form a triangle with an area of 30 square units

35. Has slope 2, and that part in the second quadrant has length 3

36. Parallel to the x-axis, below the point of intersection of the lines $x = y$ and $x + y = 4$, and forms with these lines a triangle with an area of 9 square units

37. A median of a triangle is a line segment drawn from a vertex to the midpoint of the opposite side. Find equations for the three medians of the triangle with vertices (1, 1), (3, 5), and (0, 4). Show that all three medians intersect in a point (called the centroid of the triangle).

38. If (x_1, y_1), (x_2, y_2), (x_3, y_3), and (x_4, y_4) are vertices of any quadrilateral, show that the line segments joining the midpoints of adjacent sides form a parallelogram.

39. Find the equation of the perpendicular bisector of the line segment joining (−1, 2) and (3, −4). (The perpendicular bisector is the line that cuts the line segment in half and is perpendicular to it.)

40. Find coordinates of the point that is equidistant from the three points (1, 2), (−1, 4), and (−3, 1).

41. Prove Theorem 1.1.

***42.** Let P be any point interior to an equilateral triangle (Fig. 1.36). Show that the sum of the distances of P from the three sides is always equal to the height h of the triangle.

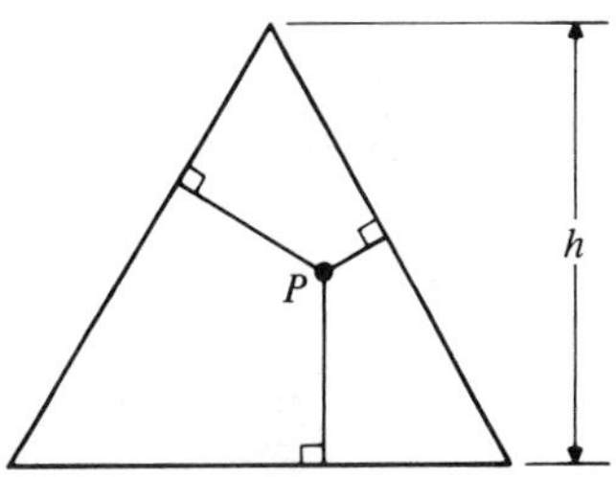

Figure 1.36

***43.** Prove that the perpendicular distance from a point (x_1, y_1) to a line $Ax + By + C = 0$ is given by

$$\frac{|Ax_1 + By_1 + C|}{\sqrt{A^2 + B^2}}.$$

The Parabola

When the y-coordinate of a point (x, y) on a curve is related to its x-coordinate by an equation of the form

$$y = ax^2 + bx + c, \tag{1.10}$$

where a, b, and c are constants ($a \neq 0$), the curve is called a *parabola*. To determine the shape of a parabola we consider a number of examples. We have already

Table 1.2

x	y
-6	42
-5	24
-4	10
-3	0
-2	-6
-1	-8
0	-6
1	0
2	10
3	24
4	42

used Table 1.1 to draw the simplest of all parabolas: $y = x^2$ (Fig. 1.27b). The values given in Table 1.2 for the parabola $y = 2x^2 + 4x - 6$ lead to the curve in Fig. 1.37. It appears from this curve that the lowest point of the parabola is at $(-1, -8)$. To prove that this is indeed the case, we rewrite the equation as

$$y = 2(x^2 + 2x) - 6 = 2(x + 1)^2 - 8.$$

This form clearly indicates that the smallest value for y is -8, and it occurs when $x = -1$.

The parabola $y = 9 - x^2$ is shown in Fig. 1.38. It is said to "open downward," whereas the parabolas shown in Figs. 1.27(b) and 1.37 "open upward."

If to the general parabola (1.10) we apply the technique used to rewrite the equation $y = 2x^2 + 4x - 6$, we obtain

$$y = a\left(x^2 + \frac{b}{a}x\right) + c = a\left(x + \frac{b}{2a}\right)^2 + \left(c - \frac{b^2}{4a}\right). \tag{1.11}$$

This is called *completing the square* of $x^2 + (b/a)x$.

Formula (1.11) for the parabola shows the following:

1. If $a > 0$, the parabola opens upward and has a minimum at the point
$$\left(-\frac{b}{2a}, c - \frac{b^2}{4a}\right).$$
2. If $a < 0$, the parabola opens downward and has a maximum at the point
$$\left(-\frac{b}{2a}, c - \frac{b^2}{4a}\right).$$
3. The parabola is symmetric about the line $x = -b/(2a)$ (Fig. 1.39); that is, for every point P on one side of the line $x = -b/(2a)$, there is a point Q on the other side that is the mirror image of P in the line.

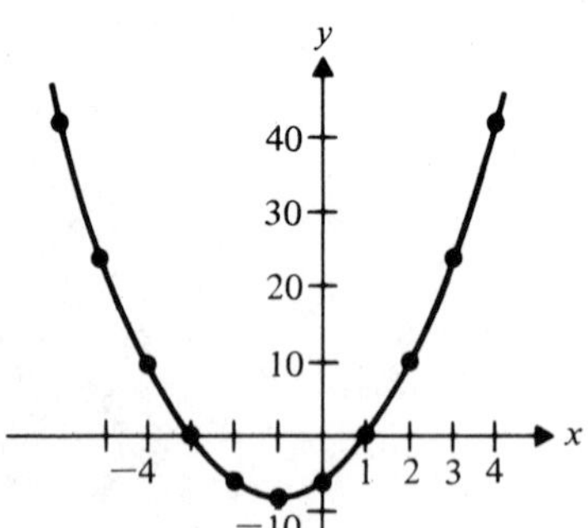

Figure 1.37

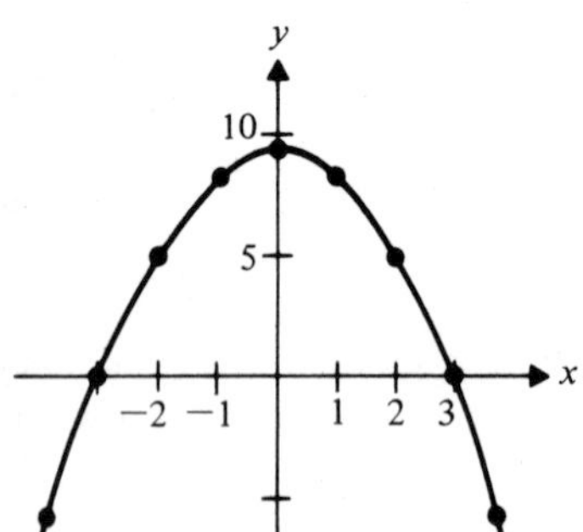

Figure 1.38

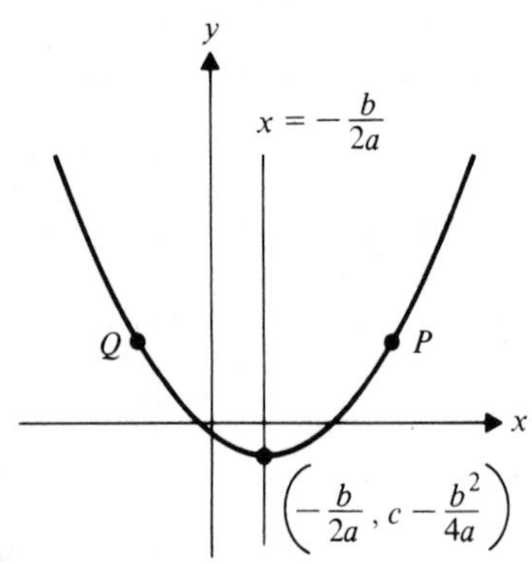

Figure 1.39

4. The parabola crosses the x-axis when
$$0 = y = a\left(x + \frac{b}{2a}\right)^2 + \left(c - \frac{b^2}{4a}\right).$$

To solve this equation for x, we first write

$$\left(x + \frac{b}{2a}\right)^2 = \frac{b^2}{4a^2} - \frac{c}{a},$$

and then take the square roots of each side:

$$x + \frac{b}{2a} = \pm\sqrt{\frac{b^2}{4a^2} - \frac{c}{a}}.$$

Finally, then, we get

$$x = -\frac{b}{2a} \pm \sqrt{\frac{b^2}{4a^2} - \frac{c}{a}} = \frac{-b \pm \sqrt{b^2 - 4ac}}{2a}. \tag{1.12}$$

This is called the *quadratic formula*. It determines the points where the parabola $y = ax^2 + bx + c$ crosses the x-axis. Equivalently, it defines roots of the quadratic equation $ax^2 + bx + c = 0$. Note that when $b^2 - 4ac > 0$, there are two solutions (Fig. 1.40a); when $b^2 - 4ac = 0$, there is only one solution (Fig. 1.40b); and when $b^2 - 4ac < 0$, there is no solution (Fig. 1.40c), i.e., no real solution.

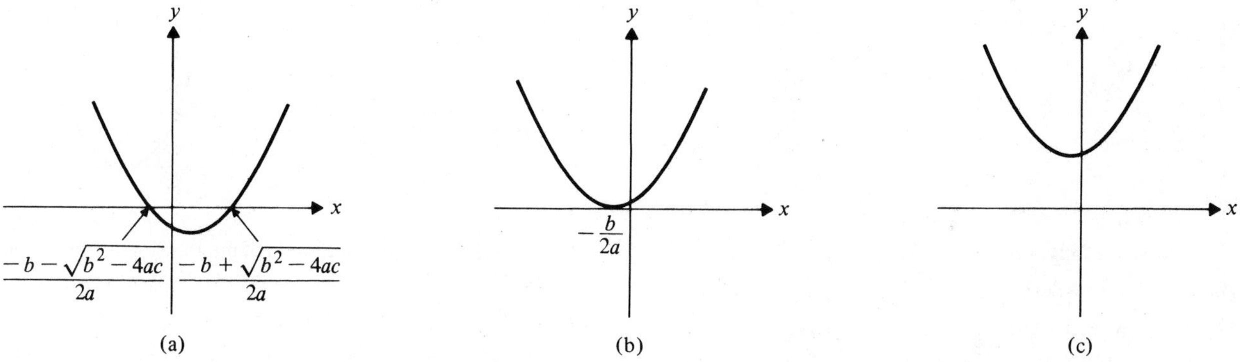

Figure 1.40

When x and y in Eq. (1.10) are interchanged, the resulting equation

$$x = ay^2 + by + c \tag{1.13}$$

is still said to define a parabola. In this case we construct a table of values by listing various values of y and calculating the corresponding values of x. For example, Table 1.3 and Fig. 1.41 and Table 1.4 and Fig. 1.42 show the parabolas $x = y^2 + 1$ and $x = -y^2 + 4y - 4$, respectively.

Table 1.3

y	−4	−3	−2	−1	0	1	2	3	4
x	17	10	5	2	1	2	5	10	17

Table 1.4

y	−1	0	1	2	3	4	5
x	−9	−4	−1	0	−1	−4	−9

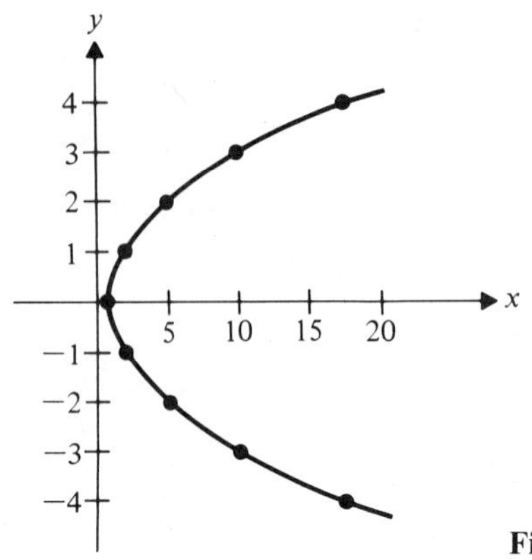

Figure 1.41

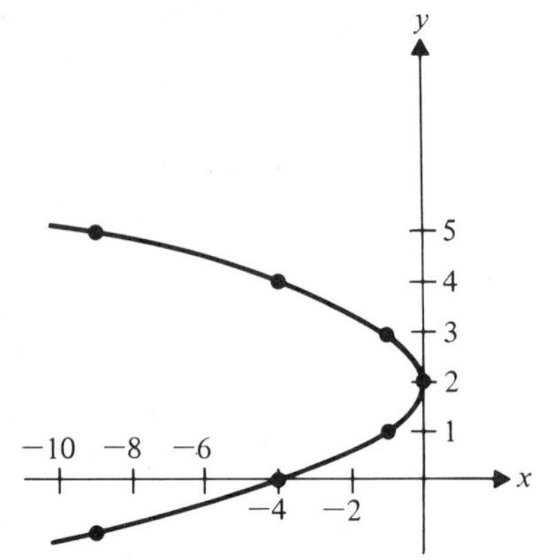

Figure 1.42

EXERCISES 1.3B

In Exercises 1–12, draw the given parabola.

1. $y = 2x^2 - 1$

2. $y = -x^2 + 4x - 3$

3. $y = x^2 - 2x + 1$

4. $3x = 4y^2 - 1$

5. $x = y^2 + 2y$

6. $2y = -x^2 + 3x + 4$

7. $x + y^2 = 1$

8. $2y^2 + x = 3y + 5$

9. $y = 4x^2 + 5x + 10$

10. $x = 10y^2$

11. $y = -x^2 + 6x - 9$

12. $x = -(4 + y)^2$

13. Find x- and y-intercepts for each of the following parabolas.

a) $y = x^2 - 2x - 5$

b) $x = 4y^2 - 8y + 4$

In Exercises 14–17, find the equation of the parabola shown.

14.

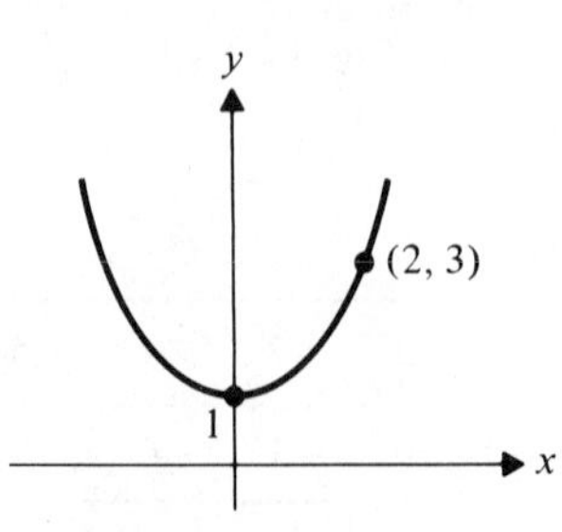

15.

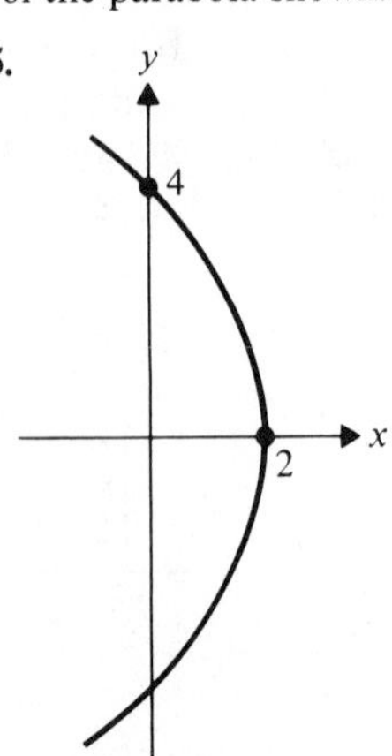

16.

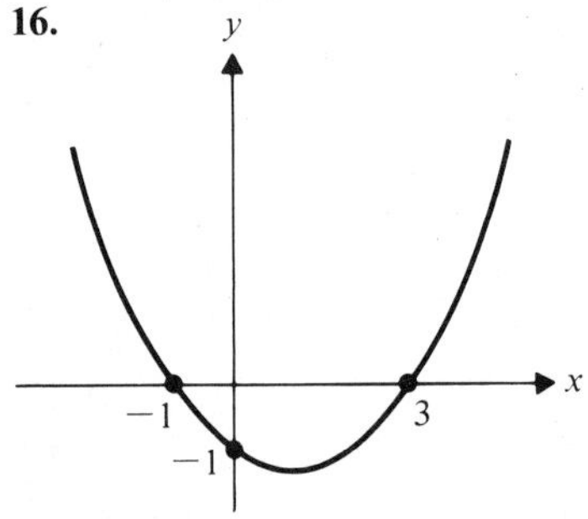

17.

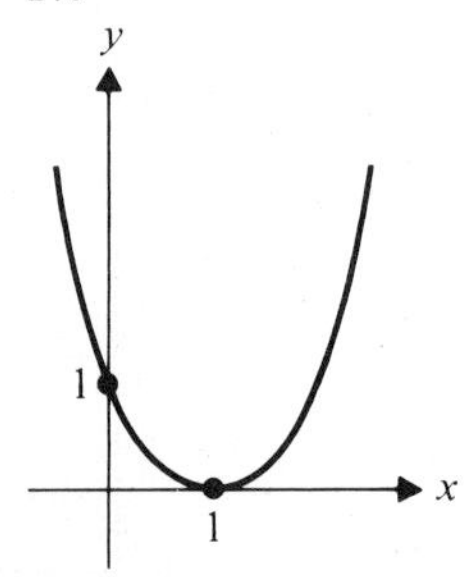

In Exercises 18–23, find all points of intersection for the given curves.

18. $y = 1 - x^2, y = x + 1$

19. $y + 2x = 0, y = 1 + x^2$

20. $y = 2x - x^2 - 6, 25 + x = 5y$

21. $x = y(y - 1), 2y = 2x + 1$

22. $x = -y^2 + 1, x = y^2 + 2y - 3$

23. $y = 6x^2 - 2, y = x^2 + x + 1$

24. When a shell is fired from an artillery gun (Fig. 1.43), it follows a parabolic path

$$y = -\frac{4.905}{v^2 \cos^2 \theta} x^2 + x \tan \theta,$$

where v is the muzzle velocity of the shell.

a) Find the range R of the shell.

b) What is the maximum height attained by the shell?

25. The cable of the suspension bridge shown in Fig. 1.44 hangs in the shape of a parabola. The towers are 200 meters apart and extend 50 meters above the roadway.

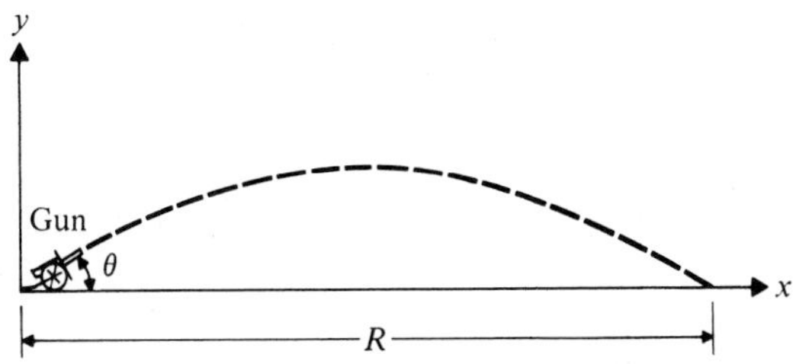

Figure 1.43

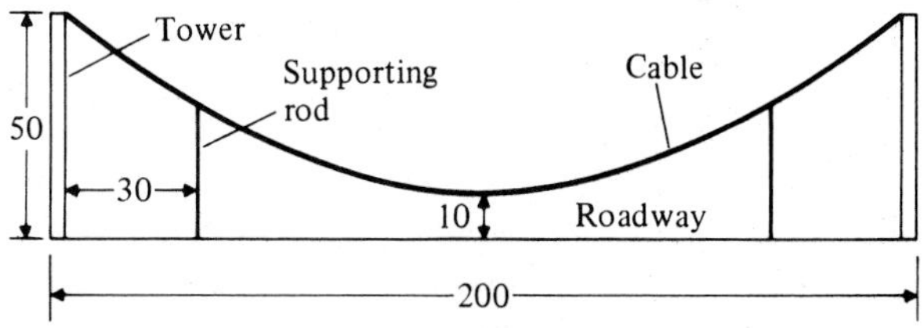

Figure 1.44

If the cable is 10 meters above the roadway at its lowest point, find the length of the supporting rods 30 meters from the towers.

26. Find points of intersection for the parabolas

$$y = (x - 2)^2 \quad \text{and} \quad 5x = y^2 + 4.$$

27. Find the height of the parabolic arch shown in Fig. 1.45.

28. Find a parabola of the type $y = ax^2 + bx + c$ passing through the points (1, 2), (−3, 10), and (3, 4).

29. Find lines through the point (3, 4) that intersect the parabola $y = x^2 - 1$ at only one point.

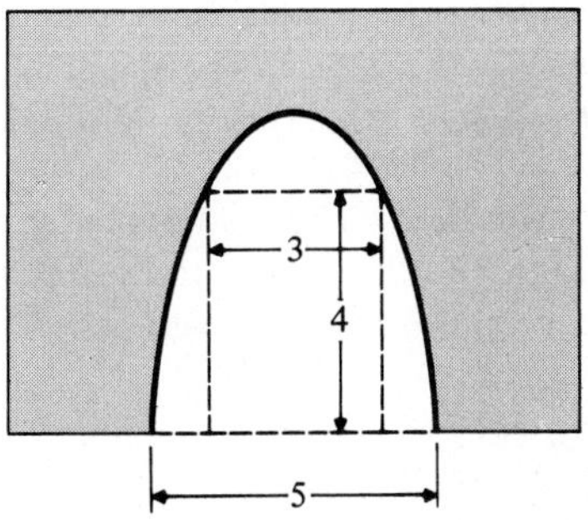

Figure 1.45

The Circle

In Section 1.2 we developed formula (1.5) for the length of the line segment joining two points. We can use this result to derive the equation of a circle. A *circle* is the set of all points that are at a constant distance (called the *radius*) from a fixed point (called the *center*). If r is the radius of the circle and (h, k) is its center (Fig. 1.46), then any point (x, y) on the circle must satisfy the equation

$$r = \sqrt{(x - h)^2 + (y - k)^2}.$$

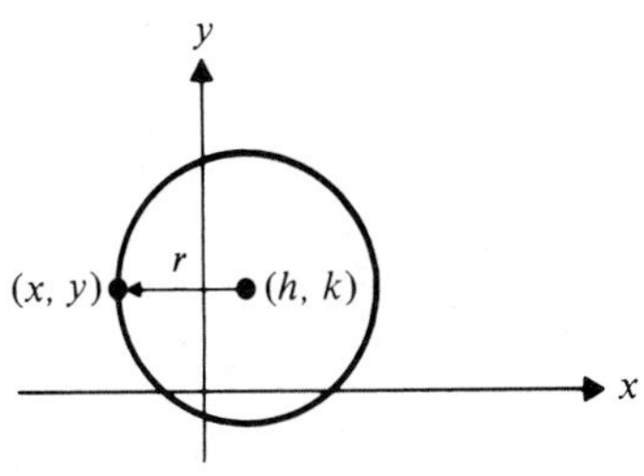

Figure 1.46

The equation of the circle with center (h, k) and radius r is therefore

$$(x - h)^2 + (y - k)^2 = r^2. \qquad (1.14)$$

For example, the radius of the circle shown in Fig. 1.47 is equal to 2; therefore its equation is

$$(x + 1)^2 + (y - 2)^2 = 4.$$

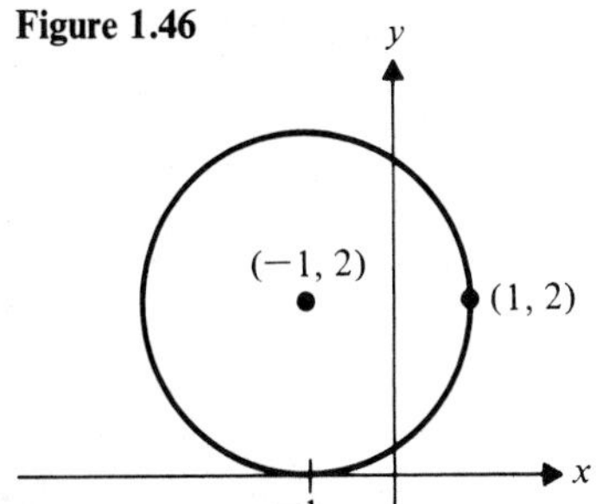

Figure 1.47

When Eq. (1.14) is expanded, we have

$$x^2 - 2hx + h^2 + y^2 - 2ky + k^2 = r^2$$

or

$$x^2 + y^2 - 2hx - 2ky = r^2 - h^2 - k^2.$$

What this shows is that the equation of a circle may be given in another form, namely,

$$x^2 + y^2 + fx + gy + e = 0. \tag{1.15}$$

In this case, the center and radius can be identified by reversing this procedure and completing the squares of $x^2 + fx$ and $y^2 + gy$. For instance, if $x^2 + y^2 + 2x - 3y - 5 = 0$, then

$$0 = (x + 1)^2 + (y - \tfrac{3}{2})^2 - 5 - 1 - \tfrac{9}{4} = (x + 1)^2 + (y - \tfrac{3}{2})^2 - \tfrac{33}{4}.$$

The center of the circle is therefore $(-1, \frac{3}{2})$ and its radius is $\sqrt{33}/2$.

When the center of a circle is the origin (0, 0), Eq. (1.14) simplifies to

$$x^2 + y^2 = r^2. \tag{1.16}$$

Example 8 Find the equation for each of the circles shown in Figs. 1.48 and 1.49.

Solution The center of the circle in Fig. 1.48 is the midpoint of the diameter joining $(-1, 1)$ and $(5, 7)$. According to Eq. (1.6), the coordinates for the center must be the averages of $(-1, 1)$ and $(5, 7)$; i.e.,

$$(h, k) = \left(\frac{5 - 1}{2}, \frac{7 + 1}{2}\right) = (2, 4).$$

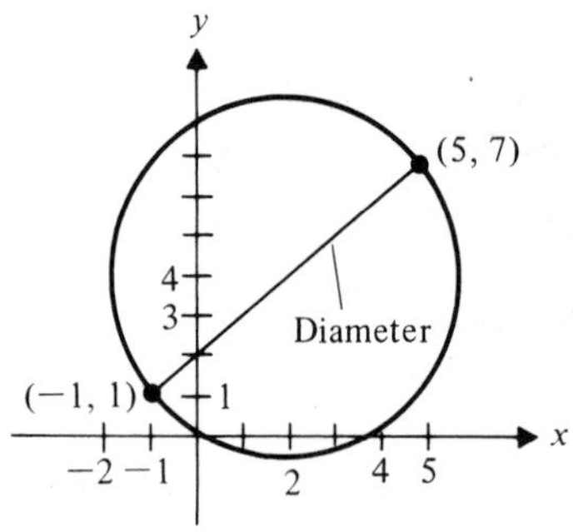

Figure 1.48

Since the radius of the circle is one-half the length of the diameter, then

$$r = \tfrac{1}{2}\sqrt{(5 + 1)^2 + (7 - 1)^2} = 3\sqrt{2},$$

and the equation of the circle is

$$(x - 2)^2 + (y - 4)^2 = 18.$$

Since the center of the circle in Fig. 1.49 must be on the y-axis, its equation must be of the form

$$x^2 + (y - k)^2 = r^2.$$

Because $(4, 0)$ is a point on the circle, these coordinates must satisfy its equation; i.e.,

$$16 + (0 - k)^2 = r^2 \quad \text{or} \quad 16 + k^2 = r^2.$$

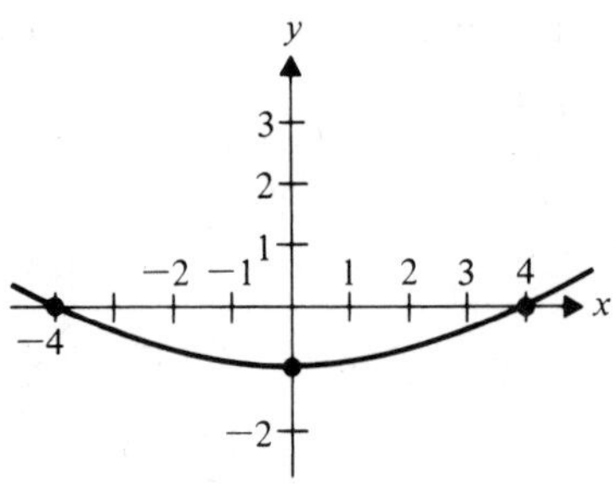

Figure 1.49

Similarly, $(0, -1)$ is on the circle; hence

$$(-1 - k)^2 = r^2 \quad \text{or} \quad 1 + 2k + k^2 = r^2.$$

If we subtract these two equations, we get

$$2k - 15 = 0,$$

from which we see that $k = \frac{15}{2}$. Consequently, $r^2 = 16 + k^2 = 16 + \frac{225}{4} = \frac{289}{4}$, and the equation of the circle is

$$x^2 + (y - \tfrac{15}{2})^2 = \tfrac{289}{4}. \ \square$$

Example 9 If P and Q are the ends of a diameter of a circle and R is any other point on the circle, prove that RP and RQ form a right angle (Fig. 1.50a).

Solution Since no coordinate system was mentioned in the problem, we are at liberty to pick our own. We choose a system that will simplify the calculations as much as possible. In particular, we choose the origin as the center of the circle and place the x-axis along PQ (Fig. 1.50b). If the radius of the circle is represented by r, then the coordinates of P and Q are $(-r, 0)$ and $(r, 0)$, respectively. If we denote the coordinates of R by (x, y), then x and y must satisfy the equation of the circle

$$x^2 + y^2 = r^2.$$

Now the angle at R is a right angle if and only if triangle PQR satisfies the Pythagorean relation; i.e., if and only if

$$|PQ|^2 = |PR|^2 + |RQ|^2$$

or

$$4r^2 = [(x + r)^2 + y^2] + [(x - r)^2 + y^2] = 2x^2 + 2y^2 + 2r^2.$$

But this equation is satisfied because x and y satisfy $x^2 + y^2 = r^2$. An alternative verification is to note that the slope of PR is $y/(x + r)$, whereas that of QR is $y/(x - r)$. Since the product of these slopes is

$$\left(\frac{y}{x + r}\right)\left(\frac{y}{x - r}\right) = \frac{y^2}{x^2 - r^2} = \frac{y^2}{-y^2} = -1,$$

PR and QR are perpendicular (Theorem 1.2). Triangle PQR is therefore right-angled at R. □

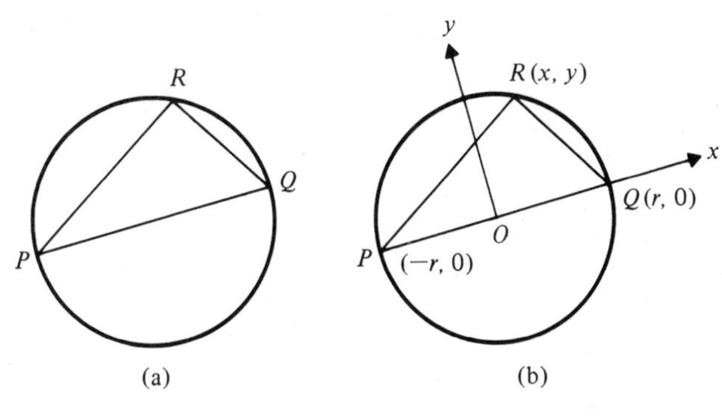

Figure 1.50

EXERCISES 1.3C

In Exercises 1–10, draw the given circle, indicating its center and radius.

1. $x^2 + y^2 = 50$
2. $(x + 5)^2 + (y - 2)^2 = 6$
3. $x^2 + 2x + y^2 = 15$
4. $x^2 + y^2 - 4y + 1 = 0$
5. $x^2 - 2x + y^2 - 2y + 1 = 0$
6. $2x^2 + 2y^2 + 6x = 25$
7. $3x^2 + 3y^2 + 4x - 2y = 6$
8. $x^2 + 4x + y^2 - 2y = 5$
9. $x^2 + y^2 - 2x - 4y + 5 = 0$
10. $x^2 + y^2 + 6x + 3y + 20 = 0$

In Exercises 11–16, find an equation for the circle shown.

11.

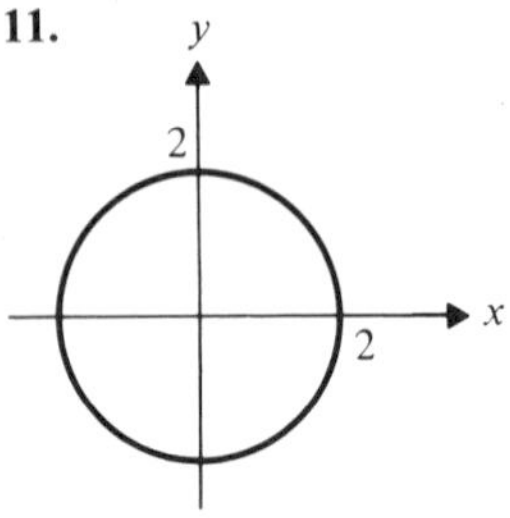

12.

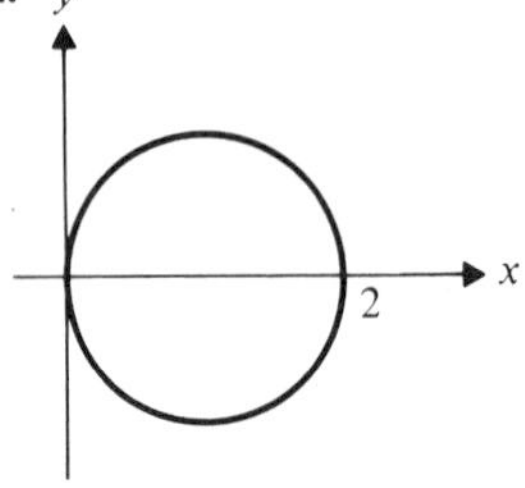

13.

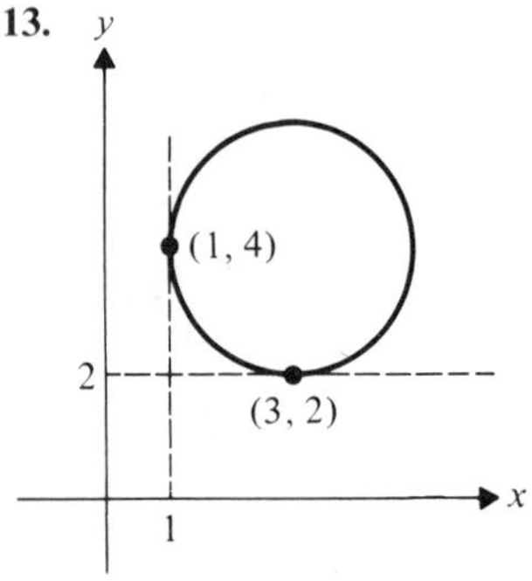

14.

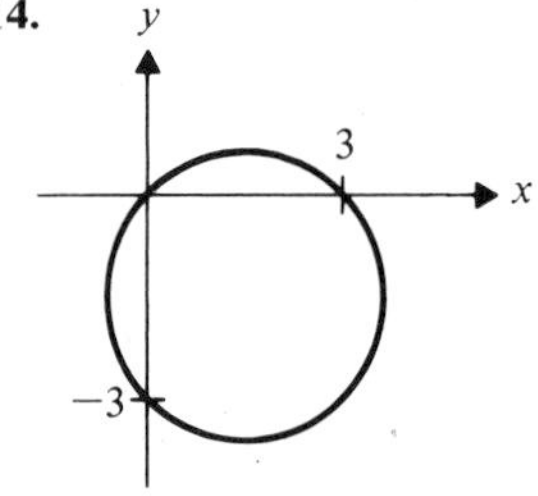

15.

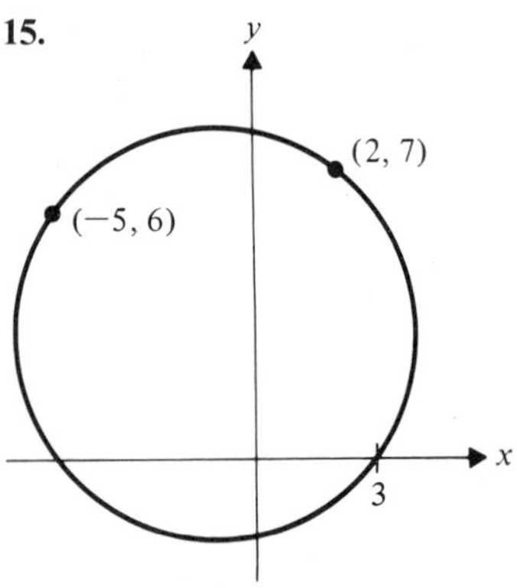

16.

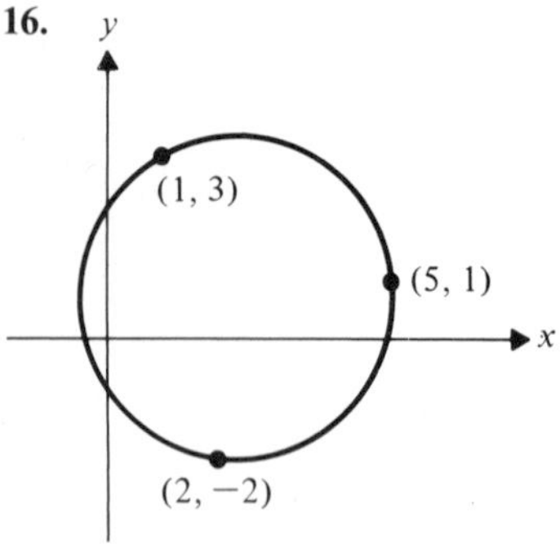

17. Find the equation of a circle that passes through the points (3, 4) and (1, −10), and has its center:

a) on the line $2x + 3y + 16 = 0$;

b) on the line $x + 7y + 19 = 0$.

Find points of intersection for the curves in Exercises 18–21.

18. $x^2 + 2x + y^2 = 4$, $y = 3x + 2$

19. $x^2 + y^2 - 4y + 1 = 0$, $2x + y = 1$

20. $x^2 + y^2 = 9$, $y = 3x^2 + 4$

21. $(x + 3)^2 + y^2 = 25$, $y^2 = 16(x + 1)$

22. Show that every equation of the form $x^2 + y^2 + fx + gy + e = 0$ represents either a circle, a point, or nothing at all.

23. Prove that the perpendicular bisector of a chord of a circle always passes through the center of the circle.

24. The *circumcircle* for a triangle is that circle which passes through all three of its vertices. Find the circumcircle for the triangle with vertices $A(1, 1)$, $B(-3, 3)$, and $C(2, 4)$ by:

a) finding its center and radius (see Exercise 23);

b) taking the equation of the circle in the form (1.15) and requiring A, B, and C to be on the circle.

25. Prove that the three altitudes of the triangle in Exercise 24 intersect in a point (called the *orthocenter* of the triangle).

***26.** Show that if a line $Ax + By + C = 0$ and a circle $(x - h)^2 + (y - k)^2 = r^2$ do not intersect, then the shortest distance between them is the smaller of the two numbers:

$$\frac{|(Ah + Bk + C) \pm r\sqrt{A^2 + B^2}|}{\sqrt{A^2 + B^2}}.$$

(*Hint:* Use the result of Exercise 43 in Section 1.3A.)

***27.** The *incircle* of a triangle is that circle which lies interior to the triangle but touches all three sides. The center of the incircle is called the *incenter*. Use the result of Exercise 43 in Section 1.3A to show that the incenter (x, y) of the triangle with vertices (0, 0), (2, 0), and (0, 1) must satisfy the equations

$$|x| = |y| = \frac{|x + 2y - 2|}{\sqrt{5}}.$$

Solve these equations for the incenter, and explain why there are four points that satisfy these equations.

1.4 FUNCTIONS AND THEIR GRAPHS

If one quantity depends on a second quantity, we say that the first quantity is a function of the second. For example, the volume V of a sphere depends on its radius r; in particular, $V = 4\pi r^3/3$. We say then that V is a function of r. If an object is dropped at time $t = 0$, the distance d that it falls is given by the formula $d = 4.905t^2$. We say that d is a function of t. Mathematically, we have the following definition.

▷ **Definition 1.7** A quantity y is said to be a function of a quantity x if there exists a rule by which we can associate exactly one value of y with each value of x. The rule that associates the value of y with each value of x is called the *function.* ◁

If we denote the rule or function in this definition by the letter f, then the value that f assigns to x is denoted by $f(x)$, and we write

$$y = f(x). \tag{1.17}$$

We should read this equation as "y is the value that f assigns to x," but we would say, generally, "y equals f of x." In our first example above, we write $V = f(r) = 4\pi r^3/3$, and the function f is the operation of cubing a number and then multiplying the result by $4\pi/3$. For $d = f(t) = 4.905t^2$, the function f is the operation of squaring a number and multiplying the result by 4.905.

We call x in Eq. (1.17) the *independent variable* (since values for x are substituted into the equation), and y the *dependent variable* (since its values depend on the assigned values of x). The *domain* of a function is the set of all specified (real) values for the independent variable. It is an essential part of a function and should always be specified. Whenever the domain of a function is not mentioned, we assume that it consists of all possible values for which $f(x)$ is a real number.

As the independent variable x takes on values in the domain, a set of values of the dependent variable is obtained. This set is called the *range* of the function. For the function $V = f(r) = 4\pi r^3/3$, which represents the volume of a sphere, the largest possible domain is $r > 0$, and the corresponding range is $V > 0$. Note that mathematically the function $f(r)$ is defined for negative as well as positive values of r (and $r = 0$); it is because of our interpretation of r as the radius of a sphere that we restrict $r > 0$. The function $d = f(t) = 4.905t^2$ represents the distance fallen by an object that is dropped at time $t = 0$. If it is dropped from a height of 20 meters, then it is clear that the range of this function is $0 \le d \le 20$. The domain that gives rise to this range is $0 \le t \le \sqrt{20/4.905}$.

Example 10 If $f(x) = \sqrt{x^2 - 3}$, calculate each of the following:

a) $f(2)$;

b) $f(-4)$;

c) $f(t)$;

d) $f(t + 4)$;

e) $\dfrac{f(x)}{f(1 - x)}$;

f) $[f(x)]^2 + 3f(x)f(-x)$.

Solution a) $f(2) = \sqrt{2^2 - 3} = 1$

b) $f(-4) = \sqrt{(-4)^2 - 3} = \sqrt{13}$

c) To find $f(t)$ we substitute t for each x in $f(x)$:

$$f(t) = \sqrt{t^2 - 3}.$$

d) For $f(t + 4)$, we replace each x in $f(x)$ by $t + 4$:

$$f(t + 4) = \sqrt{(t + 4)^2 - 3} = \sqrt{t^2 + 8t + 13}.$$

e) $$\frac{f(x)}{f(1 - x)} = \frac{\sqrt{x^2 - 3}}{\sqrt{(1 - x)^2 - 3}} = \frac{\sqrt{x^2 - 3}}{\sqrt{x^2 - 2x - 2}}$$

f) $$\begin{aligned}[f(x)]^2 + 3f(x)f(-x) &= [\sqrt{x^2 - 3}]^2 + 3\sqrt{x^2 - 3}\sqrt{(-x)^2 - 3} \\ &= x^2 - 3 + 3(x^2 - 3) \\ &= 4(x^2 - 3). \ \square\end{aligned}$$

Example 11 Find the largest possible domain for the function

$$f(x) = \sqrt{\frac{x - 4}{x + 1}}.$$

Solution The schematic diagram in Fig. 1.51 shows that the quotient $(x - 4)/(x + 1)$ is positive when $x > 4$ and when $x < -1$. Since the quotient is equal to zero when $x = 4$ and is undefined when $x = -1$, the largest possible domain for $f(x)$ consists of all values of x in the intervals $x < -1$ and $x \geq 4$. $\square$

Example 12 Does the equation $x - 4 - y^2 = 0$ define y as a function of x for $x \geq 4$?

Solution For any $x > 4$, the equation has two solutions for y:

$$y = \pm\sqrt{x - 4}.$$

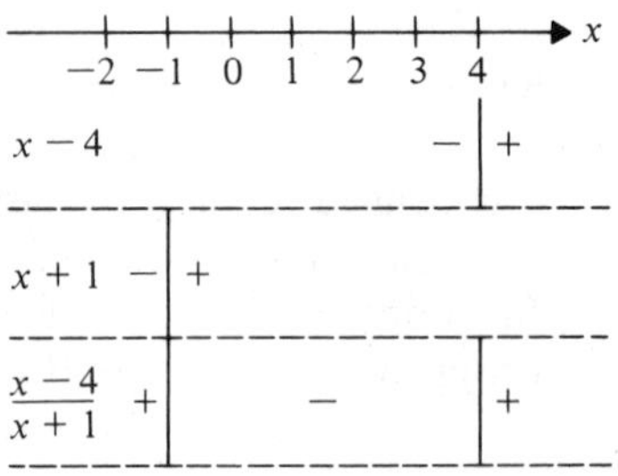

Figure 1.51

Since the equation does not therefore define a unique value of y for each value of x, it does not define y as a function of x. If we add to this equation an additional restriction such as

$$x - 4 - y^2 = 0, \qquad y \geq 0,$$

then y is now defined as a function of x, namely,

$$y = \sqrt{x - 4}.$$

Note that $x - 4 - y^2 = 0$ does, however, define x as a function of y:

$$x = y^2 + 4.$$

In this case, y is the independent variable and x the dependent variable. In other words, whenever an equation (such as $x - 4 - y^2 = 0$) is to be regarded as defining a function, it must be made clear which variable is to be considered as independent and which as dependent. $\square$

In the study of calculus and its applications we are interested in the behavior of functions; i.e., for certain values of the independent variable, what can we say about the dependent variable? The simplest and most striking method for displaying characteristics of a function is a graph. To obtain the graph of a function $f(x)$ we use a plane coordinatized with Cartesian coordinates x and y

(in short, the Cartesian xy-plane). The *graph* of the function $f(x)$ is defined to be the curve with equation $y = f(x)$, where x is limited to the domain of the function. We therefore take each value of x in the domain of the function, calculate $y = f(x)$, and plot the point (x, y) in the Cartesian plane. For example, if $f(x) = x^3$, the values in Table 1.5 lead to the curve in Fig. 1.52. This curve

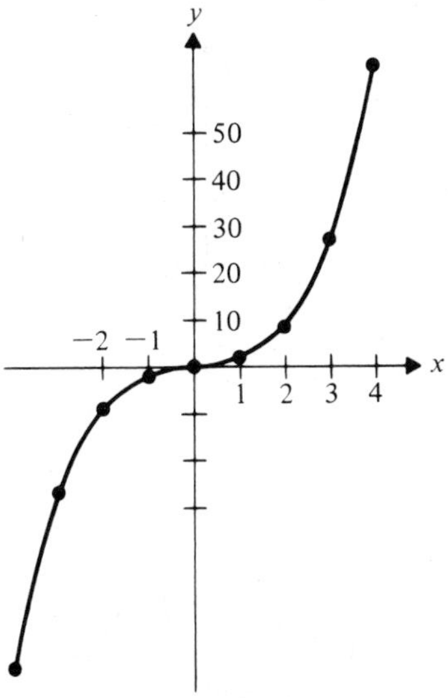

Figure 1.52

Table 1.5

x	-4	-3	-2	-1	0	1	2	3	4
y	-64	-27	-8	-1	0	1	8	27	64

is a pictorial or geometric representation of the function $f(x) = x^3$. The value of the function for given x is visually displayed as the y-coordinate of that point on the curve with x-coordinate equal to the given x. This graph clearly illustrates properties of the function that may not be quite so obvious from the algebraic definition $f(x) = x^3$—for instance:

1. $f(x)$ is positive if and only if x is positive;
2. As x increases, $f(x)$ increases;
3. For $x > 1$, $f(x)$ increases more rapidly than does x;
4. The value of $f(x)$ at $-x$ is the negative of the value of $f(x)$ at x.

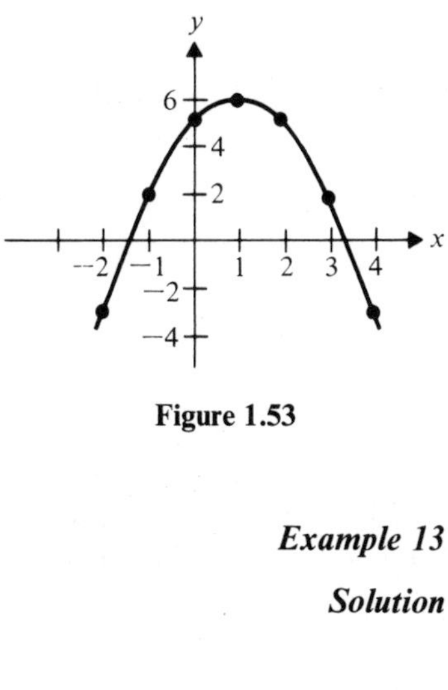

Figure 1.53

The graph of the function $f(x) = -x^2 + 2x + 5$ is the parabola $y = -x^2 + 2x + 5$ in Fig. 1.53. It indicates that $f(x)$ has a maximum value 6 when $x = 1$.

There are two distinct types of graphs for a function: a plot and a sketch. A plot is an accurate graph of the function usually obtained through an extensive table of values; this is more or less the method emphasized in Section 1.3. Although a plot of a function certainly has its advantages (and is indeed necessary in some instances), for many problems its construction is too time-consuming. Quite often we need only sketch the graph of a function. A sketch gives a general idea of the behavior of the function, i.e., a general indication of the shape of the graph. In the following examples we illustrate some useful techniques for obtaining sketches. Others will be introduced in Chapters 2, 3, and 4.

Example 13 Sketch a graph of the function $f(x) = |x^3 + 8|$.

Solution To obtain the curve $y = |x^3 + 8|$, we first sketch $y = x^3$, as in Fig. 1.54(a). Next we sketch the curve $y = x^3 + 8$ by adding 8 to every ordinate in Fig. 1.54(a). The result is shown in Fig. 1.54(b). The last step is to take the absolute value of every ordinate on the curve $y = x^3 + 8$. This changes no ordinate that is already positive, but changes the sign of any ordinate that is negative. The result and final sketch is shown in Fig. 1.54(c). □

Note that in our "building" procedure we did not attempt to sketch $y = |x^3 + 8|$ in one step. By a series of steps we began with the curve $y = x^3$, proceeded to $y = x^3 + 8$, and finally obtained $y = |x^3 + 8|$.

Example 14 Sketch a graph of the function $f(x) = 2\sqrt{x}$.

Figure 1.54

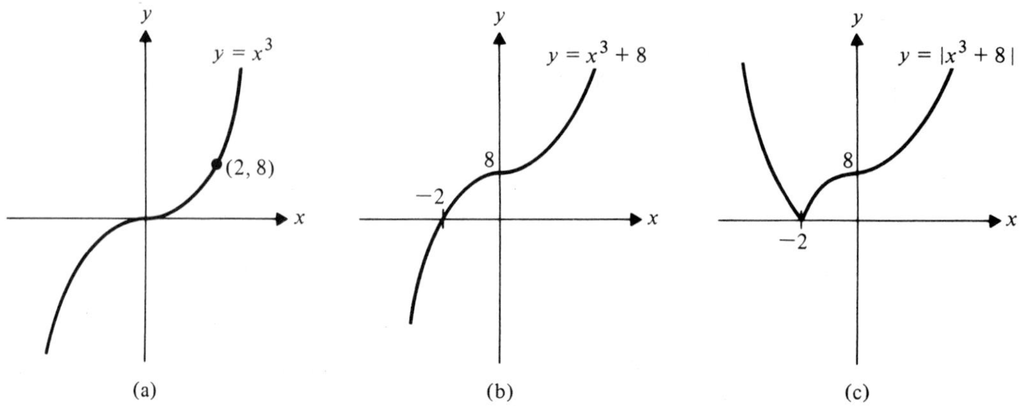

Solution If we square $y = 2\sqrt{x}$, we find that $y^2 = 4x$, which we immediately recognize as the parabola in Fig. 1.55(a). This is not the required graph, however, for this is not even the graph of a function (there being two y-values for each $x > 0$). The required graph is the upper half of the parabola, shown in Fig. 1.55(b) (because the square root in $y = 2\sqrt{x}$ demands that y be positive). □

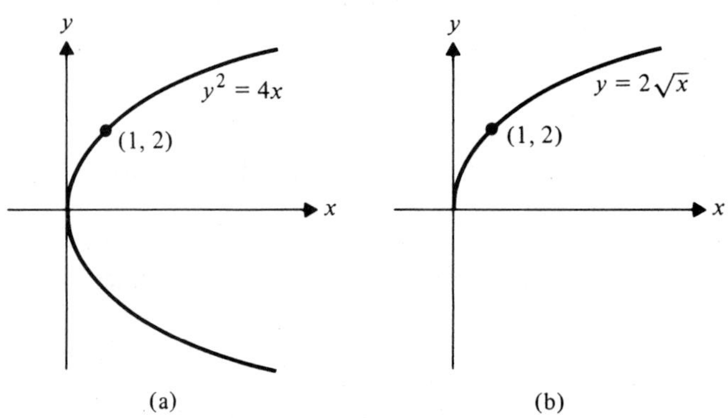

Figure 1.55

Example 15 Sketch a graph of the function $f(x) = |x| + x$.

Solution To sketch this function we first draw the straight line $y = x$ (Fig. 1.56a). The curve $y = |x|$ is obtained therefrom (shown dotted) by taking the absolute values of all ordinates. The required curve $y = |x| + x$ has as its ordinates the sums of the ordinates $|x|$ and x just sketched. Consequently, to obtain $y = |x| + x$, we add the corresponding ordinates of these two curves (Fig. 1.56b). □

This building procedure, called *addition of ordinates*, is used again in the following example.

Example 16 Sketch a graph of the function $f(x) = 2x + \sqrt{x - 4}$.

Solution If we square $y = \sqrt{x - 4}$, then $y^2 = x - 4$ or $x = y^2 + 4$. This equation describes a parabola, the top half of which is shown in Fig. 1.57(a). On the same axes we draw the straight line $y = 2x$. Addition of ordinates now gives the required graph, shown in Fig. 1.57(b). □

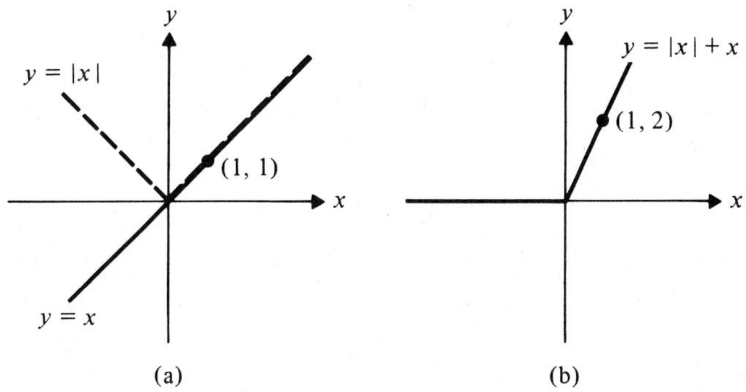

Figure 1.56

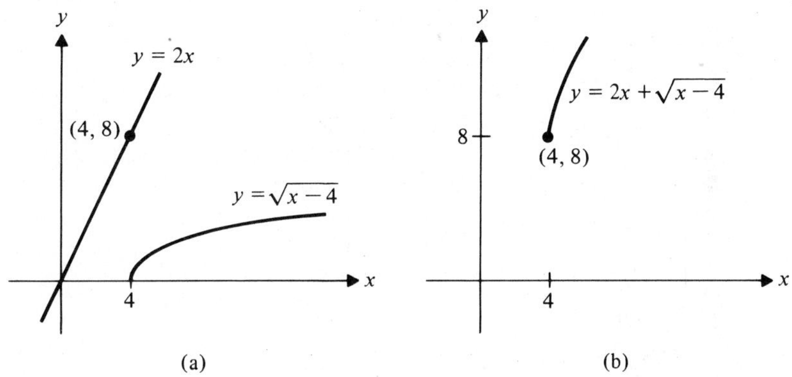

Figure 1.57

Example 17 Sketch a graph of the function $f(x) = x^2\sqrt{1-x}$.

Solution We first consider a sketch of $y = \sqrt{1-x}$, which implies that $y^2 = 1 - x$ or $x = 1 - y^2$. The top half of this parabola is shown in Fig. 1.58(a). On the same axes we sketch the parabola $y = x^2$. The required curve $y = x^2\sqrt{1-x}$ has ordinates $x^2\sqrt{1-x}$, which are products of the ordinates x^2 and $\sqrt{1-x}$ just sketched. Thus to obtain $y = x^2\sqrt{1-x}$ we multiply corresponding ordinates of these two curves (Fig. 1.58b). □

This building procedure is called *multiplication of ordinates.*

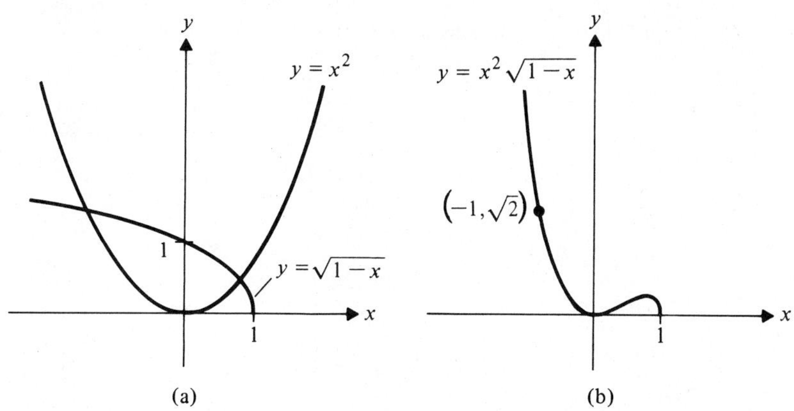

Figure 1.58

Example 18 When a ball is dropped from the top of a building 20 meters high at time $t = 0$, the distance d that it falls is a function of t given by $d = f(t) = 4.905t^2$. Sketch a graph of the function.

Solution In this example the independent and dependent variables are denoted by the letters t and d, which suggest their physical meaning (time and distance), rather than the labels x and y. With the axes labeled correspondingly as the t-axis and d-axis, we sketch that part of the parabola $d = 4.905t^2$ shown in Fig. 1.59. The remainder of the parabola has no physical significance in the context of this problem. □

Every function $f(x)$ can be represented pictorially by its graph: the curve with equation $y = f(x)$. But what about the reverse situation? Does every curve in the xy-plane represent a function $f(x)$? The curves in Fig. 1.60, which both extend between $x = a$ and $x = b$, illustrate that the answer is no. The curve in Fig. 1.60(a) represents a function, whereas the curve in Fig. 1.60(b) does not, because for values of x between a and c there are two possible values of y. In other words, a curve represents a function $f(x)$ if every vertical line that intersects the curve does so at exactly one point.

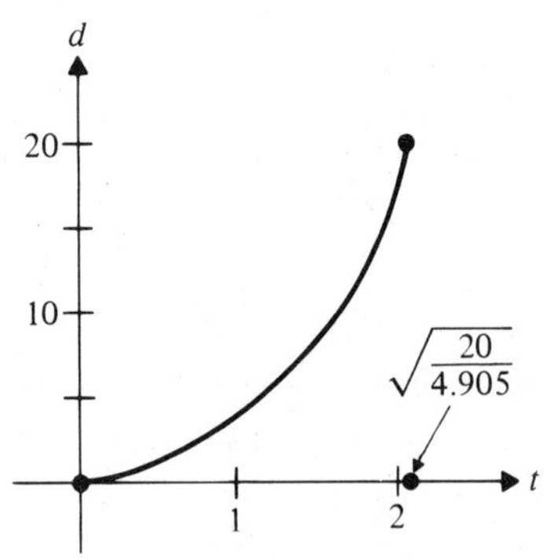

Figure 1.59

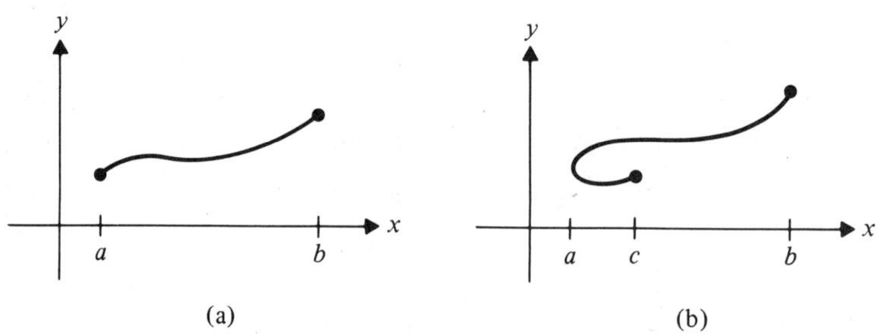

Figure 1.60

EXERCISES 1.4

In Exercises 1–6, find the largest possible domain for the given function.

1. $f(x) = \sqrt{9 - x^2}$

2. $f(x) = \dfrac{1}{x - 2}$

3. $f(x) = \dfrac{1}{x\sqrt{x^2 + 4}}$

4. $f(x) = \dfrac{1}{x\sqrt{x^2 - 4}}$

5. $f(x) = \dfrac{1}{x\sqrt{4 - x^2}}$

6. $f(x) = \sqrt{\dfrac{x^2 - 4}{9 - x^2}}$

In Exercises 7–20, sketch a graph of the given function.

7. $f(x) = x^4 - 3$

8. $f(x) = 3x^3 + 2$

9. $f(x) = x^2 - x$

10. $f(x) = -x^3 + 3x - 2$

11. $f(x) = \sqrt{5 - x^2}$

12. $f(x) = -\sqrt{5 - x^2}$

13. $f(x) = |x| + 2x$

14. $f(x) = x^{1/3}$

15. $f(x) = x\sqrt{x+1}$

16. $f(x) = x^3 + 3x^2 - 4x - 12$

17. $f(x) = -x\sqrt{1-x^2}$

18. $f(x) = x^3(x-1)^4$

19. $f(x) = |x^2 - x - 12|$

20. $f(x) = x^2\sqrt{1-x^2}$

21. What is the condition that a curve in the xy-plane represents a function $x = f(y)$?

In Exercises 22–29, sketch a graph of the given function.

22. $f(x) = \sqrt{2x - x^2}$

23. $f(x) = x + 2 + \sqrt{x}$

24. $f(x) = \sqrt{1-x^2} + \sqrt{x^2-1}$

25. $f(x) = x^2 + |x| - 2$

26. $f(x) = \sqrt{(x^2-4)^2}$

27. $f(x) = (x^2 + x - 2)^2$

28. $f(x) = \sqrt{(x^2-1)^2 - (x^2-1)}$

29. $f(x) = \sqrt{2 - \sqrt{1+x}}$

30. a) Use graphs of the functions $f(x) = x^3$ and $g(x) = 2 - x$ to show quickly that the equation

$$x^3 + x - 2 = 0$$

has only one real solution.

b) Graph functions to illustrate the number of solutions of the equation

$$\sqrt{4-x} + x^2 - x^6 = 0.$$

31. A rectangle with sides parallel to the axes is inscribed inside the curve $b^2x^2 + a^2y^2 = a^2b^2$ (Fig. 1.61). Find a formula for the area A of the rectangle in terms of x. Sketch a graph of this function.

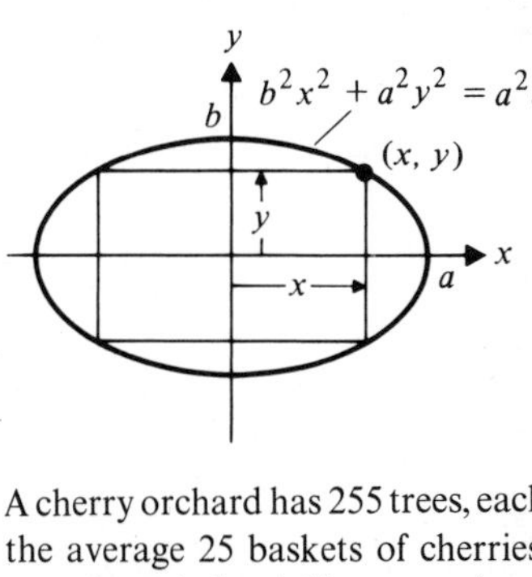

Figure 1.61

32. A cherry orchard has 255 trees, each of which produces on the average 25 baskets of cherries. For each additional tree planted, the yield per tree decreases by one-twelfth of a basket. If x represents the number of extra trees (beyond 255) and Y the yield, find Y as a function of x, and sketch its graph. How many more trees should be planted for maximum yield?

33. When two substances A and B are brought together, a chemical reaction takes place to form a new substance C. It requires 2 liters of A for each liter of B to produce 3 liters of C. The rate R at which A and B react to form C is proportional to the product of the amounts of A and B present at that instant. If the original amounts of A and B are 20 liters and 40 liters, respectively, and if x represents the amount of C present in the reaction at any given time, find a formula for R as a function of x. Sketch a graph of this function, and determine when the reaction rate is highest.

34. A man 2 m tall walks along the edge of a straight road that is 5 m wide. On the other edge of the road stands a street light 10 m high. Find a functional relationship for the length of the man's shadow in terms of his distance from the point on his side of the road directly opposite the light. Sketch a graph of the function.

35. A square plate 4 m on each side is slowly submerged in a large tank of water. One diagonal is kept vertical and lowered at a rate of $\frac{1}{2}$ m/s, entering the water at time $t = 0$. If A is the area of the submerged portion of the surface (one side only) at time t (until complete submersion occurs), find A as a function of t and sketch its graph.

36. In ordinary physics and engineering, the mass m of an object is constant, independent of how fast it is moving. In special relativity, however, m is given by the formula

$$m = \frac{m_0}{\sqrt{1 - (v^2/c^2)}},$$

where m_0 is the mass of the object when it is not moving, v is the speed of the mass, and c is a constant (the speed of light). Sketch a graph of this function, and draw any conclusions that you feel are suggested.

37. Due to construction, no passing is permitted on a 10-km stretch of highway. If cars travel at v km/h along this stretch, a safe distance between them must be maintained, and this distance increases as v increases. In particular, the highway traffic commission has determined that for speeds over 50 km/h, the distance between cars should be

$$d = \frac{3v^2}{500} \text{ meters.}$$

If it is supposed that everyone maintains the safe distance, find the number q of cars leaving the "bottleneck" per hour as a function of speed v. Sketch a graph of this function for $50 \le v \le 100$, and determine the speed that maximizes q.

***38.** A box 1 m on each side is attached to a rope. The rope passes over a small pulley 10 m from the ground and a truck pulls on the other end in a horizontal direction along a line 1 m above the ground (Fig. 1.62). We denote the positions of the truck and the bottom of the box by x and y, respectively. Find y as a function of x if the truck starts at position $x = 5$ m and stops when the top of the box touches the pulley. Assume that the length of rope between truck and box is 25 m. Sketch a graph of the function.

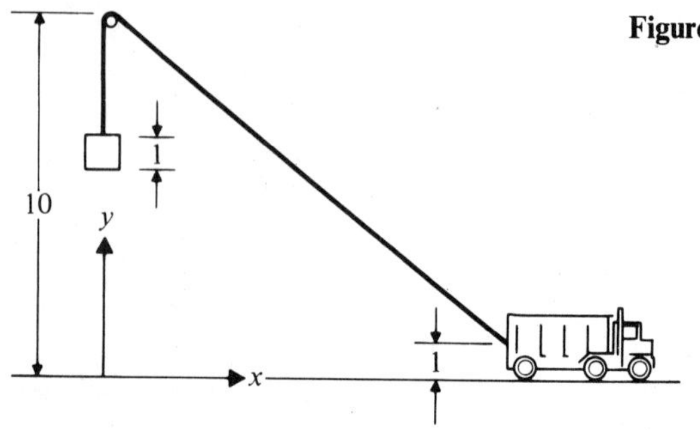

Figure 1.62

1.5 EXPLICIT AND IMPLICIT DEFINITIONS OF CURVES AND FUNCTIONS

Curves and functions can be defined explicitly, implicitly, and parametrically. We discuss explicit and implicit definitions here, and parametric definitions in Chapter 10.

▷ **Definition 1.8** A curve is said to be defined *explicitly* if its equation is given in either of the two forms

$$y = f(x) \tag{1.18a}$$

or

$$x = g(y). \quad \triangleleft \tag{1.18b}$$

In other words, a curve is defined explicitly if one of the two variables has been separated out to stand alone on the left side of the equation. Examples 13–18 have curves that are defined explicitly. The curve represented by the equation $x = y^3 - 3y^2$ is also defined explicitly. If we write $x = y^3 - 3y^2 = y^2(y - 3)$ and sketch the curves $x = y^2$ and $x = y - 3$ (Fig. 1.63a), then multiplication of abscissas gives the sketch in Fig. 1.63(b).

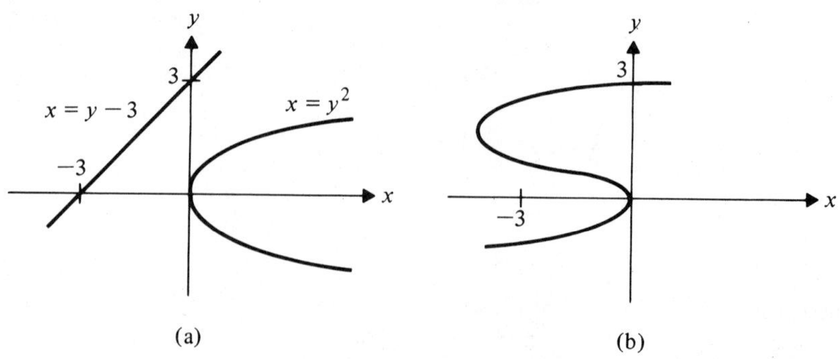

Figure 1.63

▷ **Definition 1.9** A curve is said to be defined *implicitly* if its equation is given in the form

$$F(x, y) = 0.\dagger \quad \triangleleft \qquad (1.19)$$

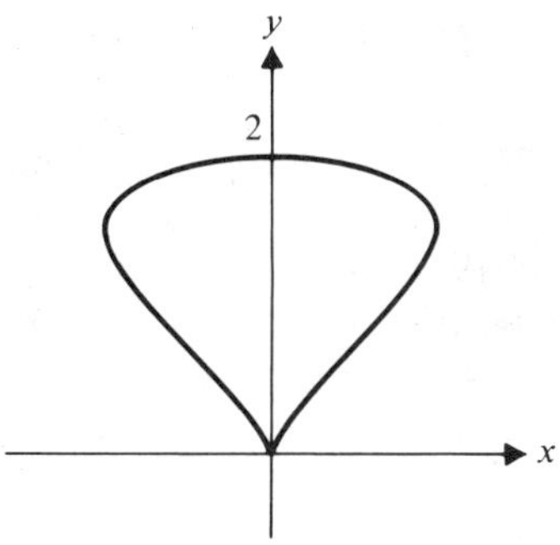

Figure 1.64

Thus a curve is defined implicitly by an equation in which neither variable has been separated out from the other. In Example 12, the equation $x - 4 - y^2 = 0$ defines a parabola, and is said to do so implicitly. An explicit definition for the parabola is $x = 4 + y^2$. If we solve for y in terms of x, rather than x in terms of y, we obtain $y = \pm\sqrt{x - 4}$. Consequently, the parabola is also defined explicitly by the equations $y = \sqrt{x - 4}$ and $y = -\sqrt{x - 4}$.

The equation of a circle is most often written in the form $(x - h)^2 + (y - k)^2 = r^2$, or $x^2 + y^2 + fx + gy + e = 0$, both of which are implicit definitions.

The curve in Fig. 1.64 is defined implicitly by the equation $y^4 - 2y^3 + x^2 = 0$. Once again we could find an explicit definition by solving for x in terms of y. The result is $x = \sqrt{2y^3 - y^4}$ and $x = -\sqrt{2y^3 - y^4}$.

Symmetries

The form of an explicit or implicit definition of a curve often suggests properties of *symmetry*, which can be very useful in obtaining a sketch of the curve. For instance, it often happens that the equation of a curve is unchanged when x is replaced by $-x$. In this case, if (x, y) is any point on the curve, then so too is $(-x, y)$; i.e., the curve is *symmetric about the y-axis*. For example, the curve $y = x^4 - x^2$ in Fig. 1.65 is symmetric about the y-axis.

If the equation of a curve is unchanged when y is replaced by $-y$, then for every point (x, y) on the curve so too must be $(x, -y)$. The curve is therefore *symmetric about the x-axis*. For example, the parabola $x = y^2$ in Fig. 1.66 is symmetric about the x-axis. Note that a curve that is symmetric about the x-axis does not represent a function.

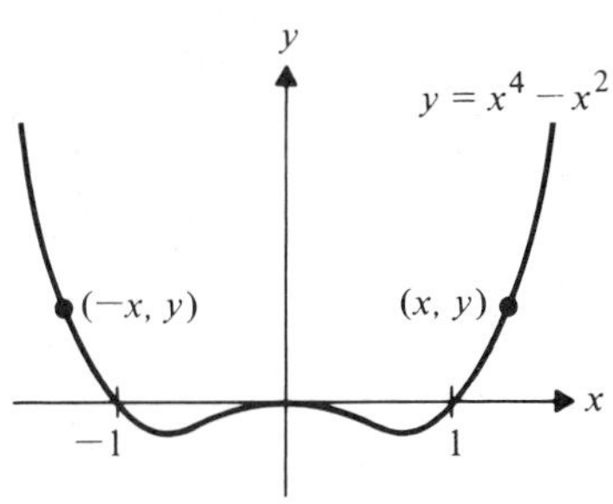

Figure 1.65

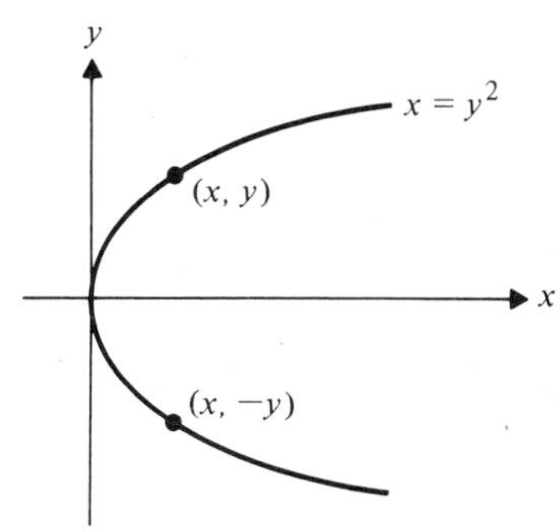

Figure 1.66

† The notation $F(x, y)$ is used to denote an expression that depends on two variables x and y. For example, the volume of a right circular cylinder depends on its radius r and height h; in particular, $V = \pi r^2 h$. In such a case we write that $V = F(r, h) = \pi r^2 h$. Similarly, the volume of a box is the product of its length l, width w, and height h, and we write $V = F(l, w, h) = lwh$.

Example 19 Sketch the curve $|x| + |y| - 1 = 0$.

Solution When x and y are both nonnegative (in which case we are in the first quadrant), then the equation reduces to $x + y = 1$, a straight-line segment shown in Fig. 1.67(a). Since the equation $|x| + |y| - 1 = 0$ is unchanged when x is replaced by $-x$ or y is replaced by $-y$, the curve is symmetric about both the x- and y-axes. This leads to the square in Fig. 1.67(b). □

Definitions 1.8 and 1.9 are also used to define functions explicitly and implicitly. We say that y is defined explicitly as a function of x by Eq. (1.18a). It is defined implicitly as a function of x by Eq. (1.19) if the curve so defined represents a function; (that is, if every vertical line that intersects the curve does so in exactly one point). Equation (1.18b) defines x explicitly as a function of y; it defines y implicitly as a function of x if every vertical line that intersects the curve does so exactly once.

Even and Odd Functions

Functions that are *even* or *odd* play a prominent role in many areas of applied mathematics. They are defined as follows.

▷ **Definition 1.10** A function $f(x)$ is said to be *even* if for each x in its domain

$$f(-x) = f(x); \tag{1.20a}$$

it is said to be *odd* if

$$f(-x) = -f(x). \quad \triangleleft \tag{1.20b}$$

Equation (1.20a) implies that the equation $y = f(x)$ for the graph of an even function is unchanged when x is replaced by $-x$; therefore the graph of an even function is symmetric about the y-axis. For example, $f(x) = x^4 - x^2$ is an even function; its graph is shown in Fig. 1.65.

Equation (1.20b) implies that if (x, y) is any point on the graph of an odd function, so too is the point $(-x, -y)$. This is illustrated by the graph of the odd function $f(x) = x^3 - 4x$ shown in Fig. 1.68. Another way to describe the graph of an odd function is to note that either half of the graph ($x < 0$ or $x > 0$) is the result of two reflections of the other half, first in the y-axis and then in the x-axis.

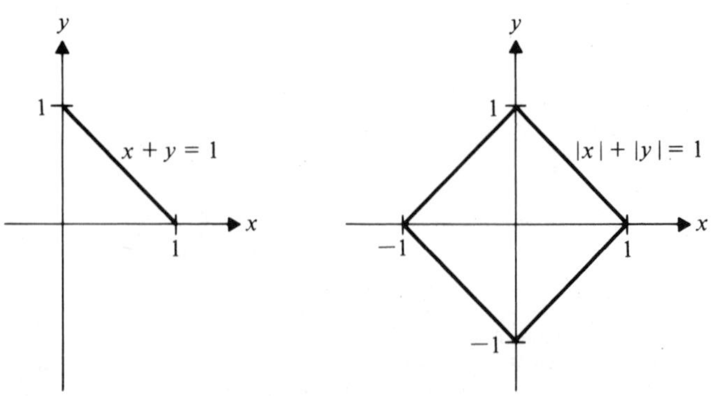

(a) **Figure 1.67** (b)

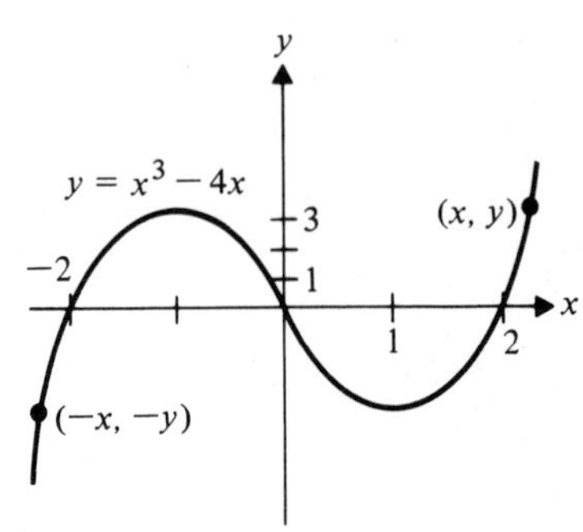

Figure 1.68

Translation of Graphs When every x in the equation of a curve is replaced by $x - c$, where c is a constant, the curve is shifted c units to the right. When every x is replaced by $x + c$, the curve is shifted c units to the left. For example, when x in the equation $x^2 + y^2 = r^2$ is replaced by $x - c$, to get $(x - c)^2 + y^2 = r^2$, the center of the circle is shifted from the origin to the point $(c, 0)$ (Fig. 1.69). When each x in the parabola $y = x^2 + x$ is replaced by $x + c$, to get

$$y = (x + c)^2 + (x + c) = (x + c)(x + c + 1),$$

the parabola is shifted c units to the left (Fig. 1.70).

Vertical shifts result when y is replaced by $y \pm c$ in the equation of a curve. For example, to sketch the curve $x = (y + 1)^3$, we could first sketch $x = y^3$ (Fig. 1.71). A shift of this curve 1 unit downward gives the curve $x = (y + 1)^3$. The curve $|x| + |y - 2| = 1$ can be sketched by shifting the curve $|x| + |y| = 1$ in Fig. 1.67(b) upward 2 units (Fig. 1.72).

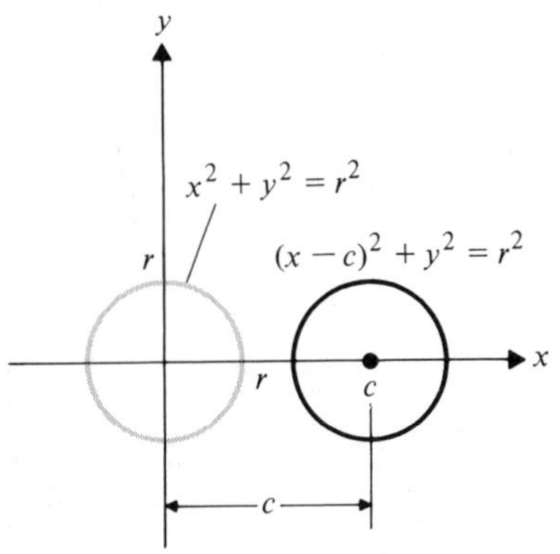

Figure 1.69

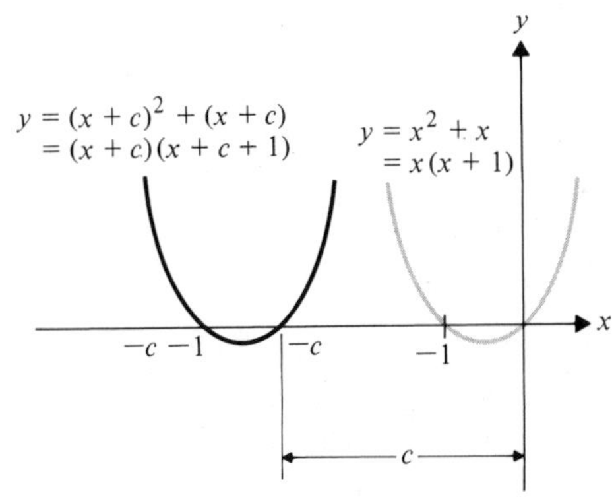

Figure 1.70

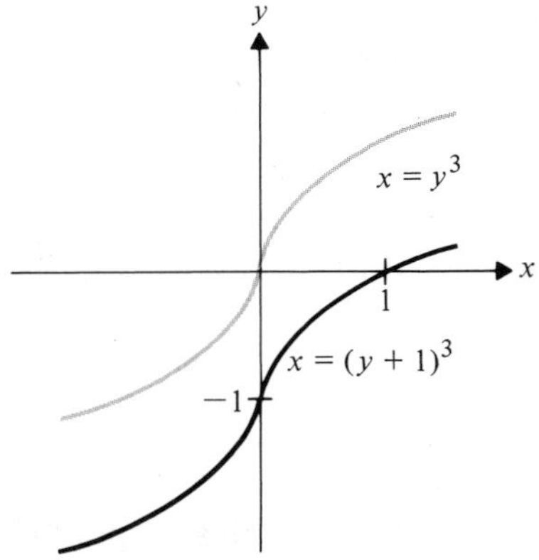

Figure 1.71

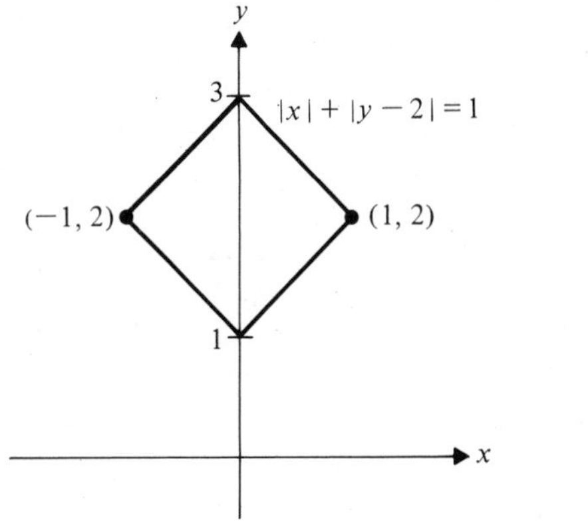

Figure 1.72

EXERCISES 1.5

In Exercises 1–20, sketch the given curve. In each case, indicate whether the curve defines y as a function of x.

1. $x + y = 4$

2. $3x = 2y + 6$

3. $x = \sqrt{2 + y}$

4. $x = -y^2 + 1$

5. $x = y^2 + 2y - 5$

6. $y = x^2 - x^6$

7. $x = 4 - |y|$

8. $4x^2 - 9y^2 = 0$

9. $y = \sqrt{12 - 4y - x^2 + 6x}$

10. $|x| - |y| = 1$

11. $y = x\sqrt{4 - x^2}$

12. $x^3 + xy^2 = 4x$

13. $y^2 = x^2(4 - x^2)$

14. $x = |4 - |y||$

15. $y^2 = x^2(4 - x)$

16. $y^2 = x^2(4 - x^2)^2$

17. $y = (x + 1)^4 - (x + 1)^2$

18. $|x| + |y + 2| = 2$

19. $y = |3 - |x + 2|| - 1$

20. $(y - 1)^2 = (x - 1)^2[1 - (x - 1)^2]$

In Exercises 21–26, determine whether the given function is even, odd, or neither even nor odd.

21. $f(x) = 1 + x^2 + 2x^4$

22. $f(x) = x^5 - x$

23. $f(x) = 12x^2 + 2x$

24. $f(x) = x^{1/5}$

25. $f(x) = \dfrac{x - 1}{x + 1}$

26. $f(x) = x(x^2 + x)$

27. Can a function be both even and odd?

28. Verify that $f(0) = 0$ for an odd function.

29. Prove each of the following.

a) The product of two even functions or two odd functions is an even function.

b) The product of an even and an odd function is an odd function.

30. a) Sketch a graph of the function $f(x) = x^{1/3}$.

b) Sketch a graph of the function

$$g(x) = \begin{cases} 0, & x < a, \\ f(x - a), & x \geq a, \end{cases} \qquad a > 0 \text{ a fixed constant.}$$

c) The Heaviside unit step function $H_a(x)$ is defined by

$$H_a(x) = \begin{cases} 0, & x < a, \\ 1, & x > a. \end{cases}$$

Sketch its graph.

d) Write $g(x)$ in terms of $f(x)$ and $H_a(x)$.

e) What in essence does multiplication of $f(x)$ by $H_a(x)$ do?

***31.** If a thermal nuclear reactor is built in the shape of a right circular cylinder of radius r and height h, then neutron diffusion theory requires r and h to satisfy an equation of the form

$$\frac{a^2}{r^2} + \frac{b^2}{h^2} = 1,$$

where a and b are positive constants. Sketch a graph of the function $r = f(h)$ defined implicitly by this equation.

Sketch the curve in each of Exercises 32–35.

***32.** $xy^5 - xy^2 = y + 1$

***33.** $x^2 + 4x^2y = x^4 + y^2$

***34.** $(x^2 + 2y - 1)^2 = y^2(1 - x^2)$

***35.** $x + y + \sqrt{2xy} = 1$

SUMMARY

In the first three sections of this chapter we developed the principles of plane analytic geometry. Analytic geometry is a marriage of geometry and algebra. On the one hand, algebraic equations are used to describe geometric curves and, on the other hand, curves are the geometric representation of equations.

The form of an equation dictates the shape of the curve, and conversely, the shape of a curve influences its equation. To illustrate this fact we discussed straight lines, circles, and parabolas. The most common equations for these curves are as follows.

Equation	Name	
$y - y_1 = m(x - x_1)$	Point-slope	Forms for the equation of a straight line
$\dfrac{x - x_1}{x_2 - x_1} = \dfrac{y - y_1}{y_2 - y_1}$	Two-point	
$y = mx + b$	Slope–y-intercept	
$y = m(x - a)$	Slope–x-intercept	
$\dfrac{x}{a} + \dfrac{y}{b} = 1$	Two-intercept	
$Ax + By + C = 0$	General	
$y = ax^2 + bx + c$	Vertical axis of symmetry	Parabola
$x = ay^2 + by + c$	Horizontal axis of symmetry	
$x^2 + y^2 + fx + gy + e = 0$		Circle
$(x - h)^2 + (y - k)^2 = r^2$		

Basic to all mathematics is the concept of a function, that is, a rule that assigns to each number x in a domain, a unique number y in the range. A function then is simply another way of saying "a quantity y depends on a quantity x." The notation $y = f(x)$ for a function immediately suggests that a function can be represented geometrically by a curve—the curve with equation $y = f(x)$—and we call this curve the graph of the function. Although graphs can always be sketched by first constructing tables of values and then joining these points smoothly, other methods such as the addition and multiplication of ordinates and symmetry principles are frequently very helpful. In Chapters 2 and 4, we will discover further curve-sketching aids, and these will all be combined for a complete discussion in Section 4.6.

Functions and curves can be defined explicitly or implicitly:

$y = f(x)$ or $x = g(y)$	Explicit definition of curve
$F(x, y) = 0$	Implicit definition of curve

REVIEW EXERCISES

In Exercises 1–10, find the largest possible domain for the given function.

1. $f(x) = \sqrt{x^2 + 5}$

2. $f(x) = \sqrt{x^2 - 5}$

3. $f(x) = \dfrac{1}{x^2 + 3x + 2}$

4. $f(x) = \dfrac{x + 4}{x^3 + 2x^2 + x}$

5. $f(x) = (x^3 - 8)^{1/3}$

6. $f(x) = x^{3/2}$

7. $f(x) = \sqrt{x^2 + 4x - 6}$

8. $f(x) = \dfrac{1}{\sqrt{2x^2 + 4x - 5}}$

9. $f(x) = \sqrt{\dfrac{2x + 1}{x - 3}} + 2$

10. $f(x) = \sqrt{x - \dfrac{1}{x}}$

In Exercises 11–20, sketch the given curve.

11. $y = 2x^2 + 3$
12. $x^2 = -y^2 + 4$
13. $y = x^{1/3} - 1$
14. $|y| = |x|$
15. $y = x^3\sqrt{1 - x^2}$
16. $y = \sqrt{-x^2 + 4x + 4}$
17. $4x^2 + y^2 = 0$
18. $x^2 - 2y = 4x - y^2$
19. $|y| = |x^3 + 1|$
20. $x = y\sqrt{y + 3}$

In Exercises 21–24, give an example of a function $y = f(x)$ with the given properties.

21. The range of the function consists of one number only.
22. The largest possible domain of the function is $-1 \leq x \leq 2$.
23. The domain of the function consists of all reals except $x = \pm 1$.
24. The domain of the function is $x \leq 0$, and the range is $y \geq 1$.
25. Is the triangle with vertices (1, 3), (2, −1), and (−2, −2) right-angled?
26. Show that the lines that successively join the points (0, 1), (3, 4), $(-\sqrt{2}, -3)$, and $(3 - \sqrt{2}, 0)$ form the sides of a rhombus. (A *rhombus* is a parallelogram with all sides of equal length.)
27. Find the distance from point (2, 1) to the line $2x + 3y = 4$.
28. a) The *greatest integer function*, denoted by $f(x) = [\![x]\!]$, is defined by $f(x) = [\![x]\!] =$ greatest integer that does not exceed x. Sketch its graph.

    b) If first-class postage is 20c per ounce, or fraction thereof, up to and including 8 ounces, sketch a graph of this cost function.

    c) Express the cost function in part (b) in terms of the greatest integer function.
29. A lighthouse is 6 kilometers off shore and a cabin on the straight shoreline is 9 kilometers from the point on the shore nearest the lighthouse (Fig. 1.73). Show that if a man rows at 3 kilometers per hour and walks at 5 kilometers per hour, and he beaches the boat at x, then the total travel time from lighthouse to cabin is

$$t = f(x) = \frac{\sqrt{x^2 + 36}}{3} + \frac{9 - x}{5}, \qquad 0 \leq x \leq 9.$$

Sketch a graph of this function.

30. Prove that the diagonals of a rhombus intersect at right angles. (See Exercise 26 for the definition of a rhombus.)

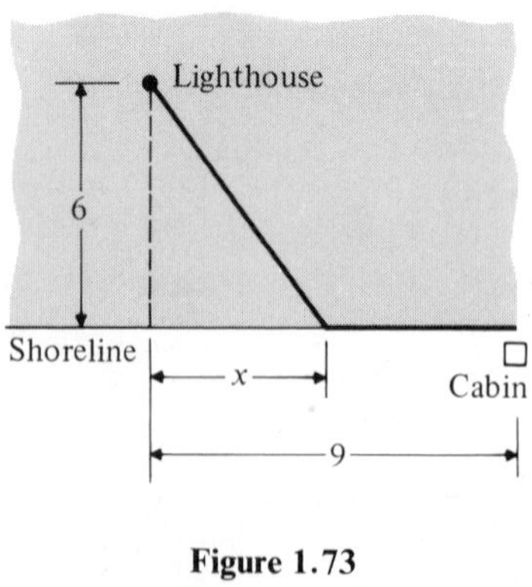

Figure 1.73

Limits and Continuity

2

2.1 LIMITS

The concept of a *limit* is crucial to calculus, for the two basic operations in calculus are differentiation and integration, each of which is defined in terms of a limit. For this reason we must have a clear understanding of limits from the beginning. In this section we present an intuitive discussion of limits of functions; in Section 2.3 we show how these ideas can be formalized mathematically.

The value of the function $f(x) = x^2 + 5$ at $x = 2$ is $f(2) = 9$. A completely different consideration is contained in the question, "What number does $f(x) = x^2 + 5$ get closer and closer to (and stay close to) as x gets closer and closer to 2?" Table 2.1 shows that *as x gets closer and closer to 2, $x^2 + 5$ gets arbitrarily close to 9 and stays close to 9.* Likewise, a graph of the function (Fig. 2.1) clearly shows that $f(x)$ approaches 9 as x approaches 2. We could rewrite the above statement to read "$x^2 + 5$ *can be made arbitrarily close to* 9 *by choosing x sufficiently close to* 2." In calculus we summarize these statements by saying "The limit of $x^2 + 5$ as x approaches 2 is 9," and we use the following notation to abbreviate this statement:

Table 2.1

x	$f(x) = x^2 + 5$	x	$f(x) = x^2 + 5$
3.0	14.0	1.0	6.0
2.1	9.41	1.9	8.61
2.01	9.0401	1.99	8.9601
2.001	9.004001	1.999	8.996001
2.0001	9.00040001	1.9999	8.99960001
2.00001	9.0000400001	1.99999	8.9999600001

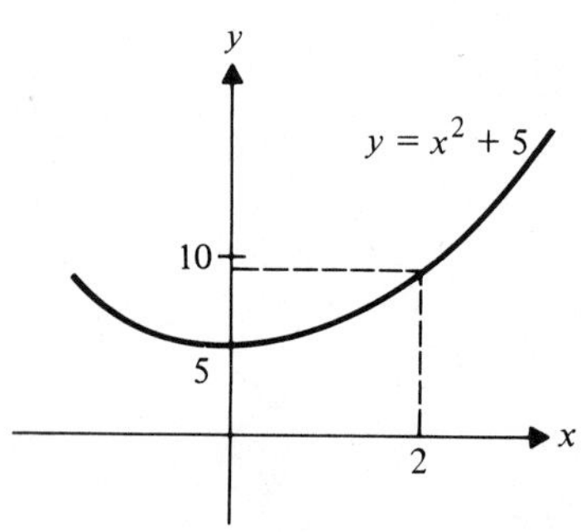

Figure 2.1

$$\lim_{x \to 2} (x^2 + 5) = 9. \tag{2.1}$$

This notation then represents either of the above statements, and we must read these statements into the symbolism. Note that limit (2.1) is not concerned with the value of $x^2 + 5$ at $x = 2$. It is concerned with the number that $x^2 + 5$ approaches as x *approaches* 2. These two numbers will not always be identical.

Generally we say that a function $f(x)$ has limit L as x approaches a, and write

$$\lim_{x \to a} f(x) = L, \tag{2.2}$$

if $f(x)$ can be made arbitrarily close to L by choosing x sufficiently close to a.

Example 1 Evaluate $\lim_{x \to 1} (x^2 + 2x + 5)$.

Solution As x gets closer and closer to 1, $x^2 + 2x + 5$ gets closer and closer to 8, and we therefore write

$$\lim_{x \to 1} (x^2 + 2x + 5) = 8. \square$$

Our discussion suggests that to calculate the somewhat more complicated limit

$$\lim_{x\to 5} \frac{x^2(3-x)}{x^3+x},$$

we should evaluate the function $x^2(3-x)/(x^3+x)$ at values of x that get closer and closer to 5. The following theorem provides a much easier method.

▶ **Theorem 2.1** If $\lim_{x\to a} f(x) = F$ and $\lim_{x\to a} g(x) = G$, then:

i) $$\lim_{x\to a} [f(x) + g(x)] = F + G; \tag{2.3}$$

ii) $$\lim_{x\to a} [f(x)g(x)] = FG; \tag{2.4}$$

iii) $$\lim_{x\to a} \frac{f(x)}{g(x)} = \frac{F}{G}, \quad \text{(provided } G \neq 0\text{).} \ ◀ \tag{2.5}$$

What this theorem says is that a limit such as $\lim_{x\to 5} [x^2(3-x)/(x^3+x)]$ can be broken down into smaller problems and reassembled later. For instance, since

$$\lim_{x\to 5} x^2 = 25, \qquad \lim_{x\to 5} (3-x) = -2, \qquad \lim_{x\to 5} x^3 = 125, \qquad \lim_{x\to 5} x = 5,$$

we may write that

$$\lim_{x\to 5} \frac{x^2(3-x)}{x^3+x} = \frac{25(-2)}{125+5} = \frac{-50}{130} = -\frac{5}{13}.$$

Although the results of Theorem 2.1 may seem evident, to prove them mathematically is not a simple task. In fact, because we have not yet given a mathematical definition for limits, a proof at this time is impossible. When we give a precise definition of limits in Section 2.3, it will then be possible to verify the theorem (see Exercises 9, 10, and 11 in Section 2.3).

Example 2 Evaluate $\lim_{x\to -2} [(x+2)/(x^2+9)]$.

Solution Since $\lim_{x\to -2} (x+2) = 0$ and $\lim_{x\to -2} (x^2+9) = 13$, part (iii) of Theorem 2.1 gives

$$\lim_{x\to -2} \frac{x+2}{x^2+9} = \frac{0}{13} = 0. \ \square$$

Example 3 Evaluate $\lim_{x\to -1} [x^2(1-x^3)/(2x^2+x+1)]$.

Solution Using Theorem 2.1, we can write that

$$\lim_{x\to -1} \frac{x^2(1-x^3)}{2x^2+x+1} = \frac{(1)(2)}{2+(-1)+1} = 1. \ \square$$

Be sure that you understand how we obtained the expression

$$\frac{(1)(2)}{2+(-1)+1}$$

for this limit. In particular, we did not set $x = -1$ in $x^2(1-x^3)/(2x^2+x+1)$.

Indeed, this is not permitted because to evaluate a limit as x approaches -1, we are not to set $x = -1$; we are to let x get closer and closer to -1. What we did do is take the limits of x^2, $(1 - x^3)$, $2x^2$, and x as x approaches -1, and then use Theorem 2.1.

The following example illustrates what can happen if we substitute $x = a$ into $f(x)$ in the evaluation of $\lim_{x \to a} f(x)$.

Example 4 Evaluate $\lim_{x \to 3} [(x^2 - 9)/(x - 3)]$.

Solution Because $\lim_{x \to 3} (x - 3) = 0$, we cannot use Theorem 2.1. Nor can we set $x = 3$ in $(x^2 - 9)/(x - 3)$ because it is inherent in the limiting procedure that we do not put $x = 3$. (Besides, if we did, we would obtain the meaningless expression 0/0.) The correct procedure is to factor $x^2 - 9$ into $(x - 3)(x + 3)$:

$$\lim_{x \to 3} \frac{x^2 - 9}{x - 3} = \lim_{x \to 3} \frac{(x - 3)(x + 3)}{x - 3} = \lim_{x \to 3} (x + 3) = 6.$$

In this calculation we have canceled the factor $x - 3$ in the numerator and denominator, an operation that would not be permissible if $x - 3$ were equal to 0; i.e., if x were equal to 3. But once again this cannot happen, because in the limiting operation we let x get closer and closer to 3, but do not put $x = 3$. □

Note in this example that although the limit is 6, there is no value of x for which the function $(x^2 - 9)/(x - 3)$ is ever equal to 6.

Example 5 Evaluate $\lim_{x \to 0} [(-2x + 3x^2)/(4x - 5x^2)]$.

Solution We have

$$\lim_{x \to 0} \frac{-2x + 3x^2}{4x - 5x^2} = \lim_{x \to 0} \frac{x(-2 + 3x)}{x(4 - 5x)} = \lim_{x \to 0} \frac{3x - 2}{4 - 5x} = -\frac{1}{2}. \ \square$$

Example 6 Evaluate

$$\lim_{x \to 0} \frac{\sqrt{1 + x} - 1}{\sqrt{x}}.$$

Solution Since the limit of the denominator as x approaches 0 is 0, we cannot use Theorem 2.1. The best procedure in this example is to rationalize the numerator (i.e., rid the numerator of the square root) by multiplying numerator and denominator by $\sqrt{1 + x} + 1$:

$$\begin{aligned} \lim_{x \to 0} \frac{\sqrt{1 + x} - 1}{\sqrt{x}} &= \lim_{x \to 0} \left\{ \frac{\sqrt{1 + x} - 1}{\sqrt{x}} \cdot \frac{\sqrt{1 + x} + 1}{\sqrt{1 + x} + 1} \right\} \\ &= \lim_{x \to 0} \frac{x}{\sqrt{x}(\sqrt{1 + x} + 1)} \\ &= \lim_{x \to 0} \frac{\sqrt{x}}{\sqrt{1 + x} + 1} = 0. \ \square \end{aligned}$$

When we write $L = \lim_{x \to a} f(x)$, we mean that $f(x)$ gets arbitrarily close to L as x gets closer and closer to a. But how is x to approach a? Does x approach a through numbers larger than a, or does it approach a through numbers smaller than a? Or does x jump back and forth between numbers larger than a and numbers smaller than a, gradually getting closer and closer to a? We have not previously mentioned "mode" of approach simply because it would have made no difference to our discussion. In each of our preceding examples, all possible modes of approach led to the same limit. In particular, Table 2.1 and Fig. 2.1 illustrate that 9 is the limit of $f(x) = x^2 + 5$ as x approaches 2 if x approaches 2 through numbers larger than 2 or through numbers smaller than 2.

Approaching a number a either through numbers larger than a or through numbers smaller than a are two modes of approach that will be extremely important to us, and we therefore give them special notations:

$\lim_{x\to a^+} f(x)$ indicates that x approaches a through numbers larger than a (often called a *right-hand limit* since x approaches a along the x-axis from the right of a);

$\lim_{x\to a^-} f(x)$ indicates that x approaches a through numbers smaller than a (called a *left-hand limit*).

Example 6 should, in fact, be designated a right-hand limit,

$$\lim_{x\to 0^+} \frac{\sqrt{1+x}-1}{\sqrt{x}} = 0,$$

since the presence of $\sqrt{x}$ in the denominator demands that x be only positive.

A natural question to ask is "What should we conclude if for a function $f(x)$

$$\lim_{x\to a^+} f(x) \neq \lim_{x\to a^-} f(x)?\text{''}$$

Our entire discussion on limits has suggested (and indeed it could be proved; see Exercise 7 in Section 2.3) that if a function has a limit as x approaches a, then it has only one such limit; that is, the limit must be the same for every possible method of approach. Consequently, if we arrive at two different results depending on the mode of approach, then we conclude that the function does not have a limit. This situation is illustrated in the following example.

Example 7 Evaluate $\lim_{x\to 0} (|x|/x)$.

Solution If $x < 0$, then $|x| = -x$, and

$$\lim_{x\to 0^-} \frac{|x|}{x} = \lim_{x\to 0^-} \frac{-x}{x} = -1.$$

If $x > 0$, then $|x| = x$, and

$$\lim_{x\to 0^+} \frac{|x|}{x} = \lim_{x\to 0^+} \frac{x}{x} = 1.$$

Since

$$\lim_{x\to 0^-} \frac{|x|}{x} \neq \lim_{x\to 0^+} \frac{|x|}{x},$$

we conclude that $\lim_{x\to 0} (|x|/x)$ does not exist. The graph of $f(x) = |x|/x$ in Fig. 2.2 clearly illustrates the situation. □

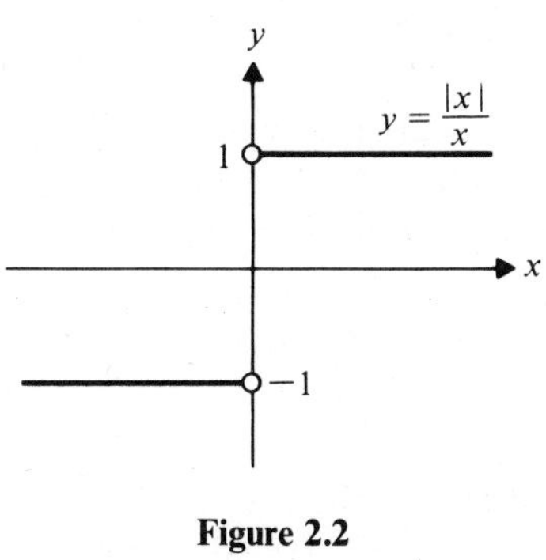

Figure 2.2

We now consider two examples of limits that are important enough to warrant special attention.

Limits at Infinity

In our first situation we ask whether the function $f(x) = (x + 2)/(2x - 1)$ approaches a number as x becomes larger and larger and larger. When x is very large, it is certainly not equal to zero, and we may divide numerator and denominator by x:

$$f(x) = \frac{1 + 2/x}{2 - 1/x}.$$

With $f(x)$ in this form we see that for very large x, $f(x)$ is approximately equal to $\frac{1}{2}$, and gets closer to $\frac{1}{2}$ as x increases. In calculus we express this fact by saying "the limit as x approaches infinity of $(x + 2)/(2x - 1)$ is $\frac{1}{2}$," and write

$$\lim_{x \to \infty} \frac{x + 2}{2x - 1} = \frac{1}{2}.$$

By using the symbolism $x \to \infty$ we are not implying that ∞ is a number; on the contrary, we stress most emphatically that ∞ is not a number. The symbolism simply means "as x gets larger and larger and larger."

Example 8 Evaluate each of the following limits.

a) $\displaystyle\lim_{x \to \infty} \frac{2x^3 - 4}{x^3 + x^2 + 2}$ b) $\displaystyle\lim_{x \to -\infty} \frac{2x^2 - 14}{3x^3 + 5x}$

Solution a) To obtain this limit we divide the numerator and denominator of the fraction by x^3:

$$\lim_{x \to \infty} \frac{2x^3 - 4}{x^3 + x^2 + 2} = \lim_{x \to \infty} \frac{2 - 4/x^3}{1 + 1/x + 2/x^3} = 2.$$

b) Once again we divide the numerator and denominator by x^3:

$$\lim_{x \to -\infty} \frac{2x^2 - 14}{3x^3 + 5x} = \lim_{x \to -\infty} \frac{2/x - 14/x^3}{3 + 5/x^2} = 0. \square$$

Example 9 Evaluate $\lim_{x \to \infty} \sin x$.

Solution Since oscillations of $\sin x$ back and forth between $+1$ and -1 never cease as x increases, there does not exist a single number that $\sin x$ approaches as x becomes very large. Thus, in spite of the fact that this function is very well behaved, it does not have a limit as x becomes very large. □

Infinite Limits

Does the function $f(x) = 1/(x - 2)^2$ approach a number as x approaches 2? Table 2.2 indicates that as x approaches 2, the value of $1/(x - 2)^2$ becomes indefinitely large; thus the function does not have a limit as x approaches 2. We usually write in such an example that

Table 2.2

x	$f(x)$	x	$f(x)$
1.9	10^2	2.1	10^2
1.99	10^4	2.01	10^4
1.999	10^6	2.001	10^6
1.9999	10^8	2.0001	10^8
1.99999	10^{10}	2.00001	10^{10}

$$\lim_{x \to 2} \frac{1}{(x - 2)^2} = \infty.$$

This statement does *not* say that the limit exists, since ∞ is not a number. It states that the *limit does not exist*, and indicates that the reason it does not exist is that values of the function become indefinitely large as x approaches 2.

Example 10 Evaluate $\lim_{x \to 1} [1/(x - 1)]$, if it exists.

Solution Since $\lim_{x\to 1^+} [1/(x-1)] = \infty$, we conclude that $\lim_{x\to 1} [1/(x-1)]$ does not exist. We also note that $\lim_{x\to 1^-} [1/(x-1)] = -\infty$. □

Right-hand limits, left-hand limits, limits at infinity, and infinite limits can all be useful in sketching graphs of functions (or curves), as is illustrated in the following examples.

Example 11 Sketch a graph of the function $f(x) = (x+3)/(x-2)$.

Solution We first note that $f(-3) = 0$ and $f(0) = -\frac{3}{2}$, giving two points $(-3, 0)$ and $(0, -\frac{3}{2})$ on the graph. Because the function is undefined at $x = 2$ (since division by zero is impossible), we take limits as x approaches 2 to investigate the behavior of the function near $x = 2$. We find that

$$\lim_{x\to 2^+} \frac{x+3}{x-2} = \infty \quad \text{and} \quad \lim_{x\to 2^-} \frac{x+3}{x-2} = -\infty;$$

this is illustrated in Fig. 2.3(a). The limits

$$\lim_{x\to \infty} \frac{x+3}{x-2} = 1^+ \quad \text{and} \quad \lim_{x\to -\infty} \frac{x+3}{x-2} = 1^-$$

indicate that as $x \to \infty$, $(x+3)/(x-2)$ approaches 1 through numbers larger than 1; and as $x \to -\infty$, $(x+3)/(x-2)$ approaches 1 through numbers smaller than 1. These limits yield the additional information shown in Fig. 2.3(b). We now join the parts of the graph smoothly to produce the graph in Fig. 2.3(c). □

We say in Fig. 2.3(c) that the line $y = 1$ is a *horizontal asymptote* for the curve and the line $x = 2$ is a *vertical asymptote.*

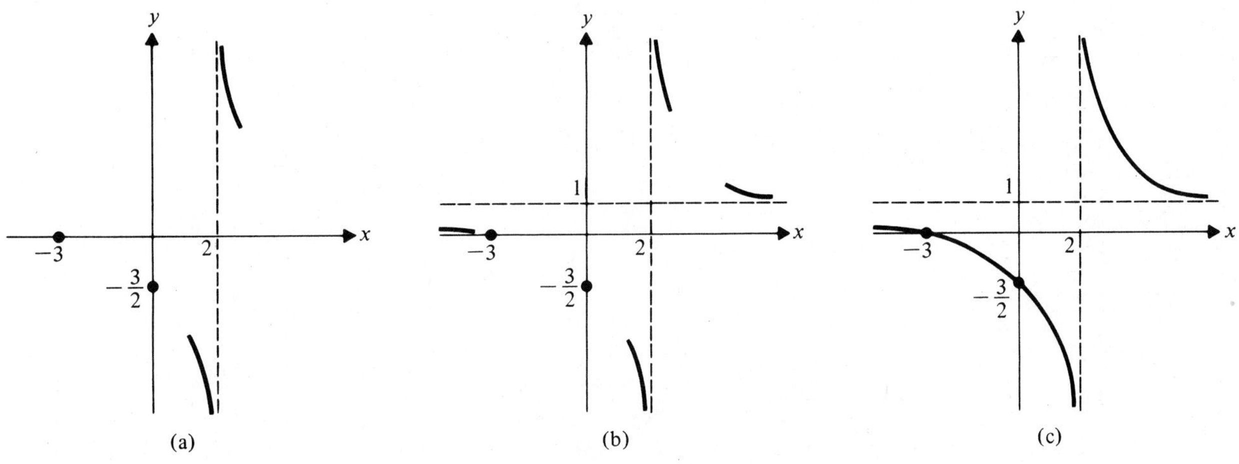

Figure 2.3

Example 12 Sketch a graph of the function $f(x) = x^2/\sqrt{x-1}$.

Solution The function is clearly undefined for $x \le 1$, and to determine its behavior near $x = 1$, we calculate that

$$\lim_{x\to 1^+} \frac{x^2}{\sqrt{x-1}} = \infty.$$

In addition, we find that

$$\lim_{x\to\infty} \frac{x^2}{\sqrt{x-1}} = \infty.$$

This information is shown in Fig. 2.4(a). We now join these parts of the curve as in Fig. 2.4(b). We cannot be certain where the low point on the curve is located, but because $f(x)$ is never equal to zero, we know that it must be above the x-axis. (In Chapter 4 we develop a method for locating the exact position of this point.) □

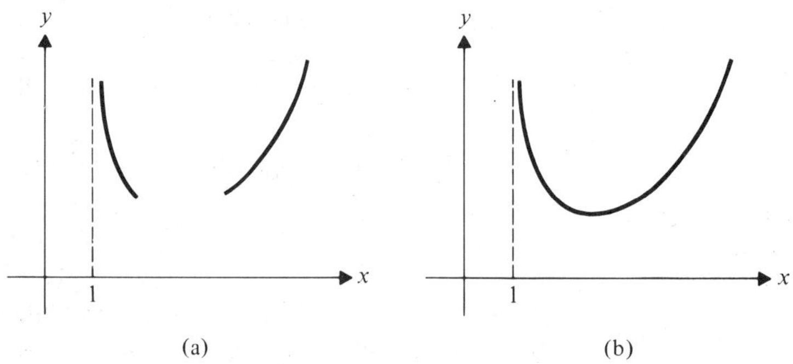

Figure 2.4

EXERCISES 2.1

In each of Exercises 1–20, find the limit if it exists.

1. $\lim_{x\to 7} \frac{x^2-5}{x+2}$
2. $\lim_{x\to -2} \frac{x^3+8}{x+5}$
3. $\lim_{x\to -5} \frac{x^2+3x+2}{x^2+25}$
4. $\lim_{x\to 0} \frac{x^4+3x^3}{3x^4-2x^3}$
5. $\lim_{x\to 2} \frac{x^3-6x^2+12x-8}{x^2-4x+4}$
6. $\lim_{x\to 1} \frac{5x}{(x-1)^3}$
7. $\lim_{x\to\infty} \frac{3x^3+4}{2x^3-5x-6}$
8. $\lim_{x\to 1/2} \frac{6x^2+7x-5}{2x-1}$
9. $\lim_{x\to\infty} \frac{2x^2+3x-5}{5x^2+6x}$
10. $\lim_{x\to\infty} \frac{x^2-2x-4}{2x^3+5}$
11. $\lim_{x\to\infty} \frac{2x^3+5}{x^2-2x-4}$
12. $\lim_{x\to 0} \frac{3x^2+2}{4x^5-2x+5}$
13. $\lim_{x\to -1} \frac{12x+5}{x^2-2x+1}$
14. $\lim_{x\to 1/4} \frac{1}{(4x-1)^4}$
15. $\lim_{x\to 1} \sqrt{\frac{2-x}{2+x}}$
16. $\lim_{x\to -\infty} \frac{x^3+2x}{x^2-3}$
17. $\lim_{x\to 5} \frac{\sqrt{1-x^2}}{3x+2}$
18. $\lim_{x\to\infty} \frac{\sqrt{x^2+1}}{2x+1}$
19. $\lim_{x\to\infty} \frac{\cos x}{x}$
20. $\lim_{x\to\infty} x\cos x$

In Exercises 21–26, sketch a graph of the given function.

21. $f(x) = \frac{1}{|x-2|}$
22. $f(x) = \frac{2}{x^2-x}$
23. $f(x) = \frac{1}{\sqrt{x-1}}$
24. $f(x) = \frac{x+1}{x-1}$
25. $f(x) = \sqrt{\frac{x-3}{x+2}}$
26. $f(x) = \frac{|x-5|}{x-5}$

In each of Exercises 27–30, evaluate the limit if it exists.

27. $\lim_{x\to 5} \frac{|x^2-25|}{x^2-25}$
28. $\lim_{x\to 2} \frac{x-2}{\sqrt{x}-\sqrt{2}}$
29. $\lim_{x\to 0} \frac{\sqrt{1-x}-\sqrt{1+x}}{x}$
30. $\lim_{x\to\infty} (\sqrt{x^2+1}-x)$

31. The weight W of an object depends on its distance d from the center of the earth. If d is less than the radius R of the earth, then W is directly proportional to d; and if d is greater than or equal to R, then W is inversely proportional to d^2. If the weight of the object on the earth's surface is W_0, find an explicit definition for W as a function of d, and sketch its graph.

32. If n is a positive integer, evaluate

$$\lim_{h\to 0}\frac{(x+h)^n - x^n}{h}.$$

Hint: Use either the binomial theorem or the result that

$$a^n - b^n = (a-b)(a^{n-1} + a^{n-2}b + \cdots + ab^{n-2} + b^{n-1}).$$

33. Evaluate

$$\lim_{h\to 0}\frac{\sqrt{x+h}-\sqrt{x}}{h}.$$

C **34.** At the present time it is impossible for us to calculate algebraically

$$\lim_{x\to 0}\frac{\sin x - x}{x^3}.$$

What can we do? One suggestion might be to use a calculator to evaluate the function $(\sin x - x)/x^3$ for various values of x that approach 0. Try this with $x = 0.1$, 0.01, 0.001, 0.0001, and 0.00001, and make any conclusion you feel might be justified. How would you feel about the use of the calculator if you knew that the value of this limit were $-\frac{1}{6}$?

35. Prove or disprove the following statement: If $f(x) < g(x)$ for all $x \neq a$, then

$$\lim_{x\to a} f(x) < \lim_{x\to a} g(x).$$

In Exercises 36–40, sketch a graph of the given function.

36. $f(x) = \dfrac{1}{x^3 - 4x}$

37. $f(x) = \dfrac{1}{x^4 + 3x^2}$

38. $f(x) = \dfrac{2x^3 - 2}{x^3 + 5x^2}$

39. $f(x) = \dfrac{|3x+1|}{x+5}$

40. $f(x) = \dfrac{1}{(x^3 - 3x^2 + 2x)^{1/3}}$

* **41.** Evaluate

$$\lim_{x\to 0}\frac{\sqrt{x+1}-\sqrt{2x+1}}{\sqrt{3x+4}-\sqrt{2x+4}}.$$

2.2 CONTINUITY

The concept of continuity is given in the following definition.

▷ **Definition 2.1** A function $f(x)$ is said to be *continuous* at $x = a$ if

$$\lim_{x\to a} f(x) = f(a). \quad \triangleleft \tag{2.6}$$

There are three conditions imposed by this definition:

1. $f(x)$ must be defined at $x = a$;
2. $\lim_{x\to a} f(x)$ must exist;
3. The numbers in (1) and (2) must be the same.

The graph in Fig. 2.5 illustrates a function that is discontinuous (i.e., is not continuous) at $x = a, b, c, d$, and e. For instance, at $x = a$ conditions (1) and (2) are satisfied but (3) is violated; at $x = b$, (2) is satisfied, but (1) and (3) are not. It appears that discontinuities of a function are characterized geometrically by separations in its graph. This is indeed true, and it is often a very informative way to illustrate the nature of a discontinuity.

When a function is defined on a closed interval $a \le x \le b$, Definition 2.1 must be rephrased in terms of right- and left-hand limits for continuity at $x = a$ and $x = b$. Specifically, we say that $f(x)$ is *continuous from the right* at $x = a$ if $\lim_{x\to a^+} f(x) = f(a)$; and that $f(x)$ is *continuous from the left* at $x = b$ if $\lim_{x\to b^-} f(x) = f(b)$.

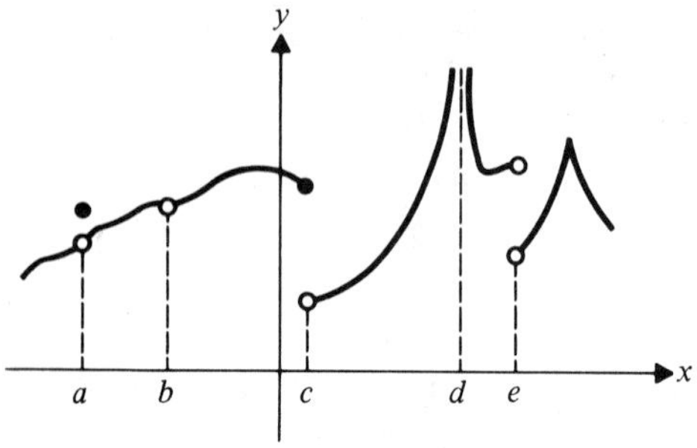

Figure 2.5

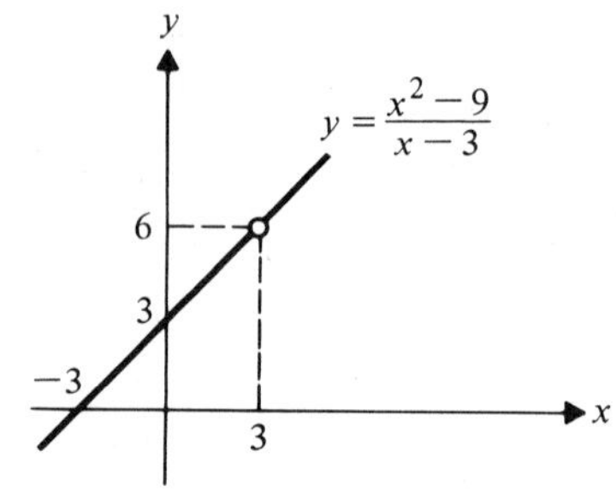

Figure 2.6

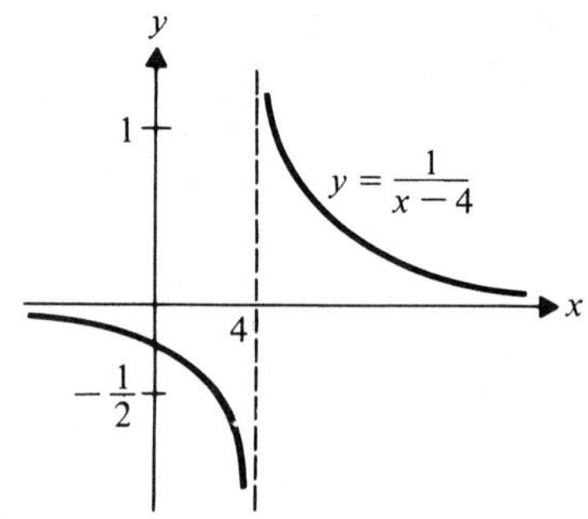

Figure 2.7

A function is said to be *continuous on an interval* if it is continuous at each point of that interval. In the event that the interval is closed, $a \le x \le b$, continuity at $x = a$ and $x = b$ is interpreted as continuity from the right and left, respectively. Geometrically, a function is continuous on an interval if a pencil can completely trace its graph without ever being lifted from the page.

Example 13 Sketch a graph of each of the following functions, indicating any discontinuities.

a) $f(x) = \dfrac{x^2 - 9}{x - 3}$

b) $f(x) = \dfrac{1}{x - 4}$

c) $f(x) = \dfrac{|x^2 - 25|}{x^2 - 25}$

Solution a) Since $f(x)$ is undefined at $x = 3$, the function is discontinuous there. For $x \neq 3$,

$$y = f(x) = \frac{(x + 3)(x - 3)}{x - 3} = x + 3.$$

Consequently, the graph of $f(x)$ is a straight line with a hole at $x = 3$ (Fig. 2.6). Note that $\lim_{x \to 3} [(x^2 - 9)/(x - 3)]$ exists and is equal to 6.

b) Since $1/(x - 4)$ is undefined at $x = 4$, the function is discontinuous there. The limits

$$\lim_{x \to 4^+} \frac{1}{x - 4} = \infty, \qquad \lim_{x \to 4^-} \frac{1}{x - 4} = -\infty,$$

$$\lim_{x \to \infty} \frac{1}{x - 4} = 0^+, \qquad \lim_{x \to -\infty} \frac{1}{x - 4} = 0^-$$

help to sketch the graph in Fig. 2.7.

c) The function $f(x) = |x^2 - 25|/(x^2 - 25)$ is undefined at $x = \pm 5$, and is therefore discontinuous at these values. When $-5 < x < 5$,

$$f(x) = \frac{25 - x^2}{x^2 - 25} = -1,$$

and when $|x| > 5$,

$$f(x) = \frac{x^2 - 25}{x^2 - 25} = 1.$$

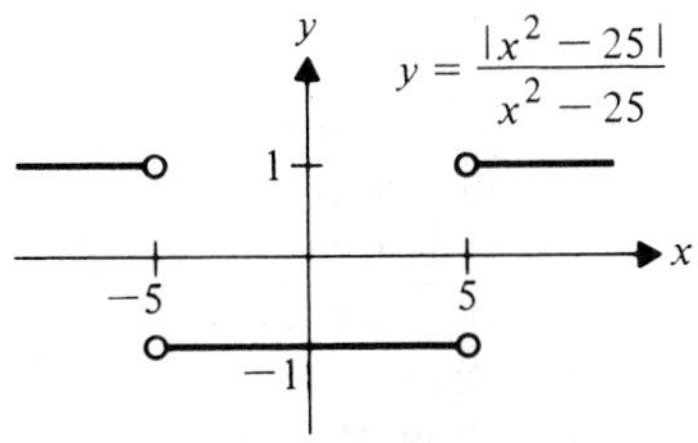

Figure 2.8

The graph is shown in Fig. 2.8. In this example, neither of the limits

$$\lim_{x \to 5} \frac{|x^2 - 25|}{x^2 - 25} \quad \text{or} \quad \lim_{x \to -5} \frac{|x^2 - 25|}{x^2 - 25}$$

exists. □

Theorem 2.1 implies the following.

▶ **Theorem 2.2** If $f(x)$ and $g(x)$ are continuous at $x = a$, then so too are the functions $f(x) + g(x)$, $f(x)g(x)$, and $f(x)/g(x)$ (provided $g(a) \neq 0$ in the case of division). ◀

It is an immediate consequence of this theorem that every function of the form

$$f(x) = a_n x^n + a_{n-1}x^{n-1} + \cdots + a_2 x^2 + a_1 x + a_0, \tag{2.7}$$

where n is a nonnegative integer, and $a_0, a_1, \ldots, a_n$ are real numbers ($a_n \neq 0$), is continuous for all x. Such a function is called a *polynomial of degree n*. Special names have become associated with some of the lower-degree polynomials because of the frequency with which they arise:

$f(x) = a_1 x + a_0$	polynomial of degree 1, linear polynomial, linear function
$f(x) = a_2 x^2 + a_1 x + a_0$	polynomial of degree 2, quadratic polynomial, quadratic function
$f(x) = a_3 x^3 + a_2 x^2 + a_1 x + a_0$	polynomial of degree 3, cubic polynomial, cubic function

A *rational function* $R(x)$ is defined as a polynomial $P(x)$ divided by a polynomial $Q(x)$:

$$R(x) = \frac{P(x)}{Q(x)}. \tag{2.8}$$

It is a corollary of Theorem 2.2 that a rational function is discontinuous only where $Q(x) = 0$.

Another useful result is contained in the following theorem.

▶ **Theorem 2.3** If $\lim_{x \to a} f(x) = L$, and $g(y)$ is a function that is continuous at $y = L$, then

$$\lim_{x \to a} g[f(x)] = g\left[\lim_{x \to a} f(x)\right] = g(L). \quad ◀ \tag{2.9}$$

What this theorem says is that the operation of a continuous function and the operation of taking a limit can be interchanged. For example, since the square root function is continuous for positive arguments, we can write that

$$\lim_{x\to 5}\sqrt{x^3+5x-2}=\sqrt{\lim_{x\to 5}(x^3+5x-2)}=\sqrt{148}.$$

For such examples it is likely that we would interchange these operations without really thinking about it, since it seems obvious (see, for example, Exercises 15, 17, and 18 in Section 2.1). In Chapter 8, however, we will encounter situations where it will not be so evident that the interchange can be made, and we will then call on Theorem 2.3 as our authority for doing so.

Discontinuous functions are encountered in many applications of mathematics. To illustrate, consider the phenomenon of radioactive disintegration, during which radioactive elements such as uranium-238 emit high-energy particles and in the process become new elements. If we analyze an ore sample that contains, along with various impurities, an amount A_0 of uranium, then disintegration of uranium nuclei into thorium-234 nuclei gradually reduces the amount of uranium in the sample. Suppose we let $A = f(t)$ represent the amount of uranium in the sample at any given time, and choose time $t = 0$ when $A = A_0$. If we were to measure carefully the amount of uranium in the sample over a very short time interval (perhaps 10^{-10} second), a graph of $A = f(t)$ might appear as shown in Fig. 2.9. Each jump (discontinuity) in this graph signals the disintegration of a uranium nucleus into a thorium nucleus. Were we to complete this graph, and assume that no two disintegrations occur at exactly the same instant, then the total number of discontinuities would be equal to the original number of uranium nuclei in the sample. In practice, we cannot hope to make such microscopic measurements. Instead, we must take measurements of the amount of uranium in the sample at much larger time intervals, and plot the data as a sequence of points, as in Fig. 2.10(a). If these points are joined with a smooth curve as in Fig. 2.10(b), we obtain a reasonable picture of A as a function of time t. This "macroscopic" view (view in the large) of $A = f(t)$ suppresses the true nature of the graph, but it would be entirely adequate if we were interested in knowing, perhaps, when disintegrations would reduce the uranium to half its original amount. Questions such as these will be answered when we continue our discussion of radioactive disintegration in Section 9.3.

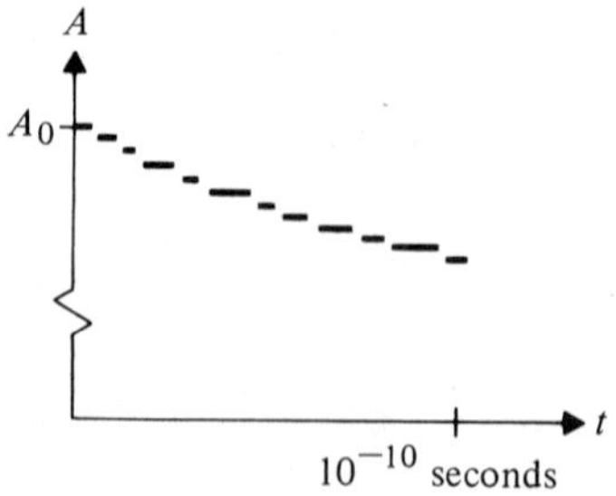

Figure 2.9

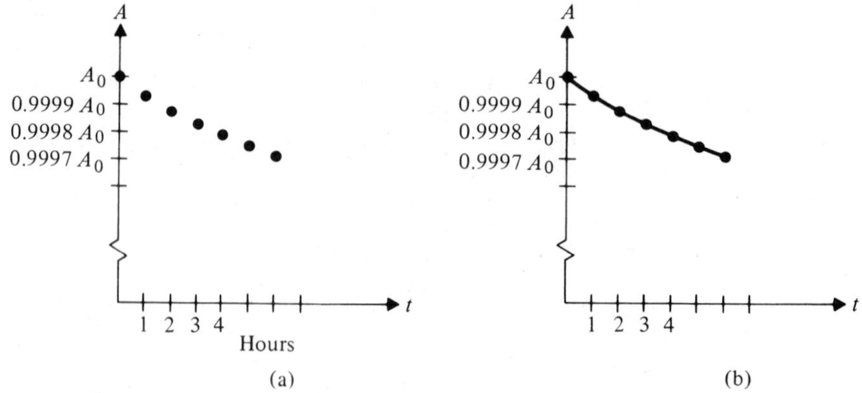

Figure 2.10

A second example is taken from the field of economics. A sugar manufacturer sells sugar in one-kilogram bags at the following prices:

40¢ per bag for the first 100 bags;

35¢ per bag for each bag after 100 up to 1000;

30¢ per bag for each bag after 1000.

If we denote the selling price of the xth bag by $f(x)$, then

$$f(x) = \begin{cases} 40, & 1 \le x \le 100, \\ 35, & 101 \le x \le 1000, \\ 30, & x \ge 1001. \end{cases}$$

This function is unlike any we have previously seen; its domain consists of only the positive integers. Its "graph," shown in Fig. 2.11, therefore consists of dots above these integers. In this example, the domain of the function is extended to all positive values of x according to

$$f(x) = \begin{cases} 40, & 0 < x \le 100, \\ 35, & 100 < x \le 1000, \\ 30, & x > 1000, \end{cases}$$

with the resulting graph shown in Fig. 2.12.

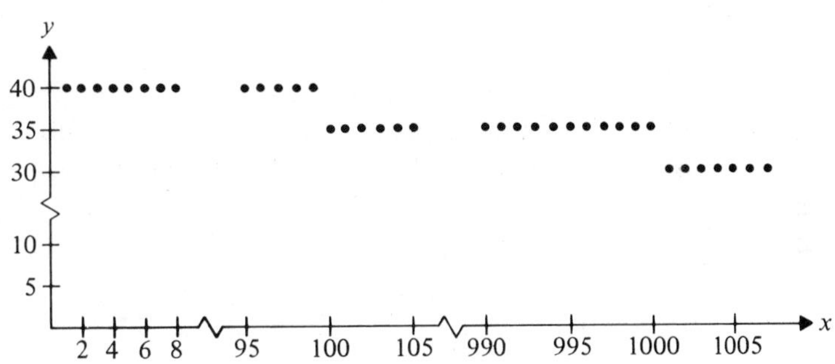

Figure 2.11

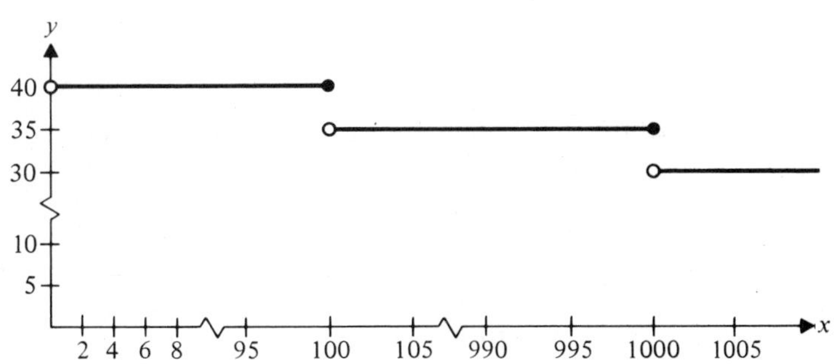

Figure 2.12

In spite of the fact that $f(x)$ is now defined for all $x > 0$, it still has significance only for x a positive integer. For instance, $f(2.5)$ would indicate the price of half a bag of sugar after two bags had already been purchased. To say that this half bag would cost 40¢ is pointless, since the manufacturer probably doesn't even sell sugar in half-kilogram bags. The advantage of extending the domain of $f(x)$ to $x > 0$ is therefore not one of physical significance. The real reason is that calculus can be applied to the extended function in order to analyze the economics of sugar sales. For the present, it is clear that the function $f(x)$ in Fig. 2.12 is discontinuous at $x = 100$ and $x = 1000$.

We can derive from $f(x)$ a function that is far more useful in economic theory: the average price $r(x)$. It is defined in this example as the average price of a one-kilogram bag of sugar when x bags are purchased. Using the fact that the average price is the total selling price divided by the total number of bags, we calculate that

$$r(x) = \begin{cases} 40, & 0 < x \leq 100, \\ \dfrac{40(100) + 35(x - 100)}{x}, & 100 < x \leq 1000, \\ \dfrac{40(100) + 35(900) + 30(x - 1000)}{x}, & x > 1000 \end{cases}$$

$$= \begin{cases} 40, & 0 < x \leq 100, \\ 35 + \dfrac{500}{x}, & 100 < x \leq 1000, \\ 30 + \dfrac{5500}{x}, & x > 1000. \end{cases}$$

Again, $r(x)$ is meaningful only for x an integer, but as shown in Fig. 2.13 we extend its domain of definition to all positive values of x. In addition, note that unlike $f(x)$, $r(x)$ has no discontinuities for $x > 0$.

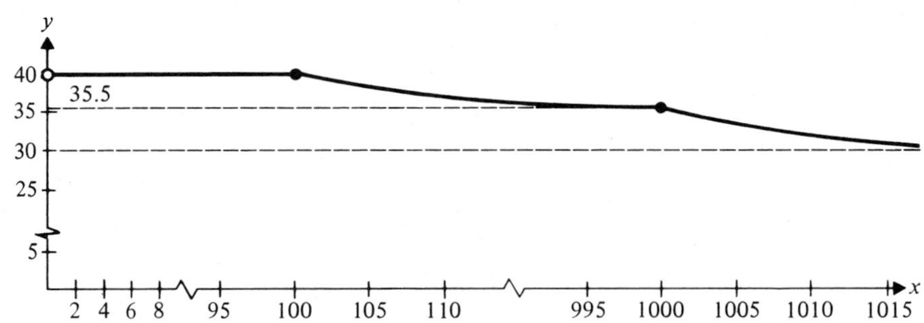

Figure 2.13

EXERCISES 2.2

In Exercises 1–12, sketch a graph of the given function, indicating any discontinuities.

1. $f(x) = \dfrac{1}{x + 2}$

2. $f(x) = \dfrac{16 - x^2}{x + 4}$

3. $f(x) = |x^2 - 5|$

4. $f(x) = \dfrac{12}{x^2 + 2}$

5. $f(x) = \dfrac{12}{x^2 + 2x}$

6. $f(x) = \dfrac{12}{x^2 + 2x + 2}$

7. $f(x) = \dfrac{3 + 2x - x^2}{x + 1}$

8. $f(x) = \dfrac{x^3 + x^2 - 2x}{x^2 - x}$

9. $f(x) = \dfrac{x^3 - 2x^2 + 5x - 10}{x - 2}$

10. $f(x) = \dfrac{x + 12}{x^2 - 9}, \ -5 \leq x \leq 4$

11. $f(x) = \dfrac{x^2 + 2x}{x^2 - 9}, \ -3 \leq x \leq 6$

12. $f(x) = \dfrac{x^3 - 27}{|x - 3|}$

13. The number of people in the city of New York is a function of time t, say, $N = f(t)$. Make two graphs of this function, one "in the small" to show birth and death situations, and a second "in the large" to get an overview of the rate of growth of the city.

14. The *signum function* (or *sign function*), denoted by sgn, is defined by

$$\operatorname{sgn} x = \begin{cases} -1, & x < 0, \\ 0, & x = 0, \\ 1, & x > 0. \end{cases}$$

Sketch its graph.

15. Sketch a graph of the function $f(x) = \operatorname{sgn}(x + 1) - \operatorname{sgn}(x - 1)$, indicating any discontinuities.

16. A toy manufacturer has two pricing policies for one of his dolls:

Policy I: \$15.00 each for the first 10 dolls, \$12.50 for each doll after 10 up to 100, \$10.00 for each doll after 100;

Policy II: \$15.00 each for the first 10 dolls, 10% off the total price on orders of 11 or more, 20% off the total price on orders of 101 or more.

Sketch average price functions for each of these policies on the same graph. Which policy is better for a retailer?

In Exercises 17–22, find a rational function whose graph fits generally the shape of the given curve.

17.

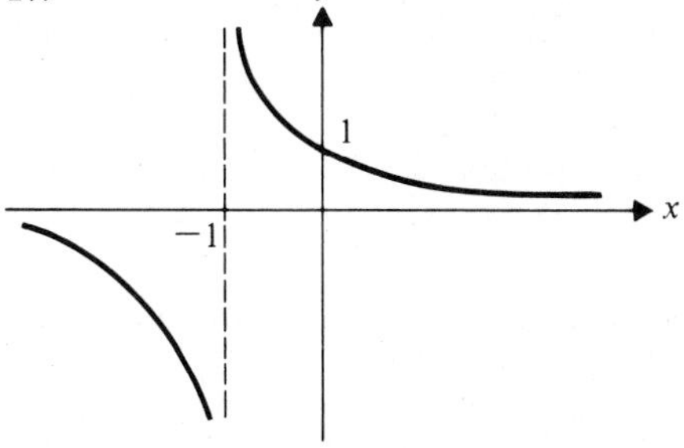

18.

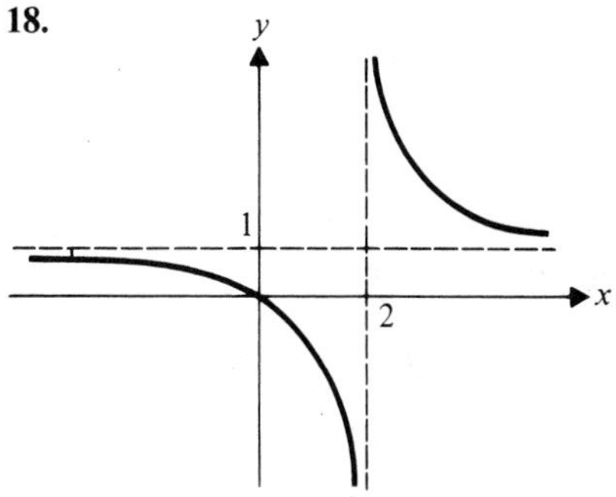

19.

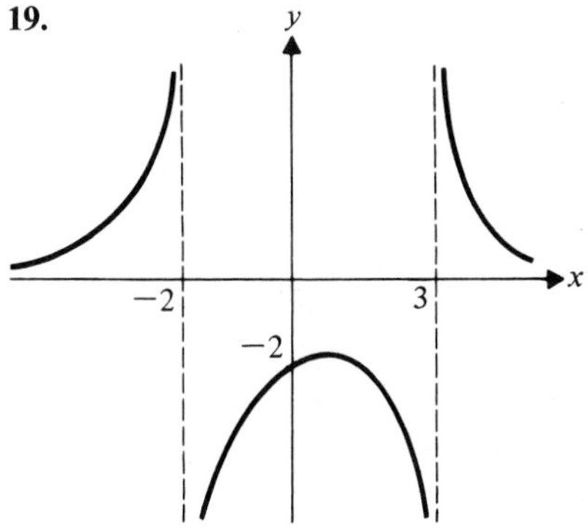

20.

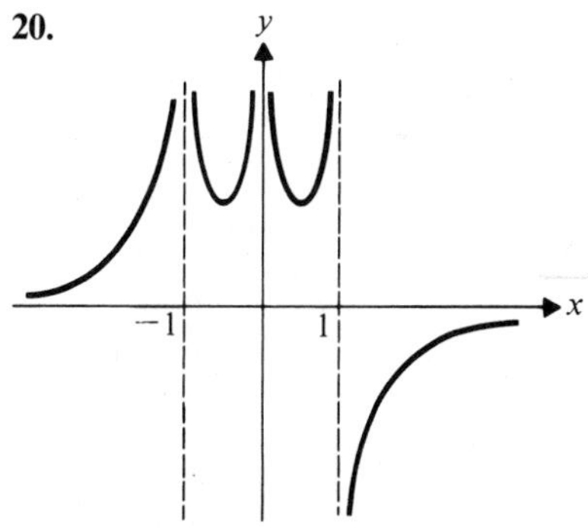

21.

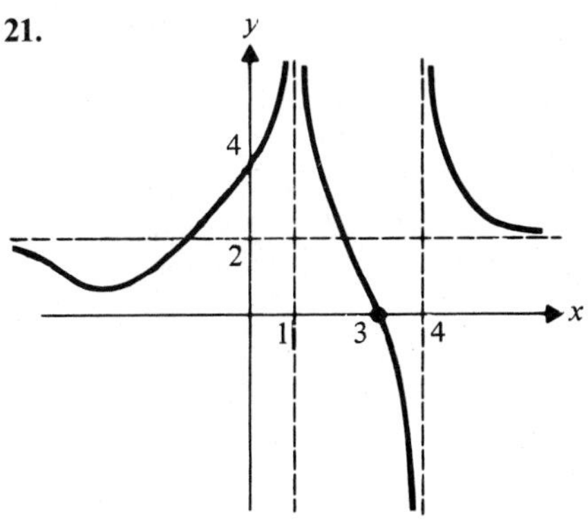

22.

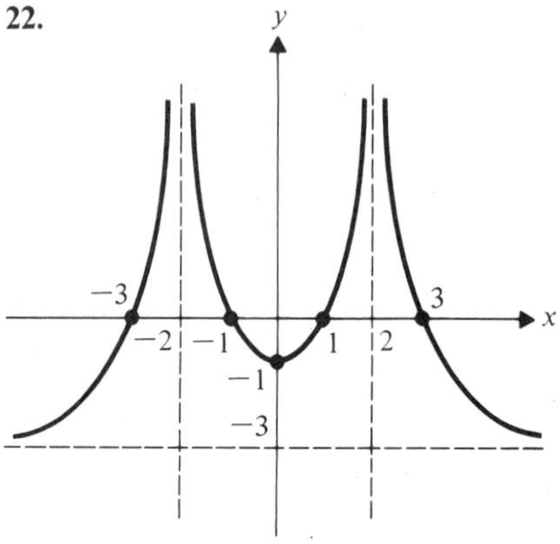

23. Three of the most important functions in physics and engineering are defined below. Sketch a graph of each.

a) The Heaviside unit step function $H_a(x)$:

$$H_a(x) = \begin{cases} 0, & x < a, \\ 1, & x > a. \end{cases}$$

b) The unit pulse function $P_{a,\varepsilon}(x)$ $(\varepsilon \neq 0)$:

$$P_{a,\varepsilon}(x) = \frac{1}{\varepsilon}\{H_a(x) - H_{a+\varepsilon}(x)\}.$$

c) The Dirac delta function $\delta_a(x)$:

$$\delta_a(x) = \lim_{\varepsilon \to 0} P_{a,\varepsilon}(x).$$

*24. A retailer of building supplies sells sand by the bag and in bulk. He quotes prices as follows:

$3.00 per 50-kilogram bag up to 20 bags,
$50.00 per ton (metric) delivered only in integer tons (i.e., 1 ton, 2 tons, etc.).

If, for example, a customer wants 2.2 tons of sand, he receives 2 tons in bulk and 4 bags. Find an algebraic expression for the average price function and sketch its graph. (*Hint:* The greatest integer function introduced in Review Exercise 28 of Chapter 1 may be useful.)

2.3 A MATHEMATICAL DEFINITION OF LIMITS

As a final consideration in this chapter, we show how our intuitive ideas of limits can be developed into a mathematical definition. We start with our intuitive statement of a limit and make a succession of paraphrases, each of which is one step closer to the precise definition. It is not our intention to cover the groundwork for a rigorous course in mathematical analysis; we wish only to demonstrate that our intuitive ideas of a limit can be stated precisely. Moreover, because the definition of a limit is at first sight quite complicated, we wish to show that it can be obtained by a fairly straightforward sequence of steps. We hope that this will be a convincing argument that the definition of a limit does indeed describe in mathematical terms the concept of a limit.

A function $f(x)$ has limit L as x approaches a if $f(x)$ gets arbitrarily close to L, and stays close to L, whenever x gets closer and closer to a.

A function $f(x)$ has limit L as x approaches a if the difference $|f(x) - L|$ gets arbitrarily close to zero, and stays close to zero, whenever x gets closer and closer to a.

A function $f(x)$ has limit L as x approaches a if given any real number $\varepsilon > 0$, no matter how small, we can make the difference $|f(x) - L|$ less than ε by choosing x sufficiently close to a.

A function $f(x)$ has limit L as x approaches a if given any $\varepsilon > 0$, we can make $|f(x) - L| < \varepsilon$ by choosing $|x - a|$ sufficiently close to zero.

Finally we have the following definition.

▷ **Definition 2.2** A function $f(x)$ has limit L as x approaches a if given any $\varepsilon > 0$, we can find a $\delta > 0$ such that

$$|f(x) - L| < \varepsilon$$

whenever $0 < |x - a| < \delta$. ◁

This definition states in precise terms our intuitive idea of a limit: that $f(x)$ can be made arbitrarily close to L by choosing x sufficiently close to a. The following two examples will help put the definition in its proper perspective. Consider proving first that

$$\lim_{x \to 3} (2x + 4) = 10.$$

We must show that given any $\varepsilon > 0$, there is a $\delta > 0$ such that

$$|(2x + 4) - 10| < \varepsilon$$

whenever $0 < |x - 3| < \delta$; i.e., we must find a δ. Since

$$|2x + 4 - 10| = |2x - 6| = 2|x - 3|,$$

we must find a δ such that

$$2|x - 3| < \varepsilon$$

whenever $0 < |x - 3| < \delta$. But this will be true if $\delta = \varepsilon/2$; in other words, if we choose x to satisfy $0 < |x - 3| < \delta = \varepsilon/2$, then

$$|(2x + 4) - 10| = 2|x - 3| < 2(\varepsilon/2) = \varepsilon.$$

The verification is now complete. We have shown that we can make $2x + 4$ as close to 10 as we want (within ε) by choosing x sufficiently close to 2 (within $\delta = \varepsilon/2$).

Verifying that

$$\lim_{x \to 2} (x^2 + 5) = 9$$

is somewhat more complicated, but the idea is the same. We must show that given any $\varepsilon > 0$, there is a $\delta > 0$ such that

$$|(x^2 + 5) - 9| < \varepsilon \quad \text{or} \quad |x^2 - 4| < \varepsilon$$

whenever $0 < |x - 2| < \delta$; i.e., we must find a δ. By working backward from the ε-inequality we can see how to choose δ. In addition, because the δ-inequality involves x in the combination $x - 2$, we rework the ε-inequality into the same combination. Since

$$|x^2 - 4| = |(x - 2)^2 + 4x - 8| = |(x - 2)^2 + 4(x - 2)|,$$

we must choose δ such that

$$|(x - 2)^2 + 4(x - 2)| < \varepsilon$$

whenever $0 < |x - 2| < \delta$. From inequality (1.4b), we can state that

$$|(x - 2)^2 + 4(x - 2)| \le |x - 2|^2 + 4|x - 2|.$$

As a result, if we can find a $\delta > 0$ such that

$$|x-2|^2 + 4|x-2| < \varepsilon$$

whenever $0 < |x-2| < \delta$, then we will have a satisfactory δ. But now we have the key. Suppose we set $z = |x-2|$ and consider the parabola $Q(z) = z^2 + 4z - \varepsilon$ shown in Fig. 2.14. It crosses the z-axis when

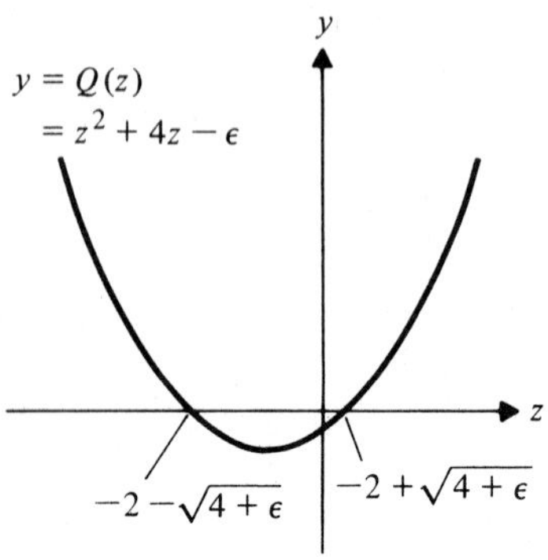

Figure 2.14

$$z^2 + 4z - \varepsilon = 0,$$

a quadratic equation with solutions

$$z = \frac{-4 \pm \sqrt{16 + 4\varepsilon}}{2} = -2 \pm \sqrt{4 + \varepsilon}.$$

The graph shows that whenever $0 < z < -2 + \sqrt{4 + \varepsilon}$, then

$$z^2 + 4z - \varepsilon < 0.$$

In other words, if $0 < z < \delta = \sqrt{4 + \varepsilon} - 2$, then

$$z^2 + 4z < \varepsilon.$$

Since $z = |x-2|$, we can say that if $0 < |x-2| < \delta = \sqrt{4 + \varepsilon} - 2$, then

$$|x-2|^2 + 4|x-2| < \varepsilon,$$

and therefore

$$|(x-2)^2 + 4(x-2)| < \varepsilon.$$

The proof is now complete. What we have shown is that we can make $x^2 + 5$ as close to 9 as we want (within ε) by choosing x sufficiently close to 2 (within $\delta = \sqrt{4 + \varepsilon} - 2$).

These examples have indicated the difference between "evaluation of" and "verification of" a limit. If you have learned from this only an appreciation for this difference, then our discussion has been worthwhile. If your appreciation goes beyond this and includes a working understanding of the definition of a limit, we could hope for no more. Results in succeeding chapters, however, will rely on our intuitive idea of a limit and make little use of the ε, δ-definition.

EXERCISES 2.3

Verify the limits in each of Exercises 1–4.

1. $\lim_{x \to 1} (x + 5) = 6$
2. $\lim_{x \to 0} (x^2 + 3) = 3$
3. $\lim_{x \to 3} (x^2 - 7x) = -12$
4. $\lim_{x \to 1} (x^2 + 3x + 5) = 9$
5. Give a mathematical definition for each of the following.
   a) $\lim_{x \to \infty} f(x) = L$
   b) $\lim_{x \to a} f(x) = \infty$
6. Verify that
$$\lim_{x \to 2} \frac{x + 2}{x - 1} = 4.$$
7. Use Definition 2.2 to prove that a function $f(x)$ cannot have two limits as x approaches a.
8. Prove that if $\lim_{x \to a} f(x) = L > 0$, then there exists an open interval I containing a in which $f(x) > 0$ except possibly at $x = a$.

In Exercises 9–11 we use Definition 2.2 to prove Theorem 2.1. In each exercise assume that $\lim_{x\to a} f(x) = F$ and that $\lim_{x\to a} g(x) = G$.

9. Show that given any $\varepsilon > 0$, there exist numbers $\delta_1 > 0$ and $\delta_2 > 0$ such that

$$|f(x) - F| < \varepsilon/2 \quad \text{whenever } 0 < |x - a| < \delta_1$$

and

$$|g(x) - G| < \varepsilon/2 \quad \text{whenever } 0 < |x - a| < \delta_2.$$

Use these results along with identity (1.4b) to prove part (i) of Theorem 2.1.

***10.** a) Verify that

$$|f(x)g(x) - FG| \le |f(x)||g(x) - G| + |G||f(x) - F|.$$

b) Show that given any $\varepsilon > 0$ there exist numbers $\delta_1 > 0$, $\delta_2 > 0$, and $\delta_3 > 0$ such that

$$|f(x)| < |F| + 1 \qquad \text{whenever } 0 < |x - a| < \delta_1,$$

$$|g(x) - G| < \frac{\varepsilon}{2(|F| + 1)} \qquad \text{whenever } 0 < |x - a| < \delta_2,$$

$$|f(x) - F| < \frac{\varepsilon}{2|G|} \qquad \text{whenever } 0 < |x - a| < \delta_3.$$

c) Use these results to prove part (ii) of Theorem 2.1.

***11.** a) Verify that when $G \neq 0$,

$$\left|\frac{f(x)}{g(x)} - \frac{F}{G}\right| \le \frac{|f(x) - F|}{|g(x)|} + \frac{|F||G - g(x)|}{|G||g(x)|}.$$

b) Show that given any $\varepsilon > 0$ there exist numbers $\delta_1 > 0$, $\delta_2 > 0$, and $\delta_3 > 0$ such that

$$|g(x)| > \frac{|G|}{2} \qquad \text{whenever } 0 < |x - a| < \delta_1,$$

$$|f(x) - F| < \frac{\varepsilon|G|}{4} \qquad \text{whenever } 0 < |x - a| < \delta_2,$$

$$|g(x) - G| < \frac{\varepsilon|G|^2}{4|F|} \qquad \text{whenever } 0 < |x - a| < \delta_3.$$

c) Now prove part (iii) of Theorem 2.1.

SUMMARY

In Section 2.1 we introduced limits of functions. For the most part our discussion was intuitive, beginning with the statement "$\lim_{x\to a} f(x) = L$ if $f(x)$ can be made arbitrarily close to L by choosing x sufficiently close to a." This idea was then extended to include the following:

Right-hand limits:	$\lim_{x\to a^+} f(x)$;
Left-hand limits:	$\lim_{x\to a^-} f(x)$;
Limits at infinity:	$\lim_{x\to\infty} f(x)$, $\lim_{x\to-\infty} f(x)$;
Infinite limits:	$\lim_{x\to a} f(x) = \infty$, $\lim_{x\to a} f(x) = -\infty$.

Keep in mind that the term "infinite limits" is somewhat of a misnomer since in both situations the limit does *not* exist.

A function $f(x)$ is said to be continuous at a point $x = a$ if it satisfies the three conditions contained in the statement

$$\lim_{x\to a} f(x) = f(a).$$

The function is said to be continuous on an interval if it is continuous at each point of that interval. Geometrically, this means that one must be able to completely trace its graph without ever lifting the pencil from the page.

We illustrated that limits can be used to sketch the graph of a function near a point of discontinuity, as well as to indicate its behavior for very large positive and negative values of x. To our list of techniques for sketching curves, which already contains such methods as the addition and multiplication of ordinates and symmetries, we therefore add the use of limits.

In Section 2.3 we developed the mathematical definition of limit, and gave two examples to illustrate how the definition can be used.

REVIEW EXERCISES

In Exercises 1–12, evaluate the given limit if it exists.

1. $\lim\limits_{x\to 3} \dfrac{x^2+5}{x^3-3}$

2. $\lim\limits_{x\to \infty} \dfrac{3x^3+2x}{x^3-10x^2+4}$

3. $\lim\limits_{x\to 6} \dfrac{x^2-36}{x-6}$

4. $\lim\limits_{x\to 0} \dfrac{x^2+2x}{3x-2x^2}$

5. $\lim\limits_{x\to -\infty} \dfrac{x+4}{2x^2-5}$

6. $\lim\limits_{x\to 1} \dfrac{x-1}{x^2-2x+1}$

7. $\lim\limits_{x\to \infty} \dfrac{\sin x}{x}$

8. $\lim\limits_{x\to 1} \dfrac{\sqrt{x}-1}{x}$

9. $\lim\limits_{x\to 0^+} \dfrac{\sqrt{x}-1}{x-1}$

10. $\lim\limits_{x\to 1} \dfrac{x^2+5}{x-1}$

11. $\lim\limits_{x\to 1} \dfrac{x^2+5}{(x-1)^2}$

12. $\lim\limits_{x\to 1/2} \dfrac{(2-4x)^3}{x(2x-1)^2}$

In Exercises 13–18, sketch a graph of the given function.

13. $f(x) = \dfrac{1}{x-2}$

14. $f(x) = \dfrac{x}{x-2}$

15. $f(x) = \dfrac{x^2}{x-2}$

16. $f(x) = \dfrac{x^2-36}{x-6}$

17. $f(x) = \left|\dfrac{x+1}{x-1}\right|$

18. $f(x) = \dfrac{2x}{x^2-3x-4}$

Differentiation

3

3.1 THE DERIVATIVE

Consider a function $y = f(x)$ that is defined for all x in the interval $x_1 \leq x \leq x_2$. If we set $y_1 = f(x_1)$ and $y_2 = f(x_2)$, then the quotient

$$\frac{y_2 - y_1}{x_2 - x_1} = \frac{f(x_2) - f(x_1)}{x_2 - x_1} \tag{3.1}$$

is called the *average rate of change* of y with respect to x in the interval from x_1 to x_2.

For example, if a car moves along a straight road coordinatized by y, its position on the road is a function of time t, say, $y = f(t)$. The quotient

$$\frac{y_2 - y_1}{t_2 - t_1} = \frac{f(t_2) - f(t_1)}{t_2 - t_1}$$

in this case is the average velocity of the car between times t_1 and t_2.

As a second example, consider an open cylindrical container placed outside a house. The depth D of water in the container during a rainstorm can be represented as a function of time t, say, $D = f(t)$. The quotient

$$\frac{D_2 - D_1}{t_2 - t_1} = \frac{f(t_2) - f(t_1)}{t_2 - t_1}$$

represents the average rate of rainfall during the time interval from t_1 to t_2.

Suppose we consider x_1 in quotient (3.1) as fixed and take a number of values of x_2 that are successively closer and closer to x_1. If we demand that x_2 get arbitrarily close to x_1, then we are really asking for the limit of quotient (3.1) as x_2 approaches x_1, and we can write this as

$$\lim_{x_2 \to x_1} \frac{y_2 - y_1}{x_2 - x_1} = \lim_{x_2 \to x_1} \frac{f(x_2) - f(x_1)}{x_2 - x_1}. \tag{3.2}$$

It would seem reasonable to define this limit, if it exists, as the *instantaneous rate of change* of y with respect to x at $x = x_1$.

For our examples,

$$\lim_{t_2 \to t_1} \frac{y_2 - y_1}{t_2 - t_1} = \lim_{t_2 \to t_1} \frac{f(t_2) - f(t_1)}{t_2 - t_1}$$

would be the instantaneous velocity of the car at time t_1, and

$$\lim_{t_2 \to t_1} \frac{D_2 - D_1}{t_2 - t_1} = \lim_{t_2 \to t_1} \frac{f(t_2) - f(t_1)}{t_2 - t_1}$$

would be the instantaneous rate of rainfall at time t_1.

Because subscripts in the notation for these instantaneous rates of change make it so cumbersome, we introduce new simplifying notation. When x in $f(x)$ is changed by a small amount, we call the change an *increment* in x and denote it by Δx. For instance, in Fig. 3.1(a), Δx is positive; and in Fig. 3.1(b), Δx is negative. If $f(x)$ is defined throughout the closed interval between x and

$x + \Delta x$, the average rate of change of y with respect to x from x to $x + \Delta x$ is

$$\frac{f(x + \Delta x) - f(x)}{(x + \Delta x) - x} = \frac{f(x + \Delta x) - f(x)}{\Delta x}.$$

The instantaneous rate of change of y with respect to x at x is

$$\lim_{\Delta x \to 0} \frac{f(x + \Delta x) - f(x)}{\Delta x}.$$

This limit is precisely what we mean by the derivative of the function $f(x)$. In other words, we have the following definition.

▷ **Definition 3.1** The *derivative* of a function $f(x)$ with respect to x is defined as

$$f'(x) = \lim_{\Delta x \to 0} \frac{f(x + \Delta x) - f(x)}{\Delta x} \tag{3.3}$$

provided the limit exists. ◁

The operation of taking the derivative is called *differentiation*. We have defined the derivative of a function algebraically as its instantaneous rate of change. Its geometric interpretation is also important.

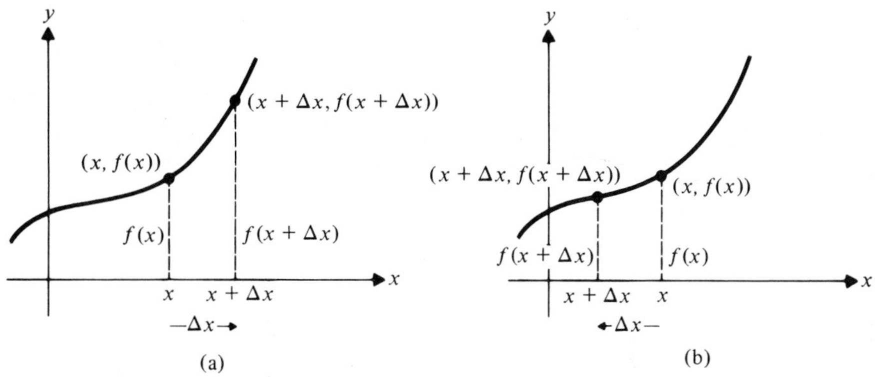

Figure 3.1

Tangent Line to a Curve

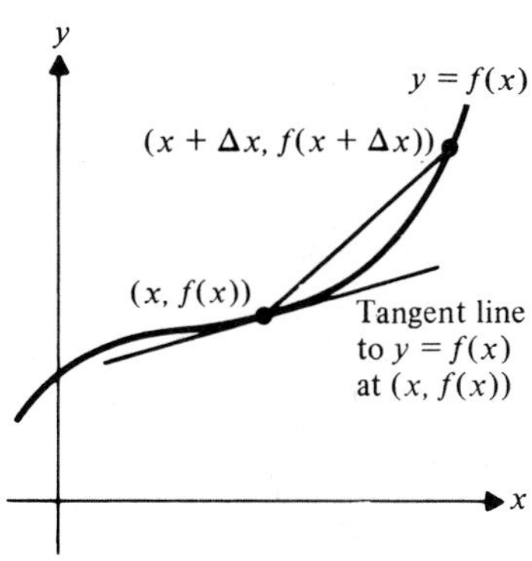

Figure 3.2

Suppose on the graph of a function $f(x)$ we join the two points $(x, f(x))$ and $(x + \Delta x, f(x + \Delta x))$ with a straight line (Fig. 3.2). If we let Δx get smaller and smaller, then the point $(x + \Delta x, f(x + \Delta x))$ moves along the curve toward $(x, f(x))$. If this line has a limiting position as Δx approaches 0, we call it the *tangent line* to the curve $y = f(x)$ at the point $(x, f(x))$. Since the quotient $[f(x + \Delta x) - f(x)]/\Delta x$ in Eq. (3.3) represents the slope of the line joining $(x, f(x))$ and $(x + \Delta x, f(x + \Delta x))$, $f'(x)$ must therefore be the slope of the tangent line to the curve at the point $(x, f(x))$. Algebraically, then, the derivative of a function is its instantaneous rate of change; geometrically, it is the slope of the tangent line to the graph of the function.

The derivative of $f(x)$ at one value of x is, in general, different from the derivative at another value of x. Geometrically, this is evident since $f'(x)$ represents the slope of the tangent line to the curve $y = f(x)$, and this will certainly vary from point to point.

Another common notation for the derivative of $y = f(x)$ with respect to x is $\dfrac{dy}{dx}$. We must be careful when using this notation not to interpret it as a quotient; it is one symbol representing an accumulation of all the operations in Eq. (3.3). It is not, therefore, to be read as "dy divided by dx." Typographically, however, it is easier to print dy/dx rather than $\dfrac{dy}{dx}$, and we will take this liberty whenever it is convenient to do so. But remember, dy/dx is not a quotient; it is one symbol representing the limit operation in Eq. (3.3).

Sometimes it is more convenient to use part of each of these notations and write

$$\frac{d}{dx} f(x).$$

In this form we understand that d/dx means to differentiate with respect to x whatever follows it, in this case $f(x)$.

Example 1 Find $f'(x)$ if $f(x) = x^2$.

Solution From Definition 3.1, we have

$$\begin{aligned} f'(x) &= \lim_{\Delta x \to 0} \frac{f(x + \Delta x) - f(x)}{\Delta x} \\ &= \lim_{\Delta x \to 0} \frac{(x + \Delta x)^2 - x^2}{\Delta x} \\ &= \lim_{\Delta x \to 0} \frac{x^2 + 2x(\Delta x) + (\Delta x)^2 - x^2}{\Delta x} \\ &= \lim_{\Delta x \to 0} \frac{(2x + \Delta x)(\Delta x)}{\Delta x} \\ &= \lim_{\Delta x \to 0} (2x + \Delta x) = 2x. \ \square \end{aligned}$$

Example 2 Find dy/dx if $y = f(x) = x^3 - 3x$.

Solution From Definition 3.1, we get

$$\begin{aligned} \frac{dy}{dx} &= \lim_{\Delta x \to 0} \frac{f(x + \Delta x) - f(x)}{\Delta x} \\ &= \lim_{\Delta x \to 0} \frac{[(x + \Delta x)^3 - 3(x + \Delta x)] - [x^3 - 3x]}{\Delta x}. \end{aligned}$$

If we expand $(x + \Delta x)^3$ and simplify the numerator, we obtain

$$\begin{aligned} \frac{dy}{dx} &= \lim_{\Delta x \to 0} \frac{(3x^2 - 3)(\Delta x) + 3x(\Delta x)^2 + (\Delta x)^3}{\Delta x} \\ &= \lim_{\Delta x \to 0} [(3x^2 - 3) + 3x(\Delta x) + (\Delta x)^2] \\ &= 3x^2 - 3. \ \square \end{aligned}$$

Example 3 Find $f'(4)$ if $f(x) = x^2 + 2x - 6$.

Solution We can find $f'(4)$ by first calculating $f'(x)$, and then evaluating the derivative at $x = 4$. By Definition 3.1,

$$\begin{aligned} f'(x) &= \lim_{\Delta x \to 0} \frac{f(x + \Delta x) - f(x)}{\Delta x} \\ &= \lim_{\Delta x \to 0} \frac{[(x + \Delta x)^2 + 2(x + \Delta x) - 6] - [x^2 + 2x - 6]}{\Delta x} \\ &= \lim_{\Delta x \to 0} \frac{x^2 + 2x\,\Delta x + (\Delta x)^2 + 2x + 2\Delta x - 6 - x^2 - 2x + 6}{\Delta x} \\ &= \lim_{\Delta x \to 0} \frac{(2x + 2)\Delta x + (\Delta x)^2}{\Delta x} \\ &= \lim_{\Delta x \to 0} [(2x + 2) + \Delta x] \\ &= 2x + 2. \end{aligned}$$

If we now set $x = 4$, we have $f'(4) = 2(4) + 2 = 10$.

Alternatively, we can calculate $f'(4)$ directly using Definition 3.1 with $x = 4$; that is,

$$\begin{aligned} f'(4) &= \lim_{\Delta x \to 0} \frac{f(4 + \Delta x) - f(4)}{\Delta x} \\ &= \lim_{\Delta x \to 0} \frac{[(4 + \Delta x)^2 + 2(4 + \Delta x) - 6] - [16 + 8 - 6]}{\Delta x} \\ &= \lim_{\Delta x \to 0} \frac{10\Delta x + (\Delta x)^2}{\Delta x} \\ &= \lim_{\Delta x \to 0} (10 + \Delta x) \\ &= 10. \ \square \end{aligned}$$

Example 4 Find dy/dx if $y = f(x) = (x - 1)/(x + 2)$.

Solution From Definition 3.1,

$$\begin{aligned} \frac{dy}{dx} &= \lim_{\Delta x \to 0} \frac{f(x + \Delta x) - f(x)}{\Delta x} \\ &= \lim_{\Delta x \to 0} \frac{1}{\Delta x} \left\{ \frac{x + \Delta x - 1}{x + \Delta x + 2} - \frac{x - 1}{x + 2} \right\}. \end{aligned}$$

If we bring the terms in braces to a common denominator to get

$$\frac{dy}{dx} = \lim_{\Delta x \to 0} \frac{1}{\Delta x} \left\{ \frac{(x + \Delta x - 1)(x + 2) - (x + \Delta x + 2)(x - 1)}{(x + \Delta x + 2)(x + 2)} \right\}$$

and then simplify the numerator, we have

$$\frac{dy}{dx} = \lim_{\Delta x \to 0} \frac{1}{\Delta x}\left\{\frac{3(\Delta x)}{(x + \Delta x + 2)(x + 2)}\right\}$$

$$= \lim_{\Delta x \to 0} \frac{3}{(x + \Delta x + 2)(x + 2)}$$

$$= \frac{3}{(x + 2)^2}. \square$$

Example 5 Find dv/dt if $v = f(t) = 1/t$.

Solution In terms of the variables v and t, Definition 3.1 takes the form

$$\frac{dv}{dt} = \lim_{\Delta t \to 0} \frac{f(t + \Delta t) - f(t)}{\Delta t}$$

$$= \lim_{\Delta t \to 0} \frac{1}{\Delta t}\left\{\frac{1}{t + \Delta t} - \frac{1}{t}\right\}$$

$$= \lim_{\Delta t \to 0} \frac{t - (t + \Delta t)}{t(t + \Delta t)\Delta t}$$

$$= \lim_{\Delta t \to 0} \frac{-1}{t(t + \Delta t)}$$

$$= -\frac{1}{t^2}. \square$$

EXERCISES 3.1

Use Definition 3.1 to calculate $f'(x)$ in each of Exercises 1–8.

1. $f(x) = x + 2$

2. $f(x) = 3x^2 + 5$

3. $f(x) = 1 + 2x - x^2$

4. $f(x) = x^3 + 2x^2$

5. $f(x) = x^4 + 4x - 12$

6. $f(x) = \dfrac{x + 4}{x - 5}$

7. $f(x) = \dfrac{x^2 + 2}{x + 3}$

8. $f(x) = x^2(x + 2)$

Find the specified rate of change in each of Exercises 9–12.

9. The rate of change of the area A of a circle with respect to its radius r

10. The rate of change of the volume V of a sphere with respect to its radius r

11. The rate of change of the volume V of a right circular cylinder with respect to its height h

12. The rate of change of the area A of a sphere with respect to its radius r

Answer Exercises 13–15 by drawing a graph of the function. Do not calculate the required derivative.

13. What is $f'(x)$ if $f(x) = mx + b$, where m and b are constants?

14. What is $f'(1)$ if $f(x) = (x - 1)^2$?

15. What is $f'(0)$ if $f(x) = x^{1/3}$?

Find $f'(x)$ in each of Exercises 16–20.

16. $f(x) = x^8$

17. $f(x) = \sqrt{x + 1}$

18. $f(x) = \dfrac{1}{(x - 2)^4}$

19. $f(x) = \dfrac{1}{\sqrt{x - 3}}$

20. $f(x) = x\sqrt{x + 1}$

Find the specified rate of change in each of Exercises 21 and 22.

21. The rate of change of the radius r of a circle with respect to its area A

22. The rate of change of the volume V of a sphere with respect to its area A

23. Find $f'(x)$ if $f(x) = |x|$.

24. A sphere of radius R has a uniform charge distribution of ρ coulombs per cubic meter. The electrostatic potential V at a distance r from the center of the sphere is defined by

$$V = f(r) = \begin{cases} \dfrac{\rho}{6\varepsilon_0}(3R^2 - r^2), & 0 \le r \le R, \\ \dfrac{R^3\rho}{3\varepsilon_0 r}, & r > R, \end{cases}$$

where ε_0 is a constant. Sketch a graph of this function. Does $f(r)$ appear to have a derivative at the surface of the sphere? That is, does $f'(R)$ exist? Prove your conjecture using Definition 3.1.

25. Repeat Exercise 24 for the magnitude E of the electrostatic field:

$$E = f(r) = \begin{cases} \dfrac{\rho r}{3\varepsilon_0}, & 0 \le r \le R, \\ \dfrac{\rho R^3}{3\varepsilon_0 r^2}, & r > R. \end{cases}$$

* **26.** Find $f'(x)$ if $f(x) = x^{1/3}$.

3.2 RULES FOR DIFFERENTIATION

Because calculus plays a key role in all branches of applied science, we will be required to differentiate many types of functions: polynomials, rational functions, trigonometric functions, exponential and logarithmic functions, etc. To use Definition 3.1 each time would be extremely laborious, and we might soon doubt the usefulness of differentiation. Fortunately, however, we can develop a number of rules for taking derivatives that eliminate the necessity of using the definition each time. We state each of these rules in the form of a theorem.

▶ **Theorem 3.1** If $f(x) = c$, where c is a constant, then $f'(x) = 0$. ◀

Proof By Definition 3.1,

$$f'(x) = \lim_{\Delta x \to 0} \frac{f(x + \Delta x) - f(x)}{\Delta x}$$

$$= \lim_{\Delta x \to 0} \frac{c - c}{\Delta x} = 0. \square$$

In short, the derivative of a constant is zero; that is,

$$\frac{d}{dx}(c) = 0. \tag{3.4}$$

▶ **Theorem 3.2** If $f(x) = x$, then $f'(x) = 1$. ◀

Proof By Definition 3.1,

$$f'(x) = \lim_{\Delta x \to 0} \frac{f(x + \Delta x) - f(x)}{\Delta x}$$

$$= \lim_{\Delta x \to 0} \frac{(x + \Delta x) - x}{\Delta x}$$

$$= \lim_{\Delta x \to 0} \frac{\Delta x}{\Delta x} = 1. \square$$

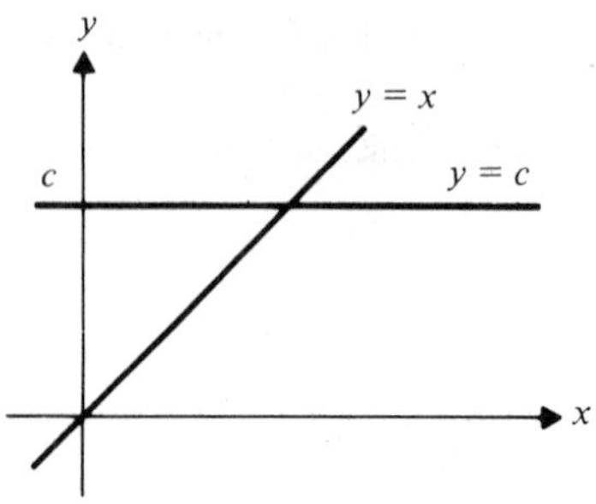

Figure 3.3

In short,

$$\frac{d}{dx}(x) = 1. \tag{3.5}$$

Note that graphs of the functions $f(x) = c$ and $f(x) = x$ in Fig. 3.3 confirm the results of Eqs. (3.4) and (3.5): The tangent line to $y = c$ always has slope zero, whereas the tangent line to $y = x$ always has slope equal to one.

▶ **Theorem 3.3** If $f(x) = x^n$, where n is a positive integer, then

$$f'(x) = nx^{n-1}. \quad ◀$$

Proof By Definition 3.1,

$$\begin{aligned}\frac{dy}{dx} = f'(x) &= \lim_{\Delta x \to 0} \frac{f(x + \Delta x) - f(x)}{\Delta x} \\ &= \lim_{\Delta x \to 0} \frac{(x + \Delta x)^n - x^n}{\Delta x}.\end{aligned}$$

If we expand $(x + \Delta x)^n$ by means of the binomial theorem, we have

$$\begin{aligned}\frac{dy}{dx} &= \lim_{\Delta x \to 0} \frac{1}{\Delta x}\Big\{x^n + nx^{n-1}\,\Delta x + \frac{n(n-1)}{2}x^{n-2}(\Delta x)^2 + \cdots \\ &\qquad + nx(\Delta x)^{n-1} + (\Delta x)^n - x^n\Big\} \\ &= \lim_{\Delta x \to 0}\Big\{nx^{n-1} + \frac{n(n-1)}{2}x^{n-2}\,\Delta x + \cdots + (\Delta x)^{n-1}\Big\} \\ &= nx^{n-1}. \ \square\end{aligned}$$

In short,

$$\frac{d}{dx}(x^n) = nx^{n-1}. \tag{3.6}$$

This is called the *power rule for differentiation.* Although we have proved the power rule only for n a positive integer, it is in fact true for every real number n. We will assume, then, that Eq. (3.6) can be used for any real number n (and will prove this when we have sufficient mathematical background to do so)—for example,

$$\frac{d}{dx}(x^3) = 3x^2 \quad \text{and} \quad \frac{d}{dx}(x^{1/3}) = \frac{1}{3}x^{-2/3}.$$

▶ **Theorem 3.4** If $g(x) = cf(x)$, where c is a constant, then

$$g'(x) = cf'(x). \quad ◀ \tag{3.7a}$$

Proof By Definition 3.1,

$$\begin{aligned} g'(x) &= \lim_{\Delta x \to 0} \frac{g(x + \Delta x) - g(x)}{\Delta x} \\ &= \lim_{\Delta x \to 0} \frac{cf(x + \Delta x) - cf(x)}{\Delta x} \\ &= c \lim_{\Delta x \to 0} \frac{f(x + \Delta x) - f(x)}{\Delta x} \\ &= cf'(x). \ \square \end{aligned}$$

Thus, for $y = f(x)$, we may write that

$$\frac{d}{dx}(cy) = c\frac{dy}{dx}. \tag{3.7b}$$

▶ **Theorem 3.5** If $h(x) = f(x) + g(x)$, then

$$h'(x) = f'(x) + g'(x). \blacktriangleleft \tag{3.8a}$$

Proof By Definition 3.1,

$$\begin{aligned} h'(x) &= \lim_{\Delta x \to 0} \frac{h(x + \Delta x) - h(x)}{\Delta x} \\ &= \lim_{\Delta x \to 0} \frac{[f(x + \Delta x) + g(x + \Delta x)] - [f(x) + g(x)]}{\Delta x} \\ &= \lim_{\Delta x \to 0} \left\{\frac{f(x + \Delta x) - f(x)}{\Delta x} + \frac{g(x + \Delta x) - g(x)}{\Delta x}\right\} \\ &= f'(x) + g'(x). \ \square \end{aligned}$$

In short, if we set $u = f(x)$ and $v = g(x)$,

$$\frac{d}{dx}(u + v) = \frac{du}{dx} + \frac{dv}{dx}, \tag{3.8b}$$

or, in words, the derivative of a sum is the sum of the derivatives.

Example 6 Find dy/dx if:

a) $y = x^4$;

b) $y = 3x^6 - x^{-2}$;

c) $y = \dfrac{x^4 - 6x^2}{3x^3}$.

Solution a) By Eq. (3.6),

$$\frac{dy}{dx} = 4x^3.$$

b) According to Eq. (3.8) we can differentiate each term separately. Because of Eq. (3.7) we can bypass the 3 in front of x^6 and the -1 in front of x^{-2} and use Eq. (3.6) to get

$$\frac{dy}{dx} = 3(6x^5) - (-2x^{-3}) = 18x^5 + 2x^{-3}.$$

c) If we write y in the form $y = (x/3) - 2x^{-1}$, then we can proceed as in (b):

$$\frac{dy}{dx} = \frac{1}{3} - 2(-x^{-2}) = \frac{1}{3} + \frac{2}{x^2}. \; \square$$

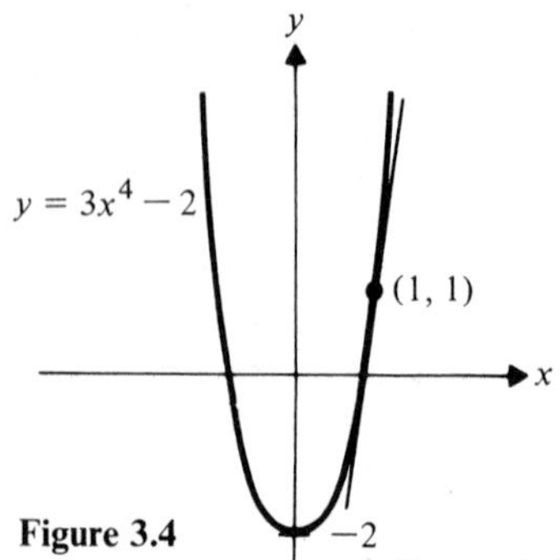

Figure 3.4

Example 7 If $f(x) = 3x^4 - 2$, evaluate $f'(1)$.

Solution Since $f'(x) = 12x^3$, it follows that $f'(1) = 12$. Geometrically, 12 is the slope of the tangent line to the curve $y = 3x^4 - 2$ at the point (1, 1) (Fig. 3.4). Algebraically, the result implies that at $x = 1$, y changes 12 times as fast as x does. $\square$

Example 8 Find the equation of the tangent line to the curve $y = f(x) = x^3 + 5x$ at the point (1, 6).

Solution Since $f'(x) = 3x^2 + 5$, the slope of the tangent line to the curve at (1, 6) is

$$f'(1) = 3(1)^2 + 5 = 8.$$

Using the point-slope form (1.8a) for a straight line, we obtain for the equation of the tangent line at (1, 6)

$$y - 6 = 8(x - 1)$$

or

$$8x - y = 2. \; \square$$

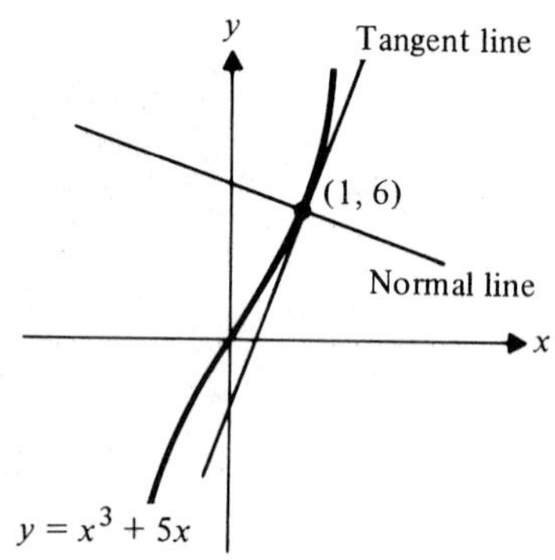

Figure 3.5

The line through (1, 6) perpendicular to the tangent line in Fig. 3.5 is called the *normal line* to the curve at (1, 6). Since two lines are perpendicular only if their slopes are negative reciprocals (see Eq. 1.9), the normal line at (1, 6) must have slope $-\frac{1}{8}$. The equation of the normal line to $y = x^3 + 5x$ at (1, 6) is therefore

$$y - 6 = -\tfrac{1}{8}(x - 1),$$

or

$$x + 8y = 49.$$

Example 9 Find points on the curve $y = x^4/4 - 2x^3/3 - x^2/2 + x$ at which the slope of the tangent line is -1.

Solution The slope of the tangent line to the curve is -1 whenever

$$-1 = \frac{dy}{dx} = x^3 - 2x^2 - x + 1,$$

or

$$0 = x^3 - 2x^2 - x + 2 = (x - 1)(x + 1)(x - 2).$$

Thus $x = 1$, -1, and 2, and the points at which the slope of the tangent line is -1 are

$$(1, \tfrac{1}{12}), \qquad (-1, -\tfrac{7}{12}), \qquad (2, -\tfrac{4}{3}). \; \square$$

EXERCISES 3.2

In each of Exercises 1–20, find $f'(x)$.

1. $f(x) = 2x^2 - 3$
2. $f(x) = 3x^3 + 4x + 5$
3. $f(x) = 10x^2 - 3x$
4. $f(x) = 4x^5 - 10x^3 + 3x$
5. $f(x) = 1/x^2$
6. $f(x) = 2/x^3$
7. $f(x) = 5x^4 - 3x^3 + 1/x$
8. $f(x) = -\dfrac{1}{2x^2} + \dfrac{3}{x^4}$
9. $f(x) = x^{10} - \dfrac{1}{x^{10}}$
10. $f(x) = 5x^4 + \dfrac{1}{4x^5}$
11. $f(x) = 5x^{-4} + \dfrac{1}{4x^{-5}}$
12. $f(x) = \sqrt{x}$
13. $f(x) = \dfrac{3}{x^2} + \dfrac{2}{\sqrt{x}}$
14. $f(x) = 1/x^{3/2} + x^{3/2}$
15. $f(x) = 2x^{1/3} - 3x^{2/3}$
16. $f(x) = \pi x^{\pi}$
17. $f(x) = (x^2 + 2)^2$
18. $f(x) = (4x^6 - x^2)/x^5$
19. $f(x) = x^{5/3} - x^{2/3} + 3$
20. $f(x) = (2x + 5)^3$

In Exercises 21–24, find equations for the tangent and normal lines to the curve at the point indicated. In each case, sketch the curve and lines.

21. $y = x^2 - 2x + 5$ at $(2, 5)$
22. $y = \sqrt{x} + 5$ at $(4, 7)$
23. $y = 2x^3 - 3x^2 - 12x$ at $(2, -20)$
24. $x = \sqrt{y + 1}$ at $(3, 8)$
25. Find the points on the curve $y = x^4/4 - 2x^3/3 - 19x^2/2 + 22x$ at which the slope of the tangent line is 2.
26. Find constants A and B in order that the function $y = f(x) = Ax + B$ satisfy the equation
$$3\frac{dy}{dx} - 4y = x + 2 \quad \text{for all } x.$$
27. A cable is stretched across a river as shown in Fig. 3.6. If the shape of the curve is approximately
$$y = f(x) = (6.94 \times 10^{-6})x^3 + 0.00472x^2 - 0.642x + 40, \quad 0 \le x \le 100,$$
find that point on the cable closest to the water.

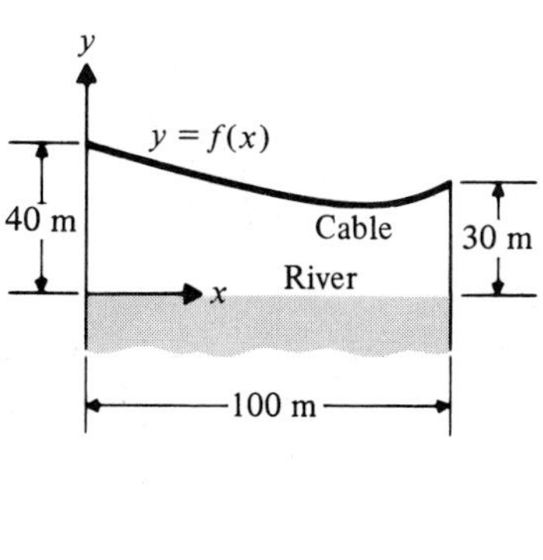

Figure 3.6

28. An electronics firm manufactures a device called a "gadget." The money that it receives from the sale of x gadgets is called total revenue and is denoted by $R(x)$. The total cost $C(x)$ of producing these x gadgets also depends on x. The total profit realized on the sale is
$$P(x) = R(x) - C(x).$$
The derivatives of these functions, $R'(x)$, $C'(x)$, and $P'(x)$ are called, respectively, marginal revenue, marginal cost, and marginal profit. Calculate these marginal functions if
$$R(x) = 75x - \frac{7x^2}{100}$$
and
$$C(x) = 50x - \frac{x^2}{30} - \frac{x^3}{60{,}000},$$
both functions being defined for $1 \le x \le 800$.
29. Sketch a graph of the function $x = f(t) = t^3 - 8t^2$. Find the value(s) of t at which the tangent to this curve is parallel to the line $x = 6t - 3$.
30. At what point(s) on the curve $y = x^3 + x^2 - 22x + 20$ will the tangent line pass through the origin?
31. At what point(s) on the parabola $y = x^2$ will the normal line pass through the point $(2, 5)$?
32. Find constants A, B, C, and D in order that $y = f(x) = Ax^3 + Bx^2 + Cx + D$ satisfy the equation
$$3\frac{dy}{dx} + 2y = x^3 - 5x \quad \text{for all } x.$$
33. Suppose the curve $y = f(x)$ has a tangent line at the point (x_0, y_0). If ϕ is the angle between the positive x-axis and this tangent line (Fig. 3.7), show that
$$\tan\phi = f'(x_0).$$

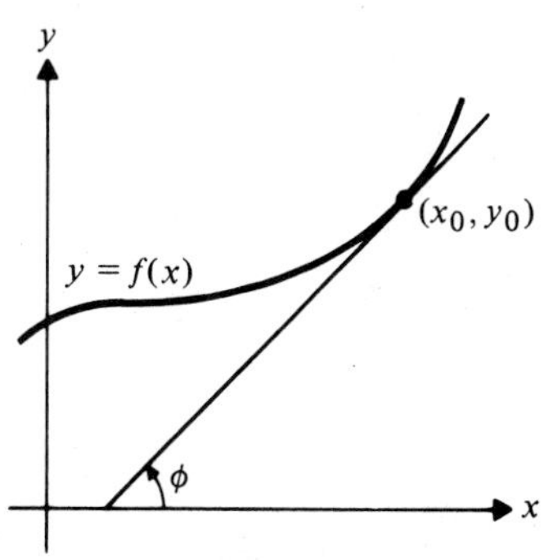

Figure 3.7

*34. If $(x) = |x^2 - x - 12|$, find $f'(x)$.

*** 35.** Show that the sum of the intercepts on the coordinate axes of the tangent line to the curve $\sqrt{x} + \sqrt{y} = \sqrt{a}$ is always equal to a.

*** 36.** Find a formula for

$$\frac{d}{dx}|x|^n, \quad n > 1 \text{ a positive integer.}$$

*** 37.** Show that if two lines have slopes m_1 and m_2, then the angle ϕ between them satisfies the equation

$$\tan\phi = \frac{m_1 - m_2}{1 + m_1 m_2}.$$

3.3 THE EXTENDED POWER RULE

Using the differentiation rules given in Section 3.2 we can differentiate any polynomial whatsoever. For instance, if $f(x) = x^3 - 3x^2 + 2x + 1$, then $f'(x) = 3x^2 - 6x + 2$. Consider, however, the polynomial $f(x) = (2x - 3)^8$, which has been conveniently factored for us. To find its derivative using the rules listed in Section 3.2, we would first expand $(2x - 3)^8$, then differentiate, and finally simplify—certainly not a pleasant task. Thus in spite of the fact that the rules of Section 3.2 permit differentiation of $(2x - 3)^8$, they are not convenient to use. Equally disturbing is the fact that although we can easily differentiate $f(x) = 1/x^2$, we cannot handle $f(x) = 1/(5x + 8)^2$, for example, with the power rule.

The following extension of the power rule permits us to differentiate a much wider class of functions (including $(2x - 3)^8$ and $1/(5x + 8)^2$).

▶ **Theorem 3.6** When $u = f(x)$ has a derivative, then

$$\frac{d}{dx}(u^n) = nu^{n-1}\frac{du}{dx}. \quad ◀ \tag{3.9}$$

This result is verified in much the same way as the original power rule of Theorem 3.3 (see Exercise 32 at the end of this section). Note that the power rule (3.6) is a special case of Eq. (3.9) when $f(x) = x$. It is important not to confuse these rules. Rule (3.6) can be used only for x^n; if anything other than x is raised to a power, Eq. (3.9) should be used. The most common error when using Eq. (3.9) is to forget the factor du/dx. Don't get caught!

Using this rule, it is now easy to differentiate $(2x - 3)^8$ and $1/(5x + 8)^2$:

$$\frac{d}{dx}(2x - 3)^8 = 8(2x - 3)^7\frac{d}{dx}(2x - 3) = 8(2x - 3)^7(2) = 16(2x - 3)^7;$$

$$\frac{d}{dx}\left(\frac{1}{(5x + 8)^2}\right) = \frac{d}{dx}(5x + 8)^{-2} = -2(5x + 8)^{-3}\frac{d}{dx}(5x + 8)$$

$$= -2(5x + 8)^{-3}(5) = \frac{-10}{(5x + 8)^3}.$$

Example 10 Find dy/dx if:

a) $y = (2x^2 + 5)^{10}$;

b) $y = \sqrt{x - x^2}$;

c) $y = \left(x + \frac{1}{x}\right)^3$;

d) $y = \frac{1}{3x^2 + 2x + 1}$.

Solution a) According to Eq. (3.9),

$$\frac{dy}{dx} = 10(2x^2 + 5)^9 \frac{d}{dx}(2x^2 + 5)$$
$$= 10(2x^2 + 5)^9(4x)$$
$$= 40x(2x^2 + 5)^9.$$

b) Again, Eq. (3.9) gives us

$$\frac{dy}{dx} = \frac{1}{2}(x - x^2)^{-1/2} \frac{d}{dx}(x - x^2)$$
$$= \frac{1}{2}(x - x^2)^{-1/2}(1 - 2x)$$
$$= \frac{1 - 2x}{2\sqrt{x - x^2}}.$$

c) $\frac{dy}{dx} = 3\left(x + \frac{1}{x}\right)^2\left(1 - \frac{1}{x^2}\right)$

d) If we write $y = (3x^2 + 2x + 1)^{-1}$, then

$$\frac{dy}{dx} = (-1)(3x^2 + 2x + 1)^{-2}(6x^2 + 2)$$
$$= -\frac{6x^2 + 2}{(3x^2 + 2x + 1)^2}. \quad \square$$

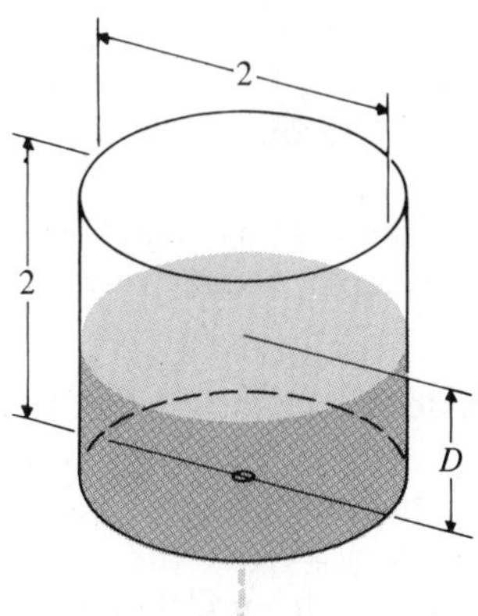

Figure 3.8

Example 11 An upright cylindrical container (Fig. 3.8) has height 2 meters and radius 1 meter. At time $t = 0$, water begins flowing out through an outlet in the bottom in such a way that the depth D of water in the container at time t seconds is given by

$$D = f(t) = (\sqrt{2} - 0.000127t)^2.$$

How fast is the water level dropping when the tank is half full and when it is just emptying?

Solution By Eq. (3.9), we have

$$f'(t) = 2(\sqrt{2} - 0.000127t)\frac{d}{dt}(\sqrt{2} - 0.000127t)$$
$$= 2(\sqrt{2} - 0.000127t)(-0.000127)$$
$$= -0.000254(\sqrt{2} - 0.000127t).$$

The tank is half full when $D = 1$, in which case

$$1 = (\sqrt{2} - 0.000127t)^2.$$

To solve this equation we first take the square root of each side,

$$\sqrt{2} - 0.000127t = \pm 1,$$

from which we get

$$t = \frac{\sqrt{2} \pm 1}{0.000127}.$$

But the tank is just emptying when $D = 0$, or

$$0 = (\sqrt{2} - 0.000127t)^2,$$

and this equation implies that

$$t = \frac{\sqrt{2}}{0.000127}.$$

It follows then that we must choose $t = (\sqrt{2} - 1)/0.000127 = 3262$ as the time when the tank is half full. At this time

$$f'(3262) = -0.000254.$$

The negative sign indicates that the water level is falling; hence we can say that the water level is dropping at 0.254 millimeter per second when the tank is half full.

When the tank is just emptying, $t = \sqrt{2}/0.000127$ and at this time $f'(t) = 0$. Physically this means that when the tank is almost empty, water just trickles out. □

In this example we have attached units to the derivative for the first time. When the units of D are meters and those of t are seconds, the units of dD/dt are meters per second (meaning meters divided by seconds). To indicate meters per second, we write "m/s." Generally, when $y = f(x)$, the units of dy/dx are units of y divided by units of x. We can see this from Definition 3.1, where the numerator of $[f(x + \Delta x) - f(x)]/\Delta x$ has units of y and the denominator has units of x. Taking the limit of this quotient as Δx approaches zero does not alter its units.

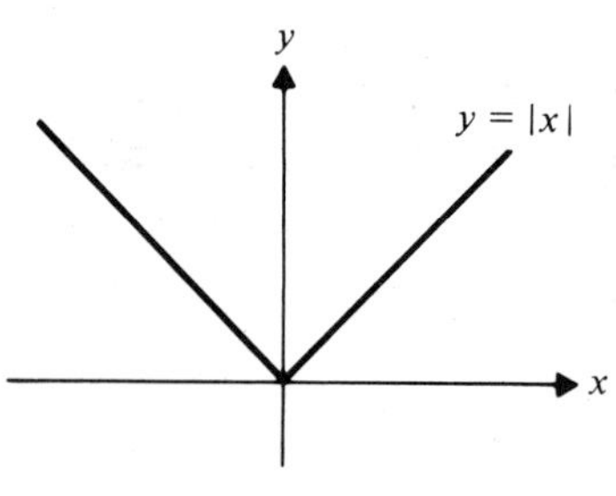

Figure 3.9

Many functions fail to have a derivative at certain isolated points. For example, consider the function $f(x) = |x|$ shown in Fig. 3.9. Geometrically, or algebraically, it is clear that for $x > 0, f'(x) = 1$, and for $x < 0, f'(x) = -1$. At $x = 0$, however, there is a problem. If we recall that the tangent line at $(0, 0)$ is the limiting position of the line joining $(0, 0)$ to other points on the curve, we must conclude that $y = |x|$ does not have a tangent line at $(0, 0)$. If the tangent line does not exist, then it can hardly have a slope. Definition 3.1 leads to the same conclusion, for if $f'(0)$ is to exist, it must be defined by

$$\lim_{\Delta x \to 0} \frac{f(0 + \Delta x) - f(0)}{\Delta x} = \lim_{\Delta x \to 0} \frac{|\Delta x|}{\Delta x}.$$

But this limit does not exist since

$$\lim_{\Delta x \to 0^-} \frac{|\Delta x|}{\Delta x} = -1 \quad \text{and} \quad \lim_{\Delta x \to 0^+} \frac{|\Delta x|}{\Delta x} = 1.$$

Consequently, $f(x) = |x|$ does not have a derivative at $x = 0$.

The same conclusion can be drawn at any point at which the graph of a function takes an abrupt change in direction. Such a point is often called a

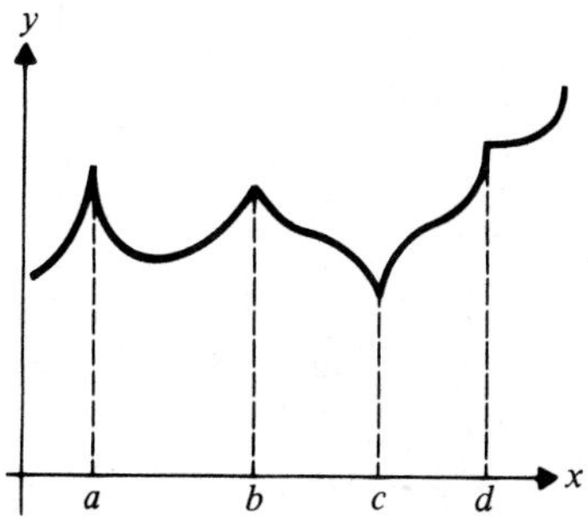

Figure 3.10

"corner." As a result, the function shown in Fig. 3.10 does not have a derivative at $x = a, b, c$, or d. It is also true that a function cannot have a derivative at a point where the function is discontinuous (see, for example, the discontinuities in Fig. 2.5). This result is an immediate consequence of the following theorem.

▶ **Theorem 3.7** If a function $f(x)$ has a derivative at $x = a$, then $f(x)$ is continuous at $x = a$. ◀

Proof The derivative $f'(a)$ is defined by the limit

$$f'(a) = \lim_{\Delta x \to 0} \frac{f(a + \Delta x) - f(a)}{\Delta x}.$$

No one will argue with the fact that

$$0 = \lim_{\Delta x \to 0} \Delta x.$$

But note that if we multiply the left-hand and right-hand sides of these two equations, we obtain

$$0 \cdot f'(a) = \left\{ \lim_{\Delta x \to 0} \Delta x \right\} \left\{ \lim_{\Delta x \to 0} \frac{f(a + \Delta x) - f(a)}{\Delta x} \right\}.$$

Consequently, using part (ii) of Theorem 2.1, we have

$$0 = \lim_{\Delta x \to 0} \{f(a + \Delta x) - f(a)\},$$

or

$$\lim_{\Delta x \to 0} f(a + \Delta x) = f(a).$$

But this is just another way of saying that

$$\lim_{x \to a} f(x) = f(a). \; \square$$

When a function $f(x)$ is defined on a closed interval $a \le x \le b$, Definition 3.1 applies only at points in the open interval $a < x < b$. For instance, it is impossible to evaluate

$$f'(a) = \lim_{\Delta x \to 0} \frac{f(a + \Delta x) - f(a)}{\Delta x}$$

since $f(x)$ is not defined for $x < a$. When a function $f(x)$ is defined only to the right of a point, we define a right-hand derivative at the point; and when $f(x)$ is defined only to the left of a point, we define its left-hand derivative.

▷ **Definition 3.2** The *right-hand derivative* of $f(x)$ with respect to x is defined as

$$f'_+(x) = \lim_{\Delta x \to 0^+} \frac{f(x + \Delta x) - f(x)}{\Delta x}, \tag{3.10a}$$

provided the limit exists. The *left-hand derivative* of $f(x)$ is

$$f'_-(x) = \lim_{\Delta x \to 0^-} \frac{f(x + \Delta x) - f(x)}{\Delta x}, \tag{3.10b}$$

if the limit exists. ◁

If a function has a derivative at a point x, its right- and left-hand derivatives both exist at x and are equal to $f'(x)$. On the other hand, a function may have both a left- and a right-hand derivative at a point but not a derivative there. An example of this is the absolute value function $f(x) = |x|$ at $x = 0$ (see Fig. 3.9).

When a function has a derivative at $x = a$, we say that it is *differentiable* at $x = a$. When it has a derivative at every point in some interval, we say that it is differentiable on that interval. In the event that the interval is closed, $a \le x \le b$, we understand that derivatives at $x = a$ and $x = b$ mean right-hand and left-hand derivatives, respectively.

EXERCISES 3.3

In each of Exercises 1–20, find $f'(x)$ in as simplified a form as possible.

1. $f(x) = (3x + 1)^3$

2. $f(x) = (2 - 6x^2)^4$

3. $f(x) = \sqrt{2x + 5}$

4. $f(x) = 1/(6x + 5)$

5. $f(x) = 3/(2x^2 + 3x + 4)$

6. $f(x) = (x^4 - 3x)^9$

7. $f(x) = 1/(x^3 - 1)$

8. $f(x) = \sqrt{-3x^2 - 6x - 1}$

9. $f(x) = \sqrt{x^5 - 10} + 1/\sqrt{3 - x}$

10. $f(x) = \sqrt[4]{4x^2 - 2}$

11. $f(x) = 1/(x^4 - x^2 + x)^5$

12. $f(x) = 1/(x + 5)^\pi$

13. $f(x) = (4x^6 - x^2)/(3x^5)$

14. $f(x) = \sqrt{1 + \sqrt{1 + x}}$

15. $f(x) = 1/(2 - 3x + 4x^2)$

16. $f(x) = (x^2 + 1)^{3/2} + \sqrt{x^2 + 1}$

17. $f(x) = \sqrt{x + \sqrt{x}}$

18. $f(x) = [x - (x - 1)^4]^4$

19. $f(x) = \left(\frac{2}{x} + \frac{5}{x^2}\right)^{1/3}$

20. $f(x) = (x^4 - 2x^2 + 3)/(4x^3)$

In Exercises 21–24, find equations for the tangent and normal lines to the curve at the point indicated. In each case, sketch the curve and lines.

21. $y = \dfrac{1}{(x - 3)^3}$ at $(2, -1)$

22. $y = \dfrac{4}{1 + x^2}$ at $(-1, 2)$

23. $y = (x^2 + 2x)^2$ at $(-1, 1)$

24. $y = -\sqrt{x + 1}$ at $(3, -2)$

25. Find points on the curve $y = (x + 3)^5$ at which the slope of the tangent line is 5.

26. Find the point of intersection of the tangent lines to the circle $x^2 + y^2 = 25$ at the points $(4, 3)$ and $(-3, 4)$.

27. Show that $y = f(x) = \frac{2}{5}x^2 - \frac{2}{15}x - \frac{4}{15} + C(x - 1)^{-1/2}$ satisfies the equation

$$\frac{dy}{dx} + \frac{y}{2(x - 1)} = x$$

for any constant C whatsoever.

28. A container in the shape of a right circular cylinder with radius 10 centimeters has a vertical axis. If water is poured into the container at the rate of 2 liters per second, how fast is the water level rising on the side of the container?

29. If $f(x) = \sqrt{x - 1} + (2 - 2x^2)^{1/4} + 3$, find $f'(x)$. For what values of x is your answer valid? Explain.

30. If a function $f(x)$ does not have a derivative at $x = a$, can its graph have a tangent line at $x = a$?

31. Which of the following functions has a right-hand derivative, a left-hand derivative, and/or a derivative at the point indicated?

a) $f(x) = |x - 5|$ at $x = 5$

b) $f(x) = x^{3/2}$ at $x = 0$

c) $f(x) = \sqrt{|x|}$ at $x = 0$

d) $f(x) = \operatorname{sgn} x = \begin{cases} -1, & x < 0, \\ 0, & x = 0, \\ 1, & x > 0, \end{cases}$ at $x = 0$

32. In this exercise we extend the proof of Theorem 3.3 to prove Theorem 3.6.

a) When $y = u^n = [f(x)]^n$, an increment Δx in x produces a change in u, which we denote by $\Delta u = f(x + \Delta x) - f(x)$. Use the binomial theorem and Definition 3.1 to show that

$$\frac{dy}{dx} = \lim_{\Delta x \to 0} \frac{1}{\Delta x}\left\{nu^{n-1}\Delta u + \frac{n(n-1)}{2}u^{n-2}(\Delta u)^2 + \cdots + (\Delta u)^n\right\}.$$

b) Factor Δu from the terms in braces and complete the proof of Theorem 3.6.

c) Theorem 3.7 must be used in part (b). Did you use it?

***33.** At what point(s) on the curve $x^2 - 16y^2 = 16$ will tangents pass through the point (2, 3)?

***34.** Generalize the result of Exercise 36 in Section 3.2 to find a formula for

$$\frac{d}{dx}|f(x)|^n, \quad n > 1 \text{ a positive integer.}$$

3.4 THE PRODUCT AND QUOTIENT RULES

In order to differentiate a function that is the product of two other functions, we use the following theorem.

▶ **Theorem 3.8** If $h(x) = f(x)g(x)$, where $f(x)$ and $g(x)$ are differentiable, then

$$h'(x) = f(x)g'(x) + f'(x)g(x). \quad ◀ \tag{3.11a}$$

Proof By Definition 3.1,

$$\begin{aligned} h'(x) &= \lim_{\Delta x \to 0} \frac{h(x + \Delta x) - h(x)}{\Delta x} \\ &= \lim_{\Delta x \to 0} \frac{f(x + \Delta x)g(x + \Delta x) - f(x)g(x)}{\Delta x}. \end{aligned}$$

To organize this quotient further, we add and subtract the quantity $f(x + \Delta x)g(x)$ in the numerator:

$$\begin{aligned} h'(x) &= \lim_{\Delta x \to 0} \frac{1}{\Delta x} \{[f(x + \Delta x)g(x + \Delta x) - f(x + \Delta x)g(x)] \\ &\qquad + [f(x + \Delta x)g(x) - f(x)g(x)]\} \\ &= \lim_{\Delta x \to 0} \left\{f(x + \Delta x)\frac{g(x + \Delta x) - g(x)}{\Delta x} + g(x)\frac{f(x + \Delta x) - f(x)}{\Delta x}\right\} \\ &= f(x)g'(x) + g(x)f'(x), \end{aligned}$$

where we have used the fact that $\lim_{\Delta x \to 0} f(x + \Delta x) = f(x)$. This follows from continuity of $f(x)$ (see Theorem 3.7). □

This result is called the *product rule for differentiation*. If we set $u = f(x)$ and $v = g(x)$, then the product rule may also be expressed in the form

$$\frac{d}{dx}(uv) = u\frac{dv}{dx} + v\frac{du}{dx}. \tag{3.11b}$$

As an immediate consequence of the product rule we have the following corollary to Theorem 3.8.

▶ **Corollary**

$$\frac{d}{dx}\left(\frac{f(x)}{f(x)}\right) = \frac{g(x)f'(x) - f(x)g'(x)}{[g(x)]^2}. \quad ◀ \tag{3.12a}$$

This result, called the *quotient rule for differentiation*, can be proved as follows:

$$\begin{aligned}\frac{d}{dx}\left(\frac{f(x)}{g(x)}\right) &= \frac{d}{dx}\{f(x)[g(x)]^{-1}\}\\ &= f(x)\frac{d}{dx}[g(x)]^{-1} + [g(x)]^{-1}\frac{d}{dx}f(x) && \text{(Using Eq. 3.11a)}\\ &= f(x)(-1)[g(x)]^{-2}g'(x) + [g(x)]^{-1}f'(x) && \text{(Using Eq. 3.9)}\\ &= -\frac{f(x)}{[g(x)]^2}g'(x) + \frac{1}{g(x)}f'(x)\\ &= \frac{g(x)f'(x) - f(x)g'(x)}{[g(x)]^2}.\end{aligned}$$

Alternatively, for $u = f(x)$ and $v = g(x)$, we may write Eq. (3.12a) in the form

$$\frac{d}{dx}\left(\frac{u}{v}\right) = \frac{v\dfrac{du}{dx} - u\dfrac{dv}{dx}}{v^2}. \tag{3.12b}$$

Example 12 Find dy/dx in as simplified a form as possible if:

a) $y = x^3\sqrt{x+1}$;

b) $y = \dfrac{\sqrt{x}}{x+1}$;

c) $y = \dfrac{x+2}{\sqrt[3]{2-x^2}}$;

d) $y = (x^3+3)^5(-x^2+5)^6$.

Solution

a) $$\frac{dy}{dx} = (x^3)\frac{1}{2}(x+1)^{-1/2} + 3x^2\sqrt{x+1} = \frac{x^3 + 6x^2(x+1)}{2\sqrt{x+1}} = \frac{x^2(7x+6)}{2\sqrt{x+1}}$$

b) $$\frac{dy}{dx} = \frac{(x+1)(1/(2\sqrt{x})) - \sqrt{x}(1)}{(x+1)^2} = \frac{(x+1)-2x}{2\sqrt{x}(x+1)^2} = \frac{1-x}{2\sqrt{x}(x+1)^2}$$

c) $$\begin{aligned}\frac{dy}{dx} &= \frac{(2-x^2)^{1/3}(1) - (x+2)(1/3)(2-x^2)^{-2/3}(-2x)}{(2-x^2)^{2/3}}\\ &= \frac{(2-x^2)^{1/3} + \dfrac{2x(x+2)}{3(2-x^2)^{2/3}}}{(2-x^2)^{2/3}}\\ &= \frac{3(2-x^2) + 2x(x+2)}{3(2-x^2)^{4/3}} = \frac{6+4x-x^2}{3(2-x^2)^{4/3}}\end{aligned}$$

d) $$\frac{dy}{dx} = ((x^3 + 3)^5)6(-x^2 + 5)^5(-2x) + ((-x^2 + 5)^6)5(x^3 + 3)^4(3x^2)$$
$$= 3x(x^3 + 3)^4(-x^2 + 5)^5[-4(x^3 + 3) + 5x(-x^2 + 5)]$$
$$= 3x(x^3 + 3)^4(-x^2 + 5)^5(-9x^3 + 25x - 12) \; \square$$

Example 13 a) Use limits to sketch a graph of the function $y = f(x) = (x - 1)/(x + 2)$.

b) Calculate dy/dx and show qualitatively that the graph in (a) agrees with your calculation.

Solution a) From the limits,

$$\lim_{x \to -2^+} f(x) = -\infty, \quad \lim_{x \to -2^-} f(x) = \infty, \quad \lim_{x \to \infty} f(x) = 1^-, \quad \lim_{x \to -\infty} f(x) = 1^+,$$

we obtain the sketch shown in Fig. 3.11.

b) Using the quotient rule, we have

$$\frac{dy}{dx} = \frac{(x + 2)(1) - (x - 1)(1)}{(x + 2)^2}$$
$$= \frac{3}{(x + 2)^2}.$$

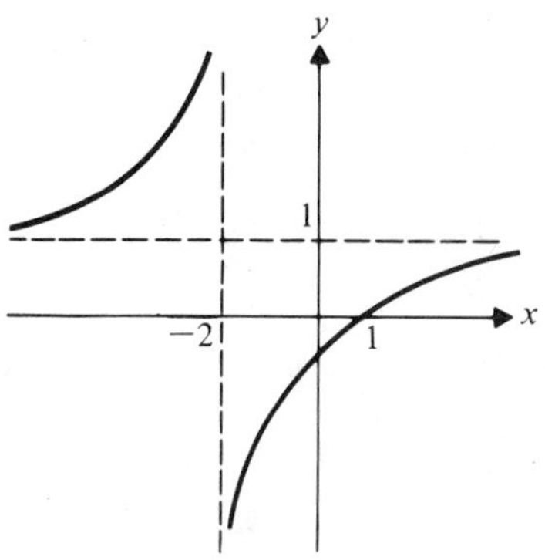

Figure 3.11

The sketch in Fig. 3.11 and dy/dx agree that:

1. The slope of the curve is always positive;
2. The slope becomes larger and larger as x approaches -2; i.e.,
$$\lim_{x \to -2} f'(x) = \infty;$$
3. The slope approaches zero as x approaches $\pm\infty$; i.e.,
$$\lim_{x \to \pm\infty} f'(x) = 0. \; \square$$

Example 14 Sketch a graph of the function $f(x) = (x - 4)/x^2$.

Solution First we note that $f(4) = 0$, and that

$$\lim_{x \to 0^+} f(x) = -\infty, \quad \lim_{x \to 0^-} f(x) = -\infty, \quad \lim_{x \to \infty} f(x) = 0^+, \quad \lim_{x \to -\infty} f(x) = 0^-.$$

This information, shown in Fig. 3.12(a), would lead us to suspect that the graph should be completed as the dashed lines indicate. To show that this is indeed the case, we calculate

$$f'(x) = \frac{x^2(1) - (x - 4)(2x)}{x^4} = \frac{8 - x}{x^3}.$$

Since

$$f'(x) < 0 \quad \text{for } x < 0,$$
$$f'(x) > 0 \quad \text{for } 0 < x < 8,$$
$$f'(x) = 0 \quad \text{for } x = 8,$$
$$f'(x) < 0 \quad \text{for } x > 8,$$

our suspicions were correct, and the final graph is shown in Fig. 3.12(b). $\square$

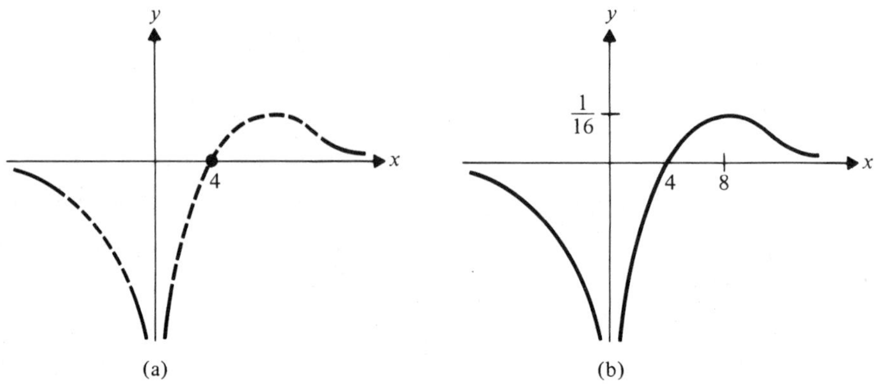

Figure 3.12

Examples 13 and 14 illustrate that derivatives can be helpful in sketching graphs. A complete discussion of the application of derivatives to curve sketching will be given in Section 4.6.

EXERCISES 3.4

In each of Exercises 1–20, find $f'(x)$ in as simplified a form as possible.

1. $f(x) = x(x^3 + 3)^4$
2. $f(x) = \dfrac{x}{3x + 2}$
3. $f(x) = \dfrac{x^2}{4x^2 - 5}$
4. $f(x) = x\sqrt{x + 1}$
5. $f(x) = \dfrac{2x^2 - 5}{3x + 4}$
6. $f(x) = \dfrac{x + 5}{2x^2 - 1}$
7. $f(x) = x^2(2x + 1)^2$
8. $f(x) = \dfrac{x}{\sqrt{2x + 1}}$
9. $f(x) = (x + 2)^2(x^2 + 3)$
10. $f(x) = \dfrac{(2x - 1)^2}{3x + 5}$
11. $f(x) = \dfrac{3x + 5}{(2x - 1)^2}$
12. $f(x) = \dfrac{x^2 + 2x + 3}{x^2 - 5x + 1}$
13. $f(x) = \dfrac{\sqrt{2x + 3}}{x}$
14. $f(x) = x^3(2 - 5x^2)^{1/3}$
15. $f(x) = \dfrac{x^3}{(2 - 5x^2)^{1/3}}$
16. $f(x) = (x + 1)^2(3x + 1)^3$
17. $f(x) = \dfrac{x(x + 1)}{1 - 3x}$
18. $f(x) = \dfrac{\sqrt{2 - 3x}}{x^2}$
19. $f(x) = \dfrac{x^{1/3}}{1 - \sqrt{x}}$
20. $f(x) = \dfrac{\sqrt{x} + 2x}{\sqrt{x} - 4}$

21. Find equations for the tangent and normal lines to the curve $y = (x + 3)/(x - 4)$ at the point $(1, -\frac{4}{3})$. Sketch the curve and lines.

22. a) Sketch a graph of the function $f(x) = x^2/(x^2 + x - 2)$.
    b) Find $f'(x)$ and show that it agrees with your sketch.

23. If the total cost of producing x items of a commodity is given by the equation
$$C(x) = ax\left(\frac{x + b}{x + c}\right),$$
where a, b, and c are constants, show that the marginal cost $C'(x)$ is
$$a\left\{1 + \frac{c(b - c)}{(x + c)^2}\right\}.$$

24. Use Definition 3.1 to prove the quotient rule.

25. Find a rule for the derivative of the product of three functions $f(x)$, $g(x)$, and $h(x)$.

26. Show that $y = f(x) = (x^3 + x + C)(4 + x)^{-1/2}$ satisfies the equation
$$\frac{dy}{dx} + \frac{y}{2(4 + x)} = \frac{3x^2 + 1}{\sqrt{4 + x}}$$
for any constant C whatsoever.

27. a) Sketch the curve $y^2 = x^2(9 - x^2)$.
    b) Find all points on this curve at which the tangent line is horizontal.

28. Find all points on the curve $y = (5 - x)/(6 + x)$ at which the tangent line passes through the origin.

29. A manufacturer's profit from the sale of x kilograms of a commodity per week is given by
$$P(x) = \frac{3x - 200}{x + 400}.$$

a) Sketch a graph of this total profit function.

b) The average profit per kilogram when x kilograms are sold is given by

$$p(x) = \frac{P(x)}{x}.$$

If a point (x, P) on the total profit curve is joined to the origin, the slope of this line is the average profit $p(x)$ for that x. Use this idea to find the sales level for highest average profit.

In each of Exercises 30–39, find $f'(x)$ in as simplified a form as possible.

30. $f(x) = \sqrt[4]{\dfrac{2-x}{2+x}}$

31. $f(x) = \left(\dfrac{x^3-1}{2x^3+1}\right)^4$

32. $f(x) = (x^3 - 2x^2)^3(x^4 - 2x)^5$

33. $f(x) = \dfrac{1}{x^3 - 3x^2 + 2x + 5}$

34. $f(x) = (x+5)^4\sqrt{1+x^3}$

35. $f(x) = \dfrac{x^3 + 3x^2 + 3x + 10}{(x+1)^3}$

36. $f(x) = x(x+5)^4\sqrt{1+x^3}$

37. $f(x) = \dfrac{x\sqrt{1-x^2}}{(3+x)^{1/3}}$

38. $f(x) = x\sqrt{1 + x\sqrt{1+x}}$

39. $f(x) = \dfrac{x^2(x^3+3)^2}{(x-2)(x+5)^2}$

* **40.** There are two points on the curve $y = x(1 + 2x - x^3)$ that share a common tangent line. Find them.

3.5 HIGHER-ORDER DERIVATIVES

If $y = f(x) = x^3 + 1/x$, then

$$\frac{dy}{dx} = 3x^2 - \frac{1}{x^2}.$$

Since dy/dx is itself a function of x, it can be differentiated with respect to x. We call the derivative of dy/dx the second derivative of y with respect to x, and denote it by d^2y/dx^2 or $f''(x)$:

$$f''(x) = \frac{d^2y}{dx^2} = \frac{d}{dx}\left(\frac{dy}{dx}\right) = 6x + \frac{2}{x^3}.$$

Similarly,

$$f'''(x) = \frac{d^3y}{dx^3} = 6 - \frac{6}{x^4}.$$

For this function we can continue the differentiation process indefinitely to produce derivatives of any positive integer order whatsoever.

Example 15 Find a formula for the nth derivative of $y = x/(x+1)$.

Solution If $n = 1$,

$$\frac{dy}{dx} = \frac{(x+1)(1) - x(1)}{(x+1)^2} = \frac{1}{(x+1)^2}.$$

If $n = 2$,

$$\frac{d^2y}{dx^2} = \frac{-2}{(x+1)^3}.$$

If $n = 3$,

$$\frac{d^3y}{dx^3} = \frac{2 \cdot 3}{(x+1)^4}.$$

If $n = 4$,

$$\frac{d^4y}{dx^4} = \frac{-2 \cdot 3 \cdot 4}{(x+1)^5}.$$

It is fairly clear that the pattern being formed is

$$\frac{d^ny}{dx^n} = \frac{(-1)^{n+1}n!}{(x+1)^{n+1}}.\dagger \quad \square$$

Example 16 Find a formula for $d^2(uv)/dx^2$ if $u = f(x)$ and $v = g(x)$.

Solution Since

$$\frac{d}{dx}(uv) = v\frac{du}{dx} + u\frac{dv}{dx},$$

then

$$\begin{aligned}\frac{d^2}{dx^2}(uv) &= \frac{d}{dx}\left(v\frac{du}{dx} + u\frac{dv}{dx}\right)\\ &= v\frac{d^2u}{dx^2} + \frac{dv}{dx}\frac{du}{dx} + u\frac{d^2v}{dx^2} + \frac{du}{dx}\frac{dv}{dx}\\ &= v\frac{d^2u}{dx^2} + 2\frac{du}{dx}\frac{dv}{dx} + u\frac{d^2v}{dx^2}. \quad \square\end{aligned}$$

Example 17 How many derivatives does $f(x) = x^{5/2}$ have at $x = 0$?

Solution Since

$$f'(x) = \tfrac{5}{2}x^{3/2}, \qquad f''(x) = (\tfrac{5}{2})(\tfrac{3}{2})x^{1/2}, \qquad f'''(x) = (\tfrac{5}{2})(\tfrac{3}{2})(\tfrac{1}{2})x^{-1/2},$$

and $f'''(0)$ is not defined, $f(x)$ has only a first and second derivative at $x = 0$. $\square$

EXERCISES 3.5

Find the indicated derivative in Exercises 1–6.

1. $f''(x)$ if $f(x) = \dfrac{x}{\sqrt{x-4}}$
2. $\dfrac{d^3s}{dt^3}$ if $s = t^3 - \dfrac{1}{(t+1)^3}$
3. $\dfrac{d^9y}{dx^9}$ if $y = x^{10}$
4. $f'''(u)$ if $f(u) = \sqrt{u^2+3}$
5. $\dfrac{d^6t}{dx^6}$ if $t = x/(2x-6)$
6. $f''''(2)$ if $f(x) = (1-x)^{2/3}$
7. When a beam of length L and uniform mass m kilograms per meter is supported as shown in Fig. 3.13, it will bend under its own weight. It can be shown that the deflection $y = f(x)$ from the horizontal must satisfy the equation

$$(EI)\frac{d^2y}{dx^2} = -\frac{mg}{2}(L-x)^2 + \frac{3}{8}mgL(L-x),$$

where E, I, and g are constants. Show that

$$y = f(x) = -\frac{mg}{48EI}(2x^4 - 5Lx^3 + 3L^2x^2)$$

satisfies this equation.

† The notation $n!$, called "n factorial," represents the product of the first n positive integers: $n! = 1(2)(3)(4)\cdots(n-1)(n)$.

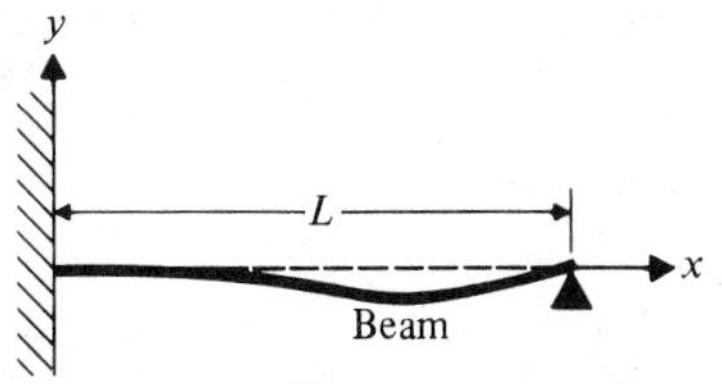

Figure 3.13

8. The steady-state temperature T in a region bounded by two concentric spheres of radii a and b $(a < b)$ must satisfy the equation

$$\frac{d^2T}{dr^2} + \frac{2}{r}\frac{dT}{dr} = 0,$$

where r is the radial distance from the common center of the spheres.

a) Show that for any constants c and d, the function

$$T = f(r) = c + \frac{d}{r}$$

satisfies the equation.

b) If temperatures on the two spheres are maintained at constant values T_a and T_b, calculate c and d in terms of a, b, T_a, and T_b.

9. Find a and b in order that the function $y = f(x) = ax^3 + bx^2$ has its second derivative equal to zero when $x = 1$ and $y = 2$.

10. Find constants a, b, c, and d in order that the function $y = f(x) = ax^3 + bx^2 + cx + d$ has its first derivative equal to 4 when $x = 1$ and $y = 0$, and its second derivative equal to 5 when $x = 2$ and $y = 4$.

11. a) Find a formula for $d^3(uv)/dx^3$ if $u = f(x)$ and $v = g(x)$.

b) On the basis of part (a) and Example 16, could you hazard a guess at a formula for $d^4(uv)/dx^4$; i.e., do you see the pattern emerging? Exercise 14 asks you to verify the correct formula.

12. If n is a positive integer, find a formula for

$$\frac{d^n}{dx^n}\left\{\frac{2x}{4 - x^2}\right\}.$$

(*Hint*: Decompose $2x/(4 - x^2)$ into two simpler fractions and proceed as in Example 15.)

13. Evaluate

$$\frac{d^{2n}}{dx^{2n}}(x^2 - 1)^n$$

for n a positive integer.

*14. If n is a positive integer, and u and v are functions of x, show (by mathematical induction) that

$$\frac{d^n}{dx^n}(uv) = \sum_{r=0}^{n} \binom{n}{r}\left(\frac{d^r u}{dx^r}\right)\left(\frac{d^{n-r} v}{dx^{n-r}}\right),$$

where $d^0u/dx^0 = u$, and $\binom{n}{r}$ are the binomial coefficients

$$\binom{n}{r} = \frac{n!}{r!(n - r)!}.$$

3.6 IMPLICIT DIFFERENTIATION

The differentiation rules given in Eqs. (3.4)–(3.12) are designed for differentiation of functions that are defined explicitly; i.e., functions in the form $y = f(x)$. In this section we show how to find the derivative of a function that is defined implicitly.

Each of the following equations defines y as a function of x implicitly:

$$y + x^2 - 2x + 5 = 0;$$

$$2y + x^2 + y^2 - 25 = 0, \qquad y \geq -1;$$

$$x + y^5 + x^2 + y = 0.$$

It is trivial to solve the first equation for the explicit definition of the function,

$$y = -x^2 + 2x - 5,$$

and obtain therefrom

$$\frac{dy}{dx} = -2x + 2.$$

We can also obtain an explicit definition of the function defined by the second equation, but not so simply. If we write

$$y^2 + 2y + (x^2 - 25) = 0,$$

and use quadratic formula (1.12), we get

$$y = \frac{-2 \pm \sqrt{4 - 4(x^2 - 25)}}{2} = -1 \pm \sqrt{26 - x^2}.$$

Since $y \geq -1$, we must choose

$$y = -1 + \sqrt{26 - x^2}.$$

It is now possible to find the derivative of y with respect to x:

$$\frac{dy}{dx} = \frac{1}{2\sqrt{26 - x^2}}(-2x) = \frac{-x}{\sqrt{26 - x^2}}.$$

The third equation is quite different, however, for it is impossible to solve this equation for an explicit definition of the function. Does this mean that it is also impossible to obtain dy/dx? The answer is no! To see this, we differentiate both sides of the equation with respect to x. Keep in mind that y is a function of x, and use Eq. (3.9) to differentiate y^5:

$$1 + 5y^4\frac{dy}{dx} + 2x + \frac{dy}{dx} = 0.$$

We can now solve this equation for dy/dx by grouping the two terms in dy/dx on one side of the equation, and transposing the remaining two terms:

$$(5y^4 + 1)\frac{dy}{dx} = -1 - 2x.$$

Division by $5y^4 + 1$ gives the required derivative,

$$\frac{dy}{dx} = -\frac{2x + 1}{5y^4 + 1}.$$

This process of differentiating an equation that implicitly defines a function is called *implicit differentiation.* It could also be used in each of the first two examples. If $y + x^2 - 2x + 5 = 0$ is differentiated with respect to x, we have

$$\frac{dy}{dx} + 2x - 2 = 0,$$

from which, as before, we get

$$\frac{dy}{dx} = -2x + 2.$$

If $2y + x^2 + y^2 - 25 = 0$ is differentiated with respect to x, we have

$$2\frac{dy}{dx} + 2x + 2y\frac{dy}{dx} = 0,$$

from which we get

$$(2 + 2y)\frac{dy}{dx} = -2x,$$

and therefore

$$\frac{dy}{dx} = \frac{-x}{y+1}.$$

Although this result appears different from the previous expression for dy/dx, when we recall that $y = -1 + \sqrt{26 - x^2}$, we find

$$\frac{dy}{dx} = \frac{-x}{-1 + \sqrt{26 - x^2} + 1} = \frac{-x}{\sqrt{26 - x^2}}.$$

These examples illustrate that if implicit differentiation is used to obtain a derivative, then the result may depend on y as well as x. Naturally if we require the derivative at a certain value of x, then the y-value used is determined by the original equation defining y implicitly as a function of x.

Example 18 Find dy/dx if y is defined implicitly as a function of x by

$$x^2y^4 - x^3y^2 + 1 = 0.$$

Solution If we differentiate (implicitly) with respect to x, we get

$$2xy^4 + 4x^2y^3\frac{dy}{dx} - 3x^2y^2 - 2x^3y\frac{dy}{dx} = 0,$$

from which we have

$$(4x^2y^3 - 2x^3y)\frac{dy}{dx} = 3x^2y^2 - 2xy^4.$$

Thus,

$$\frac{dy}{dx} = \frac{3x^2y^2 - 2xy^4}{4x^2y^3 - 2x^3y} = \frac{3xy - 2y^3}{4xy^2 - 2x^2}. \square$$

Example 19 Find dy/dx and d^2y/dx^2 if y is defined implicitly as a function of x by

$$x^5 + y^3 + y^2 = 1.$$

Solution If we differentiate with respect to x, we obtain

$$5x^4 + 3y^2\frac{dy}{dx} + 2y\frac{dy}{dx} = 0,$$

from which we then get

$$\frac{dy}{dx} = \frac{-5x^4}{3y^2 + 2y}.$$

Differentiation with respect to x now yields

$$\frac{d^2y}{dx^2} = \frac{(3y^2 + 2y)(-20x^3) + 5x^4\left(6y\frac{dy}{dx} + 2\frac{dy}{dx}\right)}{(3y^2 + 2y)^2}$$

$$= \frac{-20x^3(3y^2 + 2y) + 5x^4(6y + 2)\left(\frac{-5x^4}{3y^2 + 2y}\right)}{(3y^2 + 2y)^2}$$

$$= -\frac{20x^3(3y^2 + 2y)^2 + 25x^8(6y + 2)}{(3y^2 + 2y)^3}. \square$$

A natural question to ask at this point is "Can all equations be differentiated?" It is not difficult to show that the answer is an emphatic no! For example, if we differentiate both sides of the equation $4x = 2x$, we obtain the ludicrous result that $4 = 2$. Obviously, then, this equation cannot be differentiated with respect to x.

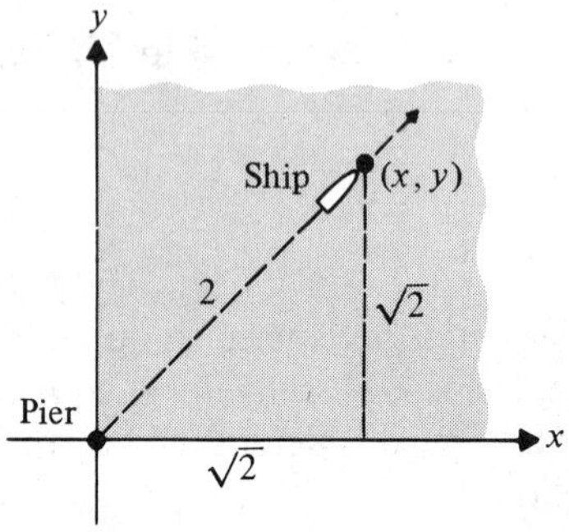

Figure 3.14

Possibly the reason differentiation fails in this example is that the equation contains only one variable (namely, x). Perhaps a more reasonable question might be "Can every equation containing two variables be differentiated?" To answer this question we consider the situation of a ship heading northeast from a pier. Let us choose east as the positive x-direction and north as the positive y-direction, both originating from the pier (Fig. 3.14). We wish to find the rate of change of the y-coordinate of the ship with respect to its x-coordinate when the ship is 2 kilometers from the pier. Obviously the answer is 1, since the x- and y-coordinates change at the same rate. But consider the following argument.

When the ship is 2 kilometers from the pier, $x = y = \sqrt{2}$, and therefore

$$x^2 + y^2 = 4.$$

If we differentiate this equation with respect to x, we get

$$2x + 2y\frac{dy}{dx} = 0, \quad \text{or} \quad \frac{dy}{dx} = -\frac{x}{y},$$

and with $x = y = \sqrt{2}$,

$$\frac{dy}{dx} = -1.$$

Differentiation of the equation $x^2 + y^2 = 4$ (which contains two variables) has again led to an erroneous result.

These two examples have certainly illustrated that not all equations can be differentiated. What then distinguishes an equation that can be differentiated from one that cannot? The answer is that an equation can be differentiated with respect to a variable only if the equation is valid for a continuous range of values of that variable. The equation $4x = 2x$ cannot be differentiated because it is true only for $x = 0$. The equation $x^2 + y^2 = 4$ in the above example cannot be differentiated because *in that example* it is valid only when the ship is 2 kilometers from the pier; i.e., only when $x = \sqrt{2}$. In another example it might be possible to differentiate $x^2 + y^2 = 4$. Each equation listed at the beginning of this section, on the other hand, was given as defining y as a function of x, and as such defines the function for some domain of values for x. The derivatives dy/dx for these functions as calculated by implicit differentiation are therefore valid at points in their respective domains (except possibly where any denominators vanish).

This is an extremely important comment, and we will return to it many times. To emphasize it once again: *We may differentiate an equation with respect to a variable only if the equation is valid for a continuous range of values of that variable.*

Earlier in this chapter we gave a proof of the power rule (3.6) in the case that n is a positive integer (and assumed its validity for all real n). With implicit differentiation it is possible to extend the proof to include the case that n is a rational number; i.e., $n = a/b$, where a and b are integers (see Exercise 36 below).

EXERCISES 3.6

Find dy/dx if y is defined implicitly as a function of x in each of Exercises 1–10.

1. $y^4 + y = 4x^3$

2. $x^4 + y^2 + y^3 = 1$

3. $xy + 2x = 4y^2 + 2$

4. $2x^3 - 3xy^4 + 5xy - 10 = 0$

5. $x + xy^5 + x^2y^3 = 3$

6. $(x + y)^2 = 2x$

7. $x(x - y) - 4y^3 = 2x + 5$

8. $\sqrt{x + y} + y^2 = 12x^2 + y$

9. $\sqrt{1 + xy} - xy = 15$

10. $\dfrac{x}{x + y} - \dfrac{y}{x} = 4$

11. Find equations for the tangent and normal lines to the curve $xy^2 + y^3 = 2$ at the point $(1, 1)$.

Find d^2y/dx^2 if y is defined implicitly as a function of x in each of Exercises 12–15.

12. $x^2 + y^3 + y = 1$

13. $2x^2 - y^3 = 4 - xy$

14. $y^2 + 2y = 5x$

15. $(x + y)^2 = x$

16. If $x^3y + xy^3 = 2$ defines y as a function of x, find dy/dx when $x = 1$.

Find dy/dx and d^2y/dx^2 if y is defined implicitly as a function of x in each of Exercises 17–20.

17. $(x + y)^2 = x^2 + y^2$

18. $x^2y^3 + 2x + 4y = 5$

19. $xy^2 - 3x^2y = x + 1$

20. $x = y\sqrt{1 - y^2}$

21. Find point(s) on the curve $xy^2 + x^2y = 16$ at which the slope of the tangent line is equal to zero.

22. Find point(s) on the curve $x^2 + y^{2/3} = 2$ at which the second derivative is equal to zero.

23. If a thermal nuclear reactor is built in the shape of a right circular cylinder of radius r and height h, then, according to neutron diffusion theory, r and h must satisfy the equation

$$\left(\frac{2.4048}{r}\right)^2 + \left(\frac{\pi}{h}\right)^2 = k = \text{constant}.$$

Find dr/dh.

24. The elasticity of a function $y = f(x)$ is defined as

$$\frac{Ey}{Ex} = \frac{x}{y}\frac{dy}{dx}.$$

Calculate elasticity for the functions defined by each of the following equations.

a) $f(x) = x\left(\dfrac{x + 1}{x + 2}\right)$ b) $x = \dfrac{400y + 200}{3 - y}$

25. Show that the elasticity of a function is equal to one if and only if the tangent line to its graph passes through the origin.

26. Find that point $P(a, b)$ on the first-quadrant part of the curve $2x^2 + 3y^2 = 14$ at which the tangent line at P is perpendicular to the line joining P and $(2, 5)$.

27. Show that the equation of the tangent line to the curve $b^2x^2 - a^2y^2 = a^2b^2$ at the point (x_0, y_0) is $b^2xx_0 - a^2yy_0 = a^2b^2$.

28. Prove that for any circle $(x - h)^2 + (y - k)^2 = r^2$,

$$\left|\frac{d^2y/dx^2}{[1 + (dy/dx)^2]^{3/2}}\right| = \frac{1}{r}.$$

29. A solution passes through a conical filter 24 centimeters deep and 16 centimeters across the top into a cylindrical vessel of diameter 12 centimeters. Find an equation relating the depth h of solution in the filter and depth H of solution in the cylinder. What is the rate of change of h with respect to H?

30. If x objects are sold at a price of $r(x)$ per object, the total revenue is given by $R(x) = xr(x)$. Find the marginal revenue $R'(x)$ if price is defined implicitly by the equation

$$x = 4a^3 - 3ar^2 + r^3,$$

where $a > 0$ is a constant, and $0 < x < 4a^3$.

31. The general polynomial of degree n is

$$a_0 + a_1x + a_2x^2 + \cdots + a_nx^n,$$

where $a_0, a_1, \ldots, a_n$ are constants. Show that two polynomials of degree n can be equal for all x,

$$a_0 + a_1x + a_2x^2 + \cdots + a_nx^n = b_0 + b_1x + b_2x^2 + \cdots + b_nx^n,$$

if and only if $a_0 = b_0, a_1 = b_1, \ldots, a_n = b_n$.

32. a) Find $f'(0)$ if $y = f(x)$ is defined implicitly as a function of x by

$$x\sqrt{1+2y} = x^2 - y.$$

b) Show that by squaring this equation

$$x^2 + 4x^2y = x^4 + y^2.$$

Differentiate this equation with respect to x to find $f'(0)$. Do you encounter any difficulties?

33. a) Use implicit differentiation to find dy/dx if $y^2 = x^2 - x^4$.

b) Can you calculate dy/dx at $x = 0$ using the result of part (a)? Sketch the curve $y^2 = x^2 - x^4$ in order to explain this difficulty.

34. a) Find dy/dx if $\sqrt{x} + \sqrt{y} = 1$, $0 \le x \le 1$, defines y implicitly as a function of x.

b) Sketch the curve $\sqrt{|x|} + \sqrt{|y|} = 1$.

35. What is dy/dx if

$$x^2 + 4x + y^2 + 6y + 15 = 0?$$

36. Verify the power rule in the case that the exponent is a rational number; i.e., prove that

$$\frac{d}{dx}u^n = nu^{n-1}\frac{du}{dx} \qquad (u > 0)$$

if $u = f(x)$ and $n = a/b$, where a and b are integers ($b \neq 0$).

37. Show that any function defined implicitly by the equation

$$\frac{x^2}{y^3} - x = C, \quad C \text{ a constant,}$$

satisfies the equation

$$3x^2\frac{dy}{dx} + y^4 = 2xy.$$

38. Show that any function defined implicitly by the equation

$$2x^2 - 3y = Cx^2y^3, \quad C \text{ a constant,}$$

satisfies the equation

$$(xy - x^3)\frac{dy}{dx} + y^2 = 0.$$

***39.** Find points on the curve $xy^2 + yx^2 = 2$ at which the slope of the tangent line is equal to one.

3.7 THE CHAIN RULE

Suppose that y is defined as a function of u by

$$y = f(u) = u^{2/3},$$

and u, in turn, is defined as a function of x by

$$u = g(x) = \frac{1}{x^2+1}.$$

These equations imply that y is a function of x; indeed,

$$y = \left(\frac{1}{x^2+1}\right)^{2/3},$$

and we can therefore calculate

$$\frac{dy}{dx} = \frac{2}{3}\left(\frac{1}{x^2+1}\right)^{-1/3}\frac{-2x}{(x^2+1)^2}.$$

If we differentiate the original functions,

$$\frac{dy}{du} = \frac{2}{3}u^{-1/3}, \qquad \frac{du}{dx} = \frac{-2x}{(x^2+1)^2},$$

then the product of these derivatives is also equal to

$$\frac{dy}{du}\frac{du}{dx} = \frac{2}{3}u^{-1/3}\frac{-2x}{(x^2+1)^2} = \frac{2}{3}\left(\frac{1}{x^2+1}\right)^{-1/3}\frac{-2x}{(x^2+1)^2}.$$

Consequently, for this example, we can write that

$$\frac{dy}{dx} = \frac{dy}{du}\frac{du}{dx}. \tag{3.13}$$

The natural question to ask now is whether this result is true in general. In other words, if

$$y = f(u) \quad \text{and} \quad u = g(x) \tag{3.14}$$

define y as a function of x,

$$y = f[g(x)], \tag{3.15}$$

does Eq. (3.13) define dy/dx correctly? We say that y is defined as a *composite function* of x by Eq. (3.15); it is composed of the two functions $f(u)$ and $g(x)$. That Eq. (3.13) is generally true is proved in the following theorem.

▶ **Theorem 3.9** (*The chain rule*). If $y = f(u)$ and $u = g(x)$ are differentiable functions, then the derivative of the composite function $y = f[g(x)]$ is

$$\frac{dy}{dx} = \frac{dy}{du}\frac{du}{dx}. \quad ◀ \tag{3.13}$$

Proof By Definition 3.1, the derivative of the composite function is

$$\frac{dy}{dx} = \lim_{\Delta x \to 0} \frac{f[g(x + \Delta x)] - f[g(x)]}{\Delta x}.$$

Now a change Δx in x will produce a change $g(x + \Delta x) - g(x)$ in u. If we denote this change by Δu, then it will in turn produce the change $f(u + \Delta u) - f(u)$ in y. We may write, therefore, that

$$\frac{dy}{dx} = \lim_{\Delta x \to 0} \frac{f(u + \Delta u) - f(u)}{\Delta x}.$$

Since $f(u)$ is differentiable, its derivative exists and is defined by the limit

$$f'(u) = \lim_{\Delta u \to 0} \frac{f(u + \Delta u) - f(u)}{\Delta u}.$$

An equivalent way to express the fact that this limit is $f'(u)$ is to say that

$$\frac{f(u + \Delta u) - f(u)}{\Delta u} = f'(u) + \varepsilon,$$

where ε must satisfy the condition $\lim_{\Delta u \to 0} \varepsilon = 0$. We may write, therefore, that

$$f(u + \Delta u) - f(u) = [f'(u) + \varepsilon]\,\Delta u,$$

and if we substitute this expression into the limit for dy/dx, we get

$$\frac{dy}{dx} = \lim_{\Delta x \to 0} \frac{[f'(u) + \varepsilon]\,\Delta u}{\Delta x} = \lim_{\Delta x \to 0} \left\{[f'(u) + \varepsilon]\frac{\Delta u}{\Delta x}\right\}.$$

But

$$\lim_{\Delta x \to 0} \frac{\Delta u}{\Delta x} = \lim_{\Delta x \to 0} \frac{g(x + \Delta x) - g(x)}{\Delta x} = \frac{du}{dx}.$$

Furthermore, because $g(x)$ is differentiable, it is continuous (Theorem 3.7), and this implies that $\Delta u \to 0$ as $\Delta x \to 0$. Consequently, $\lim_{\Delta x \to 0} \varepsilon = \lim_{\Delta u \to 0} \varepsilon = 0$, and these results give

$$\frac{dy}{dx} = f'(u)\frac{du}{dx} = \frac{dy}{du}\frac{du}{dx}. \square$$

This result is called the *chain rule* for the derivative of a composite function. It expresses the derivative of a composite function as the product of the derivatives of the functions in the composition. From the point of view of rates of change, the chain rule seems quite reasonable. It states that if a variable is defined in terms of a second variable, which is in turn defined in terms of a third variable, then the rate of change of the first variable with respect to the third is the rate of change of the first with respect to the second multiplied by the rate of change of the second with respect to the third. For example, if car A travels twice as fast as car B, and car B travels three times as fast as car C, then car A travels six times as fast as car C.

Note that the extended power rule (3.9) is a special case of the chain rule (when $f(u) = u^n$).

But why would we ever want to use the chain rule to calculate dy/dx in the first place? Wouldn't it be easier to substitute with Eqs. (3.14) to obtain the composite function (3.15), and then simply differentiate with respect to x? This would avoid the intermediate variable u altogether. This is indeed correct and can always be done. But you can imagine how difficult $f[g(x)]$ might be if $g(x)$ and $f(u)$ were at all complicated, say,

$$f(u) = \sqrt{\frac{u^2 + 2u}{u - 5}}, \qquad g(x) = \frac{x^2(x^3 + 3)^2}{(x + 5)^2(x - 2)}.$$

Example 20 Find dy/dx if $y = u^3 - 5u + 1$ and $u = \sqrt{x}/(x + 1)$.

Solution By the chain rule, we have

$$\frac{dy}{dx} = \frac{dy}{du}\frac{du}{dx} = (3u^2 - 5)\left\{\frac{(x + 1)(1/(2\sqrt{x})) - \sqrt{x}(1)}{(x + 1)^2}\right\}$$

$$= \frac{(3u^2 - 5)(1 - x)}{2\sqrt{x}(x + 1)^2}. \square$$

Example 21 Find dy/dt at $t = 1$ if $y = x^2 - 2x$ and $x = (\sqrt{t + 3})/t^2$.

Solution By the chain rule, we have

$$\frac{dy}{dt} = \frac{dy}{dx}\frac{dx}{dt} = (2x - 2)\left\{\frac{t^2(1/2)(t + 3)^{-1/2} - \sqrt{t + 3}(2t)}{t^4}\right\}.$$

When $t = 1$, $x = (\sqrt{1 + 3})/(1)^2 = 2$. If $dy/dt|_{t=1}$ denotes dy/dt evaluated at $t = 1$, then

$$\left.\frac{dy}{dt}\right|_{t=1} = (2 \cdot 2 - 2)\left\{\frac{1^2(1/2)(1 + 3)^{-1/2} - \sqrt{1 + 3}(2 \cdot 1)}{1^4}\right\} = -\frac{15}{2}. \square$$

Example 22 Find dy/dx and d^2y/dx^2 if $y = v^4 - 2$ and $v = x^3 + x$.

Solution By the chain rule, we get

$$\frac{dy}{dx} = \frac{dy}{dv}\frac{dv}{dx} = 4v^3(3x^2 + 1).$$

If we now differentiate with respect to x, we have

$$\frac{d^2y}{dx^2} = 12v^2\frac{dv}{dx}(3x^2 + 1) + 4v^3(6x)$$

$$= 12v^2(3x^2 + 1)^2 + 24v^3x. \square$$

These examples illustrate that should we use the chain rule (3.13) to calculate dy/dx when y is defined as a function of x by Eq. (3.14), then the result will usually involve both x and u. Naturally if we require the derivative at a certain value of x, then the u-value used is that value corresponding to the given x.

EXERCISES 3.7

In each of Exercises 1–8, where y is defined as a composite function of x, find dy/dx.

1. $y = t^2 + \dfrac{1}{t^3}, t = \sqrt{4 - x^2}$

2. $y = (2s - s^2)^{1/3}, s = \dfrac{1}{x^2 + 5}$

3. $y = \dfrac{v^2}{v^3 - 1}, v = x\sqrt{x^2 - 1}$

4. $y = \dfrac{u}{u + 5}, u = \dfrac{\sqrt{x - 1}}{x}$

5. $y = u^4(u^3 - 2u)^2, u = \sqrt{x - 2x^2}$

6. $y = t + \sqrt{t + \sqrt{t}}, t = \dfrac{x^2 + 1}{x^2 - 1}$

7. $y = \left(\dfrac{v^2 + 1}{1 - v^3}\right)^3, v = \dfrac{1}{x^3 + 3x^2 + 2}$

8. $y = \dfrac{\sqrt{k}}{1 + k + k^2}, k = x(x^2 + 5)^5$

Find d^2y/dx^2 if y is defined as a composite function of x in each of Exercises 9–12.

9. $y = v^2 + v, v = \dfrac{x}{x + 1}$

10. $y = (u + 1)^3 - \dfrac{1}{u}, u = x + \sqrt{x + 1}$

11. $y = \sqrt{t - 1}, t = (x + x^2)^2$

12. $y = \dfrac{s}{s + 6}, s = \dfrac{\sqrt{x}}{1 + \sqrt{x}}$

13. If $y = f(u)$, $u = g(s)$, and $s = h(x)$, show that

$$\frac{dy}{dx} = \frac{dy}{du}\frac{du}{ds}\frac{ds}{dx}.$$

14. When an electrostatic charge of $q = 5 \times 10^{-6}$ coulombs is at a distance r meters from a stationary charge $Q = 3 \times 10^{-6}$ coulombs, the magnitude of the force of repulsion of Q on q is

$$F = \frac{Qq}{4\pi\varepsilon_0 r^2} \text{ newtons,}$$

where $\varepsilon_0 = 8.85 \times 10^{-12}$. If q is moved directly away from Q at 2 meters per second, how fast is F changing when $r = 2$ meters?

15. When a mass m of 5 kilograms is r meters from the center of the earth, the magnitude of the force of attraction of the earth on m is

$$F = \frac{GmM}{r^2} \text{ newtons,}$$

where M is the mass of the earth and $G = 6.67 \times 10^{-11}$. If m is falling at 100 kilometers per hour when it is 5 kilometers above the surface of the earth, how fast is F changing? The mean density of the earth is 5.52×10^3 kilograms per cubic meter, and its mean radius is 6370 kilometers.

16. Show that if an equation $F(x, y) = 0$ can be solved for both y in terms of x, $y = f(x)$, and x in terms of y, $x = g(y)$, then

$$\frac{dy}{dx} = \frac{1}{\dfrac{dx}{dy}}.$$

17. Prove that the derivative of an odd function is an even function, and that the derivative of an even function is an odd function.

18. Find dy/dx if $y = u/\sqrt{u^2 - 1}$ and u is defined implicitly as a function of x by

$$x^2u^2 + \sqrt{u^2 - 1} = 4.$$

19. a) If $y = f(u)$ and $u = g(x)$, use the chain rule to show that

$$\frac{d^2y}{dx^2} = \frac{d^2u}{dx^2}\frac{dy}{du} + \frac{d^2y}{du^2}\left(\frac{du}{dx}\right)^2.$$

b) Use the result in (a) to calculate d^2y/dx^2 at $x = 1$ if

$$y = (u + 1)^3, \qquad u = 3x - \frac{2}{x^2}.$$

20. Given that the equations

$$y^4 + yv^3 = 3,$$

$$x^2v + 3xv^2 = 2x^3y + 1$$

define y as a function of v and v as a function of x, find dy/dx in terms of x, y, and v.

21. Find the rate of change of $y = f(x) = x^9 + x^6$ with respect to x^3.

22. Find the rate of change of $y = f(x) = \sqrt{1 - x^2}$ with respect to $x/(x + 1)$.

***23.** If $y = f(u)$ and $u = g(x)$, show that

$$\frac{d^3y}{dx^3} = \frac{d^3u}{dx^3}\frac{dy}{du} + 3\frac{d^2y}{du^2}\frac{d^2u}{dx^2}\frac{du}{dx} + \frac{d^3y}{du^3}\left(\frac{du}{dx}\right)^3.$$

3.8 TRIGONOMETRY REVIEW

Many physical systems exhibit an oscillatory nature: vibrations of a plucked guitar string, motion of a pendulum, alternating electric currents, fluctuations in room temperature as the thermostat continually engages and disengages the furnace on a cold day, and the rise and fall of tides and waves. Magnitudes of these oscillations are best described by the sine and cosine functions of trigonometry. In addition, rates at which these oscillations occur can be represented by the derivatives of these functions. In other words, the study of oscillatory systems requires a knowledge of the sine and cosine functions and their derivatives. In this section we present a brief review of the trigonometric functions and their properties, placing special emphasis on those aspects that are most useful to the study of calculus. In Section 3.9 we derive derivatives for the sine and cosine functions, postponing a discussion of derivatives of the remaining trigonometric functions until Chapter 8.

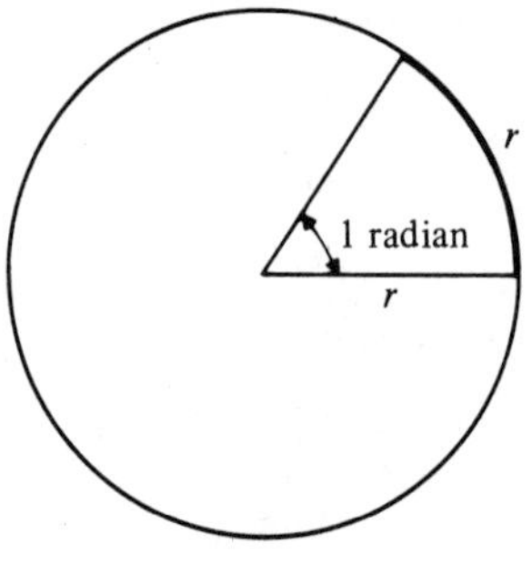

Figure 3.15

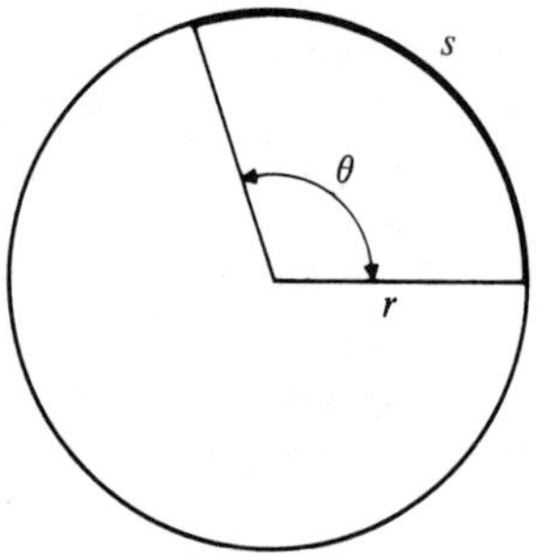

Figure 3.16

In trigonometry, angles are measured in either degrees or radians; in calculus angles are always measured in radians, never degrees. The reason for this will become evident in Section 3.9 when we differentiate the sine and cosine functions. In preparation, then, for the calculus of trigonometric functions, we will work completely in radian measure. By definition, a *radian* is that angle subtended at the center of a circle of radius r (Fig. 3.15) by an arc of equal length r. When an arc has length s (Fig. 3.16), the number of "units of radius" in this arc is s/r; therefore the angle θ subtended at the center of the circle is

$$\theta = \frac{s}{r}. \tag{3.16}$$

In particular, if s contains π units of radius ($s = \pi r$), then s represents one-half the circumference of the complete circle, and $\theta = \pi r/r = \pi$ radians. In degree measure, this angle is 180°; hence we can state that π radians is equivalent to

180°. This statement enables us to convert any angle expressed in degrees to radian measure (and vice versa). If an angle measures $\phi°$, then it is $\pi\phi/180$ radians; conversely, if it measures θ radians, then it is $180\theta/\pi°$. For example, the radian measure for $\phi = 45°$ is $45\pi/180 = \pi/4$ radians.

Elementary trigonometry is concerned with relationships among angles and lengths and, in particular, angles and sides of triangles. This naturally restricts angles to the range $0 \le \theta \le \pi$. For the purposes of calculus, however, we need to talk about angles θ, where θ is any real number—positive, negative, or zero. To do this we first define what we mean by the *standard position of an angle.* If $\theta > 0$, we draw a line segment OP through the origin O of the xy-plane (Fig. 3.17) in such a way that the positive x-axis must rotate counterclockwise through an angle θ to coincide with OP. In other words, we now regard an angle as a rotation, that is, a rotation of the positive x-axis to some terminal position. If $0 < \theta < \pi/2$, then OP lies in the first quadrant; and if $\pi/2 < \theta < \pi$, then OP is in the second quadrant. But now if $\theta > \pi$, we have a geometric representation of θ also. For example, angles of $7\pi/4$ and $9\pi/4$ are shown in Fig. 3.18. When $\theta < 0$, we regard θ as a clockwise rotation (Fig. 3.19).

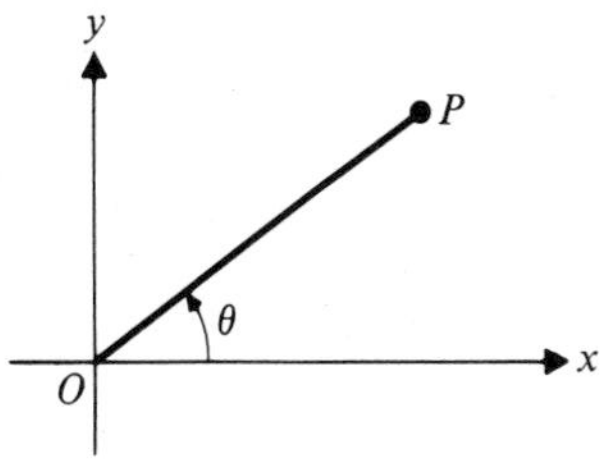

Figure 3.17

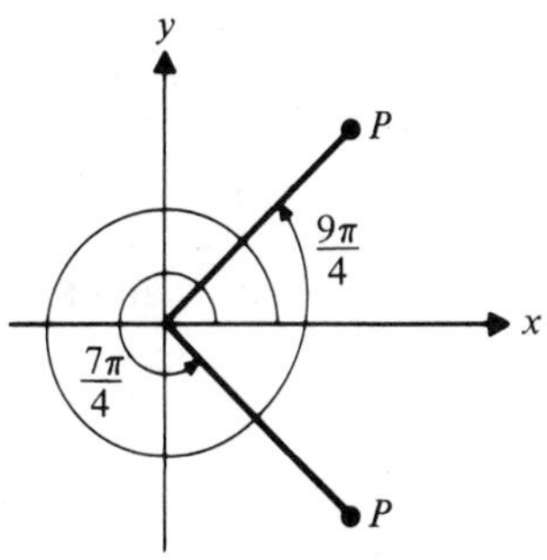

Figure 3.18

Note that for any angle θ whatsoever, the angles $\theta + 2n\pi$ for n an integer have the same terminal position of OP as θ. They are different angles, however, because the positive x-axis must encircle the origin one or more times before reaching the terminal position in the case of $\theta + 2n\pi$.

With this representation of angles as rotations, it is straightforward to define the six trigonometric functions for all angles. If θ is an angle in standard position (Fig. 3.20) and (x, y) are the coordinates of P, then we define

$$\begin{aligned} \sin\theta &= \frac{y}{r}, & \cos\theta &= \frac{x}{r}, & \tan\theta &= \frac{y}{x}, \\ \csc\theta &= \frac{r}{y}, & \sec\theta &= \frac{r}{x}, & \cot\theta &= \frac{x}{y}, \end{aligned} \tag{3.17}$$

wherever these functions are defined, and where $r = \sqrt{x^2 + y^2}$ is always assumed positive. Since x is positive in the first and fourth quadrants, and y is positive in the first and second, signs of the trigonometric functions in the various quadrants are as shown in Fig. 3.21. Furthermore, because $r \ne 0$, $\sin\theta$ and $\cos\theta$ are defined for all real θ, whereas $\tan\theta$ and $\sec\theta$ are not defined for $x = 0$, and $\csc\theta$ and $\cot\theta$ do not exist when $y = 0$.

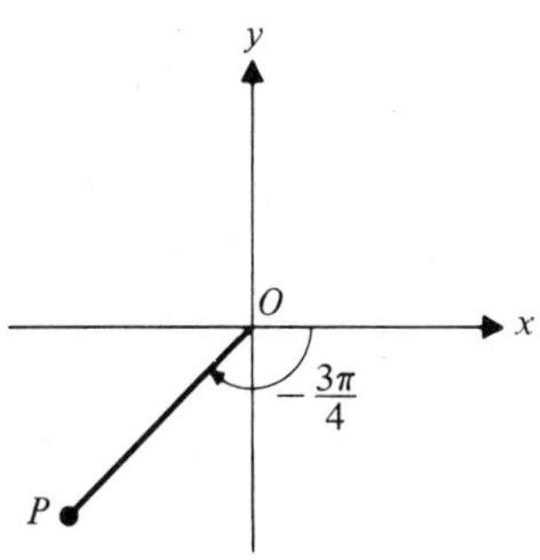

Figure 3.19

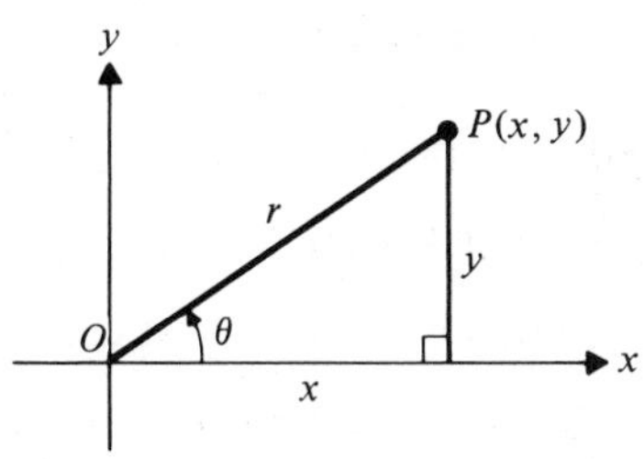

Figure 3.20

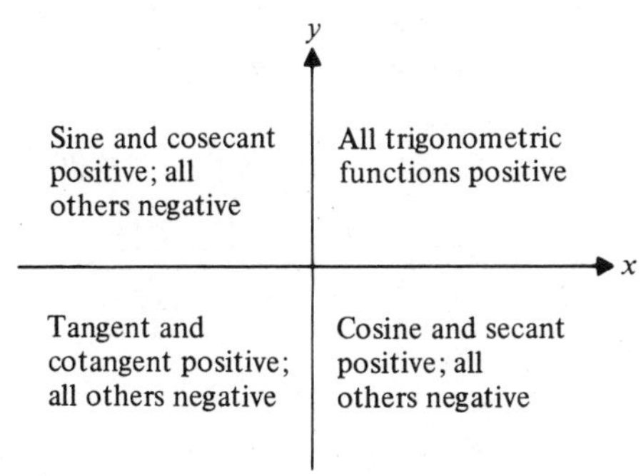

Figure 3.21

Clearly these functions are not all independent since whenever the functions are defined,

$$\csc\theta = \frac{1}{\sin\theta}, \qquad \sec\theta = \frac{1}{\cos\theta}, \qquad \cot\theta = \frac{1}{\tan\theta}. \tag{3.18}$$

In addition, the fact that $r^2 = x^2 + y^2$ leads to the identity

$$\sin^2\theta + \cos^2\theta = 1, \tag{3.19}$$

which in turn implies that

$$1 + \tan^2\theta = \sec^2\theta, \tag{3.20a}$$

$$1 + \cot^2\theta = \csc^2\theta. \tag{3.20b}$$

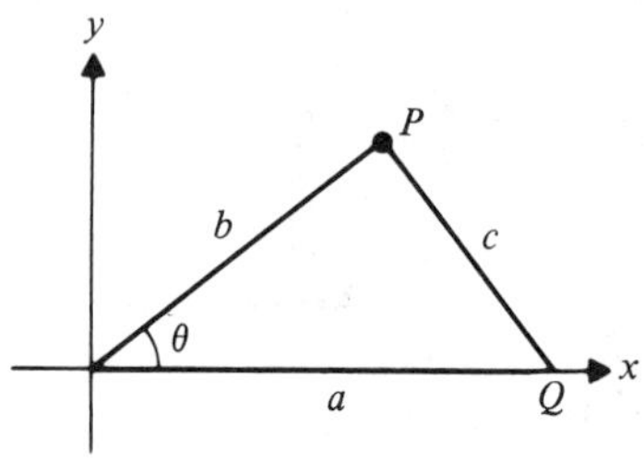

Figure 3.22

Identity (3.19) is simply a restatement of the Pythagorean relation ($r^2 = x^2 + y^2$), which allows us to express the hypotenuse of a right-angled triangle in terms of the two sides making the right angle. If the triangle is not right-angled (Fig. 3.22), it is not possible to express c in terms of a and b alone, but it is possible to express it in terms of a and b and the angle θ between them. Clearly the coordinates of P and Q in the figure are $(b\cos\theta, b\sin\theta)$ and $(a, 0)$, and therefore by formula (1.5),

$$\begin{aligned} c^2 = |PQ|^2 &= (b\cos\theta - a)^2 + (b\sin\theta)^2 \\ &= b^2(\cos^2\theta + \sin^2\theta) + a^2 - 2ab\cos\theta. \end{aligned}$$

Since $\sin^2\theta + \cos^2\theta = 1$, we have

$$c^2 = a^2 + b^2 - 2ab\cos\theta. \tag{3.21}$$

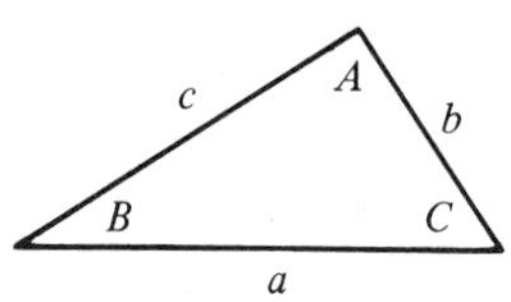

Figure 3.23

This result, called the *cosine law*, generalizes the Pythagorean relation to triangles that are not right-angled. Note that it reduces to the Pythagorean relation $c^2 = a^2 + b^2$ when $\theta = \pi/2$.

If A, B, and C are the angles in the triangle shown in Fig. 3.23, and a, b, and c are the lengths of the opposite sides, then by drawing altitudes of the triangle we can easily show that

$$\frac{\sin A}{a} = \frac{\sin B}{b} = \frac{\sin C}{c}. \tag{3.22}$$

This result, known as the *sine law*, is also useful in many problems.

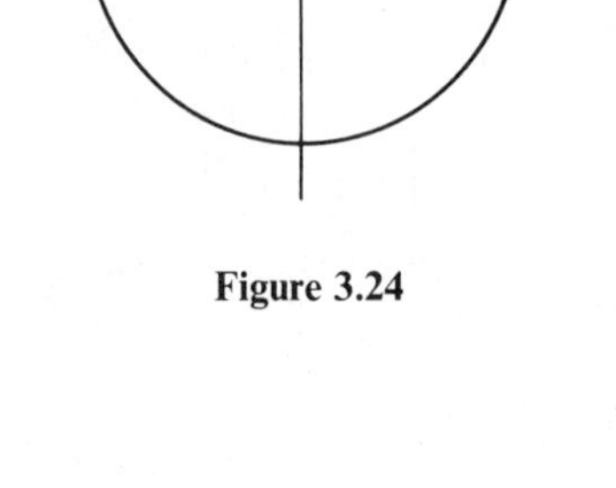

Figure 3.24

There is a large number of identities satisfied by the trigonometric functions, some of major importance, others of minor importance. It is reassuring to know that they can all be derived from four identities, called the *sum and difference formulas* for the sine and cosine functions:

$$\sin(A + B) = \sin A\cos B + \cos A\sin B, \tag{3.23a}$$

$$\sin(A - B) = \sin A\cos B - \cos A\sin B, \tag{3.23b}$$

$$\cos(A + B) = \cos A\cos B - \sin A\sin B, \tag{3.23c}$$

$$\cos(A - B) = \cos A\cos B + \sin A\sin B. \tag{3.23d}$$

Let us prove one of these, say identity (3.23d), where A and B are the angles shown in Fig. 3.24. If points P and Q lie on the circle $x^2 + y^2 = r^2$, then their coordinates are ($r\cos A$, $r\sin A$) and ($r\cos B$, $r\sin B$), respectively. According to formula (1.5), the length of PQ is therefore given by

$$\begin{aligned}|PQ|^2 &= (r\cos A - r\cos B)^2 + (r\sin A - r\sin B)^2\\ &= r^2(\cos^2 A + \sin^2 A) + r^2(\cos^2 B + \sin^2 B)\\ &\quad - 2r^2(\cos A\cos B + \sin A\sin B)\\ &= 2r^2 - 2r^2(\cos A\cos B + \sin A\sin B).\end{aligned}$$

But according to the cosine law (3.21),

$$\begin{aligned}|PQ|^2 &= |OP|^2 + |OQ|^2 - 2|OP||OQ|\cos(A - B)\\ &= r^2 + r^2 - 2r^2\cos(A - B)\\ &= 2r^2 - 2r^2\cos(A - B).\end{aligned}$$

Comparison of these two expressions for $|PQ|^2$ immediately implies identity (3.23d).

With equations (3.23), it is quite simple to derive the following identities:

$$\sin(-\theta) = -\sin\theta, \tag{3.24a}$$

$$\cos(-\theta) = \cos\theta, \tag{3.24b}$$

$$\tan(-\theta) = -\tan\theta, \tag{3.24c}$$

$$\csc(-\theta) = -\csc\theta, \tag{3.24d}$$

$$\sec(-\theta) = \sec\theta, \tag{3.24e}$$

$$\cot(-\theta) = -\cot\theta; \tag{3.24f}$$

$$\sin(\pi/2 - \theta) = \cos\theta, \tag{3.25a}$$

$$\cos(\pi/2 - \theta) = \sin\theta, \tag{3.25b}$$

$$\tan(\pi/2 - \theta) = \cot\theta, \tag{3.25c}$$

$$\csc(\pi/2 - \theta) = \sec\theta, \tag{3.25d}$$

$$\sec(\pi/2 - \theta) = \csc\theta, \tag{3.25e}$$

$$\cot(\pi/2 - \theta) = \tan\theta; \tag{3.25f}$$

$$\sin(\pi - \theta) = \sin\theta, \tag{3.26a}$$

$$\cos(\pi - \theta) = -\cos\theta, \tag{3.26b}$$

$$\tan(\pi - \theta) = -\tan\theta, \tag{3.26c}$$

$$\csc(\pi - \theta) = \csc\theta, \tag{3.26d}$$

$$\sec(\pi - \theta) = -\sec\theta, \tag{3.26e}$$

$$\cot(\pi - \theta) = -\cot\theta; \tag{3.26f}$$

$$\tan(A + B) = \frac{\tan A + \tan B}{1 - \tan A\tan B}, \tag{3.27a}$$

$$\tan(A - B) = \frac{\tan A - \tan B}{1 + \tan A \tan B}; \tag{3.27b}$$

$$\sin 2A = 2 \sin A \cos A; \tag{3.28}$$

$$\cos 2A = \cos^2 A - \sin^2 A, \tag{3.29a}$$
$$= 2\cos^2 A - 1, \tag{3.29b}$$
$$= 1 - 2\sin^2 A; \tag{3.29c}$$

$$\tan 2A = \frac{2 \tan A}{1 - \tan^2 A}; \tag{3.30}$$

$$\sin A \sin B = \tfrac{1}{2}[-\cos(A + B) + \cos(A - B)], \tag{3.31a}$$

$$\sin A \cos B = \tfrac{1}{2}[\sin(A + B) + \sin(A - B)], \tag{3.31b}$$

$$\cos A \cos B = \tfrac{1}{2}[\cos(A + B) + \cos(A - B)]; \tag{3.31c}$$

$$\sin A + \sin B = 2 \sin\left(\frac{A + B}{2}\right) \cos\left(\frac{A - B}{2}\right), \tag{3.32a}$$

$$\sin A - \sin B = 2 \cos\left(\frac{A + B}{2}\right) \sin\left(\frac{A - B}{2}\right), \tag{3.32b}$$

$$\cos A + \cos B = 2 \cos\left(\frac{A + B}{2}\right) \cos\left(\frac{A - B}{2}\right), \tag{3.32c}$$

$$\cos A - \cos B = -2 \sin\left(\frac{A + B}{2}\right) \sin\left(\frac{A - B}{2}\right). \tag{3.32d}$$

Briefly, Eqs. (3.27a and b) are a direct result of divisions of (3.23a and b) by (3.23c and d). Equations (3.24), (3.25), and (3.26) are Eqs. (3.23b and d) and (3.27b) (and their reciprocals) with $A = 0$, $\pi/2$, and π, respectively. *Double-angle formulas* (3.28), (3.29), and (3.30) can be obtained by setting $A = B$ in (3.23a and c) and (3.27a). To verify *product formulas* (3.31), appropriate combinations of (3.23) are either added or subtracted. Finally, the *sum formulas* (3.32) are essentially the corresponding equations (3.31) with a renaming of angles.

The next three examples are typical of problems that we encounter in the calculus of trigonometric functions.

Example 23 Find $R > 0$ and $0 < \phi < \pi$ such that $3 \cos\theta - 4 \sin\theta$ can be expressed in the form

$$R \sin(\theta + \phi) = 3 \cos\theta - 4 \sin\theta \quad \text{for all real } \theta.$$

Solution If we expand the left-hand side of this equation according to (3.23a), we get

$$R \sin\theta \cos\phi + R \cos\theta \sin\phi = 3 \cos\theta - 4 \sin\theta.$$

Since this equation is to be valid for all real θ, it must therefore be valid for $\theta = 0$ and $\theta = \pi/2$. Substitution of these values gives the two equations

$$R \sin\phi = 3 \quad \text{and} \quad R \cos\phi = -4.$$

If we square and add these equations, we have

$$R^2 \sin^2 \phi + R^2 \cos^2 \phi = R^2 = 3^2 + (-4)^2 = 25.$$

Consequently, $R = 5$, and

$$\sin \phi = \tfrac{3}{5}, \qquad \cos \phi = -\tfrac{4}{5}.$$

The only angle in the range $0 < \phi < \pi$ satisfying these equations is $\phi = 2.50$ radians. Thus, $3 \cos \theta - 4 \sin \theta$ can be expressed in the form

$$3 \cos \theta - 4 \sin \theta = 5 \sin (\theta + 2.50). \ \square$$

Example 24 Find all solutions of each of the following equations:

a) $2 \sin^2 \theta = 1$;

b) $\sin \theta + \cos \theta = \tfrac{1}{2}$.

Solution a) If $2 \sin^2 \theta = 1$, then

$$\sin \theta = \pm \frac{1}{\sqrt{2}}.$$

One solution of $\sin \theta = 1/\sqrt{2}$ is $\theta = \pi/4$. But according to (3.26a), $\sin(\pi - \pi/4) = \sin(\pi/4)$, and therefore $\theta = 3\pi/4$ is also a solution. Since $\sin(\theta + 2n\pi) = \sin \theta$ for any integer n, it follows that each of the angles

$$\theta = \begin{cases} \pi/4 + 2n\pi, \\ 3\pi/4 + 2n\pi, \end{cases}$$

is a solution of $\sin \theta = 1/\sqrt{2}$. A similar procedure gives solutions

$$\theta = \begin{cases} -\pi/4 + 2n\pi, \\ -3\pi/4 + 2n\pi, \end{cases}$$

of $\sin \theta = -1/\sqrt{2}$. All solutions, then, of $2 \sin^2 \theta = 1$ are contained in

$$\theta = \begin{cases} \pm\pi/4 + 2n\pi, \\ \pm 3\pi/4 + 2n\pi, \end{cases}$$

and this set of solutions can be described more compactly by

$$\theta = \pm \frac{\pi}{4} + n\pi = \left(\frac{4n \pm 1}{4}\right)\pi, \quad n \text{ an integer}$$

(see Fig. 3.25).

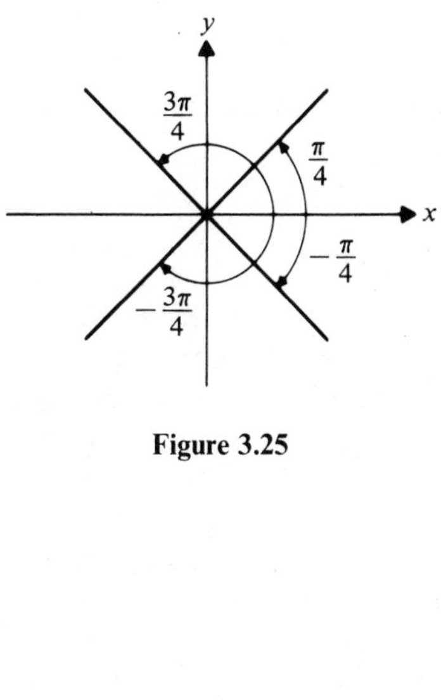

Figure 3.25

b) We begin by using (3.19) to write the equation in terms of $\sin \theta$ only:

$$\sin \theta \pm \sqrt{1 - \sin^2 \theta} = \tfrac{1}{2}.$$

If we now transpose $\sin \theta$ and square both sides, we get

$$1 - \sin^2 \theta = \tfrac{1}{4} - \sin \theta + \sin^2 \theta$$

or

$$8 \sin^2 \theta - 4 \sin \theta - 3 = 0.$$

But this is a quadratic equation in $\sin \theta$ with solutions

$$\sin \theta = \frac{4 \pm \sqrt{16 + 96}}{16} = \frac{1 \pm \sqrt{7}}{4}.$$

The acute angle for which $\sin\theta = (1 + \sqrt{7})/4$ is $\theta = 1.15$ radians. If we reason as in part (a), we would expect as solutions

$$\theta = \begin{cases} 1.15 + 2n\pi, \\ (\pi - 1.15) + 2n\pi \end{cases}$$

(since each of these angles has a sine equal to $(1 + \sqrt{7})/4$). When we check these angles in the original equation, however, we find that only $\theta = (\pi - 1.15) + 2n\pi$ are solutions. An analogous discussion for the case where $\sin\theta = (1 - \sqrt{7})/4$ leads to the solutions $\theta = -0.424 + 2n\pi$. The complete set of solutions then is

$$\theta = \begin{cases} -1.15 + (2n + 1)\pi, \\ -0.424 + 2n\pi, \end{cases} \quad n \text{ an integer.} \ \square$$

Example 25 Write the expression $\cos^4\theta$ in terms of $\cos 2\theta$ and $\cos 4\theta$.

Solution If we replace A by θ in double-angle formula (3.29b),

$$\cos 2\theta = 2\cos^2\theta - 1,$$

and solve this equation for $\cos^2\theta$, we have

$$\cos^2\theta = \frac{1 + \cos 2\theta}{2}.$$

Consequently,

$$\cos^4\theta = (\cos^2\theta)^2 = \tfrac{1}{4}(1 + \cos 2\theta)^2 = \tfrac{1}{4}(1 + 2\cos 2\theta + \cos^2 2\theta).$$

But if we now replace θ by 2θ in the identity $\cos^2\theta = (1 + \cos 2\theta)/2$, we get

$$\cos^2 2\theta = \frac{1 + \cos 4\theta}{2}.$$

Thus,

$$\cos^4\theta = \frac{1}{4}\left(1 + 2\cos 2\theta + \frac{1 + \cos 4\theta}{2}\right) = \frac{1}{8}(3 + 4\cos 2\theta + \cos 4\theta). \ \square$$

By regarding "arguments" of the trigonometric functions as angles, we have been stressing geometric properties of these functions. In particular, identities (3.23)–(3.32) have all been based on definitions of the trigonometric functions as functions of angles. These identities relate the trigonometric functions one to another. Note, however, that we have not yet made an indepth study of any one particular trigonometric function, since that belongs in the area of the calculus of trigonometric functions.

What is important about a function, be it a trigonometric function or any other kind of function, is that there is a number associated with each value of the independent variable. How we arrive at this number is irrelevant. So far as properties of the function are concerned, only its values are taken into account. Thus when we discuss properties of a trigonometric function, what is important is not that its argument can be regarded as an angle and that its values can be defined as ratios of sides of a triangle, but that we know its values. Indeed, it is sometimes unwise to regard arguments of trigonometric functions as angles. Consider, for example, the motion of a mass m suspended from a spring with spring constant k (Fig. 3.26). If $x = 0$ is the position at which the

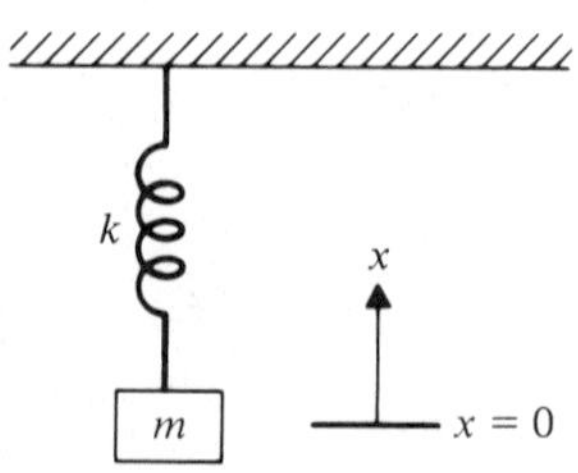

Figure 3.26

mass would hang motionless, then when vertical oscillations are initiated, the position of m as a function of time t will always be of the form

$$x = A\cos\left(\sqrt{\frac{k}{m}}\,t\right) + B\sin\left(\sqrt{\frac{k}{m}}\,t\right),$$

where A and B are constants. Clearly there are no angles associated with the motion of m, and it is therefore unnatural to attempt to interpret the argument $(\sqrt{k/m})t$ as a physical angle.

Our point, then, is that we will now consider the trigonometric functions as functions of a real variable. If it is convenient to regard the argument as an angle, then we will do so, but we are not compelled to. To emphasize this fact, we replace θ with our usual generic label for the independent variable of a function, namely, x. With this change, the trigonometric functions are $\sin x$, $\cos x$, $\tan x$, $\csc x$, $\sec x$, and $\cot x$. Their graphs are shown in Fig. 3.27.

These functions are the first examples of what are called *periodic functions*. A function $f(x)$ is said to be *periodic* if there exists a number T such that for all x,

$$f(x + T) = f(x). \tag{3.33}$$

The smallest such positive number T is called the *period* of $f(x)$. Clearly, then, $\sin x$, $\cos x$, $\csc x$, and $\sec x$ are all periodic with period 2π, whereas $\tan x$ and $\cot x$ have period π.

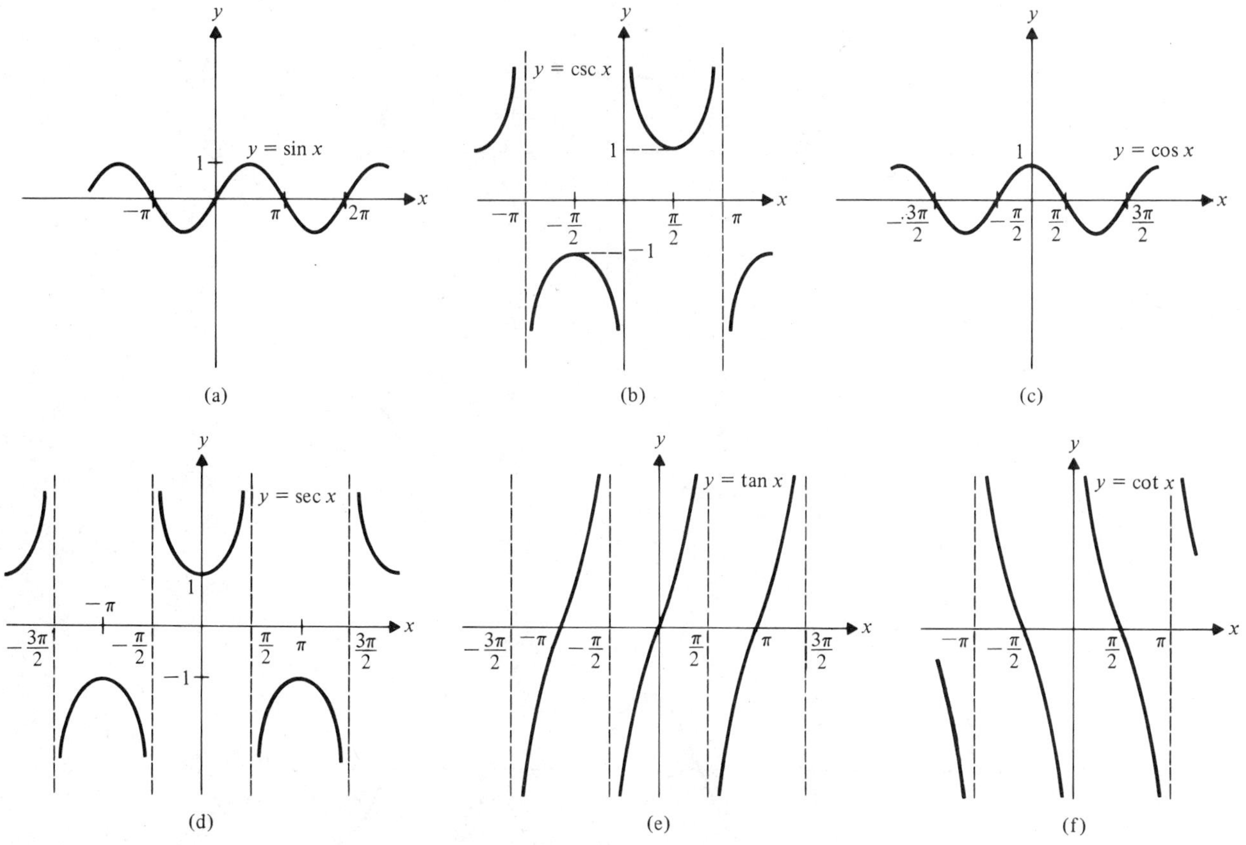

Figure 3.27

When a periodic function is added to a periodic function, the resulting function is always a periodic function. For example, the function $f(x) = 3\cos x - 4\sin x$ is the addition of two 2π-periodic functions $3\cos x$ and $-4\sin x$. In Example 23, we showed that $f(x)$ can also be expressed in the form $f(x) = 5\sin(x + 2.50)$, and from this representation it is straightforward to produce the graph shown in Fig. 3.28. The function again has period 2π. Note that the 5 in $f(x)$ indicates that oscillations occur between the limits ± 5, and the 2.50 represents a shift of the usual sine curve 2.50 units to the left.

This curve is an example of a *general sine function*,

$$f(x) = A\sin(\omega x + \phi), \tag{3.34}$$

where A and ω are positive constants and ϕ is also constant. Its graph is shown in Fig. 3.29. The number A, which represents half the range of the function, is called the *amplitude* of the oscillations. The *period* is $2\pi/\omega$ and $-\phi/\omega$ is called the *phase shift*.

Example 26 Sketch a graph of the function $f(x) = 3\sin(4x - \pi)$.

Solution An amplitude of 3 indicates that the oscillations are confined between $y = \pm 3$. If we combine this information with a period of $2\pi/4 = \pi/2$ and a phase shift of $\pi/4$ we obtain the graph shown in Fig. 3.30. □

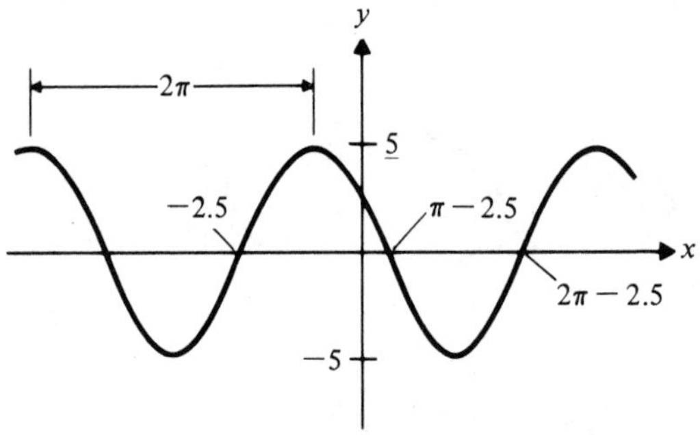

Figure 3.28

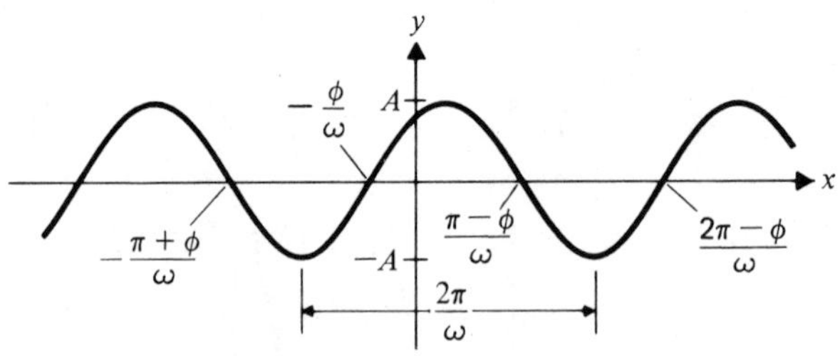

Figure 3.29

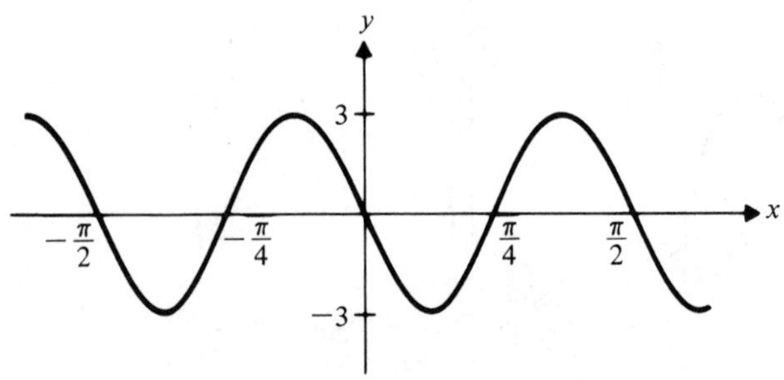

Figure 3.30

EXERCISES 3.8

In Exercises 1–10, express the given angle in radians.

1. 30°
2. 60°
3. 135°
4. −90°
5. −300°
6. 765°
7. 72°
8. −128°
9. 321°
10. −213°

In Exercises 11–20, express the given angle in degrees.

11. $\pi/3$
12. $-5\pi/4$
13. $3\pi/2$
14. 8π
15. $-5\pi/6$
16. 1
17. −3
18. 2.5
19. −3.6
20. 11

21. What angle is subtended at the center of a circle of radius 4 by an arc of length (a) 2; (b) 7; (c) 3.2?

22. When the angle in Fig. 3.20 is acute (between 0 and $\pi/2$), the coordinates of P are both positive, and no "sign" considerations enter the definitions (3.17). Often, then, only the triangle is drawn with no reference to x- and y-axes (Fig. 3.31). To define the trigonometric functions of angle x, we label the lengths of the sides of the triangle as *adjacent*, *opposite*, and *hypotenuse*. Values of the trigonometric functions are then

$$\sin x = \frac{\text{opposite}}{\text{hypotenuse}},$$

$$\cos x = \frac{\text{adjacent}}{\text{hypotenuse}},$$

$$\tan x = \frac{\text{opposite}}{\text{adjacent}},$$

with the corresponding definitions for csc x, sec x, and cot x. Use the triangles in Fig. 3.32 and 3.33 to find values of the trigonometric functions for $x = \pi/4$, $\pi/6$, and $\pi/3$. Fill these in Table 3.1, and finish the table.

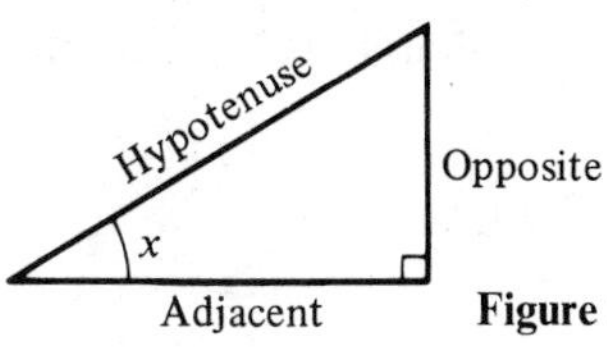

Figure 3.31

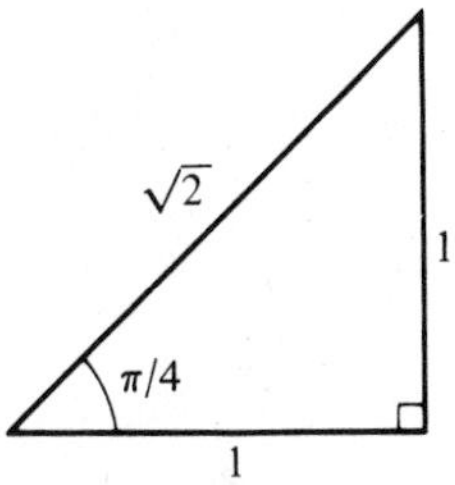

Figure 3.32

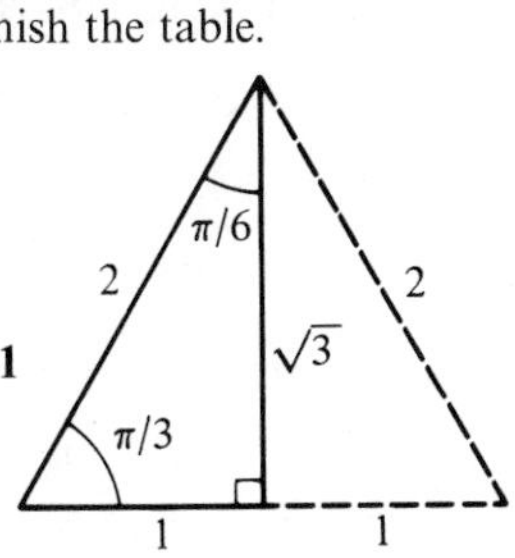

Figure 3.33

Table 3.1

	sin x	cos x	tan x	csc x	sec x	cot x
0						
$\pi/6$						
$\pi/4$						
$\pi/3$						
$\pi/2$						

For each of the triangles shown in Exercises 23–26, use the cosine law and/or the sine law to find the lengths of all sides and all interior angles.

23.

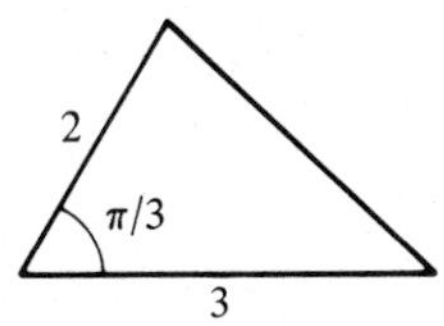

24.

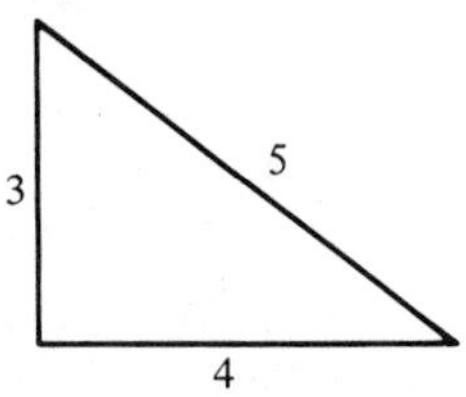

25.

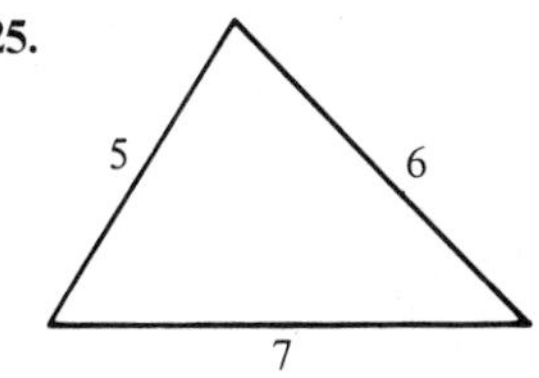

26.

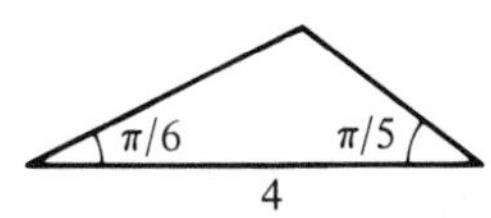

27. Use the sum and difference formulas (3.23) to prove (3.27).

28. Verify the double-angle formulas (3.28)–(3.30).

29. Use (3.23) to prove the product formulas (3.31).

30. Show that the sum formulas (3.32) can be obtained from (3.31).

In Exercises 31–40, sketch a graph of the given function.

31. $f(x) = 3 \sin x$
32. $f(x) = \sin 2x$
33. $f(x) = 3 \sin 2x$
34. $f(x) = \sin (x + \pi/4)$
35. $f(x) = 3 \sin (x + \pi/4)$
36. $f(x) = \sin (2x + \pi/4)$
37. $f(x) = 3 \sin (2x + \pi/4)$
38. $f(x) = 4 \cos (x/3)$
39. $f(x) = 2 \sin (x/2 - \pi)$
40. $f(x) = 5 \cos (\pi/2 - 3x)$

41. Show that the compound-angle formulas (3.23a, b, and c) can be derived from (3.23d).

In Exercises 42–49, find all solutions of the given equation.

42. $\sin x = \sqrt{3}/2$
43. $\cos x = 2$
44. $3 \sin x = \cos x$
45. $3 \cos^2 x = 1$
46. $\sin 2x = \sin x$
47. $\sin^2 x + 3 \sin x + 1 = 0$
48. $\sin^2 x + \sin x - 1 = 0$
49. $\sin x + \cos x = 1$

In Exercises 50–53, express each function as a general sine function, identifying its amplitude, period, and phase shift. Sketch a graph of each function.

50. $f(x) = \sin x - 2\cos x$

51. $f(x) = 4\sin 2x + \cos 2x$

52. $f(x) = -2\sin 3x - 4\cos 3x$

53. $f(x) = \sin x \cos x$

54. Which of the trigonometric functions shown in Fig. 3.27 are even? Which are odd?

Verify the identity in each of Exercises 55–59.

55. $\cos 3x = 4\cos^3 x - 3\cos x$

56. $\sin 4x = 8\cos^3 x \sin x - 4\cos x \sin x$

57. $\tan 3x = \dfrac{3\tan x - \tan^3 x}{1 - 3\tan^2 x}$

58. $\tan\left(\dfrac{x}{2}\right) = \dfrac{\sin x}{1 + \cos x}$

59. $\dfrac{1 + \tan x}{1 - \tan x} = \tan\left(x + \dfrac{\pi}{4}\right)$

60. Show that a function $f(x) = A\cos\omega x + B\sin\omega x$ can always be written in the form

$$f(x) = \sqrt{A^2 + B^2}\,\sin(\omega x + \phi),$$

where ϕ is defined by the equations

$$\sin\phi = \frac{A}{\sqrt{A^2 + B^2}} \quad \text{and} \quad \cos\phi = \frac{B}{\sqrt{A^2 + B^2}}.$$

61. In Exercise 60 can we replace the two equations defining ϕ with the single equation $\tan\phi = A/B$?

In Exercises 62–65, find all solutions of the given equation in the interval $0 \le x < 2$.

62. $\sin 4x = \cos 2x$

63. $\cos x + \cos 3x = 0$

64. $\sin 2x + \cos 3x = \sin 4x$

***65.** $\sin x + \cos x = \sqrt{3}\sin x \cos x$

***66.** Verify that if A, B, and C are the angles of a triangle, then

$$\tan A + \tan B + \tan C = \tan A \tan B \tan C.$$

(*Hint:* Expand $\tan(A + B + C)$ in terms of $\tan A$, $\tan B$, and $\tan C$.)

3.9 DIFFERENTIATION OF THE SINE AND COSINE FUNCTIONS

In order to find derivatives for the sine and cosine functions, we require the following theorem.

▶ **Theorem 3.10**

$$\lim_{\theta\to 0}\frac{\sin\theta}{\theta} = 1. \quad ◀ \tag{3.35}$$

Proof If θ is a positive acute angle as shown in Fig. 3.34, then

area of triangle BOP < area of sector BOA < area of triangle OBT.

When θ is expressed in radians, the area of a sector of a circle is $(\frac{1}{2})r^2\theta$. Consequently,

$$\tfrac{1}{2}|BP||OP| < \tfrac{1}{2}(1)^2\theta < \tfrac{1}{2}|OB||BT|,$$

or

$$\frac{1}{2}(\sin\theta)(\cos\theta) < \frac{\theta}{2} < \frac{1}{2}(1)(\tan\theta).$$

Division by $\frac{1}{2}\sin\theta$ gives

$$\cos\theta < \frac{\theta}{\sin\theta} < \frac{1}{\cos\theta},$$

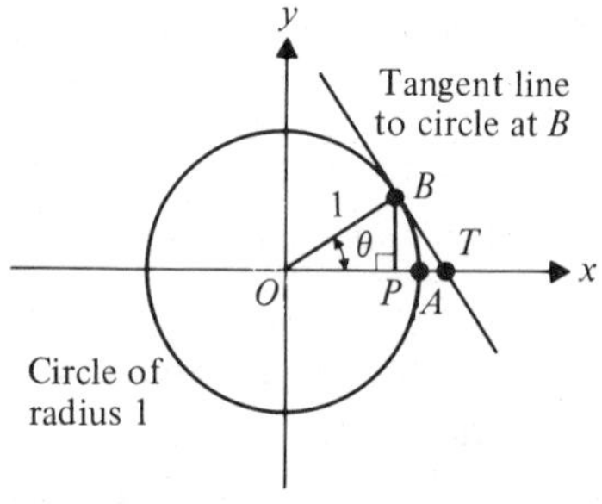

Figure 3.34

or

$$\frac{1}{\cos\theta} > \frac{\sin\theta}{\theta} > \cos\theta.$$

If we now take the limit as $\theta \to 0^+$, we obtain

$$\lim_{\theta\to 0^+} \frac{\sin\theta}{\theta} = 1.$$

If $\theta < 0$, we set $\phi = -\theta$, in which case

$$\lim_{\theta\to 0^-} \frac{\sin\theta}{\theta} = \lim_{\phi\to 0^+} \frac{\sin(-\phi)}{-\phi} = \lim_{\phi\to 0^+} \frac{\sin\phi}{\phi} = 1.$$

Therefore,

$$\lim_{\theta\to 0} \frac{\sin\theta}{\theta} = 1. \square$$

Because we expressed θ in radians in the proof of Theorem 3.10, the result in Eq. (3.35) is valid only for θ in radians. The value of the limit when θ is expressed in degrees can be found in Exercise 21 at the end of this section.

We can now prove the following theorem.

▶ **Theorem 3.11**

$$\frac{d}{dx}\sin x = \cos x \quad ◀ \tag{3.36}$$

Proof If we set $f(x) = \sin x$, then by Definition 3.1 we have

$$\begin{aligned} f'(x) &= \lim_{\Delta x\to 0} \frac{f(x+\Delta x) - f(x)}{\Delta x} \\ &= \lim_{\Delta x\to 0} \frac{\sin(x+\Delta x) - \sin x}{\Delta x} \\ &= \lim_{\Delta x\to 0} \frac{1}{\Delta x}\left\{2\cos\left(\frac{2x+\Delta x}{2}\right)\sin\left(\frac{\Delta x}{2}\right)\right\}, \end{aligned}$$

where we have used trigonometric identity (3.32b):

$$\sin A - \sin B = 2\cos\left(\frac{A+B}{2}\right)\sin\left(\frac{A-B}{2}\right).$$

Consequently,

$$\begin{aligned} f'(x) &= \lim_{\Delta x\to 0}\left\{\cos\left(x+\frac{\Delta x}{2}\right)\frac{\sin(\Delta x/2)}{\Delta x/2}\right\} \\ &= \lim_{\Delta x\to 0}\cos\left(x+\frac{\Delta x}{2}\right)\lim_{\Delta x/2\to 0}\frac{\sin(\Delta x/2)}{\Delta x/2} \\ &= \cos x. \square \end{aligned}$$

It is a simple application of the chain rule to prove the following.

▶ **Corollary** If $u = f(x)$, then

$$\frac{d}{dx}\sin u = \cos u \frac{du}{dx}. \quad ◀ \tag{3.37}$$

For example,

$$\frac{d}{dx}\sin(3x) = \cos(3x)\frac{d}{dx}(3x) = 3\cos(3x)$$

and

$$\frac{d}{dx}\sin^3(4x) = 3\sin^2(4x)\frac{d}{dx}\sin(4x)$$
$$= 12\sin^2(4x)\cos(4x).$$

Since the cosine function can always be expressed in terms of the sine function, it is straightforward to prove the following theorem.

▶ **Theorem 3.12**

$$\frac{d}{dx}\cos x = -\sin x \quad ◀ \tag{3.38}$$

Proof

$$\frac{d}{dx}\cos x = \frac{d}{dx}\sin\left(\frac{\pi}{2} - x\right) = -\cos\left(\frac{\pi}{2} - x\right) = -\sin x \ \square$$

▶ **Corollary** If $u = f(x)$, then

$$\frac{d}{dx}\cos u = -\sin u \frac{du}{dx}. \quad ◀ \tag{3.39}$$

Derivatives of the remaining trigonometric functions are developed in Chapter 8.

Example 27 Find dy/dx if y is defined as a function of x in each of the following:

a) $y = 4\sin 2x - 3\cos 5x$;
b) $y = \sqrt{1 - 3\sin 2x}$;
c) $y = \dfrac{\sin^3 x}{1 + \cos^2 x}$;
d) $x\sin y + y\sin x = 10$.

Solution a) $\dfrac{dy}{dx} = 4(\cos 2x)(2) - 3(-\sin 5x)(5) = 8\cos 2x + 15\sin 5x$

b) $\dfrac{dy}{dx} = \dfrac{1}{2}(1 - 3\sin 2x)^{-1/2}\dfrac{d}{dx}(1 - 3\sin 2x)$

$$= \frac{1}{2}(1 - 3\sin 2x)^{-1/2}(-6\cos 2x) = \frac{-3\cos 2x}{\sqrt{1 - 3\sin 2x}}$$

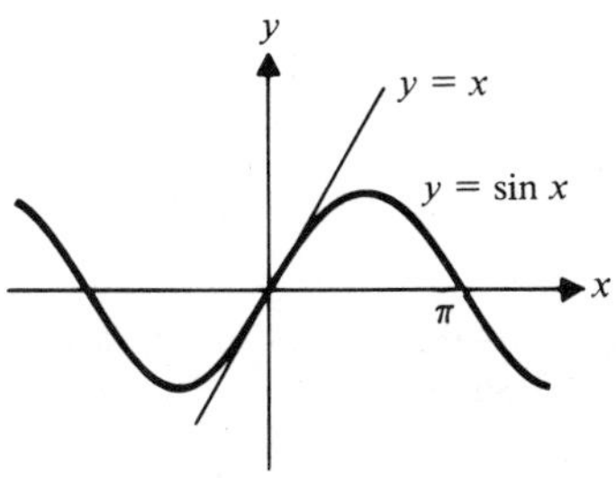

Figure 3.35

c) $$\frac{dy}{dx} = \frac{(1 + \cos^2 x)3 \sin^2 x \cos x - \sin^3 x(-2 \cos x \sin x)}{(1 + \cos^2 x)^2}$$

$$= \frac{\sin^2 x \cos x(3 + 3 \cos^2 x + 2 \sin^2 x)}{(1 + \cos^2 x)^2}$$

$$= \frac{\sin^2 x \cos x(5 + \cos^2 x)}{(1 + \cos^2 x)^2}$$

d) If we differentiate with respect to x, we have

$$\sin y + x \cos y \frac{dy}{dx} + \frac{dy}{dx} \sin x + y \cos x = 0,$$

from which

$$\frac{dy}{dx} = -\frac{y \cos x + \sin y}{x \cos y + \sin x}. \square$$

Example 28 Find the equation of the tangent line to the curve $y = \sin x$ at the point (0, 0).

Solution Since the slope of the tangent line at (0, 0) is

$$\left.\frac{dy}{dx}\right|_{x=0} = \cos(0) = 1,$$

its equation is $y = x$. Figure 3.35 illustrates that for $x > 0$, the curve $y = \sin x$ lies below the tangent line $y = x$ at $x = 0$. The reason for this is that just to the right of $x = 0$, the slope of $y = \sin x$ is less than 1 (the slope of $y = x$); therefore $y = \sin x$ must curve away from $y = x$. □

Example 29 A block of wood floating on a lake does not move horizontally, but due to the motion of the waves, it moves vertically (Fig. 3.36). Its distance away from the position it would occupy on a tranquil lake is defined by the function

$$y = f(t) = 15 \cos(2\pi t), \qquad t \geq 0,$$

where y is measured in centimeters and t is time in seconds.

a) What is the velocity of the block of wood?

b) When is the velocity of the block of wood largest in absolute value?

Solution a) Since velocity is the rate of change of position with respect to time, we have

$$v = \frac{dy}{dt} = 15[-\sin(2\pi t)](2\pi)$$

$$= -30\pi \sin(2\pi t) \text{ cm/s}.$$

b) The extreme values of $\sin(2\pi t)$ are ± 1 and these occur when $2\pi t = (n + \frac{1}{2})\pi$, n an integer. It follows, then, that the maximum value of $|v|$ is 30π, and this occurs when

$$y = 15 \cos[(n + \tfrac{1}{2})\pi] = 0,$$

that is, midway between a crest and a trough. □

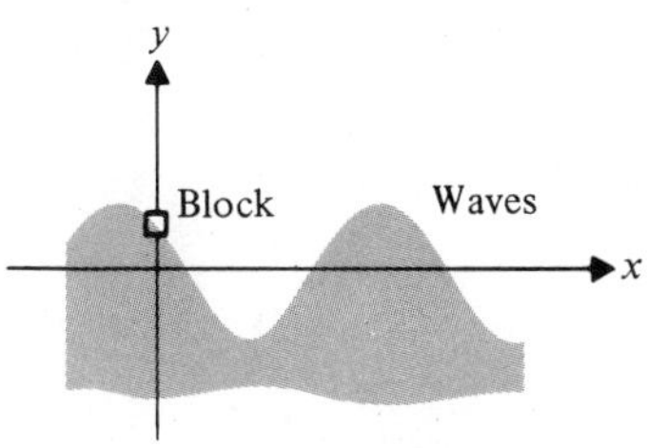

Figure 3.36

EXERCISES 3.9

If y is defined as a function of x in Exercises 1–12, find dy/dx.

1. $y = 2 \sin 3x$

2. $y = \cos x - 4 \sin 5x$

3. $y = \sin^2 x$

4. $y = \cos (2 - 3x)$

5. $y = \sin^3 x + 2 \cos x$

6. $y = \sqrt{1 + \cos (x^2)}$

7. $y = \dfrac{\sin^2 x}{1 + \cos^3 x}$

8. $y = x^2 \sin (2x + 3)$

9. $2 \sin y + 3 \cos x = 1$

10. $y = \sin (x + y)$

11. $x \cos y - y \cos x = 3$

12. $4 \sin^2 x - \cos^3 y = 1$

13. Find d^2y/dx^2 if y is defined implicitly as a function of x by

$$\sin y = x^2 + y.$$

14. The angular displacement of the pendulum shown in Fig. 3.37 at time t is given by

$$\theta = f(t) = A \cos (\omega t + \phi), \qquad t \geq 0,$$

where A, ω, and ϕ are constants. Show that $\theta = f(t)$ satisfies the equation

$$\frac{d^2\theta}{dt^2} + \omega^2\theta = 0.$$

In Exercises 15–20, evaluate the given limit if it exists.

15. $\displaystyle\lim_{x \to 0} \frac{\tan x}{x}$

16. $\displaystyle\lim_{x \to 0} \frac{\sin 2x}{x}$

17. $\displaystyle\lim_{x \to 0} \frac{(x + 1)^2 \sin x}{3x^3}$

18. $\displaystyle\lim_{x \to 0} \frac{1 - \cos x}{x}$

19. $\displaystyle\lim_{x \to \infty} \frac{\sin (2/x)}{\sin (1/x)}$

20. $\displaystyle\lim_{x \to \pi/2} \frac{\cos x}{(x - \pi/2)^2}$

21. Evaluate $\lim_{\theta \to 0} (\sin \theta/\theta)$ if θ is measured in degrees rather than radians.

22. If the voltage source labeled E in the circuit shown in Fig. 3.38 is suddenly short-circuited by opening the switch labeled S, the current I in the circuit thereafter must satisfy the equation

$$L\frac{d^2I}{dt^2} + \frac{1}{C}I = 0,$$

where L and C are the sizes of the inductor and capacitor, respectively, and t is time. Show that

$$I = f(t) = A \cos \left(\frac{t}{\sqrt{LC}}\right) + B \sin \left(\frac{t}{\sqrt{LC}}\right)$$

satisfies this equation for any constants A and B whatsoever.

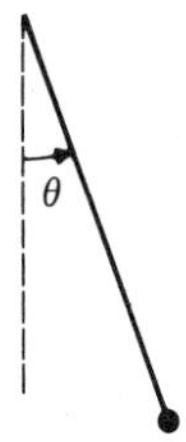

Figure 3.37

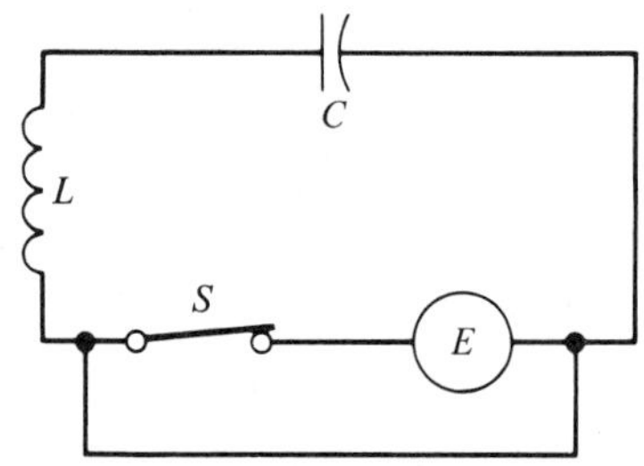

Figure 3.38

23. Find

$$\frac{d}{dx}|\sin x|.$$

***24.** If n is a positive integer, find formulas for

$$\frac{d^n}{dx^n}(\sin x) \quad \text{and} \quad \frac{d^n}{dx^n}(\cos x).$$

***25.** Illustrate graphically that the function

$$f(x) = \begin{cases} x \sin \left(\dfrac{1}{x}\right), & x \neq 0, \\ 0, & x = 0, \end{cases}$$

is continuous at $x = 0$. Show that $f'(0)$ does not exist, nor does

$$\lim_{x \to 0} f'(x).$$

SUMMARY

We defined the derivative of a function $y = f(x)$ as

$$\frac{dy}{dx} = f'(x) = \lim_{\Delta x \to 0} \frac{f(x + \Delta x) - f(x)}{\Delta x}.$$

Algebraically, it is the instantaneous rate of change of y with respect to x; geometrically, it is the slope of the tangent line to the graph of (x). To eliminate the necessity of using this definition over and over again, we derived the sum, product, quotient, and power rules:

$$\frac{d}{dx}(u + v) = \frac{du}{dx} + \frac{dv}{dx},$$

$$\frac{d}{dx}(uv) = u\frac{dv}{dx} + v\frac{du}{dx},$$

$$\frac{d}{dx}\left(\frac{u}{v}\right) = \frac{v\dfrac{du}{dx} - u\dfrac{dv}{dx}}{v^2},$$

$$\frac{d}{dx}(u^n) = nu^{n-1}\frac{du}{dx}.$$

These four simple rules are fundamental to all of calculus.

When a function $y = f(x)$ is defined implicitly by some equation $F(x, y) = 0$, we use implicit differentiation to find its derivative. We differentiate each term in the equation with respect to x, and then solve the resulting equation for dy/dx. We pointed out that care must be taken in differentiating equations. An equation can be differentiated with respect to a certain variable only if it is valid for a continuous range of values of that variable.

The chain rule defines the derivative of a composite function $y = f[g(x)]$ as the product of the derivatives of $y = f(u)$ and $u = g(x)$:

$$\frac{dy}{dx} = \frac{dy}{du}\frac{du}{dx}.$$

These rules and techniques form the basis for the rest of differential calculus. With a few more formulas for derivatives of some special functions (including the sine and cosine given in Section 3.9, and other transcendental functions discussed in Chapter 8), we will be well prepared to handle those applications of calculus that involve differentiation.

REVIEW EXERCISES

In each of Exercises 1–30, assume that y is defined as a function of x. Find dy/dx in as simplified a form as possible.

1. $y = x^3 + \dfrac{1}{x^2}$

2. $y = 3x^2 + 2x + \dfrac{1}{x}$

3. $y = x\sqrt{1 - x}$

4. $y = \dfrac{3x^2}{x^3 - 5}$

5. $x^2y + y\sqrt{1 + x} = 3$

6. $y = \sin^2 x \cos^2 x$

7. $\dfrac{x}{y} + \dfrac{y}{x} = x$

8. $y = u^2 - 2u,$
$u = (1 + 2x)^{5/3}$

9. $y = x^2\sqrt{1 - 2x^2}$

10. $y = \dfrac{x^3 + 5}{1 - 3x^2}$

11. $\dfrac{2xy}{3x + 4} = x^2 + 2$

12. $y = \sqrt{1 + \sqrt{1 + x}}$

13. $y = x^2 \sin (x^2)$

14. $y = \dfrac{\sin x}{\cos x}$

15. $x \sin y + 2xy = 4$
16. $5 \cos (x - y) = 1$
17. $y = t + \cos 2t,$
$t = x - \cos 2x$
18. $y = \sqrt{\dfrac{4 + x}{4 - x}}$
19. $y = \sin (\cos (x^2))$
20. $x^2 + y^2 = 9$
21. $x = \sqrt{1 + x \sin (y^2)}$
22. $y = v \cos^2 v,$
$v = \sqrt{1 - x^2}$
23. $\sin (x^2 y) = x$
24. $y = (x^2 + 3)^3(x^2 - 4)^4$
25. $y = \dfrac{\sin^3 x}{\cos^2 x}$
26. $x = \sqrt{y + y^2}$
27. $y = \sqrt{7 - \sqrt{7 - \sqrt{x}}}$
28. $y = \dfrac{x^2\sqrt{1 - x}}{x + 5}$
29. $x = \sqrt{\dfrac{4 + y}{4 - y}}$
30. $\dfrac{x}{x + y} = \dfrac{y}{x - y}$

In Exercises 31–34, find equations for the tangent and normal lines to the curve at the point indicated.

31. $y = x^3 + 3x - 2, \quad (1, 2)$
32. $y = \dfrac{1}{x + 5}, \quad (0, \frac{1}{5})$
33. $y = \cos 2x, \quad (\pi/2, -1)$
34. $y = (x^2 + 3x)/(2x - 5), \quad (1, -\frac{4}{3})$

In Exercises 35–37, find d^2y/dx^2 if y is defined implicitly as a function of x.

35. $x^2 - y^2 + 2(x - y) = y^3$
36. $(x - y)^2 = 3xy$
37. $\sin (x + y) = x$
38. Sketch a graph of $f(x) = \sin (x^2)$. Is it periodic?
39. Find all points on the curve $y = x^3 + x^2$ at which the tangent line passes through the origin.
40. Find that point on the curve $x = y^2 - 4$ at which the normal line will pass through the point $(-6, 7)$. Can you suggest an application of this result?
41. a) How many functions with domain $-1 \le x \le 1$ are defined implicitly by the equation $x^2 + y^2 = 1$?
b) How many continuous functions with domain $-1 \le x \le 1$ are defined implicitly by the equation?
42. What is the rate of change of the area A of an equilateral triangle with respect to its side length l?
43. In a cold climate, house temperature varies as the thermostat continually engages and disengages the furnace. Suppose that at the thermostat, the temperature T in degrees Celsius over a four-hour time interval $0 \le t \le 4$ is given by

$$T = f(t) = 20 + 3 \sin (4\pi t - \pi/2).$$

a) Sketch a graph of $f(t)$.
b) How many times is the furnace on during the four-hour period?
c) What is the maximum time rate of change of temperature?
44. The curve defined by the equation $(x^2 + y^2)^2 = x^2 - y^2$ and shown in Fig. 3.39 is called a *lemniscate*. Find the four points at which the tangent line is horizontal.

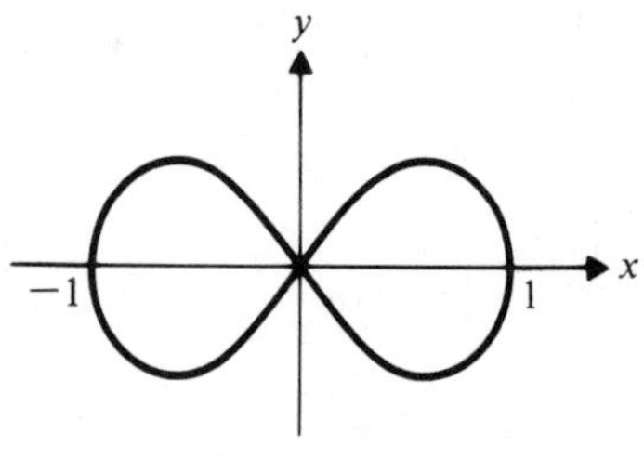

Figure 3.39

Applications of Differentiation

4

4.1 VELOCITY AND ACCELERATION

We showed in Section 3.1 that if $x = x(t)$ represents the position of a particle moving along the x-axis, then dx/dt is its instantaneous velocity v (provided, of course, that t is time):

$$v = v(t) = \frac{dx}{dt}; \qquad (4.1)$$

i.e., velocity is the rate of change of position with respect to time. The instantaneous *acceleration* a of the particle is defined as the rate of change of velocity with respect to time:

$$a = a(t) = \frac{dv}{dt} = \frac{d^2x}{dt^2}. \qquad (4.2)$$

In actual fact, $x(t)$, $v(t)$, and $a(t)$ are the x-components of the displacement, velocity, and acceleration vectors, respectively. Until we begin a complete discussion of vectors in Chapter 12, however, we shall omit the terms "vector" and "component," and simply call $x(t)$, $v(t)$, and $a(t)$ displacement, velocity, and acceleration, respectively.

When x is measured in meters and t in seconds, velocity v is measured in meters per second (m/s). Since acceleration is the time derivative of velocity, its units must be units of velocity divided by units of time, i.e., meters per second per second (m/s/s). Usually we shorten this by saying "meters per second squared," and write m/s^2.

Example 1 Discuss the motion of a particle that moves along the x-axis with displacement in meters given by

$$x(t) = t^3 - 15t^2 - 33t + 400, \qquad t \geq 0 \text{ in seconds.}$$

Solution From the equation for x, we obtain the velocity and acceleration:

$$v(t) = \frac{dx}{dt} = 3t^2 - 30t - 33 = 3(t - 11)(t + 1) \text{ m/s},$$

$$a(t) = \frac{dv}{dt} = 6t - 30 = 6(t - 5) \text{ m/s}^2.$$

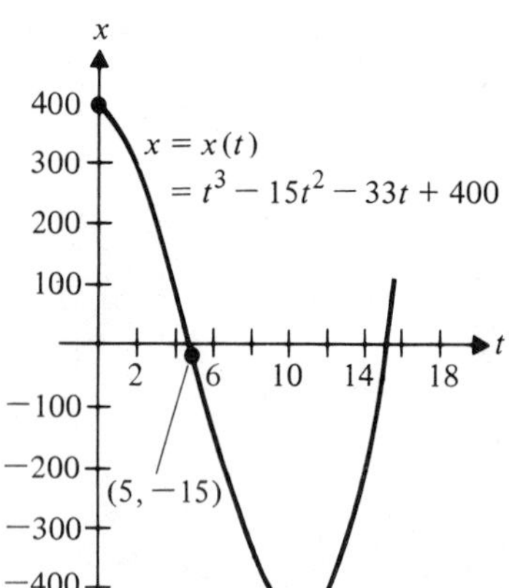

Figure 4.1

To represent the motion of the particle, we draw a graph of its displacement function $x(t)$ (Fig. 4.1). The position of the particle is displayed at any time t by the ordinate of the curve. For instance, at time $t = 0$, the particle is 400 meters to the right of the origin; and at $t = 11$ seconds, it is 447 meters to the left of the origin. Geometrically, the velocity dx/dt of the particle is the slope of the curve, and the acceleration d^2x/dt^2 is the rate of change of its slope. It is now a simple matter to trace the history of the particle. It starts 400 meters to the right of the origin with velocity -33 m/s, where negative velocity indicates that motion is to the left. The magnitude of velocity (disregarding its direction) is called *speed*, so that we can say alternatively that the particle's speed is initially 33 m/s (to the left). At this time the acceleration is -30 m/s^2, which indicates that the speed of the particle to the left is increasing; were the acceleration to continue at $-30\ \text{m/s}^2$, the particle's speed would increase 30 m/s each second.

At time $t = 5$ s, the acceleration of the particle is zero, and for $t > 5$, acceleration is to the right. It follows that the particle will have attained a maximum speed to the left at this time of $|v(5)| = 108$ m/s. Geometrically, at $(5, -15)$, the graph has attained its smallest possible slope (i.e., its most negative slope). The particle continues moving to the left until $t = 11$ s, when it instantaneously comes to rest 447 meters to the left of the origin. At this point the slope of the graph is zero (corresponding to zero velocity). The particle then moves to the right with ever-increasing speed.

Thus a graph of the position function $x(t)$ not only indicates where the particle is as a function of time; it also displays graphically the particle's speed and velocity, and the points at which its acceleration is positive and negative. The acceleration is negative to the left of the point $(5, -15)$; this is reflected in the "$\diagdown$" shape, which indicates that the velocity (slope) is decreasing. To the right of $(5, -15)$, acceleration is positive; this is shown in the "$\smile$" shape for increasing velocity (slope). □

Example 2 If a one-kilogram mass is attached to a certain spring (Fig. 4.2), pulled two centimeters below the position where it would hang motionless ($y = 0$), and released, its position as a function of time t (in seconds) is given by

$$y(t) = -\tfrac{1}{50}\cos(24t) \text{ meters} \quad (t \geq 0).$$

Discuss the motion of the mass.

Solution The position function indicates that the mass oscillates vertically about the $y = 0$ position with an amplitude of $\frac{1}{50}$ m and a period of $\pi/12$ s. Its velocity and acceleration are

$$v(t) = \tfrac{24}{50}\sin(24t) \text{ m/s}$$

and

$$a(t) = \tfrac{576}{50}\cos(24t) = -576y(t) \text{ m/s}^2.$$

The velocity of the mass is zero whenever $24t = n\pi$, n a nonnegative integer, and at these times $y = (-\frac{1}{50})\cos(n\pi) = \pm\frac{1}{50}$. The mass has zero velocity, therefore, at the extreme positions of its oscillations. The acceleration of the mass is zero every time it passes through $y = 0$.

Note that these conclusions are equally obvious from a graph of $y(t)$ (Fig. 4.3). The motion is oscillatory, the velocity is zero when $y = \pm\frac{1}{50}$, and the acceleration is zero when $y = 0$. □

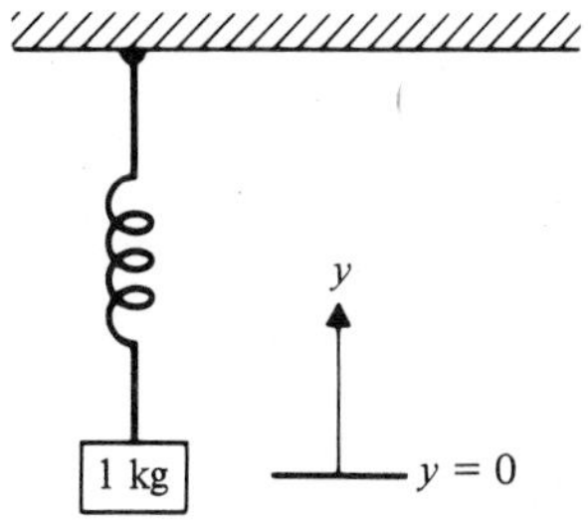

Figure 4.2

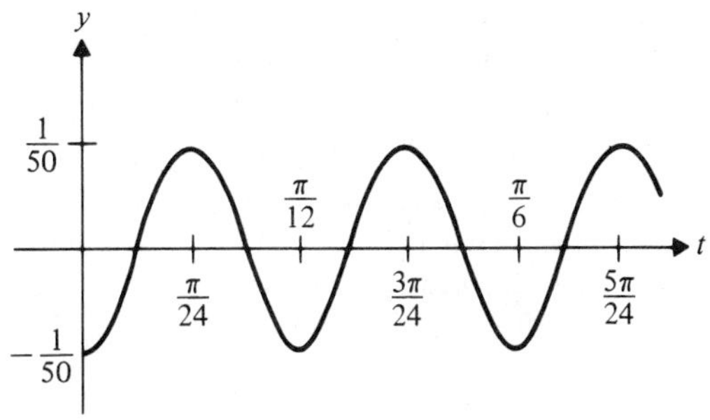

Figure 4.3

In summary, Eqs. (4.1) and (4.2) define the velocity and acceleration of a particle moving along the x-axis whenever its position is known as a function of time. A graph of the position function gives a clear picture of the particle's motion. It not only indicates position at any time, it also displays speed and velocity and the times at which these quantities are increasing and decreasing. In addition, it indicates times at which acceleration is positive and negative.

In Section 5.2 we will reverse this process and show that if we know the acceleration of a particle as a function of time, then we can find its velocity and position. This process is essential to physics and engineering, where the acceleration of a particle is determined by the forces acting on it.

EXERCISES 4.1

Find the velocity and acceleration of an object that moves along the x-axis with position as defined in Exercises 1–10. In each exercise, discuss the motion, including in your discussion a graph of the function $x(t)$. Assume that x is measured in meters and t in seconds.

1. $x(t) = 2t + 5, t \geq 5$

2. $x(t) = t^2 - 7t + 6, t \geq 0$

3. $x(t) = t^2 + 5t + 10, t \geq 1$

4. $x(t) = -2t^3 + 2t^2 + 16t - 1, t \geq 0$

5. $x(t) = t^3 - 9t^2 + 15t + 3, t \geq 0$

6. $x(t) = 3 \cos 4t, t \geq 0$

7. $x(t) = 1/t, t \geq 1$

8. $x(t) = t + 4/t, t \geq 1$

9. $x(t) = (t - 4)/t^2, t \geq 2$

10. $x(t) = t^2\sqrt{1 - t}, 0 \leq t \leq 1$

11. A particle moves along the x-axis in such a way that its position as a function of time t is given by

$$x(t) = t \sin t, \qquad 0 \leq t \leq 9.$$

Determine graphically each of the following.

a) The number of times the particle reverses direction

b) The number of times the particle has zero acceleration

c) The direction of motion at time $t = 6$

12. Can the position curve $x = x(t)$ of a realistic particle moving along the x-axis be represented by a function $x(t)$ that has a discontinuity? Explain.

In Exercises 13–16, we have defined the position function of an object moving along the x-axis. Sketch graphs of the position, velocity, and acceleration functions.

13. $x(t) = t^2 + 2t - 15, t \geq 0$

14. $x(t) = t^3 - 3t, t \geq 0$

15. $x(t) = t + 4/t, t \geq 1$

16. $x(t) = 5 \sin t, t \geq 0$

17. In many velocity and acceleration problems it is more convenient to express acceleration in terms of a derivative with respect to position (as opposed to a derivative with respect to time). Show that acceleration can be written in the form

$$a = v\frac{dv}{dx}.$$

18. When an object moves with constant acceleration a along the x-axis, its position as a function of time t must be of the form

$$x = x(t) = \tfrac{1}{2}at^2 + bt + c,$$

where b and c are constants.

a) If the object is at positions x_1 and x_2 at times t_1 and t_2, what is its average velocity over the time interval $t_1 \leq t \leq t_2$?

b) At what time in this interval is the instantaneous velocity of the object equal to the average velocity? Where is the object at this time? Is it at the midpoint of the interval between x_1 and x_2, is it closer to x_1, or is it closer to x_2? Assume in this part of the problem that $a > 0$ and that the velocity of the object is positive at time t_1.

4.2 RELATED RATES

Many interesting problems involving rates of change are of the type shown in the following two examples.

Example 3 A man 2 meters tall walks along the edge of a straight road 10 meters wide. On the other edge of the road stands a streetlight 8 meters high. If the man walks at $\frac{3}{2}$ meters per second, how fast is his shadow lengthening when he is 10 meters from the point directly opposite the light?

Example 4 One end of a rope is tied to a box. The other end is passed over a pulley 5 meters above the floor and tied to the back of a truck 1 meter above the floor. If the rope is taut and the truck moves at $\frac{1}{2}$ meter per second, how fast is the box rising when the truck is 3 meters from the plumb line through the pulley?

Each of these examples is called a *related-rate problem.* In the first, the rate at which the man walks is given, and we must find the rate at which his shadow is lengthening. In the second, the rate at which the truck moves is given, and we must find the rate at which the box is rising. To obtain a general procedure for solving such problems, we consider three simple examples before returning to Examples 3 and 4. Example 5 is a simplified version of Example 3.

Example 5 A man 2 meters tall walks directly away from a streetlight that is 8 meters high at the rate of $\frac{3}{2}$ meters per second. How fast is the length of his shadow changing?

Solution In Fig. 4.4 we have shown the rate dx/dt, which is given, and the rate ds/dt, which is required. Similar triangles enable us to relate the length s of the man's shadow to his distance x from the light,

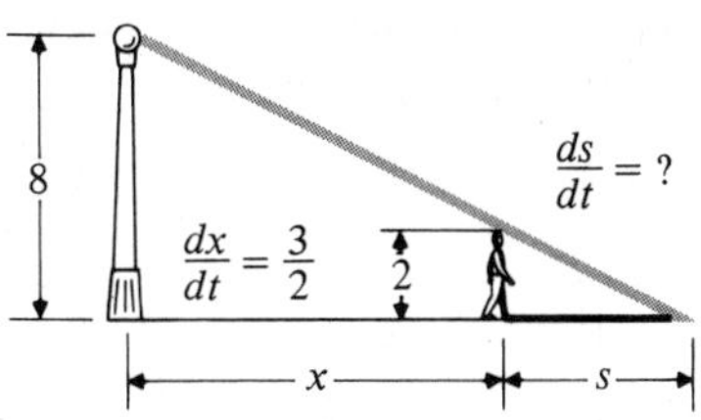

Figure 4.4

$$\frac{x+s}{s} = \frac{8}{2},$$

and this equation can be solved for s in terms of x:

$$s = \frac{x}{3}.$$

Now, s and x are each functions of time t,

$$s = f(t) \quad \text{and} \quad x = g(t),$$

(even though we have not calculated the exact form of these functions). Since the equation $s = x/3$ is valid at any time when the man is walking away from the light, we may differentiate with respect to time t to obtain

$$\frac{ds}{dt} = \frac{1}{3}\frac{dx}{dt}.$$

This equation relates the known rate $dx/dt = 3/2$ with the unknown rate ds/dt. It follows that

$$\frac{ds}{dt} = \frac{1}{3}\cdot\frac{3}{2} = \frac{1}{2},$$

and the man's shadow is therefore getting longer at the rate of $\frac{1}{2}$ meter per second. □

Example 6 One end of a ladder 20 meters long slips down the side of a house at the rate of 1 meter per second. How fast is the foot of the ladder slipping away from the house when it is 15 meters from the house?

Solution From Fig. 4.5, we may write

$$x^2 + y^2 = 20^2,$$

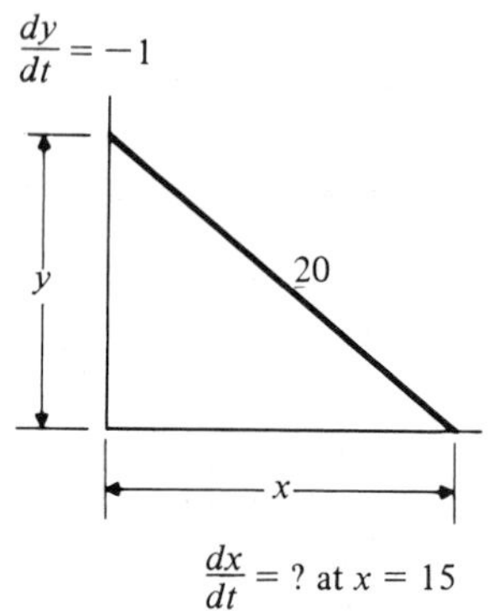

Figure 4.5

and this equation is valid at any time during which the ladder is slipping. If we differentiate, therefore, with respect to time t, we have

$$2x\frac{dx}{dt} + 2y\frac{dy}{dt} = 0.$$

Since y is decreasing at the rate of 1 meter per second, it follows that $dy/dt = -1$. If we combine this with the fact that $y = \sqrt{400 - 15^2}$ when $x = 15$, we obtain

$$15\frac{dx}{dt} + \sqrt{175}(-1) = 0.$$

Thus, when $x = 15$,

$$\frac{dx}{dt} = \frac{\sqrt{175}}{15} = \frac{\sqrt{7}}{3},$$

and the foot of the ladder is therefore slipping away at the rate of $\sqrt{7}/3$ meters per second. □

Example 7 A tank is being filled with water at the rate of 4 cubic meters per minute. How fast is the surface of the water rising when the depth is 3 meters if the tank is a right-circular cone with altitude 6 meters and base radius 3 meters?

Solution In Fig. 4.6 we have shown a cross-section of the tank. When the depth of water in the tank is D, the volume V of water is

$$V = \tfrac{1}{3}\pi r^2 D.$$

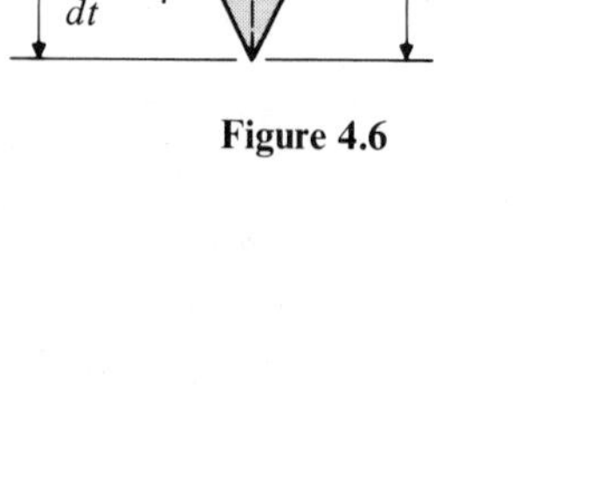

Figure 4.6

Of the three variables V, r, and D in this equation, we are concerned only with V and D, since dV/dt is given and dD/dt is what we want. This suggests that we should eliminate r, and this can be done using similar triangles. Since $r/D = 3/6$, we have $r = D/2$; therefore,

$$V = \frac{1}{3}\pi\left(\frac{D}{2}\right)^2 D = \frac{1}{12}\pi D^3.$$

Because this result is valid for all time t during the process of filling the tank, we can differentiate with respect to t:

$$\frac{dV}{dt} = \frac{1}{4}\pi D^2 \frac{dD}{dt}.$$

Since $dV/dt = 4$, we find that when $D = 3$,

$$4 = \frac{1}{4}\pi(3)^2\frac{dD}{dt} \quad \text{or} \quad \frac{dD}{dt} = \frac{16}{9\pi}.$$

The surface is therefore rising at a rate of $16/(9\pi)$ meters per minute. □

Examples 5–7 illustrate the following general procedure for solving related-rate problems.

1. Sketch a diagram illustrating all given information.
2. Find an equation valid for all time (in some interval about the instant in question) that involves only variables whose rates of change are either given or required.
3. Differentiate the equation in (2) and solve for the required rate.

Steps (1) and (3) are usually quite straightforward; step (2), on the other hand, may tax your ingenuity. To find the equation in the appropriate variables, it may be necessary to introduce and substitute for additional variables. Finding these substitutions will force you to analyze the problem very closely.

We now apply this procedure to Examples 3 and 4.

Example 3 A man 2 meters tall walks along the edge of a straight road 10 meters wide. On the other edge of the road stands a streetlight 8 meters high. If the man walks at 3/2 meters per second, how fast is his shadow lengthening when he is 10 meters from the point directly opposite the light?

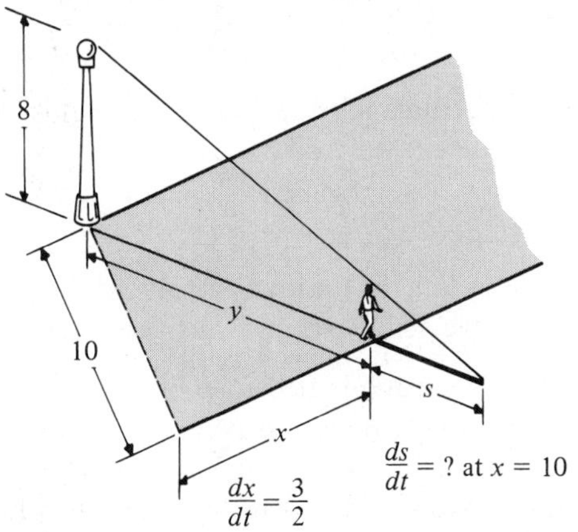

Figure 4.7

Solution By the similar triangles shown in Fig. 4.7, we have

$$\frac{y+s}{s} = \frac{8}{2} = 4,$$

or

$$y = 3s,$$

an equation that relates s to y, rather than s to x. However, since

$$y^2 = x^2 + 100,$$

we substitute to obtain

$$9s^2 = x^2 + 100.$$

The derivative of this equation with respect to time t gives

$$18s\frac{ds}{dt} = 2x\frac{dx}{dt}.$$

When $x = 10$, we obtain $s = \sqrt{100 + 100/3} = 10\sqrt{2/3}$ and, at this instant,

$$18\left(\frac{10\sqrt{2}}{3}\right)\frac{ds}{dt} = 2(10)\left(\frac{3}{2}\right).$$

Thus,

$$\frac{ds}{dt} = \frac{30}{60\sqrt{2}} = \frac{1}{2\sqrt{2}},$$

and the man's shadow is therefore lengthening at the rate of $\sqrt{2}/4$ meters per second. □

Note in this example that were we to solve for ds/dt before substituting $x = 10$ and $s = 10\sqrt{2/3}$, then we would have

$$\frac{ds}{dt} = \frac{x}{9s}\frac{dx}{dt}.$$

Since dx/dt is always equal to 3/2, and $s = \sqrt{x^2 + 100/3}$, we can write

$$\frac{ds}{dt} = \frac{x}{3\sqrt{x^2 + 100}}\left(\frac{3}{2}\right) = \frac{x}{2\sqrt{x^2 + 100}},$$

a general formula for ds/dt. The limit of this rate as x becomes very large is

$$\lim_{x\to\infty}\frac{ds}{dt} = \lim_{x\to\infty}\frac{x}{2\sqrt{x^2 + 100}} = \frac{1}{2}.$$

But for very large x, the man actually walks directly away from the light, and this answer, as we might expect, is identical to that of Example 5.

Example 4 One end of a rope is tied to a box. The other end is passed over a pulley 5 meters above the floor and tied to the back of a truck 1 meter above the floor. If the rope is taut and the truck moves at $\frac{1}{2}$ meter per second, how fast is the box rising when the truck is 3 meters from the plumb line through the pulley?

Solution If we denote the length of the rope by l (Fig. 4.8), then when the box is rising, we have

$$z + (5 - y) = l$$

and

$$z^2 = x^2 + 16.$$

Consequently,

$$(l - 5 + y)^2 = x^2 + 16,$$

and we differentiate this equation with respect to time t:

$$2(l - 5 + y)\frac{dy}{dt} = 2x\frac{dx}{dt}.$$

When $x = 3$, we obtain

$$l - 5 + y = \sqrt{9 + 16} = 5,$$

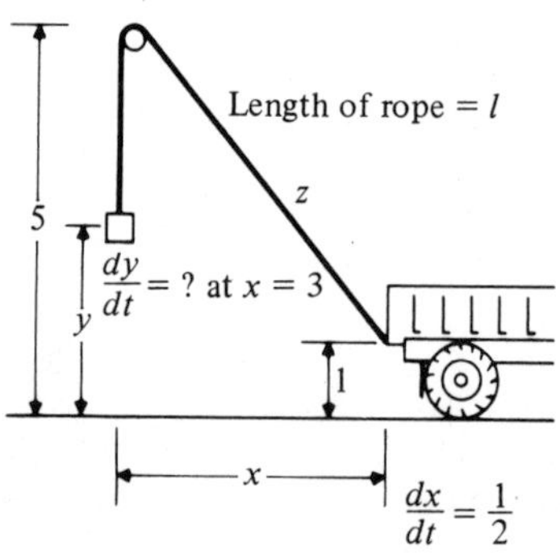

Figure 4.8

and at this instant

$$2(5)\frac{dy}{dt} = 2(3)\left(\frac{1}{2}\right).$$

Thus, $dy/dt = 3/10$ and the box is rising at a rate of 3/10 meter per second. □

EXERCISES 4.2

1. A fisherman is trolling at a rate of 2 meters per second with his lure 100 meters behind the boat and on the surface. Suddenly a fish strikes and dives vertically at a rate of 3 meters per second. If the fisherman permits the line to run freely and it always remains straight, how fast is the line being played out when the reel is 50 meters from its position at the time of the strike?

2. Boyle's law for a perfect gas states that the pressure exerted by a gas on its containing vessel is inversely proportional to the volume occupied by the gas. If when the volume is 10 liters and the pressure is 50 newtons per square meter, the volume is increasing at $\frac{1}{2}$ liter per second, find the rate of change of the pressure of the gas.

3. A woman driving 100 kilometers per hour along a straight highway notes that the shadow of a cloud is keeping pace with her. What can she conclude about the speed of the cloud?

4. Air expands adiabatically in accordance with the law $PV^{7/5} =$ constant. If, at a given time, the volume V is 100 liters and the pressure P is 40 newtons per square centimeter, at what rate is the pressure changing when the volume is decreasing at 1 liter per second?

5. Sand is poured into a right-circular cylinder of radius $\frac{1}{2}$ meter along its axis (Fig. 4.9). Once sand completely covers the bottom, a right-circular cone is formed on the top.

a) If 0.02 cubic meter of sand enters the container every minute, how fast is the top of the sand pile rising?

b) How fast is the sand traveling along the side of the cylinder?

6. A balloon has the shape of a right-circular cylinder of radius r and length l with a hemisphere at each end of radius r. The balloon is being filled at a rate of 10 milliliters per second in such a way that l increases twice as fast as r. Find the rate of change of r when $r = 8$ centimeters and $l = 20$ centimeters.

7. An oval racetrack has a straight stretch 100 meters long and two semicircles, each of radius 50 meters (Fig. 4.10). Car 1, on the infield, moves along the x-axis from O to B. It accelerates from rest at O, attains a speed of 10 meters per second at C, and maintains this speed along CB. Car 2 travels along the quarter oval $ADEB$.

a) Car 2 is at D when Car 1 is at C, and between D and B, Car 2 maintains the same rate of change of its x-coordinate as does Car 1. Find a formula for the rate of change of the y-coordinate of Car 2 between D and B.

b) How fast is the y-coordinate of Car 2 changing when it is at point E?

c) If the cars collide at B, which car will have the most damage?

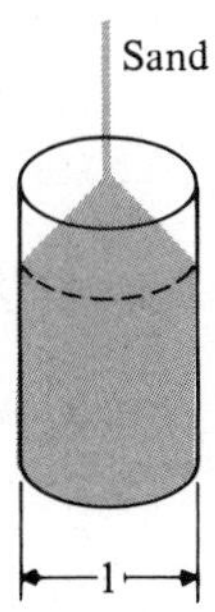

Figure 4.9

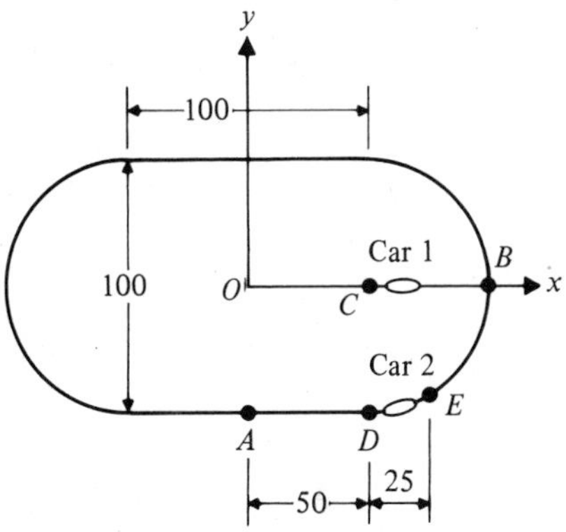

Figure 4.10

8. A ship is 1 kilometer north of a pier and is traveling N 30° E at a rate of 3 kilometers per hour. A second ship is $\frac{3}{4}$ kilometer east of the pier and is traveling east at a rate of 7 kilometers per hour. How fast are the ships separating?

9. Eight skaters form a "whip." Show that the seventh person on the whip travels twice as fast as the fourth person.

10. A solution passes through a conical filter 24 centimeters deep and 16 centimeters across the top into a cylindrical container of diameter 12 centimeters. When the depth of solution in the filter is 12 centimeters, its level is falling at the rate of 1 centimeter per minute. How fast is the level of solution rising in the cylinder at this instant?

11. A light is at the top of a pole 25 meters high and a ball is dropped at the same height from a point 10 meters from the light. How fast is the shadow of the ball moving along the ground 1 second later? The distance fallen by the ball t seconds after it has been dropped is $d = 4.905t^2$ meters.

12. A point moves along the parabola $y = x^2 - 3x$ (x and y measured in meters) in such a way that its x-coordinate changes at the rate of 2 meters per second. How fast is its distance from the point (1, 2) changing when it is at (4, 4)?

13. Repeat Exercise 12 given that the parabola is replaced by the curve $(x + y)^2 = 16x$.

14. The volume of wood in the trunk of a tree is sometimes calculated by considering it as a frustrum of a right-circular cone (Fig. 4.11).

a) Verify that the volume of the trunk is

$$V = \tfrac{1}{3}\pi h(R^2 + rR + r^2).$$

b) Suppose that at the present time the radii of the top and bottom are $r = 10$ and $R = 50$ centimeters, and the height is $h = 30$ meters. If the tree continues to grow so that ratios r/R and r/h always remain the same as they are now, and R increases at a rate of $\frac{1}{2}$ centimeter per year, how fast will the volume be changing in 2 years?

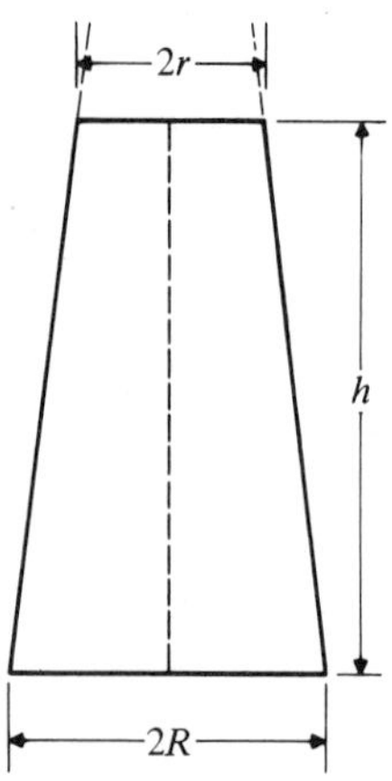

Figure 4.11

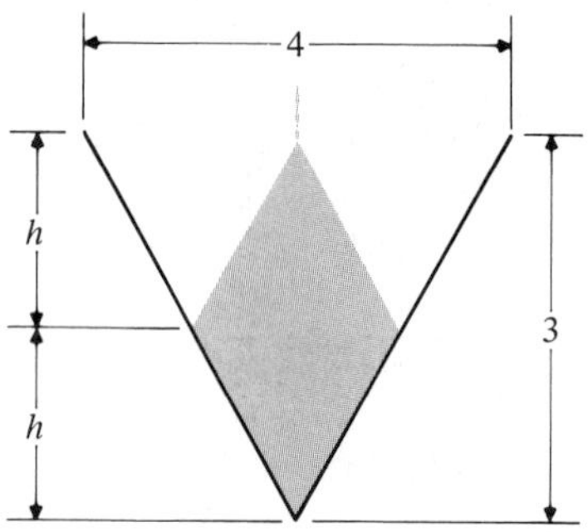

Figure 4.12

15. Sand is poured into a right-circular cone of radius 2 meters and height 3 meters along its axis (Fig. 4.12). The sand forms two cones of equal height h, one inverted on top of the other.

a) If 0.02 cubic meter of sand enters the container every minute, how fast is the top of the pile rising when it is just level with the top of the container?

b) How fast is the sand traveling along the side of the container at this instant?

16. Figure 4.13 shows a rod OB rotating counterclockwise in the xy-plane about O at a rate of 2 revolutions per second. Attached to OB is a rod AB, and A is confined to sliding horizontally along the x-axis. We are given that OB has length l meters, and AB has length L meters ($L > 2l$).

a) Find the velocity of A as a function of x and θ.

*b) Find the acceleration of A.

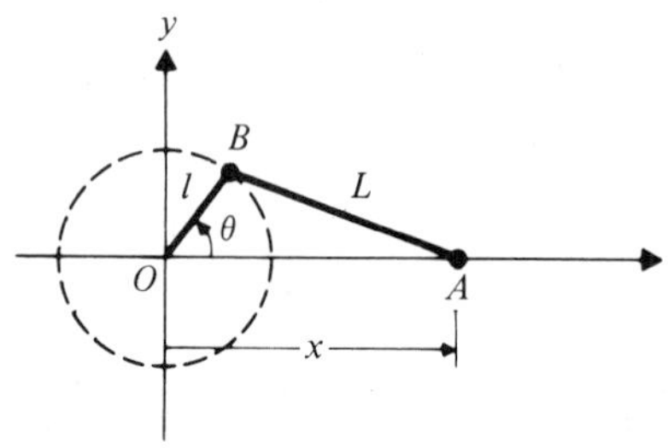

Figure 4.13

17. If to the mechanism shown in Fig. 4.13 we add a rod AC (Fig. 4.14), where C is confined to sliding vertically, find the velocity of C in terms of x, θ, and y.

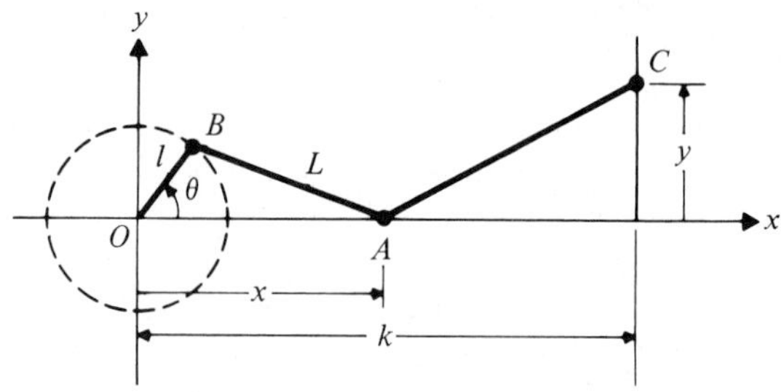

Figure 4.14

18. In Fig. 4.15 the boy's feet make 1 revolution per second around a sprocket of radius R meters. The chain travels around a sprocket of radius r meters on the back wheel, which itself has radius R' meters. If a stone imbedded in the tire becomes dislodged, how fast is it traveling when it leaves the tire?

Figure 4.15

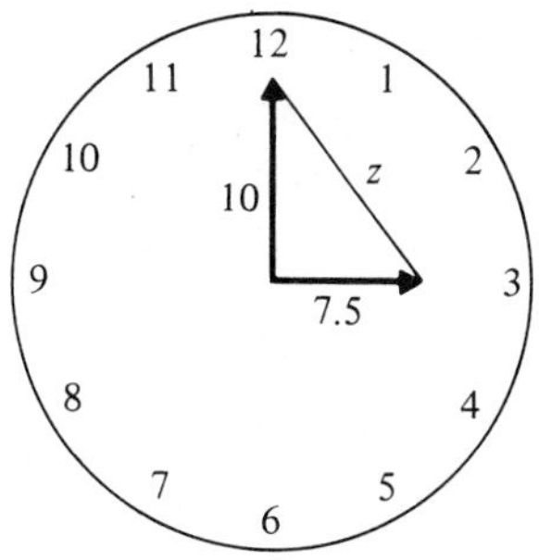

Figure 4.16

19. a) If θ is the angle formed by the minute and hour hands of a clock, what is the time rate of change of θ (in radians per minute)?

b) If the lengths of the hands on the clock are 10 centimeters and 7.5 centimeters, find the rate at which their tips approach each other at 3:00 (i.e., find dz/dt in Fig. 4.16).

c) Repeat (b) but replace time 3:00 with time 8:05.

*** 20.** If the sides of a triangle have lengths a, b, and c, its area is given by

$$A = \sqrt{s(s-a)(s-b)(s-c)},$$

where $s = (a+b+c)/2$ is one-half its perimeter. If the length of each side increases at a rate of 1 centimeter per minute, how fast is A changing when $a = 3$ centimeters, $b = 4$ centimeters, and $c = 5$ centimeters?

4.3 RELATIVE MAXIMA AND MINIMA

In Sections 4.3–4.7 we study one of the most important applications of differentiation: maxima and minima of functions of one variable. We begin by defining a critical point of a function.

▷ **Definition 4.1** A *critical point of a function* is a point in the domain of the function at which the first derivative either is equal to zero or does not exist. ◁

Specifically, x is a critical point for $f(x)$ if

$$f'(x) = 0 \tag{4.3}$$

or $f'(x)$ does not exist. Figure 4.17 illustrates that there are three types of critical points at which the first derivative vanishes, one type at $x = c$ and $x = e$, a second type at $x = d$ and $x = p$, and a third type at $x = h$ and $x = j$. For critical points at which the first derivative does not exist, there either is a corner ($x = g$, i, l, and n in Fig. 4.17), or $f(x)$ is discontinuous ($x = k$ and $x = m$). Geometrically, then, a critical point is a point at which the graph of a function has a horizontal tangent line, a vertical tangent line, or no tangent line at all.

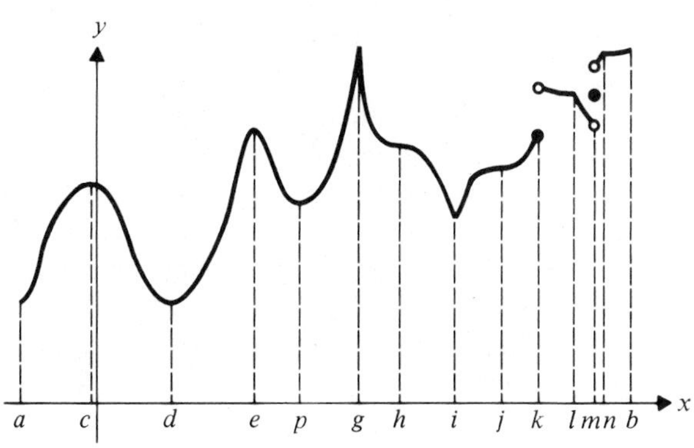

Figure 4.17

At points $x = a$, b, and c in Fig. 4.18, $f'(x)$ does not exist, but because the function is not defined at these points, they are not critical points.

Example 8 Find all critical points for each of the following functions:

a) $f(x) = x^3 - 7x^2 + 11x + 6$;

b) $f(x) = \dfrac{x^2}{x^3 - 1}$.

Solution a) For critical points we first solve

$$0 = f'(x) = 3x^2 - 14x + 11 = (3x - 11)(x - 1),$$

and obtain $x = \frac{11}{3}$ and $x = 1$. These are the only critical points since $f'(x)$ exists everywhere.

b) For critical points we calculate

$$f'(x) = \frac{(x^3 - 1)(2x) - x^2(3x^2)}{(x^3 - 1)^2} = \frac{-x(x^3 + 2)}{(x^3 - 1)^2}.$$

Clearly $f'(x) = 0$ when $x = 0$ or $x^3 + 2 = 0$, which implies that $x = -2^{1/3}$. The derivative does not exist when $x^3 - 1 = 0$; i.e., when $x = 1$. But $f(x)$ is not defined at $x = 1$ either; hence $x = 1$ is not a critical point. There are only two critical points, then: $x = 0$ and $x = -2^{1/3}$. □

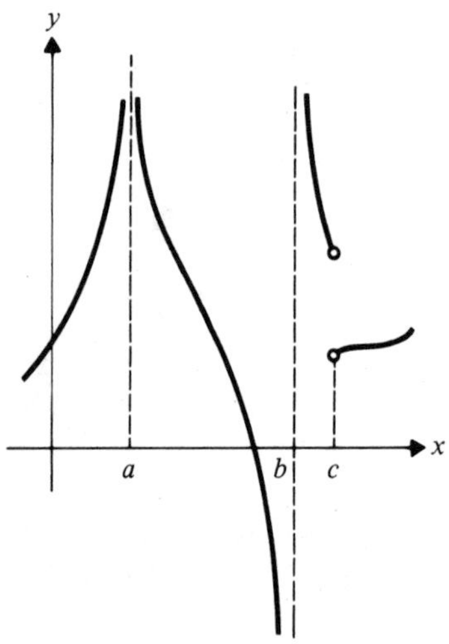

Figure 4.18

Critical points $x = c$, $x = e$, and $x = g$ in Fig. 4.17 represent "high" points on the graph of the function. We describe this property in the following definition.

▷ **Definition 4.2** A function $f(x)$ is said to have a *relative maximum* $f(x_0)$ at $x = x_0$ if there exists an open interval I containing x_0 such that for all x in I

$$f(x) \leq f(x_0). \quad ◁ \tag{4.4}$$

Because such open intervals can be drawn around $x = c$, $x = e$, and $x = g$, relative maxima occur at these points.

Critical points $x = d$, $x = p$, and $x = i$ in Fig. 4.17, where the graph has "low" points, are described in the following definition.

▷ **Definition 4.3** A function $f(x)$ is said to have a *relative minimum* $f(x_0)$ at $x = x_0$ if there exists an open interval I containing x_0 such that for all x in I

$$f(x) \geq f(x_0). \quad ◁ \tag{4.5}$$

Relative minima, therefore, occur at $x = d$, $x = p$, and $x = i$.

At critical points $x = h$ and $x = j$, where $f'(x) = 0$, the function has neither a relative maximum nor a relative minimum. It has instead *horizontal points of inflection*, which we will discuss in Section 4.5.

Figure 4.17 suggests the following result.

▶ **Theorem 4.1** Every relative maximum or minimum of a function $f(x)$ occurs at a critical point of $f(x)$. ◀

Proof To verify this result we prove that at any point at which $f'(x)$ exists and is not zero, it is impossible for $f(x)$ to have a relative extremum (i.e., a relative maximum or a relative minimum). Suppose that at $x = x_0$,

$$0 < f'(x_0) = \lim_{\Delta x \to 0} \frac{f(x_0 + \Delta x) - f(x_0)}{\Delta x}.$$

According to Exercise 8 in Section 2.3, there exists an open interval $I: a < x < b$ containing x_0 in which

$$\frac{f(x_0 + \Delta x) - f(x_0)}{\Delta x} > 0.$$

This implies that when $\Delta x > 0$ (and $x_0 + \Delta x$ is in I), $f(x_0 + \Delta x) - f(x_0)$ must also be positive, and therefore $f(x_0 + \Delta x) > f(x_0)$. But when $\Delta x < 0$ (and $x_0 + \Delta x$ is in I), $f(x_0 + \Delta x) - f(x_0)$ must be negative, and $f(x_0 + \Delta x) < f(x_0)$. Consequently, there is an interval $a < x < x_0$ in which $f(x) < f(x_0)$, and an interval $x_0 < x < b$ in which $f(x) > f(x_0)$; thus x_0 cannot be a relative extremum.

A similar proof holds when $f'(x_0) < 0$. □

Although every relative extremum of a function must occur at a critical point, not all critical points give relative extrema. They may give instead horizontal points of inflection or corners. For continuous functions, there is a very simple test to determine whether a critical point gives a relative maximum, a relative minimum, a horizontal point of inflection, or a corner. To understand this test, consider the critical points shown in Fig. 4.17 as you read the statements. A proof of the test is given in Exercise 21 of Section 4.10.

First-derivative Test for Critical Points of a Continuous Function

i) If $f'(x)$ (slope of the graph) changes from a negative quantity to a positive quantity as x increases through a critical point, then that critical point yields a relative minimum of $f(x)$.

ii) If $f'(x)$ changes from a positive quantity to a negative quantity as x increases through a critical point, then that critical point yields a relative maximum of $f(x)$.

iii) If $f'(x)$ does not change sign as x passes through a critical point at which $f'(x) = 0$, then that critical point yields a horizontal point of inflection; and if $f'(x)$ does not change sign as x passes through a critical point at which $f'(x)$ does not exist, then that critical point yields a corner that is neither a relative maximum nor a relative minimum.

Thus, to classify a critical point x_0 at which $f(x)$ is continuous, we check whether $f'(x)$ changes sign as x passes through x_0. To show that $f'(x)$ does change sign, we must find an open interval I_1 to the left of x_0 (Fig. 4.19) in which $f'(x)$ has one sign, and an open interval I_2 to the right of x_0 in which $f'(x)$ has the opposite sign. We illustrate a simple way to do this in Example 9.

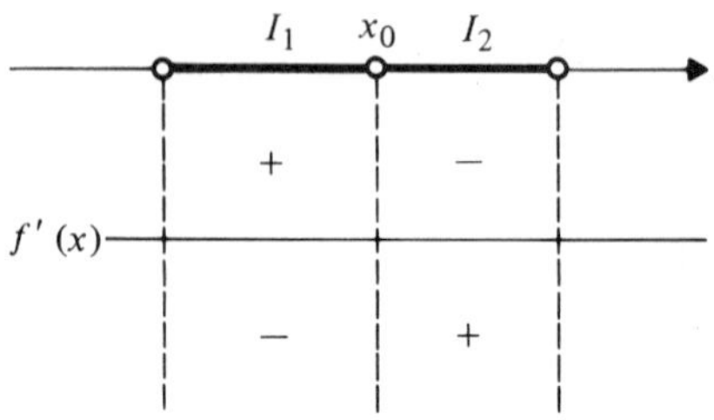

Figure 4.19

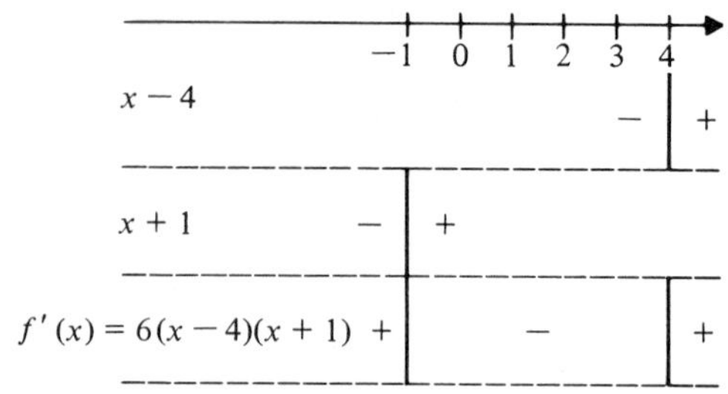

Figure 4.20

The first-derivative test applies to critical points at which $f(x)$ is continuous. What about points at which $f(x)$ is not continuous? Unfortunately, there is no simple general criterion that will determine the nature of critical points at which $f(x)$ is not continuous. Each example must be considered as its own special case.

Example 9 Find all relative extrema for each of the following functions:

a) $f(x) = 2x^3 - 9x^2 - 24x + 6$;

b) $f(x) = \dfrac{x^2}{x^3 - 1}$;

c) $f(x) = x^{5/3} - x^{2/3}$;

d) $f(x) = x + \sin x$.

Solution a) For critical points we first solve

$$0 = f'(x) = 6x^2 - 18x - 24 = 6(x-4)(x+1),$$

and obtain $x = -1$ and $x = 4$. The schematic diagram in Fig. 4.20 illustrates that as x increases through -1, $f'(x)$ changes from positive to negative; therefore, $x = -1$ yields a relative maximum of $f(-1) = 19$. Since $f'(x)$ changes from negative to positive as x increases through 4, $x = 4$ yields a relative minimum of $f(4) = -106$.

b) In Example 8 we found that

$$f'(x) = \frac{-x(x^3 + 2)}{(x^3 - 1)^2},$$

and that the critical points are $x = 0$ and $x = -2^{1/3}$. The schematic shown in Fig. 4.21 indicates that $f'(x)$ changes from negative to positive as x increases through $-2^{1/3}$; therefore, $x = -2^{1/3}$ gives a relative minimum of $f(-2^{1/3}) = -2^{2/3}/3$. Since $f'(x)$ changes from positive to negative as x increases through $x = 0$, we have a relative maximum of $f(0) = 0$ at $x = 0$.

c) For critical points we first solve

$$0 = f'(x) = \frac{5}{3}x^{2/3} - \frac{2}{3}x^{-1/3} = \frac{5x - 2}{3x^{1/3}}.$$

Clearly, $x = \frac{2}{5}$ is a critical point $[f'(\frac{2}{5}) = 0]$. In addition, because $f'(0)$ does not exist, but $f(0) = 0$, then $x = 0$ is also a critical point. From Fig. 4.22, we see that $f'(x)$ changes from negative to positive as x increases through $x = \frac{2}{5}$, which implies a relative minimum of $f(\frac{2}{5}) = (\frac{2}{5})^{5/3} - (\frac{2}{5})^{2/3} = -0.33$. Since $f(x)$ is continuous at $x = 0$, and $f'(x)$ changes from positive to negative as x increases through 0, $f(0) = 0$ is a relative maximum.

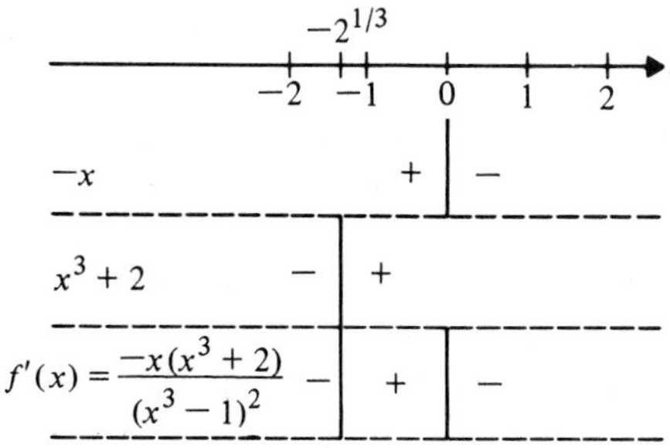

Figure 4.21

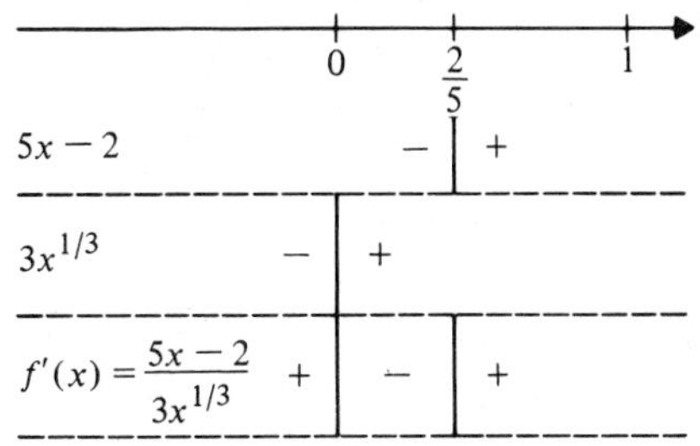

Figure 4.22

d) For critical points we first solve

$$0 = f'(x) = 1 + \cos x,$$

and obtain $x = (2n + 1)\pi$, n an integer. Since $1 + \cos x$ is always nonnegative, it follows that $f'(x)$ cannot change sign at the critical points, and each must yield a horizontal point of inflection. The function therefore has no relative extrema. □

It is perhaps evident that if we were required to sketch the graph of a function, it would be very helpful to know the function's critical points and whether these points give relative maxima, relative minima, horizontal points of inflection, or corners. This is indeed true, but before we consider the entire topic of curve sketching in Section 4.6, we introduce two further concepts in Sections 4.4 and 4.5.

EXERCISES 4.3

In Exercises 1–10, find all critical points for the given function.

1. $f(x) = x^2 - 2x + 6$

2. $f(x) = 2x^3 + 15x^2 + 24x + 1$

3. $f(x) = x^4 - \frac{4}{3}x^3 + 2x^2 - 24x$

4. $f(x) = (x - 1)^5$

5. $f(x) = \dfrac{x + 1}{x^2 + 8}$

6. $f(x) = \dfrac{x^2 + 1}{x - 1}$

7. $f(x) = \dfrac{x}{\sqrt{1 - x}}$

8. $f(x) = \dfrac{x^3}{x^4 + 1}$

9. $f(x) = \sin^2 x$

10. $f(x) = x^2 - 2\cos x$

In Exercises 11–20, find all relative maxima and minima for the given function.

11. $f(x) = \dfrac{x^3}{3} - \dfrac{x^2}{2} - 2x$

12. $f(x) = x^3 - 6x^2 + 12x + 9$

13. $f(x) = 3x^4 - 16x^3 + 18x^2 + 2$

14. $f(x) = x + \dfrac{1}{x}$

15. $f(x) = 2x^3 - 15x^2 + 6x + 4$

16. $f(x) = \dfrac{x^2 + 1}{x^2 - 1}$

17. $f(x) = (x - 1)^{2/3}$

18. $f(x) = \dfrac{(x - 1)^3}{(x + 1)^4}$

19. $f(x) = (x + 2)^3(x - 4)^3$

20. $f(x) = x + 2\sin x$

In Exercises 21–24, find all critical points for the given function.

21. $f(x) = (x + 2)^4(x - 4)^4$

22. $f(x) = x^2 + \dfrac{25x^2}{(x - 2)^2}$

23. $f(x) = \left(\dfrac{x + 8}{x}\right)\sqrt{x^2 + 100}$

24. $f(x) = \dfrac{1 + x + x^2 + x^3}{1 + x^3}$

In Exercises 25–30, find all relative extrema for the given function.

25. $f(x) = x^{5/4} - x^{1/4}$

26. $f(x) = \dfrac{x^2}{x^2 - 4}$

27. $f(x) = \dfrac{x^2}{x^3 - 1}$

28. $f(x) = \dfrac{(2x - 1)(x - 8)}{(x - 1)(x - 4)}$

29. $f(x) = \sin^2 x \cos x, 0 \le x \le 2\pi$

30. $f(x) = \sin x + \cos x$

C **31.** Verify that the function $f(x) = x^3 + \cos x$ has two critical points, one of which is $x = 0$. Use a calculator to find the other critical point correct to three decimal places.

32. The distance from the origin to any point (x, y) on the curve $C: y = f(x)$ is given by $D = \sqrt{x^2 + y^2}$. Show that if $P(x_0, y_0)$ is a point on C for which D has a relative extremum, then the line OP is perpendicular to the tangent line to C at P. Assume that $f(x)$ is differentiable, and that the curve does not pass through the origin.

33. An experiment is performed n times, resulting in n measurements $x_1, x_2, \ldots, x_n$ of the variable x. To obtain the best estimate of the value of x, we choose that value $\bar{x}$ which makes the sum of the squares of the differences between the x_i and $\bar{x}$ as small as possible; i.e., we choose $\bar{x}$ to minimize

$$S(x) = (x - x_1)^2 + (x - x_2)^2 + \cdots + (x - x_n)^2.$$

Show that $\bar{x}$ is the ordinary arithmetic mean of the x_i:

$$\bar{x} = \frac{1}{n}(x_1 + x_2 + \cdots + x_n).$$

34. A 200-gram mass is attached to a spring with constant 20 newtons per meter (Fig. 4.23). If the mass is pulled 10 centimeters below the position where it would hang motionless ($y = 0$), and given an upward velocity of 2 meters per second, then its position as a function of time t (in seconds) is given by

$$y(t) = -\tfrac{1}{10}\cos(10t) + \tfrac{1}{5}\sin(10t) \text{ meters.}$$

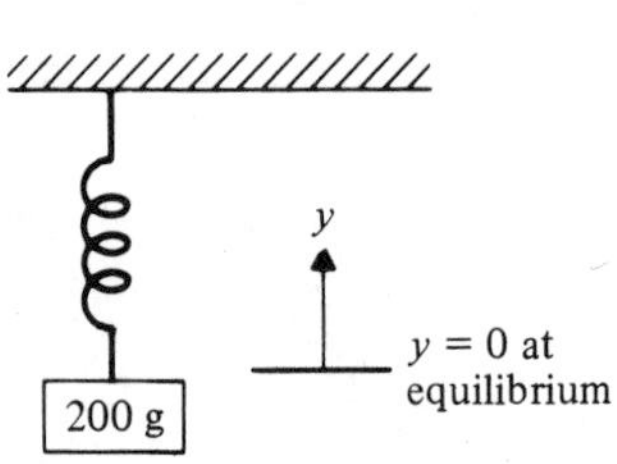

Figure 4.23

Find the maximum distance from $y = 0$ attained by the mass.

35. An inductor L, a resistor R, and a capacitor C are connected in series with a generator, producing an oscillatory voltage $E = E_0 \cos \omega t, t \ge 0$ (Fig. 4.24). If L, C, R, E_0, and ω are all constants, the steady-state current I in the circuit is given by

$$I = \frac{E_0}{Z}\cos(\omega t - \phi),$$

where

$$Z = \sqrt{R^2 + \left(\omega L - \frac{1}{\omega C}\right)^2}$$

and

$$\tan \phi = \frac{\omega L - 1/(\omega C)}{R}.$$

Find the value of ω that makes the amplitude E_0/Z of the current a maximum. What is the maximum amplitude?

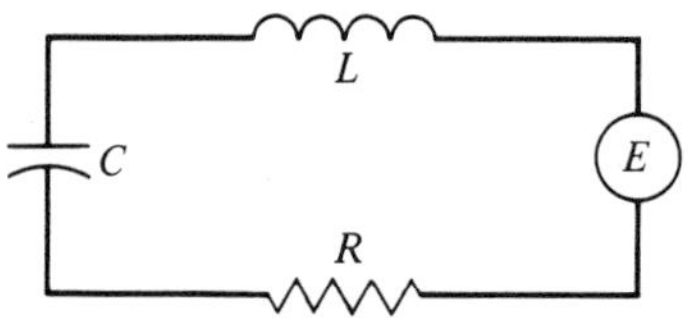

Figure 4.24

* **36.** A landing approach is to be shaped generally as shown in Fig. 4.25. The following conditions are imposed on the approach pattern:

a) The altitude must be h when descent commences;

b) Touchdown must occur at $x = 0$;

c) A constant horizontal speed U must be maintained throughout;

d) At no time must vertical acceleration in absolute value exceed some fixed positive constant k.

Find a cubic polynomial satisfying these requirements.

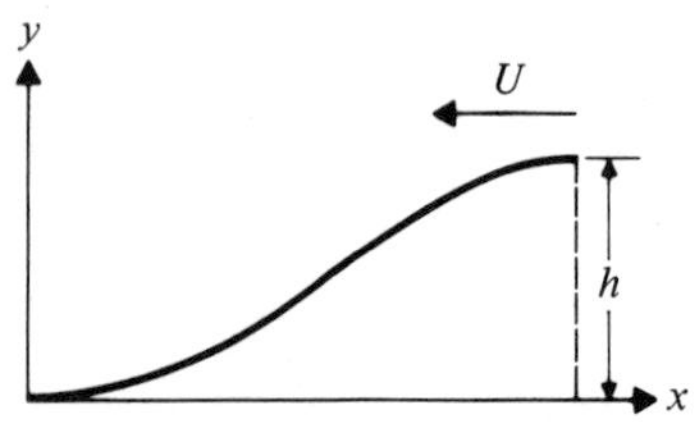

Figure 4.25

4.4 INCREASING AND DECREASING FUNCTIONS

▷ **Definition 4.4** A function $f(x)$ is said to be *increasing* on an interval I if for all $x_1 > x_2$ in I

$$f(x_1) > f(x_2). \tag{4.6}$$

A function $f(x)$ is said to be *decreasing* on I if for all $x_1 > x_2$ in I

$$f(x_1) < f(x_2). \quad ◁ \tag{4.7}$$

The function shown in Fig. 4.17 is increasing on the intervals

$$a \le x \le c, \qquad d \le x \le e, \qquad p \le x \le g, \qquad i \le x \le k, \qquad m \le x \le b,$$

and decreasing on the intervals

$$c \le x \le d, \qquad e \le x \le p, \qquad g \le x \le i, \qquad k < x < m.$$

This function illustrates that the points that separate intervals on which a function is increasing and decreasing are either relative maxima and minima or points of discontinuity of the function. The graph also indicates that the sign of $f'(x)$ determines whether a function is increasing or decreasing on an interval. A precise statement of the situation is as follows.

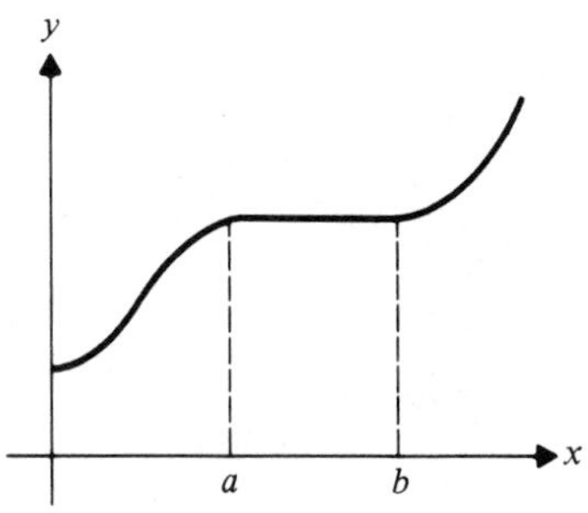

Figure 4.26

Test for Increasing and Decreasing Functions

i) A function $f(x)$ is increasing on an interval I if on I

$$f'(x) \ge 0 \tag{4.8}$$

and is equal to zero at only a finite number of points.

ii) A function $f(x)$ is decreasing on an interval I if on I

$$f'(x) \le 0 \tag{4.9}$$

and is equal to zero at only a finite number of points.

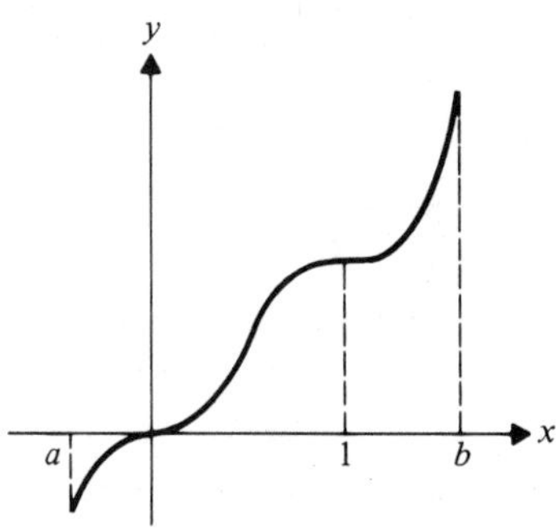

Figure 4.27

A proof of these results is outlined in Exercise 20 in Section 4.10.

In conditions (4.8) and (4.9) we permit $f'(x)$ to vanish at only a finite number of points and do not, therefore, allow it to vanish on an interval. The reason for this is illustrated in Figs. 4.26 and 4.27. In Fig. 4.26 $f'(x)$ is equal to zero on the interval $a \le x \le b$, and certainly $f(x)$ is not increasing on any interval that contains these points. The function $f(x)$ in Fig. 4.27, on the other hand, has $f'(0) = 0 = f'(1)$, and yet $f(x)$ is increasing on the interval $a \le x \le b$.

Example 10 Find intervals on which each of the following functions is increasing and decreasing:

a) $f(x) = 2x^3 + 3x^2 - 12x + 4$;

b) $f(x) = \dfrac{x}{x+1}$.

Solution a) Since $f'(x) = 6x^2 + 6x - 12 = 6(x-1)(x+2)$, we see (Fig. 4.28) that $f'(x) \ge 0$ when $x \le -2$ and when $x \ge 1$, and that $f'(x) \le 0$ when $-2 \le x \le 1$. Consequently, $f(x)$ is increasing on the intervals $x \le -2$ and $x \ge 1$, and decreasing on $-2 \le x \le 1$.

b) Because

$$f'(x) = \frac{(x+1)(1) - x(1)}{(x+1)^2} = \frac{1}{(x+1)^2},$$

$f'(x)$ is always positive, except at $x = -1$ where it does not exist. We conclude, therefore, that $f(x)$ is increasing on the intervals $x < -1$ and $x > -1$. The function is not, however, increasing for all x [since, for example, $f(0) < f(-2)$]. A graph of this function, shown in Fig. 4.29, clarifies the situation. □

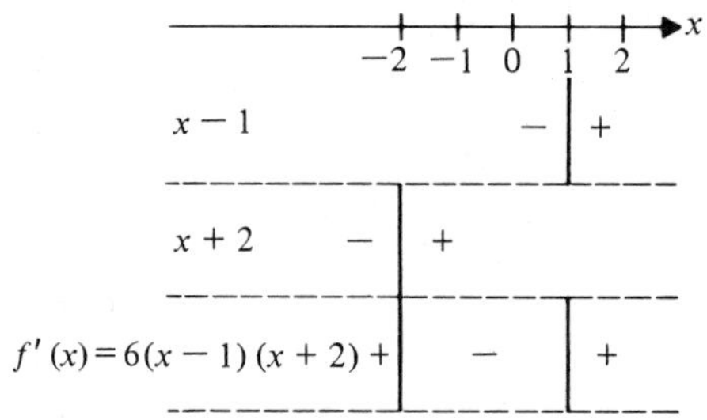

Figure 4.28

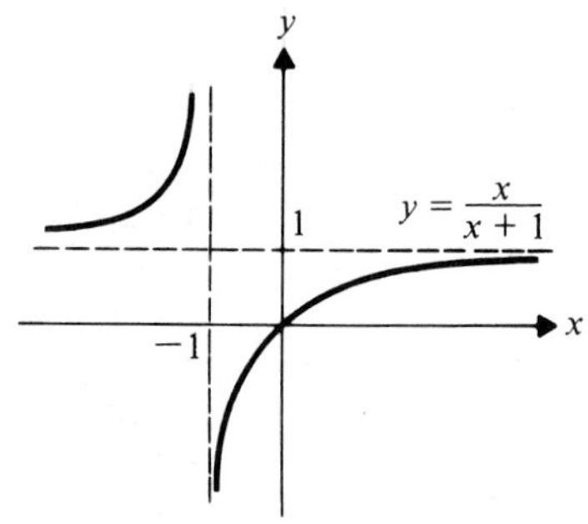

Figure 4.29

EXERCISES 4.4

In Exercises 1–8, determine intervals on which the given function is increasing and decreasing. In each case, state whether the points that separate adjoining intervals are relative maxima, relative minima, or points of discontinuity of the function.

1. $f(x) = x^2 - 3x + 4$
2. $f(x) = -2x^2 + 5x$
3. $f(x) = 2x^3 - 18x^2 + 18x + 1$
4. $f(x) = 3x^4 + 4x^3 - 24x + 2$
5. $f(x) = x^2 + \dfrac{1}{x^2}$
6. $f(x) = \dfrac{x^2 + 4}{x^2 - 1}$
7. $f(x) = \dfrac{x^3}{x + 1}$
8. $f(x) = |x^2 - 1| + 1$
9. If the price of a certain gadget is set at r, then the market will demand x gadgets per week, where

$$x = 4a^3 - 3ar^2 + r^3,$$

$a > 0$ a constant and $0 < r < 2a$. Show that the price function $r = f(x)$ defined implicitly by this equation is a decreasing function.

In Fig. 4.30(a) we have sketched the graph of a function $f(x)$ and in Fig. 4.30(b) the graph of $f'(x)$. Note that $f'(x) \geq 0$ when $f(x)$ is increasing and $f'(x) \leq 0$ when $f(x)$ is decreasing. In addition, the discontinuity in $f'(x)$ at $x = 2$ is reflected in a corner for $f(x)$. In each of Exercises 10–13, draw similar graphs for the function and its derivative.

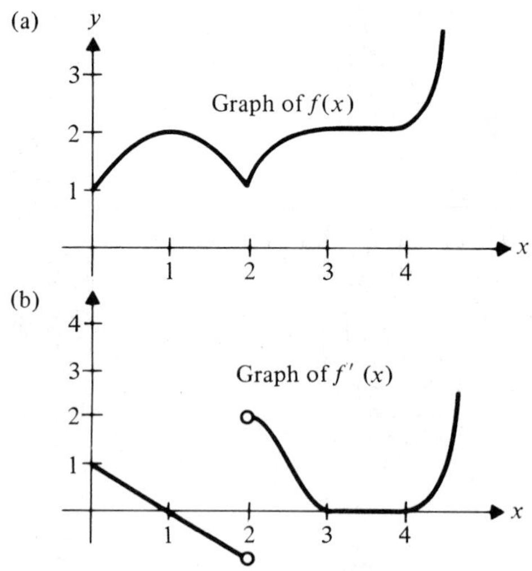

Figure 4.30

10. $f(x) = x^2 + 2x$
11. $f(x) = x^4 - x^2$
12. $f(x) = \dfrac{x - 1}{x + 2}$
13. $f(x) = \dfrac{x - 4}{x^2}$

In Exercises 14 and 15 we have given you the graph of a function $f(x)$, but not an algebraic formula for $f(x)$. In each case, use the graph to sketch a graph of $f'(x)$.

14.

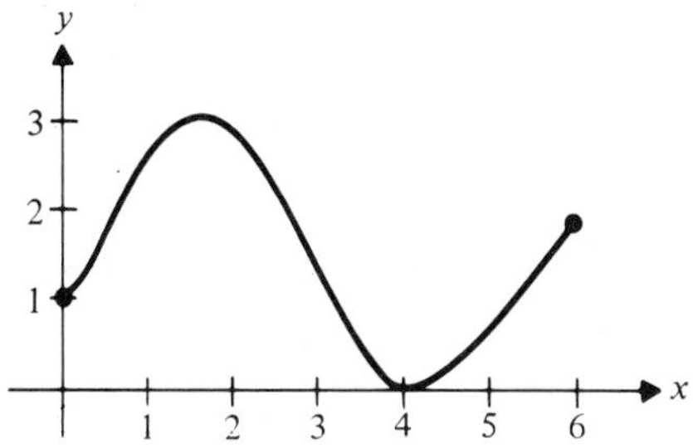

15.

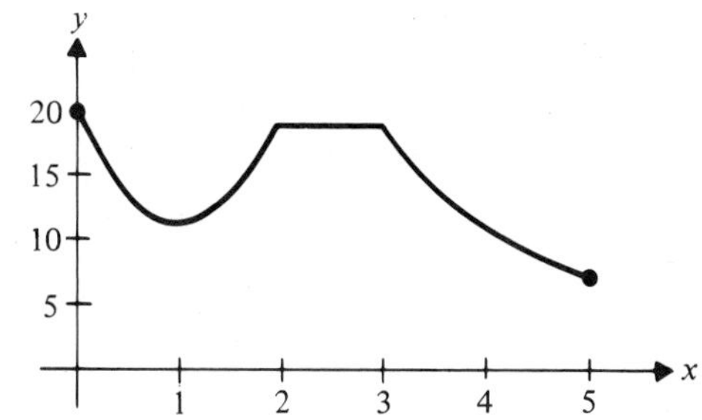

4.5 CONCAVITY AND POINTS OF INFLECTION

In Section 4.1 we illustrated that when an object moves along the x-axis with position x defined as a function of time t, $x = x(t)$, it is useful to draw a graph of this function. Suppose, for example, that $x(t) = t^4 - 8t^3 + 18t^2 - 8t$, $t \geq 0$, the graph of which is shown in Fig. 4.31. The first derivative of $x(t)$ represents the particle's velocity:

$$\begin{aligned}\frac{dx}{dt} &= 4t^3 - 24t^2 + 36t - 8 \\ &= 4(t-2)(t^2 - 4t + 1).\end{aligned}$$

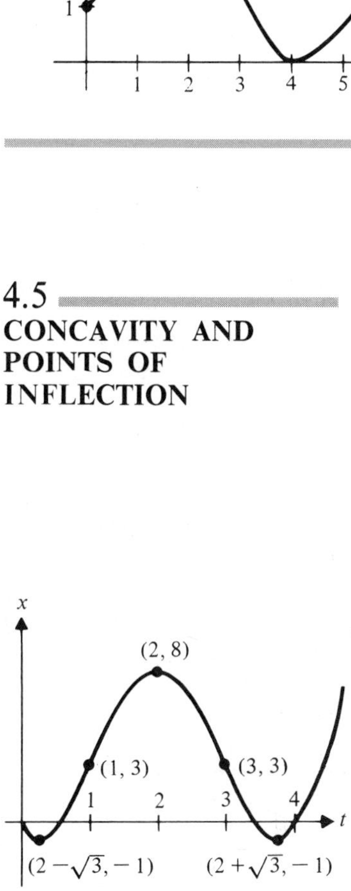

Figure 4.31

Critical points of $x(t)$ occur at $t = 2$ and $t = 2 \pm \sqrt{3}$; when $t = 2$, $x(t)$ has a relative maximum, and when $t = 2 \pm \sqrt{3}$, $x(t)$ has relative minima. At these critical points, $x'(t) = 0$; therefore, they physically represent times when the object comes to an instantaneous stop. In other words, when a function represents a physical quantity, properties of its first derivative can often be interpreted physically as well as mathematically.

In this section we discuss the mathematical significance of a function's second derivative, and illustrate that second derivatives can sometimes be interpreted physically also. Specifically, for the function shown in Fig. 4.31, we already know that its second derivative,

$$\frac{d^2x}{dt^2} = 12t^2 - 48t + 36 = 12(t-1)(t-3),$$

represents the particle's acceleration. It is clear that $x''(t) = 0$ when $t = 1$ and $t = 3$, and physically these are times when the object has zero acceleration. Note also that on the intervals $0 < t < 1$ and $t > 3$ the object's acceleration is positive and its velocity $x'(t)$ is increasing; on the other hand, for $1 < t < 3$, its acceleration is negative and its velocity is decreasing. For displacement $x(t)$ we have therefore illustrated that $x'(t)$ is increasing when $x''(t) > 0$, and that $x'(t)$ is decreasing when $x''(t) < 0$.

These results are valid not just for displacement functions, but for any function whatsoever; that is, the sign of the second derivative of a function determines when its first derivative is increasing and decreasing. The shape of the graph of a function when its first derivative is increasing and when it is decreasing is given in the following definition.

▷ **Definition 4.5** The graph of a function $f(x)$ is said to be *concave upward* on an interval I if $f'(x)$ is increasing on I, and *concave downward* on I if $f'(x)$ is decreasing on I. ◁

▷ **Definition 4.6** Points on a graph that separate intervals of opposite concavity are called *points of inflection*. ◁

The graph in Fig. 4.32 is concave upward on the intervals

$$c \le x \le d, \qquad e \le x < j, \qquad j < x \le g, \qquad i \le x \le b,$$

and concave downward on the intervals

$$a \le x \le c, \qquad d \le x \le e, \qquad g \le x < h, \qquad h < x \le i.$$

Note that $x = j$ and $x = h$ are not included in these intervals because $f'(x)$ is not defined at these points.

Points on the curve corresponding to $x = c$, d, e, g, and i are points of inflection. Note that at a point of inflection, the tangent line to the graph of the function actually crosses the graph. At the points of inflection where $x = g$ and $x = i$ in Fig. 4.32, the slope of the tangent line is zero; these points are the *horizontal points of inflection* we discussed in Section 4.3.

In Section 4.4 we stated that a function is increasing on an interval if its first derivative is greater than or equal to zero on that interval (and equal to zero at only a finite number of points); it is decreasing if its first derivative is less than or equal to zero. Furthermore, points that separate intervals on which the function is increasing and decreasing are points at which the first derivative is either equal to zero (and changes sign) or does not exist. We can use these ideas to locate analogously points of inflection and intervals on which the graph of a function is concave upward and concave downward.

Test for Concavity

i) The graph of a function $f(x)$ is concave upward on an interval I if on I

$$f''(x) \ge 0 \qquad (4.10)$$

and is equal to zero at only a finite number of points.

ii) The graph of a function $f(x)$ is concave downward on an interval I if on I

$$f''(x) \le 0 \qquad (4.11)$$

and is equal to zero at only a finite number of points.

iii) Points of inflection occur where

$$f''(x) = 0 \qquad (4.12)$$

and $f''(x)$ changes sign, or perhaps also where $f''(x)$ does not exist.

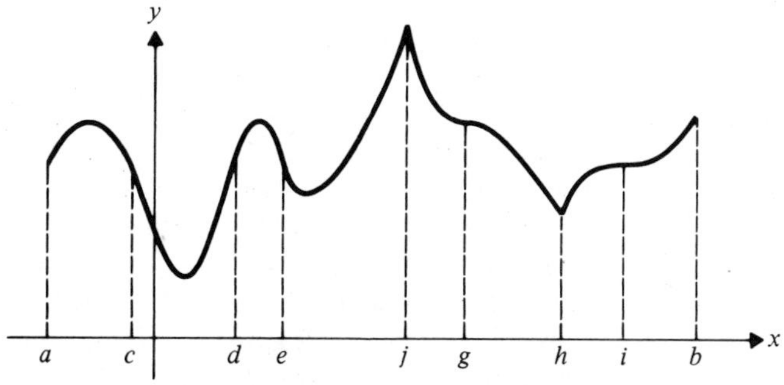

Figure 4.32

Consequently, just as the first derivative of a function is used to determine its relative extrema and intervals on which it is increasing and decreasing, the second derivative is used to find points of inflection and intervals on which the function is concave upward and concave downward.

This parallelism between the first and second derivative is reiterated in the following table.

First Derivative

1. A relative extremum occurs at a point where $f'(x) = 0$ and $f'(x)$ changes sign as x increases through the point.
2. A relative extremum may occur at a point where $f'(x)$ does not exist.
3. $f(x)$ is increasing on an interval if $f'(x) \geq 0$; $f(x)$ is decreasing on an interval if $f'(x) \leq 0$. [$f'(x)$ may be equal to zero at only a finite number of points in either case.]
4. Points that separate intervals on which a function is increasing and decreasing are relative extrema or points of discontinuity of the function.

Second Derivative

1. A point of inflection occurs at a point where $f''(x) = 0$ and $f''(x)$ changes sign as x increases through the point.
2. A point of inflection may occur at a point where $f''(x)$ does not exist.
3. $f(x)$ is concave upward on an interval if $f''(x) \geq 0$; $f(x)$ is concave downward on an interval if $f''(x) \leq 0$. [$f''(x)$ may be equal to zero at only a finite number of points in either case.]
4. Points that separate intervals on which a function is concave upward and concave downward are points of inflection or points of discontinuity of its first derivative.

The results of our discussion lead to what is often called the *second-derivative test* for determining whether a critical point at which $f'(x) = 0$ yields a relative maximum or a relative minimum.

Second-derivative Test for Critical Points

i) If $f'(x_0) = 0$ and $f''(x_0) > 0$, then $x = x_0$ yields a relative minimum.

ii) If $f'(x_0) = 0$ and $f''(x_0) < 0$, then $x = x_0$ yields a relative maximum.

iii) If $f'(x_0) = 0$ and $f''(x_0) = 0$, then no conclusion can be made.

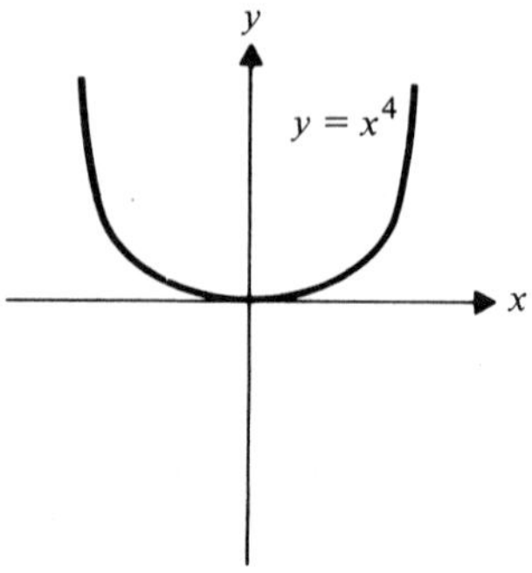

Figure 4.33

In (i), $f'(x_0) = 0$ implies that $x = x_0$ is a critical point with a horizontal tangent line; $f''(x_0) > 0$ implies that around $x = x_0$ the curve is concave upward, and $x = x_0$ therefore yields a relative minimum. Similarly, $f''(x_0) < 0$ at a critical point in (ii) implies that the curve is concave downward and therefore has a relative maximum.

To show that the test fails in case (iii), consider the three functions $f(x) = x^4$, $f(x) = -x^4$, and $f(x) = x^3$ shown in Figs. 4.33, 4.34, and 4.35, respectively. For each of these functions $f'(0) = f''(0) = 0$; yet the first has a relative minimum, the second a relative maximum, and the third a horizontal point of inflection at $x = 0$.

The second-derivative test very quickly determines the nature of a critical point x_0 when $f''(x_0) \neq 0$. For instance, the second derivative of the function $f(x) = 2x^3 - 9x^2 - 24x + 6$ in Example 9(a) is $f''(x) = 12x - 18$. Since $f''(-1) = -30$, $x = -1$ yields a relative maximum, and because $f''(4) = 30$, $x = 4$ gives a relative minimum. Should the second-derivative test fail (as it would in Example 9d), we can always use the first-derivative test.

A more refined test to determine the nature of a critical point x_0 when $f'(x_0) = f''(x_0) = 0$ is introduced in Exercise 17.

Example 11 For each of the following functions, find all points of inflection and intervals on which the graph of the function is concave upward and concave downward.

a) $f(x) = \frac{2}{3}x^3 - 6x^2 + 16x + 1$

b) $f(x) = x^{6/5} + x^{1/5}$

Solution a) For points of inflection we first solve

$$0 = f''(x) = \frac{d}{dx}(2x^2 - 12x + 16) = 4x - 12,$$

and obtain $x = 3$. Since $f''(x)$ changes sign as x passes through 3, $x = 3$ gives the point of inflection (3, 13). The curve $y = f(x)$ is concave downward for $x \leq 3$ (since $f''(x) \leq 0$) and concave upward for $x \geq 3$ ($f''(x) \geq 0$).

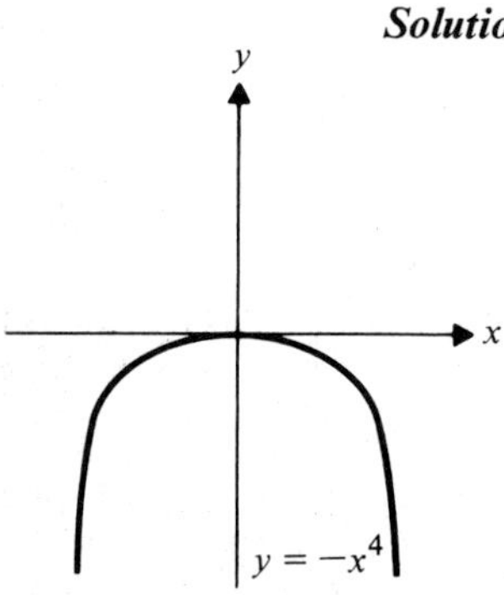

Figure 4.34

b) For points of inflection we first solve

$$0 = f''(x) = \frac{d}{dx}\left(\frac{6}{5}x^{1/5} + \frac{1}{5}x^{-4/5}\right) = \frac{6}{25}x^{-4/5} - \frac{4}{25}x^{-9/5}$$

$$= \frac{6x - 4}{25x^{9/5}},$$

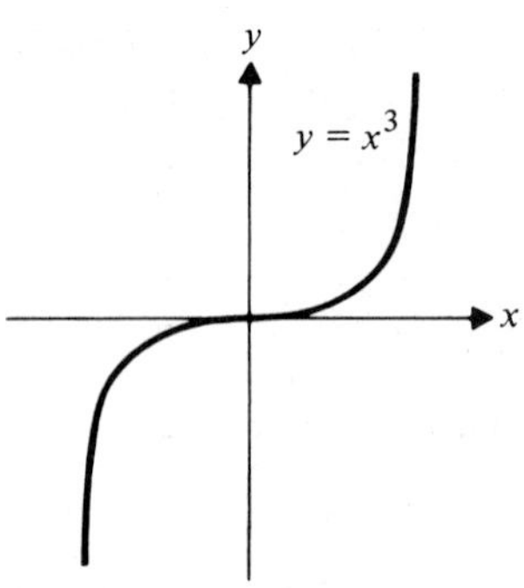

Figure 4.35

and find $x = \frac{2}{3}$. Since $f''(x)$ changes sign as x passes through $\frac{2}{3}$ (Fig. 4.36), $x = \frac{2}{3}$ gives a point of inflection with $f(\frac{2}{3}) = 1.5$.

We also note that $f'(x)$ and $f''(x)$ do not exist at $x = 0$, although $f(0) = 0$. Because $f''(x)$ changes sign as x passes through 0, (0, 0) must also be a point of inflection. Finally, the schematic shown in Fig. 4.36 indicates that the graph $y = f(x)$ is concave upward for $x < 0$ and $x \geq \frac{2}{3}$ and concave downward for $0 < x \leq \frac{2}{3}$. □

In Section 4.7 we will see the importance of critical points in the solution of applied maxima–minima problems. In some problems the significance of inflection points is equally important. We have already seen that if $x(t)$ represents the position of an object moving along the x-axis, then points of inflection

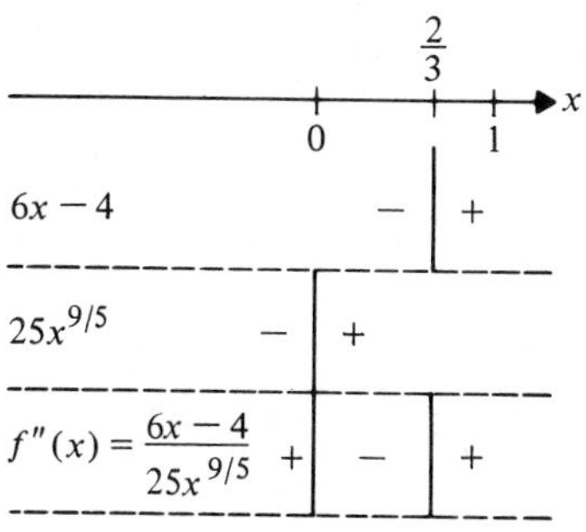

Figure 4.36

when $x''(t) = 0$ represent times at which the object has zero acceleration. As a second example, we consider how an infectious disease spreads through a population of susceptible individuals. To simplify the problem, we assume that all individuals are equally susceptible and that the disease is introduced into the population by one infective. Contact between individuals spreads the disease slowly at first, then more rapidly as more individuals become infected, and finally, when most of the population has become infected, the spread tapers off. To describe this process in more mathematical terms, we denote by N the total number of individuals in the population, and by $x(t)$ the number of infectives at any given time t. Since the disease is spread through contacts between infectives and noninfectives, it follows that the rate at which individuals become infected depends on both the number of infectives x and the number of noninfectives $(N - x)$ at any given time. In practice, it is often assumed that the rate of infection (represented by dx/dt) is directly proportional to x and $N - x$:

$$\frac{dx}{dt} = kx(N - x).$$

Given that this equation describes the rate at which the disease is spread, it is possible to determine the number of infectives $x(t)$ present at any given time. It is not a problem that we can solve completely at this point, but we can at least obtain some information about its graph.

First we note that dx/dt is always nonnegative, so that the graph of $x(t)$ is increasing. Second, since points of inflection on the graph are defined by a vanishing second derivative, we set

$$0 = x''(t) = \frac{d}{dt}\{kNx - kx^2\} = kN\frac{dx}{dt} - 2kx\frac{dx}{dt}$$
$$= k(N - 2x)\frac{dx}{dt}.$$

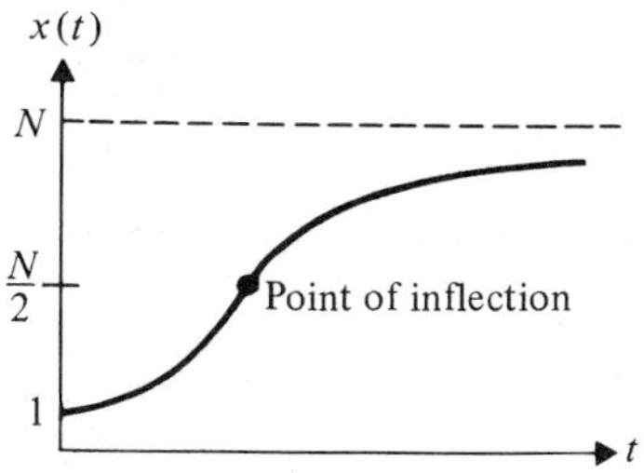

Figure 4.37

But $k \neq 0$, and $dx/dt = 0$ only when the population is either totally infected or not at all infected; therefore, a point of inflection occurs only when $N - 2x = 0$ or $x = N/2$ (i.e., when the population is half infected). Since $x''(t) > 0$ when $x < N/2$ and $x''(t) < 0$ when $x > N/2$, the graph of $x(t)$ is concave upward for $x < N/2$ and concave downward for $x > N/2$. These two pieces of information are shown in Fig. 4.37, where we have chosen a time convention of $t = 0$ when the first infective is introduced. In other words, we have completely determined the shape of the graph of $x(t)$. (What remains is the proper insertion of a time scale and this we shall do in Section 9.8.)

At this point the significance of the point of inflection at $x = N/2$ is clear. Since the slope of this curve (dx/dt) is increasing for $x < N/2$ and decreasing for $x > N/2$, it follows that when $x = N/2$, the rate of spread of disease is at its peak.

EXERCISES 4.5

Determine where the graph of the function in each of Exercises 1–4 is concave upward, is concave downward, and has points of inflection.

1. $f(x) = x^3 - 3x^2 - 3x + 5$

2. $f(x) = 3x^4 + 4x^3 - 24x + 2$

3. $f(x) = x^2 + \dfrac{1}{x^2}$

4. $f(x) = \dfrac{x^2 + 4}{x^2 - 1}$

In each of Exercises 5–8, use the second-derivative test to determine whether critical points where $f'(x) = 0$ yield relative maxima or relative minima.

5. $f(x) = x^3 - 3x^2 - 3x + 5$

6. $f(x) = x + \dfrac{1}{x}$

7. $f(x) = 3x^4 - 16x^3 + 18x^2 + 2$

8. $f(x) = x^{5/4} - x^{1/4}$

9. Is the graph of a function concave upward, concave downward, both, or neither on an interval I if on I its second derivative is always equal to zero?

10. Prove that the curve $y = 2 \cos x$ passes through all points of inflection of the curve $y = x \sin x$.

Sketch the graph of a function $f(x)$ that is defined everywhere on the interval $I: a \le x \le b$, and that possesses the properties in Exercises 11–16. In each case, assume that $a < x_0 < b$.

11. $f(x)$ is increasing on I and discontinuous at x_0

12. $f''(x) \ge 0$ on I and $f(x)$ is not concave upward on I

13. $f(x)$ is decreasing on I, $f(x)$ is continuous at x_0, but $f'(x)$ is discontinuous at x_0

14. $f(x)$ is increasing on $a \le x \le x_0$, increasing on $x_0 < x \le b$, but not increasing on I

15. $f(x)$ is concave downward on $a \le x < x_0$, concave downward on $x_0 < x \le b$, but not concave downward on I

16. $f(x)$ is increasing on $a \le x \le x_0$, concave downward on $a \le x < x_0$, decreasing on $x_0 \le x \le b$, and concave upward on $x_0 < x \le b$

17. The second-derivative test fails to classify a critical point x_0 of a function $f(x)$ as yielding a relative maximum, a relative minimum, or a horizontal point of inflection if $f''(x_0) = 0$. The following test can be used in such cases. Suppose $f(x)$ has derivatives of all orders in an open interval around x_0, and the first n derivatives all vanish at x_0,

$$0 = f'(x_0) = f''(x_0) = \cdots = f^{(n)}(x_0),^{\dagger}$$

but $f^{(n+1)}(x_0) \neq 0$. Then:

i) if n is even, $f(x)$ has a horizontal point of inflection at x_0;

ii) if n is odd and $f^{(n+1)}(x_0) > 0$, $f(x)$ has a relative minimum at x_0;

iii) if n is odd and $f^{(n+1)}(x_0) < 0$, $f(x)$ has a relative maximum at x_0.

(A proof of this result requires the use of material from Chapter 11 and is therefore delayed until that time (see Exercise 12 in Section 11.7). Note that the second-derivative test is the special case when $n = 1$. Use this test to determine whether critical points of each of the following functions yield relative maxima, relative minima, or horizontal points of inflection.)

a) $f(x) = (x^2 - 1)^3$ b) $f(x) = x^2\sqrt{1 - x}$

Would the first-derivative test be faster in classifying the critical points in these examples?

18. Show that the points of inflection of $f(x) = (k - x)/(x^2 + k^2)$ (k a constant) all lie on a straight line.

***19.** Prove that if a cubic polynomial has both a relative maximum and a relative minimum, then the point of inflection between these extrema is the midpoint of the line segment joining them.

† We use the notation $f^{(n)}(x_0)$ to represent the nth derivative of $f(x)$ evaluated at x_0.

4.6 CURVE SKETCHING

We were introduced to curve sketching in Section 1.3, when we used the basic idea of a table of values. Following this, we illustrated in Section 1.4 that the addition and multiplication of ordinates often yield rough but very quick sketches. We also saw that symmetries of a curve about the x- and y-axes can be helpful in many cases. We used limits in Section 2.2 to discover the behavior of the graph of a function near a point of discontinuity or for very large positive and negative values of the independent variable. Each of these techniques has its value and none should be neglected.

In this chapter we have defined critical points, relative extrema, increasing and decreasing functions, concavity, and points of inflection, and each of these concepts can add information to a sketch. A natural question is "If we are given a curve to sketch, which of the many tools that are now available should we use?" There is really no answer to this question, except to say that we should use those tools that give us a suitable sketch with the least amount of effort. But what is a suitable sketch for one problem may not be suitable for another. For some problems, a very rough sketch might be sufficient, in which case we might need to use only addition or multiplication of ordinates; in other problems, a fairly detailed graph might be needed, and therefore we would need further information such as relative extrema and points of inflection. In this section we illustrate how further detail can be added to a sketch by using more and more tools on a specific example. At the same time, we see how the various techniques complement one another.

Example 12 Sketch a graph of the function $f(x) = (x - 4)/x^2$.

Solution With the limits

$$\lim_{x\to 0^+} f(x) = -\infty, \quad \lim_{x\to 0^-} f(x) = -\infty, \quad \lim_{x\to \infty} f(x) = 0^+, \quad \lim_{x\to -\infty} f(x) = 0^-,$$

we begin our sketch as shown in Fig. 4.38(a). This information would lead us to suspect that the graph should be completed as shown by the dashed line. To verify this we find critical points for $f(x)$:

$$0 = f'(x) = \frac{x^2(1) - (x - 4)(2x)}{x^4} = \frac{x(8 - x)}{x^4} = \frac{8 - x}{x^3}.$$

Clearly, $x = 8$ is the only critical point of the function, and the first-derivative test indicates that it gives a relative maximum. Our sketch is now as shown in Fig. 4.38(b).

There is obviously a point of inflection to the right of $x = 8$, which we could pinpoint with $f''(x)$:

$$f''(x) = \frac{x^3(-1) - (8 - x)(3x^2)}{x^6} = \frac{2(x - 12)}{x^4}.$$

Since $f''(12) = 0$ and $f''(x)$ changes sign as x passes through 12, (12, 1/18) is the point of inflection. □

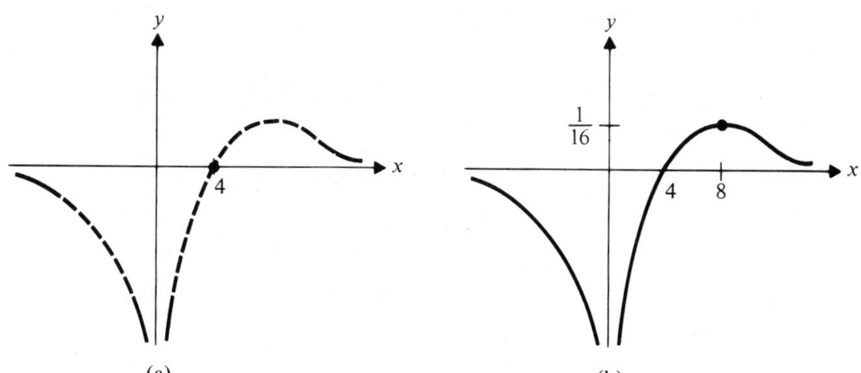

Figure 4.38

Example 13 Sketch a graph of the function $f(x) = x + \sin x$.

Solution We might try addition of ordinates for the curves $y = x$ and $y = \sin x$ shown in Fig. 4.39(a). It would, however, be difficult to decide which of the curves in Figs. 4.39(b), (c), or (d) is correct. To make this decision we consider the critical points of $f(x)$, defined by

$$0 = f'(x) = 1 + \cos x.$$

The solutions of this equation are $x = (2n + 1)\pi$, n an integer. Since $f'(x)$ does not change sign as x passes through each of these points, each yields a horizontal point of inflection, and the graph in Fig. 4.39(c) is correct. □

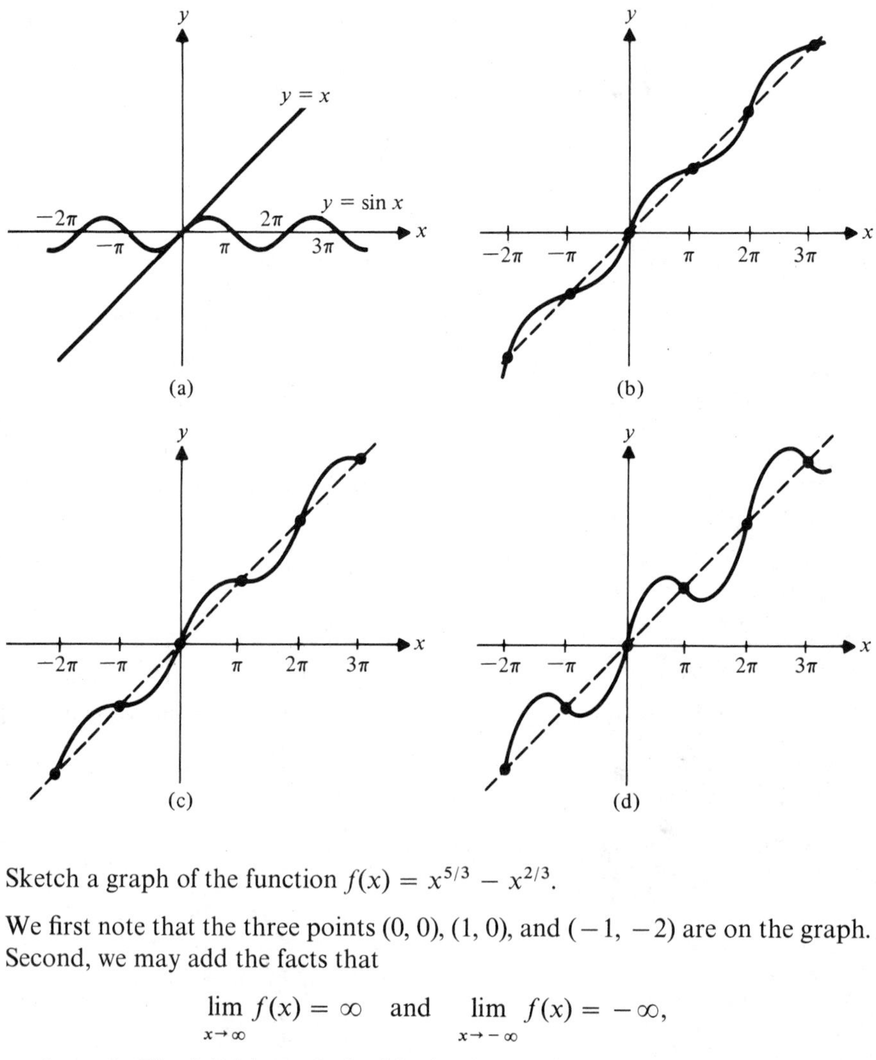

Figure 4.39

Example 14 Sketch a graph of the function $f(x) = x^{5/3} - x^{2/3}$.

Solution We first note that the three points (0, 0), (1, 0), and (−1, −2) are on the graph. Second, we may add the facts that

$$\lim_{x\to\infty} f(x) = \infty \quad \text{and} \quad \lim_{x\to-\infty} f(x) = -\infty,$$

as shown in Fig. 4.40(a). To find critical points of $f(x)$, we first solve

$$0 = f'(x) = \frac{5}{3}x^{2/3} - \frac{2}{3}x^{-1/3} = \frac{5x - 2}{3x^{1/3}}.$$

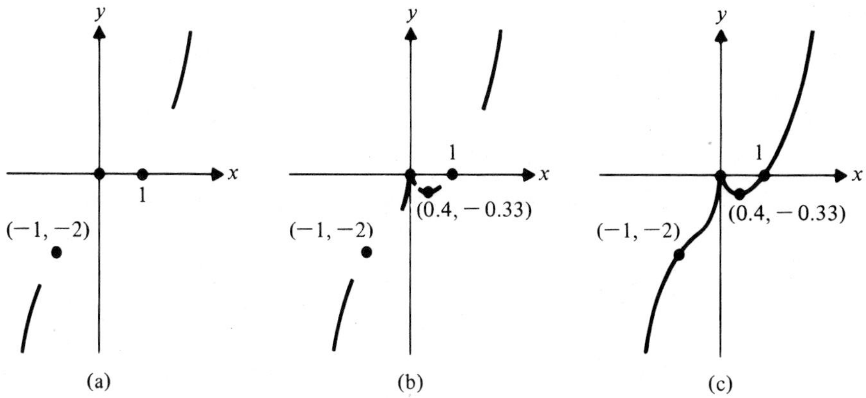

Figure 4.40

The only solution is $x = \frac{2}{5}$, and because $f'(x)$ changes from negative to positive as x increases through $\frac{2}{5}$, we have a relative minimum of $f(\frac{2}{5}) = (\frac{2}{5})^{5/3} - (\frac{2}{5})^{2/3} = -0.33$. Since $f'(x)$ does not exist at $x = 0$, this is also a critical point, and we calculate

$$\lim_{x \to 0^+} f'(x) = -\infty \quad \text{and} \quad \lim_{x \to 0^-} f'(x) = \infty.$$

This information, along with the relative minimum, is shown in Fig. 4.40(b). We now join these parts smoothly to produce the final sketch in Fig. 4.40(c). Note that we have drawn the graph concave downward for large negative x, and concave upward for large positive x. We can verify this by calculating

$$f''(x) = \left(\frac{5}{3}\right)\left(\frac{2}{3}\right)x^{-1/3} - \left(\frac{2}{3}\right)\left(-\frac{1}{3}\right)x^{-4/3} = \frac{2(5x + 1)}{9x^{4/3}}.$$

Clearly there is only one point of inflection (when $x = -\frac{1}{5}$), and the concavity is as illustrated. □

Example 15 Sketch a graph of the function $f(x) = (x + 2)^2/x^2$.

Solution We write

$$f(x) = \frac{(x + 2)^2}{x^2} = \left(\frac{x + 2}{x}\right)^2 = \left(1 + \frac{2}{x}\right)^2.$$

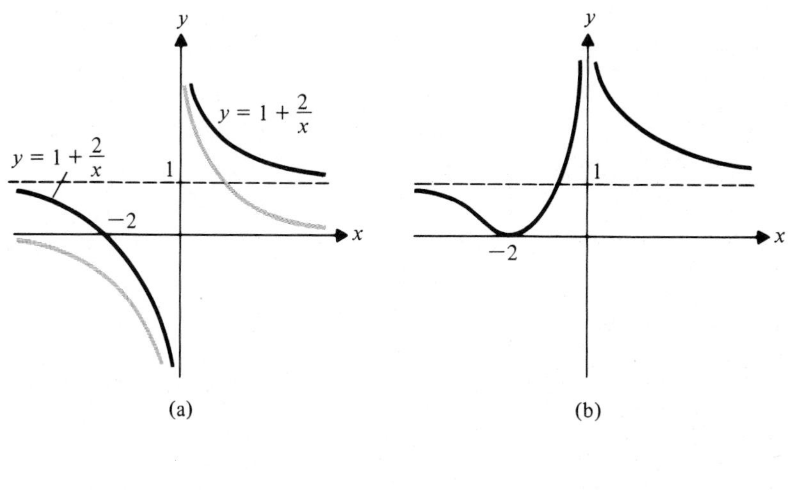

Figure 4.41

A sketch of $y = 2/x$ is shown dashed in Fig. 4.41(a), and $y = 1 + 2/x$ requires only a shift of 1 in the y-direction. If we now square all ordinates of the curve $y = 1 + 2/x$, we obtain the sketch of $y = (1 + 2/x)^2$ shown in Fig. 4.41(b). The only information that could be added is the exact position of the point of inflection to the left of $x = -2$. □

Example 16 Sketch a graph of the function $f(x) = (x^2 + 1)/(x - 1)$.

Solution If we divide $x^2 + 1$ by $x - 1$, we get

$$f(x) = (x + 1) + \frac{2}{x - 1}.$$

Sketches of $y = x + 1$ and $y = 2/(x - 1)$ are shown in Fig. 4.42(a), and by the addition of ordinates we obtain the graph in Fig. 4.42(b).

To add the positions of the relative extrema we find the critical points of $f(x)$. Since

$$f'(x) = \frac{(x - 1)(2x) - (x^2 + 1)(1)}{(x - 1)^2} = \frac{x^2 - 2x - 1}{(x - 1)^2},$$

we set

$$x^2 - 2x - 1 = 0.$$

Critical points are, therefore,

$$x = \frac{2 \pm \sqrt{4 + 4}}{2} = 1 \pm \sqrt{2}.$$

Our sketch in Fig. 4.42(b) clearly indicates that $x = 1 + \sqrt{2}$ gives a relative minimum and $x = 1 - \sqrt{2}$ a relative maximum. The final graph is shown in Fig. 4.42(c). □

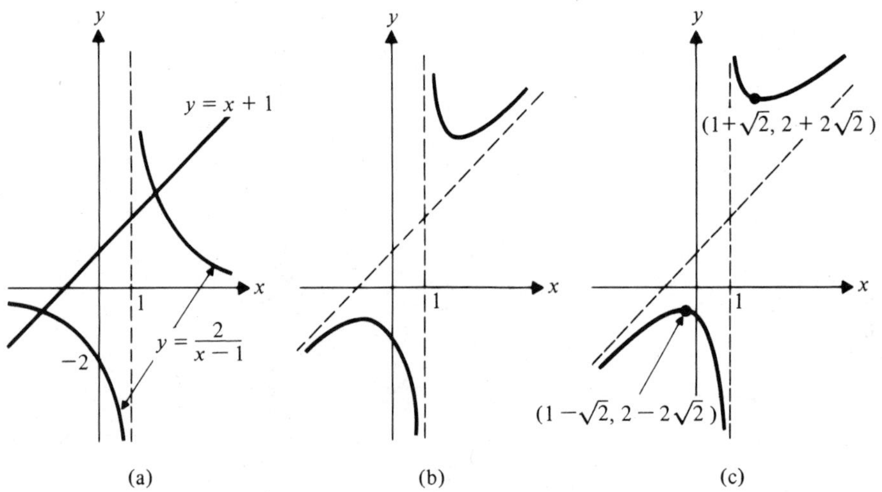

Figure 4.42

EXERCISES 4.6

In Exercises 1–10, sketch a graph of the given function.

1. $f(x) = \frac{1}{3}x^3 - \frac{1}{2}x^2 - 2x$
2. $f(x) = x^3 - 6x^2 + 12x + 9$
3. $f(x) = 3x^4 - 16x^3 + 18x^2 + 2$
4. $f(x) = x + \dfrac{1}{x}$
5. $f(x) = 2x^3 - 15x^2 + 6x + 4$
6. $f(x) = (x^2 + 1)/(x^2 - 1)$
7. $f(x) = (x - 1)^3/(x + 1)^4$
8. $f(x) = (x - 1)^{2/3}$
9. $f(x) = (x + 2)^3(x - 4)^3$
10. $f(x) = x + 2 \sin x$
11. A voltage source with constant electromotive force E and constant internal resistance r maintains a current I through a circuit with resistance R (Fig. 4.43), where

$$E = I(r + R).$$

The power (work per unit time) necessary to maintain this current in r and R is given by

$$P = I^2(r + R) = I^2r + I^2R.$$

a) If we define $P_R = I^2R$ and $P_r = I^2r$ as the power dissipated in R and r respectively, sketch P_R and P_r as functions of R.

b) Sketch a graph of $P(R) = P_R(R) + P_r(R)$.

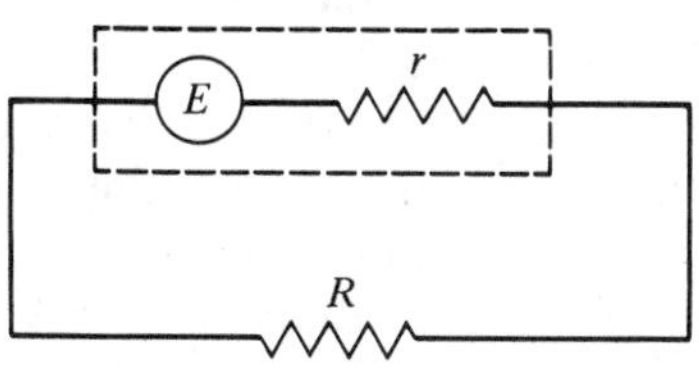

Figure 4.43

12. The rate of photosynthesis P in a leaf depends on the intensity of light I on the leaf according to

$$P = \frac{MI}{I + K} - R,$$

where $M > R$ and K are all positive constants. Sketch a graph of this function.

In Exercises 13–24, sketch a graph of the given function.

13. $f(x) = \dfrac{x}{x^2 + 3}$
14. $f(x) = x^4 + 10x^3 + 6x^2 - 64x + 5$
15. $f(x) = x^4 - 2x^3 + 2x$
16. $f(x) = \dfrac{x^2 - 8}{x - 5}$
17. $f(x) = \dfrac{x^2}{x^2 - 4}$
18. $f(x) = x^{5/4} - x^{1/4}$
19. $f(x) = |x^2 - 9| + 2$
20. $f(x) = \dfrac{(2x - 1)(x - 8)}{(x - 1)(x - 4)}$
21. $f(x) = \sin^2 x \cos x,\ 0 \le x \le 2\pi$
22. $f(x) = \sin x + \cos x$
23. $f(x) = \dfrac{x}{\sqrt{1 - x}}$
24. $f(x) = \left(\dfrac{x + 8}{x}\right)\sqrt{x^2 + 100}$
25. a) A company produces x kilograms of a commodity per day at a total cost of

$$C(x) = \frac{x^2}{300}\left(\frac{x + 100}{x + 300}\right) + 60, \qquad 1 \le x \le 200.$$

Show that the cost graph of $C(x)$ is always concave upward. Sketch the curve.

b) The average production cost per kilogram when x kilograms are produced is given by $c(x) = C(x)/x$. Sketch a graph of the average cost function $c(x)$.

c) Show that the output at which the average cost is least satisfies the equation

$$(x + 300)^2(x^2 - 18{,}000) - 60{,}000x^2 = 0.$$

Do this in two ways:

i) by finding critical points for $c(x)$;

ii) by noting that the average cost is the slope of the line joining a point $(x, C(x))$ to the origin; hence minimum average cost occurs when the tangent line to the $C(x)$ graph passes through the origin.

26. Sketch the curve $y^2 = x^2(x^2 - 4)$.

*27. Sketch a graph of the function

$$f(x) = \frac{1 + x + x^2 + x^3}{1 + x^3}.$$

4.7 ABSOLUTE MAXIMA AND MINIMA

In this section we will show that for many applied maxima–minima problems, we must extend the notion of relative maxima and minima to *absolute maxima and minima*. We define absolute extrema as follows.

▷ **Definition 4.7** The *absolute maximum* of a function $f(x)$ on an interval I is $f(x_0)$ if x_0 is in I and if for all x in I

$$f(x) \le f(x_0); \tag{4.13}$$

$f(x_0)$ is said to be the *absolute minimum* of $f(x)$ on I if for all x in I

$$f(x) \ge f(x_0). \quad \triangleleft \tag{4.14}$$

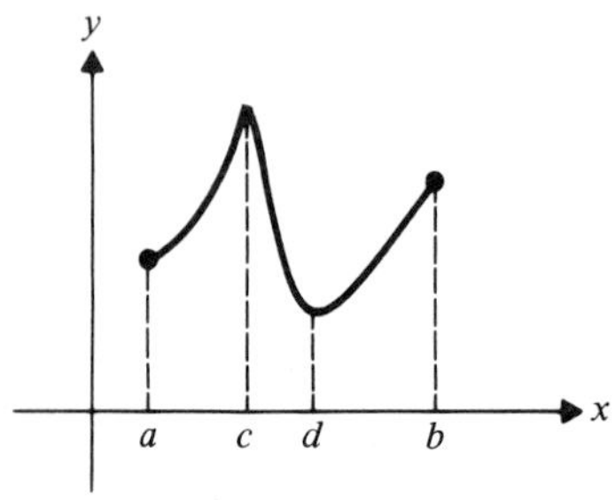

Figure 4.44

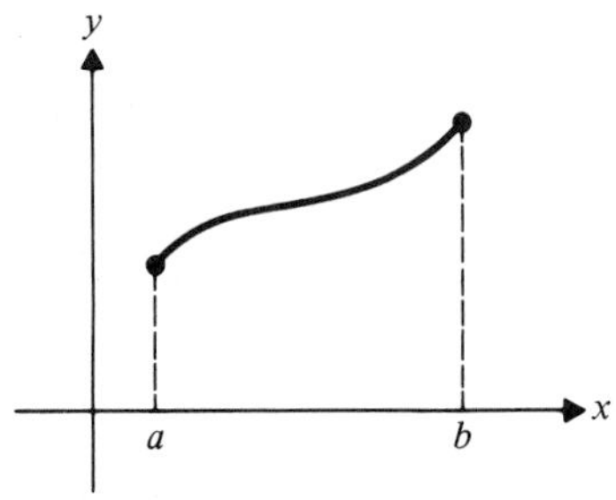

Figure 4.45

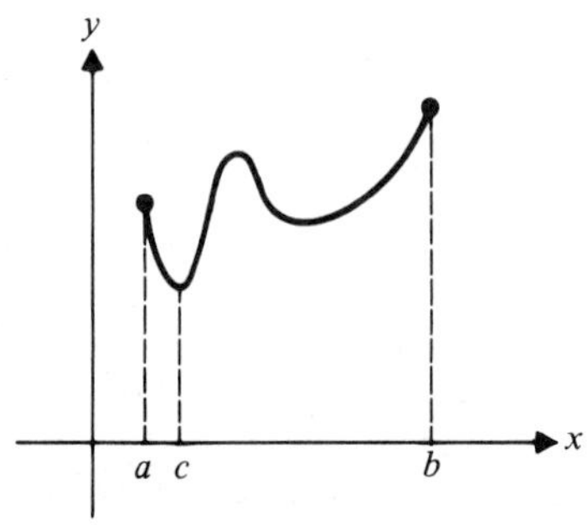

Figure 4.46

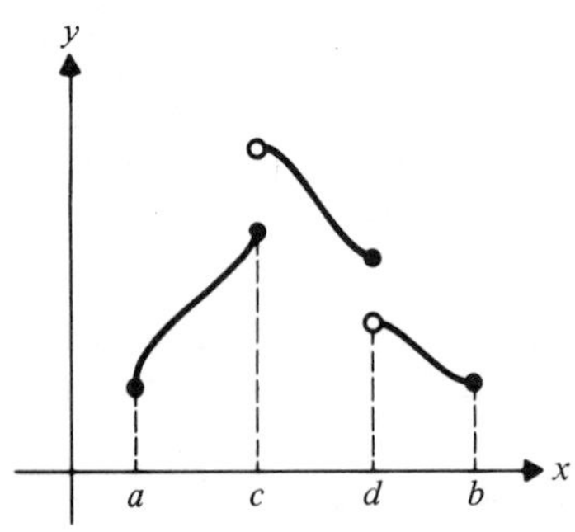

Figure 4.47

For the function shown in Fig. 4.44, the absolute maximum of $f(x)$ on the interval $a \le x \le b$ is $f(c)$, and the absolute minimum is $f(d)$. For the function in Fig. 4.45, the absolute maximum on $a \le x \le b$ is $f(b)$, and the absolute minimum is $f(a)$. For the function in Fig. 4.46, the absolute maximum on $a \le x \le b$ is $f(b)$, and the absolute minimum is $f(c)$. Note that we speak of absolute maxima and minima (absolute extrema) of a function only on some specified interval; i.e., we do not ask such questions as "What are the absolute maximum and minimum of the function $f(x)$?"

Each of the functions in Figs. 4.44–4.46 is continuous on a closed interval $a \le x \le b$. The following theorem from analysis, which we will not prove here,† asserts that every such function has absolute extrema.

▶ **Theorem 4.2** A function that is continuous on a closed interval must attain an absolute maximum and an absolute minimum on that interval. ◀

The conditions of this theorem are sufficient to guarantee existence of absolute extrema; that is, if a function is continuous on a closed interval, then it *must* have absolute extrema on that interval. If these conditions are not met, the function may or may not have absolute extrema. For instance, if the function shown in Fig. 4.44 is confined to the open interval $a < x < b$, it still attains its absolute extrema at $x = c$ and $x = d$. On the other hand, the function in Fig. 4.45 does not have an absolute maximum or an absolute minimum on the open interval $a < x < b$. The function in Fig. 4.47 is not continuous on $a \le x \le d$, and it has no absolute maximum on this interval. Nor is it continuous on $d \le x \le b$, but it does have absolute extrema on this interval.

Absolute extrema for the functions in Figs. 4.44–4.46 always occur either at a critical point of the function or at an end of the interval $a \le x \le b$. This result is, in fact, always true for continuous functions.

▶ **Theorem 4.3** A function that is continuous on a closed interval always attains its absolute extrema either at critical points or at the ends of the interval. ◀

Proof Let $f(x)$ be a continuous function on a closed interval $a \le x \le b$. To verify the theorem, we show that $f(x)$ cannot have an absolute extremum at any point x_0 in $a < x < b$ that is not critical. If $a < x_0 < b$ and x_0 is not critical,

† For a proof of this result, the interested reader should consult more advanced books.

x_0 cannot yield a relative extremum for $f(x)$ (see Theorem 4.1). Definitions 4.2 and 4.3 now imply that in every open interval around x_0, $f(x)$ takes on values that are larger than $f(x_0)$ and values that are smaller than $f(x_0)$. But then $f(x_0)$ cannot possibly be an absolute extremum for $f(x)$ on $a \leq x \leq b$. □

Because of this theorem, there is a simple procedure to determine the absolute extrema of a *continuous* function $f(x)$ on a *closed* interval $a \leq x \leq b$. We evaluate $f(x)$ at all critical points in $a \leq x \leq b$ and at $x = a$ and $x = b$; then the absolute maximum and absolute minimum of $f(x)$ on $a \leq x \leq b$ are the largest and smallest of these numbers. Note that it is not necessary to classify the critical points of $f(x)$ as yielding relative maxima, relative minima, horizontal points of inflection, or corners. We need only evaluate $f(x)$ at $x = a$, $x = b$, and its critical points. The largest and smallest of these numbers *must* be the absolute extrema.

We now show how absolute extrema can be used to solve applied problems.

Example 17 A rectangular field is to be fenced on three sides with 1000 meters of fencing (the fourth side being a straight river's edge). Find the dimensions of the field in order that the area be as large as possible.

Solution Since the area of the field is to be maximized, we first define a function representing this area. The area of a field of width w and length l (Fig. 4.48) is given by

$$A = lw.$$

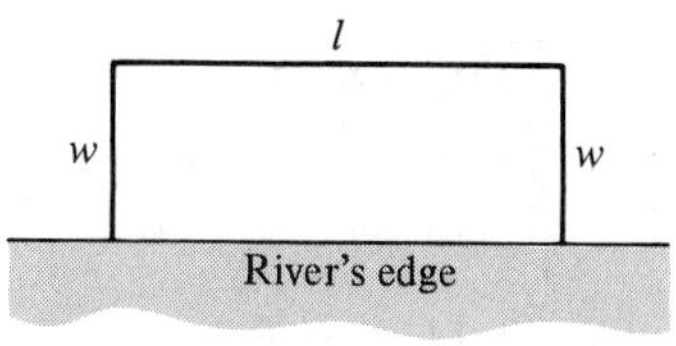

Figure 4.48

This function represents the area of a field with arbitrary length l and arbitrary width w. Our problem states that only 1000 meters of fencing are available for the three sides; therefore, l and w must satisfy the equation

$$2w + l = 1000.$$

With this equation we can express A completely in terms of w (or l):

$$A(w) = w(1000 - 2w) = 1000w - 2w^2.$$

To maximize the area of the field, we must therefore maximize the function $A(w)$. But what are the values of w under consideration? Clearly, w cannot be negative, and in order to satisfy the restriction $2w + l = 1000$, w cannot exceed 500. The physical problem has now been modeled mathematically: We must find the absolute maximum of the (continuous) function $A(w)$ on the interval $0 \leq w \leq 500$.

We first find the critical points of $A(w)$ by solving

$$0 = A'(w) = 1000 - 4w;$$

the only solution is $w = 250$. Since $A'(w)$ exists for all $0 \leq w \leq 500$, we evaluate $A(w)$ at $w = 250$ and the end points:

$$A(0) = 0,$$

$$A(500) = 0,$$

$$A(250) = 1000(250) - 2(250)^2 > 0.$$

We conclude that the largest possible area is obtained when the width of the field is 250 meters and its length is 500 meters. □

Example 18 When a projectile (such as an artillery shell) is fired up an inclined plane (Fig. 4.49), the distance that it travels before striking the plane is called its range R. It can be shown that

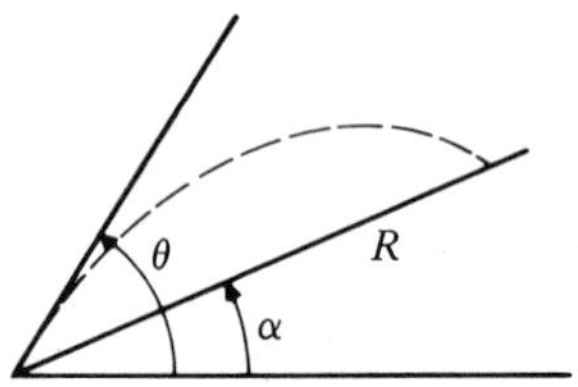

Figure 4.49

$$R(\theta) = \frac{2v^2 \cos\theta \sin(\theta - \alpha)}{g \cos^2 \alpha},$$

where v is the initial speed of the projectile, $g > 0$ is a constant, α is the angle of inclination of the plane with the horizontal, and θ is the angle of projection with the horizontal. Find the angle θ for which R is a maximum.

Solution It is evident from Fig. 4.49 that maximum R will occur for θ somewhere in the interval $\alpha \le \theta \le \pi/2$. For critical points of R we solve

$$0 = R'(\theta) = \frac{2v^2}{g \cos^2 \alpha}\{-\sin\theta \sin(\theta - \alpha) + \cos\theta \cos(\theta - \alpha)\}$$

$$= \frac{2v^2}{g \cos^2 \alpha} \cos(2\theta - \alpha).$$

Consequently, $\cos(2\theta - \alpha) = 0$, from which we get

$$2\theta - \alpha = \frac{\pi}{2} + n\pi, \quad n \text{ an integer},$$

or

$$\theta = \frac{\pi}{4} + \frac{\alpha}{2} + \frac{n\pi}{2}.$$

Only one of these values of θ lies in the specified interval, namely, $\theta = \pi/4 + \alpha/2$. Since $R(\alpha) = 0 = R(\pi/2)$ and $R(\pi/4 + \alpha/2) > 0$, it follows that maximum range will be obtained for $\theta = \pi/4 + \alpha/2$. □

Example 19 A printer contracts to print 200,000 copies of a membership card. It costs \$10 per hour to run the press, and the press produces 1000 impressions per hour. The printer is at liberty to choose the number of set types per impression (to a limit of 40), each set type costing \$2. If x set types are chosen, then each impression yields x cards. How many set types should be used?

Solution The printer will choose that number of set types which will minimize production costs. If x set types are used, then the cost for the set types themselves is $2x$ dollars. With this number of set types, the press prints $1000x$ cards per hour. In order to produce 200,000 cards, the press must therefore run for 200,000/($1000x$) hours at a cost of [10][200,000/($1000x$)] dollars. The total cost of producing the cards when x set types are used is therefore given by

$$C(x) = 2x + \frac{2000}{x}.$$

This function must be minimized on the interval $1 \le x \le 40$, since at least one set type must be used, and only a maximum of 40 are possible.

For critical points of $C(x)$, we solve

$$0 = C'(x) = 2 - \frac{2000}{x^2}.$$

The only solution of this equation in the interval $1 \le x \le 40$ is $x = 10\sqrt{10} = 31.6$. Since the printer cannot use part of a set type, we calculate

$$C(1) = 2002, \qquad C(31) = 126.52, \qquad C(32) = 126.50, \qquad C(40) = 130.$$

The printer should therefore use 32 set types. □

Example 20 a) A lighthouse is 6 kilometers off shore and a cabin on the straight shoreline is 9 kilometers from the point on the shore nearest the lighthouse. If a man rows at a rate of 3 kilometers per hour, and walks at a rate of 5 kilometers per hour, where should he beach his boat in order to get from the lighthouse to the cabin as quickly as possible?

b) Repeat the problem given that the cabin is 4 kilometers along the shoreline.

Solution a) If t_1 and t_2 are rowing and walking times, respectively, when the boat is beached at position x (Fig. 4.50), then the total travel time is given by

$$t(x) = t_1 + t_2 = \frac{\sqrt{x^2 + 36}}{3} + \frac{9 - x}{5}, \quad 0 \le x \le 9.$$

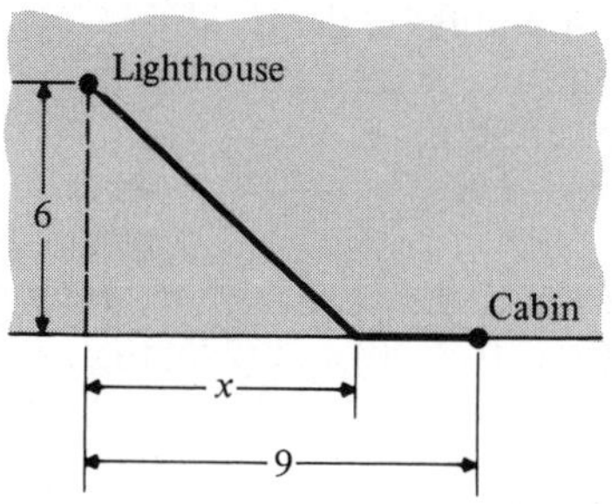

Figure 4.50

For critical points of $t(x)$ we solve

$$0 = t'(x) = \frac{x}{3\sqrt{x^2 + 36}} - \frac{1}{5},$$

or

$$5x = 3\sqrt{x^2 + 36}.$$

Thus,

$$25x^2 = 9(x^2 + 36),$$

from which we accept only the positive solution $x = \frac{9}{2}$. Since $t'(x)$ exists for all x, and

$$t(0) = \frac{\sqrt{36}}{3} + \frac{9}{5} = \frac{19}{5},$$

$$t(9) = \frac{\sqrt{81 + 36}}{3} = \sqrt{13},$$

$$t\left(\frac{9}{2}\right) = \frac{\sqrt{\frac{81}{4} + 36}}{3} + \frac{9 - \frac{9}{2}}{5} = \frac{17}{5} < \sqrt{13},$$

the boat should therefore be beached 4.5 kilometers from the cabin.

b) If the cabin is 4 kilometers along the shoreline, then

$$t(x) = \frac{\sqrt{x^2 + 36}}{3} + \frac{4 - x}{5}, \qquad 0 \le x \le 4.$$

Critical points are again given by the equation

$$\frac{x}{3\sqrt{x^2 + 36}} - \frac{1}{5} = 0,$$

but the solution $x = \frac{9}{2}$ must be rejected since it does not fall in the interval $0 \le x \le 4$. Travel time must therefore be minimized by heading directly to shore or directly to the cabin. Since

$$t(0) = \frac{\sqrt{36}}{3} + \frac{4}{5} = \frac{14}{5},$$

$$t(4) = \frac{\sqrt{16 + 36}}{3} = \frac{2\sqrt{13}}{3} < \frac{14}{5},$$

the boat should head directly toward the cabin. □

Example 21 A wall of a building is to be braced by a beam that must pass over a parallel wall 5 meters high and 2 meters from the building (Fig. 4.51). Find the length of the shortest beam that can be used.

Solution If we introduce coordinates as shown in Fig. 4.51, we see that the shortest beam is the shortest of all line segments that stretch from points on the positive x-axis to points on the positive y-axis and pass through the point (2, 5). The length of the line segment that has x- and y-intercepts of x and y is given by

$$D = \sqrt{x^2 + y^2}.$$

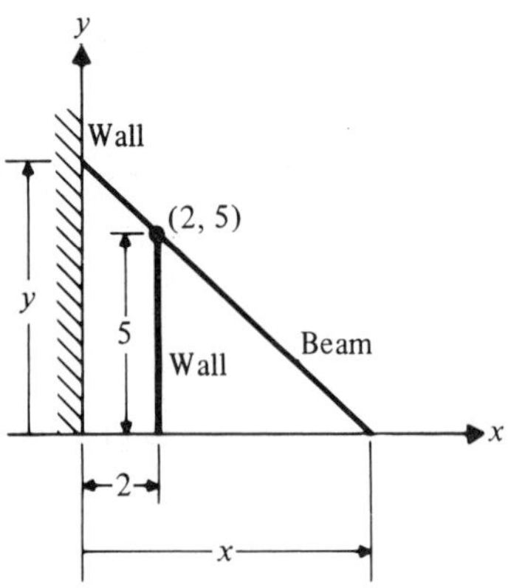

Figure 4.51

To express D in terms of only one variable, we must find an equation relating x and y. Since slopes of the line segments joining $(0, y)$ to $(x, 0)$ and $(2, 5)$ to $(x, 0)$ must be identical, we have

$$\frac{y}{-x} = \frac{-5}{x - 2}$$

or

$$y = \frac{5x}{x - 2}.$$

Thus,

$$D = \sqrt{x^2 + \frac{25x^2}{(x - 2)^2}}.$$

Since D is a minimum if and only if D^2 is a minimum, we consider minimizing

$$D^2 = f(x) = x^2 + \frac{25x^2}{(x - 2)^2}, \qquad x > 2.$$

For critical points of $f(x)$ we solve

$$0 = f'(x) = 2x + \frac{50x}{(x - 2)^2} - \frac{50x^2}{(x - 2)^3}$$

$$= \frac{2x[(x - 2)^3 + 25(x - 2) - 25x]}{(x - 2)^3}$$

$$= \frac{2x}{(x - 2)^3}[(x - 2)^3 - 50].$$

Since x cannot be equal to zero, we set

$$(x - 2)^3 = 50,$$

and this implies that $x = 2 + 50^{1/3}$. In this problem D^2 is defined on the open interval $x > 2$ rather than on a closed interval. In this case we replace evaluation of D^2 at the ends of the interval with limits as x approaches the ends of the interval; i.e., we evaluate

$$f(2 + 50^{1/3}) > 0,$$

$$\lim_{x \to 2^+} f(x) = \infty, \qquad \lim_{x \to \infty} f(x) = \infty.$$

We conclude that D^2, and therefore D, must be minimized for $x = 2 + 50^{1/3}$, and the shortest beam has length

$$\sqrt{f(2 + 50^{1/3})} = (2 + 50^{1/3})\sqrt{1 + \frac{25}{50^{2/3}}} = 9.6 \text{ meters.} \ \square$$

EXERCISES 4.7

In each of Exercises 1–5, find absolute extrema for the function on the interval.

1. $f(x) = x^3 - x^2 - 5x + 4, \ -2 \le x \le 3$
2. $f(x) = (x - 4)/(x + 1), \ 0 \le x \le 10$
3. $f(x) = x + 1/x, \ \frac{1}{2} \le x \le 5$
4. $f(x) = x - 2 \sin x, \ 0 \le x \le 4\pi$
5. $f(x) = x\sqrt{x + 1}, \ -1 \le x \le 1$
6. An agronomist wishes to fence eight rectangular plots for experimentation as shown in Fig. 4.52. If each plot must contain 9000 square meters, find the minimum amount of fencing that can be used.

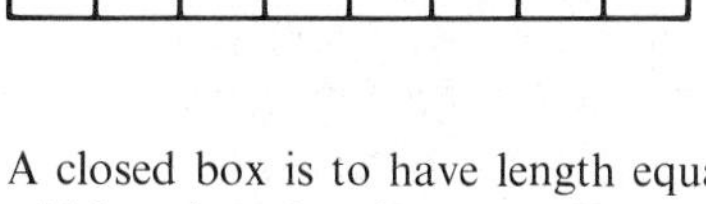

Figure 4.52

7. A closed box is to have length equal to three times its width and total surface area of 30 square meters. Find the dimensions that produce the maximum volume.
8. One end of a uniform cantilever beam of length L is built into a wall, and the other end is simply supported (Fig. 4.53). If the beam has constant mass per unit length m, its deflection y from the horizontal at a distance x from the built-in end is given by

$$(48EI)y = -mg(2x^4 - 5Lx^3 + 3L^2x^2),$$

where E and I are constants depending on the material and cross-section of the beam, and $g > 0$ is a constant. How far from the built-in end does maximum deflection occur?

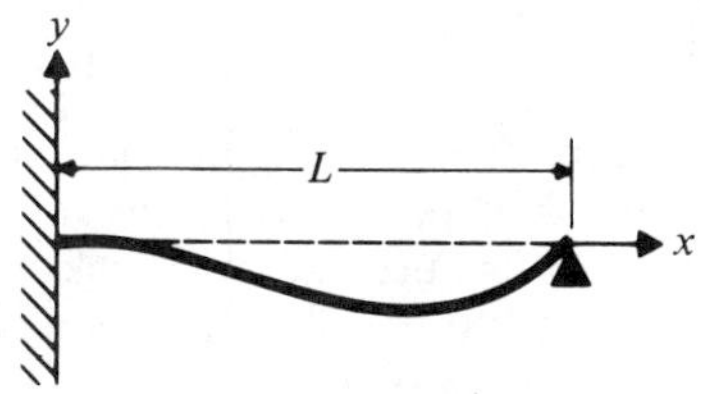

Figure 4.53

9. Soda cans are made in the shape of right-circular cylinders. Find the dimensions of the can that minimize surface area (and therefore total cost if unit material costs for top, bottom, and sides are all the same), given that the can must hold 300 milliliters.
10. A high school football field is being designed to accommodate a 400-meter track. The track runs along the length of the field and has semicircles beyond the end zones with diameters equal to the width of the field. If the end zones must be 10 meters deep, find the dimensions of the field that maximize playing area (not including the end zones).
11. A cherry orchard has 255 trees, each of which produces on the average 25 baskets of cherries. For each additional tree planted, the yield per tree decreases by one-twelfth of a basket. How many more trees will produce a maximum crop?
12. When a manufacturing company sells x objects per month, it sets the price r of each object at

$$r(x) = 100 - \frac{x^2}{10{,}000}.$$

The total cost C of producing the x objects per month is given by

$$C(x) = \frac{x^2}{10} + 2x + 20.$$

Find the number of objects the company should sell per month in order to realize maximum profits.

13. Find the point on the parabola $y = x^2$ closest to $(-2, 5)$.

14. A light source is to be placed directly above the center of a circular area of radius r (Fig. 4.54). The illumination at any point on the edge of the circle is directly proportional to the cosine of the angle θ and inversely proportional to the square of the distance d from the source. Find the height h above the circle at which illumination on the edge of the table is maximized.

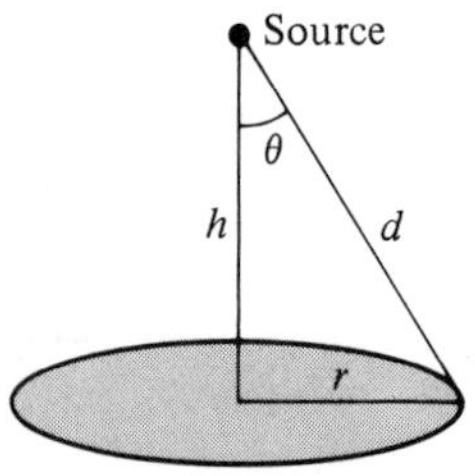

Figure 4.54

15. Among all line segments that stretch from points on the positive x-axis to points on the positive y-axis and pass through the point (2, 5), find that one which makes with the positive x- and y-axes the triangle with least possible area.

16. A long piece of metal one meter wide is to be bent in two places to form a spillway so that its cross-section is an isosceles trapezoid (Fig. 4.55). Find the angle θ at which the bends should be formed in order to obtain maximum possible flow along the spillway if the lengths of AB, BC, and CD are all one-third meter.

Figure 4.55

17. The strength of a rectangular beam varies as the product of its width and the square of its depth. Find the dimensions of the strongest rectangular beam that can be cut from a circular log of radius 25 centimeters.

18. At noon a ship S_1 is 20 kilometers north of ship S_2. If S_1 sails south at 6 kilometers per hour, and S_2 east at 8 kilometers per hour, find when the two ships are closest together.

19. A military courier is located on a desert 6 kilometers from a point P, which is the point on a long straight road nearest him. He is ordered to report to a point Q on the road. If we assume that he can travel at a rate of 14 kilometers per hour on the desert and a rate of 50 kilometers per hour on the road, find the point where he should reach the road in order to get to Q in the least possible time when:

a) Q is 3 kilometers from P;

b) Q is 1 kilometer from P.

20. Among all rectangles that can be inscribed inside the curve $b^2x^2 + a^2y^2 = a^2b^2$ and have sides parallel to the axes (Fig. 4.56), find the one with largest area.

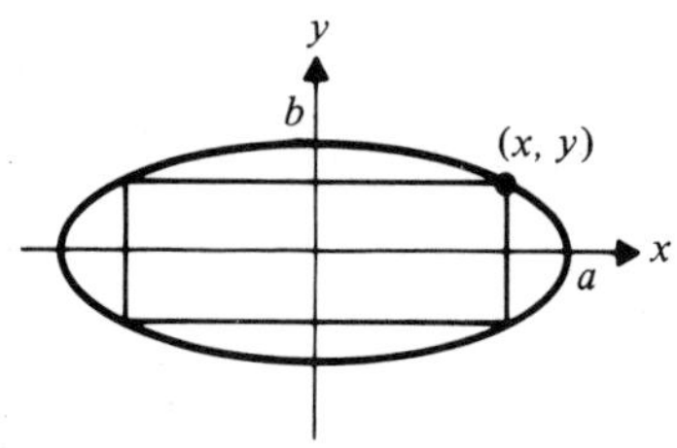

Figure 4.56

21. A rental company buys a new machine for p dollars, which it then rents to various customers. If the company keeps the machine for t years (before replacing it), the average replacement cost per year for the t years is given by p/t. During these t years, the rental company must make repairs on the machine, the number n depending on t as given by $n(t) = t^\alpha/\beta$, where $\alpha > 1$ and $\beta > 0$ are constants. If r is the average cost per repair, then the average maintenance cost per year over the life t of the machine is given by nr/t. The total yearly expense associated with the machine if it is kept for t years is therefore

$$C(t) = \frac{p}{t} + \frac{nr}{t}.$$

Find the optimum time at which to replace the machine.

22. Discuss Exercise 21 given that $\alpha = 1$ and $0 < \alpha < 1$.

23. The illumination at a point is inversely proportional to the square of the distance from the light source and directly proportional to the intensity of the source. If two light sources of intensities I_1 and I_2 are a distance d apart, at what point on the line segment joining the sources is the sum of their illuminations a minimum (relative to all other points on the line segment)?

24. A window is in the form of a rectangle surmounted by a semicircle with diameter equal to the width of the window. If the rectangle is of clear glass while the semicircle is of colored glass that transmits only half as much light per unit area as the clear glass, and if the total perimeter is fixed, find the proportions of the window that admit the most light.

25. A company wishes to construct a storage tank in the form of a rectangular parallelepiped with a square horizontal cross-section. The volume of the tank must be 86.4 cubic meters.

a) If material for the sides and top costs \$1.25 per square meter, and material for the bottom costs \$4.75 per square meter, find the dimensions that minimize material costs.

b) Repeat (a) if given that the 12 edges must be welded at a cost of \$7.50 per meter of weld.

26. An underground pipeline is to be constructed between two cities A and B (Fig. 4.57). Analysis of the substructure indicates that construction costs per kilometer in regions I ($y > 0$) and II ($y < 0$) are c_1 and c_2, respectively. Show that the total construction cost is minimized when x is chosen so that

$$c_1 \sin \theta_1 = c_2 \sin \theta_2 .$$

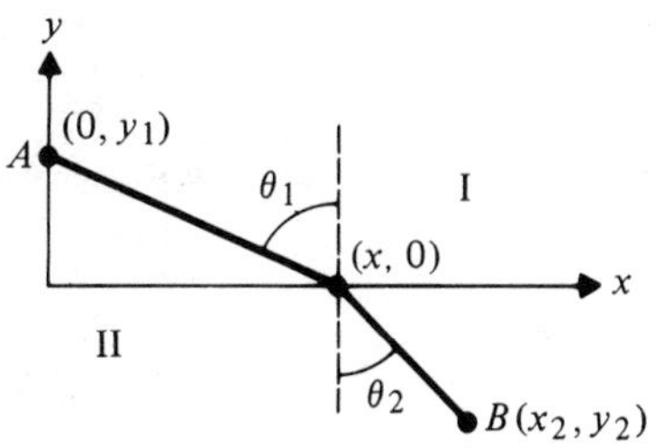

Figure 4.57

27. When an unloaded die is thrown, there is a probability of $\frac{1}{6}$ that it will come up "two." If the die is loaded, on the other hand, the probability that a "two" will appear is not $\frac{1}{6}$, but is some number p between zero and one, $0 \le p \le 1$. To find p we could roll the die a large number of times, say n, and count the number of times "two" appears, say m. It seems reasonable that an estimate for p is m/n. The mathematician defines a likelihood function, which for the present situation turns out to be

$$f(x) = \frac{n!}{m!(n-m)!} x^m (1-x)^{n-m}.$$

The value of x that maximizes $f(x)$ on the interval $0 \le x \le 1$ is called the maximum likelihood estimate of p. Show that this estimate is m/n.

28. The cost of fuel per hour for running a ship varies directly as the cube of the speed, and is \$$B$ per hour when the speed is b kilometers per hour. There are also fixed costs of \$$A$ per hour. Find the most economical speed at which to make a trip.

29. A packing company wishes to form the 1-meter by 2-meter piece of cardboard shown in Fig. 4.58(a) into a box, as shown in Fig. 4.58(b). Cuts are to be made along solid lines and folds along dotted lines, and two sides are to be taped together as shown. If the outer flaps (on top and bottom) must meet in the center but the inner flaps need not, find the dimensions of the box holding the most volume. How far apart will the inner flaps be?

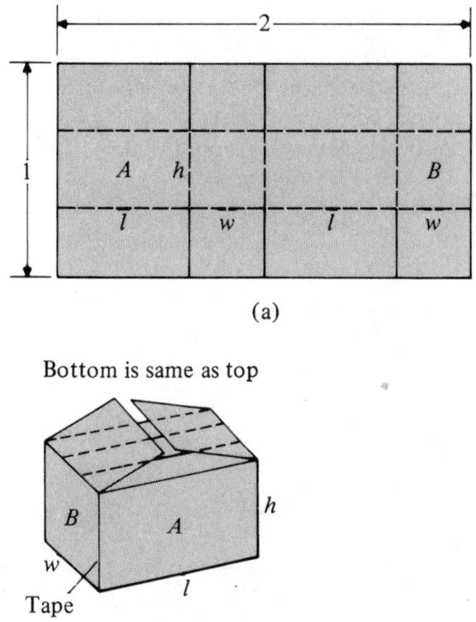

Figure 4.58

30. A paper drinking cup in the form of a right-circular cone can be made from a circular piece of paper by removing a sector and joining edges OA and OB, as shown in Fig. 4.59. If the radius of the circle is R, what choice of θ yields a cup of maximum volume?

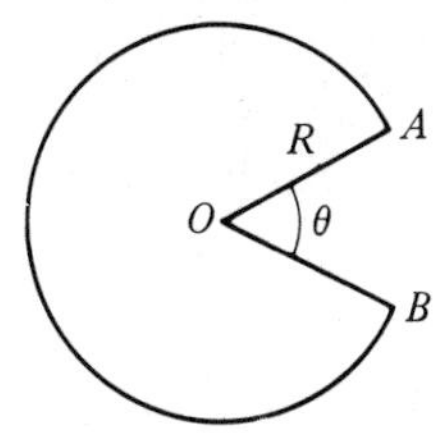

Figure 4.59

* **31.** A truck driver estimates that the cost per kilometer of operating his own rig is $40 + (50 - v)^2/10$ cents, where v is his average speed in kilometers per hour. In order to cover this cost he charges \$20 per hour for rental of his truck. For his driving he charges \$10 per hour. For a 1000-kilometer nonstop haul, what average speed would you recommend to the driver?

* **32.** Prove that the minimum distance from a point (x_1, y_1) to a line $Ax + By + C = 0$ is

$$\frac{|Ax_1 + By_1 + C|}{\sqrt{A^2 + B^2}}.$$

* **33.** A window is in the form of a rectangle surmounted by a semicircle with diameter equal to the width of the rectangle. The rectangle is of clear glass costing \$$a$ per unit area, while the semicircle is of colored glass costing \$$b$ per unit area. The colored glass transmits only a fraction p

($0 < p < 1$) as much light per unit area as the clear glass. In addition, the curved portion of the window is surmounted by a special frame at a cost of \$$c$ per unit length. If the total cost of the window must not exceed \$$A$, find the dimensions of the window that admit the most light.

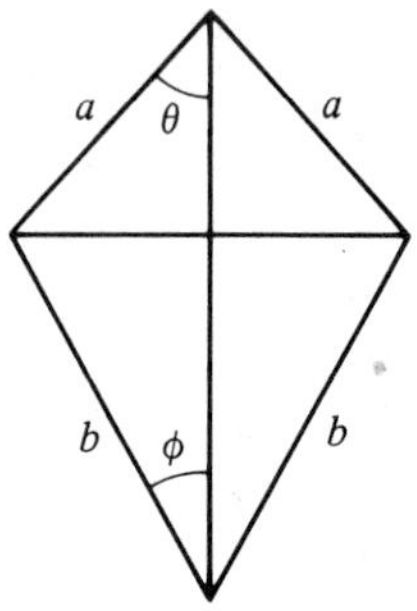

Figure 4.60

*** 34.** A kite is to be made from six pieces of wood (and some plastic) as shown in Fig. 4.60. The four outside pieces have predetermined lengths a and b ($b > a$). Yet to be cut are the two diagonal pieces. How long should they be in order to make the area of the kite as large as possible?

*** 35.** There are n red stakes securely driven into the ground in a straight line. A blue stake is to be added to the line, and each red stake is to be joined to the blue one by a rope. Where should the blue stake be placed in order that the total length of all ropes be as small as possible? (*Hint:* Consider first the cases $n = 2$, 3, and 4.)

4.8 NEWTON'S ITERATIVE PROCEDURE FOR SOLVING EQUATIONS

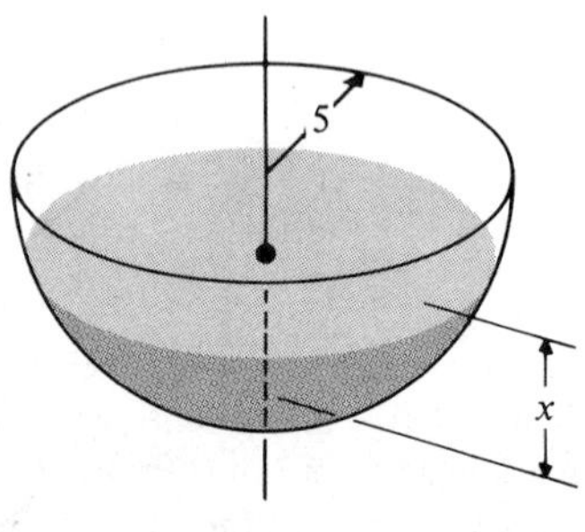

Figure 4.61

In many of the examples and exercises of Chapters 1–4, it has been necessary to solve equations, but most of these equations have had fairly obvious solutions. In practical problems, this does not happen; the equation to be solved generally has a noninteger solution that can be at best approximated. For example, consider finding the depth of water that will half fill (by volume) the hemispherical tank shown in Fig. 4.61. It can be shown (see Exercise 15 in Section 7.4) that if the water is x meters deep, the volume of water in the tank is given by

$$\frac{\pi}{3}(15x^2 - x^3).$$

Since the volume of one-half the tank is one-quarter that of a sphere of radius 5, it follows that the tank will be half full when x satisfies the equation

$$\frac{1}{4}\left\{\frac{4}{3}\pi(5)^3\right\} = \frac{\pi}{3}(15x^2 - x^3),$$

and this equation simplifies to

$$x^3 - 15x^2 + 125 = 0.$$

The question now is, "How do we solve this equation when we know that the solution we want is somewhere around $x = 3$?" One of many possible methods is called *Newton's iterative procedure*.

If we let $f(x) = x^3 - 15x^2 + 125$, then we can produce a reasonably accurate sketch of $f(x)$ between $x = 3$ and $x = 3.5$ as shown in Fig. 4.62. If we set $x_1 = 3$, then x_1 is an approximation to the solution of the equation—not a good approximation, but an approximation nonetheless. Suppose we draw the tangent line to $y = f(x)$ at $(x_1, f(x_1))$. If x_2 is the point of intersection

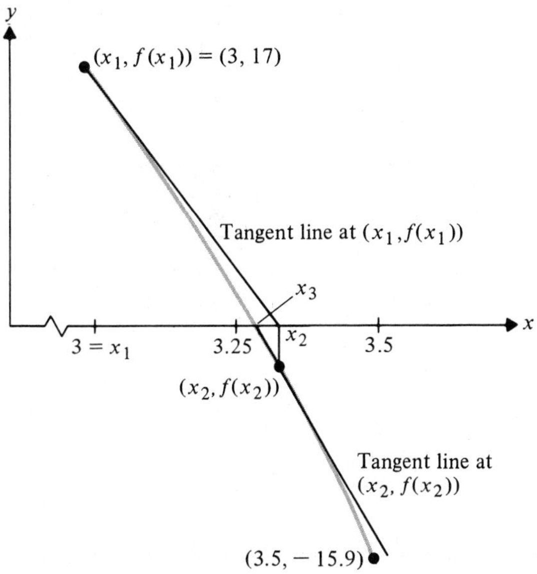

Figure 4.62

of this tangent line with the x-axis, it is clear that x_2 is a better approximation to the solution of the equation than x_1 is. If we now draw the tangent line to $y = f(x)$ at $(x_2, f(x_2))$, its intersection point x_3 with the x-axis is an even better approximation. Continuation of this process leads to a succession of numbers $x_1, x_2, x_3, \ldots$, each of which is closer to the solution of the equation $f(x) = 0$ than the preceding numbers. This procedure for finding a better and better approximation to the solution of an equation is called Newton's iterative procedure. We say that the numbers $x_1, x_2, x_3, \ldots$ converge to the root of the equation.

What we need now is an algebraic formula by which to calculate the numbers $x_1, x_2, x_3, \ldots$. If we denote the nth number by x_n, then the equation of the tangent line to $y = f(x)$ at $(x_n, f(x_n))$ is

$$y - f(x_n) = f'(x_n)(x - x_n).$$

To find the point of intersection of this tangent line with the x-axis, we set $y = 0$ and solve for x; that is,

$$-f(x_n) = f'(x_n)(x - x_n),$$

from which we get

$$x = x_n - \frac{f(x_n)}{f'(x_n)}.$$

But the point of intersection of the tangent line at $(x_n, f(x_n))$ with the x-axis is the $(n + 1)$st approximation denoted by x_{n+1}:

$$x_{n+1} = x_n - \frac{f(x_n)}{f'(x_n)}. \tag{4.15}$$

In other words, we now have an algebraic formula that defines each approximation in terms of its predecessor.

For our example, $f(x) = x^3 - 15x^2 + 125$, in which case $f'(x) = 3x^2 - 30x$ and

$$x_{n+1} = x_n - \frac{x_n^3 - 15x_n^2 + 125}{3x_n^2 - 30x_n}.$$

Calculation of the first five approximations gives

$$x_2 = x_1 - \frac{x_1^3 - 15x_1^2 + 125}{3x_1^2 - 30x_1} = 3 - \frac{3^3 - 15(3)^2 + 125}{3(3)^2 - 30(3)} = 3.26984;$$

$$x_3 = x_2 - \frac{x_2^3 - 15x_2^2 + 125}{3x_2^2 - 30x_2} = 3.26352;$$

$$x_4 = x_3 - \frac{x_3^3 - 15x_3^2 + 125}{3x_3^2 - 30x_3} = 3.2635182;$$

$$x_5 = x_4 - \frac{x_4^3 - 15x_4^2 + 125}{3x_4^2 - 30x_4} = 3.2635182.$$

Newton's iterative procedure has therefore produced 3.263518 as an approximate solution to the equation $x^3 - 15x^2 + 125 = 0$. In spite of the fact that we have written six decimals in this final answer, our analysis in no way guarantees this degree of accuracy; we have simply judged on the basis that $x_4 = x_5$ that 3.263518 might be accurate to six decimals. Discussions on the accuracy of the nth iteration of Newton's iterative procedure can be found in advanced books on numerical analysis.

Example 22 Use Newton's iterative procedure to find the only positive root of the equation

$$3x^4 + 15x^3 - 125x - 1500 = 0.$$

(This equation defines critical points for the cost function given in Exercise 25b in Section 4.7 if the volume of the tank is 100 cubic meters rather than 86.4 cubic meters.)

Solution If we set $f(x) = 3x^4 + 15x^3 - 125x - 1500$, then a quick calculation shows that $f(4) = -272$ and $f(5) = 1625$. Thus the required solution is between $x = 4$ and $x = 5$. If we set $x_1 = 4$, then the nth approximation predicted by Newton's iterative procedure is defined by

$$x_{n+1} = x_n - \frac{f(x_n)}{f'(x_n)} = x_n - \frac{3x_n^4 + 15x_n^3 - 125x_n - 1500}{12x_n^3 + 45x_n^2 - 125}.$$

We find that

$$x_2 = 4.19956, \qquad x_3 = 4.187268, \qquad x_4 = 4.1872187, \qquad x_5 = 4.1872187.$$

We estimate then that to six decimals the required solution is 4.187219. □

Example 23 Use Newton's iterative procedure to find the smallest root of the cubic equation

$$x^3 - 3x + 1 = 0.$$

Solution To find an estimate of the smallest root of this equation, we sketch a quick graph of the function $f(x) = x^3 - 3x + 1$. Critical points are defined by $0 = f'(x) = 3x^2 - 3 = 3(x^2 - 1)$. We find a relative maximum of $f(-1) = 3$ and a relative minimum of $f(1) = -1$. These extrema, together with the values $f(0) = 1$, $f(-2) = -1$, and $f(2) = 3$, give us the sketch shown in Fig. 4.63. Clearly, then,

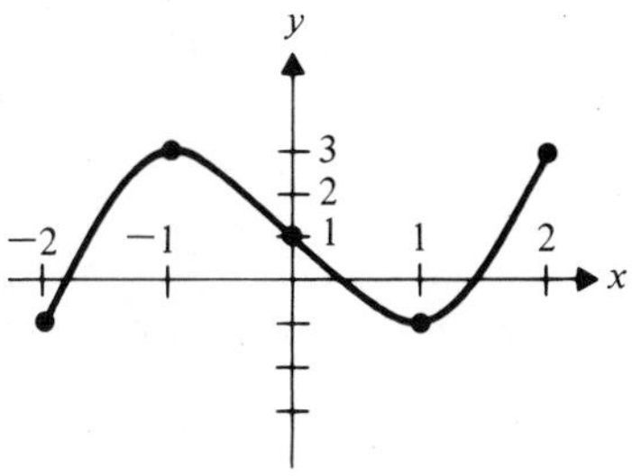

Figure 4.63

the smallest root of $f(x) = 0$ is between $x = -2$ and $x = -1$. We take $x_1 = -2$ and use Newton's iterative procedure to define the nth approximation

$$x_{n+1} = x_n - \frac{f(x_n)}{f'(x_n)}$$
$$= x_n - \frac{x_n^3 - 3x_n + 1}{3x_n^2 - 3}.$$

From this formula, we get

$$x_2 = -1.88889, \qquad x_3 = -1.87945,$$
$$x_4 = -1.879385, \qquad x_5 = -1.8793852.$$

We estimate then that the smallest root, to six decimals, is -1.879385. □

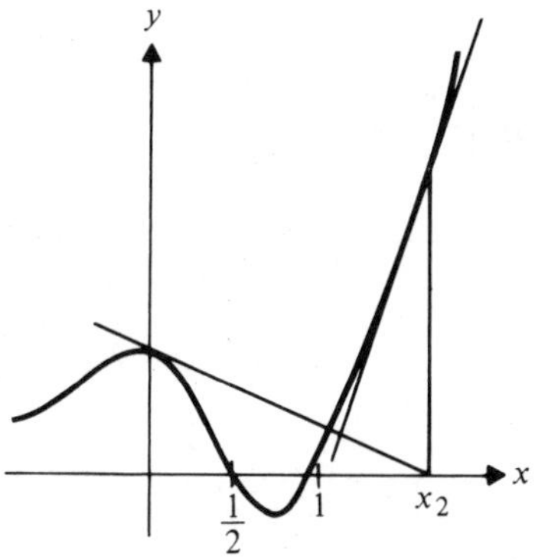

Figure 4.64

The numbers produced by Newton's iterative procedure do not always converge, and sometimes even when they do so, they do not converge to what was expected. For instance, were we to attempt to approximate the largest root in Example 23, and inadvertently chose $x_1 = 1$, we could not define the second approximation: algebraically, because $3x_1^2 - 3 = 0$, and geometrically, because at $x_1 = 1$, the tangent line is horizontal.

The function $f(x)$ in Fig. 4.64 has zeros near $x = \frac{1}{2}$ and $x = 1$. Were we to attempt to find the smaller zero using an initial approximation of $x_1 = 0$, we would find the second approximation x_2 to be larger than 1. Further iterations would then converge to the zero near $x = 1$, not the zero near $x = \frac{1}{2}$.

We conclude, therefore, that the initial approximation in Newton's iterative procedure is most important. A poor choice of x_1 may lead to numbers that either converge to the wrong root or do not converge at all. Obviously, then, we must be judicious in our choice of an initial approximation.

EXERCISES 4.8

In Exercises 1–8, use Newton's iterative procedure to find approximations to all roots of the given equation. In each case, we suggest making a sketch in order to obtain an initial approximation to each root.

C **1.** $x^2 + 3x + 1 = 0$

C **2.** $x^2 - x - 4 = 0$

C **3.** $x^3 + x - 3 = 0$

C **4.** $x^3 - x^2 + x - 22 = 0$

C **5.** $x^3 - 5x^2 - x + 4 = 0$

C **6.** $x^5 + x - 1 = 0$

C **7.** $x^4 + 3x^2 - 7 = 0$

C **8.** $\dfrac{x+1}{x-2} = x^2 + 1$

C **9.** By sketching the curves $y = x$ and $y = \cos x$, show graphically that the equation $x = \cos x$ has only one solution. Find it.

In Exercises 10–13, find all points of intersection for the given curves.

C **10.** $y = x^3$,
$y = x + 5$

C **11.** $y = (x + 1)^2$,
$y = x^3 - 4x$

C **12.** $y = x^4 - 20$,
$y = x^3 - 2x^2$

C **13.** $y = \dfrac{x}{x+1}$,
$y = x^2 + 2$

C **14.** Find all critical points for the function $f(x) = x^4 - 10x^2 - 4x + 5$.

4.9 DIFFERENTIALS

If $y = f(x)$, then the increment Δy in y corresponding to an increment Δx in x is defined by

$$\Delta y = f(x + \Delta x) - f(x). \tag{4.16}$$

In other words, Δy is the change in y resulting from the change Δx in x. For the function shown in Figs. 4.65(a) and (b), Δy is positive when Δx is positive, and Δy is negative when Δx is negative.

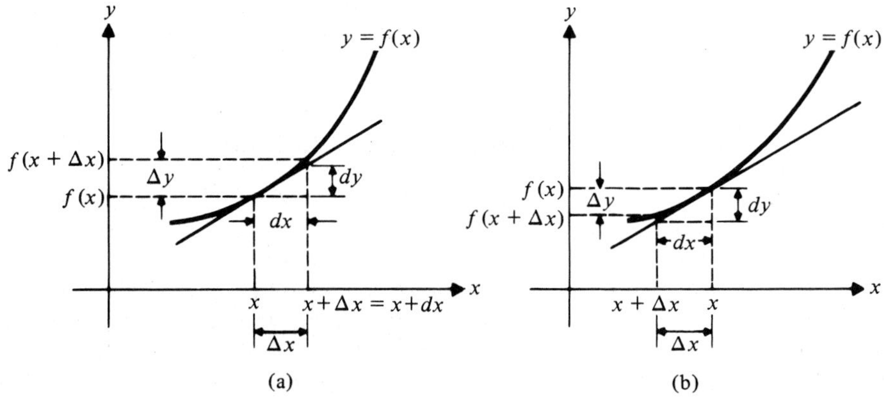

Figure 4.65

The increment Δx is given a new notation and name in the following definition.

▷ **Definition 4.8** An increment Δx in the independent variable x is denoted by

$$dx = \Delta x, \tag{4.17}$$

and when written as dx is called the *differential* of x. ◁

For the differential of the dependent variable we have the following.

▷ **Definition 4.9** The differential of $y = f(x)$, corresponding to the differential dx in x, is denoted by dy and is defined by

$$dy = f'(x)\,dx. \quad ◁ \tag{4.18}$$

When we draw a tangent line to $y = f(x)$ at the point (x, y), the slope of this line is $f'(x)$. Figures 4.65(a) and (b) suggest that, geometrically, dy is the change in y corresponding to the change dx in x if we follow the tangent line at (x, y) rather than the curve itself. They also suggest that if dx is close to zero, then dy is a good approximation for Δy; i.e.,

$$dy \simeq \Delta y \quad \text{when} \quad dx \simeq 0.$$

Note that if we were to divide both sides of Eq. (4.18) by the differential dx, then we would have

$$\frac{dy}{dx} = f'(x).$$

Now the left side of this equation has been determined as the differential of y divided by the differential of x. What we have done then is to define differentials dx and dy in such a way that the derivative notation dy/dx may now be considered as a quotient—a quotient of differentials.

We should also note that the differential dy is a function of two variables, x and dx, and that these two quantities can be specified independently of one another.

Example 24 If $f(x) = \sqrt{x^2 + 1}$, find dy when $x = 2$ and $dx = 0.1$.

Solution By Definition 4.9, we have

$$dy = f'(x)\,dx = \frac{x}{\sqrt{x^2 + 1}}\,dx.$$

When $x = 2$ and $dx = 0.1$,

$$dy = \frac{2}{\sqrt{2^2 + 1}}(0.1) = 0.089. \ \square$$

If a quantity y undergoes a change Δy, then its *relative change* is defined as

$$\frac{\Delta y}{y},$$

and its *percentage change* is given by

$$100\,\frac{\Delta y}{y}.$$

We can see why it is sometimes more important to consider relative and percentage changes rather than actual changes if we consider the volume V of a sphere. The volume depends on the radius of the sphere according to the formula $V = 4\pi r^3/3$. If the radius of the sphere is increased from 0.10 meter to 0.11 meter, then the change in the volume of the sphere is

$$\Delta V = \tfrac{4}{3}\pi(0.11)^3 - \tfrac{4}{3}\pi(0.1)^3 = 4.4\pi \times 10^{-4} \text{ cubic meter.}$$

The relative change of V is then

$$\frac{\Delta V}{V} = \frac{4.4\pi \times 10^{-4}}{\tfrac{4}{3}\pi(0.1)^3} = 0.33,$$

and the percentage change is

$$100\,\frac{\Delta V}{V} = 33\%.$$

Suppose now that the same increase of 0.01 meter is applied to a sphere with radius 100 meters. The change in the volume is now

$$\Delta V = \tfrac{4}{3}\pi(100.01)^3 - \tfrac{4}{3}\pi(100)^3 = 4.0\pi \times 10^2 \text{ cubic meters.}$$

The relative change of V is then

$$\frac{\Delta V}{V} = \frac{4.0\pi \times 10^2}{\tfrac{4}{3}\pi(100)^3} = 3.0 \times 10^{-4},$$

and the percentage change is

$$100\frac{\Delta V}{V} = 0.03\,\%.$$

Thus, although the change 400π in V when $r = 100$ is much larger than the change 0.00044π when $r = 0.1$, the relative and percentage changes are much smaller when $r = 100$. We see, then, that it may be relative or percentage changes that are significant rather than actual changes.

In the above example, when $r = 100$ meters, the change $dr = 0.01$ is certainly small compared to r. We should therefore be able to use the differential $dV = V'(r)\,dr = 4\pi r^2\,dr$ to approximate ΔV. With $r = 100$ and $dr = 0.01$, we have

$$dV = 4\pi(100)^2(0.01) = 4.0\pi \times 10^2 \text{ cubic meters.}$$

(To two significant figures, therefore, dV is equal to ΔV.) Thus relative change in V is approximately dV/V and percentage change $100\,dV/V$. For small changes then in an independent variable, the differential of the dependent variable may be used in place of its increment in the calculation of relative and percentage changes. This technique is used again in the following example.

Example 25 When a pendulum swings, the frequency (number of cycles per second) of its oscillations is given by

$$f = h(l) = 2\pi\sqrt{\frac{g}{l}},$$

where l is the length of the pendulum and $g > 0$ is a constant. If the length of the pendulum is increased by $\frac{1}{4}\%$, calculate the approximate percentage change in f.

Solution The approximate change in f is given by

$$df = h'(l)\,dl = 2\pi\sqrt{g}(-\tfrac{1}{2})l^{-3/2}\,dl = -\pi\sqrt{g}\,\frac{dl}{l^{3/2}};$$

hence the approximate percentage change in f is

$$100\frac{df}{f} = 100\left(\frac{-\pi\sqrt{g}\,dl}{l^{3/2}}\right)\left(\frac{l^{1/2}}{2\pi\sqrt{g}}\right) = -\frac{1}{2}\left(100\frac{dl}{l}\right).$$

But if l increases by $\frac{1}{4}\%$, then $100(dl/l) = \frac{1}{4}$, and

$$100\frac{df}{f} = -\frac{1}{8}.$$

The frequency therefore changes by $-\frac{1}{8}\%$, the negative sign indicating that because l increases, f decreases. □

The differential dy cannot always be used as an approximation for the actual change Δy in a function $y = f(x)$. Sometimes it cannot even be used when dx is very close to zero. For example, if $f(x) = 2x^3 + 9x^2 - 24x + 6$, then

$$dy = f'(x)\,dx = (6x^2 + 18x - 24)\,dx.$$

If x is changed from 1 to 1.01, then the approximate change in y as predicted by the differential is

$$dy = (6 + 18 - 24)(0.01) = 0.$$

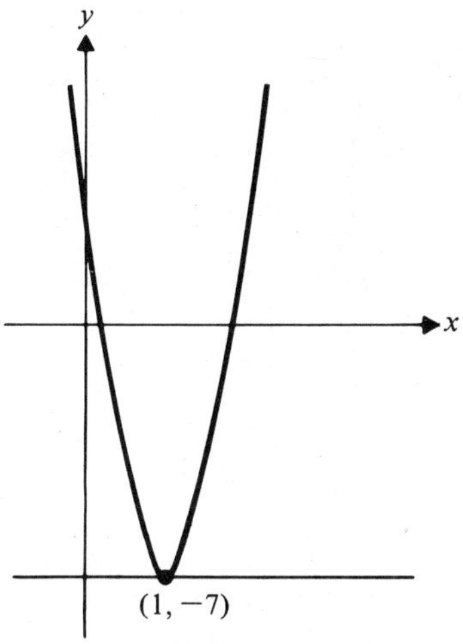

Figure 4.66

In fact, for any dx whatsoever, we find that $dy = 0$. Geometrically speaking, we can see why. Since $f'(1) = 0$, $x = 1$ is a critical point of $f(x)$ (Fig. 4.66); therefore dy, which is the tangent line approximation to Δy, will always be zero. Clearly, then, we cannot use differentials to approximate function changes at critical points.

The differential for the function $y = f(x) = x^{100}$ is

$$dy = 100x^{99}\,dx.$$

If x is changed by 1% from $x = 1$ to $x = 1.01$, then $dx = 0.01$ and

$$dy = 100(1)^{99}(0.01) = 1.0.$$

The actual change in y is

$$\Delta y = (1.01)^{100} - 1^{100} = 1.7.$$

We would hardly regard this dy as a very good approximation for Δy even though the change in x was only one percent.

This latter example begs the question, "How small in general must dx be in order that dy be a reasonable approximation for Δy?" This is not a simple question to answer. In fact, to attempt to answer it is just not worth the value of the differential as an approximating tool. We will discuss approximations in more detail in Chapter 11, where we will find far better methods than differentials for making approximations. In addition, we will then be able to answer a more fundamental question, "How good an approximation to Δy is dy?" After all, it isn't much use to say that $dy \simeq \Delta y$ if we cannot say to how many decimal places the approximation is accurate. Suffice it to say now that use of the differential dy to approximate Δy is to be regarded with some reservation. This is not to say that differentials are useless. We will see in Chapters 5–7 that differentials are indispensable to the topic of integration.

EXERCISES 4.9

In each of Exercises 1–10, find dy in terms of x and dx.

1. $y = x^2 + 3x - 2$

2. $y = \dfrac{x+1}{x-1}$

3. $y = \sqrt{x^2 - 2x}$

4. $y = \sin(x^2 + 2) - \cos x$

5. $y = x^{1/3} - x^{5/3}$

6. $y = x^3\sqrt{3 - 4x^2}$

7. $y = x^2 \sin x$

8. $y = \dfrac{x^3 - 3x^2 + 3x + 5}{x^2 - 2x + 1}$

9. $y = \sqrt{1 + \sqrt{1 - x}}$

10. $y = \dfrac{x^2(x-2)}{x^3 + 5x}$

11. The momentum M and kinetic energy K of a mass m moving with speed v are given by

$$M = mv \quad \text{and} \quad K = \tfrac{1}{2}mv^2.$$

If v is changed by 1%, what are the approximate percentage changes in M and K?

12. The magnitude of the gravitational force of attraction between two point masses m and M is given by

$$F = \frac{GmM}{r^2}, \qquad G = \text{a constant},$$

where r is the distance between the masses. If r changes by 2%, by how much does F change approximately?

13. Under adiabatic expansion, a gas obeys the law

$$PV^{7/5} = \text{a constant},$$

where P is pressure and V is volume. If the pressure is

increased by 2%, find the approximate percentage change in the volume.

14. The magnitude of the gravitational force of attraction between two point masses m and M is defined in Exercise 12. If the earth is considered a perfect sphere (radius 6.37×10^6 meters), then this law predicts a gravitational attraction of $9.81m$ newtons on a mass m on its surface. Use differentials to determine the height above the surface of the earth at which the gravitational attraction decreases to $9.80m$ newtons.

4.10 ROLLE'S THEOREM AND THE MEAN VALUE THEOREMS

In this section we prove a very important theorem in calculus: Cauchy's generalized mean value theorem. We will see in Section 4.11 and in future chapters that this theorem is exceptionally valuable for establishing other results. Our first theorem is a stepping stone to this result.

▶ **Theorem 4.4** (*Rolle's theorem*). If $f(x)$ is continuous on the closed interval $a \le x \le b$ and has a first derivative at each point in the open interval $a < x < b$, and if $f(a) = f(b)$, then there exists at least one point c in the open interval at which $f'(c) = 0$.

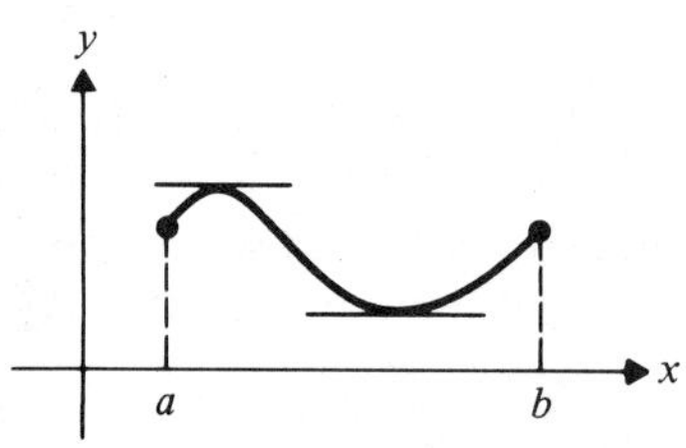

Figure 4.67

For the function $f(x)$ shown in Fig. 4.67 there are two possible choices for c.

Proof To prove the theorem we divide our discussion into three cases.

Case 1: $f(x) = f(a)$ for all x in $a < x < b$ (Fig. 4.68). In this case, $f'(x) = 0$ for all x in $a < x < b$, and c may be chosen anywhere in the interval.

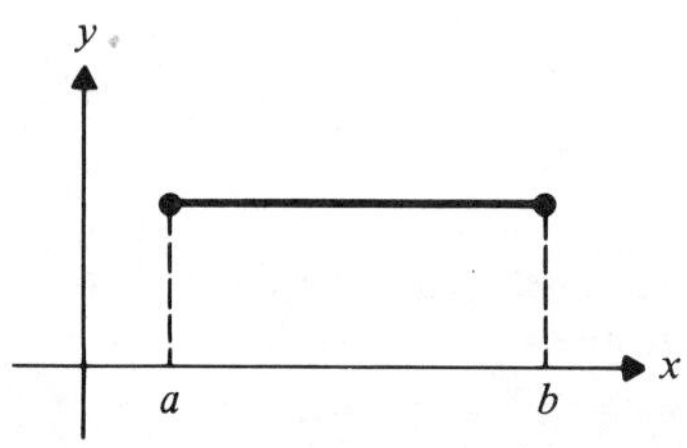

Figure 4.68

Case 2: There exists some point x_0 in $a < x < b$ for which $f(x_0) > f(a)$ (Fig. 4.69). According to Theorem 4.2, a function that is continuous on a closed interval $a \le x \le b$ must have an absolute maximum value at some point c in that interval. Furthermore, c must be at one of three places: at a critical point, at $x = a$, or at $x = b$. Since $f'(x)$ exists everywhere in $a < x < b$, and we already know that $f(x_0) > f(a)$, it follows that c must be a critical point of $f(x)$ at which $f'(c) = 0$.

Case 3: A proof similar to that in Case 2 (but using absolute minima) dispenses with the remaining case, in which $f(x) < f(a)$ for all x in $a < x < b$. □

We can now prove the following.

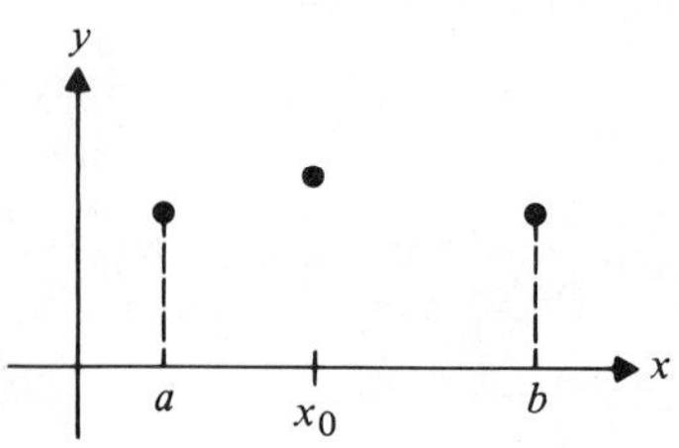

Figure 4.69

▶ **Theorem 4.5** (*Cauchy's generalized mean value theorem*). If functions $f(x)$ and $g(x)$ are continuous on the closed interval $a \le x \le b$ and differentiable on the open interval $a < x < b$, and if $g'(x) \ne 0$ on $a < x < b$, then there exists at least one point c in the open interval for which

$$\frac{f(b) - f(a)}{g(b) - g(a)} = \frac{f'(c)}{g'(c)}. \quad ◀ \tag{4.19}$$

Proof First note that $g(b) - g(a)$ cannot equal zero. If it did, then $g(a)$ would equal $g(b)$, and Rolle's theorem applied to $g(x)$ on the interval $a \le x \le b$ would imply the existence of a point c at which $g'(c) = 0$ (a contradiction). To prove the theorem, we construct a function $h(x)$ to satisfy the conditions of Rolle's theorem. Specifically, we consider

$$h(x) = f(x) - f(a) - \frac{f(b) - f(a)}{g(b) - g(a)}[g(x) - g(a)].$$

Since $f(x)$ and $g(x)$ are continuous for $a \le x \le b$, so too is $h(x)$. In addition,

$$h'(x) = f'(x) - \frac{f(b) - f(a)}{g(b) - g(a)}g'(x);$$

therefore, $h'(x)$ exists for $a < x < b$. Finally, since $h(a) = h(b) = 0$, we may conclude from Rolle's theorem that there exists a number c such that $a < c < b$ and

$$0 = h'(c) = f'(c) - \frac{f(b) - f(a)}{g(b) - g(a)}g'(c);$$

i.e.,

$$\frac{f'(c)}{g'(c)} = \frac{f(b) - f(a)}{g(b) - g(a)}. \square$$

An important special case of this result occurs when $g(x) = x$.

▶ **Theorem 4.6** (*The mean value theorem*). If $f(x)$ is continuous on the closed interval $a \le x \le b$ and has a first derivative at each point of the open interval $a < x < b$, then there exists at least one point c in the open interval for which

$$f'(c) = \frac{f(b) - f(a)}{b - a}. \quad ◀ \tag{4.20}$$

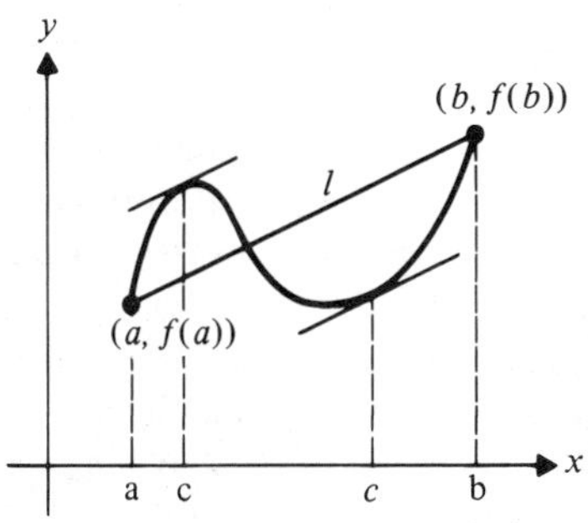

Figure 4.70

Algebraically speaking, we get very little feeling for these results, but geometrically, Theorem 4.6 seems almost obvious. Figure 4.70 illustrates that the quotient $[f(b) - f(a)]/(b - a)$ is the slope of the line l joining the points $(a, f(a))$ and $(b, f(b))$ on the graph $y = f(x)$. The mean value theorem states that there is at least one point c between a and b at which the tangent line is parallel to l. In Fig. 4.70 there are clearly two such points.

A similar interpretation of Theorem 4.5 is given in Section 10.6.

Example 26 Find all values of c satisfying the mean value theorem for the function $f(x) = x^3 - 4x$ on the interval $-1 \le x \le 3$.

Solution Since $f(x)$ is differentiable (and therefore continuous) at each point of $-1 \le x \le 3$, we can indeed apply the mean value theorem and claim the existence of at least one number c in $-1 < x < 3$ such that

$$f'(c) = \frac{f(3) - f(-1)}{3 - (-1)};$$

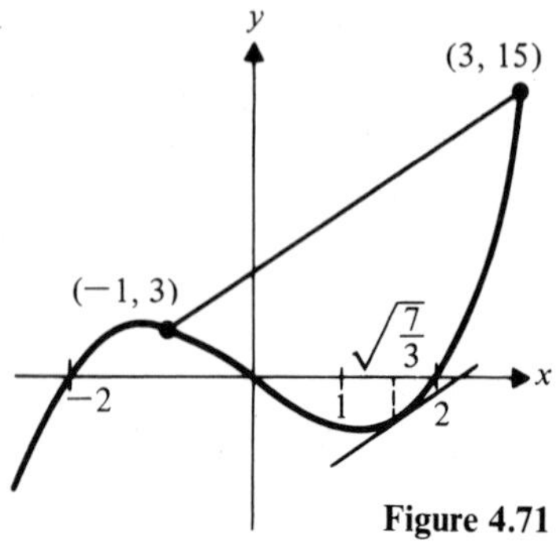

Figure 4.71

i.e.,

$$3c^2 - 4 = \frac{15 - 3}{4} = 3.$$

Consequently,

$$c = \pm\sqrt{\frac{7}{3}}.$$

Since $-\sqrt{\frac{7}{3}} < -1$, the only value of c in the interval $-1 < x < 3$ is $c = \sqrt{\frac{7}{3}}$ (Fig. 4.71). □

EXERCISES 4.10

In each of Exercises 1–10, decide whether the mean value theorem can be applied to the function on the interval. If it cannot, explain why not. If it can, find all values of c in the interval that satisfy Eq. (4.20).

1. $f(x) = x^2 + 2x,\ -3 \le x \le 2$
2. $f(x) = 4 + 3x - 2x^2,\ 1 \le x \le 3$
3. $f(x) = x + 5,\ 2 \le x \le 3$
4. $f(x) = |x|,\ -1 \le x \le 1$
5. $f(x) = |x|,\ 0 \le x \le 1$
6. $f(x) = x^3 + 2x^2 - x - 2,\ -3 \le x \le 2$
7. $f(x) = x^3 + 2x^2 - x - 2,\ -1 \le x \le 2$
8. $f(x) = (x + 2)/(x - 1),\ 2 \le x \le 4$
9. $f(x) = (x + 1)/(x + 2),\ -3 \le x \le 2$
10. $f(x) = x^2/(x + 3),\ -2 \le x \le 3$

In each of Exercises 11–14, decide whether Cauchy's generalized mean value theorem can be applied to the functions on the interval. If it cannot, explain why not. If it can, find all values of c in the interval that satisfy Eq. (4.19).

11. $f(x) = x^2,\ g(x) = x,\ 1 \le x \le 2$
12. $f(x) = x + 1,\ g(x) = |x|^{3/2},\ -1 \le x \le 1$
13. $f(x) = x^2 + 3x - 1,\ g(x) = x^3 + 5x + 4,\ 0 \le x \le 2$
14. $f(x) = x/(x + 1),\ g(x) = x/(x - 1),\ -3 \le x \le -2$
15. Prove that if a function $f(x)$ is continuous on $a \le x \le b$, and if $f'(x) > 0$ for $a < x < b$, then $f(x)$ is increasing on the interval $a \le x \le b$. (This is a somewhat restrictive version of the result contained in inequality (4.8). The general result is shown in Exercise 20.)
16. Show that if $|f'(x)| \le M$ on $a \le x \le b$, then
$$|f(b) - f(a)| \le M(b - a).$$
17. A traffic plane (Fig. 4.72) measures the time that it takes a car to travel between points A and B on the highway as 15 seconds, and radios this information ahead to a patrol car. What is the maximum speed at which the police officer can claim the car was traveling between A and B? Explain.

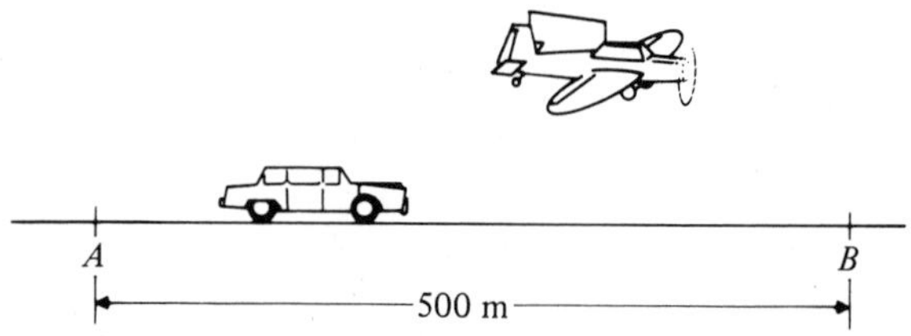

Figure 4.72

18. Show that the value c that satisfies the mean value theorem for any quadratic function $f(x) = dx^2 + ex + g$ on any interval $a \le x \le b$ whatsoever is $c = (a + b)/2$.
19. Let $f(x)$ and $g(x)$ be two functions that are differentiable at each point of the interval $a \le x \le b$. Prove that if $f(a) = g(a)$ and $f(b) = g(b)$, then there exists c in the open interval $a < x < b$ for which $f'(c) = g'(c)$.
20. We now outline a proof of the result in inequality (4.8) that if $f'(x) \ge 0$ on $a \le x \le b$ (and is equal to zero at only a finite number of points), $f(x)$ is increasing on $a \le x \le b$. Let $c_1, c_2, \ldots, c_{n-1}$ be the points in $a \le x \le b$ at which $f'(x) = 0$ and define $c_0 = a$ and $c_n = b$.
    a) Use the mean value theorem to verify that $f(x)$ is increasing on each subinterval $c_{i-1} \le x \le c_i$ $(i = 1, \ldots, n)$ (see Exercise 15).
    b) Does it now follow that $f(x)$ is increasing on $a \le x \le b$?
21. We now outline a proof of the first-derivative test for classifying critical points at which a function is continuous. Let x_0 be a critical point of $f(x)$ at which $f(x)$ is continuous, and suppose that $f'(x)$ changes from a negative quantity to a positive quantity as x increases through x_0. This implies that there exists an interval $a \le x < x_0$ in which $f'(x) < 0$ and an interval $x_0 < x \le b$ in which $f'(x) > 0$.

a) Show that $f(x)$ is decreasing on the interval $a \leq x \leq x_0$.

b) Show that $f(x)$ is increasing on the interval $x_0 \leq x \leq b$.

c) Verify now that $f(x_0)$ is a relative minimum.

d) Construct a similar proof to verify that $f(x_0)$ is a relative maximum if x_0 is critical and $f'(x)$ changes from a positive to a negative quantity as x increases through x_0.

e) Prove the remainder of the first-derivative test.

*** 22.** Verify that, for a cubic polynomial $f(x) = dx^3 + ex^2 + gx + h$ defined on any interval $a \leq x \leq b$, the values of c that satisfy Eq. (4.20) are equidistant from $x = -e/(3d)$.

4.11 INDETERMINATE FORMS AND L'HÔPITAL'S RULE

If we let x approach zero in the numerator and denominator of the limit

$$\lim_{x \to 0^+} \frac{\sqrt{1+x}-1}{\sqrt{x}},$$

we find that

$$\lim_{x \to 0^+} (\sqrt{1+x}-1) = 0 \quad \text{and} \quad \lim_{x \to 0^+} \sqrt{x} = 0.$$

Because of this we say that the original limit is of the "indeterminate form 0/0." Likewise, the limit

$$\lim_{x \to 2} \frac{x^3 - x^2 - 8x + 12}{x^2 - 4x + 4}$$

is of the indeterminate form 0/0 since as x approaches 2 both numerator and denominator approach zero.

In Section 2.1 we evaluated the first limit by rationalizing the numerator (see Example 6 there). Factorization of numerator and denominator in the second limit gives us

$$\lim_{x \to 2} \frac{x^3 - x^2 - 8x + 12}{x^2 - 4x + 4} = \lim_{x \to 2} \frac{(x-2)^2(x+3)}{(x-2)^2} = \lim_{x \to 2} (x+3) = 5.$$

These two examples are typical of the "trickery" to which we resorted in Section 2.1 in order to evaluate limits. With Cauchy's generalized mean value theorem, however, we can prove a result called L'Hôpital's rule, which makes evaluation of many limits of the indeterminate form 0/0 quite trivial.

▶ **Theorem 4.7** (*L'Hôpital's rule*). Let $f(x)$ and $g(x)$ be functions that are differentiable on an open interval I except possibly at the point $x = a$ in I. Suppose further that $g'(x) \neq 0$ in I (except possibly at a). If $\lim_{x \to a} f(x) = 0$ and $\lim_{x \to a} g(x) = 0$, and if

$$\lim_{x \to a} \frac{f'(x)}{g'(x)} = L,$$

then

$$\lim_{x \to a} \frac{f(x)}{g(x)}$$

exists and is also equal to L. ◀

Proof We define two functions $F(x)$ and $G(x)$ that are identical to $f(x)$ and $g(x)$ on I but have value zero at $x = a$:

$$F(x) = \begin{cases} f(x), & x \neq a, \\ 0, & x = a, \end{cases} \qquad G(x) = \begin{cases} g(x), & x = a, \\ 0, & x = a. \end{cases}$$

Since $F'(x) = f'(x)$ and $G'(x) = g'(x)$ for x in I except possibly at $x = a$, $F(x)$ and $G(x)$ are therefore differentiable on I except possibly at $x = a$. In addition, differentiability of a function implies continuity of the function (Theorem 3.7), so that $F(x)$ and $G(x)$ must certainly be continuous on I except possibly at $x = a$. But

$$\lim_{x \to a} F(x) = \lim_{x \to a} f(x) = 0 = F(a),$$

and the same is true for $G(x)$; hence $F(x)$ and $G(x)$ are continuous for all x in I. Consequently, $F(x)$ and $G(x)$ are identical to $f(x)$ and $g(x)$ in every respect, except that they have been assigned a value at $x = a$ to guarantee their continuity there. This extra condition permits us to apply Cauchy's generalized mean value theorem to $F(x)$ and $G(x)$ on the interval between a and x (so long as x is in I). We can conclude that there exists a number c between a and x such that

$$\frac{F(x) - F(a)}{G(x) - G(a)} = \frac{F'(c)}{G'(c)},$$

or, since $F(a) = G(a) = 0$,

$$\frac{F(x)}{G(x)} = \frac{F'(c)}{G'(c)}.$$

Since x and c are points in I, we can also write that $F(x) = f(x)$, $F'(c) = f'(c)$, $G(x) = g(x)$, $G'(c) = g'(c)$, and, therefore,

$$\frac{f(x)}{g(x)} = \frac{F(x)}{G(x)} = \frac{F'(c)}{G'(c)} = \frac{f'(c)}{g'(c)}.$$

If we now let x approach a, then c must also approach a since it is always between a and x. Consequently,

$$\lim_{x \to a} \frac{f(x)}{g(x)} = \lim_{c \to a} \frac{f'(c)}{g'(c)} = \lim_{x \to a} \frac{f'(x)}{g'(x)},$$

and if

$$L = \lim_{x \to a} \frac{f'(x)}{g'(x)},$$

it follows that

$$\lim_{x \to a} \frac{f(x)}{g(x)} = L. \ \square$$

Note that this theorem is also valid if L is replaced by ∞ or $-\infty$. The only difference in the proof is to make the same change in the last sentence.

Theorem 4.7 is also valid if $x \to a$ is replaced by either a right-hand limit, $x \to a^+$, or a left-hand limit, $x \to a^-$. The only difference in these cases is that interval I is replaced by open intervals $a < x < b$ and $b < x < a$, respectively (and the proofs are almost identical). In addition, the following theorem indicates that $x \to a$ can be replaced by $x \to \infty$ (or $x \to -\infty$).

Theorem 4.8 (*L'Hôpital's rule*). Let $f(x)$ and $g(x)$ be functions that are differentiable for some interval $x > b > 0$, and suppose that $g'(x) \neq 0$ on this interval. If $\lim_{x\to\infty} f(x) = 0$ and $\lim_{x\to\infty} g(x) = 0$, and if

$$\lim_{x\to\infty} \frac{f'(x)}{g'(x)} = L \quad (\text{or } \pm\infty),$$

then

$$\lim_{x\to\infty} \frac{f(x)}{g(x)} = L \quad (\text{or } \pm\infty).$$

In other words, L'Hôpital's rule applies to any type of limit that yields the indeterminate form 0/0 (be it $x \to a$, $x \to a^+$, $x \to a^-$, $x \to \infty$, or $x \to -\infty$). A common error made by many students when using L'Hôpital's rule is to differentiate $f(x)/g(x)$ with the quotient rule and then to take the limit of the resulting derivative. L'Hôpital's rule calls for taking the limit of $f'(x)/g'(x)$.

If we use L'Hôpital's rule on the first of the examples at the beginning of this section, we get

$$\lim_{x\to 0^+} \frac{\sqrt{1+x}-1}{\sqrt{x}} = \lim_{x\to 0^+} \frac{1/(2\sqrt{1+x})}{1/(2\sqrt{x})} = \lim_{x\to 0^+} \frac{\sqrt{x}}{\sqrt{1+x}} = 0.$$

For the second of these examples, we have

$$\lim_{x\to 2} \frac{x^3 - x^2 - 8x + 12}{x^2 - 4x + 4} = \lim_{x\to 2} \frac{3x^2 - 2x - 8}{2x - 4},$$

which is still a limit of the indeterminate form 0/0. Note that this is a conditional equation; that is, it says that the limit on the left is equal to the limit on the right, provided the limit on the right exists. If we apply L'Hôpital's rule a second time to the limit on the right, we get

$$\lim_{x\to 2} \frac{x^3 - x^2 - 8x + 12}{x^2 - 4x + 4} = \lim_{x\to 2} \frac{6x - 2}{2} = 5.$$

Example 27 Evaluate each of the following limits:

a) $\displaystyle\lim_{x\to -3} \frac{3+x}{\sqrt{3}-\sqrt{-x}}$;

b) $\displaystyle\lim_{x\to 0} \frac{\sqrt{1-x}-\sqrt{1+x}}{x}$;

c) $\displaystyle\lim_{x\to 4} \frac{x-4}{x^2 - 8x + 16}$;

d) $\displaystyle\lim_{x\to 4} \frac{x^3 - 4x^2 + 9x - 36}{x^2 + 5}$.

Solution a) Because we have the indeterminate form 0/0, we use L'Hôpital's rule to get

$$\lim_{x\to -3} \frac{3+x}{\sqrt{3}-\sqrt{-x}} = \lim_{x\to -3} \frac{1}{1/(2\sqrt{-x})} = \lim_{x\to -3} 2\sqrt{-x} = 2\sqrt{3}.$$

b) If we use L'Hôpital's rule, we have

$$\lim_{x\to 0} \frac{\sqrt{1-x}-\sqrt{1+x}}{x} = \lim_{x\to 0} \frac{-1/(2\sqrt{1-x}) - 1/(2\sqrt{1+x})}{1}$$

$$= \lim_{x\to 0} \frac{-(\sqrt{1+x}+\sqrt{1-x})}{2\sqrt{1-x}\sqrt{1+x}} = -1.$$

c) By L'Hôpital's rule,

$$\lim_{x\to 4} \frac{x-4}{x^2-8x+16} = \lim_{x\to 4} \frac{1}{2x-8}.$$

Since

$$\lim_{x\to 4^+} \frac{1}{2x-8} = \infty \quad \text{and} \quad \lim_{x\to 4^-} \frac{1}{2x-8} = -\infty,$$

we conclude that

$$\lim_{x\to 4^+} \frac{x-4}{x^2-8x+16} = \infty \quad \text{and} \quad \lim_{x\to 4^-} \frac{x-4}{x^2-8x+16} = -\infty.$$

d) This limit is not of the indeterminate form 0/0 since $\lim_{x\to 4} (x^2+5) = 21$; thus we cannot use L'Hôpital's rule. Since $\lim_{x\to 4} (x^3 - 4x^2 + 9x - 36) = 0$,

$$\lim_{x\to 4} \frac{x^3-4x^2+9x-36}{x^2+5} = 0. \;\square$$

Had we used L'Hôpital's rule in part (d), we would have obtained an incorrect answer:

$$\lim_{x\to 4} \frac{x^3-4x^2+9x-36}{x^2+5} = \lim_{x\to 4} \frac{3x^2-8x+9}{2x} = \frac{25}{8}.$$

In other words, L'Hôpital's rule is not to be used indiscriminately; it must be used only on the indeterminate forms for which it is designed. The limits

$$\lim_{x\to \infty} \frac{1+\sqrt{x-1}}{2x+5} \quad \text{and} \quad \lim_{x\to \infty} \frac{3x^3+3x}{4x^2-2}$$

are said to be of the indeterminate form ∞/∞ since the numerator and denominator of each limit become increasingly large as $x \to \infty$. Theorems 4.7 and 4.8 for L'Hôpital's rule can be adapted to this indeterminate form also; hence we calculate that

$$\lim_{x\to \infty} \frac{1+\sqrt{x-1}}{2x+5} = \lim_{x\to \infty} \frac{1/(2\sqrt{x-1})}{2} = \lim_{x\to \infty} \frac{1}{4\sqrt{x-1}} = 0$$

and

$$\lim_{x\to \infty} \frac{3x^3+3x}{4x^2-2} = \lim_{x\to \infty} \frac{9x^2+3}{8x} = \lim_{x\to \infty} \frac{18x}{8} = \infty.$$

EXERCISES 4.11

Evaluate the given limit (if it exists) in Exercises 1–25.

1. $\lim_{x\to 0} \dfrac{x^2 + 3x}{x^3 + 5x^2}$

2. $\lim_{x\to 3} \dfrac{x^2 - 9}{x - 3}$

3. $\lim_{x\to -\infty} \dfrac{x^3 + 3x - 2}{x^2 + 5x + 1}$

4. $\lim_{x\to \infty} \dfrac{2x^2 + 3x}{5x^3 + 4}$

5. $\lim_{x\to 5} \dfrac{x^2 - 10x + 25}{x^3 - 125}$

6. $\lim_{x\to 1} \dfrac{1}{(x - 1)^2}$

7. $\lim_{x\to \infty} \dfrac{\sqrt{x^2 + 1}}{2x + 5}$

8. $\lim_{x\to -\infty} \dfrac{\sin x}{x}$

9. $\lim_{x\to \infty} \dfrac{\sin (2/x)}{\sin (1/x)}$

10. $\lim_{x\to \pi/2} \dfrac{\cos x}{(x - \pi/2)^2}$

11. $\lim_{x\to 1^+} \dfrac{(1 - 1/x)^3}{\sqrt{x - 1}}$

12. $\lim_{x\to \infty} \dfrac{\sin (1/x)}{1/x^2}$

13. $\lim_{x\to 9^-} \dfrac{\sqrt{x} - 3}{\sqrt{9 - x}}$

14. $\lim_{x\to 0} \dfrac{\sqrt{5 + x} - \sqrt{5 - x}}{x}$

15. $\lim_{x\to 0} \dfrac{x - \sin x}{x^3}$

16. $\lim_{x\to a} \dfrac{x^n - a^n}{x - a}$

17. $\lim_{x\to 0} \dfrac{(1 - \cos x)^2}{3x^2}$

18. $\lim_{x\to 0} \dfrac{\tan x}{x}$

19. $\lim_{x\to 0} \dfrac{\sin 3x}{\tan 2x}$

20. $\lim_{x\to 1} \dfrac{(1 - \sqrt{2 - x})^{3/2}}{x - 1}$

21. $\lim_{x\to 0} \dfrac{\sqrt{x + 1} - \sqrt{2x + 1}}{\sqrt{3x + 4} - \sqrt{2x + 4}}$

22. $\lim_{x\to 0} \dfrac{(1 - \cos x)^2}{3x^4}$

23. $\lim_{x\to \infty} x \sin \left(\dfrac{1}{x}\right)$

24. $\lim_{x\to 2} \dfrac{(x - 2)^{10}}{(\sqrt{x} - \sqrt{2})^{10}}$

25. $\lim_{x\to 0} \left(\dfrac{4}{x^2} - \dfrac{2}{1 - \cos x}\right)$

26. The following limit arises in the calculation of the electric field intensity for a half-wave antenna:

$$\lim_{\theta\to 0} f(\theta),$$

where

$$f(\theta) = \sin \theta \left\{\frac{\sin [\pi/2 (\cos \theta - 1)]}{\cos \theta - 1} + \frac{\sin [\pi/2 (\cos \theta + 1)]}{\cos \theta + 1}\right\}.$$

Evaluate this limit by first showing that $f(\theta)$ can be written in the form

$$f(\theta) = \frac{2 \cos \left(\dfrac{\pi}{2} \cos \theta\right)}{\sin \theta},$$

and then using L'Hôpital's rule.

SUMMARY

In this chapter we discussed a number of applications of differentiation, the first of which is velocity and acceleration. When an object moves along a straight line, its velocity and acceleration are the first and second derivatives, respectively, of its displacement with respect to time. In other words, if we observe straight-line motion of an object and record its position as a function of time, then we can easily calculate the velocity and acceleration of that object at any instant. In Section 5.2 we will reverse this process and show that velocity and position can be calculated once acceleration is known as a function of time.

Changes in a number of interrelated quantities usually produce changes in the others—sometimes small, sometimes important. How the rates of change of these variables relate to each other was the subject of Section 4.2. Related-rate problems made us acutely aware of the importance of differentiating an equation with respect to a variable only if the equation is valid for a continuous range of values of that variable.

In Section 4.3 we defined a critical point of a function as a point in its domain where its first derivative either vanishes or does not exist. Geometrically, this corresponds to a point where the graph of the function has a horizontal tangent line, a vertical tangent line, or no tangent line at all. A critical point where the derivative vanishes yields either a relative maximum, a relative minimum, or a horizontal point of inflection; to determine which, we can use the first-derivative test. A critical point where the first derivative does not exist is either a relative extremum or a corner. A function is increasing (decreasing) on an interval if its graph slopes upward to the right (left), and this is characterized by a non-negative (nonpositive) derivative. It is said to be concave upward (downward) if its slope is increasing (decreasing), and consequently if its second derivative is nonnegative (nonpositive). Points that separate intervals of opposite concavity are called points of inflection. In Section 4.6 we showed how our curve sketching techniques, which now include addition and multiplication of ordinates, symmetries, limits, relative extrema, and points of inflection, complement one another in drawing graphs.

In Section 4.7 we illustrated that many applied extrema problems require absolute extrema rather than relative extrema. Absolute extrema of a continuous function on a closed interval must occur at either critical points or the ends of the interval. This fact implies that to find the absolute extrema of a continuous function $f(x)$ on a closed interval $a \le x \le b$, we evaluate $f(x)$ at its critical points and at a and b. The largest and smallest of these values are the absolute extrema of $f(x)$ on $a \le x \le b$.

On the basis of our familiarity with tangent lines, we developed Newton's iterative procedure for approximating the roots of equations. It is perhaps the most popular approximation method, principally because of its speed, simplicity, and accuracy.

Cauchy's generalized mean value theorem enabled us to develop L'Hôpital's rule in Section 4.11. With this rule we can easily calculate limits of indeterminate forms $0/0$ and ∞/∞. We will see in Section 8.9 that other indeterminate forms can also be handled using L'Hôpital's rule.

REVIEW EXERCISES

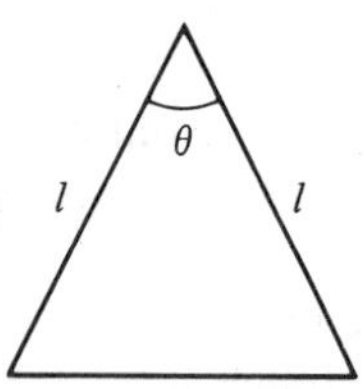

Figure 4.73

1. a) Prove that the area of the isosceles triangle shown in Fig. 4.73 is

$$A = \tfrac{1}{2}l^2 \sin \theta.$$

b) If angle θ is increasing at a rate of $\frac{1}{2}$ radian per minute, but l remains constant, how fast is the area of the triangle changing? What does your answer predict when $\theta = 0$, $\pi/2$, and π?

c) When does A change most rapidly? most slowly? that is, when is $|dA/dt|$ largest? smallest?

2. Sketch a graph of each of the following functions, indicating all relative maxima and minima and points of inflection.

a) $f(x) = 4x^3 + x^2 - 2x + 1$

b) $f(x) = \dfrac{x^2 - 2x + 4}{x^2 - 2x + 1}$

3. Of all pairs of positive numbers that add to some given constant $c > 0$, find that pair which has the largest product.

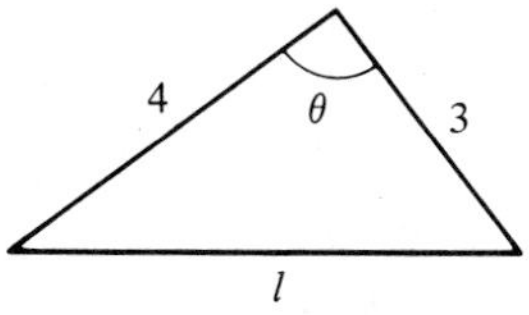

Figure 4.74

4. Solve Exercise 3 for the smallest product.

5. Of all pairs of positive numbers that multiply to some given constant $c > 0$, find that pair which has the smallest sum.

6. Solve Exercise 5 for the largest sum.

7. Two sides of the triangle shown in Fig. 4.74 maintain constant lengths of 3 centimeters and 4 centimeters but the length l of the third side decreases at the rate of 1 centimeter per minute. How fast is angle θ changing when l is 4 centimeters?

In Exercises 8–13, evaluate the given limit if it exists.

8. $\lim_{x\to 0} \dfrac{3x^2 + 2x^3}{3x^3 - 2x^2}$

9. $\lim_{x\to\infty} \dfrac{\sin 3x}{2x}$

10. $\lim_{x\to 4} \dfrac{x^2 - 16}{x - 4}$

11. $\lim_{x\to 0} \dfrac{\sin 3x}{2x}$

12. $\lim_{x\to -\infty} \dfrac{\sin (x^2)}{x^2}$

13. $\lim_{x\to 2^+} \dfrac{\sqrt{x - 2}}{\sqrt{x} - \sqrt{2}}$

C14. Use Newton's iterative procedure to find all critical points for the following functions:

a) $f(x) = x^4 + 3x^2 - 2x + 5;$

b) $f(x) = \dfrac{x^3 + 1}{3x^3 + 5x + 1}.$

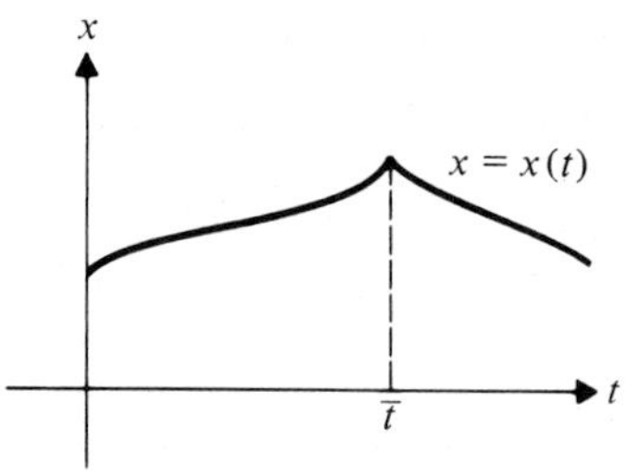

Figure 4.75

15. An object moves along the x-axis with its position defined as a function of time t by

$$x = x(t) = t^4 - \tfrac{44}{3}t^3 + 62t^2 - 84t, \qquad t \geq 0.$$

Sketch a graph of this function, indicating times when the velocity and acceleration of the object are equal to zero.

16. If the graph in Fig. 4.75 represents the position $x(t)$ of an object moving along the x-axis, what could physically cause the corner at $\bar{t}$?

17. If an object moves along the x-axis with constant acceleration, is it possible for a graph of its position function $x(t)$ (t being time) to have a point of inflection?

18. An open box is formed from a square piece of cardboard (l units long on each side) by cutting out a square at each corner and folding up the sides. What is the maximum possible volume for the box?

19. a) If the average speed of a car for a trip is 80 kilometers per hour, must it at some time have had an instantaneous speed of 80 kilometers per hour? Explain.

b) Must the car at some instant have had a speed of 83 kilometers per hour?

20. Sketch a graph of each of the following functions, indicating all relative maxima and minima and points of inflection:

a) $f(x) = \dfrac{x^3}{x^2 - 1};$

b) $f(x) = x^2 + \sin^2 x$

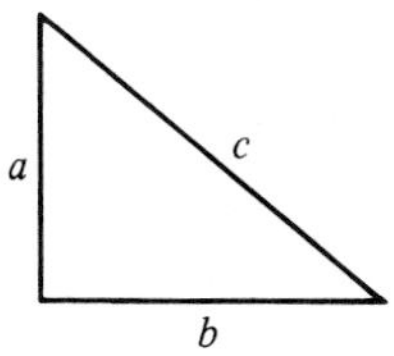

Figure 4.76

21. If at some instant of time sides a and b of the triangle shown in Fig. 4.76 form a right angle, and if these sides are increasing at equal rates, does it remain a right-angled triangle?

22. A football team presently sells tickets at prices of \$8, \$9, and \$10 per seat, depending on the position of the seat. At these prices it averages sales of 10,000 at \$10, 20,000 at \$9, and 30,000 at \$8. The team wishes to raise the price of each ticket by the same amount, but feels that for every dollar the price is raised, 10% fewer tickets of each type will be sold. What price increase per ticket will maximize revenue?

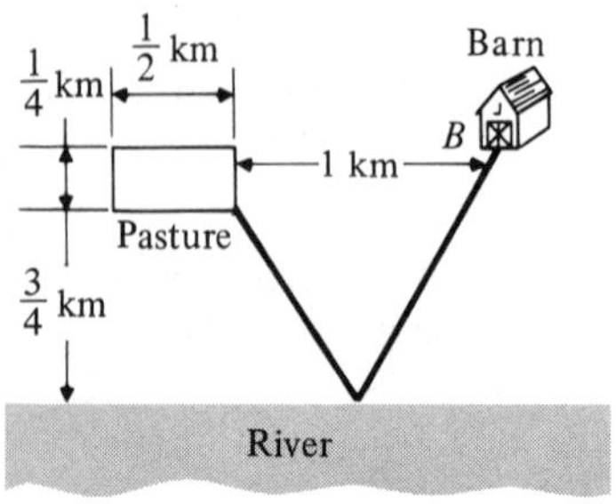

Figure 4.77

23. Repeat Exercise 22 given that the team takes into account the fact that profit from concession sales for each person at the game is 50c.

24. If a particle moves away from the origin along the positive x-axis with a constant speed of 10 meters per second, how fast is its distance from the curve $y = x^2$ changing when it is at $x = 3$ meters?

25. Each evening a cow in a pasture returns to its barn at point B (Fig. 4.77). But it always does so by first walking to the river for a drink. If the cow walks at a rate of 2 kilometers per hour and stops for a 2-minute drink, what is the minimum time it takes for the cow to get from the pasture to the barn? What is the minimum time if the cow walks twice as fast?

26. A farmer has 100 hectares to plant in corn and potatoes. Undamaged corn yields $\$p$ per hectare and potatoes $\$q$ per hectare. If the farmer plants x hectares of corn, the proportion lost to disease and pests varies directly with the area planted; i.e., proportional loss is equal to ax. The total loss of corn is therefore ax^2 hectares. Similarly, the total loss of potatoes is by^2, if the area planted in potatoes is y hectares. Find the areas that should be planted in corn and potatoes in order to minimize monetary loss. Substitute sample values for a, b, p, and q to see whether your result looks reasonable.

The Indefinite Integral or Antiderivative

5

5.1 THE REVERSE OPERATION OF DIFFERENTIATION

In Chapter 3 we introduced the derivative and learned how to differentiate functions defined explicitly and implicitly. We were then able to discuss various applications of calculus, including velocity and acceleration, related rates, maxima and minima, Newton's iterative procedure, and L'Hôpital's rule. In the first of these applications, we showed that if an object moves along the x-axis with position $x = x(t)$, then its velocity and acceleration are the first and second time derivatives of $x(t)$, respectively:

$$v = \frac{dx}{dt} \quad \text{and} \quad a = \frac{d^2x}{dt^2}.$$

For example, if $x(t) = t^3 + 3t^2$, then

$$v(t) = 3t^2 + 6t \quad \text{and} \quad a(t) = 6t + 6.$$

When engineers and physicists study problems of dynamic systems, they are generally looking for the position of an object, since it is not usually given. What they might know, however, is the acceleration of the object (perhaps through Newton's Second Law, which states that acceleration is proportional to the resultant force on the object). So the question we must now ask is, "If we know the acceleration of an object as a function of time, $a = a(t)$, can we obtain its velocity and position by reversing the differentiations?" In the above example, if we know that

$$a = \frac{dv}{dt} = 6t + 6,$$

can we find that function $v(t)$ which differentiates to give $6t + 6$? We know that to arrive at the terms $6t$ and 6 after differentiation means that $v(t)$ must have contained the terms $3t^2$ and $6t$. In other words, one possible velocity function that differentiates to give $a(t) = 6t + 6$ is $v(t) = 3t^2 + 6t$. It is not, however, the only one; $v(t) = 3t^2 + 6t + 10$ and $v(t) = 3t^2 + 6t - 22$ also have derivative $6t + 6$. In fact, for any constant C whatsoever, the derivative of

$$v(t) = 3t^2 + 6t + C$$

is $a(t) = 6t + 6$. In Theorem 5.1 we will show that this velocity function represents all functions that have $6t + 6$ as their derivative. We will also demonstrate how to evaluate the constant C.

For the present let us put $C = 0$ in this example, so that

$$v(t) = \frac{dx}{dt} = 3t^2 + 6t.$$

We now ask what the position function $x(t)$ is that differentiates to give $3t^2 + 6t$. One possibility is $x(t) = t^3 + 3t^2$, and for the same reason as above, so too is

$$x(t) = t^3 + 3t^2 + D$$

for any constant D whatsoever.

Thus by reversing the differentiation operation in this example we have proceeded from the acceleration $a(t) = 6t + 6$ to possible velocity functions $v(t)$, and then to position functions $x(t)$. This process of "backwards differentiation" or "antidifferentiation" has applications far beyond velocity and acceleration problems. In fact, these applications are so diverse that antidifferentiation is as important to calculus as differentiation is. In many cases we will find ourselves doing both: differentiating at one stage, antidifferentiating at another.

In this chapter we begin an investigation of antiderivatives that will be completed in Chapter 9.

▷ **Definition 5.1** A function $F(x)$ is called an *antiderivative* or an *indefinite integral* of $f(x)$ on an interval I if on I

$$F'(x) = f(x). \quad \triangleleft \tag{5.1}$$

For example, since

$$\frac{d}{dx}(x^4) = 4x^3,$$

we say that x^4 is an antiderivative or an indefinite integral of $4x^3$ for all x. But for any constant C, $x^4 + C$ is also an antiderivative of $4x^3$. The following theorem indicates that there are no other antiderivatives of $4x^3$ besides $x^4 + C$.

▶ **Theorem 5.1** If $F(x)$ is an indefinite integral of $f(x)$ on an interval I, then every indefinite integral of $f(x)$ on I is of the form

$$F(x) + C, \qquad C = \text{a constant.} \quad ◀$$

Proof Suppose that $F(x)$ and $G(x)$ are two antiderivatives of $f(x)$ on I. If we define a function $D(x) = G(x) - F(x)$, then on I

$$D'(x) = G'(x) - F'(x) = f(x) - f(x) = 0.$$

If x_1 and x_2 are any two points in the interval I, then certainly $D'(x) = 0$ on the interval $x_1 \le x \le x_2$. But differentiability of $D(x)$ on $x_1 \le x \le x_2$ implies continuity of $D(x)$ thereon also (Theorem 3.7). We may therefore apply the mean value theorem to $D(x)$ on the interval $x_1 \le x \le x_2$, and conclude that there exists a number c between x_1 and x_2 such that

$$D'(c) = \frac{D(x_2) - D(x_1)}{x_2 - x_1}.$$

But because $D'(c) = 0$, it follows that $D(x_1) = D(x_2)$. Since x_1 and x_2 are any two points in I, it follows that $D(x)$ must have the same value at every point in I; i.e., on I,

$$D(x) = G(x) - F(x) = C = \text{a constant.}$$

Consequently,

$$G(x) = F(x) + C,$$

and our proof is complete. □

Because of this theorem, if we find one antiderivative $F(x)$ of $f(x)$ by any means whatsoever, then we have found every antiderivative of $f(x)$, since every antiderivative can be written as $F(x)$ plus a constant C. Since every indefinite integral of $f(x)$ is of the form $F(x) + C$, we call $F(x) + C$ *the* indefinite integral of $f(x)$. The operation of taking the indefinite integral is denoted by

$$\int f(x)\,dx = F(x) + C, \tag{5.2}$$

where the differential dx indicates that the antidifferentiation is with respect to x. We call $f(x)$ the *integrand* of the indefinite integral—for example,

$$\int (2x^3 - 4x)\,dx = \frac{x^4}{2} - 2x^2 + C$$

and

$$\int \sqrt{x+1}\,dx = \tfrac{2}{3}(x+1)^{3/2} + C.$$

In the first example we reasoned as follows: An antiderivative for $2x^3$ is $x^4/2$ and an antiderivative of $4x$ is $2x^2$; therefore, the indefinite integral is $x^4/2 - 2x^2 + C$. This reasoning is indeed correct, but it has used the results of the following theorem.

▶ **Theorem 5.2** If $f(x)$ and $g(x)$ have antiderivatives on an interval I, then on I:

i) $$\int [f(x) + g(x)]\,dx = \int f(x)\,dx + \int g(x)\,dx; \tag{5.3}$$

ii) $$\int kf(x)\,dx = k\int f(x)\,dx, \qquad k = \text{a constant.} \tag{5.4}$$ ◀

Proof of this theorem is a simple exercise in differentiation and use of the definition of the indefinite integral.

One of the simplest antidifferentiation formulas, but perhaps the most important, is that for x^n,

$$\int x^n\,dx = \frac{x^{n+1}}{n+1} + C, \qquad n \neq -1. \tag{5.5}$$

We will deal with the exceptional case when $n = -1$ in Chapter 9.

Example 1 Find indefinite integrals for each of the following functions:

a) $f(x) = x^2 + 3x$;

b) $f(x) = 2x + \dfrac{1}{x^2}$;

c) $f(x) = x^3 - \dfrac{3}{x^3} + 4$.

Solution According to Theorem 5.2 we may take the antiderivative of each term separately. Using formula (5.5) on each, we have:

a) $\int (x^2 + 3x)\,dx = \tfrac{1}{3}x^3 + \tfrac{3}{2}x^2 + C$;

b) $\int\left(2x + \frac{1}{x^2}\right) dx = x^2 - \frac{1}{x} + C;$

c) $\int\left(x^3 - \frac{3}{x^3} + 4\right) dx = \frac{1}{4}x^4 + \frac{3}{2x^2} + 4x + C.$ □

Example 2 Find a curve that passes through the point (1, 5) and whose tangent line at each point (x, y) has slope $5x^4 - 3x^2 + 2$.

Solution If $y = y(x)$ is the equation of the curve, then

$$\frac{dy}{dx} = 5x^4 - 3x^2 + 2.$$

If we take antiderivatives of both sides of this equation with respect to x, we get

$$\int \frac{dy}{dx}\, dx = \int (5x^4 - 3x^2 + 2)\, dx,$$

or

$$y = x^5 - x^3 + 2x + C.$$

Since (1, 5) is a point on the curve, its coordinates must satisfy the equation of the curve:

$$5 = (1)^5 - (1)^3 + 2(1) + C.$$

Thus $C = 3$, and the required curve is

$$y = x^5 - x^3 + 2x + 3. \ \square$$

In taking the antiderivative of each side of the equation

$$\frac{dy}{dx} = 5x^4 - 3x^2 + 2$$

in this example, we added an arbitrary constant C to the right-hand side. You might question why we did not add a constant to the left-hand side. Had we done so, the result would have been

$$y + D = x^5 - x^3 + 2x + E.$$

If we then wrote

$$y = x^5 - x^3 + 2x + (E - D)$$

and defined $C = E - D$, we would have obtained exactly the same result. Hence, nothing is gained by adding an arbitrary constant to both sides; a constant on one side is sufficient.

Example 3 Discuss solutions of the following equations:

a) $\frac{dy}{dx} = 4x^2 + 7x;$

b) $\frac{dy}{dx} = -\frac{x}{3y}.$

Solution a) If we take antiderivatives with respect to x, we have

$$y = \tfrac{4}{3}x^3 + \tfrac{7}{2}x^2 + C.$$

Algebraically, this solution describes a one-parameter family of cubic polynomials (C being the parameter). Geometrically speaking, we have the one-parameter family of curves shown in Fig. 5.1. The constant C represents a vertical shift of one curve in relation to another. Note that if a vertical line is drawn at any position x, then the slope of every curve intersecting this line has exactly the same value, namely, $4x^2 + 7x$. In particular, the slope of each cubic at $x = 0$ is zero.

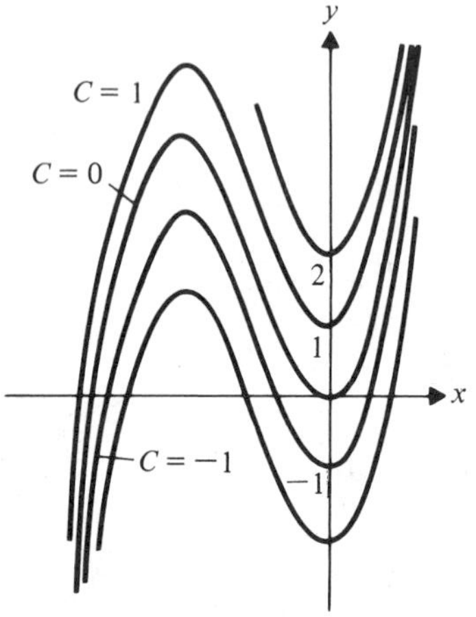

Figure 5.1

b) Because the unknown function $y = f(x)$ appears on the right-hand side of the equation, we cannot immediately evaluate antiderivatives of both sides. To remedy this, we transpose the y-term to the left-hand side and write

$$y\frac{dy}{dx} = -\frac{x}{3}.$$

If we now take antiderivatives of both sides with respect to x, we have

$$\int y\frac{dy}{dx}\,dx = \int -\frac{x}{3}\,dx,$$

or

$$\frac{y^2}{2} = -\frac{x^2}{6} + C.$$

Geometrically, this solution represents the one-parameter family of curves

$$x^2 + 3y^2 = D, \qquad (D = 6C),$$

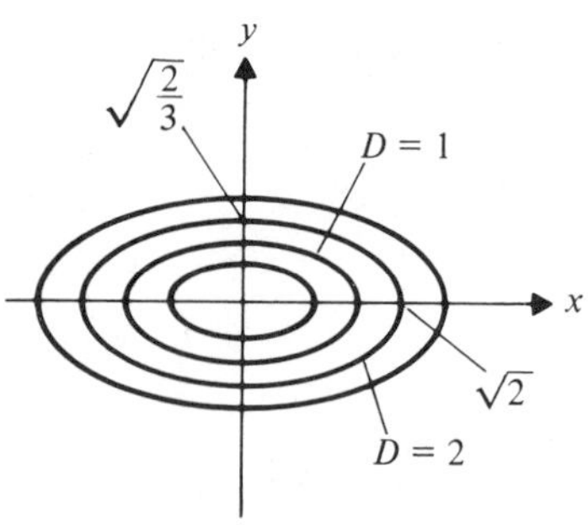

Figure 5.2

shown in Fig. 5.2. Algebraically speaking, we have shown that any differentiable function defined implicitly by the equation $x^2 + 3y^2 = D$ has a derivative satisfying the equation $dy/dx = -x/(3y)$. □

Note that what we did in Example 3(b) is to reverse the process of implicit differentiation.

EXERCISES 5.1

Evaluate the indefinite integral in each of Exercises 1–20.

1. $\int (x^3 - 2x)\,dx$
2. $\int (x^4 + 3x^2 + 5x)\,dx$
3. $\int (2x^3 - 3x^2 + 6x + 6)\,dx$
4. $\int \sin x\,dx$
5. $\int 3\cos x\,dx$
6. $\int \left(x^{10} - \frac{1}{x^3}\right)dx$
7. $\int \left(\frac{1}{x^2} - \frac{2}{x^4}\right)dx$
8. $\int \sqrt{x}\,dx$
9. $\int (x^{3/2} - x^{2/7})\,dx$
10. $\int \left(\frac{1}{x^2} + \frac{1}{2\sqrt{x}}\right)dx$
11. $\int \left(\frac{4}{x^{3/2}} + 2x^{1/3}\right)dx$
12. $\int \left(-\frac{1}{2x^2} + 3x^3\right)dx$
13. $\int \frac{1}{x^\pi}\,dx$
14. $\int (2\sqrt{x} + 3x^{3/2} - 5x^{5/2})\,dx$

15. $\int x^2(x^2 - 3)\,dx$

16. $\int \sqrt{x}(x + 1)\,dx$

17. $\int \left(\frac{x - 2}{x^3}\right) dx$

18. $\int x^2(1 + x^2)^2\,dx$

19. $\int (x^2 + 1)^3\,dx$

20. $\int \frac{(x - 1)^2}{\sqrt{x}}\,dx$

In Exercises 21–24, find the curve $y = f(x)$ that passes through the given point and whose slope at each point (x, y) is defined by the indicated derivative.

21. $dy/dx = x^2 - 3x + 2$, $(2, 1)$

22. $dy/dx = 2x^3 + 4x$, $(0, 5)$

23. $dy/dx = -2x^4 + 3x^2 + 6$, $(1, 0)$

24. $dy/dx = 2 - 4x + 8x^7$, $(1, 1)$

25. Find the equation of the curve that has a second derivative equal to $6x^2$ and passes through the points $(0, 2)$ and $(-1, 3)$.

26. Find a function $f(x)$ that has a relative maximum $f(2) = 3$ and has a second derivative equal to $-5x$.

27. Is it possible to find a function $f(x)$ that has a relative minimum $f(2) = 3$ and has a second derivative equal to $-5x$?

In each of Exercises 28–31, find a one-parameter family of functions (defined explicitly or implicitly) that satisfies the equation.

28. $\frac{dy}{dx} = x^3 - \frac{1}{x^2}$

29. $\frac{dy}{dx} = 3xy^2$

30. $\frac{dy}{dx} = \frac{(x + 1)^2}{y}$

31. $2y^2x^2\frac{dy}{dx} = 1 + x^3$

In Exercises 32–45, guess the antiderivative. If you guess incorrectly, ask yourself whether the guess can be multiplied by a suitable constant to yield the correct antiderivative. If it can, do so; if it cannot, try another guess. Keep two formulas in mind each time you guess: formula (5.5) and the power rule (3.9) for differentiation.

32. $\int \sqrt{x + 2}\,dx$

33. $\int (x + 5)^{3/2}\,dx$

34. $\int \sqrt{2 - x}\,dx$

35. $\int \frac{1}{\sqrt{x + 3}}\,dx$

36. $\int (2x - 3)^{3/2}\,dx$

37. $\int (x + 1)^5\,dx$

38. $\int (1 - 2x)^7\,dx$

39. $\int \frac{1}{(x + 4)^2}\,dx$

40. $\int \frac{1}{(1 + 3x)^6}\,dx$

41. $\int x(x^2 + 1)^3\,dx$

42. $\int x^2(2 + 3x^3)^7\,dx$

43. $\int \frac{x}{(2 + x^2)^2}\,dx$

44. $\int \cos 2x\,dx$

45. $\int \cos^3 x \sin x\,dx$

In each of Exercises 46–55, find a one-parameter family of functions (defined explicitly or implicitly) that satisfies the equation.

46. $\frac{dy}{dx} = x\sqrt{1 - x^2}$

47. $\sqrt{x - 1}\frac{dy}{dx} = \frac{1}{y}$

48. $\frac{dy}{dx} = (2x + 1)(x^2 + x)^{15}$

49. $\frac{dy}{dx} = \frac{9x^2 - 6x}{(x^3 - x^2 + 4)^4}$

50. $\frac{dy}{dx} = -\frac{2x}{1 + 3y^2}$

51. $\frac{dy}{dx} = \frac{2x + 2xy^3 - y}{x - 3x^2y^2}$

52. $(2x^2 + 3y)\frac{dy}{dx} = \frac{1 - 2xy^2}{y}$

53. $(y - 3x^3y^2\sqrt{x^2 + y^2})\frac{dy}{dx} = -x + 3x^2y^3\sqrt{x^2 + y^2}$

54. $\frac{dy}{dx} = -\frac{y + \sin y}{x \cos y + x - 2y}$

55. $(2\sqrt{2 + \cos y} + \sin y)\frac{dy}{dx} = 2\sqrt{2 + \cos y}$

56. Find a function $y = f(x)$ that satisfies the equation

$$\frac{dy}{dx} = \frac{1}{x^2}$$

and passes through the two points $(1, 1)$ and $(-1, -2)$.

57. a) Find the indefinite integral of the signum function

$$\operatorname{sgn} x = \begin{cases} 1, & x > 0, \\ 0, & x = 0, \\ -1, & x < 0, \end{cases}$$

on any interval not containing $x = 0$.

*b) Prove that sgn x cannot have an antiderivative on any interval containing $x = 0$.

*58. Find the indefinite integral of the Heaviside unit step function

$$H_a(x) = \begin{cases} 0, & x < a, \\ 1, & x > a. \end{cases}$$

Sketch a graph of the indefinite integral function.

5.2 VELOCITY AND ACCELERATION

For motion along the x-axis, velocity is the derivative of position with respect to time: $v(t) = dx/dt$. We can say, therefore, that position is the antiderivative of velocity:

$$x(t) = \int v(t)\,dt. \tag{5.6}$$

Similarly, because acceleration is the derivative of velocity,

$$v(t) = \int a(t)\,dt. \tag{5.7}$$

Thus, given the acceleration of an object moving along a straight line, we can antidifferentiate to find its velocity and antidifferentiate again for its position. Since each antidifferentiation introduces an arbitrary constant, additional information must be specified in order to evaluate these constants.

Example 4 A car pulls away from rest at a stoplight and its initial acceleration is 10 meters per second per second. This acceleration decreases linearly to zero after 10 seconds. Find the velocity and position of the car during this time interval.

Solution Let us set up a coordinate system as shown in Fig. 5.3 and choose time $t = 0$ when the car pulls away from the light. Given this time convention, we see that the acceleration of the car as a function of time t (see Fig. 5.4) is given by

$$a(t) = 10 - t, \qquad 0 \le t \le 10.$$

If $v(t)$ denotes the velocity of the car during the time interval, then

$$\frac{dv}{dt} = 10 - t,$$

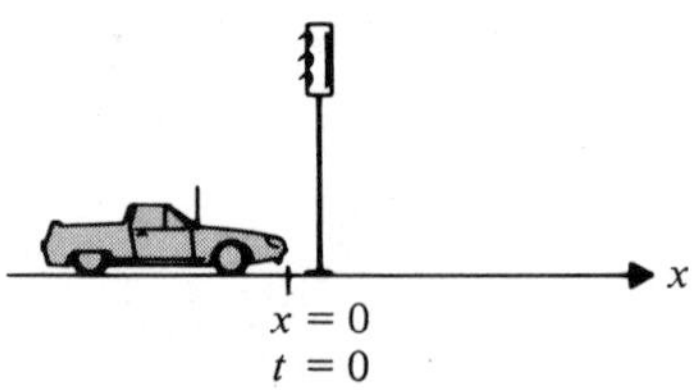

Figure 5.3

and antidifferentiation gives us

$$v(t) = 10t - \frac{t^2}{2} + C.$$

By our time convention, the velocity of the car is zero at time $t = 0$; i.e., $v(0) = 0$, and this implies that

$$0 = 10(0) - \frac{(0)^2}{2} + C.$$

Consequently,

$$v(t) = 10t - \frac{t^2}{2}\ \text{m/s},$$

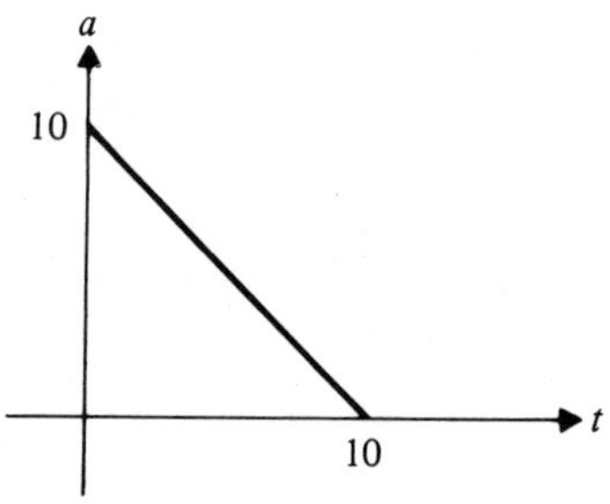

Figure 5.4

and we now know the velocity of the car during the time interval $0 \le t \le 10$.

Since the position of the car with respect to its original position at time $t = 0$ is denoted by x,

$$v = \frac{dx}{dt} = 10t - \frac{t^2}{2}.$$

Thus,

$$x(t) = 5t^2 - \frac{t^3}{6} + D.$$

Since $x(0) = 0$, D must also be zero, and the position of the car for $0 \le t \le 10$ is

$$x(t) = 5t^2 - \frac{t^3}{6} \text{ m. } \square$$

Example 5 A stone is thrown vertically upward over the edge of a cliff at a rate of 25 meters per second. When does it hit the base of the cliff if the cliff is 100 meters high?

Solution Let us measure y as positive upward, taking $y = 100$ and $t = 0$ at the instant of projection (Fig. 5.5). A law of physics states that when an object near the earth's surface is acted on by gravity, and gravity alone, it will experience an acceleration whose magnitude is 9.81 meters per second per second.† If a denotes the acceleration of the stone and v its velocity, then

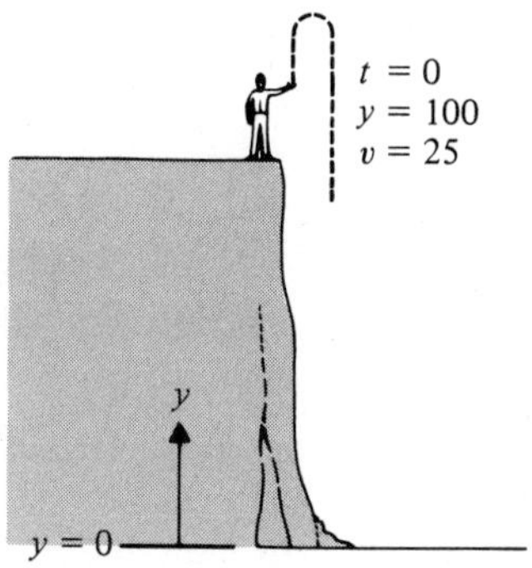

Figure 5.5

$$a = \frac{dv}{dt} = -9.81$$

(and a is negative since it is in the negative y-direction). To obtain v we take antiderivatives with respect to t:

$$v(t) = -9.81t + C.$$

By our time convention, $v(0) = 25$ so that

$$25 = -9.81(0) + C.$$

Thus,

$$v(t) = -9.81t + 25 \text{ m/s}, \qquad t \ge 0,$$

and we now know the velocity of the stone at any given instant. To find the position of the stone we set $v = dy/dt$,

$$\frac{dy}{dt} = -9.81t + 25,$$

and take antiderivatives once again:

$$y(t) = -4.905t^2 + 25t + D.$$

Since $y = 100$ when $t = 0$,

$$100 = -4.905(0)^2 + 25(0) + D$$

and

$$y(t) = -4.905t^2 + 25t + 100 \text{ m}, \qquad t \ge 0.$$

We have now found the equation that tells us exactly where the stone is at any given time. To determine when the stone strikes the base of the cliff, we set $y(t) = 0$; i.e.,

$$0 = -4.905t^2 + 25t + 100,$$

† The acceleration experienced by an object falling under the influence of the earth's gravitational pull is usually denoted by the letter g. Its value near the earth's surface is $g = \pm 9.81$ meters per second per second (the sign depends on whether position is measured positive upward or downward).

a quadratic equation with solutions

$$t = \frac{-25 \pm \sqrt{25^2 - 4(-4.905)(100)}}{-9.81} = \frac{25 \pm \sqrt{2587}}{9.81} = 7.7 \quad \text{or} \quad -2.6.$$

Since the negative root must be rejected, we find that the stone strikes the base of the cliff after 7.7 seconds. □

Example 6 Two stones are thrown vertically upward one second apart over the edge of the cliff in Example 5. The first is thrown at a rate of 25 meters per second; the second at 20 meters per second. Determine if and when they ever pass each other.

Solution Let us measure y as positive upward, taking $y = 0$ and $t = 0$ at the instant of projection of the first stone (Fig. 5.6). The acceleration of stone 1 is

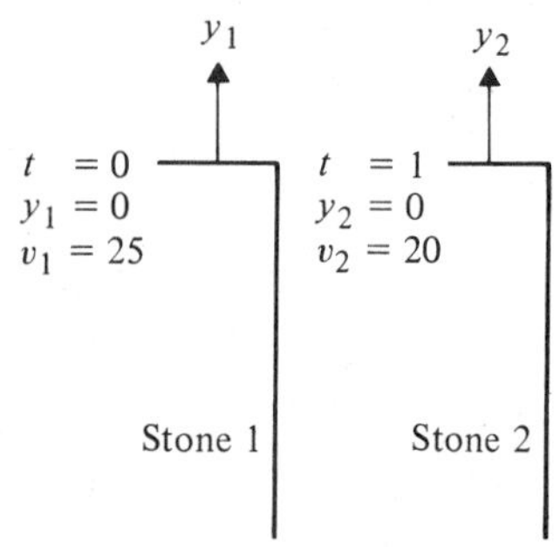

Figure 5.6

$$a_1 = -9.81 = \frac{dv_1}{dt},$$

from which we get

$$v_1(t) = -9.81t + C.$$

Since $v_1(0) = 25$, $25 = C$ and

$$v_1(t) = -9.81t + 25 \text{ m/s}, \qquad t \geq 0.$$

Thus,

$$\frac{dy_1}{dt} = -9.81t + 25,$$

from which we have

$$y_1(t) = -4.905t^2 + 25t + D.$$

Since $y_1(0) = 0$, $0 = D$ and

$$y_1(t) = -4.905t^2 + 25t \text{ m}, \qquad t \geq 0.$$

The acceleration of stone 2 is also -9.81; hence we have

$$a_2 = -9.81 = \frac{dv_2}{dt},$$

from which we get

$$v_2(t) = -9.81t + E.$$

Since $v_2(1) = 20$, $20 = -9.81(1) + E$, or $E = 29.81$. Thus,

$$v_2(t) = -9.81t + 29.81 \text{ m/s}, \qquad t \geq 1.$$

Consequently,

$$\frac{dy_2}{dt} = -9.81t + 29.81,$$

from which

$$y_2(t) = -4.905t^2 + 29.81t + F.$$

Since $y_2(1) = 0$, $0 = -4.905(1)^2 + 29.81(1) + F$, or $F = -24.905$. Thus,

$$y_2(t) = -4.905t^2 + 29.81t - 24.905 \text{ m}, \qquad t \geq 1.$$

The stones will pass each other if y_1 and y_2 are ever equal for the same time t; i.e., if

$$-4.905t^2 + 25t = -4.905t^2 + 29.81t - 24.905.$$

The solution of this equation is

$$t = \frac{24.905}{4.81} = 5.2.$$

Because the first stone does not strike the base of the cliff for 7.7 seconds (Example 5), it follows that the stones do indeed pass one another 5.2 seconds after the first stone is projected. □

An important point to note in Examples 4, 5, and 6 is that the coordinate system and time convention were specified immediately; i.e., we decided, and did so at the start of the problem, where to place the origin of our coordinate system, which direction to choose as positive, and when to choose $t = 0$. Then, and only then, were we able to specify the correct sign for acceleration. Furthermore, we determined constants arising from antidifferentiations using initial conditions expressed in terms of our coordinate system and time convention. Throughout the solutions, we were careful to refer everything to our choice of coordinates and time. Remember, then, to specify clearly the coordinate system and time convention at the beginning of a problem.

We should also note that in each of the examples we antidifferentiated with respect to time only those equations that were valid for a range of values of time. It is a common error to antidifferentiate an equation that is not an identity. For instance, we have often seen students who would reason in Example 5 as follows: Because the initial velocity is 25, $v = dy/dt = 25$ and hence, $y(t) = 25t + C$. This is not correct, however, since the equation $dy/dt = 25$ is valid only at time $t = 0$ and cannot, therefore, be antidifferentiated.

EXERCISES 5.2

In each of Exercises 1–8, we have defined acceleration $a(t)$ of an object moving along the x-axis during some time interval. Find the velocity $v(t)$ and position $x(t)$ of the object as functions of time, using the initial conditions $v(0)$ and $x(0)$ specified.

1. $a(t) = t + 2$, $0 \leq t \leq 3$; $v(0) = 0$, $x(0) = 0$
2. $a(t) = 6 - 2t$, $0 \leq t \leq 3$; $v(0) = 5$, $x(0) = 0$
3. $a(t) = 6 - 2t$, $0 \leq t \leq 4$; $v(0) = 5$, $x(0) = 0$
4. $a(t) = 120t - 12t^2$, $0 \leq t \leq 10$; $v(0) = 0$, $x(0) = 4$
5. $a(t) = t^2 + 1$, $0 \leq t \leq 5$; $v(0) = -1$, $x(0) = 1$
6. $a(t) = t^2 + 5t + 4$, $0 \leq t \leq 15$; $v(0) = -2$, $x(0) = -3$
7. $a(t) = \cos t$, $t \geq 0$; $v(0) = 0$, $x(0) = 0$
8. $a(t) = 3 \sin t$, $t \geq 0$; $v(0) = 1$, $x(0) = 4$
9. Find how far a plane will move when landing if, in t seconds after touching the ground, its speed in meters per second is given by $180 - 18t$.
10. A stone is thrown directly upward with an initial speed of 10 meters per second. How high will it rise?
11. You are standing on a bridge 25 meters above a river. If you wish to drop a stone onto a piece of floating wood, how soon before the wood reaches the appropriate spot should you drop the stone?

12. You are standing at the base of a building and wish to throw a ball to a friend on the roof 20 meters above you. With what speed must you throw the ball?

13. A car is traveling at a rate of 20 meters per second when the brakes are applied. What constant deceleration must the car experience if it is to stop before striking a tree that is 50 meters from the car at the instant the brakes are applied? Assume that the car travels in a straight line.

14. The position of an object moving along the x-axis is given by

$$x(t) = t^3 - 6t^2 + 9t - 20, \qquad t \geq 0.$$

Sketch graphs of the position, velocity, and acceleration functions for this motion. Pay particular attention to the fact that $v(t)$ represents the slope of $x = x(t)$ and $a(t)$ is the slope of $v = v(t)$.

15. Repeat Exercise 14 given that the acceleration of the object is

$$a(t) = 6t - 30, \qquad t \geq 0,$$

and $v(0) = -33$, $x(0) = 400$.

16. You are called on as an expert to testify at a traffic accident hearing. The question concerns the speed of a car that made an emergency stop with brakes locked and wheels sliding. The skid mark on the road measured 9 meters. Assuming that the deceleration of the car could not exceed the acceleration due to gravity of a freely falling body (and this is indeed a reasonable assumption), what can you say about the speed of the car before the brakes were applied? Are you testifying for the prosecution or the defense?

17. When the brakes of an automobile are applied, they produce a constant deceleration of 5 meters per second per second. What is the distance (from point of application of brakes) required to stop a car traveling at:

a) 100 kilometers per hour?

b) 50 kilometers per hour?

c) What is the ratio of these distances?

d) Repeat (a), (b), and (c) given that the reaction time of the driver to get his foot from the accelerator to the brake is $\frac{3}{4}$ second, and distances are calculated taking this reaction time into account.

18. a) Two trains, one traveling at 100 kilometers per hour and the other at 60 kilometers per hour, are headed toward one another along a straight level track. When they are 2 kilometers apart, each engineer sees the other's train and locks his wheels. If the deceleration of each train has magnitude $\frac{1}{4}$ meter per second per second, determine whether a collision occurs.

b) Repeat part (a) given that the deceleration is caused by the wheels being reversed rather than locked.

c) Illustrate graphically the difference between the situations in (a) and (b).

19. A steel bearing is dropped from the roof of a building. An observer standing in front of a window 1 meter high notes that the bearing takes $\frac{1}{8}$ second to fall from the top to the bottom of the window. The bearing continues to fall, makes a completely elastic collision with a horizontal sidewalk, and reappears at the bottom of the window 2 seconds after passing it on the way down. How tall is the building? (After a completely elastic collision, the bearing will have the same speed at a point going up as it had going down.)

20. In the theory of special relativity, Newton's Second Law ($F = ma$) is replaced by

$$F = m_0 \frac{d}{dt}\left\{\frac{v}{\sqrt{1 - (v^2/c^2)}}\right\},$$

where F is the applied force, m_0 the mass of the particle measured at rest, v its speed, and c the speed of light (a constant). Show that if we set $a = dv/dt$, then

$$F = \frac{m_0 a}{[1 - (v^2/c^2)]^{3/2}}.$$

Explain the difference between this law and Newton's Second Law.

***21.** Two stones are thrown vertically upward over the edge of a bottomless abyss, the second stone t_0 units of time after the first. The first stone has an initial speed of v' , and the second an initial speed of v_0''.

a) Show that if the two stones are ever to pass each other during their motions, two conditions must be satisfied:

$$gt_0 > v_0' - v_0'' \quad \text{and} \quad v_0' > gt_0/2,$$

where $g > 0$ is the acceleration due to gravity.

b) Show that the first condition is equivalent to the requirement that stone 1 must begin its downward trajectory before stone 2.

c) Show that the second condition is equivalent to the requirement that stone 1 must not pass its original projection point before the projection of stone 2.

5.3 CHANGE OF VARIABLE IN THE INDEFINITE INTEGRAL

It is not too difficult to guess that

$$\int (3x + 2)^7 \, dx = \tfrac{1}{24}(3x + 2)^8 + C.$$

But what could we do if this were not clear, or if the indefinite integral were not quite so simple? In this section we show how a change of variable will often make a difficult integration problem a simpler one.

Suppose in this problem we set

$$u = 3x + 2.$$

Then the function $(3x + 2)^7$ is equal to u^7. Now the differential dx is used to indicate that antidifferentiation in the problem is with respect to x. But surely there must be another reason why we choose the differential to denote this. If we regard $(3x + 2)^7 \, dx$ as a product of $(3x + 2)^7$ and dx (as it certainly appears), then perhaps we should obtain an expression for dx in terms of du. To do this we note that since $u = 3x + 2$, then

$$\frac{du}{dx} = 3, \quad \text{or} \quad dx = \frac{du}{3}.$$

If we make these substitutions, then

$$\int (3x + 2)^7 \, dx$$

is replaced by

$$\int u^7 \frac{du}{3} = \frac{1}{3} \int u^7 \, du = \frac{1}{24} u^8 + C.$$

If u is now replaced by $3x + 2$, we obtain the correct result:

$$\tfrac{1}{24}(3x + 2)^8 + C.$$

We have found then for this example an alternative method of solution that eliminates the necessity of our being clever guessers. That this method is generally applicable is justified in the following theorem.

▶ **Theorem 5.3** If by the substitution $u = g(x)$, the product $f(x)\,dx$ becomes $h(u)\,du$, and if

$$\int h(u) \, du = H(u) + C,$$

then

$$\int f(x) \, dx = H[g(x)] + C. \quad ◀$$

Proof To prove this theorem we must show that

$$\frac{d}{dx}\{H[g(x)]\} = f(x).$$

If we set $y = H(u)$, then by the chain rule, we have

$$\frac{dy}{dx} = \frac{dy}{du}\frac{du}{dx} = H'(u)\frac{du}{dx} = h(u)\frac{du}{dx}.$$

But $h(u)\,du = f(x)\,dx$ so that

$$\frac{dy}{dx} = \frac{f(x)\,dx}{dx} = f(x),$$

and the proof is complete. □

Example 7 Evaluate each of the following indefinite integrals:

a) $\int \sqrt[5]{2x+4}\,dx$;

b) $\int \frac{x}{\sqrt{x+1}}\,dx$;

c) $\int \sin^3 x \cos^2 x\,dx$.

Solution a) We can guess the solution in this case:

$$\int (2x+4)^{1/5}\,dx = \tfrac{5}{12}(2x+4)^{6/5} + C.$$

Alternatively, if we set $u = 2x + 4$, then $du = 2\,dx$ and

$$\int (2x+4)^{1/5}\,dx = \int u^{1/5}\frac{du}{2} = \frac{1}{2}\left\{\frac{5}{6}u^{6/5}\right\} + C = \frac{5}{12}(2x+4)^{6/5} + C.$$

b) If we set $u = x + 1$, then $du = dx$ and

$$\int \frac{x}{\sqrt{x+1}}\,dx = \int \frac{(u-1)}{\sqrt{u}}\,du = \int (u^{1/2} - u^{-1/2})\,du = \frac{2}{3}u^{3/2} - 2u^{1/2} + C$$

$$= \frac{2}{3}(x+1)^{3/2} - 2(x+1)^{1/2} + C.$$

We could alternatively set $u = \sqrt{x+1}$, in which case

$$du = \frac{1}{2\sqrt{x+1}}\,dx$$

and

$$\int \frac{x}{\sqrt{x+1}}\,dx = \int (u^2 - 1)(2\,du) = 2\left\{\frac{u^3}{3} - u\right\} + C$$

$$= \tfrac{2}{3}(x+1)^{3/2} - 2(x+1)^{1/2} + C.$$

c) If we set $u = \cos x$, then $du = -\sin x\,dx$ and

$$\int \sin^3 x \cos^2 x\,dx = \int \sin^2 x \cos^2 x \sin x\,dx$$

$$= \int (1 - \cos^2 x)\cos^2 x \sin x\,dx$$

$$= \int (1 - u^2)u^2(-du) = \int (u^4 - u^2)\,du$$

$$= \frac{u^5}{5} - \frac{u^3}{3} + C = \frac{1}{5}\cos^5 x - \frac{1}{3}\cos^3 x + C. \;\square$$

EXERCISES 5.3

Evaluate the indefinite integral in each of Exercises 1–24.

1. $\int (5x + 14)^9\, dx$
2. $\int \sqrt{1 - 2x}\, dx$
3. $\int \frac{1}{(3y - 12)^4}\, dy$
4. $\int \frac{5}{(5 - 42x)^{1/4}}\, dx$
5. $\int x^2(3x^3 + 10)^4\, dx$
6. $\int \frac{x}{(x^2 + 4)^2}\, dx$
7. $\int \sin^4 x \cos x\, dx$
8. $\int \frac{x^2}{(x - 2)^4}\, dx$
9. $\int z\sqrt{1 - 3z}\, dz$
10. $\int \frac{x}{\sqrt{2x + 3}}\, dx$
11. $\int \frac{1 + \sqrt{x}}{\sqrt{x}}\, dx$
12. $\int s^3\sqrt{s^2 + 5}\, ds$
13. $\int \sin^2 x \cos^3 x\, dx$
14. $\int \sqrt{1 - \cos x} \sin x\, dx$
15. $\int \frac{x^3}{(3 - x^2)^3}\, dx$
16. $\int y^2\sqrt{y - 4}\, dy$
17. $\int \frac{(1 + \sqrt{u})^{1/2}}{\sqrt{u}}\, du$
18. $\int x^8(3x^3 - 5)^6\, dx$
19. $\int \frac{1 + z^{1/4}}{\sqrt{z}}\, dz$
20. $\int \frac{x + 1}{(x^2 + 2x + 2)^{1/3}}\, dx$
21. $\int \frac{(x - 1)(x + 2)}{\sqrt{x}}\, dx$
22. $\int \frac{\cos^3 x}{(3 - 4 \sin x)^4}\, dx$
23. $\int \sqrt{1 + \sin(4t)} \cos^3(4t)\, dt$
24. $\int \sqrt{1 + \sqrt{x}}\, dx$
25. Evaluate
$$\int \sin^3 x \cos^3 x\, dx$$
by:
   a) making the substitution $u = \sin x$;
   b) making the substitution $u = \cos x$.
   c) Verify that these answers are the same.
26. Evaluate
$$\int \sqrt{\frac{x^2}{1 + x}}\, dx.$$

In each of Exercises 27–30, use the suggested change of variable to evaluate the indefinite integral.

27. $\int \frac{\sqrt{4x - x^2}}{x^3}\, dx;$ set $u = \frac{2}{x}$.
28. $\int \frac{\sqrt{x - x^2}}{x^4}\, dx;$ set $u = \frac{1}{x}$.
29. $\int \frac{x}{(5 - 4x - x^2)^{3/2}}\, dx;$ set $u^2 = \frac{5 + x}{1 - x}$.
*30. $\int \frac{1}{3(1 - x^2) - (5 + 4x)\sqrt{1 - x^2}}\, dx;$ set $u^2 = \frac{1 - x}{1 + x}$
*31. Show that the substitution $u - x = \sqrt{x^2 + x + 4}$ will replace the integral
$$\int \sqrt{x^2 + x + 4}\, dx$$
with the integral of a rational function of u.
*32. Show that the substitution $(x + 1)u = \sqrt{4 + 3x + x^2}$ will replace the integral
$$\int \frac{1}{\sqrt{4 + 3x - x^2}}\, dx$$
with the integral of a rational function of u.

5.4 AN INTRODUCTION TO DIFFERENTIAL EQUATIONS

In Example 3 we solved the equations
$$\frac{dy}{dx} = 4x^2 + 7x \quad \text{and} \quad \frac{dy}{dx} = -\frac{x}{3y}$$
for functions $y = f(x)$. When an equation such as either one of these is to be solved for an unknown function $y = f(x)$ and the equation contains derivatives of $f(x)$, it is called a *differential equation* in $y = f(x)$. In the exercises of Chapter 3 we encountered other differential equations, although at that time we referred to them simply as equations.

Differential equations is one of the most important areas of calculus. In this section we illustrate how a number of problems from completely different areas lead to the use of differential equations, and how these equations can be solved mathematically. These examples are in no way exhaustive, as a quick glance at any book on differential equations will testify. We hope only to show how calculus can be used to set up mathematical models describing physical situations, and how analysis of these models leads to conclusions about the original physical problems.

Example 8 An upright cylindrical container (Fig. 5.7) has height 2 meters and radius 1 meter. Water flows out through a hole in the bottom with speed in meters per second given by

$$v = \frac{3\sqrt{2gD}}{\ },$$

where D (in meters) is the depth of water in the container and $g = 9.81$ meters per second per second is the acceleration due to gravity. Find a formula for D as a function of time if the tank is originally full and if the area of the hole is 3 square centimeters.

Solution If V is the volume of water in the container at any time t, then

$$V = \pi(1)^2 D = \pi D.$$

The derivative

$$\frac{dV}{dt} = \pi \frac{dD}{dt}$$

represents the rate of change of the volume of water in the container, where it is clear that $dV/dt < 0$ since the volume is decreasing. But this rate of change must be equal to the rate at which the water exits through the hole. Since the area of the hole is 0.0003 square meter, and water exits with speed v meters per second, it follows that water leaves the container at a rate of $0.0003v$ cubic meter per second. Consequently,

$$-\pi \frac{dD}{dt} = 0.0003v = 0.00018\sqrt{2gD},$$

and the unknown function $D = f(t)$ must satisfy the differential equation

$$\frac{1}{\sqrt{D}} \frac{dD}{dt} = -\frac{0.00018\sqrt{2g}}{\pi}.$$

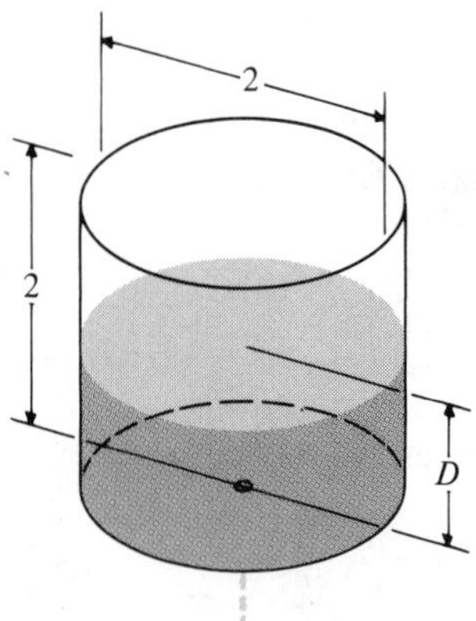

Figure 5.7

Antiderivatives with respect to t yield

$$2\sqrt{D} = -\frac{0.00018\sqrt{2g}}{\pi} t + C.$$

If we choose time $t = 0$ when the container is full, then $f(0) = 2$ and

$$2\sqrt{2} = C.$$

Thus the solution of the differential equation for $D = f(t)$ is defined implicitly by the equation

$$2\sqrt{D} = -\frac{0.00018\sqrt{2g}}{\pi} t + 2\sqrt{2}.$$

By solving this equation for D in terms of t, we obtain the explicit solution

$$D = f(t) = \left(\sqrt{2} - \frac{0.00009\sqrt{2g}}{\pi}t\right)^2.$$

What does this solution tell us about the physical situation? For one thing, we can determine how long it takes for the container to empty. If we set $D = 0$ in the implicit definition of $f(t)$, we get

$$0 = -\frac{0.00018\sqrt{2g}}{\pi}t + 2\sqrt{2}.$$

Thus,

$$t = \frac{\pi}{0.00009\sqrt{g}} = \frac{\pi}{0.00009\sqrt{9.81}} = 11{,}145,$$

and the container therefore empties in 186 minutes. □

Example 9 Find all curves for which the normal at every point passes through the origin.

Solution If $y = f(x)$ is the equation of any curve whatsoever, then the slope of the tangent line at a point (x, y) on this curve is dy/dx, and the slope of the normal line is the negative reciprocal $-dx/dy$. For a curve satisfying the given condition, the normal line must pass through the origin; hence the slope of the normal line at (x, y) must also be $(y - 0)/(x - 0)$ (Fig. 5.8). It follows that

$$\frac{y}{x} = -\frac{dx}{dy},$$

or

$$\frac{dy}{dx} = -\frac{x}{y},$$

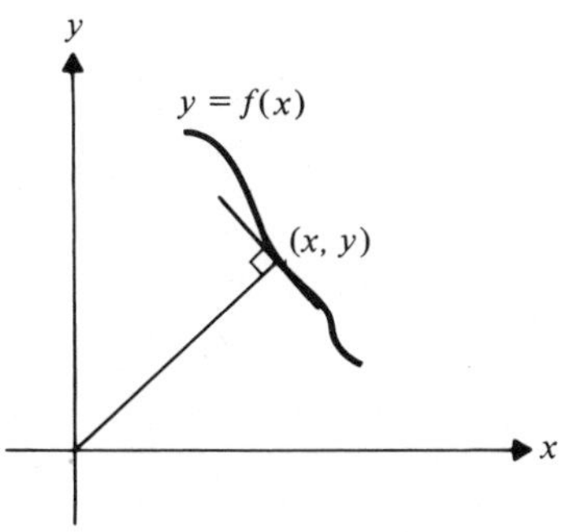

Figure 5.8

and this differential equation must be satisfied at every point on the required curve. To solve the equation we write

$$y\frac{dy}{dx} = -x$$

and antidifferentiate both sides with respect to x:

$$\frac{y^2}{2} = -\frac{x^2}{2} + C.$$

If we set $D^2 = 2C$, then

$$x^2 + y^2 = D^2;$$

therefore, the only curves satisfying the given condition are circles centered at the origin. □

Example 10 When a mothball is exposed to the air, it slowly evaporates. If the mothball remains spherical at all times, and its radius after one year is only one-half its original radius, when will it completely disappear?

Solution The size of the mothball can be measured in various ways—by its volume, its surface area, or its radius. Since we were given information about its radius, let us denote by $r = f(t)$ the radius of the mothball as a function of time t, and choose time $t = 0$ when the mothball is originally exposed to the air.

If we let R represent the original radius of the mothball, then we can say that $f(0) = R$ and $f(1) = R/2$. This is the only information about $f(t)$ that we were given, and it certainly is not enough to solve the problem.

What we need is some statement about the instantaneous rate of evaporation of the mothball. Since we were not given such a statement, we must make our own. In other words, we must make what we feel is a reasonable assumption about how fast the mothball evaporates. Since evaporation takes place at the surface, we assume that the amount of substance evaporating per unit time is proportional to the surface area of the mothball. If V and A represent volume and area, respectively, of the mothball at any time, then this assumption translates into the language of calculus as

$$\frac{dV}{dt} = kA, \qquad k = \text{a constant.}$$

But now we have four variables in the problem: t, r, A, and V. By substituting $V = \frac{4}{3}\pi r^3$ and $A = 4\pi r^2$, we can eliminate V and A:

$$\frac{d}{dt}\left\{\frac{4}{3}\pi r^3\right\} = k(4\pi r^2).$$

When differentiation is carried out on the left-hand side, we get

$$4\pi r^2 \frac{dr}{dt} = 4\pi r^2 k,$$

and this equation reduces to

$$\frac{dr}{dt} = k.$$

Now we have a differential equation for $r = f(t)$, with solution

$$r = f(t) = kt + C.$$

The conditions $f(0) = R$ and $f(1) = R/2$ imply that $C = R$ and $k = -R/2$. Consequently,

$$r = f(t) = R - \frac{Rt}{2} = R\left(1 - \frac{t}{2}\right).$$

The mothball completely disappears when $r = 0$, and this occurs when $t = 2$ years. □

Example 11 A tank originally contains 1000 liters of water in which has been dissolved 5 kilograms of salt.

a) If a brine mixture containing 2 kilograms of salt for each 100 liters of solution is poured into the tank at a rate of 10 milliliters per second, find the amount of salt in the tank as a function of time.

b) If at the same time brine is being added, the mixture in the tank is being drawn off at a rate of 5 milliliters per second, find the amount of salt in the tank as a function of time. Assume that the mixture is kept uniform by stirring.

Solution a) Suppose we let $S = f(t)$ represent the number of grams of salt in the tank at time t. If we choose time $t = 0$ at the instant the brine mixture begins entering the original solution, then $f(0) = 5000$. Since 10 milliliters of mixture enter the tank each second, and each milliliter contains 0.02 gram of salt, it follows that 0.2 grams of salt enter the tank each second. Consequently, after t seconds, $0.2t$ grams of salt have been added to the tank, and the total amount of salt in the solution is

$$S = f(t) = 5000 + 0.2t \text{ grams.}$$

b) Once we knew the rate at which salt entered the tank in (a), our solution became straightforward. To try the same procedure again, we calculate the net rate of change of salt $f(t)$ in the tank since salt is constantly being added and, at the same time, being taken out. Clearly, it is still being added at the constant rate of 0.2 gram per second. The rate at which salt leaves the tank, on the other hand, is not constant; it depends on the concentration of salt in the tank. Since the tank originally contained 10^6 milliliters of solution, and a net 5 milliliters enters each second, the amount of solution in the tank at time t is $10^6 + 5t$ milliliters. It follows that the concentration of salt in the solution at time t is

$$\frac{S}{10^6 + 5t} \text{ grams/milliliter.}$$

Because solution is being drawn off at the rate of 5 milliliters per second, the rate at which salt leaves the tank is

$$\frac{S}{10^6 + 5t}(5) \text{ grams/second.}$$

The net rate at which salt enters the tank is therefore

$$0.2 - \frac{5S}{10^6 + 5t} \text{ grams/second.}$$

But dS/dt represents the rate of change of the amount of salt in the tank; hence we may set

$$\frac{dS}{dt} = \frac{1}{5} - \frac{5S}{10^6 + 5t}.$$

To find $S = f(t)$ we must solve this differential equation subject to the condition that $f(0) = 5000$. We can clear the term $10^6 + 5t$ from the denominator if we multiply the equation by this factor:

$$(10^6 + 5t)\frac{dS}{dt} + 5S = \frac{1}{5}(10^6 + 5t).$$

But note now that the left-hand side of this differential equation is the derivative of the product of S and $10^6 + 5t$; i.e.,

$$\frac{d}{dt}\{S(10^6 + 5t)\} = (10^6 + 5t)\frac{dS}{dt} + 5S.$$

Consequently, we may express the differential equation in the form

$$\frac{d}{dt}\{S(10^6 + 5t)\} = \frac{1}{5}(10^6 + 5t)$$

and now take antiderivatives with respect to t on both sides:

$$S(10^6 + 5t) = \tfrac{1}{5}\int(10^6 + 5t)\,dt = \tfrac{1}{50}(10^6 + 5t)^2 + C.$$

Since $f(0) = 5000$, we find

$$5000(10^6) = \tfrac{1}{50}(10^6)^2 + C,$$

or

$$C = -15 \times 10^9.$$

Thus $S = f(t)$ is defined implicitly by

$$S(10^6 + 5t) = \tfrac{1}{50}(10^6 + 5t)^2 - 15 \times 10^9$$

or explicitly by

$$S = f(t) = \frac{1}{50}(10^6 + 5t) - \frac{15 \times 10^9}{10^6 + 5t} \text{ grams.}$$

A sketch of this function is shown in Fig. 5.9. □

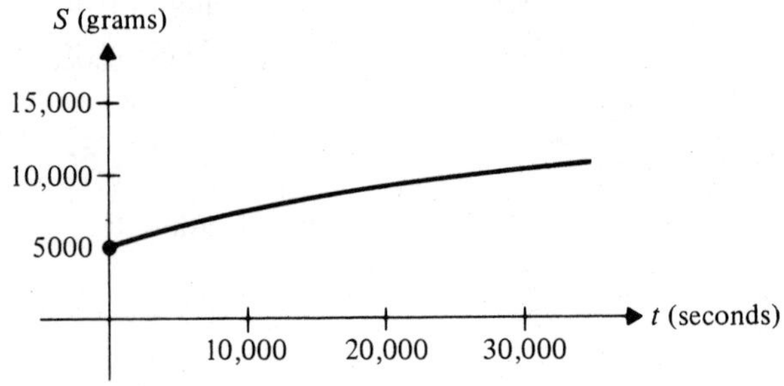

Figure 5.9

EXERCISES 5.4

1. A boy lives 6 kilometers from school. He decides to travel to school so that his speed is always proportional to the square of his distance from the school. If he is half-way to school after 1 hour, find his distance from school at any time. How long does it take him to reach school?

2. A diving board of length L and uniform mass per unit length m is fixed horizontally (Fig. 5.10). If a diver of mass M stands on the end of the board, the shape $y = f(x)$ of the board must satisfy the differential equation

$$(EI)\frac{d^2y}{dx^2} = Mg(L - x) + \frac{mg}{2}(L - x)^2,$$

where E and I are constants depending on the shape and material of the board, and $g < 0$ is the acceleration due to gravity.

a) Find $y = f(x)$.

b) What is the maximum deflection of the board from the horizontal?

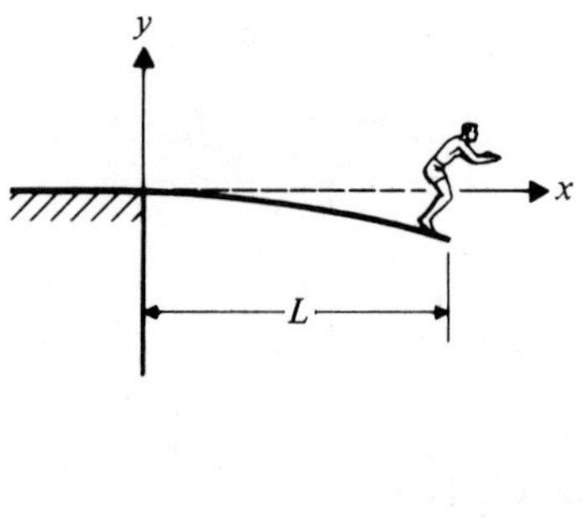

Figure 5.10

3. The water trough shown in Fig. 5.11 is 4 meters long. Its cross-section is an isosceles triangle with a $\frac{1}{2}$-meter base and a $\frac{1}{2}$-meter altitude. Water leaks out through a hole of area 1 square centimeter in the bottom with speed in meters per second given by

$$v = \sqrt{\frac{gD}{2}},$$

where D (in meters) is the depth of water in the trough, and $g > 0$ is the acceleration due to gravity. Find how long a full trough takes to empty.

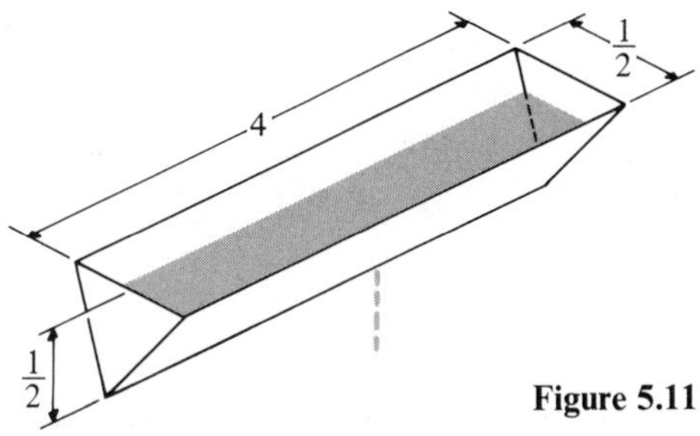

Figure 5.11

4. The y-intercept of the normal line to a curve at any point is always equal to 2. If the curve passes through the point (3, 4), find its equation.

5. A tank has 100 liters of solution containing 4 kilograms of sugar. A mixture with 10 grams of sugar per liter of solution is added at a rate of 200 milliliters per minute. At the same time, 100 milliliters of well-stirred mixture are removed each minute. Find the amount of sugar in the tank as a function of time.

6. A tank in the form of a right-circular cone of height H and radius R, with vertex below the base, is filled with water. Water escapes through a hole of cross-sectional area A at the vertex with speed $v = c\sqrt{2gD}$, where $0 < c < 1$ is a constant (called the discharge coefficient), $g > 0$ is the acceleration due to gravity, and D is the depth of water in the tank. Find a formula for the time the tank takes to empty.

7. A spring of negligible mass and constant $k > 0$ is attached to a wall at one end and a mass M at the other (Fig. 5.12). The mass is free to slide horizontally along a frictionless surface. If $x = 0$ is taken as the position of M when the spring is unstretched and M is set into motion, the differential equation describing the position $x(t)$ of M is

$$\frac{d^2x}{dt^2} = -\frac{k}{M}x.$$

If motion is initiated by imparting a speed v_0 in the positive x-direction to M at position $x = 0$, find the velocity of M as a function of position. *Hint:* Note that

$$\frac{d^2x}{dt^2} = \frac{dv}{dt} = \frac{dv}{dx}\frac{dx}{dt} = v\frac{dv}{dx}.$$

8. When a mass m falls under the influence of gravity, and gravity alone, it experiences an acceleration d^2r/dt^2 described by

$$m\frac{d^2r}{dt^2} = -\frac{GmM}{r^2},$$

where M is the mass of the earth, G is a positive constant, and r is the distance from m to the center of the earth (Fig. 5.13). If m is dropped from a height h above the surface of the earth, find its velocity when it strikes the earth. What is the maximum attainable speed of m?

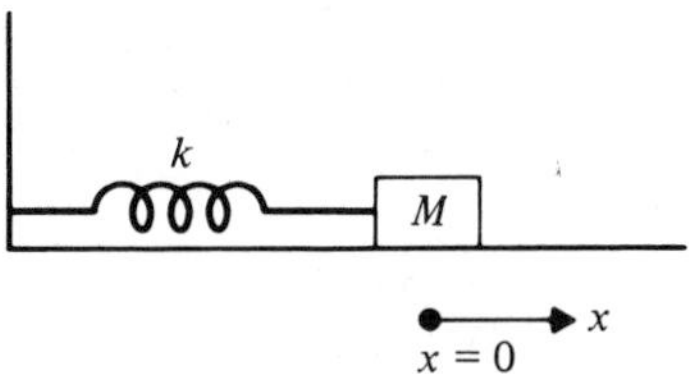

Figure 5.12

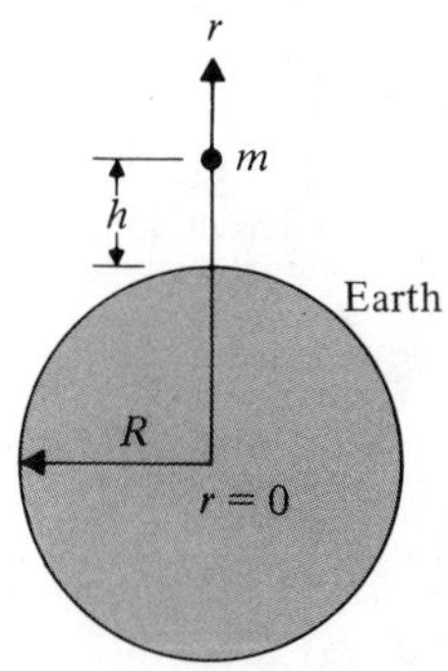

Figure 5.13

9. A container in the form of an inverted right-circular cone of radius 4 centimeters and height 10 centimeters is full of water. Water evaporates from the surface at a rate proportional to the area of the surface. If the water level drops 1 centimeter in the first 5 days, how long will it take for the water to evaporate completely?

10. Allometry is the quantitative study of the size of one part of a biological object in relation to the size of another part. For example, a researcher might be interested in the rate at which a child's eyes grow in relation to the rate of growth of his or her head. It is clear that the head grows more rapidly than the eyes, but just how much more rapidly? If $x = f(t)$ and $y = g(t)$ represent volumes of the eye and head, respectively, at any given time t, then $(1/x)\,dx/dt$ and $(1/y)\,dy/dt$ are called *relative growth rates*. Numerous experimental results have suggested that these relative growth rates are in the proportion

$$\frac{1}{y}\frac{dy}{dt} = k\frac{1}{x}\frac{dx}{dt},$$

called the *allometric law*. If we eliminate t from this equation, we obtain a differential equation for y as a function of x:

$$\frac{dy}{dx} = k\frac{y}{x}.$$

Solve this differential equation. (*Hint:* Multiply by x^k.)

11. In psychology, it is sometimes assumed that if R represents the reaction of a subject to an amount S of stimulus, then relative increases of R and S are proportional. In other words, if S is changed at rate dS/dt, and the resulting rate of change of R is dR/dt, then

$$\frac{1}{R}\frac{dR}{dt} = k\frac{1}{S}\frac{dS}{dt}.$$

(Although this is often called the Brentano–Stevens law, it is essentially the allometric law in a different setting.) Find R as a function of S.

12. Three boys carry a board of length L and uniform mass per unit length m horizontally, as shown in Fig. 5.14. The weight of the board causes it to bend, and its shape is the same on either side of the middle boy (if all three are of the same height). Between the first two boys (i.e., for $0 \leq x \leq L/2$), the shape of the board $y = f(x)$ must satisfy the differential equation

$$(EI)\frac{d^2y}{dx^2} = Ax + \frac{mg}{2}x^2,$$

where E and I are constants depending on the shape and material of the board, A is the weight supported by the boy at $x = 0$, and $g < 0$ is the acceleration due to gravity.

a) Solve this equation along with the conditions $f(0) = f(L/2) = 0$ to find $f(x)$ as a function of x, E, I, m, g, and A.

b) Find a condition that permits the determination of A.

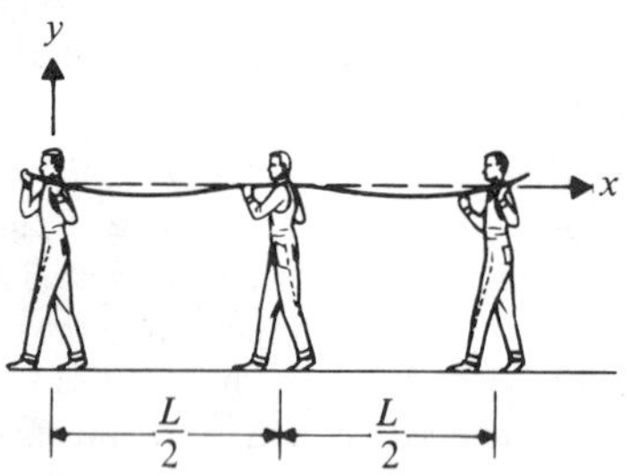

Figure 5.14

13. The steady-state temperature T in a region bounded by two concentric spheres of radii 1 meter and 2 meters must satisfy the differential equation

$$\frac{d^2T}{dr^2} + \frac{2}{r}\frac{dT}{dr} = 0,$$

where r is the radial distance from the common center of the spheres. If the temperatures on the inner and outer spheres are maintained at 10°C and 20°C, respectively, find the temperature distribution $T = f(r)$ between the spheres.

***14.** Find all curves such that if the normal is drawn at any point (x, y) on the curve, then that part of the normal between (x, y) and the x-axis is bisected by the y-axis.

***15.** a) It is sometimes assumed that the density ρ of the atmosphere is related to height h above sea level according to the differential equation

$$\frac{d\rho}{dh} = -\frac{\rho^{2-\delta}}{k\delta},$$

where $\delta > 1$ and $k > 0$ are constants. Show that if ρ_0 is density at sea level, then

$$\rho^{\delta-1} = -\frac{h}{k}\left(\frac{\delta-1}{\delta}\right) + \rho_0^{\delta-1}.$$

b) If air pressure P and density are related by the equation $P = k\rho^{\delta}$, prove that

$$P^{1-1/\delta} = P_0^{1-1/\delta} - [(1 - 1/\delta)\rho_0 P_0^{-1/\delta}]h,$$

where P_0 is air pressure at sea level.

c) Show that the effective height of the atmosphere is

$$\frac{\delta P_0}{(\delta - 1)\rho_0}.$$

SUMMARY

The indefinite integral (or antiderivative) of a function $f(x)$ is a family of functions, each of which has $f(x)$ as its first derivative. If we can find one antiderivative of $f(x)$, then every other antiderivative in the family differs from this one by at most an additive constant. Although the concept of an antiderivative is relatively simple, the actual operation of finding antiderivatives is much more complicated than that of finding derivatives. We will see just how much more complicated in Chapter 9, but to illustrate, note that given any rational function whatsoever, we can quickly find its derivative using the quotient rule. On the other hand, to find the antiderivative of even as simple a rational function as $(x + 2)/(x^2 + 3)$ is beyond our present capabilities.

Sometimes by making a change of variable, we can turn a difficult integration problem into a simpler one. We introduced this technique in Section 5.3, and will discuss it further in Chapter 9.

In many disciplines that require the use of calculus, finding antiderivatives is equally as important as finding derivatives. For instance, we illustrated that antiderivatives can be used to find the velocity and position of an object once its acceleration is specified as a function of time. In addition, many physical systems are described by differential equations, and the solution of a differential equation invariably involves one or more integrations.

REVIEW EXERCISES

In Exercises 1–20, evaluate the given indefinite integral.

1. $\int (3x^3 - 4x^2 + 5)\,dx$
2. $\int \left(\frac{1}{x^5} + 2x - \frac{1}{x^3}\right) dx$
3. $\int (2x^2 - 3x + 7x^6)\,dx$
4. $\int \left(\frac{1}{x^2} - 2\sqrt{x}\right) dx$
5. $\int \sqrt{x - 2}\,dx$
6. $\int x(1 + 3x^2)^4\,dx$
7. $\int \left(\sqrt{x} - \frac{1}{\sqrt{x}}\right) dx$
8. $\int \left(\frac{x^2 + 5}{\sqrt{x}}\right) dx$
9. $\int \frac{1}{(x + 5)^4}\,dx$
10. $\int \left(\frac{\sqrt{x}}{x^2} - \frac{15}{\sqrt{x}}\right) dx$
11. $\int \sin 3x\,dx$
12. $\int x\sqrt{1 - x^2}\,dx$
13. $\int x \cos x^2\,dx$
14. $\int x^2(1 - 2x^2)^2\,dx$
15. $\int x\sqrt{1 + x}\,dx$
16. $\int \frac{x}{\sqrt{2 - x}}\,dx$
17. $\int \frac{1}{(1 + x)^2}\,dx$
18. $\int (2 + \sqrt{x})^2\,dx$
19. $\int \frac{1}{\sqrt{x}(2 + \sqrt{x})^2}\,dx$
20. $\int \sin^4 x \cos x\,dx$

Find a one-parameter family of solutions for the given differential equation in Exercises 21–24.

21. $\frac{dy}{dx} = \frac{2x^2}{y}$
22. $\frac{dy}{dx} = y^2$
23. $x^2 \frac{dy}{dx} = y^3$
24. $\frac{dy}{dx} = \cos x \csc y$

25. Find the curve $y = f(x)$ for which $f''(x) = x^2 + 1$, and that passes through the point $(1, 1)$ with slope 4.

26. Find the curve $y = f(x)$ for which $f''(x) = 12x^2$, and that passes through the two points $(1, 4)$ and $(-1, -3)$.

27. Find the curve $y = f(x)$ for which $f''(x) = 24x^2 + 6x$, and that is tangent to the line $y = 4x + 4$ at $(1, 8)$.

28. A boy lives 6 kilometers from school. He decides to travel to school so that his speed is always proportional to the square root of his distance from the school. If he is halfway to school after 1 hour, find his distance from school at any time. How long does it take him to reach school?

29. If a ball is thrown vertically upward with a speed of 30 meters per second, how high will it rise?

30. A stone is thrown vertically downward over the edge of a bridge 50 meters above a river. If the stone strikes the water in 2.2 seconds, what was its initial speed?

In Exercises 31–36, evaluate the given indefinite integral.

31. $\displaystyle\int \frac{1}{\sqrt{1 + \sqrt{x}}}\,dx$

32. $\displaystyle\int \frac{x}{\sqrt{1 + x} + 1}\,dx$

33. $\displaystyle\int \frac{\sin x}{\sqrt{4 + 3\cos x}}\,dx$

34. $\displaystyle\int x^8(3 - 2x^3)^6\,dx$

35. $\displaystyle\int \frac{(2 + x)^4}{x^6}\,dx$

36. $\displaystyle\int \sin^3 x \cos^3 x\,dx$

37. A graph of the acceleration $a(t)$ of an object is shown in Fig. 5.15. Find its velocity $v(t)$ and position $x(t)$ in the time interval $0 \le t \le 15$ if $v(0) = 0 = x(0)$, and sketch graphs of each.

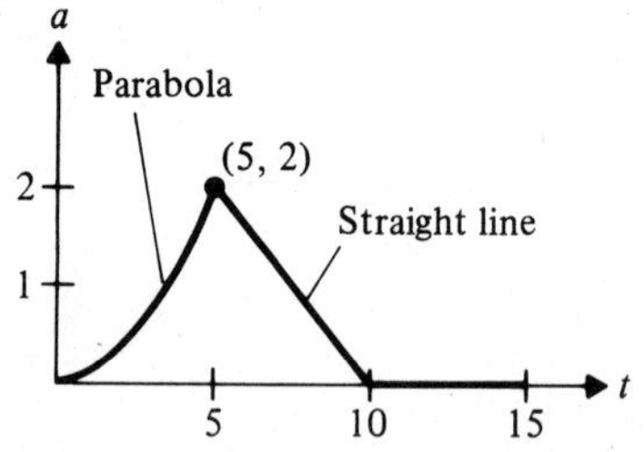

Figure 5.15

38. Find the equation of the curve that passes through the point $(1, 1)$ such that the slope of the tangent line at any point (x, y) is one-half the square of the slope of the line from the origin to (x, y).

39. The equation $y = x^3 + C$ describes a family of cubics where C represents the distance of its horizontal point of inflection above the x-axis. Find the equation of the curve that passes through the point $(1, 1)$ and intersects each of these curves at right angles.

The Definite Integral

6

6.1 SIGMA NOTATION

One of the most important notations in calculus, *sigma notation*, is used to represent the sum of a number of quantities, all of which are similar in form. For example, the six terms in the sum

$$\frac{1}{1+2^2}+\frac{2}{1+3^2}+\frac{3}{1+4^2}+\frac{4}{1+5^2}+\frac{5}{1+6^2}+\frac{6}{1+7^2}$$

are all formed in the same way: Each is an integer divided by one plus the square of the next integer. If k represents an integer, we can say that every term has the form $k/[1+(k+1)^2]$; the first term is obtained by setting $k=1$, the second by setting $k=2$, and so on. Because $k/[1+(k+1)^2]$ represents each and every term in the sum, we can describe the sum in words by saying, "Add together all the numbers obtained by setting k in $k/[1+(k+1)^2]$ equal to the positive integers between 1 and 6 inclusively." The notation used to represent this statement is

$$\sum_{k=1}^{6}\frac{k}{1+(k+1)^2}.$$

The $\sum$ symbol is the Greek letter sigma, and in this case means "sum." What is summed is expressions of the form $k/[1+(k+1)^2]$, and the "$k=1$" and "6" indicate that every integer from 1 to 6 is substituted into $k/[1+(k+1)^2]$. We call $k/[1+(k+1)^2]$ the general term of the sum, since it represents each and every term therein. The letter k is called the *index of summation* or *variable of summation*, and 1 and 6 are called the *limits of summation.*

Thus, if we see the sigma sign with limits, say $\sum_{k=5}^{15}$, we know that every integer from 5 to 15 is substituted into the general term, and all such numbers are then added together. For instance, the notation $\sum_{k=5}^{15}k^3$ represents the sum of the cubes of the integers from 5 to 15:

$$\sum_{k=5}^{15}k^3=5^3+6^3+7^3+8^3+9^3+10^3+11^3+12^3+13^3+14^3+15^3.$$

It is often necessary to sum a large number of terms, and thus very inconvenient to write them all down. In such cases, we write a few terms to indicate the pattern by which the terms are formed, three dots, and the last term. For example, to indicate the sum of the cubes of those positive integers less than or equal to 100, we would write

$$1^3+2^3+3^3+4^3+5^3+\cdots+100^3.$$

The dots indicate that all numbers between 5^3 and 100^3 are to be filled in according to the pattern suggested. Clearly, it would be preferable to write

$$\sum_{k=1}^{100}k^3,$$

in which case there can be no question as to the pattern; in addition, the notation is more compact.

There is no reason why k must be used as the index of summation; other letters can also be used. The most commonly used are i, j, k, l, m, and n.

Example 1 Write each of the following sums in sigma notation:

a) $\dfrac{1}{2\cdot 3}+\dfrac{4}{3\cdot 4}+\dfrac{9}{4\cdot 5}+\dfrac{16}{5\cdot 6}+\cdots+\dfrac{169}{14\cdot 15}$;

b) $\dfrac{16}{\sqrt{2}}+\dfrac{32}{\sqrt{3}}+\dfrac{64}{\sqrt{4}}+\dfrac{128}{\sqrt{5}}+\cdots+\dfrac{4096}{\sqrt{10}}$.

Solution a) If we use i as the index of summation, the general term is $i^2/[(i+1)(i+2)]$. Consequently,

$$\frac{1}{2\cdot 3}+\frac{4}{3\cdot 4}+\frac{9}{4\cdot 5}+\frac{16}{5\cdot 6}+\cdots+\frac{169}{14\cdot 15}=\sum_{i=1}^{13}\frac{i^2}{(i+1)(i+2)}.$$

b) If n is the index of summation, then

$$\frac{16}{\sqrt{2}}+\frac{32}{\sqrt{3}}+\frac{64}{\sqrt{4}}+\frac{128}{\sqrt{5}}+\cdots+\frac{4096}{\sqrt{10}}=\sum_{n=4}^{12}\frac{2^n}{\sqrt{n-2}}.\ \square$$

The representations for the sums in Example 1 in terms of sigma notation are not the only ones. In fact, there is an infinity of representations for each sum. Consider, for example, the sum represented by

$$\sum_{i=1}^{9}\frac{2^{i+3}}{\sqrt{i+1}}.$$

If we write the summation out in full, we find that

$$\sum_{i=1}^{9}\frac{2^{i+3}}{\sqrt{i+1}}=\frac{16}{\sqrt{2}}+\frac{32}{\sqrt{3}}+\frac{64}{\sqrt{4}}+\frac{128}{\sqrt{5}}+\cdots+\frac{4096}{\sqrt{10}},$$

the same sum as that in Example 1(b). This sum can also be represented by

$$\sum_{j=-2}^{6}\frac{2^{j+6}}{\sqrt{j+4}}\quad\text{and}\quad\sum_{m=15}^{23}\frac{2^{m-11}}{\sqrt{m-13}}.$$

We can transform any one of these representations into any other by making a change of variable of summation in much the same way that we changed variables of integration in indefinite integrals. For example, if in the summation of Example 1(b) we set $i = n - 3$, then $n = i + 3$; therefore,

$$\frac{2^n}{\sqrt{n-2}}=\frac{2^{i+3}}{\sqrt{i+1}}.$$

Since $i = 1$ when $n = 4$, and $i = 9$ when $n = 12$, it follows that

$$\sum_{n=4}^{12}\frac{2^n}{\sqrt{n-2}}=\sum_{i=1}^{9}\frac{2^{i+3}}{\sqrt{i+1}}.$$

Similarly, the changes $j = n - 6$ and $m = n + 11$ transform

$$\sum_{n=4}^{12}\frac{2^n}{\sqrt{n-2}}$$

into

$$\sum_{j=-2}^{6} \frac{2^{j+6}}{\sqrt{j+4}}$$

and

$$\sum_{m=15}^{23} \frac{2^{m-11}}{\sqrt{m-13}}.$$

Example 2 Change each of the following summations into representations that are initiated with the integer 1:

a) $\displaystyle\sum_{i=4}^{26} \frac{i^{2/3}}{i^2+i+1}$;

b) $\displaystyle\sum_{j=-3}^{102} \frac{j^2+2j+5}{\sin(j+5)}$.

Solution a) To initiate the summation at 1 that is 3 less than $i = 4$, we set $n = i - 3$. Then $i = n + 3$, and

$$\sum_{i=4}^{26} \frac{i^{2/3}}{i^2+i+1} = \sum_{n=1}^{23} \frac{(n+3)^{2/3}}{(n+3)^2+(n+3)+1}$$

$$= \sum_{n=1}^{23} \frac{(n+3)^{2/3}}{n^2+7n+13}.$$

b) In this case, we set $n = j + 4$. Then $j = n - 4$, and

$$\sum_{j=-3}^{102} \frac{j^2+2j+5}{\sin(j+5)} = \sum_{n=1}^{106} \frac{(n-4)^2+2(n-4)+5}{\sin(n+1)}$$

$$= \sum_{n=1}^{106} \frac{n^2-6n+13}{\sin(n+1)}. \quad \square$$

Every summation represented in sigma notation is of the form

$$\sum_{i=m}^{n} f(i), \tag{6.1}$$

where m and n are integers ($n > m$), and $f(i)$ is some function of the index of summation i. In Example 2(a), $f(i) = i^{2/3}/(i^2 + i + 1)$, $m = 4$, and $n = 26$; in Example 1(a), $f(i) = i^2/[(i + 1)(i + 2)]$, $m = 1$ and $n = 13$. The following properties of sigma notation are easily proved by writing out each summation.

▶ **Theorem 6.1**

i) $$\sum_{i=m}^{n} [f(i) + g(i)] = \sum_{i=m}^{n} f(i) + \sum_{i=m}^{n} g(i); \tag{6.2}$$

ii) $$\sum_{i=m}^{n} cf(i) = c\sum_{i=m}^{n} f(i), \tag{6.3}$$

if c is a constant (independent of i). ◀

Note that these properties correspond to those for the indefinite integral in Theorem 5.2.

Summation $\sum_{i=1}^{n} i$ represents the sum of the first n integers:

$$\sum_{i=1}^{n} i = 1 + 2 + 3 + 4 + \cdots + (n-1) + n.$$

If we write the terms on the right in the reverse order, we have

$$\sum_{i=1}^{n} i = n + (n-1) + (n-2) + \cdots + 4 + 3 + 2 + 1.$$

Addition of these two equations gives us

$$2\sum_{i=1}^{n} i = (n+1) + (n+1) + (n+1) + \cdots + (n+1) + (n+1)$$
$$= n(n+1).$$

Consequently,

$$\sum_{i=1}^{n} i = \frac{n(n+1)}{2}. \tag{6.4}$$

This result can be used to develop formulas for the sums of the squares, cubes, etc., of the positive integers. For example, since

$$i^3 - (i-1)^3 = 3i^2 - 3i + 1$$

for any integer i whatsoever, it follows that

$$\sum_{i=1}^{n} [i^3 - (i-1)^3] = \sum_{i=1}^{n} (3i^2 - 3i + 1).$$

But if we write the left-hand side in full, we get

$$\sum_{i=1}^{n} [i^3 - (i-1)^3] = [1^3 - 0^3] + [2^3 - 1^3]$$
$$+ [3^3 - 2^3] + \cdots + [n^3 - (n-1)^3].$$

Most of these terms cancel one another, leaving only n^3; i.e.,

$$\sum_{i=1}^{n} [i^3 - (i-1)^3] = n^3.$$

Thus,

$$n^3 = \sum_{i=1}^{n} (3i^2 - 3i + 1)$$
$$= 3\sum_{i=1}^{n} i^2 - 3\sum_{i=1}^{n} i + \sum_{i=1}^{n} 1 \qquad \text{(Using Eqs. 6.2 and 6.3)}$$
$$= 3\sum_{i=1}^{n} i^2 - 3\frac{n(n+1)}{2} + n. \qquad \text{(Using Eq. 6.4)}$$

We can now solve this equation for $\sum_{i=1}^{n} i^2$:

$$\sum_{i=1}^{n} i^2 = \frac{1}{3}\left\{n^3 + \frac{3n(n+1)}{2} - n\right\} = \frac{n(n+1)(2n+1)}{6}. \tag{6.5}$$

A similar procedure with the identity $i^4 - (i-1)^4 = 4i^3 - 6i^2 + 4i - 1$ yields

$$\sum_{i=1}^{n} i^3 = \frac{n^2(n+1)^2}{4}. \tag{6.6}$$

These results can also be established independently of one another by the method of mathematical induction†.

We make one final remark. Sigma notation is simply that—a notation. It does not evaluate sums; it merely represents them in a concise symbolism. When we encounter sigma notation, we must mentally visualize it written out in its entirety.

EXERCISES 6.1

In Exercises 1–10, express the given sum in sigma notation. Initiate the summation with the integer 1.

1. $2\cdot 3 + 3\cdot 4 + 4\cdot 5 + 5\cdot 6 + 6\cdot 7 + \cdots + 99\cdot 100$

2. $\frac{1}{2} + \frac{2}{4} + \frac{3}{8} + \frac{4}{16} + \frac{5}{32} + \cdots + \frac{10}{1024}$

3. $\frac{16}{14+15} + \frac{17}{15+16} + \frac{18}{16+17} + \cdots + \frac{199}{197+198}$

4. $1 + \sqrt{2} + \sqrt{3} + 2 + \sqrt{5} + \sqrt{6} + \sqrt{7} + \sqrt{8} + 3 + \cdots + 121$

5. $1 + \frac{1}{2} + \frac{1}{2\cdot 3} + \frac{1}{2\cdot 3\cdot 4} + \cdots + \frac{1}{2\cdot 3\cdot 4\cdot 5\cdot\cdots\cdot 16}$

6. $-2 + 3 - 4 + 5 - 6 + 7 - 8 + \cdots - 1020$

7. $\frac{2\cdot 3}{1\cdot 4} + \frac{6\cdot 7}{5\cdot 8} + \frac{10\cdot 11}{9\cdot 12} + \frac{14\cdot 15}{13\cdot 16} + \cdots + \frac{414\cdot 415}{413\cdot 416}$

8. $\frac{\tan 1}{2} + \frac{\tan 2}{1+2^2} + \frac{\tan 3}{1+3^2} + \frac{\tan 4}{1+4^2} + \cdots + \frac{\tan 225}{1+225^2}$

9. $4^3 + 5^2 + 6 + 1 + \frac{1}{8} + \frac{1}{9^2} + \cdots + \frac{1}{25^{18}}$

10. $0.9 + 0.99 + 0.999 + \cdots + 0.999999999$

In each of Exercises 11–15, verify by a change of variable of summation that the two summations are identical.

11. $\sum_{n=1}^{24} \frac{n^2}{2n+1}$ $\qquad \sum_{i=4}^{27} \frac{i^2 - 6i + 9}{2i - 5}$

12. $\sum_{k=2}^{101} \frac{3k - k^2}{\sqrt{k+5}}$ $\qquad \sum_{m=0}^{99} \frac{2 - m - m^2}{\sqrt{7+m}}$

13. $\sum_{n=5}^{20} (-1)^n \frac{2^n}{n^2+1}$ $\qquad \sum_{j=1}^{16} 16(-1)^j \frac{2^j}{j^2 + 8j + 17}$

14. $\sum_{i=0}^{37} \frac{3^{3i}}{i!}$ $\qquad \sum_{m=2}^{39} \frac{3^{3m}}{729(m-2)!}$

15. $\sum_{r=15}^{225} \frac{1}{r^2 - 10r}$ $\qquad \sum_{n=10}^{220} \frac{1}{n^2 - 25}$

In each of Exercises 16–21, use identities (6.2) and (6.3) and the results in Eqs. (6.4)–(6.6) to evaluate the sum.

16. $\sum_{n=1}^{12} (3n + 2)$

17. $\sum_{j=1}^{21} (2j^2 + 3j)$

18. $\sum_{m=1}^{n} (4m - 2)^2$

19. $\sum_{k=2}^{29} (k^3 - 3k^2)$

20. $\sum_{n=1}^{25} (n+5)(n-4)$

21. $\sum_{i=1}^{n} i(i-3)^2$

22. Verify the results contained in identities (6.2) and (6.3).

23. Find a formula for

$$\sum_{k=1}^{n} \frac{1}{k(k+1)}.$$

Hint:

$$\frac{1}{k(k+1)} = \frac{1}{k} - \frac{1}{k+1}.$$

24. If $f(x)$ is a function of x, defined for all x, simplify the sum

$$\sum_{i=1}^{n} [f(i) - f(i-1)].$$

25. Prove formula (6.6).

26. Is the following identity true or false?

$$\sum_{i=1}^{n} [f(i)g(i)] = \left\{\sum_{i=1}^{n} f(i)\right\}\left\{\sum_{i=1}^{n} g(i)\right\}$$

† The method of proof called *mathematical induction* is described in Appendix A.

27. A geometric series is a sum of terms of the form

$$a + ar + ar^2 + ar^3 + \cdots + ar^{n-1}.$$

There is a first term a, and every term thereafter is obtained by multiplying the preceding term by r (called the *common ratio*).

a) If S_n represents this sum, express S_n in sigma notation.
b) Prove that

$$S_n = \frac{a(1 - r^n)}{1 - r}.$$

Use the formula given in Exercise 27 to sum the geometric series in each of Exercises 28–31.

28. $\frac{1}{8} + \frac{1}{16} + \frac{1}{32} + \frac{1}{64} + \cdots + \frac{1}{1{,}048{,}576}$

29. $1 - \frac{1}{3} + \frac{1}{9} - \frac{1}{27} + \frac{1}{81} - \cdots - \frac{1}{19{,}683}$

30. $40(0.99) + 40(0.99)^2 + 40(0.99)^3 + \cdots + 40(0.99)^{15}$

31. $\sqrt{0.99} + 0.99 + (0.99)^{3/2} + (0.99)^2 + \cdots + (0.99)^{10}$

32. Prove that

$$\left| \sum_{i=1}^{n} f(i) \right| \le \sum_{i=1}^{n} |f(i)|.$$

*33. Express the following summation in sigma notation:

$$1 + \frac{1}{2} - \frac{1}{4} - \frac{1}{8} + \frac{1}{16} + \frac{1}{32} - \frac{1}{64} - \frac{1}{128} + \cdots + \frac{1}{4096}.$$

6.2 THE NEED FOR THE DEFINITE INTEGRAL

We now consider four problems, two of which are geometric in nature and two of which are physical. These problems are inherently different—the first being a problem on area, the second on volume, the third on blood flow, and the fourth on work—but we will see that a common method of solution exists for all four problems. It is this common theme that leads to the definition of the definite integral in Section 6.3.

Problem 1 We must find the area A shown in Fig. 6.1(a). There are only a few geometric shapes for which we can find areas (for example, rectangles, triangles, polygons, circles), and certainly the shape in Fig. 6.1(a) is not one of them. We can, however, find an approximation to the area by constructing rectangles in the following way. Between a and b we pick $n - 1$ points $x_1, x_2, x_3, \ldots, x_{n-1}$ on the x-axis such that

$$a = x_0 < x_1 < x_2 < \cdots < x_{n-1} < x_n = b.$$

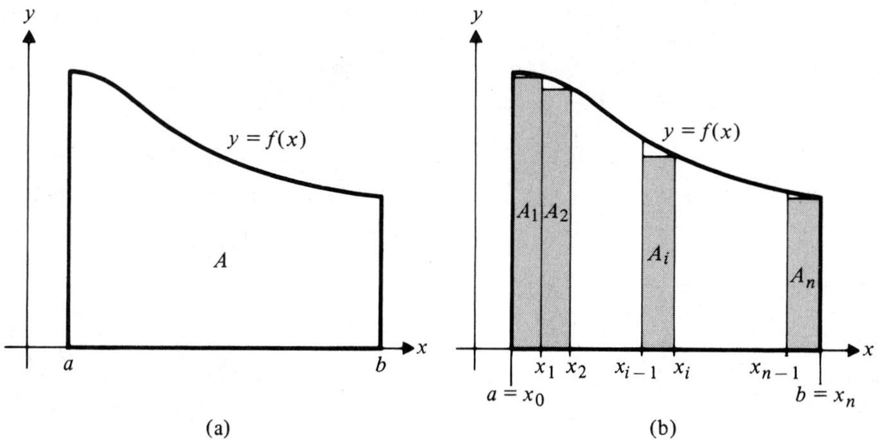

Figure 6.1

We draw vertical lines through each of these $n - 1$ points to intersect the curve $y = f(x)$ and form rectangles, as shown in Fig. 6.1(b).

If we denote by A_i the area of the ith such rectangle ($i = 1, \ldots, n$), then

$$\begin{aligned} A_i &= \text{(height of rectangle)(width of rectangle)} \\ &= f(x_i)(x_i - x_{i-1}). \end{aligned}$$

The sum of these n rectangular areas is an approximation to the required area; i.e., A is approximately equal to

$$\sum_{i=1}^{n} A_i = \sum_{i=1}^{n} f(x_i)(x_i - x_{i-1}).$$

If we let the number of these rectangles get larger and larger (and at the same time require each to have a smaller and smaller width that eventually approaches zero), it appears that the approximation will get better and better. In fact, we expect that

$$A = \lim_{n\to\infty} \sum_{i=1}^{n} A_i = \lim_{n\to\infty} \sum_{i=1}^{n} f(x_i)(x_i - x_{i-1}). \square \tag{6.7}$$

Problem 2 If the area shown in Fig. 6.1(a) is rotated about the x-axis, it traces out a volume V. This is certainly not a standard volume for which we have a formula; therefore, to calculate V we first find an approximation. To do this we take the rectangles shown in Fig. 6.1(b) (which approximate the area) and rotate them about the x-axis (Fig. 6.2). Each A_i will trace out a disc of volume V_i ($i = 1, \ldots, n$), where

$$\begin{aligned} V_i &= \text{(surface area of disc)(thickness of disc)} \\ &= \pi\text{(radius of disc)}^2\text{(thickness of disc)} \\ &= \pi[f(x_i)]^2(x_i - x_{i-1}). \end{aligned}$$

An approximation to the required volume is then

$$\sum_{i=1}^{n} V_i = \sum_{i=1}^{n} \pi[f(x_i)]^2(x_i - x_{i-1}).$$

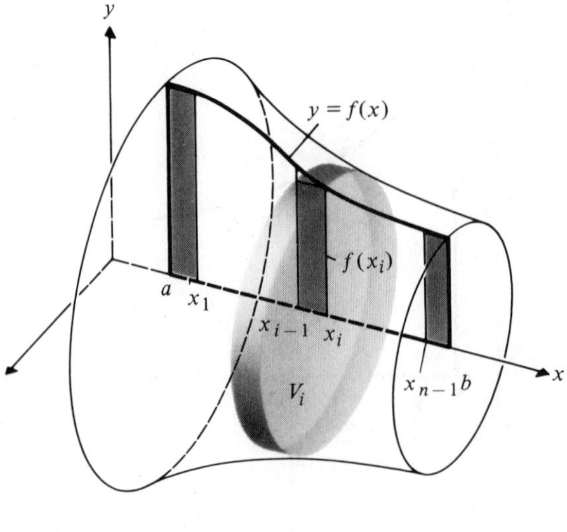

Figure 6.2

Again we feel intuitively that as the number of discs becomes larger and larger (and the width of each approaches zero), the approximation becomes better and better, and in the limit

$$V = \lim_{n\to\infty} \sum_{i=1}^{n} V_i = \lim_{n\to\infty} \sum_{i=1}^{n} \pi[f(x_i)]^2(x_i - x_{i-1}). \quad \square \tag{6.8}$$

Problem 3 When blood flows through a vein or artery, it encounters resistance due to friction with the walls of the blood vessel and due to the viscosity of the blood itself. As a result, the velocity of the blood is not constant across a cross-section of the vessel. It has been shown that for laminar blood flow in a vessel of circular cross-section (Fig. 6.3), the velocity of blood is given by

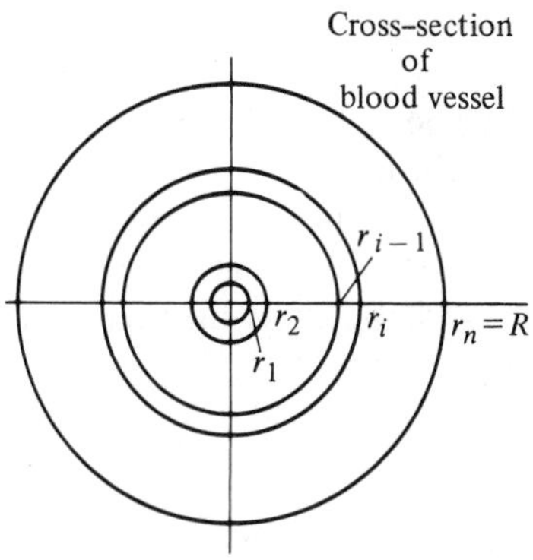

Figure 6.3

$$v = v(r) = c(R^2 - r^2), \qquad 0 \le r \le R,$$

where $c > 0$ is a constant, R is the radius of the blood vessel, and r is radial distance measured from the center of the vessel. As a result, blood flows more quickly near the center of a blood vessel than near its walls. We wish to find the rate of blood flow through the vessel; i.e., the volume of blood flowing through the cross-section per unit time.

If v were constant over the cross-section, then flow per unit time would be the product of v and the cross-sectional area. Unfortunately this is not the case, but we can still use the idea that flow is velocity multiplied by area. We divide the cross-section into rings with radii

$$0 = r_0 < r_1 < r_2 < \cdots < r_{n-1} < r_n = R.$$

Over the ith ring, the variation in v is small and v can be approximated by $v(r_i)$. The flow through the ith ring can therefore be estimated by

$$F_i = (\text{area of ring})(\text{velocity at outer radius of ring})$$
$$= (\pi r_i^2 - \pi r_{i-1}^2)v(r_i).$$

An approximation to the required flow F is the sum of these F_i:

$$\sum_{i=1}^{n} F_i = \sum_{i=1}^{n} (\pi r_i^2 - \pi r_{i-1}^2)v(r_i),$$

and it seems reasonable that as the number of rings increases, so too does the accuracy of the approximation; i.e., we anticipate that

$$F = \lim_{n\to\infty} \sum_{i=1}^{n} F_i = \lim_{n\to\infty} \sum_{i=1}^{n} (\pi r_i^2 - \pi r_{i-1}^2)v(r_i). \quad \square \tag{6.9}$$

Problem 4 One end of a spring is fixed horizontally into a wall, and the other end is free. We must find the work done in stretching the spring three centimeters by pulling on its free end (Fig. 6.4a).

Let us choose an x-axis positive to the right with $x = 0$ at the position of the free end of the spring when it is in the unstretched position (Fig. 6.4b). In order to calculate the work to stretch the spring, we must know something about the forces involved. It has been shown experimentally that the force F that must be exerted on the free end of the spring in order to maintain a stretch x in the spring is proportional to x:

$$F = F(x) = kx, \qquad k > 0.$$

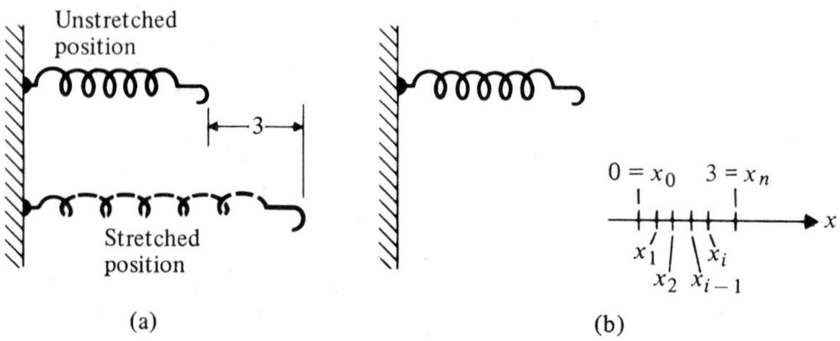

Figure 6.4

This then is the force that will perform the work. The basic definition of work W done by a constant force F acting along a straight line segment of length d is $W = Fd$. Unfortunately, our force $F(x)$ is not constant; it depends on x, and we cannot therefore simply multiply force by distance. What we can find, however, is an approximation to the required work by dividing the length between $x = 0$ and $x = 3$ into n subintervals by $n - 1$ points $x_1, x_2, \ldots, x_{n-1}$ such that

$$0 = x_0 < x_1 < x_2 < \cdots < x_{n-1} < x_n = 3.$$

When the spring is stretched between x_{i-1} and x_i, the force necessary to maintain this stretch does not vary greatly and can be approximated by $F(x_i)$. It follows then that the work necessary to stretch the spring from x_{i-1} to x_i is approximately equal to

$$W_i = F(x_i)(x_i - x_{i-1}).$$

As a result an approximation to the total work required to pull the free end of the spring from $x = 0$ to $x = 3$ is

$$\sum_{i=1}^{n} W_i = \sum_{i=1}^{n} F(x_i)(x_i - x_{i-1}).$$

Once again we expect that as n becomes indefinitely large, this approximation approaches the required work W, and

$$W = \lim_{n\to\infty} \sum_{i=1}^{n} W_i = \lim_{n\to\infty} \sum_{i=1}^{n} F(x_i)(x_i - x_{i-1}). \ \square \qquad (6.10)$$

Each of these four problems on area, volume, blood flow, and work has been attacked in the same way, and the method can be described qualitatively as follows.

The quantity to be calculated, say W, cannot be obtained for the geometric object G given (because no formula exists). As a result, n smaller objects, say G_i, are constructed. The G_i are chosen in such a way that the quantity W can be calculated (exactly or approximately) for each G_i, say W_i. Then an approximation for W is

$$\sum_{i=1}^{n} W_i.$$

If the number of G_i is then increased indefinitely, this approximation becomes more and more accurate and in the limit

$$W = \lim_{n\to\infty} \sum_{i=1}^{n} W_i.$$

It is this "limit–summation" process that we discuss throughout the remainder of the chapter. We begin in Section 6.3 with a mathematical description of the process, and by doing so, we obtain a unified approach to the whole idea. We then discover that there is a very simple way to calculate these limits. At that point we will be ready to use the technique in a multitude of applications, including the four problems given in this section.

6.3 THE DEFINITE INTEGRAL

To define the definite integral of a function $f(x)$ defined on an interval $a \leq x \leq b$ (Fig. 6.5), we first subdivide the interval into n subintervals by any $n - 1$ points:

$$a = x_0 < x_1 < x_2 < \cdots < x_i < \cdots < x_{n-1} < x_n = b.$$

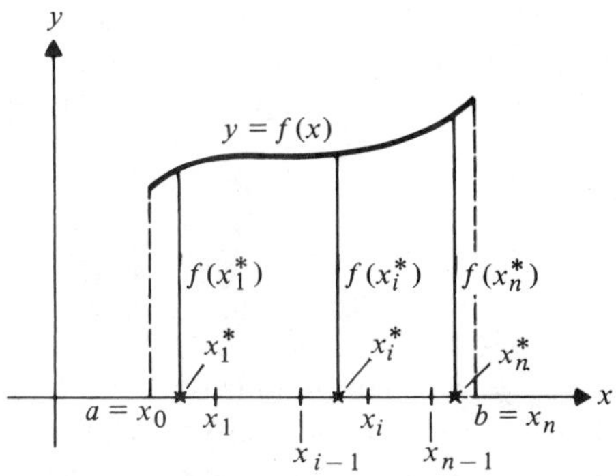

Figure 6.5

Next we choose in each subinterval $x_{i-1} \leq x \leq x_i$ any point x_i^* whatsoever, and evaluate $f(x_i^*)$. We now form the sum

$$f(x_1^*)(x_1 - x_0) + f(x_2^*)(x_2 - x_1) + \cdots + f(x_n^*)(x_n - x_{n-1}) = \sum_{i=1}^{n} f(x_i^*)(x_i - x_{i-1})$$

$$= \sum_{i=1}^{n} f(x_i^*)\,\Delta x_i,$$

where we have set

$$\Delta x_i = x_i - x_{i-1}.$$

If this sum approaches a limit as the number of subintervals becomes increasingly large and every subinterval shrinks to a point, we call the limit the definite integral of $f(x)$ with respect to x from $x = a$ to $x = b$, and we denote it by

$$\int_a^b f(x)\,dx = \lim_{n\to\infty} \sum_{i=1}^{n} f(x_i^*)\,\Delta x_i. \qquad (6.11)$$

We understand by writing $\lim_{n\to\infty}$ that each and every one of the Δx_i shrinks to a point.

If this limit were dependent on the choice of subdivision Δx_i or star-points x_i^*, then the definite integral would be of little use. We stipulate, therefore, that in order for the definite integral to exist, the limit of the sum in Eq. (6.11) must be independent of the manner of subdivision of the interval $a \le x \le b$ and choice of star-points in the subintervals. At first sight this requirement might seem rather severe since we must now check that all subdivisions and all choices of star-points lead to the same limit before concluding that the definite integral exists. Fortunately, however, the following theorem indicates that for continuous functions this is unnecessary. A proof of this theorem can be found in more advanced books.

▶ **Theorem 6.2** If a function $f(x)$ is continuous on a finite interval $a \le x \le b$, then the definite integral of $f(x)$ with respect to x from $x = a$ to $x = b$ exists. ◀

For a continuous function, then, the definite integral exists and any choice of subdivision and star-points will lead to its correct value through the limiting process (6.11). We call $f(x)$ on the left-hand side of Eq. (6.11) the integrand, and a and b the lower and upper limits of integration, respectively.

Example 3 Evaluate

$$\int_0^1 x^2\,dx.$$

Solution Since $f(x) = x^2$ is continuous on the interval $0 \le x \le 1$, the definite integral exists, and we may choose any subdivision and star-points in its evaluation. The simplest subdivision is into n equal subintervals of length $1/n$ by the points

$$x_i = \frac{i}{n}, \qquad i = 0, \ldots, n,$$

(Fig. 6.6).

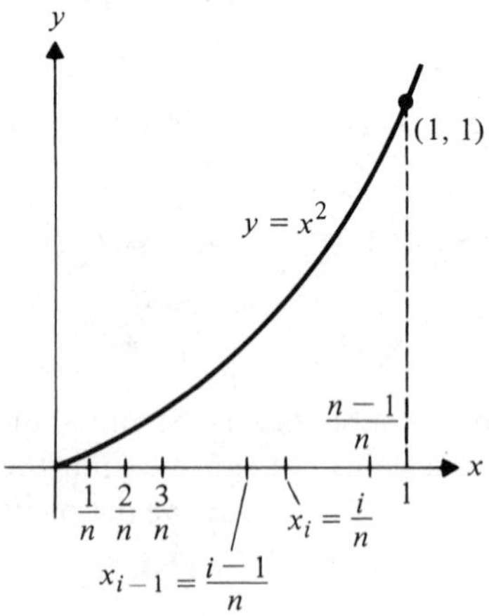

Figure 6.6

We choose for star-points the right end of each subinterval; i.e., in $x_{i-1} \le x \le x_i$ we choose $x_i^* = x_i = i/n$. Then, by definition (6.11), we have

$$\begin{aligned}\int_0^1 x^2\,dx &= \lim_{n\to\infty} \sum_{i=1}^{n} f(x_i^*)\,\Delta x_i \\ &= \lim_{n\to\infty} \sum_{i=1}^{n} \left(\frac{i}{n}\right)^2 \left(\frac{i}{n} - \frac{i-1}{n}\right) \\ &= \lim_{n\to\infty} \frac{1}{n^3} \sum_{i=1}^{n} i^2.\end{aligned}$$

If we now use Eq. (6.5) for the sum of the squares of the first n positive integers, we get

$$\int_0^1 x^2\,dx = \lim_{n\to\infty} \frac{1}{n^3} \cdot \frac{n(n+1)(2n+1)}{6} = \frac{1}{3}. \square$$

This example illustrates that even for a simple function such as $f(x) = x^2$, evaluation of the definite integral by definition (6.11) is not particularly simple. In fact, had we not known formula (6.5) for the sum of the squares of the integers, we would not have been able to complete the calculation. Imagine the magnitude of the problem were the integrand equal to $f(x) = x(x+1)^{-2/3}$. In other words, if definite integrals are to be at all useful, we must find a simpler way to evaluate them. This we do in Section 6.4, but in order to stress the definite integral as a limit–summation, we consider one more example.

Example 4 Evaluate

$$\int_{-1}^{1} (5x - 2)\, dx.$$

Solution Since $f(x) = 5x - 2$ is continuous on the interval $-1 \leq x \leq 1$, the definite integral exists, and we may choose any subdivision and star-points in its evaluation. For n equal subdivisions of length $2/n$, we use points

$$x_i = -1 + \frac{2i}{n}, \qquad i = 0, \ldots, n$$

(Fig. 6.7). If we choose the right end of each subinterval as a star-point, i.e., $x_i^* = x_i = -1 + 2i/n$, then definition (6.11) gives us

$$\begin{aligned}\int_{-1}^{1} (5x - 2)\, dx &= \lim_{n\to\infty} \sum_{i=1}^{n} f(x_i^*)\, \Delta x_i \\ &= \lim_{n\to\infty} \sum_{i=1}^{n} \left\{5\left(-1 + \frac{2i}{n}\right) - 2\right\}\left(\frac{2}{n}\right) \\ &= \lim_{n\to\infty} \sum_{i=1}^{n} \left\{\frac{20i - 14n}{n^2}\right\}.\end{aligned}$$

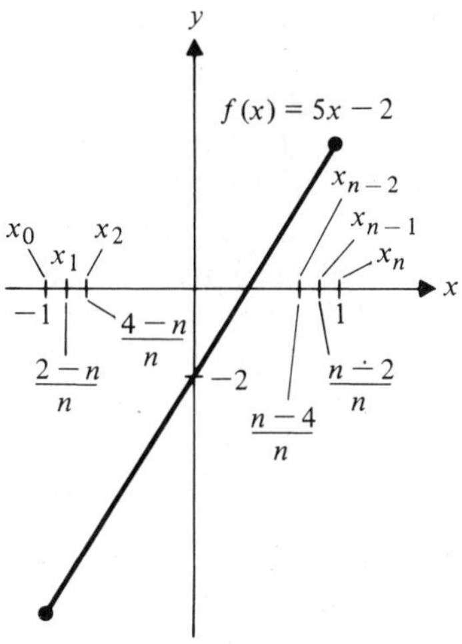

Figure 6.7

If we divide the summation into two parts, we have

$$\begin{aligned}\int_{-1}^{1} (5x - 2)\, dx &= \lim_{n\to\infty} \left\{\sum_{i=1}^{n} \frac{20i}{n^2} - \sum_{i=1}^{n} \frac{14}{n}\right\} \\ &= \lim_{n\to\infty} \left\{\frac{20}{n^2} \sum_{i=1}^{n} i - \frac{14}{n} \sum_{i=1}^{n} 1\right\} \\ &= \lim_{n\to\infty} \left\{\frac{20}{n^2} \frac{n(n+1)}{2} - \frac{14}{n} n\right\} \qquad \text{(Using Eq. 6.4)} \\ &= \lim_{n\to\infty} \left\{\frac{10 - 4n}{n}\right\} \\ &= -4. \ \square\end{aligned}$$

Note that in Example 3 the value of the definite integral is positive, and in Example 4 it is negative. The value of the definite integral in Exercise 8 is zero. In other words, the value of a definite integral can be positive, negative, or zero, depending on the limits and the integrand.

EXERCISES 6.3

In Exercises 1–8, use definition (6.11) to evaluate the given definite integral.

1. $\int_0^1 x\,dx$

2. $\int_0^2 3x\,dx$

3. $\int_0^1 (3x + 2)\,dx$

4. $\int_0^2 x^3\,dx$

5. $\int_1^2 (x^2 + 2x)\,dx$

6. $\int_{-1}^0 (-x + 1)\,dx$

7. $\int_{-1}^1 x^2\,dx$

8. $\int_{-1}^1 x^3\,dx$

***9.** Evaluate

$$\int_{-1}^1 x^{15}\,dx.$$

6.4 THE FUNDAMENTAL THEOREM OF CALCULUS

In Section 6.3 we demonstrated how to evaluate definite integrals by using definition (6.11). Note that the integrands x^2 and $5x - 2$ in Examples 3 and 4 are very simple polynomials, as are all integrands in Exercises 6.3. In spite of this, calculations were frequently laborious, and invariably dependent on identities (6.4)–(6.6). We promised a very simple technique that would replace these calculations, and this is the substance of the *fundamental theorem of calculus.* This theorem is a startling result in that we have defined two types of integral, the indefinite and the definite, and there seems to be no connection whatsoever between them. One is an operation of "backwards differentiation," and the other is a limit–summation. The fundamental theorem shows that these two concepts are very closely related, that definite integrals can be evaluated using indefinite integrals. The complete result is as follows.

▶ **Theorem 6.3** (*The fundamental theorem of calculus*). If $f(x)$ is continuous on the interval $a \le x \le b$, and $F(x)$ is an antiderivative of $f(x)$ thereon, then

$$\int_a^b f(x)\,dx = F(b) - F(a). \quad ◀ \tag{6.12}$$

Proof Since $f(x)$ is continuous on $a \le x \le b$, the definite integral of $f(x)$ from $x = a$ to $x = b$ exists and is defined by Eq. (6.11), where we are at liberty to choose the Δx_i and x_i^*. For any choice of Δx_i whatsoever, a convenient choice for the x_i^* can be found by applying the mean value theorem to $F(x)$ n times, once on each subinterval $x_{i-1} \le x \le x_i$. This is possible since $F'(x) = f(x)$ is continuous for $a \le x \le b$. The mean value theorem states that for each subinterval there exist points c_i such that $x_{i-1} < c_i < x_i$, and

$$\frac{F(x_i) - F(x_{i-1})}{x_i - x_{i-1}} = F'(c_i), \qquad i = 1, \ldots, n.$$

But $F'(c_i) = f(c_i)$, so that

$$F(x_i) - F(x_{i-1}) = f(c_i)\,\Delta x_i, \qquad i = 1, \ldots, n.$$

If we now choose $x_i^* = c_i$, then

$$f(x_i^*)\,\Delta x_i = F(x_i) - F(x_{i-1}),$$

and definition (6.11) gives us

$$\begin{aligned}\int_a^b f(x)\,dx &= \lim_{n\to\infty} \sum_{i=1}^{n} f(x_i^*)\,\Delta x_i \\ &= \lim_{n\to\infty} \sum_{i=1}^{n} \{F(x_i) - F(x_{i-1})\} \\ &= \lim_{n\to\infty} \{[F(x_1) - F(x_0)] + [F(x_2) - F(x_1)] + \cdots \\ &\qquad + [F(x_n) - F(x_{n-1})]\} \\ &= \lim_{n\to\infty} \{F(x_n) - F(x_0)\} \\ &= \lim_{n\to\infty} \{F(b) - F(a)\} \\ &= F(b) - F(a). \ \square\end{aligned}$$

We have shown then that to evaluate the definite integral of a continuous function $f(x)$ from $x = a$ to $x = b$, all we need do is find an antiderivative of $f(x)$, substitute $x = b$ and $x = a$, and subtract. If we use the notation

$$\{F(x)\}_a^b$$

to represent $F(b) - F(a)$, then

$$\int_a^b f(x)\,dx = \left\{\int f(x)\,dx\right\}_a^b. \tag{6.13}$$

For Example 3, we easily find that

$$\int_0^1 x^2\,dx = \left\{\frac{x^3}{3}\right\}_0^1 = \frac{1}{3} - 0 = \frac{1}{3}.$$

Note that had we used *the* indefinite integral $x^3/3 + C$ of x^2, we would have

$$\int_0^1 x^2\,dx = \left\{\frac{x^3}{3} + C\right\}_0^1 = \left\{\frac{1}{3} + C\right\} - \{C\} = \frac{1}{3}.$$

Because the arbitrary constant always vanishes in the evaluation of definite integrals, we omit the constant in this context.

Similarly, for Example 4, we get

$$\int_{-1}^1 (5x - 2)\,dx = \left\{\frac{5x^2}{2} - 2x\right\}_{-1}^1 = \left\{\frac{5}{2} - 2\right\} - \left\{\frac{5}{2} + 2\right\} = -4.$$

Example 5 Evaluate each of the following definite integrals:

a) $\displaystyle\int_1^2 (3x^2 - x + 4)\,dx$;

b) $\displaystyle\int_{-2}^4 \sqrt{x + 4}\,dx$.

Solution a) $\int_1^2 (3x^2 - x + 4)\,dx = \left\{x^3 - \frac{x^2}{2} + 4x\right\}_1^2$

$= \{8 - 2 + 8\} - \{1 - \frac{1}{2} + 4\} = \frac{19}{2}$

b) $\int_{-2}^4 \sqrt{x+4}\,dx = \left\{\frac{2}{3}(x+4)^{3/2}\right\}_{-2}^4$

$= \frac{2}{3}(8)^{3/2} - \frac{2}{3}(2)^{3/2} = \frac{28\sqrt{2}}{3}$ □

Four results that are fairly evident and very useful are contained in the following theorems.

▶ **Theorem 6.4** If $f(x)$ is continuous on $a \le x \le b$, and $a < c < b$, then:

i) $\int_b^a f(x)\,dx = -\int_a^b f(x)\,dx;$ (6.14)

ii) $\int_a^b f(x)\,dx = \int_a^c f(x)\,dx + \int_c^b f(x)\,dx.$ ◀ (6.15)

▶ **Theorem 6.5** If $f(x)$ and $g(x)$ are continuous on $a \le x \le b$, then:

i) $\int_a^b [f(x) + g(x)]\,dx = \int_a^b f(x)\,dx + \int_a^b g(x)\,dx;$ (6.16)

ii) $\int_a^b kf(x)\,dx = k\int_a^b f(x)\,dx,\quad k = \text{a constant}.$ ◀ (6.17)

Note that the properties in Theorem 6.5 correspond to those for the indefinite integral in Theorem 5.2.

EXERCISES 6.4

In Exercises 1–20, evaluate the given definite integral.

1. $\int_3^4 (x^3 + 3)\,dx$

2. $\int_1^3 (x^2 - 2x + 3)\,dx$

3. $\int_{-1}^1 (4x^3 + 2x)\,dx$

4. $\int_{-3}^{-1} \frac{1}{x^2}\,dx$

5. $\int_4^2 \left(x^2 + \frac{3}{x^3}\right)dx$

6. $\int_0^{\pi/2} \sin x\,dx$

7. $\int_{-1}^1 (x^2 - 1 - x^4)\,dx$

8. $\int_{-1}^{-2} \left(\frac{1}{x^2} - 2x\right)dx$

9. $\int_1^2 (x^4 + 3x^2 + 2)\,dx$

10. $\int_0^1 x(x^2 + 1)\,dx$

11. $\int_0^1 x^2(x^2 + 1)^2\,dx$

12. $\int_0^{2\pi} \cos 2x\,dx$

13. $\int_1^3 \frac{x^2 + 3}{x^2}\,dx$

14. $\int_0^1 (x^{2.2} - x^{\pi})\,dx$

15. $\int_{-1}^1 x^2(x^3 - x)\,dx$

16. $\int_3^4 \frac{(x^2 - 1)^2}{x^2}\,dx$

17. $\int_1^2 \left(\sqrt{x} - \frac{1}{\sqrt{x}}\right)dx$

18. $\int_{-2}^3 (x - 1)^3\,dx$

19. $\int_2^4 \frac{(x^2 - 1)(x^2 + 1)}{x^2}\,dx$

20. $\int_0^{\pi/4} 3\cos x\,dx$

21. Evaluate each of the following:

a) $\frac{d}{dx}\left\{\int_1^2 \left(x^2 - \frac{1}{\sqrt{x}} + \sqrt{x^2 + 1}\right)dx\right\};$

b) $\frac{d}{dx}\left\{\int_1^x \left(t^2 + \frac{1}{\sqrt{t^2 + 1}}\right)dt\right\}.$

In Exercises 22–25, evaluate the given definite integral.

22. $\int_0^5 |x|\,dx$

23. $\int_0^4 x|x+1|\,dx$

24. $\int_{-5}^5 |x|\,dx$

25. $\int_{-2}^1 x|x+1|\,dx$

26. If $v(t) = 3t^2 - 6t - 105$ represents the velocity (in meters per second) of a particle moving along the x-axis, evaluate and interpret physically:

a) $\int_0^{12} v\,dt$; b) $\int_0^{12} |v|\,dt$.

27. Assuming that a function that is continuous on a closed interval has an antiderivative on that interval, prove the mean value theorem for definite integrals: If $f(x)$ is continuous on the closed interval $a \le x \le b$, then there exists a number c between a and b such that

$$\int_a^b f(x)\,dx = (b-a)f(c).$$

6.5 CHANGE OF VARIABLE IN THE DEFINITE INTEGRAL

The fundamental theorem of calculus indicates that to evaluate the definite integral

$$\int_{-4}^{-1} \frac{x}{\sqrt{x+5}}\,dx$$

we should first find an indefinite integral for $x/\sqrt{x+5}$. To do this we set $u = x + 5$, in which case $du = dx$, and

$$\int \frac{x}{\sqrt{x+5}}\,dx = \int \frac{(u-5)}{u^{1/2}}\,du = \int (u^{1/2} - 5u^{-1/2})\,du$$
$$= \tfrac{2}{3}u^{3/2} - 10u^{1/2} + C = \tfrac{2}{3}(x+5)^{3/2} - 10(x+5)^{1/2} + C.$$

Consequently,

$$\int_{-4}^{-1} \frac{x}{\sqrt{x+5}}\,dx = \left\{\frac{2}{3}(x+5)^{3/2} - 10(x+5)^{1/2}\right\}_{-4}^{-1}$$
$$= \{\tfrac{2}{3}(4)^{3/2} - 10(4)^{1/2}\} - \{\tfrac{2}{3}(1)^{3/2} - 10(1)^{1/2}\}$$
$$= -\tfrac{16}{3}.$$

There is an alternative solution that usually turns out to be less work. We illustrate the method in the above example and then prove its general validity.

Instead of making the change of variable $u = x + 5$ in the corresponding indefinite integral, we make it in the definite integral. In this case we again replace

$$\frac{x}{\sqrt{x+5}}\,dx$$

by

$$\frac{(u-5)}{u^{1/2}}\,du.$$

In addition, we replace the limits $x = -4$ and $x = -1$ by those values of u that correspond to $x = -4$ and $x = -1$, namely, $u = 1$ and $u = 4$, respectively. We obtain then

$$\begin{aligned}\int_{-4}^{-1} \frac{x}{\sqrt{x+5}}\,dx &= \int_1^4 \frac{(u-5)}{u^{1/2}}\,du = \int_1^4 (u^{1/2} - 5u^{-1/2})\,du \\ &= \{\tfrac{2}{3}u^{3/2} - 10u^{1/2}\}_1^4 \\ &= \{\tfrac{2}{3}(4)^{3/2} - 10(4)^{1/2}\} - \{\tfrac{2}{3}(1)^{3/2} - 10(1)^{1/2}\} \\ &= -\tfrac{16}{3}.\end{aligned}$$

That this method is generally acceptable is proved in the following theorem.

▶ **Theorem 6.6** Suppose $f(x)$ is continuous on $a \le x \le b$, and we set $x = g(u)$, where $a = g(\alpha)$ and $b = g(\beta)$. Then

$$\int_a^b f(x)\,dx = \int_\alpha^\beta f[g(u)]g'(u)\,du, \tag{6.18}$$

if $g'(u)$ is continuous on $\alpha \le u \le \beta$, and if when u is between α and β, $g(u)$ is between a and b. ◀

Proof If we define

$$F(x) = \int f(x)\,dx \quad \text{and} \quad G(u) = \int f[g(u)]g'(u)\,du,$$

then we must show that

$$F(b) - F(a) = G(\beta) - G(\alpha).$$

Now the chain rule applied to $y = F(x)$, $x = g(u)$ yields

$$\frac{dy}{du} = \frac{dy}{dx}\frac{dx}{du} = f(x)g'(u);$$

i.e.,

$$\frac{d}{du}F[g(u)] = f[g(u)]g'(u).$$

Since this is an identity in u, we may integrate with respect to u from $u = \alpha$ to $u = \beta$:

$$\int_\alpha^\beta \frac{d}{du}F[g(u)]\,du = \int_\alpha^\beta f[g(u)]g'(u)\,du.$$

Thus,

$$\{F[g(u)]\}_\alpha^\beta = \{G(u)\}_\alpha^\beta,$$

or

$$G(\beta) - G(\alpha) = F[g(\beta)] - F[g(\alpha)] = F(b) - F(a). \;\square$$

Example 6 Evaluate

$$\int_2^4 \frac{\sqrt{x}}{(\sqrt{x}-1)^4}\,dx.$$

Solution If we set $u = \sqrt{x} - 1$, then $du = [1/(2\sqrt{x})]\,dx$ and

$$\int_2^4 \frac{\sqrt{x}}{(\sqrt{x}-1)^4}\,dx = \int_{\sqrt{2}-1}^1 \frac{u+1}{u^4}\,2(u+1)\,du$$

$$= 2\int_{\sqrt{2}-1}^1 \left\{\frac{1}{u^2} + \frac{2}{u^3} + \frac{1}{u^4}\right\} du$$

$$= 2\left\{-\frac{1}{u} - \frac{1}{u^2} - \frac{1}{3u^3}\right\}\Bigg|_{\sqrt{2}-1}^1$$

$$= -2\{1 + 1 + \tfrac{1}{3}\}$$

$$+ 2\left\{\frac{1}{\sqrt{2}-1} + \frac{1}{(\sqrt{2}-1)^2} + \frac{1}{3(\sqrt{2}-1)^3}\right\}$$

$$= 21.2. \ \square$$

EXERCISES 6.5

In Exercises 1–18, evaluate the given definite integral.

1. $\int_1^2 x(3x^2 - 2)^4\,dx$
2. $\int_0^1 z\sqrt{1-z}\,dz$
3. $\int_{-1}^0 \frac{x}{\sqrt{x+3}}\,dx$
4. $\int_{\pi/4}^{\pi/3} \cos^5 x \sin x\,dx$
5. $\int_1^3 x^2\sqrt{9-x^2}\,dx$
6. $\int_{-5}^6 \frac{x}{\sqrt{x^2-12}}\,dx$
7. $\int_4^5 y^2\sqrt{y-4}\,dy$
8. $\int_{1/2}^1 \sqrt{\frac{x^2}{1+x}}\,dx$
9. $\int_1^4 \frac{\sqrt{1+\sqrt{u}}}{\sqrt{u}}\,du$
10. $\int_{-2}^1 \frac{x+1}{(x^2+2x+2)^{1/3}}\,dx$
11. $\int_3^4 \frac{x^2}{(x-2)^4}\,dx$
12. $\int_0^{\pi/6} \sqrt{2 + 3\sin x}\cos x\,dx$
13. $\int_{\pi/4}^{\pi/2} \frac{\sin^3 x}{(1+\cos x)^4}\,dx$
14. $\int_1^4 \frac{(x+1)(x-1)}{\sqrt{x}}\,dx$
15. $\int_4^9 \sqrt{1+\sqrt{x}}\,dx$
16. $\int_{-1/2}^1 \sqrt{\frac{x^2}{1+x}}\,dx$
17. $\int_{-1}^1 \frac{|x|}{(x+2)^3}\,dx$
18. $\int_{-1}^1 \left|\frac{x}{(x+2)^3}\right|\,dx$

19. Show that if $f(x)$ is an odd function, then

$$\int_{-a}^a f(x)\,dx = 0;$$

and that if $f(x)$ is an even function,

$$\int_{-a}^a f(x)\,dx = 2\int_0^a f(x)\,dx.$$

(See Eqs. (1.20) for definitions of even and odd functions.)

Use the suggested substitution to evaluate the definite integral in each of Exercises 20–22.

20. $\int_1^3 \frac{1}{x^{3/2}\sqrt{4-x}}\,dx$; set $u = \frac{1}{\sqrt{x}}$.

21. $\int_{-6}^{-1} \frac{\sqrt{x^2-6x}}{x^4}\,dx$; set $u = \frac{1}{x}$.

22. $\int_{-4}^0 \frac{x}{(5-4x-x^2)^{3/2}}\,dx$; set $u^2 = \frac{1-x}{5+x}$.

*23. Show that

$$\int_0^\pi \frac{\cos^2((\pi/2)\cos\theta)}{\sin\theta}\,d\theta = \frac{\pi}{2}\int_0^{2\pi} \frac{1-\cos\phi}{\phi(2\pi-\phi)}\,d\phi.$$

We use this integral when calculating the radiated power of a half-wave antenna.

6.6 THE RELATIONSHIP BETWEEN DEFINITE AND INDEFINITE INTEGRALS

To close this chapter we emphasize again the difference between the definite integral and the indefinite integral and their relationship to one another.

The indefinite integral of a function $f(x)$ is the family of all those functions that have $f(x)$ as their first derivative. The indefinite integral of $f(x)$, then, is not a number, not a function, but a one-parameter family of functions, each of which has $f(x)$ as its derivative.

On the other hand, the definite integral of $f(x)$ from $x = a$ to $x = b$ is a number. It is the number defined by the limit–summation in Eq. (6.11).

The remarkable connection is that definite integrals are most easily evaluated using indefinite integrals. Indeed, to find the definite integral of $f(x)$ from $x = a$ to $x = b$, we subtract the numbers obtained by substituting $x = b$ and $x = a$ into any indefinite integral of $f(x)$ whatsoever. But remember, this is only an evaluation process, *not* a definition. The definite integral is a *limit of a sum*, a very special sum formed in a very special way, but nevertheless, a limit–summation.

In definition (6.11),

$$\int_a^b f(x)\,dx = \lim_{n\to\infty} \sum_{i=1}^{n} f(x_i^*)\,\Delta x_i,$$

the right-hand side defines the left-hand side; i.e., the definite integral on the left is defined to be the limit–summation on the right. In Chapter 7 when we apply the definite integral to various problems, we will look at the equation in a different way. We will be confronted repeatedly with limits of summations of the form shown in Eq. (6.11). But such a limit–summation can be interpreted immediately as a definite integral. It is the definite integral with respect to x of the function obtained by taking everything behind the summation sign and before the Δx_i, and replacing all the x_i^*'s with x's. The limits on the definite integral are determined by the Δx_i's. Once we have interpreted the limit–summation as a definite integral, we can evaluate it by means of the corresponding indefinite integral.

We make one final remark that is too often confused. A definite integral with respect to x from $x = a$ to $x = b$ is an integration along the x-axis. Although we will use the definite integral to obtain areas, volumes, work, and many other physical and geometric quantities, every definite integral with respect to x is an integration along the x-axis.

SUMMARY

The definite integral of a continuous function $f(x)$ from $x = a$ to $x = b$ is a number, one that depends on the function $f(x)$ and the limits a and b. In particular, we defined the definite integral by subdividing the interval $a \le x \le b$ into n parts by $n + 1$ points x_i $(i = 0, \ldots, n)$, and choosing a point x_i^* in each subinterval $x_{i-1} \le x \le x_i$. The definite integral is then the limit of the summation

$$\int_a^b f(x)\,dx = \lim_{n\to\infty} \sum_{i=1}^{n} f(x_i^*)\,\Delta x_i,$$

where $\Delta x_i = x_i - x_{i-1}$. Because all calculations in this limit take place on the x-axis, we regard the definite integral as an integration along the x-axis from $x = a$ to $x = b$.

To calculate definite integrals we use indefinite integrals. Indeed, the definite integral of $f(x)$ from $x = a$ to $x = b$ can be evaluated by taking any indefinite integral of $f(x)$ and subtracting its value at $x = a$ from its value at $x = b$.

We saw in Chapter 5 that a change of variable may sometimes replace a difficult indefinite integral with a simpler one. If the indefinite integral is required for evaluation of a definite integral, we suggest that the original definite integral be transformed to a new definite integral by replacing the old limits by the corresponding limits in the new variable.

REVIEW EXERCISES

In Exercises 1–20, evaluate the given definite integral.

1. $\int_0^3 (x^2 + 3x - 2)\,dx$

2. $\int_{-1}^1 (x^2 - x^4)\,dx$

3. $\int_{-1}^1 (x^3 - 3x)\,dx$

4. $\int_0^2 (x^2 - 2x)\,dx$

5. $\int_1^2 (x + 1)^2\,dx$

6. $\int_{-3}^{-2} \frac{1}{x^2}\,dx$

7. $\int_4^9 \left(\frac{1}{\sqrt{x}} - \sqrt{x}\right) dx$

8. $\int_0^\pi \cos x\,dx$

9. $\int_{-1}^1 x(x + 1)^2\,dx$

10. $\int_1^2 x^2(x^2 + 3)\,dx$

11. $\int_0^3 \sqrt{x + 1}\,dx$

12. $\int_1^5 x\sqrt{x^2 - 1}\,dx$

13. $\int_1^4 \left(\frac{\sqrt{x} + 1}{\sqrt{x}}\right) dx$

14. $\int_{-1}^0 x\sqrt{x + 1}\,dx$

15. $\int_1^2 \frac{x^2 + 1}{(x + 1)^4}\,dx$

16. $\int_{-4}^{-2} x^2\sqrt{2 - x}\,dx$

17. $\int_0^{\pi/4} \frac{\cos x}{(1 + \sin x)^2}\,dx$

18. $\int_2^3 x(1 + 2x^2)^4\,dx$

19. $\int_1^8 \frac{(1 + x^{1/3})^2}{x^{2/3}}\,dx$

20. $\int_{-4}^4 |x + 2|\,dx$

21. Use definition (6.11) to evaluate each of the following definite integrals:

a) $\int_0^2 (x - 5)\,dx$; b) $\int_0^3 (x^2 + 3)\,dx$.

22. Prove that each of the following answers is incorrect, but do so without evaluating the definite integral. Think about what the definite integral represents.

a) $\int_0^4 (x^2 + 3x)\,dx = -3$ b) $\int_{-3}^{-2} \frac{1}{x}\,dx = 5$

In Exercises 23–30, evaluate the given definite integral.

23. $\displaystyle\int_0^1 \frac{x^3}{(x^2+1)^{3/2}}\,dx$

24. $\displaystyle\int_0^{\pi/6} \frac{\cos^3 x}{\sqrt{1+\sin x}}\,dx$

25. $\displaystyle\int_{-1}^1 x^3\sqrt{1-x^2}\,dx$

26. $\displaystyle\int_{-1}^2 \left|\frac{x}{\sqrt{3+x}}\right| dx$

27. $\displaystyle\int_{-1}^2 x^2(4-x^3)^5\,dx$

28. $\displaystyle\int_1^5 \frac{6x^2+8x+2}{\sqrt{x^3+2x^2+x}}\,dx$

29. $\displaystyle\int_1^2 \frac{x-25}{\sqrt{x}-5}\,dx$

30. $\displaystyle\int_0^1 \frac{1}{\sqrt{2+x}+\sqrt{x}}\,dx$

Applications of the Definite Integral

7

7.1 AREA

At this point in our study, polygons and circles are the only shapes in the xy-plane for which we can find area. It is impossible for us to calculate the area A of a shape such as that in Fig. 7.1(a). In this section we take advantage of the fact that area for odd shapes has not yet been defined, and we make our own definition.

Figure 7.1

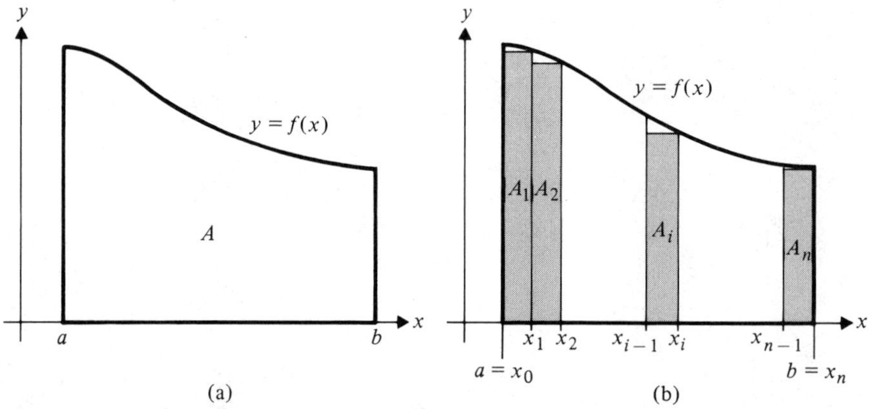

According to Problem 1 of Section 6.2 the area in Fig. 7.1(a) can be approximated by rectangles. Specifically, we divide the interval $a \le x \le b$ into n parts by points

$$a = x_0 < x_1 < x_2 < \cdots < x_{n-1} < x_n = b,$$

and construct rectangles as shown in Fig. 7.1(b). The area A_i of the ith rectangle is

$$A_i = f(x_i)(x_i - x_{i-1}) = f(x_i)\,\Delta x_i,$$

and an approximation to the required area A is therefore

$$\sum_{i=1}^{n} A_i = \sum_{i=1}^{n} f(x_i)\,\Delta x_i.$$

If we let the number of these rectangles get larger and larger, and at the same time require each to have a smaller and smaller width that eventually approaches zero, we feel that we can obtain a better and better approximation. In fact, were we to take the limit as n becomes indefinitely large, we would get what we think is A. Because we have as yet no formal definition for the area of odd-shaped figures, we take this opportunity to make our own. We define the area A in Fig. 7.1(a) as:

$$A = \lim_{n\to\infty} \sum_{i=1}^{n} f(x_i)\,\Delta x_i. \tag{7.1}$$

Area A then has been defined as the limit of a sum of rectangular areas.

But the right-hand side of definition (7.1) may also be interpreted as the definite integral of $f(x)$ with respect to x from $x = a$ to $x = b$; i.e.,

$$\int_a^b f(x)\,dx = \lim_{n\to\infty} \sum_{i=1}^{n} f(x_i)\,\Delta x_i.$$

Consequently, area A of Fig. 7.1(a) may be calculated by means of the definite integral

$$A = \int_a^b f(x)\,dx. \tag{7.2}$$

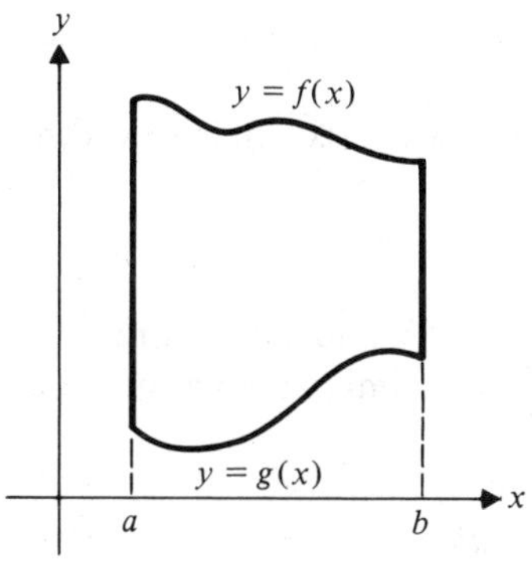

Figure 7.2

It is important to realize that we are not saying by (7.2) that a definite integral should always be thought of as an area; on the contrary, we will find that definite integrals can represent many other quantities. What we have said is that the area shown in Fig. 7.1(a) is defined by limit (7.1). It just so happens that this limit may also be interpreted as the definite integral of $f(x)$ with respect to x from $x = a$ to $x = b$; hence the area may be calculated by definite integral (7.2). Once again, the area shown in Fig. 7.1(a) is defined as limit (7.1), but it is calculated by means of definite integral (7.2). Since definite integrals can be easily evaluated using indefinite integrals, it seems that we have found a very simple way to find areas.

It is a simple matter to extend this result to the problem of finding the area shown in Fig. 7.2. With our interpretation of Eq. (7.2) we can state that the area under the curve $y = f(x)$, above the x-axis, and between the vertical lines $x = a$ and $x = b$ is given by

$$A_2 = \int_a^b f(x)\,dx.$$

Because the area under $g(x)$ is given similarly by

$$A_1 = \int_a^b g(x)\,dx,$$

it follows that the required area is

$$A = A_2 - A_1 = \int_a^b f(x)\,dx - \int_a^b g(x)\,dx = \int_a^b [f(x) - g(x)]\,dx. \tag{7.3}$$

We now have two formulas for finding areas: Eq. (7.2) for the area under a curve, above the x-axis, and between two vertical lines; and Eq. (7.3) for the area between two curves and two vertical lines. Note that (7.2) is a special case of (7.3). At this point we could solve a number of area problems using these two results, but to do so would strongly suggest that integration should be approached from a "formula" point of view; and if there is any point of view that we wish to adopt, it is completely the opposite. By the end of this chapter we hope you will have developed a clear understanding of the limit–summation process, so much so that you will be able to develop the use of integration in other situations.

To illustrate exactly how we arrive at the correct definite integral for area problems without having memorized either (7.2) or (7.3), consider again

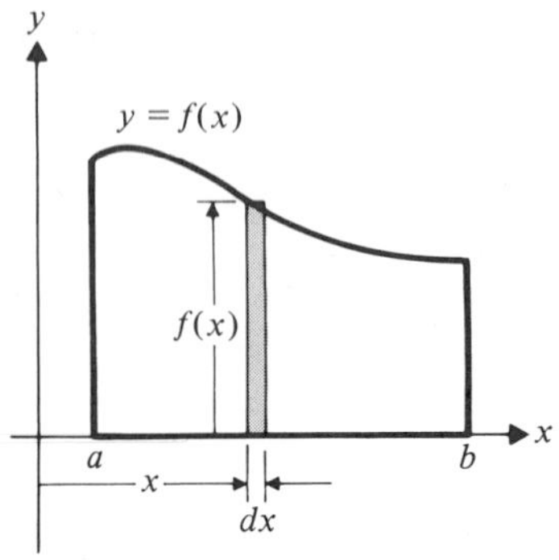

Figure 7.3

finding the area shown in Fig. 7.1(a). We draw a rectangle of width dx at position x as shown in Fig. 7.3. The area of this rectangle is

$$f(x)\,dx.$$

We now visualize this rectangle as a representative for a large number of such rectangles between a and b. We find the required area by adding together all such rectangular areas and taking the limit as their widths diminish. But this is precisely the concept of the definite integral, so that we write for the limit–summation process

$$A = \int_a^b f(x)\,dx.$$

The limits a and b identify the x-positions of the first and last rectangles, respectively.

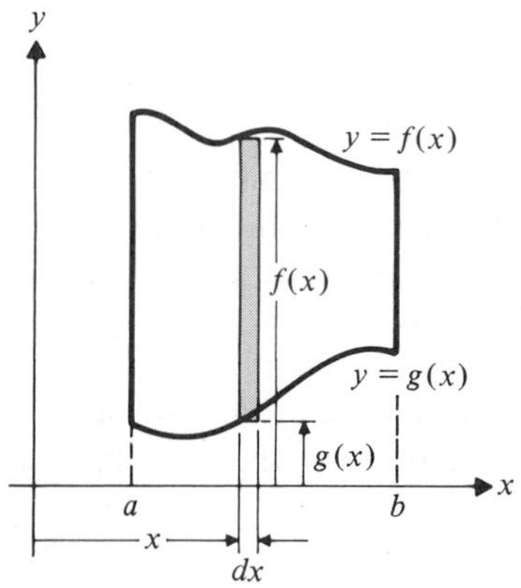

Figure 7.4

Similarly, for the area shown in Fig. 7.4, we draw a rectangle of width dx and length $f(x) - g(x)$ and, therefore, of area

$$[f(x) - g(x)]\,dx.$$

To add the areas of all such rectangles between a and b, and at the same time to take the limit as their widths approach zero, we use the limit–summation process of the definite integral:

$$A = \int_a^b [f(x) - g(x)]\,dx.$$

For area problems, then, we start with the area of a representative rectangle and proceed to the required area by summation with the definite integral.

Example 1 Find the area enclosed by the curves $y = x^2$ and $y = x^3$.

Solution The area of the representative rectangle in Fig. 7.5 is given by

$$(x^2 - x^3)\,dx;$$

hence,

$$A = \int_0^1 (x^2 - x^3)\,dx = \left\{\frac{x^3}{3} - \frac{x^4}{4}\right\}_0^1$$

$$= \frac{1}{12}. \quad \square$$

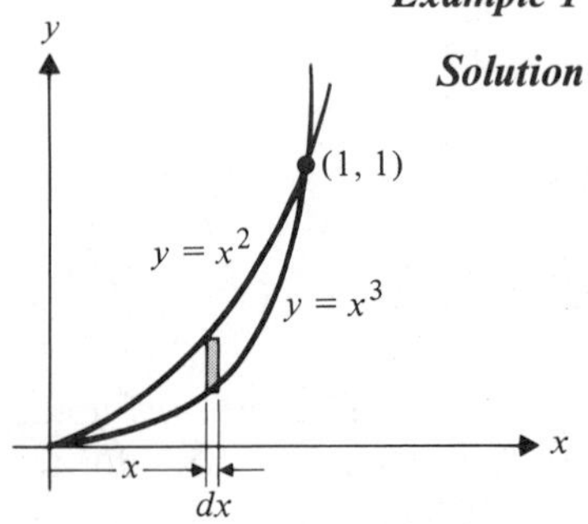

Figure 7.5

Example 2 Find the area bounded by the x-axis and the curve $y = x^3 - 3x^2 + 2x$ between $x = 0$ and $x = 2$.

Solution The area of any rectangle between $x = 0$ and $x = 1$ (Fig. 7.6) is

$$(x^3 - 3x^2 + 2x)\,dx,$$

whereas the area of a rectangle between $x = 1$ and $x = 2$ is

$$[0 - (x^3 - 3x^2 + 2x)]\,dx.$$

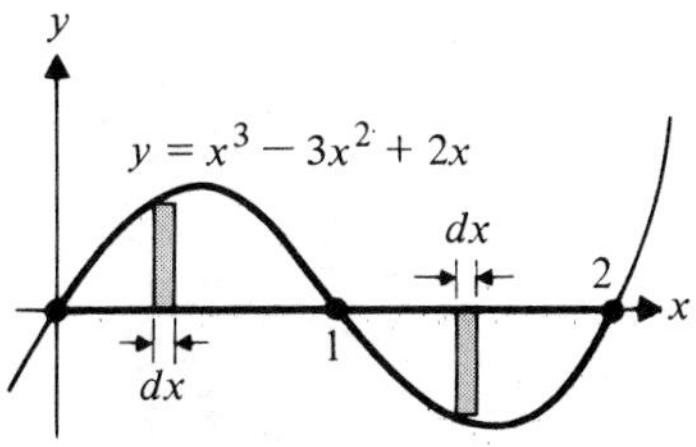

Figure 7.6

Consequently,

$$\begin{aligned} A &= \int_0^1 (x^3 - 3x^2 + 2x)\,dx + \int_1^2 (-x^3 + 3x^2 - 2x)\,dx \\ &= \left\{\frac{x^4}{4} - x^3 + x^2\right\}_0^1 + \left\{-\frac{x^4}{4} + x^3 - x^2\right\}_1^2 \\ &= \left\{\frac{1}{4} - 1 + 1\right\} + \{-4 + 8 - 4\} - \left\{-\frac{1}{4} + 1 - 1\right\} \\ &= \frac{1}{2}. \;\square \end{aligned}$$

Example 2 indicates that when the expression for the area of the representative rectangle changes, we must set up a new definite integral. Further illustration of this point is found in the following example.

Example 3 Find the area of the triangle with edges $y = x$, $y = -x/2$, and $y = 5x - 44$.

Solution The areas of representative rectangles to the left and right of $x = 8$ (Fig. 7.7) are, respectively,

$$[x - (-x/2)]\,dx \quad \text{and} \quad [x - (5x - 44)]\,dx;$$

therefore,

$$\begin{aligned} A &= \int_0^8 [x - (-x/2)]\,dx + \int_8^{11} [x - (5x - 44)]\,dx \\ &= \frac{3}{2}\int_0^8 x\,dx + \int_8^{11} (44 - 4x)\,dx \\ &= \frac{3}{4}\{x^2\}_0^8 + \{44x - 2x^2\}_8^{11} \\ &= 48 + (484 - 242) - (352 - 128) \\ &= 66. \;\square \end{aligned}$$

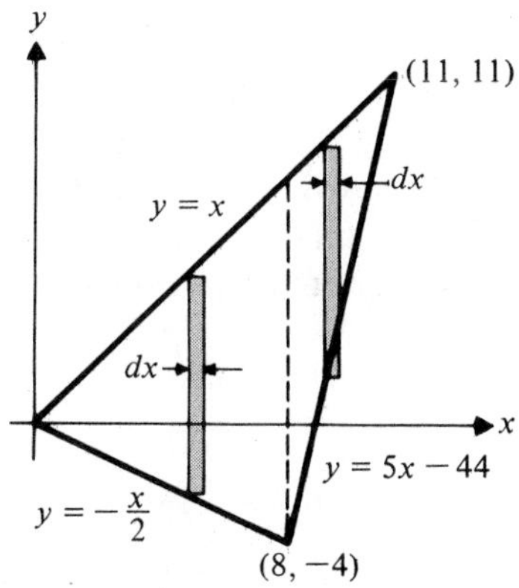

Figure 7.7

Example 4 Find the area bounded by the curves $y = \sqrt{x + 14}$, $x = \sqrt{y}$, and $y = 0$.

Solution Were we to choose vertical rectangles as in the three previous examples, we would require two definite integrals. Instead, we construct horizontal rectangles of width dy and length

$$x_2 - x_1 = \sqrt{y} - (y^2 - 14)$$

(Fig. 7.8). Then we have

$$\begin{aligned} A &= \int_0^4 (\sqrt{y} - y^2 + 14)\,dy = \left\{\frac{2}{3}y^{3/2} - \frac{1}{3}y^3 + 14y\right\}_0^4 \\ &= \frac{16}{3} - \frac{64}{3} + 56 = 40. \;\square \end{aligned}$$

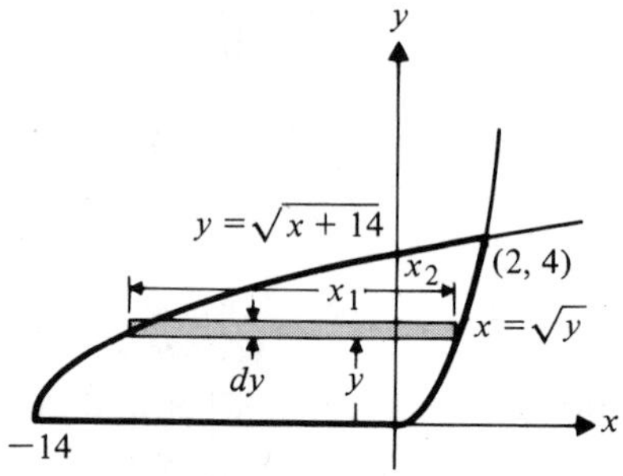

Figure 7.8

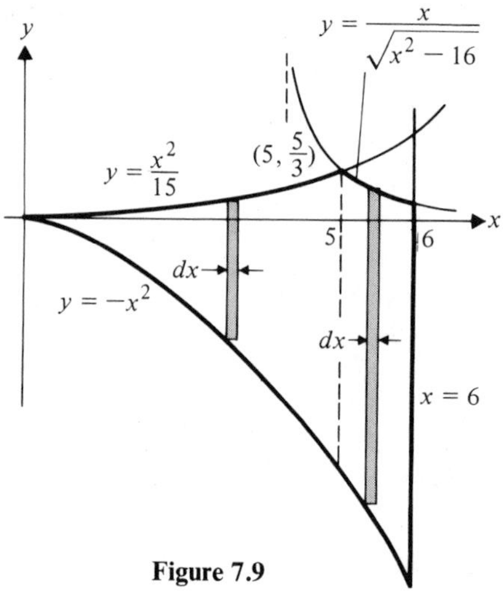

Figure 7.9

Example 4 illustrates that we can use horizontal rectangles as well as vertical rectangles or both. In any given problem it is our choice. In choosing we consider two objectives:

1. To minimize the number of integrations;
2. To obtain simple definite integrals.

For instance, by choosing one type of rectangle we may obtain only one definite integral, but it may be very difficult to evaluate. If the other type of rectangle leads to two simple definite integrals, then we would be smart to choose the two simple integrals.

Example 5 Find the area enclosed by the curves

$$y = \frac{x}{\sqrt{x^2 - 16}}, \qquad y = \frac{x^2}{15}, \qquad y = -x^2, \qquad x = 6, \qquad x \geq 0.$$

Solution From Fig. 7.9, we see that

$$A = \int_0^5 \left\{\frac{x^2}{15} - (-x^2)\right\} dx + \int_5^6 \left\{\frac{x}{\sqrt{x^2 - 16}} - (-x^2)\right\} dx$$

$$= \frac{16}{15}\int_0^5 x^2\, dx + \int_5^6 \left\{\frac{x}{\sqrt{x^2 - 16}} + x^2\right\} dx$$

$$= \frac{16}{45}\{x^3\}_0^5 + \left\{\sqrt{x^2 - 16} + \frac{x^3}{3}\right\}_5^6$$

$$= \frac{16}{45}(125) + (\sqrt{20} + 72) - \left(3 + \frac{125}{3}\right) = 76.2. \ \square$$

EXERCISES 7.1

In Exercises 1–10, find the area bounded by the given curves.

1. $y^2 = 4x$, $x^2 = 4y$

2. $y = x^3 + 8$, $y = 4x + 8$

3. $yx^2 = 4$, $y = 5 - x^2$

4. $x = y(y - 2),\ x + y = 12$

5. $x = 4y - 4y^2,\ y = x - 3,\ y = 1,\ y = 0$

6. $y = \sqrt{x + 4},\ y = (x + 4)^2/8$

7. $y = (x - 1)^5,\ x = 0,\ y = 0$

8. $y = \sin x\ (0 \le x \le \pi),\ y = 0$

9. $y = x^3 - x,\ y = 0$

10. $x + y = 1,\ x + y = 5,\ y = 2x + 1,\ y = 2x + 6$

11. a) Find the area A bounded by the curves

$$y = 16 - x^2, \qquad y = 0, \qquad x = 1, \qquad x = 3.$$

In the remainder of this problem we will approximate A by rectangles and show that the accuracy of the approximation increases as the number of rectangles increases. To do this we need a large graph of the function $f(x) = 16 - x^2$ on the interval $1 \le x \le 3$.

b) On the graph we draw two rectangles of equal width (one unit) and with heights as determined in Fig. 7.1(b). If A_2 denotes the sum of the areas of these two rectangles, what is A_2? What is the error in the approximation of A by A_2? Illustrate this error on the graph.

c) Repeat part (b) with four rectangles all of width one-half, denoting the sum of the areas of the four rectangles by A_4. Show on the graph the extra precision of A_4 over A_2.

d) Repeat part (b) with eight rectangles all of equal width, denoting the sum of the areas of the eight rectangles by A_8. Show on the graph the extra precision of A_8 over A_4.

12. Repeat Exercise 11 given the area bounded by the curves

$$y = x^3 + 1, \qquad x = 1, \qquad x = 3, \qquad y = 0.$$

In Exercises 13–16, set up, but do not evaluate, definite integral(s) for the area bounded by the curves.

13. $x = 1/\sqrt{4 - y^2},\ 4x = -y^2,\ y = -1,\ y = 1$

14. $x^2 + y^2 = 4,\ x^2 + y^2 = 4x$ (interior to both)

15. $x^2 + y^2 = 4,\ x^2 + y^2 = 6x$ (interior to both)

16. $x^2 + y^2 = 16,\ x = y^2$ (smaller area)

17. Find the area bounded by $y^2 = 4ax$ and $x^2 = 4ay$ ($a > 0$ a constant).

In Exercises 18–26, find the area bounded by the given curves.

18. $y = x/\sqrt{x + 3},\ x = 1,\ x = 6,\ y = -x^2$

19. $x = y^2 + 2,\ x = -(y - 4)^2,\ y = -x + 4,\ y = 0$

20. $y = x^3 - x,\ x + y + 1 = 0,\ x = \sqrt{y + 1}$

21. $y = \left|\dfrac{x}{(x - 2)^3}\right|,\ y = 0,\ x = -1,\ x = 1$

22. $|x|^{1/2} + |y|^{1/2} = 1$

23. $y^2 = x^2(4 - x^2)$

24. $y^2 = x^4(9 + x)$

25. $y^2 = x^2(x^2 - 4),\ x = 5$

26. $(2x - y)^2 = x^3,\ x = 4$

27. Interpret the mean value theorem for definite integrals (Exercise 27 in Section 6.4) in terms of area if $f(x) \ge 0$ on $a \le x \le b$.

* **28.** Find a point (a, b) on the curve $y = x/\sqrt{x^2 + 1}$ such that the area bounded by this curve, the x-axis, and the line $x = a$ is equal to twice the area bounded by the curve, the y-axis, and the line $y = b$.

* **29.** If 2^n $(n = 1, 2, \ldots)$ rectangles (all of equal width) are drawn in Exercise 11 to approximate A, and A_{2^n} denotes the sum of the areas of these rectangles, show that

$$A_{2^n} = A - \left\{\frac{1}{2^{n-2}} + \frac{1}{6}\left(\frac{3}{2^{n-3}} + \frac{1}{2^{2n-3}}\right)\right\},$$

and hence that

$$\lim_{n\to\infty} A_{2^n} = A.$$

* **30.** Let P be a point on the cubic curve $y = f(x) = ax^3$. Let the tangent line at P intersect $y = f(x)$ again at Q, and let A be the area of the region bounded by $y = f(x)$ and the line PQ. Let B be the area of the region defined in the same way by starting with Q instead of P. Show that B is 16 times as large as A.

* **31.** Prove that the result in Exercise 30 is valid for any cubic $y = f(x) = ax^3 + bx^2 + cx + d$.

7.2 VOLUMES OF SOLIDS OF REVOLUTION

If the area shown in Fig. 7.1(a) is revolved about the x-axis, the region through which it passes is called a *solid of revolution*. In this section we develop a definition for the volume of such solids. According to Problem 2 of Section 6.2, the volume V of the solid of revolution when the area shown in Fig. 7.1(a) is rotated

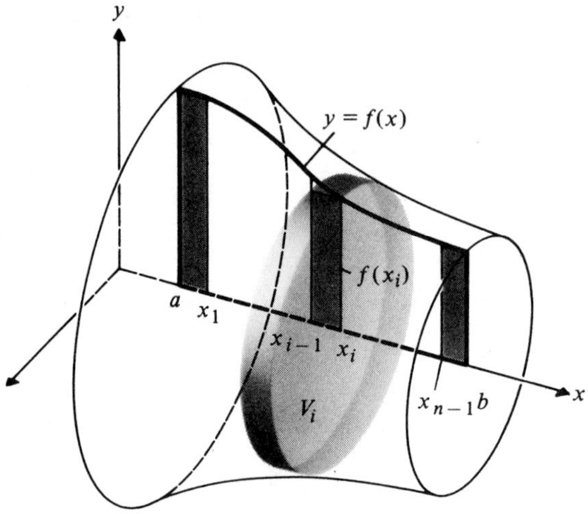

Figure 7.10

about the x-axis can be approximated by *discs*. Specifically, we divide the interval $a \leq x \leq b$ into n parts by points

$$a = x_0 < x_1 < x_2 < \cdots < x_{n-1} < x_n = b$$

and construct rectangles as in Fig. 7.1(b). If each of these rectangles is rotated about the x-axis, then n discs will be formed. Since the radius of the ith disc is $f(x_i)$ (Fig. 7.10), its volume is given by

$$\begin{aligned} V_i &= \pi[f(x_i)]^2(x_i - x_{i-1}) \\ &= \pi[f(x_i)]^2 \, \Delta x_i. \end{aligned}$$

An approximation to the required volume V is therefore

$$\sum_{i=1}^{n} V_i = \sum_{i=1}^{n} \pi[f(x_i)]^2 \, \Delta x_i.$$

If we let the number of rectangles get larger and larger and at the same time require the width of each to approach zero, it seems reasonable that we will obtain a better and better approximation. Furthermore, were we to take the limit as n becomes indefinitely large, we would get what we think is V. Because we have as yet no formal definition for such volumes, we make our own. We define

$$V = \lim_{n \to \infty} \sum_{i=1}^{n} \pi[f(x_i)]^2 \, \Delta x_i. \tag{7.4}$$

But we can interpret the right-hand side of this definition as the definite integral of $\pi[y(x)]^2$ with respect to x from $x = a$ to $x = b$; i.e.,

$$\int_a^b \pi[f(x)]^2 \, dx = \lim_{n \to \infty} \sum_{i=1}^{n} \pi[f(x_i)]^2 \, \Delta x_i.$$

Consequently, we can calculate the volume of the solid of revolution generated by rotating the area shown in Fig. 7.1(a) about the x-axis by means of the definite integral

$$V = \int_a^b \pi[f(x)]^2\, dx. \tag{7.5}$$

We point out again that the volume of the solid of revolution has been defined by limit (7.4), but for evaluation of this limit we use definite integral (7.5).

To avoid memorizing (7.5) as a formula, we use the technique introduced in Section 7.1. To find the volume obtained by rotating the area shown in Fig. 7.1(a) about the x-axis, we construct a rectangle of width dx at position x, as shown in either Fig. 7.3 or Fig. 7.11. If this rectangle is rotated about the x-axis, the volume of the disc generated is

$$\pi[f(x)]^2\, dx,$$

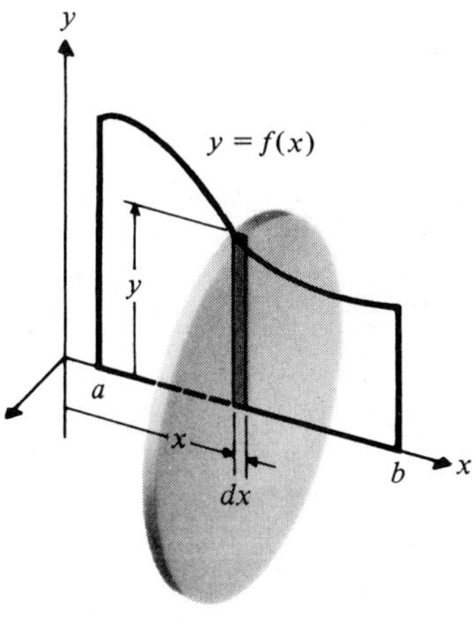

Figure 7.11

where dx is the thickness of the disc and $\pi[f(x)]^2$ is its surface area. This disc is pictured as representative of a large number of such discs between a and b. We can find the required volume by adding together the volumes of all such discs and taking the limit as their widths diminish. But this is precisely the concept of the definite integral, so that we write for the limit–summation process

$$V = \int_a^b \pi[f(x)]^2\, dx,$$

where the limits $x = a$ and $x = b$ identify the x-positions of the first and last discs, respectively. This is called the *disc method* for finding the volume of a solid of revolution.

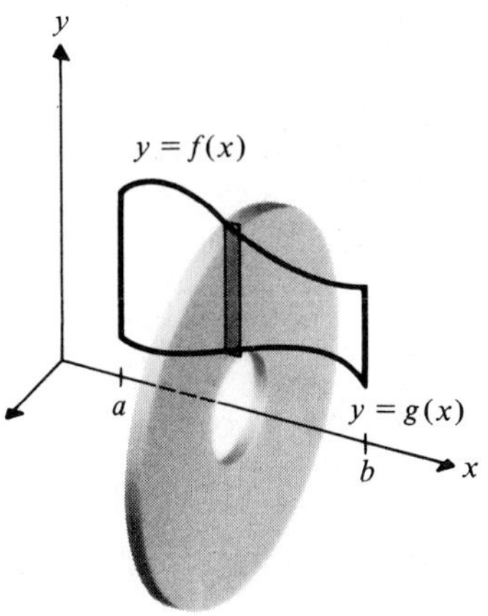

Figure 7.12

A slightly more general problem is that of finding the volume of the solid of revolution generated by rotating the area shown in Fig. 7.12 about the x-axis. If a representative rectangle of width dx at position x is rotated about the x-axis, the volume formed is a "washer." Since the outer and inner radii of this washer are $f(x)$ and $g(x)$, respectively, its volume is given by

$$\{\pi[f(x)]^2 - \pi[g(x)]^2\}\, dx.$$

To add the volumes of all such washers between a and b, and at the same time take the limit as their widths approach zero, we use the definite integral

$$V = \int_a^b \{\pi[f(x)]^2 - \pi[g(x)]^2\}\, dx. \tag{7.6}$$

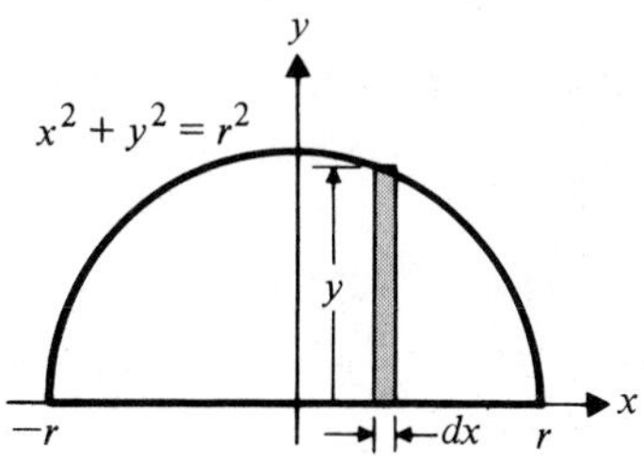

Figure 7.13

We could also have determined this result by using Eq. (7.5) as a formula and subtracting the volume obtained by rotating the area under $f(x)$ from the volume obtained by rotating the area under $g(x)$. But the key word in this approach is "formula," and memorization is something that we want to avoid as much as possible.

Example 6 Prove that the volume of a sphere of radius r is $4\pi r^3/3$.

Solution A sphere of radius r is formed if the semicircle $x^2 + y^2 \le r^2$ $(y \ge 0)$ is rotated about the x-axis. The volume of the representative disc generated by rotating the rectangle shown in Fig. 7.13 about the x-axis is

$$\pi y^2\, dx = \pi(r^2 - x^2)\, dx.$$

Since the volume formed by the left quarter circle is the same as for the right quarter circle, we calculate the volume generated by the right quarter and double the result:

$$V = 2\int_0^r \pi(r^2 - x^2)\,dx = 2\pi\left\{r^2x - \frac{x^3}{3}\right\}_0^r = \frac{4}{3}\pi r^3. \ \square$$

Example 7 Prove that the volume of a right-circular cone of base radius r and height h is $\pi r^2 h/3$.

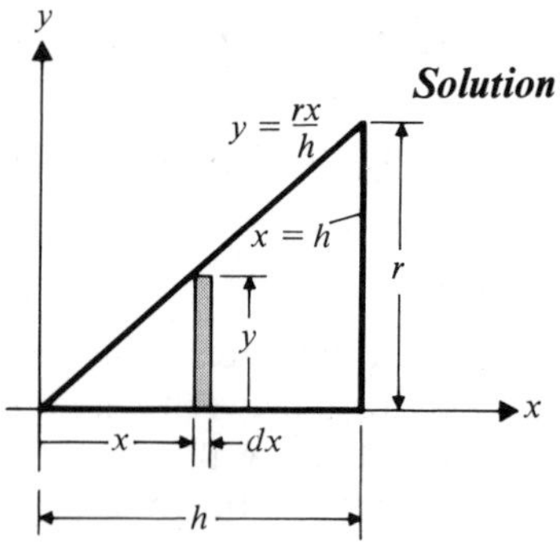

Figure 7.14

Solution The cone can be generated by rotating the triangle shown in Fig. 7.14 about the x-axis. The volume of the representative disc formed by rotating the rectangle shown is given by

$$\pi y^2\,dx = \pi\left(\frac{rx}{h}\right)^2 dx;$$

hence,

$$V = \int_0^h \frac{\pi r^2 x^2}{h^2}\,dx = \frac{\pi r^2}{h^2}\left\{\frac{x^3}{3}\right\}_0^h = \frac{1}{3}\pi r^2 h. \ \square$$

Example 8 Find the volume of the solid of revolution obtained by rotating the area bounded by the curves $y = 1 - x^2$ and $y = 4 - 4x^2$ about (a) the x-axis; (b) the line $y = -1$.

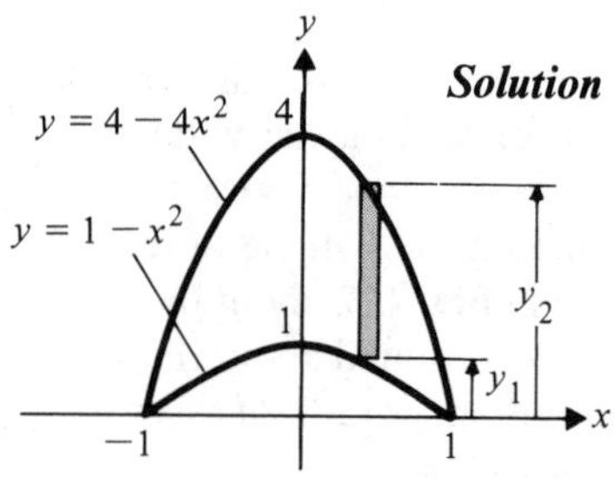

Figure 7.15

Solution

a) When the rectangle shown in Fig. 7.15 is rotated about the x-axis, the volume of the washer formed is

$$\begin{aligned}(\pi y_2^2 - \pi y_1^2)\,dx &= \pi[(4 - 4x^2)^2 - (1 - x^2)^2]\,dx \\ &= 15\pi(1 - 2x^2 + x^4)\,dx.\end{aligned}$$

Because of the symmetry of the area, we rotate only the right half and double the result:

$$V = 2\int_0^1 15\pi(1 - 2x^2 + x^4)\,dx = 30\pi\left\{x - \frac{2x^3}{3} + \frac{x^5}{5}\right\}_0^1 = 16\pi.$$

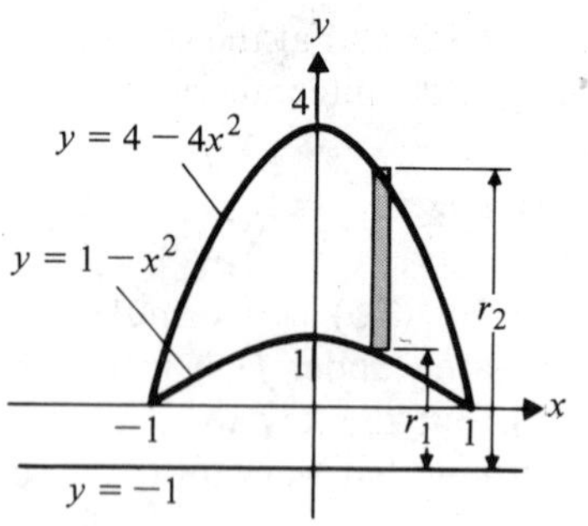

Figure 7.16

b) When the rectangle shown in Fig. 7.16 is rotated about the line $y = -1$, the inner and outer radii of the washer are r_1 and r_2, respectively. The volume of the washer formed is therefore given by

$$\begin{aligned}(\pi r_2^2 - \pi r_1^2)\,dx &= \pi[(4 - 4x^2 + 1)^2 - (1 - x^2 + 1)^2]\,dx \\ &= 3\pi(7 - 12x^2 + 5x^4)\,dx.\end{aligned}$$

Consequently,

$$V = 2\int_0^1 3\pi(7 - 12x^2 + 5x^4)\,dx = 6\pi\{7x - 4x^3 + x^5\}_0^1 = 24\pi. \ \square$$

Example 9 Find the volume of the solid of revolution obtained by rotating the area enclosed by the curves $y = x^2 - 1$ and $y = 0$ about the line $x = 5$.

Solution When the horizontal rectangle shown in Fig. 7.17 is rotated around the line $x = 5$, the volume of the washer formed is given by

$$(\pi r_2^2 - \pi r_1^2)\,dy = \pi[(5 - x_2)^2 - (5 - x_1)^2]\,dy.$$

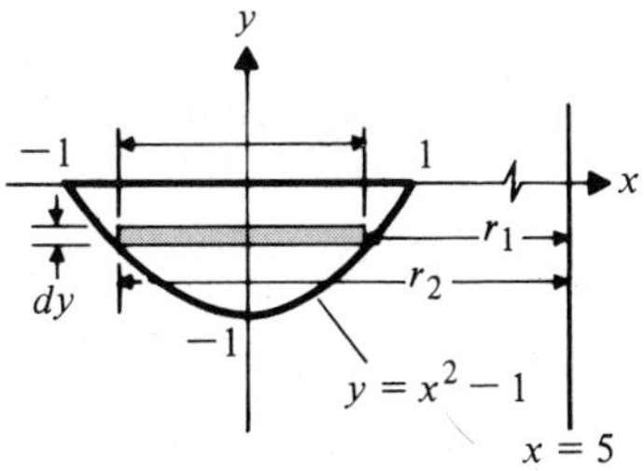

Figure 7.17

Since x_1 and x_2 are x-coordinates of points on the curve $y = x^2 - 1$, we solve this equation for $x = \pm\sqrt{y+1}$. Since $x_1 > 0$ and $x_2 < 0$, we set $x_1 = \sqrt{y+1}$ and $x_2 = -\sqrt{y+1}$. Thus the volume of the washer can be expressed as

$$\pi[(5 + \sqrt{y+1})^2 - (5 - \sqrt{y+1})^2]\,dy = 20\pi\sqrt{y+1}\,dy.$$

The required volume is therefore

$$V = \int_{-1}^{0} 20\pi\sqrt{y+1}\,dy = 20\pi\left\{\frac{2}{3}(y+1)^{3/2}\right\}_{-1}^{0} = \frac{40\pi}{3}. \square$$

Example 9 illustrates not only that horizontal as well as vertical washers can be used in the calculation of volumes of solids of revolution, but more important, that the washer method is not particularly convenient for some problems. In fact, for some problems use of the washer method is totally inappropriate. Consider, for example, the volume of the solid of revolution obtained by rotating the area shown in Fig. 7.18 about the x-axis. Each of the rectangles shown yields a washer with a volume formula different from every other. As a result, use of the washer method requires six definite integrals and only one of these is straightforward to set up. The determination of this volume seems to lend itself to the use of horizontal, rather than vertical, rectangles. To see that this is indeed true, we divide the interval $a \le y \le b$ into n parts by points

$$a = y_0 < y_1 < y_2 < \cdots < y_{n-1} < y_n = b$$

(Fig. 7.19). In each subinterval $y_{i-1} \le y \le y_i$, we find the midpoint

$$y_i^* = \frac{y_{i-1} + y_i}{2}$$

and construct a rectangle of length $f(y_i^*) - g(y_i^*)$ and width $y_i - y_{i-1}$ as shown in Fig. 7.19. When this ith rectangle is rotated around the x-axis, a

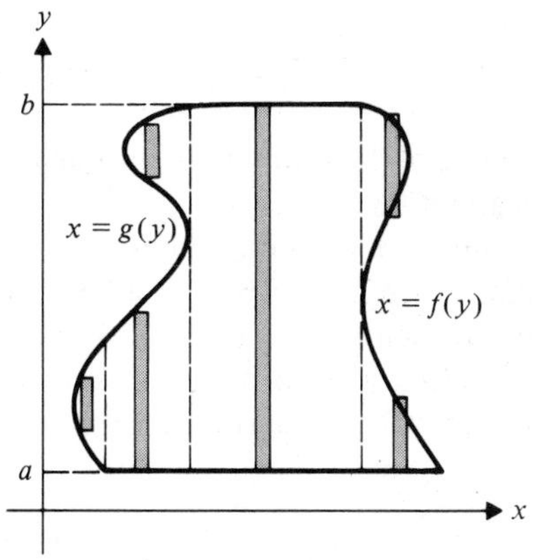

Figure 7.18

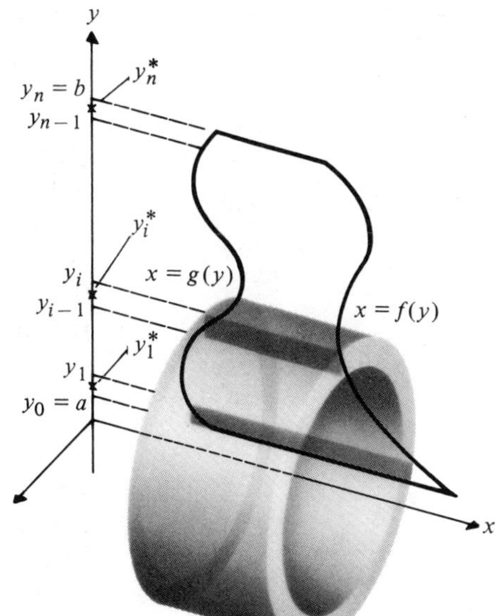

Figure 7.19

cylindrical shell is formed. Since the length of the shell is $f(y_i^*) - g(y_i^*)$, and its inner and outer radii are y_{i-1} and y_i, respectively, its volume is given by

$$\begin{aligned}(\pi y_i^2 - \pi y_{i-1}^2)&[f(y_i^*) - g(y_i^*)] \\ &= \pi(y_i + y_{i-1})(y_i - y_{i-1})[f(y_i^*) - g(y_i^*)] \\ &= 2\pi y_i^*[f(y_i^*) - g(y_i^*)]\,\Delta y_i,\end{aligned}$$

where

$$\Delta y_i = y_i - y_{i-1}.$$

If we add together the volumes of all such shells, we obtain an approximation to the required volume V:

$$\sum_{i=1}^{n} 2\pi y_i^*[f(y_i^*) - g(y_i^*)]\,\Delta y_i.$$

If we let the number of these rectangles get larger and larger and at the same time require the width of each to approach zero, it seems reasonable to expect that we will get a better and better approximation. The limit, then, as n becomes indefinitely large should yield what we think is V. We therefore define the required volume as

$$\boxed{V = \lim_{n\to\infty} \sum_{i=1}^{n} 2\pi y_i^*[f(y_i^*) - g(y_i^*)]\,\Delta y_i.} \tag{7.7}$$

But the right-hand side of definition (7.7) may be interpreted as the definite integral of $2\pi y[f(y) - g(y)]$ with respect to y from $y = a$ to $y = b$; i.e.,

$$\int_a^b 2\pi y[f(y) - g(y)]\,dy = \lim_{n\to\infty} \sum_{i=1}^{n} 2\pi y_i^*[f(y_i^*) - g(y_i^*)]\,\Delta y_i.$$

Consequently, the definite integral

$$V = \int_a^b 2\pi y[f(y) - g(y)]\,dy \tag{7.8}$$

gives the volume of the solid of revolution generated by rotating the area shown in Fig. 7.18 about the x-axis.

Once again we point out that it is not necessary to memorize Eq. (7.8) to use as a formula. Indeed, we strongly advise against memorization, and suggest that the correct definite integral can be obtained as follows. We construct a rectangle of width dy at position y as shown in Fig. 7.20(a). If this rectangle is rotated about the x-axis, the volume of the cylindrical shell generated is approximately

$$2\pi y[f(y) - g(y)]\,dy.$$

This we obtain by picturing the shell as being cut along the rectangle (Fig. 7.20b) and opened up into a slab with dimensions dy, $f(y) - g(y)$, and $2\pi y$. The thickness dy of the shell corresponds to the thickness of the slab; the length $f(y) - g(y)$ of the shell corresponds to that of the slab; and the inner circumference $2\pi y$ of the shell corresponds to the width of the slab. If we now

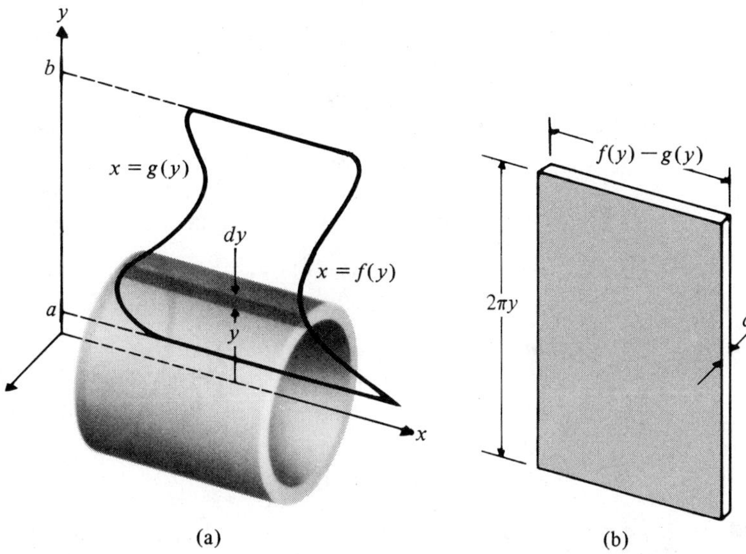

Figure 7.20

add the volumes of all such cylindrical shells, and at the same time take the limit as their widths approach zero, we obtain the required volume. But this limit–summation operation is what we mean by the definite integral; hence we write

$$V = \int_a^b 2\pi y[f(y) - g(y)]\, dy.$$

This is called the *cylindrical-shell method* for finding the volume of a solid of revolution.

We now reconsider the volume given in Example 9. If we rotate the rectangle shown in Fig. 7.21 about the line $x = 5$, the volume of the cylindrical shell is approximately

$$\begin{aligned} 2\pi(5 - x)(0 - y)\, dx &= 2\pi(5 - x)(1 - x^2)\, dx \\ &= 2\pi(x^3 - 5x^2 - x + 5)\, dx. \end{aligned}$$

The total volume is therefore

$$\begin{aligned} V &= \int_{-1}^{1} 2\pi(x^3 - 5x^2 - x + 5)\, dx = 2\pi\left\{\frac{x^4}{4} - \frac{5x^3}{3} - \frac{x^2}{2} + 5x\right\}_{-1}^{1} \\ &= 2\pi\left\{\frac{1}{4} - \frac{5}{3} - \frac{1}{2} + 5\right\} - 2\pi\left\{\frac{1}{4} + \frac{5}{3} - \frac{1}{2} - 5\right\} = \frac{40\pi}{3}. \end{aligned}$$

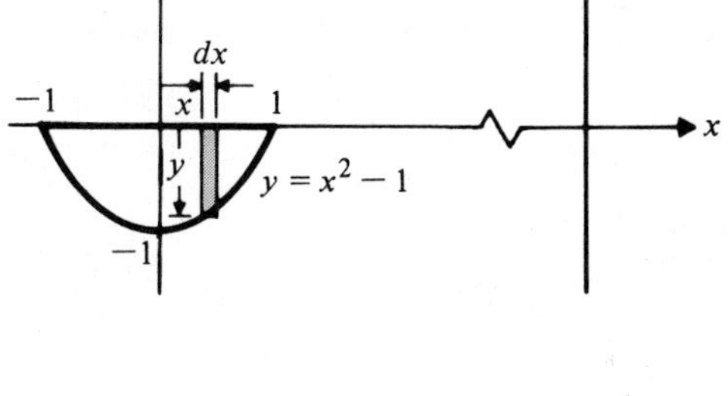

Figure 7.21

Note that even though the area in this example is symmetric about the y-axis, the volume generated by the right half is less than that generated by the left. For this reason we cannot integrate from $x = 0$ to $x = 1$ and double the result. If rotation were performed about the x-axis, we could indeed integrate over either half of the interval and double. Finally, for rotation about the y-axis we would integrate over only one half of the interval and neglect the other half (in order to eliminate duplications).

Example 10 Find the volume of the solid of revolution obtained by rotating the area bounded by the curves $y = x^3$, $y = -x$, and $y = 1$ about the line $y = -2$.

Solution If we rotate the rectangle shown in Fig. 7.22 around the line $y = -2$, the volume of the representative cylindrical shell is approximately

$$2\pi(y + 2)(x_2 - x_1)\,dy = 2\pi(y + 2)(y^{1/3} + y)\,dy.$$

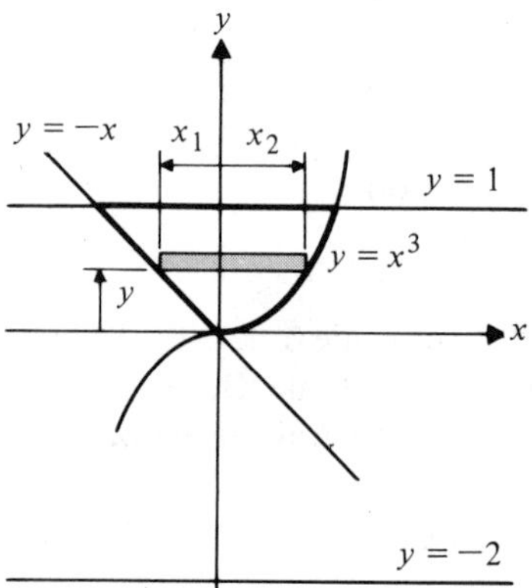

Figure 7.22

The total volume is therefore

$$\begin{aligned} V &= \int_0^1 2\pi(y + 2)(y^{1/3} + y)\,dy \\ &= 2\pi \int_0^1 (y^{4/3} + 2y^{1/3} + y^2 + 2y)\,dy \\ &= 2\pi\left\{\frac{3}{7}y^{7/3} + \frac{3}{2}y^{4/3} + \frac{1}{3}y^3 + y^2\right\}_0^1 \\ &= 2\pi\left\{\frac{3}{7} + \frac{3}{2} + \frac{1}{3} + 1\right\} \\ &= \frac{137\pi}{21}. \ \square \end{aligned}$$

EXERCISES 7.2

In each of Exercises 1–10, use the disc or washer method to find the volume of the solid of revolution obtained by rotating the area bounded by the curves about the line indicated.

1. $x^2 + y^2 = 36$, about $y = 0$

2. $y^2 = 5 - x$, $x = 0$, about $x = 0$

3. $y = x^2 + 4$, $y = 2x^2$, about $y = 0$

4. $x - y^2 = 16$, $x = 20$, about $x = 0$

5. $x - 1 = y^2$, $x = 5$, about $x = 1$

6. $x + y + 1 = 0$, $2y = x - 2$, $y = 0$, about $y = 0$

7. $y = 4x^2 - 4x$, $y = x^3$, about $y = -2$

8. $x = 2y - y^2 - 2$, $x = -5$, about $x = 0$

9. $y^2 = 5 - x$, $x = 0$, about $x = 6$

10. $y = x^2 - 2x$, $y = 2x - x^2$, about $y = 2$

In each of Exercises 11–20, use the cylindrical shell method to find the volume of the solid of revolution obtained by rotating the area bounded by the curves about the line indicated.

11. $y = 1 - x^3$, $x = 0$, $y = 0$, about $x = 0$

12. $y = -\sqrt{4 - x}$, $x = 0$, $y = 0$, about $y = 0$

13. $y = (x - 1)^2$, $y = 1$, about $x = 0$

14. $x + y = 4$, $y = 2\sqrt{x - 1}$, $y = 0$, about $y = 0$

15. $y = 3x - x^2$, $y = x^2 - 3x$, about $x = 4$

16. $y = 2 - |x|$, $y = 0$, about $y = -1$

17. $x = y^3$, $x = 2 - y^2$, $y = 0$, about $y = 1$

18. $y = x^2$, $y = -x^2$, $x = -1$, about $x = -1$

19. $y = -\sqrt{9 - x}$, $x = 0$, $y = 0$ about $y = 0$

20. $y = x$, $xy = 9$, $x + y = 10$, $(x \geq y)$, about $x = 0$

In each of Exercises 21–28, find the volume of the solid of revolution obtained by rotating the area bounded by the curves about the line indicated.

21. $(x^2 + 1)^2 y = 4$, $y = 1$, about $x = 0$

22. $y = (x - 1)^2 - 4$, $5y = 12x$, $x = 0$, $(x \geq 0)$, about $x = 0$

23. $x^2 - y^2 = 5, 9y = x^2 + 9, 9y + x^2 + 9 = 0,$ $(-3 \le x \le 3)$, about $x = 0$

24. $y = |x^2 - 1|, x = -2, x = 2, y = -1$, about $y = -1$

25. $y = x^2 - 2, y = 0$, about $y = -1$

26. $x = \sqrt{4 + 12y^2}, x - 20y = 24, y = 0$, about $y = 0$

27. $y = (x + 1)^{1/4}, y = -(x + 1)^2, x = 0$, about $x = 0$

28. $y = x^4 - 3, y = 0$, about $y = -1$

29. A tapered rod of length L has circular cross-sections. If the radii of its ends are a and b, what is the volume of the rod?

30. During one revolution an airplane propeller displaces an amount of air that can be calculated as a volume of a solid of revolution. If the area yielding the volume is that bounded by the curves $64x = y(y - 4)$ and $64x = y(4 - y)$, and is rotated about the x-axis, calculate the volume of air displaced.

31. If a sphere of radius r is sliced a distance h from its center, show that the volume of the smaller piece is

$$V = \frac{\pi}{3}(r - h)^2(2r + h).$$

32. An embankment is to be built around a circular children's wading pool, as shown in Fig. 7.23(a). Figure 7.23(b) shows a cross-section of the embankment.

a) Find a cubic polynomial $y = ax^3 + bx^2 + cx + d$ to fit the three points (0, 0), (4, 2), and (6, 0).

b) Determine the amount of fill required to build the embankment.

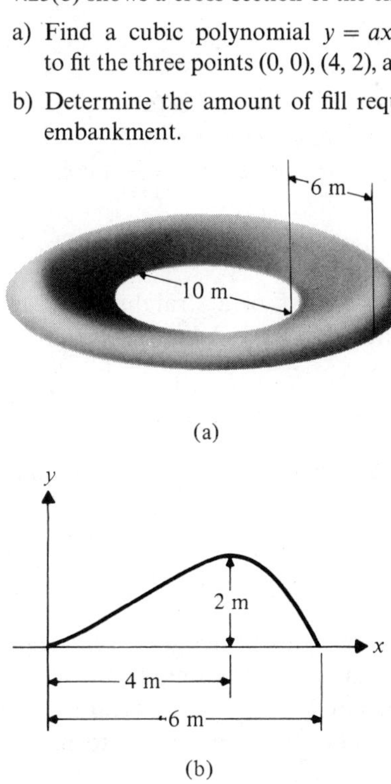

Figure 7.23

*33. Find the volume of the donut obtained by rotating the circle shown in Fig. 7.24 about the y-axis.

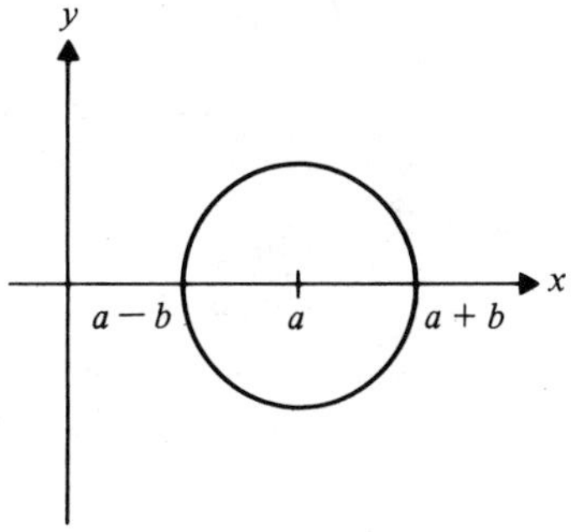

Figure 7.24

*34. A cylindrical hole is bored through the center of a sphere, the length of the hole being L. Show that no matter what the radius of the sphere, the volume of the sphere that remains is always the same, and equal to the volume of a sphere of diameter L.

*35. Water half fills a cylindrical pail of radius a and height L. When the pail is rotated about its axis of symmetry with angular speed ω (Fig. 7.25), the surface of the water assumes a parabolic shape, the cross-section of which is given by

$$y = H + \frac{\omega^2 x^2}{2g},$$

where $g > 0$ is the acceleration due to gravity and H is a constant. Find the speed ω (in terms of L, a, and g) at which water spills over the top.

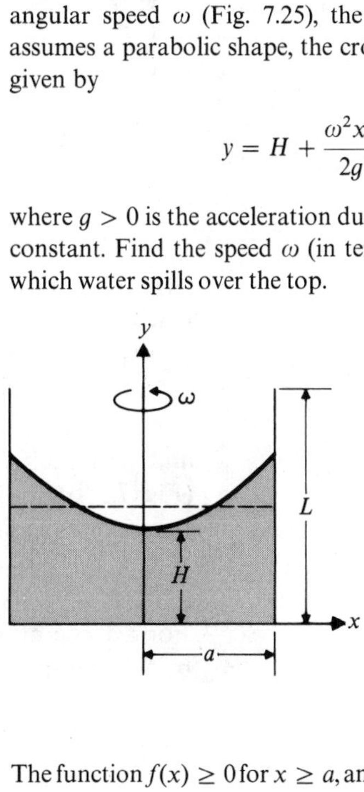

Figure 7.25

*36. The function $f(x) \ge 0$ for $x \ge a$, and is continuous on this interval. If the area bounded by $y = f(x)$, $y = 0$, $x = a$, and $x = b$ is rotated about the x-axis, the volume of the solid of revolution always turns out to be $b^3 + 3b^2 + 4$. What is $f(x)$?

7.3 LENGTHS OF CURVES

According to formula (1.5), the length of the straight-line segment joining any two points $P(x_1, y_1)$ and $Q(x_2, y_2)$ (Fig. 7.26) is given by

$$|PQ| = \sqrt{(x_2 - x_1)^2 + (y_2 - y_1)^2}.$$

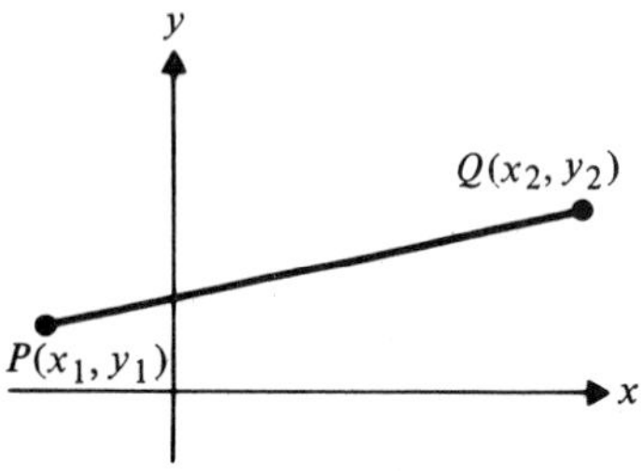

Figure 7.26

In Fig. 7.27, the arc subtends an angle of θ radians at the center of the circle, and the length of the arc is therefore

$$s = r\theta. \tag{7.9}$$

These two formulas represent the extent of our present ability to find lengths along curves. In this section we derive a result that will enable us to find lengths along many other curves.

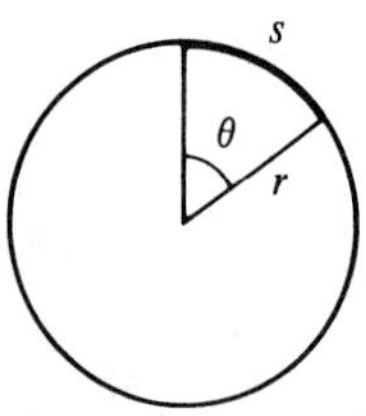

Figure 7.27

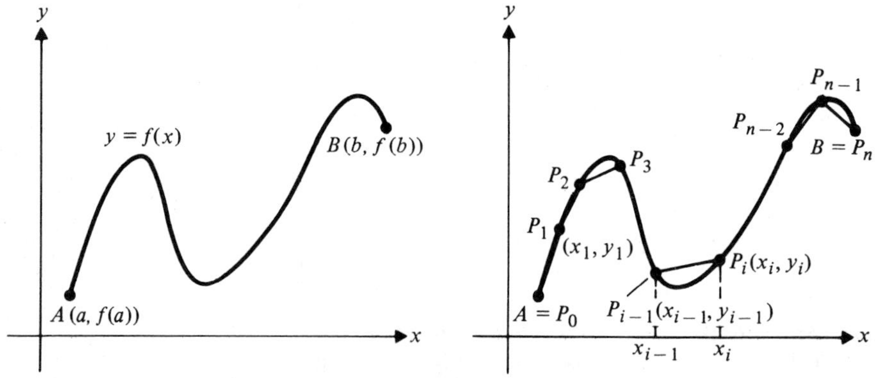

(a) (b)

Figure 7.28

Consider first the curve $C\colon y = f(x)$ shown in Fig. 7.28(a) joining points $A(a, f(a))$ and $B(b, f(b))$. To find the length L of C we begin by approximating C with a series of straight-line lengths. Specifically, we choose $n - 1$ consecutive points on C between A and B,

$$A = P_0, P_1, P_2, \ldots, P_{n-1}, P_n = B,$$

and join each P_{i-1} to P_i $(i = 1, \ldots, n)$ by means of a straight-line segment (Fig. 7.28b). If the coordinates of P_i are denoted by (x_i, y_i), then the length of the line segment joining P_{i-1} and P_i is given by

$$|P_{i-1}P_i| = \sqrt{(x_i - x_{i-1})^2 + (y_i - y_{i-1})^2}.$$

For a large number of these line segments, it is reasonable to approximate L by the sum of the lengths of the segments:

$$\sum_{i=1}^{n} |P_{i-1}P_i| = \sum_{i=1}^{n} \sqrt{(x_i - x_{i-1})^2 + (y_i - y_{i-1})^2}.$$

In fact, as we increase n and, at the same time, decrease the length of each segment, we suspect that the approximation will become correspondingly more and more accurate. The limit, then, as n becomes indefinitely large should yield L. Because we have no other way to calculate L, we take this opportunity to make our own definition. We define

$$L = \lim_{n\to\infty} \sum_{i=1}^{n} |P_{i-1}P_i|$$
$$= \lim_{n\to\infty} \sum_{i=1}^{n} \sqrt{(x_i - x_{i-1})^2 + (y_i - y_{i-1})^2}. \tag{7.10}$$

If we set $\Delta x_i = x_i - x_{i-1}$, and note that $y_{i-1} = f(x_{i-1})$ and $y_i = f(x_i)$, we have

$$L = \lim_{n\to\infty} \sum_{i=1}^{n} \sqrt{(\Delta x_i)^2 + [f(x_i) - f(x_{i-1})]^2}$$
$$= \lim_{n\to\infty} \sum_{i=1}^{n} \sqrt{1 + \left\{\frac{f(x_i) - f(x_{i-1})}{\Delta x_i}\right\}^2}\, \Delta x_i.$$

In order that (7.10) be a useful definition from a calculation point of view, we must find a convenient way to evaluate this limit–summation. To do this we assume that $f'(x)$ exists at every point in the interval $a \le x \le b$, and we apply the mean value theorem to $f(x)$ on each subinterval $x_{i-1} \le x \le x_i$ (Fig. 7.29). The theorem guarantees the existence of at least one point x_i^* between x_{i-1} and x_i such that

$$f'(x_i^*) = \frac{f(x_i) - f(x_{i-1})}{x_i - x_{i-1}}$$
$$= \frac{f(x_i) - f(x_{i-1})}{\Delta x_i}.$$

Consequently, using these points x_i^*, we can express the length of C in the form

$$L = \lim_{n\to\infty} \sum_{i=1}^{n} \sqrt{1 + [f'(x_i^*)]^2}\, \Delta x_i. \tag{7.11}$$

But the right-hand side of this equation is precisely the definition of the definite integral of the function $\sqrt{1 + [f'(x)]^2}$ with respect to x from $x = a$ to $x = b$:

$$\int_a^b \sqrt{1 + [f'(x)]^2}\, dx = \lim_{n\to\infty} \sum_{i=1}^{n} \sqrt{1 + [f'(x_i^*)]^2}\, \Delta x_i.$$

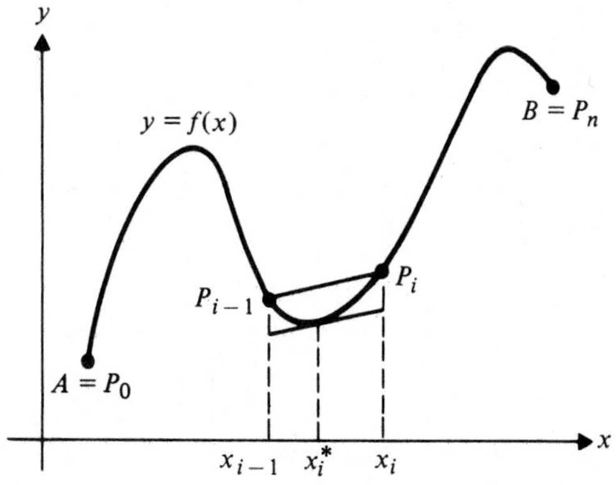

Figure 7.29

Thus we can evaluate the length of C by the definite integral

$$L = \int_a^b \sqrt{1 + \left(\frac{dy}{dx}\right)^2}\, dx. \tag{7.12}$$

In this derivation we assumed that $f'(x)$ was defined at each point in the interval $a \le x \le b$ (necessary for the mean value theorem). But in order to guarantee existence of the definite integral in (7.12), Theorem 6.2 requires continuity of the integrand $\sqrt{1 + (dy/dx)^2}$. Consequently, to ensure that the length of a curve can be calculated by means of (7.12), we assume that $f(x)$ has a continuous first derivative on the interval $a \le x \le b$.

Example 11 Find the length of the curve $y = x^{3/2}$ from $(1, 1)$ to $(2, 2\sqrt{2})$.

Solution Since

$$\frac{dy}{dx} = \frac{3}{2}x^{1/2},$$

we have (Fig. 7.30)

$$\begin{aligned} L &= \int_1^2 \sqrt{1 + \left(\frac{3}{2}x^{1/2}\right)^2}\, dx \\ &= \int_1^2 \sqrt{1 + \frac{9}{4}x}\, dx \\ &= \left\{\frac{8}{27}\left(1 + \frac{9}{4}x\right)^{3/2}\right\}_1^2 \\ &= \frac{8}{27}\left\{\left(\frac{11}{2}\right)^{3/2} - \left(\frac{13}{4}\right)^{3/2}\right\} \\ &= \frac{1}{27}(22^{3/2} - 13^{3/2}). \ \square \end{aligned}$$

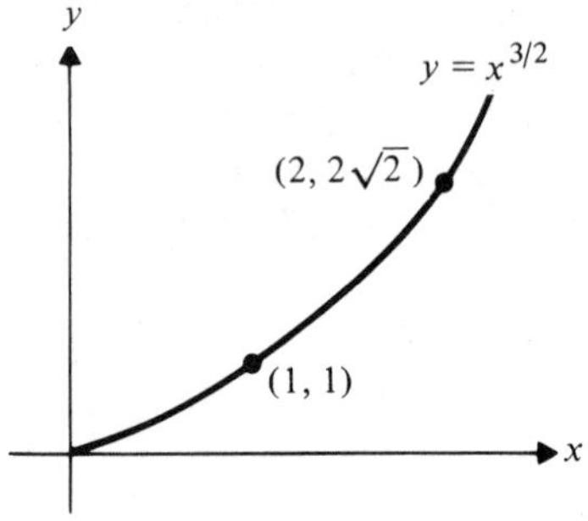

Figure 7.30

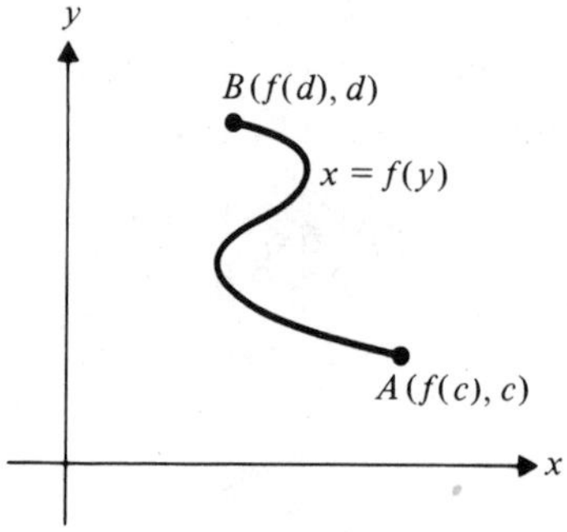

Figure 7.31

The result in (7.12) is clearly useful when the equation of the curve is expressed explicitly in the form $y = f(x)$. If the curve is defined explicitly in the form $x = f(y)$ (Fig. 7.31), a similar analysis gives

$$L = \int_c^d \sqrt{\left(\frac{dx}{dy}\right)^2 + 1}\, dy. \tag{7.13}$$

For example, we may write the equation of the curve in Example 11 in the form $x = y^{2/3}$, in which case $dx/dy = \frac{2}{3}y^{-1/3}$, and

$$\begin{aligned} L &= \int_1^{2\sqrt{2}} \sqrt{\left(\frac{2}{3}y^{-1/3}\right)^2 + 1}\, dy \\ &= \int_1^{2\sqrt{2}} \sqrt{\frac{4}{9}y^{-2/3} + 1}\, dy \\ &= \int_1^{2\sqrt{2}} \sqrt{\frac{4 + 9y^{2/3}}{9y^{2/3}}}\, dy \\ &= \int_1^{2\sqrt{2}} \frac{\sqrt{4 + 9y^{2/3}}}{3y^{1/3}}\, dy. \end{aligned}$$

If we set $u = 4 + 9y^{2/3}$, then $du = 6y^{-1/3}\,dy$, and

$$L = \frac{1}{3}\int_{13}^{22} \sqrt{u}\,\frac{du}{6} = \frac{1}{18}\left\{\frac{2}{3}u^{3/2}\right\}_{13}^{22} = \frac{1}{27}(22^{3/2} - 13^{3/2}).$$

In Sections 7.1 and 7.2 we developed a technique that eliminated the necessity for memorizing formulas for areas and volumes of solids of revolution. So far in this section we seem to have presented (7.12) and (7.13) as formulas. We now show how each of these formulas can be developed geometrically.

To find the length of the curve $C: y = f(x)$ in Fig. 7.28(a), we draw the tangent line at any point (x, y) on C, and consider at x an interval of length dx (Fig. 7.32). For very small dx we can approximate very closely the length along C inside the interval dx by the length of the tangent line inside the interval, namely,

$$\sqrt{(dx)^2 + (dy)^2}.$$

If we now picture a large number of these tangential line segments between a and b, we can find the total length of C by adding together all such lengths and taking the limit as each approaches zero. But this process is represented by the definite integral, and we therefore write

$$L = \int_{x=a}^{x=b} \sqrt{(dx)^2 + (dy)^2} \quad \text{or} \quad L = \int_{y=c}^{y=d} \sqrt{(dx)^2 + (dy)^2}.$$

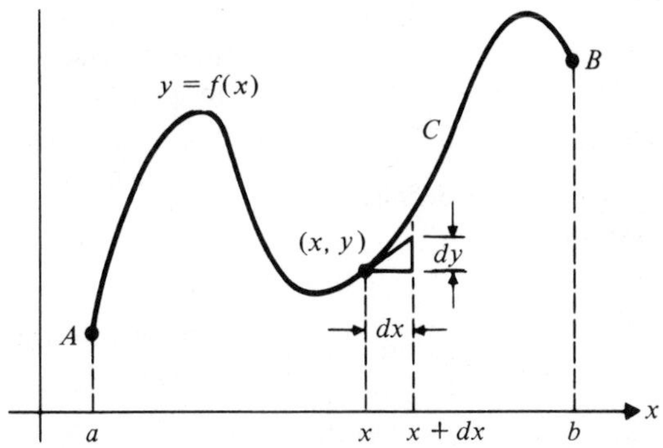

Figure 7.32

The first integral is chosen when it is more convenient to integrate with respect to x, in which case dx is taken outside the square root:

$$L = \int_a^b \sqrt{1 + \left(\frac{dy}{dx}\right)^2}\,dx \qquad \text{(Eq. 7.12);}$$

the second corresponds to a more convenient integral with respect to y, in which case dy is taken outside the square root:

$$L = \int_c^d \sqrt{\left(\frac{dx}{dy}\right)^2 + 1}\,dy \qquad \text{(Eq. 7.13).}$$

Example 12 Find the length of the curve $24xy = x^4 + 48$ between $(2, \frac{4}{3})$ and $(3, \frac{43}{24})$.

Solution Since it is easier to express y in terms of x (rather than x in terms of y), we write

$$y = \frac{x^4 + 48}{24x} = \frac{x^3}{24} + \frac{2}{x}.$$

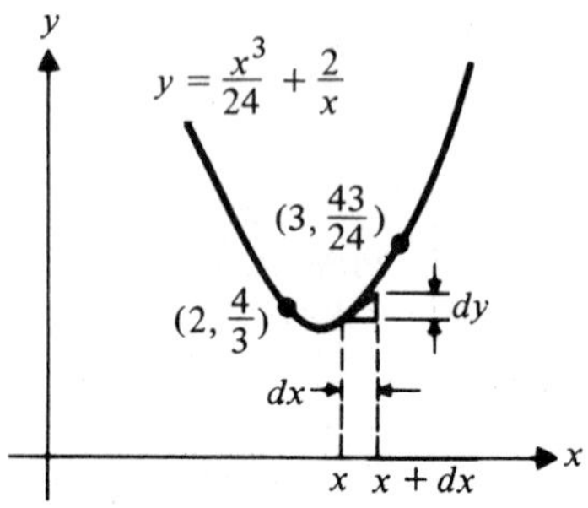

Figure 7.33

We then use the fact that small lengths along the curve are approximated by

$$\sqrt{(dx)^2 + (dy)^2} = \sqrt{1 + \left(\frac{dy}{dx}\right)^2}\, dx = \sqrt{1 + \left(\frac{x^2}{8} - \frac{2}{x^2}\right)^2}\, dx$$

$$= \sqrt{1 + \frac{x^4}{64} - \frac{1}{2} + \frac{4}{x^4}}\, dx = \sqrt{\frac{1}{64x^4}(x^8 + 32x^4 + 256)}\, dx$$

$$= \frac{1}{8x^2}\sqrt{(x^4 + 16)^2}\, dx = \frac{x^4 + 16}{8x^2}\, dx.$$

The total length along the required curve is therefore (Fig. 7.33)

$$L = \int_2^3 \frac{x^4 + 16}{8x^2}\, dx = \frac{1}{8}\int_2^3 \left(x^2 + \frac{16}{x^2}\right) dx$$

$$= \frac{1}{8}\left\{\frac{x^3}{3} - \frac{16}{x}\right\}_2^3 = \frac{9}{8}. \quad \square$$

EXERCISES 7.3

In each of Exercises 1–4, find the length of the curve between the points.

1. $3y = 2(x^2 + 1)^{3/2}$ from $(-2, 10\sqrt{5/3})$ to $(-1, 4\sqrt{2/3})$
2. $y = (x - 1)^{3/2}$ from $(2, 1)$ to $(10, 27)$
3. $36xy = x^4 + 108$ from $(3, \frac{7}{4})$ to $(4, \frac{91}{36})$
4. $y = x^5/5 + 1/(12x^3)$ from $(1, \frac{17}{60})$ to $(2, \frac{3077}{480})$

In each of Exercises 5–9 set up, but do not evaluate, a definite integral for the length of the curve between the points.

5. $y = x^2$ from $(0, 0)$ to $(1, 1)$
6. $x^2 - y^2 = 1$ from $(1, 0)$ to $(2, \sqrt{3})$
7. $x = y^2 - 2y$ from $(0, 0)$ to $(0, 2)$
8. $8y^2 = x^2(1 - x^2)$ (complete length)
9. $x^2/4 + y^2/9 = 1$ (complete length)
10. Find the length of the curve $x^{2/3} + y^{2/3} = 1$.
11. Find the length of the curve $3x = \sqrt{y}(y - 3)$ from $(-\frac{2}{3}, 1)$ to $(\frac{2}{3}, 4)$.

*12. If n is any number other than 1 or -1, and $0 < a < b$, find the length of the curve

$$y = \frac{x^{n+1}}{n+1} + \frac{1}{4(n-1)x^{n-1}}$$

between $x = a$ and $x = b$.

7.4 WORK

When a body moves a distance d along a straight line l under the action of a constant force F, which acts in the same direction as the motion (Fig. 7.34), the work done on the body by F is defined as

$$W = Fd. \tag{7.14}$$

Forces often vary in either magnitude or direction, or both, and when such forces act on a particle, calculation of the work done is not as simple as merely using Eq. (7.14). In this section we show how we can use definite integrals to calculate work done by forces that always act along the direction of motion, but do not have constant magnitude. Specifically, let us consider a particle that

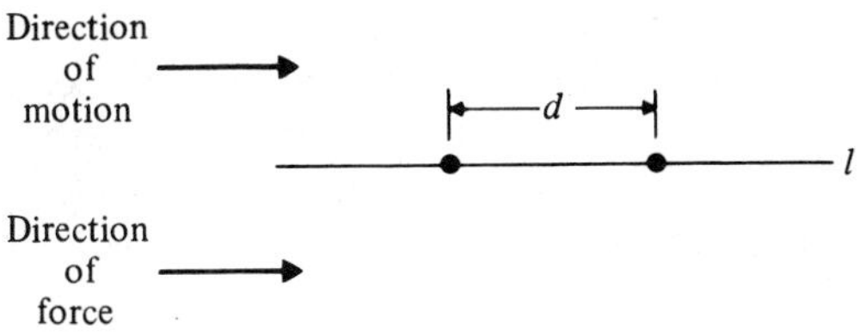

Figure 7.34

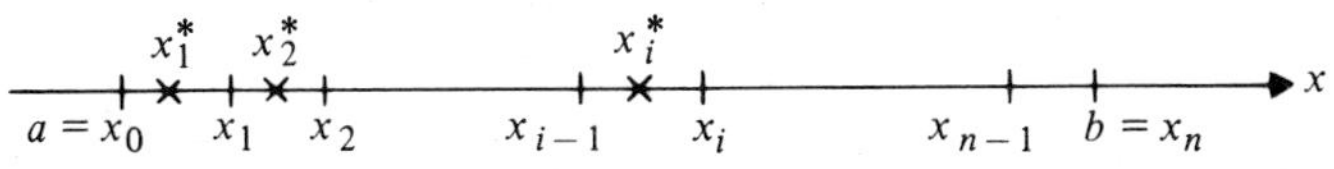

Figure 7.35

moves along the x-axis from $x = a$ to $x = b\,(b > a)$ under the action of a force F. Suppose that F always acts in the positive x-direction but its magnitude, which we denote by $F(x)$, is not constant along the x-axis. To find the work done by F as the particle moves from $x = a$ to $x = b$, we cannot simply multiply force by distance. (Where would we evaluate $F(x)$?) Instead, we divide the interval $a \le x \le b$ into n subintervals by points

$$a = x_0 < x_1 < x_2 < \cdots < x_{n-1} < x_n = b$$

(Fig. 7.35). In each subinterval $x_{i-1} \le x \le x_i$ we choose a point x_i^*. If lengths of the subintervals are small and $F(x)$ is continuous, $F(x)$ will not vary greatly over any given subinterval. In such a situation we may approximate $F(x)$ by a constant force $F(x_i^*)$ on each subinterval $x_{i-1} \le x \le x_i$. It follows that the work done by the force over the ith subinterval is approximately

$$F(x_i^*)(x_i - x_{i-1}) = F(x_i^*)\,\Delta x_i.$$

Further, an approximation to the total work done by the force as the particle moves from $x = a$ to $x = b$ is

$$\sum_{i=1}^{n} F(x_i^*)\,\Delta x_i.$$

As n becomes larger and larger and each Δx_i approaches zero, this approximation becomes better and better, and we therefore define

$$W = \lim_{n\to\infty} \sum_{i=1}^{n} F(x_i^*)\,\Delta x_i. \tag{7.15}$$

But this limit may also be interpreted as the definite integral of $F(x)$ with respect to x from $x = a$ to $x = b$. Consequently, W may be calculated with the definite integral

$$W = \int_a^b F(x)\,dx. \tag{7.16}$$

It is simple enough to interpret the various parts of Eq. (7.16), and in so doing we begin to feel the definite integral at work. The integrand $F(x)$ is the force at position x, and dx is a small distance along the x-axis. The product $F(x)\,dx$ is therefore interpreted as the (approximate) work done by the force along dx. The definite integral then adds over all dx's, beginning at $x = a$ and ending at $x = b$, to give the total work.

What is important about Eq. (7.16) is not the particular form of the definite integral but the fact that work done by a force can be evaluated by means of a definite integral. As we solve the work problems in this section we will find that the form of the definite integral varies considerably from problem to problem, but the underlying fact remains that each problem is solved with a definite integral.

Note too that in the derivation of Eq. (7.16) we assumed continuity of $F(x)$, and this guarantees existence of the definite integral.

Example 13 Find the work necessary to expand a spring from a stretch of 0.05 meter to a stretch of 0.15 meter if a force of 2.0×10^{-4} newton stretches it 0.10 meter.

Solution Let the spring be stretched in the positive x-direction, and let $x = 0$ correspond to the free end of the spring in the unstretched position (Fig. 7.36). Hooke's law states that when the spring is stretched an amount x, the restoring force in the spring is proportional to x:

$$F_s = -kx, \qquad k > 0.$$

The negative sign indicates that the force is in the negative x-direction. Since $F_s = -2.0 \times 10^{-4}$ newton when $x = 0.10$ meter,

$$-2.0 \times 10^{-4} = -k(0.10),$$

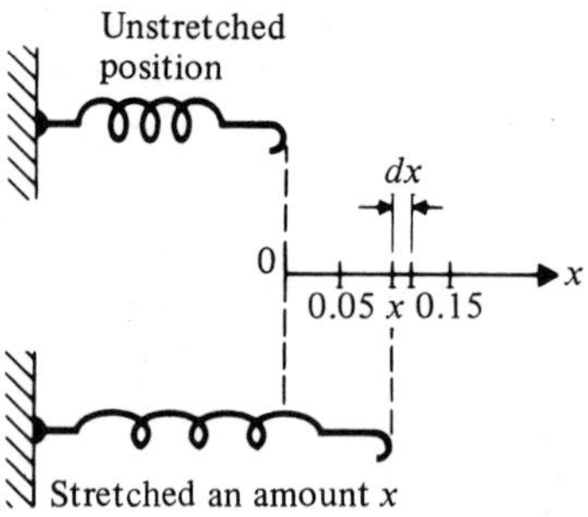

Figure 7.36

which implies that $k = 2.0 \times 10^{-3}$ N/m. The force required to counteract the restoring force of the spring is therefore

$$F(x) = 2.0 \times 10^{-3}x.$$

The work done by this force in stretching the spring from position x a further amount dx is approximately

$$2.0 \times 10^{-3}x\,dx,$$

and hence

$$W = \int_{0.05}^{0.15} 2.0 \times 10^{-3}x\,dx = 2.0 \times 10^{-3}\left\{\frac{x^2}{2}\right\}_{0.05}^{0.15} = 2.0 \times 10^{-5}.$$

The total work done is therefore 2.0×10^{-5} newton-meter (or joules). □

Example 14 A cable of length 100 meters and density 3 kilograms per meter hangs vertically from the top of a building. What work is required to lift the entire cable to the top of the building?

Solution When the cable has been lifted to a point where its lower end is y meters above its original position (Fig. 7.37), the force necessary to overcome the weight of the remaining cable is $3(9.81)(100 - y)$ newtons. The work that this force does in raising the end of the cable an additional amount dy is approximately

$$3(9.81)(100 - y)\,dy,$$

and the total work to raise the entire cable is therefore

$$W = \int_0^{100} 3(9.81)(100 - y)\,dy = 29.43\left\{100y - \frac{y^2}{2}\right\}_0^{100}$$

$$= 147{,}150 \text{ joules.} \;\square$$

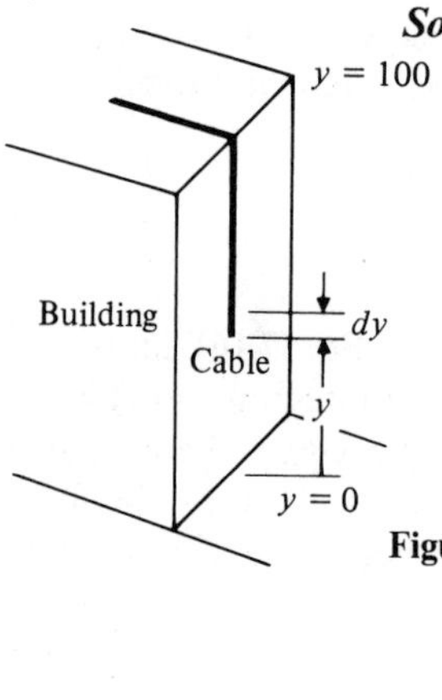

Figure 7.37

Example 15 A tank in the form of an inverted right-circular cone of depth 10 meters and radius 4 meters is full of water. Find the work required to pump the water to a level 1 meter above the top of the tank.

Solution A cross-section of the tank is shown in Fig. 7.38. Suppose we approximate the water in the tank with circular discs formed by rotating the rectangles shown about the y-axis. Since water has a density of 1000 kg/m^3, the force of gravity on the representative disc at position y is

$$-1000(9.81)\pi x^2\,dy,$$

and is negative because gravity acts in the negative y-direction. The work done by an equal and opposite force in lifting the disc to a level 1 meter above the top of the tank (a distance $11 - y$) is approximately

$$(11 - y)1000(9.81)\pi x^2\,dy.$$

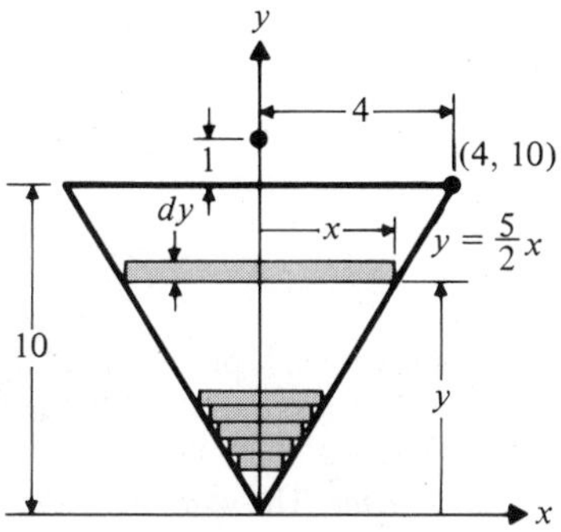

Figure 7.38

The total work to empty the tank is therefore

$$W = \int_0^{10} (11 - y)9810\pi x^2\,dy.$$

To express x in terms of y, we note that x and y are the coordinates of points on the straight line through the origin and the point (4, 10). Since the equation of this line is $y = 5x/2$, we obtain

$$\begin{aligned} W &= 9810\pi \int_0^{10} (11 - y)\left(\frac{2y}{5}\right)^2 dy \\ &= 1570\pi \int_0^{10} (11y^2 - y^3)\,dy \\ &= 1570\{\tfrac{11}{3}y^3 - \tfrac{1}{4}y^4\}_0^{10} \\ &= 5.75 \times 10^6 \text{ joules.} \ \square \end{aligned}$$

Two observations are noteworthy:

1. In Examples 13 and 14 the differential represents distance moved and the integrand represents force. In Example 15 the differential is part of the force, and the distance moved is part of the integrand. This is why we stated earlier that it is not advisable to use Eq. (7.16) as a formula since it does not always apply.
2. In each of these examples we set up the coordinate system; it was not given. We could use any coordinate system whatsoever; but once we have chosen our coordinates, we must refer everything to that system. For instance, had we used the coordinate system shown in Fig. 7.39 for Example 15, the definite integral would have been

$$W = \int_{-10}^{0} (1 - y)9810\pi[\tfrac{2}{5}(10 + y)]^2\,dy.$$

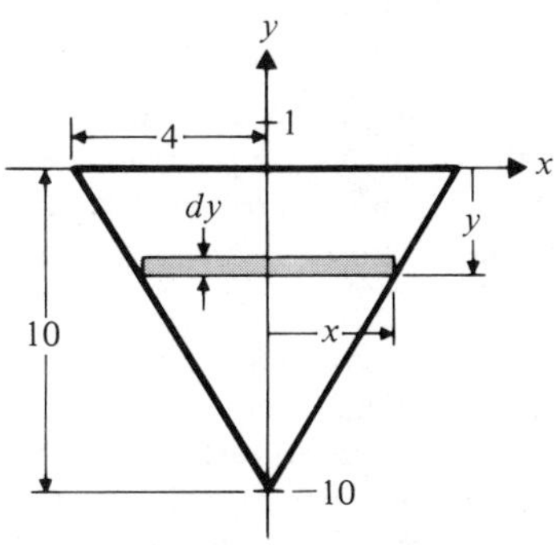

Figure 7.39

Evaluation of this definite integral again would lead to the result

$$W = 5.75 \times 10^6 \text{ joules.}$$

EXERCISES 7.4

1. A spring requires a 10-newton force to stretch it 3 centimeters. Find the work done in increasing the stretch of the spring from 5 centimeters to 7 centimeters.
2. Find the work done in increasing the stretch of the spring given in Exercise 1 from 7 centimeters to 9 centimeters.
3. A cage of mass M kilograms is to be lifted from the bottom of a mine shaft h meters deep. If the mass of the cable used to hoist the cage is m kilograms per meter, find the work done.
4. A tank filled with water has the form of a paraboloid of revolution with vertical axis (Fig. 7.40). If the depth of the tank is 12 meters and the diameter of the top is 8 meters, find the work done in pumping the water to the top of the tank.

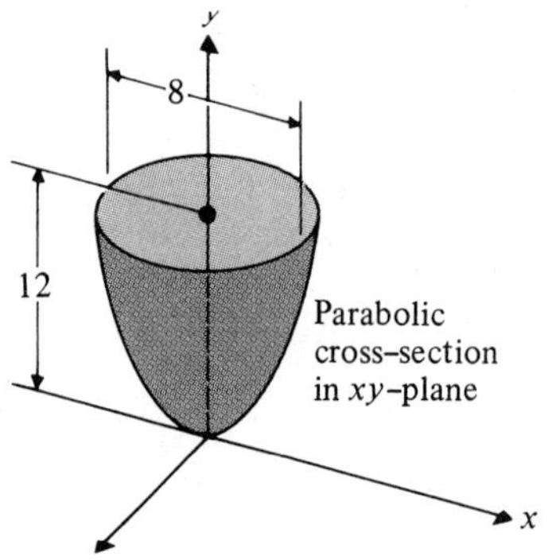

Figure 7.40

5. In Exercise 4, find the work done in emptying the tank through an outlet 2 meters above the top of the tank.
6. A 2-meter chain of mass 20 kilograms lies on the floor. If friction between floor and chain is ignored, how much work is required to lift one end of the chain 2 meters straight up?
7. The force of repulsion between two point charges of like sign, one of size q and the other of size Q, has magnitude

$$F = \frac{qQ}{4\pi\varepsilon_0 r^2},$$

where ε_0 is a constant and r is the distance between the charges. If charge Q is placed at the origin, and q is moved along the x-axis from $x = 2$ to $x = 5$, find the work done by the electrostatic force.
8. A 5-meter chain of mass 15 kilograms hangs vertically. You want to have two 2.5-meter chains hanging from the same height (Fig. 7.41). Which of the following procedures would you adopt?

   a) To cut the 5-meter chain at its midpoint and lift the middle of the chain 2.5 meters straight up.

   b) To lift the lower end of the chain 5 meters and then cut the chain.

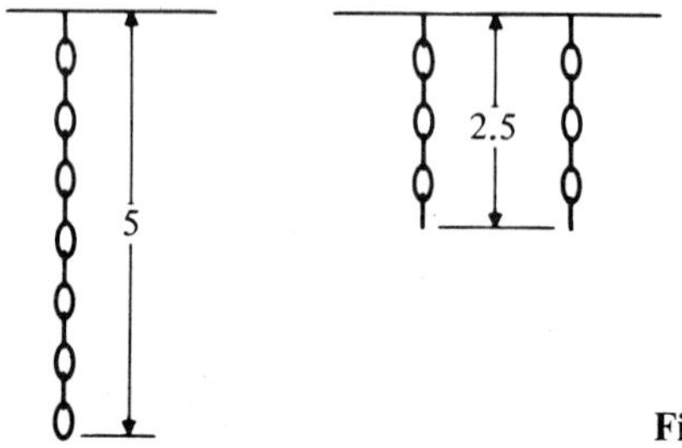

Figure 7.41

9. If the force of a crossbow (in newtons) is proportional to the draw (in meters), and it is 200 newtons at a full draw of 50 centimeters, what is the speed of a 20-gram arrow from a fully drawn crossbow? *Hint:* The work necessary to draw the bow is converted into kinetic energy K, where $K = mv^2/2$ (m = the mass of the arrow and v = the speed of the arrow).
10. Two similar springs, each 1 meter long in an unstretched position, have spring constant of k newtons per meter. The springs are joined together at P (Fig. 7.42), and their free ends are fastened to two posts 4 meters apart. What work is done in moving the midpoint P a distance of b meters to the right?

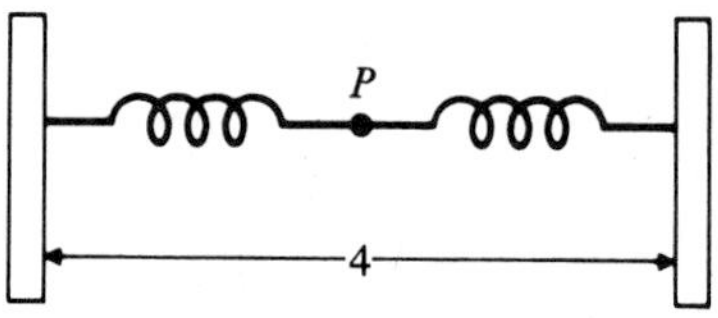

Figure 7.42

11. A rectangular swimming pool full of water is 25 meters long and 10 meters wide. The depth is 3 meters for the first 10 meters of length, decreasing linearly to 1 meter at the shallow end. How much work is required to empty the pool by pumping the water over its edge?
12. Two positive charges q_1 and q_2 are placed at positions $x = 5$ and $x = -2$ on the x-axis. A third positive charge q_3 is moved along the x-axis from $x = 1$ to $x = -1$. Find the total work done by the electrostatic forces of q_1 and q_2 on q_3.
13. In Exercise 6, if the chain is stretched out straight on the floor, and if the coefficient of friction between floor and chain is 0.01, what work is required to raise one end of the chain 2 meters? The force of friction on that part of the chain on the floor is 0.01 times the weight of the chain on the floor.

14. A bucket of water with a mass of 100 kilograms is on the ground attached to one end of a cable of density 5 kilograms per meter. The other end of the cable is attached to a windlass 100 meters above the bucket. If the bucket is raised at a constant speed, water runs out through a hole in the bottom at a constant rate to the extent that the bucket has a mass of 80 kilograms when it reaches the top. To further complicate matters, a rather fat pigeon of mass 60 kilograms lands on the bucket when it is 50 meters above the ground. He immediately begins taking a bath, splashing water over the side of the bucket at the rate of 1 kilogram per meter. Find the work done by the windlass in raising the bucket 100 meters.

15. A hemispherical tank (with a diametral plane at the top) has a radius of 5 meters.

C a) Show that if the depth of oil in the tank is h meters when the tank is one-half full (by volume), then h must satisfy the equation

$$h^3 - 15h^2 + 125 = 0.$$

Find h to four decimals.

b) Find the work required to empty the half-full tank if the oil has a density of 750 kilograms per cubic meter and the outlet is at the top of the tank.

16. A gas is confined in a cylinder, closed on one end with a piston on the other. If the gas expands adiabatically, then the pressure P and volume V of the gas obey the law

$$PV^{7/5} = C,$$

where C is a constant. The force on the piston face due to the pressure of the gas is given by $F = PA$, where A is the area of the face of the piston. Show that if the piston is moved so as to reduce the volume occupied by the gas from V_0 to $V_0/2$, then the work done by the piston is

$$\tfrac{5}{2}(2^{2/5} - 1)CV_0^{-2/5}.$$

***17.** A diving bell of mass 10,000 kilograms is attached to a chain of mass 5 kilograms per meter, and the bell sits on the bottom of the ocean 100 meters below the surface (Fig. 7.43). How much work is required to lift the bell to deck level 5 meters above the surface? Take into account the fact that when the bell and chain are below the surface, they weigh less than when above. The apparent loss in weight is equal to the weight of water displaced by the bell and chain. Assume that the bell is a perfect cube, 2 meters on each side; therefore, when completely submerged, it displaces 8 cubic meters of water. Assume also that the chain displaces 1 liter of water per meter of length.

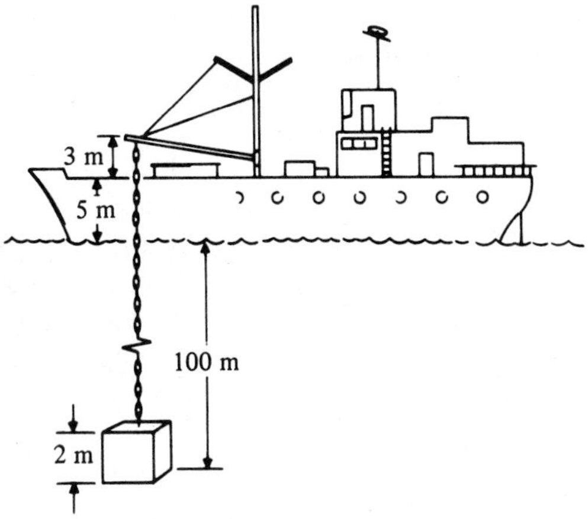

Figure 7.43

***18.** A particle is moved along the x-axis by a force $F(x)$ in the x-direction. Use Newton's second law to show that if $F(x)$ is the total resultant force on the particle, then the work done by $F(x)$ is equal to the change in kinetic energy of the particle.

***19.** Two springs with spring constants k_1 and k_2 are joined together, and then one end is fastened to a wall (Fig. 7.44). It is shown in physics that when a horizontal force is applied to the free end, the ratio of the stretches x_1 and x_2 in the two springs is inversely proportional to the ratio of their spring constants:

$$\frac{x_2}{x_1} = \frac{k_1}{k_2}.$$

Show that if the free end is moved so as to produce a total stretch in the springs of L, the amount of work done is

$$\frac{k_1 k_2 L^2}{2(k_1 + k_2)}.$$

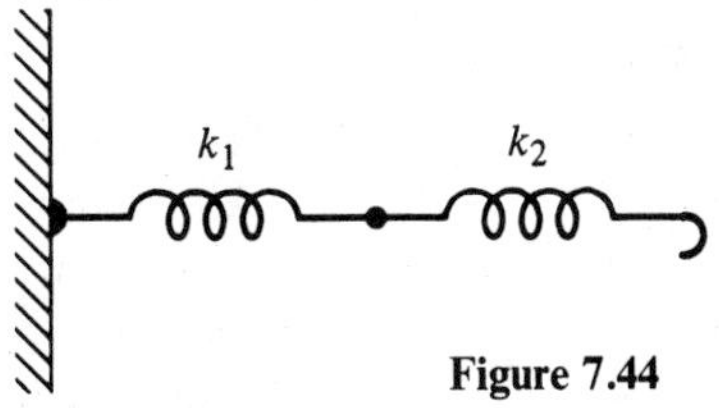

Figure 7.44

7.5 FLUID PRESSURE

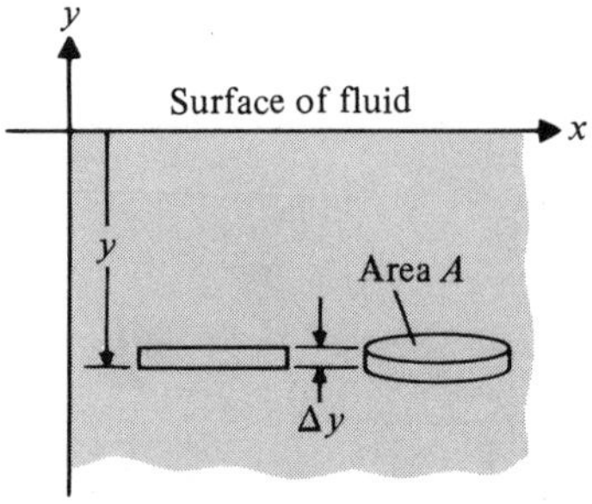

Figure 7.45

When an object is immersed in a fluid, it is acted on by forces that are often called *pressure forces.* These forces are independent of the kind of object that is immersed in the fluid, and are therefore a property of the fluid itself. The fluid forces act at every point in the fluid, and when an object is immersed in the fluid, they act on and normal to its surface. We use the concept of pressure to describe these fluid forces. We define *pressure at a point in a fluid as the magnitude of the force per unit area that would act on a surface at that point in the fluid.* Because pressure is the magnitude of the fluid force at a point, it is therefore a positive quantity. Experience indicates that pressure depends on two factors: (1) depth below the surface of the fluid and (2) the type of fluid itself. In order to discover the precise dependence, we consider a small horizontal disc of the fluid, as shown in Fig. 7.45. Suppose we denote the pressure at the bottom of the disc by $P(y)$, in which case the pressure at the top is $P(y + \Delta y)$. If the fluid is stationary, then the sum of all vertical forces on the disc must be equal to zero. Since the only forces acting on the disc are fluid forces and gravity, we obtain

$$AP(y) - AP(y + \Delta y) - \gamma(A\Delta y) = 0,$$

where A is the surface area of the disc and $\gamma > 0$ is the specific weight of the fluid (weight per unit volume). Rearrangement of this equation yields

$$\frac{P(y + \Delta y) - P(y)}{\Delta y} = -\gamma,$$

and if we take limits of both sides as $\Delta y \to 0$, we obtain

$$\frac{dP}{dy} = -\gamma.$$

This differential equation for $P(y)$ is immediately integrable:

$$P(y) = -\gamma y + C.$$

Since pressure at the surface of the fluid is equal to zero ($P(0) = 0$), C must be equal to zero; hence,

$$P = -\gamma y. \tag{7.17}$$

Pressure is therefore the product of the specific weight of the fluid and depth d below the surface of the fluid:

$$P = \gamma d, \tag{7.18}$$

where d is always assumed positive.

An illuminating interpretation of formula (7.18) is suggested by Fig. 7.46. Above a point at depth d below the surface of a fluid, we consider a column of fluid of unit cross-sectional area. The weight of this column of fluid is its volume multiplied by the specific weight of the fluid:

$$W = \gamma V.$$

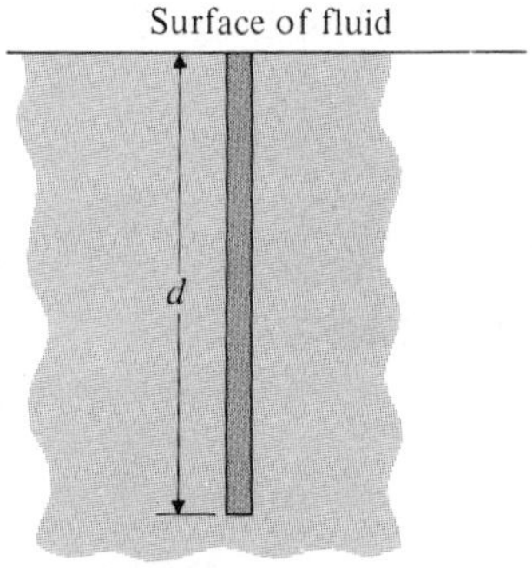

Figure 7.46

But V is the product of the length of the column, d, and the (unit) cross-sectional area; i.e., $V = (1)d = d$, and hence

$$W = \gamma d. \tag{7.19}$$

A comparison of Eqs. (7.18) and (7.19) suggests that pressure at any point is precisely the weight of a column of fluid of unit cross-sectional area above that point.

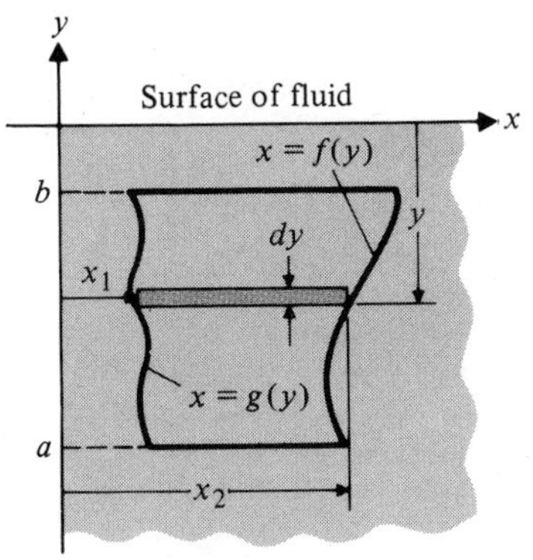

Figure 7.47

We now consider the problem of determining total force on one side of a flat vertical plate, as shown in Fig. 7.47, when it is immersed in a fluid of specific weight γ. If the area of the plate is subdivided into horizontal rectangles of width dy, then the pressure at each point on this rectangle is approximately

$$P = -\gamma y.$$

This is an approximation because slight variations in P will occur over vertical displacements within the rectangle. Since the force on the rectangle is its area multiplied by the pressure at points on the rectangle, it follows that the force on the rectangle is approximately

$$-\gamma y(x_2 - x_1)\,dy = -\gamma y[f(y) - g(y)]\,dy.$$

Total force on the plate is found by adding together the forces on all such rectangles and taking the limit as their widths approach zero. We obtain the required force, therefore, as

$$F = \int_a^b -\gamma y[f(y) - g(y)]\,dy. \tag{7.20}$$

Once again we do not suggest that Eq. (7.20) be memorized because it is intimately connected with the choice of coordinates in Fig. 7.47. For a different coordinate system, the definite integral would be correspondingly different (see Example 16). What is important is the procedure: Subdivide the area of the plate into horizontal rectangles, find the force on a representative rectangle, and finally add over all rectangles by using the definite integral.

Note that vertical rectangles cannot be used without further discussion since it is not evident how to calculate the force on such a rectangle. Consideration of vertical rectangles is given in Exercise 9.

Example 16 The vertical face of a dam is parabolic with breadth 100 meters and height 50 meters. Find the total force due to fluid pressure on the face.

Solution If we set up a coordinate system as shown in Fig. 7.48, we see that the edge of the dam has an equation of the form $y = kx^2$. Because (50, 50) is a point on this curve, it follows that $k = \frac{1}{50}$, and

$$y = \frac{x^2}{50}.$$

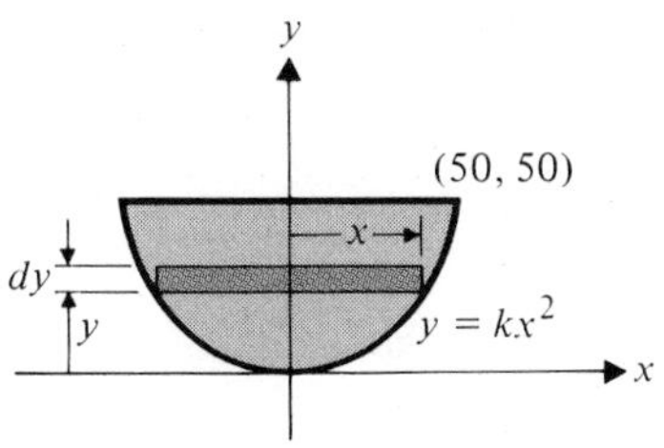

Figure 7.48

Since the area of the representative rectangle shown is $2x\,dy$, and it is at a depth $50 - y$ below the surface of the water, the force on it is approximately

$$(9.81)1000(50 - y)2x\,dy = 19{,}620(50 - y)\sqrt{50y}\,dy.$$

The total force on the dam must therefore be

$$\begin{aligned} F &= \int_0^{50} 19{,}620(50 - y)5\sqrt{2}y^{1/2}\,dy \\ &= 98{,}100\sqrt{2}\int_0^{50}(50y^{1/2} - y^{3/2})\,dy \\ &= 98{,}100\sqrt{2}\left\{\frac{100}{3}y^{3/2} - \frac{2}{5}y^{5/2}\right\}_0^{50} \\ &= 6.54 \times 10^8 \text{ newtons.} \; \square \end{aligned}$$

Example 17 A tank in the form of a right-circular cylinder of radius 2 meters and length 10 meters lies on its side. If it is half filled with oil of density ρ kilograms per cubic meter, find the force on each end of the tank.

Solution The force on the representative rectangle in Fig. 7.49 is approximately

$$9.81\rho(-y)2x\,dy = -2\gamma y\sqrt{4 - y^2}\,dy,$$

where $\gamma = 9.81\rho$. The total force on each end of the tank is therefore

$$F = \int_{-2}^{0} -2\gamma y\sqrt{4 - y^2}\,dy = 2\gamma\left\{\frac{1}{3}(4 - y^2)^{3/2}\right\}_{-2}^{0} = \frac{16}{3}\gamma \text{ newtons.} \ \square$$

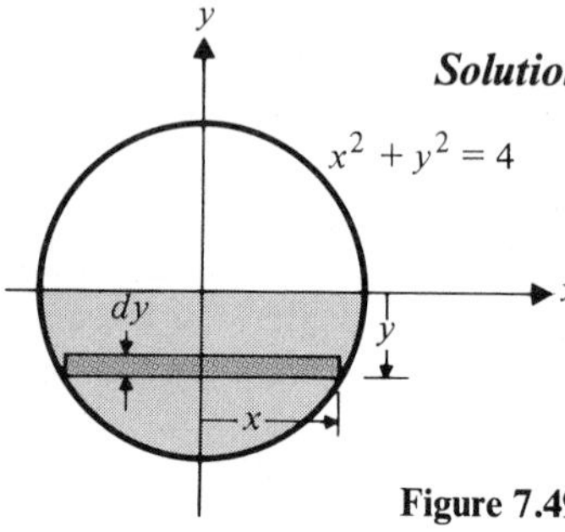

Figure 7.49

EXERCISES 7.5

1. A tropical fish tank has length 1 meter, width $\frac{1}{2}$ meter, and depth $\frac{1}{2}$ meter. Find the force due to water pressure on each of the sides and bottom when the tank is full.

2. The vertical surface of a dam exposed to the water of a lake has the shape shown in Fig. 7.50. Find the force of the water on the face of the dam.

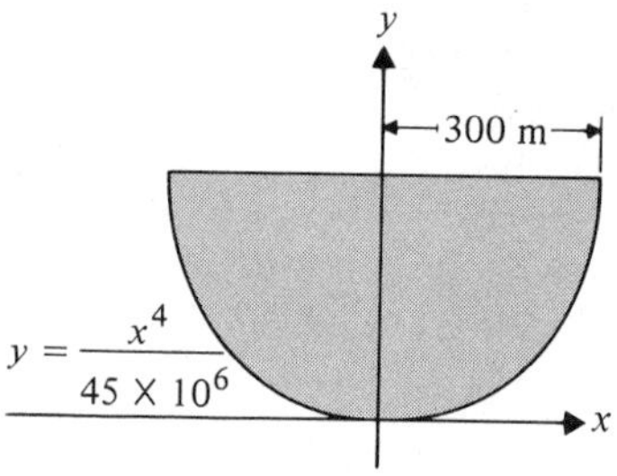

Figure 7.50

3. The vertical end of a water trough is an isosceles triangle with width 2 meters and depth 1 meter. Find the force of the water on each end when the trough is half full (by volume).

4. A square plate, 2 meters on each side, has one diagonal vertical. If it is one-half submerged in water, what is the force due to water pressure on each side of the plate?

5. A cylindrical oil tank of radius r and height h has its axis vertical. If the specific weight of the oil is γ, find the force on the bottom of the tank when it is full.

6. A rectangular swimming pool full of water is 25 meters long and 10 meters wide. The depth is 3 meters for the first 10 meters, decreasing linearly to 1 meter at the shallow end. Find the force due to the weight of the water on each of the sides and ends of the pool.

7. a) A dam across a river has the shape of a parabola 36 meters across the top and 9 meters deep at the center. Find the maximum force that the river can exert on the dam.

   b) What is the force on the face of the dam if the face is slanted at a 45° angle (Fig. 7.51)?

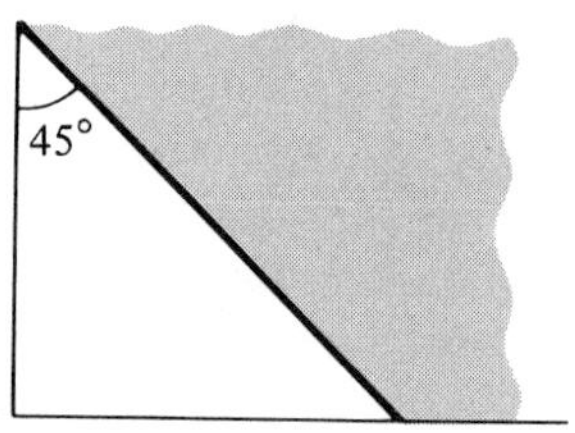

Figure 7.51

8. The bow of a landing barge consists of a rectangular flat plate A meters wide and B meters long. When the barge is stationary, this plate makes an angle of 30° with the surface of the water. Find the maximum force of the water on the bow.

9. a) Show that the force due to fluid pressure on the vertical rectangle shown in Fig. 7.52 is

$$F = \frac{\gamma}{2}h(y_1^2 - y_2^2),$$

where γ is the specific weight of the fluid.

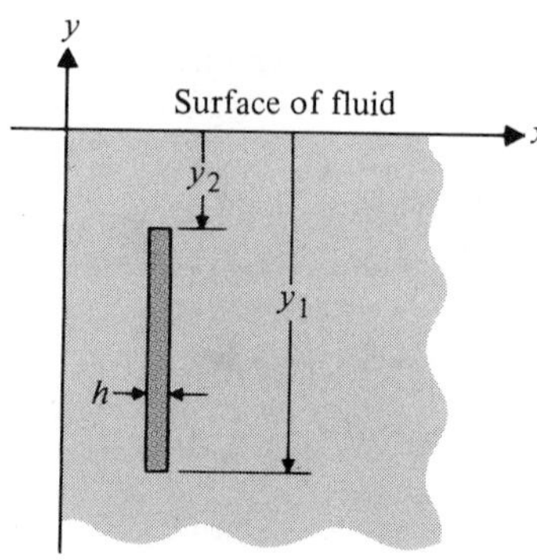

Figure 7.52

b) A flat plate in the shape of a trapezoid is submerged vertically in a fluid of specific weight γ. The plate has two parallel vertical sides of lengths 6 and 8 and a third side of length 5 that is perpendicular to the parallel sides and at a depth of 1 below the surface of the fluid. Find the force due to fluid pressure on each side of the plate, using both horizontal and vertical rectangles.

10. Set up, but do not evaluate, a definite integral(s) to find the force due to water pressure on each side of the flat vertical plate shown in Fig. 7.53.

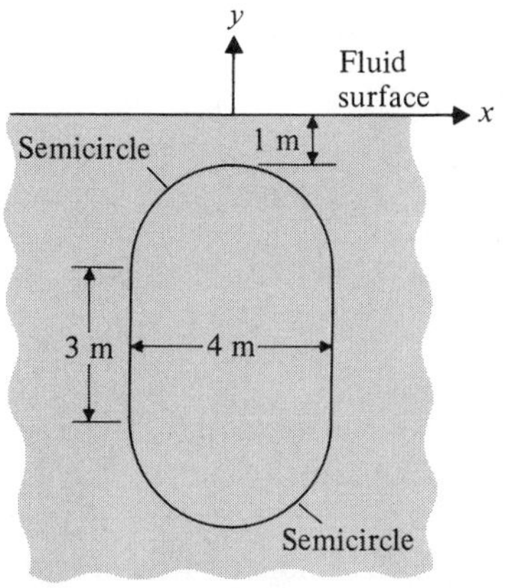

Figure 7.53

11. The base of a triangular plate, of length a, lies in the surface of a fluid of specific weight γ. The third vertex of the triangle is at a depth b below the surface (Fig. 7.54). Show that the force due to fluid pressure on each side of the plate is $\gamma ab^2/6$, no matter what the shape of the triangle.

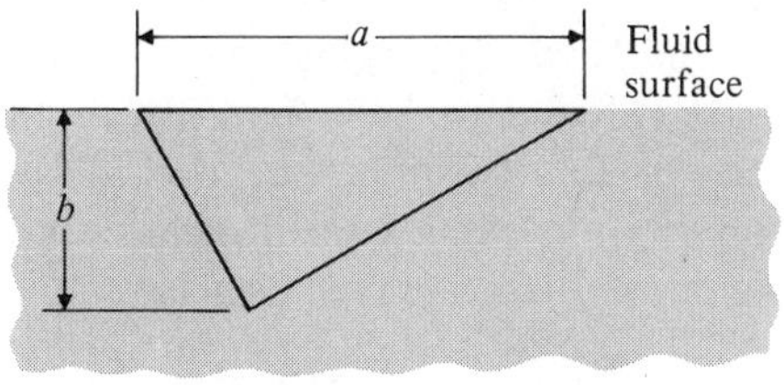

Figure 7.54

12. In Exercise 6, find the force due to water pressure on each part of the bottom of the swimming pool.

13. A water tank is to be built in the form of a rectangular box with all six sides welded along their joins (Fig. 7.55). Sides $ABCD$ and $EFGH$ are to be square, and all sides except $ABCD$ are supported from the outside. If the maximum force that $ABCD$ can withstand is 20,000 newtons, what is the largest cross-section that can be built (assuming that at some stage the tank will be full)?

14. A cylindrical oil tank of radius r and length h has its axis horizontal. If the specific weight of the oil is γ, find the force on each end of the tank when it is half full.

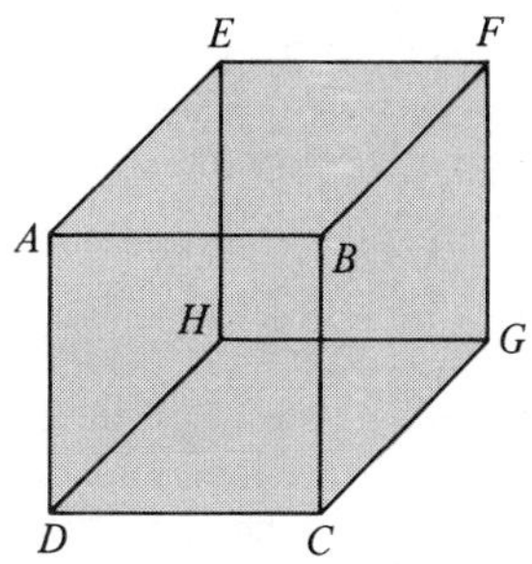

Figure 7.55

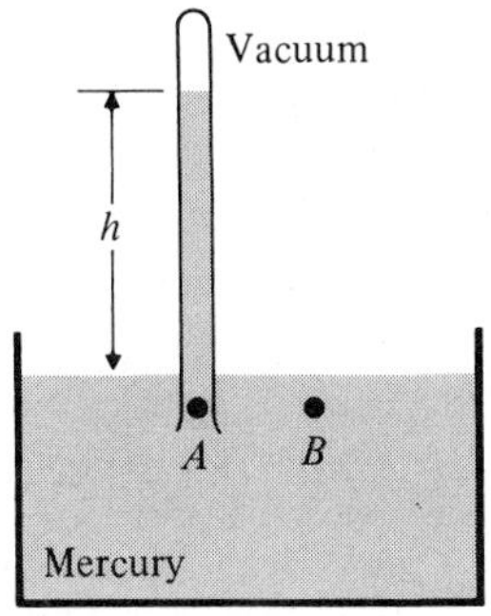

Figure 7.56

15. A right-circular cylinder of radius r and height h is immersed in a fluid of specific weight γ with its axis vertical. Show that the buoyant force on the cylinder due to the pressure of the fluid is equal to the weight of the fluid displaced. (This fact is known as *Archimedes' principle*, and is valid for an object of any shape.)

16. Archimedes' principle states that the buoyant force on an object when immersed or partially immersed in a fluid is equal to the weight of the fluid displaced by the object.

a) Show that if an object floats partially submerged in water, the percentage of the volume of the object above water is

$$100\,\frac{\rho_w - \rho_o}{\rho_w},$$

where ρ_o and ρ_w are the densities of the object and water, respectively.

b) If the densities of ice and water are 915 and 1000 kilograms per cubic meter, respectively, show that only 8.5% of the volume of an iceberg is above water.

17. If a full tube of mercury is inverted in a large container of mercury, the level of mercury in the tube will fall, but it will stabilize at a point higher than that in the container (Fig. 7.56). This is due to the fact that air pressure acts on the surface of the mercury in the container but not on the surface of the mercury in the tube. The extra column of mercury (of height h) creates a force at A that counteracts the atmospheric pressure transmitted through the mercury in the container to the tube so that the total pressure at A is equal to the total pressure at B.

a) Show that if the density of mercury is 13.6 grams per cubic centimeter, then the atmospheric pressure at the surface of the mercury in the container is

$$P = 1.33 \times 10^5 h \text{ newtons per square meter},$$

provided h is measured in meters.

b) If h is measured as 761 millimeters, find the atmospheric pressure.

18. a) A block of wood (density 0.40 gram per cubic centimeter) is cubical ($\frac{1}{4}$ meter on each side). If it floats in water, how deep is its lowest point below the surface? (Refer to Archimedes' principle in Exercise 16.)

*b) Repeat part (a) given that the block is a sphere of radius $\frac{1}{4}$ meter.

***19.** A tank is to be built in the form of a right-circular cylinder with horizontal axis. The ends are to be joined to the cylindrical side by a continuous weld. One end of the tank and the cylindrical side are supported from the outside. The remaining unsupported end can withstand a total force of 40,000 newtons less 1000 newtons for each meter of weld on that end. What is the maximum radius for the tank if it is to hold a fluid of specific weight 1.019×10^3 kilograms per cubic meter?

***20.** Find the ratio L/R such that the forces due to fluid pressure on the rectangular and semicircular parts of the plate shown in Fig. 7.57 are equal.

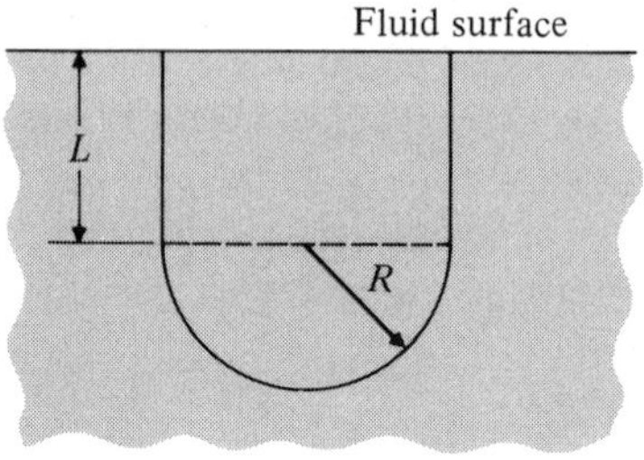

Figure 7.57

7.6 CENTERS OF MASS AND CENTROIDS

Everyone is acquainted with the action of a teeter-totter, or seesaw. Two children of unequal masses can pass many an hour rocking, provided the child with greater mass sits closer to the fulcrum. To discover the mathematics of a seesaw, we consider in Fig. 7.58 a uniform seesaw of length $2L$ balanced at its center, with a child of mass m at one end. If a second child of equal mass is placed at the other end, then the ideal seesaw situation is created. If, however, the mass of the second child is $M > m$, then this child must be placed somewhat closer to the fulcrum. To find the exact position, we must determine what might be called the "rocking power" of a mass. A little experimentation shows that when $M = 2m$, M must be placed halfway between the end and the fulcrum; when $M = 3m$, M must be placed $L/3$ away from the fulcrum; and in general, when $M = am$ ($a > 1$), M must be placed L/a away from the fulcrum. Now the rocking power of the child of mass m is constant, and for each mass $M = am$ we have found an equal and opposite rocking power if M is placed at L/a. Clearly, then, rocking power depends on both mass and distance from the fulcrum. A little thought shows that the mathematical quantity that remains constant for the various masses $M = am$ is the product of M and distance to the fulcrum; in each case, this product is $(am)(L/a) = mL$, the same product as for the child of mass m. It would appear, then, that rocking power should be defined as the product of mass and distance.

▷ **Definition 7.1** The first moment of a mass m about a point O is the product md of m and the directed distance d from O to m. ◁

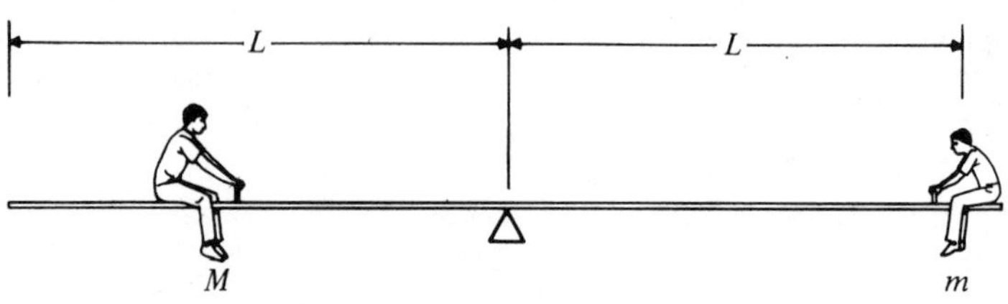

Figure 7.58

Because we use directed distances in calculating first moments, we must choose distances measured to one side of a point about which moments are being taken as positive, and distances to the other side as negative. We illustrate this in the problem shown in Fig. 7.59, where we have placed on the same seesaw five children of masses m_1, m_2, m_3, m_4, and m_5. A sixth child of mass m_6 is to be placed somewhere on the seesaw so that all six children will form the ideal seesaw.

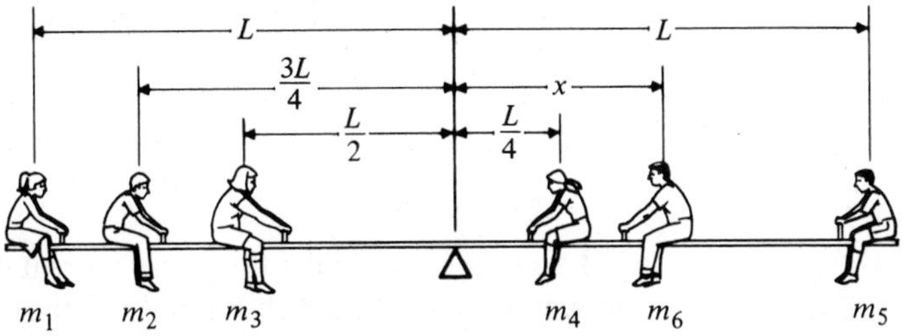

Figure 7.59

To solve this problem we let x be the directed distance from the fulcrum to the point where the sixth child should be placed. The total first moment of all six children about the fulcrum, choosing distances to the right as positive (and to the left as negative), is

$$m_1(-L) + m_2(-3L/4) + m_3(-L/2) + m_4(L/4) + m_5(L) + m_6(x).$$

We regard this as the resultant first moment of all six children attempting to turn the seesaw (clockwise if the moment is positive, counterclockwise if the moment is negative). Balance will occur if this resultant first moment is zero:

$$0 = -m_1L - \tfrac{3}{4}m_2L - \tfrac{1}{2}m_3L + \tfrac{1}{4}m_4L + m_5L + m_6x;$$

we may solve this equation for the position of m_6:

$$x = \frac{L}{4m_6}\{4m_1 + 3m_2 + 2m_3 - m_4 - 4m_5\}.$$

We now turn this problem around and place the six children at distances from the left end, as shown in Fig. 7.60. If the mass of the seesaw itself is neglected, where should the fulcrum be placed in order to create the ideal seesaw? To solve this problem, we let the distance that the fulcrum should be placed from the left end be represented by $\bar{x}$. In order for balance to occur, the total first moment of all six children about the fulcrum must vanish; hence,

$$0 = m_1(x_1 - \bar{x}) + m_2(x_2 - \bar{x}) + m_3(x_3 - \bar{x}) + m_4(x_4 - \bar{x}) + m_5(x_5 - \bar{x}) + m_6(x_6 - \bar{x}).$$

Solution of this equation gives us

$$\bar{x} = \frac{m_1x_1 + m_2x_2 + m_3x_3 + m_4x_4 + m_5x_5 + m_6x_6}{m_1 + m_2 + m_3 + m_4 + m_5 + m_6} = \frac{1}{M}\sum_{i=1}^{6} m_ix_i, \tag{7.21}$$

where $M = \sum_{i=1}^{6} m_i$ is the total mass of all six children.

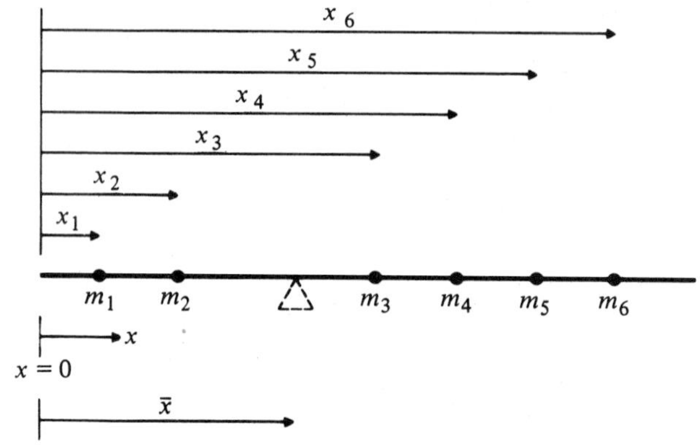

Figure 7.60

This point $\bar{x}$ at which the fulcrum creates a balancing position is called the *center of mass* for the six children. It is a point where moments of masses to the right are balanced by moments of masses to the left.

In the remainder of this section we extend the idea of a center of mass of point masses distributed along a line (the seesaw) to the center of mass of a distribution of point masses in a plane, and then to the center of mass of a continuous distribution of mass. Our first step is to define the first moment of a point mass in a plane about a line in the plane.

When a particle of mass m is located at a directed distance d from a fixed line l (Fig. 7.61), *the first moment of the mass about the line is defined as the product md.* In particular, when m is located at position (x, y) in the xy-plane, its first moments about the x- and y-axes are my and mx, respectively.

For a system of n particles of masses $m_1, m_2, \ldots, m_n$ located at points $(x_1, y_1), (x_2, y_2), \ldots, (x_n, y_n)$, respectively (Fig. 7.62), the first moments of the system about the x- and y- axes are defined as

$$\sum_{i=1}^{n} m_i y_i \quad \text{and} \quad \sum_{i=1}^{n} m_i x_i. \tag{7.22}$$

What is the physical meaning of these first moments? Do they, for instance, play the same role that first moments did for the seesaw? To see this we imagine that each point mass is embedded in a thin plastic plate in the xy-plane. The plate itself is massless and extends to include all n masses m_i. Picture now that the

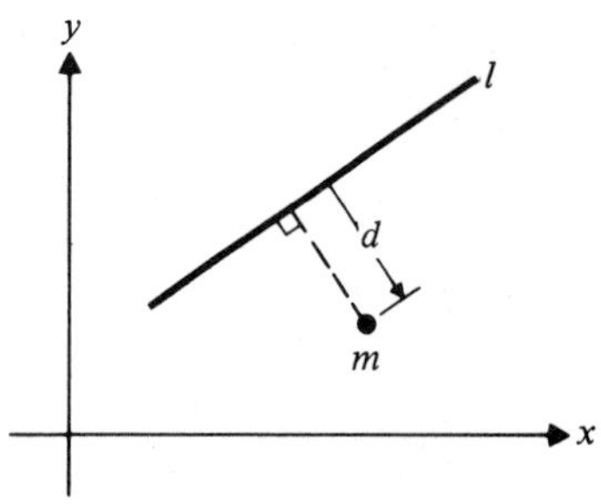

Figure 7.61

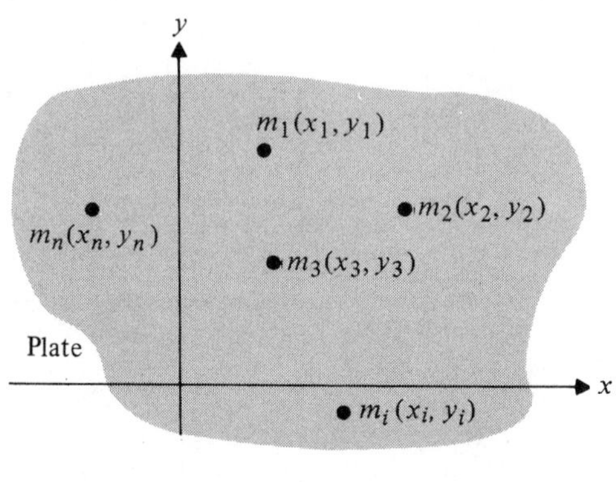

Figure 7.62

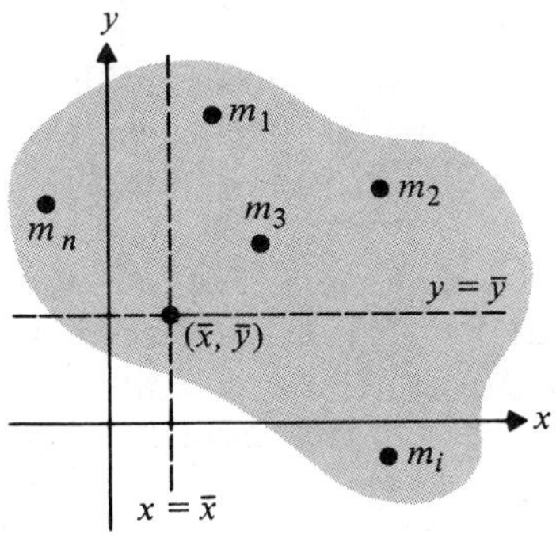

Figure 7.63

plate is horizontal, and we place a sharp edge along the y-axis. Will the plate rotate about this edge, or will it balance on it? It is fairly clear that the plate will balance if the first moment of the system about the y-axis is equal to zero, and it will rotate otherwise (the direction depending on whether the first moment is positive or negative). Similarly, the plate will balance on a sharp edge placed along the x-axis if and only if the first moment of the system about the x-axis vanishes; i.e., if and only if $\sum_{i=1}^{n} m_i y_i = 0$. In general, the plate will balance along any straight edge if and only if the first moment of the system about that edge vanishes.

We defined the center of mass for a given distribution of children on the see-saw as the point at which to place the fulcrum in order to obtain a balance. Analogously, we define the center of mass $(\bar{x}, \bar{y})$ of the distribution of point masses in Fig. 7.62 as the point at which to place a sharp point in order to obtain a balance. Remember that the plastic itself is massless and only the point masses m_i can create moments. Balance occurs at a point $(\bar{x}, \bar{y})$ if and only if balance occurs about every straight line through $(\bar{x}, \bar{y})$. In particular, balance must occur about the lines $x = \bar{x}$ and $y = \bar{y}$ parallel to the y- and x-axes (Fig. 7.63). Since balance occurs about $x = \bar{x}$ if and only if the total first moment of the system about this line vanishes, we obtain the condition that

$$0 = m_1(x_1 - \bar{x}) + m_2(x_2 - \bar{x}) + \cdots + m_n(x_n - \bar{x}),$$

from which we get

$$\bar{x} = \frac{1}{M} \sum_{i=1}^{n} m_i x_i, \tag{7.23}$$

where $M = \sum_{i=1}^{n} m_i$ is the total mass of the system. Note that this equation is exactly the same as Eq. (7.21). Similarly, for balance about $y = \bar{y}$, we find that $\bar{y}$ must be

$$\bar{y} = \frac{1}{M} \sum_{i=1}^{n} m_i y_i. \tag{7.24}$$

We have obtained a unique point $(\bar{x}, \bar{y})$ based on conditions of balance about the lines $x = \bar{x}$ and $y = \bar{y}$. Does this necessarily imply that balance will occur about every straight line through $(\bar{x}, \bar{y})$? The answer is yes, although we will leave a proof of this fact to Exercise 28.

We have now proved that every planar point mass distribution has a center of mass $(\bar{x}, \bar{y})$, where $\bar{x}$ and $\bar{y}$ are defined by Eqs. (7.23) and (7.24). Our derivation has shown that the first moment of the system about any line through $(\bar{x}, \bar{y})$ must be equal to zero. Are there any other significant features of this point? To answer this question we note that if a fictitious particle of mass M (the total mass of the system) is located at the center of mass $(\bar{x}, \bar{y})$, then its first moment about the y-axis is $M\bar{x}$. But from Eq. (7.23), we have

$$M\bar{x} = \sum_{i=1}^{n} m_i x_i, \tag{7.25}$$

and we conclude that the first moment of this fictitious particle M about the y-axis is exactly the same as the first moment of the system about the y-axis. Similarly, the first moment of M about the x-axis is $M\bar{y}$, and from Eq. (7.24), we have

$$M\bar{y} = \sum_{i=1}^{n} m_i y_i. \tag{7.26}$$

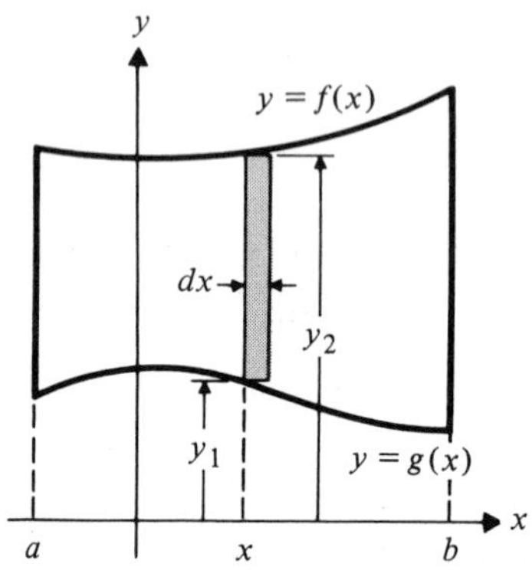

Figure 7.64

Thus the center of mass of a system of point masses m_i is a point at which a single particle of mass $M = \sum_{i=1}^{n} m_i$ has the same first moments about the x- and y-axes as the original system. It can be shown further that the first moment of M about any line is the same as the first moment of the system about that line. For a proof of this fact, see Exercise 28.

In summary, we have defined the center of mass of a system of point masses as a *balance point*. We have found as a result that the center of mass is a point at which a single particle of mass equal to the total mass of the system will have the same first moment about any line as the system itself. This result could be, and sometimes is, used as the definition of the center of mass.

We now wish to make a transition from a discrete system of particles to a continuous distribution of mass in the form of a thin plate of constant density ρ (mass per unit area) (Fig. 7.64). In order to find the mass of the plate, we proceed in exactly the same way that we did for areas: We divide the plate into vertical rectangles. The mass in a representative rectangle of width dx at position x is

$$\rho[f(x) - g(x)]\,dx.$$

To find the total mass of the plate, we add over all such rectangles of ever-diminishing widths to obtain

$$M = \int_a^b \rho[f(x) - g(x)]\,dx. \tag{7.27}$$

We proved for a discrete distribution that the center of mass was a point at which a particle of mass M will have the same first moments about the x- and y-axes as the original distribution. We use this as our definition of the center of mass of a continuous distribution. We define the center of mass of a continuous distribution of mass as that point $(\bar{x}, \bar{y})$ where a particle of mass M will have the same first moments about the x- and y-axes as the distribution. To express this statement in algebraic terms, we first note that $M\bar{x}$ is the first moment about the y-axis of a particle of mass M at $(\bar{x}, \bar{y})$. To this we must equate the first moment of the original distribution about the y-axis. Now each point in the representative rectangle in Fig. 7.64 is approximately the same distance x from the y-axis—approximately, because the rectangle does have finite width. The first moment, then, of this rectangle about the y-axis is approximately

$$x\rho[f(x) - g(x)]\,dx.$$

We find the first moment of the entire plate about the y-axis by adding together the first moments of all such rectangles and taking the limit as their widths approach zero. But this is once again the process defined by the definite integral, and we obtain, therefore, for the first moment of the plate about the y-axis

$$\int_a^b x\rho[f(x) - g(x)]\,dx.$$

Consequently,

$$M\bar{x} = \int_a^b x\rho[f(x) - g(x)]\,dx, \tag{7.28}$$

and this equation can be solved for $\bar{x}$ once M and the integral on the right have been evaluated. Equation (7.28) represents for continuous distributions what Eq. (7.23) does for discrete distributions.

To find $\bar{y}$ we must equate the product $M\bar{y}$ to the first moment of the plate about the x-axis. If we consider the representative rectangle shown in Fig. 7.64 we see that not all points therein are the same distance from the x-axis. To circumvent this problem, we consider all of the mass of the rectangle to be concentrated at its center of mass. Since the center of mass is at the midpoint of the rectangle, a point distant $[f(x) + g(x)]/2$ from the x-axis, it follows that the first moment of this rectangle about the x-axis is

$$\tfrac{1}{2}[f(x) + g(x)]\rho[f(x) - g(x)]\,dx.$$

In summary, then, the total first moment of the plate about the x-axis is the definite integral of this expression, and we write

$$M\bar{y} = \int_a^b \tfrac{1}{2}[f(x) + g(x)]\rho[f(x) - g(x)]\,dx. \qquad (7.29)$$

This equation is used to evaluate $\bar{y}$.

Equations (7.28) and (7.29) can be memorized as formulas for $\bar{x}$ and $\bar{y}$, respectively, but it is far easier to perform the above operations mentally and arrive at those equations. In addition, for various shapes of plates, we might use horizontal rectangles or even combinations of horizontal and vertical rectangles (and in such cases Eqs. 7.28 and 7.29 would have to be modified).

On the basis of definitions (7.28) and (7.29) for $(\bar{x}, \bar{y})$, we can show that the first moment of M at $(\bar{x}, \bar{y})$ about any line is the same as the first moment of the plate about that same line. This implies automatically that the plate will balance along any line through $(\bar{x}, \bar{y})$ and therefore at $(\bar{x}, \bar{y})$.

Example 18 Find the center of mass of a thin plate of constant density ρ (mass per unit area) if its edges are defined by the curves

$$y = 2 - x^2, \qquad y = 0, \qquad x = 0, \qquad x \geq 0.$$

Solution Using vertical rectangles (Fig. 7.65), we see that

$$M = \int_0^{\sqrt{2}} \rho y\,dx = \int_0^{\sqrt{2}} \rho(2 - x^2)\,dx$$

$$= \rho\left\{2x - \frac{x^3}{3}\right\}_0^{\sqrt{2}} = \frac{4\sqrt{2}}{3}\rho.$$

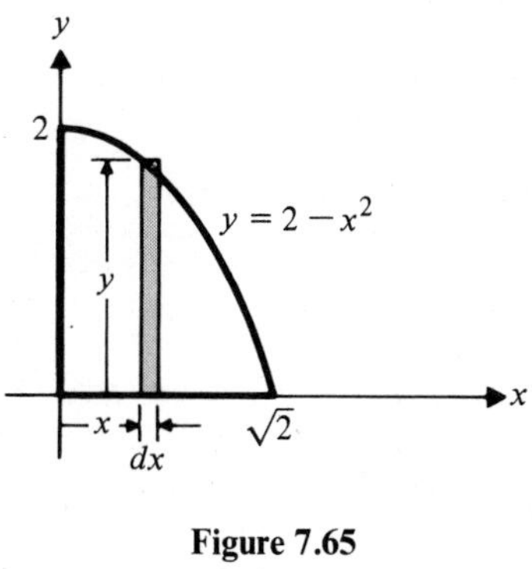

Figure 7.65

If $(\bar{x}, \bar{y})$ is the center of mass, then

$$M\bar{x} = \int_0^{\sqrt{2}} x\rho(2 - x^2)\,dx = \rho\int_0^{\sqrt{2}} (2x - x^3)\,dx = \rho\left\{x^2 - \frac{x^4}{4}\right\}_0^{\sqrt{2}} = \rho.$$

Thus,

$$\bar{x} = \rho\frac{3}{4\sqrt{2}\rho} = \frac{3}{4\sqrt{2}}.$$

To find $\bar{y}$ we calculate

$$M\bar{y} = \int_0^{\sqrt{2}} \frac{y}{2} \rho y \, dx$$
$$= \frac{\rho}{2} \int_0^{\sqrt{2}} (2 - x^2)^2 \, dx$$
$$= \frac{\rho}{2} \int_0^{\sqrt{2}} (4 - 4x^2 + x^4) \, dx$$
$$= \frac{\rho}{2} \left\{ 4x - \frac{4}{3}x^3 + \frac{1}{5}x^5 \right\}_0^{\sqrt{2}} = \frac{16\sqrt{2}}{15}\rho.$$

Thus,

$$\bar{y} = \frac{16\sqrt{2}\rho}{15} \cdot \frac{3}{4\sqrt{2}\rho} = \frac{4}{5}. \ \square$$

Example 19 Find the center of mass of a thin plate of constant density ρ (mass per unit area) if its edges are defined by the curves

$$y = 2x - x^2, \qquad y = x^2 - 4.$$

Solution Using vertical rectangles (Fig. 7.66), we see that

$$M = \int_{-1}^{2} \rho(y_2 - y_1) \, dx$$
$$= \int_{-1}^{2} \rho[(2x - x^2) - (x^2 - 4)] \, dx$$
$$= \rho \int_{-1}^{2} (4 + 2x - 2x^2) \, dx$$
$$= \rho \left\{ 4x + x^2 - \frac{2}{3}x^3 \right\}_{-1}^{2}$$
$$= 9\rho.$$

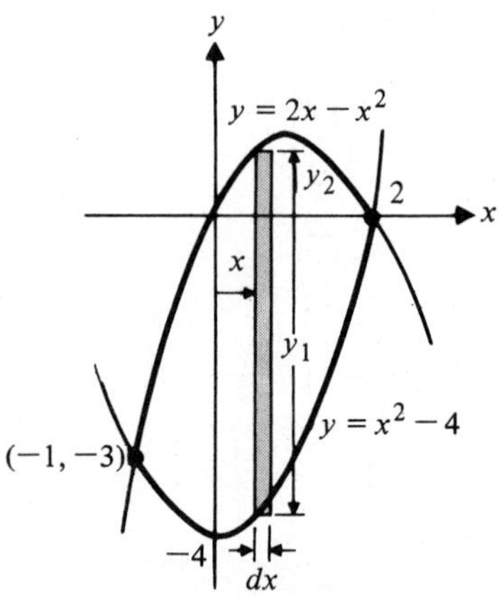

Figure 7.66

If $(\bar{x}, \bar{y})$ is the center of mass, then

$$M\bar{x} = \int_{-1}^{2} x\rho(y_2 - y_1) \, dx$$
$$= \rho \int_{-1}^{2} (4x + 2x^2 - 2x^3) \, dx$$
$$= \rho \left\{ 2x^2 + \frac{2}{3}x^3 - \frac{1}{2}x^4 \right\}_{-1}^{2} = \frac{9}{2}\rho.$$

Thus,

$$\bar{x} = \frac{9\rho}{2}\frac{1}{9\rho} = \frac{1}{2}.$$

To find $\bar{y}$ we calculate

$$\begin{aligned}
M\bar{y} &= \int_{-1}^{2} \frac{1}{2}(y_1 + y_2)\rho(y_2 - y_1)\,dx \\
&= \frac{\rho}{2}\int_{-1}^{2} (y_2^2 - y_1^2)\,dx \\
&= \frac{\rho}{2}\int_{-1}^{2} [(2x - x^2)^2 - (x^2 - 4)^2]\,dx \\
&= 2\rho\int_{-1}^{2} (-x^3 + 3x^2 - 4)\,dx \\
&= 2\rho\left\{-\frac{1}{4}x^4 + x^3 - 4x\right\}_{-1}^{2} \\
&= -\frac{27}{2}\rho.
\end{aligned}$$

Thus,

$$\bar{y} = -\frac{27\rho}{2}\frac{1}{9\rho} = -\frac{3}{2}. \quad \square$$

Example 20 Find the first moments of a thin plate of constant density ρ (mass per unit area) about the lines (a) $y = 0$, (b) $y = -2$, and (c) $x = -2$, if its edges are defined by the curves

$$x = |y|^3, \qquad x = 2 - y^2.$$

Solution

a) Since the mass is distributed symmetrically about the x-axis (Fig. 7.67), the first moment about $y = 0$ is zero.

b) Since the center of mass of the plate is on the x-axis, its first moment about the line $y = -2$ is

$$\begin{aligned}
(2)\text{ (mass of plate)} &= 2(2)(\text{mass of plate above the } x\text{-axis}) \\
&= 4\int_{0}^{1} \rho(x_2 - x_1)\,dy \\
&= 4\rho\int_{0}^{1} (2 - y^2 - y^3)\,dy \\
&= 4\rho\left\{2y - \frac{1}{3}y^3 - \frac{1}{4}y^4\right\}_{0}^{1} \\
&= \frac{17}{3}\rho.
\end{aligned}$$

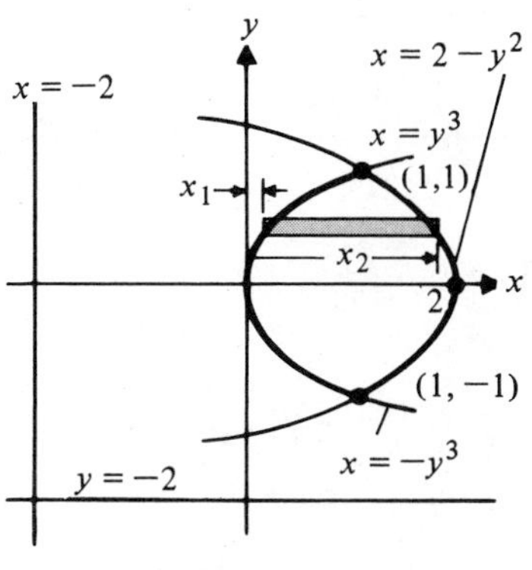

Figure 7.67

c) The first moment of the plate about the line $x = -2$ is twice the first moment of its upper half:

$$2\int_0^1 \left\{2 + \frac{x_1 + x_2}{2}\right\}\rho(x_2 - x_1)\,dy = \rho\int_0^1 [4(x_2 - x_1) + (x_2^2 - x_1^2)]\,dy$$

$$= 4\rho\int_0^1 (x_2 - x_1)\,dy + \rho\int_0^1 [(2 - y^2)^2 - (y^3)^2]\,dy$$

$$= \frac{17}{3}\rho + \rho\int_0^1 (4 - 4y^2 + y^4 - y^6)\,dy$$

$$= \frac{17}{3}\rho + \rho\left\{4y - \frac{4}{3}y^3 + \frac{1}{5}y^5 - \frac{1}{7}y^7\right\}_0^1$$

$$= \frac{881}{105}\rho. \square$$

It has become apparent through our discussions and examples that in calculating the center of mass of a thin plate of constant density ρ, ρ is really unnecessary. As a constant it is taken out of each integration and cancels in the final division. The location of the center of mass in actual fact, then, depends only on the geometric shape of the plate, and for this reason we could replace all references to mass by area. In particular, the mass of the plate M can be replaced by its area A, first moments (of mass) $M\bar{x}$ and $M\bar{y}$ can be replaced by first moments (of area) $A\bar{x}$ and $A\bar{y}$, and Eqs. (7.28) and (7.29) would then take the form

$$A\bar{x} = \int_a^b x[f(x) - g(x)]\,dx, \tag{7.30}$$

$$A\bar{y} = \int_a^b \frac{1}{2}[f(x) + g(x)][f(x) - g(x)]\,dx. \tag{7.31}$$

It is customary when using first moments of area to call $(\bar{x}, \bar{y})$ the centroid of the area, rather than the center of mass of the plate, simply because all reference to mass has been deleted. We emphasize, however, that the statements in this paragraph apply only when density is constant.

EXERCISES 7.6

Find the center of mass of the thin plate of constant density ρ (mass per unit area) shown in each of the following figures.

1.

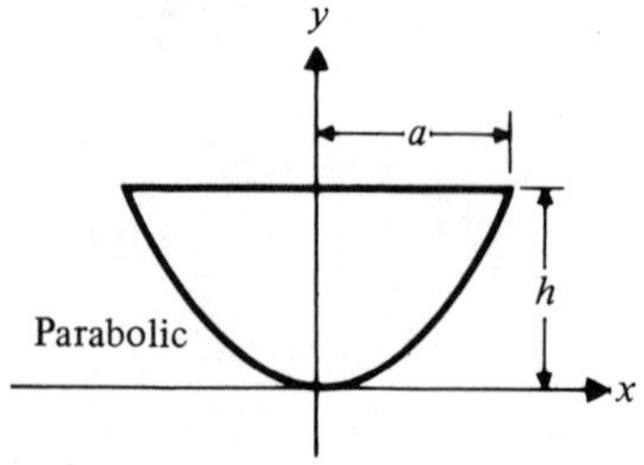

2.

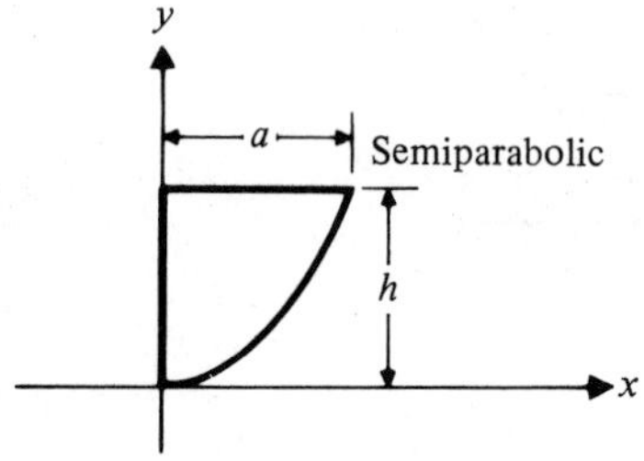

3.

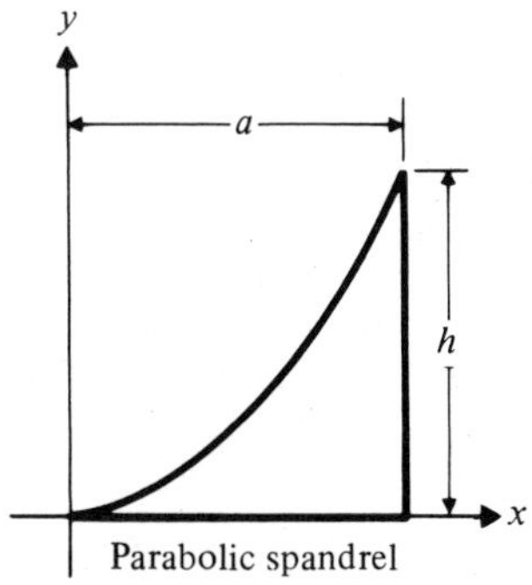

4.

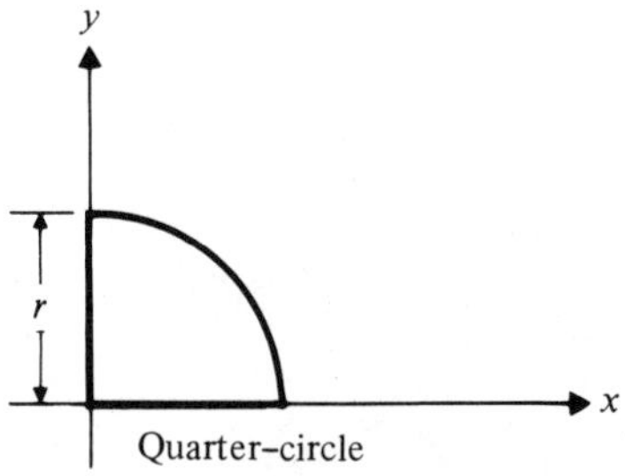

5.

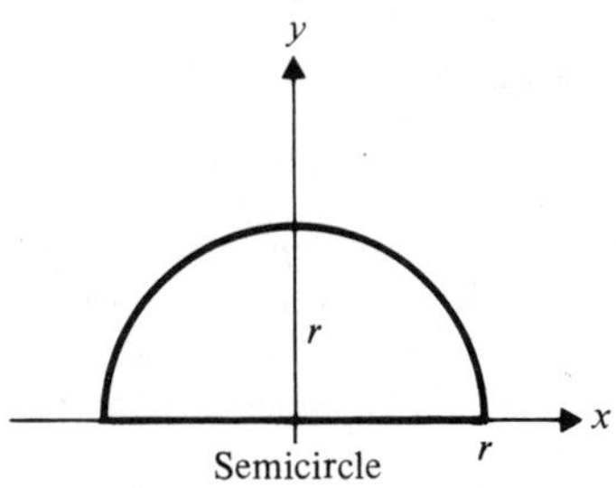

Find the centroid for the area bounded by the curves in each of Exercises 6–13.

6. $y = x^2 - 1,\ y = -x^2 - 2x - 1$

7. $y = \sqrt{|x|},\ y = 2 - x^2$

8. $y = x^3,\ x = y^3$

9. $y = x,\ y = 2x,\ 2y = x + 3$

10. $x = y^2 - 2y,\ x + y = 12$

11. $y = \sqrt{2 - x},\ x + y = 2$

12. $x = 4y - 4y^2,\ y = x - 3,\ y = 1,\ y = 0$

13. $x^3 y = 8,\ y = 9 - x^3$

14. The edges of a thin plate of constant density ρ (mass per unit area) are defined by the curves

$$y = |x|^{1/2}, \qquad y = x + 2, \qquad y = 2 - x.$$

Find its first moment about the line $x = -5$.

15. Show that the centroids of areas A and B in Fig. 7.68 are:

$$\bar{x}_A = \left(\frac{n+1}{n+2}\right)a, \qquad \bar{y}_A = \left(\frac{n+1}{4n+2}\right)b,$$

$$\bar{x}_B = \left(\frac{n+1}{2n+4}\right)a, \qquad \bar{y}_B = \left(\frac{n+1}{2n+1}\right)b.$$

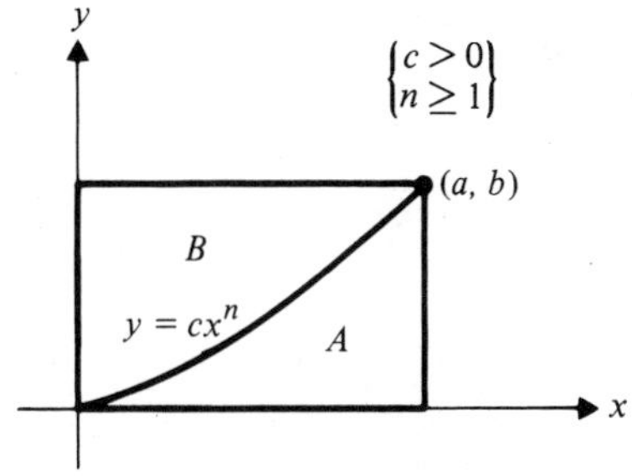

Figure 7.68

16. Find the center of mass of the seesaw shown in Fig. 7.60 if the mass of the seesaw is not neglected. Assume that it has uniform mass per unit length ρ.

Find the centroid of the area bounded by the curves in each of Exercises 17–19.

17. $x = \sqrt{y + 2},\ y = x,\ y = 0$

18. $y + x^2 = 0,\ x = y + 2,\ x + y + 2 = 0,\ y = 2$ (above $y + x^2 = 0$)

19. $y = \sqrt{2 - x},\ 15y = x^2 - 4$

20. Find the center of mass of the thin plate shown in Fig. 7.69 if it has constant density.

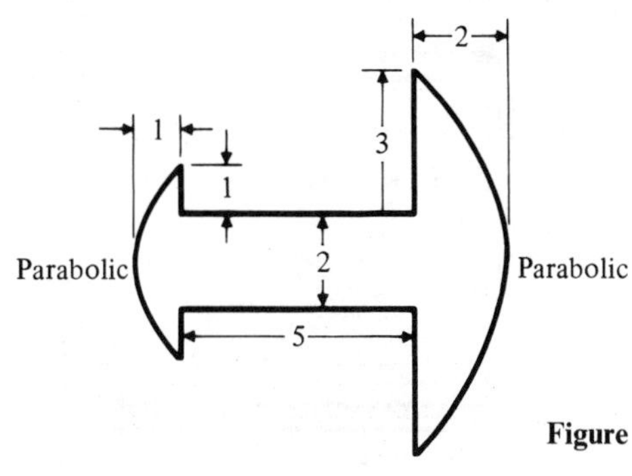

Figure 7.69

21. Find the first moment about the line $x + y = 1$ for the area defined by the curves

$$x + 2 = y^2, \qquad y = x.$$

22. If an area A can be subdivided into n subareas A_i ($i = 1, \ldots, n$) such that the centroid of each A_i is $(\bar{x}_i, \bar{y}_i)$, show that the centroid $(\bar{x}, \bar{y})$ of A is given by

$$\bar{x} = \frac{1}{A}\sum_{i=1}^{n} A_i \bar{x}_i, \qquad \bar{y} = \frac{1}{A}\sum_{i=1}^{n} A_i \bar{y}_i.$$

Use the technique suggested in Exercise 22 to find the centroid for the area in each of the following figures.

23.

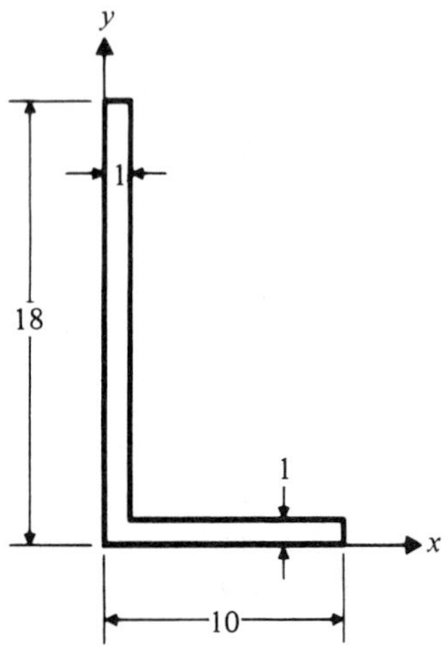

24.

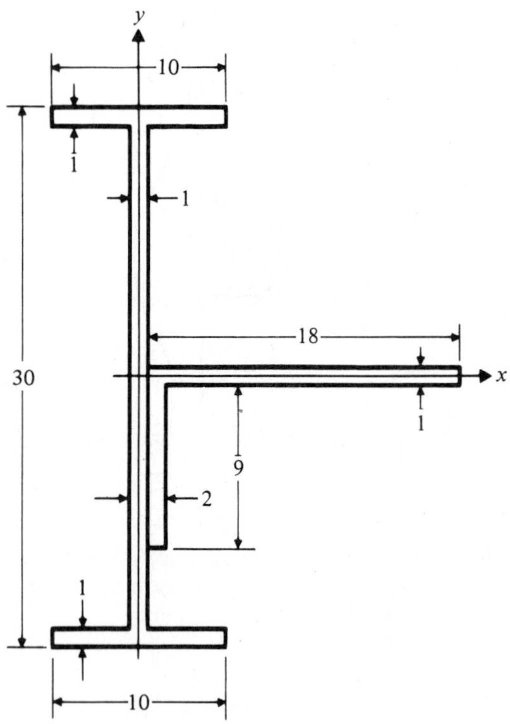

25.

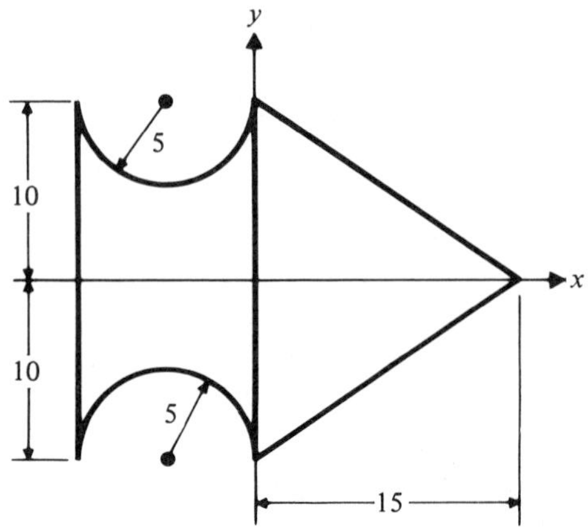

26.

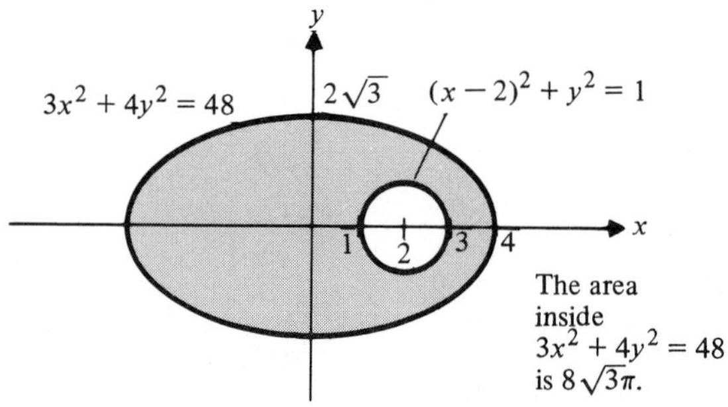

27. a) Prove the following *theorem of Pappus*: If a plane area is revolved about a coplanar axis not crossing the area, the volume generated is equal to the product of the area and the circumference of the circle described by the centroid of the area.

b) Use the result in (a) to find the volume of the donut in Exercise 33 of Section 7.2.

28. a) Show that the first moment of the system of point masses in Fig. 7.62 about any line is the same as the first moment of a point mass $M = \sum_{i=1}^{n} m_i$ at $(\bar{x}, \bar{y})$ about that line. (*Hint:* See Exercise 43 in Section 1.3A for a formula for the distance from a point to a line.)

b) Does it follow that the first moment of the system about any line through $(\bar{x}, \bar{y})$ is zero?

***29.** A thin flat plate of area A is immersed vertically in a fluid of specific weight γ. Show that the total force (due to fluid pressure) on each side of the plate is equal to the product of γ, A, and the depth of the centroid of the plate below the surface of the fluid.

***30.** Show that the centroid of a triangle with vertices (x_1, y_1), (x_2, y_2), and (x_3, y_3) is

$$\bar{x} = \frac{x_1 + x_2 + x_3}{3}, \qquad \bar{y} = \frac{y_1 + y_2 + y_3}{3}.$$

7.7 MOMENTS OF INERTIA

If a particle of mass m is located at a distance d from a fixed line (Fig. 7.61), *the moment of inertia (second moment) of the mass about the line is defined as* md^2. In particular, if m is at position (x, y) in the xy-plane, its moments of inertia about the x- and y-axes are my^2 and mx^2, respectively. For a system of n particles of masses $m_1, m_2, \ldots, m_n$ located at points $(x_1, y_1), (x_2, y_2), \ldots, (x_n, y_n)$ (Fig. 7.62), the moments of inertia of the system about the x- and y-axes are

$$\sum_{i=1}^{n} m_i y_i^2 \quad \text{and} \quad \sum_{i=1}^{n} m_i x_i^2. \tag{7.32}$$

The transition from the discrete case to a continuous distribution in the form of a thin plate of constant density ρ (mass per unit area) is not always so simple as for first moments. First consider the moment of inertia of the plate shown in Fig. 7.64 about the y-axis. The mass of the representative rectangle shown is given by

$$\rho[f(x) - g(x)]\,dx,$$

and each point of the rectangle is approximately the same distance x from the y-axis. The moment of inertia, then, of this rectangle about the y-axis is approximately

$$x^2\rho[f(x) - g(x)]\,dx.$$

The moment of inertia of the entire plate about the y-axis is found by adding together the moments of inertia of all such rectangles and taking the limit as their widths approach zero. But again this process defines a definite integral; therefore, the moment of inertia about the y-axis of the plate shown in Fig. 7.64 is

$$I = \int_a^b x^2\rho[f(x) - g(x)]\,dx. \tag{7.33}$$

Example 21 Find the moment of inertia about the y-axis of a thin plate of constant density ρ (mass per unit area) if its edges are defined by the curves

$$y = x^3, \qquad y = \sqrt{2 - x}, \qquad x = 0.$$

Solution With vertical rectangles (Fig. 7.70), we see that

$$\begin{aligned} I &= \int_0^1 x^2\rho(y_2 - y_1)\,dx \\ &= \rho\int_0^1 x^2(\sqrt{2 - x} - x^3)\,dx \\ &= \rho\int_0^1 x^2\sqrt{2 - x}\,dx - \rho\int_0^1 x^5\,dx. \end{aligned}$$

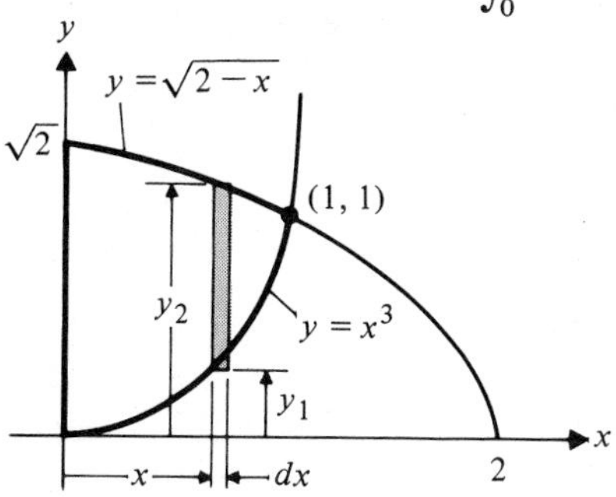

Figure 7.70

In the first integral we set $u = 2 - x$, in which case $du = -dx$ and

$$I = \rho\int_2^1 (2-u)^2 u^{1/2}(-du) - \rho\left\{\frac{1}{6}x^6\right\}_0^1$$

$$= \rho\int_1^2 (4u^{1/2} - 4u^{3/2} + u^{5/2})\,du - \frac{\rho}{6}$$

$$= \rho\left\{\frac{8}{3}u^{3/2} - \frac{8}{5}u^{5/2} + \frac{2}{7}u^{7/2}\right\}_1^2 - \frac{\rho}{6}$$

$$= \frac{256\sqrt{2} - 319}{210}\rho. \square$$

Example 22 Find the moment of inertia about the line $y = -1$ of a thin plate of constant density ρ (mass per unit area) if its edges are defined by the curves

$$x = y^2, \qquad x = 2y.$$

Solution The distance from $y = -1$ to points in the horizontal rectangle shown in Fig. 7.71 is $y + 1$; therefore,

$$I = \int_0^2 (y+1)^2 \rho(x_2 - x_1)\,dy$$

$$= \rho\int_0^2 (y+1)^2(2y - y^2)\,dy$$

$$= \rho\int_0^2 (-y^4 + 3y^2 + 2y)\,dy$$

$$= \rho\left\{-\frac{y^5}{5} + y^3 + y^2\right\}_0^2$$

$$= \frac{28}{5}\rho. \square$$

In Examples 21 and 22, and in the discussion leading to Eq. (7.33), we chose rectangles that had lengths parallel to the line about which we required the moment of inertia. This is not just coincidence; the use of perpendicular

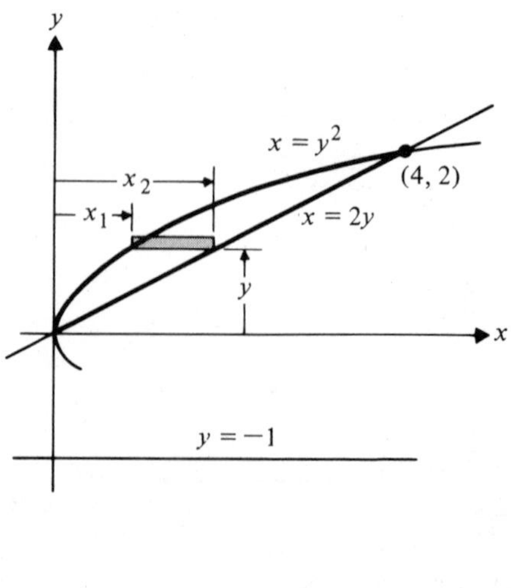

Figure 7.71

rectangles leads to unexpected difficulties. Consider, for instance, finding the moment of inertia about the x-axis of the plate given in Example 21. Should we wish to use the vertical rectangles shown in Fig. 7.70, then we first require the moment of inertia of such a rectangle about the x-axis. There is no problem with the mass of the rectangle; it is as in Example 21. But this mass must now be multiplied by the square of the distance from the x-axis to the rectangle. Unfortunately, different points in the rectangle are at different distances. One suggestion is to concentrate all of the mass of the rectangle at its center of mass and use the distance from the x-axis to the center of mass. This is *incorrect*. The center of mass is a point at which mass can be concentrated if we are discussing first moments. But we are discussing second moments. What then are we to do? If we wish to use this type of rectangle, we are forced to memorize a formula for its moment of inertia. To obtain this formula, let us consider the moment of inertia about the x-axis of the rectangle of width h and length $y_2 - y_1$ shown in Fig. 7.72. If we subdivide this rectangle into smaller rectangles of width dy, the moment of inertia of the smaller rectangle about the x-axis is approximately

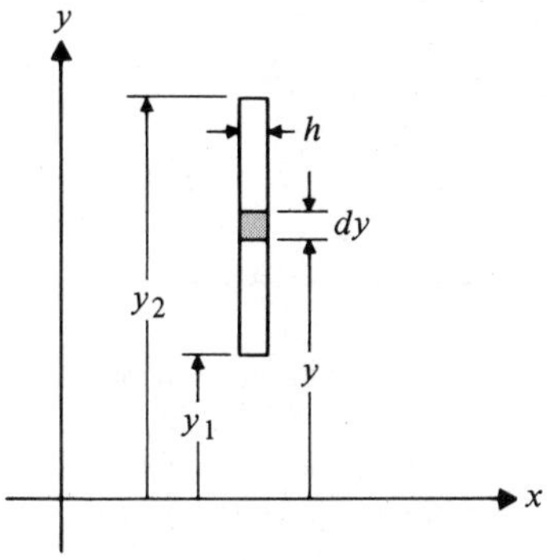

Figure 7.72

$$y^2 \rho h\, dy.$$

The moment of inertia of the large rectangle then is

$$I = \int_{y_1}^{y_2} y^2 \rho h\, dy = \rho h \int_{y_1}^{y_2} y^2\, dy = \frac{\rho h}{3}(y_2^3 - y_1^3). \tag{7.34}$$

We can now use this formula to state that the moment of inertia about the x-axis of the rectangle shown in Fig. 7.70 is

$$\frac{\rho}{3}(y_2^3 - y_1^3)\, dx = \frac{\rho}{3}[(2 - x)^{3/2} - x^9]\, dx.$$

The moment of inertia of the plate about the x-axis is therefore

$$\begin{aligned} I &= \int_0^1 \frac{\rho}{3}[(2 - x)^{3/2} - x^9]\, dx \\ &= \frac{\rho}{3}\left\{-\frac{2}{5}(2 - x)^{5/2} - \frac{1}{10}x^{10}\right\}_0^1 \\ &= \frac{16\sqrt{2} - 5}{30}\rho. \end{aligned}$$

The alternative procedure for this problem is to use horizontal rectangles and obtain two definite integrals:

$$I = \int_0^1 y^2 \rho y^{1/3}\, dy + \int_1^{\sqrt{2}} y^2 \rho (2 - y^2)\, dy.$$

In summary, we have two alternatives for determining the moments of inertia of thin plates:

1. We can choose rectangles parallel to the line about which the moment of inertia is required, in which case we use only the basic idea of a moment of inertia (mass times distance squared).

2. We can choose rectangles perpendicular to the line about which the moment of inertia is required, in which case we need either formula (7.34) or a variation of it.

As for finding centers of mass, we could, in the special case of constant density, drop all references to mass and talk about the second moments of area about a line.

EXERCISES 7.7

In each of Exercises 1–10, the curves define a thin plate with constant density ρ (mass per unit area). Find its moment of inertia about the line indicated.

1. $y = x^2$, $y = x^3$, about the y-axis

2. $y = x$, $y = 2x + 4$, $y = 0$, about the x-axis

3. $y = x^2$, $2y = x^2 + 4$, about $y = 0$

4. $y = x^2 - 4$, $y = 2x - x^2$, about $x = -2$

5. $xy\sqrt{y^2 + 12} = 1$, $x = 0$, $y = 1$, $y = \frac{1}{2}$, about $y = 0$

6. $y = |x|^{1/3}$, $y = 2 - |x|^{1/3}$, about $x = 0$

7. $x + y = 2$, $y - x = 2$, $y = 0$, about $x = -2$

8. $x = 1 - y^2$, $x = y^2 - 1$, about $x = -1$

9. $x = 1 - y^2$, $x = y^2 - 1$, about $y = 1$

10. $x = y^2$, $x + y = 2$, about $y = 3$

11. a) If I_{Ax} and I_{Bx} represent the second moments of area about the x-axis of the areas shown in Fig. 7.68, show that

$$I_{Bx} = 3nI_{Ax}.$$

b) Show that if I_{Ay} and I_{By} represent the second moments of area about the y-axis, then

$$nI_{Ay} = 3I_{By}.$$

12. The radius of gyration r of a thin plate (of constant density) about a line is defined by

$$I = Mr^2,$$

where M is the mass of the plate and I is its moment of inertia about that line. Find the radii of gyration about the x- and y-axes for the plate with edges

$$y = 2x^3, \qquad y + x^3 = 0, \qquad 2y = x + 3.$$

13. Show that the kinetic energy of a long-playing record is equal to one-half the product of the moment of inertia of the record about a line through its center (and perpendicular to its face), and the square of its angular speed.

14. Prove the *parallel axis theorem*: The moment of inertia of a thin plate (of constant density) with respect to any coplanar line is equal to the moment of inertia with respect to the parallel line through the center of mass plus the mass multiplied by the square of the distance between the lines.

***15.** Find the second moment of area of a rectangle about its diagonal.

***16.** The polar moment of inertia of a point mass m at (x, y) is defined as the product of m and the square of its distance from the origin

$$J_0 = m(x^2 + y^2).$$

For the thin plate of constant density shown in Fig. 7.73, let I_x and I_y be its moments of inertia about the x- and y-axes. Show that

$$J_0 = I_x + I_y.$$

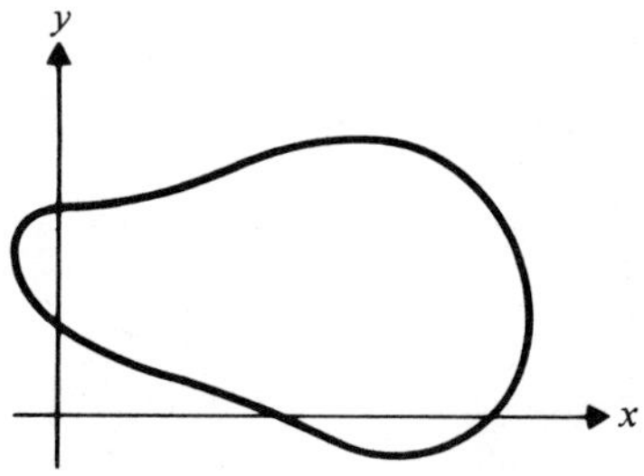

Figure 7.73

***17.** If a line $x = \tilde{x}$ is drawn through the area shown in Fig. 7.64, what integral represents the second moment of area about this line? What should be the value of $\tilde{x}$ for the smallest possible second moment?

7.8 ADDITIONAL APPLICATIONS

In Sections 7.1–7.4 we used limits of sums to define area, the volume of a solid of revolution, length along a curve, and work. We then showed that the limit–summations could be evaluated by appropriate definite integrals. In order to avoid memorizing these definite integrals as formulas, we illustrated how differentials and representative elements could be used to summarize intuitively the entire derivation. Note, however, that the heuristic approach was based on the mathematical discussion leading to the definite integral; that is, the mathematical derivation came first, and then an intuitive mnemonic was fitted to that derivation.

Having done this four times, we felt in Sections 7.5–7.7 that we could omit the mathematical discussion and proceed directly to its intuitive representation in terms of representative rectangles. It is important to realize, however, that we could give detailed discussions of limit–summations leading to Eqs. (7.20), (7.27),(7.28), (7.29), and (7.33). In this section we continue with only the intuitive representation while investigating other applications.

Volumes by Slicing If we can represent the area of parallel cross-sections of a volume as a function of one variable, we can use a definite integral to calculate the volume. In particular, when we use the disc (or washer) method to determine the volume of a solid of revolution, the parallel cross-sections are circles. In the following example, parallel cross-sections are squares.

Example 23 A uniformly tapered rod of length 2 meters has square cross-sections. If the areas of its ends are 4 and 16 square centimeters (Fig. 7.74a) what is the volume of the rod?

Solution If we define an x-coordinate perpendicular to the square cross-sections (Fig. 7.74b), then the side lengths of the cross-sections at $x = 0$ and $x = 2$ are 0.02 and 0.04 meter, respectively. Since the rod is uniformly tapered, the side length of the cross-section at x is

$$0.02 + x\left(\frac{0.04 - 0.02}{2}\right) = 0.02 + 0.01x = \frac{2 + x}{100} \text{ meters.}$$

The area of the cross-section at x is therefore given by

$$A(x) = 10^{-4}(2 + x)^2 \text{ m}^2.$$

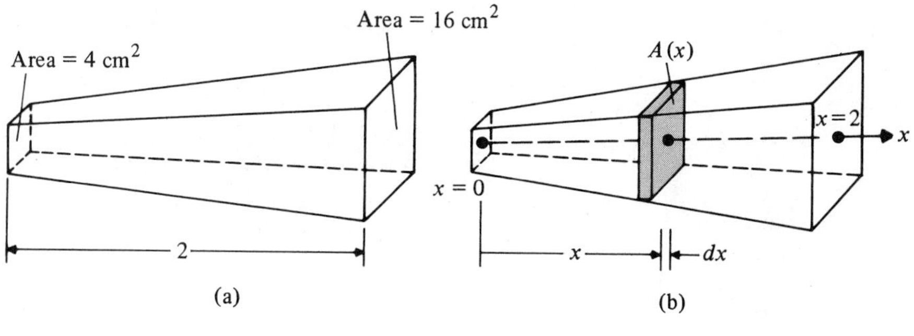

Figure 7.74

If we construct at x a "slab" of surface area $A(x)$ and width dx, the volume of the slab is given by

$$A(x)\,dx.$$

To obtain the volume of the rod we add the volumes of all such slabs between $x = 0$ and $x = 2$ and take the limit as their widths approach zero:

$$\begin{aligned} V &= \int_0^2 A(x)\,dx \\ &= \int_0^2 10^{-4}(2 + x)^2\,dx \\ &= \left\{\frac{10^{-4}}{3}(2 + x)^3\right\}_0^2 \\ &= \frac{10^{-4}}{3}(64 - 8) \\ &= \frac{56}{3} \times 10^{-4}\ \text{m}^3. \ \square \end{aligned}$$

Area of a Surface of Revolution

If a curve in the xy-plane is rotated about the x- or y-axis (or a line parallel to the x- or y-axis) in order to produce a surface, we can calculate the area of this surface using the ideas of lengths along curves that we discussed in Section 7.3.

Example 24 If that part of the parabola $y = x^2$ between $x = 0$ and $x = 1$ is rotated around the y-axis, find the area of the surface of revolution traced out by the curve (Fig. 7.75).

Solution We can approximate the length along the parabola corresponding to a change dx in x by the tangential straight-line length:

$$\begin{aligned} \sqrt{(dx)^2 + (dy)^2} &= \sqrt{1 + \left(\frac{dy}{dx}\right)^2}\,dx \\ &= \sqrt{1 + (2x)^2}\,dx \\ &= \sqrt{1 + 4x^2}\,dx. \end{aligned}$$

If this straight-line segment is rotated around the y-axis, each point follows a circular path of radius approximately equal to x; therefore, the area traced out by the line segment is approximately equal to

$$(2\pi x)(\sqrt{1 + 4x^2}\,dx).$$

We can find the total surface area by adding together all such areas, and taking the limit as the widths dx approach zero:

$$\begin{aligned} A &= \int_0^1 2\pi x\sqrt{1 + 4x^2}\,dx \\ &= 2\pi\left\{\frac{1}{12}(1 + 4x^2)^{3/2}\right\}_0^1 \\ &= \frac{\pi}{6}(5\sqrt{5} - 1). \ \square \end{aligned}$$

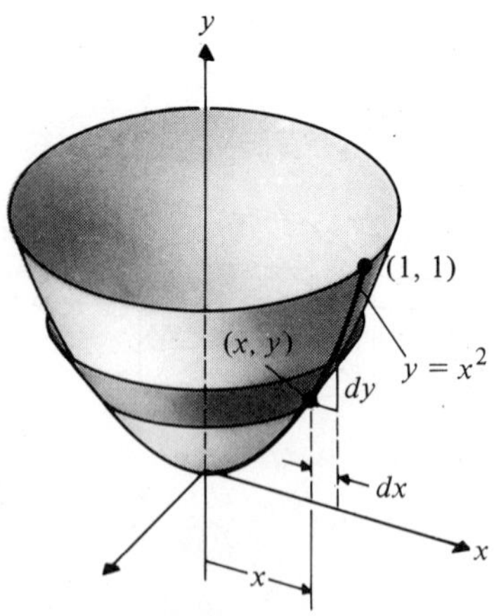

Figure 7.75

Example 25 Find the area of the surface of revolution traced out by rotating that part of the curve $y = x^3$ between $x = 1$ and $x = 2$ about the x-axis.

Solution We can approximate the length along the cubic corresponding to a change dx in x by the tangential straight-line length:

$$\begin{aligned}\sqrt{(dx)^2 + (dy)^2} &= \sqrt{1 + \left(\frac{dy}{dx}\right)^2}\, dx \\ &= \sqrt{1 + (3x^2)^2}\, dx \\ &= \sqrt{1 + 9x^4}\, dx.\end{aligned}$$

If this straight-line segment is rotated about the x-axis (Fig. 7.76), each point follows a circular path of radius approximately equal to y; therefore, the area traced out by the line segment is approximately equal to

$$(2\pi y)(\sqrt{1 + 9x^4}\, dx).$$

We can find the total surface area by adding together all such areas, and taking the limit as the widths dx approach zero:

$$\begin{aligned}A &= \int_1^2 2\pi y\sqrt{1 + 9x^4}\, dx \\ &= 2\pi \int_1^2 x^3\sqrt{1 + 9x^4}\, dx \\ &= 2\pi\left\{\frac{(1 + 9x^4)^{3/2}}{54}\right\}_1^2 \\ &= \frac{\pi}{27}(145^{3/2} - 10^{3/2}). \quad \square\end{aligned}$$

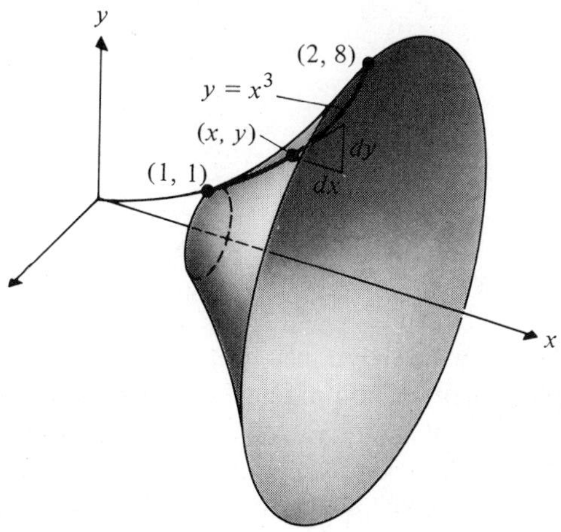

Figure 7.76

Rates of Flow In Problem 3 of Section 6.2 we considered laminar blood flow in a circular vessel. Specifically, we noted that the velocity of blood through the cross-section shown in Fig. 7.77(a) is a function of radial distance r from the center of

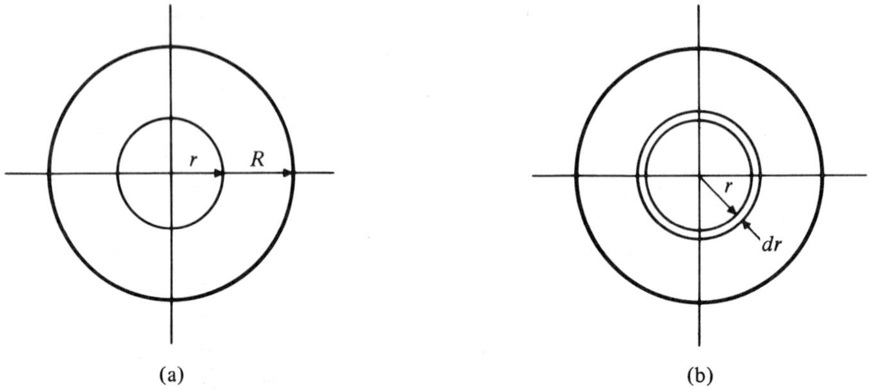

Figure 7.77

the vessel:

$$v(r) = c(R^2 - r^2), \qquad 0 \le r \le R,$$

where $c > 0$ is a constant and R is the radius of the vessel.

If we construct at radius r a thin ring of width dr (Fig. 7.77b), then the area of this ring is approximately $(2\pi r)\,dr$. Because v does not vary greatly over this ring, it follows that the amount of blood flowing through the ring per unit time is approximately $v(r)$ multiplied by the area of the ring:

$$v(r)(2\pi r\,dr).$$

We can find the total flow through the blood vessel by adding the flows through all rings and taking the limit as the widths dr of the rings approach zero:

$$\begin{aligned} F &= \int_0^R v(r)2\pi r\,dr \\ &= 2\pi \int_0^R rc(R^2 - r^2)\,dr \\ &= 2\pi c\left\{\frac{R^2 r^2}{2} - \frac{r^4}{4}\right\}_0^R \\ &= \frac{\pi c R^4}{2}. \end{aligned}$$

Repairs, Depreciation, and Replacement

In most businesses, consideration must be given to the depreciation and eventual replacement of equipment. Take, for example, a salesman who must decide when to replace his car. When he buys a new car, he knows that its value will depreciate quickly at first and less rapidly after a few years. On the other hand, the longer he keeps the car, the more costly will be the repair bills. The problem then is to decide on the optimum time at which to replace the car. Let us consider a situation in which the rate of depreciation of the car is $f(t)$, and the rate of accumulation of maintenance costs is $g(t)$. We have shown a possible $f(t)$ in Fig. 7.78, where the ordinate at time t represents the number of dollars per year at which the car depreciates at that time. If we draw a rectangle of width

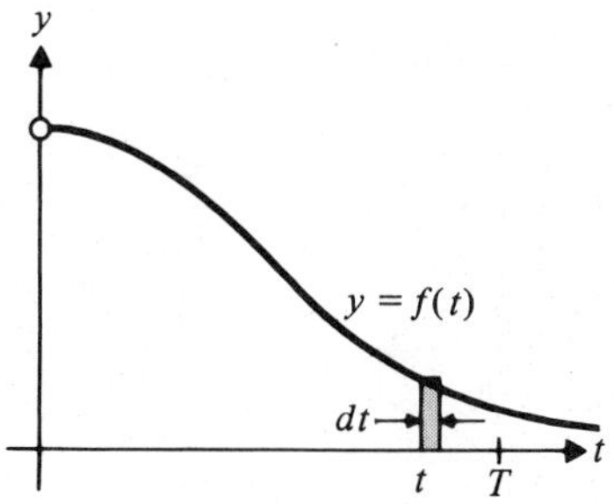

Figure 7.78

dt at time t, then the area of this rectangle, $f(t)\,dt$, represents the amount the car depreciates over the time interval dt. Consequently, the total depreciation of the car from the time it is purchased ($t = 0$) to time T is given by

$$\int_0^T f(t)\,dt.$$

In a similar way, the total maintenance cost from time $t = 0$ to $t = T$ (Fig. 7.79) is given by

$$\int_0^T g(t)\,dt.$$

The expense associated with owning and maintaining the car over a period of T years is

$$\int_0^T f(t)\,dt + \int_0^T g(t)\,dt = \int_0^T [f(t) + g(t)]\,dt,$$

and the average yearly expense is

$$C(T) = \frac{1}{T}\int_0^T [f(t) + g(t)]\,dt.$$

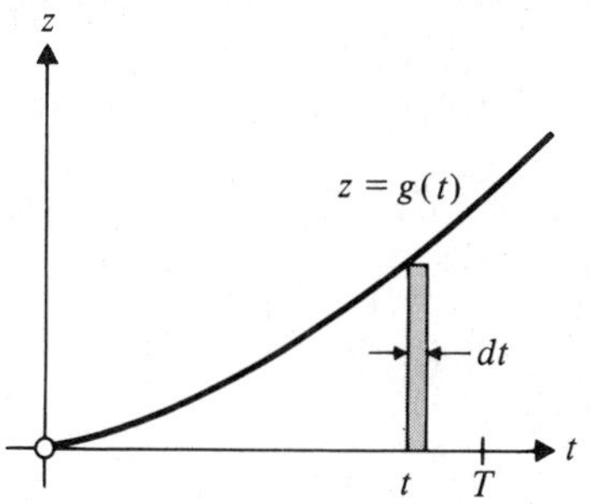

Figure 7.79

The best time to replace the car is given by that value T which minimizes this average yearly expense. In order to differentiate $C(T)$ with respect to T, we first set

$$H(T) = \int_0^T [f(t) + g(t)]\,dt,$$

and let $F(t)$ and $G(t)$ be antiderivatives of $f(t)$ and $g(t)$. Then we have

$$H(T) = \{F(t) + G(t)\}_0^T = F(T) - F(0) + G(T) - G(0);$$

hence,

$$H'(T) = F'(T) + G'(T) = f(T) + g(T).$$

For critical points of $C(T)$, then,

$$0 = C'(T) = \frac{d}{dT}\left(\frac{1}{T}H(T)\right) = -\frac{1}{T^2}H(T) + \frac{1}{T}(f(T) + g(T)),$$

or

$$\frac{H(T)}{T} = f(T) + g(T).$$

In other words, the salesman should purchase a new car when his average yearly expenses are equal to the rate at which his expenses are changing (provided this critical point leads to an absolute minimum for $C(T)$).

Example 26 Find the optimum time to replace a car if the rates of depreciation and accumulation of maintenance costs are, respectively,

$$f(t) = -\frac{V}{50}(t - 10), \qquad g(t) = \frac{V}{400}t^2,$$

where V is the value of the new car. Assume that the car must be replaced by the time it has depreciated to no value.

Solution Since the total depreciation of the car in T years is given by

$$\int_0^T f(t)\,dt = \int_0^T -\frac{V}{50}(t-10)\,dt$$
$$= -\frac{V}{50}\left\{\frac{t^2}{2} - 10t\right\}_0^T$$
$$= -\frac{V}{100}(T^2 - 20T),$$

it follows that the car will have completely depreciated when

$$V = -\frac{V}{100}(T^2 - 20T).$$

The only solution of this equation is $T = 10$ years. (Do not make the mistake of assuming that complete depreciation occurs when $f(t) = 0$; this is not always true. Complete depreciation occurs when the area under $f(t)$ is equal to the initial value of the car.)

The average yearly expense associated with the car if it is kept for T years is

$$C(T) = \frac{1}{T}\int_0^T [f(t) + g(t)]\,dt$$
$$= \frac{1}{T}\int_0^T \left\{-\frac{Vt}{50} + \frac{V}{5} + \frac{Vt^2}{400}\right\} dt$$
$$= \frac{1}{T}\left\{-\frac{VT^2}{100} + \frac{VT}{5} + \frac{VT^3}{1200}\right\}$$
$$= -\frac{VT}{100} + \frac{V}{5} + \frac{VT^2}{1200}, \qquad 0 < T \leq 10.$$

For critical points of $C(T)$, we solve

$$0 = C'(T) = -\frac{V}{100} + \frac{VT}{600},$$

which implies that $T = 6$. Since

$$C(6) = \frac{17V}{100}, \qquad C(10) = \frac{11V}{60}, \qquad \lim_{T\to 0^+} C(T) = \frac{V}{5},$$

it follows that the car should be replaced in 6 years. □

EXERCISES 7.8

1. Verify that the surface area of a sphere of radius r is $4\pi r^2$.
2. Find the area of the curved surface of a right-circular cone of radius r and height h.
3. Calculate the rate of flow of blood through a circular vessel of radius R if the velocity profile is:
   a) $v = f(r) = cR\sqrt{R^2 - r^2}$;
   b) $v = f(r) = \frac{c}{R^2}(R^2 - r^2)^2$.
4. Find the volume of the pyramid shown in Fig. 7.80.

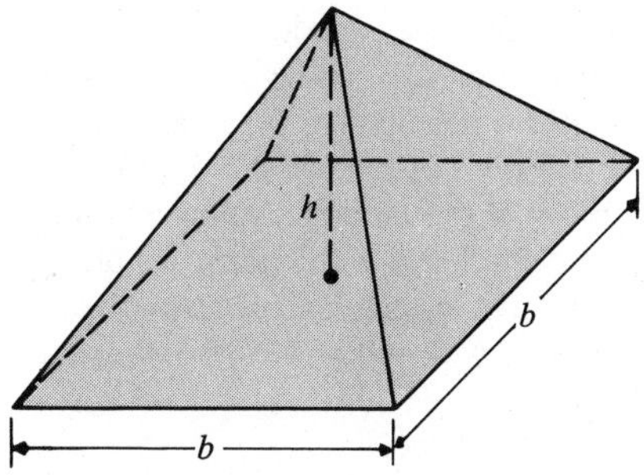

Figure 7.80

5. The amount of water consumed by a community varies throughout the day, peaking, of course, around meal hours. During the six-hour period between 12:00 noon ($t = 0$) and 6:00 P.M. ($t = 6$), we find that the number of cubic meters of water consumed per hour at time t is given by the function

$$f(t) = 5000 + 21.65t^2 - 249.7t^3 + 97.52t^4 - 9.680t^5.$$

Find the total consumption during this 6-hour period.

6. The density of bees (number of bees per unit area) at a distance x from a hive is given by

$$\rho(x) = \frac{600{,}000}{31\pi R^5}(R^3 + 2R^2x - Rx^2 - 2x^3), \qquad 0 \le x \le R,$$

where R is the maximum distance traveled by the bees.

a) What is the number of bees in the colony?

b) How many bees are within $R/2$ of the hive?

7. If the radius of the blood vessel shown in Fig. 7.77 is reduced to $R/2$ (because of arteriosclerosis), the velocity profile is

$$v(r) = c(R^2 - 4r^2), \qquad 0 \le r \le R/2.$$

What percentage of the normal flow ($\pi cR^4/2$) gets through the "hardened" vessel?

8. A tree trunk of diameter 50 centimeters (Fig. 7.81) has a wedge cut from it by two planes. The lower plane is perpendicular to the axis of the trunk, and together the planes make an angle $\pi/3$ radians (meeting along a diameter of the circular cross-section of the trunk). Find the volume of the wedge.

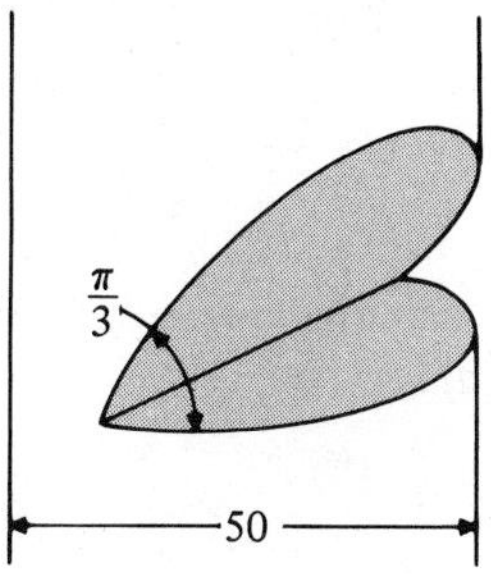

Figure 7.81

9. Find the area of the surface of revolution formed by rotating that part of the curve $24xy = x^4 + 48$ between $x = 1$ and $x = 2$ about the x-axis.

10. Find the area of the surface of revolution generated by rotating the curve $8y^2 = x^2(1 - x^2)$ about the x-axis.

11. Find the area of the surface of revolution generated by rotating the loop of the curve $9y^2 = x(3 - x)^2$ about: (a) the y-axis; (b) the x-axis.

12. The base of a solid is the circle $x^2 + y^2 = r^2$, and every plane section perpendicular to the x axis is an isosceles triangle, as shown in Fig. 7.82. Find the volume of the solid.

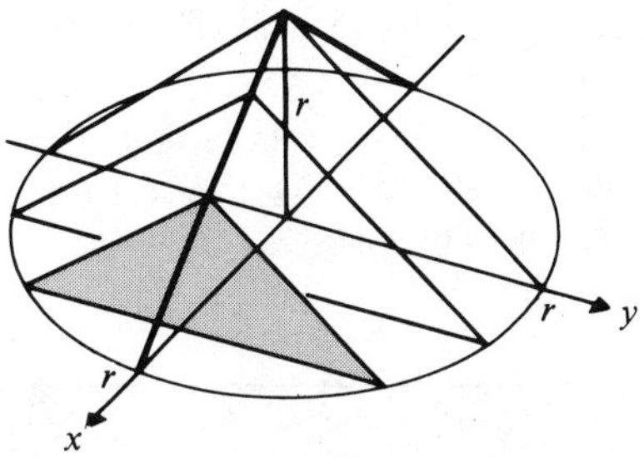

Figure 7.82

13. Electrons are fired from an electron gun at a target, as shown in Fig. 7.83. The probability that an electron will strike the target in a ring of unit area at a distance x from the center of the target is given by

$$p = f(x) = \frac{5}{3\pi R^5}(R^3 - x^3), \qquad 0 \le x \le R.$$

What percentage of a cascade of electrons will hit within a distance r from the center of the target?

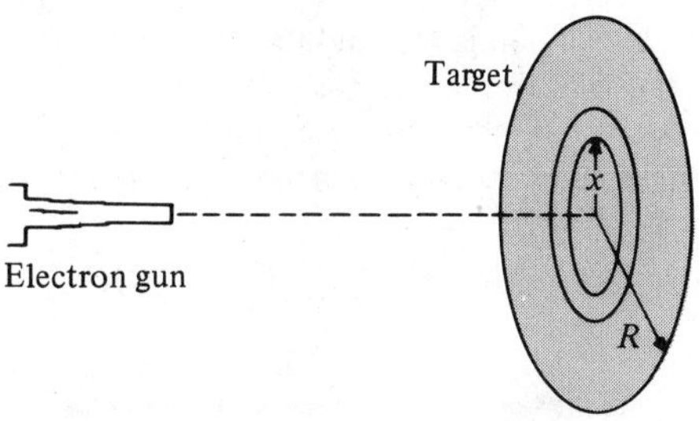

Figure 7.83

In each of Exercises 14–16, $f(t)$ and $g(t)$ represent, respectively, rate of depreciation and rate of accumulation of maintenance costs for a piece of machinery. Sketch graphs of these functions, and show that the total area under $y = y(t)$ is equal to V (the original value of the machine). Determine the optimum time at which to replace the machine. Sketch a graph of $C(T)$ (the average yearly expense) to illustrate your answer.

14. $f(t) = \begin{cases} \dfrac{V}{10} - \dfrac{V}{200}t, & 0 < t \leq 20, \\ 0, & t > 20; \end{cases} \qquad g(t) = \dfrac{Vt^2}{4000}$

15. $f(t) = \begin{cases} -\dfrac{V}{50}(t - 10), & 0 < t \leq 10, \\ 0, & t > 10; \end{cases} \qquad g(t) = \dfrac{Vt}{400}$

16. $f(t) = \begin{cases} \dfrac{V}{10} - \dfrac{6Vt^2}{15{,}625}, & 0 < t \leq 12.5, \\ 0, & t > 12.5; \end{cases} \qquad g(t) = \dfrac{Vt^2}{3000}$

17. Two right-circular cylinders, each of radius r, have axes that intersect at right angles. Find the volume common to the two cylinders.

18. What is the volume of air trapped in the attic of a house if the roof has shape as shown in Fig. 7.84? The peak of the roof is 2 meters above its base.

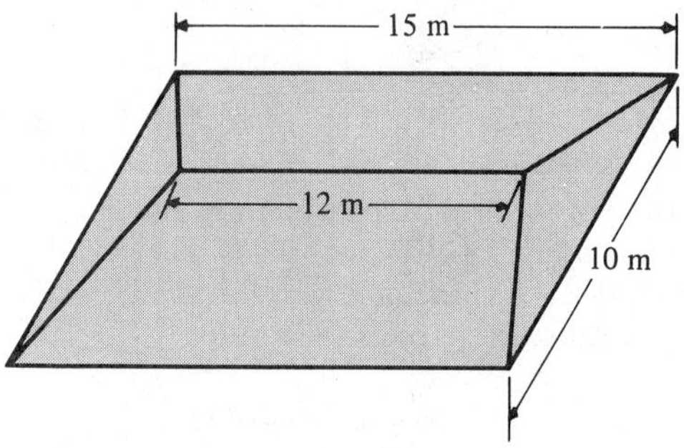

Figure 7.84

19. Suppose that at time $t = 0$ the number of people in a certain population is N_0, and the probability that each individual will live to age t is $p(t)$. For time $t > 0$, the birth rate is $r(t)$ individuals per unit time. Show that at time T the number of people in the population is given by

$$N(T) = N_0 p(T) + \int_0^T p(T - t)r(t)\,dt.$$

***20.** The depth of water in the container shown in Fig. 7.85 is originally 10 centimeters. Water evaporates from the surface at a rate proportional to the area of the surface. If the water level drops 1 centimeter in 5 days, how long will it take for the water to evaporate completely? (Note that Exercise 9 in Section 5.4 is this same problem given a container that is a right-circular cone. Now we ask you to repeat the problem with no knowledge of the shape of the container.)

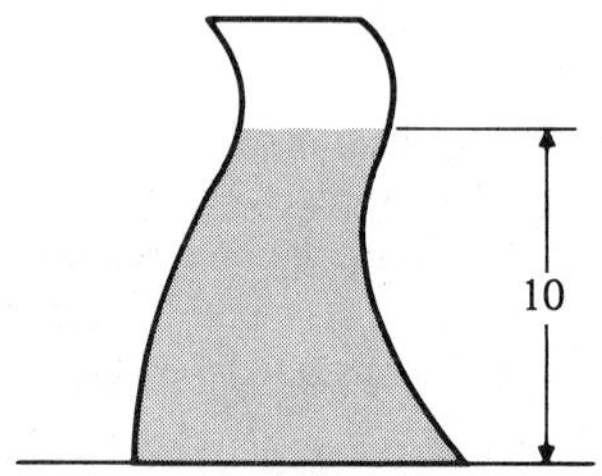

Figure 7.85

***21.** An appliance retailer must be concerned with her inventory costs. For example, given that she sells N refrigerators per year, she must decide whether to order the year's supply at one time, in which case a large storage area would be necessary, or to make periodic orders throughout the year, perhaps of $N/12$ at the first of each month. In the latter case she would incur costs due to paper work, delivery charges, etc. Let us suppose that the retailer decides to order in equal-lot sizes, x, at equally spaced intervals, N/x times per year.

a) If each order has a fixed cost of F dollars, plus f dollars for each refrigerator, what are the total yearly ordering costs?

b) Between successive deliveries, the retailer's stock dwindles from x to 0. If she assumes that the number of refrigerators in stock decreases linearly in time, and the yearly stocking cost per refrigerator is p dollars, what are the yearly stocking costs?

c) If the retailer's total yearly inventory costs are her ordering costs (in part a) plus her stocking costs (in part b), what should x be in order to minimize inventory costs?

7.9 IMPROPER INTEGRALS

According to the fundamental theorem of calculus (Theorem 6.3), we can evaluate definite integrals by indefinite integrals,

$$\int_a^b f(x)\,dx = \left\{\int f(x)\,dx\right\}_a^b,$$

provided $f(x)$ is continuous on $a \leq x \leq b$. In this section we investigate what

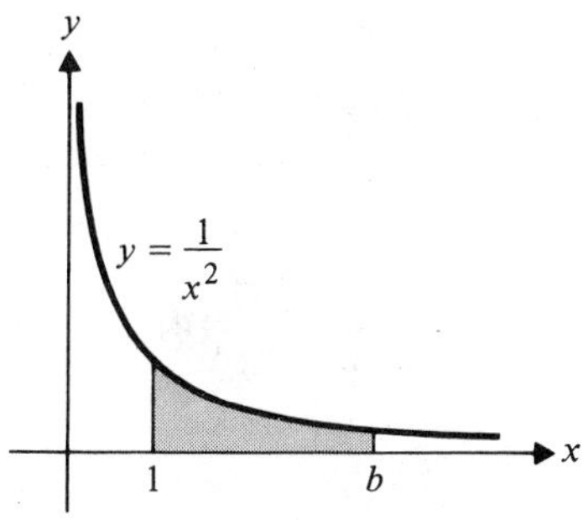

Figure 7.86

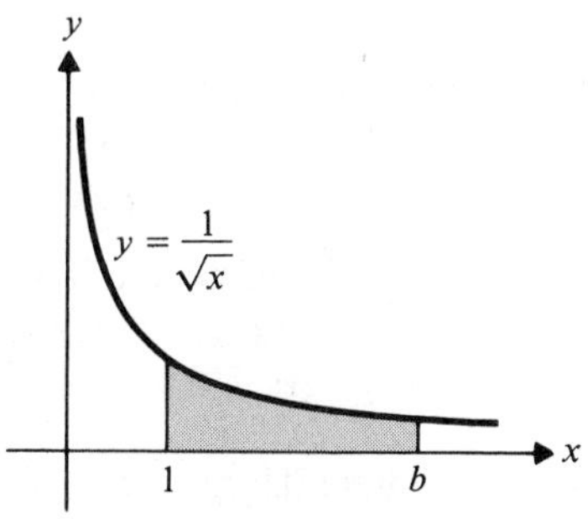

Figure 7.87

to do when $f(x)$ is not continuous on $a \leq x \leq b$, or when either a or b is infinite. Consider first whether a reasonable meaning can be given to

$$\int_1^\infty \frac{1}{x^2}\,dx \quad \text{and} \quad \int_1^\infty \frac{1}{\sqrt{x}}\,dx.$$

Both of these integrals are "improper" in the sense that their upper limits are not finite, and we therefore call them *improper integrals.* If $b > 1$, there is no difficulty with the evaluation or a possible interpretation of either

$$\int_1^b \frac{1}{x^2}\,dx \quad \text{or} \quad \int_1^b \frac{1}{\sqrt{x}}\,dx.$$

Clearly,

$$\int_1^b \frac{1}{x^2}\,dx = \left\{-\frac{1}{x}\right\}_1^b = 1 - \frac{1}{b}$$

and

$$\int_1^b \frac{1}{\sqrt{x}}\,dx = \{2\sqrt{x}\}_1^b = 2\sqrt{b} - 2.$$

We can interpret the first integral as the area under the curve $y = 1/x^2$, above the x-axis, and between the vertical lines $x = 1$ and $x = b$ (Fig. 7.86). The second integral has exactly the same interpretation but uses the curve $y = 1/\sqrt{x}$ in place of $y = 1/x^2$ (Fig. 7.87).

Table 7.1

b	$\int_1^b \frac{1}{x^2}\,dx$	$\int_1^b \frac{1}{\sqrt{x}}\,dx$
100	0.99	18
10,000	0.9999	198
1,000,000	0.999999	1,998
100,000,000	0.99999999	19,998

In Table 7.1 we have listed values of these definite integrals corresponding to various values of b. It is clear that these integrals display completely different characteristics. As b is made very, very large, the integral of $1/x^2$ is always less than one, but gets closer and closer to one as b increases. In other words,

$$\lim_{b\to\infty} \int_1^b \frac{1}{x^2}\,dx = 1.$$

Geometrically speaking, the area shown in Fig. 7.86 is always less than one but approaches one as b approaches infinity.

Contrast this with the integral of $1/\sqrt{x}$, which becomes indefinitely large as b increases:

$$\lim_{b\to\infty} \int_1^b \frac{1}{\sqrt{x}}\,dx = \infty;$$

i.e., the area shown in Fig. 7.87 can be made as large as desired by choosing b sufficiently large.

On the basis of these calculations, we can reasonably say that the improper integral of $1/x^2$ has a value of 1, whereas the improper integral of $1/\sqrt{x}$ has no value. In other words, if these improper integrals are to have values, they should be defined by the limits

$$\int_1^{\infty} \frac{1}{x^2}\,dx = \lim_{b\to\infty} \int_1^b \frac{1}{x^2}\,dx \quad \text{and} \quad \int_1^{\infty} \frac{1}{\sqrt{x}}\,dx = \lim_{b\to\infty} \int_1^b \frac{1}{\sqrt{x}}\,dx.$$

In the second case the limit does not exist, and we interpret this to mean that the improper integral does not exist (i.e., does not have a value).

In general, then, improper integrals are defined in terms of limits according to the following.

▷ **Definition 7.2** If $f(x)$ is continuous for $x \geq a$, we define

$$\int_a^{\infty} f(x)\,dx = \lim_{b\to\infty} \int_a^b f(x)\,dx, \tag{7.35}$$

provided the limit exists. ◁

If the limit exists, we say that the improper integral is equal to the limit, or that the improper integral converges to the limit. If the limit does not exist, we say that the improper integral has no value, or that it diverges.

Example 27 Determine whether each of the following improper integrals converges or diverges.

a) $\displaystyle\int_3^{\infty} \frac{1}{(x-2)^3}\,dx$

b) $\displaystyle\int_0^{\infty} \frac{x^2}{\sqrt{1+x^3}}\,dx$

c) $\displaystyle\int_1^{\infty} \frac{2x^2+6}{3x^2+5}\,dx$

Solution a)
$$\int_3^{\infty} \frac{1}{(x-2)^3}\,dx = \lim_{b\to\infty} \int_3^b \frac{1}{(x-2)^3}\,dx = \lim_{b\to\infty} \left\{-\frac{1}{2(x-2)^2}\right\}_3^b$$
$$= \lim_{b\to\infty} \left\{\frac{1}{2} - \frac{1}{2(b-2)^2}\right\} = \frac{1}{2},$$

and the improper integral therefore converges to $\frac{1}{2}$.

b)
$$\int_0^{\infty} \frac{x^2}{\sqrt{1+x^3}}\,dx = \lim_{b\to\infty} \int_0^b \frac{x^2}{\sqrt{1+x^3}}\,dx = \lim_{b\to\infty} \left\{\frac{2}{3}\sqrt{1+x^3}\right\}_0^b$$
$$= \lim_{b\to\infty} \left\{\frac{2}{3}\sqrt{1+b^3} - \frac{2}{3}\right\} = \infty,$$

and the improper integral therefore diverges.

c) In spite of the fact that we cannot (at this time) find an antiderivative for $f(x) = (2x^2+6)/(3x^2+5)$, we can still solve this problem by interpreting

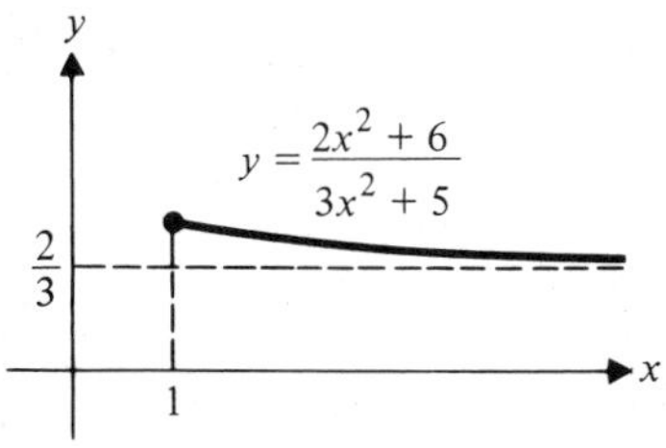

Figure 7.88

the improper integral as an area. The graph of $f(x)$ in Fig. 7.88 indicates that the curve is asymptotic to the line $y = \frac{2}{3}$. Clearly, then, the improper integral

$$\int_1^{\infty} \frac{2x^2 + 6}{3x^2 + 5}\, dx$$

can have no value since the area under the curve $y = f(x)$ is larger than the area of a rectangle of width $\frac{2}{3}$ and infinite length. □

We can define improper integrals with infinite lower limits in exactly the same way as we do improper integrals with infinite upper limits. Specifically, if $f(x)$ is continuous for $x \le b$, we define

$$\int_{-\infty}^{b} f(x)\, dx = \lim_{a \to -\infty} \int_a^b f(x)\, dx, \tag{7.36}$$

provided the limit exists. For example,

$$\int_{-\infty}^{5} \frac{1}{(x-6)^5}\, dx = \lim_{a \to -\infty} \int_a^5 \frac{1}{(x-6)^5}\, dx = \lim_{a \to -\infty} \left\{ -\frac{1}{4(x-6)^4} \right\}_a^5$$

$$= \lim_{a \to -\infty} \left\{ -\frac{1}{4} + \frac{1}{4(a-6)^4} \right\} = -\frac{1}{4}.$$

When $f(x)$ is continuous for all x, we define

$$\int_{-\infty}^{\infty} f(x)\, dx = \lim_{a \to -\infty} \int_a^0 f(x)\, dx + \lim_{b \to \infty} \int_0^b f(x)\, dx, \tag{7.37}$$

provided both limits exist. For example,

$$\int_{-\infty}^{\infty} \frac{x}{(1+x^2)^2}\, dx = \lim_{a \to -\infty} \int_a^0 \frac{x}{(1+x^2)^2}\, dx + \lim_{b \to \infty} \int_0^b \frac{x}{(1+x^2)^2}\, dx$$

$$= \lim_{a \to -\infty} \left\{ \frac{-1}{2(1+x^2)} \right\}_a^0 + \lim_{b \to \infty} \left\{ \frac{-1}{2(1+x^2)} \right\}_0^b$$

$$= \lim_{a \to -\infty} \left\{ \frac{1}{2(1+a^2)} - \frac{1}{2} \right\} + \lim_{b \to \infty} \left\{ \frac{1}{2} - \frac{1}{2(1+b^2)} \right\}$$

$$= -\frac{1}{2} + \frac{1}{2} = 0.$$

Example 28 a) Is it possible to assign a number to represent the area bounded by the curves $y = 1/x^{3/4}$, $x = 1$, and $y = 0$?

b) If the region in (a) is rotated around the x-axis, is it possible to assign a number to represent the volume of the solid so generated?

Solution a) If the region bounded by the curves is to have an area (Fig. 7.89), it must be defined by the improper integral

$$A = \int_1^\infty \frac{1}{x^{3/4}}\,dx$$

$$= \lim_{b\to\infty} \int_1^b \frac{1}{x^{3/4}}\,dx$$

$$= \lim_{b\to\infty} \{4x^{1/4}\}_1^b$$

$$= \lim_{b\to\infty} (4b^{1/4} - 4)$$

$$= \infty.$$

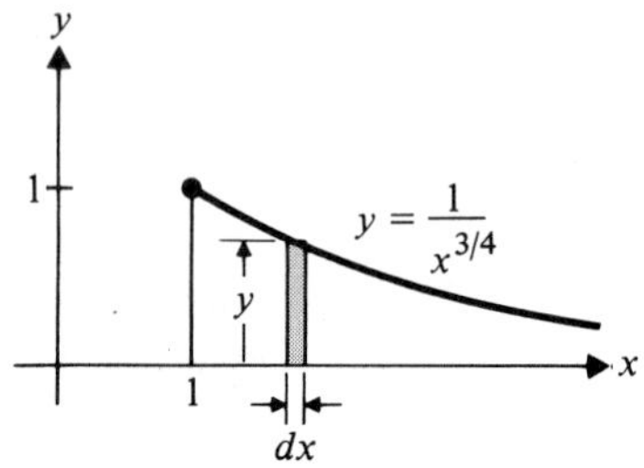

Figure 7.89

Consequently, we cannot assign an area to the region.

b) If we use circular discs to calculate the volume in the solid formed by rotating the region shown in Fig. 7.89 about the x-axis, then

$$V = \int_1^\infty \pi y^2\,dx$$

$$= \lim_{b\to\infty} \int_1^b \frac{\pi}{x^{3/2}}\,dx$$

$$= \lim_{b\to\infty} \left\{\frac{-2\pi}{x^{1/2}}\right\}_1^b$$

$$= \lim_{b\to\infty} \left\{2\pi - \frac{2\pi}{\sqrt{x}}\right\}$$

$$= 2\pi.$$

Thus we should assign to the solid a volume of 2π cubic units. □

Example 29 Find the escape velocity of a projectile from the earth's surface.

Solution Suppose the projectile is fired from the earth's surface with velocity v in the x-direction, as shown in Fig. 7.90. If the projectile is to escape the earth's gravitational pull, its initial kinetic energy must be greater than or equal to the work done against the gravitational pull as the projectile travels from the earth's surface ($x = R$) to a point where it is free of gravity ($x = \infty$). When the projectile is a distance x from the center of the earth, the force of attraction on it is

$$F(x) = -\frac{GmM}{x^2},$$

where $G > 0$ is a constant, m is the mass of the projectile, and M is the mass of the earth. (This is known as *Newton's universal law of gravitation.*) The

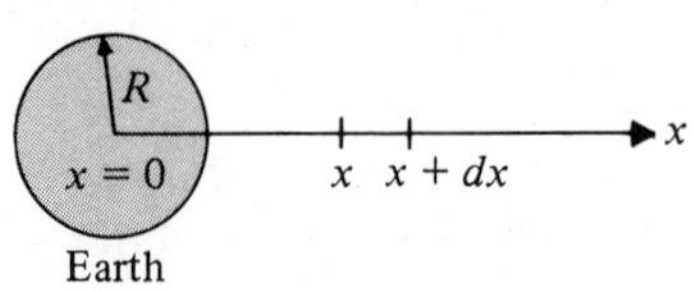

Figure 7.90

work done against gravity in traveling from $x = R$ to $x = \infty$ is therefore given by

$$\begin{aligned} W &= \int_R^\infty \frac{GmM}{x^2}\,dx \\ &= \lim_{b\to\infty} \int_R^b \frac{GmM}{x^2}\,dx \\ &= \lim_{b\to\infty} \left\{-\frac{GmM}{x}\right\}_R^b \\ &= \lim_{b\to\infty} \left\{\frac{GmM}{R} - \frac{GmM}{b}\right\} \\ &= \frac{GmM}{R}. \end{aligned}$$

Since the initial kinetic energy of the projectile is $\frac{1}{2}mv^2$, it will escape the gravitational pull of the earth if

$$\frac{1}{2}mv^2 \geq \frac{GmM}{R};$$

i.e., if

$$v \geq \sqrt{\frac{2GM}{R}},$$

and hence $\sqrt{2GM/R}$ is the escape velocity. If we take the mean radius of the earth as 6370 kilometers, its mean density as $\rho = 5.52 \times 10^3$ kilograms per cubic meter, and $G = 6.67 \times 10^{-11}$, we obtain $v = 11.2$ kilometers per second. Compare this with a 308 Winchester that fires a 150-grain bullet with a muzzle velocity of only 0.81 kilometer per second. □

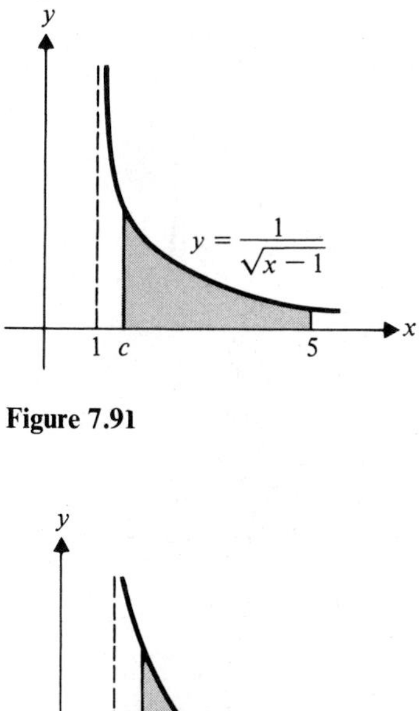

Figure 7.91

A second type of improper integral occurs when the integrand is discontinuous at a point or points in the interval of integration. For example, the integrands of the improper integrals

$$\int_1^5 \frac{1}{\sqrt{x-1}}\,dx \quad\text{and}\quad \int_1^5 \frac{1}{(x-1)^4}\,dx$$

each have infinite discontinuities at $x = 1$. If $1 < c < 5$, then it is straightforward to calculate

$$\int_c^5 \frac{1}{\sqrt{x-1}}\,dx = \{2\sqrt{x-1}\}_c^5 = 4 - 2\sqrt{c-1}$$

and

$$\int_c^5 \frac{1}{(x-1)^4}\,dx = \left\{\frac{-1}{3(x-1)^3}\right\}_c^5 = \frac{1}{3(c-1)^3} - \frac{1}{12},$$

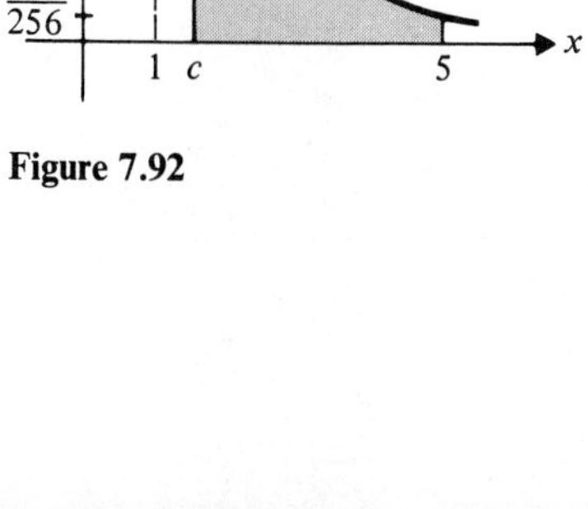

Figure 7.92

and one possible interpretation of these definite integrals is the areas shown in Figs. 7.91 and 7.92.

If we let c approach 1 from the right in the first integral, we find that

$$\lim_{c \to 1^+} \int_c^5 \frac{1}{\sqrt{x-1}}\,dx = \lim_{c \to 1^+} (4 - 2\sqrt{c-1}) = 4;$$

and in the second integral,

$$\lim_{c \to 1^+} \int_c^5 \frac{1}{(x-1)^4}\,dx = \lim_{c \to 1^+} \left\{\frac{1}{3(c-1)^3} - \frac{1}{192}\right\} = \infty.$$

It would seem reasonable then to define the area under the curve $y = 1/\sqrt{x-1}$, above the x-axis, and between the vertical lines $x = 1$ and $x = 5$ as 4 square units. On the other hand, the area bounded by $y = 1/(x-1)^4$, $y = 0$, $x = 5$, and $x = c$ becomes increasingly large as $x = c$ moves closer to $x = 1$, and we cannot therefore assign a value to this area when $c = 1$. In other words, if we define

$$\int_1^5 \frac{1}{\sqrt{x-1}}\,dx = \lim_{c \to 1^+} \int_c^5 \frac{1}{\sqrt{x-1}}\,dx$$

and

$$\int_1^5 \frac{1}{(x-1)^4}\,dx = \lim_{c \to 1^+} \int_c^5 \frac{1}{(x-1)^4}\,dx,$$

then the first improper integral has a value of 4, whereas the second has no value.

These improper integrals are examples of the general situation described in the following definition and illustrated in Fig. 7.93.

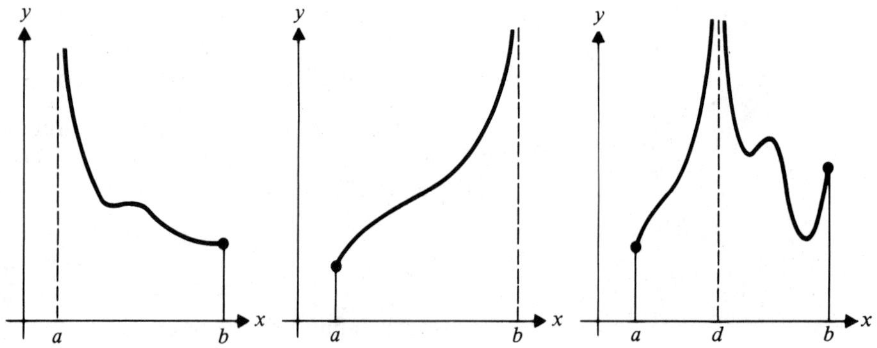

Figure 7.93

▷ **Definition 7.3** If $f(x)$ is continuous at every point in the interval $a \le x \le b$ except at $x = a$, then

$$\int_a^b f(x)\,dx = \lim_{c \to a^+} \int_c^b f(x)\,dx, \tag{7.38}$$

provided the limit exists (Fig. 7.93a). If $f(x)$ is continuous at every point in the interval $a \le x \le b$ except at $x = b$, then

$$\int_a^b f(x)\,dx = \lim_{c \to b^-} \int_a^c f(x)\,dx, \tag{7.39}$$

if the limit exists (Fig. 7.93b). If $f(x)$ is continuous at every point in the interval $a \le x \le b$ except at $x = d$ where $(a < d < b)$, then

$$\int_a^b f(x)\,dx = \lim_{c \to d^-} \int_a^c f(x)\,dx + \lim_{c \to d^+} \int_c^b f(x)\,dx, \tag{7.40}$$

provided both limits exist (Fig. 7.93c). ◁

Example 30 Determine whether each of the following improper integrals converges or diverges.

a) $\displaystyle\int_{-2}^{2} \frac{1}{(x-2)^3}\,dx$

b) $\displaystyle\int_{-1}^{1} \frac{1}{x^4}\,dx$

c) $\displaystyle\int_{1}^{5} \frac{x}{\sqrt{x-1}}\,dx$

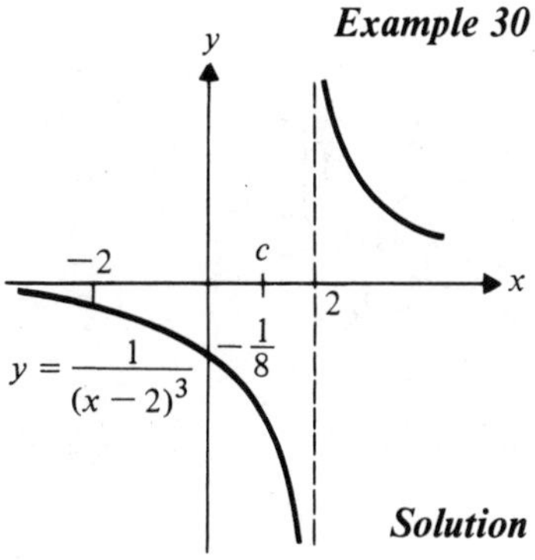

Figure 7.94

Solution a) Since the integrand is discontinuous at $x = 2$, and only $x = 2$ (Fig. 7.94), we have

$$\begin{aligned}
\int_{-2}^{2} \frac{1}{(x-2)^3}\,dx &= \lim_{c \to 2^-} \int_{-2}^{c} \frac{1}{(x-2)^3}\,dx \\
&= \lim_{c \to 2^-} \left\{ \frac{-1}{2(x-2)^2} \right\}_{-2}^{c} \\
&= \lim_{c \to 2^-} \left\{ \frac{-1}{2(c-2)^2} + \frac{1}{32} \right\} \\
&= -\infty.
\end{aligned}$$

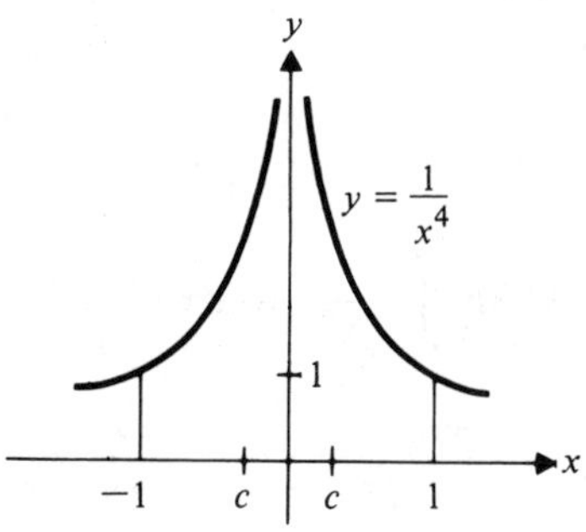

Figure 7.95

b) By Definition 7.3 (Fig. 7.95), we have

$$\begin{aligned}
\int_{-1}^{1} \frac{1}{x^4}\,dx &= \lim_{c \to 0^-} \int_{-1}^{c} \frac{1}{x^4}\,dx + \lim_{c \to 0^+} \int_{c}^{1} \frac{1}{x^4}\,dx \\
&= \lim_{c \to 0^-} \left\{ -\frac{1}{3x^3} \right\}_{-1}^{c} + \lim_{c \to 0^+} \left\{ -\frac{1}{3x^3} \right\}_{c}^{1} \\
&= \lim_{c \to 0^-} \left\{ -\frac{1}{3} - \frac{1}{3c^3} \right\} + \lim_{c \to 0^+} \left\{ \frac{1}{3c^3} - \frac{1}{3} \right\}.
\end{aligned}$$

Since neither of these limits exists, the improper integral diverges.

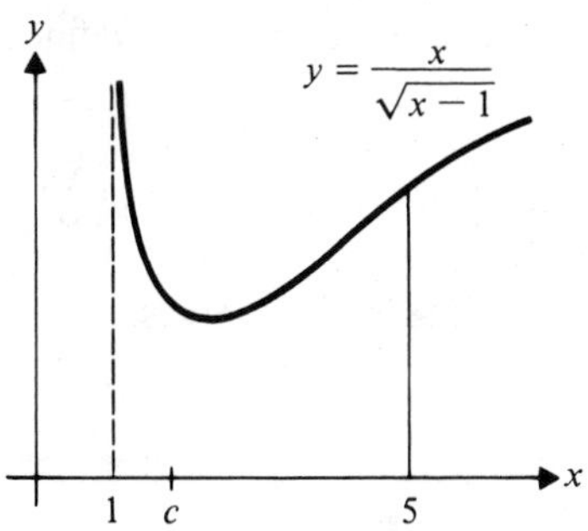

Figure 7.96

c) Since the integrand is discontinuous only at $x = 1$ (Fig. 7.96), we have

$$\int_{1}^{5} \frac{x}{\sqrt{x-1}}\,dx = \lim_{c \to 1^+} \int_{c}^{5} \frac{x}{\sqrt{x-1}}\,dx.$$

If we set $u = x - 1$, then $du = dx$ and

$$\int \frac{x}{\sqrt{x-1}}\,dx = \int \frac{u+1}{\sqrt{u}}\,du$$

$$= \int (u^{1/2} + u^{-1/2})\,du$$

$$= \frac{2}{3}u^{3/2} + 2u^{1/2} + C$$

$$= \frac{2}{3}(x-1)^{3/2} + 2\sqrt{x-1} + C.$$

Thus,

$$\int_1^5 \frac{x}{\sqrt{x-1}}\,dx = \lim_{c\to 1^+} \left\{\frac{2}{3}(x-1)^{3/2} + 2\sqrt{x-1}\right\}_c^5$$

$$= \lim_{c\to 1^+} \left\{\frac{28}{3} - \frac{2}{3}(c-1)^{3/2} - 2\sqrt{c-1}\right\}$$

$$= \frac{28}{3},$$

and the improper integral converges. □

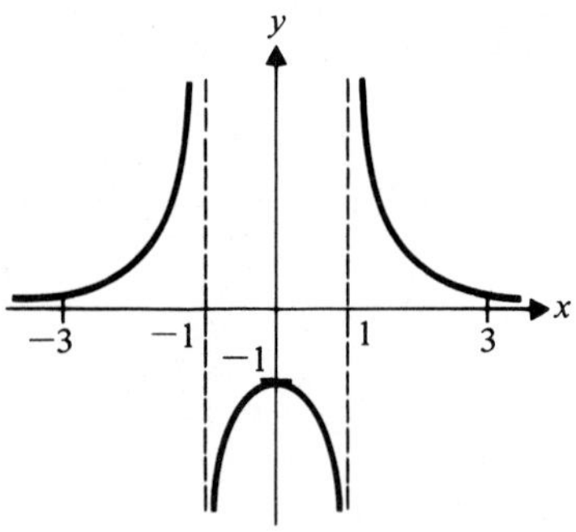

Figure 7.97

Various types of improper integrals may occur in the same problem. For example, Fig. 7.97 shows that the integral of $f(x) = 1/(x^2 - 1)$ from $x = -3$ to $x = \infty$ involves the use of five limits:

$$\int_{-3}^{\infty} \frac{1}{x^2-1}\,dx = \lim_{c\to -1^-} \int_{-3}^{c} \frac{1}{x^2-1}\,dx + \lim_{c\to -1^+} \int_{c}^{0} \frac{1}{x^2-1}\,dx$$

$$+ \lim_{c\to 1^-} \int_{0}^{c} \frac{1}{x^2-1}\,dx + \lim_{c\to 1^+} \int_{c}^{10} \frac{1}{x^2-1}\,dx$$

$$+ \lim_{b\to\infty} \int_{10}^{b} \frac{1}{x^2-1}\,dx.$$

If just one of these limits fails to exist, then the improper integral diverges.

If the integrand has a finite discontinuity, such as the function $f(x)$ in Fig. 7.98 does, then the improper integral of $f(x)$ from $x = a$ to $x = b$ is defined in terms of two limits,

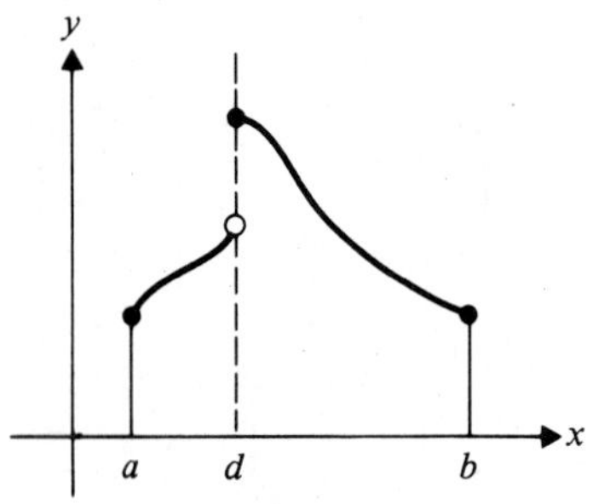

Figure 7.98

$$\int_a^b f(x)\,dx = \lim_{c\to d^-} \int_a^c f(x)\,dx + \lim_{c\to d^+} \int_c^b f(x)\,dx,$$

but there is no question in this case that the improper integral converges.

EXERCISES 7.9

In each of Exercises 1–18, determine whether the improper integral converges or diverges. Find the value for each convergent integral.

1. $\displaystyle\int_3^{\infty} \frac{1}{(x+4)^2}\,dx$

2. $\displaystyle\int_3^{\infty} \frac{1}{(x+4)^{1/3}}\,dx$

3. $\displaystyle\int_{-\infty}^{-4} \frac{x}{\sqrt{x^2-2}}\,dx$

4. $\displaystyle\int_{-\infty}^{-4} \frac{x}{(x^2-2)^4}\,dx$

5. $\displaystyle\int_{-\infty}^{\infty} \frac{10^{10}x^3}{(x^4+5)^2}\,dx$

6. $\displaystyle\int_{-\infty}^{\infty} \frac{x^3}{(x^4+5)^{1/4}}\,dx$

7. $\displaystyle\int_{0}^{1} \frac{1}{(1-x)^{5/3}}\,dx$

8. $\displaystyle\int_{0}^{1} \frac{1}{\sqrt{1-x}}\,dx$

9. $\displaystyle\int_{1}^{\infty} x\sqrt{x^2-1}\,dx$

10. $\displaystyle\int_{2}^{5} \frac{x}{\sqrt{x^2-4}}\,dx$

11. $\displaystyle\int_{-1}^{1} \frac{x}{(1-x^2)^2}\,dx$

12. $\displaystyle\int_{-\infty}^{\infty} \frac{1}{x^2}\,dx$

13. $\displaystyle\int_{0}^{\infty} \frac{1}{\sqrt{x}}\,dx$

14. $\displaystyle\int_{-\infty}^{\pi/2} \frac{x}{(x^2-4)^2}\,dx$

15. $\displaystyle\int_{4}^{\infty} \cos x\,dx$

16. $\displaystyle\int_{-\infty}^{\infty} \sin x\,dx$

17. $\displaystyle\int_{0}^{\infty} \frac{x}{\sqrt{x+3}}\,dx$

18. $\displaystyle\int_{2}^{3} \frac{x^3}{\sqrt{x^2-4}}\,dx$

19. If $f(x)$ is a continuous odd function, is it necessarily true that

$$\int_{-\infty}^{\infty} f(x)\,dx = 0?$$

20. a) Is it possible to assign a number to represent the area bounded by the curves

$$y = 1 - x^{-1/4}, \qquad y = 1, \qquad x = 1?$$

b) If the region in (a) is rotated about the line $y = 1$, is it possible to assign a number to represent its volume?

21. Repeat Exercise 20 if $y = 1 - x^{-1/4}$ is replaced by $y = 1 - x^{-2/3}$.

22. Repeat Exercise 20 if $y = 1 - x^{-1/4}$ is replaced by $y = 1 - x^{-3}$.

23. A function $f(x)$ qualifies as a *probability density function* (pdf) on the interval $x \geq 0$ if it satisfies two conditions: $f(x) \geq 0$ for $x \geq 0$, and

$$\int_{0}^{\infty} f(x)\,dx = 1.$$

a) Show that each of the following functions qualifies as a pdf.

i) $f(x) = \dfrac{6x}{(1+3x^2)^2}$ ii) $f(x) = \dfrac{2x}{(1+x)^3}$

b) If a variable x has pdf $f(x)$ defined for $x \geq 0$, then the probability that x will lie in any interval I is the definite integral of $f(x)$ over I. In particular, the probability that x will be greater than or equal to a is

$$P(x \geq a) = \int_{a}^{\infty} f(x)\,dx.$$

Calculate $P(x \geq 3)$ for each pdf in part (a).

24. Verify that

$$\int_{1}^{\infty} \frac{1}{x^p}\,dx$$

converges if $p > 1$ and diverges if $p < 1$. (The case $p = 1$ is considered in Exercise 33.)

25. The force of repulsion between two point charges of like sign, one of size q and the other of size unity, has magnitude

$$F = \frac{q}{4\pi\varepsilon_0 r^2},$$

where ε_0 is a constant and r is the distance between the charges. The potential V at any point P due to charge q is defined as the work required to bring the unit charge to P from infinity along the straight line joining q and P. Find a formula for V.

26. Verify that if $f(x)$ is continuous on $a \leq x \leq b$ except for a finite discontinuity at d (Fig. 7.98), then the improper integral of $f(x)$ from $x = a$ to $x = b$ must converge.

27. Find the length of the loop of the curve $9y^2 = x(3-x)^2$.

28. Is definition (7.37) equivalent to the following equation?

$$\int_{-\infty}^{\infty} f(x)\,dx = \lim_{a\to\infty} \int_{-a}^{a} f(x)\,dx$$

Determine whether the improper integral in each of Exercises 29–32 converges or diverges. (*Hint:* Compare each to a known convergent or divergent integral.)

29. $\displaystyle\int_{2}^{\infty} \frac{x^2}{\sqrt{x^2-1}}\,dx$

30. $\displaystyle\int_{1}^{3} \frac{x^3}{(27-x^3)^2}\,dx$

31. $\displaystyle\int_{0}^{1} \frac{x^2}{\sqrt{1-x^2}}\,dx$

32. $\displaystyle\int_{-\infty}^{-2} \frac{\sqrt{-x}}{(x^2+5)^2}\,dx$

* 33. To show that the improper integral

$$\int_{1}^{\infty} \frac{1}{x}\,dx$$

diverges, draw rectangles as shown in Fig. 7.99. For any integer $b \geq 2$, it is clear that

$$\int_{1}^{b} \frac{1}{x}\,dx > \frac{1}{2} + \frac{1}{3} + \frac{1}{4} + \cdots + \frac{1}{b}.$$

Verify that if $b = 2^n$ ($n \geq 1$ an integer),

$$\frac{1}{2} + \frac{1}{3} + \frac{1}{4} + \cdots + \frac{1}{b} \geq \frac{n}{2};$$

hence

$$\int_{1}^{2^n} \frac{1}{x}\,dx > \frac{n}{2}.$$

Now take the limit as $n \to \infty$ to complete the proof.

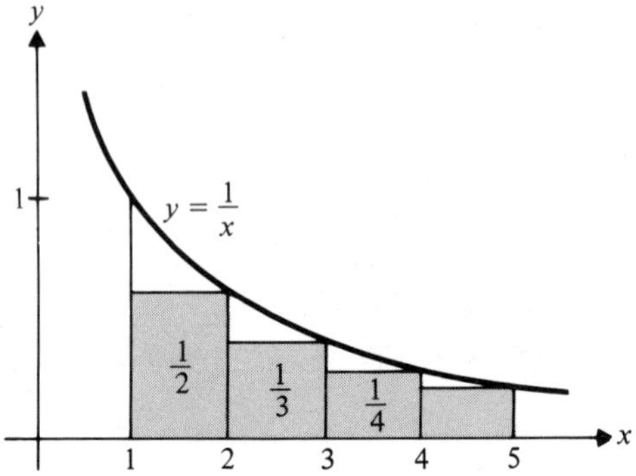

Figure 7.99

* **34.** One of the most important functions in physics and engineering is the *Dirac delta function* $\delta_a(x)$ introduced in Exercise 23 of Section 2.2 as the limit of the unit pulse function (Fig. 7.100):

$$\delta_a(x) = \lim_{\varepsilon \to 0} P_{a,\varepsilon}(x).$$

For a continuous function $f(x)$,

$$\int_{-\infty}^{\infty} f(x)\delta_a(x)\,dx = \int_{-\infty}^{\infty} f(x)\left\{\lim_{\varepsilon \to 0} P_{a,\varepsilon}(x)\right\} dx.$$

Assuming that the order of integration and the process of taking the limit as ε approaches zero can be interchanged, that is,

$$\int_{-\infty}^{\infty} f(x)\left\{\lim_{\varepsilon \to 0} P_{a,\varepsilon}(x)\right\} dx = \lim_{\varepsilon \to 0} \int_{-\infty}^{\infty} f(x)P_{a,\varepsilon}(x)\,dx,$$

show that

$$\int_{-\infty}^{\infty} f(x)\delta_a(x)\,dx = f(a).$$

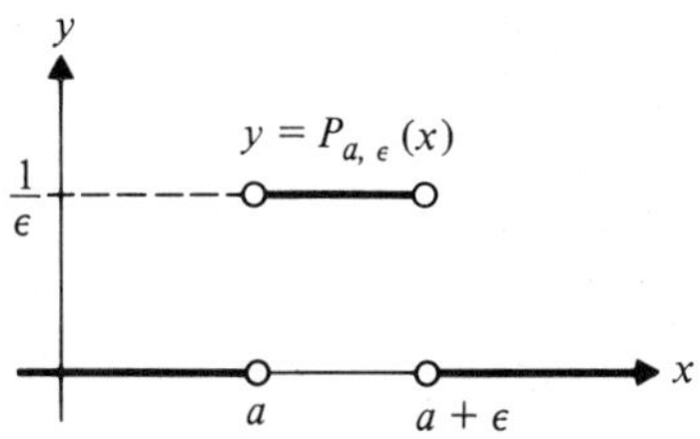

Figure 7.100

SUMMARY

In Chapter 6 we defined the definite integral of a function $f(x)$ with respect to x from $x = a$ to $x = b$ as a number, the number obtained by evaluating the limit of the sum

$$\int_a^b f(x)\,dx = \lim_{n \to \infty} \sum_{i=1}^{n} f(x_i^*)\,\Delta x_i. \tag{6.11}$$

In this chapter we have seen the advantage of using this abstract definition for the definite integral. We used limits of sums of the form in (6.11) to define area, volume of a solid of revolution, length along a curve, and work, and because of (6.11), we were able to interpret each sum as the definite integral of a function, the function obtained by replacing all x_i^*'s (or x_i's) with x's. Consequently, we can calculate area, volume of a solid of revolution, length along a curve, and work using definite integrals.

In order to avoid memorization of these definite integrals as formulas, we illustrated how differentials and representative elements could be used to summarize intuitively each derivation. In each case, however, the heuristic approach was based on the mathematical discussion leading to the definite integral; i.e., the mathematical derivation came first, and then we merely adjusted an intuitive scheme to that derivation. In the other applications (centers of mass, moments of inertia, fluid pressure, etc.), we omitted the mathematical discussion and proceeded directly to its intuitive representation in terms of representative elements. It is important to realize, however, that we could have given rigorous mathematical discussions that would have

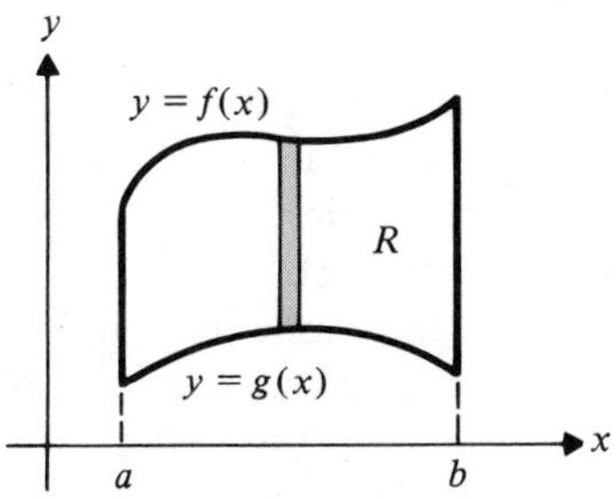

Figure 7.101

verified the use of definite integrals in each of these applications. Note also that should we use an intuitive discussion in terms of representative elements to set up a definite integral in an application, and should we be unsure of our derivation, then the only way to verify the appropriateness of the definite integral is to return to first principles; i.e., set up an approximating sum, take its limit, and determine the definite integral represented by this limit. Then, and only then, can we be sure that we have the proper definite integral for that application.

To summarize the use of representative elements in various applications, consider the region R shown in Fig. 7.101. To find the area of R we draw at position x a rectangle of length $f(x) - g(x)$ and width dx. The areas $[f(x) - g(x)]\,dx$ of all such rectangles are then added together to find the total area:

$$\text{Area} = \int_a^b [f(x) - g(x)]\,dx.$$

If R is rotated around the x-axis to form a solid of revolution, the rectangle generates a washer with volume $\{\pi[f(x)]^2 - \pi[g(x)]^2\}\,dx$, and therefore the total volume is

$$\text{Volume} = \int_a^b \{\pi[f(x)]^2 - \pi[g(x)]^2\}\,dx.$$

If R is rotated about the y-axis, the rectangle traces out a cylindrical shell of volume $2\pi x[f(x) - g(x)]\,dx$, and the total volume is

$$\text{Volume} = \int_a^b 2\pi x[f(x) - g(x)]\,dx.$$

The first moments of this rectangle about the x- and y-axes are

$$\tfrac{1}{2}[f(x) + g(x)][f(x) - g(x)]\,dx$$

and

$$x[f(x) - g(x)]\,dx,$$

and these lead to the definite integrals

$$\int_a^b \tfrac{1}{2}\{[f(x)]^2 - [g(x)]^2\}\,dx \quad \text{and} \quad \int_a^b x[f(x) - g(x)]\,dx$$

for the first moments of area of R about the x- and y-axes. The second moments of area of R about the x- and y-axes are

$$\int_a^b \tfrac{1}{3}\{[f(x)]^3 - [g(x)]^3\}\,dx \quad \text{and} \quad \int_a^b x^2[f(x) - g(x)]\,dx,$$

where in the first integral it was necessary to use formula (7.34) for the second moment of the rectangle.

The key, then, to the use of the definite integral is to divide the region into smaller elements, calculate the required quantity for a representative element, and then use a definite integral to add over all elements.

An important point to keep in mind is that whenever we require a length in the y-direction we always take "upper y minus lower y," and when we need a length in the x-direction, we use "larger x minus smaller x." It is surprising how many sign errors we can avoid if we keep this in mind when calculating *lengths*. For instance, note that this rule is always used to find the length of a rectangle (in areas, volumes, fluid pressure), the radius of a circle (in volumes, surface area), depth (fluid pressure), and some distances (work).

An improper integral is a definite integral that has an infinite limit and/or a point of discontinuity somewhere in the interval of integration. The value of an improper integral is defined by first calculating it over a finite interval in which the integrand is continuous, and then letting the limit on the integral approach either infinity or the point of discontinuity.

REVIEW EXERCISES

In each of Exercises 1–5, calculate for the region bounded by the given curves (a) its area; (b) the volumes of the solids of revolution obtained by rotating the region about the x- and y-axes; (c) its centroid; (d) its second moments of area about the x- and y-axes.

1. $y = 9 - x^2$, $y = 0$

2. $y = x^3$, $y = 2 - x$, $x = 0$

3. $x = y^2 - 1$, $x = 4 - 4y^2$

4. $2y = x$, $y + 1 = x$, $y = 0$

5. $2x + y = 2$, $y - 2x = 2$, $x = y + 1$, $y + x + 1 = 0$

In each of Exercises 6–10, find the volume of the solid of revolution obtained by rotating the area bounded by the curves about the line.

6. $y = |x| + 2$, $y = 3$, about $y = 1$

7. $x = y^4$, $x = 4$, about $x = 4$

8. $y = x^3 - x$, $y = 0$, about $y = 0$

9. $yx^2 = 1$, $x = 1$, $x = 2$, $y = 0$, about $y = -1$

10. $y = x + 2$, $y = \sqrt{4 - 2x}$, $y = 0$, about $y = 1$

In each of Exercises 11–15, find the first and second moments of area of the region bounded by the curves about the line.

11. $y = 4x^2$, $y = 2 + 2x^2$, about $y = -1$

12. $2x = \sqrt{y}$, $\sqrt{2x} = \sqrt{y - 2}$, $x = 0$, about $x = 1$

13. $3y = 2x + 6$, $2y + x = 4$, $y = 0$, about $y = -2$

14. $3y = 2x + 6$, $2y + x = 4$, $y = 0$, about $y = 2$

15. $y = x^2 - 2x$ $(y \le 0)$, $y = 2x$, $y = 4 - 2x$, about $x = 1$

16. A water tank in the form of a right-circular cylinder with radius 1 meter and height 3 meters has its axis vertical. If it is half full, how much work is required to empty the tank through an outlet at the top of the tank?

17. A car runs off a bridge into a river and submerges. If the tops of its front and rear side windows are 2 meters below the surface of the water (Fig. 7.102), what is the force due to water pressure on each window?

18. Find the surface area of the volume of the solid of revolution obtained by rotating the area bounded by the curves

$$2x + y = 2, \qquad y - 2x = 2, \qquad y = 0$$

about the x-axis.

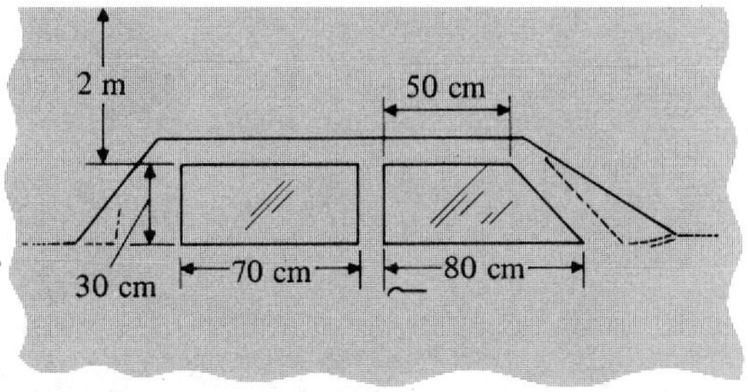

Figure 7.102

19. A tank in the form of an inverted right-circular cone of depth 10 meters and radius 4 meters is full of water. How much work is required to lower the water level in the tank by 2 meters by pumping water out through an outlet 1 meter above the top of the tank?

20. Show that if it takes W units of work to stretch a spring from equilibrium to a certain stretch, then it requires $3W$ units more work to double that stretch.

21. Newton's universal law of gravitation states that the force of attraction between two point masses m and M has magnitude

$$F = \frac{GmM}{r^2},$$

where r is the distance between the masses and G is a constant (6.67×10^{-11} Nm^2/kg^2).

a) If M represents the mass of the earth and we regard it as a point mass concentrated at its center (assume for the calculation of M that the earth is a sphere with radius 6370 kilometers and mean density 5.52×10^3 kilograms per cubic meter), show that Newton's universal law of gravitation at the earth's surface reduces to $F = mg$, where $g = 9.82$ meters per second per second.

b) Use the original law $F = GmM/r^2$ with the earth regarded as a point mass to calculate the work required to lift a mass of 10 kilograms from the earth's surface to a height of 10 kilometers.

c) Calculate the work in (b) using the constant gravitational force $F = mg$ in (a). Is there a significant difference?

In Exercises 22–29, determine whether the given improper integral converges or diverges. Evaluate each convergent integral.

22. $\displaystyle\int_1^{\infty} \frac{1}{x^{3/2}}\,dx$

23. $\displaystyle\int_0^{3} \frac{1}{\sqrt{3-x}}\,dx$

24. $\displaystyle\int_{-1}^{0} \frac{1}{(x+1)^2}\,dx$

25. $\displaystyle\int_{-2}^{2} \frac{x}{\sqrt{4-x^2}}\,dx$

26. $\displaystyle\int_{-\infty}^{\infty} \frac{1}{(x+3)^3}\,dx$

27. $\displaystyle\int_{-\infty}^{-3} \frac{1}{\sqrt{-x}}\,dx$

28. $\displaystyle\int_{-6}^{\infty} x\sqrt{x^2+4}\,dx$

29. $\displaystyle\int_{-\infty}^{\infty} \frac{x}{(x^2-1)^2}\,dx$

30. A uniformly tapered rod of length 1 meter has circular cross-sections, the radii of its ends being 1 centimeter and 2 centimeters. Find the volume of the rod by:

a) using the fact that every cross-section is circular;

b) considering the rod as a volume of a solid of revolution.

31. In Exercise 16, how much work is required to empty the tank if its axis is horizontal and the outlet is at the top of the tank?

32. The lower half of a cubical tank 2 meters on each side is occupied by water, and the upper half by oil (density 0.90 gram per cubic centimeter).

    a) What is the force on each side of the tank due to the pressure of the water and the oil?

    b) If the oil and water are stirred to create a uniform mixture, does the force on each side change? If not, explain why not. If so, by how much does it increase or decrease?

Transcendental Functions and Their Derivatives

8

In Chapter 3 we learned how to differentiate functions that involve a finite number of additions, subtractions, multiplications, divisions, and roots. Such functions fall into a wider class of functions called algebraic functions. A function $y = f(x)$ is said to be *algebraic* if it satisfies identically an equation of the form

$$P_0(x)y^n + P_1(x)y^{n-1} + \cdots + P_{n-1}(x)y + P_n(x) = 0, \tag{8.1}$$

where $P_0(x), \ldots, P_n(x)$ are polynomials in x, and n is a positive integer. Polynomials and rational functions are algebraic (see Eqs. 2.7 and 2.8). The function $f(x) = x^{1/3}$ is algebraic since it satisfies $y^3 - x = 0$. The equation $y^3 + y = x$ defines y implicitly as a function of x. We cannot find an explicit form for the function, but it is an algebraic function.

A function that is not algebraic is called *transcendental.* In this chapter we consider various transcendental functions, including the trigonometric, inverse trigonometric, exponential, logarithm, and hyperbolic functions.

8.1 THE TRIGONOMETRIC FUNCTIONS

In Section 3.8 we gave a brief review of trigonometric functions, stressing those aspects important to the study of calculus. Graphs of the trigonometric functions are shown in Fig. 3.27, where once again we point out that in calculus all angles are measured in radians.

Many functions can be built from the elementary trigonometric functions, as illustrated in the following examples.

Example 1 Sketch a graph of the function $f(x) = 4 \sin (3x + \pi/4)$.

Solution This function is an example of a general sine function (see Eq. 3.34). The coefficient 4 before the sine function indicates that the amplitude is 4; the coefficient 3 before x indicates that the period is $2\pi/3$; and the 3 and $\pi/4$ combine to show a shift of $\pi/12$ to the left. The result is shown in Fig. 8.1. □

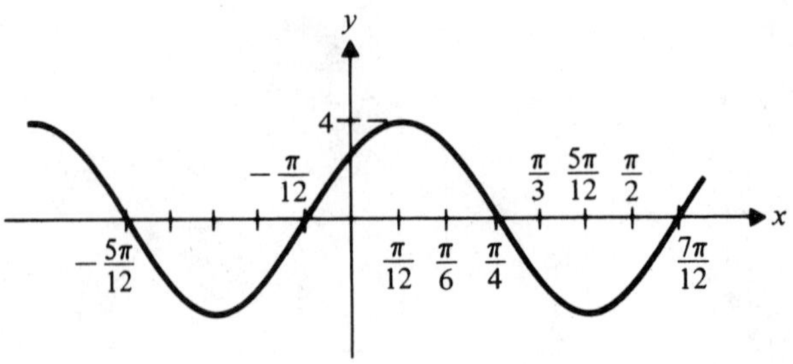

Figure 8.1

Example 2 Sketch a graph of each of the following functions:

a) $f(x) = x \sin x$;

b) $f(x) = \sqrt{1 + \csc x}$;

c) $f(x) = \sin 3x - 2 \cos 3x$.

Solution a) By multiplying the ordinates of the curves $y = x$ and $y = \sin x$ shown in Fig. 8.2(a), we obtain the curve $y = x \sin x$ in Fig. 8.2(b). Note that $y = x \sin x$ touches $y = x$ when $\sin x = 1$, and $y = -x$ when $\sin x = -1$.

b) We first sketch $y = 1 + \csc x$ by adding 1 to each ordinate of the graph in Fig. 3.27(e) and the result is shown in Fig. 8.3(a). The required graph can now be obtained by taking the square root of each nonnegative ordinate (Fig. 8.3b). Note that $y = 0$ when $x = -\pi/2 + 2n\pi = (4n - 1)\pi/2$, n an integer.

c) To sketch $y = \sin 3x - 2\cos 3x$, we could sketch $y = \sin 3x$ and $y = -2\cos 3x$ and then add the ordinates. The result would certainly be

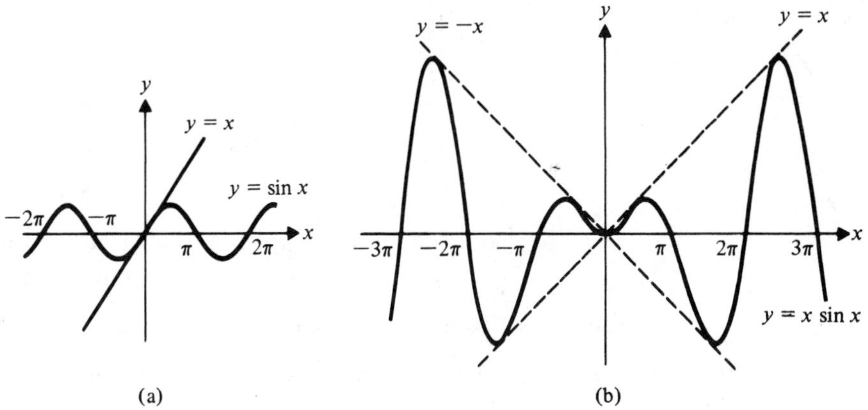

Figure 8.2

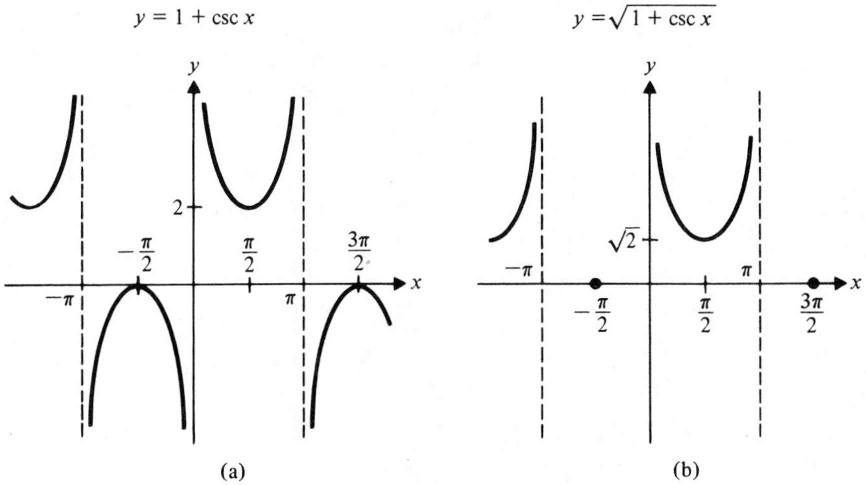

Figure 8.3

"sketchy," however. A better way is to take the approach discussed in Example 23 of Chapter 3 and find an R and an α such that

$$R \sin (3x + \alpha) = y = \sin 3x - 2 \cos 3x.$$

If we expand $\sin (3x + \alpha)$ by means of identity (3.23a), we get

$$R \sin 3x \cos \alpha + R \cos 3x \sin \alpha = \sin 3x - 2 \cos 3x.$$

Substitution of $x = \pi/6$ and $x = 0$ gives us two equations for R and α:

$$R \cos \alpha = 1 \quad \text{and} \quad R \sin \alpha = -2.$$

If we square and add these equations, we get

$$5 = R^2 \cos^2 \alpha + R^2 \sin^2 \alpha = R^2,$$

from which we have

$$R = \pm\sqrt{5}.$$

If we choose $R = \sqrt{5}$, then

$$\cos \alpha = \frac{1}{\sqrt{5}} \quad \text{and} \quad \sin \alpha = \frac{-2}{\sqrt{5}},$$

and these equations imply that $\alpha = -1.11$. Consequently, we may write

$$y = \sqrt{5} \sin (3x - 1.11),$$

the graph of which is shown in Fig. 8.4. (The same graph is obtained if we choose $R = -\sqrt{5}$, except in this case $\alpha = 2.03$.) □

Example 3 Sketch the curve defined by the equation $\cos (x + y) = 1$.

Solution Since the only angles that have a cosine equal to 1 are integer multiples of 2π, we may write equivalently that

$$x + y = 2n\pi, \qquad n = \text{an integer}.$$

This equation represents the family of straight lines all with slope -1, shown in Fig. 8.5. In other words, the original equation defines a family of curves rather than just one curve. □

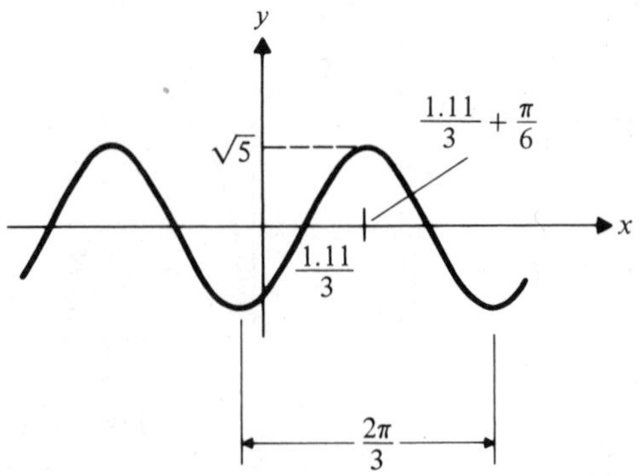

Figure 8.4

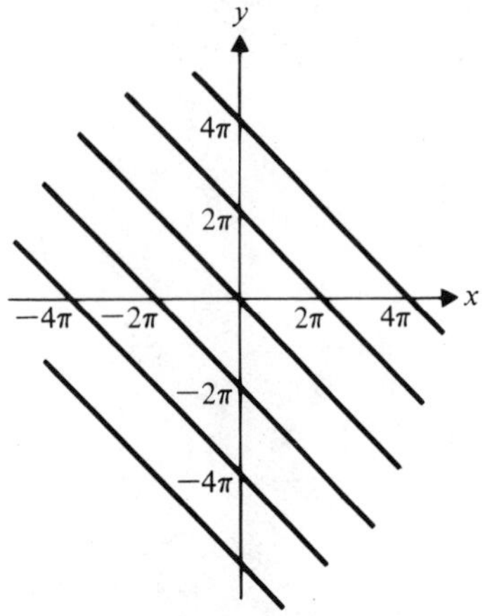

Figure 8.5

EXERCISES 8.1

Sketch the given curve in Exercises 1–10.

1. $y = 2\cos(2\pi x + 3\pi/4)$
2. $y = 4 - \tan^2 x$
3. $y = \cos^2 x - \cos x$
4. $y = |\sin x|$
5. $y = \sin|x|$
6. $|y| = |\sin x|$
7. $|y| = \sin|x|$
8. $x = \cos^2 y - \sin^2 y$
9. $y^2 = \cos x$
10. $y = 4\sin x - 3\cos x$

11. Sketch the curve defined by the equation $\sin(x - y) = -1$.

Sketch a graph of the given function in Exercises 12–17.

12. $f(x) = \sin^4 x + \cos^4 x$

13. $f(x) = \sqrt{3 - 4\sin^2 x}$

14. $f(x) = \sqrt{1 - \cos^2 x} - \sin x$

15. $f(x) = \sqrt{1 - 2\sin^2 x}$

16. $f(x) = \sin\left(\frac{1}{x}\right),\ |x| \le \frac{1}{\pi}$

17. $f(x) = x\sin\left(\frac{1}{x}\right),\ |x| \le \frac{1}{\pi}$

18. Find a single equation that represents the family of straight lines given by

$$3x + 4y = (2n + 1)\pi, \qquad n = \text{an integer.}$$

Sketch a graph of the given function in Exercises 19–20.

19. $f(\theta) = A\csc\theta + B\sec\theta,\ 0 < \theta < \pi/2$, where A and B are positive constants

20. $f(\theta) = A\csc\theta - B\cot\theta, 0 < \theta \le \pi/2$, where $A > B > 0$

21. If the curve shown in Fig. 8.6 is a general sine curve, find its equation.

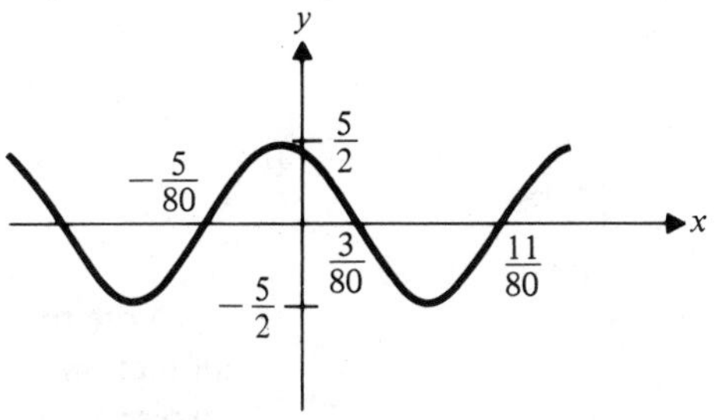

Figure 8.6

8.2 DERIVATIVES OF THE TRIGONOMETRIC FUNCTIONS

In Section 3.9 we proved that if $u(x)$ is a differentiable function of x, then

$$\frac{d}{dx}\sin u = \cos u \frac{du}{dx} \quad \text{and} \quad \frac{d}{dx}\cos u = -\sin u \frac{du}{dx}.$$

With these formulas, it is relatively easy to find the derivatives of the remaining trigonometric functions. For the tangent function, we use the chain rule and the quotient rule to calculate

$$\frac{d}{dx}\tan u = \frac{d}{du}\left(\frac{\sin u}{\cos u}\right)\frac{du}{dx} = \frac{\cos u(\cos u) - \sin u(-\sin u)}{\cos^2 u}\frac{du}{dx} = \frac{1}{\cos^2 u}\frac{du}{dx}.$$

Consequently,

$$\frac{d}{dx}\tan u = \sec^2 u \frac{du}{dx}. \tag{8.2}$$

A similar calculation gives us

$$\frac{d}{dx}\cot u = -\csc^2 u \frac{du}{dx}. \tag{8.3}$$

For the secant function, we obtain

$$\frac{d}{dx}\sec u = \frac{d}{du}\left(\frac{1}{\cos u}\right)\frac{du}{dx} = (-1)(\cos u)^{-2}(-\sin u)\frac{du}{dx} = \frac{1}{\cos u}\frac{\sin u}{\cos u}\frac{du}{dx},$$

or

$$\boxed{\frac{d}{dx}\sec u = \sec u \tan u \frac{du}{dx}.} \tag{8.4}$$

Similarly,

$$\boxed{\frac{d}{dx}\csc u = -\csc u \cot u \frac{du}{dx}.} \tag{8.5}$$

Note that we can easily find the derivative of every *cofunction* (cosine is the cofunction of sine, cotangent of tangent, and cosecant of secant) from the corresponding derivative for the function: We introduce a negative sign and replace the function by its cofunction. In other words, this procedure, along with derivatives of the sine, tangent, and secant, allows us to determine derivatives of the cosine, cotangent, and cosecant.

Example 4 Find dy/dx if y is defined as a function of x in each of the following:

a) $y = \sin 2x$;
b) $y = 4 \sec (2x^3 + 5)$;
c) $y = \tan^2 4x$;
d) $y = 2 \csc (3x^2) + 5x \sin x$;
e) $x^2 \tan y + y \sin x = 5$.

Solution a) $\dfrac{dy}{dx} = (\cos 2x)\dfrac{d}{dx}(2x) = 2 \cos 2x.$

b) $$\frac{dy}{dx} = 4 \sec (2x^3 + 5) \tan (2x^3 + 5)\frac{d}{dx}(2x^3 + 5)$$
$$= 24x^2 \sec (2x^3 + 5) \tan (2x^3 + 5).$$

c) $\dfrac{dy}{dx} = 2 \tan 4x \dfrac{d}{dx}\tan 4x = 2 \tan 4x\, (4 \sec^2 4x) = 8 \tan 4x \sec^2 4x.$

d) $$\frac{dy}{dx} = -2 \csc (3x^2) \cot (3x^2)\frac{d}{dx}(3x^2) + 5 \sin x + 5x \cos x.$$
$$= -12x \csc (3x^2) \cot (3x^2) + 5 \sin x + 5x \cos x.$$

e) If we differentiate the equation (implicitly) with respect to x, we get

$$2x \tan y + x^2 \sec^2 y\frac{dy}{dx} + \frac{dy}{dx}\sin x + y \cos x = 0;$$

therefore,

$$\frac{dy}{dx} = -\frac{2x \tan y + y \cos x}{x^2 \sec^2 y + \sin x}. \square$$

Example 5 A revolving light 2 kilometers from a straight shoreline makes 5 revolutions per minute. Find the speed of the beam of light along the shoreline at the instant when it makes an angle of $\pi/3$ with the shoreline.

Solution From Fig. 8.7, we see that $x = 2 \tan \theta$. If we differentiate this equation with respect to time t, we obtain

$$\frac{dx}{dt} = 2 \sec^2 \theta \frac{d\theta}{dt}.$$

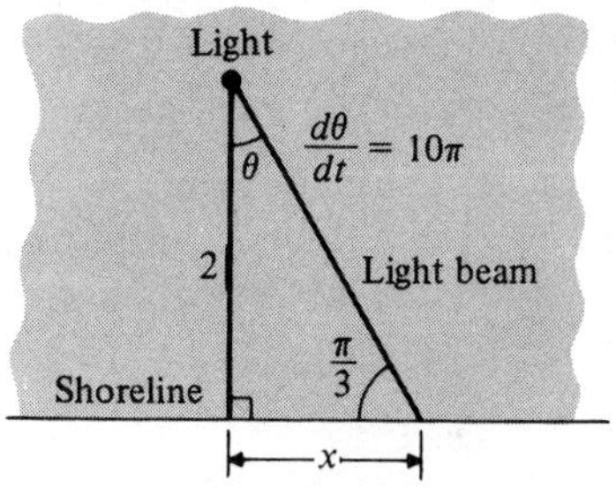

Figure 8.7

Now, $d\theta/dt = 10\pi$ radians per minute, and $\theta = \pi/6$ at the instant in question, so that at this instant,

$$\frac{dx}{dt} = 2(\sec^2 (\pi/6))(10\pi) = 2(2/\sqrt{3})^2(10\pi) = \frac{80\pi}{3}.$$

The speed of the beam is therefore $80\pi/3$ kilometers per minute along the shoreline. $\square$

Example 6 Show that the relative maxima and minima of the function $f(x) = x \sin x$ $(x > 0)$ occur just to the right of the points of intersection of the curves $y = \pm x$ and $y = x \sin x$. (In Example 2 we sketched $y = x \sin x$ by multiplying the ordinates of the two curves $y = x$ and $y = \sin x$.)

Solution For critical points of $f(x)$, we solve

$$0 = f'(x) = \sin x + x \cos x.$$

Solutions of this equation are impossible to find analytically, but we can illustrate with not too much difficulty that they exist. Solving this equation is equivalent to solving the equation

$$-x = \tan x.$$

We illustrate these solutions geometrically as the x-coordinates of the points of intersection of the curves $y = -x$ and $y = \tan x$. Sketches of these curves in Fig. 8.8 indicate that the points of intersection lie just to the right of the line $x = (2n + 1)\pi/2$, n a nonnegative integer.

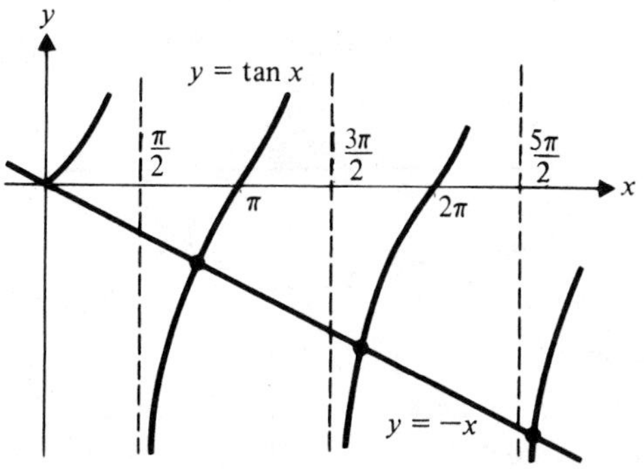

Figure 8.8

Since the curves $y = x$ and $y = x \sin x$ intersect when

$$x = x \sin x \quad \text{or} \quad \sin x = 1,$$

and this implies that

$$x = \frac{(2n + 1)}{2}\pi,$$

we have proved the conjecture. □

Example 7 Two corridors, one 3 meters wide and the other 6 meters wide, meet at right angles. Find the length of the longest beam that can be transported horizontally around the corner. (Ignore the thickness of the beam.)

Solution In Fig. 8.9 we show a number of straight lines that touch the walls of the two corridors and pass through corner C. The longest beam that can be transported horizontally around C is represented by the shortest of all such lines. If L is the length of one of these lines AB, then

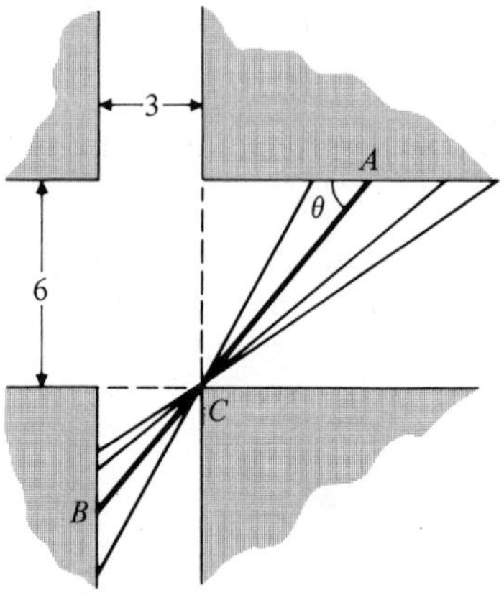

Figure 8.9

$$L = |AC| + |BC| = 6 \csc \theta + 3 \sec \theta.$$

Thus to find the longest beam that can be moved around the corner, we must find the absolute minimum for the function

$$L(\theta) = 6 \csc \theta + 3 \sec \theta, \qquad 0 < \theta < \frac{\pi}{2}.$$

For critical points of $L(\theta)$, we solve

$$0 = L'(\theta) = -6 \csc \theta \cot \theta + 3 \sec \theta \tan \theta,$$

from which we get

$$\frac{6 \cos \theta}{\sin^2 \theta} = \frac{3 \sin \theta}{\cos^2 \theta}.$$

This equation implies that $\tan^3 \theta = 2$, or $\tan \theta = 2^{1/3}$. The acute angle with tangent equal to $2^{1/3}$ is $\theta = 0.90$ radian.

Geometrically (and algebraically), it is clear that

$$\lim_{\theta \to 0^+} L(\theta) = \infty, \qquad \lim_{\theta \to \pi/2^-} L(\theta) = \infty, \quad \text{and} \quad L(0.90) > 0.$$

It follows that $\theta = 0.90$ must yield the minimum value of $L(\theta)$; therefore, the longest beam that can be transported around the corner is

$$L(0.90) = 6 \csc (0.90) + 3 \sec (0.90) = 12.5 \text{ meters.}$$ □

EXERCISES 8.2

In Exercises 1–20, y is defined as a function of x. Find dy/dx in as simplified a form as possible.

1. $y = \sec^2 x$
2. $y = 12 \sin (3x + 5)$
3. $y = \sqrt{1 + \tan x}$
4. $y = x \csc^3 x$
5. $y = \sec^2 (2x) + 3 \cos x$
6. $y = \sin (2x) \cos (2x)$
7. $y = u^3 \sec u$, $u = x \tan (x + 1)$
8. $y = \sqrt{3 - \sec v}$, $v = \tan \sqrt{x}$
9. $y = \sin (\sin x)$
10. $y = \dfrac{\sec^3 (3x)}{\csc^2 (2x)}$
11. $\tan (x + y) = y$
12. $x + \sec (xy) = 5$
13. $y = (1 + \sec^2 x)^{1/3}$

14. $y = \sin^2 t \sec^2 t,$
$t = x \tan x$

15. $y = x^3 - x^2 \cos x + 2x \sin x + 2 \cos x$

16. $y = \cos(\tan x)$

17. $y = \sin^4(x^2) - \cos^4(x^2)$

18. $x^3 y + \tan^2 y = 3x$

19. $x = y^3 \csc^3 y$

20. $y = \dfrac{1 + \tan^3(3x^2 - 4)}{x^2 \sin x}$

21. Find d^2y/dx^2 if y is defined implicitly as a function of x by

$$\sin y = 2x^2 + 3y.$$

22. A missile rises vertically from a point on the ground 1000 meters from an observer at 4000 meters per minute. How fast is the angle of elevation of the missile from the observer changing when the missile is 2000 meters high?

23. The differential equation describing the vertical oscillations of a mass m on the end of a spring (Fig. 8.10) is

$$m\frac{d^2y}{dt^2} + ky = 0,$$

where $k > 0$ is the spring constant. Show that a solution is

$$y = f(t) = A \sin\left(\sqrt{\frac{k}{m}}\,t\right) + B \cos\left(\sqrt{\frac{k}{m}}\,t\right)$$

for any constants A and B.

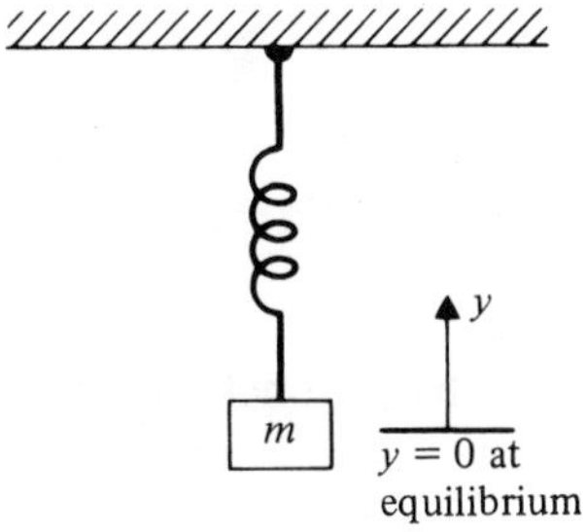

Figure 8.10

24. a) A movie camera in a plane is focused on a stationary herd of caribou. The plane flies 100 meters above the ground and heads directly over the herd at a rate of 200 kilometers per hour. At what rate is the camera rotating when the plane is 100 meters from the herd?

b) Repeat part (a) if the herd stampedes at a rate of 60 kilometers per hour in the same direction in which the plane is flying.

C **25.** Use Newton's iterative procedure to find the smallest positive root of the equation $\tan x = x$.

26. Find d^2y/dx^2 if y is defined implicitly as a function of x by

$$\tan y = x + xy.$$

27. Find all critical points of the function

$$f(x) = \cos(x + 1/x).$$

28. A bee's cell is always constructed in the shape of a regular hexagonal cylinder open on one end and a trihedral apex at the other (Fig. 8.11). It can be shown that the total area of the nine faces is given by

$$A = f(\theta) = 6xy + \frac{3}{2}x^2\left\{-\cot\theta + \frac{\sqrt{3}}{\sin\theta}\right\}.$$

Find the angle that minimizes A.

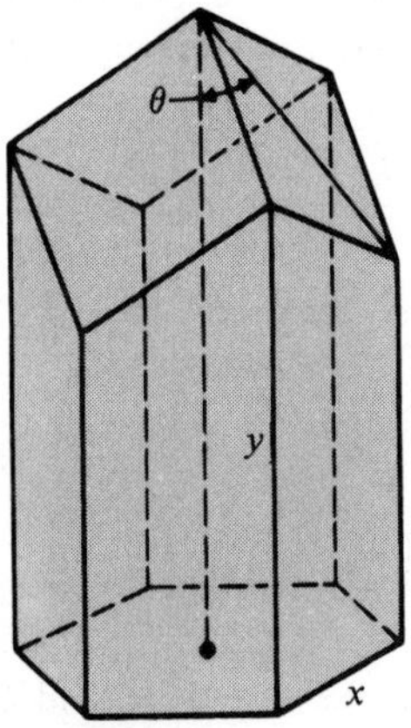

Figure 8.11

29. Find all relative extrema for $f(x) = (\tan x)/x$, and sketch a graph of the function.

* **30.** Show that the function

$$f(x) = \begin{cases} x^2 \sin(1/x), & x \neq 0, \\ 0, & x = 0, \end{cases}$$

has a derivative at $x = 0$, but $f'(x)$ is not continuous at $x = 0$.

* **31.** A submarine is sailing on the surface due east at a rate of s kilometers per hour. It is to pass 1 kilometer north of a point of land on an island at midnight. Soldiers on the island wishing to escape the enemy plan to intercept the submarine by rowing a rubber raft in a straight-line course at a rate of v kilometers per hour ($v < s$). What is the last possible instant that they can leave the island and expect to make contact with the submarine?

* **32.** Any cross-section of the reflector in a car headlight is in the form of a parabola

$$y^2 = 4x, \qquad 0 \le x \le 2,$$

with the bulb at the focus (1, 0) of the parabola. It is a principle of optics that all light rays from the bulb will be reflected by the mirror so that the angle between the incident ray and the normal to the mirror will be equal to the angle between the reflected ray and the normal. Show that all rays are reflected parallel to the x-axis.

***33.** When blood flows through a vein or artery, it encounters resistance due to friction with the walls of the blood vessel and the viscosity of the blood itself. *Poiseuille's law* for laminar blood flow states that for a circular vessel, resistance R to blood flow is proportional to the length L of the blood vessel and inversely proportional to the fourth power of its radius r:

$$R = k\frac{L}{r^4}, \qquad k = \text{a constant.}$$

Figure 8.12 shows a blood vessel of radius r_1 from A to B and a branching vessel from D to C of radius $r_2 < r_1$. The resistance encountered by the blood in flowing from A to C is given by

$$R = k\frac{L_1}{r_1^4} + k\frac{L_2}{r_2^4},$$

where L_1 and L_2 are the lengths of AD and DC, respectively.

a) If B is assumed to the right of E, show that R can be expressed in terms of θ as

$$R = f(\theta) = \frac{k}{r_1^4}(X - Y\cot\theta) + \frac{k}{r_2^4}Y\csc\theta,$$

where X and Y are the lengths of AE and CE, respectively.

b) Show that $f(\theta)$ has only one critical point $\bar{\theta}$ in the range $0 < \theta < \pi/2$, and $\bar{\theta}$ is defined by

$$\cos\bar{\theta} = \frac{r_2^4}{r_1^4}.$$

c) Verify that $\bar{\theta}$ yields a relative minimum for $f(\theta)$.

d) Show that

$$R(\bar{\theta}) = \frac{kX}{r_1^4} + \frac{kY}{r_2^4}\sqrt{1 - \left(\frac{r_2}{r_1}\right)^8} < f\left(\frac{\pi}{2}\right).$$

e) Does $\bar{\theta}$ provide an absolute minimum for $f(\theta)$?

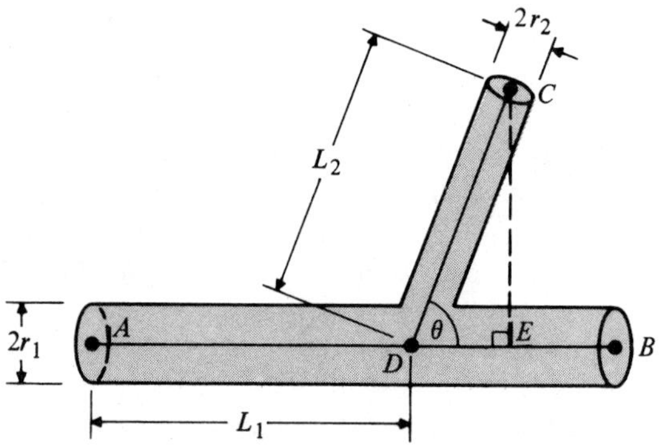

Figure 8.12

8.3 INVERSE FUNCTIONS

We now discuss the concept of inverse functions, a topic that plays a prominent role in the remainder of this chapter.

▷ **Definition 8.1** Given a function $f(x)$, we say that a function $g(x)$ is the *inverse function* of $f(x)$ if

$$x = g[f(x)] \tag{8.6}$$

and the domain of $g(x)$ is the same as the range of $f(x)$. ◁

To explain this definition we take a value of x in the domain of $f(x)$. If this x is substituted into the function, a value $f(x)$ is produced. The function $g(x)$ is the inverse of $f(x)$ if when we substitute the value $f(x)$ into $g(x)$, the result produced is the original x. Thus $g(x)$ is the inverse of $f(x)$ if it reverses the effect of $f(x)$.

For example, consider the function $f(x) = x^2$ $(x \geq 0)$, which squares nonnegative numbers. The operation that reverses this action is the positive square root function; hence the inverse function of $f(x)$ is $g(x) = \sqrt{x}$. Similarly,

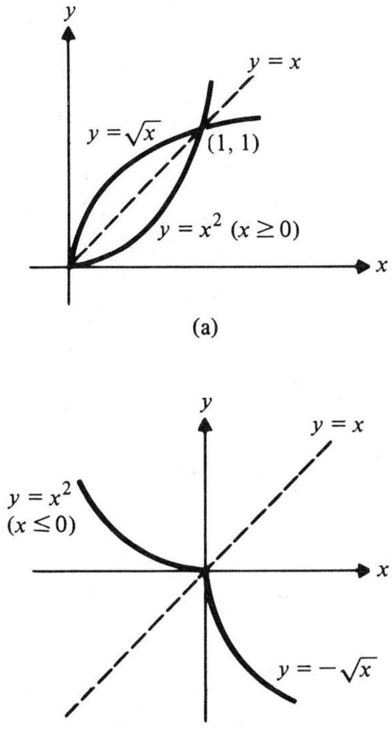

Figure 8.13

the inverse function of $f(x) = x^2$ $(x \leq 0)$ is the negative square root function $g(x) = -\sqrt{x}$, since for any $x \leq 0$

$$g[f(x)] = -\sqrt{x^2} = x.$$

(In particular, if $x = -4$, then $g[f(-4)] = -\sqrt{(-4)^2} = -4$.)

Functions

$$f(x) = x^2 \ (x \leq 0) \quad \text{and} \quad f(x) = x^2 \ (x \geq 0),$$

which are different because they have different domains, have both been shown to have inverse functions. It is natural to ask whether the function $y = x^2$, $-\infty < x < \infty$, which combines these two functions, has an inverse; that is, is there a function that "undoes" it? The answer is no! No function can decide for a given y whether y came from $\sqrt{y}$ or $-\sqrt{y}$.

Suppose we reverse the roles of $f(x)$ and $g(x)$ in this example, and ask whether the square root functions $g(x) = \sqrt{x}$ and $g(x) = -\sqrt{x}$ both defined only for $x \geq 0$ have inverses. It is not too difficult to show that the inverse of $g(x) = \sqrt{x}$ is $y = x^2$, $x \geq 0$; and the inverse of $g(x) = -\sqrt{x}$ is $y = x^2$, $x \leq 0$.

If nothing else, this example should make it clear that domains become very important when we are discussing inverse functions.

Figure 8.13, which shows graphs of the above inverse pairs, demonstrates that the graph of a function and its inverse are closely related. Indeed, it appears that graphs of these inverse pairs are symmetric about the line $y = x$; i.e., one curve is the mirror image of the other in the line $y = x$. This is not a coincidence of this example, but as we now show, graphs of inverse functions are always mirror images of one another in the line $y = x$.

Suppose that $g(x)$ is the inverse function of $f(x)$, and that (x, y) is any point on the graph of $y = f(x)$. Since $g(x)$ is the inverse of $f(x)$, the result of substituting y into $g(x)$ is x; that is, the point (y, x) is on the graph of $g(x)$. But the point (y, x) is the mirror image of (x, y) in the line $y = x$. Combine this with the fact that the domain of $g(x)$ is the range of $f(x)$, and our conjecture is verified.

This discussion suggests a very simple way to determine graphically the inverse of a function $f(x)$: Take its mirror image in the line $y = x$. Note too that if the mirror image does not represent a function, then no inverse function for $f(x)$ exists. For example, the mirror image of $y = x^2$, $-\infty < x < \infty$, in the line $y = x$ is shown in Fig. 8.14 and does not represent a function. We conclude (as before) that $y = x^2$, $-\infty < x < \infty$, does not have an inverse function.

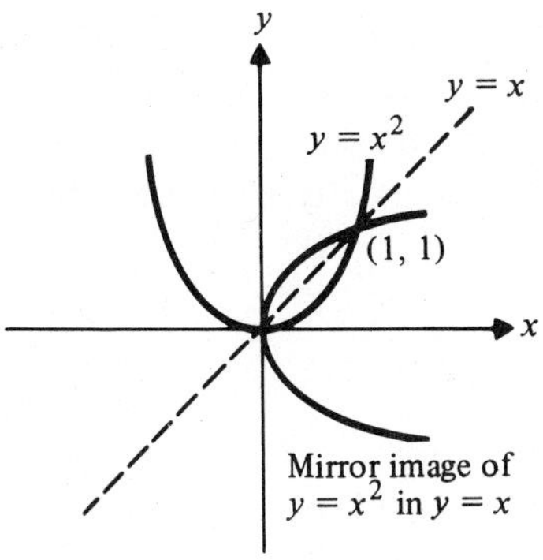

Figure 8.14

When a function $f(x)$ has an inverse, it is customary to adopt the notation $f^{-1}(x)$ to represent the inverse function. For example, when $f(x) = x^2$, $x \geq 0$, we have seen that $f^{-1}(x) = \sqrt{x}$; and when $f(x) = x^2$, $x \leq 0$, the inverse function is $f^{-1}(x) = -\sqrt{x}$. A word of warning about this notation: Do not interpret the "-1" as a power and write $f^{-1}(x) = 1/f(x)$. This is not correct. The notation f^{-1} represents a function, just as tan represents the tangent function and $\sqrt{\ }$ represents the positive square root function.

Example 8 Illustrate the inverse function of $f(x) = (x + 1)/(x - 2)$ graphically and find its equation.

Solution Graphically, the inverse function of $f(x)$ is the mirror image of the curve $y = (x + 1)/(x - 2)$ in the line $y = x$ (Fig. 8.15).

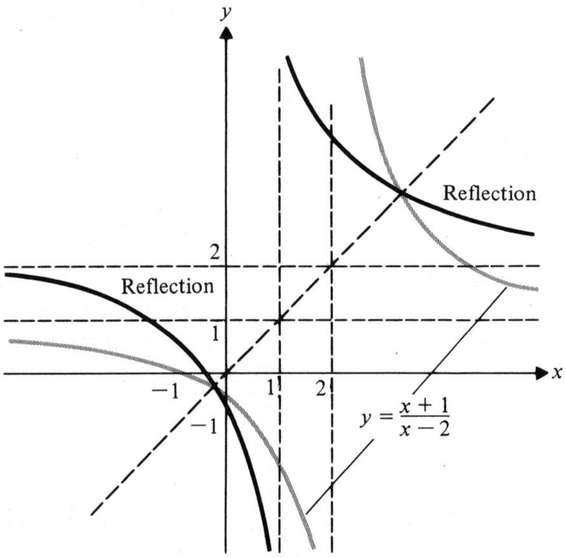

Figure 8.15

To find an algebraic definition of the inverse function, we solve the equation $y = (x + 1)/(x - 2)$ for x. First we cross-multiply,

$$x + 1 = y(x - 2);$$

then group terms in x,

$$x(y - 1) = 2y + 1;$$

and finally divide by $y - 1$,

$$x = \frac{2y + 1}{y - 1}.$$

What have we accomplished by solving for x in terms of y? If we take an x in the domain of $f(x)$, then the function produces $y = (x + 1)/(x - 2)$. Now this same x and y satisfy the equation $x = (2y + 1)/(y - 1)$ since this equation is simply a rearrangement of $y = (x + 1)/(x - 2)$. Consequently, if we substitute y into the right-hand side of the equation, we get

$$x = \frac{2y + 1}{y - 1},$$

or the original x. This means that the latter equation must define the inverse function of $f(x)$. In other words, the inverse function of $f(x)$ is

$$f^{-1}(x) = \frac{2x + 1}{x - 1}. \square$$

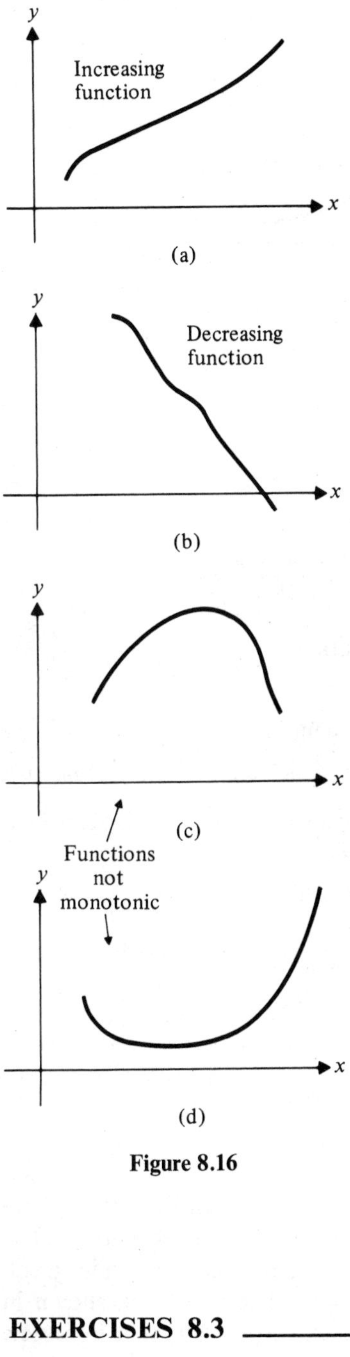

Figure 8.16

We have shown in Example 8 that to find the inverse of a function

$$f = y(x),$$

we can solve this equation for x in terms of y, and then rename variables.

To determine geometrically whether a function $f(x)$ defined on an interval I has an inverse function, we reflect its graph $y = f(x)$ in the line $y = x$. The reflected curve represents the inverse function if and only if it represents a function. But we know that a curve represents a function if every vertical line that intersects it does so at exactly one point. Furthermore, a vertical line intersects the reflected curve at more than one point if and only if the horizontal reflection of this line intersects $y = f(x)$ at more than one point. These two facts enable us to state that a function $f(x)$ has an inverse function if and only if every horizontal line that intersects it does so at exactly one point. See, for example, the functions in Figs. 8.13–8.15. In Figs. 8.13 and 8.15, horizontal lines intersect the given curves in at most one point, whereas in Fig. 8.14, horizontal lines $y = c > 0$ intersect the curve $y = x^2$ in two points.

If we examine the graphs in Fig. 8.16, we come to the conclusion that if a function is either increasing or decreasing on an interval I, but not sometimes increasing and sometimes decreasing, then it has an inverse function on that interval. Such a function is said to be *monotonic*.

It is important to note that monotonicity is a sufficient condition for the existence of an inverse function; i.e., if a function is monotonic, then it has an inverse. It is not, however, a necessary condition. The function shown in Fig. 8.17(a) is not monotonic on the interval $-1 \le x \le 1$, but it does have the inverse function shown in Fig. 8.17(b).

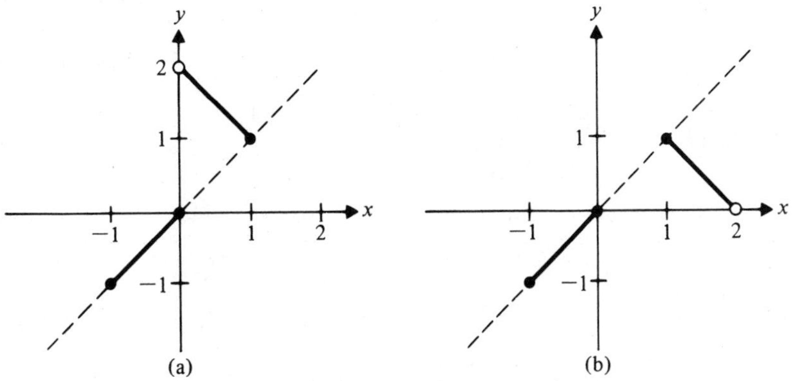

Figure 8.17

EXERCISES 8.3

In Exercises 1–14, determine graphically whether the given function has an inverse function. Find each inverse function.

1. $f(x) = 2x + 3$
2. $f(x) = \sqrt{x + 1}$
3. $f(x) = x^2 + x$
4. $f(x) = \dfrac{x + 5}{2x + 4}$
5. $f(x) = \dfrac{1}{x}$
6. $f(x) = 3x^3 + 2$

7. $f(x) = \sqrt{4 - x^2}, 0 \le x \le 2$
8. $f(x) = 2x + |x|$
9. $f(x) = x + |x|$
10. $f(x) = \sqrt{1 - x^2}$
11. $f(x) = x^4 + 2x^2 + 2, x \le 0$
12. $f(x) = x^2 - 2x + 4, x \ge 1$
13. $f(x) = \left(\frac{x + 2}{x - 2}\right)^3$
14. $f(x) = \frac{x}{3 + x^2}$
15. Prove or disprove the following statement: If $f^{-1}(x)$ is the inverse function of $f(x)$, then $f(x)$ is the inverse function of $f^{-1}(x)$.
16. If a manufacturing firm sells x objects of a certain commodity per week, it sells them at a price of r per object, and r depends on x, $r = f(x)$. In economic theory this function is usually considered decreasing, perhaps as shown in Fig. 8.18(a); hence it has an inverse function $x = f^{-1}(r)$. In this function x depends on r, indicating that if the price of the object is set at r, then the market will demand x of them per week. This function is therefore called the *demand function* (Fig. 8.18b). Find demand functions if:

    a) $r = \frac{a}{x + b} + c$, a, b, and c positive constants;

    b) $r = \frac{a}{x^2 + b} + c$, a, b, and c positive constants.

    Sketch the demand function $x = f^{-1}(r)$ and given function $r = f(x)$ in each case.

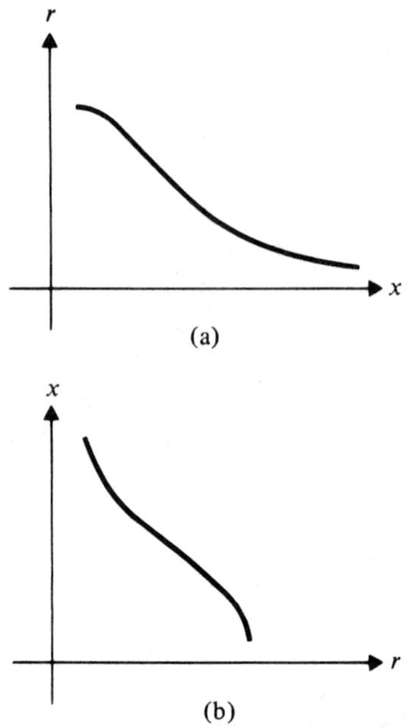

Figure 8.18

17. Show that the demand function

$$x = f(r) = 4a^3 - 3ar^2 + r^3, \qquad 0 < r < 2a$$

has an inverse function $r = f^{-1}$. What is the domain of f^{-1}? Sketch its graph.

18. Find the inverse function for

$$f(x) = \frac{x^2}{(1 + x)^2}, \qquad -1 < x \le 0.$$

8.4 THE INVERSE TRIGONOMETRIC FUNCTIONS

Associated with each of the six trigonometric functions is an inverse trigonometric function. We discuss first the inverse function associated with $f(x) = \sin x$. If $f(x) = \sin x$ is to have an inverse function, then we can obtain its graph by reflecting the curve $y = \sin x$ in the line $y = x$. The reflection, shown in Fig. 8.19, does not represent a function (since each vertical line that intersects the curve does so at an infinity of points), and hence $f(x) = \sin x$ does not have an inverse function. By restricting the range of values of the reflected curve, we can produce a (single-valued) function, and this can be done in many ways. When we choose to restrict the y-values of the reflected curve by

$$-\frac{\pi}{2} \le y \le \frac{\pi}{2}, \tag{8.7}$$

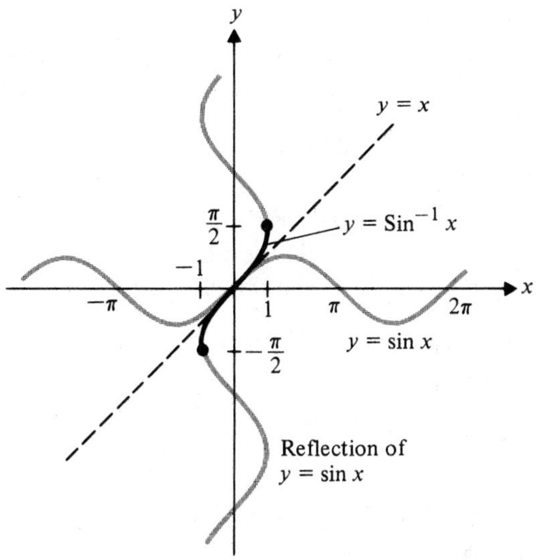

Figure 8.19

we denote the resulting function by

$$y = \text{Sin}^{-1}\, x, \tag{8.8}$$

and call it the *inverse sine function*. The values of $\text{Sin}^{-1}\, x$ defined by (8.7) are called the *principal values* of the inverse sine function. Note very carefully what we have done. We have taken the reflection of $y = \sin x$ in the line $y = x$ and constructed from it a function denoted by $\text{Sin}^{-1}\, x$. It is not, however, the inverse function of $f(x) = \sin x$, since the sine function has no inverse. Were we to restrict the domain of the sine function by $f(x) = \sin x$, $-\pi/2 \le x \le \pi/2$, then $\text{Sin}^{-1}\, x$ would indeed be the inverse of this function and vice versa (since each is the reflection of the other in the line $y = x$). It is perhaps then a misnomer to call the function $\text{Sin}^{-1}\, x$ the inverse sine function, but this has become the accepted terminology.

We now know what the inverse sine function looks like graphically, but suppose we are given a value of x in the domain ($-1 \le x \le 1$) of $\text{Sin}^{-1}\, x$ and must determine the value $y = \text{Sin}^{-1}\, x$ (to, say, four decimals). How shall we calculate y? To answer this, we note that if (x, y) is a point on $y = \text{Sin}^{-1}\, x$, then (y, x) is a point on the sine curve; i.e.,

$$y = \text{Sin}^{-1}\, x \quad \text{only if} \quad x = \sin y. \tag{8.9}$$

But in this latter equation y is an angle and x is the sine of that angle. Thus when we see $y = \text{Sin}^{-1}\, x$, we should not read "y equals inverse sine x," but rather "y is an angle whose sine is x." For instance, if $x = \frac{1}{2}$, then $y = \text{Sin}^{-1}\left(\frac{1}{2}\right)$ is read "y is an angle whose sine is $\frac{1}{2}$." Clearly the angle whose sine is $\frac{1}{2}$ is $\pi/6$, and we write $y = \text{Sin}^{-1}\left(\frac{1}{2}\right) = \pi/6$. Remember that we must choose the principal value $\pi/6$; there are many angles that have a sine equal to $\frac{1}{2}$, but the inverse sine function demands that we choose that angle in the range $-\pi/2 \le y \le \pi/2$, and state it with a number in this range. For example, $2\pi + \pi/6$ may represent the same angle with the positive x-axis as $\pi/6$, but the number $2\pi + \pi/6$ does not lie between $-\pi/2$ and $\pi/2$. In summary, the function $\sin x$ regards x as an angle and assigns to x the sine of the angle; the inverse sine function $\mathrm{Sin}^{-1} x$ regards x as the sine of an angle and assigns to x the angle with that sine.

Another notation that is commonly used for the inverse sine function is $\arcsin x$. One reason the notation $\arcsin x$ is preferable to $\mathrm{Sin}^{-1} x$ is the possible misinterpretation of $\mathrm{Sin}^{-1} x$. Sometimes students regard the "-1" as a power and write $\mathrm{Sin}^{-1} x = 1/\sin x$. This is *not* correct, a fact that we warned you about in Section 8.3. Sin^{-1} is the name of a function, just as sin is the name of the sine function, and $\sqrt{}$ is the notation for the positive square root function. The capital S in Sin^{-1} should also warn you that this is the inverse sine function.

Example 9 Simplify each of the following expressions:

a) $\mathrm{Sin}^{-1}(-\sqrt{3}/2)$;
b) $\mathrm{Sin}^{-1}(1)$;
c) $\mathrm{Sin}^{-1}(3)$;
d) $\mathrm{Sin}^{-1}(3/5) + \mathrm{Sin}^{-1}(4/5)$;
e) $\sin[\mathrm{Sin}^{-1}(\sqrt{3}/2)]$;
f) $\mathrm{Sin}^{-1}[\sin(7\pi/4)]$.

Solution a) $\mathrm{Sin}^{-1}(-\sqrt{3}/2)$ asks for the angle whose sine is equal to $-\sqrt{3}/2$. Clearly,

$$\mathrm{Sin}^{-1}(-\sqrt{3}/2) = -\pi/3.$$

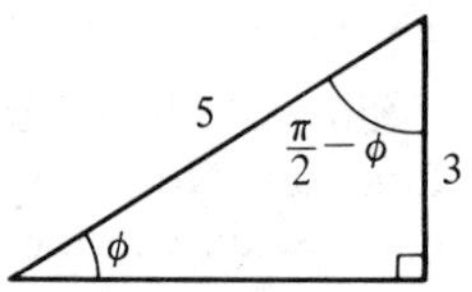

Figure 8.20

b) $\mathrm{Sin}^{-1}(1) = \pi/2$.

c) $\mathrm{Sin}^{-1}(3)$ is not defined since the domain of $\mathrm{Sin}^{-1} x$ is $-1 \le x \le 1$.

d) If $\phi = \mathrm{Sin}^{-1}(3/5)$, then ϕ is illustrated in the triangle in Fig. 8.20. Since the third side must have length 4, it follows that $\mathrm{Sin}^{-1}(4/5) = \pi/2 - \phi$, and

$$\mathrm{Sin}^{-1}\left(\frac{3}{5}\right) + \mathrm{Sin}^{-1}\left(\frac{4}{5}\right) = \phi + \frac{\pi}{2} - \phi = \frac{\pi}{2}.$$

e) $\sin[\mathrm{Sin}^{-1}(\sqrt{3}/2)] = \sin(\pi/3) = \sqrt{3}/2$.

f) $\mathrm{Sin}^{-1}\left[\sin\left(\frac{7\pi}{4}\right)\right] = \mathrm{Sin}^{-1}\left(\frac{-1}{\sqrt{2}}\right) = -\frac{\pi}{4}$. □

We do not wish to repeat the above discussion five times, once for each of the inverse functions associated with the other five trigonometric functions. Instead, we give a summarized version for the inverse cosine function and then tabulate the remaining four functions.

The reflection of the graph of the function $f(x) = \cos x$ is shown in Fig. 8.21 and does not represent a function. If we restrict the y-values on the reflected curve to

$$0 \le y \le \pi, \tag{8.10}$$

then we do obtain a function called the inverse cosine function, denoted by

$$y = \mathrm{Cos}^{-1} x. \tag{8.11}$$

The values in (8.10) are called principal values of the inverse cosine function. Note again that $\mathrm{Cos}^{-1} x$ is not the inverse function of $\cos x$, but is the inverse function of $f(x) = \cos x, 0 \le x \le \pi$.

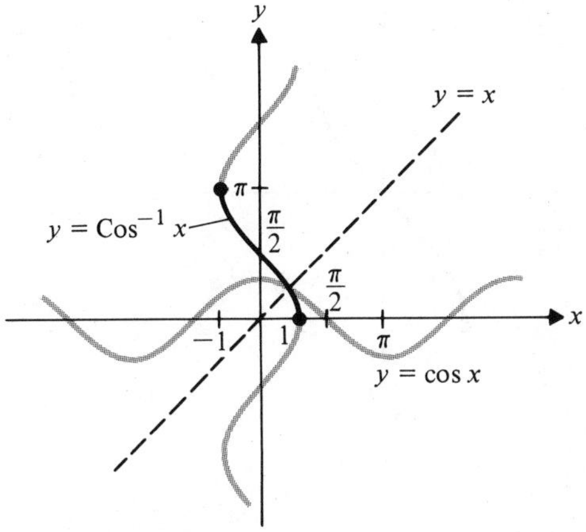

Figure 8.21

When we write $y = \mathrm{Cos}^{-1}\, x$, we do not read this as "y equals inverse cosine x," but rather as "y is an angle whose cosine is x," for if $y = \mathrm{Cos}^{-1}\, x$, then $x = \cos y$.

The remaining four inverse trigonometric functions are shown in Figs. 8.22(c)–(f). The lighter curves represent reflections in the line $y = x$ of the trigonometric functions $\tan x$, $\cot x$, $\csc x$, and $\sec x$. The heavier curves represent principal values of the inverse trigonometric functions $\mathrm{Tan}^{-1}\, x$, $\mathrm{Cot}^{-1}\, x$, $\mathrm{Csc}^{-1}\, x$, and $\mathrm{Sec}^{-1}\, x$ associated with each trigonometric function. The principal values of the six inverse trigonometric functions are listed in Table 8.1.

Table 8.1

Inverse Trigonometric Function	Principal Values
$\mathrm{Sin}^{-1}\, x$	$-\frac{\pi}{2} \le y \le \frac{\pi}{2}$
$\mathrm{Tan}^{-1}\, x$	$-\frac{\pi}{2} < y < \frac{\pi}{2}$
$\mathrm{Cos}^{-1}\, x$	$0 \le y \le \pi$
$\mathrm{Cot}^{-1}\, x$	$0 < y < \pi$
$\mathrm{Csc}^{-1}\, x$	$-\pi < y \le -\frac{\pi}{2}, 0 < y \le \frac{\pi}{2}$
$\mathrm{Sec}^{-1}\, x$	$-\pi \le y < -\frac{\pi}{2}, 0 \le y < \frac{\pi}{2}$

It would be reasonable to ask why the principal values of $\mathrm{Sec}^{-1}\, x$ were not chosen as $0 \le y < \pi/2$, $\pi/2 < y \le \pi$. Had they been chosen in this way they would have been very similar to those of $\mathrm{Cos}^{-1}\, x$ and $\mathrm{Cot}^{-1}\, x$. Likewise, why are the principal values of $\mathrm{Csc}^{-1}\, x$ not $-\pi/2 \le y < 0$, $0 < y \le \pi/2$? Our answer is that they could have been selected in this way, and some authors do indeed make this choice; it is simply a matter of preference. Each choice does, however, create corresponding changes in later work. Specifically, we show in the exercises of Section 8.5 that if the principal values of $\mathrm{Sec}^{-1}\, x$ and $\mathrm{Csc}^{-1}\, x$ had been selected in this alternative way, then the derivatives of these functions would be modified accordingly.

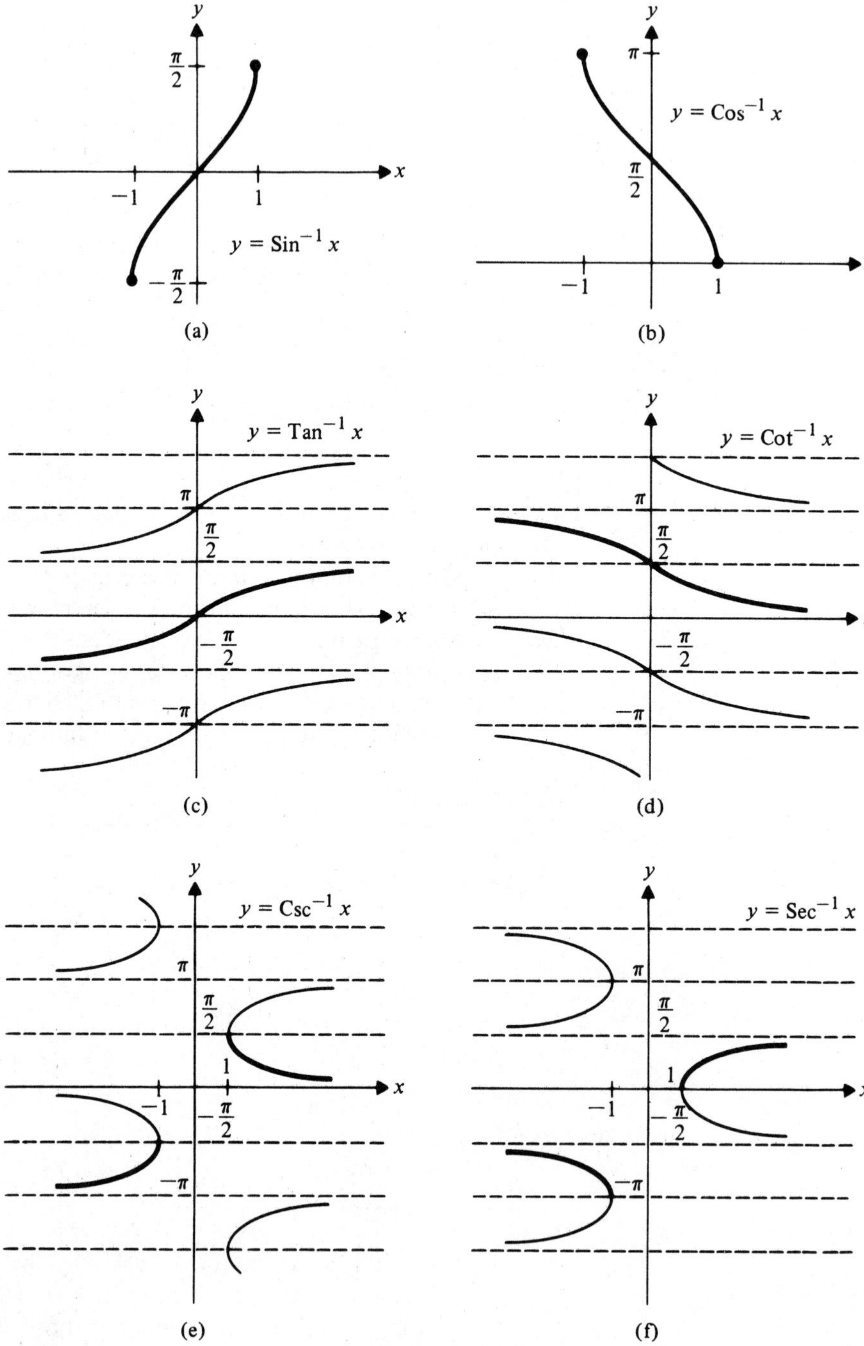

Figure 8.22

Example 10 Sketch a graph of the function $f(x) = 4 + (\mathrm{Cot}^{-1} x)^2$.

Solution A sketch of $y = (\mathrm{Cot}^{-1} x)^2$ can be obtained by squaring each ordinate in the graph of Fig. 8.22(d), the result being shown in Fig. 8.23(a). Figure 8.23(b) is the curve shown in Fig. 8.23(a) shifted vertically 4 units. □

Example 11 Does the equation $x = \mathrm{Sec}^{-1}(1/(y^2 + 2)) + 4x^2$ implicitly define y as a function of x?

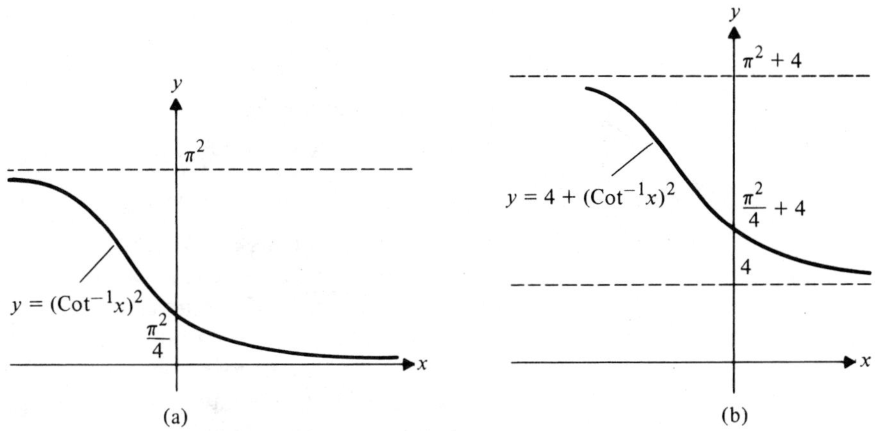

Figure 8.23

Solution Since $y^2 + 2$ is always greater than or equal to 2, it follows that $0 < 1/(y^2 + 2) \leq 1/2$. But the domain of the inverse secant function does not include these values (Fig. 8.22f). Consequently, there are no values of x and y that satisfy $x = \mathrm{Sec}^{-1}(1/(y^2 + 2)) + 4x^2$, and the equation does not define y as a function of x. □

Example 12 Show that

$$\mathrm{Sin}^{-1} x = \mathrm{Cos}^{-1} \sqrt{1 - x^2}, \qquad 0 \leq x \leq 1.$$

Solution If we set $y = \mathrm{Sin}^{-1} x$, then $x = \sin y$. Since $\sin^2 y + \cos^2 y = 1$,

$$\cos y = \pm\sqrt{1 - \sin^2 y} = \pm\sqrt{1 - x^2}.$$

But if $y = \mathrm{Sin}^{-1} x$, and $0 \leq x \leq 1$, then the principal values of $\mathrm{Sin}^{-1} x$ imply that $0 \leq y \leq \pi/2$. Consequently, $\cos y$ must be nonnegative, and

$$\cos y = \sqrt{1 - x^2}.$$

Therefore,

$$y = \mathrm{Cos}^{-1} \sqrt{1 - x^2}. \; \square$$

Example 13 Find all solutions of the equation

$$\sin^2 x + 3 \sin x = 2.$$

Solution The quadratic formula applied to the equation

$$(\sin x)^2 + 3(\sin x) - 2 = 0$$

yields

$$\sin x = \frac{-3 \pm \sqrt{9 + 8}}{2} = \frac{-3 \pm \sqrt{17}}{2}.$$

Since $\sin x$ can take on only values in the interval $-1 \leq \sin x \leq 1$, it follows that

$$\sin x = \frac{-3 + \sqrt{17}}{2}.$$

One solution of this equation is

$$x = \mathrm{Sin}^{-1}\left(\frac{\sqrt{17} - 3}{2}\right) = 0.60 \text{ radian}.$$

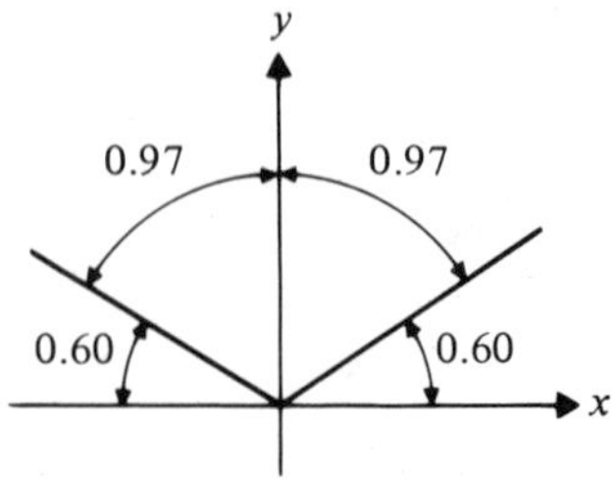

Figure 8.24

This is not the only solution, however, since the complete set of solutions is

$$2n\pi + 0.60, \qquad 2n\pi + (\pi - 0.60),$$

where n is an integer, or, equivalently (Fig. 8.24),

$$\left(\frac{4n+1}{2}\right)\pi \pm 0.97. \ \square$$

Often in problems like these $x = \mathrm{Sin}^{-1}((\sqrt{17} - 3)/2) = 0.60$ is the only solution given. We can see the reasoning behind this conclusion: There is only one principal value of the inverse sine function, and therefore there is only one solution to the equation. But the equation $\sin x = (\sqrt{17} - 3)/2$ says nothing about the inverse sine function. We have introduced it simply as a convenience. When we have obtained the solution 0.60, we must ask whether there are other possible solutions to the original equation; for this example, there are. In some problems, the principal value is the only acceptable solution. Remember, then, that if we introduce an inverse trigonometric function into a problem along with its corresponding principal values, we must ask whether there are other possibilities besides the principal values. This point is emphasized again in the following example.

Example 14 Find all solutions of the equation

$$\tan(\cos x) = \frac{1}{\sqrt{3}}.$$

Solution If we set $y = \cos x$, then $\tan y = 1/\sqrt{3}$, and one solution of this equation is $y = \mathrm{Tan}^{-1}(1/\sqrt{3}) = \pi/6$. But there are many other solutions for y, namely,

$$y = \frac{\pi}{6} + n\pi, \qquad n \text{ an integer.}$$

But $y = \cos x$, and $\cos x$ must take on values in the interval $-1 \le \cos x \le 1$. There is therefore only one possibility for n, $n = 0$; hence,

$$\cos x = \frac{\pi}{6}.$$

From the solution $x = \mathrm{Cos}^{-1}(\pi/6) = 1.02$ and Fig. 8.25, we obtain

$$x = \pm 1.02 + 2n\pi, \qquad n \text{ an integer}$$

as the complete set of solutions. $\square$

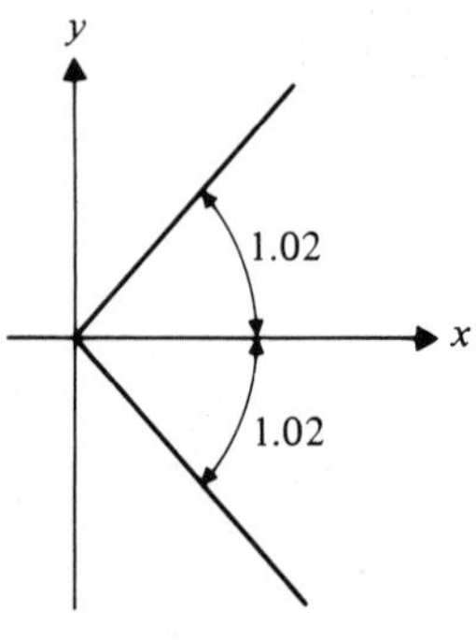

Figure 8.25

EXERCISES 8.4

Evaluate the expression in each of Exercises 1–16 (if it has a value).

1. $\mathrm{Tan}^{-1}(-1/\sqrt{3})$
2. $\mathrm{Sin}^{-1}(\frac{1}{2})$
3. $\mathrm{Sec}^{-1}(\sqrt{2})$
4. $\mathrm{Csc}^{-1}(-2/\sqrt{3})$
5. $\mathrm{Cot}^{-1}(1)$
6. $\mathrm{Cos}^{-1}(\frac{3}{2})$
7. $\mathrm{Sin}^{-1}(\pi/2)$
8. $\mathrm{Tan}^{-1}(-1)$
9. $\sin(\mathrm{Tan}^{-1}(\sqrt{3}))$
10. $\tan(\mathrm{Sin}^{-1}(\sqrt{3}))$
11. $\mathrm{Sin}^{-1}(\tan(\pi/6))$
12. $\mathrm{Tan}^{-1}(\sin(\pi/6))$
13. $\sec(\mathrm{Cos}^{-1}(\frac{1}{2}))$
14. $\mathrm{Sin}^{-1}(\sin(3\pi/4))$
15. $\sin(\mathrm{Sin}^{-1}(1/\sqrt{2}))$
16. $\mathrm{Sin}^{-1}(\cos(\mathrm{Sec}^{-1}(-\sqrt{2})))$

In Exercises 17–22, find all solutions of the given equation.

17. $\sin x = \frac{1}{2}$
18. $\tan x = -1$

19. $4\sin^2 x - 2\cos^2 x = 1$

20. $4\sin^2 x + 2\cos^2 x = 1$

21. $\cos^2 x - 3\cos x + 1 = 0$

22. $\sin^2 x - 3\sin x - 5 = 0$

Sketch a graph of the function in each of Exercises 23–26.

23. $f(x) = 2 + (\mathrm{Csc}^{-1} x)^2$

24. $f(x) = \sqrt{\mathrm{Tan}^{-1} x} + \sqrt{\mathrm{Sec}^{-1} x}$

25. $f(x) = \mathrm{Sin}^{-1}(x - 3)$

26. $f(x) = \mathrm{Sin}^{-1} x + \mathrm{Csc}^{-1} x$

27. If ϕ is the angle formed by AB and AO in Fig. 4.13, find ϕ as a function of θ.

In Exercises 28–33, find all solutions of the given equation.

28. $\sin x \tan^2 x - 3 + \tan^2 x - 3\sin x = 0$

29. $\sin x + \cos x = 1$

30. $\sec(\sin x) = -\sqrt{2}$

31. $\cos(\mathrm{Sin}^{-1} x) = 1/2$

32. $\sec(\tan x) = -\sqrt{2}$

33. $\mathrm{Cos}^{-1}(\tan(x^2 + 4)) = 5 - 2\pi$

34. Sketch a graph of each of the following functions:

a) $f(x) = \sin(\mathrm{Sin}^{-1} x)$;

b) $f(x) = \mathrm{Sin}^{-1}(\sin x)$.

35. Prove that

$$\mathrm{Tan}^{-1} A - \mathrm{Tan}^{-1} B = \mathrm{Tan}^{-1}\left(\frac{A - B}{1 + AB}\right) + n\pi$$

for some integer n. (*Hint:* You should use the expansion for $\tan(a - b)$.)

36. In Example 12 we proved the identity

$$\mathrm{Sin}^{-1} x = \mathrm{Cos}^{-1}\sqrt{1 - x^2}$$

for $0 \le x \le 1$. We wish to extend the proof so that it is valid for $-1 \le x \le 1$. If we change the inequality $0 \le y \le \pi/2$ to $-\pi/2 \le y \le \pi/2$, does the proof become valid for $-1 \le x \le 1$? Explain. Can the proof be extended to negative values of x?

37. a) Show that

$$\mathrm{Csc}^{-1} x = \mathrm{Sin}^{-1}\left(\frac{1}{x}\right), \qquad x \ge 1.$$

b) How must this identity be modified if $x \le -1$?

38. In Fig. 8.26, a ray of light is incident on a prism at angle i. The light is refracted as shown and leaves the prism at an angle of deviation ψ relative to the incident direction.

a) Show that $\gamma = \alpha + \beta$, and use this to derive the expression

$$\psi = i + \phi - \gamma.$$

b) To express ψ in terms of i and properties of the prism, we use Snell's law, which states that if the prism is in air, then the relationships between the incident and refracted light at the two faces are

$$n \sin\alpha = \sin i \quad \text{and} \quad n \sin\beta = \sin\phi,$$

where n is the refractive index of the prism. Use these equations to show that ψ can be expressed in the form

$$\psi = i - \gamma + \mathrm{Sin}^{-1}\left(\sqrt{n^2 - \sin^2 i}\,\sin\gamma - \sin i\cos\gamma\right).$$

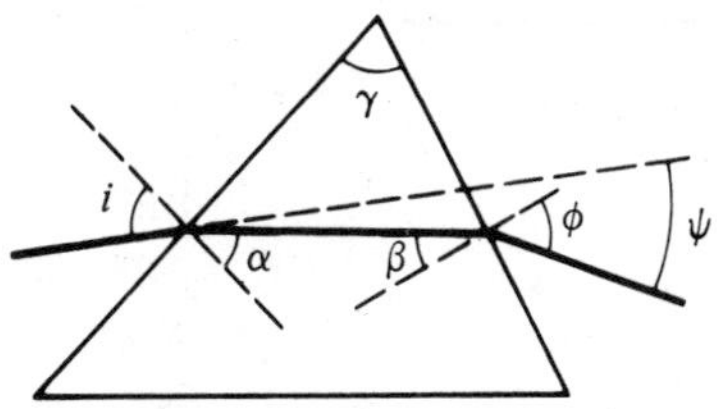

Figure 8.26

***39.** Show that for the prism in Exercise 38 the angle of deviation is a minimum when $i = \phi$. If ψ_m is the minimum angle of deviation, prove that

$$n = \frac{\sin[(\psi_m + \gamma)/2]}{\sin(\gamma/2)}.$$

In practice, this equation is used to determine the index of refraction of the prism.

8.5 DERIVATIVES OF THE INVERSE TRIGONOMETRIC FUNCTIONS

We now obtain derivatives of the six inverse trigonometric functions.

▶ **Theorem 8.1**

$$\frac{d}{dx}\mathrm{Sin}^{-1} x = \frac{1}{\sqrt{1 - x^2}} \quad ◀ \qquad (8.12)$$

Proof If we set $y = \text{Sin}^{-1} x$, then $x = \sin y$, and we can differentiate (implicitly) with respect to x:

$$1 = \cos y \frac{dy}{dx}.$$

Thus,

$$\frac{dy}{dx} = \frac{1}{\cos y}.$$

To express $\cos y$ in terms of x we can proceed in two ways:

i) From the trigonometric identity

$$\sin^2 y + \cos^2 y = 1,$$

we obtain

$$\cos y = \pm\sqrt{1 - \sin^2 y} = \pm\sqrt{1 - x^2}.$$

We know that $y = \text{Sin}^{-1} x$ and the principal values of $\text{Sin}^{-1} x$ are $-\pi/2 \leq y \leq \pi/2$. Since, therefore, y is an angle in either the first or fourth quadrant, it follows that $\cos y \geq 0$, and we must choose

$$\cos y = \sqrt{1 - x^2}.$$

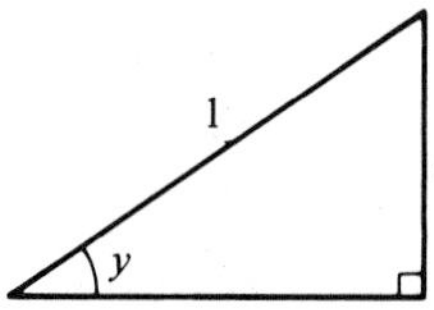

Figure 8.27

ii) With this method we use the triangle shown in Fig. 8.27 to replace the trigonometric identity. The triangle is designed to fit the equation $x = \sin y$. The third side is then $\sqrt{1 - x^2}$, in which case

$$\cos y = \sqrt{1 - x^2}.$$

But $\cos y$ is not always positive as indicated here. (For example, if y is between $\pi/2$ and π, then $\cos y < 0$.) To decide on the correct sign for $\cos y$, we use the discussion of principal values in (i), and obtain again $\cos y = \sqrt{1 - x^2}$. Finally, then, we have

$$\frac{dy}{dx} = \frac{1}{\sqrt{1 - x^2}}. \square$$

▶ **Corollary** If $u(x)$ is a differentiable function, then

$$\frac{d}{dx} \text{Sin}^{-1} u = \frac{1}{\sqrt{1 - u^2}} \frac{du}{dx}. \blacktriangleleft \tag{8.13}$$

▶ **Theorem 8.2**

$$\frac{d}{dx} \text{Cos}^{-1} x = \frac{-1}{\sqrt{1 - x^2}} \blacktriangleleft \tag{8.14}$$

Proof If we set $y = \text{Cos}^{-1} x$, then $x = \cos y$, and we can differentiate with respect to x,

$$1 = -\sin y \frac{dy}{dx},$$

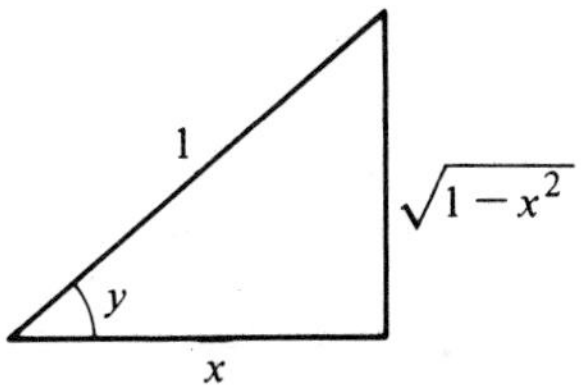

Figure 8.28

and solve for

$$\frac{dy}{dx} = \frac{-1}{\sin y}.$$

The triangle shown in Fig. 8.28 (obtained from $x = \cos y$) yields $\sin y = \sqrt{1 - x^2}$. Since the principal values of $y = \text{Cos}^{-1} x$ are $0 \le y \le \pi$, it follows that $\sin y$ is indeed nonnegative, and therefore

$$\frac{dy}{dx} = \frac{-1}{\sqrt{1 - x^2}}. \square$$

▶ **Corollary** If $u(x)$ is differentiable, then

$$\frac{d}{dx}\text{Cos}^{-1} u = \frac{-1}{\sqrt{1 - u^2}}\frac{du}{dx}. \blacktriangleleft \tag{8.15}$$

Similar derivations lead to derivatives for the remaining inverse trigonometric functions. We include derivatives of $\text{Sin}^{-1} u$ and $\text{Cos}^{-1} u$ for completeness:

$$\frac{d}{dx}\text{Sin}^{-1} u = \frac{1}{\sqrt{1 - u^2}}\frac{du}{dx}; \tag{8.13}$$

$$\frac{d}{dx}\text{Cos}^{-1} u = \frac{-1}{\sqrt{1 - u^2}}\frac{du}{dx}; \tag{8.15}$$

$$\frac{d}{dx}\text{Tan}^{-1} u = \frac{1}{1 + u^2}\frac{du}{dx}; \tag{8.16}$$

$$\frac{d}{dx}\text{Cot}^{-1} u = \frac{-1}{1 + u^2}\frac{du}{dx}; \tag{8.17}$$

$$\frac{d}{dx}\text{Sec}^{-1} u = \frac{1}{u\sqrt{u^2 - 1}}\frac{du}{dx}; \tag{8.18}$$

$$\frac{d}{dx}\text{Csc}^{-1} u = \frac{-1}{u\sqrt{u^2 - 1}}\frac{du}{dx}. \tag{8.19}$$

Note that the derivative of any inverse cofunction is the negative of the derivative of the corresponding inverse function. In other words, it is really only necessary to learn derivatives for $\text{Sin}^{-1} u$, $\text{Tan}^{-1} u$, and $\text{Sec}^{-1} u$. The derivatives for $\text{Cos}^{-1} u$, $\text{Cot}^{-1} u$, and $\text{Csc}^{-1} u$ are their negatives.

Example 15 Find dy/dx if y is defined as a function of x in each of the following:

a) $y = \text{Sec}^{-1}(x^2)$;

b) $y = 2x\,\text{Cot}^{-1}(3x)$;

c) $y = [\text{Sin}^{-1}(x^3)]^2$;

d) $y = \dfrac{\text{Sin}^{-1} x}{\text{Cos}^{-1} x}$;

e) $\text{Cos}^{-1}(xy) + x^2y + 5 = 0$.

Solution a) $$\frac{dy}{dx} = \frac{1}{x^2\sqrt{(x^2)^2 - 1}}\frac{d}{dx}(x^2) = \frac{2x}{x^2\sqrt{x^4 - 1}} = \frac{2}{x\sqrt{x^4 - 1}}$$

b) $$\frac{dy}{dx} = 2\,\mathrm{Cot}^{-1}(3x) + 2x\left(\frac{-1}{1 + 9x^2}\right)(3) = 2\,\mathrm{Cot}^{-1}(3x) - \frac{6x}{1 + 9x^2}$$

c) $$\frac{dy}{dx} = 2\,\mathrm{Sin}^{-1}(x^3)\frac{d}{dx}\mathrm{Sin}^{-1}(x^3)$$

$$= 2\,\mathrm{Sin}^{-1}(x^3)\frac{1}{\sqrt{1 - x^6}}(3x^2) = \frac{6x^2\,\mathrm{Sin}^{-1}(x^3)}{\sqrt{1 - x^6}}$$

d) $$\frac{dy}{dx} = \frac{\mathrm{Cos}^{-1}x\left(\frac{1}{\sqrt{1 - x^2}}\right) - \mathrm{Sin}^{-1}x\left(\frac{-1}{\sqrt{1 - x^2}}\right)}{(\mathrm{Cos}^{-1}x)^2} = \frac{\mathrm{Cos}^{-1}x + \mathrm{Sin}^{-1}x}{\sqrt{1 - x^2}(\mathrm{Cos}^{-1}x)^2}$$

e) If we differentiate with respect to x, we have

$$\frac{-1}{\sqrt{1 - x^2y^2}}\left(y + x\frac{dy}{dx}\right) + 2xy + x^2\frac{dy}{dx} = 0.$$

Thus,

$$\frac{dy}{dx}\left(\frac{-x}{\sqrt{1 - x^2y^2}} + x^2\right) = \frac{y}{\sqrt{1 - x^2y^2}} - 2xy,$$

from which we get

$$\frac{dy}{dx} = \frac{y - 2xy\sqrt{1 - x^2y^2}}{\sqrt{1 - x^2y^2}} \cdot \frac{\sqrt{1 - x^2y^2}}{-x + x^2\sqrt{1 - x^2y^2}} = \frac{y - 2xy\sqrt{1 - x^2y^2}}{-x + x^2\sqrt{1 - x^2y^2}}.$$ □

Example 16 Find the derivative of the function $f(x) = \mathrm{Sin}^{-1}x + \mathrm{Cos}^{-1}x$. What do you conclude?

Solution

$$f'(x) = \frac{1}{\sqrt{1 - x^2}} + \frac{-1}{\sqrt{1 - x^2}} = 0$$

Since this result is valid for the entire domain $-1 \le x \le 1$ of $f(x)$, we can antidifferentiate both sides of the equation

$$f(x) = C = \text{a constant.}$$

Because $f(0) = \mathrm{Sin}^{-1}(0) + \mathrm{Cos}^{-1}(0) = 0 + \pi/2 = \pi/2$, it follows that $C = \pi/2$. Consequently,

$$\mathrm{Sin}^{-1}x + \mathrm{Cos}^{-1}x = \frac{\pi}{2}.$$ □

Example 17 A road information sign specifying the distance to the next city is 1 meter high and sits 5 meters above road level (Fig. 8.29). If the average motorist's eye level is $\frac{3}{2}$ meters above ground level, at what distance from the sign will the letters appear tallest?

Solution The letters will appear tallest when the angle θ that they subtend at the eye (Fig. 8.30) is greatest. Since

$$\theta = \beta - \alpha = \mathrm{Tan}^{-1}\left(\frac{9/2}{x}\right) - \mathrm{Tan}^{-1}\left(\frac{7/2}{x}\right), \qquad 0 < x < \infty,$$

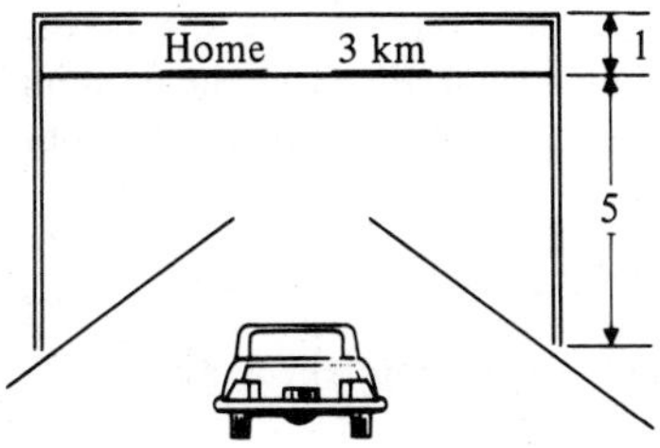

Figure 8.29

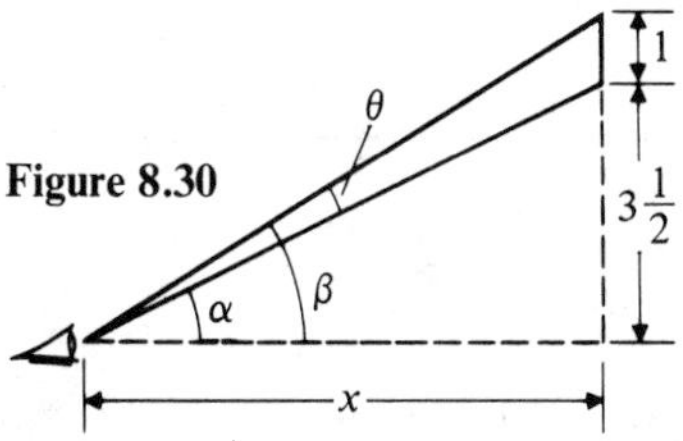

Figure 8.30

critical points of $\theta(x)$ are given by

$$0 = \frac{d\theta}{dx} = \frac{1}{1 + \left(\frac{9}{2x}\right)^2}\left(-\frac{9}{2x^2}\right) - \frac{1}{1 + \left(\frac{7}{2x}\right)^2}\left(-\frac{7}{2x^2}\right)$$

$$= \frac{-18}{4x^2 + 81} + \frac{14}{4x^2 + 49}.$$

Equivalently,

$$18(4x^2 + 49) = 14(4x^2 + 81),$$

which has only one positive solution, $x = 3\sqrt{7}/2$.

Since $\theta(x)$ approaches zero as x approaches zero and as x becomes very large, it follows that θ is maximized when the motorist is $3\sqrt{7}/2$ meters from the sign. □

EXERCISES 8.5

In Exercises 1–30, y is defined as a function of x. Find dy/dx in as simplified a form as possible.

1. $y = \text{Cos}^{-1}(2x + 3)$
2. $y = \text{Cot}^{-1}(x^2 + 2)$
3. $y = \text{Csc}^{-1}(3 - 4x)$
4. $y = \text{Tan}^{-1}(2 - x^2)$
5. $y = \text{Sec}^{-1}(3 - 2x^2)$
6. $y = x\,\text{Csc}^{-1}(x^2 + 5)$
7. $y = (x^2 + 2)\,\text{Sin}^{-1}(2x)$
8. $y = \text{Tan}^{-1}\sqrt{x + 2}$
9. $y = \text{Sin}^{-1}\sqrt{1 - x^2}$
10. $y = \text{Cot}^{-1}\sqrt{x^2 - 1}$
11. $y = (\text{Tan}^{-1}(x^2))^2$
12. $y = x^2\,\text{Sec}^{-1} x$
13. $y = \tan(3\,\text{Sin}^{-1} x)$
14. $y = \text{Cot}^{-1}((1 + x)/(1 - x))$
15. $y = \text{Csc}^{-1}(1/x)$
16. $y = \text{Sin}^{-1}[(1 - x)/(1 + x)]$
17. $y = \text{Tan}^{-1}(u^2 + 1/u)$,
    $u = \tan(x^2 + 4)$
18. $y = t\,\text{Cos}^{-1} t$,
    $t = \sqrt{1 - x^2}$
19. $y^2 \sin x + y = \text{Tan}^{-1} x$
20. $\text{Sin}^{-1}(xy) = 5x + 2y$
21. $y = \sqrt{x^2 - 1} - \text{Sec}^{-1} x$
22. $y = (1/9)\,\text{Csc}^{-1}(x/3) - \sqrt{x^2 - 9}/x^2$
23. $y = x\,\text{Cos}^{-1}(x/2) - \sqrt{4 - x^2}$
24. $y = (\text{Csc}^{-1}(3x))/x - \sqrt{9x^2 - 1}/x$
25. $y = x^2\,\text{Sec}^{-1} x - \sqrt{x^2 - 1}$
26. $y = x(\text{Cos}^{-1} x)^2 - 2x - 2\sqrt{1 - x^2}$
27. $y = (1 + 9x^2)\,\text{Cot}^{-1}(3x) + 3x$
28. $y = (x - 2)\sqrt{4x - x^2} + 4\,\text{Sin}^{-1}\left(\frac{x - 2}{2}\right)$
29. $y = (2x^2 - 1)\,\text{Sin}^{-1} x + x\sqrt{1 - x^2}$
30. $y = \text{Tan}^{-1}((\sqrt{2}x)/\sqrt{1 + x^4})$
31. Evaluate the derivative of

    $$f(x) = \text{Sec}^{-1} x + \text{Cot}^{-1}\sqrt{x^2 - 1}.$$

    What do you conclude?
32. Find all relative extrema and points of inflection for the function

    $$f(x) = x\,\text{Sin}^{-1} x + \sqrt{1 - x^2}.$$

    Sketch its graph.
33. a) Show that if the principal values of $f(x) = \text{Sec}^{-1} x$ were chosen as $0 \le y \le \pi$, $y \ne \pi/2$, then the derivative of the function would be

    $$\frac{d}{dx}\,\text{Sec}^{-1} x = \frac{1}{|x|\sqrt{x^2 - 1}}.$$

    b) What would be the derivative of $f(x) = \text{Csc}^{-1} x$ if its principal values were chosen as $-\pi/2 \le y \le \pi/2$, $y \ne 0$?

34. Find the angle between the tangent lines to the curves $y = \mathrm{Sin}^{-1} x$ and $y = \mathrm{Cos}^{-1} x$ at their point of intersection.

35. Verify the results in Eqs. (8.16)–(8.19).

36. When a force F is applied to the object of mass m shown in Fig. 8.31, two other forces act on m: the force of gravity mg ($g > 0$) directly downward, and a frictional force opposing F. It can be shown that the least force F that will overcome friction and produce motion is given by

$$F = \frac{\mu m g}{\cos\theta + \mu \sin\theta},$$

where μ is a constant called the *coefficient of static friction*. Find the angle θ for which F is minimal.

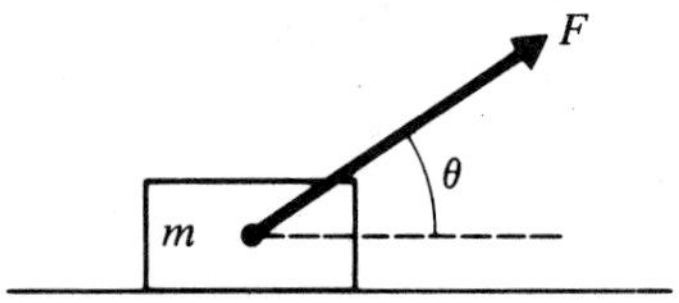

Figure 8.31

37. A runner moves counterclockwise around the track shown in Fig. 8.32 at a rate of 4 meters per second. A camera at the center of the track is placed on a swivel so that it can follow the runner. Find the rate at which the camera turns when:

a) the runner is at A; *b) the runner is at B.

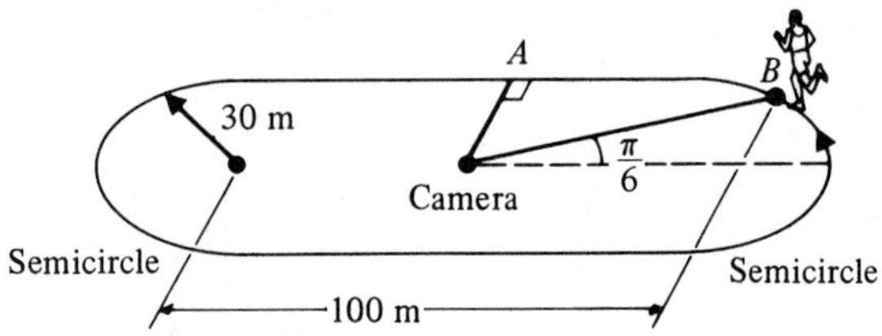

Figure 8.32

***38.** If an oscillating voltage $E = E_0 \sin \omega t$ (ω = a constant) is applied to the circuit shown in Fig. 8.33, the resulting steady-state current must satisfy the differential equation

$$L\frac{d^2I}{dt^2} + R\frac{dI}{dt} + \frac{1}{C}I = \frac{dE}{dt}.$$

Show that a solution of this differential equation is

$$I = f(t) = \frac{E_0}{Z}\sin(\omega t - \phi),$$

where

$$Z = \sqrt{R^2 + \left(\omega L - \frac{1}{\omega C}\right)^2}$$

and

$$\phi = \mathrm{Tan}^{-1}\left(\frac{\omega L - 1/\omega C}{R}\right).$$

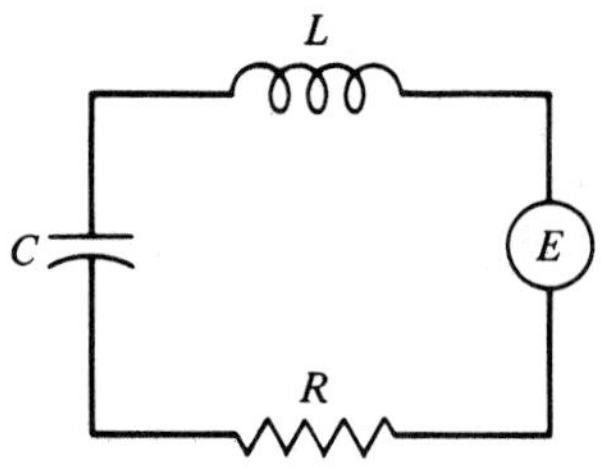

Figure 8.33

***39.** Verify that if $0 \le x < 1$, then

$$2\,\mathrm{Tan}^{-1}\sqrt{\frac{1+x}{1-x}} = \pi - \mathrm{Cos}^{-1} x.$$

8.6 THE EXPONENTIAL AND LOGARITHM FUNCTIONS

In elementary algebra we learned the basic rules for combining exponents:

$$a^b a^c = a^{b+c}; \tag{8.20a}$$

$$\frac{a^b}{a^c} = a^{b-c}; \tag{8.20b}$$

$$(a^b)^c = a^{bc}. \tag{8.20c}$$

These rules can be used to define the *exponential function*:

$$f(x) = a^x; \tag{8.21}$$

that is, a raised to the power of x for variable x.

Consider first the case when $a > 1$. The meaning of a^x when x is an integer is clear, and when $x = 1/n$, $n > 0$ an integer, $a^x = a^{1/n}$ is the nth root of a. When x is a positive rational number n/m (n and m positive integers), (8.20c) implies that

$$a^x = a^{n/m} = (a^n)^{1/m},$$

or

$$a^x = a^{n/m} = (a^{1/m})^n.$$

When x is a negative rational, we write $a^x = 1/a^{-x}$, where $-x$ is positive. These results lead to the graph of $y = a^x$ ($a > 1$) shown in Fig. 8.34.

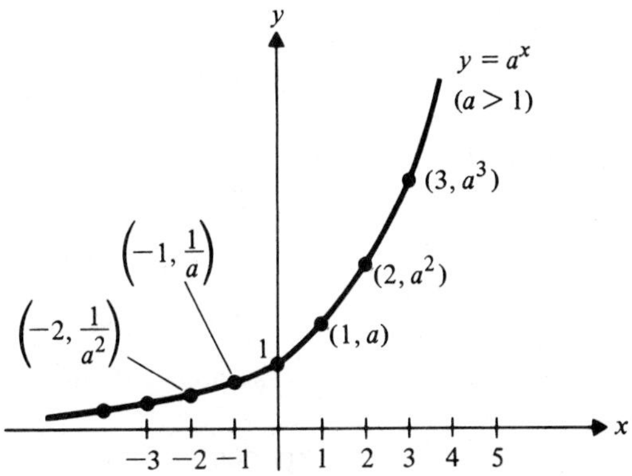

Figure 8.34

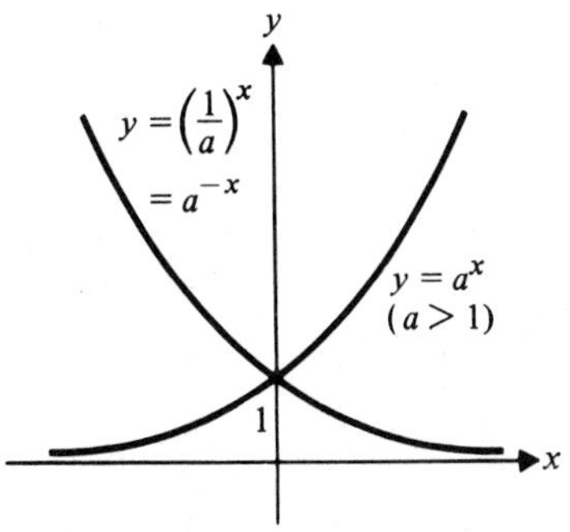

Figure 8.35

There is a difficulty with the definition of a^x when x is not rational. For example, if $x = \sqrt{2}$, then what is the value of $a^{\sqrt{2}}$? Two essentially equivalent approaches can be used to solve this problem. The first is to define $a^{\sqrt{2}}$ to be $\lim_{x\to\sqrt{2}} a^x$, where x approaches $\sqrt{2}$ through rational numbers only. It is then possible to prove that a^x is a continuous function. Alternatively, we can define a^x as a continuous function, in which case it follows that $a^{\sqrt{2}} = \lim_{x\to\sqrt{2}} a^x$. A detailed discussion of either of these approaches is not needed at this time, but we must at least realize that the definition of a^x for irrational x is not straightforward.

When one curve is obtained from another by replacing each x by $-x$, the two curves are mirror images of one another in the y-axis. The graph of $f(x) = a^{-x}$ is therefore the reflection of $f(x) = a^x$ in the y-axis (Fig. 8.35). But

$$a^{-x} = \frac{1}{a^x} = \left(\frac{1}{a}\right)^x,$$

where $0 < 1/a < 1$. In other words, the graph of the exponential function for $0 < a < 1$ is as shown in Fig. 8.35.

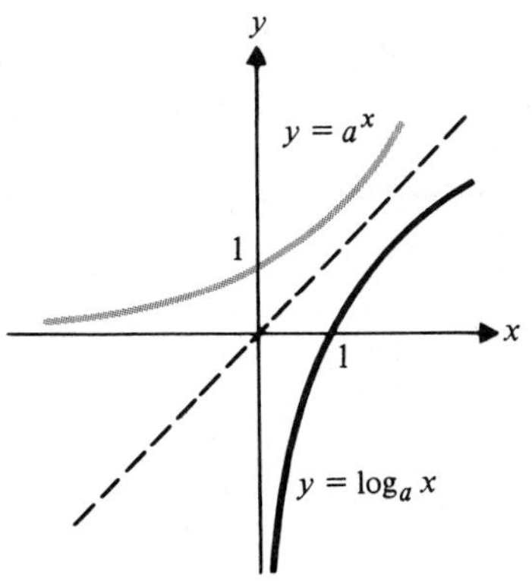

Figure 8.36

For the case in which $a = 1$, the graph of $f(x) = a^x = 1^x = 1$ is a horizontal straight line. We do not consider $f(x) = a^x$ for $a < 0$. (Can you see why?)

We now ask whether the exponential function $f(x) = a^x$ has an inverse function. The reflection of $y = a^x$ in the line $y = x$ (see Fig. 8.36 for the case in which $a > 1$) indicates that an inverse function does indeed exist. This inverse function is called the *logarithm function* to base a, and is denoted by

$$y = \log_a x. \tag{8.22}$$

Note that $\log_a x$ is defined only for $x > 0$ (i.e., we cannot take the logarithm of zero or a negative number), and the logarithm of 1 for any $a > 1$ is equal to 0.

We can also define $\log_a x$ for $0 < a < 1$, but such a logarithm is simply the negative of $\log_{1/a} x$, where $1/a > 1$ (see Exercise 33). Consequently, unless otherwise specified, we will always assume the base of logarithms to be greater than unity.

Although we have a graphical representation of the logarithm function, we still have the difficulty encountered with the inverse trigonometric functions. Given a value of $x > 0$, how shall we calculate $y = \log_a x$? Suppose that (x, y) is a point on $y = \log_a x$. Then (y, x) is a point on the exponential curve; i.e.,

$$y = \log_a x \quad \text{only if} \quad x = a^y. \tag{8.23}$$

Equations (8.23) contain the real meaning of the term *logarithm.* They state that y is the logarithm of x to base a only if x is a to the power y. In other words, the *logarithm of x to base a is the power to which a must be raised in order to produce x.* We may write from Eqs. (8.23) that

$$x = a^{\log_a x}, \qquad x > 0, \tag{8.24a}$$

$$\log_a (a^x) = x. \tag{8.24b}$$

These equations express the fact that the exponential and logarithm functions are inverses of each other.

In terms of the exponential function, exponential rules (8.20) take the following forms:

$$a^{x_1}a^{x_2} = a^{x_1+x_2}, \tag{8.25a}$$

$$\frac{a^{x_1}}{a^{x_2}} = a^{x_1-x_2}, \tag{8.25b}$$

$$(a^{x_1})^{x_2} = a^{x_1x_2}. \tag{8.25c}$$

Corresponding to these rules for the exponential function are the following rules for logarithms:

$$\log_a (x_1x_2) = \log_a x_1 + \log_a x_2, \tag{8.26a}$$

$$\log_a (x_1/x_2) = \log_a x_1 - \log_a x_2, \tag{8.26b}$$

$$\log_a ({x_1}^{x_2}) = x_2 \log_a x_1. \tag{8.26c}$$

To prove (8.26a), say, we set $z = \log_a (x_1x_2)$, in which case

$$\begin{aligned} a^z &= x_1x_2 \\ &= (a^{\log_a x_1})(a^{\log_a x_2}) \qquad \text{(Using (8.24a))} \\ &= a^{\log_a x_1 + \log_a x_2}. \qquad \text{(Using (8.25a))} \end{aligned}$$

Thus,

$$\log_a x_1 + \log_a x_2 = z = \log_a (x_1x_2).$$

We leave the proofs of (8.26b) and (8.26c) to the exercises.

Example 18 Simplify each of the following expressions:

a) $3^{\log_3 (x^2)}$;

b) $10^{-4 \log_{10} x}$;

c) $\log_a (a^{-x+3})$;

d) $\log_2 8 + \log_3 (1/27)$.

Solution a) Identity (8.24a) implies that

$$3^{\log_3 (x^2)} = x^2.$$

b) Because -4 intervenes between the logarithm and exponential operations, we cannot use (8.24a) immediately. The -4 can be relocated, however, with (8.26c):

$$10^{-4 \log_{10} x} = 10^{\log_{10} (x^{-4})} = 1/x^4 \quad (\text{if } x > 0).$$

c) Identity (8.24b) gives us

$$\log_a (a^{-x+3}) = -x + 3.$$

d) $\log_2 8 + \log_3 (1/27) = 3 - 3 = 0.$ □

Example 19 Solve each of the following equations:

a) $\log_5 x = -3$;

b) $\log_{10} x + \log_{10} (x + 1) = 0$;

c) $10^x - 12 + 10^{-x} = 0$.

Solution a) By means of (8.23), we have

$$x = 5^{-3} = 1/125.$$

b) Since $\log_{10} x + \log_{10} (x + 1) = \log_{10} [x(x + 1)]$, we can write

$$0 = \log_{10} [x(x + 1)].$$

If we now take exponentials to base 10, we get

$$10^0 = 10^{\log_{10} [x(x+1)]},$$

or

$$1 = x(x + 1).$$

This quadratic equation has solutions

$$x = \frac{-1 \pm \sqrt{1 + 4}}{2} = \frac{-1 \pm \sqrt{5}}{2}.$$

Since x must be positive (the original equation demands this), we have

$$x = \frac{-1 + \sqrt{5}}{2}.$$

c) If we multiply the equation by 10^x, then we get

$$0 = 10^{2x} - 12(10^x) + 1 = (10^x)^2 - 12(10^x) + 1.$$

But this is a quadratic equation in 10^x so that

$$10^x = \frac{12 \pm \sqrt{144 - 4}}{2} = 6 \pm \sqrt{35}.$$

Finally, then, we have

$$x = \log_{10}(6 \pm \sqrt{35}). \square$$

It is sometimes necessary to change from one base of logarithms to another. If we take logarithms to base b on both sides of identity (8.24a), we immediately obtain

$$\log_b x = (\log_a x)(\log_b a). \tag{8.27}$$

This equation defines $\log_b a$ as the conversion factor from logarithms to base a to logarithms to base b.

Example 20 Sketch a graph of the function

$$f(x) = \log_{10}\left(\frac{x-1}{x+1}\right).$$

Solution Since $f(x) = \log_{10}(1 - 2/(x+1))$, we first sketch $y = -2/(x+1)$, shown tinted in Fig. 8.37(a). A sketch of $y = 1 - 2/(x+1)$ is obtained therefrom by a shift of 1 unit upward. The final step is to take the logarithm to base 10 of each ordinate of the curve $y = 1 - 2/(x+1)$. The result is shown in Fig. 8.37(b). $\square$

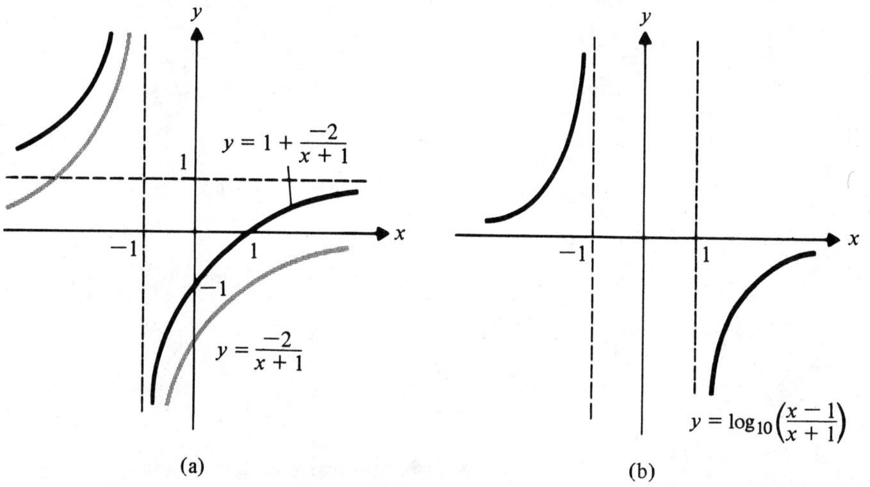

Figure 8.37

EXERCISES 8.6

In Exercises 1–24, find all values of x satisfying the given equation.

1. $\log_{10}(2 + x) = -1$
2. $10^{3x} = 5$
3. $\log_{10}(x^2 + 2x + 1) = 1$
4. $\log_{10}(x^2 + 2x + 10) = 1$
5. $10^{5-x^2} = 100$
6. $10^{1-x^2} = 100$
7. $\log_{10}(x - 3) + \log_{10} x = 1$
8. $\log_{10}(3 - x) + \log_{10} x = 1$
9. $\log_{10}[x(x - 3)] = 1$
10. $2\log_{10} x + \log_{10}(x - 1) = 2$
11. $2 = 10^{\sin x}$
12. $13 = 10^{\cos x}$
13. $14 \log_{10}(\tan x) = 1$
14. $14 \log_{10}(\mathrm{Tan}^{-1} x) = 1$
15. $\log_{10}(\sin x) = 8$
16. $\log_{10}(\sin x) = -1/8$

17. $13 = \text{Tan}^{-1}(\log_{10}(2x))$

18. $3\,\text{Cos}^{-1}(10^{\sin x}) = \pi$

19. $10^{\tan(\cos x)} = 4$

20. $10^{\cos(\tan x)} = 4$

21. $\log_a x + \log_a(x + 2) = 2$

22. $\log_a [x(x + 2)] = 2$

23. $2\cos(\log_{10}(\sin x)) = 1$

24. $\log_{10}\left(\log_{10}\left(\frac{x+3}{200x}\right) + 4\right) = -1$

Sketch a graph of the given function in Exercises 25–28.

25. $f(x) = a^{(-x^2)}$

26. $f(x) = \log_{10}(\cos x)$

27. $f(x) = \log_a |x^2 - 1|$

28. $f(x) = a^{\log_a(2x+1)}$

29. a) In the early period of reforestation, the percentage increase of timber each year is almost constant. If the original amount A_0 of a certain timber increases 3.5% the first year (and therefore 3.5% each year thereafter), find an expression for the amount of timber after t years.

b) How long does it take for timber of this type to double?

30. A new car costs \$10,000. In any year it depreciates to 75% of its value at the beginning of that year. What is the value of the car after t years?

31. Prove (8.26b) and (8.26c).

32. Is identity (8.26a) valid for any x_1 and x_2 whatsoever?

33. Show that if y is the logarithm of x to base a, then $-y$ is the logarithm of x to base $1/a$.

34. If the effective height of the earth's atmosphere (in meters) is the solution of the equation

$$10^{-6} = (1 - 2.08 \times 10^{-6}y)^{56},$$

find y. (See also Exercise 24 in Section 9.3.)

35. In acoustics, the minimum sound level detectable by the normal human ear is taken as "the tick of a watch at 20 feet under otherwise quiet conditions." This sound level, called the audible sound threshold, is said to have a decibel reading of 0. All other sounds are given decibel readings relative to this audible sound threshold. For example, a normal voice is approximately 70 decibels, the sound of a car 100 decibels, and a jet engine 160 decibels. Mathematically, the loudness of a sound is said to be L decibels if

$$L = 10\log_{10}\left(\frac{I}{I_0}\right),$$

where I is the intensity of the sound and I_0 is the intensity of sound at the audible threshold.

a) Turn this definition around and express the intensity I of a sound in terms of I_0 and its decibel reading L.

b) Intensity is a measure of energy per unit area per unit time transmitted by sound waves. Find the intensity of the normal voice, a car, and a jet engine relative to the intensity I_0.

c) If the pain threshold for sound has an intensity 10^{14} times I_0, what is its decibel reading?

d) If the intensity I_1 of one sound is 10 times the intensity I_2 of a second sound, how do their decibel readings differ?

36. The magnitude of an earthquake is measured in much the same way as a noise level. An earthquake of minimal size is taken as having value 0 on the Richter scale. Any other earthquake of intensity I is said to have magnitude R on the Richter scale if

$$R = \log_{10}\left(\frac{I}{I_0}\right),$$

where I_0 is the intensity of the minimal earthquake being used as reference.

a) Express the intensity of an earthquake in terms of I_0 and its reading on the Richter scale.

b) What is the reading on the Richter scale of earthquakes that have intensities 1.20×10^6 and 6.20×10^4 times I_0?

37. Sketch a graph of the function $f(x) = (\log_a x)^2$. Do you expect it to have any points of inflection?

38. a) If P dollars is invested at i% compounded n times per year, show that the accumulated value after t years is given by

$$A = P\left(1 + \frac{i}{100n}\right)^{nt}.$$

b) How long does it take to double an investment if interest is 8% compounded semiannually?

In Exercises 39–42, find all values of x satisfying the given equation.

39. $3a^{2x} + 3a^{-2x} = 10$

40. $2^x + 4^x = 8^x$

41. $3^{x+4} = 7^{x-1}$

42. $\log_x 2 = \log_{2x} 8$

43. Sketch a graph of the function

$$f(x) = \log_a(x + \sqrt{x^2 + 1}).$$

44. In Exercise 30, if repair costs on the car are estimated at \$50 the first year, increasing by 20% each year thereafter, set up a function $C(t)$ that represents the average yearly cost associated with owning the car for t years. (*Hint:* You may need the result of Exercise 27 in Section 6.1.)

45. A straight wire conductor has length $2L$ and circular cross-section of radius R. If the wire carries current $I > 0$, then the magnitude of the "vector" potential at a distance r from the center of the wire is given by

$$f(r) = \begin{cases} \dfrac{\mu_0 I}{4\pi}\left[\log_e\left(1 + \dfrac{4L^2}{R^2}\right) - 1 + \dfrac{r^2}{R^2}\right], & 0 \le r \le R, \\ \dfrac{\mu_0 I}{4\pi}\log_e\left(1 + \dfrac{4L^2}{r^2}\right), & r > R, \end{cases}$$

where μ_0 and $e > 1$ are positive constants.

a) Sketch a graph of this function.

b) Find the radius $r > R$ for which $f(r) = f(0)$.

In Exercises 46–48, solve the given equation for x in terms of y.

46. $y = \dfrac{a^x - a^{-x}}{2}$

47. $y = \dfrac{a^x + a^{-x}}{2}$

48. $y = \dfrac{a^x - a^{-x}}{a^x + a^{-x}}$

8.7 DERIVATIVES OF THE EXPONENTIAL AND LOGARITHM FUNCTIONS

The importance of π in mathematics and its applications is quite obvious. Another extremely important number, denoted by e, is defined as the limit

$$e = \lim_{z \to 0} (1 + z)^{1/z}. \tag{8.28}$$

An alternative definition is obtained by setting $v = 1/z$:

$$e = \lim_{v \to \infty} \left(1 + \frac{1}{v}\right)^v. \tag{8.29}$$

To actually prove the existence of either limit is extremely difficult. Substitution of various values of v into $(1 + 1/v)^v$ may suggest, however, whether a limit exists. The values in Table 8.2 are steadily increasing but because they are rapidly getting closer together, we suspect that they are indeed approaching a limit. This is true, and to twelve decimals

$$e = 2.718281828459.$$

Our first application of the number e is in the following theorem, where we find the derivative of the logarithm function.

Table 8.2

v	$\left(1 + \frac{1}{v}\right)^v$
1	2.000000
3	2.370370
5	2.488320
10	2.593742
100	2.704814
1000	2.716924
10000	2.718146
100000	2.718255
1000000	2.718282

▶ **Theorem 8.3**

$$\frac{d}{dx}\log_a x = \frac{1}{x}\log_a e \quad ◀ \tag{8.30}$$

Proof By Definition 3.1,

$$\begin{aligned} \frac{d}{dx}\log_a x &= \lim_{\Delta x \to 0} \frac{\log_a (x + \Delta x) - \log_a x}{\Delta x} \\ &= \lim_{\Delta x \to 0} \frac{1}{\Delta x}\left\{\log_a\left(\frac{x + \Delta x}{x}\right)\right\} \qquad \text{(By Eq. 8.26b)} \\ &= \lim_{\Delta x \to 0} \frac{1}{\Delta x}\left\{\log_a\left(1 + \frac{\Delta x}{x}\right)\right\} \end{aligned}$$

$$= \frac{1}{x} \lim_{\Delta x \to 0} \frac{x}{\Delta x} \left\{ \log_a \left(1 + \frac{\Delta x}{x} \right) \right\}$$

$$= \frac{1}{x} \lim_{\Delta x \to 0} \log_a \left(1 + \frac{\Delta x}{x} \right)^{x/\Delta x}. \qquad \text{(By Eq. 8.26c)}$$

Because the logarithm function is continuous, we may interchange the limit and logarithm operations (see Theorem 2.3):

$$\frac{d}{dx} \log_a x = \frac{1}{x} \log_a \left\{ \lim_{\Delta x \to 0} \left(1 + \frac{\Delta x}{x} \right)^{x/\Delta x} \right\}.$$

But

$$\lim_{\Delta x \to 0} \left(1 + \frac{\Delta x}{x} \right)^{x/\Delta x} = \lim_{\Delta x/x \to 0} \left(1 + \frac{\Delta x}{x} \right)^{x/\Delta x} = e,$$

and therefore

$$\frac{d}{dx} \log_a x = \frac{1}{x} \log_a e. \ \square$$

If we choose e as the base of logarithms (i.e., set $a = e$), derivative (8.30) simplifies to

$$\frac{d}{dx} \log_e x = \frac{1}{x}.$$

Because of the simplicity of this result, logarithms to base e are used more extensively than those to any other base. We call logarithms to base e *natural*, or *Naperian*,† *logarithms*, and usually denote them by $\ln x$ rather than $\log_e x$. We may state the following.

▶ **Corollary 1**

$$\frac{d}{dx} \ln x = \frac{1}{x} \quad ◀ \tag{8.31}$$

The chain rule also gives us the following.

▶ **Corollary 2** If $u(x)$ is a differentiable function of x, then

$$\frac{d}{dx} \log_a u = \frac{1}{u} \frac{du}{dx} \log_a e. \quad ◀ \tag{8.32}$$

▶ **Corollary 3** If $u(x)$ is differentiable,

$$\frac{d}{dx} \ln u = \frac{1}{u} \frac{du}{dx}. \quad ◀ \tag{8.33}$$

We now use implicit differentiation to find the derivative of the exponential function.

---

† Named after the Scottish mathematician John Napier (1550–1617), who first invented logarithms.

▶ **Theorem 8.4**

$$\frac{d}{dx} a^x = a^x \ln a \quad ◀ \tag{8.34}$$

Proof If we set $y = a^x$, then $x = \log_a y$, and we can differentiate (implicitly) with respect to x:

$$1 = \frac{1}{y}\frac{dy}{dx} \log_a e.$$

Thus,

$$\frac{dy}{dx} = \frac{y}{\log_a e} = \frac{a^x}{\log_a e}.$$

But if we substitute $x = b = e$ in Eq. (8.27), we find

$$\log_e e = (\log_a e)(\log_e a),$$

or

$$1 = (\log_a e)(\ln a).$$

Consequently, $1/\log_a e = \ln a$ and

$$\frac{dy}{dx} = a^x \ln a. \quad \square$$

Certainly it is again clear that should we use exponentials to base e, the differentiation formula in Theorem 8.4 simplifies to the following.

▶ **Corollary 1**

$$\frac{d}{dx} e^x = e^x \quad ◀ \tag{8.35}$$

The exponential function e^x is therefore its own derivative. We also have the following corollaries.

▶ **Corollary 2** For a differentiable function $u(x)$,

$$\frac{d}{dx} a^u = a^u \frac{du}{dx} \ln a. \quad ◀ \tag{8.36}$$

▶ **Corollary 3** For a differentiable function $u(x)$,

$$\frac{d}{dx} e^u = e^u \frac{du}{dx}. \quad ◀ \tag{8.37}$$

Example 21 Find dy/dx if y is defined as a function of x in each of the following:

a) $y = 2^{3x}$;

b) $y = \log_{10}(3x^2 + 4)$;

c) $y = x^2 e^{-2x}$;

d) $y = \dfrac{\ln x}{x}$;

e) $y = \sqrt{1 + e^x}$;

f) $y = \ln(\text{Tan}^{-1}(2x))$.

Solution a) $\dfrac{dy}{dx} = 2^{3x} \dfrac{d}{dx}(3x) \ln 2 = (3 \ln 2)2^{3x}$

b) $\dfrac{dy}{dx} = \dfrac{1}{3x^2 + 4} \dfrac{d}{dx}(3x^2 + 4) \log_{10} e = \dfrac{6x}{3x^2 + 4} \log_{10} e$

c) $\dfrac{dy}{dx} = 2xe^{-2x} - 2x^2 e^{-2x} = 2x(1 - x)e^{-2x}$

d) $\dfrac{dy}{dx} = \dfrac{x(1/x) - \ln x}{x^2} = \dfrac{1 - \ln x}{x^2}$

e) $\dfrac{dy}{dx} = \dfrac{1}{2\sqrt{1 + e^x}} \dfrac{d}{dx}(1 + e^x) = \dfrac{e^x}{2\sqrt{1 + e^x}}$

f) $\dfrac{dy}{dx} = \dfrac{1}{\text{Tan}^{-1}(2x)} \dfrac{d}{dx} \text{Tan}^{-1}(2x) = \dfrac{1}{\text{Tan}^{-1}(2x)} \dfrac{2}{1 + 4x^2}$ □

Example 22 Sketch a graph of the function $f(x) = x \ln x$.

Solution For critical points of $f(x)$, we solve

$$0 = f'(x) = \frac{x}{x} + \ln x = 1 + \ln x.$$

The only critical point is $x = 1/e$. Since $f''(x) = 1/x$, it follows that $f''(1/e) = e > 0$, and $x = 1/e$ yields a relative minimum of $f(1/e) = -1/e$. Further, because f'' is never equal to zero, there are no points of inflection on the graph. This information, along with the point (1, 0), is shown in Fig. 8.38(a). To complete the sketch we use l'Hôpital's rule to calculate

$$\lim_{x \to 0^+} x \ln x = \lim_{x \to 0^+} \frac{\ln x}{1/x} = \lim_{x \to 0^+} \frac{1/x}{-1/x^2} = \lim_{x \to 0^+} (-x) = 0^-,$$

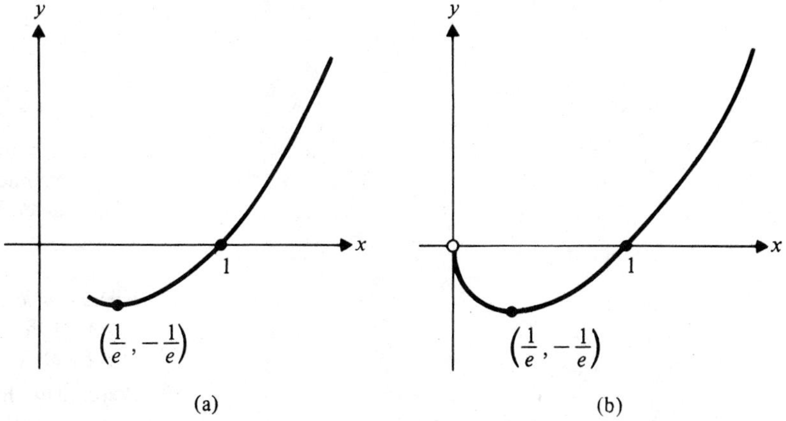

Figure 8.38

and add the extra information that

$$\lim_{x \to 0^+} f'(x) = -\infty.$$

The final graph is shown in Fig. 8.38(b). □

Recall that in Section 3.2 we proved power rule (3.6) only in the case that n was a nonnegative integer. It is now trivial to prove it for all real n, at least when $x > 0$. For in this case x^n can be written in the form $x^n = e^{n \ln x}$, and formulas (8.37) and (8.31) give us

$$\frac{d}{dx} x^n = \frac{d}{dx} e^{n \ln x} = e^{n \ln x} \frac{d}{dx}(n \ln x) = x^n\left(\frac{n}{x}\right) = nx^{n-1}.$$

A discussion of the power rule in the case that $x \le 0$ depends on the value of n (see Exercise 43).

EXERCISES 8.7

In Exercises 1–20, y is defined as a function of x. Find dy/dx in as simplified a form as possible.

1. $y = x \ln (x + 1)$
2. $y = x^2 + x^3 e^{4x}$
3. $y = \dfrac{e^{1-x}}{1 - x}$
4. $y = \sin (e^{2x})$
5. $y = \ln (\ln x)$
6. $y = e^{-2x} \sin 3x$
7. $y = \ln (x^2 e^{4x})$
8. $y = \dfrac{e^x - e^{-x}}{e^x + e^{-x}}$
9. $y = e^{-x} \ln x$
10. $y = x (\sin (\ln x) - \cos (\ln x))$
11. $y = e^{\sin u}$,
    $u = e^{1/x}$
12. $y = \ln (\cos v)$,
    $v = \sin^2 x$
13. $\ln (x + y) = x^2 y$
14. $xe^y + x^2 \ln y + y \sin x = 0$
15. $y = \ln (\sec x + \tan x)$
16. $y = x\sqrt{x^2 + 1} - \ln (x + \sqrt{x^2 + 1})$
17. $y = \ln (x + 4 + \sqrt{8x + x^2})$
18. $y = x - \frac{1}{4} \ln (1 + 5e^{4x})$
19. $e^{xy} = (x + y)^2$
20. $e^{1/x} + e^{1/y} = \dfrac{1}{x} + \dfrac{1}{y}$

In each of Exercises 21–24, find all relative extrema and points of inflection for the function, and sketch its graph.

21. $f(x) = e^{1/x}$
22. $f(x) = x^2 \ln x$
23. $f(x) = xe^{-x}$
24. $f(x) = x^2 e^{-x}$
25. Find the area of the largest rectangle that has one side on the x-axis and two vertices on the curve $y = e^{-x^2}$.
26. Find dy^2/dy^2 if y is defined implicitly as a function of x by
$$\ln (x^2 + y^2) = 2 \operatorname{Tan}^{-1} (y/x).$$
27. The function
$$f(x) = \frac{1}{\sqrt{2\pi}\sigma} e^{-(x-\mu)^2/(2\sigma^2)} \qquad (\sigma > 0, \mu \text{ constants})$$
is called the *normal probability density function*. Sketch its graph.
28. a) Use multiplication of ordinates to sketch a graph of the function $f(x) = e^{-x} \sin x$, $x \ge 0$.
    b) For what values of x does $f(x)$ have relative extrema? Does your sketch in (a) agree?
29. If a drug is injected into the bloodstream at time $t = 0$, it is sometimes assumed that the concentration of the drug is given by the function
$$f(t) = k(e^{-at} - e^{-bt}),$$
where k, a, and b are positive constants ($b > a$). Sketch a graph of this function, indicating any relative extrema and points of inflection.
30. A spring (with constant 5 newtons per meter) is attached to a fixed wall on one end and a 1-kilogram mass m on the other (Fig. 8.39). We choose a coordinate system with x positive to the right and with $x = 0$ at the center of mass

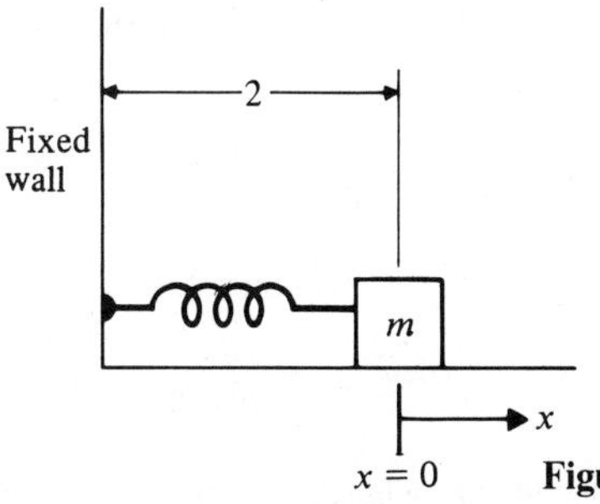

Figure 8.39

m when the spring is in the unstretched position. At time $t = 0$, m is pulled 1 meter to the left of $x = 0$ and given a speed of 3 meters per second to the right. During the subsequent motion, a frictional force equal in newtons to twice the velocity of the mass acts on m. We can show that if $x(t)$ is the position of m, then $x(t)$ must satisfy the differential equation

$$\frac{d^2x}{dt^2} + 2\frac{dx}{dt} + 5x = 0$$

and initial conditions

$$x(0) = -1, \qquad x'(0) = 3.$$

a) Verify that

$$x(t) = e^{-t}(\sin 2t - \cos 2t)$$

satisfies the differential equation and initial conditions. Sketch a graph of this function.

b) Find the maximum distance m will be from the wall.

31. Two long parallel rectangular loops lying in the same plane have lengths l and L and widths w and W, respectively (Fig. 8.40). If the loops do not overlap, and the distance between the near sides is s, the mutual inductance between the loops is given by

$$M = \frac{\mu_0 l}{2\pi} \ln \left\{ \frac{s + W}{s\left(1 + \dfrac{W}{s + w}\right)} \right\}.$$

where $\mu_0 > 0$ is a constant. Show that M is a decreasing function of s for $s > 0$.

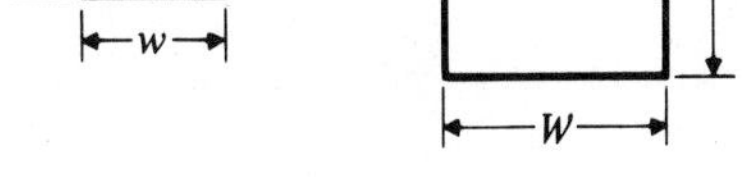

Figure 8.40

32. Two parallel wires carrying currents I (Fig. 8.41) create a magnetic field. The flux Φ of this field through the loop of dimensions h and w is given by

$$\Phi = \frac{\mu_0 hI}{2\pi} \ln \left\{ \frac{R(r + w)}{r(R + w)} \right\},$$

where $\mu_0 > 0$ is a constant. Sketch a graph of Φ first as a function of h and then as a function of w.

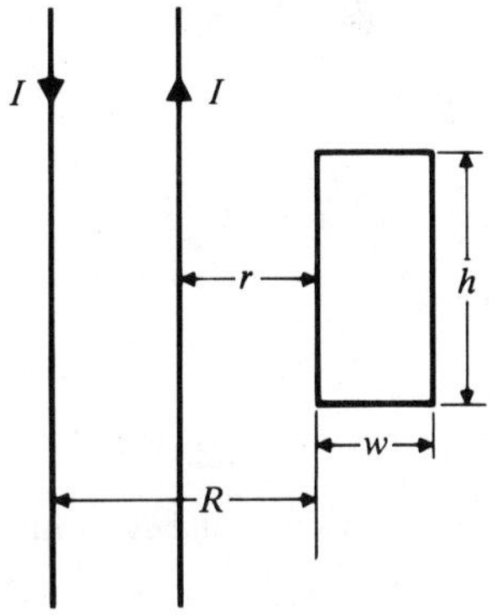

Figure 8.41

In many calculus books the natural logarithm function is defined by

$$\ln x = \int_1^x \frac{1}{t}\,dt$$

rather than the inverse function of the exponential function e^x. In Exercises 33–36, we show how this definition can be used to develop properties (8.26) of the logarithm function. Remember, then, in these exercises that you are to assume nothing about $\ln x$ other than the above definition.

33. Prove that

$$\frac{d}{dx}\ln x = \frac{1}{x},$$

and extend the result to

$$\frac{d}{dx}\ln u = \frac{1}{u}\frac{du}{dx}$$

when u is a function of x.

34. For $x_1 > 0$ a fixed constant and any $x > 0$, show that

$$\frac{d}{dx}\ln(xx_1) = \frac{1}{x}.$$

But from Exercise 33, we have that $d(\ln x)/dx = 1/x$. Use this to show that

$$\ln(x_1x_2) = \ln x_1 + \ln x_2.$$

35. Show that $\ln(1/x) = -\ln x$, and use this result and Exercise 34 to verify (8.26b) for $\ln x$.

36. By differentiating the function $\ln(x^{x_2})$ (for fixed $x_2 > 0$) with respect to x, show that

$$\frac{d}{dx}\ln(x^{x_2}) = \frac{d}{dx}(x_2 \ln x),$$

and from this that (8.26c) is valid for $\ln x$.

37. Show that the function

$$y = f(x) = Ax + B\left\{\frac{x}{2}\ln\left(\frac{x-1}{x+1}\right) + 1\right\}$$

satisfies the differential equation

$$(1 - x^2)\frac{d^2y}{dx^2} - 2x\frac{dy}{dx} + 2y = 0$$

for any constants A and B.

38. The equation

$$x = e^y - e^{-y}$$

defines y implicitly as a function of x. Find dy/dx in terms of x in two ways:

a) Solve first for an explicit definition of $y = f(x)$.

b) Differentiate implicitly with respect to x.

***39.** Snow has been falling for some time when a snowplow starts plowing a highway. The plow begins at 12:00, travels 2 kilometers during the first hour, and 1 kilometer during the second hour. Make reasonable assumptions to find out when the snow began falling.

***40.** The current I in the RC-circuit shown in Fig. 8.42 must satisfy the differential equation

$$R\frac{dI}{dt} + \frac{I}{C} = \frac{dE}{dt}.$$

If $E = E_0 \sin(\omega t)$, where E_0 and ω are constants, show that a solution is

$$I = f(t) = Ae^{-t/(RC)} + \frac{E_0}{Z}\sin(\omega t - \phi),$$

where A is any constant whatsoever, and

$$Z = \sqrt{R^2 + \frac{1}{\omega^2 C^2}}, \qquad \phi = \mathrm{Tan}^{-1}\left(-\frac{1}{\omega CR}\right).$$

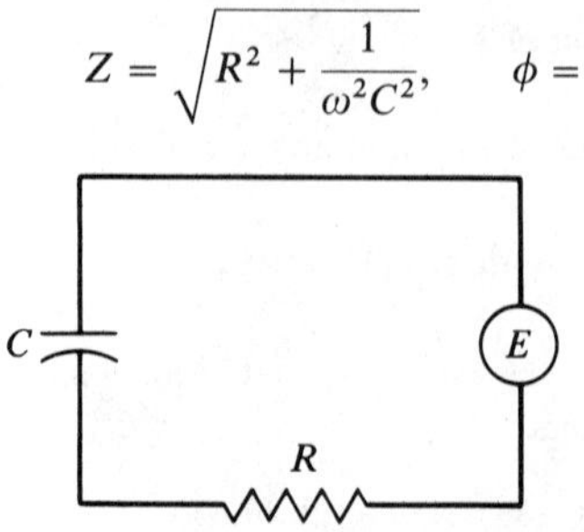

Figure 8.42

***41.** The current I in the LR-circuit shown in Fig. 8.43 must satisfy the differential equation

$$L\frac{dI}{dt} + RI = E.$$

If E is as in Exercise 40, show that a solution is

$$I = f(t) = Ae^{-Rt/L} + \frac{E_0}{Z}\sin(\omega t - \phi),$$

where A is any constant whatsoever, and

$$Z = \sqrt{R^2 + \omega^2 L^2}, \qquad \phi = \mathrm{Tan}^{-1}\left(\frac{\omega L}{R}\right).$$

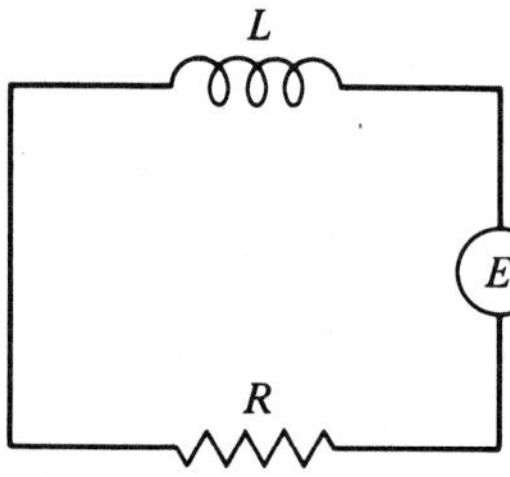

Figure 8.43

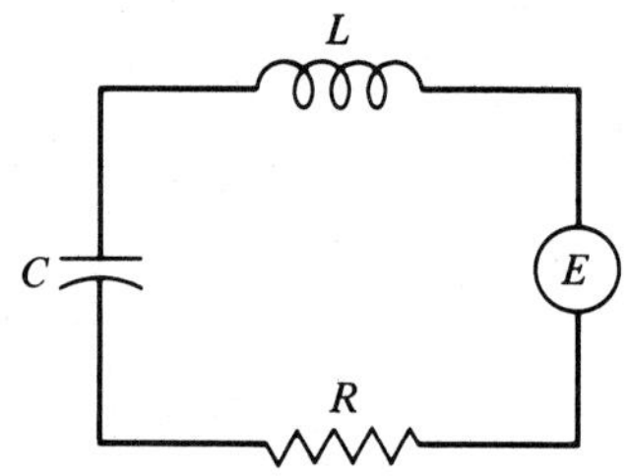

Figure 8.44

***42.** The current in the RCL-circuit shown in Fig. 8.44 must satisfy the differential equation

$$L\frac{d^2I}{dt^2} + R\frac{dI}{dt} + \frac{I}{C} = \frac{dE}{dt}.$$

If E is as in Exercise 40, show that a solution is

$$I = f(t) = e^{-Rt/(2L)}\{A\cos(vt) + B\sin(vt)\}$$

$$+ \frac{E_0}{Z}\sin(\omega t - \phi),$$

where A and B are any constants whatsoever, and

$$v = \sqrt{\frac{1}{LC} - \frac{R^2}{4L^2}},$$

$$Z = \sqrt{R^2 + \left(\omega L - \frac{1}{\omega C}\right)^2},$$

$$\phi = \mathrm{Tan}^{-1}\left(\frac{\omega L - 1/(\omega C)}{R}\right).$$

***43.** a) The function $f(x) = x^n$ is not defined for $x < 0$ if n is irrational (i.e., is not rational). Verify power rule (3.6) for $x < 0$ if n is rational and x^n is defined.

b) Discuss the derivative of $f(x) = x^n$ at $x = 0$.

8.8 LOGARITHMIC DIFFERENTIATION

Compare the three functions

$$x^n, \qquad a^x, \qquad x^x.$$

The first, in which the exponent is constant and the base is variable, is called a *power function.* The second, in which the base is constant and the exponent is variable, is called an *exponential function.* We have differentiation rules for the power function and the exponential function, namely,

$$\frac{d}{dx}x^n = nx^{n-1} \quad \text{and} \quad \frac{d}{dx}a^x = a^x \ln a,$$

respectively. The third function x^x is neither a power nor an exponential function; therefore, we cannot use either of the above differentiation formulas. To obtain the derivative of $y = f(x) = x^x$, we take natural logarithms,

$$\ln y = x \ln x,$$

and differentiate (implicitly) with respect to x:

$$\frac{1}{y}\frac{dy}{dx} = \ln x + \frac{x}{x} = \ln x + 1.$$

Consequently,

$$\frac{dy}{dx} = y(\ln x + 1),$$

or

$$\frac{d}{dx}x^x = x^x(\ln x + 1).$$

This process of taking logarithms and then differentiating is called *logarithmic differentiation.* As a second example, consider the following.

Example 23 Find $d(x^{\sin x})/dx$ when $x > 0$.

Solution If we set $y = x^{\sin x}$, then

$$\ln y = \sin x \ln x.$$

If we differentiate with respect to x, we get

$$\frac{1}{y}\frac{dy}{dx} = \cos x \ln x + \frac{1}{x}\sin x,$$

from which we have

$$\frac{dy}{dx} = x^{\sin x}\left(\cos x \ln x + \frac{1}{x}\sin x\right). \ \square$$

Logarithmic differentiation can also be used to differentiate complicated products or quotients, as in the following example.

Example 24 Find dy/dx for $-\pi < x < \pi$ if

$$y = f(x) = \frac{x^3(x^2+1)^{2/3}}{\sin^3 x}.$$

Solution When $x > 0$ we take natural logarithms of both sides of the definition of y:

$$\ln y = 3 \ln x + \tfrac{2}{3} \ln (x^2 + 1) - 3 \ln (\sin x).$$

Differentiation with respect to x gives

$$\frac{1}{y}\frac{dy}{dx} = \frac{3}{x} + \frac{2}{3}\frac{2x}{x^2+1} - \frac{3}{\sin x}\cos x,$$

and therefore

$$\frac{dy}{dx} = \frac{x^3(x^2+1)^{2/3}}{\sin^3 x}\left\{\frac{3}{x} + \frac{4x}{3(x^2+1)} - 3\cot x\right\}.$$

When $x < 0$, we write

$$y = \frac{(-x)^3(x^2+1)^{2/3}}{(-\sin x)^3}$$

and again take natural logarithms:

$$\ln y = 3\ln(-x) + \tfrac{2}{3}(x^2+1) - 3\ln(-\sin x).$$

Differentiation of this equation results in the same formula for dy/dx. □

EXERCISES 8.8

In each of Exercises 1–8, use logarithmic differentiation to find $f'(x)$.

1. $f(x) = x^{4x}$, $x > 0$
2. $f(x) = x^{4\cos x}$, $x > 0$
3. $f(x) = (x^2 + 3x^4)^3(x^2+5)^4$
4. $f(x) = \dfrac{\sqrt{x}(1+2x^2)}{\sqrt{1+x^2}}$
5. $f(x) = \left(1 + \dfrac{1}{x}\right)^x$, $x > 0$
6. $f(x) = x^2e^{4x}$
7. $f(x) = (\ln x)^{\ln x}$, $x > 1$
8. $f(x) = \left(\dfrac{1}{x}\right)^{1/x}$, $x > 0$
9. If $u(x)$ is positive for all x, find a formula for the derivative of u^u with respect to x.
10. If a company sells a certain commodity at a price r, then the market will demand x items per week, where
$$x = r^a e^{-b(r+c)}, \qquad r > \frac{a}{b},$$
and a, b, and c are positive constants.
    a) Show that the demand increases as the price decreases.
    b) Calculate the elasticity of demand defined by
$$\frac{Er}{Ex} = \frac{x}{r}\frac{dr}{dx}.$$

In Exercises 11 and 12, use logarithmic differentiation to find $f'(x)$.

11. $f(x) = \dfrac{\sqrt{x}(1-x^2)}{\sqrt{1+x^2}}$
12. $f(x) = (x^2+3x)^3(x^2+5)^4$

8.9 MORE INDETERMINATE FORMS

L'Hôpital's rule was used in Section 4.11 to evaluate limits of the indeterminate forms 0/0 and ∞/∞. With suitable restrictions on the functions $f(x)$ and $g(x)$, the rule states that if the limit of $f(x)/g(x)$ as x approaches a is of either of these indeterminate forms, then

$$\lim_{x\to a}\frac{f(x)}{g(x)} = \lim_{x\to a}\frac{f'(x)}{g'(x)},$$

provided the limit on the right exists or is equal to $\pm\infty$. For example,

$$\lim_{x\to 0}\frac{\tan x}{x}=\lim_{x\to 0}\frac{\sec^2 x}{1}=1;$$

$$\lim_{x\to\infty}\frac{\ln x}{x}=\lim_{x\to\infty}\frac{1/x}{1}=0;$$

$$\lim_{x\to\infty}\frac{e^{5x}}{x}=\lim_{x\to\infty}\frac{5e^{5x}}{1}=\infty.$$

The limits

$$\lim_{x\to\infty} xe^{-2x} \quad\text{and}\quad \lim_{x\to 0^+} x^2\ln x$$

are said to be of the indeterminate form $0\cdot\infty$. L'Hôpital's rule can again be used if we first rearrange the limits into one of the forms 0/0 or ∞/∞:

$$\lim_{x\to\infty} xe^{-2x}=\lim_{x\to\infty}\frac{x}{e^{2x}}=\lim_{x\to\infty}\frac{1}{2e^{2x}}=0;$$

$$\lim_{x\to 0^+} x^2\ln x=\lim_{x\to 0^+}\frac{\ln x}{1/x^2}=\lim_{x\to 0^+}\frac{1/x}{-2/x^3}=\lim_{x\to 0^+}\left(-\frac{x^2}{2}\right)=0.$$

Note that had we converted the second limit into the 0/0 form, then we would have had

$$\lim_{x\to 0^+} x^2\ln x=\lim_{x\to 0^+}\frac{x^2}{\dfrac{1}{\ln x}}=\lim_{x\to 0^+}\frac{2x}{\dfrac{-1}{x(\ln x)^2}}=\lim_{x\to 0^+} -2x^2(\ln x)^2.$$

Although this is correct, the limit on the right is more difficult to evaluate than the original. In other words, we must be somewhat judicious in converting a limit from the $0\cdot\infty$ indeterminate form to either 0/0 or ∞/∞.

Example 25 Evaluate each of the following limits if they exist:

a) $\lim_{x\to\pi/2} (x-\pi/2)\sec x$;

b) $\lim_{x\to 0^+} xe^{1/x}$.

Solution

a) $$\lim_{x\to\pi/2}(x-\pi/2)\sec x=\lim_{x\to\pi/2}\frac{x-\pi/2}{\cos x}=\lim_{x\to\pi/2}\frac{1}{-\sin x}=-1$$

b) $$\lim_{x\to 0^+} xe^{1/x}=\lim_{x\to 0^+}\frac{e^{1/x}}{1/x}=\lim_{x\to 0^+}\frac{e^{1/x}(-1/x^2)}{-1/x^2}=\lim_{x\to 0^+} e^{1/x}=\infty. \square$$

Various other indeterminate forms arise in the evaluation of limits, and many of these can be reduced to the 0/0 and ∞/∞ cases by introducing logarithms. In particular, the limits

$$\lim_{x\to 0^+} x^x,\qquad \lim_{x\to\infty}\left(1+\frac{1}{x}\right)^{x^2},\quad\text{and}\quad \lim_{x\to\pi/2^-}(\sec x)^{\cos x}$$

are said to display the indeterminate forms 0^0, 1^∞, and ∞^0, respectively. If we set

$$y = \lim_{x \to 0^+} x^x$$

and take natural logarithms of both sides, we have

$$\ln y = \ln \left\{ \lim_{x \to 0^+} x^x \right\}.$$

Because the logarithm function is continuous, we may interchange the limit and logarithm operations (see Theorem 2.3) to get

$$\ln y = \lim_{x \to 0^+} \{\ln x^x\} = \lim_{x \to 0^+} x \ln x = \lim_{x \to 0^+} \frac{\ln x}{1/x}.$$

We are now in a position to use L'Hôpital's rule:

$$\ln y = \lim_{x \to 0^+} \frac{1/x}{-1/x^2} = \lim_{x \to 0^+} (-x) = 0.$$

Exponentiation of both sides of $\ln y = 0$ now gives $y = e^0 = 1$; i.e.,

$$\lim_{x \to 0^+} x^x = 1.$$

For the second limit we again set

$$y = \lim_{x \to \infty} \left(1 + \frac{1}{x}\right)^{x^2}$$

and take natural logarithms:

$$\ln y = \ln \left\{ \lim_{x \to \infty} \left(1 + \frac{1}{x}\right)^{x^2} \right\} = \lim_{x \to \infty} \left\{ \ln \left(1 + \frac{1}{x}\right)^{x^2} \right\}$$

$$= \lim_{x \to \infty} \left\{ x^2 \ln \left(1 + \frac{1}{x}\right) \right\} = \lim_{x \to \infty} \left\{ \frac{\ln \left(\dfrac{x+1}{x}\right)}{1/x^2} \right\}.$$

By l'Hôpital's rule, we have

$$\ln y = \lim_{x \to \infty} \left\{ \frac{\dfrac{x}{x+1}\left(\dfrac{-1}{x^2}\right)}{-2/x^3} \right\} = \lim_{x \to \infty} \frac{x^2}{2(x+1)} = \lim_{x \to \infty} \frac{2x}{2} = \infty.$$

Consequently,

$$y = \lim_{x \to \infty} \left(1 + \frac{1}{x}\right)^{x^2} = \infty.$$

In the third limit we set

$$y = \lim_{x \to \pi/2^-} (\sec x)^{\cos x},$$

in which case

$$\ln y = \ln\left\{\lim_{x\to\pi/2^-} (\sec x)^{\cos x}\right\} = \lim_{x\to\pi/2^-} \{\cos x \ln(\sec x)\}$$

$$= \lim_{x\to\pi/2^-} \left\{\frac{\ln(\sec x)}{\sec x}\right\} = \lim_{x\to\pi/2^-} \left\{\frac{\frac{1}{\sec x}\sec x\tan x}{\sec x \tan x}\right\}$$

$$= \lim_{x\to\pi/2^-} \cos x = 0^+.$$

Thus,

$$y = \lim_{x\to\pi/2^-} (\sec x)^{\cos x} = e^0 = 1.$$

EXERCISES 8.9

Evaluate the limit (if it exists) in each of Exercises 1–12.

1. $\lim_{x\to\infty} x^2e^{-4x}$
2. $\lim_{x\to-\infty} x \sin\left(\frac{4}{x}\right)$
3. $\lim_{x\to 0} x \cot x$
4. $\lim_{x\to 0} \csc x\,(1 - \cos x)$
5. $\lim_{x\to 0^+} (\sin x)^x$
6. $\lim_{x\to 0^+} x^{\sin x}$
7. $\lim_{x\to\infty} \left(\frac{x+5}{x+3}\right)^x$
8. $\lim_{x\to 0} (1 + x)^{\cot x}$
9. $\lim_{x\to\infty} x^{1/x}$
10. $\lim_{x\to 0^+} |\ln x|^{\sin x}$
11. $\lim_{x\to 0^+} xe^{1/x}$
12. $\lim_{x\to 0} (\tan x - \csc x)$

Sketch a graph of the given function in Exercises 13–16.

13. $f(x) = x^{10}e^{-x}$
14. $f(x) = xe^{1/x}$
15. $f(x) = x^x, x > 0$
16. $f(x) = \dfrac{x^2}{\ln x}$
17. Use l'Hôpital's rule to calculate

$$\lim_{n\to\infty} \left(1 + \frac{i}{100n}\right)^n.$$

(See also Exercise 3a in Section 9.3.)

18. Evaluate

$$\lim_{x\to\infty} \left(\frac{x+a}{x+b}\right)^{cx}$$

for any constants a, b, and c.

The limits in Exercises 19–22 are said to exhibit the indeterminate form $\infty - \infty$. Calculate each limit by first transforming it into either of the indeterminate forms 0/0 or ∞/∞.

19. $\lim_{x\to\pi/2} (\sec x - \tan x)$
20. $\lim_{x\to 1} \left(\frac{x}{\ln x} - \frac{1}{x \ln x}\right)$
21. $\lim_{x\to 1} \left(\frac{x}{x-1} - \frac{1}{\ln x}\right)$
22. $\lim_{x\to 0} \left(\frac{1}{x^2} - \frac{1}{\sin^2 x}\right)$
23. The indeterminate forms 0^0, 1^∞, and ∞^0 are often evaluated by introducing logarithms. Show that the limit

$$\lim_{x\to\infty} (x - \ln x)$$

can be evaluated by introducing exponentials.

24. When an electrostatic field E is applied to a gaseous or liquid polar dielectric, a net dipole moment P per unit volume is set up, where

$$P(E) = \frac{e^E + e^{-E}}{e^E - e^{-E}} - \frac{1}{E}.$$

Show that $\lim_{E\to 0^+} P(E) = 0$.

*25. Planck's law for the energy density ψ of blackbody radiation states that

$$\psi = \psi(\lambda) = \frac{k\lambda^{-5}}{e^{c/\lambda} - 1},$$

where k and c are positive constants and λ is the wavelength of the radiation.

a) Show that

$$\lim_{\lambda \to 0^+} \psi(\lambda) = 0; \qquad \lim_{\lambda \to \infty} \psi(\lambda) = 0.$$

b) Show that a critical point of $\psi(\lambda)$ must satisfy the equation

$$(5\lambda - c)e^{c/\lambda} = 5\lambda.$$

Assuming that there is one and only one solution of this equation, sketch a graph of the function $\psi(\lambda)$.

* **26.** Find all values of a, b, and c for which

$$\lim_{x \to 0} \frac{e^{ax} - bx - \cos(x + cx^2)}{2x^3 + 5x^2} = 5.$$

8.10 HYPERBOLIC FUNCTIONS AND THEIR DERIVATIVES

Two combinations of the exponential function occur so often in physical applications that they have been given special names. Specifically, we define half the difference of e^x and e^{-x} as the hyperbolic sine function and half their sum as the hyperbolic cosine function:

$$\sinh x = \frac{e^x - e^{-x}}{2}, \qquad \cosh x = \frac{e^x + e^{-x}}{2}. \tag{8.38}$$

Note that the names of these functions, and their notations, bear a striking resemblance to those for the trigonometric functions. There are two important reasons for this. First, the hyperbolic functions $\sinh x$ and $\cosh x$ are related to the curve $x^2 - y^2 = 1$ (called the *unit hyperbola*) in much the same way as the trigonometric functions $\sin x$ and $\cos x$ are related to the unit circle $x^2 + y^2 = 1$. (We will point out two of these similarities in Examples 29 and 30.) Second, for each identity satisfied by the trigonometric functions, there is a corresponding identity satisfied by the hyperbolic functions—not the same identity, but one very similar. For example, using definitions (8.38), we have

$$\begin{aligned}(\cosh x)^2 - (\sinh x)^2 &= \left(\frac{e^x + e^{-x}}{2}\right)^2 - \left(\frac{e^x - e^{-x}}{2}\right)^2 \\ &= \tfrac{1}{4}\{(e^{2x} + 2 + e^{-2x}) - (e^{2x} - 2 + e^{-2x})\} \\ &= 1.\end{aligned}$$

Thus the hyperbolic sine and cosine functions satisfy the identity

$$\cosh^2 x - \sinh^2 x = 1, \tag{8.39}$$

which is reminiscent of the identity $\cos^2 x + \sin^2 x = 1$ for the trigonometric functions.

Just as four other trigonometric functions are defined in terms of $\sin x$ and $\cos x$, we define four corresponding hyperbolic functions as follows:

$$\begin{aligned}\tanh x &= \frac{\sinh x}{\cosh x} = \frac{e^x - e^{-x}}{e^x + e^{-x}}, \qquad \coth x = \frac{\cosh x}{\sinh x} = \frac{e^x + e^{-x}}{e^x - e^{-x}}, \\ \operatorname{sech} x &= \frac{1}{\cosh x} = \frac{2}{e^x + e^{-x}}, \qquad \operatorname{csch} x = \frac{1}{\sinh x} = \frac{2}{e^x - e^{-x}}.\end{aligned} \tag{8.40}$$

These definitions and (8.39) immediately imply that

$$1 - \tanh^2 x = \operatorname{sech}^2 x, \tag{8.41a}$$

$$\coth^2 x - 1 = \operatorname{csch}^2 x \tag{8.41b}$$

(analogous to $1 + \tan^2 x = \sec^2 x$ and $1 + \cot^2 x = \csc^2 x$, respectively).

Graphs of cosh x and sinh x (Fig. 8.45) are easily obtained by addition and subtraction of ordinates of e^x and e^{-x}. Note that cosh x is an even function, whereas sinh x is odd, and that neither function is periodic. To sketch tanh x, we use division of ordinates, together with the facts that tanh x is odd and that $\lim_{x\to\infty} \tanh x = 1$ (Fig. 8.45c). These three graphs then lead to the graphs for coth x, sech x, and csch x shown in Figs. 8.45(d)–(f).

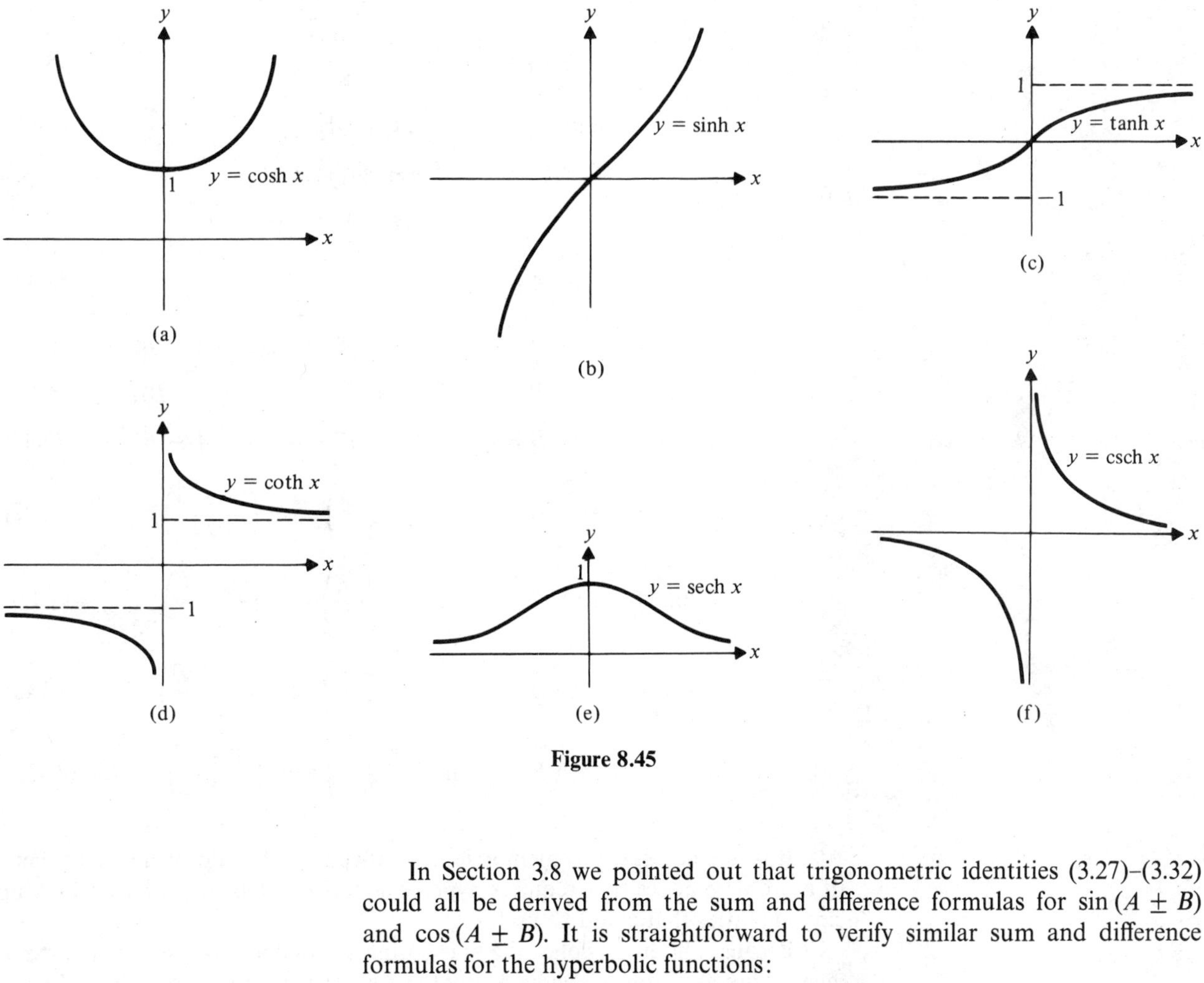

Figure 8.45

In Section 3.8 we pointed out that trigonometric identities (3.27)–(3.32) could all be derived from the sum and difference formulas for $\sin(A \pm B)$ and $\cos(A \pm B)$. It is straightforward to verify similar sum and difference formulas for the hyperbolic functions:

$$\sinh(A \pm B) = \sinh A \cosh B \pm \cosh A \sinh B, \tag{8.42a}$$

$$\cosh(A \pm B) = \cosh A \cosh B \pm \sinh A \sinh B. \tag{8.42b}$$

For example, definitions (8.38) give

$$\begin{aligned}\cosh A \cosh B - \sinh A \sinh B &= \left(\frac{e^A + e^{-A}}{2}\right)\left(\frac{e^B + e^{-B}}{2}\right) \\ &\quad - \left(\frac{e^A - e^{-A}}{2}\right)\left(\frac{e^B - e^{-B}}{2}\right) \\ &= \tfrac{1}{4}\{(e^{A+B} + e^{A-B} + e^{B-A} + e^{-A-B}) \\ &\quad - (e^{A+B} - e^{A-B} - e^{B-A} + e^{-A-B})\} \\ &= \tfrac{1}{2}\{e^{A-B} + e^{-(A-B)}\} = \cosh(A - B).\end{aligned}$$

With these formulas, we can now derive hyperbolic identities analogous to the trigonometric identities (3.27)–(3.32):

$$\tanh(A \pm B) = \frac{\tanh A \pm \tanh B}{1 \pm \tanh A \tanh B}, \tag{8.42c}$$

$$\sinh 2A = 2 \sinh A \cosh A, \tag{8.42d}$$

$$\begin{aligned}\cosh 2A &= \cosh^2 A + \sinh^2 A \\ &= 2\cosh^2 A - 1 = 1 + 2\sinh^2 A,\end{aligned} \tag{8.42e}$$

$$\tanh 2A = \frac{2 \tanh A}{1 + \tanh^2 A}, \tag{8.42f}$$

$$\sinh A \sinh B = \tfrac{1}{2}\cosh(A + B) - \tfrac{1}{2}\cosh(A - B), \tag{8.42g}$$

$$\sinh A \cosh B = \tfrac{1}{2}\sinh(A + B) + \tfrac{1}{2}\sinh(A - B), \tag{8.42h}$$

$$\cosh A \cosh B = \tfrac{1}{2}\cosh(A + B) + \tfrac{1}{2}\cosh(A - B), \tag{8.42i}$$

$$\sinh A + \sinh B = 2 \sinh\left(\frac{A + B}{2}\right) \cosh\left(\frac{A - B}{2}\right), \tag{8.42j}$$

$$\sinh A - \sinh B = 2 \cosh\left(\frac{A + B}{2}\right) \sinh\left(\frac{A - B}{2}\right), \tag{8.42k}$$

$$\cosh A + \cosh B = 2 \cosh\left(\frac{A + B}{2}\right) \cosh\left(\frac{A - B}{2}\right), \tag{8.42l}$$

$$\cosh A - \cosh B = 2 \sinh\left(\frac{A + B}{2}\right) \sinh\left(\frac{A - B}{2}\right). \tag{8.42m}$$

It is not necessary to memorize these identities; but do be aware of them. If a problem arises that requires hyperbolic functions, be prepared to look up any identities that could be useful.

Because the hyperbolic sine and cosine are defined in terms of the exponential function (for which we know the derivative), and the remaining hyperbolic functions are defined in terms of the hyperbolic sine and cosine, it follows that calculation of the derivatives of the hyperbolic functions should be straightforward. Indeed, if $u(x)$ is a differentiable function of x, then

$$\frac{d}{dx}\sinh u = \cosh u \frac{du}{dx}; \tag{8.43a}$$

$$\frac{d}{dx}\cosh u = \sinh u \frac{du}{dx}; \tag{8.43b}$$

$$\frac{d}{dx}\tanh u = \operatorname{sech}^2 u \frac{du}{dx}; \tag{8.43c}$$

$$\frac{d}{dx}\coth u = -\operatorname{csch}^2 u \frac{du}{dx}; \tag{8.43d}$$

$$\frac{d}{dx}\operatorname{sech} u = -\operatorname{sech} u \tanh u \frac{du}{dx}; \tag{8.43e}$$

$$\frac{d}{dx}\operatorname{csch} u = -\operatorname{csch} u \coth u \frac{du}{dx}. \tag{8.43f}$$

Example 26 Find dy/dx if y is defined as a function of x in each of the following:

a) $y = \operatorname{sech}(3x^2)$;
b) $y = \tanh(1 - 4x)$;
c) $y = \cos 2x \sinh 2x$;
d) $y = \cosh(\operatorname{Tan}^{-1} x^2)$.

Solution

a) $$\frac{dy}{dx} = -\operatorname{sech}(3x^2)\tanh(3x^2)\frac{d}{dx}(3x^2)$$
$$= -6x \operatorname{sech}(3x^2)\tanh(3x^2)$$

b) $$\frac{dy}{dx} = \operatorname{sech}^2(1-4x)\frac{d}{dx}(1-4x) = -4\operatorname{sech}^2(1-4x)$$

c) $$\frac{dy}{dx} = \cos 2x\,(2\cosh 2x) + (-2\sin 2x)\sinh 2x$$
$$= 2(\cos 2x \cosh 2x - \sin 2x \sinh 2x)$$

d) $$\frac{dy}{dx} = \sinh(\operatorname{Tan}^{-1} x^2)\frac{d}{dx}\operatorname{Tan}^{-1} x^2 = \sinh(\operatorname{Tan}^{-1} x^2)\frac{2x}{1+x^4} \quad \square$$

Example 27 If a uniform cable is hung between two fixed supports (Fig. 8.46), then the shape of the curve $y = f(x)$ must satisfy the differential equation

$$\frac{d^2y}{dx^2} = \frac{\rho g}{H}\sqrt{1 + \left(\frac{dy}{dx}\right)^2},$$

where ρ is the density of the cable (mass per unit length), $g > 0$ is the acceleration due to gravity, and $H > 0$ is the tension in the cable at its lowest point.

Show that a solution of this differential equation is

$$y = f(x) = \frac{H}{\rho g}\cosh\left(\frac{\rho g x}{H}\right) + C, \qquad C = \text{a constant.}$$

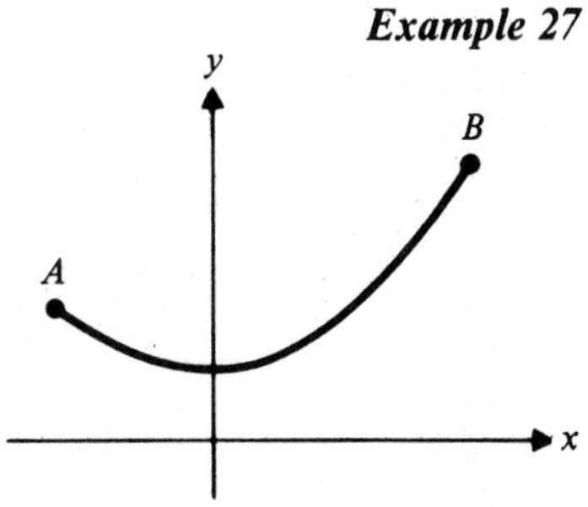

Figure 8.46

(In Exercise 32 in Section 9.5 we will derive this solution. For now we simply wish to verify that the hyperbolic cosine is indeed a solution.)

Solution Since $dy/dx = \sinh(\rho g x/H)$, it follows that

$$\frac{d^2y}{dx^2} = \frac{\rho g}{H}\cosh\left(\frac{\rho g x}{H}\right).$$

Further

$$\frac{\rho g}{H}\sqrt{1+\left(\frac{dy}{dx}\right)^2} = \frac{\rho g}{H}\sqrt{1+\sinh^2\left(\frac{\rho g x}{H}\right)}$$
$$= \frac{\rho g}{H}\sqrt{\cosh^2\left(\frac{\rho g x}{H}\right)} = \frac{\rho g}{H}\cosh\left(\frac{\rho g x}{H}\right),$$

and the proof is complete. □

What this example shows is that the many telephone and hydro wires crisscrossing the country hang in the form of a hyperbolic cosine. Engineers often call this curve a *catenary*.

Example 28 In studying wave guides, the electrical engineer often encounters the differential equation

$$\frac{d^2y}{dx^2} - ky = 0, \qquad k > 0 \text{ a constant.}$$

Show that $y = f(x) = A\cosh\sqrt{k}x + B\sinh\sqrt{k}x$ is a solution for any constants A and B.

Solution Since $dy/dx = \sqrt{k}A\sinh\sqrt{k}x + \sqrt{k}B\cosh\sqrt{k}x$, then

$$\frac{d^2y}{dx^2} = kA\cosh\sqrt{k}x + kB\sinh\sqrt{k}x = k(A\cosh\sqrt{k}x + B\sinh\sqrt{k}x) = ky.$$
□

In Examples 29 and 30 we illustrate the geometric parallel between the trigonometric sine and cosine functions and the hyperbolic sine and cosine functions.

Example 29

a) Show that every point (x, y) on the unit circle $x^2 + y^2 = 1$ can be expressed in the form $x = \cos t$, $y = \sin t$ for some real number t in the interval $0 \le t < 2\pi$.
b) Show that every point (x, y) on the right half of the unit hyperbola $x^2 - y^2 = 1$ can be expressed in the form $x = \cosh t$, $y = \sinh t$ for some real number t.

Solution

a) If t is the angle shown in Fig. 8.47, then clearly the coordinates of P are $x = \cos t$, $y = \sin t$. As angle t ranges from 0 to 2π, P traces the circle exactly once.
b) A sketch of the unit hyperbola $x^2 - y^2 = 1$ is shown in Fig. 8.48. If $x = \cosh t$ and $y = \sinh t$ are coordinates of a point P in the plane, where t is some real number, then identity (8.39) implies that $x^2 - y^2 = \cosh^2 t - \sinh^2 t = 1$. In other words, P is on the unit hyperbola. Furthermore,

since $x = \cosh t$ is always positive, P must be on the right half of the hyperbola. Finally, because the range of $x = \cosh t$ is $x \geq 1$ (Fig. 8.45a), and the range of $y = \sinh t$ is $-\infty < y < \infty$ (Fig. 8.45b), it follows that every point on the right half of the hyperbola can be obtained from some value of t.

Note that t is *not* in this case the angle formed by the positive x-axis and the line joining the origin to (x, y). □

Example 30 Show that the area bounded by the positive x-axis, the line joining the origin to the point $P(x, y)$ on the curves in Figs. 8.47 and 8.48, and the curves themselves is $t/2$.

Solution

a) Since the area of a sector of a circle is one-half the product of the square of the radius and the angle at the center of the circle, the shaded area in Fig. 8.47 is $(1/2)(1)^2(t) = t/2$.

b) The shaded area in Fig. 8.48 is the area of triangle OQP less the area beneath the hyperbola between $x = 1$ and Q:

$$\text{Area} = \frac{1}{2}(x)(y) - \int_1^x y\,dx.$$

Since the coordinates of P can be expressed in the form $x = \cosh t$ and $y = \sinh t$ (Example 29), we may set

$$\text{Area} = \frac{1}{2}\cosh t \sinh t - \int_0^t \sinh t\,(\sinh t\,dt).$$

If we now use identities (8.42d), and (8.42e) in the form

$$\sinh^2 t = (\cosh 2t - 1)/2,$$

we have

$$\text{Area} = \frac{1}{4}\sinh 2t - \frac{1}{2}\int_0^t (\cosh 2t - 1)\,dt.$$

Since the derivative of $\sinh 2t$ is $2\cosh 2t$, it follows that an antiderivative of $\cosh 2t$ is $(\frac{1}{2})\sinh 2t$ and

$$\text{Area} = \frac{1}{4}\sinh 2t - \frac{1}{2}\left\{\frac{1}{2}\sinh 2t - t\right\}_0^t = \frac{t}{2}. \;\square$$

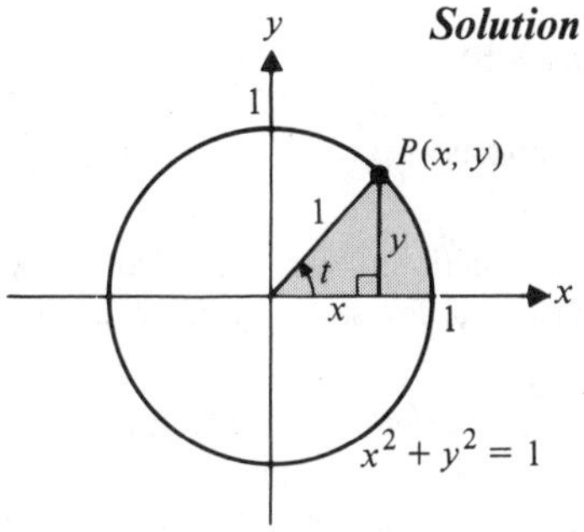

Figure 8.47

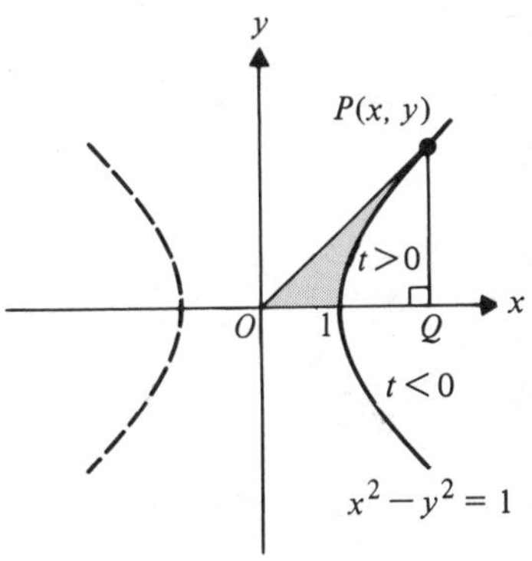

Figure 8.48

EXERCISES 8.10

In Exercises 1–10, evaluate the given expression (if it has a value). (You will need an electronic calculator with at least the exponential function or a set of tables for these exercises.)

C **1.** $3 \cosh 1$

C **2.** $\sinh(\pi/2)$

C **3.** $\tanh \sqrt{1 - \sin 3}$

C **4.** $\text{Sin}^{-1}(\text{sech}\,10)$

C **5.** $\text{Cos}^{-1}(2\,\text{csch}\,1)$

C **6.** $\coth(\sinh 5)$

C **7.** $\sqrt{\ln|\sinh(-3)|}$

C **8.** $\text{sech}(\sec(\pi/3))$

C **9.** $e^{-2\cosh e}$

C **10.** $\sinh(\text{Cot}^{-1}(-3\pi/10))$

In Exercises 11–20, y is defined as a function of x. Find dy/dx in as simplified a form as possible.

11. $y = \text{csch}(2x + 3)$

12. $y = x \sinh(x/2)$

13. $y = \sqrt{1 - \text{sech}\,x}$

14. $y = \tanh(\ln x)$

15. $\cosh(x + y) = 2x$

16. $y + \coth x = \sqrt{1 + y}$

17. $y = u \cosh u,$
$u = e^x + e^{-x}$

18. $y = \tan(\cosh t),$
$t = \cos(\tanh x)$

19. $y = \mathrm{Tan}^{-1}(\sinh x)$

20. $y = \ln\sqrt{\tanh 2x}$

21. Verify the results in identities (8.42c)–(8.42m).

22. Verify the differentiation formulas (8.43).

23. Show that $y = \sinh x$ and $y = \tanh x$ have points of inflection at $x = 0$.

24. To analyze vertical vibrations of the beam shown in Fig. 8.49, we must solve the differential equation

$$\frac{d^4y}{dx^4} - k^4y = 0, \qquad k > 0 \text{ a constant.}$$

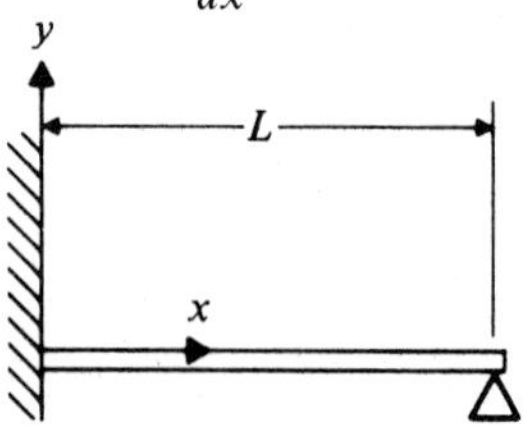

Figure 8.49

a) Show that a solution is

$$y = f(x) = A\cos(kx) + B\sin(kx) + C\cosh(kx) + D\sinh(kx)$$

for any constants A, B, C, and D.

b) If the left end ($x = 0$) is fastened horizontally and the right end ($x = L$) is pinned, then $f(x)$ must satisfy the conditions

$$f(0) = f'(0) = f(L) = f''(L) = 0.$$

Show that these restrictions imply that $C = -A$, $D = -B$, and A and B must satisfy the equations

$$A(\cos kL - \cosh kL) + B(\sin kL - \sinh kL) = 0,$$

$$A(\cos kL + \cosh kL) + B(\sin kL + \sinh kL) = 0.$$

c) Eliminate A and B between these equations to show that k must satisfy the condition

$$\tan kL = \tanh kL.$$

25. Show that $y = \operatorname{sech} x$ has points of inflection for $x = \pm\ln(\sqrt{2} + 1)$.

C **26.** A uniform hydro cable P meters long with a density of ρ kilograms per meter is hung from two supports at the same level L meters apart (Fig. 8.50). If the supports are b meters higher than the lowest point in the cable and H is the tension in the cable at the lowest point, it can be shown that

$$\rho g P = 2H \sinh\frac{\rho g L}{2H}.$$

a) If $L = 70$ meters, $\rho = 0.5$ kilogram per meter, and $P = 80$ meters, find H. [*Hint*: Set $z = \rho g/(2H)$, and use Newton's iterative procedure with $z_1 = 0.01$ to solve $Pz = \sinh(Lz)$ for z.]

b) Use the result in (a) to find b. (See Example 27 for the equation of the curve describing the cable.)

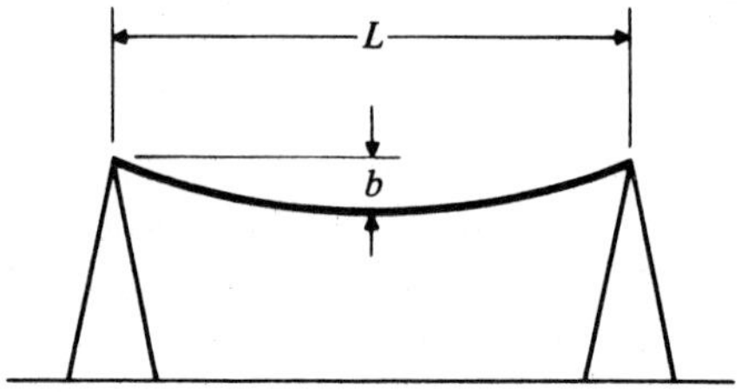

Figure 8.50

27. Each hyperbolic function has associated with it an inverse hyperbolic function.

a) Draw the inverse functions $\mathrm{Sinh}^{-1} x$, $\mathrm{Tanh}^{-1} x$, $\mathrm{Coth}^{-1} x$, and $\mathrm{Csch}^{-1} x$ for $\sinh x$, $\tanh x$, $\coth x$, and $\operatorname{csch} x$.

b) Why do $\cosh x$ and $\operatorname{sech} x$ not have inverse functions? It is customary to associate functions $\mathrm{Cosh}^{-1} x$ and $\mathrm{Sech}^{-1} x$ with $\cosh x$ and $\operatorname{sech} x$ by taking the reflections of these hyperbolic functions, and restricting their ranges to be positive numbers. Draw graphs of $\mathrm{Cosh}^{-1} x$ and $\mathrm{Sech}^{-1} x$.

c) Obtain the following derivatives for the inverse hyperbolic functions:

$$\frac{d}{dx}\mathrm{Sinh}^{-1} x = \frac{1}{\sqrt{x^2+1}};$$

$$\frac{d}{dx}\mathrm{Cosh}^{-1} x = \frac{1}{\sqrt{x^2-1}};$$

$$\frac{d}{dx}\mathrm{Tanh}^{-1} x = \frac{1}{1-x^2}, \quad |x| < 1;$$

$$\frac{d}{dx}\mathrm{Coth}^{-1} x = \frac{1}{1-x^2}, \quad |x| > 1;$$

$$\frac{d}{dx}\mathrm{Sech}^{-1} x = \frac{-1}{x\sqrt{1-x^2}};$$

$$\frac{d}{dx}\mathrm{Csch}^{-1} x = \frac{-1}{|x|\sqrt{1+x^2}}.$$

*d) Show that

$$\mathrm{Sinh}^{-1} x = \ln(x + \sqrt{x^2+1});$$

$$\mathrm{Cosh}^{-1} x = \ln(x + \sqrt{x^2-1});$$

$$\mathrm{Tanh}^{-1} x = \frac{1}{2}\ln\left(\frac{1+x}{1-x}\right), \quad |x| < 1.$$

***28.** When an object of mass m falls from rest at time $t = 0$ under the influence of gravity and an air resistance proportional to the square of velocity, its velocity v as a function of time is always defined by the equation

$$\frac{1}{2}\sqrt{\frac{\beta}{mg}}\ln\left(\frac{\sqrt{mg/\beta} - v}{\sqrt{mg/\beta} + v}\right) = -\frac{\beta t}{m}.$$

a) Show that when this equation is solved for v in terms of t, we get

$$v(t) = \sqrt{\frac{mg}{\beta}}\tanh\left(\sqrt{\frac{\beta g}{m}}\,t\right).$$

b) What is the limit of v for large t, called the *limiting velocity*?

SUMMARY

We have now completed the list of differentiation formulas. It is nice to know that the vast majority of functions that we will ever encounter in applications can be differentiated using the rules given in Chapter 3 and the formulas given in Chapter 8. At first sight there appears to be quite a lengthy list of formulas in this chapter, but only some are essential, and others follow immediately from those. Derivatives of $\sin u$, $\tan u$, and $\sec u$ are essential; derivatives of the cofunctions $\cos u$, $\cot u$, and $\csc u$ can be obtained from these by replacing each function by its cofunction and introducing a negative sign. It is also true that once we know the derivatives of $\sin u$ and $\cos u$, it takes but one line to calculate the derivative of any other trigonometric function. Derivatives of $\mathrm{Sin}^{-1} u$, $\mathrm{Tan}^{-1} u$, and $\mathrm{Sec}^{-1} u$ are essential; derivatives of the inverse cofunctions $\mathrm{Cos}^{-1} u$, $\mathrm{Cot}^{-1} u$, and $\mathrm{Csc}^{-1} u$ are the negatives of these. In Section 9.5 we will also see that for integration purposes, it is again $\mathrm{Sin}^{-1} u$, $\mathrm{Tan}^{-1} u$, and $\mathrm{Sec}^{-1} u$ that are essential. For the exponential and logarithm functions, it is true that once we know the formulas in Eqs. (8.32) and (8.36), all other derivatives are special cases. We would suggest, however, that for most applications, formulas (8.33) and (8.37) are more important.

Hyperbolic functions are special combinations of exponential functions, and as such their derivatives can always be calculated quite easily. If, on the other hand, we were involved in an area of applied mathematics that suggested continual use of hyperbolic functions and their derivatives, then formulas (8.43) would be essential.

The transcendental functions in this chapter (there are others) have been divided into compact groups: trigonometric functions and inverse trigonometric functions, the exponential function and its inverse, the logarithm function, and hyperbolic functions (and their inverses in Exercise 27 at the end of Section 8.10). The relationship between a function and its inverse is an extremely important one. For instance, in Section 9.3, our use of the inverse nature of the exponential and logarithm functions is indispensable to the solution of many differential equations. The integration technique discussed in Section 9.5 is based on the relation of an inverse trigonometric function to the corresponding trigonometric function. Evaluation of the indeterminate forms 0^0, 1^∞, and ∞^0 in Section 8.9 was based on the fact that taking logarithms and exponentials are inverse operations. By taking logarithms, we transformed each of these indeterminate forms into a limit to which we could apply l'Hôpital's rule ($0/0$ or ∞/∞). We then used exponentials to recapture the original limit.

REVIEW EXERCISES

In Exercises 1–10, find all solutions of the given equation.

1. $e^x = 5$

2. $4 \sin 2x = 1$

3. $2 \ln x = 11$

4. $\mathrm{Tan}^{-1}(3x + 2) = 5$

5. $\cos 2x = \sin x$

6. $\ln x + \ln (x + 1) = \ln 3$

7. $3\,\mathrm{Sin}^{-1}(e^{x+2}) = 2$

8. $3 \sin (e^{x+2}) = 2$

9. $\tan (x \cosh 2) = \frac{1}{4}$

10. $\sinh x = 4$

In Exercises 11–26, y is defined as a function of x. Find dy/dx in as simplified a form as possible.

11. $y = e^{2x} \sin 3x$

12. $y = \ln (3x^2 + 5)$

13. $y = \mathrm{Sin}^{-1}(2 - 3x)$

14. $y = 3 \sinh (x^2)$

15. $y = \dfrac{\mathrm{Cos}^{-1} x}{\mathrm{Sin}^{-1} x}$

16. $y = \mathrm{Tan}^{-1}\left(\dfrac{1}{x} + x\right)$

17. $y = e^{\cosh x}$

18. $y = \sqrt{\ln \sec x}$

19. $y = x^{1-x},\ x > 0$

20. $y = \ln (x + \sqrt{x^2 - 1})$

21. $e^x + e^y = 25xy$

22. $\sinh y = \sin x$

23. $x = \sin y + \cos y$

24. $\ln (\tan (x + y)) = 1$

25. $y = \ln \left(\dfrac{4 + x}{4 - x}\right)$

26. $y = x + 3 \ln (\cos x + 3 \sin x)$

27. Sketch a graph of the function $f(x) = (\ln x)^2$. Where is its point of inflection?

28. Sketch a graph of the function $f(x) = (e^x - 1)^{-1}$.

29. Find a constant m in order that the function $y = f(x) = e^{mx}$ satisfy the differential equation

$$\frac{d^2y}{dx^2} - \frac{dy}{dx} - 6y = 0.$$

30. Sketch graphs of each of the following functions:

a) $f(x) = \tan (\mathrm{Tan}^{-1} x)$;

b) $f(x) = \mathrm{Tan}^{-1}(\tan x)$.

31. If P_0, T_0, g, R, and λ are all constants, solve the following equation for y in terms of P:

$$\ln P = \frac{g}{R\lambda} \ln (T_0 - \lambda y) + \ln P_0 - \frac{g}{R\lambda} \ln T_0.$$

32. When an object of mass m falls from rest at time $t = 0$ under the influence of gravity and an air resistance proportional to velocity, its velocity v as a function of time is always defined by the equation

$$\ln \left(g - \frac{\beta v}{m}\right) = -\frac{\beta t}{m} + \ln g,$$

where $g > 0$ is the acceleration due to gravity and $\beta > 0$ is the air resistance constant.

a) Solve this equation for v in terms of t.

b) What is the limit of v for large t, called the *limiting velocity*?

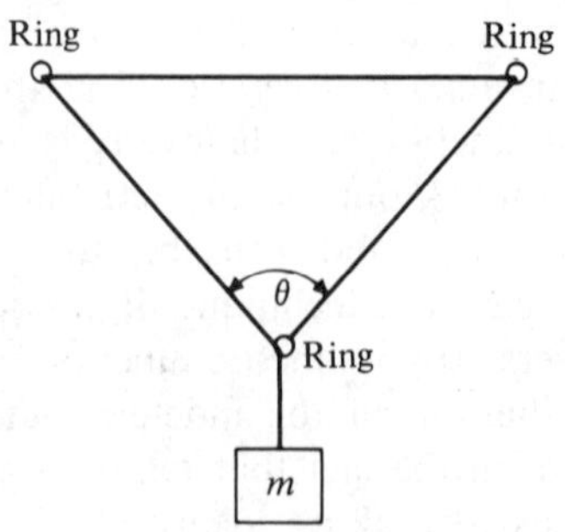

Figure 8.51

***33.** A rope with a ring at one end is passed through two fixed rings at the same level (Fig. 8.51). The end without the ring is then passed through the ring at the other end, and a mass m is attached to it. If the rope moves so as to maximize the distance from m to the line through the fixed rings, find the angle θ.

Techniques of Integration

9

9.1 INTRODUCTION

We now develop a number of techniques for evaluating indefinite integrals. Through our differentiations, we have already developed a number of integration formulas—for example,

$$\int u^n\,du = \frac{u^{n+1}}{n+1} + C, \qquad n \neq -1; \tag{9.1}$$

$$\int \frac{1}{u}\,du = \ln|u| + C; \tag{9.2}$$

$$\int e^u\,du = e^u + C; \tag{9.3}$$

$$\int a^u\,du = \frac{a^u}{\ln a} + C = a^u \log_a e + C. \tag{9.4}$$

The formulas in Eqs. (9.1), (9.3), and (9.4) need no verification; they are restatements of corresponding differentiation formulas. Equation (9.2), however, is not quite so obvious. Certainly if $u > 0$, it is clear from our differentiation of the logarithm function that

$$\int \frac{1}{u}\,du = \ln u + C.$$

On the other hand, if $u < 0$ and we set $y = -u$, then y is positive, and

$$\int \frac{1}{u}\,du = \int -\frac{1}{y}(-dy) = \int \frac{1}{y}\,dy = \ln y + C = \ln(-u) + C.$$

The two statements

$$\int \frac{1}{u}\,du = \ln u + C \quad \text{if } u > 0$$

and

$$\int \frac{1}{u}\,du = \ln(-u) + C \quad \text{if } u < 0$$

are both contained in Eq. (9.2).

We have many other potential integration formulas. For instance, each differentiation of a trigonometric and inverse trigonometric function can be written in terms of an integral (as can differentiations of hyperbolic functions). A partial list of useful formulas is as follows:

$$\int \cos u\,du = \sin u + C; \tag{9.5}$$

$$\int \sin u\, du = -\cos u + C; \tag{9.6}$$

$$\int \sec^2 u\, du = \tan u + C; \tag{9.7}$$

$$\int \csc^2 u\, du = -\cot u + C; \tag{9.8}$$

$$\int \sec u \tan u\, du = \sec u + C; \tag{9.9}$$

$$\int \csc u \cot u\, du = -\csc u + C; \tag{9.10}$$

$$\int \tan u\, du = \ln|\sec u| + C = -\ln|\cos u| + C; \tag{9.11}$$

$$\int \cot u\, du = -\ln|\csc u| + C = \ln|\sin u| + C; \tag{9.12}$$

$$\int \sec u\, du = \ln|\sec u + \tan u| + C; \tag{9.13}$$

$$\int \csc u\, du = \ln|\csc u - \cot u| + C; \tag{9.14}$$

$$\int \frac{1}{\sqrt{a^2 - u^2}}\, du = \mathrm{Sin}^{-1}\left(\frac{u}{a}\right) + C; \tag{9.15}$$

$$\int \frac{1}{a^2 + u^2}\, du = \frac{1}{a}\mathrm{Tan}^{-1}\left(\frac{u}{a}\right) + C; \tag{9.16}$$

$$\int \frac{1}{u\sqrt{u^2 - a^2}}\, du = \frac{1}{a}\mathrm{Sec}^{-1}\left(\frac{u}{a}\right) + C. \tag{9.17}$$

Formulas (9.5)–(9.10) are restatements of differentiation formulas listed in Chapter 8. Formulas (9.11)–(9.14) can be verified by differentiating the right-hand side to give the integrand on the left. We will also show how to derive these formulas in Section 9.4. Equations (9.15)–(9.17) are generalizations of differentiation formulas for inverse trigonometric functions. They can be checked by differentiation of the right-hand sides, or derived by the technique of trigonometric substitutions discussed in Section 9.5.

Many other integration formulas exist, but one of the purposes of this chapter is to develop integration techniques that will eliminate the need for excessive memorization of formulas. In other words, our emphasis will be on the development of antiderivatives rather than the memorization of formulas, and we therefore prefer to keep the list of formulas as short as possible at this stage. We suggest, however, that the integrations in Eqs. (9.1)–(9.14) are indispensable for all further work, and should therefore be committed to memory.

9.2 INTEGRATION BY SUBSTITUTION

In Chapter 5 we demonstrated how a change of variable could sometimes replace a difficult integration problem with a simpler one. In many problems, the form of the integrand strongly suggests a particular substitution. We illustrate with the following example.

Example 1 Evaluate each of the following indefinite integrals:

a) $\int x\sqrt{1 + 3x^2}\, dx$;

b) $\int \frac{e^{\sqrt{x}}}{\sqrt{x}}\, dx$;

c) $\int \frac{x}{2 - 5x^2}\, dx$.

Solution a) The complexity of the integrand lies in the sum of two terms under the square root. This suggests that we set $u = 1 + 3x^2$, in which case $du = 6x\, dx$, and

$$\int x\sqrt{1 + 3x^2}\, dx = \int u^{1/2} \frac{du}{6} = \frac{1}{6}\int u^{1/2}\, du.$$

We may now use formula (9.1):

$$\int x\sqrt{1 + 3x^2}\, dx = \frac{1}{6}\left\{\frac{2}{3}u^{3/2}\right\} + C = \frac{1}{9}(1 + 3x^2)^{3/2} + C.$$

b) The appearance of $\sqrt{x}$ in two places suggests that we set $u = \sqrt{x}$. Then $du = (1/2\sqrt{x})\, dx$ and

$$\int \frac{e^{\sqrt{x}}}{\sqrt{x}}\, dx = \int e^u (2\, du) = 2\int e^u\, du.$$

Formula (9.3) now gives us

$$\int \frac{e^{\sqrt{x}}}{\sqrt{x}}\, dx = 2e^u + C = 2e^{\sqrt{x}} + C.$$

c) To simplify the denominator of the integrand we set $u = 2 - 5x^2$, in which case $du = -10x\, dx$ and

$$\int \frac{x}{2 - 5x^2}\, dx = \int \frac{1}{u}\left(-\frac{du}{10}\right) = -\frac{1}{10}\int \frac{1}{u}\, du.$$

If we now use (9.2), we get

$$\int \frac{x}{2 - 5x^2}\, dx = -\frac{1}{10}\ln|u| + C = -\frac{1}{10}\ln|2 - 5x^2| + C. \square$$

Substitutions in this example were straightforward and led immediately to integrals that could be evaluated by means of formulas (9.1)–(9.3). Actually each of these problems can be solved by a somewhat simpler method, one that requires only that our differentiation be up to par. The method is what we call "guess and fix up." For (a) of Example 1, it is clear that to obtain $\sqrt{1 + 3x^2}$

through differentiation, one possibility might be to differentiate

$$(1 + 3x^2)^{3/2}.$$

This we use as an initial "guess" for the solution to our problem. To check whether it is correct, we quickly compare the derivative of this function,

$$\tfrac{3}{2}(1 + 3x^2)^{1/2}(6x),$$

with the original integrand. It is the same except for a factor of 9. It follows, then, that the correct indefinite integral can be obtained by "fixing up" our original guess using $\frac{1}{9}$:

$$\int x\sqrt{1 + 3x^2}\, dx = \frac{1}{9}(1 + 3x^2)^{3/2} + C.$$

For part (b), an initial guess might be

$$e^{\sqrt{x}},$$

since differentiation of e to any power gives e to that power back again. Differentiation of our initial guess gives us

$$e^{\sqrt{x}}\frac{1}{2\sqrt{x}},$$

and we need fix up only a factor of 2 to obtain

$$\int \frac{e^{\sqrt{x}}}{\sqrt{x}}\, dx = 2e^{\sqrt{x}} + C.$$

A similar line of reasoning can be used in part (c). One word of caution must be given. The "guess and fix up" method can be used only if our initial guess is correct up to a *constant*. When the difference between the derivative of our initial guess and the integrand is a factor involving x, it *cannot* be corrected. We must try another guess or use some other technique. For example, an initial guess for the problem

$$\int \sqrt{1 + 3x^2}\, dx$$

might be $(1 + 3x^2)^{3/2}$. Differentiation of this guess yields (as before) $9x\sqrt{1 + 3x^2}$. Since this differs from the integrand by a factor of $9x$, the original guess cannot be fixed up. Solution of this problem is particularly difficult and we will postpone its discussion until Section 9.7.

If the integrand contains $x^{1/n}$ (n a positive integer), it is often useful to set $u = x^{1/n}$ or, equivalently, $x = u^n$.

Example 2 Evaluate each of the following indefinite integrals:

a) $\displaystyle\int \frac{2 - \sqrt{x}}{2 + \sqrt{x}}\, dx;$

b) $\displaystyle\int \frac{x^{1/3}}{1 + x^{2/3}}\, dx;$

c) $\displaystyle\int \frac{\sqrt{x}}{1 + x^{1/3}}\, dx.$

Solution a) If we set $u = \sqrt{x}$ or $x = u^2$, then $dx = 2u\,du$ and

$$\int \frac{2 - \sqrt{x}}{2 + \sqrt{x}}\,dx = \int \frac{2 - u}{2 + u}(2u\,du) = 2\int \frac{2u - u^2}{2 + u}\,du.$$

Long division of $-u^2 + 2u$ by $u + 2$ immediately gives us

$$\int \frac{2 - \sqrt{x}}{2 + \sqrt{x}}\,dx = 2\int \left(-u + 4 - \frac{8}{u + 2}\right) du$$

$$= 2\left\{-\frac{u^2}{2} + 4u - 8\ln|u + 2|\right\} + C$$

$$= -x + 8\sqrt{x} - 16\ln(2 + \sqrt{x}) + C.$$

b) If we set $u = x^{1/3}$ or $x = u^3$, then $dx = 3u^2\,du$ and

$$\int \frac{x^{1/3}}{1 + x^{2/3}}\,dx = \int \frac{u}{1 + u^2}(3u^2\,du) = 3\int \frac{u^3}{1 + u^2}\,du.$$

Division of u^3 by $u^2 + 1$ leads to

$$\int \frac{x^{1/3}}{1 + x^{2/3}}\,dx = 3\int \left(u - \frac{u}{u^2 + 1}\right) du$$

$$= 3\left\{\frac{u^2}{2} - \frac{1}{2}\ln|u^2 + 1|\right\} + C$$

$$= \frac{3}{2}x^{2/3} - \frac{3}{2}\ln(x^{2/3} + 1) + C.$$

c) If we set $u = x^{1/6}$ or $x = u^6$, then $dx = 6u^5\,du$ and

$$\int \frac{\sqrt{x}}{1 + x^{1/3}}\,dx = \int \frac{u^3}{1 + u^2}(6u^5\,du)$$

$$= 6\int \left(u^6 - u^4 + u^2 - 1 + \frac{1}{1 + u^2}\right) du$$

$$= 6\left\{\frac{u^7}{7} - \frac{u^5}{5} + \frac{u^3}{3} - u + \mathrm{Tan}^{-1}\,u\right\} + C$$

$$= \frac{6}{7}x^{7/6} - \frac{6}{5}x^{5/6} + 2x^{1/2} - 6x^{1/6} + 6\,\mathrm{Tan}^{-1}(x^{1/6}) + C. \square$$

EXERCISES 9.2

In each of Exercises 1–24, evaluate the indefinite integral.

1. $\displaystyle\int \frac{x^2}{5 - 3x^3}\,dx$

2. $\displaystyle\int xe^{-2x^2}\,dx$

3. $\displaystyle\int \frac{x}{(x^2 + 2)^{1/3}}\,dx$

4. $\displaystyle\int \frac{e^x}{1 + e^x}\,dx$

5. $\displaystyle\int \frac{4t + 8}{t^2 + 4t + 5}\,dt$

6. $\displaystyle\int x^2\sqrt{1 - 3x^3}\,dx$

7. $\displaystyle\int (x + 1)(x^2 + 2x)^{1/3}\,dx$

8. $\displaystyle\int \frac{x^2}{(1 + x^3)^3}\,dx$

9. $\displaystyle\int \frac{\sqrt{x}}{1 - \sqrt{x}}\,dx$

10. $\displaystyle\int \frac{1 - \sqrt{x}}{\sqrt{x}}\,dx$

11. $\displaystyle\int \frac{x + 2}{x + 1}\,dx$

12. $\displaystyle\int \frac{x^2 + 2}{x^2 + 1}\,dx$

13. $\int \frac{\sin\theta}{\cos\theta - 1}\,d\theta$

14. $\int \frac{x+3}{\sqrt{2x+4}}\,dx$

15. $\int \frac{e^x}{1+e^{2x}}\,dx$

16. $\int \sin^3 2x \cos 2x\,dx$

17. $\int x^5(2x^2-5)^4\,dx$

18. $\int \frac{x^3}{(x+5)^2}\,dx$

19. $\int z^2\sqrt{3-z}\,dz$

20. $\int \tan 3x\,dx$

21. $\int \frac{1}{\sqrt{4-x^2}}\,dx$

22. $\int \frac{1}{3+x^2}\,dx$

23. $\int \frac{1}{\sqrt{4-9x^2}}\,dx$

24. $\int \frac{1}{x\sqrt{3x^2-16}}\,dx$

25. a) Is it possible to assign a number to represent the area bounded by the curves $y = 0$, $y = 1/x$, and $x = 1$?

b) If the region in (a) is rotated about the x-axis, is it possible to assign a number to represent its volume?

In Exercises 26–29, evaluate the given indefinite integral.

26. $\int \frac{\sqrt{x}}{1+x^{1/4}}\,dx$

27. $\int \frac{(x-3)^{2/3}}{(x-3)^{2/3}+1}\,dx$

28. $\int \frac{e^x-1}{e^x+1}\,dx$

29. $\int \frac{1}{x^2-4}\,dx$

30. A gas is confined in a cylinder, closed on one end with a piston on the other. If the temperature is held constant, then $PV = C$, where P is the pressure, V is the volume of the gas in the cylinder, and C is a constant. The force needed to move the piston is $F = PA$, where A is the area of the face of the piston. If the radius of the cylinder is 3 units, find the work done in moving the piston from a point 10 units to a point 5 units from the closed end.

31. Find the area bounded by the curves

$$y = \frac{x^2+1}{x+1} \quad \text{and} \quad x + 3y = 7.$$

32. Find the centroid of the area bounded by the curves

$$y = (3+x)^{3/2}, \qquad x = 1, \quad \text{and} \quad y = 0.$$

33. In Exercise 32, find the second moment of area about the line $x = 1$.

34. If $f(x)$ is a continuous even function ($f(-x) = f(x)$), prove that

$$\int_{-a}^{a} f(x)\,dx = 2\int_{0}^{a} f(x)\,dx;$$

and if $f(x)$ is a continuous odd function ($f(-x) = -f(x)$),

$$\int_{-a}^{a} f(x)\,dx = 0.$$

35. a) Show that the function $f(x) = \lambda e^{-\lambda x}$ ($\lambda > 0$ a constant) qualifies as a probability density function on the interval $x \geq 0$. (See Exercise 23 in Section 7.9 for the definition of a pdf.)

b) If the probability that $x \geq 3$ is 0.5, what is λ?

36. Evaluate

$$\int_0^{4\pi} \sqrt{1+\cos x}\,dx.$$

(*Hint:* $\cos x = 2\cos^2(x/2) - 1$.)

37. A paratrooper (and his parachute) falls from rest. The combined mass of the paratrooper and his parachute is 100 kilograms. At any instant during descent, the parachute has an air resistance force acting on it that in newtons is equal to one-half its speed (in meters per second). If we assume that the paratrooper falls vertically downward and that the parachute is already open when the jump takes place, the differential equation describing his motion is

$$200\frac{dv}{dt} = 1962 - v,$$

where v is velocity and t is time.

a) Find v as a function of t.

b) If x measures the vertical distance fallen, then $v = dx/dt$. Use this to find the position of the paratrooper as a function of time.

38. Show that the substitution $u - x = \sqrt{x^2+bx+c}$ will replace the integral

$$\int \sqrt{x^2+bx+c}\,dx$$

with the integral of a rational function of u.

39. Show that if the quadratic $c + bx - x^2$ factors as

$$c + bx - x^2 = (p+x)(q-x),$$

then either of the substitutions $(p+x)u = \sqrt{c+bx-x^2}$ or $(q-x)u = \sqrt{c+bx-x^2}$ will replace the integral

$$\int \frac{1}{\sqrt{c+bx-x^2}}\,dx$$

with the integral of a rational function of u.

40. Evaluate

$$\int \frac{1}{x(3+2x^n)}\,dx$$

for any $n > 0$. *Hint:* Find A and B in order that

$$\frac{1}{x(3+2x^n)} = \frac{A}{x} + \frac{Bx^{n-1}}{3+2x^n}.$$

***41.** Use the substitution $u - x = \sqrt{x^2 + 3x + 4}$ to evaluate

$$\int \frac{1}{(x^2 + 3x + 4)^{3/2}}\, dx.$$

***42.** If a, b, and n are constants, evaluate

$$\int x(a + bx)^n\, dx.$$

***43.** The function

$$f(x) = \frac{1}{\sqrt{2\pi}\,\sigma} e^{-(x-\mu)^2/(2\sigma^2)}, \qquad -\infty < x < \infty,$$

where μ and σ are constants, is called the *normal probability density function*. Consequently,

$$\int_{-\infty}^{\infty} f(x)\, dx = 1.$$

If the expected value of x is defined as

$$\int_{-\infty}^{\infty} xf(x)\, dx,$$

show that μ is the expected value.

***44.** Evaluate, if possible,

$$\int_{-\infty}^{\infty} e^{-|a-x|}e^{-|x|}\, dx, \qquad a = \text{a constant}.$$

9.3 EXPONENTIAL GROWTH AND DECAY

The fact that the exponential function is its own antiderivative makes it very useful in the solution of differential equations. In this section we give a number of illustrative examples.

Consider first the growth of the number of bacteria in a culture. Suppose that at some initial time (which we take as $t = 0$) the number of bacteria is N_0. Given a proper environment, the culture will grow, and by studying this growth over a limited time span, perhaps we will be able to predict the number $N(t)$ of bacteria present at any given time. If we carefully observe the culture over a very small time interval (perhaps 10^{-7} seconds) starting at time $t = 0$, we could draw a graph of $N = N(t)$ as shown in Fig. 9.1. Each jump (discontinuity) in this graph indicates the birth of a new bacterium. Such a "microscopic" analysis of $N(t)$, in which we attempt to predict the birth of each and every bacterium, is unrealistic. We are interested in a more "macroscopic" analysis, with which we might be able to predict when the number of bacteria will increase by 25% or perhaps double. In order to do so, we must take measurements of the number of bacteria at much larger time intervals than those shown in Fig. 9.1. The result would be a sequence of points such as that shown in

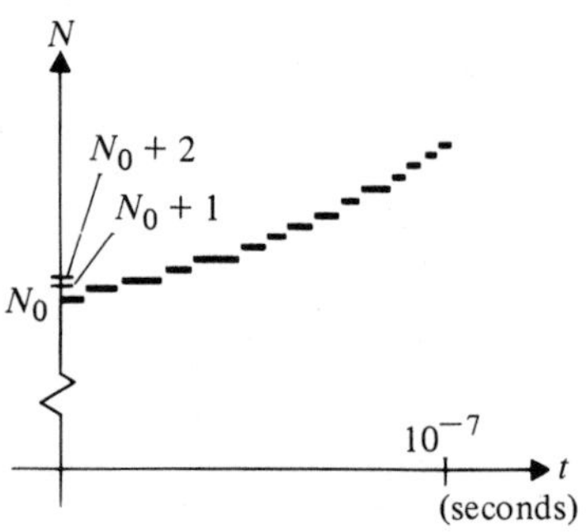

Figure 9.1

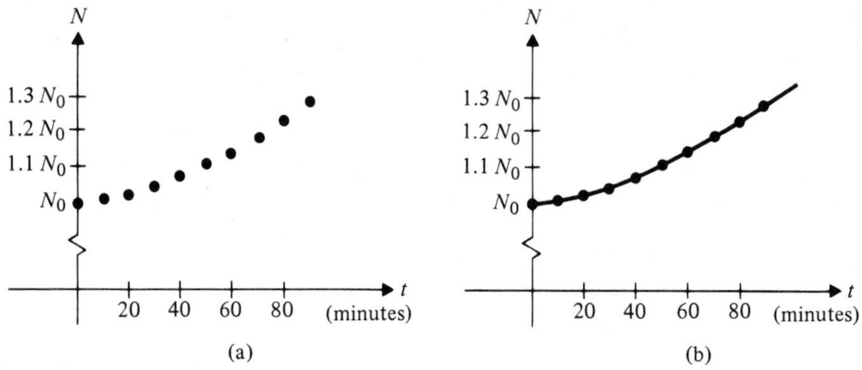

Figure 9.2

Fig. 9.2(a). If we now join these points with a smooth curve as in Fig. 9.2(b), we obtain a reasonable view of N as a function of time. Figure 9.2(b) is, however, only a graphical representation of physical measurements. It is now our problem to fit a mathematical formula to this graph; that is, to find an algebraic expression for $N(t)$. To do this often requires a great deal of experimentation with different types of curves. For this particular problem, the key is in the following measurements. At any time t we measure for a very small interval Δt the corresponding change $N(t + \Delta t) - N(t)$ in N (Fig. 9.3). If we take a number of such measurements, we find that $N(t + \Delta t) - N(t)$ is always proportional to the product of N and Δt; i.e.,

$$N(t + \Delta t) - N(t) = kN\,\Delta t, \tag{9.18}$$

where $k > 0$ is a constant. Furthermore, this is not a peculiarity of just one type of bacterium but is a characteristic of the growth pattern for many bacteria. If we divide both sides of Eq. (9.18) by Δt,

$$\frac{N(t + \Delta t) - N(t)}{\Delta t} = kN,$$

and take the limit as Δt approaches zero, we obtain

$$\frac{dN}{dt} = kN. \tag{9.19}$$

Experimental measurements have therefore led to mathematical law (9.19) for the rate of growth of bacteria in a culture. It states that the time rate of change of the number of bacteria in a culture is proportional to the number of bacteria present at that time. In other words, the more bacteria present, the faster their number increases.

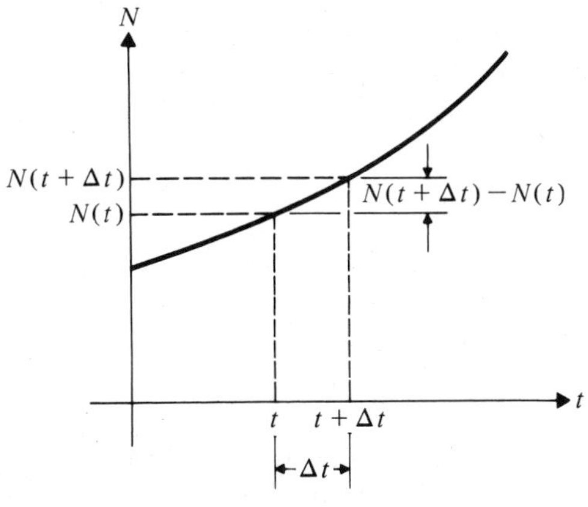

Figure 9.3

To find $N(t)$ we must solve this differential equation subject to the condition that $N(0) = N_0$. If we rewrite Eq. (9.19) in the form

$$k = \frac{1}{N}\frac{dN}{dt}$$

and take antiderivatives with respect to t, we get

$$kt + C = \ln|N| = \ln N \quad (\text{since } N > 0).$$

Exponentiation of both sides to base e gives us

$$e^{kt+C} = e^{\ln N},$$

or

$$N = e^C e^{kt} = De^{kt},$$

where we have set $D = e^C$. The initial condition implies that $N_0 = D$, and hence

$$N = N_0 e^{kt}. \tag{9.20}$$

On the basis of experimental evidence, we have shown that the number of bacteria in a culture increases exponentially. To obtain the constant k, we must specify one additional piece of information. For example, if our measurements indicate that the number of bacteria increases by 10% in 50 minutes, then $N(50) = (\frac{11}{10})N_0$; i.e.,

$$\tfrac{11}{10}N_0 = N_0 e^{50k}.$$

If we cancel the N_0 on each side and take natural logarithms, we have

$$\ln\left(\tfrac{11}{10}\right) = 50k, \quad \text{or} \quad k = \tfrac{1}{50}\ln\left(\tfrac{11}{10}\right) = 0.0019.$$

Finally, then, the number of bacteria in this particular culture is given by

$$N(t) = N_0 e^{(t/50)\ln(1.1)} = N_0 e^{0.0019t}.$$

As a second example we continue the discussion of radioactive disintegration begun in Section 2.2. Consider an ore sample that contains, along with various impurities, an amount A_0 of radioactive material (say, uranium). Disintegrations will gradually reduce this amount of uranium. To describe the rate at which these disintegrations occur, we let $A(t)$ represent the amount of uranium in the sample at any given time, and choose time $t = 0$ when $A = A_0$. If we were to take measurements and perform an analysis similar to that for bacteria in a culture (see also Section 2.2), we would obtain the following *law of radioactive disintegration*: The time rate of change of the amount of radioactive material is proportional at any instant to the amount of radioactive material present. Algebraically, therefore,

$$\frac{dA}{dt} = kA, \tag{9.21}$$

where k is a constant. Because A is decreasing, it follows that dA/dt and hence k must be negative. Except for the sign of k, this is the same differential equation as Eq. (9.19), and must have a solution of the form

$$A(t) = A_0 e^{kt}. \tag{9.22}$$

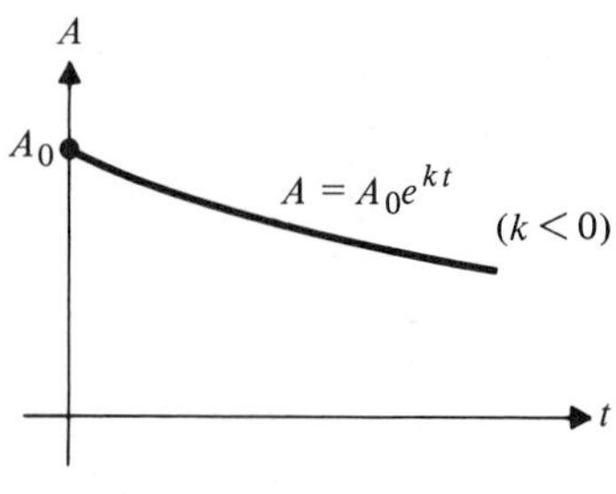

Figure 9.4

The amount of radioactive material therefore decreases exponentially, as shown in Fig. 9.4. Determination of k requires us to know A at some given time. For example, if we know that one ten-millionth of one percent of the original amount of uranium decays in 6.5 years, then we have

$$0.999999999A_0 = A_0 e^{13k/2}.$$

If we solve this equation for k, we get

$$\begin{aligned} k &= \tfrac{2}{13}\ln(0.999999999) \\ &= -1.54 \times 10^{-10}, \end{aligned}$$

and therefore

$$A = A_0 e^{-1.54 \times 10^{-10} t}.$$

The law of radioactive disintegration has an important application in the dating of once-living plants and animals. All living tissue contains two isotopes of carbon: C^{14} (carbon-14), which is radioactive, and C^{12} (carbon-12), which is stable. While the tissue is *living*, the ratio of the amount of C^{14} to that of C^{12} is essentially the same small quantity (1/10,000) for all fragments of the tissue. When the tissue *dies*, however, the ratio changes due to the fact that no more carbon is produced, and the original C^{14} present decays radioactively into an element other than C^{12}. Thus, as the dead tissue ages, the ratio of C^{14} to C^{12} decreases, and by measuring this ratio, it is possible to predict how long ago the tissue was alive. Suppose, for example, that the present ratio of C^{14} to C^{12} in a fragment of tissue is 1/100,000; i.e., one-tenth that for a living tissue. Then 90% of the original amount of C^{14} in the fragment has disintegrated. If we let $A(t)$ be the amount of C^{14} in the fragment at time t, taking $A = A_0$ to be the amount present in the living fragment at its death ($t = 0$), then $A = A_0 e^{kt}$.

The time for one-half an original sample of radioactive material to disintegrate is called its *half-life*; the half-life of C^{14} is approximately 5550 years. This implies that when $t = 5550$, A will be equal to $A_0/2$:

$$\frac{A_0}{2} = A_0 e^{5550k}.$$

If we divide by A_0 and take natural logarithms, we have

$$k = -\tfrac{1}{5550} \ln 2 = -0.000125.$$

Consequently, the amount of C^{14} in the fragment at any time t is given by

$$A = A_0 e^{-0.000125t}.$$

If T is the present time, when the amount of C^{14} is known to be 10% of its original amount, then at this time

$$0.1A_0 = A_0 e^{-0.000125T}.$$

The solution of this equation is

$$T = \frac{\ln 10}{0.000125} = 18{,}400,$$

and we conclude that the tissue died 18,400 years ago.

EXERCISES 9.3

1. Bacteria in a culture increase at a rate proportional to the number present. If the original number increases by 25% in 2 hours, when will it double?

2. If one-half of a sample of a radioactive substance decays in 15 days (called the half-life of the substance), how long does it take for 90% of the sample to decay?

3. If P dollars is invested at i% compounded n times per year, the accumulated value after one year is given by

$$A = P\left(1 + \frac{i}{100n}\right)^n$$

(see Exercise 38 in Section 8.6).

a) Calculate the maximum possible value of A if i is fixed but the number of times that the interest can be compounded is unlimited; i.e., calculate $\lim_{n\to\infty} A$. This method of calculating interest is called *continuously compounded interest*.

b) What is the accumulated value of a \$1000 investment after 10 years at 6% compounded continuously?

4. The amount of a drug injected into the body decreases at a rate proportional to the amount still present. If a dose decreases by 5% in the first hour, when will it decrease to one-half its original amount?

5. In psychology, it is sometimes assumed that if R represents the reaction of a subject to an amount S of stimulus, then dR/dS is inversely proportional to S; that is,

$$\frac{dR}{dS} = \frac{k}{S},$$

where $k > 0$ is a constant. This is called the *Weber–Fechner law*. Find R as a function of S if S_0 is the smallest stimulus detectable by the subject.

6. If the Weber–Fechner law in Exercise 5 is replaced by the Brentano–Stevens law,

$$\frac{dR}{dS} = k\frac{R}{S},$$

find $R(S)$. Can the detection threshold S_0 of Exercise 5 be used to determine the constant of integration in this case?

7. In the arctic, the snowy owl preys on the lemming, and the numbers of owls and lemmings in any given area fluctuate in time. If the number of lemmings increases, so too does the number of owls, but with some time delay. Eventually an overabundance of owls results, and a scarcity of food (lemmings) to support such a population causes a decrease in the number of owls. The lemming population consequently begins to rise and the cycle repeats. One model to describe the relationship between L, the number of lemmings, and s, the number of snowy owls, states that $s(L)$ must satisfy the differential equation

$$\frac{ds}{dL} = \frac{s(AL - B)}{L(C - Ds)},$$

where A, B, C, and D are positive constants. Solve this differential equation for an implicit definition of $s(L)$.

8. Show how much simpler it is to solve Exercise 10 in Section 5.4 now that we know how to use the logarithm and exponential functions.

9. A sugar cube 1 centimeter on each side is dropped into a cup of coffee. If the sugar dissolves in such a way that the cube always remains a cube, compare the times for the cube to completely dissolve under the following conditions:

a) Dissolving occurs at a rate proportional to the surface area of the remaining cube;

b) Dissolving occurs at a rate proportional to the amount of sugar remaining.

10. Solve Exercise 9 if the sugar is not in the form of a cube but rather in free form from the sugar bowl. Assume that the sugar consists of n spherical particles each of radius r_0 centimeters.

11. An analysis of a sample of fossil remains shows that it contains only 1.51% of the original C^{14} in the living creature. When did the creature die?

12. Glucose is administered intravenously to the bloodstream at a constant rate of R units per unit time. As the glucose is added, it is converted by the body into other substances at a rate that is proportional to the amount in the blood at that time. Show that the amount of glucose in the blood as a function of time t is given by

$$C(t) = \frac{R}{k}(1 - e^{-kt}) + C_0 e^{-kt},$$

where k is a constant and C_0 is the amount at time $t = 0$ when the intravenous feeding is initiated.

13. If water at temperature 90°C is placed in a room at constant temperature 20°C, the temperature of the water will decrease. Newton's law of cooling states that the time rate of change of the temperature T of the water is proportional to the difference between T and the temperature of the environment:

$$\frac{dT}{dt} = k(T - 20), \qquad k = \text{a constant}.$$

If the water cools to 60°C in 40 minutes, find T as a function of t.

14. Prove that if a quantity decreases at a rate proportional to its present amount, and if its percentage decrease in some interval of time is i%, then its percentage decrease in any interval of time of the same length is i%.

15. An investor knows that interest is calculated at 10% annually, but does not know how many times per year the interest is calculated. What is the shortest possible time in which the investment can double? (*Hint:* See Exercise 3.)

The exponential growth of bacteria as defined by Eq. (9.20) is unrealistic for large t in that there is no maximum for $N(t)$. Sketch the growth curve in each of Exercises 16–19 and show that each has a maximum.

16. The *logistic* growth curve

$$N(t) = \frac{N_0(1 + b)}{1 + be^{-kt}}, \qquad t \geq 0,$$

where b, k, and N_0 are positive constants.

17. The *Gompertz* growth curve

$$N(t) = Ae^{-be^{-kt}}, \qquad t \geq 0,$$

where A, b, and k are positive constants.

18. The *monomolecular* growth curve

$$N(t) = \frac{N_0}{1-b}(1 - be^{-kt}), \qquad t \geq 0,$$

where $b < 1$, k, and N_0 are positive constants.

19. The *Von Bertalanffy* growth curve

$$N(t) = N_0(1 - be^{-kt})^3, \qquad t \geq 0,$$

where $b < 1$, k, and N_0 are positive constants.

20. Sketch a graph of the sigmoid survival curve

$$s(x) = 1 - (1 - e^{-kx})^n, \qquad x \geq 0,$$

where $k > 0$ is a constant and n is a positive integer.

21. The differential equation

$$\frac{dN}{dt} = kN(t)\cos t, \qquad k > 0, \qquad t \geq 0,$$

describes a population that undergoes periodic growths and declines. If $N(0) = N_0$, find $N(t)$ and sketch its graph.

22. Your calculus professor is found dead in the lecture theatre after the final examination. The coroner arrives on the scene at 9:00 A.M. and immediately records the temperature of the body as 30.1°C. One hour later she notes that the body's temperature has dropped to 29.2°C, and that during the hour, room temperature has remained at 20°C. Assuming that the body's temperature obeys Newton's law of cooling (Exercise 13), determine when the accident took place. (Normal body temperature is 37°C.)

23. When a deep-sea diver inhales air, his body tissues absorb extra amounts of nitrogen. The nitrogen pressure P in the tissue increases as he remains under water until a maximum pressure $\bar{P}$ at depth d is reached. It is found that if he remains at depth d, the time rate of change of P is proportional to the difference $\bar{P} - P$. Show that if the diver drops very quickly to the bottom, then P is given as a function of time t by

$$P = \bar{P}(1 - e^{-kt}), \qquad k = \text{constant},$$

provided P is assumed equal to zero at time $t = 0$.

24. a) In Section 7.5 we proved that the rate of change of pressure P with respect to vertical distance y within a fluid of specific weight γ is $dP/dy = -\gamma$. Since air may be considered a fluid, air pressure above sea level ($y = 0$) must satisfy the equation

$$\frac{dP}{dy} = -\gamma = -\rho g,$$

where $g > 0$ is the acceleration due to gravity and $\rho = \rho(y)$ is air density at altitude y. If air is assumed to be a perfect gas, the pressure of a given volume V of gas must satisfy the equation

$$P = \rho RT,$$

where $T = T(y)$ is the temperature of the air and $R > 0$ is a constant. Finally, temperature varies linearly with respect to y:

$$T = T_0 - \lambda y$$

where T_0 is sea level temperature and λ is a constant (about 6.10×10^{-4}°C per 100 meters). Show that $P(y)$ must satisfy the differential equation

$$\frac{dP}{dy} = -\frac{Pg}{R(T_0 - \lambda y)},$$

hence that

$$P(y) = P_0\left(1 - \frac{\lambda y}{T_0}\right)^{g/(\lambda R)},$$

where P_0 is sea level pressure.

b) Find the altitude at which air pressure is one one-millionth that of sea level pressure, if $T_0 = 293$°K, $g = 9.81$ m/s^2, and $R = 287$ m · N/kg · °K.

25. When light passes through a material (such as water or glass), its intensity decreases with depth of penetration. The Bouguer–Lambert law for absorption of light states that the amount of light absorbed by a small thickness of the material is proportional to the product of this thickness and the intensity of light incident on it.

a) Show that if I is the intensity of light, then

$$I = I(x) = I_0 e^{-kx}, \qquad k > 0 \text{ a constant},$$

where I_0 is the intensity at the surface of the material and x is the penetration distance from the surface of the material (Fig. 9.5).

b) Give an alternative statement of the Bouguer–Lambert law in terms of rates of change.

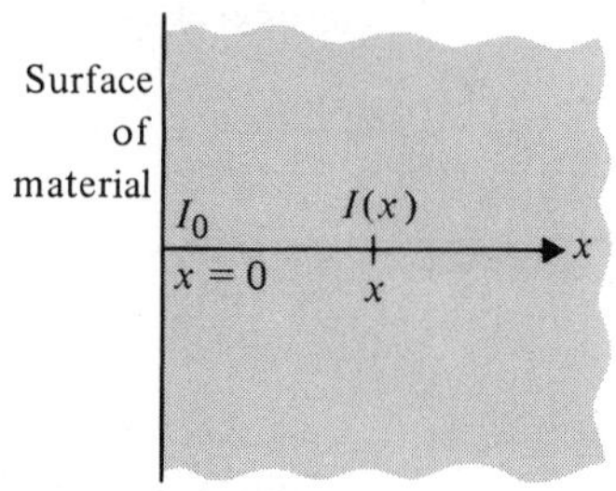

Figure 9.5

***26.** a) To perform a one-hour operation on a dog a veterinarian anesthetizes the dog with sodium pentobarbital. During the operation the dog's body breaks down the drug at a rate proportional to the amount present, and only half the original dose remains after 5 hours. If the dog weighs 20 kilograms, and 20 milligrams of sodium pentobarbital per kilogram of body weight are required to maintain surgical anesthesia, what original dose is required?

b) If after the hour, the operation is unfinished and the dog begins to stir, a further dose of anesthetic must be administered. What extra dose will guarantee surgical anesthesia for an additional 20 minutes?

9.4 TRIGONOMETRIC INTEGRALS

In this section we discuss indefinite integrals involving trigonometric functions. Our differentiation formulas for trigonometric functions suggest three pairings of these functions: sine and cosine, tangent and secant, and cotangent and cosecant. The derivative of either function in a given pair involves at most functions of that pair. Note that integration formulas (9.5)–(9.14) also substantiate this pairing: Each antiderivative gives functions in the same pairing. Because of this, a first step in every integral involving trigonometric functions is to rewrite the integrand so that each term involves only one pair. For example, to evaluate

$$\int \frac{\sec x \cot^2 x}{\sin^3 x}\,dx,$$

we would first rewrite it in the form

$$\int \frac{\cos x}{\sin^5 x}\,dx.$$

We now consider these three pairings individually.

Integrals Involving Sine and Cosine

Equations (9.5) and (9.6) are simple restatements of differentiation formulas involving the sine and cosine functions and should therefore present no problem. Frequently we encounter integrals of the form

$$\int \cos^n x \sin^m x\,dx,$$

where m and n may or may not be integers. The key to evaluation of this type of integral is recognition of the fact that if either $n = 1$ or $m = 1$, then the integration can be done mentally:

$$\int \cos^n x \sin x\,dx = -\frac{\cos^{n+1} x}{n+1} + C, \qquad n \neq -1; \tag{9.23}$$

$$\int \sin^n x \cos x\,dx = \frac{\sin^{n+1} x}{n+1} + C, \qquad n \neq -1. \tag{9.24}$$

Our knowledge of differentiation allows us to perform these integrations mentally. For instance, differentiation of $\sin^n x$ by power rule (3.9) gives

$$\frac{d}{dx}\sin^n x = n \sin^{n-1} x \cos x.$$

This result is valid for any value of n whatsoever, so that differentiation of any power of $\sin x$ leads to a power of $\sin x$ one lower than the original, multiplied by $\cos x$. Therefore, in order to reverse the procedure and antidifferentiate a power of $\sin x$, we need a factor of $\cos x$. It then becomes a matter of raising the power on $\sin x$ by one and dividing by the new power.

What we are suggesting is that (9.23) and (9.24) should not be memorized as formulas. Instead, remember that differentiation of a power of $\sin x$ leads to a $\cos x$, and therefore a power of $\sin x$ must be accompanied by $\cos x$ for integration. Of course, a simple substitution can always be made to evaluate integrals of this type. For example if in (9.24) we set $u = \sin x$, then $du = \cos x\, dx$, and

$$\int \sin^n x \cos x\, dx = \int u^n\, du = \frac{u^{n+1}}{n+1} + C = \frac{\sin^{n+1} x}{n+1} + C.$$

If neither m nor n is equal to 1, then we attempt to use trigonometric identities to rewrite the integrand with either $m = 1$ or $n = 1$. Keep in mind, then, that we constantly work toward integrals of the form in (9.23) and (9.24). We will find that trigonometric identity (3.19) and double-angle formulas (3.28) and (3.29) are very useful in this respect. In terms of x, these identities read as follows:

$$\sin^2 x + \cos^2 x = 1, \tag{9.25}$$

$$\sin 2x = 2 \sin x \cos x, \tag{9.26}$$

$$\cos 2x = 2 \cos^2 x - 1, \tag{9.27}$$

$$\cos 2x = 1 - 2 \sin^2 x. \tag{9.28}$$

Example 3 Evaluate each of the following indefinite integrals:

a) $\int \sin^2 x\, dx$;

b) $\int \sin^3 x\, dx$;

c) $\int \cos^4 x\, dx$;

d) $\int \sin 2x \cos^6 2x\, dx$;

e) $\int \sin^3 x \cos^5 x\, dx$;

f) $\int \sin^2 x \cos^2 x\, dx$;

g) $\int \frac{\cos^3 3x}{\sin^2 3x}\, dx$;

h) $\int \sqrt{\cos x} \sin^3 x\, dx$.

Solution a) If we solve double-angle formula (9.28) for $\sin^2 x$, we obtain $\sin^2 x = (1 - \cos 2x)/2$, and therefore

$$\int \sin^2 x\, dx = \frac{1}{2}\int (1 - \cos 2x)\, dx = \frac{1}{2}\left\{x - \frac{1}{2}\sin 2x\right\} + C.$$

b) Because of (9.25), we have

$$\int \sin^3 x\, dx = \int \sin x(1 - \cos^2 x)\, dx = \int (\sin x - \cos^2 x \sin x)\, dx$$

$$= -\cos x + \frac{1}{3}\cos^3 x + C.$$

c) With double-angle formula (9.27) rewritten in the form $\cos^2 x = (1 + \cos 2x)/2$, we obtain

$$\int \cos^4 x\,dx = \frac{1}{4}\int (1 + \cos 2x)^2\,dx = \frac{1}{4}\int (1 + 2\cos 2x + \cos^2 2x)\,dx.$$

We now use (9.27) again in the form $\cos 4x = 2\cos^2 2x - 1$:

$$\int \cos^4 x\,dx = \frac{1}{4}\int \left\{1 + 2\cos 2x + \frac{1 + \cos 4x}{2}\right\} dx$$
$$= \frac{1}{4}\left\{\frac{3}{2}x + \sin 2x + \frac{1}{8}\sin 4x\right\} + C.$$

d) $$\int \sin 2x \cos^6 2x\,dx = -\frac{1}{14}\cos^7 2x + C$$

e) $$\int \sin^3 x \cos^5 x\,dx = \int \sin x(1 - \cos^2 x)\cos^5 x\,dx$$
$$= \int (\cos^5 x \sin x - \cos^7 x \sin x)\,dx$$
$$= -\frac{1}{6}\cos^6 x + \frac{1}{8}\cos^8 x + C$$

f) With double-angle formulas (9.26) and (9.28), we can write

$$\int \sin^2 x \cos^2 x\,dx = \int (\sin x \cos x)^2\,dx = \frac{1}{4}\int \sin^2 2x\,dx$$
$$= \frac{1}{4}\int \left(\frac{1 - \cos 4x}{2}\right) dx = \frac{1}{8}\left\{x - \frac{1}{4}\sin 4x\right\} + C.$$

g) $$\int \frac{\cos^3 3x}{\sin^2 3x}\,dx = \int \frac{\cos 3x(1 - \sin^2 3x)}{\sin^2 3x}\,dx = \int \left\{\frac{\cos 3x}{\sin^2 3x} - \cos 3x\right\} dx$$
$$= \frac{-1}{3\sin 3x} - \frac{1}{3}\sin 3x + C$$

h) $$\int \sqrt{\cos x}\,\sin^3 x\,dx = \int \cos^{1/2} x \sin x(1 - \cos^2 x)\,dx$$
$$= \int (\cos^{1/2} x \sin x - \cos^{5/2} x \sin x)\,dx$$
$$= -\frac{2}{3}\cos^{3/2} x + \frac{2}{7}\cos^{7/2} x + C \;\square$$

Example 4 Evaluate

$$\int \sin 5x \cos 2x\,dx.$$

Solution Unlike Example 3, in which all trigonometric functions had the same argument, this integrand is the product of trigonometric functions with different arguments. Our first step is to rewrite the integrand so that we do not have such a product.

(A sum is obviously integrable but not a product.) The product formula (3.31b),

$$\sin A \cos B = \tfrac{1}{2}[\sin (A + B) + \sin (A - B)],$$

can be used to great advantage here. We can rewrite the integrand in the form

$$\int \sin 5x \cos 2x \, dx = \int \frac{1}{2}(\sin 7x + \sin 3x) \, dx,$$

and now integration is easy:

$$\int \sin 5x \cos 2x \, dx = \frac{1}{2}\left\{-\frac{1}{7}\cos 7x - \frac{1}{3}\cos 3x\right\} + C$$

$$= -\frac{1}{6}\cos 3x - \frac{1}{14}\cos 7x + C. \square$$

Integrals Involving Tangent and Secant

For integrals involving tangent and secant, we have two alternatives: to rewrite the integrand in terms of sines and cosines, or to express the integrand in terms of easily integrated combinations of the tangent and secant. It is usually more fruitful to investigate the second possibility before expressing the integrand completely in terms of sines and cosines. In this regard we note that combinations of tangent and secant that can be integrated immediately are

$$\int \tan^n x \sec^2 x \, dx = \frac{\tan^{n+1} x}{n + 1} + C, \qquad n \neq -1; \tag{9.29}$$

$$\int \sec^n x \tan x \, dx = \frac{\sec^n x}{n} + C, \qquad n \neq 0. \tag{9.30}$$

Once again these results are suggested by differentiations of powers of $\tan x$ and $\sec x$. Alternatively, these integrals can be evaluated with substitutions: $u = \tan x$ for (9.29) and $u = \sec x$ for (9.30).

To rearrange integrands into these integrable combinations, we use the identity

$$1 + \tan^2 x = \sec^2 x. \tag{9.31}$$

Example 5 Evaluate each of the following indefinite integrals:

a) $\int \tan x \, dx$;

b) $\int \sec x \, dx$;

c) $\int \tan^2 x \, dx$;

d) $\int \tan^3 x \, dx$;

e) $\int \sec^4 x \, dx$;

f) $\int \tan^4 3x \sec^2 3x \, dx$;

g) $\int \tan^3 x \sec^3 x \, dx$;

h) $\int \frac{\sec^4 x}{\tan^2 x} dx$.

Solution a) We know that

$$\int \tan x \, dx = \int \frac{\sin x}{\cos x} dx = -\ln |\cos x| + C = \ln |\sec x| + C,$$

establishing (9.11). If integration to $\ln|\cos x|$ is not obvious, try the substitution $u = \cos x$.

b) To obtain the antiderivative of $\sec x$ requires a subtle trick. We first multiply numerator and denominator by $\sec x + \tan x$:

$$\int \sec x\,dx = \int \sec x \frac{\sec x + \tan x}{\sec x + \tan x}\,dx = \int \frac{\sec^2 x + \sec x \tan x}{\sec x + \tan x}\,dx.$$

If we now set $u = \sec x + \tan x$, then $du = (\sec x \tan x + \sec^2 x)\,dx$, and

$$\int \sec x\,dx = \int \frac{1}{u}\,du = \ln|u| + C = \ln|\sec x + \tan x| + C.$$

This is formula (9.13), which we suggested be committed to memory (because it is probably as easy to remember the formula as the trick).

c) $\displaystyle\int \tan^2 x\,dx = \int (\sec^2 x - 1)\,dx = \tan x - x + C$

d)
$$\begin{aligned}\int \tan^3 x\,dx &= \int \tan x(\sec^2 x - 1)\,dx\\ &= \int (\sec^2 x \tan x - \tan x)\,dx\\ &= \frac{1}{2}\sec^2 x + \ln|\cos x| + C\end{aligned}$$

e)
$$\begin{aligned}\int \sec^4 x\,dx &= \int \sec^2 x(1 + \tan^2 x)\,dx\\ &= \int (\sec^2 x + \tan^2 x \sec^2 x)\,dx\\ &= \tan x + \frac{1}{3}\tan^3 x + C\end{aligned}$$

f) $\displaystyle\int \tan^4 3x \sec^2 3x\,dx = \frac{1}{15}\tan^5 3x + C$

g)
$$\begin{aligned}\int \tan^3 x \sec^3 x\,dx &= \int \tan x(\sec^2 x - 1)\sec^3 x\,dx\\ &= \int (\sec^5 x \tan x - \sec^3 x \tan x)\,dx\\ &= \frac{1}{5}\sec^5 x - \frac{1}{3}\sec^3 x + C\end{aligned}$$

h)
$$\begin{aligned}\int \frac{\sec^4 x}{\tan^2 x}\,dx &= \int \frac{\sec^2 x(1 + \tan^2 x)}{\tan^2 x}\,dx\\ &= \int \left\{\frac{\sec^2 x}{\tan^2 x} + \sec^2 x\right\}dx\\ &= -\frac{1}{\tan x} + \tan x + C \quad \square\end{aligned}$$

The integrands in the following example cannot be transformed by means of identity (9.31) into terms of the type (9.29) and (9.30), and we therefore express the integrands in terms of sines and cosines.

Example 6 Evaluate each of the following indefinite integrals:

a) $\displaystyle\int \frac{\tan^2 2x}{\sec^3 2x}\,dx;$

b) $\displaystyle\int \frac{1}{\sec x \tan^2 x}\,dx.$

Solution a) $\displaystyle\int \frac{\tan^2 2x}{\sec^3 2x}\,dx = \int \sin^2 2x \cos 2x\,dx = \frac{1}{6}\sin^3 2x + C$

b) $$\begin{aligned}\int \frac{1}{\sec x \tan^2 x}\,dx &= \int \frac{\cos^3 x}{\sin^2 x}\,dx\\ &= \int \frac{\cos x(1 - \sin^2 x)}{\sin^2 x}\,dx\\ &= \int \left\{\frac{\cos x}{\sin^2 x} - \cos x\right\} dx\\ &= -\frac{1}{\sin x} - \sin x + C \;\Box\end{aligned}$$

Integrals Involving Cotangent and Cosecant

Because the derivatives of the cotangent and cosecant pair

$$\frac{d}{dx}\cot x = -\csc^2 x, \qquad \frac{d}{dx}\csc x = -\csc x \cot x$$

are analogous to those of the tangent and secant pair

$$\frac{d}{dx}\tan x = \sec^2 x, \qquad \frac{d}{dx}\sec x = \sec x \tan x,$$

a discussion parallel to that concerning the tangent and secant could be made here. Aside from the fact that the cotangent replaces the tangent and the cosecant replaces the secant, there are also sign changes. Because of these minor differences, we will leave examples of integrations of the cotangent and cosecant pair to the exercises.

EXERCISES 9.4

In Exercises 1–20, evaluate the given indefinite integral.

1. $\displaystyle\int \cos^3 x \sin x\,dx$

2. $\displaystyle\int \frac{\cos x}{\sin^3 x}\,dx$

3. $\displaystyle\int \tan^5 x \sec^2 x\,dx$

4. $\displaystyle\int \csc^3 x \cot x\,dx$

5. $\displaystyle\int \cos^3 (x + 2)\,dx$

6. $\displaystyle\int \sqrt{\tan x}\,\sec^4 x\,dx$

7. $\displaystyle\int \frac{1}{\sin^4 t}\,dt$

8. $\displaystyle\int \sec^6 3x \tan 3x\,dx$

9. $\displaystyle\int \cos^2 x\,dx$

10. $\displaystyle\int \frac{\tan^3 x \sec^2 x}{\sin^2 x}\,dx$

11. $\displaystyle\int \sin^3 y \cos^2 y\,dy$

12. $\displaystyle\int \frac{\csc^2 \theta}{\cot^2 \theta}\,d\theta$

13. $\displaystyle\int \frac{\sin\theta}{1+\cos\theta}\,d\theta$

14. $\displaystyle\int \frac{\sec^2 x}{\sqrt{1+\tan x}}\,dx$

15. $\displaystyle\int \cos\theta \sin 2\theta\,d\theta$

16. $\displaystyle\int \frac{3+4\csc^2 x}{\cot^2 x}\,dx$

17. $\displaystyle\int \sin^5 x\cos^5 x\,dx$

18. $\displaystyle\int \sin^4 x\,dx$

19. $\displaystyle\int \frac{\tan^3 x}{\sec^4 x}\,dx$

20. $\displaystyle\int \frac{\csc^4 x}{\cot^3 x}\,dx$

21. Find the area between the x-axis and $y=\sin x$, $0 \le x \le \pi$.

22. Find the volume of the solid of revolution obtained by rotating the area bounded by $y=\tan x$, $y=0$, and $x=\pi/4$ about the line $y=-1$.

23. Evaluate

$$\int_0^{\pi} \sqrt{1-\sin^2 x}\,dx.$$

In Exercises 24–30, evaluate the given indefinite integral.

24. $\displaystyle\int \cot^4 z\,dz$

25. $\displaystyle\int \frac{\cos^3\theta}{3+\sin\theta}\,d\theta$

26. $\displaystyle\int \frac{\cos^4\theta}{1+\sin\theta}\,d\theta$

27. $\displaystyle\int \sin^4 x\cos^2 x\,dx$

28. $\displaystyle\int \cos 6x\cos 2x\,dx$

29. $\displaystyle\int \cos^2 2x\sin 3x\,dx$

30. $\displaystyle\int \frac{1}{\sin x\cos^2 x}\,dx$

31. The alternating current in a power line is given by

$$I=f(t)=A\cos\omega t+B\sin\omega t,$$

where A, B, and ω are constants, and t is time. The root-mean-square (rms) current I_{rms} is defined by

$$(I_{\text{rms}})^2=\frac{1}{T}\int_0^T I^2\,dt,$$

where the interval $0 \le t \le T$ represents any number of complete oscillations of the current. Show that I_{rms} is $1/\sqrt{2}$ times the amplitude of the current.

* 32. Evaluate

$$\int (1+\sin\theta+\sin^2\theta+\sin^3\theta+\cdots)\,d\theta.$$

* 33. If n is a positive even integer, show that

$$\int \sec^n x\,dx=\sum_{r=0}^{n/2-1}\binom{n/2-1}{r}\frac{\tan^{2r+1}x}{2r+1}+C.$$

* 34. A set of functions $f_1(x), f_2(x), \ldots$ is said to be orthonormal over an interval $a \le x \le b$ if

$$\int_a^b f_n(x)f_m(x)\,dx=\begin{cases}0, & m\ne n,\\ 1, & m=n.\end{cases}$$

Show that the set of functions

$$\frac{1}{\sqrt{2\pi}},\quad \frac{1}{\sqrt{\pi}}\sin x,\quad \frac{1}{\sqrt{\pi}}\cos x,$$

$$\frac{1}{\sqrt{\pi}}\sin 2x,\quad \frac{1}{\sqrt{\pi}}\cos 2x,\quad \frac{1}{\sqrt{\pi}}\sin 3x,\ldots$$

is orthonormal over the interval $0 \le x \le 2\pi$.

9.5 TRIGONOMETRIC SUBSTITUTIONS

Trigonometric substitutions are meant primarily for integrals involving the square roots

$$\sqrt{a^2+b^2x^2},\qquad \sqrt{a^2-b^2x^2},\qquad \sqrt{b^2x^2-a^2}$$

or powers thereof. The object of the substitution is to collapse the two terms under each square root into a perfect square, thereby ridding the integrand of the square root completely. For example, consider the three indefinite integrals

$$\int \frac{1}{\sqrt{1-x^2}}\,dx,\qquad \int \frac{1}{\sqrt{4-x^2}}\,dx,\qquad \int \frac{1}{\sqrt{4-9x^2}}\,dx.$$

The answer to the first problem is evident from our differentiations of the inverse trigonometric functions:

$$\int \frac{1}{\sqrt{1-x^2}}\,dx=\operatorname{Sin}^{-1}x+C.$$

The answer to the second problem can be obtained using formula (9.15) (as could the first):

$$\int \frac{1}{\sqrt{4 - x^2}}\, dx = \operatorname{Sin}^{-1}\left(\frac{x}{2}\right) + C.$$

The third integral could be evaluated by making the substitution $u = 3x$ and then using formula (9.15).

What we now show is that a method exists for evaluation of the first integral, should we forget how to differentiate the inverse sine function, and that this method is also applicable to the remaining two problems. The advantage of this method is that there is no need to memorize (9.15). If instead of the square root $\sqrt{1 - x^2}$ we had the square root $\sqrt{1 - \sin^2 \theta}$, we could immediately simplify the latter square root to $\sqrt{\cos^2 \theta} = |\cos \theta|$. This prompts us to make the substitution $x = \sin \theta$, from which we have $dx = \cos \theta \, d\theta$ and

$$\int \frac{1}{\sqrt{1 - x^2}}\, dx = \int \frac{1}{|\cos \theta|} \cos \theta \, d\theta.$$

All we need now do is decide whether $\cos \theta$ is positive or negative in order to eliminate the absolute value sign. There is a problem, however, in that the equation $x = \sin \theta$ does not really define θ; for any given x, we do not have a unique θ, but an infinite number of possibilities. We must therefore restrict the values of θ, and we do so by specifying that $\theta = \operatorname{Sin}^{-1} x$. In other words, although we have used the equation $x = \sin \theta$ to change variables, and will continue to do so, it is really the equation $\theta = \operatorname{Sin}^{-1} x$ that properly defines the change.

Because θ so defined must lie in the range $-\pi/2 \le \theta \le \pi/2$ (the principal values of the inverse sine function), it follows that $\cos \theta \ge 0$, and the absolute value sign may be dropped. Consequently,

$$\int \frac{1}{\sqrt{1 - x^2}}\, dx = \int \frac{\cos \theta}{|\cos \theta|}\, d\theta = \int d\theta = \theta + C = \operatorname{Sin}^{-1} x + C.$$

For our second example, the substitution $x = \sin \theta$ is of little use since the square root $\sqrt{4 - x^2}$ reduces to $\sqrt{4 - \sin^2 \theta}$. If, however, the square root were $\sqrt{4 - 4\sin^2 \theta}$, it would immediately change to $\sqrt{4 - 4\sin^2 \theta} = \sqrt{4\cos^2 \theta} = 2|\cos \theta|$. We can obtain this result if the substitution is modified to $x = 2 \sin \theta$ (or, more properly, $\theta = \operatorname{Sin}^{-1}(x/2)$). Since $dx = 2\cos \theta \, d\theta$, we have

$$\int \frac{1}{\sqrt{4 - x^2}}\, dx = \int \frac{1}{2|\cos\theta|} 2\cos \theta \, d\theta = \int \frac{2\cos \theta}{2\cos \theta}\, d\theta$$

$$= \int d\theta = \theta + C = \operatorname{Sin}^{-1}\left(\frac{x}{2}\right) + C.$$

To eliminate the square root in the third example, we use the substitution

$$x = \tfrac{2}{3} \sin \theta,$$

from which we get

$$dx = \tfrac{2}{3} \cos \theta \, d\theta.$$

We obtain

$$\int \frac{1}{\sqrt{4-9x^2}}\,dx = \int \frac{(2/3)\cos\theta}{\sqrt{4-4\sin^2\theta}}\,d\theta = \frac{2}{3}\int \frac{\cos\theta}{2\cos\theta}\,d\theta$$

$$= \frac{1}{3}\int d\theta = \frac{1}{3}\theta + C = \frac{1}{3}\,\mathrm{Sin}^{-1}\left(\frac{3x}{2}\right) + C.$$

Consider now the pair of indefinite integrals

$$\int \frac{1}{1+x^2}\,dx, \qquad \int \frac{1}{3+5x^2}\,dx.$$

The first obviously has the answer $\mathrm{Tan}^{-1}\,x + C$. Alternatively, we note that if instead of $1 + x^2$ we had $1 + \tan^2\theta$, the two terms in the denominator would simplify to the one term $\sec^2\theta$. We therefore substitute $x = \tan\theta$ (or, more properly, $\theta = \mathrm{Tan}^{-1}\,x$), from which we have $dx = \sec^2\theta\,d\theta$. With this substitution, we get

$$\int \frac{1}{1+x^2}\,dx = \int \frac{\sec^2\theta}{\sec^2\theta}\,d\theta = \int d\theta = \theta + C = \mathrm{Tan}^{-1}\,x + C.$$

The substitution $x = \tan\theta$ in the second example yields $3 + 5x^2 = 3 + 5\tan^2\theta$. For simplification we need a 3 in front of the $\tan^2\theta$ and not a 5. This can be accomplished by modification of our substitution to

$$x = \sqrt{\tfrac{3}{5}}\tan\theta,$$

from which we get

$$dx = \sqrt{\tfrac{3}{5}}\sec^2\theta\,d\theta.$$

Then

$$\int \frac{1}{3+5x^2}\,dx = \int \frac{\sqrt{\frac{3}{5}}\sec^2\theta}{3+3\tan^2\theta}\,d\theta = \frac{1}{\sqrt{15}}\int d\theta$$

$$= \frac{\theta}{\sqrt{15}} + C = \frac{1}{\sqrt{15}}\,\mathrm{Tan}^{-1}\left(\frac{\sqrt{5}x}{\sqrt{3}}\right) + C.$$

As a final illustrative example, consider

$$\int \frac{1}{\sqrt{x^2-4}}\,dx.$$

If instead of $\sqrt{x^2-4}$ we had $\sqrt{4\sec^2\theta - 4}$, an immediate simplification would occur. We therefore substitute $x = 2\sec\theta$, from which we obtain $dx = 2\sec\theta\tan\theta\,d\theta$. (Once again it would be better to set $\theta = \mathrm{Sec}^{-1}(x/2)$, in which case note that θ would be an angle in either the first or third quadrant.) With this change of variable, we have

$$\sqrt{x^2-4} = \sqrt{4\sec^2\theta - 4} = \sqrt{4\tan^2\theta} = 2|\tan\theta| = 2\tan\theta$$

and

$$\int \frac{1}{\sqrt{x^2-4}}\,dx = \int \frac{2\sec\theta\tan\theta}{2\tan\theta}\,d\theta = \int \sec\theta\,d\theta = \ln|\sec\theta + \tan\theta| + C.$$

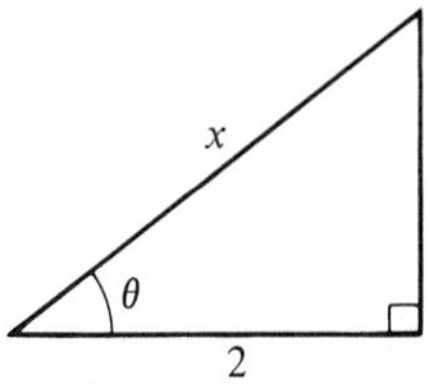

Figure 9.6

To express $\tan\theta$ in terms of x, we draw the triangle shown in Fig. 9.6 to fit the change of variable $x/2 = \sec\theta$. Since the third side is $\sqrt{x^2 - 4}$, we have

$$\int \frac{1}{\sqrt{x^2-4}}\,dx = \ln\left|\frac{x}{2} + \frac{\sqrt{x^2-4}}{2}\right| + C$$

$$= \ln|x + \sqrt{x^2-4}| + D, \qquad (D = C - \ln 2).$$

We have illustrated by simple examples that trigonometric substitutions can be useful in the evaluation of indefinite integrals involving the square roots $\sqrt{a^2 - b^2x^2}$, $\sqrt{a^2 + b^2x^2}$, and $\sqrt{b^2x^2 - a^2}$. Essentially, the method replaces the terms under the square root by a perfect square and thereby rids the integrand of the square root. To obtain the trigonometric substitution appropriate to each square root, we suggest the following procedure. We have been working basically with two trigonometric identities:

$$1 - \sin^2\theta = \cos^2\theta,$$
$$1 + \tan^2\theta = \sec^2\theta,$$
$$\sec^2\theta - 1 = \tan^2\theta.$$

To determine which trigonometric substitution is appropriate to, say, $\sqrt{a^2 - b^2x^2}$, we mentally set $a = b = 1$ and obtain $\sqrt{1 - x^2}$. We note that $1 - x^2$ resembles $1 - \sin^2\theta$, and immediately set $x = \sin\theta$. To simplify the original square root $\sqrt{a^2 - b^2x^2}$, we then modify this substitution to

$$x = \frac{a}{b}\sin\theta \qquad (\text{or } \theta = \text{Sin}^{-1}(bx/a)).$$

Similarly, for the square root $\sqrt{a^2 + b^2x^2}$, we initially set $x = \tan\theta$, thinking that with $a = b = 1$, $1 + x^2$ resembles $1 + \tan^2\theta$, and then modify to

$$x = \frac{a}{b}\tan\theta \qquad (\text{or } \theta = \text{Tan}^{-1}(bx/a)).$$

Finally, for the square root $\sqrt{b^2x^2 - a^2}$, we initially set $x = \sec\theta$ and then modify to

$$x = \frac{a}{b}\sec\theta \qquad (\text{or } \theta = \text{Sec}^{-1}(bx/a)).$$

The square roots $\sqrt{a^2 - b^2x^2}$, $\sqrt{a^2 + b^2x^2}$, and $\sqrt{b^2x^2 - a^2}$ will be replaced, respectively, by

$\sqrt{a^2 - a^2\sin^2\theta}$	$\sqrt{a^2 + a^2\tan^2\theta}$	$\sqrt{a^2\sec^2\theta - a^2}$
$= \sqrt{a^2\cos^2\theta}$	$= \sqrt{a^2\sec^2\theta}$	$= \sqrt{a^2\tan^2\theta}$
$= a\lvert\cos\theta\rvert$	$= a\lvert\sec\theta\rvert$	$= a\lvert\tan\theta\rvert$
$= a\cos\theta;$	$= a\sec\theta;$	$= a\tan\theta.$

In all cases it is our choice of principal values for the inverse trigonometric functions that enables us to neglect the absolute value.

We now consider a number of more involved examples.

Example 7 Evaluate each of the following indefinite integrals:

a) $\displaystyle\int \frac{1}{(4-x^2)^{3/2}}\,dx$;

b) $\displaystyle\int \frac{1}{x^2(9x^2+2)}\,dx$;

c) $\displaystyle\int x^2\sqrt{3-x^2}\,dx$;

d) $\displaystyle\int \frac{(16-x^2)^{3/2}}{x}\,dx$.

Solution a) If we set $x = 2\sin\theta$, then $dx = 2\cos\theta\,d\theta$ and

$$\int \frac{1}{(4-x^2)^{3/2}}\,dx = \int \frac{2\cos\theta}{8\cos^3\theta}\,d\theta = \frac{1}{4}\int \sec^2\theta\,d\theta = \frac{1}{4}\tan\theta + C.$$

From the triangle shown in Fig. 9.7, we see that

$$\int \frac{1}{(4-x^2)^{3/2}}\,dx = \frac{1}{4}\frac{x}{\sqrt{4-x^2}} + C.$$

b) If we set $x = (\sqrt{2}/3)\tan\theta$, then $dx = (\sqrt{2}/3)\sec^2\theta\,d\theta$ and

$$\begin{aligned}\int \frac{1}{x^2(9x^2+2)}\,dx &= \int \frac{(\sqrt{2}/3)\sec^2\theta}{\frac{2}{9}\tan^2\theta\, 2\sec^2\theta}\,d\theta \\ &= \frac{3}{2\sqrt{2}}\int \cot^2\theta\,d\theta \\ &= \frac{3}{2\sqrt{2}}\int (\csc^2\theta - 1)\,d\theta \\ &= \frac{3}{2\sqrt{2}}\{-\cot\theta - \theta\} + C \\ &= -\frac{3}{2\sqrt{2}}\left\{\frac{\sqrt{2}}{3x} + \mathrm{Tan}^{-1}\left(\frac{3x}{\sqrt{2}}\right)\right\} + C \qquad \text{(Fig. 9.8)} \\ &= -\frac{1}{2x} - \frac{3}{2\sqrt{2}}\mathrm{Tan}^{-1}\left(\frac{3x}{\sqrt{2}}\right) + C.\end{aligned}$$

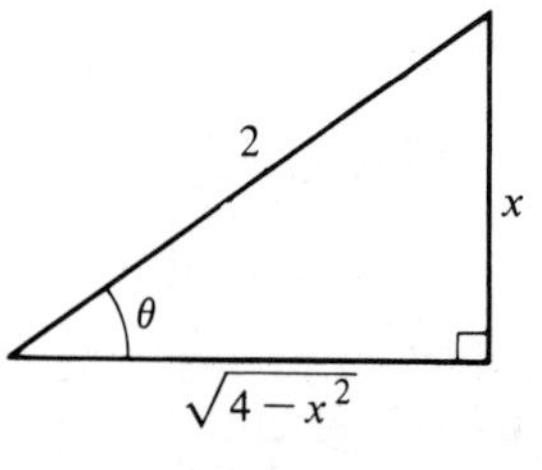

Figure 9.7

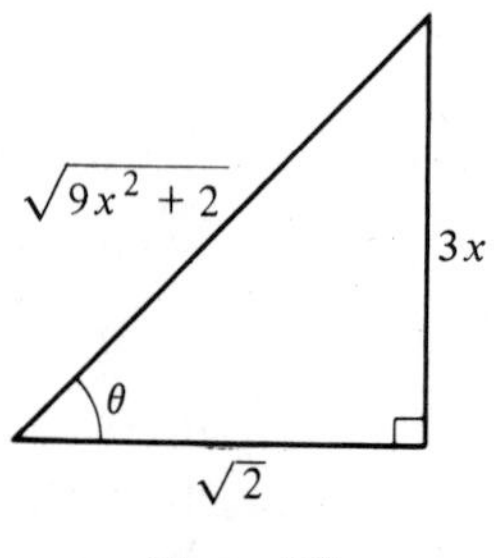

Figure 9.8

c) If we set $x = \sqrt{3}\sin\theta$, then $dx = \sqrt{3}\cos\theta\, d\theta$ and

$$\int x^2\sqrt{3-x^2}\,dx = \int 3\sin^2\theta\sqrt{3}\cos\theta\sqrt{3}\cos\theta\,d\theta = 9\int \sin^2\theta\cos^2\theta\,d\theta$$

$$= \frac{9}{4}\int \sin^2 2\theta\,d\theta = \frac{9}{4}\int\left(\frac{1-\cos 4\theta}{2}\right)d\theta$$

$$= \frac{9}{8}\left\{\theta - \frac{1}{4}\sin 4\theta\right\} + C$$

$$= \frac{9}{8}\operatorname{Sin}^{-1}\left(\frac{x}{\sqrt{3}}\right) - \frac{9}{32}(2\sin 2\theta\cos 2\theta) + C$$

$$= \frac{9}{8}\operatorname{Sin}^{-1}\left(\frac{x}{\sqrt{3}}\right) - \frac{9}{16}(2\sin\theta\cos\theta)(1-2\sin^2\theta) + C$$

$$= \frac{9}{8}\operatorname{Sin}^{-1}\left(\frac{x}{\sqrt{3}}\right) - \frac{9}{8}\left(\frac{x}{\sqrt{3}}\right)\frac{\sqrt{3-x^2}}{\sqrt{3}}\left(1-\frac{2x^2}{3}\right) + C \qquad \text{(Fig. 9.9)}$$

$$= \frac{9}{8}\operatorname{Sin}^{-1}\left(\frac{x}{\sqrt{3}}\right) - \frac{x}{8}\sqrt{3-x^2}(3-2x^2) + C.$$

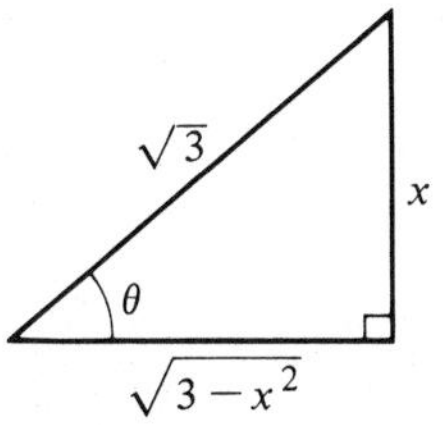

Figure 9.9

d) If we set $x = 4\sin\theta$, then $dx = 4\cos\theta\,d\theta$ and

$$\int \frac{(16-x^2)^{3/2}}{x}\,dx = \int \frac{64\cos^3\theta\, 4\cos\theta}{4\sin\theta}\,d\theta$$

$$= 64\int \frac{\cos^4\theta}{\sin\theta}\,d\theta$$

$$= 64\int \frac{\cos^2\theta(1-\sin^2\theta)}{\sin\theta}\,d\theta$$

$$= 64\int\left\{\frac{1-\sin^2\theta}{\sin\theta} - \cos^2\theta\sin\theta\right\}d\theta$$

$$= 64\int(\csc\theta - \sin\theta - \cos^2\theta\sin\theta)\,d\theta$$

$$= 64\left\{\ln|\csc\theta - \cot\theta| + \cos\theta + \frac{1}{3}\cos^3\theta\right\} + C$$

$$= 64\left\{\ln\left|\frac{4}{x} - \frac{\sqrt{16-x^2}}{x}\right| + \frac{\sqrt{16-x^2}}{4} + \frac{1}{3}\frac{(16-x^2)^{3/2}}{64}\right\} + C \qquad \text{(Fig. 9.10)}$$

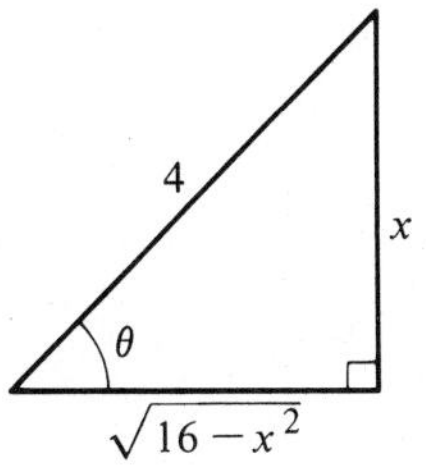

Figure 9.10

$$= 64\ln\left|\frac{4-\sqrt{16-x^2}}{x}\right| + 16\sqrt{16-x^2} + \frac{1}{3}(16-x^2)^{3/2} + C. \square$$

Example 8 Evaluate

$$\int \frac{x}{\sqrt{9 + x^2}}\,dx.$$

Solution ***Method (i)*** If we set $x = 3\tan\theta$, then $dx = 3\sec^2\theta\,d\theta$ and

$$\int \frac{x}{\sqrt{9 + x^2}}\,dx = \int \frac{3\tan\theta\, 3\sec^2\theta}{3\sec\theta}\,d\theta$$

$$= 3\int \tan\theta\sec\theta\,d\theta$$

$$= 3\sec\theta + C$$

$$= 3\frac{\sqrt{9 + x^2}}{3} + C \qquad \text{(Fig. 9.11)}$$

$$= \sqrt{9 + x^2} + C.$$

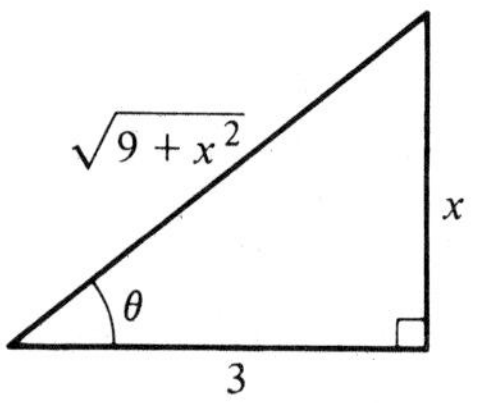

Figure 9.11

Method (ii) If we set $u = 9 + x^2$, then $du = 2x\,dx$ and

$$\int \frac{x}{\sqrt{9 + x^2}}\,dx = \int \frac{1}{\sqrt{u}}\frac{du}{2} = \frac{1}{2}\{2u^{1/2}\} + C = \sqrt{9 + x^2} + C.$$

Method (iii) If we set $u = \sqrt{9 + x^2}$, then $du = 2x/(2\sqrt{9 + x^2})\,dx$ and

$$\int \frac{x}{\sqrt{9 + x^2}}\,dx = \int du = u + C = \sqrt{9 + x^2} + C. \ \square$$

Although we have presented three methods of solution for Example 8, we suggest that with a little thought we could have guessed the answer without any substitution whatsoever.

Example 9 Verify the formula πr^2 for the area of a circle of radius r.

Solution If we consider the circle

$$x^2 + y^2 = r^2$$

shown in Fig. 9.12, we see that its area is four times that in the first quadrant:

$$A = 4\int_0^r \sqrt{r^2 - x^2}\,dx.$$

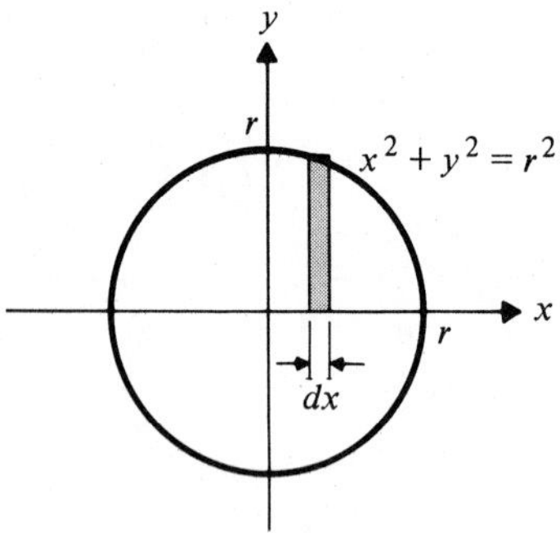

Figure 9.12

If we set $x = r\sin\theta$, then $dx = r\cos\theta\,d\theta$ and

$$A = 4\int_0^{\pi/2} r\cos\theta\, r\cos\theta\,d\theta$$

$$= 4r^2\int_0^{\pi/2}\left(\frac{1 + \cos 2\theta}{2}\right)d\theta$$

$$= 2r^2\left\{\theta + \frac{1}{2}\sin 2\theta\right\}_0^{\pi/2}$$

$$= 2r^2(\pi/2)$$

$$= \pi r^2. \ \square$$

EXERCISES 9.5

In Exercises 1–20, evaluate the given indefinite integral.

1. $\int \frac{1}{x\sqrt{2x^2-4}}\,dx$

2. $\int \frac{1}{\sqrt{9-5x^2}}\,dx$

3. $\int \frac{1}{10+x^2}\,dx$

4. $\int \frac{1}{x^2\sqrt{4-x^2}}\,dx$

5. $\int \sqrt{7-x^2}\,dx$

6. $\int x\sqrt{5x^2+3}\,dx$

7. $\int x^3\sqrt{4+x^2}\,dx$

8. $\int \frac{1}{1-x^2}\,dx$

9. $\int \frac{1}{\sqrt{x^2-5}}\,dx$

10. $\int \frac{x+5}{10x^2+2}\,dx$

11. $\int \frac{1}{x\sqrt{x^2+3}}\,dx$

12. $\int \frac{\sqrt{4-x^2}}{x}\,dx$

13. $\int \frac{x^2}{(2-9x^2)^{3/2}}\,dx$

14. $\int \frac{\sqrt{x^2-16}}{x^2}\,dx$

15. $\int \frac{1}{x^2\sqrt{2x^2+7}}\,dx$

16. $\int \frac{1}{x^3\sqrt{x^2-4}}\,dx$

17. $\int \frac{\sqrt{9-z^2}}{z^4}\,dz$

18. $\int \frac{y^3}{\sqrt{y^2+4}}\,dy$

19. $\int \frac{1}{(4x^2-9)^{3/2}}\,dx$

20. $\int \frac{x^2+2}{x^3+x}\,dx$

21. Show that if $a > 0$, then

$$\int \frac{1}{a^2-x^2}\,dx = \frac{1}{2a}\ln\left|\frac{a+x}{a-x}\right| + C.$$

22. Verify that the area inside the curve $b^2x^2 + a^2y^2 = a^2b^2$ is πab.

23. Find the area common to the circles $x^2 + y^2 = 4$ and $x^2 + y^2 = 4x$.

24. A boy initially at O (Fig. 9.13) walks along the edge of a pier (the y-axis) towing his sailboat by a string of length L.

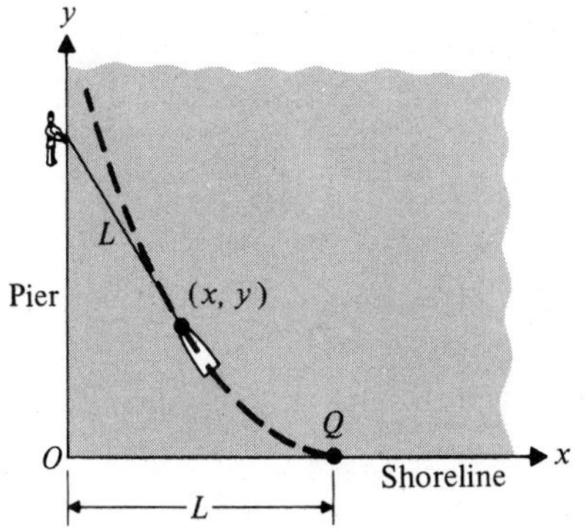

Figure 9.13

a) If the boat starts at Q and the string always remains straight, show that the equation of the curved path $y = f(x)$ followed by the boat must satisfy the differential equation

$$\frac{dy}{dx} = -\frac{\sqrt{L^2-x^2}}{x}.$$

b) Find $y = f(x)$.

25. If a thin circular plate has radius 2 and constant density ρ (mass per unit area), find its moment of inertia about any tangent line to its edge.

26. A cylindrical oil can of radius r and length h has its axis horizontal. If the specific weight of the oil is γ, find the force on each end of the can when it is full.

In Exercises 27–30, evaluate the given indefinite integral.

27. $\int \frac{2x^4-x^2}{2x^2+1}\,dx$

28. $\int (7-x^2)^{3/2}\,dx$

29. $\int \frac{1}{x-x^3}\,dx$

30. $\int \frac{1}{x^3(4x^2-1)^{3/2}}\,dx$

31. Find the length of the curve $8y^2 = x^2(1-x^2)$.

32. When a flexible cable of constant density ρ hangs between two fixed points A and B (Fig. 9.14), the shape $y = f(x)$ of the cable must satisfy the differential equation

$$\frac{d^2y}{dx^2} = \frac{\rho g}{H}\sqrt{1+(dy/dx)^2},$$

where $g > 0$ and $H > 0$ are constants. If we set $k = \rho g/H$ and $p = dy/dx$, then

$$\frac{dp}{dx} = k\sqrt{1+p^2}, \quad \text{or} \quad \frac{1}{\sqrt{1+p^2}}\frac{dp}{dx} = k.$$

It follows that

$$\int \frac{1}{\sqrt{1+p^2}}\,dp = kx + C.$$

a) Evaluate the integral shown to find $p = dy/dx$.

b) Integrate once more to find the shape of the cable. (See also Example 27 in Chapter 8.)

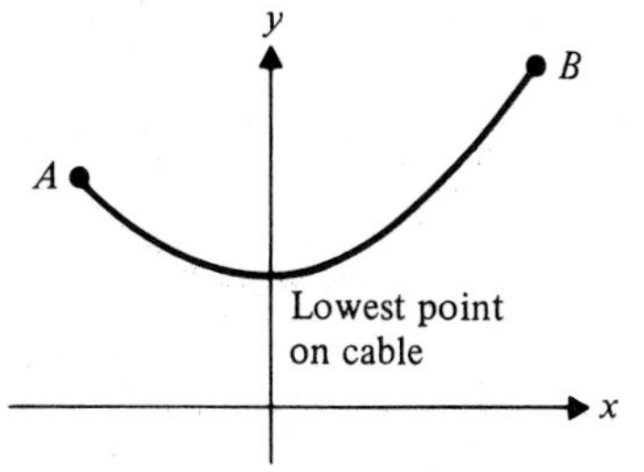

Figure 9.14

*33. Find the area inside the loop of the strophoid $y^2(a+x) = x^2(a-x)$.

***34.** a) By rationalizing the denominator, show that

$$\frac{1}{x+1+\sqrt{x^2+1}} = \frac{x+1-\sqrt{x^2+1}}{2x}.$$

b) Use (a) to prove that

$$\int_0^1 \frac{1}{x+1+\sqrt{x^2+1}}\,dx$$

$$= 1 - \frac{\sqrt{2}}{2} + \frac{1}{2}\ln\left(\frac{1+\sqrt{2}}{2}\right).$$

***35.** Show that

$$\int_{-1}^{1} \sqrt{\frac{1+x}{1-x}}\,dx = \pi.$$

9.6 COMPLETING THE SQUARE AND TRIGONOMETRIC SUBSTITUTIONS

Trigonometric substitutions reduce integrals containing the square roots $\sqrt{a^2 - b^2x^2}$, $\sqrt{a^2 + b^2x^2}$, and $\sqrt{b^2x^2 - a^2}$ to trigonometric integrals. For integrals containing the general square root

$$\sqrt{ax^2 + bx + c},$$

we can again reduce the integral to a trigonometric integral by a trigonometric substitution, if we first complete the square. For example, consider

$$\int \frac{1}{(x^2+2x+5)^{3/2}}\,dx.$$

If we complete the square of $x^2 + 2x$, we get

$$\int \frac{1}{(x^2+2x+5)^{3/2}}\,dx = \int \frac{1}{[(x+1)^2+4]^{3/2}}\,dx.$$

Had the denominator been $(x^2+4)^{3/2}$, we would have set $x = 2\tan\theta$. It is natural then for the denominator $[(x+1)^2+4]^{3/2}$ to set

$$x + 1 = 2\tan\theta,$$

in which case $dx = 2\sec^2\theta\,d\theta$ and

$$\begin{aligned}\int \frac{1}{(x^2+2x+5)^{3/2}}\,dx &= \int \frac{2\sec^2\theta}{[4\tan^2\theta+4]^{3/2}}\,d\theta \\ &= \int \frac{2\sec^2\theta}{8\sec^3\theta}\,d\theta \\ &= \frac{1}{4}\int \cos\theta\,d\theta \\ &= \frac{1}{4}\sin\theta + C \\ &= \frac{x+1}{4\sqrt{x^2+2x+5}} + C. \qquad \text{(Fig. 9.15)}\end{aligned}$$

Completion of the square in the following integral leads to

$$\int \sqrt{4 + x - x^2}\, dx = \int \sqrt{-(x - \tfrac{1}{2})^2 + \tfrac{17}{4}}\, dx.$$

If we now set

$$x - \frac{1}{2} = \frac{\sqrt{17}}{2} \sin \theta,$$

then

$$dx = \frac{\sqrt{17}}{2} \cos \theta \, d\theta$$

and

$$\begin{aligned}
\int \sqrt{4 + x - x^2}\, dx &= \int \frac{\sqrt{17}}{2} \cos \theta \frac{\sqrt{17}}{2} \cos \theta \, d\theta \\
&= \frac{17}{4} \int \left(\frac{1 + \cos 2\theta}{2}\right) d\theta \\
&= \frac{17}{8} \left\{\theta + \frac{1}{2} \sin 2\theta\right\} + C \\
&= \frac{17}{8} \{\theta + \sin \theta \cos \theta\} + C \\
&= \frac{17}{8} \operatorname{Sin}^{-1}\left(\frac{2x - 1}{\sqrt{17}}\right) + \frac{17}{8} \left\{\frac{2}{\sqrt{17}}\left(x - \frac{1}{2}\right)\right\} \\
&\quad \times \frac{\sqrt{4 + x - x^2}}{\sqrt{17}/2} + C \qquad \text{(Fig. 9.16)} \\
&= \frac{17}{8} \operatorname{Sin}^{-1}\left(\frac{2x - 1}{\sqrt{17}}\right) + \frac{1}{4}(2x - 1)\sqrt{4 + x - x^2} + C.
\end{aligned}$$

As a final illustrative example, we have

$$\int \frac{x}{\sqrt{2x^2 + 3x - 6}}\, dx = \frac{1}{\sqrt{2}} \int \frac{x}{\sqrt{x^2 + \dfrac{3x}{2} - 3}}\, dx = \frac{1}{\sqrt{2}} \int \frac{x}{\sqrt{(x + \frac{3}{4})^2 - \frac{57}{16}}}\, dx.$$

If we set

$$x + \frac{3}{4} = \frac{\sqrt{57}}{4} \sec \theta,$$

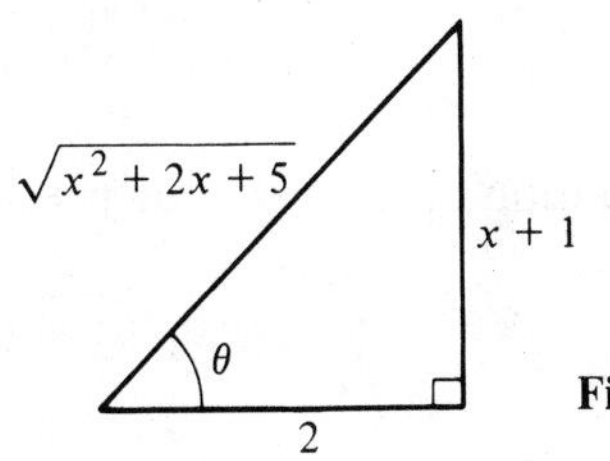

Figure 9.15

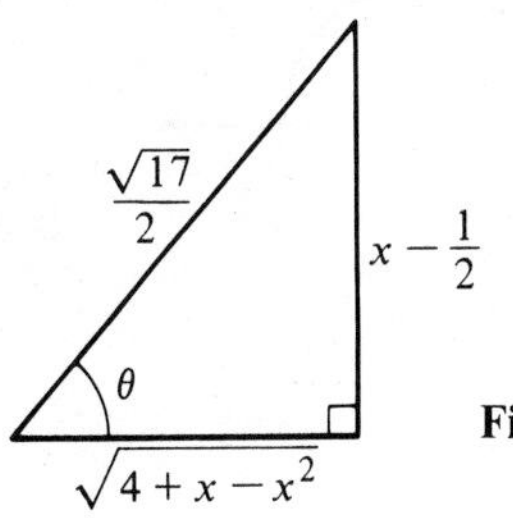

Figure 9.16

then

$$dx = \frac{\sqrt{57}}{4}\sec\theta\tan\theta\,d\theta$$

and

$$\int \frac{x}{\sqrt{2x^2+3x-6}}\,dx = \frac{1}{\sqrt{2}}\int \frac{[(\sqrt{57}/4)\sec\theta - \frac{3}{4}](\sqrt{57}/4)\sec\theta\tan\theta}{(\sqrt{57}/4)\tan\theta}\,d\theta$$

$$= \frac{1}{\sqrt{2}}\int\left\{\frac{\sqrt{57}}{4}\sec^2\theta - \frac{3}{4}\sec\theta\right\}d\theta$$

$$= \frac{\sqrt{57}}{4\sqrt{2}}\tan\theta - \frac{3}{4\sqrt{2}}\ln|\sec\theta + \tan\theta| + C$$

$$= \frac{\sqrt{57}}{4\sqrt{2}}\left\{\frac{\sqrt{x^2+(3/2)x-3}}{\sqrt{57}/4}\right\}$$

$$- \frac{3}{4\sqrt{2}}\ln\left|\frac{x+3/4}{\sqrt{57}/4} + \frac{\sqrt{x^2+(3/2)x-3}}{\sqrt{57}/4}\right| + C$$

(Fig. 9.17)

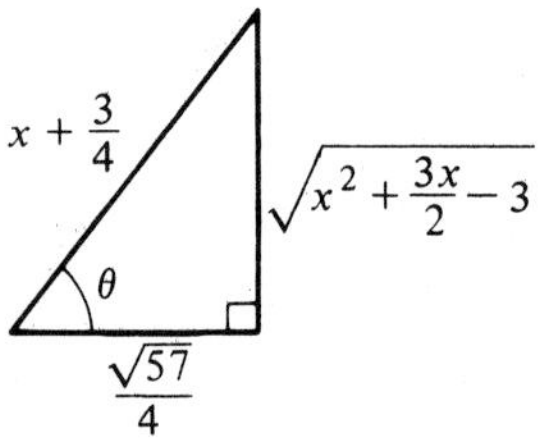

Figure 9.17

$$= \frac{1}{2}\sqrt{2x^2+3x-6}$$

$$- \frac{3}{4\sqrt{2}}\ln|4x+3+2\sqrt{2}\sqrt{2x^2+3x-6}| + D,$$

where

$$D = C + \frac{3}{4\sqrt{2}}\ln(\sqrt{57}).$$

We have shown that the technique of completing the square is the natural generalization of the technique of trigonometric substitutions. It reduces integrals containing the square root $\sqrt{ax^2+bx+c}$ to trigonometric integrals.

Example 10 Find the volume of a donut.

Solution A donut is generated (as the volume of the solid of revolution) when the circle

$$(x-a)^2 + y^2 = b^2, \qquad (a > b),$$

is rotated around the y-axis (Fig. 9.18). Cylindrical shells for the vertical rectangle shown yield the volume generated by the upper semicircle, which can be doubled to give

$$V = 2\int_{a-b}^{a+b} 2\pi xy\,dx.$$

If we solve the equation of the circle for $y = \pm\sqrt{b^2-(x-a)^2}$, then

$$V = 2\int_{a-b}^{a+b} 2\pi x\sqrt{b^2-(x-a)^2}\,dx.$$

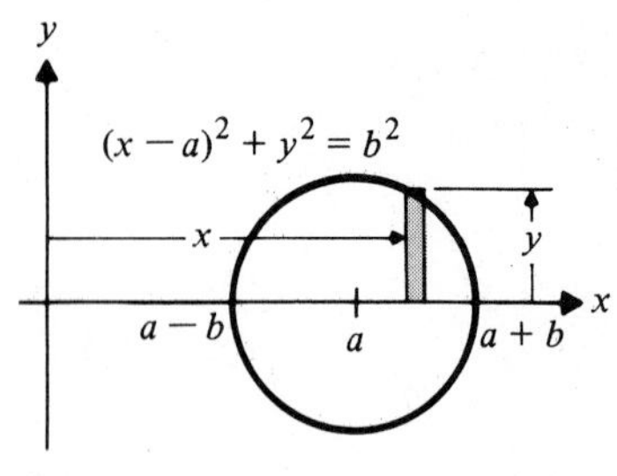

Figure 9.18

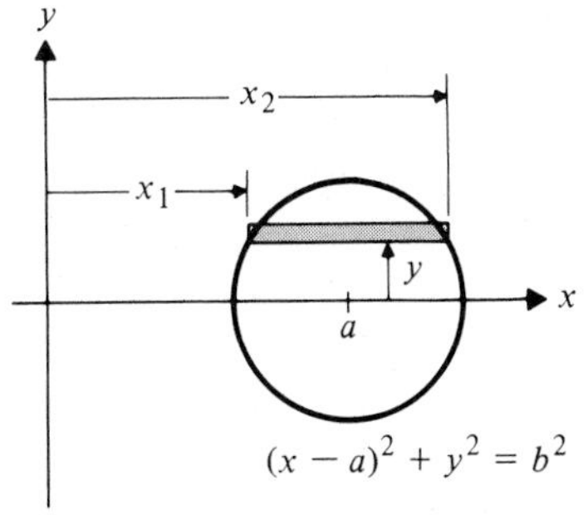

Figure 9.19

We now set $x - a = b \sin \theta$, in which case $dx = b \cos \theta \, d\theta$ and

$$V = 4\pi \int_{-\pi/2}^{\pi/2} (a + b \sin \theta) b \cos \theta \, b \cos \theta \, d\theta$$

$$= 4\pi b^2 \int_{-\pi/2}^{\pi/2} \left\{ a\left(\frac{1 + \cos 2\theta}{2}\right) + b \cos^2 \theta \sin \theta \right\} d\theta$$

$$= 4\pi b^2 \left\{ \frac{a}{2}\left(\theta + \frac{1}{2}\sin 2\theta\right) - \frac{b}{3}\cos^3 \theta \right\}_{-\pi/2}^{\pi/2}$$

$$= 4\pi b^2 \left\{ \frac{a}{2}\left(\frac{\pi}{2} + \frac{\pi}{2}\right)\right\} = 2\pi^2 ab^2.$$

Should we use horizontal rectangles, as in Fig. 9.19, the washer method yields

$$V = 2 \int_0^b (\pi x_2^2 - \pi x_1^2)\, dy.$$

We can solve the equation of the circle for $x = a \pm \sqrt{b^2 - y^2}$; hence we have

$$x_1 = a - \sqrt{b^2 - y^2},$$
$$x_2 = a + \sqrt{b^2 - y^2}.$$

Thus

$$V = 2\pi \int_0^b [(a + \sqrt{b^2 - y^2})^2 - (a - \sqrt{b^2 - y^2})^2]\, dy,$$

and this simplifies to

$$V = 2\pi \int_0^b 4a\sqrt{b^2 - y^2}\, dy.$$

If we set $y = b \sin \theta$, then $dy = b \cos \theta \, d\theta$ and

$$V = 8\pi a \int_0^{\pi/2} b \cos \theta \, b \cos \theta \, d\theta$$

$$= 8\pi ab^2 \int_0^{\pi/2} \left(\frac{1 + \cos 2\theta}{2}\right) d\theta$$

$$= 4\pi ab^2 \left\{ \theta + \frac{1}{2}\sin 2\theta \right\}_0^{\pi/2} = 4\pi ab^2(\pi/2) = 2\pi^2 ab^2. \ \square$$

EXERCISES 9.6

In Exercises 1–10, evaluate the given indefinite integral.

1. $\displaystyle\int \frac{x}{\sqrt{27 + 6x - x^2}}\, dx$

2. $\displaystyle\int \frac{1}{\sqrt{x^2 + 2x + 2}}\, dx$

3. $\displaystyle\int \frac{1}{(y^2 + 4y)^{3/2}}\, dy$

4. $\displaystyle\int \frac{1}{3x - x^2 - 4}\, dx$

5. $\displaystyle\int \frac{\sqrt{x^2 + 2x - 3}}{x + 1}\, dx$

6. $\displaystyle\int \frac{x}{(4x - x^2)^{3/2}}\, dx$

7. $\displaystyle\int \sqrt{-y^2 + 6y}\, dy$

8. $\displaystyle\int \frac{2x - 3}{x^2 + 6x + 13}\, dx$

9. $\displaystyle\int \frac{5-4x}{\sqrt{12x-4x^2-8}}\,dx$

10. $\displaystyle\int \frac{1}{x\sqrt{6+4\ln x+(\ln x)^2}}\,dx$

11. Use the substitution $z = 1/x$ to evaluate

$$\int \frac{1}{x\sqrt{x^2+6x+3}}\,dx.$$

12. One of the gates in a dam is circular with radius 1 meter. If the gate is closed and the surface of the water is 3 meters above the top of the gate, find the force due to water pressure on the gate.

13. Evaluate

$$\int \frac{1}{3x-x^2}\,dx$$

a) by completing the square, and setting $x - \frac{3}{2} = (\frac{3}{2})\sin\theta$;

b) by multiplying numerator and denominator by -1, completing the square, and setting $x - \frac{3}{2} = (\frac{3}{2})\sec\theta$.

c) Explain the difference in the two answers.

Evaluate the integrals in Exercises 14 and 15.

***14.** $\displaystyle\int \frac{1}{x\sqrt{2x-x^2}}\,dx$

***15.** $\displaystyle\int \frac{1}{(2x+5)\sqrt{2x-3}+8x-12}\,dx$

9.7 INTEGRATION BY PARTS

One of the most powerful integration techniques is that of *integration by parts*. It results from the product rule for differentiation:

$$\frac{d}{dx}(uv) = u\frac{dv}{dx} + v\frac{du}{dx}.$$

If we take antiderivatives of both sides of this equation with respect to x, we get

$$\int \frac{d}{dx}(uv)\,dx = \int u\frac{dv}{dx}\,dx + \int v\frac{du}{dx}\,dx,$$

or

$$\int u\,dv = uv - \int v\,du. \qquad (9.32)$$

Equation (9.32) is called the *integration-by-parts formula*. It says that if we are given an unknown integral (the left-hand side of Eq. 9.32), and we arbitrarily divide the function after the integral sign into two parts, one called u and the other called dv; then we can set the unknown integral equal to the right-hand side of Eq. (9.32). What this does is replace the integral

$$\int u\,dv \quad \text{with} \quad \int v\,du.$$

This will be useful if the new integral is easier to evaluate than the original. The problem then is to choose u and dv so that the new integral is simpler than the original. As a general rule we let dv be the most complicated part of the integrand (plus differential) that we can integrate mentally. The example

$$\int xe^x\,dx$$

will clarify these ideas. To use integration by parts on this problem we must define u and dv in such a way that $u\,dv = xe^x\,dx$. There are four possibilities:

$$\begin{aligned} u &= 1, & dv &= xe^x\,dx; \\ u &= x, & dv &= e^x\,dx; \\ u &= e^x, & dv &= x\,dx; \\ u &= xe^x, & dv &= dx. \end{aligned}$$

Our rule suggests the choice

$$u = x, \qquad dv = e^x\,dx,$$

in which case

$$du = dx, \qquad v = e^x.$$

Then by (9.32), we get

$$\int xe^x\,dx = xe^x - \int e^x\,dx.$$

We have therefore replaced the integration of xe^x with the integration of e^x, a definite simplification. The final solution is therefore

$$\int xe^x\,dx = xe^x - e^x + C.$$

As a second example, consider

$$\int \ln x\,dx.$$

Integration by parts leaves no choice for u and dv; they must be

$$u = \ln x, \qquad dv = dx,$$

in which case

$$du = \frac{1}{x}\,dx, \qquad v = x.$$

Then we have

$$\int \ln x\,dx = x\ln x - \int x\frac{1}{x}\,dx = x\ln x - x + C.$$

These two examples illustrate that integration by parts replaces an integral with two others: one mental integration and a second integration that is hopefully simpler than the original. We should also note that in the evaluation of the mental integration we did not include a constant of integration. For example, in the integration from $dv = e^x\,dx$ to $v = e^x$, we did not write $v = e^x + D$. Had we done so, the solution would have proceeded as follows:

$$\begin{aligned} \int xe^x\,dx &= x(e^x + D) - \int (e^x + D)\,dx \\ &= x(e^x + D) - e^x - Dx + C \\ &= xe^x - e^x + C. \end{aligned}$$

The constant D has therefore disappeared from the eventual solution. This always occurs, and consequently we make the agreement not to include an arbitrary constant in the mental integration for v.

Example 11 Evaluate each of the following indefinite integrals:

a) $\int x^2 \cos x\, dx$;

b) $\int e^x \sin x\, dx$.

Solution a) If we set

$$u = x^2, \qquad dv = \cos x\, dx,$$

then

$$du = 2x\, dx, \qquad v = \sin x,$$

and

$$\int x^2 \cos x\, dx = x^2 \sin x - \int 2x \sin x\, dx.$$

Integration by parts has therefore reduced the power on x from 2 to 1. To eliminate the x in front of the trigonometric function completely, we perform integration by parts once again, this time with

$$u = x, \qquad dv = \sin x\, dx.$$

Then,

$$du = dx, \qquad v = -\cos x,$$

and

$$\int x^2 \cos x\, dx = x^2 \sin x - 2\left\{-x \cos x - \int -\cos x\, dx\right\}$$

$$= x^2 \sin x + 2x \cos x - 2 \sin x + C.$$

b) If we set

$$u = e^x, \qquad dv = \sin x\, dx,$$

then

$$du = e^x\, dx, \qquad v = -\cos x,$$

and

$$\int e^x \sin x\, dx = -e^x \cos x - \int -e^x \cos x\, dx.$$

Integration by parts appears to have led nowhere: The integration of $e^x \cos x$ is essentially as difficult as that of $e^x \sin x$. If, however, we persevere and integrate by parts once more with

$$u = e^x, \qquad dv = \cos x\, dx,$$

then

$$du = e^x\, dx, \qquad v = \sin x,$$

and

$$\int e^x \sin x\,dx = -e^x \cos x + e^x \sin x - \int e^x \sin x\,dx.$$

This equation can now be solved for the unknown integral

$$\int e^x \sin x\,dx = \frac{1}{2} e^x(\sin x - \cos x) + C,$$

to which we have added the arbitrary constant C. □

The technique in part (b) of this example is often called *integration by reproduction.* By two applications of integration by parts, we have reproduced the unknown integral and then solved the equation for it. Another example of this method follows.

Example 12 Evaluate

$$\int \sec^3 x\,dx.$$

Solution If we set

$$u = \sec x, \qquad dv = \sec^2 x\,dx,$$

then

$$du = \sec x \tan x\,dx, \qquad v = \tan x,$$

and

$$\int \sec^3 x\,dx = \sec x \tan x - \int \sec x \tan^2 x\,dx.$$

Since $\tan^2 x = \sec^2 x - 1$, we can write

$$\begin{aligned}\int \sec^3 x\,dx &= \sec x \tan x - \int \sec x(\sec^2 x - 1)\,dx \\ &= \sec x \tan x - \int \sec^3 x\,dx + \ln|\sec x + \tan x|.\end{aligned}$$

We now solve for

$$\int \sec^3 x\,dx = \frac{1}{2}\{\sec x \tan x + \ln|\sec x + \tan x|\} + C. \;\square$$

Example 13 Evaluate

$$\int \frac{x^3}{\sqrt{9 + x^2}}\,dx.$$

Solution ***Method (i)*** If we set $u = 9 + x^2$, then $du = 2x\,dx$ and

$$\begin{aligned}\int \frac{x^3}{\sqrt{9+x^2}}\,dx &= \int \frac{u-9}{\sqrt{u}}\frac{du}{2} = \frac{1}{2}\int (u^{1/2} - 9u^{-1/2})\,du \\ &= \frac{1}{2}\left\{\frac{2}{3}u^{3/2} - 18u^{1/2}\right\} + C = \frac{1}{3}(9 + x^2)^{3/2} - 9\sqrt{9 + x^2} + C.\end{aligned}$$

Method (ii) If we set $u = \sqrt{9+x^2}$, then $du = (x/\sqrt{9+x^2})\,dx$ and

$$\int \frac{x^3}{\sqrt{9+x^2}}\,dx = \int (u^2-9)\,du$$

$$= \frac{1}{3}u^3 - 9u + C$$

$$= \frac{1}{3}(9+x^2)^{3/2} - 9\sqrt{9+x^2} + C.$$

Method (iii) If we set $x = 3\tan\theta$, then $dx = 3\sec^2\theta\,d\theta$ and

$$\int \frac{x^3}{\sqrt{9+x^2}}\,dx = \int \frac{27\tan^3\theta\,3\sec^2\theta}{3\sec\theta}\,d\theta$$

$$= 27\int \tan\theta(\sec^2\theta - 1)\sec\theta\,d\theta$$

$$= 27\int (\sec^3\theta\tan\theta - \sec\theta\tan\theta)\,d\theta$$

$$= 27\left\{\frac{1}{3}\sec^3\theta - \sec\theta\right\} + C$$

$$= 9\left\{\frac{\sqrt{9+x^2}}{3}\right\}^3 - 27\frac{\sqrt{9+x^2}}{3} + C \qquad \text{(Fig. 9.20)}$$

$$= \frac{1}{3}(9+x^2)^{3/2} - 9\sqrt{9+x^2} + C.$$

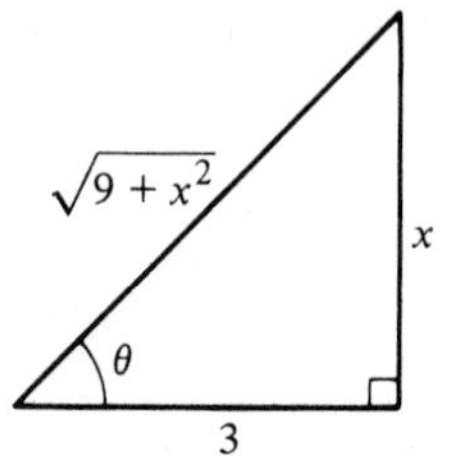

Figure 9.20

Method (iv) If we set

$$u = x^2, \qquad dv = \frac{x}{\sqrt{9+x^2}}\,dx,$$

then

$$du = 2x\,dx, \qquad v = \sqrt{9+x^2},$$

and

$$\int \frac{x^3}{\sqrt{9+x^2}}\,dx = x^2\sqrt{9+x^2} - \int 2x\sqrt{9+x^2}\,dx$$

$$= x^2\sqrt{9+x^2} - 2\left\{\frac{1}{3}(9+x^2)^{3/2}\right\} + C$$

$$= x^2\sqrt{9+x^2} - \frac{2}{3}(9+x^2)^{3/2} + C. \ \square$$

Example 13 illustrates that there may be more than one way to evaluate an indefinite integral, and the answer may appear different depending on the technique used. According to Theorem 5.1, these solutions should be identical, up to an additive constant. To illustrate that this is indeed the case, note that the solution by integration by parts can be written as

$$\begin{aligned}\sqrt{9+x^2}\{x^2 - \tfrac{2}{3}(9+x^2)\} + C &= \tfrac{1}{3}\sqrt{9+x^2}\{-18+x^2\} + C\\ &= \tfrac{1}{3}\sqrt{9+x^2}\{-27+(9+x^2)\} + C\\ &= -9\sqrt{9+x^2} + \tfrac{1}{3}(9+x^2)^{3/2} + C,\end{aligned}$$

which is the solution obtained by the other three methods.

The integration-by-parts formula in Eq. (9.32) applies to indefinite integrals. It is not difficult to show that for definite integrals

$$\int_a^b u\,dv = \left\{uv\right\}_a^b - \int_a^b v\,du. \tag{9.33}$$

Example 14 Evaluate

$$\int_0^4 \frac{x}{\sqrt{x+8}}\,dx.$$

Solution If we set

$$u = x, \qquad dv = \frac{1}{\sqrt{x+8}}\,dx,$$

then

$$du = dx, \qquad v = 2\sqrt{x+8},$$

and

$$\begin{aligned}\int_0^4 \frac{x}{\sqrt{x+8}}\,dx &= \left\{2x\sqrt{x+8}\right\}_0^4 - \int_0^4 2\sqrt{x+8}\,dx\\ &= 8\sqrt{12} - 2\left\{\frac{2}{3}(x+8)^{3/2}\right\}_0^4\\ &= \frac{16}{3}(4\sqrt{2} - 3\sqrt{3}). \ \square\end{aligned}$$

EXERCISES 9.7

In Exercises 1–18, evaluate the given indefinite integral.

1. $\int x \sin x\,dx$
2. $\int x^2 e^{2x}\,dx$
3. $\int x^4 \ln x\,dx$
4. $\int \sqrt{x} \ln(2x)\,dx$
5. $\int z \sec^2(z/3)\,dz$
6. $\int x\sqrt{3-x}\,dx$
7. $\int \mathrm{Sin}^{-1} x\,dx$
8. $\int x^2\sqrt{x+5}\,dx$
9. $\int \frac{x}{\sqrt{2+x}}\,dx$
10. $\int \frac{x^2}{\sqrt{2+x}}\,dx$
11. $\int \frac{x}{\sqrt{2+x^2}}\,dx$
12. $\int (x-1)^2 \ln x\,dx$
13. $\int e^{2x} \cos 3x\,dx$
14. $\int \mathrm{Tan}^{-1} x\,dx$
15. $\int \cos(\ln x)\,dx$
16. $\int \sin x \cos 3x\,dx$

17. $\int \frac{x^3}{\sqrt{5+3x^2}}\,dx$

18. $\int \ln(x^2+4)\,dx$

19. The face of a dam is shown in Fig. 9.21, where the curve has equation

$$y = e^{k|x|} - 1, \qquad k = \frac{1}{100}\ln 201.$$

Find the total force on the dam due to water pressure when the water level on the dam is 100 meters.

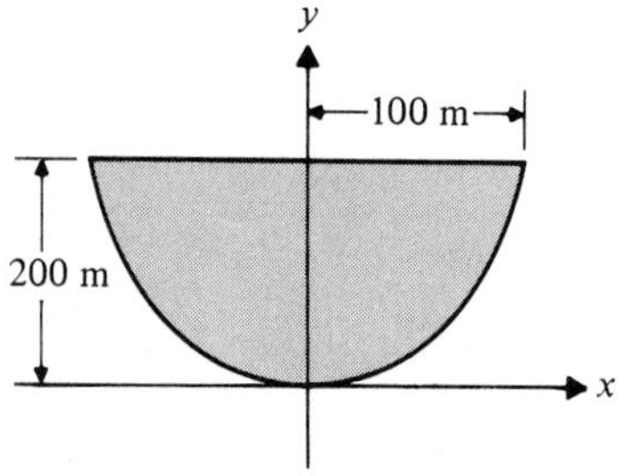

Figure 9.21

20. Consider the indefinite integral $I = \int \cos^6 x\,dx$.

a) Evaluate I using trigonometric identities.

b) If $n \geq 2$ is an integer, develop the reduction formula

$$\int \cos^n x\,dx = \frac{1}{n}\cos^{n-1} x \sin x + \frac{n-1}{n}\int \cos^{n-2} x\,dx.$$

c) Use (b) to evaluate I.

d) Verify that the answers in (a) and (c) are identical.

21. Consider the evaluation of $\int x^5 e^x\,dx$. It is clear that were we to use integration by parts the answer would eventually be of the form

$$\int x^5e^x\,dx = Ax^5e^x + Bx^4e^x + Cx^3e^x + Dx^2e^x + Exe^x + Fe^x + G.$$

Differentiate this equation, and thereby obtain A, B, C, D, E, and F (with no integration).

22. Find the length of the parabola $y = x^2$ from $(0, 0)$ to $(1, 1)$.

23. If we set $u = x^{-1}$ and $dv = dx$, then $du = -x^{-2}\,dx$ and $v = x$ and

$$\int \frac{1}{x}\,dx = \left(\frac{1}{x}\right)x - \int x\left(-\frac{1}{x^2}\,dx\right) = 1 + \int \frac{1}{x}\,dx.$$

Subtraction of $\int x^{-1}\,dx$ from each side of this equation now gives $0 = 1$. What is wrong with the argument?

24. The gamma function $\Gamma(n)$ for $n > 0$ is defined by the improper integral

$$\Gamma(n) = \int_0^\infty x^{n-1}e^{-x}\,dx.$$

Show that if n is a positive integer, then

$$\Gamma(n) = (n-1)!.$$

25. The Laplace transform of a function $F(t)$ is the function $f(s)$ defined by the improper integral

$$f(s) = \int_0^\infty e^{-st}F(t)\,dt.$$

Find Laplace transforms for each of the following functions:

a) e^{3t}; b) t^2; c) $\sin t$; d) te^{-t}.

26. Show that the function

$$f(x) = \frac{x^{\alpha-1}e^{-x/\beta}}{\Gamma(\alpha)\beta^\alpha}$$

for positive constants α and β qualifies as a probability density function on the interval $x \geq 0$. (See Exercise 23 in Section 7.9 for the definition of a pdf and Exercise 24 for the definition of $\Gamma(\alpha)$.)

In Exercises 27–34, evaluate the given indefinite integral.

27. $\int \mathrm{Tan}^{-1}\sqrt{x}\,dx$

28. $\int x^2\cos^2 x\,dx$

29. $\int \sec^5 x\,dx$

30. $\int \sqrt{x^2-4}\,dx$

31. $\int xe^x \sin x\,dx$

32. $\int \sqrt{1+3x^2}\,dx$

33. $\int \frac{x^2}{\sqrt{x^2-5}}\,dx$

34. $\int \sqrt{x^2-2x-3}\,dx$

* **35.** Evaluate $\int \sec^9 x\,dx$ by first developing the reduction formula

$$\int \sec^n x\,dx = \frac{\tan x \sec^{n-2} x}{n-1} + \frac{n-2}{n-1}\int \sec^{n-2} x\,dx.$$

* **36.** a) Use a trigonometric substitution to evaluate

$$I = \int \frac{5x+2}{(3+x^2)^4}\,dx.$$

b) Prove the following reduction formula for $n \geq 2$:

$$\int \frac{1}{(a^2+x^2)^n}\,dx = \frac{1}{2a^2(n-1)(a^2+x^2)^{n-1}} + \frac{2n-3}{2a^2(n-1)}\int \frac{1}{(a^2+x^2)^{n-1}}\,dx.$$

c) Use (b) to evaluate I.

* **37.** If n is a positive integer, define

$$I_n = \int_{-1}^{1} (1-x^2)^n\,dx.$$

a) Show that

$$(2n+1)I_n = 2nI_{n-1}, \qquad n \geq 2.$$

b) Prove that

$$I_n = \frac{2^{2n+1}(n!)^2}{(2n+1)!}, \qquad n \geq 1.$$

c) Evaluate

$$\int_0^{\pi/2} \cos^{2n+1} x\, dx.$$

9.8 PARTIAL FRACTIONS

The use of partial fractions is an integration technique applicable to integrands that are rational functions of x; i.e., to integrals

$$\int \frac{N(x)}{D(x)}\, dx, \tag{9.34}$$

where $N(x)$ and $D(x)$ are polynomials in x. Basically, the technique breaks down the fraction $N(x)/D(x)$ into simpler fractions.

Consider the integral

$$\int \frac{x^2+3x-2}{x-1}\, dx.$$

If we divide $x-1$ into x^2+3x-2, we obtain

$$\int \frac{x^2+3x-2}{x-1}\, dx = \int \left\{x+4+\frac{2}{x-1}\right\} dx = \frac{x^2}{2} + 4x + 2\ln|x-1| + C.$$

For this example the simple process of performing the long division indicated by the integrand enables us to rewrite the integrand in a form that permits immediate integration. Long division produces a polynomial quotient ($x+4$, in our example) and a remainder of degree less than that of the denominator only if initially the degree of the numerator is greater than or equal to that of the denominator. In most integrations of rational functions, it is best to perform this division first. We state the following, therefore, as a general rule.

If degree $N(x) \geq$ degree $D(x)$, divide $D(x)$ into $N(x)$ to produce a quotient polynomial $Q(x)$ and a remainder polynomial $R(x)$ of degree less than degree $D(x)$:

$$\frac{N(x)}{D(x)} = Q(x) + \frac{R(x)}{D(x)}, \qquad \deg R < \deg D.$$

Integration problem (9.34) can now be written in the form

$$\int \frac{N(x)}{D(x)}\, dx = \int Q(x)\, dx + \int \frac{R(x)}{D(x)}\, dx.$$

The first integral on the right is trivial, leaving only integration of $R(x)/D(x)$. Since this procedure can always be followed for a rational integrand, we need only consider now integration of rational functions

$$\int \frac{R(x)}{D(x)}\, dx, \quad \text{where } \deg R < \deg D.$$

We will assume that $D(x)$ is a monic polynomial; i.e., that the coefficient of the highest power of x is equal to 1. If $D(x)$ is not monic, it can always be made so by factoring the coefficient of its highest power out of every term and taking this coefficient outside the integration.

It is an established fact of algebra that every real monic polynomial, and specifically $D(x)$, can be factored into real linear factors $x - a$ and irreducible real quadratic factors $x^2 + bx + c$. We subdivide our discussion into five cases, depending on the number of linear and the number of quadratic factors.

Case 1 $D(x)$ factors into n nonrepeated linear factors.

In this case,

$$D(x) = (x - a_1)(x - a_2)\cdots(x - a_n), \tag{9.35a}$$

where the $a_1, a_2, \ldots, a_n$ are all different. We claim in this case that it is possible to rewrite $R(x)/D(x)$ in the form

$$\frac{R(x)}{D(x)} = \frac{A_1}{x - a_1} + \frac{A_2}{x - a_2} + \cdots + \frac{A_n}{x - a_n}, \tag{9.35b}$$

where the $A_1, A_2, \ldots, A_n$ are constants. The right-hand side of Eq. (9.35b) is called the *partial fraction decomposition* of $R(x)/D(x)$. Note that each term on the right of Eq. (9.35b) is immediately integrable as a logarithm, so that a decomposition of $R(x)/D(x)$ into a sum of terms of this form will certainly facilitate integration. Our only problem is to determine the unknown constants $A_1, A_2, \ldots, A_n$. We illustrate three methods for finding the constants for the example

$$\int \frac{2x^2 - x - 4}{x^3 + 2x^2 - x - 2}\,dx.$$

Since $x^3 + 2x^2 - x - 2 = (x - 1)(x + 1)(x + 2)$, the partial fraction decomposition of the integrand is

$$\frac{2x^2 - x - 4}{x^3 + 2x^2 - x - 2} = \frac{A}{x - 1} + \frac{B}{x + 1} + \frac{C}{x + 2}.$$

Method (i) If the right side of the decomposition is brought to a common denominator, we get

$$\frac{2x^2 - x - 4}{(x - 1)(x + 1)(x + 2)} = \frac{A(x + 1)(x + 2) + B(x - 1)(x + 2) + C(x - 1)(x + 1)}{(x - 1)(x + 1)(x + 2)}.$$

We now equate numerators and, at the same time, gather together terms in the various powers of x on the right-hand side:

$$2x^2 - x - 4 = (A + B + C)x^2 + (3A + B)x + (2A - 2B - C).$$

This equation can be true for a range of values of x only if the coefficients of corresponding powers of x are identical. (This result was proved in Exercise 31 in Section 3.6.) We set, therefore,

$$\begin{aligned} 2 &= A + B + C, \\ -1 &= 3A + B, \\ -4 &= 2A - 2B - C. \end{aligned}$$

The solution of this system of linear equations is

$$A = -\tfrac{1}{2}, \qquad B = \tfrac{1}{2}, \qquad C = 2.$$

Method (ii) We again bring the right-hand side of the partial fraction decomposition to a common denominator and equate numerators to get

$$2x^2 - x - 4 = A(x + 1)(x + 2) + B(x - 1)(x + 2) + C(x - 1)(x + 1),$$

but we do not multiply out the right-hand side. Instead we substitute into this equation various convenient values of x (since this equation is to be valid for a range of values of x). In particular, if we set $x = 1$, we have

$$2(1)^2 - (1) - 4 = A(2)(3),$$

which implies that $A = -\frac{1}{2}$. For $x = -1$,

$$2(-1)^2 - (-1) - 4 = B(-2)(1),$$

from which we get $B = \frac{1}{2}$. Finally, $x = -2$ gives us

$$2(-2)^2 - (-2) - 4 = C(-3)(-1),$$

from which we know that $C = 2$.

Method (iii) We mentally think of the partial fraction decomposition multiplied by $x - 1$:

$$\frac{2x^2 - x - 4}{(x+1)(x+2)} = A + \frac{B(x - 1)}{x + 1} + \frac{C(x - 1)}{x + 2}.$$

If we now set $x = 1$, we obtain

$$\frac{2(1)^2 - (1) - 4}{(2)(3)} = A,$$

which implies that $A = -\frac{1}{2}$. Effectively, A is obtained by covering up the $(x - 1)$ factor in the denominator of the integrand and setting $x = 1$. To obtain B, we cover up $x + 1$ and set $x = -1$:

$$B = \frac{2(-1)^2 - (-1) - 4}{(-2)(1)} = \frac{1}{2}.$$

For C, we cover up $x + 2$ and set $x = -2$:

$$C = \frac{2(-2)^2 - (-2) - 4}{(-3)(-1)} = 2.$$

All three methods have led to the same partial fraction decomposition, and the resulting integration is straightforward:

$$\int \frac{2x^2 - x - 4}{x^3 + 2x^2 - x - 2}\,dx = \int \left\{\frac{-\frac{1}{2}}{x - 1} + \frac{\frac{1}{2}}{x + 1} + \frac{2}{x + 2}\right\} dx$$

$$= -\frac{1}{2}\ln|x - 1| + \frac{1}{2}\ln|x + 1| + 2\ln|x + 2| + C.$$

We must point out that although method (iii) seems the fastest, methods (i) and (ii) for finding the unknown coefficients can also be used in the remaining four cases, but method (iii) applies only to the case of a nonrepeated linear factor. Finally, in methods (ii) and (iii) we should not actually be substituting values of x but rather taking limits as x approaches these values. Since the same result is obtained in either case, we will continue the substitution process.

Example 15 Evaluate

$$\int \frac{1}{9 - x^2}\,dx.$$

Solution Since

$$\frac{1}{x^2 - 9} = \frac{1}{(x-3)(x+3)} = \frac{\frac{1}{6}}{x-3} + \frac{-\frac{1}{6}}{x+3},$$

we have

$$\int \frac{1}{9 - x^2}\,dx = \int \left\{\frac{-\frac{1}{6}}{x-3} + \frac{\frac{1}{6}}{x+3}\right\} dx = -\frac{1}{6}\ln|x-3| + \frac{1}{6}\ln|x+3| + C. \square$$

(You might also try the substitution $x = 3\sin\theta$.)

Case 2 $D(x)$ factors into one linear factor repeated n times.

In this case,

$$D(x) = (x - a)^n, \tag{9.36a}$$

and it is possible to rewrite $R(x)/D(x)$ in the form

$$\frac{R(x)}{D(x)} = \frac{A_1}{x-a} + \frac{A_2}{(x-a)^2} + \cdots + \frac{A_n}{(x-a)^n}. \tag{9.36b}$$

Note that each term in this decomposition is immediately integrable as either a logarithm or a negative power of $x - a$. Once again, such a decomposition will facilitate integration.

We illustrate methods (i) and (ii) for finding the A_i in the following example (method iii is not applicable):

$$\int \frac{x^2 + 2}{x^3 - 3x^2 + 3x - 1}\,dx.$$

Since $x^3 - 3x^2 + 3x - 1 = (x-1)^3$, the partial fraction decomposition of the integrand is

$$\frac{x^2 + 2}{x^3 - 3x^2 + 3x - 1} = \frac{A}{x-1} + \frac{B}{(x-1)^2} + \frac{C}{(x-1)^3}.$$

If the right-hand side is brought to a common denominator, we have

$$\frac{x^2+2}{(x-1)^3} = \frac{A(x-1)^2 + B(x-1) + C}{(x-1)^3},$$

hence

$$x^2 + 2 = A(x-1)^2 + B(x-1) + C.$$

Method (i) of equating coefficients of like powers of x leads to the equations

$$1 = A,$$

$$0 = -2A + B,$$

$$2 = A - B + C.$$

Method (ii) with the values $x = 1, 0$, and 2 leads to the equations

$$3 = C,$$
$$2 = A - B + C,$$
$$6 = A + B + C.$$

Both sets of equations have solutions $A = 1$, $B = 2$, and $C = 3$. Thus,

$$\int \frac{x^2 + 2}{x^3 - 3x^2 + 3x - 1}\,dx = \int \left\{ \frac{1}{x - 1} + \frac{2}{(x - 1)^2} + \frac{3}{(x - 1)^3} \right\} dx$$
$$= \ln|x - 1| - \frac{2}{x - 1} - \frac{\frac{3}{2}}{(x - 1)^2} + C.$$

Example 16 Evaluate each of the following indefinite integrals:

a) $\displaystyle\int \frac{s}{s^2 + 4s + 4}\,ds;$

b) $\displaystyle\int \frac{1}{(y + 3)^4}\,dy.$

Solution a) Since

$$\frac{s}{s^2 + 4s + 4} = \frac{s}{(s + 2)^2} = \frac{A}{s + 2} + \frac{B}{(s + 2)^2},$$

then

$$s = A(s + 2) + B.$$

Clearly, $A = 1$ and $2A + B = 0$, which imply that $B = -2$. Thus,

$$\int \frac{s}{s^2 + 4s + 4}\,ds = \int \left\{ \frac{1}{s + 2} + \frac{-2}{(s + 2)^2} \right\} ds$$
$$= \ln|s + 2| + \frac{2}{s + 2} + C.$$

(You might also try the substitution $u = s + 2$.)

b)

$$\int \frac{1}{(y + 3)^4}\,dy = \frac{1}{-3(y + 3)^3} + C \;\square$$

Do not be misled in part (b) of this example to a partial fraction decomposition of the integrand. Had we done so, we would have set

$$\frac{1}{(y + 3)^4} = \frac{A}{y + 3} + \frac{B}{(y + 3)^2} + \frac{C}{(y + 3)^3} + \frac{D}{(y + 3)^4}.$$

No matter how we choose to find the constants, the solution is $A = B = C = 0$ and $D = 1$. In other words, the integrand was already in partial fraction form; it can be simplified no further.

Case 3 $D(x)$ factors into n nonrepeated irreducible quadratic factors.

In this case,

$$D(x) = (x^2 + b_1 x + c_1)(x^2 + b_2 x + c_2)\cdots(x^2 + b_n x + c_n), \quad (9.37a)$$

and it is possible to rewrite $R(x)/D(x)$ in the form

$$\frac{R(x)}{D(x)} = \frac{A_1 x + B_1}{x^2 + b_1 x + c_1} + \frac{A_2 x + B_2}{x^2 + b_2 x + c_2} + \cdots + \frac{A_n x + B_n}{x^2 + b_n x + c_n}. \quad (9.37b)$$

Each of these terms can be integrated using techniques that we have already developed. For example, if $b_1 = 0$, the first term can be broken into $A_1 x/(x^2 + c_1)$, which integrates to $(A_1/2)\ln(x^2 + c_1)$, and $B_1/(x^2 + c_1)$, for which we would set $x = \sqrt{c_1}\tan\theta$. If $b_1 \neq 0$, it is necessary to complete the square in the denominator and use a trigonometric substitution.

For the integral

$$\int \frac{1}{(x^2 + x + 1)(x^2 + x + 2)}\,dx$$

we set

$$\frac{1}{(x^2 + x + 1)(x^2 + x + 2)} = \frac{Ax + B}{x^2 + x + 1} + \frac{Cx + D}{x^2 + x + 2}$$

$$= \frac{(Ax + B)(x^2 + x + 2) + (Cx + D)(x^2 + x + 1)}{(x^2 + x + 1)(x^2 + x + 2)}.$$

Thus,

$$1 = (Ax + B)(x^2 + x + 2) + (Cx + D)(x^2 + x + 1).$$

If we identify corresponding coefficients of powers of x, we get

$$0 = A + C,$$
$$0 = A + B + C + D,$$
$$0 = 2A + B + C + D,$$
$$1 = 2B + D,$$

the solution of which is $A = 0$, $B = 1$, $C = 0$, $D = -1$. Alternatively, if we choose to set, say, $x = 0, 1, -1$, and 2, we have

$$1 = 2B + D,$$
$$1 = 4A + 4B + 3C + 3D,$$
$$1 = -2A + 2B - C + D,$$
$$1 = 16A + 8B + 14C + 7D,$$

and the solution is again $A = 0$, $B = 1$, $C = 0$, and $D = -1$. We can write then

$$\int \frac{1}{(x^2 + x + 1)(x^2 + x + 2)}\,dx = \int \left\{\frac{1}{x^2 + x + 1} + \frac{-1}{x^2 + x + 2}\right\} dx$$

$$= \int \left\{\frac{1}{(x + \frac{1}{2})^2 + \frac{3}{4}} - \frac{1}{(x + \frac{1}{2})^2 + \frac{7}{4}}\right\} dx.$$

In the first term we set $x + 1/2 = (\sqrt{3}/2)\tan\theta$, which implies that $dx = (\sqrt{3}/2)\sec^2\theta\, d\theta$ and

$$\begin{aligned}\int \frac{1}{(x+\frac{1}{2})^2+\frac{3}{4}}\,dx &= \int \frac{\sqrt{3}/2\sec^2\theta}{\frac{3}{4}\sec^2\theta}\,d\theta \\ &= \frac{2}{\sqrt{3}}\theta + C \\ &= \frac{2}{\sqrt{3}}\operatorname{Tan}^{-1}\left(\frac{2x+1}{\sqrt{3}}\right) + C.\end{aligned}$$

The second term has a similar antiderivative; hence,

$$\begin{aligned}\int \frac{1}{(x^2+x+1)(x^2+x+2)}\,dx = {}& \frac{2}{\sqrt{3}}\operatorname{Tan}^{-1}\left(\frac{2x+1}{\sqrt{3}}\right) \\ & - \frac{2}{\sqrt{7}}\operatorname{Tan}^{-1}\left(\frac{2x+1}{\sqrt{7}}\right) + C.\end{aligned}$$

Example 17 Evaluate

$$\int \frac{x}{x^4+6x^2+5}\,dx.$$

Solution Since

$$\begin{aligned}\frac{x}{x^4+6x^2+5} &= \frac{x}{(x^2+1)(x^2+5)} \\ &= \frac{Ax+B}{x^2+1} + \frac{Cx+D}{x^2+5} \\ &= \frac{(Ax+B)(x^2+5)+(Cx+D)(x^2+1)}{(x^2+1)(x^2+5)},\end{aligned}$$

we set

$$x = (Ax+B)(x^2+5) + (Cx+D)(x^2+1).$$

If coefficients of powers of x are equated, we get

$$\begin{aligned}0 &= A + C,\\ 0 &= B + D,\\ 1 &= 5A + C,\\ 0 &= 5B + D,\end{aligned}$$

the solution of which is $A = \frac{1}{4}$, $B = 0$, $C = -\frac{1}{4}$, $D = 0$. Thus,

$$\begin{aligned}\int \frac{x}{x^4+6x^2+5}\,dx &= \int \left\{\frac{x/4}{x^2+1} + \frac{-x/4}{x^2+5}\right\} dx \\ &= \frac{1}{8}\ln|x^2+1| - \frac{1}{8}\ln|x^2+5| + C. \square\end{aligned}$$

Case 4 $D(x)$ factors into one irreducible quadratic factor repeated n times.

In this case,

$$D(x) = (x^2 + bx + c)^n, \tag{9.38a}$$

and it is possible to rewrite $R(x)/D(x)$ in the form

$$\frac{R(x)}{D(x)} = \frac{A_1x + B_1}{x^2 + bx + c} + \frac{A_2x + B_2}{(x^2 + bx + c)^2} + \cdots + \frac{A_nx + B_n}{(x^2 + bx + c)^n}. \tag{9.38b}$$

Example 18 Evaluate

$$\int \frac{x^2 + x + 2}{(x^2 + 1)^2}\,dx.$$

Solution If

$$\frac{x^2 + x + 2}{(x^2 + 1)^2} = \frac{Ax + B}{x^2 + 1} + \frac{Cx + D}{(x^2 + 1)^2} = \frac{(Ax + B)(x^2 + 1) + (Cx + D)}{(x^2 + 1)^2},$$

then

$$x^2 + x + 2 = (Ax + B)(x^2 + 1) + (Cx + D).$$

Clearly,

$$\begin{aligned} 0 &= A, \\ 1 &= B, \\ 1 &= A + C, \\ 2 &= B + D, \end{aligned}$$

from which we get $C = D = 1$. Thus,

$$\begin{aligned} \int \frac{x^2 + x + 2}{(x^2 + 1)^2}\,dx &= \int \left\{ \frac{1}{x^2 + 1} + \frac{x + 1}{(x^2 + 1)^2} \right\} dx \\ &= \operatorname{Tan}^{-1} x + \frac{-\frac{1}{2}}{x^2 + 1} + \int \frac{1}{(x^2 + 1)^2}\,dx. \end{aligned}$$

If we set $x = \tan\theta$, then $dx = \sec^2\theta\, d\theta$ and

$$\begin{aligned} \int \frac{1}{(x^2 + 1)^2}\,dx &= \int \frac{\sec^2\theta}{\sec^4\theta}\,d\theta = \int \cos^2\theta\, d\theta = \int \left(\frac{1 + \cos 2\theta}{2} \right) d\theta \\ &= \frac{1}{2}\left\{ \theta + \frac{1}{2}\sin 2\theta \right\} + C \\ &= \frac{1}{2}\{\theta + \sin\theta\cos\theta\} + C \\ &= \frac{1}{2}\operatorname{Tan}^{-1} x + \frac{1}{2}\frac{x}{\sqrt{x^2 + 1}}\frac{1}{\sqrt{x^2 + 1}} + C. \qquad \text{(Fig. 9.22)} \end{aligned}$$

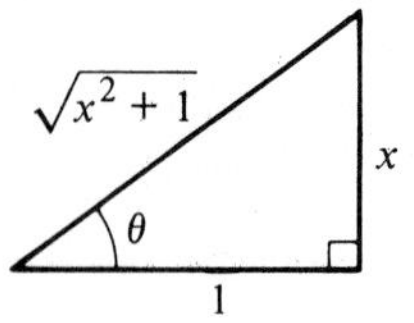

Figure 9.22

Finally, we have

$$\int \frac{x^2 + x + 2}{(x^2 + 1)^2}\,dx = \frac{3}{2}\operatorname{Tan}^{-1} x + \frac{x - 1}{2(x^2 + 1)} + C. \ \square$$

Since we had to make the substitution $x = \tan\theta$ eventually, perhaps we should have used it at the beginning of the problem (i.e., forget partial fractions and go directly to the trigonometric substitution). Try it.

Case 5 $D(x)$ factors into a combination of the factors in cases 1–4.

In this case the partial fraction decomposition of $R(x)/D(x)$ is found by considering each factor separately—be it linear or quadratic, repeated or non-repeated—and adding all such terms together.

Example 19 Evaluate each of the following indefinite integrals:

a) $\displaystyle\int \frac{x}{x^3 - 5x^2 + 8x - 4}\,dx$;

b) $\displaystyle\int \frac{1}{y^4 + y^3 + y^2 + y}\,dy$.

Solution a) If

$$\frac{x}{x^3 - 5x^2 + 8x - 4} = \frac{x}{(x-1)(x-2)^2}$$

$$= \frac{A}{x-1} + \frac{B}{x-2} + \frac{C}{(x-2)^2}$$

$$= \frac{A(x-2)^2 + B(x-1)(x-2) + C(x-1)}{(x-1)(x-2)^2},$$

then

$$x = A(x-2)^2 + B(x-1)(x-2) + C(x-1).$$

Substitution of $x = 1, 2$, and 0 gives

$$1 = A,$$

$$2 = C,$$

$$0 = 4A + 2B - C,$$

which imply that $B = -1$. Thus,

$$\int \frac{x}{x^3 - 5x^2 + 8x - 4}\,dx = \int \left\{\frac{1}{x-1} + \frac{-1}{x-2} + \frac{2}{(x-2)^2}\right\} dx$$

$$= \ln|x-1| - \ln|x-2| - \frac{2}{x-2} + C.$$

b) If

$$\frac{1}{y^4 + y^3 + y^2 + y} = \frac{1}{y(y+1)(y^2+1)} = \frac{A}{y} + \frac{B}{y+1} + \frac{Cy + D}{y^2 + 1}$$

$$= \frac{A(y+1)(y^2+1) + By(y^2+1) + (Cy+D)(y+1)y}{y(y+1)(y^2+1)},$$

then

$$1 = A(y+1)(y^2+1) + By(y^2+1) + (Cy+D)(y+1)y.$$

Substitution of $y = 0, -1, 1$, and 2 gives

$$1 = A,$$
$$1 = -2B,$$
$$1 = 4A + 2B + 2C + 2D,$$
$$1 = 15A + 10B + 12C + 6D.$$

These imply that $B = C = D = -\frac{1}{2}$, and

$$\int \frac{1}{y^4 + y^3 + y^2 + y}\,dy = \int \left\{ \frac{1}{y} + \frac{-\frac{1}{2}}{y+1} + \frac{-y/2 - \frac{1}{2}}{y^2+1} \right\} dy$$
$$= \ln|y| - \frac{1}{2}\ln|y+1| - \frac{1}{4}\ln|y^2+1|$$
$$- \frac{1}{2}\mathrm{Tan}^{-1}\, y + C. \square$$

In Section 4.5 we discussed the rate at which an infectious disease might spread through a population of susceptible individuals. If N represents the number of individuals in the population and $x(t)$ the number infected at any given time t, it is often assumed that the rate of infection is proportional to both the number of infectives and the number of noninfectives:

$$\frac{dx}{dt} = kx(N - x), \qquad k = \text{a constant}.$$

Thus $x(t)$ must satisfy this differential equation subject to an initial condition such as, perhaps, $x(0) = 1$, where therefore one infective is introduced into the population at time $t = 0$. To solve this differential equation, we write

$$\frac{1}{x(N-x)}\frac{dx}{dt} = k,$$

and determine a partial fraction decomposition for the left-hand side, namely,

$$\left\{ \frac{1/N}{x} + \frac{1/N}{N-x} \right\} \frac{dx}{dt} = k.$$

Integration of both sides of this equation with respect to time gives

$$\frac{1}{N}\{\ln|x| - \ln|N - x|\} = kt + C,$$

or

$$\ln\left\{\frac{x}{N-x}\right\} = N(kt + C).$$

If we exponentiate both sides of this result, we get

$$\frac{x}{N-x} = De^{Ft},$$

where we have substituted $D = e^{NC}$ and $F = kN$. We can now solve this equation for an explicit definition of $x(t)$:

$$x = De^{Ft}(N - x),$$

or

$$x(1 + De^{Ft}) = NDe^{Ft}.$$

Thus,

$$x(t) = \frac{NDe^{Ft}}{1 + De^{Ft}} = \frac{ND}{D + e^{-Ft}}.$$

Since $x(0) = 1$,

$$1 = \frac{ND}{1 + D},$$

which implies that $D = 1/(N - 1)$ and

$$x(t) = \frac{\dfrac{N}{N-1}}{\dfrac{1}{N-1} + e^{-Ft}} = \frac{N}{1 + (N - 1)e^{-Ft}}.$$

The shape of this curve is illustrated in Fig. 9.23(a) (see also the discussion in Section 4.5). The time scale can be affixed to the diagram once one further piece of information is given. For instance, if it is known that one-half the population will be infected in 20 days, then $x(20) = N/2$, and the time scale is as shown in Fig. 9.23(b). This added information determines F, since

$$\frac{N}{2} = \frac{N}{1 + (N - 1)e^{-20F}},$$

and this equation gives

$$F = \tfrac{1}{20} \ln (N - 1).$$

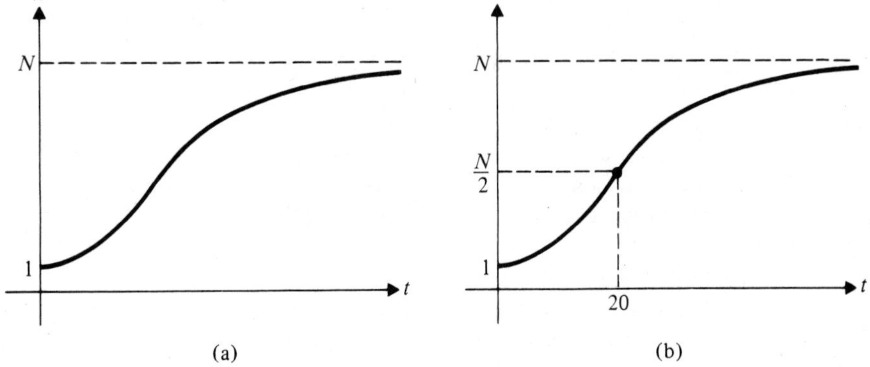

Figure 9.23

EXERCISES 9.8

In Exercises 1–16, evaluate the given indefinite integral.

1. $\int \frac{x+2}{x^2-2x+1}\,dx$

2. $\int \frac{1}{y^3+3y^2+3y+1}\,dy$

3. $\int \frac{1}{z^3+z}\,dz$

4. $\int \frac{x^2+2x-4}{x^2-2x-8}\,dx$

5. $\int \frac{x}{(x-4)^2}\,dx$

6. $\int \frac{y+1}{y^3+y^2-6y}\,dy$

7. $\int \frac{3x+5}{x^3-x^2-x+1}\,dx$

8. $\int \frac{x^3}{(x^2+2)^2}\,dx$

9. $\int \frac{1}{x^2-3}\,dx$

10. $\int \frac{y^2}{y^2+3y+2}\,dy$

11. $\int \frac{z^2+3z-2}{z^3+5z}\,dz$

12. $\int \frac{y^2+6y+4}{y^4+5y^2+4}\,dy$

13. $\int \frac{x}{x^4+7x^2+6}\,dx$

14. $\int \frac{x^2+3}{x^4+x^2-2}\,dx$

15. $\int \frac{3t+4}{t^4-3t^3+3t^2-t}\,dt$

16. $\int \frac{x^3+6}{x^4+2x^3-3x^2-4x+4}\,dx$

17. Find the centroid of the area bounded by the curves

$$(x+2)^2y = 4-x, \qquad x=0, \qquad y=0, \qquad (x, y \geq 0).$$

18. Natural factors such as space and available food supply inhibit the growth of any population. Let N represent the population of a species and M the maximum size of a stable population (for which the death rate is just equal to the birth rate). If the time rate of change of N is assumed proportional to both N and $M-N$, then N must satisfy the differential equation

$$\frac{dN}{dt} = kN(M-N).$$

If $N(0) = N_0 < M$, solve this equation for $N(t)$, and sketch its graph.

19. Two substances A and B react to form a third substance C in such a way that one gram of A reacts with one gram of B to produce two grams of C. The rate at which C is formed is proportional to the amounts of A and B present in the mixture.

a) Show that if 10 grams of A and 15 grams of B are brought together at time $t=0$, the amount $x(t)$ of C present in the mixture must satisfy the differential equation

$$\frac{dx}{dt} = k(x-20)(x-30), \qquad x(0)=0,$$

where k is a constant.

b) Solve this equation for $x(t)$.

In Exercises 20–25, evaluate the given indefinite integral.

20. $\int \frac{1}{x^5+x^4+2x^3+2x^2+x+1}\,dx$

21. $\int \frac{x^3+x+2}{x^5+2x^3+x}\,dx$

22. $\int \frac{1}{1+x^3}\,dx$

23. $\int \frac{1}{(x^2+5)(x^2+2x+3)}\,dx$

24. $\int \frac{x^4+8x^3-x^2+2x+1}{x^5+x^4+x^2+x}\,dx$

25. $\int \frac{\sin x}{\cos x\,(1+\cos^2 x)}\,dx$

26. If an integrand is a rational function of $\sin x$ and $\cos x$, it can be reduced to a rational function of t by the substitution

$$t = \tan\left(\frac{x}{2}\right).$$

Show that with this substitution

$$dx = \frac{2}{1+t^2}\,dt, \qquad \sin x = \frac{2t}{1+t^2},$$

$$\cos x = \frac{1-t^2}{1+t^2}.$$

Use the substitution in Exercise 26 to evaluate the indefinite integral in each of Exercises 27–30.

27. $\int \sec x\,dx$

28. $\int \frac{1}{3+5\sin x}\,dx$

29. $\int \frac{1}{1-2\cos x}\,dx$

*30. $\int \frac{1}{\sin x+\cos x}\,dx$

31. Show that the answer to Exercise 27 can be expressed in the usual form $\ln|\sec x+\tan x| + C$.

*32. Evaluate

$$\int \frac{x^2+x+3}{x^4+x^3+2x^2+11x-5}\,dx.$$

9.9 NUMERICAL INTEGRATION

To evaluate the definite integral of a continuous function $f(x)$ with respect to x from $x = a$ to $x = b$, we have always used the method of Theorem 6.3: Find an indefinite integral for $f(x)$, substitute the limits $x = b$ and $x = a$, and then subtract. The evaluation procedure depends on our ability to produce an indefinite integral for $f(x)$, and in this chapter we have developed a number of integration techniques to help us find antiderivatives. Unfortunately, there exist many functions for which none of these techniques will produce an antiderivative. Even worse, there are functions that have antiderivatives that cannot be expressed as a finite sum of terms. For instance, the rather innocent-looking $f(x) = e^{x^2}$ falls into the latter category. What, then, are we to do should we require the value of a definite integral for which there is no obvious antiderivative? If $f(x) \geq 0$ on $a \leq x \leq b$ (Fig. 9.24), the definite integral of $f(x)$ with respect to x can always be interpreted as the area bounded by $y = f(x)$, $y = 0$, $x = a$, and $x = b$. We have been careful to point out that it is not always wise to think of a definite integral as area, but for the purpose of our discussion here it is convenient to do so. Our problem, then, is to find the area shown in Fig. 9.24 when, because of the complexity of $f(x)$, it is very difficult, or even impossible, to find an antiderivative of $f(x)$. We now develop three numerical techniques that approximate the definite integral (area) by subdividing the interval $a \leq x \leq b$ into a number of smaller subintervals and replacing $f(x)$ with an easily integrated function on each subinterval.

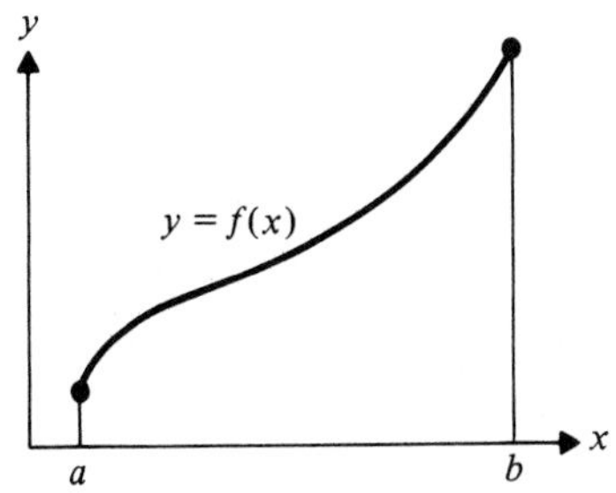

Figure 9.24

Rectangular Rule

The first method is essentially to return to definition (6.11) for the definite integral. This means that we subdivide the interval $a \leq x \leq b$ into n subintervals by points $a = x_0 < x_1 < x_2 < \cdots < x_{n-1} < x_n = b$, and pick a point x_i^* in each subinterval. For simplicity, we choose n equal subdivisions, in which case

$$x_i = a + i\left(\frac{b-a}{n}\right),$$

and select $x_i^* = x_i$. If we denote by h the length of each subdivision, then $h = \Delta x_i = (b - a)/n$, and definition (6.11) reduces to

$$\int_a^b f(x)\,dx = \lim_{n\to\infty} \sum_{i=1}^{n} f(x_i^*)\,\Delta x_i = h \lim_{n\to\infty} \sum_{i=1}^{n} f(x_i).$$

If we delete the limit from this equation, the summation becomes an approximation to the definite integral, and we write

$$\int_a^b f(x)\,dx \simeq h \sum_{i=1}^{n} f(x_i). \tag{9.39}$$

We call this the *rectangular rule* for approximating the definite integral, the reason being that we have replaced the area under $y = f(x)$ with n rectangles (Fig. 9.25). In other words, we have replaced the original function $f(x)$ by a function that is constant on each subinterval, but the constant value may vary from subinterval to subinterval. Such a function is called a *step function*.

Graphically, it is reasonable to expect that an increase in the number of subdivisions will result in an increase in the accuracy of the approximation.

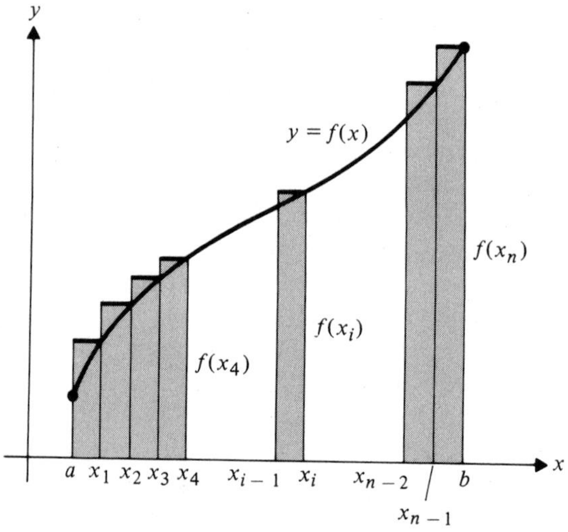

Figure 9.25

Example 20 Use a subdivision of the interval $1 \le x \le 3$ into 5, 10, and 20 equal parts to approximate the definite integral

$$\int_1^3 \sin x \, dx$$

using the rectangular rule.

Solution With 5 equal parts, $h = \frac{2}{5}$ and $x_i = 1 + 2i/5$. Consequently,

$$\int_1^3 \sin x \, dx \simeq \frac{2}{5} \sum_{i=1}^{5} \sin\left(1 + \frac{2i}{5}\right)$$

$$= \frac{2}{5}\left\{\sin\frac{7}{5} + \sin\frac{9}{5} + \sin\frac{11}{5} + \sin\frac{13}{5} + \sin 3\right\}$$

$$= 1.370.$$

With 10 equal parts, $h = \frac{2}{10} = \frac{1}{5}$ and $x_i = 1 + i/5$, so that

$$\int_1^3 \sin x \, dx \simeq \frac{1}{5} \sum_{i=1}^{10} \sin\left(1 + \frac{i}{5}\right)$$

$$= \frac{1}{5}\left\{\sin\frac{6}{5} + \sin\frac{7}{5} + \sin\frac{8}{5} + \sin\frac{9}{5} + \sin 2 + \sin\frac{11}{5} + \sin\frac{12}{5} + \sin\frac{13}{5} + \sin\frac{14}{5} + \sin 3\right\}$$

$$= 1.455.$$

With 20 equal parts, we have

$$\int_1^3 \sin x \, dx \simeq \frac{1}{10} \sum_{i=1}^{20} \sin\left(1 + \frac{i}{10}\right)$$

$$= \frac{1}{10}\left\{\sin\frac{11}{10} + \sin\frac{6}{5} + \sin\frac{13}{10} + \cdots + \sin\frac{29}{10} + \sin 3\right\}$$

$$= 1.494. \square$$

When we compare these results with the correct answer to Example 20,

$$\int_1^3 \sin x\, dx = \{-\cos x\}_1^3 = -\cos 3 + \cos 1 = 1.5302948,$$

we see that as n increases from 5 to 10 to 20, the approximation improves, but it must be increased even further to give a reasonable approximation to this integral.

Trapezoidal Rule Regarding area, it is clear in Fig. 9.25 that were we to join successive points $(x_i, f(x_i))$ on the curve $y = f(x)$ with straight lines (Fig. 9.26a), the area under this broken straight line would be a much better approximation to the area under $y = f(x)$ than that provided by the rectangular rule. Effectively, then, we now approximate the area by n trapezoids. Since the area of a trapezoid is its width multiplied by the average of its parallel lengths, it follows that the area of the ith trapezoid (Fig. 9.26b) is given by

$$h\left(\frac{f(x_i) + f(x_{i-1})}{2}\right),$$

where again $h = x_i - x_{i-1}$. As a result, the area under $y = f(x)$ can be approximated by the sum

$$h\left\{\frac{f(x_1) + f(x_0)}{2} + \frac{f(x_2) + f(x_1)}{2} + \cdots + \frac{f(x_n) + f(x_{n-1})}{2}\right\}$$

$$= \frac{h}{2}\{f(a) + 2f(x_1) + 2f(x_2) + \cdots + 2f(x_i) + \cdots + 2f(x_{n-1}) + f(b)\}$$

$$= \frac{h}{2}\left\{f(a) + 2\sum_{i=1}^{n-1} f(x_i) + f(b)\right\}.$$

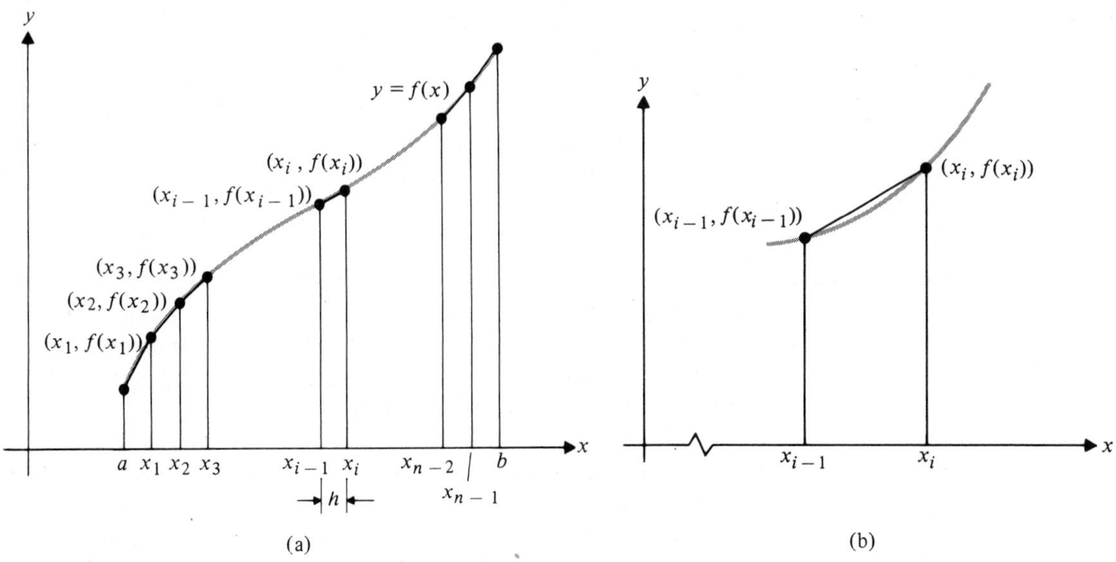

Figure 9.26

We write therefore that

$$\int_a^b f(x)\,dx \simeq \frac{h}{2}\left\{f(a) + 2\sum_{i=1}^{n-1} f(x_i) + f(b)\right\}, \tag{9.40}$$

where $x_i = a + ih = a + i(b - a)/n$, and call this the *trapezoidal rule* for approximating a definite integral. Note that if (9.40) is written in the form

$$\int_a^b f(x)\,dx \simeq h\left(\frac{f(a) - f(b)}{2}\right) + h\sum_{i=1}^{n} f(x_i),$$

the summation on the right, except for the first two terms, is the rectangular rule. In other words, the extra numerical calculation involved in using the trapezoidal rule rather than the rectangular rule is minimal, but it would appear that the accuracy is significantly increased. For this reason, the trapezoidal rule supplants the rectangular rule in most applications.

Example 21 Using the trapezoidal rule, approximate the definite integral given in Example 20.

Solution With 5 equal partitions,

$$\int_1^3 \sin x\,dx \simeq \frac{(\frac{2}{5})}{2}\left\{\sin 1 + 2\sum_{i=1}^{4} \sin\left(1 + \frac{2i}{5}\right) + \sin 3\right\}$$
$$= 1.5098.$$

With 10 equal parts,

$$\int_1^3 \sin x\,dx \simeq \frac{(\frac{1}{5})}{2}\left\{\sin 1 + 2\sum_{i=1}^{9} \sin\left(1 + \frac{i}{5}\right) + \sin 3\right\}$$
$$= 1.5252.$$

Finally, with $h = \frac{1}{10}$, we have

$$\int_1^3 \sin x\,dx \simeq \frac{(\frac{1}{10})}{2}\left\{\sin 1 + 2\sum_{i=1}^{19} \sin\left(1 + \frac{i}{10}\right) + \sin 3\right\}$$
$$= 1.5290.$$

As expected, these approximations are significantly better than corresponding results using the rectangular rule. □

Simpson's Rule

The rectangular rule replaces a function $f(x)$ with a step function; the trapezoidal rule replaces $f(x)$ with a succession of linear functions (geometrically, a broken straight line). So far as ease of integration is concerned, the next simplest function is a quadratic function. Consider, then, replacing the curve $y = f(x)$ by a succession of parabolas on the subintervals $x_{i-1} \le x \le x_i$. Now the equation of a parabola with a vertical axis of symmetry is of the form $y = ax^2 + bx + c$ with three constants a, b, and c to be determined. If this parabola is to approximate $y = f(x)$ on $x_{i-1} \le x \le x_i$, we should have the parabola pass through the end points $(x_{i-1}, f(x_{i-1}))$ and $(x_i, f(x_i))$. But this will impose only

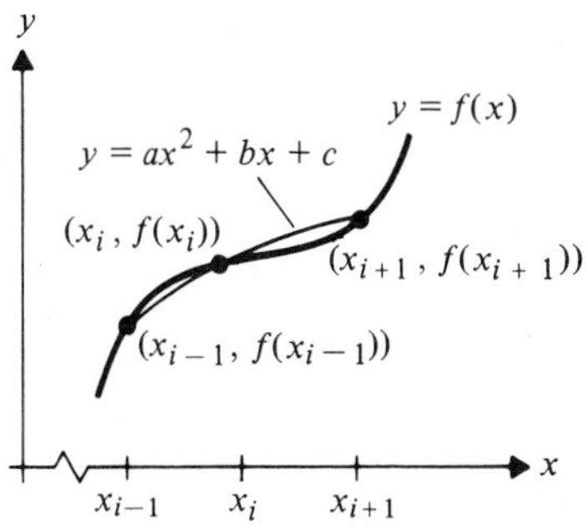

Figure 9.27

two conditions on a, b, and c, not three. To take advantage of this flexibility, we demand that the parabola also pass through the point $(x_{i+1}, f(x_{i+1}))$ (Fig. 9.27). In other words, instead of replacing $y = f(x)$ with n parabolas, one on each subinterval $x_{i-1} \leq x \leq x_i$, we replace it with $n/2$ parabolas, one on each pair of subintervals $x_{i-1} \leq x \leq x_{i+1}$. Note that this requires n to be an even integer. These three conditions imply that a, b, and c must satisfy the equations

$$\begin{aligned} ax_{i-1}^2 + bx_{i-1} + c &= f(x_{i-1}), \\ ax_i^2 \quad + bx_i \quad + c &= f(x_i), \\ ax_{i+1}^2 + bx_{i+1} + c &= f(x_{i+1}). \end{aligned}$$

We could solve these equations for a, b, and c, but perhaps this won't be necessary. Suppose for the moment that these equations do determine a, b, and c, and we continue our main discussion. With the parabola $y = ax^2 + bx + c$ replacing the curve $y = f(x)$ on the interval $x_{i-1} \leq x \leq x_{i+1}$, we approximate the area under $y = f(x)$ with that under the parabola, namely,

$$\begin{aligned} \int_{x_{i-1}}^{x_{i+1}} (ax^2 + bx + c)\,dx &= \frac{a}{3}(x_{i+1}^3 - x_{i-1}^3) + \frac{b}{2}(x_{i+1}^2 - x_{i-1}^2) + c(x_{i+1} - x_{i-1}) \\ &= \frac{x_{i+1} - x_{i-1}}{6}\{2a(x_{i+1}^2 + x_{i+1}x_{i-1} + x_{i-1}^2) \\ &\quad + 3b(x_{i+1} + x_{i-1}) + 6c\}. \end{aligned}$$

Now, if we use $x_{i+1} - x_{i-1} = 2h$, $x_{i+1} = x_i + h$, and $x_{i-1} = x_i - h$ to write everything in terms of h and x_i, we obtain

$$\begin{aligned} \int_{x_{i-1}}^{x_{i+1}} (ax^2 + bx + c)\,dx &= \frac{2h}{6}\{2a[(x_i + h)^2 + (x_i + h)(x_i - h) + (x_i - h)^2] \\ &\quad + 3b[(x_i + h) + (x_i - h)] + 6c\} \\ &= \frac{h}{3}\{a(6x_i^2 + 2h^2) + 6bx_i + 6c\}. \end{aligned}$$

But if we take the equations defining a, b, and c, and add the first and third and four times the second, we find that

$$\begin{aligned} f(x_{i-1}) + 4f(x_i) + f(x_{i+1}) &= a(x_{i-1}^2 + 4x_i^2 + x_{i+1}^2) + b(x_{i-1} + 4x_i + x_{i+1}) + 6c \\ &= a[(x_i - h)^2 + 4x_i^2 + (x_i + h)^2] \\ &\quad + b[(x_i - h) + 4x_i + (x_i + h)] + 6c \\ &= a(6x_i^2 + 2h^2) + 6bx_i + 6c. \end{aligned}$$

Thus the area under the parabola may be written in the form

$$\int_{x_{i-1}}^{x_{i+1}} (ax^2 + bx + c)\,dx = \frac{h}{3}\{f(x_{i-1}) + 4f(x_i) + f(x_{i+1})\},$$

and this expression is free of a, b, and c. When we add all such integrals over the

$n/2$ subintervals $x_{i-1} \le x \le x_{i+1}$ between $x = a$ and $x = b$, we obtain

$$\frac{h}{3}\{f(x_0) + 4f(x_1) + f(x_2)\} + \frac{h}{3}\{f(x_2) + 4f(x_3) + f(x_4)\} + \cdots + \frac{h}{3}\{f(x_{n-2}) + 4f(x_{n-1}) + f(x_n)\}$$

$$= \frac{h}{3}\{f(x_0) + 4f(x_1) + 2f(x_2) + 4f(x_3) + \cdots + 2f(x_{n-2}) + 4f(x_{n-1}) + f(x_n)\}.$$

In other words, the definite integral of $f(x)$ from $x = a$ to $x = b$ can be approximated by

$$\int_a^b f(x)\,dx \simeq \frac{h}{3}\{f(a) + 4f(x_1) + 2f(x_2) + 4f(x_3) + \cdots + 2f(x_{n-2}) + 4f(x_{n-1}) + f(b)\}, \quad (9.41)$$

where $x_i = a + ih = a + i(b - a)/n$. This result is called *Simpson's rule* for approximating a definite integral. Although this formula does not explicitly display it (except by counting the terms), do not forget that n must be an even integer.

Example 22 Approximate the definite integral given in Example 20 using Simpson's rule with $n = 10$ and $n = 20$.

Solution With $n = 10$,

$$\int_1^3 \sin x\,dx \simeq \frac{(\frac{1}{5})}{3}\left\{\sin 1 + 4\sin\frac{6}{5} + 2\sin\frac{7}{5} + 4\sin\frac{8}{5} + \cdots + 2\sin\frac{13}{5} + 4\sin\frac{14}{5} + \sin 3\right\}$$

$$= 1.5303085.$$

With $n = 20$,

$$\int_1^3 \sin x\,dx \simeq \frac{(\frac{1}{10})}{3}\left\{\sin 1 + 4\sin\frac{11}{10} + 2\sin\frac{6}{5} + 4\sin\frac{13}{10} + \cdots + 2\sin\frac{14}{5} + 4\sin\frac{29}{10} + \sin 3\right\}$$

$$= 1.5302957. \square$$

Table 9.1 lists the relative accuracies of the approximations in Examples 20–22. The correct answer for $\int_1^3 \sin x\,dx$ is 1.5302948 (to 7 decimals). It is clear that each method gives a better approximation as the value of n is increased, and that Simpson's rule is by far the most accurate. We can say this, however, only because we know the real answer. In practice, we use these rules to evaluate definite integrals that cannot be handled analytically, and we will not therefore have the correct answer with which to compare the approximation. We would still like to make some statement about the accuracy of the approximation, however, since otherwise what good is the approximation?

Table 9.1

	Rectangular rule	Trapezoidal rule	Simpson's rule
$n = 5$	1.370	1.5098	
$n = 10$	1.455	1.5252	1.5303085
$n = 20$	1.494	1.5290	1.5302957

If, for example, we use the trapezoidal rule or Simpson's rule with, say, twenty subdivisions to evaluate a definite integral, we would like to be able to state that the result is accurate to a certain number of decimals. The following two theorems give error estimates for the trapezoidal rule and for Simpson's rule.

▶ **Theorem 9.1** If $f''(x)$ exists on $a \le x \le b$, and T_n is the error in approximating the definite integral of $f(x)$ from $x = a$ to $x = b$ using the trapezoidal rule with n equal subdivisions,

$$T_n = \int_a^b f(x)\,dx - \frac{h}{2}\left\{f(a) + 2\sum_{i=1}^{n-1} f(x_i) + f(b)\right\},$$

then

$$|T_n| \le \frac{M(b-a)^3}{12n^2}, \tag{9.42}$$

where M is the maximum value of $|f''(x)|$ on $a \le x \le b$. ◀

▶ **Theorem 9.2** If $f''''(x)$ exists on $a \le x \le b$, and S_n is the error in approximating the definite integral of $f(x)$ from $x = a$ to $x = b$ by means of Simpson's rule with n equal subdivisions,

$$S_n = \int_a^b f(x)\,dx - \frac{h}{3}\left\{f(a) + f(b) + 4\sum_{i=1}^{n/2} f(x_{2i-1}) + 2\sum_{i=1}^{n/2-1} f(x_{2i})\right\},$$

then

$$|S_n| \le \frac{M(b-a)^5}{180n^4}, \tag{9.43}$$

where M is the maximum value of $|f''''(x)|$ on $a \le x \le b$. ◀

Proofs of these theorems can be found in any book on numerical analysis. Note that because of the n^4 factor in the denominator of (9.43), the accuracy of Simpson's rule increases much more rapidly than does that of the trapezoidal rule.

For the function $f(x) = \sin x$ in Examples 20–22,

$$f''(x) = -\sin x \quad \text{and} \quad f''''(x) = \sin x.$$

Consequently, in both cases we can state that $M = 1$, and therefore

$$|T_n| \le \frac{(3-1)^3}{12n^2} = \frac{2}{3n^2} \quad \text{and} \quad |S_n| \le \frac{(3-1)^5}{180n^4} = \frac{8}{45n^4}.$$

For $n = 10$ and $n = 20$, we find

$$|T_{10}| \leq \frac{2}{3(10)^2} < 0.0067 \quad \text{and} \quad |S_{10}| \leq \frac{8}{45(10)^4} < 0.000018;$$

$$|T_{20}| \leq \frac{2}{3(20)^2} < 0.0017 \quad \text{and} \quad |S_{20}| \leq \frac{8}{45(20)^4} < 0.0000012.$$

Differences between the correct value for the integral and the approximations listed in Table 9.1 corroborate these predictions.

Example 23 How many equal partitions of the interval $1 \leq x \leq 4$ will guarantee an error of less than 10^{-5} in the approximation of the definite integral

$$\int_1^4 \frac{1}{x^2}\, dx$$

using Simpson's rule?

Solution According to Theorem 9.2, the error in using Simpson's rule with n equal partitions to approximate this definite integral is

$$|S_n| \leq \frac{M(4-1)^5}{180n^4},$$

where M is the maximum value of the fourth derivative of $1/x^2$ on the interval $1 \leq x \leq 4$. Since $d^4(1/x^2)/dx^4 = 5!/x^6$, it follows that $M = 5!$ and

$$|S_n| \leq \frac{5!\,3^5}{180n^4}.$$

This error will be less than 10^{-5} if n is large enough to guarantee

$$\frac{5!\,3^5}{180n^4} < 10^{-5};$$

i.e., if

$$n > \left(\frac{5!\,3^5 10^5}{180}\right)^{1/4} = 63.44.$$

Thus if n is greater than or equal to 64, Simpson's rule will guarantee an error of less than 10^{-5} for this integral. □

The error bounds in Theorems 9.1 and 9.2 are somewhat idealistic in the sense that they are error predictions based on the use of exact numbers. For instance, they would predict that by increasing n indefinitely, we can get any degree of accuracy that we wish. Theoretically this is true, but practically speaking it is not. No matter how we choose to evaluate the summation in (9.40) or (9.41) (be it by hand, by an electronic hand calculator, or by a high-speed computer), each calculation will be rounded off to a certain number of decimals. The final sum will take into account many, many of these "approximate numbers," and must therefore be inherently inaccurate. We call this *round-off error* and it is very difficult to predict how extensive it will be. It depends on both the number and nature of the operations involved in (9.40) and (9.41).

In the approximation of a definite integral by the trapezoidal rule or Simpson's rule, there are, then, two sources of error. Theorems 9.1 and 9.2 predict errors due to the methods themselves; round-off errors may also be appreciable for large n.

We should emphasize once again that although we have used area as a convenient vehicle by which to explain the approximation of definite integrals by the rectangular, the trapezoidal, and Simpson's rules, this is not a restriction on these methods. All three can be used to approximate the definite integral of a function $f(x)$, be it positive, negative, or sometimes positive and sometimes negative on the interval of integration. The only condition that we have imposed is that $f(x)$ be continuous. In view of our discussion of improper integrals in Section 7.9, even this condition can be weakened, but we prefer not to go into detail here.

EXERCISES 9.9

In each of Exercises 1–10, use the trapezoidal rule and Simpson's rule with 10 equal subdivisions to approximate the definite integral. In each case, evaluate the integral analytically to get an idea of the accuracy of the approximation.

C 1. $\int_1^2 \frac{1}{x}\,dx$

C 2. $\int_2^3 \frac{1}{\sqrt{x+2}}\,dx$

C 3. $\int_0^1 \tan x\,dx$

C 4. $\int_0^{1/2} e^x\,dx$

C 5. $\int_{-1}^1 \sqrt{x+1}\,dx$

C 6. $\int_{-3}^{-2} \frac{1}{x^3}\,dx$

C 7. $\int_{1/2}^1 \cos x\,dx$

C 8. $\int_0^1 \frac{1}{3+x^2}\,dx$

C 9. $\int_1^3 \frac{1}{x^2+x}\,dx$

C 10. $\int_0^{1/2} xe^{x^2}\,dx$

In each of Exercises 11–14, use the trapezoidal rule and Simpson's rule with 10 equal subdivisions to approximate the definite integral.

C 11. $\int_0^2 \frac{1}{1+x^3}\,dx$

C 12. $\int_0^1 e^{x^2}\,dx$

C 13. $\int_1^2 \sqrt{1+x^4}\,dx$

C 14. $\int_{-1}^0 \sin(x^2)\,dx$

15. Show graphically that if $y = f(x)$ is concave downward on the interval $a \le x \le b$, then the trapezoidal rule will underestimate the definite integral of $f(x)$ on this interval.

16. What happens to the errors in (9.42) and (9.43) if the number of partitions is doubled?

C 17. The definite integral

$$\int_a^b e^{-x^2}\,dx$$

is very important in mathematical statistics. Use Simpson's rule with $n = 16$ to evaluate this integral for $a = 0$ and $b = 1$.

C 18. Use Simpson's rule with 10 equal intervals to find the length of the parabola $y = x^2$ between $x = 0$ and $x = 1$. Compare the answer to that of Exercise 22 in Section 9.7.

In Exercises 19–22, how many equal subdivisions of the interval of integration will guarantee an error of less than 10^{-4} in the approximation of the definite integral using (a) the trapezoidal rule? (b) Simpson's rule?

19. $\int_1^4 \frac{1}{x}\,dx$

20. $\int_0^{\pi/4} \cos x\,dx$

21. $\int_0^{1/3} e^{2x}\,dx$

22. $\int_4^5 \frac{1}{\sqrt{x+2}}\,dx$

C 23. Numerical techniques can be very important in practical applications when we know that y is a function of x, but we do not know the exact form of the function. What might be available is a set of experimental data listing values of y corresponding to various values of x (Table 9.2). Plot these points on a graph, and show how the trapezoidal rule and Simpson's rule can be used to approximate the definite integral of $y = f(x)$ with respect to x from $x = 1$ to $x = 3$, even though we do not have a formula for $f(x)$. Can you estimate the error of the approximation?

C 24. When the truss shown in Fig. 9.28 is subjected to a force F at its center, point A deflects an amount y from its equilibrium position. Forces (in kilonewtons) required to produce deflections from 0 to 5 centimeters at intervals of 0.5 centimeter are listed in Table 9.3. Plot these points on a set of axes with y horizontal and F vertical. Use Simpson's rule to approximate the work done by F in deflecting A by a full 5 centimeters.

Table 9.2

x	y
1.0	0.37
1.2	0.36
1.4	0.35
1.6	0.32
1.8	0.30
2.0	0.27
2.2	0.24
2.4	0.22
2.6	0.19
2.8	0.17
3.0	0.15

Table 9.3

y	F
0.0	0.0
0.5	1.45
1.0	2.90
1.5	4.40
2.0	5.90
2.5	7.43
3.0	9.05
3.5	10.7
4.0	13.2
4.5	15.3
5.0	18.0

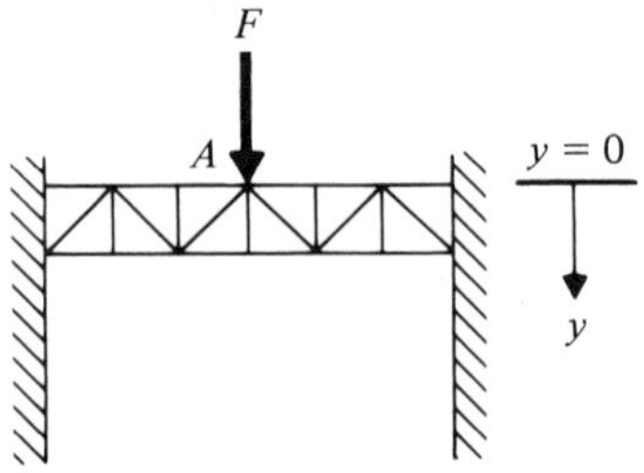

Figure 9.28

25. Show that when $f(x)$ is a cubic polynomial, evaluation of the definite integral of $f(x)$ from $x = a$ to $x = b$ by Simpson's rule always gives the *exact* answer. (*Hint:* See Theorem 9.2.) Illustrate with an example.

SUMMARY

Chapters 3 and 8 contain virtually everything that we need to know in order to differentiate the vast majority of functions that we would ever encounter in practice. In other words, differentiation is a simple, methodical procedure using derivatives of the transcendental functions (given in Chapter 8) and the rules listed in Chapter 3. Antidifferentiation, on the other hand, is far more complicated; it is not possible to state a set of rules and formulas that will suffice for most functions. Certainly there are integration formulas that we must learn, but most of those given in this chapter have been the differentiation formulas listed in Chapter 8 written in terms of integrals rather than derivatives. We have tried to stress the idea that it is important to know these obvious integration formulas, but beyond this it becomes an organizational problem; that is, organize a difficult integral into a form that utilizes these simple formulas. The three most important techniques for doing this are substitutions, integration by parts, and partial fractions. Some integrals can be approached in a number of different ways, but all methods lead to the same antiderivative.

There is a variety of substitutions for the evaluation of indefinite integrals, many suggested by the form of the integrand. For example, a term in $x^{1/n}$ (n an integer) suggests the substitution $u = x^{1/n}$. The trigonometric substitution is, perhaps, the most important. It is designed to eliminate square roots of the form $\sqrt{a^2 \pm b^2x^2}$ and $\sqrt{b^2x^2 - a^2}$ and the general root $\sqrt{ax^2 + bx + c}$.

Integration by parts is a powerful integration technique. It is used to evaluate antiderivatives of transcendental functions that are multiplied by powers of x; it leads to the method of integration by reproduction, and it can be used to develop reduction formulas.

The method of partial fractions decomposes complicated rational functions into simple fractions that are either immediately integrable or amenable to other methods such as trigonometric substitutions.

Even with these techniques, there are many functions that either cannot be antidifferentiated at all or can be antidifferentiated only with extreme difficulty. To evaluate definite integrals of such functions, numerical techniques such as the trapezoidal rule and Simpson's rule are essential.

REVIEW EXERCISES

In Exercises 1–50, evaluate the given indefinite integral.

1. $\int \sqrt{2 - x}\, dx$

2. $\int \frac{1}{(x + 3)^2}\, dx$

3. $\int \frac{x^2 + 3}{x}\, dx$

4. $\int \frac{x^2 + 3}{x + 1}\, dx$

5. $\int \frac{x^2 + 3}{x^2 + 1}\, dx$

6. $\int \frac{x}{\sqrt{x + 3}}\, dx$

7. $\int \sin^2 x \cos^3 x\, dx$

8. $\int x \sin x\, dx$

9. $\int \tan^2 (2x)\, dx$

10. $\int \frac{x}{x^2 + 2x - 3}\, dx$

11. $\int \frac{1}{\sqrt{4 - 3x^2}}\, dx$

12. $\int \frac{2 - \sqrt{x}}{\sqrt{x} + 5}\, dx$

13. $\int \frac{x}{3x^2 + 4}\, dx$

14. $\int \frac{e^x}{\sqrt{1 - e^{2x}}}\, dx$

15. $\int x^2 \ln x\, dx$

16. $\int \frac{x}{(x^2 + 1)^2}\, dx$

17. $\int \frac{x^2}{(x^2 + 1)^2}\, dx$

18. $\int \frac{x^3}{(x^2 + 1)^2}\, dx$

19. $\int \frac{x + 1}{x^3 - 4x}\, dx$

20. $\int \left(\frac{x + 1}{x - 1}\right)^2 dx$

21. $\int \frac{x^2}{(1 + 3x^3)^4}\, dx$

22. $\int \mathrm{Cos}^{-1}\, x\, dx$

23. $\int \sin x \cos 2x\, dx$

24. $\int \sin x \cos 5x\, dx$

25. $\int e^{3x} \cos 2x\, dx$

26. $\int \frac{1}{\sqrt{x^2 + 4x - 5}}\, dx$

27. $\int \frac{1}{x^2 + 4x - 5}\, dx$

28. $\int x^3 \sqrt{4 - x^2}\, dx$

29. $\int \frac{\cos 2x}{1 - \sin 2x}\, dx$

30. $\int \frac{6x}{4 - x^2}\, dx$

31. $\int \frac{1}{x\sqrt{\ln x}}\, dx$

32. $\int \frac{1}{x^2 + 4x - 4}\, dx$

33. $\int \frac{\sin x}{1 + \cos^2 x}\, dx$

34. $\int \frac{1}{x^4 + x^3}\, dx$

35. $\int x \sec^2 (3x)\, dx$

36. $\int \frac{1}{\sqrt{16 - 3x + x^2}}\, dx$

37. $\int \frac{\sqrt{x^2 - 4}}{x^2}\, dx$

38. $\int x^2 \operatorname{Tan}^{-1} x\, dx$

39. $\int \frac{x^2}{x^3 + 3x^2 + 3x + 1}\, dx$

40. $\int \frac{\ln x}{x}\, dx$

41. $\int \frac{x^2}{1 + 4x^6}\, dx$

42. $\int \frac{1}{x(9 + x^2)^2}\, dx$

43. $\int \frac{x^2 + 2}{x^3 + 5x^2 + 4x}\, dx$

44. $\int \frac{x^2 + 2}{x^3 + 4x^2 + 4x}\, dx$

45. $\int \frac{x^2 + 2}{x^3 + x^2 + 4x}\, dx$

46. $\int \frac{3x^2 + 2x + 4}{x^3 + x^2 + 4x}\, dx$

47. $\int x \operatorname{Sin}^{-1} x\, dx$

48. $\int \sqrt{\cot x}\, \csc^4 x\, dx$

49. $\int \ln(\sqrt{x} + 1)\, dx$

50. $\int \frac{1}{(4x - x^2)^{3/2}}\, dx$

Use Simpson's rule with ten equal partitions to approximate the definite integral in each of Exercises 51–54.

C51. $\int_1^2 \frac{\sin x}{x}\, dx$

C52. $\int_0^1 \sqrt{\sin x}\, dx$

C53. $\int_2^4 \frac{1}{\ln x}\, dx$

C54. $\int_{-1}^3 \frac{1}{1 + e^x}\, dx$

55. If 10% of a sample of radioactive material decays in 10 days, what is its half-life?

In Exercises 56–71, evaluate the given indefinite integral.

56. $\int \frac{1}{x^{1/3} - \sqrt{x}}\, dx$

57. $\int \ln(1 + x^2)\, dx$

58. $\int \frac{x}{x^4 + 16}\, dx$

59. $\int \csc^3 x\, dx$

60. $\int \frac{1}{(3x - x^2)^{3/2}}\, dx$

61. $\int \frac{1}{x^3\sqrt{x^2 - 9}}\, dx$

62. $\int \sin \sqrt{x}\, dx$

63. $\int \sin(\ln x)\, dx$

64. $\int x \cos x \sin 3x\, dx$

65. $\int \frac{x^4 + 3x^2 + 1}{x(x^2 + 1)^2}\, dx$

66. $\int \frac{1}{1 + \cos 2x}\, dx$

67. $\int \frac{x^4 + 3x^2 - 2x + 5}{x^2 - 3x + 7}\, dx$

68. $\int \sin^2 x \cos 3x\, dx$

69. $\int \frac{1}{x^3(4 - x^2)^{3/2}}\, dx$

70. $\int \sqrt{1 - x^2}\, \operatorname{Sin}^{-1} x\, dx$

71. $\int \frac{1}{x + \sqrt{x^2 + 4}}\, dx$

72. a) Use the substitution $z^2 = (1 + x)/(1 - x)$ to show that

$$\int \sqrt{\frac{1 + x}{1 - x}}\, dx = 2 \operatorname{Tan}^{-1} \sqrt{\frac{1 + x}{1 - x}} - \sqrt{1 - x^2} + C.$$

b) Evaluate the integral by multiplying the numerator and denominator of the integrand by $\sqrt{1 + x}$. Verify that this answer is the same as that in (a).

Conic Sections, Polar Coordinates, and Parametric Equations

10

10.1 CONIC SECTIONS

In Section 1.3 we introduced analytic geometry by discussing straight lines, circles, and parabolas. These curves and their representations as equations demonstrate how geometric objects can be represented algebraically, and conversely, how algebraic equations can be interpreted geometrically. Parabolas are part of a larger group of curves referred to as *conic sections* (or *conics*, for short). The group also includes curves called ellipses and hyperbolas; circles and straight

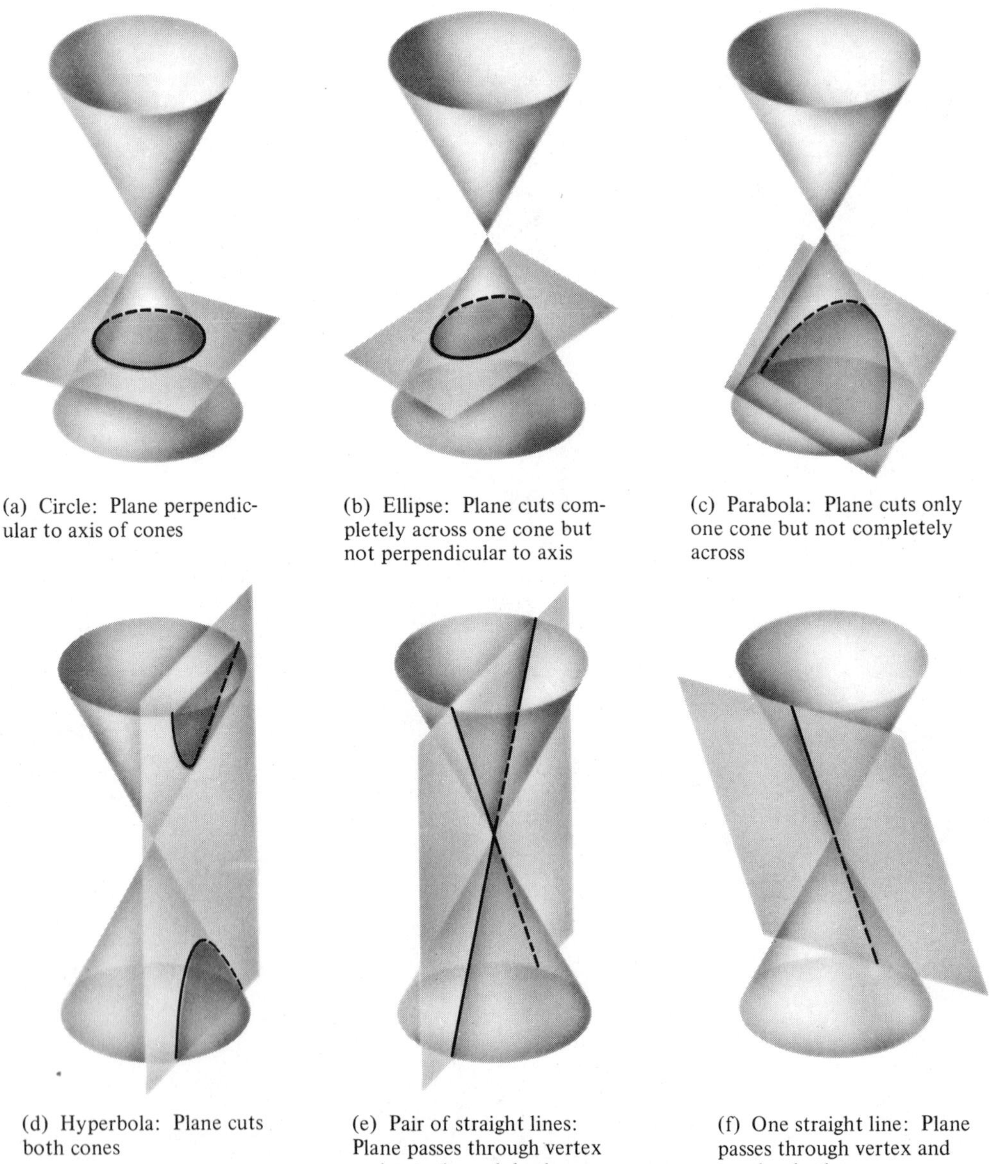

Figure 10.1 (a) Circle: Plane perpendicular to axis of cones (b) Ellipse: Plane cuts completely across one cone but not perpendicular to axis (c) Parabola: Plane cuts only one cone but not completely across (d) Hyperbola: Plane cuts both cones (e) Pair of straight lines: Plane passes through vertex and cuts through both cones (f) One straight line: Plane passes through vertex and touches both cones

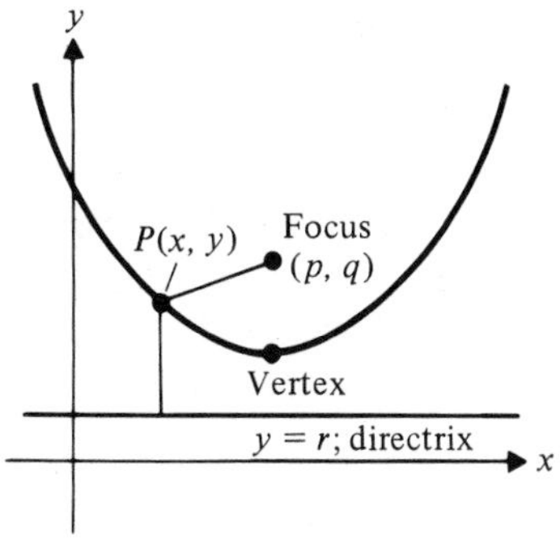

Figure 10.2

lines are often referred to as degenerate conic sections. Each of these curves can be visualized as the curve of intersection of a plane with a pair of right-circular cones. In particular, if a plane cuts completely across one of the cones but is not perpendicular to the axis of the cones (Fig. 10.1b), the resulting curve is called an *ellipse*. If the plane cuts through both cones but does not pass through the vertex of the cones (Fig. 10.1d), the curve is called a *hyperbola*. Parabolas, circles, and straight lines are obtained by other orientations of the plane. This visualization of ellipses, hyperbolas, and parabolas certainly suggests why they are referred to as conic sections, but because of the three-dimensional nature of a cone, an analysis of conics from this point of view is not yet possible. However, each of these curves is contained in a plane, and our knowledge of plane analytic geometry should permit us to analyze each conic completely. In this section we develop other definitions for conics that rely only on plane analytic geometry.

The Parabola

▷ **Definition 10.1** A *parabola* is the curve traced out by a point that moves in a plane so that its distances from a fixed point (called the *focus*) and a fixed line (called the *directrix*) are always equal. ◁

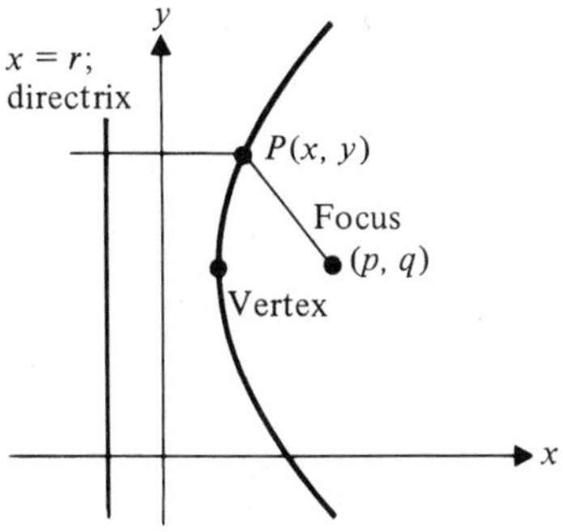

Figure 10.3

For example, suppose the focus is the point (p, q) and the directrix is a line $y = r$ parallel to the x-axis (Fig. 10.2). If $P(x, y)$ is any point on the parabola, then its distance from (p, q) must be equal to its distance from $y = r$:

$$\sqrt{(x - p)^2 + (y - q)^2} = |y - r|. \tag{10.1}$$

If we square this equation and rearrange terms, we obtain

$$y = \frac{1}{2(q - r)}\{(x - p)^2 + (q^2 - r^2)\}. \tag{10.2}$$

We can rewrite this equation in our accustomed form $y = ax^2 + bx + c$ for a parabola, but the present form provides more information. First, the line $x = p$ through the focus and perpendicular to the directrix is the line of symmetry for the parabola. Second, the parabola opens upward if $q > r$, in which case the focus is above the directrix, and opens downward if $q < r$. Finally, the vertex of the parabola is found by setting $x = p$, in which case $y = (q + r)/2$, halfway between the focus and the directrix.

A similar analysis shows that when the directrix is parallel to the y-axis (Fig. 10.3), the equation of the parabola is of the form $x = ay^2 + by + c$.

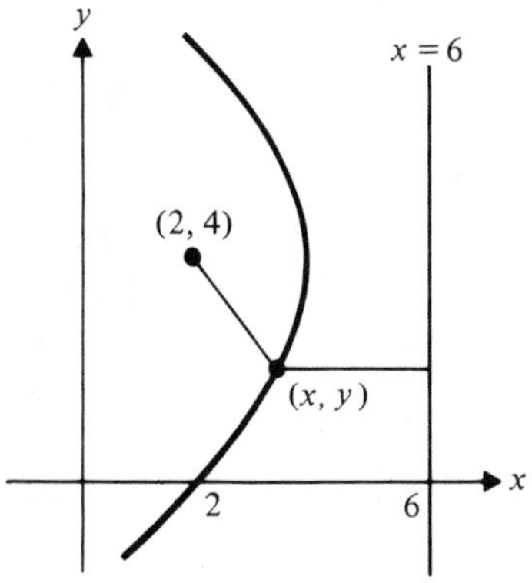

Figure 10.4

Given the focus and directrix (parallel to a coordinate axis) of a parabola, we can easily find its equation: We follow the algebraic steps leading from (10.1) to (10.2). Conversely, given the equation of a parabola in the form $y = ax^2 + bx + c$ or $x = ay^2 + by + c$, we can identify its focus and directrix (see Exercises 52 and 53).

Example 1 Find the equation of the parabola that has focus (2, 4) and directrix $x = 6$.

Solution If (x, y) is any point on the parabola (Fig. 10.4), the fact that its distance from (2, 4) is equal to its distance from $x = 6$ is expressed as

$$\sqrt{(x - 2)^2 + (y - 4)^2} = 6 - x.$$

If we square this equation and simplify, the result is $x = (16 + 8y - y^2)/8$. □

The Ellipse

▷ **Definition 10.2** An *ellipse* is the curve traced out by a point that moves in a plane so that the sum of its distances from two fixed points (called *foci*) remains constant. ◁

The equation of an ellipse is simplest when the foci lie on either the x-axis or the y-axis and are equidistant from the origin. For example, suppose the foci are $(c, 0)$ and $(-c, 0)$ (Fig. 10.5), and that the sum of the distances from these foci is $2a$ $(a > c \geq 0)$. If $P(x, y)$ is any point on the ellipse, Definition 10.2 implies that

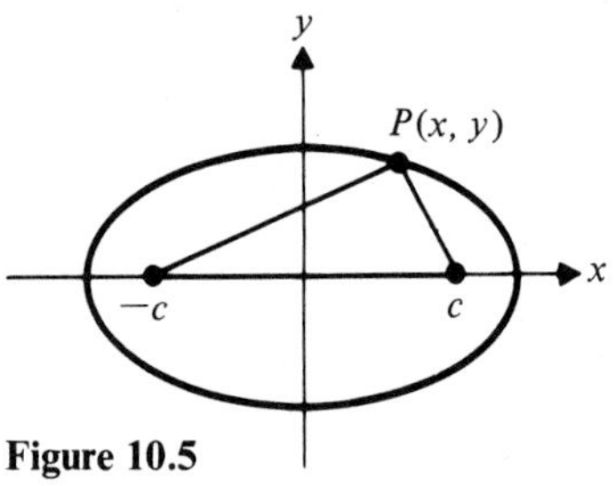

Figure 10.5

$$\sqrt{(x + c)^2 + y^2} + \sqrt{(x - c)^2 + y^2} = 2a. \tag{10.3}$$

If we transpose the second term and square both sides, we obtain

$$(x + c)^2 + y^2 = 4a^2 - 4a\sqrt{(x - c)^2 + y^2} + (x - c)^2 + y^2,$$

and this equation simplifies to

$$a^2 - cx = a\sqrt{(x - c)^2 + y^2}.$$

Squaring once again leads to

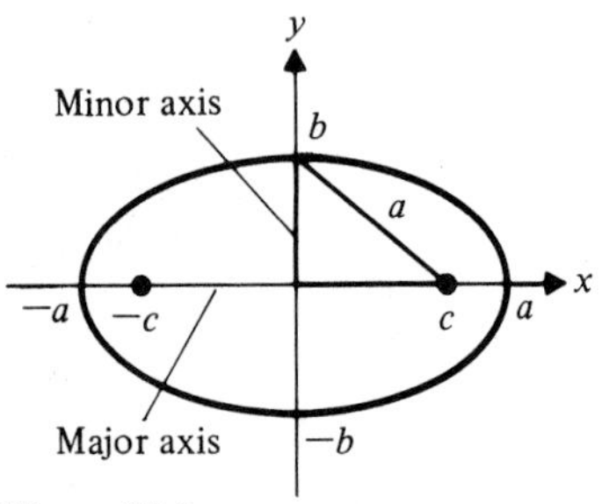

Figure 10.6

$$\frac{x^2}{a^2} + \frac{y^2}{a^2 - c^2} = 1. \tag{10.4}$$

It is customary to denote the y-intercepts of an ellipse by $\pm b$ $(b > 0)$ (Fig. 10.6), in which case $b^2 = a^2 - c^2$, and the equation of the ellipse becomes

$$\frac{x^2}{a^2} + \frac{y^2}{b^2} = 1, \quad \text{or} \quad b^2x^2 + a^2y^2 = a^2b^2. \tag{10.5}$$

The line segment across the ellipse and through the foci is called the *major axis* of the ellipse; it has length $2a$. The midpoint of the major axis is called the *center* of the ellipse. The line segment across the ellipse, through its center, and perpendicular to the major axis is called the *minor axis*; it has length $2b$. Note that the line segment joining either end of the minor axis to a focus (Fig. 10.6) has length a, and the triangle formed specifies the relationship among a, b, and c, namely, $a^2 = b^2 + c^2$.

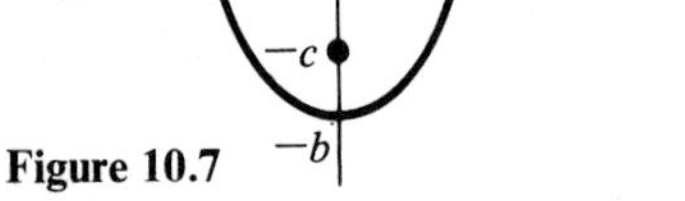

Figure 10.7

A similar analysis shows that when the foci of the ellipse are on the y-axis, equidistant from the origin, the equation of the ellipse is again in the form (10.5). In this case, however, $2b$ represents the length of the major axis (Fig. 10.7), and $b^2 = a^2 + c^2$.

What we should remember is that an equation of the form (10.5) always specifies an ellipse. The foci are on the longer axis and can be located using $c^2 = |a^2 - b^2|$. The length of the major axis represents the sum of the distances from any point on the ellipse to the foci.

Example 2 Sketch the ellipse $16x^2 + 9y^2 = 144$, indicating its foci.

Solution If we write the ellipse in the form $x^2/9 + y^2/16 = 1$, its x- and y-intercepts are clearly ± 3 and ± 4. A sketch of the ellipse is therefore as shown in Fig. 10.8. The foci must lie on the y-axis at distances $\pm c = \pm\sqrt{4^2 - 3^2} = \pm\sqrt{7}$ from the origin. □

If $a = b$ in Eq. (10.5), then $x^2 + y^2 = a^2$, and this is the equation for a circle with radius a and center at the origin. But if $a = b$, the distance from the origin

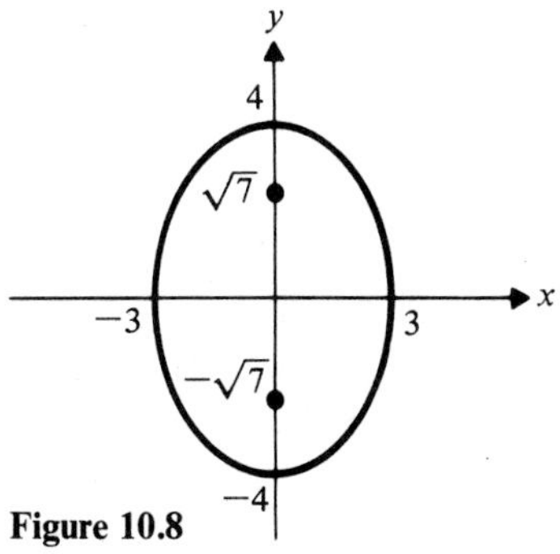

Figure 10.8

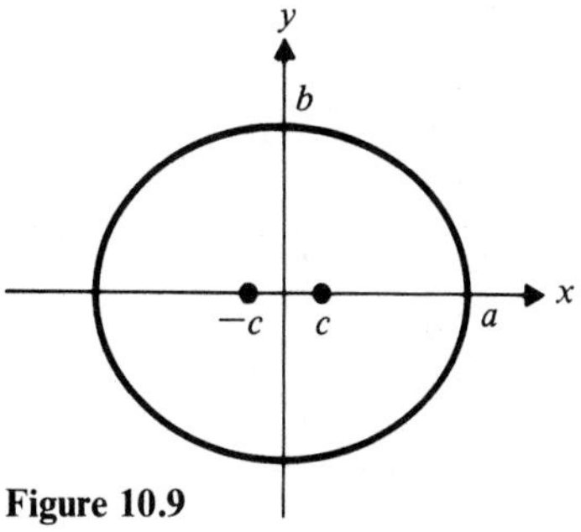

Figure 10.9

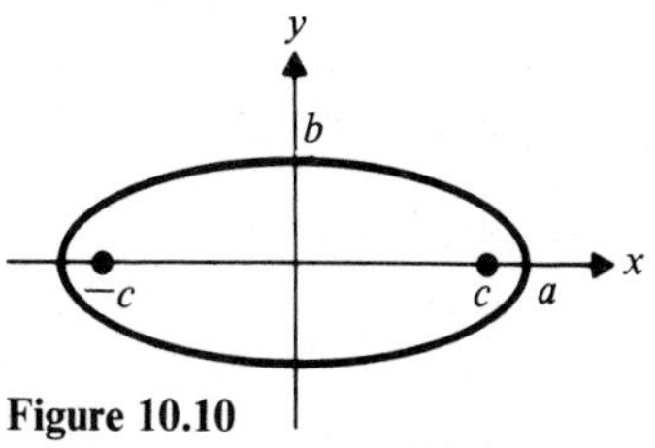

Figure 10.10

to each focus must be $c = 0$. In other words, a circle may be regarded as a degenerate ellipse whose foci are at one and the same point. It is also true that when c is very small compared to half the length of the major axis (Fig. 10.9), the ellipse is shaped very much like a circle; on the other hand, when these lengths are almost equal (Fig. 10.10), the ellipse is long and narrow.

When the center of an ellipse is at point (h, k) and the foci lie on either the line $x = h$ or the line $y = k$ (Fig. 10.11), the equation for the ellipse is only somewhat more complicated than (10.5). A calculation similar to that leading from (10.3) to (10.5) gives

$$\frac{(x-h)^2}{a^2} + \frac{(y-k)^2}{b^2} = 1 \qquad (10.6)$$

(see Exercise 55). Alternatively, the curves in Fig. 10.11 are those in Fig. 10.5 and 10.7 translated h units in the x-direction and k units in the y-direction. According to Section 1.5, equations for the translated ellipses can be obtained by replacing x and y in $x^2/a^2 + y^2/b^2 = 1$ by $x - h$ and $y - k$, respectively.

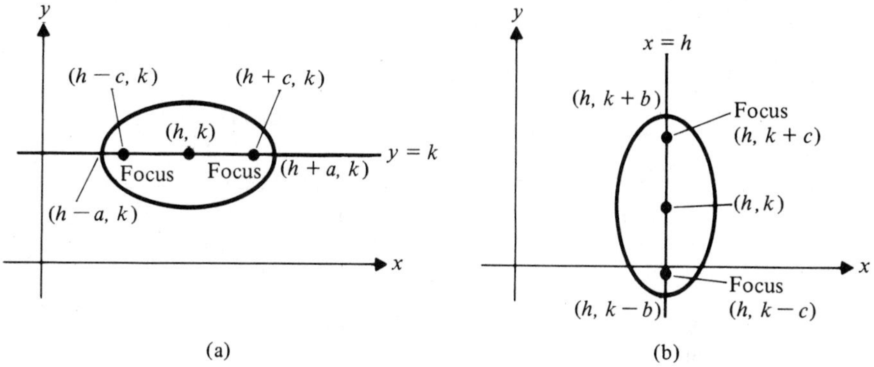

Figure 10.11

Example 3 Sketch the ellipse $16x^2 + 25y^2 - 160x + 50y - 1175 = 0$.

Solution If we complete the squares on the x- and y-terms, we obtain

$$16(x-5)^2 + 25(y+1)^2 = 1600,$$

or

$$\frac{(x-5)^2}{100} + \frac{(y+1)^2}{64} = 1.$$

The center of the ellipse is $(5, -1)$, and the lengths of its major and minor axes are 20 and 16, respectively (Fig. 10.12). □

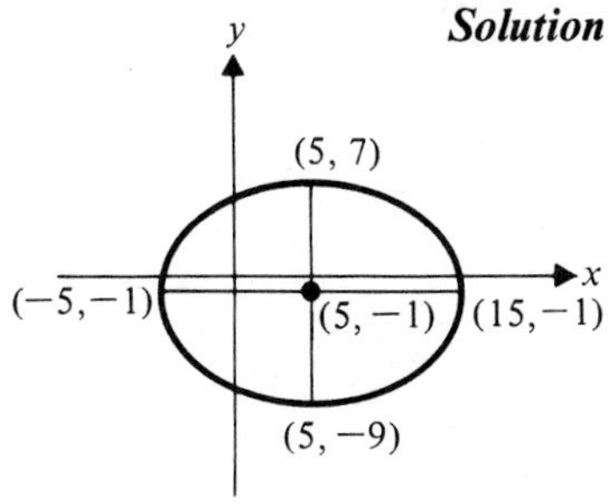

Figure 10.12

The Hyperbola

▷ **Definition 10.3** A *hyperbola* is the path traced out by a point that moves in a plane so that the difference of its distances from two fixed points (called *foci*) remains constant. ◁

Like the ellipse, the simplest hyperbolas have foci on either the x- or y-axis, equidistant from the origin. For example, if the foci are $(\pm c, 0)$ (Fig. 10.13) and

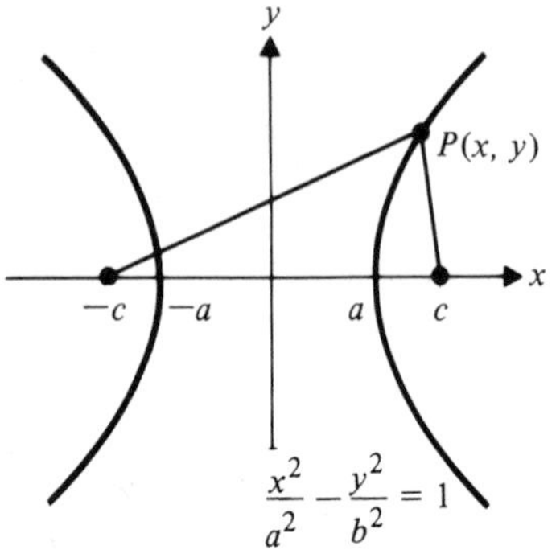

Figure 10.13

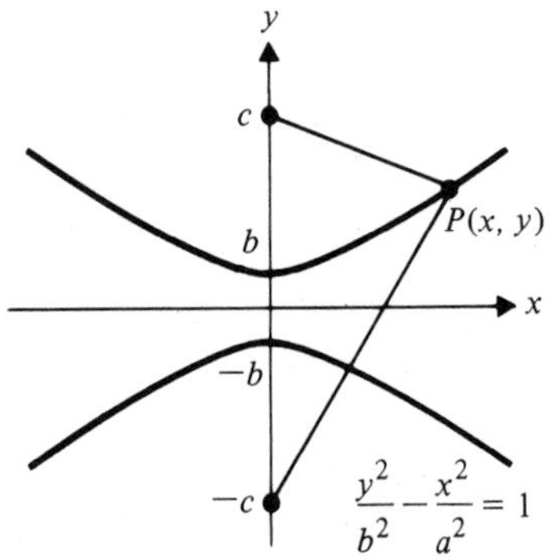

Figure 10.14

the difference in the distances from $P(x, y)$ to these foci is $2a$, then Definition 10.3 implies that

$$\left|\sqrt{(x + c)^2 + y^2} - \sqrt{(x - c)^2 + y^2}\right| = 2a. \tag{10.7}$$

This equation can be simplified by a calculation similar to that used for (10.3), the result being

$$\frac{x^2}{a^2} - \frac{y^2}{b^2} = 1, \tag{10.8}$$

where $b^2 = c^2 - a^2$.

When the foci are on the y-axis (Fig. 10.14), the equation of the hyperbola becomes

$$\frac{y^2}{b^2} - \frac{x^2}{a^2} = 1, \tag{10.9}$$

where $2b$ is the constant difference in the distances from a point (x, y) to the foci, and $a^2 = c^2 - b^2$.

If we solve Eqs. (10.8) and (10.9) for y in terms of x, we obtain

$$y = \pm\frac{b}{a}\sqrt{x^2 - a^2} \quad \text{and} \quad y = \frac{b}{a}\sqrt{x^2 + a^2}.$$

These expressions indicate that for very large positive or negative x, y is essentially equal to $\pm(b/a)x$. In other words, the hyperbolas are asymptotic to the lines $y = \pm(b/a)x$ (Fig. 10.15). Note also that we can draw triangles indicating the relationship among a, b, and c in the same way that we did for ellipses in Fig. 10.6. The line segment joining the vertices of a hyperbola is called the *transverse axis* of the hyperbola; it has length $2a$ (Fig. 10.15a) or $2b$ (Fig. 10.15b). The midpoint of the transverse axis is called the *center* of the hyperbola. The line segment perpendicular to the transverse axis, through its center, and of length $2b$ (Fig. 10.15a) or $2a$ (Fig. 10.15b) is called the *conjugate axis.*

What we should remember is that an equation of the form (10.8) or (10.9) specifies a hyperbola. The foci lie on the extension of the transverse axis and can be located using $c^2 = a^2 + b^2$. The length of the transverse axis represents the difference of the distances from any point on the hyperbola to the foci.

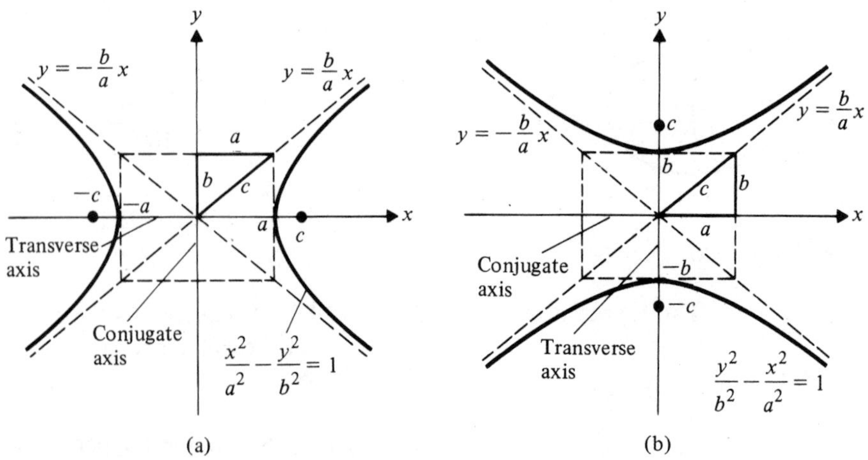

Figure 10.15

Example 4 Sketch the hyperbola $16x^2 - 9y^2 = 144$, indicating its foci.

Solution If we express the hyperbola in the form $x^2/9 - y^2/16 = 1$, its x-intercepts are ± 3. With asymptotes $y = \pm(\frac{4}{3})x$, we obtain the sketch shown in Fig. 10.16. The foci must lie on the x-axis at distances $\pm c = \pm\sqrt{4^2 + 3^2} = \pm 5$ from the origin. □

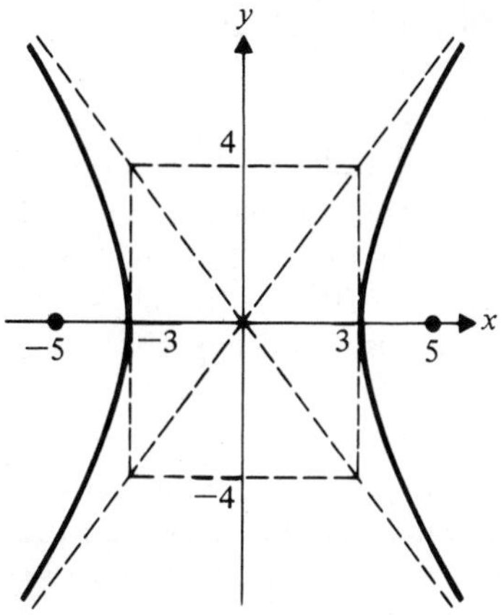

Figure 10.16

When the center of a hyperbola is at point (h, k) and its foci are on the lines $x = h$ or $y = k$, Eqs. (10.8) and (10.9) are modified in exactly the same way as Eq. (10.5) was modified for an ellipse: We replace each x by $x - h$ and each y by $y - k$ (see Exercise 56). Consequently, equations for the hyperbolas in Figs. 10.17(a) and (b) are

$$\frac{(x-h)^2}{a^2} - \frac{(y-k)^2}{b^2} = 1 \quad \text{and} \quad \frac{(y-k)^2}{b^2} - \frac{(x-h)^2}{a^2} = 1, \qquad (10.10)$$

respectively.

Example 5 Sketch the hyperbola $x^2 - y^2 + 4x + 10y = 5$.

Solution If we complete the squares on the x- and y-terms, we have

$$(x+2)^2 - (y-5)^2 = -16,$$

or

$$\frac{(y-5)^2}{16} - \frac{(x+2)^2}{16} = 1.$$

The center of the hyperbola is $(-2, 5)$, and the length of its transverse axis (along $x = -2$) is 8. If we solve the equation of the hyperbola for y, we obtain

$$y = 5 \pm \sqrt{(x+2)^2 + 16};$$

the asymptotes of the hyperbola are then

$$y = 5 \pm (x + 2).$$

The hyperbola can now be sketched as shown in Fig. 10.18. □

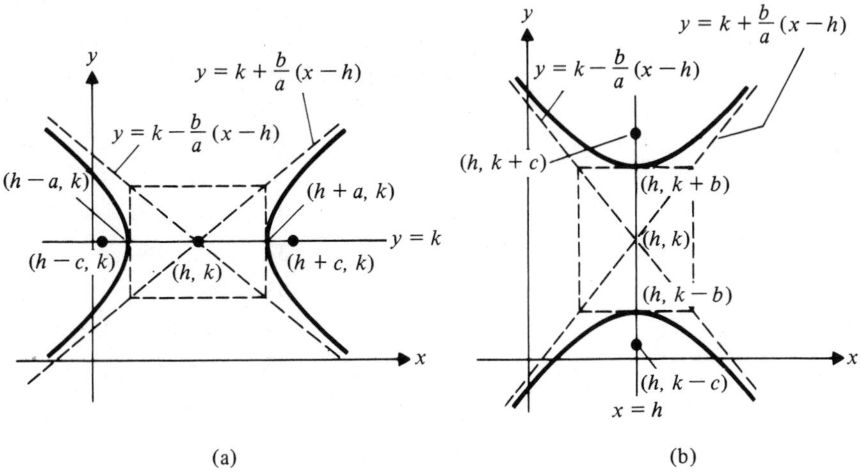

Figure 10.17

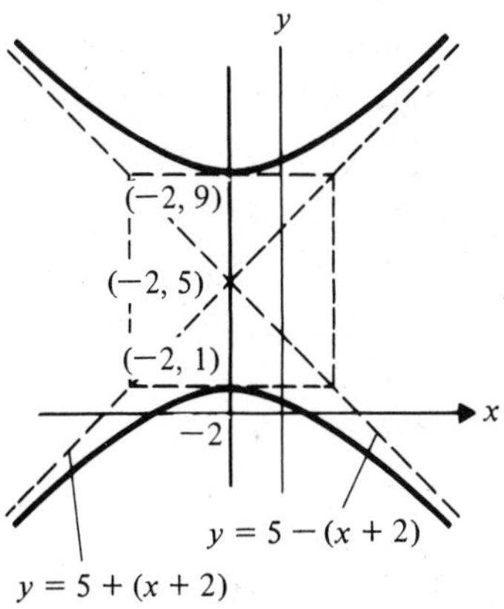

Figure 10.18

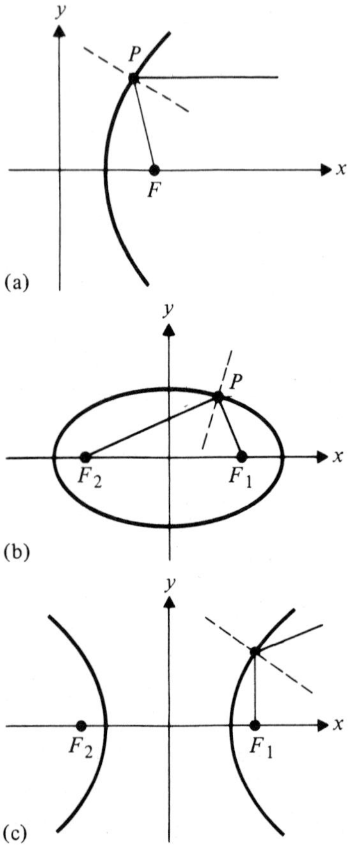

Figure 10.19

If P is a point on a conic section, the *focal radii* at P are the lines joining P to the foci (Fig. 10.19). As a result, a parabola has one focal radius, and an ellipse and hyperbola each have two. One of the properties of conics that makes them so useful is the fact that the normal line to the conic at any point bisects the angle between the focal radii. For the parabola, the normal bisects the angle between the focal radius and the line through P parallel to the axis of symmetry of the parabola. These facts are verified in Exercises 59 and 60. To obtain one physical significance of these results, suppose each conic in Fig. 10.19 is rotated about the x-axis to form a surface of revolution, which we regard as a mirror. It is a law of physics that when a ray of light strikes a reflecting surface, the angle between incident light and the normal to the surface is always equal to the angle between reflected light and the normal. Consequently, if a beam of light travels in the negative x-direction and strikes the parabolic mirror shown in Fig. 10.19(a), all light is reflected toward the focus. Conversely, if F is a source of light, all light striking the mirror is reflected parallel to the x-axis. If either focus of the ellipse shown in Fig. 10.19(b) is a light source, all light striking the elliptic mirror is reflected toward the other focus. Similarly, if either focus of the hyperbola shown in Fig. 10.19(c) is a source, all light striking the mirror is reflected in a direction that would make it seem to originate at the other focus. Conversely, if light that is directed at one focus first strikes the mirror, it is reflected toward the other focus.

It is not difficult to realize that this is precisely why we have parabolic reflectors in automobile headlights and searchlights, parabolic and hyperbolic reflectors in telescopes, and elliptic ceilings in whispering rooms.

Conic sections (parabolas, ellipses, and hyperbolas) have been defined in terms of distances; the parabola uses a focus and a directrix, and the ellipse and hyperbola two foci. In Section 10.5 we show that ellipses and hyperbolas can also be defined in terms of a focus and a directrix.

EXERCISES 10.1

In Exercises 1–14, identify each curve as a straight line, circle, parabola, ellipse, hyperbola, or none of these.

1. $2x + 3y = y^2$

2. $x^2 + y^2 - 3x + 2y = 25$

3. $2x - y = 3$

4. $x^2 + y^3 = 3x + 2$

5. $5x^2 = 11 - 2y^2$

6. $2x^2 - 3y^2 + 5 = 0$

7. $y^2 - x + 3y = 14 - x^2$

8. $x^2 + 2x = 3y + 4$

9. $y^2 + x^2 - 2x + 6y + 15 = 0$

10. $x^2 + 2y^2 + 24 = 0$

11. $5 + y^2 = 3x^2$

12. $x^2 + 2y^2 = 24$

13. $y^3 = 3x + 4$

14. $3 - x = 4y$

In Exercises 15–36, sketch each curve. Identify foci for each ellipse and hyperbola.

15. $y = 2x^2 - 1$

16. $\dfrac{x^2}{25} + \dfrac{y^2}{36} = 1$

17. $x^2 - \dfrac{y^2}{16} = 1$

18. $3x = 4y^2 - 1$

19. $\dfrac{y^2}{4} - \dfrac{x^2}{25} = 1$

20. $7x^2 + 3y^2 = 16$

21. $x + y^2 = 1$

22. $2y^2 + x = 3y + 5$

23. $9x^2 + 289y^2 = 2601$

24. $y^2 = 10(2 - x^2)$

25. $3x^2 - 4y^2 = 25$

26. $y^2 - x^2 = 5$

27. $y = -x^2 + 6x - 9$

28. $2x^2 - 3y^2 = 5$

29. $3x^2 + 6y^2 = 21$

30. $x^2 + 16y^2 = 2$

31. $y^2 - 3x^2 = 1$

32. $x = -(4 + y)^2$

33. $x^2 + 2x + 4y^2 - 16y + 13 = 0$

34. $x^2 - 6x - 4y^2 - 24y = 11$

35. $9x^2 + y^2 - 18x - 6y = 0$

36. $9x^2 - 16y^2 - 18x - 64y = 91$

37. Find the equation of a hyperbola that passes through the point (1, 2) and has asymptotes $y = \pm 4x$.

38. Find the equation of an ellipse through the points $(-2, 4)$ and (3, 1).

39. Find the width of the elliptic arch shown in Fig. 10.20.

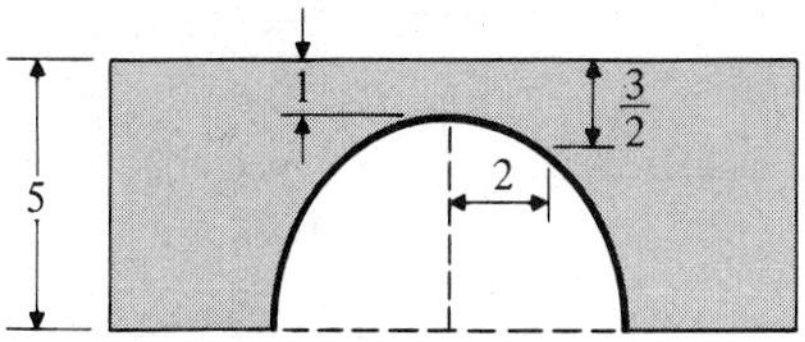

Figure 10.20

40. Find the height of the parabolic arch shown in Fig. 10.21.

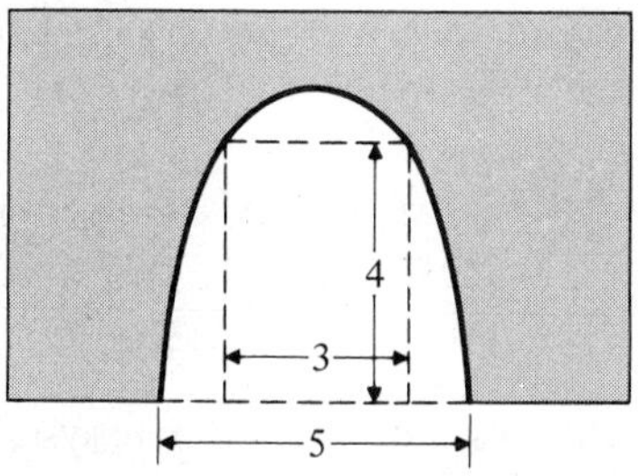

Figure 10.21

41. Explain how an ellipse can be drawn with a piece of string, two tacks, and a pencil.

42. Find the equation of the ellipse traced out by a point that moves so that the sum of its distances from $(\pm 4, 0)$ is always equal to 10:
    a) by using Eq. (10.5) with suitable values for a and b;
    b) by establishing and simplifying an equation similar to (10.3).

43. Find the equation of the hyperbola traced out by a point that moves so that the difference of its distances from $(0, \pm 3)$ is always equal to 1:
    a) by using Eq. (10.9) with suitable values for a and b;
    b) by establishing and simplifying an equation similar to (10.7).

44. Show that the equation of every straight line, every circle, and every conic discussed in this section can be obtained by appropriate choices of constants A, C, D, E, and F in the equation

$$Ax^2 + Cy^2 + Dx + Ey + F = 0.$$

45. Show that the equation of the tangent line to the ellipse $b^2x^2 + a^2y^2 = a^2b^2$ at a point (x_0, y_0) is $b^2xx_0 + a^2yy_0 = a^2b^2$.

46. Show that the equation of the tangent line to the hyperbola $b^2x^2 - a^2y^2 = a^2b^2$ at a point (x_0, y_0) is $b^2xx_0 - a^2yy_0 = a^2b^2$.

47. Find that point P on the first quadrant of the ellipse $2x^2 + 3y^2 = 14$ where the tangent line at P is perpendicular to the line joining P and (2, 5).

48. Find the area inside the ellipse $b^2x^2 + a^2y^2 = a^2b^2$.

49. Among all rectangles that can be inscribed inside the ellipse $b^2x^2 + a^2y^2 = a^2b^2$ and have sides parallel to the axes, find that one with the largest possible area.

50. A *prolate spheroid* is the solid of revolution obtained by rotating an ellipse about its major axis. An *oblate spheroid* is obtained by rotating an ellipse about its minor axis. Find volumes for the prolate and oblate spheroids generated by the ellipse $b^2x^2 + a^2y^2 = a^2b^2$ if $a > b$.

51. A sharp noise originating at one focus F_1 of an ellipse is reflected by the ellipse toward the other focus F_2. Explain why all reflected noise arrives at F_2 at exactly the same time.

52. Use Eq. (10.2) to show that when a parabola is written in the form $y = ax^2 + bx + c$, the following formulas identify its focus (p, q) and directrix $y = r$:

$$p = -\frac{b}{2a}; \qquad q = \frac{1}{4a}(1 + 4ac - b^2);$$

$$r = \frac{1}{4a}(-1 + 4ac - b^2).$$

53. Show that formulas for the position (p, q) of the focus and directrix $x = r$ for a parabola of the type $x = ay^2 + by + c$ are given by the equations

$$p = \frac{1}{4a}(1 + 4ac - b^2); \qquad q = -\frac{b}{2a};$$

$$r = \frac{1}{4a}(-1 + 4ac - b^2).$$

54. Use the formulas in Exercises 52 and 53 to identify the focus and directrix for any parabolas in Exercises 15–36.

55. Show that when the center of an ellipse is at point (h, k) and its foci are on the line $x = h$ or $y = k$, Definition 10.2 leads to Eq. (10.6).

56. Show that when the center of a hyperbola is at point (h, k) and its foci are on the line $x = h$ or $y = k$. Definition 10.3 leads to Eqs. (10.10).

57. Prove that the normal line to the parabola $x = ay^2$ shown in Fig. 10.22 bisects the angle between the focal radius FP and the line PQ parallel to the x-axis. (*Hint:* Draw the tangent line PR at P and show that $|PF| = |RF|$.)

*58. When a beam of light traveling in the negative x-direction (Fig. 10.23) strikes a parabolic mirror with cross-section $x = ay^2 + c$, all light rays are reflected to the focus F of the mirror. Show that all photons that pass simultaneously through $x = d$ arrive at F at the same time.

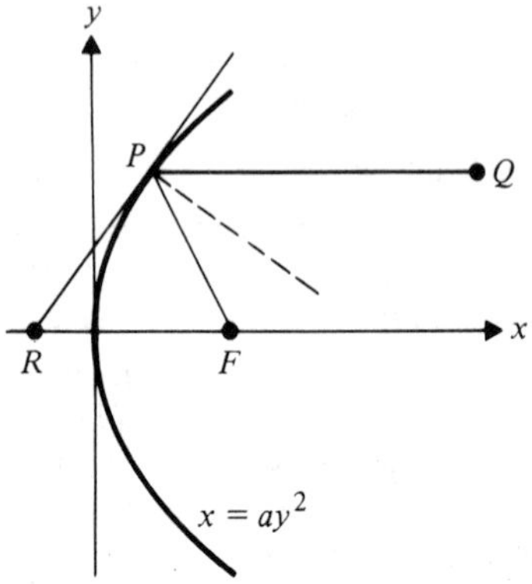

Figure 10.22

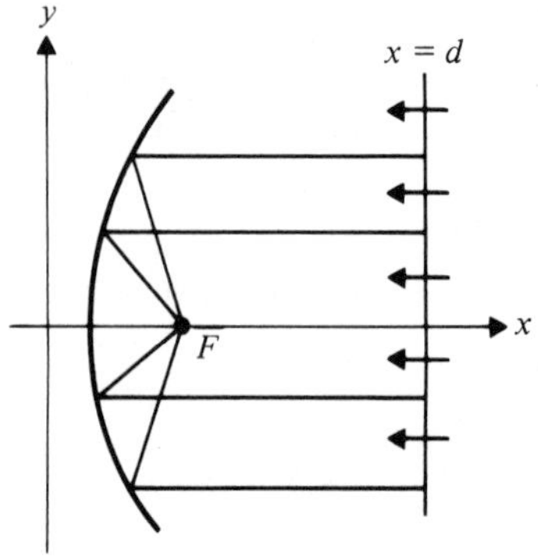

Figure 10.23

***59.** Prove that the normal line to an ellipse or hyperbola bisects the angle between the focal radii.

***60.** Prove that the normal line to the parabola $y = ax^2 + bx + c$ at any point P bisects the angle between the focal radius and the line through P parallel to the y-axis.

***61.** A line segment through the focus of a parabola with ends on the parabola is called a *focal chord*. It is an established result that tangents to a parabola at the ends of a focal chord are perpendicular to each other and intersect on the directrix. Prove this for the parabola $y = ax^2$.

10.2 POLAR COORDINATES

In this section we introduce polar coordinates, a new coordinate system for the plane. We will see both here and in subsequent chapters that many problems that have difficult solutions using Cartesian coordinates may become much simpler to solve using polar coordinates.

Polar coordinates are defined by choosing a point O in the plane called the *pole* and a half-line originating at O called the *polar axis* (Fig. 10.24). If P is a point in the plane, we join O and P. One of the polar coordinates of P is the angle θ (in radians) through which the polar axis must be rotated to coincide with line segment OP. Counterclockwise rotations are regarded as positive, and clockwise rotations as negative. In Fig. 10.25, position OQ is reached through a positive rotation of $\pi/6$ radians; therefore, for point Q, $\theta = \pi/6$. But clearly we could arrive at this position in many other ways. We could, for instance, rotate the polar axis counterclockwise through any number of complete revolutions, bringing it back to its original position, and then rotate a further $\pi/6$ radians. Alternatively, we could rotate in a clockwise direction any number of complete revolutions, and then a further $-11\pi/6$ radians. In other words, polar coordinate θ for Q could be any of the values $\pi/6 + 2n\pi$, n an integer. Possible values of θ for point R in Fig. 10.25 are $3\pi/4 + 2n\pi$.

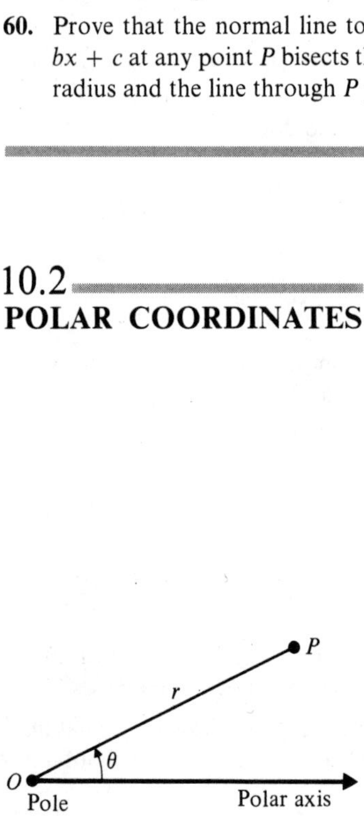

Figure 10.24

The other polar coordinate of a point P is defined as the directed distance r along the line OP from O to P (Fig. 10.24). For point Q in Fig. 10.25, $r = 2$, and for R, $r = 1$. Polar coordinates r and θ for a point are written in the form (r, θ) so that possible polar coordinates for Q and R are $(2, \pi/6 + 2n\pi)$ and $(1, 3\pi/4 + 2n\pi)$.

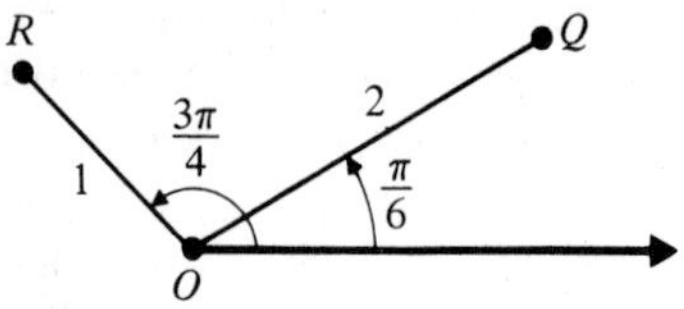

Figure 10.25

We have defined r as directed distance, which is certainly to suggest that r can be negative. By negative r we mean that distance should not be measured along the ray at angle θ, but instead along this ray extended in the opposite direction; i.e., along the ray at angle $\theta + \pi$. In other words, we could also specify

Q in Fig. 10.25 by rotating $\theta = 7\pi/6$ radians and then proceeding a distance $r = -2$ along this ray. Consequently, we also accept as polar coordinates of Q $(-2, 7\pi/6)$; and in addition to these, we must accept $(-2, 7\pi/6 + 2n\pi)$. Similarly, for point R, $(-1, 7\pi/4 + 2n\pi)$ are possible polar coordinates besides $(1, 3\pi/4 + 2n\pi)$, already mentioned.

This situation is not at all like that for Cartesian coordinates, where each point has only one set of coordinates (x, y), and every ordered pair of real numbers specifies one point. With polar coordinates, every ordered pair of real numbers (r, θ) still represents one and only one point, but every point has an infinity of possible representations. We should point out that in some applications this is not a desirable situation. For instance, when polar coordinates are used to define the polar representation of complex numbers, it is customary to restrict r to be nonnegative. This eliminates half the possible sets of polar coordinates of a point. In the branch of mathematics called *tensor analysis*, it is necessary that polar coordinates assign exactly one pair of coordinates to each point. This can be accomplished in any region that does not contain the pole by demanding, for instance, that

$$r > 0, \qquad -\pi < \theta \leq \pi. \tag{10.11}$$

Even in Section 10.4, where we will use polar coordinates to find areas, we will see that we must be particular about our choice of r and θ. For now, however, no advantage is to be gained by imposing restrictions on r and θ, and we therefore accept the fact that if (r, θ) are polar coordinates of a point, then so too are $(r, \theta + 2n\pi)$ and $(-r, \theta + (2n + 1)\pi)$ for any integer n. Note also that polar coordinates for the pole are $(0, \theta)$ for any θ whatsoever.

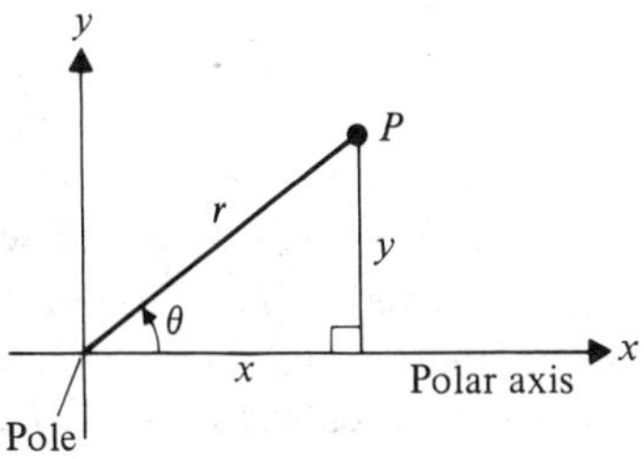

Figure 10.26

If we introduce into a plane both a system of Cartesian coordinates (x, y) and a system of polar coordinates (r, θ), then relations will exist between the two systems. Suppose, for example, that the pole of the polar coordinates and the origin of the Cartesian coordinates are chosen as the same point, and that the polar axis is chosen as the positive x-axis (Fig. 10.26). In this case, the Cartesian and polar coordinates of any point P are related by

$$x = r \cos \theta, \qquad y = r \sin \theta. \tag{10.12}$$

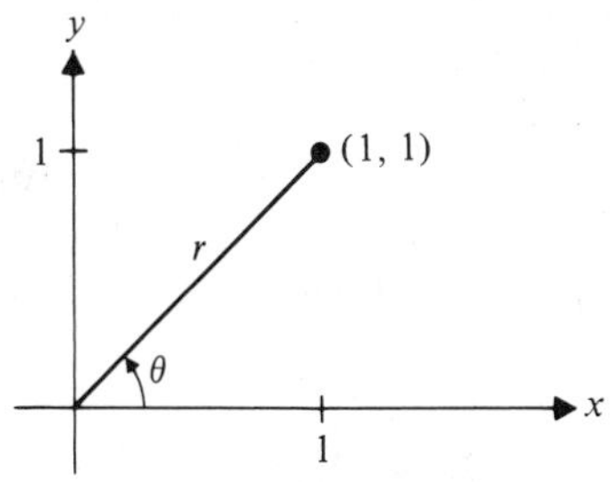

Figure 10.27

These equations define the Cartesian coordinates of a point in terms of its polar coordinates; i.e., given its polar coordinates (r, θ) we can calculate its Cartesian coordinates (x, y) by means of (10.12). For example, if the polar coordinates of a point are $(3, 2)$, then its Cartesian coordinates are

$$x = 3 \cos 2 = -1.25, \qquad y = 3 \sin 2 = 2.73.$$

It is also true that Eqs. (10.12) define the polar coordinates of a point in terms of its Cartesian coordinates, not explicitly, but implicitly. For instance, if the Cartesian coordinates of a point are $(1, 1)$ (Fig. 10.27), its polar coordinates must satisfy

$$1 = r \cos \theta, \qquad 1 = r \sin \theta.$$

If we square these equations and add, we have

$$1 + 1 = r^2 \cos^2 \theta + r^2 \sin^2 \theta = r^2,$$

and therefore $r = \pm\sqrt{2}$. When $r = \sqrt{2}$, $\cos \theta = \sin \theta = 1/\sqrt{2}$, from which we get $\theta = \pi/4 + 2n\pi$. When $r = -\sqrt{2}$, $\cos \theta = \sin \theta = -1/\sqrt{2}$, in which case

$\theta = 5\pi/4 + 2n\pi$. Thus the polar coordinates of the point are $(\sqrt{2}, \pi/4 + 2n\pi)$ and $(-\sqrt{2}, 5\pi/4 + 2n\pi)$.

We have indicated that Eqs. (10.12) define r and θ implicitly in terms of x and y, but obviously it is preferable to have explicit definitions. There is no problem expressing r explicitly in terms of x and y,

$$r = \pm\sqrt{x^2 + y^2}, \tag{10.13}$$

but the case for θ is not so simple. If we substitute expression (10.13) for r into Eqs. (10.12), we obtain

$$\cos\theta = \frac{\pm x}{\sqrt{x^2 + y^2}}, \qquad \sin\theta = \frac{\pm y}{\sqrt{x^2 + y^2}}. \tag{10.14}$$

Except for the pole, these two equations determine all possible values of θ for given x and y. What we would like to do is obtain one equation, if possible, that defines θ. If we divide the second of these equations by the first, we have

$$\tan\theta = \frac{y}{x}. \tag{10.15}$$

This equation does define all possible values of θ, but unfortunately it does not determine which values of θ accompany positive r and which accompany negative r. For example, if Cartesian coordinates of a point are $(-1, -1)$, (10.15) becomes

$$\tan\theta = \frac{-1}{-1} = 1,$$

and solutions of this equation are $\theta = \pi/4 + 2n\pi$ and $\theta = 5\pi/4 + 2n\pi$. The question is, Which of these angles should be grouped with $r = \sqrt{2}$ and which should be grouped with $r = -\sqrt{2}$ for polar coordinates of $(-1, -1)$? A diagram showing the point $(-1, -1)$ (Fig. 10.28) clearly indicates that $r = \sqrt{2}$ corresponds to $\theta = 5\pi/4 + 2n\pi$ and $r = -\sqrt{2}$ corresponds to $\theta = \pi/4 + 2n\pi$. But Eq. (10.15) by itself does not provide this correspondence. It appears, then, that there is no simple way to express θ explicitly in terms of x and y.

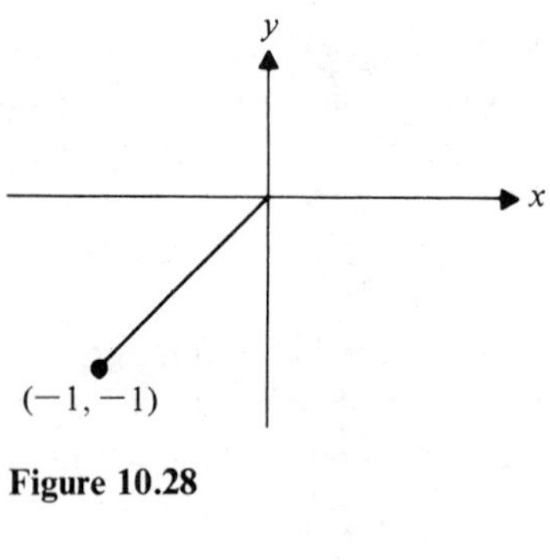

Figure 10.28

We suggest a slight modification of this procedure in order to determine θ from given x and y. First we calculate angle θ defined by

$$\theta = \mathrm{Tan}^{-1}\left(\frac{y}{x}\right). \tag{10.16}$$

This equation always defines a possible value of θ, and therefore all possible values can be listed as

$$\mathrm{Tan}^{-1}\left(\frac{y}{x}\right) + 2n\pi, \qquad \mathrm{Tan}^{-1}\left(\frac{y}{x}\right) + (2n + 1)\pi.$$

To decide which of these sets of values of θ corresponds to positive r and which corresponds to negative r, we make a diagram such as Fig. 10.28. We illustrate this procedure in the following example.

Example 6 Find polar coordinates for points with the following Cartesian coordinates:

a) $(1, 2)$;
b) $(-2, 3)$;
c) $(3, -1)$;
d) $(-2, -4)$.

Solution a) $r = \pm\sqrt{1^2 + 2^2} = \pm\sqrt{5}$. One possible value of θ is $\text{Tan}^{-1}\left(\frac{2}{1}\right) = 1.11$. Since the point is in the first quadrant (Fig. 10.29), the polar coordinates are $(\sqrt{5}, 1.11 + 2n\pi)$ and $(-\sqrt{5}, 1.11 + \pi + 2n\pi) = (-\sqrt{5}, 4.25 + 2n\pi)$.

b) $r = \pm\sqrt{(-2)^2 + 3^2} = \pm\sqrt{13}$. One possible value of θ is $\text{Tan}^{-1}\left(-\frac{3}{2}\right) = -0.98$. Since the point is in the second quadrant, the polar coordinates are $(\sqrt{13}, 2.16 + 2n\pi)$ and $(-\sqrt{13}, -0.98 + 2n\pi)$.

c) $r = \pm\sqrt{3^2 + (-1)^2} = \pm\sqrt{10}$. One possible value for θ is $\text{Tan}^{-1}\left(-\frac{1}{3}\right) = -0.32$. With the point in the fourth quadrant, possible polar coordinates are $(\sqrt{10}, -0.32 + 2n\pi)$ and $(-\sqrt{10}, 2.82 + 2n\pi)$.

d) $r = \pm\sqrt{(-2)^2 + (-4)^2} = \pm 2\sqrt{5}$. With $\text{Tan}^{-1}(2) = 1.11$ and the point in the third quadrant, the polar coordinates are $(2\sqrt{5}, 4.25 + 2n\pi)$ and $(-2\sqrt{5}, 1.11 + 2n\pi)$. □

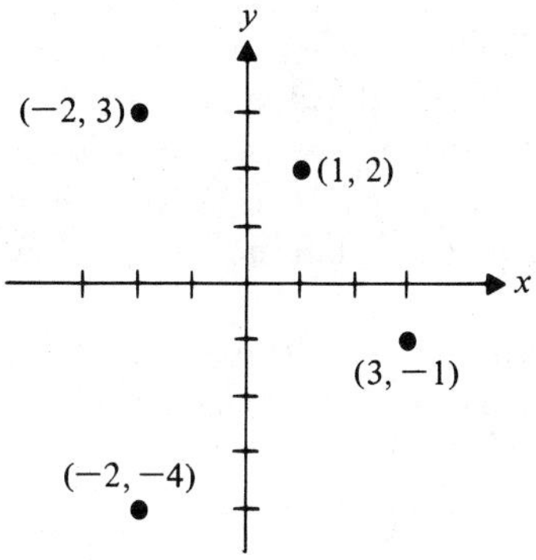

Figure 10.29

The results in Eqs. (10.12)–(10.16) are valid only when the pole and origin coincide and the polar axis and positive x-axis are identical. Should we decide on a different arrangement, we would modify these relations accordingly. For example, if the pole is situated at the point with Cartesian coordinates (h, k) and the polar axis is as shown in Fig. 10.30, Eqs. (10.12) are replaced by

$$x = h + r\cos(\theta + \alpha), \qquad y = k + r\sin(\theta + \alpha), \tag{10.17}$$

(see Exercise 13).

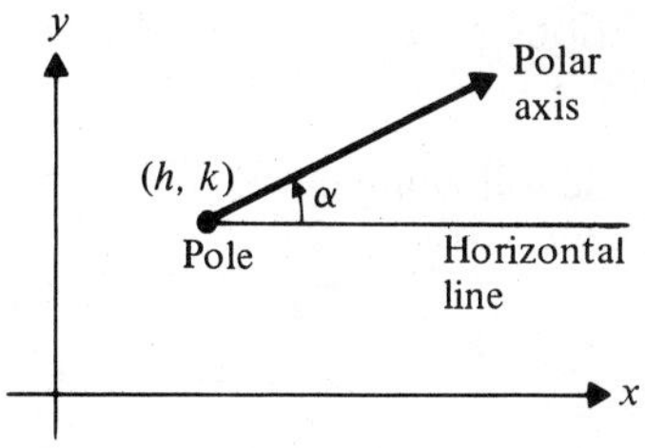

Figure 10.30

The usual choice of pole and polar axis is that shown in Fig. 10.26, and unless otherwise stipulated we will assume this to be the case.

EXERCISES 10.2

In Exercises 1–8, plot the point having the given set of Cartesian coordinates, and find all possible polar coordinates.

1. $(1, -1)$

2. $(-1, \sqrt{3})$

3. $(4, 3)$

4. $(-2\sqrt{3}, 2)$

5. $(2, 6)$

6. $(-1, -4)$

7. $(7, -5)$

8. $(-5, 2)$

In Exercises 9–12, plot the point having the given set of polar coordinates, and find its Cartesian coordinates.

9. $(2, \pi/4)$

10. $(6, -\pi/6)$

11. $(7, 1)$

12. $(3, -2.4)$

13. Verify that polar and Cartesian coordinates as shown in Fig. 10.30 are related by Eqs. (10.17).

10.3 CURVES IN POLAR COORDINATES

A curve is defined explicitly in Cartesian coordinates by equations $y = f(x)$ or $x = g(y)$ and implicitly by $F(x, y) = 0$. A point is on a curve if and only if its Cartesian coordinates (x, y) satisfy the equation of the curve.

Analogously, we say that a curve is defined explicitly in polar coordinates if its equation is expressed in either of the forms

$$r = f(\theta) \quad \text{or} \quad \theta = g(r), \tag{10.18}$$

and is defined implicitly when its equation is given in the form

$$F(r, \theta) = 0. \tag{10.19}$$

Unlike the Cartesian coordinates, however, a point is on a curve if at least one of its sets of polar coordinates (r, θ) satisfies the equation. It follows, then, that all sets of polar coordinates for a point need not satisfy the equation. For example, the point with polar coordinates $(\frac{3}{2}, \pi/3)$ is on the curve $r = 1 + \cos\theta$, since these coordinates clearly satisfy the equation. In addition, the polar coordinates $(\frac{3}{2}, \pi/3 + 2n\pi)$ of the same point also satisfy the equation, but its polar coordinates $(-\frac{3}{2}, 4\pi/3 + 2n\pi)$ do not. Likewise, the pole (or origin) is on this curve, and its polar coordinates $(0, (2n + 1)\pi)$ satisfy $r = 1 + \cos\theta$, but no other polar coordinates for the pole satisfy the equation.

Often we are required to transform the equation of a curve from Cartesian coordinates to polar coordinates, and vice versa. To transform from Cartesian to polar is simple: We merely replace each x with $r\cos\theta$ and each y with $r\sin\theta$. For example, the equation $x^2 + y^2 = 9$ describes a circle centered at the origin with radius 3. In polar coordinates, its equation is

$$9 = (r\cos\theta)^2 + (r\sin\theta)^2 = r^2\cos^2\theta + r^2\sin^2\theta = r^2.$$

Consequently, $r = \pm 3$, and either of these equations describes the circle. We would generally use the positive value, in this case $r = 3$.

Example 7 Find equations in polar coordinates for each of the following curves:

a) $2x + 3y = 3$;

b) $x^2 - 2x + y^2 = 0$;

c) $x^2 + y^2 = \sqrt{x^2 + y^2} - 4x$.

Solution a) If $2x + 3y = 3$, then

$$3 = 2r\cos\theta + 3r\sin\theta, \quad \text{or} \quad r = \frac{3}{2\cos\theta + 3\sin\theta}.$$

b) If $x^2 - 2x + y^2 = 0$, then

$$0 = -2x + (x^2 + y^2) = -2r\cos\theta + r^2 = r(r - 2\cos\theta).$$

Thus,

$$r = 0 \quad \text{or} \quad r = 2\cos\theta.$$

Since $r = 0$ defines the origin, and this point also satisfies $r = 2\cos\theta$ (for $\theta = \pi/2$), it follows that we need only write $r = 2\cos\theta$.

c) If $x^2 + y^2 = \sqrt{x^2 + y^2} - 4x$, then

$$r^2 = r - 4r\cos\theta = r(1 - 4\cos\theta).$$

Thus,

$$0 = r^2 - r(1 - 4\cos\theta) = r(r - 1 + 4\cos\theta),$$

from which we have

$$r = 0 \quad \text{or} \quad r = 1 - 4\cos\theta.$$

Again the origin satisfies the second of these equations, and we obtain the required equation in polar coordinates as $r = 1 - 4\cos\theta$. □

To transform equations of curves in polar coordinates to equations in Cartesian coordinates can sometimes be more difficult, principally because we have no substitution for θ. If, however, the equation involves $\cos\theta$ and/or $\sin\theta$, we can substitute from Eqs. (10.14).

Example 8 Find equations in Cartesian coordinates for each of the following curves:

a) $r = 1 + \cos\theta$;

b) $r^2 \cos 2\theta = 1$;

c) $r^2 = 9 \sin 2\theta$.

Solution a) If $r = 1 + \cos\theta$, then

$$\pm\sqrt{x^2 + y^2} = 1 + \frac{\pm x}{\sqrt{x^2 + y^2}},$$

and this equation simplifies to

$$(x^2 + y^2 - x)^2 = x^2 + y^2.$$

b) If $r^2 \cos 2\theta = 1$, we use double-angle formula (3.29b) to rewrite the equation in terms of $\cos\theta$ rather than $\cos 2\theta$:

$$\begin{aligned} 1 = r^2(2\cos^2\theta - 1) &= (x^2 + y^2)\left\{(2)\frac{x^2}{x^2 + y^2} - 1\right\} \\ &= (x^2 + y^2)\frac{x^2 - y^2}{x^2 + y^2} \\ &= x^2 - y^2. \end{aligned}$$

c) This time we use double-angle formula (3.28) to get

$$x^2 + y^2 = 18 \sin\theta\cos\theta = 18\frac{\pm y}{\sqrt{x^2 + y^2}}\frac{\pm x}{\sqrt{x^2 + y^2}},$$

or

$$(x^2 + y^2)^2 = 18xy. \;\square$$

Examples 7 and 8 illustrate that equations for some curves become much simpler when expressed in polar coordinates. It is also true that some curves are more easily sketched using a polar rather than a Cartesian representation. For instance, consider the curve $r = 1 + \cos\theta$ of Example 8. To sketch this curve we could construct a table of values (Table 10.1) and obtain the set of points shown in Fig. 10.31(a). These points could then be joined to produce the curve, and indeed the correct curve, in Fig. 10.31(b). It is called a *cardioid*.

Table 10.1

θ	$-\frac{3\pi}{4}$	$-\frac{\pi}{2}$	$-\frac{\pi}{4}$	0	$\frac{\pi}{4}$	$\frac{\pi}{2}$	$\frac{3\pi}{4}$	π
r	0.29	1	1.71	2	1.71	1	0.29	0

We note, however, that this is not the only way to join the eight points in Fig. 10.31(a). For instance, the curve in Fig. 10.31(c) also agrees with these eight points and bears little resemblance to the cardioid. It is true that this could be rectified by constructing an increasingly larger table of values, but we offer an alternative method that makes use of our building procedures for sketching curves.

We draw a θ axis horizontally and an r axis perpendicular to it. On this set of axes we sketch a graph of the function $r = f(\theta) = 1 + \cos\theta$ (Fig. 10.32). This is

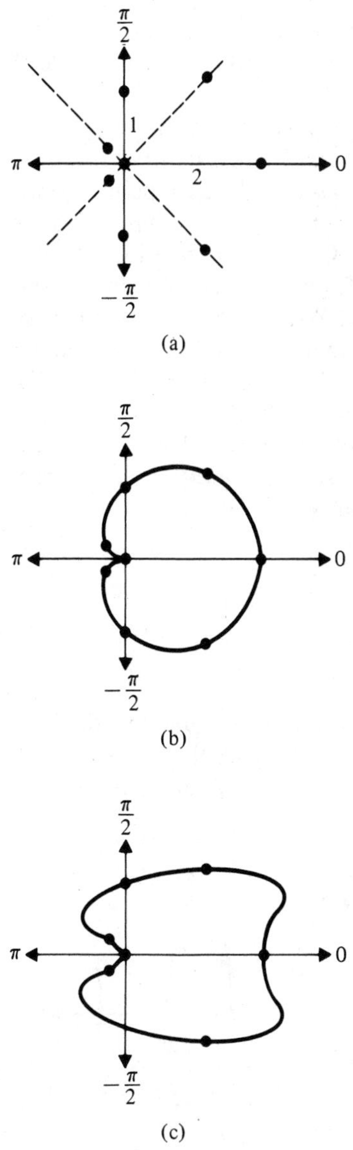

Figure 10.31

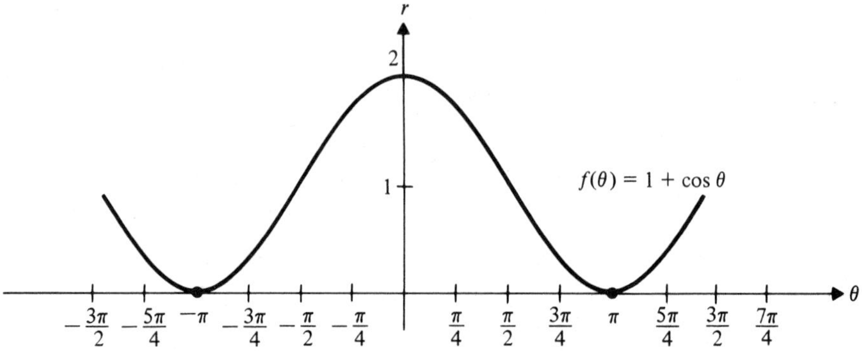

Figure 10.32

not the required curve; it is a sketch of the function $f(\theta)$, illustrating values of r for various values of θ. We might say that this sketch represents an "infinite table of values."

To sketch the cardioid $r = 1 + \cos\theta$, we now read from this sketch pairs of polar coordinates (r, θ), and interpret r as directed distance and θ as rotation. For example, we might start by selecting from this sketch the eight points shown in Fig. 10.31(a). But using Fig. 10.32 it is a simple procedure to join these points correctly. Take, for instance, the pair of points $(2, 0)$ and $(1.71, \pi/4)$. Figure 10.32 indicates that as the angle of rotation θ increases from 0 to $\pi/4$, the distance r from the origin (pole) steadily decreases from 2 to 1.71. We obtain, then, that part of Fig. 10.31(b) joining $(2, 0)$ to $(1.71, \pi/4)$. For that part of the curve joining $(1.71, \pi/4)$ and $(1, \pi/2)$, Fig. 10.32 shows that as θ increases from $\pi/4$ to $\pi/2$, r decreases steadily from 1.71 to 1. Continuation of this process quickly yields the cardioid shown in Fig. 10.31(b). Note that because $f(\theta) = 1 + \cos\theta$ is 2π-periodic, only those values in the interval $-\pi < \theta \leq \pi$ need be considered. Values outside this interval retrace previous points.

Example 9 Sketch each of the following curves:

a) $r^2 = 9 \sin 2\theta$;

b) $r = 1 - 4\cos\theta$;

c) $r = \theta, \theta \geq 0$.

Solution a) If $r^2 = 9\sin 2\theta$, then $r = \pm 3\sqrt{\sin 2\theta}$. A sketch of the function $\sin 2\theta$ is shown in Fig. 10.33(a), and from this we obtain the sketch of $f(\theta) = 3\sqrt{\sin 2\theta}$ in Fig. 10.33(b). The curve $r = 3\sqrt{\sin 2\theta}$ can now be obtained by reading

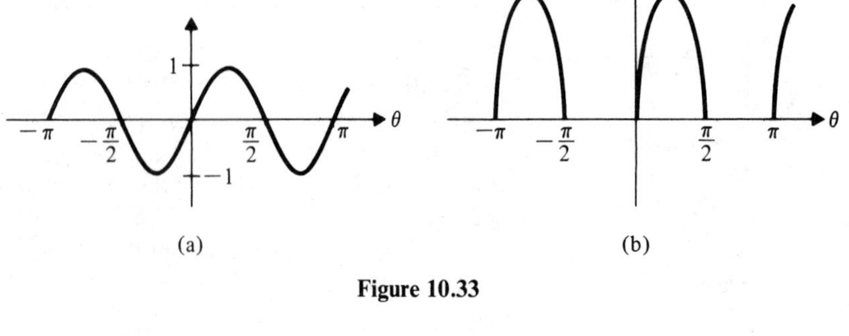

Figure 10.33

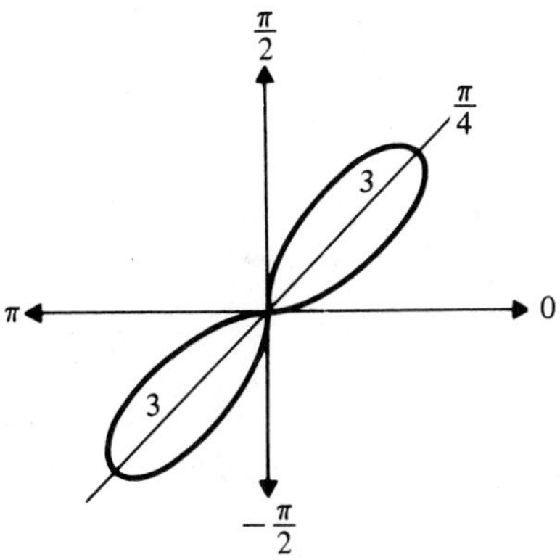

Figure 10.34

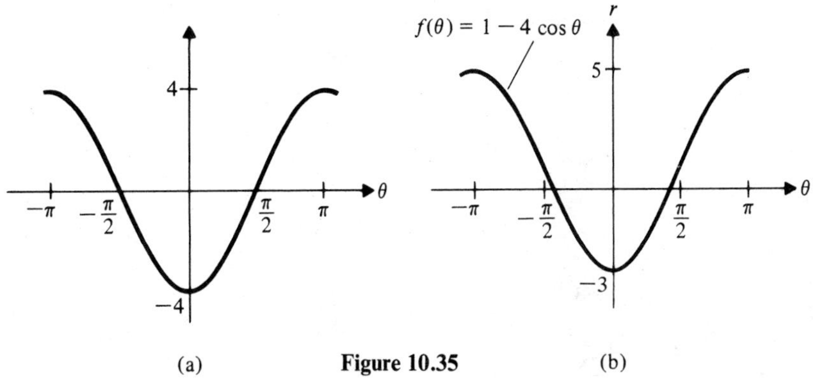

Figure 10.35

polar coordinate pairs (r, θ) from this graph, the result being shown in Fig. 10.34.

The alternative $r = -3\sqrt{\sin 2\theta}$ simply retraces the same curve. It is called a *lemniscate.*

b) A sketch of the function $-4 \cos \theta$ is shown in Fig. 10.35(a), and a vertical shift of one unit yields the sketch of $f(\theta) = 1 - 4\cos\theta$ shown in Fig. 10.35(b). This leads to the curve $r = 1 - 4\cos\theta$ shown in Fig. 10.36, called a *limaçon.* Note that at the pole the radial lines $\theta = \mathrm{Cos}^{-1}\left(\frac{1}{4}\right)$ and $\theta = \pi - \mathrm{Cos}^{-1}\left(\frac{1}{4}\right)$ are tangent to the limaçon.

c) The sketch of $f(\theta) = \theta$ in Fig. 10.37 leads to the spiral in Fig. 10.38. □

To find the points of intersection of two curves whose equations are given in Cartesian coordinates, we solve the two equations simultaneously for all real solutions. Each such solution represents a distinct point of intersection. For curves whose equations are given in polar coordinates, the situation is somewhat different because we have multiple names for points. To find the points of intersection when the curves are given in polar coordinates, we again solve the equations simultaneously for all real solutions. Each such solution represents a point of intersection, but because points have many sets of polar coordinates, some of these solutions may represent the same point. In addition, it may also happen that one set of polar coordinates for a point of intersection satisfies one equation, whereas a different set satisfies the other equation. Particularly troublesome in this respect is the origin, which has so many sets of polar coordinates. The best way to handle these difficulties is to sketch the two curves.

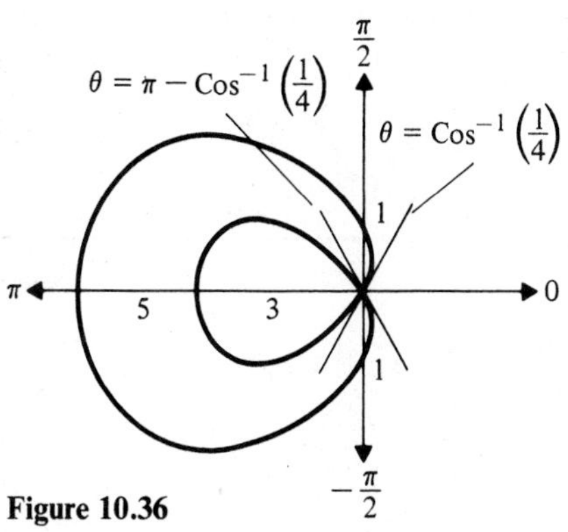

Figure 10.36

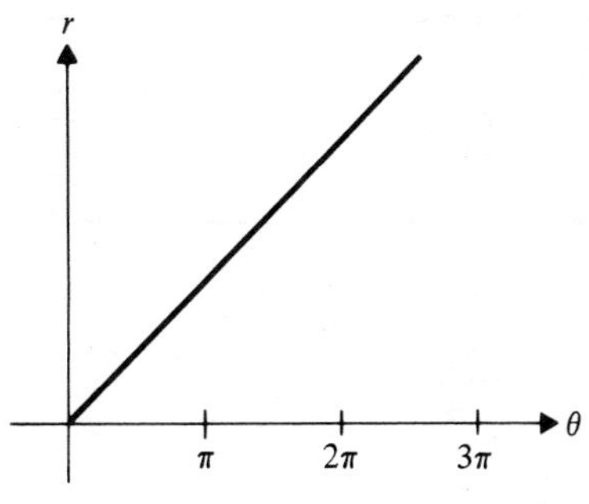

Figure 10.37

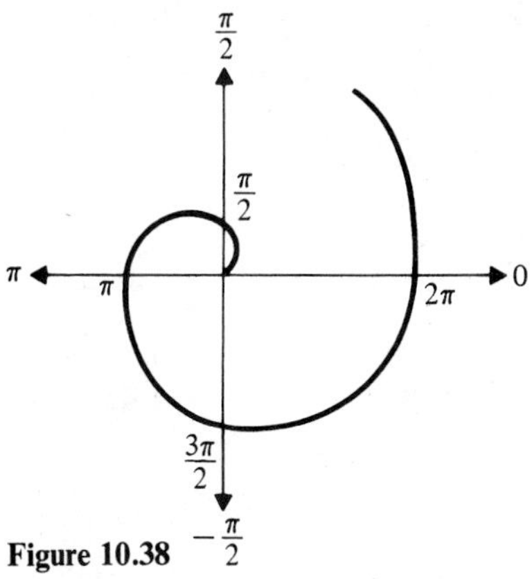

Figure 10.38

Example 10 Find all points of intersection for each of the following pairs of curves:

a) $r = \sin\theta$, $r = 1 - \sin\theta$;

b) $r = 1$, $r = 2\sin 2\theta$.

Solution a) If we set $\sin\theta = 1 - \sin\theta$, then $\sin\theta = \frac{1}{2}$. All solutions of this equation are defined by

$$\theta = \begin{cases} \dfrac{\pi}{6} + 2n\pi, \\ \dfrac{5\pi}{6} + 2n\pi, \end{cases} \qquad n = \text{an integer}.$$

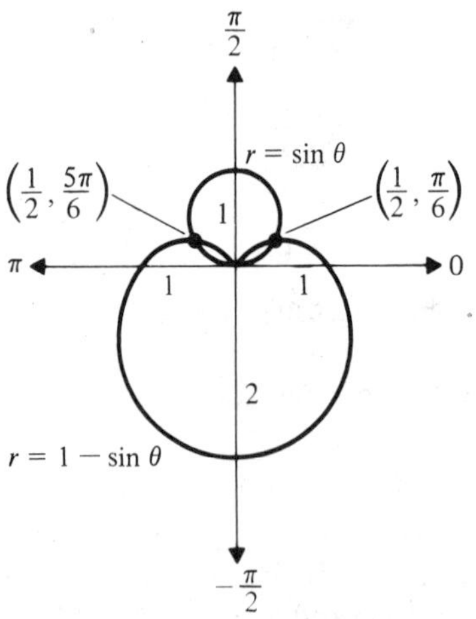

Figure 10.39

Sketches of the curves in Fig. 10.39 indicate that these give the points of intersection $(\frac{1}{2}, \pi/6)$ and $(\frac{1}{2}, 5\pi/6)$. The figure also illustrates that the origin is a point of intersection of these two curves. We did not obtain this result from the above discussion because different values of θ yield $r = 0$ in $r = \sin\theta$ and $r = 1 - \sin\theta$. For $r = \sin\theta$ to give $r = 0$, we must choose $\theta = n\pi$, whereas to obtain $r = 0$ for $r = 1 - \sin\theta$, we must set $\theta = (2n + 1)\pi/2$. Thus both curves pass through the pole, but the pole cannot be obtained by simultaneously solving the equations of the curves.

b) For points of intersection we set $1 = 2\sin 2\theta$, or $\sin 2\theta = \frac{1}{2}$. All solutions of this equation are given by

$$2\theta = \begin{cases} \dfrac{\pi}{6} + 2n\pi, \\ \dfrac{5\pi}{6} + 2n\pi, \end{cases} \quad \text{or} \quad \theta = \begin{cases} \dfrac{\pi}{12} + n\pi, \\ \dfrac{5\pi}{12} + n\pi \end{cases} \qquad n = \text{an integer}.$$

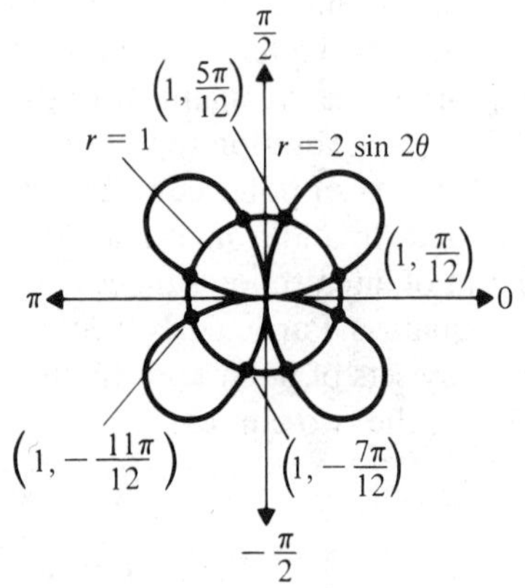

Figure 10.40

Sketches of the curves in Fig. 10.40 indicate that these solutions yield four points of intersection in the first and third quadrants:

$$(1, \pi/12), \qquad (1, 5\pi/12), \qquad (1, -7\pi/12), \qquad (1, -11\pi/12).$$

But clearly there are four other points of intersection in the second and fourth quadrants, which by symmetry are

$$(1, 7\pi/12), \qquad (1, 11\pi/12), \qquad (1, -\pi/12), \qquad (1, -5\pi/12).$$

These could be obtained by taking the alternative equation $r = -1$ for the circle and solving it simultaneously with $r = 2\sin 2\theta$. Certainly this is not an obvious procedure, and it strongly supports our recommendation that any discussion on points of intersection should always be accompanied by sketches of the curves. □

EXERCISES 10.3

In each of Exercises 1–10, find an equation for the curve in polar coordinates, and draw a sketch.

1. $x + 2y = 5$
2. $y = -x$
3. $x^2 + y^2 = 3$
4. $x^2 - 2x + y^2 - 2y + 1 = 0$
5. $y = 4x^2$
6. $x^2 + 2y^2 = 3$
7. $x^2 + y^2 = x$
8. $x^2 + y^2 = \sqrt{x^2 + y^2} -$
9. $(x^2 + y^2)^2 = x$
10. $y = 1/x^2$

In each of Exercises 11–20, find the equation of the curve in Cartesian coordinates, and draw a sketch.

11. $r = 5$

12. $\theta = 1$

13. $r = 3 \sin \theta$

14. $r^2 = 4 \sin 2\theta$

15. $r = 3 + 3 \sin \theta$

16. $r = 2 \sin 2\theta$

17. $r^2 = -4 \cos 2\theta$

18. $r = 3 - 4 \cos \theta$

19. $r = 5 \csc \theta$

20. $r = \cot^2 \theta \csc \theta$

21. Sketch each of the following three curves:

a) $r = 2 + 2 \sin \theta$;

b) $r = 2 + 4 \sin \theta$;

c) $r = 4 + 2 \sin \theta$.

In Exercises 22–25, find all points of intersection for the given curves.

22. $r = 2$
$r^2 = 8 \cos 2\theta$

23. $r = \cos \theta$,
$r = 1 + \cos \theta$

24. $r = 1 + \cos \theta$,
$r = 2 - 2 \cos \theta$

25. $r = 1$,
$r = 2 \cos 2\theta$

26. Show that if $f(\theta)$ is an even function, then the curve $r = f(\theta)$ is symmetric about the $\theta = 0$ line (or x-axis). Illustrate with two examples.

Sketch the given curve in Exercises 27–34.

27. $r = \sin 3\theta$

28. $r = \cos 2\theta$

29. $r = \sin 4\theta$

30. $r^2 = \theta$

31. $r = e^{\theta}$

32. $r = 2 \sin (\theta/2)$

33. $r = -2 \cos (\theta/2)$

34. $r = 1 + \cos (\theta + \pi/6)$

35. a) The electrostatic charge distribution consisting of a charge $q > 0$ at the point with polar coordinates $(s, 0)$ and a charge $-q$ at (s, π) is called a *dipole*. The lines of force for the dipole are defined by the equation $r = A \sin^2 \theta$, where each value of the constant $A > 0$ defines a particular line of force. Sketch lines of force for $A = 1, 2$, and 3.

b) The equipotential lines for the dipole are defined by $r^2 = B \cos \theta$, where $B > 0$ is a constant. Sketch equipotential lines for $B = 1, 2$, and 3.

36. Sketch the bifolium $(x^2 + y^2)^2 = x^2 y$.

37. Any curve of the form $r^2 = a^2 \cos 2\theta$ or $r^2 = a^2 \sin 2\theta$ ($a > 0$ constant) is called a lemniscate.

a) Sketch all such curves.

b) Find equations for the lemniscates in Cartesian coordinates.

38. Any curve of the form $r = a(1 \pm \cos \theta)$ or $r = a(1 \pm \sin \theta)$ ($a > 0$ constant) is called a cardioid.

a) Sketch all such curves.

b) Find equations for the cardioids in Cartesian coordinates.

39. Any curve of the form $r = b \pm a \cos \theta$ or $r = b \pm a \sin \theta$ ($a, b > 0$ constants) is called a limaçon.

a) Sketch these curves in the three cases that $a < b$, $a = b$, and $a > b$.

b) Find equations for the limaçons in Cartesian coordinates.

c) Compare limaçons when $a = b$ with cardioids as in Exercise 38.

40. Any curve of the form $r = a \sin (n\theta)$ or $r = a \cos (n\theta)$ ($a > 0$ constant, and $n > 0$ an integer) is called a *rose*. Show that if n is odd, the rose has n "petals," and if n is even, it has $2n$ petals.

10.4 AREAS IN POLAR COORDINATES

In Section 7.1 we used definite integrals to find areas bounded by curves whose equations could be conveniently expressed in Cartesian coordinates. For the area shown in Fig. 10.41, we write

$$\text{Area} = \int_a^b [f(x) - g(x)]\, dx,$$

where $[f(x) - g(x)]\, dx$ is interpreted as the area of a representative rectangle, and the integral adds over all such rectangles from $x = a$ to $x = b$.

In this section we indicate how definite integrals in polar coordinates can be used to calculate areas bounded by curves. To do this we require the formula

$$\tfrac{1}{2} r^2 (\theta_2 - \theta_1) \qquad (10.20)$$

for the area of the shaded sector of the circle shown in Fig. 10.42. This result is

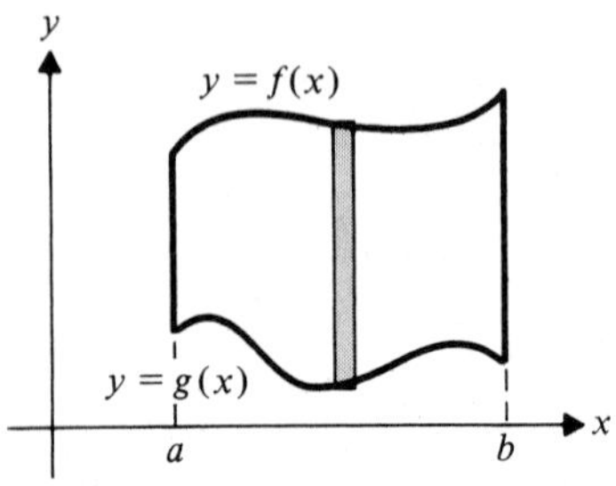

Figure 10.41

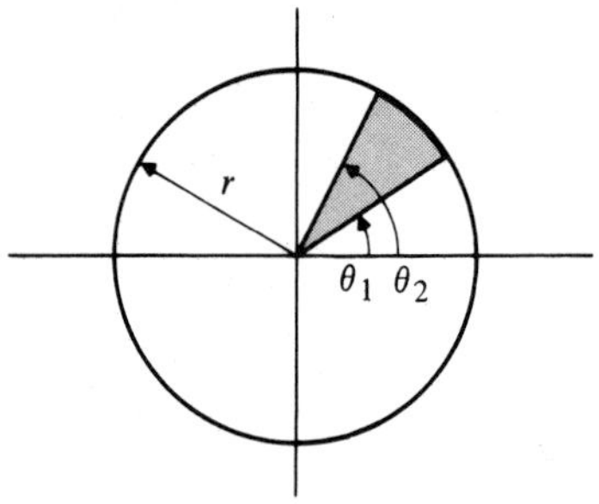

Figure 10.42

an immediate consequence of the fact that the area of the sector is the fractional part $(\theta_2 - \theta_1)/(2\pi)$ of the area πr^2 of the circle.

Consider finding the area shown in Fig. 10.43(a) bounded by the radial lines $\theta = \alpha$ and $\theta = \beta$ and the curve $r = f(\theta)$. We subdivide the area by means of $n + 1$ radial lines $\theta = \theta_i$, where

$$\alpha = \theta_0 < \theta_1 < \theta_2 < \cdots < \theta_{n-1} < \theta_n = \beta.$$

On that part of the curve $r = f(\theta)$ between $\theta = \theta_{i-1}$ and $\theta = \theta_i$, we pick any point with polar coordinates $(f(\theta_i^*), \theta_i^*)$ (Fig. 10.43b). If between the lines $\theta = \theta_{i-1}$ and $\theta = \theta_i$ we draw the arc of a circle with center at the pole and radius $f(\theta_i^*)$, a sector is formed with area

$$\Delta A_i = \tfrac{1}{2}[f(\theta_i^*)]^2 \Delta\theta_i,$$

where $\Delta\theta_i = \theta_i - \theta_{i-1}$. Since this sector approximates that part of the required area between the radial lines $\theta = \theta_i$ and $\theta = \theta_{i-1}$, we can say that an approximation to the required area is

$$\sum_{i=1}^{n} \Delta A_i = \sum_{i=1}^{n} \frac{1}{2} [f(\theta_i^*)]^2 \Delta\theta_i.$$

By increasing the number of sectors indefinitely, and at the same time requiring each of the $\Delta\theta_i$ to approach zero, we obtain a better and better approximation, and in the limit

$$\text{Area} = \lim_{n\to\infty} \sum_{i=1}^{n} \frac{1}{2} [f(\theta_i^*)]^2 \Delta\theta_i.$$

But this limit is precisely the definition of the definite integral of the function $\frac{1}{2}[f(\theta)]^2$ with respect to θ from $\theta = \alpha$ to $\theta = \beta$, and we write therefore that

$$\text{Area} = \int_{\alpha}^{\beta} \frac{1}{2} [f(\theta)]^2 \, d\theta. \tag{10.21}$$

In order to develop this formula in any given problem, rather than memorize it, we use the procedure adopted for definite integrals in Cartesian coordinates. We draw at angle θ a representative sector of angular width $d\theta$ and radius r, as

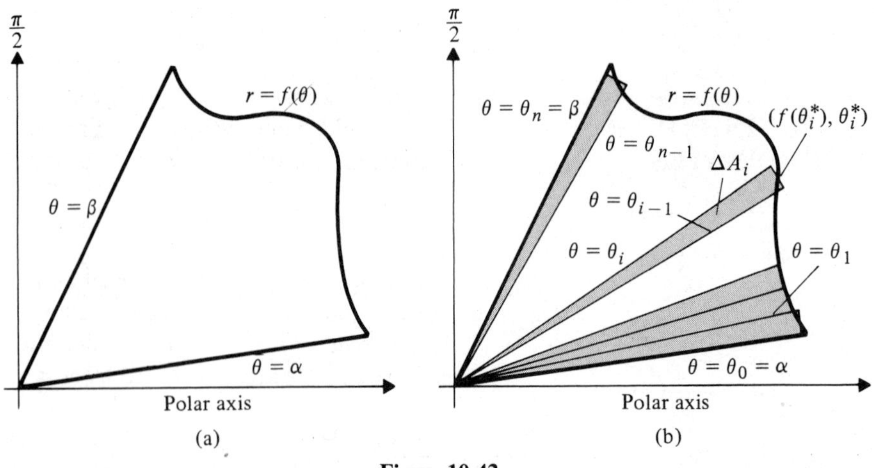

Figure 10.43

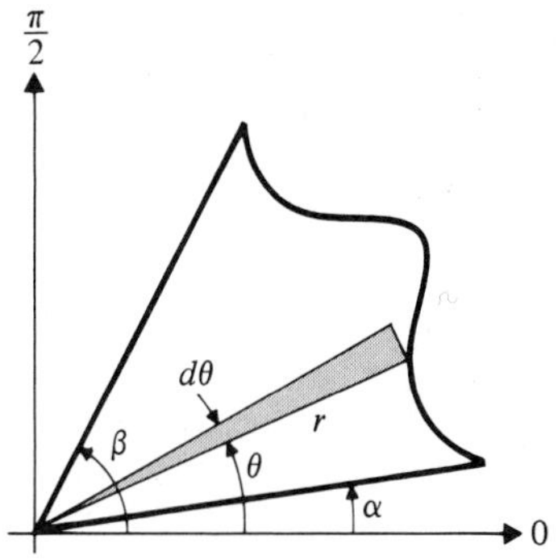

Figure 10.44

shown in Fig. 10.44. The area of this sector is

$$\tfrac{1}{2}r^2\, d\theta,$$

where r depends on θ, $r = f(\theta)$. If areas of all such sectors from angle α to angle β are added together, and the limit is taken as their widths approach zero, the required area is obtained. But this is the process defined by the definite integral, and we therefore write Eq. (10.21) for the area. Definite integral (10.21) is certain to exist if we assume that $f(\theta)$ is continuous for $\alpha \le \theta \le \beta$.

For example, consider finding the area inside the cardioid $r = 1 + \sin\theta$ (Fig. 10.45). The area of the representative sector shown is

$$\tfrac{1}{2}r^2\, d\theta = \tfrac{1}{2}(1 + \sin\theta)^2\, d\theta,$$

and we must add over all sectors interior to the cardioid. Since areas on either side of the y-axis are identical, we calculate the area to the right and double the result. To find the area to the right of the y-axis, we must identify angular positions of the first and last sectors. The first sector is at the pole, and the equation of the cardioid indicates that $r = 0$ when $\sin\theta = -1$. But which of the values $\theta = -\pi/2 + 2n\pi$ shall we choose? Similarly, the last sector occurs when $r = 2$, in which case θ could be any of $\theta = \pi/2 + 2n\pi$. Again, which shall we choose? If we choose $\alpha = -\pi/2$ and $\beta = \pi/2$, then all values of θ in the interval $-\pi/2 \le \theta \le \pi/2$ yield points on the right half of the cardioid with no duplications. Consequently,

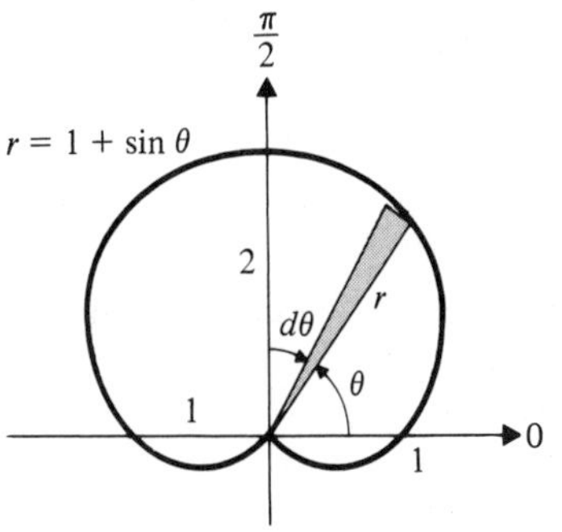

Figure 10.45

$$\text{Area} = 2\int_{-\pi/2}^{\pi/2} \frac{1}{2}(1 + \sin\theta)^2\, d\theta = \int_{-\pi/2}^{\pi/2} (1 + 2\sin\theta + \sin^2\theta)\, d\theta$$

$$= \int_{-\pi/2}^{\pi/2} \left\{1 + 2\sin\theta + \frac{1 - \cos 2\theta}{2}\right\} d\theta$$

$$= \left\{\frac{3\theta}{2} - 2\cos\theta - \frac{\sin 2\theta}{4}\right\}_{-\pi/2}^{\pi/2} = \frac{3\pi}{2}.$$

Example 11 Find the area common to the circles $x^2 + y^2 = 4$ and $x^2 + y^2 = 4x$.

Solution If A_s is the shaded area shown in Fig. 10.46, the area common to the circles is the area of either circle less twice A_s:

$$\text{Area} = \pi(2)^2 - 2A_s.$$

Equations for the circles in polar coordinates are $r = 2$ and $r = 4\cos\theta$, and they intersect in the points with polar coordinates $(2, \pm\pi/3)$. The area of the representative element is the difference in the areas of two sectors:

$$\tfrac{1}{2}(4\cos\theta)^2\, d\theta - \tfrac{1}{2}(2)^2\, d\theta = 2(4\cos^2\theta - 1)\, d\theta.$$

Since all sectors in A_s can be identified by values of θ between 0 and $\pi/3$, it follows that the required area is

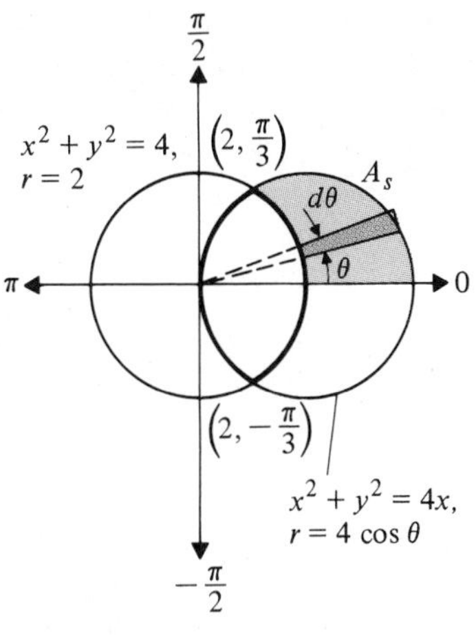

Figure 10.46

$$\text{Area} = 4\pi - 2\int_0^{\pi/3} 2(4\cos^2\theta - 1)\, d\theta = 4\pi - 4\int_0^{\pi/3} \{2(1 + \cos 2\theta) - 1\}\, d\theta$$

$$= 4\pi - 4\{\theta + \sin 2\theta\}_0^{\pi/3} = 4\pi - 4\left\{\frac{\pi}{3} + \frac{\sqrt{3}}{2}\right\} = \frac{8\pi}{3} - 2\sqrt{3}. \square$$

This area can also be found by using a definite integral with respect to x or y, but the integration requires a trigonometric substitution (see Exercise 23 in Section 9.5).

EXERCISES 10.4

In Exercises 1–10, find the area enclosed by the given curve.

1. $r = 3 \sin \theta$

2. $r = -6 \cos \theta$

3. $r = 2 \sin 2\theta$

4. $r^2 = 2 \sin 2\theta$

5. $r^2 = -\cos \theta$

6. $r = 2 - 2 \cos \theta$

7. $r = 4 - 4 \cos \theta$

8. $r = 4 - 2 \cos \theta$

9. $r = \sin 3\theta$

10. $r = 2 (\cos \theta + \sin \theta)$

In Exercises 11–21, find the indicated area.

11. Outside $r = 3$ but inside $r = 6 \sin \theta$

12. Inside both $r = 1$ and $r = 1 - \sin \theta$

13. Inside $r = 2 \sin 2\theta$ but outside $r = 1$

14. Inside both $r = 2 + 2 \cos \theta$ and $r = 2 - 2 \cos \theta$

15. Inside both $r = \sin \theta$ and $r = \cos \theta$

16. Inside both $r = \cos \theta$ and $r = 1 - \cos \theta$

17. Inside $r = |1 - 4 \cos \theta|$

18. Inside $r = 4 + 3 \sin \theta$ and outside $r = 2$

19. Inside the smaller loop of $r = 1 - 4 \cos \theta$

20. Inside the bifolium $r = \sin \theta \cos^2 \theta$

21. Bounded by $\theta = 0$ and $r = \theta, 0 \leq \theta \leq \pi$

***22.** a) Show that in polar coordinates the *strophoid*

$$y^2 = (x^2)\frac{a - x}{a + x}, \qquad a > 0 \text{ a constant,}$$

takes the form

$$r = a \cos 2\theta \sec \theta.$$

b) Sketch the curve and find the area inside its loop.

10.5 CONIC SECTIONS IN POLAR COORDINATES

In Section 10.1 we defined parabolas using a focus and directrix and ellipses and hyperbolas using two foci. In this section we show that all three conics can be defined using a focus and directrix, and that in polar coordinates one equation represents all three conics.

Let F be a fixed point (called the *focus*) and l be a fixed line (called the *directrix*) that does not pass through F (Fig. 10.47a). We propose to find the equation of the curve traced out by a point that moves so that its undirected distances from F and l always remain in a constant ratio ε (called the *eccentricity*). To do this we set up polar coordinates with F as pole and polar axis directed away from l but perpendicular to l (Fig. 10.47b). Suppose that (r, θ) are the polar coordinates for any point P on the required curve. If we choose only positive values for r, the fact that the ratio of the distances from F and l to P is equal to ε is expressed as

$$\frac{r}{d + r \cos \theta} = \varepsilon. \qquad (10.22)$$

When this equation is solved for r, we have

$$r = \frac{\varepsilon d}{1 - \varepsilon \cos \theta}, \qquad (10.23)$$

and this is the equation of the curve traced out by the point. Certainly this curve should be a parabola when $\varepsilon = 1$. We now verify this, and show that the curve is an ellipse when $\varepsilon < 1$ and a hyperbola when $\varepsilon > 1$. To do this we transform the

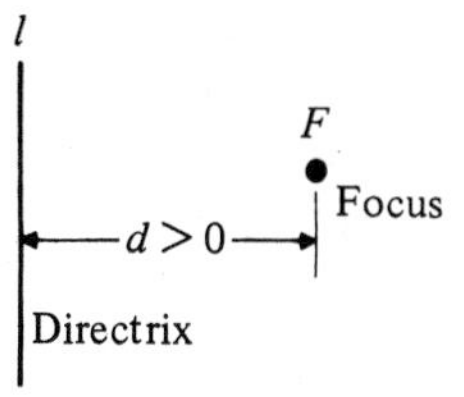

(a)

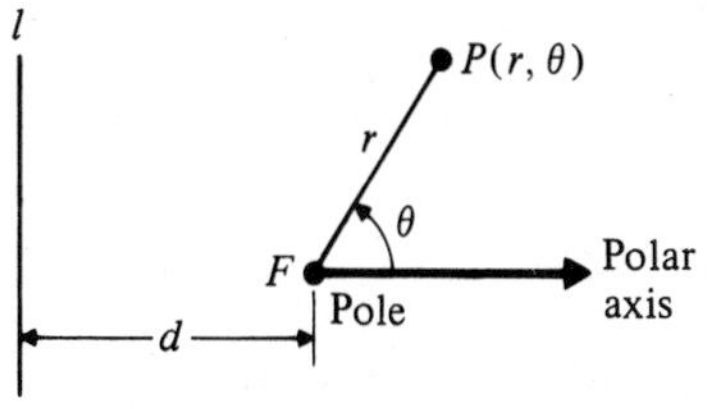

(b)

Figure 10.47

equation into the usual Cartesian coordinates $x = r\cos\theta$ and $y = r\sin\theta$. From Eqs. (10.13) and (10.14), we obtain

$$\sqrt{x^2 + y^2} = \frac{\varepsilon d}{1 - \dfrac{\varepsilon(x)}{\sqrt{x^2 + y^2}}},$$

and this equation simplifies to

$$x^2 + y^2 = \varepsilon^2(d + x)^2.$$

When $\varepsilon = 1$, the x^2-terms cancel, and the equation reduces to that for a parabola:

$$x = \frac{1}{2d}(y^2 - d^2).$$

When $\varepsilon \neq 1$, we write

$$x^2 + y^2 = \varepsilon^2(d^2 + 2dx + x^2),$$

or

$$(1 - \varepsilon^2)x^2 - 2d\varepsilon^2 x + y^2 = \varepsilon^2 d^2.$$

When $\varepsilon < 1$, we divide by $1 - \varepsilon^2$,

$$x^2 - \frac{2d\varepsilon^2}{1 - \varepsilon^2}x + \frac{y^2}{1 - \varepsilon^2} = \frac{\varepsilon^2 d^2}{1 - \varepsilon^2},$$

and complete the square on the x-terms:

$$\left(x - \frac{d\varepsilon^2}{1 - \varepsilon^2}\right)^2 + \frac{y^2}{1 - \varepsilon^2} = \frac{\varepsilon^2 d^2}{1 - \varepsilon^2} + \frac{d^2\varepsilon^4}{(1 - \varepsilon^2)^2} = \left(\frac{\varepsilon d}{1 - \varepsilon^2}\right)^2.$$

According to (10.6), this is the equation for an ellipse with center at position $x = d\varepsilon^2/(1 - \varepsilon^2)$ on the x-axis. Since one focus is at the origin, it follows that the other must also be on the x-axis at $x = 2d\varepsilon^2/(1 - \varepsilon^2)$ (Fig. 10.48).

A similar calculation shows that when $\varepsilon > 1$, the hyperbola

$$\left(x + \frac{d\varepsilon^2}{\varepsilon^2 - 1}\right)^2 - \frac{y^2}{\varepsilon^2 - 1} = \left(\frac{\varepsilon d}{\varepsilon^2 - 1}\right)^2$$

(Fig. 10.49) is obtained. (To obtain the left half of the hyperbola we must permit r in (10.23) to be negative as well as positive.)

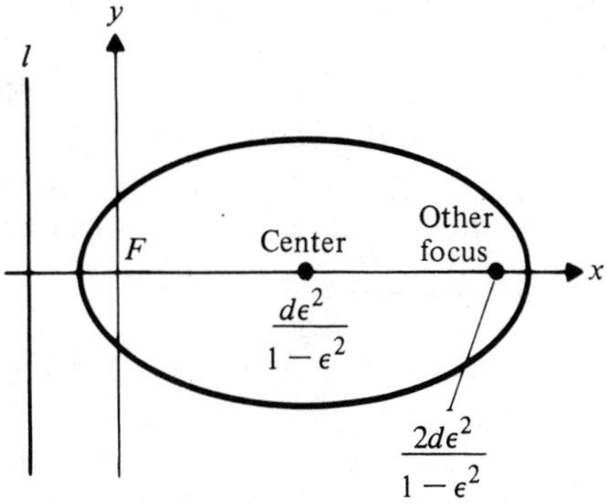

Figure 10.48

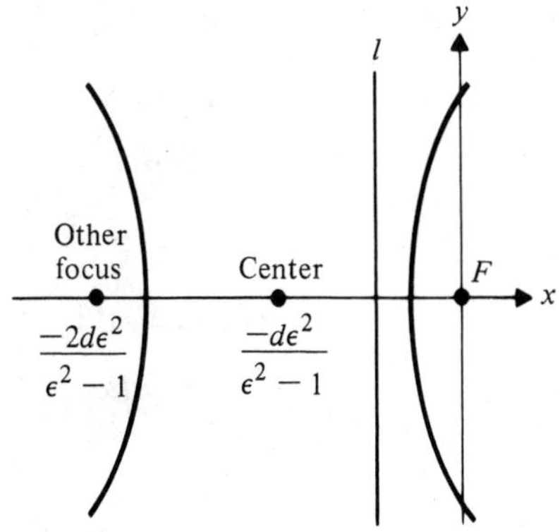

Figure 10.49

We have shown that Eq. (10.23) defines an ellipse when $0 < \varepsilon < 1$, a parabola when $\varepsilon = 1$, and a hyperbola when $\varepsilon > 1$, and this provides a unifying approach to conic sections. Instead of approaching parabolas with a focus and a directrix and ellipses and hyperbolas with two foci, we now know that it is possible to approach all three using a focus and a directrix. In fact, many authors use this unified approach to introduce conic sections.

It is clear that Eq. (10.23) can only yield a parabola that opens to the right and has its focus on the x-axis and an ellipse and hyperbola with foci on the x-axis. To obtain parabolas that open to the left, or up, or down, and ellipses and hyperbolas with foci on the y-axis, we must change the position of the directrix. For instance, conic sections with directrices as shown in Figs. 10.50 (a), (b), and (c) have equations

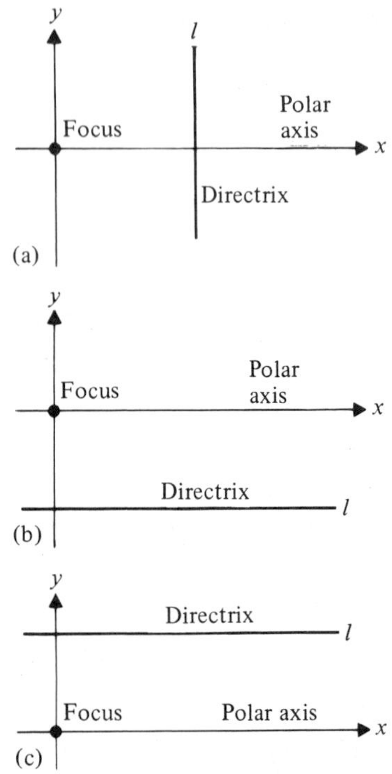

Figure 10.50

$$r = \frac{\varepsilon d}{1 + \varepsilon \cos \theta}, \qquad r = \frac{\varepsilon d}{1 - \varepsilon \sin \theta}, \qquad r = \frac{\varepsilon d}{1 + \varepsilon \sin \theta}, \tag{10.24}$$

respectively. The first leads to parabolas opening to the left; the second and third yield parabolas that open upward and downward and ellipses and hyperbolas with foci on the y-axis. Note, however, that in each case one focus of the conic section is chosen as the pole. In other words, the simplicity of Eqs. (10.23) and (10.24) to describe conic sections is a direct consequence of the fact that the pole is at a focus, and the directrix is either parallel or perpendicular to the polar axis.

Example 12 Sketch the curve $r = 15/(3 + 2 \cos \theta)$.

Solution If we write the equation in form (10.24),

$$r = \frac{5}{1 + (\frac{2}{3}) \cos \theta},$$

the eccentricity $\varepsilon = \frac{2}{3}$ indicates that the curve is an ellipse. Both foci lie on the x-axis, and one is at the origin. The ends of the major axis occur when $\theta = 0$ and π, and for these values $r = 3$ and 15 (Fig. 10.51). It now follows that the center of the ellipse is at $x = -6$, and its remaining focus is at $x = -12$. If $b > 0$ denotes the maximum y-value on the ellipse (when $x = -6$), then $b^2 = a^2 - c^2 = (9)^2 - (6)^2 = 45$. Consequently, $b = 3\sqrt{5}$, and the ellipse is as shown in Fig. 10.51. This information now permits us to write the equation of the ellipse in Cartesian coordinates:

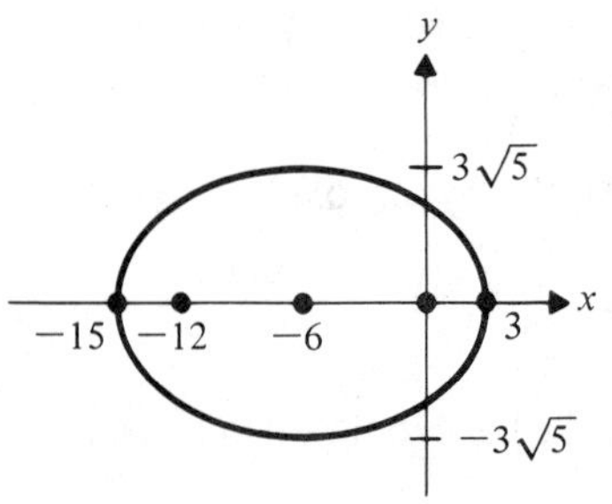

Figure 10.51

$$\frac{(x + 6)^2}{81} + \frac{y^2}{45} = 1. \ \square$$

An alternative solution is to substitute $r = \pm\sqrt{x^2 + y^2}$ and $\cos \theta = \pm x/\sqrt{x^2 + y^2}$ into the polar equation for the conic, simplify the equation to $(x + 6)^2/81 + y^2/45 = 1$, and then sketch the ellipse from this equation. We illustrate this method in the following example.

Example 13 Sketch the curve $r = 2/(3 - 4 \sin \theta)$.

Solution If we set $r = \pm\sqrt{x^2 + y^2}$ and $\sin \theta = \pm y/\sqrt{x^2 + y^2}$, then

$$\pm\sqrt{x^2 + y^2} = \frac{2}{3 - 4\dfrac{\pm y}{\sqrt{x^2 + y^2}}} = \frac{2\sqrt{x^2 + y^2}}{3\sqrt{x^2 + y^2} \mp 4y}.$$

Division by $\sqrt{x^2 + y^2}$ and cross-multiplication gives

$$3\sqrt{x^2 + y^2} \mp 4y = \pm 2,$$

or

$$3\sqrt{x^2 + y^2} = \pm(4y + 2).$$

If we now square, we have

$$9(x^2 + y^2) = 4 + 16y + 16y^2,$$

which simplifies to

$$\frac{(y + \frac{8}{7})^2}{\frac{36}{49}} - \frac{x^2}{\frac{4}{7}} = 1.$$

This equation describes a hyperbola with center at $(0, -\frac{8}{7})$ and y-intercepts equal to $-\frac{8}{7} \pm \frac{6}{7} = -\frac{2}{7}, -2$. Its asymptotes can be obtained by solving the equation for y:

$$y = -\frac{8}{7} \pm \frac{6}{7}\sqrt{1 + \frac{7x^2}{4}}.$$

The asymptotes are

$$y = -\frac{8}{7} \pm \frac{6}{7}\sqrt{\frac{7x^2}{4}} = -\frac{8}{7} \pm \frac{3x}{\sqrt{7}},$$

and we can draw the sketch shown in Fig. 10.52. □

Example 14 Find a polar representation for the ellipse

$$\frac{(x - 1)^2}{4} + \frac{y^2}{9} = 1.$$

Solution To find a polar representation for the ellipse (Fig. 10.53), we could use the usual polar coordinates defined by $x = r\cos\theta$ and $y = r\sin\theta$, but the resulting equation would not be a simple one. (Try it.) We know that a simple polar representation must result if the pole is chosen at a focus of the ellipse and the polar axis either parallel or perpendicular to the directrix. For this ellipse, foci are on the line $x = 1$ at distances $\pm c = \pm\sqrt{9 - 4} = \pm\sqrt{5}$ from the x-axis. Let us choose the pole at position $(1, \sqrt{5})$ and the polar axis parallel to the x-axis (and therefore parallel to the directrix). According to Eq. (10.17), the polar and Cartesian coordinates are related by $x = 1 + r\cos\theta$ and $y = \sqrt{5} + r\sin\theta$. If we substitute

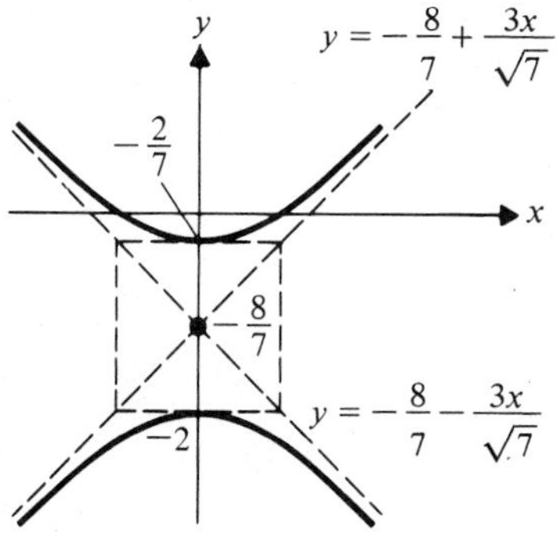

Figure 10.52

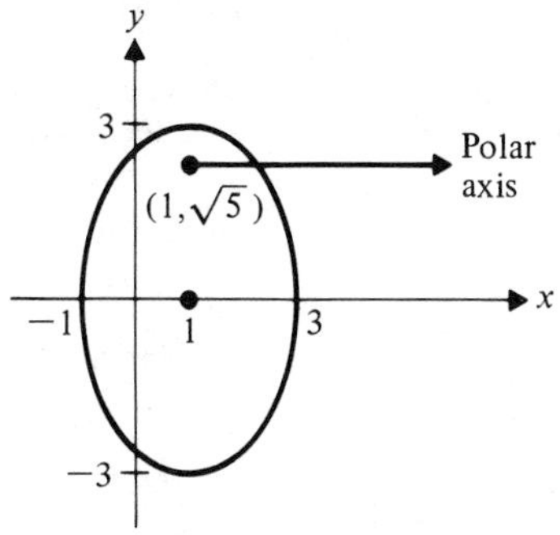

Figure 10.53

these into the equation for the ellipse, we have

$$\frac{r^2\cos^2\theta}{4}+\frac{(\sqrt{5}+r\sin\theta)^2}{9}=1,$$

and this equation can be rewritten as a quadratic equation in r:

$$r^2(9\cos^2\theta+4\sin^2\theta)+r(8\sqrt{5}\sin\theta)-16=0.$$

Solutions for r are

$$\begin{aligned} r&=\frac{-8\sqrt{5}\sin\theta\pm\sqrt{320\sin^2\theta+64(9\cos^2\theta+4\sin^2\theta)}}{2(9\cos^2\theta+4\sin^2\theta)}\\ &=\frac{-8\sqrt{5}\sin\theta\pm\sqrt{576(\cos^2\theta+\sin^2\theta)}}{2(9\cos^2\theta+4\sin^2\theta)}\\ &=\frac{4(\pm 3-\sqrt{5}\sin\theta)}{9\cos^2\theta+4\sin^2\theta}.\end{aligned}$$

If we choose $+3$, rather than -3, in the numerator (which corresponds to the solution with positive r), we obtain

$$\begin{aligned} r&=\frac{4(3-\sqrt{5}\sin\theta)}{9(1-\sin^2\theta)+4\sin^2\theta}=\frac{4(3-\sqrt{5}\sin\theta)}{9-5\sin^2\theta}\\ &=\frac{4(3-\sqrt{5}\sin\theta)}{(3-\sqrt{5}\sin\theta)(3+\sqrt{5}\sin\theta)}=\frac{4}{3+\sqrt{5}\sin\theta}.\ \square\end{aligned}$$

EXERCISES 10.5

In each of Exercises 1–10, sketch the conic section.

1. $r=\dfrac{3}{1+\cos\theta}$

2. $r=\dfrac{16}{3+5\cos\theta}$

3. $r=\dfrac{4}{3-3\sin\theta}$

4. $r=\dfrac{16}{5+3\cos\theta}$

5. $r=\dfrac{4}{3-4\sin\theta}$

6. $r=\dfrac{4}{4-3\sin\theta}$

7. $r=\dfrac{1}{2-2\cos\theta}$

8. $r=\dfrac{1}{2+\sin\theta}$

9. $r=\dfrac{\sec\theta}{3+6\sec\theta}$

10. $r=\dfrac{4\csc\theta}{7\csc\theta-2}$

In Exercises 11–16, find the Cartesian equation for the given curve.

11. $r=\dfrac{3}{1-\sin\theta}$

12. $r=\dfrac{1}{3+\cos\theta}$

13. $r=\dfrac{1}{1+2\cos\theta}$

14. $r=\dfrac{2}{1-3\cos\theta}$

15. $r=\dfrac{4}{6-3\sin\theta}$

16. $r=\dfrac{4}{5+5\cos\theta}$

In Exercises 17–20, find a polar equation in one of the four forms (10.23) or (10.24) for the given conic.

17. $x^2-16y^2=16$

18. $4x^2+9y^2=36$

19. $(x-1)^2+\dfrac{(y+1)^2}{4}=1$

20. $x^2-9(y-2)^2=9$

21. a) Show that the eccentricity ε of an ellipse or hyperbola is always equal to the distance from its center to either focus, divided by half the length of the major or transverse axis.

b) Discuss the eccentricities of the ellipses in Figs. 10.9 and 10.10.

22. A circle has been described as a degenerate ellipse in the sense that its two foci coincide. What happens to the eccentricity of an ellipse if the distance between its foci approaches zero? (*Hint*: See Exercise 21.)

10.6 PARAMETRIC EQUATIONS

In Section 1.5 we defined a curve in Cartesian coordinates explicitly by equations $y = f(x)$ and $x = g(y)$ and implicitly by $F(x, y) = 0$. A third method is described as follows.

▷ **Definition 10.4** A curve is said to be defined parametrically if it is given in the form

$$x = x(t), \qquad y = y(t), \qquad \alpha \le t \le \beta. \quad \triangleleft \tag{10.25}$$

Variable t is called a parameter and is simply a connecting link between x and y. Each possible value of t is substituted into Eqs. (10.25), and the pair $(x, y) = (x(t), y(t))$ is used as a point on the curve.

Example 15 Sketch the curve defined parametrically by $x = 3 - t$, $y = t^2 - 2$.

Solution We could construct a table of values (Table 10.2) by substituting various values of t and calculating corresponding values of x and y. This would lead to the points in Fig. 10.54(a), which we would then join as shown in Fig. 10.54(b). Alternatively, if we solve the first equation for $t = 3 - x$, and substitute into the second, we have

$$y = (3 - x)^2 - 2 = x^2 - 6x + 7,$$

an explicit definition of the curve. The original equations are therefore parametric equations for this parabola. □

Table 10.2

t	-3	-2	-1	0	1	2	3
x	6	5	4	3	2	1	0
y	7	2	-1	-2	-1	2	7

Example 16 Discuss the curve defined parametrically by

$$x = 2\cos\theta, \quad y = 2\sin\theta, \qquad 0 \le \theta < 2\pi.$$

Solution If these equations are squared and added, we obtain

$$x^2 + y^2 = 4\cos^2\theta + 4\sin^2\theta = 4,$$

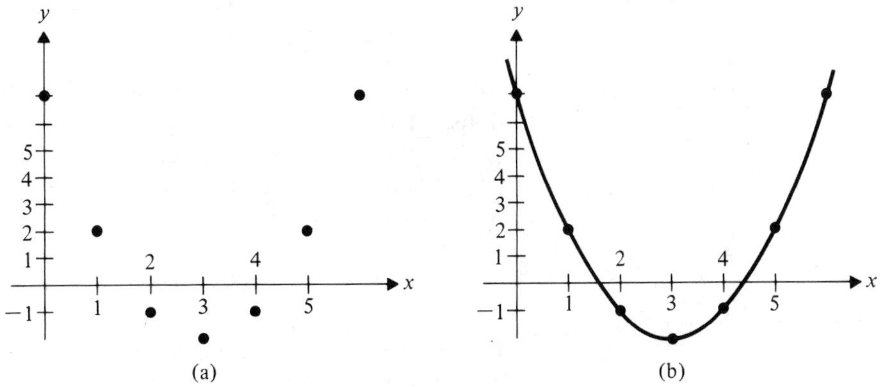

Figure 10.54

an implicit definition of the curve. The original equations therefore define this circle parametrically. In this case we can actually interpret the parameter θ as the angle (expressed in radians) shown in Fig. 10.55. The values $0 \le \theta < 2\pi$ of θ guarantee that the parametric equations describe the complete circle. Additional values of θ would only duplicate points on this circle. □

Example 17 Sketch the curve defined parametrically by

$$x = t - t^3, \qquad y = t + t^3.$$

Solution It is possible once again to eliminate the parameter, for

$$x + y = 2t \quad \text{and} \quad y - x = 2t^3.$$

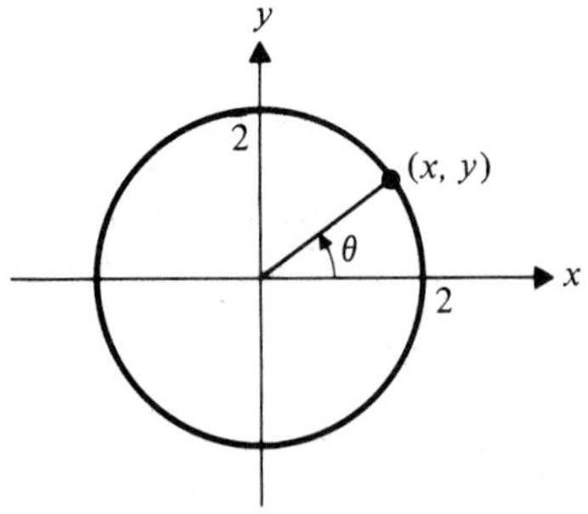

Figure 10.55

Consequently,

$$y - x = \frac{(x + y)^3}{4},$$

an implicit definition of the curve, but not a particularly simple one. An explicit definition is impossible. It seems then for this example that parametric equations are the easiest with which to work. We start by noting that when $t = 0$ and $t = 1$, we obtain the points (0, 0) and (0, 2) on the curve. As t increases from $t = 0$ to $t = 1$, y steadily increases from $y = 0$ to $y = 2$. At the same time, x begins at $x = 0$, increases to some maximum value, and then decreases back to $x = 0$. This information leads to the part of the curve joining (0, 0) and (0, 2) in Fig. 10.56. As t increases beyond $t = 1$, y steadily increases and x steadily decreases. This completes the graph above the x-axis. We now note that if t is replaced by $-t$ in the parametric equations, the signs of both x and y are reversed. In other words, to obtain that part of the curve for $t < 0$, we take each point (x, y) on that part for $t > 0$, and construct the point $(-x, -y)$. This yields the lower half of the curve. □

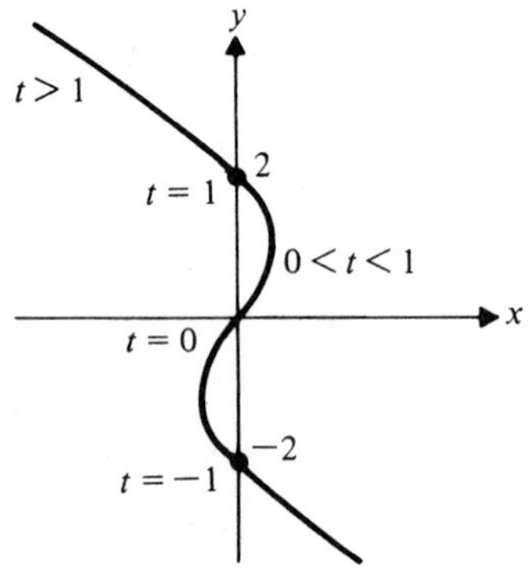

Figure 10.56

Parametric equations can frequently be used in problems concerning the motion of objects. For example, suppose a stone is thrown horizontally over the edge of a cliff from a height of 100 meters above a river (Fig. 10.57). If the initial speed of the stone is 30 meters per second, the path followed by the stone is described parametrically by

$$x = 30t, \qquad y = 100 - 4.905t^2, \qquad 0 \le t \le \sqrt{\frac{100}{4.905}},$$

where t is time measured in seconds and x and y are in meters. For these equations, t has been chosen equal to zero at the instant of projection, and $t = \sqrt{100/4.905}$ is the time at which the stone strikes the river.

Elimination of the parameter t indicates that the stone follows a parabolic path, explicitly defined by

$$y = 100 - 4.905\left(\frac{x}{30}\right)^2 = 100 - 0.00545x^2.$$

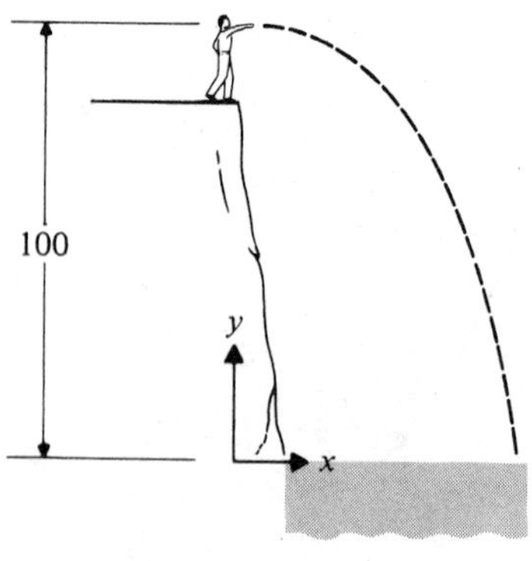

Figure 10.57

In spite of the fact that we might prefer the explicit definition for the path followed by the stone, were we to discuss the motion of the stone (as regards, say, its velocity and acceleration), we would find the parametric equations indispensable.

We say that Eqs. (10.25) define y parametrically as a function of x if the curve so defined represents a function; i.e., if every vertical line that intersects the curve does so in exactly one point. The parametric equations in Example 15 therefore define a function parametrically, but in Examples 16 and 17, they do not. Note, however, that each of the curves in these last two examples can be divided into subcurves that do represent functions. For instance, in Example 17, each of the subcurves

$$\begin{matrix} x = t - t^3, \\ y = t + t^3, \end{matrix} \quad t \le -\frac{1}{\sqrt{3}}; \qquad \begin{matrix} x = t - t^3, \\ y = t + t^3, \end{matrix} \quad -\frac{1}{\sqrt{3}} \le t \le \frac{1}{\sqrt{3}};$$

$$\begin{matrix} x = t - t^3, \\ y = t + t^3, \end{matrix} \quad t \ge \frac{1}{\sqrt{3}}$$

defines a function, and does so parametrically, and together all three curves make up the original curve.

When Eqs. (10.25) do define y parametrically as a function of x, it is straightforward to find the derivative of the function. If we denote this function by $y = f(x)$, the chain rule applied to $y = f(x)$, $x = x(t)$ gives us

$$\frac{dy}{dt} = \frac{dy}{dx}\frac{dx}{dt}.$$

Provided then that $dx/dt \ne 0$, we may solve for

$$\frac{dy}{dx} = \frac{\dfrac{dy}{dt}}{\dfrac{dx}{dt}}. \tag{10.26}$$

This formula is called the *parametric rule* for differentiation of a function defined parametrically by Eqs. (10.25); it defines the derivative of y with respect to x in terms of derivatives of the given functions $x(t)$ and $y(t)$ with respect to t.

For example, if y is defined parametrically as a function of x by

$$x(t) = \frac{1}{t^2 + 1}, \qquad y(t) = \frac{2}{t(t^2 + 1)}, \qquad t \ge 0,$$

then

$$\frac{dy}{dx} = \frac{\dfrac{dy}{dt}}{\dfrac{dx}{dt}} = \frac{\dfrac{-2(3t^2 + 1)}{(t^3 + t)^2}}{\dfrac{-2t}{(t^2 + 1)^2}} = \frac{3t^2 + 1}{t^3}.$$

We can also calculate d^2y/dx^2 in spite of the fact that dy/dx is in terms of t (rather than x). Once again it is the chain rule that comes to our rescue:

$$\frac{d^2y}{dx^2} = \frac{d}{dx}\left(\frac{dy}{dx}\right) = \frac{d}{dt}\left(\frac{dy}{dx}\right)\frac{dt}{dx}$$

$$= \frac{\dfrac{d}{dt}\left(\dfrac{dy}{dx}\right)}{\dfrac{dx}{dt}} = \frac{\dfrac{t^3(6t) - (3t^2 + 1)(3t^2)}{t^6}}{\dfrac{-2t}{(t^2 + 1)^2}} = \frac{3(t^2 + 1)^3}{2t^5}.$$

If functions $x(t)$ and $y(t)$ in (10.25) are complicated, a useful device in sketching the curve is to first sketch graphs of these functions. We can then read pairs of values $(x(t), y(t))$ from these two graphs. We illustrate this technique in the following example.

Example 18 Sketch the strophoid

$$x = \frac{1-t^2}{1+t^2}, \qquad y = \frac{t(1-t^2)}{1+t^2}$$

and find the points at which its tangent line is horizontal.

Solution To sketch the strophoid we first sketch graphs of the functions $x(t)$ and $y(t)$. For $x(t)$, we write $x = -1 + 2/(1+t^2)$, and first draw $2/(1+t^2)$ (tinted in Fig. 10.58). The graph of $x(t)$ is this curve shifted downward one unit. Since $y(t) = tx(t)$, we sketch a graph of $y(t)$ by multiplication of ordinates (Fig. 10.59).

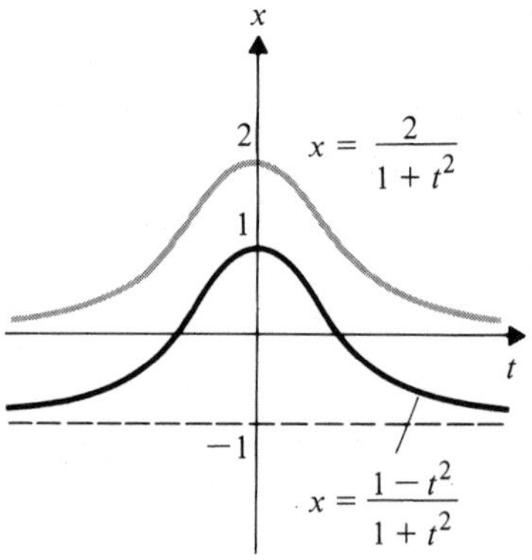

Figure 10.58

To sketch the strophoid we now read pairs of values of x and y from these graphs. When $t = 0$, we begin with the point $(1, 0)$. As t increases from $t = 0$ to $t = 1$, x steadily decreases from $x = 1$ to $x = 0$; y at the same time begins at $y = 0$, increases to some maximum value, and then returns to $y = 0$. This gives that part of the strophoid in the first quadrant (Fig. 10.60). As t increases beyond $t = 1$, x becomes negative and approaches -1 while y becomes very large negatively. This defines that part of the strophoid in the third quadrant. Negative values of t determine the other half of the curve.

To find the two points at which the tangent line is horizontal, we set

$$0 = \frac{dy}{dx} = \frac{\frac{dy}{dt}}{\frac{dx}{dt}} = \frac{\frac{(1+t^2)(1-3t^2)-(t-t^3)(2t)}{(1+t^2)^2}}{\frac{(1+t^2)(-2t)-(1-t^2)(2t)}{(1+t^2)^2}} = \frac{1-4t^2-t^4}{-4t}.$$

But this equation implies that

$$t^4 + 4t^2 - 1 = 0.$$

If we regard this as a quadratic equation in t^2 and use the quadratic formula, we have

$$t^2 = \frac{-4 \pm \sqrt{16+4}}{2} = -2 \pm \sqrt{5}.$$

Since t^2 must be nonnegative, it follows that $t^2 = \sqrt{5} - 2$, and therefore

$$t = \pm\sqrt{\sqrt{5}-2}.$$

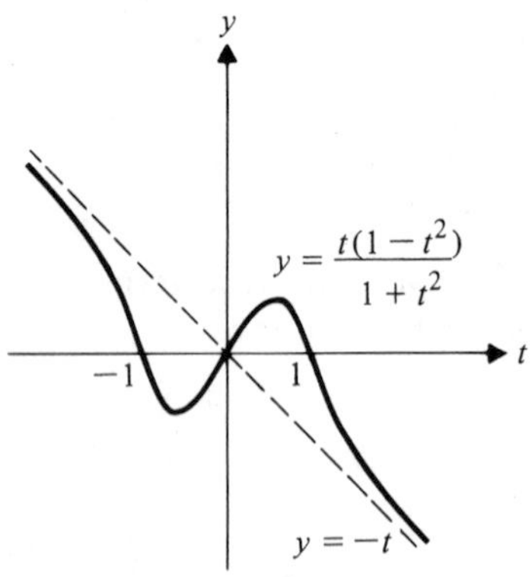

Figure 10.59

For these values of t,

$$x = \frac{1-(\sqrt{5}-2)}{1+(\sqrt{5}-2)} = \frac{3-\sqrt{5}}{\sqrt{5}-1}\cdot\frac{\sqrt{5}+1}{\sqrt{5}+1} = \frac{\sqrt{5}-1}{2} = 0.62,$$

and

$$y = tx = (\pm\sqrt{\sqrt{5}-2})\left(\frac{\sqrt{5}-1}{2}\right) = \pm 0.30.$$

The points with a horizontal tangent line are therefore $(0.62, \pm 0.30)$. □

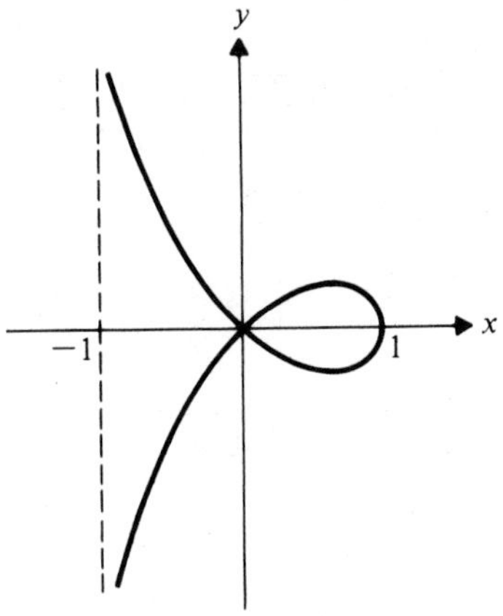

Figure 10.60

When a curve C is defined parametrically by (10.25), its length is given by the definite integral

$$L = \int_{\alpha}^{\beta} \sqrt{\left(\frac{dx}{dt}\right)^2 + \left(\frac{dy}{dt}\right)^2}\, dt. \tag{10.27}$$

To verify this we first note that any curve defined parametrically by (10.25) can be divided into subcurves (Fig. 10.61) in such a way that each subcurve represents a function $y = f(x)$. Suppose we denote these subcurves by $C_1, C_2, \ldots, C_n$ and let P_i be the point joining C_i to C_{i+1}. According to Eq. (7.12), the length of C_i is given by

$$L_i = \int_{x_{i-1}}^{x_i} \sqrt{1 + \left(\frac{dy}{dx}\right)^2}\, dx$$

when $x_i > x_{i-1}$ and by

$$L_i = \int_{x_i}^{x_{i-1}} \sqrt{1 + \left(\frac{dy}{dx}\right)^2}\, dx$$

when $x_{i-1} > x_i$. Let t_i be the value of t yielding P_i. If we set $x = x(t)$, $y = y(t)$ in these integrals and use (10.26), the first integral becomes

$$L_i = \int_{t_{i-1}}^{t_i} \sqrt{1 + \left(\frac{dy/dt}{dx/dt}\right)^2}\, dx = \int_{t_{i-1}}^{t_i} \sqrt{\frac{\left(\frac{dx}{dt}\right)^2 + \left(\frac{dy}{dt}\right)^2}{\left(\frac{dx}{dt}\right)^2}}\, dx$$

$$= \int_{t_{i-1}}^{t_i} \sqrt{\left(\frac{dx}{dt}\right)^2 + \left(\frac{dy}{dt}\right)^2}\, \frac{dt}{dx}\, dx = \int_{t_{i-1}}^{t_i} \sqrt{\left(\frac{dx}{dt}\right)^2 + \left(\frac{dy}{dt}\right)^2}\, dt.$$

The second integral when $x_{i-1} > x_i$ becomes

$$L_i = \int_{t_i}^{t_{i-1}} \sqrt{1 + \left(\frac{dy/dt}{dx/dt}\right)^2}\, dx = \int_{t_i}^{t_{i-1}} \sqrt{\frac{\left(\frac{dx}{dt}\right)^2 + \left(\frac{dy}{dt}\right)^2}{\left(\frac{dx}{dt}\right)^2}}\, dx$$

$$= \int_{t_i}^{t_{i-1}} \sqrt{\left(\frac{dx}{dt}\right)^2 + \left(\frac{dy}{dt}\right)^2}\left(-\frac{dt}{dx}\right) dx \qquad \left(\text{since } \frac{dx}{dt} < 0\right)$$

$$= \int_{t_{i-1}}^{t_i} \sqrt{\left(\frac{dx}{dt}\right)^2 + \left(\frac{dy}{dt}\right)^2}\, dt.$$

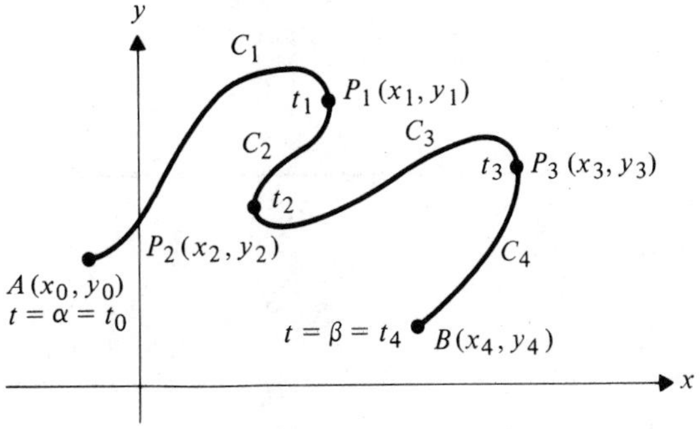

Figure 10.61

The total length of the curve is therefore

$$L = \sum_{i=1}^{n} L_i = \sum_{i=1}^{n} \int_{t_{i-1}}^{t_i} \sqrt{\left(\frac{dx}{dt}\right)^2 + \left(\frac{dy}{dt}\right)^2}\, dt = \int_{\alpha}^{\beta} \sqrt{\left(\frac{dx}{dt}\right)^2 + \left(\frac{dy}{dt}\right)^2}\, dt.$$

Example 19 Find the length of the curve

$$x = e^t \cos t, \qquad y = e^t \sin t, \qquad 0 \le t \le \pi/2.$$

Solution According to (10.27),

$$L = \int_0^{\pi/2} \sqrt{(e^t \cos t - e^t \sin t)^2 + (e^t \sin t + e^t \cos t)^2}\, dt$$

$$= \int_0^{\pi/2} \sqrt{2}e^t\, dt = \sqrt{2}\{e^t\}_0^{\pi/2} = \sqrt{2}(e^{\pi/2} - 1). \square$$

When a curve has a polar equation of the form $r = f(\theta)$, $\alpha \le \theta \le \beta$, parametric equations for the curve are

$$x = f(\theta)\cos\theta, \qquad y = f(\theta)\sin\theta, \qquad \alpha \le \theta \le \beta. \tag{10.28}$$

It is not difficult to show that with θ as parameter, Eqs. (10.26) and (10.27) become

$$\frac{dy}{dx} = \frac{f'(\theta)\sin\theta + f(\theta)\cos\theta}{f'(\theta)\cos\theta - f(\theta)\sin\theta} \tag{10.29}$$

and

$$L = \int_a^{\beta} \sqrt{r^2 + \left(\frac{dr}{d\theta}\right)^2}\, d\theta. \tag{10.30}$$

Example 20 Find the length of the cardioid $r = 1 - \cos\theta$.

Solution According to (10.30) (Fig. 10.62),

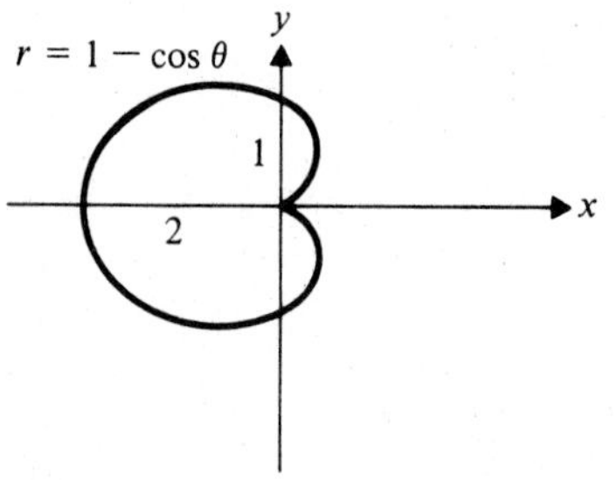

Figure 10.62

$$L = \int_0^{2\pi} \sqrt{(1 - \cos\theta)^2 + (\sin\theta)^2}\, d\theta$$

$$= \int_0^{2\pi} \sqrt{2}\sqrt{1 - \cos\theta}\, d\theta$$

$$= \sqrt{2}\int_0^{2\pi} \sqrt{1 - (1 - 2\sin^2(\theta/2))}\, d\theta$$

$$= 2\int_0^{2\pi} \sin(\theta/2)\, d\theta = 2\{-2\cos(\theta/2)\}_0^{2\pi} = 8. \square$$

EXERCISES 10.6

In Exercises 1–8, sketch the given curve.

1. $x = 2 + t$,
   $y = 3t - 1$
2. $x = t^2 + 3t + 4$,
   $y = 1 - t$
3. $x = 1 + 2\cos t$,
   $y = 2 + 2\sin t$, $\quad 0 \le t < 2\pi$
4. $x = t + 1/t$,
   $y = t - 1/t$

5. $x = 2\cos t,\ y = 4\sin t,\quad 0 \le t < 2\pi$

6. $x = t - t^2,\ y = t + t^2$

7. $x = t^2 + 1,\ y = t^3 + 3$

8. $x = 2\cot\theta,\ y = 2\sin^2\theta,\quad -\frac{\pi}{2} \le \theta \le \frac{\pi}{2},\quad \theta \ne 0$

In Exercises 9–16, assume that y is defined parametrically as a function of x, and find dy/dx.

9. $x(t) = t^3 + 3t - 2,\ y(t) = t^2 - 1$

10. $x(u) = \dfrac{u}{u-1},\ y(u) = \dfrac{u^2}{u^2-1}$

11. $x(v) = (v^3 + 2)\sqrt{v-1},\ y(v) = 2v^3 + 3$

12. $x(t) = \sqrt{\dfrac{2+t}{2-t}},\ y(t) = \sqrt{\dfrac{2-t}{2+t}}$

13. $x(s) = s^{3/2} - s^{2/3},\ y(s) = s^2 + 2s$

14. $x(t) = (2t+3)^4,\ y(t) = \dfrac{t}{t+6}$

15. $x(u) = \left(\dfrac{1+u}{1-u}\right)^{1/3},\ y(u) = \left(\dfrac{1-u}{1+u}\right)^4$

16. $x(t) = \sqrt{-t^2 + 3t + 5},\ y(t) = \dfrac{1}{t^2 + 2t - 5}$

In Exercises 17–20, find dy/dx and d^2y/dx^2 if y is defined parametrically as a function of x.

17. $x(t) = t^2 + \dfrac{1}{t},\ y(t) = t^2 - \dfrac{1}{t}$

18. $x(t) = \sqrt{t-1},\ y(t) = \sqrt{t+1}$

19. $x(u) = 2u + 5,\ y(u) = 7 - 14u$

20. $x(v) = v^2 + 2v + 3,\ y(v) = 2v - 4$

21. Find equations for the tangent and normal lines to the curve

$$x = t + \frac{1}{t}, \qquad y = t - \frac{1}{t}$$

at the point corresponding to $t = 4$.

22. Find point(s) on the curve

$$x = \frac{t^3}{3} - 3t, \qquad y = \frac{3t^2}{2} + t$$

where the slope of the tangent line to the curve is equal to 1.

23. Is there any difference between the two curves

$$y = 2x^2 - 1 \quad \text{and} \quad \begin{cases} x = \cos t, \\ y = 2\cos^2 t - 1? \end{cases}$$

24. What curve is described by the parametric equations

$$x = h + a\cos\theta, \qquad y = k + b\sin\theta,$$
$$0 \le \theta < 2\pi?$$

25. Find parametric equations for a circle with center (h, k) and radius r.

26. Show that the straight line through two points $P_1(x_1, y_1)$ and $P_2(x_2, y_2)$ has parametric equations

$$x = x_1 + (x_2 - x_1)t, \qquad y = y_1 + (y_2 - y_1)t.$$

27. Sketch the curves defined parametrically by:

a) $x = \sec\theta,\ y = \tan\theta,\ -\pi/2 < \theta < \pi/2$;

b) $x = \cosh\phi,\ y = \sinh\phi$;

c) $x = (t + 1/t)/2,\ y = (t - 1/t)/2,\ t \ge 0$.

Are these curves related?

In Exercises 28–31, find parametric equations for the given curve.

28. $y = \dfrac{x+1}{x-2}$

29. $x + y^3 + xy = 5y^2$

30. $x^2 + y^2 + 2x - 4y = 0$

31. $4 - x^2 + 2y^2 = 0$

32. Two particles move along straight lines l_1 and l_2 defined parametrically by

$$l_1: \begin{matrix} x = 1 - t, \\ y = t, \end{matrix} \quad t \ge 0; \qquad l_2: \begin{matrix} x = 4t - 5, \\ y = 2t - 1, \end{matrix} \quad t \ge 0;$$

t being time. When are they closest together?

33. If $x = x(t)$ and $y = y(t)$ define y as a function of x, show that

$$\frac{d^2y}{dx^2} = \frac{\dot{x}\ddot{y} - \dot{y}\ddot{x}}{\dot{x}^3},$$

where $\cdot$ indicates differentiation with respect to t.

34. Use (10.27) and (10.28) to verify (10.30).

In Exercises 35–38, find the length of the given curve.

35. $x = 3 + 4\cos t,\ y = -2 + 4\sin t,\ 0 \le t \le 2\pi$

36. $x = e^{-t}\sin t,\ y = e^{-t}\cos t,\ 0 \le t \le 1$

37. The cardioid $r = a(1 - \cos\theta)$

38. $x = t + \ln t,\ y = t - \ln t,\ 1 \le t \le 2$

39. Set up, but do not evaluate, a definite integral representing the length of the ellipse $x = a\cos\theta,\ y = b\sin\theta,\ 0 \le \theta \le 2\pi$.

40. Use (10.26) and (10.28) to verify (10.29).

41. Find the slope of the tangent line to the curve $r = 3/(1 - \sin\theta)$ at the point with polar coordinates $(6, \pi/6)$.

Do this in two ways:

a) by using (10.29);

b) by finding the equation of the curve in Cartesian coordinates, and calculating dy/dx.

42. Suppose $x(t)$ and $y(t)$ in Eqs. (10.25) are continuous on $\alpha \le t \le \beta$ and have derivatives on $\alpha < t < \beta$, and that $x'(t) \ne 0$ on $\alpha < t < \beta$. Show that Cauchy's generalized mean value theorem (Theorem 4.5) implies that there exists a number c between α and β such that

$$\frac{y(\beta) - y(\alpha)}{x(\beta) - x(\alpha)} = \frac{y'(c)}{x'(c)}.$$

Interpret this result geometrically.

43. The equations $x = t^2 + 2t - 1$, $y = t + 5$, $1 \le t \le 4$ define a curve parametrically. Find the parametric equations that describe the same curve, but have values of the parameter in the intervals (a) $0 \le t \le 3$; (b) $0 \le t \le 1$.

44. A particle travels around the circle $x^2 + y^2 = 4$ counterclockwise at constant speed, making 2 revolutions each second. If the particle starts at point (2, 0) at time $t = 0$, find the parametric equations for its position in terms of t.

45. a) A stone is embedded in the tread of a tire and the tire rolls (without slipping) along the x-axis (Fig. 10.63). If the stone starts at the origin, the path that it follows is called a cycloid. Show that if θ is the angle through which the stone has turned, the parametric equations for the cycloid are

$$x = R(\theta - \sin\theta), \qquad y = R(1 - \cos\theta).$$

b) Find the area under one arch of the cycloid.

c) Find the length of one arch of the cycloid.

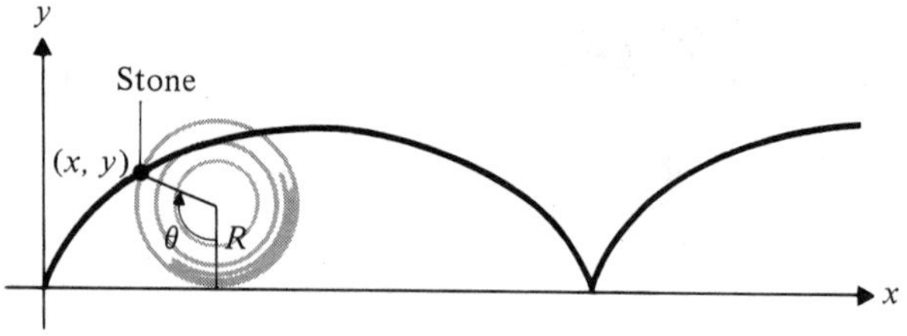

Figure 10.63

46. In Exercise 45, if the stone is embedded in the side of the tire rather than the tread, its path is called a *trochoid* (Fig. 10.64). Show that if the distance from the center of the tire to the stone is b, the parametric equations for the trochoid are

$$x = R\theta - b\sin\theta, \qquad y = R - b\cos\theta.$$

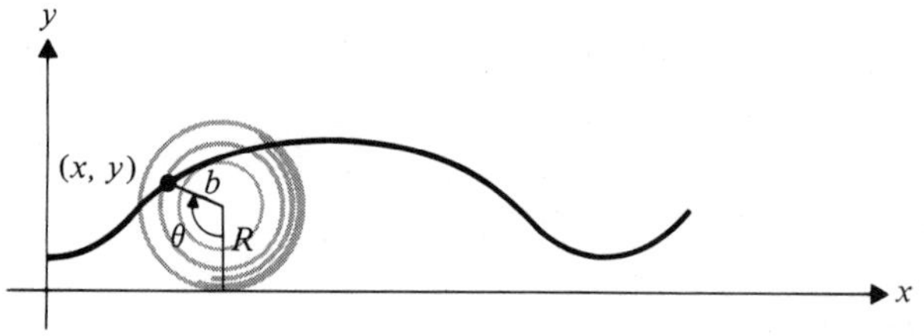

Figure 10.64

In Exercises 47–49, sketch the given curve.

47. $x = \cos\theta$, $y = \sin 3\theta$ (Curve of Lissajous)

48. $x = 3t/(1 + t^3)$, $y = 3t^2/(1 + t^3)$ (Folium of Descartes)

49. $x = \cos^3\theta$, $y = \sin^3\theta$ (Astroid or hypocycloid of four cusps)

SUMMARY

Polar coordinates provide an alternative way to identify the position of points in a plane. They use distance from a point, called the pole, and rotation of a half line, called the polar axis. Many curves that have complex equations in Cartesian coordinates can be represented very simply using polar coordinates. Particularly simple are multileaved roses, cardioids, lemniscates, limaçons, and some circles. Polar coordinates also provide a unified approach to conic sections. Each conic can be described as the curve traced out by a point that moves so that the ratio ε of its distances from a fixed point and a fixed line remains constant. The conic is either an ellipse, a parabola, or a hyperbola depending on whether $0 < \varepsilon < 1$, $\varepsilon = 1$, and $\varepsilon > 1$, respectively. Ellipses and hyperbolas can also be defined as curves traced out by a point that moves so that the sum and difference of its distances from two fixed points remains constant.

When a curve $r = f(\theta)$ in polar coordinates encloses a region R, the definite integral

$$\int_\alpha^\beta \frac{1}{2}[f(\theta)]^2\,d\theta$$

for appropriate choices of α and β can be used to find the area of R. The quantity $(\frac{1}{2})[f(\theta)]^2\,d\theta$ is regarded as the area of a sector of angular width $d\theta$ and radius r, and the definite integral adds over all sectors between angles $\theta = \alpha$ and $\theta = \beta$.

In Chapter 1 we defined curves and functions explicitly and implicitly. In this chapter we added a third description—the parametric definition

$$x = x(t), \qquad y = y(t), \qquad \alpha \le t \le \beta.$$

When such a curve defines a function, its derivative can be calculated using the parametric rule for differentiation:

$$\frac{dy}{dx} = \frac{\dfrac{dy}{dt}}{\dfrac{dx}{dt}};$$

second and higher-order derivatives can be calculated using the chain rule. The length of such a curve is defined by the definite integral

$$\int_\alpha^\beta \sqrt{\left(\frac{dx}{dt}\right)^2 + \left(\frac{dy}{dt}\right)^2}\,dt.$$

If a curve has polar equation $r = f(\theta)$, $\alpha \le \theta \le \beta$, a set of parametric equations for the curve is

$$x(\theta) = f(\theta)\cos\theta, \qquad y(\theta) = f(\theta)\sin\theta, \qquad \alpha \le \theta \le \beta.$$

In such a case, the above integral for length of the curve becomes

$$\int_\alpha^\beta \sqrt{r^2 + \left(\frac{dr}{d\theta}\right)^2}\,d\theta.$$

REVIEW EXERCISES

In Exercises 1–30, sketch the given curve.

1. $r = \cos\theta$

2. $r = -\sin\theta$

3. $r = \dfrac{3}{1 + 2\sin\theta}$

4. $r = \dfrac{3}{2 + \sin\theta}$

5. $r = \dfrac{1}{1 + \sin\theta}$

6. $r = \dfrac{3}{2 - 2\cos\theta}$

7. $r + 1 = 2\sin\theta$

8. $r^2 = 4\cos\theta$

9. $x = 2\cos t,\ y = 3\sin t,\quad 0 \le t < 2\pi$

10. $x = 4 + t,\ y = 5 - 3t^2$

11. $x = 1 + 4\sin t,\ y = -2 + 4\cos t,\quad 0 \le t \le \pi$

12. $x = \sin^2 t,\ y = \cos^2 t,\quad 0 \le t < 2\pi$

13. $(x^2 + y^2)^3 = x$

14. $x^2 + y^2 = 2\sqrt{x^2 + y^2}$

15. $r^2 = 4\cos^2\theta$

16. $r = 3\cos 2\theta$

17. $x + y = \sqrt{x^2 + y^2}$
18. $x^2 + y^2 = x + y$
19. $r = 4 \cos 3\theta$
20. $r = \sin^2 \theta - \cos^2 \theta$
21. $x^2 + y^2 - 3x + 2y = 1$
22. $x^2 - 3y^2 + 4 = 0$
23. $2x^2 + 3y^2 - 6y = 0$
24. $y^2 + x + 2y = 3$
25. $r(2\cos\theta - \sin\theta) = 3$
26. $r = \cos(\theta/2)$
27. $r = \sin^2 \theta$
28. $x^2 = y^2 + y$
29. $x = e^{-t}$, $y = \ln t$, $t > 0$
30. $x = \sin \theta$, $y = \sin 2\theta$

In Exercises 31–35, find the indicated area.

31. Inside $r = 2 + 2\cos\theta$
32. Inside $r = 4$ and outside $r = 4 \sin 2\theta$
33. Common to $r = 2$ and $r^2 = 9 \cos 2\theta$
34. Inside $r = \sin^2 \theta$
35. Common to $r = 1 + \sin\theta$ and $r = 2 - 2\sin\theta$

In Exercises 36 and 37, assume that y is defined as a function of x, and find dy/dx and d^2y/dx^2.

36. $x(t) = t^3 + 2t$,
    $y(t) = 3t - t^3$
37. $x(u) = 2 \sin u$,
    $y(u) = 3 \cos u$

In Exercises 38–40, find the length of the given curve.

38. $r = 2 + 2\cos\theta$
39. $x = t^2, y = t^3, 0 \le t \le 1$
40. $x = e^t \cos t, y = e^t \sin t, 0 \le t \le \pi/2$
41. Find the equation of the tangent line to the curve $r = 2 - 2\sin\theta$ at the point with polar coordinates $(1, \pi/6)$.

Infinite Sequences and Series

11

11.1 INFINITE SEQUENCES OF CONSTANTS

In Section 4.8 we used Newton's iterative procedure to develop a set of numbers $x_1, x_2, x_3, \ldots$, to approximate the root of an equation $f(x) = 0$. The first number is chosen as some initial approximation to the solution of the equation, and then each of the other numbers defined by the formula

$$x_{n+1} = x_n - \frac{f(x_n)}{f'(x_n)}$$

is a better and better approximation. This formula associates a number x_n with each positive integer n, and we call such a rule an *infinite sequence of constants*. In general, we have the following definition.

▷ **Definition 11.1** An *infinite sequence of constants* is a function $f(n)$ whose domain is the set of positive integers. ◁

The word "infinite" simply indicates that an infinity of numbers is defined by the sequence (since there is an infinity of positive integers), but indicates nothing about the nature of the numbers. Often we write the numbers $f(n)$ in a line separated by commas,

$$f(1), f(2), \ldots, f(n), \ldots \tag{11.1a}$$

and refer to this array as the sequence, rather than the law by which it is formed. Because this notation is somewhat cumbersome, we adopt a notation similar to that used for the sequence defined by Newton's iterative procedure. We set $c_1 = f(1)$, $c_2 = f(2), \ldots, c_n = f(n), \ldots$, and write for (11.1a)

$$c_1, c_2, c_3, \ldots, c_n, \ldots. \tag{11.1b}$$

The first number c_1 is called the first term of the sequence, c_2 the second term, and for general n, c_n is called the nth term (or general term) of the sequence.

It is not necessary that the first term of a sequence correspond to the integer 1. We could define a sequence as a function whose domain is the set of all integers larger than or equal to some fixed integer N (which could be positive, negative, or zero). Indeed, in Section 11.6 we will find it convenient to initiate the assignment with $N = 0$. For now we prefer to choose $N = 1$, in which case we have the natural situation where the first term of the sequence corresponds to $n = 1$, the second term to $n = 2$, etc.

Example 1 The general terms of four sequences are:

a) $\dfrac{1}{2^{n-1}}$;

b) $\dfrac{n}{n+1}$;

c) $(-1)^n|n-3|$;

d) $(-1)^{n+1}$.

Write out the first six terms of each sequence.

Solution The first six terms of each sequence are as follows:

a) $1, \frac{1}{2}, \frac{1}{4}, \frac{1}{8}, \frac{1}{16}, \frac{1}{32}$;

b) $\frac{1}{2}, \frac{2}{3}, \frac{3}{4}, \frac{4}{5}, \frac{5}{6}, \frac{6}{7}$;

c) $-2, 1, 0, 1, -2, 3$;

d) $1, -1, 1, -1, 1, -1$. □

Example 2 A sequence is defined recursively by

$$c_1 = 1, \qquad c_{n+1} = 5 + \sqrt{2 + c_n}, \qquad n \geq 1.$$

List its first four terms.

Solution The first four terms of the sequence are

$$c_1 = 1,$$
$$c_2 = 5 + \sqrt{2 + c_1} = 5 + \sqrt{3},$$
$$c_3 = 5 + \sqrt{2 + c_2} = 5 + \sqrt{7 + \sqrt{3}},$$
$$c_4 = 5 + \sqrt{2 + c_3} = 5 + \sqrt{7 + \sqrt{7 + \sqrt{3}}}. \ \square$$

Note the difference between a sequence that is defined explicitly (Example 1) and one that is defined recursively (Example 2 and the sequence defined by Newton's iterative procedure). To find the 100th term of a sequence defined explicitly, we simply substitute $n = 100$ into the explicit formula. To find the 100th term of a sequence that is defined recursively, we must know the 99th; but to find the 99th, we must know the 98th; to find the 98th, we need the 97th; and so on down the line. In other words, to find a term in the sequence, we must first find every term that precedes it. Obviously it is much more convenient to have an explicit definition of c_n in terms of n, but to find an explicit definition for a recursive sequence is usually very difficult.

Example 3 The nth term of a sequence is the nth prime integer (greater than 1) when all such primes are listed in ascending size. List its first ten terms.

Solution The first ten terms of the sequence are

$$2, \quad 3, \quad 5, \quad 7, \quad 11, \quad 13, \quad 17, \quad 19, \quad 23, \quad 29. \ \square$$

Since no formula for the prime integers is known, it is impossible to obtain an explicit definition for this sequence.

If the general term of a sequence is known explicitly ($1/2^{n-1}$ in Example 1a), any term in the sequence is obtained by substituting the appropriate value of n. In other words, the general term specifies every term in the sequence. We therefore use the general term to abbreviate the notation for a sequence by writing the general term in braces and using this to represent the sequence. Specifically, we write

$$\left\{\frac{1}{2^{n-1}}\right\}_1^\infty = 1, \frac{1}{2}, \frac{1}{4}, \frac{1}{8}, \ldots, \frac{1}{2^{n-1}}, \ldots,$$

where 1 and ∞ indicate that the first term corresponds to the integer $n = 1$, and that there is an infinite number of terms in the sequence. In general,

$$\{c_n\}_1^\infty = c_1, c_2, c_3, \ldots, c_n, \ldots. \tag{11.2}$$

If (as is the case in this section) the first term of the sequence corresponds to the integer $n = 1$, we abbreviate the notation further and simply write $\{c_n\}$ in place of $\{c_n\}_1^{\infty}$.

Since a sequence $\{c_n\}$ is a function whose domain is the set of positive integers, we can represent $\{c_n\}$ graphically. The sequences of Example 1 are shown in Fig. 11.1.

We now define a number called the *limit* of a sequence. Intuitively, a number L is called the limit of a sequence $\{c_n\}$ if, as we go farther and farther out in the sequence, the terms get arbitrarily close to L, and stay close to L. If such a number L exists, we write

$$L = \lim_{n\to\infty} c_n,$$

and say that the sequence $\{c_n\}$ *converges* to L. If no such number exists, we say that the sequence does not have a limit, or that the sequence *diverges.*

For the sequences of Example 1, it is evident that:

a) $\lim_{n\to\infty} \dfrac{1}{2^{n-1}} = 0;$

b) $\lim_{n\to\infty} \dfrac{n}{n+1} = 1;$

c) $\lim_{n\to\infty} (-1)^n|n-3|$ does not exist;

d) $\lim_{n\to\infty} (-1)^{n+1}$ does not exist;

and in Example 3,

$$\lim_{n\to\infty} c_n \text{ does not exist.}$$

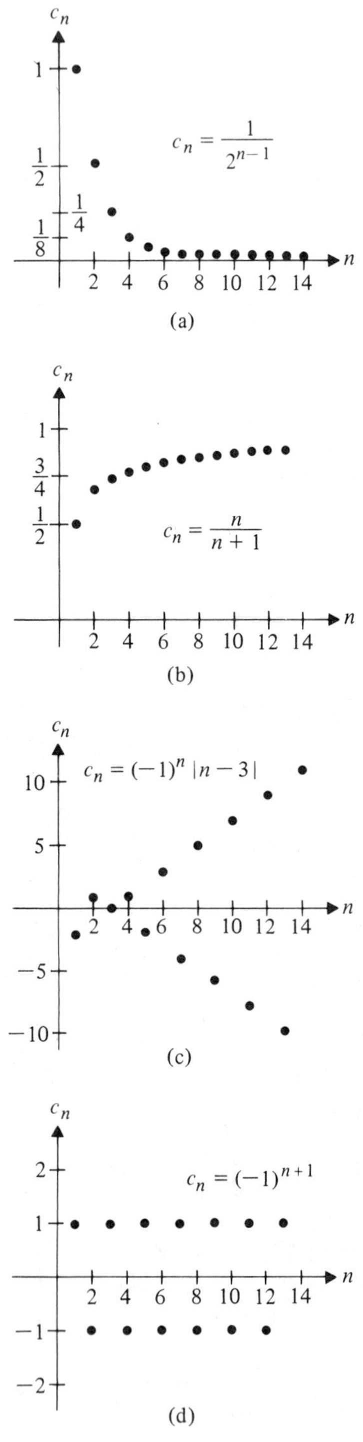

Figure 11.1

Note how the points on the graphs in Figs. 11.1(a) and (b) cluster around the limits 0 and 1 as n gets larger and larger. No such clustering occurs in the remaining two figures. Note also that sequence $\{(-1)^{n+1}\}$ in Fig. 11.1(d) is perfectly well behaved and, in spite of this, it has no limit. In other words, divergence of a sequence should not be confused with wild, uncontrolled behavior. Many well behaved sequences do not have limits.

It is not at all clear whether the recursive sequence of Example 2 has a limit. Likewise for the recursive sequence $\{x_n\}$ defined by Newton's iterative procedure, we pointed out in Section 4.8 that sometimes the sequence does not converge, but gave no conditions that would ensure that it does. What we really need to do now, then, is to develop criteria by which to determine whether a given sequence has a limit. The following two definitions lead to Theorem 11.1, which can be very useful in determining whether a recursive sequence has a limit.

▷ **Definition 11.2** A sequence $\{c_n\}$ is said to be:

i) *increasing* if $c_{n+1} > c_n$ for all $n \geq 1$; (11.3a)

ii) *nondecreasing* if $c_{n+1} \geq c_n$ for all $n \geq 1$; (11.3b)

iii) *decreasing* if $c_{n+1} < c_n$ for all $n \geq 1$; (11.4a)

iv) *nonincreasing* if $c_{n+1} \leq c_n$ for all $n \geq 1$. (11.4b)

If a sequence satisfies any one of these four properties, it is said to be *monotonic.* ◁

▷ **Definition 11.3** A sequence $\{c_n\}$ is said to have an *upper bound* U (be bounded above by U) if

$$c_n \leq U \tag{11.5}$$

for all $n \geq 1$. It has a *lower bound* V (is bounded below by V) if

$$c_n \geq V \tag{11.6}$$

for all $n \geq 1$. ◁

We note that if U is an upper bound for a sequence, any number greater than U is also an upper bound. If V is a lower bound, so too is any number smaller than V.

For the sequences of Example 1, we find that $\{1/2^{n-1}\}$ is decreasing and has an upper bound $U = 1$ and a lower bound $V = 0$; $\{n/(n+1)\}$ is increasing and has an upper bound $U = 5$ and a lower bound $V = -2$; $\{(-1)^n|n-3|\}$ is not monotonic and has neither an upper nor a lower bound; $\{(-1)^{n+1}\}$ is not monotonic and has an upper bound $U = 1$ and a lower bound $V = -3$.

For example 2, it is evident that a lower bound is $V = 1$. It can be shown by mathematical induction† that the sequence is increasing and has an upper bound $U = 10$.

The sequence of Example 3 is increasing and has a lower bound $V = 2$, but has no upper bound.

We now state the following.

▶ **Theorem 11.1** A bounded monotonic sequence has a limit. ◀

By this theorem we can now conclude that the recursive sequence of Example 2 has a limit. To evaluate the limit, say L, we let n become infinite on each side of the equation defining the sequence recursively:

$$\lim_{n\to\infty} c_{n+1} = \lim_{n\to\infty} (5 + \sqrt{2 + c_n}).$$

It is important to note that this cannot be done until the conditions of Theorem 11.1 have been checked. From this equation we conclude that

$$L = 5 + \sqrt{2 + L}.$$

If we transpose the 5 and square both sides of the equation, we obtain the quadratic equation

$$L^2 - 11L + 23 = 0,$$

with solutions

$$L = \frac{11 \pm \sqrt{29}}{2}.$$

Only the positive square root satisfies the original equation defining L, so that $L = (11 + \sqrt{29})/2$. The negative square root can also be eliminated on the grounds that all terms beyond the first are greater than 5.

Example 4 Find the limit of the sequence

$$c_1 = 2, \qquad c_{n+1} = \frac{1}{4 - c_n}, \qquad n \geq 1$$

if it exists.

† The method of proof called *mathematical induction* is described in Appendix A.

Solution The first four terms of the sequence are

$$c_1 = 2,$$

$$c_2 = \frac{1}{4-2} = \frac{1}{2},$$

$$c_3 = \frac{1}{4-1/2} = \frac{2}{7},$$

$$c_4 = \frac{1}{4-2/7} = \frac{7}{26}.$$

It would appear that the sequence is decreasing since $c_2 < c_1$, $c_3 < c_2$, and $c_4 < c_3$, and that 2 and 0 are upper and lower bounds. To prove these statements we use mathematical induction. Since $0 \le c_1 \le 2$, we suppose that k is some integer for which $0 \le c_k \le 2$. Then

$$0 \ge -c_k \ge -2,$$

and if we add 4 to each side of these inequalities, we have

$$4 \ge 4 - c_k \ge 2.$$

Consequently,

$$\frac{1}{4} \le \frac{1}{4 - c_k} \le \frac{1}{2};$$

therefore

$$\tfrac{1}{4} \le c_{k+1} \le \tfrac{1}{2}.$$

But then certainly $0 \le c_{k+1} \le 2$, and we have shown by mathematical induction that $0 \le c_n \le 2$ for all n.

To show that $\{c_n\}$ is decreasing, we must show that $c_{n+1} < c_n$ for all n. Clearly this is true for $n = 1$, and we therefore suppose that k is some integer for which $c_{k+1} < c_k$. Then

$$-c_{k+1} > -c_k$$

and

$$4 - c_{k+1} > 4 - c_k.$$

Consequently,

$$\frac{1}{4 - c_{k+1}} < \frac{1}{4 - c_k},$$

and therefore $c_{k+2} < c_{k+1}$. By mathematical induction, then, $\{c_n\}$ is decreasing.

By Theorem 11.1 we can now state that $\{c_n\}$ has a limit L, and L may be obtained by taking the limit of the recursive definition:

$$\lim_{n\to\infty} c_{n+1} = \lim_{n\to\infty} \frac{1}{4 - c_n}.$$

It follows that

$$L = \frac{1}{4 - L},$$

and of the two solutions $2 \pm \sqrt{3}$ of this equation, we must choose

$$L = 2 - \sqrt{3}. \ \square$$

In the field of numerical analysis, recursive sequences commonly arise in the form of iterative procedures. We have already encountered Newton's iterative procedure as one example. When this method is applied to the equation

$$x^3 - 3x + 1 = 0$$

with an initial approximation of $x_1 = 0$ to find the root between 0 and 1, the sequence obtained is

$$x_1 = 0, \qquad x_{n+1} = x_n - \frac{f(x_n)}{f'(x_n)} = x_n - \frac{x_n^3 - 3x_n + 1}{3x_n^2 - 3}, \qquad n \geq 1.$$

The first five terms in this sequence are

$$x_1 = 0, \qquad x_2 = 1/3, \qquad x_3 = 0.347222,$$
$$x_4 = 0.34729635, \qquad x_5 = 0.34729636,$$

which leads us to conclude that an approximation to the required root is $x = 0.3472964$. But this conclusion is based on the assumption that the recursive sequence is convergent and that it converges to the required root of the equation. We really have no basis for this assumption other than the fact that the first five terms of the sequence are getting closer together. Assumptions of this sort can often lead to bizarre results, and should therefore be avoided (see, for example, Exercise 38). In order that we may confidently accept Newton's iterative procedure, we discuss conditions that guarantee convergence of its sequence in Exercise 50.

Another way to solve for this same root is to rewrite the equation in the form

$$x = \tfrac{1}{3}(x^3 + 1),$$

and use this to define the recursive sequence

$$x_1 = 0, \qquad x_{n+1} = \tfrac{1}{3}(x_n^3 + 1), \qquad n \geq 1.$$

It is straightforward to show that this sequence is bounded above by 1 and below by 0 and that it is increasing. According to Theorem 11.1, it therefore has a limit L, which must lie in the interval $0 \leq L \leq 1$. To find L, we set

$$\lim_{n \to \infty} x_{n+1} = \frac{1}{3} \lim_{n \to \infty} (x_n^3 + 1),$$

or

$$L = \tfrac{1}{3}(L^3 + 1).$$

But this implies that L is the required root of the original equation. In other words, we have defined a recursive sequence $\{x_n\}$ in such a way that its limit is the required solution of $x^3 - 3x + 1 = 0$. We can now be sure that successive terms in this sequence will be better and better approximations to the solution of $x^3 - 3x + 1 = 0$. The first nine terms of the sequence are:

$$\begin{aligned} x_1 &= 0, & x_6 &= 0.34729353,\\ x_2 &= 1/3, & x_7 &= 0.34729601,\\ x_3 &= 0.345679, & x_8 &= 0.34729631,\\ x_4 &= 0.34710219, & x_9 &= 0.34729635,\\ x_5 &= 0.34727295, \end{aligned}$$

which leads us again to the conclusion that an approximation to the root of

$x^3 - 3x + 1 = 0$ between 0 and 1 is $x = 0.3472964$ (but this time with some measure of confidence).

This method of finding the root of an equation is often called the *method of successive substitutions*. It is discussed further in Exercise 49.

So far our approach to limits has been intuitive, but results in the remaining sections of this chapter require us to be more mathematical. To define limits precisely, we start with our intuitive description of a limit and make a succession of paraphrases, each of which is one step closer to a mathematical definition:

A sequence $\{c_n\}$ has limit L if its terms get arbitrarily close to L, and stay close to L, as n gets larger and larger.

A sequence $\{c_n\}$ has limit L if the difference $|c_n - L|$ gets arbitrarily close to zero, and stays close to zero, as n gets larger and larger.

A sequence $\{c_n\}$ has limit L if given any real number $\varepsilon > 0$, no matter how small, we can make the difference $|c_n - L|$ less than ε by choosing n sufficiently large.

Finally, we arrive at the following definition.

▷ **Definition 11.4** A sequence $\{c_n\}$ has *limit* L if for any given $\varepsilon > 0$, there exists an integer N such that for all $n > N$

$$|c_n - L| < \varepsilon. \quad \triangleleft$$

This definition puts in precise terms, and in the simplest possible way, our intuitive idea of a limit. (For those who have studied Section 2.3, note the similarity between Definition 11.4 and Definition 2.2.)

In this chapter we are primarily concerned with evaluating the limit L of a sequence $\{c_n\}$, rather than actually proving that L is the limit. We cannot, however, simply forget Definition 11.4, since proofs of many theorems in this chapter require an understanding of this definition. Take, for example, the following theorem, which we have implicitly assumed throughout our discussion of limits.

▶ **Theorem 11.2** A sequence can have only one limit. ◀

A proof of this theorem is easily formulated using Definition 11.4.

Proof Suppose on the contrary that a sequence $\{c_n\}$ has two distinct limits L_1 and L_2, where $L_2 > L_1$, and let $L_2 - L_1 = \delta$. Since $\lim_{n\to\infty} c_n = L_1$, if we set $\varepsilon = \delta/3$, then according to Definition 11.4, there exists an integer N_1 such that for all $n > N_1$

$$|c_n - L_1| < \varepsilon = \delta/3;$$

i.e., for $n > N_1$, all terms in the sequence are within a distance $\delta/3$ of L_1.

But similarly, there exists an N_2 such that for $n > N_2$, all terms in the sequence are within a distance $\varepsilon = \delta/3$ of L_2:

$$|c_n - L_2| < \varepsilon = \delta/3.$$

But this is impossible (see Fig. 11.2), since L_1 and L_2 are a distance δ apart. This contradiction therefore implies that $\{c_n\}$ cannot have two limits; it can have at most one. □

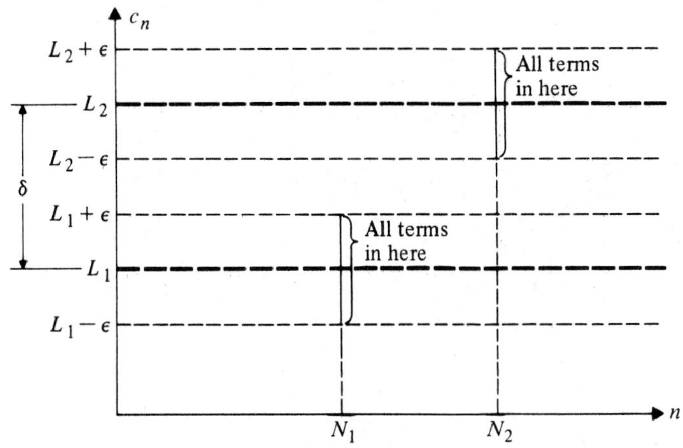

Figure 11.2

The following theorem, which states that convergent sequences can be added, subtracted, multiplied, and divided, can also be proved using Definition 11.4. (Only the first two parts, however, are straightforward. See Exercises 39, 51, and 52.)

▶ **Theorem 11.3** If sequences $\{c_n\}$ and $\{d_n\}$ have limits C and D, then:
i) $\{kc_n\}$ has limit kC if k is a constant;
ii) $\{c_n + d_n\}$ has limit $C + D$;
iii) $\{c_n d_n\}$ has limit CD;
iv) $\{c_n/d_n\}$ has limit C/D provided that $D \neq 0$, and none of the $d_n = 0$. ◀

EXERCISES 11.1

In each of Exercises 1–14, state (without proof) whether the sequence is monotonic, and whether it has an upper bound, a lower bound, and a limit.

1. $\{1/n\}$
2. $\{3^n + 1\}$
3. $\{3\}$
4. $\{(3/4)^{n+1}\}$
5. $\{(\frac{4}{3})^{n+1}\}$
6. $\{(-\frac{15}{16})^{n+5}\}$
7. $\{\sin (n\pi/2)\}$
8. $\{n/(n^2 + n + 2)\}$
9. $\{(-1)^n/n\}$
10. $\{\mathrm{Tan}^{-1}\, n\}$
11. $\{\ln (n)/(n^2 + 1)\}$
12. $\{(-1)^n\sqrt{n^2 + 1}\}$
13. $c_1 = 2,\ c_{n+1} = \dfrac{2}{c_n - 1},\ n \geq 1$
14. $c_1 = 4,\ c_{n+1} = -c_n/n^2,\ n \geq 1$

In each of Exercises 15–18, discuss, with all necessary proofs, whether the sequence is monotonic, and whether it has an upper bound, a lower bound, and a limit.

15. $\left\{\dfrac{n^2 + 5n - 4}{n^2 + 2n - 2}\right\}$
16. $c_1 = 3,\ c_{n+1} = \sqrt{5 + c_n},\ n \geq 1$
17. $c_1 = 3,\ c_{n+1} = 1 + \sqrt{6 + c_n},\ n \geq 1$
18. $\{ne^{-n}\}$

Find an explicit formula for the general term of the sequence in Exercises 19–22. In each case assume that the remaining terms follow the pattern suggested by the given terms.

19. $\frac{1}{2}, \frac{3}{4}, \frac{7}{8}, \frac{15}{16}, \frac{31}{32}, \ldots$
20. $4, \frac{7}{4}, \frac{10}{9}, \frac{13}{16}, \frac{16}{25}, \frac{19}{36}, \ldots$
21. $1, 0, 1, 0, 1, 0, \ldots$
22. $\dfrac{\ln 2}{\sqrt{2}}, -\dfrac{\ln 3}{\sqrt{3}}, \dfrac{\ln 4}{\sqrt{4}}, -\dfrac{\ln 5}{\sqrt{5}}, \ldots$
23. If all terms in a sequence $\{c_n\}$ are less than U, and $L = \lim_{n\to\infty} c_n$ exists, is it necessary that $L < U$?
24. The equilateral triangle in Fig. 11.3(a) has perimeter P. If each side of the triangle is divided into three equal parts and an equilateral triangle is drawn on each side, as shown in Fig. 11.3(b), what is the perimeter P_1 of this figure? If each side of Fig. 11.3(b) is now subdivided into three equal portions and equilateral triangles are constructed, as shown in Fig. 11.3(c), what is the perimeter

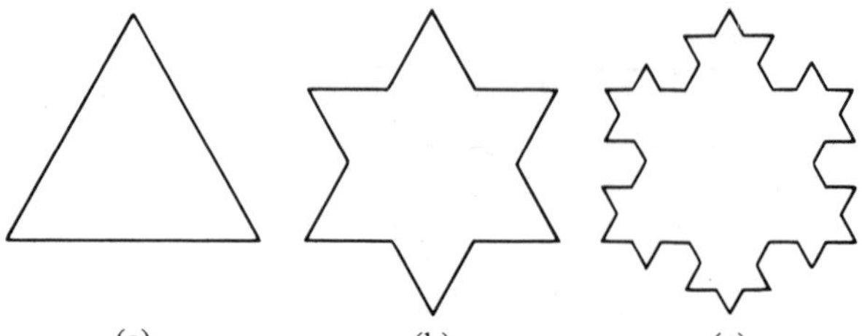

Figure 11.3

P_2 of this figure? If this subdivision process is continued indefinitely, what is the perimeter P_n after the nth subdivision? What is $\lim_{n\to\infty} P_n$?

C **25.** a) Use Newton's iterative procedure with $x_1 = 1$ to approximate the root of $x^4 - 15x + 2 = 0$ between 0 and 1.

b) Use the method of successive substitutions with $x_1 = 1$ and $x_{n+1} = (2 + x_n^4)/15$ to approximate the root in (a).

c) Use Newton's iterative procedure to approximate the root between 2 and 3.

d) What happens if the sequence in (b) is used to approximate the root between 2 and 3 with $x_1 = 2$ and $x_1 = 3$?

26. A superball is dropped from the top of a building 20 meters high. Each time it strikes the ground, it rebounds to 99% of the height from which it fell.

a) If d_n denotes the distance traveled by the ball between the nth and $(n + 1)$st bounce, find a formula for d_n.

b) If t_n denotes the time between the nth and $(n + 1)$st bounce, find a formula for t_n.

27. A dog sits at a farmhouse, carefully watching for his master to return from the fields. When the farmer is one kilometer from the farmhouse, the dog immediately takes off for the farmer. When he reaches the farmer, he turns and runs back to the farmhouse, whereupon he again turns and runs to the farmer. The dog continues this frantic action until the farmer reaches the farmhouse. If the dog runs twice as fast as the farmer, find the distance d_n run by the dog from the point when he reaches the farmer for the nth time to the point when he reaches the farmer for the $(n + 1)$st time. (Ignore any accelerations of the dog in the turns.)

28. Verify that the sequence $\{n^{1/n}\}$ is decreasing for $n \geq 3$ by showing that the function $f(x) = x^{1/x}$ has a negative derivative for $x \geq 3$. Illustrate this result graphically. What is the limit of this sequence?

29. A sequence $\{c_n\}$ is defined by $c_n = 3n/(n^2 - 5n + 1)$.

a) Consider the following proof that the sequence is decreasing: $c_{n+1} < c_n$ if and only if

$$\frac{3(n+1)}{(n+1)^2 - 5(n+1) + 1} < \frac{3n}{n^2 - 5n + 1}.$$

But this is true if and only if

$$3(n+1)(n^2 - 5n + 1) < 3n(n^2 - 3n - 3),$$

or

$$n^3 - 4n^2 - 4n + 1 < n^3 - 3n^2 - 3n;$$

i.e., if and only if

$$1 < n^2 + n.$$

Since this inequality is true for all $n \geq 1$, the sequence is decreasing. Do you see any error in this proof?

b) Calculate the first five terms in the sequence. What do you now conclude about the argument in (a)?

For the sequence in each of Exercises 30–33 discuss, with all necessary proofs, whether the sequence is monotonic, and whether it has bounds and a limit.

30. $c_1 = 1,\ c_{n+1} = \dfrac{1}{3 - c_n},\ n \geq 1$

31. $c_1 = 2,\ c_{n+1} = c_n^2 + 4c_n - 5,\ n \geq 1$

32. $c_1 = 1,\ c_{n+1} = \dfrac{4c_n}{4 + c_n},\ n \geq 1$

33. $c_1 = 1,\ c_2 = 2,\ c_{n+1} = c_n c_{n-1},\ n \geq 2$

Show how L'Hôpital's rule can be used to evaluate the limit for the sequence in each of Exercises 34–37.

34. $\left\{\dfrac{\ln(n)}{\sqrt{n}}\right\}$

35. $\left\{\dfrac{n^3 + 1}{e^n}\right\}$

36. $\left\{n \sin\left(\dfrac{4}{n}\right)\right\}$

37. $\left\{\left(\dfrac{n+5}{n+3}\right)^n\right\}$

38. a) Prove that if a sequence converges, the differences between successive terms in the sequence must approach zero.

b) The following example illustrates that the converse is not true. Show that the differences between successive terms of the sequence $\{\ln(n)\}$ approach zero, but the sequence itself diverges.

39. Prove parts (i) and (ii) of Theorem 11.3.

40. Find bounds for the sequence

$$c_1 = 1, \qquad c_{n+1} = \frac{1 + c_n}{1 + 2c_n}, \qquad n \geq 1.$$

41. Determine whether the following sequence is monotonic, has bounds, and has a limit.

$$c_1 = -30, \qquad c_2 = -20, \qquad c_{n+1} = 5 + \tfrac{1}{2}c_n + \tfrac{1}{3}c_{n-1}, \qquad n \geq 2.$$

42. If A_n is the area of the figure with perimeter P_n in Exercise 24, find a formula for A_n.

43. Prove that if $\lim_{n\to\infty} c_n = L$, there exists an integer N such that for all $n > N$,

$$c_n < L + 1.$$

44. A sequence $\{c_n\}$ has only positive terms. Prove that if $\lim_{n\to\infty} c_n = L < 1$, there exists an integer N such that all $n > N$,

$$c_n < \frac{L+1}{2}.$$

45. The *Fibonacci sequence* (used in computer programming) is defined by

$$c_1 = 1, \qquad c_2 = 1, \qquad c_{n+1} = c_n + c_{n-1}, \qquad n \geq 2.$$

a) Write out the first ten terms of this sequence.

b) Prove that

$$c_n^2 - c_{n-1}c_{n+1} = (-1)^{n+1}, \qquad n \geq 2.$$

c) Verify that an explicit formula for c_n is

$$c_n = \frac{1}{\sqrt{5}}\left\{\left(\frac{1+\sqrt{5}}{2}\right)^n - \left(\frac{1-\sqrt{5}}{2}\right)^n\right\}.$$

46. Consider the sequence

$$c_1 = 1, \qquad c_{n+1} = \frac{1}{2}\left(c_n + \frac{a}{c_n}\right), \qquad n \geq 1,$$

where a is any number greater than 1.

a) Show that $\sqrt{a} < c_n < a$ for $n \geq 2$.

b) Prove that the sequence is decreasing for $n \geq 2$.

c) Verify that the sequence converges to $\sqrt{a}$. (This result establishes an iterative procedure for finding the square root of real numbers. The sequence converges to $\sqrt{a}$ for $0 < a < 1$ also.)

d) Show that this sequence can be obtained from Newton's iterative procedure for the solution of the equation $x^2 = a$.

47. Consider the sequence $\{e_n\}$, where

$$e_n = \left(1 + \frac{1}{n}\right)^n.$$

a) Use the binomial theorem to show that e_{n+1} can be expressed in the form

$$e_{n+1} = 1 + \sum_{r=1}^{n+1} \frac{1}{r!}(1)\left(1 - \frac{1}{n+1}\right)\left(1 - \frac{2}{n+1}\right)\cdots\left(1 - \frac{r-1}{n+1}\right).$$

b) Use the result in (a) to prove that the sequence is increasing and bounded. This sequence therefore has a limit that, according to Eq. (8.29), is e.

* **48.** Prove that the sequence

$$c_1 = d, \qquad c_{n+1} = \sqrt{a + 2bc_n}, \qquad n \geq 1$$

(a, b, and d all positive constants) is increasing if and only if $d < b + \sqrt{a + b^2}$. What happens when $d = b + \sqrt{a + b^2}$?

49. a) Show that if α is the only root of the equation $x = f(x)$ and Fig. 11.4(a) is a graph of $f(x)$, then Fig. 11.4(b) exhibits geometrically the sequence of approximations of α determined by the method of successive substitutions.

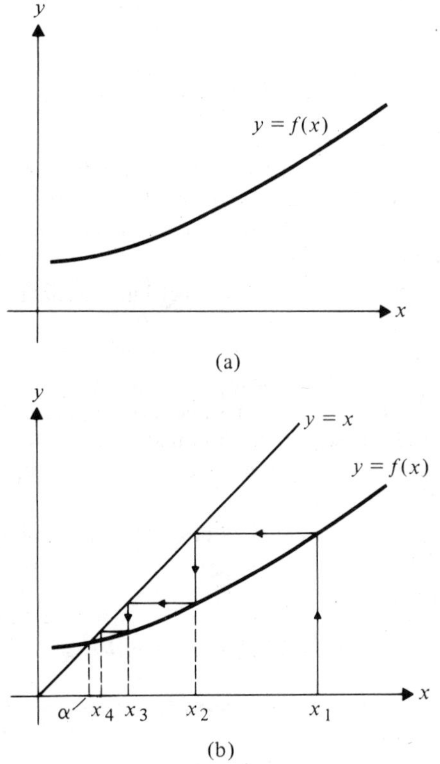

Figure 11.4

b) Illustrate graphically how the sequence defined by successive substitutions converges to the root $x = \alpha$ for the equation $x = f(x)$ if $f(x)$ is as shown in Fig. 11.5.

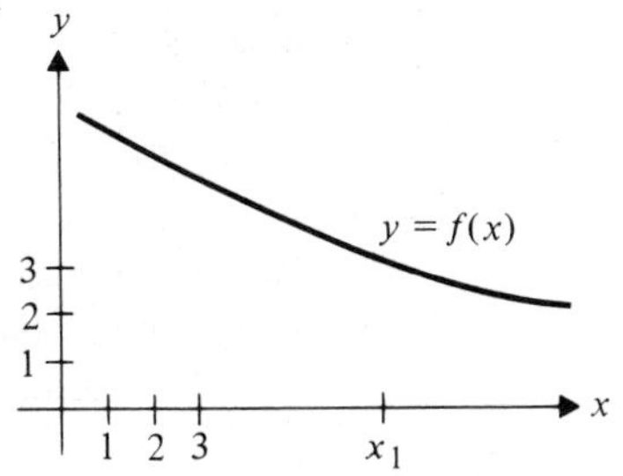

Figure 11.5

c) Illustrate graphically that the method fails for the function $f(x)$ shown in Fig. 11.6.

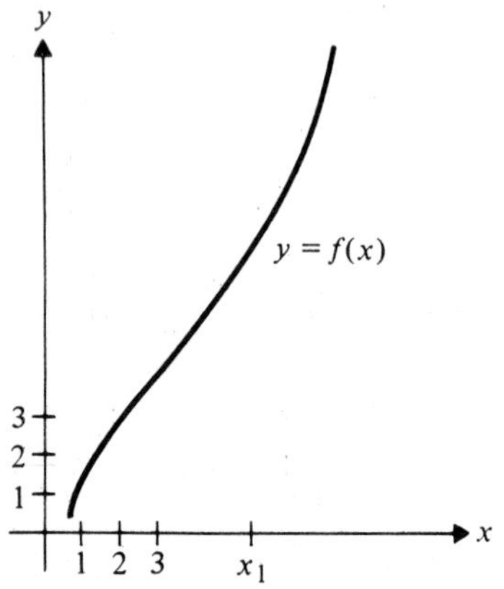

Figure 11.6

*d) Prove that if $x = f(x)$ has a root $x = \alpha$, the method of successive substitutions with an initial approximation of x_1 will always converge to α if $|f'(x)| \leq a < 1$ on the interval $|x - \alpha| \leq |x_1 - \alpha|$.

***50.** Suppose that $''(x)$ exists on an open interval containing a root $x = \alpha$ of the equation $f(x) = 0$. Use the result of Exercise 49 to prove that if $f'(\alpha) \neq 0$, Newton's iterative sequence will always converge to α provided the initial approximation x_1 is chosen sufficiently close to α. (*Hint:* First use Exercise 49(d) to show that Newton's sequence converges to α if on the interval $|x - \alpha| \leq |x_1 - \alpha|$, $|ff''/(f')^2| \leq a < 1$.)

***51.** In this exercise we outline a proof of part (iii) of Theorem 11.3.

a) Verify that

$$|c_n d_n - CD| \leq |c_n||d_n - D| + |D||c_n - C|.$$

b) Show that given any $\varepsilon > 0$, there exist positive integers N_1, N_2, and N_3 such that

$$|c_n| < |C| + 1 \qquad \text{whenever } n > N_1,$$

$$|d_n - D| < \frac{\varepsilon}{2(|C| + 1)} \qquad \text{whenever } n > N_2,$$

$$|c_n - C| < \frac{\varepsilon}{2|D|} \qquad \text{whenever } n > N_3.$$

c) Use these results to prove part (iii) of Theorem 11.3.

***52.** In this exercise we outline a proof of part (iv) of Theorem 11.3.

a) Verify that when $d_n \neq 0$ and $D \neq 0$,

$$\left|\frac{c_n}{d_n} - \frac{C}{D}\right| \leq \frac{|c_n - C|}{|d_n|} + \frac{|C||d_n - D|}{|D||d_n|}.$$

b) Show that given any $\varepsilon > 0$, there exist positive integers N_1, N_2, and N_3 such that

$$|d_n| > \frac{|D|}{2} \qquad \text{whenever } n > N_1,$$

$$|c_n - C| < \frac{\varepsilon|D|}{4} \qquad \text{whenever } n > N_2,$$

$$|d_n - D| < \frac{\varepsilon|D|^2}{4|C|} \qquad \text{whenever } n > N_3.$$

c) Now prove part (iv) of Theorem 11.3.

***53.** Find an explicit formula for the recursive sequence

$$c_1 = 1, \qquad c_2 = 2, \qquad c_{n+1} = \tfrac{1}{2}(c_n + c_{n-1}), \qquad n \geq 2.$$

11.2 INFINITE SERIES OF CONSTANTS

If $\{c_n\}$ is an infinite sequence of constants, an expression of the form

$$c_1 + c_2 + c_3 + \cdots + c_n + \cdots \tag{11.7}$$

is called an *infinite series* (or series). We often use sigma notation to represent a series:

$$\sum_{n=1}^{\infty} c_n = c_1 + c_2 + c_3 + \cdots + c_n + \cdots. \tag{11.8}$$

By initiating the sigma notation with $n = 1$, we once again have the convenient situation in which $n = 1$ yields the first term of the series, $n = 2$ the second term, etc. We usually abbreviate the notation further by writing $\sum c_n$ in place of

$\sum_{n=1}^{\infty} c_n$ whenever the lower limit of summation is chosen as unity. When the lower limit is not unity, it is included in the sigma notation.

From Example 1 we may form the following series.

Example 5 Write out the first six terms of each of the following series:

a) $\displaystyle\sum_{n=1}^{\infty} \frac{1}{2^{n-1}}$;

b) $\displaystyle\sum_{n=1}^{\infty} \frac{n}{n+1}$;

c) $\displaystyle\sum_{n=1}^{\infty} (-1)^n |n-3|$;

d) $\displaystyle\sum_{n=1}^{\infty} (-1)^{n+1}$.

Solution a) $\displaystyle\sum_{n=1}^{\infty} \frac{1}{2^{n-1}} = 1 + \frac{1}{2} + \frac{1}{4} + \frac{1}{8} + \frac{1}{16} + \frac{1}{32} + \cdots$

b) $\displaystyle\sum_{n=1}^{\infty} \frac{n}{n+1} = \frac{1}{2} + \frac{2}{3} + \frac{3}{4} + \frac{4}{5} + \frac{5}{6} + \frac{6}{7} + \cdots$

c) $\displaystyle\sum_{n=1}^{\infty} (-1)^n |n-3| = -2 + 1 - 0 + 1 - 2 + 3 - \cdots$

d) $\displaystyle\sum_{n=1}^{\infty} (-1)^{n+1} = 1 - 1 + 1 - 1 + 1 - 1 + \cdots$ □

Example 6 An *(infinite) geometric series* is defined as a series of the form

$$a + ar + ar^2 + ar^3 + \cdots,$$

where each term is obtained by multiplying its predecessor by the same constant r. The number $a \neq 0$ is called the first term of the series and r the common ratio. Write the geometric series in sigma notation.

Solution In sigma notation, we have

$$\sum_{n=1}^{\infty} ar^{n-1} = a + ar + ar^2 + ar^3 + \cdots. \ \square \tag{11.9}$$

Example 7 The series

$$\sum_{n=1}^{\infty} \frac{1}{n} = 1 + \frac{1}{2} + \frac{1}{3} + \frac{1}{4} + \cdots \tag{11.10}$$

is called the *harmonic series*. What is the sum of its first 20 terms?

Solution The sum of the first 20 terms of the harmonic series is

$$1 + \frac{1}{2} + \frac{1}{3} + \frac{1}{4} + \cdots + \frac{1}{20} = 3.5977. \ \square$$

It is quite convenient to write expressions like (11.7) and say that they are infinite series, but because we can add only finitely many numbers, (11.7) is as yet meaningless. We now remedy this by defining what we mean by an infinite series. We form a sequence $\{S_n\}$ defined by

$$\begin{aligned} S_1 &= c_1, \\ S_2 &= c_1 + c_2, \\ S_3 &= c_1 + c_2 + c_3, \\ &\vdots \\ S_n &= c_1 + c_2 + c_3 + \cdots + c_n, \\ &\vdots \end{aligned}$$

called the *sequence of partial sums* of the series $\sum c_n$. The nth term of the sequence $\{S_n\}$ represents the sum of the first n terms of the series. Suppose as n gets very large, the terms S_n of the sequence of partial sums get arbitrarily close to a number S and stay close to S. This implies that the more and more terms of the series that we add together, the closer and closer the sum will get to S. It seems reasonable, then, to call S the sum of the series. We therefore make the following definition.

▷ **Definition 11.5** Let $S_n = \sum_{k=1}^{n} c_k$ be the nth partial sum of a series $\sum_{n=1}^{\infty} c_n$. If the sequence of partial sums $\{S_n\}$ has limit S,

$$S = \lim_{n\to\infty} S_n,$$

we call S the sum of the series, and write

$$\sum_{n=1}^{\infty} c_n = S;$$

if $\{S_n\}$ does not have a limit, we say that the series does not have a sum. ◁

If a series has sum S, we say that the series *converges* to S, which means that its sequence of partial sums converges to S. If a series does not have a sum, we say that the series *diverges*, which means that its sequence of partial sums diverges. We now consider the series of Examples 5–7.

Example 8 For what values of r does the geometric series $\sum_{n=1}^{\infty} ar^{n-1}$ converge?

Solution If $\{S_n\}$ is the sequence of partial sums for the series, then

$$S_n = a + ar + ar^2 + \cdots + ar^{n-1}.$$

If we write

$$rS_n = ar + ar^2 + \cdots + ar^n,$$

it follows that $S_n - rS_n = a - ar^n$, and hence for $r \neq 1$,

$$S_n = \frac{a(1 - r^n)}{1 - r}. \tag{11.11a}$$

Furthermore, if $r = 1$,

$$S_n = na.$$

Since $\lim_{n\to\infty} r^n = 0$ for $|r| < 1$, we obtain the following result:

$$\lim_{n\to\infty} S_n = \begin{cases} \dfrac{a}{1-r}, & |r| < 1, \\ \text{does not exist}, & |r| \geq 1. \end{cases}$$

The infinite geometric series therefore has sum

$$\sum_{n=1}^{\infty} ar^{n-1} = \frac{a}{1-r} \tag{11.11b}$$

if $|r| < 1$, but otherwise it diverges. □

Example 9 Discuss convergence of the series in Example 5.

Solution

a) Series $\sum 1/2^{n-1}$ is a geometric series with $a = 1$ and $r = \frac{1}{2}$; hence according to formula (11.11b),

$$\sum_{n=1}^{\infty} \frac{1}{2^{n-1}} = \frac{1}{1-\frac{1}{2}} = 2.$$

b) Since every term of series $\sum n/(n+1)$ is greater than or equal to $\frac{1}{2}$, it follows that $S_n \geq n(\frac{1}{2}) = n/2$. Because the sequence of partial sums is therefore (increasing and) unbounded, it cannot possibly have a limit. The series does not therefore have a sum.

c) Examination of the first few partial sums of series $\sum (-1)^n|n-3|$ leads to the result that for $n > 1$,

$$S_n = \begin{cases} \dfrac{n-4}{2}, & \text{if } n \text{ is even}, \\ \dfrac{1-n}{2}, & \text{if } n \text{ is odd}. \end{cases}$$

Since the sequence $\{S_n\}$ does not have a limit, the series diverges.

d) Series $\sum (-1)^{n+1}$ is a geometric series with $r = -1$, and does not therefore have a sum. □

Example 10 Does the harmonic series $\sum 1/n$ converge?

Solution We note that

$$S_1 = 1,$$

$$S_2 = 1 + \frac{1}{2} = \frac{3}{2},$$

$$S_4 = S_2 + \frac{1}{3} + \frac{1}{4} > \frac{3}{2} + \frac{1}{4} + \frac{1}{4} = \frac{4}{2},$$

$$S_8 = S_4 + \frac{1}{5} + \frac{1}{6} + \frac{1}{7} + \frac{1}{8} > \frac{4}{2} + \frac{1}{8} + \frac{1}{8} + \frac{1}{8} + \frac{1}{8} = \frac{5}{2},$$

$$S_{16} = S_8 + \frac{1}{9} + \frac{1}{10} + \cdots + \frac{1}{16} > \frac{5}{2} + \frac{1}{16} + \frac{1}{16} + \cdots + \frac{1}{16} = \frac{6}{2}.$$

This procedure can be continued indefinitely, and shows that the sequence of partial sums is unbounded. The harmonic series therefore diverges. □

In each of the above examples, we used Definition 11.5 for the sum of a series to determine whether the series converges or diverges; i.e., we actually formed the sequence of partial sums $\{S_n\}$ in order to consider its limit. For most examples, it is too difficult, or even impossible, to evaluate S_n in closed form, and in such cases consideration of the limit of sequence $\{S_n\}$ is impractical. Consequently, we must develop alternative ways to decide on the convergence of a series. This we do in Sections 11.3 and 11.4.

First, however, we mention some fairly simple but important results on convergence of series. Suppose a series $\sum c_n$ has each of its terms multiplied by a constant k to produce the series $\sum kc_n$. It is evident from the sequences of partial sums that if $\sum c_n$ converges to S, then $\sum kc_n$ converges to kS; if $\sum c_n$ diverges, then $\sum kc_n$ diverges also (unless $k = 0$). Second, if a series $\sum c_n$ has a finite number of its terms altered in any fashion whatsoever, the new series converges if and only if the original series converges. The new series may converge to a different sum, but it will converge if the original series converges. Finally, if $\sum c_n$ and $\sum d_n$ are convergent with sums C and D, then $\sum (c_n + d_n)$ converges to $C + D$.

Our first convergence test is a corollary to the following theorem.

▶ **Theorem 11.4** If a series $\sum_{n=1}^{\infty} c_n$ converges, then $\lim_{n\to\infty} c_n = 0$. ◀

Proof Series $\sum c_n$ has a sum S if and only if its sequence of partial sums $\{S_n\}_2^{\infty}$ has limit S. It is also true that sequence $\{S_{n-1}\}_2^{\infty}$ has limit S. According to Theorem 11.3, sequence $\{S_n - S_{n-1}\}_2^{\infty}$ has limit $S - S = 0$. Since $S_n - S_{n-1} = c_n$, we have shown that $\lim_{n\to\infty} c_n = 0$. □

Theorem 11.4 states that a necessary condition for a series $\sum c_n$ to converge is that the sequence $\{c_n\}$ of its terms must approach zero. Unfortunately, this is not the kind of theorem for which we are searching. We wish theorems stating sufficient conditions to guarantee convergence or divergence of a series. Theorem 11.4 begins with convergence of a series and makes a conclusion from it. We can, however, immediately take the contrapositive of the theorem and obtain the following.

▶ **Corollary** (nth *term test*). If $\lim_{n\to\infty} c_n \neq 0$, or does not exist, then the series $\sum_{n=1}^{\infty} c_n$ diverges. ◀

This is our first convergence test: the nth term test. It is in fact a test for divergence rather than convergence, stating that if $\lim_{n\to\infty} c_n$ exists and is equal to anything but zero, or the limit does not exist, then the series $\sum c_n$ diverges. Note well that the *nth term test never indicates that a series converges.* Even if $\lim_{n\to\infty} c_n = 0$, we can conclude nothing about the convergence or divergence of $\sum c_n$; it may converge or it may diverge. For example, the harmonic series $\sum 1/n$ and the geometric series $\sum 1/2^{n-1}$ both satisfy the condition $\lim_{n\to\infty} c_n = 0$, yet one series diverges and the other converges. The nth term test therefore sometimes indicates that a series diverges, but it never indicates that a series converges.

In particular, both series $\sum(-1)^n|n-3|$ and $\sum(-1)^{n+1}$ of Example 9 diverge by the nth term test.

To understand series it is crucial that we distinguish clearly among three entities: the series itself, its sequence of partial sums, and its sequence of terms. For any series $\sum c_n$ we can form its sequence of terms $\{c_n\}$ and its sequence of partial sums $\{S_n\}$. Each of these sequences may give information about the sum of the series $\sum c_n$, but in very different ways.

The sum of the series is defined by its sequence of partial sums in that $\sum c_n$ has a sum if and only if $\{S_n\}$ has a limit. The sequence of partial sums therefore tells us definitely whether the series converges or diverges (provided, however, that we can evaluate S_n in closed form).

The sequence of terms, on the other hand, may or may not tell us whether the series diverges. If $\{c_n\}$ has no limit or has a limit other than zero, we know that the series does not have a sum. If $\{c_n\}$ has limit zero, we obtain no information about convergence of the series $\sum c_n$ and must continue our examination.

EXERCISES 11.2

In Exercises 1–10, determine whether the given series converges or diverges. Find the sum of each convergent series. To get a feeling for a series, it is sometimes helpful to write out its first few terms. Try it.

1. $\displaystyle\sum_{n=1}^{\infty}\frac{n+1}{2n}$

2. $\displaystyle\sum_{n=1}^{\infty}\frac{2^n}{5^{n+1}}$

3. $\displaystyle\sum_{n=1}^{\infty}\cos\left(\frac{n\pi}{2}\right)$

4. $\displaystyle\sum_{n=1}^{\infty}\left(\frac{n}{n+1}\right)^n$

5. $\displaystyle\sum_{n=1}^{\infty}\frac{7^{2n+3}}{3^{2n-2}}$

6. $\displaystyle\sum_{n=1}^{\infty}\frac{7^{n+3}}{3^{2n-2}}$

7. $\displaystyle\sum_{n=1}^{\infty}\sqrt{\frac{n^2-1}{n^2+1}}$

8. $\displaystyle\sum_{n=1}^{\infty}\frac{\cos(n\pi)}{2^n}$

9. $\displaystyle\sum_{n=1}^{\infty}\frac{4^n+3^n}{3^n}$

10. $\displaystyle\sum_{n=1}^{\infty}\text{Tan}^{-1}\,n$

In Exercises 11–14, we have given a repeating decimal. Express each as a geometric series and use formula (11.11b) to express the decimal as a rational number; i.e., as an integer divided by an integer.

11. 0.666666...

12. 0.1313131313...

13. 1.347346346346...

14. 43.020502050205...

Complete the statement in each of Exercises 15–17, and give a short proof to substantiate your claim.

15. If $\sum a_n$ and $\sum b_n$ converge, then $\sum(a_n+b_n)$

16. If $\sum a_n$ converges and $\sum b_n$ diverges, then $\sum(a_n+b_n)$

17. If $\sum a_n$ and $\sum b_n$ diverge, then $\sum(a_n+b_n)$

In Exercises 18–21, determine whether the series converges or diverges. Find the sum of each convergent series.

18. $\displaystyle\sum_{n=1}^{\infty}\frac{2^n+3^n}{4^n}$

19. $\displaystyle\sum_{n=1}^{\infty}\frac{3^n-1}{2^n}$

20. $\displaystyle\sum_{n=1}^{\infty}\frac{n^2+2^{2n}}{4^n}$

21. $\displaystyle\sum_{n=1}^{\infty}\frac{2^n+4^n-8^n}{2^{3n}}$

22. Find the sum of the series

$$\sum_{n=1}^{\infty}\frac{1}{n(n+1)}.$$

(*Hint:* Use partial fractions on the nth term and find the sequence of partial sums.)

23. In Exercise 26 of Section 11.1, find the total distance traveled by the superball before it comes to rest.

24. In Exercise 26 of Section 11.1, find the time taken for the superball to come to rest.

25. In Exercise 27 of Section 11.1, what distance does the dog run from the time when he sees the farmer until the farmer reaches the farmhouse?

26. In Exercise 42 of Section 11.1, find a simplified formula for the area A_n. What is $\lim_{n\to\infty} A_n$?

27. In Example 8 we showed that the nth partial sum of the geometric series $1+r+r^2+r^3+\cdots$ is $S_n=(1-r^n)/(1-r)$. If $\bar{S}_n$ denotes the nth partial sum of the series

$$1+2r+3r^2+4r^3+\cdots,$$

show that

$$\bar{S}_n-S_n=r(\bar{S}_n-nr^{n-1}).$$

Solve this equation for $\bar{S}_n$ and take the limit as n becomes infinite to show that

$$\sum_{n=1}^{\infty}nr^{n-1}=\frac{1}{(1-r)^2},\qquad |r|<1.$$

28. Two people flip a single coin to see who can first flip a head. Show that the probability that the first person to do so wins the game is represented by the series

$$\frac{1}{2} + \frac{1}{8} + \frac{1}{32} + \cdots + \frac{1}{2^{2n-1}} + \cdots.$$

What is the sum of this series?

29. Two people throw a single die to see who can first throw a six. Find the probability that the person who first throws wins the game.

30. One of Zeno's paradoxes describes a race between Achilles and a tortoise. Zeno claimed that if the tortoise is given a head start, then no matter how fast Achilles runs, he can never catch the tortoise. He reasoned as follows: In order to catch the tortoise, Achilles must first make up the length of the head start. But while he is running this distance, the tortoise "runs" a further distance. While Achilles makes up this distance, the tortoise covers a further distance, and so on. It follows that Achilles is always making up distance covered by the tortoise, and therefore can never catch the tortoise.

If the tortoise is given a head start of length L and Achilles runs $c > 1$ times as fast as the tortoise, use infinite series to show that Achilles does, in fact, catch the tortoise and that the distance he covers in doing so is $cL(c-1)^{-1}$.

31. Find the time between 1:05 and 1:10 when the minute and hour hands of a clock point in the same direction:

a) by reasoning with infinite series as in Exercise 30;

b) by finding expressions for the angular displacements of the hands as functions of time.

32. Repeat Exercise 31 for the instant between 10:50 and 10:55 when the hands coincide.

33. A child has a large number of identical cubical blocks, which she stacks as shown in Fig. 11.7. The top block protrudes $\frac{1}{2}$ its length over the second block, which protrudes $\frac{1}{4}$ its length over the third block, which protrudes $\frac{1}{6}$ its length over the fourth block, etc. Show that the center of mass of the top n blocks lies directly over the edge of the $(n+1)$st block. Now deduce that if a sufficient number of blocks is piled, the top block can be made to protrude as far over the bottom block as desired without the stack falling.

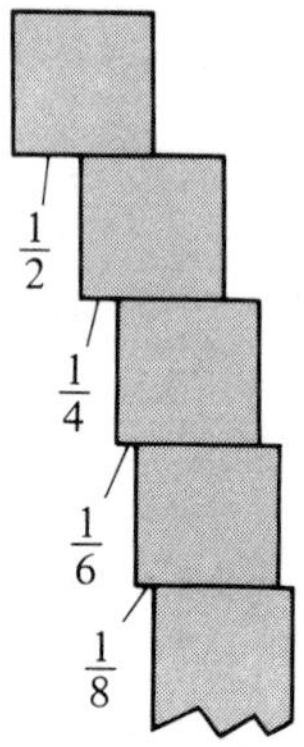

Figure 11.7

34. It is customary to assume that when a drug is administered to the human body it will be eliminated exponentially; i.e., if A represents the amount of drug in the body, then

$$A = A_0 e^{-kt},$$

where $k > 0$ is a constant and A_0 is the amount injected at time $t = 0$. Suppose that n successive injections of amount A_0 are administered at equally spaced time intervals T, the first injection at time $t = 0$.

a) Show that the amount of drug in the body at time t between the nth and $(n+1)$st injection is given by

$$A_n(t) = A_0 e^{-kt}\left\{\frac{1 - e^{knT}}{1 - e^{kT}}\right\}, \qquad (n-1)T < t < nT.$$

b) Sketch graphs of these functions on one set of axes.

c) What is the amount of drug in the body immediately after the nth injection for very large n; i.e., what is

$$\lim_{n\to\infty} A_n[(n-1)T]?$$

35. Prove the following result: If $\sum_{n=1}^{\infty} c_n$ converges, then its terms can be grouped in any manner, and the resulting series will be convergent with the same sum as the original series.

11.3 NONNEGATIVE SERIES

In order to obtain further convergence and divergence tests, we consider two classes of series:

1. Series with terms that are all nonnegative;
2. Series with both positive and negative terms.

We might consider a third classification of series with terms that are all nonpositive, but such a series is simply the negative of the first type listed above.

▷ **Definition 11.6** A series $\sum_{n=1}^{\infty} c_n$ is said to be *nonnegative* if each term is nonnegative: $c_n \geq 0$. ◁

For example, the harmonic series $\sum 1/n$ and the geometric series $\sum 1/2^{n-1}$ are both nonnegative series. We have already seen that both series have the property that $\lim_{n\to\infty} c_n = 0$, yet the harmonic series diverges and the geometric series converges. It appears that whether a nonnegative series does or does not have a sum depends on how quickly its sequence of terms $\{c_n\}$ approaches zero. In essence, then, the study of convergence of nonnegative series is an investigation into the question "How fast must the sequence of terms of a nonnegative series approach zero in order that the series have a sum?" This is not a simple problem, and we will give only a partial answer through the comparison and ratio tests. The comparison test is contained in the following theorem.

▶ **Theorem 11.5** (*Comparison test*). If $0 \leq c_n \leq a_n$ for all n and $\sum_{n=1}^{\infty} a_n$ converges, then $\sum_{n=1}^{\infty} c_n$ converges. If $c_n \geq d_n \geq 0$ for all n and $\sum_{n=1}^{\infty} d_n$ diverges, then $\sum_{n=1}^{\infty} c_n$ diverges. ◀

Proof Since $a_n \geq 0$ and $\sum a_n$ converges, its sequence of partial sums is nondecreasing and bounded (Theorem 11.1). Because $0 \leq c_n \leq a_n$ for all n, it follows that the sequence of partial sums of $\sum c_n$ is nondecreasing and bounded also. This sequence therefore has a limit and series $\sum c_n$ converges. A similar argument can be made for the divergent case. □

Essentially, Theorem 11.5 states that if the terms of a nonnegative series $\sum c_n$ are smaller than those of a known convergent series, then $\sum c_n$ must converge; if they are larger than a known nonnegative divergent series, then $\sum c_n$ must diverge.

In order for the comparison test to be useful, we require a group of known convergent and divergent series to which we may compare other series. Already at our disposal are geometric series,

$$\sum_{n=1}^{\infty} ar^{n-1} = a + ar + ar^2 + \cdots,$$

which converge if $|r| < 1$ and diverge if $|r| \geq 1$. A second extremely useful series is

$$\sum_{n=1}^{\infty} \frac{1}{n^p} = 1 + \frac{1}{2^p} + \frac{1}{3^p} + \frac{1}{4^p} + \cdots, \tag{11.12}$$

called the *p-series*. The special case of $p = 1$ is the harmonic series and has been shown to diverge. If $p < 1$,

$$\frac{1}{n^p} \geq \frac{1}{n},$$

and since the harmonic series diverges, the comparison test implies that the p-series diverges for $p < 1$.

We can use the curve $y = 1/x^p$ in Fig. 11.8(a) to prove that the p-series converges for $p > 1$. Area under this curve, above the x-axis, and to the right of $x = 1$ is defined by the improper integral

$$\int_1^{\infty} \frac{1}{x^p}\,dx = \lim_{b\to\infty} \int_1^b \frac{1}{x^p}\,dx = \lim_{b\to\infty} \left\{\frac{1}{-(p-1)x^{p-1}}\right\}_1^b$$
$$= \frac{1}{1-p}\left\{\lim_{b\to\infty} \frac{1}{b^{p-1}} - 1\right\} = \frac{1}{p-1},$$

and is therefore finite.

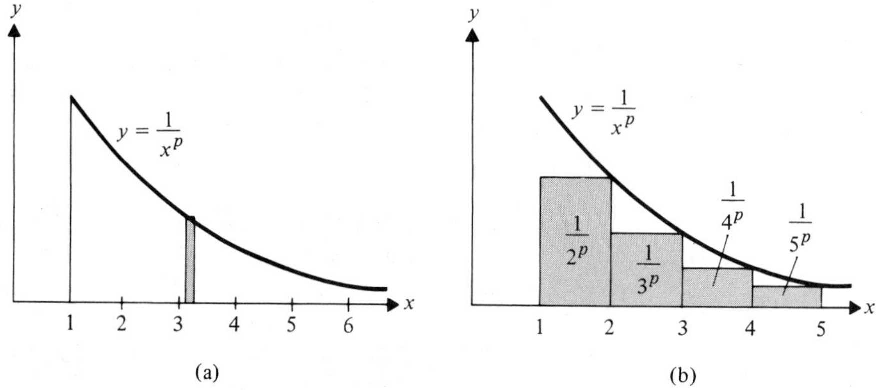

Figure 11.8

If $\{S_n\}$ denotes the sequence of partial sums of the p-series for $p > 1$, then

$$S_n = 1 + \frac{1}{2^p} + \frac{1}{3^p} + \cdots + \frac{1}{n^p},$$

and this sequence is clearly increasing. To show that it has an upper bound, we construct rectangles beneath the curve $y = 1/x^p$ as shown in Fig. 11.8(b). Since the areas of these rectangles are

$$\frac{1}{2^p},\ \frac{1}{3^p},\ \frac{1}{4^p},\ldots,$$

it follows that

$$S_n = 1 + \frac{1}{2^p} + \frac{1}{3^p} + \cdots + \frac{1}{n^p}$$
$$< 1 + \int_1^n \frac{1}{x^p}\,dx < 1 + \int_1^{\infty} \frac{1}{x^p}\,dx = 1 + \frac{1}{p-1} = \frac{p}{p-1}.$$

We have shown that an upper bound for sequence $\{S_n\}$ is $p/(p-1)$, and by Theorem 11.1, $\{S_n\}$ has a limit. The p-series therefore converges for $p > 1$.

We now have a fairly wide catalogue of known convergent and divergent series. Geometric series converge for $|r| < 1$ and diverge for $|r| \geq 1$; p-series converge for $p > 1$ and diverge for $p \leq 1$.

Example 11 Determine whether each of the following series converges or diverges:

a) $\sum_{n=1}^{\infty} \frac{n+4}{n^2}$;

b) $\sum_{n=1}^{\infty} \frac{2n^2-1}{15n^4+14}$;

c) $\sum_{n=1}^{\infty} \frac{2n^2+1}{15n^4-14}$;

d) $\sum_{n=1}^{\infty} \frac{\ln(n)}{n}$.

Solution a) For series $\sum (n+4)/n^2$, we note that

$$\frac{n+4}{n^2} > \frac{n}{n^2} = \frac{1}{n}.$$

Since $\sum 1/n$ diverges, so too does $\sum (n+4)/n^2$ (by the comparison test).

b) For series $\sum (2n^2-1)/(15n^4+14)$, we find that

$$\frac{2n^2-1}{15n^4+14} < \frac{2n^2}{15n^4} = \frac{2}{15n^2}.$$

Since $\sum 2/15n^2 = \frac{2}{15}\sum 1/n^2$ converges (p-series with $p = 2$), so too does the original series (by the comparison test).

c) For this series we note that the inequality

$$\frac{2n^2+1}{15n^4-14} \le \frac{3}{n^2}$$

is valid if and only if

$$45n^4 - 42 \ge 2n^4 + n^2.$$

But this inequality is valid if and only if

$$n^2(43n^2 - 1) \ge 42,$$

which is obviously true for $n \ge 1$. Consequently,

$$\frac{2n^2+1}{15n^4-14} \le \frac{3}{n^2},$$

and since $\sum 3/n^2 = 3\sum 1/n^2$ converges, so too does the original series (by the comparison test).

d) For this series, we note that for $n \ge 3$,

$$\frac{\ln(n)}{n} > \frac{1}{n}.$$

Since $\sum_{n=3}^{\infty} 1/n$ diverges (harmonic series with first two terms changed to zero), so too does $\sum_{n=3}^{\infty} \ln(n)/n$ (by the comparison test). Thus

$$\sum_{n=1}^{\infty} \frac{\ln(n)}{n}$$

diverges. □

Two observations about Example 11 are worthwhile:

1. To use the comparison test we must first have a suspicion as to whether the given series converges or diverges in order to discuss the correct inequality; that is, if we suspect that the given series converges, we search for a convergent series for the right-hand side of the inequality $\leq$, and if we suspect that the given series diverges, we search for a divergent series for the right-hand side of the opposite inequality $\geq$.
2. Recall that when the terms of a nonnegative series approach zero, whether the series has a sum depends on how fast these terms approach zero. The only difference in the nth terms of Examples 11(b) and (c) is the position of the negative sign. For very large n this difference becomes negligible since each nth term can, for large n, be closely approximated by $2/15n^2$. We might then expect similar analyses for these examples, and yet they are quite different. In addition, it is natural to ask where we obtained the factor 3 in Example 11(c). The answer is, "by trial and error."

Each of these observations points out a definite weakness in the comparison test, a problem that can be eliminated in many examples with the following.

▶ **Theorem 11.6** (*Limit comparison test*). If $0 \leq c_n$ and $0 < b_n$ and

$$\lim_{n\to\infty} \frac{c_n}{b_n} = l, 0 < l < \infty,$$

then series $\sum_{n=1}^{\infty} c_n$ converges if $\sum_{n=1}^{\infty} b_n$ converges, and diverges if $\sum_{n=1}^{\infty} b_n$ diverges. ◀

Proof Suppose that series $\sum b_n$ converges. Because sequence $\{c_n/b_n\}$ converges to l, we can say that for all n greater than or equal to some integer N,

$$\frac{c_n}{b_n} < l + 1, \quad \text{or} \quad c_n < (l + 1)b_n$$

(see Exercise 43 in Section 11.1). Since the series $\sum_{n=N}^{\infty} (l+1)b_n = (l+1)\sum_{n=N}^{\infty} b_n$ converges, so also must $\sum_{n=N}^{\infty} c_n$ (by the comparison test). Hence, $\sum_{n=1}^{\infty} c_n$ converges. A similar argument can be made for the divergent case. □

If $\lim_{n\to\infty} c_n/b_n = l$, then for large n we can say that $c_n \simeq lb_n$. It follows that if $\{b_n\}$ approaches zero, $\{c_n\}$ approaches zero $1/l$ times as fast as $\{b_n\}$. Theorem 11.6 implies then that if the sequence of nth terms of two nonnegative series approach zero at proportional rates, the series converge or diverge together.

To utilize the limit comparison test we again must find a series $\sum b_n$ so that the limit of ratio c_n/b_n is finite and greater than zero. To obtain a series $\sum b_n$ it is sufficient in many examples to simply answer the question, "What does the given series really look like for very large n?" In both Examples 11(b) and (c) we see that for large n

$$c_n \simeq \frac{2n^2}{15n^4} = \frac{2}{15n^2}.$$

Consequently, we calculate in Example 11(c) that

$$l = \lim_{n\to\infty} \frac{\dfrac{2n^2 + 1}{15n^4 - 14}}{(2/15n^2)} = 1.$$

Since $\sum 2/15n^2$ converges, so too does $\sum (2n^2 + 1)/(15n^4 - 14)$.

Example 12 Determine whether each of the following series converges or diverges:

a) $\displaystyle\sum_{n=1}^{\infty} \frac{\sqrt{n^2 + 2n - 1}}{n^{5/2} + 15n - 3}$;

b) $\displaystyle\sum_{n=1}^{\infty} \frac{2^n + 1}{3^n + 5}$.

Solution

a) For this series we note that for large n,

$$\frac{\sqrt{n^2 + 2n - 1}}{n^{5/2} + 15n - 3} \simeq \frac{n}{n^{5/2}} = \frac{1}{n^{3/2}}.$$

We calculate therefore that

$$l = \lim_{n\to\infty} \frac{\dfrac{\sqrt{n^2 + 2n - 1}}{n^{5/2} + 15n - 3}}{\dfrac{1}{n^{3/2}}} = 1.$$

Since $\sum 1/n^{3/2}$ converges, so too does the given series (by the limit comparison test).

b) For this series we note that for large n,

$$\frac{2^n + 1}{3^n + 5} \simeq \left(\frac{2}{3}\right)^n.$$

We calculate therefore that

$$l = \lim_{n\to\infty} \frac{\dfrac{2^n + 1}{3^n + 5}}{\left(\dfrac{2}{3}\right)^n} = 1.$$

Since $\sum (\frac{2}{3})^n$ converges (a geometric series with $r = \frac{2}{3}$), so too does the given series (by the limit comparison test). □

A second useful convergence test is the limit ratio test, which determines whether a series resembles a geometric series.

► **Theorem 11.7** (*Limit ratio test*). Suppose $c_n > 0$ and $\lim_{n\to\infty} c_{n+1}/c_n = L$. Then:

i) $\sum_{n=1}^{\infty} c_n$ converges if $L < 1$;
ii) $\sum_{n=1}^{\infty} c_n$ diverges if $L > 1$ (or if $\lim_{n\to\infty} c_{n+1}/c_n = \infty$);
iii) $\sum_{n=1}^{\infty} c_n$ may converge or diverge if $L = 1$. ◄

Proof

i) If the limit of sequence $\{c_{n+1}/c_n\}$ is less than 1, there exists an integer N such that for all $n \geq N$,

$$\frac{c_{n+1}}{c_n} < \frac{L + 1}{2}$$

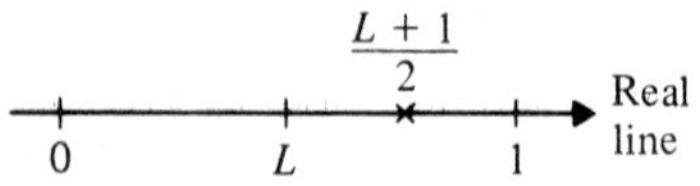

Figure 11.9

(see Exercise 44 in Section 11.1 and Fig. 11.9). As a result,

$$c_{N+1} < \left(\frac{L+1}{2}\right)c_N;$$

$$c_{N+2} < \left(\frac{L+1}{2}\right)c_{N+1} < \left(\frac{L+1}{2}\right)^2 c_N;$$

$$c_{N+3} < \left(\frac{L+1}{2}\right)c_{N+2} < \left(\frac{L+1}{2}\right)^3 c_N;$$

etc. Hence,

$$c_N + c_{N+1} + c_{N+2} + \cdots < c_N + \left(\frac{L+1}{2}\right)c_N + \left(\frac{L+1}{2}\right)^2 c_N + \cdots.$$

Since the right-hand side of this inequality is a geometric series with common ratio $(L+1)/2 < 1$, it follows (by the comparison test) that $\sum_{n=N}^{\infty} c_n$ converges. Therefore, $\sum_{n=1}^{\infty} c_n$ converges also.

ii) If $\lim_{n\to\infty} c_{n+1}/c_n = L > 1$ (or $\lim_{n\to\infty} c_{n+1}/c_n = \infty$), there exists an integer N such that for all $n \geq N$,

$$\frac{c_{n+1}}{c_n} > \frac{L+1}{2} > 1.$$

This clearly implies that for all $n > N$, $c_n > c_N$, and therefore

$$\lim_{n\to\infty} c_n \neq 0.$$

Hence $\sum c_n$ diverges by the nth term test.

iii) To show that the limit ratio test fails when $L = 1$, consider the two p-series $\sum 1/n$ and $\sum 1/n^2$. For each series, $L = 1$, yet the first series diverges and the second converges. □

Results of the limit ratio test certainly seem reasonable. If $\lim_{n\to\infty} c_{n+1}/c_n = L$, then for large n, each term of series $\sum c_n$ is essentially L times the term before it; i.e., the series resembles a geometric series with common ratio L. We would expect convergence of the series if $L < 1$ and divergence if $L > 1$. We might also anticipate some indecision about the $L = 1$ case (depending on how this limit is reached) since this case corresponds to the common ratio which separates convergent and divergent geometric series.

Example 13 Determine whether each of the following series converges or diverges:

a) $\displaystyle\sum_{n=1}^{\infty} \frac{2^n}{n^4}$;

b) $\displaystyle\sum_{n=1}^{\infty} \frac{n^{100}}{1\cdot 3\cdot 5\cdot \cdots \cdot (2n-1)}$;

c) $\displaystyle\sum_{n=1}^{\infty} \frac{n^n}{n!}$.†

† Recall that $n!$ is the notation representing the product of the first n positive integers:

$$n! = (1)(2)(3)\cdots(n).$$

Solution a) For series $\sum 2^n/n^4$,

$$L = \lim_{n\to\infty} \frac{\dfrac{2^{n+1}}{(n+1)^4}}{\dfrac{2^n}{n^4}} = 2,$$

and the series therefore diverges (by the limit ratio test).

b) For this series,

$$L = \lim_{n\to\infty} \frac{\dfrac{(n+1)^{100}}{1\cdot 3\cdot 5\cdot \cdots \cdot(2n+1)}}{\dfrac{n^{100}}{1\cdot 3\cdot 5\cdot \cdots \cdot(2n-1)}} = \lim_{n\to\infty}\left(\frac{n+1}{n}\right)^{100}\frac{1}{2n+1} = 0,$$

and the series therefore converges (by the limit ratio test).

c) For series $\sum n^n/n!$,

$$L = \lim_{n\to\infty} \frac{\dfrac{(n+1)^{n+1}}{(n+1)!}}{\dfrac{n^n}{n!}} = \lim_{n\to\infty}\left(\frac{n+1}{n}\right)^n = e > 1$$

(see Eq. 8.29), and the series diverges (by the limit ratio test). □

We now have at our disposal three tests that can be used to determine whether a series converges or diverges:

1. nth term test applicable to any series;
2. Comparison and limit comparison test applicable to nonnegative series;
3. Limit ratio test applicable to series of positive terms.

Each test in this section has the advantage that it avoids the sequence of partial sums $\{S_n\}$, and therefore eliminates the necessity of calculating S_n in closed form. On the other hand, this proves to have one disadvantage. Although each of the comparison, limit comparison, and limit ratio tests may indicate that a series has a sum, the tests in no way suggest the value of the sum. The problem of actually calculating the sum is often more difficult than showing that it exists in the first place. In Section 11.5, we begin a discussion on calculating the sum of a series.

There are many other tests for determining whether nonnegative series converge or diverge, two of which are discussed in Exercises 32 and 36. Other tests can be found in books devoted to infinite series.

EXERCISES 11.3

In Exercises 1–30, determine whether the given series converges or diverges.

1. $\displaystyle\sum_{n=1}^{\infty} \frac{e^n}{n^4}$

2. $\displaystyle\sum_{n=2}^{\infty} \frac{1}{n^3-1}$

3. $\displaystyle\sum_{n=1}^{\infty} \frac{1}{n!}$

4. $\displaystyle\sum_{n=1}^{\infty} \frac{1}{(2n-1)(2n+1)}$

5. $\displaystyle\sum_{n=1}^{\infty} \frac{n-5}{n^2+3n-2}$

6. $\displaystyle\sum_{n=1}^{\infty} \frac{n^3}{2^n}$

7. $\displaystyle\sum_{n=1}^{\infty} \frac{1}{n^n}$

8. $\displaystyle\sum_{n=1}^{\infty} \frac{(n-1)(n-2)}{n^2 2^n}$

9. $\sum_{n=1}^{\infty} \frac{1}{\sqrt{n(n+1)}}$

10. $\sum_{n=1}^{\infty} \frac{(2n)!}{(n!)^2}$

11. $\sum_{n=1}^{\infty} \frac{1}{2^n} \sin\left(\frac{\pi}{n}\right)$

12. $\sum_{n=1}^{\infty} \frac{e^{-n}}{\sqrt{n+\pi}}$

13. $\sum_{n=1}^{\infty} \frac{2^n+n}{3^n+1}$

14. $\sum_{n=1}^{\infty} n(\frac{3}{4})^n$

15. $\sum_{n=1}^{\infty} \frac{1+1/n}{e^n}$

16. $\sum_{n=1}^{\infty} \frac{\ln(n)}{n+1}$

17. $\sum_{n=1}^{\infty} \frac{1}{n^2} \operatorname{Sin}^{-1}\left(\frac{1}{n}\right)$

18. $\sum_{n=1}^{\infty} \frac{n^4+3}{5^{n/2}}$

19. $\sum_{n=1}^{\infty} \frac{\operatorname{Tan}^{-1} n}{n^2+1}$

20. $\sum_{n=2}^{\infty} \frac{1}{n^2 \ln(n)}$

21. $\sum_{n=1}^{\infty} \frac{\sqrt{n+1}}{n^{n+1/2}}$

22. $\sum_{n=1}^{\infty} \frac{3^{-n}+2^{-n}}{4^{-n}+5^{-n}}$

23. $\sum_{n=1}^{\infty} \frac{2 \cdot 4 \cdot \cdots \cdot (2n)}{4 \cdot 7 \cdot \cdots \cdot (3n+1)}$

24. $\sum_{n=1}^{\infty} \frac{n^{n-1}}{3^{n-1}(n-1)!}$

25. $\sum_{n=2}^{\infty} \frac{1+\ln^2 n}{n \ln^2 n}$

26. $\sum_{n=1}^{\infty} \frac{2 \cdot 4 \cdot \cdots \cdot (2n)}{3 \cdot 5 \cdot \cdots \cdot (2n+1)} \left(\frac{1}{n^2}\right)$

27. $\sum_{n=1}^{\infty} \frac{n^n}{(n+1)^{n+1}}$

28. $\sum_{n=1}^{\infty} \frac{(n+1)^n}{n^{n+1}}$

29. $\sum_{n=1}^{\infty} \frac{2^n+n^2 3^n}{4^n}$

30. $\sum_{n=1}^{\infty} \frac{n^2 2^n - n}{n^3+1}$

31. Prove that the sequence $\{n!/n^n\}$ has limit zero. (*Hint:* Consider $\sum_{n=1}^{\infty} n!/n^n$.)

32. The *integral test* for a series all of whose terms are positive is stated as follows: Suppose $c_n = f(n)$ and $f(x)$ is a continuous, positive, decreasing function for $x \geq 1$. Then the series $\sum_{n=1}^{\infty} c_n$ converges if the improper integral $\int_1^{\infty} f(x)\,dx$ converges, and diverges if the improper integral diverges. Prove the integral test.

In Exercises 33–35, use the integral test to determine whether the given series converges or diverges.

33. $\sum_{n=1}^{\infty} \frac{1}{n^p}$

34. $\sum_{n=1}^{\infty} n^2 e^{-n}$

35. $\sum_{n=2}^{\infty} \frac{1}{n \ln(n)}$

36. The *limit root test* for a nonnegative series is stated as follows: Suppose $c_n \geq 0$ and $R = \lim_{n\to\infty} (c_n)^{1/n}$. Then:
   i) $\sum_{n=1}^{\infty} c_n$ converges if $R < 1$;
   ii) $\sum_{n=1}^{\infty} c_n$ diverges if $R > 1$ (or if $\lim_{n\to\infty} (c_n)^{1/n} = \infty$);
   iii) $\sum_{n=1}^{\infty} c_n$ may converge or diverge if $R = 1$.

   Prove the limit root test.

In Exercises 37–39, use the limit root test to determine whether the given series converges or diverges.

37. $\sum_{n=1}^{\infty} (\frac{2}{3})^n$

38. $\sum_{n=1}^{\infty} \left(\frac{n}{n+1}\right)^{n^2}$

39. $\sum_{n=2}^{\infty} \frac{1}{(\ln(n))^n}$

40. For what values of a is the series

$$\sum_{n=1}^{\infty} \frac{(n!)^2}{(an)!} \qquad (a \geq 2 \text{ an integer})$$

convergent?

*41. Use the integral test (Exercise 32) to discuss convergence of the series

$$\sum_{n=2}^{\infty} \frac{1}{(\ln(n))^p}.$$

11.4 SERIES WITH POSITIVE AND NEGATIVE TERMS

We noted in Section 11.3 that the comparison, limit comparison, and limit ratio tests apply only to series with nonnegative or positive terms. Series with both positive and negative terms are somewhat more complicated. Certainly the nth term test can be applied to series with positive and negative terms, but remember that it may imply divergence but never convergence. Thus far, then, we have no tests for convergence of a series with positive and negative terms (except the definition of the sum in terms of its sequence of partial sums). It is natural to ask whether there is any possibility of utilizing the comparison, limit comparison, and limit ratio tests. The answer is yes, and the way is through the following definition.

▷ **Definition 11.7** A series $\sum_{n=1}^{\infty} c_n$ is said to be absolutely convergent if the series of absolute values $\sum_{n=1}^{\infty} |c_n|$ converges. ◁

At first glance it might seem that absolute convergence is a strange concept indeed. What possible good could it do to consider the series of absolute values, a series quite different from the original series? The fact is that if we can prove that the series of absolute values converges, it automatically follows that the original series converges also (see Theorem 11.8). And since the series of absolute values has all nonnegative or positive terms, we can use the comparison, limit comparison, or limit ratio test to consider its convergence.

▶ **Theorem 11.8** If a series converges absolutely, then it converges. ◀

Proof Let $\{S_n\}$ be the sequence of partial sums of the absolutely convergent series $\sum c_n$. We define sequences $\{P_n\}$ and $\{N_n\}$ where P_n is the sum of all positive terms in S_n, and N_n is the sum of the absolute values of all negative terms in S_n. Then

$$S_n = P_n - N_n.$$

The sequence of partial sums for the series of absolute values $\sum |c_n|$ is

$$\{P_n + N_n\},$$

and this sequence must be nondecreasing and bounded. Because each of the sequences $\{P_n\}$ and $\{N_n\}$ is nondecreasing and a part of $\{P_n + N_n\}$, it follows that each is bounded, and therefore has a limit, say P and N, respectively. As a result, sequence $\{P_n - N_n\} = \{S_n\}$ has limit $P - N$, and series $\sum c_n$ converges to $P - N$. □

Example 14 Determine whether each of the following series converges:

a) $\displaystyle\sum_{n=1}^{\infty} \frac{(-1)^n n}{(n+1)2^n}$;

b) $$1 - \frac{1}{2^2} - \frac{1}{3^3} + \frac{1}{4^4} + \frac{1}{5^5} + \frac{1}{6^6} - \frac{1}{7^7} - \frac{1}{8^8} - \frac{1}{9^9} - \frac{1}{10^{10}} + \frac{1}{11^{11}} + \cdots + \frac{1}{15^{15}} - \cdots.$$

Solution a) The series of absolute values is $\sum n/[(n+1)2^n]$. Since

$$l = \lim_{n\to\infty} \frac{\dfrac{n}{(n+1)2^n}}{\dfrac{1}{2^n}} = 1,$$

and $\sum (\frac{1}{2})^n$ is convergent (a geometric series with $r = \frac{1}{2}$), it follows that the series of absolute values converges (by the limit comparison test). The original series therefore converges absolutely.

b) The series of absolute values is $\sum 1/n^n$. Since

$$\frac{1}{n^n} \le \frac{1}{n^2}$$

and $\sum 1/n^2$ converges, so too does $\sum 1/n^n$ (by the comparison test). The original series therefore converges absolutely. □

In Example 14(a), absolute convergence of series $\sum (-1)^n n/[(n+1)2^n]$ implies convergence of the series itself, but it is customary to omit such a statement. It is important to realize, however, that it is the original series $\sum (-1)^n n/[(n+1)2^n]$ that is being analyzed, and its convergence is guaranteed by Theorem 11.8.

We must now ask whether series can converge without converging absolutely. If there are such series, and indeed there are, we must devise new convergence tests. We describe these series as follows.

▷ **Definition 11.8** A series that converges but does not converge absolutely is said to *converge conditionally.* ◁

The most important type of series with both positive and negative terms is an alternating series. As the name suggests, *an alternating series has terms that are alternately positive and negative.* For example,

$$\sum_{n=1}^{\infty} \frac{(-1)^{n+1}}{n} = 1 - \frac{1}{2} + \frac{1}{3} - \frac{1}{4} + \frac{1}{5} - \cdots$$

is an alternating series (called the *alternating harmonic series*).

Given an alternating series to examine for convergence, we first test for absolute convergence. Should this fail, we test for conditional convergence as follows.

▶ **Theorem 11.9** (*Alternating series test*). An alternating series $\sum_{n=1}^{\infty} c_n$ converges if the sequence of absolute values of the terms $\{|c_n|\}$ is nonincreasing and has limit zero. ◀

Proof Suppose $c_1 > 0$, in which case all odd terms are positive and all even terms are negative. If $\{S_n\}$ is the sequence of partial sums of $\sum c_n$, S_{2n} can be expressed in the two forms

$$S_{2n} = (c_1 + c_2) + (c_3 + c_4) + \cdots + (c_{2n-1} + c_{2n}),$$

$$S_{2n} = c_1 + (c_2 + c_3) + (c_4 + c_5) + \cdots + (c_{2n-2} + c_{2n-1}) + c_{2n}.$$

Since $\{|c_n|\}$ is nonincreasing, each term in the parentheses of the first expression is nonnegative. Consequently, the subsequence $\{S_{2n}\}$ of $\{S_n\}$ is nondecreasing. Since each term in the parentheses in the second expression is nonpositive, as is c_{2n}, it follows that $S_{2n} \le c_1$ for all n. By Theorem 11.1, then, the sequence $\{S_{2n}\}$ has a limit, say, S.

The subsequence of odd partial sums $\{S_{2n-1}\}$ is such that $S_{2n-1} = S_{2n} - c_{2n}$. Since $\{S_{2n}\}$ has limit S and $\{c_{2n}\}$ has limit zero, it follows that $\{S_{2n-1}\}$ has limit S also. Consequently, $\{S_n\}$ has limit S, and series $\sum c_n$ converges. A similar proof holds if $c_1 < 0$. □

Example 15 Determine whether each of the following series converges absolutely, converges conditionally, or diverges:

a) $\displaystyle\sum_{n=1}^{\infty} \frac{(-1)^{n+1}}{n}$;

b) $\displaystyle\sum_{n=1}^{\infty} (-1)^n \frac{\sqrt{n^2+5n}}{n^{3/2}}$;

c) $\displaystyle\sum_{n=1}^{\infty} (-1)^{n+1} \frac{4^n}{n^5 3^n}$.

Solution a) The alternating harmonic series $\sum (-1)^{n+1}/n$ is not absolutely convergent since $\sum 1/n$ diverges. Since the sequence of absolute values of the terms $\{1/n\}$ is decreasing with limit zero, series $\sum (-1)^{n+1}/n$ converges conditionally.

b) For the alternating series $\sum (-1)^n(\sqrt{n^2+5n}/n^{3/2})$, we first consider the series of absolute values $\sum \sqrt{n^2+5n}/n^{3/2}$. Since

$$l = \lim_{n\to\infty} \frac{\dfrac{\sqrt{n^2+5n}}{n^{3/2}}}{\dfrac{1}{n^{1/2}}} = 1$$

and $\sum 1/n^{1/2}$ diverges, so too does $\sum \sqrt{n^2+5n}/n^{3/2}$ (by the limit comparison test). The original series does not therefore converge absolutely. The sequence $\{\sqrt{n^2+5n}/n^{3/2}\}$ of absolute values of the terms of the series is nonincreasing if

$$\frac{\sqrt{(n+1)^2+5(n+1)}}{(n+1)^{3/2}} \le \frac{\sqrt{n^2+5n}}{n^{3/2}},$$

or

$$\begin{aligned} n^3(n^2+7n+6) &\le (n^2+5n)(n+1)^3 \\ &= n^5 + 8n^4 + 18n^3 + 16n^2 + 5n; \end{aligned}$$

i.e., if

$$n^4 + 12n^3 + 16n^2 + 5n \ge 0,$$

which is obviously valid. Since further $\lim_{n\to\infty} \sqrt{n^2+5n}/n^{3/2} = 0$, alternating series $\sum (-1)^n(\sqrt{n^2+5n}/n^{3/2})$ converges conditionally.

c) If we apply the limit ratio test to the series of absolute values $\sum 4^n/(n^5 3^n)$, we have

$$L = \lim_{n\to\infty} \frac{\dfrac{4^{n+1}}{(n+1)^5 3^{n+1}}}{\dfrac{4^n}{n^5 3^n}} = \lim_{n\to\infty} \frac{4}{3}\left(\frac{n}{n+1}\right)^5 = \frac{4}{3}.$$

The original alternating series does not therefore converge absolutely. But the fact that $L = \frac{4}{3}$ tells us more. It implies that for large n, each term in the series of absolute values is approximately $\frac{4}{3}$ times the term that precedes it, and therefore

$$\lim_{n\to\infty} \frac{4^n}{n^5 3^n} = \infty.$$

Consequently,

$$\lim_{n\to\infty} (-1)^{n+1} \frac{4^n}{n^5 3^n}$$

cannot possibly exist, and the original series diverges by the nth term test. □

We have noted several times that the essential question for convergence of a nonnegative series is, "Do the terms approach zero quickly enough to guarantee convergence of the series?" With a series that has infinitely many positive and negative terms, this question is improper. Such a series may converge because of a partial canceling effect; for example, a negative term may offset the effect of a large positive term. This kind of process may produce a convergent series even though the series would be divergent if all terms were replaced by their absolute values. In particular, the alternating harmonic series converges (conditionally) because of this canceling effect, whereas the harmonic series itself, which has no cancellations, diverges.

In Sections 11.2–11.4, we have obtained a number of tests for determining whether a series converges or diverges. To test a series for convergence, we suggest the following procedure:

1. Try the nth term test for divergence.
2. If $\{c_n\}$ has limit zero and the series is nonnegative or positive, try the comparison, limit comparison, or limit ratio test.
3. If $\{c_n\}$ has limit zero and the series contains both positive and negative terms, try for absolute convergence as in (2). If this fails and the series is alternating, try for conditional convergence with the alternating series test.

Each of the comparison, limit comparison, limit ratio, and alternating series tests requires conditions to be satisfied for all terms of the series. Specifically, the comparison and limit comparison tests require $c_n \geq 0$ for all n; the limit ratio test requires $c_n > 0$; and the alternating series test requires $\{|c_n|\}$ to be nonincreasing and $\{c_n\}$ to be alternately positive and negative. None of these requirements is essential for all n; in fact, so long as they are satisfied for all terms in the series beyond some point, say for n greater than or equal to some integer N, the particular test may be used on the series $\sum_{n=N}^{\infty} c_n$. The original series $\sum_{n=1}^{\infty} c_n$ then converges if and only if $\sum_{n=N}^{\infty} c_n$ converges.

EXERCISES 11.4

In Exercises 1–10, determine whether the series converges absolutely, converges conditionally, or diverges.

1. $\sum_{n=1}^{\infty} (-1)^n \frac{n}{n^3+1}$
2. $\sum_{n=1}^{\infty} (-1)^n \frac{n}{n^2+1}$
3. $\sum_{n=1}^{\infty} \frac{\cos(n\pi/2)}{2n^2}$
4. $\sum_{n=1}^{\infty} (-1)^n \frac{n^3}{3^n}$
5. $\sum_{n=1}^{\infty} \frac{(-1)^{n+1}}{\sqrt{n}}$
6. $\sum_{n=1}^{\infty} (-1)^n \frac{3^n}{n^3}$

7. $\sum_{n=1}^{\infty} (-1)^n \dfrac{n}{n^2+n+2}$

8. $\sum_{n=1}^{\infty} \dfrac{n \sin (n\pi/4)}{2^n}$

9. $\sum_{n=1}^{\infty} (-1)^{n+1} \left(\dfrac{n}{n+1}\right)$

10. $\sum_{n=1}^{\infty} (-1)^{n+1} \dfrac{\sqrt{3n-2}}{n}$

11. Show that the sequence $\{x^n/n!\}$ converges to zero for any x whatsoever. (*Hint:* Consider the series $\sum_{n=1}^{\infty} x^n/n!$.)

In Exercises 12–15, determine whether the series converges absolutely, converges conditionally, or diverges.

12. $\sum_{n=1}^{\infty} (-1)^n \left(\dfrac{n}{n+1}\right)^n$

13. $\sum_{n=1}^{\infty} (-1)^n \dfrac{\sqrt{n^2+3}}{n^2+5}$

14. $\sum_{n=2}^{\infty} (-1)^{n-1} \dfrac{\ln (n)}{n}$

15. $\sum_{n=1}^{\infty} \dfrac{\cos (n\pi/10) \operatorname{Cot}^{-1} n}{n^3+5n}$

16. Discuss convergence of the series

$$\sum_{n=1}^{\infty} \frac{\sin (nx)}{n^2}.$$

17. Prove that if $\sum c_n$ converges absolutely, then $\sum c_n^p$ converges absolutely for all integers $p > 1$.

* **18.** Discuss convergence of the series

$$\sum_{n=1}^{\infty} (-1)^n \frac{n^n}{(n+1)^{n+1}}.$$

11.5 CALCULATION OF THE SUM OF A SERIES

We have considered only one-half the convergence problem for infinite series of constants. The comparison, limit comparison, limit ratio, and alternating series tests may determine whether a series converges or diverges, but they do not determine the sum of the series in the case of a convergent series. This part of the problem, as has been suggested before, can sometimes be far more difficult.

If the convergent series is a geometric series, no problem exists for we can use formula (11.11b) to find its sum. It may also happen that the nth partial sum S_n of the series can be calculated in closed form, in which case the sum of the series is $\lim_{n\to\infty} S_n$. Cases of this latter type are rare and quite often artificial. These two methods exhaust our present capabilities for obtaining the sum of a convergent series, but we will find additional methods in Sections 11.7 and 11.8.

In many applications of infinite series we need only a reasonable approximation to the sum of a series, and we therefore turn our attention to the problem of estimating the sum of a series. The easiest method for estimating the sum S of a series $\sum c_n$ is simply to choose the partial sum S_N for some fixed N as an approximation; i.e., simply truncate the series after N terms and choose

$$S \simeq S_N = c_1 + c_2 + c_3 + \cdots + c_N.$$

But an approximation is of value only if we can make some definitive statement about its accuracy. In truncating the series, we have neglected the infinity of terms $\sum_{n=N+1}^{\infty} c_n$, and the accuracy of the approximation is therefore determined by the size of $\sum_{n=N+1}^{\infty} c_n$: The smaller it is, the better the approximation. The problem is that we do not know the exact value of $\sum_{n=N+1}^{\infty} c_n$; if we did, there would be no need to approximate the sum of the original series in the first place. What we must do then is estimate the sum $\sum_{n=N+1}^{\infty} c_n$. Consider, for example, using the first seven terms of the convergent series $\sum_{n=1}^{\infty} 1/n!$ to approximate its sum. A quick calculation gives $S_7 = 1.7182539$. The remainder of the terms have sum

$$\begin{aligned}
\sum_{n=8}^{\infty} \frac{1}{n!} &= \frac{1}{8!} + \frac{1}{9!} + \frac{1}{10!} + \frac{1}{11!} + \cdots \\
&= \frac{1}{8!}\left\{1 + \frac{1}{9} + \frac{1}{9 \cdot 10} + \frac{1}{9 \cdot 10 \cdot 11} + \cdots\right\} \\
&< \frac{1}{8!}\left\{1 + \frac{1}{9} + \frac{1}{9^2} + \frac{1}{9^3} + \cdots\right\} \\
&= \frac{1}{8!} \cdot \frac{1}{1 - \frac{1}{9}} \qquad \text{(Using Eq. 11.11b)} \\
&= \frac{9}{8 \cdot 8!} \\
&< 0.0000280.
\end{aligned}$$

We may write, therefore, that

$$1.7182539 < \sum_{n=1}^{\infty} \frac{1}{n!} < 1.7182819,$$

or, to four decimals,

$$\sum_{n=1}^{\infty} \frac{1}{n!} = 1.7183.$$

In practical situations, it frequently happens that we must decide how many terms of a series to take in order to guarantee a certain degree of accuracy. This type of problem is illustrated in the following example.

Example 16 How many terms in the convergent series $\sum_{n=1}^{\infty} n/[(n + 1)3^n]$ will ensure a truncation error of less than 10^{-5}?

Solution If this series is truncated after the Nth term, the error is

$$\begin{aligned}
\sum_{n=N+1}^{\infty} \frac{n}{(n + 1)3^n} &= \frac{N + 1}{(N + 2)3^{N+1}} + \frac{N + 2}{(N + 3)3^{N+2}} + \cdots \\
&< \frac{1}{3^{N+1}} + \frac{1}{3^{N+2}} + \cdots \\
&= \frac{\frac{1}{3^{N+1}}}{1 - \frac{1}{3}} \qquad \text{(Using Eq. 11.11b)} \\
&= \frac{1}{2 \cdot 3^N}.
\end{aligned}$$

Consequently, the error is guaranteed to be less than 10^{-5} if N satisfies the inequality

$$\frac{1}{2 \cdot 3^N} < 10^{-5},$$

or

$$2 \cdot 3^N > 10^5.$$

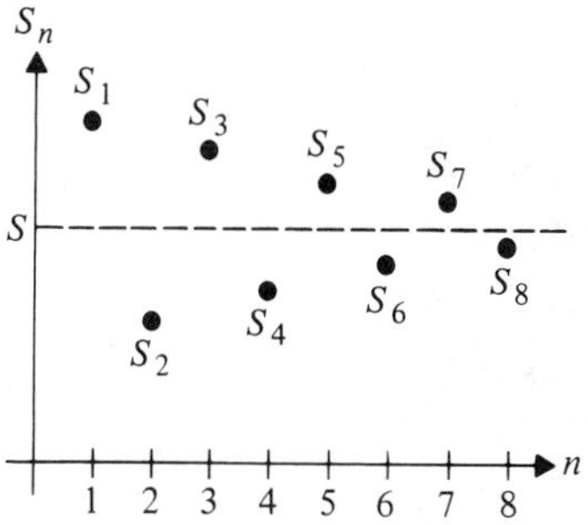

Figure 11.10

A quick calculation shows that this is true if $N \geq 10$, and therefore 10 or more terms will yield the required accuracy. □

It is very simple to obtain an estimate of the accuracy of a truncated alternating series $\sum c_n$ if the sequence $\{|c_n|\}$ is nonincreasing with limit zero. In the proof of Theorem 11.9, we showed that if $c_1 > 0$, the subsequence $\{S_{2n}\}$ of partial sums is nondecreasing and therefore approaches the sum of the series $\sum c_n$ from below (Fig. 11.10). In a similar way, we can show that subsequence $\{S_{2n-1}\}$ is nonincreasing and approaches the sum of the series from above. It therefore follows that the sum $\sum c_n$ must be somewhere between any two terms of the subsequences $\{S_{2n}\}$ and $\{S_{2n-1}\}$. In particular, it must be between any two successive partial sums S_N and S_{N+1}. Thus if $\sum c_n$ is approximated by S_N, the maximum possible error is $S_{N+1} - S_N = c_{N+1}$, the next term.

Example 17 Use the first twenty terms of the series $\sum_{n=1}^{\infty} (-1)^{n+1}/n^3$ to estimate its sum. Obtain an error estimate.

Solution The sum of the first twenty terms of the series is 0.901485. Since the series is alternating, the maximum possible error in this estimate is the twenty-first term $1/21^3 < 0.000108$. Thus,

$$0.901485 < \sum_{n=1}^{\infty} \frac{(-1)^{n+1}}{n^3} < 0.901593. \ \square$$

The value 0.901485 for S_{20} in this example is correct to six decimals. Since roundoff error would have crept into the calculation of $\sum_{n=1}^{20} (-1)^{n+1}/n^3$, more than six decimals were carried in arriving at this result.

EXERCISES 11.5

In Exercises 1–8, use the number of terms indicated to find an approximation to the sum of the series. In each case, obtain an estimate of the truncation error.

C **1.** $\sum_{n=1}^{\infty} \frac{1}{n2^n}$ (10 terms)

C **2.** $\sum_{n=1}^{\infty} \frac{(-1)^n}{n^4}$ (20 terms)

C **3.** $\sum_{n=1}^{\infty} \frac{1}{n^n}$ (5 terms)

C **4.** $\sum_{n=2}^{\infty} \frac{(-1)^{n+1}}{n^3 3^n}$ (3 terms)

C **5.** $\sum_{n=1}^{\infty} \frac{1}{2^n} \sin\left(\frac{\pi}{n}\right)$ (15 terms)

C **6.** $\sum_{n=2}^{\infty} \frac{2^n - 1}{3^n + n}$ (20 terms)

C **7.** $\sum_{n=2}^{\infty} \frac{2^n + 1}{3^n + n}$ (20 terms)

C **8.** $\sum_{n=1}^{\infty} \frac{(-1)^n}{n}$ (100 terms)

How many terms in the series in Exercises 9–11 will guarantee an approximation to the sum with a truncation error of less than 10^{-4}?

9. $\sum_{n=1}^{\infty} \frac{(-1)^n}{n^2}$

10. $\sum_{n=1}^{\infty} \frac{1}{n^2 4^n}$

11. $\sum_{n=1}^{\infty} \frac{2^n}{n!}$

12. This exercise shows that we must be very careful in predicting the accuracy of a result. Consider the series

$$S = 3.1251001 - 0.00009018\left\{1 + \frac{1}{10} + \frac{1}{10^2} + \frac{1}{10^3} + \cdots\right\}.$$

a) Show that the sum of this series is exactly $S = 3.1249999$. To two decimals, then, the value of S is 3.12.

b) Verify that the first four partial sums of the series are

$$S_1 = 3.1251001,$$
$$S_2 = 3.12500992,$$
$$S_3 = 3.125000902,$$
$$S_4 = 3.1250000002.$$

c) If $E_n = S_n - S$ are the differences between the sum of the series and its first four partial sums, show that

$$E_1 = 0.0001002,$$
$$E_2 = 0.00001002,$$
$$E_3 = 0.000001002,$$
$$E_4 = 0.0000001002.$$

What can you say about the accuracy of S_1, S_2, S_3, and S_4 as approximations to S?

d) If S is approximated by any of S_1, S_2, S_3, or S_4 to two decimals, the result is 3.13, not 3.12 as in (a). Thus in spite of the accuracy predicted in (c), S_1, S_2, S_3, and S_4 do not predict S correctly to two decimals. Do they predict S correctly to three or four decimals?

11.6 POWER SERIES AND INTERVALS OF CONVERGENCE

An infinite sequence of functions is the assignment of functions to integers,

$$\{f_n(x)\}_0^\infty = f_0(x),\ f_1(x),\ f_2(x), \ldots, \tag{11.13}$$

where you will note that we have initiated our assignment with the integer $n = 0$. The reason for this will become apparent shortly. From such a sequence we may form an infinite series of functions:

$$\sum_{n=0}^{\infty} f_n(x) = f_0(x) + f_1(x) + f_2(x) + \cdots. \tag{11.14}$$

When $f_n(x)$ is a constant multiplied by x^n,

$$f_n(x) = a_n x^n, \qquad a_n = \text{a constant},$$

series (11.14) takes the form

$$\sum_{n=0}^{\infty} a_n x^n = a_0 + a_1 x + a_2 x^2 + \cdots + a_n x^n + \cdots, \tag{11.15}$$

and is called a *power series in x*.

For power series the variable of summation n no longer identifies terms in the series; it is actually one step out of phase: $n = 0$ identifies the first term, $n = 1$ the second term, etc. Instead, the value of n now identifies the power of x. In this section we will often write $\sum a_n x^n$ in place of $\sum_{n=0}^{\infty} a_n x^n$ whenever the lower limit is $n = 0$.

For each value of x that we substitute into a power series, we obtain a series of constants. For example, if into the power series

$$\sum_{n=0}^{\infty} x^n = 1 + x + x^2 + x^3 + \cdots$$

we substitute $x = \frac{1}{2}$, we obtain the series of constants,

$$\sum_{n=0}^{\infty} \left(\frac{1}{2}\right)^n = 1 + \frac{1}{2} + \frac{1}{2^2} + \frac{1}{2^3} + \cdots,$$

a geometric series that converges to 2. When we substitute $x = -3$, we obtain the divergent geometric series

$$\sum_{n=0}^{\infty} (-3)^n = 1 - 3 + 9 - 27 + \cdots.$$

▷ **Definition 11.9** The totality of values of x for which a power series converges is called its *interval of convergence.* ◁

Obviously the interval of convergence for a power series $\sum a_n x^n$ always includes the value $x = 0$, but what other possibilities are there? Just by the choice of words "interval of convergence" in Definition 11.9, we seem to suggest that the values of x for which a power series converges will form some kind of interval. This is indeed true, as we will soon see.

Example 18 Find the interval of convergence for the power series $\sum_{n=0}^{\infty} x^n$.

Solution Since the series

$$\sum_{n=0}^{\infty} x^n = 1 + x + x^2 + x^3 + \cdots$$

is a geometric series with common ratio x, its interval of convergence is $-1 < x < 1$. □

Example 19 Find intervals of convergence for each of the following power series:

a) $\displaystyle\sum_{n=1}^{\infty} \frac{(-1)^{n+1}}{n} x^n$;

b) $\displaystyle\sum_{n=1}^{\infty} \frac{x^n}{n!}$;

c) $\displaystyle\sum_{n=1}^{\infty} n!x^n$.

Solution a) Consider the series of absolute values $\sum |x|^n/n$. If we use the limit ratio test, we have

$$L = \lim_{n\to\infty} \frac{\dfrac{|x|^{n+1}}{n+1}}{\dfrac{|x|^n}{n}} = \lim_{n\to\infty} |x| \frac{n}{n+1} = |x|.$$

Since $L = |x|$, it follows that the series of absolute values converges if $|x| < 1$ and diverges if $|x| > 1$. The original series therefore converges absolutely for $-1 < x < 1$. Furthermore, when $|x| > 1$, so too is L, and it follows that the terms in the sequence $\{|x|^n/n\}$ must become indefinitely large. But then

$$\lim_{n\to\infty} \frac{(-1)^{n+1}}{n} x^n$$

cannot possibly exist, and the series diverges (by the nth term test). (If $x < -1$, terms in the sequence $\{(-1)^{n+1}x^n/n\}$ are all negative and "approach negative infinity"; if $x > 1$, the terms are alternately positive and negative with increasing absolute values.) When $x = 1$, the power series reduces to the alternating harmonic series

$$\sum_{n=1}^{\infty} \frac{(-1)^{n+1}}{n} = 1 - \frac{1}{2} + \frac{1}{3} - \frac{1}{4} + \cdots,$$

which converges conditionally; and when $x = -1$, it reduces to

$$\sum_{n=1}^{\infty} \frac{(-1)^{n+1}}{n}(-1)^n = -1 - \frac{1}{2} - \frac{1}{3} - \frac{1}{4} - \cdots,$$

which diverges. The interval of convergence is therefore $-1 < x \leq 1$.

b) For the series of absolute values $\sum |x|^n/n!$, we calculate that

$$L = \lim_{n\to\infty} \frac{\dfrac{|x|^{n+1}}{(n+1)!}}{\dfrac{|x|^n}{n!}} = |x| \lim_{n\to\infty} \frac{1}{n+1} = 0 \quad \text{for all } x.$$

The power series therefore converges absolutely for all x.

c) Consider again the series of absolute values $\sum n!|x|^n$. Since

$$L = \lim_{n\to\infty} \frac{(n+1)!|x|^{n+1}}{n!|x|^n} = |x| \lim_{n\to\infty} (n+1) = \infty \quad \text{for all } x \neq 0,$$

the original power series does not converge absolutely for any $x \neq 0$. Furthermore, because $L = \infty$, it follows (as in part (a)) that $\lim_{n\to\infty} n!x^n$ does not exist. The original series thus diverges for all $x \neq 0$, and converges only for $x = 0$. □

Example 19 has illustrated that there are at least three possible types of intervals of convergence for a power series $\sum a_n x^n$:

1. The power series converges only for $x = 0$;
2. The power series converges absolutely for all x;
3. There exists a number $R > 0$ such that the power series converges absolutely for $|x| < R$, diverges for $|x| > R$, and may or may not converge for $x = \pm R$.

These are in fact the only possibilities for an interval of convergence, and this is proved in Exercises 32 and 33. In (3) we call R the *radius of convergence* of the power series. It is one-half the length of the interval of convergence; in other words, the distance we may proceed in either direction along the x-axis from $x = 0$ and expect absolute convergence of the power series (with the possible exceptions of $x = \pm R$). In order to have a radius of convergence associated with every power series, we say in (1) and (2) that $R = 0$ and $R = \infty$, respectively.

Every power series $\sum a_n x^n$ now has a radius of convergence R. If $R = 0$, the power series converges only for $x = 0$; if $R = \infty$, the power series converges absolutely for all x; and if $0 < R < \infty$, the power series converges absolutely for $|x| < R$, diverges for $|x| > R$, and may or may not converge for $x = \pm R$. For many power series the radius of convergence can be calculated according to the following theorem.

▶ **Theorem 11.10** The radius of convergence of a power series $\sum_{n=0}^{\infty} a_n x^n$ is defined by

$$R = \lim_{n\to\infty} \left| \frac{a_n}{a_{n+1}} \right|, \tag{11.16}$$

provided the limit exists or is equal to infinity. ◀

Proof If the limit ratio test is applied to the series of absolute values $\sum |a_n x^n|$,

$$L = \lim_{n\to\infty} \frac{|a_{n+1}x^{n+1}|}{|a_n x^n|} = |x| \lim_{n\to\infty} \left| \frac{a_{n+1}}{a_n} \right|.$$

Now we have assumed that either $\lim_{n\to\infty} |a_n/a_{n+1}|$ exists or is equal to infinity. There are three possibilities:

i) If $\lim_{n\to\infty} |a_n/a_{n+1}| = 0$, then $\lim_{n\to\infty} |a_{n+1}/a_n| = \infty$. But then $L = \infty$, and the power series diverges for all $x \neq 0$. In other words,

$$R = 0 = \lim_{n\to\infty} \left| \frac{a_n}{a_{n+1}} \right|.$$

ii) If $\lim_{n\to\infty} |a_n/a_{n+1}| = \infty$, then $\lim_{n\to\infty} |a_{n+1}/a_n| = 0$. But then $L = 0$, and the power series converges absolutely for all x. Consequently,

$$R = \infty = \lim_{n\to\infty} \left| \frac{a_n}{a_{n+1}} \right|.$$

iii) If $\lim_{n\to\infty} |a_n/a_{n+1}| = R$, then $\lim_{n\to\infty} |a_{n+1}/a_n| = 1/R$. In this case, $L = |x|/R$. Since the power series converges absolutely for $L < 1$ and diverges for $L > 1$, it follows that absolute convergence occurs for $|x| < R$ and divergence for $|x| > R$. But then R is the radius of convergence of the power series. □

With Theorem 11.10, the limit ratio test can be avoided in determining intervals of convergence for many power series. For the series in Example 19 we would proceed as follows.

a) For power series $\sum_{n=1}^{\infty} ((-1)^{n+1}/n)x^n$,

$$R = \lim_{n\to\infty} \left| \frac{a_n}{a_{n+1}} \right| = \lim_{n\to\infty} \left| \frac{\dfrac{(-1)^{n+1}}{n}}{\dfrac{(-1)^{n+2}}{n+1}} \right| = \lim_{n\to\infty} \left(\frac{n+1}{n} \right) = 1.$$

When $x = 1$, the power series reduces to the alternating harmonic series

$$\sum_{n=1}^{\infty} \frac{(-1)^{n+1}}{n} = 1 - \frac{1}{2} + \frac{1}{3} - \frac{1}{4} + \cdots,$$

which converges conditionally; and when $x = -1$, it reduces to

$$\sum_{n=1}^{\infty} \frac{(-1)^{n+1}}{n}(-1)^n = -1 - \frac{1}{2} - \frac{1}{3} - \frac{1}{4} - \cdots,$$

which diverges. The interval of convergence is therefore $-1 < x \leq 1$.

b) For power series $\sum_{n=1}^{\infty} x^n/n!$,

$$R = \lim_{n\to\infty} \left|\frac{a_n}{a_{n+1}}\right| = \lim_{n\to\infty} \frac{\dfrac{1}{n!}}{\dfrac{1}{(n+1)!}} = \lim_{n\to\infty} (n+1) = \infty,$$

and the power series converges absolutely for all x.

c) For power series $\sum_{n=1}^{\infty} n!x^n$,

$$R = \lim_{n\to\infty} \left|\frac{a_n}{a_{n+1}}\right| = \lim_{n\to\infty} \frac{n!}{(n+1)!} = \lim_{n\to\infty} \frac{1}{n+1} = 0,$$

and the power series converges only for $x = 0$.

Example 20 Find the interval of convergence for power series $\sum_{n=0}^{\infty} nx^n/5^{2n}$.

Solution Since

$$R = \lim_{n\to\infty} \left|\frac{a_n}{a_{n+1}}\right| = \lim_{n\to\infty} \frac{\dfrac{n}{5^{2n}}}{\dfrac{n+1}{5^{2n+2}}} = 25,$$

the power series converges absolutely for $-25 < x < 25$. At $x = 25$, the power series reduces to $\sum_{n=0}^{\infty} n$, which diverges; and at $x = -25$, it reduces to $\sum_{n=0}^{\infty} n(-1)^n$, which also diverges. The interval of convergence is therefore $-25 < x < 25$. □

Example 21 Find the interval of convergence for power series $\sum_{n=1}^{\infty} x^{2n+1}/(n^2 2^n)$.

Solution Because the sequence $\{a_n/a_{n+1}\}$ is not defined for this power series, we cannot immediately find its radius of convergence according to Theorem 11.10. Instead we write

$$\sum_{n=1}^{\infty} \frac{x^{2n+1}}{n^2 2^n} = x \sum_{n=1}^{\infty} \frac{(x^2)^n}{n^2 2^n}$$

and set $y = x^2$:

$$\sum_{n=1}^{\infty} \frac{x^{2n+1}}{n^2 2^n} = \sqrt{y} \sum_{n=1}^{\infty} \frac{y^n}{n^2 2^n}.$$

According to Theorem 11.10, the radius of convergence of this series in y is

$$\lim_{n\to\infty} \frac{\dfrac{1}{n^2 2^n}}{\dfrac{1}{(n+1)^2 2^{n+1}}} = \lim_{n\to\infty} 2\left(\frac{n+1}{n}\right)^2 = 2.$$

Since $x = \sqrt{y}$, it follows that the radius of convergence of the power series in x is $R = \sqrt{2}$. When $x = \sqrt{2}$, the series reduces to $\sum_{n=1}^{\infty} \sqrt{2}/n^2 = \sqrt{2} \sum_{n=1}^{\infty} 1/n^2$, which converges; and when $x = -\sqrt{2}$, it reduces to the negative of this series. The interval of convergence is therefore $-\sqrt{2} \le x \le \sqrt{2}$. □

The sum of a power series is identified in the following definition.

▷ **Definition 11.10** A function $f(x)$ is said to be the sum of a power series $\sum_{n=0}^{\infty} a_n x^n$ if at each point x in the interval of convergence of the power series, the sum of the series is the same as the value of $f(x)$ at that x. ◁

We write

$$f(x) = \sum_{n=0}^{\infty} a_n x^n \tag{11.17}$$

to signify that $f(x)$ is the sum of the power series. For example, the sum of the geometric series

$$\sum_{n=0}^{\infty} x^n = 1 + x + x^2 + x^3 + \cdots$$

is $1/(1 - x)$ for $|x| < 1$, and we write

$$\frac{1}{1 - x} = 1 + x + x^2 + x^3 + \cdots, \qquad |x| < 1.$$

Note that we affix to the equation the interval of convergence of the power series, thus identifying for which values of x the function $1/(1 - x)$ is the sum of the series. The function $1/(1 - x)$ is defined for all x (except $x = 1$), but it represents the sum of the series only for x in the range $-1 < x < 1$.

In Section 11.2 we defined the sum of an infinite series of constants as the limit of its sequence of partial sums. In the case of power series (11.17), the sequence of partial sums is a sequence of polynomials:

$$\begin{aligned}
S_1(x) &= a_0, \\
S_2(x) &= a_0 + a_1 x, \\
S_3(x) &= a_0 + a_1 x + a_2 x^2, \\
&\;\;\vdots \\
S_n(x) &= a_0 + a_1 x + a_2 x^2 + \cdots + a_{n-1} x^{n-1}.
\end{aligned}$$

By Eq. (11.17) we are saying that for each x in the interval of convergence of the power series,

$$f(x) = \lim_{n \to \infty} S_n(x).$$

For example, the nth partial sum of the geometric series $\sum_{n=0}^{\infty} x^n$ is

$$\begin{aligned}
S_n(x) &= 1 + x + x^2 + \cdots + x^{n-1} \\
&= \begin{cases} n, & x = 1, \\ \dfrac{1 - x^n}{1 - x}, & x \neq 1 \end{cases}
\end{aligned}$$

(see Example 8). Algebraically it is clear that for large n, $S_n(x)$ approaches $1/(1 - x)$, provided $|x| < 1$, and we write therefore

$$\frac{1}{1 - x} = 1 + x + x^2 + x^3 + \cdots, \qquad |x| < 1.$$

In Fig. 11.11 we show the first five partial sums of this geometric series together with the sum, just to illustrate how the partial sums of the series approximate the sum more closely for increasing n. Note how these figures illustrate the following facts about the geometric series and its partial sums:

1. The partial sums $S_n(x)$ are defined for all x and the sum $1/(1 - x)$ is defined for all $x \neq 1$, but the curves $y = S_n(x)$ approach $y = 1/(1 - x)$ only for $|x| < 1$.
2. For $x > 0$, the partial sums $S_n(x)$ approach $1/(1 - x)$ from below; for $x < 0$, the $S_n(x)$ oscillate about $1/(1 - x)$, but gradually approach $1/(1 - x)$.

Figure 11.11

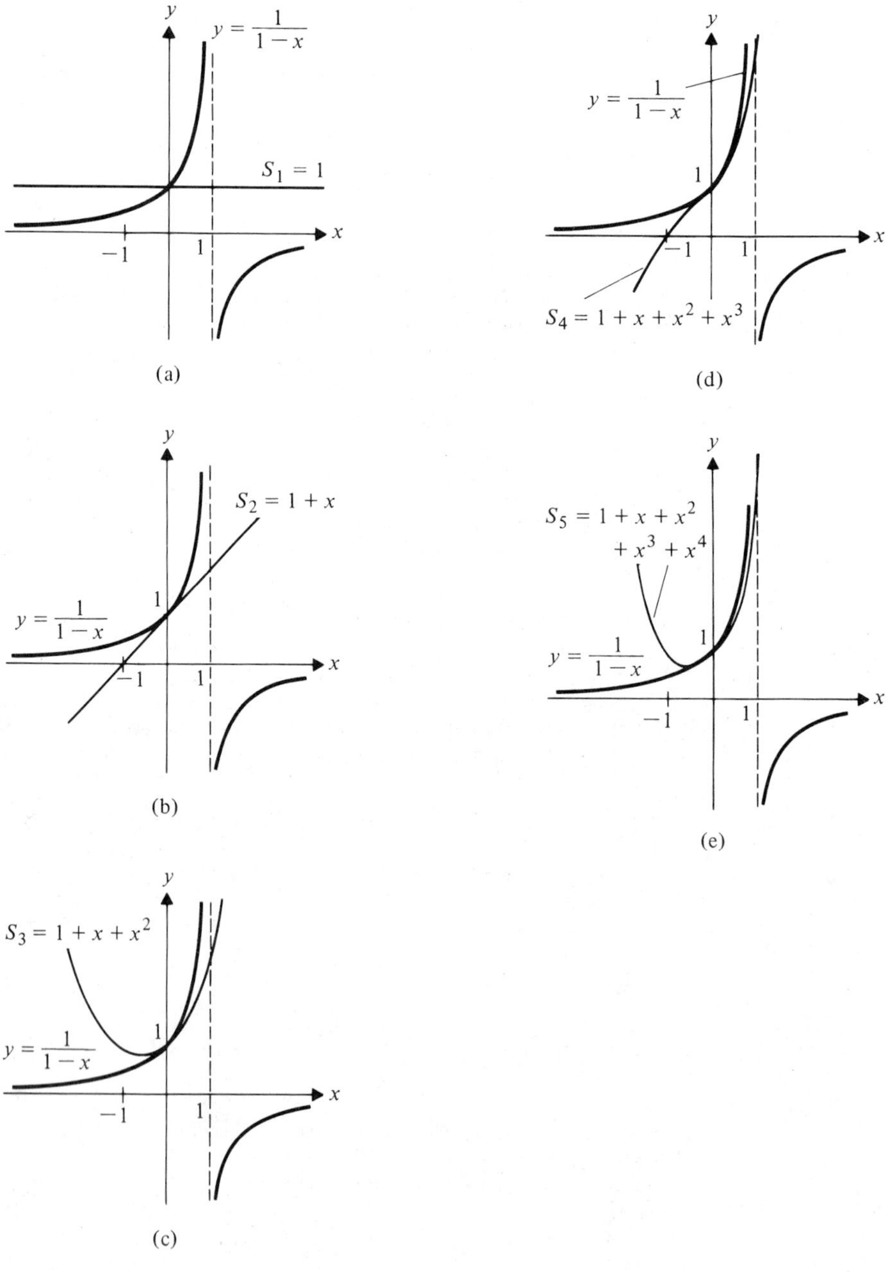

Example 22 Find the sum of the series

$$\sum_{n=0}^{\infty} \frac{(-1)^n}{2^n} x^{(2n+1)/2}.$$

Solution We write

$$\sum_{n=0}^{\infty} \frac{(-1)^n}{2^n} x^{(2n+1)/2} = x^{1/2} \sum_{n=0}^{\infty} \frac{(-1)^n}{2^n} x^n = x^{1/2} \sum_{n=0}^{\infty} \left(-\frac{x}{2}\right)^n$$

and note that power series $\sum_{n=0}^{\infty} (-x/2)^n$ is a geometric series with sum $1/(1 + x/2)$, provided $|x/2| < 1$. Consequently,

$$\sum_{n=0}^{\infty} \frac{(-1)^n}{2^n} x^{(2n+1)/2} = \sqrt{x}\, \frac{1}{1 + \frac{x}{2}} = \frac{2\sqrt{x}}{x + 2}, \qquad 0 \le x < 2. \square$$

For many power series there is no known function that is equal to the sum of the power series over its interval of convergence. In such cases we continue to write Eq. (11.17), but say that the power series defines the value of $f(x)$ at each x. For instance, a very important series in engineering and physics is

$$\sum_{n=0}^{\infty} \frac{(-1)^n}{2^{2n}(n!)^2} x^{2n},$$

which converges absolutely for all x. It arises so often in applications that it is given a special name, the Bessel function of the first kind of order zero, and is denoted by $J_0(x)$. In other words, we write

$$J_0(x) = \sum_{n=0}^{\infty} \frac{(-1)^n}{2^{2n}(n!)^2} x^{2n},$$

and say that the power series defines the value of $J_0(x)$ for each x.

We have called (11.15) a power series in x. It is also said to be a power series about the point 0 (meaning the point 0 on the x-axis), where we note that for any interval of convergence whatsoever, 0 is always at its center. This suggests that power series about other points might be considered, and this is indeed the case. The general power series about a point a on the x-axis is

$$\sum_{n=0}^{\infty} a_n(x - a)^n = a_0 + a_1(x - a) + a_2(x - a)^2 + \cdots \qquad (11.18)$$

and is also said to be a power series in $x - a$.

A power series in $x - a$ has an interval of convergence and a radius of convergence analogous to a power series in x. In particular, every power series in $x - a$ has a radius of convergence R such that if $R = 0$, the power series converges only for $x = a$; if $R = \infty$, the power series converges absolutely for all x; and if $0 < R < \infty$, the power series converges absolutely for $|x - a| < R$, diverges for $|x - a| > R$, and may or may not converge for $x = a \pm R$. The radius of convergence is again given by Eq. (11.16), provided the limit exists or is equal to infinity. For power series in $x - a$, then, the point a is the center of the interval of convergence.

Example 23 Find the interval of convergence of power series

$$2^2(x + 2) + 2^3(x + 2)^2 + 2^4(x + 2)^3 + \cdots + 2^{n+1}(x + 2)^n + \cdots.$$

Solution By writing the power series in sigma notation,

$$\sum_{n=1}^{\infty} 2^{n+1}(x+2)^n,$$

we can use Eq. (11.16) to calculate its radius of convergence,

$$R = \lim_{n\to\infty}\left|\frac{a_n}{a_{n+1}}\right| = \lim_{n\to\infty}\frac{2^{n+1}}{2^{n+2}} = \frac{1}{2}.$$

Since this is a power series about -2, the ends of the interval of convergence are $x = -\frac{5}{2}$ and $x = -\frac{3}{2}$. At $x = -\frac{5}{2}$, the power series reduces to $\sum_{n=1}^{\infty} 2(-1)^n$, which diverges; and at $x = -\frac{3}{2}$, it reduces to $\sum_{n=1}^{\infty} 2$, which also diverges. The interval of convergence is therefore $-\frac{5}{2} < x < -\frac{3}{2}$. □

EXERCISES 11.6

Find the interval of convergence for the power series in each of Exercises 1–22.

1. $\sum_{n=1}^{\infty} \frac{x^n}{n}$
2. $\sum_{n=1}^{\infty} n^2 x^n$
3. $\sum_{n=0}^{\infty} \frac{x^n}{(n+1)^3}$
4. $\sum_{n=0}^{\infty} n^2 3^n x^n$
5. $\sum_{n=0}^{\infty} \frac{1}{2^n}(x-1)^n$
6. $\sum_{n=4}^{\infty} (-1)^n n^3 (x+3)^n$
7. $\sum_{n=1}^{\infty} \frac{(x+2)^n}{\sqrt{n}}$
8. $\sum_{n=2}^{\infty} 2^n\left(\frac{n-1}{n+2}\right)^2 (x-4)^n$
9. $\sum_{n=1}^{\infty} \frac{x^{2n}}{n^2}$
10. $\sum_{n=0}^{\infty} (-1)^n x^{3n}$
11. $\sum_{n=0}^{\infty} \frac{n-1}{n+1}(2x)^n$
12. $\sum_{n=0}^{\infty} \frac{1}{\sqrt{n+1}} x^{3n+1}$
13. $\sum_{n=0}^{\infty} \frac{(-1)^n}{3^n} x^{2n+1}$
14. $\sum_{n=1}^{\infty} \frac{(-e)^n}{n^2} x^n$
15. $\frac{x}{9} + \frac{4}{3^4}x^2 + \frac{9}{3^6}x^3 + \cdots + \frac{n^2}{3^{2n}}x^n + \cdots$
16. $x + 2^2x^2 + 3^3x^3 + 4^4x^4 + \cdots + n^n x^n + \cdots$
17. $\frac{1}{36}(x+10)^6 + \frac{1}{49}(x+10)^7 + \frac{1}{64}(x+10)^8 + \cdots + \frac{1}{n^2}(x+10)^n + \cdots$
18. $3x + 8(3x)^2 + 27(3x)^3 + \cdots + n^3(3x)^n + \cdots$
19. $\frac{3x^2}{4} + \frac{9x^4}{9} + \frac{27x^6}{16} + \cdots + \frac{3^n x^{2n}}{(n+1)^2} + \cdots$
20. $1 + \frac{x^3}{5} + \frac{x^6}{25} + \cdots + \frac{x^{3n}}{5^n} + \cdots$
21. $\sum_{n=2}^{\infty} \frac{x^n}{\ln(n)}$
22. $\sum_{n=2}^{\infty} \frac{x^n}{n^2 \ln(n)}$

Find the radius of convergence of the power series in each of Exercises 23–25.

23. $\sum_{n=1}^{\infty} \frac{(n!)^3}{(3n)!} x^n$
24. $\sum_{n=1}^{\infty} \frac{2\cdot 4\cdot 6\cdot \cdots \cdot (2n)}{3\cdot 5\cdot 7\cdot \cdots \cdot (2n+1)} x^n$
25. $\sum_{n=1}^{\infty} \frac{[1\cdot 3\cdot 5\cdot \cdots \cdot (2n+1)]^2}{2^{2n}(2n)!} x^n$

In Exercises 26–29, find the sum of the given power series.

26. $\sum_{n=0}^{\infty} \frac{1}{4^n} x^{3n}$
27. $\sum_{n=1}^{\infty} (-e)^n x^n$
28. $\sum_{n=1}^{\infty} \frac{1}{3^{2n}}(x-1)^n$
29. $\sum_{n=2}^{\infty} (x+5)^{2n}$

30. If m is a nonnegative integer, the Bessel function of order m of the first kind is defined by the power series
$$J_m(x) = \sum_{n=0}^{\infty} \frac{(-1)^n}{2^{2n+m} n!(n+m)!} x^{2n+m}, \qquad (0! = 1).$$
 a) Write out the first five terms of $J_0(x)$, $J_1(x)$, and $J_m(x)$.
 b) Find the interval of convergence for each $J_m(x)$.
31. The hypergeometric series is
$$1 + \frac{\alpha\beta}{\gamma}x + \frac{\alpha(\alpha+1)\beta(\beta+1)}{2!\gamma(\gamma+1)}x^2 + \frac{\alpha(\alpha+1)(\alpha+2)\beta(\beta+1)(\beta+2)}{3!\gamma(\gamma+1)(\gamma+2)}x^3 + \cdots,$$

where α, β, and γ are all constants.

a) Write this series in sigma notation.

b) What is the radius of convergence of the hypergeometric series if γ is not zero or a negative integer?

32. Show that if a power series $\sum a_n x^n$ converges for $x_1 \neq 0$, then it converges absolutely for all x in the interval $|x| < |x_1|$.

*** 33.** Prove that the interval of convergence of a power series $\sum a_n x^n$ must be one of the three possibilities listed on p. 462. Use Exercise 32 and the *completeness property* for real numbers, which states: If a nonempty set S of real numbers has an upper bound, then it has a least upper bound; i.e., there exists a number l such that all numbers in S are less than or equal to l and there is no upper bound smaller than l.

11.7 TAYLOR'S REMAINDER FORMULA AND TAYLOR SERIES

If $f(x)$ is the sum of a power series $\sum_{n=0}^{\infty} a_n(x-a)^n$, we write

$$f(x) = \sum_{n=0}^{\infty} a_n(x-a)^n. \tag{11.19}$$

In Section 11.6 we considered this equation with the power series given and the sum to be determined. An alternative view is to consider the function $f(x)$ as given and the power series as an expansion of the function about the point a. For example, given the function $f(x) = (2x+3)^{-1}$, it is quite simple to use properties of geometric series to find a power series expansion in x for $f(x)$. If we write

$$f(x) = \frac{1}{2x+3} = \frac{1}{3\left(1 + \dfrac{2x}{3}\right)}$$

and interpret the right-hand side as one-third the sum of a geometric series with first term equal to 1 and common ratio $-2x/3$, then

$$f(x) = \frac{1}{3}\sum_{n=0}^{\infty}\left(-\frac{2x}{3}\right)^n, \qquad \left|-\frac{2x}{3}\right| < 1$$

$$= \sum_{n=0}^{\infty} \frac{(-1)^n 2^n}{3^{n+1}} x^n, \qquad |x| < \frac{3}{2}.$$

It is not clear at this time why we should want to expand a function in the form of an infinite series; in Section 11.9, when we consider applications of series, we will find such expansions extremely useful.

In this section we establish two results. First we give conditions that guarantee that a function $f(x)$ has a power series expansion about a point a and that this power series converges to $f(x)$. Second, we show that if a function has a power series expansion about a point, then it has only one such series. The first result is an immediate consequence of the following theorem.

▶ **Theorem 11.11** (*Taylor's remainder formula*). If $f(x)$ and its first $n-1$ derivatives are continuous on the closed interval between a and x, and if $f(x)$ has an nth derivative on the open interval between a and x, then there exists a point c between a and x such that

$$f(x) = f(a) + f'(a)(x - a) + \frac{f''(a)}{2!}(x - a)^2 + \frac{f'''(a)}{3!}(x - a)^3 + \cdots$$
$$+ \frac{f^{(n-1)}(a)}{(n-1)!}(x - a)^{n-1} + \frac{f^{(n)}(c)}{n!}(x - a)^n.\dagger \quad (11.20)$$

This result can be proved by applying Rolle's theorem (Theorem 4.4) to the function

$$F(z) = f(x) - f(z) - f'(z)(x - z) - \frac{f''(z)}{2!}(x - z)^2 - \cdots - \frac{f^{(n-1)}(z)}{(n-1)!}(x - z)^{n-1}$$
$$- \left(\frac{x - z}{x - a}\right)^n \left\{ f(x) - f(a) - f'(a)(x - a) - \frac{f''(a)}{2!}(x - a)^2 - \cdots \right.$$
$$\left. - \frac{f^{(n-1)}(a)}{(n-1)!}(x - a)^{n-1} \right\}$$

defined on the interval $a \le z \le x$ (or $x \le z \le a$).

Note that when $n = 1$, Taylor's remainder formula reduces to the mean value theorem of Section 4.10. In other words, the mean value theorem is a special case of Taylor's remainder formula.

Consider now a function $f(x)$ that has derivatives of all orders on the closed interval between a and x. By Taylor's remainder formula, we may write for each positive integer n

$$f(x) = f(a) + f'(a)(x - a) + \frac{f''(a)}{2!}(x - a)^2 + \cdots$$
$$+ \frac{f^{(n-1)}(a)}{(n-1)!}(x - a)^{n-1} + R_n, \quad (11.21a)$$

where R_n is the remainder,

$$R_n = R_n(a, x) = \frac{f^{(n)}(c)}{n!}(x - a)^n, \quad (11.21b)$$

and point c (which may vary with n) is always between a and x. We have written $R_n(a, x)$ to emphasize the fact that the remainder depends on the point of expansion a and the value of x being considered. We call R_n the remainder for the simple reason that if we were to use

$$f(x) \simeq f(a) + f'(a)(x - a) + \cdots + \frac{f^{(n-1)}(a)}{(n-1)!}(x - a)^{n-1}$$

as an approximation to $f(x)$, then R_n would represent the remainder or error of the approximation. We will give a detailed discussion of approximations in Section 11.9 but for now this gives us an indication of why Theorem 11.11 is called Taylor's remainder formula.

---

† We use the notation $f^{(n)}(c)$ to represent the nth derivative of $f(x)$ evaluated at $x = c$.

Thus if $f(x)$ has derivatives of all orders between a and x, we have Taylor's remainder formula for each $n \geq 1$. If, further,

$$\lim_{n \to \infty} R_n = 0,$$

then Eq. (11.21) implies that the power series

$$f(a) + f'(a)(x - a) + \frac{f''(a)}{2!}(x - a)^2 + \cdots$$

converges to $f(x)$. We may therefore write that

$$f(x) = f(a) + f'(a)(x - a) + \frac{f''(a)}{2!}(x - a)^2 + \cdots, \tag{11.22}$$

and this equation is valid for all those x's for which $\lim_{n\to\infty} R_n(a, x) = 0$. This power series in $x - a$ is called the *Taylor series* of $f(x)$ about the point a. We have shown then that *if the sequence of remainders* $\{R_n\}$ *for a function* $f(x)$ *exists and has limit zero, the Taylor series converges to* $f(x)$ *for that* x.

When $a = 0$, the Taylor series takes the form

$$f(x) = f(0) + f'(0)x + \frac{f''(0)}{2!}x^2 + \cdots, \tag{11.23}$$

a power series in x called *the Maclaurin series for* $f(x)$.

To obtain the Taylor or Maclaurin series for a function $f(x)$ is theoretically quite simple: Evaluate all derivatives of $f(x)$ at the point of expansion. To show that the series converges to $f(x)$ is a different matter. This can be a very difficult task, principally because we must show that sequence $\{R_n\}$ has limit zero, and R_n is defined in terms of some unknown point c. This problem is usually circumvented by finding some maximum value for R_n and showing that this maximum value approaches zero. We illustrate this procedure in the following three examples.

Example 24 Find the Maclaurin series for $\sin x$ and show that it converges to $\sin x$ for all x.

Solution If $f(x) = \sin x$, then:

$$f(0) = \sin x_{|x=0} = 0;$$

$$f'(0) = \cos x_{|x=0} = 1;$$

$$f''(0) = -\sin x_{|x=0} = 0;$$

$$f'''(0) = -\cos x_{|x=0} = -1;$$

$$f''''(0) = \sin x_{|x=0} = 0; \quad \text{etc.}$$

Taylor's remainder formula for $\sin x$ and $a = 0$ yields

$$\sin x = x - \frac{x^3}{3!} + \frac{x^5}{5!} - \frac{x^7}{7!} + \cdots + (n\text{th term}) + R_n,$$

where

$$R_n = \frac{d^n}{dx^n}(\sin x)_{|x=c}\frac{x^n}{n!}.$$

But $d^n(\sin x)/dx^n$ is either $\pm\sin x$ or $\pm\cos x$, so that

$$\left| \frac{d^n}{dx^n} (\sin x)_{|x=c} \right| \leq 1;$$

hence,

$$|R_n| \leq \frac{|x|^n}{n!}.$$

But according to Exercise 11 in Section 11.4 $\lim_{n\to\infty} (|x|^n/n!) = 0$ for any x whatsoever, and therefore

$$\lim_{n\to\infty} R_n = 0.$$

The Maclaurin series for $\sin x$ therefore converges to $\sin x$ for all x, and we may write

$$\sin x = x - \frac{x^3}{3!} + \frac{x^5}{5!} - \frac{x^7}{7!} + \cdots$$

$$= \sum_{n=0}^{\infty} (-1)^n \frac{x^{2n+1}}{(2n+1)!}, \qquad -\infty < x < \infty. \square$$

Example 25 Find the Maclaurin series for e^x and show that it converges to e^x for all x.

Solution Since

$$\frac{d^n}{dx^n} (e^x)_{|x=0} = e^x{}_{|x=0} = 1,$$

Taylor's remainder formula for e^x and $a = 0$ gives

$$e^x = 1 + x + \frac{x^2}{2!} + \frac{x^3}{3!} + \cdots + \frac{x^{n-1}}{(n-1)!} + R_n,$$

where

$$R_n = \frac{d^n}{dx^n} (e^x)_{|x=c} \frac{x^n}{n!} = e^c \frac{x^n}{n!}.$$

Now, if $x < 0$, then $x < c < 0$, and

$$|R_n| < e^0 \frac{|x|^n}{n!},$$

which approaches zero as n becomes infinite (see again Exercise 11 in Section 11.4). If $x > 0$, then $0 < c < x$, and

$$|R_n| < e^x \frac{|x|^n}{n!},$$

which again has limit zero as n approaches infinity. Thus for any x whatsoever, $\lim_{n\to\infty} R_n = 0$, and the Maclaurin series for e^x converges to e^x:

$$e^x = 1 + x + \frac{x^2}{2!} + \frac{x^3}{3!} + \cdots$$

$$= \sum_{n=0}^{\infty} \frac{x^n}{n!}, \qquad -\infty < x < \infty. \square$$

Example 26 Find the Taylor series for $\ln x$ about the point 1 and show that it converges to $\ln x$ for $\frac{1}{2} \le x \le 2$.

Solution If $f(x) = \ln x$, then:

$$f(1) = \ln x_{|x=1} = 0;$$

$$f'(1) = \frac{1}{x}_{|x=1} = 1;$$

$$f''(1) = \frac{-1}{x^2}_{|x=1} = -1;$$

$$f'''(1) = \frac{2}{x^3}_{|x=1} = 2;$$

$$f''''(1) = \frac{-3!}{x^4}_{|x=1} = -3!; \quad \text{etc.}$$

Taylor's remainder formula for $\ln x$ and $a = 1$ yields

$$\ln x = (x-1) + \frac{-1}{2!}(x-1)^2 + \frac{2!}{3!}(x-1)^3 + \cdots$$
$$+ \frac{(-1)^n(n-2)!}{(n-1)!}(x-1)^{n-1} + R_n,$$

where

$$R_n = \frac{d^n}{dx^n}(\ln x)_{|x=c}\frac{(x-1)^n}{n!} = (-1)^{n+1}\frac{(n-1)!}{c^n}\frac{(x-1)^n}{n!}$$
$$= \frac{(-1)^{n+1}}{n}\left(\frac{x-1}{c}\right)^n,$$

and c is between 1 and x.

If $1 < x \le 2$, then $1 < c < x$ and

$$|R_n| < \frac{1}{n}\frac{(1)}{1},$$

and therefore,

$$\lim_{n\to\infty} R_n = 0.$$

If $\frac{1}{2} \le x < 1$, then $0 < 1 - x \le \frac{1}{2}$. Combine this with $x < c < 1$, and we can state that $0 < (1-x)/c < 1$. Again then

$$|R_n| < \frac{1}{n} \quad \text{and} \quad \lim_{n\to\infty} R_n = 0.$$

Thus for $\frac{1}{2} \le x \le 2$, $\{R_n\}$ approaches zero, and the Taylor series converges to $\ln x$ for these values of x:

$$\ln x = (x-1) - \frac{1}{2}(x-1)^2 + \frac{1}{3}(x-1)^3 + \cdots$$
$$= \sum_{n=1}^{\infty}\frac{(-1)^{n+1}}{n}(x-1)^n, \qquad \frac{1}{2} \le x \le 2. \ \square$$

This series actually converges to $\ln x$ on the larger interval $0 < x \leq 2$, but a proof using Taylor's remainder formula is not possible. We will give an alternative proof in Example 30.

We have shown that if the sequence of Taylor's remainders $\{R_n\}$ for a function $f(x)$ and a point a has limit zero, then $f(x)$ has at least one power series expansion about a, namely its Taylor series,

$$f(x) = \sum_{n=0}^{\infty} \frac{f^{(n)}(a)}{n!}(x-a)^n, \qquad f^{(0)}(a) = f(a), \tag{11.24}$$

and this series converges to $f(x)$. We now show that this is the *only* power series expansion of $f(x)$ about a. To do this we require the following theorem.

▶ **Theorem 11.12** If a function $f(x)$ has a power series expansion $f(x) = \sum_{n=0}^{\infty} a_n(x-a)^n$ with nonzero radius of convergence R, then each of the following power series has radius of convergence R:

$$f'(x) = \sum_{n=0}^{\infty} na_n(x-a)^{n-1}, \tag{11.25}$$

$$\int f(x)\,dx = \sum_{n=0}^{\infty} \frac{a_n}{n+1}(x-a)^{n+1} + C. \tag{11.26}$$

In other words, power series expansions for $f'(x)$ and $\int f(x)\,dx$ can be obtained by term-by-term differentiation and integration of the power series of $f(x)$, and all three series have the same radius of convergence. ◀

Because of the difficulty in proving this theorem, and in order to preserve the continuity of our discussion, we prefer to omit a proof. Note that the theorem is stated in terms of radii of convergence rather than intervals of convergence. This is due to the fact that in differentiating a power series we may lose the end points of the original interval of convergence, and in integrating we may pick them up.

We can now prove that there is at most one power series expansion of a function about a given point.

▶ **Theorem 11.13** If a power series expansion of a function has a nonzero radius of convergence, then it is the Taylor series. ◀

Proof Suppose the power series expansion

$$f(x) = \sum_{n=0}^{\infty} a_n(x-a)^n = a_0 + a_1(x-a) + a_2(x-a)^2 + \cdots$$

has a nonzero radius of convergence. Substitution of $x = a$ gives

$$f(a) = a_0.$$

If we differentiate the power series according to Theorem 11.12, we get

$$f'(x) = a_1 + 2a_2(x-a) + 3a_3(x-a)^2 + \cdots,$$

and if we substitute $x = a$, we then have

$$f'(a) = a_1.$$

If we differentiate the power series for $f'(x)$, we obtain

$$f''(x) = 2a_2 + 3 \cdot 2a_3(x - a) + 4 \cdot 3a_4(x - a)^2 + \cdots,$$

and if we substitute $x = a$, we have

$$f''(a) = 2a_2 \quad \text{or} \quad a_2 = \frac{f''(a)}{2!}.$$

Continued differentiation and substitution leads to the result that for all n,

$$a_n = \frac{f^{(n)}(a)}{n!}. \square$$

An immediate consequence of this theorem is as follows.

▶ **Corollary** If two power series $\sum_{n=0}^{\infty} a_n(x - a)^n$ and $\sum_{n=0}^{\infty} b_n(x - a)^n$ with positive radii of convergence have identical sums,

$$\sum_{n=0}^{\infty} a_n(x - a)^n = \sum_{n=0}^{\infty} b_n(x - a)^n,$$

then $a_n = b_n$ for all n. ◀

Our theory of power series expansions of functions is now complete. Given a function $f(x)$ and a point a, there can be only one power series expansion for $f(x)$ about a: its Taylor series. To show that the Taylor series does indeed have sum $f(x)$, it is sufficient to show that the sequence of Taylor's remainders $\{R_n\}$ has limit zero. In this case we write

$$f(x) = \sum_{n=0}^{\infty} \frac{f^{(n)}(a)}{n!}(x - a)^n. \tag{11.27}$$

Equation (11.27) says, then, that the right-hand side is the Taylor series of $f(x)$ and that the Taylor series converges to $f(x)$. This second point is most important because there do exist functions $f(x)$ whose Taylor series exist but do not converge to $f(x)$ (see Exercise 14). We therefore write (11.27) only when we know that the Taylor series of $f(x)$ exists and that it converges to $f(x)$.

Only one point of consideration remains. Although we have a formula for the calculation of the Taylor series of a function $f(x)$, determination of $f^{(n)}(a)$ could prove to be very complicated. In addition, if a reasonable formula for $f^{(n)}(a)$ cannot be found, how will we ever prove that the sequence of Taylor's remainders has limit zero? Fortunately, in many of the more common situations, special devices exist that enable us to find Taylor series without actually calculating $f^{(n)}(a)$. Indeed, if by any method whatsoever we can obtain a power series representation of $f(x)$, then we can be assured by Theorem 11.13 that it is the Taylor series. In Section 11.8 we illustrate some of these techniques.

Although we have not given a proof of Theorem 11.12, it is a very important result, and we will call on it many times in Sections 11.8 and 11.9. One important application of the theorem is in finding sums of power series. For instance, consider finding the sum of power series $\sum_{n=1}^{\infty} nx^{n-1}$. If we denote the sum of this series by $S(x)$,

$$S(x) = \sum_{n=1}^{\infty} nx^{n-1},$$

and use Theorem 11.12 to find the antiderivative of $S(x)$, we have

$$\int S(x)\,dx = \sum_{n=1}^{\infty} x^n + C.$$

But the series on the right is a geometric series with known sum $x/(1-x)$, provided $|x| < 1$, and we may write therefore that

$$\int S(x)\,dx = \frac{x}{1-x} + C, \qquad |x| < 1.$$

If we now differentiate this equation, we obtain

$$S(x) = \frac{(1-x)(1) - x(-1)}{(1-x)^2} = \frac{1}{(1-x)^2},$$

and therefore

$$\sum_{n=1}^{\infty} nx^{n-1} = \frac{1}{(1-x)^2}, \qquad |x| < 1.$$

EXERCISES 11.7

In Exercises 1–3, use Taylor's remainder formula to find Taylor series for the function $f(x)$ about the point indicated. In each case, find the interval on which the Taylor series converges to $f(x)$.

1. $f(x) = \cos x$, about $x = 0$
2. $f(x) = e^{5x}$, about $x = 0$
3. $f(x) = \sin(10x)$, about $x = 0$
4. Prove the corollary to Theorem 11.13.
5. Prove that if a power series with positive radius of convergence has sum zero, $\sum a_n(x-a)^n = 0$, then $a_n = 0$ for all n.

In Exercises 6–10, use Theorem 11.12 to find the sum of the given power series.

6. $\sum_{n=1}^{\infty} n^2x^{n-1}$
7. $\sum_{n=0}^{\infty} \frac{(-1)^n}{2n+1}x^{2n+1}$
8. $\sum_{n=1}^{\infty} \frac{1}{n}x^n$
9. $\sum_{n=1}^{\infty} (n^2+2n)x^n$
10. $\sum_{n=0}^{\infty} \left(\frac{n+1}{n+2}\right)x^n$
11. In Section 4.5 we stated the second-derivative test for determining whether a critical point x_0 at which $f'(x_0) = 0$ yields a relative maximum or a relative minimum. Use Taylor's remainder formula to verify this result when $f'(x)$ and $f''(x)$ are continuous on an open interval containing x_0.

*12. Extend the result of Exercise 11 to verify the extrema test of Exercise 17 in Section 4.5.

13. There is an integral form for the remainder $R_n(a, x)$ in Taylor's remainder formula that is sometimes more useful than the derivative form in Theorem 11.11: If $f(x)$ and its first n derivatives are continuous on the closed interval between a and x, then

$$f(x) = f(a) + f'(a)(x-a) + \frac{f''(a)}{2!}(x-a)^2 + \cdots + \frac{f^{(n-1)}(a)}{(n-1)!}(x-a)^{n-1} + R_n(a, x),$$

where

$$R_n(a, x) = \frac{1}{(n-1)!}\int_a^x (x-t)^{n-1} f^{(n)}(t)\,dt.$$

In this exercise we outline a proof of the result.

a) Show that

$$f(x) = f(a) + \int_a^x f'(t)\,dt.$$

b) Use integration by parts with $u = f'(t)$, $du = f''(t)\,dt$, $dv = dt$, and $v = t - x$ on the integral in (a) to obtain

$$f(x) = f(a) + f'(a)(x-a) + \int_a^x (x-t)f''(t)\,dt.$$

c) Use integration by parts with $u = f''(t)$, $du = f'''(t)\,dt$, $dv = (x - t)\,dt$, and $v = -\frac{1}{2}(x - t)^2$ to obtain

$$f(x) = f(a) + f'(a)(x - a) + \frac{f''(a)}{2!}(x - a)^2 + \frac{1}{2!}\int_a^x (x - t)^2 f'''(t)\,dt.$$

d) Continue this process to obtain the integral form for Taylor's remainder formula.

***14.** a) Sketch a graph of the function

$$f(x) = \begin{cases} e^{-1/x^2}, & x \neq 0, \\ 0, & x = 0. \end{cases}$$

b) Use L'Hôpital's rule to show that for every positive integer n,

$$\lim_{x \to 0} \frac{e^{-1/x^2}}{x^n} = 0.$$

c) Prove, by mathematical induction, that $f^{(n)}(0) = 0$ for $n \geq 1$.

d) What is the Maclaurin series for $f(x)$?

e) For what values of x does the Maclaurin series of $f(x)$ converge to $f(x)$?

11.8 TAYLOR SERIES EXPANSIONS OF FUNCTIONS

In Section 11.7 we pointed out that if by any method whatsoever we can find a power series representation for a function $f(x)$, then we can be assured by Theorem 11.13 that it is the Taylor series for $f(x)$. In this section, we illustrate various techniques for producing power series. Our first example uses geometric series.

Example 27 Find (a) the Maclaurin series for $(4 + 5x)^{-1}$; (b) the Taylor series about 5 for $(13 - 2x)^{-1}$.

Solution a) If we write

$$\frac{1}{4 + 5x} = \frac{1}{4\left(1 + \frac{5x}{4}\right)}$$

and interpret the right-hand side as one-quarter of the sum of a geometric series with first term equal to one and common ratio $-5x/4$, then

$$\frac{1}{4 + 5x} = \frac{1}{4}\left\{1 + \left(-\frac{5x}{4}\right) + \left(-\frac{5x}{4}\right)^2 + \cdots\right\}, \qquad \left|-\frac{5x}{4}\right| < 1,$$

$$= \sum_{n=0}^{\infty} \frac{(-1)^n 5^n}{4^{n+1}} x^n, \qquad |x| < \frac{4}{5}.$$

b) By a similar procedure, we have

$$\frac{1}{13 - 2x} = \frac{1}{3 - 2(x - 5)}$$

$$= \frac{1}{3(1 - \frac{2}{3}(x - 5))}$$

$$= \frac{1}{3}\left\{1 + \frac{2}{3}(x - 5) + \left(\frac{2}{3}\right)^2 (x - 5)^2 + \cdots\right\}, \qquad \left|\frac{2}{3}(x - 5)\right| < 1,$$

$$= \sum_{n=0}^{\infty} \frac{2^n}{3^{n+1}}(x - 5)^n, \qquad |x - 5| < \frac{3}{2}. \;\square$$

Note that in both of these examples, the properties of geometric series gave us not only the required series, but also their intervals of convergence.

Addition and Subtraction of Power Series In Section 11.2 we noted that convergent series of constants can be added or subtracted to give convergent series. It follows, therefore, that if $f(x) = \sum a_n(x - a)^n$ and $g(x) = \sum b_n(x - a)^n$, then

$$f(x) \pm g(x) = \sum_{n=0}^{\infty} (a_n \pm b_n)(x - a)^n, \tag{11.28}$$

and these results are valid for every x that is common to the intervals of convergence of the two series. We use this result in the following example.

Example 28 Find the Maclaurin series for $f(x) = 5x/(x^2 - 3x - 4)$.

Solution If we decompose $f(x)$ into its partial fractions,

$$f(x) = \frac{5x}{x^2 - 3x - 4} = \frac{4}{x - 4} + \frac{1}{x + 1},$$

and expand each of these terms in its Maclaurin series, we have

$$\frac{4}{x - 4} = \frac{-1}{1 - \dfrac{x}{4}} = -\left\{1 + \frac{x}{4} + \frac{x^2}{4^2} + \cdots\right\}, \qquad |x| < 4,$$

$$\frac{1}{1 + x} = 1 - x + x^2 - x^3 + \cdots, \qquad |x| < 1.$$

Addition of these series gives the required Maclaurin series for $f(x)$:

$$\begin{aligned}\frac{5x}{x^2 - 3x - 4} &= \left\{-1 - \frac{x}{4} - \frac{x^2}{4^2} - \frac{x^3}{4^3} - \cdots\right\} + \{1 - x + x^2 - x^3 + \cdots\}\\ &= \left(-1 - \frac{1}{4}\right)x + \left(1 - \frac{1}{4^2}\right)x^2 + \left(-1 - \frac{1}{4^3}\right)x^3\\ &\quad + \left(1 - \frac{1}{4^4}\right)x^4 + \cdots\\ &= \sum_{n=1}^{\infty} \left((-1)^n - \frac{1}{4^n}\right)x^n, \quad |x| < 1. \ \square\end{aligned}$$

Differentiation and Integration of Power Series Perhaps the most important technique for generating power series expansions is to differentiate or integrate known expansions according to Theorem 11.12.

Example 29 Find Maclaurin series for each of the following:

a) $\cos x$;

b) $\dfrac{1}{(1 - x)^3}$.

Solution a) Since

$$\sin x = x - \frac{x^3}{3!} + \frac{x^5}{5!} - \frac{x^7}{7!} + \cdots, \qquad -\infty < x < \infty$$

(see Example 24), term-by-term differentiation gives

$$\cos x = 1 - \frac{x^2}{2!} + \frac{x^4}{4!} - \frac{x^6}{6!} + \cdots$$

$$= \sum_{n=0}^{\infty} \frac{(-1)^n}{(2n)!} x^{2n}, \qquad -\infty < x < \infty.$$

b) Term-by-term differentiation of the series

$$\frac{1}{1-x} = 1 + x + x^2 + x^3 + \cdots + x^n + \cdots, \qquad |x| < 1$$

gives

$$\frac{1}{(1-x)^2} = 1 + 2x + 3x^2 + \cdots + nx^{n-1} + \cdots, \qquad |x| < 1.$$

Another differentiation yields

$$\frac{2}{(1-x)^3} = 2 + 3 \cdot 2x + 4 \cdot 3x^2 + \cdots + n(n-1)x^{n-2} + \cdots, \quad |x| < 1.$$

Consequently, we have

$$\frac{1}{(1-x)^3} = 1 + 3x + 6x^2 + \cdots + \frac{n(n-1)}{2} x^{n-2} + \cdots$$

$$= \sum_{n=2}^{\infty} \frac{n(n-1)}{2} x^{n-2}$$

$$= \frac{1}{2} \sum_{n=0}^{\infty} (n+2)(n+1)x^n, \qquad |x| < 1. \ \square$$

Example 30 Find the Taylor series about 1 for ln x.

Solution Since ln x is an antiderivative of $1/x$, we first expand $1/x$ in a Taylor series about 1:

$$\frac{1}{x} = \frac{1}{(x-1)+1} = 1 - (x-1) + (x-1)^2 - (x-1)^3 + \cdots, \qquad |x-1| < 1.$$

If we integrate this series term by term, we have

$$\ln|x| = \left\{x - \frac{1}{2}(x-1)^2 + \frac{1}{3}(x-1)^3 - \frac{1}{4}(x-1)^4 + \cdots\right\} + C.$$

Substitution of $x = 1$ implies that $0 = 1 + C$, and hence

$$\ln|x| = (x-1) - \frac{1}{2}(x-1)^2 + \frac{1}{3}(x-1)^3 - \cdots$$

$$= \sum_{n=1}^{\infty} \frac{(-1)^{n+1}}{n}(x-1)^n.$$

According to Theorem 11.12, the radius of convergence of this series is $R = 1$ so that the series certainly converges to $\ln|x|$ for $0 < x < 2$. But when $x = 2$, the series reduces to the alternating harmonic series, which converges conditionally. Consequently, we can extend the interval of convergence to include $x = 2$ and write

$$\ln x = \sum_{n=1}^{\infty} \frac{(-1)^{n+1}}{n}(x-1)^n, \qquad 0 < x \leq 2.$$

Convergence of this series at $x = 2$ does not, by itself, imply convergence to $\ln 2$, as this equation suggests. It is, however, true, and this is a direct application of the following result: If the Taylor series $\sum_{n=0}^{\infty} a_n(x-a)^n$ of a function $f(x)$ converges at the endpoint $x = a + R$ of its interval of convergence, and if $f(x)$ is continuous at $a + R$, then the Taylor series evaluated at $a + R$ converges to $f(a + R)$. The same result is valid at the other endpoint $x = a - R$. □

Comparison of the discussions in Examples 26 and 30 clearly indicates the advantage of avoiding the use of Taylor's remainder formula.

Multiplication of Power Series

In Example 28 we added the Maclaurin series

$$\frac{4}{x-4} = -\left\{1 + \frac{x}{4} + \frac{x^2}{4^2} + \cdots\right\}, \qquad |x| < 4$$

and

$$\frac{1}{1+x} = 1 - x + x^2 - x^3 + \cdots, \qquad |x| < 1$$

to obtain the Maclaurin series for $5x/(x^2 - 3x - 4)$:

$$\frac{5x}{x^2 - 3x - 4} = \frac{4}{x-4} + \frac{1}{x+1} = \sum_{n=1}^{\infty}\left((-1)^n - \frac{1}{4^n}\right)x^n.$$

An alternative procedure might be to multiply these two series since

$$\frac{5x}{x^2 - 3x - 4} = 5x\left\{\frac{1}{x-4}\right\}\left\{\frac{1}{x+1}\right\}$$

$$= \frac{5x}{-4}\left\{1 + \frac{x}{4} + \frac{x^2}{4^2} + \frac{x^3}{4^3} + \cdots\right\}\{1 - x + x^2 - x^3 + \cdots\}.$$

The rules of algebra would demand that we multiply every term of the first series by every term of the second. If we do this and group all products with like powers of x, we obtain

$$\frac{5x}{x^2 - 3x - 4} = \frac{5x}{-4}\left\{1 + \left(-1 + \frac{1}{4}\right)x\right.$$

$$\left. + \left(1 - \frac{1}{4} + \frac{1}{4^2}\right)x^2 + \left(-1 + \frac{1}{4} - \frac{1}{4^2} + \frac{1}{4^3}\right)x^3 + \cdots\right\}.$$

It is clear that the coefficient of x^n in the braces is a finite geometric series (to which we can apply Eq. 11.11a):

$$(-1)^n\left\{1-\frac{1}{4}+\frac{1}{4^2}-\frac{1}{4^3}+\cdots+\frac{(-1)^n}{4^n}\right\}=(-1)^n\left\{\frac{1-(-\frac{1}{4})^{n+1}}{1+\frac{1}{4}}\right\}$$

$$=(-1)^n\frac{4}{5}[1-(-\tfrac{1}{4})^{n+1}]$$

Consequently,

$$\frac{5x}{x^2-3x-4}=\frac{5x}{-4}\sum_{n=0}^{\infty}(-1)^n\frac{4}{5}[1-(-\tfrac{1}{4})^{n+1}]x^n$$

$$=\sum_{n=0}^{\infty}(-1)^{n+1}\left\{1-\frac{(-1)^{n+1}}{4^{n+1}}\right\}x^{n+1}$$

$$=\sum_{n=0}^{\infty}\left((-1)^{n+1}-\frac{1}{4^{n+1}}\right)x^{n+1}$$

$$=\sum_{n=1}^{\infty}\left((-1)^n-\frac{1}{4^n}\right)x^n.$$

For this example, then, multiplication as well as addition of power series leads to the Maclaurin series. Clearly, addition of power series is much simpler for this example, but we have at least demonstrated that power series can be multiplied together. That this is generally possible is stated in the following theorem.

▶ **Theorem 11.14** If $f(x)=\sum_{n=0}^{\infty}a_n(x-a)^n$ and $g(x)=\sum_{n=0}^{\infty}b_n(x-a)^n$, then

$$f(x)g(x)=\sum_{n=0}^{\infty}c_n(x-a)^n, \tag{11.29a}$$

where

$$c_n=\sum_{i=0}^{n}a_ib_{n-i}=a_0b_n+a_1b_{n-1}+\cdots+a_{n-1}b_1+a_nb_0, \tag{11.29b}$$

and this series converges absolutely at every point at which the series for $f(x)$ and $g(x)$ converge absolutely. ◀

Example 31 Find the Maclaurin series for $f(x)=[1/(x-1)]\ln(1-x)$.

Solution If we integrate the Maclaurin series

$$\frac{1}{1-x}=1+x+x^2+x^3+\cdots,\qquad |x|<1,$$

we find

$$-\ln|1-x|=\left\{x+\frac{x^2}{2}+\frac{x^3}{3}+\frac{x^4}{4}+\cdots\right\}+C.$$

By setting $x = 0$, we obtain $C = 0$ and

$$\ln(1-x) = -x - \frac{x^2}{2} - \frac{x^3}{3} - \frac{x^4}{4} - \cdots.$$

(We have dropped the absolute values since the radius of convergence of the series is 1.) We now form the Maclaurin series for $f(x)$:

$$\begin{aligned}\frac{1}{x-1}\ln(1-x) &= \frac{-1}{1-x}\ln(1-x)\\ &= \{1 + x + x^2 + x^3 + \cdots\}\left\{x + \frac{x^2}{2} + \frac{x^3}{3} + \cdots\right\}\\ &= x + \left(1 + \frac{1}{2}\right)x^2 + \left(1 + \frac{1}{2} + \frac{1}{3}\right)x^3 + \cdots\\ &= \sum_{n=1}^{\infty}\left(1 + \frac{1}{2} + \frac{1}{3} + \cdots + \frac{1}{n}\right)x^n.\end{aligned}$$

Since both of the multiplied series converge absolutely on the interval $|x| < 1$, the Maclaurin series for $(x-1)^{-1}\ln(1-x)$ is also valid on this interval. □

Example 32 Find the first three nonzero terms in the Maclaurin series for $\tan x$.

Solution If $\tan x = \sum a_n x^n$, then by setting $\tan x = \sin x/\cos x$, we have

$$\sin x = \cos x \sum a_n x^n.$$

We now substitute the Maclaurin series for $\sin x$ and $\cos x$:

$$\left\{x - \frac{x^3}{3!} + \frac{x^5}{5!} - \frac{x^7}{7!} + \cdots\right\} = \left\{1 - \frac{x^2}{2!} + \frac{x^4}{4!} - \frac{x^6}{6!} + \cdots\right\}\{a_0 + a_1 x + a_2 x^2 + \cdots\}.$$

According to the corollary of Theorem 11.13, two power series can be identical only if their corresponding coefficients are equal. We therefore multiply the right-hand side and equate coefficients:

$$1: 0 = a_0;$$

$$x: 1 = a_1;$$

$$x^2: 0 = a_2 - \frac{a_0}{2!}, \quad \text{which implies } a_2 = 0;$$

$$x^3: -\frac{1}{3!} = a_3 - \frac{a_1}{2!}, \quad \text{which implies } a_3 = \frac{1}{2!} - \frac{1}{3!} = \frac{1}{3};$$

$$x^4: 0 = a_4 - \frac{a_2}{2!} + \frac{a_0}{4!}, \quad \text{from which } a_4 = 0;$$

$$x^5: \frac{1}{5!} = a_5 - \frac{a_3}{2!} + \frac{a_1}{4!}, \quad \text{from which } a_5 = \frac{1}{5!} + \frac{1}{6} - \frac{1}{4!} = \frac{2}{15}.$$

The first three nonzero terms in the Maclaurin series for tan x are therefore

$$\tan x = x + \frac{1}{3}x^3 + \frac{2}{15}x^5 + \cdots. \quad \square$$

Binomial Expansion One of the most widely used power series is the *binomial expansion.* We are well acquainted with the binomial theorem, which predicts the product $(a + b)^m$ for any positive integer m:

$$(a + b)^m = \sum_{n=0}^{m} \binom{m}{n} a^n b^{m-n}. \tag{11.30}$$

With the usual definition of the binomial coefficients,

$$\binom{m}{n} = \frac{m!}{(m-n)!n!} = \frac{m(m-1)(m-2)\cdots(m-n+1)}{n!},$$

the binomial theorem becomes

$$(a + b)^m = a^m + ma^{m-1}b + \frac{m(m-1)}{2!}a^{m-2}b^2 + \cdots + mab^{m-1} + b^m. \tag{11.31}$$

Even when m is not a positive integer, this form for the binomial theorem remains almost intact. To show this, we consider the power series

$$1 + \sum_{n=1}^{\infty} \frac{m(m-1)(m-2)\cdots(m-n+1)}{n!}x^n$$

for any real number m except a nonnegative integer. The radius of convergence of this power series is

$$R = \lim_{n\to\infty} \left| \frac{m(m-1)(m-2)\cdots(m-n+1)}{n!} \cdot \frac{(n+1)!}{m(m-1)(m-2)\cdots(m-n)} \right|$$

$$= \lim_{n\to\infty} \left| \frac{n+1}{m-n} \right| = 1.$$

The power series therefore converges absolutely for $|x| < 1$. Whether the series converges at the end points $x = \pm 1$ depends on the value of m. For the time being, we will be content to work on the interval $|x| < 1$, and at the end of the discussion, we will state the complete result. Let us denote the sum of the series by

$$f(x) = 1 + \sum_{n=1}^{\infty} \frac{m(m-1)\cdots(m-n+1)}{n!}x^n, \qquad |x| < 1.$$

If we differentiate this series term by term (according to Theorem 11.12),

$$f'(x) = \sum_{n=1}^{\infty} \frac{m(m-1)\cdots(m-n+1)}{(n-1)!}x^{n-1}, \qquad |x| < 1,$$

and then multiply both sides by x, we have

$$xf'(x) = \sum_{n=1}^{\infty} \frac{m(m-1)\cdots(m-n+1)}{(n-1)!}x^n, \qquad |x| < 1.$$

If we add these results, we obtain

$$\begin{aligned}(1+x)f'(x) &= \sum_{n=1}^{\infty} \frac{m(m-1)\cdots(m-n+1)}{(n-1)!}x^{n-1} \\ &\quad + \sum_{n=1}^{\infty} \frac{m(m-1)\cdots(m-n+1)}{(n-1)!}x^{n} \\ &= \sum_{n=0}^{\infty} \frac{m(m-1)\cdots(m-n)}{n!}x^{n} + \sum_{n=1}^{\infty} \frac{m(m-1)\cdots(m-n+1)}{(n-1)!}x^{n} \\ &= m + \sum_{n=1}^{\infty} \frac{m(m-1)\cdots(m-n+1)}{(n-1)!}\left\{\frac{m-n}{n}+1\right\}x^{n} \\ &= m\left\{1 + \sum_{n=1}^{\infty} \frac{m(m-1)\cdots(m-n+1)}{n!}x^{n}\right\} \\ &= mf(x).\end{aligned}$$

Consequently, the function $f(x)$ must satisfy the differential equation

$$\frac{f'(x)}{f(x)} = \frac{m}{1+x}.$$

Integration immediately gives

$$\ln|f(x)| = m \ln|1+x| + C,$$

or

$$f(x) = D(1+x)^m.$$

To evaluate D, we note that from the original definition of $f(x)$ as the sum of the power series, $f(0) = 1$, and this implies that $D = 1$. Thus,

$$f(x) = (1+x)^m,$$

and we may write finally that

$$(1+x)^m = 1 + \sum_{n=1}^{\infty} \frac{m(m-1)(m-2)\cdots(m-n+1)}{n!}x^n, \qquad |x| < 1 \tag{11.32a}$$

$$= 1 + mx + \frac{m(m-1)}{2!}x^2 + \frac{m(m-1)(m-2)}{3!}x^3 + \cdots, \qquad |x| < 1. \tag{11.32b}$$

This is called the binomial expansion of $(1+x)^m$. We have verified the binomial expansion for m any real number except a nonnegative integer, but in the case of a nonnegative integer, the series terminates after $m+1$ terms and is therefore valid for these values of m also. We mentioned earlier that the binomial expansion may also converge at the end points $x = \pm 1$ depending on the value of m. The complete result states that (11.32) is valid for

$-\infty < x < \infty$	if m is a nonnegative integer,
$-1 < x < 1$	if $m \le -1$,
$-1 < x \le 1$	if $-1 < m < 0$,
$-1 \le x \le 1$	if $m > 0$ but not an integer.

It is not difficult to generalize this result to expand $(a + b)^m$ for real m. If $|b| < |a|$, we write

$$(a + b)^m = a^m\left(1 + \frac{b}{a}\right)^m$$

and now expand the bracketed term by means of (11.32):

$$(a + b)^m = a^m\left\{1 + m\left(\frac{b}{a}\right) + \frac{m(m-1)}{2!}\left(\frac{b}{a}\right)^2 + \cdots\right\} \qquad |b| < |a|,$$

$$= a^m + ma^{m-1}b + \frac{m(m-1)}{2!}a^{m-2}b^2 + \cdots \qquad |b| < |a|,$$

which as we predicted is Eq. (11.31) except that the series does not terminate.

Example 33 Use the binomial expansion to find the Maclaurin series for $1/(1 - x)^3$.

Solution By (11.32b), we have

$$\frac{1}{(1-x)^3} = (1 - x)^{-3}$$

$$= 1 + (-3)(-x) + \frac{(-3)(-4)}{2!}(-x)^2 + \frac{(-3)(-4)(-5)}{3!}(-x)^3 + \cdots$$

$$= 1 + 3x + \frac{3\cdot 4}{2}x^2 + \frac{4\cdot 5}{2}x^3 + \frac{5\cdot 6}{2}x^4 + \cdots$$

$$= \frac{1}{2}\sum_{n=0}^{\infty}(n + 1)(n + 2)x^n, \qquad |x| < 1. \square$$

This result was also obtained in Example 29 by differentiation of the Maclaurin series for $1/(1 - x)$.

Example 34 Find the Maclaurin series for $\text{Sin}^{-1}\, x$.

Solution By the binomial expansion, we have

$$\frac{1}{\sqrt{1-x^2}} = 1 + \left(-\frac{1}{2}\right)(-x^2) + \frac{(-\frac{1}{2})(-\frac{3}{2})}{2!}(-x^2)^2$$

$$+ \frac{(-\frac{1}{2})(-\frac{3}{2})(-\frac{5}{2})}{3!}(-x^2)^3 + \cdots$$

$$= 1 + \frac{1}{2}x^2 + \frac{3}{2^2 2!}x^4 + \frac{3\cdot 5}{2^3 3!}x^6 + \frac{3\cdot 5\cdot 7}{2^4 4!}x^8 + \cdots, \qquad |x| < 1.$$

Integration of this series gives

$$\text{Sin}^{-1}\, x = \left\{x + \frac{1}{2\cdot 3}x^3 + \frac{3}{2^2 2!5}x^5 + \frac{3\cdot 5}{2^3 3!7}x^7 + \frac{3\cdot 5\cdot 7}{2^4 4!9}x^9 + \cdots\right\} + C.$$

Evaluation of both sides of this equation at $x = 0$ requires $C = 0$. According to Theorem 11.12, the radius of convergence of this series must be 1, and we can write

$$\operatorname{Sin}^{-1} x = x + \sum_{n=1}^{\infty} \frac{1 \cdot 3 \cdot 5 \cdot \cdots \cdot (2n-1)}{2^n n!(2n+1)} x^{2n+1}$$

$$= x + \sum_{n=1}^{\infty} \frac{1 \cdot 2 \cdot 3 \cdot 4 \cdot 5 \cdot \cdots \cdot (2n-2)(2n-1)(2n)}{2 \cdot 4 \cdot \cdots \cdot (2n) 2^n n!(2n+1)} x^{2n+1}$$

$$= x + \sum_{n=1}^{\infty} \frac{(2n)!}{(2n+1)2^{2n}(n!)^2} x^{2n+1}, \qquad |x| < 1.$$

(In this case, integration of the series for $(1 - x^2)^{-1/2}$ does not pick up the end points $x = \pm 1$, although this is a very difficult fact to prove.) □

EXERCISES 11.8

In Exercises 1–20, find the power series expansion of the function about the point indicated.

1. $f(x) = \dfrac{1}{3x + 2}$, about $x = 0$

2. $f(x) = \dfrac{1}{4 + x^2}$, about $x = 0$

3. $f(x) = \dfrac{1}{x + 3}$, about $x = 2$

4. $f(x) = \cos(x^2)$, about $x = 0$

5. $f(x) = \dfrac{1}{\sqrt{1 + x}}$, about $x = 0$

6. $f(x) = e^{5x}$, about $x = 0$

7. $f(x) = \cosh x = \dfrac{e^x + e^{-x}}{2}$, about $x = 0$

8. $f(x) = \sinh x = \dfrac{e^x - e^{-x}}{2}$, about $x = 0$

9. $f(x) = \ln(1 + 2x)$, about $x = 0$

10. $f(x) = (1 + 3x)^{3/2}$, about $x = 0$

11. $f(x) = \dfrac{1}{x}$, about $x = 4$

12. $f(x) = x^4 + 3x^2 - 2x + 1$, about $x = 0$

13. $f(x) = x^4 + 3x^2 - 2x + 1$, about $x = -2$

14. $f(x) = \dfrac{1}{(x + 2)^3}$, about $x = 0$

15. $f(x) = \dfrac{1}{x^2 + 8x + 15}$, about $x = 0$

16. $f(x) = e^x$, about $x = 3$

17. $f(x) = \operatorname{Tan}^{-1} x$, about $x = 0$

18. $f(x) = \sqrt{x + 3}$, about $x = 0$

19. $f(x) = \dfrac{x^2}{(1 + x^2)^2}$, about $x = 0$

20. $f(x) = x(1 - x)^{1/3}$, about $x = 0$

Find the first four nonzero terms in the Maclaurin series for the function in each of Exercises 21–23.

21. $\tan 2x$

22. $\sec x$

23. $e^x \sin x$

24. Find the Maclaurin series for $\cos^2 x$.

In Exercises 25–28, find the Maclaurin series for the given function.

25. $f(x) = \dfrac{1}{x^6 - 3x^3 - 4}$

26. $f(x) = \operatorname{Sin}^{-1}(x^2)$

27. $f(x) = \dfrac{2x^2 + 4}{x^2 + 4x + 3}$

28. $f(x) = \ln\left[\dfrac{1 + x/\sqrt{2}}{1 - x/\sqrt{2}}\right]$

Verify the results in Exercises 29–32.

29. $\displaystyle\sum_{n=0}^{\infty} \frac{1}{n!} = e$

30. $\displaystyle\sum_{n=0}^{\infty} \frac{(-1)^n}{(2n)!} = \cos 1$

31. $\displaystyle\sum_{n=1}^{\infty} \frac{(-1)^{n+1}}{n} = \ln 2$

32. $\displaystyle\sum_{n=1}^{\infty} \frac{n}{2^n} = 2$

33. If during a working day, one person drinks from a fountain every 30 seconds (on the average), then the probability that exactly n people drink in a time interval of length t seconds is given by the *Poisson distribution*:

$$P_n(t) = \frac{1}{n!}\left(\frac{t}{30}\right)^n e^{-t/30}.$$

Calculate $\sum_{n=0}^{\infty} P_n(t)$ and interpret the result.

34. Find the Maclaurin series of the error function erf(x) defined by

$$\text{erf}(x) = \frac{2}{\sqrt{\pi}}\int_0^x e^{-t^2}\,dt.$$

35. Find Maclaurin series for the Fresnel integrals $C(x)$ and $S(x)$ defined by

$$C(x) = \int_0^x \cos(\pi t^2/2)\,dt, \qquad S(x) = \int_0^x \sin(\pi t^2/2)\,dt.$$

Show that Bessel functions of the first kind (defined in Exercise 30 in Section 11.6) satisfy the properties in Exercises 36 and 37.

36. $2mJ_m(x) - xJ_{m-1}(x) = xJ_{m+1}(x)$

37. $J_{m-1}(x) - J_{m+1}(x) = 2J'_m(x)$

38. If the function $(1 - 2\mu x + x^2)^{-1/2}$ is expanded in a Maclaurin series in x,

$$\frac{1}{\sqrt{1 - 2\mu x + x^2}} = \sum_{n=0}^{\infty} P_n(\mu)x^n,$$

the coefficients $P_n(\mu)$ are called the *Legendre polynomials.* Find $P_0(\mu)$, $P_1(\mu)$, $P_2(\mu)$, and $P_3(\mu)$.

39. a) If we define $f(x) = x/(e^x - 1)$ at $x = 0$ as $f(0) = 1$, it turns out that $f(x)$ has a Maclaurin series expansion with positive radius of convergence. When this expansion is expressed in the form

$$\frac{x}{e^x - 1} = 1 + B_1 x + \frac{B_2}{2!}x^2 + \frac{B_3}{3!}x^3 + \cdots,$$

the coefficients $B_1, B_2, B_3, \ldots$ are called the *Bernoulli numbers.* Write this equation in the form

$$x = (e^x - 1)\left(1 + B_1 x + \frac{B_2}{2!}x^2 + \cdots\right),$$

and substitute the Maclaurin series for e^x to find the first five Bernoulli numbers.

*b) Show that the odd Bernoulli numbers all vanish for $n \geq 3$.

***40.** By substituting power series expansions for $e^{xt/2}$ and $e^{-x/(2t)}$ in terms of powers of t and $1/t$, respectively, show that

$$e^{x(t-1/t)/2} = \sum_{n=0}^{\infty} J_n(x)t^n.$$

For a definition of $J_n(x)$, see Exercise 30 in Section 11.6.

11.9 APPLICATIONS OF TAYLOR SERIES AND TAYLOR'S REMAINDER FORMULA

We suggested in Section 11.7 that if Taylor's remainder $R_n(a, x)$ is truncated from the remainder formula (11.20), then an approximation to $f(x)$ is

$$f(x) \simeq f(a) + f'(a)(x - a) + \frac{f''(a)}{2!}(x - a)^2 + \cdots + \frac{f^{(n-1)}(a)}{(n-1)!}(x - a)^{n-1}.$$

Approximations

The remainder,

$$R_n(a, x) = \frac{f^{(n)}(c)}{n!}(x - a)^n$$

(where c is between a and x), therefore represents the accuracy of the approximation: the smaller R_n, the better the approximation. In this section we use Taylor series and Taylor's remainder formula in a number of situations that require approximations.

Consider using the first three nonzero terms of the Maclaurin series for $\sin x$ to approximate $\sin x$ on the interval $-\frac{1}{2} \leq x \leq \frac{1}{2}$. Taylor's remainder formula (with $a = 0$) states that

$$\sin x = x - \frac{x^3}{3!} + \frac{x^5}{5!} + R_6,$$

where

$$R_6 = \frac{d^6}{dx^6}(\sin x)_{|x=c}\frac{x^6}{6!} = \frac{-\sin c}{6!}x^6$$

and c is between 0 and x. If we knew the value of c, we would know precisely the error in replacing the complete Taylor series of $\sin x$ by its first six terms $x - x^3/3! + x^5/5!$. Unfortunately, Taylor's remainder formula only tells us that c is somewhere between 0 and x. What we can do, however, is find a maximum possible value for R_6 considering all values of x between $-\frac{1}{2}$ and $\frac{1}{2}$. No matter what the value of c, $|\sin c| < 1$, and therefore it follows that

$$|R_6| < \frac{|x|^6}{6!} \leq \frac{(\frac{1}{2})^6}{6!} < 0.000022.$$

Consequently, if we use $x - x^3/3! + x^5/5!$ to approximate $\sin x$ for any x in the interval $|x| \leq \frac{1}{2}$, the approximation must be accurate to within 0.000022.

Actually, we can do better than this. Since the coefficient of x^6 in the Maclaurin series for $\sin x$ is $a_6 = 0$, we can write

$$\sin x = x - \frac{x^3}{3!} + \frac{x^5}{5!} + R_7,$$

where

$$|R_7| = \left|\frac{d^7}{dx^7}(\sin x)_{|x=c}\frac{x^7}{7!}\right| \leq \frac{(\frac{1}{2})^7}{7!} < 0.0000016.$$

In other words, approximation of $\sin x$ by $x - x^3/3! + x^5/5!$ on $|x| \leq \frac{1}{2}$ is accurate to within 0.0000016.

Problems of this type often take the form shown in the following example.

Example 35 How many terms in the Maclaurin series of $\ln(1 + x)$ will guarantee a truncation error of less than 10^{-6} for any x in the interval $0 \leq x \leq \frac{1}{2}$?

Solution It is straightforward to calculate that the nth derivative of $\ln(1 + x)$ is

$$\frac{d^n}{dx^n}\ln(1 + x) = \frac{(-1)^{n+1}(n-1)!}{(x+1)^n}, \qquad n \geq 1,$$

and therefore Taylor's remainder formula with $a = 0$ states that

$$\ln(1 + x) = x - \frac{x^2}{2} + \frac{x^3}{3} - \frac{x^4}{4} + \cdots + \frac{(-1)^n}{n-1}x^{n-1} + R_n,$$

where

$$R_n = \frac{(-1)^{n+1}(n-1)!}{(c+1)^n}\frac{x^n}{n!} = \frac{(-1)^{n+1}}{n(c+1)^n}x^n.$$

Since c is between 0 and x and $0 \leq x \leq \frac{1}{2}$, we can state that

$$|R_n| < \frac{(\frac{1}{2})^n}{n(1)^n} = \frac{1}{n2^n}.$$

The truncated Maclaurin series will have an error less than 10^{-6}, then, if

$$\frac{1}{n2^n} < 10^{-6}, \quad \text{or} \quad n2^n > 10^6.$$

Evaluation of $n2^n$ for a few values of n quickly indicates that the smallest value of n for which this is true is $n = 16$. Consequently, if $\ln(1 + x)$ is approximated by

$$\ln(1 + x) \simeq x - \frac{x^2}{2} + \frac{x^3}{3} - \frac{x^4}{4} + \cdots + \frac{x^{15}}{15}$$

on the interval $0 \le x \le \frac{1}{2}$, the truncation error will be less than 10^{-6}.

Because the Maclaurin series for $\ln(1 + x)$ is an alternating series, we could also use the results of Section 11.5. The maximum possible error when

$$\ln(1 + x) = x - \frac{x^2}{2} + \frac{x^3}{3} - \frac{x^4}{4} + \cdots + \frac{(-1)^{n+1}}{n}x^n + \cdots$$

is truncated after the term $(-1)^n x^{n-1}/(n - 1)$ is the value of the next term, $(-1)^{n+1}x^n/n$. If we consider the values of x in the interval $0 \le x \le \frac{1}{2}$, this error is maximized when $x = \frac{1}{2}$; i.e., the magnitude of the maximum possible truncation error is again

$$\left|\frac{(-1)^{n+1}(\frac{1}{2})^n}{n}\right| = \frac{1}{n2^n}. \quad \square$$

Truncated power series are often used to evaluate definite integrals of functions that do not have antiderivatives. For instance, to evaluate

$$\int_0^{1/2} \frac{\sin x}{x}\, dx,$$

where $\sin x/x$ is defined as 1 at $x = 0$, we expand the integrand in a Maclaurin series:

$$\frac{1}{x}\sin x = \frac{1}{x}\left\{x - \frac{x^3}{3!} + \frac{x^5}{5!} - \cdots\right\} = 1 - \frac{x^2}{3!} + \frac{x^4}{5!} - \cdots, \qquad -\infty < x < \infty.$$

According to Theorem 11.12,

$$\int \frac{\sin x}{x}\, dx = \left\{x - \frac{x^3}{3 \cdot 3!} + \frac{x^5}{5 \cdot 5!} - \cdots\right\} + C,$$

and therefore

$$\int_0^{1/2} \frac{\sin x}{x}\, dx = \frac{1}{2} - \frac{(\frac{1}{2})^3}{3 \cdot 3!} + \frac{(\frac{1}{2})^5}{5 \cdot 5!} - \cdots.$$

If we approximate this alternating series with its first three terms, the truncation error is the fourth term. Since

$$\frac{1}{2} - \frac{(\frac{1}{2})^3}{3 \cdot 3!} + \frac{(\frac{1}{2})^5}{5 \cdot 5!} = 0.4931076$$

and

$$\frac{(\frac{1}{2})^7}{7 \cdot 7!} < 0.000000222,$$

we obtain

$$0.4931074 < \int_0^{1/2} \frac{\sin x}{x}\, dx < 0.4931076.$$

Consequently, using only three terms of the Maclaurin series for $\sin x/x$, we can say that to five decimals

$$\int_0^{1/2} \frac{\sin x}{x}\, dx = 0.49311.$$

Example 36 A very important function in engineering and physics is the error function erf(x) defined by

$$\operatorname{erf}(x) = \frac{2}{\sqrt{\pi}} \int_0^x e^{-x^2}\, dx.$$

Calculate erf(1) correct to three decimals.

Solution If we replace the integrand by its Maclaurin series, we have

$$\begin{aligned} \frac{\sqrt{\pi}}{2}\operatorname{erf}(1) &= \int_0^1 \left\{1 - x^2 + \frac{(-x^2)^2}{2!} + \frac{(-x^2)^3}{3!} + \cdots\right\} dx \\ &= \left\{x - \frac{x^3}{3} + \frac{x^5}{5\cdot 2!} - \frac{x^7}{7\cdot 3!} + \cdots\right\}_0^1 \\ &= 1 - \frac{1}{3} + \frac{1}{5\cdot 2!} - \frac{1}{7\cdot 3!} + \frac{1}{9\cdot 4!} - \cdots. \end{aligned}$$

Since

$$\frac{2}{\sqrt{\pi}}\left\{1 - \frac{1}{3} + \frac{1}{5\cdot 2!} - \frac{1}{7\cdot 3!} + \frac{1}{9\cdot 4!} - \frac{1}{11\cdot 5!}\right\} = 0.84259367$$

and the seventh nonzero term is

$$\frac{2}{\sqrt{\pi}}\frac{1}{13\cdot 6!} < 0.00012056,$$

it follows that

$$0.84259367 < \operatorname{erf}(1) < 0.84271423,$$

and to three decimals

$$\operatorname{erf}(1) = 0.843. \ \square$$

In Section 9.9 we developed three numerical techniques for the evaluation of definite integrals of functions $f(x)$ that have no obvious antiderivatives: the rectangular rule, the trapezoidal rule, and Simpson's rule. Each method divides the interval of integration into a number of subintervals and approximates $f(x)$ by a more elementary function on each subinterval. The rectangular rule replaces $f(x)$ by a step function, the trapezoidal rule replaces $f(x)$ by a succession of linear functions, and Simpson's rule uses a sequence of quadratic functions.

In this section we have shown that another possibility is to replace $f(x)$ by a truncated power series (a polynomial) over the entire interval of integration.

Differential Equations Many differential equations arising in physics and engineering have solutions that can be expressed only in terms of infinite series. One such equation is Bessel's differential equation of order zero for a function $y = f(x)$:

$$xy'' + y' + xy = 0,$$

where $y' = dy/dx$ and $y'' = d^2y/dx^2$. Before considering this rather difficult differential equation, we introduce the ideas through a somewhat easier example. It is a simple substitution to verify that for any constant C, $y = f(x) = Ce^{2x} - \frac{1}{2} - x/4$ is a solution of the differential equation

$$y' - 2y = x,$$

and in Chapter 16 we will show how to derive this solution in a very straightforward way. Our purpose here is to show that it can be obtained by assuming that the solution of the differential equation can be represented as an infinite series,

$$y = f(x) = \sum_{n=0}^{\infty} a_n x^n = a_0 + a_1 x + a_2 x^2 + \cdots.$$

If we substitute this form for $f(x)$ into the differential equation, we have

$$\{a_1 + 2a_2 x + 3a_3 x^2 + 4a_4 x^3 + \cdots\} - 2\{a_0 + a_1 x + a_2 x^2 + \cdots\} = x,$$

and if we gather together terms in the various powers of x, we obtain

$$0 = (a_1 - 2a_0) + (2a_2 - 2a_1 - 1)x + (3a_3 - 2a_2)x^2 + (4a_4 - 2a_3)x^3 + \cdots.$$

Since the power series on the right always has sum zero, its coefficients must all vanish (see Exercise 5 in Section 11.7), and therefore we must set

$$\begin{aligned} a_1 - 2a_0 &= 0; \\ 2a_2 - 2a_1 - 1 &= 0; \\ 3a_3 - 2a_2 &= 0; \\ 4a_4 - 2a_3 &= 0 \ldots . \end{aligned}$$

These equations imply that

$$\begin{aligned} a_1 &= 2a_0; \\ a_2 &= \frac{1}{2}(1 + 2a_1) = \frac{1}{2}(1 + 4a_0); \\ a_3 &= \frac{2}{3}a_2 = \frac{2}{3!}(1 + 4a_0); \\ a_4 &= \frac{2}{4}a_3 = \frac{2^2}{4!}(1 + 4a_0); \ldots ; \\ a_n &= \frac{2^{n-2}}{n!}(1 + 4a_0). \end{aligned}$$

Thus,

$$\begin{aligned} f(x) &= a_0 + 2a_0 x + \frac{1}{2}(1 + 4a_0)x^2 + \cdots + \frac{2^{n-2}}{n!}(1 + 4a_0)x^n + \cdots \\ &= a_0 + 2a_0 x + \frac{1}{4}(1 + 4a_0)\left\{\frac{2^2x^2}{2!} + \frac{2^3x^3}{3!} + \cdots + \frac{2^nx^n}{n!} + \cdots\right\}. \end{aligned}$$

But the Maclaurin series for e^{2x} is

$$e^{2x} = 1 + (2x) + \frac{(2x)^2}{2!} + \frac{(2x)^3}{3!} + \cdots,$$

and therefore,

$$\begin{aligned} f(x) &= a_0 + 2a_0 x + \frac{1}{4}(1 + 4a_0)\{e^{2x} - 1 - 2x\} \\ &= -\frac{1}{4} - \frac{x}{2} + \frac{1}{4}(1 + 4a_0)e^{2x} \\ &= Ce^{2x} - \frac{1}{4} - \frac{x}{2}. \end{aligned}$$

Certainly this is a laborious way to derive the solution of this particular differential equation, but it clearly illustrates the steps in a solution obtained through power series. We now consider Bessel's differential equation of order zero: $xy'' + y' + xy = 0$. If we assume that the solution has a power series expansion $y = f(x) = \sum a_n x^n$ and substitute, we obtain

$$\begin{aligned} 0 &= x \sum_{n=0}^{\infty} n(n-1)a_n x^{n-2} + \sum_{n=0}^{\infty} na_n x^{n-1} + x \sum_{n=0}^{\infty} a_n x^n \\ &= \sum_{n=0}^{\infty} n(n-1)a_n x^{n-1} + \sum_{n=0}^{\infty} na_n x^{n-1} + \sum_{n=0}^{\infty} a_n x^{n+1}. \end{aligned}$$

In order to combine all terms in x, all terms in x^2, etc., we change the index of summation in the last term:

$$\begin{aligned} 0 &= \sum_{n=0}^{\infty} n(n-1)a_n x^{n-1} + \sum_{n=0}^{\infty} na_n x^{n-1} + \sum_{n=2}^{\infty} a_{n-2} x^{n-1} \\ &= a_1 + \sum_{n=2}^{\infty} \{n(n-1)a_n + na_n + a_{n-2}\}x^{n-1}. \end{aligned}$$

But the only way a power series can be equal to zero is for all of its coefficients to be equal to zero; i.e.,

$$a_1 = 0; \qquad n(n-1)a_n + na_n + a_{n-2} = 0, \quad n \geq 2.$$

Thus,

$$a_n = -\frac{a_{n-2}}{n^2}, \qquad n \geq 2,$$

a recursive relation defining the unknown coefficient a_n of x^n in terms of the coefficient a_{n-2} of x^{n-2}. Since $a_1 = 0$, it follows that

$$0 = a_1 = a_3 = a_5 = \cdots.$$

For $n = 2$,

$$a_2 = -\frac{a_0}{2^2}.$$

For $n = 4$,

$$a_4 = -\frac{a_2}{4^2} = \frac{a_0}{2^2 4^2} = \frac{a_0}{2^4(2!)^2}.$$

For $n = 6$,

$$a_6 = -\frac{a_4}{6^2} = \frac{-a_0}{2^4(2!)^2 6^2} = -\frac{a_0}{2^6(3!)^2}.$$

The solution obtained is therefore

$$f(x) = a_0 - \frac{a_0}{2^2}x^2 + \frac{a_0}{2^4(2!)^2}x^4 - \frac{a_0}{2^6(3!)^2}x^6 + \cdots$$
$$= a_0 \sum_{n=0}^{\infty} \frac{(-1)^n}{2^{2n}(n!)^2} x^{2n}.$$

The function defined by the infinite series

$$J_0(x) = \sum_{n=0}^{\infty} \frac{(-1)^n}{2^{2n}(n!)^2} x^{2n}, \qquad -\infty < x < \infty$$

is called the *zero-order Bessel function of the first kind.*

EXERCISES 11.9

In Exercises 1–4, find a maximum possible error in using the given terms of the Taylor series to approximate the function on the interval specified.

1. $\cos x \simeq 1 - \frac{x^2}{2!} + \frac{x^4}{4!}$ for $|x| \leq \frac{1}{10}$

2. $e^x \simeq 1 + x + \frac{x^2}{2!} + \frac{x^3}{3!} + \frac{x^4}{4!}$ for $0 \leq x < 0.01$

3. $\mathrm{Tan}^{-1}\, x \simeq x - \frac{x^3}{3} + \frac{x^5}{5} - \frac{x^7}{7} + \cdots - \frac{x^{99}}{99}$ for $|x| < 1$

4. $\ln x \simeq (x-1) - \frac{1}{2}(x-1)^2 + \frac{1}{3}(x-1)^3 - \frac{1}{4}(x-1)^4$ for $|x - 1| \leq \frac{1}{2}$

Evaluate the integral in Exercises 5–8 correct to three decimals.

C 5. $\int_0^1 \frac{\sin x}{x}\, dx$

C 6. $\int_0^{1/2} \cos(x^2)\, dx$

C 7. $\int_0^{2/3} \frac{1}{x^4 + 1}\, dx$

C 8. $\int_{-1}^{1} x^{11} \sin x\, dx$

Find a series solution in powers of x for the differential equation in Exercises 9–12.

9. $y' + 3y = 4$

10. $y'' + y' = 0$

11. $xy' - 4y = 3x$

12. $4xy'' + 2y' + y = 0$

13. In special relativity theory, the kinetic energy K of an object moving with speed v is defined by

$$K = c^2(m - m_0),$$

where c is a constant (the speed of light), m_0 is the rest mass of the object, and m is its mass when moving with speed v. The masses m and m_0 are related by

$$m = \frac{m_0}{\sqrt{1 - v^2/c^2}}.$$

Use the binomial expansion to show that

$$K = \frac{1}{2}m_0 v^2 + m_0 c^2\left\{\frac{3}{8}\frac{v^4}{c^4} + \frac{5}{16}\frac{v^6}{c^6} + \cdots\right\},$$

and hence, to a first approximation, kinetic energy is defined by the classical expression $m_0 v^2/2$.

Find a series solution in powers of x for the differential equation in Exercises 14 and 15.

14. $xy'' + y = 0$

15. $y'' + y = 0$

In Exercises 16–18, determine where the Maclaurin series for the function may be truncated in order to guarantee the accuracy indicated.

16. $\sin(x/3)$ on $|x| \le 4$ with error less than 10^{-3}

17. $\dfrac{1}{\sqrt{1+x^3}}$ on $0 < x < \frac{1}{2}$ with error less than 10^{-4}

18. $\ln(1-x)$ on $|x| < \frac{1}{3}$ with error less than 10^{-2}

19. The ellipse $x^2/a^2 + y^2/b^2 = 1$ can be represented parametrically by

$$x = a\cos t, \qquad y = b\sin t, \qquad 0 \le t < 2\pi.$$

a) Show that the length of the circumference of the ellipse is defined by the definite integral

$$L = 4b\int_0^{\pi/2} \sqrt{1 - k^2\sin^2 t}\,dt, \qquad k^2 = 1 - \frac{a^2}{b^2}.$$

b) Use the binomial expansion to show that

$$L = 2\pi b\left\{1 - \frac{k^2}{4} - \frac{3k^4}{64} - \cdots\right\}$$

so that to a first approximation, L is the circumference of a circle of radius b.

20. In determining the radiated power from a half-wave antenna, it is necessary to evaluate

$$\int_0^{2\pi} \frac{1-\cos\theta}{\theta}\,d\theta.$$

Find a two-decimal approximation for this integral.

21. Planck's law for the energy density Ψ of blackbody radiation of wavelength λ states that

$$\Psi(\lambda) = \frac{8\pi ch\lambda^{-5}}{e^{ch/(\lambda kT)} - 1},$$

where $h > 0$ is Planck's constant, c is the (constant) speed of light, and T is temperature, also assumed constant. Show that for long wavelengths Planck's law reduces to the Rayleigh–Jeans law:

$$\Psi(\lambda) = \frac{8\pi kT}{\lambda^4}.$$

SUMMARY

An infinite sequence of constants is a function whose domain is the set of positive integers, and as such could be denoted by $f(n)$. It therefore assigns to the positive integers $1, 2, 3, \ldots$, numbers $f(1), f(2), f(3), \ldots$. In this chapter we have preferred an alternative notation $c_1, c_2, c_3, \ldots$ for the terms of a sequence, and correspondingly the notation $\{c_n\}_1^\infty$ or $\{c_n\}$ for the sequence as a whole.

In most applications of sequences, the prime consideration is whether the sequence has a limit. If the sequence has its terms defined explicitly, then our ability to take limits of continuous functions ("limits at infinity" in Chapter 2 and L'Hôpital's rule in Chapters 4 and 8) can be very helpful. If the sequence is defined recursively, existence of the limit can sometimes be established by showing that the sequence is monotonic and bounded.

An expression of the form

$$\sum_{n=1}^{\infty} c_n = c_1 + c_2 + \cdots + c_n + \cdots$$

is called an infinite series. To define what is meant by such an expression we define its sequence of partial sums $\{S_n\}$, where S_n is the sum of the first n terms of the series. If this sequence has a limit, we call it the sum of the series; if the sequence does not have a limit, we say that the series does not have a sum. Unfortunately, for most series it is impossible to find a simple formula for S_n, and therefore analysis of the limit of the sequence $\{S_n\}$ is out of the question. To remedy this, we developed various convergence tests that avoided the sequence $\{S_n\}$: nth term, comparison, limit comparison, limit ratio, and alternating series tests. Note the number of sequences that have now been associated with a series $\sum c_n$:

$\{S_n\}$	sequence of partial sums for the definition of a sum;
$\{c_n\}$	sequence of terms for nth term test;
$\{c_n/b_n\}$	sequence for limit comparison test;
$\{c_{n+1}/c_n\}$	sequence for limit ratio test;
$\{\lvert c_n \rvert\}$	sequence for alternating series test.

Depending on the limits of these sequences (if they exist), we may be able to infer something about convergence of the series.

Each of these tests suffers from the fact that it does not determine the sum of the series in the case that the series converges. It often becomes necessary then to approximate the sum of a series by using some specific partial sum; i.e., the series is truncated at some stage, and the sum of the terms to this stage is used as an approximation to the full sum. Remember, however, that there is little point in making such an approximation unless it is accompanied by some definitive statement about the maximum possible error incurred by truncation.

From infinite sequences and series of constants we proceeded to a discussion of infinite sequences and series of functions—in particular, power series. Our first point of view was that a power series was given, and the sum was to be determined. We saw that every power series $\sum a_n(x - a)^n$ has a radius of convergence R and an associated interval of convergence. If $R = 0$, the interval of convergence consists of only one point $x = a$; if $R = \infty$, the power series converges absolutely for all x; and if $0 < R < \infty$, the interval of convergence must be one of four possibilities: $a - R < x < a + R$, $a - R \le x < a + R$, $a - R < x \le a + R$, or $a - R \le x \le a + R$. The radius of convergence is given by $\lim_{n\to\infty} |a_n/a_{n+1}|$ provided the limit exists or is equal to infinity. If at each point in the interval of convergence of the power series the value of a function $f(x)$ is the same as the sum of the series, we write $f(x) = \sum a_n(x - a)^n$ and call $f(x)$ the sum of the series.

The opposite viewpoint is to consider a function $f(x)$ and a point a as given, and ask whether $f(x)$ has a power series expansion about a. We saw that there can be at most one power series expansion of $f(x)$ about a with a positive radius of convergence, and this series must be its Taylor series. In other words, if a function is to have a power series expansion about a point, it must be the Taylor series. One way to verify that $f(x)$ does indeed have a Taylor series about a and that this series converges to $f(x)$ is to show that the sequence of Taylor's remainders $\{R_n(a, x)\}$ exists and has limit zero. Fortunately, however, there exist alternative methods for finding Taylor series such as addition, multiplication, differentiation, and integration of known series.

When a Taylor series is truncated, Taylor's remainder $R_n(a, x)$ represents the truncation error and, in spite of the fact that R_n is expressed in terms of some unknown point c, it is often possible to calculate a maximum value for the error.

Sometimes $R_n(a, x)$ can be avoided altogether. For instance, if the Taylor series is an alternating series, then the maximum possible truncation error is the value of the next term.

REVIEW EXERCISES

In Exercises 1–4, discuss, with all necessary proofs, whether the sequence is monotonic and has an upper bound, a lower bound, and a limit.

1. $\left\{\dfrac{n^2 - 5n + 3}{n^2 + 5n + 4}\right\}$

2. $c_1 = 1, c_{n+1} = \frac{1}{2}\sqrt{c_n^2 + 1}, n \geq 1$

3. $\left\{\dfrac{\text{Tan}^{-1}(1/n)}{n^2 + 1}\right\}$

4. $c_1 = 7, c_{n+1} = 15 + \sqrt{c_n - 2}, n \geq 1$

C **5.** Use Newton's iterative procedure and the method of successive substitutions to approximate the root of the equation

$$x = \left(\frac{x + 5}{x + 4}\right)^2$$

between $x = 1$ and $x = 2$.

6. For what values of k does the sequence

$$c_1 = k, \qquad c_{n+1} = c_n^2, \qquad n \geq 1$$

converge?

7. Find an explicit definition for the sequence

$$c_1 = 1, \qquad c_{n+1} = \sqrt{1 + c_n^2}, \qquad n \geq 1.$$

8. Use the derivative of the function $f(x) = (\ln x)/x$ to prove that the sequence $\{\ln(n)/n\}$ is decreasing for $n \geq 3$.

In Exercises 9–28, determine whether the series converges or diverges. In the case of a convergent series that has both positive and negative terms, indicate whether it converges absolutely or conditionally.

9. $\displaystyle\sum_{n=1}^{\infty} \frac{n^2 - 3n + 2}{n^3 + 4n}$

10. $\displaystyle\sum_{n=1}^{\infty} \frac{n^2 + 5n + 3}{n^4 - 2n + 5}$

11. $\displaystyle\sum_{n=1}^{\infty} \frac{5^{2n}}{n!}$

12. $\displaystyle\sum_{n=1}^{\infty} \frac{n^2 + 3}{n3^n}$

13. $\displaystyle\sum_{n=1}^{\infty} \frac{(\ln(n))^2}{\sqrt{n}}$

14. $\displaystyle\sum_{n=1}^{\infty} (-1)^n\left(\frac{n + 1}{n^2}\right)$

15. $\displaystyle\sum_{n=1}^{\infty} (-1)^n\left(\frac{n + 1}{n^3}\right)$

16. $\displaystyle\sum_{n=1}^{\infty} \text{Cos}^{-1}\left(\frac{1}{n}\right)$

17. $\displaystyle\sum_{n=1}^{\infty} \frac{1}{n}\,\text{Cos}^{-1}\left(\frac{1}{n}\right)$

18. $\displaystyle\sum_{n=1}^{\infty} \frac{1}{n^2}\,\text{Cos}^{-1}\left(\frac{1}{n}\right)$

19. $\displaystyle\sum_{n=1}^{\infty} \frac{2 \cdot 4 \cdot 6 \cdot \cdots \cdot (2n)}{n!}$

20. $\displaystyle\sum_{n=1}^{\infty} \frac{3 \cdot 6 \cdot 9 \cdot \cdots \cdot (3n)}{(2n)!}$

21. $\sum_{n=1}^{\infty} \sqrt{\frac{n^2+1}{n^2+5}}$

22. $\sum_{n=1}^{\infty} (-1)^{n+1}\left(1+\frac{1}{n}\right)^3$

23. $\sum_{n=1}^{\infty} \frac{1}{n^2} \sin(n)$

24. $\sum_{n=1}^{\infty} \frac{10^n}{5^{3n+2}}$

25. $\sum_{n=1}^{\infty} (-1)^n \frac{\ln(n)}{n}$

26. $\sum_{n=1}^{\infty} \frac{1}{e^{n\pi}}$

27. $\sum_{n=1}^{\infty} \frac{2^n + 2^{-n}}{3^n}$

28. $\sum_{n=1}^{\infty} \frac{1}{\sqrt{n}} \cos(n\pi)$

In Exercises 29–36, find the interval of convergence for the given power series.

29. $\sum_{n=0}^{\infty} \frac{n+1}{n^2+1} x^n$

30. $\sum_{n=1}^{\infty} \frac{1}{n^2 2^n} x^n$

31. $\sum_{n=0}^{\infty} (n+1)^3 x^n$

32. $\sum_{n=1}^{\infty} \frac{x^n}{n^n}$

33. $\sum_{n=0}^{\infty} \frac{1}{4^n} (x-2)^n$

34. $\sum_{n=2}^{\infty} \sqrt{\frac{n+1}{n-1}} (x+3)^n$

35. $\sum_{n=1}^{\infty} n 3^n x^{2n}$

36. $\sum_{n=1}^{\infty} \frac{2^n}{n} x^{3n}$

In Exercises 37–45, find the power series expansion of the function about the point indicated.

37. $f(x) = \sqrt{1+x^2}$, about $x = 0$

38. $f(x) = e^{x+5}$, about $x = 0$

39. $f(x) = \cos(x + \pi/4)$, about $x = 0$

40. $f(x) = x \ln(2x+1)$, about $x = 0$

41. $f(x) = \sin x$, about $x = \pi/4$

42. $f(x) = x/(x^2 + 4x + 3)$, about $x = 0$

43. $f(x) = e^x$, about $x = 3$

44. $f(x) = (x+1)\ln(x+1)$, about $x = 0$

45. $f(x) = x^3 e^{x^2}$, about $x = 0$

46. How many terms in the Maclaurin series for $f(x) = e^{-x^2}$ will guarantee a truncation error of less than 10^{-5} for all x in the interval $0 \le x \le 2$?

47. Find a power series solution in powers of x for the differential equation

$$y'' - 4y = 0.$$

48. Find the Maclaurin series for $f(x) = \sqrt{1 + \sin x}$ valid for $-\pi/2 \le x \le \pi/2$ by first showing that $f(x)$ can be written in the form

$$f(x) = \sin(x/2) + \cos(x/2).$$

Why is the restriction $-\pi/2 \le x \le \pi/2$ necessary?

C **49.** On a calculator, take the cosine of 1 (radian). Take the cosine of this result, and again, and again, and again, ... What happens? Interpret what is going on.

50. A certain experiment is to be performed until it is successful. The probability that it will be successful in any given attempt is p $(0 < p < 1)$, and therefore the probability that it will fail is $q = 1 - p$. The expected number of times that the experiment must be performed in order to be successful can be shown to be represented by the infinite series

$$\sum_{n=1}^{\infty} npq^{n-1} = \sum_{n=1}^{\infty} np(1 - p)^{n-1}.$$

a) What is the sum of this series?

b) If p is the probability that a single die will come up 6, is the answer to (a) what you would expect?

Vectors and Three-dimensional Analytic Geometry

12

12.1 RECTANGULAR COORDINATES IN SPACE

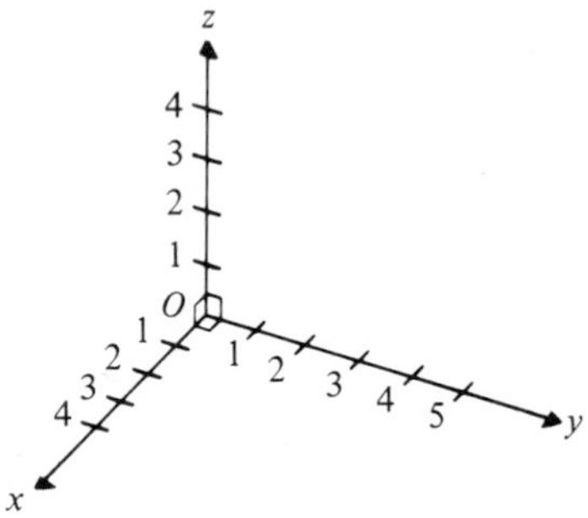

Figure 12.1

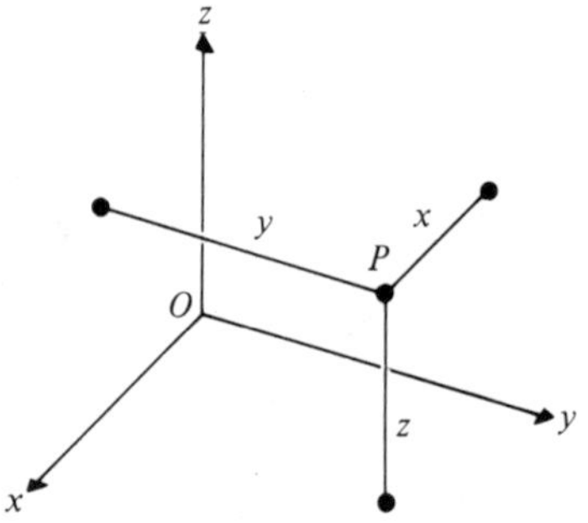

Figure 12.2

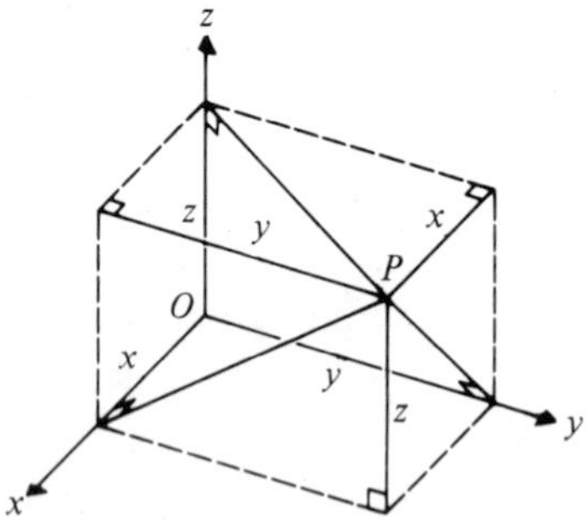

Figure 12.3

The coordinate of a point on the real line is defined as its directed distance from a fixed point called the *origin*. The Cartesian coordinates of a point in a plane are its directed distances from two fixed lines called the *coordinate axes*. In space, Cartesian coordinates will be directed distances from three fixed planes called the *coordinate planes*. In particular, we draw through a point O, called the *origin*, three mutually perpendicular lines called the x-, y-, and z-axes (Fig. 12.1). Each of the axes is coordinatized with some unit distance (which need not be the same for all three axes). These three coordinate axes determine three planes called *coordinate planes*: The xy-coordinate plane is that plane containing the x- and y-axes, the yz-coordinate plane contains the y- and z-axes, and the xz-coordinate plane contains the x- and z-axes.

If P is any point in space, we draw lines from P perpendicular to the three coordinate planes (Fig. 12.2). The directed distance from the yz-coordinate plane to P is parallel to the x-axis, and is called the x-coordinate of P. Similarly, the y- and z-coordinates are defined as the directed distances from the xz- and xy-coordinate planes to P. These three coordinates of P, written (x, y, z), are called the *Cartesian* or *rectangular coordinates* of P. It is fairly clear that if we draw through P lines perpendicular to the axes, then the coordinates of P are also the directed distances from O to the points of intersection of these perpendiculars with the axes (Fig. 12.3).

By either definition, each point in space has a unique ordered set of Cartesian coordinates (x, y, z); and conversely, every ordered triple of real numbers (x, y, z) is the set of coordinates for one and only one point in space. For example, the points with coordinates $(1, 1, 1)$, $(2, -3, 4)$, $(3, 4, -1)$, and $(-2, 5, 3)$ are shown in Fig. 12.4.

The coordinate systems in Figs. 12.1–12.4 are called *right-handed coordinate systems*, because if we curl the fingers on our right hand from the positive x-direction toward the positive y-direction, then the thumb points in the positive z-direction (Fig. 12.5). The coordinate system in Fig. 12.6, on the other hand, is a *left-handed coordinate system*, since the thumb of the left hand points in the

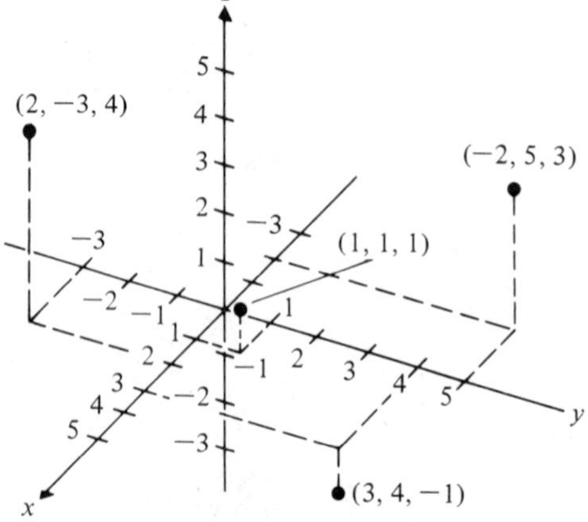

Figure 12.4

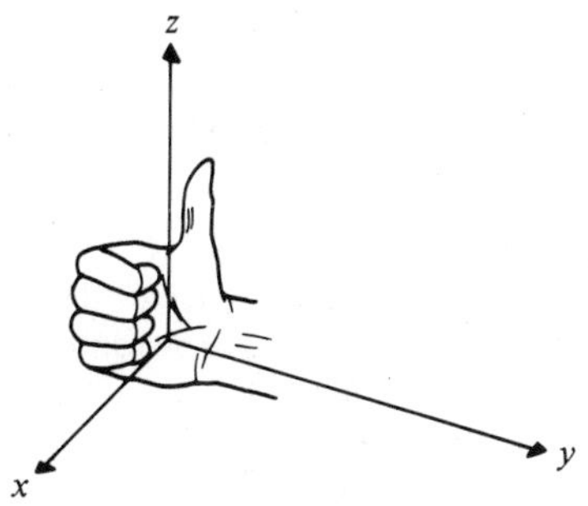

Figure 12.5

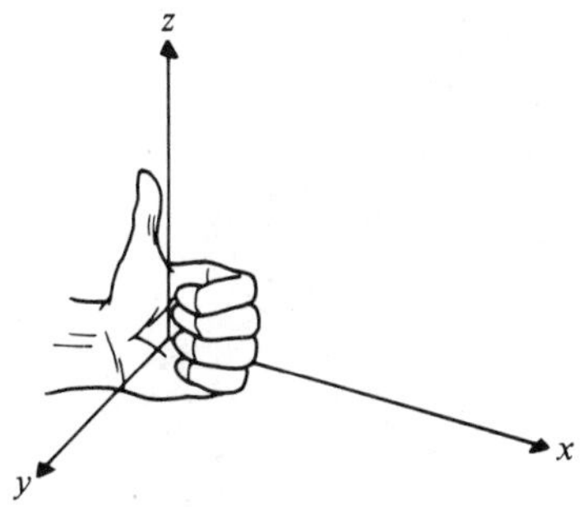

Figure 12.6

positive z-direction when the fingers of this hand are curled from the positive x-direction to the positive y-direction. We will always use right-handed systems in this book.

Suppose we construct for any two points P_1 and P_2 with coordinates (x_1, y_1, z_1) and (x_2, y_2, z_2), respectively, a box with sides parallel to the coordinate planes and with line segment P_1P_2 as diagonal (Fig. 12.7). Because triangles P_1AB and P_1BP_2 are right-angled, we can write that

$$\begin{aligned}|P_1P_2|^2 &= |P_1B|^2 + |BP_2|^2\\ &= |P_1A|^2 + |AB|^2 + |BP_2|^2\\ &= (x_2 - x_1)^2 + (y_2 - y_1)^2 + (z_2 - z_1)^2.\end{aligned}$$

In other words, the length of the line segment joining two points $P_1(x_1, y_1, z_1)$ and $P_2(x_2, y_2, z_2)$ is

$$|P_1P_2| = \sqrt{(x_2 - x_1)^2 + (y_2 - y_1)^2 + (z_2 - z_1)^2}. \tag{12.1}$$

This is the analogue of formula (1.5) for the length of a line segment joining two points in the xy-plane.

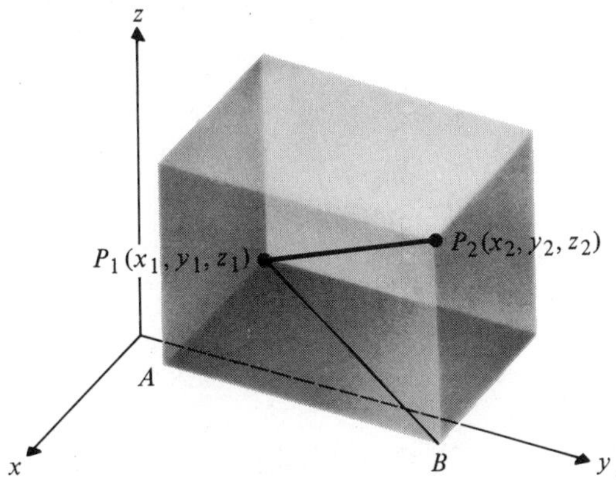

Figure 12.7

EXERCISES 12.1

1. In a space with Cartesian coordinates x, y, and z, show the points $(1, 2, 1)$, $(-1, 3, 2)$, $(1, -2, 4)$, $(3, 4, -5)$, $(-1, -2, -3)$, $(-2, -5, 4)$, $(8, -3, -6)$, and $(-4, 3, -5)$.
2. Find the length of the line segment joining the points $(1, -2, 5)$ and $(-3, 2, 4)$.
3. Prove that the triangle with vertices $(2, 0, 4\sqrt{2})$, $(3, -1, 5\sqrt{2})$, and $(4, -2, 4\sqrt{2})$ is right-angled and isosceles.
4. A cube has sides of length 2 units. What are the coordinates of its corners if one corner is at the origin, three of its faces lie in the coordinate planes, and one corner has all three coordinates positive.
5. Show that the (undirected, perpendicular) distances from a point (x, y, z) to the x-, y-, and z-axes are, respectively,

$$\sqrt{y^2 + z^2}, \quad \sqrt{x^2 + z^2}, \quad \sqrt{x^2 + y^2}.$$

In Exercises 6–9, find the (undirected) distances from (a) the origin; (b) the x-axis; (c) the y-axis; (d) the z-axis to the point.

6. $(2, 3, -4)$
7. $(1, -5, -6)$
8. $(4, 3, 0)$
9. $(-2, 1, -3)$
10. Prove that the three points $(1, 3, 5)$, $(-2, 0, 3)$, and $(7, 9, 9)$ are collinear.

11. Find that point in the xy-plane that is equidistant from the two points (1, 3, 2) and (2, 4, 5) and has a y-coordinate equal to three times its x-coordinate.

12. Find an equation describing all points that are equidistant from the two points $(-3, 0, 4)$ and (2, 1, 5). What does this equation describe geometrically?

13. a) If $(\sqrt{3}-3, 2+2\sqrt{3}, 2\sqrt{3}-1)$ and $(2\sqrt{3}, 4, \sqrt{3}-2)$ are two vertices of an equilateral triangle, and if the third vertex lies on the z-axis, find the third vertex.

b) Can you find a third vertex on the x-axis?

14. A birdhouse is built from a box $\frac{1}{2}$ meter on each side with a roof as shown in Fig. 12.8. If the distance from each corner of the roof to the peak is $\frac{3}{4}$ meter, find the coordinates of the nine corners of the house. (The sides of the box are parallel to the coordinate planes.)

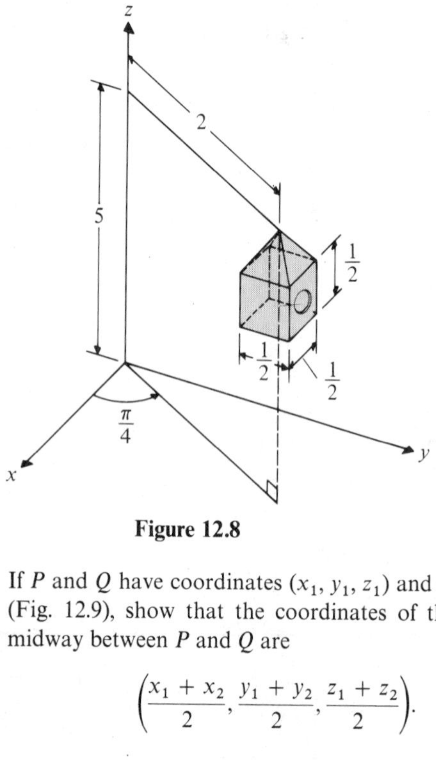

Figure 12.8

15. If P and Q have coordinates (x_1, y_1, z_1) and (x_2, y_2, z_2) (Fig. 12.9), show that the coordinates of the point R midway between P and Q are

$$\left(\frac{x_1+x_2}{2}, \frac{y_1+y_2}{2}, \frac{z_1+z_2}{2}\right).$$

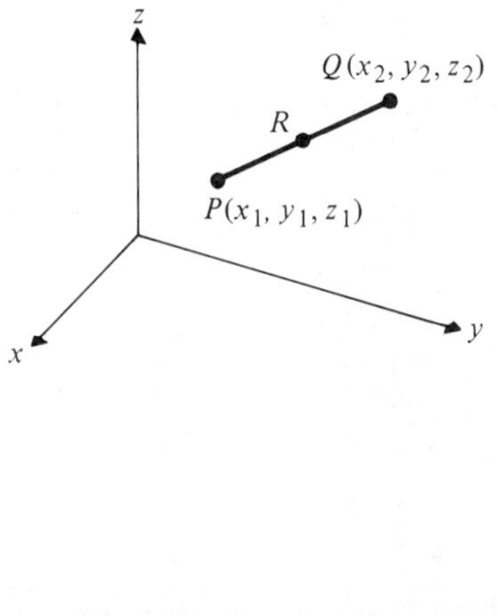

Figure 12.9

16. a) Find the midpoint of the line segment joining the points $P(1, -1, -3)$ and $Q(3, 2, -4)$.

b) If the line segment joining P and Q is extended its own length beyond Q to a point R, find the coordinates of R.

17. The four-sided object shown in Fig. 12.10 is a *tetrahedron.* If the four vertices of the tetrahedron are as shown, prove that the three lines joining the midpoints of opposite edges (one of which is PQ) meet at a point that bisects each of them.

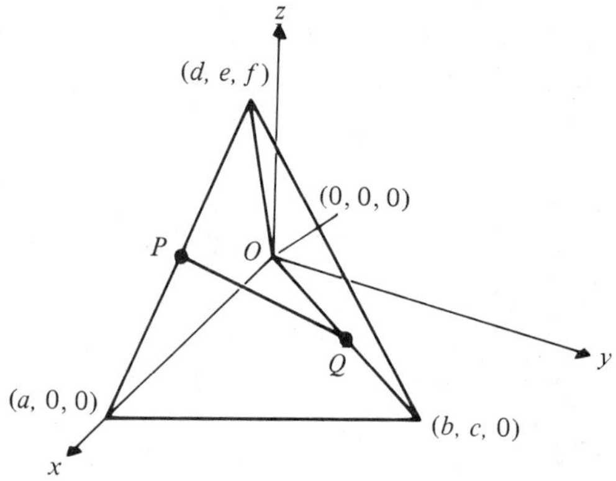

Figure 12.10

***18.** Generalize the result of Exercise 15 to prove that if a point R divides the length PQ so that

$$\frac{\text{length } PR}{\text{length } RQ} = \frac{r_1}{r_2},$$

where r_1 and r_2 are positive integers, then the coordinates of R are

$$x = \frac{r_1x_2 + r_2x_1}{r_1 + r_2}, \qquad y = \frac{r_1y_2 + r_2y_1}{r_1 + r_2},$$

$$z = \frac{r_1z_2 + r_2z_1}{r_1 + r_2}.$$

***19.** A man 2 meters tall walks along the edge of a straight road 10 meters wide (Fig. 12.11). On the other edge of the road stands a streetlight 8 meters high. A building runs parallel to the road and 1 meter from it. If Cartesian coordinates are set up as shown (with the x- and y-axes in the plane of the road), find the coordinates of the tip of the man's shadow when he is 10 meters from the point directly opposite the light.

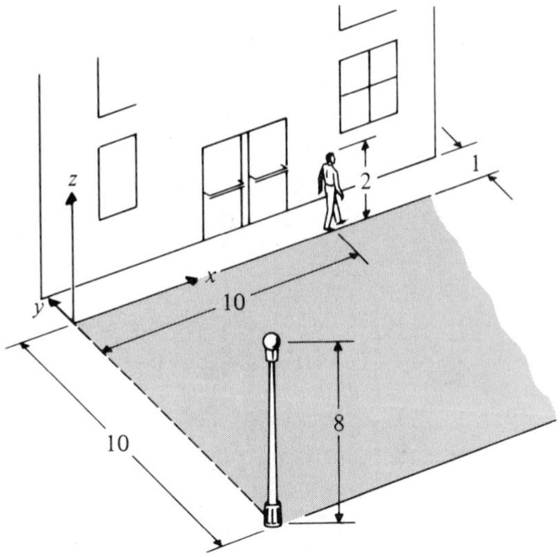

Figure 12.11

12.2 CURVES AND SURFACES

Figure 12.12

An equation in the x- and y-coordinates of points in the xy-plane usually specifies a curve. For example, the equation $x^2 + y^2 = 4$ describes a circle of radius 2 centered at the origin (Fig. 12.12). We now ask what is defined by an equation in the Cartesian coordinates (x, y, z) of points in space. Some equations are obvious. For example, the equation $z = 0$ describes all points in the xy-plane since all such points have a z-coordinate equal to zero. Similarly, $y = 2$ describes all points in the plane parallel to and two units to the right of the xz-plane. What does the equation $x^2 + y^2 = 4$ describe? In other words, regarded as a restriction on the x-, y-, and z-coordinates of points in space, rather than a restriction on the x- and y-coordinates of points in the xy-plane, what does it represent? Because the equation says nothing about z, there is no restriction whatsoever on z. In other words, the z-coordinate can take on all possible values, but the x- and y-coordinates must be restricted by $x^2 + y^2 = 4$. If we consider those points in the xy-plane ($z = 0$) that satisfy $x^2 + y^2 = 4$, we obtain the circle shown in Fig. 12.12. In space, each of these points has coordinates $(x, y, 0)$, where x and y still satisfy $x^2 + y^2 = 4$ (Fig. 12.13). If we now take any point Q that is either directly above or directly below a point $P(x, y, 0)$ on this circle, it will have exactly the same x- and y-coordinates as P; only its z-coordinate will differ. Thus the x- and y-coordinates of Q also satisfy $x^2 + y^2 = 4$. Since we can do this for any point P on the circle, it follows that $x^2 + y^2 = 4$ describes the right-circular cylinder of radius 2 and infinite extent shown in Fig. 12.13.

By reasoning similar to that used above, we can show that the equation $2x + y = 2$ describes the plane in Fig. 12.14 parallel to the z-axis and standing on the straight line $2x + y = 2$, $z = 0$ in the xy-plane.

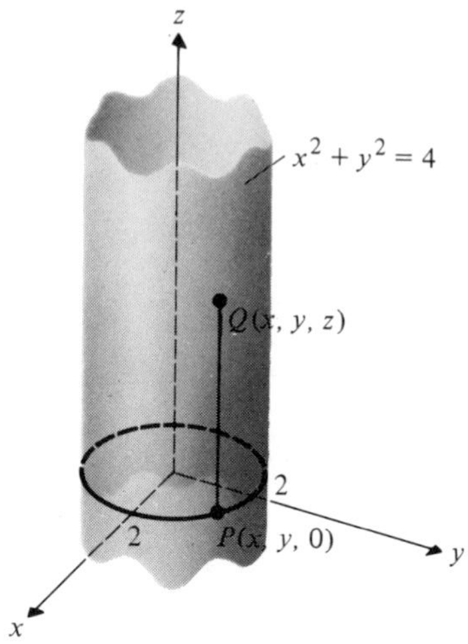

Figure 12.13

Finally, consider the equation $x^2 + y^2 + z^2 = 9$. Since $\sqrt{x^2 + y^2 + z^2}$ is the distance from the origin to a point with coordinates (x, y, z), this equation describes all points that are three units away from the origin. In other words, $x^2 + y^2 + z^2 = 9$ describes the points on a sphere of radius 3 centered at the origin.

It appears, then, that one equation in the coordinates (x, y, z) of points in space specifies a surface. The shape of the surface is determined by the form of the equation. If one equation in the coordinates (x, y, z) specifies a surface, it is easy to see what two simultaneous equations specify. For instance, suppose we ask for all points in space whose coordinates satisfy both of the equations

$$x^2 + y^2 = 4, \qquad z = 1.$$

By itself, $x^2 + y^2 = 4$ describes the cylinder in Fig. 12.13. The equation $z = 1$ describes all points in a plane parallel to the xy-plane and one unit above it. To ask for all points that satisfy $x^2 + y^2 = 4$ and $z = 1$ simultaneously is to ask for all points that lie on both surfaces. Consequently, the equations $x^2 + y^2 = 4$, $z = 1$ describe the curve of intersection of the two surfaces—the circle in Fig. 12.15.

The equation $x = 0$ describes the yz-plane; the equation $y = 0$ describes the xz-plane. If we put the two equations together, $x = 0$ and $y = 0$, we obtain all points that lie on both the yz-plane and the xz-plane; i.e., the z-axis. In other words, equations for the z-axis are $x = 0$, $y = 0$.

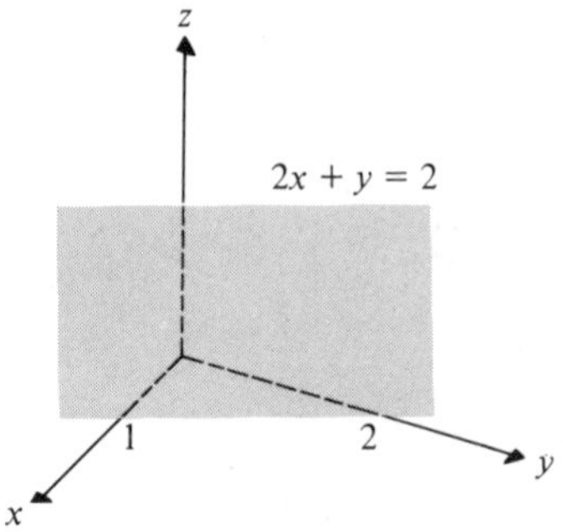

Figure 12.14

Finally, $x^2 + y^2 + z^2 = 9$ is the equation of a sphere of radius 3 centered at the origin, and $y = 2$ is the equation of a plane parallel to the xz-plane and two units to the right. Together, the equations $x^2 + y^2 + z^2 = 9$, $y = 2$ describe the curve of intersection of the two surfaces—the circle in Fig. 12.16. Note that by substituting $y = 2$ into the equation of the sphere, we can write alternatively that $x^2 + z^2 = 5$, $y = 2$. This pair of equations is equivalent to the original pair in the sense that all points that satisfy $x^2 + y^2 + z^2 = 9$, $y = 2$ also satisfy $x^2 + z^2 = 5$, $y = 2$, and vice versa. This new pair of equations provides an alternative way of visualizing the curve. Again $y = 2$ is the plane of Fig. 12.16, but $x^2 + z^2 = 5$ describes a right-circular cylinder of radius $\sqrt{5}$ and infinite extent around the y-axis (Fig. 12.17). Our discussion has shown that the cylinder and plane intersect in the same curve as the sphere and plane.

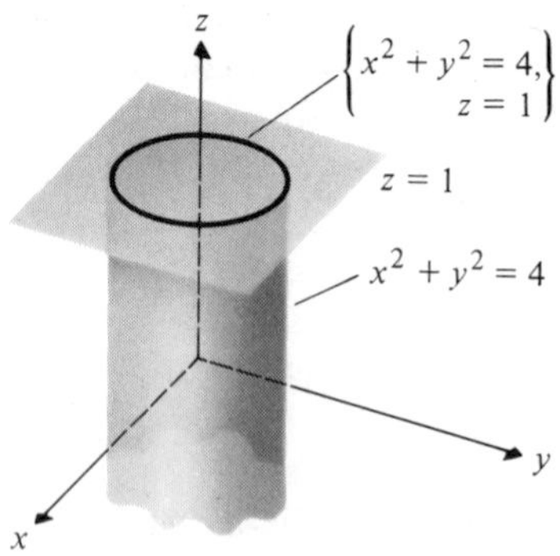

Figure 12.15

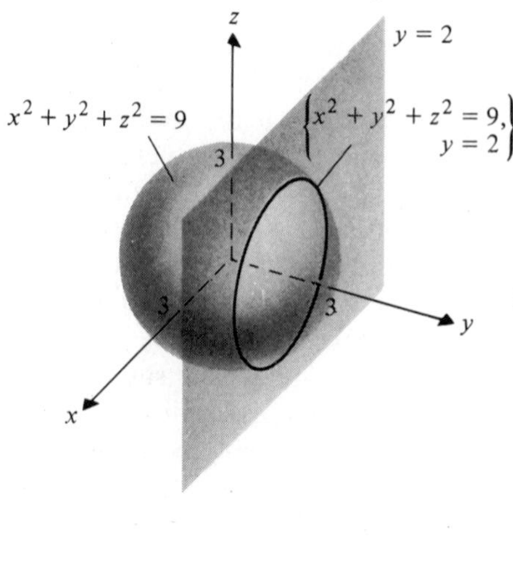

Figure 12.16

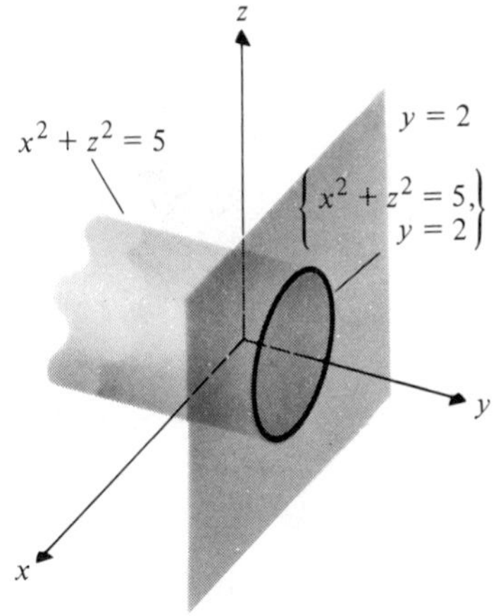

Figure 12.17

In summary, we have illustrated that one equation in the coordinates (x, y, z) of a point specifies a surface; two simultaneous equations specify a curve, the curve of intersection of the two surfaces (provided, of course, that the surfaces do intersect).

In Chapters 4–10 we learned to appreciate the value of curve sketching in the xy-plane. Sometimes a sketch serves as a device by which we can interpret algebraic statements geometrically (such as the mean value theorem or the interpretation of a critical point of a function as a point where the tangent line to the graph of the function is horizontal, vertical, or does not exist). Sometimes it plays an integral part in the solution of a problem (such as when the definite integral is used to find areas, volumes, etc.). Sometimes a sketch is a complete solution to a problem (such as to determine whether a given function has an inverse). We will find that sketching surfaces can be just as useful for multivariable calculus in Chapters 13–15. Unfortunately, surface sketching is more difficult than curve sketching, principally because we are trying to draw a space diagram on a page. Our ability to sketch curves, as we will see, can be of immense help.

One of the most helpful techniques for sketching a surface is to intersect the surface with various planes—in particular, the coordinate planes. From these cross-sections of the surface, it is sometimes possible to visualize the entire surface. For example, if we intersect the surface $z = x^2 + y^2$ with the yz-plane, we obtain the parabola $z = y^2$, $x = 0$. Similarly, the parabola $z = x^2$, $y = 0$ is the intersection curve with the xz-plane. These curves, shown in Fig. 12.18(a), would lead us to suspect that the surface $z = x^2 + y^2$ might be shaped as shown in Fig. 12.18(b). To verify this, we intersect the surface with a plane $z = k$ (k a constant), giving the curve

$$\begin{array}{l} z = x^2 + y^2, \\ z = k, \end{array} \quad \text{or} \quad \begin{array}{r} x^2 + y^2 = k, \\ z = k. \end{array}$$

These latter equations indicate that cross-sections of $z = x^2 + y^2$ with planes $z = k$ are circles centered on the z-axis with radii $\sqrt{k}$ that increase as k increases. This certainly confirms the sketch in Fig. 12.18(b).

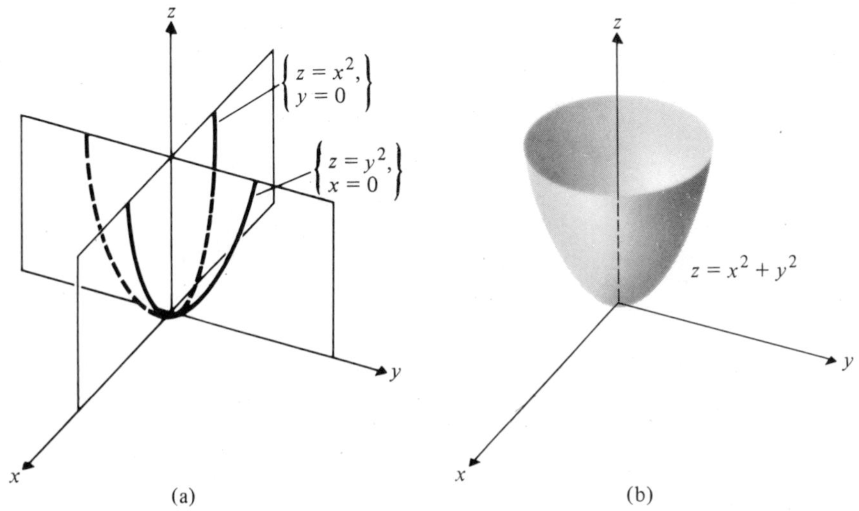

Figure 12.18

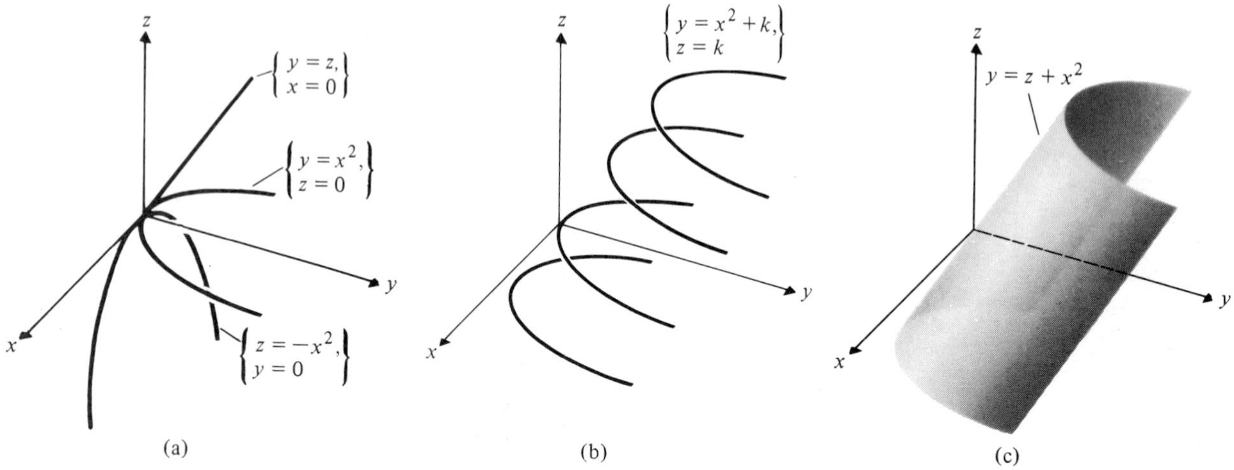

Figure 12.19

Intersections of the surface $y = z + x^2$ with the xy-, xz-, and yz-coordinate planes give two parabolas and a straight line, shown in Fig. 12.19(a). These really do not seem to help us visualize the surface. If, however, we intersect the surface with planes $z = k$, we obtain the parabolas

$$\begin{aligned} y &= z + x^2, \\ z &= k \end{aligned} \quad \text{or} \quad \begin{aligned} y &= x^2 + k, \\ z &= k. \end{aligned}$$

These parabolas, shown in Fig. 12.19(b), indicate that the surface $y = z + x^2$ should be drawn as in Fig. 12.19(c).

We can sometimes "build" surfaces in much the same way that we "built" curves in Chapter 1. For the surface $z = 1 - x^2 - y^2$, we first draw the surface $z = x^2 + y^2$ in Fig. 12.18(b). To sketch $z = -(x^2 + y^2)$, we turn $z = x^2 + y^2$ upside down (Fig. 12.20a), and finally we see that $z = 1 - x^2 - y^2$ is $z = -(x^2 + y^2)$ shifted upward one unit (Fig. 12.20b).

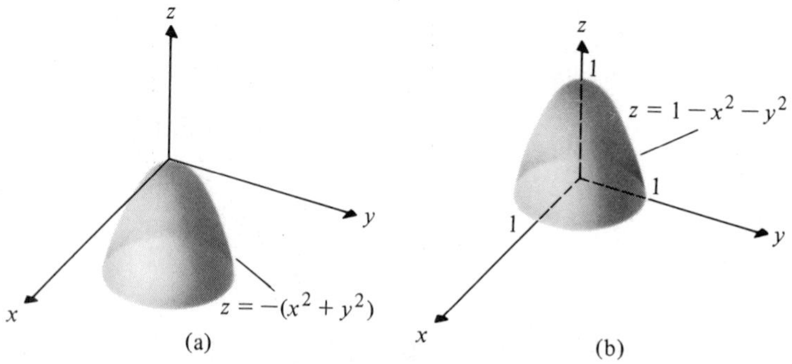

Figure 12.20

Example 1 Sketch the surface defined by each of the following equations:

a) $z = \sqrt{4x + 2y - x^2 - y^2 - 4}$;

b) $y = 1 + \sqrt{x^2 + z^2}$.

Solution a) If we square the equation, and at the same time complete the squares on $-x^2 + 4x$ and $-y^2 + 2y$, we have

$$z^2 = -(x-2)^2 - (y-1)^2 + 1,$$

or

$$(x-2)^2 + (y-1)^2 + z^2 = 1.$$

Because $\sqrt{(x-2)^2 + (y-1)^2 + z^2}$ is the distance from a point (x, y, z) to $(2, 1, 0)$, this equation states that (x, y, z) must always be a unit distance from $(2, 1, 0)$; i.e., the equation $(x-2)^2 + (y-1)^2 + z^2 = 1$ defines a sphere of radius 1 centered at $(2, 1, 0)$ (Fig. 12.21a). Because the original equation requires z to be nonnegative, the required surface is the upper half of this sphere—the hemisphere shown in Fig. 12.21(b).

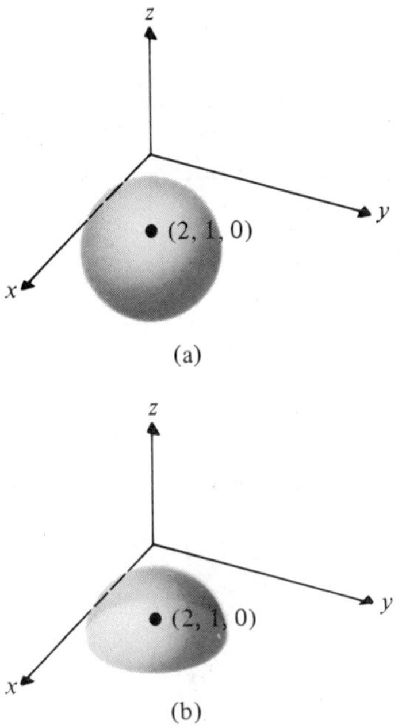

Figure 12.21

b) If we intersect the surface $y = \sqrt{x^2 + z^2}$ with the xy-plane, we obtain the broken straight line $y = |x|$, $z = 0$ shown in Fig. 12.22(a). Intersections of the surface with planes $y = k$ (k a constant) give

$$\begin{aligned} y &= \sqrt{x^2 + z^2}, \\ y &= k, \end{aligned} \quad \text{or} \quad \begin{aligned} x^2 + z^2 &= k^2, \\ y &= k. \end{aligned}$$

These define circles of radii k in the planes $y = k$ (Fig. 12.22b). Consequently, $y = \sqrt{x^2 + z^2}$ defines the right-circular cone in Fig. 12.22(c). The surface $y = 1 + \sqrt{x^2 + z^2}$ can now be obtained by shifting the cone one unit in the y-direction (Fig. 12.22d). □

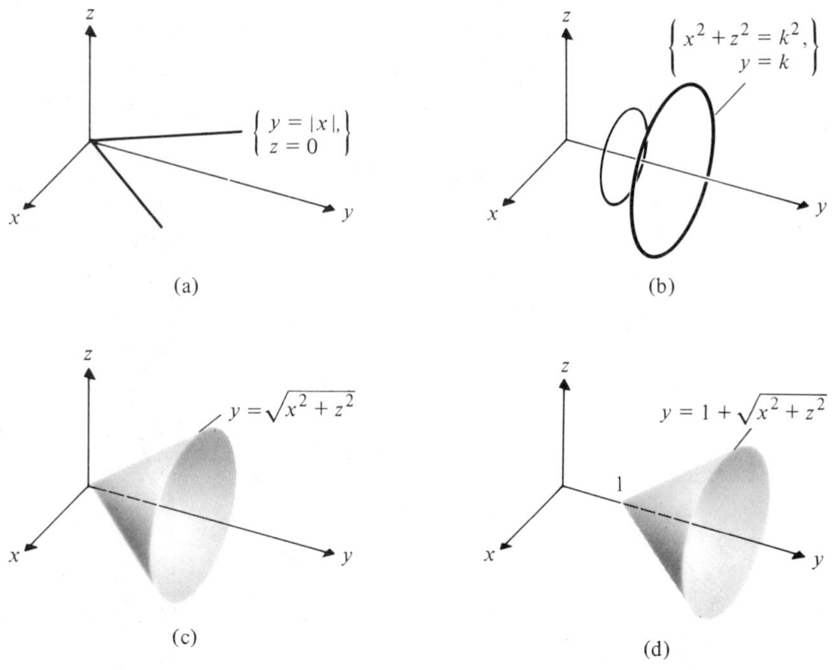

Figure 12.22

Cylinders When one of the coordinates is missing from the equation of a surface, cross-sections of the surface with planes perpendicular to the axis of the missing variable are all identical. Such a surface is said to be a *cylinder*. For example, z is missing from the equation $x^2 + y^2 = 4$, and we saw in Fig. 12.13 that this is the equation of a right-circular cylinder around the z-axis. Every cross-section of this surface with a plane $z = k$ is a circle of radius 2 centered on the z-axis. The equation $z = x^2$ is free of y. Each cross-section of this surface with a plane $y = k$ is the parabola $z = x^2$ in the plane $y = k$. Consequently, $z = x^2$ is the equation for the parabolic cylinder shown in Fig. 12.23.

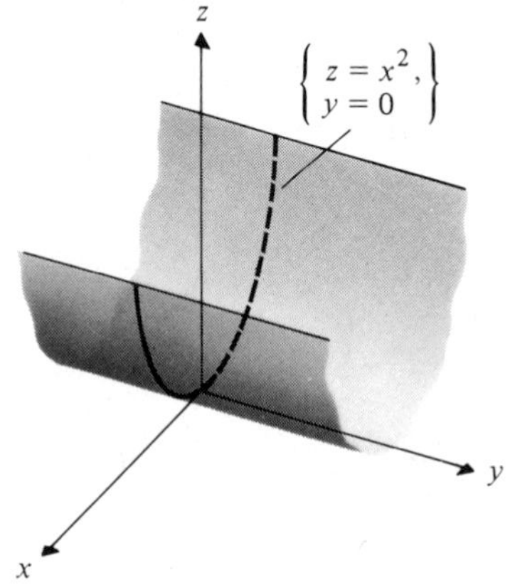

Figure 12.23

EXERCISES 12.2

In Exercises 1–20, sketch the surface defined by the given equation.

1. $2y + 3z = 6$
2. $2x - 3y = 0$
3. $y = x^2 + 2$
4. $z = x^3$
5. $y^2 + z^2 = 1$
6. $x^2 + y^2 + z^2 = 4$
7. $x^2 + 4y^2 = 1$
8. $y^2 - z^2 = 4$
9. $z = 2(x^2 + y^2)$
10. $x = \sqrt{y^2 + z^2}$
11. $x = \sqrt{1 - y^2}$
12. $z = 2 - x$
13. $x^2 = y^2$
14. $x = z^2 + 2$
15. $z = y + 3$
16. $4z = 3\sqrt{x^2 + y^2}$
17. $x^2 - 2x + z^2 = 0$
18. $yz = 1$
19. $x^2 + y^2 + (z - 1)^2 = 3$
20. $z + 5 = 4(x^2 + y^2)$

In Exercises 21–30, sketch the curve defined by the given equations.

21. $x + 2y = 6,$ $y - 2z = 3$
22. $x^2 + y^2 = 2,$ $z = 4$
23. $x^2 + y^2 = 2,$ $x + z = 1$
24. $z = x^2 + y^2,$ $x^2 + y^2 = 5$
25. $z + 2x^2 = 1,$ $y = z$
26. $z = \sqrt{x^2 + y^2},$ $y = x$
27. $z = y,$ $y = x^2$
28. $z = \sqrt{4 - x^2 - y^2},$ $x^2 + y^2 - 2y = 0$
29. $z = x^2,$ $z = y^2$
30. $x^2 + z^2 = 1,$ $y^2 + z^2 = 1$

In Exercises 31–40, sketch whatever is defined by the equation or equations.

31. $x = 0, y = 5$
32. $(x - 2)^2 + y^2 + z^2 = 0$
33. $x + y = 15, y - x = 4$
34. $\sqrt{x} + \sqrt{y} = 1, z = x$
35. $z = |x|$
36. $z = 1 - (x^2 + y^2)^{1/3}$
37. $x = \ln (y^2 + z^2)$
38. $z = x^2, y = z^2$
39. $x = 2, y = 4, z^2 - 1 = 0$
40. $z = |x - y|$

12.3 SURFACES OF REVOLUTION AND QUADRIC SURFACES

Surfaces of Revolution

When a plane curve is rotated about a fixed line lying in the plane of the curve, the surface generated is called a *surface of revolution*. In the case that the curve lies in one of the three coordinate planes and the fixed line is a coordinate axis, it is straightforward to find the equation of the surface of revolution. To illustrate, we rotate the curve

$$y = f(x), \qquad z = 0 \tag{12.2}$$

in Fig. 12.24 about the x-axis to form the surface of revolution shown in Fig. 12.25. Given any point $P(x^*, y^*, 0)$ on the original curve, we see that an entire circle of points is generated on the surface of revolution. If $Q(x, y, z)$ are the coordinates of any point on this circle, then clearly $x^* = x$, and the length of QR is the same as the length of PR; i.e.,

$$y^* = \sqrt{y^2 + z^2} \quad \text{when } y^* \geq 0,$$

and

$$-y^* = \sqrt{y^2 + z^2} \quad \text{when } y^* < 0.$$

Since x^* and y^* must satisfy Eqs. (12.2), it follows that $y^* = f(x^*)$; hence

$$f(x^*) = \sqrt{y^2 + z^2} \quad \text{when } f(x^*) \geq 0,$$

and

$$-f(x^*) = \sqrt{y^2 + z^2} \quad \text{when } f(x^*) < 0.$$

Because $x^* = x$, we may write

$$f(x) = \sqrt{y^2 + z^2} \quad \text{when } f(x) \geq 0,$$

and

$$-f(x) = \sqrt{y^2 + z^2} \quad \text{when } f(x) < 0.$$

These equations can be combined into the single equation

$$[f(x)]^2 = y^2 + z^2. \tag{12.3}$$

This equation must be satisfied by the coordinates of every point on the surface of revolution and it must therefore be the equation of the surface.

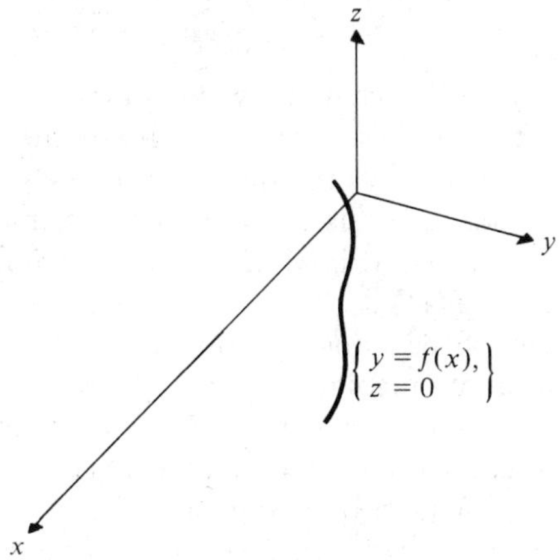

Figure 12.24

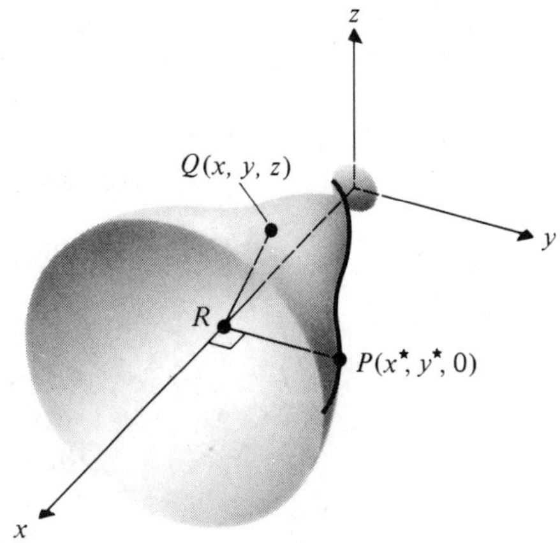

Figure 12.25

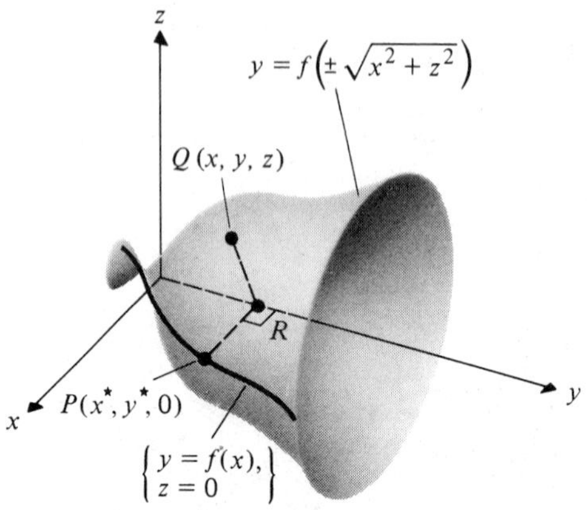

Figure 12.26

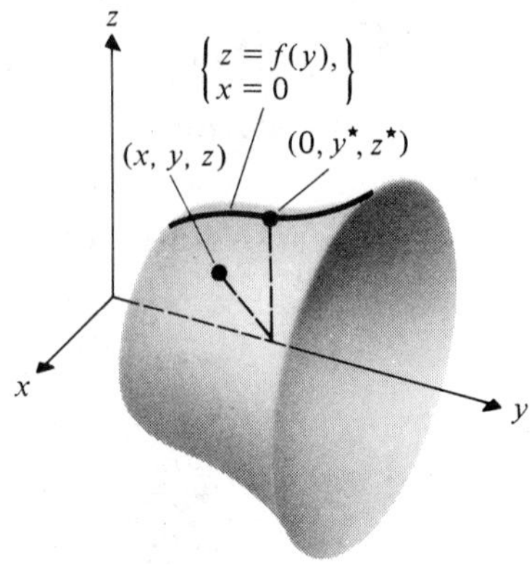

Figure 12.27

The equation of the surface of revolution obtained by rotating the curve $y = f(x)$, $z = 0$ in Fig. 12.26 about the y-axis can be obtained in a similar fashion, but the result is not quite so simple. In this case $y^* = y$, and because the length of QR is the same as that of PR,

$$x^* = \sqrt{x^2 + z^2} \quad \text{when } x^* \geq 0,$$

and

$$-x^* = \sqrt{x^2 + z^2} \quad \text{when } x^* < 0.$$

Since x^* and y^* satisfy $y^* = f(x^*)$,

$$y^* = f(\sqrt{x^2 + z^2}) \quad \text{when } x^* \geq 0,$$

and

$$y^* = f(-\sqrt{x^2 + z^2}) \quad \text{when } x^* < 0.$$

If we now set $y^* = y$, we obtain

$$y = \begin{cases} f(\sqrt{x^2 + z^2}), & x \geq 0 \\ f(-\sqrt{x^2 + z^2}), & x < 0 \end{cases}. \tag{12.4}$$

Unfortunately we cannot always combine these two results into a single equation. Of course, if x is always positive on the original curve $y = f(x)$ (or always negative), then only one equation is needed.

A similar analysis shows that when the curve $z = f(y)$, $x = 0$ is rotated about the y-axis (Fig. 12.27), the equation of the surface of revolution is $x^2 + z^2 = [f(y)]^2$.

Many of the surfaces with which we are already familiar can be regarded as surfaces of revolution: The sphere $x^2 + y^2 + z^2 = R^2$ is generated by rotating the circle $x^2 + y^2 = R^2$, $z = 0$ about the x- (or y-) axis (Fig. 12.28); the right-circular cone $z = \sqrt{x^2 + y^2}$ is generated by rotating the straight line $z = x$, $y = 0$ $(x \geq 0)$ about the z-axis (Fig. 12.29); and the right-circular cylinder $x^2 + y^2 = R^2$ is formed when the line $x = R$, $y = 0$ is rotated about the z-axis (Fig. 12.30).

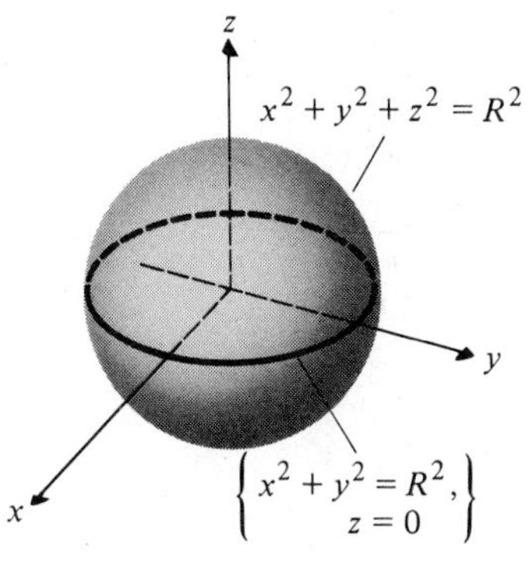

Figure 12.28

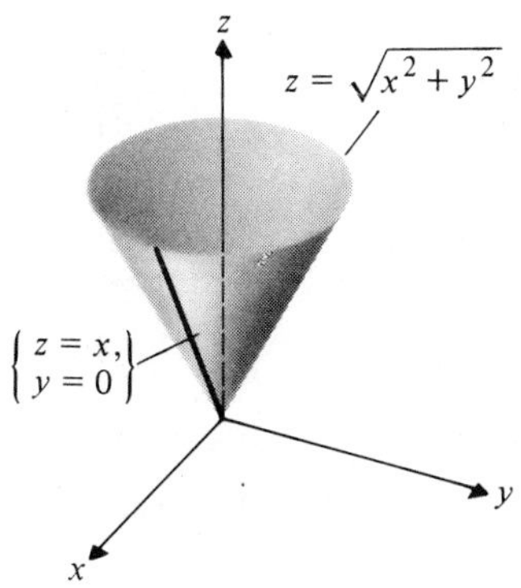

Figure 12.29

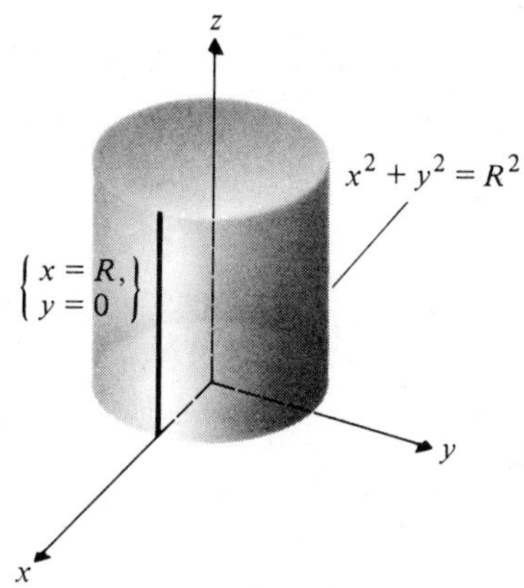

Figure 12.30

Example 2 Find the equation of the surface of revolution obtained by rotating the parabola $y = z^2 + 1$, $x = 0$ about the z-axis.

Solution If $P(0, y^*, z^*)$ is any point on the parabola, then because $|QR| = |PR|$, $\sqrt{x^2 + y^2} = y^*$ (and note that y^* is always positive). Since $y^* = (z^*)^2 + 1$, and $z^* = z$, we obtain

$$\sqrt{x^2 + y^2} = (z^*)^2 + 1 = z^2 + 1.$$

Thus $x^2 + y^2 = (z^2 + 1)^2$ is the equation of the surface of revolution (Fig. 12.31). □

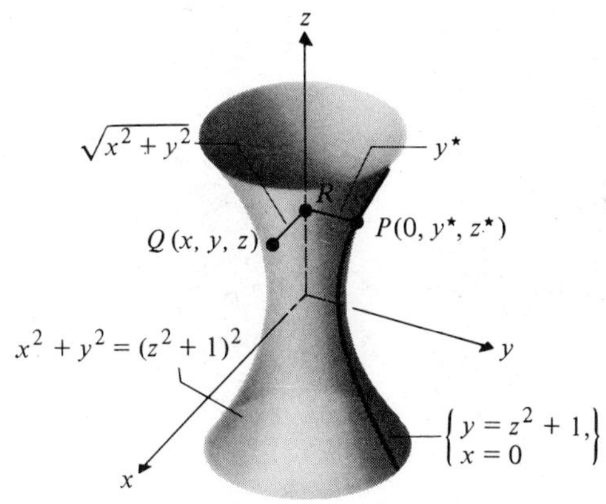

Figure 12.31

Example 3 Sketch the surface defined by the equation $y^2 - z^2 = 1 + x^2$.

Solution If we write $y^2 = 1 + x^2 + z^2$, then this surface can be interpreted as a surface of revolution. It is the surface generated by rotating the curve $y^2 = 1 + x^2$, $z = 0$ about the y-axis. But this curve is the hyperbola $y^2 - x^2 = 1$, $z = 0$, shown in Fig. 12.32(a). By rotating this curve about the y-axis, we obtain the surface shown in Fig. 12.32(b). □

Figure 12.32

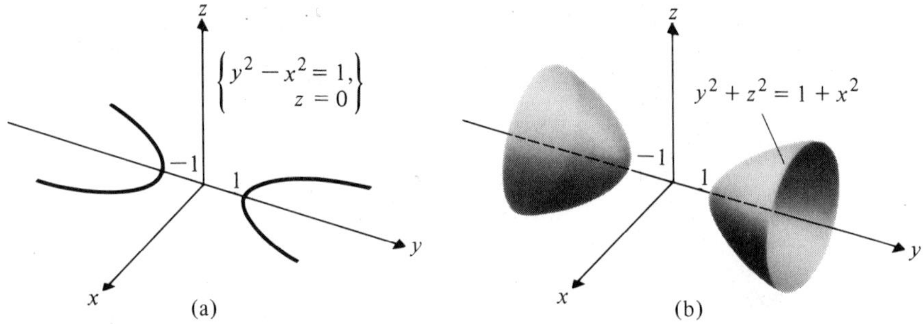

Quadric Surfaces

A *quadric surface* is a surface whose equation is quadratic in x, y, and z, the most general such equation being

$$Ax^2 + By^2 + Cz^2 + Dxy + Eyz + Fxz + Gx + Hy + Iz + J = 0. \quad (12.5)$$

For the most part, we will encounter those quadric surfaces whose equations are of the forms

$$Ax^2 + By^2 + Cz^2 = J \quad \text{or} \quad Ax^2 + By^2 = Iz,$$

or these equations with x, y, and z interchanged. Surfaces with these equations fall into nine major classes depending on whether the constants are positive, negative, or zero. They are illustrated in Figs. 12.33–12.41.

ELLIPTIC CYLINDER

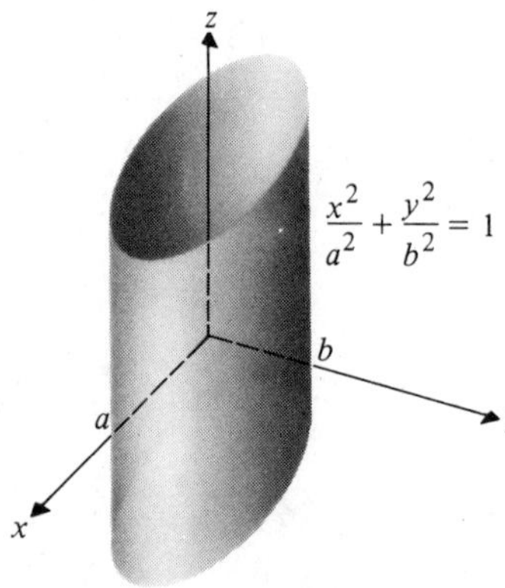

Figure 12.33

HYPERBOLIC CYLINDER

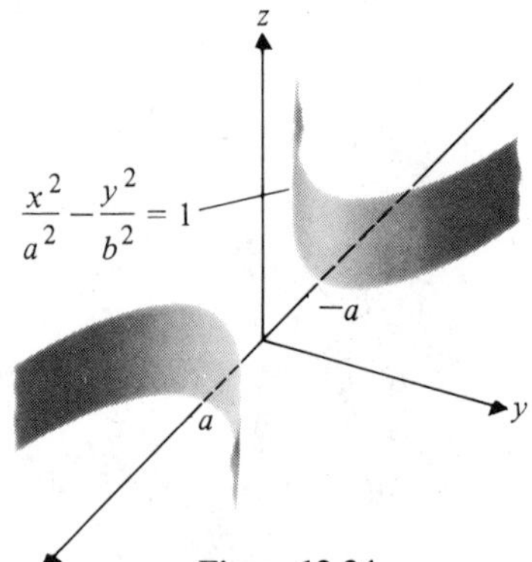

Figure 12.34

PARABOLIC CYLINDER

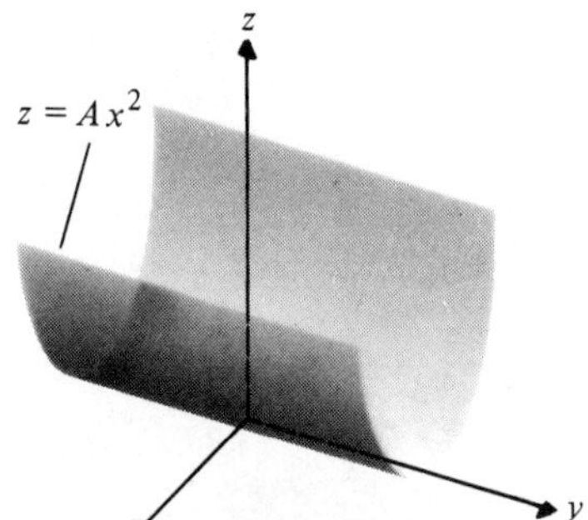

Figure 12.35

ELLIPSOID

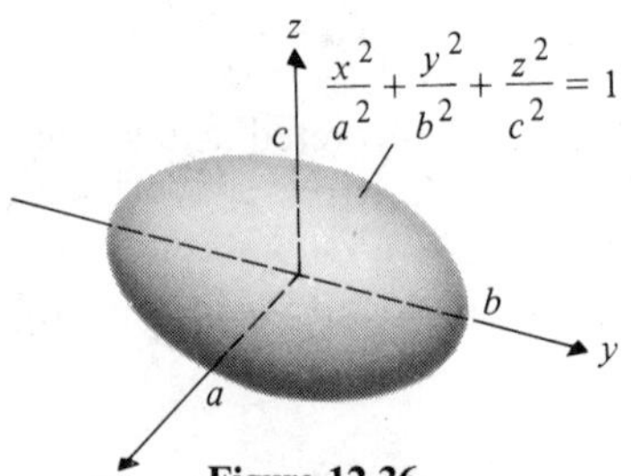

Figure 12.36

ELLIPTIC PARABOLOID

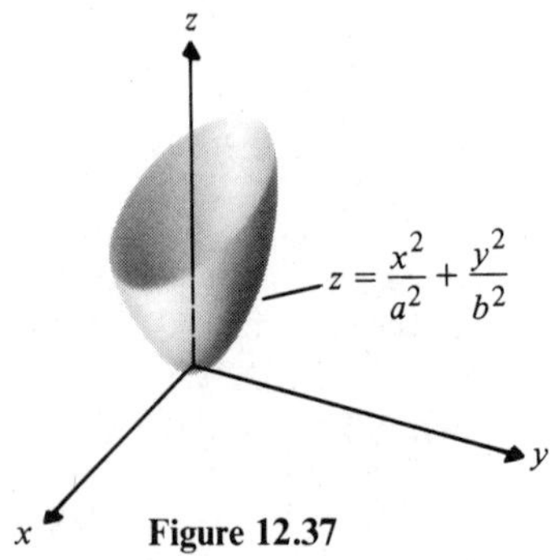

Figure 12.37

ELLIPTIC CONE

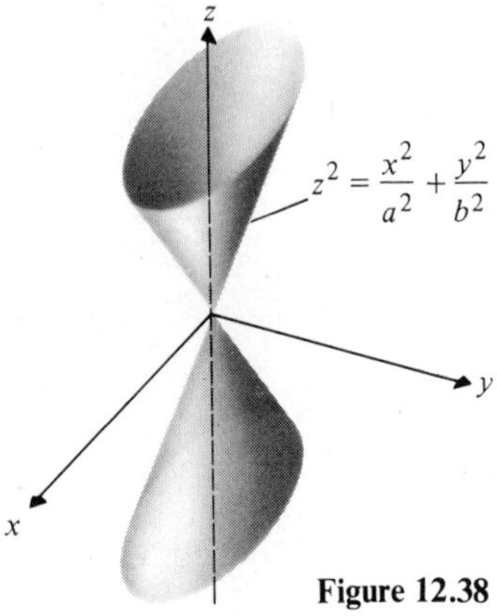

Figure 12.38

HYPERBOLIC PARABOLOID

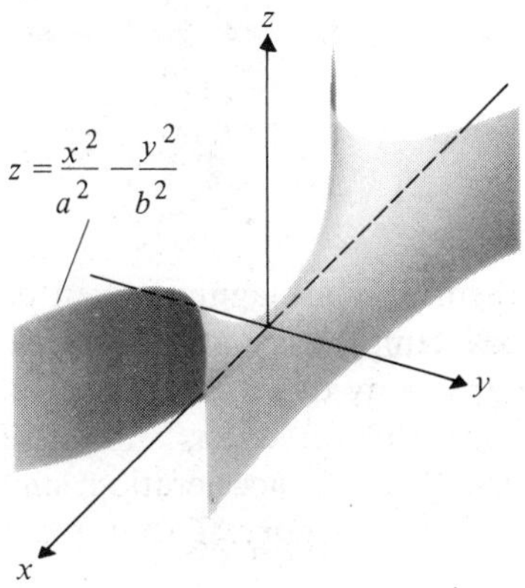

Figure 12.39

ELLIPTIC HYPERBOLOID
of one sheet

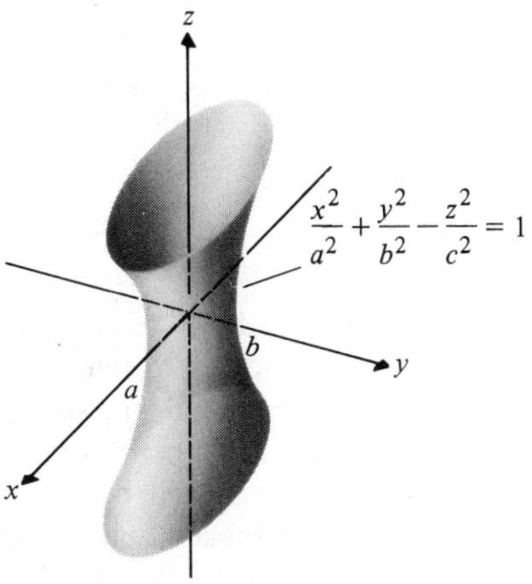

Figure 12.40

ELLIPTIC HYPERBOLOID
of two sheets

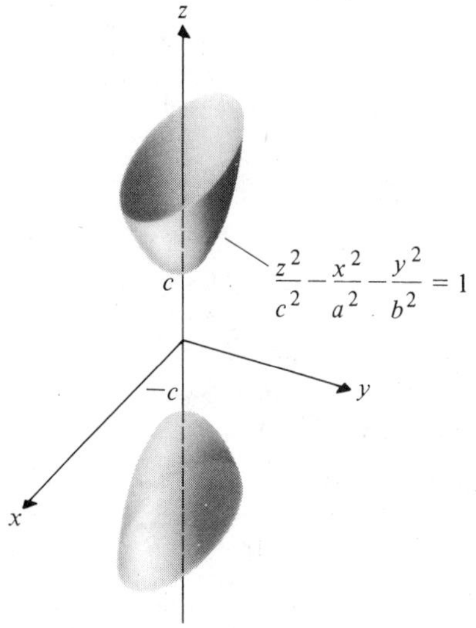

Figure 12.41

Note that six of the nine surfaces refer to ellipses, because in each case certain cross-sections of the surface are ellipses. If $a = b$ in any one of these surfaces, then the elliptical cross-sections become circular, and the surface can then be regarded as a surface of revolution. For instance, if $a = b$ in the elliptic hyperboloid of one sheet, then we can write that

$$\frac{1}{a^2}(x^2 + y^2) = 1 + \frac{z^2}{c^2}.$$

But this is the equation for the surface of revolution obtained by rotating the hyperbola

$$\frac{x^2}{a^2} = 1 + \frac{z^2}{c^2}, \qquad y = 0$$

around the z-axis.

EXERCISES 12.3

In Exercises 1–10, find the equation for the surface of revolution obtained by rotating the curve about the line indicated. In each case, sketch the curve and surface.

1. $y = 2x$, $z = 0$, about the y-axis
2. $y = \sin x$, $z = 0$, about the x-axis
3. $z^2 = x$, $y = 0$, about the x-axis
4. $x + y = 1$, $z = 0$, about the x-axis
5. $x = y^3$, $z = 0$, about the x-axis
6. $x = y^3$, $z = 0$, about the y-axis
7. $z = y - y^2$, $x = 0$, about the y-axis
8. $z = e^{-x^2}$, $y = 0$, about the z-axis
9. $z = 1 - |x|$, $y = 0$, about the z axis
10. $(y - a)^2 + z^2 = b^2$, $x = 0$ $(b < a)$, about the z-axis

Sketch the surface in each of Exercises 11–16, interpreting it as a surface of revolution.

11. $x^2 + z^2 = y^2$
12. $x^2 + z^2 = y^2 + 1$
13. $y^2 + z^2 = x^6$
14. $x^2 + y^2 + 4z^2 = 1$
15. $9z^2 = x^2 + y^2 + 1$
16. $(y^2 + z^2)^2 = x + 1$

Sketch the quadric surface in each of Exercises 17–25.

17. $z^2 + 4y^2 = 1$

18. $y - z^2 = 0$

19. $x^2 - z^2 = 4$

20. $x^2 + y^2/4 + z^2/9 = 1$

21. $z = x^2/4 + y^2/25$

22. $x^2 = z^2 + 9y^2$

23. $z = y^2/16 - x^2/4$

24. $x^2 + y^2/4 - z^2/25 = 1$

25. $z^2 - 9x^2 - 16y^2 = 1$

26. Show that when the curve $y = f(x)$, $z = 0$ is rotated about the line $y = a$, $z = 0$, the equation of the surface of revolution is

$$(y - a)^2 + z^2 = [f(x) - a]^2.$$

12.4 VECTORS

Physical quantities that have associated with them only a magnitude can be represented by real numbers. Some examples are temperature, density, area, moment of inertia, speed, and pressure. There are many quantities, however, that have associated with them both magnitude and direction, and these quantities cannot be described by real numbers. Velocity, acceleration, and forces are perhaps the most notable concepts in this category. To represent such quantities mathematically, we introduce *vectors*.

▷ **Definition 12.1** A *vector* is defined as a directed line segment. ◁

To denote a vector we use a letter in boldface type, such as **v**. In Figs. 12.42(a), (b), and (c) we show two vectors **u** and **v** along a line, three vectors **u**, **v**, and **w** in a plane, and three vectors **u**, **v**, and **w** in space, respectively. It is customary to place an arrowhead on a vector and call this end the *tip* of the vector. The other end is called the *tail* of the vector, and the direction of the vector is from tail to tip. A vector then has both *direction* and *length*.

Definition 12.1 for a vector says nothing about its point of application; i.e., where its tail should be placed. This means that we may place the tail anywhere we wish. This suggests the following definition for equality of vectors.

▷ **Definition 12.2** Two vectors are equal if and only if they have the same length and direction. Their points of application are irrelevant. ◁

For example, vectors **u** and **v** in Fig. 12.43 have exactly the same length and direction, and are therefore one and the same. Although vector **w** in the same figure is parallel to **u** and **v** and has the same length, it points in the opposite direction and is not, therefore, the same as **u** and **v**.

Components of Vectors

We realized back in Chapter 1 that to solve geometric problems, it is often helpful to represent them algebraically. In fact, our entire development of single-variable calculus has hinged on our ability to represent a curve by an algebraic equation and also to draw the curve described by an equation. Since vectors are directed line segments along a line, in a plane, or in space, and into each of these we can introduce Cartesian coordinates, surely we must be able to find an algebraic representation for vectors. Suppose we denote by

$\overrightarrow{\mathbf{PQ}}$ the vector from P to Q in Fig. 12.44. If P and Q have coordinates (x_1, y_1, z_1) and (x_2, y_2, z_2) in the coordinate system shown, then the length of $\overrightarrow{\mathbf{PQ}}$ is

$$\text{length of } \overrightarrow{\mathbf{PQ}} = \sqrt{(x_2 - x_1)^2 + (y_2 - y_1)^2 + (z_2 - z_1)^2}. \tag{12.6}$$

Note also that if we start at point P, proceed $x_2 - x_1$ units in the x-direction, then $y_2 - y_1$ units in the y-direction, and finally $z_2 - z_1$ units in the z-direction, then we arrive at Q. In other words, the three numbers $x_2 - x_1$, $y_2 - y_1$, and $z_2 - z_1$ characterize both the direction and the length of the vector joining P to Q. Because of this we make the following agreement.

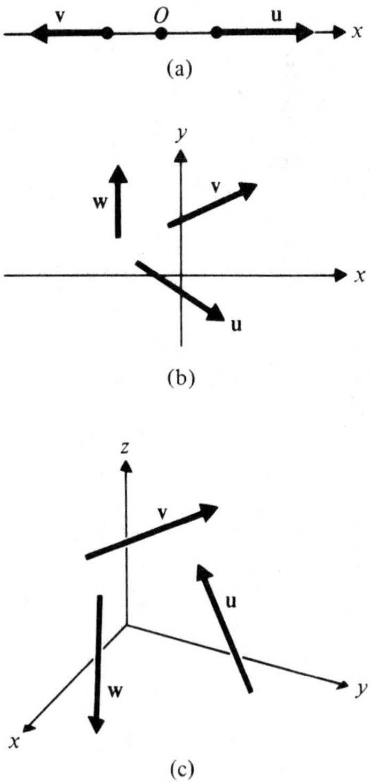

Figure 12.42

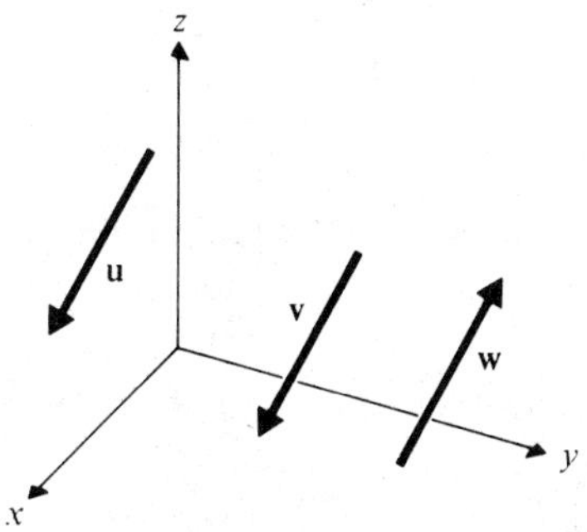

Figure 12.43

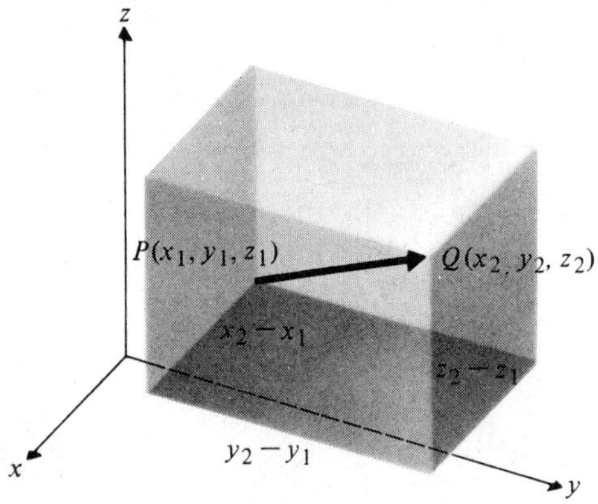

Figure 12.44

▷ **Definition 12.3** If the tail of a vector $\mathbf{v}$ is at $P(x_1, y_1, z_1)$ and its tip is at $Q(x_2, y_2, z_2)$, then $\mathbf{v}$ shall be represented by the triple of numbers $x_2 - x_1$, $y_2 - y_1$, $z_2 - z_1$. In such a case we enclose the numbers in parentheses and write

$$\mathbf{v} = (x_2 - x_1, y_2 - y_1, z_2 - z_1). \quad \triangleleft \tag{12.7}$$

The equals sign in (12.7) means "is represented by." The number $x_2 - x_1$ is called the *x-component* of $\mathbf{v}$, $y_2 - y_1$ the *y-component*, and $z_2 - z_1$ the *z-component*. Vectors in the xy-plane have only an x- and a y-component:

$$\mathbf{v} = (x_2 - x_1, y_2 - y_1),$$

where (x_1, y_1) and (x_2, y_2) are the coordinates of the tail and tip of $\mathbf{v}$. Vectors along the x-axis have only an x-component $x_2 - x_1$, where x_1 and x_2 are the coordinates of the tail and tip of $\mathbf{v}$.

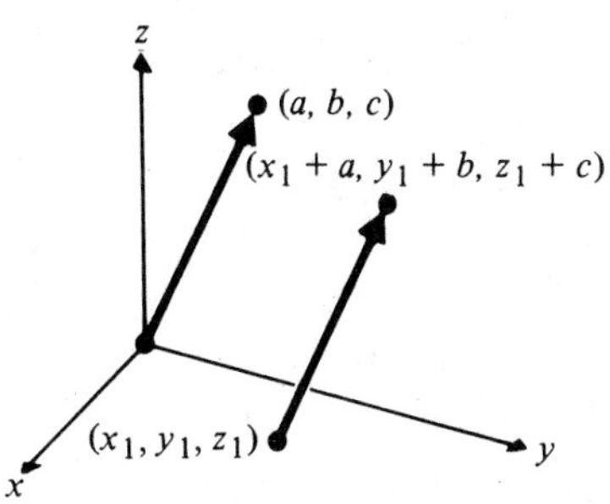

Figure 12.45

We now have an algebraic representation for vectors. Each vector has associated with it a set of components that can be found by subtracting the coordinates of its tail from the coordinates of its tip. Conversely, given a set of real numbers (a, b, c), there is one and only one vector with these numbers as components. We can visualize this vector by placing its tail at the origin and its tip at the point with coordinates (a, b, c) (Fig. 12.45). Alternatively, we can place the tail of the vector at any point (x_1, y_1, z_1) and its tip at the point $(x_1 + a, y_1 + b, z_1 + c)$.

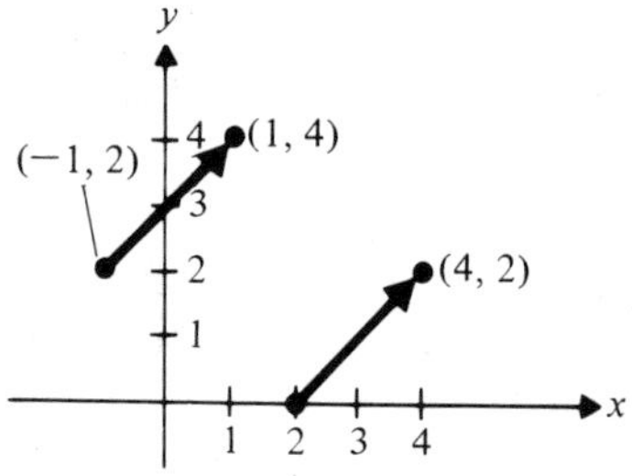

Figure 12.46

It is worth emphasizing once again that the same components of a vector are obtained for any point of application whatsoever. For example, the two vectors in Fig. 12.46 are identical, and in both cases the components (2, 2) are obtained by subtracting the coordinates of the tail from the tip. What we are saying is that Definition 12.2 for equality of vectors can be stated algebraically as follows.

▶ **Theorem 12.1** Two vectors are equal if and only if they have the same components. ◀

Example 4 Find the components of a vector in the xy-plane that has length 5 and its tail at the origin, and makes an angle of $\pi/6$ radians with the positive x-axis.

Solution Figure 12.47 illustrates that there are two such vectors, **u** and **v**. From the triangles shown, it is clear that

$$|OP| = 5\cos(\pi/6) = 5\sqrt{3}/2 \quad \text{and} \quad |PQ| = |PR| = 5\sin(\pi/6) = \tfrac{5}{2}.$$

Consequently, Q and R have coordinates $Q = (5\sqrt{3}/2, \frac{5}{2})$ and $R = (5\sqrt{3}/2, -\frac{5}{2})$, and

$$\mathbf{u} = \left(\frac{5\sqrt{3}}{2}, \frac{5}{2}\right), \qquad \mathbf{v} = \left(\frac{5\sqrt{3}}{2}, -\frac{5}{2}\right). \ \square$$

Example 5 Find the components of the vector in the xy-plane that has its tail at the point (4, 5), has length 3, and points directly toward the point (2, −3).

Solution In Fig. 12.48, $|PR| = 8$, $|QR| = 2$, and therefore $|PQ| = \sqrt{8^2 + 2^2} = 2\sqrt{17}$. Because of the similar triangles, we can write that

$$\frac{|ST|}{|PS|} = \frac{|QR|}{|PQ|} \quad \text{or} \quad |ST| = \frac{3(2)}{2\sqrt{17}} = \frac{3}{\sqrt{17}}.$$

Similarly,

$$|PT| = |PS|\frac{|PR|}{|PQ|} = \frac{3(8)}{2\sqrt{17}} = \frac{12}{\sqrt{17}}.$$

Since $|ST|$ and $|PT|$ represent the differences in the x- and y-coordinates of P and S (except for signs), the components of $\overrightarrow{\mathbf{PS}}$ are $(-3/\sqrt{17}, -12/\sqrt{17})$. □

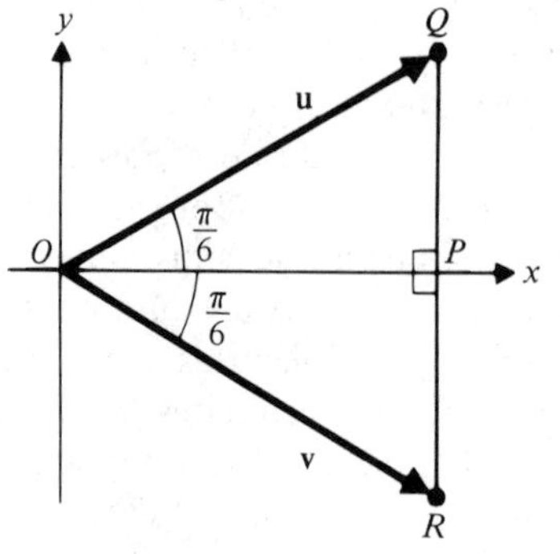

Figure 12.47

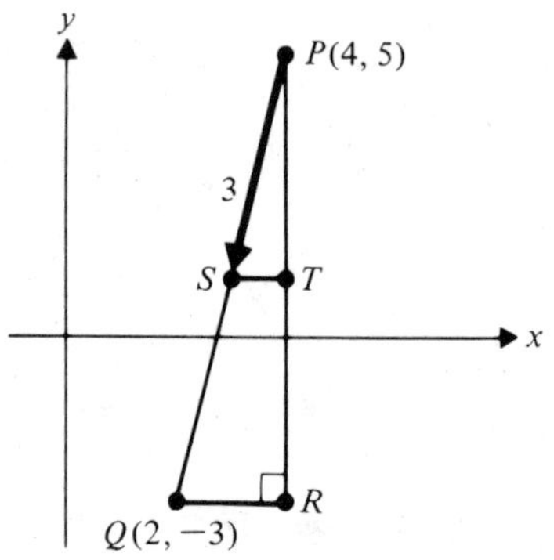

Figure 12.48

Unit Vectors and Scalar Multiplication

If the x-, y-, and z-components of a vector $\mathbf{v}$ are (v_x, v_y, v_z), often called the *Cartesian components* of $\mathbf{v}$, then these components represent the differences in the coordinates of its tip and tail. But then according to Eq. (12.6), the length of the vector, which we denote by $|\mathbf{v}|$, is

$$|\mathbf{v}| = \sqrt{v_x^2 + v_y^2 + v_z^2}. \tag{12.8}$$

In words, the length of a vector is the square root of the sum of the squares of its components.

▷ **Definition 12.4** A vector $\mathbf{v}$ is said to be a unit vector if it has length equal to one unit; i.e., $\mathbf{v}$ is a unit vector if

$$v_x^2 + v_y^2 + v_z^2 = 1. \quad \triangleleft \tag{12.9}$$

To indicate that a vector is a unit vector, we place a circumflex ˆ above it: $\hat{\mathbf{v}}$.

Example 6 What is the length of the vector from $(1, -1, 0)$ to $(2, -3, -5)$?

Solution Since the components of the vector are $(1, -2, -5)$, its length is

$$\sqrt{(1)^2 + (-2)^2 + (-5)^2} = \sqrt{30}. \ \square$$

We now have vectors, which are directed line segments, and real numbers, which we call scalars. We know that scalars can be added, subtracted, multiplied, and divided, but can we do the same with vectors, and can we combine vectors and scalars? In the remainder of this section we show how to add and subtract vectors and multiply vectors by scalars; in Sections 12.5 and 12.6 we define two ways to multiply vectors. Each of these operations can be approached either algebraically or geometrically. The geometric approach uses the geometric properties of vectors, namely, length and direction; the algebraic approach uses components of vectors. Neither method is suitable for all situations. Sometimes an idea is more easily introduced with a geometric approach; sometimes an algebraic approach is more suitable. We will choose whichever we feel expresses the idea more clearly. But, whenever we take a geometric approach, we will be careful to follow it up with the algebraic equivalent; conversely, when an algebraic approach is taken, we will always illustrate the geometric significance of the results.

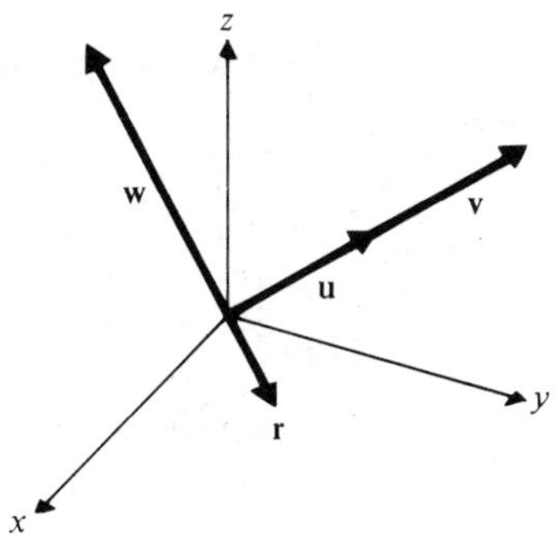

Figure 12.49

To introduce multiplication of a vector by a scalar, consider the vectors $\mathbf{u}$ and $\mathbf{v}$ in Fig. 12.49, both of which have their tails at the origin; $\mathbf{v}$ is in the same direction as $\mathbf{u}$ but is twice as long as $\mathbf{u}$. In such a situation we would like to say that $\mathbf{v}$ is equal to $2\mathbf{u}$ and write $\mathbf{v} = 2\mathbf{u}$. Vector $\mathbf{w}$ is in the opposite direction to $\mathbf{r}$ and is three times as long as $\mathbf{r}$, and we would like to denote this vector by $\mathbf{w} = (-3)\mathbf{r}$. These examples suggest the following meaning for multiplication of a vector by a scalar.

▷ **Definition 12.5** If $\lambda > 0$ is a scalar and $\mathbf{v}$ is a vector, then $\lambda\mathbf{v}$ is the vector that is in the same direction as $\mathbf{v}$ and λ times as long as $\mathbf{v}$; if $\lambda < 0$, then $\lambda\mathbf{v}$ is the vector that is in the opposite direction to $\mathbf{v}$ and $|\lambda|$ times as long as $\mathbf{v}$. ◁

This is a geometric definition of scalar multiplication; it describes the length and direction of $\lambda\mathbf{v}$. We now show that the components of $\lambda\mathbf{v}$ are λ

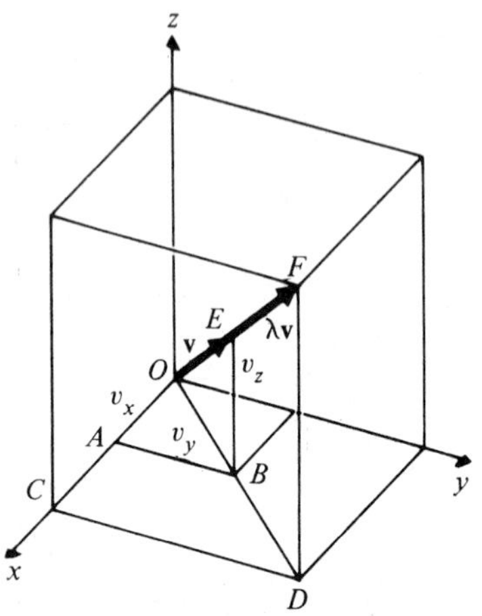

Figure 12.50

times the components of **v**. In Fig. 12.50 we show a box with faces parallel to the coordinate planes and $\lambda\mathbf{v}$ as diagonal, and have given vector **v** components (v_x, v_y, v_z). From the pairs of similar triangles OAB and OCD and OBE and ODF, we can write that

$$\frac{|OC|}{v_x} = \frac{|CD|}{v_y} = \frac{|OD|}{|OB|} = \frac{|DF|}{v_z}.$$

If we equate each of these ratios to a constant μ, then we have

$$|OC| = \mu v_x, \qquad |CD| = \mu v_y, \qquad |DF| = \mu v_z,$$

where $|OC|$, $|CD|$, and $|DF|$ are the components of $\lambda\mathbf{v}$. Since the length of $\lambda\mathbf{v}$ is λ times that of **v**, $|\lambda\mathbf{v}| = \lambda|\mathbf{v}|$, and by (12.8),

$$\sqrt{(\mu v_x)^2 + (\mu v_y)^2 + (\mu v_z)^2} = \lambda\sqrt{v_x^2 + v_y^2 + v_z^2}.$$

But this equation implies that $\lambda = \mu$, and therefore the components of $\lambda\mathbf{v}$ are $(\lambda v_x, \lambda v_y, \lambda v_z)$; i.e.,

$$\boxed{\lambda\mathbf{v} = \lambda(v_x, v_y, v_z) = (\lambda v_x, \lambda v_y, \lambda v_z).} \tag{12.10}$$

To multiply a vector by a scalar, then, we multiply each component by the scalar.

Example 7 Find components for the unit vector in the same direction as $\mathbf{v} = (2, -2, 1)$.

Solution The length of **v** is $|\mathbf{v}| = \sqrt{(2)^2 + (-2)^2 + 1^2} = 3$. According to our definition of multiplication of a vector by a scalar, the vector $\frac{1}{3}\mathbf{v}$ must have length 1 ($\frac{1}{3}$ that of **v**) and the same direction as **v**. Consequently, a unit vector in the same direction as **v** is

$$\hat{\mathbf{v}} = \tfrac{1}{3}\mathbf{v} = \tfrac{1}{3}(2, -2, 1) = (\tfrac{2}{3}, -\tfrac{2}{3}, \tfrac{1}{3}). \ \square$$

This example illustrates that a unit vector in the same direction as a given vector **v** is

$$\boxed{\hat{\mathbf{v}} = \frac{\mathbf{v}}{|\mathbf{v}|}.} \tag{12.11}$$

Example 8 Find components for the vector of length 4 in the direction opposite that of $\mathbf{v} = (1, 2, -3)$.

Solution Since $|\mathbf{v}| = \sqrt{1 + 4 + 9} = \sqrt{14}$, a unit vector in the same direction as **v** is

$$\hat{\mathbf{v}} = \frac{1}{\sqrt{14}}\mathbf{v}.$$

The vector of length 4 in the opposite direction to **v** must therefore be

$$(-4)\hat{\mathbf{v}} = \left(\frac{-4}{\sqrt{14}}\right)\mathbf{v} = \left(\frac{-4}{\sqrt{14}}\right)(1, 2, -3) = \left(-\frac{4}{\sqrt{14}}, -\frac{8}{\sqrt{14}}, \frac{12}{\sqrt{14}}\right). \ \square$$

Note that with the operation of scalar multiplication, we can simplify the solution of Example 5. The vector that points from (4, 5) to (2, −3) is $\mathbf{v} = (-2, -8)$, and therefore the unit vector in this direction is

$$\hat{\mathbf{v}} = \frac{1}{\sqrt{4+64}}(-2, -8) = \frac{1}{2\sqrt{17}}(-2, -8) = \frac{1}{\sqrt{17}}(-1, -4).$$

The required vector of length 3 must be

$$3\hat{\mathbf{v}} = \frac{3}{\sqrt{17}}(-1, -4) = \left(-\frac{3}{\sqrt{17}}, -\frac{12}{\sqrt{17}}\right).$$

Addition and Subtraction of Vectors

In Fig. 12.51 we show two parallel vectors **u** and **v** and have placed the tail of **v** on the tip of **u**. It would seem natural to denote the vector that has its tail at the tail of **u** and its tip at the tip of **v** by $\mathbf{u} + \mathbf{v}$. For instance, if **u** and **v** were equal, then we would simply be saying that $\mathbf{u} + \mathbf{u} = 2\mathbf{u}$. We use this idea to define addition of vectors even when the vectors are not parallel.

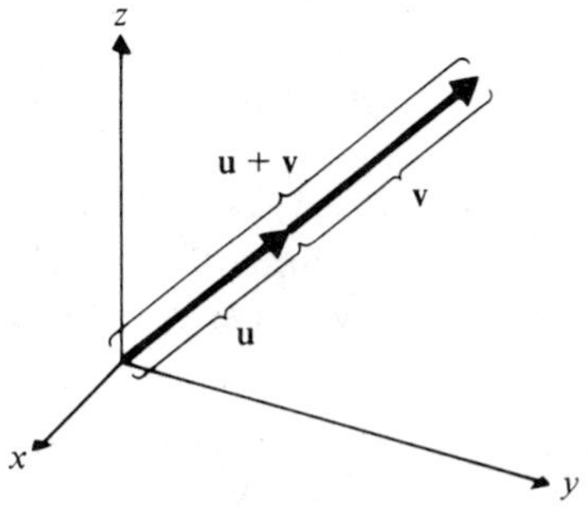

Figure 12.51

▷ **Definition 12.6** The sum of two vectors **u** and **v**, denoted by $\mathbf{u} + \mathbf{v}$, is the vector from the tail of **u** to the tip of **v** when the tail of **v** is placed on the tip of **u**. ◁

Because the three vectors **u**, **v**, and $\mathbf{u} + \mathbf{v}$ then form a triangle (Fig. 12.52), we call this *triangular addition* of vectors.

Note that were we to place tails of **u** and **v** both at the same point (Fig. 12.53), and complete the parallelogram with **u** and **v** as sides, then the diagonal of this parallelogram would also represent the vector $\mathbf{u} + \mathbf{v}$. This is an equivalent method for geometrically finding $\mathbf{u} + \mathbf{v}$, and it is called *parallelogram addition* of vectors.

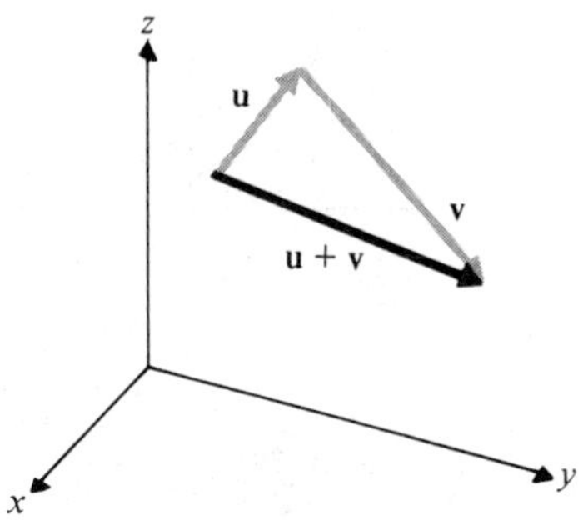

Figure 12.52

Algebraically, vectors are added component by component; i.e., if $\mathbf{u} = (u_x, u_y, u_z)$ and $\mathbf{v} = (v_x, v_y, v_z)$, then

$$\boxed{\mathbf{u} + \mathbf{v} = (u_x + v_x, u_y + v_y, u_z + v_z).} \tag{12.12}$$

To verify this we simply note that the differences in the coordinates of P and Q in Fig. 12.54 are (u_x, u_y, u_z), and the differences in those of Q and R are (v_x, v_y, v_z). Consequently, the differences in the coordinates of P and R must be $(u_x + v_x, u_y + v_y, u_z + v_z)$.

It is not difficult to show (see Exercise 26) that vector addition and scalar multiplication obey the following rules:

$$\mathbf{u} + \mathbf{v} = \mathbf{v} + \mathbf{u}; \tag{12.13a}$$

$$(\mathbf{u} + \mathbf{v}) + \mathbf{w} = \mathbf{u} + (\mathbf{v} + \mathbf{w}); \tag{12.13b}$$

$$\lambda(\mathbf{u} + \mathbf{v}) = \lambda\mathbf{u} + \lambda\mathbf{v}; \tag{12.13c}$$

$$(\lambda + \mu)\mathbf{v} = \lambda\mathbf{v} + \mu\mathbf{v}. \tag{12.13d}$$

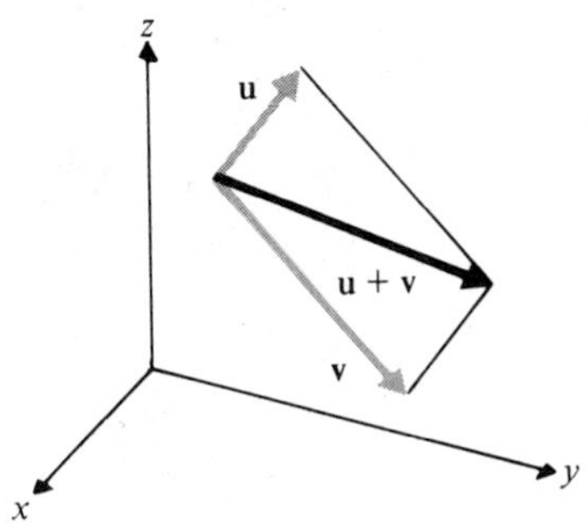

Figure 12.53

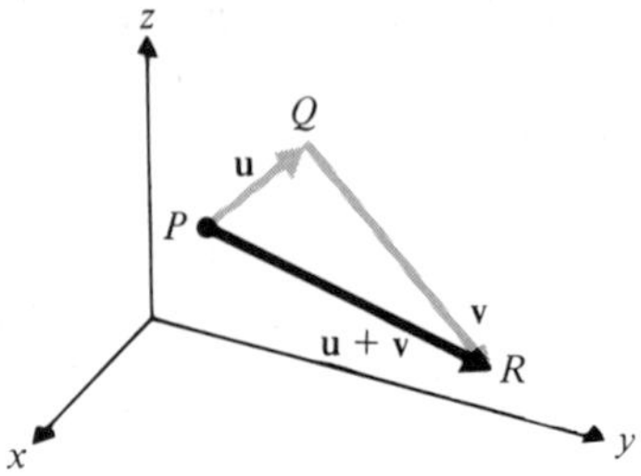

Figure 12.54

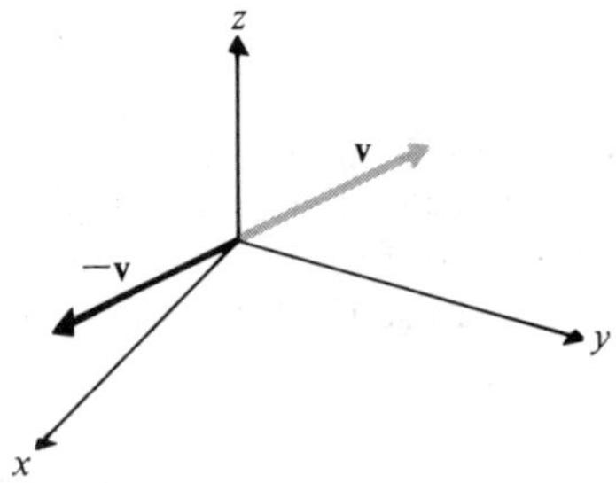

Figure 12.55

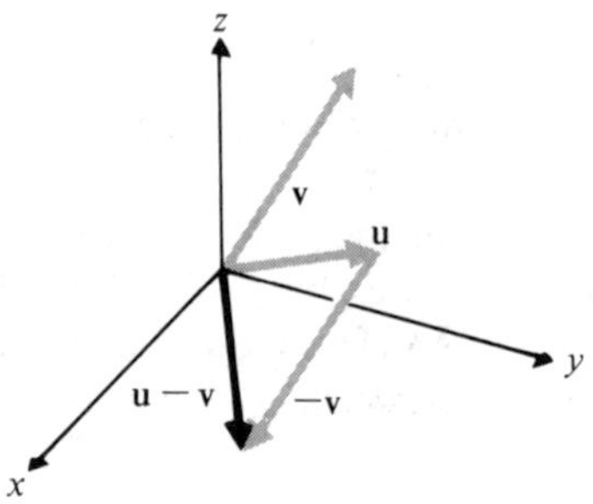

Figure 12.56

If we denote the vector $(-1)\mathbf{v}$ by $-\mathbf{v}$, then the components of $-\mathbf{v}$ are the negatives of those of $\mathbf{v}$:

$$-\mathbf{v} = (-1)\mathbf{v} = (-1)(v_x, v_y, v_z) = (-v_x, -v_y, -v_z). \tag{12.14}$$

This vector has the same length as $\mathbf{v}$, but is opposite in direction to $\mathbf{v}$ (Fig. 12.55). When $\mathbf{v}$ is added to $-\mathbf{v}$, the resultant vector has components that are all zero:

$$\mathbf{v} + (-\mathbf{v}) = (v_x, v_y, v_z) + (-v_x, -v_y, -v_z) = (0, 0, 0).$$

This vector, called the *zero vector*, is denoted by $\mathbf{0}$, and has the property that

$$\mathbf{v} + \mathbf{0} = \mathbf{0} + \mathbf{v} = \mathbf{v} \tag{12.15}$$

for any vector $\mathbf{v}$ whatsoever.

To subtract a vector $\mathbf{v}$ from $\mathbf{u}$, we add $-\mathbf{v}$ to $\mathbf{u}$.

▷ **Definition 12.7** The difference $\mathbf{u} - \mathbf{v}$ between two vectors $\mathbf{u}$ and $\mathbf{v}$ is the vector

$$\mathbf{u} - \mathbf{v} = \mathbf{u} + (-\mathbf{v}). \quad \triangleleft \tag{12.16}$$

In Fig. 12.56 $\mathbf{u} - \mathbf{v}$ is determined by a triangle, and in Fig. 12.57 by a parallelogram. Alternatively, if we denote by $\mathbf{r}$ the vector joining the tip of $\mathbf{v}$ to the tip of $\mathbf{u}$ as in Fig. 12.58, then by triangle addition, we have $\mathbf{v} + \mathbf{r} = \mathbf{u}$. Addition of $-\mathbf{v}$ to each side of this equation gives

$$-\mathbf{v} + \mathbf{v} + \mathbf{r} = -\mathbf{v} + \mathbf{u} \quad \text{or} \quad \mathbf{0} + \mathbf{r} = -\mathbf{v} + \mathbf{u}.$$

Thus $\mathbf{r} = \mathbf{u} - \mathbf{v}$, and $\mathbf{u} - \mathbf{v}$ is the vector joining the tip of $\mathbf{v}$ to the tip of $\mathbf{u}$. Definition 12.7 implies that vectors are subtracted component by component:

$$\boxed{\mathbf{u} - \mathbf{v} = (u_x, u_y, u_z) - (v_x, v_y, v_z) = (u_x - v_x, u_y - v_y, u_z - v_z).} \tag{12.17}$$

Example 9 If $\mathbf{u} = (1, 1, 1)$, $\mathbf{v} = (-2, 3, 0)$, and $\mathbf{w} = (-10, 10, -2)$, find:

a) $3\mathbf{u} + 2\mathbf{v} - \mathbf{w}$;

b) $2\mathbf{u} - 4\mathbf{v} + \mathbf{w}$;

c) $|\mathbf{u}|\mathbf{v} + \dfrac{4}{|\mathbf{v}|}\mathbf{w}$.

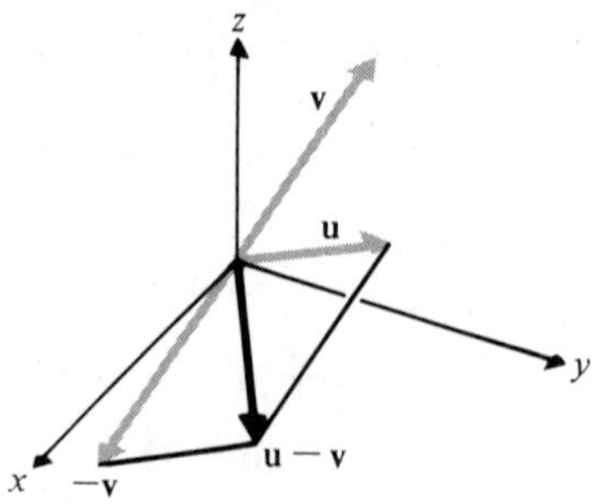

Figure 12.57

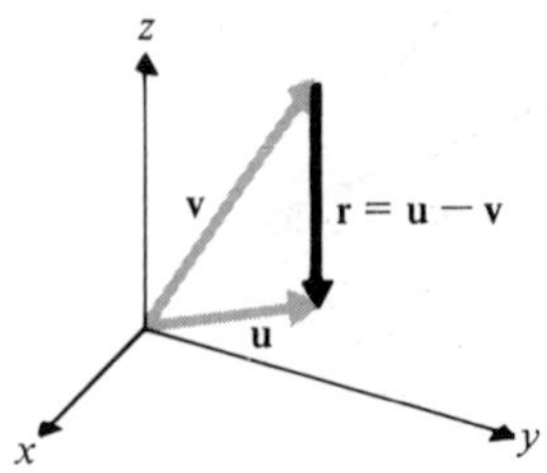

Figure 12.58

Solution a) $3\mathbf{u} + 2\mathbf{v} - \mathbf{w} = 3(1, 1, 1) + 2(-2, 3, 0) - (-10, 10, -2)$
$= (3, 3, 3) + (-4, 6, 0) + (10, -10, 2)$
$= (9, -1, 5)$

b) $2\mathbf{u} - 4\mathbf{v} + \mathbf{w} = 2(1, 1, 1) - 4(-2, 3, 0) + (-10, 10, -2)$
$= (0, 0, 0) = \mathbf{0}$

c) Since $|\mathbf{u}| = \sqrt{1^2 + 1^2 + 1^2} = \sqrt{3}$ and $|\mathbf{v}| = \sqrt{(-2)^2 + 3^2} = \sqrt{13}$,

$$|\mathbf{u}|\mathbf{v} + \frac{4}{|\mathbf{v}|}\mathbf{w} = \sqrt{3}(-2, 3, 0) + \frac{4}{\sqrt{13}}(-10, 10, -2)$$
$$= \left(-2\sqrt{3} - \frac{40}{\sqrt{13}}, 3\sqrt{3} + \frac{40}{\sqrt{13}}, -\frac{8}{\sqrt{13}}\right). \square$$

Forces We have already mentioned that quantities such as temperature, area, and density have associated with them only a magnitude and can therefore be represented by real numbers (scalars). There are many quantities, however, that have associated with them both magnitude and direction, and these can be described by vectors. The most notable of this group are forces. When we speak of a force, we mean a push or pull of some size in some specific direction. For example, when the boy shown in Fig. 12.59(a) pulls his wagon, he exerts a force in the direction indicated by the handle. Suppose that he pulls with a force of 10 newtons and that the angle between the handle and the horizontal is $\pi/4$ radians. To represent this force as a vector $\mathbf{F}_1$, we choose a coordinate system as shown in Fig. 12.59(b), and make the agreement that the *length of* $\mathbf{F}_1$ *will correspond to the magnitude of the force.* Since $\mathbf{F}_1$ represents a force of 10 newtons, it follows that the length of $\mathbf{F}_1$ will be 10 units. Furthermore, because $\mathbf{F}_1$ makes an angle of $\pi/4$ radians with the positive x- and y-axes, the difference in the x-coordinates (and the y-coordinates) of its tip and tail must be $10 \cos(\pi/4) = 5\sqrt{2}$. The components of $\mathbf{F}_1$ are therefore $\mathbf{F}_1 = (5\sqrt{2}, 5\sqrt{2})$. If the boy's young sister drags her feet on the ground, then she effectively exerts a force $\mathbf{F}_2$ in the negative x-direction. If the magnitude of this force is 3 newtons, then its vector representation is $\mathbf{F}_2 = (-3, 0)$. Finally, if the combined weight of the wagon and the girl is 200 newtons, then the force $\mathbf{F}_3$ of gravity on the wagon and its load is $\mathbf{F}_3 = (0, -200)$. In mechanics we replace the individual forces $\mathbf{F}_1$, $\mathbf{F}_2$, and $\mathbf{F}_3$ by a single force that has the same effect on the

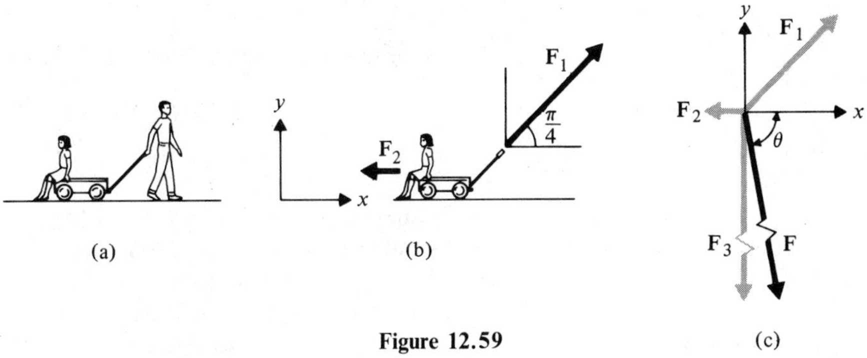

Figure 12.59

wagon as all three forces combined. This force, called the *resultant* of $\mathbf{F}_1$, $\mathbf{F}_2$, and $\mathbf{F}_3$, is represented by the vector $\mathbf{F}$, which is the sum of the vectors $\mathbf{F}_1$, $\mathbf{F}_2$, and $\mathbf{F}_3$:

$$\begin{aligned}\mathbf{F} &= \mathbf{F}_1 + \mathbf{F}_2 + \mathbf{F}_3\\ &= (5\sqrt{2}, 5\sqrt{2}) + (-3, 0) + (0, -200)\\ &= (5\sqrt{2} - 3, 5\sqrt{2} - 200).\end{aligned}$$

The magnitude of this force corresponds to the length of $\mathbf{F}$,

$$|\mathbf{F}| = \sqrt{(5\sqrt{2} - 3)^2 + (5\sqrt{2} - 200)^2} = 193.0,$$

and must therefore be 193.0 newtons. Its direction is shown in Fig. 12.59(c), where

$$\theta = \text{Tan}^{-1}\left(\frac{200 - 5\sqrt{2}}{5\sqrt{2} - 3}\right) = 1.55 \text{ radians.}$$

Example 10 If two water skiers S_1 and S_2 exert forces of 500 and 600 newtons, respectively, on the boat shown in Fig. 12.60(a), what is the resultant force on the boat?

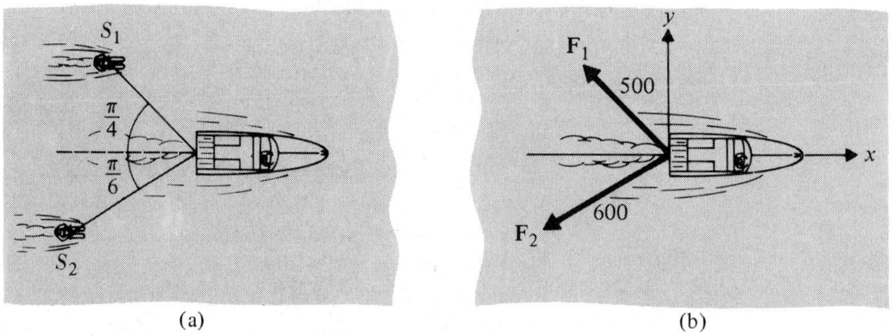

Figure 12.60

Solution If we introduce the coordinate system in Fig. 12.60(b), then the lengths of the force vectors $\mathbf{F}_1$ and $\mathbf{F}_2$ exerted by S_1 and S_2 are 500 and 600 units, respectively. The components of $\mathbf{F}_1$ and $\mathbf{F}_2$ are easily seen to be

$$\begin{aligned}\mathbf{F}_1 &= \left(-\frac{500}{\sqrt{2}}, \frac{500}{\sqrt{2}}\right) & \mathbf{F}_2 &= \left(-\frac{600\sqrt{3}}{2}, -\frac{600}{2}\right)\\ &= (-250\sqrt{2}, 250\sqrt{2}); & &= (-300\sqrt{3}, -300).\end{aligned}$$

The resultant force $\mathbf{F}$ on the boat is then the sum of $\mathbf{F}_1$ and $\mathbf{F}_2$:

$$\begin{aligned}\mathbf{F} = \mathbf{F}_1 + \mathbf{F}_2 &= (-250\sqrt{2} - 300\sqrt{3}, 250\sqrt{2} - 300)\\ &= 50(-5\sqrt{2} - 6\sqrt{3}, 5\sqrt{2} - 6). \square\end{aligned}$$

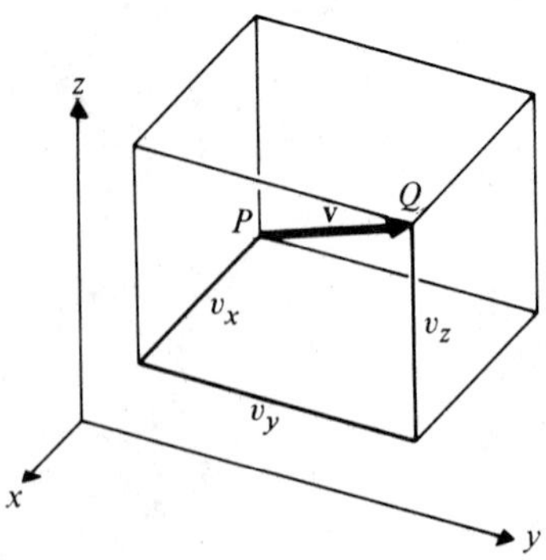

Figure 12.61

By the x-, y-, and z-components (v_x, v_y, v_z) of a vector $\mathbf{v}$, we mean that if we start at a point P (Fig. 12.61) and proceed v_x units in the x-direction, v_y units in the y-direction, and v_z units in the z-direction to a point Q, then $\mathbf{v}$ is the directed line segment joining P and Q. To phrase this another way, we introduce three special vectors parallel to the coordinate axes. We define $\hat{\mathbf{i}}$ as

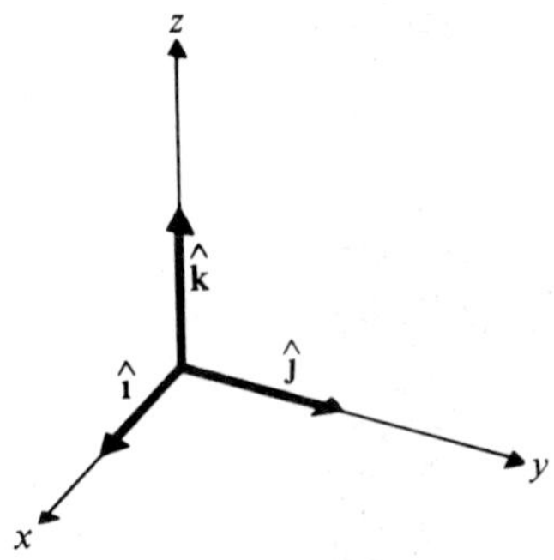

Figure 12.62

a unit vector in the positive x-direction, $\hat{\mathbf{j}}$ as a unit vector in the positive y-direction, and $\hat{\mathbf{k}}$ as a unit vector in the positive z-direction. We have shown these vectors with their tails at the origin in Fig. 12.62, and it is clear that their components are

$$\hat{\mathbf{i}} = (1, 0, 0), \qquad \hat{\mathbf{j}} = (0, 1, 0), \qquad \hat{\mathbf{k}} = (0, 0, 1). \tag{12.18}$$

But note, then, that we can write the vector $\mathbf{v} = (v_x, v_y, v_z)$ in the form

$$\begin{aligned}\mathbf{v} &= (v_x, 0, 0) + (0, v_y, 0) + (0, 0, v_z)\\ &= v_x(1, 0, 0) + v_y(0, 1, 0) + v_z(0, 0, 1)\\ &= v_x\hat{\mathbf{i}} + v_y\hat{\mathbf{j}} + v_z\hat{\mathbf{k}}.\end{aligned}$$

In other words, every vector in space can be written as a linear combination of the three vectors $\hat{\mathbf{i}}$, $\hat{\mathbf{j}}$, and $\hat{\mathbf{k}}$ (i.e., as a constant times $\hat{\mathbf{i}}$ plus a constant times $\hat{\mathbf{j}}$ plus a constant times $\hat{\mathbf{k}}$). Furthermore, the constants multiplying $\hat{\mathbf{i}}$, $\hat{\mathbf{j}}$, and $\hat{\mathbf{k}}$ are precisely the Cartesian components of the vector. This result is equally obvious geometrically. In Fig. 12.61, we have shown the vector $\mathbf{v}$ from P to Q. If we define points A and B as shown in Fig. 12.63, then clearly

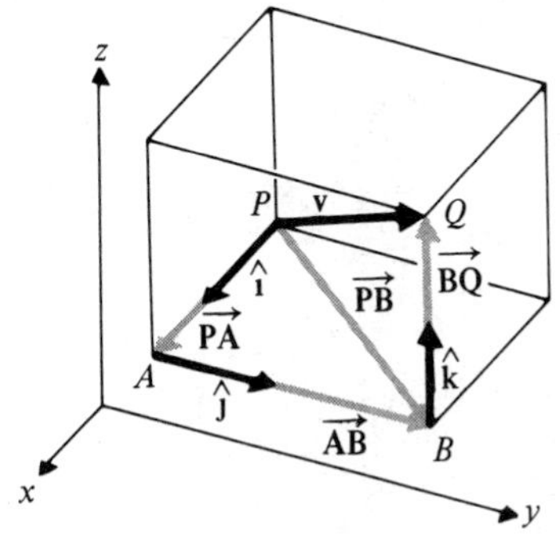

Figure 12.63

$$\mathbf{v} = \overrightarrow{\mathbf{PQ}} = \overrightarrow{\mathbf{PB}} + \overrightarrow{\mathbf{BQ}} = \overrightarrow{\mathbf{PA}} + \overrightarrow{\mathbf{AB}} + \overrightarrow{\mathbf{BQ}}.$$

But because $\overrightarrow{\mathbf{PA}}$ is a vector in the positive x-direction and has length v_x, it follows that $\overrightarrow{\mathbf{PA}} = v_x\hat{\mathbf{i}}$. Similarly, $\overrightarrow{\mathbf{AB}} = v_y\hat{\mathbf{j}}$ and $\overrightarrow{\mathbf{BQ}} = v_z\hat{\mathbf{k}}$, and therefore

$$\mathbf{v} = v_x\hat{\mathbf{i}} + v_y\hat{\mathbf{j}} + v_z\hat{\mathbf{k}}. \tag{12.19}$$

To say then that v_x, v_y, and v_z are the x-, y-, and z-components of a vector $\mathbf{v}$ is equivalent to saying that $\mathbf{v}$ can be written in the form (12.19).

Some books refer to v_x, v_y, and v_z as the *scalar components* of the vector $\mathbf{v}$, and the vectors $v_x\hat{\mathbf{i}}$, $v_y\hat{\mathbf{j}}$, and $v_z\hat{\mathbf{k}}$ as the *vector components* of $\mathbf{v}$. In this book, the word "component" will always mean "scalar component." We point this out only to keep you from being confused between the terminology used here and that used in other texts.

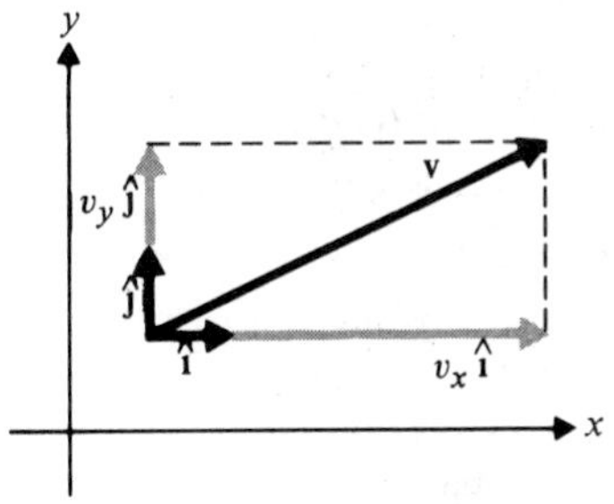

Figure 12.64

Vectors in the xy-plane have only an x- and a y-component, and can therefore be written in terms of $\hat{\mathbf{i}}$ and $\hat{\mathbf{j}}$. If $\mathbf{v} = (v_x, v_y)$, then we write equivalently that $\mathbf{v} = v_x\hat{\mathbf{i}} + v_y\hat{\mathbf{j}}$ (Fig. 12.64). Vectors along the x-axis have only an x-component and can therefore be written in the form $\mathbf{v} = v_x\hat{\mathbf{i}}$.

We use this new notation in the following example.

Example 11 The force $\mathbf{F}$ exerted on a point charge q_1 coulombs by a charge q_2 coulombs is defined by Coulomb's law as

$$\mathbf{F} = \frac{q_1 q_2}{4\pi\varepsilon_0 r^2}\,\hat{\mathbf{r}} \text{ newtons,}$$

where $\varepsilon_0 > 0$ is a constant, r is the distance in meters between the charges, and $\hat{\mathbf{r}}$ is a unit vector in the direction from q_2 to q_1 (Fig. 12.65). When q_1 and q_2 are both positive charges or both negative charges, then $\mathbf{F}$ is repulsive, and when one is positive and the other is negative, $\mathbf{F}$ is attractive. In particular, suppose that charges of 2 and -2 coulombs are placed at $(0, 0, 0)$ and $(3, 0, 0)$, respectively, and a third charge of 1 coulomb is placed at $(1, 1, 1)$. According to Coulomb's law, the 2-coulomb charge will exert a repulsive force on the 1-coulomb charge, and the -2-coulomb charge an attractive force. Find the resultant of these two forces on the 1-coulomb charge.

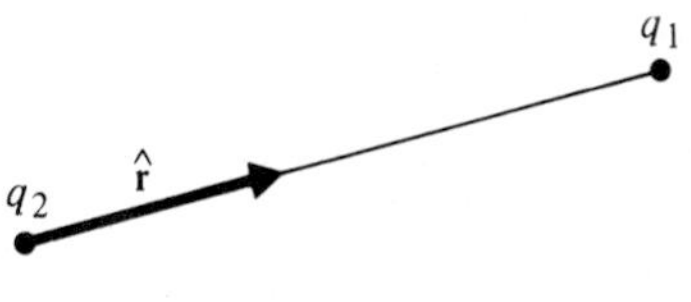

Figure 12.65

Solution If $\mathbf{F}_1$ is the force on the 1-coulomb charge by the -2-coulomb charge (Fig. 12.66), then

$$\mathbf{F}_1 = \frac{(1)(-2)}{4\pi\varepsilon_0 r^2}\hat{\mathbf{r}},$$

where the distance between the charges is $r = \sqrt{(-2)^2 + 1^2 + 1^2} = \sqrt{6}$. The vector from (3, 0, 0) to (1, 1, 1) is $(-2, 1, 1)$, and therefore

$$\hat{\mathbf{r}} = \frac{1}{\sqrt{6}}(-2\hat{\mathbf{i}} + \hat{\mathbf{j}} + \hat{\mathbf{k}}).$$

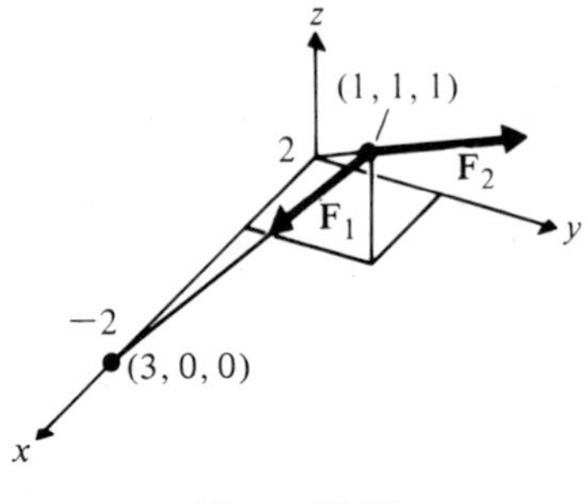

Figure 12.66

Consequently,

$$\mathbf{F}_1 = \frac{-2}{4\pi\varepsilon_0(6)}\cdot\frac{1}{\sqrt{6}}(-2\hat{\mathbf{i}} + \hat{\mathbf{j}} + \hat{\mathbf{k}}) = \frac{1}{12\sqrt{6}\pi\varepsilon_0}(2\hat{\mathbf{i}} - \hat{\mathbf{j}} - \hat{\mathbf{k}}).$$

Similarly, the force $\mathbf{F}_2$ exerted by the charge at the origin is

$$\mathbf{F}_2 = \frac{(1)(2)}{4\pi\varepsilon_0(3)}\cdot\frac{1}{\sqrt{3}}(\hat{\mathbf{i}} + \hat{\mathbf{j}} + \hat{\mathbf{k}}) = \frac{1}{6\sqrt{3}\pi\varepsilon_0}(\hat{\mathbf{i}} + \hat{\mathbf{j}} + \hat{\mathbf{k}}).$$

The resultant of these forces is

$$\mathbf{F} = \mathbf{F}_1 + \mathbf{F}_2 = \frac{1}{12\sqrt{6}\pi\varepsilon_0}\{(2 + 2\sqrt{2})\hat{\mathbf{i}} + (2\sqrt{2} - 1)\hat{\mathbf{j}} + (2\sqrt{2} - 1)\hat{\mathbf{k}}\}\ N. \square$$

EXERCISES 12.4

If $\mathbf{u} = (1, 3, 6)$, $\mathbf{v} = (-2, 0, 4)$, and $\mathbf{w} = (4, 3, -2)$, find components for the vector in each of Exercises 1–10.

1. $3\mathbf{u} - 2\mathbf{v}$
2. $2\mathbf{w} + 3\mathbf{v}$
3. $\mathbf{w} - 3\mathbf{u} - 3\mathbf{v}$
4. $\hat{\mathbf{v}}$
5. $2\hat{\mathbf{w}} - 3\mathbf{v}$
6. $|\mathbf{v}|\mathbf{v} - 2|\hat{\mathbf{v}}|\mathbf{w}$
7. $(15 - 2|\mathbf{w}|)(\mathbf{u} + \mathbf{v})$
8. $|3\mathbf{u}|\mathbf{v} - |-2\mathbf{v}|\mathbf{u}$
9. $|2\mathbf{u} + 3\mathbf{v} - \mathbf{w}|\hat{\mathbf{w}}$
10. $\dfrac{\mathbf{v} - \mathbf{w}}{|\mathbf{v} + \mathbf{w}|}$

If $\mathbf{u} = 2\hat{\mathbf{i}} + \hat{\mathbf{j}}$ and $\mathbf{v} = -\hat{\mathbf{i}} + 3\hat{\mathbf{j}}$, find the components of the vector in each of Exercises 11–14 and illustrate each geometrically.

11. $\mathbf{u} + \mathbf{v}$
12. $\mathbf{u} - \mathbf{v}$
13. $2\hat{\mathbf{u}}$
14. $\hat{\mathbf{v}} + \hat{\mathbf{u}}$

In Exercises 15–24, find the Cartesian components for the spatial vector described. In each case, draw the vector.

15. From (1, 3, 2) to $(-1, 4, 5)$
16. With length 5 in the positive x-direction
17. With length 2 in the negative z-direction
18. With tail at (1, 1, 1), length 3, and pointing toward the point (1, 3, 5)
19. With positive y-component, length 1, and parallel to the line through (1, 3, 6) and $(-2, 1, 4)$
20. In the same direction as the vector from $(1, 0, -1)$ to $(3, 2, -4)$, but only half as long
21. With positive and equal x- and y-components, length 10, and z-component equal to 4
22. Has its tail at the origin, makes angles of $\pi/3$ and $\pi/4$ radians with the positive x- and y-axes, respectively, and has length $\frac{5}{2}$
23. From $(1, 3, -2)$ to the midpoint of the line segment joining $(2, 4, -3)$ and (1, 5, 6)
24. Has its tail at the origin, makes equal angles with the positive coordinate axes, has all positive components, and has length 2
25. If P, Q, and R are the points with coordinates $(3, 2, -1)$, (0, 1, 4), and $(6, 5, -2)$, respectively, find the coordinates of a point S in order that $\overrightarrow{\mathbf{PQ}} = \overrightarrow{\mathbf{RS}}$.
26. Prove that vector addition and scalar multiplication have the properties in Eqs. (12.13).
27. Draw all spatial vectors of length 1 that have equal x- and y-components and tails at the origin.

28. Draw all spatial vectors of length 2 that have their tails at the origin and make an angle of $\pi/4$ with the positive z-axis.

29. If $\mathbf{u} = 3\hat{\imath} + 2\hat{\jmath} - 4\hat{\mathbf{k}}$, and $\mathbf{v} = \hat{\imath} + 6\hat{\jmath} + 5\hat{\mathbf{k}}$, find scalars λ and ρ so that the vector $\mathbf{w} = 5\hat{\imath} - 18\hat{\jmath} - 32\hat{\mathbf{k}}$ can be written in the form $\mathbf{w} = \lambda\mathbf{u} + \rho\mathbf{v}$.

30. Find a vector $\mathbf{T}$ of length 3 along the tangent line to the curve $y = x^2$ at the point (2, 4) (Fig. 12.67).

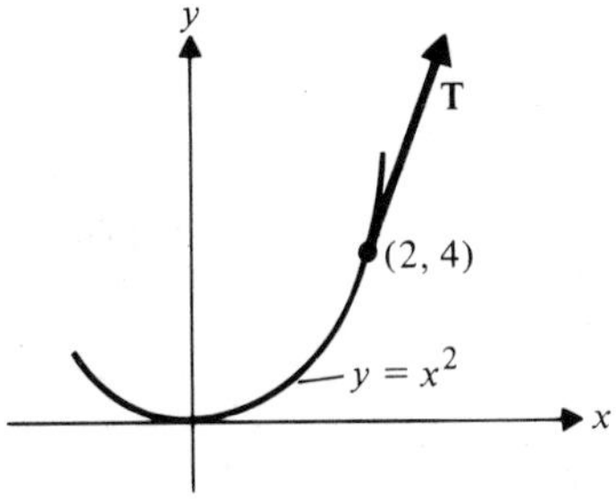

Figure 12.67

31. Use Coulomb's law (see Example 11) to find the force on a charge of 2 coulombs at the origin due to equal charges of 3 coulombs at the points (1, 1, 2) and (2, −1, −2).

32. Newton's universal law of gravitation states that the force $\mathbf{F}$ of attraction on a mass of m kilograms by a mass of M kilograms is

$$\mathbf{F} = \frac{GmM}{r^2}\hat{\mathbf{r}} \text{ newtons,}$$

where $G = 6.67 \times 10^{-11}$ is a constant, r is the distance in meters between the masses, and $\hat{\mathbf{r}}$ is a unit vector in the direction from m to M. If point masses, each of 5 kilograms, are situated at (5, 1, 3) and (−1, 2, 1), what is the resultant force on a mass of 10 kilograms at (2, 2, 2)?

33. Illustrate geometrically the triangle inequality for vectors

$$|\mathbf{u} + \mathbf{v}| \leq |\mathbf{u}| + |\mathbf{v}|.$$

Prove the result algebraically.

Vectors $\mathbf{u}$, $\mathbf{v}$, and $\mathbf{w}$ are said to be linearly dependent if there exist three scalars a, b, and c, not all zero, such that

$$a\mathbf{u} + b\mathbf{v} + c\mathbf{w} = \mathbf{0}.$$

If this equation can only be satisfied with $a = b = c = 0$, the vectors are said to be linearly independent. In Exercises 34–37, determine whether the vectors are linearly dependent or linearly independent.

34. $\mathbf{u} = (1, 1, 1)$, $\mathbf{v} = (2, 1, 3)$, $\mathbf{w} = (4, 2, 6)$

35. $\mathbf{u} = (1, 1, 1)$, $\mathbf{v} = (2, 1, 3)$, $\mathbf{w} = (1, 6, 4)$

36. $\mathbf{u} = (-1, 3, -5)$, $\mathbf{v} = (2, 4, -1)$, $\mathbf{w} = (3, 11, -7)$

37. $\mathbf{u} = (4, 2, 6)$, $\mathbf{v} = (1, 3, -2)$, $\mathbf{w} = (7, 1, 4)$

38. Use vectors to show that the line joining the midpoints of two sides of a triangle is parallel to the third side and its length is one-half the length of the third side (Fig. 12.68).

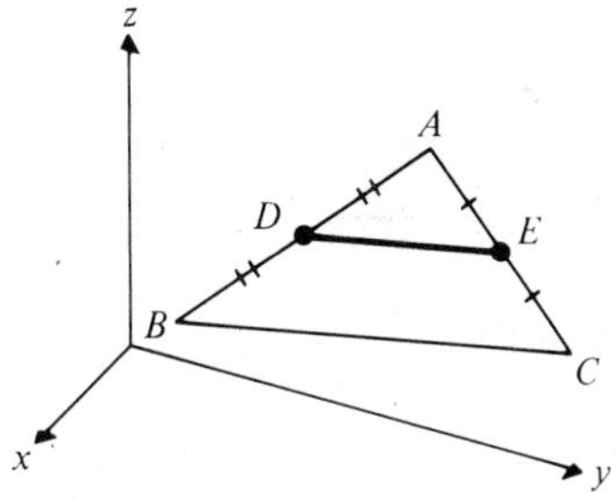

Figure 12.68

39. Use vectors to show that the medians of a triangle (Fig. 12.69) all meet in a point with coordinates

$$\left(\frac{x_1 + x_2 + x_3}{3}, \frac{y_1 + y_2 + y_3}{3}, \frac{z_1 + z_2 + z_3}{3}\right).$$

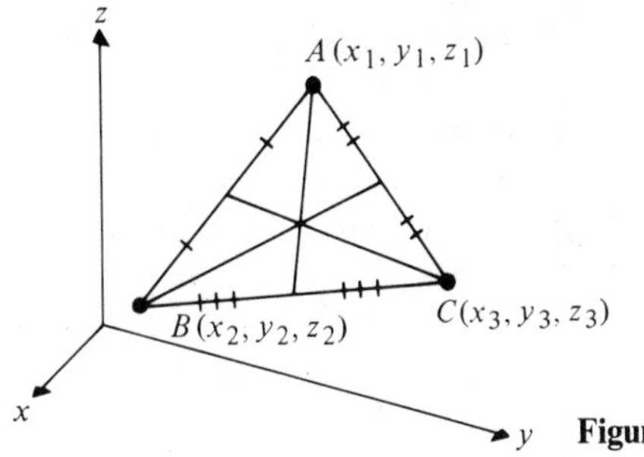

Figure 12.69

40. If n point masses m_i are located at points (x_i, y_i) in the xy-plane, then Eqs. (7.25) and (7.26) define the center of mass $(\bar{x}, \bar{y})$ of the system. Show that these two scalar equations are represented by the one vector equation

$$M(\bar{x}, \bar{y}) = \sum_{i=1}^{n} m_i\mathbf{r}_i$$

where $\mathbf{r}_i$ is the vector joining the origin to the point (x_i, y_i).

*41. If $\mathbf{u} = u_x\hat{\imath} + u_y\hat{\jmath}$ and $\mathbf{v} = v_x\hat{\imath} + v_y\hat{\jmath}$, show that every vector $\mathbf{w}$ in the xy-plane can be written in the form

$$\mathbf{w} = \lambda\mathbf{u} + \rho\mathbf{v}$$

provided $u_xv_y - u_yv_x \neq 0$.

*42. Two springs (with constants k_1 and k_2 and unstretched lengths l) are fixed at points A and B (Fig. 12.70). They are joined to a sleeve that slides along the x-axis. Find the resultant force of the springs on the sleeve at any point between O and C.

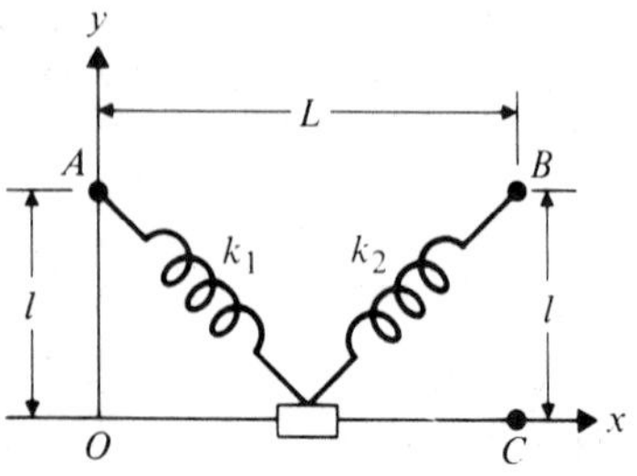

Figure 12.70

* **43.** In Fig. 12.71(a), the xy-plane is the interface between two materials that both transmit light. If a ray of light strikes the surface in a direction defined by the unit vector $\hat{\mathbf{u}} = (u_x, u_y, u_z)$, then some of the light is reflected along a vector $\hat{\mathbf{v}}$, and some is refracted along $\hat{\mathbf{w}}$. The three vectors $\hat{\mathbf{u}}$, $\hat{\mathbf{v}}$, and $\hat{\mathbf{w}}$ all lie in a plane that is perpendicular to the xy-plane.

a) If the angle of incidence i (Fig. 12.71b) is equal to the angle of reflection ϕ, find components for $\hat{\mathbf{v}}$ in terms of those of $\hat{\mathbf{u}}$.

b) If the angle of refraction θ is related to the angle of incidence by $n_1 \sin i = n_2 \sin \theta$, where n_1 and n_2 are the indices of refraction of the two materials, find components for $\hat{\mathbf{w}}$.

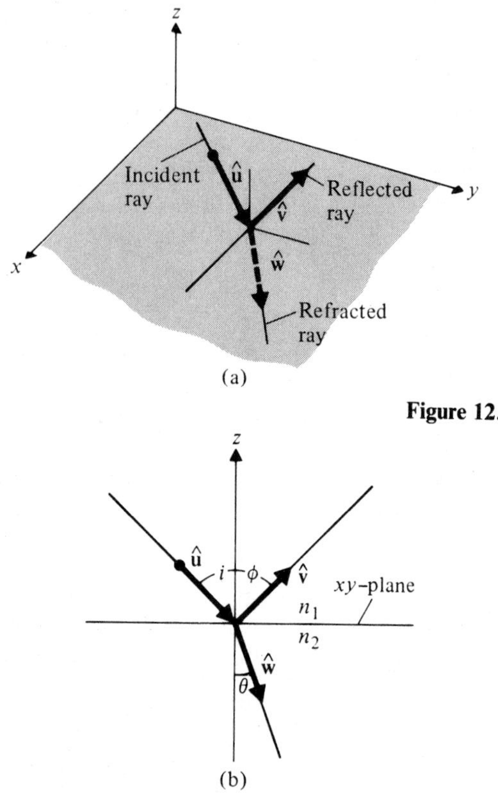

Figure 12.71

12.5 THE SCALAR PRODUCT

In Section 12.4 we learned how to multiply a vector by a scalar and how to add and subtract vectors. The next question naturally is, "Can vectors be multiplied?" The answer is yes, and in fact we will define two products for vectors, one of which yields a scalar and the other a vector. The first is defined algebraically as follows.

▷ **Definition 12.8** The *scalar product* (*dot product* or *inner product*) of two vectors $\mathbf{u}$ and $\mathbf{v}$ with Cartesian components (u_x, u_y, u_z) and (v_x, v_y, v_z) is defined as

$$\mathbf{u} \cdot \mathbf{v} = u_x v_x + u_y v_y + u_z v_z. \quad \triangleleft \tag{12.20}$$

If $\mathbf{u}$ and $\mathbf{v}$ have only x- and y-components, Definition 12.8 reduces to

$$\mathbf{u} \cdot \mathbf{v} = u_x v_x + u_y v_y, \tag{12.21}$$

and if they have only x-components, it becomes

$$\mathbf{u} \cdot \mathbf{v} = u_x v_x. \tag{12.22}$$

It is straightforward to check that

$$\hat{\imath}\cdot\hat{\imath} = \hat{\jmath}\cdot\hat{\jmath} = \hat{\mathbf{k}}\cdot\hat{\mathbf{k}} = 1, \tag{12.23a}$$

$$\hat{\imath}\cdot\hat{\jmath} = \hat{\jmath}\cdot\hat{\mathbf{k}} = \hat{\imath}\cdot\hat{\mathbf{k}} = 0, \tag{12.23b}$$

and that the scalar product is commutative and distributive:

$$\mathbf{u}\cdot\mathbf{v} = \mathbf{v}\cdot\mathbf{u}, \tag{12.24a}$$

$$\mathbf{u}\cdot(\lambda\mathbf{v} + \rho\mathbf{w}) = \lambda(\mathbf{u}\cdot\mathbf{v}) + \rho(\mathbf{u}\cdot\mathbf{w}). \tag{12.24b}$$

Example 12 If $\mathbf{u} = (-2, 1, 3)$ and $\mathbf{v} = (3, -2, -1)$, evaluate each of the following:

a) $\mathbf{u}\cdot\mathbf{v}$;

b) $3\mathbf{u}\cdot(2\mathbf{u} - 4\mathbf{v})$.

Solution a) $\mathbf{u}\cdot\mathbf{v} = (-2)(3) + (1)(-2) + (3)(-1) = -11$

b) $$\begin{aligned} 3\mathbf{u}\cdot(2\mathbf{u} - 4\mathbf{v}) &= (-6, 3, 9)\cdot(-16, 10, 10) \\ &= (-6)(-16) + (3)(10) + (9)(10) \\ &= 216. \ \square \end{aligned}$$

Note that by taking the scalar product of a vector $\mathbf{v} = (v_x, v_y, v_z)$ with itself, we obtain

$$\mathbf{v}\cdot\mathbf{v} = v_x^2 + v_y^2 + v_z^2 = |\mathbf{v}|^2. \tag{12.25}$$

Because Definition 12.8 for the scalar product of two vectors **u** and **v** is phrased in terms of the components of **u** and **v**, and these components depend on the coordinate system used, it follows that this definition also depends on the fact that we have used Cartesian coordinates. Were we to use a different set of coordinates (such as polar coordinates), then the definition of $\mathbf{u}\cdot\mathbf{v}$ in terms of components in that coordinate system would be quite different. For this reason we must find a geometric definition for the scalar product (which is therefore independent of coordinate systems). This we do in the following.

▶ **Theorem 12.2** If two nonzero vectors **u** and **v** are placed tail to tail, and θ is the angle between these vectors $(0 \le \theta \le \pi)$, then

$$\mathbf{u}\cdot\mathbf{v} = |\mathbf{u}|\,|\mathbf{v}|\cos\theta. \ ◀ \tag{12.26}$$

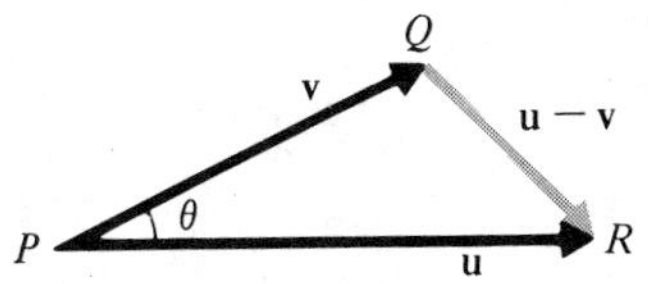

Figure 12.72

Proof The cosine law applied to the triangle in Fig. 12.72 gives

$$|\overrightarrow{\mathbf{QR}}|^2 = |\overrightarrow{\mathbf{PQ}}|^2 + |\overrightarrow{\mathbf{PR}}|^2 - 2|\overrightarrow{\mathbf{PQ}}|\,|\overrightarrow{\mathbf{PR}}|\cos\theta,$$

or

$$|\mathbf{u} - \mathbf{v}|^2 = |\mathbf{v}|^2 + |\mathbf{u}|^2 - 2|\mathbf{v}|\,|\mathbf{u}|\cos\theta.$$

Consequently,

$$|\mathbf{u}|\,|\mathbf{v}|\cos\theta = \tfrac{1}{2}\{|\mathbf{u}|^2 + |\mathbf{v}|^2 - |\mathbf{u} - \mathbf{v}|^2\},$$

and if (u_x, u_y, u_z) and (v_x, v_y, v_z) are the Cartesian components of **u** and **v**, then

$$\begin{aligned} |\mathbf{u}|\,|\mathbf{v}|\cos\theta &= \tfrac{1}{2}\{(u_x^2 + u_y^2 + u_z^2) + (v_x^2 + v_y^2 + v_z^2) \\ &\quad - [(u_x - v_x)^2 + (u_y - v_y)^2 + (u_z - v_z)^2]\} \\ &= u_x v_x + u_y v_y + u_z v_z. \ \square \end{aligned}$$

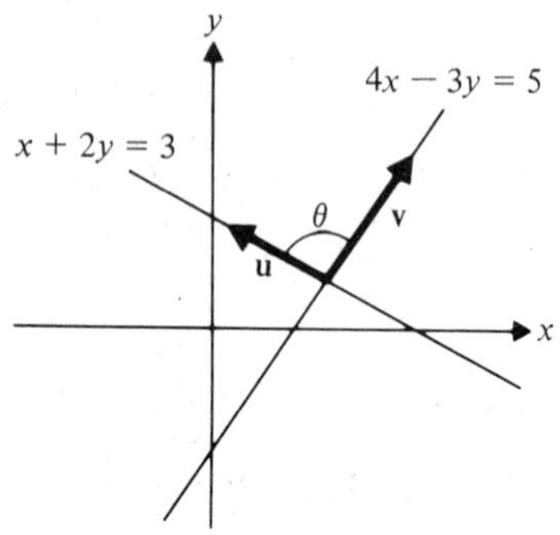

Figure 12.73

An immediate consequence of this result is the following.

Corollary Two nonzero vectors **u** and **v** are perpendicular if and only if

$$\mathbf{u} \cdot \mathbf{v} = 0. \tag{12.27}$$

For example, the vectors $\mathbf{u} = (1, 2, -1)$ and $\mathbf{v} = (4, 2, 8)$ are perpendicular since $\mathbf{u} \cdot \mathbf{v} = (1)(4) + (2)(2) + (-1)(8) = 0$.

Expression (12.26) doesn't just tell us whether or not the angle between two vectors is $\pi/2$ radians; it can also be used to determine the exact angle between any two given vectors **u** and **v**, simply by solving for θ:

$$\cos \theta = \frac{\mathbf{u} \cdot \mathbf{v}}{|\mathbf{u}|\,|\mathbf{v}|}. \tag{12.28}$$

Since the principal values of the inverse cosine function lie between 0 and π, precisely the range for θ, we can write that

$$\theta = \operatorname{Cos}^{-1}\left(\frac{\mathbf{u} \cdot \mathbf{v}}{|\mathbf{u}|\,|\mathbf{v}|}\right). \tag{12.29}$$

Example 13 Find the angle between the vectors $\mathbf{u} = (2, -3, 1)$ and $\mathbf{v} = (5, 2, 4)$.

Solution Since $\mathbf{u} \cdot \mathbf{v} = |\mathbf{u}|\,|\mathbf{v}| \cos \theta$, where θ is the angle between **u** and **v**,

$$\cos \theta = \frac{\mathbf{u} \cdot \mathbf{v}}{|\mathbf{u}|\,|\mathbf{v}|} = \frac{10 - 6 + 4}{\sqrt{4 + 9 + 1}\sqrt{25 + 4 + 16}} = \frac{8}{3\sqrt{70}}.$$

Thus,

$$\theta = \operatorname{Cos}^{-1}\left(\frac{8}{3\sqrt{70}}\right) = 1.25. \ \square$$

Example 14 Find the angle between the lines $x + 2y = 3$, $4x - 3y = 5$.

Solution Since the slope of $x + 2y = 3$ is $-\frac{1}{2}$, a vector along this line is $\mathbf{u} = (-2, 1)$. Similarly, a vector along $4x - 3y = 5$ is $\mathbf{v} = (3, 4)$. If θ is the angle between these vectors (and therefore between the lines (Fig. 12.73)), then $\mathbf{u} \cdot \mathbf{v} = |\mathbf{u}|\,|\mathbf{v}| \cos \theta$. Consequently,

$$\cos \theta = \frac{\mathbf{u} \cdot \mathbf{v}}{|\mathbf{u}|\,|\mathbf{v}|} = \frac{-6 + 4}{\sqrt{5}\sqrt{25}} = \frac{-2}{5\sqrt{5}},$$

and $\theta = \operatorname{Cos}^{-1}[-2/(5\sqrt{5})] = 1.75$. The acute angle between the lines is $\pi - 1.75 = 1.39$. $\square$

Planes Scalar products can be used to find equations for planes in space. A plane can be characterized in various ways: by two intersecting lines, by a line and a point not on the line, or by three noncollinear points. For our present purposes, we use the fact that given a point $P(x_1, y_1, z_1)$ and a vector (A, B, C) (Fig. 12.74), there is one and only one plane through P that is perpendicular to (A, B, C). To find the equation of this plane we note that if $Q(x, y, z)$ is any other point in the plane, then the vector $\overrightarrow{\mathbf{PQ}} = (x - x_1, y - y_1, z - z_1)$ lies in the plane.

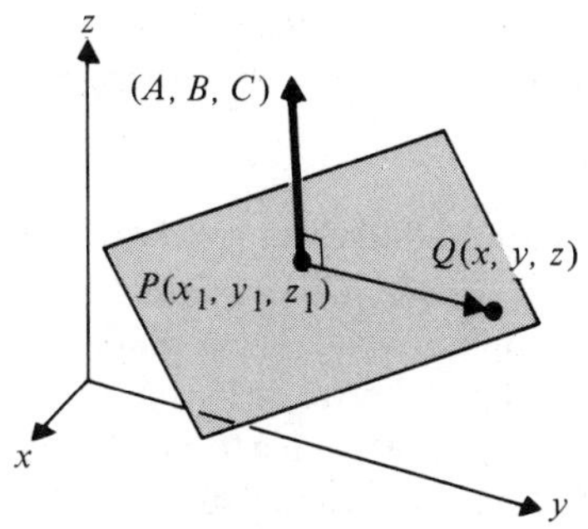

Figure 12.74

But it must then be perpendicular to (A, B, C); hence by the corollary to Theorem 12.2,

$$(A, B, C) \cdot (x - x_1, y - y_1, z - z_1) = 0.$$

Because this equation must be satisfied by every point (x, y, z) in the plane (and at the same time is not satisfied by any point not in the plane), it must be the equation of the plane through P and perpendicular to (A, B, C). If we expand the scalar product, we obtain the equation of the plane in the form

$$A(x - x_1) + B(y - y_1) + C(z - z_1) = 0.$$

This result is worth stating as a theorem.

▶ **Theorem 12.3** An equation for the plane through (x_1, y_1, z_1) perpendicular to the vector (A, B, C) is

$$A(x - x_1) + B(y - y_1) + C(z - z_1) = 0. \quad ◀ \tag{12.30}$$

Equation (12.30) can also be written in the form

$$Ax + By + Cz + D = 0, \tag{12.31}$$

where $D = -(Ax_1 + By_1 + Cz_1)$, and this equation is said to be *linear* in x, y, and z. We have shown then that every plane has a linear equation, and the coefficients A, B, and C of x, y, and z in the equation are the components of a vector (A, B, C) that is perpendicular to the plane. Instead of saying that (A, B, C) is perpendicular to the plane $Ax + By + Cz + D = 0$, we often say that (A, B, C) is *normal* to the plane or that (A, B, C) is a *normal vector* to the plane.

We can also prove the converse result that every linear equation (12.31) describes a plane by reversing the argument leading to Theorem 12.3. Given an equation of the form (12.31), it is always possible to find a point (x_1, y_1, z_1) such that $D = -(Ax_1 + By_1 + Cz_1)$. Equation (12.31) can then be written

$$\begin{aligned} 0 &= Ax + By + Cz - Ax_1 - By_1 - Cz_1 \\ &= A(x - x_1) + B(y - y_1) + C(z - z_1) \\ &= (A, B, C) \cdot (x - x_1, y - y_1, z - z_1). \end{aligned}$$

But a point (x, y, z) satisfies this equation only if the vector $(x - x_1, y - y_1, z - z_1)$ joining (x_1, y_1, z_1) to (x, y, z) is perpendicular to (A, B, C); i.e., (x, y, z) must lie in the plane through (x_1, y_1, z_1) perpendicular to (A, B, C).

Example 15 Find an equation for the plane through the point $(4, -3, 5)$ and normal to the vector $(4, -8, 3)$.

Solution According to (12.30), the equation of the plane is

$$4(x - 4) - 8(y + 3) + 3(z - 5) = 0,$$

or

$$4x - 8y + 3z = 55. \quad \square$$

Example 16 Determine whether the planes $x + 2y - 4z = 10$ and $2x + 4y - 8z = 11$ are parallel.

Solution Normal vectors to these planes are $\mathbf{N}_1 = (1, 2, -4)$ and $\mathbf{N}_2 = (2, 4, -8)$. Since $\mathbf{N}_2 = 2\mathbf{N}_1$, the normal vectors are in the same direction, and therefore the planes must be parallel. □

Resolution of Vectors into Perpendicular Directions

In many applications of mathematics, we are called on to find distances between geometric objects. For example, we already have formulas (1.5) and (12.1) for the distances between points in a plane and points in space. In a plane we might want to find the distance from a point to a line or between two parallel lines. In space, there are many other distances that might prove useful: distance from a point to a plane or line, between parallel planes, between nonintersecting lines, etc. Each of these distances can be calculated in a number of ways, but the easiest method is to use components of vectors.

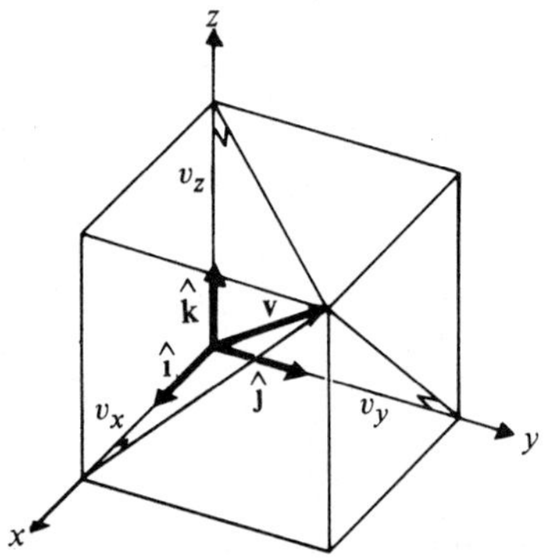

Figure 12.75

We have defined what is meant by the Cartesian components of a vector. In particular, the Cartesian components (v_x, v_y, v_z) of a vector $\mathbf{v}$ are those scalars that multiply $\hat{\mathbf{i}}$, $\hat{\mathbf{j}}$, and $\hat{\mathbf{k}}$ so that $\mathbf{v} = v_x\hat{\mathbf{i}} + v_y\hat{\mathbf{j}} + v_z\hat{\mathbf{k}}$. They can be represented in terms of scalar products as

$$v_x = \mathbf{v} \cdot \hat{\mathbf{i}}, \qquad v_y = \mathbf{v} \cdot \hat{\mathbf{j}}, \qquad v_z = \mathbf{v} \cdot \hat{\mathbf{k}}. \tag{12.32}$$

Geometrically they can be found by drawing $\mathbf{v}$, $\hat{\mathbf{i}}$, $\hat{\mathbf{j}}$, and $\hat{\mathbf{k}}$ all at the origin and dropping perpendiculars from the tip of $\mathbf{v}$ to the x-, y-, and z-axes (Fig. 12.75). We now generalize our definition of a component along an axis to define the component of a vector in any direction whatsoever.

▷ **Definition 12.9** To define the component of a vector $\mathbf{v}$ in a direction $\mathbf{u}$, we place $\mathbf{v}$ and $\mathbf{u}$ tail to tail at a point P and draw the perpendicular from the tip of $\mathbf{v}$ to the line containing $\mathbf{u}$ (Fig. 12.76). The directed distance PR is called the component of $\mathbf{v}$ in the direction $\mathbf{u}$. If R is on the same side of P as the tip of $\mathbf{u}$, PR is taken as positive; and if R is on that side of P opposite to the tip of $\mathbf{u}$, PR is negative. ◁

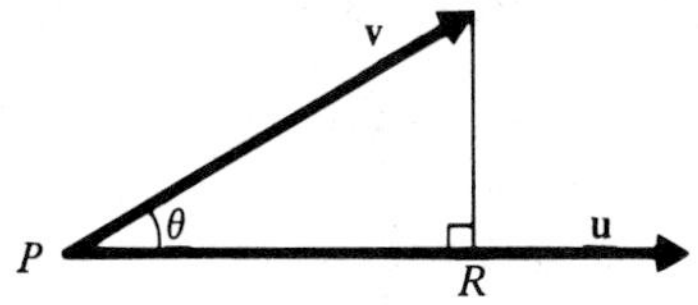

Figure 12.76

In Fig. 12.76 the component of $\mathbf{v}$ in the direction $\mathbf{u}$ is positive, and in Fig. 12.77 the component of $\mathbf{v}$ in the direction $\mathbf{u}$ is negative. Note that the length of $\mathbf{u}$ is irrelevant; it is only the direction of $\mathbf{u}$ that determines the component of $\mathbf{v}$ in the direction $\mathbf{u}$. If θ is the angle between $\mathbf{u}$ and $\mathbf{v}$, then

$$PR = |\mathbf{v}| \cos\theta.$$

The right-hand side of this equation looks very much like the scalar product of $\mathbf{v}$ and a vector that makes an angle θ with $\mathbf{v}$. It lacks only the length of this second vector. Clearly, $\mathbf{u}$ is a vector that makes an angle θ with $\mathbf{v}$, but unfortunately we *cannot* write $PR = |\mathbf{v}|\,|\mathbf{u}| \cos\theta$, since the length of $\mathbf{u}$ need not be 1. If, however, $\hat{\mathbf{u}}$ is the unit vector in the same direction as $\mathbf{u}$, then we can write

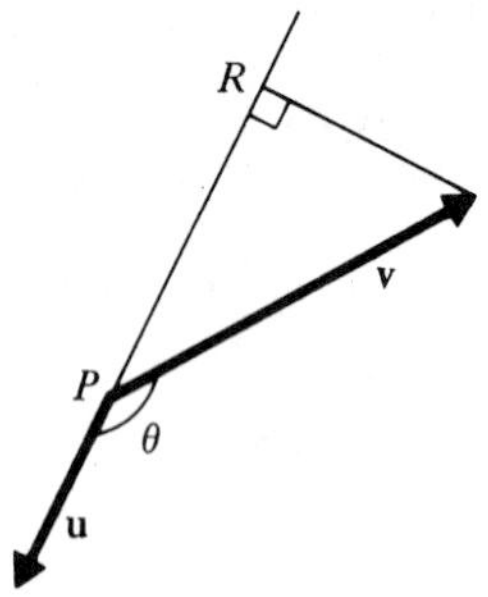

Figure 12.77

$$PR = |\mathbf{v}|\,|\hat{\mathbf{u}}| \cos\theta = \mathbf{v} \cdot \hat{\mathbf{u}}. \tag{12.33}$$

In other words, we have the following theorem.

▶ **Theorem 12.4** The component of a vector $\mathbf{v}$ in a direction $\mathbf{u}$ is $\mathbf{v} \cdot \hat{\mathbf{u}}$, where $\hat{\mathbf{u}}$ is the unit vector in the direction $\mathbf{u}$. ◀

This result agrees with (12.32) for the x-, y-, and z-components of $\mathbf{v}$.

But what good does it do to have a component of a vector in every direction? For one thing, we can represent a vector in other ways besides (12.19). For instance, consider the two perpendicular unit vectors $\hat{\mathbf{e}}_1$ and $\hat{\mathbf{e}}_2$ in Fig. 12.78. The components of vector $\mathbf{v} = 3\hat{\mathbf{i}} + 4\hat{\mathbf{j}}$ in the directions of $\hat{\mathbf{e}}_1$ and $\hat{\mathbf{e}}_2$ are

$$\mathbf{v}\cdot\hat{\mathbf{e}}_1 = \frac{3}{\sqrt{2}} + \frac{4}{\sqrt{2}} = \frac{7}{\sqrt{2}}$$

and

$$\mathbf{v}\cdot\hat{\mathbf{e}}_2 = -\frac{3}{\sqrt{2}} + \frac{4}{\sqrt{2}} = \frac{1}{\sqrt{2}}.$$

Geometrically it is clear that we can express $\mathbf{v}$ as the sum

$$\mathbf{v} = \frac{7}{\sqrt{2}}\hat{\mathbf{e}}_1 + \frac{1}{\sqrt{2}}\hat{\mathbf{e}}_2.$$

In other words, $\mathbf{v}$, and every other vector in the plane, can be expressed as a linear combination of the perpendicular vectors $\hat{\mathbf{e}}_1$ and $\hat{\mathbf{e}}_2$ (instead of $\hat{\mathbf{i}}$ and $\hat{\mathbf{j}}$). We say that $\mathbf{v}$ has been *resolved* into components along $\hat{\mathbf{e}}_1$ and $\hat{\mathbf{e}}_2$. We will find this idea of resolving vectors in terms of perpendicular vectors other than $\hat{\mathbf{i}}$, $\hat{\mathbf{j}}$, and $\hat{\mathbf{k}}$ very important in Section 12.11.

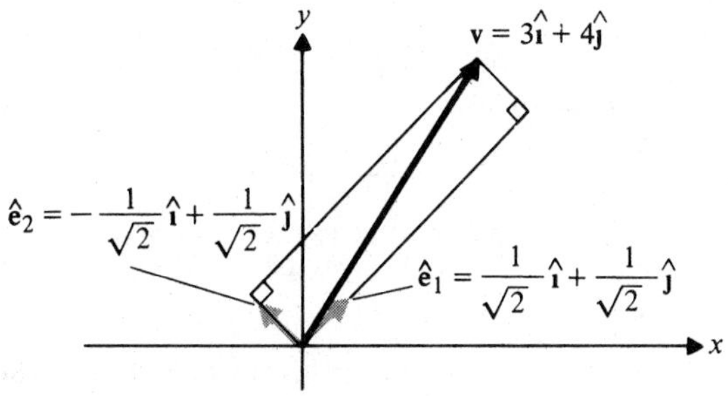

Figure 12.78

Distances Between Points, Lines, and Planes

We stated earlier that components of vectors in arbitrary directions can be used to calculate distances between various geometric objects. The first problem that we consider is the distance from a point $P(x, y)$ to a line l: $Ax + By + C = 0$ in the xy-plane (Fig. 12.79). There are many ways to calculate this distance, and by this distance we mean the perpendicular (or, equivalently, the shortest) distance d from P to l. One way is to use the formula $|Ax + By + C|/\sqrt{A^2 + B^2}$ developed in Exercise 43 in Section 1.3A. A second way is to find the function that describes the distance from P to any point on l, and then determine the absolute minimum for the function. The easiest method is to use components of vectors. If Q is any point on l, then d is simply the component of $\overrightarrow{\mathbf{PQ}}$ in the direction $\overrightarrow{\mathbf{PR}}$; i.e.,

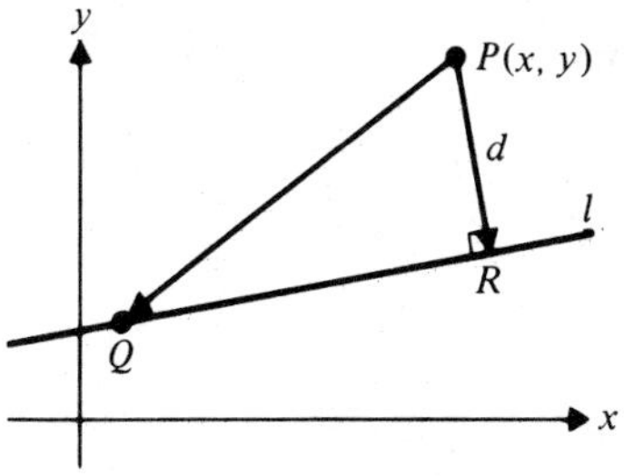

Figure 12.79

$$d = \overrightarrow{\mathbf{PQ}}\cdot\widehat{\mathbf{PR}},$$

where $\widehat{\mathbf{PR}}$ is the unit vector in the direction of $\overrightarrow{\mathbf{PR}}$. To illustrate, suppose that P has coordinates (3, 5) and l is the line $2y = x + 2$ (Fig. 12.80). Since $Q(0, 1)$ is a point on l, the vector $\overrightarrow{\mathbf{PQ}}$ has Cartesian components $(-3, -4)$. The distance from P to l is the component of $\overrightarrow{\mathbf{PQ}}$ in the direction $\overrightarrow{\mathbf{PR}}$. Now we do not know the coordinates of $\mathbf{R}$ (else the problem would be finished), but we really do not need them. What we want is the unit vector along $\overrightarrow{\mathbf{PR}}$. Since the slope of l is $\frac{1}{2}$, a vector along l is (2, 1). A vector perpendicular to l is therefore $(1, -2)$ (and

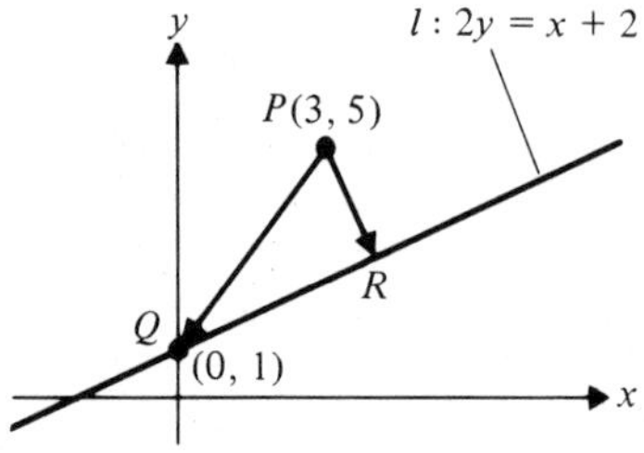

Figure 12.80

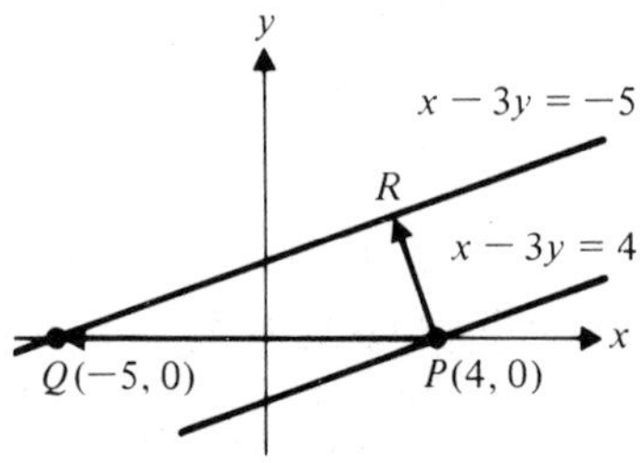

Figure 12.81

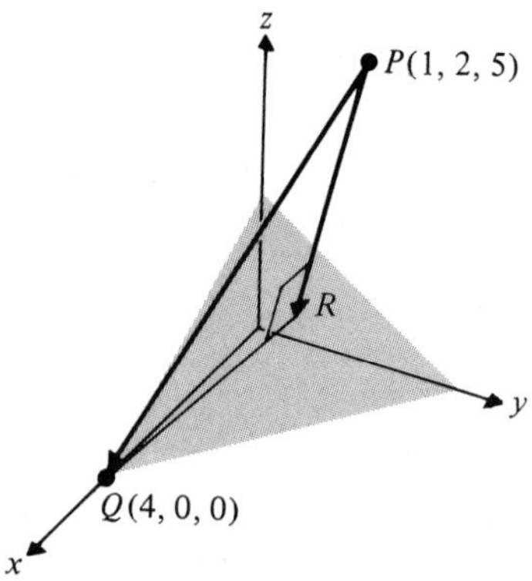

Figure 12.82

note that the scalar product of (2, 1) and (1, −2) vanishes). Consequently, the unit vector along $\overrightarrow{\mathbf{PR}}$ is

$$\widehat{\mathbf{PR}} = \frac{1}{\sqrt{5}}(1, -2) \quad \text{and} \quad d = (-3, -4) \cdot \frac{1}{\sqrt{5}}(1, -2) = \frac{5}{\sqrt{5}} = \sqrt{5}.$$

The following examples illustrate that the same technique may be applied to other distances.

Example 17 Find the distance between the lines $x - 3y = 4$ and $x - 3y = -5$.

Solution The required distance can be calculated as the distance from $P(4, 0)$ (a point on $x - 3y = 4$) to $x - 3y = -5$ (Fig. 12.81). For $Q(-5, 0)$ on $x - 3y = -5$, $\overrightarrow{\mathbf{PQ}} = (-9, 0)$. The required distance is the component of $\overrightarrow{\mathbf{PQ}}$ in the direction $\overrightarrow{\mathbf{PR}}$. Since (3, 1) is a vector along either of the lines (each has slope $\frac{1}{3}$), a vector perpendicular to the lines is (−1, 3). Consequently,

$$d = \overrightarrow{\mathbf{PQ}} \cdot \widehat{\mathbf{PR}} = (-9, 0) \cdot \frac{1}{\sqrt{10}}(-1, 3) = \frac{9}{\sqrt{10}}. \quad \square$$

Example 18 Find the distance from the point (1, 2, 5) to the plane $x + y + 2z = 4$.

Solution If Q is any point in the plane, say (4, 0, 0), then the distance d from $P(1, 2, 5)$ to the plane is the component of $\overrightarrow{\mathbf{PQ}}$ in the direction $\overrightarrow{\mathbf{PR}}$ normal to the plane; i.e., $d = \overrightarrow{\mathbf{PQ}} \cdot \widehat{\mathbf{PR}}$ (Fig. 12.82). Since (1, 1, 2) is a vector normal to the plane,

$$\overrightarrow{\mathbf{PR}} = \frac{1}{\sqrt{6}}(-1, -1, -2)$$

and

$$d = (3, -2, -5) \cdot \frac{1}{\sqrt{6}}(-1, -1, -2) = \frac{9}{\sqrt{6}}. \quad \square$$

Work In Section 7.4 we described work as the product of force and distance. Now that we have represented forces by vectors, we can be much more precise with our definition. In particular, if a particle moves along the line in Fig. 12.83 from P to Q, then vector $\overrightarrow{\mathbf{PQ}}$ represents its displacement. If a constant force $\mathbf{F}$

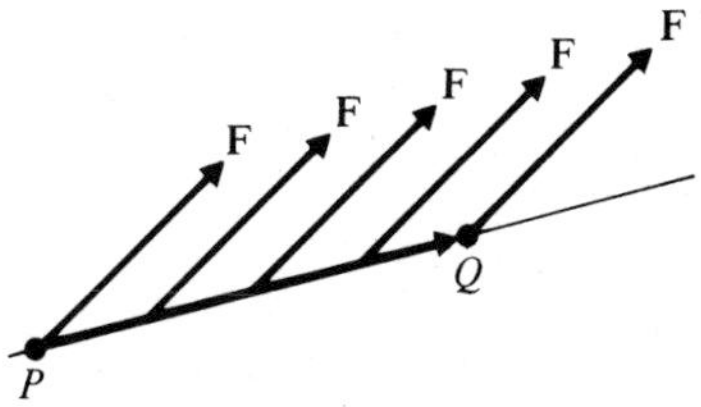

Figure 12.83

acts on the particle during this motion, then the work done by $\mathbf{F}$ is defined as

$$W = \mathbf{F} \cdot \overrightarrow{\mathbf{PQ}}. \tag{12.34}$$

It is important to keep in mind exactly when this definition of work can be used: for constant forces acting along straight lines, and by constant $\mathbf{F}$, we mean that $\mathbf{F}$ is constant in both magnitude and direction. Note that when $\mathbf{F}$ and $\overrightarrow{\mathbf{PQ}}$ are both in the same direction, then the angle between the vectors is zero. In this case,

$$W = |\mathbf{F}|\,|\overrightarrow{\mathbf{PQ}}| \cos(0) = |\mathbf{F}|\,|\overrightarrow{\mathbf{PQ}}|,$$

and this is essentially the equation dealt with in Section 7.4. When $\mathbf{F}$ and $\overrightarrow{\mathbf{PQ}}$ are not in the same direction, then

$$W = |\mathbf{F}|\,|\overrightarrow{\mathbf{PQ}}| \cos\theta.$$

Since $|\mathbf{F}|\cos\theta$ is the component of $\mathbf{F}$ along $\overrightarrow{\mathbf{PQ}}$, this equation simply states that when $\mathbf{F}$ and $\overrightarrow{\mathbf{PQ}}$ are not in the same direction, $|\overrightarrow{\mathbf{PQ}}|$ should be multiplied by the component of $\mathbf{F}$ in the direction of $\overrightarrow{\mathbf{PQ}}$.

Example 19 If the boy shown in Fig. 12.84 pulls the wagon handle with a force of 10 newtons at an angle of $\pi/4$ radians with the horizontal, how much work does he do in walking a distance of 20 meters in a straight line?

Solution The force $\mathbf{F}$ exerted by the boy has magnitude $|\mathbf{F}| = 10$, and points in a direction that makes an angle of $\pi/4$ radians with the displacement vector. If $\mathbf{d}$ is the displacement vector, then

$$W = \mathbf{F} \cdot \mathbf{d} = |\mathbf{F}|\,|\mathbf{d}| \cos\theta = 10(20)\cos\left(\frac{\pi}{4}\right) = \frac{200}{\sqrt{2}} = 100\sqrt{2} \text{ joules. } \square$$

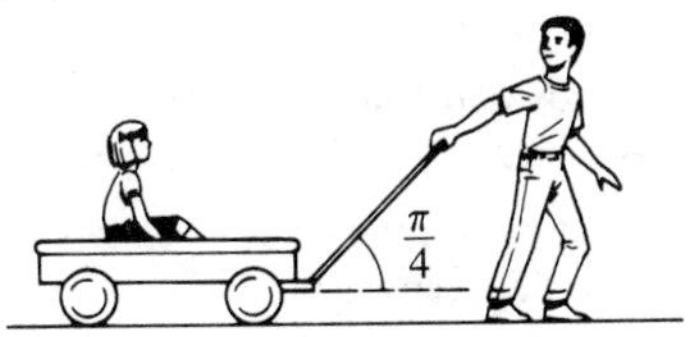

Figure 12.84

When motion is along a straight line, but $\mathbf{F}$ is *not* constant in direction or magnitude or both, we must use integration. The following example illustrates such a situation.

Example 20 The spring in Fig. 12.85 is fixed at A and moves the sleeve frictionlessly along the rod from B to C. If the spring is unstretched when the sleeve is at C, find the work done by the spring.

Solution If we set up a coordinate system as shown, then at position D, the force $\mathbf{F}$ exerted by the spring on the sleeve has magnitude

$$|\mathbf{F}| = k\left(\sqrt{(1-x)^2 + \tfrac{1}{4}} - \tfrac{1}{2}\right),$$

where k is the spring constant. Since the spring is always stretched during the motion, the direction of $\mathbf{F}$ is along the vector $\overrightarrow{\mathbf{DA}}$. Clearly, then, $\mathbf{F}$ changes in both magnitude and direction as the sleeve moves from B to C. For a small displacement dx at position x, the amount of work done by $\mathbf{F}$ is (approximately)

$$\mathbf{F} \cdot (dx\,\hat{\mathbf{i}}) = |\mathbf{F}|\,dx\,(\cos\theta) = k\left(\sqrt{(1-x)^2 + \tfrac{1}{4}} - \tfrac{1}{2}\right) dx \frac{1-x}{\sqrt{(1-x)^2 + \frac{1}{4}}}$$

$$= k(1-x)\left\{1 - \frac{\frac{1}{2}}{\sqrt{(1-x)^2 + \frac{1}{4}}}\right\} dx.$$

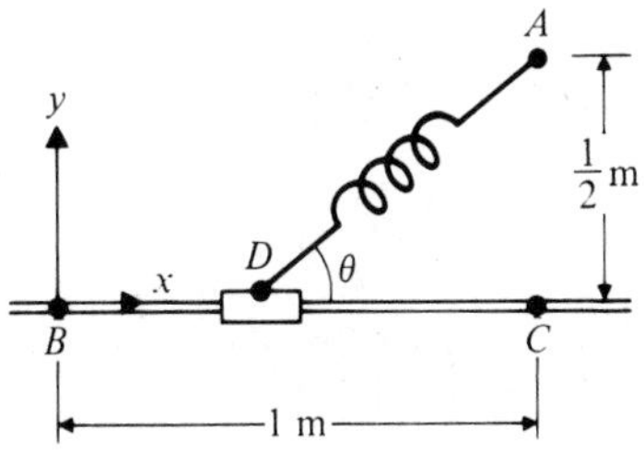

Figure 12.85

The total work done by the spring as the sleeve moves from B to C must therefore be

$$W = \int_0^1 k(1-x)\left\{1 - \frac{1}{2\sqrt{(1-x)^2 + \frac{1}{4}}}\right\} dx$$

$$= k\int_0^1 \left\{1 - x - \frac{1-x}{2\sqrt{(1-x)^2 + \frac{1}{4}}}\right\} dx$$

$$= k\left\{x - \frac{x^2}{2} + \frac{1}{2}\sqrt{(1-x)^2 + \frac{1}{4}}\right\}_0^1$$

$$= \frac{k}{4}(3 - \sqrt{5})\text{ joules. } \square$$

EXERCISES 12.5

If $\mathbf{u} = 2\hat{\mathbf{i}} - 3\hat{\mathbf{j}} + \hat{\mathbf{k}}$, $\mathbf{v} = \hat{\mathbf{j}} - \hat{\mathbf{k}}$, and $\mathbf{w} = 6\hat{\mathbf{i}} - 2\hat{\mathbf{j}} + 3\hat{\mathbf{k}}$, evaluate the scalar or find the components of the vector in Exercises 1–10.

1. $\mathbf{u} \cdot \mathbf{v}$
2. $(\mathbf{v} \cdot \mathbf{w})\mathbf{u}$
3. $(2\mathbf{u} - 3\mathbf{v}) \cdot \mathbf{w}$
4. $2\hat{\mathbf{i}} \cdot \hat{\mathbf{u}}$
5. $|2\mathbf{u}|\mathbf{v} \cdot \mathbf{w}$
6. $(3\mathbf{u} - 4\mathbf{w}) \cdot (2\hat{\mathbf{i}} + 3\mathbf{u} - 2\mathbf{v})$
7. $\mathbf{w} \cdot \hat{\mathbf{w}}$
8. $\dfrac{(105\mathbf{u} + 240\mathbf{v}) \cdot (105\mathbf{u} + 240\mathbf{v})}{|105\mathbf{u} + 240\mathbf{v}|^2}$
9. $|\mathbf{u} - \mathbf{v} + \hat{\mathbf{k}}|(\hat{\mathbf{j}} + \mathbf{w}) \cdot \hat{\mathbf{k}}$
10. $\mathbf{u} \cdot \mathbf{v} + \mathbf{v} \cdot \mathbf{w} - (\mathbf{u} + \mathbf{w}) \cdot \mathbf{v}$

In Exercises 11–14, determine whether the vectors are perpendicular.

11. $(1, 2), (3, 5)$
12. $(2, 4), (-8, 4)$
13. $(1, 3, 6), (-2, 1, -4)$
14. $(2, 3, -6), (-6, 6, 1)$

In Exercises 15–20, find the angle between the vectors.

15. $(3, 4), (2, -5)$
16. $(1, 6), (-4, 7)$
17. $(4, 2, 3), (1, 5, 6)$
18. $(3, 1, -1), (-2, 1, 4)$
19. $(2, 0, 5), (0, 3, 0)$
20. $(1, 3, -2), (-2, -6, 4)$

Find an equation for the plane in Exercises 21 and 22.

21. Through the point $(1, -1, 3)$ and normal to the vector $(4, 3, -2)$
22. Through the point $(2, 1, 5)$ and normal to the vector joining $(2, 1, 5)$ and $(4, 2, 3)$

In Exercises 23–28, find the indicated distance.

23. From the point $(1, 2)$ to the line $x - y = 1$
24. From the point $(3, 2)$ to the line $2x + 3y = 18$
25. From the point $(-1, 4)$ to the line $4x + 3y = 8$
26. From the point $(1, 3, 4)$ to the plane $x + y - 2z = 0$
27. Between the lines $x + 4 = y$ and $3x + 7 = 3y$
28. Between the planes $2x + 3y - z = 15$ and $4x + 6y - 2z = 7$.
29. Verify the results in Eqs. (12.24).
30. Show that a vector perpendicular to the line $Ax + By + C = 0$ in the xy-plane is (A, B).
31. The acute angle between two intersecting planes is defined as the acute angle between their normals. Find the acute angle between the planes $x - 2y + 4z = 6$ and $2x + y = z + 4$.
32. Use Eq. (12.25) to prove the parallelogram law:
$$|\mathbf{u} + \mathbf{v}|^2 + |\mathbf{u} - \mathbf{v}|^2 = 2|\mathbf{u}|^2 + 2|\mathbf{v}|^2.$$
Why is this called the parallelogram law?
33. The angles between a vector $\mathbf{v} = (v_x, v_y, v_z)$ and the vectors $\hat{\mathbf{i}}$, $\hat{\mathbf{j}}$, and $\hat{\mathbf{k}}$ are called the direction angles α, β, and γ of $\mathbf{v}$. Show that
$$\alpha = \operatorname{Cos}^{-1}\left(\frac{\mathbf{v} \cdot \hat{\mathbf{i}}}{|\mathbf{v}|}\right) = \operatorname{Cos}^{-1}\left(\frac{v_x}{|\mathbf{v}|}\right); \qquad \beta = \operatorname{Cos}^{-1}\left(\frac{\mathbf{v} \cdot \hat{\mathbf{j}}}{|\mathbf{v}|}\right) = \operatorname{Cos}^{-1}\left(\frac{v_y}{|\mathbf{v}|}\right);$$
$$\gamma = \operatorname{Cos}^{-1}\left(\frac{\mathbf{v} \cdot \hat{\mathbf{k}}}{|\mathbf{v}|}\right) = \operatorname{Cos}^{-1}\left(\frac{v_z}{|\mathbf{v}|}\right).$$

34. Find direction angles for each of the following vectors:

a) $(1, 2, -3)$;

b) $(0, 1, -3)$;

c) $(-1, -2, 6)$.

35. If $\mathbf{F}$ is a constant force, show that the work done by $\mathbf{F}$ around any closed polygon is zero.

In Exercises 36 and 37, verify that $\hat{\mathbf{v}}$ and $\hat{\mathbf{w}}$ are perpendicular, and then resolve $\mathbf{u}$ into components along $\hat{\mathbf{v}}$ and $\hat{\mathbf{w}}$; i.e., find λ and ρ so that $\mathbf{u} = \lambda\hat{\mathbf{v}} + \rho\hat{\mathbf{w}}$.

36. $\mathbf{u} = (2, 1)$; $\hat{\mathbf{v}} = (1/\sqrt{2}, 1/\sqrt{2})$, $\hat{\mathbf{w}} = (1/\sqrt{2}, -1/\sqrt{2})$

37. $\mathbf{u} = 3\hat{\mathbf{i}} - 2\hat{\mathbf{j}}$; $\hat{\mathbf{v}} = (\hat{\mathbf{i}} - 2\hat{\mathbf{j}})/\sqrt{5}$, $\hat{\mathbf{w}} = (2\hat{\mathbf{i}} + \hat{\mathbf{j}})/\sqrt{5}$

In Exercises 38 and 39, verify that $\hat{\mathbf{u}}$, $\hat{\mathbf{v}}$, and $\hat{\mathbf{w}}$ are mutually perpendicular, and then resolve $\mathbf{r}$ into components along $\hat{\mathbf{u}}$, $\hat{\mathbf{v}}$, and $\hat{\mathbf{w}}$.

38. $\mathbf{r} = (1, 3, -4)$; $\hat{\mathbf{u}} = (2, 1, 0)/\sqrt{5}$, $\hat{\mathbf{v}} = (-1, 2, 3)/\sqrt{14}$, $\hat{\mathbf{w}} = (3, -6, 5)/\sqrt{70}$

39. $\mathbf{r} = 2\hat{\mathbf{i}} - \hat{\mathbf{k}}$; $\hat{\mathbf{u}} = (\hat{\mathbf{i}} + \hat{\mathbf{j}} + \hat{\mathbf{k}})/\sqrt{3}$, $\hat{\mathbf{v}} = (\hat{\mathbf{i}} + \hat{\mathbf{j}} - 2\hat{\mathbf{k}})/\sqrt{6}$, $\hat{\mathbf{w}} = (\hat{\mathbf{i}} - \hat{\mathbf{j}})/\sqrt{2}$

40. If $\mathbf{u} = (3, 2)$, $\mathbf{v} = (1, -3)$, and $\mathbf{w} = (6, 2)$, verify that $\mathbf{v}$ and $\mathbf{w}$ are perpendicular, and find scalars λ and ρ so that $\mathbf{u} = \lambda\mathbf{v} + \rho\mathbf{w}$.

41. If $\mathbf{u} = (1, 0, 1)$, $\mathbf{v} = (1, 1, -1)$, $\mathbf{w} = (-1, 2, 1)$, and $\mathbf{r} = (-2, -3, 4)$, verify that $\mathbf{u}$, $\mathbf{v}$, and $\mathbf{w}$ are mutually perpendicular, and find scalars λ, ρ, and μ so that $\mathbf{r} = \lambda\mathbf{u} + \rho\mathbf{v} + \mu\mathbf{w}$.

42. Find the equation of a plane normal to $\hat{\mathbf{i}} - 2\hat{\mathbf{j}} + 3\hat{\mathbf{k}}$ and two units from the point $(1, 2, 3)$.

43. If a, b, and c (all positive constants) are the intercepts of a plane with the x-, y-, and z-axes, and p is the length of the perpendicular from the origin to the plane, show that

$$\frac{1}{p^2} = \frac{1}{a^2} + \frac{1}{b^2} + \frac{1}{c^2}.$$

44. Repeat Example 20 if the spring has an unstretched length l that is less than the length of AC.

45. Two positive charges q_1 and q_2 are placed at positions $(5, 5)$ and $(-2, 3)$ in the xy-plane, respectively. A third positive charge q_3 is moved along the x-axis from $x = 1$ to $x = -1$. Find the total work done by the electrostatic forces of q_1 and q_2 on q_3.

46. When the rocket shown in Fig. 12.86 passes close to the spherical asteroid, it is attracted to the asteroid by a gravitational force with magnitude

$$\frac{GmM}{r^2},$$

where m and M are the masses of the rocket and asteroid, respectively, r is the distance from the rocket to the center of the asteroid, and $G > 0$ is Newton's gravitational constant. Find how much work the rocket must do against this force in order to follow the straight-line path from A to B.

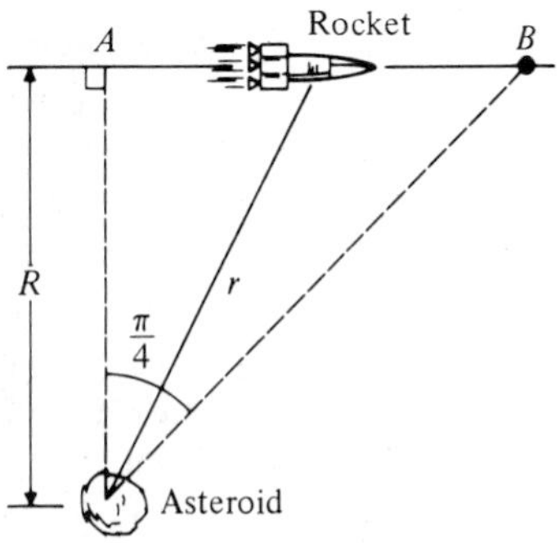

Figure 12.86

***47.** Prove that the (undirected) distance from a point $P(x', y', z')$ to the plane $Ax + By + Cz + D = 0$ is

$$\frac{|Ax' + By' + Cz' + D|}{\sqrt{A^2 + B^2 + C^2}}.$$

***48.** Prove that the (undirected) distance between two parallel planes $Ax + By + Cz + D_1 = 0$ and $Ax + By + Cz + D_2 = 0$ is

$$\frac{|D_1 - D_2|}{\sqrt{A^2 + B^2 + C^2}}.$$

12.6 THE VECTOR PRODUCT

In many applications of vectors we need a vector perpendicular to two given vectors. For instance, suppose we must find the equation for the plane passing through the three given points P_1, P_2, and P_3 in Fig. 12.87. Clearly, $\overrightarrow{\mathbf{P}_1\mathbf{P}_2}$ and $\overrightarrow{\mathbf{P}_1\mathbf{P}_3}$ are vectors that lie in the plane, and to find a normal to the plane we therefore require a vector perpendicular to both $\overrightarrow{\mathbf{P}_1\mathbf{P}_2}$ and $\overrightarrow{\mathbf{P}_1\mathbf{P}_3}$.

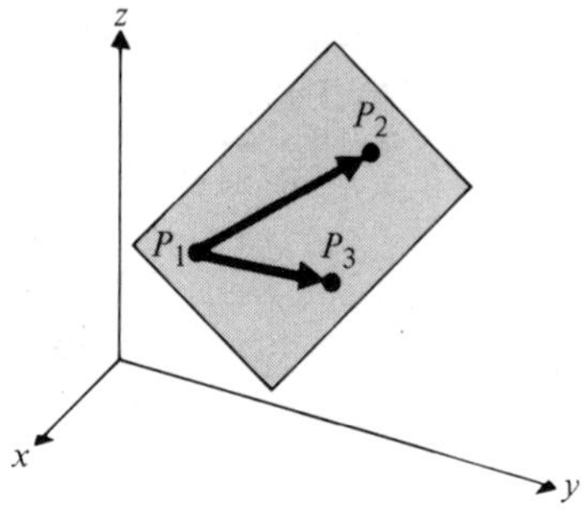

Figure 12.87

Consider, then, finding a vector $\mathbf{r} = (a, b, c)$ perpendicular to two given vectors $\mathbf{u} = (u_x, u_y, u_z)$ and $\mathbf{v} = (v_x, v_y, v_z)$. The corollary to Theorem 12.2 requires a, b, and c to satisfy

$$0 = \mathbf{r} \cdot \mathbf{u} = au_x + bu_y + cu_z,$$

$$0 = \mathbf{r} \cdot \mathbf{v} = av_x + bv_y + cv_z.$$

When we solve these equations, we find that there is an infinite number of solutions all represented by

$$a = s(u_y v_z - u_z v_y), \qquad b = s(u_z v_x - u_x v_z), \qquad c = s(u_x v_y - u_y v_x),$$

where s is any real number. In other words, any vector of the form

$$\mathbf{r} = s(u_y v_z - u_z v_y, u_z v_x - u_x v_z, u_x v_y - u_y v_x)$$

is perpendicular to the vectors $\mathbf{u} = (u_x, u_y, u_z)$ and $\mathbf{v} = (v_x, v_y, v_z)$. When we choose $s = 1$, the resulting vector is called the vector product of $\mathbf{u}$ and $\mathbf{v}$.

▷ **Definition 12.10** The *vector product* (*cross product* or *outer product*) of two vectors $\mathbf{u}$ and $\mathbf{v}$ with Cartesian components (u_x, u_y, u_z) and (v_x, v_y, v_z) is defined as

$$\mathbf{u} \times \mathbf{v} = (u_y v_z - u_z v_y)\hat{\mathbf{i}} + (u_z v_x - u_x v_z)\hat{\mathbf{j}} + (u_x v_y - u_y v_x)\hat{\mathbf{k}}. \quad \triangleleft \quad (12.35)$$

To eliminate the need for memorizing the exact placing of the six components of $\mathbf{u}$ and $\mathbf{v}$ in this definition, we borrow the notation for determinants from linear algebra. A brief discussion of determinants and their properties is given in Appendix B. We set up a 3×3 determinant with $\hat{\mathbf{i}}$, $\hat{\mathbf{j}}$, and $\hat{\mathbf{k}}$ across the top row and the components of $\mathbf{u}$ and $\mathbf{v}$ across the second and third rows:

$$\begin{vmatrix} \hat{\mathbf{i}} & \hat{\mathbf{j}} & \hat{\mathbf{k}} \\ u_x & u_y & u_z \\ v_x & v_y & v_z \end{vmatrix}.$$

In actual fact, this is not a determinant, since three of the entries are vectors and six are scalars. If we ignore this fact, and apply the rules for the expansion of a 3×3 determinant along its first row (namely, $\hat{\mathbf{i}}$ times the 2×2 determinant obtained by deleting the row and column containing $\hat{\mathbf{i}}$, minus $\hat{\mathbf{j}}$ times the 2×2 determinant obtained by deleting the row and column containing $\hat{\mathbf{j}}$, plus $\hat{\mathbf{k}}$ times the 2×2 determinant obtained by deleting the row and column containing $\hat{\mathbf{k}}$), we obtain

$$\begin{vmatrix} \hat{\mathbf{i}} & \hat{\mathbf{j}} & \hat{\mathbf{k}} \\ u_x & u_y & u_z \\ v_x & v_y & v_z \end{vmatrix} = \begin{vmatrix} u_y & u_z \\ v_y & v_z \end{vmatrix}\hat{\mathbf{i}} - \begin{vmatrix} u_x & u_z \\ v_x & v_z \end{vmatrix}\hat{\mathbf{j}} + \begin{vmatrix} u_x & u_y \\ v_x & v_y \end{vmatrix}\hat{\mathbf{k}}.$$

But the definition for the value of a 2×2 determinant is

$$\begin{vmatrix} a & b \\ c & d \end{vmatrix} = ad - bc.$$

Consequently,

$$\begin{vmatrix} \hat{\mathbf{i}} & \hat{\mathbf{j}} & \hat{\mathbf{k}} \\ u_x & u_y & u_z \\ v_x & v_y & v_z \end{vmatrix} = (u_y v_z - u_z v_y)\hat{\mathbf{i}} - (u_x v_z - u_z v_x)\hat{\mathbf{j}} + (u_x v_y - u_y v_x)\hat{\mathbf{k}},$$

and this is the same as the right-hand side of (12.35). We may write, therefore, as a memory-saving device that

$$\boxed{\mathbf{u} \times \mathbf{v} = \begin{vmatrix} \hat{\mathbf{i}} & \hat{\mathbf{j}} & \hat{\mathbf{k}} \\ u_x & u_y & u_z \\ v_x & v_y & v_z \end{vmatrix},} \tag{12.36}$$

so long as we evaluate the right-hand side using the general rules for the expansion of a determinant along its first row.

For example, if $\mathbf{u} = (1, -1, 2)$ and $\mathbf{v} = (2, 3, -5)$, then

$$\mathbf{u} \times \mathbf{v} = \begin{vmatrix} \hat{\mathbf{i}} & \hat{\mathbf{j}} & \hat{\mathbf{k}} \\ 1 & -1 & 2 \\ 2 & 3 & -5 \end{vmatrix} = (5-6)\hat{\mathbf{i}} - (-5-4)\hat{\mathbf{j}} + (3+2)\hat{\mathbf{k}} = -\hat{\mathbf{i}} + 9\hat{\mathbf{j}} + 5\hat{\mathbf{k}}.$$

It is straightforward to verify that

$$\hat{\mathbf{i}} \times \hat{\mathbf{j}} = \hat{\mathbf{k}}, \qquad \hat{\mathbf{j}} \times \hat{\mathbf{k}} = \hat{\mathbf{i}}, \qquad \hat{\mathbf{k}} \times \hat{\mathbf{i}} = \hat{\mathbf{j}}, \tag{12.37}$$

and that the cross-product is anticommutative and distributive:

$$\mathbf{u} \times \mathbf{v} = -\mathbf{v} \times \mathbf{u}, \tag{12.38a}$$

$$\mathbf{u} \times (\lambda\mathbf{v} + \rho\mathbf{w}) = \lambda(\mathbf{u} \times \mathbf{v}) + \rho(\mathbf{u} \times \mathbf{w}). \tag{12.38b}$$

Our preliminary analysis indicated that $\mathbf{u} \times \mathbf{v}$ is perpendicular to both $\mathbf{u}$ and $\mathbf{v}$. The following theorem relates the length of $\mathbf{u} \times \mathbf{v}$ to the lengths of $\mathbf{u}$ and $\mathbf{v}$.

▶ **Theorem 12.5** If θ is the angle between two vectors $\mathbf{u}$ and $\mathbf{v}$, then

$$|\mathbf{u} \times \mathbf{v}| = |\mathbf{u}|\,|\mathbf{v}| \sin\theta. \tag{12.39}$$ ◀

Proof Since θ is an angle between 0 and π, $\sin\theta$ must be positive, and we can write from Eq. (12.28) that

$$\sin\theta = \sqrt{1 - \cos^2\theta} = \sqrt{1 - \left(\frac{\mathbf{u} \cdot \mathbf{v}}{|\mathbf{u}|\,|\mathbf{v}|}\right)^2} = \frac{1}{|\mathbf{u}|\,|\mathbf{v}|}\sqrt{|\mathbf{u}|^2|\mathbf{v}|^2 - (\mathbf{u} \cdot \mathbf{v})^2}.$$

Consequently,

$$|\mathbf{u}|\,|\mathbf{v}| \sin\theta = \sqrt{|\mathbf{u}|^2|\mathbf{v}|^2 - (\mathbf{u} \cdot \mathbf{v})^2}.$$

If $\mathbf{u} = (u_x, u_y, u_z)$ and $\mathbf{v} = (v_x, v_y, v_z)$, then

$$\begin{aligned}
|\mathbf{u}|^2|\mathbf{v}|^2 \sin^2\theta &= (u_x^2 + u_y^2 + u_z^2)(v_x^2 + v_y^2 + v_z^2) - (u_x v_x + u_y v_y + u_z v_z)^2 \\
&= u_x^2(v_x^2 + v_y^2 + v_z^2) + u_y^2(v_x^2 + v_y^2 + v_z^2) \\
&\quad + u_z^2(v_x^2 + v_y^2 + v_z^2) - (u_x^2 v_x^2 + u_y^2 v_y^2 + u_z^2 v_z^2 \\
&\quad + 2u_x v_x u_y v_y + 2u_x v_x u_z v_z + 2u_y v_y u_z v_z) \\
&= (u_y^2 v_z^2 - 2u_y v_y u_z v_z + u_z^2 v_y^2) + (u_x^2 v_z^2 - 2u_x v_x u_z v_z + u_z^2 v_x^2) \\
&\quad + (u_x^2 v_y^2 - 2u_x v_x u_y v_y + u_y^2 v_x^2) \\
&= (u_y v_z - u_z v_y)^2 + (u_x v_z - u_z v_x)^2 + (u_x v_y - u_y v_x)^2 \\
&= |\mathbf{u} \times \mathbf{v}|^2,
\end{aligned}$$

or

$$|\mathbf{u} \times \mathbf{v}| = |\mathbf{u}|\,|\mathbf{v}| \sin\theta. \;\square$$

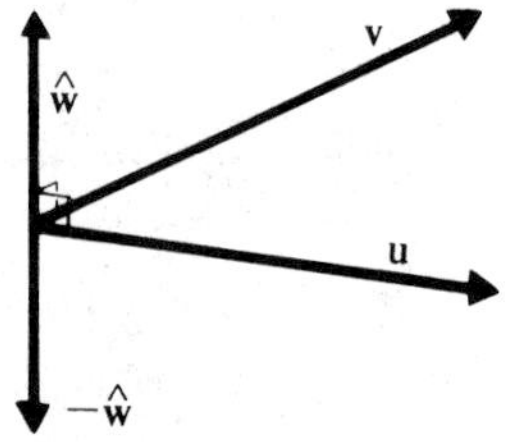

Figure 12.88

We now know that $\mathbf{u} \times \mathbf{v}$ is perpendicular to $\mathbf{u}$ and $\mathbf{v}$, and has length $|\mathbf{u}|\,|\mathbf{v}| \sin\theta$, where θ is the angle between $\mathbf{u}$ and $\mathbf{v}$. Figure 12.88 illustrates that there are only two directions that are perpendicular to $\mathbf{u}$ and $\mathbf{v}$, and one is the negative of the other. Let us denote by $\hat{\mathbf{w}}$ the unit vector along that direction which is perpendicular to $\mathbf{u}$ and $\mathbf{v}$ and is determined by the right-hand rule (curl the fingers of the right hand from $\mathbf{u}$ toward $\mathbf{v}$ and the thumb points in the direction $\hat{\mathbf{w}}$).

We now show that $\mathbf{u} \times \mathbf{v}$ always points in the direction determined by the right-hand rule (i.e., in the direction $\hat{\mathbf{w}}$ rather than $-\hat{\mathbf{w}}$). To see this we place $\mathbf{u}$ and $\mathbf{v}$ tail to tail and establish a coordinate system with this common point as origin and the positive x-axis along $\mathbf{u}$ (Fig. 12.89a). Let the plane determined by $\mathbf{u}$ and $\mathbf{v}$ be the xy-plane. In this coordinate system, $\mathbf{u}$ has only an x-component, $\mathbf{u} = u_x\hat{\mathbf{i}}$ $(u_x > 0)$, and $\mathbf{v}$ has only x- and y-components, $\mathbf{v} = v_x\hat{\mathbf{i}} + v_y\hat{\mathbf{j}}$. The cross-product of $\mathbf{u}$ and $\mathbf{v}$ is therefore

$$\mathbf{u} \times \mathbf{v} = \begin{vmatrix} \hat{\mathbf{i}} & \hat{\mathbf{j}} & \hat{\mathbf{k}} \\ u_x & 0 & 0 \\ v_x & v_y & 0 \end{vmatrix} = u_x v_y \hat{\mathbf{k}}.$$

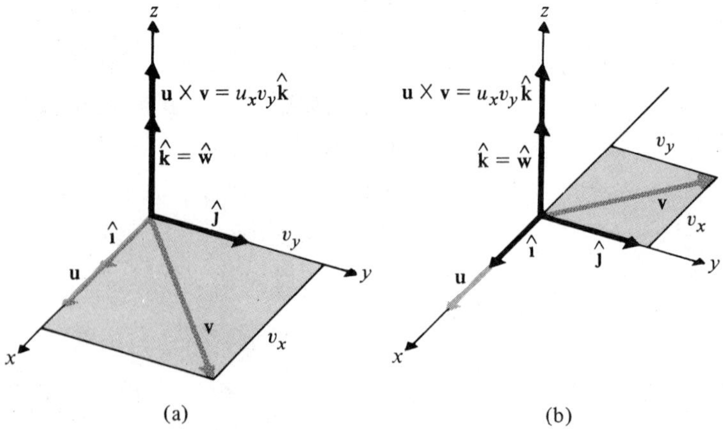

Figure 12.89

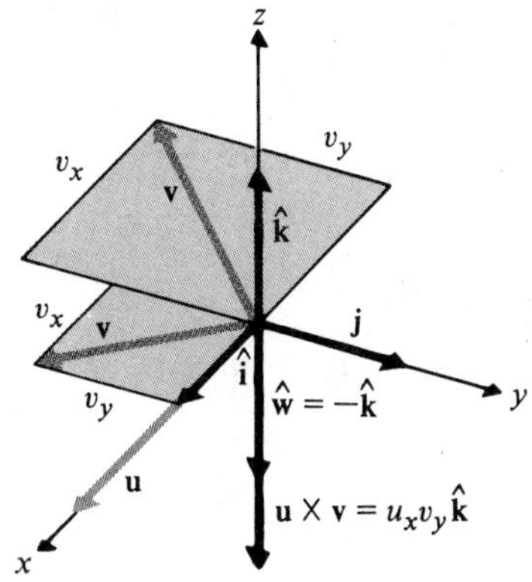

Figure 12.90

For **v** in Fig. 12.89(a), v_y is clearly positive and therefore $u_x v_y$, the component of $\mathbf{u} \times \mathbf{v}$, is also positive. But then $\mathbf{u} \times \mathbf{v}$ is indeed determined by the right-hand rule since $\hat{\mathbf{w}} = \hat{\mathbf{k}}$.

When the tip of **v** lies in the second quadrant of the xy-plane (Fig. 12.89b), then $\mathbf{u} \times \mathbf{v}$ is once again in the positive z-direction, the direction of $\hat{\mathbf{w}}$. When the tip of **v** is in the third or fourth quadrant (Fig. 12.90), $v_y < 0$, and $\mathbf{u} \times \mathbf{v}$ therefore has a negative z-component. But in this case $\hat{\mathbf{w}} = -\hat{\mathbf{k}}$, and the direction of $\mathbf{u} \times \mathbf{v}$ is once again determined by the right-hand rule.

What we have now established is the following coordinate-free definition for the vector product of two vectors **u** and **v**:

$$\mathbf{u} \times \mathbf{v} = (|\mathbf{u}|\,|\mathbf{v}| \sin \theta)\hat{\mathbf{w}}. \tag{12.40}$$

The unit vector $\hat{\mathbf{w}}$ defines the direction of $\mathbf{u} \times \mathbf{v}$ and the factor $|\mathbf{u}|\,|\mathbf{v}| \sin \theta$ is its length.

The fact that the vector product $\mathbf{u} \times \mathbf{v}$ is perpendicular to both **u** and **v** makes it a powerful tool in many applications.

Example 21 Find the equation of the plane through the points $(1, 2, 3)$, $(-2, 0, 4)$, and $(5, 2, -1)$.

Solution If the plane passes through $P(1, 2, 3)$, $Q(-2, 0, 4)$, and $R(5, 2, -1)$, then $\overrightarrow{\mathbf{PQ}} = (-3, -2, 1)$ and $\overrightarrow{\mathbf{PR}} = (4, 0, -4)$ are vectors that lie in the plane. It follows that a vector normal to the plane is

$$\overrightarrow{\mathbf{PQ}} \times \overrightarrow{\mathbf{PR}} = \begin{vmatrix} \hat{\mathbf{i}} & \hat{\mathbf{j}} & \hat{\mathbf{k}} \\ -3 & -2 & 1 \\ 4 & 0 & -4 \end{vmatrix} = (8-0)\hat{\mathbf{i}} - (12-4)\hat{\mathbf{j}} + (0+8)\hat{\mathbf{k}} = 8(\hat{\mathbf{i}} - \hat{\mathbf{j}} + \hat{\mathbf{k}}).$$

Since $\hat{\mathbf{i}} - \hat{\mathbf{j}} + \hat{\mathbf{k}}$ must therefore be normal to the plane, the equation of the plane must be

$$(1, -1, 1) \cdot (x - 1, y - 2, z - 3) = 0,$$

or

$$x - y + z = 2. \ \square$$

Lines In Section 12.2 we indicated that space curves can be described by two simultaneous equations in x, y, and z, and that such a representation describes the curve as the intersection of two surfaces. In this section we use vectors to discuss straight lines. A straight line in space is uniquely characterized by a point on it and a vector parallel to it. For instance, there is one and only one line l through the point $P_0(x_0, y_0, z_0)$ and in the direction $\mathbf{v} = (v_x, v_y, v_z)$ in Fig. 12.91. If $P(x, y, z)$ is any point on this line, then the vector **r** joining the origin to P has components $\mathbf{r} = (x, y, z)$. Now this vector can be expressed as the sum of $\mathbf{r}_0$, the vector from O to P_0 and $\overrightarrow{\mathbf{P}_0\mathbf{P}}$:

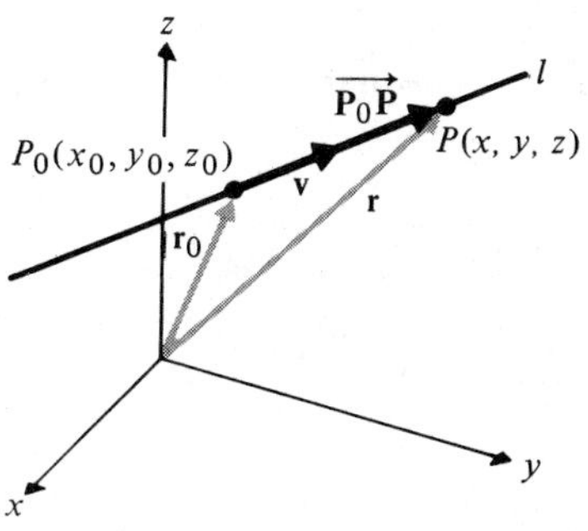

Figure 12.91

$$\mathbf{r} = \mathbf{r}_0 + \overrightarrow{\mathbf{P}_0\mathbf{P}}.$$

But $\overrightarrow{\mathbf{P}_0\mathbf{P}}$ is in the same direction as **v**; therefore it must be some scalar multiple of **v**, $\overrightarrow{\mathbf{P}_0\mathbf{P}} = t\mathbf{v}$. Consequently, the vector joining O to any point on l can be written in the form

$$\mathbf{r} = \mathbf{r}_0 + t\mathbf{v}, \tag{12.41}$$

for an appropriate value of t. Because the components of $\mathbf{r} = (x, y, z)$ are also the coordinates of the point (x, y, z) on l, this equation is called the *vector equation* of the line l through (x_0, y_0, z_0) in the direction $\mathbf{v}$. If we substitute components for $\mathbf{r}$, $\mathbf{r}_0$, and $\mathbf{v}$, then

$$\begin{aligned}(x, y, z) &= (x_0, y_0, z_0) + t(v_x, v_y, v_z)\\ &= (x_0 + tv_x, y_0 + tv_y, z_0 + tv_z).\end{aligned}$$

Since two vectors are equal if and only if corresponding components are identical, we can write that

$$x = x_0 + v_x t, \tag{12.42a}$$

$$y = y_0 + v_y t, \tag{12.42b}$$

$$z = z_0 + v_z t. \tag{12.42c}$$

These three scalar equations are equivalent to vector equation (12.41); they are called *parametric equations* for line l. They illustrate once again that a line in space is characterized by a point (x_0, y_0, z_0) on it and a vector (v_x, v_y, v_z) along it. Each value of t substituted into (12.42) yields a point (x, y, z) on l, and conversely, every point on l is represented by some value of t. For instance, $t = 0$ yields P_0, and $t = 1$ gives the point at the tip of $\mathbf{v}$ in Fig. 12.91.

If none of v_x, v_y, and v_z is equal to zero, we can solve Eqs. (12.42) for t and equate the three expressions to obtain

$$\frac{x - x_0}{v_x} = \frac{y - y_0}{v_y} = \frac{z - z_0}{v_z}. \tag{12.43}$$

These are called *symmetric equations* for the line l through (x_0, y_0, z_0) parallel to $\mathbf{v} = (v_x, v_y, v_z)$. There are only two independent equations in (12.43), which therefore substantiates our previous result that a curve (in this case, a line) can be described by two equations in x, y, and z. We could, for instance, write

$$v_x(y - y_0) = v_y(x - x_0) \quad \text{and} \quad v_z(y - y_0) = v_y(z - z_0),$$

or

$$v_y x - v_x y = v_y x_0 - v_x y_0, \qquad v_y z - v_z y = v_y z_0 - v_z y_0.$$

Since the first of these is linear in x and y and the second is linear in y and z, each describes a plane. This means that the line has been described as the curve of intersection of two planes.

Example 22 Find, if possible, vector, parametric, and symmetric equations for the line through the points $(-1, 2, 1)$ and $(3, -2, 1)$.

Solution A vector along the line is $(3, -2, 1) - (-1, 2, 1) = (4, -4, 0)$, and so too is $(1, -1, 0)$. A vector equation for the line is

$$\mathbf{r} = (-1, 2, 1) + t(1, -1, 0) = (t - 1, -t + 2, 1).$$

Parametric equations are therefore

$$x = t - 1, \qquad y = -t + 2, \qquad z = 1.$$

Because the z-component of every vector along the line is zero, we cannot write full symmetric equations for the line. By eliminating t between the x- and y-equations, however, we can write

$$x + 1 = \frac{y - 2}{-1}, \qquad z = 1.$$

If we set $x + 1 = 2 - y$, $z = 1$, or $x + y = 1$, $z = 1$, we represent the line as the intersection of the planes $x + y = 1$ and $z = 1$ (Fig. 12.92). □

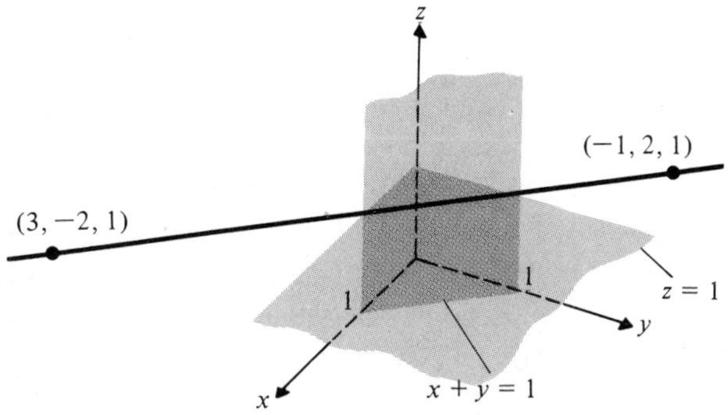

Figure 12.92

Example 23 Find the equation of the plane containing the origin and the line $2x + y - z = 4$, $x + z = 5$.

Solution We can easily find two more points on the plane. For instance, if $x = 0$, then the equations of the line require $z = 5$ and $y = 9$; and if $z = 0$, then $x = 5$ and $y = -6$. Thus $P(0, 9, 5)$ and $Q(5, -6, 0)$, as well as $O(0, 0, 0)$, are points on the plane. It follows then that $\overrightarrow{\mathbf{OP}} = (0, 9, 5)$ and $\overrightarrow{\mathbf{OQ}} = (5, -6, 0)$ are vectors in the plane, and a vector normal to the plane is

$$\overrightarrow{\mathbf{OP}} \times \overrightarrow{\mathbf{OQ}} = \begin{vmatrix} \hat{\mathbf{i}} & \hat{\mathbf{j}} & \hat{\mathbf{k}} \\ 0 & 9 & 5 \\ 5 & -6 & 0 \end{vmatrix} = (30, 25, -45).$$

So too, then, is $(6, 5, -9)$, and the equation of the plane is

$$0 = (6, 5, -9) \cdot (x - 0, y - 0, z - 0) = 6x + 5y - 9z. \ \square$$

Example 24 Find symmetric equations for the line $x + y - 2z = 6, 2x - 3y + 4z = 10$.

Solution To find symmetric equations, we require a vector parallel to the line and a point on it. By setting $x = 0$ and solving $y - 2z = 6$, $-3y + 4z = 10$, we obtain $y = -22$, $z = -14$. Consequently, $(0, -22, -14)$ is a point on the line. To find a vector along the line, we could find another point on the line, say, by setting $z = 0$ and solving $x + y = 6$, $2x - 3y = 10$ for $x = \frac{28}{5}$, $y = \frac{2}{5}$. A vector along the line is therefore $(\frac{28}{5}, \frac{2}{5}, 0) - (0, -22, -14) = (\frac{28}{5}, \frac{112}{5}, 14)$, and so too is $(\frac{5}{14})(\frac{28}{5}, \frac{112}{5}, 14) = (2, 8, 5)$.

Alternatively, we know that $(1, 1, -2)$ and $(2, -3, 4)$ are vectors that are normal to the planes $x + y - 2z = 6$ and $2x - 3y + 4z = 10$, and a vector along the line of intersection of the planes must be perpendicular to both of these vectors (Fig. 12.93). Consequently, a vector along the line of intersection is

$$\begin{vmatrix} \hat{\mathbf{i}} & \hat{\mathbf{j}} & \hat{\mathbf{k}} \\ 1 & 1 & -2 \\ 2 & -3 & 4 \end{vmatrix} = (-2, -8, -5).$$

Symmetric equations for the line are therefore

$$\frac{x}{2} = \frac{y+22}{8} = \frac{z+14}{5}. \square$$

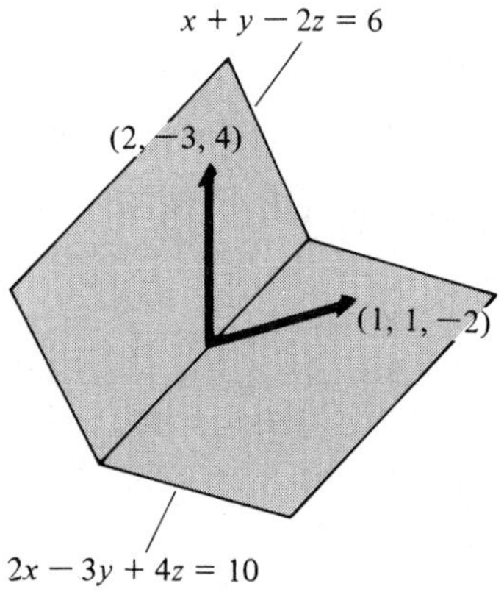

Figure 12.93

Example 25 Find the distance from the point (1, 3, 6) to the line

$$l{:}\quad x - 1 = \frac{y-2}{2} = \frac{z-4}{3}.$$

Solution Clearly $Q(1, 2, 4)$ is a point on the line, and therefore the required distance d is the component of $\overrightarrow{\mathbf{PQ}}$ in the direction $\overrightarrow{\mathbf{PR}}$ (Fig. 12.94). By Eq. (12.33), then, $d = \overrightarrow{\mathbf{PQ}} \cdot \widehat{\mathbf{PR}}$. To find $\widehat{\mathbf{PR}}$ we need a vector in the direction $\overrightarrow{\mathbf{PR}}$. This is not nearly as simple as finding $\overrightarrow{\mathbf{PR}}$ for a line in the xy-plane. Since $S(0, 0, 1)$ is also a point on l, $\overrightarrow{\mathbf{QS}} = (-1, -2, -3)$ is a vector along l. But then

$$\overrightarrow{\mathbf{QS}} \times \overrightarrow{\mathbf{PQ}} = \begin{vmatrix} \hat{\mathbf{i}} & \hat{\mathbf{j}} & \hat{\mathbf{k}} \\ -1 & -2 & -3 \\ 0 & -1 & -2 \end{vmatrix} = \hat{\mathbf{i}} - 2\hat{\mathbf{j}} + \hat{\mathbf{k}}$$

is perpendicular to both $\overrightarrow{\mathbf{PQ}}$ and $\overrightarrow{\mathbf{QS}}$. It now follows that a vector in the direction $\overrightarrow{\mathbf{PR}}$ must be

$$(\overrightarrow{\mathbf{QS}} \times \overrightarrow{\mathbf{PQ}}) \times \overrightarrow{\mathbf{QS}} = \begin{vmatrix} \hat{\mathbf{i}} & \hat{\mathbf{j}} & \hat{\mathbf{k}} \\ 1 & -2 & 1 \\ -1 & -2 & -3 \end{vmatrix} = 8\hat{\mathbf{i}} + 2\hat{\mathbf{j}} - 4\hat{\mathbf{k}}.$$

Finally, then,

$$\widehat{\mathbf{PR}} = \frac{(8, 2, -4)}{\sqrt{64+4+16}} = \frac{(4, 1, -2)}{\sqrt{21}}$$

and

$$d = (0, -1, -2) \cdot \frac{(4, 1, -2)}{\sqrt{21}} = \frac{\sqrt{21}}{7}.$$

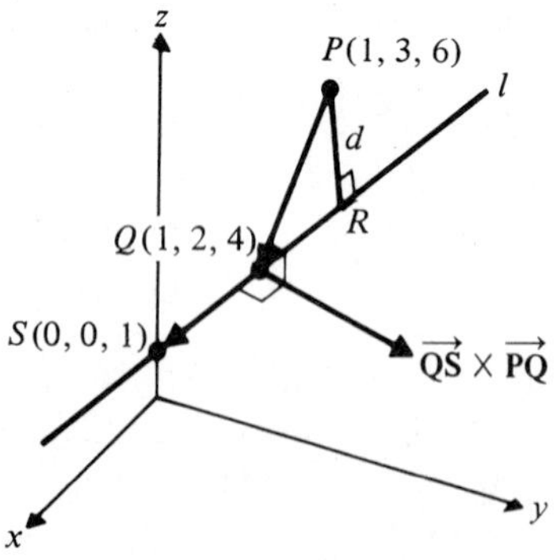

Figure 12.94

An alternative procedure is to note in Fig. 12.95 that $d = |\mathbf{PQ}| \sin\theta$. But the right-hand side of this equation looks like the length of the cross-product of $\overrightarrow{\mathbf{PQ}}$ and a vector that makes an angle θ with $\overrightarrow{\mathbf{PQ}}$. Since $\overrightarrow{\mathbf{RQ}}$ makes an angle θ with $\overrightarrow{\mathbf{PQ}}$, we could write that

$$d = |\overrightarrow{\mathbf{PQ}}||\widehat{\mathbf{RQ}}| \sin\theta = |\overrightarrow{\mathbf{PQ}} \times \widehat{\mathbf{RQ}}|.$$

Now $\overrightarrow{\mathbf{PQ}} = (0, -1, -2)$ and a vector along l is $(1, 2, 3)$. Thus $\widehat{\mathbf{RQ}} = (-1, -2, -3)/\sqrt{14}$ and

$$\overrightarrow{\mathbf{PQ}} \times \widehat{\mathbf{RQ}} = \begin{vmatrix} \hat{\mathbf{i}} & \hat{\mathbf{j}} & \hat{\mathbf{k}} \\ 0 & -1 & -2 \\ -1/\sqrt{14} & -2/\sqrt{14} & -3/\sqrt{14} \end{vmatrix} = \frac{1}{\sqrt{14}}(-1, 2, -1).$$

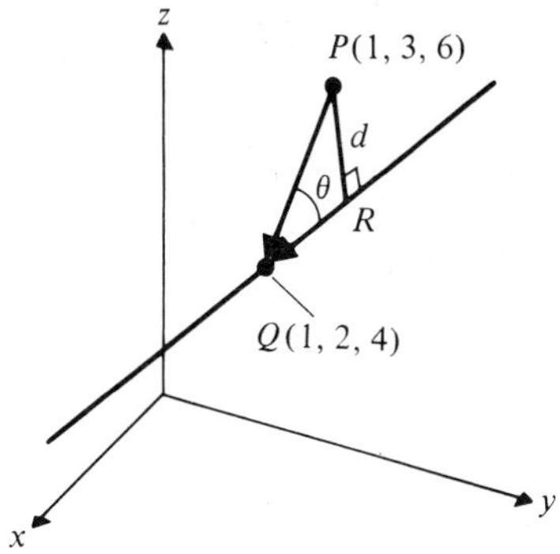

Figure 12.95

Finally,

$$d = \frac{1}{\sqrt{14}}\sqrt{1+4+1} = \frac{\sqrt{21}}{7}. \square$$

This example suggests the following formula for the area of a triangle.

▶ **Theorem 12.6** If A, B, and C are the vertices of a triangle, then

$$\text{Area of } \Delta ABC = \tfrac{1}{2}|\overrightarrow{\mathbf{AB}} \times \overrightarrow{\mathbf{AC}}|. \blacktriangleleft \qquad (12.44)$$

Proof The area of triangle ABC in Fig. 12.96 is

$$\tfrac{1}{2}|\overrightarrow{\mathbf{AC}}||\overrightarrow{\mathbf{BD}}| = \tfrac{1}{2}|\overrightarrow{\mathbf{AC}}||\overrightarrow{\mathbf{AB}}|\sin\theta = \tfrac{1}{2}|\overrightarrow{\mathbf{AB}} \times \overrightarrow{\mathbf{AC}}|. \square$$

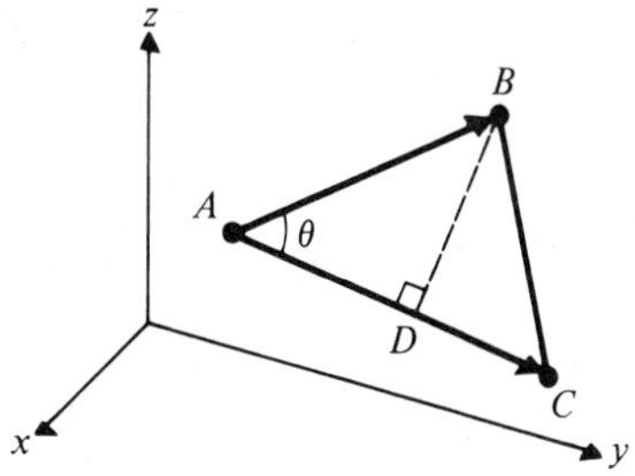

Figure 12.96

For the triangle with vertices $A(1, 1, 1)$, $B(2, -3, 2)$, and $C(4, 1, 5)$ in Fig. 12.97,

$$\overrightarrow{\mathbf{AB}} \times \overrightarrow{\mathbf{AC}} = \begin{vmatrix} \hat{\mathbf{i}} & \hat{\mathbf{j}} & \hat{\mathbf{k}} \\ 1 & -4 & 1 \\ 3 & 0 & 4 \end{vmatrix} = (-16, -1, 12).$$

The area of the triangle is therefore

$$\tfrac{1}{2}|(-16, -1, 12)| = \tfrac{1}{2}\sqrt{256+1+144} = \tfrac{1}{2}\sqrt{401}.$$

The following corollary is a direct consequence.

▶ **Corollary** The area of a parallelogram with coterminal sides $\overrightarrow{\mathbf{AB}}$ and $\overrightarrow{\mathbf{AC}}$ (Fig. 12.98) is

$$\text{Area} = |\overrightarrow{\mathbf{AB}} \times \overrightarrow{\mathbf{AC}}|. \blacktriangleleft \qquad (12.45)$$

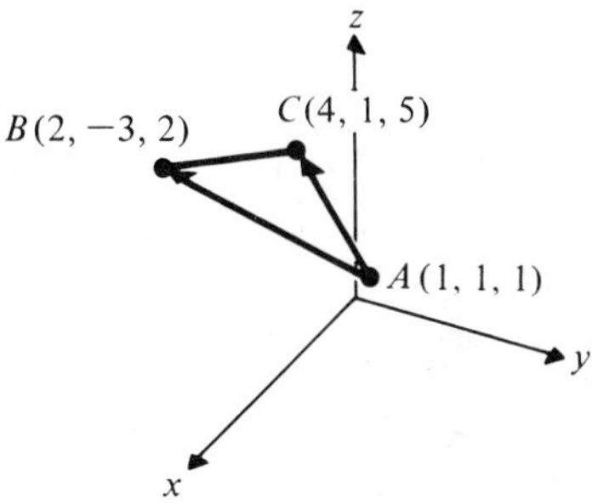

Figure 12.97

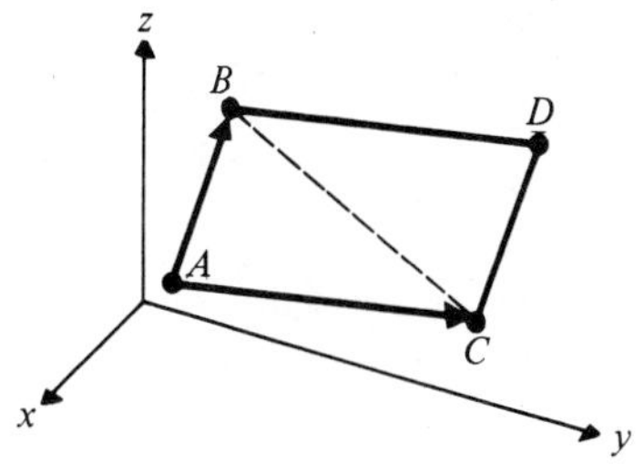

Figure 12.98

EXERCISES 12.6

If $\mathbf{u} = (3, 1, 4)$, $\mathbf{v} = (-1, 2, 0)$, and $\mathbf{w} = (-2, -3, 5)$, evaluate the scalar or find the components of the vector in each of Exercises 1–10.

1. $\mathbf{v} \times \mathbf{w}$
2. $(-3\mathbf{u}) \times (2\mathbf{v})$
3. $\mathbf{u} \cdot (\mathbf{v} \times \mathbf{w})$
4. $\hat{\mathbf{u}} \times \hat{\mathbf{w}}$
5. $(3\mathbf{u}) \times \mathbf{w} + \mathbf{u} \times \mathbf{v}$
6. $\mathbf{u} \times (3\mathbf{v} - \mathbf{w})$
7. $\dfrac{\mathbf{w} \times \mathbf{u}}{|\mathbf{u} \times \mathbf{v}|}$
8. $\mathbf{u} \times \mathbf{w} - \mathbf{u} \times \mathbf{v} + \mathbf{u} \times (2\mathbf{u} + \mathbf{v})$
9. $(\mathbf{u} \times \mathbf{v}) \times \mathbf{w}$
10. $\mathbf{u} \times (\mathbf{v} \times \mathbf{w})$

In Exercises 11–14, find components for the given vector.

11. Perpendicular to the vectors (1, 3, 5) and (−2, 1, 4)

12. Perpendicular to the y-axis and the vector joining the points $(2, 4, -3)$ and $(1, 5, 6)$
13. Perpendicular to the plane containing the lines $x = 1 + 3t$, $y = 2 - 4t$, $z = -1 + 6t$, and $x = 1 + 5t$, $y = 2 - t$, $z = -1$
14. Perpendicular to the triangle with vertices $(-1, 0, 3)$, $(5, 1, 2)$, and $(-6, 2, 4)$
15. What is the area of the triangle in Exercise 14?
16. Verify the results in Eqs. (12.38).
17. Find the area of the parallelogram with vertices $(1, 2, 3)$, $(4, 3, 7)$, $(-1, 3, 6)$, and $(2, 4, 10)$.

In Exercises 18–27, find vector, parametric, and symmetric (if possible) equations for the straight line.

18. Through the point $(1, -1, 3)$ and parallel to the vector $(2, 4, -3)$
19. Through the point $(-1, 3, 6)$ and parallel to the vector $(2, -3, 0)$
20. Through the points $(2, -3, 4)$ and $(5, 2, -1)$
21. Through the points $(-2, 3, 3)$ and $(-2, -3, -3)$
22. Through the points $(1, 3, 4)$ and $(1, 3, 5)$
23. Through the point $(1, -3, 5)$ and parallel to the line

$$\frac{x}{5} = \frac{y - 2}{3} = \frac{z + 4}{-2}$$

24. Through the point $(2, 0, 3)$ and parallel to the line $x = 4 + t$, $y = 2$, $z = 6 - 2t$
25. Through the point of intersection of the lines

$$\frac{x - 1}{2} = \frac{y + 4}{-3} = \frac{z - 2}{5};$$

$$\frac{x - 1}{6} = \frac{y + 4}{3} = \frac{z - 2}{4};$$

and parallel to the line joining the points $(1, 3, -2)$ and $(2, -2, 1)$

26. $2x - y = 5$, $3x + 4y + z = 10$
27. Through the point $(-2, 3, 1)$ and parallel to the line $x + y = 3$, $2x - y + z = -2$

In Exercises 28–35, find an equation for the given plane.

28. Containing the points $(1, 3, 2)$, $(-2, 0, -2)$, $(1, 4, 3)$
29. Containing the point $(2, -4, 3)$ and the line $(x - 1)/3 = (y + 5)/4 = z + 2$
30. Containing the lines $x = 2y = (z + 1)/4$ and $x = t$, $y = 2t$, $z = 6t - 1$
31. Containing the lines $(x - 1)/6 = y/8 = (z + 2)/2$ and $(x + 1)/3 = (y - 2)/4 = z + 5$
32. Containing the line $x - y + 2z = 4$, $2x + y + 3z = 6$ and the point $(1, -2, 4)$
33. Containing the two lines

$$x + 2y + 4z = 21,$$
$$x - y + 6z = 13,$$

and

$$x = 2 + 3t, \quad y = 4, \quad z = -3 + 5t$$

34. Containing the two lines

$$3x + 4y = -6, \quad x + 2y + z = 2, \qquad \text{and} \qquad 2y + 3z = 19, \quad 3x - 2y - 9z = -58$$

35. Containing the line $x + y - 4z = 6$, $2x + 3y + 5z = 10$ and:
    a) perpendicular to the xy-plane;
    b) perpendicular to the xz-plane;
    c) perpendicular to the yz-plane.
36. Does the line $(x - 3)/2 = y - 2 = (z + 1)/4$ lie in the plane $x - y + 2z = -1$?
37. Show that the lines joining the midpoints of the sides of any quadrilateral form a parallelogram.
38. Show that if a plane has nonzero intercepts a, b, and c on the x-, y-, and z-axes, then its equation is

$$\frac{x}{a} + \frac{y}{b} + \frac{z}{c} = 1.$$

39. Find equations for the four faces of the tetrahedron in Fig. 12.10.
40. Find equations for the nine planes forming the sides, bottom, and roof of the birdhouse shown in Fig. 12.8.
41. Show that the cross-product is not associative; i.e., in general,

$$\mathbf{u} \times (\mathbf{v} \times \mathbf{w}) \neq (\mathbf{u} \times \mathbf{v}) \times \mathbf{w}.$$

42. Show that the three lines

$$\frac{x - 4}{3} = \frac{y - 8}{4} = \frac{z + 7}{-4};$$

$$\frac{x + 5}{3} = \frac{y + 2}{2} = \frac{z + 1}{2};$$

$$x = 1, \quad y = 5 + t, \quad z = -6 - 3t$$

form a triangle, and find its area.

If a force $\mathbf{F}$ acts at a point Q (Fig. 12.99), then the moment of $\mathbf{F}$ about a point P is defined as the vector $\mathbf{M} = \mathbf{r} \times \mathbf{F}$, where $\mathbf{r} = \overrightarrow{PQ}$. In Exercises 43–47, calculate the moment for the force about the given point.

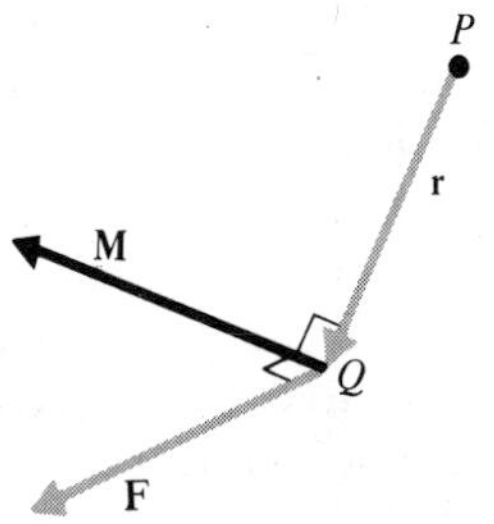

Figure 12.99

43. $\mathbf{F} = 2\hat{\mathbf{i}} + 3\hat{\mathbf{j}} - 4\hat{\mathbf{k}}$ at $(1, 3, 2)$ about $(-1, 4, 2)$
44. $\mathbf{F} = \hat{\mathbf{i}} + 2\hat{\mathbf{j}}$ at $(1, 1, 0)$ about $(2, 1, -5)$
45. $\mathbf{F} = -\hat{\mathbf{i}} + 3\hat{\mathbf{k}}$ at $(0, 0, 0)$ about $(-1, 3, 0)$
46. $\mathbf{F} = 3\hat{\mathbf{i}} - \hat{\mathbf{j}} + 4\hat{\mathbf{k}}$ at $(1, 1, 1)$ about $(2, 2, 2)$
47. $\mathbf{F} = 6\hat{\mathbf{i}}$ at $(0, 1, 3)$ about $(2, 0, 0)$
48. a) If $\mathbf{u} \neq \mathbf{0}$, show that if both the conditions $\mathbf{u} \cdot \mathbf{v} = \mathbf{u} \cdot \mathbf{w}$ and $\mathbf{u} \times \mathbf{v} = \mathbf{u} \times \mathbf{w}$ are satisfied, then $\mathbf{v} = \mathbf{w}$.

    b) Show that if one of the conditions in (a) is satisfied, but the other is not, then $\mathbf{v}$ cannot be equal to $\mathbf{w}$.

49. The scalar $\mathbf{u} \cdot \mathbf{v} \times \mathbf{w}$ is called the *scalar triple product* of $\mathbf{u}$, $\mathbf{v}$, and $\mathbf{w}$.

    a) Find $\mathbf{u} \cdot \mathbf{v} \times \mathbf{w}$ if $\mathbf{u} = (6, -1, 0)$, $\mathbf{v} = (1, 3, 4)$, and $\mathbf{w} = (-2, -1, 4)$.

    b) Prove that
    $$\mathbf{u} \cdot \mathbf{v} \times \mathbf{w} = \mathbf{u} \times \mathbf{v} \cdot \mathbf{w}.$$

    c) Show that $|\mathbf{u} \cdot \mathbf{v} \times \mathbf{w}|$ can be interpreted as the volume of the parallelepiped with $\mathbf{u}$, $\mathbf{v}$, and $\mathbf{w}$ as coterminal sides (Fig. 12.100).

    d) Verify that three nonzero vectors $\mathbf{u}$, $\mathbf{v}$, and $\mathbf{w}$ all lie in the same plane if and only if $\mathbf{u} \cdot \mathbf{v} \times \mathbf{w} = 0$.

In Exercises 50 and 51, prove the identity.

50. $(\mathbf{u} \times \mathbf{v}) \cdot (\mathbf{w} \times \mathbf{r}) = (\mathbf{u} \cdot \mathbf{w})(\mathbf{v} \cdot \mathbf{r}) - (\mathbf{u} \cdot \mathbf{r})(\mathbf{v} \cdot \mathbf{w})$

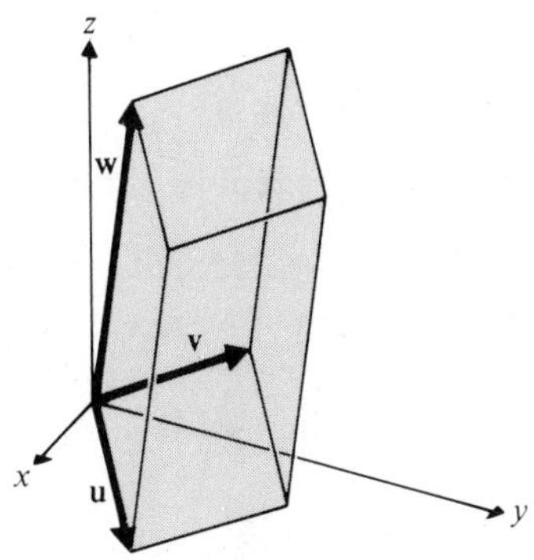

Figure 12.100

51. $\mathbf{u} \times (\mathbf{v} \times \mathbf{w}) = (\mathbf{u} \cdot \mathbf{w})\mathbf{v} - (\mathbf{u} \cdot \mathbf{v})\mathbf{w}$
52. Verify that the equation of the plane passing through the three points $P_1(x_1, y_1, z_1)$, $P_2(x_2, y_2, z_2)$, and $P_3(x_3, y_3, z_3)$ can be written in the form
$$\overrightarrow{\mathbf{P_1P}} \cdot \overrightarrow{\mathbf{P_1P_2}} \times \overrightarrow{\mathbf{P_1P_3}} = 0,$$
where $P(x, y, z)$ is any point in the plane.
53. Use vectors to prove the sine law
$$\frac{\sin A}{a} = \frac{\sin B}{b} = \frac{\sin C}{c}$$
for the triangle in Fig. 12.101. (*Hint:* Note that $\overrightarrow{\mathbf{PQ}} + \overrightarrow{\mathbf{QR}} + \overrightarrow{\mathbf{RP}} = \mathbf{0}$. Cross this equation with $\overrightarrow{\mathbf{PQ}}$ and take lengths.)

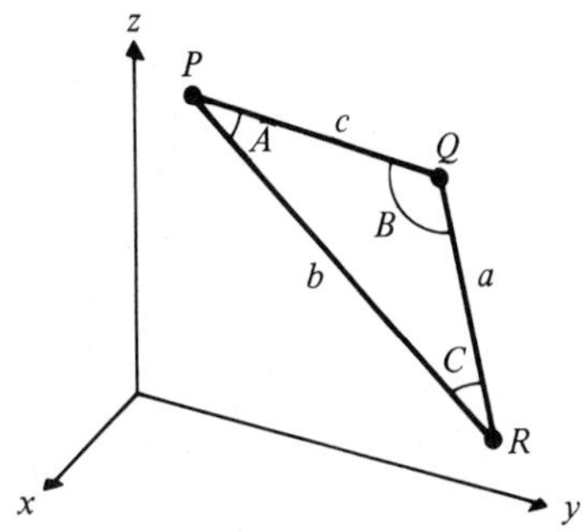

Figure 12.101

54. The region S_{xy} in the xy-plane bounded by the straight lines $x = 0$, $x = 1$, $y = 0$, $y = 1$ is a rectangle with unit area.

    a) Show that the region in the plane $y = z$ that projects onto S_{xy} is also a rectangle, but with area $\sqrt{2}$.

    b) Show that the region in the plane $x + y - 2z = 0$ that projects onto S_{xy} is a parallelogram with area $\sqrt{6}/2$.

    *c) Generalize the results of (a) and (b) to show that if S is the area in a plane $Ax + By + Cz + D = 0$ $(C \neq 0)$ that projects onto S_{xy}, then the area of S is $\sec \gamma$, where γ is the acute angle between the given plane and the xy-plane.

*55. Find the shortest distance between the lines
$$\frac{x - 1}{2} = \frac{y + 3}{3} = 4 - z;$$
$$x = -1 + t, \quad y = 2t, \quad z = 3 - 2t.$$

12.7 DIFFERENTIATION AND INTEGRATION OF VECTORS

In Sections 12.5 and 12.6 we used vectors to represent forces and moments; we will also use them to describe many other physical quantities, such as position, velocity, acceleration, electric and magnetic fields, and fluid flow. In applications such as these, vectors seldom have constant components; instead they have components that are either functions of position, or functions of some parameter, or both. For instance, the spring force in Example 20 varies in both magnitude and direction as the sleeve moves from B to C. Consequently, components of the vector $\mathbf{F}$ representing this force are functions of position x between B and C:

$$\mathbf{F} = F_x\hat{\mathbf{i}} + F_y\hat{\mathbf{j}} = F_x(x)\hat{\mathbf{i}} + F_y(x)\hat{\mathbf{j}}.$$

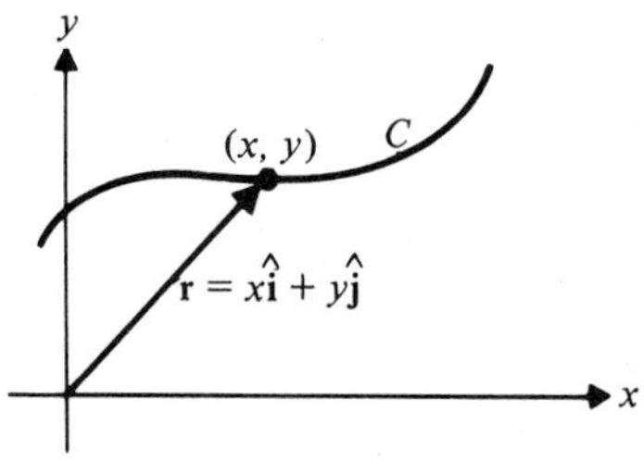

Figure 12.102

When a particle moves along a curve C in the xy-plane defined parametrically by

$$C:\quad x = x(t),\quad y = y(t),\qquad \alpha \le t \le \beta,$$

its position (x, y) relative to the origin is represented by the vector $\mathbf{r} = x\hat{\mathbf{i}} + y\hat{\mathbf{j}}$ (Fig. 12.102). This vector is called the *position*, or *displacement*, *vector* of the particle relative to the origin. We will have more to say about it in Section 12.11. Note, however, that if we substitute from the parametric equations for C, we have

$$\mathbf{r} = x(t)\hat{\mathbf{i}} + y(t)\hat{\mathbf{j}},$$

which indicates that the displacement vector has components that are functions of the parameter t.

In this section we consider the general situation in which the components v_x, v_y, and v_z of a vector $\mathbf{v}$ are functions of some parameter t,

$$\mathbf{v} = v_x(t)\hat{\mathbf{i}} + v_y(t)\hat{\mathbf{j}} + v_z(t)\hat{\mathbf{k}}, \tag{12.46}$$

and show how the operations of differentiation and integration can be applied to such a vector. In Sections 12.9 and 12.10 we will use these results to discuss the geometry of curves, and this will pave the way for an analysis of the motion of particles in Section 12.11.

Because $\mathbf{v}$ in (12.46) has components that are functions of t, we say that $\mathbf{v}$ itself is a function of t, and write $\mathbf{v} = \mathbf{v}(t)$. Each of the component functions $v_x(t)$, $v_y(t)$, and $v_z(t)$ has a domain, and their common domain is called the domain of the vector function $\mathbf{v}(t)$. Given that this domain is some interval $\alpha \le t \le \beta$, we express (12.46) more fully in the form

$$\mathbf{v} = \mathbf{v}(t) = v_x(t)\hat{\mathbf{i}} + v_y(t)\hat{\mathbf{j}} + v_z(t)\hat{\mathbf{k}},\qquad \alpha \le t \le \beta. \tag{12.47}$$

To differentiate and integrate vector functions, we first require the concept of a limit.

▷ **Definition 12.11** If $\mathbf{v}(t) = v_x(t)\hat{\mathbf{i}} + v_y(t)\hat{\mathbf{j}} + v_z(t)\hat{\mathbf{k}}$, then

$$\lim_{t\to t_0} \mathbf{v}(t) = \left(\lim_{t\to t_0} v_x(t)\right)\hat{\mathbf{i}} + \left(\lim_{t\to t_0} v_y(t)\right)\hat{\mathbf{j}} + \left(\lim_{t\to t_0} v_z(t)\right)\hat{\mathbf{k}}, \tag{12.48}$$

provided each of the limits on the right exists. ◁

This definition states that to take the limit of a vector function, we take the limit of each component separately. As an illustration, consider the following example.

Example 26 If $\mathbf{v} = \mathbf{v}(t) = (t^2 + 1)\hat{\mathbf{i}} + 3t\hat{\mathbf{j}} - (\sin t)\hat{\mathbf{k}}$, calculate $\lim_{t\to 5} \mathbf{v}(t)$.

Solution According to Definition 12.11,

$$\lim_{t\to 5} \mathbf{v}(t) = \left(\lim_{t\to 5} (t^2 + 1)\right)\hat{\mathbf{i}} + \left(\lim_{t\to 5} (3t)\right)\hat{\mathbf{j}} - \left(\lim_{t\to 5} \sin t\right)\hat{\mathbf{k}}$$
$$= 26\hat{\mathbf{i}} + 15\hat{\mathbf{j}} - (\sin 5)\hat{\mathbf{k}}. \square$$

In the remainder of this section we define continuity, derivatives, and antiderivatives for vector functions. Each definition is an exact duplicate of the corresponding definition for a scalar function $y(t)$, except that $y(t)$ is replaced by $\mathbf{v}(t)$. We then show that the vector definition can be rephrased in terms of components of the vector. We begin with continuity in the following definition.

▷ **Definition 12.12** A vector function $\mathbf{v}(t)$ is said to be *continuous* at a point t_0 if

$$\mathbf{v}(t_0) = \lim_{t\to t_0} \mathbf{v}(t). \quad \triangleleft \tag{12.49}$$

It is a simple matter to prove this theorem.

▶ **Theorem 12.7** A vector function is continuous at a point if and only if its components are continuous at that point. ◀

Proof If $\mathbf{v}(t) = v_x(t)\hat{\mathbf{i}} + v_y(t)\hat{\mathbf{j}} + v_z(t)\hat{\mathbf{k}}$, then according to (12.49), $\mathbf{v}(t)$ is continuous at t_0 if and only if

$$v_x(t_0)\hat{\mathbf{i}} + v_y(t_0)\hat{\mathbf{j}} + v_z(t_0)\hat{\mathbf{k}} = \lim_{t\to t_0} \{v_x(t)\hat{\mathbf{i}} + v_y(t)\hat{\mathbf{j}} + v_z(t)\hat{\mathbf{k}}\}.$$

Definition 12.11 implies that we can write this condition in the form

$$v_x(t_0)\hat{\mathbf{i}} + v_y(t_0)\hat{\mathbf{j}} + v_z(t_0)\hat{\mathbf{k}} = \left(\lim_{t\to t_0} v_x(t)\right)\hat{\mathbf{i}} + \left(\lim_{t\to t_0} v_y(t)\right)\hat{\mathbf{j}} + \left(\lim_{t\to t_0} v_z(t)\right)\hat{\mathbf{k}}.$$

But because two vectors are equal if and only if their components are equal, we can say that $\mathbf{v}(t)$ is continuous at t_0 if and only if

$$v_x(t_0) = \lim_{t\to t_0} v_x(t); \qquad v_y(t_0) = \lim_{t\to t_0} v_y(t); \qquad v_z(t_0) = \lim_{t\to t_0} v_z(t);$$

i.e., $\mathbf{v}(t)$ is continuous at t_0 if and only if its components are continuous at t_0. $\square$

For example, the vector function

$$\mathbf{v}(t) = (t - 1)\hat{\mathbf{i}} + (1/t)\hat{\mathbf{j}} + (t^2 - 1)^{-1}\hat{\mathbf{k}}$$

is discontinuous for $t = 0$ (since $v_y(0)$ is not defined) and for $t = \pm 1$ (since $v_z(\pm 1)$ is not defined).

The derivative of a scalar function $y(t)$ is its instantaneous rate of change

$$\frac{dy}{dt} = \lim_{\Delta t\to 0} \frac{y(t + \Delta t) - y(t)}{\Delta t}.$$

The derivative of a vector function is also a rate of change defined by a similar limit.

▷ **Definition 12.13** The derivative of a vector function $\mathbf{v}(t)$ is defined as

$$\frac{d\mathbf{v}}{dt} = \lim_{\Delta t \to 0} \frac{\mathbf{v}(t + \Delta t) - \mathbf{v}(t)}{\Delta t}, \tag{12.50}$$

provided the limit exists. ◁

In practice, we seldom use the definition of a derivative to calculate dy/dt for a scalar function $y(t)$; formulas such as the power, product, quotient, and chain rules are more convenient. It would be helpful to have corresponding formulas for derivatives of vector functions. The following theorem shows that to differentiate a vector function we simply differentiate its Cartesian components.

▶ **Theorem 12.8** If $\mathbf{v}(t) = v_x(t)\hat{\mathbf{i}} + v_y(t)\hat{\mathbf{j}} + v_z(t)\hat{\mathbf{k}}$, and $d\mathbf{v}/dt$ exists, then

$$\frac{d\mathbf{v}}{dt} = \frac{dv_x}{dt}\hat{\mathbf{i}} + \frac{dv_y}{dt}\hat{\mathbf{j}} + \frac{dv_z}{dt}\hat{\mathbf{k}}. \blacktriangleleft \tag{12.51}$$

Proof If we substitute the components of $\mathbf{v}(t + \Delta t)$ and $\mathbf{v}(t)$ into Definition 12.13, then we have

$$\begin{aligned}\frac{d\mathbf{v}}{dt} &= \lim_{\Delta t \to 0} \left\{\frac{[v_x(t+\Delta t)\hat{\mathbf{i}} + v_y(t+\Delta t)\hat{\mathbf{j}} + v_z(t+\Delta t)\hat{\mathbf{k}}] - [v_x(t)\hat{\mathbf{i}} + v_y(t)\hat{\mathbf{j}} + v_z(t)\hat{\mathbf{k}}]}{\Delta t}\right\} \\ &= \lim_{\Delta t \to 0} \left\{\left[\frac{v_x(t+\Delta t) - v_x(t)}{\Delta t}\right]\hat{\mathbf{i}} + \left[\frac{v_y(t+\Delta t) - v_y(t)}{\Delta t}\right]\hat{\mathbf{j}} + \left[\frac{v_z(t+\Delta t) - v_z(t)}{\Delta t}\right]\hat{\mathbf{k}}\right\}\end{aligned}$$

(according to Eq. 12.17), and

$$\begin{aligned}\frac{d\mathbf{v}}{dt} &= \left(\lim_{\Delta t \to 0} \frac{v_x(t + \Delta t) - v_x(t)}{\Delta t}\right)\hat{\mathbf{i}} + \left(\lim_{\Delta t \to 0} \frac{v_y(t + \Delta t) - v_y(t)}{\Delta t}\right)\hat{\mathbf{j}} \\ &\quad + \left(\lim_{\Delta t \to 0} \frac{v_z(t + \Delta t) - v_z(t)}{\Delta t}\right)\hat{\mathbf{k}}\end{aligned}$$

(according to Eq. 12.48). If $d\mathbf{v}/dt$ exists, then each of the three limits on the right must exist, and we can write that

$$\frac{d\mathbf{v}}{dt} = \frac{dv_x}{dt}\hat{\mathbf{i}} + \frac{dv_y}{dt}\hat{\mathbf{j}} + \frac{dv_z}{dt}\hat{\mathbf{k}}. \square$$

This proof is also reversible, giving the following corollary to Theorem 12.8.

▶ **Corollary** If the derivatives dv_x/dt, dv_y/dt, and dv_z/dt of the components of a vector function $\mathbf{v}(t) = v_x(t)\hat{\mathbf{i}} + v_y(t)\hat{\mathbf{j}} + v_z(t)\hat{\mathbf{k}}$ exist at a point, then so too does $d\mathbf{v}/dt$. ◀

Theorem 12.8 gives us a working rule for differentiating vector functions: To differentiate a vector function, we differentiate its Cartesian components.

Example 27 If $\mathbf{v}(t) = t^2\hat{\mathbf{i}} + (3t^3 - 2t)\hat{\mathbf{j}} + 5\hat{\mathbf{k}}$, find $\mathbf{v}'(3)$.

Solution According to (12.51),

$$\frac{d\mathbf{v}}{dt} = 2t\hat{\mathbf{i}} + (9t^2 - 2)\hat{\mathbf{j}}.$$

Consequently,

$$\mathbf{v}'(3) = 6\hat{\mathbf{i}} + 79\hat{\mathbf{j}}. \square$$

The sum rule (3.8) for differentiation of scalar functions has its counterpart in the sum rule for vector functions,

$$\frac{d}{dt}(\mathbf{u} + \mathbf{v}) = \frac{d\mathbf{u}}{dt} + \frac{d\mathbf{v}}{dt} \tag{12.52}$$

(see Exercise 22). There are three types of products associated with vectors: the product of a scalar and a vector, the dot product of two vectors, and the cross product of two vectors. Corresponding to each, we have a product rule for differentiation, but all resemble the product rule for scalar functions.

▶ **Theorem 12.9** If $f(t)$ is a differentiable function and $\mathbf{u}(t)$ and $\mathbf{v}(t)$ are differentiable vector functions, then:

$$\frac{d}{dt}(f\mathbf{v}) = \frac{df}{dt}\mathbf{v} + f\frac{d\mathbf{v}}{dt}, \tag{12.53}$$

$$\frac{d}{dt}(\mathbf{u} \cdot \mathbf{v}) = \mathbf{u} \cdot \frac{d\mathbf{v}}{dt} + \frac{d\mathbf{u}}{dt} \cdot \mathbf{v}, \tag{12.54}$$

$$\frac{d}{dt}(\mathbf{u} \times \mathbf{v}) = \mathbf{u} \times \frac{d\mathbf{v}}{dt} + \frac{d\mathbf{u}}{dt} \times \mathbf{v}. \blacktriangleleft \tag{12.55}$$

For a proof of these results, see Exercise 23.

Example 28 If $f(t) = t^2 + 2t + 3$, $\mathbf{u}(t) = t\hat{\mathbf{i}} + t^2\hat{\mathbf{j}} - 3\hat{\mathbf{k}}$, and $\mathbf{v}(t) = t(\hat{\mathbf{i}} + \hat{\mathbf{j}} + \hat{\mathbf{k}})$, use (12.53)–(12.55) to evaluate:

a) $\dfrac{d}{dt}(f\mathbf{u})$; b) $\dfrac{d}{dt}(\mathbf{u} \cdot \mathbf{v})$; c) $\dfrac{d}{dt}(\mathbf{u} \times \mathbf{v})$.

Solution a) With (12.53),

$$\frac{d}{dt}(f\mathbf{u}) = \frac{df}{dt}\mathbf{u} + f\frac{d\mathbf{u}}{dt} = (2t + 2)(t\hat{\mathbf{i}} + t^2\hat{\mathbf{j}} - 3\hat{\mathbf{k}}) + (t^2 + 2t + 3)(\hat{\mathbf{i}} + 2t\hat{\mathbf{j}})$$

$$= (3t^2 + 4t + 3)\hat{\mathbf{i}} + (4t^3 + 6t^2 + 6t)\hat{\mathbf{j}} - 6(t + 1)\hat{\mathbf{k}}.$$

b) With (12.54),

$$\frac{d}{dt}(\mathbf{u} \cdot \mathbf{v}) = \mathbf{u} \cdot \frac{d\mathbf{v}}{dt} + \frac{d\mathbf{u}}{dt} \cdot \mathbf{v}$$

$$= (t\hat{\mathbf{i}} + t^2\hat{\mathbf{j}} - 3\hat{\mathbf{k}}) \cdot (\hat{\mathbf{i}} + \hat{\mathbf{j}} + \hat{\mathbf{k}}) + (\hat{\mathbf{i}} + 2t\hat{\mathbf{j}}) \cdot (t\hat{\mathbf{i}} + t\hat{\mathbf{j}} + t\hat{\mathbf{k}})$$

$$= (t + t^2 - 3) + (t + 2t^2) = 3t^2 + 2t - 3.$$

c) With (12.55),

$$\begin{aligned}\frac{d}{dt}(\mathbf{u}\times\mathbf{v}) &= \mathbf{u}\times\frac{d\mathbf{v}}{dt}+\frac{d\mathbf{u}}{dt}\times\mathbf{v} = (t\hat{\mathbf{i}}+t^2\hat{\mathbf{j}}-3\hat{\mathbf{k}})\times(\hat{\mathbf{i}}+\hat{\mathbf{j}}+\hat{\mathbf{k}})\\ &\quad + (\hat{\mathbf{i}}+2t\hat{\mathbf{j}})\times(t\hat{\mathbf{i}}+t\hat{\mathbf{j}}+t\hat{\mathbf{k}})\\ &= \begin{vmatrix}\hat{\mathbf{i}} & \hat{\mathbf{j}} & \hat{\mathbf{k}}\\ t & t^2 & -3\\ 1 & 1 & 1\end{vmatrix} + \begin{vmatrix}\hat{\mathbf{i}} & \hat{\mathbf{j}} & \hat{\mathbf{k}}\\ 1 & 2t & 0\\ t & t & t\end{vmatrix}\\ &= [(t^2+3)\hat{\mathbf{i}}-(3+t)\hat{\mathbf{j}}+(t-t^2)\hat{\mathbf{k}}]\\ &\quad + [2t^2\hat{\mathbf{i}}-t\hat{\mathbf{j}}+(t-2t^2)\hat{\mathbf{k}}]\\ &= (3t^2+3)\hat{\mathbf{i}}-(3+2t)\hat{\mathbf{j}}+(2t-3t^2)\hat{\mathbf{k}}.\ \square\end{aligned}$$

If vector functions can be differentiated, then they can be antidifferentiated. Formally we make the following statement.

▷ **Definition 12.14** A vector function $\mathbf{V}(t)$ is said to be an antiderivative or indefinite integral of $\mathbf{v}(t)$ on the interval $\alpha \le t \le \beta$ if

$$\frac{d\mathbf{V}}{dt} = \mathbf{v}(t) \quad \text{for } \alpha \le t \le \beta. \quad ◁ \tag{12.56}$$

For example, an antiderivative of $\mathbf{v}(t) = 2t\hat{\mathbf{i}} - \hat{\mathbf{j}} + 3t^2\hat{\mathbf{k}}$ is

$$\mathbf{V}(t) = t^2\hat{\mathbf{i}} - t\hat{\mathbf{j}} + t^3\hat{\mathbf{k}}.$$

If we add to $\mathbf{V}(t)$ in (12.56) any vector with constant components, denoted by $\mathbf{C}$, then $\mathbf{V}(t) + \mathbf{C}$ is also an antiderivative of $\mathbf{v}(t)$. We call this vector *the* antiderivative or *the* indefinite integral of $\mathbf{v}(t)$, and write

$$\int \mathbf{v}(t)\,dt = \mathbf{V}(t) + \mathbf{C}. \tag{12.57}$$

For our example, then,

$$\int (2t\hat{\mathbf{i}} - \hat{\mathbf{j}} + 3t^2\hat{\mathbf{k}})\,dt = t^2\hat{\mathbf{i}} - t\hat{\mathbf{j}} + t^3\hat{\mathbf{k}} + \mathbf{C}.$$

Because vectors can be differentiated component by component, it follows that they may also be integrated component by component; i.e., if $\mathbf{v}(t) = v_x(t)\hat{\mathbf{i}} + v_y(t)\hat{\mathbf{j}} + v_z(t)\hat{\mathbf{k}}$, then

$$\int \mathbf{v}(t)\,dt = \left(\int v_x(t)\,dt\right)\hat{\mathbf{i}} + \left(\int v_y(t)\,dt\right)\hat{\mathbf{j}} + \left(\int v_z(t)\,dt\right)\hat{\mathbf{k}}. \tag{12.58}$$

Example 29 Find the antiderivative of $\mathbf{v}(t) = \sqrt{t-1}\,\hat{\mathbf{i}} + e^t\hat{\mathbf{j}} + 6t^2\hat{\mathbf{k}}$.

Solution According to Eq. (12.58),

$$\begin{aligned}\int \mathbf{v}(t)\,dt &= \left(\frac{2}{3}(t-1)^{3/2} + C_1\right)\hat{\mathbf{i}} + (e^t + C_2)\hat{\mathbf{j}} + (2t^3 + C_3)\hat{\mathbf{k}}\\ &= \frac{2}{3}(t-1)^{3/2}\hat{\mathbf{i}} + e^t\hat{\mathbf{j}} + 2t^3\hat{\mathbf{k}} + \mathbf{C},\end{aligned}$$

where $\mathbf{C} = C_1\hat{\mathbf{i}} + C_2\hat{\mathbf{j}} + C_3\hat{\mathbf{k}}$ is a constant vector. □

EXERCISES 12.7

In Exercises 1–5, find the largest possible domain for the given vector function.

1. $\mathbf{v}(t) = t^2\hat{\mathbf{i}} + \sqrt{t-1}\,\hat{\mathbf{j}} + \hat{\mathbf{k}}$
2. $\mathbf{v}(t) = (\sin t)\hat{\mathbf{i}} + (\cos t)\hat{\mathbf{j}} - t^3\hat{\mathbf{k}}$
3. $\mathbf{v}(t) = (\mathrm{Sin}^{-1}\, t)\hat{\mathbf{i}} - t^2\hat{\mathbf{j}} + (t+1)\hat{\mathbf{k}}$
4. $\mathbf{v}(t) = \ln(t+4)(\hat{\mathbf{i}} + \hat{\mathbf{j}})$
5. $\mathbf{v}(t) = e^t\hat{\mathbf{i}} + (\cos^2 t)\hat{\mathbf{j}} - (e^t \cos t)\hat{\mathbf{k}}$

If $f(t) = t^2 + 3$, $g(t) = 2t^3 - 3t$, $\mathbf{u}(t) = t\hat{\mathbf{i}} - t^2\hat{\mathbf{j}} + 2t\hat{\mathbf{k}}$, and $\mathbf{v}(t) = \hat{\mathbf{i}} - 2t\hat{\mathbf{j}} + 3t^2\hat{\mathbf{k}}$, find the scalar or the components of the vector in Exercises 6–21.

6. $\dfrac{d\mathbf{u}}{dt}$ 7. $\dfrac{d}{dt}(f(t)\mathbf{v}(t))$ 8. $\dfrac{d}{dt}(g(t)\mathbf{u}(t))$

9. $\dfrac{d}{dt}(\mathbf{u} \times \mathbf{v})$ 10. $\dfrac{d}{dt}(\mathbf{u} \times t\mathbf{v})$ 11. $\dfrac{d}{dt}(2\mathbf{u} \cdot \mathbf{v})$

12. $\dfrac{d}{dt}(3\mathbf{u} + 4\mathbf{v})$ 13. $\int \mathbf{u}(t)\, dt$ 14. $\dfrac{d}{dt}(f(t)\mathbf{u} + g(t)\mathbf{v})$

15. $\int 4\mathbf{v}(t)\, dt$ 16. $\dfrac{d}{dt}(t(\mathbf{u} \times \mathbf{v}))$ 17. $\int (f(t)\mathbf{u}(t))\, dt$

18. $\int (3g(t)\mathbf{v}(t) + \mathbf{u}(t))\, dt$ 19. $\int (f(t)\mathbf{u} \cdot \mathbf{v})\, dt$

20. $\mathbf{u} \times \dfrac{d\mathbf{v}}{dt} - f(t)\mathbf{u} \cdot \dfrac{d\mathbf{v}}{dt}\mathbf{v}$ 21. $\mathbf{u} \cdot \dfrac{d\mathbf{v}}{dt} - \mathbf{v} \cdot \int \mathbf{u}(t)\, dt$

22. Prove (12.52).

23. Verify the results in Eqs. (12.53)–(12.55).

24. Prove that for differentiable functions $\mathbf{u}(t)$, $\mathbf{v}(t)$, and $\mathbf{w}(t)$,
$$\frac{d}{dt}(\mathbf{u} \cdot \mathbf{v} \times \mathbf{w}) = \frac{d\mathbf{u}}{dt} \cdot \mathbf{v} \times \mathbf{w} + \mathbf{u} \cdot \frac{d\mathbf{v}}{dt} \times \mathbf{w} + \mathbf{u} \cdot \mathbf{v} \times \frac{d\mathbf{w}}{dt}.$$

25. Prove that if a differentiable function $\mathbf{v}(t)$ has constant length, then at any point at which $d\mathbf{v}/dt \neq \mathbf{0}$, $d\mathbf{v}/dt$ is perpendicular to $\mathbf{v}$.

26. If $\mathbf{v} = \mathbf{v}(s)$ is a differentiable vector function and $s = s(t)$ is a differentiable scalar function, prove that
$$\frac{d\mathbf{v}}{dt} = \frac{d\mathbf{v}}{ds}\frac{ds}{dt}.$$
This result is called the *chain rule* for differentiation of vector functions.

*27. Show that the following definition for the limit of a vector function is equivalent to Definition 12.11: A vector function $\mathbf{v}(t)$ is said to have limit $\mathbf{V}$ as t approaches t_0 if given any $\varepsilon > 0$, there exists a $\delta > 0$ such that
$$|\mathbf{v}(t) - \mathbf{V}| < \varepsilon$$
whenever $0 < |t - t_0| < \delta$.

12.8 PARAMETRIC AND VECTOR REPRESENTATIONS OF CURVES

In Section 12.2 we presented curves in space as the intersection of two surfaces. For example, each of the equations
$$x^2 + y^2 + z^2 = 9, \qquad y = 2$$
describes a surface (the first a sphere and the second a plane), and together they describe the curve of intersection of the surfaces—the circle in Fig. 12.103.

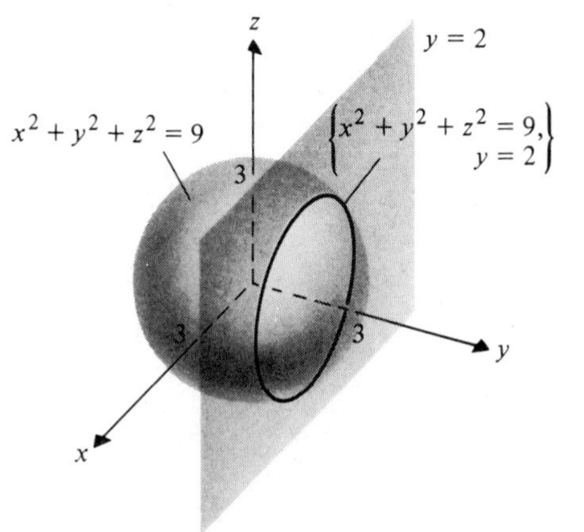

Figure 12.103

Parametric Representation of Curves

In many applications it is more convenient to have a curve defined "parametrically."

▷ **Definition 12.15** A curve in space is defined parametrically by three functions:

$$C: \quad x = x(t), \qquad y = y(t), \qquad z = z(t), \qquad \alpha \leq t \leq \beta. \quad \triangleleft \quad (12.59)$$

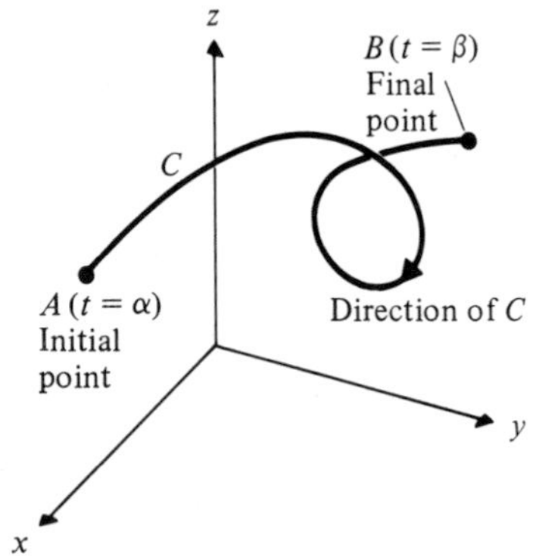

Figure 12.104

Each value of t in the interval $\alpha \leq t \leq \beta$ is substituted into the three functions, and the triple $(x, y, z) = (x(t), y(t), z(t))$ is used as a point on the curve. Definition 12.15 clearly corresponds to parametric Definition 10.4 for a plane curve.

It is customary to assign a direction to a curve by calling that point on C corresponding to $t = \alpha$ the initial point and that point corresponding to $t = \beta$ the final point, and the direction of C is that direction along C from initial point to final point (Fig. 12.104). Note in particular that the direction of a curve always corresponds to the direction in which the parameter t *increases* along the curve. Because of this, whenever we describe a curve in nonparametric form but with a specified direction, we must be careful in setting up parametric equations to ensure that the parameter increases in the appropriate direction.

When a curve is described as the curve of intersection of two surfaces, we can obtain parametric equations for the curve by specifying one of x, y, or z as a function of t, and then solving the equations for the other two as functions of t. Considerable ingenuity is sometimes required in arriving at an initial function of t. We illustrate this in the following example.

Example 30 Find parametric equations for each of the following curves:

a) $z - 1 = x^2 + y^2$, $x - y = 0$ directed so that z increases when x and y are positive;

b) $x + 2y + z = 4$, $2x + y + 3z = 6$ directed so that y increases along the curve;

c) $x^2 + (y - 1)^2 = 4$, $z = x$ directed so that y increases when x is positive.

Solution a) The curve of intersection of the paraboloid of revolution $z = 1 + x^2 + y^2$ and the plane $y = x$ is shown in Fig. 12.105. If we choose x as the parameter along the curve by setting $x = t$, then

$$x = t, \qquad y = t, \qquad z = 1 + 2t^2$$

(and note that z increases when x and y are positive).

b) The straight-line intersection of the two planes is shown in Fig. 12.106. If we choose y as the parameter by setting $y = t$, then

$$x + z = 4 - 2t, \qquad 2x + 3z = 6 - t.$$

The solution of these equations for x and z in terms of t gives the parametric equations

$$x = 6 - 5t, \qquad y = t, \qquad z = -2 + 3t.$$

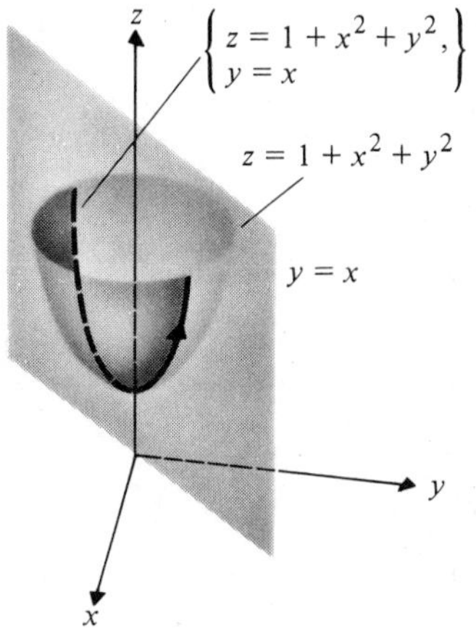

Figure 12.105

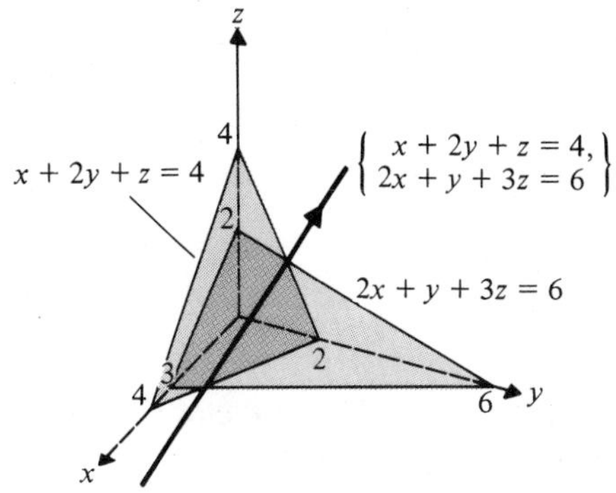

Figure 12.106

c) The curve of intersection of the right-circular cylinder $x^2 + (y - 1)^2 = 4$ and the plane $z = x$ is shown in Fig. 12.107. If we set $x = 2\cos t$, then $y = 1 \pm 2\sin t$. A set of parametric equations for the curve is therefore

$$x = 2\cos t, \qquad y = 1 + 2\sin t, \qquad z = 2\cos t, \qquad 0 \le t < 2\pi.$$

Any range of values of t of length 2π will trace the curve exactly once. Note that had we chosen the equations

$$x = 2\cos t, \qquad y = 1 - 2\sin t, \qquad z = 2\cos t, \qquad 0 \le t < 2\pi,$$

we would have generated the same set of points traced in the opposite direction. □

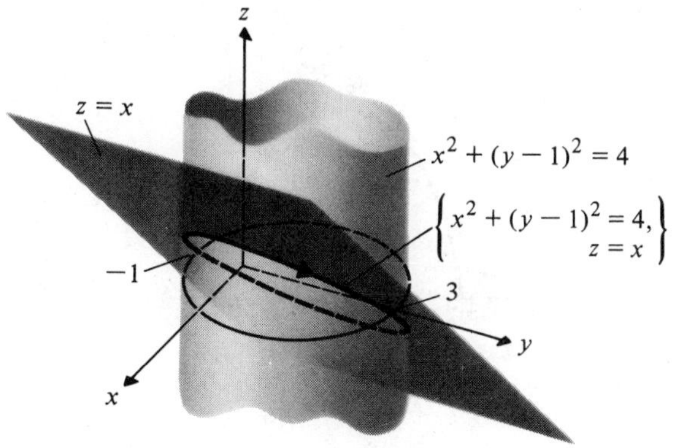

Figure 12.107

▷ **Definition 12.16** A curve C: $x = x(t)$, $y = y(t)$, $z = z(t)$, $\alpha \le t \le \beta$, is said to be *continuous* if each of the functions $x(t)$, $y(t)$, and $z(t)$ is continuous for $\alpha \le t \le \beta$. ◁

Geometrically, this implies that the curve is at no point separated. Each of the curves in Example 30 is therefore continuous.

Vector Representation of Curves The *position*, or *displacement*, *vector* of a point $P(x, y, z)$ in space is the vector

$$\mathbf{r} = (x, y, z) = x\hat{\mathbf{i}} + y\hat{\mathbf{j}} + z\hat{\mathbf{k}}.$$

If we consider only points that lie on a curve defined parametrically by (12.59), then for these points we can write that

$$\mathbf{r} = \mathbf{r}(t) = x(t)\hat{\mathbf{i}} + y(t)\hat{\mathbf{j}} + z(t)\hat{\mathbf{k}}, \qquad \alpha \le t \le \beta. \tag{12.60}$$

As t varies from $t = \alpha$ to $t = \beta$, the tip of this vector traces the curve C from initial point to final point (Fig. 12.108). We call (12.60) the *vector representation* of a curve.

Example 31 Sketch the curve with position vector

$$\mathbf{r} = \mathbf{r}(t) = (2\cos t)\hat{\mathbf{i}} + (3\sin t)\hat{\mathbf{j}}, \qquad 0 \le t \le \pi.$$

Solution When we set $x = 2\cos t$ and $y = 3\sin t$, it is clear that $x^2/4 + y^2/9 = 1$. The position vector $\mathbf{r} = \mathbf{r}(t)$ therefore describes points on an ellipse in the xy-plane. As t varies from 0 to π, x varies from 2 to -2, and y from 0 to 3 to 0 again. This means that only the top half of the ellipse is defined by $\mathbf{r} = \mathbf{r}(t)$ (Fig. 12.109). □

Example 32 Sketch the curve with position vector

$$\mathbf{r} = \mathbf{r}(t) = t\hat{\mathbf{i}} + t^2\hat{\mathbf{j}} + t\hat{\mathbf{k}}, \qquad t \ge 0.$$

Solution When we set $x = t$, $y = t^2$, and $z = t$, then $z = x$ and $y = x^2$. These imply that $\mathbf{r} = \mathbf{r}(t)$ describes points on the curve of intersection of the surfaces $y = x^2$ and $z = x$ (Fig. 12.110). Because $t \ge 0$, only that half of the curve of intersection in the first octant is defined by $\mathbf{r} = \mathbf{r}(t)$. □

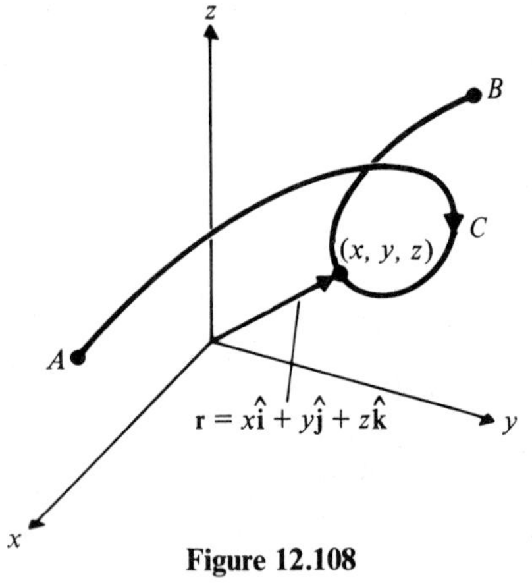

Figure 12.108

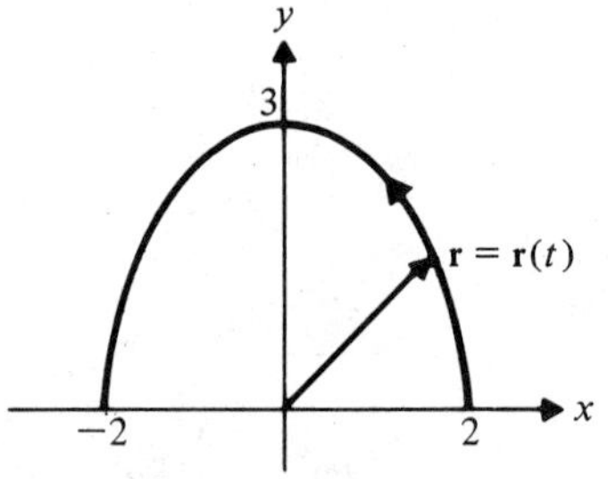

Figure 12.109

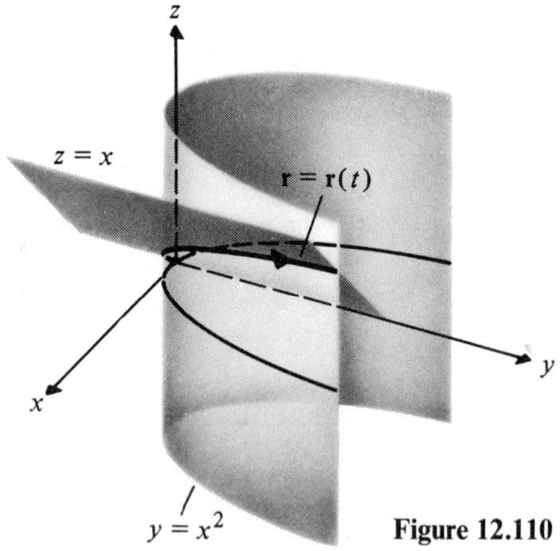

Figure 12.110

EXERCISES 12.8

In Exercises 1–10, find parametric and vector representations for the curve. In each case, sketch the curve.

1. $x + 2y + 3z = 6$,
$y - 2z = 3$,
directed so that z increases along the curve

2. $x^2 + y^2 = 2$,
$z = 4$,
directed so that y increases in the first octant

3. $x^2 + y^2 = 2$,
$x + y + z = 1$,
directed so that y decreases when x is positive

4. $z = x^2 + y^2$,
$x^2 + y^2 = 5$,
directed clockwise as viewed from the origin

5. $z + 2x^2 = 1$,
$y = z$,
directed so that x decreases along the curve

6. $z = \sqrt{x^2 + y^2}$,
$y = x$,
directed so that y increases when x is positive

7. $z = x + y$,
$y = x^2$,
directed so that x increases along the curve

8. $z = \sqrt{4 - x^2 - y^2}$,
$x^2 + y^2 - 2y = 0$,
directed so that z decreases when x is positive

9. $x = \sqrt{z}$,
$z = y^2$,
directed away from the origin in the first octant

10. $z = \sqrt{x^2 + y^2}$,
$y = x^2$,
directed so that y decreases in the first octant

In Exercises 11–15, sketch the curve with the given position vector.

11. $\mathbf{r}(t) = t\hat{\mathbf{i}} + t\hat{\mathbf{j}} + t^2\hat{\mathbf{k}}, \quad t \geq 0$

12. $\mathbf{r}(t) = (2\cos t)\hat{\mathbf{i}} + (2\sin t)\hat{\mathbf{j}} + 3t\hat{\mathbf{k}}, 0 \leq t \leq 4\pi$

13. $\mathbf{r}(t) = (t - 2)\hat{\mathbf{i}} + (2 - 3t)\hat{\mathbf{j}} + 5t\hat{\mathbf{k}}$

14. $\mathbf{r}(t) = (t^2 - t)\hat{\mathbf{i}} + t\hat{\mathbf{j}} + 5\hat{\mathbf{k}}$

15. $\mathbf{r}(t) = (\cos t)\hat{\mathbf{i}} + (\sin t)\hat{\mathbf{j}} + (\cos t)\hat{\mathbf{k}}, 0 \leq t \leq \pi$

12.9 TANGENT VECTORS AND LENGTHS OF CURVES

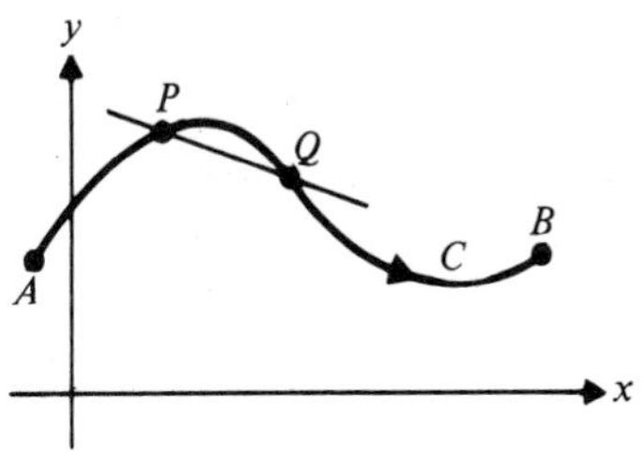

Figure 12.111

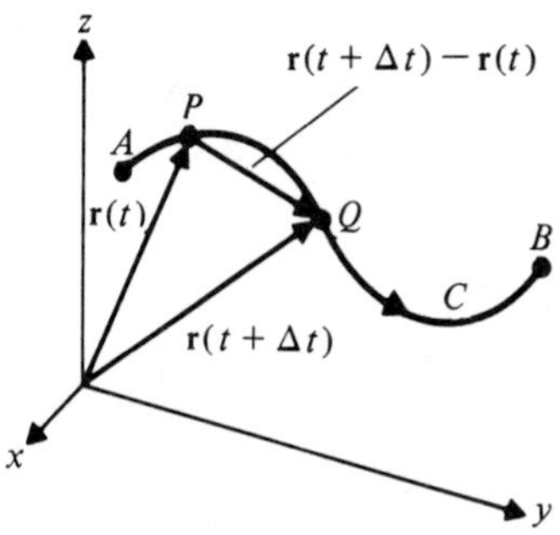

Figure 12.112

If C is a curve in the xy-plane (Fig. 12.111), then the tangent line to C at P is defined as the limiting position of the line PQ as Q moves along C toward P (see Section 3.1). We take the same approach in defining tangent vectors to curves in an arbitrary plane or in space. On a curve C defined by (12.59), we let P and Q be the points corresponding to the parameter values t and $t + \Delta t$. The position vectors of P and Q are then

$$\mathbf{r}(t) = x(t)\hat{\mathbf{i}} + y(t)\hat{\mathbf{j}} + z(t)\hat{\mathbf{k}}$$

and

$$\mathbf{r}(t + \Delta t) = x(t + \Delta t)\hat{\mathbf{i}} + y(t + \Delta t)\hat{\mathbf{j}} + z(t + \Delta t)\hat{\mathbf{k}}$$

(Fig. 12.112), and the vector $\overrightarrow{\mathbf{PQ}}$ joining P to Q is clearly equal to

$$\overrightarrow{\mathbf{PQ}} = \mathbf{r}(t + \Delta t) - \mathbf{r}(t).$$

If we let Δt approach zero, then the point Q moves along C toward P, and the direction of $\overrightarrow{\mathbf{PQ}}$ becomes closer and closer to what seems to be a reasonable definition of the tangent direction to C at P. Perhaps then we should define $\lim_{\Delta t \to 0} [\mathbf{r}(t + \Delta t) - \mathbf{r}(t)]$ as a tangent vector to C at P. Unfortunately, the limit vector has length zero, and therefore

$$\lim_{\Delta t \to 0} [\mathbf{r}(t + \Delta t) - \mathbf{r}(t)] = \mathbf{0}.$$

If, however, we divide $\mathbf{r}(t + \Delta t) - \mathbf{r}(t)$ by Δt, then the resulting vector

$$\frac{\mathbf{r}(t + \Delta t) - \mathbf{r}(t)}{\Delta t}$$

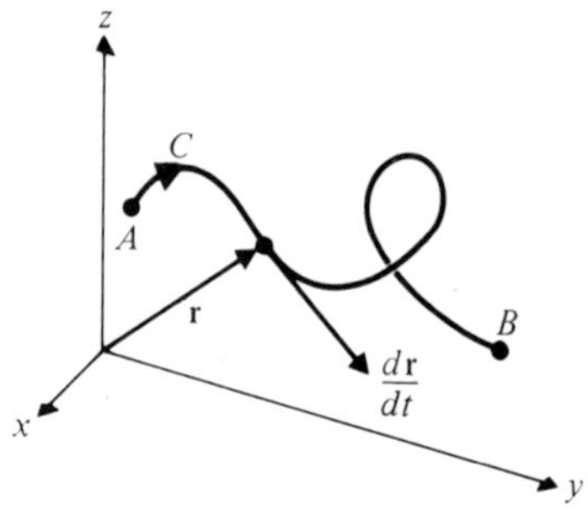

Figure 12.113

is not equal to $\overrightarrow{\mathbf{PQ}}$, but it does have the same direction as $\overrightarrow{\mathbf{PQ}}$. Consider, then, taking the limit of this vector as Δt approaches zero:

$$\lim_{\Delta t \to 0} \frac{\mathbf{r}(t + \Delta t) - \mathbf{r}(t)}{\Delta t}.$$

If the limit vector exists, then it will be tangent to C at P. But according to Eq. (12.50), this limit defines the derivative $d\mathbf{r}/dt$,

$$\frac{d\mathbf{r}}{dt} = \lim_{\Delta t \to 0} \frac{\mathbf{r}(t + \Delta t) - \mathbf{r}(t)}{\Delta t} = \frac{dx}{dt}\hat{\mathbf{i}} + \frac{dy}{dt}\hat{\mathbf{j}} + \frac{dz}{dt}\hat{\mathbf{k}}, \tag{12.61}$$

provided each of the derivatives dx/dt, dy/dt, and dz/dt exists. We have just established the following result.

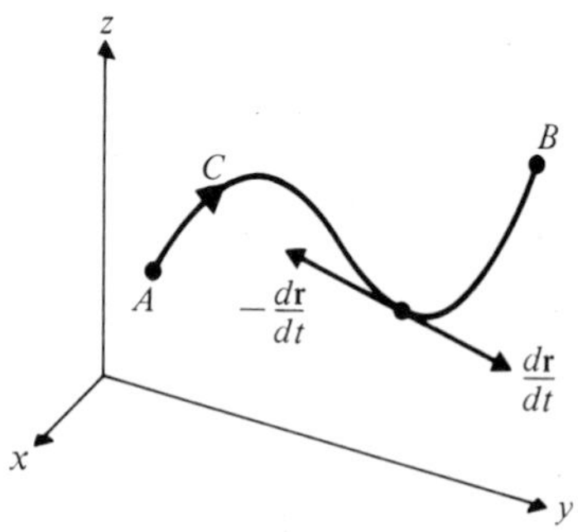

Figure 12.114

▶ **Theorem 12.10** If $\mathbf{r} = \mathbf{r}(t) = x(t)\hat{\mathbf{i}} + y(t)\hat{\mathbf{j}} + z(t)\hat{\mathbf{k}}$, $\alpha \le t \le \beta$, is the vector representation of a curve C, then at any point on C at which $x'(t)$, $y'(t)$, and $z'(t)$ all exist and do not simultaneously vanish,

$$\mathbf{T} = \frac{d\mathbf{r}}{dt} = \frac{dx}{dt}\hat{\mathbf{i}} + \frac{dy}{dt}\hat{\mathbf{j}} + \frac{dz}{dt}\hat{\mathbf{k}} \tag{12.62}$$

is a tangent vector to C (Fig. 12.113). ◀

It is obvious that at any point on a curve there are two tangent directions. One of these has been shown to be $d\mathbf{r}/dt$; the other must be defined by $-d\mathbf{r}/dt$ (Fig. 12.114). How can we tell which one is $d\mathbf{r}/dt$? A closer analysis of the limit in (12.61) indicates the following (see Exercise 15).

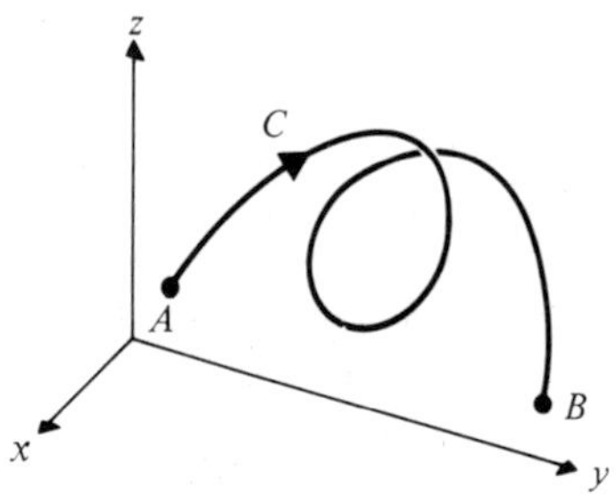

Figure 12.115

▶ **Corollary** The tangent vector $d\mathbf{r}/dt$ to a curve C: $x = x(t)$, $y = y(t)$, $z = z(t)$, $\alpha \le t \le \beta$, always points in the direction in which the parameter t increases along C. ◀

▷ **Definition 12.17** A curve C: $x = x(t)$, $y = y(t)$, $z = z(t)$, $\alpha \le t \le \beta$, is said to be smooth if the derivatives $x'(t)$, $y'(t)$, and $z'(t)$ are all continuous for $\alpha \le t \le \beta$. ◁

Since $x'(t)$, $y'(t)$, and $z'(t)$ are the components of a tangent vector to C, this definition implies that along a smooth curve, small changes in t produce small changes in the direction of the tangent vector. In other words, the tangent vector turns gradually, or "smoothly." The curve in Fig. 12.115 is smooth; that in Fig. 12.116 is not because abrupt changes in the direction of the curve occur at P and Q. According to the following definition, this curve is piecewise smooth.

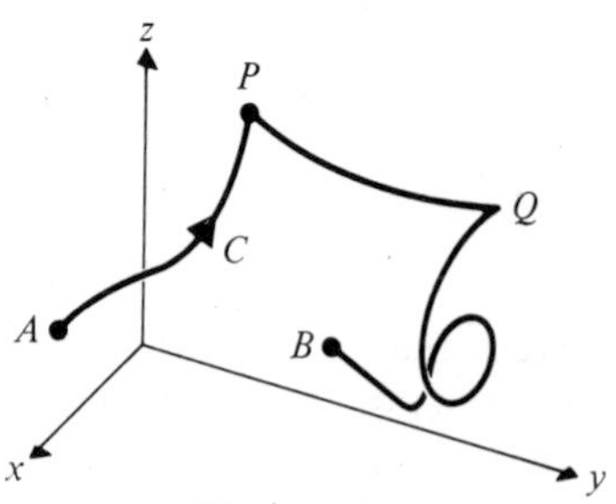

Figure 12.116

▷ **Definition 12.18** A curve is said to be *piecewise smooth* if it is continuous and can be divided into a finite number of smooth subcurves. ◁

Example 33 For the curve in Example 32, find a tangent vector at the point (3, 9, 3).

Solution A tangent vector to this curve at any point on the curve is

$$\frac{d\mathbf{r}}{dt} = \frac{dx}{dt}\hat{\mathbf{i}} + \frac{dy}{dt}\hat{\mathbf{j}} + \frac{dz}{dt}\hat{\mathbf{k}} = \hat{\mathbf{i}} + 2t\hat{\mathbf{j}} + \hat{\mathbf{k}}.$$

Since $t = 3$ yields the point $(3, 9, 3)$, a tangent vector at this point is $\mathbf{r}'(3) = \hat{\mathbf{i}} + 6\hat{\mathbf{j}} + \hat{\mathbf{k}}$. □

Example 34 Find a tangent vector at the point $(2, 0, 3)$ to the helix

$$x = 2\cos t, \qquad y = 2\sin t, \qquad z = \frac{3t}{2\pi}, \qquad t \geq 0.$$

Is the helix smooth?

Solution A tangent vector to the helix at any point is

$$\frac{d\mathbf{r}}{dt} = \frac{dx}{dt}\hat{\mathbf{i}} + \frac{dy}{dt}\hat{\mathbf{j}} + \frac{dz}{dt}\hat{\mathbf{k}} = (-2\sin t)\hat{\mathbf{i}} + (2\cos t)\hat{\mathbf{j}} + \left(\frac{3}{2\pi}\right)\hat{\mathbf{k}}.$$

Since $t = 2\pi$ yields the point $(2, 0, 3)$, a tangent vector at this point is $\mathbf{r}'(2\pi) = 2\hat{\mathbf{j}} + (3/(2\pi))\hat{\mathbf{k}}$ (see Fig. 12.117). Since $x'(t)$, $y'(t)$, and $z'(t)$ are all continuous functions, the helix is indeed smooth. □

Unit Tangent Vectors

When a curve in the xy-plane is defined parametrically by

$$C: \quad x = x(t), \qquad y = y(t), \qquad \alpha \leq t \leq \beta, \tag{12.63}$$

a tangent vector to C is

$$\mathbf{T} = \frac{d\mathbf{r}}{dt} = \frac{dx}{dt}\hat{\mathbf{i}} + \frac{dy}{dt}\hat{\mathbf{j}}, \tag{12.64}$$

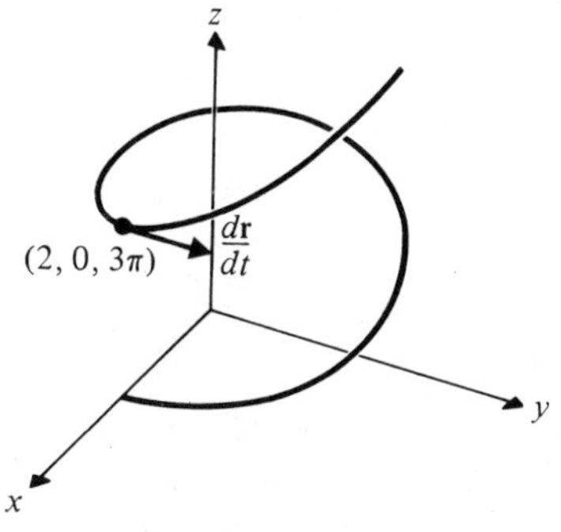

Figure 12.117

and this tangent vector points in the direction in which t increases along C. To produce a unit tangent vector to C at any point, we divide $\mathbf{T}$ by its length:

$$\hat{\mathbf{T}} = \frac{\mathbf{T}}{|\mathbf{T}|} = \frac{d\mathbf{r}/dt}{|d\mathbf{r}/dt|}. \tag{12.65}$$

We now show that if the length along C is used as the parameter by which to specify its points, then division by $|\mathbf{T}|$ is unnecessary.

In Section 7.3 we showed that small lengths along a plane curve C can be approximated by straight-line lengths along tangent lines to the curve, and that the total length of a smooth curve from A to B (Fig. 12.118) is

$$L = \int_A^B \sqrt{(dx)^2 + (dy)^2}.$$

With parametric equations (12.63) we can write this formula as a definite integral with respect to t (see also Eq. 10.27):

$$L = \int_\alpha^\beta \sqrt{\left(\left(\frac{dx}{dt}\right)^2 + \left(\frac{dy}{dt}\right)^2\right)(dt)^2} = \int_\alpha^\beta \sqrt{\left(\frac{dx}{dt}\right)^2 + \left(\frac{dy}{dt}\right)^2}\, dt. \tag{12.66}$$

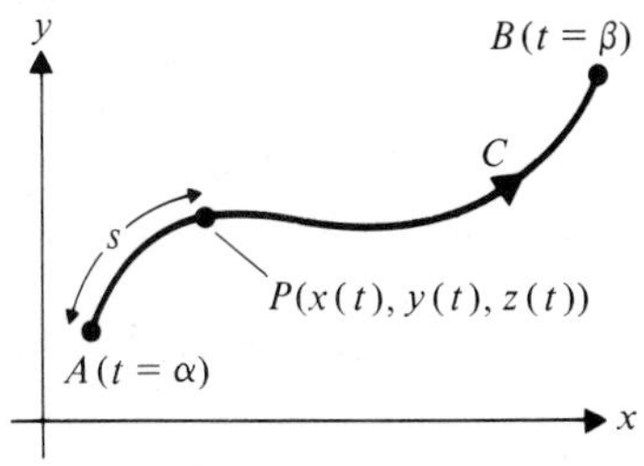

Figure 12.118

Furthermore, if we denote by $s = s(t)$ the length of that part of C from its initial point A (where $t = \alpha$) to any point $P(x(t), y(t))$ on C (Fig. 12.118), then $s(t)$ is defined by the integral

$$s(t) = \int_\alpha^t \sqrt{\left(\frac{dx}{dt}\right)^2 + \left(\frac{dy}{dt}\right)^2}\, dt. \tag{12.67}$$

It follows, then, that the derivative of $s(t)$ is

$$\frac{ds}{dt} = \sqrt{\left(\frac{dx}{dt}\right)^2 + \left(\frac{dy}{dt}\right)^2}. \tag{12.68}$$

But according to (12.64), $d\mathbf{r}/dt$ is a tangent vector to C with the same length:

$$\left|\frac{d\mathbf{r}}{dt}\right| = \sqrt{\left(\frac{dx}{dt}\right)^2 + \left(\frac{dy}{dt}\right)^2} = \frac{ds}{dt}. \tag{12.69}$$

When this equation is multiplied by dt, it gives

$$|d\mathbf{r}| = \sqrt{(dx)^2 + (dy)^2} = ds. \tag{12.70}$$

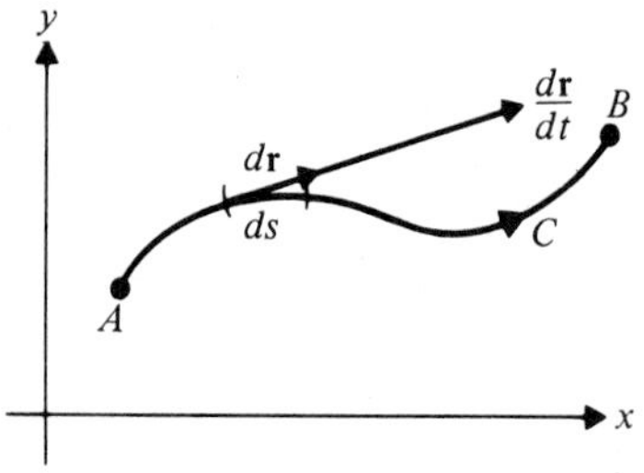

Figure 12.119

This equation states that ds is the length of the tangent vector $d\mathbf{r} = dx\hat{\mathbf{i}} + dy\hat{\mathbf{j}}$, and therefore ds is a measure of length along the tangent line to C. In spite of this we often think of ds as a measure of small lengths along C itself (Fig. 12.119), and that ds is approximated by the tangential straight-line length

$$|d\mathbf{r}| = \sqrt{(dx)^2 + (dy)^2}.$$

Note too that if we use length s along C as a parameter, then the chain rule applied to $\mathbf{r} = \mathbf{r}(s)$, $s = s(t)$ gives

$$\frac{d\mathbf{r}}{dt} = \frac{d\mathbf{r}}{ds}\frac{ds}{dt}. \tag{12.71}$$

(The chain rule is proved in Exercise 26 of Section 12.7.) Consequently, Eq. (12.69) implies that

$$\frac{d\mathbf{r}}{ds} = \frac{d\mathbf{r}/dt}{ds/dt} = \frac{d\mathbf{r}/dt}{|d\mathbf{r}/dt|}. \tag{12.72}$$

What we have shown, then, is that if we choose length along a curve C as the parameter by which to specify points on the curve ($C\colon x = x(s)$, $y = y(s)$), then the vector

$$\hat{\mathbf{T}} = \frac{d\mathbf{r}}{ds} = \frac{dx}{ds}\hat{\mathbf{i}} + \frac{dy}{ds}\hat{\mathbf{j}} \tag{12.73}$$

is a unit tangent vector to C. In addition, the corollary to Theorem 12.10 implies that $d\mathbf{r}/ds$ points in the direction in which s increases along C. This suggests perhaps that we should always set up parametric equations for a curve with length along the curve as parameter. Theoretically this is quite acceptable, but practically it is impossible. For most curves we have enough difficulty just finding a set of parametric equations, let alone finding that set with length along the curve as parameter. If we then use a parameter t other than length along the curve, a unit tangent vector is calculated according to (12.65).

These results can be extended to space curves as well. When a smooth curve C has parametric equations (12.59), Eq. (12.65) still defines a unit tangent vector to C, but because C is a space curve, $d\mathbf{r}/dt$ is calculated according to (12.62).

Corresponding to formula (12.67) for length along a curve in the xy-plane, length along a smooth curve in space is defined by the definite integral

$$s(t) = \int_\alpha^t \sqrt{\left(\frac{dx}{dt}\right)^2 + \left(\frac{dy}{dt}\right)^2 + \left(\frac{dz}{dt}\right)^2}\, dt. \tag{12.74}$$

These two results imply that

$$\left|\frac{d\mathbf{r}}{dt}\right| = \sqrt{\left(\frac{dx}{dt}\right)^2 + \left(\frac{dy}{dt}\right)^2 + \left(\frac{dz}{dt}\right)^2} = \frac{ds}{dt}, \tag{12.75}$$

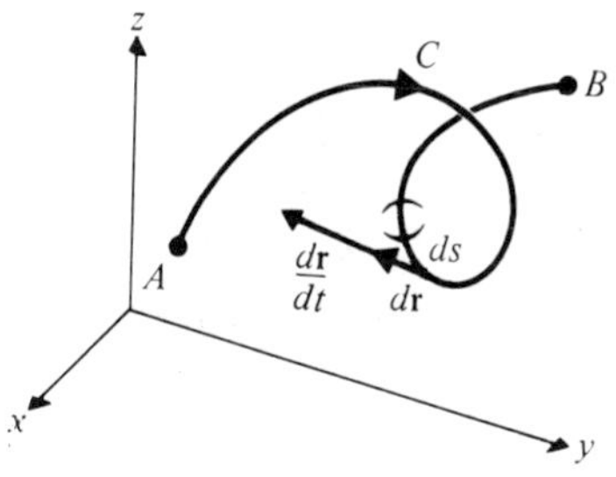

Figure 12.120

the three-space analogue of (12.69). Once again we are led to the fact that when s is used as parameter along C, then

$$\hat{\mathbf{T}} = \frac{d\mathbf{r}}{ds} = \frac{dx}{ds}\hat{\mathbf{i}} + \frac{dy}{ds}\hat{\mathbf{j}} + \frac{dz}{ds}\hat{\mathbf{k}} \tag{12.76}$$

is a unit tangent vector to C. In addition, multiplication of (12.75) by dt yields

$$|d\mathbf{r}| = ds = \sqrt{(dx)^2 + (dy)^2 + (dz)^2}, \tag{12.77}$$

indicating that small lengths ds along C (Fig. 12.120) are defined in terms of small lengths $|d\mathbf{r}|$ along the tangent line to C.

Example 35 Find a unit tangent vector to the curve

$$C:\ x = \sin t, \qquad y = 2\cos t, \qquad z = 2t/\pi, \qquad t \geq 0$$

at the point $(0, -2, 2)$.

Solution A tangent vector to C at any point is

$$\frac{d\mathbf{r}}{dt} = \frac{dx}{dt}\hat{\mathbf{i}} + \frac{dy}{dt}\hat{\mathbf{j}} + \frac{dz}{dt}\hat{\mathbf{k}} = \cos t\hat{\mathbf{i}} - 2\sin t\hat{\mathbf{j}} + \left(\frac{2}{\pi}\right)\hat{\mathbf{k}}.$$

Since $t = \pi$ yields the point $(0, -2, 2)$, a tangent vector at this point is $\mathbf{r}'(\pi) = -\hat{\mathbf{i}} + (2/\pi)\hat{\mathbf{k}}$. A unit tangent vector is then

$$\hat{\mathbf{T}} = \frac{-\hat{\mathbf{i}} + (2/\pi)\hat{\mathbf{k}}}{\sqrt{1 + 4/\pi^2}} = \frac{-\pi\hat{\mathbf{i}} + 2\hat{\mathbf{k}}}{\sqrt{4 + \pi^2}}. \square$$

Example 36 Find the length of that part of the curve $x = y^{2/3}$, $x = z^{2/3}$ between the points $(0, 0, 0)$ and $(4, 8, 8)$.

Solution If we use $x = t$, $y = t^{3/2}$, $z = t^{3/2}$, $0 \leq t \leq 4$, as parametric equations for the curve, then

$$L = \int_0^4 \sqrt{\left(\frac{dx}{dt}\right)^2 + \left(\frac{dy}{dt}\right)^2 + \left(\frac{dz}{dt}\right)^2}\, dt = \int_0^4 \sqrt{1 + \left(\frac{3}{2}\sqrt{t}\right)^2 + \left(\frac{3}{2}\sqrt{t}\right)^2}\, dt$$

$$= \int_0^4 \sqrt{1 + \frac{9t}{2}}\, dt = \left\{\frac{4}{27}\left(1 + \frac{9t}{2}\right)^{3/2}\right\}_0^4 = \frac{4}{27}(19\sqrt{19} - 1). \square$$

EXERCISES 12.9

In Exercises 1–5, express each curve in vector form and find the unit tangent vector $\hat{\mathbf{T}}$ at each point on the curve.

1. $x = \sin t$, $y = \cos t$, $z = t$, $-\infty < t < \infty$

2. $x = t$, $y = t^2$, $z = t^3$, $t \geq 1$

3. $x = (t-1)^2$, $y = (t+1)^2$, $z = -t$, $-3 \leq t \leq 4$

4. $x + y = 5$, $x^2 - y = z$ from $(5, 0, 25)$ to $(0, 5, -5)$

5. $x + y + z = 4$, $x^2 + y^2 = 4$, $y \geq 0$ from $(2, 0, 2)$ to $(-2, 0, 6)$

In Exercises 6–10, find $\hat{\mathbf{T}}$ at the point indicated.

6. $x = 4\cos t$, $y = 6\sin t$, $z = 2\sin t$, $-\infty < t < \infty$; $(2\sqrt{2}, 3\sqrt{2}, \sqrt{2})$

7. $x = 2 - 5t, \quad y = 1 + t, \quad z = 6 + 4t, \quad -\infty < t < \infty$; $(7, 0, 2)$

8. $x^2 + y^2 + z^2 = 4$, $z = \sqrt{x^2 + y^2}$, directed so that x increases when y is positive; $(1, 1, \sqrt{2})$

9. $x = y^2 + 1, z = x + 5$, directed so that y increases along the curve; $(5, 2, 10)$

10. $x^2 + (y - 1)^2 = 4$, $z = x$, directed so that z decreases when y is negative; $(2, 1, 2)$

In Exercises 11–14, find the length of the curve. In each case, sketch the curve.

11. $x = 2 \cos t, y = 2 \sin t, z = 3t, 0 \le t \le 2\pi$

12. $x = 2 - 5t, y = 1 + t, z = 6 + 4t, -1 \le t \le 0$

13. $x = t^3, y = t^2, z = t^3, 0 \le t \le 1$

14. $x = t, y = t^{3/2}, z = 4t^{3/2}, 1 \le t \le 4$

15. Show that the tangent vector $d\mathbf{r}/dt$ to a curve (12.60) always points in the direction in which t increases along the curve.

16. a) What happens when Eq. (12.62) is used to determine a tangent vector to the curve $x = t^2$, $y = t^3$, $z = t^2$, $-\infty < t < \infty$, at the origin?

b) Can you devise a way in which to obtain a tangent vector?

12.10 NORMAL VECTORS, CURVATURE, AND RADIUS OF CURVATURE

Normal Vectors

In discussing curves we distinguish between two types of properties: intrinsic and not intrinsic. An intrinsic property is one that is independent of the parameter used to specify the curve; a property that is not intrinsic is parameter-dependent. To illustrate, the tangent vector $\mathbf{T} = d\mathbf{r}/dt$ in (12.62) is not intrinsic; a change of parameter results in a change in the length of $\mathbf{T}$. The unit tangent vector $\hat{\mathbf{T}}$, on the other hand, is intrinsic; there is only one unit tangent vector in the direction of the curve. Length of a curve from its initial point to an arbitrary point is an intrinsic property; a change of parameter along the curve does not affect length between points.

Because length along a curve is an intrinsic property, it is customary in theoretical discussions to use it as the parameter by which to specify points on the curve. When C is a smooth curve in the xy-plane, parametric equations for C in terms of length s along C take the form

$$C: \quad x = x(s), \qquad y = y(s), \qquad 0 \le s \le L. \tag{12.78}$$

Figure 12.121

The normal line at a point P on such a curve C is that line which is perpendicular to the tangent line to C at P (Fig. 12.121). Any vector along this normal line is said to be a normal vector to the curve at P. Since the unit tangent vector to C at P is

$$\hat{\mathbf{T}} = \frac{d\mathbf{r}}{ds} = \frac{dx}{ds}\hat{\mathbf{i}} + \frac{dy}{ds}\hat{\mathbf{j}},$$

it follows that

$$\hat{\mathbf{N}} = -\frac{dy}{ds}\hat{\mathbf{i}} + \frac{dx}{ds}\hat{\mathbf{j}} \tag{12.79}$$

is a unit normal vector to C at P (note that $\hat{\mathbf{T}} \cdot \hat{\mathbf{N}} = 0$). Because there is only one direction normal to C at P, every normal vector to C at P must be some multiple $\lambda\hat{\mathbf{N}}$ of $\hat{\mathbf{N}}$.

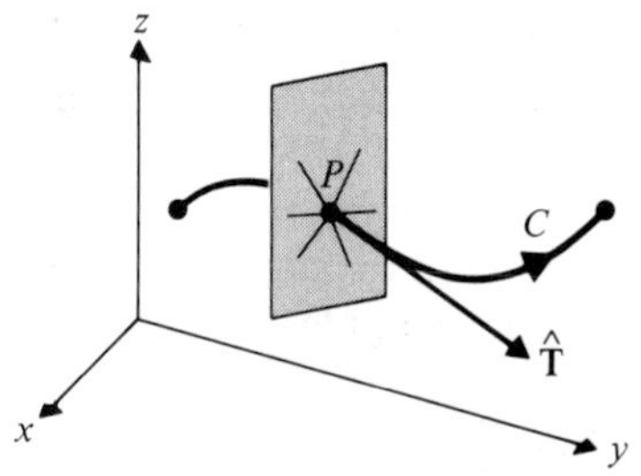

Figure 12.122

The situation is quite different for space curves (Fig. 12.122). If $\hat{\mathbf{T}}$ is the unit tangent vector to a smooth curve C, then there is an entire plane of normal vectors to C at P. In the following discussion, we single out two normal vectors called the *principal normal* and the *binormal*. Suppose that C is defined parametrically by

$$C: \quad x = x(s), \qquad y = y(s), \qquad z = z(s), \qquad 0 \le s \le L, \tag{12.80}$$

and that $\hat{\mathbf{T}}$ is the unit tangent vector to C defined by (12.76). Because $\hat{\mathbf{T}}$ has length unity,

$$1 = \hat{\mathbf{T}} \cdot \hat{\mathbf{T}}.$$

If we use Eq. (12.54) to differentiate this equation with respect to s, we have

$$0 = \frac{d\hat{\mathbf{T}}}{ds} \cdot \hat{\mathbf{T}} + \hat{\mathbf{T}} \cdot \frac{d\hat{\mathbf{T}}}{ds} = 2\left(\hat{\mathbf{T}} \cdot \frac{d\hat{\mathbf{T}}}{ds}\right).$$

But if neither of the vectors $\hat{\mathbf{T}}$ nor $d\hat{\mathbf{T}}/ds$ is equal to zero, then the fact that their scalar product is equal to zero implies that they are perpendicular. In other words,

$$\mathbf{N} = \frac{d\hat{\mathbf{T}}}{ds} \tag{12.81}$$

is a normal vector to C at any point. The unit normal vector in this direction

$$\hat{\mathbf{N}} = \frac{\mathbf{N}}{|\mathbf{N}|} = \frac{d\hat{\mathbf{T}}/ds}{|d\hat{\mathbf{T}}/ds|} \tag{12.82}$$

is called the *principal normal* (vector) to C (Fig. 12.123).

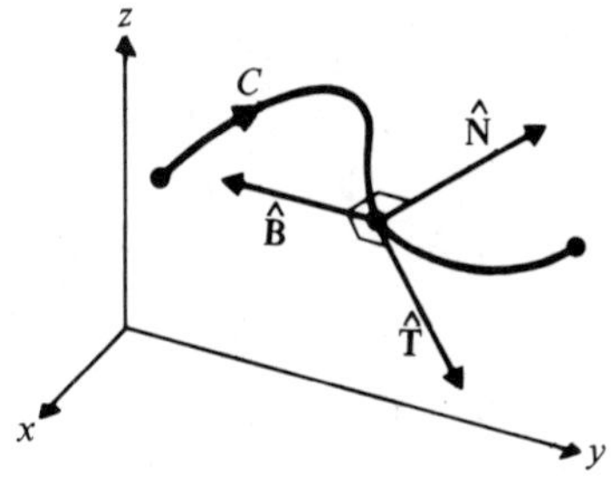

Figure 12.123

Because $\hat{\mathbf{N}}$ is defined in terms of intrinsic properties $\hat{\mathbf{T}}$ and s for a curve, it must also be an intrinsic property. It follows, then, that no matter what parameter is used to specify points on a curve, $\hat{\mathbf{N}}$ is always the same. But how do we find $\hat{\mathbf{N}}$ when a curve C is specified in terms of a parameter other than length along C, say, in the form

$$C: \quad x = x(t), \qquad y = y(t), \qquad z = z(t), \qquad \alpha \le t \le \beta? \tag{12.83}$$

If $s = s(t)$ is length along C (measured from $t = \alpha$), then by the chain rule

$$\frac{d\hat{\mathbf{T}}}{ds} = \frac{d\hat{\mathbf{T}}}{dt}\frac{dt}{ds},$$

where dt/ds must be positive since both s and t increase along C. Consequently,

$$\hat{\mathbf{N}} = \frac{d\hat{\mathbf{T}}/ds}{|d\hat{\mathbf{T}}/ds|} = \frac{(d\hat{\mathbf{T}}/dt)(dt/ds)}{|(d\hat{\mathbf{T}}/dt)(dt/ds)|} = \frac{(d\hat{\mathbf{T}}/dt)(dt/ds)}{|d\hat{\mathbf{T}}/dt|\,dt/ds} = \frac{d\hat{\mathbf{T}}/dt}{|d\hat{\mathbf{T}}/dt|}. \tag{12.84}$$

In other words, for any parametrization of C whatsoever, the vector $d\hat{\mathbf{T}}/dt$ always points in the direction of the principal normal, and to find $\hat{\mathbf{N}}$, we simply find the unit vector in the direction of $d\hat{\mathbf{T}}/dt$.

In the study of space curves, a second normal vector to C, called the *binormal*, is defined by

$$\hat{\mathbf{B}} = \hat{\mathbf{T}} \times \hat{\mathbf{N}}. \tag{12.85}$$

Since the cross product of two vectors is always perpendicular to each of the vectors, it follows that the binormal is perpendicular to both $\hat{\mathbf{T}}$ and $\hat{\mathbf{N}}$, and must therefore indeed be a normal vector to C (Fig. 12.123).

We have singled out three vectors at each point P on a curve C: a unit tangent vector $\hat{\mathbf{T}}$ and two unit normal vectors $\hat{\mathbf{N}}$ and $\hat{\mathbf{B}}$. As P moves along C, these vectors constantly change direction but always have unit length.

Example 37 Find $\hat{\mathbf{T}}$, $\hat{\mathbf{N}}$, and $\hat{\mathbf{B}}$ for the curve $x = t$, $y = t^2$, $z = t^2$, $t \geq 0$.

Solution The unit tangent vector $\hat{\mathbf{T}}$ is defined by

$$\hat{\mathbf{T}} = \frac{d\mathbf{r}/dt}{|d\mathbf{r}/dt|} = \frac{(1, 2t, 2t)}{\sqrt{1 + 4t^2 + 4t^2}} = \frac{(1, 2t, 2t)}{\sqrt{1 + 8t^2}}.$$

The principal normal $\hat{\mathbf{N}}$ lies along the vector $\mathbf{N} = d\hat{\mathbf{T}}/dt$, and according to Eq. (12.53), we can write that

$$\begin{aligned}
\mathbf{N} = \frac{d\hat{\mathbf{T}}}{dt} &= \frac{d}{dt}\left(\frac{1}{\sqrt{1 + 8t^2}}\right)(1, 2t, 2t) + \frac{1}{\sqrt{1 + 8t^2}}\frac{d}{dt}(1, 2t, 2t) \\
&= \frac{-8t}{(1 + 8t^2)^{3/2}}(1, 2t, 2t) + \frac{1}{\sqrt{1 + 8t^2}}(0, 2, 2) \\
&= \frac{1}{(1 + 8t^2)^{3/2}}\{-8t(1, 2t, 2t) + (1 + 8t^2)(0, 2, 2)\} \\
&= \frac{(-8t, 2, 2)}{(1 + 8t^2)^{3/2}}.
\end{aligned}$$

The principal normal is therefore

$$\hat{\mathbf{N}} = \frac{\mathbf{N}}{|\mathbf{N}|} = \frac{(-8t, 2, 2)}{\sqrt{64t^2 + 4 + 4}} = \frac{(-4t, 1, 1)}{\sqrt{2 + 16t^2}}.$$

The binormal is

$$\begin{aligned}
\hat{\mathbf{B}} = \hat{\mathbf{T}} \times \hat{\mathbf{N}} &= \frac{(1, 2t, 2t)}{\sqrt{1 + 8t^2}} \times \frac{(-4t, 1, 1)}{\sqrt{2 + 16t^2}} \\
&= \frac{1}{\sqrt{1 + 8t^2}\sqrt{2 + 16t^2}}\begin{vmatrix} \hat{\mathbf{i}} & \hat{\mathbf{j}} & \hat{\mathbf{k}} \\ 1 & 2t & 2t \\ -4t & 1 & 1 \end{vmatrix} \\
&= \frac{1}{\sqrt{2}\sqrt{1 + 8t^2}\sqrt{1 + 8t^2}}(0, -1 - 8t^2, 1 + 8t^2) \\
&= \frac{1 + 8t^2}{\sqrt{2}(1 + 8t^2)}(0, -1, 1) \\
&= \frac{(0, -1, 1)}{\sqrt{2}}.
\end{aligned}$$

The significance of the fact that the binormal has constant direction can be seen from a sketch of the curve. Because the parametric equations imply that $y = x^2$ and $z = y$, the curve is the curve of intersection of these two surfaces (Fig. 12.124). Since the curve lies in the plane $-y + z = 0$, and a normal vector to this plane is $(0, -1, 1)$, it follows that $(0, -1, 1)$ is always normal to the curve. But this is precisely the direction of $\hat{\mathbf{B}}$. In other words, constant $\hat{\mathbf{B}}$ implies that the curve lies in a plane that has $\hat{\mathbf{B}}$ as normal. (This result is in fact always true, although this is difficult to prove.) □

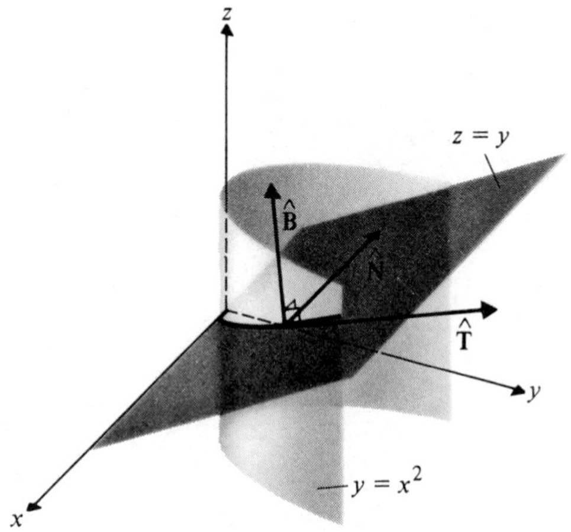

Figure 12.124

Example 38 Show that for a smooth curve C: $x = x(t)$, $y = y(t)$, $\alpha \le t \le \beta$ in the xy-plane, the principal normal is

$$\hat{\mathbf{N}} = \operatorname{sgn}\left(\frac{dy}{dt}\frac{d^2x}{dt^2} - \frac{dx}{dt}\frac{d^2y}{dt^2}\right)\frac{(dy/dt, -dx/dt)}{\sqrt{(dx/dt)^2 + (dy/dt)^2}},$$

where the signum function sgn (u) is defined by

$$\operatorname{sgn}(u) = \begin{cases} 1, & \text{if } u > 0, \\ 0, & \text{if } u = 0, \\ -1, & \text{if } u < 0. \end{cases}$$

Solution The unit tangent vector to C is

$$\hat{\mathbf{T}} = \frac{(dx/dt, dy/dt)}{\sqrt{(dx/dt)^2 + (dy/dt)^2}}.$$

For simplicity in notation, we use a dot "·" above a variable to indicate that the variable is differentiated with respect to t. For example, $\dot{x} = dx/dt$ and $\ddot{x} = d^2x/dt^2$. With this notation,

$$\hat{\mathbf{T}} = \frac{(\dot{x}, \dot{y})}{\sqrt{\dot{x}^2 + \dot{y}^2}}.$$

By Eq. (12.84), $\hat{\mathbf{N}} = (d\hat{\mathbf{T}}/dt)/|d\hat{\mathbf{T}}/dt|$, where

$$\frac{d\hat{\mathbf{T}}}{dt} = \frac{d}{dt}\frac{(\dot{x}, \dot{y})}{\sqrt{\dot{x}^2 + \dot{y}^2}} = \frac{d}{dt}\left(\frac{1}{\sqrt{\dot{x}^2 + \dot{y}^2}}\right)(\dot{x}\hat{\mathbf{i}} + \dot{y}\hat{\mathbf{j}}) + \frac{1}{\sqrt{\dot{x}^2 + \dot{y}^2}}(\ddot{x}\hat{\mathbf{i}} + \ddot{y}\hat{\mathbf{j}})$$

$$= \left(\frac{-\dot{x}\ddot{x} - \dot{y}\ddot{y}}{(\dot{x}^2 + \dot{y}^2)^{3/2}}\right)(\dot{x}\hat{\mathbf{i}} + \dot{y}\hat{\mathbf{j}}) + \frac{1}{\sqrt{\dot{x}^2 + \dot{y}^2}}(\ddot{x}\hat{\mathbf{i}} + \ddot{y}\hat{\mathbf{j}})$$

$$= \frac{1}{(\dot{x}^2 + \dot{y}^2)^{3/2}}\{-(\dot{x}\ddot{x} + \dot{y}\ddot{y})(\dot{x}\hat{\mathbf{i}} + \dot{y}\hat{\mathbf{j}}) + (\dot{x}^2 + \dot{y}^2)(\ddot{x}\hat{\mathbf{i}} + \ddot{y}\hat{\mathbf{j}})\}$$

$$= \frac{1}{(\dot{x}^2 + \dot{y}^2)^{3/2}}\{(-\dot{x}^2\ddot{x} - \dot{x}\dot{y}\ddot{y} + \dot{x}^2\ddot{x} + \dot{y}^2\ddot{x})\hat{\mathbf{i}}$$
$$+ (-\dot{x}\dot{y}\ddot{x} - \dot{y}^2\ddot{y} + \dot{x}^2\ddot{y} + \dot{y}^2\ddot{y})\hat{\mathbf{j}}\}$$

$$= \frac{1}{(\dot{x}^2 + \dot{y}^2)^{3/2}}\{\dot{y}(\dot{y}\ddot{x} - \dot{x}\ddot{y})\hat{\mathbf{i}} + \dot{x}(\dot{x}\ddot{y} - \dot{y}\ddot{x})\hat{\mathbf{j}}\}$$

$$= \frac{\dot{y}\ddot{x} - \dot{x}\ddot{y}}{(\dot{x}^2 + \dot{y}^2)^{3/2}}(\dot{y}\hat{\mathbf{i}} - \dot{x}\hat{\mathbf{j}}).$$

If $\dot{y}\ddot{x} - \dot{x}\ddot{y}$ is positive, then

$$\hat{\mathbf{N}} = \frac{\dot{y}\hat{\mathbf{i}} - \dot{x}\hat{\mathbf{j}}}{\sqrt{\dot{x}^2 + \dot{y}^2}};$$

whereas if $\dot{y}\ddot{x} - \dot{x}\ddot{y}$ is negative, then

$$\hat{\mathbf{N}} = \frac{-\dot{y}\hat{\mathbf{i}} + \dot{x}\hat{\mathbf{j}}}{\sqrt{\dot{x}^2 + \dot{y}^2}}.$$

In other words,

$$\hat{\mathbf{N}} = \operatorname{sgn}(\dot{y}\ddot{x} - \dot{x}\ddot{y})\left(\frac{\dot{y}\hat{\mathbf{i}} - \dot{x}\hat{\mathbf{j}}}{\sqrt{\dot{x}^2 + \dot{y}^2}}\right). \quad \square$$

Curvature and Radius of Curvature

When length s along a smooth curve C is used as the parameter by which to identify points on the curve (Eqs. 12.80), the vector $\hat{\mathbf{T}} = d\mathbf{r}/ds$ is a unit tangent vector to C. Suppose we differentiate $\hat{\mathbf{T}}$ with respect to s to form $d\hat{\mathbf{T}}/ds = d^2\mathbf{r}/ds^2$. Since $\hat{\mathbf{T}}$ has constant length unity, only its direction can change; therefore, the derivative $d\hat{\mathbf{T}}/ds$ must be a measure of the rate of change of the direction of $\hat{\mathbf{T}}$. Since $\hat{\mathbf{T}}$ is really our way of specifying the direction of the curve itself, we can also say that $d\hat{\mathbf{T}}/ds$ is a measure of how fast the direction of C changes. But exactly how does $d\hat{\mathbf{T}}/ds$ measure this directional rate of change? According to Eq. (12.81), $d\hat{\mathbf{T}}/ds$ is always normal to the curve; at every point on C, it lies along the principal normal to C. In Fig. 12.125 we show a number of circles in the xy-plane, all of which are tangent to the y-axis at the origin. Parametric equations for the circle with center $(R, 0)$ and radius R in terms of length s along the circle (as measured from $(2R, 0)$) are

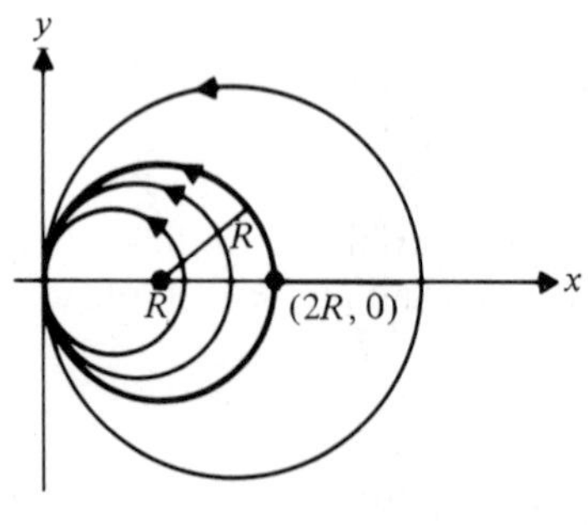

Figure 12.125

$$x = R + R\cos(s/R), \qquad y = R\sin(s/R), \qquad 0 \le s < 2\pi R.$$

Consequently,

$$\hat{\mathbf{T}} = \frac{d\mathbf{r}}{ds} = -\sin\left(\frac{s}{R}\right)\hat{\mathbf{i}} + \cos\left(\frac{s}{R}\right)\hat{\mathbf{j}}$$

and

$$\frac{d\hat{\mathbf{T}}}{ds} = -\frac{1}{R}\cos\left(\frac{s}{R}\right)\hat{\mathbf{i}} - \frac{1}{R}\sin\left(\frac{s}{R}\right)\hat{\mathbf{j}}.$$

At the origin, $s = \pi R$, and

$$\frac{d\hat{\mathbf{T}}}{ds}_{|s=\pi R} = \frac{1}{R}\hat{\mathbf{i}}.$$

Thus for each of the circles in Fig. 12.125, the vector $d\hat{\mathbf{T}}/ds$ has exactly the same direction. Yet the rate of change of the direction of $\hat{\mathbf{T}}$ is not the same for each circle; the direction changes more rapidly as the radius of the circle decreases. We must conclude, therefore, that it cannot be the direction of $d\hat{\mathbf{T}}/ds$ that measures the rate of change of $\hat{\mathbf{T}}$. Since a vector has only length and direction, it must be the length of $d\hat{\mathbf{T}}/ds$ that measures this rate of change. The circles in Fig. 12.125 certainly support this claim; the length of $d\hat{\mathbf{T}}/ds$ is $1/R$, and this quantity increases as the radii of the circles decrease. This agrees with the fact that the rate at which $\hat{\mathbf{T}}$ turns increases as R decreases. According to the following definition, we call $|d\hat{\mathbf{T}}/ds|$ *curvature* and $1/|d\hat{\mathbf{T}}/ds|$ *radius of curvature.*

▷ **Definition 12.19** If $x = x(s)$, $y = y(s)$, $z = z(s)$, $0 \le s \le L$, are parametric equations for a smooth curve in terms of length s along the curve, we define the *curvature* of the curve at a point as

$$\kappa(s) = \left|\frac{d\hat{\mathbf{T}}}{ds}\right|, \tag{12.86}$$

its *radius of curvature* as

$$\rho(s) = \frac{1}{\kappa(s)}, \tag{12.87}$$

and its circle of curvature as that circle in the plane of $\hat{\mathbf{T}}$ and $\hat{\mathbf{N}}$ with center at $\mathbf{r}(s) + \rho(s)\hat{\mathbf{N}}$ and radius $\rho(s)$. ◁

The circle of curvature is illustrated in Fig. 12.126.

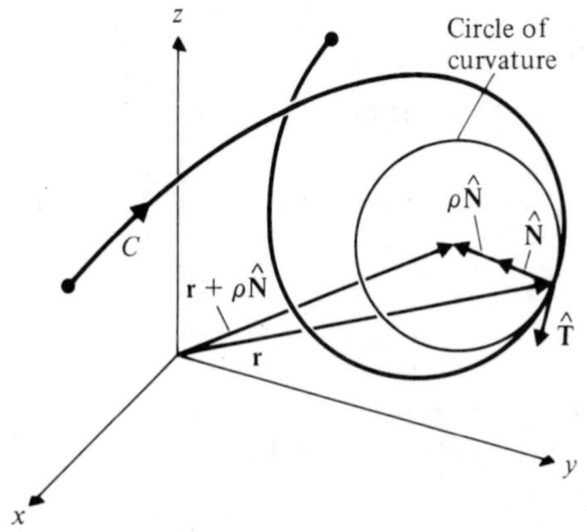

Figure 12.126

For the circles in Fig. 12.125 we have already shown that $d\hat{\mathbf{T}}/ds = R^{-1}\hat{\mathbf{i}}$. Consequently, for these circles, the curvature is always R^{-1}, and the radius of curvature is R, the radius of the circle. In other words, for a circle, the circle of curvature is the circle itself, its radius of curvature is its radius, and its curvature is the inverse of its radius. For the case when a curve is not a circle, we show in Exercise 27 that at any point on the curve the circle of curvature is in some sense the "best-fitting" circle to the curve at that point.

Because curvature and radius of curvature have been defined in terms of intrinsic properties $\hat{\mathbf{T}}$ and s for a curve, they must also be intrinsic properties. It follows, then, that no matter what parameter is used to specify points on a curve, curvature and radius of curvature are always the same. The following theorem shows how to calculate κ and ρ when the curve is specified in terms of a parameter other than length along the curve.

▶ **Theorem 12.11** When a smooth curve is defined parametrically by

$$C: \quad x = x(t), \qquad y = y(t), \qquad z = z(t), \qquad \alpha \le t \le \beta,$$

its curvature $\kappa(t)$ is defined by

$$\kappa(t) = \frac{|\dot{\mathbf{r}} \times \ddot{\mathbf{r}}|}{|\dot{\mathbf{r}}|^3}, \tag{12.88}$$

where $\dot{\mathbf{r}} = d\mathbf{r}/dt$ and $\ddot{\mathbf{r}} = d^2\mathbf{r}/dt^2$. ◀

Proof If $s(t)$ is length along C (measured from $t = \alpha$), then by the chain rule

$$\kappa = \left|\frac{d\hat{\mathbf{T}}}{ds}\right| = \left|\frac{d\hat{\mathbf{T}}}{dt}\frac{dt}{ds}\right| = \frac{|d\hat{\mathbf{T}}/dt|}{|ds/dt|} = \frac{|d\hat{\mathbf{T}}/dt|}{ds/dt} = \frac{|d\hat{\mathbf{T}}/dt|}{|\dot{\mathbf{r}}|}$$

(see Eq. 12.75). Now, we can write $\dot{\mathbf{r}}$ in the form $\dot{\mathbf{r}} = |\dot{\mathbf{r}}|\hat{\mathbf{T}}$; therefore, using (12.53), we have

$$\begin{aligned}
\ddot{\mathbf{r}} &= \left(\frac{d}{dt}|\dot{\mathbf{r}}|\right)\hat{\mathbf{T}} + |\dot{\mathbf{r}}|\frac{d\hat{\mathbf{T}}}{dt} \\
&= \left(\frac{d}{dt}|\dot{\mathbf{r}}|\right)\hat{\mathbf{T}} + (|\dot{\mathbf{r}}|\,|d\hat{\mathbf{T}}/dt|)\frac{d\hat{\mathbf{T}}/dt}{|d\hat{\mathbf{T}}/dt|} \\
&= \left(\frac{d}{dt}|\dot{\mathbf{r}}|\right)\hat{\mathbf{T}} + \left(|\dot{\mathbf{r}}|\left|\frac{d\hat{\mathbf{T}}}{dt}\right|\right)\hat{\mathbf{N}}.
\end{aligned}$$

If we take the cross product of this vector with $\dot{\mathbf{r}}$, we get

$$\begin{aligned}
\dot{\mathbf{r}} \times \ddot{\mathbf{r}} &= \left(\frac{d}{dt}|\dot{\mathbf{r}}|\right)\dot{\mathbf{r}} \times \hat{\mathbf{T}} + \left(|\dot{\mathbf{r}}|\left|\frac{d\hat{\mathbf{T}}}{dt}\right|\right)\dot{\mathbf{r}} \times \hat{\mathbf{N}} \\
&= \left(|\dot{\mathbf{r}}|\left|\frac{d\hat{\mathbf{T}}}{dt}\right|\right)\dot{\mathbf{r}} \times \hat{\mathbf{N}}. \qquad \text{(Since } \dot{\mathbf{r}} \text{ is parallel to } \hat{\mathbf{T}})
\end{aligned}$$

Because $\dot{\mathbf{r}}$ is perpendicular to $\hat{\mathbf{N}}$, it follows that $|\dot{\mathbf{r}} \times \hat{\mathbf{N}}| = |\dot{\mathbf{r}}|\,|\hat{\mathbf{N}}| \sin(\pi/2) = |\dot{\mathbf{r}}|$, and therefore

$$|\dot{\mathbf{r}} \times \ddot{\mathbf{r}}| = \left(|\dot{\mathbf{r}}|\left|\frac{d\hat{\mathbf{T}}}{dt}\right|\right)|\dot{\mathbf{r}}|.$$

Consequently,

$$\left|\frac{d\hat{\mathbf{T}}}{dt}\right| = \frac{|\dot{\mathbf{r}} \times \ddot{\mathbf{r}}|}{|\dot{\mathbf{r}}|^2}$$

and

$$\kappa = \kappa(t) = \frac{|\dot{\mathbf{r}} \times \ddot{\mathbf{r}}|}{|\dot{\mathbf{r}}|^3}. \square$$

For the radius of curvature, we have the following.

▶ **Corollary** When a smooth curve is defined in terms of an arbitrary parameter t,

$$\rho(t) = \frac{|\dot{\mathbf{r}}|^3}{|\dot{\mathbf{r}} \times \ddot{\mathbf{r}}|}. \blacktriangleleft \tag{12.89}$$

Example 39 Find the curvature and radius of curvature for the curve in Example 37.

Solution According to (12.88),

$$\kappa(t) = \frac{|\dot{\mathbf{r}} \times \ddot{\mathbf{r}}|}{|\dot{\mathbf{r}}|^3} = \frac{|(1, 2t, 2t) \times (0, 2, 2)|}{|(1, 2t, 2t)|^3}$$

$$= \frac{1}{(1 + 4t^2 + 4t^2)^{3/2}} \left\| \begin{matrix} \hat{\mathbf{i}} & \hat{\mathbf{j}} & \hat{\mathbf{k}} \\ 1 & 2t & 2t \\ 0 & 2 & 2 \end{matrix} \right\| = \frac{1}{(1 + 8t^2)^{3/2}} |(0, -2, 2)|$$

$$= \frac{2\sqrt{2}}{(1 + 8t^2)^{3/2}};$$

$$\rho(t) = \frac{1}{\kappa(t)} = \frac{(1 + 8t^2)^{3/2}}{2\sqrt{2}}.$$

Note in particular that as t increases, so too does ρ, a fact that is certainly supported by Fig. 12.124. □

Example 40 Show that for a smooth curve $y = y(x)$ in the xy-plane,

$$\kappa(x) = \frac{|y''|}{[1 + (y')^2]^{3/2}}.$$

Solution When we use x as parameter along the curve $y = y(x)$, parametric equations are $x = x$, $y = y(x)$. Then,

$$\dot{\mathbf{r}} = (1, y'(x)),$$
$$\ddot{\mathbf{r}} = (0, y''(x)),$$

and

$$\dot{\mathbf{r}} \times \ddot{\mathbf{r}} = \begin{vmatrix} \hat{\mathbf{i}} & \hat{\mathbf{j}} & \hat{\mathbf{k}} \\ 1 & y' & 0 \\ 0 & y'' & 0 \end{vmatrix} = y''\hat{\mathbf{k}}.$$

Thus,

$$\kappa(x) = \frac{|\dot{\mathbf{r}} \times \ddot{\mathbf{r}}|}{|\dot{\mathbf{r}}|^3} = \frac{|y''|}{|(1, y')|^3} = \frac{|y''|}{[1 + (y')^2]^{3/2}}. \square$$

EXERCISES 12.10

In Exercises 1–5, find $\hat{\mathbf{N}}$ and $\hat{\mathbf{B}}$ at each point on the curve.

1. $x = \sin t,\ y = \cos t,\ z = t,\ -\infty < t < \infty$
2. $x = t,\ y = t^2,\ z = t^3,\ t \geq 1$
3. $x = (t-1)^2,\ y = (t+1)^2,\ z = -t,\ -3 \leq t \leq 4$
4. $x + y = 5,\ x^2 - y = z$, from $(5, 0, 25)$ to $(0, 5, -5)$
5. $z = x,\ x^2 + y^2 = 4,\ y \geq 0$, from $(2, 0, 2)$ to $(-2, 0, -2)$

In Exercises 6–10, find $\hat{\mathbf{N}}$ and $\hat{\mathbf{B}}$ at the point indicated.

6. $x = 4\cos t,\quad y = 6\sin t,\quad z = 2\sin t,\quad -\infty < t < \infty;$ $(2\sqrt{2}, 3\sqrt{2}, \sqrt{2})$
7. $x = 2 - 5t,\quad y = 1 + t,\quad z = 6 + 4t^3,\quad -\infty < t < \infty;$ $(7, 0, 2)$
8. $x^2 + y^2 + z^2 = 4,\ z = \sqrt{x^2 + y^2}$, directed so that x increases when y is positive; $(1, 1, \sqrt{2})$
9. $x = y^2 + 1,\ z = x + 5$, directed so that y increases along the curve; $(5, 2, 10)$
10. $x^2 + (y-1)^2 = 4,\ x = z$, directed so that z decreases when y is negative; $(2, 1, 2)$

In Exercises 11–18, find the curvature and the radius of curvature of the curve (if they exist). In each case, sketch the curve.

11. $(x - h)^2 + (y - k)^2 = R^2,\ z = 0$, directed counterclockwise
12. $x = x_0 + at,\ y = y_0 + bt,\ z = z_0 + ct,\ -\infty < t < \infty$ (x_0, y_0, z_0, a, b, c all constants)
13. $x = t,\ y = t^2,\ z = 0,\ t \geq 0$
14. $x = e^t \cos t,\ y = e^t \sin t,\ z = t,\ -\infty < t < \infty$
15. $x = t,\ y = t^3,\ z = t^2,\ t \geq 0$
16. $x = 2\cos t,\ y = 2\sin t,\ z = 2\sin t,\ 0 \leq t < 2\pi$
17. $x = t + 1,\ y = t^2 - 1,\ z = t + 1,\ -\infty < t < \infty$
18. $x = t^2,\ y = t^4,\ z = 2t,\ -1 \leq t \leq 5$
19. At what points on the ellipse $b^2x^2 + a^2y^2 = a^2b^2 (a > b)$ is the curvature a maximum and a minimum?
20. Show that curvature for a smooth curve $x = x(t),\ y = y(t),\ \alpha \leq t \leq \beta$, in the xy-plane can be expressed in the form
$$\kappa(t) = \frac{\left|\dfrac{dy}{dt}\dfrac{d^2x}{dt^2} - \dfrac{dx}{dt}\dfrac{d^2y}{dt^2}\right|}{\left[\left(\dfrac{dx}{dt}\right)^2 + \left(\dfrac{dy}{dt}\right)^2\right]^{3/2}}.$$
21. Show that the only curves for which curvature is identically equal to zero are straight lines.
22. What happens to curvature at a point of inflection on the graph of a function $y = y(x)$?
23. Let C be the curve $x = t,\ y = t^2$ in the xy-plane.
    a) At each point on C calculate the unit tangent vector $\hat{\mathbf{T}}$ and the principal normal $\hat{\mathbf{N}}$. What is $\hat{\mathbf{B}}$? (See Example 38 for $\hat{\mathbf{N}}$.)
    b) $\mathbf{F} = t^2\hat{\mathbf{i}} + t^4\hat{\mathbf{j}}$ is a vector that is defined at each point P on C. Denote by F_T and F_N the components of $\mathbf{F}$ in the directions $\hat{\mathbf{T}}$ and $\hat{\mathbf{N}}$ at P. Find F_T and F_N as functions of t.
    c) Express $\mathbf{F}$ in terms of $\hat{\mathbf{T}}$ and $\hat{\mathbf{N}}$.
24. Repeat Exercise 23 for the curve C: $x = 2\cos t,\ y = 2\sin t$, and the vector $\mathbf{F} = x^2\hat{\mathbf{i}} + y^2\hat{\mathbf{j}}$.
25. The vectors $\hat{\mathbf{T}}$, $\hat{\mathbf{N}}$, and $\hat{\mathbf{B}}$ were calculated at each point on the curve $x = t,\ y = t^2,\ z = t^2$ in Example 37, If $\mathbf{F} = t^2\hat{\mathbf{i}} + 2t\hat{\mathbf{j}} - 3\hat{\mathbf{k}}$ is a vector defined along C, find the components of $\mathbf{F}$ in the directions $\hat{\mathbf{T}}$, $\hat{\mathbf{N}}$, and $\hat{\mathbf{B}}$. Express $\mathbf{F}$ in terms of $\hat{\mathbf{T}}$, $\hat{\mathbf{N}}$, and $\hat{\mathbf{B}}$.
26. Calculate $\hat{\mathbf{T}}$, $\hat{\mathbf{N}}$, and $\hat{\mathbf{B}}$ for the curve $x = \cos t,\ y = \sin t,\ z = t$. Express the vector $\mathbf{F} = x\hat{\mathbf{i}} + xy^2\hat{\mathbf{j}} + \hat{\mathbf{k}}$ in terms of $\hat{\mathbf{T}}$, $\hat{\mathbf{N}}$, and $\hat{\mathbf{B}}$.
27. In this exercise we discuss our claim that the circle of curvature is the "best-fitting" circle to the curve at a point.
    a) First, is it true that the circle of curvature at a point on a curve passes through that point?
    b) Second, show that the circle of curvature and curve share the same tangent line at their common point.
    c) Finally, verify that the circle of curvature and curve have the same curvature at their common point.
28. If ϕ is the angle between $\hat{\mathbf{i}}$ and $\hat{\mathbf{T}}$ for a curve in the xy-plane (Fig. 12.127), show that
$$\kappa(s) = \left|\frac{d\phi}{ds}\right|.$$

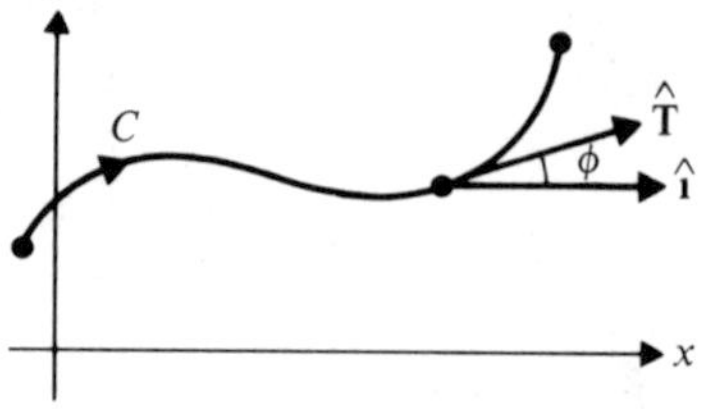

Figure 12.127

*29. In differential geometry the rates of change $d\hat{\mathbf{T}}/ds$, $d\hat{\mathbf{N}}/ds$, and $d\hat{\mathbf{B}}/ds$ of $\hat{\mathbf{T}}$, $\hat{\mathbf{N}}$, and $\hat{\mathbf{B}}$ with respect to length along a curve are represented by the Frenet–Serret formulas. We develop these equations in this exercise.

a) The first Frenet–Serret formula is

$$\frac{d\hat{\mathbf{T}}}{ds} = \kappa\hat{\mathbf{N}}.$$

Verify this result.

b) The second Frenet–Serret formula is

$$\frac{d\hat{\mathbf{B}}}{ds} = -\tau\hat{\mathbf{N}},$$

where τ is called the torsion of the curve. To verify this result, differentiate the equations $\hat{\mathbf{B}} = \hat{\mathbf{T}} \times \hat{\mathbf{N}}$ and $\hat{\mathbf{B}} \cdot \hat{\mathbf{B}} = 1$ to show that $d\hat{\mathbf{B}}/ds$ is perpendicular to $\hat{\mathbf{T}}$ and $\hat{\mathbf{B}}$ and therefore parallel to $\hat{\mathbf{N}}$.

c) The third formula is

$$\frac{d\hat{\mathbf{N}}}{ds} = \tau\hat{\mathbf{B}} - \kappa\hat{\mathbf{T}}.$$

Verify this result by showing that $\hat{\mathbf{N}} = \hat{\mathbf{B}} \times \hat{\mathbf{T}}$ and then calculating $d\hat{\mathbf{N}}/ds$.

12.11 DISPLACEMENT, VELOCITY, AND ACCELERATION

In Sections 4.1 and 5.2 we introduced the concepts of displacement, velocity, and acceleration for moving objects, but indicated that our terminology at that time was somewhat loose. In particular, we stated that if $x = x(t)$ represents the position of a particle moving along the x-axis, then the instantaneous velocity of the particle is

$$v = \frac{dx}{dt}, \tag{12.90}$$

provided, of course, that t is time, and the acceleration of the particle is

$$a = \frac{dv}{dt} = \frac{d^2x}{dt^2}. \tag{12.91}$$

We illustrated by examples that given any one of $x(t)$, $v(t)$, or $a(t)$ and sufficient initial conditions, it is always possible to find the other two. There was nothing wrong with the calculations in the examples—they were correct—but our terminology was not quite correct. We now rectify this situation and give precise definitions of velocity and acceleration.

Suppose a particle moves along some curve C in space (under perhaps the influence of various forces), and that C is defined as a function of time t by the parametric equation

$$C:\quad x = x(t), \qquad y = y(t), \qquad z = z(t), \qquad t \geq 0. \tag{12.92}$$

The position of the particle can then be described as a function of time by its position or displacement vector:

$$\mathbf{r} = \mathbf{r}(t) = x(t)\hat{\mathbf{i}} + y(t)\hat{\mathbf{j}} + z(t)\hat{\mathbf{k}}, \qquad t \geq 0. \tag{12.93}$$

The *velocity* $\mathbf{v}$ of the particle at any time t is defined as the time rate of change of its displacement vector:

$$\mathbf{v} = \frac{d\mathbf{r}}{dt}. \tag{12.94}$$

Velocity, then, is a vector, and because of Theorem 12.8, the components of velocity are the derivatives of the components of displacement:

$$\mathbf{v} = \frac{d\mathbf{r}}{dt} = \frac{dx}{dt}\hat{\mathbf{i}} + \frac{dy}{dt}\hat{\mathbf{j}} + \frac{dz}{dt}\hat{\mathbf{k}}. \tag{12.95}$$

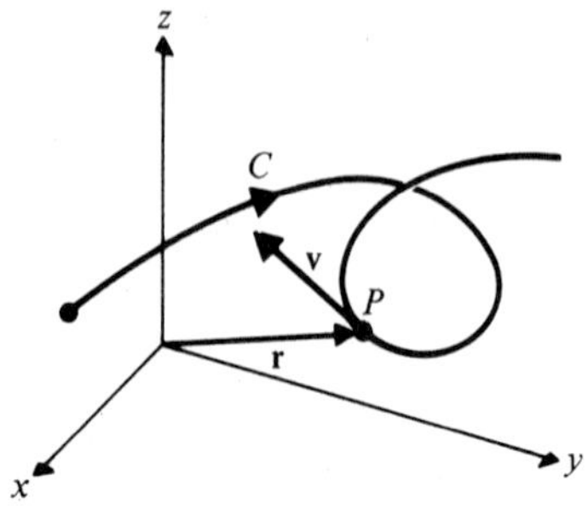

Figure 12.128

But according to Theorem 12.10, the vector $d\mathbf{r}/dt$ is tangent to the curve C (Fig. 12.128). In other words, if a particle is at position P, and we draw its velocity vector with tail at P, then $\mathbf{v}$ will be tangent to the trajectory.

In some applications it is the length or magnitude of velocity that is important, not its direction. This quantity, called *speed*, is therefore defined by

$$|\mathbf{v}| = \sqrt{\left(\frac{dx}{dt}\right)^2 + \left(\frac{dy}{dt}\right)^2 + \left(\frac{dz}{dt}\right)^2}. \tag{12.96}$$

Equation (12.75) implies that if $s(t)$ is length along the trajectory C (where $s(0) = 0$), then $|\mathbf{v}| = ds/dt$. In other words, speed is the time rate of change of distance traveled along C.

It is important to understand this difference between velocity and speed. Velocity is the time derivative of displacement; speed is the time derivative of distance traveled. Velocity is a vector; speed is a scalar—the magnitude of velocity.

The *acceleration* of the particle as it moves along the curve C in Eq. (12.92) is defined as the rate of change of velocity with respect to time:

$$\mathbf{a} = \frac{d\mathbf{v}}{dt} = \frac{d^2\mathbf{r}}{dt^2} = \frac{d^2x}{dt^2}\hat{\mathbf{i}} + \frac{d^2y}{dt^2}\hat{\mathbf{j}} + \frac{d^2z}{dt^2}\hat{\mathbf{k}}. \tag{12.97}$$

Acceleration, then, is also a vector; it is the derivative of velocity, and therefore its components are the derivatives of the components of the velocity vector. Alternatively, it is the second derivative of displacement and has components that are the second derivatives of the displacement vector.

In the special case in which C is a curve in the xy-plane, the definitions of displacement, velocity, speed, and acceleration become, respectively,

$$\mathbf{r} = x(t)\hat{\mathbf{i}} + y(t)\hat{\mathbf{j}}, \tag{12.98}$$

$$\mathbf{v} = \frac{d\mathbf{r}}{dt} = \frac{dx}{dt}\hat{\mathbf{i}} + \frac{dy}{dt}\hat{\mathbf{j}}, \tag{12.99}$$

$$|\mathbf{v}| = \sqrt{\left(\frac{dx}{dt}\right)^2 + \left(\frac{dy}{dt}\right)^2}, \tag{12.100}$$

$$\mathbf{a} = \frac{d\mathbf{v}}{dt} = \frac{d^2\mathbf{r}}{dt^2} = \frac{d^2x}{dt^2}\hat{\mathbf{i}} + \frac{d^2y}{dt^2}\hat{\mathbf{j}}. \tag{12.101}$$

For motion along the x-axis,

$$\mathbf{r} = x(t)\hat{\mathbf{i}}, \tag{12.102}$$

$$\mathbf{v} = \frac{d\mathbf{r}}{dt} = \frac{dx}{dt}\hat{\mathbf{i}}, \tag{12.103}$$

$$|\mathbf{v}| = \left|\frac{dx}{dt}\right|, \tag{12.104}$$

$$\mathbf{a} = \frac{d\mathbf{v}}{dt} = \frac{d^2\mathbf{r}}{dt^2} = \frac{d^2x}{dt^2}\hat{\mathbf{i}}. \tag{12.105}$$

If we compare Eqs. (12.103) and (12.105) with (12.90) and (12.91), we see that for motion along the x-axis, $x(t)$, $v(t)$, and $a(t)$ are the components of the displacement, velocity, and acceleration vectors, respectively. Because these are the only components of $\mathbf{r}(t)$, $\mathbf{v}(t)$, and $\mathbf{a}(t)$, it follows that consideration of the components of the vectors is equivalent to consideration of the vectors themselves. For one-dimensional motion, then, we can drop the vector notation and work with components (and this is precisely the procedure that we followed in Sections 4.1 and 5.2).

The most important law governing the motion of objects is *Newton's second law*. It states that if an object of mass m is subjected to a force $\mathbf{F}$, then the time rate of change of its momentum must be equal to $\mathbf{F}$:

$$\mathbf{F} = \frac{d}{dt}(m\mathbf{v}). \tag{12.106}$$

In most cases, the mass of the object is constant, and this equation then yields its acceleration:

$$\mathbf{F} = m\frac{d\mathbf{v}}{dt} = m\mathbf{a}. \tag{12.107}$$

If $\mathbf{F}$ is known as a function of time t, $\mathbf{F} = \mathbf{F}(t)$, then (12.107) defines the acceleration of the object as a function of time,

$$\mathbf{a}(t) = \frac{1}{m}\mathbf{F}(t),$$

and integration of this equation leads to expressions for the velocity $\mathbf{v}(t)$ and position $\mathbf{r}(t)$ as functions of time.

Example 41 Find the velocity and acceleration of a particle and describe its motion, if its position as a function of time is given by

$$x = 2t + \sqrt{2}, \qquad y = \sqrt{4t^2 + 4\sqrt{2}t + 1}, \qquad z = 0, \qquad t \geq 0.$$

Solution If $\mathbf{r} = x(t)\hat{\mathbf{i}} + y(t)\hat{\mathbf{j}}$, then the velocity of the particle is

$$\mathbf{v} = \frac{dx}{dt}\hat{\mathbf{i}} + \frac{dy}{dt}\hat{\mathbf{j}} = 2\hat{\mathbf{i}} + \left(\frac{4t + 2\sqrt{2}}{\sqrt{4t^2 + 4\sqrt{2}t + 1}}\right)\hat{\mathbf{j}},$$

and its acceleration is

$$\mathbf{a} = \frac{d^2x}{dt^2}\hat{\mathbf{i}} + \frac{d^2y}{dt^2}\hat{\mathbf{j}} = \frac{-4}{(4t^2 + 4\sqrt{2}t + 1)^{3/2}}\hat{\mathbf{j}}.$$

Because $t = (x - \sqrt{2})/2$, an explicit definition of the path is

$$y = \sqrt{4\left(\frac{x - \sqrt{2}}{2}\right)^2 + 4\sqrt{2}\left(\frac{x - \sqrt{2}}{2}\right) + 1} = \sqrt{x^2 - 1}.$$

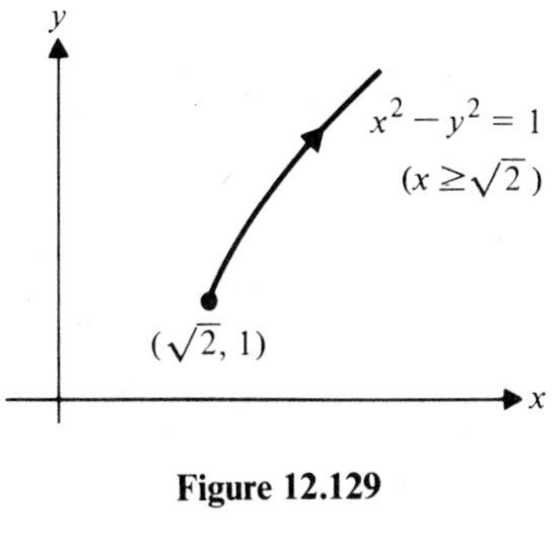

Figure 12.129

Since $t \geq 0$, it follows that $x \geq \sqrt{2}$, and the path is that part of the hyperbola $x^2 - y^2 = 1$ shown in Fig. 12.129. Note that $dx/dt = 2$, so that the x-component of the velocity is always equal to 2. This is also reflected in the fact that the acceleration has no x-component. Since the y-component of acceleration is always negative, the y-component of velocity must be decreasing in time. □

Example 42 A particle starts at time $t = 0$ from position (1, 1) with a speed of 2 meters per second in the negative y-direction. It is subjected to an acceleration that is given as a function of time by

$$\mathbf{a}(t) = \frac{1}{\sqrt{t+1}}\hat{\mathbf{i}} + 6t\hat{\mathbf{j}}$$

in meters per second per second. Find its velocity and position as functions of time.

Solution If $\mathbf{a} = d\mathbf{v}/dt = (1/\sqrt{t+1})\hat{\mathbf{i}} + 6t\hat{\mathbf{j}}$, then

$$\mathbf{v} = 2\sqrt{t+1}\,\hat{\mathbf{i}} + 3t^2\hat{\mathbf{j}} + \mathbf{C},$$

where $\mathbf{C}$ is some constant vector. Because the initial velocity of the particle is 2 m/s in the negative y-direction, $\mathbf{v}(0) = -2\hat{\mathbf{j}}$. Consequently,

$$-2\hat{\mathbf{j}} = 2\hat{\mathbf{i}} + \mathbf{C},$$

which implies that $\mathbf{C} = -2\hat{\mathbf{i}} - 2\hat{\mathbf{j}}$. The velocity, then, of the particle at any time $t \geq 0$ is

$$\mathbf{v}(t) = (2\sqrt{t+1} - 2)\hat{\mathbf{i}} + (3t^2 - 2)\hat{\mathbf{j}}\text{ m/s}.$$

Because $\mathbf{v} = d\mathbf{r}/dt$, integration gives

$$\mathbf{r} = (\tfrac{4}{3}(t+1)^{3/2} - 2t)\hat{\mathbf{i}} + (t^3 - 2t)\hat{\mathbf{j}} + \mathbf{D}.$$

Since the particle starts from position (1, 1), $\mathbf{r}(0) = \hat{\mathbf{i}} + \hat{\mathbf{j}}$, and

$$\hat{\mathbf{i}} + \hat{\mathbf{j}} = \tfrac{4}{3}\hat{\mathbf{i}} + \mathbf{D}, \quad \text{or} \quad \mathbf{D} = -\tfrac{1}{3}\hat{\mathbf{i}} + \hat{\mathbf{j}}.$$

The displacement of the particle is therefore

$$\mathbf{r}(t) = (\tfrac{4}{3}(t+1)^{3/2} - 2t - \tfrac{1}{3})\hat{\mathbf{i}} + (t^3 - 2t + 1)\hat{\mathbf{j}}\text{ m.} \quad \square$$

Example 43 A shell is fired from an artillery gun with a speed of v_0 meters per second at an angle θ with the horizontal (Fig. 12.130). If gravity is assumed to be the only force acting on the shell:

a) Find the position of the projectile as a function of time t.
b) Find the range of the shell; i.e., find the horizontal distance traveled by the shell.
c) Find the maximum height attained by the shell.
d) Show that the path of the shell is a parabola.

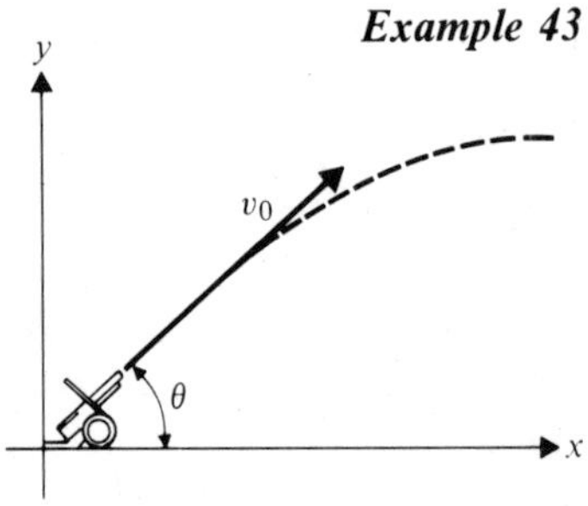

Figure 12.130

Solution a) If m is the mass of the shell, then the force of gravity on the shell is

$$\mathbf{F} = -9.81m\hat{\mathbf{j}}.$$

According to Newton's second law, the acceleration of the shell is defined by

$$-9.81m\hat{\mathbf{j}} = m\mathbf{a};$$

i.e.,

$$\mathbf{a} = -9.81\hat{\mathbf{j}}.$$

Since $\mathbf{a} = d\mathbf{v}/dt$, we can write that

$$\frac{d\mathbf{v}}{dt} = -9.81\hat{\mathbf{j}},$$

and integration of this equation gives

$$\mathbf{v}(t) = -9.81t\hat{\mathbf{j}} + \mathbf{C}.$$

If we choose time $t = 0$ at the instant the shell is fired, then $\mathbf{v}(0) = v_0 \cos\theta\hat{\mathbf{i}} + v_0 \sin\theta\hat{\mathbf{j}}$, and hence

$$v_0 \cos\theta\hat{\mathbf{i}} + v_0 \sin\theta\hat{\mathbf{j}} = \mathbf{C}.$$

Because $\mathbf{v} = d\mathbf{r}/dt$, the position vector of the shell is

$$\mathbf{r}(t) = -4.905t^2\hat{\mathbf{j}} + \mathbf{C}t + \mathbf{D}.$$

Since $\mathbf{r}(0) = \mathbf{0}$, we must set $\mathbf{D} = \mathbf{0}$, and

$$\begin{aligned} \mathbf{r}(t) &= -4.905t^2\hat{\mathbf{j}} + v_0 \cos\theta t\hat{\mathbf{i}} + v_0 \sin\theta t\hat{\mathbf{j}} \\ &= (tv_0 \cos\theta)\hat{\mathbf{i}} + (-4.905t^2 + tv_0 \sin\theta)\hat{\mathbf{j}}. \end{aligned}$$

b) The shell strikes the ground when the y-component of $\mathbf{r}$ vanishes; i.e., when

$$\begin{aligned} 0 &= -4.905t^2 + tv_0 \sin\theta \\ &= t(-4.905t + v_0 \sin\theta). \end{aligned}$$

Clearly, $t = (v_0 \sin\theta)/4.905$, and at this time, the x-component of $\mathbf{r}$ is the range

$$v_0 \cos\theta\left(\frac{v_0 \sin\theta}{4.905}\right) = \frac{v_0^2 \sin\theta\cos\theta}{4.905} = \frac{v_0^2 \sin 2\theta}{9.81} \text{ meters.}$$

c) The shell will be at its maximum height when its y-component of velocity is equal to zero; i.e., when

$$0 = -9.81t + v_0 \sin\theta,$$

or

$$t = \frac{v_0 \sin\theta}{9.81}.$$

At this time the y-component of the position vector is

$$-4.905\left(\frac{v_0 \sin\theta}{9.81}\right)^2 + v_0 \sin\theta\left(\frac{v_0 \sin\theta}{9.81}\right) = \frac{v_0^2 \sin^2\theta}{19.62} \text{ meters.}$$

d) Since $x(t) = tv_0 \cos\theta$ and $y(t) = -4.905t^2 + tv_0 \sin\theta$, it follows that

$$\begin{aligned} y &= -4.905\left(\frac{x}{v_0 \cos\theta}\right)^2 + v_0 \sin\theta\left(\frac{x}{v_0 \cos\theta}\right) \\ &= \frac{-4.905}{v_0^2 \cos^2\theta}x^2 + x\tan\theta, \end{aligned}$$

a parabola. □

Tangential and Normal Components of Velocity and Acceleration

For some types of motion it is inconvenient to express velocity and acceleration in terms of Cartesian components; sometimes it is an advantage to resolve these vectors into tangential and normal components. When the trajectory C of a particle is specified as a function of time t by (12.92), its velocity $\mathbf{v} = d\mathbf{r}/dt$ is tangent to C, and we can therefore write

$$\mathbf{v} = |\mathbf{v}|\hat{\mathbf{T}}. \tag{12.108}$$

In other words, the tangential component of velocity is speed, and $\mathbf{v}$ has no normal component. Differentiation of this equation gives the particle's acceleration:

$$\begin{aligned}\mathbf{a} = \frac{d\mathbf{v}}{dt} &= \left(\frac{d}{dt}|\mathbf{v}|\right)\hat{\mathbf{T}} + |\mathbf{v}|\frac{d\hat{\mathbf{T}}}{dt}\\ &= \left(\frac{d}{dt}|\mathbf{v}|\right)\hat{\mathbf{T}} + \left(|\mathbf{v}|\left|\frac{d\hat{\mathbf{T}}}{dt}\right|\right)\frac{d\hat{\mathbf{T}}/dt}{|d\hat{\mathbf{T}}/dt|}\\ &= \left(\frac{d}{dt}|\mathbf{v}|\right)\hat{\mathbf{T}} + \left(|\mathbf{v}|\left|\frac{d\hat{\mathbf{T}}}{dt}\right|\right)\hat{\mathbf{N}}.\end{aligned} \tag{12.109}$$

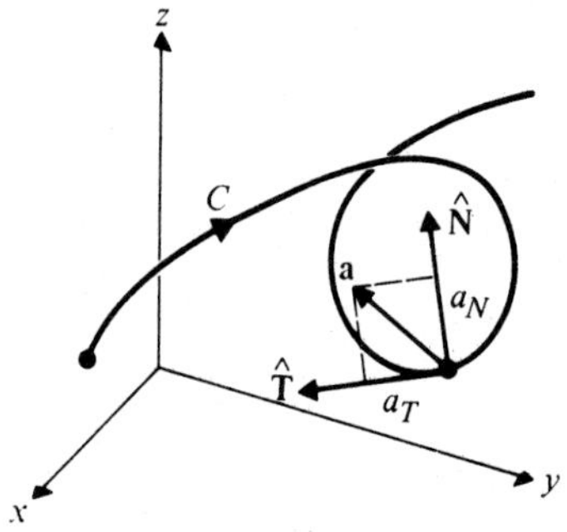

Figure 12.131

We have therefore expressed $\mathbf{a}$ in terms of the unit tangent vector $\hat{\mathbf{T}}$ to C and the principal normal $\hat{\mathbf{N}}$ (Fig. 12.131). We call $d(|\mathbf{v}|)/dt$ and $|\mathbf{v}||d\hat{\mathbf{T}}/dt|$ the tangential and normal components of acceleration, respectively. If a_T and a_N denote these components, then we can write that

$$\mathbf{a} = a_T\hat{\mathbf{T}} + a_N\hat{\mathbf{N}}, \tag{12.110a}$$

where

$$a_T = \mathbf{a}\cdot\hat{\mathbf{T}} = \frac{d}{dt}|\mathbf{v}|, \qquad a_N = \mathbf{a}\cdot\hat{\mathbf{N}} = |\mathbf{v}|\left|\frac{d\hat{\mathbf{T}}}{dt}\right|. \tag{12.110b}$$

Note that the tangential component of acceleration is the time rate of change of speed. Since acceleration is the rate of change of velocity, the normal component of acceleration must determine the rate of change of the direction of $\mathbf{v}$.

To calculate a_N using (12.110b) is often quite complicated. A far easier formula results if we take the scalar product of $\mathbf{a}$ as defined by (12.110a) with itself,

$$\begin{aligned}\mathbf{a}\cdot\mathbf{a} &= (a_T\hat{\mathbf{T}} + a_N\hat{\mathbf{N}})\cdot(a_T\hat{\mathbf{T}} + a_N\hat{\mathbf{N}})\\ &= a_T^2\hat{\mathbf{T}}\cdot\hat{\mathbf{T}} + 2a_Ta_N\hat{\mathbf{T}}\cdot\hat{\mathbf{N}} + a_N^2\hat{\mathbf{N}}\cdot\hat{\mathbf{N}}\\ &= a_T^2 + a_N^2,\end{aligned}$$

since $\hat{\mathbf{T}}\cdot\hat{\mathbf{N}} = 0$ and $\hat{\mathbf{T}}\cdot\hat{\mathbf{T}} = \hat{\mathbf{N}}\cdot\hat{\mathbf{N}} = 1$. Consequently,

$$a_N^2 = \mathbf{a}\cdot\mathbf{a} - a_T^2 = |\mathbf{a}|^2 - a_T^2,$$

and because a_N is always positive (see Eq. 12.110b),

$$a_N = \sqrt{|\mathbf{a}|^2 - a_T^2}. \tag{12.111}$$

Example 44 If the trajectory of a particle is defined by

$$x = t^2 + 1, \qquad y = 2t^2 - 1, \qquad z = t^2 + 5t, \qquad t \geq 0,$$

where t is time, find the tangential and normal components of the particle's velocity and acceleration.

Solution The velocity and acceleration of the particle have Cartesian components

$$\mathbf{v} = 2t\hat{\mathbf{i}} + 4t\hat{\mathbf{j}} + (2t + 5)\hat{\mathbf{k}}, \qquad \mathbf{a} = 2\hat{\mathbf{i}} + 4\hat{\mathbf{j}} + 2\hat{\mathbf{k}}.$$

The tangential component of the particle's velocity is its speed:

$$|\mathbf{v}| = \sqrt{(2t)^2 + (4t)^2 + (2t + 5)^2} = \sqrt{24t^2 + 20t + 25}.$$

According to (12.110b), the tangential component of the acceleration is

$$a_T = \frac{d}{dt}|\mathbf{v}| = \frac{24t + 10}{\sqrt{24t^2 + 20t + 25}}.$$

With (12.111), the normal component of acceleration is

$$a_N = \sqrt{|\mathbf{a}|^2 - a_T^2} = \left\{(4 + 16 + 4) - \frac{(24t + 10)^2}{24t^2 + 20t + 25}\right\}^{1/2}$$

$$= \left\{\frac{24(24t^2 + 20t + 25) - (24t + 10)^2}{24t^2 + 20t + 25}\right\}^{1/2}$$

$$= \frac{10\sqrt{5}}{\sqrt{24t^2 + 20t + 25}}. \quad \square$$

Example 45 Prove that the tangential component of the acceleration of a particle is equal to zero if and only if its speed is constant.

Solution The tangential component of the acceleration of a particle is $a_T = d(|\mathbf{v}|)/dt$ and is equal to zero, therefore, if and only if

$$\frac{d}{dt}|\mathbf{v}| = 0.$$

But this is valid if and only if

$$|\mathbf{v}| = \text{a constant.} \quad \square$$

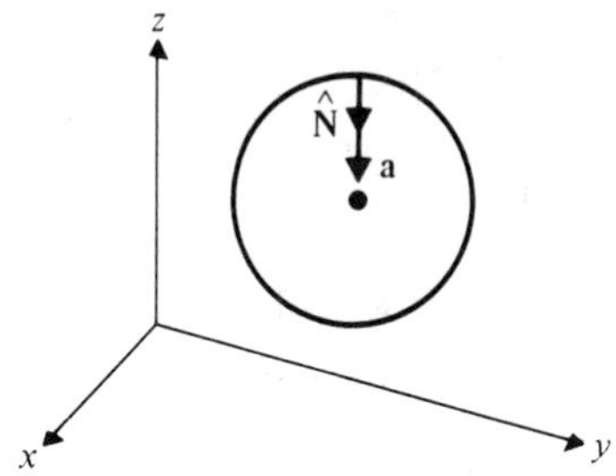

Figure 12.132

What this example shows is that when a particle moves with constant speed, its acceleration always points along the principal normal to the curve. In the special case in which the trajectory is a circle (Fig. 12.132), the principal normal always points directly toward the center of the circle; therefore, the acceleration always points toward the center of the circle also. Understand, however, that this is true only when speed is constant. If speed is not constant for circular motion, then a_T is not equal to zero; therefore, acceleration is not directed toward the center of the circle.

EXERCISES 12.11

In Exercises 1–5, find the velocity, speed, and acceleration of a particle if the given equations represent its position as a function of time.

1. $x(t) = \sqrt{t^2 + 1},\ y(t) = t\sqrt{t^2 + 1},\ t \geq 0$

2. $x(t) = t + 1/t,\ y(t) = t - 1/t,\ t \geq 1$

3. $x(t) = \sin t,\ y(t) = 3\cos t,\ z(t) = \sin t,\ 0 \leq t \leq 10\pi$

4. $x(t) = t^2 + 1,\ y(t) = 2te^t,\ z(t) = 1/t^2,\ 1 \leq t \leq 5$

5. $x(t) = e^{-t^2},\ y(t) = t \ln t,\ z(t) = 5,\ t \geq 1$

In Exercises 6 and 7, a particle at (1, 2, −1) starts from rest at time $t = 0$. Find its position as a function of time if the given function defines its acceleration.

6. $\mathbf{a}(t) = 3t^2\hat{\mathbf{i}} + (t + 1)\hat{\mathbf{j}} - 4t^3\hat{\mathbf{k}},\ t \geq 0$

7. $\mathbf{a}(t) = 3\hat{\mathbf{i}} + \hat{\mathbf{j}}/(t + 1)^3,\ t \geq 0$

In Exercises 8 and 9, find the tangential and normal components of acceleration for a particle moving with position defined by the given functions (where t is time).

8. $x(t) = t$, $y(t) = t^2 + 1$, $t \geq 0$

9. $x(t) = \cos t$, $y(t) = \sin t$, $z = t$, $t \geq 0$

10. Show that the normal component of acceleration of a particle can be expressed in the form

$$a_N = \frac{|\mathbf{v}|^2}{\rho} = \kappa|\mathbf{v}|^2.$$

11. The kinetic energy of an object of mass m moving with velocity $\mathbf{v}$ is defined as

$$K = \tfrac{1}{2}m|\mathbf{v}|^2.$$

Find kinetic energy for the particle in each of Exercises 1–5 if its mass is 2 grams. Assume that x, y, and z are measured in meters and t in seconds.

12. A particle starts at the origin and moves along the curve $4y = x^2$ to the point (4, 4).

a) If the y-component of its acceleration is always equal to 2 and the y-component of its velocity is initially zero, find the x-component of its acceleration.

b) If the x-component of its acceleration is equal to $24t^2$ (t being time) and the x-component of its velocity is initially zero, find the y-component of its acceleration.

13. A particle moves along the curve

$$x(t) = t, \qquad y(t) = t^3 - 3t^2 + 2t, \qquad 0 \leq t \leq 5$$

in the xy-plane (where t is time). Is there any point at which its velocity is parallel to its displacement?

14. A particle moves along the curve $y = x^3 - 2x + 3$ so that its x-component of velocity is always equal to 5. Find its acceleration.

15. If a particle starts at time $t = 0$ from rest at position (3, 4) and experiences an acceleration $\mathbf{a} = -5t^4\hat{\imath} - (2t^3 + 1)\hat{\jmath}$, find its speed at $t = 2$.

16. A particle travels counterclockwise around the circle $(x - h)^2 + (y - k)^2 = R^2$ in Fig. 12.133. Show that the speed of the particle at any time is $|\mathbf{v}| = \omega R$, where $\omega = d\theta/dt$ is called the *angular speed* of the particle.

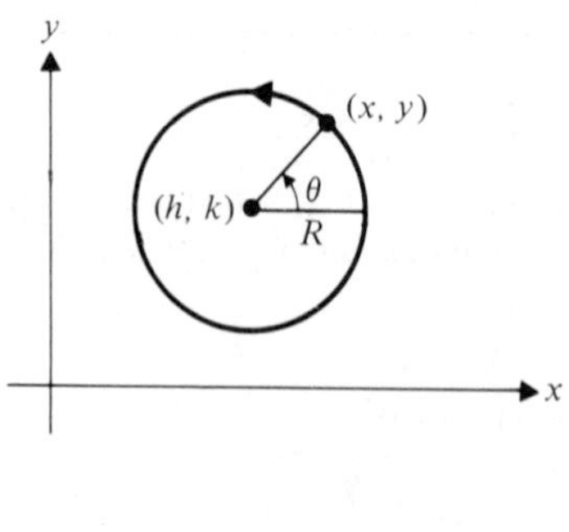

Figure 12.133

17. A particle travels around the circle $x^2 + y^2 = 4$ counterclockwise at a constant speed, making 2 revolutions each second. If x and y are measured in meters, what is the velocity of the particle when it is at the point $(1, -\sqrt{3})$?

18. a) Show that if an object moves with constant speed in a circular path of radius R, the magnitude of its acceleration is $|\mathbf{a}| = |\mathbf{v}|^2/R$.

b) If a satellite moves with constant speed in a circular orbit 200 kilometers above the earth's surface, what is its speed? [*Hint:* Use Newton's universal law of gravitation (see Exercise 32 in Section 12.4) to determine the acceleration $\mathbf{a}$ of the satellite. Assume that the earth is a sphere with radius 6370 kilometers and density 5.52×10^3 kilograms per cubic meter.]

19. Two particles move along curves C_1 and C_2 (Fig. 12.134). If at some instant of time the particles are at positions P_1 and P_2, then the vector $\overrightarrow{\mathbf{P_1P_2}}$ is the displacement of P_2 with respect to P_1. Clearly, $\overrightarrow{\mathbf{OP}}_1 + \overrightarrow{\mathbf{P_1P_2}} = \overrightarrow{\mathbf{OP}}_2$. Show that when this equation is differentiated with respect to time, we have

$$\mathbf{v}_{P_1/O} + \mathbf{v}_{P_2/P_1} = \mathbf{v}_{P_2/O},$$

where $\mathbf{v}_{P_1/O}$ and $\mathbf{v}_{P_2/O}$ are the velocities of P_1 and P_2 with respect to the origin, and $\mathbf{v}_{P_2/P_1}$ is the velocity of P_2 with respect to P_1.

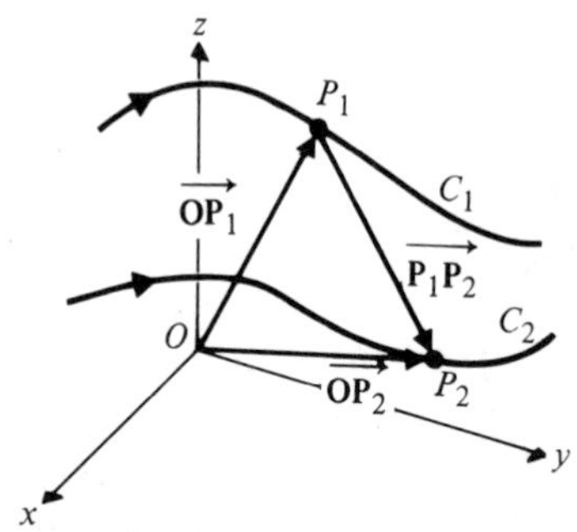

Figure 12.134

20. A plane flies on course N30°E with an air speed of 650 kilometers per hour (i.e., the speed of the plane relative to the air is 650). If the air is moving at a rate of 40 kilometers per hour directly east, find the ground velocity and speed of the plane. (*Hint:* Use Exercise 19.)

21. a) In Fig. 12.135, a cannon is fired up an inclined plane. If the speed at which the ball is ejected from the cannon is S, show that the range R of the ball is given by

$$R = \frac{2S^2 \cos\theta \sin(\theta - \alpha)}{g \cos^2 \alpha},$$

where $g > 0$ is the acceleration due to gravity.

b) What angle θ will make R a maximum?

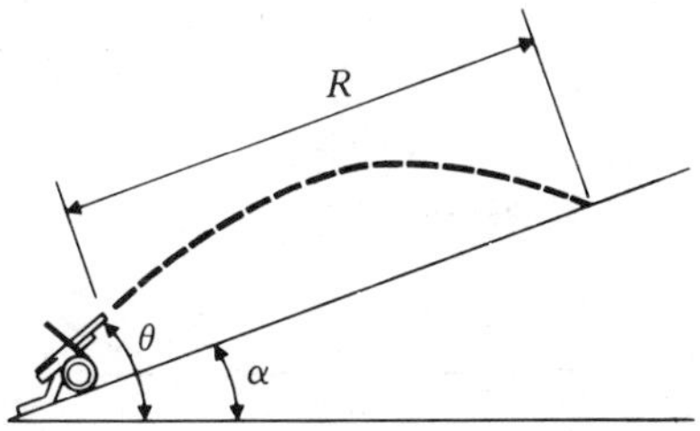

Figure 12.135

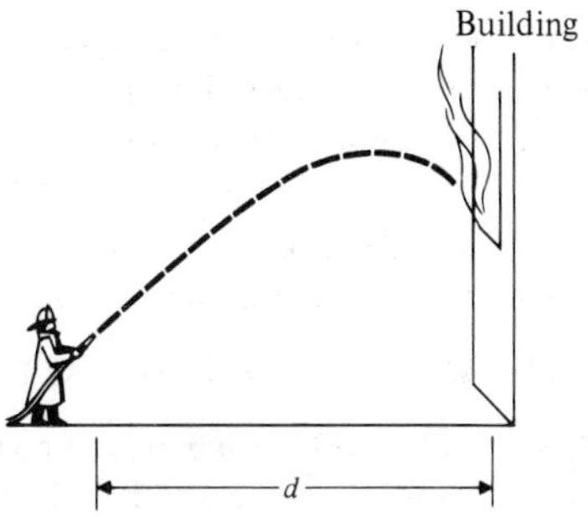

Figure 12.136

22. What constant acceleration must a particle experience if it is to travel from (1, 2, 3) to (4, 5, 7) along the straight line joining the points, starting from rest, and covering the distance in 2 units of time?

23. Calculate the normal component a_N of the acceleration of the particle in Example 44 using (12.110b).

24. A particle moves along the curve

$$x(t) = 2 + \sqrt{1 - t^2}, \qquad y(t) = t, \qquad 0 \le t \le \tfrac{1}{2},$$

where t is time. Is there a time at which its acceleration is perpendicular to its velocity?

25. a) Show that motion along a straight line results in both of the following situations:

i) The initial velocity is zero, and the acceleration is constant.

ii) The initial velocity is nonzero, and the acceleration is constant and parallel to the initial velocity.

b) Can we generalize the results of (a), and state that constant acceleration produces straight-line motion? Illustrate.

26. A particle starts from position $\mathbf{r}_0 = (x_0, y_0, z_0)$ at time $t = t_0$ with velocity $\mathbf{v}_0$. If it experiences constant acceleration $\mathbf{a}$, show that

$$\mathbf{r} = \mathbf{r}_0 + \mathbf{v}_0(t - t_0) + \tfrac{1}{2}\mathbf{a}(t - t_0)^2.$$

27. A ladder 8 meters long has its upper end against a vertical wall and its lower end on a horizontal floor. Suppose that the lower end slips away from the wall at a constant speed of 1 meter per second.

a) Find the velocity and acceleration of the middle point of the ladder when the foot of the ladder is 3 meters from the wall.

b) How fast does the middle point of the ladder strike the floor?

28. Water issues from the nozzle of a fire hose at speed S (Fig. 12.136). Show that the maximum height attainable by the water on the building is given by $(S^4 - g^2d^2)/(2gS^2)$, where $g > 0$ is the acceleration due to gravity.

29. A boy stands on a cliff 50 meters high that overlooks a river 85 meters wide (Fig. 12.137). If he can throw a stone at 25 meters per second, can he throw it across the river?

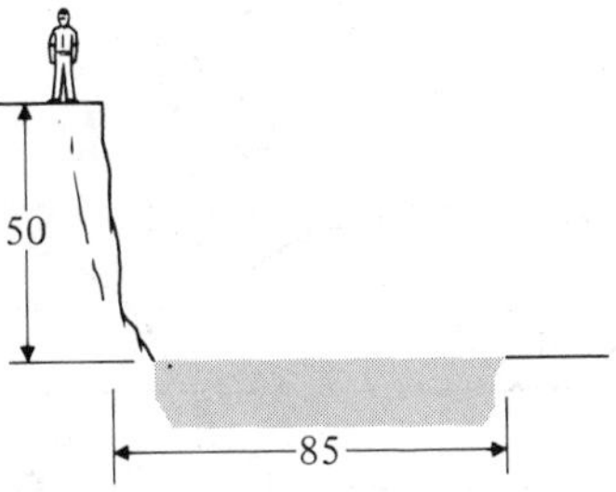

Figure 12.137

30. A golfer can drive a maximum of 300 meters on a level fairway. From the tee shown in Fig. 12.138, can he expect to clear the stream?

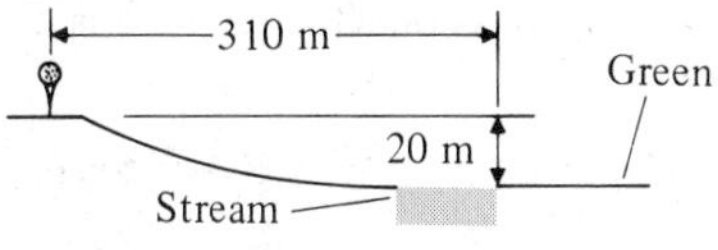

Figure 12.138

31. A particle is confined to move in a circular path of radius R in the xy-plane if and only if its position vector is $\mathbf{r} = x\hat{\imath} + y\hat{\jmath}$, where

$$x = h + R\cos\omega(t), \qquad y = k + R\sin\omega(t),$$

$\omega(t)$ being some function of time t. Show that the acceleration of the particle is directed radially toward the center of the circular path if and only if $\omega(t) = At + B$, where A and B are constants. Furthermore, in this case verify that the speed of the particle is constant.

32. If $\mathbf{r}$ is the position vector of a particle with mass m moving under the action of a force $\mathbf{F}$, the torque (or moment) of $\mathbf{F}$ about the origin is $\mathbf{M} = \mathbf{r} \times \mathbf{F}$. The angular momentum of m about O is defined as $\mathbf{H} = \mathbf{r} \times m\mathbf{v}$. Use Newton's second law in the form (12.106) to show that $\mathbf{M} = d\mathbf{H}/dt$.

33. A stone is embedded in the tread of a tire and the tire rolls (without slipping) along the x-axis (Fig. 12.139). If the stone starts at the origin, the path that it traces is called a *cycloid*.

a) Show that if θ is the angle through which the stone has turned, then parametric equations for the cycloid are

$$x = R(\theta - \sin\theta), \qquad y = R(1 - \cos\theta).$$

b) Verify that if the center of the tire moves at constant speed S, with $t = 0$ when the stone is at the origin, then $\theta = St/R$.

c) Find the velocity, speed, and acceleration of the stone at any time.

d) What are the normal and tangential components of the stone's acceleration?

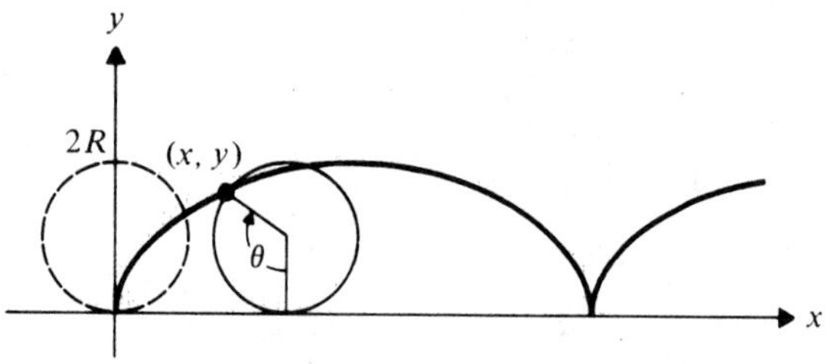

Figure 12.139

34. In Exercise 33, if the stone is embedded in the side of the tire instead of the tread, its path is called a *trochoid* (Fig. 12.140).

a) Show that if the distance from the center of the tire to the stone is b, then parametric equations for the trochoid are

$$x = R\theta - b\sin\theta, \qquad y = R - b\cos\theta.$$

b) Find the velocity, speed, and acceleration of the stone if the tire rolls so that its center has constant speed S.

c) What are the normal and tangential components of the stone's acceleration?

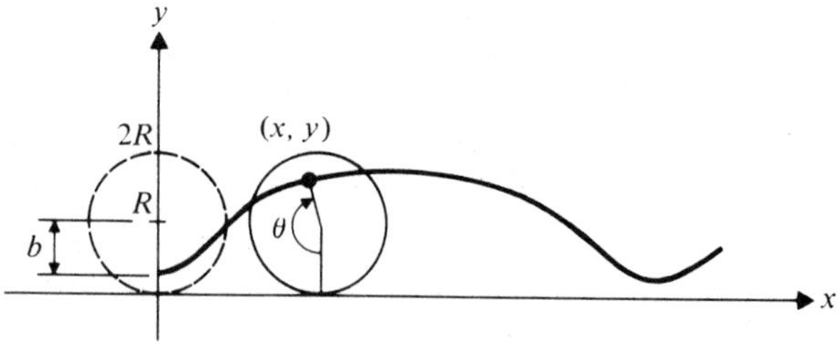

Figure 12.140

35. The two circles C_1 and C_2 in Fig. 12.141 represent cross-sections of two cylinders. The left cylinder remains stationary while the right one rolls (without slipping) around the left one, and the cylinders always remain in contact. If the right cylinder picks up a speck of dirt at the origin, the path that the dirt traces out during one revolution is a *cardioid*.

a) Show that parametric equations for the cardioid are

$$x = R(2\cos\theta - \cos 2\theta), \qquad y = R(2\sin\theta - \sin 2\theta).$$

b) Verify that if the point of contact moves at a constant speed S, with $t = 0$ when the speck of dirt is picked up, then $\theta = St/R$.

c) Find the velocity, speed, and acceleration of the speck of dirt.

d) What are the normal and tangential components of the dirt's acceleration?

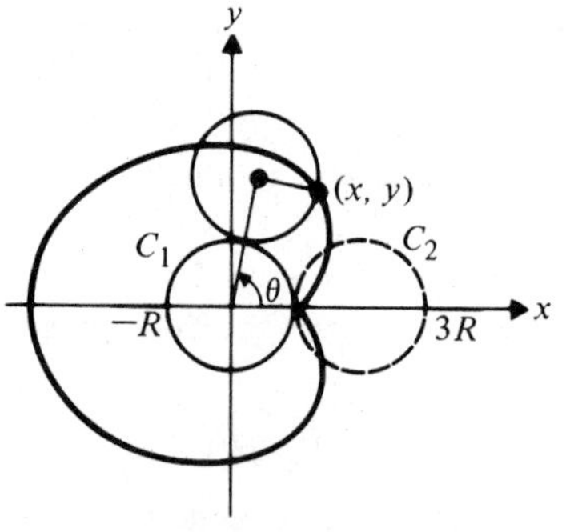

Figure 12.141

SUMMARY

We have now established the groundwork for multivariable calculus: We discussed curves and surfaces in space and introduced vectors. Our approach to three-dimensional analytic geometry paralleled that to plane analytic geometry in Chapter 1. We described points by Cartesian coordinates (x, y, z) and then illustrated that an equation $F(x, y, z) = 0$ in these coordinates usually defines a surface. A second equation $G(x, y, z) = 0$ also defines a surface, and the pair of simultaneous equations $F(x, y, z) = 0$, $G(x, y, z) = 0$ describes the curve of intersection of the two surfaces (if the two surfaces do intersect). It is often more useful to have parametric equations for a curve, and these can be obtained by specifying one of x, y, or z as a function of a parameter t and solving the given equations for the other two in terms of t: $x = x(t)$, $y = y(t)$, $z = z(t)$.

The most common surfaces that we encountered were planes and quadric surfaces. Every plane has an equation of the form $Ax + By + Cz + D = 0$; conversely, every such equation describes a plane (provided A, B, and C are not all zero). A plane is uniquely defined by a vector that is perpendicular to it [and (A, B, C) is one such vector] and a point on it. Quadric surfaces are surfaces whose equations are quadratic in x, y, and z, the most important of which were sketched in Figs. 12.33–12.41. Many surfaces, including some of the quadric surfaces, have cross-sections that are circles around one of the axes, and in such cases the surface can be visualized as the surface of revolution generated by rotating a plane curve about the axis.

Every straight line in space is characterized by a vector along it and a point on it. (Contrast this with the above characterization of a plane.) If (a, b, c) are the components of a vector along a line and (x_0, y_0, z_0) are the coordinates of a point on it, then the vector, symmetric, and parametric equations for the line are, respectively,

$$(x, y, z) = (x_0, y_0, z_0) + t(a, b, c);$$

$$\frac{x - x_0}{a} = \frac{y - y_0}{b} = \frac{z - z_0}{c};$$

$$x = x_0 + at,$$

$$y = y_0 + bt,$$

$$z = z_0 + ct.$$

Geometrically, vectors are defined as directed line segments; algebraically, they are represented by ordered sets of real numbers (v_x, v_y, v_z), called their Cartesian components. Vectors can be added or subtracted geometrically with triangles or parallelograms; algebraically, they are added and subtracted component by component. Vectors can also be multiplied by scalars to give parallel vectors of different lengths.

We defined two products of vectors: the scalar product and the vector product. The scalar product of two vectors $\mathbf{u} = (u_x, u_y, u_z)$ and $\mathbf{v} = (v_x, v_y, v_z)$ is defined as

$$\mathbf{u} \cdot \mathbf{v} = u_x v_x + u_y v_y + u_z v_z = |\mathbf{u}|\,|\mathbf{v}| \cos \theta,$$

where $|\mathbf{u}| = \sqrt{u_x^2 + u_y^2 + u_z^2}$ is the length of $\mathbf{u}$ and θ is the angle between $\mathbf{u}$ and $\mathbf{v}$. If the components of $\mathbf{u}$ and $\mathbf{v}$ are known, then this definition defines the angle θ between the vectors. The scalar product has many uses: finding components of vectors in arbitrary directions, calculating distances between geometric objects, and finding mechanical work, among others.

The vector product of two vectors $\mathbf{u}$ and $\mathbf{v}$ is

$$\mathbf{u} \times \mathbf{v} = \begin{vmatrix} \hat{\mathbf{i}} & \hat{\mathbf{j}} & \hat{\mathbf{k}} \\ u_x & u_y & u_z \\ v_x & v_y & v_z \end{vmatrix} = |\mathbf{u}|\,|\mathbf{v}| \sin \theta \hat{\mathbf{w}},$$

where $\hat{\mathbf{w}}$ is the unit vector perpendicular to **u** and **v** determined by the right-hand rule. Because of the perpendicularity property, the vector product is indispensable in finding vectors perpendicular to other vectors. We used this fact when finding a vector along the line of intersection of two planes, a vector perpendicular to the plane containing three given points, and distances between geometric objects. It can also be used to find areas of triangles and parallelograms.

If a curve is represented vectorially in the form $\mathbf{r}(t) = x(t)\hat{\mathbf{i}} + y(t)\hat{\mathbf{j}} + z(t)\hat{\mathbf{k}}$, then a unit vector tangent to the curve at any point is

$$\hat{\mathbf{T}} = \frac{d\mathbf{r}/dt}{|d\mathbf{r}/dt|}.$$

Two unit vectors normal to the curve are the principal normal $\hat{\mathbf{N}}$ and the binormal $\hat{\mathbf{B}}$:

$$\hat{\mathbf{N}} = \frac{d\hat{\mathbf{T}}/dt}{|d\hat{\mathbf{T}}/dt|}; \qquad \hat{\mathbf{B}} = \hat{\mathbf{T}} \times \hat{\mathbf{N}}.$$

These three vectors form a moving triad of mutually perpendicular unit vectors along the curve. The curvature of a curve, defined by $\kappa(t) = |\dot{\mathbf{r}} \times \ddot{\mathbf{r}}|/|\dot{\mathbf{r}}|^3$, measures the rate at which the curve changes direction: The larger κ, the faster the curve turns. The reciprocal of curvature $\rho = \kappa^{-1}$ is called radius of curvature. It is the radius of that circle which best approximates the curve at any point.

If parametric equations for a curve represent the position of a particle and t is time, then the velocity and acceleration of the particle are, respectively,

$$\mathbf{v} = \frac{d\mathbf{r}}{dt}; \qquad \mathbf{a} = \frac{d\mathbf{v}}{dt} = \frac{d^2\mathbf{r}}{dt^2};$$

and its speed is the magnitude of velocity, $|\mathbf{v}|$.

Tangential and normal components of velocity and acceleration of the particle are defined by

$$\mathbf{v} = |\mathbf{v}|\hat{\mathbf{T}}; \qquad \mathbf{a} = a_T\hat{\mathbf{T}} + a_N\hat{\mathbf{N}};$$

where

$$a_T = \frac{d}{dt}|\mathbf{v}| \quad \text{and} \quad a_N = |\mathbf{v}|\left|\frac{d\hat{\mathbf{T}}}{dt}\right| = \sqrt{|\mathbf{a}|^2 - a_T^2}.$$

What these results say is that velocity is always tangent to the trajectory of the particle, and its acceleration always lies in the plane of the velocity vector and the principal normal.

REVIEW EXERCISES

In each of Exercises 1–10, find the value of the scalar or the components of the vector if $\mathbf{u} = (1, 3, -2)$, $\mathbf{v} = (2, 4, -1)$, $\mathbf{w} = (0, 2, 1)$, and $\mathbf{r} = (2, 0, -1)$.

1. $2\mathbf{u} - 3\mathbf{w} + \mathbf{r}$
2. $\mathbf{u} \cdot \mathbf{v} \times \mathbf{w}$
3. $(3\mathbf{u} \times 4\mathbf{v}) - \mathbf{w}$
4. $3\mathbf{u} \times (4\mathbf{v} - \mathbf{w})$
5. $|\mathbf{u}|\mathbf{v} - |\mathbf{v}|\mathbf{r}$
6. $(\mathbf{u} + \mathbf{v}) \cdot (\mathbf{r} - \mathbf{w})$

7. $(\mathbf{u} + \mathbf{v}) \times (\mathbf{r} - \mathbf{w})$
8. $(\mathbf{u} \times \mathbf{v}) \times (\mathbf{r} \times \mathbf{w})$
9. $(\mathbf{u} \cdot \mathbf{v})\mathbf{r} - 3(\mathbf{v} \cdot \mathbf{w})\mathbf{u}$
10. $\dfrac{2\mathbf{r}}{\mathbf{v} \cdot \mathbf{w}} + 3(\mathbf{v} + \mathbf{u})$

In Exercises 11–26, sketch whatever the equation, or equations, describe in space.

11. $x - y + 2z = 6$
12. $x^2 + z^2 = 1$
13. $x = \sqrt{y^2 + z^2}$
14. $x - y = 5,\ 2x + y = 6$
15. $x^2 + y^2 + z^2 = 6z + 10$
16. $x^2 + y^2 + z^2 = 6z - 10$
17. $x + y = 5,\ 2x - 3y + 6z = 1,\ y = z$
18. $x = t^2,\ y = t,\ z = t^3$
19. $x = t,\ y = t^3 + 1$
20. $\dfrac{x-1}{3} = \dfrac{y+5}{2} = z$
21. $z = 4 - x^2 - 2y^2$
22. $y^2 + z^2 = 1,\ y = z$
23. $y^2 + z^2 = 1,\ x = z$
24. $x^2 + y^2 = z^2 + 1$
25. $x = y^2,\ x = z^2$
26. $z^2 = x^2 - y^2$

In each of Exercises 27–30, find equations for the given line.

27. Through the points $(-2, 3, 0)$ and $(1, -2, 4)$
28. Through $(6, 6, 2)$ and perpendicular to the plane $5x - 2y + z = 4$
29. Parallel to the line $x - y = 5,\ 2x + 3y + 6z = 4$, and through the origin
30. Perpendicular to the line $x = t + 2,\ y = 3 - 2t,\ z = 4 + t$, intersecting this line, and through the point $(1, 3, 2)$

In Exercises 31–34, find the equation for the given plane.

31. Through the points $(1, 3, 2)$, $(2, -1, 0)$, and $(6, 1, 3)$
32. Through the point $(1, 2, -1)$ and perpendicular to the line $y = z,\ x + y = 4$
33. Containing the line $x - y + z = 3,\ 3x + 4y = 6$, and the point $(2, 2, 2)$
34. Containing the lines $x = 3t,\ y = 1 + 2t,\ z = 4 - t$, and $x = y = z$

In Exercises 35–39, find the distance.

35. Between the points $(1, 3, -2)$ and $(6, 4, 1)$
36. From the point $(6, 2, 1)$ to the plane $6x + 2y - z = 4$
37. From the line $x - y + z = 2,\ 2x + y + z = 4$, to the plane $x - y = 5$
38. From the line $x - y + z = 2,\ 2x + y + z = 4$, to the plane $3x + 6y = 4$
39. From the point $(6, 2, 3)$ to the line $x - y + z = 6,\ 2x + y + 4z = 1$
40. Find the area of the triangle with vertices $(1, 1, 1)$, $(-2, 1, 0)$, and $(6, 3, -2)$.
41. If the points in Exercise 40 are three vertices of a parallelogram, what are the possibilities for the fourth vertex? What are the areas of these parallelograms?

In Exercises 42 and 43, find the unit tangent vector $\hat{\mathbf{T}}$, the principal normal $\hat{\mathbf{N}}$, and the binormal $\hat{\mathbf{B}}$ for the curve.

42. $x = 2 \sin t,\ y = 2 \cos t,\ z = t$
43. $x = t^3,\ y = 2t^2,\ z = t + 4$
44. If a particle has a trajectory defined by $x = t,\ y = t^2,\ z = t^2$, where t is time, find its velocity, speed, and acceleration at any time. What are the normal and tangential components of its velocity and acceleration?

45. A force $\mathbf{F} = x^{-2}(2\mathbf{i} + 3\mathbf{j})$ acts on a particle moving from $x = 1$ to $x = 4$ along the x-axis. How much work does it do?

46. A ball rolls off a table (1 meter high) with a speed of 0.5 meter per second (Fig. 12.142).

a) With what speed does it strike the floor?

b) What is its displacement vector relative to the point where it left the table when it strikes the floor?

*c) If it rebounds in the direction shown but loses 20% of its speed in the bounce, find the position of its second bounce.

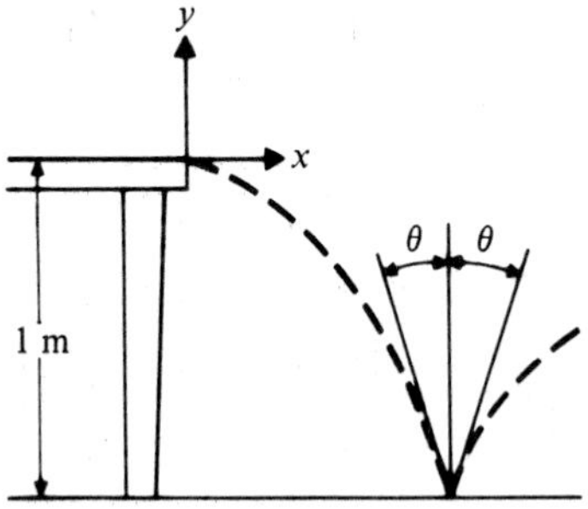

Figure 12.142

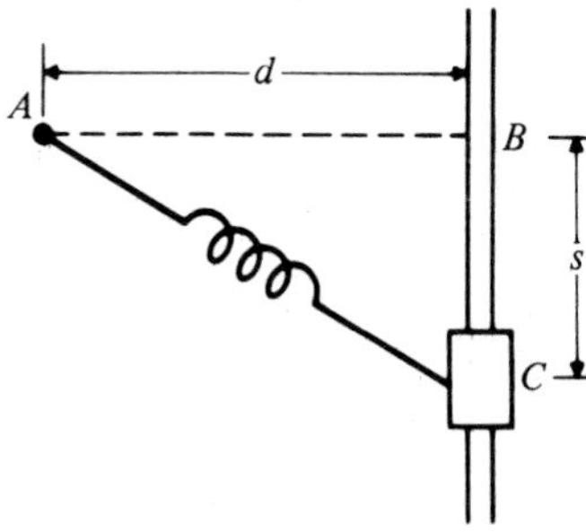

Figure 12.143

47. In Fig. 12.143 a spring (with constant k) is fixed at A and attached to a sleeve C. The sleeve is free to slide without friction on a vertical rod, and when the spring is horizontal (at B), it is unstretched. If the sleeve is slowly lowered, there will be a position at which the vertical component of the spring force on C is balanced by the force of gravity on the sleeve (ignoring the weight of the spring itself). If the mass of the sleeve is m, find an equation determining s in terms of d, m, k, and the acceleration $g > 0$ due to gravity.

48. Find the Cartesian components of the spring force $\mathbf{F}$ on the sleeve in Example 20.

Differential Calculus of Multivariable Functions

13

13.1 MULTIVARIABLE FUNCTIONS

Few quantities depend on only one variable; most depend on a multitude of interrelated variables. In order to study the properties of such complicated relationships, we begin in this chapter the study of functions of more than one variable. Much of the theory and many of our examples involve functions of two or three variables, because in these cases we can give geometric as well as analytic explanations. If the situation is completely analogous for functions of more variables, then it is likely that no mention of this fact will be made; on the other hand, if the situation is different for a higher number of variables, we will be careful to point out these differences.

If a variable T depends on other variables x, y, z, and t, we write $T = f(x, y, z, t)$ and speak of T as a function of x, y, z, and t. For example, T might be temperature, x, y, and z might be the coordinates of points in some region of space, and t might be time. The stopping distance D of a car depends on many factors: the initial speed s, the reaction time t of the driver to move from the accelerator to the brake, the texture T of the road, the moisture level M on the road, etc. We write $D = f(s, t, T, M, \ldots)$ to represent this functional dependence. The function $P = f(I, R) = I^2R$ represents the power necessary to maintain a current I through a wire with resistance R.

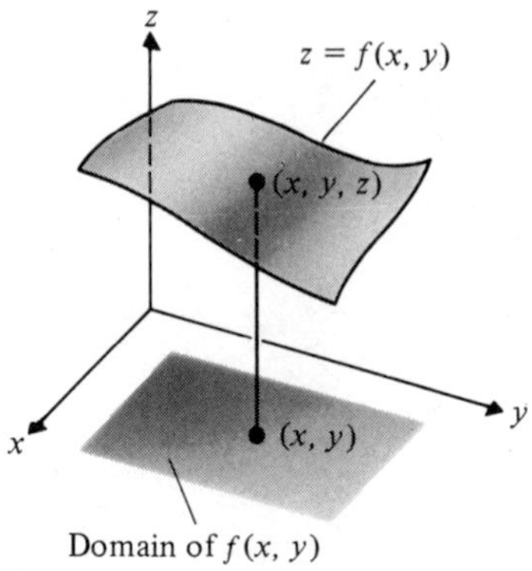

Figure 13.1

More precisely, a variable z is said to be a function of two independent variables x and y if x and y are not related and each pair of values of x and y determines a unique value of z. We write $z = f(x, y)$ to indicate that z is a function of x and y. Each possible pair of values x and y of the independent variables can be represented geometrically as a point (x, y) in the xy-plane. The totality of all points for which $f(x, y)$ is defined will form a region in the xy-plane called the *domain* of the function. Figure 13.1, for example, illustrates a rectangular domain. If for each point (x, y) in the domain we plot a point $f(x, y)$ units above the xy-plane, we obtain a surface like the one shown in Fig. 13.1. Each point on this surface has coordinates (x, y, z) that satisfy the equation

$$z = f(x, y), \tag{13.1}$$

and therefore (13.1) is the equation of the surface. Functions $f(x, y)$ of two independent variables, then, can be represented pictorially as surfaces in space.

It is clear that functions of more than two independent variables cannot be represented pictorially as surfaces. For example, if $u = f(x, y, z)$ is a function of three independent variables, values (x, y, z) of these independent variables can be represented geometrically as points in space. Our above procedure [for $z = f(x, y)$] was to construct an axis for the dependent variable (z) perpendicular to the axes for the independent variables (x and y). To do this for $f(x, y, z)$ would require a u-axis perpendicular to the x-, y-, and z-axes, a somewhat difficult task geometrically. We can certainly think of $u = f(x, y, z)$ as defining a surface in four-dimensional $xyzu$ space, but visually we are stymied.

There are alternative methods for representing multivariable functions such as level curves and surfaces, and these representations are sometimes quite useful.

We make one last remark. Although every function $f(x, y)$ of two independent variables can be represented geometrically as a surface, not every surface represents a function $f(x, y)$. A given surface does represent a function

$f(x, y)$ if and only if every vertical line (in the z-direction) that intersects the surface does so in exactly one point.

Example 1 Sketch the surface defined by the function $f(x, y) = x^2 + 4y^2$.

Solution To sketch the surface $z = x^2 + 4y^2$, we note that if the surface is intersected with a plane $z = k > 0$, then the ellipse $x^2 + 4y^2 = k$, $z = k$ is obtained. As k increases, the ellipse enlarges. In other words, cross-sections of this surface are ellipses that expand with increasing z. If we intersect the surface with the yz-plane ($x = 0$), we obtain the parabola $z = 4y^2$, $x = 0$. Similarly, intersection of the surface with the xz-plane gives the parabola $z = x^2$, $y = 0$. These facts lead to Fig. 13.2. □

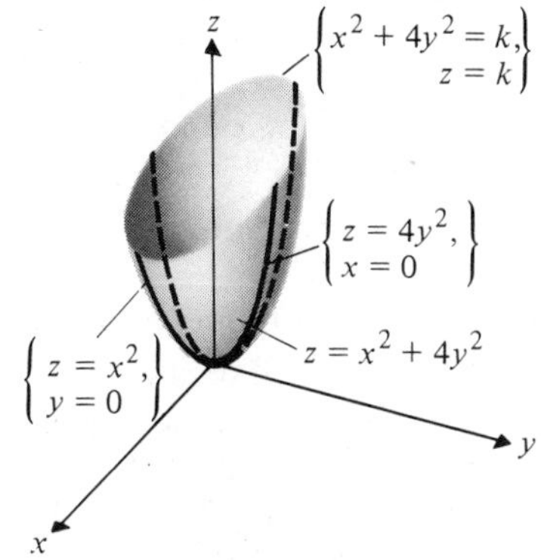

Figure 13.2

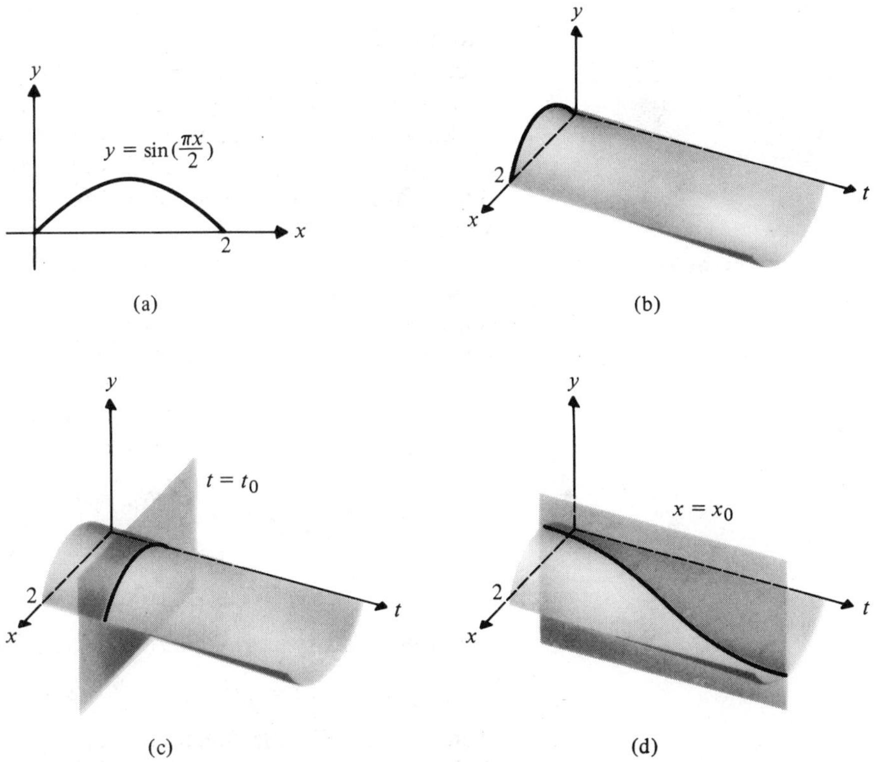

Figure 13.3

Example 2 The ends of a certain taut string are fixed at $x = 0$ and $x = 2$ on the x-axis. At time $t = 0$, the string is given a displacement in the y-direction of $y = \sin(\pi x/2)$ (Fig. 13.3a). If the string is then released, its displacement thereafter is given by

$$y = f(x, t) = \sin(\pi x/2)\cos(8\pi t).$$

Physically, this function need only be considered for $t \geq 0$ and $0 \leq x \leq 2$, and is pictured graphically as a surface in Fig. 13.3(b). Interpret physically the intersections of this surface with planes $t = t_0$ (t_0 = a constant) and $x = x_0$ (x_0 = a constant).

Solution If we intersect the surface $y = f(x, t)$ with a plane $t = t_0$ ($t_0 \geq 0$) (Fig. 13.3c), the curve of intersection is a picture of the position of the string at time t_0.

If we intersect the surface with a plane $x = x_0$ $(0 \leq x_0 \leq 2)$ (Fig. 13.3d), we obtain a graphical history of the displacement of the particle at position x_0. □

The functions in Examples 1 and 2 were defined explicitly, that is, with the dependent variable on the left-hand side of the equation and the independent variables on the right. Generally, then, z is defined explicitly as a function of x and y when an equation is specified in the form

$$z = f(x, y). \tag{13.2}$$

A variable z is said to be defined implicitly as a function of x and y by an equation

$$F(x, y, z) = 0 \tag{13.3}$$

if for each point (x, y) in some domain of the xy-plane, there is one and only one z for which (x, y, z) satisfies Eq. (13.3). In other words, Eq. (13.3) defines z as a function of x and y if every vertical line that intersects the surface defined by the equation does so in exactly one point. For example, the equation $x^2 + y^2 + z^2 = 1$ defines the sphere of radius 1 centered at the origin in Fig. 13.4. Clearly, this surface does not define z as a function of x and y. Alternatively, we could solve $x^2 + y^2 + z^2 = 1$ for z,

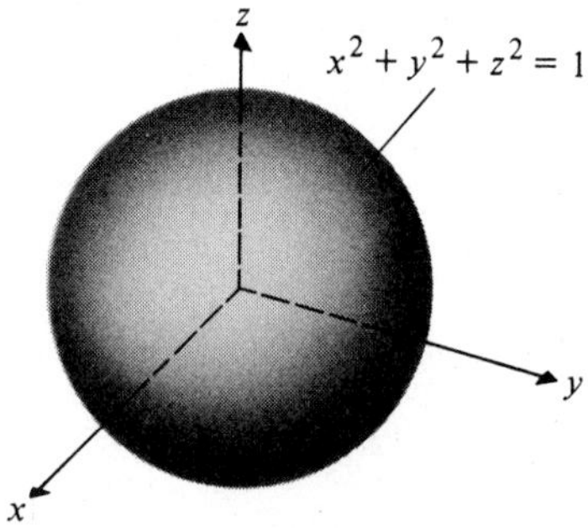

Figure 13.4

$$z = \pm\sqrt{1 - x^2 - y^2},$$

but because there are two possible solutions corresponding to a point (x, y), we again conclude that z is not defined as a function of x and y by the equation $x^2 + y^2 + z^2 = 1$.

Suppose that we are given three equations in five variables u, v, w, x, and y as follows:

$$\begin{aligned} F(u, v, w, x, y) &= 0, \\ G(u, v, w, x, y) &= 0, \\ H(u, v, w, x, y) &= 0. \end{aligned} \tag{13.4}$$

If we can solve these equations for u, v, and w in terms of x and y,

$$u = f(x, y), \qquad v = g(x, y), \qquad w = h(x, y), \tag{13.5}$$

then we say that the functions (13.5) are defined implicitly by Eqs. (13.4). Although it may be impossible to determine the functions (13.5) explicitly, we still regard Eqs. (13.4) as defining the functions implicitly if for each point (x, y) in some domain of the xy-plane, there exists exactly one set of numbers (u, v, w) for which the set (u, v, w, x, y) satisfies (13.4). Usually we expect m equations in n variables $(n > m)$ to implicitly define m dependent variables as functions of $n - m$ independent variables.

EXERCISES 13.1

1. If $f(x, y) = x^3y + x \sin y$, evaluate:

a) $f(1, 2)$;
b) $f(-2, -2)$;
c) $f(x^2 + y, x - y^2)$;
d) $f(x + h, y) - f(x, y)$.

2. If $f(x, y, z) = x^2y^2 - x^4 + 4zx^2$, show that $f(a + b, a - b, ab) = 0$.

In Exercises 3–6, find and illustrate geometrically the largest possible domain for the given function.

3. $f(x, y) = \sqrt{4 - x^2 - y^2}$

4. $f(x, y) = \ln(1 - x^2 + y^2)$

5. $f(x, y) = \text{Sin}^{-1}(x^2y + 1)$

6. $f(x, y, z) = \dfrac{1}{x^2 + y^2 + z^2}$

7. Give an example of an equation $F(x, y, z) = 0$ that implicitly defines each of x, y, and z as a function of the other two.

8. For what values of x and y is the function

$$f(x, y) = \frac{12xy - x^2y^2}{2(x + y)}$$

equal to zero? Illustrate these values as points in the xy-plane. What is the largest possible domain of the function?

9. What is the condition that a surface in xyz-space represent a function $x = f(y, z)$?

In each of Exercises 10–23, sketch the surface defined by the function.

10. $f(x, y) = y^2$

11. $f(x, y) = 4 - x - 2y$

12. $f(x, y) = x^2 + y^2$

13. $f(x, y) = \sqrt{x^2 + y^2}$

14. $f(x, y) = y + x$

15. $f(x, y) = 1 - x^3$

16. $f(x, y) = 2(x^2 + y^2)$

17. $f(x, y) = 1 - x^2 - 4y^2$

18. $f(x, y) = xy$

19. $f(x, y) = y - x^2$

20. $f(x, y) = e^{-x^2-y^2}$

21. $f(x, y) = |x - y|$

22. $f(x, y) = x^2 - y^2$

23. $f(x, y) = \sqrt{1 + x^2 - y^2}$

24. A closed box is to have a total surface area of 30 square meters. Find a formula for the volume of the box in terms of its length l and width w.

25. a) A company wishes to construct a storage tank in the form of a rectangular box. If the material for the sides and top costs \$1.25 per square meter and the material for the bottom costs \$4.75 per square meter, find the cost of building the tank as a function of its length l, width w, and height h.

b) If the tank must hold 1000 cubic meters, find the construction cost in terms of l and w.

c) Repeat (a) and (b) if the 12 edges of the tank must be welded at a cost of \$7.50 per meter of weld.

26. A rectangular box is inscribed inside the ellipsoid $x^2/a^2 + y^2/b^2 + z^2/c^2 = 1$ with sides parallel to the coordinate planes and corners on the ellipsoid. Find a formula for the volume of the box in terms of x and y.

27. a) A silo is to be built in the shape of a right-circular cylinder surmounted by a right-circular cone. If the radius of each is 6 meters, find a formula for the volume V in the silo as a function of the heights H and h of the cylinder and the cone, respectively.

b) If the total surface area of the silo must be 200 square meters (not including the base), find V as a function of h. (The area of the curved surface of a cone is $\pi r\sqrt{r^2 + h^2}$.)

28. A long piece of metal 1 meter wide is bent in two places A and B (Fig. 13.5) to form a channel with three straight sides. Find a formula for the cross-sectional area of the channel in terms of x, θ, and ϕ.

Figure 13.5

29. A uniform circular rod has flat ends at $x = 0$ and $x = \pi$ on the x-axis. The round side of the rod is insulated and the faces at $x = 0$ and $x = \pi$ are both kept at temperature 0°C for time $t > 0$. If the initial temperature (at time $t = 0$) of the rod is given by $100 \sin x$ $(0 \le x \le \pi)$, then the temperature thereafter is

$$T = f(x, t) = 100e^{-kt} \sin x \qquad (k > 0 \text{ constant}).$$

a) Sketch the surface $T = f(x, t)$.

b) Interpret physically the curves of intersection of this surface with planes $x = x_0$ and $t = t_0$.

30. A cow's daily diet consists of three foods: hay, grain, and supplements. The animal is always given 11 kilograms of hay per day, 50% of which is digestive material and 12% of which is protein. Grain is 74% digestive and 8.8% protein. The supplements are 62% digestive material and 34% protein. Hay costs \$27.50 for 1000 kilograms, whereas grain and supplements cost \$110 and \$175, respectively, for 1000 kilograms. A healthy cow's daily diet must contain between 9.5 and 11.5 kilograms of digestive material and between 1.9 and 2.0 kilograms of protein. Find a formula for the cost per day C of feeding a cow in terms of the number of kilograms of grain G and supplements S fed to the cow daily. What is the domain of this function?

13.2 LIMITS AND CONTINUITY

The concepts of limit and continuity for multivariable functions are exactly the same as for functions of one variable; on the other hand, the work involved with the application of these concepts is much more complicated for multivariable functions.

Intuitively, a function $f(x, y)$ is said to have limit L as x and y approach x_0 and y_0, if $f(x, y)$ gets arbitrarily close to L, and stays close to L, as x and y get arbitrarily close to x_0 and y_0. To give a mathematical definition of limits, it is convenient to represent pairs of independent variables as points (x, y) in the xy-plane. We have, then, the following definition.

▷ **Definition 13.1** A function $f(x, y)$ has limit L as (x, y) approaches (x_0, y_0), written

$$\lim_{(x, y)\to(x_0, y_0)} f(x, y) = L, \tag{13.6}$$

if given any $\varepsilon > 0$, we can find a $\delta > 0$ such that

$$|f(x, y) - L| < \varepsilon$$

whenever $0 < \sqrt{(x - x_0)^2 + (y - y_0)^2} < \delta$ and (x, y) is in the domain of $f(x, y)$. ◁

In other words, $f(x, y)$ has limit L as (x, y) approaches (x_0, y_0) if $f(x, y)$ can be made arbitrarily close to L (within ε) by choosing points (x, y) sufficiently close to (x_0, y_0) (within a circle of radius δ). Note the similarity of this definition to that for the limit of a function $f(x)$ of one variable in Section 2.3.

It is clear that

$$\lim_{(x, y)\to(2, 1)} (x^2 + 2xy - 5) = 3,$$

but the limit

$$\lim_{(x, y)\to(0, 0)} \frac{x^2 - y^2}{x^2 + y^2}$$

presents a problem, since both numerator and denominator approach zero as x and y approach zero.

To conclude that $\lim_{x\to a} f(x) = L$, the limit must be L no matter how x approaches a—be it through numbers larger than a, through numbers smaller than a, or through any other approach. For limit (13.6), the limit of $f(x, y)$ must also be L for all possible ways of approaching (x_0, y_0). But in this case there might be a multitude of ways of approaching (x_0, y_0). We might be able to approach (x_0, y_0) along straight lines with various slopes, along parabolas, along cubics, etc. Definition 13.1 implies, then, that the limit exists only if it is independent of the manner of approach. It is assumed, however, that we approach (x_0, y_0) only through points (x, y) that lie in the domain of definition of the function.

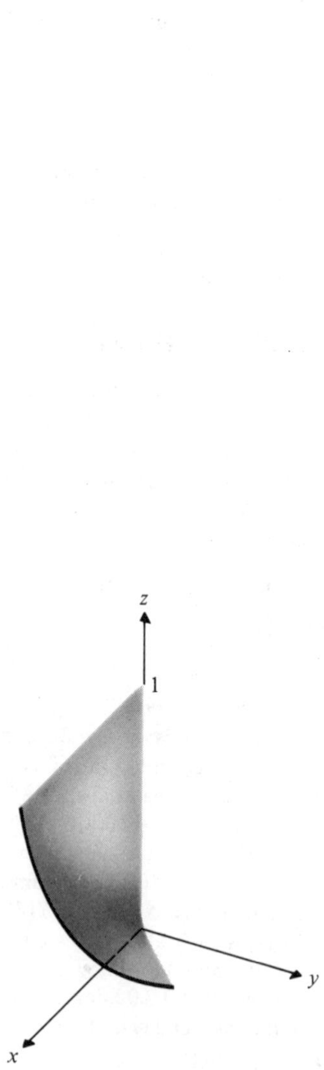

Figure 13.6

For the second example above, suppose we approach the origin along the straight line $y = mx$. Then, along this line,

$$\lim_{(x, y)\to(0, 0)} \frac{x^2 - y^2}{x^2 + y^2} = \lim_{x\to 0} \frac{x^2 - m^2x^2}{x^2 + m^2x^2} = \frac{1 - m^2}{1 + m^2}.$$

Because this result depends on m, we have shown that as (x, y) approaches $(0, 0)$ along various straight lines, the function $(x^2 - y^2)/(x^2 + y^2)$ approaches different numbers. We conclude, therefore, that the function does not have a limit as (x, y) approaches $(0, 0)$. In Fig. 13.6 we show this function to illustrate our conclusion.

Example 3 Evaluate

$$\lim_{(x, y)\to(0, 0)} \frac{x^2 - y^2}{x + y}$$

if it exists.

Solution Because points on the line $y = -x$ are not within the domain of definition of the function, we can write that

$$\lim_{(x, y)\to(0, 0)} \frac{x^2 - y^2}{x + y} = \lim_{(x, y)\to(0, 0)} \frac{(x - y)(x + y)}{x + y} = \lim_{(x, y)\to(0, 0)} (x - y) = 0. \square$$

The concept of continuity for multivariable functions is contained in the following definition.

▷ **Definition 13.2** A function $f(x, y)$ is said to be continuous at a point (x_0, y_0) if

$$\lim_{(x, y)\to(x_0, y_0)} f(x, y) = f(x_0, y_0). \triangleleft \tag{13.7}$$

Just as Definition 2.1 for the continuity of a function of only one variable contains three conditions, so too does Definition 13.2. It demands that:

1. $f(x, y)$ be defined at (x_0, y_0);
2. $\lim_{(x, y)\to(x_0, y_0)} f(x, y)$ exist;
3. the numbers in (1) and (2) be the same.

Geometrically, a function $f(x, y)$ is continuous at a point (x_0, y_0) if the surface $z = f(x, y)$ does not have a separation at the point $(x_0, y_0, f(x_0, y_0))$. The function $f(x, y) = (x^2 - y^2)/(x^2 + y^2)$ in Fig. 13.6 is discontinuous at $(0, 0)$.

EXERCISES 13.2

In each of Exercises 1–20, evaluate the limit if it exists.

1. $\lim_{(x, y)\to(2, -3)} \frac{x^2 - 1}{x + y}$
2. $\lim_{(x, y)\to(1, 1)} \frac{x^3 + 2y^3}{x^3 + 4y^3}$
3. $\lim_{(x, y)\to(3, 2)} \frac{2x - 3y}{x + y}$
4. $\lim_{(x, y, z)\to(2, 3, -1)} \frac{xyz}{x^2 + y^2 + z^2}$
5. $\lim_{(x, y)\to(1, 0)} \frac{x}{y}$
6. $\lim_{(x, y, z)\to(0, \pi/2, 1)} \operatorname{Tan}^{-1} (x/(yz))$
7. $\lim_{(x, y, z)\to(0, \pi/2, 1)} \operatorname{Tan}^{-1} (yz/x)$
8. $\lim_{(x, y, z)\to(0, \pi/2, 1)} \operatorname{Tan}^{-1} |yz/x|$
9. $\lim_{(x, y)\to(3, 4)} \frac{|x^2 - y^2|}{x^2 - y^2}$

10. $\displaystyle\lim_{(x,y)\to(3,4)} \frac{|x^2+y^2|}{x^2+y^2}$

11. $\displaystyle\lim_{(x,y)\to(2,1)} \frac{x^2-y^2}{x-y}$

12. $\displaystyle\lim_{(x,y)\to(2,2)} \frac{x^2-y^2}{x-y}$

13. $\displaystyle\lim_{(x,y)\to(0,0)} \frac{x^2-y^2}{x-y}$

14. $\displaystyle\lim_{(x,y)\to(0,0)} \frac{x-y}{x+y}$

15. $\displaystyle\lim_{(x,y,z)\to(0,0,0)} \frac{x^2-y^2}{y^2+z^2+1}$

16. $\displaystyle\lim_{(x,y)\to(2,1)} \frac{(x-2)^2(y+1)}{x-2}$

17. $\displaystyle\lim_{(x,y,z)\to(1,1,1)} |2x-y-z|$

18. $\displaystyle\lim_{(x,y)\to(0,0)} \frac{3x^3-y^3}{2x^3+4y^3}$

19. $\displaystyle\lim_{(x,y)\to(0,0)} \operatorname{Sec}^{-1}\left(\frac{-1}{x^2+y^2}\right)$

20. $\displaystyle\lim_{(x,y)\to(0,0)} \operatorname{Sec}^{-1}(x^2+y^2)$

In Exercises 21–26, find all points of discontinuity for the given function.

21. $f(x, y) = \dfrac{x^2-1}{x+y}$

22. $f(x, y) = \dfrac{xy}{x^2+y^2}$

23. $f(x, y) = \dfrac{1}{1-x^2-y^2}$

24. $f(x, y, z) = \dfrac{1}{xyz}$

25. $f(x, y) = |x - y|$

26. $f(x, y) = \dfrac{x+y}{x^2y+xy^2}$

27. Evaluate

$$\lim_{(x,y)\to(a,a)} \{\cos(x+y) - \sqrt{1-\sin^2(x+y)}\},$$

where $0 \le a \le \pi/2$.

In each of Exercises 28–31, evaluate the limit if it exists.

28. $\displaystyle\lim_{(x,y)\to(0,0)} \frac{x^4+y^2}{x^4-y^2}$

[*Hint:* Approach (0, 0) along parabolas.]

29. $\displaystyle\lim_{(x,y)\to(1,1)} \frac{x^2-2x+y^2+2y-2}{x^2-y^2-2x+2y}$

[*Hint:* Approach (1, 1) along straight lines.]

30. $\displaystyle\lim_{(x,y)\to(1,0)} \frac{\sqrt{x+y}-\sqrt{x-y}}{y}$

31. $\displaystyle\lim_{(x,y)\to(0,0)} \frac{\sin(x^2+y^2)}{x^2+y^2}$

32. a) Does the limit

$$\lim_{(x,y)\to(1,1)} \frac{\sin(x-y)}{x-y}$$

exist? Explain. Is the function continuous at (1, 1)?

b) If we define the function everywhere by giving it a value along the line $y = x$,

$$\frac{\sin(x-y)}{x-y} = \begin{cases} \dfrac{\sin(x-y)}{x-y}, & y \neq x, \\ 1, & y = x, \end{cases}$$

does the limit of this function exist at (1, 1)? Is this function continuous at (1, 1)?

33. Give a mathematical definition for

$$\lim_{(x,y,z)\to(x_0,y_0,z_0)} f(x, y, z) = L.$$

34. Is the following statement true or false? If a function $f(x, y)$ is undefined at every point on a curve C, then for any point (x_0, y_0) on C,

$$\lim_{(x,y)\to(x_0,y_0)} f(x, y)$$

does not exist. Explain. Give an example.

35. Let

$$f(x, y) = \begin{cases} \dfrac{x^2y^2}{x^4+y^4}, & \text{if } (x, y) \neq (0, 0), \\ 0, & \text{if } (x, y) = (0, 0). \end{cases}$$

a) Show that $f(x, y)$ is continuous in each variable separately at (0, 0). In other words, show that $f(x, 0)$ and $f(0, y)$ are continuous at $x = 0$ and $y = 0$.

b) Show that $f(x, y)$ is not continuous at (0, 0).

***36.** Prove that:

a) $\displaystyle\lim_{(x,y)\to(0,0)} (xy + 5) = 5;$

b) $\displaystyle\lim_{(x,y)\to(1,1)} (x^2 + 2xy + 5) = 8.$

13.3 PARTIAL DERIVATIVES

We now define partial derivatives of multivariable functions and interpret these derivatives algebraically and geometrically.

▷ **Definition 13.3** The *partial derivative* of a function $f(x, y)$ with respect to x is

$$\frac{\partial f}{\partial x} = \lim_{\Delta x \to 0} \frac{f(x + \Delta x, y) - f(x, y)}{\Delta x}, \tag{13.8}$$

and the partial derivative with respect to y is

$$\frac{\partial f}{\partial y} = \lim_{\Delta y \to 0} \frac{f(x, y + \Delta y) - f(x, y)}{\Delta y}. \quad \triangleleft \tag{13.9}$$

It is evident from (13.8) that the partial derivative of $f(x, y)$ with respect to x is simply the ordinary derivative of $f(x, y)$ with respect to x, where y is considered a constant. Similarly, $\partial f/\partial y$ is the ordinary derivative of $f(x, y)$ with respect to y, holding x constant.

For the partial derivative of a function of more than two independent variables, we again permit one variable to vary, but hold all others constant. For example, the partial derivative of $f(x, y, z, t, \ldots)$ with respect to z is

$$\frac{\partial f}{\partial z} = \lim_{\Delta z \to 0} \frac{f(x, y, z + \Delta z, t, \ldots) - f(x, y, z, t, \ldots)}{\Delta z}. \tag{13.10}$$

Hence, we differentiate with respect to z while treating $x, y, t, \ldots$ as constants.

Other notations for the partial derivative are common. In particular, for $\partial f/\partial x$ when $z = f(x, y)$, there are also

$$\frac{\partial z}{\partial x}, \; f_x, \; z_x, \; \left.\frac{\partial f}{\partial x}\right)_y \quad \text{and} \quad \left.\frac{\partial z}{\partial x}\right)_y,$$

the last two indicating that the variable y is held constant when differentiation with respect to x is performed.

Example 4 Find $\partial z/\partial x$ and $\partial z/\partial y$ if $z = f(x, y) = x^2y^3 + e^{xy}$.

Solution For this function,

$$\frac{\partial z}{\partial x} = 2xy^3 + ye^{xy} \quad \text{and} \quad \frac{\partial z}{\partial y} = 3x^2y^2 + xe^{xy}. \; \square$$

Example 5 Find $\partial f/\partial x$ at the point (1, 2, 3) if $f(x, y, z) = x^2/y^4 + 3xz + 4$.

Solution Since $\partial f/\partial x = 2x/y^4 + 3z$,

$$\left.\frac{\partial f}{\partial x}\right|_{(1,2,3)} = 2(1)/2^4 + 3(3) = \frac{73}{8}. \; \square$$

For a function $y = f(x)$ of one variable, we defined differentials dx and dy in such a way that the derivative dy/dx could be thought of as a quotient. This will *not* be done for functions of more than one variable. Although we will write the partial derivative $\dfrac{\partial f}{\partial x}$ in the form $\partial f/\partial x$ (for typographical reasons), we will *never* consider it as a quotient.

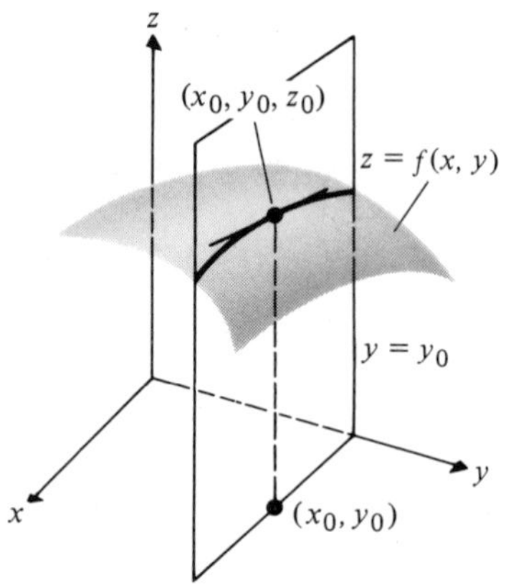

Figure 13.7

Algebraically, the partial derivative $\partial f/\partial x$ represents the rate of change of $f(x, y, \ldots)$ with respect to x when all other variables in $f(x, y, \ldots)$ are held constant. For instance, $V = \pi r^2 h/3$ represents the volume of a right-circular cone with height h and radius r, and therefore $\partial V/\partial r = 2\pi r h/3$ represents the rate of change of the volume of the cone as the base radius changes and the height remains fixed.

We can interpret the partial derivative of a function geometrically when the function can be interpreted geometrically, namely, when there are only two independent variables. Consider, then, a function $f(x, y)$ that is represented geometrically as a surface $z = f(x, y)$ in Fig. 13.7. If we intersect this surface with a plane $y = y_0 =$ a constant, we obtain a curve with equations

$$y = y_0, \qquad z = f(x, y_0).$$

Because this curve lies in the plane $y = y_0$, we can talk about its tangent line at the point (x_0, y_0, z_0), where $z_0 = f(x_0, y_0)$. The slope of this tangent must be the derivative of z with respect to x, but because y is being held constant at y_0, it must be the partial derivative of z with respect to x. In other words, the slope of the tangent line to the curve in Fig. 13.7 at the point (x_0, y_0, z_0) is $\partial f/\partial x_{|(x_0, y_0)}$. Similarly, the partial derivative $\partial f/\partial y$ evaluated at (x_0, y_0) represents the slope of the tangent line to the curve of intersection of $z = f(x, y)$ and the plane $x = x_0$ at the point (x_0, y_0, z_0).

EXERCISES 13.3

In each of Exercises 1–20, evaluate $\partial f/\partial x$ and $\partial f/\partial y$.

1. $f(x, y) = x^3y^2 + 2xy$
2. $f(x, y) = 3xy - 4x^4y^4$
3. $f(x, y) = x^4/y^3$
4. $f(x, y) = x/(x + y) - x/y$
5. $f(x, y) = x/(2x^2 + y)$
6. $f(x, y) = \sin(xy)$
7. $f(x, y) = x\cos(x + y)$
8. $f(x, y) = \sqrt{x^2 + y^2}$
9. $f(x, y) = x\sqrt{x^2 - y^2}$
10. $f(x, y) = \tan(2x^2 + y^2)$
11. $f(x, y) = e^{x+y}$
12. $f(x, y) = e^{xy}$
13. $f(x, y) = xye^{xy}$
14. $f(x, y) = \ln(x^2 + y^2)$
15. $f(x, y) = (x + 1)\ln(xy)$
16. $f(x, y) = \sin(ye^x)$
17. $f(x, y) = \mathrm{Tan}^{-1}(x/y)$
18. $f(x, y) = \sqrt[3]{1 - \cos^3(x^2y)}$
19. $f(x, y) = \dfrac{\sin x}{\cos y}$
20. $f(x, y) = \ln(\sec\sqrt{x + y})$

In Exercises 21–30, evaluate the indicated derivative.

21. $\partial f/\partial x$ if $f(x, y, z) = xyze^{x^2+y^2}$
22. $\partial f/\partial z$ if $f(x, z) = \mathrm{Tan}^{-1}(1/(x^2 + z^2))$
23. $\partial f/\partial y$ at $(1, 1, 0)$ if $f(x, y, z) = xy(x^2 + y^2 + z^2)^{1/3}$
24. $\partial f/\partial x$ at $(1, -1, 1, -1)$ if $f(x, y, z, t) = zt/(x^2 + y^2 - t^2)$
25. $\partial f/\partial t$ if $f(x, y, t) = x\sqrt{t^2 - y^2}/t^2 + (x/y)\,\mathrm{Sec}^{-1}(t/3)$
26. $\partial f/\partial x$ if $f(x, y, z) = \mathrm{Cot}^{-1}(1 + x + y + z)$
27. $\partial f/\partial y$ at $(1, 2, 3)$ if $f(x, y, t) = \mathrm{Sin}^{-1}(xyt)/\mathrm{Cos}^{-1}(xyt)$
28. $\partial f/\partial x$ if $f(x, y, z) = x^3/y + x\sin(yz/x)$
29. $\partial f/\partial t$ if $f(x, y, z, t) = xyz\ln(x^2 + y^2 + z^2)$
30. $\partial f/\partial z$ if $f(x, z) = (z^2/2)\,\mathrm{Sin}^{-1}(x/z) + (x/2)\sqrt{z^2 - x^2}$
31. If $z = f(x, y) = x^3y/(x - y)$, show that

$$x\frac{\partial z}{\partial x} + y\frac{\partial z}{\partial y} = 3z.$$

32. If $u = f(x, y, z) = (x^4 + y^4 + z^4)/(xyz)$, show that

$$x\frac{\partial f}{\partial x} + y\frac{\partial f}{\partial y} + z\frac{\partial f}{\partial z} = f.$$

33. If $u = f(x, y, z) = (x^2 + y^2)\cos[(y + z)/x]$, show that

$$x\frac{\partial f}{\partial x} + y\frac{\partial f}{\partial y} + z\frac{\partial f}{\partial z} = 2f.$$

34. The equation of continuity for three-dimensional, unsteady flow of a compressible fluid is

$$\frac{\partial \rho}{\partial t} + \frac{\partial}{\partial x}(\rho u) + \frac{\partial}{\partial y}(\rho v) + \frac{\partial}{\partial z}(\rho w) = 0,$$

where $\rho(x, y, z, t)$ is the density of the fluid, and $u\hat{\mathbf{i}} + v\hat{\mathbf{j}} + w\hat{\mathbf{k}}$ is the velocity of the fluid at position (x, y, z) and time t. Determine whether the continuity equation is satisfied if:

a) $\rho =$ constant, $u = (2x^2 - xy + z^2)t$,
$v = (x^2 - 4xy + y^2)t$,
$w = (-2xy - yz + y^2)t$;

b) $\rho = xy + zt$, $u = x^2y + t$, $v = y^2z - 2t^2$,
$w = 5x + 2z$.

35. In complex variable theory, two functions $u(x, y)$ and $v(x, y)$ are said to be harmonic conjugates in a region R if in R they satisfy the Cauchy–Riemann equations

$$\frac{\partial u}{\partial x} = \frac{\partial v}{\partial y}, \qquad \frac{\partial v}{\partial x} = -\frac{\partial u}{\partial y}.$$

Show that the following pairs of functions are harmonic conjugates:

a) $u(x, y) = -3xy^2 + y + x^3$,
$v(x, y) = 3x^2y - y^3 - x + 5$;

b) $u(x, y) = (x^2 + x + y^2)/(x^2 + y^2)$,
$v(x, y) = -y/(x^2 + y^2)$;

c) $u(x, y) = e^x(x\cos y - y\sin y)$,
$v(x, y) = e^x(x\sin y + y\cos y)$.

36. If r and θ are polar coordinates, then the Cauchy–Riemann equations in Exercise 35 for functions $u(r, \theta)$ and $v(r, \theta)$ take the form

$$\frac{\partial u}{\partial r} = \frac{1}{r}\frac{\partial v}{\partial \theta}, \qquad \frac{1}{r}\frac{\partial u}{\partial \theta} = -\frac{\partial v}{\partial r}, \qquad r \neq 0.$$

Show that the following pairs of functions satisfy these equations:

a) $u(r, \theta) = (r^2 + r\cos\theta)/(1 + r^2 + 2r\cos\theta)$,
$v(r, \theta) = r\sin\theta/(1 + r^2 + 2r\cos\theta)$;

b) $u(r, \theta) = \sqrt{r}\cos(\theta/2)$, $v(r, \theta) = \sqrt{r}\sin(\theta/2)$;

c) $u(r, \theta) = \ln r$, $v(r, \theta) = \theta$.

13.4 GRADIENTS

Suppose a function $f(x, y, z)$ is defined at each point in some region of space, and that at each point of the region all three partial derivatives

$$\frac{\partial f}{\partial x}, \quad \frac{\partial f}{\partial y}, \quad \frac{\partial f}{\partial z}$$

exist. For example, if $f(x, y, z)$ represents the present temperature at each point in the room in which you are working, then these derivatives represent the rates of change of temperature in directions parallel to the x-, y-, and z-axes, respectively. There is a particular combination of these derivatives that will prove very useful in later work. This combination is contained in the following definition.

▷ **Definition 13.4** If a function $f(x, y, z)$ has partial derivatives $\partial f/\partial x$, $\partial f/\partial y$, and $\partial f/\partial z$ at each point in some region D of space, then at each point in D we define a vector called the *gradient* of $f(x, y, z)$, written grad f or ∇f, by

$$\text{grad } f = \nabla f = \frac{\partial f}{\partial x}\hat{\mathbf{i}} + \frac{\partial f}{\partial y}\hat{\mathbf{j}} + \frac{\partial f}{\partial z}\hat{\mathbf{k}}. \quad ◁ \tag{13.11}$$

For a function $f(x, y)$ of only two independent variables, we have

$$\nabla f = \frac{\partial f}{\partial x}\hat{\mathbf{i}} + \frac{\partial f}{\partial y}\hat{\mathbf{j}}. \tag{13.12}$$

Example 6 If $f(x, y, z) = x^2yz - 2x/y$, find ∇f at $(1, -1, 3)$.

Solution Since

$$\nabla f = (2xyz - 2/y)\hat{\mathbf{i}} + (x^2z + 2x/y^2)\hat{\mathbf{j}} + (x^2y)\hat{\mathbf{k}},$$

then

$$\nabla f_{|(1,-1,3)} = -4\hat{\mathbf{i}} + 5\hat{\mathbf{j}} - \hat{\mathbf{k}}. \square$$

Example 7 If $f(x, y, z) = \text{Tan}^{-1}(xy/z)$, what is ∇f?

Solution
$$\nabla f = \left\{\frac{1}{1 + (xy/z)^2}\left(\frac{y}{z}\right)\right\}\hat{\mathbf{i}} + \left\{\frac{1}{1 + (xy/z)^2}\left(\frac{x}{z}\right)\right\}\hat{\mathbf{j}} + \left\{\frac{1}{1 + (xy/z)^2}\left(\frac{-xy}{z^2}\right)\right\}\hat{\mathbf{k}}$$
$$= \frac{yz}{z^2 + x^2y^2}\hat{\mathbf{i}} + \frac{xz}{z^2 + x^2y^2}\hat{\mathbf{j}} - \frac{xy}{z^2 + x^2y^2}\hat{\mathbf{k}}$$
$$= (yz\hat{\mathbf{i}} + xz\hat{\mathbf{j}} - xy\hat{\mathbf{k}})/(z^2 + x^2y^2). \square$$

Gradients arise in a multitude of applications in applied mathematics—heat conduction, electromagnetic theory, and fluid flow, to name a few—and two of the properties that make them so indispensable are discussed in detail in Sections 13.8 and 13.9. Examples 8 and 9 suggest these properties, but we make no attempt at a complete discussion here. For the moment we simply want you to be familiar with the definition of gradients and be able to calculate them.

Example 8 The equation $F(x, y, z) = 0$, where $F(x, y, z) = x^2 + y^2 + z^2 - 4$, defines a sphere of radius 2 centered at the origin. Show that the gradient vector ∇F at any point on the sphere is perpendicular to the sphere.

Solution If $P(x, y, z)$ is any point on the sphere (Fig. 13.8), then the position vector $\mathbf{r} = x\hat{\mathbf{i}} + y\hat{\mathbf{j}} + z\hat{\mathbf{k}}$ from the origin to P is clearly perpendicular to the sphere. On the other hand,

$$\nabla F = 2x\hat{\mathbf{i}} + 2y\hat{\mathbf{j}} + 2z\hat{\mathbf{k}} = 2\mathbf{r}.$$

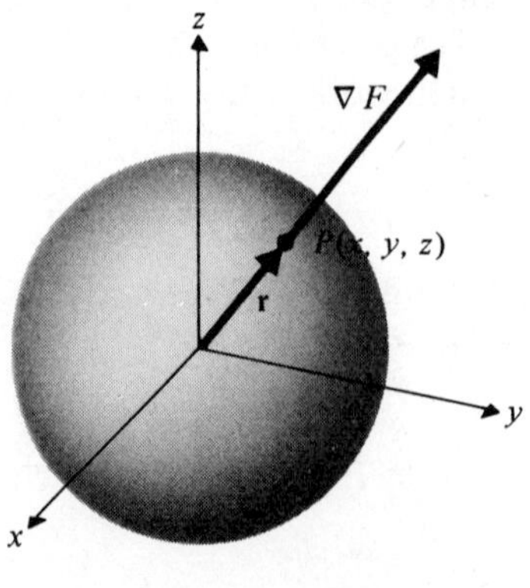

Figure 13.8

Consequently, at any point P on the sphere, ∇F is also perpendicular to the sphere. □

This example suggests that gradients may be useful in finding perpendiculars to surfaces (and, as we will see, perpendiculars to curves).

Example 9 The function $f(x, y) = 2x^2 - 4x + 3y^2 + 2y + 6$ is defined at every point in the xy-plane. If we start at the origin $(0, 0)$ and move along the positive x-axis, the rate of change of the function is $f_x(0, 0) = -4$; if we move along the y-axis, the rate of change is $f_y(0, 0) = 2$. Calculate the rate of change of $f(x, y)$ at $(0, 0)$ if we move toward the point $(1, 1)$ along the line $y = x$, and show that it is equal to the component of $\nabla f_{|(0,0)}$ in the direction $\mathbf{v} = (1, 1)$.

Solution The difference in the values of $f(x, y)$ at any point (x, y) and $(0, 0)$ is $f(x, y) - f(0, 0) = (2x^2 - 4x + 3y^2 + 2y + 6) - (6) = 2x^2 - 4x + 3y^2 + 2y$. If we divide this by the length of the line joining $(0, 0)$ and (x, y), we obtain

$$\frac{f(x, y) - f(0, 0)}{\sqrt{x^2 + y^2}} = \frac{2x^2 - 4x + 3y^2 + 2y}{\sqrt{x^2 + y^2}}.$$

The limit of this quotient as (x, y) approaches $(0, 0)$ along the line $y = x$ should yield the required rate of change. We therefore set $y = x$ and take the limit as x approaches zero through positive numbers:

$$\lim_{x \to 0^+} \frac{2x^2 - 4x + 3x^2 + 2x}{\sqrt{x^2 + x^2}} = \lim_{x \to 0^+} \frac{x(5x - 2)}{\sqrt{2}x} = -\sqrt{2}.$$

This quantity, then, is the rate of change of $f(x, y)$ at $(0, 0)$ along the line $y = x$ toward the point $(1, 1)$.

On the other hand,

$$\nabla f_{|(0,0)} = (4x - 4, 6y + 2)_{|(0,0)} = (-4, 2),$$

and the component of this vector in the direction $\mathbf{v} = (1, 1)$ is

$$\nabla f_{|(0,0)} \cdot \hat{\mathbf{v}} = (-4, 2) \cdot \frac{(1, 1)}{\sqrt{2}} = \frac{-4 + 2}{\sqrt{2}} = -\sqrt{2}. \ \square$$

This example indicates that gradients may be useful in calculating rates of change of functions in directions other than those parallel to the coordinate axes.

Example 10 The electrostatic potential at a point (x, y, z) in space due to a charge q fixed at the origin is given by

$$V = \frac{q}{4\pi\varepsilon_0 r},$$

where $r = \sqrt{x^2 + y^2 + z^2}$. If a second charge Q is placed at (x, y, z), it experiences a force $\mathbf{F}$, where

$$\mathbf{F} = \frac{qQ}{4\pi\varepsilon_0 r^3}\mathbf{r},$$

where $\mathbf{r} = x\hat{\mathbf{i}} + y\hat{\mathbf{j}} + z\hat{\mathbf{k}}$. Show that $\mathbf{F} = -Q\nabla V$.

Solution
$$\nabla V = \nabla\left(\frac{q}{4\pi\varepsilon_0 r}\right) = \frac{q}{4\pi\varepsilon_0}\nabla\left(\frac{1}{\sqrt{x^2+y^2+z^2}}\right)$$
$$= \frac{q}{4\pi\varepsilon_0}\left\{\frac{-x}{(x^2+y^2+z^2)^{3/2}}\hat{\mathbf{i}} + \frac{-y}{(x^2+y^2+z^2)^{3/2}}\hat{\mathbf{j}} + \frac{-z}{(x^2+y^2+z^2)^{3/2}}\hat{\mathbf{k}}\right\}$$
$$= \frac{-q}{4\pi\varepsilon_0 r^3}(x\hat{\mathbf{i}} + y\hat{\mathbf{j}} + z\hat{\mathbf{k}}) = -\frac{q}{4\pi\varepsilon_0 r^3}\mathbf{r} = -\frac{\mathbf{F}}{Q} \quad \square$$

EXERCISES 13.4

In Exercises 1–10, find the gradient of the given function.

1. $f(x, y, z) = x^2y + xz + yz^2$
2. $f(x, y, z) = x^2yz$
3. $f(x, y, z) = x^2y/z - 2xz^6$
4. $f(x, y) = x^2y + xy^2$
5. $f(x, y) = \sin(x + y)$
6. $f(x, y, z) = \text{Tan}^{-1}(xyz)$
7. $f(x, y) = \text{Tan}^{-1}(y/x)$
8. $f(x, y, z) = e^{x+y+z}$
9. $f(x, y) = 1/(x^2 + y^2)$
10. $f(x, y, z) = 1/\sqrt{x^2 + y^2 + z^2}$

In Exercises 11–15, find the gradient of the function at the indicated point.

11. $f(x, y) = xy + x + y$ at $(1, 3)$
12. $f(x, y, z) = \cos(x + y + z)$ at $(-1, 1, 1)$
13. $f(x, y, z) = (x^2 + y^2 + z^2)^2$ at $(0, 3, 6)$
14. $f(x, y) = e^{-x^2-y^2}$ at $(2, 2)$
15. $f(x, y) = xy\ln(x + y)$ at $(4, -2)$
16. The equation $F(x, y, z) = Ax + By + Cz + D = 0$ defines a plane in space. Show that at any point on this plane the vector ∇F is perpendicular to the plane.
17. Use the result of Exercise 16 to illustrate that a vector along the line
$$F(x, y, z) = 2x + 3y - 2z + 4 = 0,$$
$$G(x, y, z) = x - y + 3z + 6 = 0$$
is $\nabla F \times \nabla G$. Find parametric equations for the line.
18. Prove that if $f(x, y, z)$ and $g(x, y, z)$ both have gradients, then
$$\nabla(fg) = f\,\nabla g + g\,\nabla f.$$
19. Repeat Example 9 for each of the following functions:
    a) $f(x, y) = x^2 + y^2$;
    b) $f(x, y) = 2x^3 - 3y$.
20. The equation $F(x, y) = x^3 + xy + y^4 - 5 = 0$ implicitly defines a curve in the xy-plane. Show that at any point on this curve, ∇F is a normal vector to the curve.

Sketch the surface defined by the equation in each of Exercises 21 and 22. At what points on these surfaces is ∇F not defined?

21. $F(x, y, z) = z - \sqrt{x^2 + y^2} = 0$
22. $F(x, y, z) = z - |x - y| = 0$
23. If $z = f(x, y) = 1 - x^2 - y^2$, find ∇f. Find the point (x, y) at which $\nabla f = \mathbf{0}$, and illustrate graphically the nature of the surface $z = f(x, y)$ at this point.
24. If the gradient of a function $f(x, y)$ is
$$\nabla f = (2xy - y)\hat{\mathbf{i}} + (x^2 - x)\hat{\mathbf{j}},$$
what is $f(x, y)$?
25. Repeat Exercise 24 if
$$\nabla f = (2x/y + 1)\hat{\mathbf{i}} + (-x^2/y^2 + 2)\hat{\mathbf{j}}.$$
26. If the gradient of a function $f(x, y, z)$ is
$$\nabla f = yz\hat{\mathbf{i}} + (xz + 2yz)\hat{\mathbf{j}} + (xy + y^2)\hat{\mathbf{k}},$$
what is $f(x, y, z)$?
27. Repeat Exercise 26 if
$$\nabla f = (x\hat{\mathbf{i}} + y\hat{\mathbf{j}} + z\hat{\mathbf{k}})/\sqrt{x^2 + y^2 + z^2}.$$
28. If $f(x, y)$ and $g(x, y)$ have first partial derivatives in a region R of the xy-plane, and if in R, $\nabla f = \nabla g$, how are $f(x, y)$ and $g(x, y)$ related?
29. If $\nabla f = \mathbf{0}$ for all points in some region R of space, what can we say about $f(x, y, z)$ in R?
*30. Show that if the equation $F(x, y) = 0$ implicitly defines a curve C in the xy-plane, then at any point on C, the vector ∇F is perpendicular to C.

13.5 HIGHER-ORDER PARTIAL DERIVATIVES

If $f(x, y) = x^3y^2 + ye^x$, then

$$\frac{\partial f}{\partial x} = 3x^2y^2 + ye^x$$

and

$$\frac{\partial f}{\partial y} = 2x^3y + e^x.$$

Since each of these partial derivatives is a function of x and y, we can take further partial derivatives. The partial derivative of $\partial f/\partial x$ with respect to x is called the second partial derivative of $f(x, y)$ with respect to x, and is written

$$\frac{\partial}{\partial x}\left(\frac{\partial f}{\partial x}\right) = \frac{\partial^2 f}{\partial x^2} = 6xy^2 + ye^x.$$

Similarly, we have three more second partial derivatives:

$$\frac{\partial}{\partial y}\left(\frac{\partial f}{\partial x}\right) = \frac{\partial^2 f}{\partial y\,\partial x} = 6x^2y + e^x,$$

$$\frac{\partial}{\partial x}\left(\frac{\partial f}{\partial y}\right) = \frac{\partial^2 f}{\partial x\,\partial y} = 6x^2y + e^x,$$

$$\frac{\partial}{\partial y}\left(\frac{\partial f}{\partial y}\right) = \frac{\partial^2 f}{\partial y^2} = 2x^3.$$

Note that the second partial derivatives $\partial^2 f/\partial x\,\partial y$ and $\partial^2 f/\partial y\,\partial x$ are identical. This is not a peculiarity of this function, but according to the following theorem it is what we should expect normally.

▶ **Theorem 13.1** If $f(x, y)$, $\partial f/\partial x$, $\partial f/\partial y$, $\partial^2 f/\partial x\,\partial y$, and $\partial^2 f/\partial y\,\partial x$ are all continuous at a point, then at that point

$$\frac{\partial^2 f}{\partial x\,\partial y} = \frac{\partial^2 f}{\partial y\,\partial x}. \quad ◀$$

For most functions with which we will be concerned, this theorem can be extended to say that a mixed partial derivative may be calculated in any order whatsoever. For example, if we require $\partial^{10} f/\partial x^3\,\partial y^7$, where $f(x, y) = \ln(y^y) + xy^{10}$, it is far more advantageous to use

$$\frac{\partial^{10} f}{\partial x^3\,\partial y^7} = \frac{\partial^{10} f}{\partial y^7\,\partial x^3} = \frac{\partial^7}{\partial y^7}\left(\frac{\partial^3 f}{\partial x^3}\right) = 0.$$

Example 11 Show that $f(x, y) = \mathrm{Tan}^{-1}(2xy/(x^2 - y^2))$ satisfies the equation

$$\frac{\partial^2 f}{\partial x^2} + \frac{\partial^2 f}{\partial y^2} = 0.$$

Solution Since

$$\frac{\partial f}{\partial x} = \frac{1}{1 + \left(\dfrac{2xy}{x^2 - y^2}\right)^2} \left\{\frac{(x^2 - y^2)(2y) - 2xy(2x)}{(x^2 - y^2)^2}\right\}$$

$$= \frac{(x^2 - y^2)^2}{(x^2 - y^2)^2 + 4x^2y^2} \left\{\frac{2y(x^2 - y^2 - 2x^2)}{(x^2 - y^2)^2}\right\}$$

$$= \frac{-2y(x^2 + y^2)}{x^4 - 2x^2y^2 + y^4 + 4x^2y^2} = \frac{-2y(x^2 + y^2)}{(x^2 + y^2)^2} = \frac{-2y}{x^2 + y^2},$$

then

$$\frac{\partial^2 f}{\partial x^2} = \frac{2y}{(x^2 + y^2)^2}(2x) = \frac{4xy}{(x^2 + y^2)^2}.$$

Similarly,

$$\frac{\partial f}{\partial y} = \frac{1}{1 + \left(\dfrac{2xy}{x^2 - y^2}\right)^2} \left\{\frac{(x^2 - y^2)(2x) - 2xy(-2y)}{(x^2 - y^2)^2}\right\}$$

$$= \frac{(x^2 - y^2)^2}{(x^2 + y^2)^2} \left\{\frac{2x(x^2 - y^2 + 2y^2)}{(x^2 - y^2)^2}\right\} = \frac{2x}{x^2 + y^2}.$$

Consequently,

$$\frac{\partial^2 f}{\partial y^2} = \frac{-2x}{(x^2 + y^2)^2}(2y) = \frac{-4xy}{(x^2 + y^2)^2},$$

and addition of these two expressions for $\partial^2 f/\partial x^2$ and $\partial^2 f/\partial y^2$ completes the proof. □

The equation

$$\frac{\partial^2 f}{\partial x^2} + \frac{\partial^2 f}{\partial y^2} = 0 \tag{13.13}$$

for a function $f(x, y)$ is one of the most important equations in applied mathematics. It is called Laplace's equation in two variables (x and y). Laplace's equation for a function $f(x, y, z)$ of three variables is

$$\frac{\partial^2 f}{\partial x^2} + \frac{\partial^2 f}{\partial y^2} + \frac{\partial^2 f}{\partial z^2} = 0. \tag{13.14}$$

A function is said to be *harmonic* in a region R if it satisfies Laplace's equation in R and has continuous second partial derivatives in R. In particular, the function $f(x, y)$ in Example 11 is harmonic in any region that does not contain points on the lines $y = \pm x$. In Exercises 29–31, we illustrate various areas of applied mathematics that make use of Laplace's equation.

EXERCISES 13.5

In Exercises 1–20, find the indicated derivative.

1. $\partial^2 f/\partial x^2$ if $f(x, y) = x^2y^2 - 2x^3y$
2. $\partial^3 f/\partial y^3$ if $f(x, y) = 2x/y + 3x^3y^4$
3. $\partial^2 f/\partial z^2$ if $f(x, y, z) = \sin(xyz)$
4. $\partial^2 f/\partial y\,\partial z$ if $f(x, y, z) = xyze^{x+y+z}$
5. $\partial^2 f/\partial y\,\partial x$ if $f(x, y) = \sqrt{x^2 + y^2}$
6. $\partial^3 f/\partial x^2\,\partial y$ if $f(x, y) = e^{x+y} - x^2/y^2$
7. $\partial^3 f/\partial y^3$ at (1, 3) if $f(x, y) = 3x^3y^3 - 3x/y$
8. $\partial^3 f/\partial x\,\partial y\,\partial z$ at (1, 0, −1) if $f(x, y, z) = x^2y^2 + x^2z^2 + y^2z^2$
9. $\partial^2 f/\partial x^2$ if $f(x, y) = \sqrt{1 - x^2 - y^2}$
10. $\partial^2 f/\partial z^2$ if $f(x, y, z) = \ln\sqrt{x^2 + y^2 + z^2}$
11. $\partial^3 f/\partial x^2\,\partial y$ if $f(x, y, z) = x^2e^y + y^2e^x$
12. $\partial^2 f/\partial x^2$ if $f(x, y) = \text{Tan}^{-1}(y/x)$
13. $\partial^3 f/\partial x\,\partial y^2$ if $f(x, y, z) = \cot(x^2 + y^2 + z^2)$
14. $\partial^2 f/\partial x\,\partial y$ at (−2, −2) if $f(x, y) = \text{Sin}^{-1}(x^2 + y^2)^{-1}$
15. $\partial^{10} f/\partial x^7\,\partial y^3$ if $f(x, y) = x^7e^xy^2 + 1/y^6$
16. $\partial^8 f/\partial x^8$ if $f(x, y, z) = x^8y^9z^{10}$
17. $\partial^6 f/\partial x^2\,\partial y^2\,\partial z^2$ if $f(x, y, z) = 1/x^2 + 1/y^2 + 1/z^2$
18. $\partial^4 f/\partial x^3\,\partial y$ if $f(x, y, z) = \cos(x + y^3)$
19. $\partial^4 f/\partial x\,\partial y\,\partial z\,\partial t$ if $f(x, y, z, t) = \sqrt{x^2 + y^2 + z^2 - t^2}$
20. $\partial^2 f/\partial x\,\partial y$ if $f(x, y) = \text{Sec}^{-1}(xy)$
21. If $z = x^2 + xy + y^2\sin(x/y)$, show that
$$x\frac{\partial z}{\partial x} + y\frac{\partial z}{\partial y} = 2z$$
$$= x^2\frac{\partial^2 z}{\partial x^2} + 2xy\frac{\partial^2 z}{\partial x\,\partial y} + y^2\frac{\partial^2 z}{\partial y^2}.$$
22. If $u = x + y + ze^{y/x}$, show that
$$x^2\frac{\partial^2 u}{\partial x^2} + y^2\frac{\partial^2 u}{\partial y^2} + z^2\frac{\partial^2 u}{\partial z^2} + 2xy\frac{\partial^2 u}{\partial x\,\partial y}$$
$$+ 2yz\frac{\partial^2 u}{\partial y\,\partial z} + 2xz\frac{\partial^2 u}{\partial x\,\partial z} = 0.$$

In Exercises 23–28, find a region (if possible) in which the function is harmonic.

23. $f(x, y) = x^2 - y^2 + 2xy + y$
24. $f(x, y) = \ln(x^2 + y^2)$
25. $f(x, y) = x^3y^2 - 3xy$
26. $f(x, y, z) = 3x^2yz - y^3z + xy$
27. $f(x, y, z) = 1/\sqrt{x^2 + y^2 + z^2}$
28. $f(x, y, z) = x^3y^3z^3$
29. a) Show that the electrostatic potential function $V(x, y, z) = q/(4\pi\varepsilon_0 r)$ in Example 10 satisfies Laplace's equation (13.14).
    b) If $V(x, y, z)$ represents the electrostatic potential at a point (x, y, z) due to a system of n point charges at points (x_i, y_i, z_i) $(i = 1, \ldots, n)$, does $V(x, y, z)$ also satisfy Laplace's equation?
30. The gravitational potential at a point (x, y, z) in space due to a uniform spherical mass distribution (mass M) at the origin is defined as
$$V = \frac{GM}{r},$$
where $G > 0$ is a constant and $r = \sqrt{x^2 + y^2 + z^2}$. Show that $V(x, y, z)$ satisfies Laplace's equation (13.14).
31. In Fig. 13.9 we show a plate bounded by the lines $x = 0$, $y = 0$, $x = 1$, and $y = 1$. Temperature along the first three sides is kept at 0°C, while that along $y = 1$ varies according to
$$f(x) = \sin(3\pi x) - 2\sin(4\pi x), \qquad 0 \le x \le 1.$$

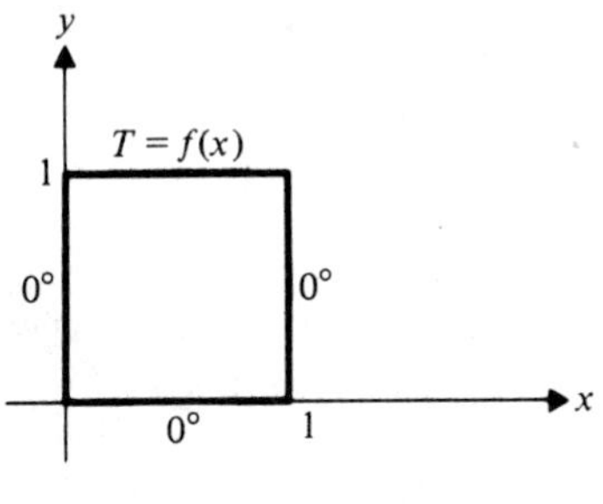

Figure 13.9

The temperature at any point interior to the plate is then
$$T(x, y) = C(e^{3\pi y} - e^{-3\pi y})\sin(3\pi x) + D(e^{4\pi y} - e^{-4\pi y})\sin(4\pi x),$$
where $C = (e^{3\pi} - e^{-3\pi})^{-1}$ and $D = -2(e^{4\pi} - e^{-4\pi})^{-1}$. Show that $T(x, y)$ is harmonic in the region $0 < x < 1$, $0 < y < 1$, and that it also satisfies the boundary conditions
$$T(0, y) = 0, \quad T(x, 0) = 0,$$
$$T(1, y) = 0, \quad T(x, 1) = f(x).$$
32. If $\partial^2 z/\partial x^2$ is positive for $x_1 \le x \le x_2$ and $y = y_0$, what can you say geometrically about the surface $z = f(x, y)$?

33. Two functions $u(x, y)$ and $v(x, y)$ are said to be harmonic conjugates if they satisfy the Cauchy–Riemann equations

$$\frac{\partial u}{\partial x} = \frac{\partial v}{\partial y} \quad \text{and} \quad \frac{\partial v}{\partial x} = -\frac{\partial u}{\partial y}.$$

(See also Exercise 35 in Section 13.3.) Show that if $u(x, y)$ and $v(x, y)$ are harmonic conjugates and have continuous second partial derivatives in a region R, then each is a harmonic function in R.

34. If the ends of a taut string are fastened at $x = 0$ and $x = L$ on the x-axis and if the string vibrates in the y-direction, then its displacement $y(x, t)$ must satisfy the one-dimensional wave equation

$$\frac{\partial^2 y}{\partial t^2} = \frac{T}{\rho}\frac{\partial^2 y}{\partial x^2},$$

where T is the constant tension in the string and ρ is its constant density. Determine whether the wave equation is satisfied by:

a) $y(x, t) = \sin\left(\frac{5\pi x}{L}\right)\cos\left(\frac{5\pi}{L}\sqrt{\frac{T}{\rho}}\,t\right)$;

b) $y(x, t) = \sum_{n=1}^{50} a_n \sin\left(\frac{n\pi x}{L}\right)\cos\left(\frac{n\pi}{L}\sqrt{\frac{T}{\rho}}\,t\right)$, where a_n are constants.

35. A uniform circular rod has flat ends at $x = 0$ and $x = L$ on the x-axis. If the round side of the rod is insulated, then heat flows in only the x-direction. If, further, the face at $x = L$ is insulated and the face at $x = 0$ is kept at temperature 100°C, then the temperature $T(x, t)$ at any point in the rod must satisfy the one-dimensional heat conduction problem

$$\frac{\partial T}{\partial t} = k\frac{\partial^2 T}{\partial x^2}, \qquad k = \text{a constant};$$

$$T(0, t) = 100;$$

$$T_x(L, t) = 0.$$

Show that for any positive integer n,

$$T(x, t) = 100 - \frac{2L}{(2n-1)\pi} e^{-k[(2n-1)\pi/(2L)]^2 t} \sin\left\{\left(\frac{2n-1}{2}\right)\frac{\pi x}{L}\right\}$$

satisfies the heat conduction problem.

36. a) Show that the function $u(x, y) = x^2 - y^2$ is harmonic in the entire xy-plane.

b) Use the Cauchy–Riemann equations in Exercise 33 to find a function $v(x, y)$ so that u and v are harmonic conjugates.

37. Repeat Exercise 36 if $u(x, y) = e^x \cos y + x$.

13.6 CHAIN RULES FOR PARTIAL DERIVATIVES

If $y = f(u)$ and $u = g(x)$, then the chain rule for the derivative dy/dx of the composite function $f[g(x)]$ is

$$\frac{dy}{dx} = \frac{dy}{du}\frac{du}{dx}. \tag{13.15}$$

Equation (13.15) can be extended only in terms of more intermediate variables, say $y = f(u)$, $u = g(s)$, $s = h(x)$, in which case

$$\frac{dy}{dx} = \frac{dy}{du}\frac{du}{ds}\frac{ds}{dx}. \tag{13.16}$$

For multivariable functions, the variations in the chain rule are countless. We will discuss two examples in considerable detail, and then show schematically how to obtain the correct chain rule for every possible situation.

Suppose z is a function of u and v and each of u and v is a function of x and y,

$$z = f(u, v), \qquad u = g(x, y), \qquad v = h(x, y), \tag{13.17}$$

and it is required to find $\partial z/\partial x$. By the substitutions

$$z = f[g(x, y), h(x, y)], \tag{13.18}$$

we could express z as a function of x and y, and the partial derivative $\partial z/\partial x$ could then be calculated. However, if the functions in (13.17) are at all complicated, you can imagine how difficult the composite function in (13.18) might be to differentiate. As a result, we search for an alternative procedure for calculating $\partial z/\partial x$, namely, the appropriate chain rule. It is contained in the following theorem.

▶ **Theorem 13.2** Let $u = g(x, y)$ and $v = h(x, y)$ be continuous and have first partial derivatives with respect to x at a point (x, y), and let $z = f(u, v)$ have continuous first partial derivatives inside a circle centered at the point $(u, v) = (g(x, y), h(x, y))$. Then

$$\frac{\partial z}{\partial x} = \frac{\partial z}{\partial u}\frac{\partial u}{\partial x} + \frac{\partial z}{\partial v}\frac{\partial v}{\partial x}. \quad ◀ \tag{13.19}$$

Proof This result can be proved in much the same way as chain rule (3.13) was proved in Section 3.7. By Definition 13.3,

$$\frac{\partial z}{\partial x} = \lim_{\Delta x \to 0} \frac{f[g(x + \Delta x, y), h(x + \Delta x, y)] - f[g(x, y), h(x, y)]}{\Delta x}.$$

Now the increment Δx in x produces changes in u and v, which we denote by

$$\Delta u = g(x + \Delta x, y) - g(x, y), \qquad \Delta v = h(x + \Delta x, y) - h(x, y).$$

If we write u and v whenever $g(x, y)$ and $h(x, y)$ are evaluated at (x, y), and substitute for $g(x + \Delta x, y)$ and $h(x + \Delta x, y)$ in the definition for $\partial z/\partial x$, then

$$\begin{aligned}
\frac{\partial z}{\partial x} &= \lim_{\Delta x \to 0} \frac{f(u + \Delta u, v + \Delta v) - f(u, v)}{\Delta x} \\
&= \lim_{\Delta x \to 0} \frac{[f(u + \Delta u, v + \Delta v) - f(u, v + \Delta v)] + [f(u, v + \Delta v) - f(u, v)]}{\Delta x} \\
&= \lim_{\Delta x \to 0} \left\{\frac{f(u + \Delta u, v + \Delta v) - f(u, v + \Delta v)}{\Delta x} + \frac{f(u, v + \Delta v) - f(u, v)}{\Delta x}\right\}.
\end{aligned}$$

We assumed that the derivative

$$\frac{\partial z}{\partial v} = \lim_{\Delta v \to 0} \frac{f(u, v + \Delta v) - f(u, v)}{\Delta v}$$

exists at (u, v), and an equivalent way to express the fact that this limit exists is to say that

$$\frac{f(u, v + \Delta v) - f(u, v)}{\Delta v} = \frac{\partial z}{\partial v} + \varepsilon_1,$$

where ε_1 must satisfy the condition that $\lim_{\Delta v \to 0} \varepsilon_1 = 0$. We can write, therefore, that

$$f(u, v + \Delta v) - f(u, v) = [z_v(u, v) + \varepsilon_1]\,\Delta v.$$

Similarly, we can write that

$$f(u + \Delta u, v + \Delta v) - f(u, v + \Delta v) = [z_u(u, v + \Delta v) + \varepsilon_2]\,\Delta u,$$

where $\lim_{\Delta u \to 0} \varepsilon_2 = 0$ (provided Δv is sufficiently small). When these expressions are substituted into the limit for $\partial z/\partial x$, we have

$$\frac{\partial z}{\partial x} = \lim_{\Delta x \to 0} \left\{ [z_u(u, v + \Delta v) + \varepsilon_2] \frac{\Delta u}{\Delta x} + [z_v(u, v) + \varepsilon_1] \frac{\Delta v}{\Delta x} \right\}.$$

We now examine each part of this limit. Clearly,

$$\lim_{\Delta x \to 0} \frac{\Delta u}{\Delta x} = \frac{\partial u}{\partial x} \quad \text{and} \quad \lim_{\Delta x \to 0} \frac{\Delta v}{\Delta x} = \frac{\partial v}{\partial x}.$$

In addition, because $g(x, y)$ and $h(x, y)$ are continuous, $\Delta u \to 0$ and $\Delta v \to 0$ as $\Delta x \to 0$. Consequently,

$$\lim_{\Delta x \to 0} \varepsilon_1 = \lim_{\Delta v \to 0} \varepsilon_1 = 0 \quad \text{and} \quad \lim_{\Delta x \to 0} \varepsilon_2 = \lim_{\Delta u \to 0} \varepsilon_2 = 0.$$

Finally, because $\partial z/\partial u$ is continuous,

$$\lim_{\Delta x \to 0} z_u(u, v + \Delta v) = \lim_{\Delta v \to 0} z_u(u, v + \Delta v) = z_u(u, v).$$

When all these results are taken into account, we have

$$\frac{\partial z}{\partial x} = z_u(u, v) \frac{\partial u}{\partial x} + z_v(u, v) \frac{\partial v}{\partial x} = \frac{\partial z}{\partial u}\frac{\partial u}{\partial x} + \frac{\partial z}{\partial v}\frac{\partial v}{\partial x}. \quad \square$$

Chain rule (13.19) defines $\partial z/\partial x$ in terms of derivatives of the given functions in (13.17). We could be more explicit by indicating which variable is being held constant in each of the five derivatives:

$$\left.\frac{\partial z}{\partial x}\right)_y = \left.\frac{\partial z}{\partial u}\right)_v \left.\frac{\partial u}{\partial x}\right)_y + \left.\frac{\partial z}{\partial v}\right)_u \left.\frac{\partial v}{\partial x}\right)_y. \tag{13.20}$$

If we are discussing rates of change, this result seems quite reasonable. The problem is to find the rate of change of z with respect to x holding y constant. The first term $(\partial z/\partial u)(\partial u/\partial x)$ accounts for the rate of change of z with respect to those x's that affect z through u. The second term $(\partial z/\partial v)(\partial v/\partial x)$ accounts for the rate of change of z with respect to those x's that affect z through v. The total rate of change is then the sum of the two parts.

Consider now the functional situation

$$z = f(u, v), \qquad u = g(x, y, s), \qquad v = h(x, y, s), \qquad x = p(t),$$
$$y = q(t), \qquad s = r(t). \tag{13.21}$$

By the substitutions

$$z = f[g(p(t), q(t), r(t)), h(p(t), q(t), r(t))], \tag{13.22}$$

we can express z as a function of t alone, and can therefore pose the problem of calculating dz/dt. If we reason as in the last paragraph, the appropriate chain rule for dz/dt must account for all t's affecting z through u and v. We obtain, then,

$$\frac{dz}{dt} = \frac{\partial z}{\partial u}\frac{du}{dt} + \frac{\partial z}{\partial v}\frac{dv}{dt},$$

where we have written du/dt and dv/dt because u and v can be expressed entirely in terms of t:

$$u = g[p(t), q(t), r(t)], \qquad v = h[p(t), q(t), r(t)].$$

Chain rules for each of du/dt and dv/dt [similar to (13.19)] yield

$$\frac{du}{dt} = \frac{\partial u}{\partial x}\frac{dx}{dt} + \frac{\partial u}{\partial y}\frac{dy}{dt} + \frac{\partial u}{\partial s}\frac{ds}{dt}, \qquad \frac{dv}{dt} = \frac{\partial v}{\partial x}\frac{dx}{dt} + \frac{\partial v}{\partial y}\frac{dy}{dt} + \frac{\partial v}{\partial s}\frac{ds}{dt}.$$

Finally, then,

$$\frac{dz}{dt} = \frac{\partial z}{\partial u}\left\{\frac{\partial u}{\partial x}\frac{dx}{dt} + \frac{\partial u}{\partial y}\frac{dy}{dt} + \frac{\partial u}{\partial s}\frac{ds}{dt}\right\} + \frac{\partial z}{\partial v}\left\{\frac{\partial v}{\partial x}\frac{dx}{dt} + \frac{\partial v}{\partial y}\frac{dy}{dt} + \frac{\partial v}{\partial s}\frac{ds}{dt}\right\}, \quad (13.23)$$

which expresses dz/dt in terms of derivatives of the given functions in (13.21).

These two examples illustrate that finding chain rules for complicated composite functions can be very difficult. Fortunately, there is an amazingly simple method that gives the correct chain rule in every situation. The method is not designed to help you understand the chain rule, but to find it quickly. We suggest that you test your understanding by developing a few chain rules in the exercises with a discussion such as in the second example above, and then check your result by the quicker method.

In the first example we represent the functional situation described in (13.17) as:

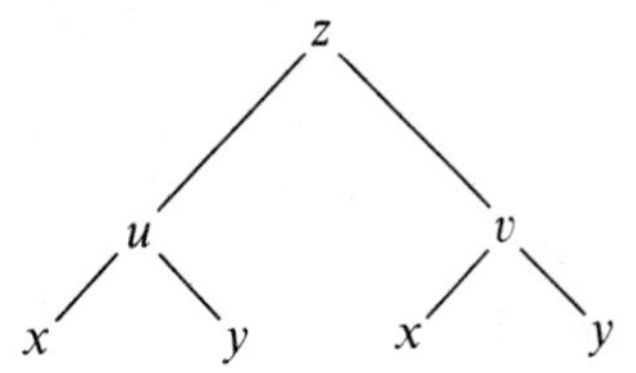

At the top of the diagram is the dependent variable z, which we wish to differentiate. In the line below z are the variables u and v in terms of which z is initially defined. In the line below u and v are x's and y's illustrating that each of u and v is defined in terms of x and y. To obtain the partial derivative $\partial z/\partial x$, we take all possible paths in this diagram from z to x. The two paths are from z through u to x, and from z through v to x. For each straight-line segment in a given path, we differentiate the upper variable with respect to the lower variable and multiply together all such derivatives in that path. The products are then added together to form the complete chain rule. In particular, for the path through u we form the product

$$\frac{\partial z}{\partial u}\frac{\partial u}{\partial x},$$

and for the path through v,

$$\frac{\partial z}{\partial v}\frac{\partial v}{\partial x}.$$

The complete chain rule is then the sum of these products,

$$\frac{\partial z}{\partial x} = \frac{\partial z}{\partial u}\frac{\partial u}{\partial x} + \frac{\partial z}{\partial v}\frac{\partial v}{\partial x},$$

and this result agrees with (13.19). We can also use the schematic to indicate which variables are to be held constant in the derivatives on the right (as in Eq. 13.20). We simply agree that all other variables on the same level are to be held constant.

For the second example in Eqs. (13.21) the schematic is:

There are six possible paths from z to t so that the chain rule for dz/dt must have six terms. We find

$$\frac{dz}{dt} = \frac{\partial z}{\partial u}\frac{\partial u}{\partial x}\frac{dx}{dt} + \frac{\partial z}{\partial u}\frac{\partial u}{\partial y}\frac{dy}{dt} + \frac{\partial z}{\partial u}\frac{\partial u}{\partial s}\frac{ds}{dt} + \frac{\partial z}{\partial v}\frac{\partial v}{\partial x}\frac{dx}{dt} + \frac{\partial z}{\partial v}\frac{\partial v}{\partial y}\frac{dy}{dt} + \frac{\partial z}{\partial v}\frac{\partial v}{\partial s}\frac{ds}{dt},$$

and this agrees with (13.23). Note too that if when forming a derivative from the schematic we see that there are two or more lines emanating from a variable, then we obtain a partial derivative; if there is only one line, then we have an ordinary derivative.

Example 12 Find chain rules for

$$\left.\frac{\partial z}{\partial x}\right)_y \quad \text{and} \quad \left.\frac{\partial z}{\partial y}\right)_x$$

if

$$z = f(r, s, x), \qquad r = g(x, y), \qquad s = h(x, y).$$

Solution From this schematic

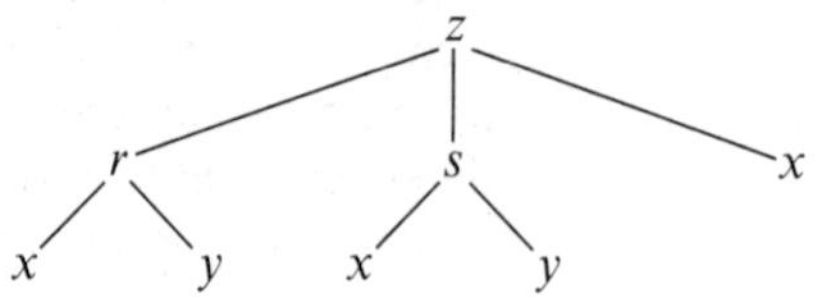

we obtain

$$\left.\frac{\partial z}{\partial y}\right)_x = \left.\frac{\partial z}{\partial r}\right)_{s,x}\left.\frac{\partial r}{\partial y}\right)_x + \left.\frac{\partial z}{\partial s}\right)_{r,x}\left.\frac{\partial s}{\partial y}\right)_x,$$

$$\left.\frac{\partial z}{\partial x}\right)_y = \left.\frac{\partial z}{\partial r}\right)_{s,x}\left.\frac{\partial r}{\partial x}\right)_y + \left.\frac{\partial z}{\partial s}\right)_{r,x}\left.\frac{\partial s}{\partial x}\right)_y + \left.\frac{\partial z}{\partial x}\right)_{r,s}. \quad \square$$

In this example it is essential that we indicate which variables to hold constant in the partial derivatives. If we were to omit these designations, then in the second result we would have a term $\partial z/\partial x$ on both sides of the equation and might be tempted to cancel them. This cannot be done, however, since they are entirely different derivatives. The term $\partial z/\partial x)_y$ indicates the derivative of z with respect to x holding y constant if z were expressed entirely in terms of x and y; the term $\partial z/\partial x)_{r,s}$ indicates the derivative of the given function $f(r, s, x)$ with respect to x holding r and s constant.

Example 13 Find dz/dt if

$$z = x^3y^2 + x \sin y + tx, \qquad x = 2t + \frac{1}{t}, \qquad y = t^2e^t.$$

Solution From this schematic

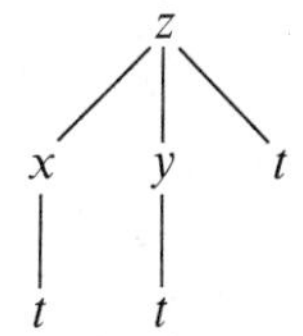

we have

$$\frac{dz}{dt} = \frac{\partial z}{\partial x}\frac{dx}{dt} + \frac{\partial z}{\partial y}\frac{dy}{dt} + \frac{\partial z}{\partial t}$$
$$= (3x^2y^2 + \sin y + t)(2 - 1/t^2) + (2x^3y + x\cos y)(2te^t + t^2e^t) + x. \square$$

When a chain rule is used to calculate a derivative, the result usually involves all intermediate variables. For instance, the derivative dz/dt in Example 13 involves not only t, but the intermediate variables x and y as well. Were dz/dt required at $t = 1$, values of x and y for $t = 1$ would be calculated $x(1) = \frac{5}{2}$ and $y(1) = e$ and all three values substituted to obtain

$$\frac{dz}{dt}_{|t=1} = [3(3)^2(e)^2 + \sin(e) + 1](2 - 1)$$
$$+ [2(3)^3e + (3)\cos(e)](2e + e) + 3$$
$$= 1378.6.$$

Example 14 Find $\partial^2 z/\partial x^2$ if

$$z = s^2t + 2\sin t, \qquad s = xy - y, \qquad t = x^2 + \frac{y}{x}.$$

Solution From this schematic

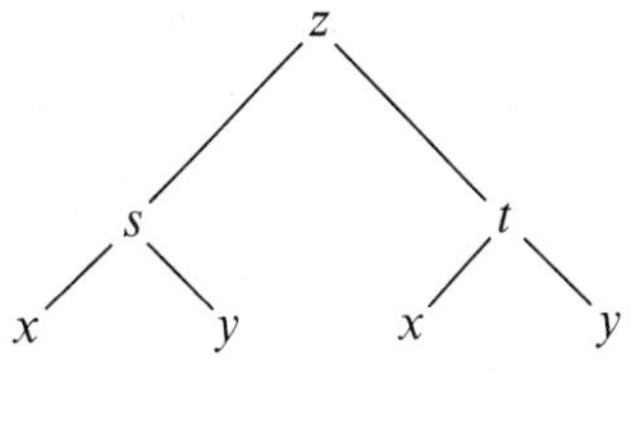

we have

$$\frac{\partial z}{\partial x} = \frac{\partial z}{\partial s}\frac{\partial s}{\partial x} + \frac{\partial z}{\partial t}\frac{\partial t}{\partial x}$$

$$= (2st)(y) + (s^2 + 2\cos t)(2x - y/x^2).$$

Now $\partial z/\partial x$ is a function of s, t, x, and y, and therefore in order to find

$$\frac{\partial^2 z}{\partial x^2} = \frac{\partial}{\partial x}\left(\frac{\partial z}{\partial x}\right),$$

we form the following schematic for $\partial z/\partial x$:

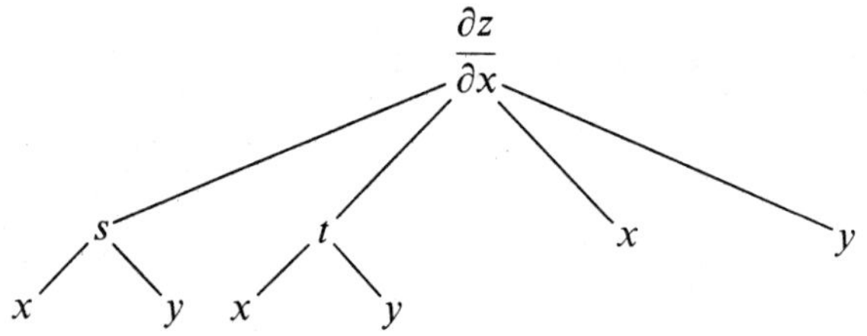

From this schematic, we obtain

$$\frac{\partial^2 z}{\partial x^2} = \frac{\partial}{\partial s}\left(\frac{\partial z}{\partial x}\right)\frac{\partial s}{\partial x} + \frac{\partial}{\partial t}\left(\frac{\partial z}{\partial x}\right)\frac{\partial t}{\partial x} + \frac{\partial}{\partial x}\left(\frac{\partial z}{\partial x}\right)_{s,t,y}$$

$$= [2ty + 2s(2x - y/x^2)](y) + [2sy - 2\sin t(2x - y/x^2)](2x - y/x^2)$$

$$+ (s^2 + 2\cos t)(2 + 2y/x^3). \ \square$$

Example 15 The temperature T at points in the atmosphere depends on both position (x, y, z) and time t: $T = T(x, y, z, t)$. When a weather balloon is released to take temperature readings, it is not free to take readings at just any point, but only at those points along the path that the winds force the balloon to follow. This path is a curve in space that can be defined parametrically by

$$C: \quad x = x(t), \qquad y = y(t), \qquad z = z(t), \qquad t \ge 0,$$

t again being time. If we substitute from the equations for C into the temperature function, then T becomes a function of t alone,

$$T = T[x(t), y(t), z(t), t],$$

and this function of time describes the temperature at points along the path of the balloon. For the derivative of this function with respect to t, the schematic

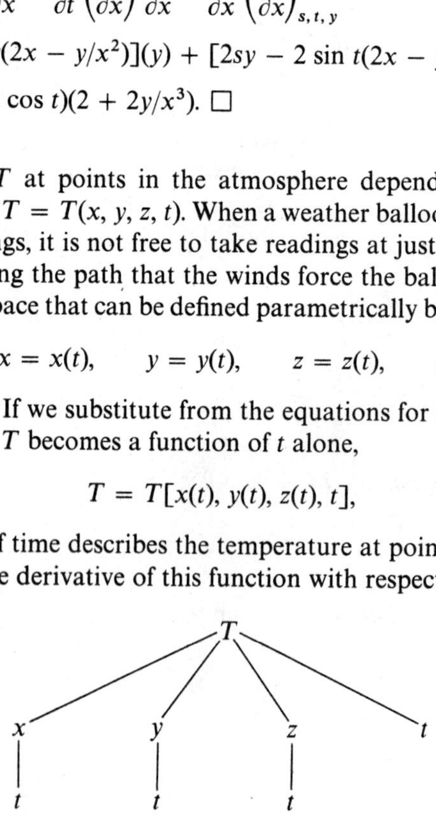

yields

$$\frac{dT}{dt} = \frac{\partial T}{\partial x}\frac{dx}{dt} + \frac{\partial T}{\partial y}\frac{dy}{dt} + \frac{\partial T}{\partial z}\frac{dz}{dt} + \frac{\partial T}{\partial t}.$$

The question we pose is "What is the physical difference between dT/dt and $\partial T/\partial t$?"

Solution The temperature at a point in space is independent of the observer measuring it; hence $T[x(t), y(t), z(t), t]$ is the temperature at points on C as measured by both the balloon and any observer fixed in the xyz-reference system. If, however, these two observers calculate at some point (x, y, z) on C the rate of change of temperature with respect to time, they calculate different results. The observer fixed in the xyz-reference system (not restricted to moving along C) calculates the rate of change of T with respect to t as the derivative of the function $T(x, y, z, t)$ partially with respect to t holding x, y, and z constant; i.e., the fixed observer calculates $\partial T/\partial t$ as the rate of change of temperature in time. The balloon, on the other hand, has no alternative but to take temperature readings as it moves along C; thus its measurement of T as a function of t is

$$T[x(t), y(t), z(t), t].$$

Therefore, when the balloon calculates the time variation of temperature, it is calculating dT/dt. It follows, then, that the terms

$$\frac{\partial T}{\partial x}\frac{dx}{dt} + \frac{\partial T}{\partial y}\frac{dy}{dt} + \frac{\partial T}{\partial z}\frac{dz}{dt}$$

describe that part of dT/dt caused by the motion of the balloon through space. □

Many important applications of the chain rule occur in the field of partial differential equations. To illustrate, we consider the one-dimensional wave equation

$$\frac{\partial^2 y}{\partial t^2} = c^2 \frac{\partial^2 y}{\partial x^2}, \qquad c = \text{a constant},$$

for $y = f(x, t)$. To solve it, we must find all functions $y = f(x, t)$ such that the second partial derivative of $f(x, t)$ with respect to t is equal to c^2 times the second partial derivative of $f(x, t)$ with respect to x. In general, partial differential equations are far more difficult to solve than ordinary differential equations, but chain rules offer a powerful technique to simplify some equations.

We saw in Chapter 9 that it is sometimes advantageous to make a change of variable in an integration problem. In Chapter 16 we will see that a change of variable sometimes facilitates the solution of an ordinary differential equation. Perhaps, then, it would be possible to make a change of independent variable, or variables, which would simplify the wave equation. In particular, suppose we introduce new independent variables u and v defined by

$$u = x + ct, \qquad v = x - ct,$$

and transform the wave equation from a partial differential equation in $y = f(x, t)$ as a function of x and t to a partial differential equation in $y = F(u, v)$

as a function of u and v. To do this, we first find an expression for $\partial y/\partial t$ in terms of derivatives of $y = F(u, v)$ with respect to u and v. The schematic

represents y as a function of u and v, which in turn are functions of x and t:

$$y = F(u, v), \qquad u = x + ct, \qquad v = x - ct.$$

From this schematic, we see that

$$\frac{\partial y}{\partial t} = \frac{\partial y}{\partial u}\frac{\partial u}{\partial t} + \frac{\partial y}{\partial v}\frac{\partial v}{\partial t} = c\frac{\partial y}{\partial u} - c\frac{\partial y}{\partial v},$$

and we have an expression for the derivative of $y = f(x, t)$ with respect to t in terms of derivatives of $y = F(u, v)$ with respect to u and v. To find an expression for $\partial^2 y/\partial t^2$, we first write

$$\frac{\partial^2 y}{\partial t^2} = \frac{\partial}{\partial t}\left(\frac{\partial y}{\partial t}\right) = \frac{\partial}{\partial t}\left(c\frac{\partial y}{\partial u} - c\frac{\partial y}{\partial v}\right),$$

and note that the term in parentheses is a function of u and v. We therefore form the schematic

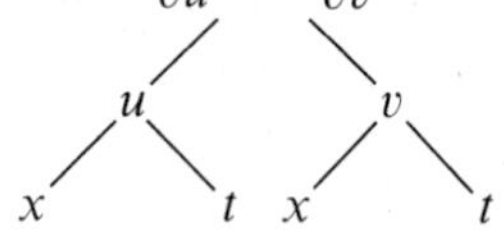

and find

$$\begin{aligned}\frac{\partial^2 y}{\partial t^2} &= \frac{\partial}{\partial t}\left(c\frac{\partial y}{\partial u} - c\frac{\partial y}{\partial v}\right)\\ &= \frac{\partial}{\partial u}\left(c\frac{\partial y}{\partial u} - c\frac{\partial y}{\partial v}\right)\frac{\partial u}{\partial t} + \frac{\partial}{\partial v}\left(c\frac{\partial y}{\partial u} - c\frac{\partial y}{\partial v}\right)\frac{\partial v}{\partial t}\\ &= c\left(\frac{\partial^2 y}{\partial u^2} - \frac{\partial^2 y}{\partial u\,\partial v}\right)c + c\left(\frac{\partial^2 y}{\partial v\,\partial u} - \frac{\partial^2 y}{\partial v^2}\right)(-c) = c^2\left(\frac{\partial^2 y}{\partial u^2} - 2\frac{\partial^2 y}{\partial u\,\partial v} + \frac{\partial^2 y}{\partial v^2}\right).\end{aligned}$$

A similar procedure gives

$$\frac{\partial^2 y}{\partial x^2} = \frac{\partial^2 y}{\partial u^2} + 2\frac{\partial^2 y}{\partial u\,\partial v} + \frac{\partial^2 y}{\partial v^2}.$$

If we substitute these expressions into the wave equation, we obtain

$$c^2\left(\frac{\partial^2 y}{\partial u^2} - 2\frac{\partial^2 y}{\partial u\,\partial v} + \frac{\partial^2 y}{\partial v^2}\right) = c^2\left(\frac{\partial^2 y}{\partial u^2} + 2\frac{\partial^2 y}{\partial u\,\partial v} + \frac{\partial^2 y}{\partial v^2}\right),$$

which reduces to

$$\frac{\partial^2 y}{\partial u\,\partial v} = 0.$$

This equation, then, is the wave equation expressed in terms of the independent variables u and v instead of x and t. If we write this equation in the form

$$\frac{\partial}{\partial u}\left(\frac{\partial y}{\partial v}\right) = 0,$$

then it is clear that $\partial y/\partial v$ cannot depend on u; i.e.,

$$\frac{\partial y}{\partial v} = h(v),$$

where $h(v)$ is any differentiable function of v whatsoever. Integration of this equation with respect to v now gives

$$y = F(u, v) = \int h(v)\,dv + G(u) = H(v) + G(u),$$

where $H(v)$ and $G(u)$ are arbitrary differentiable functions of u and v. Finally, by setting $u = x + ct$ and $v = x - ct$, we obtain the solution of the one-dimensional wave equation

$$y = f(x, t) = H(x - ct) + G(x + ct).$$

In other words, if we take any function $H(v)$ and any function $G(u)$ and replace each v by $x - ct$ and each u by $x + ct$, then the function $H(x - ct) + G(x + ct)$ will satisfy the one-dimensional wave equation.

EXERCISES 13.6

In Exercises 1–10, we have defined a general functional situation and a specific example. Find the chain rule for the indicated derivative in the general situation, and then use that result to calculate the same derivative in the specific example.

1. dz/dt if $z = f(x, t)$, $x = g(t)$; $z = xt^2/(x + t)$, $x = e^{3t}$
2. $\left.\frac{\partial z}{\partial t}\right)_s$ if $z = f(x, y)$, $x = g(s, t)$, $y = h(s, t)$; $z = x^2e^y + y \ln x$, $x = s^2 \cos t$, $y = 4 \operatorname{Sec}^{-1}(t^2 + 2s)$
3. $\left.\frac{\partial u}{\partial s}\right)_t$ if $u = f(x, y, z)$, $x = g(s, t)$, $y = h(s, t)$, $z = k(s, t)$; $u = \sqrt{x^2 + y^2 + z^2}$, $x = 2st$, $y = s^2 + t^2$, $z = st$
4. dz/du if $z = f(x, y, v)$, $x = g(u)$, $y = h(u)$, $v = k(u)$; $z = x^2yv^3$, $x = u^3 + 2u$, $y = \ln(u^2 + 1)$, $v = ue^u$
5. $\left.\frac{\partial u}{\partial r}\right)_t$ if $u = f(x, y, s)$, $x = g(t)$, $y = h(r)$, $s = k(r, t)$; $u = \sqrt{x^2 + y^2s}$, $x = t/(t + 5)$, $y = \operatorname{Sin}^{-1}(r^2 + 5)$, $s = \tan(rt)$
6. $\left.\frac{\partial z}{\partial t}\right)_r$ if $z = f(x)$, $x = g(y)$, $y = h(r, t)$; $z = 3^{x+2}$, $x = y^2 + 5$, $y = \csc(r^2 + t)$
7. $\left.\frac{\partial u}{\partial x}\right)_y$ if $u = f(x, y, z)$, $z = g(x, y)$; $u = y/\sqrt{x^2 + y^2 + z^2}$, $z = x/y$
8. $\left.\frac{\partial x}{\partial y}\right)_z$ if $x = f(r, s, t)$, $r = g(y)$, $s = h(y, z)$, $t = k(y, z)$; $x = s^2r^2t^2$, $r = y^{-5}$, $s = 1/(y^2 + z^2)$, $t = 1/y^2 + 1/z^2$
9. $\left.\frac{\partial z}{\partial t}\right)_s$ if $z = f(x, y)$, $x = g(r)$, $y = h(r)$, $r = k(s, t)$; $z = e^{x+y}$, $x = 2r + 5$, $y = 2r - 5$, $r = t \ln(s^2 + t^2)$
10. dz/dt if $z = f(x, y, u)$, $x = g(v)$, $u = h(x, y)$, $v = k(t)$, $y = p(t)$; $z = x^2 + y^2 + u^2$, $x = v^3 - 3v^2$, $u = 1/(x^2 - y^2)$, $v = e^t$, $y = e^{4t}$

In Exercises 11–15, find the indicated derivative.

11. $\left(\dfrac{\partial^2 z}{\partial t^2}\right)_s$ if $z = x^2y^2 + xe^y$, $x = s + t^2$, $y = s - t^2$

12. d^2x/dt^2 if $x = y^2 + yt - t^2$, $y = t^2e^t$

13. $\left(\dfrac{\partial^2 u}{\partial s^2}\right)_t$ if $u = x^2 + y^2 + z^2 + xyz$, $x = s^2 + t^2$, $y = s^2 - t^2$, $z = st$

14. d^2z/dv^2 if $z = \sin(xy)$, $x = 3\cos v$, $y = 4\sin v$

15. $\partial^2 u/\partial x\,\partial y$ if $u = y/\sqrt{x^2 + y^2 + z^2}$, $z = x/y$

16. Suppose that u is a differentiable function of r and $r = \sqrt{x^2 + y^2 + z^2}$. Show that

$$\left(\frac{\partial u}{\partial x}\right)^2 + \left(\frac{\partial u}{\partial y}\right)^2 + \left(\frac{\partial u}{\partial z}\right)^2 = \left(\frac{du}{dr}\right)^2.$$

17. Consider a gas that is moving through some region D of space. If we follow a particular particle of the gas, it will trace out some curved path (Fig. 13.10)

$$C: \quad x = x(t), \qquad y = y(t), \qquad z = z(t), \qquad t \geq 0.$$

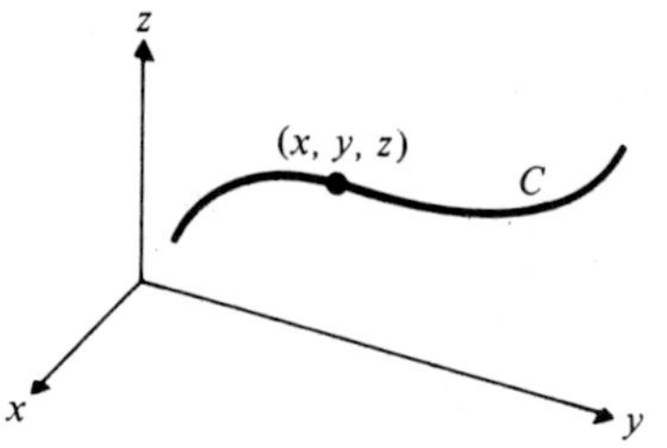

Figure 13.10

Suppose that the density of the gas at any point in the region D at time t is denoted by $\rho(x, y, z, t)$. We can write, then, that along C

$$\rho = \rho[x(t), y(t), z(t), t].$$

a) Obtain the chain rule defining $d\rho/dt$ in terms of $\partial\rho/\partial t$ and the derivatives of x, y, and z with respect to t.

b) Explain the physical difference between $d\rho/dt$ and $\partial\rho/\partial t$.

18. The radius and height of a right-circular cone are 10 and 20 centimeters, respectively. If the radius is increasing at a rate of 1 centimeter per minute and the height is decreasing at a rate of 2 centimeters per minute, how fast is the volume changing? Do you need to use multivariable calculus to solve this problem?

19. If two sides of a triangle have lengths x and y and the angle between them is θ, then the area of the triangle is $A = (\frac{1}{2})xy \sin\theta$. How fast is the area changing when x is 1 meter, y is 2 meters, and θ is $\frac{1}{3}$ radian, if x and y are each increasing at the rate of $\frac{1}{2}$ meter per second and θ is decreasing at the rate of $\frac{1}{10}$ radian per second?

20. When a rocket rises from the earth's surface, its mass decreases because fuel is being consumed at the rate of 50 kilograms per second. Use Newton's universal law of gravitation (see Example 29 in Section 7.9) to determine how fast the force of gravity of the earth on the rocket is changing when the rocket is 100 kilometers above the earth's surface and traveling at the rate of 2 kilometers per second. Assume that the mass of the rocket at this height is 12×10^6 kilograms.

21. If $z = f(u, v)$, $u = g(x, y)$, and $v = h(x, y)$, find the chain rule for the second derivative $\partial^2 z/\partial x^2$.

22. The gradient of $f(x, y)$ is a vector with Cartesian components $\partial f/\partial x$ and $\partial f/\partial y$. If $f(x, y)$ is expressed in terms of polar coordinates r and θ, then it can be shown that ∇f has polar components

$$\nabla f = \frac{\partial f}{\partial r}\hat{\mathbf{e}}_r + \frac{1}{r}\frac{\partial f}{\partial \theta}\hat{\mathbf{e}}_\theta$$

(see Exercise 28). Use this formula to find the polar components of ∇f for each of the following functions:

a) $f(x, y) = x^2 + y^2$;

b) $f(x, y) = x^2 - y^2$;

c) $f(x, y) = 2x + 3y + 4$.

23. If $f(s)$ and $g(t)$ are differentiable functions, show that

$$\nabla f(x^2 - y^2) \cdot \nabla g(xy) = 0.$$

24. If $f(s)$ is a differentiable function, show that $f(x - y)$ satisfies the equation

$$\frac{\partial f}{\partial y} = -\frac{\partial f}{\partial x}.$$

25. Show that if variables x and y are replaced by s and t in the partial differential equation

$$3\frac{\partial u}{\partial x} + 4\frac{\partial u}{\partial y} = 5,$$

where $x = 3s + t$ and $y = 4s + t$, then u must satisfy

$$\frac{\partial u}{\partial s} = 5.$$

Solve this equation and thereby obtain $u = f(x, y)$.

26. If the variables x and y in the function $u = f(x, y)$ are replaced by s and t, where $x = s + t$ and $y = s - t$, show that

$$\left(\frac{\partial u}{\partial x}\right)^2 + \left(\frac{\partial u}{\partial y}\right)^2 = \frac{1}{2}\left(\frac{\partial u}{\partial s}\right)^2 + \frac{1}{2}\left(\frac{\partial u}{\partial t}\right)^2.$$

27. Use the change of variables $x = s + t$, $y = s - t$ to solve the partial differential equation

$$\frac{\partial^2 u}{\partial x^2} - \frac{\partial^2 u}{\partial y^2} = 0.$$

28. If $u = f(x, y)$, then the gradient of u is a vector with Cartesian components $\partial f/\partial x$ and $\partial f/\partial y$. Show that if u is written in terms of polar coordinates r and θ, then

$$\nabla u = \frac{\partial u}{\partial r}\hat{\mathbf{e}}_r + \frac{1}{r}\frac{\partial u}{\partial \theta}\hat{\mathbf{e}}_\theta,$$

where

$$\hat{\mathbf{e}}_r = \cos\theta\hat{\mathbf{i}} + \sin\theta\hat{\mathbf{j}}, \qquad \hat{\mathbf{e}}_\theta = -\sin\theta\hat{\mathbf{i}} + \cos\theta\hat{\mathbf{j}}.$$

(*Hint:* You will need the derivatives $\partial r/\partial x = \cos\theta$, $\partial r/\partial y = \sin\theta$, $\partial\theta/\partial x = -\sin\theta/r$, $\partial\theta/\partial y = \cos\theta/r$. Verify these results using formulas (10.13) and (10.14).)

* **29.** Transform the two-dimensional Laplace equation (13.13) into polar coordinates.

* **30.** It is postulated in one of the theories of "traffic flow" that the average speed u at a point x on a straight highway (along the x-axis) is related to the concentration k of traffic by the differential equation

$$u\frac{\partial u}{\partial x} + \frac{\partial u}{\partial t} = -c^2 k^n \frac{\partial k}{\partial x},$$

where t is time, and $c > 0$ and n are constants.

a) Use the chain rules for $\partial u/\partial x$ and $\partial u/\partial t$ in the functional situation $u = f(k)$ and $k = g(x, t)$ to show that

$$\frac{du}{dk}\left(u\frac{\partial k}{\partial x} + \frac{\partial k}{\partial t}\right) + c^2 k^n \frac{\partial k}{\partial x} = 0.$$

b) The equation of continuity for vehicular traffic flow states that

$$\frac{\partial k}{\partial t} + \frac{\partial (ku)}{\partial x} = 0.$$

Use these last two equations to obtain the differential equation relating speed and concentration:

$$\frac{du}{dk} = -ck^{(n-1)/2}.$$

c) Solve this differential equation for $u = f(k)$.

* **31.** A function $f(x, y, z)$ is said to be positively homogeneous of degree n if for $t > 0$

$$f(tx, ty, tz) = t^n f(x, y, z).$$

a) Show that if such a function has first partial derivatives, then

$$x\frac{\partial f}{\partial x} + y\frac{\partial f}{\partial y} + z\frac{\partial f}{\partial z} = nf(x, y, z).$$

b) Find and prove a similar result for the second partial derivatives of $f(x, y, z)$.

* **32.** Two equal masses m are connected by springs having equal spring constant k so that the masses are free to slide on a frictionless table (Fig. 13.11). The walls A and B are fixed.

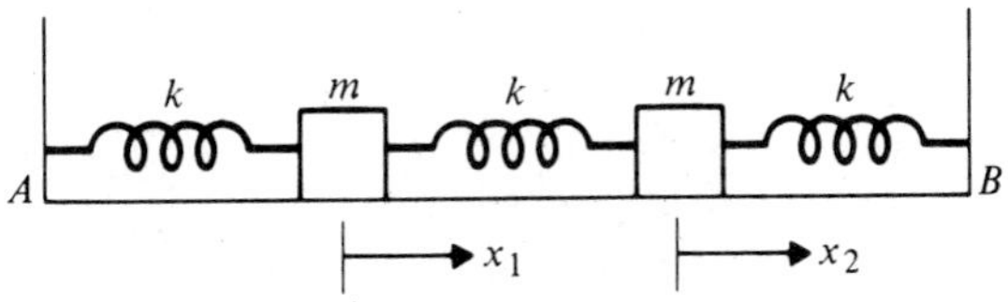

Figure 13.11

a) Use Newton's second law to show that the differential equations for the motions of the masses are

$$m\ddot{x}_1 = k(x_2 - 2x_1), \qquad m\ddot{x}_2 = k(x_1 - 2x_2),$$

where x_1 and x_2 are the displacements of the masses from their equilibrium positions, $\ddot{x}_1 = d^2x_1/dt^2$ and $\ddot{x}_2 = d^2x_2/dt^2$.

b) The Euler–Lagrange equations from the field of theoretical mechanics for this system are

$$\frac{d}{dt}\left(\frac{\partial L}{\partial \dot{x}_1}\right) - \frac{\partial L}{\partial x_1} = 0, \qquad \frac{d}{dt}\left(\frac{\partial L}{\partial \dot{x}_2}\right) - \frac{\partial L}{\partial x_2} = 0,$$

where L is defined as the kinetic energy of the two masses less the energy stored in the springs. Show that

$$L(x_1, x_2, \dot{x}_1, \dot{x}_2) = \frac{m}{2}(\dot{x}_1^2 + \dot{x}_2^2) - k(x_1^2 + x_2^2 - x_1x_2).$$

c) Obtain the equations in (a) from the Euler–Lagrange equations in (b).

* **33.** A bead slides from rest at the origin on a frictionless wire in a vertical plane to the point (x_0, y_0) under the influence of gravity (Fig. 13.12). It can be shown that the time for the bead to traverse the path is

$$t = \frac{1}{\sqrt{2g}}\int_0^{x_0}\sqrt{\frac{1 + (y')^2}{y}}\,dx,$$

where g is the acceleration due to gravity and $y' = dy/dx$.

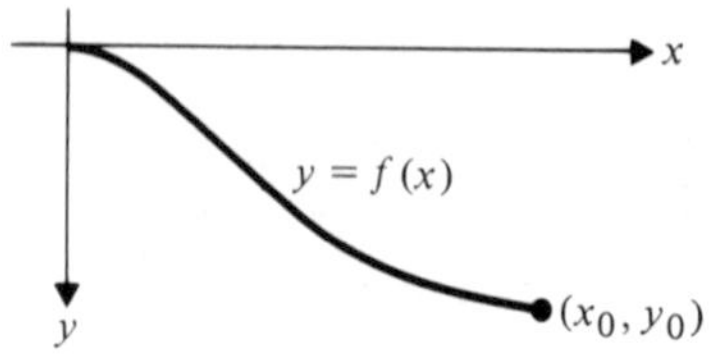

Figure 13.12

The problem of finding that shape of wire which makes t as small as possible is often called the *brachistochrone problem.* It is shown in the calculus of variations that $y = f(x)$ must satisfy the equation

$$\frac{d}{dx}\left(\frac{\partial F}{\partial y'}\right) - \frac{\partial F}{\partial y} = 0,$$

where

$$F(y, y') = \sqrt{\frac{1 + (y')^2}{y}}.$$

a) Show that $f(x)$ must satisfy the differential equation

$$1 + (y')^2 + 2yy'' = 0.$$

b) Show that the curve that satisfies the equation in (a) is the cycloid defined parametrically by

$$x = a(\theta - \sin\theta), \quad y = a(1 - \cos\theta),$$
$$a = \text{a constant}.$$

* **34.** Show that if a solution $f(x, y, z)$ of the three-dimensional Laplace equation (13.14) can be expressed in the form $u = g(r)$, where $r = \sqrt{x^2 + y^2 + z^2}$, then $f(x, y, z)$ must be of the form

$$f(x, y, z) = \frac{C}{\sqrt{x^2 + y^2 + z^2}} + D,$$

where C and D are constants.

* **35.** Suppose the second-order partial differential equation

$$p\frac{\partial^2 z}{\partial x^2} + q\frac{\partial^2 z}{\partial x\,\partial y} + r\frac{\partial^2 z}{\partial y^2} = F\left(x, y, z, \frac{\partial z}{\partial x}, \frac{\partial z}{\partial y}\right)$$

(p, q, and r constants) is subjected to the change of variables

$$s = ax + by, \qquad t = cx + dy, \qquad a, b, c, d = \text{constants}.$$

Show that the partial differential equation in s and t is

$$P\frac{\partial^2 z}{\partial s^2} + Q\frac{\partial^2 z}{\partial s\,\partial t} + R\frac{\partial^2 z}{\partial t^2} = G\left(s, t, z, \frac{\partial z}{\partial s}, \frac{\partial z}{\partial t}\right),$$

where

$$Q^2 - 4PR = (q^2 - 4pr)(ad - bc)^2.$$

13.7 IMPLICIT DIFFERENTIATION

In Section 3.6 we introduced the technique of implicit differentiation in order to obtain the derivative of a function $y = f(x)$ defined implicitly by an equation

$$F(x, y) = 0. \tag{13.24}$$

Essentially, the technique is to differentiate all terms in the equation with respect to x, considering all the while that y is a function of x. For example, if y is defined implicitly by

$$x^2y^3 + 3xy = 3x + 2,$$

then (by implicit differentiation)

$$2xy^3 + 3x^2y^2\frac{dy}{dx} + 3y + 3x\frac{dy}{dx} = 3.$$

We can now solve for

$$\frac{dy}{dx} = \frac{3 - 2xy^3 - 3y}{3x^2y^2 + 3x}.$$

With the chain rule we can actually present a formula for dy/dx. Since Eq. (13.24), when written in the form

$$F[x, f(x)] = 0,$$

must be valid for all x in the domain of the function $f(x)$, we can differentiate it with respect to x. From the schematic

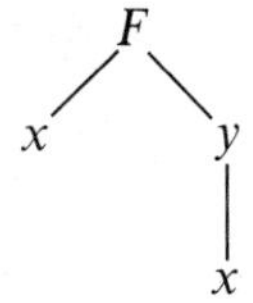

the derivative of the left side of the equation is

$$\frac{dF}{dx} = \frac{\partial F}{\partial x} + \frac{\partial F}{\partial y}\frac{dy}{dx}.$$

If we equate this to the derivative of the right side of the equation, we find

$$F_x + F_y \frac{dy}{dx} = 0,$$

or

$$\frac{dy}{dx} = -\frac{F_x}{F_y}. \tag{13.25}$$

For the above example, Eq. (13.25) gives

$$\frac{dy}{dx} = -\frac{2xy^3 + 3y - 3}{3x^2y^2 + 3x},$$

and this result is identical to that obtained by implicit differentiation.

Similarly, if the equation

$$F(x, y, z) = 0 \tag{13.26}$$

defines z implicitly as a function of x and y, the schematic

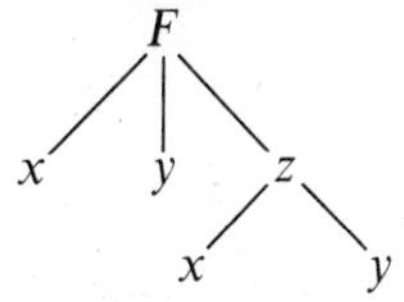

immediately yields

$$\frac{\partial F}{\partial x} + \frac{\partial F}{\partial z}\frac{\partial z}{\partial x} = 0, \qquad \frac{\partial F}{\partial y} + \frac{\partial F}{\partial z}\frac{\partial z}{\partial y} = 0.$$

From these we obtain the results

$$\frac{\partial z}{\partial x} = -\frac{F_x}{F_z}, \qquad \frac{\partial z}{\partial y} = -\frac{F_y}{F_z}. \tag{13.27}$$

We do not suggest that the formulas in Eqs. (13.25) and (13.27) be memorized. On the contrary, we intend to obtain results in this section that include (13.25) and (13.27) as special cases.

We have already mentioned (Section 13.1) that m equations in n variables usually specify m dependent variables as functions of $n - m$ independent variables. As a specific example, three equations in five variables,

$$\begin{aligned} F(x, y, u, v, w) &= 0, \\ G(x, y, u, v, w) &= 0, \\ H(x, y, u, v, w) &= 0, \end{aligned} \tag{13.28}$$

might specify u, v, and w as functions of x and y for some domain of values of x and y (and do so implicitly). It might even be possible to solve the system and obtain explicit definitions of the functions

$$\begin{aligned} u &= f(x, y), \\ v &= g(x, y), \\ w &= h(x, y). \end{aligned} \tag{13.29}$$

We would like to pose the problem of finding the six first-order partial derivatives of u, v, and w with respect to x and y supposing that it is undesirable or even impossible to obtain the explicit form of the functions. To do this, we note that were the results (13.29) known and substituted into (13.28), then

$$\begin{aligned} F[x, y, f(x, y), g(x, y), h(x, y)] &= 0, \\ G[x, y, f(x, y), g(x, y), h(x, y)] &= 0, \\ H[x, y, f(x, y), g(x, y), h(x, y)] &= 0, \end{aligned}$$

would be identities in x and y. As a result we could differentiate each equation with respect to x, obtaining from the schematic

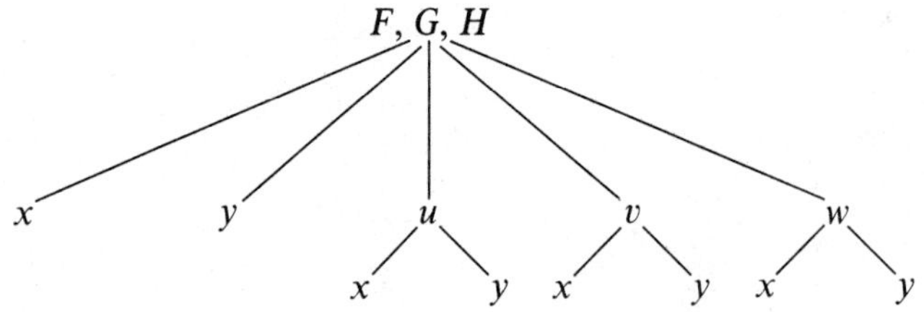

the following:

$$\begin{aligned} \frac{\partial F}{\partial x} + \frac{\partial F}{\partial u}\frac{\partial u}{\partial x} + \frac{\partial F}{\partial v}\frac{\partial v}{\partial x} + \frac{\partial F}{\partial w}\frac{\partial w}{\partial x} &= 0, \\ \frac{\partial G}{\partial x} + \frac{\partial G}{\partial u}\frac{\partial u}{\partial x} + \frac{\partial G}{\partial v}\frac{\partial v}{\partial x} + \frac{\partial G}{\partial w}\frac{\partial w}{\partial x} &= 0, \\ \frac{\partial H}{\partial x} + \frac{\partial H}{\partial u}\frac{\partial u}{\partial x} + \frac{\partial H}{\partial v}\frac{\partial v}{\partial x} + \frac{\partial H}{\partial w}\frac{\partial w}{\partial x} &= 0, \end{aligned} \tag{13.30}$$

or

$$
\begin{aligned}
F_u \frac{\partial u}{\partial x} + F_v \frac{\partial v}{\partial x} + F_w \frac{\partial w}{\partial x} &= -F_x,\\
G_u \frac{\partial u}{\partial x} + G_v \frac{\partial v}{\partial x} + G_w \frac{\partial w}{\partial x} &= -G_x,\\
H_u \frac{\partial u}{\partial x} + H_v \frac{\partial v}{\partial x} + H_w \frac{\partial w}{\partial x} &= -H_x.
\end{aligned}
\tag{13.31}
$$

We have in (13.31) three equations in the three unknowns $\partial u/\partial x$, $\partial v/\partial x$, and $\partial w/\partial x$, and because the equations are linear in the unknowns, solutions can be obtained with Cramer's rule.† In particular,

$$
\frac{\partial u}{\partial x} = \frac{\begin{vmatrix} -F_x & F_v & F_w \\ -G_x & G_v & G_w \\ -H_x & H_v & H_w \end{vmatrix}}{\begin{vmatrix} F_u & F_v & F_w \\ G_u & G_v & G_w \\ H_u & H_v & H_w \end{vmatrix}} = -\frac{\begin{vmatrix} F_x & F_v & F_w \\ G_x & G_v & G_w \\ H_x & H_v & H_w \end{vmatrix}}{\begin{vmatrix} F_u & F_v & F_w \\ G_u & G_v & G_w \\ H_u & H_v & H_w \end{vmatrix}}. \tag{13.32}
$$

The two determinants on the right of (13.32) involve only derivatives of the given functions F, G, and H, and we have therefore obtained a method for finding $\partial u/\partial x$ that avoids solving (13.28) for (13.29). We could list similar formulas for the remaining five derivatives, but first we introduce some simplifying notation.

▷ **Definition 13.5** The *Jacobian* of functions F, G, and H with respect to variables u, v, and w is denoted by $\partial(F, G, H)/\partial(u, v, w)$ and is defined as the determinant

$$
\frac{\partial(F, G, H)}{\partial(u, v, w)} = \begin{vmatrix} F_u & F_v & F_w \\ G_u & G_v & G_w \\ H_u & H_v & H_w \end{vmatrix} = \begin{vmatrix} \dfrac{\partial F}{\partial u} & \dfrac{\partial F}{\partial v} & \dfrac{\partial F}{\partial w} \\[2ex] \dfrac{\partial G}{\partial u} & \dfrac{\partial G}{\partial v} & \dfrac{\partial G}{\partial w} \\[2ex] \dfrac{\partial H}{\partial u} & \dfrac{\partial H}{\partial v} & \dfrac{\partial H}{\partial w} \end{vmatrix}. \tag{13.33}
$$
◁

With this notation we can write (13.32) in the form

$$
\frac{\partial u}{\partial x} = -\frac{\partial(F, G, H)/\partial(x, v, w)}{\partial(F, G, H)/\partial(u, v, w)}. \tag{13.34}
$$

The remaining derivatives of $v = g(x, y)$ and $w = h(x, y)$ with respect to x can be obtained from Eqs. (13.31) by Cramer's rule also:

$$
\frac{\partial v}{\partial x} = -\frac{\partial(F, G, H)/\partial(u, x, w)}{\partial(F, G, H)/\partial(u, v, w)}, \qquad \frac{\partial w}{\partial x} = -\frac{\partial(F, G, H)/\partial(u, v, x)}{\partial(F, G, H)/\partial(u, v, w)}. \tag{13.35}
$$

† Cramer's rule is discussed in Appendix B.

A similar procedure yields

$$\frac{\partial u}{\partial y} = -\frac{\partial(F, G, H)/\partial(y, v, w)}{\partial(F, G, H)/\partial(u, v, w)}, \qquad \frac{\partial v}{\partial y} = -\frac{\partial(F, G, H)/\partial(u, y, w)}{\partial(F, G, H)/\partial(u, v, w)},$$

$$\frac{\partial w}{\partial y} = -\frac{\partial(F, G, H)/\partial(u, v, y)}{\partial(F, G, H)/\partial(u, v, w)}. \tag{13.36}$$

The formulas in (13.34)–(13.36) apply only to the situation in which Eqs. (13.28) define u, v, and w as functions of x and y. It is, however, fairly evident how to construct formulas in other situations. Each formula has a Jacobian divided by a Jacobian (and do not forget the negative sign). In the denominator, it is the Jacobian of the functions defining the original equations with respect to the dependent variables. The only difference in the Jacobian in the numerator is that the dependent variable that is being differentiated is replaced by the independent variable with respect to which differentiation is being performed.

The results in Eqs. (13.34)–(13.36) are valid provided, of course, that the Jacobian

$$\frac{\partial(F, G, H)}{\partial(u, v, w)} \neq 0.$$

In actual fact, it is this condition that guarantees that Eqs. (13.28) do define u, v, and w as functions of x and y in the first place.

As a second example, if

$$F(x, y, s, t) = 0,$$
$$G(x, y, s, t) = 0$$

define x and y as functions of s and t, then

$$\frac{\partial x}{\partial s} = -\frac{\partial(F, G)/\partial(s, y)}{\partial(F, G)/\partial(x, y)}, \qquad \frac{\partial x}{\partial t} = -\frac{\partial(F, G)/\partial(t, y)}{\partial(F, G)/\partial(x, y)},$$

$$\frac{\partial y}{\partial s} = -\frac{\partial(F, G)/\partial(x, s)}{\partial(F, G)/\partial(x, y)}, \qquad \frac{\partial y}{\partial t} = -\frac{\partial(F, G)/\partial(x, t)}{\partial(F, G)/\partial(x, y)}.$$

We now show that the results in Eqs. (13.25) and (13.27) are special cases of these considerations. If Eq. (13.24) defines y implicitly as a function of x, then

$$\frac{dy}{dx} = -\frac{\partial(F)/\partial(x)}{\partial(F)/\partial(y)} = -\frac{F_x}{F_y}.$$

If Eq. (13.26) defines z implicitly as a function of x and y, then

$$\frac{\partial z}{\partial x} = -\frac{\partial(F)/\partial(x)}{\partial(F)/\partial(z)} = -\frac{F_x}{F_z}; \qquad \frac{\partial z}{\partial y} = -\frac{\partial(F)/\partial(y)}{\partial(F)/\partial(z)} = -\frac{F_y}{F_z}.$$

Example 16 If $x^2y^2z^3 + zx \sin y = 5$ defines z as a function of x and y, find $\partial z/\partial x$.

Solution If we set $F(x, y, z) = x^2y^2z^3 + zx \sin y - 5 = 0$, then

$$\frac{\partial z}{\partial x} = -\frac{\partial(F)/\partial(x)}{\partial(F)/\partial(z)} = -\frac{F_x}{F_z} = -\frac{2xy^2z^3 + z \sin y}{3x^2y^2z^2 + x \sin y}. \quad \square$$

Example 17 The equations $x = r\cos\theta$, $y = r\sin\theta$ define Cartesian coordinates x and y in terms of polar coordinates r and θ. At the same time they can be used to implicitly define r and θ in terms of x and y. Find $\partial\theta/\partial x$.

Solution If we set

$$F(x, y, r, \theta) = x - r\cos\theta = 0,$$

$$G(x, y, r, \theta) = y - r\sin\theta = 0,$$

then

$$\frac{\partial\theta}{\partial x} = -\frac{\partial(F, G)/\partial(r, x)}{\partial(F, G)/\partial(r, \theta)} = -\frac{\begin{vmatrix} F_r & F_x \\ G_r & G_x \end{vmatrix}}{\begin{vmatrix} F_r & F_\theta \\ G_r & G_\theta \end{vmatrix}} = -\frac{\begin{vmatrix} -\cos\theta & 1 \\ -\sin\theta & 0 \end{vmatrix}}{\begin{vmatrix} -\cos\theta & r\sin\theta \\ -\sin\theta & -r\cos\theta \end{vmatrix}} = -\frac{\sin\theta}{r}.$$

□

We have already mentioned that partial derivatives are not to be considered quotients. In particular, note that had we thought of $\partial\theta/\partial x$ in Example 17 as a quotient, we might have been tempted to write

$$\frac{\partial\theta}{\partial x} = \frac{1}{\partial x/\partial\theta} = \frac{1}{-r\sin\theta},$$

which certainly does not agree with the result obtained.

Example 18 In the theory of thermodynamics, the variables pressure P, temperature T, volume V, and internal energy U are related by two equations of state:

$$F(P, T, V, U) = 0, \qquad G(P, T, V, U) = 0.$$

The second law of thermodynamics implies that if U and P are regarded as functions of T and V, then the functions $U(T, V)$ and $P(T, V)$ must satisfy the equation

$$\frac{\partial U}{\partial V} - T\frac{\partial P}{\partial T} + P = 0.$$

Show that if in the equations of state U and V are chosen as independent variables rather than T and V, then the second law takes the form

$$\frac{\partial T}{\partial V} + T\frac{\partial P}{\partial U} - P\frac{\partial T}{\partial U} = 0.$$

Solution When U and V are taken as independent variables,

$$\frac{\partial T}{\partial V} = -\frac{\partial(F, G)/\partial(V, P)}{\partial(F, G)/\partial(T, P)},$$

$$\frac{\partial P}{\partial U} = -\frac{\partial(F, G)/\partial(T, U)}{\partial(F, G)/\partial(T, P)},$$

$$\frac{\partial T}{\partial U} = -\frac{\partial(F, G)/\partial(U, P)}{\partial(F, G)/\partial(T, P)}.$$

Consequently,

$$\frac{\partial T}{\partial V} + T\frac{\partial P}{\partial U} - P\frac{\partial T}{\partial U}$$

$$= -\frac{1}{\partial(F,G)/\partial(T,P)}\left\{\frac{\partial(F,G)}{\partial(V,P)} + T\frac{\partial(F,G)}{\partial(T,U)} - P\frac{\partial(F,G)}{\partial(U,P)}\right\}.$$

To show that this expression must be equal to zero, we note that when T and V are independent variables, the second law implies that

$$\begin{aligned} 0 &= -\frac{\partial(F,G)/\partial(V,P)}{\partial(F,G)/\partial(U,P)} + T\frac{\partial(F,G)/\partial(U,T)}{\partial(F,G)/\partial(U,P)} + P \\ &= -\frac{1}{\partial(F,G)/\partial(U,P)}\left\{\frac{\partial(F,G)}{\partial(V,P)} - T\frac{\partial(F,G)}{\partial(U,T)} - P\frac{\partial(F,G)}{\partial(U,P)}\right\} \\ &= -\frac{1}{\partial(F,G)/\partial(U,P)}\left\{\frac{\partial(F,G)}{\partial(V,P)} + T\frac{\partial(F,G)}{\partial(T,U)} - P\frac{\partial(F,G)}{\partial(U,P)}\right\}. \end{aligned}$$

Hence

$$\frac{\partial T}{\partial V} + T\frac{\partial P}{\partial U} - P\frac{\partial T}{\partial U} = 0. \ \square$$

EXERCISES 13.7

In Exercises 1–4, y is defined implicitly as a function of x. Find dy/dx.

1. $x^3y^2 - 2xy + 5 = 0$

2. $(x + y)^2 = 2x$

3. $x(x - y) - 4y^3 = 2e^{xy} + 6$

4. $\sin(x + y) + y^2 = 12x^2 + y$

In Exercises 5–8, z is defined implicitly as a function of x and y. Find $\partial z/\partial x$ and $\partial z/\partial y$.

5. $x^2 \sin z - ye^z = 2x$

6. $x^2z^2 + yz + 3x = 4$

7. $z \sin^2 y + y \sin^2 x = z^3$

8. $\mathrm{Tan}^{-1}(yz) = xz$

In Exercises 9–13, find the required derivative. Assume that the system of equations does define the function(s) indicated.

9. $\partial u/\partial x$ and $\partial v/\partial y$ if

$$x^2 - y^2 + u^2 + 2v^2 = 1,$$
$$x^2 + y^2 = 2 + u^2 + v^2$$

10. $\partial x/\partial t$ if

$$\sin(x + t) - \sin(x - t) = z$$

11. $\left.\dfrac{\partial \phi}{\partial x}\right)_{y,z}$ if

$$x = r\sin\phi\cos\theta,$$
$$y = r\sin\phi\sin\theta,$$
$$z = r\cos\phi$$

12. dz/dx if

$$x^2 + y^2 - z^2 + 2xy = 1,$$
$$x^3 + y^3 - 5y = 4$$

13. $\left.\dfrac{\partial u}{\partial y}\right)_x$ if

$$xyu + vw = 4,$$
$$y^2 + u^2 - u^2v = y,$$
$$yw + xu + v + 4 = 0$$

14. Given that the equations

$$x^2 - y\cos(uv) + z^2 = 0,$$
$$x^2 + y^2 - \sin(uv) + 2z^2 = 2,$$
$$xy - \sin u \cos v + z = 0$$

define x, y, and z as functions of u and v, find $(\partial x/\partial u)_v$ at the values $x = 1$, $y = 1$, $u = \pi/2$, $v = 0$, and $z = 0$.

15. If the equation $F(x, y, z) = 0$ defines each of x, y, and z as a function of the other two, show that

$$\left(\frac{\partial z}{\partial x}\right)_y \left(\frac{\partial x}{\partial y}\right)_z \left(\frac{\partial y}{\partial z}\right)_x = -1.$$

16. If $z = e^x \cos y$, where x and y are functions of t defined by

$$x^3 + e^x - t^2 - t = 1,$$

$$yt^2 + y^2t - t + y = 0,$$

find dz/dt.

17. If the equations

$$x = u^2 - v^2,$$

$$y = 2uv$$

define u and v as functions of x and y, find $\partial^2 u/\partial x^2$.

18. If the equations

$$x^2 - 2y^2s^2t - 2st^2 = 1,$$

$$x^2 + 2y^2s^2t + 5st^2 = 1$$

define s and t as functions of x and y, find $\partial^2 t/\partial y^2$.

19. Show that if V and P are chosen as independent variables in Example 18, then the second law of thermodynamics takes the form

$$T - P\frac{\partial T}{\partial P} + \frac{\partial(T, U)}{\partial(V, P)} = 0.$$

***20.** a) Suppose the equations

$$F(u, v, s, t) = 0, \qquad G(u, v, s, t) = 0$$

define u and v as functions of s and t, and the equations

$$H(s, t, x, y) = 0, \qquad I(s, t, x, y) = 0$$

define s and t as functions of x and y. Show that

$$\frac{\partial(u, v)}{\partial(s, t)}\frac{\partial(s, t)}{\partial(x, y)} = \frac{\partial(u, v)}{\partial(x, y)}.$$

b) If the equations

$$F(u, v, x, y) = 0, \qquad G(u, v, x, y) = 0$$

define u and v as functions of x and y, and also define x and y as functions of u and v, show that

$$\frac{\partial(u, v)}{\partial(x, y)} = \frac{1}{\partial(x, y)/\partial(u, v)}.$$

***21.** Suppose the system of m linear equations in n unknowns $(n > m)$

$$\sum_{j=1}^{n} a_{ij}x_j = c_i, \qquad i = 1, \ldots, m,$$

defines $x_1, x_2, \ldots, x_m$ as functions of $x_{m+1}, x_{m+2}, \ldots, x_n$. Show that if $1 \le i \le m$ and $m + 1 \le j \le n$, then

$$\frac{\partial x_i}{\partial x_j} = -\frac{D_{ij}}{D},$$

where $D = |a_{ij}|_{m\times m}$, and D_{ij} is the same as the determinant D except that the ith column is replaced by the jth column of $[a_{ij}]_{m\times n}$.

13.8 DIRECTIONAL DERIVATIVES

If a function $f(x, y, z)$ is defined throughout some region of space, then at any point (x_0, y_0, z_0) we can calculate its partial derivatives $\partial f/\partial x$, $\partial f/\partial y$, and $\partial f/\partial z$. These derivatives define the rates of change of $f(x, y, z)$ at (x_0, y_0, z_0) in directions parallel to the x-, y-, and z-axes. But what if we want the rate of change of $f(x, y, z)$ at (x_0, y_0, z_0) in some arbitrary direction defined by a vector $\mathbf{v}$ (Fig. 13.13)? By the rate of change of $f(x, y, z)$ in the direction $\mathbf{v}$, we mean the rate of change with respect to distance as measured along a line through (x_0, y_0, z_0) in the direction $\mathbf{v}$. Let us define s as a measure of directed distance along this line, taking $s = 0$ at (x_0, y_0, z_0) and positive s in the direction of $\mathbf{v}$. What we want, then, is the derivative of $f(x, y, z)$ with respect to s at $s = 0$.

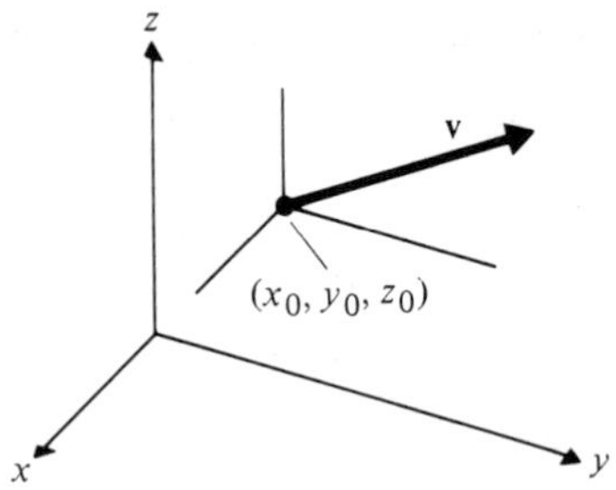

Figure 13.13

To express $f(x, y, z)$ in terms of s, we use parametric equations of the line through (x_0, y_0, z_0) along $\mathbf{v}$. If $\hat{\mathbf{v}} = (v_x, v_y, v_z)$ is a unit vector in the direction of $\mathbf{v}$, then parametric equations for this line (see Eqs. 12.42) are

$$x = x_0 + v_x s, \qquad y = y_0 + v_y s, \qquad z = z_0 + v_z s. \tag{13.37}$$

From the schematic

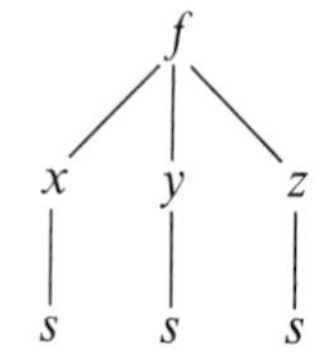

we obtain

$$\begin{aligned}\frac{df}{ds} &= \frac{\partial f}{\partial x}\frac{dx}{ds} + \frac{\partial f}{\partial y}\frac{dy}{ds} + \frac{\partial f}{\partial z}\frac{dz}{ds} \\ &= \frac{\partial f}{\partial x} v_x + \frac{\partial f}{\partial y} v_y + \frac{\partial f}{\partial z} v_z,\end{aligned}$$

where all partial derivatives of $f(x, y, z)$ are to be evaluated at (x_0, y_0, z_0). This derivative df/ds is called the *directional derivative* of $f(x, y, z)$ in the direction $\mathbf{v}$ at the point (x_0, y_0, z_0), and is usually given the alternative notation

$$D_{\mathbf{v}} f = \frac{\partial f}{\partial x} v_x + \frac{\partial f}{\partial y} v_y + \frac{\partial f}{\partial z} v_z. \tag{13.38}$$

Now, v_x, v_y, and v_z are the components of the unit vector $\hat{\mathbf{v}}$ in the direction of $\mathbf{v}$, and $\partial f/\partial x$, $\partial f/\partial y$, and $\partial f/\partial z$ are the components of the gradient of $f(x, y, z)$. We can write, therefore, that

$$D_{\mathbf{v}} f = \nabla f \cdot \hat{\mathbf{v}}. \tag{13.39}$$

Consequently, the derivative (rate of change) of a function in any given direction is the scalar product of the gradient of the function and a unit vector in the required direction—in other words, the following theorem.

▶ **Theorem 13.3** The directional derivative of a function in any direction is the component of the gradient of the function in that direction. ◀

Example 19 Find $D_{\mathbf{v}} f$ at (4, 0, 16) if $f(x, y, z) = x^3 e^y + xz$ and $\mathbf{v}$ is the vector from (4, 0, 16) to $(-2, 1, 4)$.

Solution Since

$$\begin{aligned}\nabla f_{|(4,0,16)} &= \{(3x^2 e^y + z)\hat{\mathbf{i}} + x^3 e^y \hat{\mathbf{j}} + x\hat{\mathbf{k}}\}_{|(4,0,16)} \\ &= 64\hat{\mathbf{i}} + 64\hat{\mathbf{j}} + 4\hat{\mathbf{k}}\end{aligned}$$

and

$$\hat{\mathbf{v}} = \frac{\mathbf{v}}{|\mathbf{v}|} = \frac{(-6, 1, -12)}{\sqrt{36 + 1 + 144}} = \frac{-1}{\sqrt{181}}(6, -1, 12),$$

we have

$$D_{\mathbf{v}}f = -(64, 64, 4) \cdot \frac{(6, -1, 12)}{\sqrt{181}} = -\frac{368}{\sqrt{181}}. \square$$

The directional derivative gives us insight into some of the properties of the gradient vector. In particular, we have this theorem.

▶ **Theorem 13.4** The gradient ∇f of a function $f(x, y, z)$ defines the direction in which the function increases most rapidly, and the maximum rate of change is $|\nabla f|$. ◀

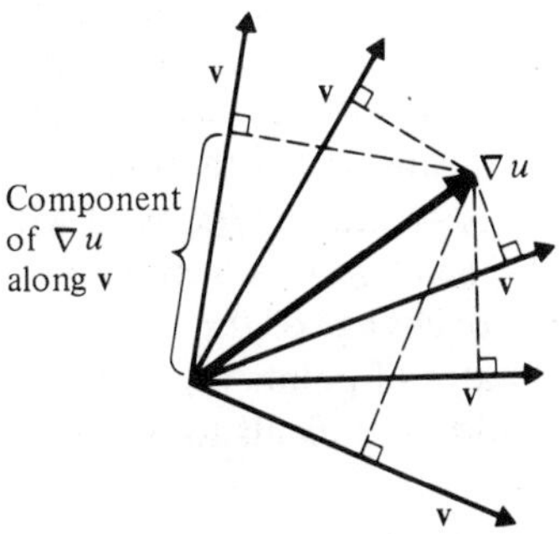

Figure 13.14

Proof Theorem 13.3 states that the directional derivative of $f(x, y, z)$ in a direction $\mathbf{v}$ is the component of ∇f in that direction. Figure 13.14, which shows components of ∇f in various directions, makes it clear that $D_{\mathbf{v}}f$ is greatest when $\mathbf{v}$ is parallel to ∇f. Alternatively, if θ is the angle between $\mathbf{v}$ and ∇f, then

$$D_{\mathbf{v}}f = \nabla f \cdot \hat{\mathbf{v}} = |\nabla f||\hat{\mathbf{v}}| \cos\theta = |\nabla f| \cos\theta.$$

Obviously $D_{\mathbf{v}}f$ is a maximum when $\cos\theta$ is a maximum—i.e., when $\cos\theta = 1$ or $\theta = 0$—and this occurs when $\mathbf{v}$ is parallel to ∇f. Finally, when $\mathbf{v}$ is parallel to ∇f, we have $D_{\mathbf{v}}f = |\nabla f|$. $\square$

Note that for any function $f(x, y, z)$,

$$D_{\hat{\mathbf{i}}}f = \frac{\partial f}{\partial x}, \qquad D_{\hat{\mathbf{j}}}f = \frac{\partial f}{\partial y}, \qquad D_{\hat{\mathbf{k}}}f = \frac{\partial f}{\partial z}.$$

In other words, the partial derivatives of a function are its directional derivatives along the coordinate directions.

Example 20 Find the direction at the point $(1, 2, -3)$ in which the function $f(x, y, z) = x^2y + xyz$ increases most rapidly.

Solution According to Theorem 13.4, $f(x, y, z)$ increases most rapidly in the direction

$$\nabla f_{|(1,2,-3)} = (2xy + yz, x^2 + xz, xy)_{|(1,2,-3)} = (-2, -2, 2). \square$$

You might feel that because the definition of the directional derivative $D_{\mathbf{v}}f$ does not involve a limit process, it is some strange new type of differentiation. To show that this is not the case, let us return to the calculation of the derivative of $f(x, y, z)$ at (x_0, y_0, z_0) in the direction $\mathbf{v}$ shown in Fig. 13.13. With parametric equations (13.37) for the line through (x_0, y_0, z_0) along $\mathbf{v}$, the value of $f(x, y, z)$ at any point (x, y, z) along this line is $f(x_0 + v_x s, y_0 + v_y s, z_0 + v_z s)$. If we take the difference between this value and $f(x_0, y_0, z_0)$ and divide by the distance s between (x_0, y_0, z_0) and (x, y, z), then the limit of this expression as $s \to 0^+$ should define the derivative of $f(x, y, z)$ at (x_0, y_0, z_0) in the direction $\mathbf{v}$; i.e.,

$$D_{\mathbf{v}}f = \lim_{s \to 0^+} \frac{f(x_0 + v_x s, y_0 + v_y s, z_0 + v_z s) - f(x_0, y_0, z_0)}{s}. \tag{13.40}$$

We can show that this limit (and this is perhaps the form we might have expected the derivative to take) also leads to the result contained in (13.39).

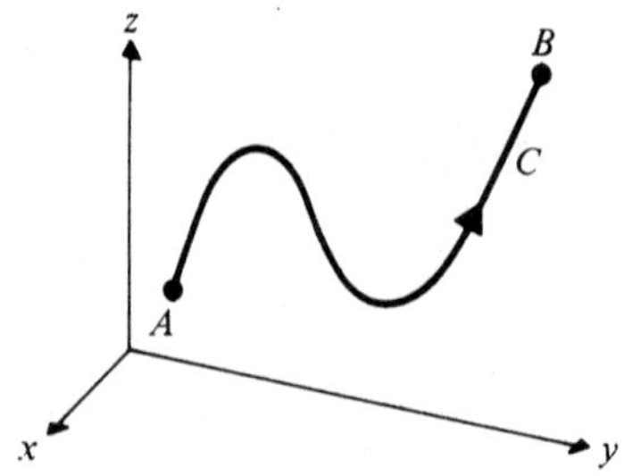

Figure 13.15

Consider a curve C in space that is defined parametrically by

$$C: \quad x = x(t), \qquad y = y(t), \qquad z = z(t), \qquad \alpha \le t \le \beta$$

(Fig. 13.15). Imagine that C is the path traced out by some particle as it moves through space under the action of some system of forces, and suppose that $f(x, y, z)$ is a function defined along C. Perhaps the particle is a weather balloon and $f(x, y, z)$ is temperature at points along its trajectory C, or barometric pressure, or humidity, or wind speed, etc. In these applications we might be asked, "What is the rate of change of $f(x, y, z)$ along C?" The experienced worker immediately knows what is being asked for, but the neophyte finds difficulty with the language. He or she understands that "rate of change" requires a derivative and that it is the derivative of $f(x, y, z)$, but cannot decide what "along C" means. Does it mean to differentiate with respect to x, or perhaps with respect to y or z? Does it perhaps mean to differentiate $f(x, y, z)$ with respect to t, since we can write $f[x(t), y(t), z(t)]$? What it really means is "with respect to distance traveled along C"; i.e., "what is the rate of change of $f(x, y, z)$ with respect to distance as measured along C"? If we then use s as a measure of distance along C (taking $s = 0$ at A), then "the rate of change of $f(x, y, z)$ along C" is synonymous with "df/ds." Since the coordinates of points (x, y, z) on C can be regarded as functions of s (although it would be difficult to find these functions explicitly), the chain rule gives

$$\begin{aligned}\frac{df}{ds} &= \frac{\partial f}{\partial x}\frac{dx}{ds} + \frac{\partial f}{\partial y}\frac{dy}{ds} + \frac{\partial f}{\partial z}\frac{dz}{ds} \\ &= \left(\frac{\partial f}{\partial x}, \frac{\partial f}{\partial y}, \frac{\partial f}{\partial z}\right) \cdot \left(\frac{dx}{ds}, \frac{dy}{ds}, \frac{dz}{ds}\right) \\ &= \nabla f \cdot \frac{d\mathbf{r}}{ds}.\end{aligned}$$

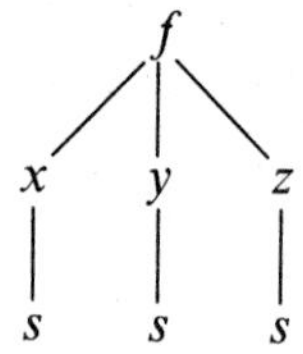

In Section 12.9 we saw that $d\mathbf{r}/ds$ is a unit tangent vector $\hat{\mathbf{T}}$ to C. Consequently,

$$\frac{df}{ds} = \nabla f \cdot \hat{\mathbf{T}}.$$

But this equation states that df/ds is the directional derivative of $f(x, y, z)$ along the tangent direction to the curve C. In other words, although "rate of change of $f(x, y, z)$ along C" means "rate of change of $f(x, y, z)$ with respect to distance as measured along C," we can calculate this rate of change using the directional derivative of $f(x, y, z)$ in the tangent direction along C.

Example 21 Find the rate of change of the function $f(x, y, z) = x^2y - xz$ along the curve $y = x^2$, $z = x$ in the direction of decreasing x at the point $(2, 4, 2)$.

Solution Since parametric equations for the curve are C: $x = -t$, $y = t^2$, $z = -t$, a tangent vector to C at any point is $\mathbf{T} = (-1, 2t, -1)$. At $(2, 4, 2)$, $t = -2$, and this tangent vector becomes $\mathbf{T} = (-1, -4, -1)$. A unit tangent vector to C at $(2, 4, 2)$ in the direction of decreasing x is therefore

$$\hat{\mathbf{T}} = \frac{(-1, -4, -1)}{\sqrt{18}} = \frac{-1}{3\sqrt{2}}(1, 4, 1).$$

The rate of change of $f(x, y, z)$ in this direction is

$$\nabla f \cdot \hat{\mathbf{T}} = (2xy - z, x^2, -x)_{|(2,4,2)} \cdot \frac{(1, 4, 1)}{-3\sqrt{2}}$$

$$= \frac{-1}{3\sqrt{2}}(14, 4, -2) \cdot (1, 4, 1) = -\frac{28}{3\sqrt{2}}. \quad \square$$

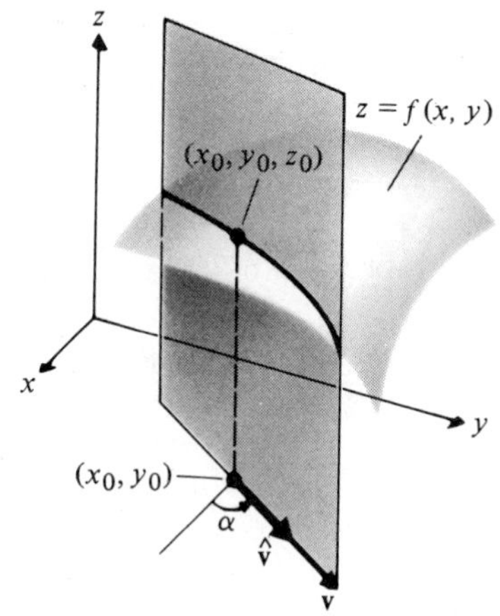

Figure 13.16

It is worthwhile at this point to discuss directional derivatives for functions $f(x, y)$ of two independent variables. Such a function can be represented graphically as a surface $z = f(x, y)$ (Fig. 13.16). For a direction $\mathbf{v}$ at (x_0, y_0) in the xy-plane,

$$D_{\mathbf{v}} f = \nabla f \cdot \hat{\mathbf{v}},$$

where ∇f is evaluated at (x_0, y_0). Algebraically, this is the derivative (rate of change) of $f(x, y)$ in the direction $\mathbf{v}$. Geometrically, this is the rate of change of the height z of the surface as we move along the curve of intersection of the surface and a vertical plane containing the vector $\mathbf{v}$, or the slope of this curve. Each direction $\mathbf{v}$ at (x_0, y_0) defines an angle α with a line through (x_0, y_0) parallel to the positive x-axis, and for this direction

$$\hat{\mathbf{v}} = \cos\alpha\hat{\mathbf{i}} + \sin\alpha\hat{\mathbf{j}}.$$

We can write, then,

$$D_{\mathbf{v}} f = \nabla f \cdot \hat{\mathbf{v}} = \left(\frac{\partial f}{\partial x}\hat{\mathbf{i}} + \frac{\partial f}{\partial y}\hat{\mathbf{j}}\right) \cdot (\cos\alpha\hat{\mathbf{i}} + \sin\alpha\hat{\mathbf{j}})$$

$$= \frac{\partial f}{\partial x}\cos\alpha + \frac{\partial f}{\partial y}\sin\alpha. \tag{13.41}$$

If $D_{\mathbf{v}} f$ represents the slope of the curve of intersection of the surface and the vertical plane through $\mathbf{v}$, then $D_{\mathbf{v}}(D_{\mathbf{v}} f)$ represents the rate of change of this slope. Now

$$D_{\mathbf{v}}(D_{\mathbf{v}} f) = \nabla(D_{\mathbf{v}} f) \cdot \hat{\mathbf{v}},$$

and

$$\nabla(D_{\mathbf{v}} f) = \left(\frac{\partial^2 f}{\partial x^2}\cos\alpha + \frac{\partial^2 f}{\partial x\,\partial y}\sin\alpha\right)\hat{\mathbf{i}} + \left(\frac{\partial^2 f}{\partial y\,\partial x}\cos\alpha + \frac{\partial^2 f}{\partial y^2}\sin\alpha\right)\hat{\mathbf{j}}.$$

Thus,

$$D_{\mathbf{v}}(D_{\mathbf{v}} f) = \left(\frac{\partial^2 f}{\partial x^2}\cos\alpha + \frac{\partial^2 f}{\partial x\,\partial y}\sin\alpha\right)\cos\alpha + \left(\frac{\partial^2 f}{\partial y\,\partial x}\cos\alpha + \frac{\partial^2 f}{\partial y^2}\sin\alpha\right)\sin\alpha$$

$$= \frac{\partial^2 f}{\partial x^2}\cos^2\alpha + 2\frac{\partial^2 f}{\partial x\,\partial y}\cos\alpha\sin\alpha + \frac{\partial^2 f}{\partial y^2}\sin^2\alpha. \tag{13.42}$$

We call $D_{\mathbf{v}}(D_{\mathbf{v}} f)$ the second directional derivative of $f(x, y)$ at (x_0, y_0) in the direction $\mathbf{v}$. If it is positive, then the curve of intersection is concave upward, whereas if it is negative, the curve is concave downward. We will find these results useful in Section 13.10 when we discuss relative extrema of a function of two independent variables.

EXERCISES 13.8

In Exercises 1–10, calculate the directional derivative of the function at the point and in the direction indicated.

1. $f(x, y, z) = 2x^2 - y^2 + z^2$ at $(1, 2, 3)$ in the direction of the vector from $(1, 2, 3)$ to $(3, 5, 0)$
2. $f(x, y, z) = x^2y + xz$ at $(-1, 1, -1)$ in the direction of the vector that joins $(3, 2, 1)$ to $(3, 1, -1)$
3. $f(x, y) = xe^y + y$ at $(3, 0)$ in the direction of the vector from $(3, 0)$ to $(-2, -4)$
4. $f(x, y, z) = \ln(xy + yz + xz)$ at $(1, 1, 1)$ in the direction from $(1, 1, 1)$ toward the point $(-1, -2, 3)$
5. $f(x, y) = \mathrm{Tan}^{-1}(xy)$ at $(1, 2)$ along the line $y = 2x$ in the direction of increasing x
6. $f(x, y) = \sin(x + y)$ at $(2, -2)$ along the line $3x + 4y = -2$ in the direction of decreasing y
7. $f(x, y, z) = x^3y \sin z$ at $(3, -1, -2)$ along the line $x = 3 + t$, $y = -1 + 4t$, $z = -2 + 2t$, in the direction of decreasing x
8. $f(x, y, z) = x^2y + y^2z + z^2x$ at $(1, -1, 0)$ along the line $x + 2y + 1 = 0$, $x - y + 2z = 2$, in the direction of decreasing z
9. $f(x, y) = 2x - 3y$ at $(1, 1)$ along the curve $y = x^2$ in the direction of increasing x
10. $f(x, y, z) = x^2y + xy^3z$ at $(2, -1, 2)$ along the curve $x^2 - y^2 = 3$, $z = x$, in the direction of increasing x

In Exercises 11–16, find the direction in which the function increases most rapidly at the point indicated. What is the rate of change in that direction?

11. $f(x, y, z) = x^4yz - xy^3 + z$ at $(1, 1, -3)$
12. $f(x, y) = 2xy + \ln(xy)$ at $(2, \frac{1}{2})$
13. $f(x, y, z) = 1/\sqrt{x^2 + y^2 + z^2}$ at $(1, -3, 2)$
14. $f(x, y, z) = -1/\sqrt{x^2 + y^2 + z^2}$ at $(1, -3, 2)$
15. $f(x, y, z) = \mathrm{Tan}^{-1}(xyz)$ at $(3, 2, -4)$
16. $f(x, y) = xye^{xy}$ at $(1, 1)$
17. In what direction is the rate of change of $f(x, y, z) = xyz$ smallest at the point $(2, -1, 3)$?
18. In the derivation of (13.38) why was it necessary to use a unit vector $\hat{\mathbf{v}}$ to determine parametric equations for the line through (x_0, y_0, z_0) along $\mathbf{v}$? In other words, why could we not use the components of $\mathbf{v}$ itself to write parametric equations for the line?
19. Find points on the curve $C: x = t$, $y = 1 - 2t$, $z = t$ at which the rate of change of $f(x, y, z) = x^2 + xyz$ along C vanishes.
20. Repeat Exercise 19 for the curve $C: z = x$, $x = y^2$, and the function $f(x, y, z) = x^2 - y^2 + z^2$.

13.9 TANGENT LINES AND TANGENT PLANES

Tangent Lines to Curves

One equation in the coordinates x, y, and z of points in space,

$$F(x, y, z) = 0, \tag{13.43}$$

usually defines a surface. When each of the equations

$$F(x, y, z) = 0, \qquad G(x, y, z) = 0 \tag{13.44}$$

defines a surface, then together they define the curve of intersection of the two surfaces. Theoretically, we can find parametric equations for the curve by setting x equal to some function of a parameter t, say $x = x(t)$, and then solving Eqs. (13.44) for y and z in terms of t: $y = y(t)$ and $z = z(t)$. The parametric definition, therefore, takes the form

$$x = x(t), \qquad y = y(t), \qquad z = z(t), \qquad \alpha \le t \le \beta, \tag{13.45}$$

where α and β specify the endpoints of the curve. Practical difficulties arise in choosing $x(t)$ and solving for $y(t)$ and $z(t)$. For some examples, it might be more convenient to specify $y(t)$ and solve for $x(t)$ and $z(t)$ or, alternatively, to specify

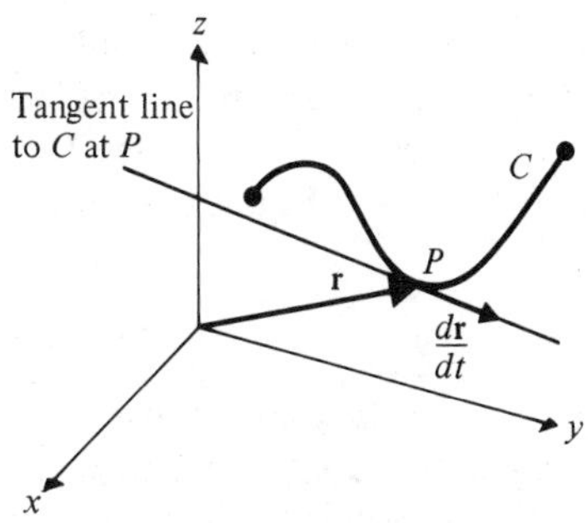

Figure 13.17

$z(t)$ and solve for $x(t)$ and $y(t)$. We considered examples of such conversions in Section 12.8.

In Section 12.9 we indicated that if a curve C is defined parametrically by (13.45), then a tangent vector to C at any point P is

$$\frac{d\mathbf{r}}{dt} = \frac{dx}{dt}\hat{\mathbf{i}} + \frac{dy}{dt}\hat{\mathbf{j}} + \frac{dz}{dt}\hat{\mathbf{k}} \tag{13.46}$$

(Fig. 13.17). The tangent line to C at P is the line through P having direction $d\mathbf{r}/dt$. If (x_0, y_0, z_0) are the coordinates of P and t_0 is the value of t yielding P, then the vector equation for the tangent line at P is

$$(x, y, z) = (x_0, y_0, z_0) + u\frac{d\mathbf{r}}{dt}_{|t=t_0} \tag{13.47a}$$

(see Eq. 12.41). Parametric equations for the tangent line are therefore

$$x = x_0 + x'(t_0)u, \qquad y = y_0 + y'(t_0)u, \qquad z = z_0 + z'(t_0)u, \tag{13.47b}$$

and in the case that none of $x'(t_0)$, $y'(t_0)$, and $z'(t_0)$ vanishes, we can also write symmetric equations for the tangent line:

$$\frac{x - x_0}{x'(t_0)} = \frac{y - y_0}{y'(t_0)} = \frac{z - z_0}{z'(t_0)}. \tag{13.47c}$$

Example 22 Find equations for the tangent line to the curve

$$C:\quad x = 2\cos t, \qquad y = 4\sin t, \qquad z = 2t/\pi$$

at $P(\sqrt{2}, 2\sqrt{2}, \frac{1}{2})$.

Solution Since $t = \pi/4$ at P, a tangent vector to C at P is

$$\frac{d\mathbf{r}}{dt}_{|t=\pi/4} = (-2\sin t, 4\cos t, 2/\pi)_{|t=\pi/4} = (-\sqrt{2}, 2\sqrt{2}, 2/\pi).$$

Symmetric equations for the tangent line are therefore

$$\frac{x - \sqrt{2}}{-\sqrt{2}} = \frac{y - 2\sqrt{2}}{2\sqrt{2}} = \frac{z - \frac{1}{2}}{2/\pi}. \quad \square$$

Example 23 Find equations for the tangent line to the curve

$$C:\quad x = 2t^3 + t^2 + 4t, \qquad y = t^2 - 3t + 5, \qquad z = t^2 - 1$$

at the point $P(0, 5, -1)$.

Solution Since $t = 0$ at P, a tangent vector to C at P is

$$\frac{d\mathbf{r}}{dt}_{|t=0} = (6t^2 + 2t + 4, 2t - 3, 2t)_{|t=0} = (4, -3, 0).$$

Because the z-component of this vector is equal to zero, we cannot use the full symmetric equations (13.47c). We can, however, use them for the x- and y-coordinates of points on the line and write

$$\frac{x}{4} = \frac{y - 5}{-3}, \quad \text{or} \quad 3x + 4y = 20.$$

Since the z-component of $d\mathbf{r}/dt$ vanishes, $d\mathbf{r}/dt$ must be parallel to the xy-plane, and so too must the tangent line to C at P. But then every point on the line has the same z-coordinate, and because the z-coordinate of P is -1, it follows that all points on the tangent line have z-coordinate equal to -1. Equations for the tangent line are therefore $3x + 4y = 20$, $z = -1$. □

Tangent Planes to Surfaces

We now consider the problem of finding the equation for the tangent plane at a point P on a surface S (Fig. 13.18). We define the tangent plane as that plane which contains all tangent lines at P to curves in S through P (provided, of course, that such a plane exists). Suppose that the surface is defined by the equation

$$F(x, y, z) = 0, \tag{13.48}$$

and that

$$C: \quad x = x(t), \qquad y = y(t), \qquad z = z(t), \qquad \alpha \le t \le \beta,$$

is any curve in S through P. Since C is in S, the equation

$$F[x(t), y(t), z(t)] = 0$$

is valid for all t in $\alpha \le t \le \beta$. If $F(x, y, z)$ has continuous first partial derivatives, and $x(t)$, $y(t)$, and $z(t)$ are all differentiable, we may differentiate this equation using the chain rule:

$$\frac{\partial F}{\partial x}\frac{dx}{dt} + \frac{\partial F}{\partial y}\frac{dy}{dt} + \frac{\partial F}{\partial z}\frac{dz}{dt} = 0.$$

This equation, which holds at all points on C, and in particular at P, can be expressed vectorially as

$$0 = \left(\frac{\partial F}{\partial x}, \frac{\partial F}{\partial y}, \frac{\partial F}{\partial z}\right) \cdot \left(\frac{dx}{dt}, \frac{dy}{dt}, \frac{dz}{dt}\right) = \nabla F \cdot \frac{d\mathbf{r}}{dt}.$$

But if the scalar product of two vectors vanishes, the vectors are perpendicular (see Eq. 12.27). Consequently, ∇F is perpendicular to the tangent vector $d\mathbf{r}/dt$ to C at P. Since C is an arbitrary curve in S, it follows that ∇F at P is perpendicular

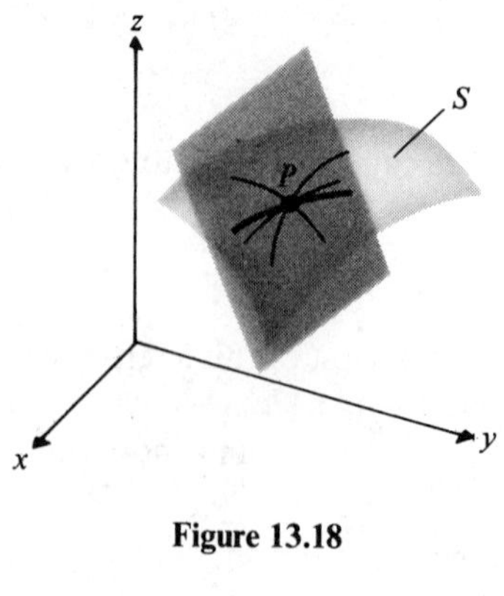

Figure 13.18

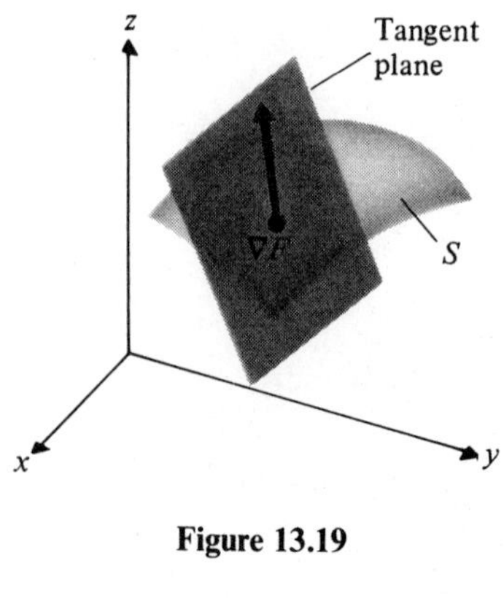

Figure 13.19

to every tangent line to every curve C in S at P. In other words, ∇F at P must be perpendicular to the tangent plane to S at P (Fig. 13.19). If the coordinates of P are (x_0, y_0, z_0), then the equation of the tangent plane to S at P is

$$\begin{aligned} 0 &= \nabla F_{|P} \cdot (x - x_0, y - y_0, z - z_0) \\ &= F_x(x_0, y_0, z_0)(x - x_0) + F_y(x_0, y_0, z_0)(y - y_0) + F_z(x_0, y_0, z_0)(z - z_0) \end{aligned} \tag{13.49}$$

(see Eq. 12.30).

Example 24 Find the equation of the tangent plane to the surface $xyz^3 + yz^2 = 4$ at the point $(1, 2, 1)$.

Solution A vector perpendicular to the tangent plane is

$$\nabla(xyz^3 + yz^2 - 4)_{|(1,2,1)} = (yz^3, xz^3 + z^2, 3xyz^2 + 2yz)_{|(1,2,1)} = (2, 2, 10).$$

But then the vector $(1, 1, 5)$ must also be perpendicular to the tangent plane, and the equation of the plane is therefore

$$0 = (1, 1, 5) \cdot (x - 1, y - 2, z - 1) = x + y + 5z - 8. \quad \square$$

We have shown in this section that if the equation $F(x, y, z) = 0$ defines a surface S, and if there is a tangent plane to S at a point P, then the vector $\nabla F_{|P}$ is normal to the tangent plane (Fig. 13.19). It is customary to state in this situation that $\nabla F_{|P}$ is normal to the surface itself at P, rather than to the tangent plane to the surface. This fact proves to be another of the important properties of the gradient vector, and is worth stating as a theorem.

▶ **Theorem 13.5** If the equation $F(x, y, z) = 0$ defines a surface S, and $F(x, y, z)$ has continuous first partial derivatives, then at any point on S, the vector ∇F is perpendicular to S. ◀

A geometric application of this fact is contained in the following example.

Example 25 Find equations for the tangent line at the point $(2, 2, 1)$ to the curve $C: x^2 + y^2 + z^2 = 9$, $x^2 + y^2 - 8z^2 = 0$.

Solution Equation (13.46) indicates that to find a tangent vector to C we should first have parametric equations for C. These can be obtained by first solving each equation for $x^2 + y^2$ and equating the results:

$$9 - z^2 = 8z^2.$$

This equation implies that $z = \pm 1$, the positive result being required here. On C, then, $x^2 + y^2 = 8$, and parametric equations for C are

$$x = 2\sqrt{2}\cos t, \qquad y = 2\sqrt{2}\sin t, \qquad z = 1, \qquad 0 \le t < 2\pi.$$

According to (13.46), a tangent vector to C at $(2, 2, 1)$ is

$$\left(\frac{dx}{dt}, \frac{dy}{dt}, \frac{dz}{dt}\right)_{|(2,2,1)} = (-2\sqrt{2}\sin t, 2\sqrt{2}\cos t, 0)_{|t=\pi/4} = (-2, 2, 0).$$

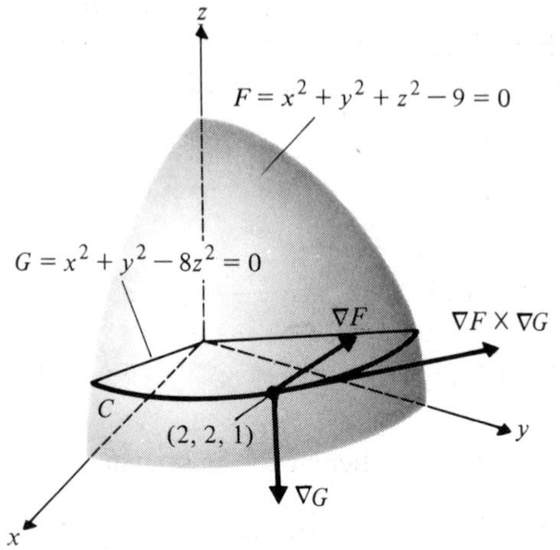

Figure 13.20

The tangent line therefore has equations

$$\frac{x-2}{1}=\frac{y-2}{-1}, \qquad z=1, \quad \text{or} \quad x+y=4, \qquad z=1.$$

The fact that gradients can be used to find normals to surfaces suggests an alternative solution. It is clear from Fig. 13.20 that if we define $F(x, y, z) = x^2 + y^2 + z^2 - 9 = 0$, then ∇F evaluated at (2, 2, 1) is perpendicular not only to the surface $F(x, y, z) = 0$, but also to the curve C. Similarly, if $G(x, y, z) = x^2 + y^2 - 8z^2 = 0$, then ∇G at (2, 2, 1) is also perpendicular to C. Since a vector along the tangent line to C at (2, 2, 1) will be perpendicular to both of these vectors, it follows that a vector along the tangent line must be

$$\begin{aligned}(\nabla F \times \nabla G)_{|(2,2,1)} &= \{(2x, 2y, 2z) \times (2x, 2y, -16z)\}_{|(2,2,1)}\\ &= (4, 4, 2) \times (4, 4, -16)\\ &= 8\begin{vmatrix}\hat{\mathbf{i}} & \hat{\mathbf{j}} & \mathbf{k}\\ 2 & 2 & 1\\ 1 & 1 & -4\end{vmatrix}\\ &= 8(-9, 9, 0)\\ &= -72(1, -1, 0)\end{aligned}$$

Once again, then, we have obtained (1, −1, 0) as a tangent vector to the curve, and equations for the tangent line can be written down as before. □

Example 25 illustrates that when a curve is defined as the intersection of two surfaces $F(x, y, z) = 0$, $G(x, y, z) = 0$ (Fig. 13.21), then a vector tangent to the curve is

$$\mathbf{T} = \nabla F \times \nabla G. \tag{13.50}$$

Thus to find a tangent vector to a curve we use (13.46) when the curve is defined parametrically. When the curve is defined as the intersection of two surfaces, we can either find parametric equations and use (13.46) or use (13.50). Note too that

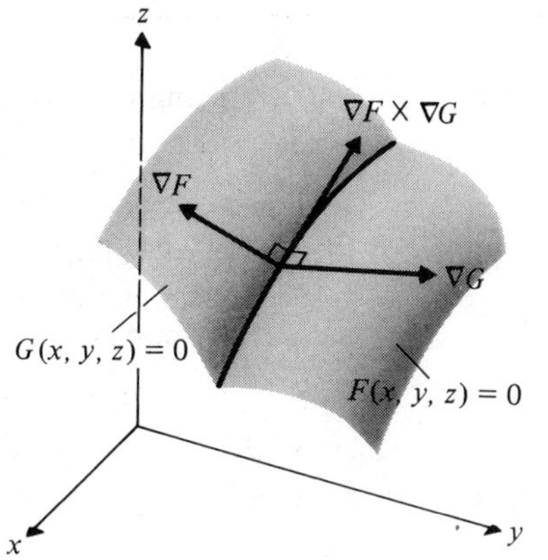

Figure 13.21

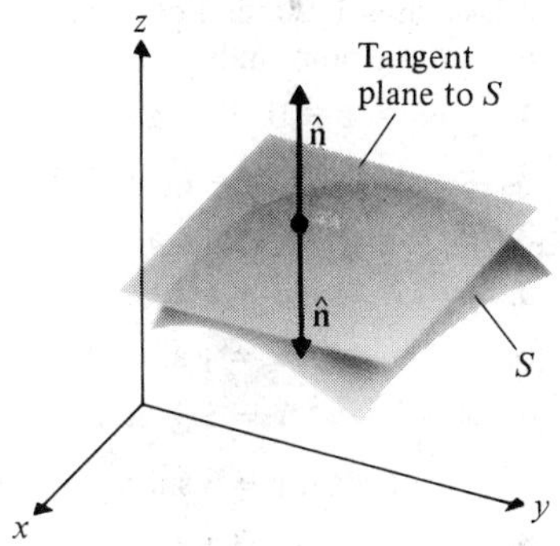

Figure 13.22

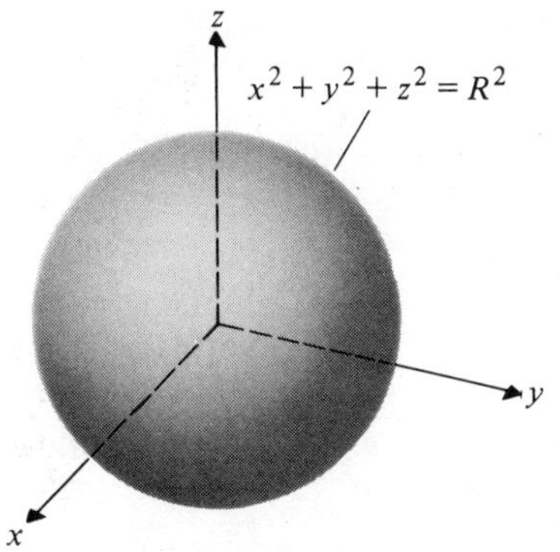

Figure 13.23

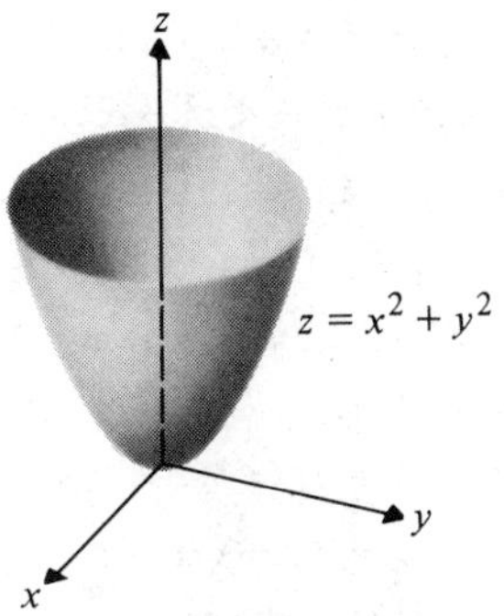

Figure 13.24

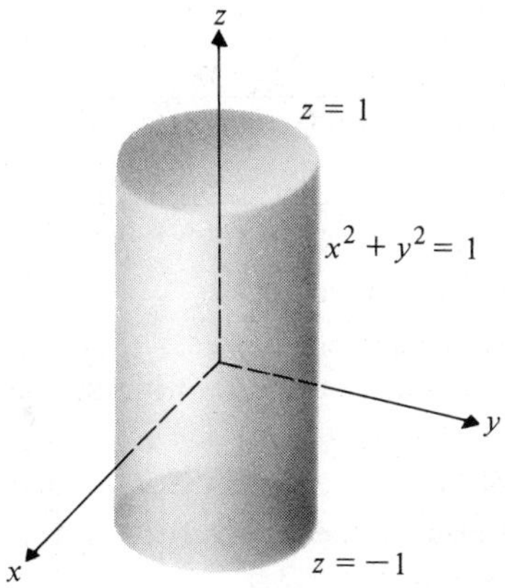

Figure 13.25

in order to find tangent lines to curves, it is not necessary to have a direction assigned to the curves.

At each point on a surface S at which S has a tangent plane (Fig. 13.22), we have defined a normal vector to S as a vector normal to the tangent plane to S. If we denote by $\hat{\mathbf{n}}$ a unit normal vector to S, then the direction of $\hat{\mathbf{n}}$ clearly varies as we move from point to point on S. We say that $\hat{\mathbf{n}}$ is a function of position (x, y, z) on S. Furthermore, at each point at which S has a unit normal vector, it has two such vectors, one in the opposite direction to the other. We say that a surface S is *smooth* if it can be assigned a unit normal $\hat{\mathbf{n}}$ that varies continuously on S. What this means geometrically is that for small changes in position, the unit normal $\hat{\mathbf{n}}$ will undergo small changes in direction. The sphere in Fig. 13.23 is smooth, as is the paraboloid in Fig. 13.24. The surface bounding the cylindrical volume in Fig. 13.25 is not smooth; a unit normal that varies continuously over the surface cannot be assigned at points on the circles $x^2 + y^2 = 1, z = \pm 1$. This surface can, however, be divided into a finite number of subsurfaces, each of which is smooth. In particular, we choose the three subsurfaces S_1: $z = 1$, $x^2 + y^2 \leq 1$; S_2: $z = -1$, $x^2 + y^2 \leq 1$; S_3: $x^2 + y^2 = 1$, $-1 \leq z \leq 1$. Such a surface is said to be *piecewise smooth.*

EXERCISES 13.9

In Exercises 1–20, find equations for the tangent line to the curve at the point indicated.

1. $y = x^2, z = 0$ at $(-2, 4, 0)$
2. $x = t, y = t^2, z = t^3$ at $(1, 1, 1)$
3. $x = \cos t, y = \sin t, z = \cos t$ at $(1, 0, 1)$
4. $y = x^2, z = x$ at $(-2, 4, -2)$
5. $x^2 = y, z + x = y$ at $(1, 1, 0)$
6. $x = 2 - t^2, y = 3 + 2t, z = t$ at $(1, 5, 1)$
7. $x = 2 \cos t, y = 3 \sin t, z = 5$ at $(\sqrt{2}, -3/\sqrt{2}, 5)$
8. $x^2y^3 + xy = 68$ at $(1, 4)$
9. $x + y + z = 4, x - y = 2$ at $(0, -2, 6)$
10. $x = e^{-t} \cos t, y = e^{-t} \sin t, z = t$ at $(1, 0, 0)$
11. $x = t^2 + 1, y = 2t - 4, y = t^3 + 3$ at $(2, -6, 2)$
12. $y^2 + z^2 = 6, x + z = 1$ at $(2, -\sqrt{5}, -1)$
13. $x^2 + y^2 + z^2 = 4, z^2 = x^2 + y^2$ at $(1, 1, -\sqrt{2})$
14. $x = t, y = 1, z = \sqrt{1 + t^2}$ at $(4, 1, \sqrt{17})$
15. $x = 1 + \cos t, y = 2 - \sin t, z = \sqrt{4 + t}$ at $(2, 2, 2)$
16. $x = z^2 + z^3, y = z - z^4$ at $(12, -14, 2)$
17. $x = y^2 + 3y^3 - 2y + 5, z = 0$ at $(7, 1, 0)$
18. $2x^2 + y^2 + 2y = 3, z = x + 1$ at $(0, 1, 1)$
19. $x = t^2, y = t, z = \sqrt{t + t^4}$ at $(1, 1, \sqrt{2})$
20. $x = t \sin t, y = t \cos t, z = 2t$ at $(0, 2\pi, 4\pi)$

In Exercises 21–26, find the equation of the tangent plane to the surface at the point indicated.

21. $z = \sqrt{x^2 + y^2}$ at $(1, 1, \sqrt{2})$
22. $x = x^2 - y^3z$ at $(2, -1, -2)$
23. $x^2y + y^2z + z^2x + 3 = 0$ at $(2, -1, -1)$
24. $x + y + z = 4$ at $(1, 1, 2)$
25. $x = y \sin(\pi z/2)$ at $(-1, -1, 1)$
26. $x^2 + y^2 + 2y = 1$ at $(1, 0, 3)$
27. Show that the curve
$$x = 2(t^3 + 2)/3, \qquad y = 2t^2, \qquad z = 3t - 2$$
intersects the surface
$$x^2 + 2y^2 + 3z^2 = 15$$
at right angles at the point $(2, 2, 1)$.
28. Verify that the curve
$$x^2 - y^2 + z^2 = 1, \qquad xy + xz = 2$$
is tangent to the surface
$$xyz - x^2 - 6y + 6 = 0$$
at the point $(1, 1, 1)$.
29. Show that the equation of the tangent plane to a surface $S: z = f(x, y)$ at a point (x_0, y_0, z_0) on S can be written in the form
$$z - z_0 = (x - x_0)f_x(x_0, y_0) + (y - y_0)f_y(x_0, y_0).$$

In Exercises 30–32, find the directional derivative of the function for the point and direction indicated.

30. $f(x, y, z) = 2x^2 + y^2z^2$ at $(3, 1, 0)$ along the curve $x + y + z = 4$, $x - y + z = 2$ in the direction of increasing x
31. $f(x, y, z) = xyz + xy + xz + yz$ at $(1, -2, 5)$ perpendicular to the surface
$$z = x^2 + y^2$$
32. $f(x, y, z) = x^2 + y^2 - z^2$ at $(3, 4, 5)$ along the curve $x^2 + y^2 - z^2 = 0, 2x^2 + 2y^2 - z^2 = 25$ in the direction of decreasing x
33. If $F(x, y) = 0$ defines a curve implicitly in the xy-plane, prove that at any point on the curve ∇F is perpendicular to the curve.
34. Suppose that the equations
$$F(x, y, z, t) = 0,$$
$$G(x, y, z, t) = 0,$$
$$H(x, y, z, t) = 0$$
implicitly define parametric equations for a curve C (t being the parameter). If $P(x_0, y_0, z_0)$ is a point on C, show that equations for the tangent line to C at P can be written in the form
$$\frac{x - x_0}{\partial(F, G, H)/\partial(t, y, z)|_P} = \frac{y - y_0}{\partial(F, G, H)/\partial(x, t, z)|_P} = \frac{z - z_0}{\partial(F, G, H)/\partial(x, y, t)|_P},$$
provided none of the Jacobians vanish.

*35. Find all points on the paraboloid $z = x^2 + y^2 - 1$ at which the normal to the surface coincides with the line joining the origin to the point.

*36. Show that the sum of the intercepts on the x-, y-, and z-axes of the tangent plane to the surface $\sqrt{x} + \sqrt{y} + \sqrt{z} = \sqrt{a}$ at any point is a.

13.10 RELATIVE MAXIMA AND MINIMA

We now study relative extrema of functions of more than one independent variable. Most of the discussion will be confined to functions $f(x, y)$ of two independent variables because we can discuss the concepts geometrically as well as algebraically. Unfortunately, not all the results are easily extended to functions of more than two independent variables, and we will therefore be careful to point out these limitations.

Before beginning the discussion, we briefly review maxima–minima results for functions $f(x)$ of one variable. We do this because maxima–minima theory for multivariable functions is essentially the same as that for single-variable functions. In fact, every definition that we make and every result that we discuss in this section has its counterpart in single-variable theory. Hence, a synopsis of single-variable results is in order. Unfortunately, proving results in the multivariable case is considerably more complicated than in the single-variable case, but if we can keep central ideas foremost in our minds and constantly make comparisons with single-variable calculus, we find that the discussions are not nearly as difficult as they might otherwise be.

Critical points of a function $f(x)$ are points at which $f'(x)$ is either equal to zero or does not exist. Geometrically, this means points at which the graph of $f(x)$ has a horizontal tangent line, a vertical tangent line, or no tangent line at all. Critical points for continuous functions are classified as yielding either relative maxima, relative minima, horizontal points of inflection, or just corners. If x_0 is a critical point for a continuous function $f(x)$, two tests can be used to determine the nature of the critical point. The first-derivative test states that if $f'(x)$ changes from a positive quantity to a negative quantity as x increases through x_0, then x_0 gives a relative maximum; if $f'(x)$ changes from negative to positive, then a relative minimum is obtained; and if $f'(x)$ does not change sign, then x_0 gives either a horizontal point of inflection or just a corner. The second-derivative test indicates the nature of a critical point at which $f'(x_0) = 0$ whenever $f''(x_0) \neq 0$. If $f''(x_0) > 0$, then a relative minimum is obtained, and if $f''(x_0) < 0$, a relative maximum is found. The second-derivative test fails to classify a critical point for which either $f'(x_0)$ does not exist or $f'(x_0) = f''(x_0) = 0$.

We begin our study of extrema theory for multivariable functions by defining critical points for functions of two independent variables.

▷ **Definition 13.6** A point (x_0, y_0) in the domain of a function $f(x, y)$ is said to be a critical point of $f(x, y)$ if

$$\left.\frac{\partial f}{\partial x}\right|_{(x_0, y_0)} = 0, \qquad \left.\frac{\partial f}{\partial y}\right|_{(x_0, y_0)} = 0, \tag{13.51}$$

or if one (or both) of these partial derivatives does not exist at (x_0, y_0). ◁

There are two ways of interpreting critical points of $f(x, y)$ geometrically. In Section 13.3, we interpreted $\partial f/\partial x$ at (x_0, y_0) as the slope of the tangent line to the curve of intersection of the surface $z = f(x, y)$ and the plane $y = y_0$, and $\partial f/\partial y$ as the slope of the tangent line to the curve of intersection with $x = x_0$. It follows, then, that (x_0, y_0) is critical if both curves have horizontal tangent lines or if either curve has a vertical tangent line or no tangent line at all. Alternatively, recall that the equation of the tangent plane to the surface $z = f(x, y)$ at

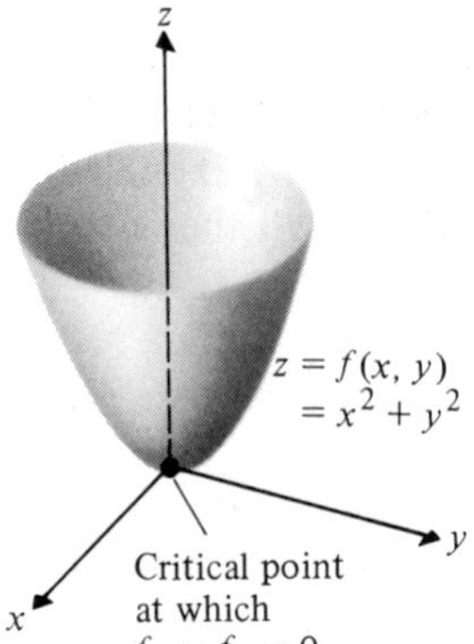

Figure 13.26

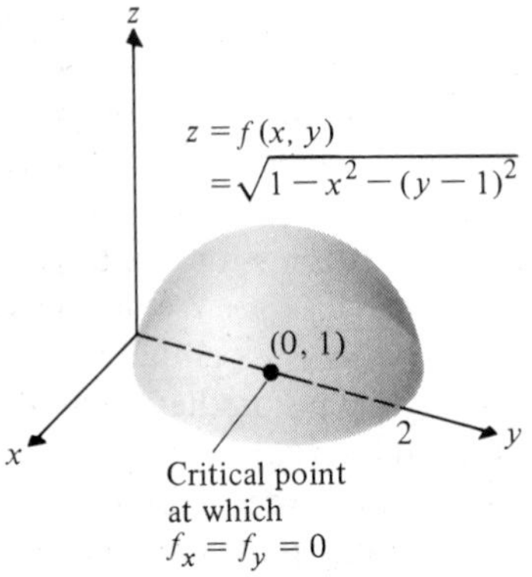

Figure 13.27

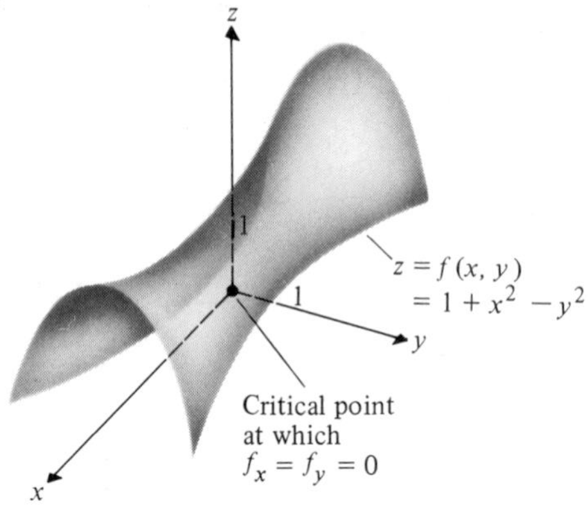

Figure 13.28

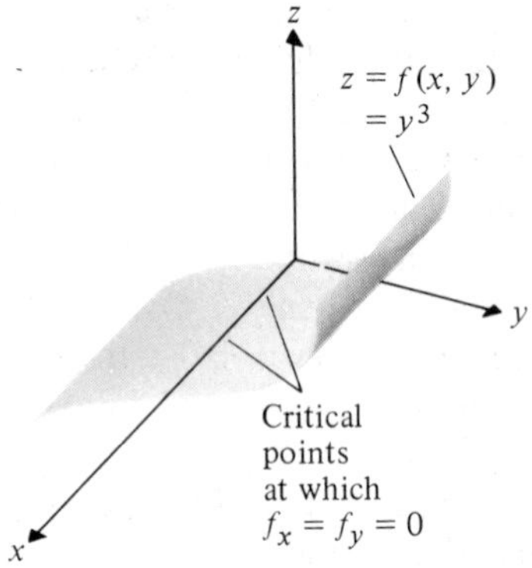

Figure 13.29

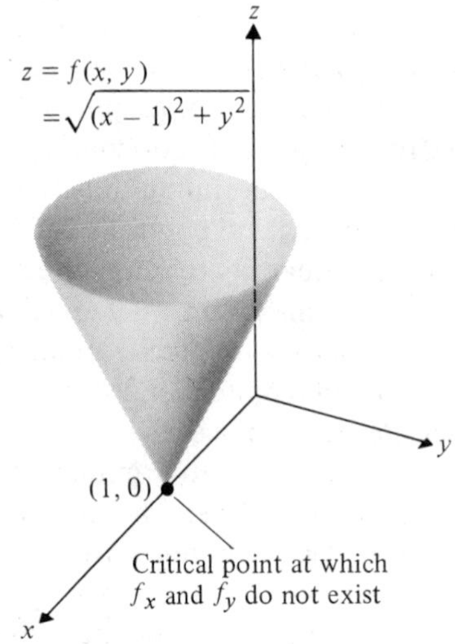

Figure 13.30

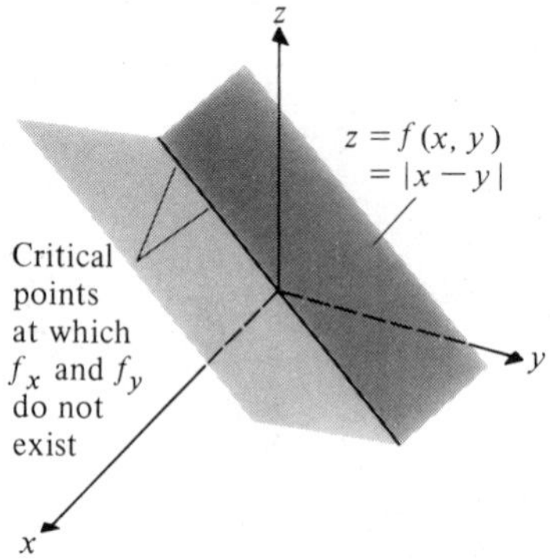

Figure 13.31

(x_0, y_0) is

$$z - z_0 = f_x(x_0, y_0)(x - x_0) + f_y(x_0, y_0)(y - y_0)$$

(see Exercise 29 in Section 13.9). If both partial derivatives vanish, then the tangent plane is horizontal with equation $z = z_0$. For example, at each of the critical points in Figs. 13.26–13.29, $\partial f/\partial x = \partial f/\partial y = 0$ and the tangent plane is horizontal. The remaining functions in Figs. 13.30–13.34 have critical points at which either $\partial f/\partial x$ or $\partial f/\partial y$ or both do not exist. In Figs. 13.30–13.33, the surfaces do not have tangent planes at critical points, and in Fig. 13.34, the tangent plane is vertical at each critical point. Consequently, (x_0, y_0) is a critical point of a function $f(x, y)$ if at (x_0, y_0) the surface $z = f(x, y)$ has a horizontal tangent plane, a vertical tangent plane, or no tangent plane at all.

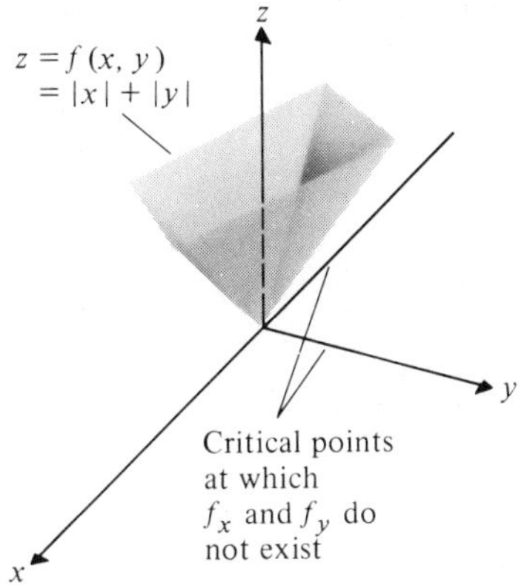

Figure 13.32

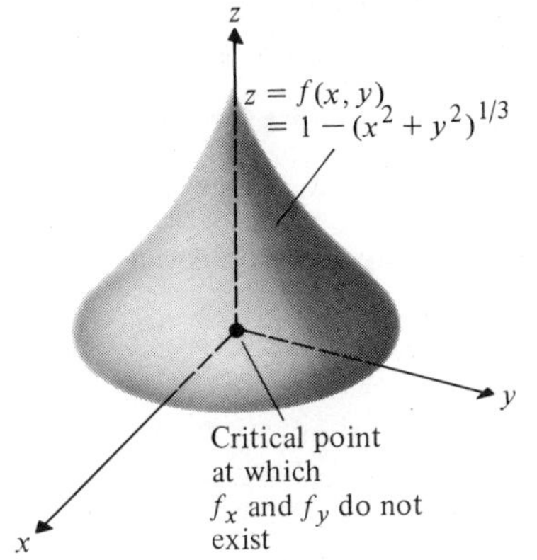

Figure 13.33

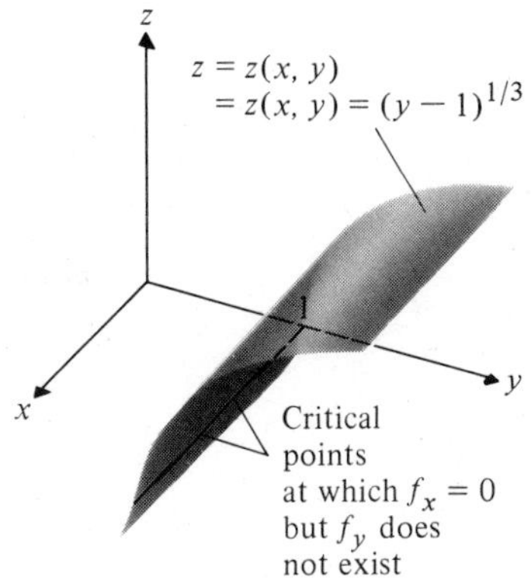

Figure 13.34

Example 26 Find all critical points for the function

$$f(x, y) = x^2y - 2xy^2 + 3xy + 4.$$

Solution For critical points, we first solve

$$0 = \frac{\partial f}{\partial x} = 2xy - 2y^2 + 3y = y(2x - 2y + 3),$$

$$0 = \frac{\partial f}{\partial y} = x^2 - 4xy + 3x = x(x - 4y + 3).$$

To satisfy these two equations simultaneously, there are four possibilities:

1. $x = 0$, $y = 0$, which gives the critical point $(0, 0)$;
2. $y = 0$, $x - 4y + 3 = 0$, which gives the critical point $(-3, 0)$;
3. $x = 0$, $2x - 2y + 3 = 0$, which gives the critical point $(0, \frac{3}{2})$;
4. $2x - 2y + 3 = 0$, $x - 4y + 3 = 0$, which gives the critical point $(-1, \frac{1}{2})$.

Since $\partial f/\partial x$ and $\partial f/\partial y$ are defined for all x and y, these are the only critical points. □

Critical points for functions of more than two independent variables can be defined algebraically, but because we have no geometric representation for such functions, there is no geometric interpretation for their critical points. For example, if $f(x, y, z, t)$ is a function of independent variables x, y, z, and t, then (x_0, y_0, z_0, t_0) is a critical point of $f(x, y, z, t)$ if all four of its first-order partial derivatives vanish at (x_0, y_0, z_0, t_0),

$$\frac{\partial f}{\partial x}_{|(x_0, y_0, z_0, t_0)} = \frac{\partial f}{\partial y}_{|(x_0, y_0, z_0, t_0)} = \frac{\partial f}{\partial z}_{|(x_0, y_0, z_0, t_0)} = \frac{\partial f}{\partial t}_{|(x_0, y_0, z_0, t_0)} = 0, \qquad (13.52)$$

or if at least one of the partial derivatives does not exist at the point. Note that because the partial derivatives of a function are the components of the gradient of the function, we can say that a critical point of a function is a point at which its gradient is either equal to zero or undefined.

Example 27 Find all critical points for the function

$$f(x, y, z) = xyz\sqrt{x^2 + y^2 + z^2}.$$

Solution For critical points, we consider the equations

$$0 = \frac{\partial f}{\partial x} = yz\sqrt{x^2 + y^2 + z^2} + \frac{x^2yz}{\sqrt{x^2 + y^2 + z^2}}$$

$$= \frac{yz}{\sqrt{x^2 + y^2 + z^2}}(2x^2 + y^2 + z^2),$$

$$0 = \frac{\partial f}{\partial y} = xz\sqrt{x^2 + y^2 + z^2} + \frac{xy^2z}{\sqrt{x^2 + y^2 + z^2}}$$

$$= \frac{xz}{\sqrt{x^2 + y^2 + z^2}}(x^2 + 2y^2 + z^2),$$

$$0 = \frac{\partial f}{\partial z} = xy\sqrt{x^2 + y^2 + z^2} + \frac{xyz^2}{\sqrt{x^2 + y^2 + z^2}}$$

$$= \frac{xy}{\sqrt{x^2 + y^2 + z^2}}(x^2 + y^2 + 2z^2).$$

The partial derivatives are clearly undefined for $x = y = z = 0$, and therefore the origin $(0, 0, 0)$ is a critical point. If x, y, and z are not all zero, then the terms in parentheses cannot vanish, and we must set

$$yz = 0, \qquad xz = 0, \qquad xy = 0.$$

If any two of x, y, and z vanish, but the third does not, then these equations are satisfied. In other words, every point on the x-axis, every point on the y-axis, and every point on the z-axis is critical. □

We now turn our attention to the classification of critical points of a function $f(x, y)$ of two independent variables. The critical points $(0, 1)$ in Fig. 13.27 and $(0, 0)$ in Fig. 13.33 yield "high" points on the surfaces. We describe this property in the following definition.

▷ **Definition 13.7** A function $f(x, y)$ is said to have a *relative maximum* $f(x_0, y_0)$ at a point (x_0, y_0) if there exists a circle in the xy-plane centered at (x_0, y_0) such that for all points (x, y) inside this circle

$$f(x, y) \leq f(x_0, y_0). \quad ◁ \tag{13.53}$$

The "low" points on the surfaces at $(0, 0)$ in Fig. 13.26 and $(1, 0)$ in Fig. 13.30 are relative minima according to the following.

▷ **Definition 13.8** A function $f(x, y)$ is said to have a *relative minimum* $f(x_0, y_0)$ at a point (x_0, y_0) if there exists a circle in the xy-plane centered at

(x_0, y_0) such that for all points (x, y) inside this circle

$$f(x, y) \geq f(x_0, y_0). \quad \triangleleft \tag{13.54}$$

Note that every critical point in Fig. 13.31 yields a relative minimum of $f(x, x) = 0$, as does the critical point (0, 0) in Fig. 13.32.

▷ **Definition 13.9** If a critical point of a function $f(x, y)$ at which $\partial f/\partial x = \partial f/\partial y = 0$ yields neither a relative maximum nor a relative minimum, it is said to yield a *saddle point.* ◁

The critical point (0, 0) in Fig. 13.28 therefore gives a saddle point, as does each of the critical points in Fig. 13.29. Saddle points for surfaces $z = f(x, y)$ are clearly the analogues of horizontal points of inflection for curves $y = f(x)$. In both cases the derivative(s) of the function vanishes but there is neither a relative maximum nor a relative minimum.

The remaining critical points (all critical points in Fig. 13.32, except (0, 0), and those in Fig. 13.34) are the counterparts of corners for the graph of a function $f(x)$. They are points at which one or both of the partial derivatives of $f(x, y)$ do not exist, but like corners for $f(x)$, they do not necessarily yield relative extrema.

Our discussion has made it abundantly clear that:

a) At a relative maximum or minimum of $f(x, y)$ either $\partial f/\partial x$ and $\partial f/\partial y$ both vanish, or one or both of the partial derivatives do not exist;

b) Saddle points may also occur where $\partial f/\partial x = \partial f/\partial y = 0$, and points where the derivatives do not exist may fail to yield relative extrema.

In other words, every relative extremum of $f(x, y)$ occurs at a critical point; but critical points do not always give relative extrema.

Given the problem of determining all relative maxima and minima of a function $f(x, y)$, we should first find its critical points. But how do we decide whether these critical points yield relative maxima, relative minima, saddle points, or none of these? We do not have a result that is equivalent to the first-derivative test for functions of one variable, but we do have a test that corresponds to the second-derivative test. For functions of two independent variables the situation is more complicated, however, since there are three second-order partial derivatives, but the idea of the test is essentially the same. It determines whether certain curves are concave upward or concave downward at the critical point. The complete result is contained in the following theorem.

▶ **Theorem 13.6** Suppose (x_0, y_0) is a critical point of $f(x, y)$ at which $\partial f/\partial x$ and $\partial f/\partial y$ both vanish. Suppose further that f_x, f_y, f_{xx}, f_{xy}, and f_{yy} are all continuous at (x_0, y_0). Define

$$A = f_{xx}(x_0, y_0), \qquad B = f_{xy}(x_0, y_0), \qquad C = f_{yy}(x_0, y_0).$$

If

i) $B^2 - AC < 0$ and $A < 0$, then $f(x, y)$ has a relative maximum at (x_0, y_0);

ii) $B^2 - AC < 0$ and $A > 0$, then $f(x, y)$ has a relative minimum at (x_0, y_0);

iii) $B^2 - AC > 0$, then $f(x, y)$ has a saddle point at (x_0, y_0);

iv) $B^2 - AC = 0$, then the test fails. ◀

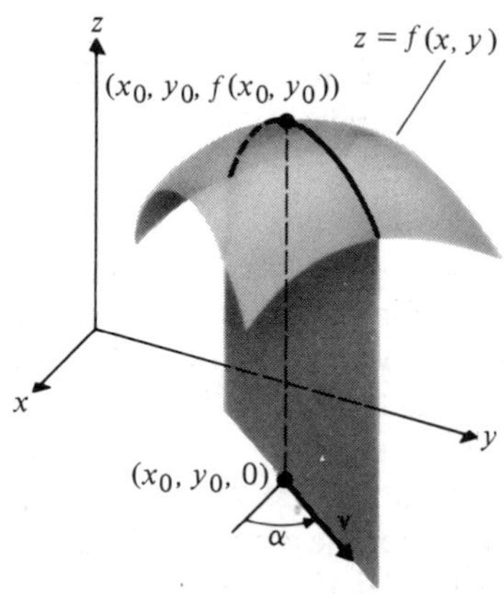

Figure 13.35

Proof (i) Suppose we intersect the surface $z = f(x, y)$ with a plane parallel to the z-axis, through the point $(x_0, y_0, 0)$, and making an angle α with the line through $(x_0, y_0, 0)$ parallel to the positive x-axis (Fig. 13.35). The slope of the curve of intersection of these surfaces at the point $(x_0, y_0, f(x_0, y_0))$ is given by the directional derivative

$$D_{\mathbf{v}} f_{|(x_0, y_0)} = \nabla f_{|(x_0, y_0)} \cdot \hat{\mathbf{v}} = \nabla f_{|(x_0, y_0)} \cdot (\cos \alpha, \sin \alpha)$$
$$= \frac{\partial f}{\partial x}_{|(x_0, y_0)} \cos \alpha + \frac{\partial f}{\partial y}_{|(x_0, y_0)} \sin \alpha$$

(see Eq. 13.41). Since (x_0, y_0) is a critical point at which $\nabla f = \mathbf{0}$, it follows that

$$D_{\mathbf{v}} f_{|(x_0, y_0)} = 0 \quad \text{for all } \alpha.$$

In Fig. 13.35 we have illustrated the critical point as a relative maximum. But how do we verify that this is indeed the case? If we can show that each and every curve of intersection of the surface with a vertical plane through $(x_0, y_0, 0)$ is concave downward at $(x_0, y_0, 0)$, then (x_0, y_0) must give a relative maximum. But to discuss concavity of a curve we require the second derivative—in this case, the second directional derivative of $f(x, y)$. According to (13.42), the second directional derivative of $f(x, y)$ at (x_0, y_0) in the direction $\hat{\mathbf{v}} = (\cos \alpha, \sin \alpha)$ is

$$D_{\mathbf{v}}(D_{\mathbf{v}} f) = \frac{\partial^2 f}{\partial x^2}_{|(x_0, y_0)} \cos^2 \alpha + 2 \frac{\partial^2 f}{\partial x\, \partial y}_{|(x_0, y_0)} \cos \alpha \sin \alpha + \frac{\partial^2 f}{\partial y^2}_{|(x_0, y_0)} \sin^2 \alpha$$
$$= A \cos^2 \alpha + 2B \cos \alpha \sin \alpha + C \sin^2 \alpha,$$

where we understand here that $D_{\mathbf{v}}(D_{\mathbf{v}} f)$ is implicitly suffixed by (x_0, y_0). In order, therefore, to verify that (x_0, y_0) gives a relative maximum, it is sufficient to show that $D_{\mathbf{v}}(D_{\mathbf{v}} f)$ is negative for each value of α in the interval $0 \le \alpha < 2\pi$. However, because $D_{\mathbf{v}}(D_{\mathbf{v}} f)$ is unchanged if α is replaced by $\alpha + \pi$, it is sufficient to verify that $D_{\mathbf{v}}(D_{\mathbf{v}} f)$ is negative for $0 \le \alpha < \pi$.

For any of these values of α except $\pi/2$, we can write

$$D_{\mathbf{v}}(D_{\mathbf{v}} f) = \cos^2 \alpha\, (A + 2B \tan \alpha + C \tan^2 \alpha),$$

and if we set $u = \tan \alpha$,

$$D_{\mathbf{v}}(D_{\mathbf{v}} f) = \cos^2 \alpha\, (A + 2Bu + Cu^2).$$

It is evident that $D_{\mathbf{v}}(D_{\mathbf{v}} f) < 0$ for all $\alpha \ne \pi/2$ if and only if

$$Q(u) = A + 2Bu + Cu^2 < 0 \quad \text{for} \quad -\infty < u < \infty.$$

Were we to sketch a graph of the quadratic $Q(u)$, we would see that it would cross the u-axis where

$$u = \frac{-2B \pm \sqrt{4B^2 - 4AC}}{2C} = \frac{-B \pm \sqrt{B^2 - AC}}{C}.$$

But if we know, as stated in Theorem 13.6(i), that $B^2 - AC < 0$, then there are no real solutions of this equation, and therefore $Q(u)$ never crosses the u-axis. Since $Q(0) = A < 0$, it follows that $Q(u) < 0$ for all u. We have shown, then, that

$$D_{\mathbf{v}}(D_{\mathbf{v}} f) < 0 \quad \text{for all } \alpha \neq \pi/2.$$

When $\alpha = \pi/2$, $D_{\mathbf{v}}(D_{\mathbf{v}} f) = C$. Since $B^2 - AC < 0$ and $A < 0$, it follows that $C < 0$ also. Consequently, if $B^2 - AC < 0$ and $A < 0$, then $D_{\mathbf{v}}(D_{\mathbf{v}} f) < 0$ for all α, and (x_0, y_0) yields a relative maximum.

ii) If $B^2 - AC < 0$ and $A > 0$, a similar argument leads to the conclusion that (x_0, y_0) yields a relative minimum; the only difference is that inequalities are reversed.

iii) If $B^2 - AC > 0$, then $Q(u)$ has real distinct zeros, in which case $Q(u)$ is sometimes negative and sometimes positive. This means that the curve of intersection is sometimes concave upward and sometimes concave downward, and the point (x_0, y_0) therefore gives a saddle point.

iv) If $B^2 - AC = 0$, the classification of the point determined by (x_0, y_0) depends on which of A, B, and C vanish, if any. □

To illustrate that we can obtain a relative maximum, a relative minimum, or a saddle point for a critical point at which $B^2 - AC = 0$, consider the three functions $f(x, y) = -y^2$, $f(x, y) = y^2$, and $f(x, y) = y^3$ in Figs. 13.36–13.38. The point (0, 0) is a critical point for each function, and at this point $B^2 - AC = 0$. Yet (0, 0) yields a relative maximum for $f(x, y) = -y^2$, a relative minimum for $f(x, y) = y^2$, and a saddle point for $f(x, y) = y^3$. In fact, every point on the x-axis is a relative maximum for $f(x, y) = -y^2$, a relative minimum for $f(x, y) = y^2$, and a saddle point for $f(x, y) = y^3$.

Example 28 Find and classify the critical points of each of the following functions as yielding relative maxima, relative minima, saddle points, or none of these:

a) $f(x, y) = 4xy - x^4 - y^4$;

b) $f(x, y) = y^3 + x^2 - 6xy + 3x + 6y$;

c) $f(x, y) = \dfrac{12xy - x^2y^2}{2(x + y)}$.

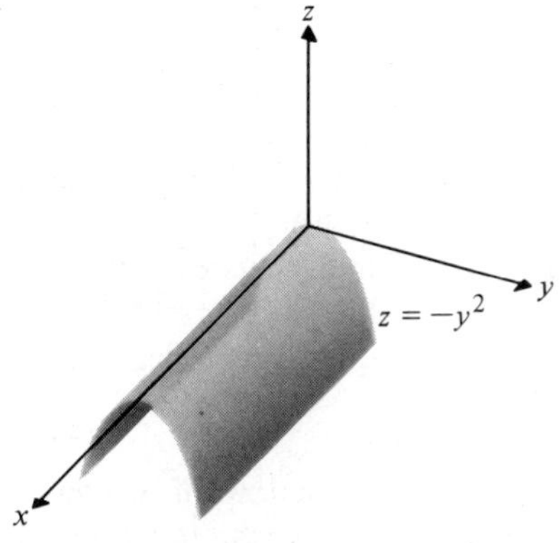

Figure 13.36

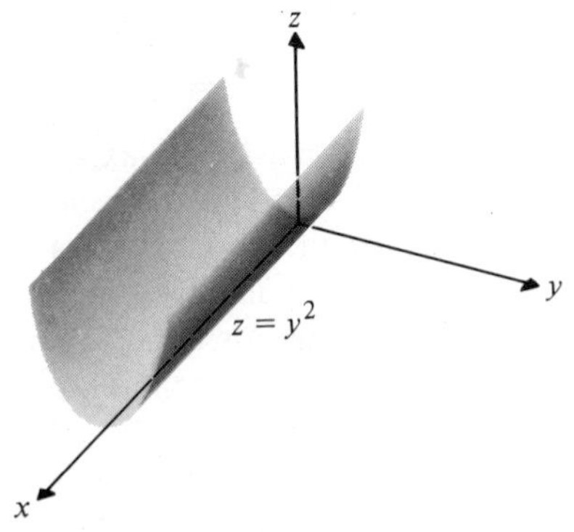

Figure 13.37

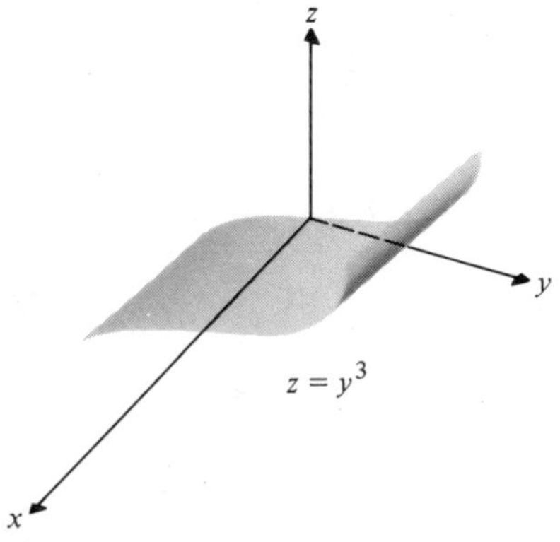

Figure 13.38

Solution a) Critical points of $f(x, y)$ are given by

$$0 = \frac{\partial f}{\partial x} = 4y - 4x^3, \qquad 0 = \frac{\partial f}{\partial y} = 4x - 4y^3.$$

Solutions of these equations are $(0, 0)$, $(1, 1)$, and $(-1, -1)$. We now calculate

$$\frac{\partial^2 f}{\partial x^2} = -12x^2, \qquad \frac{\partial^2 f}{\partial x\, \partial y} = 4, \qquad \frac{\partial^2 f}{\partial y^2} = -12y^2.$$

At $(0, 0)$, $B^2 - AC = 16 > 0$, and therefore $(0, 0)$ yields a saddle point. At $(1, 1)$, $B^2 - AC = -128$ and $A = -12$, and therefore $(1, 1)$ gives a relative maximum. At $(-1, -1)$, $B^2 - AC = -128$ and $A = -12$, and $(-1, -1)$ also gives a relative maximum.

b) Critical points for $f(x, y)$ are given by

$$0 = \frac{\partial f}{\partial x} = 2x - 6y + 3, \qquad 0 = \frac{\partial f}{\partial y} = 3y^2 - 6x + 6.$$

Solutions of these equations are $(\frac{27}{2}, 5)$ and $(\frac{3}{2}, 1)$. The second derivatives of $f(x, y)$ are

$$\frac{\partial^2 f}{\partial x^2} = 2, \qquad \frac{\partial^2 f}{\partial x\, \partial y} = -6, \qquad \frac{\partial^2 f}{\partial y^2} = 6y.$$

At $(\frac{27}{2}, 5)$, $B^2 - AC = -24$ and $A = 2$, and therefore $(\frac{27}{2}, 5)$ yields a relative minimum. At $(\frac{3}{2}, 1)$, $B^2 - AC = 24$, and therefore $(\frac{3}{2}, 1)$ gives a saddle point.

c) For critical points, we consider

$$0 = \frac{\partial f}{\partial x} = \frac{y^2(12 - x^2 - 2xy)}{2(x + y)^2}, \qquad 0 = \frac{\partial f}{\partial y} = \frac{x^2(12 - y^2 - 2xy)}{2(x + y)^2},$$

or

$$0 = y^2(12 - x^2 - 2xy), \qquad 0 = x^2(12 - y^2 - 2xy).$$

Solutions of these equations are $(2, 2)$ and $(-2, -2)$ (and the solution $(0, 0)$ has been rejected since $f(x, y)$ is not defined there). To classify these critical points by means of the second-derivative test would lead to some very messy calculations; thus we will show that the classification can be achieved without the test. First we note that $f(x, y) = 0$ whenever

$$0 = 12xy - x^2y^2 = xy(12 - xy);$$

i.e., along the x-axis, the y-axis, and the hyperbola $xy = 12$ (except $(0, 0)$; see Fig. 13.39). Further, between the axes and that part of the hyperbola in the first quadrant, the function is always positive. Finally, since $(2, 2)$ is the only critical point in this region, it follows that $(2, 2)$ could only yield a relative maximum. A similar argument in the third quadrant indicates that $(-2, -2)$ gives a relative minimum. □

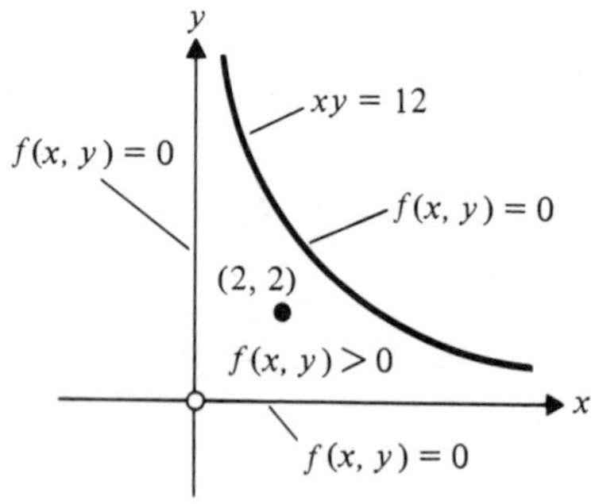

Figure 13.39

This completes our discussion of relative extrema of functions of two independent variables. Our next step should be to extend the theory to functions of more than two variables. It is a simple matter to give definitions of relative maxima and minima for such functions; they are almost identical to Definitions

13.7 and 13.8. On the other hand, to develop a theorem for functions of more than two independent variables that is analogous to Theorem 13.6 is beyond the scope of this book. We refer the interested reader to more advanced books.

EXERCISES 13.10

In Exercises 1–12, find all critical points for the function and classify each as yielding a relative maximum, a relative minimum, a saddle point, or none of these.

1. $f(x, y) = x^2 + 2xy + 2y^2 - 6y$
2. $f(x, y) = 3xy - x^3 - y^3$
3. $f(x, y) = x^3 - 3x + y^2 + 2y$
4. $f(x, y) = x^2y^2 + 3x$
5. $f(x, y) = xy - x^2 + y^2$
6. $f(x, y) = x \sin y$
7. $f(x, y) = xye^{-(x^2+y^2)}$
8. $f(x, y) = x^2 - 2xy + y^2$
9. $f(x, y) = (x^2 + y^2)^{2/3}$
10. $f(x, y) = x^4y^3$
11. $f(x, y) = 2xy^2 + 3xy + x^2y^3$
12. $f(x, y) = |x| + y^2$

In Exercises 13–16, find all critical points for the function.

13. $f(x, y, z) = x^2 + y^2 - z^2 + 3x - 2y + 5$
14. $f(x, y, z, t) = x^2y^2z^2 + t^2x^2 + 3x$
15. $f(x, y, z) = xyz + x^2yz - y$
16. $f(x, y, z) = xyze^{x^2+y^2+z^2}$
17. Suppose that $f(x, y)$ is harmonic in the region D: $x^2 + y^2 < 1$. Show that $f(x, y)$ cannot have a relative maximum or minimum at any point in D at which either f_{xx} or f_{xy} does not vanish.
18. a) Plot the ten points in the following table to show that they are reasonably close to being collinear.

x	y
1	6.05
2	8.32
3	10.74
4	13.43
5	15.90
6	18.38
7	20.93
8	23.32
9	24.91
10	28.36

b) If $y = mx + b$ is the equation of a straight line that we might use to approximate the function $y = f(x)$ described by these points, then the following sum is a measure of how well the line fits the data:

$$S = S(m, b) = \sum_{i=1}^{10} (\bar{y}_i - mx_i - b)^2,$$

where $(x_i, \bar{y}_i)$ are the data points in the table. What does this sum represent geometrically?

c) To find the best straight line to fit the data, the least-squares method suggests that m and b be chosen to minimize S. Show that S has only one critical point (m, b), which is defined by the linear equations

$$\left(\sum_{i=1}^{10} x_i^2\right)m + \left(\sum_{i=1}^{10} x_i\right)b = \sum_{i=1}^{10} x_i\bar{y}_i,$$

$$\left(\sum_{i=1}^{10} x_i\right)m + 10b = \sum_{i=1}^{10} \bar{y}_i.$$

d) Solve these equations for m and b.

19. a) Plot the sixteen points in the following table.

x	y
3.00	31.5
3.25	30.4
3.50	29.2
3.75	28.1
4.00	26.9
4.25	26.4
4.50	25.3
4.75	25.2
5.00	25.1
5.25	25.2
5.50	25.4
5.75	26.3
6.00	27.0
6.25	28.2
6.50	29.3
6.75	29.9

Do they seem to follow a parabolic path?

b) If $y = ax^2 + bx + c$ is the equation of a parabola that is to approximate the function $y = f(x)$ described by these points, then the following sum is a measure of the accuracy of the fit:

$$S = S(a, b, c) = \sum_{i=1}^{16} (\bar{y}_i - ax_i^2 - bx_i - c)^2,$$

where $(x_i, \bar{y}_i)$ are the points in the table. To find the best possible fit, the least-squares method says to choose a, b, and c to minimize S. Show that S has only one critical point (a, b, c), which is defined by the linear equations

$$(\textstyle\sum x_i^4)a + (\sum x_i^3)b + (\sum x_i^2)c = \sum x_i^2 \bar{y}_i,$$
$$(\textstyle\sum x_i^3)a + (\sum x_i^2)b + (\sum x_i)c = \sum x_i \bar{y}_i,$$
$$(\textstyle\sum x_i^2)a + (\sum x_i)b + \quad 16c = \sum \bar{y}_i,$$

where all summations range from $i = 1$ to $i = 16$.

c) Solve these equations for a, b, and c.

20. Find and classify the critical points of $f(x, y) = x^4 + 3xy^2 + y^2$ as yielding relative maxima, relative minima, or saddle points.

13.11 ABSOLUTE MAXIMA AND MINIMA

So far as applications of maxima and minima are concerned, absolute maxima and minima are more important than relative maxima and minima. In this section and in Section 13.12 we discuss the theory of absolute extrema and consider a number of applications. Once again we begin with functions $f(x, y)$ of two independent variables and base our discussion on the theory of absolute extrema for functions of one variable.

We learned in Section 4.7 that a function $f(x)$ that is continuous on a finite interval $a \leq x \leq b$ must have an absolute maximum and an absolute minimum on that interval. Furthermore, these absolute extrema must occur at either critical points or at the ends $x = a$ and $x = b$ of the interval. Consequently, to find the absolute extrema of a function $f(x)$, we evaluate $f(x)$ at all critical points, at $x = a$, and at $x = b$; the largest of these numbers is the absolute maximum of $f(x)$ on $a \leq x \leq b$, and the smallest is the absolute minimum.

The procedure is much the same for a function $f(x, y)$ that is continuous on a region R that is finite and includes all the points on its boundary. First, however, we define exactly what we mean by absolute extrema of $f(x, y)$ and consider a number of simple examples. We will then be able to make general statements about the nature of all absolute extrema, and proceed to the important area of applications.

▷ **Definition 13.10** The absolute maximum of a function $f(x, y)$ on a region R is $f(x_0, y_0)$ if (x_0, y_0) is in R and

$$f(x, y) \leq f(x_0, y_0) \tag{13.55}$$

for all (x, y) in R. The absolute minimum of $f(x, y)$ on R is $f(x_0, y_0)$ if (x_0, y_0) is in R and

$$f(x, y) \geq f(x_0, y_0) \tag{13.56}$$

for all (x, y) in R. ◁

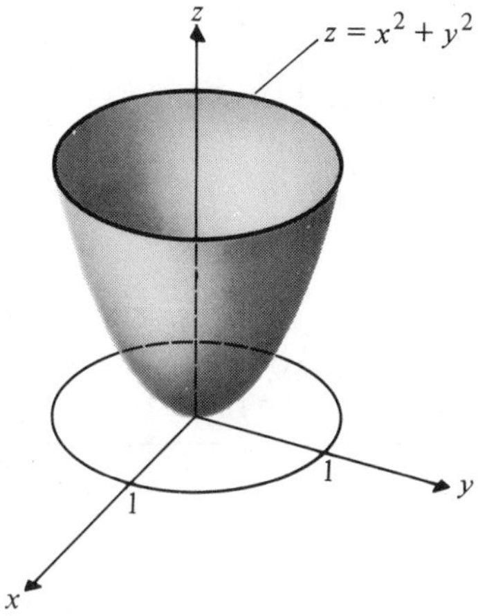

Figure 13.40

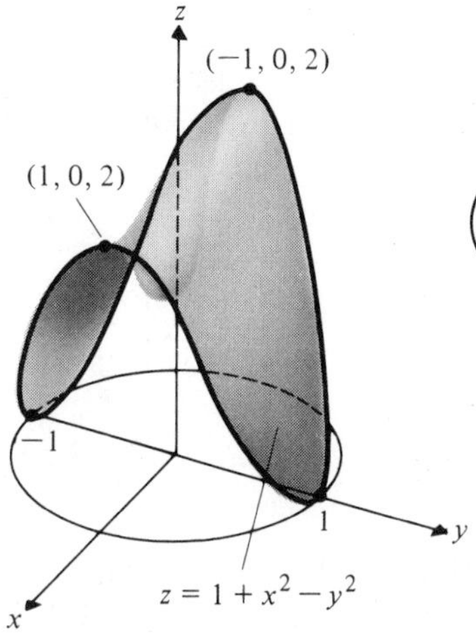

Figure 13.41

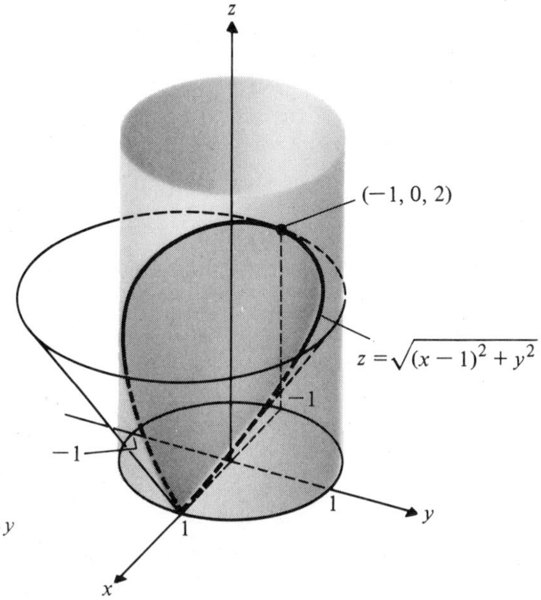

Figure 13.42

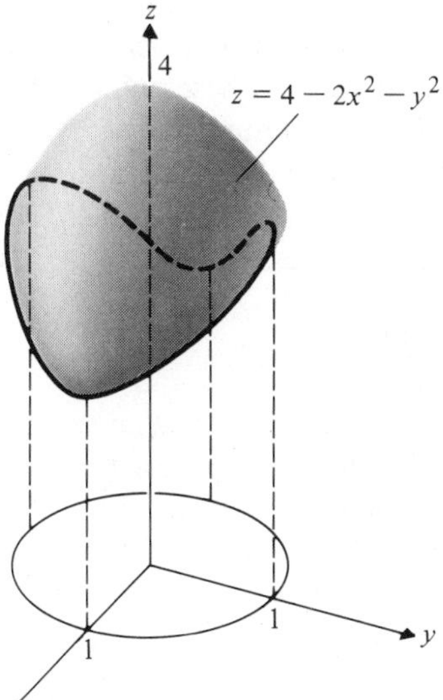

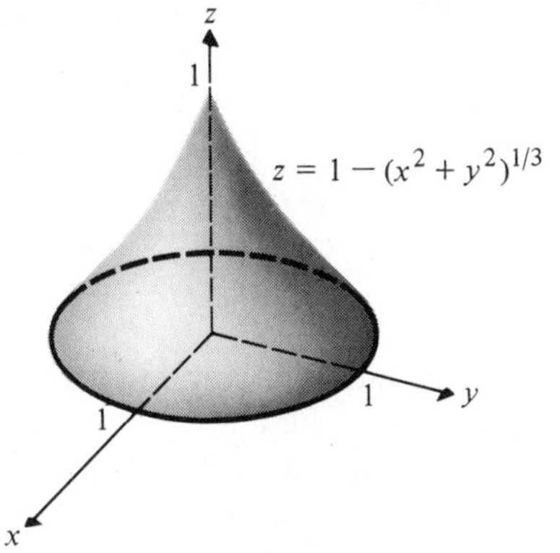

Figure 13.44

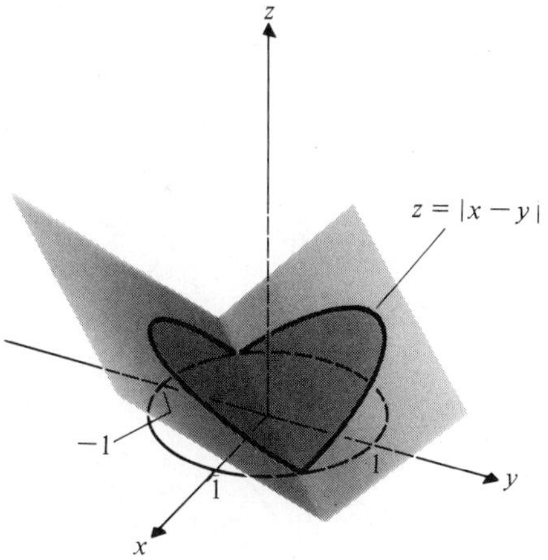

Figure 13.45

x

Figure 13.43

In Figs. 13.40–13.45, we have shown six functions defined on the circle R: $x^2 + y^2 \leq 1$. The absolute maxima and minima of these functions for this region are shown in Table 13.1.

For each of the functions in these figures, *absolute extrema occur at either a critical point or a point on the boundary of R.* This result is in fact true for any *continuous function defined on a finite region that includes all the points on its boundary.* Although this result may seem fairly obvious geometrically, to prove it analytically is very difficult; thus we will be content to assume its validity and carry on from there.

Table 13.1

Function $f(x, y)$	Position of absolute maximum	Value of absolute maximum	Position of absolute minimum	Value of absolute minimum
$x^2 + y^2$	Every point on $x^2 + y^2 = 1$	1	$(0, 0)$	0
$1 + x^2 - y^2$	$(\pm 1, 0)$	2	$(0, \pm 1)$	0
$\sqrt{(x-1)^2 + y^2}$	$(-1, 0)$	2	$(1, 0)$	0
$4 - 2x^2 - y^2$	$(0, 0)$	4	$(\pm 1, 0)$	2
$1 - (x^2 + y^2)^{1/3}$	$(0, 0)$	1	Every point on $x^2 + y^2 = 1$	0
$\lvert x - y\rvert$	$(\pm 1/\sqrt{2}, \mp 1/\sqrt{2})$	$\sqrt{2}$	Every point on $y = x$, $-1/\sqrt{2} \le x \le 1/\sqrt{2}$	0

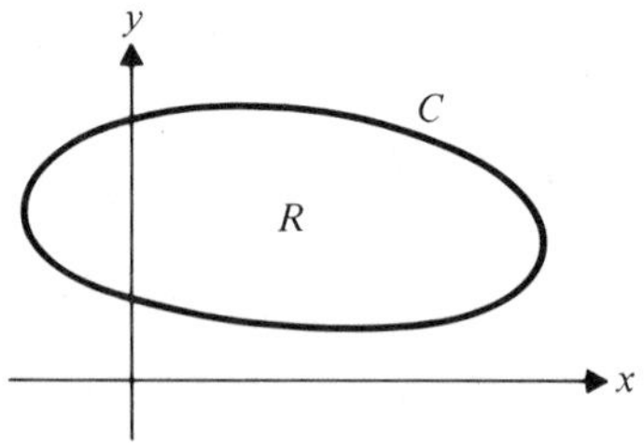

Figure 13.46

It is very simple to look at the surface defined by a function $f(x, y)$ and pick off its absolute extrema, but in practice this just does not happen. Usually we must determine the absolute extrema algebraically from the function itself. Suppose, then, that a continuous function $f(x, y)$ is given and we are required to find its absolute extrema on a finite region R (which includes its boundary points). The above discussion has indicated that the extrema must occur either at critical points or on the boundary of R. Consequently, we should first determine all critical points of $f(x, y)$ in R, and evaluate $f(x, y)$ at each of these points. These values should now be compared to the maximum and minimum values of $f(x, y)$ on the boundary of R. But how do we find the maximum and minimum values of $f(x, y)$ on the boundary? If the boundary of R is denoted by C (Fig. 13.46), and if C has parametric equations $x = x(t)$, $y = y(t)$, $\alpha \le t \le \beta$, then on C we can express $f(x, y)$ in terms of t, and t alone:

$$f[x(t), y(t)], \qquad \alpha \le t \le \beta.$$

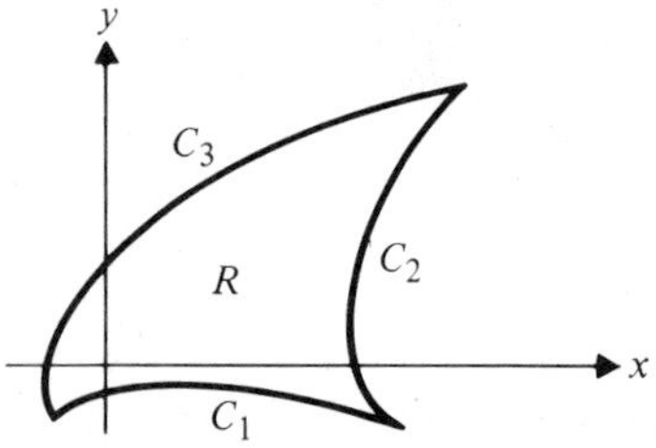

Figure 13.47

To find the maximum and minimum values of $f(x, y)$ on C is now an absolute extrema problem for a function of one variable. The function $f[x(t), y(t)]$ should therefore be evaluated at each of its critical points and at $t = \alpha$ and $t = \beta$.

If the boundary of R consists of a number of curves (Fig. 13.47), then this boundary procedure must be performed for each part. In other words, on each part of the boundary we express $f(x, y)$ as a function of one variable, and then evaluate this function at its critical points and at the ends of that part of the boundary to which it applies.

The absolute maximum of $f(x, y)$ on R is then the largest of all values of $f(x, y)$ evaluated at the critical points inside R, the critical points on the boundary of R, and the endpoints of each part of the boundary. The absolute minimum of $f(x, y)$ on R is the smallest of all these values.

Recall that to find the absolute extrema of a function $f(x)$, continuous on $a \le x \le b$, we evaluate $f(x)$ at all critical points and at the boundary points $x = a$ and $x = b$. The procedure that we have established here for $f(x, y)$ is much the same—the difference is that for $f(x, y)$, the boundary consists not of two points, but of entire curves. Evaluation of $f(x, y)$ on the boundary therefore reduces to one or more extrema problems for functions of one variable. Note too that for $f(x, y)$ (or $f(x)$), it is not necessary to determine the nature of the critical points; it is necessary only to evaluate $f(x, y)$ at these points.

Example 29 Find the maximum value of the function $z = f(x, y) = 4xy - x^4 - 2y^2$ on the region R: $-2 \leq x \leq 2$, $-2 \leq y \leq 2$.

Solution Critical points of $f(x, y)$ are given by

$$0 = \frac{\partial f}{\partial x} = 4y - 4x^3, \qquad 0 = \frac{\partial f}{\partial y} = 4x - 4y.$$

Solutions of these equations are $(0, 0)$, $(1, 1)$, and $(-1, -1)$, and the values of $f(x, y)$ at these critical points are

$$f(0, 0) = 0, \qquad f(1, 1) = 1, \qquad f(-1, -1) = 1.$$

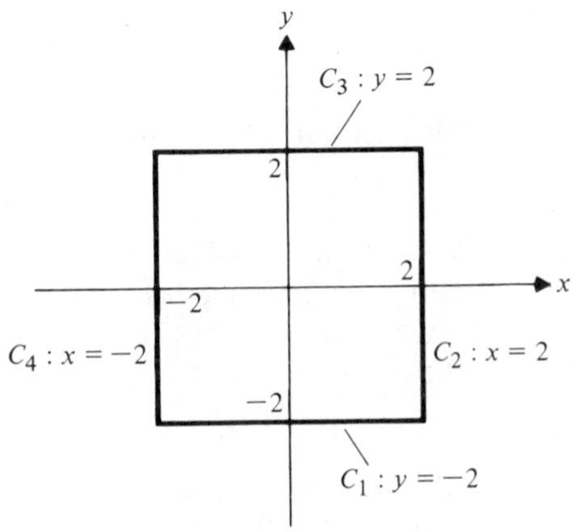

Figure 13.48

We denote the four parts of the boundary of R by C_1, C_2, C_3, and C_4 (Fig. 13.48). On C_1, $y = -2$, in which case

$$z = -8x - x^4 - 8, \qquad -2 \leq x \leq 2.$$

For critical points of this function, we solve

$$0 = \frac{dz}{dx} = -8 - 4x^3.$$

The only solution is $x = -2^{1/3}$, at which the value of z is

$$z = 8 \cdot 2^{1/3} - 2^{4/3} - 8 = -0.44.$$

On C_2, $x = 2$, in which case

$$z = 8y - 16 - 2y^2, \qquad -2 \leq y \leq 2.$$

Critical points are defined by

$$0 = \frac{dz}{dy} = 8 - 4y.$$

The only solution $y = 2$ defines one of the corners of the square, and at this point

$$z = -8.$$

On C_3, $y = 2$ and

$$z = 8x - x^4 - 8, \qquad -2 \le x \le 2.$$

For critical points, we solve

$$0 = \frac{dz}{dx} = 8 - 4x^3.$$

At the single point $x = 2^{1/3}$,

$$z = 8 \cdot 2^{1/3} - 2^{4/3} - 8 = -0.44.$$

On the final curve C_4, $x = -2$ and

$$z = -8y - 16 - 2y^2, \qquad -2 \le y \le 2.$$

Critical points are given by

$$0 = \frac{dz}{dy} = -8 - 4y.$$

The solution $y = -2$ defines another corner of the square at which

$$z = -8.$$

We have now evaluated $f(x, y)$ at all critical points inside R and at all critical points on the four parts of the boundary of R. It remains only to evaluate $f(x, y)$ at the corners of the square. Two corners have already been accounted for; the other two give

$$f(2, -2) = -40, \qquad f(-2, 2) = -40.$$

The largest value of $f(x, y)$ produced is 1, and this is therefore the maximum value of $f(x, y)$ on R. □

Example 30 The temperature at each point (x, y) in a semicircular plate defined by $x^2 + y^2 \le 1$, $y \ge 0$ is given by

$$T(x, y) = 16x^2 - 24xy + 40y^2.$$

Find the hottest and coldest points in the plate.

Solution For critical points of $T(x, y)$, we solve

$$0 = \frac{\partial T}{\partial x} = 32x - 24y, \qquad 0 = \frac{\partial T}{\partial y} = -24x + 80y.$$

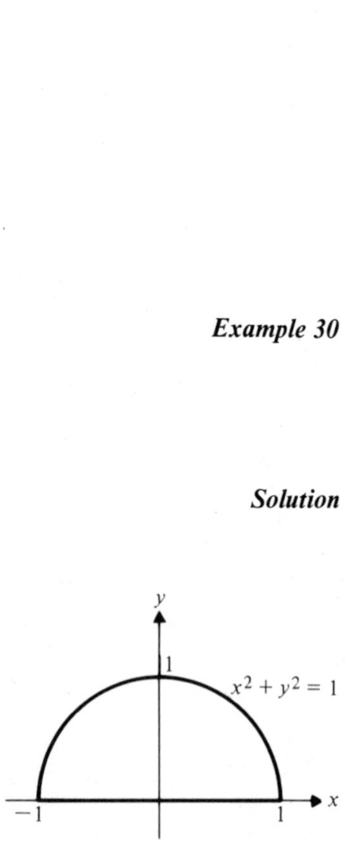

Figure 13.49

The only solution of these equations is $(0, 0)$ at which $T = 0$. On the upper edge of the plate (Fig. 13.49), we set $x = \cos t$, $y = \sin t$, $0 \le t \le \pi$, in which case

$$T = 16\cos^2 t - 24\cos t \sin t + 40\sin^2 t, \qquad 0 \le t \le \pi.$$

For critical points of this function, we solve

$$0 = \frac{dT}{dt} = -32\cos t \sin t - 24(-\sin^2 t + \cos^2 t) + 80\sin t \cos t$$

$$= 24(\sin 2t - \cos 2t).$$

If we divide by cos $2t$ (since $\cos 2t = 0$ does not lead to a solution of this equation), we have

$$\tan 2t = 1.$$

The only solutions of this equation in the interval $0 \le t \le \pi$ are $t = \pi/8$ and $t = 5\pi/8$. When $t = \pi/8$, $T = 11.0$; and when $t = 5\pi/8$, $T = 45.0$. At the ends of this part of the boundary, $t = 0, \pi$; and $T(1, 0) = 16$, $T(-1, 0) = 16$. On the lower edge of the plate, $y = 0$, in which case

$$T = 16x^2, \qquad -1 \le x \le 1.$$

The only critical point of this function is $x = 0$, at which $T = 0$. The hottest point in the plate is therefore $(\cos(5\pi/8), \sin(5\pi/8)) = (-0.38, 0.92)$, where the temperature is 45°, and the coldest point is (0, 0) with temperature 0°. □

Example 31 Find the point on the plane $2x - 3y - 4z = 25$ closest to the point (3, 2, 1).

Solution The distance D from (3, 2, 1) to any point (x, y, z) in space is defined by

$$D^2 = (x-3)^2 + (y-2)^2 + (z-1)^2.$$

Because we want the minimum distance from (3, 2, 1) to points in the plane $2x - 3y - 4z = 25$, we must minimize D but consider only those points (x, y, z) that satisfy the equation of the plane (Fig. 13.50). At the moment, D^2 is a function of three variables x, y, and z, but they are not all independent because of the planar restriction. If we solve the equation of the plane for z in terms of x and y and substitute, then

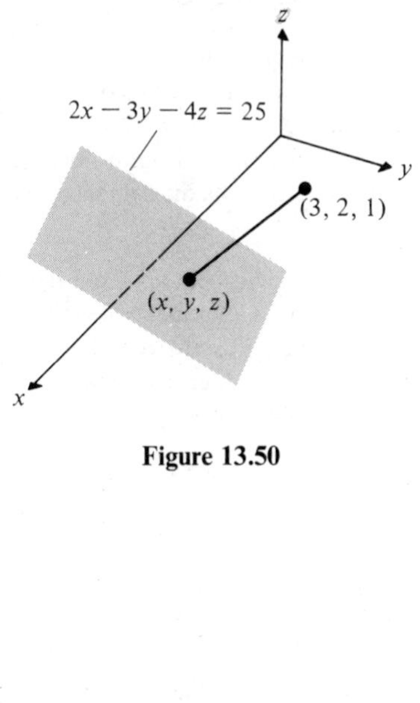

Figure 13.50

$$D^2 = f(x, y) = (x-3)^2 + (y-2)^2 + \left(\frac{2x - 3y - 25}{4} - 1\right)^2$$

$$= (x-3)^2 + (y-2)^2 + \left(\frac{2x - 3y - 29}{4}\right)^2,$$

where x and y are independent variables. Now D is minimized when D^2 is minimized, and we will therefore find that point (x, y) which minimizes D^2. First we locate the critical points of D^2, by solving

$$0 = \frac{\partial f}{\partial x} = 2(x-3) + \left(\frac{2x - 3y - 29}{4}\right)$$

and

$$0 = \frac{\partial f}{\partial y} = 2(y-2) - \frac{3}{2}\left(\frac{2x - 3y - 29}{4}\right).$$

These reduce to

$$10x - 3y = 53, \qquad -6x + 25y = -55.$$

The only solution is $x = 5$, $y = -1$. In this problem we do not have a finite region R in the xy-plane in which D^2 is defined. It is geometrically clear, however, that as x and y become very large, so too does D^2. It follows, therefore, that the one critical point obtained, namely $(5, -1)$, must yield the absolute minimum of D^2. Consequently, the point closest to (3, 2, 1) in the plane $2x - 3y - 4z = 25$ is $(5, -1, -3)$. □

EXERCISES 13.11

In Exercises 1–5, find the maximum and minimum values of the function on the region indicated.

1. $f(x, y) = x^2 + y^3$ on $R: x^2 + y^2 \leq 1$

2. $f(x, y) = x^2 + x + 3y^2 + y$ on the region R bounded by $y = x + 1$, $y = 1 - x$, $y = x - 1$, $y = -x - 1$

3. $f(x, y) = 3x + 4y$ on the region R bounded by the lines $x + y = 1$, $x + y = 4$, $y + 1 = x$, $y - 1 = x$

4. $f(x, y) = x^2y + xy^2 + y$ on $R: -1 \leq x \leq 1, -1 \leq y \leq 1$

5. $f(x, y) = 3x^2 + 2xy - y^2 + 5$ on $R: 4x^2 + 9y^2 \leq 36$

6. Find the point on the plane $x + y - 2z = 6$ closest to the origin.

7. Find the shortest distance from $(-1, 1, 2)$ to the plane $2x - 3y + 6z = 14$.

8. Find the point on the surface $z = x^2 + y^2$ closest to the point $(1, 1, 0)$.

9. Find the point on that part of the plane $x + y + 2z = 4$ in the first octant that is closest to the point $(3, 3, 1)$. For this question assume that the curves of intersection of the plane with the coordinate planes are part of the surface.

10. The electrostatic potential at each point in the region $0 \leq x \leq 1, 0 \leq y \leq 1$, is given by

$$V(x, y) = 48xy - 32x^3 - 24y^2.$$

Find the maximum and minimum potentials in the region.

11. When a rectangular box is sent through the mail, the post office demands that the length of the box plus twice the sum of its height and width be no more than 250 centimeters. Find the dimensions of the box satisfying this requirement that encloses the largest possible volume.

12. An open tank in the form of a rectangular parallelepiped is to be built to hold 1000 liters of acid. If the cost of lining the base of the tank is three times that of the sides, what dimensions will minimize the cost of lining the tank?

13. Find the point on the curve

$$x^2 - xy + y^2 - z^2 = 1,$$
$$x^2 + y^2 = 1$$

closest to the origin.

14. Find the dimensions of the box with largest possible volume that can be fitted inside the ellipsoid $x^2/a^2 + y^2/b^2 + z^2/c^2 = 1$, assuming that its edges are parallel to the coordinate axes.

15. Find the maximum and minimum values of the function $f(x, y, z) = xyz$ on the sphere $x^2 + y^2 + z^2 = 1$.

16. If P is the perimeter of a triangle with sides of lengths x, y, and z, then the area of the triangle is

$$A = \sqrt{\frac{P}{2}\left(\frac{P}{2} - x\right)\left(\frac{P}{2} - y\right)\left(\frac{P}{2} - z\right)},$$

where $P = x + y + z$. Show that A is maximized for fixed P when the triangle is equilateral.

17. Show that for any triangle with interior angles A, B, and C,

$$\sin(A/2)\sin(B/2)\sin(C/2) \leq \tfrac{1}{8}.$$

Hint: Find the maximum value of the function

$$f(A, B, C) = \sin(A/2)\sin(B/2)\sin(C/2).$$

18. A silo is in the shape of a right-circular cylinder surmounted by a right-circular cone. If the radius of each is 6 meters and the total surface area must be 200 square meters (not including the base), what heights for the cone and cylinder will yield the maximum enclosed volume?

19. A long piece of metal 1 meter wide is bent at A and B, as shown in Fig. 13.51, to form a channel with three straight sides. If the bends occur equidistant from the ends, where should they be made in order to obtain the maximum possible flow of fluid along the channel?

***20.** Find the maximum and minimum values of the function $f(x, y, z) = xy + xz$ on the region $x^2 + y^2 + z^2 \leq 1$.

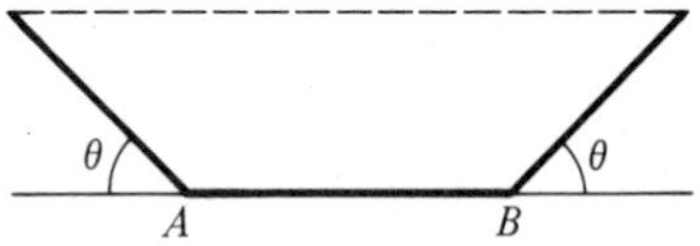

Figure 13.51

***21.** Find the maximum and minimum values of the function $f(x, y, z) = x^2yz$ on the region $x^2 + y^2 \leq 1, 0 \leq z \leq 1$.

***22.** A cow's daily diet consists of three foods: hay, grain, and supplements. The cow is always given 11 kilograms of hay per day, 50% of which is digestive material and 12% of which is protein. Grain is 74% digestive material and 8.8% protein, whereas supplements are 62% digestive material and 34% protein. The cost of hay is \$27.50 for 1000 kilograms, and grain and supplements cost \$110 and

\$175, respectively, for 1000 kilograms. A healthy cow's daily diet must contain between 9.5 and 11.5 kilograms of digestive material and between 1.9 and 2.0 kilograms of protein. Determine the daily amounts of grain and supplements that the cow should be fed in order that total food cost be kept to a minimum.

13.12 LAGRANGE MULTIPLIERS

Many applied maxima and minima problems result in *constraint problems.* In particular, Examples 30 and 31 contain such problems. In Example 30, to find extreme values of T on the edge of the plate we maximized and minimized $T(x, y) = 16x^2 - 24x + 40y^2$ subject to first the constraint $x^2 + y^2 = 1$, and then the constraint $y = 0$. Our method there was to substitute from the constraint equation into $T(x, y)$ in order to obtain a function of one variable. In Example 31, to find the minimum distance from (3, 2, 1) to the plane $2x - 3y - 4z = 25$, we minimized $D^2 = (x - 3)^2 + (y - 2)^2 + (z - 1)^2$ subject to the constraint $2x - 3y - 4z = 25$. Again we substituted from the constraint to obtain D^2 as a function of two independent variables.

A natural question to ask is whether problems of this type can be solved without substituting from the constraint equation, for if the constraint equation is at all complicated, substitution may be very difficult or even impossible. To show that there is indeed an alternative, consider the situation in which a function $f(x, y, z)$ is to be maximized or minimized subject to two constraints:

$$F(x, y, z) = 0, \tag{13.57a}$$

$$G(x, y, z) = 0. \tag{13.57b}$$

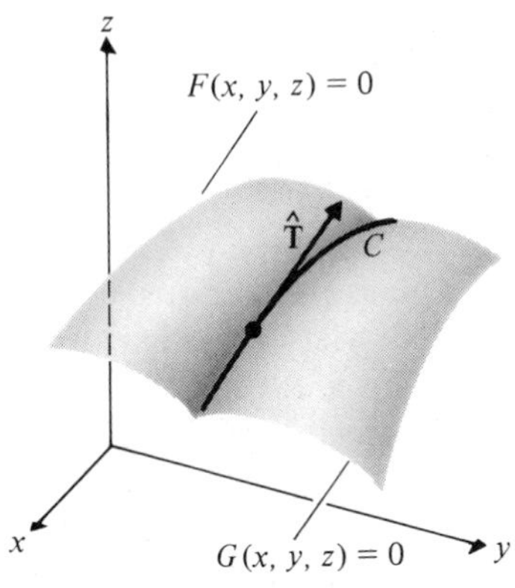

Figure 13.52

Algebraically, we are to find extreme values of $f(x, y, z)$, considering only those values of x, y, and z that satisfy the two equations (13.57). Geometrically, we can interpret each of these conditions as specifying a surface, so that we are seeking extreme values of $f(x, y, z)$, considering only those points on the curve of intersection C of the surfaces $F(x, y, z) = 0$ and $G(x, y, z) = 0$ (Fig. 13.52). Extreme values of $f(x, y, z)$ along C will occur either at critical points of the function or at the ends of the curve. But what derivative or derivatives of $f(x, y, z)$ are we talking about when we say critical points? Since we are concerned only with values of $f(x, y, z)$ on C, we must mean the derivative of $f(x, y, z)$ along C; i.e., the directional derivative in the tangent direction to C. If $\mathbf{T}$, then, is a tangent vector to C, critical points of $f(x, y, z)$ along C are given by

$$0 = D_{\mathbf{T}}f = \nabla f \cdot \hat{\mathbf{T}},$$

or at points where the directional derivative is undefined. According to Eq. (13.50), a tangent vector to C is $\mathbf{T} = \nabla F \times \nabla G$, and hence a unit tangent vector is

$$\hat{\mathbf{T}} = \frac{\nabla F \times \nabla G}{|\nabla F \times \nabla G|}.$$

The directional derivative of $f(x, y, z)$ at points along C is therefore given by

$$D_{\mathbf{T}}f = \nabla f \cdot \frac{\nabla F \times \nabla G}{|\nabla F \times \nabla G|}.$$

It follows, then, that critical points of $f(x, y, z)$ are points (x, y, z) that satisfy the equation

$$\nabla f \cdot \nabla F \times \nabla G = 0,$$

or points at which the left side is not defined. But the scalar triple product of three vectors vanishes if and only if the vectors are coplanar (see Exercise 49 in Section 12.6). Provided, then, that ∇F and ∇G are not parallel, we can write ∇f in terms of them; i.e., there exist scalars λ and μ such that

$$\nabla f = (-\lambda)\nabla F + (-\mu)\nabla G,$$

or

$$\nabla f + \lambda \nabla F + \mu \nabla G = \mathbf{0}. \tag{13.58}$$

This vector equation is equivalent to the three scalar equations

$$\frac{\partial f}{\partial x} + \lambda \frac{\partial F}{\partial x} + \mu \frac{\partial G}{\partial x} = 0, \tag{13.59a}$$

$$\frac{\partial f}{\partial y} + \lambda \frac{\partial F}{\partial y} + \mu \frac{\partial G}{\partial y} = 0, \tag{13.59b}$$

$$\frac{\partial f}{\partial z} + \lambda \frac{\partial F}{\partial z} + \mu \frac{\partial G}{\partial z} = 0, \tag{13.59c}$$

and these equations must be satisfied at a critical point at which the directional derivative of $f(x, y, z)$ vanishes. Note too that at a point at which the directional derivative of $f(x, y, z)$ does not exist, one of the partial derivatives in these equations does not exist. In other words, we have shown that critical points of $f(x, y, z)$ are points that satisfy Eqs. (13.59) or points at which the equations are undefined. These equations, however, contain five unknowns: x, y, z, λ, and μ. To complete the system we add Eqs. (13.57) since they must also be satisfied by a critical point. Equations (13.57) and (13.59) therefore yield a system of five equations in the five unknowns: x, y, z, λ, and μ; the first three unknowns (x, y, z) define a critical point of $f(x, y, z)$ along C. The advantage of this system of equations lies in the fact that the differentiations in (13.59) involve only the given functions (and no substitutions from the constraint equations are necessary). What we have sacrificed is a system of three equations in the three unknowns (x, y, z) for a system of five equations in the five unknowns (x, y, z, λ, μ).

Let us not forget that the original problem was to find extreme values for the function $f(x, y, z)$ subject to constraints (13.57). What we have shown so far is that critical points at which the directional derivative of $f(x, y, z)$ vanishes can be found by solving Eqs. (13.57) and (13.59). In addition, critical points at which the directional derivative of $f(x, y, z)$ does not exist are points at which Eqs. (13.59) are not defined. What remains is to evaluate $f(x, y, z)$ at all critical points and at the ends of C. *If C is a closed curve* (i.e., if C rejoins itself), then $f(x, y, z)$ *need be evaluated only at the critical points.*

Through the directional derivative and tangent vectors to curves, we have shown that Eqs. (13.57) and (13.59) define critical points of a function $f(x, y, z)$ that is subject to two constraints: $F(x, y, z) = 0$ and $G(x, y, z) = 0$. But what about other situations? Let us say, for example, that we require extreme values of a function $f(x, y, z, t)$ subject to a single constraint $F(x, y, z, t) = 0$. How shall we find critical points of this function? Fortunately, as we now show, there is a very simple method that yields Eqs. (13.57) and (13.59), and this method generalizes to other situations also.

To find critical points of $f(x, y, z)$ subject to constraints (13.57), we define a function

$$L(x, y, z, \lambda, \mu) = f(x, y, z) + \lambda F(x, y, z) + \mu G(x, y, z),$$

and regard it as a function of five independent variables x, y, z, λ, and μ. To find critical points of this function, we would first solve the equations obtained by setting each of the partial derivatives of L equal to zero:

$$0 = \frac{\partial L}{\partial x} = \frac{\partial f}{\partial x} + \lambda \frac{\partial F}{\partial x} + \mu \frac{\partial G}{\partial x},$$

$$0 = \frac{\partial L}{\partial y} = \frac{\partial f}{\partial y} + \lambda \frac{\partial F}{\partial y} + \mu \frac{\partial G}{\partial y},$$

$$0 = \frac{\partial L}{\partial z} = \frac{\partial f}{\partial z} + \lambda \frac{\partial F}{\partial z} + \mu \frac{\partial G}{\partial z},$$

$$0 = \frac{\partial L}{\partial \lambda} = F(x, y, z),$$

$$0 = \frac{\partial L}{\partial \mu} = G(x, y, z).$$

In addition, we would consider points at which the partial derivatives of L do not exist. Clearly, this means points (x, y, z) at which any of the partial derivatives of $f(x, y, z)$, $F(x, y, z)$, and $G(x, y, z)$ do not exist. But these are precisely Eqs. (13.57) and (13.59). We have shown, then, that finding critical points (x, y, z) of $f(x, y, z)$ subject to $F(x, y, z) = 0$ and $G(x, y, z) = 0$ is equivalent to finding critical points (x, y, z, λ, μ) of $L(x, y, z, \lambda, \mu)$. The two unknowns λ and μ that accompany a critical point (x, y, z) of $f(x, y, z)$ are called *Lagrange multipliers*. They are not a part of the solution (x, y, z) to the original problem, but have been introduced as a convenience by which to arrive at that solution. The function $L(x, y, z, \lambda, \mu)$ is often called the Lagrangian of the problem.

The method for other constraint problems should now be evident. Given a function $f(x, y, z, t, \ldots)$ of n variables to maximize or minimize subject to m constraints

$$F_1(x, y, z, t, \ldots) = 0, \qquad F_2(x, y, z, t, \ldots) = 0, \ldots, F_m(x, y, z, t, \ldots) = 0, \tag{13.60}$$

we introduce m Lagrange multipliers $\lambda_1, \lambda_2, \ldots, \lambda_m$ into a Lagrangian of $n + m$ independent variables $x, y, z, t, \ldots, \lambda_1, \lambda_2, \ldots, \lambda_m$:

$$L(x, y, z, t, \ldots, \lambda_1, \lambda_2, \ldots, \lambda_m) = f(x, y, z, t, \ldots) + \lambda_1 F_1(x, y, z, t, \ldots) + \cdots + \lambda_m F_m(x, y, z, t, \ldots). \tag{13.61}$$

Critical points $(x, y, z, t, \ldots)$ of $f(x, y, z, t, \ldots)$ are then determined by the equations defining critical points of $L(x, y, z, t, \ldots, \lambda_1, \lambda_2, \ldots, \lambda_m)$, namely,

$$0 = \frac{\partial L}{\partial x} = \frac{\partial f}{\partial x} + \lambda_1 \frac{\partial F_1}{\partial x} + \cdots + \lambda_m \frac{\partial F_m}{\partial x}, \tag{13.62a}$$

$$0 = \frac{\partial L}{\partial y} = \frac{\partial f}{\partial y} + \lambda_1 \frac{\partial F_1}{\partial y} + \cdots + \lambda_m \frac{\partial F_m}{\partial y}, \tag{13.62b}$$

$$\vdots \qquad\qquad \vdots \qquad\qquad \vdots$$

$$0 = \frac{\partial L}{\partial \lambda_1} = F_1(x, y, z, t, \ldots), \tag{13.62c}$$

$$0 = \frac{\partial L}{\partial \lambda_2} = F_2(x, y, z, t, \ldots), \tag{13.62d}$$

$$\vdots \qquad\qquad \vdots$$

$$0 = \frac{\partial L}{\partial \lambda_m} = F_m(x, y, z, t, \ldots). \tag{13.62e}$$

To use Lagrange multipliers in Example 31, we define the Lagrangian

$$\begin{aligned} L(x, y, z, \lambda) &= D^2 + \lambda(2x - 3y - 4z - 25) \\ &= (x - 3)^2 + (y - 2)^2 + (z - 1)^2 + \lambda(2x - 3y - 4z - 25). \end{aligned}$$

Critical points of $L(x, y, z, \lambda)$ are defined by

$$0 = \frac{\partial L}{\partial x} = 2(x - 3) + 2\lambda,$$

$$0 = \frac{\partial L}{\partial y} = 2(y - 2) - 3\lambda,$$

$$0 = \frac{\partial L}{\partial z} = 2(z - 1) - 4\lambda,$$

$$0 = \frac{\partial L}{\partial \lambda} = 2x - 3y - 4z - 25.$$

The solution of this linear system is $(x, y, z, \lambda) = (5, -1, -3, -2)$, yielding as before the critical point $(5, -1, -3)$ of D^2.

Example 32 Find the maximum value of the function

$$f(x, y, z) = 2x^2y^2 + 2y^2z^2 + 3z,$$

considering only those values of x, y, and z that satisfy the equations

$$z = x^2 + y^2, \qquad x^2 + 3y^2 = 1.$$

Solution For this example we illustrate the difference between a solution that utilizes Lagrange multipliers and one that does not. The solution that does not use Lagrange multipliers requires us first to express $f(x, y, z)$ in terms of a single independent variable. We can do this by solving the constraint equations for

$x^2 = 1 - 3y^2$ and $y^2 = (1 - z)/2$ and substituting into $f(x, y, z)$:

$$\begin{aligned} f(x, y, z) &= 2(1 - 3y^2)y^2 + 2y^2z^2 + 3z \\ &= 2\left(1 - \frac{3}{2} + \frac{3z}{2}\right)\left(\frac{1}{2} - \frac{z}{2}\right) + 2z^2\left(\frac{1}{2} - \frac{z}{2}\right) + 3z \\ &= \frac{1}{2}(-2z^3 - z^2 + 10z - 1). \end{aligned}$$

Since $z = x^2 + y^2 = (1 - 3y^2) + y^2 = 1 - 2y^2$, and y must be restricted to $|y| \leq 1/\sqrt{3}$, it follows that the only possible values for z are $\frac{1}{3} \leq z \leq 1$. We have shown, then, that maximizing $f(x, y, z)$ subject to the two constraints $z = x^2 + y^2$ and $x^2 + 3y^2 = 1$ is equivalent to maximizing

$$F(z) = \frac{1}{2}(-2z^3 - z^2 + 10z - 1), \qquad \frac{1}{3} \leq z \leq 1.$$

For critical points of $F(z)$, we solve

$$0 = \frac{dF}{dz} = \frac{1}{2}(-6z^2 - 2z + 10),$$

from which we get

$$z = \frac{2 \pm \sqrt{4 + 240}}{-12} = \frac{-1 \pm \sqrt{61}}{6}.$$

These two critical points must be rejected as not lying within the interval $\frac{1}{3} \leq z \leq 1$, and the maximum value of $F(z)$ must therefore occur at either $z = \frac{1}{3}$ or $z = 1$. Since $F(\frac{1}{3}) = \frac{29}{27}$ and $F(1) = 3$, it follows that the maximum value of $F(z)$ and therefore of $f(x, y, z)$ is 3.

To maximize $f(x, y, z)$ using Lagrange multipliers, we define the Lagrangian

$$L(x, y, z, \lambda, \mu) = 2x^2y^2 + 2y^2z^2 + 3z + \lambda(x^2 + y^2 - z) + \mu(x^2 + 3y^2 - 1).$$

Critical points of L and therefore of $f(x, y, z)$ are defined by the equations

$$0 = \frac{\partial L}{\partial x} = 4xy^2 + 2\lambda x + 2\mu x = 2x(2y^2 + \lambda + \mu),$$

$$0 = \frac{\partial L}{\partial y} = 4x^2y + 4yz^2 + 2\lambda y + 6\mu y = 2y(2x^2 + 2z^2 + \lambda + 3\mu),$$

$$0 = \frac{\partial L}{\partial z} = 4y^2z + 3 - \lambda,$$

$$0 = \frac{\partial L}{\partial \lambda} = x^2 + y^2 - z,$$

$$0 = \frac{\partial L}{\partial \mu} = x^2 + 3y^2 - 1.$$

If we choose $x = 0$ to satisfy the first equation, then the remaining equations imply that $y = \pm 1/\sqrt{3}$, $z = \frac{1}{3}$ (and $\lambda = \frac{31}{9}$, $\mu = -\frac{11}{9}$). If we choose $y = 0$ to satisfy the second equation, then the remaining equations require $x = \pm 1$, $z = 1$ (and $\lambda = 3$, $\mu = -3$). The only other way to satisfy the first two equations is to set

$$2y^2 + \lambda + \mu = 0,$$
$$2x^2 + 2z^2 + \lambda + 3\mu = 0.$$

If we multiply the first of these by 3 and subtract the second, we have

$$\begin{aligned} 0 &= 6y^2 - 2x^2 - 2z^2 + 2\lambda \\ &= 6y^2 - 2x^2 - 2(x^2 + y^2)^2 + 2\lambda \\ &= 6y^2 - 2x^2 - 2(x^4 + 2x^2y^2 + y^4) + 2\lambda \\ &= 6y^2 - 2(1 - 3y^2) - 2(1 - 3y^2)^2 \\ &\quad -4y^2(1 - 3y^2) - 2y^4 + 2\lambda \\ &= 2(\lambda - 2 + 10y^2 - 4y^4). \end{aligned}$$

But from the equation for $\partial L/\partial z$, we can also write

$$\begin{aligned} 0 &= 3 - \lambda + 4y^2(x^2 + y^2) \\ &= 3 - \lambda + 4y^2(1 - 3y^2 + y^2) \\ &= 3 - \lambda + 4y^2 - 8y^4. \end{aligned}$$

These two equations in y and λ imply that

$$2 - 10y^2 + 4y^4 = 3 + 4y^2 - 8y^4, \quad \text{or} \quad 12y^4 - 14y^2 - 1 = 0.$$

Thus,

$$y^2 = \frac{14 \pm \sqrt{196 + 48}}{24} = \frac{7 \pm \sqrt{61}}{12},$$

where we reject the negative solution. But substitution of this result into the constraint $x^2 + 3y^2 = 1$ requires x^2 to be negative. Consequently, only four critical points (x, y, z) are obtained: $(0, \pm 1/\sqrt{3}, \frac{1}{3})$ and $(\pm 1, 0, 1)$. Since the curve defined by the constraints is closed, we need evaluate $f(x, y, z)$ only at these critical points:

$$f(0, \pm 1/\sqrt{3}, \tfrac{1}{3}) = 2(\tfrac{1}{3})(\tfrac{1}{9}) + 3(\tfrac{1}{3}) = \tfrac{29}{27},$$
$$f(\pm 1, 0, 1) = 3.$$

The maximum value of $f(x, y, z)$ is again 3.

Note that the endpoints in the first solution are critical points in the Lagrangian solution. This emphasizes the fact that, as we pointed out earlier, when the curve is closed and we use a Lagrangian, it is necessary to evaluate the function to be maximized or minimized only at its critical points. □

Example 33 Find the maximum and minimum values of the function $f(x, y, z) = xyz$ subject to the constraint $x^2 + y^2 + z^2 = 1$.

Solution If we define the Lagrangian

$$L(x, y, z, \lambda) = xyz + \lambda(x^2 + y^2 + z^2 - 1),$$

then critical points of L, and therefore of $f(x, y, z)$, are defined by the equations

$$0 = \frac{\partial L}{\partial x} = yz + 2\lambda x,$$

$$0 = \frac{\partial L}{\partial y} = xz + 2\lambda y,$$

$$0 = \frac{\partial L}{\partial z} = xy + 2\lambda z,$$

$$0 = \frac{\partial L}{\partial \lambda} = x^2 + y^2 + z^2 - 1.$$

If we multiply the first equation by y and the second by x, and equate the resulting expressions for $2\lambda xy$, we have

$$y^2 z = x^2 z.$$

Consequently, either $z = 0$ or $y = \pm x$.

Case I: $z = 0$. In this case the equations reduce to

$$\lambda x = 0, \qquad \lambda y = 0, \qquad xy = 0, \qquad x^2 + y^2 = 1.$$

The first implies that either $x = 0$ or $\lambda = 0$. If $x = 0$, then $y = \pm 1$, and we have two critical points $(0, \pm 1, 0)$. If $\lambda = 0$, then the third equation requires $x = 0$ or $y = 0$. We therefore obtain two additional critical points $(\pm 1, 0, 0)$.

Case II: $y = x$. In this case the equations reduce to

$$xz + 2\lambda x = 0, \qquad x^2 + 2\lambda z = 0, \qquad 2x^2 + z^2 = 1.$$

The first implies that either $x = 0$ or $z = -2\lambda$. If $x = 0$, then $z = \pm 1$, and we have the two critical points $(0, 0, \pm 1)$. If $z = -2\lambda$, then the last two equations imply that $x = \pm 1/\sqrt{3}$, and we obtain the four critical points

$$(\pm 1/\sqrt{3}, \pm 1/\sqrt{3}, 1/\sqrt{3}) \quad \text{and} \quad (\pm 1/\sqrt{3}, \pm 1/\sqrt{3}, -1/\sqrt{3}).$$

Case III: $y = -x$. This case is similar to that for $y = x$, and leads to the additional four critical points

$$(\pm 1/\sqrt{3}, \mp 1/\sqrt{3}, 1/\sqrt{3}) \quad \text{and} \quad (\pm 1/\sqrt{3}, \mp 1/\sqrt{3}, -1/\sqrt{3}).$$

Because $x^2 + y^2 + z^2 = 1$ is a surface without a boundary, we complete the problem by evaluating $f(x, y, z)$ at each of the critical points:

$$f(\pm 1, 0, 0,) = f(0, \pm 1, 0) = f(0, 0, \pm 1) = 0,$$

$$f(\pm 1/\sqrt{3}, \pm 1/\sqrt{3}, 1/\sqrt{3}) = f(\pm 1/\sqrt{3}, \mp 1/\sqrt{3}, -1/\sqrt{3}) = \sqrt{3}/9,$$

$$f(\pm 1/\sqrt{3}, \pm 1/\sqrt{3}, -1/\sqrt{3}) = f(\pm 1/\sqrt{3}, \mp 1/\sqrt{3}, 1/\sqrt{3}) = -\sqrt{3}/9.$$

The maximum and minimum values of $f(x, y, z)$ on $x^2 + y^2 + z^2 = 1$ are therefore $\sqrt{3}/9$ and $-\sqrt{3}/9$. □

To compare the Lagrangian solution in this example to that without a Lagrange multiplier, see Exercise 15 in Section 13.11.

Example 34 When a thermonuclear reactor is built in the form of a right-circular cylinder, neutron diffusion theory requires its radius and height to satisfy the equation

$$\left(\frac{2.4048}{r}\right)^2 + \left(\frac{\pi}{h}\right)^2 = k = \text{constant}.$$

Find r and h in terms of k if the reactor is to occupy as small a volume as possible.

Solution The volume of a right-circular cylinder is $V = \pi r^2 h$, and were there no constraints on r and h, this function would be considered for all points in the first quadrant of the rh-plane. However, r and h must satisfy a constraint that geometrically can be interpreted as a curve in the rh-plane. What we must do then is minimize $V = \pi r^2 h$, considering only those points (r, h) on the curve defined by the constraint. Clearly there is only one independent variable in the problem—either r or h, but not both. If we choose r as the independent variable, then we note from the constraint that as h becomes very large, r approaches $2.4048/\sqrt{k}$. Since there is no upper bound on r, we can state that the values of r to be considered in the minimization of V are $r > 2.4048/\sqrt{k}$.

To find critical points of V we introduce the Lagrangian

$$L(r, h, \lambda) = \pi r^2 h + \lambda\left\{\left(\frac{2.4048}{r}\right)^2 + \left(\frac{\pi}{h}\right)^2 - k\right\},$$

and first solve the equations

$$0 = \frac{\partial L}{\partial r} = 2\pi r h + \lambda\left(\frac{-2(2.4048)^2}{r^3}\right),$$

$$0 = \frac{\partial L}{\partial h} = \pi r^2 + \lambda\left(\frac{-2\pi^2}{h^3}\right),$$

$$0 = \frac{\partial L}{\partial \lambda} = \left(\frac{2.4048}{r}\right)^2 + \left(\frac{\pi}{h}\right)^2 - k.$$

If we solve each of the first two equations for λ and equate the resulting expressions, we have

$$\frac{\pi r^4 h}{2.4048^2} = \frac{r^2 h^3}{2\pi}.$$

Since neither r nor h can be zero, we divide by $r^2 h$:

$$\frac{\pi r^2}{2.4048^2} = \frac{h^2}{2\pi},$$

from which

$$r = \frac{2.4048h}{\sqrt{2}\pi}.$$

Substitution of this result into the constraint equation gives

$$\left(\frac{\sqrt{2}\pi}{h}\right)^2 + \left(\frac{\pi}{h}\right)^2 = k,$$

and this equation can be solved for $h = \pi\sqrt{3/k}$. This gives

$$r = \frac{2.4048}{\sqrt{2\pi}} \frac{\pi\sqrt{3}}{\sqrt{k}}$$
$$= 2.4048\sqrt{3/(2k)}.$$

We have obtained therefore only one critical point (r, h) at which the derivatives of L vanish. The only values of r and h at which the derivatives of L do not exist are $r = 0$ and $h = 0$, but these must be rejected since the constraint requires both r and h to be positive.

To finish the problem we note that

$$\lim_{r\to\infty} V = \infty, \qquad \lim_{r\to 2.4048/\sqrt{k}^+} V = \lim_{h\to\infty} V = \infty.$$

It follows, therefore, that the single critical point at which $r = 2.4048\sqrt{3/(2k)}$ and $h = \pi\sqrt{3/k}$ must give the absolute minimum value of $V(r, h)$. □

EXERCISES 13.12

In Exercises 1–8, use Lagrange multipliers to find the maximum and minimum values of the function subject to the given constraints. In each case, interpret the constraints geometrically.

1. $f(x, y) = x^2 + y$ subject to $x^2 + y^2 = 4$
2. $f(x, y, z) = 5x - 2y + 3z + 4$ subject to $x^2 + 2y^2 + 4z^2 = 9$
3. $f(x, y) = x + y$ subject to $(x - 1)^2 + y^2 = 1$
4. $f(x, y, z) = x^3 + y^3 + z^3$ subject to $x^2 + y^2 + z^2 = 9$
5. $f(x, y, z) = xyz$ subject to $x^2 + 2y^2 + 3z^2 = 12$
6. $f(x, y, z) = x^2y + z$ subject to $x^2 + y^2 = 1$, $z = y$
7. $f(x, y, z) = x^2 + y^2 + z^2$ subject to $x^2 + y^2 + z^2 = 2z$, $x + y + z = 1$
8. $f(x, y, z) = xyz - x^2z$ subject to $x^2 + y^2 = 1$, $z = \sqrt{x^2 + y^2}$
9. Use Lagrange multipliers to solve Exercise 7 in Section 13.11.
10. Use Lagrange multipliers to solve Exercise 8 in Section 13.11.
11. Use Lagrange multipliers to solve Exercise 14 in Section 13.11.
12. Use Lagrange multipliers to solve Exercise 16 in Section 13.11.
13. Use Lagrange multipliers to solve Exercise 17 in Section 13.11.
14. Use Lagrange multipliers to solve Exercise 18 in Section 13.11.
15. Suppose that $F(x, y) = 0$ and $G(x, y) = 0$ define two curves C_1 and C_2 in the xy-plane. Let $P(x_0, y_0)$ and $Q(X_0, Y_0)$ be the points on C_1 and C_2 that minimize the distance between C_1 and C_2. If C_1 and C_2 have tangent lines at P and Q, show that the line PQ is perpendicular to these tangent lines.

In Exercises 16–18, use Lagrange multipliers to find the maximum and minimum values of the given function.

16. $f(x, y) = 3x^2 + 2xy - y^2 + 5$ for $4x^2 + 9y^2 \leq 36$
17. $f(x, y) = x^2y + xy^2 + y$ for $-1 \leq x \leq 1$, $-1 \leq y \leq 1$
18. $f(x, y, z) = xy + xz$ for $x^2 + y^2 + z^2 \leq 1$
19. Find the maximum value of $f(x, y, z) = x^2yz - xzy^2$ subject to the constraints $x^2 + y^2 = 1$, $z = \sqrt{x^2 + y^2}$.

*20. To find the point on the curve

$$x^2 - xy + y^2 - z^2 = 1, \qquad x^2 + y^2 = 1$$

closest to the origin, we must minimize the function

$$f(x, y, z) = x^2 + y^2 + z^2$$

subject to the constraints defined by the equations of the curve. Show that this can be done by:

a) using two Lagrange multipliers;
b) expressing $f(x, y, z)$ in terms of x and y alone, $u = 1 - xy$, and minimizing this function subject to

$x^2 + y^2 = 1$ (with, therefore, one Lagrange multiplier);

c) expressing $f(x, y, z)$ in terms of x alone, $u = 1 \pm x\sqrt{1 - x^2}$, and minimizing these functions on appropriate intervals;

d) writing $x = \cos t, y = \sin t$ along the curve, expressing $f(x, y, z)$ in terms of t, $u = 1 - \sin t \cos t$, and minimizing this function on appropriate intervals.

***21.** Find the maximum value of $f(x, y, z) = (xy + x^2)/(z^2 + 1)$ subject to the constraint $x^2(4 - x^2) = y^2$.

13.13 DIFFERENTIALS

If $y = f(x)$ is a function of one variable, the differential of y, defined by $dy = f'(x)\,dx$, was found to be an approximation to the increment $\Delta y = f(x + dx) - f(x)$ for small dx. In particular, dy is the change in y corresponding to the change dx in x if we follow the tangent line to the curve at (x, y) instead of the curve itself.

We take the same approach in defining differentials for multivariable functions. First consider a function $f(x, y)$ of two independent variables that can be represented geometrically as a surface with equation $z = f(x, y)$ (Fig. 13.53). If we change the values of x and y by amounts $\Delta x = dx$ and $\Delta y = dy$, then the corresponding change in z is

$$\Delta z = f(x + dx, y + dy) - f(x, y).$$

Geometrically, this is the difference in the heights of the surface at the points $(x + dx, y + dy)$ and (x, y). If we draw the tangent plane to the surface at (x, y), then very near (x, y) the height of the tangent plane approximates the height of the surface (Fig. 13.54). In particular, the height of the tangent plane at $(x + dx, y + dy)$ for small dx and dy approximates the height of the surface. Therefore, we define the differential dz as the change in z corresponding to the changes dx and dy in x and y if we follow the tangent plane at (x, y) instead of the surface itself. To find dz in terms of dx and dy, we note that the vector joining

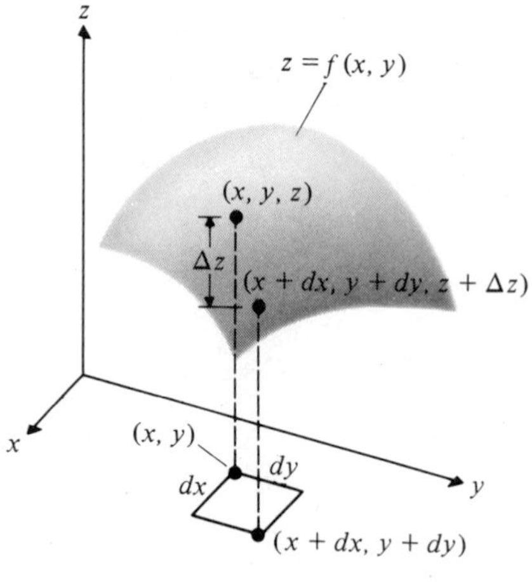

Figure 13.53

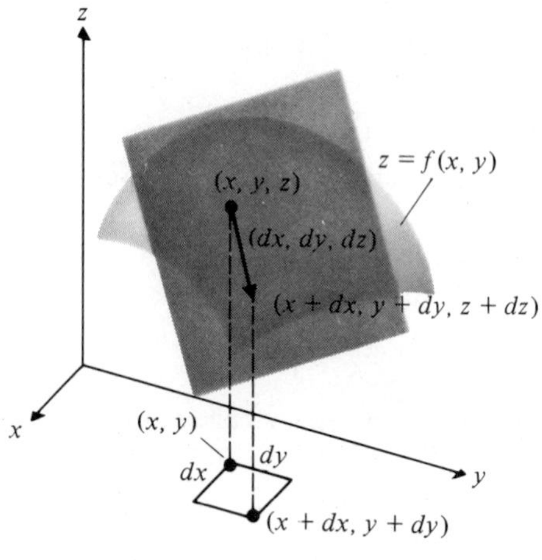

Figure 13.54

the points (x, y, z) and $(x + dx, y + dy, z + dz)$ has components (dx, dy, dz), and this vector lies in the tangent plane. Since a normal vector to the tangent plane is

$$\nabla(z - f(x, y)) = (-f_x, -f_y, 1),$$

it follows that the vectors $(-f_x, -f_y, 1)$ and (dx, dy, dz) must be perpendicular. Consequently,

$$0 = (-f_x, -f_y, 1) \cdot (dx, dy, dz) = -\frac{\partial f}{\partial x}\,dx - \frac{\partial f}{\partial y}\,dy + dz,$$

and hence

$$dz = \frac{\partial f}{\partial x}\,dx + \frac{\partial f}{\partial y}\,dy. \tag{13.63}$$

Note that if y is held constant in the function $f(x, y)$, then $dy = 0$ and Eq. (13.63) for dz reduces to the definition of the differential of a function of one variable.

In Section 4.9 we expressed reservations about using the differential dy as an approximation for the change Δy in a function $f(x)$. For the same reasons, we must be judicious in our use of dz as an approximation for Δz. Indeed, we have stated that dz is an approximation for Δz for small dx and dy, but the difficulty is deciding how small is small and how good is the approximation. In addition, note that if (x, y) is a critical point of the function $f(x, y)$, then either $dz = 0$ for all dx and dy, or dz is undefined. In other words, dz cannot be used to approximate Δz at a critical point.

Example 35 If the radius of a right-circular cone is changed from 10 centimeters to 10.1 centimeters and the height is changed from 1 meter to 0.99 meter, use differentials to approximate the change in its volume.

Solution The volume of a cone of radius r and height h is given by the formula $V = \pi r^2 h/3$. The differential of this function is

$$dV = \frac{\partial V}{\partial r}\,dr + \frac{\partial V}{\partial h}\,dh = \frac{2}{3}\pi rh\,dr + \frac{1}{3}\pi r^2\,dh.$$

If $r = 10$, $dr = 0.1$, $h = 100$, and $dh = -1$, then

$$dV = \frac{2}{3}\pi(10)(100)(0.1) + \frac{1}{3}\pi(10)^2(-1) = \frac{100\pi}{3} \text{ cubic centimeters. } \square$$

Equation (13.63) suggests the following definition for the differential of a function of more than two independent variables.

▷ **Definition 13.11** If $u = f(x, y, z, t, \ldots, w)$, then the differential of

$$f(x, y, z, t, \ldots, w)$$

is defined as

$$du = \frac{\partial f}{\partial x}\,dx + \frac{\partial f}{\partial y}\,dy + \frac{\partial f}{\partial z}\,dz + \frac{\partial f}{\partial t}\,dt + \cdots + \frac{\partial f}{\partial w}\,dw. \triangleleft \tag{13.64}$$

Example 36 The area of the triangle in Fig. 13.55 is given by the formula $A = (\frac{1}{2})ab \sin \theta$. If when $\theta = \pi/3$, a and b are changed by $\frac{1}{3}\%$ and θ by $\frac{1}{2}\%$ use differentials to find the approximate percentage change in A.

Solution Since

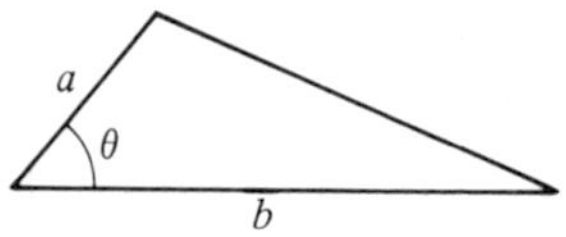

Figure 13.55

$$dA = \frac{\partial A}{\partial a}\,da + \frac{\partial A}{\partial b}\,db + \frac{\partial A}{\partial \theta}\,d\theta$$

$$= \frac{1}{2} b \sin \theta \, da + \frac{1}{2} a \sin \theta \, db + \frac{1}{2} ab \cos \theta \, d\theta,$$

the approximate percentage change in A is

$$100\left(\frac{dA}{A}\right) = \frac{100}{A}\left\{\frac{1}{2} b \sin \theta \, da + \frac{1}{2} a \sin \theta \, db + \frac{1}{2} ab \cos \theta \, d\theta\right\}$$

$$= 100\left\{\frac{da}{a} + \frac{db}{b} + \cot \theta \, d\theta\right\}.$$

Since a and b are changed by $\frac{1}{3}\%$ and θ by $\frac{1}{2}\%$,

$$100\left(\frac{da}{a}\right) = 100\left(\frac{db}{b}\right) = \frac{1}{3}$$

and

$$100\left(\frac{d\theta}{\theta}\right) = \frac{1}{2}.$$

Thus,

$$100\left(\frac{dA}{A}\right) = \frac{1}{3} + \frac{1}{3} + \frac{\theta}{2} \cot \theta = \frac{2}{3} + \frac{\theta}{2} \cot \theta,$$

and when $\theta = \pi/3$,

$$100\left(\frac{dA}{A}\right) = \frac{2}{3} + \frac{1}{2}\left(\frac{\pi}{3}\right)\left(\frac{1}{\sqrt{3}}\right) = 0.97.$$

The approximate percentage change in A is therefore 1%. □

EXERCISES 13.13

In each of Exercises 1–10, find the differential of the function.

1. $f(x, y) = x^2y - \sin y$
2. $f(x, y) = \mathrm{Tan}^{-1}(xy)$
3. $f(x, y, z) = xyz - x^3e^z$
4. $f(x, y, z) = \sin(xyz) - x^2y^2z^2$
5. $f(x, y, z) = \ln(x^2 + y^2 + z^2)$
6. $f(x, y) = \mathrm{Sin}^{-1}(xy)$
7. $f(x, y) = \mathrm{Sin}^{-1}(x + y) + \mathrm{Cos}^{-1}(x + y)$
8. $f(x, y, z, t) = xy + yz + zt + xt$
9. $f(x, y, z, w) = xy \tan(zw)$
10. $f(x, y, z, t) = e^{x^2+y^2+z^2-t^2}$
11. A right-circular cone has radius 10 centimeters and height 20 centimeters. If its radius is increased by 1 millimeter and its height is decreased by 3 millimeters, use differen-

tials to find the approximate change in its volume. Compare this with the actual change in volume.

12. When the ellipse $b^2x^2 + a^2y^2 = a^2b^2$ is rotated about the x-axis, the volume V of the spheroid is $4\pi ab^2/3$. If a and b are each changed by 1%, use differentials to find the approximate percentage change in V.

13. When two resistors of resistances R_1 and R_2 are connected in parallel, their effective resistance R is

$$R = \frac{R_1R_2}{R_1 + R_2}.$$

Show that if R_1 and R_2 are both increased by a small percentage c, then the percentage increase of R is also c.

SUMMARY

We began the study of multivariable functions in this chapter, concentrating our attention on differentiation and its applications. We introduced two types of derivatives for a multivariable function: partial derivatives and directional derivatives. In actual fact, partial derivatives are directional derivatives in directions parallel to the coordinate axes. The directional derivative of a function $f(x, y, z)$ in the direction $\mathbf{v}$ is given by the formula $D_{\mathbf{v}}f = \nabla f \cdot \hat{\mathbf{v}}$, where $\hat{\mathbf{v}}$ is the unit vector in the direction of $\mathbf{v}$, and the gradient ∇f is evaluated at the point at which $D_{\mathbf{v}}f$ is required. This formula leads to the fact that the gradient $\nabla f(x, y, z)$ points in the direction in which $f(x, y, z)$ increases most rapidly, and $|\nabla f|$ is the (maximum) rate of change of $f(x, y, z)$ in the direction ∇f. A second property of gradient vectors (which is related to the first) is that if $F(x, y, z) = 0$ is the equation of a surface, then at any point on the surface ∇F is perpendicular to the surface. This property, along with the fact that perpendicularity to a surface is synonymous with perpendicularity to its tangent plane, enables us to find equations for tangent planes to surfaces and tangent lines to curves.

We illustrated various ways to calculate partial derivatives of a multivariable function, depending on whether the function is defined explicitly, implicitly, or as a composite function. Since partial derivatives are ordinary derivatives with other variables held constant, there is no difficulty calculating partial derivatives when the function is defined explicitly; we simply use the rules from single-variable calculus. When the partial derivative of a composite function is required, we use a schematic diagram illustrating functional dependences to develop the appropriate chain rule. Partial derivatives for functions defined implicitly are calculated using Jacobians.

Critical points of a multivariable function are points at which all of its first partial derivatives vanish or at which one or more of these partial derivatives does not exist. Critical points can yield relative maxima, relative minima, saddle points, or none of these. For functions of two independent variables, a second-derivative test exists that may determine whether a critical point at which the partial derivatives vanish yields a relative maximum, a relative minimum, or a saddle point. This test is analogous to that for functions of one variable.

A continuous function of two independent variables defined on a region that includes its boundary always takes on a maximum value and a minimum value. To find these values we evaluate the function at each of its critical points and compare these numbers to the maximum and minimum values of the function on its boundary. Finding the extreme values on the boundary involves one or more extrema problems for a function of one variable, the number of such problems depending on the complexity of the boundary.

We suggest two methods for finding the extreme values of a function when the variables of the function are subject to constraints. We either solve the constraint equations for dependent variables and express the given function in terms of independent variables or we use Lagrange multipliers. Lagrange multipliers eliminate the necessity for solving the constraint equations, but they do, on the other hand, give a larger system of equations to solve for critical points.

REVIEW EXERCISES

In Exercises 1–20, find the indicated derivative.

1. $\partial f/\partial x$ if $f(x, y) = x^2/y^3 - \mathrm{Sin}^{-1}(xy)$

2. $\partial^2 f/\partial y^2$ if $f(x, y, z) = \ln(x^2 + y^2 + z^2)$

3. $\partial^3 f/\partial x^2\,\partial y$ if $f(x, y, z, t) = x^3 e^y - xzt^2 - \sin(x + y + z + t)$

4. $\partial z/\partial x$ if $z^2 x + \mathrm{Tan}^{-1} z + y = 3x$

5. $\partial u/\partial y$ if $u \cos y + y \cos(xu) + z^2 = 5x$

6. df/dt if $f(x, y) = x^2 + y^2 - e^{xy}$, $x = t^3 + 3t$, $y = t \ln t$

7. dy/dx if $x = y^3 + 3y^2 - 2y + 4$

8. $\partial u/\partial x)_y$ if $u^2 + v^2 - xy = 5$, $3u - 2v + x^2 u = 2v^3$

9. $\partial^2 f/\partial u\,\partial v$ if $f(u, v) = u^2/\sqrt{v} - v/\sqrt{u}$

10. df/dt if $f(x, y) = xy - x^2 - y^2$, $x = te^t$, $y = te^{-t}$

11. $\partial z/\partial t)_u$ if $z = x^2 - y^2$, $x = 2u - 3v^2 + 3uvt$, $y = u \cos(vt)$, $v = t^2 - 2t$

12. $\partial r/\partial x)_y$ if $x = r \cos\theta$, $y = r \sin\theta$

13. $\partial\theta/\partial x)_{y,z}$ if $x = r \sin\phi \cos\theta$, $y = r \sin\phi \sin\theta$, $z = r \cos\phi$

14. $\partial u/\partial r)_\theta$ if $u = x^2 - y^2 x^3$, $x = r \cos\theta$, $y = r \sin\theta$

15. $\partial^2 u/\partial r^2)_\theta$ if $u = x^2 - y^2 x^3$, $x = r \cos\theta$, $y = r \sin\theta$

16. d^2u/dt^2 if $u = x/z^2 - z/x^2$, $x = t^3 - 3$, $z = 1/t^3$

17. dz/dt if $z = y - xy^2 + x$, and $x^2 - y^2 + xt = 2t$, $xy = 4t^2$

18. $\partial^2 z/\partial x^2$ if $xz - x^2 z^3 + y^2 = 3$

19. dy/dx if $yx - x^2 z^2 + 5x = 3$, $2xz - 3x^2 y^2 = 4z^4$

20. $\partial u/\partial t)_v$ if $u = xyt^2 - 3\,\mathrm{Sin}^{-1}(xy)$, $x = v^2 t^2 - 2t$, $y = v \tan t$

21. If $u = (x^2 + y^2)(1 + \sin(x/z))$, show that

$$x\frac{\partial u}{\partial x} + y\frac{\partial u}{\partial y} + z\frac{\partial u}{\partial z} = 2u.$$

22. If $u = 2x^2 - 3y^2 + xy$, show that

$$x^2\frac{\partial^2 u}{\partial x^2} + 2xy\frac{\partial^2 u}{\partial x\,\partial y} + y^2\frac{\partial^2 u}{\partial y^2} = 2u.$$

23. If $f(s)$ is a differentiable function, show that $f(3x - 2y)$ satisfies the equation

$$2\frac{\partial f}{\partial x} + 3\frac{\partial f}{\partial y} = 0.$$

24. If $f(s, t)$ has continuous first partial derivatives, show that $f(x^2 - y^2, y^2 - x^2)$ satisfies the equation

$$y\frac{\partial f}{\partial x} + x\frac{\partial f}{\partial y} = 0.$$

In Exercises 25–30, find the directional derivative of the function at the point and in the direction indicated.

25. $f(x, y) = x^2 \sin y$ at $(3, -1)$ in the direction $\mathbf{v} = (2, 4)$

26. $f(x, y, z) = x^2 + y^2 + z^2$ at $(1, 0, 1)$ in the direction from $(1, 0, 1)$ to $(2, -1, 3)$

27. $f(x, y, z) = z \operatorname{Tan}^{-1}(x + y)$ at $(-1, 2, 5)$ in the direction $\mathbf{v}$ perpendicular to the surface $z = x^2 + y^2$ with positive z-component

28. $f(x, y, z) = x^2 + y - 2z$ at $(1, -1, 2)$ along the line $x - y + z = 4$, $2x + 4y + 2 = 0$ in the direction of increasing x

29. $f(x, y) = \ln(x + y)$ at $(3, 10)$ along the curve $y = x^2 + 1$ in the direction of decreasing y

30. $f(x, y, z) = 2xyz - x^2 - z^2$ at $(0, 1, 1)$ along the curve $x^2 + y^2 + z^2 = 2$, $y = z$ in the direction of increasing x

In Exercises 31–33, find the equation of the tangent plane to the surface at the point.

31. $z = x^2 + y^2$ at $(1, 3, 10)$

32. $x^2 + z^3 = y^2$ at $(-1, 3, 2)$

33. $x^2 + y^2 = z^2 + 1$ at $(1, 0, 0)$

In Exercises 34–36, find equations for the tangent line to the curve at the point.

34. $x = t^2 + 1$, $y = t^2 - 1$, $z = t^3 + 5t$ at $(2, 0, 6)$

35. $x + y + z = 0$, $2x - 3y - 6z = 11$ at $(1, 1, -2)$

36. $z = xy$, $x^2 + y^2 = 2$ at $(1, 1, 1)$

In Exercises 37–40, find all critical points for the function and classify each as yielding a relative maximum, a relative minimum, or a saddle point.

37. $f(x, y) = x^3 + 3y^2 - 6x + 4$

38. $f(x, y) = ye^x$

39. $f(x, y) = x^2 - xy + y^2 + x - 4y$

40. $f(x, y) = (x^2 + y^2 - 1)^2$

41. If $f(x, y) = (x^2 + y^2)F(x, y)$, where $F(x, y) = x^3/y - y^3/x$, verify that

$$\frac{\partial^2 f}{\partial x^2} + \frac{\partial^2 f}{\partial y^2} = (x^2 + y^2)\left(\frac{\partial^2 F}{\partial x^2} + \frac{\partial^2 F}{\partial y^2}\right) + 12F(x, y).$$

42. Find the maximum and minimum values of the function $f(x, y) = xy$ on the circle $x^2 + y^2 \le 1$.

43. Find the maximum and minimum values of the function $f(x, y, z) = 2x + 3y - 4z$ on the sphere $x^2 + y^2 + z^2 \le 2$.

44. Find the points on the curve $x^2 + x^4 + y^2 = 1$ closest to and farthest from the origin.

45. Find the point(s) on the surface $z^2 = 1 + xy$ closest to the origin.

46. Generalize Review Exercise 26 in Chapter 4 to incorporate a third crop, say sunflowers, with a yield of \$$r$ per hectare and a proportional loss cz.

47. If the equation $u = f(x - ut)$ defines u implicitly as a function of x and t, show that

$$\frac{\partial u}{\partial t} + u\frac{\partial u}{\partial x} = 0.$$

Multiple Integrals

14

14.1 INTRODUCTION

The definite integral of a function $f(x)$ of one variable is defined as the limit of a sum of the form

$$f(x_1^*)\,\Delta x_1 + f(x_2^*)\,\Delta x_2 + \cdots + f(x_n^*)\,\Delta x_n \tag{14.1}$$

as n approaches infinity (and every Δx_i approaches zero). We have seen that definite integrals can be used to calculate such quantities as area, volume, work, fluid force, and moments. In spite of the fact that some of these are two- and three-dimensional concepts, we have been careful to emphasize that a definite integral with respect to x is an integration along the x-axis, and a definite integral with respect to y is an integration along the y-axis. In other words, independent of how we interpret the result of the integration, a definite integral is a "limit–summation" *along a line.*

Generalization of these limiting sums to functions of two and three independent variables leads to definitions of double and triple integrals.

14.2 DOUBLE INTEGRALS AND DOUBLE ITERATED INTEGRALS

Suppose a function $f(x, y)$ is defined in some region R of the xy-plane that has finite area (Fig. 14.1). We divide R into n subregions of areas $\Delta A_1, \Delta A_2, \ldots, \Delta A_n$, in any manner whatsoever. In each subregion ΔA_i ($i = 1, \ldots, n$) we choose an arbitrary point (x_i^*, y_i^*) and form the sum

$$f(x_1^*, y_1^*)\,\Delta A_1 + f(x_2^*, y_2^*)\,\Delta A_2 + \cdots + f(x_n^*, y_n^*)\,\Delta A_n = \sum_{i=1}^{n} f(x_i^*, y_i^*)\,\Delta A_i. \tag{14.2}$$

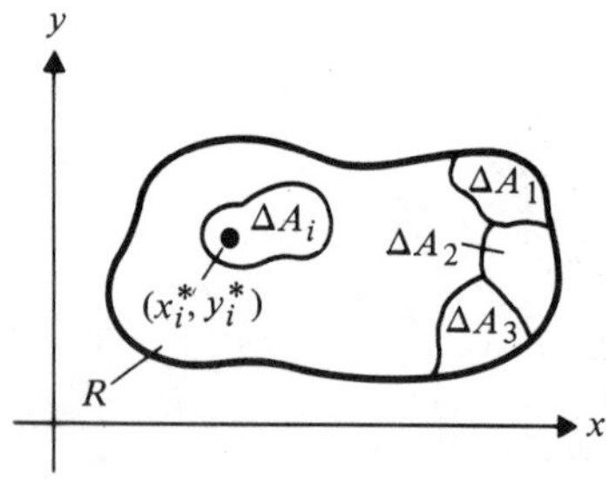

Figure 14.1

If this sum approaches a limit as the number of subregions becomes increasingly large and every subregion shrinks to a point, we call the limit the *double integral* of $f(x, y)$ over the region R, and denote it by

$$\iint_R f(x, y)\,dA = \lim_{n \to \infty} \sum_{i=1}^{n} f(x_i^*, y_i^*)\,\Delta A_i, \tag{14.3}$$

where it is understood that every $\Delta A_i \to 0$ as $n \to \infty$.

If this limit were dependent on the choice of subdivision ΔA_i or choice of star-points (x_i^*, y_i^*), then double integrals would be of little use. We therefore demand that the limit of the sum be independent of the manner of subdivision of R and choice of star-points in the subregions. At first sight this requirement might seem rather severe since we must now check that all subdivisions and all choices of star-points lead to the same limit before concluding that the double integral exists. Fortunately, however, the following theorem indicates that for a large class of functions this is unnecessary.

▶ **Theorem 14.1** Let C be a closed, piecewise-smooth curve that encloses a region R with finite area. If $f(x, y)$ is a continuous function inside and on C, then the double integral of $f(x, y)$ over R exists. ◀

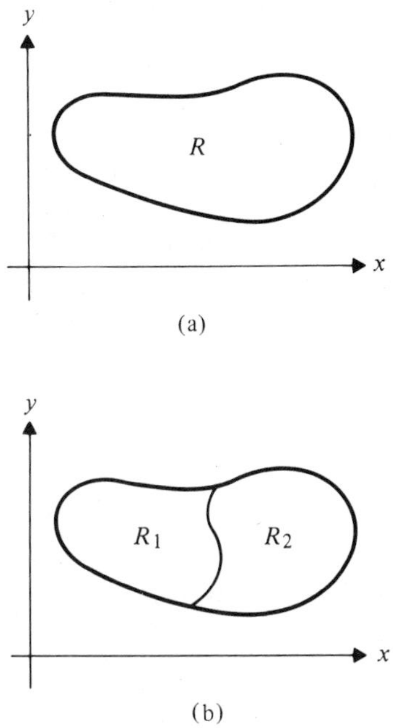

Figure 14.2

For a continuous function, then, the double integral exists and any choice of subdivision and star-points will lead to its correct value through the limiting process (14.3). Note that continuity was also the condition that guaranteed existence of the definite integral in Theorem 6.3.

We cannot overemphasize the fact that a double integral is simply the limit of a sum. Moreover, any limit of the form (14.3) may be interpreted as the double integral of the function $f(x, y)$ over the region defined by the ΔA_i.

The following properties of double integrals are easily proved using definition (14.3):

1. If the double integral of $f(x, y)$ over R exists and c is a constant, then

$$\iint_R cf(x, y)\, dA = c \iint_R f(x, y)\, dA. \tag{14.4}$$

2. If double integrals of $f(x, y)$ and $g(x, y)$ over R exist, then

$$\iint_R [f(x, y) + g(x, y)]\, dA = \iint_R f(x, y)\, dA + \iint_R g(x, y)\, dA. \tag{14.5}$$

3. If a region R is subdivided by a piecewise-smooth curve into two parts R_1 and R_2 that have at most boundary points in common (Fig. 14.2), and the double integral of $f(x, y)$ over R exists, then

$$\iint_R f(x, y)\, dA = \iint_{R_1} f(x, y)\, dA + \iint_{R_2} f(x, y)\, dA. \tag{14.6}$$

4. The area of a region R is

$$\text{Area of } R = \iint_R dA. \tag{14.7}$$

In spite of the fact that double integrals are defined as "limits of sums," we certainly do not intend to evaluate them by using this definition. Just as definite integrals are evaluated with indefinite integrals, we will find that double integrals can be calculated using double iterated integrals.

Double Iterated Integrals

We have already seen that a function $f(x, y)$ of two independent variables has two first-order partial derivatives, one with respect to x holding y constant, and one with respect to y holding x constant. We now reverse this process and define "partial" integration of $f(x, y)$ with respect to x and y. Quite naturally, we define *a* partial indefinite integral of $f(x, y)$ with respect to x as *an* antiderivative of $f(x, y)$ with respect to x holding y constant. For example, since

$$\frac{\partial}{\partial x}(x^3 + x^2 y) = 3x^2 + 2xy,$$

$x^3 + x^2 y$ is an antiderivative with respect to x of $3x^2 + 2xy$. But so too is $x^3 + x^2 y + y$. In fact for any differentiable function $C(y)$ of y whatsoever, $x^3 + x^2 y + C(y)$ is an antiderivative of $3x^2 + 2xy$ with respect to x. Since this expression represents all antiderivatives of $3x^2 + 2xy$, we call it *the* indefinite integral of $3x^2 + 2xy$ with respect to x, and write

$$\int (3x^2 + 2xy)\, dx = x^3 + x^2 y + C(y).$$

Similarly, the indefinite integral of $3x^2 + 2xy$ with respect to y is

$$\int (3x^2 + 2xy)\, dy = 3x^2y + xy^2 + D(x),$$

where $D(x)$ is an arbitrary differentiable function of x.

In this chapter we will be concerned only with partial definite integrals. Limits on a partial definite integral with respect to x must not depend on x, but may depend on y. In general, then, a partial definite integral with respect to x will be of the form

$$\int_{g(y)}^{h(y)} f(x, y)\, dx; \tag{14.8}$$

similarly, a partial definite integral with respect to y will be of the form

$$\int_{g(x)}^{h(x)} f(x, y)\, dy. \tag{14.9}$$

Each of these partial definite integrals is evaluated by substituting the limits into a corresponding partial indefinite integral. For example,

$$\begin{aligned}\int_{x^2}^{x+2} (2y + xe^y)\, dy &= \{y^2 + xe^y\}_{x^2}^{x+2}\\ &= \{(x+2)^2 + xe^{x+2}\} - \{(x^2)^2 + xe^{x^2}\}\\ &= (x+2)^2 - x^4 + x(e^{x+2} - e^{x^2}).\end{aligned}$$

Once the antidifferentiation in (14.8) has been completed and the limits substituted, the result will be a function of y alone. It would then be possible to integrate this function with respect to y between any two limits, say from $y = c$ to $y = d$:

$$\int_c^d \left\{ \int_{g(y)}^{h(y)} f(x, y)\, dx \right\} dy.$$

In practice we omit the braces and simply write

$$\int_c^d \int_{g(y)}^{h(y)} f(x, y)\, dx\, dy, \tag{14.10}$$

but we understand that in the evaluation we proceed from the inner integral to the outer. This is called a *double iterated integral* first with respect to x and then with respect to y (or, more concisely, with respect to x and y). In a similar way we can define double iterated integrals with respect to y and x. They will always be of the form

$$\int_a^b \int_{g(x)}^{h(x)} f(x, y)\, dy\, dx. \tag{14.11}$$

Example 1 Evaluate each of the following double iterated integrals:

a) $\displaystyle\int_0^1 \int_x^{x-1} (x^2 + e^y)\, dy\, dx;$

b) $\displaystyle\int_{-1}^1 \int_0^y xye^{x^2}\, dx\, dy.$

Solution a) $$\int_0^1\int_x^{x-1}(x^2+e^y)\,dy\,dx=\int_0^1\{x^2y+e^y\}_x^{x-1}\,dx$$
$$=\int_0^1\{x^2(x-1)+e^{x-1}-x^3-e^x\}\,dx$$
$$=\int_0^1\{-x^2+e^{x-1}-e^x\}\,dx$$
$$=\left\{-\frac{x^3}{3}+e^{x-1}-e^x\right\}_0^1$$
$$=\left\{-\frac{1}{3}+1-e\right\}-\{e^{-1}-1\}$$
$$=\frac{5}{3}-e-e^{-1}.$$

b) $$\int_{-1}^1\int_0^y xye^{x^2}\,dx\,dy=\int_{-1}^1\left\{\frac{1}{2}ye^{x^2}\right\}_0^y dy=\frac{1}{2}\int_{-1}^1(ye^{y^2}-y)\,dy$$
$$=\frac{1}{2}\left\{\frac{1}{2}e^{y^2}-\frac{y^2}{2}\right\}_{-1}^1=\frac{1}{2}\left\{\frac{e}{2}-\frac{1}{2}-\frac{e}{2}+\frac{1}{2}\right\}=0.\ \square$$

EXERCISES 14.2

In each of Exercises 1–30, evaluate the double iterated integral.

1. $\displaystyle\int_{-1}^2\int_y^{y+2}(x^2-xy)\,dx\,dy$
2. $\displaystyle\int_{-3}^3\int_{-\sqrt{18-2y^2}}^{\sqrt{18-2y^2}}x\,dx\,dy$
3. $\displaystyle\int_0^1\int_{x^2}^x(2xy+3y^2)\,dy\,dx$
4. $\displaystyle\int_{-1}^0\int_y^2(1+y)^2\,dx\,dy$
5. $\displaystyle\int_3^4\int_0^{\pi/2}x\sin y\,dy\,dx$
6. $\displaystyle\int_1^2\int_1^y e^{x+y}\,dx\,dy$
7. $\displaystyle\int_{-1}^1\int_{-x}^5(x^2+y^2)\,dy\,dx$
8. $\displaystyle\int_{-1}^1\int_x^{2x}(xy+x^3y^3)\,dy\,dx$
9. $\displaystyle\int_0^1\int_x^1(x+y)^4\,dy\,dx$
10. $\displaystyle\int_1^2\int_x^{2x}\frac{1}{(x+y)^3}\,dy\,dx$
11. $\displaystyle\int_0^1\int_0^{3x}\sqrt{x+y}\,dy\,dx$
12. $\displaystyle\int_{-1}^1\int_1^e\frac{y}{x}\,dx\,dy$
13. $\displaystyle\int_1^4\int_{\sqrt{x}}^{x^2}(x^2+2xy-3y^2)\,dy\,dx$
14. $\displaystyle\int_0^2\int_{x^2}^{2x^2}x\cos y\,dy\,dx$
15. $\displaystyle\int_0^1\int_1^{\tan x}\frac{1}{1+y^2}\,dy\,dx$
16. $\displaystyle\int_0^1\int_0^{y^3}\frac{1}{1+y^2}\,dx\,dy$
17. $\displaystyle\int_2^3\int_0^1\frac{x}{\sqrt{1-y^2}}\,dy\,dx$
18. $\displaystyle\int_0^2\int_{-x}^x(8-2x^2)^{3/2}\,dy\,dx$
19. $\displaystyle\int_0^1\int_0^x\frac{1}{\sqrt{1-y^2}}\,dy\,dx$
20. $\displaystyle\int_{-9}^0\int_0^{x^2\sqrt{9+x}}dy\,dx$
21. $\displaystyle\int_0^2\int_{\sqrt{4-x^2}}^2 y^2\,dy\,dx$

22. $\displaystyle\int_{-1}^{0}\int_{y}^{0} x\sqrt{x^2+y^2}\,dx\,dy$

23. $\displaystyle\int_{2}^{3}\int_{1}^{2x} \frac{1}{(xy+x^2)^2}\,dy\,dx$

24. $\displaystyle\int_{0}^{1}\int_{0}^{\mathrm{Cos}^{-1}x} x\cos y\,dy\,dx$

25. $\displaystyle\int_{0}^{1}\int_{\sqrt{y^2+y}}^{\sqrt{2y}} x^3\sqrt{x^2-y^2}\,dx\,dy$

26. $\displaystyle\int_{0}^{1}\int_{\sqrt{2y}}^{\sqrt{y^2+y}} x^3\sqrt{x^2-y^2}\,dx\,dy$

27. $\displaystyle\int_{-2}^{0}\int_{x^4}^{4x^2} \sqrt{y-x^4}\,dy\,dx$

28. $\displaystyle\int_{-2}^{0}\int_{y}^{0} \frac{x}{\sqrt{x^2+y^2}}\,dx\,dy$

29. $\displaystyle\int_{-1}^{2}\int_{-1}^{y^3} \sqrt{1+y}\,dx\,dy$

*** 30.** $\displaystyle\int_{0}^{1}\int_{0}^{x} \sqrt{x^2+y^2}\,dy\,dx$

14.3 EVALUATION OF DOUBLE INTEGRALS BY DOUBLE ITERATED INTEGRALS

According to Theorem 14.1, if a function $f(x, y)$ is continuous on a finite region R with a piecewise-smooth boundary, then double integral (14.3) exists, and its evaluation by means of that limit is independent of both the manner of subdivision of R into areas ΔA_i and choice of star-points (x_i^*, y_i^*). We now show that if we make particular choices of ΔA_i, double integrals can be evaluated by means of double iterated integrals in x and y.

Consider first a rectangle R with edges parallel to the x- and y-axes as shown in Fig. 14.3. We divide R into smaller rectangles by a network of $n + 1$ vertical lines and $m + 1$ horizontal lines identified by abscissa

$$a = x_0 < x_1 < x_2 < \cdots < x_{n-1} < x_n = b$$

and ordinates

$$c = y_0 < y_1 < y_2 < \cdots < y_{m-1} < y_m = d.$$

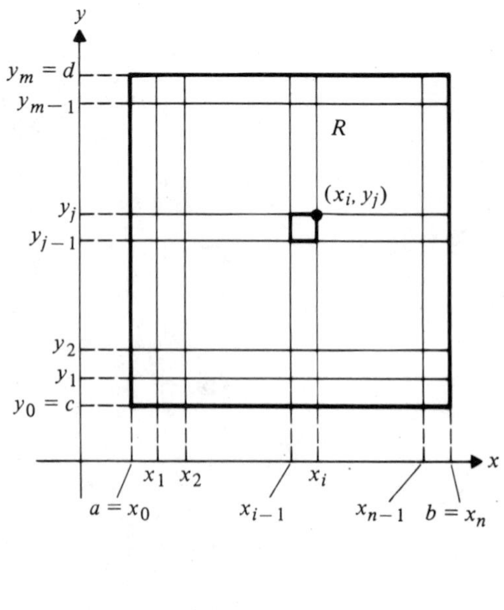

Figure 14.3

If the (i, j)th rectangle is that rectangle bounded by the lines $x = x_{i-1}$, $x = x_i$, $y = y_{j-1}$, and $y = y_j$, then its area is $\Delta x_i \, \Delta y_j$, where $\Delta x_i = x_i - x_{i-1}$ and $\Delta y_j = y_j - y_{j-1}$. We choose as star-point in the (i, j)th rectangle the upper right corner: $(x_i^*, y_j^*) = (x_i, y_j)$. With this rectangular subdivision of R and choice of star-points, definition (14.3) for the double integral of $f(x, y)$ over R takes the form

$$\iint_R f(x, y)\, dA = \lim_{\substack{n \to \infty \\ m \to \infty}} \sum_{i=1}^{n} \sum_{j=1}^{m} f(x_i, y_j)\, \Delta x_i \, \Delta y_j, \tag{14.12}$$

where it is understood that every $\Delta x_i \to 0$ and $\Delta y_j \to 0$ as $n \to \infty$ and $m \to \infty$. Suppose we choose to perform first the limit on m and then the limit on n, and write therefore

$$\iint_R f(x, y)\, dA = \lim_{n \to \infty} \sum_{i=1}^{n} \left\{ \lim_{m \to \infty} \sum_{j=1}^{m} f(x_i, y_j)\, \Delta y_j \right\} \Delta x_i.$$

Since x_i is constant in the limit with respect to y, the y-limit is precisely the definition of the definite integral of $f(x_i, y)$ with respect to y from $y = c$ to $y = d$; i.e.,

$$\lim_{m \to \infty} \sum_{j=1}^{m} f(x_i, y_j)\, \Delta y_j = \int_c^d f(x_i, y)\, dy.$$

Consequently,

$$\iint_R f(x, y)\, dA = \lim_{n \to \infty} \sum_{i=1}^{n} \left\{ \int_c^d f(x_i, y)\, dy \right\} \Delta x_i.$$

Because the term in braces is a function of x_i alone, we can interpret this limit as a definite integral with respect to x:

$$\iint_R f(x, y)\, dA = \int_a^b \left\{ \int_c^d f(x, y)\, dy \right\} dx = \int_a^b \int_c^d f(x, y)\, dy\, dx, \tag{14.13}$$

a double iterated integral. By reversing the order of taking limits, we can show similarly that the double integral can be evaluated by using a double iterated integral with respect to x and y:

$$\iint_R f(x, y)\, dA = \int_c^d \int_a^b f(x, y)\, dx\, dy. \tag{14.14}$$

We have shown, then, that for the special case of a rectangle R with sides parallel to the axes, a double integral over R can be evaluated by using double iterated integrals. Conversely, every double iterated integral with constant limits represents a double integral over a rectangle. The double iterated integral simply indicates that a rectangular subdivision has been chosen to evaluate the double integral.

We have just stated that the choice of a double iterated integral to evaluate a double integral implies that the region of integration has been subdivided into small rectangles. We now show that the x- and y-integrations themselves can be interpreted geometrically. These interpretations will simplify the transition to more difficult regions of integration.

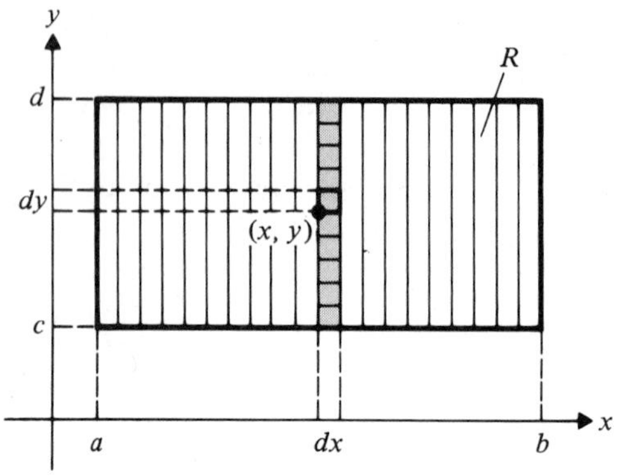

Figure 14.4

In the subdivision of R into rectangles, suppose we denote the dimensions of a representative rectangle at position (x, y) by dx and dy (Fig. 14.4). In the inner integral

$$\int_c^d f(x, y)\, dy\, dx$$

of Eq. (14.13), x is held constant and integration is performed in the y-direction. This (partial) definite integral is therefore interpreted as summing over the rectangles in the vertical strip of width dx at position x. The limits $y = c$ and $y = d$ identify the initial and terminal positions of this vertical strip. It is important to note that we are not adding the areas of the rectangles of dimensions dx and dy in the strip. On the contrary, each rectangle of area $dy\, dx$ is multiplied by the value of $f(x, y)$ for that rectangle,

$$f(x, y)\, dy\, dx,$$

and it is these quantities that are added.

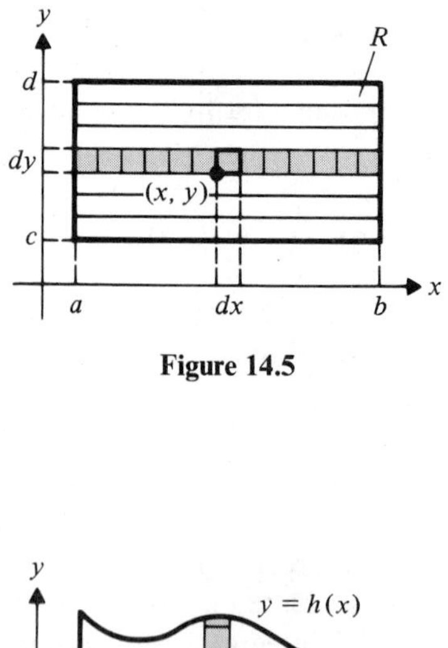

Figure 14.5

The x-integration in Eq. (14.13) is interpreted as adding over all strips starting at $x = a$ and ending at $x = b$. The limits on x, therefore, identify the positions of the first and last strips.

Although our diagram illustrates finite rectangles of dimensions dx and dy and finite strips of width dx, we must keep in mind that the integrations take limits as these dimensions approach zero.

Analogously, the double iterated integral in (14.14) is interpreted as adding over horizontal strips, as shown in Fig. 14.5. The inner limits indicate where each strip starts and stops, and the outer limits indicate the positions of the first and last strips.

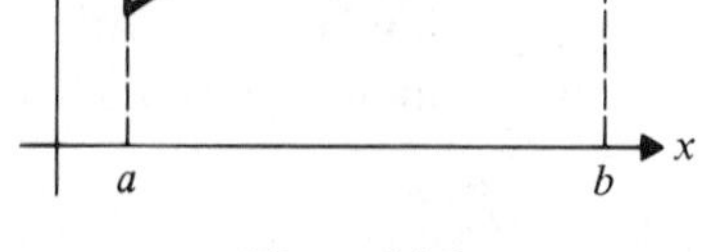

Figure 14.6

The transition now to general regions is quite straightforward. For the double integral of $f(x, y)$ over the region shown in Fig. 14.6, we use a double iterated integral with respect to y and x. The y-integration adds the quantities $f(x, y)\, dy\, dx$ over rectangles in a vertical strip. We write

$$\int_{g(x)}^{h(x)} f(x, y)\, dy\, dx,$$

where $g(x)$ and $h(x)$ indicate that each vertical strip starts on the curve $y = g(x)$

and ends on the curve $y = h(x)$. The x-integration now adds over all strips, beginning at $x = a$ and ending at $x = b$:

$$\iint_R f(x, y)\, dA = \int_a^b \int_{g(x)}^{h(x)} f(x, y)\, dy\, dx. \tag{14.15}$$

A double iterated integral in the reverse order is not convenient for this region because horizontal strips neither all start on the same curve nor all end on the same curve.

For the region shown in Fig. 14.7, we obtain easily

$$\iint_R f(x, y)\, dA = \int_c^d \int_{g(y)}^{h(y)} f(x, y)\, dx\, dy. \tag{14.16}$$

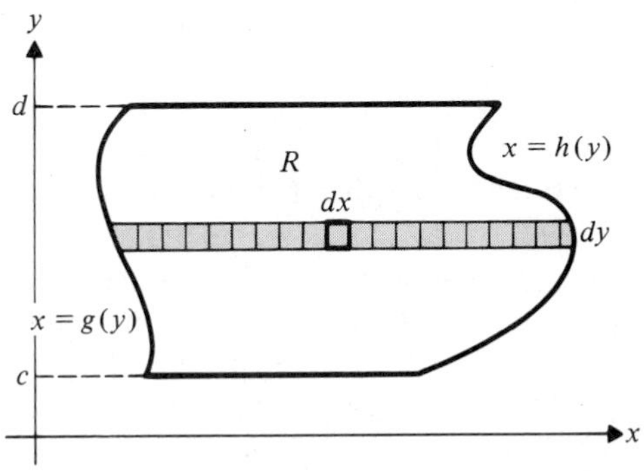

Figure 14.7

The limits on double iterated integrals have been interpreted schematically as follows:

$$\int_{\text{position of first horizontal strip}}^{\text{position of last horizontal strip}} \int_{\text{where each and every horizontal strip starts}}^{\text{where each and every horizontal strip stops}} f(x, y)\, dx\, dy;$$

$$\int_{\text{position of first vertical strip}}^{\text{position of last vertical strip}} \int_{\text{where each and every vertical strip starts}}^{\text{where each and every vertical strip stops}} f(x, y)\, dy\, dx.$$

With these interpretations on the limits, you can see how important it is to have a well-labeled diagram.

Example 2 Evaluate the double integral of $f(x, y) = xy^2 + x^2$ over the region bounded by the curves $y = x^2$ and $x = y^2$.

Solution If we use vertical strips (Fig. 14.8), we have

$$\iint_R (xy^2 + x^2)\, dA = \int_0^1 \int_{x^2}^{\sqrt{x}} (xy^2 + x^2)\, dy\, dx = \int_0^1 \left\{\frac{xy^3}{3} + x^2 y\right\}_{x^2}^{\sqrt{x}} dx$$

$$= \int_0^1 \left\{\frac{1}{3}x^{5/2} + x^{5/2} - \frac{x^7}{3} - x^4\right\} dx = \left\{\frac{8x^{7/2}}{21} - \frac{x^8}{24} - \frac{x^5}{5}\right\}_0^1$$

$$= \frac{8}{21} - \frac{1}{24} - \frac{1}{5} = \frac{39}{280}. \square$$

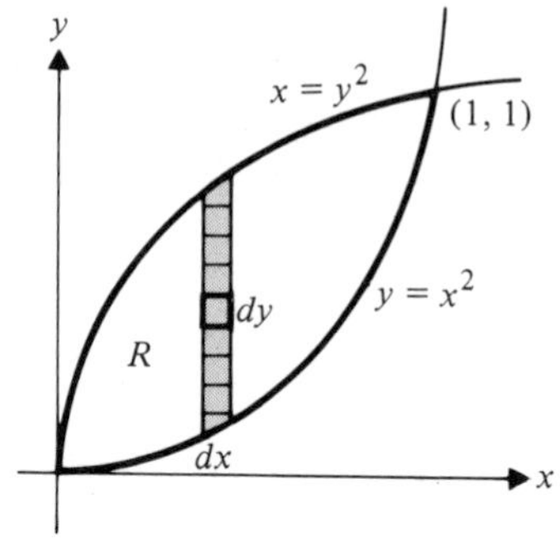

Figure 14.8

There are two distinct parts to every double integral: first, the function $f(x, y)$ being integrated, which becomes the integrand; second, the region R over which integration is being performed, and this region determines the limits on the corresponding double iterated integral. Note that we do not use $f(x, y)$ to determine limits on the double iterated integral; the region determines the limits. Conversely, if we are given a double iterated integral, then we know that it represents the double integral of its integrand over some region, and the region is completely defined by the limits on the iterated integral. This point is emphasized in the following example.

Example 3 Evaluate the double iterated integral

$$\int_0^2 \int_y^2 e^{x^2}\,dx\,dy.$$

Solution The function $f(x, y) = e^{x^2}$ does not have an elementary antiderivative with respect to x, and it is therefore impossible to evaluate the double iterated integral as it now stands. But the double iterated integral represents the double integral of $f(x, y)$ over some region R in the xy-plane. To find R we note that the inner integral indicates horizontal strips that all start on the line $x = y$ and stop on the line $x = 2$ (Fig. 14.9a). The outer limits state that the first and last strips are at $y = 0$ and $y = 2$, respectively. This defines R as the triangle bounded by the straight lines $y = x$, $x = 2$, and $y = 0$ (Fig. 14.9b). If we now reverse the order of integration and use vertical strips, we have

$$\int_0^2 \int_y^2 e^{x^2}\,dx\,dy = \iint_R e^{x^2}\,dA = \int_0^2 \int_0^x e^{x^2}\,dy\,dx = \int_0^2 \{ye^{x^2}\}_0^x\,dx$$

$$= \int_0^2 xe^{x^2}\,dx = \left\{\frac{1}{2}e^{x^2}\right\}_0^2 = \frac{e^4 - 1}{2}. \;\square$$

This example also points out that an iterated integral in one order may be much easier to evaluate than the corresponding iterated integral in the opposite order.

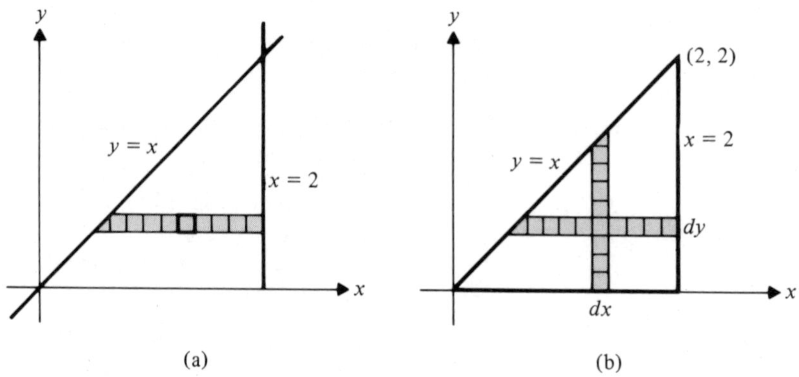

Figure 14.9

EXERCISES 14.3

In each of Exercises 1–12, evaluate the double integral over the region indicated.

1. $\iint_R (x^2 + y^2)\,dA,$

   where R is bounded by $y = x^2$, $x = y^2$

2. $\iint_R (4 - x^2 - y)\,dA,$

   where R is bounded by $x = \sqrt{4 - y}$, $x = 0$, $y = 0$

3. $\iint_R (x + y)\,dA,$

   where R is bounded by $x = y^3 + 2$, $x = 1$, $y = 1$

4. $\iint_R xy^2\,dA,$

   where R is bounded by $x + y + 1 = 0$, $x + y^2 = 1$

5. $\iint_R xe^y\,dA,$

   where R is bounded by $y = x$, $y = 0$, $x = 1$

6. $\iint_R (x + y)\,dA,$

   where R is bounded by $x^2 + y^2 = 9$

7. $\iint_R x^2y\,dA,$

   where R is bounded by $y = \sqrt{x + 4}$, $y = 0$, $x + y = 2$

8. $\iint_R (xy + y^2 - 3x^2)\,dA,$

   where R is bounded by $y = |x|$, $y = 1$, $y = 2$

9. $\iint_R (1 - x)^2\,dA,$

   where R is bounded by $x + y = 1$, $x + y = -1$, $x - y = 1$, $y - x = 1$

10. $\iint_R (x + y)\,dA,$

    where R is bounded by $x = y^2$, $x^2 - y^2 = 12$

11. $\iint_R x\,dA,$

    where R is bounded by $y = 3x$, $y = x$, $x + y = 4$

12. $\iint_R y^2\,dA,$

    where R is bounded by $x = 0$, $y = 1$, $y = \frac{1}{2}$, $x = 1/\sqrt{y^4 + 12y^2}$

In Exercises 13–16, evaluate the given double iterated integral.

13. $\int_0^2 \int_0^{\sqrt{4-x^2}} (4 - y^2)^{3/2}\,dy\,dx$

14. $\int_0^1 \int_y^1 \sin(x^2)\,dx\,dy$

15. $\int_{-2}^0 \int_{-y}^2 y(x^2 + y^2)^8\,dx\,dy$

16. $\int_{-2}^0 \int_{-2}^x \frac{x}{\sqrt{x^2 + y^2}}\,dy\,dx$

The average value of a function $f(x, y)$ over a region R is defined by

$$\bar{f} = \frac{1}{\text{Area}(R)} \iint_R f(x, y)\,dA.$$

In Exercises 17–20, find the average value of the function over the region.

17. $f(x, y) = xy$ over the region in the first quadrant bounded by $x = 0$, $y = 0$, $y = \sqrt{1 - x^2}$

18. $f(x, y) = x + y$ over the region bounded by $y = x$, $y = 0$, $y = \sqrt{2 - x}$

19. $f(x, y) = x$ over the region between $y = \sin x$ and $y = 0$ for $0 \le x \le 2\pi$

20. $f(x, y) = e^{x+y}$ over the region bounded by $y = x + 1$, $y = x - 1$, $y = 1 - x$, $y = -1 - x$

In Exercises 21–28, evaluate the double integral over the region indicated.

21. $\iint_R x^2\,dA,$

    where R is bounded by $x^2 + y^2 = 4$

22. $\iint_R x^7 \cos(x^2)\,dA,$

    where R is bounded by $y = 4 - |x|$, $y = x^2$

23. $\iint_R 6x^5\,dA,$

    where R is the region under $x + 5y = 16$, above $y = x - 4$, and bounded by $x = (y - 2)^2$

24. $\iint_R (6 - x - 2y)\, dA,$

where R is bounded by $x^2 + y^2 = 4$

25. $\iint_R \sqrt{1 + y}\, dA,$

where R is bounded by $x = -1, y = 2, x = y^3$

26. $\iint_R ye^x\, dA,$

where R is bounded by $y = x, x + y = 2, y = 0$

27. $\iint_R (x^2 + y^2)\, dA,$

where R is bounded by $x^2 + y^2 = 9$

28. $\iint_R y\sqrt{x^2 + y^2}\, dA,$

where R is bounded by $y = x, x = -1, y = 0$

29. Evaluate the double iterated integral

$$\int_0^1 \int_0^1 |x - y|\, dy\, dx.$$

30. Evaluate the double integral

$$\iint_R |y - 2x^2 + 1|\, dA,$$

where R is the square bounded by $x = \pm 1, y = \pm 1$.

14.4 AREAS AND VOLUMES OF SOLIDS OF REVOLUTION

Areas

Because Eq. (14.7) represents the area of a region R as a double integral, and double integrals are evaluated by means of double iterated integrals, it follows that areas can be calculated using double iterated integrals. In particular, to find the area of the region shown in Fig. 14.10, we subdivide R into rectangles of dimensions dx and dy and therefore of area $dA = dy\, dx$. The areas of these rectangles are then added in the y-direction to give the area of a vertical strip

$$\int_{g(x)}^{h(x)} dy\, dx,$$

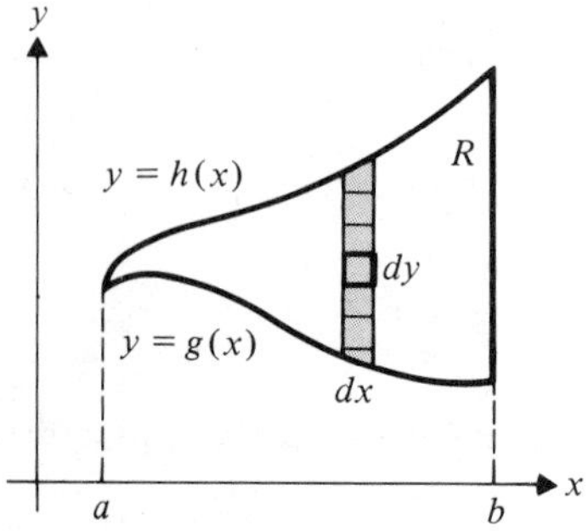

Figure 14.10

where the limits indicate that every vertical strip starts on the curve $y = g(x)$ and ends on the curve $y = h(x)$. Finally, the areas of the vertical strips are added together to give the total area:

$$\text{Area} = \int_a^b \int_{g(x)}^{h(x)} dy\, dx,$$

where a and b indicate the x-positions of the first and last strips.

Example 4 Find the area bounded by the curves $xy = 2, x = 2\sqrt{y}, y = 4$.

Solution If we choose horizontal strips for this area (Fig. 14.11), we have

$$\text{Area} = \int_1^4 \int_{2/y}^{2\sqrt{y}} dx\, dy = \int_1^4 \left\{2\sqrt{y} - \frac{2}{y}\right\} dy$$

$$= \left\{\frac{4}{3} y^{3/2} - 2 \ln |y|\right\}_1^4 = \frac{28}{3} - 2 \ln 4.$$

For vertical strips (Fig. 14.12), we require two iterated integrals because to the left of the line $x = 2$, strips begin on the hyperbola $xy = 2$, whereas to the right of $x = 2$, they begin on the parabola $x = 2\sqrt{y}$ We obtain

$$\text{Area} = \int_{1/2}^{2}\int_{2/x}^{4} dy\,dx + \int_{2}^{4}\int_{x^2/4}^{4} dy\,dx = \int_{1/2}^{2}\left\{4 - \frac{2}{x}\right\}dx + \int_{2}^{4}\left\{4 - \frac{x^2}{4}\right\}dx$$

$$= \{4x - 2\ln|x|\}_{1/2}^{2} + \left\{4x - \frac{x^3}{12}\right\}_{2}^{4} = \frac{28}{3} - 2\ln 4. \square$$

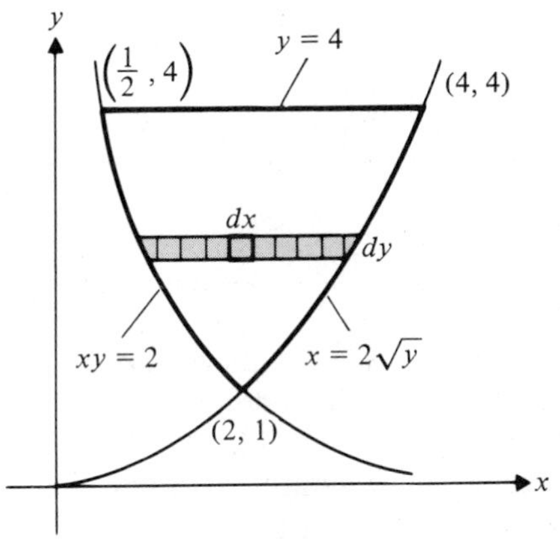

Figure 14.11

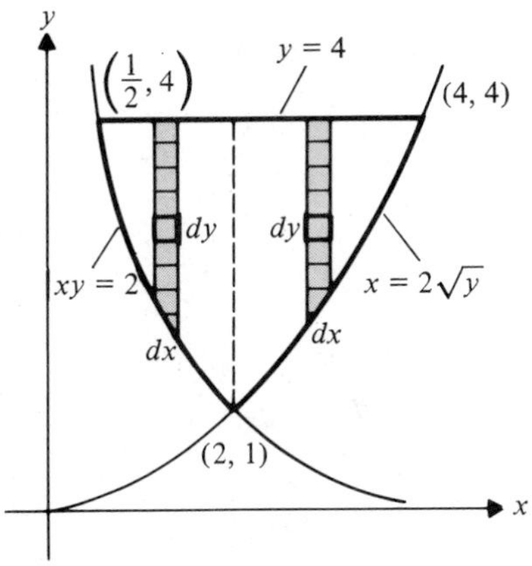

Figure 14.12

If we compare finding areas by definite integrals (Section 7.1) and finding the same areas by double integrals, it is clear that no great advantage is derived by using double integrals. In fact, it is probably more work because we must perform two, rather than one, integrations, although the first integration is trivial. The advantage of double integrals is therefore not in finding area; it is in finding volume of solids of revolution, center of mass, moment of inertia, and fluid pressure, among other applications.

Volumes of Solids of Revolution

If the area in Fig. 14.13 is rotated around the x-axis, the volume of the resulting solid of revolution can be evaluated by using the washer method introduced in Section 7.2:

$$\text{Volume} = \int_a^b \{\pi[h(x)]^2 - \pi[g(x)]^2\}\,dx. \tag{14.17}$$

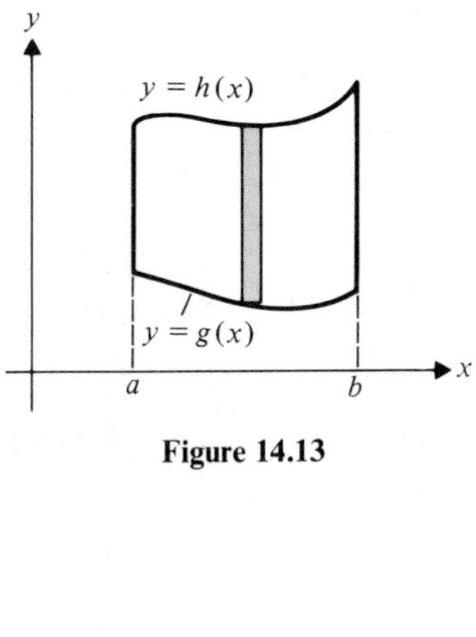

Figure 14.13

If this area is rotated around the y-axis, the volume generated is calculated by using the cylindrical shell method:

$$\text{Volume} = \int_a^b 2\pi x[h(x) - g(x)]\,dx. \tag{14.18}$$

Thus once we have chosen to use vertical rectangles, the axis of revolution determines whether we use either the washer or cylindrical shell method. We now show that with double integrals one method works for both problems.

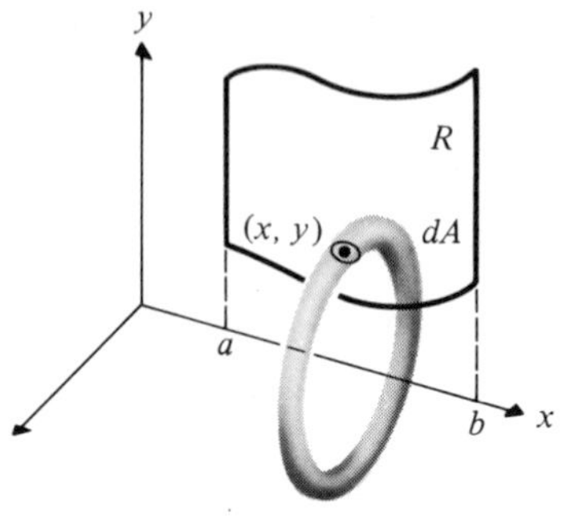

Figure 14.14

To rotate this area around the x-axis we subdivide R into small areas dA (Fig. 14.14). If the area dA at a point (x, y) is rotated about the x-axis, it generates a "ring" with cross-sectional area dA. Since (x, y) travels a distance $2\pi y$ in traversing the ring, it follows that the volume in the ring is approximately $2\pi y\, dA$. To find the total volume obtained by rotating R about the x-axis, we add the volumes of all such rings and take the limit as the areas shrink to points. But this is precisely what we mean by the double integral of $2\pi y$ over the region R, and we therefore write

$$\text{Volume} = \iint_R 2\pi y\, dA. \tag{14.19}$$

On the other hand, if dA is rotated about the y-axis, it again forms a ring, but with approximate volume $2\pi x\, dA$. The total volume, then, when R is rotated about the y-axis is

$$\text{Volume} = \iint_R 2\pi x\, dA. \tag{14.20}$$

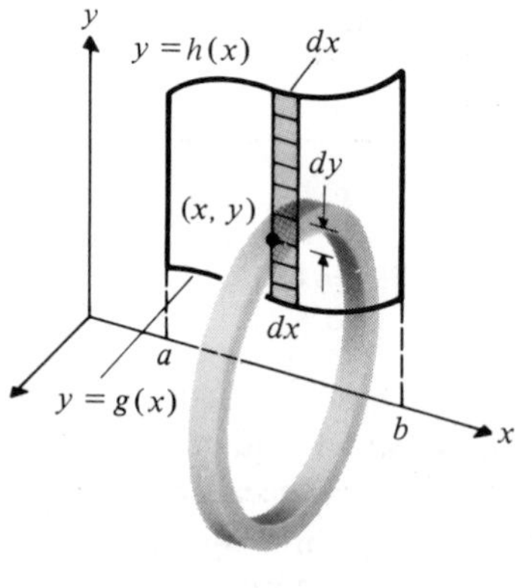

Figure 14.15

Since double iterated integrals are used to evaluate double integrals, it follows that we can set up double iterated integrals to find the volumes represented by Eqs. (14.19) and (14.20). The decision to use a double iterated integral with respect to y and x implies a subdivision of R into rectangles of dimensions dx and dy (Fig. 14.15). The volume of the ring formed when this rectangle is rotated around the x-axis is $2\pi y\, dy\, dx$. If we choose to integrate first with respect to y, we are adding over all rectangles in a vertical strip

$$\int_{g(x)}^{h(x)} 2\pi y\, dy\, dx,$$

where the limits indicate that all vertical strips start on the curve $y = g(x)$ and end on the curve $y = h(x)$. This integral is the volume generated by rotating the vertical strip around the x-axis. Integration now with respect to x adds over all strips to give the required volume

$$\text{Volume} = \int_a^b \int_{g(x)}^{h(x)} 2\pi y\, dy\, dx. \tag{14.21}$$

Note that when we actually do perform the inner integration, we get

$$\text{Volume} = \int_a^b \{\pi y^2\}_{g(x)}^{h(x)}\, dx = \int_a^b \{\pi[h(x)]^2 - \pi[g(x)]^2\}\, dx,$$

and this is the result contained in Eq. (14.17).

When R is rotated around the y-axis, the rectangular area $dy\, dx$ generates a ring of volume $2\pi x\, dy\, dx$. Addition over the rectangles in a vertical strip

$$\int_{g(x)}^{h(x)} 2\pi x\, dy\, dx$$

gives the volume generated by rotating the strip about the y-axis. Finally, integration with respect to x adds over all strips to give

$$\text{Volume} = \int_a^b \int_{g(x)}^{h(x)} 2\pi x\, dy\, dx. \tag{14.22}$$

Note this time that the inner integration leads to

$$\text{Volume} = \int_a^b \{2\pi xy\}_{g(x)}^{h(x)}\,dx = \int_a^b 2\pi x[h(x) - g(x)]\,dx,$$

the same result as in Eq. (14.18).

The advantage, then, in using double integrals to find volumes of solids of revolution is that it requires only one idea, that of rings. The first integration then leads to the concepts of washers or cylindrical shells.

Example 5 Find the volumes of the solids of revolution if the area bounded by the curves $y = 2x - x^2$, $y = x^2 - 2x$ is rotated about:

a) the y-axis;

b) $x = -3$;

c) $y = 2$.

Solution a) If we use vertical strips (Fig. 14.16), then

$$\text{Volume} = \int_0^2 \int_{x^2-2x}^{2x-x^2} 2\pi x\,dy\,dx = 2\pi \int_0^2 \{xy\}_{x^2-2x}^{2x-x^2}\,dx$$

$$= 2\pi \int_0^2 x\{(2x - x^2) - (x^2 - 2x)\}\,dx$$

$$= 4\pi \int_0^2 (2x^2 - x^3)\,dx = 4\pi \left\{\frac{2x^3}{3} - \frac{x^4}{4}\right\}_0^2 = \frac{16\pi}{3}.$$

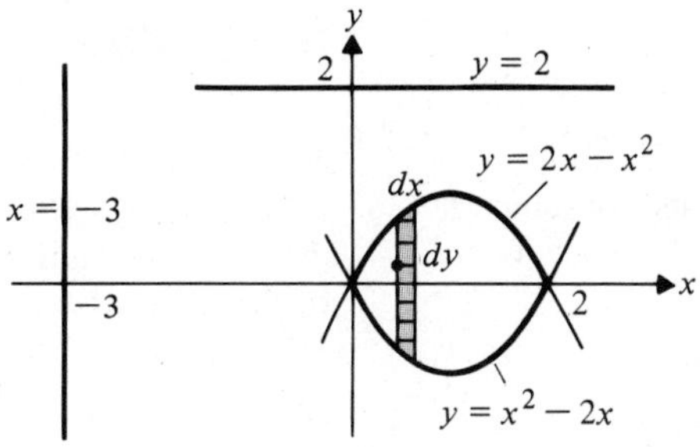

Figure 14.16

b) In this case the radius of the ring formed by rotating the rectangle about $x = -3$ is $x + 3$, and therefore

$$\text{Volume} = \int_0^2 \int_{x^2-2x}^{2x-x^2} 2\pi(x + 3)\,dy\,dx = 2\pi \int_0^2 \{y(x + 3)\}_{x^2-2x}^{2x-x^2}\,dx$$

$$= 2\pi \int_0^2 (x + 3)(4x - 2x^2)\,dx = 4\pi \int_0^2 (6x - x^2 - x^3)\,dx$$

$$= 4\pi \left\{3x^2 - \frac{x^3}{3} - \frac{x^4}{4}\right\}_0^2 = \frac{64\pi}{3}.$$

c) When the rectangle is rotated around $y = 2$, the radius of the ring is $2 - y$, and hence

$$\text{Volume} = \int_0^2 \int_{x^2-2x}^{2x-x^2} 2\pi(2 - y)\, dy\, dx = 2\pi \int_0^2 \left\{ -\frac{1}{2}(2 - y)^2 \right\}_{x^2-2x}^{2x-x^2} dx$$

$$= \pi \int_0^2 \{(2 - x^2 + 2x)^2 - (2 - 2x + x^2)^2\}\, dx = 8\pi \int_0^2 (2x - x^2)\, dx$$

$$= 8\pi \left\{ x^2 - \frac{x^3}{3} \right\}_0^2 = \frac{32\pi}{3}. \square$$

EXERCISES 14.4

In each of Exercises 1–10, use a double integral to find the area bounded by the curves.

1. $y = 4x^2,\ x = 4y^2$
2. $y = x^2,\ y = 5x + 6$
3. $x = y^2,\ x = 3y - 2$
4. $y = x^3 + 8,\ y = 4x + 8$
5. $y = 4/x^2,\ y = 5 - x^2$
6. $y = xe^{-x},\ y = x,\ x = 2$
7. $x = 4y - 4y^2,\ y = x - 3,\ y = 1,\ y = 0$
8. $x = y(y - 2),\ x + y = 12$
9. $y = x^3 - x^2 - 2x + 2,\ y = 2$
10. $x + y = 1,\ x + y = 5,\ y = 2x + 1,\ y = 2x + 6$

In Exercises 11–20, use a double integral to find the volume of the solid of revolution obtained by rotating the area bounded by the curves about the line indicated.

11. $y = -\sqrt{4 - x},\ x = 0,\ y = 0$ about $y = 0$
12. $4x^2 + 9y^2 = 36$ about $y = 0$
13. $y = (x - 1)^2,\ y = 1$ about $x = 0$
14. $y = x^2 + 4,\ y = 2x^2$ about $y = 0$
15. $x - 1 = y^2,\ x = 5$ about $x = 1$
16. $x = y^3,\ y = \sqrt{2 - x},\ y = 0$ about $y = 1$
17. $y = 4x^2 - 4x,\ y = x^3$ about $y = -2$
18. $x = 3y - y^2,\ x = y^2 - 3y$ about $y = 4$
19. $x = 2y - y^2 - 2,\ x = -5$ about $x = 1$
20. $x + y = 4,\ y = 2\sqrt{x - 1},\ y = 0$ about $y = -1$

In Exercises 21–30, use a double integral to find the area bounded by the curves.

21. $y = 2x^3,\ y = 4x + 8,\ y = 0$
22. $y = x/\sqrt{x + 3},\ x = 1,\ x = 6,\ y = -x^2$
23. $x = y^2 + 2,\ x = -(y - 4)^2,\ y = 4 - x,\ y = 0$
24. $y = x^3 - x,\ x + y + 1 = 0,\ x = \sqrt{y + 1}$
25. $y^2 = x^2(4 - x^2)$
26. $x^2 + y^2 = 4,\ x^2 + y^2 = 4x$ (interior to both)
27. $x = 1/\sqrt{4 - y^2},\ 4x + y^2 = 0,\ y + 1 = 0,\ y - 1 = 0$
28. $y^2 = x^4(9 + x)$
29. $y = (x^2 + 1)/(x + 1),\ x + 3y = 7$
30. $(x + 2)^2 y = 4 - x,\ x = 0,\ y = 0\ (x, y \geq 0)$

In Exercises 31–35, use a double integral to find the volume of the solid of revolution obtained by rotating the area bounded by the curves about the line indicated.

31. $y = 4/(x^2 + 1)^2,\ y = 1$ about $x = 0$
32. $y = x^2 - 2,\ y = 0$ about $y = -1$
33. $y = |x^2 - 1|,\ x = -2,\ x = 2,\ y = -1$ about $y = -2$
34. $x = \sqrt{4 + 12y^2},\ x - 20y = 24,\ y = 0$ about $y = 0$
35. $y = (x + 1)^{1/4},\ y = -(x + 1)^2,\ x = 0$ about $x = 0$
36. Find the area common to the two circles
$$x^2 + y^2 = 4, \qquad x^2 + y^2 = 6x.$$
*37. Prove that the area above the line $y = h$ and under the circle $x^2 + y^2 = r^2$ $(r > h)$ is given by
$$A = \pi r^2/2 - h\sqrt{r^2 - h^2} - r^2\,\mathrm{Sin}^{-1}(h/r).$$

14.5 FLUID PRESSURE

In Section 7.5 we defined pressure at a point in a fluid as the magnitude of the force per unit area that would act on any surface placed at that point. We discovered that at a depth $d > 0$ below the surface of a fluid, pressure is given by

$$P = \gamma d, \tag{14.23}$$

where $\gamma > 0$ is the specific weight of the fluid (magnitude of the weight per unit volume). With these ideas and the definite integral, we were able to calculate fluid forces on flat surfaces in the fluid. In particular, the magnitude of the total force on each side of the vertical surface in Fig. 14.17 is given by the definite integral

$$\text{Force} = \int_a^b -\gamma y[h(y) - g(y)]\, dy. \tag{14.24}$$

Although horizontal rectangles are convenient for this problem, it is clear that they are not reasonable for the surface in Fig. 14.18.

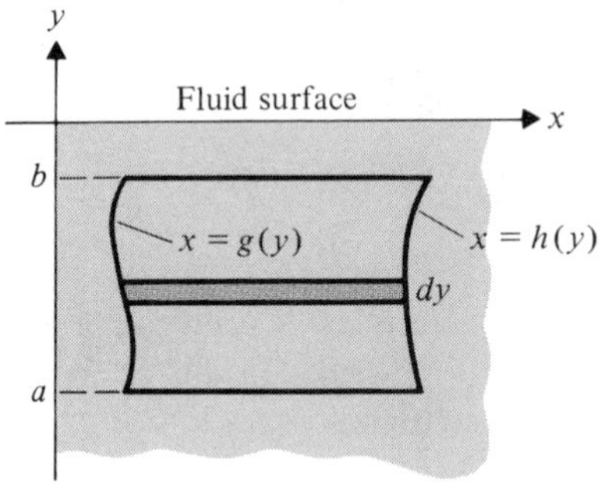

Figure 14.17

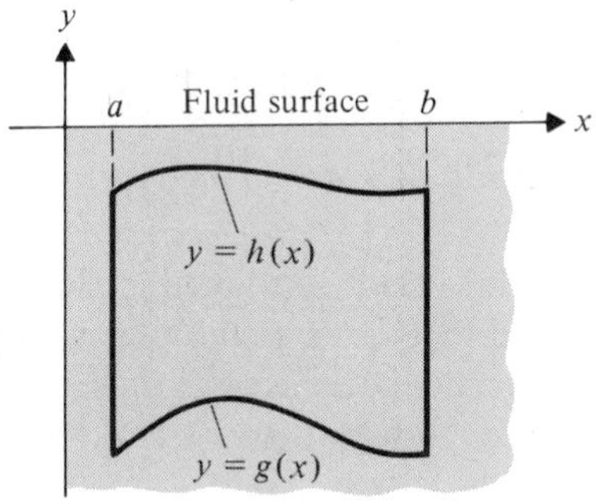

Figure 14.18

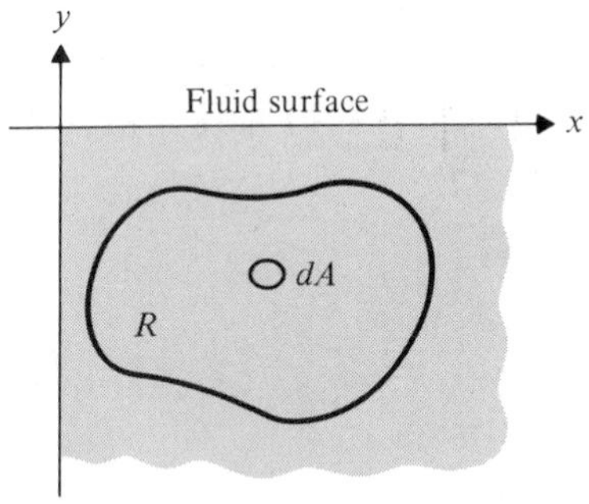

Figure 14.19

Double integrals, on the other hand, can be applied with equal ease to both surfaces. To see this we consider first the total force on each side of the surface shown in Fig. 14.19. If the surface is divided into small areas dA, then the force on dA is $P\, dA$, where P is pressure at that depth. The total force on R is the sum of the forces on all such areas in R as the areas dA shrink to a point. But once again this is the concept of the double integral, and we therefore write

$$\text{Force} = \iint_R P\, dA. \tag{14.25}$$

To set up a double iterated integral in order to evaluate this double integral, we use our interpretation of the double iterated integral as a limit of a sum in which the areas dA have been chosen as rectangles. In particular, for the surface in Fig. 14.17 we draw rectangles of dimensions dx and dy, as shown in Fig. 14.20. The force on this rectangle is its area $dx\, dy$ multiplied by the pressure $-\gamma y$ at that depth, $-\gamma y\, dx\, dy$. Addition of these quantities over all rectangles in a horizontal strip gives the force on the strip,

$$\int_{g(y)}^{h(y)} -\gamma y\, dx\, dy,$$

where the limits indicate that all horizontal strips start on the curve $x = g(y)$

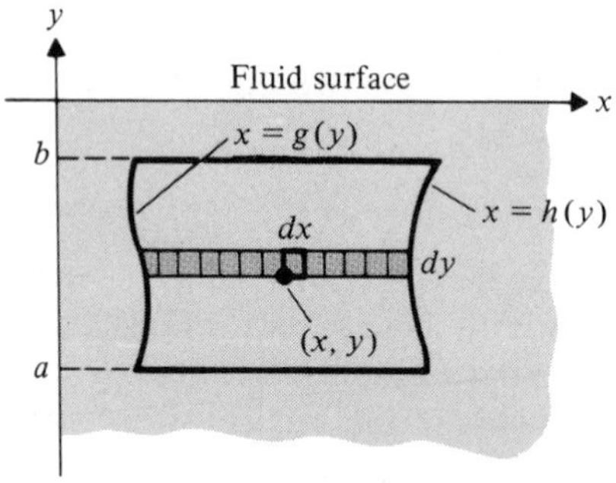

Figure 14.20

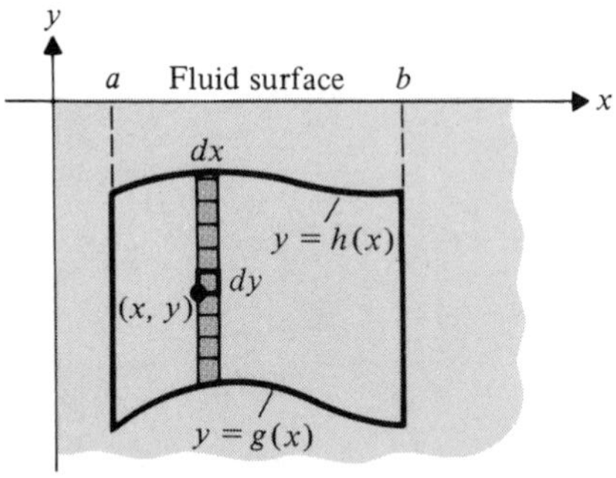

Figure 14.21

and stop on the curve $x = h(y)$. Integration with respect to y now adds over all horizontal strips to give the total force on the surface:

$$\text{Force} = \int_a^b \int_{g(y)}^{h(y)} -\gamma y \, dx \, dy. \tag{14.26}$$

Note that when we perform the inner integration, we get

$$\text{Force} = \int_a^b \{-\gamma y x\}_{g(y)}^{h(y)} \, dy = \int_a^b -\gamma y[h(y) - g(y)] \, dy,$$

and this is the result contained in Eq. (14.24).

For the surface in Fig. 14.18 we again draw rectangles of dimensions dx and dy and calculate the force on such a rectangle, $-\gamma y \, dy \, dx$. In this case it is more convenient to add over rectangles in a vertical strip to give the force on the strip (Fig. 14.21):

$$\int_{g(x)}^{h(x)} -\gamma y \, dy \, dx.$$

The force on the entire surface can now be found by adding over all vertical strips:

$$\text{Force} = \int_a^b \int_{g(x)}^{h(x)} -\gamma y \, dy \, dx. \tag{14.27}$$

Example 6 The face of a dam is parabolic with breadth 100 meters and height 50 meters. Find the magnitude of the total force due to fluid pressure on the face.

Solution If we use the coordinate system shown in Fig. 14.22, then the edge of the dam has an equation of the form $y = kx^2$. Since (50, 50) is a point on this curve, it follows that $k = \frac{1}{50}$. Because the force on the left half of the dam is the same as that on the right half, we can integrate for the right half and double the result; i.e.,

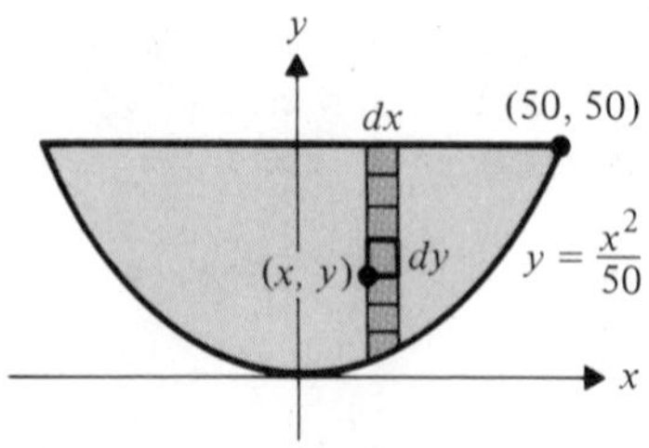

Figure 14.22

$$\text{Force} = 2\int_0^{50} \int_{x^2/50}^{50} 1000(9.81)(50 - y)dy \, dx$$

$$= 19{,}620 \int_0^{50} \left\{50y - \frac{y^2}{2}\right\}_{x^2/50}^{50} dx = 19{,}620 \int_0^{50} \left\{1250 - x^2 + \frac{x^4}{5000}\right\} dx$$

$$= 19{,}620\left\{1250x - \frac{x^3}{3} + \frac{x^5}{25{,}000}\right\}_0^{50} = 6.54 \times 10^8 \text{ newtons. } \square$$

Example 7 A tank in the form of a right-circular cylinder of radius $\frac{1}{2}$ meter and length 10 meters has its axis horizontal. If it is full of water, find the force due to water pressure on each end of the tank.

Solution Since the force on that part of the end to the left of the y-axis (Fig. 14.23) is identical to the force on that part to the right, we can write that

$$\text{Force} = 2\int_{-1/2}^{1/2}\int_{0}^{\sqrt{1/4-y^2}} 1000(9.81)(\tfrac{1}{2} - y)\,dx\,dy$$

$$= 9810\int_{-1/2}^{1/2}\int_{0}^{\sqrt{1/4-y^2}} dx\,dy - 19{,}620\int_{-1/2}^{1/2}\int_{0}^{\sqrt{1/4-y^2}} y\,dx\,dy.$$

The first double iterated integral represents the area of one-half the end of the tank. Consequently,

$$\text{Force} = 9810\{\tfrac{1}{2}\pi(\tfrac{1}{2})^2\} - 19{,}620\int_{-1/2}^{1/2} y\sqrt{\tfrac{1}{4} - y^2}\,dy$$

$$= \frac{4905\pi}{4} - 19{,}620\{-\tfrac{1}{3}(\tfrac{1}{4} - y^2)^{3/2}\}_{-1/2}^{1/2} = \frac{4905\pi}{4}\text{ newtons.} \quad \square$$

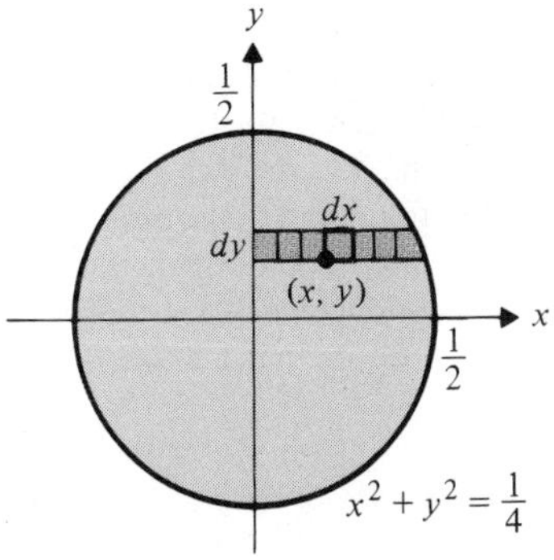

Figure 14.23

EXERCISES 14.5

In Exercises 1–8, the surface described is submerged vertically in a fluid of specific weight γ. Find the force due to fluid pressure on one side of the surface.

1. An equilateral triangle of side length 2 with one side lying in the surface

2. A parabolic segment of base 12 and height 4 with the base in the surface

3. A square of side length 3 with one diagonal vertical and the uppermost vertex in the surface of the fluid

4. A triangle of side lengths 5, 5, and 8, with the longest side uppermost, horizontal, and 3 units below the fluid surface

5. A triangle of side lengths 5, 5, and 8, with the longest side below the opposite vertex, horizontal, and 6 units below the fluid surface

6. A trapezoid with vertical parallel sides of lengths 6 and 8, and a third side perpendicular to the parallel sides, of length 5, and in the surface of the fluid

7. A triangle of side lengths 3, 3, and 4, with the longest side vertical, and the uppermost vertex 1 unit below the surface of the fluid

8. A semicircle of radius 5 with the (diameter) base in the surface of the fluid

9. The vertical end of a water trough is an isosceles triangle with width 2 meters and depth 1 meter. Find the force of the water on each end when the trough is one-half filled (by volume) with water.

10. A dam across a river has the shape of a parabola 36 meters across the top and 9 meters deep at the center. Find the maximum force due to water pressure on the dam.

In Exercises 11–15, the surface described is submerged vertically in a fluid of specific weight γ. Find the force due to fluid pressure on one side of the surface.

11. A circle of radius 2 with center 3 units below the surface of the fluid

12. A rectangle of side lengths 2 and 5, with one diagonal vertical and the uppermost vertex in the surface of the fluid

13. An ellipse with major and minor axes of lengths 8 and 6, and with the major axis horizontal and 5 units below the surface of the fluid

14. A parallelogram of side lengths 4 and 5, with one of the longer sides horizontal and in the surface of the fluid, and two sides making an angle of $\pi/6$ radians with the surface

***15.** A triangle of side lengths 2, 3, and 4, with the longest side vertical, the side of length 2 above the side of length 3, and the uppermost vertex 1 unit below the surface of the fluid

16. An oil can is in the form of a right-circular cylinder of radius r and height h. If the axis of the can is horizontal, and the can is full of oil with specific weight γ, find the force due to fluid pressure on each end.

17. Find the force due to water pressure on each side of the flat vertical plate shown in Fig. 14.24.

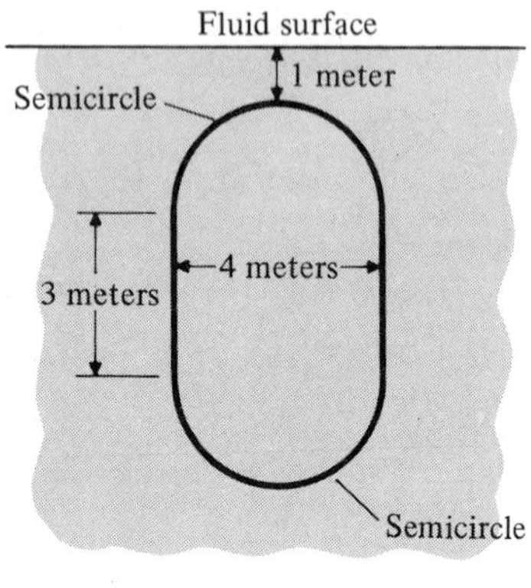

Figure 14.24

18. A square plate of side length 2 meters has one side on the bottom of a swimming pool 3 meters deep. The plate is inclined at an angle of $\pi/4$ radians with the bottom of the pool so that its horizontal upper edge is $3 - \sqrt{2}$ meters below the surface. Find the force due to water pressure on each side of the plate.

* 19. A thin triangular piece of wood with sides of length 2, 2, and 3 meters floats in a pond. A piece of rope is tied to the vertex opposite the longest side. A rock is then attached to the other end of the rope and lowered into the water. When the rock sits on the bottom (and the rope is taut), the longest side of the wood still floats in the surface of the water, but the opposite vertex is 1 meter below the surface. Find the force due to water pressure on each side of the piece of wood.

14.6 CENTERS OF MASS AND MOMENTS OF INERTIA

We now show how double integrals can be used to replace definite integrals in calculating first moments, centers of mass, and moments of inertia of thin plates. Consider, then, a thin plate of density ρ (mass per unit area) such as that shown in Fig. 14.25. Note that unlike our discussion in Section 7.6 where we assumed ρ constant, we have made no such assumption here. In other words, density could be a function of position, $\rho = \rho(x, y)$.

The center of mass $(\bar{x}, \bar{y})$ of the plate is a point at which a particle of mass M (equal to the total mass of the plate) has the same first moments about the x- and y-axes as the plate itself. If we divide the plate into small areas dA, then the mass in dA is $\rho\, dA$. Addition over all such areas in R as each dA shrinks to a point gives the mass of the plate

$$M = \iint_R \rho\, dA. \tag{14.28}$$

Since the first moment of the mass in dA about the y-axis is $x\rho\, dA$, it follows that the first moment of the entire plate about the y-axis is

$$\iint_R x\rho\, dA.$$

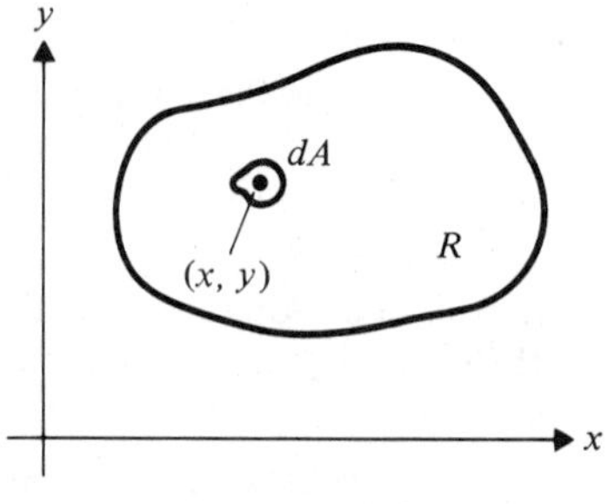

Figure 14.25

But this must be equal to the first moment of the particle of mass M at $(\bar{x}, \bar{y})$ about the y-axis, and hence

$$M\bar{x} = \iint_R x\rho\, dA. \tag{14.29}$$

This equation can be solved for $\bar{x}$ once the double integral on the right and M have been calculated.

Similarly, $\bar{y}$ is determined by the equation

$$M\bar{y} = \iint_R y\rho\, dA, \tag{14.30}$$

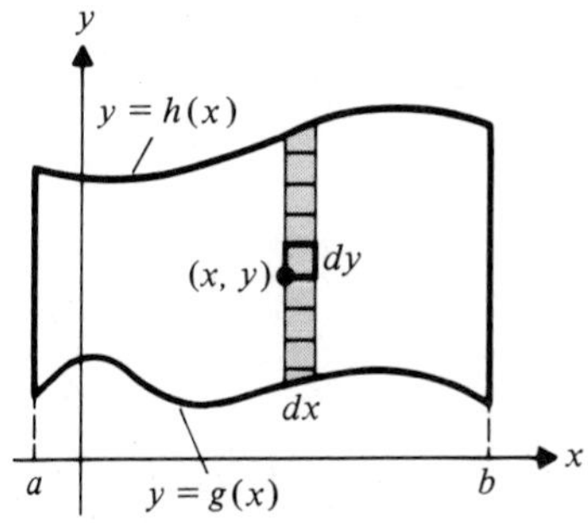

Figure 14.26

where the double integral on the right is the first moment of the plate about the x-axis.

In any given problem, the double integrals in (14.28)–(14.30) are evaluated by means of double iterated integrals. For example, if we divide the plate in Fig. 14.26 into rectangles of dimensions dx and dy and use vertical strips, then we have

$$M = \int_a^b \int_{g(x)}^{h(x)} \rho \, dy \, dx, \tag{14.31}$$

$$M\bar{x} = \int_a^b \int_{g(x)}^{h(x)} x\rho \, dy \, dx, \tag{14.32}$$

$$M\bar{y} = \int_a^b \int_{g(x)}^{h(x)} y\rho \, dy \, dx. \tag{14.33}$$

Once again we point out that Eqs. (14.31)–(14.33) should not be memorized as formulas. Indeed, each can be derived as needed. For instance, to obtain Eq. (14.33) we reason that the first moment of the mass in a rectangle of dimensions dx and dy at position (x, y) about the x-axis is $y\rho \, dy \, dx$. Addition over the rectangles in a vertical strip gives the first moment of the strip about the x-axis:

$$\int_{g(x)}^{h(x)} y\rho \, dy \, dx,$$

and integration with respect to x now adds over all strips to give the first moment of the entire plate about the x-axis. Note that if ρ is constant and the inner integrations are performed in each of Eqs. (14.31)–(14.33), then

$$M = \int_a^b \{\rho y\}_{g(x)}^{h(x)} \, dx = \int_a^b \rho[h(x) - g(x)] \, dx,$$

$$M\bar{x} = \int_a^b \{\rho xy\}_{g(x)}^{h(x)} \, dx = \int_a^b \rho x[h(x) - g(x)] \, dx,$$

$$M\bar{y} = \int_a^b \left\{\rho \frac{y^2}{2}\right\}_{g(x)}^{h(x)} dx = \int_a^b \frac{\rho}{2}\{[h(x)]^2 - [g(x)]^2\} \, dx.$$

These are precisely the results contained in Eqs. (7.27)–(7.29) (with different names for the curves), but the simplicity of the discussion leading to the double iterated integrals certainly demonstrates its advantage over use of the definite integral described in Section 7.6.

Example 8 Find the center of mass of a thin plate of constant density ρ if its edges are defined by the curves $y = 2x - x^2$ and $y = x^2 - 4$.

Solution For vertical strips as shown in Fig. 14.27,

$$\begin{aligned} M &= \int_{-1}^{2} \int_{x^2-4}^{2x-x^2} \rho \, dy \, dx \\ &= \rho \int_{-1}^{2} \{(2x - x^2) - (x^2 - 4)\} \, dx \\ &= \rho\left\{x^2 - \frac{2x^3}{3} + 4x\right\}_{-1}^{2} = 9\rho. \end{aligned}$$

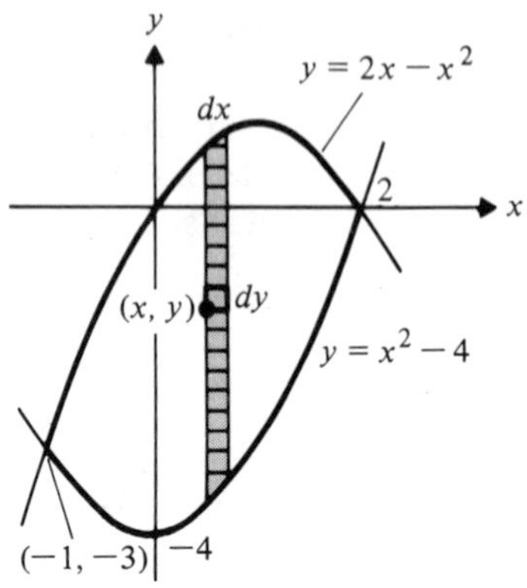

Figure 14.27

If the center of mass of the plate is $(\bar{x}, \bar{y})$, then

$$M\bar{x} = \int_{-1}^{2}\int_{x^2-4}^{2x-x^2} x\rho\, dy\, dx = \rho\int_{-1}^{2} x\{(2x - x^2) - (x^2 - 4)\}\, dx$$

$$= \rho\left\{\frac{2x^3}{3} - \frac{x^4}{2} + 2x^2\right\}_{-1}^{2} = \frac{9\rho}{2}.$$

Thus,

$$\bar{x} = \frac{9\rho}{2}\cdot\frac{1}{9\rho} = \frac{1}{2}.$$

Since

$$M\bar{y} = \int_{-1}^{2}\int_{x^2-4}^{2x-x^2} y\rho\, dy\, dx = \rho\int_{-1}^{2}\left\{\frac{y^2}{2}\right\}_{x^2-4}^{2x-x^2} dx$$

$$= \frac{\rho}{2}\int_{-1}^{2}(-4x^3 + 12x^2 - 16)\, dx$$

$$= \frac{\rho}{2}\{-x^4 + 4x^3 - 16x\}_{-1}^{2} = -\frac{27\rho}{2},$$

we find

$$\bar{y} = -\frac{27\rho}{2}\cdot\frac{1}{9\rho} = -\frac{3}{2}. \square$$

Example 9 Find the first moment of area about the line $y = -2$ for the region bounded by the curves $x = |y|^3$ and $x = 2 - y^2$.

Solution ***Method 1*** The first moment about $y = -2$ of a rectangle of dimensions dx and dy at position (x, y) is $(y + 2)\, dx\, dy$ (Fig. 14.28). For the entire plate, then, the required first moment is

$$\int_{-1}^{0}\int_{-y^3}^{2-y^2}(y + 2)\, dx\, dy + \int_{0}^{1}\int_{y^3}^{2-y^2}(y + 2)\, dx\, dy$$

$$= \int_{-1}^{0}\{x(y + 2)\}_{-y^3}^{2-y^2}\, dy + \int_{0}^{1}\{x(y + 2)\}_{y^3}^{2-y^2}\, dy$$

$$= \int_{-1}^{0}(y^4 + y^3 - 2y^2 + 2y + 4)\, dy + \int_{0}^{1}(-y^4 - 3y^3 - 2y^2 + 2y + 4)\, dy$$

$$= \left\{\frac{y^5}{5} + \frac{y^4}{4} - \frac{2y^3}{3} + y^2 + 4y\right\}_{-1}^{0} + \left\{-\frac{y^5}{5} - \frac{3y^4}{4} - \frac{2y^3}{3} + y^2 + 4y\right\}_{0}^{1}$$

$$= \frac{17}{3}.$$

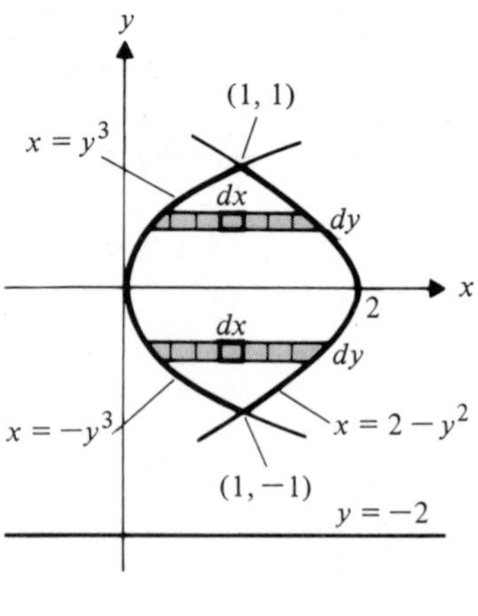

Figure 14.28

Method 2 By symmetry, the centroid of the region is somewhere along the x-axis. Hence, the required first moment is $2A$, where A is the area of the region and 2 is the distance from $y = -2$ to the centroid. Since the area of the region is equally distributed about the x-axis, we obtain the required first moment as

$$2(2)(\text{Area of plate above } x\text{-axis}) = 4\int_0^1\int_{y^3}^{2-y^2} dx\,dy = 4\int_0^1 (2 - y^2 - y^3)\,dy$$

$$= 4\left\{2y - \frac{y^3}{3} - \frac{y^4}{4}\right\}_0^1 = \frac{17}{3}. \square$$

To calculate moments of inertia (second moments) of thin plates about lines parallel to the x- and y-axes is equally as easy as calculating first moments if we use double integrals. In particular, the mass in area dA in Fig. 14.25 is $\rho\,dA$; thus its moments of inertia about the x- and y-axes are, respectively, $y^2\rho\,dA$ and $x^2\rho\,dA$. The moments of inertia of the entire plate about the x- and y-axes are therefore given by the double integrals

$$I_x = \iint_R y^2\rho\,dA \quad\text{and}\quad I_y = \iint_R x^2\rho\,dA. \tag{14.34}$$

For a plate such as that shown in Fig. 14.29, we would evaluate these double integrals by means of double iterated integrals with respect to x and y:

$$I_x = \int_a^b\int_{g(y)}^{h(y)} y^2\rho\,dx\,dy, \tag{14.35}$$

$$I_y = \int_a^b\int_{g(y)}^{h(y)} x^2\rho\,dx\,dy. \tag{14.36}$$

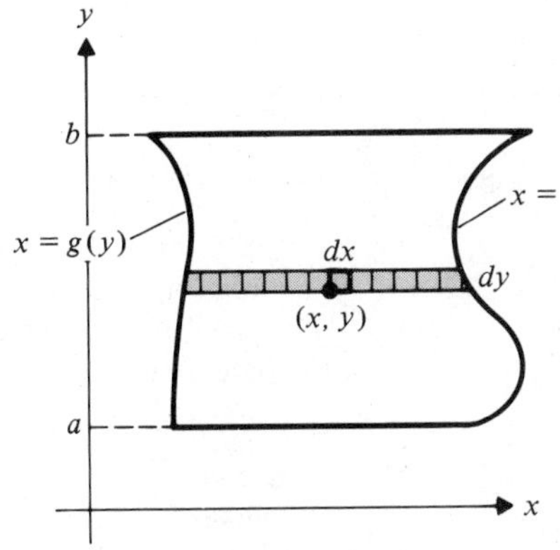

Figure 14.29

Example 10 Find the moment of inertia about the x-axis of a thin plate of constant density ρ if its edges are defined by the curves $y = x^3$, $y = \sqrt{2-x}$, and $x = 0$.

Solution With vertical strips as shown in Fig. 14.30, the moment of inertia about the x-axis is

$$\int_0^1\int_{x^3}^{\sqrt{2-x}} y^2\rho\,dy\,dx = \rho\int_0^1\left\{\frac{y^3}{3}\right\}_{x^3}^{\sqrt{2-x}} dx = \frac{\rho}{3}\int_0^1 \{(2-x)^{3/2} - x^9\}\,dx$$

$$= \frac{\rho}{3}\left\{-\frac{2}{5}(2-x)^{5/2} - \frac{x^{10}}{10}\right\}_0^1 = \frac{16\sqrt{2}-5}{30}\rho. \square$$

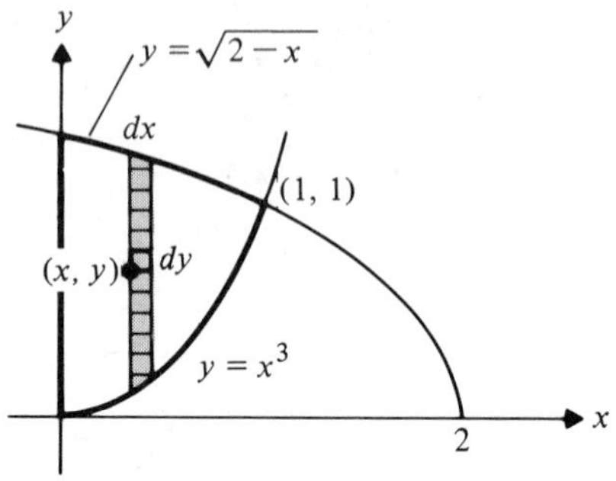

Figure 14.30

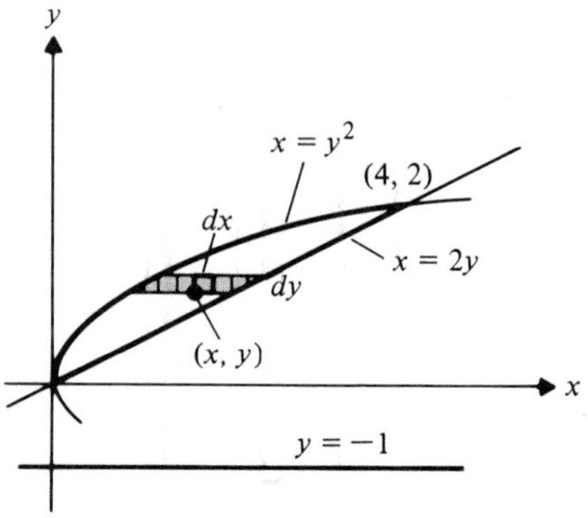

Figure 14.31

Example 11 Find the second moment of area about the line $y = -1$ of the area bounded by the curves $x = y^2$ and $x = 2y$.

Solution The second moment of area about the line $y = -1$ (Fig. 14.31) is

$$\int_0^2 \int_{y^2}^{2y} (y+1)^2 \, dx \, dy = \int_0^2 \{x(y+1)^2\}_{y^2}^{2y} \, dy = \int_0^2 (y+1)^2(2y - y^2) \, dy$$

$$= \int_0^2 (-y^4 + 3y^2 + 2y) \, dy = \left\{ -\frac{y^5}{5} + y^3 + y^2 \right\}_0^2$$

$$= \frac{28}{5}. \square$$

EXERCISES 14.6

In each of Exercises 1–10, find the centroid of the area bounded by the curves.

1. $x = y + 2, x = y^2$
2. $y = 8 - 2x^2, y + x^2 = 4$
3. $y = x^2 - 1, y + (x + 1)^2 = 0$
4. $x + y = 5, xy = 4$
5. $y = e^x, y = 0, x = 0, x = 1$
6. $y = \sqrt{4 - x^2}, y = x, x = 0$
7. $y = 1/(x - 1), y = 1, y = 2, x = 0$
8. $x = 4y - 4y^2, x = y + 3, y = 1, y = 0$
9. $y = |x^2 - 1|, y = 2$
10. $y = x, y = 2x, 2y = x + 3$

In Exercises 11–15, find the second moment of area of the region bounded by the curves about the line indicated.

11. $y = x^2, y = x^3$ about the y-axis
12. $y = x, y = 2x + 4, y = 0$ about the x-axis
13. $y = x^2, 2y = x^2 + 4$ about $y = 0$
14. $y = x^2 - 4, y = 2x - x^2$ about $x = -2$
15. $x = 1/\sqrt{y^4 + 12y^2}, x = 0, y = \frac{1}{2}, y = 1$ about $y = 0$
16. Find the first moment about the line $y = -2$ of a thin plate of constant density ρ (mass per unit area) if its edges are defined by the curves $y = 2 - 2x^2$ and $y = x^2 - 1$.

In Exercises 17–22, find the centroid of the area bounded by the curves.

17. $x = \sqrt{y + 2}, y = x, y = 0$
18. $y + x^2 = 0, x = y + 2, x + y + 2 = 0, y = 2$ (above $y + x^2 = 0$)
19. $y^2 = x^4(1 - x^2)$ (right loop)
20. $3x^2 + 4y^2 = 48, (x - 2)^2 + y^2 = 1$
21. $y = \ln x, y + \sqrt{x - 1} = 0, x = 2$
22. $y = \sqrt{2 - x}, 15y = x^2 - 4$

In Exercises 23–25, find the second moment of area of the region bounded by the curves about the line indicated.

23. $4x^2 + 9y^2 = 36$ about $y = -2$
24. $x = y^2$, $x + y = 2$ about $x = -1$
25. $y = \sqrt{a^2 - x^2}$, $y = a$, $x = a$ $(a > 0)$ about the x-axis
26. Find the first moment of area about the line $x + y = 1$ for the area bounded by the curves $x = y^2 - 2$ and $y = x$.
27. Prove the *theorem of Pappus*: If a plane area is revolved about a coplanar axis not crossing the area, the volume generated is equal to the product of the area and the circumference of the circle described by the centroid of the area.
28. A thin flat plate of area A is immersed vertically in a fluid of specific weight γ. Show that the total force (due to fluid pressure) on each side of the plate is equal to the product of A, γ, and the depth of the centroid of the plate below the surface of the fluid. Use this result to find the forces in some of the problems in Exercises 14.5, say 1, 3, 4, 5, 7, 11, 12, 13, 16, and 17. For those problems involving triangles, recall the result of Exercise 30 in Section 7.6.
29. Prove the *parallel axis theorem*: The moment of inertia of a thin plate (of constant density) with respect to any coplanar line is equal to the moment of inertia with respect to the parallel line through the center of mass plus the mass multiplied by the square of the distance between the lines.

14.7 SURFACE AREA

In Section 7.3, to find the length of a curve we approximated the curve by tangent line segments. To find the area of a surface we follow a similar procedure by approximating the surface with tangential planes. In particular, consider finding the area of a smooth surface S given that every vertical line that intersects the surface does so in exactly one point (Fig. 14.32). If S_{xy} is the area in the xy-plane onto which S projects, we divide S_{xy} into n subareas ΔA_i in any fashion whatsoever, and choose a point (x_i, y_i) in each ΔA_i. At the point (x_i, y_i, z_i) on the surface S that projects onto (x_i, y_i), we draw the tangent plane to S. Suppose we now project ΔA_i upward onto S and onto the tangent plane at (x_i, y_i, z_i) and denote these projected areas by ΔS_i and ΔS_{Ti}, respectively. Now

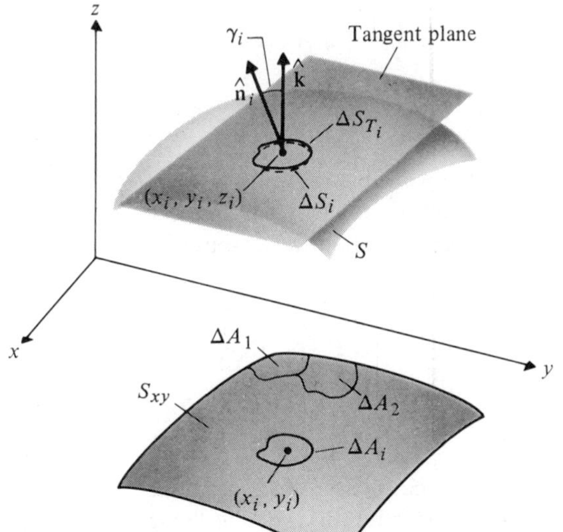

Figure 14.32

ΔS_{Ti} is an approximation to ΔS_i and, as long as ΔA_i is small, a reasonably good approximation. In fact, the smaller ΔA_i, the better the approximation. We therefore define the area of S as

$$\text{Area of } S = \lim_{n\to\infty} \sum_{i=1}^{n} \Delta S_{Ti}, \tag{14.37}$$

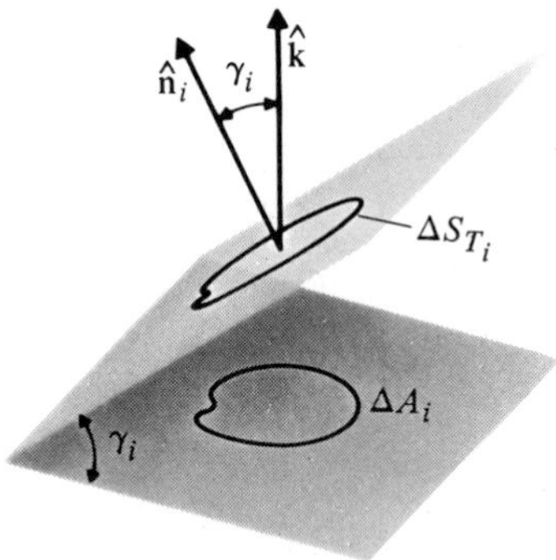

Figure 14.33

where in taking the limit we must make sure that each and every ΔA_i shrinks to a point. We have therefore defined area on a curved surface in terms of flat areas on tangent planes to the surface. The advantage of this definition is that we can calculate ΔS_{Ti} in terms of ΔA_i. To see how, we denote by $\hat{\mathbf{n}}_i$ the unit normal vector to S at (x_i, y_i, z_i) with positive z-component and by γ_i the acute angle between $\hat{\mathbf{n}}_i$ and $\hat{\mathbf{k}}$. Now ΔS_{Ti} projects onto ΔA_i and γ_i is the acute angle between the planes containing ΔA_i and ΔS_{Ti} (Fig. 14.33). It follows that ΔA_i and ΔS_{Ti} are related by

$$\Delta A_i = \cos\gamma_i \,\Delta S_{Ti} \tag{14.38}$$

(see Exercise 54 in Section 12.6). Note that if ΔS_{Ti} is horizontal, then $\gamma_i = 0$ and $\Delta S_{Ti} = \Delta A_i$; and if ΔS_{Ti} tends toward the vertical ($\gamma_i \to \pi/2$), then ΔS_{Ti} becomes very large for fixed ΔA_i.

Because the surface projects in a one-to-one fashion onto the area S_{xy} in the xy-plane, we can take the equation for S in the form $z = f(x, y)$. A vector normal to S at any point is therefore

$$\nabla(z - f(x, y)) = \left(-\frac{\partial f}{\partial x}, -\frac{\partial f}{\partial y}, 1\right);$$

hence

$$\hat{\mathbf{n}} = \frac{\left(-\frac{\partial f}{\partial x}, -\frac{\partial f}{\partial y}, 1\right)}{\sqrt{1 + \left(\frac{\partial f}{\partial x}\right)^2 + \left(\frac{\partial f}{\partial y}\right)^2}} = \frac{\left(-\frac{\partial z}{\partial x}, -\frac{\partial z}{\partial y}, 1\right)}{\sqrt{1 + \left(\frac{\partial z}{\partial x}\right)^2 + \left(\frac{\partial z}{\partial y}\right)^2}}.$$

Since $\hat{\mathbf{n}}_i \cdot \hat{\mathbf{k}} = |\hat{\mathbf{n}}_i||\hat{\mathbf{k}}| \cos\gamma_i = \cos\gamma_i$, it follows that

$$\cos\gamma_i = \hat{\mathbf{n}}_i \cdot \hat{\mathbf{k}} = \frac{1}{\sqrt{1 + z_x^2(x_i, y_i) + z_y^2(x_i, y_i)}}.$$

When we substitute this expression into (14.38), we obtain the result that area ΔS_{Ti} on the tangent plane to $z = f(x, y)$ at the point (x_i, y_i, z_i) is related to its projection ΔA_i in the xy-plane according to

$$\Delta S_{Ti} = \sqrt{1 + z_x^2(x_i, y_i) + z_y^2(x_i, y_i)}\,\Delta A_i. \tag{14.39}$$

Definition (14.37) for the area of S can now be written in the form

$$\text{Area of } S = \lim_{n\to\infty} \sum_{i=1}^{n} \sqrt{1 + z_x^2(x_i, y_i) + z_y^2(x_i, y_i)}\,\Delta A_i, \tag{14.40}$$

where the summation is carried out over all areas ΔA_i in S_{xy} as each ΔA_i shrinks to a point. But this is precisely the definition of the double integral of the function $\sqrt{1 + (\partial z/\partial x)^2 + (\partial z/\partial y)^2}$ over the region S_{xy}. In other words, on the basis of

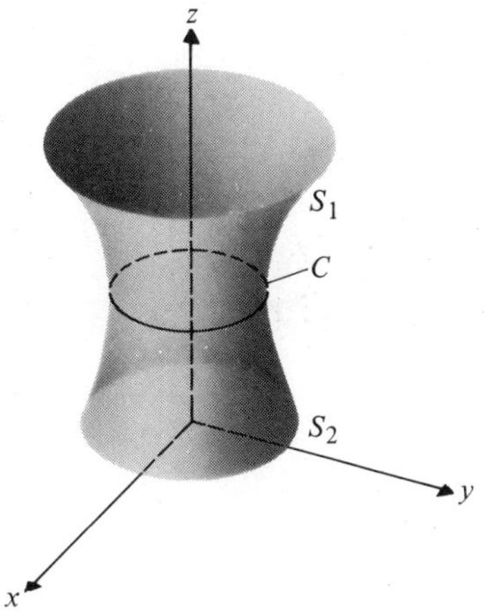

Figure 14.34

definition (14.37), areas on surfaces can be calculated according to

$$\text{Area of } S = \int_{S_{xy}} \sqrt{1 + \left(\frac{\partial z}{\partial x}\right)^2 + \left(\frac{\partial z}{\partial y}\right)^2}\, dA \tag{14.41}$$

Note the analogy between Eqs. (14.41) and (7.12). In Eq. (7.12) we think of $\sqrt{1 + (dy/dx)^2}\, dx$ as a small length along a curve $C: y = f(x)$ that projects onto the length dx along the x-axis. In actual fact, $\sqrt{1 + (dy/dx)^2}\, dx$ is along the tangent line to C, but we think of it as along C itself (see Section 12.9). The total length of C is then found by adding over all projections of C from $x = a$ to $x = b$. Similarly, in Eq. (14.41) we think of $\sqrt{1 + (\partial z/\partial x)^2 + (\partial z/\partial y)^2}\, dA$ as a small area on a surface $S: z = f(x, y)$ that projects onto the area dA in the xy-plane. It is, in fact, a small area on the tangent plane to S, but it is usually easier to think of it as being on S itself. The total area of S is then the addition over all projections S_{xy} of S.

Note too that when S is smooth, $\partial z/\partial x$ and $\partial z/\partial y$ are continuous, thus guaranteeing existence of the double integral (14.41). If S is piecewise smooth, we simply divide it into smooth subsurfaces and integrate over each piece separately.

This discussion has been based on the assumption that S projects one-to-one onto some region S_{xy} in the xy-plane. If this condition is not met, one possibility is to subdivide S into parts, each of which projects one-to-one onto the xy-plane. The total area of S is then the sum of the areas of its parts. For instance, to find the area of the surface S shown in Fig. 14.34, we could subdivide S into S_1 and S_2 along the curve C. The area of S is then the sum of the areas of S_1 and S_2, each of which projects one-to-one onto the xy-plane.

Alternatively, we could note that there is nothing sacred about projecting S onto the xy-plane. We could develop similar results if S projects one-to-one onto either the yz- or the xz-plane. If S_{yz} and S_{xz} represent these projections, then

$$\text{Area of } S = \iint_{S_{yz}} \sqrt{1 + \left(\frac{\partial x}{\partial y}\right)^2 + \left(\frac{\partial x}{\partial z}\right)^2}\, dA, \tag{14.42a}$$

$$\text{Area of } S = \iint_{S_{xz}} \sqrt{1 + \left(\frac{\partial y}{\partial x}\right)^2 + \left(\frac{\partial y}{\partial z}\right)^2}\, dA. \tag{14.42b}$$

Example 12 Find the area of that part of the plane $x + 2y + 3z = 6$ in the first octant.

Solution This area projects one-to-one onto the triangular area S_{xy} in the xy-plane bounded by the lines $x = 0$, $y = 0$, $x + 2y = 6$ (Fig. 14.35). Since $z = 2 - x/3 - 2y/3$,

$$\text{Area} = \iint_{S_{xy}} \sqrt{1 + \left(\frac{\partial z}{\partial x}\right)^2 + \left(\frac{\partial z}{\partial y}\right)^2}\, dA = \iint_{S_{xy}} \sqrt{1 + \left(-\frac{1}{3}\right)^2 + \left(-\frac{2}{3}\right)^2}\, dA$$

$$= \frac{\sqrt{14}}{3} \iint_{S_{xy}} dA = \frac{\sqrt{14}}{3} (\text{Area of } S_{xy}) = \frac{\sqrt{14}}{3} \left(\frac{1}{2}(3)(6)\right) = 3\sqrt{14}. \square$$

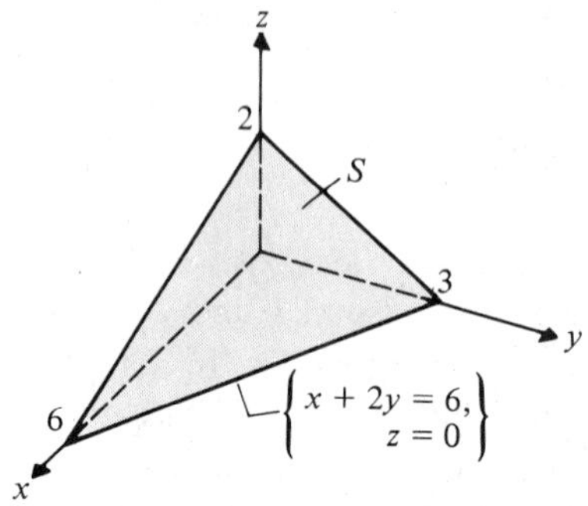

Figure 14.35

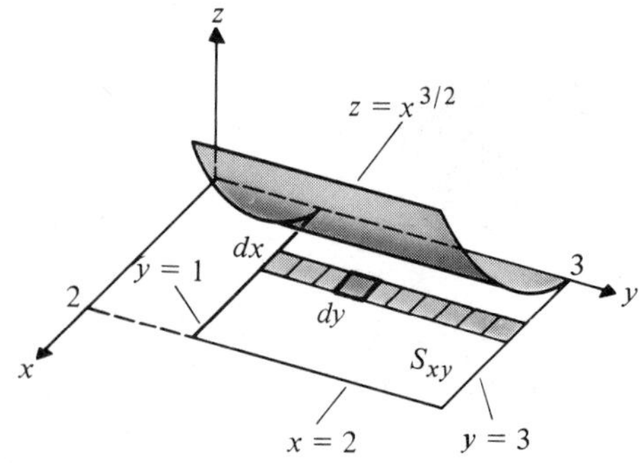

Figure 14.36

Example 13 Find the area of the surface $z = x^{3/2}$ that projects onto the rectangle in the xy-plane bounded by the straight lines $x = 0$, $x = 2$, $y = 1$, and $y = 3$.

Solution Since the surface projects one-to-one onto the rectangle (Fig. 14.36), we find that

$$\begin{aligned}
\text{Area} &= \iint_{S_{xy}} \sqrt{1 + \left(\frac{\partial z}{\partial x}\right)^2 + \left(\frac{\partial z}{\partial y}\right)^2}\, dA = \iint_{S_{xy}} \sqrt{1 + \left(\frac{3}{2}x^{1/2}\right)^2}\, dA \\
&= \frac{1}{2}\int_0^2 \int_1^3 \sqrt{4 + 9x}\, dy\, dx = \frac{1}{2}\int_0^2 \{y\sqrt{4 + 9x}\}_1^3\, dx = \int_0^2 \sqrt{4 + 9x}\, dx \\
&= \left\{\frac{2}{27}(4 + 9x)^{3/2}\right\}_0^2 = \frac{2}{27}(22\sqrt{22} - 8). \ \square
\end{aligned}$$

Example 14 Find the area of the cone $y = \sqrt{x^2 + z^2}$ to the left of the plane $y = 1$.

Solution ***Method 1*** The surface projects one-to-one onto the interior of the circle $x^2 + z^2 = 1$ in the xz-plane (Fig. 14.37). Since the area of the surface is four times that in the first octant, if we let S_{xz} be the quarter-circle $x^2 + z^2 \leq 1$, $x \geq 0$, $z \geq 0$, then

$$\begin{aligned}
\text{Area} &= 4\iint_{S_{xz}} \sqrt{1 + \left(\frac{\partial y}{\partial x}\right)^2 + \left(\frac{\partial y}{\partial z}\right)^2}\, dA \\
&= 4\iint_{S_{xz}} \sqrt{1 + \left(\frac{x}{\sqrt{x^2 + z^2}}\right)^2 + \left(\frac{z}{\sqrt{x^2 + z^2}}\right)^2}\, dA \\
&= 4\iint_{S_{xz}} \sqrt{1 + \frac{x^2}{x^2 + z^2} + \frac{z^2}{x^2 + z^2}}\, dA = 4\iint_{S_{xz}} \sqrt{2}\, dA \\
&= 4\sqrt{2}\,(\text{Area of } S_{xz}) = 4\sqrt{2}\,\{\tfrac{1}{4}\pi(1)^2\} = \sqrt{2}\pi.
\end{aligned}$$

Method 2 Suppose instead that we project that part of the surface in the first octant onto the triangle S_{xy} in Fig. 14.38. We write the equation of this part of the

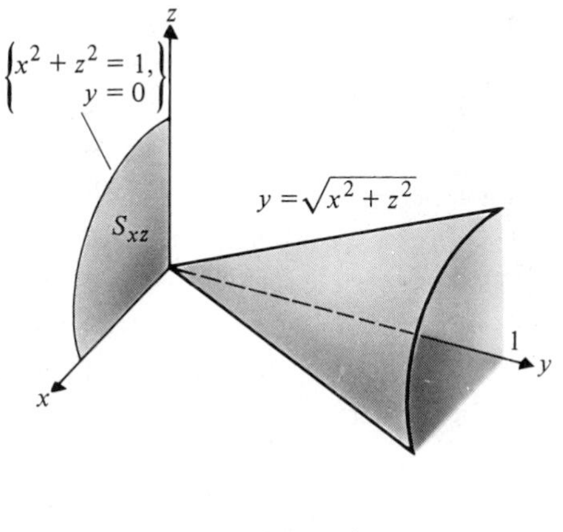

Figure 14.37

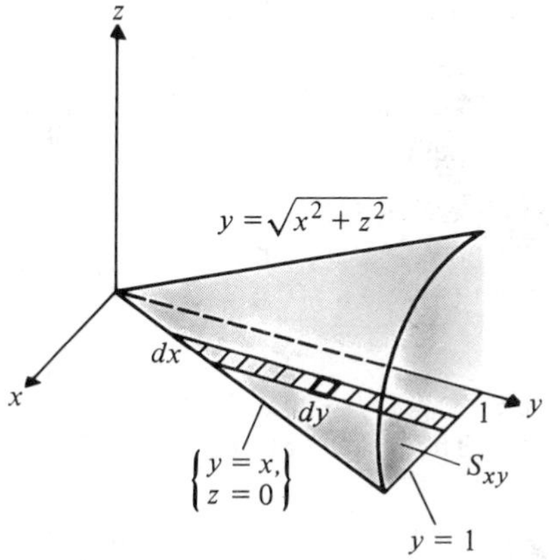

Figure 14.38

surface in the form $z = \sqrt{y^2 - x^2}$ and calculate

$$\begin{aligned}\text{Area} &= 4 \iint_{S_{xy}} \sqrt{1 + \left(\frac{\partial z}{\partial x}\right)^2 + \left(\frac{\partial z}{\partial y}\right)^2}\, dA \\ &= 4 \iint_{S_{xy}} \sqrt{1 + \left(\frac{-x}{\sqrt{y^2 - x^2}}\right)^2 + \left(\frac{y}{\sqrt{y^2 - x^2}}\right)^2}\, dA \\ &= 4 \iint_{S_{xy}} \sqrt{1 + \frac{x^2}{y^2 - x^2} + \frac{y^2}{y^2 - x^2}}\, dA = 4\sqrt{2} \iint_{S_{xy}} \frac{y}{\sqrt{y^2 - x^2}}\, dA.\end{aligned}$$

To evaluate this double integral, it is advantageous to integrate first with respect to y:

$$\begin{aligned}\text{Area} &= 4\sqrt{2} \int_0^1 \int_x^1 \frac{y}{\sqrt{y^2 - x^2}}\, dy\, dx = 4\sqrt{2} \int_0^1 \{\sqrt{y^2 - x^2}\}_x^1\, dx \\ &= 4\sqrt{2} \int_0^1 \sqrt{1 - x^2}\, dx.\end{aligned}$$

If we now set $x = \sin\theta$, then $dx = \cos\theta\, d\theta$, and

$$\begin{aligned}\text{Area} &= 4\sqrt{2} \int_0^{\pi/2} \cos\theta \cos\theta\, d\theta = 4\sqrt{2} \int_0^{\pi/2} \left\{\frac{1 + \cos 2\theta}{2}\right\} d\theta \\ &= 2\sqrt{2} \left\{\theta + \frac{1}{2}\sin 2\theta\right\}_0^{\pi/2} = \sqrt{2}\pi. \square\end{aligned}$$

EXERCISES 14.7

In Exercises 1–6, find the required area.

1. The area of $2x + 3y + 6z = 1$ in the first octant
2. The area of $x + 2y - 3z + 4 = 0$ for which $x \leq 0, y \leq 0$, and $z \geq 0$
3. The area of $z = 1 - 4\sqrt{x^2 + y^2}$ above the xy-plane
4. The area of $z = \sqrt{2xy}$ cut out by the planes $x = 1, x = 2, y = 1, y = 3$

5. The area in the first octant cut out from the surface $z = x + y$ by the plane $x + 2y = 4$
6. The area of $z = x^{3/2} + y^{3/2}$ in the first octant cut off by the plane $x + y = 1$

In Exercises 7–12, set up, but do not evaluate, double iterated integrals to find the required area.

7. The area of $x^2 + y^2 + z^2 = 2$ inside the cone $z = \sqrt{x^2 + y^2}$
8. The area of $4x = y^2 + z^2$ cut off by $x = 4$
9. The area in the first octant cut from $y = xz$ by the cylinder $x^2 + z^2 = 1$
10. The area of $z = (x^2 + y^2)^2$ below $z = 4$
11. The area of $y = 1 - x^2 - 3z^2$ to the right of the xz-plane
12. The area of $z = \ln(1 + x + y)$ in the first octant cut off by $y = 1 - x^2$
13. Find the area of the surface $z = \ln x$ that projects onto the rectangle in the xy-plane bounded by the lines $x = 1$, $x = 2$, $y = 0$, $y = 2$.
14. Verify that the area of the curved portion of a right-circular cone of radius r and height h is $\pi r\sqrt{r^2 + h^2}$.

In Exercises 15–19, set up, but do not evaluate, double iterated integrals to find the required area.

15. The area of $y = x^2 + z^2$ cut off by $y + z = 1$
16. The area of $y = z^2 + x$ inside $x^2 + y^2 = 1$
17. The area of $y^2 = z + x^2$ inside $x^2 + y^2 = 4$
18. The area of $z = x^3 + y^3$ that is in the first octant and between the planes $x + y = 1$ and $x + y = 2$
19. The area of $x^2 + y^2 = z^2 + 1$ between the planes $z = 1$ and $z = 4$

14.8 DOUBLE ITERATED INTEGRALS IN POLAR COORDINATES

So far we have used only double iterated integrals in x and y to evaluate double integrals. But for some problems this is not convenient. For instance, the double integral of a continuous function $f(x, y)$ over the region R in Fig. 14.39 requires three double iterated integrals in x and y. In other words, a subdivision of R into rectangles by coordinate lines x = constant and y = constant is simply not convenient for this region. We will see that for such an area, polar coordinates can be very useful.

Polar coordinates with the origin as pole and the positive x-axis as polar axis are defined by

$$x = r\cos\theta, \qquad y = r\sin\theta$$

(see Section 10.2). We wish to obtain double iterated integrals in polar coordinates that represent the double integral of $f(x, y)$ over the region R in Fig. 14.39. To do this we return to definition (14.3) for a double integral, and choose a subdivision of R into subareas convenient to polar coordinates. When using Cartesian coordinates we drew coordinate lines x = constant and y = constant. When using polar coordinates we draw coordinate curves r = constant and θ = constant. In particular, we subdivide R by a network of $n + 1$ circles $r = r_i$, where

$$a = r_0 < r_1 < r_2 < \cdots < r_{n-1} < r_n = b,$$

and $m + 1$ radial lines $\theta = \theta_j$, where

$$c = \theta_0 < \theta_1 < \theta_2 < \cdots < \theta_{m-1} < \theta_m = d$$

(Fig. 14.40). If ΔA_{ij} represents the area bounded by the circles $r = r_{i-1}$ and $r = r_i$

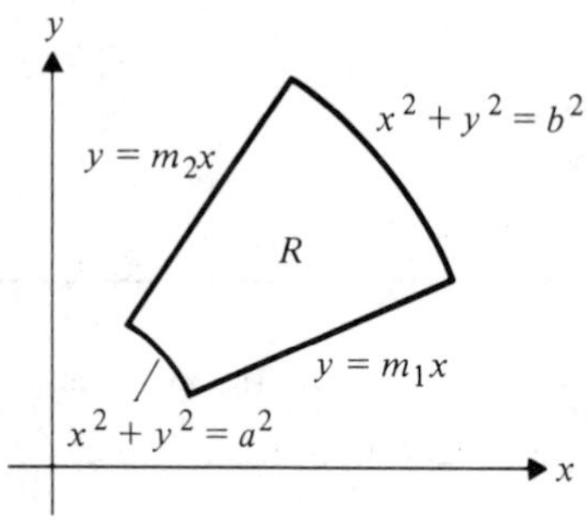

Figure 14.39

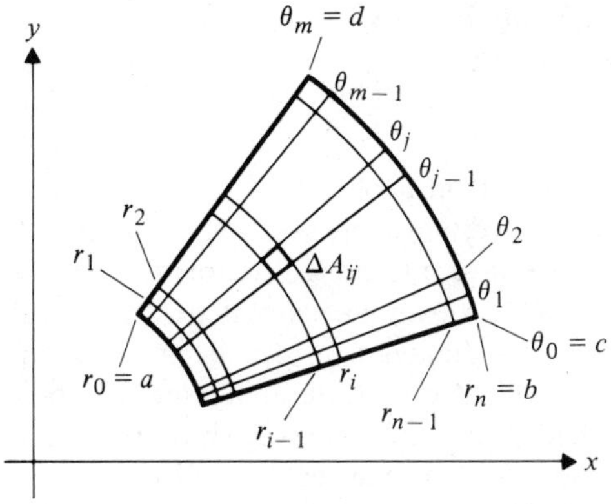

Figure 14.40

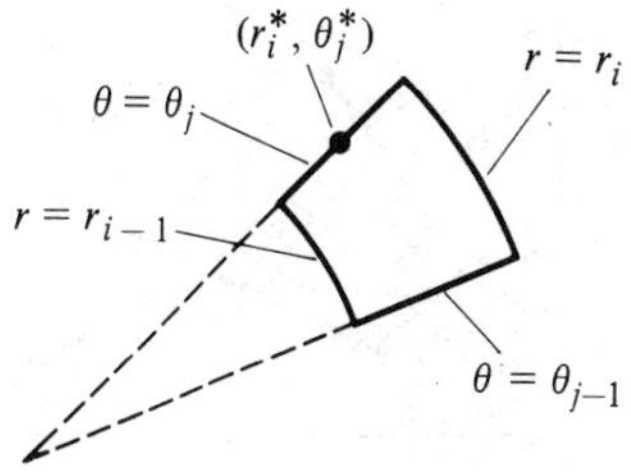

Figure 14.41

and the radial lines $\theta = \theta_{j-1}$ and $\theta = \theta_j$ (Fig. 14.41), then it is straightforward to show that

$$\Delta A_{ij} = \frac{1}{2}(r_i^2 - r_{i-1}^2)(\theta_j - \theta_{j-1}).$$

If we set $\Delta r_i = r_i - r_{i-1}$ and $\Delta\theta_j = \theta_j - \theta_{j-1}$, then

$$\begin{aligned}\Delta A_{ij} &= \frac{1}{2}(r_i + r_{i-1})(r_i - r_{i-1})(\theta_j - \theta_{j-1}) \\ &= \left[\frac{r_i + r_{i-1}}{2}\right]\Delta r_i \Delta\theta_j.\end{aligned} \tag{14.43}$$

Our next task in using definition (14.3) for the double integral of $f(x, y)$ over R is to choose a star-point in each ΔA_{ij}. If we select

$$(r_i^*, \theta_j^*) = \left(\frac{r_i + r_{i-1}}{2}, \theta_j\right),$$

then

$$\Delta A_{ij} = r_i^* \,\Delta r_i \,\Delta\theta_j,$$

and by definition (14.3),

$$\begin{aligned}\iint_R f(x, y)\, dA &= \lim_{\substack{n\to\infty \\ m\to\infty}} \sum_{j=1}^{m} \sum_{i=1}^{n} f(r_i^* \cos\theta_j^*, r_i^* \sin\theta_j^*)\,\Delta A_{ij} \\ &= \lim_{\substack{n\to\infty \\ m\to\infty}} \sum_{j=1}^{m} \sum_{i=1}^{n} f(r_i^* \cos\theta_j^*, r_i^* \sin\theta_j^*) r_i^* \,\Delta r_i \,\Delta\theta_j,\end{aligned}$$

where it is understood that every $\Delta r_i \to 0$ and every $\Delta\theta_j \to 0$ as $n \to \infty$ and $m \to \infty$. If we take the limit first on n and then on m (as in Section 14.3), we obtain the double iterated integral

$$\iint_R f(x, y)\, dA = \int_c^d \int_a^b f(r\cos\theta, r\sin\theta) r \, dr\, d\theta. \tag{14.44}$$

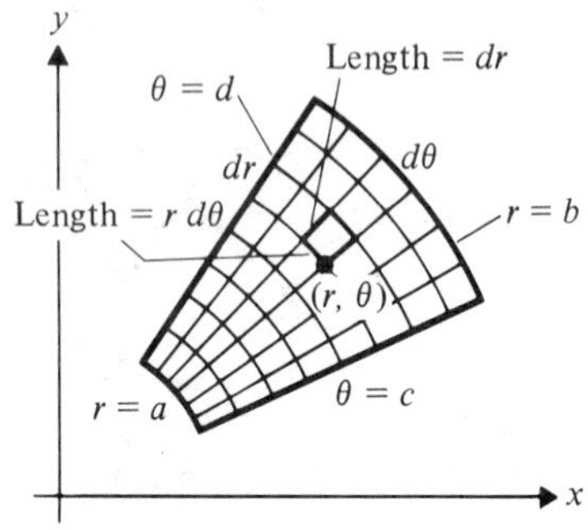

Figure 14.42

Reversing the order of taking limits reverses the order of the iterated integrals:

$$\iint_R f(x, y)\, dA = \int_a^b \int_c^d f(r\cos\theta, r\sin\theta) r\, d\theta\, dr. \tag{14.45}$$

For the region R of Fig. 14.40, then, there are two double iterated integrals in polar coordinates representing the double integral of $f(x, y)$ over R.

We have interpreted double iterated integrals in Cartesian coordinates as integrations over horizontal or vertical strips. Double iterated integrals in polar coordinates can also be interpreted geometrically. Take, for instance, Eq. (14.44). A double iterated integral in polar coordinates implies a subdivision of the region R into areas as shown in Fig. 14.41. Let us denote small variations in r and θ for a representative piece of area at position (r, θ) by dr and $d\theta$ (Fig. 14.42). If dr and $d\theta$ are very small (as is implied in the definition of the double integral), then this piece of area is almost rectangular with an approximate area of $(r\, d\theta)\, dr$. In polar coordinates, then, we think of dA in (14.44) as being replaced by

$$dA = r\, dr\, d\theta. \tag{14.46}$$

Each such area at (r, θ) is multiplied by the value of $f(x, y)$ at (r, θ) to give the product

$$f(r\cos\theta, r\sin\theta) r\, dr\, d\theta.$$

The inner integral

$$\int_a^b f(r\cos\theta, r\sin\theta) r\, dr\, d\theta$$

with respect to r holds θ constant and is therefore interpreted as a summation over the small areas in a wedge $d\theta$ from $r = a$ to $r = b$. The θ-integration then adds over all wedges starting at $\theta = c$ and ending at $\theta = d$. Limits on θ therefore identify the positions of the first and last wedges.

If the order of integration is reversed (Eq. 14.45), then the inner integral

$$\int_c^d f(r\cos\theta, r\sin\theta) r\, d\theta\, dr$$

holds r constant. We interpret this as an addition over the small areas in a ring dr, where the limits indicate that each and every ring starts on the curve $\theta = c$ and ends on the curve $\theta = d$. The outer r-integration is an addition over all rings with the first ring at $r = a$ and the last at $r = b$.

Double iterated integrals in polar coordinates for more general regions are now quite simple. For the region R of Fig. 14.43,

$$\iint_R f(x, y)\, dA = \int_\alpha^\beta \int_{g(\theta)}^{h(\theta)} f(r\cos\theta, r\sin\theta) r\, dr\, d\theta.$$

Example 15 Find the area inside the circle $x^2 + y^2 = 4x$ and outside the circle $x^2 + y^2 = 1$.

Solution If R is the region bounded by these circles and above the x-axis (Fig. 14.44), then the required area is

$$2\iint_R dA.$$

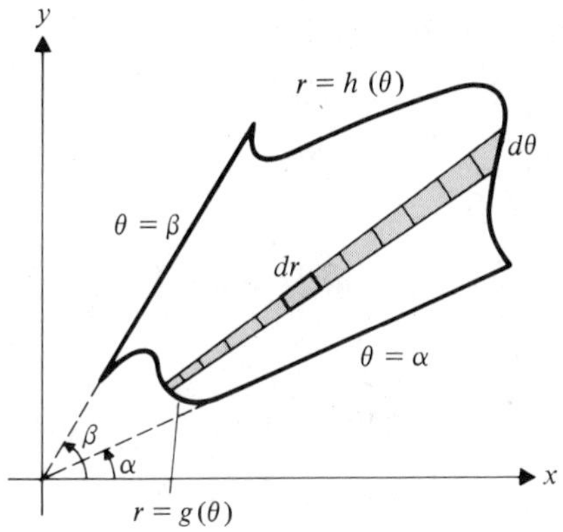

Figure 14.43

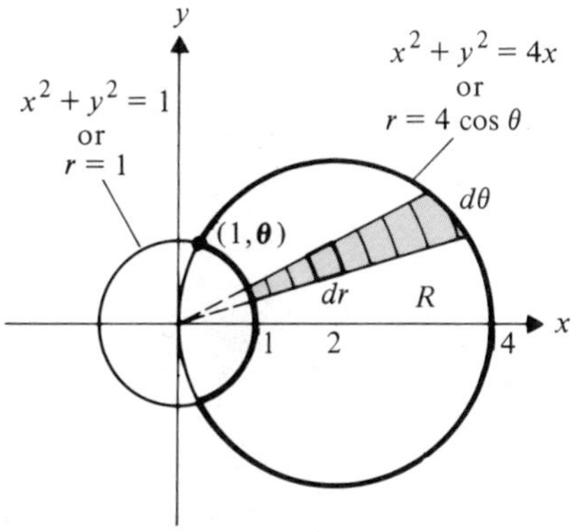

Figure 14.44

Since the curves intersect in the first quadrant at a point where $\theta = \boldsymbol{\theta} = \mathrm{Cos}^{-1}(\frac{1}{4})$, then

$$\begin{aligned}\text{Area} &= 2\int_0^{\boldsymbol{\theta}}\int_1^{4\cos\theta} r\,dr\,d\theta = 2\int_0^{\boldsymbol{\theta}}\left\{\frac{r^2}{2}\right\}_1^{4\cos\theta} d\theta = \int_0^{\boldsymbol{\theta}}(16\cos^2\theta - 1)\,d\theta \\ &= \int_0^{\boldsymbol{\theta}}\left\{16\left(\frac{1+\cos 2\theta}{2}\right) - 1\right\}d\theta = \int_0^{\boldsymbol{\theta}}(7 + 8\cos 2\theta)\,d\theta = \{7\theta + 4\sin 2\theta\}_0^{\boldsymbol{\theta}} \\ &= 7\boldsymbol{\theta} + 4\sin 2\boldsymbol{\theta} = 7\,\mathrm{Cos}^{-1}(\tfrac{1}{4}) + 8\cos\boldsymbol{\theta}\sin\boldsymbol{\theta} \\ &= 7\,\mathrm{Cos}^{-1}(\tfrac{1}{4}) + 8(\tfrac{1}{4})\sqrt{1 - \tfrac{1}{16}} = 7\,\mathrm{Cos}^{-1}(\tfrac{1}{4}) + \sqrt{15}/2. \;\square\end{aligned}$$

Example 16 Evaluate the double iterated integral

$$\int_0^1\int_0^{\sqrt{-x^2+x}} y^2\,dy\,dx.$$

Solution The limits identify the region of integration as the interior of the semicircle shown in Fig. 14.45. The integrand suggests an interpretation of the integral as the second moment of area of this semicircle about the x-axis. Since the semicircle R in Fig. 14.46 has exactly the same second moment about the x-axis, we can state that

$$\begin{aligned}\int_0^1\int_0^{\sqrt{-x^2+x}} y^2\,dy\,dx &= \iint_R y^2\,dA = \int_0^{\pi}\int_0^{1/2}(r^2\sin^2\theta)r\,dr\,d\theta = \int_0^{\pi}\frac{\sin^2\theta}{64}\,d\theta \\ &= \frac{1}{128}\int_0^{\pi}(1 - \cos 2\theta)\,d\theta = \frac{1}{128}\left\{\theta - \frac{\sin 2\theta}{2}\right\}_0^{\pi} = \frac{\pi}{128}. \;\square\end{aligned}$$

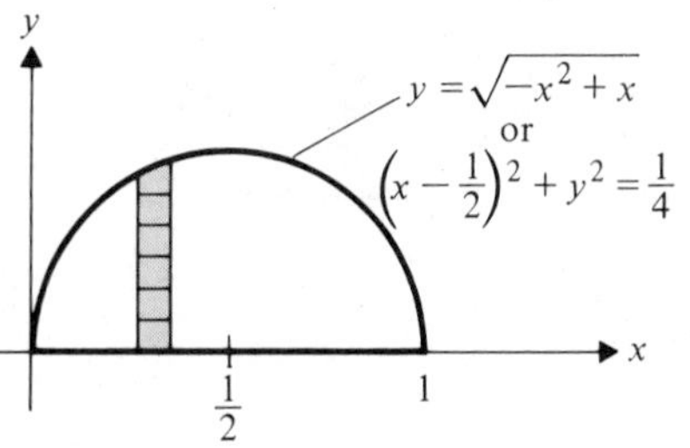

Figure 14.45

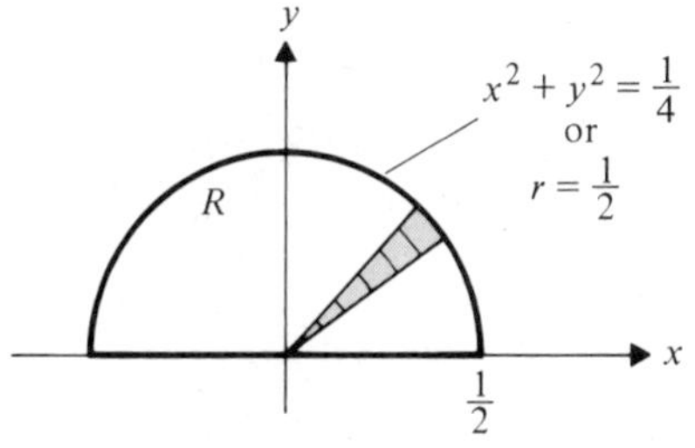

Figure 14.46

Example 17 Find the centroid of the region R in Fig. 14.47.

Solution Evidently, $\bar{x} = 0$, and the area of the region is $A = (\pi b^2 - \pi a^2)/2$. Since

$$A\bar{y} = \iint_R y\, dA = \int_0^\pi \int_a^b (r \sin\theta) r\, dr\, d\theta = \int_0^\pi \frac{1}{3}(b^3 - a^3)\sin\theta\, d\theta$$

$$= \frac{1}{3}(b^3 - a^3)\{-\cos\theta\}_0^\pi = \frac{2}{3}(b^3 - a^3),$$

it follows that

$$\bar{y} = \frac{2}{3}(b^3 - a^3)\frac{2}{\pi(b^2 - a^2)} = \frac{4}{3\pi}\frac{b^2 + ab + a^2}{a + b}. \square$$

Example 18 Find the area of that portion of the sphere $x^2 + y^2 + z^2 = 2$ inside the cone $z = \sqrt{x^2 + y^2}$.

Solution If S is that portion of the sphere that is inside the cone and also in the first octant (Fig. 14.48), then the required area is four times that of S; i.e.,

$$\text{Area} = 4\iint_{S_{xy}} \sqrt{1 + \left(\frac{\partial z}{\partial x}\right)^2 + \left(\frac{\partial z}{\partial y}\right)^2}\, dA,$$

where S_{xy} is the projection of S on the xy-plane. The curve of intersection of the cone and the sphere has equations

$$\begin{aligned} x^2 + y^2 + z^2 &= 2, \\ z &= \sqrt{x^2 + y^2}, \end{aligned} \qquad \text{or equivalently} \qquad \begin{aligned} z &= 1, \\ x^2 + y^2 &= 1. \end{aligned}$$

Consequently, S_{xy} is the interior of the quarter-circle $x^2 + y^2 \le 1$, $x \ge 0$, $y \ge 0$. On S,

$$\frac{\partial z}{\partial x} = -\frac{x}{z} \quad \text{and} \quad \frac{\partial z}{\partial y} = -\frac{y}{z},$$

so that

$$\text{Area} = 4\iint_{S_{xy}} \sqrt{1 + \frac{x^2}{z^2} + \frac{y^2}{z^2}}\, dA = 4\iint_{S_{xy}} \sqrt{\frac{x^2 + y^2 + z^2}{z^2}}\, dA$$

$$= 4\iint_{S_{xy}} \sqrt{\frac{2}{z^2}}\, dA = 4\sqrt{2}\iint_{S_{xy}} \frac{1}{\sqrt{2 - x^2 - y^2}}\, dA.$$

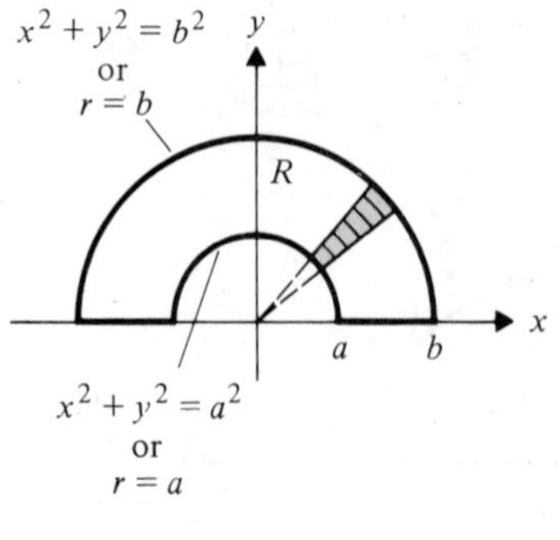

Figure 14.47

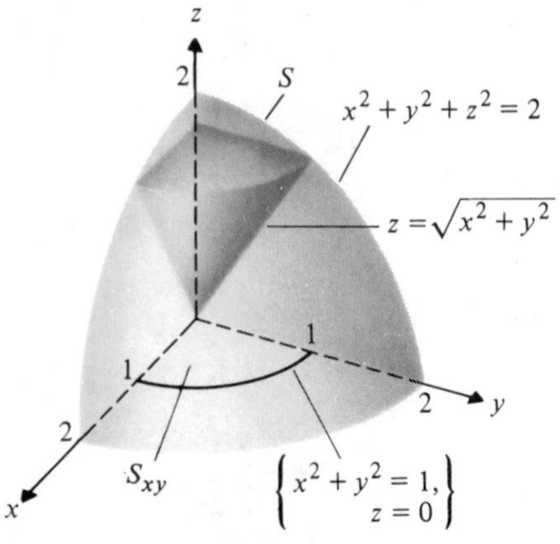

Figure 14.48

If we now use polar coordinates to evaluate this double integral, we have

$$\text{Area} = 4\sqrt{2}\int_0^{\pi/2}\int_0^1 \frac{1}{\sqrt{2-r^2}}\, r\,dr\,d\theta = 4\sqrt{2}\int_0^{\pi/2}\{-\sqrt{2-r^2}\}_0^1\,d\theta$$
$$= 4\sqrt{2}(\sqrt{2}-1)\int_0^{\pi/2} d\theta = 4\sqrt{2}(\sqrt{2}-1)\frac{\pi}{2} = 2\sqrt{2}\pi(\sqrt{2}-1). \square$$

EXERCISES 14.8

In each of Exercises 1–5, evaluate the double integral of the function over the region R.

1. $f(x, y) = e^{x^2+y^2}$, where R is bounded by $x^2 + y^2 = a^2$
2. $f(x, y) = x$, where R is bounded by $x = \sqrt{2y - y^2}$, $x = 0$
3. $f(x, y) = \sqrt{x^2 + y^2}$, where R is bounded by $y = \sqrt{9 - x^2}$, $y = x$, $x = 0$
4. $f(x, y) = 1/\sqrt{x^2 + y^2}$, where R is the region outside $x^2 + y^2 = 4$ and inside $x^2 + y^2 = 4x$
5. $f(x, y) = \sqrt{1 + 2x^2 + 2y^2}$, where R is bounded by $x^2 + y^2 = 1, x^2 + y^2 = 4$

Evaluate the double iterated integral in Exercises 6 and 7.

6. $\displaystyle\int_0^1\int_0^{\sqrt{1-x^2}} \sqrt{x^2 + y^2}\,dy\,dx$
7. $\displaystyle\int_{-\sqrt{2}}^0\int_{-y}^{\sqrt{4-y^2}} x^2\,dx\,dy$

In each of Exercises 8–12, find the area bounded by the curves.

8. Outside $x^2 + y^2 = 9$ and inside $x^2 + y^2 = 2\sqrt{3}y$
9. $r = 9(1 + \cos\theta)$
10. $r = \cos 3\theta$
11. Common to $r = 2$ and $r^2 = 9\cos 2\theta$
12. Common to $r = 1 + \sin\theta$ and $r = 2 - 2\sin\theta$
13. Find the centroid of the area bounded by the curves $y = x$, $y = -x$, and $x = \sqrt{2 - y^2}$.
14. Find the second moment of area for a circular plate of radius R about any diameter.
15. A water tank in the form of a right-circular cylinder with radius R and length h has its axis horizontal. If it is full, what is the force due to water pressure on each end?

In Exercises 16–18, find the area of the surface described.

16. The area of $z = x^2 + y^2$ below $z = 4$
17. The area of $x^2 + y^2 + z^2 = 4$ inside $x^2 + y^2 = 1$
18. The area of $z = xy$ inside $x^2 + y^2 = 9$
19. Prove that the area of a sphere of radius R is $4\pi R^2$.

In Exercises 20 and 21, find the volume of the solid of revolution obtained by rotating the area bounded by the curve about the line indicated.

20. $r = \cos^2\theta$ about the x-axis
21. $r = 1 + \sin\theta$ about the y-axis
22. A circular plate of radius R (Fig. 14.49) has a uniform charge distribution of ρ coulombs per square meter. If P is a point directly above the center of the plate and dA is a small area on the plate, then the potential at P due to dA is given by

$$\frac{1}{4\pi\varepsilon_0}\frac{\rho\,dA}{s},$$

where s is the distance from P to dA.

a) Show that, in terms of polar coordinates, the potential at P due to the entire plate is

$$V = \frac{\rho}{4\pi\varepsilon_0}\int_{-\pi}^{\pi}\int_0^R \frac{r}{\sqrt{r^2 + d^2}}\,dr\,d\theta.$$

b) Evaluate the double iterated integral to find V.

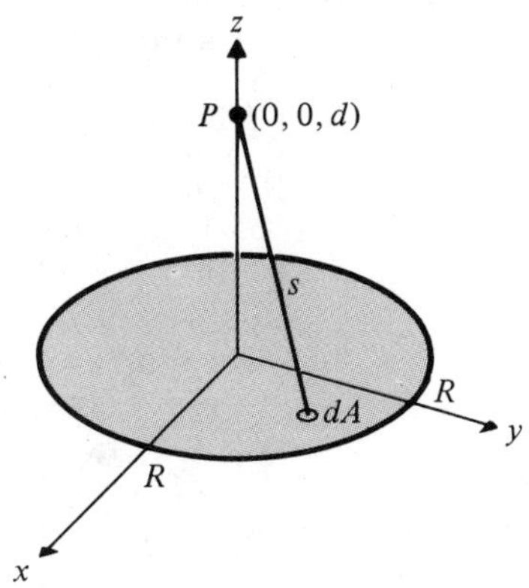

Figure 14.49

In Exercises 23–25, find the area bounded by the curves.

23. $(x^2 + y^2)^2 = 2xy$
24. Inside both $r = 6\cos\theta$ and $r = 4 - 2\cos\theta$
25. $r = \cos^2\theta\sin\theta$
26. Find the centroid of the cardioid $r = 1 + \cos\theta$.

27. Find the second moment of area about the x-axis for the area bounded by $r^2 = 9 \cos 2\theta$.

28. Figure 14.50 illustrates a piece of an artery or vein with circular cross-section (radius R). The speed of blood flowing through the blood vessel is not uniform because of the viscosity of the blood and friction at the walls. Poiseuille's law states that for laminar blood flow, the speed of blood at a distance r from the center of the blood vessel is given by

$$v = \frac{P}{4nL}(R^2 - r^2),$$

where P is the pressure difference between the ends of the blood vessel, L is the length of the vessel, and n is the viscosity of the blood. Find the amount of blood flowing over a cross-section of the blood vessel per unit time.

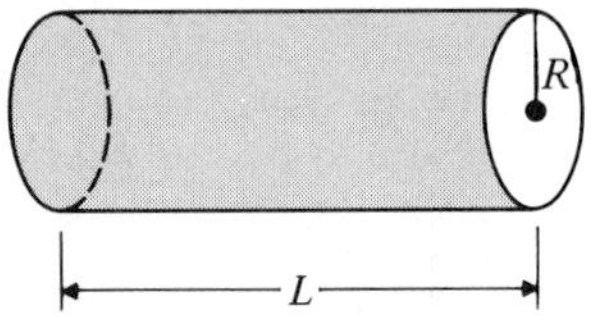

Figure 14.50

***29.** Find the area of that portion of the surface $x^2 + z^2 = a^2$ cut out by $x^2 + y^2 = a^2$.

***30.** A very important integral in statistics is

$$I = \int_0^\infty e^{-x^2}\,dx.$$

To evaluate this integral we set

$$I = \int_0^\infty e^{-y^2}\,dy,$$

and then multiply these two equations. Do this to prove that $I = \sqrt{\pi}/2$.

***31.** A circular pasture has radius R. A cow is tied to a stake at the edge of the pasture. What length of rope will permit the cow to graze on exactly half the pasture?

14.9 TRIPLE INTEGRALS AND TRIPLE ITERATED INTEGRALS

Triple integrals are defined in much the same way as double integrals. Suppose $f(x, y, z)$ is a function defined in some region V of space that has finite volume (Fig. 14.51). We divide V into n subregions of volumes $\Delta V_1, \Delta V_2, \ldots, \Delta V_n$ in any manner whatsoever, and in each subregion ΔV_i ($i = 1, \ldots, n$) we choose an arbitrary point (x_i^*, y_i^*, z_i^*). We then form the sum

$$f(x_1^*, y_1^*, z_1^*)\,\Delta V_1 + f(x_2^*, y_2^*, z_2^*)\,\Delta V_2 + \cdots + f(x_n^*, y_n^*, z_n^*)\,\Delta V_n$$

$$= \sum_{i=1}^{n} f(x_i^*, y_i^*, z_i^*)\,\Delta V_i. \qquad (14.47)$$

If this sum approaches a limit as the number of subregions becomes increasingly large and every subregion shrinks to a point, we call the limit the triple integral of $f(x, y, z)$ over the region V and denote it by

$$\iiint_V f(x, y, z)\,dV = \lim_{n\to\infty} \sum_{i=1}^{n} f(x_i^*, y_i^*, z_i^*)\,\Delta V_i. \qquad (14.48)$$

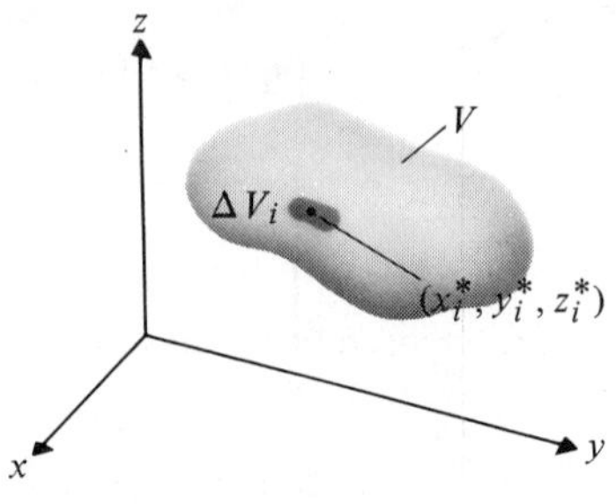

Figure 14.51

As in the case of double integrals, we require that this limit be independent of the manner of subdivision of V and the choice of star-points in the subregions. This is guaranteed for continuous functions by the following theorem.

▶ **Theorem 14.2** Let S be a piecewise-smooth surface that encloses a region V with finite volume. If $f(x, y, z)$ is continuous inside and on S, then the triple integral of $f(x, y, z)$ over V exists. ◀

Properties analogous to those in Eqs. (14.4)–(14.7) hold for triple integrals, although we will not list the first three here. Corresponding to Eq. (14.7), the volume of a region V is given by the triple integral

$$\text{Volume of } V = \iiint_V dV. \tag{14.49}$$

We will evaluate triple integrals by triple iterated integrals. If we use Cartesian coordinates there are six possible triple iterated integrals of a function $f(x, y, z)$, corresponding to the six permutations of the product of the differentials dx, dy, and dz:

$$dz\,dy\,dx, \qquad dz\,dx\,dy, \qquad dx\,dz\,dy, \qquad dx\,dy\,dz, \qquad dy\,dx\,dz, \qquad dy\,dz\,dx.$$

The general triple iterated integral of $f(x, y, z)$ with respect to z, y, and x is of the form

$$\int_a^b \int_{g_1(x)}^{g_2(x)} \int_{h_1(x,y)}^{h_2(x,y)} f(x, y, z)\,dz\,dy\,dx. \tag{14.50}$$

Because the first integration with respect to z holds x and y constant, the limits on z may therefore depend on x and y. Similarly, the second integration with respect to y holds x constant, and the limits may be functions of x.

Example 19 Evaluate the triple iterated integral

$$\int_0^1 \int_0^{x^2} \int_{xy}^{x+y} xyz\,dz\,dy\,dx.$$

Solution

$$\begin{aligned}
\int_0^1 \int_0^{x^2} \int_{xy}^{x+y} xyz\,dz\,dy\,dx &= \int_0^1 \int_0^{x^2} \left\{\frac{xyz^2}{2}\right\}_{xy}^{x+y} dy\,dx \\
&= \frac{1}{2}\int_0^1 \int_0^{x^2} \{xy(x+y)^2 - xy(xy)^2\}\,dy\,dx \\
&= \frac{1}{2}\int_0^1 \int_0^{x^2} (x^3y + 2x^2y^2 + xy^3 - x^3y^3)\,dy\,dx \\
&= \frac{1}{2}\int_0^1 \left\{\frac{x^3y^2}{2} + \frac{2x^2y^3}{3} + \frac{xy^4}{4} - \frac{x^3y^4}{4}\right\}_0^{x^2} dx \\
&= \frac{1}{24}\int_0^1 (6x^7 + 8x^8 + 3x^9 - 3x^{11})\,dx \\
&= \frac{1}{24}\left\{\frac{3x^8}{4} + \frac{8x^9}{9} + \frac{3x^{10}}{10} - \frac{x^{12}}{4}\right\}_0^1 = \frac{19}{270}. \ \square
\end{aligned}$$

Because of the analogy between double and triple integrals, we accept without proof that triple integrals can be evaluated by using triple iterated integrals. We must, however, examine how triple iterated integrals effect

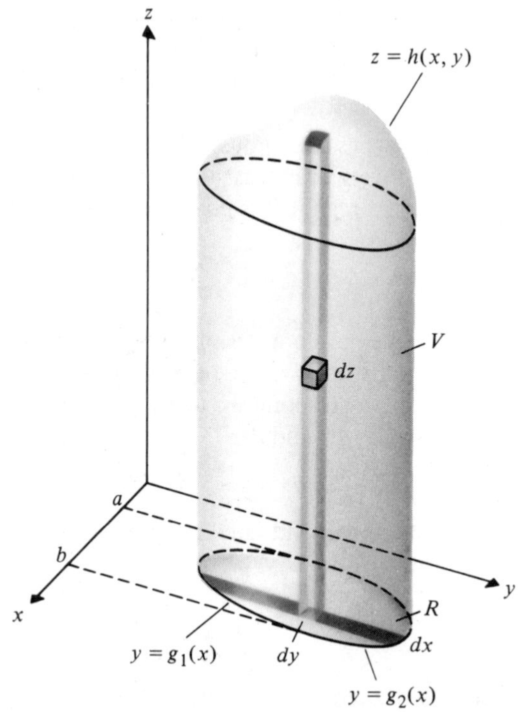

Figure 14.52

summations represented by triple integrals, for it is only by thoroughly understanding this process that we can obtain limits for triple iterated integrals.

In Section 14.3 we discussed in considerable detail the evaluation of double integrals by means of double iterated integrals. In particular, we showed that double iterated integrals in Cartesian coordinates represent the subdivision of an area into small rectangles by coordinate lines x = constant and y = constant. The first integration creates a summation over rectangles in a strip, and the second integration adds over all strips. It is fairly straightforward to generalize these ideas to triple integrals.

Consider evaluating the triple integral

$$\iiint_V f(x, y, z)\, dV$$

over the region V in Fig. 14.52 bounded above by the surface $z = h(x, y)$, below by the area R in the xy-plane, and on the sides by a cylindrical wall standing on the curve bounding R.

The choice of a triple iterated integral in Cartesian coordinates to evaluate this triple integral implies a subdivision of V into small rectangular parallelepipeds (boxes for short) by means of coordinate planes x = constant, y = constant, and z = constant. The dimensions of a representative box at position (x, y, z) in V are denoted by dx, dy, and dz, with resulting volume $dx\, dy\, dz$. If we decide on a triple iterated integral with respect to z, y, and x, then the first integration on z holds x and y constant. This integration therefore adds the quantities

$$f(x, y, z)\, dz\, dy\, dx$$

over boxes in a vertical column of cross-sectional dimensions dx and dy. Lower and upper limits on z identify where each and every column starts and stops, and must consequently be 0 and $h(x, y)$:

$$\int_0^{h(x,y)} f(x, y, z)\, dz\, dy\, dx.$$

Since this integration produces a function of x and y alone, the remaining integration with respect to y and x is essentially a double iterated integral in the xy-plane. These integrations must account for all columns in V and therefore the area in the xy-plane over which this double iterated integral is performed is the area R upon which all columns in V stand. Since the y-integration adds inside a strip in the y-direction and limits identify where all strips start and stop, they must therefore be $g_1(x)$ and $g_2(x)$. We have now

$$\int_{g_1(x)}^{g_2(x)} \int_0^{h(x,y)} f(x, y, z)\, dz\, dy\, dx.$$

Finally, the x-integration adds over all strips and the limits are $x = a$ and $x = b$:

$$\iiint_V f(x, y, z)\, dV = \int_a^b \int_{g_1(x)}^{g_2(x)} \int_0^{h(x,y)} f(x, y, z)\, dz\, dy\, dx.$$

Suppose now that V is the region bounded above by the surface $z = h_2(x, y)$ and below by $z = h_1(x, y)$ (Fig. 14.53). In this case, the limits on the first integration with respect to z are $h_1(x, y)$ and $h_2(x, y)$ since each and every column starts on the surface $z = h_1(x, y)$ and ends on the surface $z = h_2(x, y)$:

$$\int_{h_1(x,y)}^{h_2(x,y)} f(x, y, z)\, dz\, dy\, dx.$$

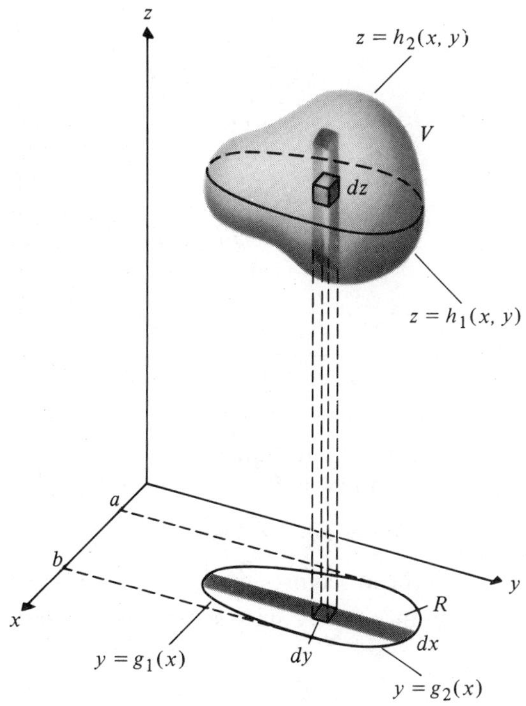

Figure 14.53

For the volume of Fig. 14.52 we interpreted the final two integrations as a double iterated integral in the xy-plane over the area R from which all columns emanated. For the present volume we interpret R as the area in the xy-plane onto which all columns project. We obtain then

$$\iiint_V f(x, y, z)\, dV = \int_a^b \int_{g_1(x)}^{g_2(x)} \int_{h_1(x, y)}^{h_2(x, y)} f(x, y, z)\, dz\, dy\, dx.$$

Schematically we have obtained the following interpretation for the limits of triple iterated integrals in Cartesian coordinates:

$$\int_{\text{position of first strip}}^{\text{position of last strip}} \int_{\text{where each and every strip starts}}^{\text{where each and every strip stops}} \int_{\text{where each and every column starts}}^{\text{where each and every column stops}} f(x, y, z) \begin{Bmatrix} dz\, dy\, dx \\ dz\, dx\, dy \\ dx\, dz\, dy \\ dx\, dy\, dz \\ dy\, dx\, dz \\ dy\, dz\, dx \end{Bmatrix}.$$

Example 20 Set up the six triple iterated integrals in Cartesian coordinates for the triple integral of a function $f(x, y, z)$ over the region V in the first octant bounded by the surfaces

$$y^2 + z^2 = 1, \qquad y = x, \qquad z = 0, \qquad x = 0.$$

Solution The triple integral of $f(x, y, z)$ over V is given by each of the following triple iterated integrals (see Fig. 14.54):

$$\int_0^1 \int_x^1 \int_0^{\sqrt{1-y2}} f(x, y, z)\, dz\, dy\, dx,$$

$$\int_0^1 \int_0^y \int_0^{\sqrt{1-y^2}} f(x, y, z)\, dz\, dx\, dy,$$

$$\int_0^1 \int_0^{\sqrt{1-z^2}} \int_0^y f(x, y, z)\, dx\, dy\, dz,$$

$$\int_0^1 \int_0^{\sqrt{1-y^2}} \int_0^y f(x, y, z)\, dx\, dz\, dy,$$

$$\int_0^1 \int_0^{\sqrt{1-x^2}} \int_x^{\sqrt{1-z^2}} f(x, y, z)\, dy\, dz\, dx,$$

$$\int_0^1 \int_0^{\sqrt{1-z^2}} \int_x^{\sqrt{1-z^2}} f(x, y, z)\, dy\, dx\, dz. \; \square$$

Example 21 Set up triple iterated integrals with respect to z, y, and x for the triple integral of the function $f(x, y, z) = x^2 y \sin z$ over the region V bounded by the surfaces

$$z = \sqrt{y}, \qquad y + z = 2, \qquad x = 0, \qquad z = 0, \qquad x = 2.$$

Solution The problem requires triple iterated integrals first with respect to z, then with respect to y and x. Since some columns end on the parabolic cylinder $z = \sqrt{y}$

(Fig. 14.55) and others on the plane $y + z = 2$, we require two iterated integrals:

$$\iiint_V x^2 y \sin z \, dV = \int_0^2 \int_0^1 \int_0^{\sqrt{y}} x^2 y \sin z \, dz \, dy \, dx$$

$$+ \int_0^2 \int_1^2 \int_0^{2-y} x^2 y \sin z \, dz \, dy \, dx. \square$$

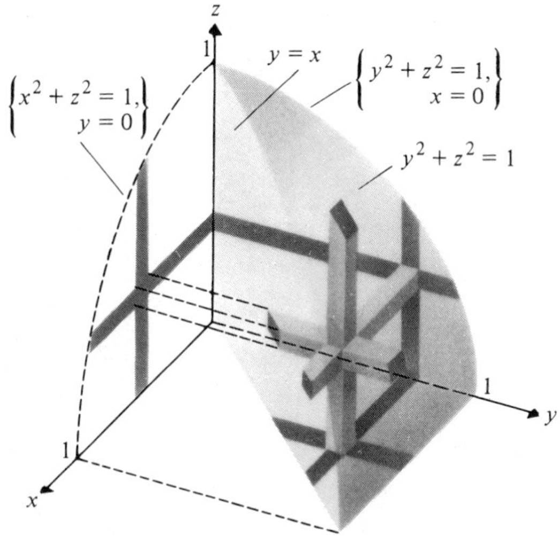

Figure 14.54

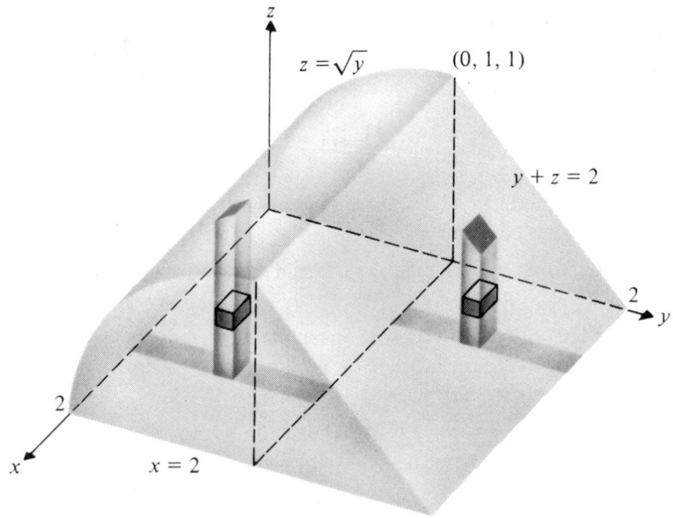

Figure 14.55

EXERCISES 14.9

In each of Exercises 1–10, evaluate the triple integral over the region indicated.

1. $\iiint_V (x^2 z + ye^x) \, dV$,
   where V is bounded by $x = 0$, $x = 1$, $y = 1$, $y = 2$, $z = 0$, $z = 1$

2. $\iiint_V x \, dV$,
   where V is bounded by $x = 0$, $y = 0$, $z = 0$, $x + y + z = 4$

3. $\iiint_V \sin(y + z) \, dV$,
   where V is bounded by $z = 0$, $y = 2x$, $y = 0$, $x = 1$, $z = x + 2y$

4. $\iiint_V xy \, dV$,
   where V is enclosed by $z = \sqrt{1 - x^2 - y^2}$, $z = 0$

5. $\iiint_V dV$,
   where V is bounded by $x^2 + y^2 = 1$, $x^2 + z^2 = 1$

6. $\iiint_V (x^2 + 2z) \, dV$,
   where V is bounded by $z = 0$, $y + z = 4$, $y = x^2$

7. $\iiint_V x^2 y^2 z^2 \, dV$,
   where V is bounded by $z = 1 + y$, $y + z = 1$, $x = 1$, $x = 0$, $z = 0$

8. $\iiint_V xyz \, dV$,
   where V is the first octant volume cut out by $z = x^2 + y^2$, $z = \sqrt{x^2 + y^2}$

9. $\iiint_V dV$,
   where V is bounded by $z = x^2$, $y + z = 4$, $y = 0$

10. $\iiint_V (x + y + z)\, dV$,

where V is bounded by $x = 0$, $x = 1$, $z = 0$, $y + z = 2$, $y = z$

11. Set up the six triple iterated integrals in Cartesian coordinates for the triple integral of a function $f(x, y, z)$ over the volume enclosed by the surfaces $y = 1 - x^2$, $z = 0$, and $y = z$.

The average value of a function $f(x, y, z)$ over a region V is defined by

$$\bar{f} = \frac{1}{\text{Volume }(V)} \iiint_V f(x, y, z)\, dV.$$

In each of Exercises 12–14, find the average value of the function over the region.

12. $f(x, y, z) = xy$ over the region bounded by the surfaces $x = 0$, $y = 0$, $z = 0$, $x + y + z = 1$

13. $f(x, y, z) = x + y + z$ over the region in the first octant bounded by the surfaces $z = 9 - x^2 - y^2$, $z = 0$, and for which $0 \le x \le 1$, $0 \le y \le 1$

14. $f(x, y, z) = x^2 + y^2 + z^2$ over the region bounded by the surfaces $x = 0$, $x = 1$, $y + z = 2$, $y = 2$, $z = 2$

In Exercises 15–18, evaluate the triple integral over the region indicated.

15. $\iiint_V (xy + z)\, dV$,

where V is bounded by $y + z = 1$, $z = 2y$, $z = y$, $x = 0$, $x = 3$

16. $\iiint_V (y + x^2)\, dV$,

where V is bounded by $x + z^2 = 1$, $z = x + 1$, $y = 1$, $y = -1$

17. $\iiint_V dV$,

where V is bounded by $z = x^2 + y^2$, $z = 4 - x^2 - y^2$

18. $\iiint_V dV$,

where V is bounded by $z = 0$, $x^2 + y^2 = 1$, $x + y + z = 2$

19. Set up, but do not evaluate, triple iterated integrals to evaluate the triple integral of the function $f(x, y, z) = x^2 + y^2 + z^2$ over the volume bounded by the surfaces $x^2 + y^2 = z^2 + 1$, $2z = \sqrt{x^2 + y^2}$, and $z = 0$.

14.10 VOLUMES

Because the volume of a region V is represented by triple integral (14.49) and triple integrals are evaluated by means of triple iterated integrals, it follows that volumes can be evaluated with triple iterated integrals. For example, to evaluate the volume of the region shown in Fig. 14.53 using a triple iterated integral in x, y, and z, we subdivide V into boxes of dimensions dx, dy, and dz and therefore of volume

$$dz\, dy\, dx.$$

Integration with respect to z adds these volumes in the z-direction to give the volume of a vertical column (Fig. 14.56):

$$\int_{h_1(x, y)}^{h_2(x, y)} dz\, dy\, dx.$$

The limits indicate that all columns start on the surface $z = h_1(x, y)$ and end on the surface $z = h_2(x, y)$. Integration with respect to y now adds the volumes of columns that project onto a strip in the y-direction:

$$\int_{g_1(x)}^{g_2(x)} \int_{h_1(x, y)}^{h_2(x, y)} dz\, dy\, dx,$$

where the limits indicate that all strips start on the curve $y = g_1(x)$ and end on the curve $y = g_2(x)$. Evidently this integration yields the volume of a slab as

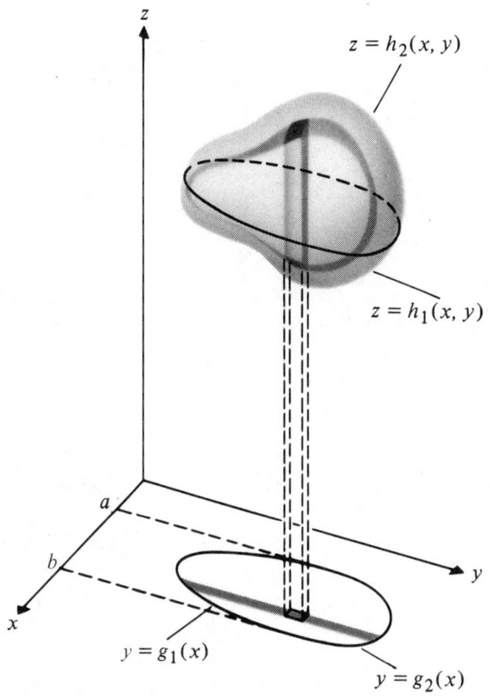

Figure 14.56

shown in Fig. 14.56. Finally, integration with respect to x adds the volumes of all such slabs in V:

$$\int_a^b \int_{g_1(x)}^{g_2(x)} \int_{h_1(x,y)}^{h_2(x,y)} dz\, dy\, dx,$$

where a and b designate the positions of the first and last strips.

Example 22 Find the volume bounded by the planes $z = x + y$, $y = 2x$, $z = 0$, $x = 0$, $y = 2$.

Solution If we use vertical columns (Fig. 14.57), we have

$$\text{Volume} = \int_0^1 \int_{2x}^2 \int_0^{x+y} dz\, dy\, dx = \int_0^1 \int_{2x}^2 (x + y)\, dy\, dx = \int_0^1 \left\{xy + \frac{y^2}{2}\right\}_{2x}^2 dx$$

$$= 2\int_0^1 (1 + x - 2x^2)\, dx = 2\left\{x + \frac{x^2}{2} - \frac{2x^3}{3}\right\}_0^1 = \frac{5}{3}. \ \square$$

Example 23 Find the volume in the first octant cut off from the cylinder $x^2 + z^2 = 4$ by the plane $y + z = 6$.

Solution With columns in the y-direction (Fig. 14.58), we have

$$\text{Volume} = \int_0^2 \int_0^{\sqrt{4-x^2}} \int_0^{6-z} dy\, dz\, dx = \int_0^2 \int_0^{\sqrt{4-x^2}} (6 - z)\, dz\, dx$$

$$= \int_0^2 \left\{6z - \frac{z^2}{2}\right\}_0^{\sqrt{4-x^2}} dx$$

$$= \int_0^2 \left\{6\sqrt{4 - x^2} - \frac{1}{2}(4 - x^2)\right\} dx.$$

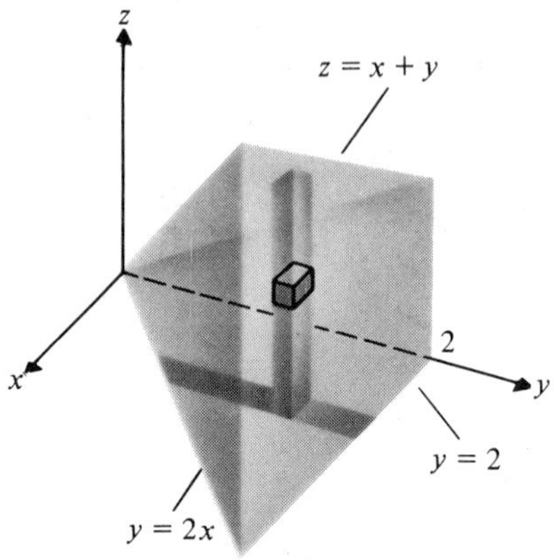

Figure 14.57

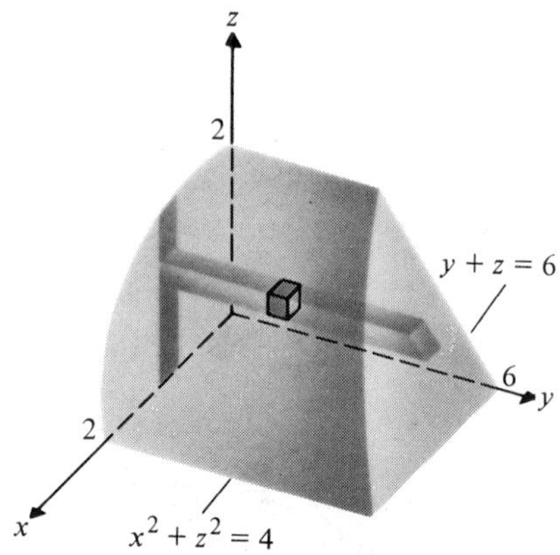

Figure 14.58

In the first term we set $x = 2 \sin \theta$, from which we get $dx = 2 \cos \theta \, d\theta$, and

$$\begin{aligned}\text{Volume} &= 6\int_0^{\pi/2} (2\cos\theta)\, 2\cos\theta \, d\theta + \left\{-2x + \frac{x^3}{6}\right\}_0^2 \\ &= 12\int_0^{\pi/2} (1 + \cos 2\theta)\, d\theta - \frac{8}{3} \\ &= 12\left\{\theta + \frac{1}{2}\sin 2\theta\right\}_0^{\pi/2} - \frac{8}{3} = 6\pi - \frac{8}{3}. \ \square\end{aligned}$$

Had we used iterated integrals with respect to z, y, and x in this example, we would have had the two integrals

$$\text{Volume} = \int_0^2 \int_0^{6-\sqrt{4-x^2}} \int_0^{\sqrt{4-x^2}} dz\, dy\, dx + \int_0^2 \int_{6-\sqrt{4-x^2}}^{6} \int_0^{6-y} dz\, dy\, dx.$$

EXERCISES 14.10

In each of Exercises 1–17, find the volume of the region bounded by the surfaces.

1. $y = x^2,\ y = 1,\ z = 0,\ z = 4$
2. $x = z^2,\ z = x^2,\ y = 0,\ y = 2$
3. $x = 3z,\ z = 3x,\ y = 1,\ y = 0,\ x = 2$
4. $x + y + z = 6,\ y = 4 - x^2,\ z = 0,\ y = 0$
5. $z = x^2 + y^2,\ y = x^2,\ y = 4,\ z = 0$
6. $x + y + z = 4,\ y = 3z,\ x = 0,\ y = 0$
7. $x^2 + y^2 = 4,\ y^2 + z^2 = 4$
8. $y = x^2 - 1,\ y = 1 - x^2,\ x + z = 1,\ z = 0$
9. $z = 16 - x^2 - 4y^2,\ x + y = 1,\ z = 16,\ x = 0,\ y = 0$ (in the first octant)
10. $z = x^2 + y^2,\ x = 1,\ z = 0,\ x = y,\ x = 2y$
11. $z = 1 - x^2 - y^2,\ z = 0$
12. $x - z = 0,\ x + z = 3,\ y + z = 1,\ z = y + 1,\ z = 0$
13. $x + y + z = 2,\ x^2 + y^2 = 1,\ z = 0$
14. $y + z = 1,\ z = 2y,\ z = y,\ x = 0,\ x + y + z = 4$
15. $x^2 + 4y^2 = z,\ x^2 + 4y^2 = 12 - 2z$
16. $y = 1 - z^2,\ y = z^2 - 1,\ x = 1 - z^2,\ x = z^2 - 1$
17. $x + 3y + 2z = 6,\ z = 0,\ y = x,\ y = 2x$
18. Find the volume in the first octant bounded by the plane $2x + y + z = 2$ and inside the cylinder $y^2 + z^2 = 1$.
19. A pyramid has a square base with side length b and has height h at its center.
    a) Find its volume by taking cross-sections parallel to the base. (See Section 7.8.)
    b) Find its volume using triple integrals.

***20.** The bottom and sides of a boat are defined by the surface equation

$$x = 10(1 - y^2 - z^2), \qquad 0 \le x \le 10,$$

where all dimensions are in meters. Find the volume of water displaced by the boat when the water level on the side of the boat is d meters below the top of the boat. Archimedes' principle states that the buoyant force on an object when immersed or partially immersed in a fluid is equal to the weight of the fluid displaced by the object. Find the maximum weight of the boat and its contents just before sinking.

14.11 CENTERS OF MASS AND MOMENTS OF INERTIA

In this section we discuss centers of mass and moments of inertia for three-dimensional objects of density $\rho(x, y, z)$ (mass per unit volume). If we divide the object occupying region V in Fig. 14.59 into small volumes dV, then the amount of mass in dV is $\rho\, dV$. The triple integral

$$M = \iiint_V \rho\, dV \tag{14.51}$$

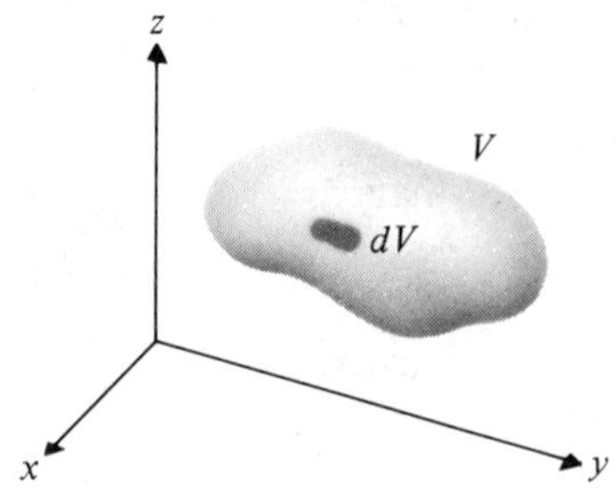

Figure 14.59

adds the masses of all such volumes (of ever-decreasing size) to produce the total mass M of the object.

Corresponding to Eqs. (14.29) and (14.30) for first moments of planar masses about the coordinate axes are the following formulas for first moments of the object about the coordinate planes:

$$\text{First moment of object about } yz\text{-plane} = \iiint_V x\rho\, dV, \tag{14.52}$$

$$\text{First moment of object about } xz\text{-plane} = \iiint_V y\rho\, dV, \tag{14.53}$$

$$\text{First moment of object about } xy\text{-plane} = \iiint_V z\rho\, dV. \tag{14.54}$$

The center of mass of the object is defined as that point $(\bar{x}, \bar{y}, \bar{z})$ at which a particle of mass M would have the same first moments about the coordinate planes as the object itself. Since the first moments of M at $(\bar{x}, \bar{y}, \bar{z})$ about the coordinate planes are $M\bar{x}$, $M\bar{y}$, and $M\bar{z}$, it follows that we can use the equations

$$M\bar{x} = \iiint_V x\rho\, dV, \tag{14.55}$$

$$M\bar{y} = \iiint_V y\rho\, dV, \tag{14.56}$$

$$M\bar{z} = \iiint_V z\rho\, dV \tag{14.57}$$

to solve for $(\bar{x}, \bar{y}, \bar{z})$ once M and the integrals on the right have been evaluated.

If we use a triple iterated integral with respect to z, y, and x to evaluate (14.57), say, for the object in Fig. (14.56), then

$$\iiint_V z\rho\, dV = \int_a^b \int_{g_1(x)}^{g_2(x)} \int_{h_1(x, y)}^{h_2(x, y)} z\rho\, dz\, dy\, dx.$$

The quantity $z\rho\,dz\,dy\,dx$ is the first moment about the xy-plane of the mass in an elemental box of dimensions dx, dy, and dz. The z-integration then adds these moments over boxes in the z-direction to give the first moment about the xy-plane of the mass in a vertical column:

$$\int_{h_1(x,y)}^{h_2(x,y)} z\rho\,dz\,dy\,dx.$$

The y-integration then adds the first moments of columns that project onto a strip in the y-direction:

$$\int_{g_1(x)}^{g_2(x)} \int_{h_1(x,y)}^{h_2(x,y)} z\rho\,dz\,dy\,dx.$$

This quantity therefore represents the first moment about the xy-plane of the slab shown in Fig. 14.56. Finally, the x-integration adds first moments of all such slabs to give the total first moment of V about the xy-plane:

$$\int_a^b \int_{g_1(x)}^{g_2(x)} \int_{h_1(x,y)}^{h_2(x,y)} z\rho\,dz\,dy\,dx.$$

Example 24 Find the center of mass of an object of constant density ρ if it is bounded by the surfaces $z = 1 - y^2$, $x = 0$, $z = 0$, $x = 2$.

Solution From the symmetry of the object (Fig. 14.60), we see that $\bar{x} = 1$ and $\bar{y} = 0$. Now

$$\begin{aligned} M &= 2\int_0^2\int_0^1\int_0^{1-y^2} \rho\,dz\,dy\,dx = 2\rho\int_0^2\int_0^1 (1 - y^2)\,dy\,dx \\ &= 2\rho\int_0^2 \left\{y - \frac{y^3}{3}\right\}_0^1 dx = \frac{4\rho}{3}\int_0^2 dx = \frac{8\rho}{3}; \end{aligned}$$

and

$$\begin{aligned} M\bar{z} &= 2\int_0^2\int_0^1\int_0^{1-y^2} z\rho\,dz\,dy\,dx = 2\rho\int_0^2\int_0^1 \left\{\frac{z^2}{2}\right\}_0^{1-y^2} dy\,dx \\ &= \rho\int_0^2\int_0^1 (1 - 2y^2 + y^4)\,dy\,dx \\ &= \rho\int_0^2 \left\{y - \frac{2y^3}{3} + \frac{y^5}{5}\right\}_0^1 dx = \frac{8\rho}{15}\int_0^2 dx = \frac{16\rho}{15}. \end{aligned}$$

Thus,

$$\bar{z} = \frac{16\rho}{15}\cdot\frac{3}{8\rho} = \frac{2}{5}. \quad \square$$

In view of our discussion on first moments in this section and on moments of inertia of thin plates in Section 14.6, it should not be necessary to give a full treatment to moments of inertia of three-dimensional objects. Instead, we simply note that the distances from a point (x, y, z) to the x-, y-, and z-axes are, re-

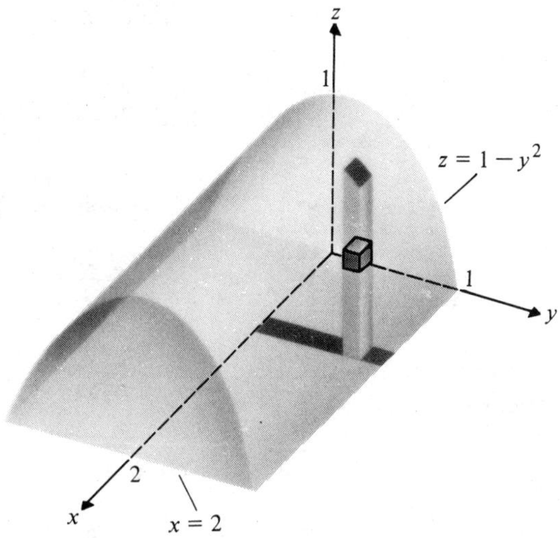

Figure 14.60

spectively, $\sqrt{y^2 + z^2}$, $\sqrt{x^2 + z^2}$, and $\sqrt{x^2 + y^2}$. It follows, then, that if an object of density $\rho(x, y, z)$ occupies a region V of space:

$$\text{Moment of inertia of object about } x\text{-axis} = \iiint_V (y^2 + z^2)\rho \, dV, \quad (14.58)$$

$$\text{Moment of inertia of object about } y\text{-axis} = \iiint_V (x^2 + z^2)\rho \, dV, \quad (14.59)$$

$$\text{Moment of inertia of object about } z\text{-axis} = \iiint_V (x^2 + y^2)\rho \, dV. \quad (14.60)$$

Example 25 Find the moment of inertia of a right-circular cylinder of constant density ρ about its axis.

Solution Let the length and radius of the cylinder be h and r and choose the coordinate system shown in Fig. 14.61. The required moment of inertia about the z-axis

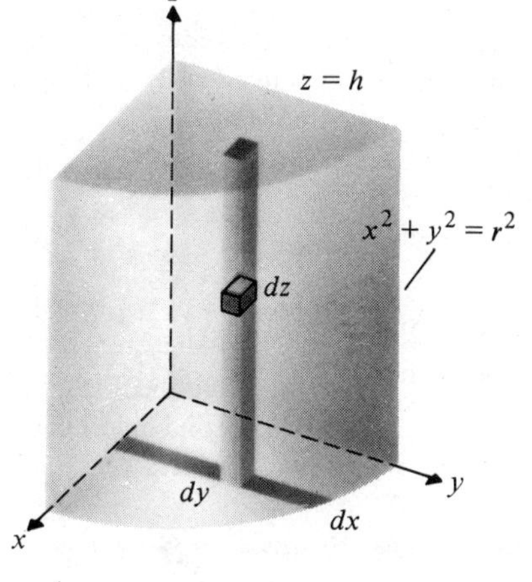

Figure 14.61

is four times the moment of inertia of that part of the cylinder in the first octant. Hence,

$$I = 4\int_0^r\int_0^{\sqrt{r^2-x^2}}\int_0^h (x^2+y^2)\rho\,dz\,dy\,dx = 4\rho h\int_0^r\int_0^{\sqrt{r^2-x^2}}(x^2+y^2)\,dy\,dx$$

$$= 4\rho h\int_0^r\left\{x^2y+\frac{y^3}{3}\right\}_0^{\sqrt{r^2-x^2}}dx = 4\rho h\int_0^r\left\{x^2\sqrt{r^2-x^2}+\frac{1}{3}(r^2-x^2)^{3/2}\right\}dx$$

$$= \frac{4\rho h}{3}\int_0^r\{r^2\sqrt{r^2-x^2}+2x^2\sqrt{r^2-x^2}\}\,dx.$$

To evaluate this definite integral we set $x = r\sin\theta$, which implies that $dx = r\cos\theta\,d\theta$, and

$$I = \frac{4\rho h}{3}\int_0^{\pi/2}\{r^2 r\cos\theta + 2r^2\sin^2\theta\, r\cos\theta\}\, r\cos\theta\,d\theta$$

$$= \frac{4\rho hr^4}{3}\int_0^{\pi/2}\{\cos^2\theta + 2\sin^2\theta\cos^2\theta\}\,d\theta$$

$$= \frac{4\rho hr^4}{3}\int_0^{\pi/2}\left\{\frac{1+\cos 2\theta}{2}+\frac{1-\cos 4\theta}{4}\right\}d\theta$$

$$= \frac{4\rho hr^4}{3}\left\{\frac{3\theta}{4}+\frac{\sin 2\theta}{4}-\frac{\sin 4\theta}{16}\right\}_0^{\pi/2} = \frac{\rho hr^4\pi}{2}. \square$$

EXERCISES 14.11

In Exercises 1–5, the surfaces bound a solid object of constant density. Find its center of mass.

1. $z = x^2 + y^2, z = 0, x = 0, y = 0, x = 1, y = 1$
2. $x + y + z = 1, x = 0, y = 0, z = 0$
3. $z = x^2, y + z = 4, y = 0$
4. $y = 4 - x^2, z = 0, y = z$
5. $x + y + z = 4, y = 3z, x = 0, y = 0$

In Exercises 6–10, the surfaces bound a solid object of constant density ρ. Find its moment of inertia about the line indicated.

6. $x = 0, y = 0, z = 0, x = 1, y = 1, z = 1$ about the x-axis
7. $z = 2x, z = 0, y = 0, y = 2, x = 3$ about the y-axis
8. $x + y + z = 2, y = 0, x = 0, 0 \le z \le \sqrt{1-y}$ about the x-axis
9. $z = xy, x^2 + y^2 = 1, z = 0$ (first octant) about the z-axis
10. $y + z = 2,\ x + z = 2,\ x = 0,\ y = 0,\ z = 0$ about the z-axis
11. Find the first moment about the xy-plane of a solid of constant density ρ if it is bounded by the surfaces $x = z$, $x + z = 0$, $z = 2$, $y = 0$, and $y = 2$.

In Exercises 12–14, the surfaces bound a solid object of constant density. Find its center of mass.

12. $y = x^3, x = y^2, z = 1 + x^2 + y^2, z = -x^2 - y^2$
13. $z = x^2, x + z = 2, z = y, y = 0$
14. $y + z = 0, y - z = 0, x + z = 0, x - z = 0, z = 2$

In Exercises 15–17, the surfaces bound a solid object of constant density ρ. Find its moment of inertia about the line indicated.

15. $z = 4 - x^2,\ x + z + 2 = 0,\ y = 0,\ y = 2$ about the y-axis
16. $x^2 + z^2 = a^2, x^2 + y^2 = a^2$ about the x-axis
17. $x + y - z = 0, x = 3y, 3y = 2x, x = 3, z = 0$ about the z-axis
18. Find the first moment about the plane $x + y + z = 1$ of a solid object of constant density ρ if it is bounded by the surfaces $x + 2y + 4z = 12$, $x = 0$, $y = 0$, and $z = 0$.

14.12 TRIPLE ITERATED INTEGRALS IN CYLINDRICAL COORDINATES

In Section 14.8 we saw that polar coordinates are sometimes more convenient to use in evaluating double integrals than Cartesian coordinates. It should come as no surprise, then, that other coordinate systems can simplify the evaluation of triple integrals. Two of the most common are cylindrical and spherical coordinate systems.

Cylindrical coordinates are useful in problems involving an axis of symmetry. They are based on a Cartesian coordinate along the axis of symmetry and polar coordinates in a plane perpendicular to the axis of symmetry. If the z-axis is the axis of symmetry and polar coordinates are defined in the xy-plane with the origin as pole and the positive x-axis as polar axis, then cylindrical coordinates and Cartesian coordinates are related by the equations

$$x = r\cos\theta, \qquad y = r\sin\theta, \qquad z = z \tag{14.61}$$

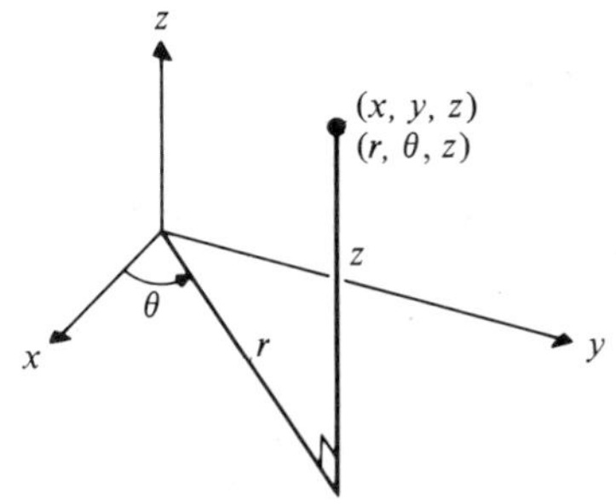

Figure 14.62

(see Fig. 14.62). Recall that r can be expressed in terms of x and y by

$$r^2 = x^2 + y^2, \tag{14.62a}$$

and θ is defined implicitly by the equations

$$\cos\theta = \frac{\pm x}{\sqrt{x^2+y^2}}, \qquad \sin\theta = \frac{\pm y}{\sqrt{x^2+y^2}}. \tag{14.62b}$$

To use cylindrical coordinates in the evaluation of triple integrals, we must express equations of surfaces in terms of these coordinates. But this is very simple, for if $F(x, y, z) = 0$ is the equation of a surface in Cartesian coordinates, then to express this equation in cylindrical coordinates we substitute from Eqs. (14.61): $F(r\cos\theta, r\sin\theta, z) = 0$. For example, the right-circular cylinder $x^2 + y^2 = 9$, which has the z-axis as its axis of symmetry, has the very simple equation $r = 3$ in cylindrical coordinates. The right-circular cone $z = \sqrt{x^2 + y^2}$ also has the z-axis as its axis of symmetry, and its equation in cylindrical coordinates takes the simple form $z = r$.

Suppose that we are to evaluate the triple integral of a continuous function $f(x, y, z)$,

$$\iiint_V f(x, y, z)\,dV,$$

over some region V of space. The choice of a triple iterated integral in cylindrical coordinates implies a subdivision of V into small volumes dV by means of coordinate surfaces $r =$ constant, $\theta =$ constant, and $z =$ constant (Fig. 14.63). Surfaces $r =$ constant are right-circular cylinders coaxial with the z-axis; surfaces $\theta =$ constant are planes containing the z-axis and therefore perpendicular to the xy-plane; and surfaces $z =$ constant are planes parallel to the xy-plane.

If we denote small variations in r, θ, and z for the element dV by dr, $d\theta$, and dz (Fig. 14.64), then the volume of the element is approximately $(r\,dr\,d\theta)\,dz$, where $r\,dr\,d\theta$ is the polar cross-sectional area parallel to the xy-plane. Hence in cylindrical coordinates we set

$$dV = r\,dz\,dr\,d\theta. \tag{14.63}$$

The integrand $f(x, y, z)$ is expressed in cylindrical coordinates as $f(r\cos\theta, r\sin\theta, z)$. It remains only to affix appropriate limits to the triple

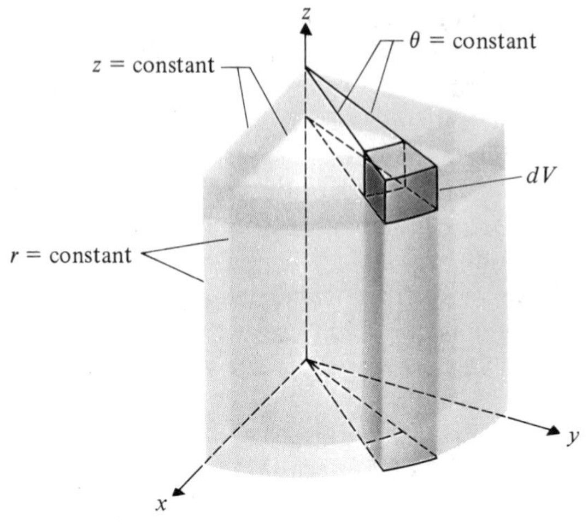

Figure 14.63

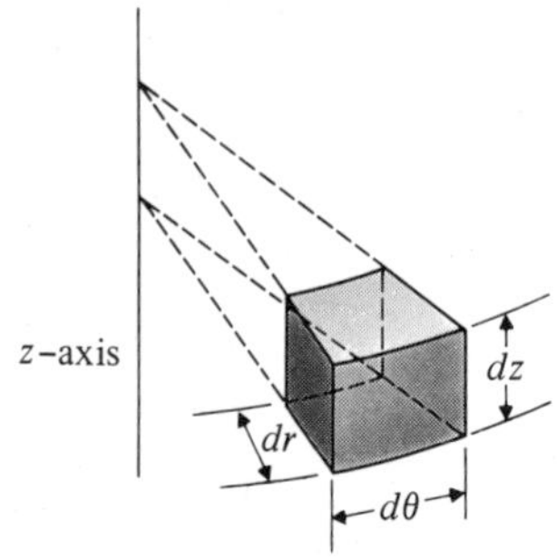

Figure 14.64

iterated integral, and these will, of course, depend on which of the six possible iterated integrals in cylindrical coordinates we choose. The most commonly used triple iterated integral is with respect to z, r, and θ, and in this case the z-integration adds the quantities

$$f(r\cos\theta, r\sin\theta, z)r\,dz\,dr\,d\theta$$

in a vertical column, where r and θ are constant. The limits therefore identify the surfaces on which each and every column starts and stops, and will generally depend on r and θ:

$$\int_{h_1(r,\theta)}^{h_2(r,\theta)} f(r\cos\theta, r\sin\theta, z)r\,dz\,dr\,d\theta.$$

The remaining integrations with respect to r and θ perform additions over the area in the xy-plane onto which all vertical columns project. Since r and θ are simply polar coordinates, the r-integration adds over small areas in a wedge and the θ-integration adds over all wedges. The triple iterated integral with respect to z, r, and θ will therefore have the form

$$\int_a^b \int_{g_1(\theta)}^{g_2(\theta)} \int_{h_1(r,\theta)}^{h_2(r,\theta)} f(r\cos\theta, r\sin\theta, z)r\,dz\,dr\,d\theta.$$

We comment on the geometric aspects of these additions more fully in the following examples.

Example 26 Find the volume inside both the sphere $x^2 + y^2 + z^2 = 2$ and the cylinder $x^2 + y^2 = 1$.

Solution The required volume is eight times the first octant volume shown in Fig. 14.65. If we use cylindrical coordinates, the volume of an elemental piece is $r\,dz\,dr\,d\theta$. A z-integration adds these pieces to give the volume in a vertical column:

$$\int_0^{\sqrt{2-r^2}} r\,dz\,dr\,d\theta,$$

where the limits indicate that for the volume in the first octant all columns start on the xy-plane (where $z = 0$) and end on the sphere (where $r^2 + z^2 = 2$). An

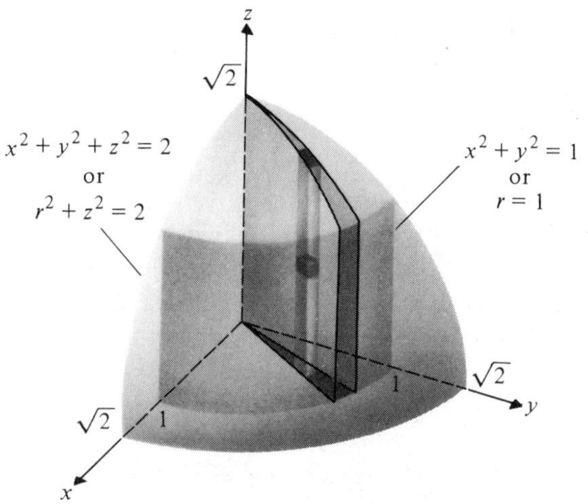

Figure 14.65

r-integration now adds the volumes of all columns that stand on a wedge:

$$\int_0^1 \int_0^{\sqrt{2-r^2}} r\, dz\, dr\, d\theta,$$

where the limits indicate that all wedges start at the origin (where $r = 0$) and end on the curve $x^2 + y^2 = 1$ (or $r = 1$) in the xy-plane. This integration therefore yields the volume of a slice (Fig. 14.65). Finally, the θ-integration adds the volumes of all such slices

$$\int_0^{\pi/2} \int_0^1 \int_0^{\sqrt{2-r^2}} r\, dz\, dr\, d\theta,$$

where the limits 0 and $\pi/2$ identify the positions of the first and last wedges, respectively, in the first quadrant. We obtain the required volume, then, as

$$\begin{aligned} 8 \int_0^{\pi/2} \int_0^1 \int_0^{\sqrt{2-r^2}} r\, dz\, dr\, d\theta &= 8 \int_0^{\pi/2} \int_0^1 r\sqrt{2 - r^2}\, dr\, d\theta \\ &= 8 \int_0^{\pi/2} \left\{ -\frac{1}{3}(2 - r^2)^{3/2} \right\}_0^1 d\theta \\ &= \frac{8}{3}(2\sqrt{2} - 1) \int_0^{\pi/2} d\theta = \frac{4\pi}{3}(2\sqrt{2} - 1). \ \square \end{aligned}$$

Had we used a triple iterated integral with respect to z, y, and x in Example 26, we would have obtained

$$\text{Volume} = 8 \int_0^1 \int_0^{\sqrt{1-x^2}} \int_0^{\sqrt{2-x^2-y^2}} dz\, dy\, dx.$$

To appreciate the value of cylindrical coordinates, try to evaluate this triple iterated integral.

As further evidence of the value of cylindrical coordinates, redo Examples 23 and 25 using cylindrical coordinates.

Example 27 Evaluate the triple iterated integral

$$I = \int_0^1 \int_0^{\sqrt{1-y^2}} \int_0^{x^2+y^2} y^2\, dz\, dx\, dy.$$

Solution The first two integrations with respect to z and x are quite straightforward, giving

$$I = \frac{1}{3}\int_0^1 y^2(1 + 2y^2)\sqrt{1 - y^2}\,dy.$$

To evaluate this definite integral we could make a trigonometric substitution $y = \sin\theta$. (Try it.) Alternatively, we could consider using cylindrical coordinates on the original triple iterated integral. We first obtain the region over which integration is being performed. The limits on z indicate that all columns begin on the xy-plane and end on the paraboloid $z = x^2 + y^2$. The limits on x indicate that in the xy-plane all strips begin on the y-axis ($x = 0$) and end on the curve $x = \sqrt{1 - y^2}$ (i.e., the circle $x^2 + y^2 = 1$). The first and last strips are at $y = 0$ and $y = 1$, respectively. These facts determine the region of integration as that region in the first octant bounded by the paraboloid $z = x^2 + y^2$ and the right-circular cylinder $x^2 + y^2 = 1$ (Fig. 14.66a). If we use a triple iterated integral with respect to z, r, and θ (Fig. 14.66b), then

$$I = \int_0^{\pi/2}\int_0^1\int_0^{r^2} r^2\sin^2\theta\, r\,dz\,dr\,d\theta = \int_0^{\pi/2}\int_0^1 r^5\sin^2\theta\,dr\,d\theta = \int_0^{\pi/2}\frac{1}{6}\sin^2\theta\,d\theta$$

$$= \frac{1}{6}\int_0^{\pi/2}\left\{\frac{1 - \cos 2\theta}{2}\right\}d\theta = \frac{1}{12}\left\{\theta - \frac{1}{2}\sin 2\theta\right\}_0^{\pi/2} = \frac{\pi}{24}. \quad \square$$

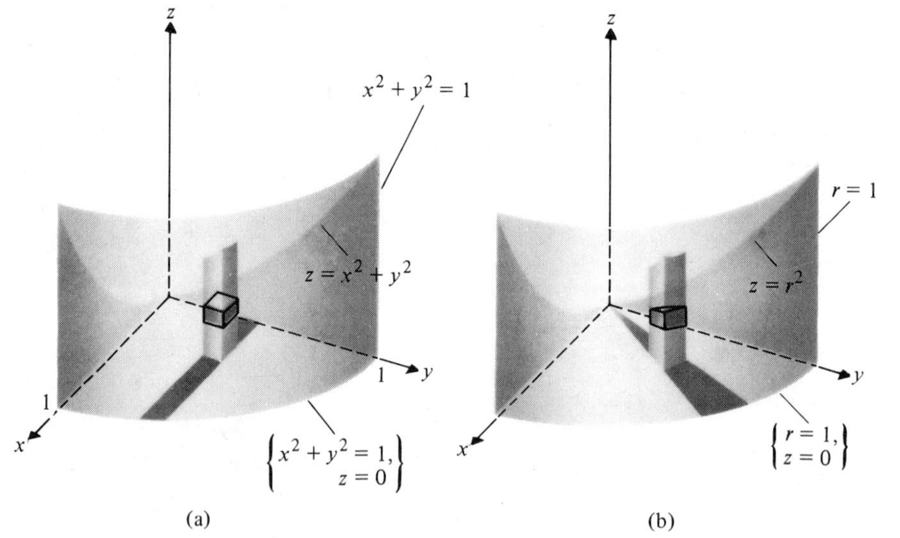

Figure 14.66

EXERCISES 14.12

In each of Exercises 1–10, find the equation for the surface in cylindrical coordinates. Sketch each surface and indicate whether the surface is symmetric about the z-axis.

1. $x^2 + y^2 + z^2 = 4$

2. $x^2 + y^2 = 1$

3. $y^2 + z^2 = 6$

4. $x + y = 5$

5. $z = 2\sqrt{x^2 + y^2}$

6. $z = x^2$

7. $x^2 + 4y^2 = 4$

8. $4z = x^2 + y^2$

9. $y = x$

10. $x^2 + y^2 = 1 + z^2$

In each of Exercises 11–15, find the volume bounded by the surfaces.

11. $z = \sqrt{x^2 + y^2}, x^2 + y^2 = 4, z = 0$

12. $z = \sqrt{2 - x^2 - y^2}, z = x^2 + y^2$

13. $z = xy, x^2 + y^2 = 1, z = 0$

14. $z = x^2 + y^2, z = 4 - x^2 - y^2$

15. $x + y + z = 2, x^2 + y^2 = 1, z = 0$

16. Find the volume inside the sphere $x^2 + y^2 + z^2 = 4$ but outside the cylinder $x^2 + y^2 = 1$.

17. Find the center of mass of a uniform hemispherical solid.

18. Set up the six triple iterated integrals in cylindrical coordinates for the triple integral of a function $f(x, y, z)$ over the volume bounded by the surfaces $z = 1 + x^2 + y^2$, $x^2 + y^2 = 9$, and $z = 0$.

19. Find the moment of inertia of a uniform right-circular cylinder of radius R and height h:

a) about its axis;

b) about a line through the center of its base and perpendicular to its axis.

20. Find the moment of inertia of a uniform sphere of radius R about any line through its center.

In Exercises 21–24, evaluate the given triple iterated integral.

21. $$\int_0^3 \int_0^{\sqrt{9-x^2}} \int_0^{\sqrt{x^2+y^2}} dz\,dy\,dx$$

22. $$\int_0^9 \int_0^{\sqrt{81-y^2}} \int_0^{\sqrt{81-x^2-y^2}} \frac{1}{\sqrt{x^2+y^2}}\,dz\,dx\,dy$$

23. $$\int_0^4 \int_0^{\sqrt{4y-y^2}} \int_0^{y+x^2} dz\,dx\,dy$$

24. $$\int_0^{\sqrt{3}/2} \int_{5-\sqrt{21-y^2}}^{\sqrt{1-y^2}} \int_0^{x^2+y^2} y\,dz\,dx\,dy$$

25. Find the center of mass of the uniform solid bounded by the surfaces

$$x^2 + y^2 = 2x, \qquad z = \sqrt{x^2 + y^2}, \qquad z = 0.$$

26. Find the moment of inertia of a uniform right-circular cone of radius R and height h about its axis.

27. A casting is in the form of a sphere of radius b with two cylindrical holes of radius $a < b$ such that the axes of the holes pass through the center of the sphere and intersect at right angles. What volume of metal is required for the casting?

In each of Exercises 28–31, find the volume bounded by the surfaces.

28. $x^2 + y^2 - z^2 = 1, 4z^2 = x^2 + y^2$

29. $z = x^2 + y^2, z = 0, (x^2 + y^2)^2 = x^2 - y^2$

30. $x^2 + y^2 + z^2 = 4,\ x^2 + y^2 + z^2 = 16,\ z = \sqrt{x^2 + y^2}$ (smaller piece)

31. $x^2 + y^2 + z^2 = 1,\ y = x,\ x = 2y,\ z = 0$ (in the first octant)

* **32.** Use cylindrical coordinates to find the volume of the torus

$$(\sqrt{x^2 + y^2} - a)^2 + z^2 = b^2, \qquad b < a.$$

* **33.** A tumbler in the form of a right-circular cylinder of radius R and height h is full of water. As the axis of the tumbler is tilted from the vertical, water pours over the side. Find the volume of water remaining in the tumbler as a function of the angle between the vertical and the axis of the tumbler.

14.13 TRIPLE ITERATED INTEGRALS IN SPHERICAL COORDINATES

Spherical coordinates are useful in solving problems concerning figures that are symmetric about a point. If the origin is that point, then spherical coordinates $(\mathscr{R}, \theta, \phi)$ are related to Cartesian coordinates (x, y, z) by the equations

$$x = \mathscr{R} \sin \phi \cos \theta,$$
$$y = \mathscr{R} \sin \phi \sin \theta, \qquad (14.64)$$
$$z = \mathscr{R} \cos \phi,$$

which are illustrated in Fig. 14.67. As is the case for polar and cylindrical coordinates, without restrictions on $\mathscr{R}$, θ, and ϕ, each point in space has many

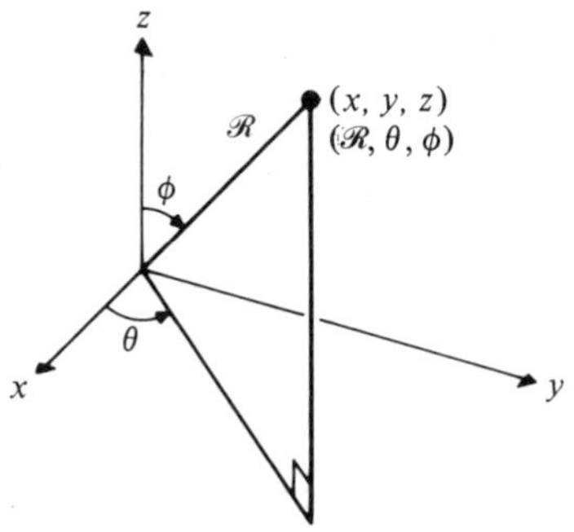

Figure 14.67

sets of spherical coordinates. The positive value of its spherical coordinate $\mathscr{R}$ is given by

$$\mathscr{R} = \sqrt{x^2 + y^2 + z^2}. \tag{14.65}$$

The θ-coordinates in cylindrical and spherical coordinates are identical, so that no explicit formula for θ in terms of x, y, and z exists. The ϕ-coordinate is the angle between the positive z-axis and the line joining the origin to the point (x, y, z), and that value of ϕ in the range $0 \le \phi \le \pi$ is determined by the formula

$$\phi = \mathrm{Cos}^{-1}\left(\frac{z}{\sqrt{x^2 + y^2 + z^2}}\right). \tag{14.66}$$

To transform equations $F(x, y, z) = 0$ of surfaces from Cartesian to spherical coordinates, we substitute from Eqs. (14.64):

$$F(\mathscr{R} \sin\phi \cos\theta, \mathscr{R} \sin\phi \sin\theta, \mathscr{R} \cos\phi) = 0.$$

For example, the sphere $x^2 + y^2 + z^2 = 4$ is symmetric about its center, and its equation in spherical coordinates is simply $\mathscr{R} = 2$. For the right-circular cone $z = \sqrt{x^2 + y^2}$, we write

$$\mathscr{R} \cos\phi = \sqrt{\mathscr{R}^2 \sin^2\phi \cos^2\theta + \mathscr{R}^2 \sin^2\phi \sin^2\theta} = \mathscr{R} \sin\phi.$$

Consequently, $\tan\phi = 1$ or $\phi = \pi/4$; i.e., $\phi = \pi/4$ is the equation of the cone in spherical coordinates.

Suppose that we are to evaluate the triple integral of a function $f(x, y, z)$,

$$\iiint_V f(x, y, z)\, dV,$$

over some region V of space. The choice of a triple iterated integral in spherical coordinates implies a subdivision of V into small volumes by means of coordinate surfaces $\mathscr{R} =$ constant, $\theta =$ constant, and $\phi =$ constant (Fig. 14.68). Surfaces $\mathscr{R} =$ constant are spheres centered at the origin; surfaces $\theta =$ constant are planes containing the z-axis; and surfaces $\phi =$ constant are right-circular cones symmetric about the z-axis with the origin as apex.

If we denote small variations in $\mathscr{R}$, θ, and ϕ for the element dV by $d\mathscr{R}$, $d\theta$, and $d\phi$ (Fig. 14.69) and approximate dV by a rectangular parallelepiped with dimensions $d\mathscr{R}$, $\mathscr{R}\, d\phi$, and $\mathscr{R} \sin\phi\, d\theta$, then

$$dV = (\mathscr{R} \sin\phi\, d\theta)(\mathscr{R}\, d\phi)\, d\mathscr{R} = \mathscr{R}^2 \sin\phi\, d\mathscr{R}\, d\theta\, d\phi. \tag{14.67}$$

The integrand $f(x, y, z)$ is expressed in spherical coordinates as

$$f(\mathscr{R} \sin\phi \cos\theta, \mathscr{R} \sin\phi \sin\theta, \mathscr{R} \cos\phi).$$

It remains only to affix appropriate limits to the triple iterated integral, and these limits will depend on which of the six possible triple iterated integrals in spherical coordinates that we choose. If we use a triple iterated integral with respect to $\mathscr{R}$, ϕ, and θ, then it will be of the form

$$\iiint_V f(x, y, z)\, dV = \int_a^b \int_{g_1(\theta)}^{g_2(\theta)} \int_{h_1(\theta,\phi)}^{h_2(\theta,\phi)}$$
$$\times\ f(\mathscr{R} \sin\phi \cos\theta, \mathscr{R} \sin\phi \sin\theta, \mathscr{R} \cos\phi)\mathscr{R}^2 \sin\phi\, d\mathscr{R}\, d\phi\, d\theta.$$

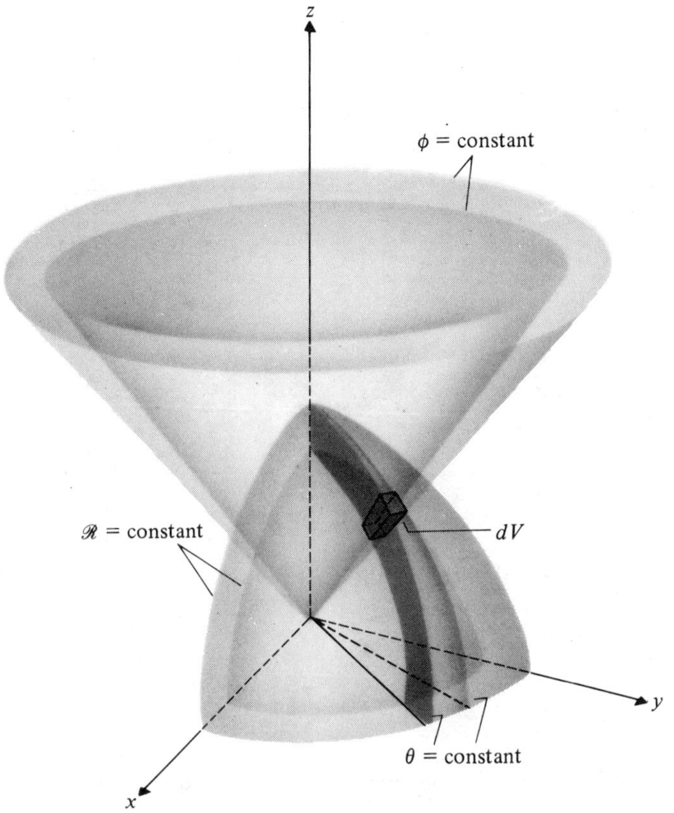

Figure 14.68

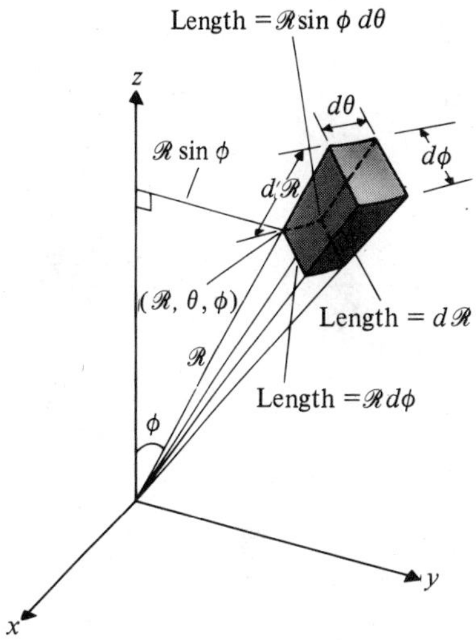

Figure 14.69

The geometric interpretations of the additions represented by these integrations with respect to $\mathscr{R}$, ϕ, and θ are left to the examples.

Example 28 Find the volume of a sphere.

Solution The equation of a sphere of radius R centered at the origin is $x^2 + y^2 + z^2 = R^2$ or, in spherical coordinates, $\mathscr{R} = R$. The volume of this sphere is eight times the first octant volume shown in Fig. 14.70. If we use spherical coordinates, the volume of an elemental piece is

$$\mathscr{R}^2 \sin \phi \, d\mathscr{R} \, d\phi \, d\theta.$$

An $\mathscr{R}$-integration adds these volumes for constant ϕ and θ to give the volume in a "spike,"

$$\int_0^R \mathscr{R}^2 \sin \phi \, d\mathscr{R} \, d\phi \, d\theta,$$

where the limits indicate that all spikes start at the origin (where $\mathscr{R} = 0$) and end on the sphere (where $\mathscr{R} = R$). A ϕ-integration now adds the volumes of spikes for constant θ. This yields the volume of a slice

$$\int_0^{\pi/2} \int_0^R \mathscr{R}^2 \sin \phi \, d\mathscr{R} \, d\phi \, d\theta,$$

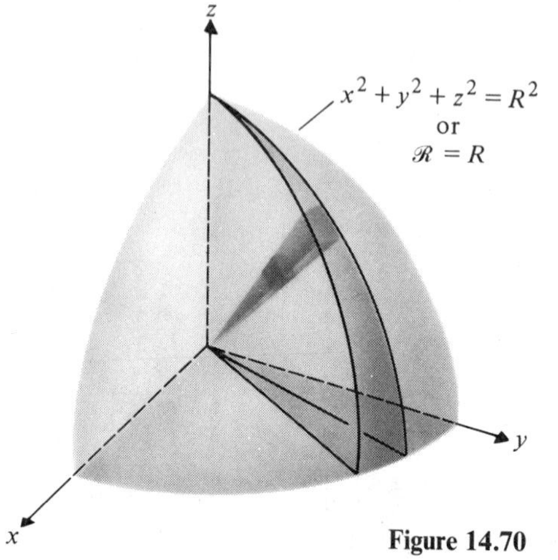

Figure 14.70

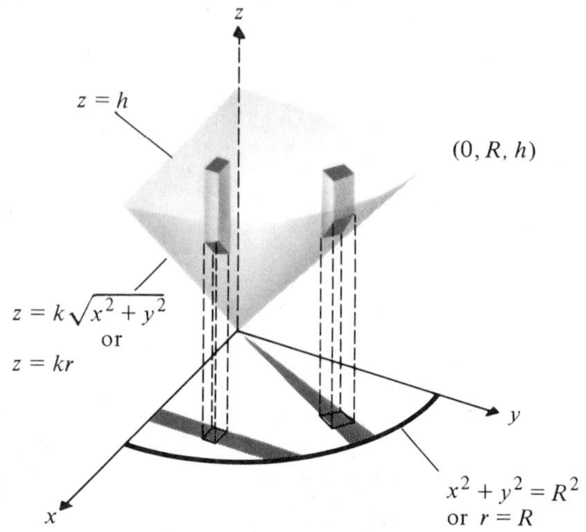

Figure 14.71

where the limits indicate that all slices in the first octant start on the z-axis (where $\phi = 0$) and end on the xy-plane (where $\phi = \pi/2$). Finally, the θ-integration adds the volumes of all such slices

$$\int_0^{\pi/2} \int_0^{\pi/2} \int_0^R \mathscr{R}^2 \sin\phi \, d\mathscr{R} \, d\phi \, d\theta,$$

where the limits 0 and $\pi/2$ identify the positions of the first and last slices, respectively, in the first octant. We obtain the required volume as

$$\begin{aligned} 8\int_0^{\pi/2} \int_0^{\pi/2} \int_0^R \mathscr{R}^2 \sin\phi \, d\mathscr{R} \, d\phi \, d\theta &= 8\int_0^{\pi/2} \int_0^{\pi/2} \left\{\frac{\mathscr{R}^3}{3} \sin\phi\right\}_0^R d\phi \, d\theta \\ &= \frac{8R^3}{3} \int_0^{\pi/2} \{-\cos\phi\}_0^{\pi/2} \, d\theta \\ &= \frac{8R^3}{3} \{\theta\}_0^{\pi/2} = \frac{4}{3}\pi R^3. \ \square \end{aligned}$$

Example 29 Find the center of mass of a solid object of constant density if it is in the shape of a right-circular cone.

Solution Let the cone have altitude h and base radius R. Then its mass is

$$M = \frac{1}{3}\pi R^2 h\rho,$$

where ρ is the density of the object. If we place axes as shown in Fig. 14.71, then $\bar{x} = \bar{y} = 0$; i.e., the center of mass is on the axis of symmetry of the cone. To find $\bar{z}$ we offer three solutions.

Method 1 If we use Cartesian coordinates, the equation of the surface of the cone is of the form $z = k\sqrt{x^2 + y^2}$. Since $(0, R, h)$ is a point on the cone, $h = kR$, and therefore $k = h/R$. Now

$$M\bar{z} = 4\int_0^R\int_0^{\sqrt{R^2-x^2}}\int_{k\sqrt{x^2+y^2}}^h z\rho\,dz\,dy\,dx = 4\rho\int_0^R\int_0^{\sqrt{R^2-x^2}}\left\{\frac{z^2}{2}\right\}_{k\sqrt{x^2+y^2}}^h dy\,dx$$

$$= 2\rho\int_0^R\int_0^{\sqrt{R^2-x^2}}\{h^2 - k^2(x^2+y^2)\}\,dy\,dx$$

$$= 2\rho\int_0^R\left\{h^2y - k^2x^2y - \frac{k^2y^3}{3}\right\}_0^{\sqrt{R^2-x^2}} dx$$

$$= 2\rho\int_0^R\left\{h^2\sqrt{R^2-x^2} - k^2x^2\sqrt{R^2-x^2} - \frac{k^2}{3}(R^2-x^2)^{3/2}\right\} dx.$$

If we set $x = R\sin\theta$, then $dx = R\cos\theta\,d\theta$, and

$$M\bar{z} = 2\rho\int_0^{\pi/2}\left\{h^2R\cos\theta - k^2(R^2\sin^2\theta)R\cos\theta - \frac{k^2}{3}R^3\cos^3\theta\right\}R\cos\theta\,d\theta$$

$$= 2\rho R^2\int_0^{\pi/2}\left\{\frac{h^2}{2}(1+\cos 2\theta) - \frac{k^2R^2}{8}(1-\cos 4\theta)\right.$$

$$\left. - \frac{k^2R^2}{12}\left(1 + 2\cos 2\theta + \frac{1+\cos 4\theta}{2}\right)\right\} d\theta$$

$$= \rho R^2\left\{h^2\left(\theta + \frac{\sin 2\theta}{2}\right) - \frac{k^2R^2}{4}\left(\theta - \frac{\sin 4\theta}{4}\right)\right.$$

$$\left. - \frac{k^2R^2}{6}\left(\frac{3\theta}{2} + \sin 2\theta + \frac{\sin 4\theta}{8}\right)\right\}_0^{\pi/2}$$

$$= \rho R^2\left\{\frac{\pi h^2}{2} - \frac{k^2R^2\pi}{8} - \frac{k^2R^2\pi}{8}\right\} = \frac{\pi R^2h^2\rho}{4}.$$

Thus,

$$\bar{z} = \frac{\pi R^2h^2\rho}{4}\cdot\frac{3}{\pi R^2h\rho} = \frac{3h}{4}.$$

Method 2 If we use cylindrical coordinates, the equation of the surface of the cone is $z = kr$. From Fig. 14.71, we see that

$$M\bar{z} = 4\int_0^{\pi/2}\int_0^R\int_{kr}^h z\rho\,r\,dz\,dr\,d\theta = 4\rho\int_0^{\pi/2}\int_0^R\left\{\frac{rz^2}{2}\right\}_{kr}^h dr\,d\theta$$

$$= 2\rho\int_0^{\pi/2}\int_0^R r(h^2 - k^2r^2)\,dr\,d\theta = 2\rho\int_0^{\pi/2}\left\{\frac{r^2h^2}{2} - \frac{k^2r^4}{4}\right\}_0^R d\theta$$

$$= \rho\left(R^2h^2 - \frac{k^2R^4}{2}\right)\{\theta\}_0^{\pi/2} = \rho\left(R^2h^2 - \frac{k^2R^4}{2}\right)\frac{\pi}{2} = \frac{\pi R^2h^2\rho}{4}.$$

Again, then, $\bar{z} = 3h/4$.

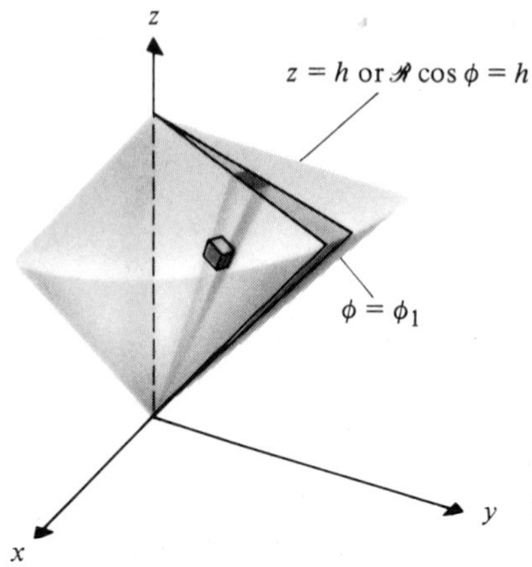

Figure 14.72

Method 3 If we use spherical coordinates, then the equation of the surface of the cone is

$$\phi = \mathrm{Cos}^{-1} \frac{h}{\sqrt{h^2 + R^2}} = \phi_1.$$

From Fig. 14.72, we see that

$$M\bar{z} = 4 \int_0^{\pi/2} \int_0^{\phi_1} \int_0^{h \sec \phi} (\mathcal{R} \cos \phi)\, \rho \mathcal{R}^2 \sin \phi \, d\mathcal{R}\, d\phi\, d\theta$$

$$= 4\rho \int_0^{\pi/2} \int_0^{\phi_1} \left\{ \frac{\mathcal{R}^4}{4} \sin \phi \cos \phi \right\}_0^{h \sec \phi} d\phi\, d\theta$$

$$= \rho h^4 \int_0^{\pi/2} \int_0^{\phi_1} \frac{\sin \phi}{\cos^3 \phi} \, d\phi\, d\theta = \rho h^4 \int_0^{\pi/2} \left\{ \frac{1}{2 \cos^2 \phi} \right\}_0^{\phi_1} d\theta$$

$$= \frac{\rho h^4}{2} \left\{ \frac{1}{\cos^2 \phi_1} - 1 \right\} \{\theta\}_0^{\pi/2} = \frac{\pi \rho h^4}{4} \left\{ \frac{h^2 + R^2}{h^2} - 1 \right\} = \frac{\pi h^2 R^2 \rho}{4}.$$

Again, $\bar{z} = 3h/4$. □

EXERCISES 14.13

In each of Exercises 1–7, find the equation of the surface in spherical coordinates. Sketch each surface.

1. $x^2 + y^2 + z^2 = 4$
2. $x^2 + y^2 = 1$
3. $z = \sqrt{x^2 + y^2/3}$
4. $4z = x^2 + y^2$
5. $y = x$
6. $x^2 + y^2 = 1 + z^2$
7. $z = -2\sqrt{x^2 + y^2}$

In each of Exercises 8–12, find the volume bounded by the surfaces.

8. $z = \sqrt{x^2 + y^2}$, $z = \sqrt{1 - x^2 - y^2}$
9. $z = 1$, $z = \sqrt{4 - x^2 - y^2}$
10. $x^2 + y^2 + z^2 = 1$, $y = x$, $y = 2x$, $z = 0$ (in first octant)
11. Inside $x^2 + y^2 + z^2 = 2$ but outside $x^2 + y^2 = 1$
12. $z = 2\sqrt{x^2 + y^2}$, $x^2 + y^2 = 4$, $z = 0$
13. Find the center of mass of a uniform hemispherical solid.
14. Find the moment of inertia of a uniform solid sphere of radius R about any line through its center.
15. Find the first moment about the yz-plane of the uniform solid in the first octant bounded by the surfaces $x^2 + y^2 + z^2 = 4$, $x^2 + y^2 + z^2 = 9$, $y = 0$, and $y = \sqrt{3}x$.
16. A solid sphere of radius R and center at the origin has a continuous charge distribution throughout. If the density of the charge is $\rho(x, y, z) = k\sqrt{x^2 + y^2 + z^2}$ coulombs per cubic meter (i.e., is proportional to the distance from the center of the sphere), find the total charge in the sphere.

17. Set up the six triple iterated integrals in spherical coordinates for the triple integral of a function $f(x, y, z)$ over the volume in the first octant under the sphere $x^2 + y^2 + z^2 = 2$ and inside the cylinder $x^2 + y^2 = 1$.

Evaluate the triple iterated integral in Exercises 18 and 19.

18. $$\int_0^9 \int_0^{\sqrt{81-y^2}} \int_0^{\sqrt{81-x^2-y^2}} \frac{1}{x^2+y^2+z^2}\, dz\, dx\, dy$$

19. $$\int_0^1 \int_0^{\sqrt{1-x^2}} \int_{\sqrt{x^2+y^2}}^{\sqrt{2-x^2-y^2}} dz\, dy\, dx$$

20. a) Use Archimedes' principle to determine the density of a spherical ball if it floats half submerged in water.

b) What force is required to keep the ball with its center at a depth of one-half the radius of the ball?

21. Find the volume bounded by the surface

$$(x^2 + y^2 + z^2)^2 = 2z(x^2 + y^2).$$

22. A sphere of radius R carries a uniform charge distribution of ρ coulombs per cubic meter (Fig. 14.73). If P is a point on the z-axis (distance $d > R$ from the center of the sphere) and dV is a small element of volume of the sphere, then the potential at P due to dV is given by

$$\frac{1}{4\pi\varepsilon_0}\frac{\rho\, dV}{s},$$

where s is the distance from P to dV.

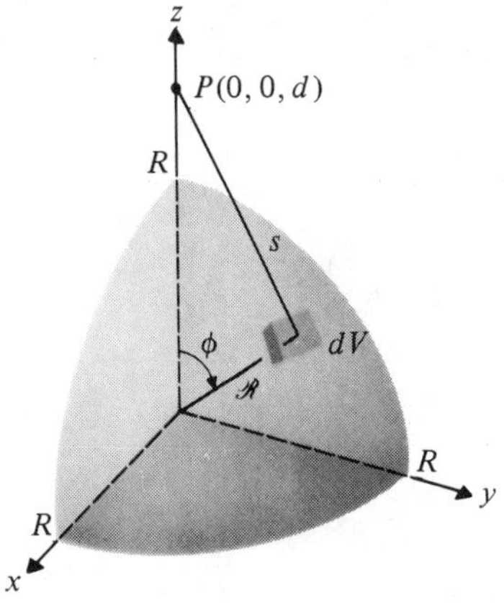

Figure 14.73

a) Show that in terms of spherical coordinates the potential at P due to the entire sphere is

$$V = \frac{\rho}{4\pi\varepsilon_0}\int_{-\pi}^{\pi}\int_0^{\pi}\int_0^R \frac{\mathcal{R}^2 \sin\phi}{\sqrt{\mathcal{R}^2 + d^2 - 2\mathcal{R}d\cos\phi}} \times d\mathcal{R}\, d\phi\, d\theta.$$

b) Because this iterated integral is very difficult to evaluate, we replace ϕ with the variable

$$s = \sqrt{\mathcal{R}^2 + d^2 - 2d\mathcal{R}\cos\phi}.$$

Show that with this change

$$V = \frac{\rho}{4\pi\varepsilon_0 d}\int_{-\pi}^{\pi}\int_0^R \int_{d-\mathcal{R}}^{d+\mathcal{R}} \mathcal{R}\, ds\, d\mathcal{R}\, d\theta.$$

c) Evaluate the integral in (b) to verify that

$$V = \frac{1}{4\pi\varepsilon_0}\frac{Q}{d},$$

where Q is the total charge on the sphere.

23. A sphere of constant density ρ and radius R is located at the origin (Fig. 14.74). If a mass m is situated at a point P on the z-axis (distance $d > R$ from the center of the sphere) and dV is a small element of volume of the sphere, then according to Newton's universal law of gravitation, the z-component of the force on m due to the mass in dV is given by

$$-\frac{Gm\rho\, dV\cos\psi}{s^2},$$

where $G > 0$ is a constant and s is the distance between P and dV.

a) Show that in spherical coordinates the total force on m due to the entire sphere has z-component

$$F_z = -\frac{Gm\rho}{2d}\int_{-\pi}^{\pi}\int_0^{\pi}\int_0^R \left(\frac{s^2 + d^2 - \mathcal{R}^2}{s^3}\right) \times \mathcal{R}^2 \sin\phi\, d\mathcal{R}\, d\phi\, d\theta.$$

b) Use the transformation in Exercise 22(b) to write F_z in the form

$$F_z = -\frac{Gm\rho}{2d^2}\int_{-\pi}^{\pi}\int_0^R \int_{d-\mathcal{R}}^{d+\mathcal{R}} \mathcal{R}\left(\frac{s^2 + d^2 - \mathcal{R}^2}{s^2}\right) \times ds\, d\mathcal{R}\, d\theta,$$

and show that

$$F_z = -\frac{GmM}{d^2},$$

where M is the total mass of the sphere.

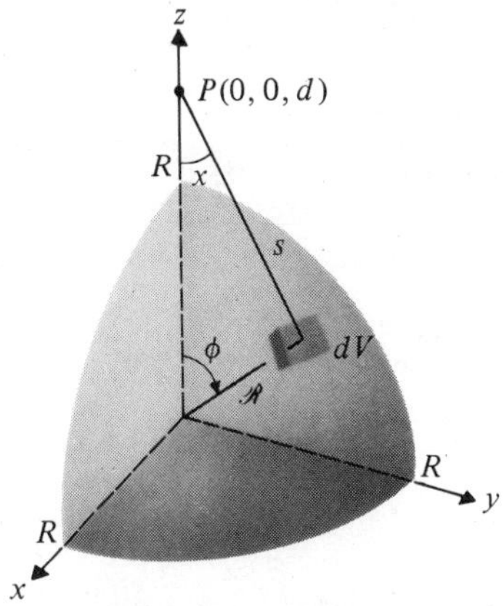

Figure 14.74

14.14 DERIVATIVES OF DEFINITE INTEGRALS

If a function $f(x, y)$ of two independent variables is integrated with respect to y from $y = a$ to $y = b$, the result depends on x. Suppose we denote this function by $F(x)$:

$$F(x) = \int_a^b f(x, y)\, dy. \tag{14.68}$$

To calculate the derivative $F'(x)$ of this function, we should first integrate with respect to y and then differentiate with respect to x. The following theorem indicates that differentiation can be done first and integration later.

▶ **Theorem 14.3** If the partial derivative $\partial f/\partial x$ of $f(x, y)$ is continuous on a rectangle $a \le y \le b$, $c \le x \le d$, then for $c < x < d$

$$\frac{d}{dx}\int_a^b f(x, y)\, dy = \int_a^b \frac{\partial f(x, y)}{\partial x}\, dy. \quad ◀ \tag{14.69}$$

Proof If we define

$$g(x) = \int_a^b \frac{\partial f(x, y)}{\partial x}\, dy$$

as the right-hand side of Eq. (14.69), then this function is defined for $c \le x \le d$. In fact, because $\partial f/\partial x$ is continuous, so too must $g(x)$ be continuous. We can therefore integrate $g(x)$ with respect to x from $x = c$ to any value of x in the interval $c \le x \le d$:

$$\int_c^x g(x)\, dx = \int_c^x \int_a^b \frac{\partial f(x, y)}{\partial x}\, dy\, dx.$$

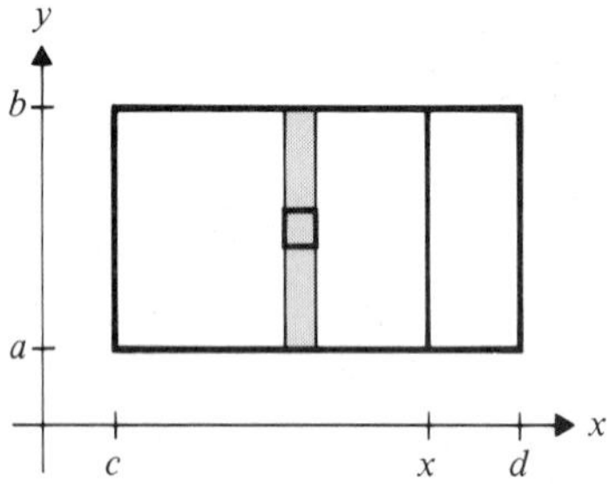

Figure 14.75

This double iterated integral represents the double integral of $\partial f/\partial x$ over the rectangle shown in Fig. 14.75, and if we reverse the order of integration, we have

$$\int_c^x g(x)\, dx = \int_a^b \int_c^x \frac{\partial f(x, y)}{\partial x}\, dx\, dy = \int_a^b \{f(x, y)\}_c^x\, dy$$

$$= \int_a^b \{f(x, y) - f(c, y)\}\, dy = \int_a^b f(x, y)\, dy - \int_a^b f(c, y)\, dy.$$

Because the second integral on the right is independent of x, if we differentiate this equation with respect to x, we get

$$\frac{d}{dx}\int_c^x g(x)\, dx = \frac{d}{dx}\int_a^b f(x, y)\, dy.$$

But elementary calculus tells us that

$$\frac{d}{dx}\int_c^x g(x)\, dx = g(x);$$

thus $g(x)$ is also given by

$$g(x) = \frac{d}{dx}\int_a^b f(x, y)\, dy.$$

This completes the proof. □

The limits on the integral in (14.68) need not be numerical constants; so far as the integration with respect to y is concerned, x is constant, and therefore a and

b could be functions of x:

$$F(x) = \int_{a(x)}^{b(x)} f(x, y)\, dy.$$

Indeed, this is precisely what does occur in the evaluation of double integrals by means of double iterated integrals. To differentiate $F(x)$ now is more complicated than in Theorem 14.3 because we must also account for the x's in $a(x)$ and $b(x)$. The chain rule can be used to develop a formula for $F'(x)$.

▶ **Theorem 14.4** (*Leibnitz's rule*). If the partial derivative $\partial f/\partial x$ is continuous on the area bounded by the curves $y = a(x)$, $y = b(x)$, $x = c$, and $x = d$, then

$$\frac{d}{dx}\int_{a(x)}^{b(x)} f(x, y)\, dy$$
$$= \int_{a(x)}^{b(x)} \frac{\partial f(x, y)}{\partial x}\, dy + f[x, b(x)]\frac{db}{dx} - f[x, a(x)]\frac{da}{dx}. \blacktriangleleft \qquad (14.70)$$

Proof If $F(x) = \int_{a(x)}^{b(x)} f(x, y)\, dy$, then the schematic

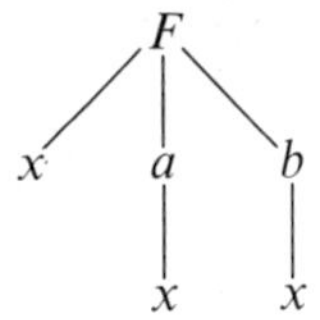

gives the chain rule

$$F'(x) = \left.\frac{\partial F}{\partial x}\right)_{a, b} + \left.\frac{\partial F}{\partial b}\right)_{a, x}\frac{db}{dx} + \left.\frac{\partial F}{\partial a}\right)_{b, x}\frac{da}{dx}.$$

The first term is precisely the situation covered in Theorem 14.3, and therefore

$$\left.\frac{\partial F}{\partial x}\right)_{a, b} = \int_a^b \frac{\partial f(x, y)}{\partial x}\, dy.$$

In the proof of Theorem 14.3 we also indicated that it is a fact from elementary calculus that for constant a

$$\frac{d}{db}\int_a^b f(y)\, dy = f(b).$$

It immediately follows that

$$\frac{\partial}{\partial b}\int_a^b f(x, y)\, dy = f(x, b).$$

In other words,

$$\left.\frac{\partial F}{\partial b}\right)_{a, x} = f(x, b).$$

Furthermore,

$$\left.\frac{\partial F}{\partial a}\right)_{b, x} = \frac{\partial}{\partial a}\int_a^b f(x, y)\, dy = -\frac{\partial}{\partial a}\int_b^a f(x, y)\, dy = -f(x, a).$$

Substitution of these facts into the chain rule now gives Leibnitz's rule. □

Our derivation of Leibnitz's rule for the differentiation of a definite integral that depends on a parameter (x in this case) shows that the first term accounts for those x's in the integrand, and the second and third terms for the x's in the upper and lower limits. The following geometric interpretation of Leibnitz's rule emphasizes this same point.

Suppose the function $f(x, y)$ has only positive values so that the surface $z = f(x, y)$ lies completely above the xy-plane (Fig. 14.76). The equations $y = a(x)$ and $y = b(x)$ describe cylindrical walls standing on the curves $y = a(x)$, $z = 0$ and $y = b(x)$, $z = 0$ in the xy-plane. Were we to slice through the surface $z = f(x, y)$ with a plane $x =$ constant, then an area would be defined in this plane bounded on the top by $z = f(x, y)$, on the sides by $y = a(x)$ and $y = b(x)$, and on the bottom by the xy-plane. This area is clearly defined by the definite integral

$$\int_{a(x)}^{b(x)} f(x, y)\, dy$$

in Leibnitz's rule, and as x varies so too does the area. Note, in particular, that as the plane varies, the area changes, not only because the height of the surface $z = f(x, y)$ changes but also because the width of the area varies (i.e., the two cylindrical walls are not a constant distance apart). The first term in Leibnitz's rule accounts for the vertical variation, whereas the remaining two terms represent variations due to the fluctuating width.

Example 30 If $F(x) = \int_x^{2x} (y^3 \ln y + x^3 e^y)\, dy$, find $F'(x)$.

Solution With Leibnitz's rule, we have

$$\begin{aligned} F'(x) &= \int_x^{2x} (3x^2 e^y)\, dy + \{(2x)^3 \ln(2x) + x^3 e^{2x}\}(2) - \{x^3 \ln x + x^3 e^x\}(1) \\ &= \{3x^2 e^y\}_x^{2x} + 16x^3(\ln 2 + \ln x) + 2x^3 e^{2x} - x^3 \ln x - x^3 e^x \\ &= 3x^2 e^{2x} - 3x^2 e^x + (16 \ln 2)x^3 + 15x^3 \ln x + 2x^3 e^{2x} - x^3 e^x \\ &= 3x^2 e^x(e^x - 1) + x^3(16 \ln 2 + 15 \ln x + 2e^{2x} - e^x). \ \square \end{aligned}$$

In Exercise 6 you are asked to evaluate the integral defining $F(x)$ in this example and then to differentiate the resulting function with respect to x. It will be clear, then, that for this example Leibnitz's rule simplifies the calculations considerably.

Example 31 Evaluate

$$\int_0^1 \frac{y^x - 1}{\ln y}\, dy \quad \text{for } x > -1.$$

Solution To attempt to find an antiderivative for $(y^x - 1)/\ln y$ is impossible. If, however, we set

$$F(x) = \int_0^1 \frac{y^x - 1}{\ln y}\, dy$$

and use Leibnitz's rule, we have

$$F'(x) = \int_0^1 \frac{\partial}{\partial x}\left\{\frac{y^x - 1}{\ln y}\right\} dy = \int_0^1 \frac{y^x \ln y}{\ln y}\, dy = \int_0^1 y^x\, dy = \left\{\frac{y^{x+1}}{x+1}\right\}_0^1 = \frac{1}{x+1}.$$

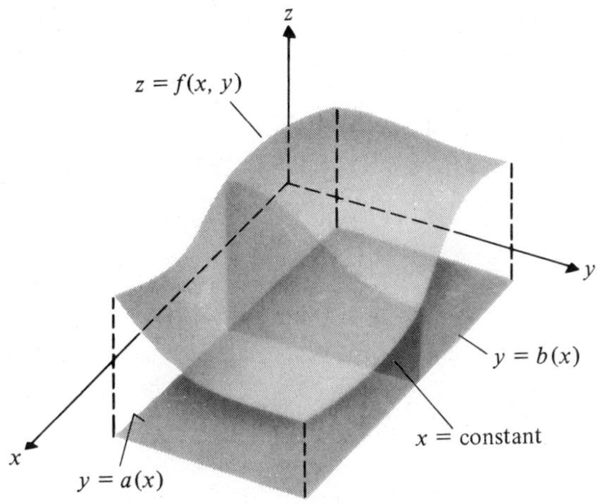

Figure 14.76

It follows, therefore, that $F(x)$ must be of the form

$$F(x) = \ln (x + 1) + C.$$

But from the definition of $F(x)$ as an integral, it is clear that $F(0) = 0$, and hence $C = 0$. Thus,

$$\int_0^1 \frac{y^x - 1}{\ln y}\, dy = \ln (x + 1). \ \square$$

Note that this problem originally had nothing whatsoever to do with Leibnitz's rule. But by introducing the rule, we were able to find a simple formula for $F'(x)$, and this immediately led to $F(x)$. This can be a very useful technique for evaluating definite integrals that depend on a parameter.

Example 32 When a shell is fired from the artillery gun shown in Fig. 14.77, the barrel recoils along a well-lubricated guide and its motion is braked by a battery of heavy springs. We set up a coordinate system where $x = 0$ represents the firing position of the gun when no stretch or compression exists in the springs. Suppose that when the gun is fired, the horizontal component of the force causing recoil is $g(t)$ (t = time). If the mass of the gun is m and the effective spring constant for the battery of springs is k, then Newton's second law states that

$$m \frac{d^2x}{dt^2} = g(t) - kx,$$

or

$$m \frac{d^2x}{dt^2} + kx = g(t).$$

Figure 14.77

This differential equation can be solved by a method called *variation of parameters*, and the solution is

$$x(t) = A \cos \sqrt{\frac{k}{m}}\, t + B \sin \sqrt{\frac{k}{m}}\, t + \frac{1}{\sqrt{mk}} \int_0^t g(u) \sin \left(\sqrt{\frac{k}{m}}\, (t - u) \right) du,$$

where A and B are arbitrary constants. We do not discuss differential equations until Chapter 16, but with Leibnitz's rule it is possible to verify that this function is indeed a solution. Do so.

Solution To differentiate the integral we rewrite Leibnitz's rule in terms of the variables of this problem:

$$\frac{d}{dt}\int_{a(t)}^{b(t)} f(t, u)\, du = \int_{a(t)}^{b(t)} \frac{\partial f(t, u)}{\partial t}\, du + f[t, b(t)]\frac{db}{dt} - f[t, a(t)]\frac{da}{dt}.$$

If we now apply this formula to the definite integral in $x(t)$ where

$$f(t, u) = g(u) \sin\left[\sqrt{\frac{k}{m}}(t - u)\right],$$

then

$$\frac{dx}{dt} = -\sqrt{\frac{k}{m}} A \sin \sqrt{\frac{k}{m}}\, t + \sqrt{\frac{k}{m}} B \cos \sqrt{\frac{k}{m}}\, t$$

$$+ \frac{1}{\sqrt{mk}}\left\{\int_0^t g(u)\sqrt{\frac{k}{m}} \cos\left(\sqrt{\frac{k}{m}}(t - u)\right) du + g(t) \sin\left(\sqrt{\frac{k}{m}}(t - t)\right)\right\}$$

$$= \sqrt{\frac{k}{m}}\left\{-A \sin \sqrt{\frac{k}{m}}\, t + B \cos \sqrt{\frac{k}{m}}\, t\right\} + \frac{1}{m}\int_0^t g(u) \cos\left(\sqrt{\frac{k}{m}}(t - u)\right) du.$$

We now use Leibnitz's rule once more to find d^2x/dt^2:

$$\frac{d^2x}{dt^2} = \sqrt{\frac{k}{m}}\left\{-\sqrt{\frac{k}{m}} A \cos \sqrt{\frac{k}{m}}\, t - \sqrt{\frac{k}{m}} B \sin \sqrt{\frac{k}{m}}\, t\right\}$$

$$+ \frac{1}{m}\left\{\int_0^t -g(u)\sqrt{\frac{k}{m}} \sin\left(\sqrt{\frac{k}{m}}(t - u)\right) du + g(t) \cos\left(\sqrt{\frac{k}{m}}(t - t)\right)\right\}$$

$$= -\frac{k}{m}\left\{A \cos \sqrt{\frac{k}{m}}\, t + B \sin \sqrt{\frac{k}{m}}\, t\right\}$$

$$- \sqrt{\frac{k}{m^3}}\int_0^t g(u) \sin\left(\sqrt{\frac{k}{m}}(t - u)\right) du + \frac{1}{m} g(t)$$

$$= -\frac{k}{m}\left\{A \cos \sqrt{\frac{k}{m}}\, t + B \sin \sqrt{\frac{k}{m}}\, t + \frac{1}{\sqrt{mk}}\int_0^t g(u) \sin\left(\sqrt{\frac{k}{m}}(t - u)\right) du\right\}$$

$$+ \frac{1}{m} g(t)$$

$$= -\frac{k}{m} x(t) + \frac{1}{m} g(t).$$

In other words, the function $x(t)$ satisfies the differential equation

$$m\frac{d^2x}{dt^2} = -kx(t) + g(t),$$

and the proof is complete. □

This example illustrates that Leibnitz's rule is essential to the manipulation of solutions of differential equations that are represented as definite integrals.

EXERCISES 14.14

In each of Exercises 1–5, use Leibnitz's rule to find the derivative of $F(x)$. Check your result by evaluating the integral and then differentiating.

1. $F(x) = \int_0^3 (x^2y^2 + 3xy)\,dy$

2. $F(x) = \int_1^x (x^2/y^2 + e^y)\,dy$

3. $F(x) = \int_{x-1}^{x^2} (x^3y + y^2 + 1)\,dy$

4. $F(x) = \int_{x^2}^{x^3-1} (x + y \ln y)\,dy$

5. $F(x) = \int_0^x \frac{y - x}{y + x}\,dy$

6. Find $F'(x)$ in Example 30 by evaluating the definite integral and then calculating $F'(x)$.

7. Use the result of Example 31 to prove that

$$\int_0^1 \frac{x^p - x^q}{\ln x}\,dx = \ln\left(\frac{p+1}{q+1}\right),$$

provided $p > -1$ and $q > -1$.

8. Find $F'(x)$ if

$$F(x) = \int_{\sin x}^{e^x} \sqrt{1 + y^3}\,dy.$$

In each of Exercises 9–11, use Leibnitz's rule to verify that the function $y(x)$ satisfies the differential equation.

9. $y(x) = \frac{1}{x^2}\int_0^x t^2 f(t)\,dt;\ x\frac{dy}{dx} + 2y = xf(x)$

10. $y(x) = \frac{1}{2}\int_0^x f(t)(e^{x-t} - e^{t-x})\,dt;\ \frac{d^2y}{dx^2} - y = f(x)$

11. $y(x) = \frac{1}{\sqrt{2}}\int_0^x e^{2(t-x)} \sin[\sqrt{2}(x-t)] f(t)\,dt;$

$$\frac{d^2y}{dx^2} + 4\frac{dy}{dx} + 6y = f(x)$$

12. a) What is the domain of the function

$$F(x) = \int_0^9 \ln(1 - x^2y^2)\,dy?$$

What is $F(0)$?

b) Find $F'(x)$ by Leibnitz's rule. What is $F'(0)$?

c) Show that the graph of the function $F(x)$ is concave downward for all x in its domain of definition.

***13.** Laplace's equation for a function $u(r, \theta)$ in polar coordinates is

$$\frac{\partial^2 u}{\partial r^2} + \frac{1}{r}\frac{\partial u}{\partial r} + \frac{1}{r^2}\frac{\partial^2 u}{\partial \theta^2} = 0.$$

If the values of $u(r, \theta)$ are specified on the circle $r = R$ as $u(R, \phi)$, $-\pi < \phi \le \pi$, then Poisson's integral formula states that the value of $u(r, \theta)$ interior to this circle is defined by

$$u(r, \theta) = \frac{R^2 - r^2}{2\pi}\int_{-\pi}^{\pi} \frac{u(R, \phi)}{R^2 + r^2 - 2rR\cos(\theta - \phi)}\,d\phi.$$

Show that this function does indeed satisfy Laplace's equation.

SUMMARY

The definite integral of a function $f(x)$ from $x = a$ to $x = b$ is a limit of a sum

$$\int_a^b f(x)\,dx = \lim_{n\to\infty} \sum_{i=1}^{n} f(x_i^*)\,\Delta x_i.$$

In this chapter we extended this idea to define double integrals of functions $f(x, y)$ over regions in the xy-plane and triple integrals of functions $f(x, y, z)$

over regions of space. Each is once again the limit of a sum:

$$\iint_R f(x, y)\, dA = \lim_{n \to \infty} \sum_{i=1}^{n} f(x_i^*, y_i^*)\, \Delta A_i,$$

$$\iiint_V f(x, y, z)\, dV = \lim_{n \to \infty} \sum_{i=1}^{n} f(x_i^*, y_i^*, z_i^*)\, \Delta V_i.$$

To evaluate double integrals we use double iterated integrals in Cartesian or polar coordinates. Which is the more useful in a given problem depends on the shape of the region R and the form of the function $f(x, y)$. For instance, circles centered at the origin and straight lines through the origin are represented very simply in polar coordinates, and therefore a region R with these curves as boundaries immediately suggests the use of polar coordinates. On the other hand, curves that can be described in the form $y = f(x)$, where $f(x)$ is a polynomial, a rational function, or a transcendental function, often suggest using double iterated integrals in Cartesian coordinates. Each integration in a double iterated integral can be interpreted geometrically, and through these interpretations it is a simple matter to find appropriate limits for the integrals. In particular, for a double iterated integral in Cartesian coordinates, the inner integration is over the rectangles in a strip (horizontal or vertical), and the outer integration adds over all strips. In polar coordinates, the inner integral is inside either a wedge or a ring, and the outer integral adds over all wedges or rings.

To evaluate triple integrals we use triple iterated integrals in Cartesian, cylindrical, or spherical coordinates. Once again the limits on these integrals can be determined by interpreting the summations geometrically. For example, the first integration in a triple iterated integral in Cartesian coordinates is always over the boxes in a column, the second over the rectangles inside a strip, and the third over all strips.

We used double integrals to find plane areas, volumes of solids of revolution, centroids, moments of inertia, fluid forces, and areas of surfaces. We dealt with these same applications (with the exception of surface area) in Chapter 7 using the definite integral, but with some difficulty: Volumes require two methods, shells and washers; centroids require an averaging formula for the first moment of a rectangle that had its length perpendicular to the axis about which a moment is required; moments of inertia need a "one-third cubed formula" for rectangles with lengths perpendicular to the axis about which the moment of inertia is required; and fluid forces require horizontal rectangles. On the other hand, double integrals eliminate these difficulties but, more important, provide a unified approach to all applications.

In Section 14.14 we used double integrals to verify Leibnitz's rule for differentiating definite integrals that depend on a parameter.

REVIEW EXERCISES

In each of Exercises 1–21, evaluate the integral over the region indicated.

1. $\iint_R (2x + y)\, dA$, where R is bounded by $y = x$, $y = 0$, $x = 2$

2. $\iiint_V xyz\, dV$, where V is bounded by $y = z$, $x = 0$, $y = 3$, $x = 1$, $z = 0$

3. $\iint_R x^3y^2\,dA$, where R is bounded by $x = 1, x = -1, y = 2, y = -2$

4. $\iiint_V (x^2 - y^3)\,dV$, where V is bounded by $z = xy, z = 0, x = 1, y = 1$

5. $\iiint_V (x^2 - y^2)\,dV$, where V is the region in Exercise 4

6. $\iint_R y\,dA$, where R is bounded by $y = (x - 1)^2, y = x + 1$

7. $\iint_R xy^2\,dA$, where R is bounded by $x = 2 - 2y^2, x = -y^2$

8. $\iint_R x^2y\,dA$, where R is the region in Exercise 7

9. $\iiint_V (x^2 + y^2 + z^2)\,dV$, where V is bounded by $z = x, z = -x, y = 0, y = 1, z = 2$

10. $\iint_R (xy - x^2y^2)\,dA$, where R is bounded by $y = 2x^2, y = 4 - 2x^2$

11. $\iint_R x \sin y\,dA$, where R is bounded by $x = \sqrt{1 - y}, x = 0, y = 0$

12. $\iiint_V (x + y + z)\,dV$, where V is bounded by $z = 1 - x^2 - y^2, z = 0$

13. $\iint_R xe^y\,dA$, where R is bounded by $x = 0, y = 5, y = 2x + 1$

14. $\iiint_V dV$, where V is bounded by $z^2 = x^2 + y^2, x^2 + y^2 = 4$

15. $\iint_R (x + y)\,dA$, where R is bounded by $y = 1 - x^2, y = 1 - \sqrt{x}, y = x - 1$

16. $\iiint_V (x^2 + y^2 + z^2)\,dV$, where V is bounded by $z = \sqrt{1 - x^2 - y^2}, z = 0$

17. $\iint_R \frac{x}{x + y}\,dA$, where R is bounded by $y = x - 1, x = 2, y = 0$

18. $\iint_R (x^2 + y^2)\,dA$, where R is bounded by $x^2 + y^2 = 2x$

19. $\iiint_V \frac{x^2}{z^2}\,dV$, where V is bounded by $z = \sqrt{4 - x^2 - y^2}, z = 1$

20. $\iint_R \frac{1}{x^2 + y^2}\,dA$, where R is bounded by $y = x, x = 1, x = 2, y = 0$

21. $\iint_R (x^2 - y^2)\,dA$, where R is bounded by $y = x, y = x - 1, y = 5 - 2x, y = 14 - 2x$

22. Find the area bounded by the curve $x^2(4 - x^2) = y^2$.

23. Find the centroid of the area bounded by the curves $4x = 4 - y^2, x = \sqrt{8 - 2y^2}$.

24. Find the volume bounded by the surfaces $y + z = 2, y - z = 2, y = x^2 + 1$.

25. If the viewing glass on a deep-sea diver's helmet is circular with diameter 10 centimeters, what is the force on the glass when the center of the glass is 50 meters below the surface?

26. Find the area of that part of $z = 1 - 4x^2 - 4y^2$ above the xy-plane.

27. Find the area bounded by the curve $x^4(4 - x^2) = y^2$.

28. Find the volumes of the solids of revolution when the area in Exercise 22 is rotated about the x- and y-axes.

29. Find the moment of inertia about the z-axis of the solid bounded by the surfaces $z = \sqrt{x^2 + y^2}$, $z = 2 - \sqrt{x^2 + y^2}$ if its density ρ is constant.

30. Find the second moment of area about the y-axis of the area bounded by the curve $x^3 + y^3 = 1$.

31. Find the center of mass of a uniform solid bounded by the surfaces $z = 1 + x^2 + y^2$, $x^2 + y^2 = 1$, $z = 0$.

32. Find the average value of the function $f(x, y) = x^2 + y^2$ over the region bounded by the curve $x^2 + y^2 = 4$.

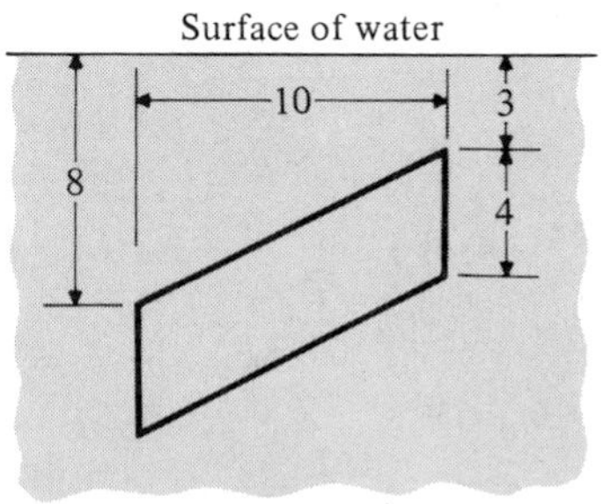

Figure 14.78

33. Find the force due to water pressure on each side of the vertical parallelogram shown in Fig. 14.78. All measurements are in meters.

34. Find the moment of inertia about the z-axis of a uniform solid bounded by the surfaces $z = 0$, $z = 1$, $x^2 + y^2 = 1 + z^2$.

35. Find the average value of the function $f(x, y, z) = x + y + z$ over the region bounded by the surfaces $-x + 2y + 2z = 4$, $x + y + z = 2$, $y = 0$, $z = 0$.

36. Find the centroid of the bifolium $r = \sin\theta\cos^2\theta$ (r, θ being polar coordinates).

37. Show that

$$y(x) = e^{-3x}(C_1 \cos x + C_2 \sin x) + \int_0^x f(t)e^{3(t-x)} \sin(x - t)\, dt$$

is a solution of the differential equation

$$\frac{d^2y}{dx^2} + 6\frac{dy}{dx} + 10y = f(x)$$

for any constants C_1 and C_2.

38. Find the center of mass of the uniform solid in the first octant common to the cylinders $x^2 + z^2 = 1$, $y^2 + z^2 = 1$.

39. Find the area of that part of $z = \ln(x^2 + y^2)$ between the cylinders $x^2 + y^2 = 1$ and $x^2 + y^2 = 4$.

40. Find the volume bounded by the surface $\sqrt{x^2 + z^2} = y(2 - y)$.

Vector Calculus

15

15.1 VECTOR FIELDS OR VECTOR FUNCTIONS

In Sections 12.7 and 12.8 we considered vectors whose components are functions of a single variable. In particular, if an object moves along a curve C and C is defined parametrically by $x = x(t)$, $y = y(t)$, $z = z(t)$, $t \geq 0$, t being time, then the components of the position, velocity, and acceleration vectors are functions of time:

$$\mathbf{r} = \mathbf{r}(t) = x(t)\hat{\mathbf{i}} + y(t)\hat{\mathbf{j}} + z(t)\hat{\mathbf{k}},$$

$$\mathbf{v} = \frac{d\mathbf{r}}{dt} = \frac{dx}{dt}\hat{\mathbf{i}} + \frac{dy}{dt}\hat{\mathbf{j}} + \frac{dz}{dt}\hat{\mathbf{k}},$$

$$\mathbf{a} = \frac{d\mathbf{v}}{dt} = \frac{d^2x}{dt^2}\hat{\mathbf{i}} + \frac{d^2y}{dt^2}\hat{\mathbf{j}} + \frac{d^2z}{dt^2}\hat{\mathbf{k}}.$$

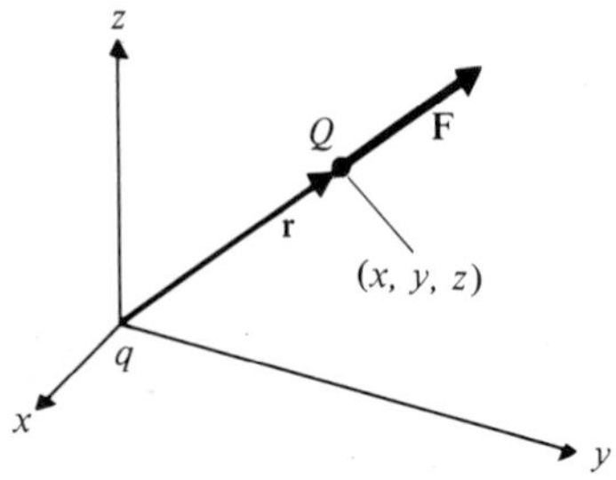

Figure 15.1

If the object has constant mass m and is subjected to a force $\mathbf{F}$, which is given as a function of time t, $\mathbf{F} = \mathbf{F}(t)$, then Newton's second law defines the acceleration of the object as $\mathbf{a} = \mathbf{F}(t)/m$. This equation can then be integrated to yield the velocity and position of the object as functions of time. Unfortunately, what often happens is that we do not know $\mathbf{F}$ as a function of time. Instead, we know that if the object were at such and such a position, then the force on it would be such and such; i.e., we know $\mathbf{F}$ as a function of position. For example, suppose a positive charge q is placed at the origin in space (Fig. 15.1), and a second positive charge Q is placed at position (x, y, z). According to Coulomb's law, the force on Q due to q is

$$\mathbf{F} = \frac{qQ}{4\pi\varepsilon_0|\mathbf{r}|^3}\mathbf{r} = \frac{qQ}{4\pi\varepsilon_0(x^2 + y^2 + z^2)^{3/2}}(x\hat{\mathbf{i}} + y\hat{\mathbf{j}} + z\hat{\mathbf{k}}).$$

If we allow Q to move under the influence of this force, we will not know $\mathbf{F}$ as a function of time, but rather as a function of position. This makes Newton's second law much more difficult to deal with, but it is in fact the normal situation. Most forces are represented as a function of position rather than time. Besides electrostatic forces, consider, for instance, spring forces, gravitational forces, and fluid forces—all of these are functions of position. Forces that are functions of position are examples of vectors that are functions of position, and we call such vectors *vector fields* or *vector functions.* To be precise, we say that $\mathbf{F}$ is a vector field or vector function in a region D if $\mathbf{F}$ assigns a vector to each point in D. If D is a region of space, then $\mathbf{F}$ assigns a vector $\mathbf{F}(x, y, z)$ to each point (x, y, z) in D. If P, Q, and R are the components of $\mathbf{F}(x, y, z)$, then each of these will be a function of x, y, and z:

$$\mathbf{F} = \mathbf{F}(x, y, z) = P(x, y, z)\hat{\mathbf{i}} + Q(x, y, z)\hat{\mathbf{j}} + R(x, y, z)\hat{\mathbf{k}}. \tag{15.1}$$

If D is a region of the xy-plane, then

$$\mathbf{F} = \mathbf{F}(x, y) = P(x, y)\hat{\mathbf{i}} + Q(x, y)\hat{\mathbf{j}}. \tag{15.2}$$

Vector fields are essential to the study of most areas in the physical sciences. Moments ($\mathbf{M} = \mathbf{r} \times \mathbf{F}$) in mechanics (see Exercise 43 in Section 12.6) depend on the position of **F**, and are therefore vector fields. The electric field intensity **E**, the electric displacement **D**, the magnetic induction **B**, and the current density **J** are all important vector fields in electromagnetic theory. The heat flux vector **q** is the basis for the study of heat conduction.

We have also encountered vector fields that are of geometric importance. For example, the gradient of a scalar function $f(x, y, z)$ is a vector function

$$\text{grad } f = \nabla f = \frac{\partial f}{\partial x}\hat{\mathbf{i}} + \frac{\partial f}{\partial y}\hat{\mathbf{j}} + \frac{\partial f}{\partial z}\hat{\mathbf{k}}.$$

It assigns the vector ∇f to each point (x, y, z) in some region of space. We have already seen that ∇f points in the direction in which $f(x, y, z)$ increases most rapidly, and its magnitude $|\nabla f|$ is the rate of increase in that direction. In addition, we know that if $f(x, y, z) = c$ is the equation of a surface that passes through a point (x, y, z), then ∇f at that point is normal to the surface.

Often we write

$$\nabla f = \left(\frac{\partial}{\partial x}\hat{\mathbf{i}} + \frac{\partial}{\partial y}\hat{\mathbf{j}} + \frac{\partial}{\partial z}\hat{\mathbf{k}}\right) f$$

and regard

$$\nabla = \frac{\partial}{\partial x}\hat{\mathbf{i}} + \frac{\partial}{\partial y}\hat{\mathbf{j}} + \frac{\partial}{\partial z}\hat{\mathbf{k}}$$

as a vector differential operator, called the *del operator*. It operates on a scalar function to produce a vector function, its gradient. As an operator ∇ should never stand alone, but should always be followed by something on which to operate. Because of this ∇ should not itself be considered a vector, in spite of the fact that it has the form of a vector. It is a differential operator, and must therefore operate on something. In the remainder of this section we use the del operator to define two extremely useful operations on vector fields: the *divergence* and the *curl*.

▷ **Definition 15.1** If $\mathbf{F}(x, y, z) = P(x, y, z)\hat{\mathbf{i}} + Q(x, y, z)\hat{\mathbf{j}} + R(x, y, z)\hat{\mathbf{k}}$ is a vector function in a region D, then the *divergence* of **F** is a scalar function in D defined by

$$\text{div } \mathbf{F} = \nabla \cdot \mathbf{F} = \frac{\partial P}{\partial x} + \frac{\partial Q}{\partial y} + \frac{\partial R}{\partial z}, \tag{15.3}$$

provided that the partial derivatives exist at each point in D. ◁

It is clear why we use the notation $\nabla \cdot \mathbf{F}$, in spite of the fact that ∇ is not a vector in the true sense of the word.

Example 1 Calculate $\nabla \cdot \mathbf{F}$ if:

a) $\mathbf{F} = 2xy\hat{\mathbf{i}} + z\hat{\mathbf{j}} + x^2 \cos(yz)\hat{\mathbf{k}}$;

b) $\mathbf{F} = \dfrac{qQ}{4\pi\varepsilon_0 |\mathbf{r}|^3}\mathbf{r}$, where $\mathbf{r} = x\hat{\mathbf{i}} + y\hat{\mathbf{j}} + z$.

Solution a) $\nabla\cdot\mathbf{F} = \dfrac{\partial}{\partial x}(2xy) + \dfrac{\partial}{\partial y}(z) + \dfrac{\partial}{\partial z}(x^2\cos(yz)) = 2y - x^2y\sin(yz)$.

b) For the derivative of the x-component of $\mathbf{F}$ with respect to x, we calculate

$$\frac{\partial}{\partial x}\left\{\frac{qQx}{4\pi\varepsilon_0(x^2+y^2+z^2)^{3/2}}\right\} = \frac{qQ}{4\pi\varepsilon_0}\left\{\frac{1}{(x^2+y^2+z^2)^{3/2}} - \frac{3x^2}{(x^2+y^2+z^2)^{5/2}}\right\}$$
$$= \frac{qQ}{4\pi\varepsilon_0}\left\{\frac{-2x^2+y^2+z^2}{(x^2+y^2+z^2)^{5/2}}\right\}.$$

With similar results for the remaining two derivatives, we obtain

$$\nabla\cdot\mathbf{F} = \frac{qQ}{4\pi\varepsilon_0}\left\{\frac{-2x^2+y^2+z^2}{(x^2+y^2+z^2)^{5/2}} + \frac{x^2-2y^2+z^2}{(x^2+y^2+z^2)^{5/2}} + \frac{x^2+y^2-2z^2}{(x^2+y^2+z^2)^{5/2}}\right\}$$
$$= 0. \square$$

Any physical interpretation of $\nabla\cdot\mathbf{F}$ will depend on the interpretation of $\mathbf{F}$. The following discussion describes the interpretation of the divergence of a certain vector field in the theory of fluid flow. If a gas flows through a region D of space, then it will flow with some velocity $\mathbf{v}$ past point $P(x, y, z)$ in D at time t (Fig. 15.2). Consider a unit area A around P perpendicular to $\mathbf{v}$. If at time t the density of gas at P is ρ, then the vector $\rho\mathbf{v}$ represents the mass of gas flowing through A per unit time. At each point P in D we have defined a vector $\rho\mathbf{v}$; its direction tells us the direction of gas flow, and its length indicates the mass of gas flowing in that direction. For changing conditions, each of ρ and $\mathbf{v}$ would depend not only on position (x, y, z) but also on time t:

$$\rho\mathbf{v} = \rho(x, y, z, t)\mathbf{v}(x, y, z, t).$$

The vector $\rho\mathbf{v}$, then, is a vector field that also depends on time.

In fluid dynamics it is shown that at each point in D, $\rho\mathbf{v}$ must satisfy

$$\nabla\cdot(\rho\mathbf{v}) = -\frac{\partial\rho}{\partial t},$$

called the *equation of continuity* (see also Section 15.11). It is this equation that gives us an interpretation of the divergence of $\rho\mathbf{v}$. Density is the mass per unit volume of gas. If $\partial\rho/\partial t$ is positive, then density is increasing. This means that more mass must be entering unit volume than leaving it. Similarly, if $\partial\rho/\partial t$ is negative, more mass is leaving than entering. Since $\nabla\cdot(\rho\mathbf{v})$ is the negative of $\partial\rho/\partial t$, it follows that $\nabla\cdot(\rho\mathbf{v})$ must be a measure of how much more

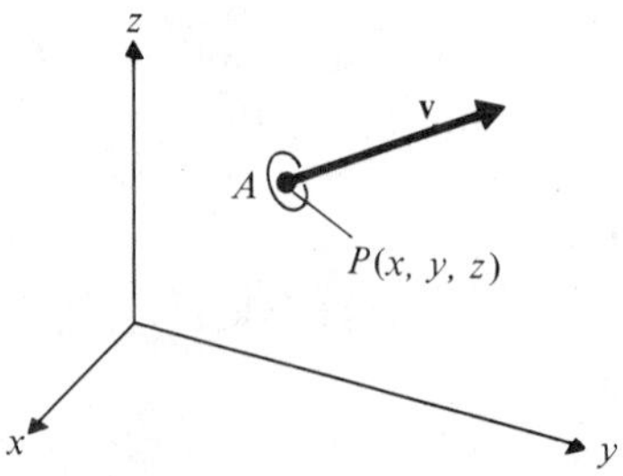

Figure 15.2

gas is leaving unit volume than entering. We can see, then, that the word "divergence" is appropriately chosen for this application.

The curl of a vector field is defined as follows.

▷ **Definition 15.2** If $\mathbf{F}(x, y, z) = P(x, y, z)\hat{\mathbf{i}} + Q(x, y, z)\hat{\mathbf{j}} + R(x, y, z)\hat{\mathbf{k}}$ is a vector function in a region D, then the *curl* of $\mathbf{F}$ is a vector function in D defined by

$$\operatorname{curl}\mathbf{F} = \nabla \times \mathbf{F} = \begin{vmatrix} \hat{\mathbf{i}} & \hat{\mathbf{j}} & \hat{\mathbf{k}} \\ \dfrac{\partial}{\partial x} & \dfrac{\partial}{\partial y} & \dfrac{\partial}{\partial z} \\ P & Q & R \end{vmatrix}$$

$$= \left(\frac{\partial R}{\partial y} - \frac{\partial Q}{\partial z}\right)\hat{\mathbf{i}} + \left(\frac{\partial P}{\partial z} - \frac{\partial R}{\partial x}\right)\hat{\mathbf{j}} + \left(\frac{\partial Q}{\partial x} - \frac{\partial P}{\partial y}\right)\hat{\mathbf{k}}, \tag{15.4}$$

provided that the partial derivatives exist at each point in D. ◁

Note that in representing curl $\mathbf{F}$ in the form of a determinant we must agree to expand the determinant along the first row.

Example 2 Calculate the curl of the vector functions in Example 1.

Solution a)

$$\nabla \times \mathbf{F} = \begin{vmatrix} \hat{\mathbf{i}} & \hat{\mathbf{j}} & \hat{\mathbf{k}} \\ \dfrac{\partial}{\partial x} & \dfrac{\partial}{\partial y} & \dfrac{\partial}{\partial z} \\ 2xy & z & x^2\cos(yz) \end{vmatrix}$$

$$= (-x^2 z\sin(yz) - 1)\hat{\mathbf{i}} + (-2x\cos(yz))\hat{\mathbf{j}} + (-2x)\hat{\mathbf{k}}$$

b)

$$\nabla \times \mathbf{F} = \frac{qQ}{4\pi\varepsilon_0}\begin{vmatrix} \hat{\mathbf{i}} & \hat{\mathbf{j}} & \hat{\mathbf{k}} \\ \dfrac{\partial}{\partial x} & \dfrac{\partial}{\partial y} & \dfrac{\partial}{\partial z} \\ \dfrac{x}{|\mathbf{r}|^3} & \dfrac{y}{|\mathbf{r}|^3} & \dfrac{z}{|\mathbf{r}|^3} \end{vmatrix}$$

The x-component of $\nabla \times \mathbf{F}$ is $qQ/(4\pi\varepsilon_0)$ multiplied by

$$\frac{\partial}{\partial y}\left(\frac{z}{|\mathbf{r}|^3}\right) - \frac{\partial}{\partial z}\left(\frac{y}{|\mathbf{r}|^3}\right) = \frac{\partial}{\partial y}\left(\frac{z}{(x^2+y^2+z^2)^{3/2}}\right) - \frac{\partial}{\partial z}\left(\frac{y}{(x^2+y^2+z^2)^{3/2}}\right)$$

$$= \frac{-3yz}{(x^2+y^2+z^2)^{5/2}} + \frac{3yz}{(x^2+y^2+z^2)^{5/2}} = 0.$$

Similar results for the y- and z-components give $\nabla \times \mathbf{F} = \mathbf{0}$. □

If a vector $\rho\mathbf{v}$ is defined as above for gas flow through a region D, then the curl of $\rho\mathbf{v}$ describes the tendency for the motion of the gas to be circular rather than flowing in a straight line (see Section 15.11). This suggests why the term "curl" is used. With this interpretation, the following definition seems reasonable.

▷ **Definition 15.3** A vector function $\mathbf{F}$ is said to be *irrotational* in a region D if in D

$$\nabla \times \mathbf{F} = \mathbf{0}. \quad \triangleleft \tag{15.5}$$

Applications of divergence and curl extend far beyond the topic of fluid dynamics. Both concepts are indispensable in many areas of applied mathematics such as electromagnetism, continuum mechanics, and heat conduction, to name a few.

The del operator ∇ operates on a scalar function to produce the gradient of the scalar function and on a vector function to give the divergence and the curl of the vector function. The following list of properties for the del operator is straightforward to verify (see Exercise 23).

If f and g are scalar functions that have first partial derivatives in a region D, and if $\mathbf{F}$ and $\mathbf{G}$ are vector functions in D with components that have first partial derivatives, then:

$$\nabla(f + g) = \nabla f + \nabla g, \tag{15.6}$$

$$\nabla \cdot (\mathbf{F} + \mathbf{G}) = \nabla \cdot \mathbf{F} + \nabla \cdot \mathbf{G}, \tag{15.7}$$

$$\nabla \times (\mathbf{F} + \mathbf{G}) = \nabla \times \mathbf{F} + \nabla \times \mathbf{G}, \tag{15.8}$$

$$\nabla(fg) = f\nabla g + g\nabla f, \tag{15.9}$$

$$\nabla \cdot (f\mathbf{F}) = \nabla f \cdot \mathbf{F} + f\nabla \cdot \mathbf{F}, \tag{15.10}$$

$$\nabla \times (f\mathbf{F}) = \nabla f \times \mathbf{F} + f\nabla \times \mathbf{F}, \tag{15.11}$$

$$\nabla \cdot (\mathbf{F} \times \mathbf{G}) = \mathbf{G} \cdot (\nabla \times \mathbf{F}) - \mathbf{F} \cdot (\nabla \times \mathbf{G}), \tag{15.12}$$

$$\nabla \times (\nabla f) = \mathbf{0}, \tag{15.13}$$

$$\nabla \cdot (\nabla \times \mathbf{F}) = 0. \tag{15.14}$$

For properties (15.13) and (15.14), we assume that f and $\mathbf{F}$ have continuous second-order partial derivatives in D. A typical way to verify these identities is to reduce each side of the identity to the same quantity. For example, to verify (15.12), we set $\mathbf{F} = P\hat{\mathbf{i}} + Q\hat{\mathbf{j}} + R\hat{\mathbf{k}}$ and $\mathbf{G} = L\hat{\mathbf{i}} + M\hat{\mathbf{j}} + N$. Then

$$\mathbf{F} \times \mathbf{G} = (QN - RM)\hat{\mathbf{i}} + (RL - PN)\hat{\mathbf{j}} + (PM - QL)\hat{\mathbf{k}};$$

thus,

$$\begin{aligned}
\nabla \cdot \mathbf{F} \times \mathbf{G} &= \frac{\partial}{\partial x}(QN - RM) + \frac{\partial}{\partial y}(RL - PN) + \frac{\partial}{\partial z}(PM - QL) \\
&= Q\frac{\partial N}{\partial x} + N\frac{\partial Q}{\partial x} - R\frac{\partial M}{\partial x} - M\frac{\partial R}{\partial x} + R\frac{\partial L}{\partial y} + L\frac{\partial R}{\partial y} - P\frac{\partial N}{\partial y} \\
&\quad - N\frac{\partial P}{\partial y} + P\frac{\partial M}{\partial z} + M\frac{\partial P}{\partial z} - Q\frac{\partial L}{\partial z} - L\frac{\partial Q}{\partial z} \\
&= L\left(\frac{\partial R}{\partial y} - \frac{\partial Q}{\partial z}\right) + M\left(\frac{\partial P}{\partial z} - \frac{\partial R}{\partial x}\right) + N\left(\frac{\partial Q}{\partial x} - \frac{\partial P}{\partial y}\right) \\
&\quad + P\left(\frac{\partial M}{\partial z} - \frac{\partial N}{\partial y}\right) + Q\left(\frac{\partial N}{\partial x} - \frac{\partial L}{\partial z}\right) + R\left(\frac{\partial L}{\partial y} - \frac{\partial M}{\partial x}\right).
\end{aligned}$$

On the other hand,

$$\mathbf{G}\cdot(\nabla\times\mathbf{F})-\mathbf{F}\cdot(\nabla\times\mathbf{G})=\mathbf{G}\cdot\begin{vmatrix}\hat{\mathbf{i}} & \hat{\mathbf{j}} & \hat{\mathbf{k}}\\ \partial/\partial x & \partial/\partial y & \partial/\partial z\\ P & Q & R\end{vmatrix}-\mathbf{F}\cdot\begin{vmatrix}\hat{\mathbf{i}} & \hat{\mathbf{j}} & \hat{\mathbf{k}}\\ \partial/\partial x & \partial/\partial y & \partial/\partial z\\ L & M & N\end{vmatrix}$$

$$=(L,M,N)\cdot\left(\frac{\partial R}{\partial y}-\frac{\partial Q}{\partial z},\frac{\partial P}{\partial z}-\frac{\partial R}{\partial x},\frac{\partial Q}{\partial x}-\frac{\partial P}{\partial y}\right)$$

$$-(P,Q,R)\cdot\left(\frac{\partial N}{\partial y}-\frac{\partial M}{\partial z},\frac{\partial L}{\partial z}-\frac{\partial N}{\partial x},\frac{\partial M}{\partial x}-\frac{\partial L}{\partial y}\right)$$

$$=L\left(\frac{\partial R}{\partial y}-\frac{\partial Q}{\partial z}\right)+M\left(\frac{\partial P}{\partial z}-\frac{\partial R}{\partial x}\right)+N\left(\frac{\partial Q}{\partial x}-\frac{\partial P}{\partial y}\right)$$

$$-P\left(\frac{\partial N}{\partial y}-\frac{\partial M}{\partial z}\right)-Q\left(\frac{\partial L}{\partial z}-\frac{\partial N}{\partial x}\right)$$

$$-R\left(\frac{\partial M}{\partial x}-\frac{\partial L}{\partial y}\right),$$

and this is the same expression as for $\nabla\cdot(\mathbf{F}\times\mathbf{G})$.

Given a scalar function $f(x, y, z)$ it is straightforward to calculate its gradient ∇f. Conversely, given the gradient of a function ∇f, it is possible to find the function $f(x, y, z)$. For example, if the vector function

$$\mathbf{F}=(3x^2yz+z^2)\hat{\mathbf{i}}+(x^3z+2y)\hat{\mathbf{j}}+(x^3y+2xz+1)\hat{\mathbf{k}}$$

is known to be the gradient of some function $f(x, y, z)$, then

$$\frac{\partial f}{\partial x}\hat{\mathbf{i}}+\frac{\partial f}{\partial y}\hat{\mathbf{j}}+\frac{\partial f}{\partial z}\hat{\mathbf{k}}=(3x^2yz+z^2)\hat{\mathbf{i}}+(x^3z+2y)\hat{\mathbf{j}}+(x^3y+2xz+1)\hat{\mathbf{k}}.$$

Since two vectors are equal if and only if they have identical components, we can say that

$$\frac{\partial f}{\partial x}=3x^2yz+z^2,\qquad \frac{\partial f}{\partial y}=x^3z+2y,\qquad \frac{\partial f}{\partial z}=x^3y+2xz+1.$$

The first of these implies that $f(x, y, z)$ must be of the form

$$f(x,y,z)=x^3yz+xz^2+v(y,z)$$

for some function $v(y, z)$. To determine $v(y, z)$, we substitute this $f(x, y, z)$ into the second equation,

$$x^3z+\frac{\partial v}{\partial y}=x^3z+2y,$$

and this implies that

$$\frac{\partial v}{\partial y}=2y.$$

Consequently,

$$v(y,z)=y^2+w(z)$$

for some $w(z)$, and therefore

$$f(x, y, z) = x^3yz + xz^2 + y^2 + w(z).$$

We now know both the x- and y-dependence of $f(x, y, z)$. To find $w(z)$ we substitute into the equation for $\partial f/\partial z$:

$$x^3y + 2xz + \frac{dw}{dz} = x^3y + 2xz + 1.$$

This equation requires

$$\frac{dw}{dz} = 1,$$

from which we have

$$w(z) = z + C, \qquad C = \text{a constant}.$$

Thus,

$$f(x, y, z) = x^3yz + xz^2 + y^2 + z + C,$$

and this represents all functions that have a gradient equal to the given vector $\mathbf{F}$.

A much more difficult question is to determine whether a given vector function $\mathbf{F}(x, y, z)$ is the gradient of some scalar function $f(x, y, z)$. In the vast majority of cases, we can say that if the above procedure fails, then the answer is no. For instance, if $\mathbf{F} = x^2y\hat{\mathbf{i}} + xy\hat{\mathbf{j}} + z\hat{\mathbf{k}}$, and we attempt to find a function $f(x, y, z)$ so that $\nabla f = \mathbf{F}$, then

$$\frac{\partial f}{\partial x} = x^2y, \qquad \frac{\partial f}{\partial y} = xy, \qquad \frac{\partial f}{\partial z} = z.$$

The first implies that

$$f(x, y, z) = \frac{x^3y}{3} + v(y, z),$$

and when this is substituted into the second, we get

$$\frac{x^3}{3} + \frac{\partial v}{\partial y} = xy, \quad \text{or} \quad \frac{\partial v}{\partial y} = xy - \frac{x^3}{3}.$$

But this is an impossible situation since v is to be a function of y and z only. How then could its derivative depend on x? Although this type of argument will suffice in most examples, it is really not a satisfactory mathematical answer. The following theorem gives a test by which to determine whether a given vector function is the gradient of some scalar function.

▶ **Theorem 15.1** Let D be the interior of a rectangular parallelepiped with faces parallel to the coordinate planes in which the components $P(x, y, z)$, $Q(x, y, z)$, and $R(x, y, z)$ of $\mathbf{F} = P\hat{\mathbf{i}} + Q\hat{\mathbf{j}} + R\hat{\mathbf{k}}$ have continuous first derivatives. Then there exists a function $f(x, y, z)$ defined in D such that $\nabla f = \mathbf{F}$ if and only if in D

$$\nabla \times \mathbf{F} = \mathbf{0}. \blacktriangleleft \tag{15.15}$$

In short, $\mathbf{F} = \nabla f$ if and only if $\nabla \times \mathbf{F} = \mathbf{0}$. It is obvious that if $\mathbf{F} = \nabla f$, then $\nabla \times \mathbf{F} = \mathbf{0}$, for this is simply the result of Eq. (15.13). To prove the converse result requires Stokes's theorem from Section 15.10, and a proof must therefore be delayed until that time.

In the special case in which $\mathbf{F}$ is a vector field in the xy-plane, D becomes a rectangle with sides parallel to the coordinate axes. The equation $\nabla \times \mathbf{F} = \mathbf{0}$ is still the condition for existence of a function $f(x, y)$ such that $\nabla f = \mathbf{F} = P\hat{\mathbf{i}} + Q\hat{\mathbf{j}}$, but the condition reduces to

$$\mathbf{0} = \begin{vmatrix} \hat{\mathbf{i}} & \hat{\mathbf{j}} & \hat{\mathbf{k}} \\ \partial/\partial x & \partial/\partial y & \partial/\partial z \\ P & Q & 0 \end{vmatrix} = \left(\frac{\partial Q}{\partial x} - \frac{\partial P}{\partial y}\right)\hat{\mathbf{k}},$$

or

$$\frac{\partial Q}{\partial x} = \frac{\partial P}{\partial y}. \tag{15.16}$$

Example 3 Find, if possible, a function $f(x, y)$ such that

$$\nabla f = \left(\frac{x^3 - 2y^2}{x^3 y}\right)\hat{\mathbf{i}} + \left(\frac{y^2 - x^3}{x^2 y^2}\right)\hat{\mathbf{j}}.$$

Solution We first note that

$$\frac{\partial}{\partial x}\left(\frac{y^2 - x^3}{x^2 y^2}\right) = \frac{\partial}{\partial x}\left(\frac{1}{x^2} - \frac{x}{y^2}\right) = \frac{-2}{x^3} - \frac{1}{y^2}$$

and

$$\frac{\partial}{\partial y}\left(\frac{x^3 - 2y^2}{x^3 y}\right) = \frac{\partial}{\partial y}\left(\frac{1}{y} - \frac{2y^2}{x^3}\right) = \frac{-1}{y^2} + \frac{-2}{x^3}.$$

Since the components of $\mathbf{F}$ are undefined whenever $x = 0$ or $y = 0$, we can state that in any rectangle D, with sides parallel to the coordinate axes, that does not contain points on either of the axes, there is a function $f(x, y)$ such that $\nabla f = \mathbf{F}$. To find $f(x, y)$, we set

$$\frac{\partial f}{\partial x} = \frac{1}{y} - \frac{2y}{x^3}, \qquad \frac{\partial f}{\partial y} = \frac{1}{x^2} - \frac{x}{y^2}.$$

From the first, we have

$$f(x, y) = \frac{x}{y} + \frac{y}{x^2} + v(y),$$

which substituted into the second gives us

$$-\frac{x}{y^2} + \frac{1}{x^2} + \frac{dv}{dy} = \frac{1}{x^2} - \frac{x}{y^2}.$$

Consequently,

$$\frac{dv}{dy} = 0 \quad \text{or} \quad v(y) = C,$$

and therefore

$$f(x, y) = \frac{x}{y} + \frac{y}{x^2} + C. \ \square$$

EXERCISES 15.1

In Exercises 1–20, calculate the required quantity.

1. ∇f if $f(x, y, z) = 3x^2y - y^3z^2$
2. ∇f if $f(x, y, z) = (x^2 + y^2 + z^2)^{-1/2}$
3. ∇f if $f(x, y) = \text{Tan}^{-1}(y/x)$
4. ∇f at $(1, 2)$ if $f(x, y) = x^3y - 2x \cos y$
5. ∇f at $(1, -1, 4)$ if $f(x, y, z) = e^{xyz}$
6. $\nabla \cdot \mathbf{F}$ if $\mathbf{F}(x, y, z) = 2xe^y\hat{\mathbf{i}} + 3x^2z\hat{\mathbf{j}} - 2x^2yz\hat{\mathbf{k}}$
7. $\nabla \cdot \mathbf{F}$ if $\mathbf{F}(x, y) = x \ln y\hat{\mathbf{i}} - y^3e^x\hat{\mathbf{j}}$
8. $\nabla \cdot \mathbf{F}$ if $\mathbf{F}(x, y, z) = \sin(x^2 + y^2 + z^2)\hat{\mathbf{i}} + \cos(y + z)\hat{\mathbf{j}}$
9. $\nabla \cdot \mathbf{F}$ if $\mathbf{F}(x, y) = e^x\hat{\mathbf{i}} + e^y\hat{\mathbf{j}}$
10. $\nabla \cdot \mathbf{F}$ at $(1, 1, 1)$ if $\mathbf{F}(x, y, z) = x^2y^3\hat{\mathbf{i}} - 3xy\hat{\mathbf{j}} + z^2\hat{\mathbf{k}}$
11. $\nabla \cdot \mathbf{F}$ at $(-1, 3)$ if $\mathbf{F}(x, y) = (x + y)^2(\hat{\mathbf{i}} + \hat{\mathbf{j}})$
12. $\nabla \cdot \mathbf{F}$ if $\mathbf{F}(x, y, z) = (x\hat{\mathbf{i}} + y\hat{\mathbf{j}} + z\hat{\mathbf{k}})/\sqrt{x^2 + y^2 + z^2}$
13. $\nabla \cdot \mathbf{F}$ if $\mathbf{F}(x, y) = \text{Cot}^{-1}(xy)\hat{\mathbf{i}} + \text{Tan}^{-1}(xy)\hat{\mathbf{j}}$
14. $\nabla \times \mathbf{F}$ if $\mathbf{F}(x, y, z) = x^2z\hat{\mathbf{i}} + 12xyz\hat{\mathbf{j}} + 32y^2z^4\hat{\mathbf{k}}$
15. $\nabla \times \mathbf{F}$ if $\mathbf{F}(x, y) = xe^y\hat{\mathbf{i}} - 2xy^2\hat{\mathbf{j}}$
16. $\nabla \times \mathbf{F}$ if $\mathbf{F}(x, y, z) = x^2\hat{\mathbf{i}} + y^2\hat{\mathbf{j}} + z^2\hat{\mathbf{k}}$
17. $\nabla \times \mathbf{F}$ at $(1, -1, 1)$ if $\mathbf{F}(x, y, z) = xz^3\hat{\mathbf{i}} - 2x^2yz\hat{\mathbf{j}} + 2yz^4\hat{\mathbf{k}}$
18. $\nabla \times \mathbf{F}$ at $(2, 0)$ if $\mathbf{F}(x, y) = y\hat{\mathbf{i}} - x\hat{\mathbf{j}}$
19. $\nabla \times \mathbf{F}$ if $\mathbf{F}(x, y, z) = \ln(x + y + z)(\hat{\mathbf{i}} + \hat{\mathbf{j}} + \hat{\mathbf{k}})$
20. $\nabla \times \mathbf{F}$ if $\mathbf{F}(x, y) = \text{Sec}^{-1}(x + y)\hat{\mathbf{i}} + \text{Csc}^{-1}(y + x)\hat{\mathbf{j}}$
21. If $\mathbf{F} = x^2y\hat{\mathbf{i}} - 2xz\hat{\mathbf{j}} + 2yz\hat{\mathbf{k}}$, find $\nabla \times (\nabla \times \mathbf{F})$.
22. a) Verify that Laplace's equation (13.14) can be expressed in the form $\nabla \cdot \nabla f = 0$.

    b) Show that $f(x, y, z) = (x^2 + y^2 + z^2)^{-1/2}$ satisfies Laplace's equation.
23. Prove properties (15.6)–(15.11), (15,13), and (15.14).
24. A gas is moving through some region D of space. If we follow a particular particle of the gas, it will trace out some curved path

    $$C:\quad x = x(t), \qquad y = y(t), \qquad z = z(t), \qquad t = \text{time}.$$

    If $\rho = \rho(x, y, z, t)$ is the density of gas at any point in D at time t, then along C we can express density in terms of t only, $\rho = \rho[x(t), y(t), z(t), t]$. Show that along C

    $$\frac{d\rho}{dt} = \frac{\partial \rho}{\partial t} + \nabla\rho \cdot \frac{d\mathbf{r}}{dt},$$

    where $\mathbf{r} = x\hat{\mathbf{i}} + y\hat{\mathbf{j}} + z\hat{\mathbf{k}}$.

In each of Exercises 25–29, find all functions $f(x, y)$ such that ∇f is equal to the given vector function.

25. $\mathbf{F}(x, y) = 2xy\hat{\mathbf{i}} + x^2\hat{\mathbf{j}}$
26. $\mathbf{F}(x, y) = (3x^2y^2 + 3)\hat{\mathbf{i}} + (2x^3y + 2)\hat{\mathbf{j}}$
27. $\mathbf{F}(x, y) = e^y\hat{\mathbf{i}} + (xe^y + 4y^2)\hat{\mathbf{j}}$
28. $\mathbf{F}(x, y) = (x + y)^{-1}(\hat{\mathbf{i}} + \hat{\mathbf{j}})$
29. $\mathbf{F}(x, y) = -xy(1 - x^2y^2)^{-1/2}(y\hat{\mathbf{i}} + x\hat{\mathbf{j}})$

In each of Exercises 30–35, find all functions $f(x, y, z)$ such that ∇f is equal to the given vector function.

30. $\mathbf{F}(x, y, z) = x\hat{\mathbf{i}} + y\hat{\mathbf{j}} + z\hat{\mathbf{k}}$
31. $\mathbf{F}(x, y, z) = yz\hat{\mathbf{i}} + xz\hat{\mathbf{j}} + (yx - 3)\hat{\mathbf{k}}$
32. $\mathbf{F}(x, y, z) = (1 + x + y + z)^{-1}(\hat{\mathbf{i}} + \hat{\mathbf{j}} + \hat{\mathbf{k}})$
33. $\mathbf{F}(x, y, z) = (2x/y^2 + 1)\hat{\mathbf{i}} - (2x^2/y^3)\hat{\mathbf{j}} - 2z\hat{\mathbf{k}}$
34. $\mathbf{F}(x, y, z) = (1 + x^2y^2)^{-1}(y\hat{\mathbf{i}} + x\hat{\mathbf{j}}) + z\hat{\mathbf{k}}$
35. $\mathbf{F}(x, y, z) = (3x^2y + yz + 2xz^2)\hat{\mathbf{i}} + (xz + x^3 + 3z^2 - 6y^2z)\hat{\mathbf{j}} + (2x^2z + 6yz - 2y^3 + xy)\hat{\mathbf{k}}$
36. a) Find constants a, b, and c in order that the vector function

    $$\mathbf{F} = (x^2 + 2y + az)\hat{\mathbf{i}} + (bx - 3y - z)\hat{\mathbf{j}} + (4x + cy + 2z)\hat{\mathbf{k}}$$

    be irrotational.

    b) If $\mathbf{F}$ is irrotational, find a scalar function $f(x, y, z)$ such that $\nabla f = \mathbf{F}$.
37. A vector function $\mathbf{F}(x, y, z)$ is said to be solenoidal if $\nabla \cdot \mathbf{F} = 0$.

    a) Are either of $\mathbf{F} = (2x^2 + 8xy^2z)\hat{\mathbf{i}} + (3x^3y - 3xy)\hat{\mathbf{j}} - (4y^2z^2 + 2x^3z)\hat{\mathbf{k}}$ or $xyz^2\mathbf{F}$ solenoidal?

    b) Show that $\nabla f \times \nabla g$ is solenoidal for arbitrary functions $f(x, y, z)$ and $g(x, y, z)$, which have continuous second partial derivatives.
38. Associated with every electric field is a scalar function $V(x, y, z)$ called potential. It is defined by $\mathbf{E} = -\nabla V$, where $\mathbf{E}$ is a vector function called the electric field intensity. In addition, if a point charge Q is placed at a point (x, y, z) in the electric field, then the force $\mathbf{F}$ on Q is $\mathbf{F} = Q\mathbf{E}$.

    a) If the force on Q due to a charge q at the origin is

    $$\mathbf{F} = \frac{qQ}{4\pi\varepsilon_0|\mathbf{r}|^3}\mathbf{r},$$

    where $\mathbf{r} = x\hat{\mathbf{i}} + y\hat{\mathbf{j}} + z\hat{\mathbf{k}}$ and ε_0 is a constant, find $V(x, y, z)$ for the field due to q.

    b) If the entire xy-plane is given a uniform charge density σ units of charge per unit area, it is found that the force on a charge Q placed z units above the plane is

    $$\mathbf{F} = \frac{Q\sigma}{2\varepsilon_0}\hat{\mathbf{k}}.$$

Find the potential V for the electric field due to this charge distribution.

39. Show that if $\mathbf{v} = \boldsymbol{\omega} \times \mathbf{r}$, where $\boldsymbol{\omega}$ is a constant vector, and $\mathbf{r} = x\hat{\mathbf{i}} + y\hat{\mathbf{j}} + z\hat{\mathbf{k}}$, then $\boldsymbol{\omega} = (\frac{1}{2})\nabla \times \mathbf{v}$.

40. Show that if a function $f(x, y, z)$ satisfies Laplace's equation (13.14), then its gradient is both irrotational and solenoidal.

41. Theorem 15.1 indicates that a vector field $\mathbf{F}$ is the gradient of some scalar field if $\nabla \times \mathbf{F} = \mathbf{0}$. Sometimes a given vector field $\mathbf{F}$ is the curl of another field $\mathbf{v}$: $\mathbf{F} = \nabla \times \mathbf{v}$. The following theorem indicates when this is the case: Let D be the interior of a sphere in which the components of a vector field $\mathbf{F}$ have continuous first partial derivatives. Then there exists a vector field $\mathbf{v}$ defined in D such that $\mathbf{F} = \nabla \times \mathbf{v}$ if and only if $\nabla \cdot \mathbf{F} = 0$ in D. In other words, $\mathbf{F}$ is the curl of a vector field if and only if $\mathbf{F}$ is solenoidal.

a) Show that if $\mathbf{F} = P\hat{\mathbf{i}} + Q\hat{\mathbf{j}} + R\hat{\mathbf{k}}$ is solenoidal, then the components of $\mathbf{v} = L\hat{\mathbf{i}} + M\hat{\mathbf{j}} + N\hat{\mathbf{k}}$ would have to satisfy the equations

$$P = \frac{\partial N}{\partial y} - \frac{\partial M}{\partial z}, \qquad Q = \frac{\partial L}{\partial z} - \frac{\partial N}{\partial x}, \qquad R = \frac{\partial M}{\partial x} - \frac{\partial L}{\partial y}.$$

*b) Show that the vector function $\mathbf{v}(x, y, z)$ defined by

$$\mathbf{v}(x, y, z) = \int_0^1 t\mathbf{F}(tx, ty, tz) \times (x, y, z)\, dt$$

satisfies these equations. In other words, this formula defines a possible vector $\mathbf{v}$. Is it unique?

*c) Show that if $\mathbf{F}$ satisfies the property that

$$\mathbf{F}(tx, ty. tz) = t^n\mathbf{F}(x, y, z),$$

then

$$\mathbf{v}(x, y, z) = \frac{1}{n + 2}\mathbf{F} \times \mathbf{r}, \qquad \mathbf{r} = x\hat{\mathbf{i}} + y\hat{\mathbf{j}} + z\hat{\mathbf{k}}.$$

In Exercises 42–44, verify that the vector field is solenoidal, and then use the formulas in Exercise 41 to find a vector field $\mathbf{v}$ such that $\mathbf{F} = \nabla \times \mathbf{v}$.

42. $\mathbf{F} = x\hat{\mathbf{i}} + y\hat{\mathbf{j}} - 2z\hat{\mathbf{k}}$

43. $\mathbf{F} = (1 + x)\hat{\mathbf{i}} - (x + z)\hat{\mathbf{k}}$

44. $\mathbf{F} = 2x^2\hat{\mathbf{i}} - y^2\hat{\mathbf{j}} + (-4xz + 2yz)\hat{\mathbf{k}}$

15.2 LINE INTEGRALS

Just as definite integrals, double integrals, and triple integrals are defined as limits of sums, so too are line and surface integrals. The only difference is that line integrals are applied to functions defined along curves, and surface integrals involve functions defined on surfaces.

A curve C in space is defined parametrically by three functions

$$C:\quad \begin{aligned} x &= x(t),\\ y &= y(t), \qquad \alpha \le t \le \beta,\\ z &= z(t), \end{aligned} \tag{15.17}$$

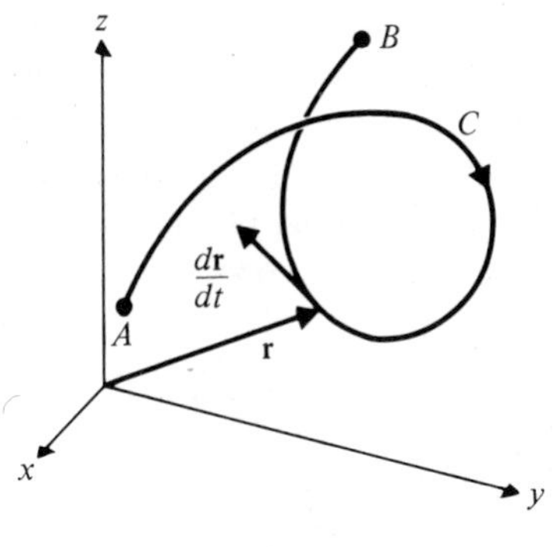

Figure 15.3

where α and β specify initial and final points A and B of the curve, respectively (Fig. 15.3). The direction of a curve is from initial to final point, and in Section 12.8 we agreed to parametrize a curve using only parameters that increase in the direction of the curve.

The curve is said to be continuous if each of the functions $x(t)$, $y(t)$, and $z(t)$ is continuous (implying that C is at no point separated). It is said to be smooth if each of these functions has a continuous first derivative; geometrically, this means that the tangent vector $d\mathbf{r}/dt = x'(t)\hat{\mathbf{i}} + y'(t)\hat{\mathbf{j}} + z'(t)\hat{\mathbf{k}}$ turns gradually or smoothly along C. A continuous curve that is not smooth but can be divided into a finite number of smooth subcurves is said to be piecewise smooth.

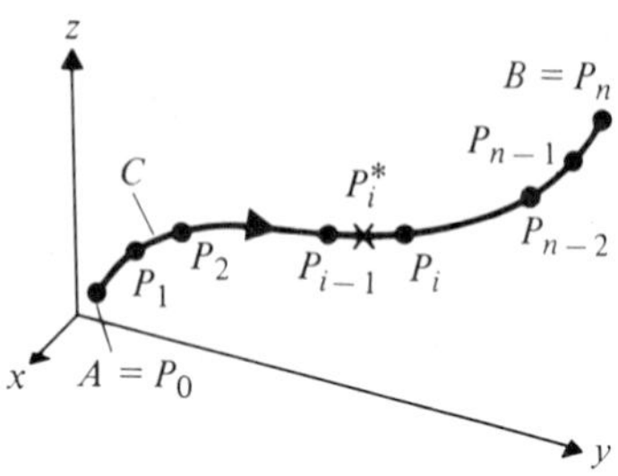

Figure 15.4

Suppose a function $f(x, y, z)$ is defined along a curve C joining A to B (Fig. 15.4). We divide C into n subcurves of lengths $\Delta s_1, \Delta s_2, \ldots, \Delta s_n$ by any $n-1$ consecutive points $A = P_0, P_1, P_2, \ldots, P_{n-1}, P_n = B$, whatsoever. On each subcurve of length Δs_i $(i = 1, \ldots, n)$ we choose an arbitrary point $P_i^*(x_i^*, y_i^*, z_i^*)$. We then form the sum

$$f(x_1^*, y_1^*, z_1^*)\,\Delta s_1 + f(x_2^*, y_2^*, z_2^*)\,\Delta s_2 + \cdots + f(x_n^*, y_n^*, z_n^*)\,\Delta s_n = \sum_{i=1}^{n} f(x_i^*, y_i^*, z_i^*)\,\Delta s_i. \quad (15.18)$$

If this sum approaches a limit as the number of subcurves becomes increasingly large and the length of every subcurve approaches zero, we call the limit the line integral of $f(x, y, z)$ along the curve C, and denote it by

$$\boxed{\int_C f(x, y, z)\,ds = \lim_{n\to\infty} \sum_{i=1}^{n} f(x_i^*, y_i^*, z_i^*)\,\Delta s_i,} \quad (15.19)$$

where it is understood that every $\Delta s_i \to 0$ as $n \to \infty$. A more appropriate name might be curvilinear integral, rather than line integral, but line integral has become the accepted terminology. We must simply regard the word "line" as meaning "curved line" rather than "straight line."

For definition (15.19) to be at all useful we must demand that the limit be independent of the manner of subdivision of C and choice of star-points on the subcurves. Theorem 15.2 indicates that for continuous functions defined on smooth curves, this requirement is indeed satisfied.

▶ **Theorem 15.2** Let $f(x, y, z)$ be continuous on a smooth curve C of finite length, C: $x = x(t)$, $y = y(t)$, $z = z(t)$, $\alpha \le t \le \beta$. Then the line integral of $f(x, y, z)$ along C exists and can be evaluated by means of the following definite integral:

$$\int_C f(x, y, z)\,ds = \int_\alpha^\beta f[x(t), y(t), z(t)]\sqrt{\left(\frac{dx}{dt}\right)^2 + \left(\frac{dy}{dt}\right)^2 + \left(\frac{dz}{dt}\right)^2}\,dt. \quad ◀ \quad (15.20)$$

It is not necessary to memorize (15.20) as a formula, since it is the result obtained by expressing x, y, z, and ds in terms of t and interpreting that result as a definite integral with respect to t. To be more explicit, recall from Eq. (12.77) that when length along a curve is measured from its initial point, then an infinitesimal length ds along C corresponding to an increment dt in t is given by

$$ds = \sqrt{(dx)^2 + (dy)^2 + (dz)^2} = \sqrt{\left\{\left(\frac{dx}{dt}\right)^2 + \left(\frac{dy}{dt}\right)^2 + \left(\frac{dz}{dt}\right)^2\right\}(dt)^2}$$
$$= \sqrt{\left(\frac{dx}{dt}\right)^2 + \left(\frac{dy}{dt}\right)^2 + \left(\frac{dz}{dt}\right)^2}\,dt.$$

If we substitute this into the left-hand side of (15.20) and at the same time use the equations for C to express $f(x, y, z)$ in terms of t, then

$$\int_C f(x, y, z)\,ds = \int_C f[x(t), y(t), z(t)]\sqrt{\left(\frac{dx}{dt}\right)^2 + \left(\frac{dy}{dt}\right)^2 + \left(\frac{dz}{dt}\right)^2}\,dt.$$

But if we now interpret the right-hand side of this equation as the definite integral of $f[x(t), y(t), z(t)]\sqrt{(dx/dt)^2 + (dy/dt)^2 + (dz/dt)^2}$ with respect to t and affix limits $t = \alpha$ and $t = \beta$ that identify end points of C, we obtain (15.20).

To evaluate a line integral, then, we simply express $f(x, y, z)$ and ds in terms of some parameter along C and evaluate the resulting definite integral. If equations for C are given in the form C: $y = y(x)$, $z = z(x)$, $x_A \le x \le x_B$, then x is a convenient parameter, and Eq. (15.20) takes the form

$$\int_C f(x, y, z)\,ds = \int_{x_A}^{x_B} f[x, y(x), z(x)]\sqrt{1 + \left(\frac{dy}{dx}\right)^2 + \left(\frac{dz}{dx}\right)^2}\,dx. \qquad (15.21)$$

Similar expressions exist if either y or z is a convenient parameter.

When C is piecewise smooth rather than smooth, the line integral along C is found by evaluating the line integral along each smooth subcurve and adding the results.

Example 4 Evaluate the line integral of $f(x, y, z) = 8x + 6xy + 30z$ from $A(0, 0, 0)$ to $B(1, 1, 1)$:

a) along the straight line joining A to B with parametrization

$$C_1:\quad x = t, \qquad y = t, \qquad z = t, \qquad 0 \le t \le 1;$$

b) along the straight line in (a) with parametrization

$$C_1:\quad x = -1 + \frac{t}{2}, \qquad y = -1 + \frac{t}{2}, \qquad z = -1 + \frac{t}{2}, \qquad 2 \le t \le 4;$$

c) along the curve (Fig. 15.5)

$$C_2:\quad x = t, \qquad y = t^2, \qquad z = t^3, \qquad 0 \le t \le 1.$$

Solution a)

$$\int_{C_1} (8x + 6xy + 30z)\,ds = \int_0^1 (8t + 6t^2 + 30t)\sqrt{(1)^2 + (1)^2 + (1)^2}\,dt$$

$$= \sqrt{3}\int_0^1 (38t + 6t^2)\,dt = \sqrt{3}\{19t^2 + 2t^3\}_0^1$$

$$= 21\sqrt{3}.$$

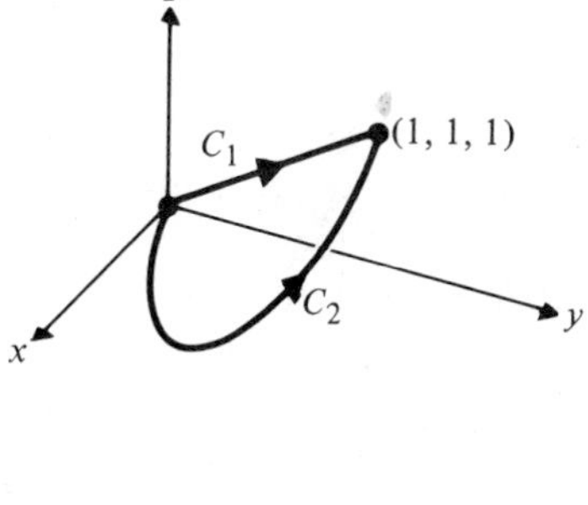

Figure 15.5

b)

$$\int_{C_1} (8x + 6xy + 30z)\, ds = \int_2^4 \{8(-1 + t/2) + 6(-1 + t/2)^2 + 30(-1 + t/2)\} \times \sqrt{(1/2)^2 + (1/2)^2 + (1/2)^2}\, dt$$

$$= \frac{\sqrt{3}}{2} \int_2^4 \{38(-1 + t/2) + 6(-1 + t/2)^2\}\, dt$$

$$= \frac{\sqrt{3}}{2} \{38(-1 + t/2)^2 + 4(-1 + t/2)^3\}_2^4$$

$$= \frac{\sqrt{3}}{2} \{38 + 4\} = 21\sqrt{3}.$$

c)

$$\int_{C_2} (8x + 6xy + 30z)\, ds = \int_0^1 (8t + 6t^3 + 30t^3)\sqrt{(1)^2 + (2t)^2 + (3t^2)^2}\, dt$$

$$= \int_0^1 (8t + 36t^3)\sqrt{1 + 4t^2 + 9t^4}\, dt$$

$$= \{\tfrac{2}{3}(1 + 4t^2 + 9t^4)^{3/2}\}_0^1 = \tfrac{2}{3}(14\sqrt{14} - 1). \square$$

Parts (a) and (b) of this example suggest that the value of a line integral does not depend on the particular parametrization of the curve used in its evaluation. This is indeed true, and should perhaps be expected since definition (15.20) makes no reference whatsoever to parametrization of the curve. For a proof of this fact see Exercise 18. Different parameters will normally lead to different definite integrals, but they will all give exactly the same value for the line integral. Parts (a) and (c) illustrate that a line integral does depend on the curve joining the points A and B; i.e., the value of the line integral changes if the curve C joining A and B changes.

Example 5 Evaluate the line integral of $f(x, y) = x^2 + y^2$ once clockwise around the circle $x^2 + y^2 = 4$, $z = 0$.

Solution If we use the parametrization

$$C:\quad x = 2\cos t, \qquad y = -2\sin t, \qquad z = 0, \qquad 0 \le t \le 2\pi$$

(Fig. 15.6), then

$$\int_C (x^2 + y^2)\, ds = \int_0^{2\pi} (4)\sqrt{\left(\frac{dx}{dt}\right)^2 + \left(\frac{dy}{dt}\right)^2}\, dt$$

$$= 4\int_0^{2\pi} \sqrt{(-2\sin t)^2 + (2\cos t)^2}\, dt = 8\int_0^{2\pi} dt = 16\pi. \square$$

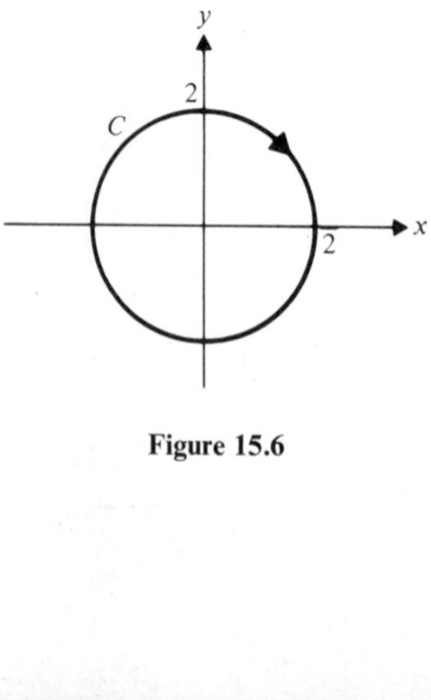

Figure 15.6

The value of this line integral, as well as many other line integrals in the xy-plane, can be given a geometric interpretation. Suppose a function $f(x, y)$ is positive along a curve C in the xy-plane. If at each point of C we draw a vertical line of height $z = f(x, y)$, then a vertical wall is constructed as shown in Fig. 15.7. Since ds is an elemental piece of length along C, the quantity $f(x, y)\, ds$ can be interpreted as approximately the area of the vertical wall projecting

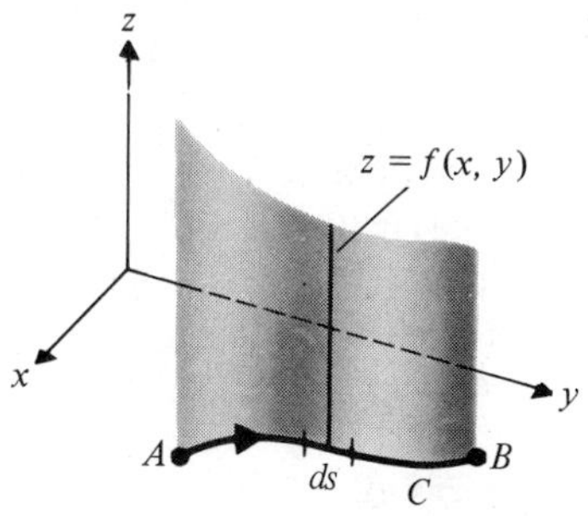

Figure 15.7

onto ds. Because the line integral

$$\int_C f(x, y)\, ds,$$

like all integrals, is a limit–summation process, we interpret the value of this line integral as the total area of the vertical wall. Correct as this interpretation is, it really is of little use in the evaluation of line integrals and, in addition, the interpretation is valid only if the curve C along which the line integral is performed is contained in a plane.

Because a line integral is a limit–summation, it should be obvious that the line integral

$$\int_C ds \tag{15.22}$$

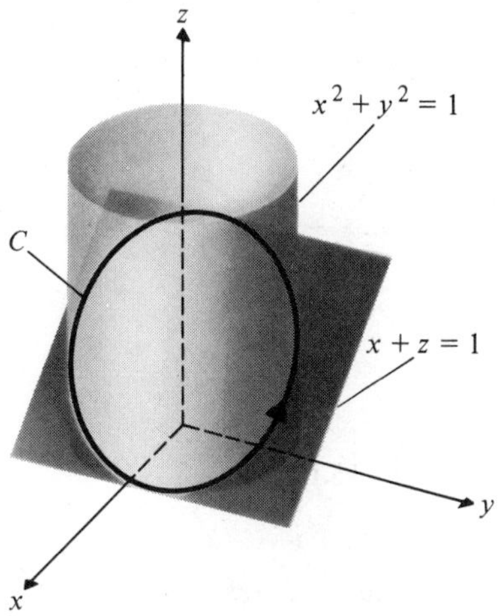

Figure 15.8

represents the length of the curve C. If C is a curve in the xy-plane, we substitute $ds = \sqrt{(dx)^2 + (dy)^2}$, and if C is a curve in space, then $ds = \sqrt{(dx)^2 + (dy)^2 + (dz)^2}$. This agrees with the results of Eqs. (12.70) and (12.77).

We make one last point about notation. A curve is said to be *closed* if its initial and final points are the same. For example, the curve of intersection of the cylinder $x^2 + y^2 = 1$ and the plane $x + z = 1$ is a closed curve (Fig. 15.8). To indicate that a line integral is being evaluated around a closed curve, we usually draw a circle on the integral sign, as follows:

$$\oint_C f(x, y, z)\, ds.$$

When C is a closed curve in the xy-plane and C does not cross itself, we indicate the direction along C by an arrowhead on the circle. For the curves shown in Fig. 15.9 we write

$$\oint_{C_1} f(x, y)\, ds \quad \text{and} \quad \oint_{C_2} f(x, y)\, ds.$$

Figure 15.9

EXERCISES 15.2

In Exercises 1–6, evaluate the given line integral.

1. $\int_C x\, ds$,
where C is the curve $y = x^2$, $z = 0$ from $(0, 0, 0)$ to $(1, 1, 0)$

2. $\oint_C (x^2 + y^2)\, ds$,
once (counterclockwise) around the square C in the xy-plane with vertices $(\pm 1, \pm 1)$

3. $\oint_C (2 + x - 2xy)\, ds$,
once (clockwise) around the circle $x^2 + y^2 = 4$, $z = 0$

4. $\int_C (x^2 + yz)\, ds$,
along the straight line from $(1, 2, -1)$ to $(3, 2, 5)$

5. $\int_C xy\, ds$,
where C is the first octant part of $x^2 + y^2 = 1$, $x^2 + z^2 = 1$ from $(1, 0, 0)$ to $(0, 1, 1)$

6. $\int_C x^2yz\, ds$,
where C is the curve $z = x + y$, $x + y + z = 1$ from $(1, -\frac{1}{2}, \frac{1}{2})$ to $(-3, \frac{7}{2}, \frac{1}{2})$

7. Prove that the length of the circumference of a circle is 2π multiplied by the radius.

8. A spring has six coils in the form of the helix

$$C:\quad x = 3\cos t, \qquad y = 3\sin t, \qquad z = 3t/(4\pi), \qquad 0 \le t \le 12\pi,$$

all dimensions in centimeters. Find the length of the spring.

9. Use the parametric equations $x = \cos^3\theta$, $y = \sin^3\theta$, $0 \le \theta < 2\pi$, to sketch the astroid $x^{2/3} + y^{2/3} = 1$ in the xy-plane. At each point (x, y) on the astroid, a vertical line is drawn with height $z = x^2 + y^2$, thus forming a cylindrical wall. Find the area of the wall.

In Exercises 10–12, evaluate the line integral.

10. $\int_C xz\, ds$,
along the first octant part of $y = x^2$, $z + y = 1$ from $(0, 0, 1)$ to $(1, 1, 0)$

11. $\int_C (x + y)^5\, ds$,
along $C: x = t + 1/t$, $y = t - 1/t$ from $(2, 0)$ to $(\frac{17}{4}, \frac{15}{4})$

12. $\int_C x\sqrt{y + z}\, ds$,
where C is that part of the curve $3x + 2y + 3z = 6$, $x - 2y + 4z = 5$ from $(1, 0, 1)$ to $(0, \frac{9}{14}, \frac{11}{7})$

13. a) If the curve C in Fig. 15.10 is rotated around the y-axis, show that the area of the surface that it traces out is represented by the line integral

$$\int_C 2\pi x\, ds.$$

b) If C is rotated around the x-axis, what line integral represents the area of the surface traced out?

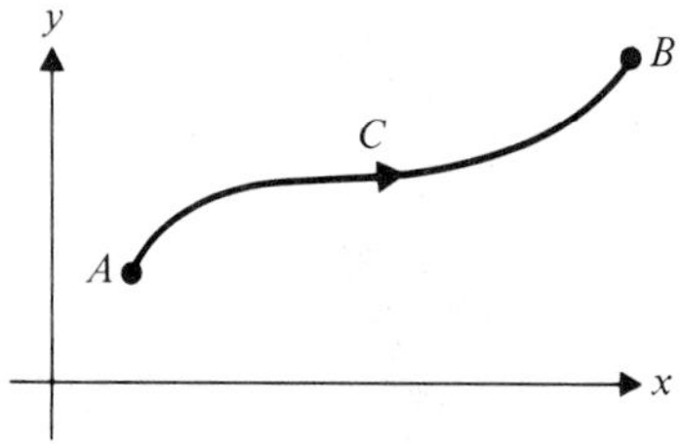

Figure 15.10

Use the method in Exercise 13 to find the area of the surface traced out when the curve in each of Exercises 14–16 is rotated about the line indicated.

14. $y = x^3$, $1 \le x \le 2$, about $y = 0$

15. $24xy = x^4 + 48$, $1 \le x \le 2$, about $x = 0$

16. $8y^2 = x^2(1 - x^2)$ about $y = 0$

17. Find the length of the parabola $y = x^2$ from $(0, 0)$ to $(1, 1)$.

18. Show that the value of a line integral is independent of the parameter used to specify the curve.

*19. Find the surface area of the torus $(\sqrt{x^2 + y^2} - a)^2 + z^2 = b^2$, $(a > b)$.

15.3 LINE INTEGRALS INVOLVING VECTOR FUNCTIONS

There are many ways in which $f(x, y, z)$ in the line integral

$$\int_C f(x, y, z)\, ds \tag{15.23}$$

can arise. According to Eq. (15.22) we must choose $f(x, y, z) = 1$ in order to find the length of the curve C; Exercise 13 in Section 15.2 indicates that for the

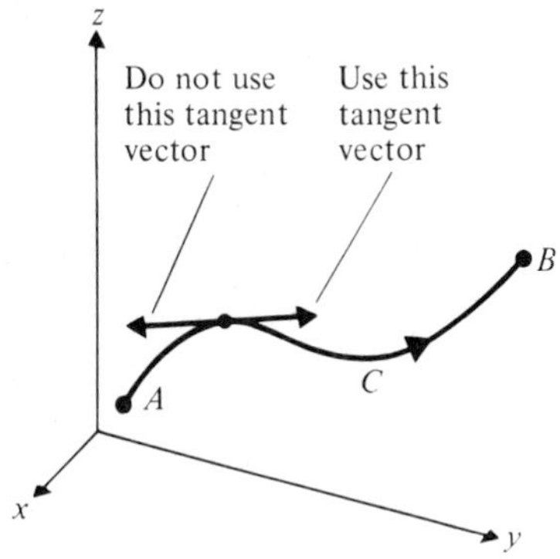

Figure 15.11

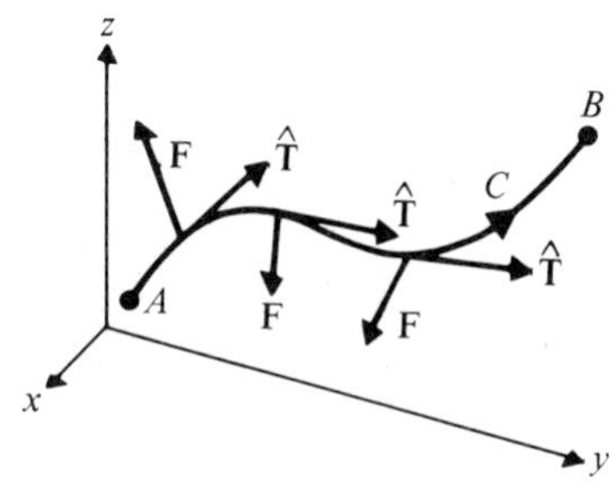

Figure 15.12

areas of the surfaces traced out when a curve in the xy-plane is rotated about the y- and x-axes, we must choose $f(x, y)$ equal to $2\pi x$ and $2\pi y$, respectively.

The most important and common type of line integral occurs when $f(x, y, z)$ is specified as the tangential component of some given vector function $\mathbf{F}(x, y, z)$ defined along C; i.e., $f(x, y, z)$ itself is not given, but $\mathbf{F}$ is, and to find $f(x, y, z)$ we must calculate the tangential component of $\mathbf{F}$ along C. By the tangential component of $\mathbf{F}(x, y, z)$ along C we mean the component of $\mathbf{F}$ along that tangent vector to C which points in the same direction as C (Fig. 15.11). In Section 12.9 we saw that if s is a measure of length along a curve C from A to B (Fig. 15.12), and if s is chosen equal to zero at A, then a unit tangent vector pointing in the direction of motion along C is

$$\hat{\mathbf{T}} = \frac{d\mathbf{r}}{ds}. \tag{15.24}$$

Consequently, if $f(x, y, z)$ is the tangential component of $\mathbf{F}(x, y, z)$ along C, then

$$f(x, y, z) = \mathbf{F} \cdot \hat{\mathbf{T}} = \mathbf{F} \cdot \frac{d\mathbf{r}}{ds}, \tag{15.25}$$

and we can write that

$$\int_C f(x, y, z)\, ds = \int_C \mathbf{F} \cdot \hat{\mathbf{T}}\, ds = \int_C \mathbf{F} \cdot \frac{d\mathbf{r}}{ds}\, ds = \int_C \mathbf{F} \cdot d\mathbf{r}. \tag{15.26}$$

If the components of the vector function $\mathbf{F}(x, y, z)$ are

$$\mathbf{F}(x, y, z) = P(x, y, z)\hat{\mathbf{i}} + Q(x, y, z)\hat{\mathbf{j}} + R(x, y, z)\hat{\mathbf{k}}, \tag{15.27}$$

then

$$\int_C \mathbf{F} \cdot d\mathbf{r} = \int_C (P\, dx + Q\, dy + R\, dz),$$

and if the parentheses are omitted, we have

$$\int_C \mathbf{F} \cdot d\mathbf{r} = \int_C P\, dx + Q\, dy + R\, dz. \tag{15.28}$$

This discussion has shown that when the integrand $f(x, y, z)$ of a line integral is specified as the tangential component of $\mathbf{F} = P\hat{\mathbf{i}} + Q\hat{\mathbf{j}} + R\hat{\mathbf{k}}$ along C, the product $f(x, y, z)\, ds$ can be replaced by the sum of products $P\, dx + Q\, dy + R\, dz$:

$$\int_C f(x, y, z)\, ds = \int_C \mathbf{F} \cdot d\mathbf{r} = \int_C P\, dx + Q\, dy + R\, dz. \tag{15.29}$$

According to the results of Section 15.2, evaluation of this line integral can be accomplished by expressing $P\, dx + Q\, dy + R\, dz$ in terms of any parametric representation of C and evaluating the resulting definite integral.

Example 6 Evaluate

$$\int_C \frac{z}{y}\, dx + (x^2 + y^2 + z^2)\, dz,$$

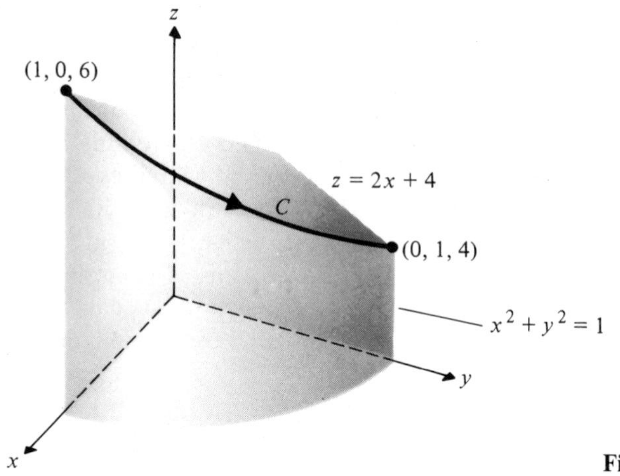

Figure 15.13

where C is the first octant intersection of $x^2 + y^2 = 1$ and $z = 2x + 4$ joining $(1, 0, 6)$ to $(0, 1, 4)$.

Solution If we choose the parametrization

$$x = \cos t, \qquad y = \sin t, \qquad z = 2\cos t + 4, \qquad 0 \le t \le \pi/2,$$

for C (Fig. 15.13), then

$$\int_C \frac{z}{y}\,dx + (x^2 + y^2 + z^2)\,dz$$

$$= \int_0^{\pi/2} \left\{\left(\frac{2\cos t + 4}{\sin t}\right)(-\sin t\,dt) + (\cos^2 t + \sin^2 t + 4\cos^2 t + 16\cos t + 16)(-2\sin t\,dt)\right\}$$

$$= -2\int_0^{\pi/2} \{\cos t + 2 + 17\sin t + 4\cos^2 t\sin t + 16\cos t\sin t\}\,dt$$

$$= -2\left\{\sin t + 2t - 17\cos t - \frac{4\cos^3 t}{3} + 8\sin^2 t\right\}_0^{\pi/2} = -2\pi - \frac{164}{3}. \;\square$$

Example 7 Evaluate

$$\oint_C y^2\,dx + x^2\,dy,$$

where C is the closed curve shown in Fig. 15.14.

Solution If we start at the origin and denote the three straight line paths by

$$C_1\colon\; y = 0,\; 0 \le x \le 1; \qquad C_2\colon\; x = 1,\; 0 \le y \le 1;$$

$$C_3\colon\; y = x = 1 - t,\; 0 \le t \le 1,$$

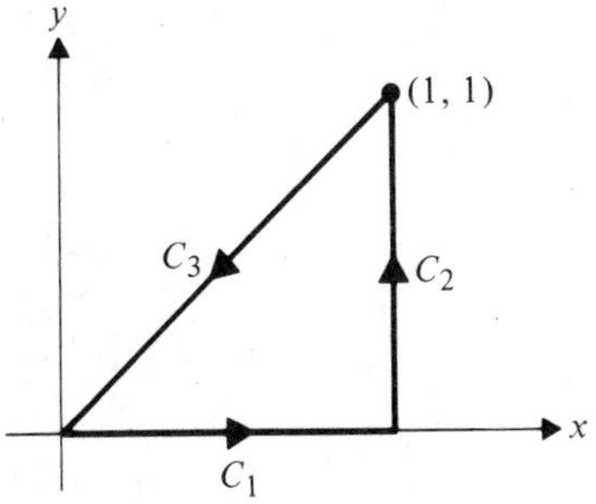

Figure 15.14

then

$$\oint_C y^2\,dx + x^2\,dy = \int_{C_1} y^2\,dx + x^2\,dy + \int_{C_2} y^2\,dx + x^2\,dy + \int_{C_3} y^2\,dx + x^2\,dy$$

$$= \int_0^1 0\,dx + x^2 0 + \int_0^1 y^2 0 + 1\,dy$$

$$+ \int_0^1 (1-t)^2(-dt) + (1-t)^2(-dt)$$

$$= \{y\}_0^1 + \left\{\frac{2}{3}(1-t)^3\right\}_0^1 = 1 - \frac{2}{3} = \frac{1}{3}. \ \square$$

Example 8 If a curve C has initial and final points A and B, then the curve that traces the same points but has initial and final points B and A is denoted by $-C$. Show that

$$\int_{-C} \mathbf{F} \cdot d\mathbf{r} = -\int_C \mathbf{F} \cdot d\mathbf{r}. \tag{15.30}$$

Solution When (15.17) are parametric equations for C, then parametric equations for $-C$ are

$$-C: \quad x = x(t), \qquad y = y(t), \qquad z = z(t), \qquad \beta \geq t \geq \alpha.$$

To obtain an increasing parameter along $-C$, we set $u = -t$, in which case

$$-C: \quad x = x(-u), \qquad y = y(-u), \qquad z = z(-u), \qquad -\beta \leq u \leq -\alpha.$$

If $\mathbf{F} = P\hat{\mathbf{i}} + Q\hat{\mathbf{j}} + R\hat{\mathbf{k}}$, then the value of the line integral along $-C$ can be expressed as a definite integral with respect to u:

$$\int_{-C} \mathbf{F} \cdot d\mathbf{r} = \int_{-C} P\,dx + Q\,dy + R\,dz$$

$$= \int_{-\beta}^{-\alpha} \left\{ P[x(-u), y(-u), z(-u)]\frac{dx}{du} \right.$$

$$\left. + Q[x(-u), y(-u), z(-u)]\frac{dy}{du} + R[x(-u), y(-u), z(-u)]\frac{dz}{du} \right\} du.$$

If we now change variables of integration by setting $t = -u$, then

$$\frac{dx}{du} = \frac{dx}{dt}\frac{dt}{du} = -\frac{dx}{dt},$$

and similarly for dy/du and dz/du. Consequently,

$$\begin{aligned}\int_{-C} \mathbf{F}\cdot d\mathbf{r} &= \int_{\beta}^{\alpha} \left\{P[x(t), y(t), z(t)]\left(-\frac{dx}{dt}\right) + Q[x(t), y(t), z(t)]\left(-\frac{dy}{dt}\right)\right. \\ &\quad \left. + R[x(t), y(t), z(t)]\left(-\frac{dz}{dt}\right)\right\}(-dt) \\ &= -\int_{\alpha}^{\beta} \left\{P[x(t), y(t), z(t)]\frac{dx}{dt} + Q[x(t), y(t), z(t)]\frac{dy}{dt}\right. \\ &\quad \left. + R[x(t), y(t), z(t)]\frac{dz}{dt}\right\} dt \\ &= -\int_{C} \mathbf{F}\cdot d\mathbf{r}. \square\end{aligned}$$

According to this example, when the direction along a curve is reversed the value of a line integral of the form (15.28) along the new curve is the negative of its value along the original curve.

Line integrals of form (15.28) are singled out for special consideration because they repeatedly arise in physical problems. For example, suppose **F** represents a force, and we consider finding the work done by this force as a particle moves along a curve C from A to B. We begin by dividing C into n subcurves of lengths Δs_i as shown in Fig. 15.15. If $\mathbf{F}(x, y, z)$ is continuous along C, then along any given Δs_i, $\mathbf{F}(x, y, z)$ does not vary greatly (provided, of course, that Δs_i is small). If we approximate $\mathbf{F}(x, y, z)$ along Δs_i by its value $\mathbf{F}(x_i, y_i, z_i)$ at the final point (x_i, y_i, z_i) of Δs_i, then an approximation to the work done by **F** along Δs_i is $\mathbf{F}(x_i, y_i, z_i)\cdot\hat{\mathbf{T}}(x_i, y_i, z_i)\,\Delta s_i$. An approximation to the

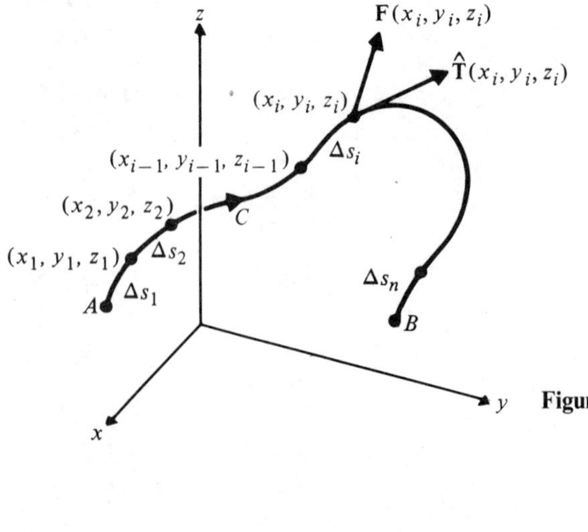

Figure 15.15

total work done by **F** along C is therefore

$$\sum_{i=1}^{n} \mathbf{F}(x_i, y_i, z_i) \cdot \hat{\mathbf{T}}(x_i, y_i, z_i)\, \Delta s_i.$$

To obtain the exact value of the work done by **F** along C we should take the limit of this sum as the number of subdivisions becomes larger and larger and each Δs_i approaches zero. But this is precisely the definition of the line integral of $\mathbf{F} \cdot \hat{\mathbf{T}}$, and we therefore write

$$W = \int_C \mathbf{F} \cdot \hat{\mathbf{T}}\, ds = \int_C \mathbf{F} \cdot d\mathbf{r}. \tag{15.31}$$

This interpretation of a line integral as the work done by a force **F** is extremely important, and we will return to it in Section 15.5.

Example 9 The force of repulsion between two positive point charges, one of size q and the other of size unity, has magnitude $q/(4\pi\varepsilon_0 r^2)$, where ε_0 is a constant and r is the distance between the charges. The potential V at any point P due to the charge q is defined as the work required to bring the unit charge to P from an infinite distance along the straight line joining q and P. Find V.

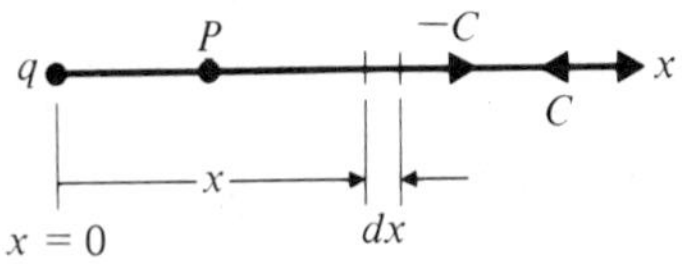

Figure 15.16

Solution If that part of the line joining q and P from infinity to P is denoted by C (Fig. 15.16), then

$$V = \int_C \mathbf{F} \cdot d\mathbf{r} = -\int_{-C} \mathbf{F} \cdot d\mathbf{r}$$

(see Eq. 15.30), where **F**, the force necessary to overcome the electrostatic repulsion, is given by

$$\mathbf{F} = \frac{-q}{4\pi\varepsilon_0 x^2}\hat{\mathbf{i}}.$$

Along $-C$, $d\mathbf{r} = dx\hat{\mathbf{i}}$, and therefore V can be evaluated by the (improper) definite integral

$$V = -\int_r^{\infty} \frac{-q}{4\pi\varepsilon_0 x^2}\, dx = -\left\{\frac{q}{4\pi\varepsilon_0 x}\right\}_r^{\infty} = \frac{q}{4\pi\varepsilon_0 r}. \;\square$$

If the vector function **F** in Eq. (15.27) has an x-component that is only a function of x, $P = P(x)$, and if the curve C is a portion of the x-axis from $x = a$ to $x = b$, then

$$\int_C \mathbf{F} \cdot d\mathbf{r} = \int_a^b P(x)\, dx.$$

This equation indicates that definite integrals with respect to x can be regarded as line integrals along the x-axis.

EXERCISES 15.3

In Exercises 1–10, evaluate the given line integral.

1. $\int_C x\,dx + x^2y\,dy,$
where C is the curve $y = x^3$, $z = 0$ from $(-1, -1, 0)$ to $(2, 8, 0)$

2. $\int_C x\,dx + yz\,dy + x^2\,dz,$
where C is the curve $y = x$, $z = x^2$ from $(-1, -1, 1)$ to $(2, 2, 4)$

3. $\int_C x\,dx + (x + y)\,dy,$
where C is the curve $x = 1 + y^2$ from $(2, 1)$ to $(2, -1)$

4. $\int_C x^2\,dx + y^2\,dy + z^2\,dz,$
where C is the curve $x + y = 1$, $x + z = 1$ from $(-2, 3, 3)$ to $(1, 0, 0)$

5. $\int_C (y + 2x^2z)\,dx,$
where C is the curve $x = y^2$, $z = x^2$ from $(4, -2, 16)$ to $(1, 1, 1)$

6. $\oint_C x^2y\,dx + (x - y)\,dy,$
once counterclockwise around the curve bounding the area described by the curves $x = 1 - y^2$, $y = x + 1$

7. $\int_C y^2\,dx + x^2\,dy,$
where C is the semicircle $x = \sqrt{1 - y^2}$ from $(0, 1)$ to $(0, -1)$

8. $\int_C y\,dx + x\,dy + z\,dz,$
where C is the curve $z = x^2 + y^2$, $x + y = 1$ from $(1, 0, 1)$ to $(-1, 2, 5)$

9. $\oint_C x^2y\,dy + z\,dx,$
where C is the curve $x^2 + y^2 = 1$, $x + y + z = 1$ directed so that x decreases when y is positive

10. $\oint_C y^2\,dx + x^2\,dy,$
once clockwise around the curve $C: |x| + |y| = 1$

11. Find the work done by the force $\mathbf{F} = x^2y\hat{\imath} + x\hat{\jmath}$ as a particle moves from $(1, 0)$ to $(6, 5)$ along the straight line joining these points.

12. Consider the line integral
$$\int_C xy\,dx + x^2\,dy,$$
where C is the quarter-circle $x^2 + y^2 = 9$ from $(3, 0)$ to $(0, 3)$. Show that for each of the following parametrizations of C the value of the line integral is the same:
a) $x = 3\cos t$, $y = 3\sin t$, $0 \le t \le \pi/2$;
b) $x = \sqrt{9 - y^2}$, $0 \le y \le 3$.

13. Evaluate the line integral
$$\int xy\,dx + x\,dy$$
from $(-5, 3, 0)$ to $(4, 0, 0)$ along each of the following three curves:
a) C_1: the straight line joining the points;
b) $C_2: x = 4 - y^2$, $z = 0$;
c) $C_3: 3y = x^2 - 16$, $z = 0$.

14. Find the work done by a force $\mathbf{F} = x\hat{\imath} + y\hat{\jmath}$ on a particle as it moves once counterclockwise around the ellipse $b^2x^2 + a^2y^2 = a^2b^2$, $z = 0$.

In Exercises 15–19, evaluate the given line integral.

15. $\int_C \frac{1}{yz}\,dx,$
where C is the curve $z = \sqrt{1 - x^2}$, $y = \sqrt{1 - x^2}$ from $(1/\sqrt{2}, 1/\sqrt{2}, 1/\sqrt{2})$ to $(-1/\sqrt{2}, 1/\sqrt{2}, 1/\sqrt{2})$

16. $\oint_C (x^2 + 2y^2)\,dy,$
twice clockwise around the circle $(x - 2)^2 + y^2 = 1$, $z = 0$

17. $\int_C y\,dx - y(x - 1)\,dy + y^2z\,dz,$
where C is the first octant intersection of $x^2 + y^2 + z^2 = 4$ and $(x - 1)^2 + y^2 = 1$ from $(2, 0, 0)$ to $(0, 0, 2)$

18. $\int_C x^2y\,dx + y\,dy + \sqrt{1 - x^2}\,dz,$
where C is the curve $y - 2z^2 = 1$, $z = x + 1$ from $(0, 3, 1)$ to $(1, 9, 2)$

19. $\int_C x\,dx + xy\,dy + 2\,dz,$
where C is the curve $x + 2y + z = 4$, $4x + 3y + 2z = 13$ from $(2, -1, 4)$ to $(3, 1, -1)$

20. Evaluate the line integral

$$\int_C \frac{x^3}{(1+x^4)^3}\,dx + y^2e^y\,dy + \frac{z}{\sqrt{1+z^2}}\,dz,$$

where C is the broken line joining successively the points $(0, -1, 1)$, $(1, -1, 1)$, $(1, 0, 1)$, and $(1, 0, 2)$.

21. One end of a spring (with spring constant k) is fixed at point D in Fig. 15.17. The other end is moved along the x-axis from A to B. If the spring is stretched an amount l at A, find the work done against the spring.

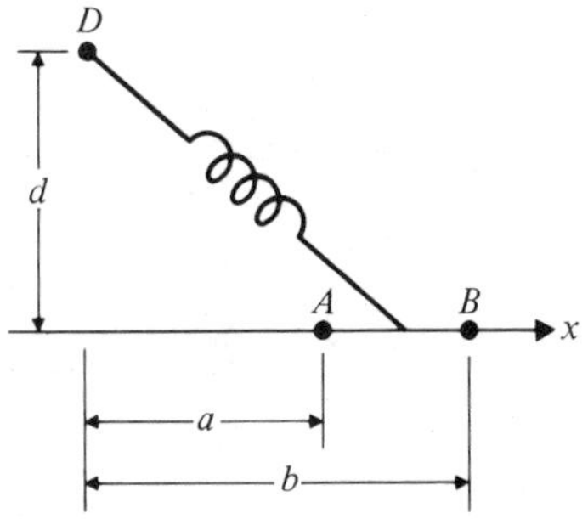

Figure 15.17

22. Two positive charges q_1 and q_2 are placed at positions $(5, 5)$ and $(-2, 3)$, respectively, in the xy-plane. A third positive charge q_3 is moved along the x-axis from $x = 1$ to $x = -1$. Find the total work done by the electrostatic forces of q_1 and q_2 on q_3.

23. Suppose a gas is flowing through a region D of space. At each point $P(x, y, z)$ in D and time t, the gas will have a certain velocity $\mathbf{v}(x, y, z, t)$. If C is a closed curve in D, then the line integral

$$\Gamma = \oint_C \mathbf{v} \cdot d\mathbf{r}$$

is called the circulation of the flow for the curve C. If C is the circle $x^2 + y^2 = r^2$, $z = 1$ (directed clockwise as viewed from the origin), calculate Γ for each of the following flow vectors:

a) $\mathbf{v}(x, y, z, t) = (x\hat{\mathbf{i}} + y\hat{\mathbf{j}} + z\hat{\mathbf{k}})/(x^2 + y^2 + z^2)^{3/2}$;

b) $\mathbf{v}(x, y, z, t) = -y\hat{\mathbf{i}} + x\hat{\mathbf{j}}$.

24. We have shown that given a line integral (15.28) it is always possible to write it uniquely in form (15.23), where $f = \mathbf{F} \cdot \hat{\mathbf{T}}$. Show that the converse is not true; i.e., given $f(x, y, z)$, there does not exist a unique $\mathbf{F}(x, y, z)$ such that $\mathbf{F} \cdot d\mathbf{r} = f(x, y, z)\,ds$.

25. The cycloid $x = a(\theta - \sin\theta)$, $y = a(1 - \cos\theta)$ (Fig. 15.18) is the curve traced out by a fixed point on the circumference of a circle of radius a rolling along the x-axis (θ being the angle through which the point has rotated). Suppose the point is acted on by a force of unit magnitude directed toward the center of the rolling circle.

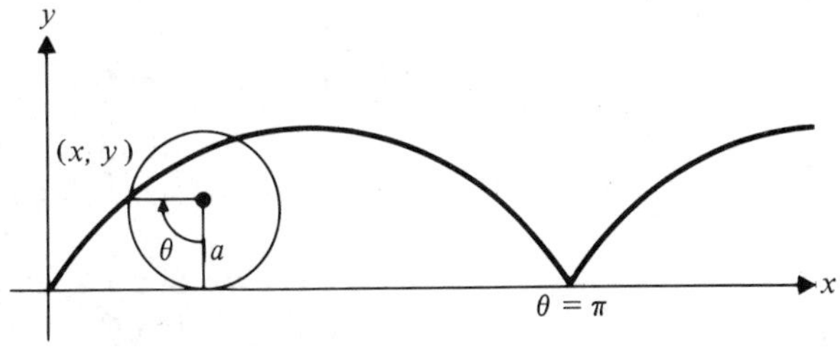

Figure 15.18

a) Find the work done by the force as the point moves from $\theta = 0$ to $\theta = \pi$.

b) How much of the work in (a) is done by the vertical component of the force?

15.4 INDEPENDENCE OF PATH

In Sections 15.2 and 15.3 we illustrated that the value of a line integral joining two points usually depends on the curve joining the points. In this section we show that certain line integrals have the same value for all curves joining the same two points. We formalize this idea in the following definition.

▷ **Definition 15.4** A line integral $\int \mathbf{F} \cdot d\mathbf{r}$ is said to be independent of path in a region D if for each pair of points A and B in D, the value of the line integral

$$\int_C \mathbf{F} \cdot d\mathbf{r}$$

is the same for all piecewise-smooth paths C in D from A to B. ◁

The value of such a line integral for given $\mathbf{F}$ will then depend only on the end points A and B. Note that we speak of independence of path only for the special class of line integrals of the form $\int \mathbf{F} \cdot d\mathbf{r}$. The question we must now ask is "How do we determine whether a given line integral is independent of path?" One answer is contained in the following theorem.

▶ **Theorem 15.3** Suppose $P(x, y, z)$, $Q(x, y, z)$, and $R(x, y, z)$ are continuous functions in some region D. The line integral

$$\int \mathbf{F} \cdot d\mathbf{r} = \int P\,dx + Q\,dy + R\,dz$$

is independent of path in D if and only if there exists a function $\phi(x, y, z)$ defined in D such that

$$\nabla\phi = \mathbf{F} = P\hat{\mathbf{i}} + Q\hat{\mathbf{j}} + R\hat{\mathbf{k}}. \quad ◀ \tag{15.32}$$

Essentially, then, a line integral is independent of path if $\mathbf{F}$ is the gradient of some scalar function.

Proof Suppose first of all that in D there exists a function $\phi(x, y, z)$ such that $\nabla\phi = P\hat{\mathbf{i}} + Q\hat{\mathbf{j}} + R\hat{\mathbf{k}}$. If

$$C:\quad x = x(t), \qquad y = y(t), \qquad z = z(t), \qquad \alpha \le t \le \beta$$

is any smooth curve in D from A to B, then

$$\int_C \mathbf{F} \cdot d\mathbf{r} = \int_C P\,dx + Q\,dy + R\,dz = \int_\alpha^\beta \left\{\frac{\partial\phi}{\partial x}\frac{dx}{dt} + \frac{\partial\phi}{\partial y}\frac{dy}{dt} + \frac{\partial\phi}{\partial z}\frac{dz}{dt}\right\} dt.$$

But the term in braces is precisely the chain rule for the derivative of the composite function $\phi[x(t), y(t), z(t)]$, and we can therefore write that

$$\begin{aligned}\int_C \mathbf{F} \cdot d\mathbf{r} &= \int_\alpha^\beta \frac{d\phi}{dt}\,dt = \{\phi[x(t), y(t), z(t)]\}_\alpha^\beta \\ &= \phi[x(\beta), y(\beta), z(\beta)] - \phi[x(\alpha), y(\alpha), z(\alpha)] \\ &= \phi(x_B, y_B, z_B) - \phi(x_A, y_A, z_A).\end{aligned}$$

(The same result is obtained even when C is piecewise smooth rather than smooth.) But because this last expression does not depend on the curve C taken from A to B, it follows that the line integral is independent of path in D.

Conversely, suppose now that the line integral

$$\int \mathbf{F} \cdot d\mathbf{r} = \int P\,dx + Q\,dy + R\,dz$$

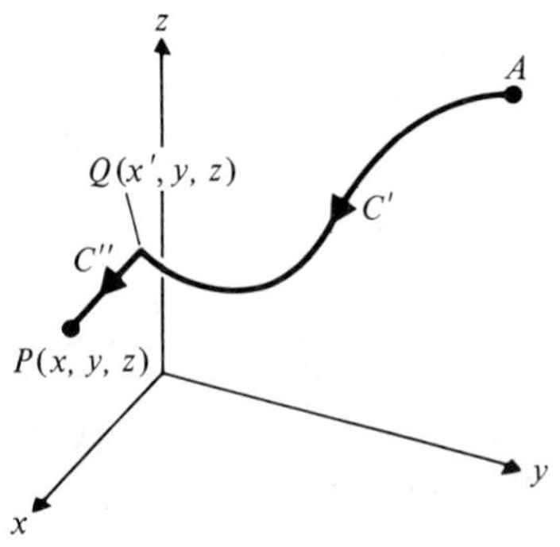

Figure 15.19

is independent of path in D, and A is chosen as some fixed point in D. If $P(x, y, z)$ is any other point in D (Fig. 15.19), and C is a piecewise-smooth curve in D from A to P, then the line integral

$$\phi(x, y, z) = \int_C \mathbf{F} \cdot d\mathbf{r}$$

defines a single-valued function $\phi(x, y, z)$ in D, and the value of $\phi(x, y, z)$ is the same for all piecewise-smooth curves from A to P. Consider a curve C composed of two parts: a straight-line portion C'' parallel to the x-axis from a fixed point $Q(x', y, z)$ to $P(x, y, z)$, and any other piecewise-smooth curve C' in D from A to Q. Then

$$\phi(x, y, z) = \int_{C'} \mathbf{F} \cdot d\mathbf{r} + \int_{C''} \mathbf{F} \cdot d\mathbf{r}.$$

Now along C'', y and z are both constant, and therefore

$$\phi(x, y, z) = \int_{C'} \mathbf{F} \cdot d\mathbf{r} + \int_{x'}^{x} P(t, y, z)\, dt.$$

The partial derivative of this function with respect to x is

$$\frac{\partial \phi}{\partial x} = \frac{\partial}{\partial x} \int_{C'} \mathbf{F} \cdot d\mathbf{r} + \frac{\partial}{\partial x} \int_{x'}^{x} P(t, y, z)\, dt,$$

but because $Q(x', y, z)$ is fixed,

$$\frac{\partial}{\partial x} \int_{C'} \mathbf{F} \cdot d\mathbf{r} = 0.$$

Consequently,

$$\frac{\partial \phi}{\partial x} = \frac{\partial}{\partial x} \int_{x'}^{x} P(t, y, z)\, dt = P(x, y, z).$$

By choosing other curves with straight-line portions parallel to the y- and z-axes, we can also show that $\partial\phi/\partial y = Q$ and $\partial\phi/\partial z = R$. Thus $\mathbf{F} = \nabla\phi$. □

This theorem points out that it is very simple to evaluate a line integral that is independent of path: We take the difference in the values of $\phi(x, y, z)$ at B and at A. This is much easier than substituting from a parametric representation of the curve and evaluating a definite integral.

Example 10 Evaluate $\int_C 2xy\, dx + x^2\, dy + 2z\, dz$, where C is the first octant intersection of $x^2 + y^2 = 1$ and $z = 2x + 4$ from $(0, 1, 4)$ to $(1, 0, 6)$.

Solution Since $\nabla(x^2y + z^2) = 2xy\hat{\mathbf{i}} + x^2\hat{\mathbf{j}} + 2z\hat{\mathbf{k}}$, the line integral is independent of path everywhere, and

$$\int_C 2xy\, dx + x^2\, dy + 2z\, dz = \left\{x^2y + z^2\right\}_{(0,1,4)}^{(1,0,6)} = 36 - 16 = 20. \ \square$$

The following corollary is an immediate consequence of Theorem 15.3.

▶ **Corollary** If $\int \mathbf{F} \cdot d\mathbf{r}$ is independent of path in D, then

$$\oint_C \mathbf{F} \cdot d\mathbf{r} = 0 \tag{15.33}$$

for every closed path in D. ◀

Theorem 15.3 states that a necessary and sufficient condition for line integral (15.28) to be independent of path is the existence of a function $\phi(x, y, z)$ such that $\nabla\phi = \mathbf{F}$. For most problems it is quite obvious whether such a function $\phi(x, y, z)$ exists; for a few, however, it is not at all clear whether $\phi(x, y, z)$ does or does not exist. Since much time could be wasted searching for $\phi(x, y, z)$ (which in fact does not exist), it would be helpful to have a test that states *a priori* whether $\phi(x, y, z)$ does indeed exist. Such a test is contained in Theorem 15.1. It states that $\mathbf{F}$ is the gradient of some scalar function $\phi(x, y, z)$ if $\nabla \times \mathbf{F} = \mathbf{0}$. When this result is combined with Theorem 15.3, we obtain this important theorem.

▶ **Theorem 15.4** Let D be the interior of a rectangular parallelepiped with faces parallel to the coordinate planes in which $P(x, y, z)$, $Q(x, y, z)$, and $R(x, y, z)$ have continuous first derivatives. Then the line integral $\int \mathbf{F} \cdot d\mathbf{r} = \int P\,dx + Q\,dy + R\,dz$ is independent of path in D if and only if $\nabla \times \mathbf{F} = \mathbf{0}$ in D. ◀

We have in Theorem 15.4 a simple test to determine whether a given line integral is independent of path: We see whether the curl of $\mathbf{F}$ is zero. Evaluation of a line integral that is independent of path still requires the function $\phi(x, y, z)$, but it is at least nice to know that ϕ exists before searching for it.

Theorems 15.1, 15.3, and 15.4 have identified an important equivalence. If $\mathbf{F} = P(x, y, z)\hat{\mathbf{i}} + Q(x, y, z)\hat{\mathbf{j}} + R(x, y, z)\hat{\mathbf{k}}$ is a vector function defined in the interior of a rectangular parallelepiped D with faces parallel to the coordinate planes, and P, Q, and R have continuous first derivatives in D, then the following properties are equivalent:

1. $\int \mathbf{F} \cdot d\mathbf{r}$ is independent of path in D;
2. $\mathbf{F} = \nabla\phi$ for some function $\phi(x, y, z)$ defined in D;
3. $\nabla \times \mathbf{F} = \mathbf{0}$ in D.

Theorem 15.1 states that (2) and (3) are equivalent; Theorem 15.3 verifies the equivalence of (1) and (2); and these two imply the equivalence of (1) and (3) (Theorem 15.4).

For line integrals in the xy-plane, this equivalence is still valid except that D becomes a rectangle with sides parallel to the coordinate axes, and $\nabla \times \mathbf{F} = \mathbf{0}$ can be stated more simply as

$$\frac{\partial Q}{\partial x} = \frac{\partial P}{\partial y} \tag{15.34}$$

(see Exercise 12).

Example 11 Evaluate $\int_C 2xye^z\,dx + (x^2e^z + y)\,dy + (x^2ye^z - z)\,dz$ along the straight line C from $(0, 1, 2)$ to $(2, 1, -8)$.

Solution ***Method 1*** Parametric equations for the straight line are

$$C:\quad x = 2t, \qquad y = 1, \qquad z = 2 - 10t, \qquad 0 \le t \le 1.$$

If I is the value of the line integral, then

$$\begin{aligned}
I &= \int_0^1 \{2(2t)(1)e^{2-10t}(2\,dt) + [(2t)^2e^{2-10t} + 1](0) \\
&\quad + [(2t)^2(1)e^{2-10t} - 2 + 10t](-10\,dt)\} \\
&= \int_0^1 \{8e^2(t - 5t^2)e^{-10t} + 20 - 100t\}\,dt \\
&= 8e^2\left\{(t - 5t^2)\frac{e^{-10t}}{-10}\right\}_0^1 + \frac{4e^2}{5}\int_0^1 (1 - 10t)e^{-10t}\,dt + 10\{2t - 5t^2\}_0^1 \\
&= 8e^2\left\{\frac{2}{5}e^{-10}\right\} - 30 + \frac{4}{5}e^2\left\{(1 - 10t)\frac{e^{-10t}}{-10}\right\}_0^1 + \frac{2e^2}{25}\int_0^1 -10e^{-10t}\,dt \\
&= \frac{16}{5}e^{-8} - 30 + \frac{4}{5}e^2\left\{\frac{9}{10}e^{-10} + \frac{1}{10}\right\} + \frac{2e^2}{25}\left\{e^{-10t}\right\}_0^1 = 4e^{-8} - 30.
\end{aligned}$$

Method 2 It is evident that

$$\nabla\left(x^2ye^z + \frac{y^2}{2} - \frac{z^2}{2}\right) = 2xye^z\hat{\mathbf{i}} + (x^2e^z + y)\hat{\mathbf{j}} + (x^2ye^z - z)\hat{\mathbf{k}},$$

and hence the line integral is independent of path. Its value is therefore

$$I = \left\{x^2ye^z + \frac{y^2}{2} - \frac{z^2}{2}\right\}_{(0,1,2)}^{(2,1,-8)} = \left\{4e^{-8} + \frac{1}{2} - 32\right\} - \left\{\frac{1}{2} - 2\right\} = 4e^{-8} - 30.$$

Method 3 Since

$$\begin{aligned}
\nabla \times (2xye^z\hat{\mathbf{i}} + (x^2e^z + y)\hat{\mathbf{j}} + (x^2ye^z - z)\hat{\mathbf{k}}) &= \begin{vmatrix} \hat{\mathbf{i}} & \hat{\mathbf{j}} & \hat{\mathbf{k}} \\ \partial/\partial x & \partial/\partial y & \partial/\partial z \\ 2xye^z & x^2e^z + y & x^2ye^z - z \end{vmatrix} \\
&= (x^2e^z - x^2e^z)\hat{\mathbf{i}} \\
&\quad + (2xye^z - 2xye^z)\hat{\mathbf{j}} \\
&\quad + (2xe^z - 2xe^z)\hat{\mathbf{k}} = \mathbf{0},
\end{aligned}$$

the line integral is independent of path. Thus there exists a function $\phi(x, y, z)$ such that

$$\nabla\phi = 2xye^z\hat{\mathbf{i}} + (x^2e^z + y)\hat{\mathbf{j}} + (x^2ye^z - z)\hat{\mathbf{k}}$$

or

$$\frac{\partial\phi}{\partial x} = 2xye^z, \qquad \frac{\partial\phi}{\partial y} = x^2e^z + y, \qquad \frac{\partial\phi}{\partial z} = x^2ye^z - z.$$

Integration of the first of these equations yields

$$\phi(x, y, z) = x^2ye^z + K(y, z).$$

Substitution of this function into the left-hand side of the second equation gives

$$x^2e^z + \frac{\partial K}{\partial y} = x^2e^z + y,$$

which implies that

$$\frac{\partial K}{\partial y} = y.$$

Consequently,

$$K(y, z) = \frac{y^2}{2} + L(z),$$

and we know both the x- and y-dependence of ϕ:

$$\phi(x, y, z) = x^2ye^z + \frac{y^2}{2} + L(z).$$

To obtain the z-dependence contained in $L(z)$, we substitute into the left-hand side of the third equation to get

$$x^2ye^z + \frac{dL}{dz} = x^2ye^z - z,$$

from which we have

$$\frac{dL}{dz} = -z.$$

Hence, $L(z) = -z^2/2 + C$ (C a constant), and

$$\phi(x, y, z) = x^2ye^z + \frac{y^2}{2} - \frac{z^2}{2} + C.$$

Finally, then, we have

$$I = \left\{x^2ye^z + \frac{y^2}{2} - \frac{z^2}{2}\right\}_{(0,1,2)}^{(2,1,-8)} = 4e^{-8} - 30. \square$$

Method 1 is one of "brute force." The function $x^2ye^z + y^2/2 - z^2/2$ in Method 2 was obtained by observation. Method 3 is the systematic procedure suggested in Section 15.1 for finding the function $\phi(x, y, z)$.

Example 12 Evaluate

$$I = \int_C \left(\frac{x^3 - 2y^2}{x^3y}\right) dx + \left(\frac{y^2 - x^3}{x^2y^2}\right) dy + 2z^2\, dz,$$

where C is the curve $y = x^2$, $z = x - 1$ from $(1, 1, 0)$ to $(2, 4, 1)$.

Solution If we set

$$\mathbf{F} = \left(\frac{x^3 - 2y^2}{x^3y}\right)\hat{\mathbf{i}} + \left(\frac{y^2 - x^3}{x^2y^2}\right)\hat{\mathbf{j}} + 2z^2\hat{\mathbf{k}},$$

it is straightforward to show that $\nabla \times \mathbf{F} = \mathbf{0}$, and therefore the line integral is independent of path provided $x \neq 0$ and $y \neq 0$. Thus there exists a function $\phi(x, y, z)$ such that $\nabla\phi = \mathbf{F}$. If we write

$$\mathbf{F} = \left(\frac{1}{y} - \frac{2y}{x^3}\right)\hat{\mathbf{i}} + \left(\frac{1}{x^2} - \frac{x}{y^2}\right)\hat{\mathbf{j}} + 2z^2\hat{\mathbf{k}},$$

then it is evident that

$$\phi(x, y, z) = \frac{x}{y} + \frac{y}{x^2} + \frac{2z^3}{3}.$$

Thus,

$$I = \left\{\frac{x}{y} + \frac{y}{x^2} + \frac{2z^3}{3}\right\}_{(1,1,0)}^{(2,4,1)} = \left\{\frac{1}{2} + 1 + \frac{2}{3}\right\} - \left\{1 + 1\right\} = \frac{1}{6}. \square$$

Example 13 In thermodynamics the state of a gas is described by four variables—pressure P, temperature T, internal energy U, and volume V. These variables are related by two equations of state,

$$F(P, T, U, V) = 0 \quad \text{and} \quad G(P, T, U, V) = 0,$$

so that two of the variables are independent and two are dependent. If U and V are chosen as independent variables, then $T = T(U, V)$ and $P = P(U, V)$. It is an experimental law called the second law of thermodynamics that the line integral

$$\int_C \frac{1}{T}\,dU + \frac{P}{T}\,dV$$

is independent of path in the UV-plane. Show that the second law can be expressed in the differential form

$$T\frac{\partial P}{\partial U} - P\frac{\partial T}{\partial U} + \frac{\partial T}{\partial V} = 0.$$

Solution According to Eq. (15.34), the line integral is independent of path if and only if

$$\frac{\partial}{\partial U}\left(\frac{P}{T}\right) = \frac{\partial}{\partial V}\left(\frac{1}{T}\right),$$

or

$$0 = \frac{T\dfrac{\partial P}{\partial U} - P\dfrac{\partial T}{\partial U}}{T^2} + \frac{1}{T^2}\frac{\partial T}{\partial V};$$

that is,

$$0 = T\frac{\partial P}{\partial U} - P\frac{\partial T}{\partial U} + \frac{\partial T}{\partial V},$$

and the proof is complete.

Since the above line integral is independent of path, there exists a function $S(U, V)$ such that

$$\frac{\partial S}{\partial U} = \frac{1}{T}, \qquad \frac{\partial S}{\partial V} = \frac{P}{T},$$

and the value of the line integral is given by

$$\int_C \frac{1}{T}\, dU + \frac{P}{T}\, dV = S(B) - S(A),$$

where C joins A and B. This function, called *entropy*, plays a key role in the field of thermodynamics. □

EXERCISES 15.4

In each of Exercises 1–10, show that the line integral is independent of path, and evaluate it.

1. $\int_C xy^2\, dx + x^2 y\, dy$,
where C is the curve $y = x^2, z = 0$ from $(0, 0, 0)$ to $(1, 1, 0)$

2. $\int_C (3x^2 + y)\, dx + x\, dy$,
where C is the straight line from $(2, 1, 5)$ to $(-3, 2, 4)$

3. $\int_C 2xe^y\, dx + (x^2 e^y + 3)\, dy$,
where C is the curve $y = \sqrt{1 - x^2}, z = 0$ from $(1, 0, 1)$ to $(-1, 0, 0)$

4. $\int_C 3x^2 yz\, dx + x^3 z\, dy + (x^3 y - 4z)\, dz$,
where C is the curve $x^2 + y^2 + z^2 = 3, y = x$ from $(-1, -1, 1)$ to $(1, 1, -1)$

5. $\int_C -\frac{y}{z} \sin x\, dx + \frac{1}{z} \cos x\, dy - \frac{y}{z^2} \cos x\, dz$,
where C is the helix $x = 2\cos t, y = 2\sin t, z = t$ from $(2, 0, 2\pi)$ to $(2, 0, 4\pi)$

6. $\oint_C y\cos x\, dx + \sin x\, dy$,
once clockwise around the circle $x^2 + y^2 - 2x + 4y = 7, z = 0$

7. $\int_C x^2\, dx + y^2\, dy + z^2\, dz$,
where C is the curve $x + y = 1, x + z = 1$ from $(-2, 3, 3)$ to $(1, 0, 0)$

8. $\int_C y\, dx + x\, dy + z\, dz$,
where C is the curve $z = x^2 + y^2, x + y = 1$ from $(1, 0, 1)$ to $(-1, 2, 5)$

9. $\int_C \frac{1}{y}\, dx - \frac{x}{y^2}\, dy + dz$,
where C is the curve $y = x^2 + 1, x + y + z = 2$ from $(0, 1, 1)$ to $(3, 10, -11)$

10. $\int_C 3x^2 y^3\, dx + 3x^3 y^2\, dy$,
where C is the curve $y = e^x$ from $(0, 1)$ to $(1, e)$

11. Show that if $f(x)$, $g(y)$, and $h(z)$ have continuous first derivatives, then the line integral

$$\int_C f(x)\, dx + g(y)\, dy + h(z)\, dz$$

is independent of path.

12. For line integrals in the xy-plane, Theorem 15.4 is stated as follows: Let D be the interior of a rectangle in the xy-plane with edges parallel to the coordinate axes in which $P(x, y)$ and $Q(x, y)$ have continuous first derivatives. Then the line integral

$$\int \mathbf{F} \cdot d\mathbf{r} = \int P\, dx + Q\, dy$$

is independent of path in D if and only if in D, $\nabla \times \mathbf{F} = \mathbf{0}$. Show that this condition requires P and Q to satisfy

$$\frac{\partial P}{\partial y} = \frac{\partial Q}{\partial x}.$$

In Exercises 13–18, evaluate the given line integral.

13. $\int_C zye^{xy}\, dx + zxe^{xy}\, dy + (e^{xy} - 1)\, dz$,
where C is the curve $y = x^2, z = x^3$ from $(1, 1, 1)$ to $(2, 4, 8)$

14. $\oint_C y(\tan x + x\sec^2 x)\, dx + x\tan x\, dy + dz$,
once around the circle $x^2 + y^2 = 1, z = 0$

15. $\int_C \left(\frac{1+y^2}{x^3}\right) dx - \left(\frac{y+x^2y}{x^2}\right) dy + z\, dz,$
where C is the broken line joining, successively, (1, 0, 0), (25, 2, 3), (5, 2, 1)

16. $\oint_C \frac{zy\, dx - xz\, dy + xy\, dz}{y^2},$
where C is the closed curve $x^2 + z^2 = 1$, $y + z = 2$

17. $\int_C -\frac{1}{x^2} \operatorname{Tan}^{-1} y\, dx + \frac{1}{x + xy^2}\, dy,$
where C is the curve $x = y^2 + 1$ from $(2, -1)$ to $(10, 3)$

18. $\int_C \frac{1}{(x-3)^2(y+5)}\, dx + \frac{1}{(x-3)(y+5)^2}\, dy + \frac{1}{z+4}\, dz,$
where C is the curve $x = y = z$ from (0, 0, 0) to (2, 2, 2)

19. Evaluate

$$\oint_C \frac{-y\, dx + x\, dy}{x^2 + y^2}$$

a) once counterclockwise around the circle $x^2 + y^2 = 1$, $z = 0$;

b) once counterclockwise around the circle $(x-2)^2 + y^2 = 1$, $z = 0$.

20. Evaluate

$$\oint_C (2xye^{x^2y} + x^2y)\, dx + x^2e^{x^2y}\, dy$$

once clockwise around the ellipse $x^2 + 4y^2 = 4$, $z = 0$.

21. The second law of thermodynamics states that the line integral

$$I = \int T^{-1}(dU + P\, dV)$$

is independent of path in the UV-plane (see Example 13).

a) The equations of state for an ideal gas are

$$PV = nRT, \qquad U = f(T),$$

where n and R are constants and f is some given function. Because of these, it is more convenient to choose T and V as independent variables and to express P and U in terms of T and V. If this is done, show that

$$I = \int kT^{-1}\, dT + nRV^{-1}\, dV,$$

where $k = dU/dT$.

b) Since the line integral is independent of path, there exists a function $S(T, V)$, called entropy, such that

$$\int_C \frac{k}{T}\, dT + \frac{nR}{V}\, dV = S(B) - S(A),$$

where C is any curve joining points A and B. Show in the case that k is constant that

$$S = k \ln T + nR \ln V + S_0, \qquad S_0 = \text{a constant}.$$

*22. A spring has one end fixed at point P. The other end is moved along the curve $y = f(x)$ from $(x_0, f(x_0))$ to $(x_1, f(x_1))$. If the initial and final stretches in the spring are a and b $(b > a)$, what work is done against the spring?

15.5 ENERGY AND CONSERVATIVE FORCE FIELDS

Suppose a force field $\mathbf{F}(x, y, z)$ acts throughout some region D of space. In physics and engineering, many force fields satisfy the following definition.

▷ **Definition 15.5** A force field $\mathbf{F}(x, y, z)$ is said to be conservative in a region D if the line integral $\int \mathbf{F} \cdot d\mathbf{r}$ is independent of path in D. ◁

Since the value of $\int_C \mathbf{F} \cdot d\mathbf{r}$ can be interpreted as the work done by $\mathbf{F}$ along C, a force field is conservative if the work done by it is independent of path taken from one point to another. According to the results of Section 15.4, we can also state that a force field $\mathbf{F}$ is conservative if and only if $\nabla \times \mathbf{F} = \mathbf{0}$ or there exists a function $\phi(x, y, z)$ such that $\mathbf{F} = \nabla\phi$.

It is customary to associate a potential energy function $U(x, y, z)$ with a conservative force field $\mathbf{F}$. This function assigns to each point (x, y, z) a potential energy in such a way that if a particle moves from point A to point B, then the difference in potential energy $U(A) - U(B)$ is precisely the work done by **F**; i.e., if C is any curve joining A and B, then

$$U(A) - U(B) = \int_C \mathbf{F} \cdot d\mathbf{r}. \tag{15.35}$$

If $\int \mathbf{F} \cdot d\mathbf{r} > 0$, then the potential energy at A is greater than the potential energy at B; if $\int \mathbf{F} \cdot d\mathbf{r} < 0$, then the potential energy at B is greater than that at A. To find $U(x, y, z)$, we use the fact that because **F** is conservative, there exists a function $\phi(x, y, z)$ such that $\mathbf{F} = \nabla\phi$, and

$$\int_C \mathbf{F} \cdot d\mathbf{r} = \phi(B) - \phi(A).$$

It follows, then, that $U(x, y, z)$ must satisfy the equation

$$U(A) - U(B) = \phi(B) - \phi(A),$$

or

$$U(A) + \phi(A) = U(B) + \phi(B).$$

Since A and B are arbitrary points in D, this last equation states that the value of the function $U(x, y, z) + \phi(x, y, z)$ is the same at every point in the force field,

$$U(x, y, z) + \phi(x, y, z) = C = \text{a constant}.$$

Thus,

$$U(x, y, z) = -\phi(x, y, z) + C. \tag{15.36}$$

Equation (15.36) shows that the force field **F** defines a potential energy function $U(x, y, z)$ up to an additive constant. (This seems reasonable in that ϕ itself is defined only to an additive constant.) Because $U = -\phi + C$ and $\mathbf{F} = \nabla\phi$, we can also regard U as being defined by the equation

$$\mathbf{F} = -\nabla U. \tag{15.37}$$

The advantage of this equation is that it defines U directly, not through the function ϕ.

For a conservative force field **F**, then, we define a potential energy function $U(x, y, z)$ by Eq. (15.37). If a particle moves from A to B, then the work done by **F** is

$$W = U(A) - U(B); \tag{15.38}$$

in other words, work done by a conservative force field is equal to loss in potential energy.

On the other hand, if a particle moves under the action of a force **F** (and only **F**), be it conservative or nonconservative, then it does so according to Newton's second law,

$$\mathbf{F} = \frac{d}{dt}(m\mathbf{v}) = m\frac{d\mathbf{v}}{dt},$$

where m is the mass of the particle (assumed constant), $\mathbf{v}$ is its velocity, and t is time. The action of **F** will produce motion along some curve C, and the work done by **F** along this curve from A to B is

$$W = \int_C \mathbf{F} \cdot d\mathbf{r} = \int_\alpha^\beta m\frac{d\mathbf{v}}{dt} \cdot \frac{d\mathbf{r}}{dt}\,dt = \int_\alpha^\beta m\frac{d\mathbf{v}}{dt} \cdot \mathbf{v}\,dt = \int_\alpha^\beta \frac{d}{dt}\left\{\frac{1}{2}m\mathbf{v} \cdot \mathbf{v}\right\}dt$$

$$= \left\{\frac{1}{2}m\mathbf{v} \cdot \mathbf{v}\right\}_\alpha^\beta = \left\{\frac{1}{2}m|\mathbf{v}|^2\right\}_\alpha^\beta.$$

Thus if $K(x, y, z)$ represents kinetic energy of the particle, the work done by **F** is equal to the gain in kinetic energy of the particle,

$$W = K(B) - K(A) \tag{15.39}$$

(and this is true for any force **F** so long as **F** is the total resultant force producing motion).

For a conservative force field **F** we have two expressions (15.38) and (15.39) for the work done as a particle moves from one point to another under the action of **F**. If we equate them, we have

$$U(A) - U(B) = K(B) - K(A),$$

or

$$U(A) + K(A) = U(B) + K(B). \tag{15.40}$$

We have shown then that if a particle moves under the action of a conservative force field *only*, the sum of the kinetic and potential energies at B must be the same as the sum of the kinetic and potential energies at A. In other words, if E is the total energy of the particle, kinetic plus potential, then

$$E(A) = E(B). \tag{15.41}$$

Since B can be any point along the path of the particle, it follows that when a particle moves under the action of a conservative force field, and only a conservative force field, then at every point along its trajectory

$$E = \text{a constant}. \tag{15.42}$$

This is the *law of conservation of energy* for a conservative force field.

Example 14 Show that the electrostatic force due to a point charge is conservative, and determine a potential energy function for the field.

Solution The electrostatic force on a charge Q due to a charge q is

$$\mathbf{F} = \frac{qQ}{4\pi\varepsilon_0|\mathbf{r}|^3}\mathbf{r},$$

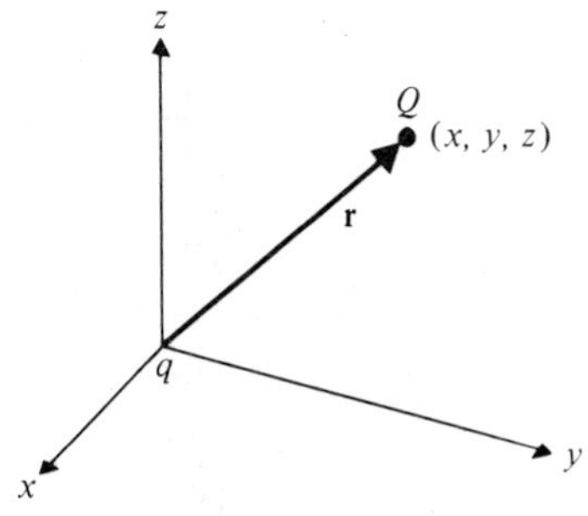

Figure 15.20

where **r** is the vector from q to Q. If we choose a Cartesian coordinate system with q at the origin (Fig. 15.20), then

$$\mathbf{F} = \frac{qQ}{4\pi\varepsilon_0(x^2 + y^2 + z^2)^{3/2}}(x\hat{\mathbf{i}} + y\hat{\mathbf{j}} + z\hat{\mathbf{k}}).$$

Since

$$\nabla\left\{\frac{-qQ}{4\pi\varepsilon_0(x^2 + y^2 + z^2)^{1/2}}\right\} = \mathbf{F},$$

the force field is conservative, and possible potential energy functions are

$$U(x, y, z) = \frac{qQ}{4\pi\varepsilon_0(x^2 + y^2 + z^2)^{1/2}} + C = \frac{qQ}{4\pi\varepsilon_0 r} + C, \qquad r = |\mathbf{r}|.$$

In electrostatics it is customary to choose $U(x, y, z)$ equal to zero at infinity, in which case $C = 0$, and

$$U(x, y, z) = \frac{qQ}{4\pi\varepsilon_0 r}.$$

In addition, if V is defined as the potential (energy) per unit test charge Q, then

$$V(x, y, z) = \frac{U}{Q} = \frac{q}{4\pi\varepsilon_0 r}.$$

This result agrees with that in Example 9. □

EXERCISES 15.5

In Exercises 1–5, determine whether the force field $\mathbf{F}$ is conservative. Identify each conservative force field, and find a potential energy function.

1. $\mathbf{F}(x, y, z) = \dfrac{q_1 q_2}{4\pi\varepsilon_0} \dfrac{x\hat{\mathbf{i}} + y\hat{\mathbf{j}} + z\hat{\mathbf{k}}}{(x^2 + y^2 + z^2)^{3/2}}$

2. $\mathbf{F}(x, y) = mx\hat{\mathbf{i}} + xy\hat{\mathbf{j}}$, m = constant

3. $\mathbf{F}(x) = -kx\hat{\mathbf{i}}$, k = constant

4. $\mathbf{F}(x, y, z) = -mg\hat{\mathbf{k}}$, m and g = constants

5. $\mathbf{F}(x, y, z) = GMm \dfrac{x\hat{\mathbf{i}} + y\hat{\mathbf{j}} + z\hat{\mathbf{k}}}{(x^2 + y^2 + z^2)^{3/2}}$, G, M, and m = constants

6. Suppose that $\mathbf{F}(x, y, z)$ is a conservative force field in some region D, and $U(x, y, z)$ is a potential energy function associated with $\mathbf{F}$. The surfaces

$$U(x, y, z) = C = \text{constant}$$

are called *equipotential surfaces*. Through each point P in D there is one and only one such equipotential surface for $\mathbf{F}$. Show that at P the force $\mathbf{F}$ is normal to the equipotential surface through P.

7. Draw the equipotential surfaces for the forces in Exercises 1, 4, and 5.

***8.** A force field $\mathbf{F}(x, y, z)$ is said to be radially symmetric about the origin if it can be written in the form

$$\mathbf{F}(x, y, z) = f(\sqrt{x^2 + y^2 + z^2})\mathbf{r}, \qquad \mathbf{r} = x\hat{\mathbf{i}} + y\hat{\mathbf{j}} + z\hat{\mathbf{k}}$$

for some function f. We often write in such a case that

$$\mathbf{F}(x, y, z) = f(r)\mathbf{r}, \quad \text{where} \quad r = |\mathbf{r}| = \sqrt{x^2 + y^2 + z^2}.$$

a) Use Theorem 15.4 to show that such a force is conservative in suitably defined regions (provided $f(r)$ has a continuous first derivative).

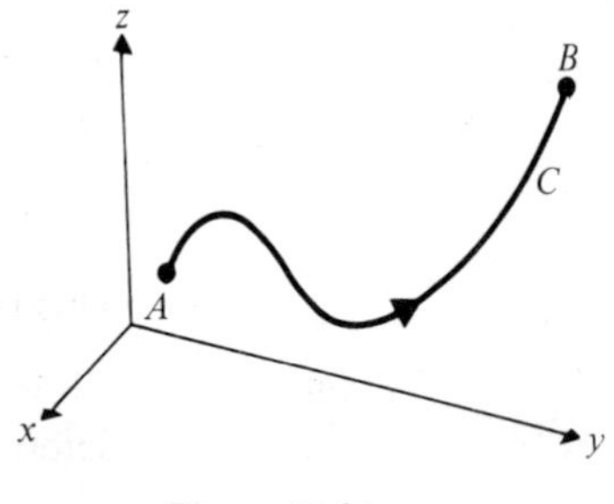

Figure 15.21

b) If A and B are the points in Fig. 15.21 joined by the curve C, show that

$$\int_C \mathbf{F} \cdot d\mathbf{r} = \int_a^b rf(r)\, dr,$$

where the limits a and b on the definite integral are the distances from the origin to A and B, respectively.

c) Have we discussed any radially symmetric force fields in this chapter?

15.6 GREEN'S THEOREM

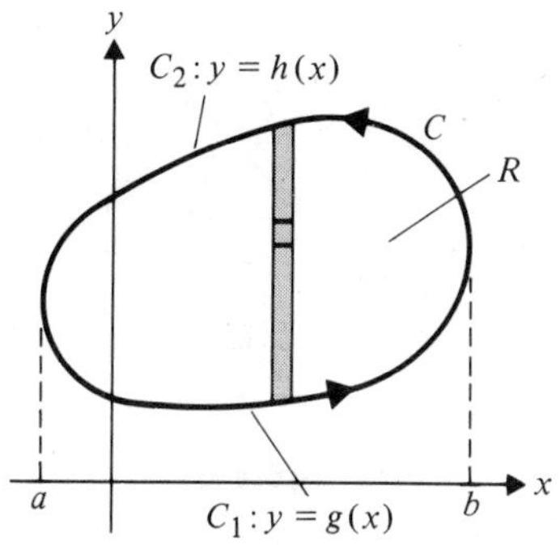

Figure 15.22

Line integrals in the xy-plane are of the form $\int_C f(x, y)\,ds$, and in the special case that $f(x, y)$ is the tangential component of some vector function $\mathbf{F}(x, y) = P(x, y)\hat{\mathbf{i}} + Q(x, y)\hat{\mathbf{j}}$ along C, they take the form

$$\int_C \mathbf{F} \cdot d\mathbf{r} = \int_C P(x, y)\,dx + Q(x, y)\,dy. \tag{15.43}$$

We now show that when C is a closed curve, line integral (15.43) can sometimes be replaced by a double integral. The precise result is contained in the following theorem.

▶ **Theorem 15.5** (*Green's theorem*). Let C be a non–self-intersecting, piecewise-smooth, closed curve in the xy-plane enclosing a region R (Fig. 15.22). Let $P(x, y)$ and $Q(x, y)$ have continuous first partial derivatives inside and on C. Then,

$$\oint_C P\,dx + Q\,dy = \iint_R \left(\frac{\partial Q}{\partial x} - \frac{\partial P}{\partial y}\right) dA. \blacktriangleleft \tag{15.44}$$

Proof We first consider a fairly simple region R. In particular, suppose every line parallel to the x- and y-axes that intersects C does so in at most two points (Fig. 15.22). Then C can be subdivided into an upper and a lower part,

$$C_2: \quad y = h(x) \quad \text{and} \quad C_1: \quad y = g(x).$$

If we first consider the second term on the right of Eq. (15.44), we have

$$\iint_R -\frac{\partial P}{\partial y}\,dA = \int_a^b \int_{g(x)}^{h(x)} -\frac{\partial P}{\partial y}\,dy\,dx = \int_a^b \left\{-P\right\}_{g(x)}^{h(x)} dx$$
$$= \int_a^b \{P[x, g(x)] - P[x, h(x)]\}\,dx.$$

On the other hand, the first term on the left of (15.44) is

$$\oint_C P\,dx = \int_{C_1} P\,dx + \int_{C_2} P\,dx = \int_{C_1} P\,dx - \int_{-C_2} P\,dx,$$

and if we use x as a parameter along C_1 and $-C_2$, then

$$\oint_C P\,dx = \int_a^b P[x, g(x)]\,dx - \int_a^b P[x, h(x)]\,dx$$
$$= \int_a^b \{P[x, g(x)] - P[x, h(x)]\}\,dx.$$

We have shown therefore that

$$\oint_C P\,dx = \iint_R -\frac{\partial P}{\partial y}\,dA.$$

By subdividing C into two parts of the type $x = g(y)$ and $x = h(y)$ (where

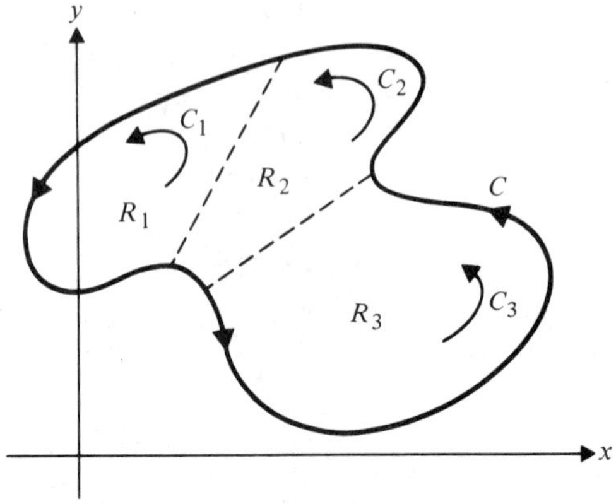

Figure 15.23

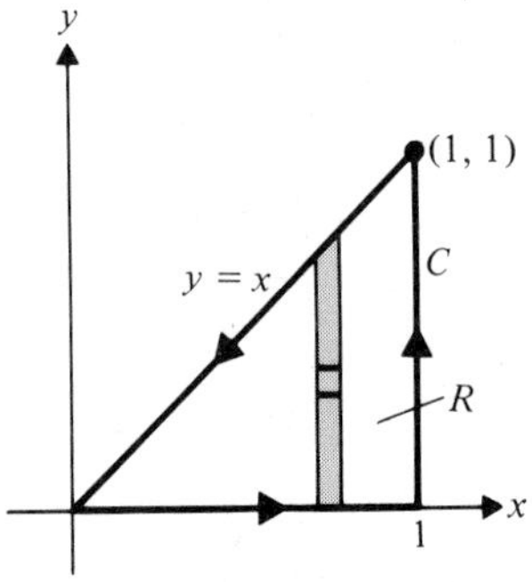

Figure 15.24

$g(y) \leq h(y)$), we can also show that

$$\oint_C Q\,dy = \iint_R \frac{\partial Q}{\partial x}\,dA.$$

Addition of these results gives Green's theorem for this C and R.

Now consider a more general region R such as that shown in Fig. 15.23, which can be decomposed into n subregions R_i, each of which satisfies the condition that lines parallel to the coordinate axes intersect its boundary in at most two points. For each subregion R_i, Green's theorem gives

$$\oint_{C_i} P\,dx + Q\,dy = \iint_{R_i} \left(\frac{\partial Q}{\partial x} - \frac{\partial P}{\partial y}\right) dA.$$

If these results are added, we get

$$\sum_{i=1}^{n} \oint_{C_i} P\,dx + Q\,dy = \sum_{i=1}^{n} \iint_{R_i} \left(\frac{\partial Q}{\partial x} - \frac{\partial P}{\partial y}\right) dA.$$

Now, R is composed of the subregions R_i; thus the right-hand side of this equation is precisely the double integral over R. Figure 15.23 illustrates that when the line integrals over the C_i are added, contributions from ancillary (interior) curves cancel in pairs, leaving the line integral around C. □

We omit a proof for even more general regions that cannot be divided into a finite number of these subregions. The interested reader should consult more advanced books.

Example 15 Evaluate the line integral of Example 7.

Solution By Green's theorem (see Fig. 15.24), we have

$$\oint_C y^2\,dx + x^2\,dy = \iint_R (2x - 2y)\,dA = 2\int_0^1 \int_0^x (x - y)\,dy\,dx$$

$$= 2\int_0^1 \left\{xy - \frac{y^2}{2}\right\}_0^x dx = \int_0^1 x^2\,dx = \left\{\frac{x^3}{3}\right\}_0^1 = \frac{1}{3}. \;\square$$

Example 16 Evaluate

$$\oint_C (5x^2 + ye^x + y)\,dx + (e^x + e^y)\,dy,$$

where C is the circle $(x - 1)^2 + y^2 = 1$.

Solution By Green's theorem (see Fig. 15.25), we have

$$\oint_C (5x^2 + ye^x + y)\,dx + (e^x + e^y)\,dy = \iint_R (e^x - e^x - 1)\,dA = -\iint_R dA$$
$$= -\,(\text{Area of } R) = -\pi(1)^2 = -\pi. \ \square$$

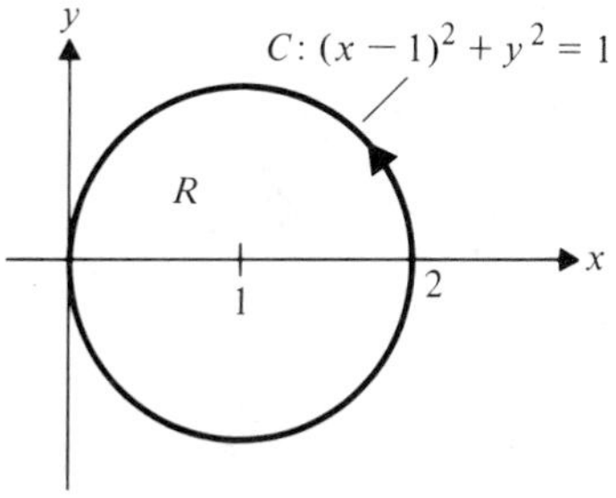

Figure 15.25

Example 17 Show that the area of a region R is defined by each of the line integrals

$$\oint_C x\,dy = \oint_C -\,y\,dx = \frac{1}{2}\oint_C x\,dy - y\,dx,$$

where C is the boundary of R.

Solution By Green's theorem, we have

$$\oint_C x\,dy = \iint_R 1\,dA = \text{Area of } R$$

and

$$\oint_C -\,y\,dx = \iint_R 1\,dA = \text{Area of } R.$$

The third expression for area is the average of these two equations. $\square$

EXERCISES 15.6

In Exercises 1–11, use Green's theorem (if possible) to evaluate the given line integral.

1. $\oint_C y^2\,dx + x^2\,dy,$
where C is the circle $x^2 + y^2 = 1$

2. $\oint_C (x^2 + 2y^2)\,dy,$
where C is the curve $(x - 2)^2 + y^2 = 1$

3. $\oint_C x^2e^y\,dx + (x + y)\,dy,$
where C is the square with vertices $(\pm 1, \pm 1)$

4. $\oint_C xy^3\,dx + x^2\,dy$,
where C is the curve enclosing the area bounded by $x = \sqrt{1+y^2}$, $x = 2$

5. $\oint_C (x^3 + y^3)\,dx + (x^3 - y^3)\,dy$,
where C is the curve enclosing the area bounded by $x = y^2 - 1$, $x = 1 - y^2$

6. $\oint_C 2\,\mathrm{Tan}^{-1}(y/x)\,dx + \ln(x^2 + y^2)\,dy$,
where C is the circle $(x-4)^2 + (y-1)^2 = 2$

7. $\oint_C (3x^2y^3 + y)\,dx + (3x^3y^2 + 2x)\,dy$,
where C is the boundary of the area enclosed by $x + y = 1$, $x = -1$, $y = -1$

8. $\oint_C (x^3 + y^3)\,dx + (x^3 - y^3)\,dy$,
where C is the curve $2|x| + |y| = 1$

9. $\oint_C (x^2y^2 + 3x)\,dx + (2xy - y)\,dy$,
where C is the boundary of the area enclosed by $x = 1 - y^2$ $(x \geq 0)$, $y = x + 1$, $y + x + 1 = 0$

10. $\oint_C (xy^2 + 2x)\,dx + (x^2y + y + x^2)\,dy$,
where C is the boundary of the area enclosed by $y^2 - x^2 = 4$, $x = 0$, $x = 3$

11. $\oint_C \dfrac{-y\,dx + x\,dy}{x^2 + y^2}$,
where C is the circle $x^2 + y^2 = 1$

12. If a vector field $\mathbf{F}$ is defined by $\mathbf{F}(x, y) = P(x, y)\hat{\mathbf{i}} + Q(x, y)\hat{\mathbf{j}}$, show that Green's theorem can be expressed vectorially in the form

$$\oint_C \mathbf{F} \cdot d\mathbf{r} = \iint_R (\nabla \times \mathbf{F}) \cdot \hat{\mathbf{k}}\,dA.$$

13. If a curve C is traced out in the direction defined by Green's theorem, it can be shown that a normal vector to C that always points to the outside of C is $\mathbf{n} = (dy, -dx)$. Show that if a vector field $\mathbf{F}$ is defined by $\mathbf{F}(x, y) = P(x, y)\hat{\mathbf{i}} + Q(x, y)\hat{\mathbf{j}}$, then Green's theorem can be written vectorially in the form

$$\oint_C \mathbf{F} \cdot \hat{\mathbf{n}}\,ds = \iint_R \nabla \cdot \mathbf{F}\,dA.$$

14. a) If C is the straight-line segment joining points $P_1(x_1, y_1)$ and $P_2(x_2, y_2)$ in the xy-plane, show that

$$\frac{1}{2}\int_C x\,dy - y\,dx = \frac{1}{2}(x_1y_2 - x_2y_1).$$

b) Let $P_1(x_1, y_1), P_2(x_2, y_2), \ldots, P_n(x_n, y_n)$ be the coordinates of a polygon labeled in the counterclockwise direction (Fig. 15.26). Use the result in (a) and Example 17 to derive the following formula for the area A of the polygon:

$$A = \tfrac{1}{2}(x_1y_2 - x_2y_1) + \tfrac{1}{2}(x_2y_3 - x_3y_2) + \cdots + \tfrac{1}{2}(x_{n-1}y_n - x_ny_{n-1}) + \tfrac{1}{2}(x_ny_1 - x_1y_n).$$

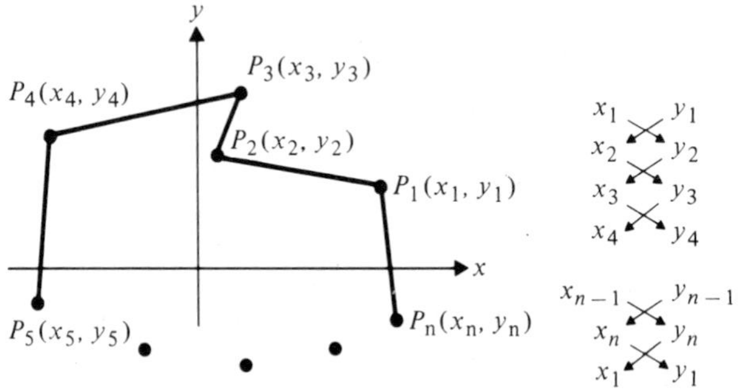

Figure 15.26

c) Show that if the coordinates of the vertices are arranged in a vertical column with (x_1, y_1) repeated at the bottom (Fig. 15.26), then

$$A = \tfrac{1}{2}\{[\text{sum of downward products to the right } (\searrow)] - [\text{sum of downward products to the left } (\swarrow)]\}.$$

In Exercises 15–18, use the result of Exercise 14 to find the area of the polygon with the given points as successive vertices.

15. $(1, 2), (-3, 2), (4, 1)$

16. $(2, -2), (1, -3), (-2, 1), (5, 6)$

17. $(3, 0), (1, 1), (2, 5), (-4, -4)$

18. $(0, 4), (-1, 0), (-2, 0), (-3, -4), (0, -5), (6, -2), (3, 0), (2, 2)$

In Exercises 19–23, evaluate the given line integral.

19. $\oint_C (2xye^{x^2y} + 3x^2y)\,dx + x^2e^{x^2y}\,dy$,
where C is the ellipse $x^2 + 4y^2 = 4$

20. $\oint_C (3x^2y^3 - x^2y)\,dx + (xy^2 + 3x^3y^2)\,dy$,
where C is the circle $x^2 + y^2 = 9$

21. $\oint_C -x^3y^2\,dx + x^2y^3\,dy$,
where C is the right loop of $(x^2 + y^2)^{3/2} = x^2 - y^2$

22. $\int_C (x - y)(dx + dy)$,
where C is the semicircular part of $x^2 + y^2 = 4$ above $y = x$ from $(-\sqrt{2}, -\sqrt{2})$ to $(\sqrt{2}, \sqrt{2})$

23. $\int_C (e^y - y \sin x)\,dx + (\cos x + xe^y)\,dy$,
where C is the curve $x = 1 - y^2$ from $(0, -1)$ to $(0,1)$

24. The result of this exercise is useful when the curve C in Green's theorem contains a point (or points) at which either P or Q fails to have continuous first partial derivatives (see Exercise 25).

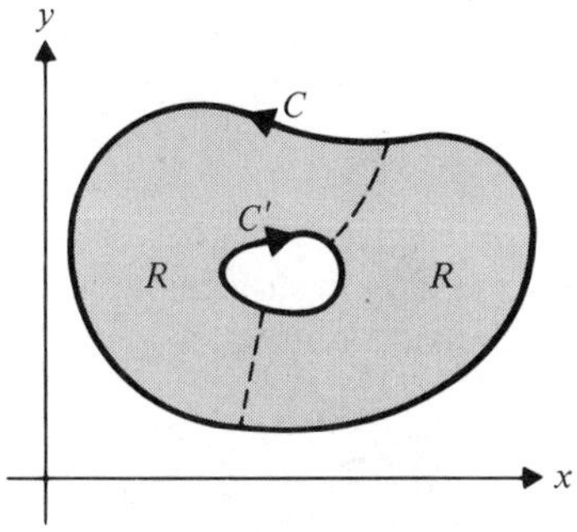

Figure 15.27

a) Suppose a piecewise-smooth curve C (Fig. 15.27) contains in its interior another piecewise-smooth curve C', and that $P(x, y)$ and $Q(x, y)$ have continuous first partial derivatives in the region R consisting of C, C', and the area between them. Prove that

$$\oint_C P\,dx + Q\,dy + \oint_{C'} P\,dx + Q\,dy = \iint_R \left(\frac{\partial Q}{\partial x} - \frac{\partial P}{\partial y}\right) dA.$$

(*Hint*: Join C and C' by two curves such as those in Fig. 15.27.)

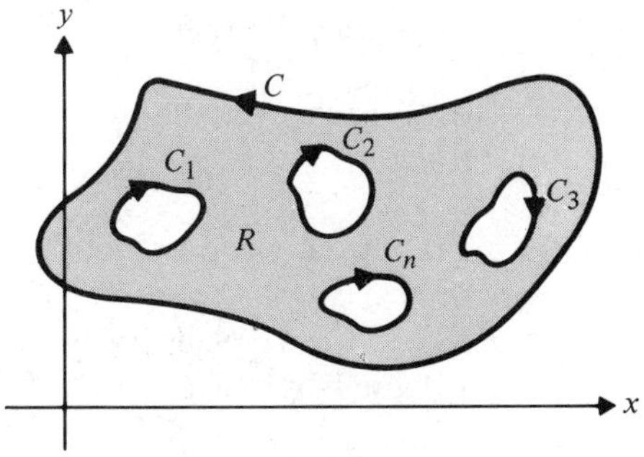

Figure 15.28

b) Extend this result to show that when C' is replaced by n distinct curves C_i (Fig. 15.28), and P and Q have continuous first partial derivatives on and between C and the C_i, then

$$\oint_C P\,dx + Q\,dy + \oint_{C_1} P\,dx + Q\,dy + \cdots + \oint_{C_n} P\,dx + Q\,dy = \iint_R \left(\frac{\partial Q}{\partial x} - \frac{\partial P}{\partial y}\right) dA.$$

c) What can we conclude in (a) and (b) if $\partial Q/\partial x = \partial P/\partial y$ in R?

25. Use the result of Exercise 24(a) to evaluate

$$\oint_C \frac{y\,dx - (x-1)\,dy}{(x-1)^2 + y^2},$$

where C is the circle $x^2 + y^2 = 4$.

In Exercises 26–28, assume that $P(x, y)$ and $Q(x, y)$ have continuous second partial derivatives in the region R consisting of C and its interior (Fig. 15.29). Let the vector $\hat{\mathbf{n}} = (dy/ds, -dx/ds)$ be the outward pointing normal to C, and let

$$\frac{\partial P}{\partial n} = \nabla P \cdot \hat{\mathbf{n}}, \qquad \frac{\partial Q}{\partial n} = \nabla Q \cdot \hat{\mathbf{n}}$$

be the directional derivatives of P and Q in the direction of $\hat{\mathbf{n}}$.

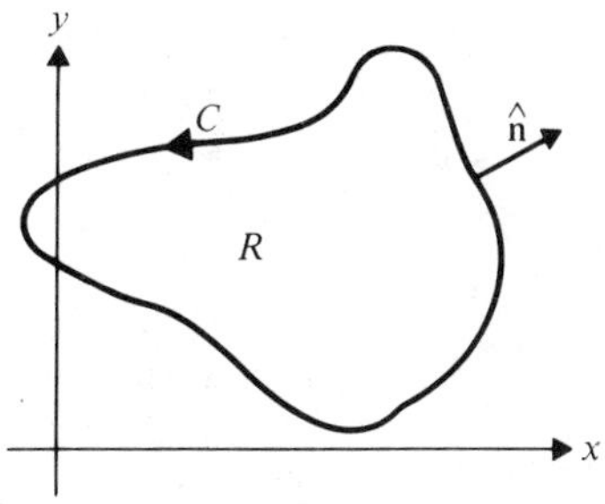

Figure 15.29

26. Show that

$$\oint_C \frac{\partial P}{\partial n}\,ds = \iint_R \nabla^2 P\,dA,$$

where

$$\nabla^2 P = \frac{\partial^2 P}{\partial x^2} + \frac{\partial^2 P}{\partial y^2}.$$

(*Hint*: See Exercise 13.) What do you conclude if $P(x, y)$ satisfies Laplace's equation in R?

27. Show that

$$\oint_C P\frac{\partial Q}{\partial n}\,ds = \iint_R P\nabla^2 Q\,dA + \iint_R \nabla P \cdot \nabla Q\,dA.$$

[*Hint*: Use identity (15.10). This result is often called Green's first identity (in the plane).]

28. Prove that

$$\oint_C \left(P\frac{\partial Q}{\partial n} - Q\frac{\partial P}{\partial n}\right) ds = \iint_R (P\nabla^2 Q - Q\nabla^2 P)\, dA.$$

This is often called Green's second identity (in the plane).

***29.** Find all possible values of the line integral

$$\oint_C \frac{-y\,dx + x\,dy}{x^2 + y^2}$$

for curves in the xy-plane not passing through the origin.

15.7 SURFACE INTEGRALS

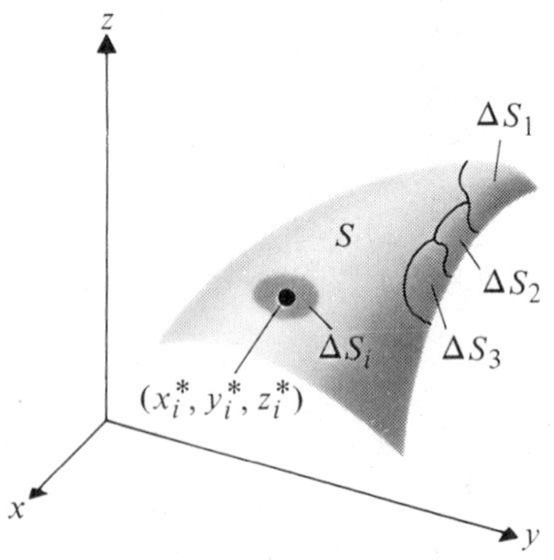

Figure 15.30

Consider a function $f(x, y, z)$ defined on some surface S (Fig. 15.30). We divide S into n subsurfaces of areas $\Delta S_1, \Delta S_2, \ldots, \Delta S_n$ in any manner whatsoever. On each subsurface ΔS_i $(i = 1, \ldots, n)$ we choose an arbitrary point (x_i^*, y_i^*, z_i^*), and form the sum

$$f(x_1^*, y_1^*, z_1^*)\,\Delta S_1 + f(x_2^*, y_2^*, z_2^*)\,\Delta S_2 + \cdots + f(x_n^*, y_n^*, z_n^*)\,\Delta S_n$$

$$= \sum_{i=1}^{n} f(x_i^*, y_i^*, z_i^*)\,\Delta S_i. \tag{15.45}$$

If this sum approaches a limit as the number of subsurfaces becomes increasingly large and every subsurface shrinks to a point, we call the limit the surface integral of $f(x, y, z)$ over the surface S, and denote it by

$$\iint_S f(x, y, z)\, dS = \lim_{n\to\infty} \sum_{i=1}^{n} f(x_i^*, y_i^*, z_i^*)\,\Delta S_i \tag{15.46}$$

(where it is understood that every $\Delta S_i \to 0$ as $n \to \infty$).

Surface integrals, like all integrals, are limit–summations. We think of dS as a small piece of area on S, and each dS is multiplied by the value of $f(x, y, z)$ for that area. All such products are then added together and the limit taken as the pieces of area shrink to points.

The following theorem guarantees the existence of the surface integral of a continuous function over a smooth surface.

▶ **Theorem 15.6** If a function $f(x, y, z)$ is continuous on a smooth surface S of finite area, then the surface integral of $f(x, y, z)$ over S exists. If S projects in a one-to-one fashion onto a region S_{xy} in the xy-plane, then the surface integral of $f(x, y, z)$ over S can be evaluated by means of the following double integral:

$$\iint_S f(x, y, z)\, dS = \iint_{S_{xy}} f[x, y, g(x, y)]\sqrt{1 + \left(\frac{\partial z}{\partial x}\right)^2 + \left(\frac{\partial z}{\partial y}\right)^2}\, dA, \tag{15.47}$$

where $z = g(x, y)$ is the equation of S. ◀

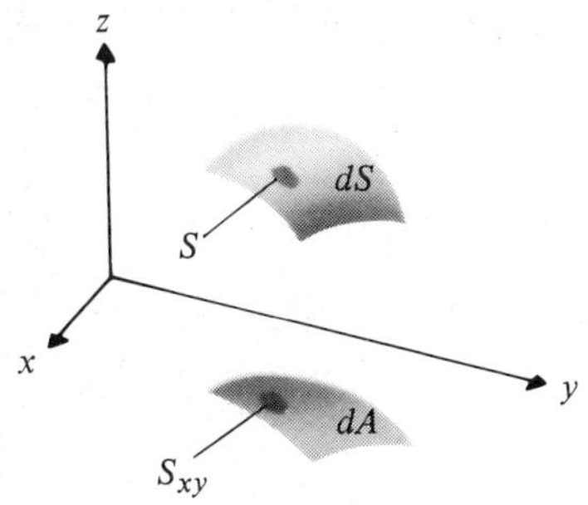

Figure 15.31

It is not necessary to memorize (15.47) as a formula, since it is the result obtained by expressing z and dS in terms of x and y and interpreting the result as a double integral over the projection of S in the xy-plane. Recall from Section 14.7 that when a surface S can be represented in the form $z = g(x, y)$, a small area dS on S is related to its projection dA in the xy-plane according to the formula

$$dS = \sqrt{1 + \left(\frac{\partial z}{\partial x}\right)^2 + \left(\frac{\partial z}{\partial y}\right)^2}\, dA \tag{15.48}$$

(Fig. 15.31). If we substitute this into the left-hand side of (15.47) and at the same time use $z = g(x, y)$ to express $f(x, y, z)$ in terms of x and y, then

$$\iint_S f(x, y, z)\, dS = \iint_S f[x, y, g(x, y)]\sqrt{1 + \left(\frac{\partial z}{\partial x}\right)^2 + \left(\frac{\partial z}{\partial y}\right)^2}\, dA.$$

But if we now interpret the right-hand side of this equation as the double integral of $f[x, y, g(x, y)]\sqrt{1 + (\partial z/\partial x)^2 + (\partial z/\partial y)^2}$ over the projection S_{xy} of S onto the xy-plane, we obtain (15.47).

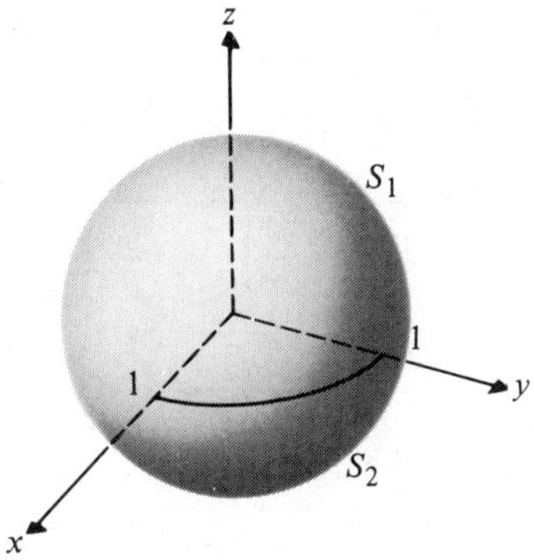

Figure 15.32

Note the analogy between Eqs. (15.47) and (15.20). Equation (15.20) states that the line integral on the left can be evaluated by means of the definite integral on the right. Equation (15.47) states that the surface integral on the left can be evaluated by means of the double integral on the right.

If S does not project one-to-one onto an area in the xy-plane, then one possibility is to subdivide S into parts, each of which projects one-to-one onto the xy-plane. The total surface integral over S is then the sum of the surface integrals over the parts. For example, if we require the surface integral of a function $f(x, y, z)$ over the sphere S: $x^2 + y^2 + z^2 = 1$ (Fig. 15.32), then we could subdivide S into two hemispheres,

$$S_1:\quad z = \sqrt{1 - x^2 - y^2},$$

$$S_2:\quad z = -\sqrt{1 - x^2 - y^2},$$

each of which projects onto

$$S_{xy}:\quad x^2 + y^2 \leq 1.$$

Then,

$$\iint_S f(x, y, z)\, dS = \iint_{S_1} f(x, y, z)\, dS + \iint_{S_2} f(x, y, z)\, dS.$$

A second possibility is to project S onto either the xz-plane or the yz-plane. If S does project one-to-one onto areas S_{xz} and S_{yz} in these planes, then

$$\iint_S f(x, y, z)\, dS = \iint_{S_{xz}} f[x, g(x, z), z]\sqrt{1 + \left(\frac{\partial y}{\partial x}\right)^2 + \left(\frac{\partial y}{\partial z}\right)^2}\, dA \tag{15.49}$$

and

$$\iint_S f(x, y, z)\, dS = \iint_{S_{yz}} f[g(y, z), y, z]\sqrt{1 + \left(\frac{\partial x}{\partial y}\right)^2 + \left(\frac{\partial x}{\partial z}\right)^2}\, dA. \tag{15.50}$$

We should also note that the area of a surface S is defined by the surface integral

$$\iint_S dS, \tag{15.51}$$

and if S projects one-to-one onto S_{xy}, then

$$\text{Area of } S = \iint_{S_{xy}} \sqrt{1 + \left(\frac{\partial z}{\partial x}\right)^2 + \left(\frac{\partial z}{\partial y}\right)^2}\, dA. \tag{15.52}$$

This is precisely formula (14.41).

The results in Eqs. (15.47)–(15.52) were based on a smooth surface S. If S is piecewise smooth, rather than smooth, we simply subdivide S into smooth parts and apply each of these results to the smooth parts. The surface integral over S is then the summation of the surface integrals over its parts.

Example 18 Evaluate $\iint_S (x + y + z)\, dS$, where S is that part of the plane $x + 2y + 4z = 8$ in the first octant.

Solution The surface S projects one-to-one onto the triangle S_{xy} in the xy-plane shown in Fig. 15.33. Since $z = (8 - x - 2y)/4$ on S,

$$\begin{aligned}
\iint_S (x + y + z)\, dS &= \iint_{S_{xy}} \left(x + y + 2 - \frac{x}{4} - \frac{y}{2}\right) \sqrt{1 + \left(\frac{\partial z}{\partial x}\right)^2 + \left(\frac{\partial z}{\partial y}\right)^2}\, dA \\
&= \frac{1}{4} \iint_{S_{xy}} (3x + 2y + 8) \sqrt{1 + \left(-\frac{1}{4}\right)^2 + \left(-\frac{1}{2}\right)^2}\, dA \\
&= \frac{\sqrt{21}}{16} \int_0^4 \int_0^{8-2y} (3x + 2y + 8)\, dx\, dy \\
&= \frac{\sqrt{21}}{16} \int_0^4 \left\{\frac{3x^2}{2} + (2y + 8)x\right\}_0^{8-2y} dy \\
&= \frac{\sqrt{21}}{8} \int_0^4 (80 - 24y + y^2)\, dy \\
&= \frac{\sqrt{21}}{8} \left\{80y - 12y^2 + \frac{y^3}{3}\right\}_0^4 = \frac{56\sqrt{21}}{3}. \quad \square
\end{aligned}$$

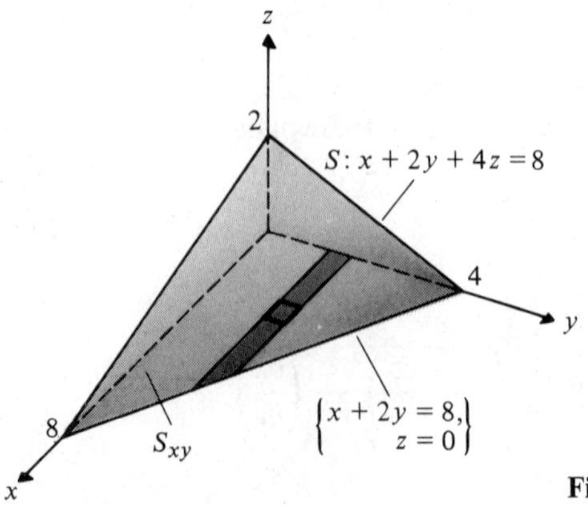

Figure 15.33

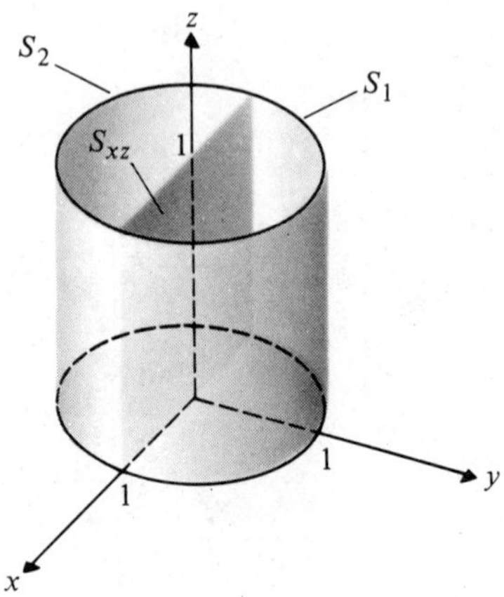

Figure 15.34

Example 19 Evaluate $\iint_S x^2z\,dS$, where S is the surface $x^2 + y^2 = 1$ for $0 \le z \le 1$.

Solution If we divide S into two parts (Fig. 15.34), we have

$$S_1:\ y = \sqrt{1 - x^2}, \qquad 0 \le z \le 1,$$
$$S_2:\ y = -\sqrt{1 - x^2}, \qquad 0 \le z \le 1,$$

and each projects one-to-one onto the rectangle

$$S_{xz}:\ y = 0, \qquad -1 \le x \le 1, \qquad 0 \le z \le 1,$$

in the xz-plane. For each subsurface,

$$\sqrt{1 + \left(\frac{\partial y}{\partial x}\right)^2 + \left(\frac{\partial y}{\partial z}\right)^2} = \sqrt{1 + \frac{x^2}{1 - x^2}} = \frac{1}{\sqrt{1 - x^2}};$$

therefore,

$$\iint_S x^2z\,dS = \iint_{S_1} x^2z\,dS + \iint_{S_2} x^2z\,dS$$
$$= \iint_{S_{xz}} \frac{x^2z}{\sqrt{1 - x^2}}\,dA + \iint_{S_{xz}} \frac{x^2z}{\sqrt{1 - x^2}}\,dA$$
$$= 2\iint_{S_{xz}} \frac{x^2z}{\sqrt{1 - x^2}}\,dA = 2\int_{-1}^{1}\int_0^1 \frac{x^2z}{\sqrt{1 - x^2}}\,dz\,dx$$
$$= \int_{-1}^{1} \frac{x^2}{\sqrt{1 - x^2}}\,dx.$$

If we set $x = \sin\theta$, then $dx = \cos\theta\,d\theta$ and

$$\iint_S x^2z\,dS = \int_{-\pi/2}^{\pi/2} \frac{\sin^2\theta}{\cos\theta}\cos\theta\,d\theta = \int_{-\pi/2}^{\pi/2} \left\{\frac{1 - \cos 2\theta}{2}\right\} d\theta$$
$$= \frac{1}{2}\left\{\theta - \frac{\sin 2\theta}{2}\right\}_{-\pi/2}^{\pi/2} = \frac{\pi}{2}. \ \square$$

EXERCISES 15.7

In Exercises 1–8, evaluate the given surface integral.

1. $\iint_S (x^2y + z)\,dS$,
where S is the first octant part of $2x + 3y + z = 6$

2. $\iint_S (x^2 + y^2)z\,dS$,
where S is that part of $z = x + y$ cut out by $x = 0$, $y = 0$, and $x + y = 1$

3. $\iint_S xyz\,dS$,
where S is the surface of the cube $0 \le x \le 1, 0 \le y \le 1$, $0 \le z \le 1$

4. $\iint_S xy\,dS$,
where S is the first octant part of $z = \sqrt{x^2 + y^2}$ cut out by $x^2 + y^2 = 1$

5. $\iint_S \frac{1}{\sqrt{z - y + 1}}\,dS$,
where S is the surface defined by $2z = x^2 + 2y, 0 \le x \le 1, 0 \le y \le 1$

6. $\iint_S \sqrt{4y + 1}\,dS$,
where S is the first octant part of $y = x^2$ cut out by $2x + y + z = 1$

7. $\iint_S z^2\,dS$,
where S is the sphere $x^2 + y^2 + z^2 = 4$

8. $\iint_S (x+y)\,dS$,
where S is the surface bounding the volume enclosed by $x=0$, $y=0$, $z=0$, $6x-3y+2z=6$

9. Set up double iterated integrals for the surface integral of a function $f(x, y, z)$ over the surface defined by $z = 4 - x^2 - 4y^2$ $(x, y, z \geq 0)$ if the surface is projected onto the xy-, the xz-, and the yz-coordinate planes.

10. Use a surface integral to find the area of the curved portion of a right-circular cone of radius R and height h.

In Exercises 11–14, evaluate the given surface integral.

11. $\iint_S xyz^3\,dS$,
where S is the surface defined by $x = y^2$, $0 \leq x \leq 4$, $0 \leq z \leq 1$

12. $\iint_S xyz\,dS$,
where S is the surface defined by $2y = \sqrt{9-x}$, $x \geq 0$, $0 \leq z \leq 3$

13. $\iint_S \frac{1}{\sqrt{2az - z^2}}\,dS$,
where S is that part of $x^2 + y^2 + (z-a)^2 = a^2$ inside the cylinder $x^2 + y^2 = ay$, underneath the plane $z = a$, and in the first octant

14. $\iint_S z\,dS$,
where S is that part of the surface $x^2 + y^2 - z^2 = 1$ between the planes $z = 0$ and $z = 1$

15. Show that if a surface S defined implicitly by the equation $F(x, y, z) = 0$ projects in a one-to-one fashion onto S_{xy} in the xy-plane, then

$$\iint_S f(x, y, z)\,dS = \iint_{S_{xy}} f[x, y, g(x, y)] \frac{|\nabla F|}{|\partial F/\partial z|}\,dA.$$

16. a) Find the area cut from the cones $z^2 = x^2 + y^2$ by the cylinder $x^2 + y^2 = 2x$.
b) Find the area cut from the cylinder $x^2 + y^2 = 2x$ by the cones $z^2 = x^2 + y^2$.

15.8 SURFACE INTEGRALS INVOLVING VECTOR FUNCTIONS

The most important and common type of surface integral occurs when $f(x, y, z)$ in (15.46) is specified as the normal component of some given vector function $\mathbf{F}(x, y, z)$ defined on S; in other words, $f(x, y, z)$ itself is not given, but $\mathbf{F}$ is, and to find $f(x, y, z)$ we must calculate the component of $\mathbf{F}$ normal to S. If $\hat{\mathbf{n}}$ is the unit normal vector to S, then

$$f(x, y, z) = \mathbf{F} \cdot \hat{\mathbf{n}} \tag{15.53}$$

and

$$\iint_S f(x, y, z)\,dS = \iint_S \mathbf{F} \cdot \hat{\mathbf{n}}\,dS. \tag{15.54}$$

It is important to remember that there are two unit normals to a surface at each point (Fig. 15.35), and we must therefore be careful to choose the appropriate one in each calculation.

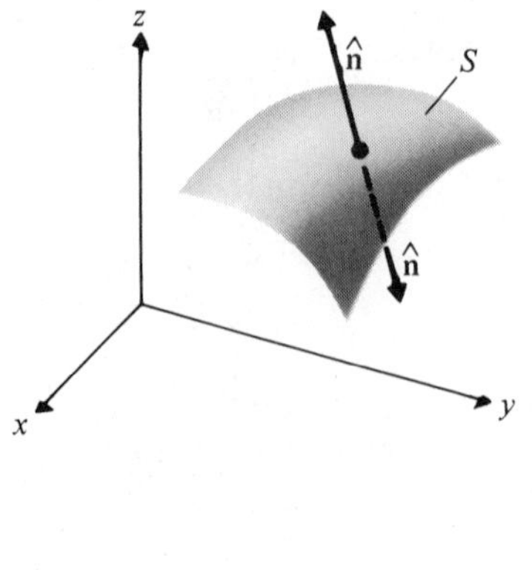

Figure 15.35

Example 20 Evaluate $\iint_S \mathbf{F}\cdot\hat{\mathbf{n}}\,dS$, where $\mathbf{F} = x^2y\hat{\mathbf{i}} + xz\hat{\mathbf{j}}$ and $\hat{\mathbf{n}}$ is the upper normal to the surface $S: z = 4 - x^2 - y^2, z \geq 0$.

Solution Since a normal vector to S is

$$\nabla(z - 4 + x^2 + y^2) = (2x, 2y, 1),$$

it follows that

$$\hat{\mathbf{n}} = \frac{(2x, 2y, 1)}{\sqrt{4x^2 + 4y^2 + 1}}.$$

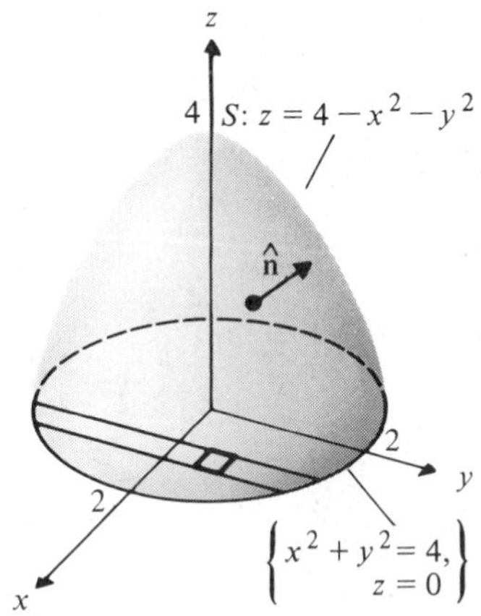

Figure 15.36

Thus,

$$\mathbf{F}\cdot\hat{\mathbf{n}} = \frac{2x^3y + 2xyz}{\sqrt{4x^2 + 4y^2 + 1}}.$$

If we project S onto $S_{xy}: x^2 + y^2 \leq 4$ in the xy-plane (Fig. 15.36), then

$$\begin{aligned}\iint_S \mathbf{F}\cdot\hat{\mathbf{n}}\,dS &= \iint_{S_{xy}} \frac{2x^3y + 2xyz}{\sqrt{4x^2 + 4y^2 + 1}}\sqrt{1 + \left(\frac{\partial z}{\partial x}\right)^2 + \left(\frac{\partial z}{\partial y}\right)^2}\,dA\\ &= \iint_{S_{xy}} \frac{2x^3y + 2xy(4 - x^2 - y^2)}{\sqrt{4x^2 + 4y^2 + 1}}\sqrt{1 + (-2x)^2 + (-2y)^2}\,dA\\ &= 2\iint_{S_{xy}} (4xy - xy^3)\,dA\\ &= 2\int_{-2}^{2}\int_{-\sqrt{4-x^2}}^{\sqrt{4-x^2}} (4xy - xy^3)\,dy\,dx\\ &= 2\int_{-2}^{2}\left\{2xy^2 - \frac{xy^4}{4}\right\}_{-\sqrt{4-x^2}}^{\sqrt{4-x^2}} dx\\ &= 0. \ \square\end{aligned}$$

Example 21 Evaluate

$$\oiint_S \mathbf{F}\cdot\hat{\mathbf{n}}\,dS,$$

where $\mathbf{F} = x\hat{\mathbf{i}} + y\hat{\mathbf{j}} + z\hat{\mathbf{k}}$ and $\hat{\mathbf{n}}$ is the unit outward-pointing normal to the surface S enclosing the volume bounded by $x^2 + y^2 = 4$, $z = 0$, $z = 2$.

Solution We can divide S into four parts (Fig. 15.37):

$$\begin{aligned}&S_1: \ z = 0, \quad x^2 + y^2 \leq 4;\\ &S_2: \ z = 2, \quad x^2 + y^2 \leq 4;\\ &S_3: \ y = \sqrt{4 - x^2}, \quad 0 \leq z \leq 2;\\ &S_4: \ y = -\sqrt{4 - x^2}, \quad 0 \leq z \leq 2.\end{aligned}$$

On $S_1, \hat{\mathbf{n}} = -\hat{\mathbf{k}}$; on $S_2, \hat{\mathbf{n}} = \hat{\mathbf{k}}$; and on S_3 and S_4,

$$\hat{\mathbf{n}} = \frac{\nabla(x^2 + y^2 - 4)}{|\nabla(x^2 + y^2 - 4)|} = \frac{(2x, 2y, 0)}{\sqrt{4x^2 + 4y^2}} = \frac{(x, y, 0)}{2}.$$

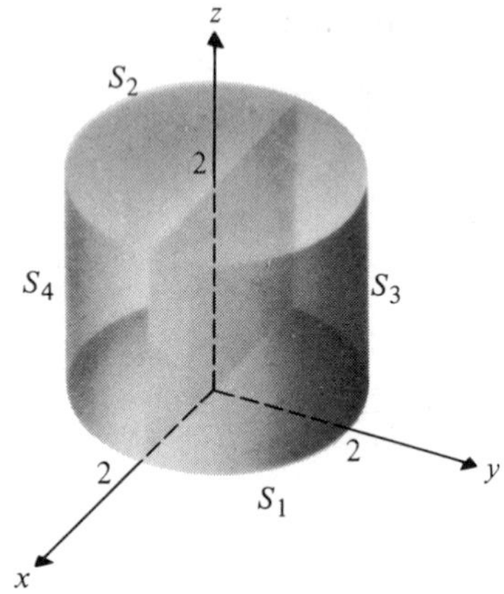

Figure 15.37

We know that S_1 and S_2 project onto S_{xy}: $x^2 + y^2 \leq 4$ in the xy-plane, and S_3 and S_4 project onto the rectangle S_{xz}: $-2 \leq x \leq 2$, $0 \leq z \leq 2$ in the xz-plane. Consequently,

$$\begin{aligned}\oiint_S \mathbf{F}\cdot\hat{\mathbf{n}}\,dS &= \iint_{S_1} -z\,dS + \iint_{S_2} z\,dS + \iint_{S_3}\left(\frac{x^2+y^2}{2}\right)dS + \iint_{S_4}\left(\frac{x^2+y^2}{2}\right)dS\\ &= 0 + \iint_{S_{xy}} 2\sqrt{1}\,dA + \frac{1}{2}\iint_{S_{xz}}(x^2+4-x^2)\sqrt{1+\left(\frac{-x}{\sqrt{4-x^2}}\right)^2}\,dA\\ &\quad + \frac{1}{2}\iint_{S_{xz}}(x^2+4-x^2)\sqrt{1+\left(\frac{x}{\sqrt{4-x^2}}\right)^2}\,dA\\ &= 2\iint_{S_{xy}} dA + 4\iint_{S_{xz}}\frac{2}{\sqrt{4-x^2}}\,dA\\ &= 2(\text{Area of } S_{xy}) + 8\int_{-2}^{2}\int_0^2 \frac{1}{\sqrt{4-x^2}}\,dz\,dx\\ &= 2(4\pi) + 16\int_{-2}^{2}\frac{1}{\sqrt{4-x^2}}\,dx.\end{aligned}$$

If we set $x = 2\sin\theta$, then $dx = 2\cos\theta\,d\theta$ and

$$\oiint_S \mathbf{F}\cdot\hat{\mathbf{n}}\,dS = 8\pi + 16\int_{-\pi/2}^{\pi/2}\frac{1}{2\cos\theta}2\cos\theta\,d\theta = 8\pi + 16\Big\{\theta\Big\}_{-\pi/2}^{\pi/2} = 24\pi. \ \square$$

Note that the notation $\oiint$ is similar to that for line integrals for the surface integral over a closed surface (a closed surface being one that encloses a volume).

Example 22 If a spherical object is submerged in a fluid of specific weight γ, it experiences a buoyant force due to fluid pressure. Show that the magnitude of this force is exactly the weight of the fluid displaced by the object.

Solution Suppose the object is represented by the sphere $x^2 + y^2 + z^2 = R^2$ and the surface of the fluid by the plane $z = h$ $(h > R)$ (Fig. 15.38). If dS is a small area on the surface, then the force due to fluid pressure on dS has magnitude $P\,dS$, where P is pressure, and this force acts normal to the surface of the sphere. If $\hat{\mathbf{n}}$ is the unit inward-pointing normal to the sphere, then the force on dS is

$$(P\,dS)\hat{\mathbf{n}}.$$

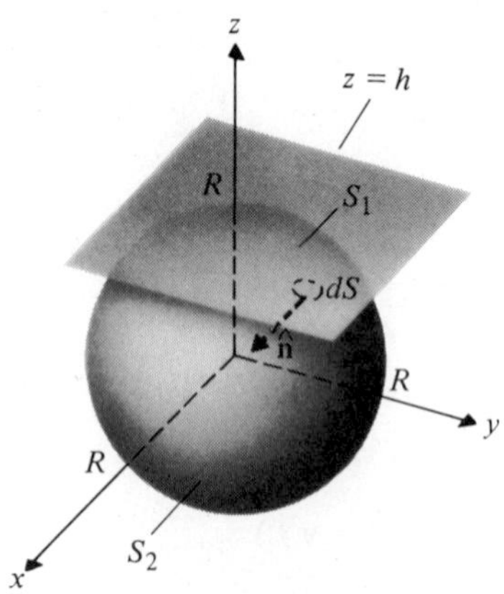

Figure 15.38

Clearly, the resultant force on the sphere will be in the z-direction, the x- and y-components canceling because of the symmetry of the sphere. The z-component of the force on dS is $(P\,dS)\hat{\mathbf{n}}\cdot\hat{\mathbf{k}}$; thus the magnitude of the resultant force on the sphere is

$$\oiint_S P\hat{\mathbf{k}}\cdot\hat{\mathbf{n}}\,dS.$$

If we subdivide S into two parts

$$S_1:\ z = \sqrt{R^2 - x^2 - y^2},$$
$$S_2:\ z = -\sqrt{R^2 - x^2 - y^2},$$

both of which project onto the circle $S_{xy}: x^2 + y^2 \leq R^2$ in the xy-plane, then a normal to each surface is

$$\nabla(x^2 + y^2 + z^2 - R^2) = (2x, 2y, 2z).$$

On S_1 and S_2, then,

$$\hat{\mathbf{n}} = \frac{-(x, y, z)}{\sqrt{x^2 + y^2 + z^2}} = -\frac{1}{R}(x, y, z).$$

Further, on both S_1 and S_2,

$$\sqrt{1 + \left(\frac{\partial z}{\partial x}\right)^2 + \left(\frac{\partial z}{\partial y}\right)^2} = \sqrt{1 + \frac{x^2}{R^2 - x^2 - y^2} + \frac{y^2}{R^2 - x^2 - y^2}}$$

$$= \frac{R}{\sqrt{R^2 - x^2 - y^2}}.$$

Finally, then, we have

$$\oiint_S P\hat{\mathbf{k}}\cdot\hat{\mathbf{n}}\,dS = \iint_{S_1} \gamma(h - z)\left(-\frac{z}{R}\right) dS + \iint_{S_2} \gamma(h - z)\left(-\frac{z}{R}\right) dS$$

$$= -\gamma \iint_{S_{xy}} (h - \sqrt{R^2 - x^2 - y^2})\frac{\sqrt{R^2 - x^2 - y^2}}{R}\frac{R}{\sqrt{R^2 - x^2 - y^2}}\,dA$$

$$+ \gamma \iint_{S_{xy}} (h + \sqrt{R^2 - x^2 - y^2})\frac{\sqrt{R^2 - x^2 - y^2}}{R}\frac{R}{\sqrt{R^2 - x^2 - y^2}}\,dA$$

$$= 2\gamma \iint_{S_{xy}} \sqrt{R^2 - x^2 - y^2}\,dA = 2\gamma \int_{-\pi}^{\pi}\int_0^R \sqrt{R^2 - r^2}\,r\,dr\,d\theta$$

$$= 2\gamma \int_{-\pi}^{\pi} \left\{-\frac{1}{3}(R^2 - r^2)^{3/2}\right\}_0^R d\theta = 2\gamma\left\{\frac{R^3}{3}\right\}(2\pi) = \frac{4}{3}\pi R^3\gamma.$$

Thus the buoyant force is equal to the weight of fluid displaced by the object. □

A proof of this result for arbitrary shapes will be given in Section 15.9.

EXERCISES 15.8

In Exercises 1–13, evaluate the given surface integral.

1. $\iint_S (x\hat{\mathbf{i}} + z\hat{\mathbf{j}})\cdot\hat{\mathbf{n}}\,dS$,
where S is the first octant part of $x + y + z = 3$, and $\hat{\mathbf{n}}$ is the unit normal to S with positive z-component

2. $\iint_S (yz^2\hat{\mathbf{i}} + ye^x\hat{\mathbf{j}} + x\hat{\mathbf{k}})\cdot\hat{\mathbf{n}}\,dS$,
where S is defined by $y = x^2, 0 \leq y \leq 4, 0 \leq z \leq 1$, and $\hat{\mathbf{n}}$ is the unit normal to S with positive y-component

3. $\iint_S (x\hat{\mathbf{i}} + y\hat{\mathbf{j}} + z\hat{\mathbf{k}})\cdot\hat{\mathbf{n}}\,dS$,
where S is the hemisphere $z = \sqrt{1 - x^2 - y^2}$, and $\hat{\mathbf{n}}$ is its upper normal

4. $\iint_S (yz\hat{\mathbf{i}} + zx\hat{\mathbf{j}} + xy\hat{\mathbf{k}})\cdot\hat{\mathbf{n}}\,dS$,
where S is that part of the surface $z = x^2 + y^2$ cut out by the planes $x = 1, x = -1, y = 1, y = -1$, and $\hat{\mathbf{n}}$ is the unit lower normal to S

5. $\oiint_S (z\hat{\mathbf{i}} - x\hat{\mathbf{j}} + y\hat{\mathbf{k}})\cdot\hat{\mathbf{n}}\,dS$,
where S is the surface enclosing the volume defined by $z = \sqrt{4 - x^2 - y^2}, z = 0$, and $\hat{\mathbf{n}}$ is the unit outer normal to S

6. $\iint_S (x\hat{\mathbf{i}} + y\hat{\mathbf{j}})\cdot\hat{\mathbf{n}}\,dS$,
where S is that part of the surface $z = \sqrt{x^2 + y^2}$ below

$z = 1$ and $\hat{\mathbf{n}}$ is the unit normal to S with a negative z-component

7. $\iint_S (xyz\hat{\mathbf{i}} - x\hat{\mathbf{j}} + z\hat{\mathbf{k}}) \cdot \hat{\mathbf{n}}\, dS$,
where S is the smaller part of $x^2 + y^2 = 9$ cut out by $z = 0$, $z = 2$, $y = |x|$, and $\hat{\mathbf{n}}$ is the unit normal to S with positive y-component

8. $\iint_S (x^2y\hat{\mathbf{i}} + xy\hat{\mathbf{j}} + z\hat{\mathbf{k}}) \cdot \hat{\mathbf{n}}\, dS$,
where S is defined by $z = 2 - x^2 - y^2$, $z \geq 0$, and $\hat{\mathbf{n}}$ is the unit normal to S with negative z-component

9. $\oiint_S (yz\hat{\mathbf{i}} + xz\hat{\mathbf{j}} + xy\hat{\mathbf{k}}) \cdot \hat{\mathbf{n}}\, dS$,
where S is the surface enclosing the volume defined by $x = 0$, $x = 2$, $z = 0$, $z = y$, $y + z = 2$, and $\hat{\mathbf{n}}$ is the unit outer normal to S

10. $\iint_S (x\hat{\mathbf{i}} + y\hat{\mathbf{j}}) \cdot \hat{\mathbf{n}}\, dS$,
where S is the surface $x^2 + y^2 + z^2 = 4$, $z \geq 1$, and $\hat{\mathbf{n}}$ is the unit upper normal to S

11. $\oiint_S (x^2\hat{\mathbf{i}} + y^2\hat{\mathbf{j}} + z^2\hat{\mathbf{k}}) \cdot \hat{\mathbf{n}}\, dS$,
where S is the sphere $x^2 + y^2 + z^2 = a^2$, and $\hat{\mathbf{n}}$ is the unit outer normal to S

12. $\iint_S (y\hat{\mathbf{i}} - x\hat{\mathbf{j}} + \hat{\mathbf{k}}) \cdot \hat{\mathbf{n}}\, dS$,
where S is the smaller surface cut from the sphere $x^2 + y^2 + z^2 = 1$ by the plane $y + z = 1$, and $\hat{\mathbf{n}}$ is the unit upper normal to S

13. $\oiint_S \mathbf{F} \cdot \hat{\mathbf{n}}\, dS$,
where $\mathbf{F} = (z^2 - x)\hat{\mathbf{i}} - xy\hat{\mathbf{j}} + 3z\hat{\mathbf{k}}$, S is the surface enclosing the volume defined by $z = 4 - y^2$, $x = 0$, $x = 3$, $z = 0$, and $\hat{\mathbf{n}}$ is the unit outer normal to S

14. Evaluate

$$\iint_S (y\hat{\mathbf{i}} - x\hat{\mathbf{j}} + z\hat{\mathbf{k}}) \cdot \hat{\mathbf{n}}\, dS,$$

where:

a) S is that part of $z = 9 - x^2 - y^2$ cut out by $z = 2y$;

b) S is that part of $z = 2y$ cut out by $z = 9 - x^2 - y^2$;

and $\hat{\mathbf{n}}$ is the unit upper normal to S in each case. (*Hint*: Use polar coordinates with pole at $(0, -1)$.)

15. Show that if a surface S, defined implicitly by the equation $G(x, y, z) = 0$, projects in a one-to-one fashion onto the area S_{xy} in the xy-plane, then

$$\iint_S \mathbf{F} \cdot \hat{\mathbf{n}}\, dS = \pm \iint_{S_{xy}} \frac{\mathbf{F} \cdot \nabla G}{|\partial G/\partial z|}\, dA.$$

16. A circular tube $S: x^2 + z^2 = 1, 0 \leq y \leq 2$ is a model for a part of an artery. Blood flows through the artery and the force per unit area at any point on the arterial wall is given by

$$\mathbf{F} = e^{-y}\hat{\mathbf{n}} + \frac{1}{y^2 + 1}\hat{\mathbf{j}},$$

where $\hat{\mathbf{n}}$ is the unit outer normal to the arterial wall. Blood diffuses through the wall in such a way that if dS is a small area on S, the amount of diffusion through dS in one second is $\mathbf{F} \cdot \hat{\mathbf{n}}\, dS$. Find the total amount of blood leaving the entire arterial wall per second.

17. A beam of light traveling in the positive y-direction has circular cross-section $x^2 + z^2 \leq a^2$. It strikes a surface S: $x^2 + y^2 + z^2 = a^2$, $y \geq a/2$. The intensity of the beam is given by

$$\mathbf{I} = \frac{e^{-t}}{y^2}\hat{\mathbf{j}}, \qquad t \text{ being time.}$$

The absorption of light by a small area dS on S in time dt is $\mathbf{I} \cdot \hat{\mathbf{n}}\, dS\, dt$, where $\hat{\mathbf{n}}$ is the unit normal to S at dS.

a) Find the total absorption over S in time dt.

b) Find the total absorption over S from time $t = 0$ to time $t = 5$.

15.9 THE DIVERGENCE THEOREM

In this section and in Section 15.10 we show that relationships may exist between line integrals and surface integrals and between surface integrals and triple integrals. In Section 15.11 we indicate how useful these relationships can be in the fields of electromagnetic theory and fluid dynamics.

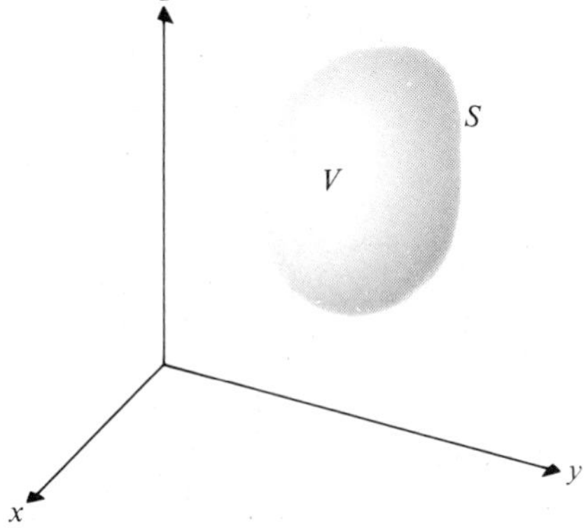

Figure 15.39

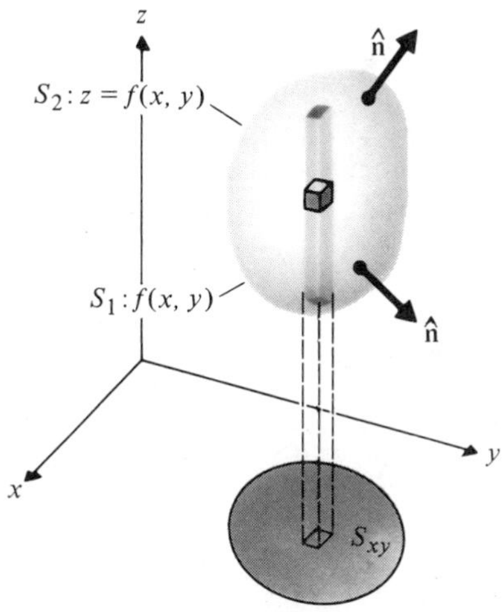

Figure 15.40

The divergence theorem relates certain surface integrals over surfaces that enclose volumes to triple integrals over the enclosed volume. More precisely, we have the following.

▶ **Theorem 15.7** (*Divergence theorem*). Let S be a piecewise-smooth surface enclosing a region V (Fig. 15.39). Let $\mathbf{F}(x, y, z) = L(x, y, z)\hat{\mathbf{i}} + M(x, y, z)\hat{\mathbf{j}} + N(x, y, z)\hat{\mathbf{k}}$ be a vector function whose components L, M, and N have continuous first partial derivatives inside and on S. If $\hat{\mathbf{n}}$ is the unit outer normal to S, then

$$\oiint_S \mathbf{F} \cdot \hat{\mathbf{n}}\, dS = \iiint_V \nabla \cdot \mathbf{F}\, dV, \tag{15.55a}$$

or

$$\oiint_S (L\hat{\mathbf{i}} + M\hat{\mathbf{j}} + N\hat{\mathbf{k}}) \cdot \hat{\mathbf{n}}\, dS = \iiint_V \left(\frac{\partial L}{\partial x} + \frac{\partial M}{\partial y} + \frac{\partial N}{\partial z}\right) dV. \blacktriangleleft \tag{15.55b}$$

Proof We consider first of all a surface S for which any line parallel to any coordinate axis intersects S in at most two points (Fig. 15.40). We can then divide S into an upper and a lower portion, $S_2 : z = f_2(x, y)$ and $S_1 : z = f_1(x, y)$, both of which have the same projection S_{xy} in the xy-plane. We consider the third term in the surface integral on the left-hand side of Eq. (15.55b):

$$\oiint_S N\hat{\mathbf{k}} \cdot \hat{\mathbf{n}}\, dS = \iint_{S_1} N\hat{\mathbf{k}} \cdot \hat{\mathbf{n}}\, dS + \iint_{S_2} N\hat{\mathbf{k}} \cdot \hat{\mathbf{n}}\, dS.$$

On S_1,

$$\hat{\mathbf{n}} = \frac{\left(\dfrac{\partial f_1}{\partial x}, \dfrac{\partial f_1}{\partial y}, -1\right)}{\sqrt{1 + \left(\dfrac{\partial f_1}{\partial x}\right)^2 + \left(\dfrac{\partial f_1}{\partial y}\right)^2}},$$

and on S_2,

$$\hat{\mathbf{n}} = \frac{-\left(\dfrac{\partial f_2}{\partial x}, \dfrac{\partial f_2}{\partial y}, -1\right)}{\sqrt{1 + \left(\dfrac{\partial f_2}{\partial x}\right)^2 + \left(\dfrac{\partial f_2}{\partial y}\right)^2}}.$$

Consequently,

$$\begin{aligned}
\oiint_S N\hat{\mathbf{k}}\cdot\hat{\mathbf{n}}\,dS &= \iint_{S_1} \frac{-N}{\sqrt{1 + \left(\dfrac{\partial f_1}{\partial x}\right)^2 + \left(\dfrac{\partial f_1}{\partial y}\right)^2}}\,dS + \iint_{S_2} \frac{N}{\sqrt{1 + \left(\dfrac{\partial f_2}{\partial x}\right)^2 + \left(\dfrac{\partial f_2}{\partial y}\right)^2}}\,dS \\
&= \iint_{S_{xy}} \frac{-N[x, y, f_1(x, y)]}{\sqrt{1 + \left(\dfrac{\partial f_1}{\partial x}\right)^2 + \left(\dfrac{\partial f_1}{\partial y}\right)^2}} \sqrt{1 + \left(\frac{\partial f_1}{\partial x}\right)^2 + \left(\frac{\partial f_1}{\partial y}\right)^2}\,dA \\
&\quad + \iint_{S_{xy}} \frac{N[x, y, f_2(x, y)]}{\sqrt{1 + \left(\dfrac{\partial f_2}{\partial x}\right)^2 + \left(\dfrac{\partial f_2}{\partial y}\right)^2}} \sqrt{1 + \left(\frac{\partial f_2}{\partial x}\right)^2 + \left(\frac{\partial f_2}{\partial y}\right)^2}\,dA \\
&= \iint_{S_{xy}} \{N[x, y, f_2(x, y)] - N[x, y, f_1(x, y)]\}\,dA.
\end{aligned}$$

On the other hand,

$$\begin{aligned}
\iiint_V \frac{\partial N}{\partial z}\,dV &= \iint_{S_{xy}} \left\{\int_{f_1(x,y)}^{f_2(x,y)} \frac{\partial N}{\partial z}\,dz\right\} dA = \iint_{S_{xy}} \left\{N\right\}_{f_1(x,y)}^{f_2(x,y)} dA \\
&= \iint_{S_{xy}} \{N[x, y, f_2(x, y)] - N[x, y, f_1(x, y)]\}\,dA.
\end{aligned}$$

We have shown then that

$$\oiint_S N\hat{\mathbf{k}}\cdot\hat{\mathbf{n}}\,dS = \iiint_V \frac{\partial N}{\partial z}\,dV.$$

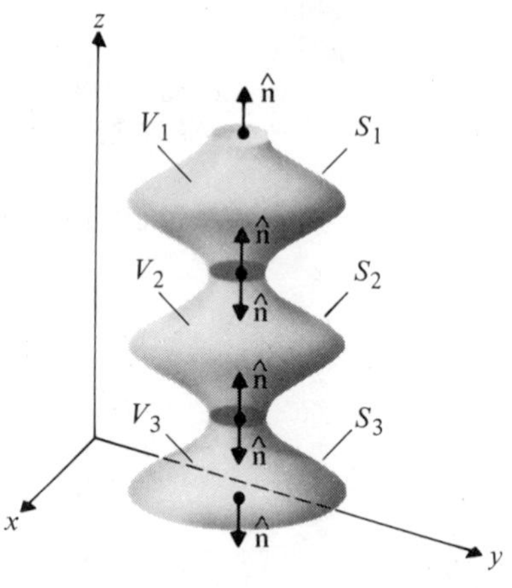

Figure 15.41

Projections of S onto the xz- and yz-planes lead in a similar way to

$$\oiint_S M\hat{\mathbf{j}} \cdot \hat{\mathbf{n}}\, dS = \iiint_V \frac{\partial M}{\partial y}\, dV \quad \text{and} \quad \oiint_S L\hat{\mathbf{i}} \cdot \hat{\mathbf{n}}\, dS = \iiint_V \frac{\partial L}{\partial x}\, dV.$$

By adding these three results, we obtain the divergence theorem for $\mathbf{F}$ and S.

The proof can be extended to general surfaces for which lines parallel to the coordinate axes intersect the surfaces in more than two points. Indeed, most volumes V bounded by surfaces S can be subdivided into n subvolumes V_i whose bounding surfaces S_i do satisfy this condition (Fig. 15.41). For each such subvolume the divergence theorem is now known to apply

$$\oiint_{S_i} \mathbf{F} \cdot \hat{\mathbf{n}}\, dS = \iiint_{V_i} \nabla \cdot \mathbf{F}\, dV, \qquad i = 1, \ldots, n.$$

If these n equations are then added together, we have

$$\sum_{i=1}^{n} \oiint_{S_i} \mathbf{F} \cdot \hat{\mathbf{n}}\, dS = \sum_{i=1}^{n} \iiint_{V_i} \nabla \cdot \mathbf{F}\, dV.$$

The right side is precisely the triple integral of $\nabla \cdot \mathbf{F}$ over V since the V_i constitute V. Figure 15.41 illustrates that when the surface integrals over the S_i are added, contributions from the auxiliary (interior) surfaces cancel in pairs, and the remaining surface integrals add to give the surface integral of $\mathbf{F} \cdot \hat{\mathbf{n}}$ over S. □

We omit a proof for even more general surfaces that cannot be subdivided into a finite number of subsurfaces of this type. The interested reader should consult more advanced books.

Example 23 Verify the divergence theorem for $\mathbf{F} = x^2\hat{\mathbf{i}} + yz\hat{\mathbf{j}} + x\hat{\mathbf{k}}$ if V is the volume bounded by the surfaces $x + y + z = 1$, $x = 0$, $y = 0$, $z = 0$.

Solution The surface S bounding V must be subdivided into four parts:

$$S_1\colon\ z = 0, \qquad S_2\colon\ x = 0, \qquad S_3\colon\ y = 0, \qquad S_4\colon\ x + y + z = 1,$$

as shown in Fig. 15.42. The surface integral of $\mathbf{F} \cdot \hat{\mathbf{n}}$ over S is then the sum of

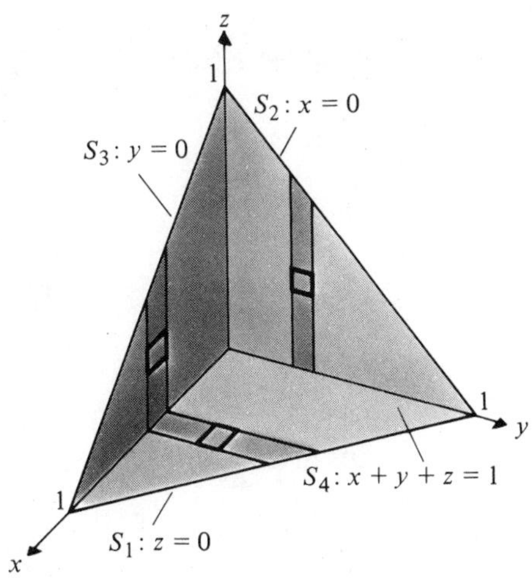

Figure 15.42

the surface integrals over these four parts:

$$\iint_{S_1} \mathbf{F}\cdot\hat{\mathbf{n}}\,dS = \iint_{S_1} \mathbf{F}\cdot(-\hat{\mathbf{k}})\,dS = \iint_{S_1} -x\,dS$$

$$= -\iint_{S_{xy}} x\sqrt{1}\,dA$$

$$= -\int_0^1\int_0^{1-x} x\,dy\,dx = -\int_0^1 \Big\{xy\Big\}_0^{1-x} dx$$

$$= -\int_0^1 x(1-x)\,dx$$

$$= -\left\{\frac{x^2}{2} - \frac{x^3}{3}\right\}_0^1$$

$$= -\frac{1}{6};$$

$$\iint_{S_2} \mathbf{F}\cdot\hat{\mathbf{n}}\,dS = \iint_{S_2} \mathbf{F}\cdot(-\hat{\mathbf{\imath}})\,dS = \iint_{S_2} -x^2\,dS = 0;$$

$$\iint_{S_3} \mathbf{F}\cdot\hat{\mathbf{n}}\,dS = \iint_{S_3} \mathbf{F}\cdot(-\hat{\mathbf{\jmath}})\,dS = \iint_{S_3} -yz\,dS = 0;$$

$$\iint_{S_4} \mathbf{F}\cdot\hat{\mathbf{n}}\,dS = \iint_{S_4} \mathbf{F}\cdot\frac{(1, 1, 1)}{\sqrt{3}}\,dS = \frac{1}{\sqrt{3}}\iint_{S_4}(x^2 + yz + x)\,dS$$

$$= \frac{1}{\sqrt{3}}\iint_{S_{xy}}\{x^2 + y(1 - x - y) + x\}\sqrt{1 + (-1)^2 + (-1)^2}\,dA$$

$$= \int_0^1\int_0^{1-x}\{(x^2 + x) + y(1 - x) - y^2\}\,dy\,dx$$

$$= \int_0^1\left\{(x^2 + x)y + (1 - x)\frac{y^2}{2} - \frac{y^3}{3}\right\}_0^{1-x} dx$$

$$= \int_0^1\left\{x - x^3 + \frac{1}{6}(1 - x)^3\right\}dx$$

$$= \left\{\frac{x^2}{2} - \frac{x^4}{4} - \frac{1}{24}(1 - x)^4\right\}_0^1$$

$$= \frac{7}{24}.$$

Adding these results gives us

$$\oiint_S \mathbf{F}\cdot\hat{\mathbf{n}}\,dS = -\frac{1}{6} + 0 + 0 + \frac{7}{24} = \frac{1}{8}.$$

Alternatively, we have

$$
\begin{aligned}
\iiint_V \nabla \cdot \mathbf{F}\, dV &= \iiint_V (2x + z)\, dV \\
&= \int_0^1 \int_0^{1-x} \int_0^{1-x-y} (2x + z)\, dz\, dy\, dx \\
&= \int_0^1 \int_0^{1-x} \left\{2xz + \frac{z^2}{2}\right\}_0^{1-x-y} dy\, dx \\
&= \frac{1}{2}\int_0^1 \int_0^{1-x} \{4x(1 - x - y) + (1 - x - y)^2\}\, dy\, dx \\
&= \frac{1}{2}\int_0^1 \left\{4x\left(y - xy - \frac{y^2}{2}\right) - \frac{1}{3}(1 - x - y)^3\right\}_0^{1-x} dx \\
&= \frac{1}{2}\int_0^1 \left\{2x(1 - x)^2 + \frac{1}{3}(1 - x)^3\right\} dx \\
&= \frac{1}{2}\int_0^1 \left\{2x - 4x^2 + 2x^3 + \frac{1}{3}(1 - x)^3\right\} dx \\
&= \frac{1}{2}\left\{x^2 - \frac{4x^3}{3} + \frac{x^4}{2} - \frac{1}{12}(1 - x)^4\right\}_0^1 \\
&= \frac{1}{8}.
\end{aligned}
$$

We have therefore verified the divergence theorem for this **F** and S. □

Example 24 Use the divergence theorem to evaluate the surface integral of Example 21.

Solution By the divergence theorem (see Fig. 15.37), we have

$$
\begin{aligned}
\oiint_S \mathbf{F} \cdot \hat{\mathbf{n}}\, dS &= \iiint_V \nabla \cdot \mathbf{F}\, dV \\
&= \iiint_V (1 + 1 + 1)\, dV \\
&= 3 \iiint_V dV \\
&= 3(\text{Volume of } V) = 3(4\pi)(2) = 24\pi. \;\square
\end{aligned}
$$

Example 25 Evaluate

$$\oiint_S (x^3\hat{\mathbf{i}} + y^3\hat{\mathbf{j}} + z^3\hat{\mathbf{k}}) \cdot \hat{\mathbf{n}}\, dS,$$

where $\hat{\mathbf{n}}$ is the unit inner normal to the surface bounding the volume defined by the surfaces $y = x$, $z = 0$, $z = 2$, $y = 0$, $x = 4$.

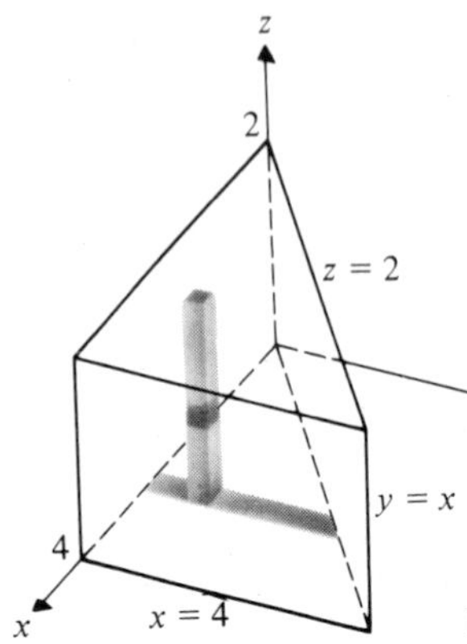

Figure 15.43

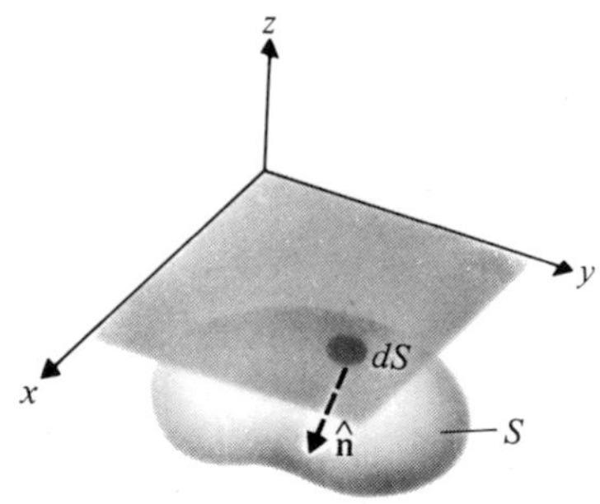

Figure 15.44

Solution By the divergence theorem (see Fig. 15.43), we have

$$\begin{aligned}\oiint_S (x^3\hat{\mathbf{i}} + y^3\hat{\mathbf{j}} + z^3\hat{\mathbf{k}})\cdot\hat{\mathbf{n}}\,dS &= -\iiint_V \nabla\cdot(x^3\hat{\mathbf{i}} + y^3\hat{\mathbf{j}} + z^3\hat{\mathbf{k}})\,dV \\ &= -\iiint_V (3x^2 + 3y^2 + 3z^2)\,dV \\ &= -3\int_0^4\int_0^x\int_0^2 (x^2 + y^2 + z^2)\,dz\,dy\,dx \\ &= -3\int_0^4\int_0^x \left\{(x^2 + y^2)z + \frac{z^3}{3}\right\}_0^2 dy\,dx \\ &= -2\int_0^4\int_0^x \{3(x^2 + y^2) + 4\}\,dy\,dx \\ &= -2\int_0^4 \{y^3 + (3x^2 + 4)y\}_0^x\,dx \\ &= -2\int_0^4 (4x^3 + 4x)\,dx = -2\{x^4 + 2x^2\}_0^4 \\ &= -576. \;\square\end{aligned}$$

Example 26 Prove Archimedes' principle, which states that when an object is submerged in a fluid, it experiences a buoyant force equal to the weight of the fluid displaced.

Solution Suppose the surface of the fluid is taken as the xy-plane ($z = 0$), and the object occupies a region V with bounding surface S (Fig. 15.44). The force due to fluid pressure P on a small area dS on S is $(P\,dS)\hat{\mathbf{n}}$, where $\hat{\mathbf{n}}$ is the unit inner normal to S at dS. If γ is the specific weight of the fluid, then $P = -\gamma z$, and the force on dS is

$$(-\gamma z\,dS)\hat{\mathbf{n}}.$$

The resultant buoyant force is in the positive z-direction (the x- and y-components canceling), so that we require only the z-component of this force,

$$(-\gamma z\,dS)\hat{\mathbf{n}}\cdot\hat{\mathbf{k}}.$$

The total buoyant force must therefore have z-component

$$\oiint_S (-\gamma z)\hat{\mathbf{n}}\cdot\hat{\mathbf{k}}\,dS = \oiint_S (-\gamma z\hat{\mathbf{k}})\cdot\hat{\mathbf{n}}\,dS = \oiint_S (\gamma z\hat{\mathbf{k}})\cdot(-\hat{\mathbf{n}})\,dS,$$

where $-\hat{\mathbf{n}}$ is the unit outer normal to S. If we now use the divergence theorem, we have

$$\oiint_S (-\gamma z)\hat{\mathbf{n}}\cdot\hat{\mathbf{k}}\,dS = \iiint_V \nabla\cdot(\gamma z\hat{\mathbf{k}})\,dV = \iiint_V \gamma\,dV = \gamma(\text{Volume of } V),$$

and this is the weight of the fluid displaced by the object. □

EXERCISES 15.9

In each of Exercises 1–8, use the divergence theorem to evaluate the surface integral.

1. $\oiint_S (x\hat{\mathbf{i}} + y\hat{\mathbf{j}} - 2z\hat{\mathbf{k}})\cdot\hat{\mathbf{n}}\,dS$, where S is the surface bounding the volume defined by $z = 2x^2 + y^2$, $x^2 + y^2 = 3$, $z = 0$

2. $\oiint_S (x^2\hat{\mathbf{i}} + y^2\hat{\mathbf{j}} + z^2\hat{\mathbf{k}})\cdot\hat{\mathbf{n}}\,dS$, where S is the sphere $x^2 + y^2 + z^2 = a^2$, and $\hat{\mathbf{n}}$ is the unit outer normal to S

3. $\oiint_S (yz\hat{\mathbf{i}} + xz\hat{\mathbf{j}} + xy\hat{\mathbf{k}})\cdot\hat{\mathbf{n}}\,dS$, where S is the surface enclosing the volume defined by $x = 0$, $x = 2$, $z = 0$, $z = y$, $y + z = 2$, and $\hat{\mathbf{n}}$ is the unit outer normal to S

4. $\oiint_S \mathbf{F}\cdot\hat{\mathbf{n}}\,dS$, where $\mathbf{F} = (z^2 - x)\hat{\mathbf{i}} - xy\hat{\mathbf{j}} + 3z\hat{\mathbf{k}}$, S is the surface enclosing the volume defined by $z = 4 - y^2$, $x = 0$, $x = 3$, $z = 0$, and $\hat{\mathbf{n}}$ is the unit outer normal to S

5. $\oiint_S \mathbf{F}\cdot\hat{\mathbf{n}}\,dS$, where $\mathbf{F} = (x^2y\hat{\mathbf{i}} + y^2z\hat{\mathbf{j}} + z^2x\hat{\mathbf{k}})/2$, S is the surface bounding the volume in the first octant defined by $x = 0$, $y = 0$, $z = 1$, $z = 0$, $x^2 + y^2 = 1$, and $\hat{\mathbf{n}}$ is the unit inner normal to S

6. $\oiint_S (x\hat{\mathbf{i}} + y\hat{\mathbf{j}} + 2z\hat{\mathbf{k}})\cdot\hat{\mathbf{n}}\,dS$, where S is the surface bounding the volume defined by $z = 2x^2 + y^2$, $x^2 + y^2 = 3$, $z = 0$, and $\hat{\mathbf{n}}$ is the unit outer normal to S

7. $\oiint_S (z\hat{\mathbf{i}} - x\hat{\mathbf{j}} + y\hat{\mathbf{k}})\cdot\hat{\mathbf{n}}\,dS$, where S is the surface enclosing the volume defined by $z = \sqrt{4 - x^2 - y^2}$, $z = 0$, and $\hat{\mathbf{n}}$ is the unit outer normal to S

8. $\oiint_S (2x^2y\hat{\mathbf{i}} - y^2\hat{\mathbf{j}} + 4xz^2\hat{\mathbf{k}})\cdot\hat{\mathbf{n}}\,dS$, where S is the surface enclosing the volume in the first octant defined by $y^2 + z^2 = 9$, $x = 2$, and $\hat{\mathbf{n}}$ is the unit outer normal to S

9. Show that if $\hat{\mathbf{n}}$ is the unit outer normal to a surface S, then the region enclosed by S has volume

$$V = \frac{1}{3}\oiint_S \mathbf{r}\cdot\hat{\mathbf{n}}\,dS, \qquad \mathbf{r} = x\hat{\mathbf{i}} + y\hat{\mathbf{j}} + z\hat{\mathbf{k}}.$$

10. If $\hat{\mathbf{n}}$ is the unit outer normal to a surface S that encloses a region V, show that the area of S can be expressed in the form

$$\text{Area}(S) = \iiint_V \nabla\cdot\hat{\mathbf{n}}\,dV.$$

In Exercises 11–13, evaluate the given surface integral.

11. $\oiint_S \mathbf{F}\cdot\hat{\mathbf{n}}\,dS$, where $\mathbf{F} = (x + y)\hat{\mathbf{i}} + y^3\hat{\mathbf{j}} + x^2z\hat{\mathbf{k}}$, and $\hat{\mathbf{n}}$ is the unit outer normal to the surface S enclosing the volume defined by $x^2 + y^2 - z^2 = 1$, $2z^2 = x^2 + y^2$

12. $\oiint_S \mathbf{F}\cdot\hat{\mathbf{n}}\,dS$, where $\mathbf{F} = (x + y)^2\hat{\mathbf{i}} + x^2y\hat{\mathbf{j}} - x^2z\hat{\mathbf{k}}$, and $\hat{\mathbf{n}}$ is the unit inner normal to the surface S enclosing the volume defined by $z^2 = (1 - x^2 - 2y^2)^2$

13. $\iint_S \mathbf{F}\cdot\hat{\mathbf{n}}\,dS$, where $\mathbf{F} = (y^3 + x^2y)\hat{\mathbf{i}} + (x^3 - xy^2)\hat{\mathbf{j}} + z\hat{\mathbf{k}}$, and $\hat{\mathbf{n}}$ is the unit upper normal to the surface S: $z = \sqrt{1 - x^2 - y^2}$

14. How would you prove Archimedes' principle in the case that an object is only partially submerged? (See Example 26.)

15. Is Green's theorem related to the divergence theorem? (See Exercise 13 in Section 15.6.)

16. If V is a region bounded by a closed surface S, and $\mathbf{B} = \nabla \times \mathbf{A}$, show that

$$\oiint_S \mathbf{B}\cdot\hat{\mathbf{n}}\,dS = 0.$$

In Exercises 17–19, assume that $P(x, y, z)$ and $Q(x, y, z)$ have continuous first and second partial derivatives in the region V consisting of a closed surface S and its interior. Let $\hat{\mathbf{n}}$ be the unit outer normal to S.

17. Show that

$$\oiint_S \nabla P\cdot\hat{\mathbf{n}}\,dS = \iiint_V \nabla^2 P\,dV,$$

where

$$\nabla^2 P = \frac{\partial^2 P}{\partial x^2} + \frac{\partial^2 P}{\partial y^2} + \frac{\partial^2 P}{\partial z^2}.$$

What do you conclude if $P(x, y, z)$ satisfies Laplace's equation in V?

18. Show that

$$\oiint_S P\nabla Q\cdot\ \ dS = \iiint_V (P\nabla^2 Q + \nabla P\cdot\nabla Q)\,dV.$$

This result is called *Green's first identity*.

19. Prove that

$$\oiint_S (P\nabla Q - Q\nabla P)\cdot\hat{\mathbf{n}}\,dS = \iiint_V (P\nabla^2 Q - Q\nabla^2 P)\,dV,$$

(Green's second identity).

20. Compare Exercises 17–19 with Exercises 26–28 of Section 15.6.

***21.** Let S be a closed surface, and $\hat{\mathbf{n}}$ be the unit outer normal to S. If $\mathbf{r}_0 = x_0\hat{\mathbf{i}} + y_0\hat{\mathbf{j}} + z_0\hat{\mathbf{k}}$ is the position vector of some fixed point P_0, show that

$$\oiint_S \frac{\mathbf{r}-\mathbf{r}_0}{|\mathbf{r}-\mathbf{r}_0|^3}\cdot\hat{\mathbf{n}}\,dS = \begin{cases} 0, & \text{if } S \text{ does not enclose } P_0,\\ 4\pi, & \text{if } S \text{ does enclose } P_0.\end{cases}$$

22. In this problem we use the result of Exercise 21 to prove Gauss's law for electrostatic fields. Let S be a closed surface containing n point charges q_i $(i = 1, \ldots, n)$ at points $\mathbf{r}_i = (x_i, y_i, z_i)$ in its interior. According to Coulomb's law, the electric field $\mathbf{E}$ at a point $\mathbf{r} = (x, y, z)$ due to this charge distribution is defined by $\mathbf{E} = -\nabla V$, where V, the potential at $\mathbf{r}$, is

$$V(x, y, z) = \sum_{i=1}^{n} \frac{q_i}{4\pi\varepsilon_0|\mathbf{r}-\mathbf{r}_i|}.$$

a) Verify that

$$\mathbf{E} = \sum_{i=1}^{n} \frac{q_i(\mathbf{r}-\mathbf{r}_i)}{4\pi\varepsilon_0|\mathbf{r}-\mathbf{r}_i|^3}.$$

b) Now use the result of Exercise 21 to show that if $\hat{\mathbf{n}}$ is the unit outer normal to S, then

$$\oiint_S \mathbf{E}\cdot\hat{\mathbf{n}}\,dS = \frac{Q}{\varepsilon_0},$$

where Q is the total charge inside S. This result is Gauss's law.

15.10 STOKES'S THEOREM

Stokes's theorem relates certain line integrals around closed curves to surface integrals over surfaces that have the curves as boundaries.

▶ **Theorem 15.8** (*Stokes's theorem*). Let C be a closed, piecewise-smooth, non–self-intersecting curve and let S be a piecewise-smooth (orientable) surface with C as boundary (Fig. 15.45). Let $\mathbf{F}(x, y, z) = P(x, y, z)\hat{\mathbf{i}} + Q(x, y, z)\hat{\mathbf{j}} + R(x, y, z)\hat{\mathbf{k}}$ be a vector function whose components P, Q, and R have continuous first partial derivatives in a region that contains

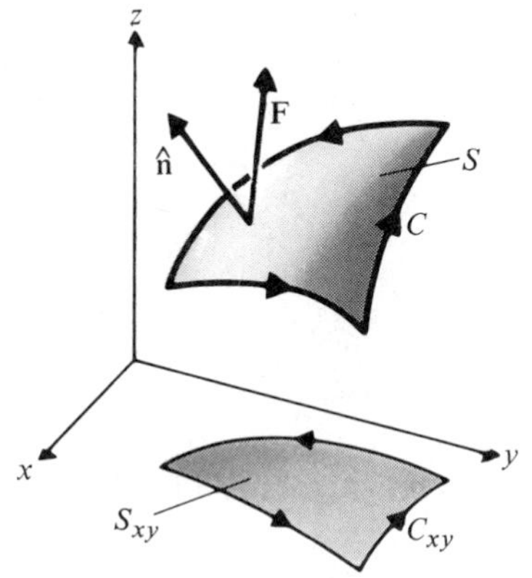

Figure 15.45

S and C in its interior. Then

$$\oint_C \mathbf{F}\cdot d\mathbf{r} = \iint_S (\nabla \times \mathbf{F})\cdot\hat{\mathbf{n}}\, dS, \tag{15.56a}$$

or

$$\oint_C P\,dx + Q\,dy + R\,dz$$
$$= \iint_S \left\{\left(\frac{\partial R}{\partial y} - \frac{\partial Q}{\partial z}\right)\hat{\mathbf{i}} + \left(\frac{\partial P}{\partial z} - \frac{\partial R}{\partial x}\right)\hat{\mathbf{j}} + \left(\frac{\partial Q}{\partial x} - \frac{\partial P}{\partial y}\right)\hat{\mathbf{k}}\right\}\cdot\hat{\mathbf{n}}\, dS, \tag{15.56b}$$

where $\hat{\mathbf{n}}$ is the unit normal to S chosen in the following way: If when moving along C the surface S is on the left-hand side, then $\hat{\mathbf{n}}$ must be chosen as the unit normal on that side of S. On the other hand, if when moving along C, the surface is on the right, then $\hat{\mathbf{n}}$ must be chosen on the opposite side of S. ◀

For a complete statement of Stokes's theorem, S must be an "orientable" surface, a property we have not yet defined. All surfaces discussed in this text are orientable, with the exception of the Möbius strip mentioned in Exercise 13. In general terms, a surface is orientable if it has two distinct sides, and it is this property that permits us to assign a unit normal to S in an unambiguous way.

Proof We first consider a surface S that projects in a one-to-one fashion onto each of the three coordinate planes. Because S projects one-to-one onto some region S_{xy} in the xy-plane, we can take the equation for S in the form $z = f(x, y)$. If the direction along C is as indicated in Fig. 15.45, and C_{xy} is the projection of C on the xy-plane, then

$$\oint_C P\,dx = \oint_{C_{xy}} P[x, y, f(x, y)]\,dx.$$

If we use Green's theorem on this line integral around C_{xy}, we have

$$\oint_C P\,dx = \iint_{S_{xy}} -\left\{\frac{\partial P}{\partial y} + \frac{\partial P}{\partial z}\frac{\partial z}{\partial y}\right\} dA.$$

On the other hand, since a unit normal to S is

$$\hat{\mathbf{n}} = \frac{\left(-\frac{\partial z}{\partial x}, -\frac{\partial z}{\partial y}, 1\right)}{\sqrt{1 + \left(\frac{\partial z}{\partial x}\right)^2 + \left(\frac{\partial z}{\partial y}\right)^2}},$$

it also follows that

$$\iint_S \left(\frac{\partial P}{\partial z}\hat{\mathbf{j}} - \frac{\partial P}{\partial y}\hat{\mathbf{k}}\right)\cdot\hat{\mathbf{n}}\,dS = \iint_S \frac{-\frac{\partial P}{\partial z}\frac{\partial z}{\partial y} - \frac{\partial P}{\partial y}}{\sqrt{1 + \left(\frac{\partial z}{\partial x}\right)^2 + \left(\frac{\partial z}{\partial y}\right)^2}}\,dS$$

$$= \iint_{S_{xy}} \frac{-\frac{\partial P}{\partial z}\frac{\partial z}{\partial y} - \frac{\partial P}{\partial y}}{\sqrt{1 + \left(\frac{\partial z}{\partial x}\right)^2 + \left(\frac{\partial z}{\partial y}\right)^2}}\sqrt{1 + \left(\frac{\partial z}{\partial x}\right)^2 + \left(\frac{\partial z}{\partial y}\right)^2}\,dA$$

$$= \iint_{S_{xy}} -\left\{\frac{\partial P}{\partial y} + \frac{\partial P}{\partial z}\frac{\partial z}{\partial y}\right\}dA.$$

We have shown, then, that

$$\oint_C P\,dx = \iint_S \left(\frac{\partial P}{\partial z}\hat{\mathbf{j}} - \frac{\partial P}{\partial y}\hat{\mathbf{k}}\right)\cdot\hat{\mathbf{n}}\,dS.$$

By projecting C and S onto the xz- and yz-planes, we can show similarly that

$$\oint_C R\,dz = \iint_S \left(\frac{\partial R}{\partial y}\hat{\mathbf{i}} - \frac{\partial R}{\partial x}\hat{\mathbf{j}}\right)\cdot\hat{\mathbf{n}}\,dS$$

and

$$\oint_C Q\,dy = \iint_S \left(\frac{\partial Q}{\partial x}\hat{\mathbf{k}} - \frac{\partial Q}{\partial z}\hat{\mathbf{i}}\right)\cdot\hat{\mathbf{n}}\,dS.$$

Addition of these three results gives Stokes's theorem for $\mathbf{F}$ and S.

The proof can be extended to general curves and surfaces that do not project in a one-to-one fashion onto all three coordinate planes. Most surfaces S with bounding curves C can be subdivided into n subsurfaces S_i with bounding curves C_i that do satisfy this condition (Fig. 15.46). For each such subsurface, Stokes's theorem applies

$$\oint_{C_i} \mathbf{F}\cdot d\mathbf{r} = \iint_{S_i} (\nabla \times \mathbf{F})\cdot\hat{\mathbf{n}}\,dS, \qquad i = 1, \ldots, n.$$

If these n equations are now added together, we have

$$\sum_{i=1}^{n} \oint_{C_i} \mathbf{F}\cdot d\mathbf{r} = \sum_{i=1}^{n} \iint_{S_i} (\nabla \times \mathbf{F})\cdot\hat{\mathbf{n}}\,dS.$$

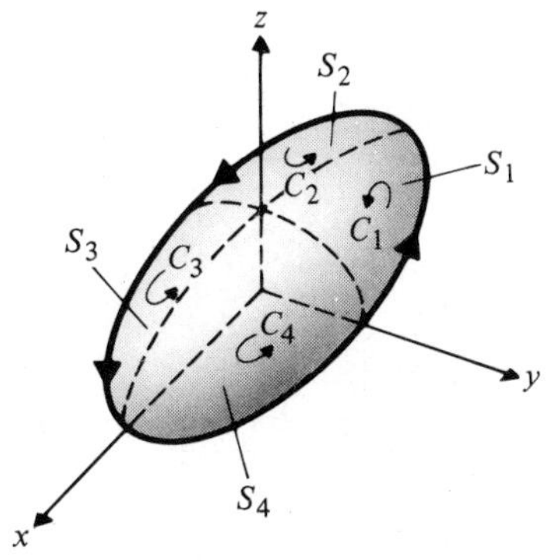

Figure 15.46

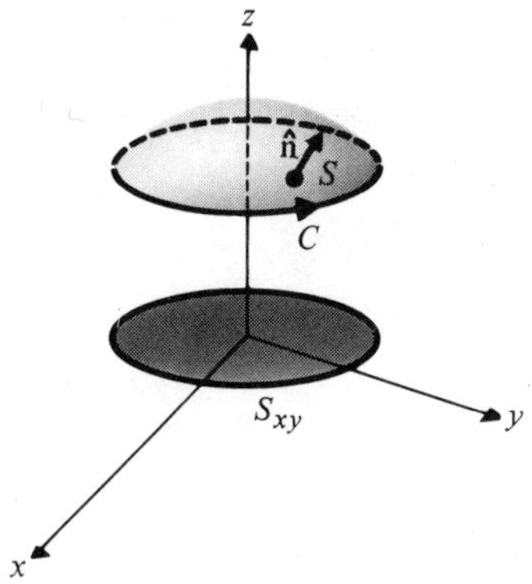

Figure 15.47

Since the S_i constitute S, the right-hand side of this equation is the surface integral of $(\nabla \times \mathbf{F}) \cdot \hat{\mathbf{n}}$ over S. Figure 15.46 illustrates that when the line integrals over the C_i are added, the contributions from the auxiliary (interior) curves cancel in pairs, and the remaining line integrals give the line integral of $\mathbf{F} \cdot d\mathbf{r}$ along C. □

For general surfaces that cannot be divided into a finite number of subsurfaces of this type, the reader should consult a more advanced book.

Note that Green's theorem is a special case of Stokes's theorem. For if $\mathbf{F} = P(x, y)\hat{\mathbf{i}} + Q(x, y)\hat{\mathbf{j}}$, and C is a closed curve in the xy-plane, then by Stokes's theorem

$$\oint_C P\,dx + Q\,dy = \iint_S \left(\frac{\partial Q}{\partial x} - \frac{\partial P}{\partial y}\right)\hat{\mathbf{k}} \cdot \hat{\mathbf{n}}\,dS,$$

where S is any surface for which C is the boundary. If we choose S as that part of the xy-plane bounded by C, then $\hat{\mathbf{n}} = \hat{\mathbf{k}}$ and

$$\oint_C P\,dx + Q\,dy = \iint_S \left(\frac{\partial Q}{\partial x} - \frac{\partial P}{\partial y}\right) dA.$$

Note too that if D is a rectangular parallelepiped with faces parallel to the coordinate planes, and if $\nabla \times \mathbf{F} = \mathbf{0}$ in D, then

$$\oint_C \mathbf{F} \cdot d\mathbf{r} = 0$$

for every closed, piecewise-smooth curve in D. But this implies that the line integral is independent of path in D, and therefore there exists a function $\phi(x, y, z)$ such that $\mathbf{F} = \nabla\phi$. This is a proof of the sufficiency part of Theorem 15.1.

Example 27 Verify Stokes's theorem if $\mathbf{F} = x^2\hat{\mathbf{i}} + x\hat{\mathbf{j}} + xyz\hat{\mathbf{k}}$, and S is that part of the sphere $x^2 + y^2 + z^2 = 4$ above the plane $z = 1$.

Solution If we choose $\hat{\mathbf{n}}$ as the upper normal to S, then C, the boundary of S, must be traversed in the direction shown in Fig. 15.47. (If $\hat{\mathbf{n}}$ is chosen as the lower normal, then C must be traversed in the opposite direction.) Since parametric equations for C are

$$x = \sqrt{3}\cos t, \qquad y = \sqrt{3}\sin t, \qquad z = 1, \qquad 0 \le t \le 2\pi,$$

then

$$\begin{aligned}
\oint_C \mathbf{F}\cdot d\mathbf{r} &= \oint_C x^2\,dx + x\,dy + xyz\,dz \\
&= \int_0^{2\pi} \{3\cos^2 t(-\sqrt{3}\sin t\,dt) + \sqrt{3}\cos t(\sqrt{3}\cos t\,dt)\} \\
&= \int_0^{2\pi} \left\{-3\sqrt{3}\cos^2 t\sin t + \frac{3}{2}(1+\cos 2t)\right\} dt \\
&= \left\{\sqrt{3}\cos^3 t + \frac{3t}{2} + \frac{3\sin 2t}{4}\right\}_0^{2\pi} \\
&= 3\pi.
\end{aligned}$$

Alternatively,

$$\begin{aligned}
\iint_S (\nabla\times\mathbf{F})\cdot\hat{\mathbf{n}}\,dS &= \iint_S (xz\hat{\mathbf{i}} - yz\hat{\mathbf{j}} + \hat{\mathbf{k}})\cdot\frac{(2x, 2y, 2z)}{\sqrt{4x^2+4y^2+4z^2}}\,dS \\
&= \iint_S \frac{x^2z - y^2z + z}{\sqrt{x^2+y^2+z^2}}\,dS \\
&= \iint_{S_{xy}} \frac{z(x^2-y^2+1)}{2}\sqrt{1 + \left(\frac{\partial z}{\partial x}\right)^2 + \left(\frac{\partial z}{\partial y}\right)^2}\,dA \\
&= \frac{1}{2}\iint_{S_{xy}} z(x^2-y^2+1)\sqrt{1+(-x/z)^2+(-y/z)^2}\,dA \\
&= \frac{1}{2}\iint_{S_{xy}} z(x^2-y^2+1)\frac{\sqrt{x^2+y^2+z^2}}{z}\,dA \\
&= \iint_{S_{xy}} (x^2-y^2+1)\,dA.
\end{aligned}$$

If we use polar coordinates to evaluate this double integral over S_{xy}: $x^2+y^2\le 3$, we have

$$\begin{aligned}
\iint_S (\nabla\times\mathbf{F})\cdot\hat{\mathbf{n}}\,dS &= \int_{-\pi}^{\pi}\int_0^{\sqrt{3}} (r^2\cos^2\theta - r^2\sin^2\theta + 1)r\,dr\,d\theta \\
&= \int_{-\pi}^{\pi} \left\{\frac{r^4}{4}(\cos^2\theta - \sin^2\theta) + \frac{r^2}{2}\right\}_0^{\sqrt{3}} d\theta \\
&= \int_{-\pi}^{\pi} \left\{\frac{9}{4}\cos 2\theta + \frac{3}{2}\right\} d\theta \\
&= \left\{\frac{9}{8}\sin 2\theta + \frac{3\theta}{2}\right\}_{-\pi}^{\pi} = 3\pi. \;\square
\end{aligned}$$

Example 28 Evaluate

$$\oint_C 2xy^3\,dx + 3x^2y^2\,dy + (2z + x)\,dz,$$

where C is the broken line joining $A(2, 0, 0)$ to $B(0, 1, 0)$ to $D(0, 0, 1)$ to A.

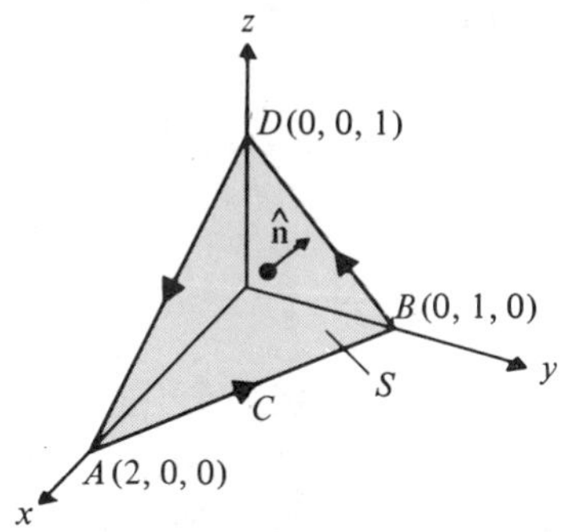

Figure 15.48

Solution By Stokes's theorem,

$$\oint_C 2xy^3\,dx + 3x^2y^2\,dy + (2z + x)\,dz = \iint_S \nabla \times (2xy^3, 3x^2y^2, 2z + x)\cdot\hat{\mathbf{n}}\,dS,$$

where S is any surface with C as boundary. If we choose S as the flat triangle bounded by C (Fig. 15.48), then a normal vector to S is

$$\overrightarrow{\mathbf{BD}} \times \overrightarrow{\mathbf{BA}} = \begin{vmatrix} \hat{\mathbf{i}} & \hat{\mathbf{j}} & \hat{\mathbf{k}} \\ 0 & -1 & 1 \\ 2 & -1 & 0 \end{vmatrix} = (1, 2, 2),$$

and therefore

$$\hat{\mathbf{n}} = \frac{(1, 2, 2)}{3}.$$

Since

$$\nabla \times (2xy^3, 3x^2y^2, 2z + x) = \begin{vmatrix} \hat{\mathbf{i}} & \hat{\mathbf{j}} & \hat{\mathbf{k}} \\ \partial/\partial x & \partial/\partial y & \partial/\partial z \\ 2xy^3 & 3x^2y^2 & 2z + x \end{vmatrix} = -\hat{\mathbf{j}},$$

it follows that

$$\nabla \times (2xy^3, 3x^2y^2, 2z + x)\cdot\hat{\mathbf{n}} = -\hat{\mathbf{j}}\cdot\frac{(1, 2, 2)}{3} = -\frac{2}{3}$$

and

$$\oint_C 2xy^3\,dx + 3x^2y^2\,dy + (2z + x)\,dz = \iint_S -\frac{2}{3}\,dS = -\frac{2}{3}(\text{Area of } S).$$

But from Eq. (12.44), the area of triangle S is

$$\frac{1}{2}|\overrightarrow{\mathbf{BD}} \times \overrightarrow{\mathbf{BA}}| = \frac{1}{2}|(1, 2, 2)| = \frac{3}{2}.$$

Finally, then,

$$\oint_C 2xy^3\,dx + 3x^2y^2\,dy + (2z + x)\,dz = -\frac{2}{3}\left(\frac{3}{2}\right) = -1. \quad \square$$

EXERCISES 15.10

In each of Exercises 1–10, use Stokes's theorem to evaluate the line integral.

1. $\oint_C x^2y\,dx + y^2z\,dy + z^2x\,dz$, where C is the curve $z = x^2 + y^2$, $x^2 + y^2 = 4$, directed counterclockwise as viewed from the origin

2. $\oint_C y^2\,dx + xy\,dy + xz\,dz$, where C is the curve $x^2 + y^2 = 2y$, $y = z$, directed so that y increases when x is positive

3. $\oint_C \mathbf{F}\cdot d\mathbf{r}$, where $\mathbf{F} = (xyz + 2yz)\hat{\mathbf{i}} + xz\hat{\mathbf{j}} + 2xy\hat{\mathbf{k}}$, and C is the curve $z = 1$, $x^2 + y^2 + z^2 = 4$, directed clockwise as viewed from the origin

4. $\oint_C \mathbf{F}\cdot d\mathbf{r}$, where $\mathbf{F} = (2xy + y)\hat{\mathbf{i}} + (x^2 + xy - 3y)\hat{\mathbf{j}} + 2xz\hat{\mathbf{k}}$, and C is the curve $z = \sqrt{x^2 + y^2}$, $z = 4$

5. $\oint_C x^2\,dx + y^2\,dy + (x^2 + y^2)\,dz$, where C is the boundary of that part of the plane $x + y + z = 1$ in the first octant, directed counterclockwise as viewed from the origin

6. $\oint_C y\,dx + x\,dy + (x^2 + y^2 + z^2)\,dz$, where C is the curve $x^2 + y^2 = 1$, $z = xy$, directed clockwise as viewed from the point $(0, 0, 1)$

7. $\oint_C zy^2\,dx + xy\,dy + (y^2 + z^2)\,dz$, where C is the curve $x^2 + z^2 = 9$, $y = \sqrt{x^2 + z^2}$, directed counterclockwise as viewed from the origin

8. $\oint_C y\,dx + z\,dy + x\,dz$, where C is the curve $x + y = 2b$, $x^2 + y^2 + z^2 = 2b(x + y)$, directed clockwise as viewed from the origin

9. $\oint_C y^2\,dx + (x + y)\,dy + yz\,dz$, where C is the curve $x^2 + y^2 = 2$, $x + y + z = 2$, directed clockwise as viewed from the origin

10. $\oint_C (x + y)^2\,dx + (x + y)^2\,dy + yz^3\,dz$, where C is the curve $z = \sqrt{x^2 + y^2}$, $(x - 1)^2 + y^2 = 1$

11. Evaluate the line integral
$$\oint_C 2x^2y\,dx - yz\,dy + xz\,dz,$$
where C is the curve $x^2 + y^2 + z^2 = 4$, $z = \sqrt{3(x^2 + y^2)}$, directed clockwise as viewed from the origin, in four ways:
    a) directly as a line integral;
    b) using Stokes's theorem with S as that part of $z = \sqrt{4 - x^2 - y^2}$ bounded by C;
    c) using Stokes's theorem with S as that part of $z = \sqrt{3(x^2 + y^2)}$ bounded by C;
    d) using Stokes's theorem with S as that part of $z = \sqrt{3}$ bounded by C.

12. Let S_1 be that part of $x^2 + y^2 + z^2 = 1$ above the xy-plane and S_2 be that part of $z = 1 - x^2 - y^2$ above the xy-plane. Show that if $\hat{\mathbf{n}}_1$ and $\hat{\mathbf{n}}_2$ are the unit upper normals to these surfaces, and $\mathbf{F}$ is a vector field defined on both S_1 and S_2, then
$$\iint_{S_1} (\nabla \times \mathbf{F})\cdot\hat{\mathbf{n}}_1\,dS = \iint_{S_2} (\nabla \times \mathbf{F})\cdot\hat{\mathbf{n}}_2\,dS.$$

13. In this exercise we construct a nonorientable surface. Take a rectangular strip of paper and label its corners consecutively as A, B, C, and D (Fig. 15.49). Join edge AB to CD by joining A to C and B to D (necessitating a half twist in the paper). This surface is called a *Möbius strip.* Show that one can start at any point on the strip with a unit normal vector and move the vector continuously around the strip without ever touching its edge in such a way as to end up back at the same point with the unit normal pointing in the opposite direction.

Figure 15.49

15.11 FLUX AND CIRCULATION

In many branches of engineering and physics we encounter the concepts of *flux* and *circulation.* In this section we discuss the relationships between flux and divergence and between circulation and curl. We find a physical setting previously mentioned in Section 15.1 most useful in developing these ideas.

Fluid Flow

Consider a gas flowing through a region D of space. At time t and point $P(x, y, z)$ in D, gas will be flowing through P with some velocity $\mathbf{v}(x, y, z, t)$. If A is a unit area around P perpendicular to $\mathbf{v}$ (Fig. 15.50) and $\rho(x, y, z, t)$ is the density of the gas at P, then the amount of gas crossing A per unit time is $\rho\mathbf{v}$. At every point P in D, then, the vector $\rho\mathbf{v}$ is such that its direction $\mathbf{v}$ gives the velocity of gas flow, and its magnitude $\rho|\mathbf{v}|$ describes the mass of gas flowing in that direction per unit time.

Consider now some surface S in D (Fig. 15.51). If $\hat{\mathbf{n}}$ is a unit normal to S, then $\rho\mathbf{v}\cdot\hat{\mathbf{n}}$ is the component of $\rho\mathbf{v}$ normal to the surface S. If dS is an element of area on S, then $\rho\mathbf{v}\cdot\hat{\mathbf{n}}\,dS$ describes the mass of gas flowing through dS per unit time. Consequently,

$$\iint_S \rho\mathbf{v}\cdot\hat{\mathbf{n}}\,dS$$

is the mass of gas flowing through S per unit time. This quantity is called the *flux* for the surface S. It is the mass of gas flowing through S per unit time.

If S is a closed surface (Fig. 15.52) and $\hat{\mathbf{n}}$ is the unit outer normal to S, then

$$\oiint_S \rho\mathbf{v}\cdot\hat{\mathbf{n}}\,dS$$

is the mass of gas flowing out of the surface S per unit time. If this flux (for closed S) is positive, then there is a net outward flow of gas through S (i.e., more gas is leaving the volume bounded by S than is entering); if the flux is negative, the net flow is inward.

If we apply the divergence theorem to the flux integral over the closed surface S, we have

$$\oiint_S \rho\mathbf{v}\cdot\hat{\mathbf{n}}\,dS = \iiint_V \nabla\cdot(\rho\mathbf{v})\,dV. \tag{15.57}$$

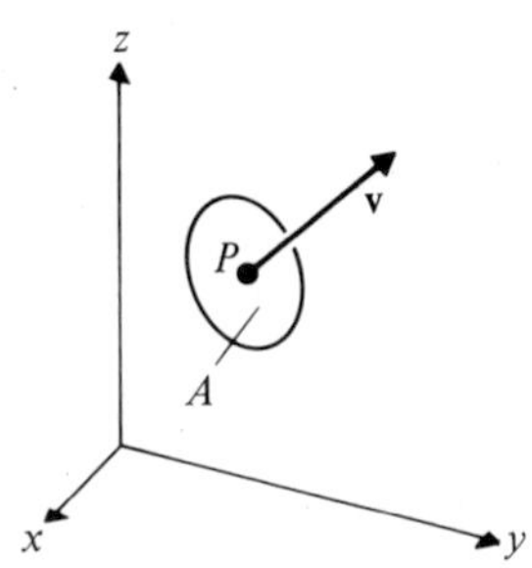

Figure 15.50

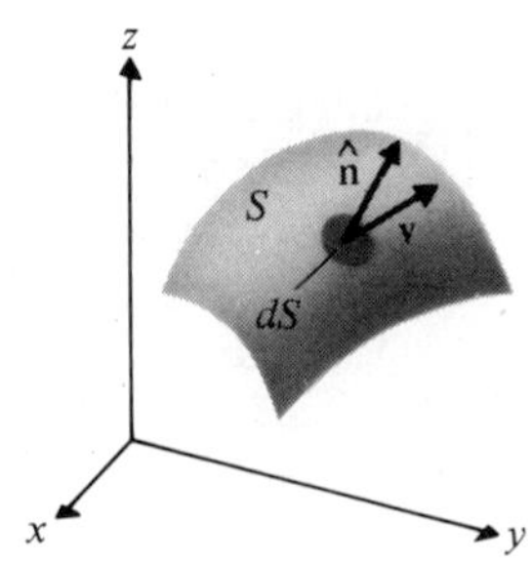

Figure 15.51

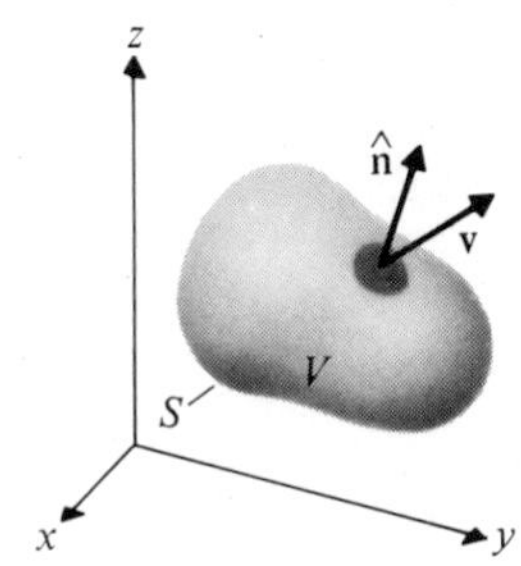

Figure 15.52

Now the flux (on the left-hand side of this equation) is the mass of gas per unit time leaving S. In order for the right-hand side to represent the same quantity, $\nabla \cdot (\rho \mathbf{v})$ must be interpreted as the mass of gas leaving unit volume per unit time, because then $\nabla \cdot (\rho \mathbf{v})\, dV$ represents the mass per unit time leaving dV, and the triple integral is the mass per unit time leaving V.

We have obtained, therefore, an interpretation of the divergence of $\rho \mathbf{v}$. The divergence of $\rho \mathbf{v}$ is the flux per unit volume per unit time at a point: the mass of gas leaving unit volume in unit time. We can use this idea of flux to derive the "equation of continuity" for fluid flow. The triple integral

$$\iiint_V \frac{\partial \rho}{\partial t}\, dV = \frac{\partial}{\partial t} \iiint_V \rho\, dV$$

measures the time rate of change of mass in a volume V. If this triple integral is positive, then there is a net inward flow of mass; if it is negative, the net flow is outward. We conclude, therefore, that this triple integral must be the negative of the flux for the volume V; i.e.,

$$\iiint_V \frac{\partial \rho}{\partial t}\, dV = - \iiint_V \nabla \cdot (\rho \mathbf{v})\, dV,$$

or

$$\iiint_V \left(\nabla \cdot (\rho \mathbf{v}) + \frac{\partial \rho}{\partial t} \right) dV = 0.$$

If $\nabla \cdot (\rho \mathbf{v})$ and $\partial \rho / \partial t$ are continuous functions, then this equation can hold for arbitrary volume V only if

$$\nabla \cdot (\rho \mathbf{v}) + \frac{\partial \rho}{\partial t} = 0. \tag{15.58}$$

This equation, called the *equation of continuity*, expresses conservation of mass.

If C is a closed curve in the flow region D, then the circulation of the flow for the curve C is defined by

$$\Gamma = \oint_C \mathbf{v} \cdot d\mathbf{r}. \tag{15.59}$$

To obtain an intuitive feeling for Γ, we consider two very simple two-dimensional flows. First suppose $\mathbf{v} = x\hat{\mathbf{i}} + y\hat{\mathbf{j}}$, so that all particles of gas flow along radial lines directed away from the origin (Fig. 15.53). In this case, the line integral defining Γ is independent of path and $\Gamma = 0$ for any curve whatsoever.

Second, suppose $\mathbf{v} = -y\hat{\mathbf{i}} + x\hat{\mathbf{j}}$, so that all particles of the gas flow counterclockwise around circles centered at the origin (Fig. 15.54). In this case Γ does not generally vanish. In particular, if C is the circle $x^2 + y^2 = r^2$, then $\mathbf{v}$ and $d\mathbf{r}$ are parallel, and

$$\Gamma = \oint_C \mathbf{v} \cdot d\mathbf{r} = \oint_C |\mathbf{v}|\, ds = \oint_C \sqrt{y^2 + x^2}\, ds = \oint_C r\, ds = 2\pi r^2.$$

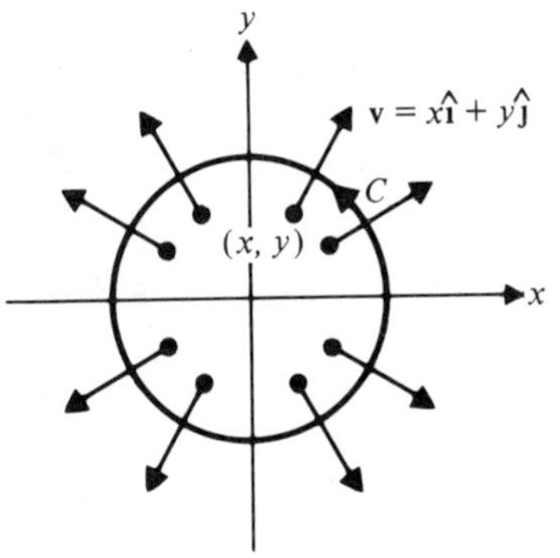

Figure 15.53

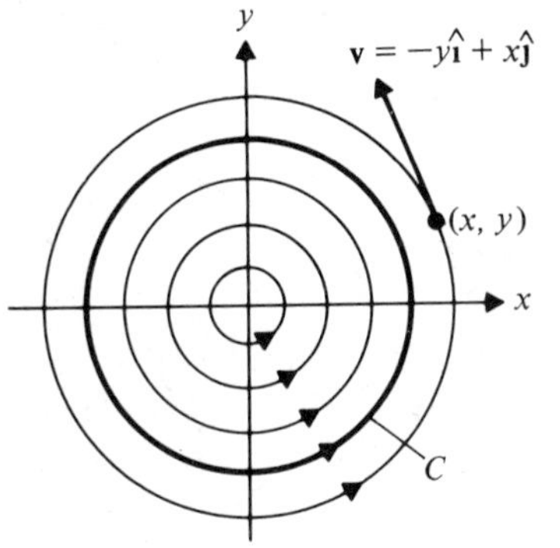

Figure 15.54

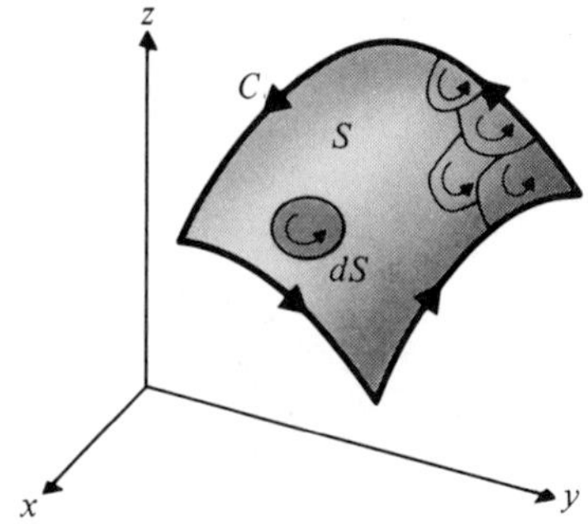

Figure 15.55

These two flow patterns indicate perhaps that circulation is a measure of the tendency for the flow to be circulatory. If we apply Stokes's theorem to the circulation integral for the closed curve C, we have

$$\oint_C \mathbf{v} \cdot d\mathbf{r} = \iint_S (\nabla \times \mathbf{v}) \cdot \hat{\mathbf{n}} \, dS, \qquad (15.60)$$

where S is any surface in the flow with boundary C. If the right-hand side of this equation is to represent the circulation for C also, then $(\nabla \times \mathbf{v}) \cdot \hat{\mathbf{n}} \, dS$ must be interpreted as the circulation for the curve bounding dS (or simply for dS itself). Then the addition process of the surface integral (Fig. 15.55) gives the circulation around C, the circulation around all internal boundaries canceling. But if $(\nabla \times \mathbf{v}) \cdot \hat{\mathbf{n}} \, dS$ is the circulation for dS, then it follows that $(\nabla \times \mathbf{v}) \cdot \hat{\mathbf{n}}$ must be the circulation for unit area perpendicular to $\hat{\mathbf{n}}$. Thus $\nabla \times \mathbf{v}$ describes the circulatory nature of the flow $\mathbf{v}$, its component $(\nabla \times \mathbf{v}) \cdot \hat{\mathbf{n}}$ in any direction describes the circulation for unit area perpendicular to $\hat{\mathbf{n}}$.

Electromagnetic Theory The concepts of flux and circulation also play a prominent role in electromagnetic theory. For example, suppose a dielectric contains a charge distribution of density ρ (charge per unit volume). This charge produces an electric field represented by the electric displacement vector $\mathbf{D}$. If S is a surface in the dielectric, then the flux of $\mathbf{D}$ through S is defined as

$$\iint_S \mathbf{D} \cdot \hat{\mathbf{n}} \, dS,$$

and, in the particular case in which S is closed, as

$$\oint\!\!\!\oint_S \mathbf{D} \cdot \hat{\mathbf{n}} \, dS.$$

Gauss's law states that this flux integral must be exactly equal to the total charge enclosed by S. If V is the region enclosed by S, we can write

$$\oint\!\!\!\oint_S \mathbf{D} \cdot \hat{\mathbf{n}} \, dS = \iiint_V \rho \, dV. \qquad (15.61)$$

On the other hand, if we apply the divergence theorem to the flux integral, we have

$$\oiint_S \mathbf{D} \cdot \hat{\mathbf{n}}\, dS = \iiint_V \nabla \cdot \mathbf{D}\, dV.$$

Consequently,

$$\iiint_V \nabla \cdot \mathbf{D}\, dV = \iiint_V \rho\, dV,$$

or

$$\iiint_V (\nabla \cdot \mathbf{D} - \rho)\, dV = 0.$$

If $\nabla \cdot \mathbf{D}$ and ρ are continuous functions, then the only way this equation can hold for arbitrary volume V in the dielectric is if

$$\nabla \cdot \mathbf{D} = \rho. \tag{15.62}$$

This is the first of Maxwell's equations for electromagnetic fields.

Another of Maxwell's equations can be obtained using Stokes's theorem. The flux through a surface S of a magnetic field $\mathbf{B}$ is defined by

$$\iint_S \mathbf{B} \cdot \hat{\mathbf{n}}\, dS.$$

If $\mathbf{B}$ is a changing field, then an induced electric field intensity $\mathbf{E}$ is created. Faraday's induction law states that the time rate of change of the flux of $\mathbf{B}$ through S must be equal to the negative of the line integral of $\mathbf{E}$ around the boundary C of S:

$$\oint_C \mathbf{E} \cdot d\mathbf{r} = -\frac{\partial}{\partial t} \iint_S \mathbf{B} \cdot \hat{\mathbf{n}}\, dS. \tag{15.63}$$

But Stokes's theorem applied to the line integral also gives

$$\oint_C \mathbf{E} \cdot d\mathbf{r} = \iint_S (\nabla \times \mathbf{E}) \cdot \hat{\mathbf{n}}\, dS.$$

It follows, therefore, that if S is stationary,

$$\iint_S (\nabla \times \mathbf{E}) \cdot \hat{\mathbf{n}}\, dS = \iint_S -\frac{\partial \mathbf{B}}{\partial t} \cdot \hat{\mathbf{n}}\, dS,$$

or

$$\iint_S \left(\nabla \times \mathbf{E} + \frac{\partial \mathbf{B}}{\partial t}\right) \cdot \hat{\mathbf{n}}\, dS = 0.$$

Once again, this equation can hold for arbitrary surfaces S, if $\nabla \times \mathbf{E}$ and $\partial \mathbf{B}/\partial t$ are continuous, only if

$$\nabla \times \mathbf{E} = -\frac{\partial \mathbf{B}}{\partial t}. \tag{15.64}$$

SUMMARY

When a vector is a function of position, it becomes susceptible to the operations of differentiation and integration. In this chapter we developed various ways of differentiating and integrating vector functions, beginning with the operations of divergence and curl. The divergence of a vector field is a scalar field, and the curl of a vector field is another vector field. Both are extremely useful in applied mathematics. Mathematically, the curl appeared in our discussion of independence of path for line integrals and in Stokes's theorem; we saw its physical importance in our study of fluid flow and electromagnetic theory. We introduced the divergence of a vector field in our discussion of the divergence theorem and in the same applications as those for the curl.

The line integral of a function $f(x, y, z)$ along a curve C is defined in the same way as a definite integral, a double integral, or a triple integral—that is, the limit of a sum,

$$\int_C f(x, y, z)\, ds = \lim_{n\to\infty} \sum_{i=1}^{n} f(x_i^*, y_i^*, z_i^*)\, \Delta s_i.$$

The most important type of line integral occurs when $f(x, y, z)$ is the tangential component of a vector function $\mathbf{F} = P\hat{\mathbf{i}} + Q\hat{\mathbf{j}} + R\hat{\mathbf{k}}$ defined along C, and in this case we write

$$\int_C f(x, y, z)\, ds = \int_C \mathbf{F} \cdot d\mathbf{r} = \int_C P\, dx + Q\, dy + R\, dz.$$

We developed three methods for evaluating line integrals:

1. Express all parts of the line integral in terms of any parameter along C and evaluate the resulting definite integral. All line integrals can be evaluated in this way, but often methods (2) and (3) lead to much simpler calculations.
2. If a line integral is independent of path, then we can evaluate it by taking the difference in values of the function ϕ (where $\nabla\phi = \mathbf{F}$) at the ends of the curve.
3. If C is a closed curve, we can sometimes use Stokes's theorem to replace a line integral with a simpler surface integral. In this regard, Green's theorem is a special case of Stokes's theorem.

If the line integral of a force field $\mathbf{F}$ is independent of path, the force field is said to be conservative. Associated with every conservative force field is a potential function U such that the work done by $\mathbf{F}$ along a curve C from A to B is equal to the difference in U at A and B. In addition, motion of an object in a conservative force field is always characterized by an exchange of potential energy for kinetic energy in such a way that the sum of the two energies is always a constant value.

Surface integrals are also limits of sums,

$$\iint_S f(x, y, z)\, dS = \lim_{n\to\infty} \sum_{i=1}^{n} f(x_i^*, y_i^*, z_i^*)\, \Delta S_i,$$

and the most important type of surface integral occurs when $f(x, y, z)$ is the normal component of a vector function $\mathbf{F}$ on S:

$$\iint_S f(x, y, z)\, dS = \iint_S \mathbf{F} \cdot \hat{\mathbf{n}}\, dS.$$

We suggested two methods for the evaluation of surface integrals:

1. Project S onto some region R in one of the coordinate planes, express all parts of the integral in terms of coordinates in that plane, and evaluate the resulting double integral over R.
2. If S is closed, it could be advantageous to replace a surface integral with the triple integral of $\nabla \cdot \mathbf{F}$ over the volume bounded by S (the divergence theorem).

REVIEW EXERCISES

In Exercises 1–10, calculate the required quantity.

1. ∇f if $f(x, y, z) = x^2y^3 - xy + z$
2. $\nabla \cdot \mathbf{F}$ if $\mathbf{F}(x, y) = x^3y\hat{\mathbf{i}} - (x^2/y)\hat{\mathbf{j}}$
3. $\nabla \times \mathbf{F}$ if $\mathbf{F}(x, y) = \sin(xy)\hat{\mathbf{i}} + \cos(xy)\hat{\mathbf{j}} + xy\hat{\mathbf{k}}$
4. $\nabla \times \mathbf{F}$ if $\mathbf{F}(x, y, z) = (x + y + z)(\hat{\mathbf{i}} + \hat{\mathbf{j}} + \hat{\mathbf{k}})$
5. ∇f if $f(x, y, z) = \ln(x^2 + y^2 + z^2)$
6. $\nabla \cdot \mathbf{F}$ if $\mathbf{F}(x, y, z) = ye^x\hat{\mathbf{i}} + ze^y\hat{\mathbf{j}} + xe^z\hat{\mathbf{k}}$
7. $\nabla \times \mathbf{F}$ if $\mathbf{F}(x, y, z) = xyz\hat{\mathbf{j}}$
8. ∇f if $f(x, y, z) = \mathrm{Sin}^{-1}(x + y)$
9. $\nabla \cdot \mathbf{F}$ if $\mathbf{F}(y, z) = yz\hat{\mathbf{i}} - (y^2 + z^2)\hat{\mathbf{j}} + y^2z^2\hat{\mathbf{k}}$
10. $\nabla \times \mathbf{F}$ if $\mathbf{F}(x, y, z) = \mathrm{Cot}^{-1}(xyz)\hat{\mathbf{i}}$

In Exercises 11–30, evaluate the given integral.

11. $\int_C y\, ds,$

    where C is the curve $y = x^3$ from $(-1, -1)$ to $(2, 8)$

12. $\iint_S (x^2 + yz)\, dS,$

    where S is that part of $x + y + z = 2$ in the first octant

13. $\iint_S (x\hat{\mathbf{i}} + y\hat{\mathbf{j}}) \cdot \hat{\mathbf{n}}\, dS,$

    where S is that part of $z = x^2 + y^2$ bounded by the surfaces $x = \pm 1$, $y = \pm 1$, and $\hat{\mathbf{n}}$ is the lower normal

14. $\iint_S (x\hat{\mathbf{i}} + y\hat{\mathbf{j}}) \cdot \hat{\mathbf{n}}\, dS,$

    where S is that part of $z = x^2 + y^2$ below $z = 1$, and $\hat{\mathbf{n}}$ is the lower normal

15. $\oint_C x\, dx + y\, dy - z^2\, dz,$

    where C is the curve $x^2 + y^2 = 1$, $y = z$

16. $\int_C xy\,dx + xz\,dz$,
where C is the curve $z = \sqrt{2 - x^2 - y^2}$, $y = \sqrt{1 + x^2}$ from $(1/\sqrt{2}, \sqrt{3/2}, 0)$ to $(-1/\sqrt{2}, \sqrt{3/2}, 0)$

17. $\oint_C 2xy^3\,dx + (3x^2y^2 + 2xy)\,dy$,
where C is the curve $(x - 1)^2 + y^2 = 1$

18. $\oint_C 2xy^3\,dx + (3x^2y^2 + x^2)\,dy$,
where C is the curve $(x - 1)^2 + y^2 = 1$

19. $\oiint_S (x^2\hat{\mathbf{i}} + y^2\hat{\mathbf{j}} + z^2\hat{\mathbf{k}}) \cdot \hat{\mathbf{n}}\,dS$,
where S is the surface bounding the volume enclosed by $y = z$, $y + z = 2$, $x = 0$, $x = 1$, $z = 0$, and $\hat{\mathbf{n}}$ is the outer normal to S

20. $\iint_S (x^2 + y^2)\,dS$,
where S is that part of $x^2 + y^2 + z^2 = 6$ inside $z = x^2 + y^2$

21. $\iint_S (x^2 + y^2)\hat{\mathbf{i}} \cdot \hat{\mathbf{n}}\,dS$,
where S is that part of $x^2 + y^2 + z^2 = 6$ inside $z = x^2 + y^2$ and $\hat{\mathbf{n}}$ is the upper normal to S

22. $\oint_C (x^2\hat{\mathbf{i}} + y\hat{\mathbf{j}} - xz\hat{\mathbf{k}}) \cdot d\mathbf{r}$,
where C is the curve $x^2 + y^2 = 1$, $z = x + 1$, directed clockwise as viewed from the origin

23. $\oint_C (xy\hat{\mathbf{i}} + z\hat{\mathbf{j}} - x^2\hat{\mathbf{k}}) \cdot d\mathbf{r}$,
where C is the curve in Exercise 22

24. $\oint_C y\,dx + 2x\,dy - 3z^2\,dz$,
where C is the curve $y = \sqrt{1 + z^2 - x^2}$, $x^2 + z^2 = 1$, directed counterclockwise as viewed from the origin

25. $\oint_C (xy + 4x^3y^2)\,dx + (z + 2x^4y)\,dy + (z^5 + x^2z^2)\,dz$,
where C is the curve $x^2 + z^2 = 4$, $x^2 + y^2 = 4$, $y = z$, directed counterclockwise as viewed from a point far up the z-axis

26. $\iint_S (x^2yz\hat{\mathbf{i}} - x^2yz\hat{\mathbf{j}} - xyz^2\hat{\mathbf{k}}) \cdot \hat{\mathbf{n}}\,dS$,
where S is that part of $z = 1 - \sqrt{x^2 - y^2}$ above the xy-plane, and $\hat{\mathbf{n}}$ is the upper normal

27. $\iint_S dS$,
where S is that part of $z = x^2 - y^2$ inside $x^2 + y^2 = 4$

28. $\iint_S y\,dS$,
where S is that part of $x = y^2 + 1$ in the first octant, which is under $x + z = 2$

29. $\oint_C (ye^{xy} + xy^2e^{xy})\,dx + (xe^{xy} + x^2ye^{xy} + x^3y)\,dy$,
where C is the curve $x^2 + y^2 = 2y$, $z = 0$

30. $\iint_S (x\hat{\mathbf{i}} + y\hat{\mathbf{j}}) \cdot \hat{\mathbf{n}}\,dS$,
where S is that part of $z^2 - x^2 - y^2 = 1$ between the planes $z = 0$ and $z = 2$, and $\hat{\mathbf{n}}$ is the lower normal

31. Let S be that part of the sphere $x^2 + y^2 + z^2 = 1$ that lies above the parabolic cylinder $2z = x^2$. Set up, but do not evaluate, double iterated integrals to calculate the surface integral

$$\iint_S x^2y^2z^2\,dS$$

by projecting S onto (a) the xy-coordinate plane; (b) the yz-coordinate plane; and (c) the xz-coordinate plane.

32. If $\mathbf{r} = x\hat{\mathbf{i}} + y\hat{\mathbf{j}} + z\hat{\mathbf{k}}$, show that

$$\nabla(|\mathbf{r}|^n) = n|\mathbf{r}|^{n-2}\mathbf{r}.$$

33. Verify that

$$\nabla \times (\nabla \times \mathbf{F}) = \nabla(\nabla \cdot \mathbf{F}) - \nabla^2\mathbf{F},$$

where

$$\nabla^2 = \partial^2/\partial x^2 + \partial^2/\partial y^2 + \partial^2/\partial z^2.$$

Differential Equations

16

16.1 INTRODUCTION

A differential equation is an equation that must be solved for an unknown function. What distinguishes a differential equation from other equations is the fact that it contains derivatives of the unknown function. For example, each of the following equations is a differential equation in y as a function of x:

$$\frac{dy}{dx} + \frac{2}{3}y = 9.81, \tag{16.1}$$

$$\frac{d^2y}{dx^2} = k\sqrt{1 + \left(\frac{dy}{dx}\right)^2}, \tag{16.2}$$

$$x\frac{d^2y}{dx^2} + \frac{dy}{dx} + xy = 0, \tag{16.3}$$

$$\frac{d^4y}{dx^4} - k^4y = 0. \tag{16.4}$$

Equation (16.1) is used to determine the position of a paratrooper who falls under the influence of gravity and an air resistance proportional to velocity (see Example 8); Eq. (16.2) describes the shape of a hanging cable (Exercise 11 in Section 16.4); Eq. (16.3), called Bessel's differential equation of order zero, is found in heat flow and vibration problems; and Eq. (16.4) is used to determine the deflection of beams.

▷ **Definition 16.1** The *order* of a differential equation is the order of the highest derivative in the equation. ◁

Of the four differential equations (16.1)–(16.4), the first is first order, the second and third are second order, and the last is fourth order.

We have considered quite a number of differential equations in Chapters 5, 8, and 9. In Section 5.4 where we introduced the idea of a differential equation for the first time, we concentrated on describing physical situations by means of differential equations; in Chapter 8 we verified that particular combinations of transcendental functions satisfied certain differential equations; and in Chapter 9 we used our integration techniques to solve various equations. Almost all these differential equations were based on applications, many from physics and engineering, but some also from geometry and other fields such as ecology, chemistry, and psychology. Because of this association with applications, the differential equations were almost always accompanied by subsidiary conditions called *initial* or *boundary conditions*. For example, if a mass m falls from rest under gravity and is acted on by a force due to air resistance that is proportional to its instantaneous velocity, then the differential equation that describes its velocity $v(t)$ as a function of time t is

$$m\frac{dv}{dt} = -kv + mg, \qquad k = \text{a constant}. \tag{16.5}$$

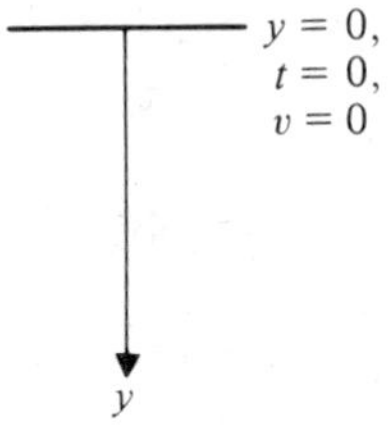

Figure 16.1

This is simply a statement of Newton's second law, where dv/dt is the vertical component of the acceleration of m, and $-kv + mg$ is the vertical component of the total force on m due to gravity (mg) and air resistance ($-kv$). If distance y is chosen as positive downward (Fig. 16.1), both g and k are positive. Furthermore, if we choose time $t = 0$ at the instant m is dropped, then because m falls from rest, the condition $v(0) = 0$ must be added to the differential equation. In other words, the real problem is to find the solution of the differential equation that also satisfies the initial condition:

$$m\frac{dv}{dt} = -kv + mg, \qquad v(0) = 0. \tag{16.6}$$

This is the form in which applied mathematicians find differential equations—the differential equation is accompanied by subsidiary conditions that express physical requirements of the solution. It is not difficult to show that the solution of Eq. (16.6) is

$$v(t) = \frac{mg}{k} - \frac{mg}{k}e^{-kt/m}. \tag{16.7}$$

(All we need do to verify this is to substitute the function into the differential equation to see that it does indeed satisfy the equation. It is clear that it does satisfy the initial condition.)

If we change the initial condition to $v(0) = v_0$, so that the initial velocity of m has vertical component v_0, then the solution becomes

$$v(t) = \frac{mg}{k} - \left(\frac{mg}{k} - v_0\right)e^{-kt/m}. \tag{16.8}$$

In other words, every solution of differential equation (16.5) can be written in the form

$$v(t) = \frac{mg}{k} + Ce^{-kt/m}, \tag{16.9}$$

and when we impose the initial condition $v(0) = v_0$, then $C = v_0 - mg/k$.

In a similar way, Eq. (16.4) is normally accompanied by four boundary conditions such as perhaps

$$y(0) = y(L) = 0;$$
$$y''(0) = y''(L) = 0.^{\dagger}$$

It can be shown that every solution of Eq. (16.4) can be expressed in the form

$$y(x) = C_1e^{kx} + C_2e^{-kx} + C_3\sin(kx) + C_4\cos(kx), \tag{16.10}$$

where C_1, C_2, C_3, and C_4 are arbitrary constants, and when the boundary

† In this chapter it is frequently convenient to use the notation y', y'', y''', ... to represent dy/dx, d^2y/dx^2, d^3y/dx^3, etc. In this notation $y''(a)$ is the second derivative of y evaluated at $x = a$. In addition, we denote the solution of a differential equation in y as a function of x by $y(x)$.

conditions are applied, these constants must satisfy the four equations

$$0 = C_1 + C_2 + C_4,$$
$$0 = C_1e^{kL} + C_2e^{-kL} + C_3 \sin (kL) + C_4 \cos (kL),$$
$$0 = C_1 + C_2 - C_4,$$
$$0 = C_1e^{kL} + C_2e^{-kL} - C_3 \sin (kL) - C_4 \cos (kL).$$

We have stated that every solution of (16.5) can be written in the form (16.9), and every solution of (16.4) can be expressed as (16.10). Note that (16.9) contains one arbitrary constant whereas (16.10) has four, but in both cases the number of arbitrary constants is exactly the same as the order of the differential equation. We might suspect that there is a general result developing that states that every solution of an nth-order differential equation can be expressed as a function involving n arbitrary constants. For many differential equations this is indeed true, but unfortunately it is not true for all equations. As an illustration, consider the equation

$$\frac{d^2y}{dx^2} = \left(\frac{dy}{dx}\right)^2. \tag{16.11}$$

In Example 6 we will apply standard techniques for solving differential equations to obtain the solution $y(x) = C_1 - \ln (C_2 + x)$, which contains two arbitrary constants C_1 and C_2. This two-parameter family of solutions does not, however, contain all solutions of the differential equation, for no choice of C_1 and C_2 will give the perfectly acceptable solution $y(x) \equiv 1$. This solution is not particularly interesting, but it is nonetheless a solution that is not contained within the two-parameter family. Such a solution is called a *singular solution.* We have illustrated that a solution that contains the same number of arbitrary constants as the order of the differential equation may or may not contain all solutions of the differential equation. In spite of this unfortunate circumstance, there do exist large classes of differential equations for which a solution with the same number of arbitrary constants as the order of the equation does indeed represent all possible solutions. Because of this we make the following definition.

▷ **Definition 16.2** A *general solution* of a differential equation is a solution that contains the same number of arbitrary constants as the order of the differential equation. ◁

Consequently, in order for a function to be a general solution of a differential equation, it must first be a solution, and second, contain the requisite number of arbitrary constants.

Once again we emphasize that a general solution of a differential equation may or may not contain all possible solutions. We will point out various important cases in which a general solution does indeed contain all solutions.

Example 1 Show that $y(x) = C_1 \cos 2x + C_2 \sin 2x$ is a general solution of the differential equation

$$\frac{d^2y}{dx^2} + 4y = 0.$$

Solution If we substitute this function into the left-hand side of the equation, we have

$$\frac{d^2y}{dx^2} + 4y = \{-4C_1 \cos 2x - 4C_2 \sin 2x\} + 4\{C_1 \cos 2x + C_2 \sin 2x\} = 0,$$

and the function is indeed a solution. Because it contains two arbitrary constants, the order of the differential equation, it is a general solution. □

▷ **Definition 16.3** A *particular solution* of a differential equation is a solution that contains no arbitrary constants. ◁

It follows, therefore, that particular solutions can be obtained by assigning specific values to the arbitrary constants in a general solution. For example, $y(x) = 5 - \ln(3 + x)$ is a particular solution of differential equation (16.11), as is $y(x) = -\ln x$, both being obtained from $y(x) = C_1 - \ln(C_2 + x)$. On the other hand, the singular solution $y(x) = 10$ is also a particular solution, but it cannot be obtained from this general solution.

Example 2 Find a particular solution of the differential equation

$$5\frac{d^3y}{dx^3} + 3\frac{d^2y}{dx^2} + 2y = 4.$$

Solution In Section 16.10 we will develop systematic techniques for finding general and particular solutions for differential equations such as this. But clearly those techniques are not needed here, since a simple glance tells us that $y(x) = 2$ is a solution. □

Many differential equations are immediately solvable (or, as we often say, immediately integrable). For example, to solve a differential equation of the form

$$\frac{dy}{dx} = M(x), \tag{16.12}$$

where $M(x)$ is given, we simply integrate both sides of the equation with respect to x to obtain the general solution

$$y(x) = \int M(x)\,dx + C. \tag{16.13}$$

We have called this *the* general solution rather than *a* general solution because in this case the general solution does indeed contain all solutions of the differential equation.

This result is easily extended to the nth-order equation

$$\frac{d^ny}{dx^n} = M(x), \qquad n \text{ a positive integer.} \tag{16.14}$$

We integrate successively n times to obtain the general solution

$$y(x) = \int \cdots \int M(x)\,dx \cdots dx + C_1 + C_2 x + \cdots + C_n x^{n-1}. \tag{16.15}$$

EXERCISES 16.1

In Exercises 1–10, show that the function is a general solution of the differential equation.

1. $y(x) = 2 + Ce^{-x^2}; \dfrac{dy}{dx} + 2xy - 4x = 0$

2. $y(x) = \dfrac{x}{1 + Cx}; \dfrac{dy}{dx} = y^2/x^2$

3. $y(x) = \dfrac{x^3}{2} + Cx^3e^{1/x^2}; x^3\dfrac{dy}{dx} + (2 - 3x^2)y = x^3$

4. $y(x) = C_1 \sin 3x + C_2 \cos 3x; \dfrac{d^2y}{dx^2} + 9y = 0$

5. $y(x) = \dfrac{C_1^2e^{2x} + 1}{2C_1e^x} + C_2; \left(\dfrac{d^2y}{dx^2}\right)^2 = 1 + \left(\dfrac{dy}{dx}\right)^2$

6. $y(x) = C_1e^{2x} \cos\left(\dfrac{x}{\sqrt{2}}\right) + C_2e^{2x} \sin\left(\dfrac{x}{\sqrt{2}}\right);$

$2\dfrac{d^2y}{dx^2} - 8\dfrac{dy}{dx} + 9y = 0$

7. $y(x) = C_1 \cos 2x + C_2 \sin 2x + C_3 \cos x + C_4 \sin x;$

$\dfrac{d^4y}{dx^4} + 5\dfrac{d^2y}{dx^2} + 4y = 0$

8. $y(x) = \left(C_1 + C_2x - \dfrac{x^2}{4}\right)e^{4x};$

$2\dfrac{d^2y}{dx^2} - 16\dfrac{dy}{dx} + 32y = -e^{4x}$

9. $y(x) = C_1 \cos(2 \ln x) + C_2 \sin(2 \ln x) + \dfrac{1}{4};$

$x^2\dfrac{d^2y}{dx^2} + x\dfrac{dy}{dx} + 4y = 1$

10. $y(x) = C_1\dfrac{\sin x}{\sqrt{x}} + C_2\dfrac{\cos x}{\sqrt{x}};$

$x^2\dfrac{d^2y}{dx^2} + x\dfrac{dy}{dx} + \left(x^2 - \dfrac{1}{4}\right)y = 0$

In Exercises 11–14, find a particular solution of the differential equation in Exercise 4 that satisfies the given conditions.

11. $y(0) = 1, y'(0) = 6$

12. $y(0) = 2, y(\pi/2) = 3$

13. $y(\pi/12) = 0, y'(\pi/12) = 1$

14. $y(1) = 1, y(2) = 2$

In Exercises 15–20, find a general solution of the differential equation.

15. $\dfrac{dy}{dx} = 6x^2 + 2x$

16. $\dfrac{dy}{dx} = \dfrac{1}{9 + x^2}$

17. $\dfrac{d^2y}{dx^2} = 2x + e^x$

18. $\dfrac{d^2y}{dx^2} = x \ln x$

19. $\dfrac{d^3y}{dx^3} = \dfrac{1}{3x^5}$

20. $\dfrac{dy}{dx} = y$

21. a) A boy initially at 0 (Fig. 16.2) walks along the edge of a swimming pool (the y-axis) towing his sailboat by a string of length L. If the boat starts at Q and the string always remains straight, show that the equation of the curved path $y = y(x)$ followed by the boat must satisfy the differential equation

$$\frac{dy}{dx} = -\frac{\sqrt{L^2 - x^2}}{x}.$$

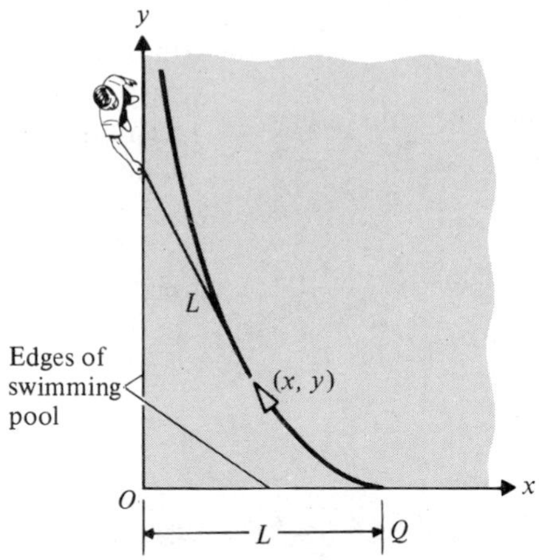

Figure 16.2

b) Solve this differential equation for $y(x)$.

22. Show that $y(x) = -(x^2 + C)^{-1}$ is a general solution of the differential equation

$$\frac{dy}{dx} = 2xy^2.$$

Find a singular solution.

23. Show that $y(x) = 1 - (x^3 + C)^{-1}$ is a general solution of the differential equation

$$\frac{dy}{dx} = 3x^2(y - 1)^2.$$

Find a singular solution.

24. a) Verify that $y(x) = Ce^{2x}$ is a general solution of the differential equation

$$\frac{dy}{dx} = 2y.$$

b) Draw the one-parameter family of curves defined by this general solution.

c) Show proof that there is a particular solution that passes through any given point (x_0, y_0), and that this solution can be obtained by choosing C appropriately.

25. a) Verify that a general solution of the differential equation

$$2x\frac{dy}{dx} = y$$

is defined implicitly by the equation $y^2 = Cx$.

b) Draw the one-parameter family of curves defined by this equation.

c) Show that, with the exception of points on the y-axis, there is a particular solution that passes through any given point (x_0, y_0), and that this solution can be obtained by specifying C appropriately.

26. a) Draw the one-parameter family of curves defined by the general solution in Exercise 22.

b) Show that, with the exception of points on the x-axis, there is a solution passing through any given point in the xy-plane.

27. Consider the differential equation

$$\frac{dy}{dx} = \frac{1}{x^2}.$$

a) Find a solution that satisfies the condition $y(1) = 1$.

b) Find a solution that satisfies the condition $y(-1) = 2$.

c) Find a solution that satisfies the conditions in both (a) and (b).

d) Draw the curves defined by the general solution to explain why this is possible.

16.2 SEPARABLE DIFFERENTIAL EQUATIONS

In this section and the next, we consider first-order differential equations that can be written in the form

$$\frac{dy}{dx} = F(x, y). \tag{16.16}$$

Since any function $F(x, y)$ can always be considered as the quotient of two other functions,

$$F(x, y) = \frac{M(x, y)}{N(x, y)},$$

Eq. (16.16) can also be written in the equivalent form

$$N(x, y)\,dy = M(x, y)\,dx. \tag{16.17}$$

Depending on the form of $F(x, y)$ (or $M(x, y)$ and $N(x, y)$) various methods can be used to obtain the unknown function $y(x)$. Two of the more important techniques are considered here and in Section 16.3; others are discussed in the exercises.

Differential equation (16.16) is said to be separable if it can be expressed in the form

$$\frac{dy}{dx} = \frac{M(x)}{N(y)}, \tag{16.18}$$

i.e., if dy/dx is equal to a function of x divided by a function of y. Equivalently, a differential equation is said to be separable if it can be written in the form

$$N(y)\,dy = M(x)\,dx. \tag{16.19}$$

When a differential equation is written in this way, it is said to be separated—separated in the sense that the x- and y-variables appear on opposite sides of the equation. For a separated equation we can write therefore that

$$N(y)\frac{dy}{dx} = M(x), \tag{16.20}$$

and if we integrate both sides with respect to x, we have

$$\int N(y)\frac{dy}{dx}\,dx = \int M(x)\,dx + C. \tag{16.21}$$

Cancellation of differentials on the left leads to the solution

$$\int N(y)\,dy = \int M(x)\,dx + C. \tag{16.22}$$

What we mean by saying that (16.21) and (16.22) represent general solutions of (16.19) is that any function defined *implicitly* by (16.21) or (16.22) is a solution of (16.19). For example, if we divide the differential equation

$$3x^3y^2\,dx + (xy^3 - xy^2)\,dy = 0$$

by xy^2, it becomes separated:

$$(1 - y)\,dy = 3x^2\,dx.$$

According to (16.21) we should divide by dx and integrate both sides with respect to x:

$$\int (1 - y)\frac{dy}{dx}\,dx = \int 3x^2\,dx = x^3 + C.$$

The antidifferentiation on the left must be interpreted as "implicit antidifferentiation," asking for that function which when differentiated with respect to x gives $(1 - y)\,dy/dx$. Since an antiderivative is $y - y^2/2$, a general solution of the differential equation is

$$y - \frac{y^2}{2} = x^3 + C.$$

Were we to use (16.22) after separation (instead of (16.21)), we would write

$$\int (1 - y)\,dy = \int 3x^2\,dx + C,$$

and integrate for a general solution

$$y - \frac{y^2}{2} = x^3 + C.$$

As indicated above, by saying that $y - y^2/2 = x^3 + C$ represents a general solution of the original differential equation, we mean that any function defined implicitly by $y - y^2/2 = x^3 + C$ is a solution.

Example 3 Find a general solution for each of the following differential equations:

a) $2x^3y^2\,dx = xy^3\,dy$;

b) $\dfrac{dy}{dx} = \dfrac{y \sin x + y^3 \sin x}{(1 + y^2)^2}$.

Solution a) If we divide the differential equation by xy^2, we obtain

$$y\,dy = 2x^2\,dx,$$

which is separated. A general solution is therefore defined implicitly by

$$\int y\,dy = \int 2x^2\,dx + C,$$

or

$$\frac{y^2}{2} = \frac{2x^3}{3} + C.$$

Note that $y(x) = 0$ is also a solution, but it cannot be obtained by specifying C. In other words, $y(x) = 0$ is a singular solution. We removed this solution when we divided the original equation by xy^2.

b) Since

$$\frac{dy}{dx} = \frac{y \sin x(1 + y^2)}{(1 + y^2)^2} = \frac{y \sin x}{1 + y^2},$$

the differential equation can be separated:

$$\sin x\,dx = \frac{1 + y^2}{y}\,dy = \left(\frac{1}{y} + y\right)dy.$$

A general solution is therefore defined implicitly by

$$\int \sin x\,dx + C = \int \left(\frac{1}{y} + y\right) dy,$$

or

$$-\cos x + C = \ln|y| + \frac{y^2}{2}. \square$$

Example 4 Find the function (in explicit form) that satisfies the differential equation

$$e^{2x+y}\,dx - 2e^{x-y}\,dy = 0, \qquad y(0) = 1.$$

Solution If we divide the equation by e^{x+y}, we have

$$e^x\,dx = 2e^{-2y}\,dy.$$

A general solution is defined implicitly by

$$e^x = -e^{-2y} + C.$$

The condition $y(0) = 1$ requires $e^0 = -e^{-2} + C$, which implies that $C = 1 + e^{-2}$. The required solution is therefore defined implicitly by

$$e^x = -e^{-2y} + e^{-2} + 1,$$

and from this equation, we have

$$e^{-2y} = e^{-2} + 1 - e^x, \quad \text{or} \quad y(x) = -\tfrac{1}{2}\ln(e^{-2} + 1 - e^x). \;\square$$

EXERCISES 16.2

In Exercises 1–10, find a general solution of the differential equation.

1. $y^2\,dx - x^2\,dy = 0$ **2.** $\dfrac{dy}{dx} + 2xy = 4x$

3. $2xy\,dx + (x^2 + 1)\,dy = 0$

4. $\dfrac{dy}{dx} = 3y + 2$

5. $3(y^2 + 2)\,dx = 4y(x - 1)\,dy$

6. $(x^2y + x^2)\,dx + (xy^2 - y^2)\,dy = 0$

7. $\dfrac{dy}{dx} = -\dfrac{\cos y}{\sin x}$

8. $(x^2ye^x - y)\,dx + xy^3\,dy = 0$

9. $(x^2y^2 \sec x \tan x + xy^2 \sec x)\,dx + xy^3\,dy = 0$

10. $\dfrac{dy}{dx} = \dfrac{1 + y^2}{1 + x^2}$

In Exercises 11–15, find a solution of the differential equation that also satisfies the initial condition.

11. $2y\,dx + (x + 1)\,dy = 0$, $y(1) = 2$

12. $(xy + y)\,dx - (xy - x)\,dy = 0$, $y(1) = 2$

13. $\dfrac{dy}{dx} = e^{x+y}$, $y(0) = 0$

14. $\dfrac{dy}{dx} = 2x(1 + y^2)$, $y(2) = 4$

15. $\dfrac{dy}{dx} = \dfrac{\sin^2 y}{\cos^2 x}$, $y(0) = \pi/2$

16. A girl lives 6 kilometers from school. She decides to travel to school so that her speed is always proportional to the square of her distance from the school.

a) Find her distance from school at any time.

b) When does she reach school?

17. Find a general solution for the differential equation

$$\frac{dy}{dx} = -\frac{1 + y^3}{xy^2 + x^3y^2}.$$

18. When a container of water at temperature 80°C is placed in a room at temperature 20°C, Newton's law of cooling states that the time rate of change of the temperature of the water is proportional to the difference between the temperature of the water and room temperature. If the water cools to 60°C in 2 minutes, find a formula for its temperature as a function of time.

19. The amount of a drug such as penicillin injected into the body is used up at a rate proportional to the amount still present. If a dose decreases by 5% in the first hour, when will it decrease to one-half its original amount?

20. When a substance such as glucose is administered intravenously into the bloodstream, it is used up by the body at a rate proportional to the amount present at that time. If it is added at a constant rate of R units per unit time, and A_0 is the amount present in the bloodstream when the intravenous feeding begins, find a formula for the amount $A(t)$ in the bloodstream at any time.

21. When a deep-sea diver inhales air, his body tissues absorb extra amounts of nitrogen. The nitrogen pressure P in the tissue increases as he remains under water until a maximum pressure $\bar{P}$ at depth d is reached. It is found that if he remains at depth d, the time rate of change of P is proportional to the difference $\bar{P} - P$. If the pressure is zero when he enters the water and he drops quickly to the bottom (depth d), find P as a function of time.

22. When two substances A and B are brought together in one solution, they react to form a third substance C in such a way that one gram of A reacts with one gram of B to produce two grams of C. The rate at which C is formed is proportional to the amounts of A and B still present in the solution. If 10 grams of A and 15 grams of B are originally brought together, find a formula for the amount of C present in the mixture at any time.

23. A first-order differential equation in $y(x)$ is said to be *homogeneous* if it can be written in the form

$$\frac{dy}{dx} = f\left(\frac{y}{x}\right).$$

Show that the change of dependent variable $v = y/x$ yields a differential equation in $v(x)$ that is always separable.

In each of Exercises 24–29, show that the differential equation is homogeneous, and use the change of variable in Exercise 23 to find a general solution.

24. $(y^2 - x^2)\,dx + xy\,dy = 0$

25. $2x\,dy - 2y\,dx = \sqrt{x^2 + 4y^2}\,dx$

26. $\dfrac{dy}{dx} = \dfrac{y + x}{y - x}$

27. $x\,dy - y\,dx = x\cos(y/x)\,dx$

28. $\dfrac{dy}{dx} = \dfrac{x^2 e^{-y/x} + y^2}{xy}$

29. $(x^2 y + y^3)\,dx + x^3\,dy = 0$

30. Find in explicit form a function that satisfies the differential equation

$$\frac{dy}{dx} = \csc y$$

and the following initial conditions: (a) $y(0) = \pi/4$; (b) $y(0) = 7\pi/4$.

31. Snow has been falling for some time when a snowplow starts plowing a highway. The plow begins at 12:00 and travels 2 kilometers during the first hour and 1 kilometer during the second hour. Make reasonable assumptions to find out when the snow started falling.

32. In order to perform a one-hour operation on a dog a veterinarian anesthetizes the dog with sodium pentobarbital. During the operation the dog's body breaks down the drug at a rate proportional to the amount still present, and only half an original dose remains after 5 hours. If the dog has a mass of 20 kilograms, and 20 milligrams of sodium pentobarbital per kilogram of body mass are required to maintain surgical anesthesia, what original dose is required?

33. If a first-order differential equation can be written in the form

$$\frac{dy}{dx} = f(ax + by),$$

where a and b are constants, show that the change of dependent variable $v = ax + by$ always gives a differential equation in $v(x)$ that is separable.

Use the method of Exercise 33 to find a general solution of the differential equation in Exercises 34–37.

34. $\dfrac{dy}{dx} = x + y$

35. $\dfrac{dy}{dx} = (x + y)^2$

36. $\dfrac{dy}{dx} = \dfrac{1}{2x + 3y}$

37. $\dfrac{dy}{dx} = \sin^2(x - y)$

38. Solve the differential equation

$$(x^3 y^4 + 2xy^4)\,dx + (x - xy^6)\,dy = 0, \qquad y(1) = 1.$$

39. Explain what the cancellation of differentials in proceeding from (16.21) to (16.22) really means.

16.3 LINEAR FIRST-ORDER DIFFERENTIAL EQUATIONS

A first-order differential equation that can be written in the form

$$\frac{dy}{dx} + P(x)y = Q(x) \tag{16.23}$$

is said to be linear. We will explain the significance of the adjective "linear" in Section 16.7. To illustrate how to solve such differential equations, consider the equation

$$\frac{dy}{dx} + \frac{1}{x}y = 1.$$

If we multiply both sides by x, we have

$$x\frac{dy}{dx} + y = x.$$

But note now that the left side is the derivative of the product xy; i.e.,

$$\frac{d}{dx}(xy) = x\frac{dy}{dx} + y.$$

In other words, we can write the differential equation in the form

$$\frac{d}{dx}(xy) = x,$$

and integration immediately gives a general solution

$$xy = \frac{x^2}{2} + C.$$

This is the principle behind all linear first-order equations: Multiply the equation by a function of x in order that the left side can be expressed as the derivative of a product. To show that this is always possible, we turn now to the general equation (16.23). If the equation is multiplied by a function $\mu(x)$,

$$\mu\frac{dy}{dx} + \mu P(x)y = \mu Q(x), \tag{16.24}$$

then this equation is equivalent to (16.23) in the sense that $y(x)$ is a solution of (16.23) if and only if it is a solution of (16.24). We ask whether it is possible to find μ so that the left side of (16.24) can be written as the derivative of the product μy; i.e., can we find $\mu(x)$ so that

$$\mu\frac{dy}{dx} + \mu P(x)y = \frac{d}{dx}(\mu y)?$$

If we expand the right side, μ must satisfy

$$\mu\frac{dy}{dx} + \mu P(x)y = \mu\frac{dy}{dx} + y\frac{d\mu}{dx},$$

from which we get

$$\mu P = \frac{d\mu}{dx} \quad \text{or} \quad \frac{d\mu}{\mu} = P\,dx.$$

Thus μ must satisfy a separated differential equation, one solution of which is

$$\ln|\mu| = \int P(x)\,dx, \quad \text{or} \quad \mu = \pm e^{\int P(x)\,dx}.$$

We have shown then that if (16.23) is multiplied by the factor

$$e^{\int P(x)\,dx} \qquad (\text{or by } -e^{\int P(x)\,dx}),$$

then the differential equation becomes

$$e^{\int P(x)\,dx}\frac{dy}{dx} + P(x)ye^{\int P(x)\,dx} = Q(x)e^{\int P(x)\,dx},$$

and the left side can be expressed as the derivative of a product:

$$\frac{d}{dx}\{ye^{\int P(x)\,dx}\} = Q(x)e^{\int P(x)\,dx}.$$

Integration now gives a general solution

$$ye^{\int P(x)\,dx} = \int Q(x)e^{\int P(x)\,dx}\,dx + C. \tag{16.25}$$

The quantity $e^{\int P(x)\,dx}$ is called an *integrating factor* for Eq. (16.23) because when Eq. (16.23) is multiplied by this factor, it becomes immediately integrable.

In summary, if a linear differential equation (16.23) is multiplied by the function $e^{\int P(x)\,dx}$, then the left side of the equation becomes the derivative of the product of y and this function. In this form the equation can immediately be integrated to give a general solution. Given a specific linear equation to solve, we have three choices on how to proceed:

1. Memorize (16.25) as a formula.
2. Memorize that an integrating factor is $e^{\int P(x)\,dx}$.
3. Develop the integrating factor in every example.

We feel that (2) is a reasonable compromise in that we must memorize something, but not too much, and we must do some calculations, but not too many.

Example 5 Find a general solution for each of the following differential equations:

a) $\dfrac{dy}{dx} + xy = x;$

b) $(y - x\sin x)\,dx + x\,dy = 0;$

c) $\cos x\dfrac{dy}{dx} + y\sin x = 1.$

Solution a) An integrating factor for this linear equation is

$$e^{\int x\,dx} = e^{x^2/2}.$$

If we multiply the equation by this integrating factor, we have

$$e^{x^2/2}\frac{dy}{dx} + yxe^{x^2/2} = xe^{x^2/2},$$

or

$$\frac{d}{dx}\{ye^{x^2/2}\} = xe^{x^2/2}.$$

Integration yields

$$ye^{x^2/2} = \int xe^{x^2/2}\,dx = e^{x^2/2} + C,$$

or

$$y(x) = 1 + Ce^{-x^2/2}.$$

b) If we write the differential equation in the form

$$\frac{dy}{dx} + \frac{y}{x} = \sin x,$$

then we see that it is linear first order. An integrating factor is therefore

$$e^{\int 1/x\,dx} = e^{\ln|x|} = |x|.$$

If we multiply the differential equation by this factor, we get

$$|x|\frac{dy}{dx} + \frac{|x|}{x}y = |x|\sin x.$$

If $x > 0$, then we write

$$x\frac{dy}{dx} + \frac{x}{x}y = x\sin x,$$

whereas if $x < 0$,

$$-x\frac{dy}{dx} - \frac{x}{x}y = -x\sin x.$$

In either case, however, the equation simplifies to

$$x\frac{dy}{dx} + y = x\sin x,$$

or

$$\frac{d}{dx}(xy) = x\sin x.$$

Integration now gives

$$xy = \int x\sin x\,dx = -x\cos x + \sin x + C.$$

Finally, then,

$$y(x) = -\cos x + \frac{\sin x}{x} + \frac{C}{x}.$$

c) An integrating factor for the linear equation

$$\frac{dy}{dx} + y\tan x = \frac{1}{\cos x}$$

is

$$e^{\int \tan x\,dx} = e^{\ln|\sec x|} = |\sec x|.$$

For either $\sec x < 0$ or $\sec x > 0$, we obtain

$$\frac{1}{\cos x}\frac{dy}{dx} + y\frac{\sin x}{\cos^2 x} = \frac{1}{\cos^2 x},$$

or

$$\frac{d}{dx}\left\{\frac{y}{\cos x}\right\} = \sec^2 x.$$

Integration now yields

$$\frac{y}{\cos x} = \tan x + C,$$

and hence

$$y(x) = \sin x + C \cos x. \square$$

EXERCISES 16.3

In each of Exercises 1–12, find a general solution for the differential equation.

1. $\dfrac{dy}{dx} + 2xy = 4x$

2. $\dfrac{dy}{dx} + \dfrac{2}{x}y = 6x^3$

3. $(2y - x)\,dx + dy = 0$

4. $\dfrac{dy}{dx} + y \cot x = 5e^{\cos x}$

5. $(x^2 + 2xy)\,dx + (x^2 + 1)\,dy = 0$

6. $(x + 1)\dfrac{dy}{dx} - 2y = 2(x + 1)$

7. $\dfrac{1}{x}\dfrac{dy}{dx} - \dfrac{y}{x^2} = \dfrac{1}{x^3}$

8. $(y + e^{2x})\,dx = dy$

9. $\dfrac{dy}{dx} + y = 2 \cos x$

10. $x^3\dfrac{dy}{dx} + (2 - 3x^2)y = x^3$

11. $\dfrac{dy}{dx} + \dfrac{y}{x \ln x} = x^2$

12. $(-2y \cot 2x - 1 + 2x \cot 2x + 2 \csc 2x)\,dx + dy = 0$

In Exercises 13–15, solve the differential equation.

13. $\dfrac{dy}{dx} + 3x^2 y = x^2,\ y(1) = 2$

14. $(-e^x \sin x + y)\,dx + dy = 0,\ y(0) = -1$

15. $\dfrac{dy}{dx} + \dfrac{x^3 y}{x^4 + 1} = x^7,\ y(0) = 1$

16. Find a general solution for the differential equation

$$(y^3 - x)\,dy = y\,dx.$$

17. A differential equation of the form

$$\frac{dy}{dx} + P(x)y = y^n Q(x)$$

is called a *Bernoulli equation*. Show that the change of dependent variable $z = y^{1-n}$ gives

$$\frac{dz}{dx} + (1 - n)Pz = (1 - n)Q,$$

a linear first-order equation in $z(x)$.

In Exercises 18–22, use the change of variable in Exercise 17 to find a general solution for the differential equation.

18. $\dfrac{dy}{dx} + y = y^2 e^x$

19. $\dfrac{dy}{dx} + \dfrac{y}{x} = \dfrac{y^2}{x^2}$

20. $\dfrac{dy}{dx} - y + (x^2 + 2x)y^2 = 0$

21. $x\,dy + y\,dx = x^3 y^5\,dx$

22. $\dfrac{dy}{dx} + y \tan x = y^4 \sin x$

23. Repeat Exercise 20 in Section 16.2 if the glucose is added at a rate $R(t)$, which is a function of time t. Find a formula for $A(t)$ in terms of $R(t)$.

24. A tank has 100 liters of solution containing 4 kilograms of sugar. A mixture with 10 grams of sugar per liter of solution is added at a rate of 200 milliliters per minute. At the same time, 100 milliliters of well-stirred mixture are

removed each minute. Find the amount of sugar in the tank as a function of time.

25. Repeat Exercise 24 if 300 milliliters of mixture are removed each minute.

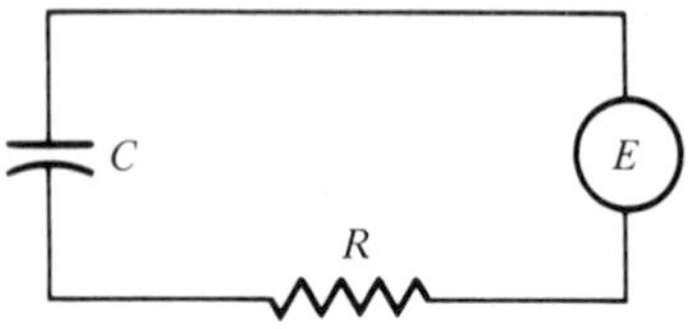

Figure 16.3

26. a) The current I in the RC-circuit shown in Fig. 16.3 must satisfy the differential equation

$$R\frac{dI}{dt} + \frac{I}{C} = \frac{dE}{dt}.$$

If $E(t) = E_0 \sin(\omega t)$, $t \geq 0$, where E_0 and ω are constants, solve this differential equation for $I(t)$, and show that the solution can be written in the form

$$I(t) = Ae^{-t/(RC)} + \frac{E_0}{Z}\sin(\omega t - \phi),$$

where A is an arbitrary constant and

$$Z = \sqrt{R^2 + \frac{1}{\omega^2 C^2}}, \qquad \phi = \mathrm{Tan}^{-1}\left(-\frac{1}{\omega CR}\right).$$

b) What is the value of A if the current in the circuit at time $t = 0$ when $E(t)$ is connected is I_0?

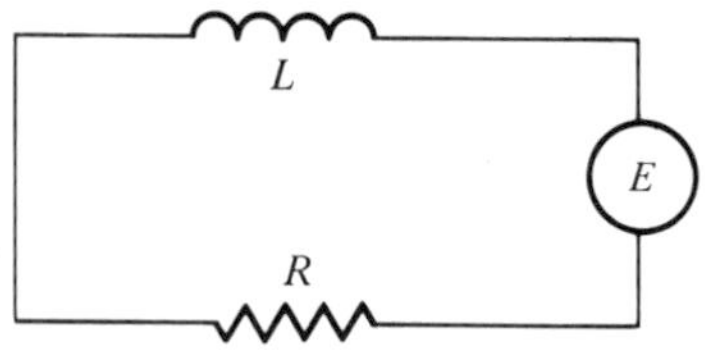

Figure 16.4

27. a) The current I in the LR-circuit shown in Fig. 16.4 must satisfy the differential equation

$$L\frac{dI}{dt} + RI = E.$$

If $E(t)$ is as in Exercise 26, solve this differential equation for $I(t)$, and show that the solution can be written in the form

$$I(t) = Ae^{-Rt/L} + \frac{E_0}{Z}\sin(\omega t - \phi),$$

where A is an arbitrary constant and

$$Z = \sqrt{R^2 + \omega^2 L^2}, \qquad \phi = \mathrm{Tan}^{-1}\left(\frac{\omega L}{R}\right).$$

b) What is the value of A if the current in the circuit at time $t = 0$ when $E(t)$ is connected is I_0?

***28.** Repeat Exercise 21 in Section 16.2 given that the diver descends slowly to the bottom. Assume that his descent is at a constant rate over a time interval of length T. Assume also that the maximum pressure $\overline{P}$ is proportional to the depth below the surface.

16.4 SECOND-ORDER EQUATIONS REDUCIBLE TO TWO FIRST-ORDER EQUATIONS

We now consider two types of second-order differential equations that can be reduced to a pair of first-order equations.

Type I: Dependent Variable Missing

If a second-order differential equation in $y(x)$ is explicitly independent of y, then it is of the form

$$F(x, y', y'') = 0. \tag{16.26}$$

In such a case we set

$$v = \frac{dy}{dx} \quad \text{and} \quad \frac{d^2y}{dx^2} = \frac{dv}{dx}. \tag{16.27}$$

If we substitute these into (16.26), we obtain

$$F(x, v, v') = 0,$$

a first-order differential equation in $v(x)$. If we can solve this equation for $v(x)$, we can then integrate $dy/dx = v(x)$ for $y(x)$.

Example 6 Find a general solution for each of the following differential equations:

a) $xy'' - y' = 0$;

b) $y'' = (y')^2$.

Solution a) Since y is explicitly missing, we substitute $v = y'$ and $y'' = v'$:

$$x\frac{dv}{dx} - v = 0.$$

Variables are now separable,

$$\frac{dv}{v} = \frac{dx}{x},$$

and a solution for $v(x)$ is

$$\ln|v| = \ln|x| + C \quad \text{or} \quad v = Dx \qquad (D = \pm e^C).$$

Because $v = dy/dx$,

$$\frac{dy}{dx} = Dx,$$

and we can integrate for

$$\begin{aligned} y(x) &= \frac{D}{2}x^2 + E \\ &= Fx^2 + E \qquad (F = D/2). \end{aligned}$$

b) Since y is explicitly missing, we substitute $v = y'$ and $y'' = v'$ to get

$$\frac{dv}{dx} = v^2.$$

Variables are again separable,

$$\frac{dv}{v^2} = dx,$$

and a solution for $v(x)$ is

$$-\frac{1}{v} = x + C.$$

Consequently,

$$v = \frac{dy}{dx} = -\frac{1}{x + C}.$$

Integration now yields

$$y(x) = -\ln|x + C| + D. \ \square$$

Type II: Independent Variable Missing If a second-order differential equation in $y(x)$ is explicitly independent of x, then it is of the form

$$F(y, y', y'') = 0. \tag{16.28}$$

In such a case we set

$$v = \frac{dy}{dx} \quad \text{and} \quad \frac{d^2y}{dx^2} = \frac{dv}{dx} = \frac{dv}{dy}\frac{dy}{dx} = v\frac{dv}{dy}. \tag{16.29}$$

When we substitute these into (16.28), we obtain

$$F\left(y, v, v\frac{dv}{dy}\right) = 0,$$

a first-order differential equation in $v(y)$. If we can solve this equation for $v(y)$, we can separate $dy/dx = v(y)$ and integrate for $y(x)$.

Example 7 Find a general solution for each of the following differential equations:

a) $yy'' + (y')^2 = 1$;

b) $y'' = (y')^2$.

Solution a) Since x is explicitly missing, we substitute

$$v = y' \quad \text{and} \quad y'' = \frac{dv}{dx} = \frac{dv}{dy}\frac{dy}{dx} = v\frac{dv}{dy}$$

to get

$$yv\frac{dv}{dy} + v^2 = 1.$$

If variables are separated, we have

$$\frac{v\,dv}{v^2 - 1} = -\frac{dy}{y},$$

and a general solution for $v(y)$ is

$$\tfrac{1}{2}\ln|v^2 - 1| = -\ln|y| + C.$$

Thus,

$$|v^2 - 1| = \frac{e^{2C}}{y^2},$$

from which we have

$$\frac{dy}{dx} = v = \frac{\pm\sqrt{D + y^2}}{y} \qquad (D = \pm e^{2C}).$$

We separate variables again to get

$$\frac{y\,dy}{\sqrt{y^2 + D}} = \pm dx,$$

and obtain an implicit definition of the solution $y(x)$,

$$\sqrt{y^2 + D} = \pm x + E.$$

b) Since x is explicitly missing, we again substitute from (16.29) to obtain

$$v\frac{dv}{dy} = v^2.$$

If variables are separated,

$$\frac{dv}{v} = dy,$$

then a solution for $v(y)$ is

$$\ln|v| = y + C.$$

Thus,

$$|v| = e^{y+C}$$

and

$$\frac{dy}{dx} = v = De^y \qquad (D = \pm e^C).$$

We separate variables again,

$$e^{-y}\,dy = D\,dx,$$

and find an implicit definition of the solution $y(x)$:

$$-e^{-y} = Dx + E. \square$$

In each of Examples 6 and 7 we solved the differential equation $y'' = (y')^2$ since both the independent variable x and the dependent variable y are missing. Although the solutions appear different, each is easily derivable from the other.

Let us summarize the results of this section. Substitutions (16.27) for differential equation (16.26), with the dependent variable missing, replace the second-order differential equation with two first-order equations: a first-order equation in $v(x)$, followed by a first-order equation in $y(x)$. Contrast this with the method for Eq. (16.28) with the independent variable missing. Substitutions (16.29) again replace the second-order equation with two first-order equations. However, the first first-order equation is in $v(y)$, so that for this equation, y is the independent variable rather than the dependent variable. The second first-order equation is again one for $y(x)$.

EXERCISES 16.4

In each of Exercises 1–10, find a general solution for the differential equation.

1. $xy'' + y' = 4x$
2. $2yy'' = 1 + (y')^2$
3. $y'' = y' + 2x$
4. $x^2y'' = (y')^2$
5. $y'' \sin x + y' \cos x = \sin x$
6. $y'' = [1 + (y')^2]^{3/2}$
7. $y'' + 4y = 0$
8. $y'' = yy'$
9. $y'' + (y')^2 = 1$
10. $(y'')^2 = 1 + (y')^2$
11. When a flexible cable of uniform density hangs between two fixed points A and B (Fig. 16.5), the differential

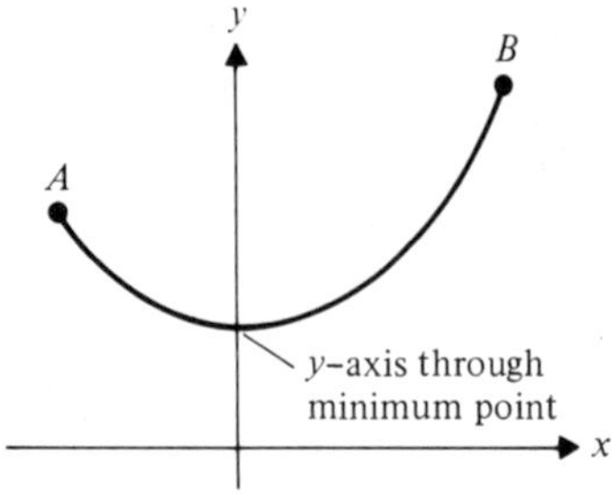

Figure 16.5

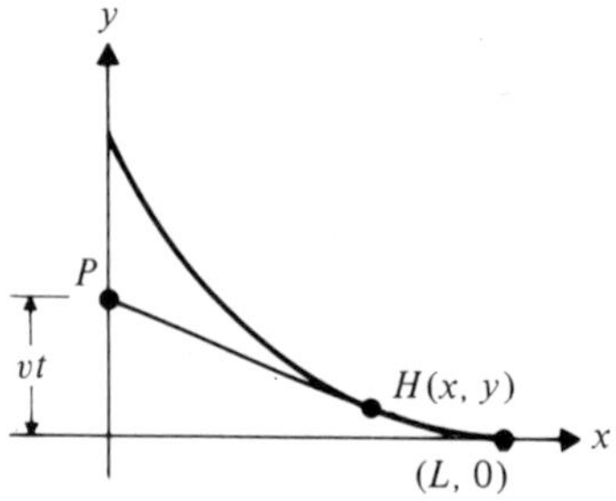

Figure 16.6

equation that describes the shape of the cable is

$$\frac{d^2y}{dx^2} = k\sqrt{1 + \left(\frac{dy}{dx}\right)^2},$$

where k is a constant. Solve this equation for the curve $y = y(x)$.

* **12.** A hawk at position $(L, 0)$ (Fig. 16.6) spots a pigeon at the origin flying with speed v in the positive y-direction. The hawk immediately takes off after the pigeon with speed $V > v$, always heading directly toward the pigeon. After a time t, the pigeon will be at position $P(0, vt)$. If the equation of the pursuit curve of the hawk is $y = y(x)$, then during time t, the hawk will have traveled distance Vt along this curve. But distance along this curve can be calculated by means of the definite integral

$$\int_x^L \sqrt{1 + \left(\frac{dy}{dx}\right)^2}\, dx.$$

a) Show that $y(x)$ must satisfy the integrodifferential equation

$$x\frac{dy}{dx} - y = \frac{v}{V}\int_L^x \sqrt{1 + \left(\frac{dy}{dx}\right)^2}\, dx.$$

b) Differentiate this equation to obtain the second-order differential equation

$$x\frac{d^2y}{dx^2} = \frac{v}{V}\sqrt{1 + \left(\frac{dy}{dx}\right)^2}.$$

c) Solve this differential equation for the pursuit curve of the hawk.

d) Show that the hawk catches the pigeon a distance $vVL/(V^2 - v^2)$ up the y-axis.

16.5 NEWTONIAN MECHANICS

One of the most important applications of differential equations is in the study of moving particles and objects. In classical mechanics, motion is governed by Newton's second law (12.107), which states that when an object of constant mass m is subjected to a force $\mathbf{F}$, the resultant acceleration is described by

$$\mathbf{F} = m\mathbf{a}. \tag{16.30}$$

When $\mathbf{F}$ is given, this is an algebraic equation giving acceleration $\mathbf{a}$. If we substitute $\mathbf{a} = d\mathbf{v}/dt$, we obtain a first-order differential equation

$$\mathbf{F} = m\frac{d\mathbf{v}}{dt} \tag{16.31}$$

for velocity $\mathbf{v}$ as a function of time t. If we substitute $\mathbf{a} = d^2\mathbf{r}/dt^2$, we obtain a second-order differential equation for position $\mathbf{r}$ as a function of time:

$$\mathbf{F} = m\frac{d^2\mathbf{r}}{dt^2}. \tag{16.32}$$

In practice it is seldom this simple. Often $\mathbf{F}$ is not given as a function of time, but as a function of position, or velocity, or some other variable. In such cases we may have to change the dependent or independent variable, or both, in order to solve the differential equation.

In this section we will be concerned with motion in one direction only. If $r(t)$, $v(t)$, $a(t)$, and F are the components of $\mathbf{r}(t)$, $\mathbf{v}(t)$, $\mathbf{a}(t)$, and $\mathbf{F}$ in this direction, then Eqs. (16.30)–(16.32) take the forms

$$F = ma, \tag{16.33}$$

$$F = m\frac{dv}{dt}, \tag{16.34}$$

$$F = m\frac{d^2r}{dt^2}, \tag{16.35}$$

respectively.

Example 8 A paratrooper and his parachute have a mass of 100 kilograms. At the instant his parachute opens, the paratrooper is traveling vertically downward at 10 meters per second. Air resistance on the chute varies directly as the instantaneous velocity and is 800 newtons when the paratrooper's velocity is 12 meters per second. Find the position and velocity of the paratrooper as functions of time.

Solution Let us measure y as positive downward, taking $y = 0$ and time $t = 0$ at the instant the paratrooper opens his parachute (Fig. 16.7). If F_a is the component of the force of air resistance, then

$$F_a = -kv.$$

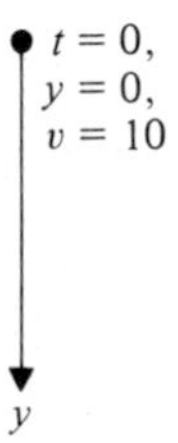

Figure 16.7

Since $F_a = -800$ when $v = 12$, it follows that $-800 = -12k$ or $k = 200/3$. Since the total force on the paratrooper during the fall has component $100g - 200v/3$ ($g = 9.81$), Newton's second law gives

$$100\frac{dv}{dt} = 100g - \frac{200}{3}v, \qquad t \geq 0.$$

Note that the force on the right is not a function of t so that the equation is not immediately integrable. It is however separable,

$$\frac{dv}{2v - 3g} = -\frac{dt}{3},$$

and a general solution for $v(t)$ is defined implicitly by

$$\frac{1}{2}\ln|2v - 3g| = -\frac{t}{3} + C.$$

An explicit solution is found by solving this equation for v:

$$v(t) = \frac{3g}{2} + De^{-2t/3} \qquad (D = \pm e^{2C}/2).$$

Since $v = 10$ when $t = 0$,

$$10 = \frac{3g}{2} + D \quad \text{or} \quad D = 10 - \frac{3g}{2}.$$

The velocity of the paratrooper as a function of time is therefore

$$v(t) = \frac{3g}{2} + \left(10 - \frac{3g}{2}\right)e^{-2t/3} \text{ m/s}.$$

Because $v(t) = dy/dt$, we set

$$\frac{dy}{dt} = \frac{3g}{2} + \left(10 - \frac{3g}{2}\right)e^{-2t/3},$$

and this equation is immediately integrable to

$$y(t) = \frac{3g}{2}t - \frac{3}{2}\left(10 - \frac{3g}{2}\right)e^{-2t/3} + D.$$

Since $y(0) = 0$,

$$0 = -\frac{3}{2}\left(10 - \frac{3g}{2}\right) + D \quad \text{or} \quad D = 15 - \frac{9g}{4}.$$

Consequently, the distance fallen by the paratrooper as a function of time is given by

$$y(t) = \frac{3g}{2}t - \left(15 - \frac{9g}{4}\right)e^{-2t/3} + \left(15 - \frac{9g}{4}\right) = 14.7t + 7.07(e^{-2t/3} - 1) \text{ m}.$$

Note that the velocity of the paratrooper does not increase indefinitely. In fact, as time passes a limiting velocity is approached:

$$\lim_{t\to\infty} v(t) = \lim_{t\to\infty}\left\{\frac{3g}{2} + \left(10 - \frac{3g}{2}\right)e^{-2t/3}\right\} = \frac{3g}{2} = 14.7 \text{ m/s}.$$

This is called the terminal velocity of the paratrooper; it is a direct result of the assumption that air resistance is proportional to instantaneous velocity. □

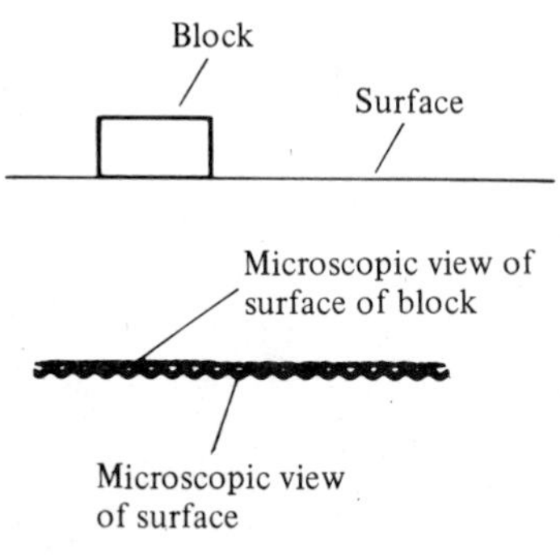

Figure 16.8

Whenever we move an object such as the block shown in Fig. 16.8 over a surface, there is resistance to the motion. This resistance, called *friction*, is due to the fact that the interface between the block and the surface is not smooth; each surface is inherently rough, and this roughness retards the motion of one surface over the other. In effect, a force opposing the motion of the block is created, and this force is called the *force of friction*. Many experiments have been performed to obtain a functional representation for this force. It turns out that when the block in Fig. 16.8 slides along a horizontal surface, the magnitude of the force of friction opposing the motion is given by

$$|\mathbf{F}| = \mu mg, \tag{16.36}$$

where m is the mass of the block, $g > 0$ is the acceleration due to gravity, and μ is a constant called the *coefficient of kinetic friction*. In other words, the force of friction is directly proportional to the weight mg of the block. We caution the reader that this result is valid for the situation shown in Fig. 16.8, but it may not be valid for other configurations.

Example 9 A block of mass 2 kilograms is given an initial speed of 5 meters per second along a horizontal surface. If the coefficient of kinetic friction between the block and surface is $\mu = 0.25$, how far does the block slide before stopping?

Solution Let us measure x as positive in the direction of motion (Fig. 16.9), taking $x = 0$ and $t = 0$ at the instant the block is released. The x-component of the force of friction on the block is then

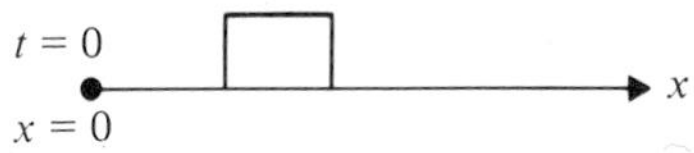

Figure 16.9

$$F = -0.25(2)g = -\frac{g}{2} \qquad (g = 9.81).$$

According to Newton's second law,

$$2\frac{dv}{dt} = -\frac{g}{2},$$

from which we get

$$v(t) = -\frac{g}{4}t + C.$$

Since $v(0) = 5$, it follows that $C = 5$, and

$$v(t) = -\frac{g}{4}t + 5 \text{ m/s}.$$

But $v = dx/dt$, and hence

$$\frac{dx}{dt} = -\frac{g}{4}t + 5.$$

Integration gives

$$x(t) = -\frac{g}{8}t^2 + 5t + D.$$

Because we chose $x = 0$ at time $t = 0$, D must also be zero, and

$$x(t) = -\frac{g}{8}t^2 + 5t \text{ m}.$$

The block comes to rest when $v = 0$; i.e., when

$$-\frac{g}{4}t + 5 = 0 \quad \text{or} \quad t = \frac{20}{g}.$$

The position of the block at this time is

$$x = -\frac{g}{8}\left(\frac{20}{g}\right)^2 + 5\left(\frac{20}{g}\right) = 5.10.$$

The block therefore slides 5.10 meters before stopping. □

EXERCISES 16.5

1. A car of mass 1500 kilograms starts from rest at an intersection. The motor exerts a constant force of 3000 newtons, and air friction causes a resistive force whose magnitude in newtons is equal to the speed of the car in meters per second. Find the speed of the car and its distance from the intersection after 10 seconds.

2. You are called on as an expert to testify in a traffic accident hearing. The question concerns the speed of a car that made an emergency stop with its brakes locked and wheels sliding. The skid mark on the road measured 9 meters. If we assume that the coefficient of kinetic friction between the tires and the road was less than one, what can

you say about the speed of the car before the brakes were applied? Are you testifying for the prosecution or the defense?

3. A boat and its contents have a mass of 250 kilograms. Water exerts a resistive force on the motion of the boat that is proportional to the instantaneous speed of the boat and is 200 newtons when the speed is 30 kilometers per hour.

a) If the boat starts from rest and the motor exerts a constant force of 250 newtons in the direction of motion, find the speed of the boat as a function of time.

b) What is the limiting speed of the boat?

4. a) In Example 8 it was necessary to solve the equation

$$\frac{1}{2}\ln|2v - 3g| = -\frac{t}{3} + C$$

for $v = v(t)$ and use the condition $v(0) = 10$ to evaluate C. We chose first to solve for $v(t)$ and then to evaluate the constant. Show the details of this analysis.

b) Instead, first use the condition $v(0) = 10$ to evaluate C, and then solve the equation for $v = v(t)$.

5. A body of weight W falls from rest under gravity. It is acted on by a resistive force that is proportional to the square root of the speed at any instant. If the magnitude of this resistive force is F^* when the speed is v^*, find an expression for its terminal speed.

6. A spring (with constant k) is attached on one end to a wall and on the other end to a mass M (Fig. 16.10). The mass is set into motion along the x-axis by pulling it a distance x_0 to the right of the position it would occupy were the spring in the unstretched position and given speed v_0 to the left. During the subsequent motion, there is a frictional force between M and the horizontal surface with coefficient of kinetic friction equal to μ.

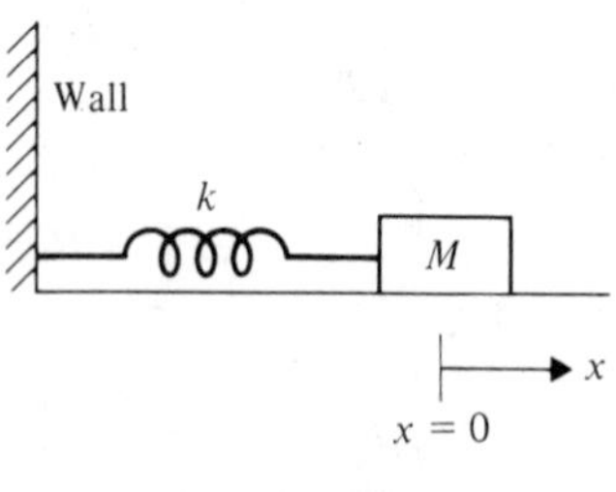

Figure 16.10

a) Show that the differential equation describing the motion of M is

$$M\frac{d^2x}{dt^2} = -kx + \mu Mg, \qquad x(0) = x_0,$$

$$v(0) = -v_0,$$

if we take $t = 0$ at the instant that motion is initiated. When is this equation valid?

b) Since t is explicitly missing from the equation in (a), show that it can be rewritten in the form

$$Mv\frac{dv}{dx} = -kx + \mu Mg, \qquad v(x_0) = -v_0,$$

and that therefore

$$\frac{k}{2}(x_0^2 - x^2) = \frac{M}{2}(v^2 - v_0^2) + \mu Mg(x_0 - x).$$

Interpret each of the terms in this equation physically.

* c) If x' represents the position at which M comes to rest for the first time, use the equation in (b) to determine x' as a function of μ, M, g, k, x_0, and v_0. Discuss the possibilities of x' being positive, negative, and zero.

7. When a body falls in air, it is acted on by gravity and also by a force due to air resistance that is proportional to its instantaneous speed.

a) If the body is initially projected upward, find its velocity as a function of time, and sketch a graph of the function.

b) Repeat (a) if the body is projected downward with velocity less than its terminal velocity.

c) Repeat (b) if the initial velocity is greater than its terminal velocity.

C **8.** A 1-kilogram rock is thrown vertically upward with a speed of 20 meters per second. Air resistance to its motion when measured in newtons has magnitude equal to one-tenth its speed in meters per second. Find the maximum height attained by the rock.

C **9.** A stone of mass 100 grams is thrown vertically upward with a speed of 20 meters per second. Air exerts a resistive force on the stone proportional to its instantaneous speed, and has a magnitude of $\frac{1}{10}$ newton when the speed of the stone is 10 meters per second. Find an equation defining the time when the stone will return to its projection point. Solve this equation numerically, and compare with the time taken if air resistance is neglected.

* **10.** A huge cannon fires a projectile with initial velocity v_0 directly toward the moon (Fig. 16.11). Newton's universal law of gravitation states that when the projectile is at a distance r above the earth's surface, the force of attraction of the earth on the projectile has magnitude

$$\frac{GMm}{(r + R)^2},$$

where G is a constant, M is the mass of the earth, and m is

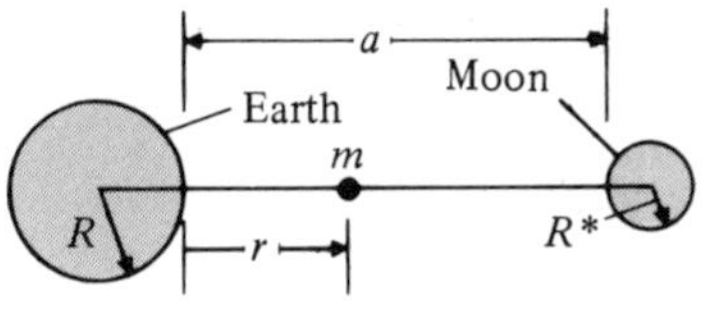

Figure 16.11

the mass of the projectile. At this point the moon's gravitational attraction has magnitude

$$\frac{GM^*m}{(a + R^* - r)^2}$$

where M^* is the mass of the moon.

a) Show that if only the above two forces of gravitational attraction are considered to act on the projectile, then the differential equation describing its motion is

$$\frac{d^2r}{dt^2} = -\frac{gR^2}{(r + R)^2} + \frac{g^*R^{*2}}{(a + R^* - r)^2},$$

where $g > 0$ and $g^* > 0$ are gravitational accelerations on the surfaces of the earth and the moon.

b) Prove that the velocity of the projectile at a distance r above the surface of the earth is defined by

$$v^2 = \frac{2gR^2}{r + R} + \frac{2g^*R^{*2}}{a + R^* - r} + v_0^2 - 2gR - \frac{2g^*R^{*2}}{a + R^*}.$$

* **11.** Newton's second law states that if an object of variable mass $m(t)$ is subjected to a force $\mathbf{F}(t)$, then

$$\frac{d}{dt}(m\mathbf{v}) = \mathbf{F},$$

where t is time and $\mathbf{v}$ is the velocity of the object. A uniform chain of length 3 meters and mass 6 kilograms is held by one end so that the other end just touches the floor. If the chain is released, find the velocity of the falling chain as a function of the length of chain still falling. How fast does the end hit the floor?

16.6 POPULATION DYNAMICS

The rate of growth of a country's population depends on many factors: the birth rate, death rate, immigration rate, size of the country, availability of food, state of technology, natural disasters, etc. It follows, therefore, that any attempt to develop a mathematical formula that would account for all possible factors would be impossible. Some factors, on the other hand, are much more important to the growth of a population than others. Perhaps then we could develop a mathematical model that would initially account for the important factors, and that could be refined at a later stage to incorporate minor factors. Our problem, then, is to develop a model that predicts populations of countries with some reasonable amount of accuracy.

On the basis of the census figures of a country such as the United States, it is possible to calculate the number of births per year in the population. If we divide this by the existing size of the population, we get what is called the relative birth rate b of the population (or the individual birth rate). It is an average figure, but it is found that for many populations the relative birth rate is reasonably constant over long periods of time. In a similar way, it is possible to define a relative death rate d (death due to natural causes), and the constant $k = b - d$ is called the relative growth rate of the population. It is the increase in population per individual per year that can be attributed solely to births and deaths. On the other hand, if $N = N(t)$ represents the size of the population at any time t (in years), then

$$\frac{1}{N}\frac{dN}{dt}$$

is the total change in population per individual per year, and should therefore account for all factors that affect the size of the population. If we were to decide that the most important factor in determining the size of a population were its birth and death rates, we would write

$$\frac{1}{N}\frac{dN}{dt} = k,$$

or

$$\frac{dN}{dt} = kN. \tag{16.37}$$

This is a mathematical model (called the *Malthusian model*) that attempts to describe the size of a population in terms only of its birth and death rates.

Because (16.37) is a separable differential equation

$$\frac{1}{N}\,dN = k\,dt,$$

we obtain a general solution

$$\ln|N| = kt + D.$$

Because N is always nonnegative, we can state that

$$N = Ce^{kt} \qquad (C = e^{D}). \tag{16.38}$$

Table 16.1

Year	1790	1800	1810	1820	1830	1840	1850	1860	1870	1880
Population (millions)	3.93	5.31	7.24	9.64	12.87	17.07	23.19	31.44	38.56	50.16

Year	1890	1900	1910	1920	1930	1940	1950	1960	1970
Population (millions)	62.95	76.00	91.97	105.71	122.78	131.67	150.70	179.32	203.19

In Table 16.1 and Fig. 16.12 we show approximate census figures for the United States from 1790 to 1970. Note that the population figure in 1870 is uncharacteristic of the previous growth from 1790 to 1860 (and history makes the reason clear). Suppose we attempt to fit the exponential curve (16.38) to the population between 1790 and 1860. One way to do this is to demand that $N(1790) = 3.93$ and $N(1860) = 31.44$:

$$3.93 = Ce^{1790k}, \qquad 31.44 = Ce^{1860k}.$$

The solution of these equations for C and k is $C = 3.17 \times 10^{-23}$, $k = 0.0297$. Consequently,

$$N = 3.17 \times 10^{-23} e^{0.0297t},$$

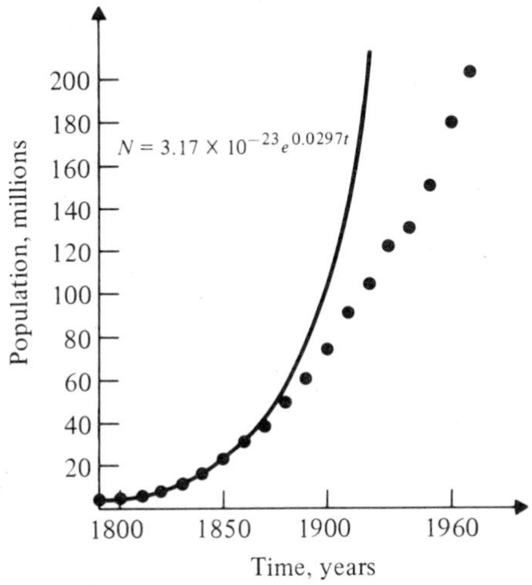

Figure 16.12

and this curve is shown in Fig. 16.12. It is indeed an excellent description of the United States population in the years between 1790 and 1860.

It is equally obvious, however, that this particular exponential function is not an adequate representation of the population beyond 1860. In fact no exponential function can describe the census points in Table 16.1. This implies that after 1860 other factors besides simply pure birth and death processes become important in determining the size of the population. One possible explanation is that because (16.37) predicts exponential growth, it must assume that there are sufficient resources (area, food, water, etc.) to sustain any level of population. This is unrealistic in the long term because there is a limit to the size of the population that any environment can support. This limit, called the *carrying capacity* C of the environment, is such that as the size of the population draws closer to C its growth rate must slow down. One way to incorporate this into model (16.37) is to set

$$\frac{dN}{dt} = kN\left(1 - \frac{N}{C}\right) \tag{16.39}$$

(called the *logistic model*). For small N, $1 - N/C$ is approximately equal to unity, and N will experience exponential growth, but as N approaches C, this factor diminishes and the growth rate decreases.

To solve this differential equation, we again separate variables,

$$k\,dt = \frac{C\,dN}{N(C-N)} = \left\{\frac{1}{N} + \frac{1}{C-N}\right\} dN,$$

and obtain a general solution

$$kt + E = \ln N - \ln(C - N).$$

We now solve this equation for $N(t)$ by first exponentiating

$$\frac{N}{C-N} = De^{kt} \qquad (D = e^E),$$

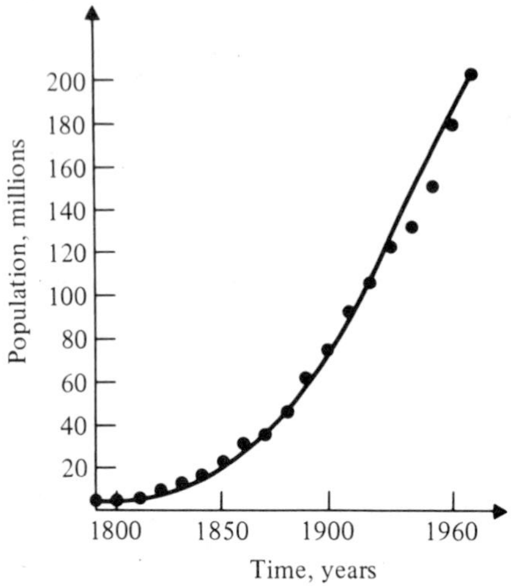

Figure 16.13

and cross-multiplying to get

$$N = (C - N)De^{kt}.$$

Thus,

$$N(t) = \frac{CDe^{kt}}{1 + De^{kt}} = \frac{C}{1 + Fe^{-kt}} \qquad (F = 1/D), \tag{16.40}$$

and this is called the *logistic growth function* (or *logistic curve*). If we use the census figures of 1790, 1880, and 1970 to determine the three constants F, k, and C, we obtain

$$N(t) = \frac{257}{1 + 3.54 \times 10^{25} e^{-0.0305t}}.$$

This function and the census figures of Table 16.1 are shown in Fig. 16.13. Note that this logistic function predicts an upper limit of 257 million people for the United States, not very realistic in light of today's population. It has resulted from assuming that the population of the United States can reasonably be described by a logistic function that passes through the census points in 1790, 1880, and 1970. This implies, then, that such a representation is still not an adequate description of the United States population; other factors besides birth and death rates and carrying capacity must be incorporated. Some of these will be introduced in the exercises.

EXERCISES 16.6

1. Bacteria in a culture increase at a rate proportional to the number present. If the original number of bacteria increases by 50% in one-half hour, how long does it take to quintuple the original number?

2. Two rodents begin a colony. During the next ten months they are permitted to reproduce and their numbers are recorded in the following table.

Time t (months)	0	2	5	7	10
Number N of rodents	2	5	15	35	110

a) Find a formula for $N(t)$ based on the Malthusian model and use the conditions $N(0) = 2$ and $N(2) = 5$ to determine the unknown constants in the function.

b) Repeat (a) but use the conditions $N(0) = 2$ and $N(10) = 110$ to evaluate the constants.

c) Sketch both functions in (a) and (b) and the data points all on the same graph.

3. If we modify the Malthusian model for predicting the population of a country to incorporate a constant immigration rate of λ individuals per unit time, then the model becomes

$$\frac{dN}{dt} = kN + \lambda.$$

Solve this differential equation for $N = N(t)$.

4. Draw the logistic curve (16.40), indicating any relative extrema and points of inflection. Assume that $N(0) = N_0 < C/2$.

5. An initial population of 100 inhabitants enters an area with a carrying capacity of 100,000. In the first year, the population increases to 120. Assume that the population follows logistic growth.

a) Determine the population as an explicit function of time.

b) How many years will it take the population to reach 95,000?

6. a) A country institutes an immigration policy that states that the immigration rate shall be proportional to the difference between the carrying capacity C and the present size N of the population. Show that a mathematical model to describe $N(t)$ is

$$\frac{dN}{dt} = -\frac{k}{C}(N - C)(N + C\lambda/k),$$

where k is the relative growth rate of the population attributed only to the birth and death rates, and $\lambda > 0$ is a constant.

b) Solve this differential equation, and show that

$$N(t) = \frac{C - (FC\lambda/k)e^{-(k+\lambda)t}}{1 + Fe^{-(k+\lambda)t}}.$$

7. a) For some species, if the population drops below a certain level, the species will become extinct. Given that m is the minimum viable population for such a species, one way to incorporate this condition into the logistic model is to write

$$\frac{dN}{dt} = kN\left(1 - \frac{N}{C}\right)\left(1 - \frac{m}{N}\right).$$

Solve this differential equation subject to the condition $N(0) = N_0$ to obtain

$$N(t) = \frac{[(C - N_0)/(N_0 - m)]me^{-[(C-m)/C]kt} + C}{1 + [(C - N_0)/(N_0 - m)]e^{-[(C-m)/C)]kt}}.$$

b) Use the result in (a) to show that if $N_0 < m$, then extinction must occur.

The logistic model (16.39) incorporates the term $1 - N/C$ in order to decrease dN/dt for N approaching C. In Exercises 8–10, we show that there are other ways of doing this.

8. The Gompertz model multiplies the right-hand side of the Malthusian model by the factor e^{-lt} ($l > 0$ a constant):

$$\frac{dN}{dt} = kNe^{-lt}.$$

Solve this differential equation for $N(t)$, and sketch the curve $N = N(t)$. Does it predict an upper limit for $N(t)$?

9. The monomolecular model multiplies the right-hand side of the Malthusian model by the factor $be^{-kt}/(1 - be^{-kt})$ ($0 < b < 1$ a constant):

$$\frac{dN}{dt} = kN\frac{be^{-kt}}{1 - be^{-kt}}.$$

Solve this differential equation for $N(t)$, and sketch the curve $N = N(t)$. Does it predict an upper limit for $N(t)$?

10. The Von Bertalanffy model is much like the monomolecular model,

$$\frac{dN}{dt} = kN\frac{3be^{-kt}}{1 - be^{-kt}}.$$

Solve this differential equation for $N(t)$, and sketch the curve $N = N(t)$. Does it predict an upper limit for $N(t)$?

***11.** Verify that if the constants C, k, and F in the logistic function (16.40) are evaluated by using the census figures of 1790, 1880, and 1970 for the United States, then

$$C = 257, \quad F = 3.54 \times 10^{25}, \quad \text{and} \quad k = 0.0305.$$

16.7 LINEAR DIFFERENTIAL EQUATIONS

Throughout Chapters 5, 8, and 9 and Sections 16.1–16.6, we have stressed the use of differential equations in solving applied problems. We have considered examples from such diverse areas as engineering, geometry, ecology, and psychology, hoping thereby to illustrate how valuable differential equations can be in modeling physical situations mathematically. Perhaps the most important type of differential equation is the *linear differential equation.* Attesting to the importance of linear differential equations is the number of texts that has been written on this topic alone; therefore, our discussion of linear equations in the next five sections can at best be an introduction.

▷ **Definition 16.4** A differential equation of the form

$$a_0(x)\frac{d^ny}{dx^n} + a_1(x)\frac{d^{n-1}y}{dx^{n-1}} + a_2(x)\frac{d^{n-2}y}{dx^{n-2}} + \cdots + a_{n-1}(x)\frac{dy}{dx} + a_n(x)y = F(x), \tag{16.41}$$

where $a_0(x) \not\equiv 0$, is called a *linear nth-order differential equation.* ◁

Note in particular that none of the derivatives of $y(x)$ are multiplied together, nor are they squared or cubed or taken to any other power, nor do they appear as the argument of any transcendental function. All we see is a function of x multiplying y, plus a function of x multiplying the first derivative of y, plus a function of x multiplying the second derivative of y, and so on.

If $n = 1$, Eq. (16.41) reduces to

$$a_0(x)\frac{dy}{dx} + a_1(x)y = F(x),$$

and at any point at which $a_0(x) \neq 0$, we can divide to obtain

$$\frac{dy}{dx} + \frac{a_1(x)}{a_0(x)}y = \frac{F(x)}{a_0(x)}.$$

If we set $P(x) = a_1(x)/a_0(x)$ and $Q(x) = F(x)/a_0(x)$, we have

$$\frac{dy}{dx} + P(x)y = Q(x);$$

i.e., every linear first-order differential equation can be expressed in this form. We discussed equations of this type in Section 16.3, where it was shown that such equations have a general solution

$$y(x) = e^{-\int P(x)\,dx}\left\{\int Q(x)e^{\int P(x)\,dx}\,dx + C\right\}.$$

In other words, we already know how to solve linear first-order differential equations, and therefore our discussion in the next five sections is directed primarily at second- and higher-order equations. Keep in mind, however, that all results are also valid for first-order linear equations.

Because Eq. (16.41) is so cumbersome, we introduce notation to simplify its representation. In particular, if we use the notation $D = d/dx$, $D^2 = d^2/dx^2$, etc., we can write

$$a_0(x)D^ny + a_1(x)D^{n-1}y + a_2(x)D^{n-2}y + \cdots + a_{n-1}(x)Dy + a_n(x)y = F(x), \tag{16.42}$$

or

$$\{a_0(x)D^n + a_1(x)D^{n-1} + a_2(x)D^{n-2} + \cdots + a_{n-1}(x)D + a_n(x)\}y = F(x). \tag{16.43}$$

The quantity in braces is called a *differential operator*; it operates on whatever follows it—in this case, y. It is a "differential" operator because it operates by taking derivatives. Because the operator involves only x's and D's, we denote it by

$$\phi(x, D) = a_0(x)D^n + a_1(x)D^{n-1} + a_2(x)D^{n-2} + \cdots + a_{n-1}(x)D + a_n(x). \tag{16.44}$$

The general linear nth-order differential equation can then be represented very simply by

$$\phi(x, D)y = F(x). \tag{16.45}$$

For example, the linear second-order differential equation

$$xy'' + y' + xy = 0$$

is called Bessel's differential equation of order zero. In operator notation we write

$$(xD^2 + D + x)y = 0,$$

or

$$\phi(x, D)y = 0,$$

where $\phi(x, D) = xD^2 + D + x$.

For the differential equation

$$y'' + 2y' - 3y = e^{-x},$$

we write

$$\phi(D)y = e^{-x},$$

where $\phi(D) = D^2 + 2D - 3$.

We now indicate the meaning of the term "linear." Suppose that L is an operator that operates on each function $y(x)$ in some set S. For example, L might be the operation that multiplies each function by 5, or perhaps squares each function, or perhaps differentiates each function. It is said to be a linear operator if it satisfies the following definition.

▷ **Definition 16.5** An operator L is said to be *linear* on a set of functions S if for any two functions $y_1(x)$ and $y_2(x)$ in S, and any constant c,

$$L(y_1 + y_2) = Ly_1 + Ly_2, \tag{16.46a}$$

$$L(cy_1) = c(Ly_1). \quad \triangleleft \tag{16.46b}$$

Many of the operations in calculus are therefore linear. For instance, taking limits is a linear operation, as is differentiation, antidifferentiation, and taking definite integrals. On the other hand, taking the square root of a positive function is not a linear operation, since

$$L(y_1 + y_2) = \sqrt{y_1 + y_2} \neq L(y_1) + L(y_2).$$

It is not difficult to show that the differential operator $\phi(x, D)$ in (16.44) is linear; i.e.,

$$\phi(x, D)(y_1 + y_2) = \phi(x, D)y_1 + \phi(x, D)y_2,$$
$$\phi(x, D)(cy_1) = c[\phi(x, D)y_1],$$

and because of this, differential equation (16.45) is also said to be linear. The following two differential equations are examples of equations that are not linear:

$$\frac{d^2y}{dx^2} + y^2 = x, \qquad \frac{d^2y}{dx^2}\frac{dy}{dx} = e^x.$$

In particular, if we substitute $y_1(x) + y_2(x)$ into the left-hand side of the first equation, we obtain

$$\frac{d^2}{dx^2}(y_1 + y_2) + (y_1 + y_2)^2 = \frac{d^2y_1}{dx^2} + \frac{d^2y_2}{dx^2} + y_1^2 + 2y_1y_2 + y_2^2.$$

If we substitute $y_1(x)$, then $y_2(x)$, and then add the results, we find a different expression:

$$\frac{d^2y_1}{dx^2} + y_1^2 + \frac{d^2y_2}{dx^2} + y_2^2.$$

EXERCISES 16.7

In each of Exercises 1–10, prove either that the operator L is linear or that it is not linear. In each case assume that the set S of functions on which L operates is the set of all functions on which L can operate. For instance, in Exercise 6, assume that S is the set of all functions $y(x)$ that have a first derivative dy/dx.

1. L multiplies functions $y(x)$ by 5
2. L multiplies functions $y(x)$ by $15x$
3. L adds the fixed function $z(x)$ to functions $y(x)$
4. L takes the limit of functions $y(x)$ as x approaches 3
5. L takes the limit of functions $y(x)$ as x approaches infinity
6. L takes the derivative of functions $y(x)$ with respect to x
7. L takes the third derivative of functions $y(x)$ with respect to x
8. L takes the antiderivative of functions $y(x)$ with respect to x
9. L takes the definite integral of functions $y(x)$ with respect to x from $x = -1$ to $x = 4$
10. L takes the cube root of functions $y(x)$

In Exercises 11–20, determine whether the given differential equation is linear or nonlinear. Write those equations that are linear in operator notation (16.45).

11. $2x\dfrac{d^2y}{dx^2} + x^3y = x^2 + 5$
12. $2x\dfrac{d^2y}{dx^2} + x^3y = x^2 + 5y$
13. $2x\dfrac{d^2y}{dx^2} + x^3y = x^2 + 5y^2$
14. $x\dfrac{d^3y}{dx^3} + 3x\dfrac{d^2y}{dx^2} - 2\dfrac{dy}{dx} + y = 10 \sin x$

15. $x\dfrac{d^3y}{dx^3} + 3x\dfrac{d^2y}{dx^2} - 2\dfrac{dy}{dx} + y^2 = 10 \sin x$

16. $y\dfrac{d^3y}{dx^3} + 3x\dfrac{d^2y}{dx^2} - 2\dfrac{dy}{dx} + y = 10 \sin x$

17. $y'' - 3y' - 2y = 9 \sec^2 x$

18. $yy'' + 3y' - 2y = e^x$

19. $\sqrt{1 + y'} + x^2 = 4$

20. $y'''' + y'' - y = \ln x$

21. The Laplace transform of a function $y(t)$ is defined as the function

$$L(y) = \int_0^\infty e^{-st}y(t)\,dt,$$

provided the improper integral converges. Show that if S is the set of all functions that have a Laplace transform, then the operation of taking the Laplace transform is linear on S.

22. The finite Fourier cosine transform of a function $y(x)$ is defined as

$$L(y) = \int_0^{2\pi} y(x) \cos nx\,dx, \qquad n \text{ a nonnegative integer,}$$

provided that the definite integral exists. Show that if S is the set of all functions that have a finite Fourier cosine transform, then the operation of taking the transform is linear on S.

16.8 HOMOGENEOUS LINEAR DIFFERENTIAL EQUATIONS

Two classes of linear differential equations present themselves: those for which $F(x) \equiv 0$, and those for which $F(x) \not\equiv 0$. In this section and the next we consider equations for which $F(x) \equiv 0$; we will discuss the more difficult class, in which $F(x) \not\equiv 0$, in Section 16.10. First let us name each of these classes of linear differential equations.

▷ **Definition 16.6** A linear differential equation $\phi(x, D)y = F(x)$ is said to be *homogeneous* if $F(x) \equiv 0$, and *nonhomogeneous* otherwise. ◁

The fundamental idea behind the solution of all linear differential equations is the following theorem.

▶ **Theorem 16.1** (*Superposition principle*). If $y_1(x), y_2(x), \ldots, y_m(x)$ are solutions of a homogeneous linear differential equation

$$\phi(x, D)y = 0,$$

then so too is any linear combination thereof,

$$C_1y_1(x) + C_2y_2(x) + \cdots + C_my_m(x)$$

(for arbitrary constants $C_1, C_2, \ldots, C_m$). ◀

Proof The proof requires only linearity of the operator $\phi(x, D)$, for

$$\begin{aligned}\phi(x, D)[C_1y_1(x) + \cdots + C_my_m(x)] &= C_1[\phi(x, D)y_1] + \cdots + C_m[\phi(x, D)y_m] \\ &= 0 + 0 + \cdots + 0 = 0. \square\end{aligned}$$

Solutions of a linear differential equation that are linearly combined to produce other solutions are said to be superposed—consequently, the name "superposition principle" for Theorem 16.1. For example, it is straightforward to verify that $y_1(x) = e^{4x}$ and $y_2(x) = e^{-3x}$ are solutions of the homogeneous equation $y'' - y' + 12y = 0$. The superposition principle then states that for any constants C_1 and C_2, the function $y(x) = C_1 y_1(x) + C_2 y_2(x) = C_1 e^{4x} + C_2 e^{-3x}$ must also be a solution. But because $y(x)$ contains two arbitrary constants, it is not only a solution, it is a general solution of the differential equation.

Similarly, superposition of the three solutions e^x, xe^x, and $x^2 e^x$ of the linear differential equation $y''' - 3y'' + 3y' - y = 0$ gives a general solution $y(x) = C_1 e^x + C_2 x e^x + C_3 x^2 e^x$.

Now we begin to see the importance of the superposition principle. If we can find n solutions $y_1(x), y_2(x), \ldots, y_n(x)$ of an nth-order homogeneous linear differential equation, then a general solution is

$$y(x) = C_1 y_1(x) + C_2 y_2(x) + \cdots + C_n y_n(x).$$

In other words, all that we need do is find n solutions; the superposition principle will do the rest. There is a problem, however, if we take things a little too literally. For instance, $y_1(x) = e^{4x}$ and $y_2(x) = 10e^{4x}$ are both solutions of $y'' - y' + 12y = 0$. By superposition, so too then is $y(x) = C_1 y_1 + C_2 y_2 = C_1 e^{4x} + 10C_2 e^{4x}$. But is it a general solution? The answer is no, because we could write $y(x) = (C_1 + 10C_2)e^{4x}$, and by setting $C_3 = C_1 + 10C_2$, we have $y(x) = C_3 e^{4x}$. Superposition of the solutions e^{4x} and $10e^{4x}$ has not therefore led to a general solution, and the reason is that they are essentially the same solution: $y_2(x)$ is $y_1(x)$ multiplied by a constant. Superposition does not therefore lead to a solution with two arbitrary constants.

In a similar way, $y_1(x) = e^x$, $y_2(x) = xe^x$, and $y_3(x) = 2e^x - 3xe^x$ are all solutions of $y''' - 3y'' + 3y' - y = 0$, and therefore so is $y(x) = C_1 y_1 + C_2 y_2 + C_3 y_3$. But because we can write

$$\begin{aligned} y(x) &= C_1 e^x + C_2 x e^x + C_3(2e^x - 3xe^x) \\ &= (C_1 + 2C_3)e^x + (C_2 - 3C_3)xe^x \\ &= C_4 e^x + C_5 x e^x, \end{aligned}$$

$y(x)$ is not a general solution. This is a direct result of the fact that $y_3(x)$ is a linear combination of the solutions $y_1(x)$ and $y_2(x)$; it is twice $y_1(x)$ minus three times $y_2(x)$.

Our problem seems to come down to this: If we have n solutions of an nth-order homogeneous linear differential equation, how can we determine whether superposition leads to a (general) solution that contains n arbitrary constants? Our examples have suggested that if any one of the solutions is a linear combination of the others, then a general solution is not obtained, and this is indeed true. If one of the solutions is a linear combination of the others, we say that the n solutions are *linearly dependent*; if no solution is a linear combination of the others, we say that the n solutions are *linearly independent*. We summarize our results in the following theorem.

▶ **Theorem 16.2** If $y_1(x), y_2(x), \ldots, y_n(x)$ are n linearly independent solutions of an nth-order homogeneous linear differential equation, then

$y(x) = C_1 y_1(x) + C_2 y_2(x) + \cdots + C_n y_n(x)$ is a general solution of the differential equation. ◀

What we should now do is devise a test to determine whether a set of n solutions is linearly independent or linearly dependent. For most examples, no test is really necessary; it is obvious whether one of the solutions can be written as a linear combination of the others. For those rare occasions when it is not obvious, Exercise 10 describes a test that can be used to determine whether functions are linearly independent.

In summary, the superposition principle states that solutions of a homogeneous linear differential equation can be superposed to produce other solutions. If n linearly independent solutions of an nth-order equation are superposed, a general solution is obtained. This is the importance of the superposition principle. We need not devise a method that will take us directly to a general solution; we need a method for finding n linearly independent solutions—the superposition principle will do the rest. Unfortunately, for completely general coefficients $a_i(x)$ in $\phi(x, D)$ (see Eq. 16.44), it is impossible to give a method that will always yield n linearly independent solutions of a homogeneous equation. There is, however, one special case of great practical importance in which it is always possible to produce n linearly independent solutions in a very simple way. This special case occurs when the coefficients $a_i(x)$ are all constants a_i, and this is the subject of Section 16.9.

Example 10 If $y_1(x) = \cos 3x$ and $y_2(x) = \sin 3x$ are solutions of the differential equation $y'' + 9y = 0$, find a general solution.

Solution Since $y_1(x)$ and $y_2(x)$ are linearly independent solutions (one is not a constant times the other), a general solution can be obtained by superposition:

$$y(x) = C_1 \cos 3x + C_2 \sin 3x. \square$$

Example 11 Given that $y_1(x) = e^{2x} \cos x$ and $y_2(x) = e^{2x} \sin x$ are solutions of the homogeneous linear differential equation $y'' - 4y' + 5y = 0$, find that solution which satisfies the conditions $y(\pi/4) = 1$, $y(\pi/3) = 2$.

Solution By superposition, a general solution of the differential equation is

$$y(x) = C_1 e^{2x} \cos x + C_2 e^{2x} \sin x = e^{2x}(C_1 \cos x + C_2 \sin x).$$

To satisfy the conditions $y(\pi/4) = 1$ and $y(\pi/3) = 2$, we have

$$1 = e^{\pi/2}(C_1/\sqrt{2} + C_2/\sqrt{2}), \qquad 2 = e^{2\pi/3}(C_1/2 + \sqrt{3}C_2/2).$$

Thus C_1 and C_2 are defined by the pair of equations

$$C_1 + C_2 = \sqrt{2}e^{-\pi/2}, \qquad C_1 + \sqrt{3}C_2 = 4e^{-2\pi/3},$$

the solution of which is

$$C_2 = \frac{4e^{-2\pi/3} - \sqrt{2}e^{-\pi/2}}{\sqrt{3} - 1} = 0.2713, \qquad C_1 = \sqrt{2}e^{-\pi/2} - C_2 = 0.0227.$$

The required solution is therefore

$$y(x) = e^{2x}(0.0227 \cos x + 0.2713 \sin x). \quad \square$$

We pointed out in Section 16.1 that a general solution of a differential equation might not contain all solutions of the equation. This is not true for linear differential equations. It can be shown that if $y(x)$ is a general solution of a linear differential equation on an interval I, then every solution of the differential equation on I can be obtained by specifying particular values for the arbitrary constants in $y(x)$. Here then is a very important class of differential equations for which a general solution contains all possible solutions.

EXERCISES 16.8

In Exercises 1–8, show that the given functions are solutions of the differential equation. Check that the differential equation is linear and homogeneous, and then find a general solution.

1. $y'' + y' - 6y = 0$; $y_1(x) = e^{2x}$, $y_2(x) = e^{-3x}$
2. $y' + y \tan x = 0$; $y_1(x) = \cos x$
3. $y'''' + 5y'' + 4y = 0$; $y_1(x) = \cos 2x$, $y_2(x) = \sin 2x$, $y_3(x) = \cos x$, $y_4(x) = \sin x$
4. $2y'' - 16y' + 32y = 0$; $y_1(x) = 3e^{4x}$, $y_2(x) = -2xe^{4x}$
5. $y''' - 3y'' + 2y' = 0$; $y_1(x) = 10$, $y_2(x) = 3e^x$, $y_3(x) = 4e^{2x}$
6. $2y'' - 8y' + 9y = 0$; $y_1(x) = e^{2x} \cos (x/\sqrt{2})$, $y_2(x) = e^{2x} \sin (x/\sqrt{2})$
7. $x^2y'' + xy' + (x^2 - \frac{1}{4})y = 0$; $y_1(x) = \sin x/\sqrt{x}$, $y_2(x) = \cos x/\sqrt{x}$
8. $x^2y'' + xy' + 4y = 0$; $y_1(x) = \cos (2 \ln x)$, $y_2(x) = \sin (2 \ln x)$
9. Show that $y_1(x) = -2/(x + 1)$ and $y_2(x) = -2/(x + 2)$ are both solutions of the differential equation

$$y'' = yy'.$$

Is $y(x) = y_1(x) + y_2(x)$ a solution? Explain.

10. We stated in this section that n functions $y_1(x), \ldots, y_n(x)$ are linearly dependent if at least one of the functions can be expressed as a linear combination of the others; they are linearly independent if none of the functions is a linear combination of the others. Another way of saying this is as follows: Functions $y_1(x), \ldots, y_n(x)$ are linearly dependent on an interval I if there exist constants $C_1, \ldots, C_n$, not all zero, such that on I

$$C_1y_1(x) + \cdots + C_ny_n(x) \equiv 0.$$

If this equation can be satisfied only if $C_1 = C_2 = \cdots = C_n = 0$, the functions are linearly independent. In this exercise we give a test to determine whether functions are linearly independent or dependent. If $y_1(x), \ldots, y_n(x)$ have derivatives up to and including order $n - 1$ on the interval I, we define the Wronskian of the functions as the $n \times n$ determinant:

$$W(y_1, \ldots, y_n) = \begin{vmatrix} y_1 & y_2 & \cdots & y_n \\ y_1' & y_2' & \cdots & y_n' \\ y_1'' & y_2'' & \cdots & y_n'' \\ \vdots & \vdots & & \vdots \\ y_1^{(n-1)} & y_2^{(n-1)} & \cdots & y_n^{(n-1)} \end{vmatrix}.$$

Show that if $y_1, \ldots, y_n$ are linearly dependent on I, then

$$W(y_1, \ldots, y_n) \equiv 0 \quad \text{on } I.$$

It follows then that if there is at least one point in I at which $W(y_1, \ldots, y_n) \neq 0$, the functions $y_1, \ldots, y_n$ are linearly independent on I.

In each of Exercises 11–15, use the method of Exercise 10 to determine whether the functions are linearly dependent or independent on the given interval.

11. $\{1, x, x^2\}$ on $-\infty < x < \infty$
12. $\{x, 2x - 3x^2, x^2\}$ on $-\infty < x < \infty$
13. $\{\sin x, \cos x\}$ on $0 \leq x \leq 2\pi$
14. $\{x, xe^x, x^2e^x\}$ on $0 \leq x \leq 1$
15. $\{x \sin x, e^{2x}\}$ on $-\infty < x < \infty$

16.9 HOMOGENEOUS LINEAR DIFFERENTIAL EQUATIONS WITH CONSTANT COEFFICIENTS

We now consider homogeneous linear differential equations

$$a_0D^ny + a_1D^{n-1}y + a_2D^{n-2}y + \cdots + a_{n-1}Dy + a_ny = 0, \tag{16.47}$$

where the coefficients $a_0, a_1, \ldots, a_n$ are all constants. In operator notation we write

$$\phi(D)y = 0, \tag{16.48a}$$

where

$$\phi(D) = a_0D^n + a_1D^{n-1} + \cdots + a_{n-1}D + a_n. \tag{16.48b}$$

The superposition principle states that a general solution of Eq. (16.47) is $y(x) = C_1y_1(x) + \cdots + C_ny_n(x)$, provided $y_1(x), \ldots, y_n(x)$ are any n linearly independent solutions of the equation. Our problem then is to devise a technique for finding n linearly independent solutions, and to illustrate a possible procedure, we first consider three examples of second-order equations.

Our first example is

$$y'' + 2y' - 3y = 0.$$

It is not unreasonable to expect that perhaps for some value of m, $y(x) = e^{mx}$ might be a solution of this equation. After all, the equation says that the second derivative of the function must be equal to three times the function minus twice its first derivative. Since the exponential function reproduces itself when differentiated, perhaps m can be chosen to produce this combination. To see whether this is possible, we substitute $y = e^{mx}$ into the differential equation, and find that if $y = e^{mx}$ is to be a solution, then

$$m^2e^{mx} + 2me^{mx} - 3e^{mx} = 0;$$

this implies that

$$0 = m^2 + 2m - 3 = (m + 3)(m - 1).$$

Thus $y = e^{mx}$ is a solution if m is chosen as either 1 or -3; i.e., $y_1 = e^x$ and $y_2 = e^{-3x}$ are solutions of the differential equation. Since they are linearly independent, a general solution is $y(x) = C_1e^x + C_2e^{-3x}$.

For our second example, we take

$$y'' + 2y' + y = 0.$$

Since exponentials worked in the first example, we once again try a solution of the form $y(x) = e^{mx}$. If we substitute into the differential equation, we obtain

$$m^2e^{mx} + 2me^{mx} + e^{mx} = 0,$$

which implies that

$$0 = m^2 + 2m + 1 = (m + 1)^2.$$

Thus $y_1 = e^{-x}$ is a solution, but unfortunately it is the only solution that we obtain as a result of our guess. We need a second linearly independent solution $y_2(x)$ in order to obtain a general solution. Clearly, no other exponential will work. Perhaps if we multiplied e^{-x} by another function, we might find a second solution; in other words, perhaps there is a solution of the form

$y(x) = v(x)e^{-x}$ for some $v(x)$. To see, we again substitute into the differential equation to get

$$0 = \{v''e^{-x} - 2v'e^{-x} + ve^{-x}\} + 2\{v'e^{-x} - ve^{-x}\} + ve^{-x} = v''e^{-x}.$$

Consequently, $v'' = 0$, and this implies that $v(x) = Ax + B$, for any constants A and B. In particular, if $A = 1$ and $B = 0$, $v(x) = x$, and $y_2(x) = v(x)e^{-x} = xe^{-x}$ is also a solution of the differential equation. By superposition, then, we find that a general solution is $y(x) = C_1e^{-x} + C_2xe^{-x} = (C_1 + C_2x)e^{-x}$. Note that if we had set $A = 0$ and $B = 1$, then $v(x) = 1$, and the solution $y(x) = v(x)e^{-x}$ would have been $y_1(x)$. Further, if we had simply set $y(x) = v(x)e^{-x} = (Ax + B)e^{-x}$, we would have found the general solution.

Our third example is

$$y'' + 2y' + 10y = 0.$$

As in the previous two examples, if we assume a solution $y = e^{mx}$, then

$$m^2e^{mx} + 2me^{mx} + 10e^{mx} = 0,$$

or

$$0 = m^2 + 2m + 10.$$

The solutions of this quadratic equation are the complex numbers

$$m = \frac{-2 \pm \sqrt{4 - 40}}{2} = -1 \pm 3i.$$

This means that there is no real exponential $y = e^{mx}$ that will satisfy the differential equation. If, however, we form complex exponentials $y_1(x) = e^{(-1+3i)x}$ and $y_2(x) = e^{(-1-3i)x}$ and superpose these solutions, then

$$y(x) = Ae^{(-1+3i)x} + Be^{(-1-3i)x}$$

must also be a solution. When we use Euler's identity for complex exponentials, $e^{i\theta} = \cos\theta + i\sin\theta$, then we can write $y(x)$ in the form

$$\begin{aligned} y(x) &= Ae^{-x}e^{3xi} + Be^{-x}e^{-3xi} \\ &= Ae^{-x}(\cos 3x + i\sin 3x) + Be^{-x}(\cos 3x - i\sin 3x) \\ &= e^{-x}[(A + B)\cos 3x + i(A - B)\sin 3x] \\ &= e^{-x}(C_1\cos 3x + C_2\sin 3x), \end{aligned}$$

where $C_1 = A + B$ and $C_2 = i(A - B)$. In other words, the function $y(x) = e^{-x}(C_1\cos 3x + C_2\sin 3x)$ is a general solution of the differential equation, and it has been derived from the complex roots $m = -1 \pm 3i$ of the equation $m^2 + 2m + 10 = 0$. Note that what multiplies x in the exponential is the real part of these complex numbers, and what multiplies x in the trigonometric functions is the imaginary part. In Exercise 17 we show that this solution can also be derived without complex numbers, but we feel that in general the use of complex numbers is the best method.

In each of these examples we guessed $y = e^{mx}$ as a possible solution. We then substituted into the differential equation to obtain an algebraic equation for m. In each case the equation for m was precisely

$$\phi(m) = 0;$$

i.e., we took the operator $\phi(D)$, replaced D by m, and set the polynomial equal to zero. This is not a peculiarity of these examples for it is straightforward to show that for any homogeneous linear equation (16.48), if we assume a solution of the form $y = e^{mx}$, then m must satisfy the equation $\phi(m) = 0$. We name this equation in the following definition.

▷ **Definition 16.7** With every linear differential equation that has constant coefficients $\phi(D)y = F(x)$, we associate an equation

$$\phi(m) = 0 \tag{16.49}$$

called the *auxiliary equation.* ◁

To summarize, in each of the examples we assumed a solution $y = e^{mx}$ and found that m had to satisfy the auxiliary equation $\phi(m) = 0$. From the roots of the auxiliary equation we obtained solutions of the differential equation, and superposition then led to a general solution. It appears that we have found what could be a general procedure for determining a general solution of every homogeneous linear differential equation with constant coefficients. This is indeed true so that every such differential equation can be solved in exactly the same way. But if the procedure is the same in every case, surely we can set down rules that eliminate the necessity of tediously repeating these steps in every example. This we do in the following theorem.

▶ **Theorem 16.3** If $\phi(m) = 0$ is the auxiliary equation associated with the homogeneous linear differential equation $\phi(D)y = 0$, then there are two possibilities:

i) $\phi(m) = 0$ has a real root m of multiplicity k. Then a solution of the differential equation is

$$(C_1 + C_2 x + \cdots + C_k x^{k-1})e^{mx}. \tag{16.50a}$$

ii) $\phi(m) = 0$ has a pair of complex conjugate roots $a \pm bi$ each of multiplicity k. Then a solution of the differential equation is

$$e^{ax}[(C_1 + C_2 x + \cdots + C_k x^{k-1}) \cos bx + (D_1 + D_2 x + \cdots + D_k x^{k-1}) \sin bx]. \tag{16.50b}$$

A general solution of the differential equation is obtained by superposing all solutions in (i) and (ii). ◀

For a proof of this theorem see Exercise 20. Let us now apply the theorem to our previous examples. The auxiliary equation for $y'' + 2y' - 3y = 0$ is

$$0 = m^2 + 2m - 3 = (m + 3)(m - 1)$$

with solutions $m = 1$ and -3. If we now use part (i) of Theorem 16.3 with two real roots, each of multiplicity 1, a general solution of the differential equation is

$$y(x) = C_1 e^x + C_2 e^{-3x}.$$

The auxiliary equation for $y'' + 2y' + y = 0$ is

$$0 = m^2 + 2m + 1 = (m + 1)^2$$

with solutions $m = -1$ and -1. Part (i) of Theorem 16.3 with a single real root of multiplicity 2 gives the general solution

$$y(x) = (C_1 + C_2 x)e^{-x}.$$

The auxiliary equation for $y'' + 2y' + 10y = 0$ is

$$0 = m^2 + 2m + 10$$

with solutions $m = -1 \pm 3i$. Part (ii) of Theorem 16.3 with a pair of complex conjugate roots, each of multiplicity 1, gives as the general solution

$$y(x) = e^{-x}(C_1 \cos 3x + C_2 \sin 3x).$$

Example 12 Find a general solution for the differential equation

$$y''' - 3y'' - 4y' + 6y = 0.$$

Solution The auxiliary equation is

$$0 = m^3 - 3m^2 - 4m + 6 = (m - 1)(m^2 - 2m - 6)$$

with solutions $m = 1$ and $1 \pm \sqrt{7}$. A general solution of the differential equation is therefore

$$y(x) = C_1 e^x + C_2 e^{(1+\sqrt{7})x} + C_3 e^{(1-\sqrt{7})x}. \ \square$$

Example 13 Find a general solution for $y''' - y = 0$.

Solution The auxiliary equation is

$$0 = m^3 - 1 = (m - 1)(m^2 + m + 1)$$

with solutions $m = 1$ and $-(1/2) \pm (\sqrt{3}/2)i$. A general solution of the differential equation is therefore

$$y(x) = C_1 e^x + e^{-x/2}[C_2 \cos(\sqrt{3}x/2) + C_3 \sin(\sqrt{3}x/2)]. \ \square$$

Example 14 If the roots of the auxiliary equation $\phi(m) = 0$ are

$$3, \quad 3, \quad 3, \quad \pm 2i, \quad -2, \quad 1 \pm \sqrt{3}, \quad -4 \pm i, \quad -4 \pm i,$$

find a general solution of the differential equation $\phi(D)y = 0$.

Solution A general solution is

$$\begin{aligned} y(x) = {} & (C_1 + C_2 x + C_3 x^2)e^{3x} + C_4 \cos 2x + C_5 \sin 2x + C_6 e^{-2x} \\ & + C_7 e^{(1+\sqrt{3})x} + C_8 e^{(1-\sqrt{3})x} + e^{-4x}[(C_9 + C_{10}x)\cos x \\ & + (C_{11} + C_{12}x)\sin x]. \ \square \end{aligned}$$

EXERCISES 16.9

In Exercises 1–12, find a general solution for the given homogeneous linear differential equation.

1. $y'' + y' - 6y = 0$
2. $2y'' - 16y' + 32y = 0$
3. $2y'' + 16y' + 82y = 0$
4. $y'' + 2y' - 2y = 0$
5. $y'' - 4y' + 5y = 0$
6. $y''' - 3y'' + y' - 3y = 0$
7. $y'''' + 2y'' + y = 0$
8. $y''' - 6y'' + 12y' - 8y = 0$
9. $3y''' - 12y'' + 18y' - 12y = 0$
10. $y'''' + 5y'' + 4y = 0$
11. $y''' - 3y'' + 2y' = 0$
12. $y'''' + 16y = 0$

In Exercises 13–16, find a homogeneous linear differential equation that has the function as general solution.

13. $y(x) = C_1e^x + (C_2 + C_3x)e^{-4x}$
14. $y(x) = e^{-2x}(C_1 \cos 4x + C_2 \sin 4x)$
15. $y(x) = C_1 + C_2e^{\sqrt{3}x} + C_3e^{-\sqrt{3}x}$
16. $y(x) = e^x(C_1 + C_2x) \cos \sqrt{2}x + e^x(C_3 + C_4x) \sin \sqrt{2}x$
17. Show that if we assume that $y(x) = e^{ax} \sin bx$ is a solution of the differential equation $y'' + 2y' + 10y = 0$, then a and b must be equal to -1 and ± 3, respectively. Verify that for this a and b, $y(x) = e^{ax} \cos bx$ is also a solution, and therefore a general solution is
$$y(x) = e^{-x}(C_1 \cos 3x + C_2 \sin 3x).$$
18. The equation
$$y''' + ay'' + by' + cy = 0,$$
where a, b, and c are constants, has solution
$$y(x) = C_1e^{-x} + e^{-2x}(C_2 \sin 4x + C_3 \cos 4x).$$
Find a, b, and c.
19. Show that if p is constant and $f(x)$ is differentiable, then
$$D\{e^{px}f(x)\} = e^{px}\{(D + p)f(x)\}.$$
Now use mathematical induction to prove that if $f(x)$ is k times differentiable, then
$$D^k\{e^{px}f(x)\} = e^{px}\{(D + p)^kf(x)\}.$$
Finally verify that
$$\phi(D)\{e^{px}f(x)\} = e^{px}\{\phi(D + p)f(x)\},$$
a result called the operator shift theorem.
20. a) If m_0 is a real root of multiplicity k for the auxiliary equation $\phi(m) = 0$, show that the operator $\phi(D)$ can be expressed in the form
$$\phi(D) = (D - m_0)^k\psi(D),$$
where $\psi(D)$ is a polynomial in D. Now use the operator shift theorem of Exercise 19 to verify that $(C_1 + C_2x + \cdots + C_kx^{k-1})e^{m_0x}$ is a solution of $\phi(D)y = 0$.

    b) If $a \pm bi$ are complex conjugate roots each of multiplicity k for the auxiliary equation $\phi(m) = 0$, show that $\phi(D)$ can be expressed in the form
$$\phi(D) = (D - a - bi)^k(D - a + bi)^k\psi(D),$$
where $\psi(D)$ is a polynomial in D. Now use the operator shift theorem to verify that (16.50b) is a solution of $\phi(D)y = 0$.

*21. If M, β, and k are all positive constants, find a general solution for the linear differential equation
$$M\frac{d^2x}{dt^2} + \beta\frac{dx}{dt} + kx = 0.$$

16.10 NONHOMOGENEOUS LINEAR DIFFERENTIAL EQUATIONS WITH CONSTANT COEFFICIENTS

The general nonhomogeneous linear differential equation with constant coefficients is

$$\phi(D)y = F(x), \tag{16.51a}$$

where

$$\phi(D) = a_0D^n + a_1D^{n-1} + \cdots + a_{n-1}D + a_n. \tag{16.51b}$$

It is natural to ask whether we can use the results of Section 16.9 concerning homogeneous equations with constant coefficients to solve nonhomogeneous problems. Fortunately the answer is yes, as shown by the following definition.

▷ **Definition 16.8** With every nonhomogeneous linear differential equation with constant coefficients

$$\phi(D)y = F(x),$$

we associate a homogeneous equation

$$\phi(D)y = 0, \tag{16.52}$$

called the homogeneous (reduced, or complimentary) equation associated with $\phi(D)y = F(x)$. ◁

We now prove the following theorem.

▶ **Theorem 16.4** A general solution of the linear differential equation $\phi(D)y = F(x)$ is $y(x) = y_h(x) + y_p(x)$, where $y_h(x)$ is a general solution of the associated homogeneous equation, and $y_p(x)$ is any particular solution of the given equation. ◀

Proof Since $\phi(D)$ is a linear operator,

$$\phi(D)(y_h + y_p) = \phi(D)y_h + \phi(D)y_p = 0 + F(x) = F(x),$$

so that $y_h + y_p$ is indeed a solution of the given differential equation. Because $y_h(x)$ is a general solution of the associated homogeneous equation, it contains the requisite number of arbitrary constants for $y(x)$ to be a general solution of $\phi(D)y = F(x)$. □

We note in passing that Theorem 16.4 is also valid for linear differential equations with variable coefficients.

Theorem 16.4 indicates that the discussion of nonhomogeneous differential equations can be divided into two separate parts. First we find a general solution $y_h(x)$ of the associated homogeneous equation (16.52), and this can be done using the results of Section 16.9. To this we add any particular solution $y_p(x)$ of (16.51). We present two methods for finding a particular solution: (1) the method of undetermined coefficients, and (2) the method of operators. Both methods apply in general only to differential equations in which $F(x)$ is a power (x^n, n a nonnegative integer), an exponential (e^{px}), a sine ($\sin px$), a cosine ($\cos px$), and/or any sums or products thereof.

Method of Undetermined Coefficients for a Particular Solution

As stated above, the method of undetermined coefficients is to be used only when $F(x)$ in Eq. (16.51) is of the form x^n, e^{px}, $\sin px$, $\cos px$, and/or sums or products thereof. For example, if

$$y'' + y' - 6y = e^{4x},$$

then the method essentially says that Ae^{4x} is the simplest function that could conceivably yield e^{4x} when substituted into the left-hand side of the differential equation. Consequently, it is natural to assume $y_p = Ae^{4x}$ and attempt to determine the unknown coefficient A. Substitution of this function into the differential equation gives

$$16Ae^{4x} + 4Ae^{4x} - 6Ae^{4x} = e^{4x}.$$

If we divide by e^{4x}, then

$$14A = 1 \quad \text{and} \quad A = \tfrac{1}{14}.$$

A particular solution is therefore $y_p = e^{4x}/14$.

Before stating a general rule, we illustrate a few more possibilities in the following example.

Example 15 Find a particular solution of $y'' + y' - 6y = F(x)$ if:

a) $F(x) = 6x^2 + 2x + 3$;

b) $F(x) = 2 \sin 2x$;

c) $F(x) = xe^{-x} - e^{-x}$.

Solution a) Since terms in x^2, x, and constants yield terms in x^2, x, and constants when substituted into the left-hand side of the differential equation, we attempt to find a particular solution of the form $y_p = Ax^2 + Bx + C$. Substitution into the differential equation gives

$$(2A) + (2Ax + B) - 6(Ax^2 + Bx + C) = 6x^2 + 2x + 3,$$

or

$$(-6A)x^2 + (2A - 6B)x + (2A + B - 6C) = 6x^2 + 2x + 3.$$

But this equation can hold for all values of x only if coefficients of corresponding powers of x are identical (see Exercise 31 in Section 3.6). Equating coefficients then gives us

$$\begin{aligned} -6A &= 6, \\ 2A - 6B &= 2, \\ 2A + B - 6C &= 3. \end{aligned}$$

These imply that $A = -1$, $B = -2/3$, $C = -17/18$, and

$$y_p = -x^2 - \frac{2x}{3} - \frac{17}{18}.$$

b) Since terms in $\sin 2x$ and $\cos 2x$ yield terms in $\sin 2x$ when substituted into the left-hand side of the differential equation, we assume that $y_p = A \sin 2x + B \cos 2x$. Substitution into the differential equation gives

$$(-4A \sin 2x - 4B \cos 2x) + (2A \cos 2x - 2B \sin 2x) \\ -6(A \sin 2x + B \cos 2x) = 2 \sin 2x,$$

or

$$(-10A - 2B) \sin 2x + (2A - 10B) \cos 2x = 2 \sin 2x.$$

Equating coefficients gives us

$$-10A - 2B = 2,$$
$$2A - 10B = 0.$$

These imply that $A = -5/26$, $B = -1/26$, and hence

$$y_p = -\frac{1}{26}(5 \sin 2x + \cos 2x).$$

c) Since terms in xe^{-x} and e^{-x} yield terms in xe^{-x} and e^{-x} when substituted into the left-hand side of the differential equation, we assume that $y_p = Axe^{-x} + Be^{-x}$. Substitution into the differential equation gives

$$(Axe^{-x} - 2Ae^{-x} + Be^{-x}) + (-Axe^{-x} + Ae^{-x} - Be^{-x}) - 6(Axe^{-x} + Be^{-x}) = xe^{-x} - e^{-x},$$

or

$$(-6A)xe^{-x} + (-A - 6B)e^{-x} = xe^{-x} - e^{-x}.$$

Equating coefficients gives us

$$-6A = 1,$$
$$-A - 6B = -1.$$

These imply that $A = -1/6$, $B = 7/36$, and hence $y_p = -(1/6)xe^{-x} + (7/36)e^{-x}$. □

The following rule encompasses each part of this example.

Rule 1 If a term of $F(x)$ consists of a power (x^n), an exponential (e^{px}), a sine ($\sin px$), a cosine ($\cos px$), or any product thereof, assume as a part of y_p a constant multiplied by that term plus a constant multiplied by any linearly independent function arising from it by differentiation.

For Example 15(a), since $F(x)$ contains the term $6x^2$, we assume y_p contains Ax^2. Differentiation of Ax^2 yields a term in x and a constant so that we form $y_p = Ax^2 + Bx + C$. No new terms for y_p are obtained from the terms $2x$ and 3 in $F(x)$.

For Example 15(b), we assume that y_p contains $A \sin 2x$ to account for the term $2 \sin 2x$ in $F(x)$. Differentiation of $A \sin 2x$ gives a linearly independent term in $\cos 2x$ so that we form $y_p = A \sin 2x + B \cos 2x$.

For Example 15(c), since $F(x)$ contains the term xe^{-x}, we assume that y_p contains Axe^{-x}. Differentiation of Axe^{-x} yields a term in e^{-x} so that we form $y_p = Axe^{-x} + Be^{-x}$. No new terms for y_p are obtained from the term $-e^{-x}$ in $F(x)$.

Example 16 What is the form of the particular solution predicted by Rule 1 for the differential equation

$$y'' + 15y' - 6y = x^2e^{4x} + x + x \cos x?$$

Solution Rule 1 suggests that

$$y_p = Ax^2e^{4x} + Bxe^{4x} + Ce^{4x} + Dx + E + Fx\cos x + Gx\sin x + H\cos x + I\sin x. \square$$

Unfortunately, exceptions to Rule 1 do occur. For the differential equation $y'' + y = \cos x$, Rule 1 would predict $y_p = A\cos x + B\sin x$. If we substitute this into the differential equation we obtain the absurd identity $0 = \cos x$, and certainly no equations to solve for A and B. This result could have been predicted had we first calculated $y_h(x)$. The auxiliary equation $m^2 + 1 = 0$ has solutions $m = \pm i$ so that $y_h(x) = C_1\cos x + C_2\sin x$. Since y_p as suggested by Rule 1 is precisely y_h with different names for the constants, then certainly $y_p'' + y_p = 0$. Suppose as an alternative we multiply this y_p by x, and assume that $y_p = Ax\cos x + Bx\sin x$. Substitution into the differential equation now gives

$$-2A\sin x + 2B\cos x = \cos x.$$

Identification of coefficients requires $A = 0$, $B = 1/2$, and $y_p = (1/2)x\sin x$.

This example suggests that if y_p predicted by Rule 1 is already contained in y_h, then a modification of y_p is necessary. A precise statement of the situation is given in the following rule.

Rule 2 Suppose that a term in $F(x)$ is of the form $x^nf(x)$ (n a non-negative integer). Suppose further that $f(x)$ can be obtained from $y_h(x)$ by specifying values for the arbitrary constants. If this term in y_h results from a root of the auxiliary equation of multiplicity k, then corresponding to $x^nf(x)$, assume as a part of y_p the term $Ax^k(x^nf(x)) = Ax^{n+k}f(x)$, plus a constant multiplied by any linearly independent function arising from it by differentiation.

To use this rule we first require $y_h(x)$. Then, and only then, can we decide on the form of $y_p(x)$. As an illustration, consider the following example.

Example 17 Find a general solution for $y''' - y = x^3e^x$.

Solution In Example 13 we solved the auxiliary equation for $m = 1$ and $-1/2 \pm (\sqrt{3}/2)i$, from which we got

$$y_h(x) = C_1e^x + e^{-x/2}[C_2\cos(\sqrt{3}x/2) + C_3\sin(\sqrt{3}x/2)].$$

Now x^3e^x is x^3 times e^x, and e^x can be obtained from y_h by specifying $C_1 = 1$ and $C_2 = C_3 = 0$. Since this term results from the root $m = 1$ of multiplicity 1, we assume that y_p contains $Ax(x^3e^x) = Ax^4e^x$. Differentiation of this function gives terms in x^3e^x, x^2e^x, xe^x, and e^x. We therefore take

$$y_p = Ax^4e^x + Bx^3e^x + Cx^2e^x + Dxe^x.$$

(We do not include a term in e^x since it is already in y_h.) Substitution into the differential equation and simplification gives

$$(12A)x^3e^x + (36A + 9B)x^2e^x + (24A + 18B + 6C)xe^x + (6B + 6C + 3D)e^x = x^3e^x.$$

Equating coefficients gives us

$$\begin{aligned} 12A \qquad\qquad\qquad &= 1, \\ 36A + \ \ 9B \qquad\qquad &= 0, \\ 24A + 18B + 6C \qquad &= 0, \\ 6B + 6C + 3D &= 0. \end{aligned}$$

These imply that $A = 1/12$, $B = -1/3$, $C = 2/3$, $D = -2/3$, and

$$y_p = \tfrac{1}{12}x^4e^x - \tfrac{1}{3}x^3e^x + \tfrac{2}{3}x^2e^x - \tfrac{2}{3}xe^x.$$

A general solution of the differential equation is therefore

$$y(x) = C_1e^x + e^{-x/2}[C_2 \cos(\sqrt{3}x/2) + C_3 \sin(\sqrt{3}x/2)] + \frac{x^4e^x}{12} - \frac{x^3e^x}{3} + \frac{2x^2e^x}{3} - \frac{2xe^x}{3}. \square$$

Example 18 If the roots of the auxiliary equation $\phi(m) = 0$ for the differential equation $\phi(D)y = x^2 - 2\sin x + xe^{-2x}$ are $m = \pm i, -2, -2, -2, 4$, and 4, what is the form of y_p predicted by the method of undetermined coefficients?

Solution From the roots of the auxiliary equation, we can form

$$y_h(x) = C_1 \cos x + C_2 \sin x + (C_3 + C_4x + C_5x^2)e^{-2x} + (C_6 + C_7x)e^{4x}.$$

Corresponding to the term x^2 in $F(x)$, Rule 1 requires that y_p contain $Ax^2 + Bx + C$. Because $-2 \sin x$ can be obtained from $y_h(x)$ by specifying $C_2 = -2$, $C_1 = C_3 = C_4 = C_5 = C_6 = C_7 = 0$, and this term results from the roots $m = \pm i$, each of multiplicity 1, Rule 2 suggests that y_p contain $Dx \sin x + Ex \cos x$. (We do not include terms in $\sin x$ and $\cos x$ since they are already in y_h.) Finally, xe^{-2x} is x times e^{-2x}, and this function can be obtained from $y_h(x)$ by setting $C_3 = 1$, and $C_1 = C_2 = C_4 = C_5 = C_6 = C_7 = 0$. Because this term results from the root $m = -2$ of multiplicity 3, y_p must contain $Fx^4e^{-2x} + Gx^3e^{-2x}$ (but not terms in x^2e^{-2x}, xe^{-2x}, and e^{-2x} since they are in y_h). The total particular solution is therefore

$$y_p = Ax^2 + Bx + C + Dx \sin x + Ex \cos x + Fx^4e^{-2x} + Gx^3e^{-2x}. \square$$

Operator Method for a Particular Solution

This method, like that of undetermined coefficients, is designed only for functions $F(x)$ in Eq. (16.51) of the form x^n, e^{px}, $\sin px$, $\cos px$, and/or sums or products thereof. Essentially the operator method says that if

$$\phi(D)y = F(x), \tag{16.53}$$

then

$$y = \frac{1}{\phi(D)}F(x). \tag{16.54}$$

But there is a problem. What does it mean to say that

$$\frac{1}{\phi(D)} = \frac{1}{a_0D^n + a_1D^{n-1} + \cdots + a_{n-1}D + a_n} \tag{16.55}$$

operates on $F(x)$? The operator method then depends on our explaining how $1/\phi(D)$ operates on $F(x)$. The simplest $\phi(D)$ is $\phi(D) = D$. In this case the differential equation is

$$Dy = F(x),$$

and the solution is

$$y(x) = \int F(x)\,dx.$$

If the solution is also to be represented in operator notation by

$$y(x) = \frac{1}{D}F(x),$$

then we must define the operator $1/D$ by

$$\frac{1}{D}F(x) = \int F(x)\,dx. \tag{16.56}$$

If $1/D$ means to integrate, then $1/D^2$ must mean to integrate twice, $1/D^3$ to integrate three times, and so on.

We have stated that the method of operators is applicable in general only when $F(x)$ consists of powers, exponentials, and/or sines and cosines. Consider first the case in which $F(x)$ is a power x^n, in which case Eqs. (16.53) and (16.54) become

$$\phi(D)y = x^n \tag{16.57}$$

and

$$y = \frac{1}{\phi(D)}x^n. \tag{16.58}$$

If we forget for the moment that D is a differential operator, and simply regard $1/\phi(D)$ as a rational function of a variable D, then we can express $1/\phi(D)$ as an infinite series of the form

$$\frac{1}{\phi(D)} = \frac{1}{D^k}\{b_0 + b_1D + b_2D^2 + \cdots\} \tag{16.59}$$

for some nonnegative integer k. (We will show how in a moment.) But this suggests that we write (16.58) in the form

$$y = \frac{1}{D^k}\{b_0 + b_1D + b_2D^2 + \cdots\}x^n. \tag{16.60}$$

If we now reinterpret D as d/dx, then the operator on the right will produce a polynomial in x. (Note that $D^mx^n = 0$ if $m > n$.) It turns out that if we ignore all arbitrary constants that result from the integrations (when $k \geq 1$), this polynomial is a solution of (16.57). In other words, when $F(x) = x^n$, a particular solution of (16.57) is

$$y_p = \frac{1}{D^k}\{b_0 + b_1D + b_2D^2 + \cdots\}x^n. \tag{16.61}$$

We now show how to expand $1/\phi(D)$ as a series of the form (16.59). Only two situations arise and each of these can be illustrated with a simple example. First suppose that $\phi(D) = D^2 + 4D + 5$. For $1/\phi(D)$ we write

$$\frac{1}{\phi(D)} = \frac{1}{D^2 + 4D + 5} = \frac{1}{5\left(1 + \dfrac{4D + D^2}{5}\right)},$$

and interpret the right-hand side (less the 5 outside the parentheses) as the sum of an infinite geometric series with common ratio $-(4D + D^2)/5$. When we write this series out, we have

$$\frac{1}{D^2 + 4D + 5} = \frac{1}{5}\left\{1 - \left(\frac{4D + D^2}{5}\right) + \left(\frac{4D + D^2}{5}\right)^2 - \cdots\right\}$$

$$= \frac{1}{5}\left\{1 - \frac{4}{5}D + \frac{11}{25}D^2 + \cdots\right\},$$

which is of the required form (16.59) with $k = 0$.

Power k in (16.59) is positive only when $\phi(D)$ has no constant term. For example, if $\phi(D) = D^4 - 2D^3 + 3D^2$, then we factor out D^2, and proceed as above:

$$\frac{1}{\phi(D)} = \frac{1}{D^2(3 - 2D + D^2)} = \frac{1}{3D^2\left(1 - \dfrac{2D - D^2}{3}\right)}$$

$$= \frac{1}{3D^2}\left\{1 + \left(\frac{2D - D^2}{3}\right) + \left(\frac{2D - D^2}{3}\right)^2 + \cdots\right\}$$

$$= \frac{1}{3D^2}\left\{1 + \frac{2}{3}D + \frac{1}{9}D^2 + \cdots\right\}.$$

Example 19 Find a particular solution of the differential equation

$$y'' + 6y' + 4y = x^2 + 4.$$

Solution We write

$$y_p = \frac{1}{D^2 + 6D + 4}(x^2 + 4) = \frac{1}{4\left(1 + \dfrac{6D + D^2}{4}\right)}(x^2 + 4)$$

$$= \frac{1}{4}\left\{1 - \left(\frac{6D + D^2}{4}\right) + \left(\frac{6D + D^2}{4}\right)^2 - \cdots\right\}(x^2 + 4).$$

Since $D^n(x^2 + 4) = 0$ if $n > 2$, we require only the constant term and terms in D and D^2:

$$y_p = \frac{1}{4}\left\{1 - \frac{3}{2}D + 2D^2 + \cdots\right\}(x^2 + 4)$$

$$= \frac{1}{4}(x^2 + 4) - \frac{3}{8}(2x) + \frac{1}{2}(2) = \frac{x^2}{4} - \frac{3x}{4} + 2. \square$$

Example 20 Find a particular solution of the differential equation

$$y''' + 2y' = x^2 - x.$$

Solution A particular solution is

$$y_p = \frac{1}{D^3 + 2D}(x^2 - x) = \frac{1}{D(2 + D^2)}(x^2 - x)$$

$$= \frac{1}{2D\left(1 + \frac{D^2}{2}\right)}(x^2 - x)$$

$$= \frac{1}{2D}\left\{1 - \frac{D^2}{2} + \cdots\right\}(x^2 - x)$$

$$= \frac{1}{2D}\left\{x^2 - x - \frac{1}{2}(2)\right\},$$

and since $1/D$ means to integrate with respect to x, we have

$$y_p = \frac{1}{2}\left\{\frac{x^3}{3} - \frac{x^2}{2} - x\right\} = \frac{x^3}{6} - \frac{x^2}{4} - \frac{x}{2}. \ \square$$

In summary, to evaluate (16.54) when $F(x)$ is a power x^n (or a polynomial), we expand $1/\phi(D)$ in a series of the form (16.59) and perform the indicated differentiations (and integrations if $k > 0$).

For all other cases of $F(x)$, we make use of a theorem called the *inverse operator shift theorem.* This theorem states that

$$\frac{1}{\phi(D)}\{e^{px}f(x)\} = e^{px}\frac{1}{\phi(D + p)}f(x) \tag{16.62}$$

(see Exercise 16). The theorem enables us to shift the exponential e^{px} past the operator $1/\phi(D)$, but in so doing, the operator must be modified to $1/\phi(D + p)$. Equation (16.62) immediately yields $y_p(x)$ whenever $f(x)$ is a power x^n, n a nonnegative integer (or a polynomial). This is illustrated in the following example.

Example 21 Find a particular solution for each of the following differential equations:

a) $y'' + 3y' + 10y = x^2e^{-x}$;
b) $y'' - 4y' + 4y = e^{2x}$;
c) $y''' + 2y'' + 3y' - y = e^{2x}$.

Solution a) Equation (16.62) gives

$$y_p = \frac{1}{D^2 + 3D + 10}x^2e^{-x} = e^{-x}\frac{1}{(D - 1)^2 + 3(D - 1) + 10}x^2$$

$$= e^{-x}\frac{1}{D^2 + D + 8}x^2,$$

and we now proceed as in Example 19:

$$y_p = e^{-x} \frac{1}{8\left(1 + \dfrac{D + D^2}{8}\right)} x^2$$

$$= \frac{e^{-x}}{8}\left\{1 - \left(\frac{D + D^2}{8}\right) + \left(\frac{D + D^2}{8}\right)^2 - \cdots\right\}x^2$$

$$= \frac{e^{-x}}{8}\left\{1 - \frac{D}{8} - \frac{7D^2}{64} + \cdots\right\}x^2 = \frac{e^{-x}}{8}\left\{x^2 - \frac{x}{4} - \frac{7}{32}\right\}.$$

b) Once again we use the inverse operator shift theorem to get

$$y_p = \frac{1}{D^2 - 4D + 4} e^{2x} = e^{2x} \frac{1}{(D + 2)^2 - 4(D + 2) + 4}(1)$$

$$= e^{2x}\frac{1}{D^2}(1) = \frac{x^2}{2}e^{2x},$$

since $1/D^2$ means to integrate twice.

c) For this differential equation,

$$y_p = \frac{1}{D^3 + 2D^2 + 3D - 1} e^{2x} = e^{2x} \frac{1}{(D+2)^3 + 2(D+2)^2 + 3(D+2) - 1}(1)$$

$$= e^{2x}\frac{1}{D^3 + 8D^2 + 23D + 21}(1) = e^{2x}\frac{1}{21\left(1 + \dfrac{23D + 8D^2 + D^3}{21}\right)}(1)$$

$$= \frac{e^{2x}}{21}\{1 + \cdots\}(1) = \frac{e^{2x}}{21}. \square$$

Example 22 Find a general solution for $y''' - 3y'' + 3y' - y = 2x^2e^x$.

Solution The auxiliary equation is

$$0 = m^3 - 3m^2 + 3m - 1 = (m - 1)^3$$

with solutions $m = 1$, 1, and 1. Thus,

$$y_h(x) = (C_1 + C_2x + C_3x^2)e^x.$$

A particular solution is

$$y_p(x) = \frac{2}{(D - 1)^3} x^2e^x = 2e^x \frac{1}{(D + 1 - 1)^3} x^2 = 2e^x \frac{1}{D^3} x^2 = \frac{x^5}{30} e^x.$$

A general solution of the differential equation is therefore

$$y(x) = (C_1 + C_2x + C_3x^2)e^x + \frac{x^5}{30}e^x. \square$$

Example 23 Find a particular solution for the differential equation

$$y''' + 3y'' - 4y = xe^{-2x} + x^2.$$

Solution

$$y_p = \frac{1}{D^3 + 3D^2 - 4}(xe^{-2x} + x^2)$$

$$= e^{-2x}\frac{1}{(D-2)^3 + 3(D-2)^2 - 4}x + \frac{1}{D^3 + 3D^2 - 4}x^2$$

$$= e^{-2x}\frac{1}{D^3 - 3D^2}x - \frac{1}{4\left(1 - \dfrac{3D^2 + D^3}{4}\right)}x^2$$

$$= e^{-2x}\frac{1}{D^2}\frac{1}{D-3}x - \frac{1}{4}\left\{1 + \left(\frac{3D^2 + D^3}{4}\right) + \cdots\right\}x^2$$

$$= e^{-2x}\frac{1}{D^2}\frac{1}{-3\left(1 - \dfrac{D}{3}\right)}x - \frac{1}{4}\left(x^2 + \frac{3}{2}\right)$$

$$= \frac{e^{-2x}}{-3}\frac{1}{D^2}\left\{1 + \frac{D}{3} + \cdots\right\}x - \frac{x^2}{4} - \frac{3}{8}$$

$$= \frac{e^{-2x}}{-3}\frac{1}{D^2}\left\{x + \frac{1}{3}\right\} - \frac{x^2}{4} - \frac{3}{8}$$

$$= \frac{e^{-2x}}{-3}\left\{\frac{x^3}{6} + \frac{x^2}{6}\right\} - \frac{x^2}{4} - \frac{3}{8}. \quad \square$$

When $F(x)$ in (16.54) is of the form $x^n \sin px$ or $x^n \cos px$, n a nonnegative integer and p a constant, we introduce complex exponentials. Specifically, because

$$x^n e^{ipx} = x^n(\cos px + i \sin px),$$

we can write

$$x^n \cos px = \text{Real part of } x^n e^{ipx} = \text{Re}\{x^n e^{ipx}\}, \tag{16.63a}$$

$$x^n \sin px = \text{Imaginary part of } x^n e^{ipx} = \text{Im}\{x^n e^{ipx}\}. \tag{16.63b}$$

To operate on either of these functions by $1/\phi(D)$, we interchange the operations of $1/\phi(D)$ and taking the real and imaginary parts:

$$\frac{1}{\phi(D)}\{x^n \cos px\} = \frac{1}{\phi(D)}\text{Re}\{x^n e^{ipx}\} = \text{Re}\left\{\frac{1}{\phi(D)}(x^n e^{ipx})\right\}, \tag{16.64a}$$

$$\frac{1}{\phi(D)}\{x^n \sin px\} = \frac{1}{\phi(D)}\text{Im}\{x^n e^{ipx}\} = \text{Im}\left\{\frac{1}{\phi(D)}(x^n e^{ipx})\right\}. \tag{16.64b}$$

We can now proceed by using the inverse operator shift theorem (16.62).

Example 24 Find a particular solution for each of the following differential equations:

a) $y'' + y = \sin 2x$;

b) $y'' + 4y = x^2 \cos x$;

c) $y'' + 9y = \sin 3x$;

d) $y'' + 4y = x \sin 2x$;

e) $y'' + 2y' + 4y = e^{-x} \sin \sqrt{3}x$.

Solution a) Equation (16.64b) gives

$$y_p = \frac{1}{D^2 + 1} \sin 2x = \frac{1}{D^2 + 1} \{\text{Im}(e^{2ix})\} = \text{Im}\left\{\frac{1}{D^2 + 1} e^{2ix}\right\}.$$

We now use the inverse operator shift theorem (16.62) to get

$$y_p = \text{Im}\left\{e^{2ix} \frac{1}{(D + 2i)^2 + 1} (1)\right\}$$

$$= \text{Im}\left\{e^{2ix} \frac{1}{D^2 + 4iD - 3} (1)\right\} = \text{Im}\left\{e^{2ix} \frac{1}{-3\left(1 - \dfrac{4iD + D^2}{3}\right)} (1)\right\}$$

$$= \text{Im}\left\{\frac{e^{2ix}}{-3}\right\} = -\frac{1}{3} \sin 2x.$$

b) $$y_p = \frac{1}{D^2 + 4} x^2 \cos x = \frac{1}{D^2 + 4} \text{Re}(x^2 e^{ix}) = \text{Re}\left\{\frac{1}{D^2 + 4} x^2 e^{ix}\right\}$$

$$= \text{Re}\left\{e^{ix} \frac{1}{(D + i)^2 + 4} x^2\right\} = \text{Re}\left\{e^{ix} \frac{1}{D^2 + 2iD + 3} x^2\right\}$$

$$= \text{Re}\left\{e^{ix} \frac{1}{3\left(1 + \dfrac{2iD + D^2}{3}\right)} x^2\right\}$$

$$= \text{Re}\left\{\frac{e^{ix}}{3}\left[1 - \left(\frac{2iD + D^2}{3}\right) + \left(\frac{2iD + D^2}{3}\right)^2 - \cdots\right] x^2\right\}$$

$$= \frac{1}{3} \text{Re}\left\{e^{ix}\left[1 - \frac{2iD}{3} - \frac{7D^2}{9} + \cdots\right] x^2\right\} = \frac{1}{3} \text{Re}\left\{e^{ix}\left[x^2 - \frac{4ix}{3} - \frac{14}{9}\right]\right\}$$

$$= \frac{1}{3}\left\{x^2 \cos x + \frac{4x}{3} \sin x - \frac{14}{9} \cos x\right\}.$$

c) For the differential equation $y'' + 9y = \sin 3x$, we have

$$y_p = \frac{1}{D^2 + 9} \sin 3x = \frac{1}{D^2 + 9} \text{Im}(e^{3ix}) = \text{Im}\left\{\frac{1}{D^2 + 9} e^{3ix}\right\}$$

$$= \text{Im}\left\{e^{3ix} \frac{1}{(D + 3i)^2 + 9} (1)\right\} = \text{Im}\left\{e^{3ix} \frac{1}{D^2 + 6iD} (1)\right\}$$

$$= \text{Im}\left\{e^{3ix} \frac{1}{D} \frac{1}{D + 6i} (1)\right\} = \text{Im}\left\{\frac{e^{3ix}}{6i} \frac{1}{D} \frac{1}{1 + D/6i} (1)\right\}$$

$$= \text{Im}\left\{-\frac{i}{6} e^{3ix} \frac{1}{D}\left(1 - \frac{D}{6i} + \cdots\right)(1)\right\} = \text{Im}\left\{-\frac{i}{6} e^{3ix} \frac{1}{D} (1)\right\}$$

$$= \text{Im}\left\{-\frac{i}{6} e^{3ix} x\right\} = -\frac{x}{6} \cos 3x.$$

d) For the differential equation $y'' + 4y = x \sin 2x$, we get

$$y_p = \frac{1}{D^2+4} x \sin 2x = \frac{1}{D^2+4} \operatorname{Im}(xe^{2ix}) = \operatorname{Im}\left\{\frac{1}{D^2+4} xe^{2ix}\right\}$$

$$= \operatorname{Im}\left\{e^{2ix} \frac{1}{(D+2i)^2+4} x\right\} = \operatorname{Im}\left\{e^{2ix} \frac{1}{D^2+4iD} x\right\}$$

$$= \operatorname{Im}\left\{e^{2ix} \frac{1}{D} \frac{1}{D+4i} x\right\} = \operatorname{Im}\left\{\frac{1}{4i} e^{2ix} \frac{1}{D}\left(1 - \frac{D}{4i} + \cdots\right)x\right\}$$

$$= \operatorname{Im}\left\{-\frac{i}{4} e^{2ix} \frac{1}{D}\left(x - \frac{1}{4i}\right)\right\} = \operatorname{Im}\left\{-\frac{i}{4} e^{2ix}\left(\frac{x^2}{2} + \frac{ix}{4}\right)\right\}$$

$$= -\frac{x^2}{8} \cos 2x + \frac{x}{16} \sin 2x.$$

e) For the differential equation $y'' + 2y' + 4y = e^{-x} \sin \sqrt{3}x$, we have

$$y_p = \frac{1}{D^2+2D+4} e^{-x} \sin \sqrt{3}x = \frac{1}{D^2+2D+4} \operatorname{Im}(e^{-x}e^{\sqrt{3}ix})$$

$$= \operatorname{Im}\left\{\frac{1}{D^2+2D+4} e^{(-1+\sqrt{3}i)x}\right\}$$

$$= \operatorname{Im}\left\{e^{(-1+\sqrt{3}i)x} \frac{1}{(D-1+\sqrt{3}i)^2 + 2(D-1+\sqrt{3}i)+4} (1)\right\}$$

$$= \operatorname{Im}\left\{e^{(-1+\sqrt{3}i)x} \frac{1}{D^2+2\sqrt{3}iD} (1)\right\} = \operatorname{Im}\left\{e^{(-1+\sqrt{3}i)x} \frac{1}{D} \frac{1}{D+2\sqrt{3}i} (1)\right\}$$

$$= \operatorname{Im}\left\{e^{(-1+\sqrt{3}i)x} \frac{1}{D} \frac{1}{2\sqrt{3}i}\right\} = \operatorname{Im}\left\{-\frac{i}{2\sqrt{3}} xe^{(-1+\sqrt{3}i)x}\right\}$$

$$= -\frac{xe^{-x}}{2\sqrt{3}} \cos \sqrt{3}x. \ \square$$

Example 25 Find a general solution of $y'' + 6y' + y = \sin 3x$.

Solution The auxiliary equation is $m^2 + 6m + 1 = 0$ with solutions $m = -3 \pm 2\sqrt{2}$. Consequently,

$$y_h(x) = C_1 e^{(-3+2\sqrt{2})x} + C_2 e^{(-3-2\sqrt{2})x}.$$

A particular solution is

$$y_p = \frac{1}{D^2+6D+1} \operatorname{Im}(e^{3ix}) = \operatorname{Im}\left\{\frac{1}{D^2+6D+1} e^{3ix}\right\}$$

$$= \operatorname{Im}\left\{e^{3ix} \frac{1}{(D+3i)^2 + 6(D+3i)+1} (1)\right\}$$

$$= \operatorname{Im}\left\{e^{3ix} \frac{1}{D^2+(6+6i)D+(-8+18i)} (1)\right\} = \operatorname{Im}\left\{e^{3ix} \frac{1}{-8+18i}\right\}$$

$$= \operatorname{Im}\left\{e^{3ix} \frac{1}{-8+18i} \frac{-8-18i}{-8-18i}\right\} = -\frac{1}{194}(4 \sin 3x + 9 \cos 3x).$$

Finally, then, a general solution of the differential equation is

$$y(x) = C_1e^{(-3+2\sqrt{2})x} + C_2e^{(-3-2\sqrt{2})x} - \frac{1}{194}(4\sin 3x + 9\cos 3x). \square$$

Examples 19–25 have illustrated that with identity (16.62) and the concept of series, we can obtain a particular solution for (16.53) whenever $F(x)$ is x^n, e^{px}, $\sin px$, $\cos px$, and/or any sums or products thereof. In summary:

When $F(x) = x^n$, use series (16.59) for $1/\phi(D)$.

In any other situation, use (16.62) and then series (16.59) for $1/\phi(D + p)$.

We make one final comment. We introduced complex exponentials e^{ipx} to handle terms involving $\sin px$ and $\cos px$. This is not the only way to treat trigonometric functions, since there do exist other methods that completely avoid complex numbers. Unfortunately, these methods require memorization of somewhat involved identities, and because of this, we prefer the use of complex numbers.

EXERCISES 16.10

In each of Exercises 1–11, find a particular solution for the differential equation. Do so by both the method of operators and the method of undetermined coefficients. Find a general solution of each equation.

1. $2y'' - 16y' + 32y = -e^{4x}$

2. $y'' + 2y' - 2y = x^2e^{-x}$

3. $y''' - 3y'' + y' - 3y = 3xe^x + 2$

4. $y'''' + 2y'' + y = \cos 2x$

5. $y''' - 6y'' + 12y' - 8y = 2e^{2x}$

6. $y'''' + 5y'' + 4y = e^{-2x}$

7. $y''' - 3y'' + 2y' = x^2 + e^{-x}$

8. $2y'' + 16y' + 82y = -2e^{2x}\sin x$

9. $y'' + y' - 6y = x + \cos x$

10. $y'' - 4y' + 5y = x\cos x$

11. $3y''' - 12y'' + 18y' - 12y = x^2 + 3x - 4$

In each of Exercises 12–15, state the form for the particular solution predicted by the method of undetermined coefficients. Do not evaluate the coefficients.

12. $y''' + 9y'' + 27y' + 27y = xe^{3x} + 2x\cos x$

13. $y''' + 4y'' + y' + 4y = xe^x \sin x$

14. $2y''' - 6y'' - 12y' + 16y = xe^x + 2x^3 - 4\cos x$

15. $2y'' - 4y' + 10y = 5e^x \sin 2x$

16. Use the operator shift theorem of Exercise 19 in Section 16.9 to verify the inverse operator shift theorem (16.62).

In Exercises 17 and 18, find a general solution for the differential equation.

17. $y'' + 2y' - 4y = \cos^2 x$

18. $2y'' - 4y' + 3y = \cos x \sin 2x$

In Exercises 19 and 20, find a solution of the differential equation.

19. $y'' - 3y' + 2y = 8x^2 + 12e^{-x}$, $y(0) = 0$, $y'(0) = 2$

20. $y'' + 9y = x(\sin 3x + \cos 3x)$, $y(0) = y'(0) = 0$

21. If J, k, and w are positive constants, find a general solution for

$$J\frac{d^4y}{dx^4} + ky = w.$$

22. The second-order linear differential equation

$$x^2\frac{d^2y}{dx^2} + ax\frac{dy}{dx} + by = F(x), \qquad a, b \text{ constants},$$

is called the *Cauchy linear equation*. Because of the x^2- and x-factors, it does not have constant coefficients, and is therefore not immediately amenable to the techniques of this chapter. Show that if we make a change of independent variable $x = e^z$, then

$$x\frac{dy}{dx} = \frac{dy}{dz}, \qquad x^2\frac{d^2y}{dx^2} = \frac{d^2y}{dz^2} - \frac{dy}{dz},$$

and that as a result the Cauchy equation is transformed

into a linear differential equation in $y(z)$ with constant coefficients.

In Exercises 23 and 24, use the technique of Exercise 22 to find a general solution for the differential equation.

23. $\dfrac{d^2u}{dr^2} + \dfrac{1}{r}\dfrac{du}{dr} - \dfrac{u}{r^2} = 0,\ r > 0$

24. $x^2y'' + xy' + 4y = 1,\ x > 0$

***25.** If M, β, k, A, and ω are all positive constants, find a particular solution of the linear differential equation

$$M\frac{d^2x}{dt^2} + \beta\frac{dx}{dt} + kx = A\sin\omega t.$$

16.11 APPLICATIONS OF LINEAR DIFFERENTIAL EQUATIONS

In Fig. 16.14 we have shown a mass M suspended vertically from a spring. If M is given an initial motion in the vertical direction (and the vertical direction only), then we expect M to oscillate up and down for some time to come. In this section we show how to describe these oscillations mathematically.

Vibrating Mass–Spring Systems

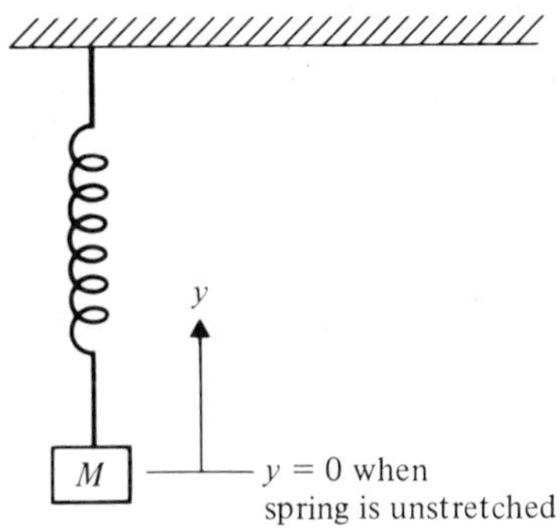

Figure 16.14

In order to describe the position of M as a function of time t, we must choose a vertical coordinate system. There are two natural places to choose the origin $y = 0$, one being the position of M when the spring is unstretched. Suppose we do this and choose y as positive upward. When M is a distance y away from the origin, the restoring force of the spring has y-component $-ky$ ($k > 0$). In addition, if $g < 0$ is the acceleration due to gravity, then the force of gravity on M has y-component Mg. Finally, suppose the oscillations take place in a medium that exerts a damping force proportional to the instantaneous velocity of M. This damping force must therefore have a y-component of the form $-\beta(dy/dt)$, where $\beta > 0$ is a constant. The total force on M therefore has y-component

$$-ky + Mg - \beta\frac{dy}{dt},$$

and Newton's second law states that the acceleration d^2y/dt^2 of M must satisfy the equation

$$-ky + Mg - \beta\frac{dy}{dt} = M\frac{d^2y}{dt^2}.$$

Consequently, the differential equation that determines the position $y(t)$ of M relative to the unstretched position of the spring is

$$M\frac{d^2y}{dt^2} + \beta\frac{dy}{dt} + ky = Mg. \tag{16.65}$$

The alternative possibility for describing oscillations is to attach M to the spring and slowly lower M until it reaches an equilibrium position. At this position, the restoring force of the spring is exactly equal to the force of gravity on the mass, and the mass, left by itself, will remain motionless. If $s > 0$ is the amount of stretch in the spring at equilibrium, and $g < 0$ is the acceleration due to gravity, then at equilibrium

$$ks + Mg = 0. \tag{16.66}$$

Suppose we take the equilibrium position as $x = 0$ and x as positive upward (Fig. 16.15). When M is a distance x away from its equilibrium position, the restoring force on M has y-component $k(s - x)$. The y-component of the force of gravity remains as Mg, and that of the damping force is $-\beta(dx/dt)$. Newton's second law therefore implies that

$$M\frac{d^2x}{dt^2} = k(s - x) + Mg - \beta\frac{dx}{dt},$$

or

$$M\frac{d^2x}{dt^2} + \beta\frac{dx}{dt} + kx = Mg + ks.$$

But according to (16.66), $Mg + ks = 0$, and hence

$$M\frac{d^2x}{dt^2} + \beta\frac{dx}{dt} + kx = 0. \tag{16.67}$$

This is the differential equation describing the displacement $x(t)$ of M relative to the equilibrium position of M.

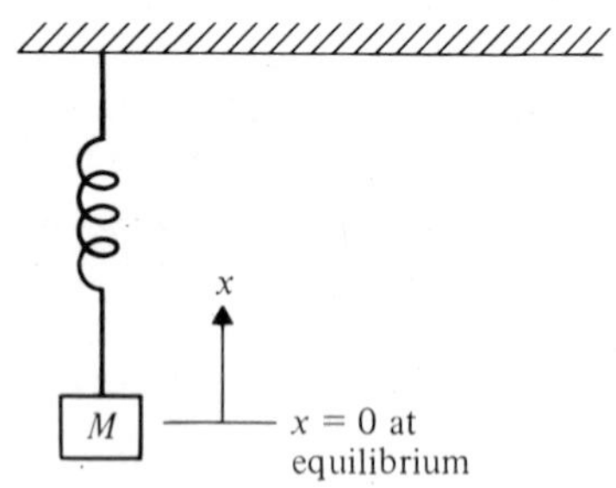

Figure 16.15

Note that both equations (16.65) and (16.67) are linear second-order differential equations with constant coefficients. The advantage of (16.67) is that it is homogeneous as well, and this is simply due to a convenient choice of dependent variable (x as opposed to y). Physically, we are saying that there are two parts to the spring force $k(s - x)$: a part ks and a part $-kx$. Gravity is always acting on M, and that part ks of the spring force is counteracting it in an attempt to restore the spring to its unstretched position. Because these forces always cancel, we might just as well eliminate both of them from our discussion. This would leave us $-kx$, and we therefore interpret $-kx$ *as the spring force attempting to restore the mass to its equilibrium position.*

If we choose Eq. (16.67) to describe the motion of M (and this equation is usually chosen over Eq. 16.65), we must remember three things: x is measured from equilibrium, $-kx$ is the spring force attempting to restore M to its equilibrium position, and gravity has been taken into account.

There are three basic ways to initiate the motion. First, we can move the mass away from its equilibrium position and then release it, giving it an initial displacement but no initial velocity. Second, we can strike the mass at the equilibrium position, imparting an initial velocity but no initial displacement. And finally, we can give the mass both an initial displacement and an initial velocity. Each of these methods will add two initial conditions to the differential equation.

To be complete we note that when (in addition to the forces already mentioned) there is an externally applied force acting on the mass that is represented as a function of time by $F(t)$, then Eq. (16.67) is modified to

$$M\frac{d^2x}{dt^2} + \beta\frac{dx}{dt} + kx = F(t). \tag{16.68}$$

Perhaps, for example, M contains some iron, and $F(t)$ is due to a magnet directly below M that constantly exerts an attractive force on M.

Example 26 A 2-kilogram mass is suspended vertically from a spring with constant 16 newtons per meter. The mass is raised 10 centimeters above its equilibrium position and then released. If damping is ignored, find the amplitude, period, and frequency of the motion.

Solution If we choose $x = 0$ at the equilibrium position of the mass and x positive upward (Fig. 16.15), then the differential equation for the motion $x(t)$ of the mass is

$$2\frac{d^2x}{dt^2} = -16x,$$

or

$$\frac{d^2x}{dt^2} + 8x = 0,$$

along with the initial conditions

$$x(0) = 1/10, \qquad x'(0) = 0.$$

The auxiliary equation is $m^2 + 8 = 0$ with solutions $m = \pm 2\sqrt{2}i$. Consequently,

$$x(t) = C_1 \cos(2\sqrt{2}t) + C_2 \sin(2\sqrt{2}t).$$

The initial conditions require

$$1/10 = C_1, \qquad 0 = 2\sqrt{2}C_2.$$

Thus,

$$x(t) = \frac{1}{10} \cos (2\sqrt{2}t) \text{ meters.}$$

The amplitude of the oscillations is 1/10 meter, the period is $2\pi/(2\sqrt{2}) = \pi/\sqrt{2}$ seconds, and the frequency is $\sqrt{2}/\pi$ cycles per second. A graph of this function (Fig. 16.16) illustrates the oscillations of the mass about its equilibrium position. This is called *simple harmonic motion.* □

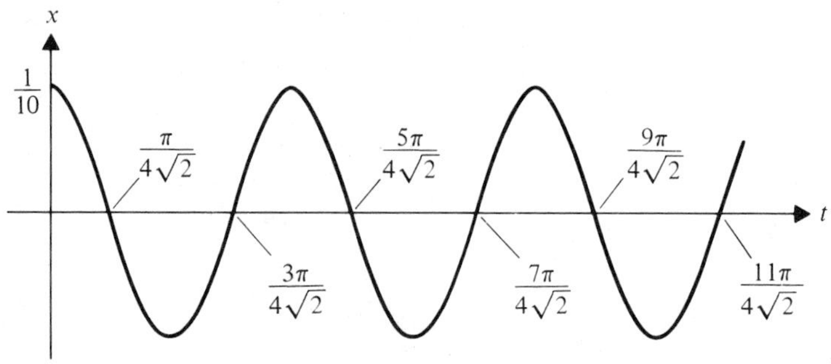

Figure 16.16

Example 27 A 100-gram mass is suspended vertically from a spring with constant 5 newtons per meter. The mass is pulled 5 centimeters below its equilibrium position and given a velocity of 2 meters per second upward. If, during the motion, the mass is acted on by a damping force in newtons numerically equal to one-twentieth the instantaneous velocity in meters per second, find the position of the mass at any time.

Solution If we choose $x = 0$ at the equilibrium position of the mass and x positive upward (Fig. 16.15), then the differential equation for the motion $x(t)$ of the mass is

$$\frac{1}{10}\frac{d^2x}{dt^2} = -5x - \frac{1}{20}\frac{dx}{dt},$$

or

$$2\frac{d^2x}{dt^2} + \frac{dx}{dt} + 100x = 0,$$

along with the initial conditions

$$x(0) = -1/20, \qquad x'(0) = 2.$$

The auxiliary equation is $2m^2 + m + 100 = 0$ with solutions

$$m = \frac{-1 \pm \sqrt{1 - 800}}{4} = \frac{-1 \pm \sqrt{799}i}{4}.$$

Consequently,

$$x(t) = e^{-t/4}[C_1 \cos (\sqrt{799}t/4) + C_2 \sin (\sqrt{799}t/4)].$$

The initial conditions require

$$-1/20 = C_1, \qquad 2 = -C_1/4 + \sqrt{799}C_2/4,$$

from which we get

$$C_2 = \frac{159\sqrt{799}}{15980}.$$

Finally, then,

$$\begin{aligned} x(t) &= e^{-t/4}\left[-\frac{1}{20}\cos\left(\frac{\sqrt{799}t}{4}\right) + \frac{159\sqrt{799}}{15980}\sin\left(\frac{\sqrt{799}t}{4}\right)\right] \\ &= e^{-0.25t}[-0.05\cos(7.07t) + 0.28\sin(7.07t)] \text{ meters.} \end{aligned}$$

The graph of this function in Fig. 16.17 clearly indicates how the amplitude of the oscillations decreases in time. □

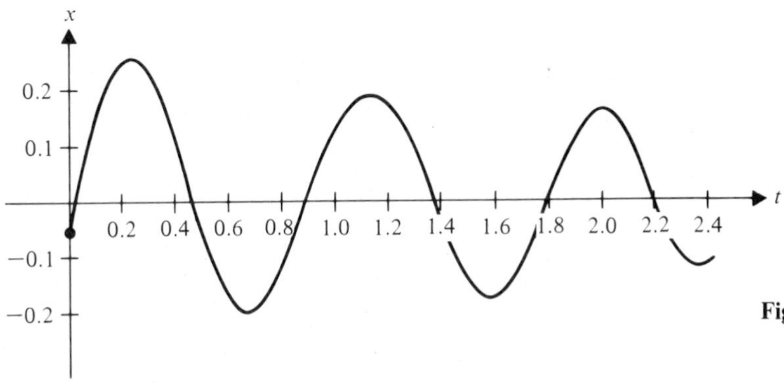

Figure 16.17

LCR Circuits If a resistance R, an inductance L, and a capacitance C are connected in series with an electromotive force $E(t)$ (Fig. 16.18) and the switch is closed, current will flow in the circuit and charge will build up in the capacitor. If at any time t, Q is the charge on the capacitor and I is the current in the loop, then Kirchhoff's law requires that

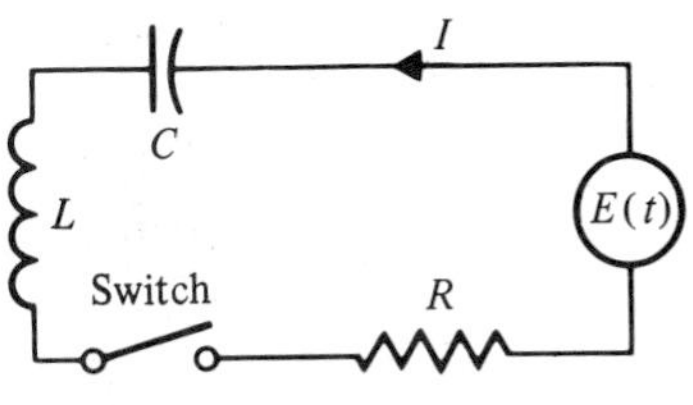

Figure 16.18

$$L\frac{dI}{dt} + RI + \frac{Q}{C} = E(t), \tag{16.69}$$

where $L\,dI/dt$, RI, and Q/C represent the voltage drops across the inductor, the resistor, and the capacitor, respectively. If we substitute $I = dQ/dt$, then

$$L\frac{d^2Q}{dt^2} + R\frac{dQ}{dt} + \frac{1}{C}Q = E(t), \tag{16.70}$$

a second-order linear differential equation for $Q(t)$.

Alternatively, if we differentiate this equation, we obtain

$$L\frac{d^2I}{dt^2} + R\frac{dI}{dt} + \frac{1}{C}I = E'(t), \tag{16.71}$$

a second-order linear differential equation for $I(t)$.

Example 28 At time $t = 0$, a 25-ohm resistor, a 2-henry inductor, and a 0.01-farad capacitor are connected in series with a generator producing an alternating voltage of $10 \sin (5t)$, $t \geq 0$ (Fig. 16.19). Find the charge on the capacitor and the current in the circuit if the capacitor is uncharged when the circuit is closed.

Solution The differential equation for the charge Q on the capacitor is (see Eq. 16.70)

$$2\frac{d^2Q}{dt^2} + 25\frac{dQ}{dt} + 100Q = 10 \sin (5t),$$

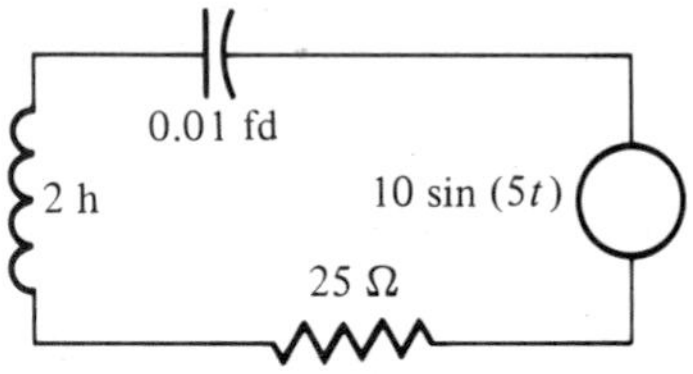

Figure 16.19

to which we add the initial conditions

$$Q(0) = 0, \qquad Q'(0) = I(0) = 0.$$

The auxiliary equation is $2m^2 + 25m + 100 = 0$ with solutions

$$m = \frac{-25 \pm \sqrt{625 - 800}}{4} = \frac{-25 \pm 5\sqrt{7}i}{4}.$$

Consequently, the general solution of the homogeneous equation is

$$Q_h(t) = e^{-25t/4}[C_1 \cos (5\sqrt{7}t/4) + C_2 \sin (5\sqrt{7}t/4)].$$

To find a particular solution of the nonhomogeneous equation by undetermined coefficients, we set

$$Q_p(t) = A \sin (5t) + B \cos (5t).$$

Substitution of this function into the differential equation gives

$$2\{-25A \sin (5t) - 25B \cos (5t)\} + 25\{5A \cos (5t) - 5B \sin (5t)\} + 100\{A \sin (5t) + B \cos (5t)\} = 10 \sin (5t).$$

This equation requires A and B to satisfy

$$50A - 125B = 10, \qquad 125A + 50B = 0,$$

the solution of which is $A = 4/145$, $B = -2/29$.

A particular solution is therefore

$$Q_p(t) = \frac{2}{145}[-5 \cos (5t) + 2 \sin (5t)]$$

and

$$\begin{aligned} Q(t) &= Q_h(t) + Q_p(t) \\ &= e^{-25t/4}\left[C_1 \cos \left(\frac{5\sqrt{7}t}{4}\right) + C_2 \sin \left(\frac{5\sqrt{7}t}{4}\right)\right] \\ &\quad + \frac{2}{145}[2 \sin (5t) - 5 \cos (5t)]. \end{aligned}$$

The initial conditions require

$$0 = C_1 - \frac{10}{145}, \qquad 0 = -\frac{25}{4}C_1 + \frac{5\sqrt{7}}{4}C_2 + \frac{20}{145},$$

and these imply that $C_1 = 2/29$, $C_2 = 34/(145\sqrt{7})$. Consequently,

$$Q(t) = \frac{e^{-25t/4}}{145\sqrt{7}}\left[10\sqrt{7}\cos\left(\frac{5\sqrt{7}t}{4}\right) + 34\sin\left(\frac{5\sqrt{7}t}{4}\right)\right]$$

$$+ \frac{2}{145}[2\sin(5t) - 5\cos(5t)] \text{ coulombs.}$$

The current in the circuit is

$$I(t) = \frac{dQ}{dt} = \left(-\frac{25}{4}\right)\frac{e^{-25t/4}}{145\sqrt{7}}\left[10\sqrt{7}\cos\left(\frac{5\sqrt{7}t}{4}\right) + 34\sin\left(\frac{5\sqrt{7}t}{4}\right)\right]$$

$$+ \frac{e^{-25t/4}}{145\sqrt{7}}\left[-\frac{175}{2}\sin\left(\frac{5\sqrt{7}t}{4}\right) + \frac{85\sqrt{7}}{2}\cos\left(\frac{5\sqrt{7}t}{4}\right)\right]$$

$$+ \frac{2}{145}[10\cos(5t) + 25\sin(5t)]$$

$$= -\frac{e^{-25t/4}}{29\sqrt{7}}\left[4\sqrt{7}\cos\left(\frac{5\sqrt{7}t}{4}\right) + 60\sin\left(\frac{5\sqrt{7}t}{4}\right)\right]$$

$$+ \frac{2}{29}[2\cos(5t) + 5\sin(5t)] \text{ amperes.}$$

The solution $Q(t)$ contains two parts. The first two terms (containing the exponential $e^{-25t/4}$) are $Q_h(t)$ with the constants C_1 and C_2 determined by the initial conditions; the last two terms are $Q_p(t)$. We point this out because the two parts display completely different characteristics. For small t, both parts of $Q(t)$ are present and contribute significantly, but for very large t, the first two terms become negligible. In other words, after a long time, the charge $Q(t)$ on the capacitor will be defined essentially by $Q_p(t)$. We call $Q_p(t)$ the steady-state part of the solution, and the two other terms in $Q(t)$ are called the transient part of the solution. Similarly, the first two terms in $I(t)$ are called the transient part of the current and the last two terms the steady-state part of the current.

Finally, note that the frequency of the steady-state part of either $Q(t)$ or $I(t)$ is exactly that of the forcing voltage $E(t)$. □

The similarity between differential equations (16.68) and (16.70) cannot go unmentioned:

$$M\frac{d^2x}{dt^2} + \beta\frac{dx}{dt} + kx = F(t),$$

$$L\frac{d^2Q}{dt^2} + R\frac{dQ}{dt} + \frac{1}{C}Q = E(t).$$

Each of the coefficients M, β, and k for the mechanical system has its analogue L, R, and $1/C$ in the electrical system. This suggests that LCR circuits might be used to model complicated physical systems subject to vibrations, and conversely, that mass–spring systems might represent complicated electrical systems.

EXERCISES 16.11

1. A 1-kilogram mass is suspended vertically from a spring with constant 16 newtons per meter. The mass is pulled 10 centimeters below its equilibrium position, and then released. Find the position of the mass, relative to its equilibrium position, at any time if:

a) damping is ignored;

b) a damping force in newtons equal to one-tenth the instantaneous velocity in meters per second acts on the mass;

c) a damping force in newtons equal to ten times the instantaneous velocity in meters per second acts on the mass.

2. A 200-gram mass suspended vertically from a spring with constant 10 newtons per meter is set into vibration by an external force in newtons given by $4 \sin(10t)$, $t \geq 0$. During the motion a damping force in newtons equal to $\frac{3}{5}$ the velocity of the mass in meters per second acts on the mass. Find the position of the mass as a function of time t.

3. A 0.001-farad capacitor and a 2-henry inductor are connected in series with a 20-volt battery. If there is no charge on the capacitor before the battery is connected, find the current in the circuit as a function of time.

4. At time $t = 0$, a 0.02-farad capacitor, a 100-ohm resistor, and a 1-henry inductor are connected in series. If the charge on the capacitor is initially 5 coulombs, find its charge as a function of time.

5. A 5-henry inductor and a 20-ohm resistor are connected in series with a generator supplying an oscillating voltage of $10 \sin(2t)$, $t \geq 0$. What are the transient and steady-state currents in the circuit?

6. At time $t = 0$, a mass M is attached to the end of a hanging spring with constant k, and then released. Assuming that friction is negligible, find the subsequent displacement of the mass as a function of time.

7. A $\frac{1}{2}$-kilogram mass sits on a table attached to a spring with constant 18 newtons per meter (Fig. 16.20). The mass is pulled so as to stretch the spring 5 centimeters and then released.

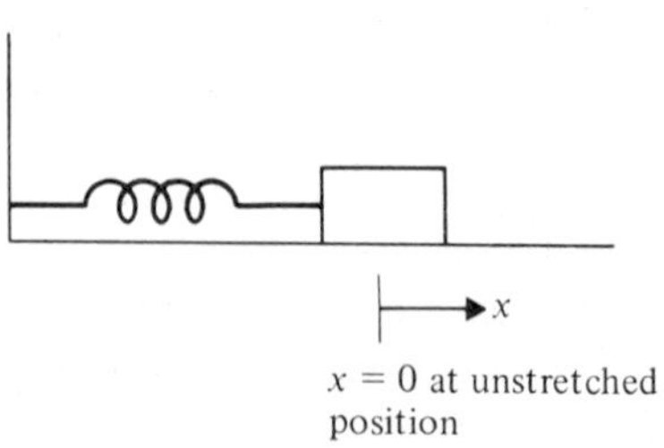

Figure 16.20

a) If friction between the mass and the table creates a force of $\frac{1}{2}$ newton that opposes motion, show that the differential equation determining motion is

$$\frac{d^2x}{dt^2} + 36x = 1, \qquad x(0) = 0.05, \qquad x'(0) = 0.$$

b) Find where the mass comes to rest for the first time. Will it move from this position?

8. Repeat Exercise 7 given that the mass is pulled 25 centimeters to the right.

9. At time $t = 0$, an uncharged 0.1-farad capacitor is connected in series with a 0.5-henry inductor and a 3-ohm resistor. If the current in the circuit at this instant is 1 ampere, find the maximum charge that the capacitor will experience.

10. Differential equation (16.67) describes the motion of a mass M at the end of a spring, taking damping proportional to velocity into account. Show each of the following.

a) If $\beta = 0$,

$$x(t) = C_1 \cos(\sqrt{k/M}\,t) + C_2 \sin(\sqrt{k/M}\,t),$$

called *simple harmonic motion.*

b) If $\beta \neq 0$ and $\beta^2 - 4kM < 0$,

$$x(t) = e^{-\beta t/(2M)}[C_1 \cos \omega t + C_2 \sin \omega t],$$

where

$$\omega = \frac{\sqrt{4kM - \beta^2}}{2M},$$

called *damped oscillatory motion.*

c) If $\beta \neq 0$ and $\beta^2 - 4kM > 0$,

$$x(t) = e^{-\beta t/(2M)}[C_1 e^{\omega t} + C_2 e^{-\omega t}]$$

where

$$\omega = \frac{\sqrt{\beta^2 - 4kM}}{2M},$$

called *overdamped motion.*

d) If $\beta \neq 0$ and $\beta^2 - 4kM = 0$,

$$x(t) = (C_1 + C_2 t)e^{-\beta t/(2M)},$$

called *critically damped motion.*

11. A vertical spring having constant 64 newtons per meter has a 1-kilogram mass attached to it. An external force $F(t) = 2 \sin 8t$, $t \geq 0$, is applied to the mass.

If the mass is at rest at its equilibrium position at time $t = 0$, and damping is negligible, find the position of the mass as a function of time. What happens to the oscillations as time progresses? This phenomenon is called *resonance*; it occurs when the frequency of the forcing function is identical to the natural frequency of the undamped system.

12. A $\frac{25}{9}$-henry inductor, a 0.04-farad capacitor, and a generator with a voltage of $15 \cos (3t)$ are connected in series at time $t = 0$. Find the current in the circuit as a function of time. Does resonance occur?

13. A mass M is suspended from a vertical spring with constant k. If an external force $F(t) = A \cos \omega t$, $t \geq 0$, is applied to the mass, find the value of ω that will cause resonance.

14. A mass M is suspended from a vertical spring with constant k. When an external force $F(t) = A \cos \omega t$, $t \geq 0$, is applied to the mass, and a damping force proportional to velocity acts on the mass during its subsequent oscillations, the differential equation governing the motion of M is

$$M \frac{d^2x}{dt^2} + \beta \frac{dx}{dt} + kx = A \cos \omega t.$$

a) Show that steady-state oscillations of M are defined by

$$x(t) = \frac{A}{(k - M\omega^2)^2 + \beta^2\omega^2} \times [(k - M\omega^2) \cos \omega t + \beta\omega \sin \omega t].$$

b) Verify that this function can be expressed in the form

$$x(t) = \frac{A}{\sqrt{(k - M\omega^2)^2 + \beta^2\omega^2}} \sin (\omega t + \phi),$$

where

$$\sin \phi = \frac{k - M\omega^2}{\sqrt{(k - M\omega^2)^2 + \beta^2\omega^2}},$$

$$\cos \phi = \frac{\beta\omega}{\sqrt{(k - M\omega^2)^2 + \beta^2\omega^2}}.$$

c) If resonance is said to occur when the amplitude of the steady-state oscillations is a maximum, what value of ω will yield resonance? What is the maximum amplitude of $x(t)$?

d) How does resonance differ for damped oscillations in this case and for undamped oscillations in Exercise 13?

15. a) A cube L meters on each side and with a mass of M kilograms floats half submerged in water. If it is pushed down slightly and then released, oscillations take place. Use Archimedes' principle to find the differential equation governing these oscillations. Assume no damping forces due to the viscosity of the water.

b) What is the frequency of the oscillations?

***16.** A sphere of radius R floats half submerged in water. It is set into vibration by pushing it down slightly and then releasing it. If y denotes the instantaneous distance of its diametral plane below the surface, show that

$$\frac{d^2y}{dt^2} = \frac{-3g}{2R^3}\left(R^2y - \frac{y^3}{3}\right),$$

where $g > 0$ is the acceleration due to gravity.

SUMMARY

A differential equation is an equation that contains an unknown function and some of its derivatives, and the equation must be solved for this function. Depending on the form of the equation, various techniques may be used to find the solution. From this point of view, solving differential equations is much like evaluating antiderivatives: We must first recognize the technique appropriate to the particular problem at hand, and then proceed through the mechanics of the technique. It is important, then, to immediately recognize the type of differential equation under consideration.

Broadly speaking, we divided differential equations into two main groups: first-order equations together with simple second-order equations, and linear differential equations. A first-order differential equation in $y(x)$ is said to be separable if it can be written in the form $N(y)\,dy = M(x)\,dx$, and a general solution for such an equation is defined implicitly by

$$\int N(y)\,dy = \int M(x)\,dx + C.$$

A differential equation of the form $dy/dx + P(x)y = Q(x)$ is said to be linear first order. If this equation is multiplied by the integrating factor $e^{\int P(x)\,dx}$, then the left side becomes the derivative of $ye^{\int P(x)\,dx}$, and the equation is immediately integrable.

If a second-order differential equation in $y(x)$ has either the dependent variable or the independent variable explicitly missing, it can be reduced to a pair of first-order equations. This is accomplished in the former case by setting $v = y'$ and $v' = y''$, and in the latter by setting $v = y'$ and $y'' = v\,dv/dy$.

A differential equation in $y(x)$ is said to be linear if it is of the form

$$a_0 \frac{d^n y}{dx^n} + a_1 \frac{d^{n-1} y}{dx^{n-1}} + \cdots + a_{n-1} \frac{dy}{dx} + a_n y = F(x).$$

The general solution of such an equation is composed of two parts: $y(x) = y_h(x) + y_p(x)$. The function $y_h(x)$ is the general solution of the associated homogeneous equation obtained by replacing $F(x)$ by 0; $y_p(x)$ is any particular solution of the given equation whatsoever. In the special case that the a_i are constants, it is always possible to find $y_h(x)$. This is done by calculating all solutions of the auxiliary equation

$$a_0 m^n + a_1 m^{n-1} + \cdots + a_{n-1} m + a_n = 0,$$

a polynomial equation in m, and then using the rules of Theorem 16.3.

When the a_i are constants and $F(x)$ is a polynomial, an exponential, a sine, a cosine, or any sums or products thereof, we can also find $y_p(x)$, either by undetermined coefficients or by operators. The method of undetermined coefficients is simply an intelligent way of guessing at $y_p(x)$ on the basis of $F(x)$ and $y_h(x)$. Note once again that $y_h(x)$ must be calculated before using the method of undetermined coefficients. The method of operators, on the other hand, is based on formal algebraic manipulations and the inverse operator shift theorem. It does not require prior calculation of $y_h(x)$ and, in its simplest form, uses complex numbers.

REVIEW EXERCISES

In each of Exercises 1–20, find a general solution for the differential equation.

1. $x^2\,dy - y\,dx = 0$

2. $(x + 1)\,dx - xy\,dy = 0$

3. $\dfrac{dy}{dx} + 3xy = 2x$

4. $\dfrac{dy}{dx} + 4y = x^2$

5. $\dfrac{d^2y}{dx^2} + 4\dfrac{dy}{dx} + 3y = 2$

6. $\dfrac{d^2y}{dx^2} + 3\dfrac{dy}{dx} + 4y = 2$

7. $yy' = \sqrt{1 + y^2}$

8. $\dfrac{d^2y}{dx^2} + \dfrac{1}{x}\dfrac{dy}{dx} = x$

9. $y'' + 6y' + 3y = xe^x$

10. $\dfrac{dy}{dx} + 2xy = 2x^3$

11. $y^2y'' = y'$

12. $y'' - 4y' + 4y = \sin x$

13. $y'' - 4y' + 4y = x^2e^{2x}$

14. $y'' + 4y = \sin 2x$

15. $y'' + 4y' = x^2$

16. $2xy^2\dfrac{dy}{dx} + (x + 1)^2y^3 = 0$

17. $\dfrac{d^3y}{dx^3} + 3\dfrac{d^2y}{dx^2} + 3\dfrac{dy}{dx} + y = 2e^{-x}$

18. $y'' + 2y' + 4y = e^{-x}\cos\sqrt{3}x$

19. $\dfrac{dy}{dx} = y\tan x + \cos x$

20. $(2y^2 + 3x)\,dy + dx = 0$

In Exercises 21–24, find the solution of the differential equation satisfying the given condition(s).

21. $y^2\,dx + (x + 1)\,dy = 0$, $y(0) = 3$

22. $y'' - 8y' - 9y = 2x + 4$, $y(0) = 3$, $y'(0) = 7$

23. $y'' + 9y = e^x$, $y(0) = 0$, $y(\pi/2) = 4$

24. $y' + \dfrac{2}{x}y = \sin x$, $y(1) = 1$

25. The quantity of radioactive material present in a sample decays at a rate proportional to the amount of the material in the sample (see Section 9.3). If one quarter of a sample decays in 5 years, how long does it take for 90% of the sample to decay?

26. a) A piece of wood rises from the bottom of a container of oil 1 meter deep. If the wood has a mass of 0.5 gram and a volume of 1.0 cubic centimeter, and the density of oil is 0.9 gram per cubic centimeter, show that Archimedes' principle predicts a buoyant force due to fluid pressure of 8.829×10^{-3} newton. What is the force on the piece of wood due to gravity and fluid pressure?

b) The viscosity of the oil opposes motion by exerting a force equal (in newtons) to twice its velocity (in meters per second). Find the distance traveled by the wood as a function of time, assuming that it starts from rest on the bottom.

27. A 100-gram mass is suspended vertically from a spring with constant 1 newton per meter. The mass is pulled 4 centimeters above its equilibrium position and then released. Find the position of the mass at any time if:

a) damping is ignored;

b) a damping force in newtons equal to $\frac{1}{5}$ the instantaneous velocity of the mass in meters per second acts on the mass;

c) a damping force in newtons equal to $\sqrt{2/5}$ the instantaneous velocity of the mass in meters per second acts on the mass.

* 28. A 10-gram stone is dropped over the side of a bridge 50 meters above a river 10 meters deep. Air resistance is negligible, but as the stone sinks in the water, its motion is retarded by a force in newtons equal to one-fifth the speed of the stone in meters per second. If the stone loses 10% of its speed in penetrating the surface of the water, find:

a) its velocity as a function of time;

b) its position relative to the drop point as a function of time;

c) the time it takes to reach the bottom of the river from the point at which it was dropped.

Assume that the water is stationary and that buoyancy due to Archimedes' principle may be neglected.

***29.** Repeat Exercise 28 but assume that, for the purpose of Archimedes' principle, the volume of the stone is 3 cubic centimeters.

Appendix A

Mathematical Induction

Mathematical induction is a type of proof used to verify propositions that depend on integers. It is best introduced through a simple example.

Suppose we are required to verify that the formula for the sum of the first n positive integers is

$$1 + 2 + 3 + \cdots + n = \frac{n(n + 1)}{2}. \tag{A.1}$$

It is a simple matter to check that this result is valid for the first few integers:

For $n = 1$, the sum is 1, and the formula gives $1(1 + 1)/2 = 1$.
For $n = 2$, the sum is $1 + 2 = 3$; the formula gives $2(2 + 1)/2 = 3$.
For $n = 3$, the sum is $1 + 2 + 3 = 6$; the formula gives $3(3 + 1)/2 = 6$.

(Actually, to use induction, it is only necessary to verify the result at this stage for $n = 1$.)

We now suppose that k is some integer for which the result is valid; i.e., we suppose that

$$1 + 2 + 3 + \cdots + k = \frac{k(k + 1)}{2}.$$

On the basis of this supposition, the sum of the first $k + 1$ integers is

$$\begin{aligned} 1 + 2 + \cdots + (k + 1) &= (1 + 2 + \cdots + k) + (k + 1) \\ &= \frac{k(k + 1)}{2} + (k + 1) \\ &= \frac{1}{2}[k(k + 1) + 2(k + 1)] \\ &= \frac{(k + 1)(k + 2)}{2}. \end{aligned}$$

But this is precisely (A.1) with n replaced by $k + 1$. In other words, if (A.1) is valid for some integer k, then it must also be valid for the next integer $k + 1$.

We now put the above two facts together to verify that (A.1) is valid for all $n \geq 1$. We have proved that it is valid for $n = 1, 2$, and 3. But if it is true for 3, then it must also be true for 4. But if it is true for 4, then it is true for 5. If it is true for 5, it is true for 6; and so on, and so on. Consequently, (A.1) must be valid for every positive integer n.

This example contains the two essential parts of every inductive proof:

1. Verification of the result for the smallest integer;
2. Verification of the fact that if the result is valid for some integer k, then it must be valid for the next integer $k + 1$.

Once these two facts are established, the above argument is expounded to verify that the result is valid for the complete set of integers. This is the principle of mathematical induction, which we state formally as follows.

Principle of Mathematical Induction Suppose that with each integer n greater than or equal to some fixed integer N, there is associated a proposition P_n. Then P_n is true for all $n \geq N$ provided:

a) P_N is valid;

b) the validity of P_k implies the validity of P_{k+1}.

All inductive arguments must therefore contain these two essential parts. Part (a) is usually quite simple to establish; part (b), on the other hand, can sometimes be quite difficult. Be sure you understand what we are doing in part (b). We are proving that under the assumption that P_k is true, then P_{k+1} must also be true. It is the principle of mathematical induction that puts (a) and (b) together to state that P_n is true for all $n \geq N$.

Example 1 Verify that

$$1^2 + 2^2 + 3^2 + \cdots + n^2 = \frac{n(n+1)(2n+1)}{6}, \qquad n \geq 1.$$

Solution When $n = 1$, the left-hand side of this equation has value 1, and the right-hand side is $1(2)(3)/6 = 1$. The required result is therefore true for $n = 1$. Next we suppose that k is some integer for which the result is valid; i.e., we suppose that

$$1^2 + 2^2 + 3^2 + \cdots + k^2 = \frac{k(k+1)(2k+1)}{6}$$

(and we must write this down because we are going to need it later on). Our objective now is to verify that the result is valid for $k + 1$; i.e., we must verify that

$$1^2 + 2^2 + \cdots + (k+1)^2 = \frac{(k+1)(k+2)[2(k+1)+1]}{6}.$$

If we begin on the left, we have

$$\begin{aligned}
1^2 + 2^2 + \cdots + (k+1)^2 &= (1^2 + 2^2 + \cdots + k^2) + (k+1)^2 \\
&= \frac{k(k+1)(2k+1)}{6} + (k+1)^2 \\
&= \frac{(k+1)}{6}[k(2k+1) + 6(k+1)] \\
&= \frac{(k+1)}{6}(2k^2 + 7k + 6) \\
&= \frac{(k+1)(k+2)(2k+3)}{6} \\
&= \frac{(k+1)(k+2)[2(k+1)+1]}{6},
\end{aligned}$$

which is the right-hand side of the required result. We have therefore verified that the result is valid for $k + 1$. Consequently, by mathematical induction the formula is correct for all $n \geq 1$. □

Example 2 Prove that 7 divides $23^{3n} - 1$ for all positive integers n.

Solution When $n = 1$, $23^{3n} - 1 = 23^3 - 1 = 12{,}166$, and this is divisible by 7. The result is therefore true for $n = 1$. Next, suppose that k is some integer for which the result is valid; i.e., suppose that 7 divides $23^{3k} - 1$. We must now verify that 7 divides $23^{3(k+1)} - 1$. Since

$$\begin{aligned} 23^{3(k+1)} - 1 = 23^{3k+3} - 1 &= (23^{3k})(23^3) - 1 \\ &= [(23^{3k})(23^3) - 23^3] + (23^3 - 1) \\ &= 23^3(23^{3k} - 1) + (23^3 - 1), \end{aligned}$$

and $23^{3k} - 1$ and $23^3 - 1$ are both divisible by 7, it follows that $23^{3(k+1)} - 1$ must also be divisible by 7. The result is therefore true for $k + 1$ and, by mathematical induction is true for all $n \geq 1$. □

Example 3 Verify that if n lines are drawn in the plane, no two of which are identical, then the maximum number of points of intersection of these lines is $n(n - 1)/2$.

Solution If we draw two lines in the plane, there can be at most one point of intersection, and this is precisely what is predicted by $n(n - 1)/2$ when $n = 2$. The result is therefore correct for $n = 2$. Suppose that the proposition is true for some integer k; i.e., suppose that when k lines are drawn in the plane, the maximum number of points of intersection is $k(k - 1)/2$. We must now verify that when any $k + 1$ lines are drawn, the maximum number of points of intersection is $(k + 1)k/2$. To do this, we remove one of the $k + 1$ lines leaving k. But for these k lines the maximum number of points of intersection is $k(k - 1)/2$. If the line that was removed is now replaced, it can add at most k more points of intersection, one with each of the k lines. The maximum number of points of intersection of the $k + 1$ lines is therefore

$$\frac{k(k-1)}{2} + k = \frac{k}{2}(k - 1 + 2) = \frac{k(k+1)}{2},$$

and this is the result for $k + 1$. By mathematical induction, then, the result is valid for all $n \geq 2$. □

The following example is typical of inductive proofs for sequences discussed in Section 11.1.

Example 4 Verify that the sequence $c_1 = 1$, $c_{n+1} = 5 + \sqrt{2 + c_n}$, $n \geq 1$ is increasing and that 10 is an upper bound.

Solution

a) The sequence is increasing if $c_{n+1} > c_n$ for all $n \geq 1$. Since $c_2 = 5 + \sqrt{2 + 1} = 5 + \sqrt{3} > c_1$, the result is valid for $n = 1$. Suppose that k is some integer for which $c_{k+1} > c_k$. To verify that the result is valid for $k + 1$, we must show that $c_{k+2} > c_{k+1}$ (or equivalently that $5 + \sqrt{2 + c_{k+1}} > 5 + \sqrt{2 + c_k}$). But this is easily shown since $c_{k+1} > c_k$ implies that

$$2 + c_{k+1} > 2 + c_k,$$

and from this,

$$\sqrt{2 + c_{k+1}} > \sqrt{2 + c_k}.$$

Consequently,

$$5 + \sqrt{2 + c_{k+1}} > 5 + \sqrt{2 + c_k},$$

or

$$c_{k+2} > c_{k+1}.$$

By mathematical induction, then, $c_{n+1} > c_n$ for all $n \geq 1$.

b) An upper bound for the sequence is 10 if $c_n \leq 10$ for all $n \geq 1$. Certainly this is true for $n = 1$ since $c_1 = 1 < 10$. Suppose that k is some integer for which $c_k \leq 10$. To verify that $c_{k+1} \leq 10$ we proceed as follows. Since $c_k \leq 10$,

$$2 + c_k \leq 12.$$

Consequently,

$$\sqrt{2 + c_k} \leq \sqrt{12}$$

and

$$c_{k+1} = 5 + \sqrt{2 + c_k} \leq 5 + \sqrt{12} < 10.$$

By mathematical induction, then, 10 is an upper bound for the sequence. □

A variation of the principle of mathematical induction is given in Exercise A14.

EXERCISES A

In each of Exercises 1–5, use mathematical induction to establish the formula for positive integer n.

1. $1^3 + 2^3 + 3^3 + \cdots + n^3 = n^2(n + 1)^2/4$
2. $1 + 3 + 6 + 10 + \cdots + n(n + 1)/2 = n(n + 1)(n + 2)/6$
3. $\dfrac{1}{1 \cdot 2} + \dfrac{1}{2 \cdot 3} + \dfrac{1}{3 \cdot 4} + \cdots + \dfrac{1}{n(n + 1)} = \dfrac{n}{n + 1}$
4. $\dfrac{1}{1 \cdot 2 \cdot 3} + \dfrac{1}{2 \cdot 3 \cdot 4} + \dfrac{1}{3 \cdot 4 \cdot 5} + \cdots + \dfrac{1}{n(n + 1)(n + 2)} = \dfrac{n(n + 3)}{4(n + 1)(n + 2)}$
5. $a + ar + ar^2 + ar^3 + \cdots + ar^{n-1} = a(1 - r^n)/(1 - r), (r \neq 1)$
6. Prove that $x - y$ divides $x^n - y^n$ for $n \geq 1$.
7. Verify that $x + y$ divides $x^{2n+1} + y^{2n+1}$ for $n \geq 0$.
8. Prove that 576 divides $5^{2n+2} - 24n - 25$ for $n \geq 1$.
9. Prove that the sequence $c_1 = 3$, $c_{n+1} = 1 + \sqrt{6 + c_n}$, $n \geq 1$, is increasing and that 5 is an upper bound.
10. Verify that for the sequence $c_1 = 1$, $c_{n+1} = 1/(3 - c_n)$, $n \geq 1$:
    a) $0 \leq c_n \leq 1$ for all n;
    b) $c_{n+1} < c_n$ for all n.
    c) Can you establish (b) without first establishing (a)?
11. Verify that the sum of the interior angles of a polygon with n sides is $(n - 2)\pi$ radians.
12. Show that if n is a positive integer so too is
$$(n^3 + 6n^2 + 2n)/3.$$
13. Prove that if the proposition
$$1 + 3 + 5 + \cdots + (2n - 1) = n^2 + 4$$
is valid for some integer k, then it is also valid for $k + 1$. Is the result valid for all $n \geq 1$?

14. A variation of the principle of mathematical induction is the following: Suppose that with each integer n greater than or equal to some fixed integer N, there is associated a proposition P_n. Then P_n is true for all $n \geq N$ provided:

a) P_N and P_{N+1} are valid;

b) the validity of P_{k-1} and P_k implies the validity of P_{k+1}.

Use this variation to verify that the following sequence is increasing and that 40 is an upper bound:

$$c_1 = -30, \quad c_2 = -20, \quad c_{n+1} = 5 + \tfrac{1}{2}c_n + \tfrac{1}{3}c_{n-1}, \quad n \geq 2.$$

15. Prove that:

a) $$1 + \cos\theta + \cos 2\theta + \cdots + \cos n\theta = \frac{1 - \cos\theta + \cos n\theta - \cos(n+1)\theta}{2 - 2\cos\theta};$$

b) $$\sin\theta + \sin 3\theta + \cdots + \sin(2n-1)\theta = \frac{\sin^2 n\theta}{\sin\theta}.$$

16. Verify that

$$\left(1 - \frac{1}{4}\right)\left(1 - \frac{1}{9}\right)\left(1 - \frac{1}{16}\right)\cdots\left(1 - \frac{1}{n^2}\right) = \frac{n+1}{2n}.$$

17. Use induction to show that for $n \geq 2$, there exist constants a_r and b_r, $r = 0, 1, 2, \ldots, n$ such that

$$\sin^n x = \sum_{r=0}^{n} (a_r \cos rx + b_r \sin rx).$$

Determinants

Appendix B

In appearance, a *determinant of order n* is n^2 numbers arranged in n rows and n columns and enclosed by two vertical lines. Thus,

$$\begin{vmatrix} 2 & 1 \\ 3 & 4 \end{vmatrix}, \quad \begin{vmatrix} -1 & 0 & 3 \\ 2 & 1 & 6 \\ 7 & -1 & 4 \end{vmatrix}, \quad \begin{vmatrix} 0 & 0 & 1 & 5 \\ 2 & -1 & 3 & 6 \\ -1 & -2 & -3 & 4 \\ 6 & 7 & 8 & 10 \end{vmatrix}$$

are determinants of order 2, 3, and 4, respectively. The general determinant of order n is usually written in the form

$$D = \begin{vmatrix} a_{11} & a_{12} & a_{13} & \cdots & a_{1n} \\ a_{21} & a_{22} & a_{23} & \cdots & a_{2n} \\ \vdots & \vdots & \vdots & & \vdots \\ a_{n1} & a_{n2} & a_{n3} & \cdots & a_{nn} \end{vmatrix}. \qquad \text{(B.1)}$$

The element in the ith row and jth column is called the (i, j)th element and is denoted by a_{ij}. For brevity we often write

$$D = |a_{ij}|_{n \times n} \qquad \text{(B.2)}$$

to identify the general determinant of order n.

We wish to assign a value to every determinant, and to do this we first define what is meant by a minor and a cofactor of an element in a determinant.

▷ **Definition B.1** The *minor* M_{ij} of a_{ij} in $D = |a_{ij}|_{n \times n}$ is the determinant of order $n - 1$ obtained by deleting the ith row and jth column of D. ◁

For example, if

$$D = \begin{vmatrix} -1 & 0 & 3 \\ 2 & 1 & 6 \\ 7 & -1 & 4 \end{vmatrix},$$

then

$$M_{11} = \begin{vmatrix} 1 & 6 \\ -1 & 4 \end{vmatrix}, \quad M_{23} = \begin{vmatrix} -1 & 0 \\ 7 & -1 \end{vmatrix}, \quad M_{31} = \begin{vmatrix} 0 & 3 \\ 1 & 6 \end{vmatrix}.$$

▷ **Definition B.2** The *cofactor* A_{ij} of a_{ij} in $D = |a_{ij}|_{n \times n}$ is $(-1)^{i+j}M_{ij}$. ◁

If

$$D = \begin{vmatrix} -1 & 0 & 3 \\ 2 & 1 & 6 \\ 7 & -1 & 4 \end{vmatrix},$$

then

$$A_{11} = (-1)^{1+1}M_{11} = \begin{vmatrix} 1 & 6 \\ -1 & 4 \end{vmatrix},$$

$$A_{23} = (-1)^{2+3}M_{23} = -\begin{vmatrix} -1 & 0 \\ 7 & -1 \end{vmatrix}.$$

The following two rules now define the value of every determinant:

1. The value of a determinant $D = |a_{11}|$ of order 1 is a_{11}.

2. The value of a determinant $D = |a_{ij}|_{n \times n}$ of order n is obtained by choosing any line (row or column) and adding elements in that line, each multiplied by its cofactor.

If we select row i, then

$$D = a_{i1}A_{i1} + a_{i2}A_{i2} + \cdots + a_{in}A_{in} = \sum_{j=1}^{n} a_{ij}A_{ij}; \tag{B.3a}$$

if we select column j, we have

$$D = a_{1j}A_{1j} + a_{2j}A_{2j} + \cdots + a_{nj}A_{nj} = \sum_{i=1}^{n} a_{ij}A_{ij}. \tag{B.3b}$$

Rule (2) defines a determinant of order n in terms of n determinants of order $n - 1$ (the cofactors); each of these determinants is defined in terms of determinants of order $n - 2$, and the process is continued until only determinants of order 1 are involved.

It is not clear that the *value of a determinant is independent of the line chosen in its evaluation*, but this is indeed the case. In other words, rules (1) and (2) define a unique value for every determinant.

The following result is essential to the speedy evaluation of determinants.

▶ **Theorem B.1** The value of a determinant of order 2 is

$$D = \begin{vmatrix} a_{11} & a_{12} \\ a_{21} & a_{22} \end{vmatrix} = a_{11}a_{22} - a_{12}a_{21}. \blacktriangleleft \tag{B.4}$$

Proof If we expand D along its first row, we have

$$D = a_{11}A_{11} + a_{12}A_{12} = a_{11}(-1)^{1+1}M_{11} + a_{12}(-1)^{1+2}M_{12}$$
$$= a_{11}(a_{22}) - a_{12}(a_{21}). \square$$

Example 1 Evaluate

$$\begin{vmatrix} 2 & 3 \\ -4 & 6 \end{vmatrix}.$$

Solution By Theorem B.1, we have

$$\begin{vmatrix} 2 & 3 \\ -4 & 6 \end{vmatrix} = (2)(6) - (3)(-4) = 24. \square$$

Example 2 Evaluate

$$D = \begin{vmatrix} 3 & -2 & 6 \\ 1 & 3 & 4 \\ 2 & -1 & 2 \end{vmatrix}.$$

Solution If we expand along the first column, we have

$$D = 3(-1)^2 \cdot \begin{vmatrix} 3 & 4 \\ -1 & 2 \end{vmatrix} + 1(-1)^3 \cdot \begin{vmatrix} -2 & 6 \\ -1 & 2 \end{vmatrix} + 2(-1)^4 \cdot \begin{vmatrix} -2 & 6 \\ 3 & 4 \end{vmatrix}$$

$$= 3(6 + 4) - (-4 + 6) + 2(-8 - 18) = -24. \square$$

Using Eq. (B.3) to evaluate determinants is not always particularly easy. For instance, even for a determinant of order 5, it is necessary to evaluate 60 determinants of order 2. As a result, we now prove two theorems that are used to simplify determinants.

▶ **Theorem B.2** If any line of a determinant D has its elements multiplied by c, the new determinant has value cD. ◀

Proof If the ith row of $D = |a_{ij}|_{n \times n}$ is multiplied by c, and the resulting determinant is expanded along this row, its value is

$$\sum_{j=1}^{n} ca_{ij}A_{ij} = c\sum_{j=1}^{n} a_{ij}A_{ij} = cD. \square$$

▶ **Theorem B.3** If a multiple of one line of a determinant D is added to a parallel line, the resulting determinant also has value D. ◀

Proof Suppose c times the ith row is added to the jth row of $D = |a_{ij}|_{n \times n}$ to form

$$E = \begin{vmatrix} a_{11} & a_{12} & \cdots & a_{1n} \\ a_{21} & a_{22} & \cdots & a_{2n} \\ \vdots & \vdots & & \vdots \\ a_{i1} & a_{i2} & \cdots & a_{in} \\ \vdots & \vdots & & \vdots \\ a_{j1} + ca_{i1} & a_{j2} + ca_{i2} & \cdots & a_{jn} + ca_{in} \\ \vdots & \vdots & & \vdots \\ a_{n1} & a_{n2} & \cdots & a_{nn} \end{vmatrix}.$$

If we expand E along the jth row, we have

$$E = \sum_{k=1}^{n} (a_{jk} + ca_{ik})A_{jk} = \sum_{k=1}^{n} a_{jk}A_{jk} + c\sum_{k=1}^{n} a_{ik}A_{jk}.$$

Now $\Sigma_{k=1}^{n} a_{jk}A_{jk} = D$, and $\Sigma_{k=1}^{n} a_{ik}A_{jk}$ is the value of the following determinant, which has identical ith and jth rows:

$$\begin{vmatrix} a_{11} & a_{12} & \cdots & a_{1n} \\ \vdots & \vdots & & \vdots \\ a_{i1} & a_{i2} & \cdots & a_{in} \\ \vdots & \vdots & & \vdots \\ a_{i1} & a_{i2} & \cdots & a_{in} \\ \vdots & \vdots & & \vdots \\ a_{n1} & a_{n2} & \cdots & a_{nn} \end{vmatrix} \begin{matrix} \\ \\ \leftarrow i\text{th row} \\ \\ \leftarrow j\text{th row} \\ \\ \\ \end{matrix}.$$

In Exercise B.16 we show that a determinant with two identical parallel lines has value zero, and therefore $E = D$. $\square$

This theorem is the key to the evaluation of determinants. We use it to replace a determinant with an equivalent determinant that has a large number of zero elements. This makes evaluation by cofactors much simpler.

Example 3 Evaluate the determinant

$$D = \begin{vmatrix} 1 & 2 & -3 & 4 \\ 2 & 4 & 0 & -1 \\ 3 & 6 & 1 & 2 \\ 4 & 0 & 1 & 5 \end{vmatrix}.$$

Solution Instead of immediately expanding the determinant according to Eq. (B.3), we use Theorem B.3 to create zeros in column 3. We do this by first adding 3 times row 3 to row 1, and second, adding -1 times row 3 to row 4. The result is

$$D = \begin{vmatrix} 10 & 20 & 0 & 10 \\ 2 & 4 & 0 & -1 \\ 3 & 6 & 1 & 2 \\ 1 & -6 & 0 & 3 \end{vmatrix}.$$

We now expand D along the third column,

$$D = (1)(-1)^{3+3} \cdot \begin{vmatrix} 10 & 20 & 10 \\ 2 & 4 & -1 \\ 1 & -6 & 3 \end{vmatrix},$$

and factor 10 from the first row and 2 from the second column,

$$D = 20 \begin{vmatrix} 1 & 1 & 1 \\ 2 & 2 & -1 \\ 1 & -3 & 3 \end{vmatrix}.$$

Finally, we expand D along the first row to obtain

$$D = 20[1(3) - 1(7) + 1(-8)] = -240. \square$$

Example 4 Evaluate

$$D = \begin{vmatrix} 3 & 0 & 2 & -1 & 4 \\ 6 & 3 & 7 & -2 & 8 \\ 10 & 3 & 1 & 0 & 6 \\ 2 & 1 & 3 & -4 & 1 \\ 6 & -2 & -3 & -5 & 2 \end{vmatrix}.$$

Solution By adding multiples of column 4 to columns 1, 3, and 5, we can create zeros in the first row:

$$D = \begin{vmatrix} 0 & 0 & 0 & -1 & 0 \\ 0 & 3 & 3 & -2 & 0 \\ 10 & 3 & 1 & 0 & 6 \\ -10 & 1 & -5 & -4 & -15 \\ -9 & -2 & -13 & -5 & -18 \end{vmatrix}.$$

Expansion along the first row gives

$$D = \begin{vmatrix} 0 & 3 & 3 & 0 \\ 10 & 3 & 1 & 6 \\ -10 & 1 & -5 & -15 \\ -9 & -2 & -13 & -18 \end{vmatrix}.$$

We now add -1 times column 2 to column 3 to get

$$D = \begin{vmatrix} 0 & 3 & 0 & 0 \\ 10 & 3 & -2 & 6 \\ -10 & 1 & -6 & -15 \\ -9 & -2 & -11 & -18 \end{vmatrix}.$$

Expansion along row 1 gives

$$D = 3(-1) \cdot \begin{vmatrix} 10 & -2 & 6 \\ -10 & -6 & -15 \\ -9 & -11 & -18 \end{vmatrix}.$$

We now factor -1 from rows 2 and 3, 2 from row 1, and 3 from column 3, and then expand along row 1 to get

$$D = -18 \cdot \begin{vmatrix} 5 & -1 & 1 \\ 10 & 6 & 5 \\ 9 & 11 & 6 \end{vmatrix} = -18[5(-19) + 1(15) + 1(56)] = 432. \ \square$$

Solution of Linear Equations by Cramer's Rule

To solve a pair of linear equations in two unknowns such as

$$\begin{aligned} 2x + 3y &= 6, \\ x - 4y &= -1, \end{aligned}$$

we eliminate one of the unknowns, say y, solve the resulting equation for x, and then substitute this value of x into either of the original equations to obtain y. The result for the pair above is $x = 21/11$, $y = 8/11$. A similar procedure can be followed for three linear equations in three unknowns:

$$\begin{aligned} 2x - 3y + 4z &= 6, \\ x + 4y - 2z &= 7, \\ 3x - 2y + z &= -2. \end{aligned}$$

First one variable is eliminated, say x, to obtain two equations in the two unknowns y and z. These are then solved for y and z and substituted into one of the original equations to find x, and the complete solution is obtained. The solution is $x = 3/7$, $y = 128/35$, $z = 141/35$.

With determinants, a formula can be presented for the solution of linear equations; this result is called *Cramer's rule*. We illustrate it for the second example above and then demonstrate its general validity. The coefficients of the unknowns are arranged in a determinant called the determinant of the system of equations. For the above system of three equations, it is

$$D = \begin{vmatrix} 2 & -3 & 4 \\ 1 & 4 & -2 \\ 3 & -2 & 1 \end{vmatrix}.$$

We now define three other determinants D_x, D_y, and D_z. They are obtained by replacing the first, second, and third columns in D by the coefficients on the right-hand sides of the equations:

$$D_x = \begin{vmatrix} 6 & -3 & 4 \\ 7 & 4 & -2 \\ -2 & -2 & 1 \end{vmatrix}, \quad D_y = \begin{vmatrix} 2 & 6 & 4 \\ 1 & 7 & -2 \\ 3 & -2 & 1 \end{vmatrix}, \quad D_z = \begin{vmatrix} 2 & -3 & 6 \\ 1 & 4 & 7 \\ 3 & -2 & -2 \end{vmatrix}.$$

Cramer's rule says that

$$x = \frac{D_x}{D}, \qquad y = \frac{D_y}{D}, \qquad z = \frac{D_z}{D}.$$

We check the first of these and leave it to the reader to verify the other two. If we expand each of D_x and D along its first row, we have

$$x = \frac{D_x}{D} = \frac{6(0) + 3(3) + 4(-6)}{2(0) + 3(7) + 4(-14)} = \frac{-15}{-35} = \frac{3}{7}.$$

For our first example of two linear equations in x and y, Cramer's rule would give

$$x = \frac{D_x}{D} = \frac{\begin{vmatrix} 6 & 3 \\ -1 & -4 \end{vmatrix}}{\begin{vmatrix} 2 & 3 \\ 1 & -4 \end{vmatrix}} = \frac{-21}{-11} = \frac{21}{11},$$

$$y = \frac{D_y}{D} = \frac{\begin{vmatrix} 2 & 6 \\ 1 & -1 \end{vmatrix}}{-11} = \frac{-8}{-11} = \frac{8}{11}.$$

To prove Cramer's rule we require the following theorem.

▶ **Theorem B.4** If the elements in any line of a determinant are multiplied by the cofactors of corresponding elements of a distinct parallel line, the resulting sum is zero. ◀

Proof Suppose we take $D = |a_{ij}|_{n \times n}$, and construct a determinant E from D by replacing its kth row by its ith row:

$$E = \begin{vmatrix} a_{11} & a_{12} & \cdots & a_{1n} \\ \vdots & \vdots & & \vdots \\ a_{i1} & a_{i2} & \cdots & a_{in} \\ \vdots & \vdots & & \vdots \\ a_{i1} & a_{i2} & \cdots & a_{in} \\ \vdots & \vdots & & \vdots \\ a_{n1} & a_{n2} & \cdots & a_{nn} \end{vmatrix} \begin{matrix} \\ \\ \leftarrow i\text{th row} \\ \\ \leftarrow k\text{th row} \\ \\ \\ \end{matrix}.$$

If we expand this determinant along its kth row, the net effect is to multiply the elements of row i by the cofactors of row k:

$$E = \sum_{j=1}^{n} a_{ij} A_{kj}.$$

However, because E has two identical parallel lines, its value must be zero (see Exercise B.16). Thus,

$$\sum_{j=1}^{n} a_{ij} A_{kj} = 0 \quad \text{whenever } i \neq k. \tag{B.5a}$$

A similar proof for columns gives

$$\sum_{i=1}^{n} a_{ij} A_{ik} = 0 \quad \text{whenever } j \neq k. \ \square \tag{B.5b}$$

We now prove the following theorem.

▶ **Theorem B.5** (*Cramer's rule*). A system of n linear equations in n unknowns $x_1, x_2, \ldots, x_n$,

$$\begin{aligned} a_{11}x_1 + a_{12}x_2 + \cdots + a_{1n}x_n &= c_1, \\ a_{21}x_1 + a_{22}x_2 + \cdots + a_{2n}x_n &= c_2, \\ \vdots \qquad\qquad\qquad & \ \ \vdots \quad\ \vdots \\ a_{n1}x_1 + a_{n2}x_2 + \cdots + a_{nn}x_n &= c_n, \end{aligned} \tag{B.6}$$

can be represented compactly in the form

$$\sum_{j=1}^{n} a_{ij} x_j = c_i, \qquad i = 1, \ldots, n. \tag{B.7}$$

From the determinant $D = |a_{ij}|_{n \times n}$ of the system, we define n other determinants D_k in the following way: D_k is formed by replacing the kth column of D by the column of constants $c_1, c_2, \ldots, c_n$. The solution of (B.6) is then

$$x_k = \frac{D_k}{D}, \qquad k = 1, \ldots, n, \tag{B.8}$$

provided $D \neq 0$. ◀

Proof We multiply the first equation in (B.6) by the cofactor A_{1k}, the second equation by A_{2k}, and so on, until the last equation is multiplied by A_{nk}. Symbolically, this is represented by multiplying the ith equation by A_{ik} to get

$$\sum_{j=1}^{n} a_{ij} A_{ik} x_j = c_i A_{ik}, \qquad i = 1, \ldots, n.$$

We now add all these equations together:

$$\sum_{i=1}^{n} \sum_{j=1}^{n} a_{ij} A_{ik} x_j = \sum_{i=1}^{n} c_i A_{ik}.$$

The right-hand side of this equation is the expansion of D_k along its kth column. Consequently,

$$D_k = \sum_{j=1}^{n} \left(\sum_{i=1}^{n} a_{ij} A_{ik} \right) x_j.$$

But according to Theorem B.4 the summation in parentheses is zero unless $j = k$; and when $j = k$, the result is

$$D_k = \left(\sum_{i=1}^{n} a_{ik} A_{ik} \right) x_k = D x_k,$$

or

$$x_k = \frac{D_k}{D}. \ \square$$

Example 5 Use Cramer's rule to solve

$$\begin{aligned} 3x - 2y \quad\quad &= -1, \\ x + 4y - 2z &= 6, \\ 3y + 4z &= 7. \end{aligned}$$

Solution We have

$$D = \begin{vmatrix} 3 & -2 & 0 \\ 1 & 4 & -2 \\ 0 & 3 & 4 \end{vmatrix} = 3(22) + 2(4) = 74;$$

$$D_x = \begin{vmatrix} -1 & -2 & 0 \\ 6 & 4 & -2 \\ 7 & 3 & 4 \end{vmatrix} = -1(22) + 2(38) = 54;$$

$$D_y = \begin{vmatrix} 3 & -1 & 0 \\ 1 & 6 & -2 \\ 0 & 7 & 4 \end{vmatrix} = 3(38) + 1(4) = 118;$$

$$D_z = \begin{vmatrix} 3 & -2 & -1 \\ 1 & 4 & 6 \\ 0 & 3 & 7 \end{vmatrix} = 3(10) - 1(-11) = 41.$$

Hence, $x = 54/74 = 27/37$, $y = 118/74 = 59/37$, $z = 41/74$. □

EXERCISES B

In Exercises 1–9, evaluate the given determinant.

1. $\begin{vmatrix} 3 & 2 \\ -1 & 4 \end{vmatrix}$

2. $\begin{vmatrix} 1 & 0 \\ 0 & 1 \end{vmatrix}$

3. $\begin{vmatrix} -2 & -4 \\ -6 & -8 \end{vmatrix}$

4. $\begin{vmatrix} 1 & 2 & 3 \\ 2 & 4 & 6 \\ -1 & 3 & 0 \end{vmatrix}$

5. $\begin{vmatrix} -2 & 0 & 5 \\ 1 & 3 & 6 \\ -7 & 8 & 10 \end{vmatrix}$

6. $\begin{vmatrix} 10 & 20 & 30 \\ 16 & 32 & 64 \\ -1 & 2 & -3 \end{vmatrix}$

7. $\begin{vmatrix} 1 & 1 & 1 & -3 \\ 2 & 1 & 3 & 6 \\ 7 & -8 & 9 & 10 \\ 3 & 4 & -2 & 1 \end{vmatrix}$

8. $\begin{vmatrix} 3 & -2 & 1 & 6 \\ 4 & 5 & 2 & -1 \\ 0 & 0 & 3 & 2 \\ -1 & 3 & 4 & -1 \end{vmatrix}$

9. $\begin{vmatrix} 0 & 1 & 2 & 3 & 4 \\ -1 & 0 & 5 & 6 & 7 \\ -2 & -5 & 0 & 8 & 9 \\ -3 & -6 & -8 & 0 & 10 \\ -4 & -7 & -9 & -10 & 0 \end{vmatrix}$

In Exercises 10–15, use Cramer's rule to solve the system of equations.

10. $-3x + 4y = 2,$
$x - 2y = 6$

11. $2x - 3y = -10,$
$x + y = 0$

12. $4r - 2s = 6,$
$3r - s = -1$

13. $2x - 3y + z = 2,$
$6x - y + 2z = 4,$
$x - y = 1$

14.
$$\begin{aligned} 3z - 2y + x &= 6, \\ z + y + 4x &= 2, \\ -z - x + 2y &= -1 \end{aligned}$$

15.
$$\begin{aligned} 2x - 3y + 4z + w &= 1, \\ x - 3y \qquad + 2w &= 6, \\ 3y + 4z - w &= 2, \\ 3x - y + z \qquad &= 0 \end{aligned}$$

16. a) Use mathematical induction to verify that when two parallel lines of a determinant D are interchanged, the value of the new determinant is $-D$.

b) Use part (a) to prove that a determinant with two identical parallel lines has value zero.

17. A determinant is said to be *skew-symmetric* if its elements satisfy the property $a_{ij} + a_{ji} = 0$. (The determinant in Exercise 9 is skew-symmetric.) Show that a skew-symmetric determinant of odd order has value zero.

18. Solve the system of equations

$$\begin{aligned} 2x + 3y - 4z + w &= 0, \\ x + y - 2z + 3w &= 0, \\ 2x - 3y + z - 2w &= 0, \\ x + y - z + w &= 0. \end{aligned}$$

19. Can we use Cramer's rule to solve the following system?

$$\begin{aligned} 2x + 3y - 4z + w &= 0, \\ x + y - 2z + 3w &= 0, \\ 2x - 3y + z - 2w &= 0, \\ 5x + y - 5z + 2w &= 0 \end{aligned}$$

20. a) Show that the equation of the straight line in the xy-plane through the two points (x_1, y_1) and (x_2, y_2) can be expressed in the form

$$\begin{vmatrix} x & y & 1 \\ x_1 & y_1 & 1 \\ x_2 & y_2 & 1 \end{vmatrix} = 0.$$

b) Use the result in (a) to find the equation of the line through $(1, 1)$ and $(-2, 3)$.

21. a) Show that the equation of the circle in the xy-plane through the three points (x_1, y_1), (x_2, y_2), and (x_3, y_3) can be expressed in the form

$$\begin{vmatrix} x^2 + y^2 & x & y & 1 \\ x_1^2 + y_1^2 & x_1 & y_1 & 1 \\ x_2^2 + y_2^2 & x_2 & y_2 & 1 \\ x_3^2 + y_3^2 & x_3 & y_3 & 1 \end{vmatrix} = 0.$$

b) Use the result in (a) to find the circle through $(2, 1)$, $(-3, -3)$, and $(7, -5)$.

22. A parabola of the form $y = ax^2 + bx + c$ is to pass through the three points $(1, 0)$, $(2, 11)$, and $(-2, 1)$. Find a determinant that implicitly defines its equation.

23. Find a condition in the form of a determinant that serves as a test to determine whether a circle can be passed through four given points, no three of which are collinear. (*Hint*: See Exercise 21.)

24. Evaluate

$$\begin{vmatrix} a+b & a & a & \cdots & a \\ a & a+b & a & \cdots & a \\ a & a & a+b & \cdots & a \\ \vdots & \vdots & \vdots & & \vdots \\ a & a & a & \cdots & a+b \end{vmatrix}_{n \times n}.$$

25. a) Evaluate

$$\begin{vmatrix} a & b & b & b & \cdots & & b \\ b & a & b & b & \cdots & & b \\ b & b & a & b & \cdots & & b \\ b & b & b & a & \cdots & & b \\ \vdots & \vdots & \vdots & \vdots & & & \vdots \\ b & b & b & b & \cdots & a & b \\ b & b & b & b & \cdots & b & a \end{vmatrix}_{n \times n}.$$

b) Use the result in (a) to solve the system of equations

$$\begin{aligned} ax_1 + bx_2 + bx_3 + \cdots + bx_n &= 1, \\ bx_1 + ax_2 + bx_3 + \cdots + bx_n &= 1, \\ bx_1 + bx_2 + ax_3 + \cdots + bx_n &= 1, \\ \vdots \qquad \vdots \qquad \vdots \qquad & \vdots \quad \vdots \\ bx_1 + bx_2 + bx_3 + \cdots + ax_n &= 1. \end{aligned}$$

Answers to Even-numbered Exercises

CHAPTER 1

Exercises 1.1

2. $x < 3$ **4.** $0 < x < 3/5$ **6.** $-3 \le x \le 3$
8. $-5/3 < x < 3$ **10.** $-1/\sqrt{2} \le x \le 1/\sqrt{2}$
12. $-11/2 \le x \le -1/2$ **14.** $x \le -3, x \ge 3$
16. $x \le -1, x \ge 1$ **18.** $x < 1/2, x > 11/9$
20. $-2 - \sqrt{10} < x < -2 + \sqrt{10}$ **22.** $x < -1 - \sqrt{7/2}$, $x > -1 + \sqrt{7/2}$ **24.** All points **28.** $x < -2$, $-4/\sqrt{5} < x < 4/\sqrt{5}, x > 2$ **30.** $\sqrt{3} - 1 \le x \le \sqrt{5} - 1$, $-\sqrt{5} - 1 \le x \le -\sqrt{3} - 1$ **32.** $x < -\sqrt{37}$, $-3 < x < 3, x > \sqrt{37}$ **34.** $x < -11/7, -1 < x < 3$
36. $x \ge 5$ **38.** $x < (3 - \sqrt{101})/2, 5 < x < (3 + \sqrt{101})/2$
40. $(1 - \sqrt{5})/2 \le x \le (1 + \sqrt{5})/2$
42. $|x + 1| + |x - 2| > 4, x < -3/2, x > 5/2$
44. $\big||x - 2| - |x + 6|\big| < 3, -7/2 < x < -1/2$
46. (a) $-11 < x < -1$; (b) $x < -11, -2 < x < -1$; (c) $x < -11, -5 < x < -1$; (d) $-11 < x < -5$, $-2 < x < -1$

Exercises 1.2

4.

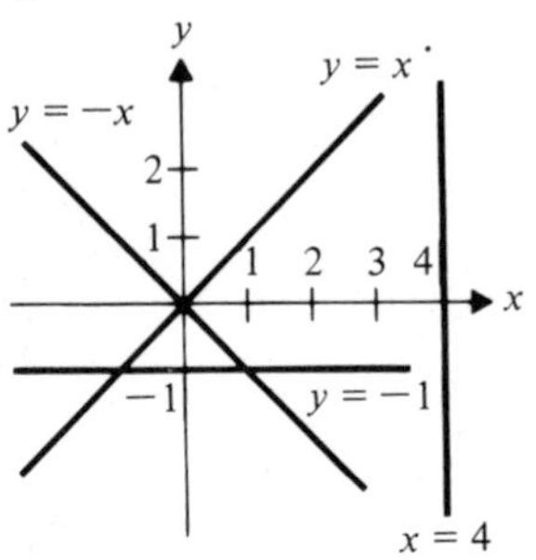

6. (a) Yes; (b) no
8. $(2, \sqrt{21})$
12. (a) $(0, 10), (-15/2, 0)$; (b) $(0, 10), (15\sqrt{3}/4, -15/4)$; (c) $(0, -10), (-15\sqrt{2}/4, -15\sqrt{2}/4)$; (d) $(-5\sqrt{3}, -5)$, $(5.75, 4.82)$

Exercises 1.3A

2.

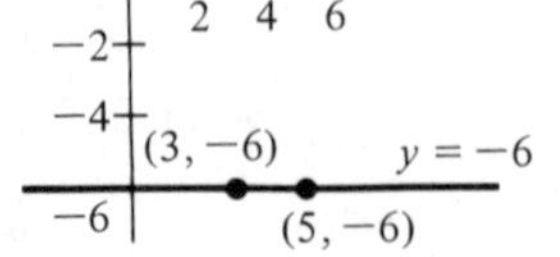

4.

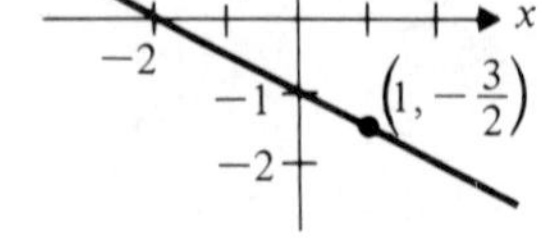

6.

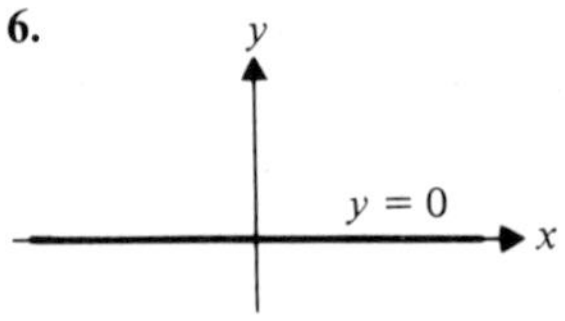

8.

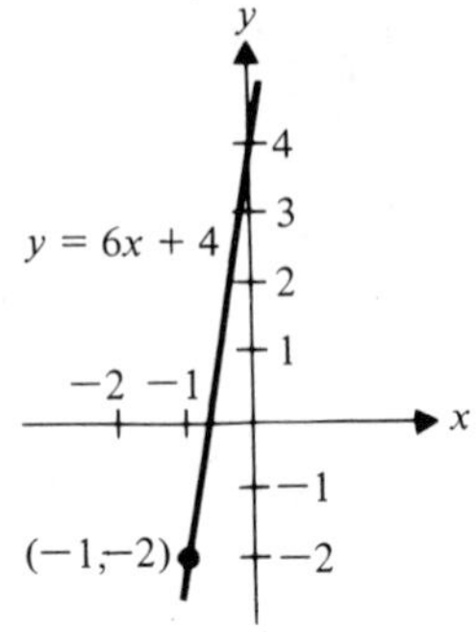

10.

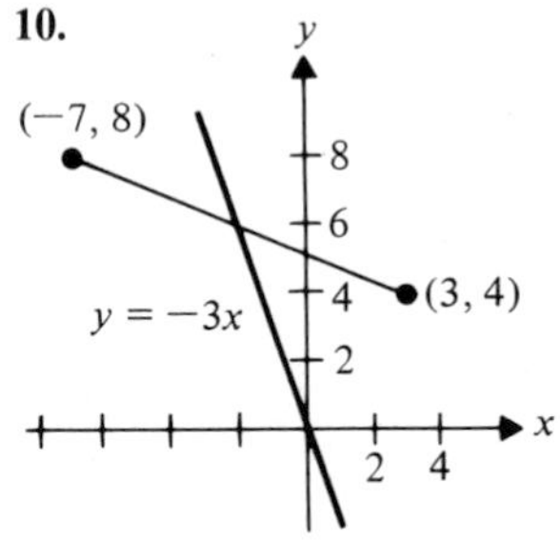

14. No **16.** Yes **18.** Yes **20.** (1, 2)
22. $(-10, -14)$ **24.** No **26.** No **30.** $x + y = 2$
32. $3y = 2x$ **34.** $5x + 3y = 30$ **36.** $y = -1$
40. $(-11/10, 19/10)$

Exercises 1.3B

2.

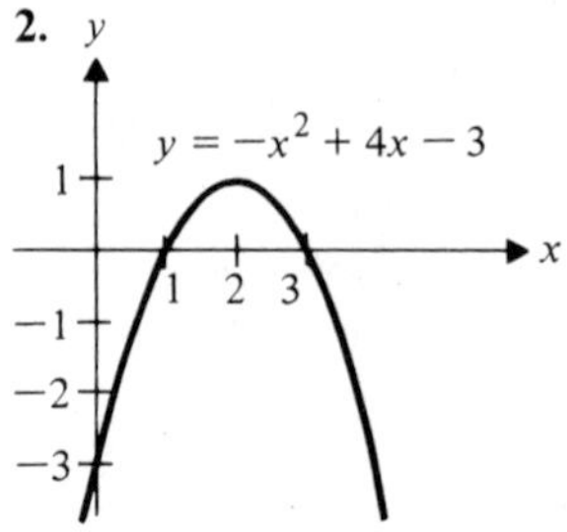

4.

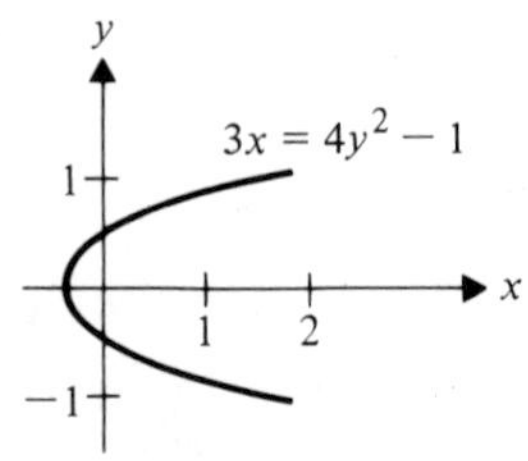

6.

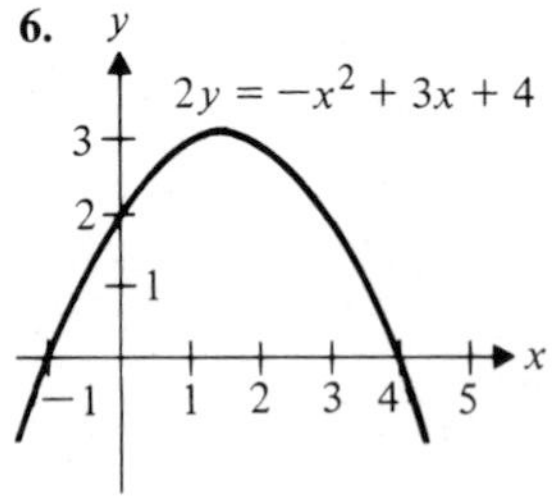

8.

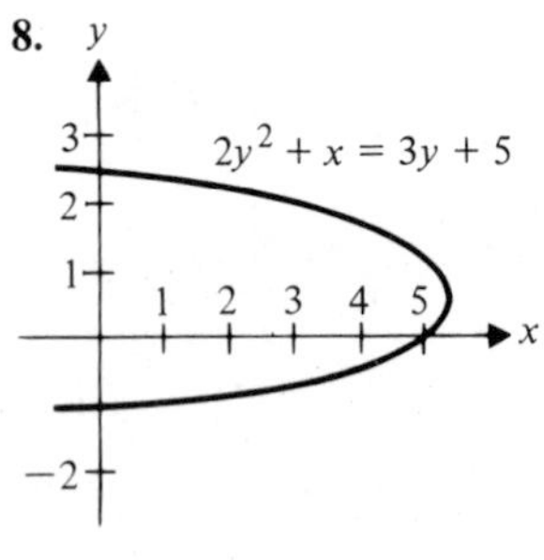

10.

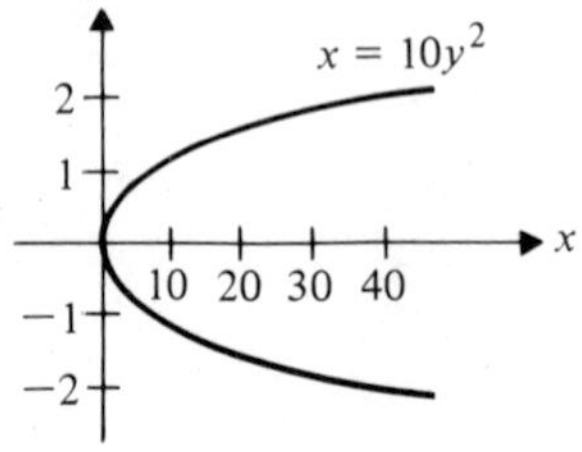

12.

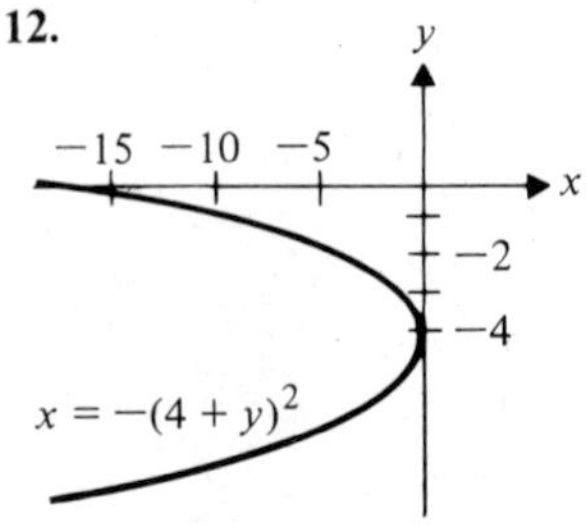

14. $y = x^2/2 + 1$ **16.** $y = (x^2 - 2x - 3)/3$ **18.** $(0, 1)$, $(-1, 0)$ **20.** No points **22.** $(0, 1), (-3, -2)$
24. (a) $(v^2/4.905) \sin\theta \cos\theta$; (b) $(v^2/19.62) \sin^2\theta$
26. $(1, 1), (4, 4)$ **28.** $y = (x^2 - 2x + 5)/2$

Exercises 1.3C

2.

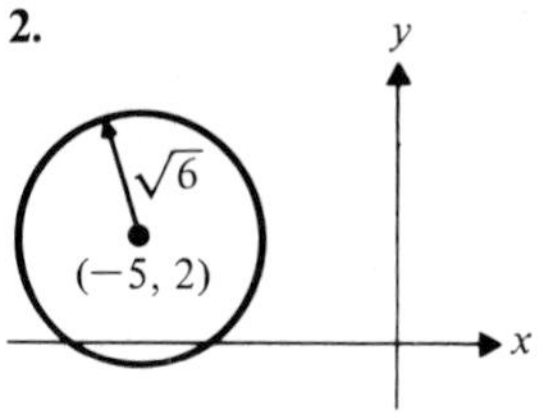

4.

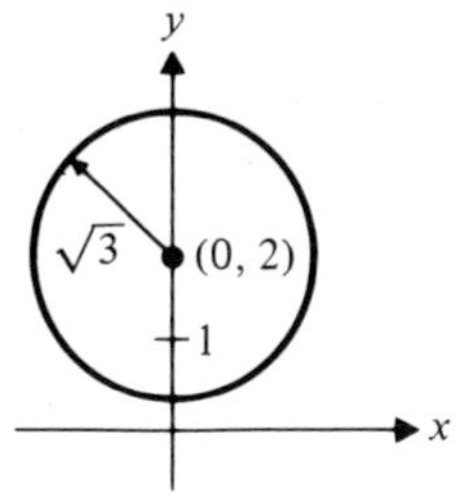

6.

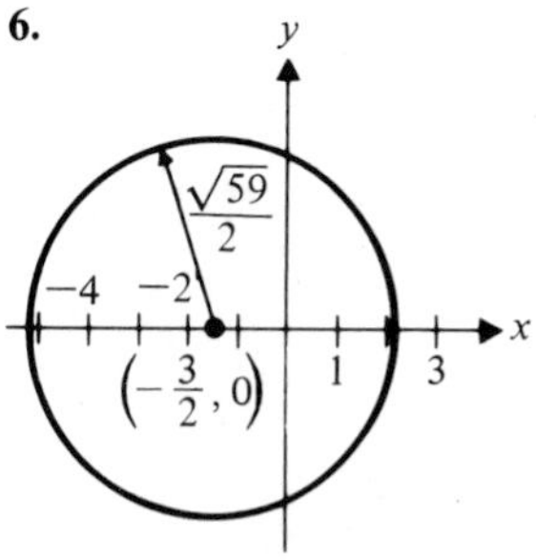

8.

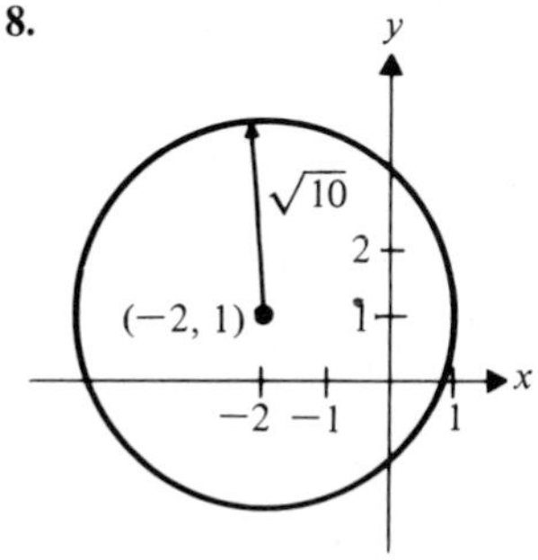

10. No points satisfy the equation. **12.** $(x - 1)^2 + y^2 = 1$
14. $(x - 3/2)^2 + (y + 3/2)^2 = 9/2$
16. $3x^2 + 3y^2 - 14x - 4y = 4$ **18.** $(0, 2), (-7/5, -11/5)$
20. No points **24.** $7(x^2 + y^2) + 6x - 44y + 24 = 0$

Exercises 1.4

2. $x \neq 2$ **4.** $|x| > 2$ **6.** $2 \leq |x| < 3$

8.

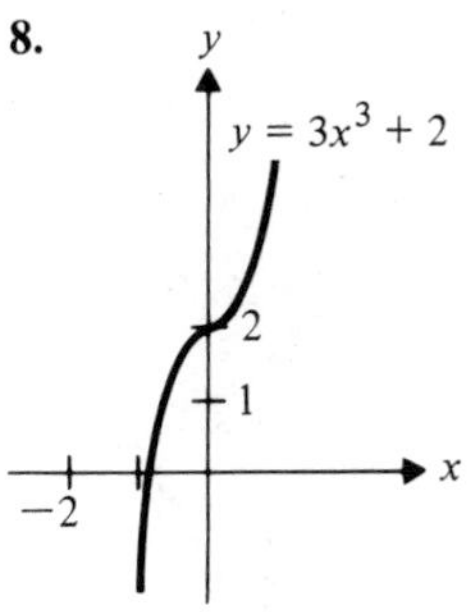

10.

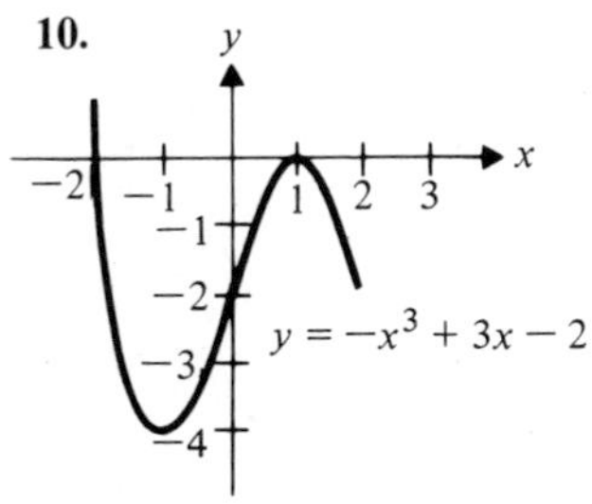

12.

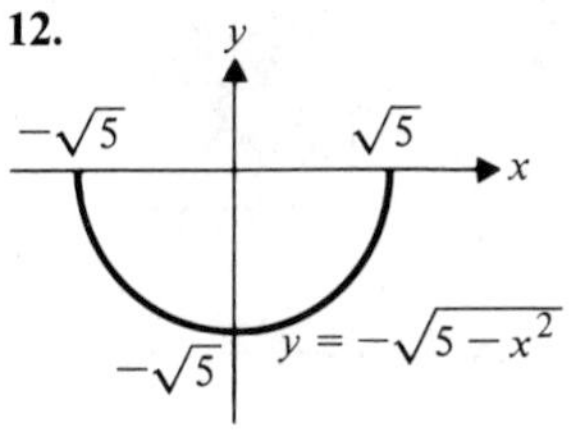

14.

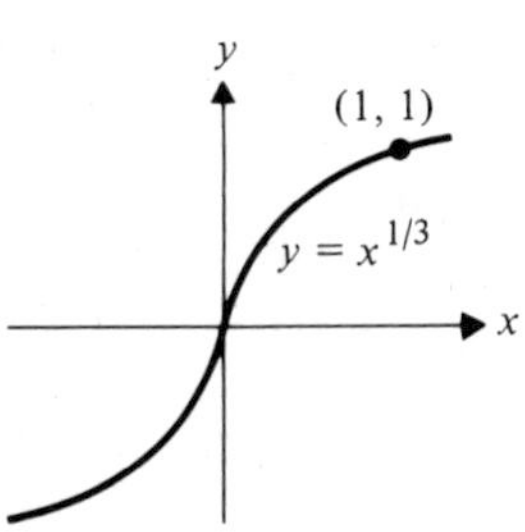

16.

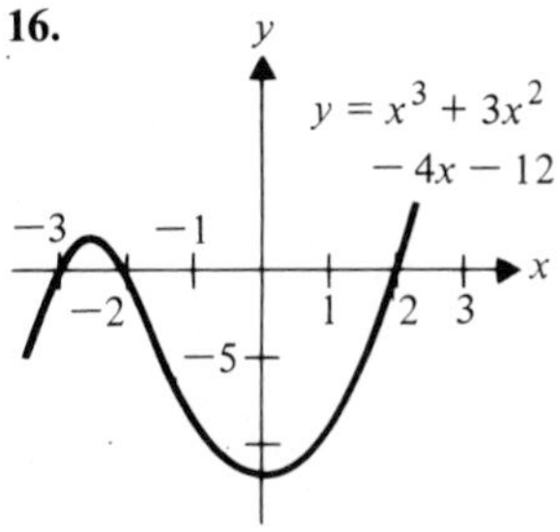

18.

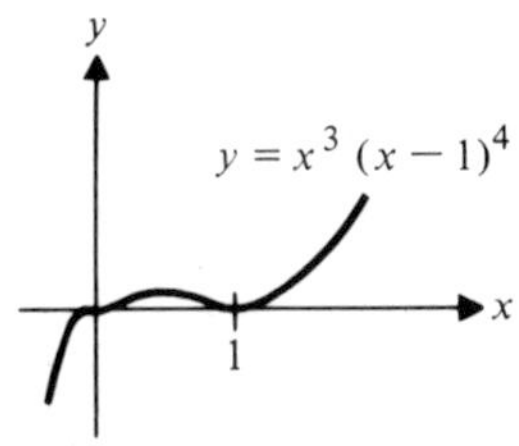

20.

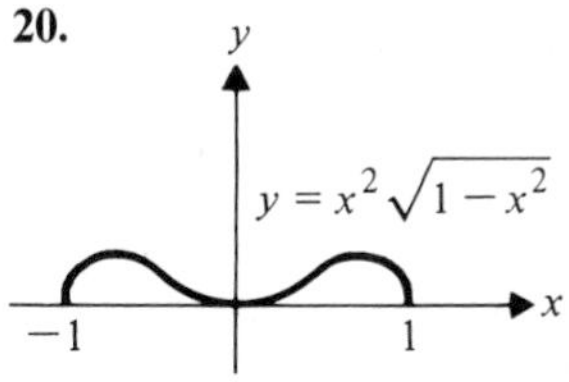

22.

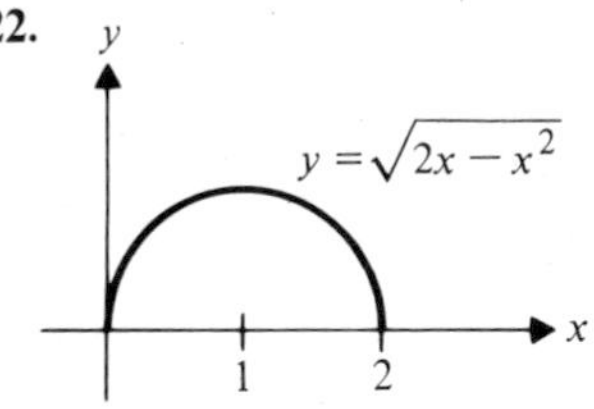

24.

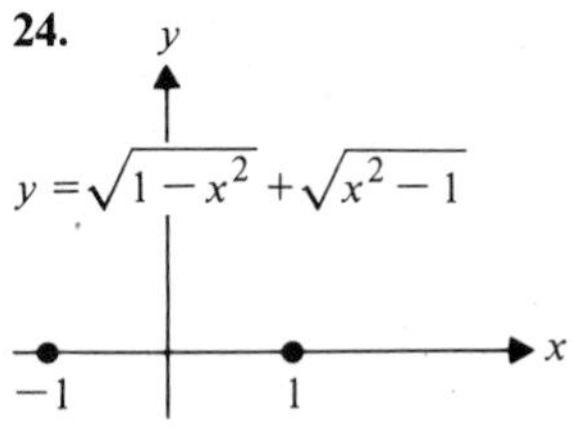

26.

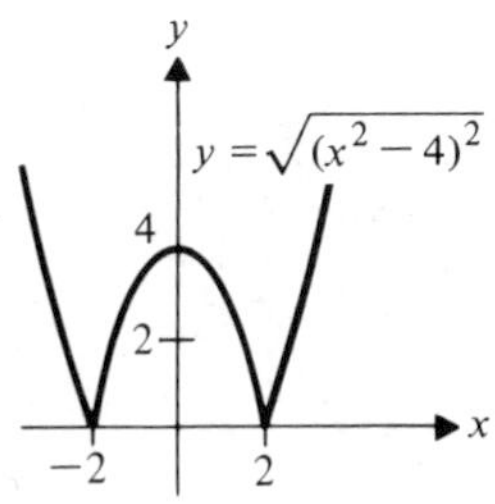

28.

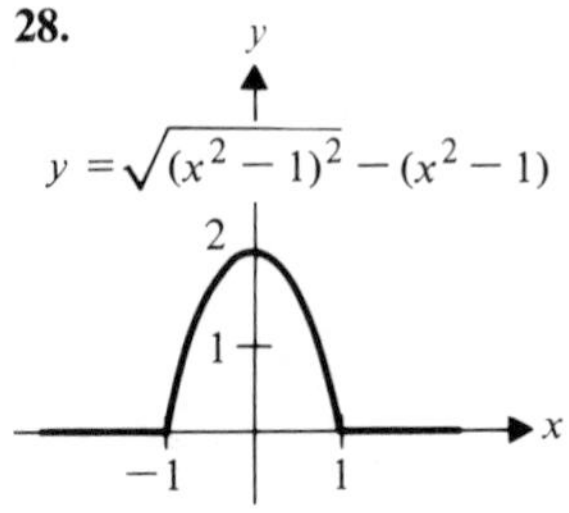

30. (b) Two solutions

32.

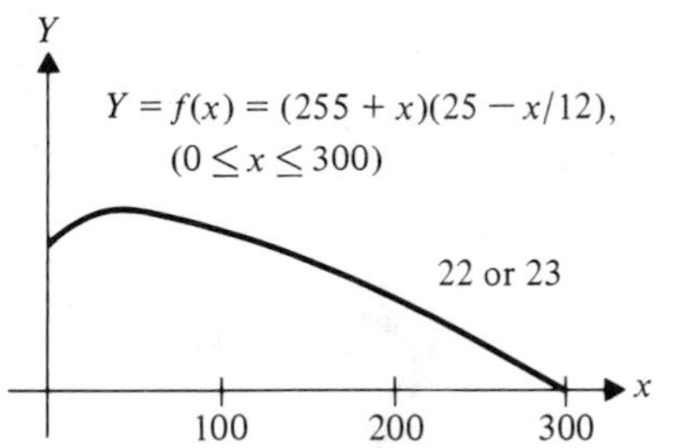

34.

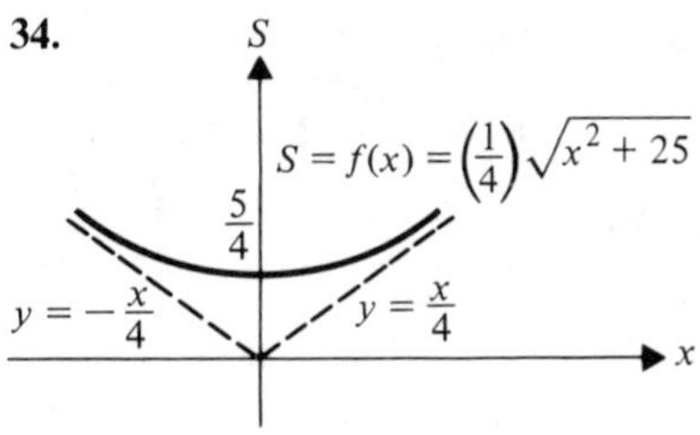

36.

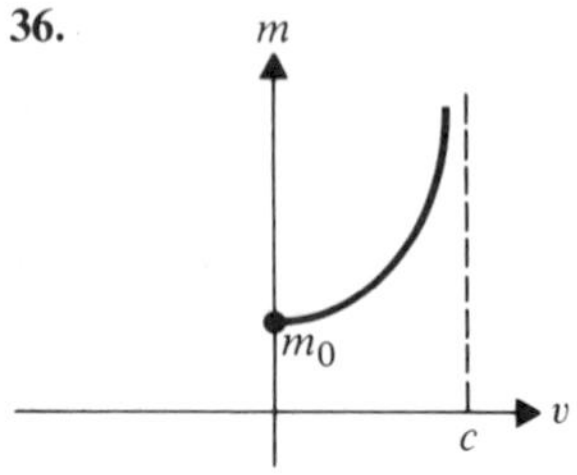

Exercises 1.5

2. Function

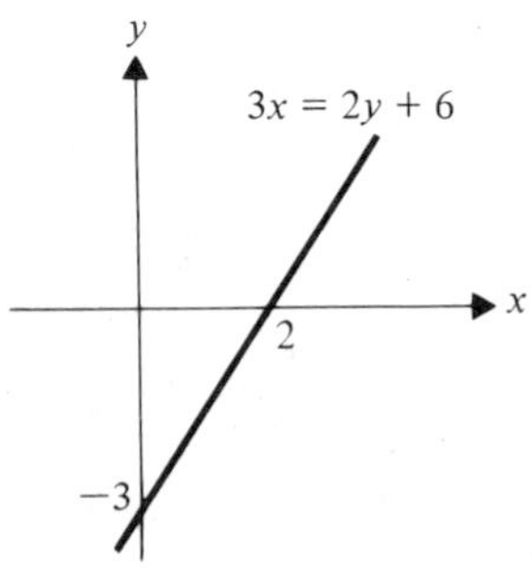

4. Not a function

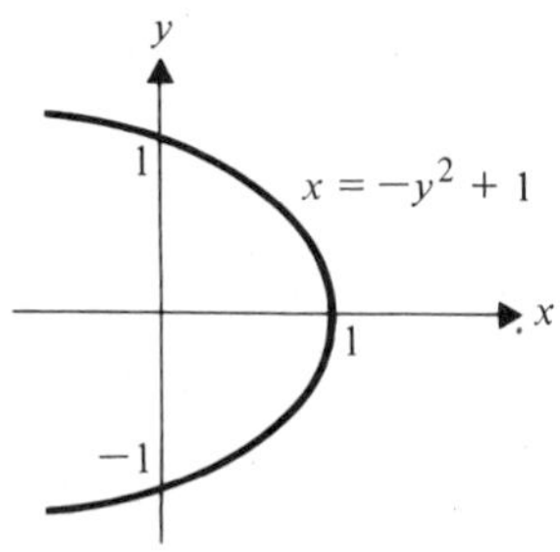

6. Function

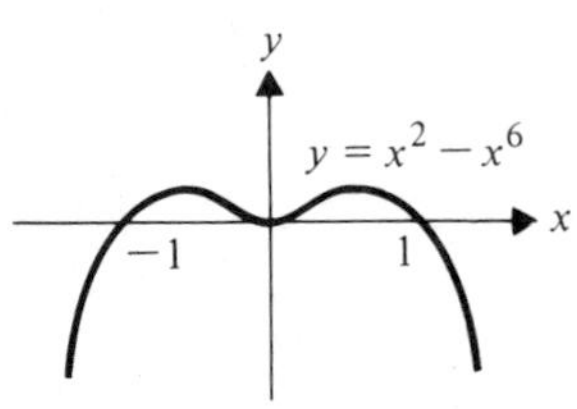

8. Not a function

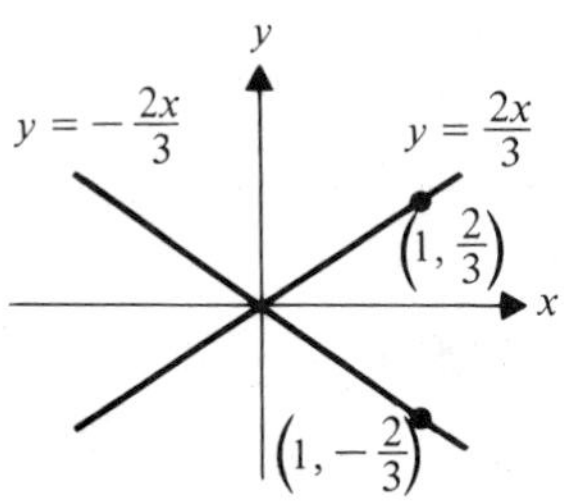

10. Not a function

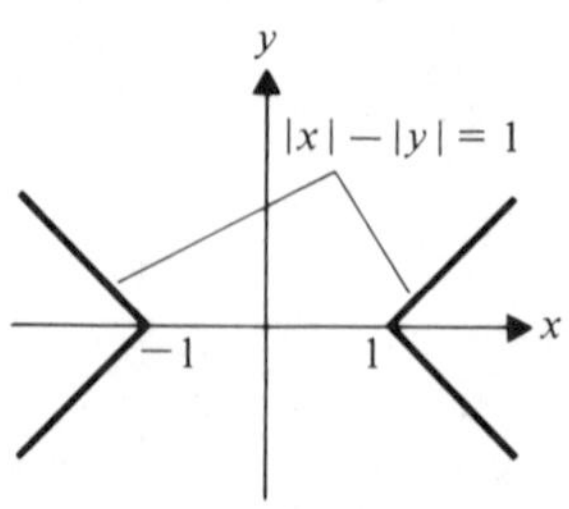

12. Not a function

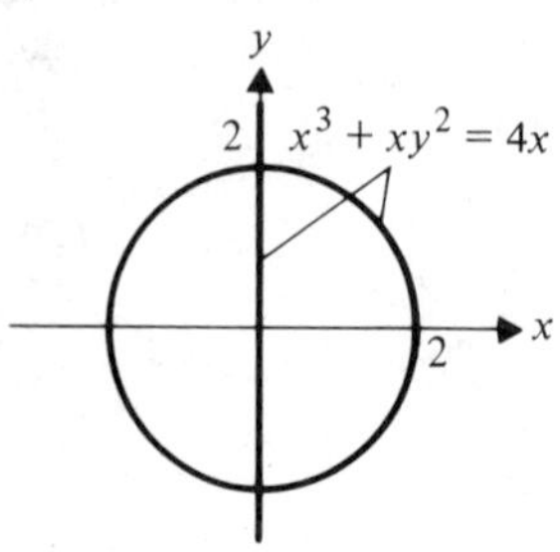

14. Not a function

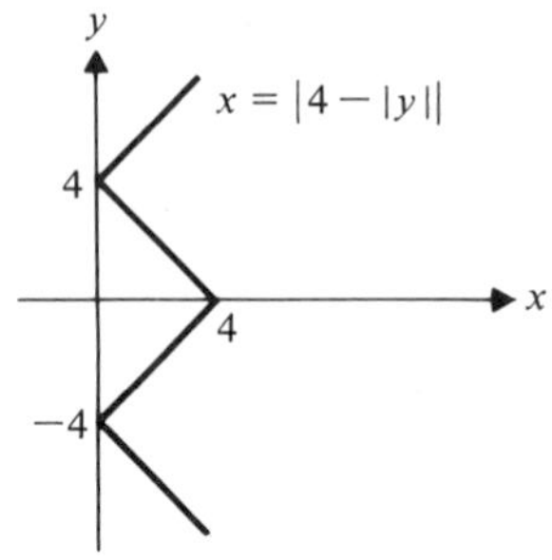

16. Not a function

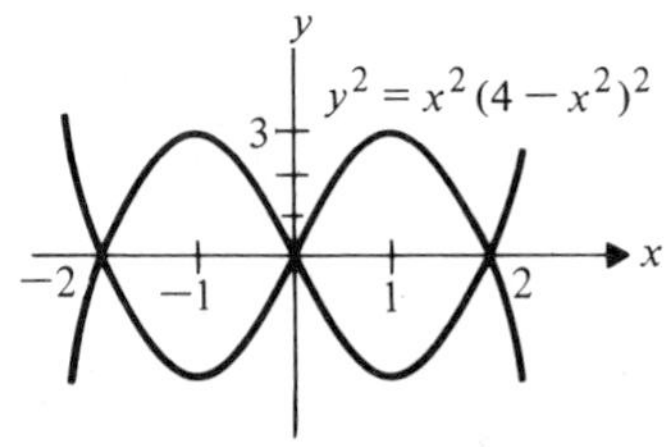

18. Not a function

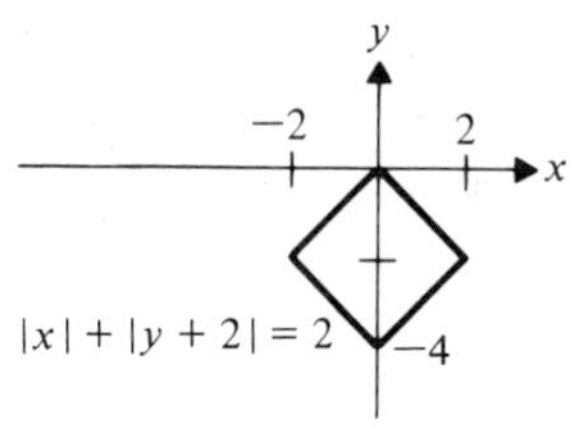

20. Not a function

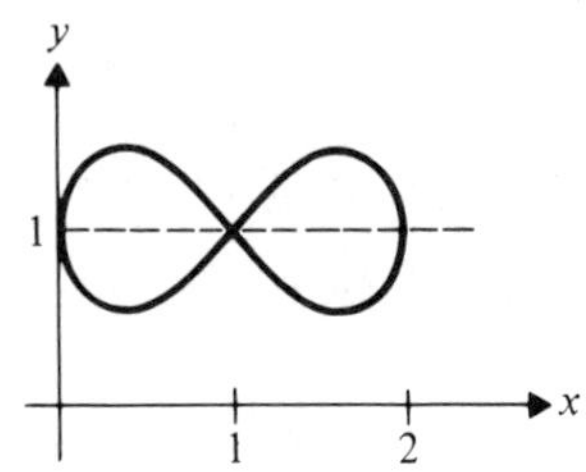

22. Odd **24.** Odd **26.** Neither

30. (a)

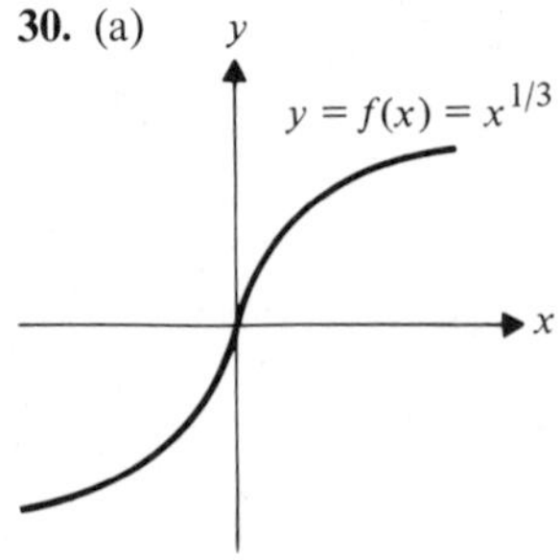

(b)

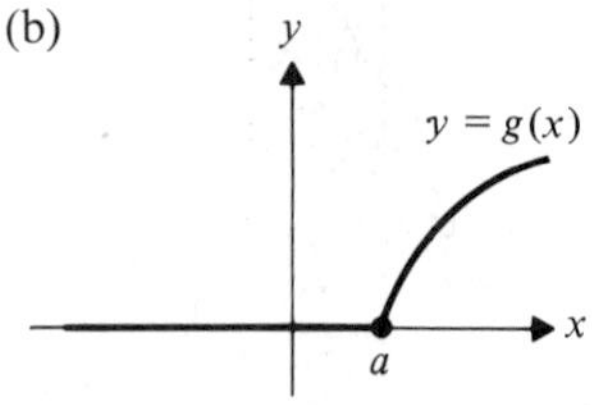

(c)

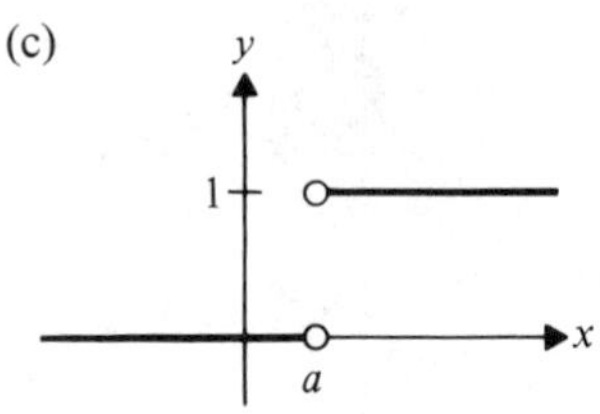

(d) $g(x) = \begin{cases} f(x-a)H_a(x), & x \neq a \\ 0, & x = a \end{cases}$

Review Exercises

2. $|x| \geq \sqrt{5}$ **4.** $x \neq 0, -1$ **6.** $x \geq 0$
8. $|x+1| > \sqrt{14}/2$ **10.** $-1 \leq x < 0, x \geq 1$

12.

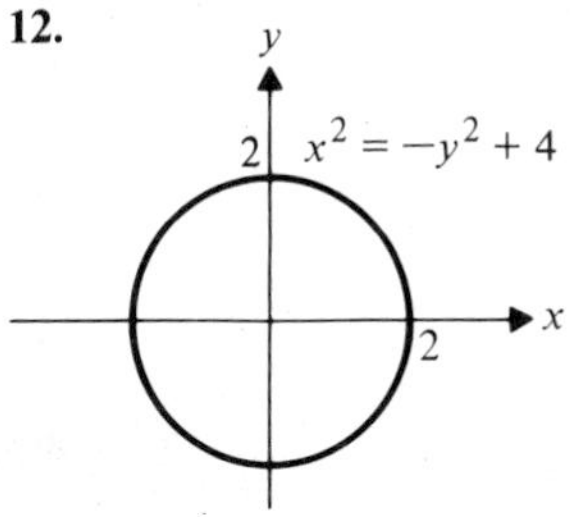

14.

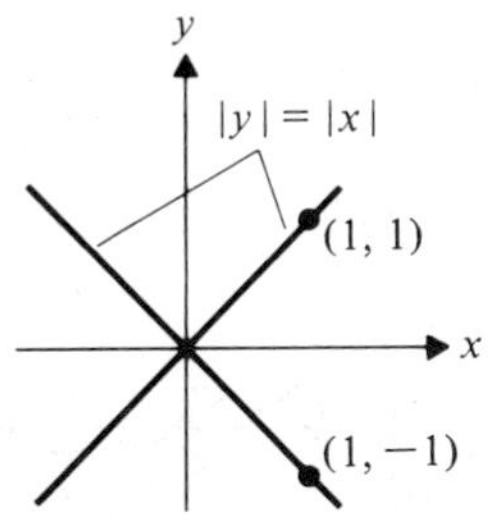

16.

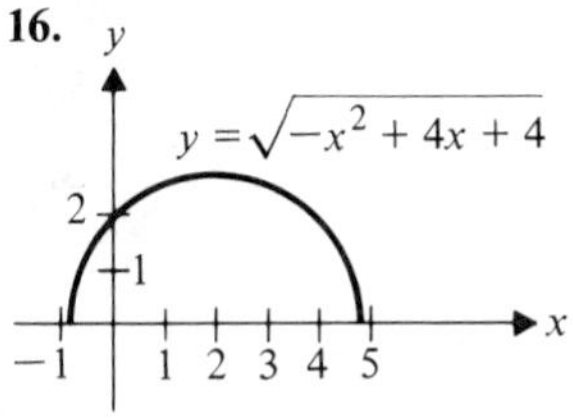

18.

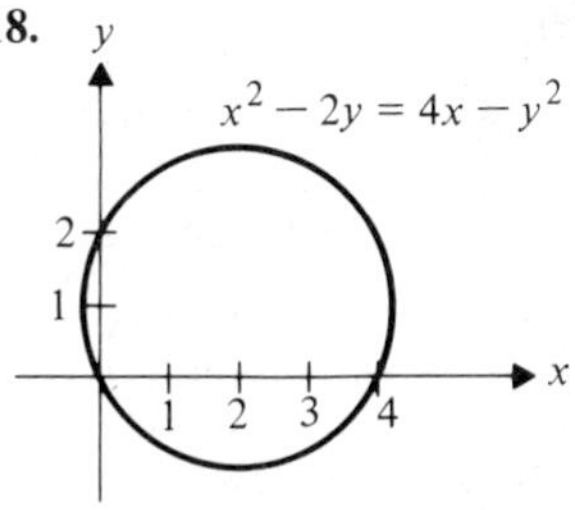

20.

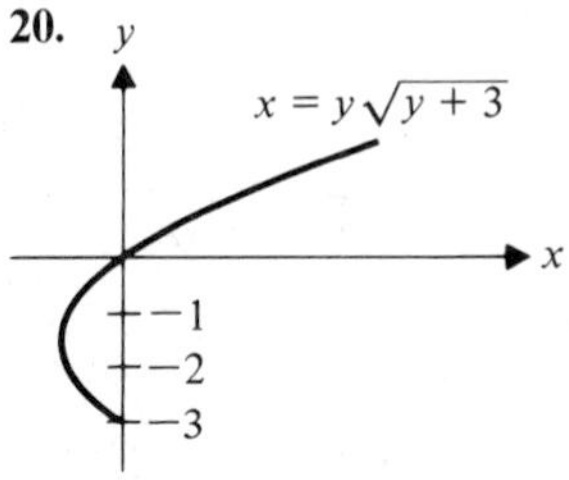

22. $f(x) = \sqrt{(x+1)/(2-x)}$
24. $f(x) = \sqrt{-x} + 1$
28. (a)

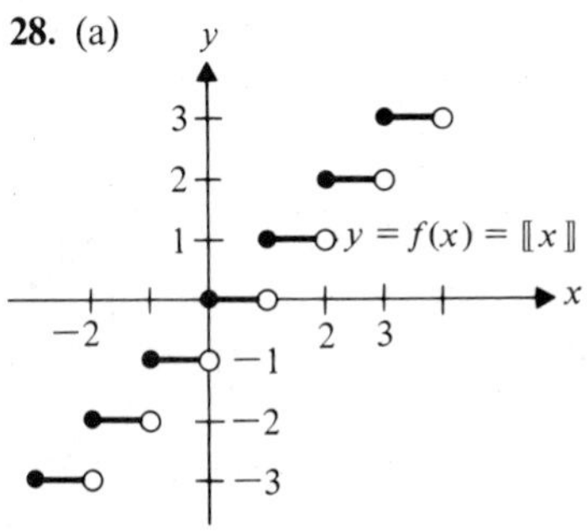

(b)

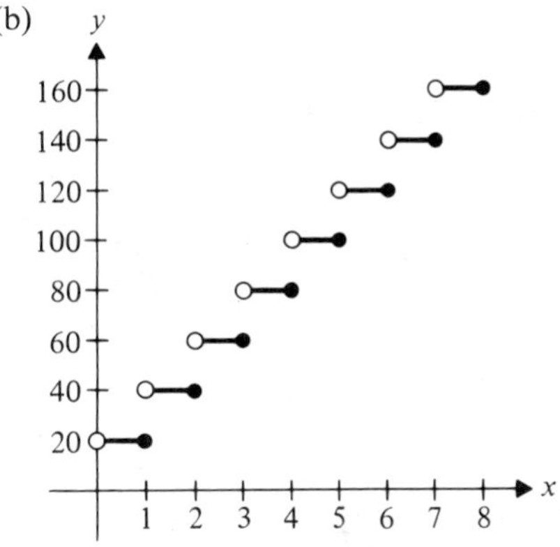

(c) $20[\![x+1]\!]$, $x \neq$ integer; $20x$, $x =$ integer; $0 < x \leq 8$

CHAPTER 2

Exercises 2.1

2. 0 **4.** −3/2 **6.** Does not exist **8.** 13/2 **10.** 0
12. 2/5 **14.** Does not exist **16.** Does not exist
18. 1/2 **20.** Does not exist

22.

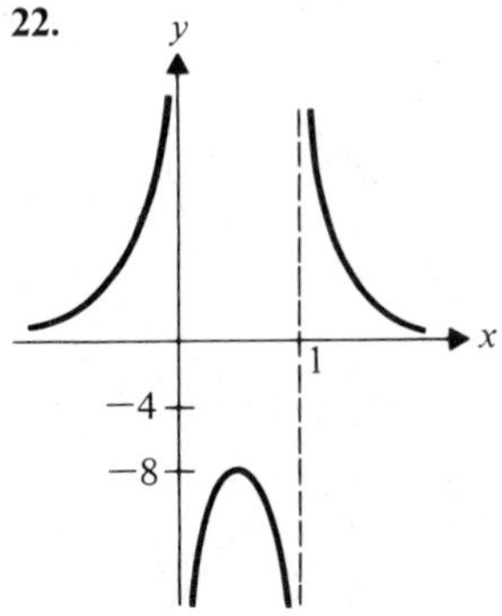

24.

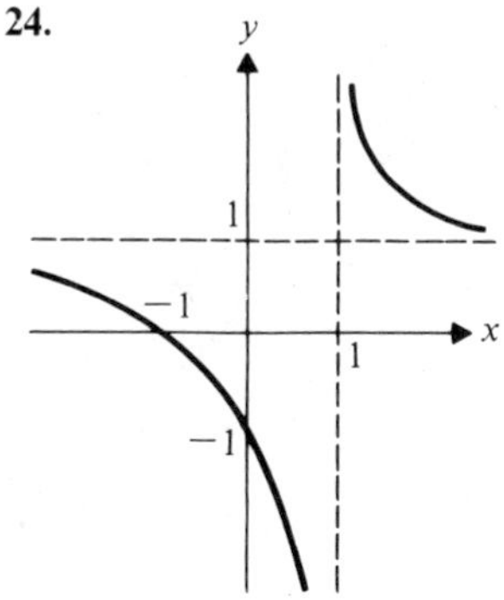

26.

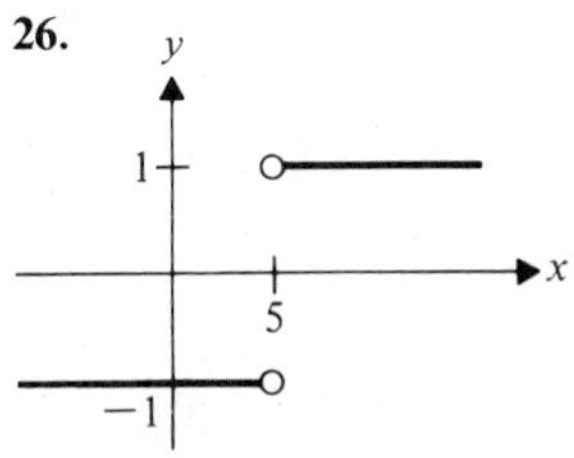

28. $2\sqrt{2}$ **30.** 0 **32.** nx^{n-1}

36.

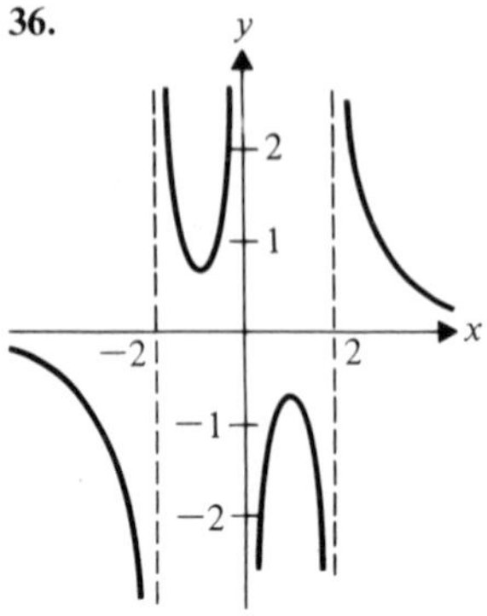

38.

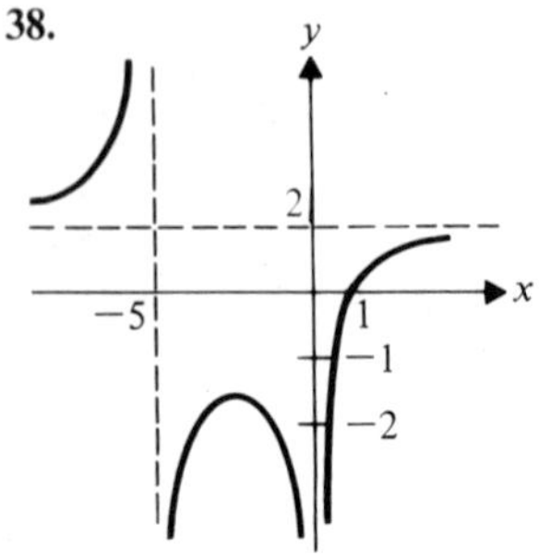

40.

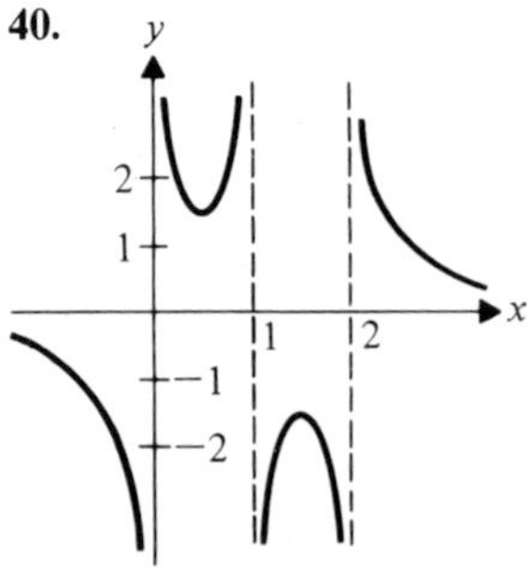

12.

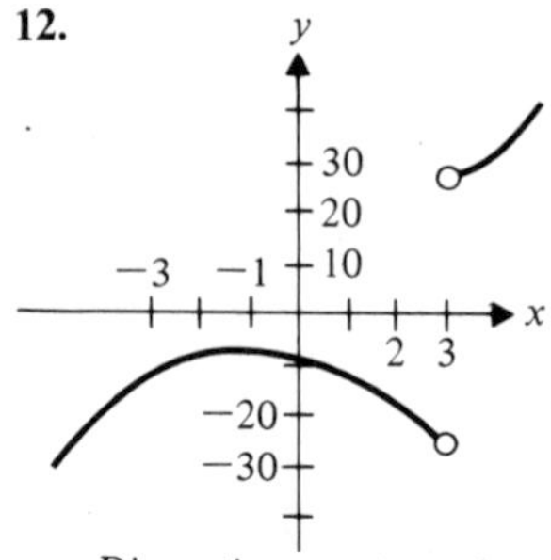

Discontinuous at $x = 3$

14.

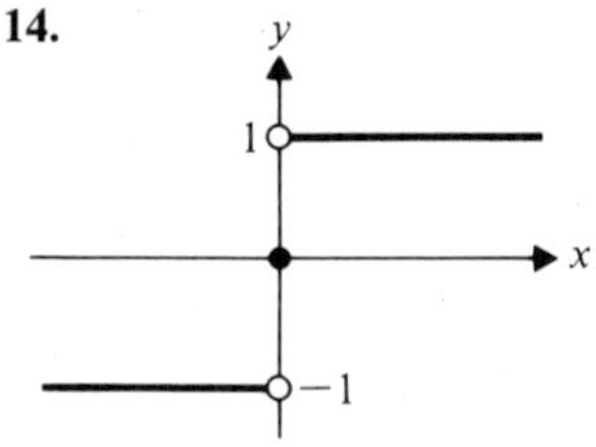

16.

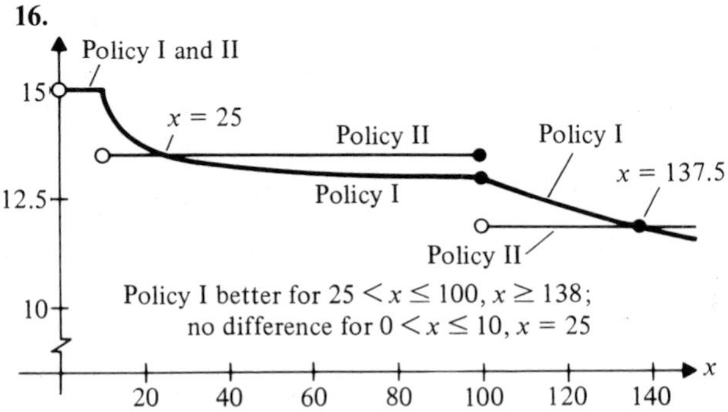

18. $x/(x-2)$ **20.** $-x^{-2}(x+1)^{-2}(x-1)^{-1}$
22. $-3(x^2-1)(x^2-9)(x^2+16/27)(x^2-4)^{-2}(x^2+1)^{-1}$

Exercises 2.2

2.

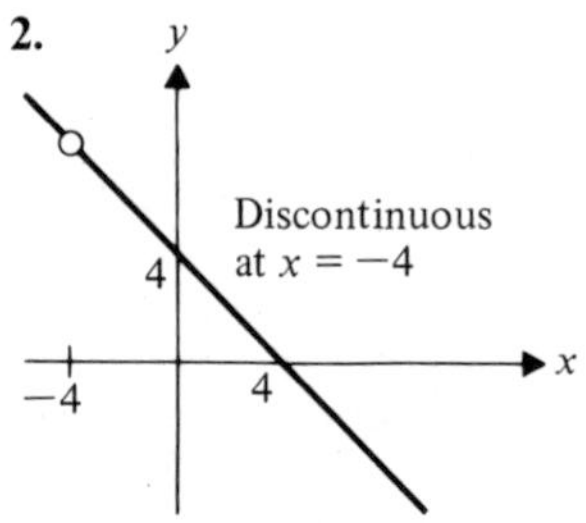

4.

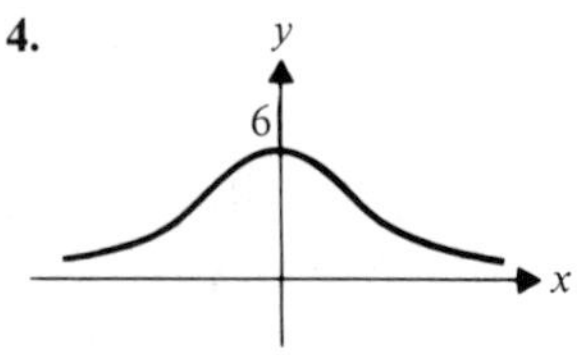

6.

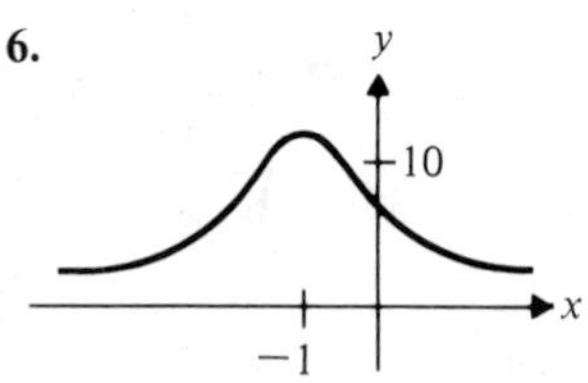

8.

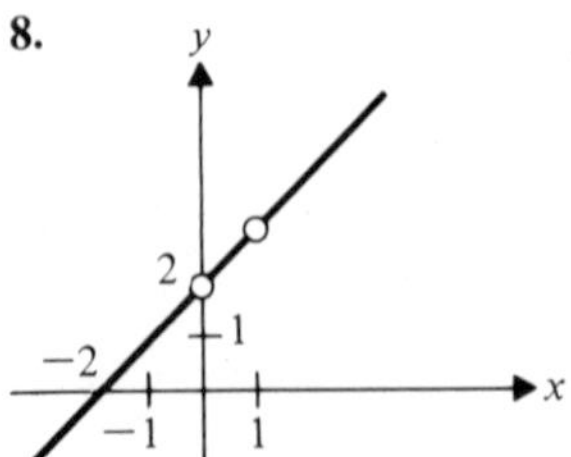

Discontinuous at $x = 0, 1$

10.

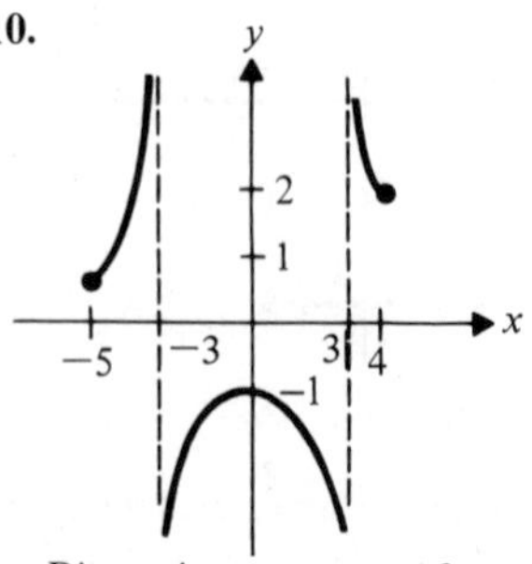

Discontinuous at $x = \pm 3$

Review Exercises

2. 3 **4.** 2/3 **6.** Does not exist **8.** 0
10. Does not exist **12.** 0

14.

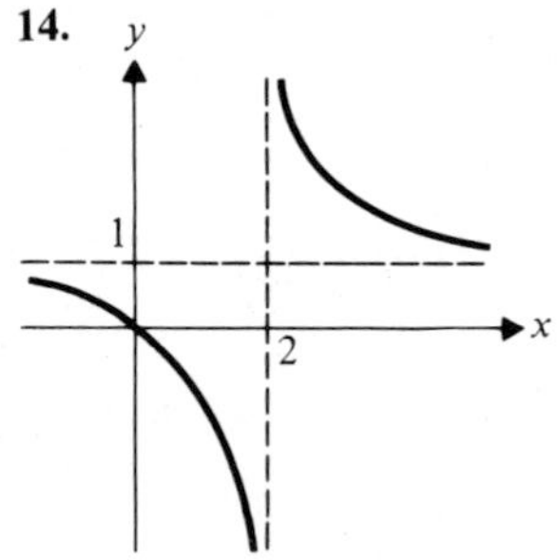

16.

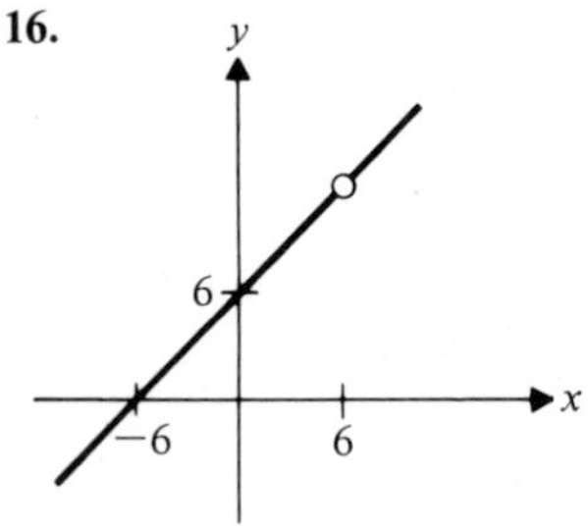

18.

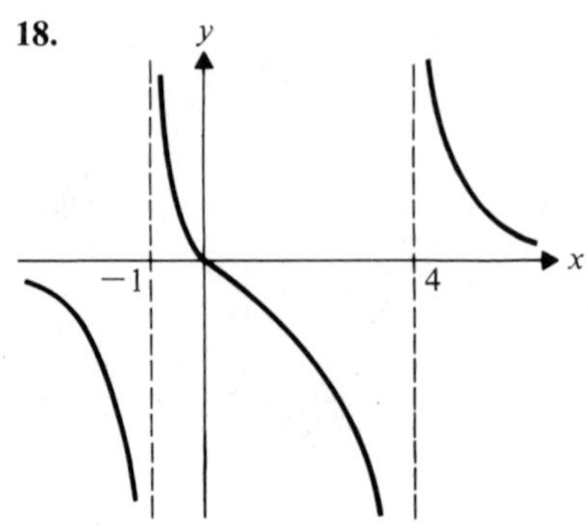

CHAPTER 3

Exercises 3.1

2. $6x$ **4.** $3x^2 + 4x$ **6.** $-9/(x-5)^2$ **8.** $3x^2 + 4x$ **10.** $4\pi r^2$ **12.** $8\pi r$ **14.** 0 **16.** $8x^7$ **18.** $-4/(x-2)^5$ **20.** $(1/2)(3x+2)/\sqrt{x+1}$ **22.** $(1/4)\sqrt{A/\pi}$ **24.** Yes

Exercises 3.2

2. $9x^2 + 4$ **4.** $20x^4 - 30x^2 + 3$ **6.** $-6/x^4$ **8.** $1/x^3 - 12/x^5$ **10.** $20x^3 - 5/(4x^6)$ **12.** $1/(2\sqrt{x})$ **14.** $-(3/2)/x^{5/2} + (3/2)x^{1/2}$ **16.** $\pi^2 x^{\pi-1}$ **18.** $4 + 3/x^4$ **20.** $6(2x+5)^2$ **22.** $4y = x + 24, 4x + y = 23$ **24.** $y = 6x - 10, x + 6y = 51$ **26.** $A = -1/4$, $B = -11/16$ **28.** $75 - 7x/50, 50 - x/15 - x^2/20{,}000$, $25 - 11x/150 + x^2/20{,}000$ **30.** $(2, -12)$ **32.** $A = 1/2$, $B = -9/4, C = 17/4, D = -51/8$

Exercises 3.3

2. $-48x(2-6x^2)^3$ **4.** $-6/(6x+5)^2$ **6.** $9(4x^3-3)(x^4-3x)^8$ **8.** $-3(x+1)/\sqrt{-3x^2-6x-1}$ **10.** $2x(4x^2-2)^{-3/4}$ **12.** $-\pi/(x+5)^{\pi+1}$ **14.** $(1/4)/[\sqrt{1+x}\sqrt{1+\sqrt{1+x}}]$ **16.** $x(3x^2+4)/\sqrt{x^2+1}$ **18.** $4[x-(x-1)^4]^3[1-4(x-1)^3]$ **20.** $(x^4+2x^2-9)/(4x^4)$ **22.** $y = 2x + 4, x + 2y = 3$ **24.** $x + 4y + 5 = 0, y = 4x - 14$ **26.** $(1, 7)$ **28.** $20/\pi$ cm/s **30.** Yes, if its tangent line is vertical.

Exercises 3.4

2. $2/(3x+2)^2$ **4.** $(3x+2)/[2\sqrt{x+1}]$ **6.** $-(2x^2+20x+1)/(2x^2-1)^2$ **8.** $(x+1)/(2x+1)^{3/2}$ **10.** $(2x-1)(6x+23)/(3x+5)^2$ **12.** $(17-4x-7x^2)/(x^2-5x+1)^2$ **14.** $x^2(18-55x^2)/[3(2-5x^2)^{2/3}]$ **16.** $(x+1)(15x+11)(3x+1)^2$ **18.** $(9x-8)/[2x^3\sqrt{2-3x}]$ **20.** $(x-8\sqrt{x}-2)/[\sqrt{x}(\sqrt{x}-4)^2]$ **22.** (b) $x(x-4)/(x^2+x-2)^2$ **28.** $(5+\sqrt{55}, -\sqrt{55}/(11+\sqrt{55}))$, $(5-\sqrt{55}, \sqrt{55}/(11-\sqrt{55}))$ **30.** $-1/[(2+x)^{5/4}(2-x)^{3/4}]$ **32.** $x^2(x^3-2x^2)^2(x^4-2x)^4(29x^4-52x^3-28x+44)$ **34.** $(x+5)^3(11x^3+15x^2+8)/[2\sqrt{1+x^3}]$ **36.** $(x+5)^3(13x^4+25x^3+10x+10)/[2\sqrt{1+x^3}]$ **38.** $(7x^2+6x+4\sqrt{1+x})/[4\sqrt{1+x}\sqrt{1+x\sqrt{1+x}}]$

Exercises 3.5

2. $6 + 60/(t+1)^6$ **4.** $-9u/(u^2+3)^{5/2}$ **6.** $-56/81$ **8.** (b) $(bT_b - aT_a)/(b-a), ab(T_a - T_b)/(b-a)$ **10.** $5/4, -5, 41/4, -13/2$ **12.** $n!/(2-x)^{n+1} + (-1)^{n+1}n!/(2+x)^{n+1}$

Exercises 3.6

2. $-4x^3/(2y+3y^2)$ **4.** $(6x^2-3y^4+5y)/(12xy^3-5x)$ **6.** $(1-x-y)/(x+y)$ **8.** $(48x\sqrt{x+y}-1)/[1+(4y-2)\sqrt{x+y}]$ **10.** $-(6x+5y)/(5x+2y)$ **12.** $-[2(3y^2+1)^2+24x^2y]/(3y^2+1)^3$ **14.** $-(25/4)/(y+1)^3$ **16.** -1 **18.** $-(2xy^3+2)/(3x^2y^2+4), [-2y^3(3x^2y^2+4)^2 + 12xy^2(2xy^3+2)(3x^2y^2+4) - 6x^2y(2xy^3+2)^2]/(3x^2y^2+4)^3$ **20.** $\sqrt{1-y^2}/(1-2y^2)$, $(3y-2y^3)/(1-2y^2)^3$ **22.** $(1, \pm 1), (-1, \pm 1)$ **24.** (a) $(x^2+4x+2)/(x^2+3x+2)$; (b) $(2y+1)(3-y)/(7y)$ **26.** $(1, 2)$ **30.** $(3r^3-6ar^2+x)/(3r^2-6ar)$ **32.** (a) -1 **34.** (a) $-\sqrt{y/x}$

Exercises 3.7

2. $(4/3)x(s-1)/[(x^2+5)^2(2s-s^2)^{2/3}]$ **4.** $(5/2)(2-x)/[x^2\sqrt{x-1}(u+5)^2]$ **6.** $[1+(2\sqrt{t}+1)/(4\sqrt{t}\sqrt{t+\sqrt{t}})][-4x/(x^2-1)^2]$ **8.** $(1-k-3k^2)(x^2+5)^4(11x^2+5)/[2\sqrt{k}(1+k+k^2)^2]$ **10.** $(6u^4+6u^3-2)(2\sqrt{x+1}+1)^2/[4u^3(x+1)] - [3(u^2+u)^2+1]/[4u^2(x+1)^{3/2}]$ **12.** $-3[2\sqrt{x}+(s+6)(1+3x+4\sqrt{x})]/[2x^{3/2}(1+\sqrt{x})^4(s+6)^3]$ **14.** -0.067 N/s **18.** $2xu/[(u^2-1)(1+2x^2\sqrt{u^2-1})]$ **20.** $3yv^2(2xv+3v^2-6x^2y)/[6x^3yv^2+(x^2+6xv)(4y^3+v^3)]$ **22.** $-x(x+1)^2/\sqrt{1-x^2}$

Exercises 3.8

2. $\pi/3$ **4.** $-\pi/2$ **6.** $17\pi/4$ **8.** -2.23 **10.** -3.72 **12.** $-225°$ **14.** $1440°$ **16.** $57.3°$ **18.** $143.2°$ **20.** $630.3°$

22.

	$\sin x$	$\cos x$	$\tan x$	$\csc x$	$\sec x$	$\cot x$
0	0	1	0	†	1	†
$\pi/6$	$1/2$	$\sqrt{3}/2$	$1/\sqrt{3}$	2	$2/\sqrt{3}$	$\sqrt{3}$
$\pi/4$	$1/\sqrt{2}$	$1/\sqrt{2}$	1	$\sqrt{2}$	$\sqrt{2}$	1
$\pi/3$	$\sqrt{3}/2$	$1/2$	$\sqrt{3}$	$2/\sqrt{3}$	2	$1/\sqrt{3}$
$\pi/2$	1	0	†	1	†	0

† indicates "is not defined."

24. $\pi/2, 0.64, 0.93$

26. $2.57, 2.19, 19\pi/30$

32.

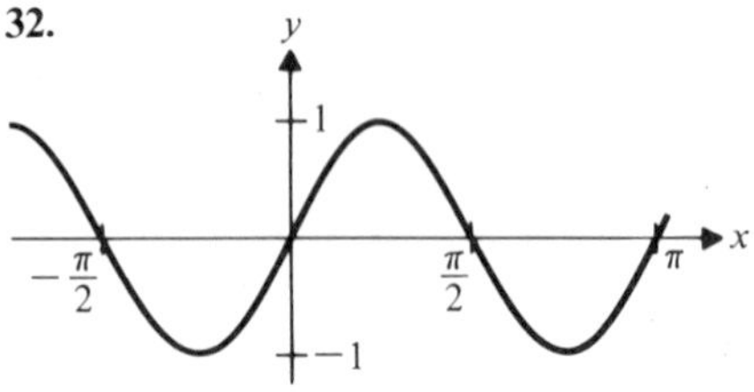

34.

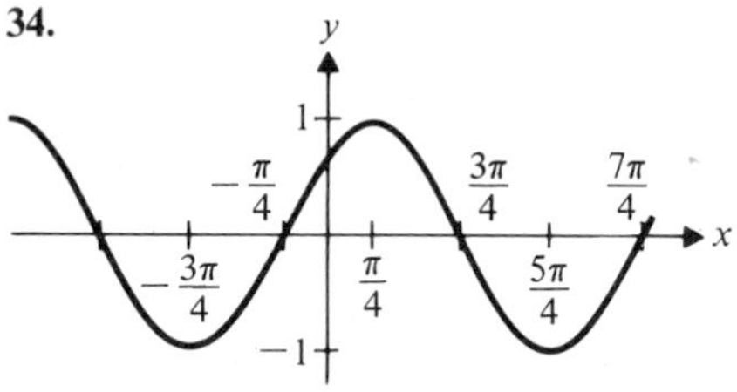

36.

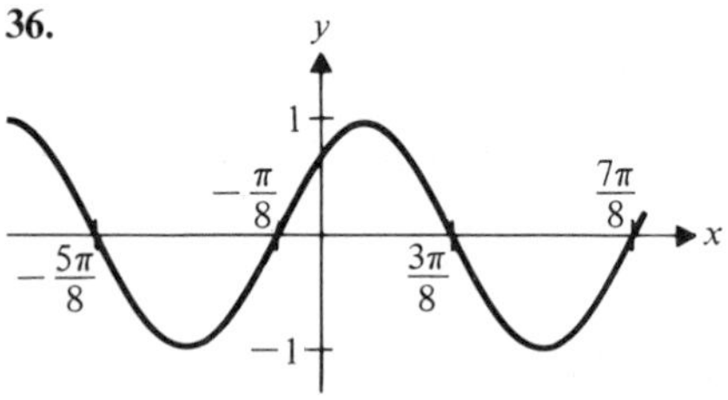

38.

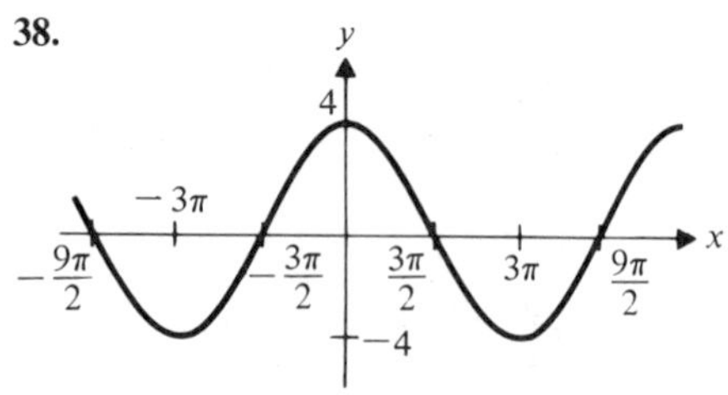

40.

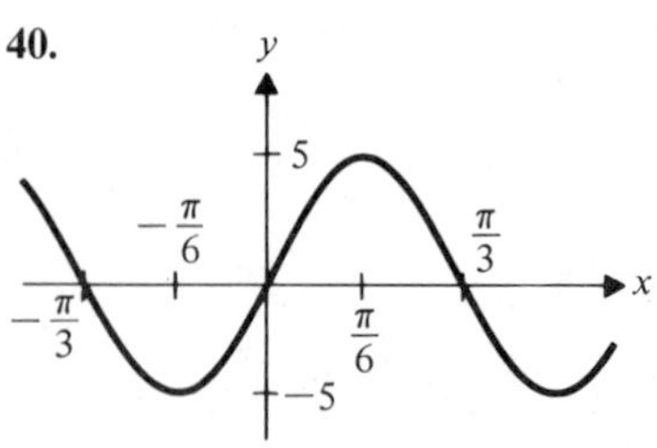

42. $(6n + 1)\pi/3$, $(6n + 2)\pi/3$

44. $0.32 + n\pi$

46. $n\pi$, $(6n \pm 1)\pi/3$

48. $0.67 + 2n\pi$, $-0.67 + (2n + 1)\pi$

50. $\sqrt{5}\sin(x - 1.11)$; amplitude $= \sqrt{5}$, period $= 2\pi$, phase shift $= 1.11$

52. $2\sqrt{5}\sin(3x - 2.03)$; amplitude $= 2\sqrt{5}$, period $= 2\pi/3$, phase shift $= 0.68$

54. $\cos x$, $\sec x$ even; $\sin x$, $\csc x$, $\tan x$, $\cot x$ odd

62. $\pi/4$, $\pi/12$, $5\pi/12$

64. $\pi/6$, $\pi/2$

Exercises 3.9

2. $-\sin x - 20\cos 5x$ **4.** $3\sin(2 - 3x)$
6. $-x\sin(x^2)/\sqrt{1 + \cos(x^2)}$
8. $2x[x\cos(2x + 3) + \sin(2x + 3)]$
10. $\cos(x + y)/[1 - \cos(x + y)]$
12. $-8\sin x\cos x/(3\cos^2 y\sin y)$ **16.** 2 **18.** 0
20. Does not exist

Review Exercises

2. $6x + 2 - 1/x^2$ **4.** $-3x(x^3 + 10)/(x^3 - 5)^2$
6. $(1/2)\sin 4x$ **8.** $(20/3)(u - 1)(1 + 2x)^{2/3}$
10. $3x(10 + x - x^3)/(1 - 3x^2)^2$
12. $1/[4\sqrt{1 + x}\sqrt{1 + \sqrt{1 + x}}]$ **14.** $\sec^2 x$ **16.** 1
18. $4(4 + x)^{-1/2}(4 - x)^{-3/2}$ **20.** $-x/y$
22. $(-x/v)(\cos^2 v - v\sin 2v)$ **24.** $14x^3(x^2 + 3)^2(x^2 - 4)^3$
26. $2x/(2y + 1)$ **28.** $x(20 - 23x - 3x^2)/[2\sqrt{1 - x}(x + 5)^2]$ **30.** $(x - y)/(x + y)$
32. $x + 25y = 5$, $125x - 5y + 1 = 0$ **34.** $23x + 9y = 11$, $27x - 69y = 119$ **36.** $-42(x^2 - 5xy + y^2)/(5x - 2y)^3$
38. No **40.** $(-3, 1)$; finding the shortest distance from $(-6, 7)$ to the curve **42.** $\sqrt{31}/2$ **44.** $(\sqrt{6}/4, \pm\sqrt{2}/4)$, $(-\sqrt{6}/4, \pm\sqrt{2}/4)$

CHAPTER 4

Exercises 4.1

2. $2t - 7$ m/s, 2 m/s^2 **4.** $-6t^2 + 4t + 16$ m/s, $-12t + 4$ m/s^2 **6.** $-12\sin 4t$ m/s, $-48\cos 4t$ m/s^2
8. $1 - 4/t^2$ m/s, $8/t^3$ m/s^2 **10.** $(4t - 5t^2)/(2\sqrt{1 - t})$ m/s, $(1/4)(15t^2 - 24t + 8)/(1 - t)^{3/2}$ m/s^2 **12.** No
14.

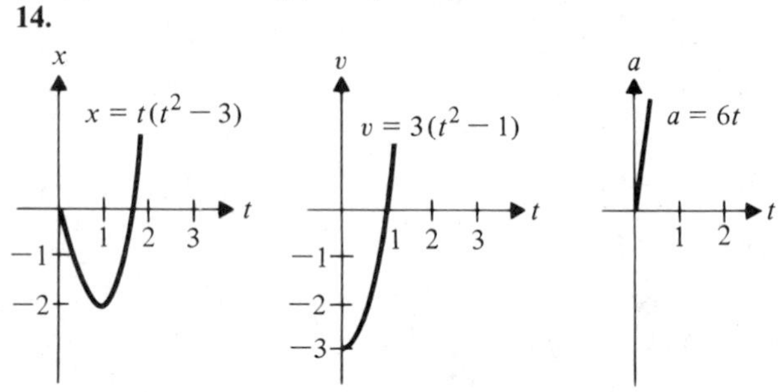

16.

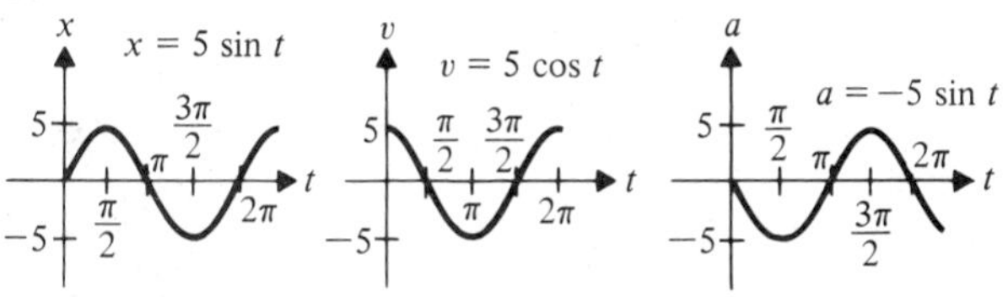

18. (a) $b + a(t_1 + t_2)/2$; (b) $(t_1 + t_2)/2$, closer to x_1

Exercises 4.2

2. -2.5 N/m^2/s **4.** 5600 N/m^2/s **6.** $5/(352\pi)$ cm/s
8. $3(11 + 4\sqrt{3})/10$ km/hr **10.** 4/9 cm/min
12. $2\sqrt{13}$ m/s **14.** (b) 0.30 m^3/year
16. (a) $4\pi x l \sin\theta/(l\cos\theta - x)$ m/s **18.** $2\pi RR'/r$ m/s

Exercises 4.3

2. $-1, -4$ **4.** 1 **6.** $1 \pm \sqrt{2}$ **8.** $0, \pm 3^{1/4}$ **10.** 0
12. No relative extrema **14.** $f(1) = 2$ rel. min.; $f(-1) = -2$ rel. max. **16.** $f(0) = -1$ rel. max.
18. $f(7) = 27/512$ rel. max.
20. $f(2\pi/3 + 2n\pi) = \sqrt{3} + 2\pi(3n + 1)/3$ rel. max.; $f(4\pi/3 + 2n\pi) = -\sqrt{3} + 2\pi(3n + 2)/3$ rel. min.
22. $0, 2 + 50^{1/3}$ **24.** 1 **26.** $f(0) = 0$ rel. max.
28. $f(-2) = 25/9$ rel. max.; $f(2) = 9$ rel. min.
30. $f(\pi/4 + 2n\pi) = \sqrt{2}$ rel. max.; $f(5\pi/4 + 2n\pi) = -\sqrt{2}$ rel. min. **34.** $10\sqrt{5}$ cm

Exercises 4.4

2. Increasing on $x \le 5/4$; decreasing on $x \ge 5/4$
4. Increasing on $x \ge 1$; decreasing on $x \le 1$
6. Increasing on $x < -1$ and $-1 < x \le 0$; decreasing on $0 \le x < 1$ and $x > 1$ **8.** Increasing on $-1 \le x \le 0$ and $x \ge 1$; decreasing on $x \le -1$ and $0 \le x \le 1$

10.

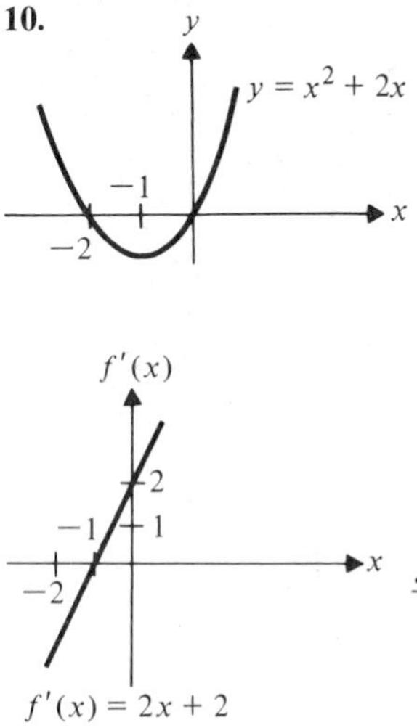

12.

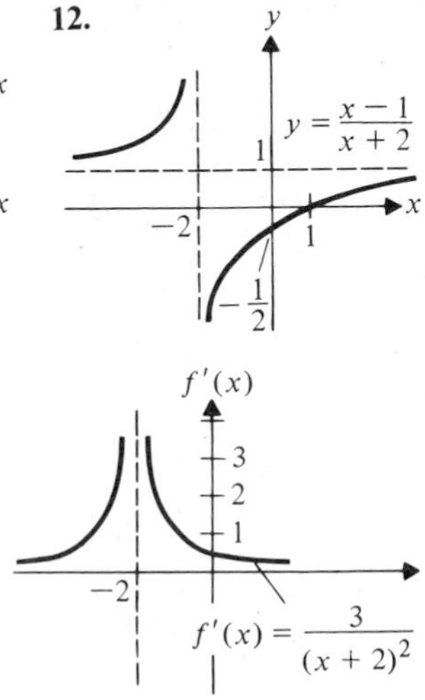

14.

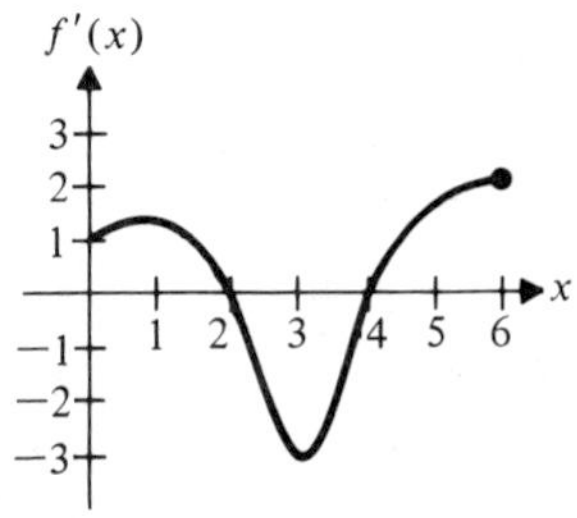

Exercises 4.5

2. Concave upward on $x \le -2/3$ and $x \ge 0$; concave downward on $-2/3 \le x \le 0$; points of inflection at (0, 2) and $(-2/3, 470/27)$
4. Concave upward on $x < -1$ and $x > 1$; concave downward on $-1 < x < 1$
6. Relative minimum at $x = 1$; relative maximum at $x = -1$
8. Relative minimum at $x = 1/5$

Exercises 4.6

2.

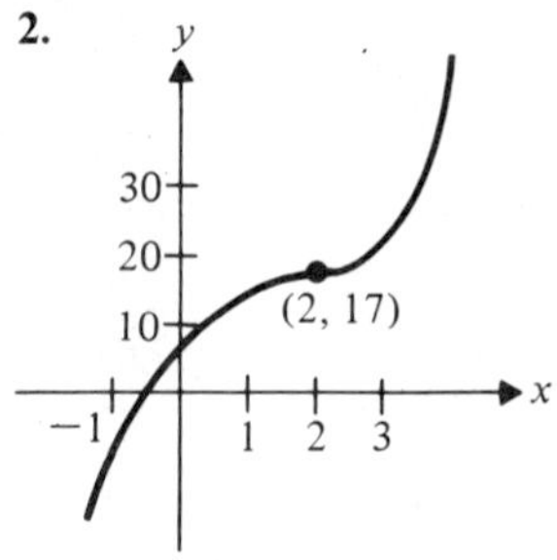

4.

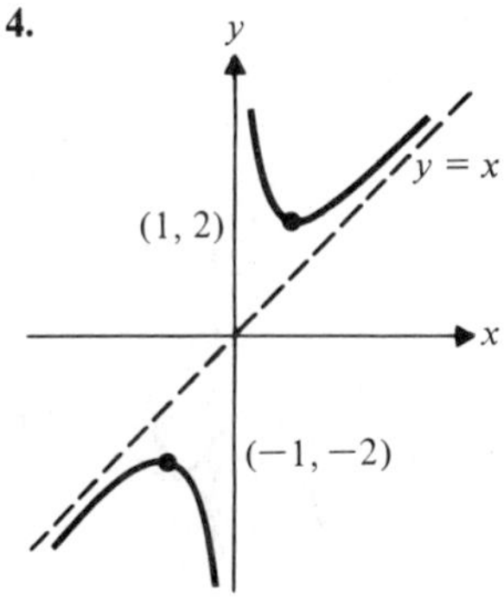

6.

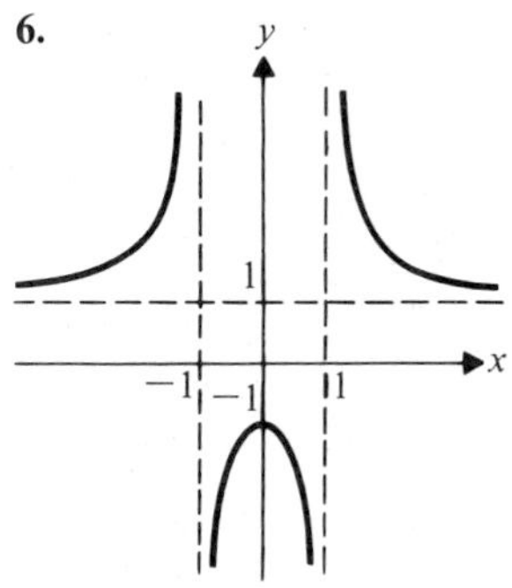

8.

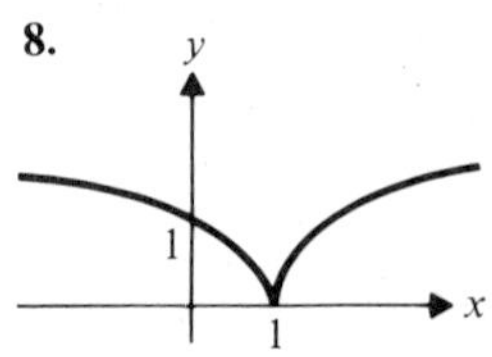

10.

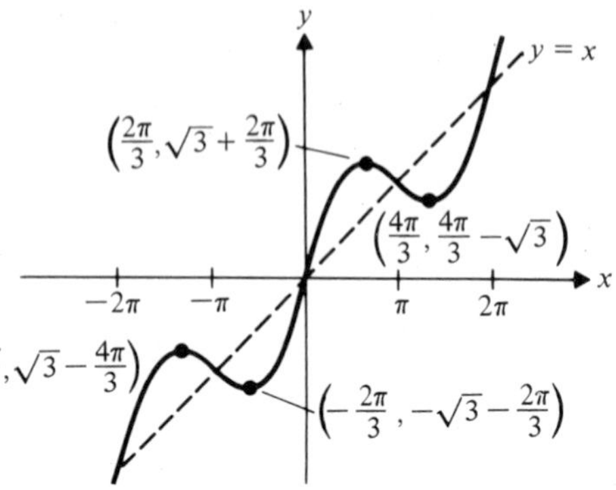

12.

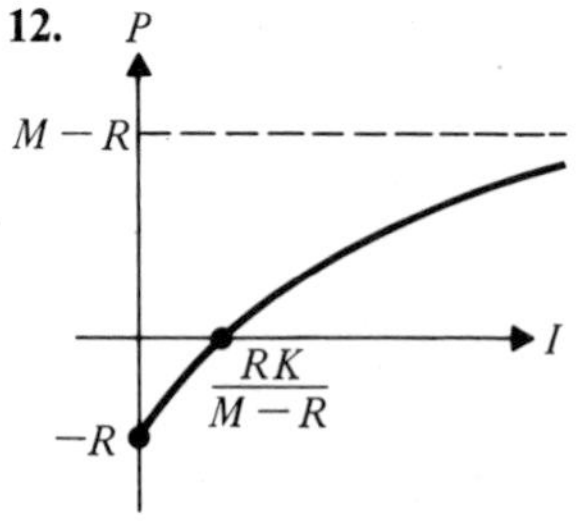

14.

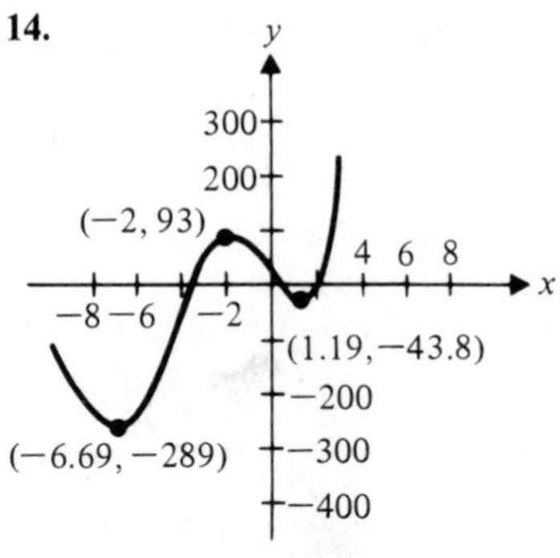

16.

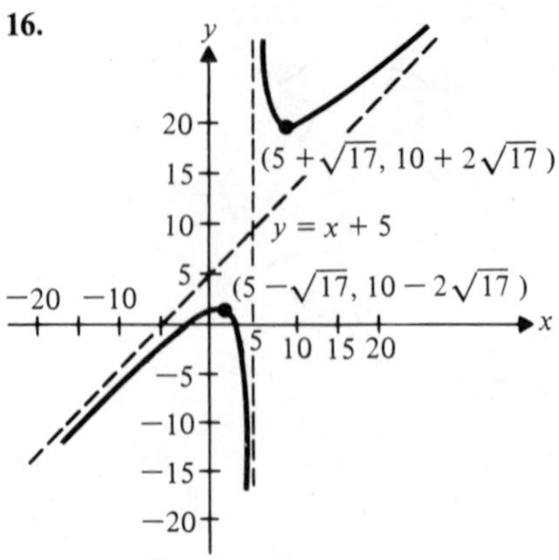

18.

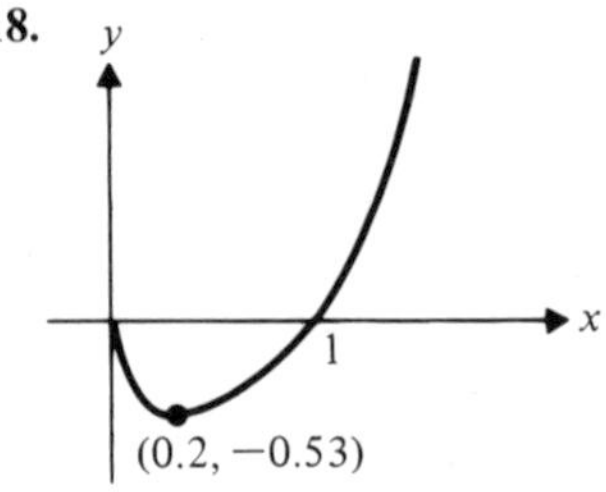

20.

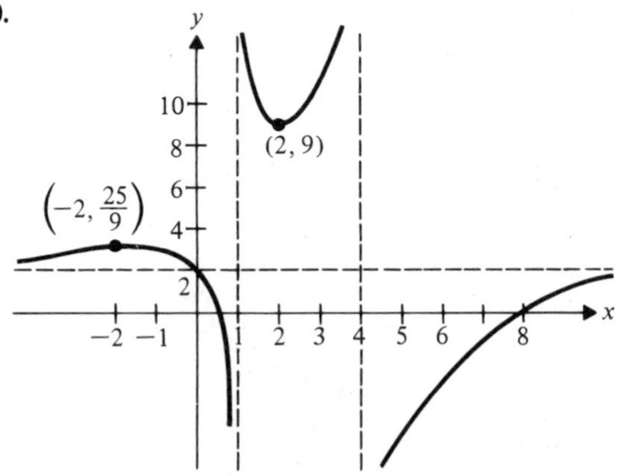

22.

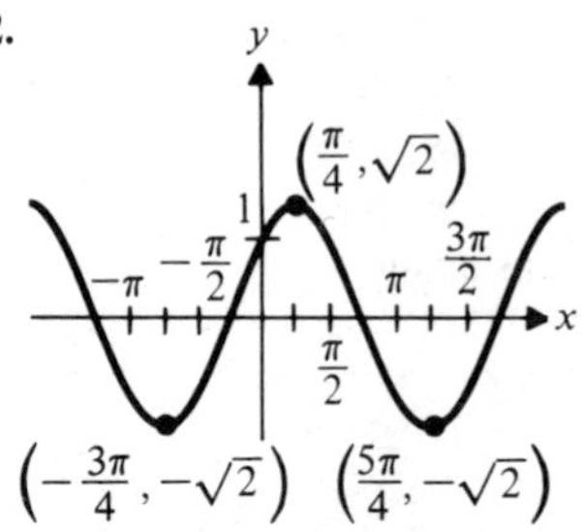

24.

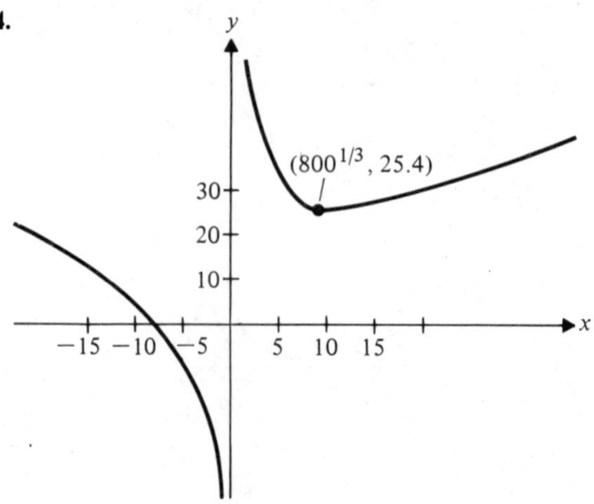

26.

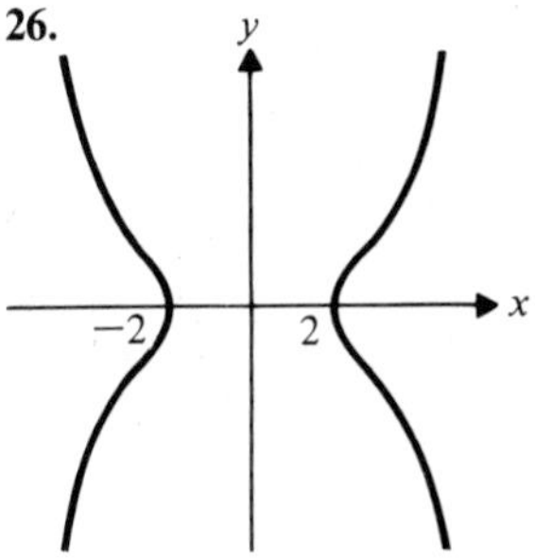

Exercises 4.7

2. $-4, 6/11$ **4.** $\pi/3 - \sqrt{3}, 11\pi/3 + \sqrt{3}$ **6.** $720\sqrt{10}$ m **8.** $(15 - \sqrt{33})L/16$ **10.** Rectangular length = 90 m, width = $180/\pi$ m **12.** 328 **14.** $r/\sqrt{2}$ **16.** $\pi/3$ **18.** 1:12 P.M. **20.** Lengths of $\sqrt{2}a$ and $\sqrt{2}b$ in the x- and y-directions **22.** For $0 < \alpha \le 1$, keep the machine as long as possible. **24.** Height/width = $(4 + \pi)/8$ **28.** $b(A/(2B))^{1/3}$ km/hr **30.** $2\pi(1 - \sqrt{6}/3)$

Exercises 4.8

2. 2.561553, −1.561553 **4.** 3.044723 **6.** 0.754878 **8.** 2.485584 **10.** (1.904161, 6.904161) **12.** (2.130189, 0.5907579), (−1.726688, −11.11094) **14.** −2.128419, −0.201640, 2.330059

Exercises 4.9

2. $-2dx/(x - 1)^2$ **4.** $[2x \cos(x^2 + 2) + \sin x]\,dx$ **6.** $x^2(9 - 16x^2)\,dx/\sqrt{3 - 4x^2}$ **8.** $[1 - 12/(x - 1)^3]\,dx$ **10.** $2(x^2 + 5x - 5)\,dx/(x^2 + 5)^2$ **12.** −4% **14.** 3250 m

Exercises 4.10

2. 2 **4.** Cannot be applied since $f'(0)$ does not exist **6.** $(-2 \pm \sqrt{19})/3$ **8.** $1 + \sqrt{3}$ **10.** $-3 + \sqrt{6}$ **12.** Cannot be applied since $g'(0) = 0$ **14.** $(1 + \sqrt{6})/(1 - \sqrt{6})$

Exercises 4.11

2. 6 **4.** 0 **6.** ∞ **8.** 0 **10.** Does not exist **12.** ∞ **14.** $1/\sqrt{5}$ **16.** na^{n-1} **18.** 1 **20.** 0 **22.** 1/12 **24.** 2^{15} **26.** 0

Review Exercises

2. (a)

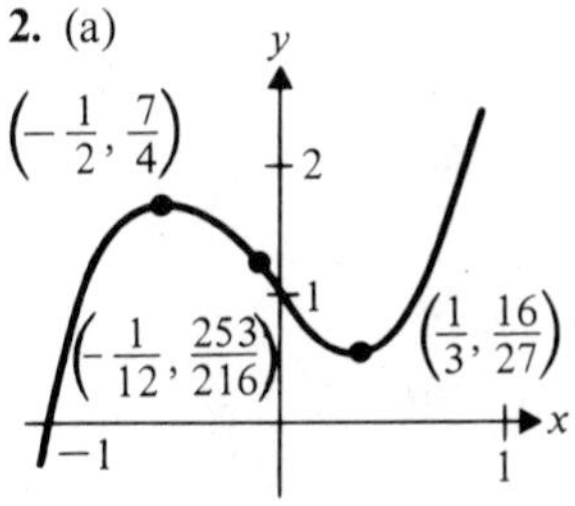

(b)

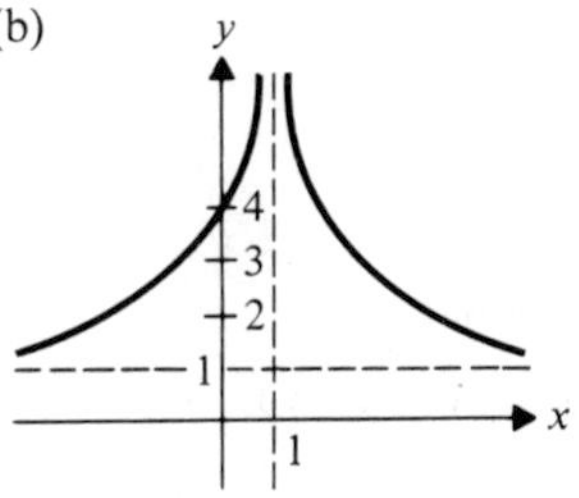

4. No solution **6.** No solution **8.** −3/2 **10.** 8 **12.** 0 **14.** (a) 0.312908; (b) 1.051888 **18.** $2l^3/27$ **20.** (a)

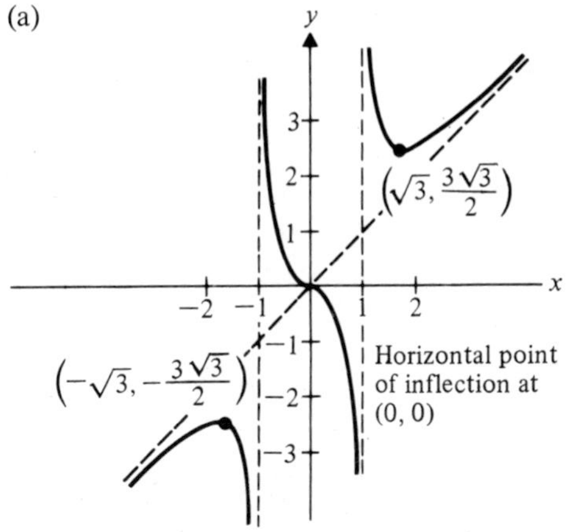

(b)

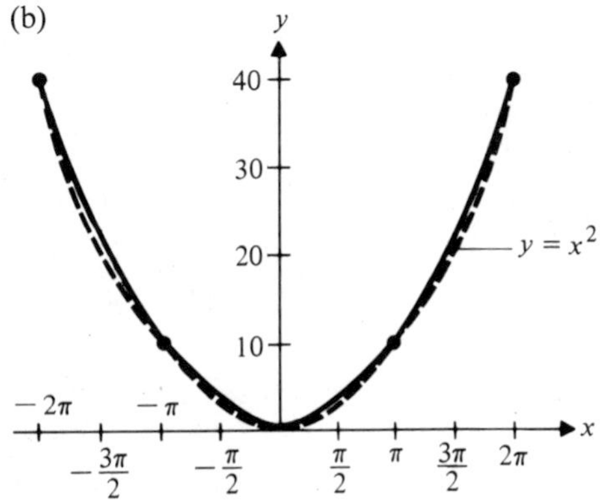

22. 67¢ **24.** $4\sqrt{5}$ m/s **26.** $x = 100bq/(ap + bq)$ hectares, $y = 100ap/(ap + bq)$ hectares

CHAPTER 5

Exercises 5.1

2. $x^5/5 + x^3 + 5x^2/2 + C$ **4.** $-\cos x + C$ **6.** $x^{11}/11 + 1/(2x^2) + C$ **8.** $(2/3)x^{3/2} + C$ **10.** $-1/x + \sqrt{x} + C$ **12.** $1/(2x) + (3/4)x^4 + C$ **14.** $(4/3)x^{3/2} + (6/5)x^{5/2} - (10/7)x^{7/2} + C$ **16.** $(2/5)x^{5/2} + (2/3)x^{3/2} + C$ **18.** $x^3/3 + 2x^5/5 + x^7/7 + C$ **20.** $(2/5)x^{5/2} - (4/3)x^{3/2} + 2\sqrt{x} + C$ **22.** $y = x^4/2 + 2x^2 + 5$ **24.** $y = 2x - 2x^2 + x^8$ **26.** $f(x) = -5x^3/6 + 10x - 31/3$ **28.** $y = x^4/4 + 1/x + C$ **30.** $y^2 = 2(x + 1)^3/3 + C$ **32.** $(2/3)(x + 2)^{3/2} + C$ **34.** $-(2/3)(2 - x)^{3/2} + C$ **36.** $(1/5)(2x - 3)^{5/2} + C$ **38.** $(-1/16)(1 - 2x)^8 + C$ **40.** $(-1/15)/(1 + 3x)^5 + C$ **42.** $(1/72)(2 + 3x^3)^8 + C$ **44.** $(1/2)\sin 2x + C$ **46.** $y = (-1/3)(1 - x^2)^{3/2} + C$

48. $y = (1/16)(x^2 + x)^{16} + C$ **50.** $y + y^3 = C - x^2$
52. $x^2y^2 + y^3 = x + C$ **54.** $x \sin y + xy = y^2 + C$
56. $y = -3 - 1/x, x < 0; y = 2 - 1/x, x > 0$

Exercises 5.2

2. $6t - t^2 + 5, 3t^2 - t^3/3 + 5t$ **4.** $60t^2 - 4t^3$, $20t^3 - t^4 + 4$ **6.** $t^3/3 + 5t^2/2 + 4t - 2$, $t^4/12 + 5t^3/6 + 2t^2 - 2t - 3$ **8.** $4 - 3\cos t$, $4 + 4t - 3\sin t$ **10.** 5.1 m **12.** $|v| \geq 19.8$ m/s

14.

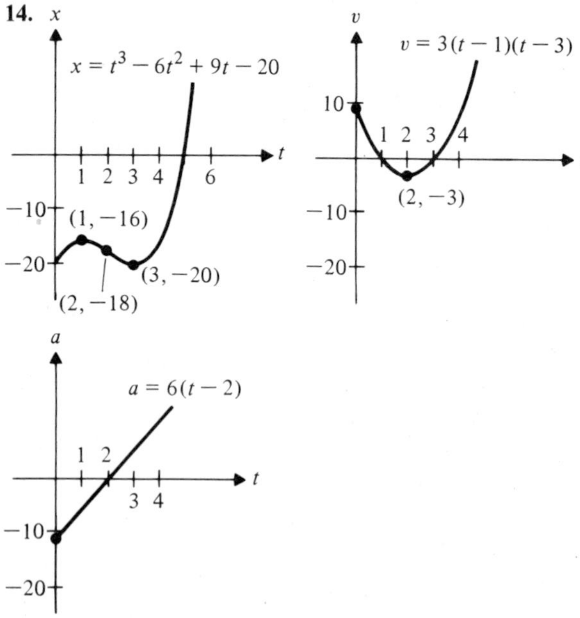

16. Less than or equal to 47.8 km/hr, defense
18. (a) Yes; (b) no

Exercises 5.3

2. $(-1/3)(1 - 2x)^{3/2} + C$ **4.** $(-10/63)(5 - 42x)^{3/4} + C$
6. $(-1/2)/(x^2 + 4) + C$
8. $-1/(x - 2) - 2/(x - 2)^2 - (4/3)/(x - 2)^3 + C$
10. $(1/6)(2x + 3)^{3/2} - (3/2)\sqrt{2x + 3} + C$
12. $(1/5)(s^2 + 5)^{5/2} - (5/3)(s^2 + 5)^{3/2} + C$
14. $(2/3)(1 - \cos x)^{3/2} + C$
16. $(2/7)(y - 4)^{7/2} + (16/5)(y - 4)^{5/2} + (32/3)(y - 4)^{3/2} + C$
18. $(1/729)(3x^3 - 5)^9 + (5/324)(3x^3 - 5)^8 + (25/567)(3x^3 - 5)^7 + C$ **20.** $(3/4)(x^2 + 2x + 2)^{2/3} + C$
22. $(7/192)/(3 - 4\sin x)^3 + (3/64)/(3 - 4\sin x)^2 - (1/64)/(3 - 4\sin x) + C$
24. $(4/5)(1 + \sqrt{x})^{5/2} - (4/3)(1 + \sqrt{x})^{3/2} + C$
26. $(2/3)(1 + x)^{3/2} - 2\sqrt{1 + x} + C$ if $x \geq 0$; $(-2/3)(1 + x)^{3/2} + 2\sqrt{1 + x} + C$ if $-1 < x \leq 0$
28. $(-2/5)(1/x - 1)^{5/2} - (2/3)(1/x - 1)^{3/2} + C$

Exercises 5.4

2. (a) $(24EI)^{-1}gx^2[4M(3L - x) + m(6L^2 - 4Lx + x^2)]$; (b) $(24EI)^{-1}gL^3(8M + 3mL)$ **4.** $x^2 + (y - 2)^2 = 13$
6. $\pi R^2\sqrt{2H}/(5cA\sqrt{g})$ **8.** $-\sqrt{2GMh/(R^2 + Rh)}$, $\sqrt{2GM/R}$ **10.** $y = Cx^k$
12. (a) $(192EI)^{-1}[8mgx^4 + 32Ax^3 - (8AL^2 + mgL^3)x]$; (b) $-3mgL/16$

Review Exercises

2. $(-1/4)/x^4 + x^2 + (1/2)/x^2 + C$
4. $-1/x - (4/3)x^{3/2} + C$ **6.** $(1/30)(1 + 3x^2)^5 + C$
8. $(2/5)x^{5/2} + 10\sqrt{x} + C$ **10.** $-2/\sqrt{x} - 30\sqrt{x} + C$
12. $(-1/3)(1 - x^2)^{3/2} + C$
14. $x^3/3 - (4/5)x^5 + (4/7)x^7 + C$
16. $(2/3)(2 - x)^{3/2} - 4\sqrt{2 - x} + C$
18. $4x + (8/3)x^{3/2} + x^2/2 + C$ **20.** $(1/5)\sin^5 x + C$
22. $y = -1/(x + C)$ **24.** $\cos y = C - \sin x$
26. $y = x^4 + 7x/2 - 1/2$ **28.** $[(\sqrt{3} - \sqrt{6})t + \sqrt{6}]^2$ km, $\sqrt{6}/(\sqrt{6} - \sqrt{3})$ hours **30.** 11.9 m/s
32. $(2/3)(1 + x)^{3/2} - x + C$
34. $-(3/56)(3 - 2x^3)^7 + (1/32)(3 - 2x^3)^8 - (1/216)(3 - 2x^3)^9 + C$
36. $(1/4)\sin^4 x - (1/6)\sin^6 x + C$ **38.** $y = 2x/(1 + x)$

CHAPTER 6

Exercises 6.1

2. $\sum_{k=1}^{10} k/2^k$ **4.** $\sum_{k=1}^{14641} \sqrt{k}$ **6.** $\sum_{k=1}^{1019} (-1)^k(k + 1)$
8. $\sum_{k=1}^{225} (\tan k)/(1 + k^2)$ **10.** $\sum_{k=1}^{9} (10^k - 1)/10^k$
16. 258 **18.** $4n(4n^2 - 1)/3$ **20.** 5350 **24.** $f(n) - f(0)$
26. False **28.** $(1 - 2^{-18})/4$ **30.** $3960(1 - (0.99)^{15})$

Exercises 6.3

2. 6 **4.** 4 **6.** 3/2 **8.** 0

Exercises 6.4

2. 20/3 **4.** 2/3 **6.** 1 **8.** $-7/2$ **10.** 3/4 **12.** 0
14. $5/16 - 1/(\pi + 1)$ **16.** 125/12 **18.** $-65/4$
20. $3/\sqrt{2}$ **22.** 25/2 **24.** 25 **26.** (a) 36 m; (b) 1114 m

Exercises 6.5

2. 4/15 **4.** 7/384 **6.** Does not exist
8. $\sqrt{3/2} - 2\sqrt{2/3}$ **10.** $3(5^{2/3} - 2^{2/3})/4$
12. $\sqrt{2}(7\sqrt{7} - 8)/18$ **14.** 52/5 **16.** $(16 - 9\sqrt{2})/6$
18. 5/18 **20.** $1/\sqrt{3}$ **22.** $-8\sqrt{5}/45$

Review Exercises

2. 4/15 **4.** $-4/3$ **6.** 1/6 **8.** 0 **10.** 66/5

12. $16\sqrt{6}$ **14.** $-4/15$ **16.** $(2112\sqrt{6} - 704)/105$
18. 120,852.5 **20.** 20 **24.** $(33\sqrt{6} - 56)/60$
26. $(24\sqrt{3} - 14\sqrt{2} - 8\sqrt{5})/3$ **28.** $24\sqrt{5} - 8$
30. $(3\sqrt{3} - 2\sqrt{2} - 1)/3$

CHAPTER 7

Exercises 7.1

2. 8 **4.** 343/6 **6.** 8/3 **8.** 2 **10.** 20/3
12. (a) 22; (b) 37, 15; (c) 29, 7; (d) 203/8, 27/8
14. $4\int_1^2 \sqrt{4 - x^2}\, dx$
16. $2\int_0^{(\sqrt{65}-1)/2} (\sqrt{16 - y^2} - y^2)\, dy$ **18.** 235/3
20. 7/6 **22.** 2/3 **24.** 23,328/35 **26.** 128/5

Exercises 7.2

2. $80\sqrt{5}\pi/3$ **4.** $5888\pi/15$ **6.** π **8.** $1088\pi/15$
10. $32\pi/3$ **12.** 8π **14.** $14\pi/3$ **16.** $40\pi/3$ **18.** $\pi/3$
20. $344\pi/3$ **22.** $775\pi/6$ **24.** $272\pi/15$ **26.** $68\pi/9$
28. $16\pi(13 + 2^{17/4} - 3^{9/4})/45$ **30.** $4\pi/3$
32. (a) $x^2(6 - x)/16$; (b) $1161\pi/10\ m^3$

Exercises 7.3

2. $(85^{3/2} - 13^{3/2})/27$ **4.** 3011/480
6. $\int_1^2 \sqrt{(2x^2 - 1)/(x^2 - 1)}\, dx$
8. $\sqrt{2}\int_0^1 (3 - 2x^2)/\sqrt{1 - x^2}\, dx$ **10.** 6

Exercises 7.4

2. 8/15 J **4.** 1.18×10^7 J **6.** 196.2 J **8.** (a) 184 J; (b) 184 J **10.** kb^2 J **12.** $q_3(q_1 - 8q_2)/(48\pi\varepsilon_0)$
14. 3.51×10^5 J

Exercises 7.5

2. 6.78×10^{10} N **4.** 9.25×10^3 N **6.** 7.60×10^5 N, 4.41×10^5 N, 4.91×10^4 N **8.** $4905AB^2/2$ N
10. $9\gamma\int_0^2 (3 + 2\sqrt{4 - x^2})\, dx$ **12.** 2.943×10^6 N, 2.969×10^6 N **14.** $2r^3\gamma/3$ **18.** (a) 10.2 cm

Exercises 7.6

2. $(3a/8, 3h/5)$ **4.** $(4r/(3\pi), 4r/(3\pi))$ **6.** $(-1/2, -1/2)$
8. (0, 0) **10.** (33/5, 1/2) **12.** (177/85, 9/17) **14.** $25\rho/3$
16. $(M + 2\rho L)^{-1}(\sum_{i=1}^{6} m_i x_i + 2\rho L^2)$ **18.** (0, 192/205)
20. On axis of symmetry, 1637/350 from smaller end
24. (12/5, −3/5) **26.** $\bar{x} = -2/(8\sqrt{3} - 1)$, $\bar{y} = 0$

Exercises 7.7

2. $32\rho/3$ **4.** $603\rho/10$ **6.** $2\rho/15$ **8.** $344\rho/105$
10. $1143\rho/20$ **12.** $\sqrt{44/45}$, $\sqrt{2}/3$

Exercises 7.8

2. $\pi r\sqrt{r^2 + h^2}$ **4.** $b^2h/3$ **6.** (a) 20,000; (b) 7000
8. $31{,}250/\sqrt{3}\ cm^3$ **10.** $\pi/2$ **12.** $r^3(3\pi - 4)/6$
14. 15 **16.** 16.5 **18.** $140\ m^3$

Exercises 7.9

2. Diverges **4.** −1/16,464 **6.** Diverges **8.** 2
10. $\sqrt{21}$ **12.** Diverges **14.** Diverges **16.** Diverges
18. $17\sqrt{5}/3$ **20.** (a) No; (b) no **22.** (a) 1/2; (b) $\pi/5$
28. No **30.** Diverges **32.** Converges

Review Exercises

2. (a) 5/4; (b) $46\pi/21$, $14\pi/15$; (c) (28/75, 92/105); (d) 73/60, 1/4 **4.** (a) 1/2; (b) $\pi/3$, π; (c) (1, 1/3); (d) 1/12, 7/12 **6.** $10\pi/3$ **8.** $16\pi/105$ **10.** $43\pi/12$
12. −5/6, 3/5 **14.** −28/3, 14 **16.** 1.04×10^5 J
18. $4\sqrt{5}\pi$ **22.** 2 **24.** Diverges **26.** Diverges
28. Diverges **30.** $7\pi/30{,}000\ m^3$ **32.** (a) 36,297 N; (b) increases by 981 N

CHAPTER 8

Exercises 8.1

2.

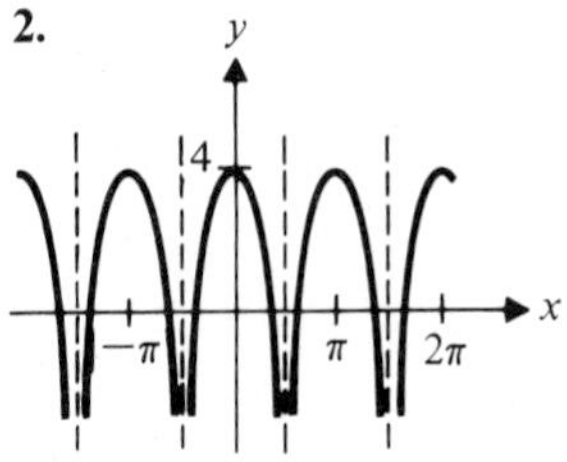

4.

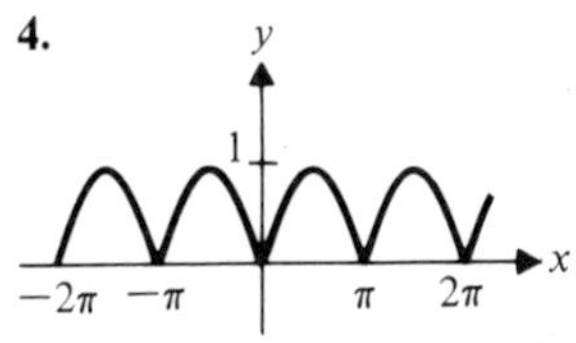

6.

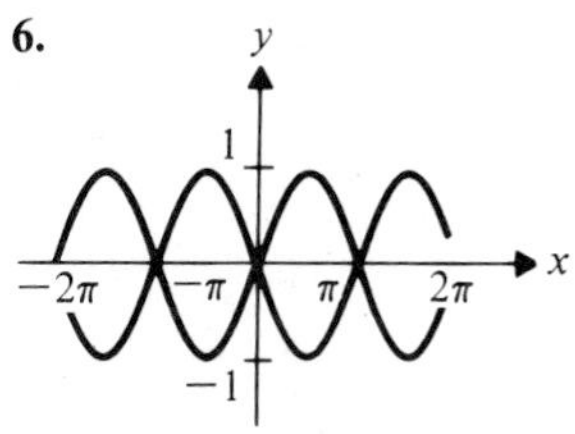

8.

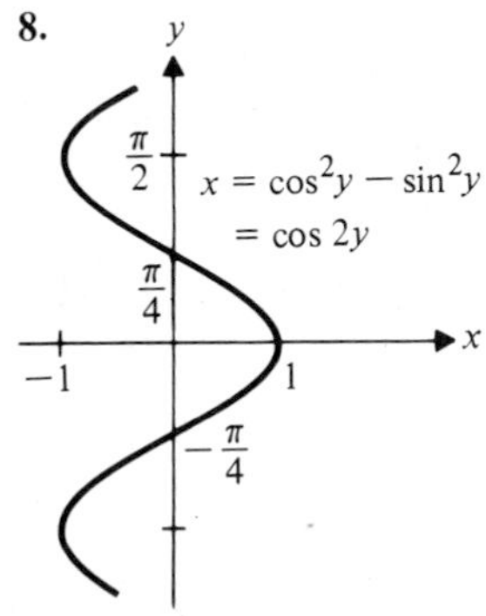

10.

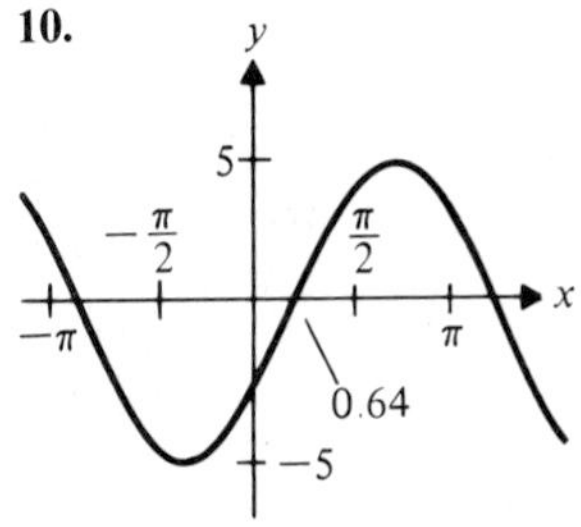

12.

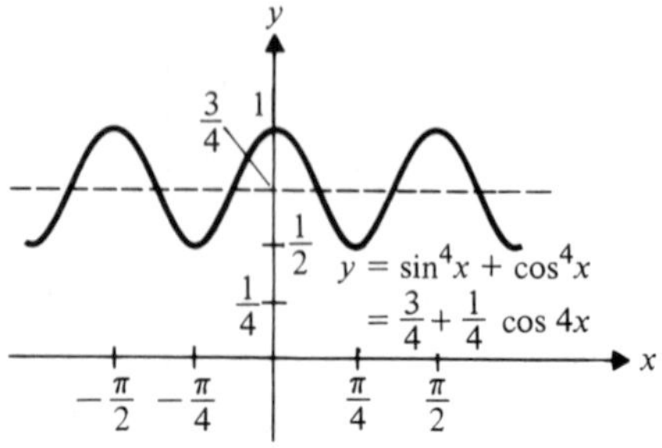

14.

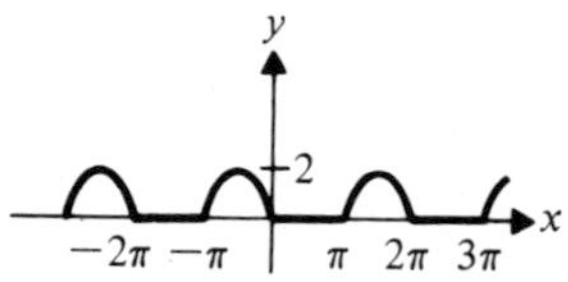

16.

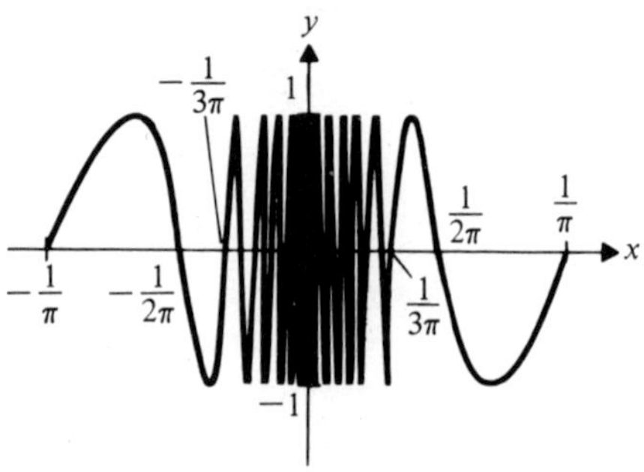

18. $\cos(3x + 4y) = -1$

20.

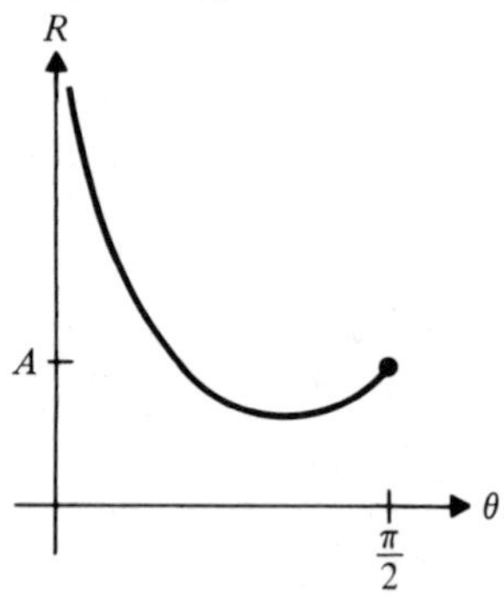

Exercises 8.2

2. $36 \cos(3x + 5)$ **4.** $\csc^3 x(1 - 3x \cot x)$
6. $2 \cos(4x)$ **8.** $-\sec v \tan v \sec^2 \sqrt{x}/[4\sqrt{x}\sqrt{3 - \sec v}]$
10. $\sec^3(3x)[9 \tan(3x) + 4 \cot(2x)]/\csc^2(2x)$
12. $-[1 + y \tan(xy) \sec(xy)]/[x \tan(xy) \sec(xy)]$
14. $2 \tan t \sec^2 t(x \sec^2 x + \tan x)$ **16.** $-\sec^2 x \sin(\tan x)$
18. $3(1 - x^2y)/(x^3 + 2 \tan y \sec^2 y)$
20. $[18x^2 \tan^2(3x^2 - 4) \sec^2(3x^2 - 4) - (2 + x \cot x)(1 + \tan^3(3x^2 - 4))]/(x^3 \sin x)$

22. 4/5 rad/min **24.** (a) 5/18 rad/s; (b) 7/36 rad/s
26. $2(y + 1)[\sec^2 y - x - (y + 1) \sec^2 y \tan y]/(\sec^2 y - x)^3$ **28.** 0.96 rad

Exercises 8.3

2.

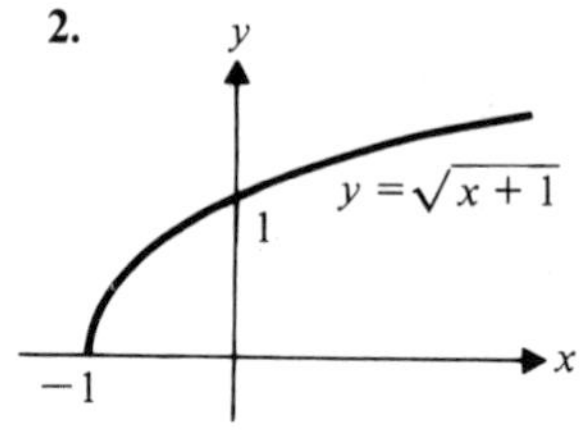

$f^{-1}(x) = x^2 - 1 (x \geq 0)$

4.

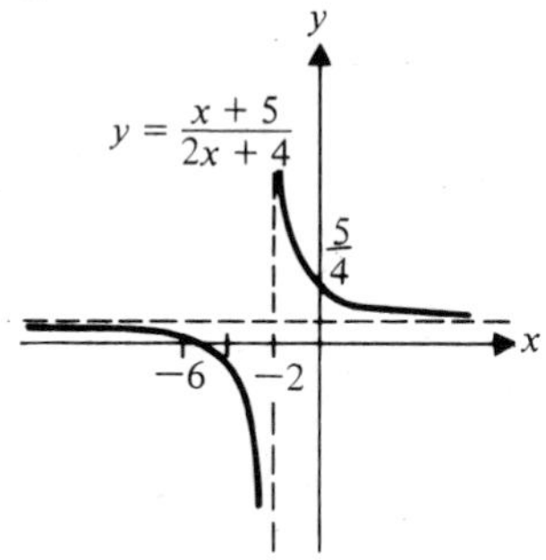

$f^{-1}(x) = \dfrac{5 - 4x}{2x - 1}$

6.

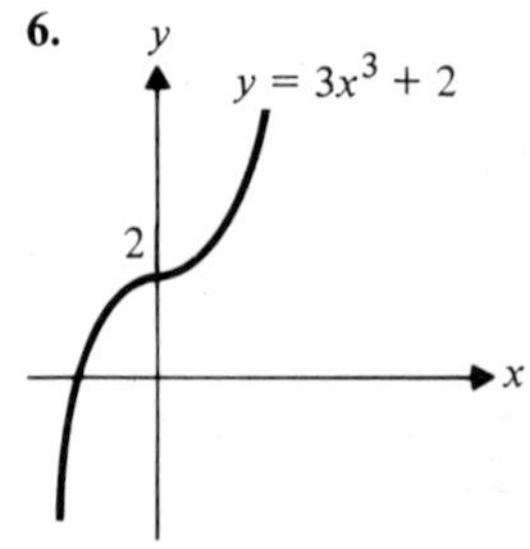

$f^{-1}(x) = \left(\dfrac{x - 2}{3}\right)^{1/3}$

8.

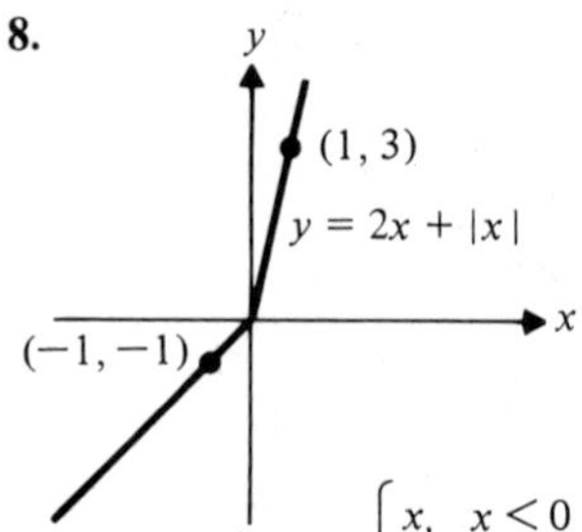

$f^{-1}(x) = \begin{cases} x, & x < 0 \\ \dfrac{x}{3}, & x \geq 0 \end{cases}$

10.

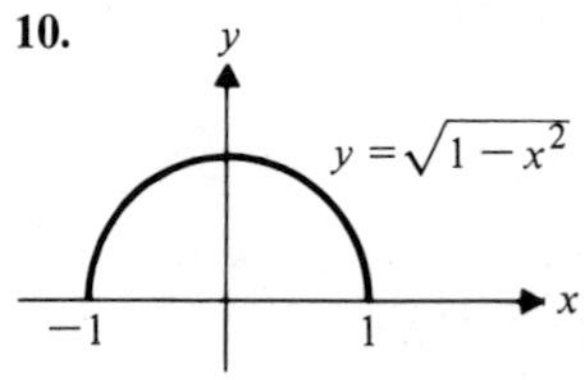

No inverse function

12.

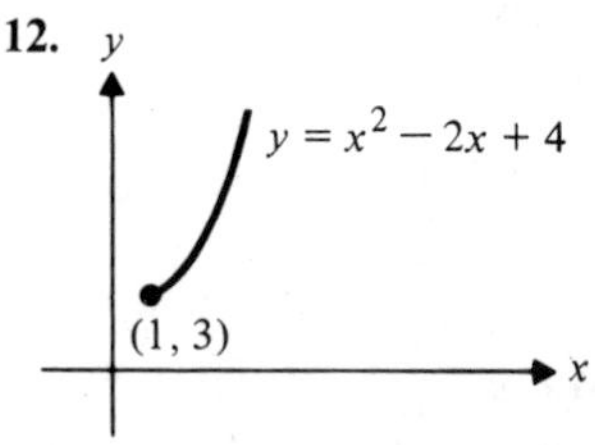

$f^{-1}(x) = 1 + \sqrt{x - 3}$

14.

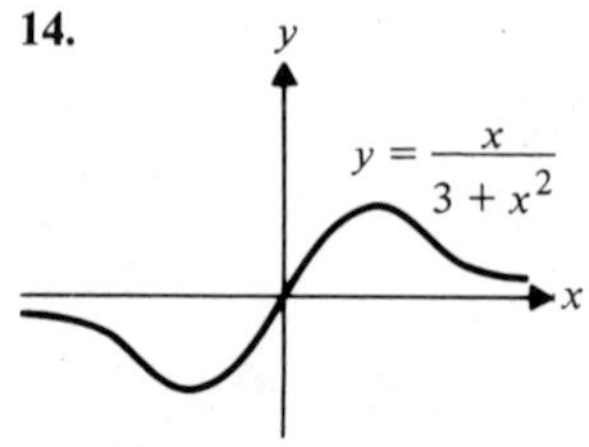

No inverse function

16. (a)

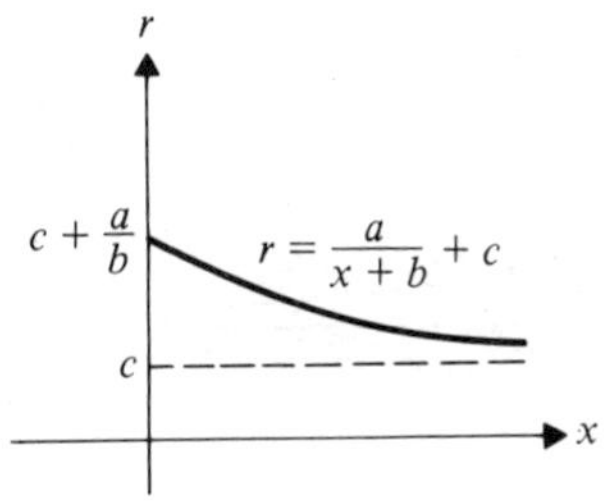

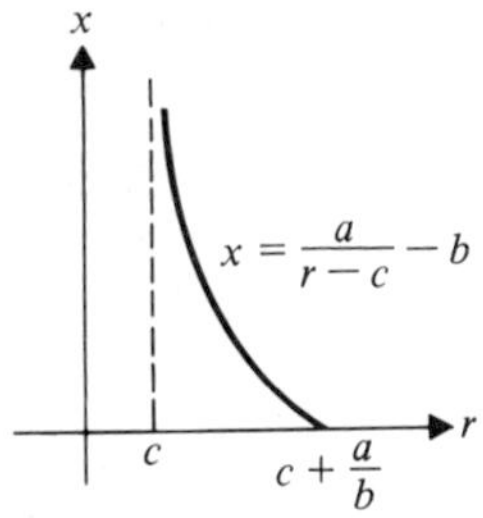

(b)

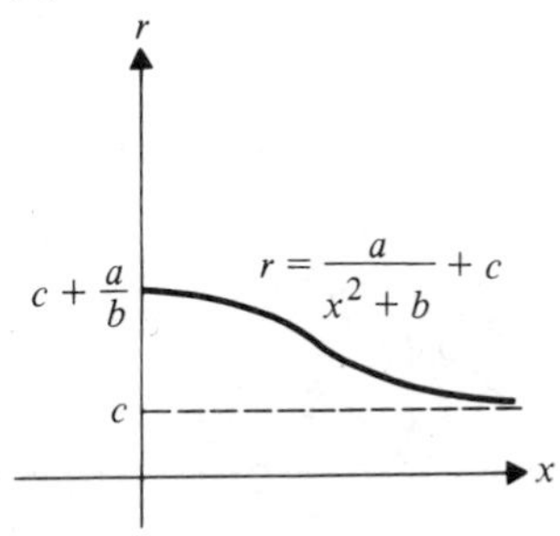

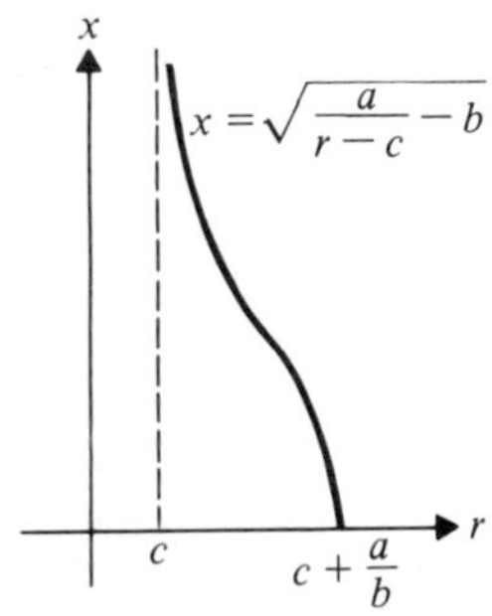

18. $f^{-1}(x) = -\sqrt{x}/(\sqrt{x} + 1)$

Exercises 8.4

2. $\pi/6$ **4.** $-2\pi/3$ **6.** Does not exist **8.** $-\pi/4$
10. Does not exist **12.** 0.464 **14.** $\pi/4$ **16.** $-\pi/4$
18. $(4n - 1)\pi/4$ **20.** No solutions **22.** No solutions

24.

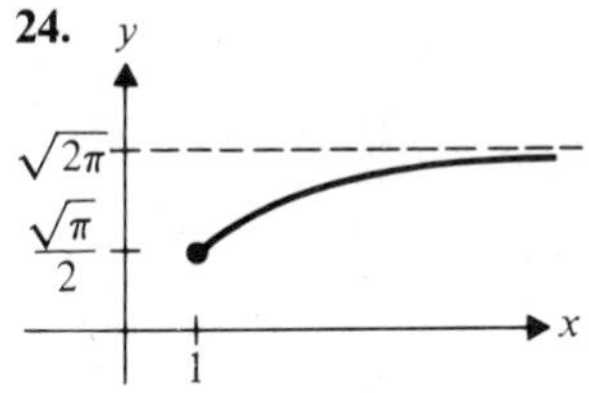

26.

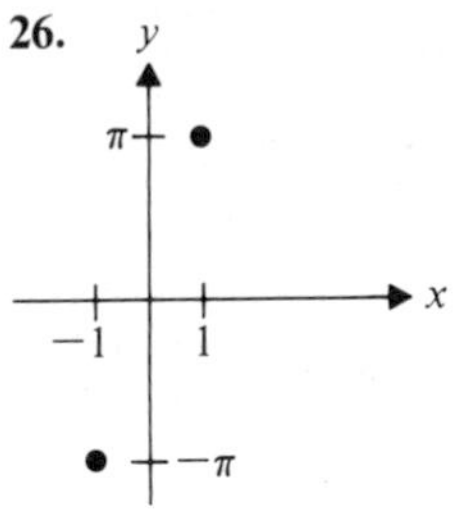

28. $(3n \pm 1)\pi/3$ **30.** No solutions
32. $\text{Tan}^{-1}(2n\pi \pm 3\pi/4) + m\pi$

34. (a)

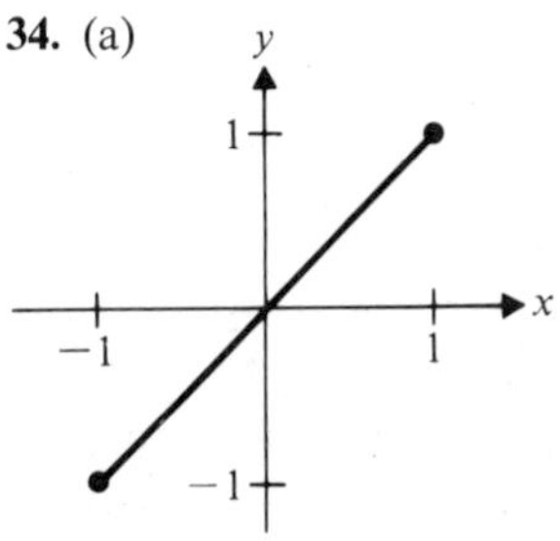

(b)

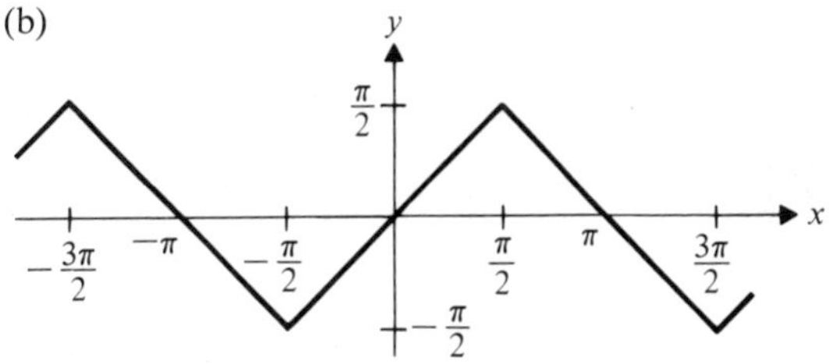

36. No; no

Exercises 8.5

2. $-2x/[1 + (x^2 + 2)^2]$ **4.** $-2x/[1 + (2 - x^2)^2]$
6. $\text{Csc}^{-1}(x^2 + 5) - 2x^2/[(x^2 + 5)\sqrt{(x^2 + 5)^2 - 1}]$
8. $1/[2(x + 3)\sqrt{x + 2}]$ **10.** $-1/(x\sqrt{x^2 - 1})$
12. $2x\,\text{Sec}^{-1}x + x/\sqrt{x^2 - 1}$ **14.** $-1/(1 + x^2)$
16. $-1/[\sqrt{x}(x + 1)]$
18. $x(t - \sqrt{1 - t^2}\,\text{Cos}^{-1}t)/[\sqrt{1 - x^2}\sqrt{1 - t^2}]$
20. $(5\sqrt{1 - x^2y^2} - y)/(x - 2\sqrt{1 - x^2y^2})$
22. $-18/[x^3\sqrt{x^2 - 9}]$
24. $(-1/x^2)\,\text{Csc}^{-1}(3x) - 2/(x^2\sqrt{9x^2 - 1})$
26. $(\text{Cos}^{-1}x)^2 - 2 + 2x(1 - \text{Cos}^{-1}x)/\sqrt{1 - x^2}$
28. $2\sqrt{4x - x^2}$ **30.** $\sqrt{2}(1 - x^2)/[(1 + x^2)\sqrt{1 + x^4}]$
32.

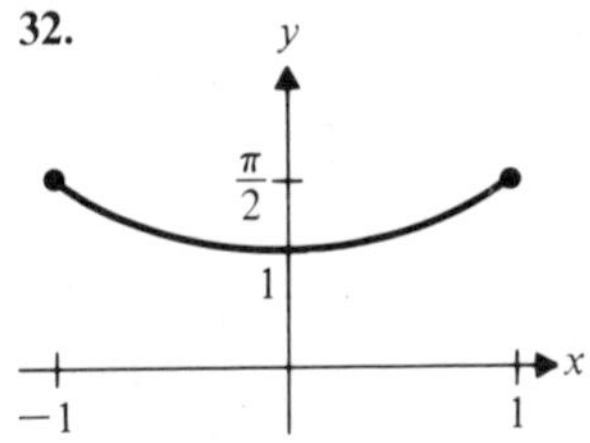

Relative minimum of $f(0) = 1$

34. 1.23 **36.** $\text{Tan}^{-1}\mu$

Exercises 8.6

2. $(1/3)\log_{10}5$ **4.** $0, -2$ **6.** No solutions
8. No solutions **10.** 5 **12.** No solutions **14.** 2.42
16. $(4n + 1)\pi/2 \pm 0.723$ **18.** $(4n - 1)\pi/2 \pm 1.27$
20. $m\pi + \text{Tan}^{-1}(2n\pi \pm 0.925)$ **22.** $-1 \pm \sqrt{1 + a^2}$
24. -3.08
26.

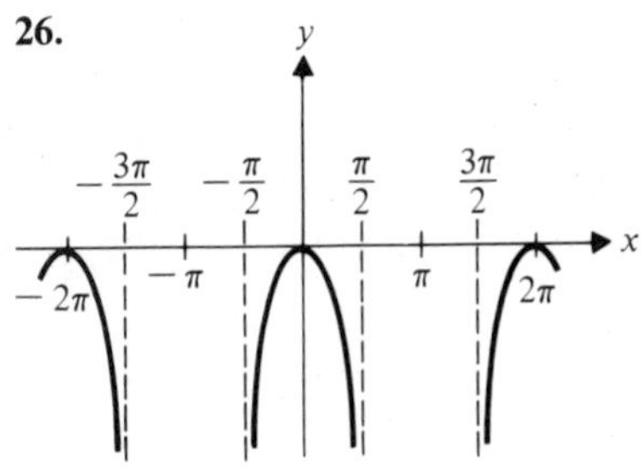

28.

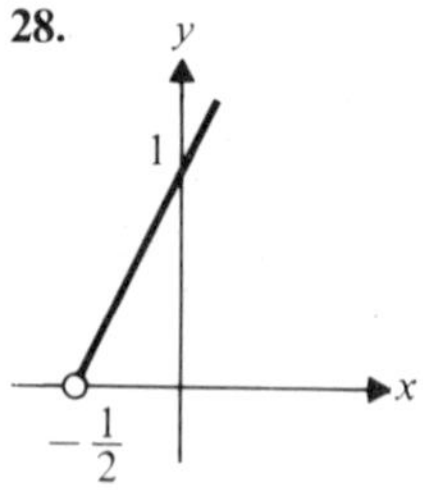

30. $10{,}000(3/4)^t$ **32.** $x_1 > 0$ and $x_2 > 0$ **34.** 105 km
36. (a) $I_0 10^R$; (b) 6.08, 4.79 **38.** (b) 9 years **40.** 0.694
42. $\sqrt{2}$ **44.** $[10{,}000(1 - (3/4)^t) + 250((6/5)^t - 1)]/t$
46. $\log_a(y + \sqrt{y^2 + 1})$ **48.** $(1/2)\log_a((1 + y)/(1 - y))$

Exercises 8.7

2. $2x + (4x^3 + 3x^2)e^{4x}$ **4.** $2e^{2x}\cos(e^{2x})$
6. $e^{-2x}(3\cos 3x - 2\sin 3x)$ **8.** $4(e^x + e^{-x})^{-2}$
10. $2\sin(\ln x)$ **12.** $-\tan v \sin 2x$
14. $-(y\cos x + 2x\ln y + e^y)/(xe^y + \sin x + x^2/y)$
16. $2x^2/\sqrt{x^2 + 1}$ **18.** $(1 + 5e^{4x})^{-1}$
20. $(y^2/x^2)(e^{1/x} - 1)/(1 - e^{1/y})$
22.

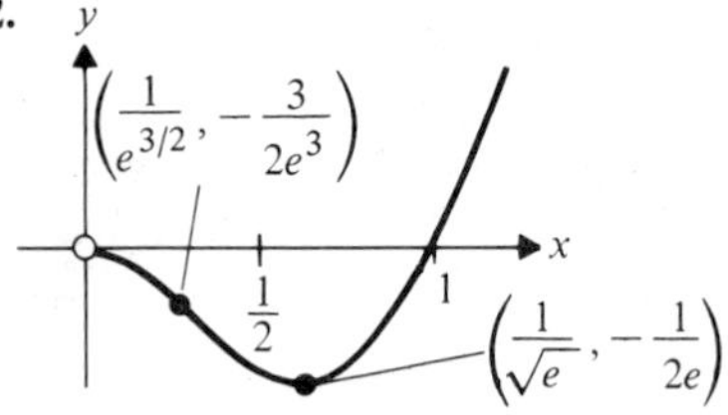

24.

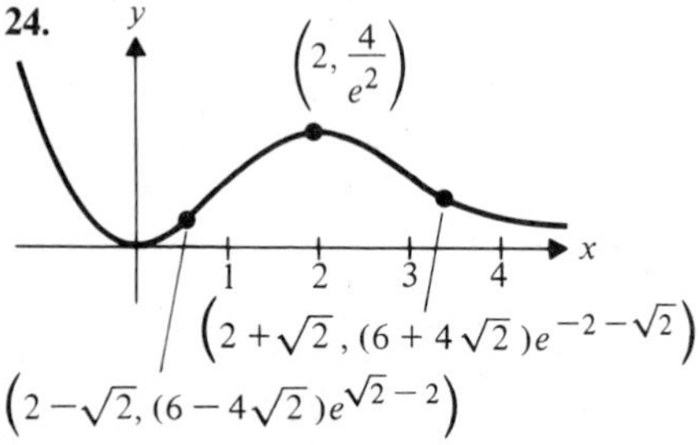

26. $2(x^2 + y^2)/(x - y)^3$
28. (a) (b) $(4n + 1)\pi/4$

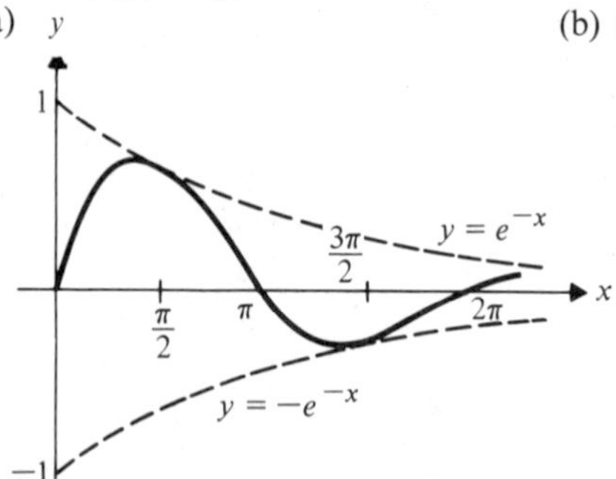

30. (a) (b) 2.491 m

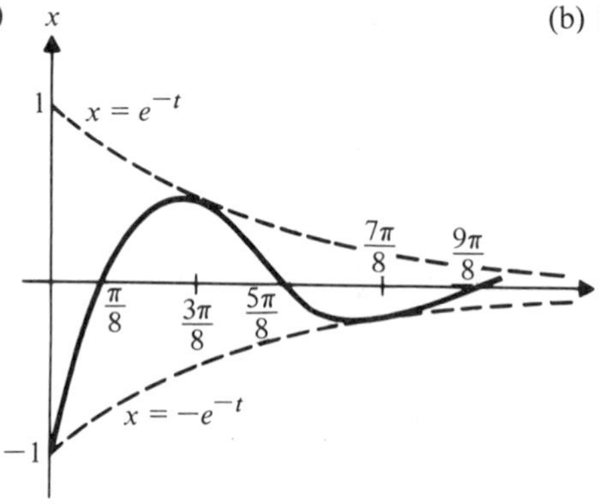

32.

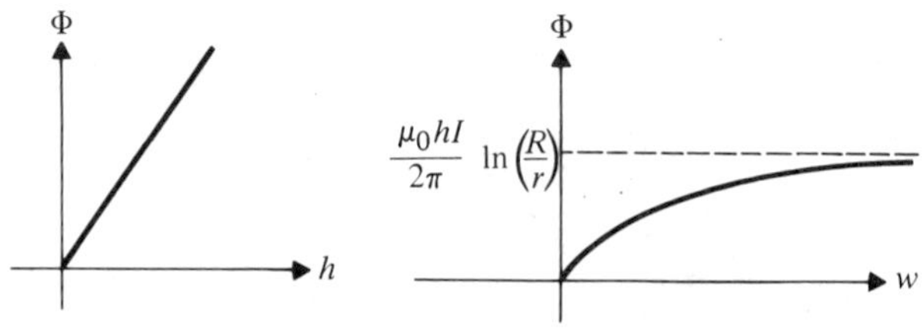

38. $1/\sqrt{x^2 + 4}$

Exercises 8.8

2. $4x^{4\cos x}(\cos x/x - \sin x \ln x)$
4. $(1 + 9x^2 + 6x^4)/[2\sqrt{x}(1 + x^2)^{3/2}]$ **6.** $2xe^{4x}(2x + 1)$
8. $(\ln x - 1)(1/x)^{2+1/x}$ **10.** (b) $(a - br)^{-1}$
12. $(x^2 + 3x)^2(x^2 + 5)^3(14x^3 + 33x^2 + 30x + 45)$

Exercises 8.9

2. 4 **4.** 0 **6.** 1 **8.** e **10.** 1 **12.** Does not exist

14.

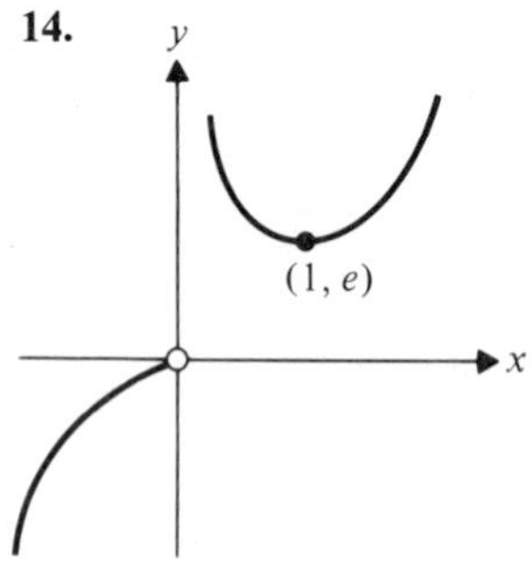

16.

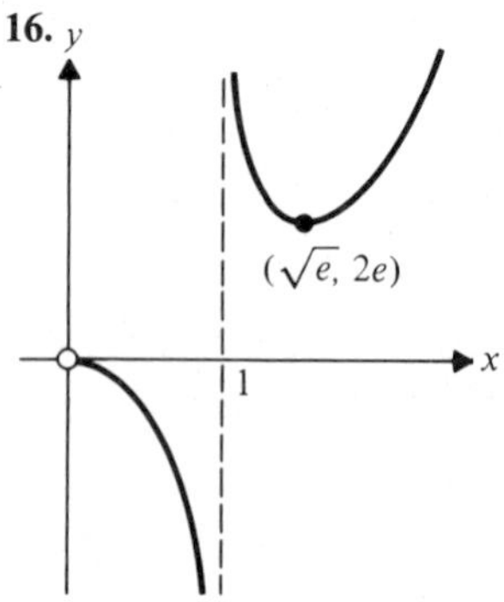

18. $e^{c(a-b)}$ **20.** 2 **22.** $-1/3$

Exercises 8.10

2. 2.30 **4.** 9.08×10^{-5} **6.** 1.00 **8.** 0.266 **10.** 5.07
12. $\sin h(x/2) + (x/2)\cosh(x/2)$ **14.** $(1/x)\operatorname{sech}^2(\ln x)$

16. $2\sqrt{1+y}\operatorname{csch}^2 x/(2\sqrt{1+y}-1)$
18. $-\sinh t \operatorname{sech}^2 x \sec^2(\cosh t)\sin(\tanh x)$
20. $2 \operatorname{csch} 4x$ **26.** (a) 189.3 N; (b) 17.0 m

Review Exercises

2. $(4n+1)\pi/4 \pm 0.659$ **4.** -1.79 **6.** $(\sqrt{13}-1)/2$
8. $\ln[(4n+1)\pi/2 \pm 0.84] - 2, n \geq 0$ **10.** 2.09
12. $6x/(3x^2+5)$ **14.** $6x\cosh(x^2)$
16. $(x^2-1)/(x^4+3x^2+1)$ **18.** $(1/2)\tan x/\sqrt{\ln\sec x}$
20. $1/\sqrt{x^2-1}$ **22.** $\cos x \operatorname{sech} y$ **24.** -1
26. $10\cos x/(\cos x + 3\sin x)$
28.

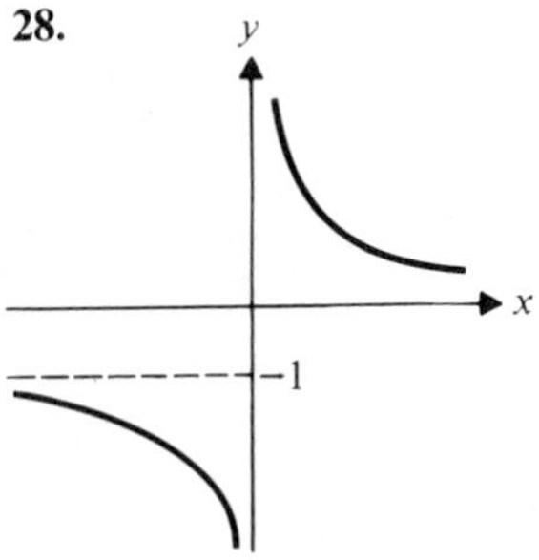

30. (a)

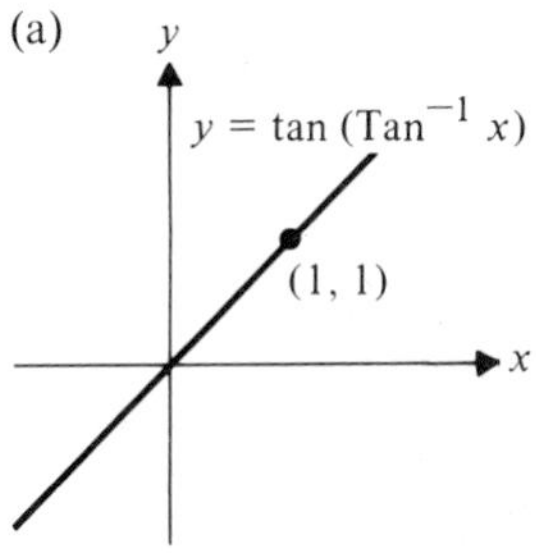

(b)

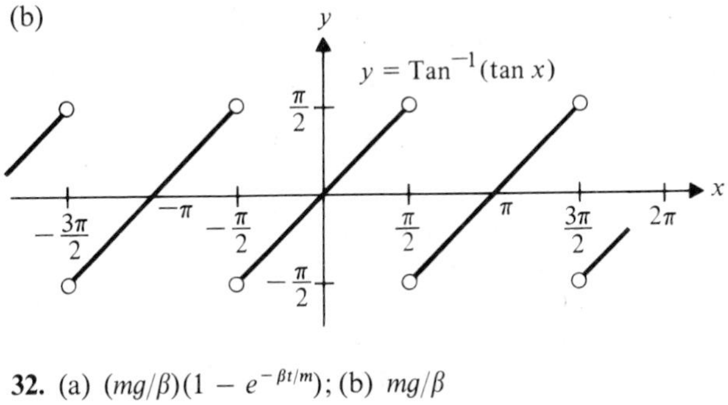

32. (a) $(mg/\beta)(1-e^{-\beta t/m})$; (b) mg/β

CHAPTER 9

Exercises 9.2

2. $(-1/4)e^{-2x^2}+C$ **4.** $\ln(1+e^x)+C$
6. $(-2/27)(1-3x^3)^{3/2}+C$ **8.** $(-1/6)(1+x^3)^{-2}+C$
10. $2\sqrt{x}-x+C$ **12.** $\operatorname{Tan}^{-1}x+x+C$
14. $(1/6)(2x+4)^{3/2}+\sqrt{2x+4}+C$
16. $(1/8)\sin^4 2x + C$
18. $(1/2)(x+5)^2 - 15x + 75\ln|x+5| + 125/(x+5) + C$
20. $(1/3)\ln|\sec 3x| + C$ **22.** $(1/\sqrt{3})\operatorname{Tan}^{-1}(x/\sqrt{3})+C$
24. $(1/4)\operatorname{Sec}^{-1}(\sqrt{3}x/4)+C$
26. $(4/5)x^{5/4} - x + (4/3)x^{3/4} - 2\sqrt{x} + 4x^{1/4} - 4\ln(x^{1/4}+1)+C$ **28.** $2\ln(e^x+1)-x+C$
30. $C\ln 2$ **32.** $(-1/7, 5/2)$ **36.** $8\sqrt{2}$
40. $(1/(3n))\ln|x^n/(3+2x^n)|+C$

Exercises 9.3

2. 49.83 days **4.** 13.51 hours **6.** CS^k; no
10. (a) Finite time; (b) infinitely long
16.

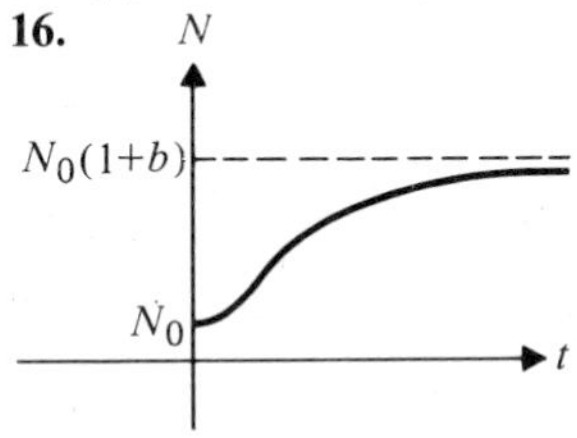

18.

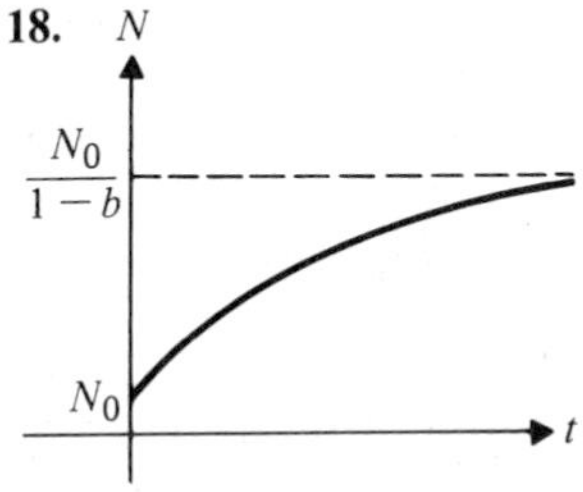

20.

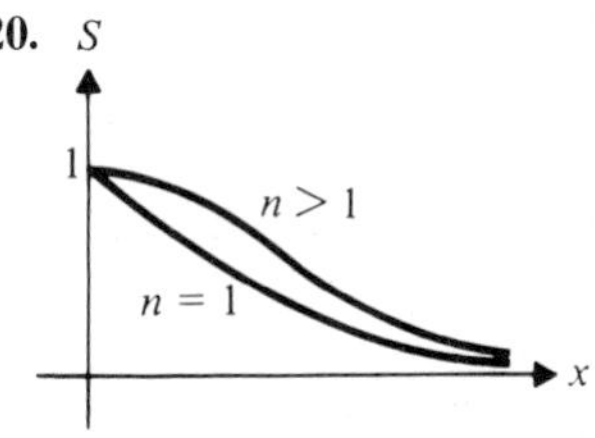

22. 3:25 A.M. **24.** (b) 118.3 km

Exercises 9.4

2. $(-1/2)\sin^{-2}x+C$ **4.** $(-1/3)\csc^3 x + C$
6. $(2/3)\tan^{3/2}x + (2/7)\tan^{7/2}x + C$
8. $(1/18)\sec^6 3x + C$ **10.** $(1/4)\sec^4 x + C$
12. $\tan\theta + C$ **14.** $2\sqrt{1+\tan x}+C$
16. $7\tan x - 3x + C$
18. $(1/32)(12x - 8\sin 2x + \sin 4x) + C$
20. $(1/2)\tan^2 x + \ln|\tan x| + C$ **22.** $\pi(1-\pi/4+\ln 2)$
24. $(-1/3)\cot^3 z + \cot z + z + C$
26. $\theta/2 + (1/4)\sin 2\theta + (1/3)\cos^3\theta + C$
28. $(1/8)\sin 4x + (1/16)\sin 8x + C$
30. $\sec x + \ln|\csc x - \cot x| + C$

Exercises 9.5

2. $(1/\sqrt{5})\operatorname{Sin}^{-1}(\sqrt{5}x/3)+C$ **4.** $-\sqrt{4-x^2}/(4x)+C$
6. $(1/15)(5x^2+3)^{3/2}+C$
8. $(1/2)\ln|(1+x)/(1-x)|+C$
10. $(1/20)\ln(5x^2+1)+(\sqrt{5}/2)\operatorname{Tan}^{-1}(\sqrt{5}x)+C$

12. $\sqrt{4-x^2} + 2\ln|(2-\sqrt{4-x^2})/x| + C$
14. $\ln|x+\sqrt{x^2-16}| - \sqrt{x^2-16}/x + C$
16. $(1/16)\,\mathrm{Sec}^{-1}(x/2) + \sqrt{x^2-4}/(8x^2) + C$
18. $(1/3)(y^2+4)^{3/2} - 4\sqrt{y^2+4} + C$
20. $2\ln|x| - (1/2)\ln(1+x^2) + C$
24. $L\ln((L+\sqrt{L^2-x^2})/x) - \sqrt{L^2-x^2}$ **26.** $\pi\gamma r^3$
28. $(147/8)\,\mathrm{Sin}^{-1}(x/\sqrt{7}) + (x/8)(35-2x^2)\sqrt{7-x^2} + C$
30. $-6\,\mathrm{Sec}^{-1}(2x) - (12x^2-1)/(2x^2\sqrt{4x^2-1}) + C$
32. (a) $(e^{kx}-e^{-kx})/2$; (b) $(e^{kx}+e^{-kx})/(2k) + C$

Exercises 9.6

2. $\ln|\sqrt{x^2+2x+2} + x + 1| + C$
4. $(-2/\sqrt{7})\,\mathrm{Tan}^{-1}((2x-3)/\sqrt{7}) + C$
6. $(x/2)/\sqrt{4x-x^2} + C$
8. $\ln(x^2+6x+13) - (9/2)\,\mathrm{Tan}^{-1}((x+3)/2) + C$
10. $\ln|\sqrt{6+4\ln x+(\ln x)^2} + \ln x + 2| + C$
12. $39{,}240\pi$ N

Exercises 9.7

2. $(x^2/2 - x/2 + 1/4)e^{2x} + C$
4. $(2/3)x^{3/2}\ln(2x) - (4/9)x^{3/2} + C$
6. $(-2/5)(2+x)(3-x)^{3/2} + C$
8. $(2x^2/3)(x+5)^{3/2} - (8x/15)(x+5)^{5/2} + (16/105)(x+5)^{7/2} + C$
10. $2x^2\sqrt{x+2} - (8x/3)(x+2)^{3/2} + (16/15)(x+2)^{5/2} + C$
12. $(1/3)(x-1)^3\ln x - x^3/9 + x^2/2 - x + (1/3)\ln x + C$
14. $x\,\mathrm{Tan}^{-1}x - (1/2)\ln(1+x^2) + C$
16. $(1/8)(3\sin x\sin 3x + \cos x\cos 3x) + C$
18. $x\ln(x^2+4) - 2x + 4\,\mathrm{Tan}^{-1}(x/2) + C$
20. (a) $5x/16 + (1/4)\sin 2x + (3/64)\sin 4x - (1/48)\sin^3 2x + C$; (c) $(1/6)\sin x\cos^5 x + (5/24)\sin x\cos^3 x + (5/16)\sin x\cos x + 5x/16 + C$
22. $(2\sqrt{5} + \ln(2+\sqrt{5}))/4$
28. $x^3/6 + (x^2/4 - 1/8)\sin 2x + (x/4)\cos 2x + C$
30. $(x/2)\sqrt{x^2-4} - 2\ln|x+\sqrt{x^2-4}| + C$
32. $(1/2)[x\sqrt{1+3x^2} + (1/\sqrt{3})\ln|\sqrt{3}x + \sqrt{1+3x^2}|] + C$
34. $(1/2)(x-1)\sqrt{x^2-2x-3} - 2\ln|x-1+\sqrt{x^2-2x-3}| + C$

Exercises 9.8

2. $(-1/2)(y+1)^{-2} + C$
4. $x + (10/3)\ln|x-4| + (2/3)\ln|x+2| + C$
6. $(-1/6)\ln|y| - (2/15)\ln|y+3| + (3/10)\ln|y-2| + C$
8. $(1/2)\ln(x^2+2) + (x^2+2)^{-1} + C$
10. $y - 4\ln|y+2| + \ln|y+1| + C$
12. $\ln(y^2+1) + \mathrm{Tan}^{-1}y - \ln(y^2+4) + C$
14. $(-1/(3\sqrt{2}))\,\mathrm{Tan}^{-1}(x/\sqrt{2}) + (2/3)\ln|(x-1)/(x+1)| + C$
16. $(-5/27)\ln|x-1| - (7/9)(x-1)^{-1} + (32/27)\ln|x+2| + (2/9)(x+2)^{-1} + C$
18.

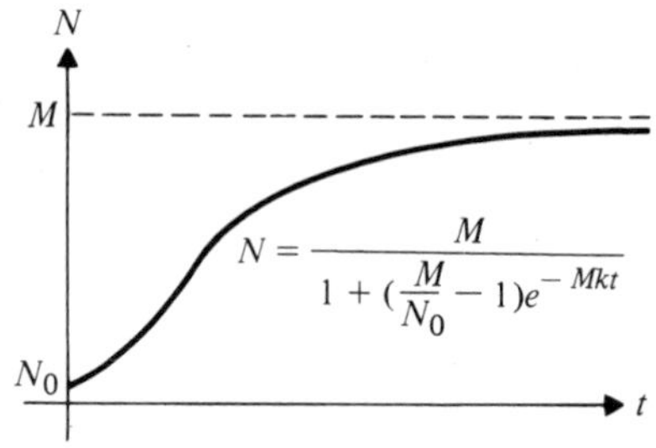

20. $(1/4)\ln|x+1| - (1/8)\ln(x^2+1) + (1/2)\,\mathrm{Tan}^{-1}x + (1/4)(x+1)/(x^2+1) + C$
22. $(1/3)\ln|x+1| + (1/\sqrt{3})\,\mathrm{Tan}^{-1}((2x-1)/\sqrt{3}) - (1/6)\ln(x^2-x+1) + C$
24. $\ln|x| + \ln(x^2-x+1) - 2\ln|x+1| - 3/(x+1) + (2/\sqrt{3})\,\mathrm{Tan}^{-1}((2x-1)/\sqrt{3}) + C$
28. $(1/4)\ln|3\tan(x/2)+1| - (1/4)\ln|\tan(x/2)+3| + C$

Exercises 9.9

2. 0.47215, 0.47214 **4.** 0.64886, 0.64872
6. −0.069570, −0.069445 **8.** 0.30220, 0.30230
10. 0.14221, 0.14201 **12.** 1.4672, 1.4627
14. 0.31117, 0.31026 **16.** 1/4 previous; 1/16 previous error **18.** 1.4789 **20.** (a) 21; (b) 4 **22.** (a) 3; (b) 2
24. 395.4 J

Review Exercises

2. $-1/(x+3) + C$ **4.** $x^2/2 - x + 4\ln|x+1| + C$
6. $(2/3)(x+3)^{3/2} - 6\sqrt{x+3} + C$
8. $-x\cos x + \sin x + C$
10. $(3/4)\ln|x+3| + (1/4)\ln|x-1| + C$
12. $14\sqrt{x} - x - 70\ln(5+\sqrt{x}) + C$
14. $\mathrm{Sin}^{-1}(e^x) + C$ **16.** $-(1/2)/(x^2+1) + C$
18. $(1/2)\ln(x^2+1) + (1/2)/(x^2+1) + C$
20. $x + 4\ln|x-1| - 4/(x-1) + C$
22. $x\,\mathrm{Cos}^{-1}x - \sqrt{1-x^2} + C$
24. $(-1/12)\cos 6x + (1/8)\cos 4x + C$
26. $\ln|x+2+\sqrt{x^2+4x-5}| + C$
28. $(1/5)(4-x^2)^{5/2} - (4/3)(4-x^2)^{3/2} + C$
30. $-3\ln|4-x^2| + C$
32. $(1/(2\sqrt{2}))\ln|(x+2-2\sqrt{2})/\sqrt{x^2+4x-4}| + C$
34. $\ln|x| + 1/x - 1/(2x^2) - \ln|x+1| + C$
36. $\ln|2\sqrt{x^2-3x+16} + 2x - 3| + C$
38. $-x^2/6 + (x^3/3)\,\mathrm{Tan}^{-1}x + (1/6)\ln(x^2+1) + C$
40. $(\ln x)^2/2 + C$
42. $(1/81)\ln|x| - (1/162)\ln(x^2+9) + (1/18)/(x^2+9) + C$
44. $(1/2)\ln|x| + (1/2)\ln|x+2| + 3/(x+2) + C$
46. $\ln|x| + \ln(x^2+x+4) + C$
48. $-(2/3)\cot^{3/2}x - (2/7)\cot^{7/2}x + C$
50. $(x-2)/(4\sqrt{4x-x^2}) + C$ **52.** 0.64041 **54.** 1.2647
56. $-2\sqrt{x} - 3x^{1/3} - 6x^{1/6} - 6\ln|1-x^{1/6}| + C$

58. $(1/8)\ \mathrm{Tan}^{-1}(x^2/4) + C$ **60.** $(4x - 6)/(9\sqrt{3x - x^2}) + C$
62. $2 \sin\sqrt{x} - 2\sqrt{x} \cos\sqrt{x} + C$
64. $(1/32) \sin 4x + (1/8) \sin 2x - (x/8) \cos 4x - (x/4) \cos 2x + C$ **66.** $(1/2) \tan x + C$
68. $(1/6) \sin 3x - (1/20) \sin 5x - (1/4) \sin x + C$
70. $(1/4)(\mathrm{Sin}^{-1} x)^2 + (x/2)\sqrt{1 - x^2}\ \mathrm{Sin}^{-1} x - x^2/4 + C$
72. (b) $\mathrm{Sin}^{-1} x - \sqrt{1 - x^2} + C$

CHAPTER 10

Exercises 10.1

2. Circle **4.** None of these **6.** Hyperbola
8. Parabola **10.** None of these **12.** Ellipse
14. Straight line

16.

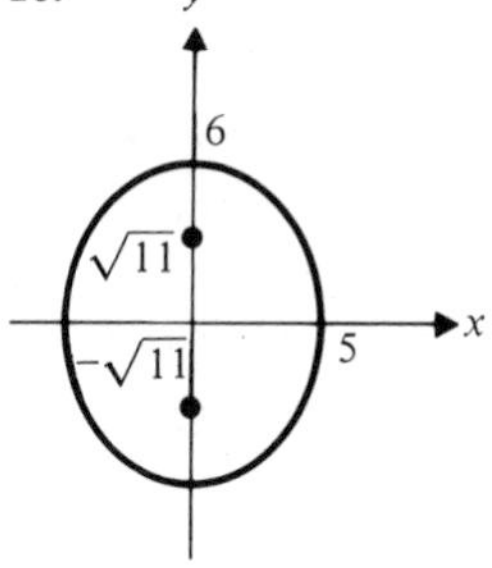

18.

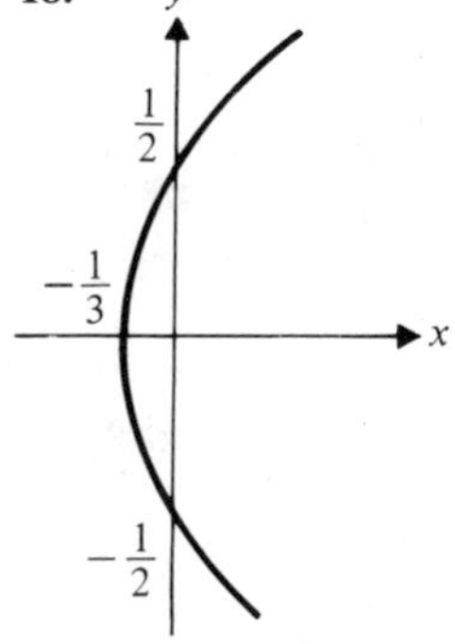

20.

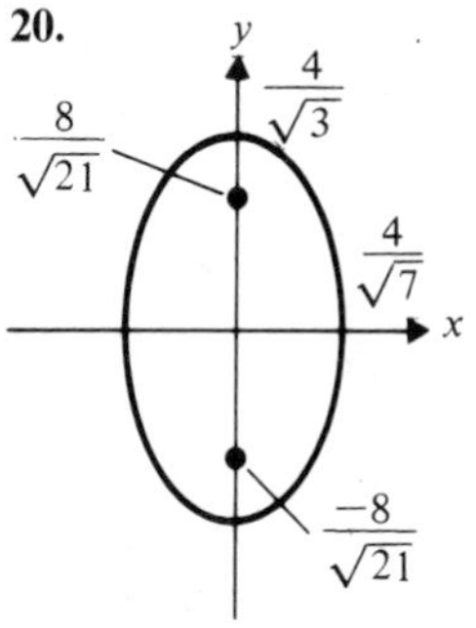

22.

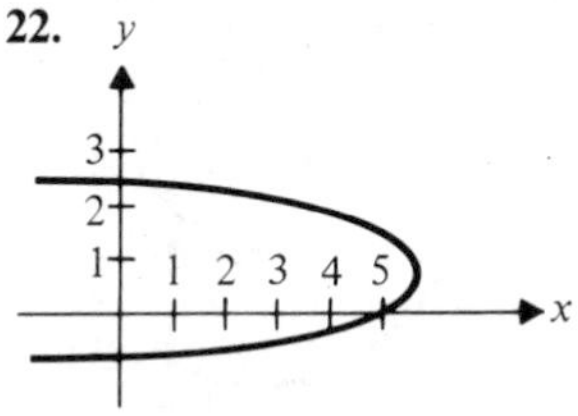

24.

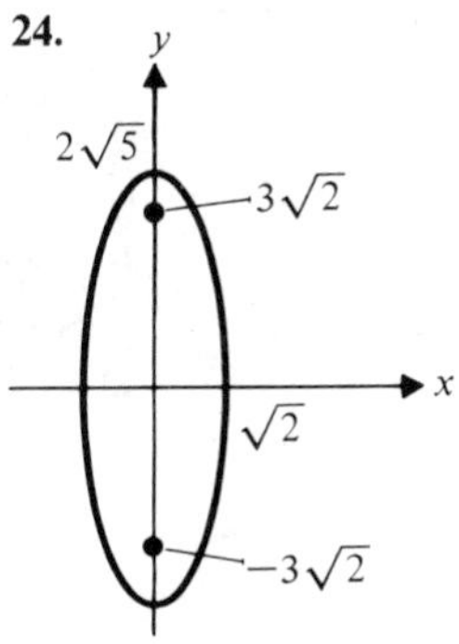

26.

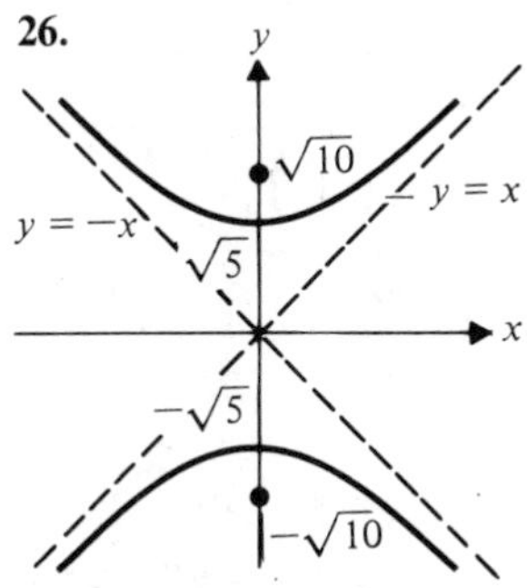

28.

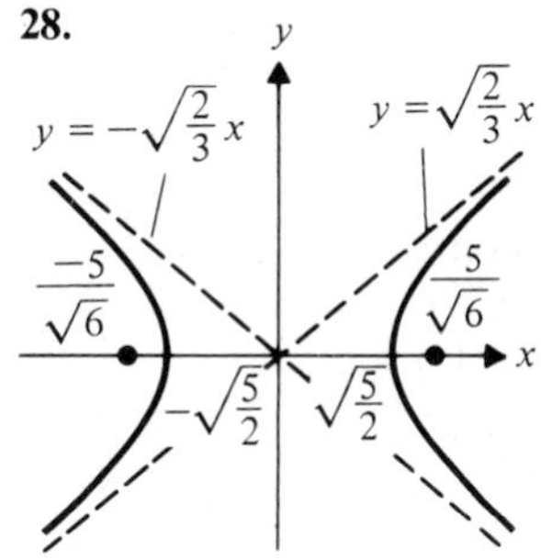

30.

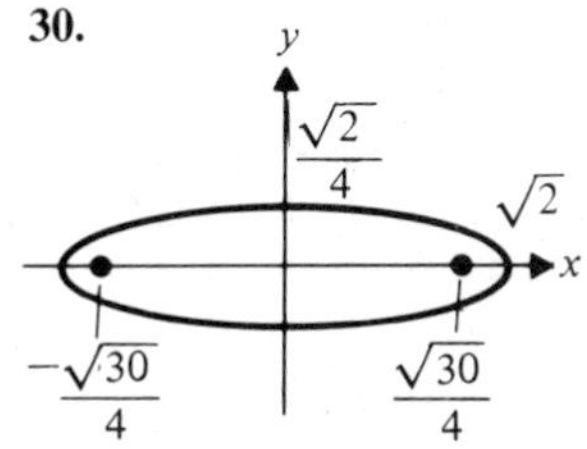

32.

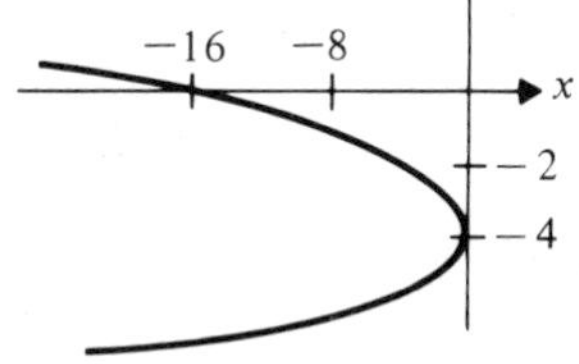

34.

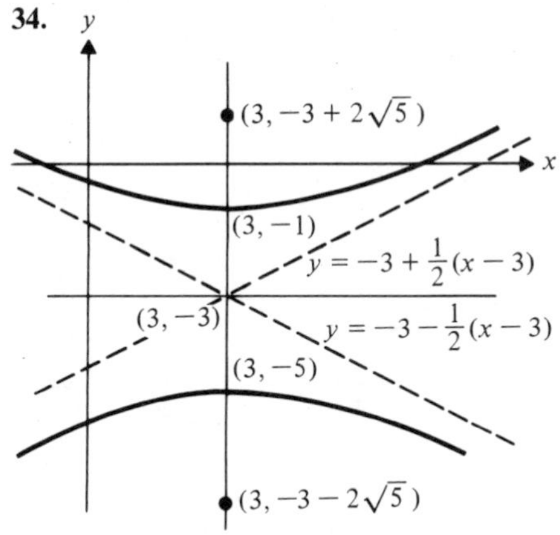

36.

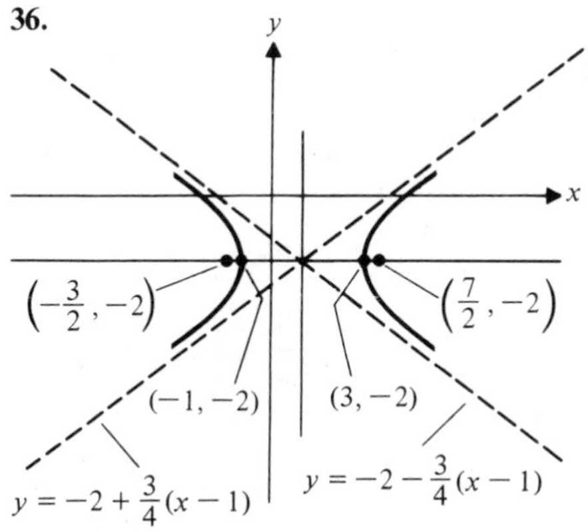

38. $3x^2 + y^2 = 28$ **40.** $25/4$ **42.** $9x^2 + 25y^2 = 225$
48. πab **50.** $4\pi ab^2/3, 4\pi a^2 b/3$
54. In Exercise 15, focus is $(0, -7/8)$, directrix is $y = -9/8$; in Exercise 18, focus is $(-7/48, 0)$, directrix is $x = -25/48$; in Exercise 21, focus is $(3/4, 0)$, directrix is $x = 5/4$; in Exercise 22, focus is $(6, 3/4)$, directrix is $x = 25/4$; in Exercise 27, focus is $(3, -1/4)$, directrix is $y = 1/4$; in Exercise 32, focus is $(-1/4, -4)$, directrix is $x = 1/4$.

Exercises 10.2

2. $(2, 2\pi/3 + 2n\pi), (-2, -\pi/3 + 2n\pi)$
4. $(4, 5\pi/6 + 2n\pi), (-4, -\pi/6 + 2n\pi)$
6. $(\sqrt{17}, -1.82 + 2n\pi), (-\sqrt{17}, 1.33 + 2n\pi)$
8. $(\sqrt{29}, 2.76 + 2n\pi), (-\sqrt{29}, -0.38 + 2n\pi)$
10. $(3\sqrt{3}, -3)$ **12.** $(-2.21, -2.03)$

Exercises 10.3

2.
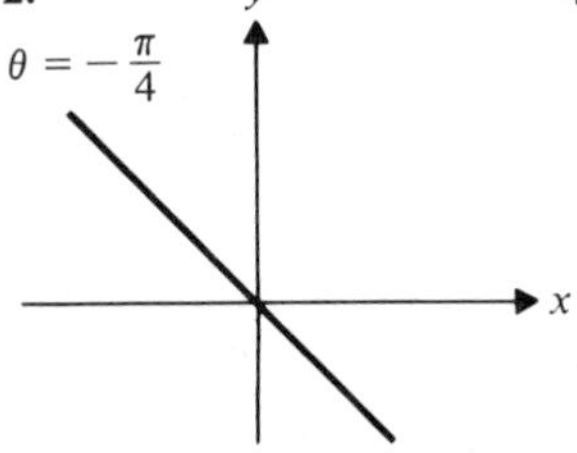

4.
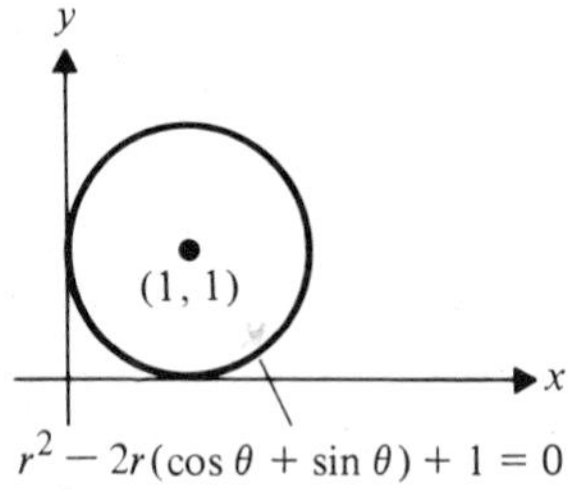

6.
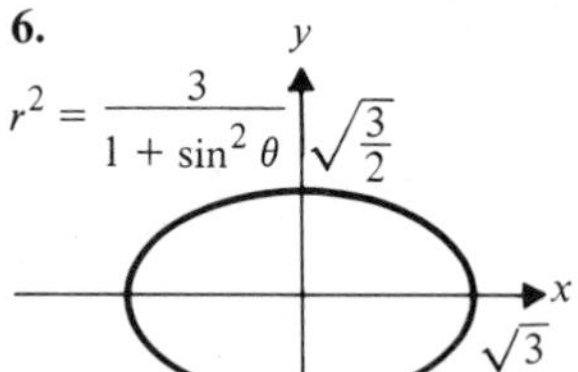

8.
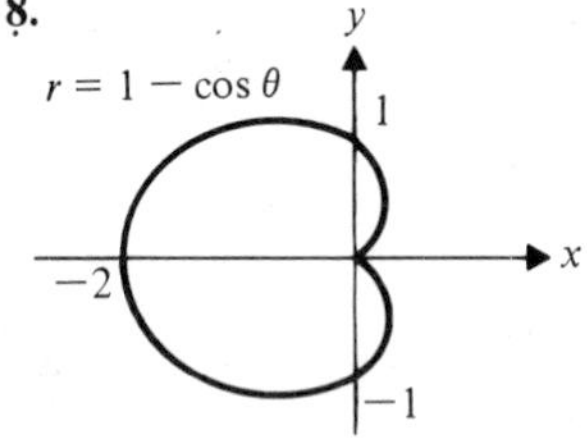

10.
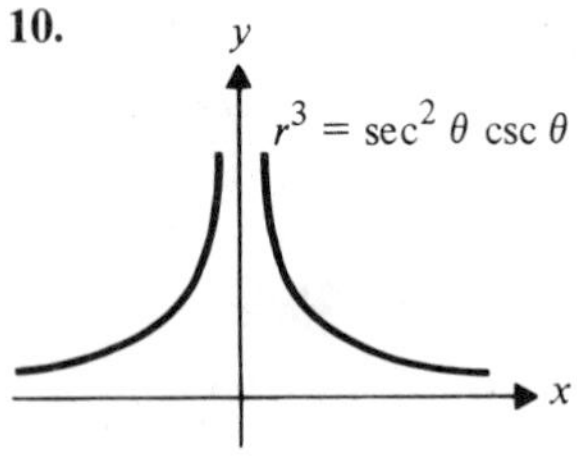

12.
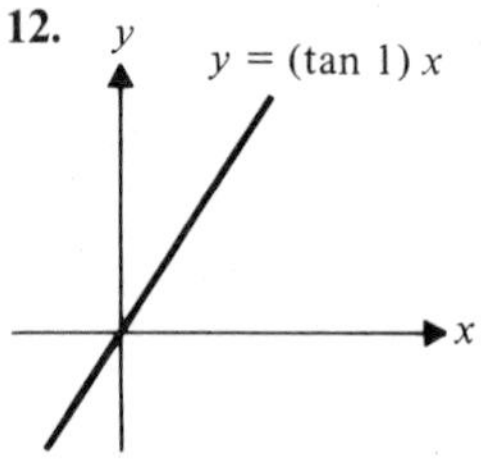

14.
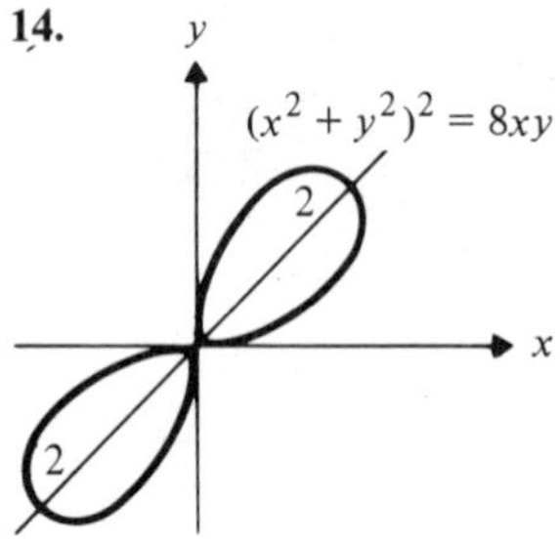

16.
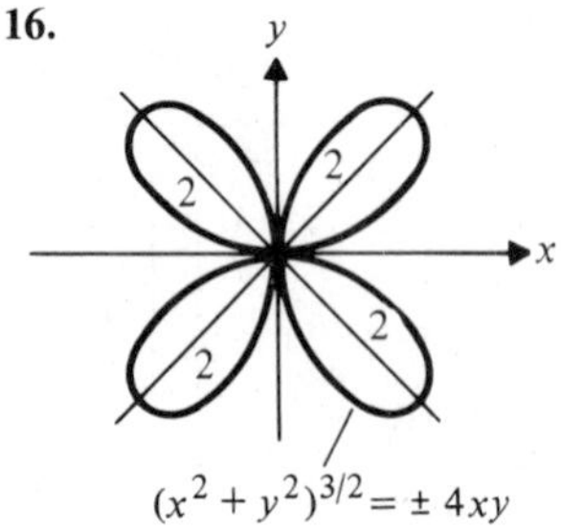

18.
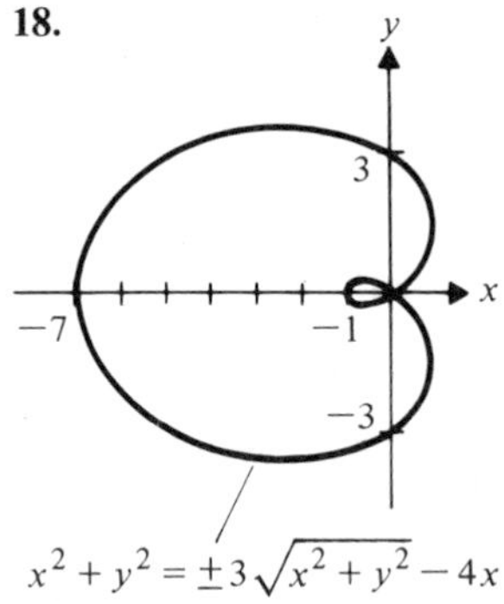

20.
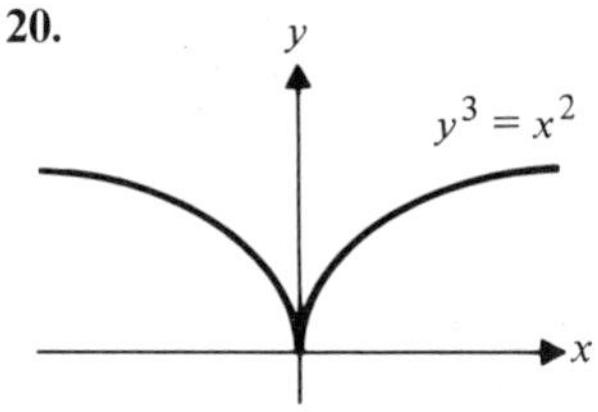

22. $(2, \pm\pi/6), (2, \pm 5\pi/6)$ **24.** $(0, \theta), (4/3, \pm 1.23)$

28.
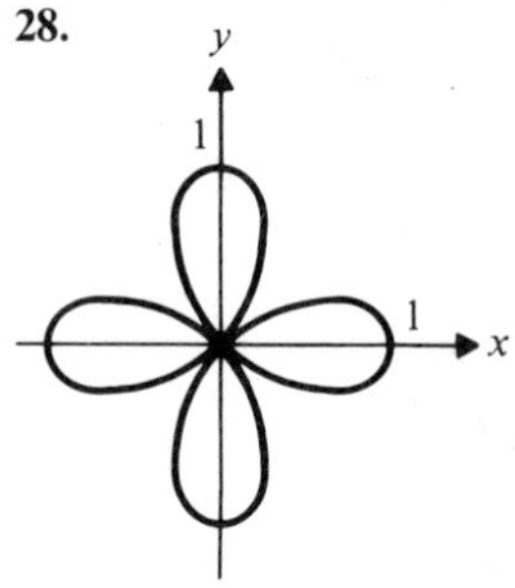

30.
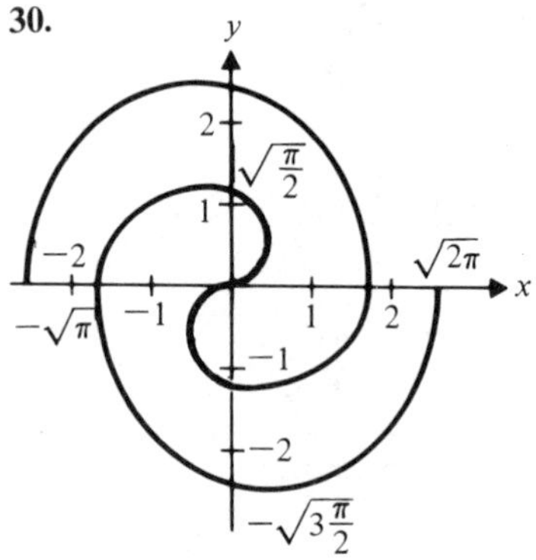

32.
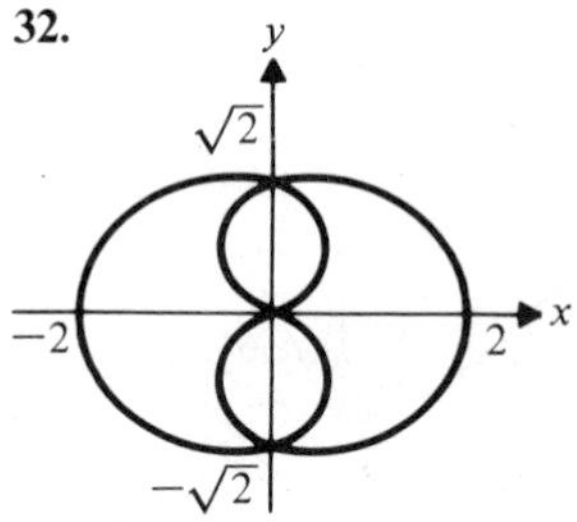

34.
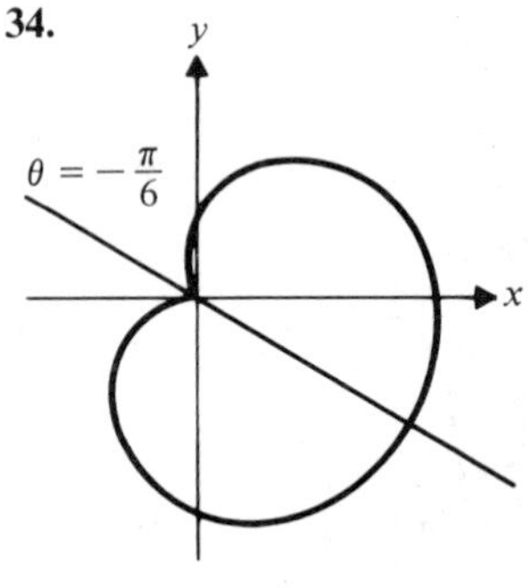

36.
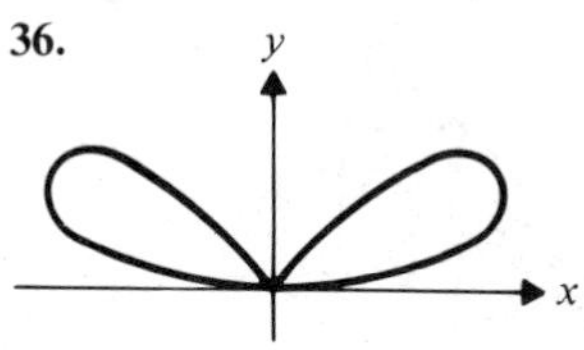

38. (a)

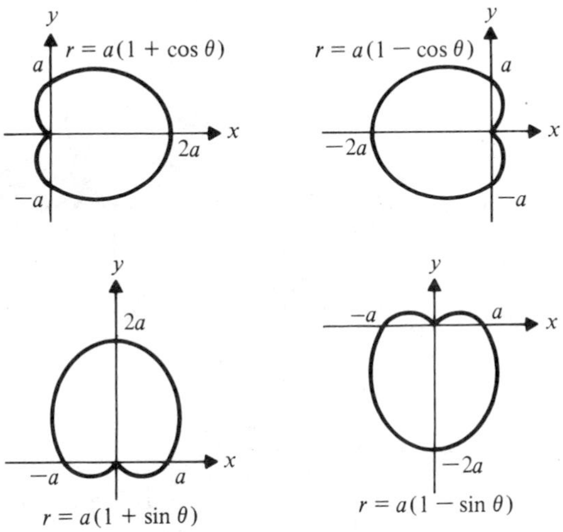

(b) $x^2 + y^2 = a(\pm\sqrt{x^2 + y^2} + x)$,
$x^2 + y^2 = a(\pm\sqrt{x^2 + y^2} - x)$,
$x^2 + y^2 = a(\pm\sqrt{x^2 + y^2} + y)$,
$x^2 + y^2 = a(\pm\sqrt{x^2 + y^2} - y)$

Exercises 10.4

2. 9π **4.** 2 **6.** 6π **8.** 18π **10.** 2π **12.** $(5\pi - 8)/4$
14. $6\pi - 16$ **16.** $(7\pi - 12\sqrt{3})/12$
18. $33[\pi + 2\,\mathrm{Sin}^{-1}(2/3)]/4 + 7\sqrt{5}$ **20.** $\pi/32$

Exercises 10.5

2.

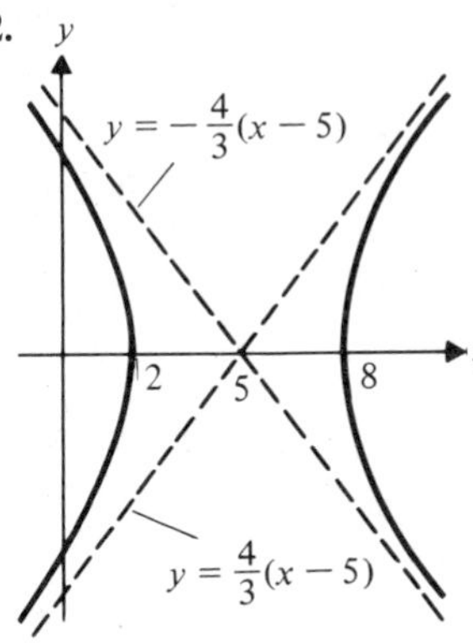

4.

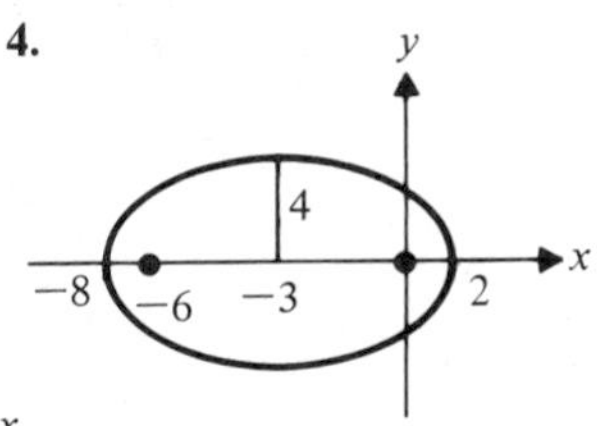

6.

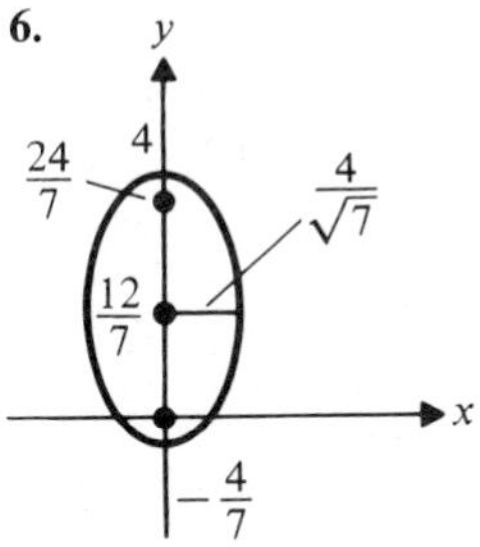

8.

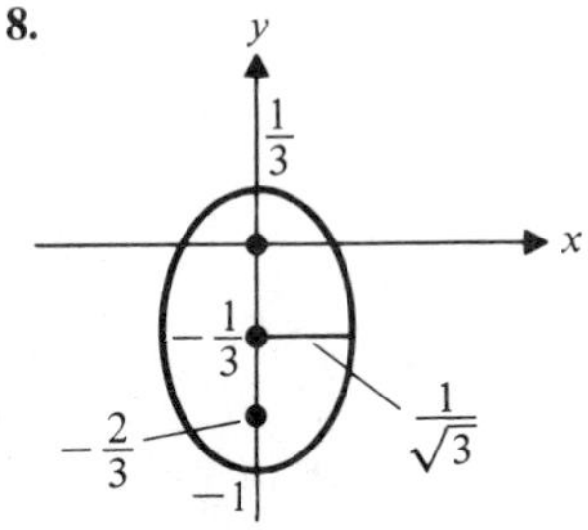

10.

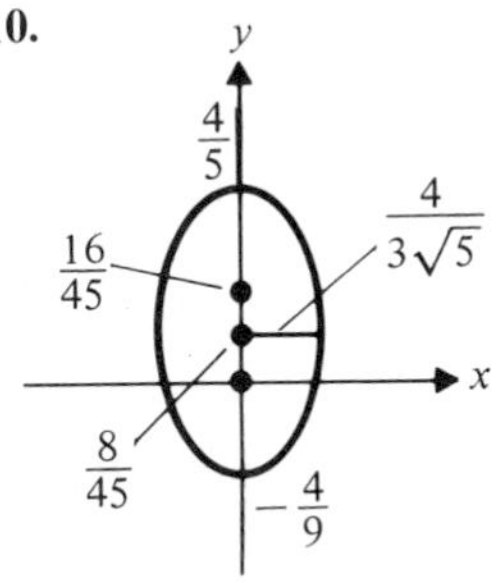

12. $8x^2 + 2x + 9y^2 = 1$ **14.** $8x^2 + 12x - y^2 + 4 = 0$
16. $40x + 25y^2 = 16$
18. With $x = \sqrt{5} + r\cos\theta$, $y = r\sin\theta$, the equation becomes $r = 4/(3 + \sqrt{5}\cos\theta)$.
20. With $x = \sqrt{10} + r\cos\theta$, $y = 2 + r\sin\theta$, the equation becomes $r = 1/(3 - \sqrt{10}\cos\theta)$. **22.** $\varepsilon \to 0$

Exercises 10.6

2.

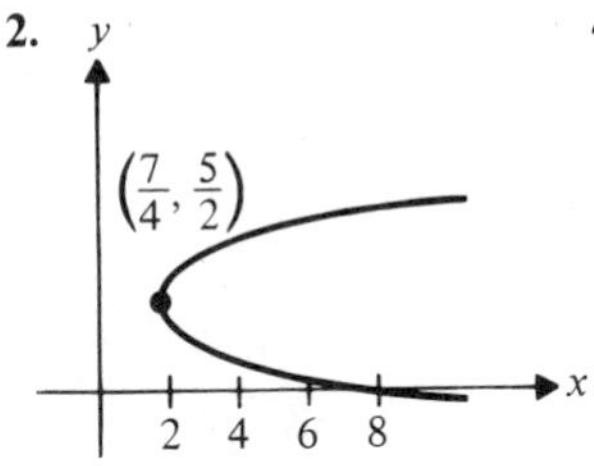

4.

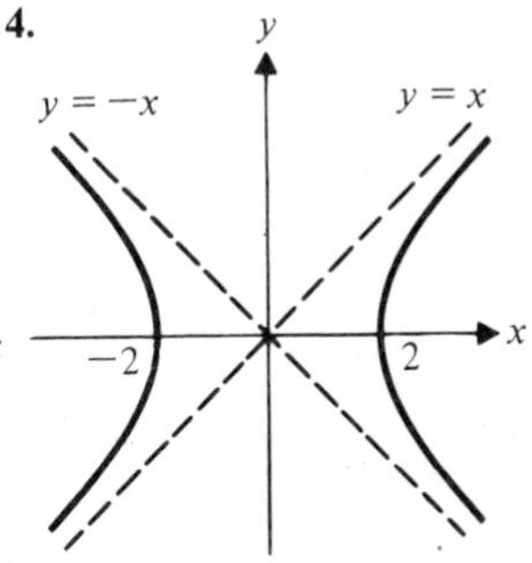

6.

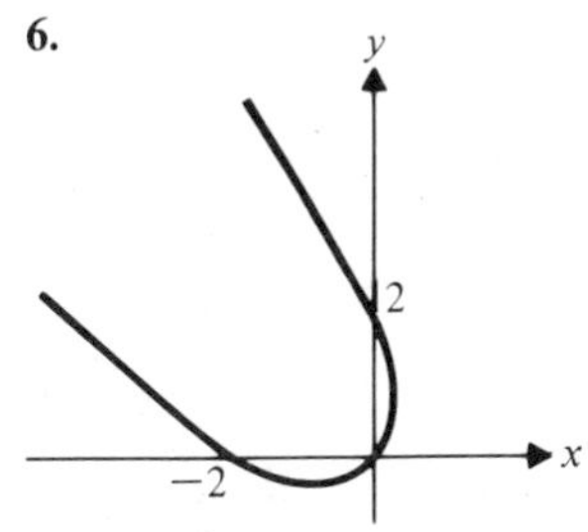

8.

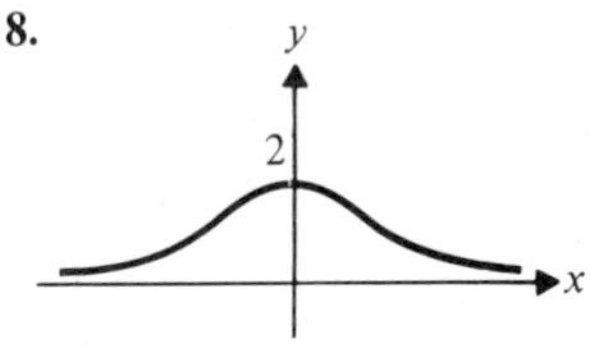

10. $2u/(u + 1)^2$ **12.** $-1/x^2$ **14.** $3/[4(t + 6)^2(2t + 3)^3]$
16. $-4(t + 1)\sqrt{-t^2 + 3t + 5}/[(3 - 2t)(t^2 + 2t - 5)^2]$
18. x/y, $2/y^3$ **20.** $1/(v + 1)$, $-(1/2)/(v + 1)^3$
22. $(8/3, 1/2)$, $(28/3, 28)$
24. Ellipse $b^2(x - h)^2 + a^2(y - k)^2 = a^2b^2$
28. $x = t$, $y = (t + 1)/(t - 2)$
30. $x = -1 + \sqrt{5}\cos t$, $y = 2 + \sqrt{5}\sin t$, $0 \le t < 2\pi$
32. 31/26 **36.** $\sqrt{2}(e - 1)/e$ **38.** 1.73
44. $x = 2\cos(4\pi t)$, $y = 2\sin(4\pi t)$, $t \ge 0$

48.

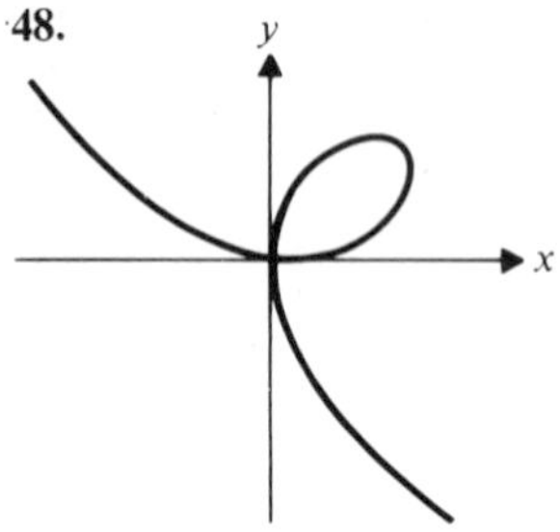

18.

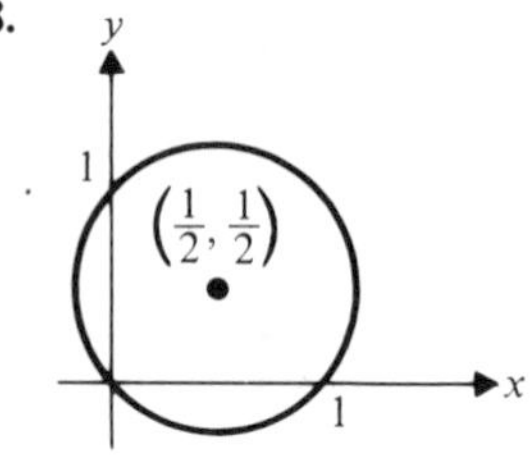

20.

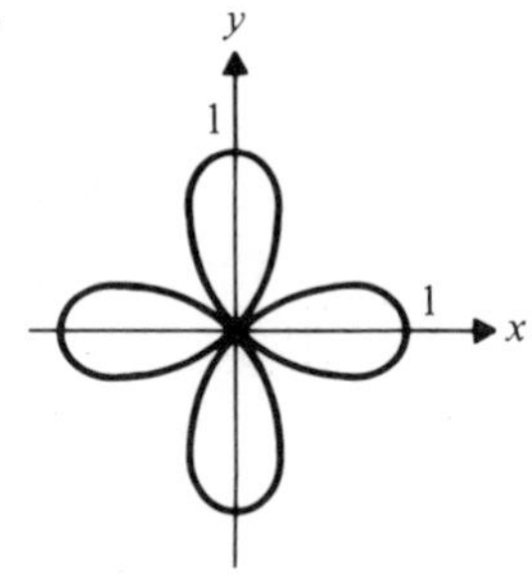

Review Exercises

2.

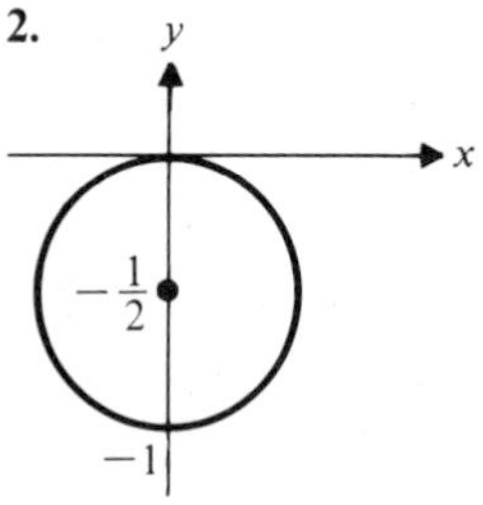

4.

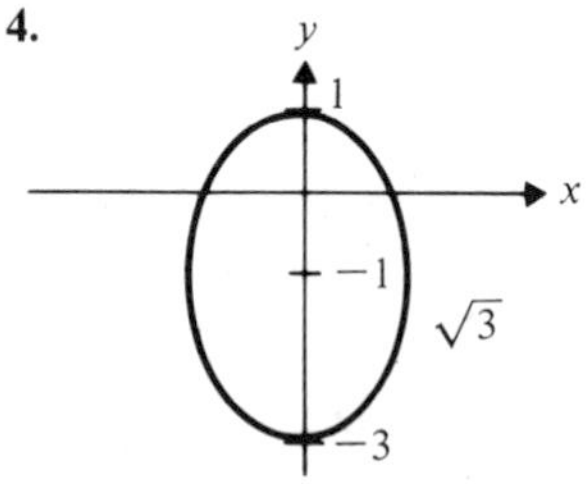

22.

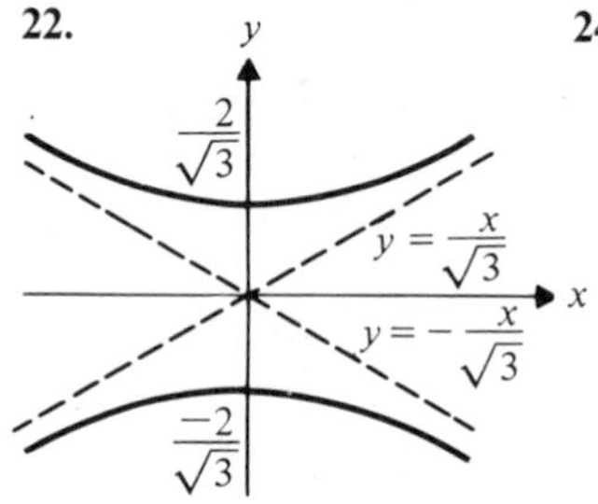

24.

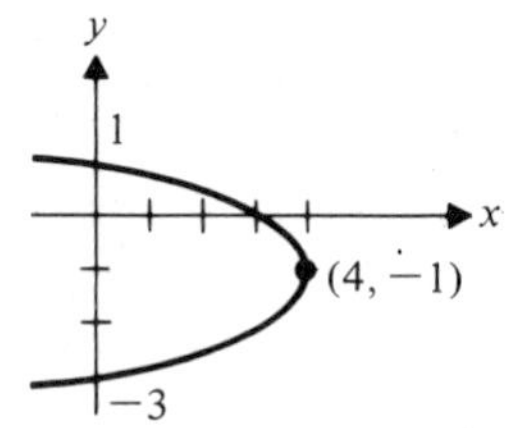

6.

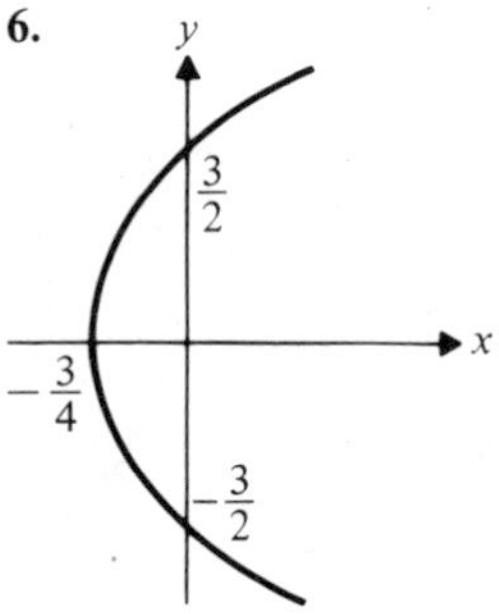

8.

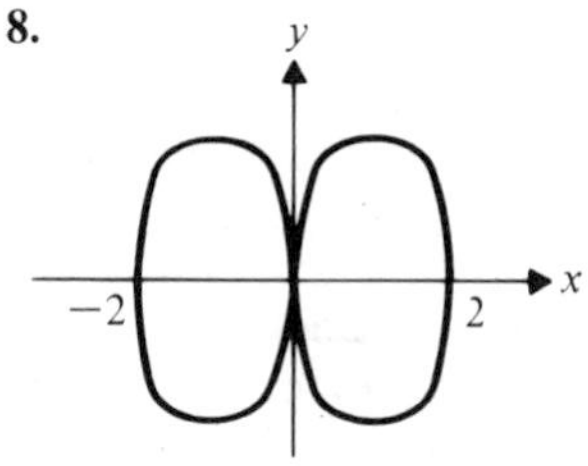

26.

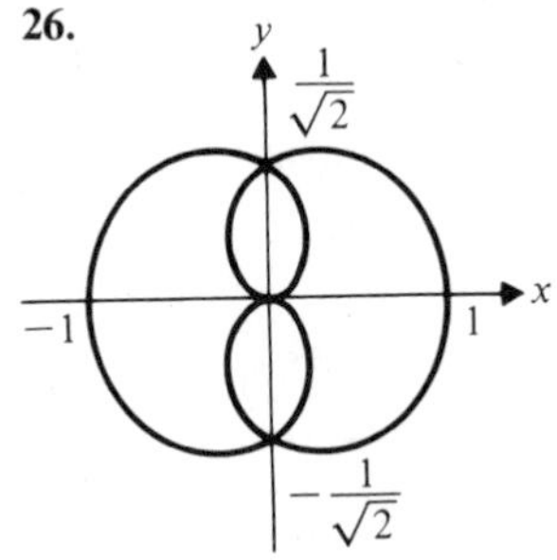

28.

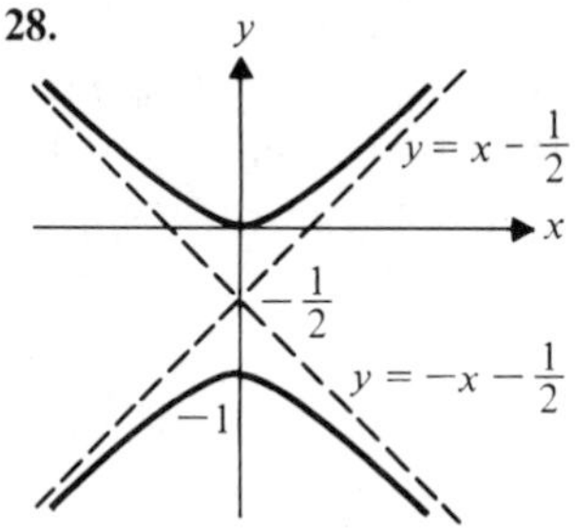

10.

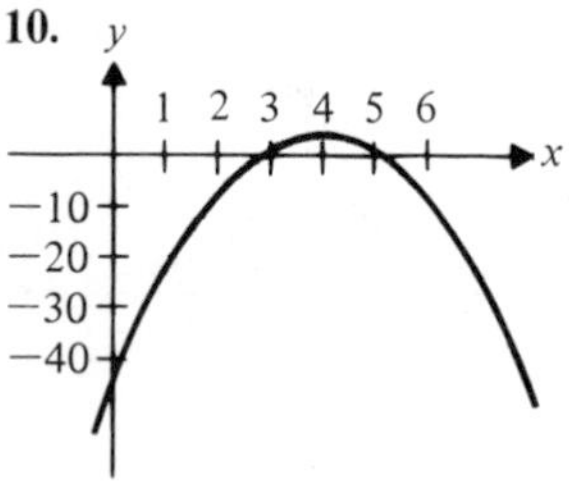

12.

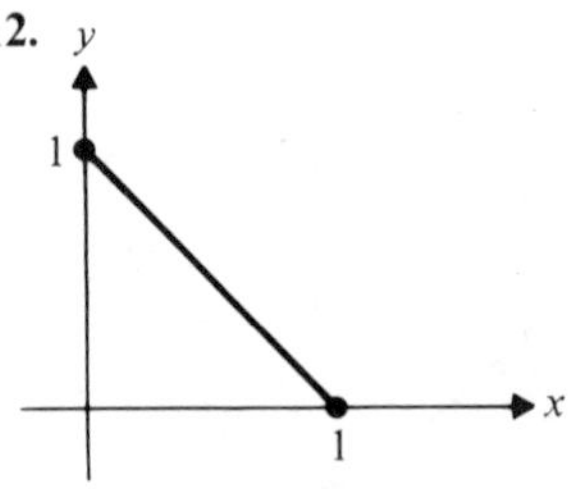

30.

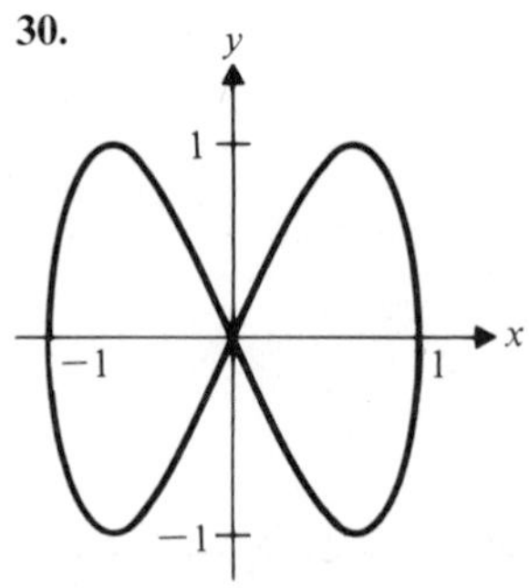

14.

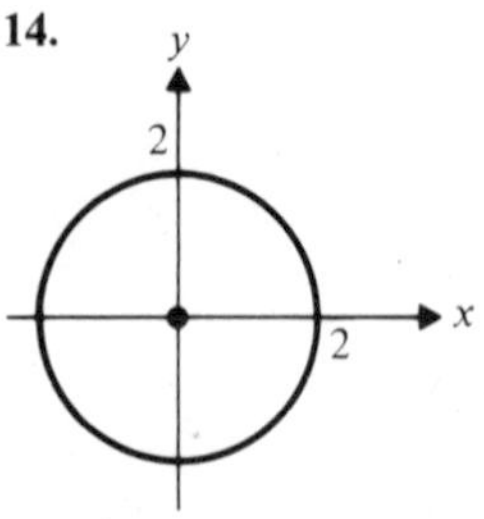

16.

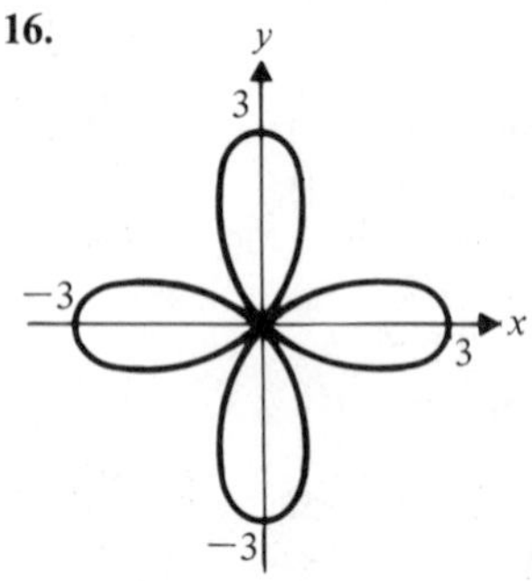

32. 8π **34.** $3\pi/8$ **36.** $(3 - 3t^2)/(2 + 3t^2)$, $-30t/(2 + 3t^2)^3$ **38.** 16 **40.** $\sqrt{2}(e^{\pi/2} - 1)$

CHAPTER 11

Exercises 11.1

2. Increasing, $V = 4$ **4.** Decreasing, $U = 9/16$, $V = 0$, $L = 0$ **6.** Not monotonic, $U = 1$, $V = -1$, $L = 0$
8. Decreasing, $U = 1/4$, $V = 0$, $L = 0$ **10.** Increasing, $U = \pi/2$, $V = \pi/4$, $L = \pi/2$ **12.** Not monotonic
14. Not monotonic, $U = 4$, $V = -4$, $L = 0$
16. Decreasing, $U = 3$, $V = 0$, $L = (1 + \sqrt{21})/2$
18. Decreasing, $U = e^{-1}$, $V = 0$, $L = 0$ **20.** $(3n + 1)/n^2$
22. $(-1)^{n+1} \ln(n + 1)/\sqrt{n + 1}$
24. $4P/3$, $16P/9$, $(4/3)^nP$, does not exist
26. (a) $40(0.99)^n$ m; (b) $4(0.981)^{-1/2}(0.99)^{n/2}$ s **28.** 1
30. Decreasing, $U = 1$, $V = 0$, $L = (3 - \sqrt{5})/2$
32. Decreasing, $U = 1$, $V = 0$, $L = 0$ **34.** 0 **36.** 4
40. $V = 1/2$, $U = 1$
42. $(\sqrt{3}P^2/36)[1 + (1/3) + (4/3^3) + (4^2/3^5) + \cdots + (4^{n-1}/3^{2n-1})]$

Exercises 11.2

2. 2/15 **4.** Diverges **6.** 21,609/2 **8.** $-1/3$
10. Diverges **12.** 13/99 **14.** 430,162/9999
16. Diverges **18.** 4 **20.** Diverges **22.** 1 **24.** 804 s
26. $(\sqrt{3}P^2/180)[8 - 3(4/9)^n]$, $2\sqrt{3}P^2/45$ **28.** 2/3
32. 600/11 minutes after 10:00 **34.** (c) $A_0e^{kT}/(e^{kT} - 1)$

Exercises 11.3

2. Converges **4.** Converges **6.** Converges
8. Converges **10.** Diverges **12.** Converges
14. Converges **16.** Diverges **18.** Converges
20. Converges **22.** Diverges **24.** Converges
26. Converges **28.** Diverges **30.** Diverges
34. Converges **38.** Converges **40.** $a \geq 2$

Exercises 11.4

2. Converges conditionally **4.** Converges absolutely
6. Diverges **8.** Converges absolutely
10. Converges conditionally **12.** Diverges
14. Converges conditionally **16.** Converges absolutely for all x

Exercises 11.5

2. -0.9470 **4.** -0.0127 **6.** 1.07 **8.** -0.7 **10.** 4

Exercises 11.6

2. $-1 < x < 1$ **4.** $-1/3 < x < 1/3$ **6.** $-4 < x < -2$
8. $7/2 < x < 9/2$ **10.** $-1 < x < 1$ **12.** $-1 \leq x < 1$
14. $-1/e \leq x \leq 1/e$ **16.** $x = 0$ **18.** $-1/3 < x < 1/3$
20. $-5^{1/3} < x < 5^{1/3}$ **22.** $-1 \leq x \leq 1$ **24.** 1
26. $4/(4 - x^3)$, $|x| < 4^{1/3}$ **28.** $(x - 1)/(10 - x)$, $-8 < x < 10$ **30.** (b) $-\infty < x < \infty$

Exercises 11.7

2. $\sum_{n=0}^{\infty} \frac{5^n}{n!} x^n$, $-\infty < x < \infty$ **6.** $(x + 1)/(1 - x)^3$, $|x| < 1$
8. $-\ln(1 - x)$, $-1 \leq x < 1$
10. $x^{-2} \ln(1 - x) + 1/(x - x^2)$, $|x| < 1$

Exercises 11.8

2. $\sum_{n=0}^{\infty} \frac{(-1)^n}{4^{n+1}} x^{2n}$, $|x| < 2$ **4.** $\sum_{n=0}^{\infty} \frac{(-1)^n}{(2n)!} x^{4n}$, $-\infty < x < \infty$
6. $\sum_{n=0}^{\infty} \frac{5^n}{n!} x^n$, $-\infty < x < \infty$
8. $\sum_{n=0}^{\infty} \frac{1}{(2n + 1)!} x^{2n+1}$, $-\infty < x < \infty$
10. $1 + 9x/2 + 27x^2/8 + \sum_{n=3}^{\infty} \frac{(-1)^n(2n - 5)!3^{n+1}}{2^{2n-3}n!(n - 3)!} x^n$, $-1/3 \leq x \leq 1/3$ **12.** $x^4 + 3x^2 - 2x + 1$
14. $\sum_{n=0}^{\infty} \frac{(-1)^n(n + 1)(n + 2)}{2^{n+4}} x^n$, $|x| < 2$
16. $\sum_{n=0}^{\infty} \frac{e^3}{n!} (x - 3)^n$, $-\infty < x < \infty$
18. $\sqrt{3}[1 + x/6] + \sum_{n=2}^{\infty} \frac{4\sqrt{3}(-1)^{n-1}(2n - 3)!}{12^n n!(n - 2)!} x^n$, $|x| \leq 3$
20. $x - x^2/3 - \sum_{n=3}^{\infty} \frac{(2)(5)\cdots(3n - 7)}{3^{n-1}(n - 1)!} x^n$, $|x| \leq 1$
22. $1 + x^2/2 + 5x^4/24 + 61x^6/720$
24. $1 + \sum_{n=1}^{\infty} \frac{(-1)^n 2^{2n-1}}{(2n)!} x^{2n}$, $-\infty < x < \infty$
26. $x^2 + \sum_{n=1}^{\infty} \frac{(2n)!}{(2n + 1)2^{2n}(n!)^2} x^{4n+2}$, $|x| < 1$
28. $\sqrt{2} \sum_{n=0}^{\infty} \frac{1}{2^n(2n + 1)} x^{2n+1}$, $|x| < \sqrt{2}$
34. $(2/\sqrt{\pi}) \sum_{n=0}^{\infty} \frac{(-1)^n}{n!(2n + 1)} x^{2n+1}$, $-\infty < x < \infty$
38. 1, μ, $(3\mu^2 - 1)/2$, $(5\mu^3 - 3\mu)/2$

Exercises 11.9

2. 8.5×10^{-13} **4.** 0.00625 **6.** 0.497
8. 0.133 **10.** $a_0 + a_1 \sum_{n=1}^{\infty} \frac{(-1)^{n+1}}{n!} x^n = C + De^{-x}$
12. $a_0 \sum_{n=0}^{\infty} \frac{(-1)^n}{(2n)!} x^n$ **14.** $a_1 \sum_{n=1}^{\infty} \frac{(-1)^{n+1}}{n!(n - 1)!} x^n$
16. $x/3 - x^3/(3^3 3!) + x^5/(3^5 5!) - x^7/(3^7 7!)$
18. $-x - x^2/2 - x^3/3$ **20.** 2.44

Review Exercises

2. Decreasing, $U = 1$, $V = 0$, $L = 1/\sqrt{3}$
4. Increasing, $V = 7$, $U = 100$, $L = (31 + \sqrt{53})/2$

6. $|k| \leq 1$ **10.** Converges **12.** Converges
14. Converges conditionally **16.** Diverges
18. Converges **20.** Converges **22.** Diverges
24. Converges **26.** Converges
28. Converges conditionally **30.** $-2 \leq x \leq 2$
32. $-\infty < x < \infty$ **34.** $-4 < x < -2$
36. $-2^{-1/3} \leq x < 2^{-1/3}$ **38.** $e^5 \sum_{n=0}^{\infty} \frac{1}{n!} x^n$, $-\infty < x < \infty$
40. $\sum_{n=2}^{\infty} \frac{(-1)^n 2^{n-1}}{n-1} x^n$, $-1/2 < x \leq 1/2$
42. $(1/2) \sum_{n=0}^{\infty} (-1)^{n+1}[1 - 1/3^n]x^n$, $|x| < 1$
44. $x + \sum_{n=2}^{\infty} \frac{(-1)^n}{n(n-1)} x^n$, $-1 < x \leq 1$ **46.** $\sum_{n=0}^{18} \frac{(-1)^n}{n!} x^{2n}$
48. $\sum_{n=0}^{\infty} \frac{(-1)^n}{2^{2n}(2n)!} x^{2n} + \sum_{n=0}^{\infty} \frac{(-1)^n}{2^{2n+1}(2n+1)!} x^{2n+1}$
50. (a) $1/p$; (b) 6

CHAPTER 12

Exercises 12.1

2. $\sqrt{33}$ **4.** (0, 0, 0), (2, 0, 0), (0, 2, 0), (0, 0, 2), (2, 2, 0), (2, 0, 2), (0, 2, 2), (2, 2, 2) **6.** $\sqrt{29}$, 5, $2\sqrt{5}$, $\sqrt{13}$
8. 5, 3, 4, 5 **12.** $10x + 2y + 2z = 5$, plane
14. $(\sqrt{2}, \sqrt{2}, 5)$, $(\sqrt{2} \pm 1/4, \sqrt{2} \pm 1/4, 5 - \sqrt{7}/4)$, $(\sqrt{2} \pm 1/4, \sqrt{2} \pm 1/4, 9/2 - \sqrt{7}/4)$
16. (a) $(2, 1/2, -7/2)$; (b) $(5, 5, -5)$

Exercises 12.2

2.
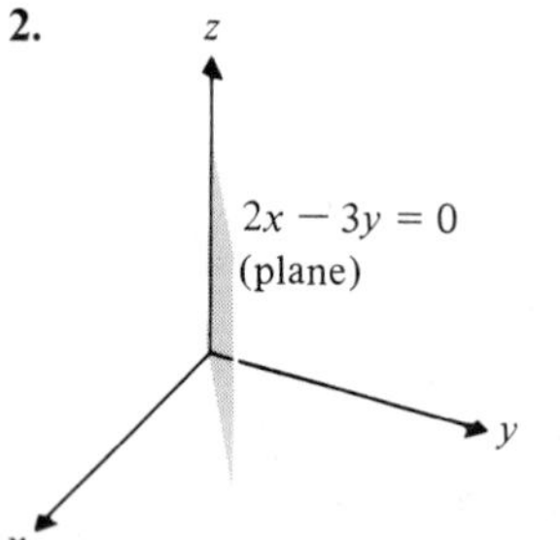

4.
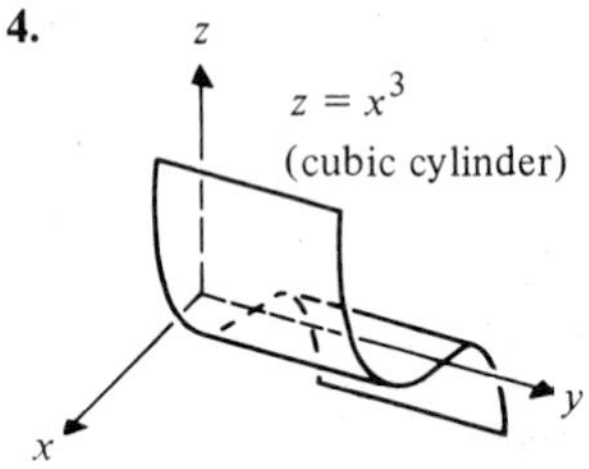

6.
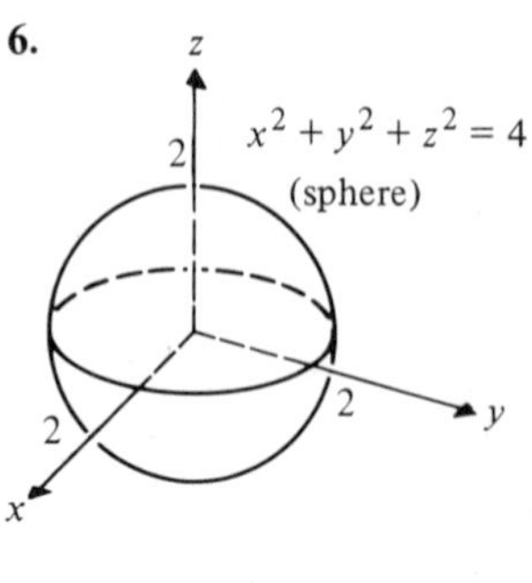

8.
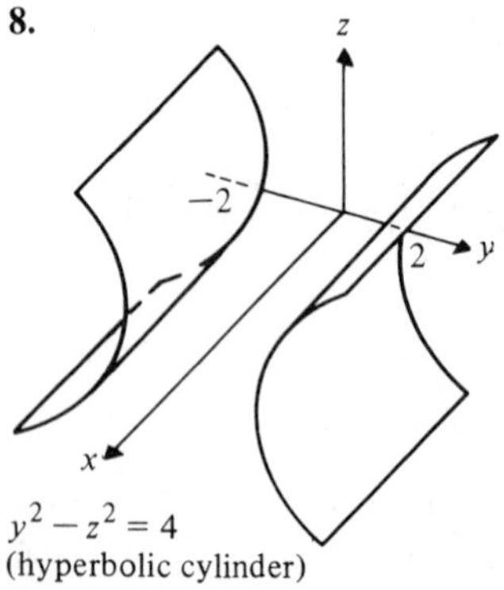

10.
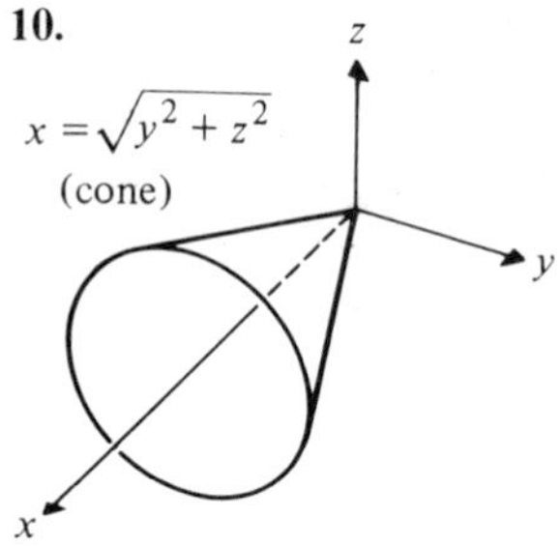

12.
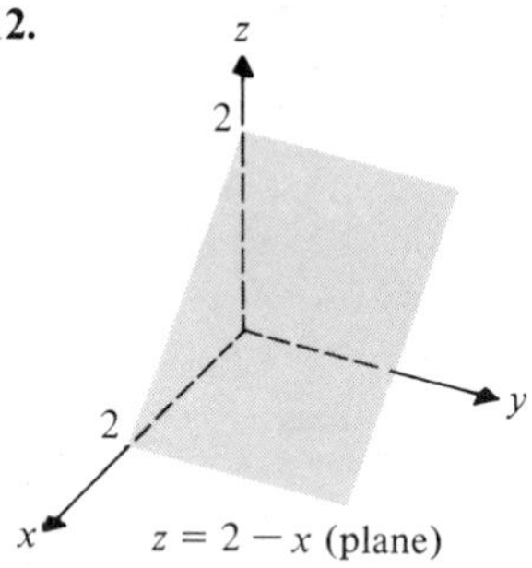

14.
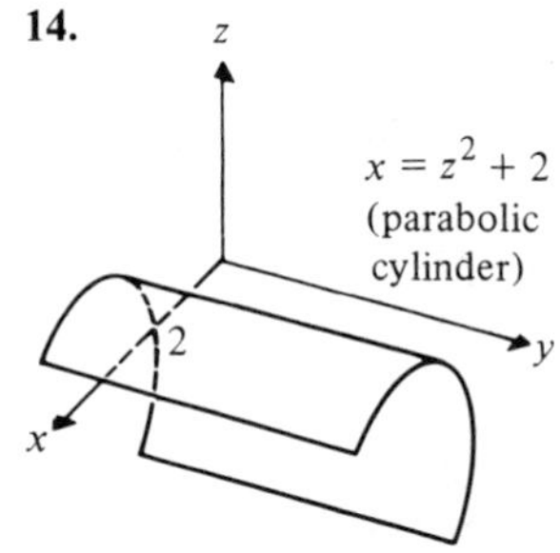

16.
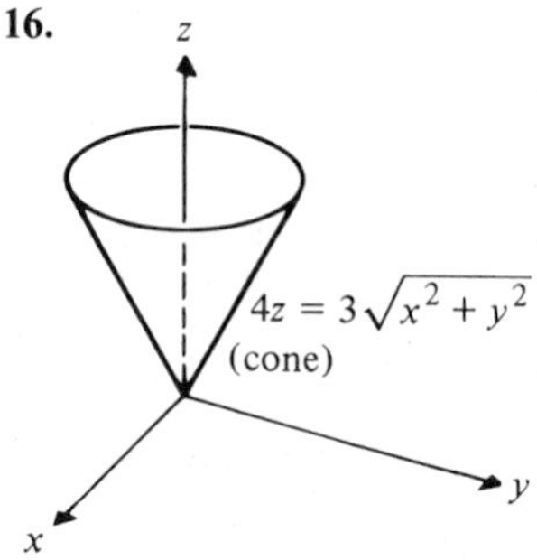

18.
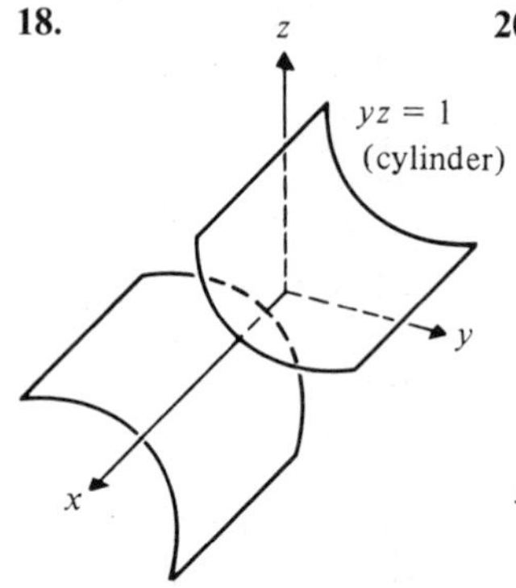

20.
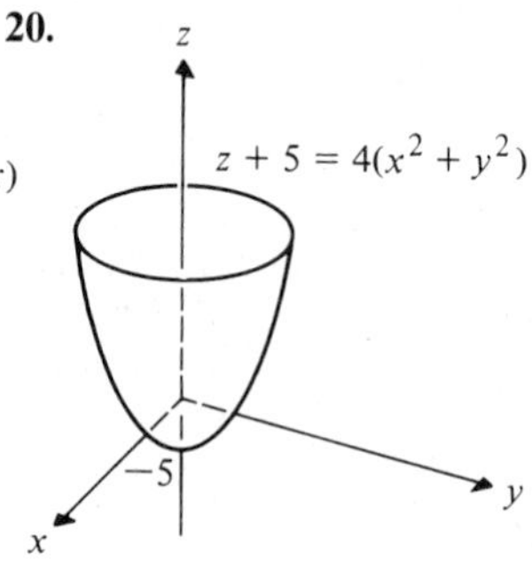

22.
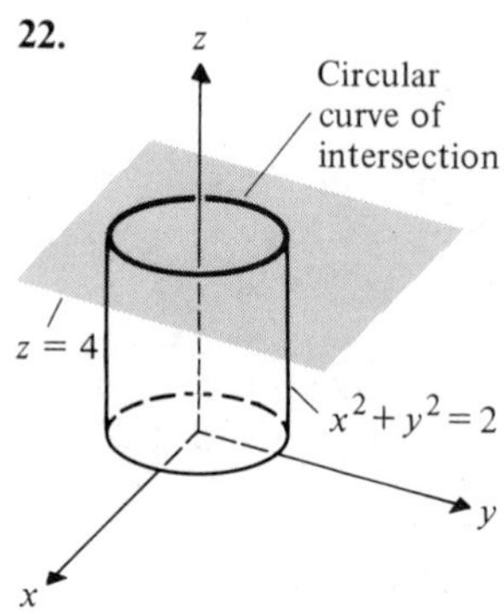

24.
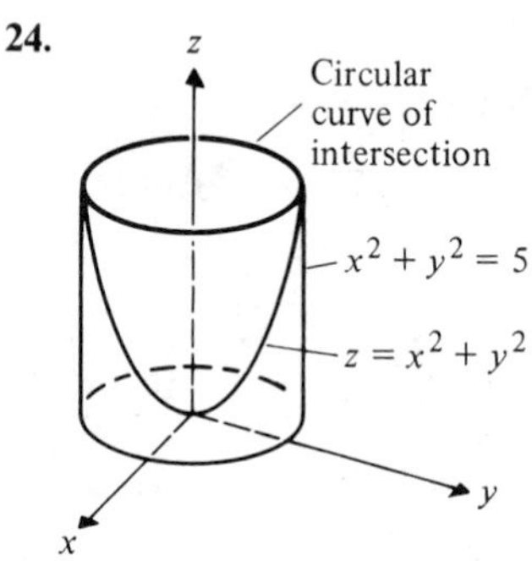

26.

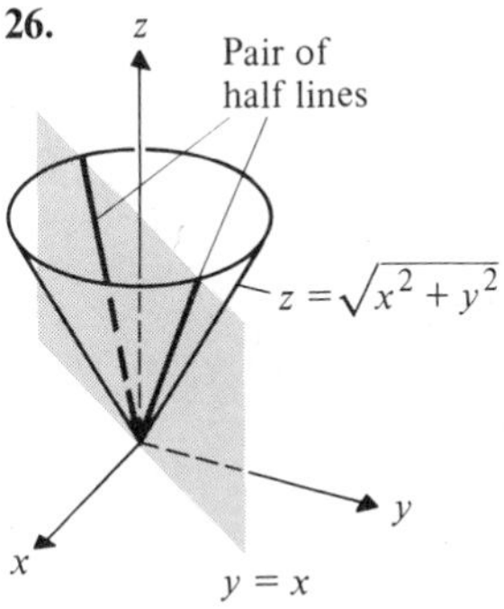

28.

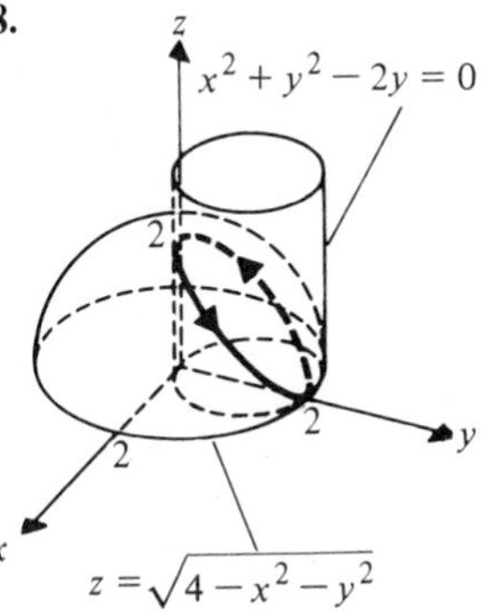

30.

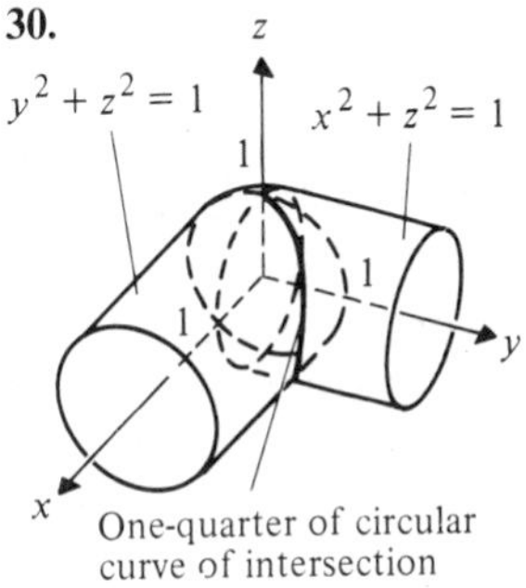

32.

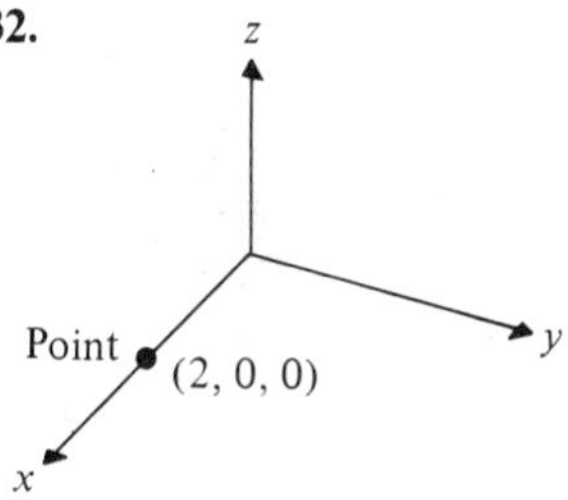

34.

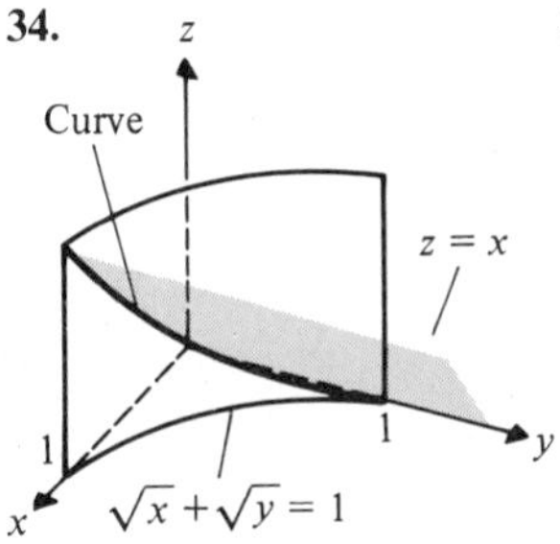

36.

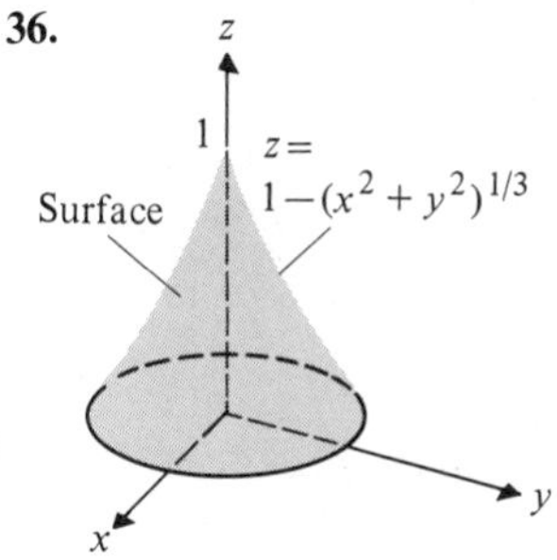

38.

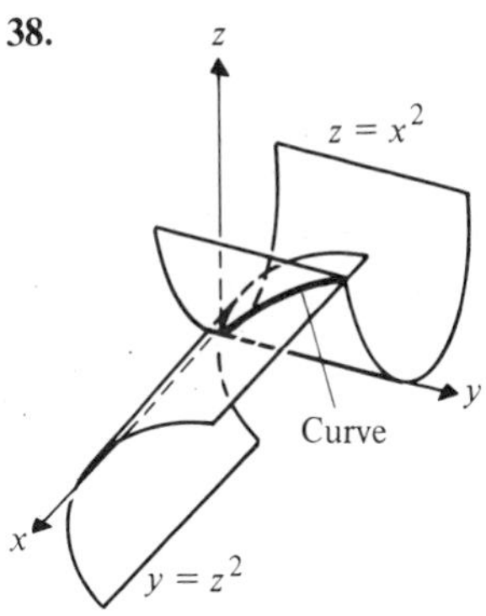

40.

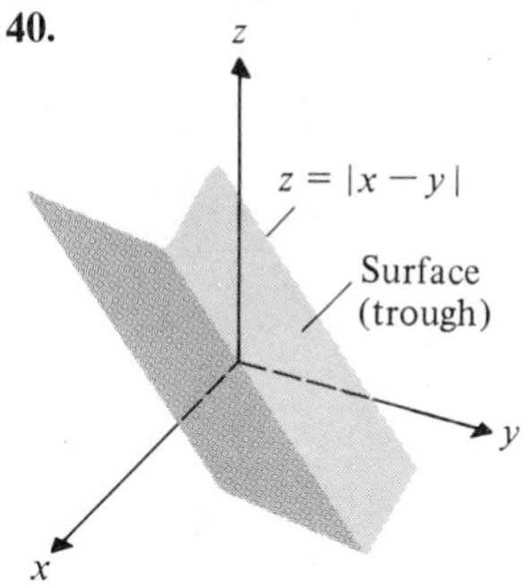

Exercises 12.3

2.

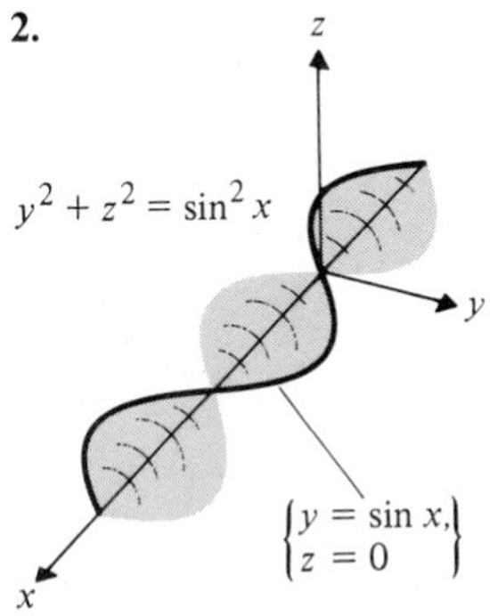

4.

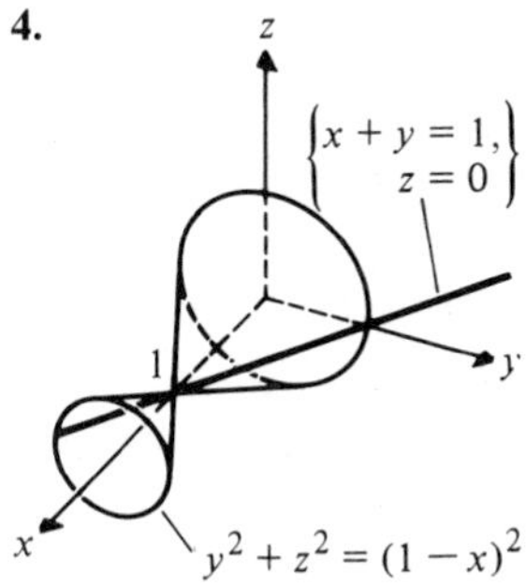

6.

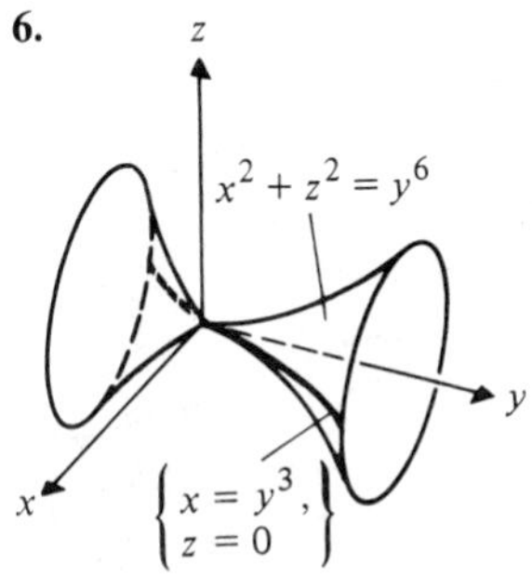

8.

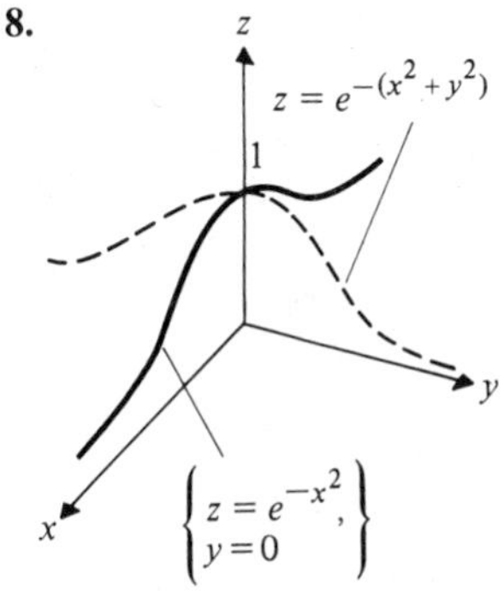

10.

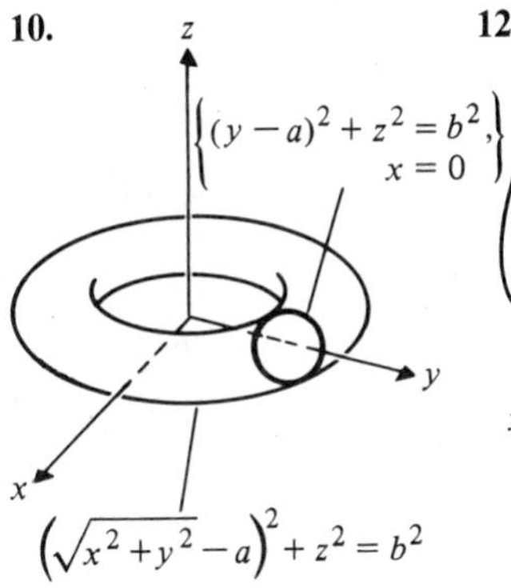

12.

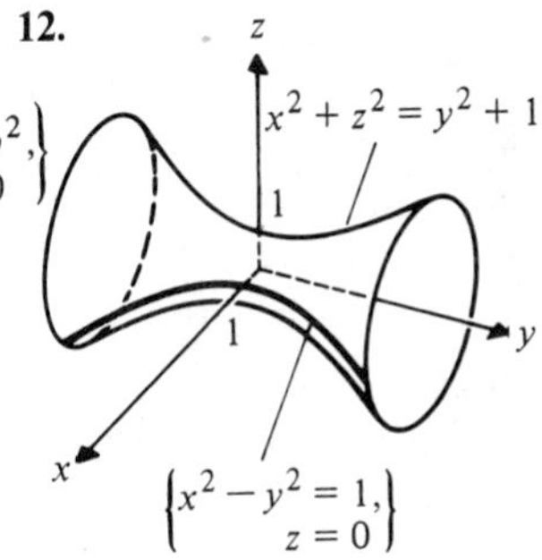

14.

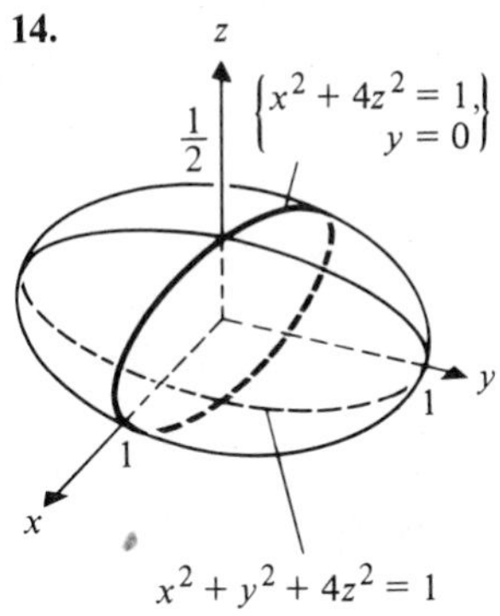

16.

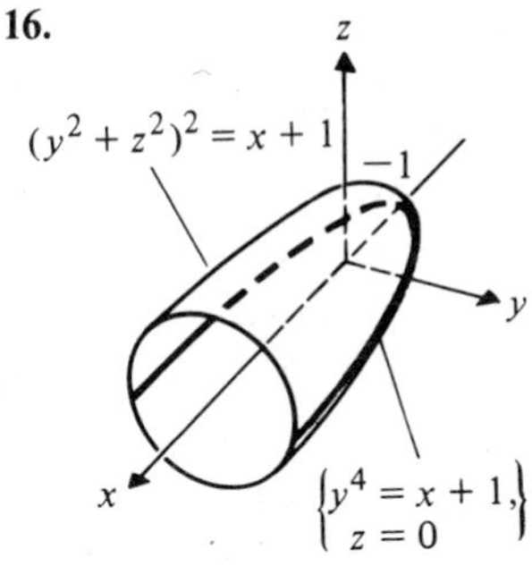

18.

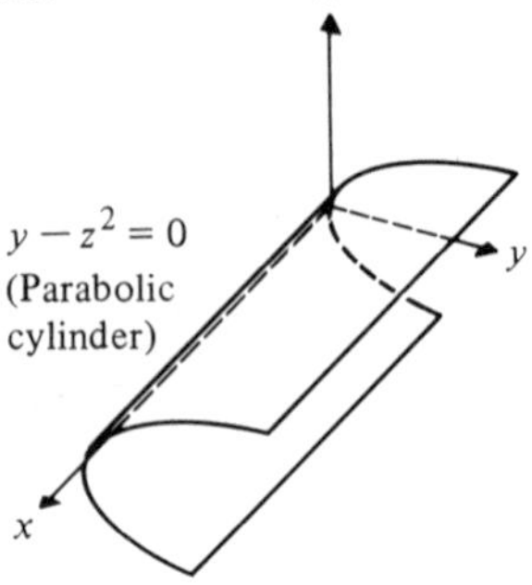

20.

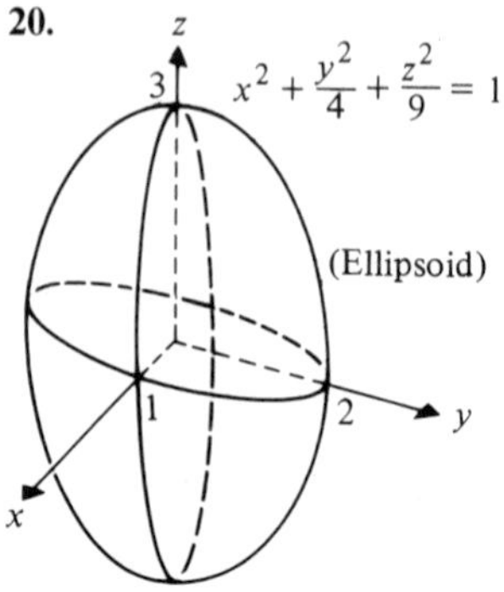

22.

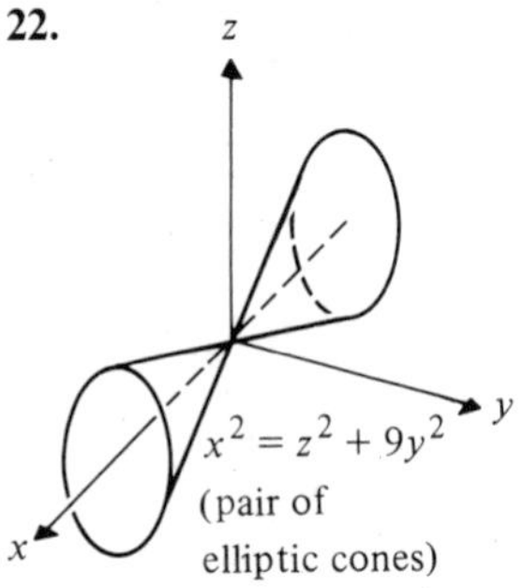

24.

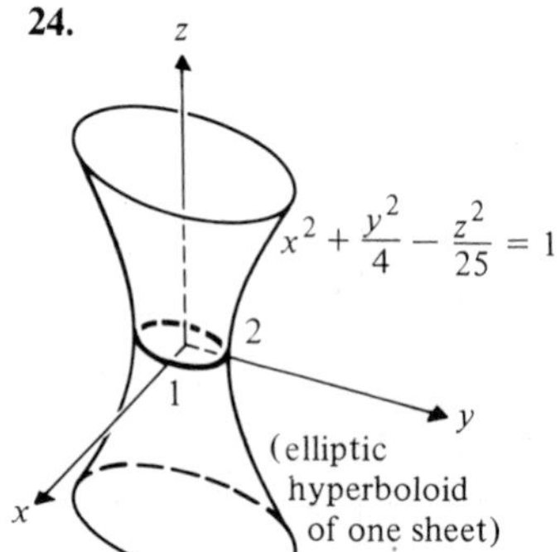

Exercises 12.4

2. $(2, 6, 8)$ **4.** $(-1/\sqrt{5}, 0, 2/\sqrt{5})$
6. $(-4\sqrt{5} - 8, -6, 8\sqrt{5} + 4)$
8. $(-6\sqrt{46} - 4\sqrt{5}, -12\sqrt{5}, 12\sqrt{46} - 24\sqrt{5})$
10. $(-6/\sqrt{17}, -3/\sqrt{17}, 6/\sqrt{17})$ **12.** $3\hat{\mathbf{i}} - 2\hat{\mathbf{j}}$
14. $(2/\sqrt{5} - 1/\sqrt{10})\hat{\mathbf{i}} + (1/\sqrt{5} + 3/\sqrt{10})\hat{\mathbf{j}}$ **16.** $(5, 0, 0)$
18. $(0, 3/\sqrt{5}, 6/\sqrt{5})$ **20.** $(1, 1, -3/2)$
22. $(5/4, 5\sqrt{2}/4, 5/4)$ **24.** $(2/\sqrt{3}, 2/\sqrt{3}, 2/\sqrt{3})$
30. $(3/\sqrt{17}, 12/\sqrt{17})$ **32.** $5G(-3/\sqrt{10} + 30\sqrt{11}/121, -10\sqrt{11}/121, -1/\sqrt{10} + 10\sqrt{11}/121)$ N
34. Linearly dependent **36.** Linearly dependent

Exercises 12.5

2. $(-10, 15, -5)$ **4.** $4/\sqrt{14}$ **6.** -178 **8.** 1 **10.** 0
12. Yes **14.** Yes **16.** 0.68 **18.** 2.20 **20.** π
22. $2x + y - 2z + 5 = 0$ **24.** $6/\sqrt{13}$ **26.** $4/\sqrt{6}$
28. $23\sqrt{14}/28$ **34.** (a) 1.30, 1.01, 2.50; (b) $\pi/2$, 1.25, 2.82; (c) 1.73, 1.89, 0.36 **36.** $3/\sqrt{2}, 1/\sqrt{2}$ **38.** $\sqrt{5}, -\sqrt{14}/2, -\sqrt{70}/2$ **40.** $-3/10, 11/20$
42. $x - 2y + 3z = 6 \pm 2\sqrt{14}$ **44.** $k[1 + l(1 - \sqrt{5})]/2$ J
46. $(\sqrt{2} - 1)GMm/(\sqrt{2}R)$

Exercises 12.6

2. $(48, 24, -42)$ **4.** $(2\sqrt{247})^{-1}(17, -23, -7)$
6. $(-41, 11, 28)$ **8.** $(17, -23, -7)$ **10.** $(-13, 19, 5)$
12. $\lambda(9, 0, 1)$ **14.** $\lambda(3, -1, 17)$
18. $(x, y, z) = (1, -1, 3) + t(2, 4, -3)$; $x = 1 + 2t, y = -1 + 4t, z = 3 - 3t$; $(x - 1)/2 = (y + 1)/4 = (z - 3)/(-3)$
20. $(x, y, z) = (2, -3, 4) + t(3, 5, -5)$; $x = 2 + 3t, y = -3 + 5t, z = 4 - 5t$; $(x - 2)/3 = (y + 3)/5 = (z - 4)/(-5)$
22. $(x, y, z) = (1, 3, 4) + t(0, 0, 1)$; $x = 1, y = 3, z = 4 + t$
24. $(x, y, z) = (2, 0, 3) + t(1, 0, -2)$; $x = 2 + t, y = 0, z = 3 - 2t$; $x - 2 = (z - 3)/(-2), y = 0$
26. $(x, y, z) = (0, -5, 30) + t(1, 2, -11)$; $x = t, y = -5 + 2t, z = 30 - 11t$; $x = (y + 5)/2 = (z - 30)/(-11)$
28. $x + 3y - 3z = 4$ **30.** $10x + 4y - 3z = 3$
32. $8x + 13y + 9z = 18$ **34.** $7x + 6y - 5z + 34 = 0$
36. No **40.** $x = \sqrt{2} \pm 1/4, y = \sqrt{2} \pm 1/4, z = 9/2 - \sqrt{7}/4$, $\sqrt{7}x - z = \sqrt{14} - 5$, $\sqrt{7}x + z = \sqrt{14} + 5$, $\sqrt{7}y - z = \sqrt{14} - 5$, $\sqrt{7}y + z = \sqrt{14} + 5$ **42.** $\sqrt{154}$ **44.** $(-10, 5, -2)$
46. $(-5, 1, 4)$

Exercises 12.7

2. $-\infty < t < \infty$ **4.** $t > -4$ **6.** $\hat{\mathbf{i}} - 2t\hat{\mathbf{j}} + 2\hat{\mathbf{k}}$
8. $2t(4t^2 - 3)\hat{\mathbf{i}} + t^2(9 - 10t^2)\hat{\mathbf{j}} + 4t(4t^2 - 3)\hat{\mathbf{k}}$
10. $3t^2(4 - 5t^2)\hat{\mathbf{i}} + 4t(1 - 3t^2)\hat{\mathbf{j}} - 3t^2\hat{\mathbf{k}}$
12. $3\hat{\mathbf{i}} - 2(3t + 4)\hat{\mathbf{j}} + 6(1 + 4t)\hat{\mathbf{k}}$
14. $9t^2\hat{\mathbf{i}} + (6t - 20t^3)\hat{\mathbf{j}} + (30t^4 - 21t^2 + 6)\hat{\mathbf{k}}$
16. $(-15t^4 + 12t^2)\hat{\mathbf{i}} + (4t - 12t^3)\hat{\mathbf{j}} - 3t^2\hat{\mathbf{k}}$
18. $(3t^4/2 - 4t^2)\hat{\mathbf{i}} + (17t^3/3 - 12t^5/5)\hat{\mathbf{j}} + (3t^6 - 27t^4/4 + t^2)\hat{\mathbf{k}} + \mathbf{C}$
20. $(-14t^4 - 6t^3 - 42t^2 + 4t)\hat{\mathbf{i}} + (28t^5 + 84t^3 - 6t^2)\hat{\mathbf{j}} - (126t^4 + 42t^6 + 2t)\hat{\mathbf{k}}$

Exercises 12.8

2.

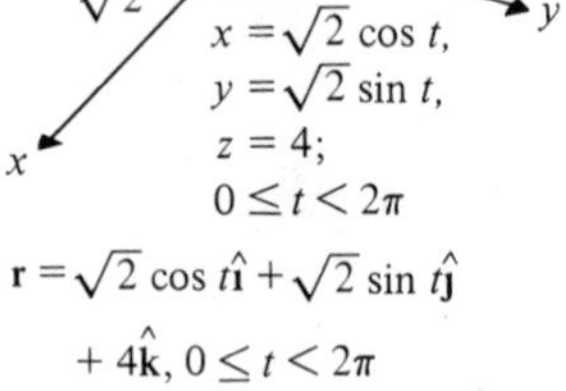

$\mathbf{r} = \sqrt{2}\cos t\hat{\mathbf{i}} + \sqrt{2}\sin t\hat{\mathbf{j}} + 4\hat{\mathbf{k}}, 0 \le t < 2\pi$

4.

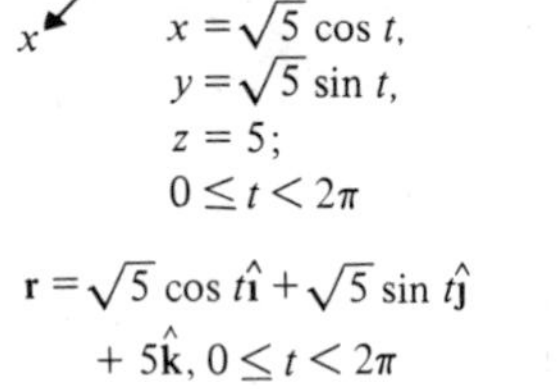

$\mathbf{r} = \sqrt{5}\cos t\hat{\mathbf{i}} + \sqrt{5}\sin t\hat{\mathbf{j}} + 5\hat{\mathbf{k}}, 0 \le t < 2\pi$

6.

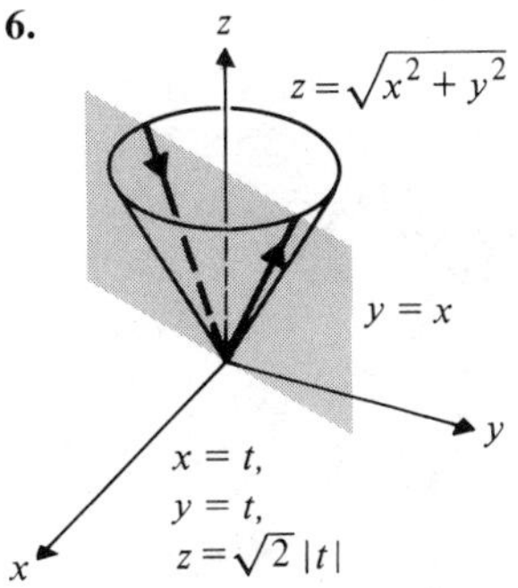

$\mathbf{r} = t\hat{\mathbf{i}} + t\hat{\mathbf{j}} + \sqrt{2}\,|t|\hat{\mathbf{k}}$

8.

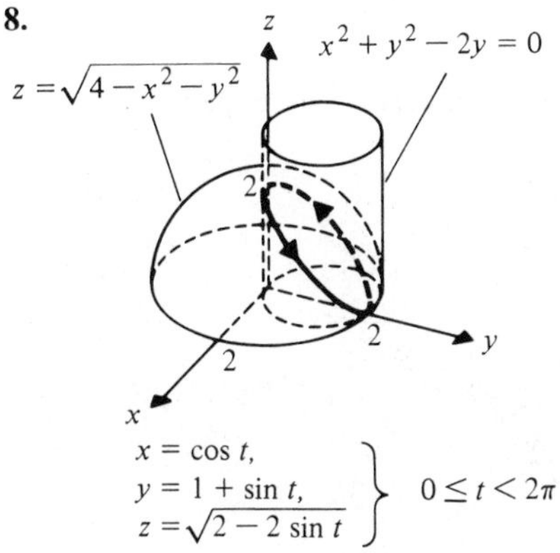

$\mathbf{r} = \cos t\hat{\mathbf{i}} + (1 + \sin t)\hat{\mathbf{j}}$

$+ \sqrt{2 - 2\sin t}\,\hat{\mathbf{k}}, 0 \le t < 2\pi$

10.

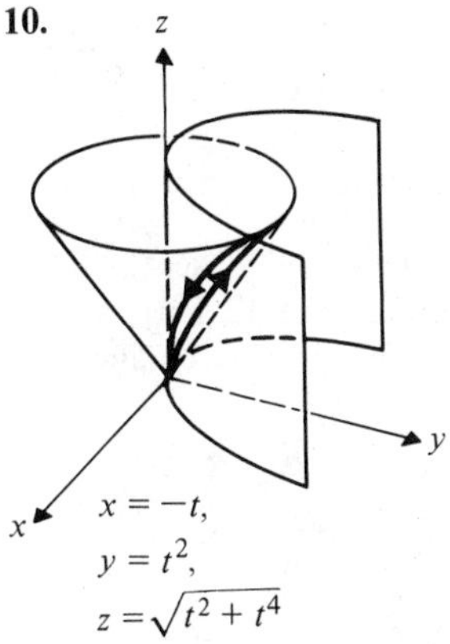

$\mathbf{r} = -t\hat{\mathbf{i}} + t^2\hat{\mathbf{j}} + \sqrt{t^2 + t^4}\,\hat{\mathbf{k}}$

12.

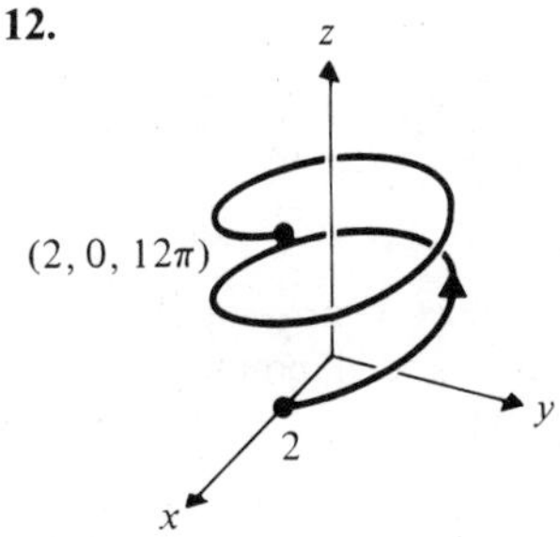

14.

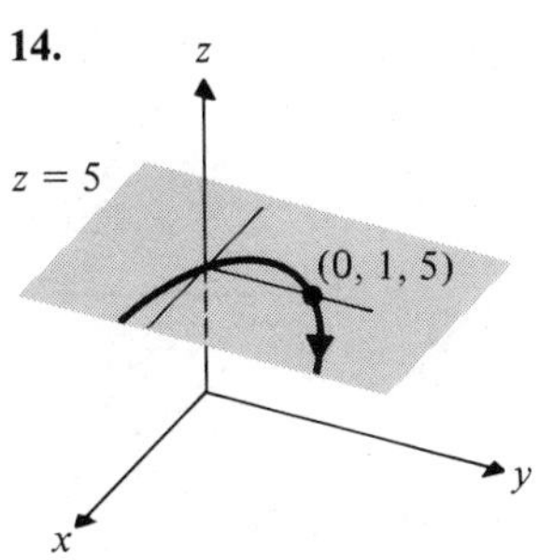

Exercises 12.9

2. $\mathbf{r} = t\hat{\mathbf{i}} + t^2\hat{\mathbf{j}} + t^3\hat{\mathbf{k}}$; $\hat{\mathbf{T}} = (\hat{\mathbf{i}} + 2t\hat{\mathbf{j}} + 3t^2\hat{\mathbf{k}})/\sqrt{1 + 4t^2 + 9t^4}$
4. $\mathbf{r} = -t\hat{\mathbf{i}} + (5 + t)\hat{\mathbf{j}} + (t^2 - t - 5)\hat{\mathbf{k}}, -5 \le t \le 0$;
$\hat{\mathbf{T}} = [-\hat{\mathbf{i}} + \hat{\mathbf{j}} + (2t - 1)\hat{\mathbf{k}}]/\sqrt{4t^2 - 4t + 3}$
6. $(-2\hat{\mathbf{i}} + 3\hat{\mathbf{j}} + \hat{\mathbf{k}})/\sqrt{14}$ **8.** $(\hat{\mathbf{i}} - \hat{\mathbf{j}})/\sqrt{2}$ **10.** $-\hat{\mathbf{j}}$
12. $\sqrt{42}$ **14.** $(616\sqrt{616} - 157\sqrt{157})/459$
16. (a) $(0, 0, 0)$; (b) $2\hat{\mathbf{i}} + 2\hat{\mathbf{k}}$

Exercises 12.10

2. $[-(2t + 9t^3)\hat{\mathbf{i}} + (1 - 9t^4)\hat{\mathbf{j}} + (3t + 6t^3)\hat{\mathbf{k}}]/\sqrt{1 + 13t^2 + 54t^4 + 117t^6 + 81t^8}$; $(3t^2\hat{\mathbf{i}} - 3t\hat{\mathbf{j}} + \hat{\mathbf{k}})/\sqrt{1 + 9t^4 + 9t^2}$ **4.** $[(2t - 1)\hat{\mathbf{i}} + (1 - 2t)\hat{\mathbf{j}} + 2\hat{\mathbf{k}}]/\sqrt{8t^2 - 8t + 6}$; $(\hat{\mathbf{i}} + \hat{\mathbf{j}})/\sqrt{2}$ **6.** $-(5\hat{\mathbf{i}} + 3\hat{\mathbf{j}} + \hat{\mathbf{k}})/\sqrt{35}$; $(-\hat{\mathbf{j}} + 3\hat{\mathbf{k}})/\sqrt{10}$ **8.** $-(\hat{\mathbf{i}} + \hat{\mathbf{j}})/\sqrt{2}$; $-\hat{\mathbf{k}}$
10. $-(\hat{\mathbf{i}} + \hat{\mathbf{k}})/\sqrt{2}$; $(\hat{\mathbf{i}} - \hat{\mathbf{k}})/\sqrt{2}$ **12.** $\kappa = 0$, ρ undefined
14. $\kappa = 2e^t\sqrt{1 + e^{2t}}/(1 + 2e^{2t})^{3/2}$,
$\rho = (1 + 2e^{2t})^{3/2}/(2e^t\sqrt{1 + e^{2t}})$
16. $\kappa = 1/[\sqrt{2}(1 + \cos^2 t)^{3/2}]$, $\rho = \sqrt{2}(1 + \cos^2 t)^{3/2}$
18. $\kappa = \sqrt{1 + 36t^4 + 16t^6}/[2(1 + t^2 + 4t^6)^{3/2}]$,
$\rho = 2(1 + t^2 + 4t^6)^{3/2}/\sqrt{1 + 36t^4 + 16t^6}$ **22.** 0
24. (a) $-\sin t\hat{\mathbf{i}} + \cos t\hat{\mathbf{j}}$; $-(\cos t\hat{\mathbf{i}} + \sin t\hat{\mathbf{j}})$; $\hat{\mathbf{k}}$;
(b) $2\sin 2t(\sin t - \cos t)$, $-4(\cos^3 t + \sin^3 t)$;
(c) $2\sin 2t(\sin t - \cos t)\hat{\mathbf{T}} - 4(\cos^3 t + \sin^3 t)\hat{\mathbf{N}}$
26. $(-\sin t\hat{\mathbf{i}} + \cos t\hat{\mathbf{j}} + \hat{\mathbf{k}})/\sqrt{2}$; $-(\cos t\hat{\mathbf{i}} + \sin t\hat{\mathbf{j}})$;
$(\sin t\hat{\mathbf{i}} - \cos t\hat{\mathbf{j}} + \hat{\mathbf{k}})/\sqrt{2}$; $(1 - \cos t\sin t + \cos^2 t\sin^2 t)\hat{\mathbf{T}}/\sqrt{2} - \cos t\,(\cos t + \sin^3 t)\hat{\mathbf{N}} +$
$(-\cos^2 t\sin^2 t + \cos t\sin t + 1)\hat{\mathbf{B}}/\sqrt{2}$

Exercises 12.11

2. $[(t^2 - 1)\hat{\mathbf{i}} + (t^2 + 1)\hat{\mathbf{j}}]/t^2$; $\sqrt{2 + 2t^4}/t^2$; $2(\hat{\mathbf{i}} - \hat{\mathbf{j}})/t^3$
4. $2t\hat{\mathbf{i}} + 2e^t(t + 1)\hat{\mathbf{j}} - 2\hat{\mathbf{k}}/t^3$; $2\sqrt{t^2 + e^{2t}(t + 1)^2 + 1/t^6}$;
$2\hat{\mathbf{i}} + 2e^t(t + 2)\hat{\mathbf{j}} + 6\hat{\mathbf{k}}/t^4$
6. $(t^4/4 + 1)\hat{\mathbf{i}} + (t^3 + 3t^2 + 12)\hat{\mathbf{j}}/6 - (t^5/5 + 1)\hat{\mathbf{k}}$
8. $4t/\sqrt{1 + 4t^2}$; $2/\sqrt{1 + 4t^2}$ **12.** (a) 0; (b) $56t^6$
14. $150x\hat{\mathbf{j}}$ **18.** (b) 7.79 km/s
20. $365\hat{\mathbf{i}} + 325\sqrt{3}\,\hat{\mathbf{j}}$ km/hr; 670.9 km/hr **22.** $(3/2, 3/2, 2)$
24. $(3, 0)$ **30.** Yes
34. (b) $(S/R)[(R - b\cos\theta)\hat{\mathbf{i}} + b\sin\theta\hat{\mathbf{j}}]$;
$(S/R)\sqrt{R^2 - 2bR\cos\theta + b^2}$; $(bS^2/R^2)(\sin\theta\hat{\mathbf{i}} + \cos\theta\hat{\mathbf{j}})$;
(c) $(bS^2/R^2)|R\cos\theta - b|/\sqrt{R^2 - 2bR\cos\theta + b^2}$;
$(bS^2/R)\sin\theta/\sqrt{R^2 - 2bR\cos\theta + b^2}$

Review Exercises

2. -8 **4.** $(39, -33, -30)$ **6.** -2 **8.** $(-16, -24, -4)$
10. $(67/7, 21, -65/7)$

12.

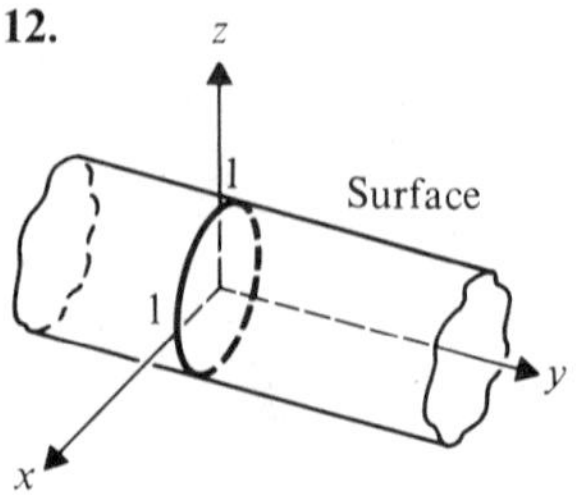

14.

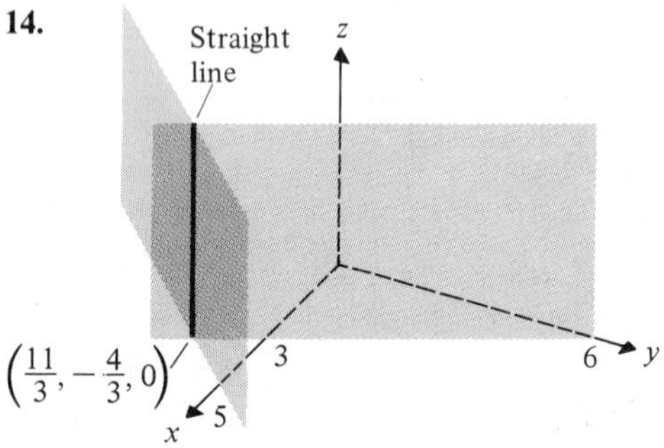

16. No points

18.

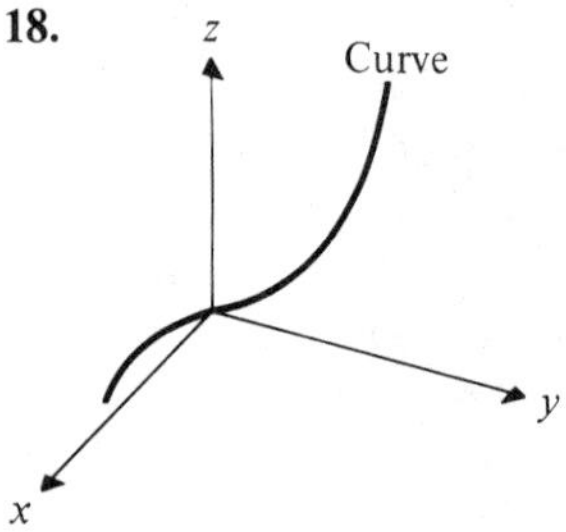

20.

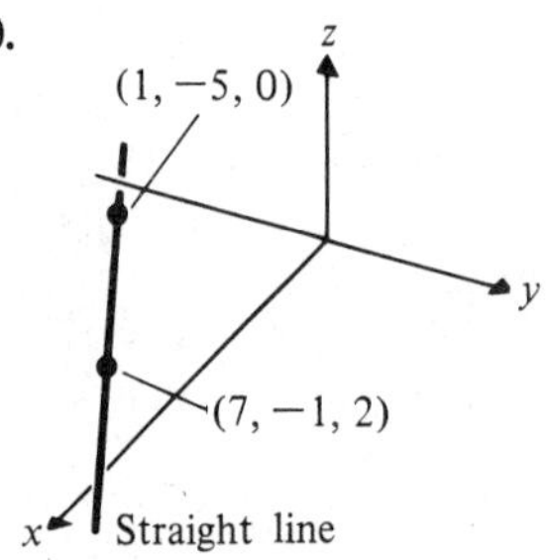

22.

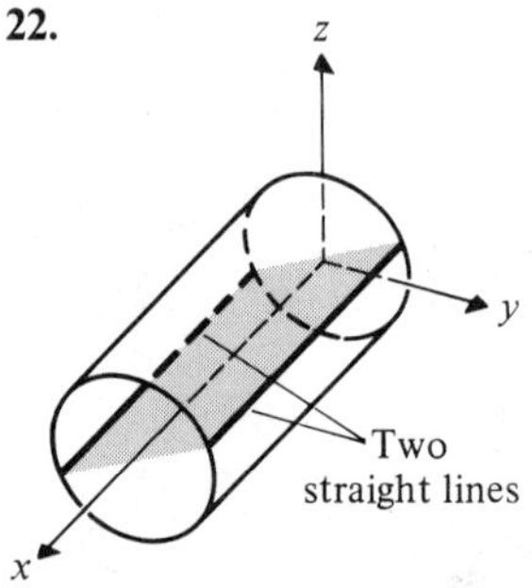

24.

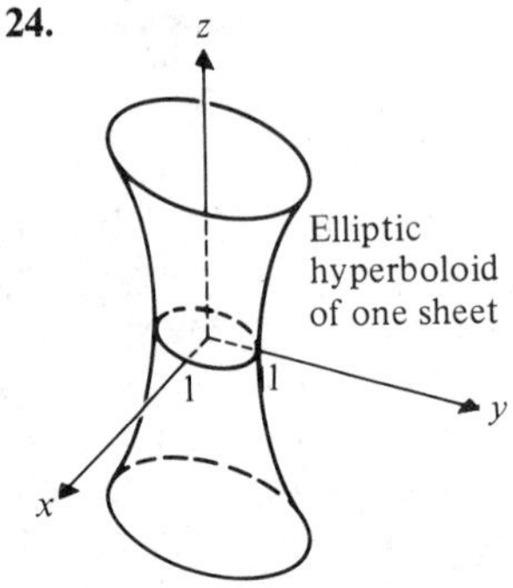

26.

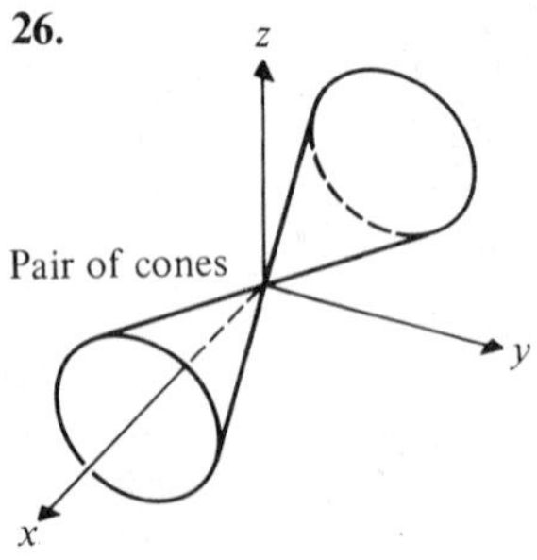

28. $x = 6 + 5t$, $y = 6 - 2t$, $z = 2 + t$
30. $x = 1 + u$, $y = 3 + 2u$, $z = 2 + 3u$ **32.** $x = y + z$
34. $3x - 4y + z = 0$ **36.** $35/\sqrt{41}$ **38.** $2\sqrt{5}/15$
40. $\sqrt{59}$ **42.** $(2 \cos t\hat{\mathbf{i}} - 2 \sin t\hat{\mathbf{j}} + \hat{\mathbf{k}})/\sqrt{5}$; $-\sin t\hat{\mathbf{i}} - \cos t\hat{\mathbf{j}}$; $(\cos t\hat{\mathbf{i}} - \sin t\hat{\mathbf{j}} - 2\hat{\mathbf{k}})/\sqrt{5}$
44. $\hat{\mathbf{i}} + 2t\hat{\mathbf{j}} + 2t\hat{\mathbf{k}}$; $\sqrt{1 + 8t^2}$; $2(\hat{\mathbf{j}} + \hat{\mathbf{k}})$; 0, $\sqrt{1 + 8t^2}$; $2\sqrt{2}/\sqrt{1 + 8t^2}$, $8t/\sqrt{1 + 8t^2}$
46. (a) 4.46 m/s; (b) $0.226\hat{\mathbf{i}} - \hat{\mathbf{j}}$ m
48. $k(1 - x)[1 - 1/\sqrt{1 + 4(1 - x)^2}]\hat{\mathbf{i}} + (k/2)[1 - 1/\sqrt{1 + 4(1 - x)^2}]\hat{\mathbf{j}}$ N

CHAPTER 13

Exercises 13.1

4. All points between the branches of the hyperbola $x^2 - y^2 = 1$

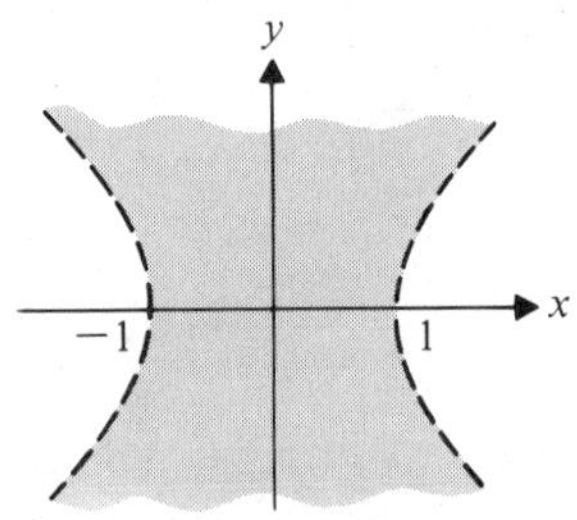

6. All points in space except (0, 0, 0)

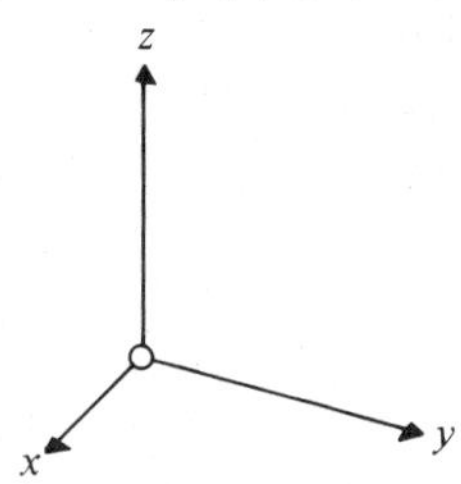

8. Function zero for all points on the x- and y-axes (except (0, 0)), and all points on the hyperbola $xy = 12$. Largest domain is all points in the xy-plane except those on line $y = -x$.

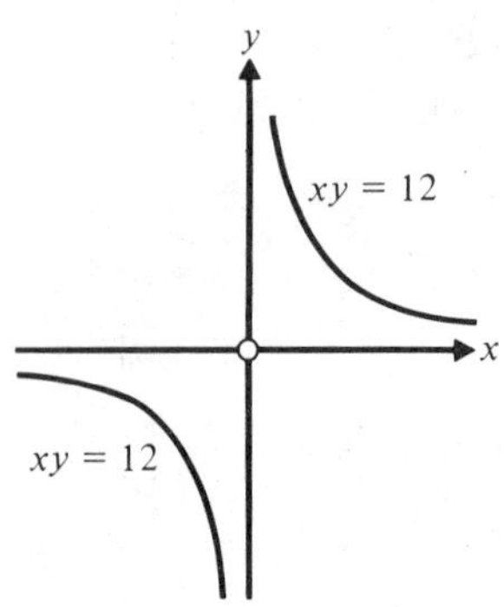

10.
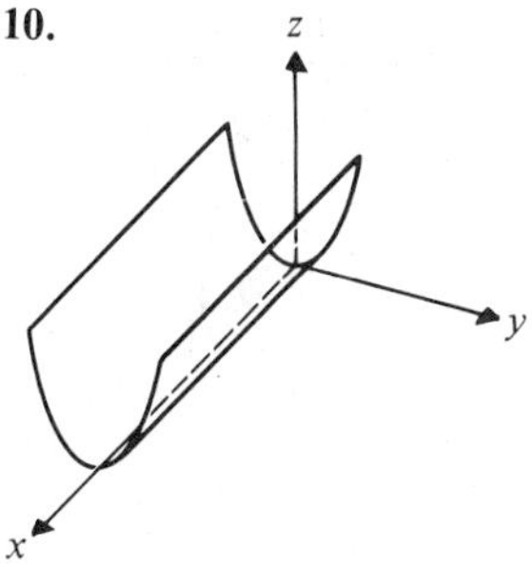

12.
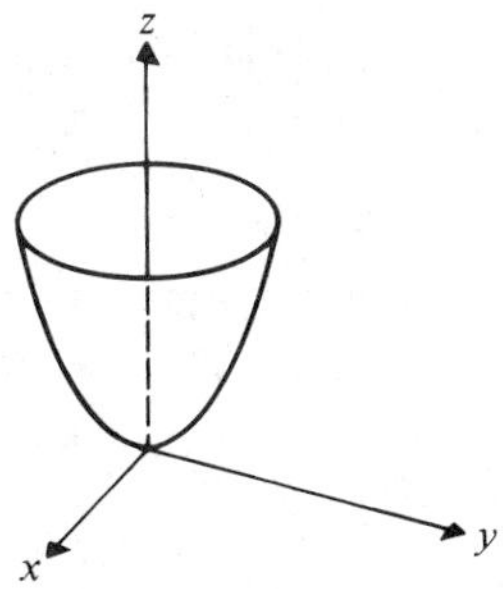

14.
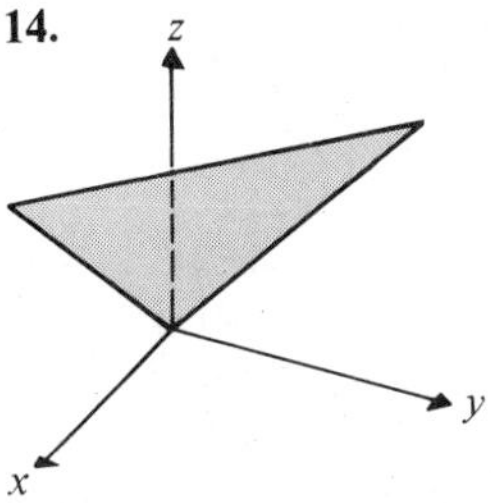

16.
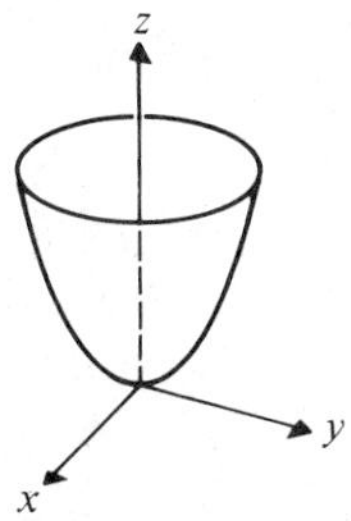

18.
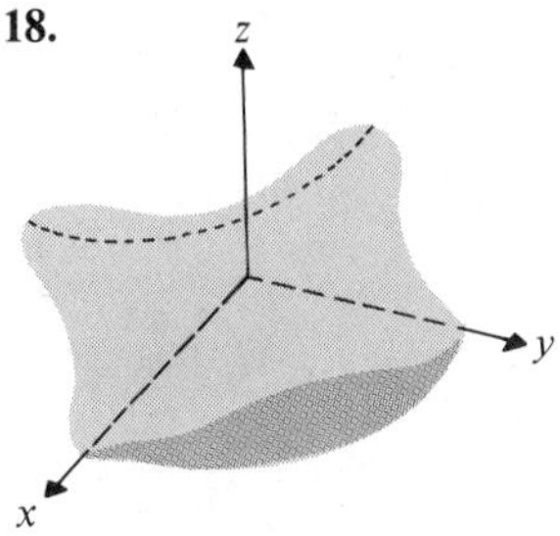

20.
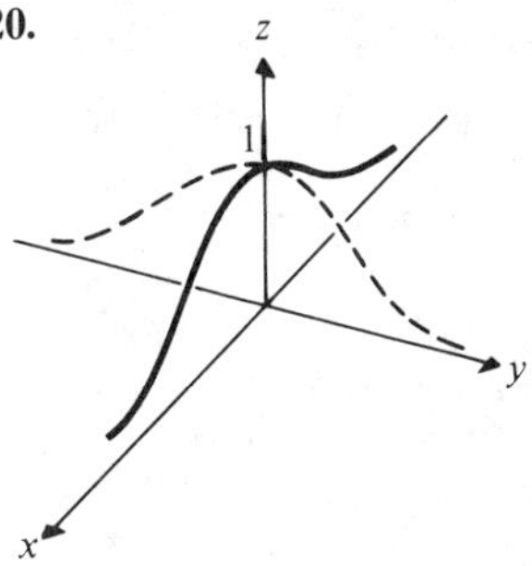

22.
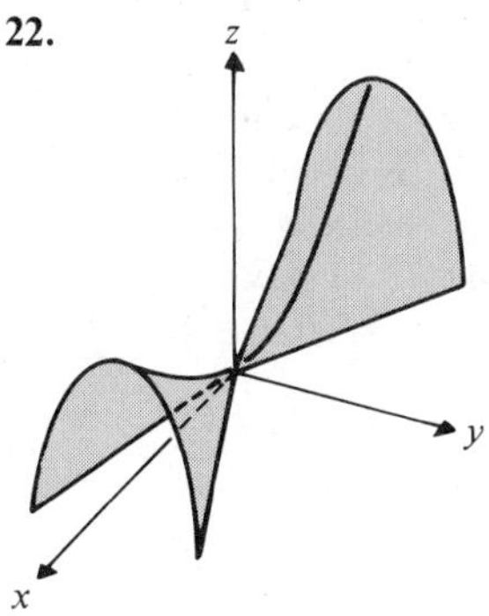

24. $lw(15 - lw)/(w + l)$ **26.** $8cxy\sqrt{1 - x^2/a^2 - y^2/b^2}$
28. $x \sin\theta(1 - x - x \sin\theta \csc\phi) + (x^2/2)\sin\theta(\cos\theta + \sin\theta\cot\phi)$
30. $30.25 + 11G + 17.5S$ cents. The domain consists of all points (G, S) satisfying the inequalities $200 \le 37G + 31S \le 300$, $145 \le 22G + 85S \le 170$.

Exercises 13.2

2. $3/5$ **4.** $-3/7$ **6.** 0 **8.** $\pi/2$ **10.** 1 **12.** 4
14. Does not exist **16.** 0 **18.** Does not exist
20. Does not exist **22.** $(0, 0)$ **24.** $x = 0; y = 0; z = 0$
26. $x = 0; y = 0; y = -x$ **28.** Does not exist **30.** 1
32. (a) 1, no; (b) 1, yes **34.** False

Exercises 13.3

2. $3y - 16x^3y^4, 3x - 16x^4y^3$
4. $-x(x + 2y)/[y(x + y)^2], x^2(x + 2y)/[y^2(x + y)^2]$
6. $y\cos(xy), x\cos(xy)$ **8.** $x/\sqrt{x^2 + y^2}, y/\sqrt{x^2 + y^2}$
10. $4x\sec^2(2x^2 + y^2), 2y\sec^2(2x^2 + y^2)$ **12.** ye^{xy}, xe^{xy}
14. $2x/(x^2 + y^2), 2y/(x^2 + y^2)$
16. $ye^x\cos(ye^x), e^x\cos(ye^x)$
18. $2xy\cos^2(x^2y)\sin(x^2y)/[1 - \cos^3(x^2y)]^{2/3}$, $x^2\cos^2(x^2y)\sin(x^2y)/[1 - \cos^3(x^2y)]^{2/3}$
20. $\tan\sqrt{x + y}/(2\sqrt{x + y}), \tan\sqrt{x + y}/(2\sqrt{x + y})$
22. $-2z/[1 + (x^2 + z^2)^2]$ **24.** 2
26. $-1/[1 + (1 + x + y + z)^2]$
28. $3x^2/y + \sin(yz/x) - (yz/x)\cos(yz/x)$
30. $z\,\mathrm{Sin}^{-1}(x/z)$ if $z > 0$; $z\,\mathrm{Sin}^{-1}(x/z) + xz/\sqrt{z^2 - x^2}$ if $z < 0$ **34.** (a) Yes; (b) no

Exercises 13.4

2. $2xyz\hat{\mathbf{i}} + x^2z\hat{\mathbf{j}} + x^2y\hat{\mathbf{k}}$ **4.** $(2xy + y^2)\hat{\mathbf{i}} + (x^2 + 2xy)\hat{\mathbf{j}}$
6. $(1 + x^2y^2z^2)^{-1}(yz\hat{\mathbf{i}} + xz\hat{\mathbf{j}} + xy\hat{\mathbf{k}})$
8. $e^{x+y+z}(\hat{\mathbf{i}} + \hat{\mathbf{j}} + \hat{\mathbf{k}})$
10. $-(x^2 + y^2 + z^2)^{-3/2}(x\hat{\mathbf{i}} + y\hat{\mathbf{j}} + z\hat{\mathbf{k}})$
12. $-(\sin 1)(\hat{\mathbf{i}} + \hat{\mathbf{j}} + \hat{\mathbf{k}})$ **14.** $-4e^{-8}(\hat{\mathbf{i}} + \hat{\mathbf{j}})$
22. ∇F not defined along the line $y = x$

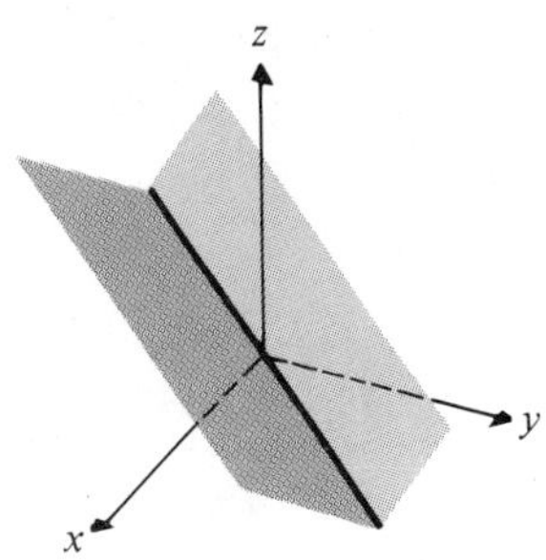

24. $x^2y - xy + C$
26. $xyz + y^2z + C$
28. $f(x, y) = g(x, y) + C$

Exercises 13.5

2. $-12x/y^4 + 72x^3y$ **4.** $x(1 + z + y + yz)e^{x+y+z}$
6. $e^{x+y} + 4/y^3$ **8.** 0
10. $(x^2 + y^2 - z^2)/(x^2 + y^2 + z^2)^2$
12. $2xy/(x^2 + y^2)^2$ **14.** $127/(756\sqrt{7})$ **16.** $8!\,y^9z^{10}$

18. $3y^2 \cos(x + y^3)$ **20.** $-xy/(x^2y^2 - 1)^{3/2}$
24. Entire plane with (0, 0) deleted **26.** All space
28. Not harmonic **32.** Curve of intersection of surface with plane $y = y_0$ is concave upward for $x_1 \le x \le x_2$.
34. (a) Yes; (b) yes **36.** (b) $2xy + C$

Exercises 13.6

2. $(2xe^y + y/x)(-s^2 \sin t) + (x^2e^y + \ln x)\times [8t/((t^2 + 2s)\sqrt{(t^2 + 2s)^2 - 1})]$
4. $xv^2[2yv(3u^2 + 2) + 2xuv/(u^2 + 1) + 3xye^u(u + 1)]$
6. $-2(\ln 3)y3^{x+2} \csc(r^2 + t) \cot(r^2 + t)$
8. $-2rst[5st/y^6 + 2yrt/(y^2 + z^2)^2 + 2rs/y^3]$
10. $6xve^t(v - 2)[1 - 2u/(x^2 - y^2)^2] + 8ye^{4t}[1 + 2u/(x^2 - y^2)^2]$
12. $2(t^2 + 2t)^2e^{2t} + (t^3 + 6t^2 + 6t + 2yt^2 + 4y + 8yt)e^t - 2$
14. $-(3y \sin v - 4x \cos v)^2 \sin(xy) - (24 \sin v \cos v + 3y \cos v + 4x \sin v) \cos(xy)$
18. $200\pi/3$ cc/min; no **20.** -7.11×10^4 N/s
22. (a) $2r\hat{\mathbf{e}}_r$; (b) $2r(\cos 2\theta\hat{\mathbf{e}}_r - \sin 2\theta\hat{\mathbf{e}}_\theta)$; (c) $(2 \cos\theta + 3 \sin\theta)\hat{\mathbf{e}}_r + (3 \cos\theta - 2 \sin\theta)\hat{\mathbf{e}}_\theta$

Exercises 13.7

2. $1/(x + y) - 1$
4. $(24x - \cos(x + y))/(2y - 1 + \cos(x + y))$
6. $-(2xz^2 + 3)/(2x^2z + y)$, $-z/(2x^2z + y)$
8. $z(1 + y^2z^2)/(y - x - xy^2z^2)$, $z/(x + xy^2z^2 - y)$
10. $[\cos(x + t) + \cos(x - t)]/[\cos(x - t) - \cos(x + t)]$
12. $(x + y)(3y^2 - 3x^2 - 5)/(3y^2z - 5z)$ **14.** 0
16. $e^x[(2t + 1) \cos y/(3x^2 + e^x) + (y^2 + 2yt - 1) \sin y/(t^2 + 2yt + 1)]$ **18.** $-2t/(9y^2)$

Exercises 13.8

2. $1/\sqrt{5}$ **4.** $-2/\sqrt{17}$ **6.** 1/5 **8.** $-13/\sqrt{29}$
10. $-40/\sqrt{6}$ **12.** (1, 4) **14.** (1, −3, 2) **16.** (1, 1)
20. (0, 0, 0), (1/4, ±1/2, 1/4)

Exercises 13.9

2. $x = 1 + u, y = 1 + 2u, z = 1 + 3u$
4. $x = -2 + t, y = 4 - 4t, z = -2 + t$
6. $x = 1 - 2u, y = 5 + 2u, z = 1 + u$
8. $132x + 49y = 328, z = 0$ **10.** $x = 1 - u, y = u, z = u$
12. $x = 2 + t, y = -\sqrt{5} + (1/\sqrt{5})t, z = -1 - t$
14. $x = 4 + \sqrt{17}u, y = 1, z = \sqrt{17} + 4u$
16. $x = 12 + 16t, y = -14 - 31t, z = 2 + t$
18. $x = t, y = 1, z = 1 + t$ **20.** $x = 2\pi u, y = 2\pi + u, z = 4\pi + 2u$ **22.** $3x + 6y + z + 2 = 0$
24. $x + y + z = 4$ **26.** $x + y = 1$ **30.** $6\sqrt{2}$ **32.** 0

Exercises 13.10

2. (0, 0) saddle; (1, 1) max. **4.** None **6.** $(0, n\pi)$ saddle
8. (x, x) min. **10.** $(x, 0)$ saddle; $(0, y)$ $y > 0$ min.; $(0, y)$ $y < 0$ max. **12.** (0, 0) min.; $(0, y)$ $y \neq 0$, none of these
14. None **16.** All points on the coordinate axes
18. (d) 2.45, 3.54 **20.** (0, 0) min.; (−1/3, ±2/9) saddle

Exercises 13.11

2. 4, −1/3 **4.** 3, −3 **6.** (1, 1, −2) **8.** (1/2, 1/2, 1/2)
10. 2, −32 **12.** $100(2/3)^{1/3} \times 100(2/3)^{1/3} \times 100(3/2)^{2/3}$ cm
14. $2a/\sqrt{3} \times 2b/\sqrt{3} \times 2c/\sqrt{3}$
18. $12/\sqrt{5}$ m, $(50\sqrt{5} - 27\pi)/(3\sqrt{5}\pi)$ m

Exercises 13.12

2. $(8 \pm 9\sqrt{13})/2$ **4.** ±27 **6.** $\pm\sqrt{32/27}$
8. $(\pm\sqrt{2} - 1)/2$ **16.** 33.12, −0.12 **18.** $\pm 1/\sqrt{2}$

Exercises 13.13

2. $y(1 + x^2y^2)^{-1}\,dx + x(1 + x^2y^2)^{-1}\,dy$
4. $(yz \cos(xyz) - 2xy^2z^2)\,dx + (xz \cos(xyz) - 2x^2yz^2)\,dy + (xy \cos(xyz) - 2x^2y^2z)\,dz$
6. $(1 - x^2y^2)^{-1/2}(y\,dx + x\,dy)$
8. $(y + t)\,dx + (x + z)\,dy + (y + t)\,dz + (z + x)\,dt$
10. $2e^{x^2+y^2+z^2-t^2}(x\,dx + y\,dy + z\,dz - t\,dt)$ **12.** 3%

Review Exercises

2. $2(x^2 - y^2 + z^2)/(x^2 + y^2 + z^2)^2$
4. $(3 - z^2)(1 + z^2)/(1 + 2xz + 2xz^3)$
6. $3(2x - ye^{xy})(t^2 + 1) + (2y - xe^{xy})(\ln t + 1)$
8. $(y + 3yv^2 - 2xuv)/(2u + 6uv^2 + 3v + vx^2)$
10. $e^t(t + 1)(y - 2x) + e^{-t}(1 - t)(x - 2y)$ **12.** $\cos\theta$
14. $(2x - 3x^2y^2) \cos\theta - 2yx^3 \sin\theta$
16. $6(6x^4t^6z + 9x^5 + x^4t^9z^2 + 2xt^9z^5 - 9t^{12}z^5 - 6xt^6z^4 - 4x^5t^3z - 2x^2t^3z^4)/(x^4t^8z^4)$
18. $xz[(1 - 3xz^2)(1 - 6xz^2 + 6x^2z^4) + (1 - 2xz^2)(1 - 3xz^2 + 6x^2z^4)]/(x - 3x^2z^2)^3$
20. $2y(v^2t - 1)(t^2 - 3/\sqrt{1 - x^2y^2}) + xv \sec^2 t(t^2 - 3/\sqrt{1 - x^2y^2}) + 2xyt$ **26.** $\sqrt{6}$
28. $9/\sqrt{14}$ **30.** 2 **32.** $x + 3y - 6z + 4 = 0$
34. $x = 2 + u, y = u, z = 6 + 4u$ **36.** $x = 1 + t, y = 1 - t, z = 1$ **38.** None **40.** (0, 0) max.; all points on $x^2 + y^2 = 1$ are relative minima **42.** 1/2, −1/2
44. $(\pm\sqrt{(\sqrt{5} - 1)/2}, 0)$; $(0, \pm 1)$
46. $x = 100bcqr/(acpr + abpq + bcqr)$, $y = 100acpr/(acpr + abpq + bcqr)$, $z = 100abpq/(acpr + abpq + bcqr)$

CHAPTER 14

Exercises 14.2

2. 0 **4.** 3/4 **6.** $e^2(1 - e)^2/2$ **8.** 0 **10.** 5/144 **12.** 0
14. −0.54 **16.** $(1 - \ln 2)/2$ **18.** $128\sqrt{2}/5$

20. 11664/35 **22.** $(1 - 2\sqrt{2})/12$ **24.** 1/3 **26.** 8/189
28. $2(1 - \sqrt{2})$

Exercises 14.3

2. 128/15 **4.** $-621/140$ **6.** 0 **8.** 0 **10.** 304/15
12. $\sqrt{13} - 7/2$ **14.** $(1 - \cos 1)/2$
16. $2(1 - \sqrt{2})$ **18.** 101/70 **20.** $(e^2 - 1)/(2e)$ **22.** 0
24. 24π **26.** $e^2 - 2e - 1$ **28.** $(1 - 2\sqrt{2})/12$
30. 44/15

Exercises 14.4

2. 343/6 **4.** 8 **6.** $1 + 3e^{-2}$ **8.** 343/6 **10.** 20/3
12. 16π **14.** $1024\pi/15$ **16.** $26\pi/15$ **18.** 45π
20. $34\pi/3$ **22.** 235/3 **24.** 7/6 **26.** $8\pi/3 - 2\sqrt{3}$
28. 23328/35 **30.** $2 - \ln 3$ **32.** $16\pi(1 + \sqrt{2})/15$
34. $68\pi/9$ **36.** 5.38

Exercises 14.5

2. $256\gamma/5$ **4.** 48γ **6.** $370\gamma/3$ **8.** $250\gamma/3$
10. 7.63×10^6 N **12.** $5\sqrt{29}\gamma$ **14.** 10γ **16.** $\pi\gamma r^3$
18. 9.00×10^4 N

Exercises 14.6

2. (0, 24/5) **4.** $\bar{x} = \bar{y} = 9/(15 - 16 \ln 2)$
6. $(8(2 - \sqrt{2})/(3\pi), 8\sqrt{2}/(3\pi))$ **8.** (177/85, 9/17)
10. (4/3, 5/3) **12.** 32/3 **14.** 603/10 **16.** $48\rho/5$
18. (0, 192/205) **20.** $(-2/(8\sqrt{3} - 1), 0)$
22. $(-61/28, 807/700)$ **24.** 4761/140 **26.** $81\sqrt{2}/40$

Exercises 14.7

2. $4\sqrt{14}/3$ **4.** $4(5\sqrt{3} - 2\sqrt{6} - 3 + \sqrt{2})/3$
6. $(247\sqrt{13} + 64)/1215$
8. $2\int_0^4 \int_0^{\sqrt{16-y^2}} \sqrt{4 + y^2 + z^2}\, dz\, dy$
10. $4\int_0^{\sqrt{2}} \int_0^{\sqrt{2-x^2}} \sqrt{1 + 16(x^2 + y^2)^3}\, dy\, dx$
12. $\int_0^1 \int_0^{1-x^2} \sqrt{1 + 2/(1 + x + y)^2}\, dy\, dx$
16. $\sqrt{2} \int_{-1/\sqrt{2}}^{1/\sqrt{2}} \int_x^{\sqrt{1-x^2}} \sqrt{2 + 1/(y - x)}\, dy\, dx$
18. $\int_0^1 \int_{1-x}^{2-x} \sqrt{1 + 9x^4 + 9y^4}\, dy\, dx +$
$\int_1^2 \int_0^{2-x} \sqrt{1 + 9x^4 + 9y^4}\, dy\, dx$

Exercises 14.8

2. 2/3 **4.** $4(3\sqrt{3} - \pi)/3$ **6.** $\pi/6$ **8.** $(3\sqrt{3} - \pi)/2$
10. $\pi/4$ **12.** $15\pi/4 - 19\sqrt{2}/3 - (9/2)\,\text{Sin}^{-1}(1/3)$
14. $\pi R^4/4$ **16.** $\pi(17\sqrt{17} - 1)/6$ **18.** $2\pi(10\sqrt{10} - 1)/3$
20. $4\pi/21$ **22.** $\rho(\sqrt{R^2 + d^2} - d)/(2\varepsilon_0)$ **24.** $9\pi - 12\sqrt{3}$
26. (5/6, 0) **28.** $\pi P R^4/(8nL)$

Exercises 14.9

2. 32/3 **4.** 0 **6.** 1024/21 **8.** 1/96 **10.** 11/6
12. 1/20 **14.** 13/3 **16.** 729/70 **18.** 2π

Exercises 14.10

2. 2/3 **4.** 704/15 **6.** 8 **8.** 8/3 **10.** 19/96 **12.** 7/3
14. 5/18 **16.** 64/15 **18.** $\pi/4 - 1/3$

Exercises 14.11

2. (1/4, 1/4, 1/4) **4.** (0, 16/7, 8/7) **6.** $2\rho/3$
8. $773\rho/2520$ **10.** $64\rho/15$
12. (6772/11,847, 7300/14,001, 1/2) **14.** (0, 0, 3/2)
16. $128\rho a^5/45$ **18.** $51\sqrt{3}\rho$

Exercises 14.12

2.

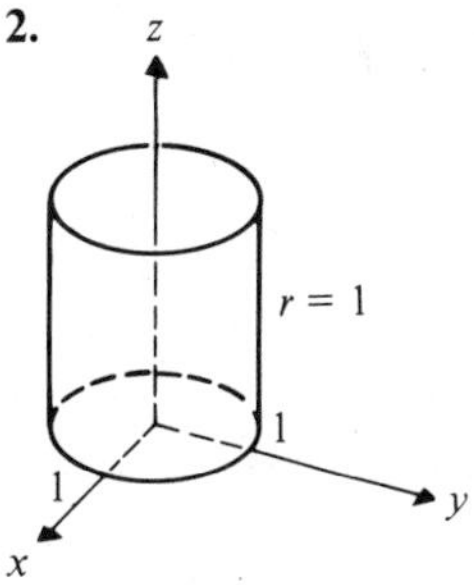

4.

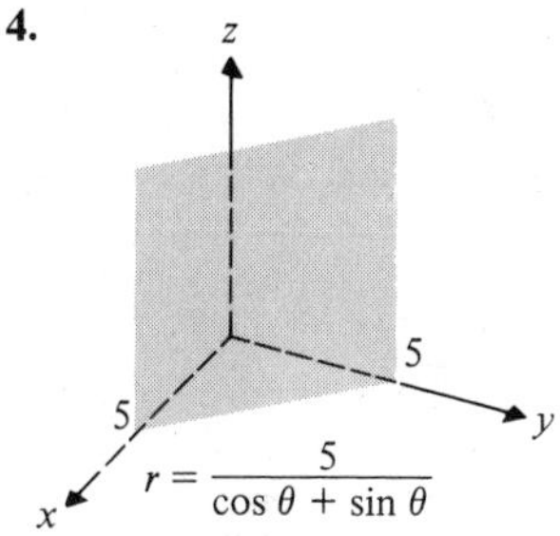

6.

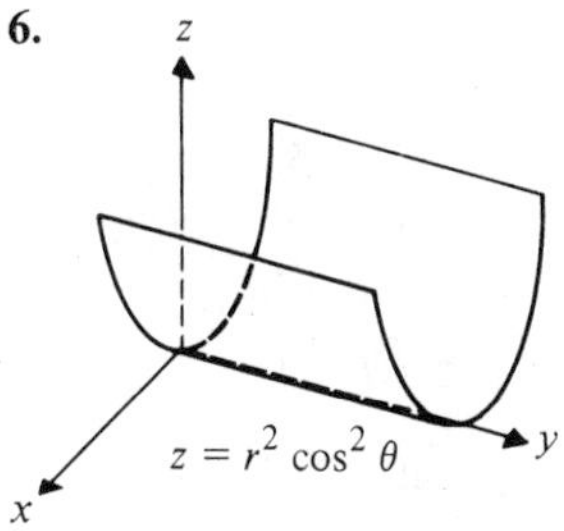

8.

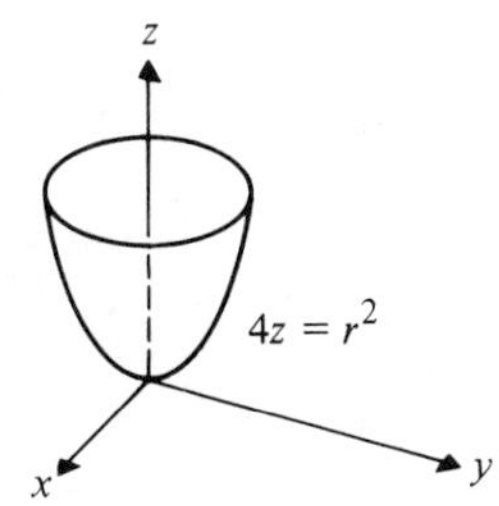

10.

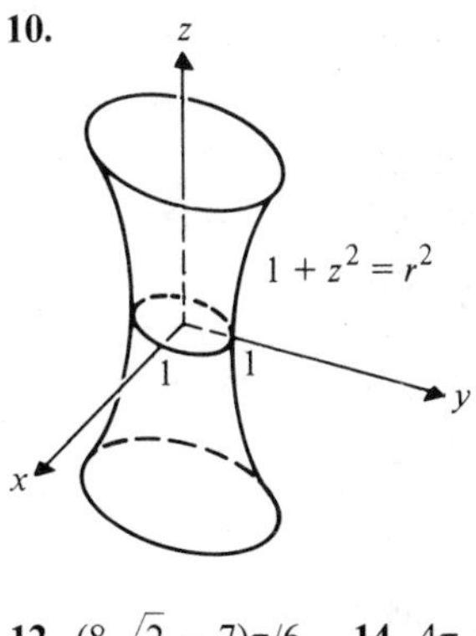

12. $(8\sqrt{2} - 7)\pi/6$ **14.** 4π **16.** $4\sqrt{3}\pi$
18. $\int_0^{2\pi} \int_0^3 \int_0^{1+r^2} f(r\cos\theta, r\sin\theta, z)\, r\, dz\, dr\, d\theta$;
$\int_0^3 \int_0^{2\pi} \int_0^{1+r^2} f(r\cos\theta, r\sin\theta, z) r\, dz\, d\theta\, dr$;
$\int_0^3 \int_0^{1+r^2} \int_0^{2\pi} f(r\cos\theta, r\sin\theta, z) r\, d\theta\, dz\, dr$;
$\int_0^1 \int_0^3 \int_0^{2\pi} f(r\cos\theta, r\sin\theta, z) r\, d\theta\, dr\, dz +$
$\int_1^{10} \int_{\sqrt{z-1}}^3 \int_0^{2\pi} f(r\cos\theta, r\sin\theta, z) r\, d\theta\, dr\, dz$;

$\int_0^{2\pi}\int_0^1\int_0^3 f(r\cos\theta, r\sin\theta, z)r\,dr\,dz\,d\theta +$
$\int_0^{2\pi}\int_1^{10}\int_{\sqrt{z-1}}^3 f(r\cos\theta, r\sin\theta, z)r\,dr\,dz\,d\theta$;
$\int_0^1\int_0^{2\pi}\int_0^3 f(r\cos\theta, r\sin\theta, z)r\,dr\,d\theta\,dz +$
$\int_1^{10}\int_0^{2\pi}\int_{\sqrt{z-1}}^3 f(r\cos\theta, r\sin\theta, z)r\,dr\,d\theta\,dz$
20. $8\pi\rho R^5/15$ **22.** $81\pi^2/8$ **24.** 0.084 **26.** $\pi h\rho R^4/10$
28. $4\sqrt{3}\pi/9$ **30.** $56\pi(2-\sqrt{2})/3$

Exercises 14.13

2. $\mathscr{R}\sin\phi = 1$ (see the figure given for Exercise 2 in Section 14.12) **4.** $\mathscr{R} = 4\cot\phi\csc\phi$ (see the figure given for Exercise 8 in Section 14.12) **6.** $\mathscr{R}^2 = -\sec 2\phi$ (see the figure given for Exercise 10 in Section 14.12) **8.** $(2-\sqrt{2})\pi/3$
10. $[4\,\mathrm{Tan}^{-1}(2) - \pi]/12$ **12.** $32\pi/3$ **14.** $8\pi\rho R^5/15$
16. $k\pi R^4$ coulombs **18.** $9\pi/2$
20. (a) $\rho_b = \rho_w/2$; (b) $11\pi\rho_w gR^3/24$

Exercises 14.14

2. $2x - 1 + e^x$
4. $4x^3 - 3x^2 - 1 + 3x^2(x^3-1)\ln(x^3-1) - 2x^3\ln(x^2)$
8. $e^x\sqrt{1+e^{3x}} - \cos x\sqrt{1+\sin^3 x}$
12. (a) $|x| < 1/9$; 0; (b) $18/x + (1/x^2)\ln((1-9x)/(1+9x))$, $x \neq 0$; $F'(0) = 0$

Review Exercises

2. 81/16 **4.** 1/40 **6.** 36/5 **8.** 0 **10.** −4544/945
12. $\pi/6$ **14.** $32\pi/3$ **16.** $2\pi/5$ **18.** $3\pi/2$ **20.** $(\pi\ln 2)/4$
22. 32/3 **24.** 16/15 **26.** $\pi(17\sqrt{17}-1)/96$
28. $128\pi/15$, $4\pi^2$ **30.** 1 **32.** 2 **34.** $14\rho\pi/15$
36. (0, 1/8) **38.** $(9\pi/64, 9\pi/64, 3/8)$ **40.** $16\pi/15$

CHAPTER 15

Exercises 15.1

2. $-(x^2+y^2+z^2)^{-3/2}(x\hat{\mathbf{i}} + y\hat{\mathbf{j}} + z\hat{\mathbf{k}})$
4. $(6 - 2\cos 2)\hat{\mathbf{i}} + (1 + 2\sin 2)\hat{\mathbf{j}}$ **6.** $2(e^y - x^2y)$
8. $2x\cos(x^2+y^2+z^2) - \sin(y+z)$
10. 1 **12.** $2/\sqrt{x^2+y^2+z^2}$
14. $4y(16z^4 - 3x)\hat{\mathbf{i}} + x^2\hat{\mathbf{j}} + 12yz\hat{\mathbf{k}}$ **16.** 0 **18.** $-2\hat{\mathbf{k}}$
20. $-2\hat{\mathbf{k}}/[(x+y)\sqrt{(x+y)^2-1}]$
26. $x^3y^2 + 3x + 2y + C$ **28.** $\ln|x+y| + C$
30. $(x^2+y^2+z^2)/2 + C$ **32.** $\ln|1+x+y+z| + C$
34. $\mathrm{Tan}^{-1}(xy) + z^2/2 + C$ **36.** (a) 4, 2, −1;
(b) $x^3/3 + 2xy + 4xz - 3y^2/2 - yz + z^2 + C$
38. (a) $q/(4\pi\varepsilon_0|\mathbf{r}|) + C$; (b) $-\sigma z/(2\varepsilon_0) + C$
42. $yz\hat{\mathbf{i}} - xz\hat{\mathbf{j}}$ **44.** $(1/4)[(4xyz - 3y^2z)\hat{\mathbf{i}} + (2xyz - 6x^2z)\hat{\mathbf{j}} + (2x^2y + xy^2)\hat{\mathbf{k}}]$

Exercises 15.2

2. 32/3 **4.** $50\sqrt{10}/3$ **6.** $37\sqrt{2}/3$ **8.** $9\sqrt{1+16\pi^2}$ cm
10. 37/80 **12.** 0.78 **14.** $(\pi/27)(145\sqrt{145} - 10\sqrt{10})$
16. $\pi/2$

Exercises 15.3

2. 51/4 **4.** −15 **6.** −99/140 **8.** 10 **10.** 0 **12.** 9
14. 0 **16.** -8π **18.** $(768 + 5\pi)/20$
20. $67/32 + \sqrt{5} - \sqrt{2} - 5/e$
22. $(4\pi\varepsilon_0)^{-1}[q_1q_3(1/\sqrt{41} - 1/\sqrt{61}) + q_2q_3(1/\sqrt{18} - 1/\sqrt{10})]$

Exercises 15.4

2. −43 **4.** −2 **6.** 0 **8.** 10 **10.** e^3 **14.** 0 **16.** 0
18. $8/105 + \ln(3/2)$ **20.** 2π

Exercises 15.5

2. Not conservative **4.** mgz

Exercises 15.6

2. 4π **4.** $2\sqrt{3}/5$ **6.** 0 **8.** −3/8
10. $4(13\sqrt{13} - 8)/3$ **16.** 24 **18.** 77/2 **20.** $81\pi/2$
22. -4π

Exercises 15.7

2. $2\sqrt{3}/15$ **4.** $\sqrt{2}/8$ **6.** $(-61 + 44\sqrt{2})/5$ **8.** −3
10. $\pi R\sqrt{R^2+h^2}$ **12.** $3(145^{5/2} - 361)/5120$
14. $\pi(3\sqrt{3} - 1)/3$ **16.** (a) $2\sqrt{2}\pi$; (b) 16

Exercises 15.8

2. $2e^2 - 10e^{-2}$ **4.** 0 **6.** $2\pi/3$ **8.** -2π **10.** $10\pi/3$
12. $\sqrt{2}\pi/4$ **14.** (a) 30π; (b) -20π **16.** $2\pi(1 - e^{-2})$

Exercises 15.9

2. 0 **4.** 16 **6.** 27π **8.** 180 **12.** 0

Exercises 15.10

2. 0 **4.** $\pm 16\pi$ **6.** 0 **8.** $-2\sqrt{2}\pi b^2$ **10.** $\pm 2\pi$

Review Exercises

2. $3x^2y + x^2/y^2$ **4.** 0 **6.** $ye^x + ze^y + xe^z$
8. $(\hat{\mathbf{i}} + \hat{\mathbf{j}})/\sqrt{1-(x+y)^2}$ **10.** $x(-y\hat{\mathbf{j}} + z\hat{\mathbf{k}})/(1 + x^2y^2z^2)$
12. $2\sqrt{3}$ **14.** π **16.** $\sqrt{2/3}$ **18.** 2π
20. $8\pi(18 - 7\sqrt{6})/3$ **22.** 0 **24.** 0 **26.** $-\pi/120$
28. $(25\sqrt{5} - 11)/120$ **30.** $8\pi/3$

CHAPTER 16

Exercises 16.1

12. $-3\sin 3x + 2\cos 3x$
14. $[(2\cos 3 - \cos 6)\sin 3x + (\sin 6 - 2\sin 3)\cos 3x]/\sin 3$

16. $(1/3)\,\mathrm{Tan}^{-1}(x/3) + C$
18. $(x^3/6)\ln x - 5x^3/36 + C_1 x + C_2$ **20.** Ce^x **22.** $y = 0$
24. (b)

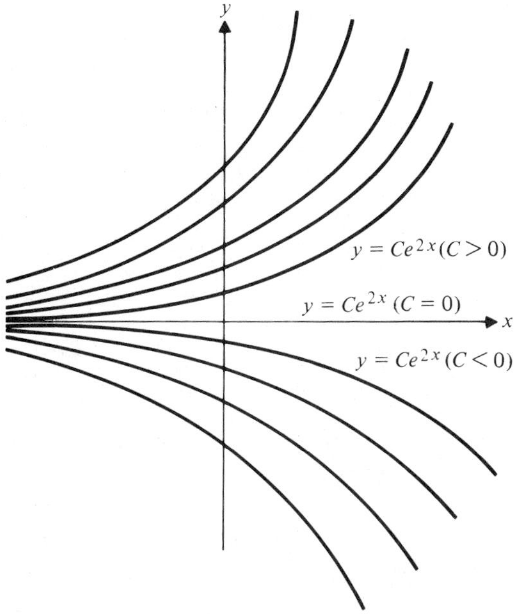

26. (a)

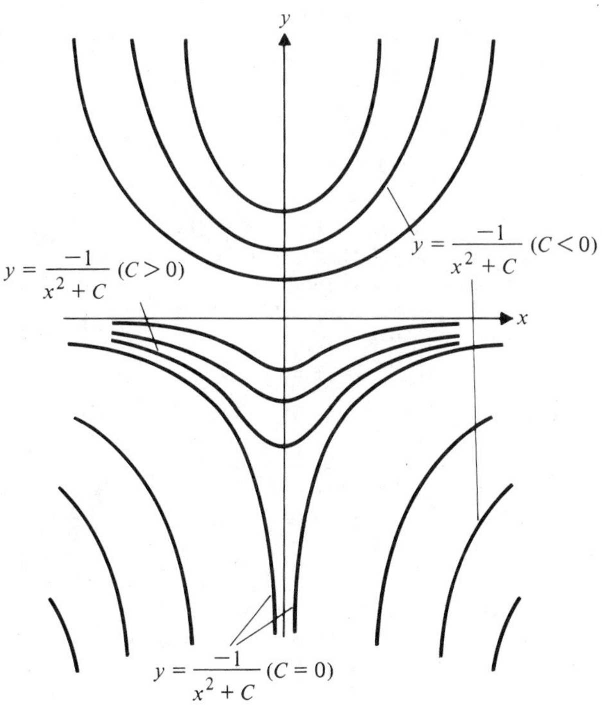

Exercises 16.2

2. $y(x) = 2 + Ce^{-x^2}$ **4.** $y(x) = Ce^{3x} - 2/3$
6. $y^2 + x^2 + 2(x - y) + 2\ln|y + 1| + 2\ln|x - 1| = C$
8. $y(x) = [C + 3\ln|x| + 3e^x(1 - x)]^{1/3}$
10. $y(x) = (x + C)/(1 - Cx)$ **12.** $xy = 2e^{y-x-1}$
14. $y(x) = [\tan(x^2 - 4) + 4]/[1 - 4\tan(x^2 - 4)]$
16. (a) $6(1 - 6kt)^{-1}$ km; (b) never **18.** $20 + 60e^{-0.203t}$
20. $(R/k)(1 - e^{-kt}) + A_0 e^{-kt}$
22. $60(1 - e^{kt})/(2 - 3e^{kt})$ grams **24.** $x^2(x^2 - 2y^2) = C$
26. $x^2 + 2xy - y^2 = C$ **28.** $(y - x)e^{y/x} = x\ln|x| + Cx$
30. (a) $y(x) = \mathrm{Cos}^{-1}(1/\sqrt{2} - x)$;
(b) $y(x) = 2\pi - \mathrm{Cos}^{-1}(1/\sqrt{2} - x)$ **32.** 459.5 milligrams
34. $y(x) = Ce^x - x - 1$ **36.** $6y - 3\ln|4x + 6y + 3| = C$
38. $y^3(x^3 + 6x - 5) = 1 + y^6$

Exercises 16.3

2. $y(x) = x^4 + C/x^2$ **4.** $y\sin x = C - 5e^{\cos x}$
6. $y(x) = C(x + 1)^2 - 2x - 2$ **8.** $y(x) = Ce^x + e^{2x}$
10. $y(x) = x^3/2 + Cx^3 e^{1/x^2}$
12. $y(x) = x + \cos 2x + C\sin 2x$
14. $y(x) = e^x(2\sin x - \cos x)/5 - 4e^{-x}/5$
16. $xy = C + y^4/4$ **18.** $y(x) = e^{-x}/(C - x)$
20. $x^2 y + Cye^{-x} = 1$
22. $y(x) = ((3/4)\cos x + C\sec^3 x)^{-1/3}$
24. $1000 + t + 3 \times 10^6/(1000 + t)$ grams
26. (b) $I_0 - E_0/(\omega C Z^2)$

Exercises 16.4

2. $2\sqrt{Cy - 1} = \pm Cx + D$
4. $y(x) = Cx - C^2\ln|x + C| + D$
6. $(x + C)^2 + (y + D)^2 = 1$ **8.** $D + x/2 = \{-1/y$; or $C\,\mathrm{Tan}^{-1}(Cy)$; or $(1/(2C))\ln|(y - C)/(y + C)|\}$
10. $y(x) = (1/2)(Ce^x + C^{-1}e^{-x}) + D$

Exercises 16.5

2. Less than 47.8 km/hr; defense **8.** 18.0 m

Exercises 16.6

2. (a) $2e^{[(1/2)\ln(5/2)]t}$; (b) $2e^{[(1/10)\ln 55]t}$
4.

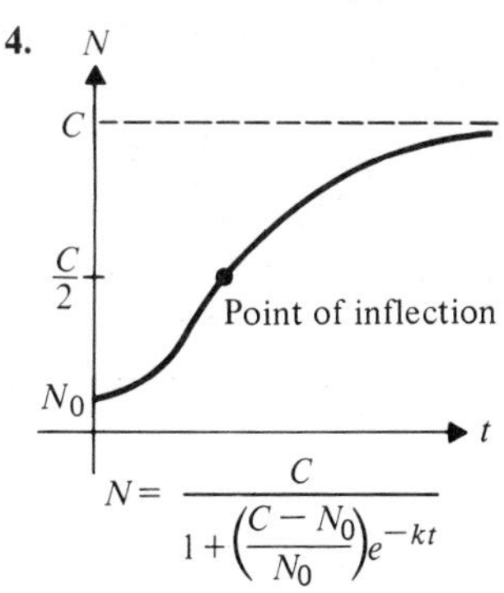

8.

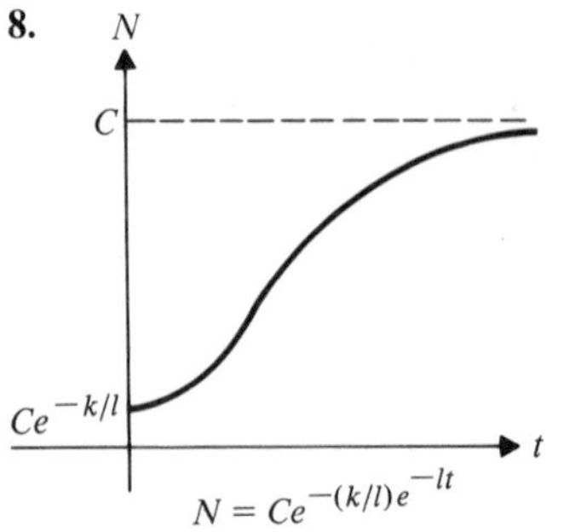

10.

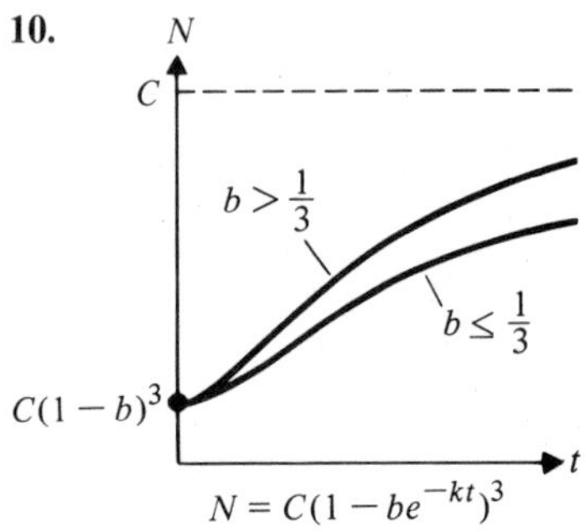

Exercises 16.7

2. Linear **4.** Linear **6.** Linear **8.** Linear
10. Not linear **12.** Linear **14.** Linear **16.** Not linear
18. Not linear **20.** Linear

Exercises 16.8

2. $C \cos x$ **4.** $C_1e^{4x} + C_2xe^{4x}$
6. $e^{2x}[C_1 \cos(x/\sqrt{2}) + C_2 \sin(x/\sqrt{2})]$
8. $C_1 \cos(2 \ln x) + C_2 \sin(2 \ln x)$ **12.** Dependent
14. Independent

Exercises 16.9

2. $(C_1 + C_2x)e^{4x}$ **4.** $C_1e^{(-1+\sqrt{3})x} + C_2e^{(-1-\sqrt{3})x}$
6. $C_1e^{3x} + C_2 \cos x + C_3 \sin x$
8. $(C_1 + C_2x + C_3x^2)e^{2x}$
10. $C_1 \cos x + C_2 \sin x + C_3 \cos 2x + C_4 \sin 2x$
12. $e^{\sqrt{2}x}(C_1 \cos(\sqrt{2}x) + C_2 \sin(\sqrt{2}x)) + e^{-\sqrt{2}x}(C_3 \cos(\sqrt{2}x) + C_4 \sin(\sqrt{2}x))$
14. $y'' + 4y' + 20y = 0$
16. $y'''' - 4y''' + 10y'' - 12y' + 9y = 0$ **18.** 5, 24, 20

Exercises 16.10

2. $-(x^2/3 + 2/9)e^{-x}$; $C_1e^{(-1+\sqrt{3})x} + C_2e^{-(1+\sqrt{3})x} - (x^2/3 + 2/9)e^{-x}$
4. $(1/9)\cos 2x$; $(C_1 + C_2x)\cos x + (C_3 + C_4x)\sin x + (1/9)\cos 2x$
6. $e^{-2x}/40$; $C_1 \cos x + C_2 \sin x + C_3 \cos 2x + C_4 \sin 2x + e^{-2x}/40$
8. $e^{2x}(\cos x - 5 \sin x)/312$; $e^{-4x}(C_1 \cos 5x + C_2 \sin 5x) + e^{2x}(\cos x - 5 \sin x)/312$
10. $[(2x + 1)\cos x - 2(x + 1)\sin x]/16$; $e^{2x}(C_1 \cos x + C_2 \sin x) + [(2x + 1)\cos x - 2(x + 1)\sin x]/16$
12. $Axe^{3x} + Be^{3x} + Cx \cos x + Dx \sin x + E \cos x + F \sin x$
14. $Ax^2e^x + Bxe^x + Cx^3 + Dx^2 + Ex + F + G \cos x + H \sin x$
18. $e^x(C_1 \cos(x/\sqrt{2}) + C_2 \sin(x/\sqrt{2})) + (4 \cos 3x - 5 \sin 3x)/246 + (\sin x + 4 \cos x)/34$
20. $-(1/108)\sin 3x + (x^2/12)(\sin 3x - \cos 3x) + (x/36)(\sin 3x + \cos 3x)$

Exercises 16.11

2. $e^{-15t/4}[(12/65)\cos(5\sqrt{23}t/4) + (20/(13\sqrt{23}))\sin(5\sqrt{23}t/4)] - (4/65)[3 \cos(10t) + 2 \sin(10t)]$ m
4. $-0.0253e^{-99.50t} + 5.03e^{-0.50t}C$
6. $(M|g|/k)\cos(\sqrt{k/M}t)$ from equilibrium
8. (b) $x = -7/36$ m; yes
12. $(9/10)[-\sin(3t) + 3t \cos(3t)]$ A; yes
14. (c) $\sqrt{k/M - \beta^2/(2M^2)}$; $(2MA/\beta)\sqrt{4kM - \beta^2}$

Review Exercises

2. $y = \pm\sqrt{2(x + \ln|x|) + C}$
4. $y = Ce^{-4x} + (8x^2 - 4x + 1)/32$
6. $y = e^{-3x/2}[C_1 \cos(\sqrt{7}x/2) + C_2 \sin(\sqrt{7}x/2)] + 1/2$
8. $y = x^3/9 + C_1 \ln|x| + C_2$ **10.** $y = Ce^{-x^2} + x^2 - 1$
12. $y = (C_1 + C_2x)e^{2x} + (3 \sin x + 4 \cos x)/25$
14. $y = C_1 \cos(2x) + C_2 \sin(2x) - (x/4)\cos(2x)$
16. $xy^2 = Ce^{-2x - x^2/2}$
18. $e^{-x}[C_1 \cos(\sqrt{3}x) + C_2 \sin(\sqrt{3}x)] + (\sqrt{3}x/6)e^{-x} \sin(\sqrt{3}x)$
20. $x = Ce^{-3y} + (2/27)(-9y^2 + 6y - 2)$
22. $y = (424/405)e^{9x} + (11/5)e^{-x} - 2x/9 - 20/81$
24. $y = -\cos x + (2 \sin x)/x + (2 \cos x + 1 - \cos 1 - 2 \sin 1)/x^2$
26. (a) 3.924×10^{-3} N; (b) $(1.962 \times 10^{-3})t + (4.905 \times 10^{-7})(e^{-4000t} - 1)$ m

Appendix B

2. 1 **4.** 0 **6.** -640 **8.** 78 **10.** $x = -14$, $y = -10$
12. $r = -4$, $s = -11$ **14.** $x = -5/18$, $y = 11/18$, $z = 5/2$ **18.** $x = y = z = w = 0$ **20.** (b) $2x + 3y = 5$
22.
$$\begin{vmatrix} y & x^2 & x & 1 \\ 0 & 1 & 1 & 1 \\ 11 & 4 & 2 & 1 \\ 1 & 4 & -2 & 1 \end{vmatrix} = 0$$
24. $b^{n-1}(b + na)$

Index

Principal Values of Inverse Trigonometric Functions

$y = \mathrm{Sin}^{-1} x \qquad -\pi/2 \le y \le \pi/2$

$y = \mathrm{Tan}^{-1} x \qquad -\pi/2 < y < \pi/2$

$y = \mathrm{Cos}^{-1} x \qquad 0 \le y \le \pi$

$y = \mathrm{Cot}^{-1} x \qquad 0 < y < \pi$

$$y = \mathrm{Csc}^{-1} x \qquad \begin{cases} -\pi < y \le -\pi/2 & \text{when } x < 0 \\ 0 < y \le \pi/2 & \text{when } x > 0 \end{cases}$$

$$y = \mathrm{Sec}^{-1} x \qquad \begin{cases} -\pi \le y < -\pi/2 & \text{when } x < 0 \\ 0 \le y < \pi/2 & \text{when } x > 0 \end{cases}$$

Derivative Formulas

$$\frac{d}{dx}(u + v) = \frac{du}{dx} + \frac{dv}{dx}$$

$$\frac{d}{dx}(uv) = u\frac{dv}{dx} + v\frac{du}{dx}$$

$$\frac{d}{dx}\left(\frac{u}{v}\right) = \frac{v\dfrac{du}{dx} - u\dfrac{dv}{dx}}{v^2}$$

$$\frac{d}{dx}(u^n) = nu^{n-1}\frac{du}{dx}$$

$$\frac{d}{dx}\sin u = \cos u\frac{du}{dx}$$

$$\frac{d}{dx}\cos u = -\sin u\frac{du}{dx}$$

$$\frac{d}{dx}\tan u = \sec^2 u\frac{du}{dx}$$

$$\frac{d}{dx}\cot u = -\csc^2 u\frac{du}{dx}$$

$$\frac{d}{dx}\sec u = \sec u\tan u\frac{du}{dx}$$

$$\frac{d}{dx}\csc u = -\csc u\cot u\frac{du}{dx}$$

$$\frac{d}{dx}\mathrm{Sin}^{-1} u = \frac{1}{\sqrt{1 - u^2}}\frac{du}{dx}$$

$$\frac{d}{dx}\mathrm{Cos}^{-1} u = \frac{-1}{\sqrt{1 - u^2}}\frac{du}{dx}$$

$$\frac{d}{dx}\mathrm{Tan}^{-1} u = \frac{1}{1 + u^2}\frac{du}{dx}$$

$$\frac{d}{dx}\mathrm{Cot}^{-1} u = \frac{-1}{1 + u^2}\frac{du}{dx}$$

$$\frac{d}{dx}\mathrm{Sec}^{-1} u = \frac{1}{u\sqrt{u^2 - 1}}\frac{du}{dx}$$

$$\frac{d}{dx}\mathrm{Csc}^{-1} u = \frac{-1}{u\sqrt{u^2 - 1}}\frac{du}{dx}$$

$$\frac{d}{dx}a^u = a^u \ln a\frac{du}{dx}$$

$$\frac{d}{dx}e^u = e^u\frac{du}{dx}$$

$$\frac{d}{dx}\log_a u = \frac{1}{u}\log_a e\frac{du}{dx}$$

$$\frac{d}{dx}\ln u = \frac{1}{u}\frac{du}{dx}$$